AstroAmerica's

DAILY EPHEMERIS

OF THE PLANETS' PLACES

2000 – 2020

Midnight GMT

Includes Chiron

Compiled and formatted by David R. Roell

Astrology Classics

ISBN: 1-933303-19-0
Daily Ephemeris, 2000-2020

Published by
Astrology Classics

The publication division of
The Astrology Center of America
207 Victory Lane, Bel Air MD 21014

On the net at www.**AstroAmerica.com**

Dedicated to those killed and maimed in wars and natural disasters, 2000-2020.

Introductory

Longitudes and declinations shown in this book are for
MIDNIGHT GMT.

The **Mean Node** retrogrades 19° 20′ per 365 day year, or about 3.124′ per day.

Julian Day is calculated from noon, GMT, which means that a midnight ephemeris (such as this one) will always show it ending as a half-day. If you divide the Julian Day for January 1, 2000, by 365.25, you get a number of years a few minutes shy of exactly 6712, or, to be precise, January 1, 4713 BC (there was no year 0). This system was invented by Joseph Scaliger in 1583. It is based on the Indiction cycle of 15 years (used in dating medieval documents), multiplied times the Metonic cycle of 18 years (lunar), times the Solar cycle of 28 (in the Julian calendar, the number of years to complete one leap year cycle). The product of these numbers is 7560. Scaliger found that all three cycles were last together on January 1, 4713 BC, hence his choice of that date. Which, so far as he - or anyone else - knew, predated all historical dates.

Ayanamsa: Lahiri, the standard ayanamsa, as used in India. There are many minor variations derived from it. To convert from the tropical zodiac to a Lahiri-based sidereal zodiac, convert tropical longitudes to 360° notation (as shown below) and subtract the ayanamsa for the month. With this you may then determine the Nakshatra.

Aspectarian: The aspectarian includes Ptolemaic aspects, sign ingresses, stations and parallels and counter parallels of declination. Conjunctions, squares and oppositions of Sun and Moon, as well as lunar sign ingresses, are not given in the Aspectarian, as they are shown elsewhere. Aspects to Chiron are not included.

Signs			Planets		Aspects		Phases (lunations)	
0°	♈	Aries	☉	Sun	☌	Conjunction 0°	●	New
30°	♉	Taurus	☽	Moon	✶	Sextile 60°	◐	First Quarter
60°	♊	Gemini	☿	Mercury	□	Square 90°	○	Full
90°	♋	Cancer	♀	Venus	△	Trine 120°	◑	Last Quarter
120°	♌	Leo	♂	Mars	☍	Opposition 180°	✹	Lunar Eclipse
150°	♍	Virgo	♃	Jupiter	∥	Parallel	✪	Solar Eclipse
180°	♎	Libra	♄	Saturn	⋕	Counter-parallel		
210°	♏	Scorpio	♅	Uranus				
240°	♐	Sagittarius	♆	Neptune				
270°	♑	Capricorn	♇	Pluto				
300°	♒	Aquarius	☊	Node (north)				
330°	♓	Pisces	⚷	Chiron				

Nakshatras *(in sidereal zodiac)*

1. Ashvini	00° 00′	00 ♈ 00′	15. Swati	186° 40′	06 ♎ 40′
2. Bharani	13° 20′	13 ♈ 20′	16. Vishakha	200° 00′	20 ♎ 00′
3. Krittika	26° 40′	26 ♈ 40′	17. Anuradna	213° 20′	03 ♏ 20′
4. Rohini	40° 00′	10 ♉ 00′	18. Jyeshtha	226° 40′	16 ♏ 40′
5. Mrigashirsha	53° 20′	23 ♉ 20′	19. Mula	240° 00′	00 ♐ 00′
6. Ardra	66° 40′	06 ♊ 40′	20. Purva Ashadha	253° 20′	13 ♐ 20′
7. Punarvasu	80° 00′	20 ♊ 00′	21. Uttara Ashadha	266° 40′	26 ♐ 40′
8. Pushya	93° 20′	03 ♋ 20′	*(Abhijit, 276° 40′ - 280° 54.13′)*		
9. Ashlesha	106° 40′	16 ♋ 40′	22. Shravana	280° 00′	10 ♑ 00′
10. Magha	120° 00′	00 ♌ 00′	23. Dhanishtha	293° 20′	23 ♑ 20′
11. Purva Phalguni	133° 20′	13 ♌ 20′	24. Shatabhisha	306° 40′	06 ♒ 40′
12. Uttara Phalguni	146° 40′	26 ♌ 40′	25. Purva Bhadrapada	320° 00′	20 ♒ 00′
13. Hasta	160° 00′	10 ♍ 00′	26. Uttara Bhadrapada	333° 20′	03 ♓ 20′
14. Chitra	173° 20′	23 ♍ 20′	27. Revati	346° 40′	16 ♓ 40′

21 04:45 00♌26 ✹ Total Lunar Eclipse (mag 1.330)

January 2000

Day	S. T.	☉	☽	☿	♀	♂	♃	♄	♅	♆	♇	☊ True
	h m s	° ' "	° ' "	° '	° '	° '	° '	° '	° '	° '	° '	° '
01 Sa	06 39 51	09♑51 33	07♏17 11	01♑07	00♐58	27♒35	25♈14	10♉R24	14♒47	03♒11	11♐26	03♌R55
02 Su	06 43 48	10 52 43	19 18 36	02 40	02 10	28 21	25 16	10 23	14 50	03 13	11 28	03 52
03 Mo	06 47 44	11 53 54	01♐12 55	04 14	03 23	29 08	25 19	10 22	14 53	03 15	11 30	03 49
04 Tu	06 51 41	12 55 04	13 03 19	05 48	04 36	29 54	25 22	10 21	14 56	03 17	11 32	03 45
05 We	06 55 38	13 56 15	24 52 27	07 22	05 48	00♓41	25 25	10 20	14 59	03 19	11 34	03 42
06 Th	06 59 34	14 57 25	06♑42 35	08 57	07 01	01 27	25 28	10 19	15 02	03 21	11 37	03 40
07 Fr	07 03 31	15 58 36	18 35 41	10 32	08 14	02 14	25 32	10 19	15 06	03 23	11 39	03 39
08 Sa	07 07 27	16 59 47	00♒33 37	12 07	09 27	03 00	25 35	10 18	15 09	03 26	11 41	03D 39
09 Su	07 11 24	18 00 57	12 38 17	13 43	10 40	03 47	25 39	10 18	15 12	03 28	11 43	03 40
10 Mo	07 15 20	19 02 07	24 51 46	15 20	11 53	04 33	25 43	10 18	15 15	03 30	11 45	03 41
11 Tu	07 19 17	20 03 16	07♓16 20	16 56	13 06	05 20	25 47	10 17	15 18	03 32	11 46	03 43
12 We	07 23 13	21 04 25	19 54 36	18 34	14 19	06 06	25 52	10 17	15 22	03 35	11 48	03 45
13 Th	07 27 10	22 05 34	02♈49 17	20 11	15 32	06 53	25 56	10D 17	15 25	03 37	11 50	03 46
14 Fr	07 31 07	23 06 42	16 03 06	21 49	16 45	07 39	26 01	10 17	15 28	03 39	11 52	03 47
15 Sa	07 35 03	24 07 49	29 38 16	23 28	17 58	08 26	26 06	10 18	15 31	03 41	11 54	03R 47
16 Su	07 38 60	25 08 55	13♉36 00	25 07	19 12	09 12	26 11	10 18	15 35	03 44	11 56	03 46
17 Mo	07 42 56	26 10 01	27 55 51	26 46	20 25	09 59	26 17	10 19	15 38	03 46	11 58	03 45
18 Tu	07 46 53	27 11 06	12♊35 15	28 26	21 38	10 45	26 22	10 19	15 41	03 48	12 00	03 44
19 We	07 50 49	28 12 11	27 29 14	00♒07	22 52	11 31	26 28	10 20	15 45	03 50	12 01	03 43
20 Th	07 54 46	29 13 14	12♋30 43	01 48	24 05	12 18	26 34	10 21	15 48	03 53	12 03	03 42
21 Fr	07 58 42	00♒14 17	27 31 15	03 30	25 19	13 04	26 40	10 22	15 52	03 55	12 05	03 41
22 Sa	08 02 39	01 15 19	12♌22 13	05 12	26 32	13 50	26 46	10 23	15 55	03 57	12 06	03 41
23 Su	08 06 36	02 16 21	26 55 58	06 54	27 46	14 37	26 52	10 24	15 58	03 59	12 08	03 40
24 Mo	08 10 32	03 17 22	11♍06 55	08 37	28 59	15 23	26 59	10 25	16 02	04 02	12 10	03 40
25 Tu	08 14 29	04 18 22	24 52 01	10 21	00♑13	16 09	27 05	10 27	16 05	04 04	12 11	03 39
26 We	08 18 25	05 19 22	08♎10 41	12 05	01 26	16 56	27 12	10 28	16 09	04 06	12 13	03 39
27 Th	08 22 22	06 20 21	21 04 29	13 49	02 40	17 42	27 19	10 30	16 12	04 09	12 15	03 38
28 Fr	08 26 18	07 21 20	03♏36 29	15 34	03 54	18 28	27 27	10 31	16 16	04 11	12 16	03 38
29 Sa	08 30 15	08 22 18	15 50 46	17 18	05 07	19 14	27 34	10 33	16 19	04 13	12 18	03D 38
30 Su	08 34 11	09 23 15	27 51 56	19 03	06 21	20 00	27 41	10 35	16 22	04 15	12 19	03 38
31 Mo	08 38 08	10 24 12	09♐44 39	20 48	07 35	20 46	27 49	10 37	16 26	04 18	12 21	03 39

Data for	01-01-2000
Julian Day	2451544.50
Ayanamsa	23 51 10
SVP	05♓15 21
☽ ☊ Mean	05♌04 R

● ◐ PHASES ○ ◑

06	18:14	●	15♑44
14	13:34	◐	23♈41
21	04:41	✹	00♌26
28	07:57	◑	07♏42

LAST ASPECT ☽ Day	h m	INGRESS Day	h m	
02	19:29	02	21:33	♐
05	01:06	05	10:24	♑
07	14:00	07	22:53	♒
10	01:41	10	09:59	♓
12	02:22	12	18:48	♈
14	17:47	15	00:38	♉
16	21:50	17	03:25	♊
18	22:21	19	04:01	♋
20	22:37	21	03:59	♌
23	01:30	23	05:08	♍
24	07:48	25	09:10	♎
27	12:00	27	17:02	♏
29	07:11	30	04:18	♐

DECLINATION

Day	☉	☽	☿	♀	♂	♃	♄	♅	♆	♇
01 Sa	23S 04	09S 00	24S 23	18S 19	13S 19	08N 35	12N 37	17S 02	19S 13	11S 24
02 Su	23 00	12 41	24 27	18 35	13 03	08 36	12 37	17 01	19 13	11 24
03 Mo	22 54	15 50	24 30	18 51	12 46	08 38	12 37	17 00	19 12	11 24
04 Tu	22 49	18 20	24 32	19 06	12 29	08 39	12 37	16 59	19 12	11 24
05 We	22 43	20 02	24 33	19 21	12 11	08 40	12 37	16 58	19 11	11 24
06 Th	22 36	20 53	24 32	19 35	11 54	08 42	12 37	16 57	19 11	11 24
07 Fr	22 29	20 47	24 30	19 48	11 37	08 43	12 37	16 56	19 10	11 24
08 Sa	22 21	19 45	24 26	20 02	11 19	08 45	12 37	16 55	19 10	11 25
09 Su	22 14	17 48	24 21	20 14	11 02	08 47	12 37	16 54	19 09	11 25
10 Mo	22 05	15 02	24 14	20 27	10 44	08 48	12 37	16 53	19 09	11 25
11 Tu	21 56	11 33	24 06	20 38	10 26	08 50	12 38	16 52	19 08	11 25
12 We	21 47	07 30	23 57	20 49	10 08	08 52	12 38	16 51	19 08	11 25
13 Th	21 38	03 03	23 46	21 00	09 50	08 54	12 38	16 50	19 07	11 25
14 Fr	21 28	01N 40	23 34	21 10	09 32	08 56	12 38	16 49	19 07	11 25
15 Sa	21 17	06 24	23 20	21 19	09 14	08 58	12 39	16 48	19 06	11 25
16 Su	21 06	10 57	23 04	21 28	08 56	09 00	12 39	16 47	19 06	11 25
17 Mo	20 55	15 00	22 47	21 36	08 38	09 02	12 40	16 46	19 05	11 25
18 Tu	20 43	18 13	22 29	21 43	08 20	09 05	12 40	16 45	19 05	11 25
19 We	20 31	20 17	22 09	21 50	08 01	09 07	12 41	16 44	19 04	11 26
20 Th	20 19	20 57	21 47	21 56	07 43	09 09	12 41	16 43	19 03	11 26
21 Fr	20 06	20 06	21 24	22 02	07 24	09 12	12 42	16 42	19 03	11 26
22 Sa	19 53	17 52	21 00	22 07	07 06	09 14	12 42	16 41	19 02	11 26
23 Su	19 39	14 30	20 34	22 11	06 47	09 17	12 43	16 40	19 02	11 26
24 Mo	19 25	10 23	20 06	22 15	06 29	09 19	12 44	16 39	19 01	11 26
25 Tu	19 11	05 50	19 37	22 18	06 10	09 22	12 44	16 38	19 01	11 26
26 We	18 56	01 09	19 06	22 21	05 51	09 25	12 45	16 37	19 00	11 26
27 Th	18 41	03S 26	18 34	22 23	05 33	09 27	12 46	16 36	19 00	11 26
28 Fr	18 26	07 44	18 00	22 24	05 14	09 30	12 47	16 35	18 59	11 26
29 Sa	18 10	11 38	17 25	22 24	04 55	09 33	12 48	16 34	18 59	11 26
30 Su	17 54	14 59	16 48	22 24	04 36	09 36	12 48	16 33	18 58	11 26
31 Mo	17 38	17 40	16 10	22 23	04 17	09 39	12 49	16 32	18 58	11 26

ASPECTARIAN

01 05:35 ☽ ✶ ☉
06:11 ☽ ☍ ♄
12:37 ☉ △ ♄
15:00 ☽ □ ♅
15:15 ☽ ∥ ♇
23:29 ☽ # ♄
02 02:18 ☽ ∥ ♂
19:29 ☽ □ ♂
21:16 ♀ ✶ ♆
03 04:07 ☽ ✶ ♆
04:53 ☽ ☌ ♀
10:15 ☽ ∥ ♅
12:27 ♂ # ♄
20:55 ☽ ☌ ♇
04 03:01 ♂ ♓
03:50 ☽ ✶ ♅
09:03 ♀ ∥ ♆
10:40 ☽ ∥ ♆
10:55 ☽ ∥ ♀
05 01:06 ☽ △ ♃
12:36 ☽ ✶ ♂
06 05:14 ☽ ☌ ☿
07:18 ☽ △ ♄
20:42 ☿ △ ♄
07 14:00 ☽ □ ♃
16:41 ♂ ∥ ♇
19:57 ☽ ∥ ♀
08 05:44 ☽ ☌ ♆
08:35 ☽ ∥ ♆
19:22 ☽ □ ♄
19:39 ☽ ✶ ♀
22:11 ☽ ✶ ♇
09 05:06 ☽ ☌ ♅
08:40 ☽ ∥ ♅
21:15 ♀ ☌ ♇
10 01:41 ☽ ✶ ♃
17:07 ☽ # ♄
20:03 ☽ ☌ ♂
11 00:55 ☽ ∥ ♇
05:48 ☽ ✶ ♄
07:32 ☽ ∥ ♂
08:38 ☽ □ ♇
12:20 ☽ □ ♀
16:20 ☽ # ♃
21:05 ☽ ✶ ☿
12 02:24 ☽ ✶ ☉
05:00 ♄ SD
21:32 ♀ ✶ ♅
13 01:29 ☽ ✶ ♆
16:28 ☽ △ ♇
22:57 ☽ ✶ ♅
14 01:23 ☽ △ ♀
11:41 ☽ □ ☿
17:47 ☽ ☌ ♃
21:51 ♀ ∥ ☉
15 07:03 ☽ □ ♆
13:28 ☽ ∥ ♃
13:53 ☽ # ♂
16:04 ☽ ✶ ♂
18:23 ☽ ☌ ♄
19:28 ♂ # ♃
16 01:19 ☿ ☌ ☉
02:42 ☽ # ♇
03:22 ☽ □ ♅
09:43 ☽ ∥ ♄
16:26 ☿ □ ♃
20:51 ☽ △ ☉
21:50 ☽ △ ☿
17 02:51 ☉ □ ♃
09:38 ☽ △ ♆
10:28 ♂ ✶ ♄
12:19 ☽ # ♅
20:51 ☽ □ ♂
23:02 ☽ ☍ ♇
18 05:03 ☽ △ ♅
08:17 ☽ # ♆
15:55 ☽ ☍ ♀
22:20 ☿ ♒
22:21 ☽ ✶ ♃
19 03:52 ☽ # ☉
16:07 ♂ □ ♇
16:23 ☿ ∥ ♀
20:32 ☽ ✶ ♄
23:38 ☽ △ ♂
20 18:23 ☉ ♒
22:37 ☽ □ ♃
23:58 ☽ # ☉
21 06:06 ☿ ☌ ♆
10:19 ☽ ☍ ♆
10:51 ☽ ☍ ☿
12:59 ☽ # ♆
20:45 ☽ □ ♄
23:34 ☽ △ ♇
22 04:58 ♀ △ ♃
05:49 ☽ ☍ ♅
09:15 ☽ # ♅
23:54 ☽ △ ♃
23 01:30 ☽ △ ♀
10:49 ☽ ∥ ♄
18:13 ☽ # ♇
22:48 ☽ △ ♄
24 01:48 ☽ □ ♇
05:40 ☽ ∥ ♃
07:48 ☽ ☍ ♂
18:08 ☉ ☌ ♆
19:54 ♀ ♑
22:09 ☽ # ♂
25 01:18 ☿ □ ♄
10:30 ☽ □ ♀
16:33 ☽ △ ♆
17:07 ☉ ∥ ♆
18:21 ☽ △ ☉
26 01:56 ☿ ✶ ♇
04:18 ☿ ∥ ♆
07:27 ☽ ✶ ♇
08:18 ☽ △ ☿
13:39 ☿ ∥ ☉
14:48 ☽ △ ♅
27 10:47 ☽ ∥ ♂
12:00 ☽ ☍ ♃
28 00:37 ☽ ✶ ♀
01:07 ☽ □ ♆
09:55 ☿ ☌ ♅
10:44 ☽ # ♃
13:32 ☽ ☍ ♄
22:46 ☽ ∥ ♇
29 00:56 ☽ □ ♅
03:23 ☽ □ ☿
07:11 ☽ △ ♂
07:59 ☽ # ♄
30 09:55 ☿ ∥ ♅
12:34 ☽ ∥ ☿
12:56 ☽ ✶ ♆
13:14 ☽ ∥ ♅
23:40 ☽ ∥ ☉
31 01:28 ☽ ✶ ☉
05:15 ☉ □ ♄
05:17 ☽ ☌ ♇
13:39 ☽ ✶ ♅
14:42 ☽ ∥ ♆
23:58 ☽ □ ♂

⚷ Chiron

01 Dec.	18 S 08
01	11♐34
04	11 54
07	12 14
10	12 34
13	12 53
16	13 12
19	13 30
22	13 48
25	14 05
28	14 21
31	14 37

February 2000

05 12:50 16♒02 ☉ Partial Solar Eclipse (mag 0.579)

Day	S. T. h m s	☉ ° ' "	☽ ° ' "	☿ ° '	♀ ° '	♂ ° '	♃ ° '	♄ ° '	♅ ° '	♆ ° '	♇ ° '	☊ True ° '
01 Tu	08 42 05	11♒ 25 08	21♐ 33 23	22♒ 33	08♑ 49	21♓ 33	27♈ 57	10♉ 39	16♒ 29	04♒ 20	12♐ 22	03♌ 40
02 We	08 46 01	12 26 03	03♑ 22 11	24 17	10 02	22 19	28 05	10 41	16 33	04 22	12 24	03 41
03 Th	08 49 58	13 26 58	15 14 36	26 01	11 16	23 05	28 13	10 44	16 36	04 24	12 25	03 43
04 Fr	08 53 54	14 27 51	27 13 34	27 44	12 30	23 51	28 21	10 46	16 40	04 27	12 26	03 44
05 Sa	08 57 51	15 28 43	09♒ 21 27	29 26	13 44	24 37	28 30	10 49	16 43	04 29	12 28	03 45
06 Su	09 01 47	16 29 35	21 40 02	01♓ 06	14 58	25 23	28 38	10 51	16 47	04 31	12 29	03R 44
07 Mo	09 05 44	17 30 25	04♓ 10 33	02 45	16 12	26 09	28 47	10 54	16 50	04 33	12 30	03 44
08 Tu	09 09 40	18 31 13	16 53 50	04 21	17 26	26 55	28 56	10 57	16 54	04 36	12 31	03 42
09 We	09 13 37	19 32 01	29 50 22	05 55	18 40	27 40	29 05	11 00	16 57	04 38	12 33	03 41
10 Th	09 17 34	20 32 46	13♈ 00 33	07 25	19 54	28 26	29 14	11 03	17 01	04 40	12 34	03 39
11 Fr	09 21 30	21 33 31	26 24 41	08 51	21 07	29 12	29 23	11 06	17 04	04 42	12 35	03 37
12 Sa	09 25 27	22 34 13	10♉ 02 58	10 13	22 21	29 58	29 32	11 10	17 08	04 45	12 36	03 36
13 Su	09 29 23	23 34 55	23 55 21	11 29	23 35	00♈ 44	29 42	11 13	17 11	04 47	12 37	03 35
14 Mo	09 33 20	24 35 34	08♊ 01 16	12 40	24 49	01 29	29 51	11 16	17 15	04 49	12 38	03D 35
15 Tu	09 37 16	25 36 12	22 19 25	13 44	26 03	02 15	00♉ 01	11 20	17 18	04 51	12 39	03 36
16 We	09 41 13	26 36 48	06♋ 47 25	14 40	27 17	03 01	00 11	11 24	17 22	04 53	12 40	03 37
17 Th	09 45 09	27 37 22	21 21 38	15 28	28 31	03 46	00 21	11 27	17 25	04 55	12 41	03R 37
18 Fr	09 49 06	28 37 55	05♌ 57 12	16 08	29 45	04 32	00 31	11 31	17 29	04 58	12 42	03 36
19 Sa	09 53 03	29 38 26	20 28 19	16 38	00♒ 59	05 18	00 41	11 35	17 32	05 00	12 43	03 35
20 Su	09 56 59	00♓ 38 55	04♍ 48 54	16 59	02 14	06 03	00 51	11 39	17 35	05 02	12 44	03 32
21 Mo	10 00 56	01 39 23	18 53 26	17 09	03 28	06 48	01 02	11 43	17 39	05 04	12 45	03 28
22 Tu	10 04 52	02 39 49	02♎ 37 38	17R 10	04 42	07 34	01 12	11 48	17 42	05 06	12 45	03 23
23 We	10 08 49	03 40 14	15 59 04	17 00	05 56	08 19	01 23	11 52	17 46	05 08	12 46	03 18
24 Th	10 12 45	04 40 37	28 57 21	16 40	07 10	09 05	01 34	11 56	17 49	05 10	12 47	03 13
25 Fr	10 16 42	05 40 59	11♏ 33 56	16 11	08 24	09 50	01 45	12 01	17 52	05 12	12 47	03 10
26 Sa	10 20 38	06 41 20	23 51 49	15 34	09 38	10 35	01 55	12 05	17 56	05 14	12 48	03 07
27 Su	10 24 35	07 41 39	05♐ 55 10	14 49	10 52	11 20	02 07	12 10	17 59	05 16	12 49	03D 06
28 Mo	10 28 32	08 41 57	17 48 46	13 57	12 06	12 06	02 18	12 15	18 02	05 18	12 49	03 07
29 Tu	10 32 28	09 42 13	29 37 46	13 00	13 20	12 51	02 29	12 20	18 06	05 20	12 50	03 09

Data for	02-01-2000
Julian Day	2451575.50
Ayanamsa	23 51 15
SVP	05 ♓ 15 18
☽ ☊ Mean	03 ♌ 26 R

● ◐ PHASES ○ ◑

05	13:03	☉	16♒02
12	23:22	◐	23♉33
19	16:27	○	00♍20
27	03:53	◑	07♐51

LAST ASPECT ☽ INGRESS

Day	h m	Day	h m	
01	13:08	01	17:10	♑
04	02:16	04	05:31	♒
06	13:34	06	16:02	♓
08	19:46	09	00:18	♈
11	05:19	11	06:21	♉
12	23:22	13	10:23	♊
15	05:52	15	12:46	♋
17	12:52	17	14:12	♌
18	19:06	19	15:54	♍
20	20:59	21	19:21	♎
23	03:15	24	01:58	♏
25	12:18	26	12:10	♐
28	00:28	29	00:45	♑

DECLINATION

Day	☉	☽	☿	♀	♂	♃	♄	♅	♆	♇
01 Tu	17S 21	19S 37	15S 31	22S 22	03S 58	09N 42	12N 50	16S 31	18S 57	11S 25
02 We	17 04	20 43	14 51	22 19	03 39	09 45	12 51	16 30	18 57	11 25
03 Th	16 47	20 54	14 09	22 17	03 21	09 48	12 52	16 29	18 56	11 25
04 Fr	16 29	20 08	13 27	22 13	03 02	09 51	12 53	16 28	18 56	11 25
05 Sa	16 12	18 25	12 43	22 09	02 43	09 55	12 54	16 27	18 55	11 25
06 Su	15 54	15 50	11 59	22 04	02 24	09 58	12 55	16 26	18 55	11 25
07 Mo	15 35	12 28	11 15	21 58	02 05	10 01	12 57	16 25	18 54	11 25
08 Tu	15 17	08 30	10 30	21 52	01 46	10 05	12 58	16 24	18 53	11 25
09 We	14 58	04 05	09 45	21 45	01 27	10 08	12 59	16 23	18 53	11 25
10 Th	14 38	00N 35	09 00	21 38	01 08	10 11	13 00	16 22	18 52	11 25
11 Fr	14 19	05 19	08 16	21 30	00 49	10 15	13 01	16 21	18 52	11 25
12 Sa	13 59	09 52	07 33	21 21	00 30	10 18	13 03	16 19	18 51	11 24
13 Su	13 40	13 58	06 51	21 12	00 11	10 22	13 04	16 18	18 51	11 24
14 Mo	13 19	17 22	06 10	21 02	00N 08	10 26	13 05	16 17	18 50	11 24
15 Tu	12 59	19 45	05 32	20 51	00 27	10 29	13 07	16 16	18 50	11 24
16 We	12 39	20 54	04 56	20 40	00 46	10 33	13 08	16 15	18 49	11 24
17 Th	12 18	20 39	04 23	20 28	01 04	10 37	13 09	16 14	18 49	11 24
18 Fr	11 57	19 00	03 53	20 16	01 23	10 40	13 11	16 13	18 48	11 24
19 Sa	11 36	16 07	03 27	20 03	01 42	10 44	13 12	16 12	18 48	11 23
20 Su	11 15	12 17	03 05	19 49	02 01	10 48	13 14	16 11	18 47	11 23
21 Mo	10 53	07 50	02 47	19 35	02 19	10 52	13 15	16 10	18 47	11 23
22 Tu	10 31	03 05	02 34	19 20	02 38	10 55	13 17	16 09	18 46	11 23
23 We	10 10	01S 41	02 26	19 05	02 57	10 59	13 18	16 08	18 46	11 23
24 Th	09 48	06 14	02 22	18 49	03 15	11 03	13 20	16 07	18 45	11 23
25 Fr	09 26	10 23	02 24	18 32	03 34	11 07	13 21	16 06	18 45	11 22
26 Sa	09 03	14 00	02 30	18 15	03 52	11 11	13 23	16 05	18 44	11 22
27 Su	08 41	16 58	02 41	17 58	04 11	11 15	13 25	16 04	18 44	11 22
28 Mo	08 18	19 11	02 56	17 40	04 29	11 19	13 26	16 03	18 43	11 22
29 Tu	07 56	20 33	03 15	17 21	04 47	11 23	13 28	16 02	18 43	11 22

ASPECTARIAN

01 02:22 ☽ ⚹ ☿
13:08 ☽ △ ♃
23:01 ☉ ⚹ ♆
02 13:07 ♀ △ ♄
14:52 ☽ △ ♄
15:04 ☽ ☌ ♀
03 16:48 ☽ ⚹ ♂
04 02:16 ☽ □ ♃
02:17 ☉ ∥ ♅
09:31 ☿ ⚹ ♃
14:22 ☽ ☌ ♆
18:04 ☽ ∥ ♆
18:10 ☿ # ♄
05 02:52 ☽ □ ♄
06:06 ☽ ⚹ ♇
08:10 ☿ ♓
14:28 ☽ ☌ ♅
19:00 ☽ ∥ ♅
23:25 ☽ ∥ ☉
06 07:14 ☉ ☌ ♅
13:34 ☽ ⚹ ♃
18:30 ☿ ∥ ♆
20:52 ☽ ☌ ☿
20:56 ☽ # ♄
07 06:44 ☽ ∥ ♆
09:37 ☽ ∥ ☿
12:47 ☽ ⚹ ♄
14:59 ☽ # ♃
15:47 ☽ □ ♇
08 01:06 ☽ ⚹ ♀
12:38 ☿ # ♃
19:46 ☽ ☌ ♂
09 08:48 ☽ ⚹ ♆
14:37 ☽ ∥ ♂
23:11 ☽ △ ♇
10 02:34 ☽ # ♂
07:15 ☽ ⚹ ♅
13:38 ☽ □ ♀
14:39 ☽ ⚹ ☉
11 05:19 ☽ ☌ ♃
13:18 ☽ # ☿
14:41 ☽ □ ♆
12 00:19 ☽ ⚹ ☿
01:04 ♂ ♈
01:57 ☽ ☌ ♄
02:30 ☽ ∥ ♃
08:41 ☽ # ♇
12:21 ☽ □ ♅
18:24 ☽ ∥ ♄
18:29 ☿ ⚹ ♄
22:14 ☽ # ☉
23:23 ☽ △ ♀
13 12:18 ☽ ⚹ ♂
15:48 ☽ # ♅
18:33 ☽ △ ♆
23:24 ☿ □ ♇
14 07:47 ☽ ☍ ♇
08:29 ☽ □ ☿
13:26 ☽ # ♆
15:34 ☽ △ ♅
15:49 ☉ # ♄
21:39 ♃ ♉
15 05:52 ☽ △ ☉
12:56 ☽ ⚹ ♃
17:24 ☽ □ ♂
17:34 ☽ # ♀
16 07:38 ☽ ⚹ ♄
13:47 ☽ △ ☿
17 04:34 ☽ # ♀
12:52 ☽ ☍ ♀
14:57 ☽ □ ♃
21:32 ☽ △ ♂
22:21 ☽ ☍ ♆
18 01:54 ☽ # ♆
04:43 ♀ ♒
09:13 ☽ □ ♄
11:08 ☽ △ ♇
14:07 ♂ ⚹ ♆
17:04 ♀ □ ♃
19:06 ☽ ☍ ♅
23:20 ☽ # ♅
19 08:33 ☉ ♓
14:02 ☉ ∥ ♆
17:15 ☽ △ ♃
18:31 ☽ ∥ ♄
20 05:00 ☽ # ♇
05:59 ☉ ⚹ ♃
06:19 ☽ # ☉
08:08 ☽ ∥ ♃
11:39 ☽ △ ♄
13:26 ☽ □ ♇
20:59 ☽ ☍ ☿
21 01:23 ☉ # ♃
12:48 ☿ SR
20:42 ☿ # ♂
22 02:04 ☽ ∥ ♂
02:38 ☽ # ☿
04:02 ☽ △ ♀
04:24 ☽ △ ♆
08:07 ♀ ☌ ♆
09:19 ☽ ☍ ♂
18:09 ☽ ⚹ ♇
23 03:17 ☽ △ ♅
03:44 ☽ ∥ ☿
06:57 ☽ # ♂
24 05:00 ☽ ☍ ♃
05:03 ♀ ∥ ♆
11:43 ☽ △ [illegible]
11:46 ☽ □ ♆
17:16 ☽ □ ♀
18:41 ☽ ∥ ☉
25 00:52 ☽ ☍ ♄
04:38 ☽ # ♃
06:11 ☽ ∥ ♇
08:33 ☽ △ ☿
12:18 ☽ □ ♅
19:34 ☽ # ♄
26 16:07 ☽ ∥ ♅
22:41 ☽ ⚹ ♆
27 08:39 ☽ ∥ ♀
11:00 ☽ ⚹ ♀
11:39 ☽ △ ♂
13:55 ☽ ☌ ♇
16:43 ☽ □ ☿
18:17 ☽ ∥ ♆
23:29 ♀ ⚹ ♂
28 00:28 ☽ ⚹ ♅
02:58 ♀ □ ♄
14:07 ♀ ⚹ ♇
14:50 ♃ # ♇
23:36 ♂ △ ♇
29 04:09 ☿ □ ♇
05:55 ☽ △ ♃
13:00 [illegible] [illegible] ♄
22:21 ☽ ⚹ ☉

⚷ Chiron

01	Dec.	18 S 17
03	14♐53	
06	15 07	
09	15 21	
12	15 35	
15	15 47	
18	15 59	
21	16 10	
24	16 20	
27	16 29	

March 2000

Day	S. T.	☉	☽	☿	♀	♂	♃	♄	♅	♆	♇	☊ True
	h m s	° ' "	° ' "	° '	° '	° '	° '	° '	° '	° '	° '	° '
01 We	10 36 25	10♓42 28	11♑27 13	12♓R00	14♒34	13♈36	02♉40	12♉24	18♒09	05♒22	12♐50	03♌11
02 Th	10 40 21	11 42 41	23 21 54	10 57	15 49	14 21	02 52	12 29	18 12	05 24	12 51	03 13
03 Fr	10 44 18	12 42 53	05♒26 02	09 54	17 03	15 06	03 03	12 34	18 16	05 26	12 51	03R 14
04 Sa	10 48 14	13 43 03	17 43 04	08 52	18 17	15 51	03 15	12 40	18 19	05 28	12 52	03 13
05 Su	10 52 11	14 43 11	00♓15 29	07 53	19 31	16 36	03 27	12 45	18 22	05 30	12 52	03 10
06 Mo	10 56 07	15 43 17	13 04 37	06 56	20 45	17 21	03 38	12 50	18 25	05 31	12 52	03 05
07 Tu	11 00 04	16 43 22	26 10 30	06 04	21 59	18 06	03 50	12 55	18 28	05 33	12 53	02 59
08 We	11 04 01	17 43 25	09♈32 02	05 18	23 13	18 50	04 02	13 01	18 32	05 35	12 53	02 52
09 Th	11 07 57	18 43 25	23 07 08	04 37	24 28	19 35	04 14	13 06	18 35	05 37	12 53	02 44
10 Fr	11 11 54	19 43 24	06♉53 18	04 02	25 42	20 20	04 27	13 12	18 38	05 39	12 53	02 38
11 Sa	11 15 50	20 43 21	20 47 56	03 34	26 56	21 05	04 39	13 18	18 41	05 40	12 54	02 33
12 Su	11 19 47	21 43 15	04♊48 43	03 12	28 10	21 49	04 51	13 23	18 44	05 42	12 54	02 30
13 Mo	11 23 43	22 43 07	18 53 46	02 58	29 24	22 34	05 03	13 29	18 47	05 44	12 54	02 28
14 Tu	11 27 40	23 42 57	03♋01 35	02 49	00♓38	23 19	05 16	13 35	18 50	05 46	12 54	02D 28
15 We	11 31 36	24 42 45	17 10 56	02D 47	01 52	24 03	05 28	13 41	18 53	05 47	12 54	02R 28
16 Th	11 35 33	25 42 30	01♌20 23	02 51	03 07	24 47	05 41	13 47	18 56	05 49	12R 54	02 27
17 Fr	11 39 30	26 42 13	15 28 04	03 00	04 21	25 32	05 54	13 53	18 59	05 50	12 54	02 25
18 Sa	11 43 26	27 41 54	29 31 24	03 15	05 35	26 16	06 06	13 59	19 02	05 52	12 54	02 22
19 Su	11 47 23	28 41 33	13♍27 08	03 36	06 49	27 01	06 19	14 05	19 05	05 54	12 54	02 16
20 Mo	11 51 19	29 41 09	27 11 36	04 01	08 03	27 45	06 32	14 11	19 08	05 55	12 54	02 07
21 Tu	11 55 16	00♈40 44	10♎41 20	04 31	09 17	28 29	06 45	14 18	19 10	05 57	12 53	01 57
22 We	11 59 12	01 40 16	23 53 40	05 05	10 31	29 13	06 58	14 24	19 13	05 58	12 53	01 46
23 Th	12 03 09	02 39 47	06♏47 11	05 44	11 45	29 57	07 11	14 30	19 16	06 00	12 53	01 36
24 Fr	12 07 05	03 39 16	19 22 02	06 26	12 59	00♉41	07 24	14 37	19 19	06 01	12 53	01 27
25 Sa	12 11 02	04 38 43	01♐39 53	07 12	14 14	01 25	07 37	14 43	19 21	06 02	12 52	01 19
26 Su	12 14 58	05 38 08	13 43 47	08 02	15 28	02 09	07 50	14 50	19 24	06 04	12 52	01 15
27 Mo	12 18 55	06 37 32	25 37 53	08 55	16 42	02 53	08 03	14 56	19 27	06 05	12 51	01 13
28 Tu	12 22 52	07 36 54	07♑27 06	09 50	17 56	03 37	08 17	15 03	19 29	06 07	12 51	01D 13
29 We	12 26 48	08 36 14	19 16 48	10 49	19 10	04 21	08 30	15 10	19 32	06 08	12 51	01 14
30 Th	12 30 45	09 35 32	01♒12 23	11 50	20 24	05 05	08 44	15 16	19 35	06 09	12 50	01 15
31 Fr	12 34 41	10 34 48	13 19 05	12 54	21 38	05 49	08 57	15 23	19 37	06 10	12 50	01R 14

Data for	03-01-2000
Julian Day	2451604.50
Ayanamsa	23 51 19
SVP	05 ♓ 15 13
☽ ☊ Mean	01 ♌ 54 R

● ◐ PHASES ○ ◑

Day	Time	Phase	Position
06	05:17	●	15♓57
13	06:59	◐	23♊01
20	04:44	○	29♍53
28	00:21	◑	07♑38

LAST ASPECT Day	h m	☽ INGRESS Day	h m	Sign
01	04:38	02	13:14	♒
04	01:12	04	23:30	♓
06	05:17	07	06:55	♈
09	02:35	09	12:02	♉
11	11:32	11	15:46	♊
13	06:59	13	18:52	♋
15	13:43	15	21:43	♌
17	18:07	18	00:49	♍
20	04:44	20	04:57	♎
22	10:26	22	11:17	♏
23	23:53	24	20:43	♐
26	11:27	27	08:51	♑
28	23:44	29	21:35	♒

DECLINATION

Day	☉	☽	☿	♀	♂	♃	♄	♅	♆	♇
01 We	07S 33	21S 01	03S 38	17S 02	05N 05	11N 27	13N 30	16S 01	18S 42	11S 21
02 Th	07 10	20 32	04 03	16 43	05 24	11 31	13 32	16 00	18 42	11 21
03 Fr	06 47	19 07	04 30	16 23	05 42	11 35	13 33	15 59	18 42	11 21
04 Sa	06 24	16 46	04 58	16 02	06 00	11 40	13 35	15 58	18 41	11 21
05 Su	06 01	13 36	05 28	15 41	06 18	11 44	13 37	15 57	18 41	11 20
06 Mo	05 38	09 43	05 57	15 20	06 36	11 48	13 39	15 56	18 40	11 20
07 Tu	05 14	05 19	06 26	14 58	06 53	11 52	13 40	15 55	18 40	11 20
08 We	04 51	00 34	06 54	14 36	07 11	11 56	13 42	15 54	18 39	11 20
09 Th	04 28	04N16	07 21	14 14	07 29	12 01	13 44	15 53	18 39	11 19
10 Fr	04 04	08 58	07 46	13 51	07 46	12 05	13 46	15 52	18 38	11 19
11 Sa	03 41	13 14	08 09	13 27	08 04	12 09	13 48	15 51	18 38	11 19
12 Su	03 17	16 49	08 30	13 03	08 21	12 13	13 50	15 50	18 38	11 19
13 Mo	02 53	19 26	08 49	12 39	08 38	12 18	13 52	15 49	18 37	11 18
14 Tu	02 30	20 53	09 06	12 15	08 56	12 22	13 54	15 48	18 37	11 18
15 We	02 06	21 00	09 20	11 50	09 13	12 26	13 56	15 48	18 36	11 18
16 Th	01 42	19 46	09 32	11 25	09 30	12 31	13 58	15 47	18 36	11 17
17 Fr	01 19	17 18	09 42	11 00	09 47	12 35	14 00	15 46	18 36	11 17
18 Sa	00 55	13 49	09 49	10 34	10 03	12 39	14 01	15 45	18 35	11 17
19 Su	00 31	09 36	09 54	10 08	10 20	12 44	14 03	15 44	18 35	11 17
20 Mo	00 07	04 56	09 57	09 42	10 37	12 48	14 05	15 43	18 34	11 16
21 Tu	00N16	00 07	09 58	09 15	10 53	12 52	14 08	15 42	18 34	11 16
22 We	00 40	04S 36	09 57	08 49	11 09	12 57	14 10	15 41	18 34	11 16
23 Th	01 04	09 00	09 53	08 22	11 26	13 01	14 12	15 41	18 33	11 15
24 Fr	01 27	12 54	09 48	07 54	11 42	13 05	14 14	15 40	18 33	11 15
25 Sa	01 51	16 11	09 41	07 27	11 58	13 10	14 16	15 39	18 33	11 15
26 Su	02 14	18 43	09 32	06 59	12 14	13 14	14 18	15 38	18 32	11 15
27 Mo	02 38	20 24	09 21	06 31	12 29	13 19	14 20	15 37	18 32	11 14
28 Tu	03 01	21 10	09 08	06 03	12 45	13 23	14 22	15 37	18 32	11 14
29 We	03 25	21 00	08 54	05 35	13 00	13 27	14 24	15 36	18 31	11 14
30 Th	03 48	19 53	08 38	05 07	13 16	13 32	14 26	15 35	18 31	11 13
31 Fr	04 11	17 50	08 20	04 38	13 31	13 36	14 28	15 34	18 31	11 13

ASPECTARIAN

01 01:01 ☽ ✶ ☿
01:58 ☽ △ ♄
04:38 ☽ □ ♂
15:12 ☿ ☌ ☉
02 19:13 ☽ □ ♃
20:19 ☉ ✶ ♄
03 00:00 ☽ ☌ ♆
03:26 ☉ □ ♇
04:58 ☽ ∥ ♆
14:06 ☽ □ ♄
14:34 ☽ ✶ ♇
20:09 ☽ ✶ ♂
04 00:39 ♀ ☌ ♅
01:09 ☽ ☌ ♅
01:12 ☽ ☌ ♀
05:11 ♀ ∥ ♅
06:37 ☽ ∥ ♅
06:48 ☽ ∥ ♀
14:18 ☉ # ♂
23:49 ☽ # ♄
05 06:06 ☽ ✶ ♃
11:46 ☽ # ♃
13:20 ☽ ☌ ☿
14:23 ☽ ∥ ♇
15:12 ☿ ∥ ☉
23:33 ☽ ✶ ♄
23:37 ☽ □ ♇
06 16:09 ☽ # ♂
18:38 ☽ ∥ ☿
07 00:21 ☽ ∥ ☉
13:06 ♂ ✶ ♅
16:56 ☽ ✶ ♆
08 05:57 ☽ △ ♇
15:59 ☽ ✶ ♅
17:26 ☽ ☌ ♂
09 00:51 ☽ # ☉
02:35 ☽ ✶ ♀
11:05 ☿ ✶ ♃
17:04 ☽ # ☿
17:17 ☽ ∥ ♂
19:14 ☽ ✶ ☿
19:41 ☽ ☌ ♃
21:50 ☽ □ ♆
10 01:36 ☿ # ♂
04:24 ♀ # ♄
10:59 ☽ ☌ ♄
12:47 ☽ # ♇
17:25 ☽ ∥ ♃
20:21 ☽ □ ♅
23:51 ☽ ✶ ☉
11 01:09 ☽ # ♀
03:24 ☽ ∥ ♄
11:32 ☽ □ ♀
16:50 ☽ # ♅
21:19 ☽ □ ☿
12 01:31 ☽ △ ♆
13:47 ☽ ☍ ♇
15:20 ☽ # ♆
23:48 ☽ △ ♅
13 06:35 ☽ ✶ ♂
11:36 ♀ ♓
18:15 ♀ # ♃
19:33 ☽ △ ♀
23:39 ☽ △ ☿
14 03:51 ☽ ✶ ♃
18:01 ☽ ✶ ♄
20:40 ☿ SD
15 11:505 ♇ SR
12:17 ☽ □ ♂
13:43 ☽ △ ☉
18:23 ☿ ☌ ♀
16 07:21 ♀ ∥ ♇
07:29 ☽ □ ♃
07:36 ☽ ☍ ♆
09:18 ☿ # ♂
12:47 ☽ # ♆
17:16 ♃ □ ♆
19:38 ☽ △ ♇
21:16 ☽ □ ♄
17 06:00 ☽ ☍ ♅
11:26 ☽ # ♅
18:07 ☽ △ ♂
22:46 ☽ ∥ ♄
18 06:33 ☽ ☍ ☿
06:53 ☽ ∥ ♃
11:25 ☽ ☍ ♀
11:29 ☽ △ ♃
12:20 ♀ ✶ ♃
14:51 ☽ # ♇
17:17 ♀ # ♂
20:18 ☽ ∥ ♂
20:50 ☽ # ♀
22:24 ☽ # ☿
23:02 ☽ □ ♇
19 01:06 ☽ △ ♄
11:05 ☿ ∥ ♀
20 07:35 ☉ ♈
15:29 ☽ △ ♆
23:20 ☽ ∥ ☉
21 02:09 ☽ # ☉
03:58 ☽ ✶ ♇
15:26 ☽ △ ♅
22 09:14 ♂ # ♇
10:26 ☽ ☍ ♂
20:43 ☽ ∥ ♀
21:54 ☽ △ ☿
22:30 ☽ □ ♆
23 00:45 ☽ ☍ ♃
01:25 ♂ ♉
05:04 ☽ ∥ ☿
10:26 ☽ △ ♀
13:24 ☽ ∥ ♇
14:47 ☽ ☍ ♄
15:32 ☽ # ♂
21:53 ♀ □ ♇
23:54 ☽ □ ♅
24 01:12 ☽ # ♃
09:07 ☽ # ♄
19:40 ☽ ∥ ♅
25 06:25 ☽ △ ☉
08:40 ☽ ✶ ♆
10:33 ♀ ✶ ♄
11:46 ☽ □ ☿
16:35 ☿ ✶ ♃
22:02 ☽ ∥ ♆
22:16 ☽ ☌ ♇
26 03:52 ☽ □ ♀
10:39 ☉ ✶ ♆
11:27 ☽ ✶ ♅
27 15:42 ☽ △ ♂
28 01:43 ☽ △ ♃
05:16 ☽ ✶ ☿
15:34 ☽ △ ♄
23:44 ☽ ✶ ♀
30 08:13 ☽ □ ♂
09:52 ☽ ☌ ♆
15:14 ☽ □ ♃
17:106 ☽ ∥ ♆
18:08 ☽ ✶ ☉
22:27 ☿ □ ♇
23:03 ☽ ✶ ♇
31 04:05 ☽ □ ♄
11:51 ♂ ∥ ♃
12:12 ♂ □ ♆
12:20 ☽ ☌ ♅
12:37 ♀ # ☉
19:20 ☽ ∥ ♅

⚷ Chiron

01 Dec.	18 S 12
01	16♐38
04	16 45
07	16 52
10	16 58
13	17 03
16	17 07
19	17 10
22	17 12
25	17 14
28	17 14R
31	17 14

27 20:08 17♐14 R

April 2000

Day	S.T. h m s	☉ ° ' "	☽ ° ' "	☿ ° '	♀ ° '	♂ ° '	♃ ° '	♄ ° '	♅ ° '	♆ ° '	♇ ° '	☊ True ° '
01 Sa	12 38 38	11♈34 03	25♒41 29	14♓01	22♓52	06♉32	09♉11	15♉30	19♒40	06♒12	12♐R49	01♌R12
02 Su	12 42 34	12 33 15	08♓23 12	15 10	24 06	07 16	09 24	15 37	19 42	06 13	12 49	01 06
03 Mo	12 46 31	13 32 26	21 26 29	16 21	25 20	08 00	09 38	15 44	19 44	06 14	12 48	00 59
04 Tu	12 50 27	14 31 35	04♈51 42	17 34	26 34	08 43	09 51	15 51	19 47	06 15	12 48	00 49
05 We	12 54 24	15 30 42	18 37 08	18 49	27 48	09 27	10 05	15 58	19 49	06 16	12 47	00 38
06 Th	12 58 21	16 29 46	02♉39 10	20 06	29 03	10 10	10 19	16 05	19 51	06 17	12 46	00 26
07 Fr	13 02 17	17 28 49	16 52 48	21 26	00♈17	10 54	10 32	16 12	19 54	06 18	12 46	00 16
08 Sa	13 06 14	18 27 50	01♊12 29	22 47	01 31	11 37	10 46	16 19	19 56	06 19	12 45	00 08
09 Su	13 10 10	19 26 48	15 33 03	24 10	02 45	12 21	11 00	16 26	19 58	06 20	12 44	00 02
10 Mo	13 14 07	20 25 44	29 50 23	25 34	03 59	13 04	11 14	16 33	20 00	06 21	12 43	29♋58
11 Tu	13 18 03	21 24 38	14♋01 44	27 01	05 13	13 47	11 28	16 41	20 02	06 22	12 42	29 56
12 We	13 21 60	22 23 30	28 05 39	28 29	06 27	14 30	11 42	16 48	20 04	06 23	12 42	29 56
13 Th	13 25 56	23 22 19	12♌01 33	29 59	07 41	15 13	11 56	16 55	20 06	06 24	12 41	29 54
14 Fr	13 29 53	24 21 06	25 49 16	01♈30	08 55	15 56	12 10	17 03	20 08	06 25	12 40	29 52
15 Sa	13 33 50	25 19 50	09♍28 30	03 04	10 09	16 40	12 24	17 10	20 10	06 25	12 39	29 47
16 Su	13 37 46	26 18 33	22 58 35	04 39	11 23	17 23	12 38	17 17	20 12	06 26	12 38	29 39
17 Mo	13 41 43	27 17 13	06♎18 19	06 15	12 37	18 05	12 52	17 25	20 14	06 27	12 37	29 29
18 Tu	13 45 39	28 15 51	19 26 13	07 53	13 50	18 48	13 06	17 32	20 16	06 27	12 36	29 16
19 We	13 49 36	29 14 27	02♏20 50	09 33	15 04	19 31	13 20	17 40	20 17	06 28	12 35	29 03
20 Th	13 53 32	00♉13 01	15 01 09	11 14	16 18	20 14	13 34	17 47	20 19	06 29	12 34	28 49
21 Fr	13 57 29	01 11 34	27 27 03	12 58	17 32	20 57	13 48	17 55	20 21	06 29	12 33	28 37
22 Sa	14 01 25	02 10 05	09♐39 26	14 42	18 46	21 39	14 02	18 02	20 22	06 30	12 32	28 27
23 Su	14 05 22	03 08 34	21 40 20	16 29	20 00	22 22	14 16	18 10	20 24	06 30	12 31	28 21
24 Mo	14 09 19	04 07 01	03♑32 55	18 17	21 14	23 05	14 31	18 17	20 25	06 31	12 29	28 17
25 Tu	14 13 15	05 05 27	15 21 20	20 07	22 28	23 47	14 45	18 25	20 27	06 31	12 28	28 16
26 We	14 17 12	06 03 51	27 10 25	21 58	23 42	24 30	14 59	18 32	20 28	06 32	12 27	28D 16
27 Th	14 21 08	07 02 13	09♒05 32	23 51	24 56	25 12	15 13	18 40	20 30	06 32	12 26	28 17
28 Fr	14 25 05	08 00 34	21 12 06	25 46	26 10	25 55	15 27	18 48	20 31	06 33	12 24	28R 17
29 Sa	14 29 01	08 58 53	03♓35 25	27 43	27 24	26 37	15 42	18 55	20 32	06 33	12 23	28 15
30 Su	14 32 58	09 57 11	16 20 03	29 41	28 37	27 19	15 56	19 03	20 34	06 33	12 22	28 11

Data for	04-01-2000
Julian Day	2451635.50
Ayanamsa	23 51 22
SVP	05 ♓ 15 07
☽ ☊ Mean	00 ♌ 15 R

● ◐ PHASES ○ ◑

Day	Time	Phase	Position
04	18:13	●	15♈16
11	13:30	◐	21♋58
18	17:42	○	28♎59
26	19:31	◑	06♒51

LAST ASPECT ☽ Day	h m	INGRESS Day	h m	Sign
31	12:20	01	8:13	♓
03	07:45	03	15:22	♈
05	02:04	05	19:29	♉
07	08:25	07	21:58	♊
09	16:01	10	00:16	♋
12	00:44	12	03:16	♌
13	21:14	14	07:19	♍
15	13:45	16	12:36	♎
18	17:42	18	19:36	♏
20	10:37	21	04:58	♐
22	21:26	23	16:48	♑
25	18:13	26	05:43	♒
28	10:45	28	17:06	♓

DECLINATION

Day	☉	☽	☿	♀	♂	♃	♄	♅	♆	♇
01 Sa	04N34	14S56	08S01	04S10	13N46	13N41	14N30	15S33	18S30	11S13
02 Su	04 58	11 16	07 40	03 41	14 01	13 45	14 33	15 33	18 30	11 12
03 Mo	05 21	06 58	07 18	03 12	14 16	13 50	14 35	15 32	18 30	11 12
04 Tu	05 44	02 14	06 54	02 43	14 31	13 54	14 37	15 31	18 30	11 12
05 We	06 06	02N45	06 29	02 14	14 45	13 58	14 39	15 31	18 29	11 11
06 Th	06 29	07 40	06 03	01 45	14 59	14 03	14 41	15 30	18 29	11 11
07 Fr	06 52	12 15	05 35	01 16	15 14	14 07	14 43	15 29	18 29	11 11
08 Sa	07 14	16 11	05 06	00 47	15 28	14 12	14 45	15 29	18 29	11 10
09 Su	07 37	19 08	04 35	00 17	15 42	14 16	14 48	15 28	18 28	11 10
10 Mo	07 59	20 54	04 04	00N12	15 55	14 20	14 50	15 27	18 28	11 10
11 Tu	08 21	21 19	03 31	00 41	16 09	14 25	14 52	15 27	18 28	11 10
12 We	08 43	20 23	02 57	01 10	16 22	14 29	14 54	15 26	18 28	11 09
13 Th	09 05	18 12	02 22	01 40	16 36	14 33	14 56	15 25	18 28	11 09
14 Fr	09 26	14 59	01 46	02 09	16 49	14 38	14 58	15 25	18 27	11 09
15 Sa	09 48	10 59	01 08	02 38	17 02	14 42	15 01	15 24	18 27	11 08
16 Su	10 09	06 28	00 30	03 08	17 15	14 47	15 03	15 24	18 27	11 08
17 Mo	10 30	01 43	00N10	03 37	17 27	14 51	15 05	15 23	18 27	11 08
18 Tu	10 51	03S 03	00 50	04 06	17 40	14 55	15 07	15 23	18 27	11 07
19 We	11 12	07 36	01 32	04 35	17 52	15 00	15 09	15 22	18 26	11 07
20 Th	11 33	11 44	02 14	05 04	18 04	15 04	15 11	15 22	18 26	11 07
21 Fr	11 53	15 18	02 57	05 33	18 16	15 08	15 14	15 21	18 26	11 06
22 Sa	12 14	18 09	03 41	06 01	18 28	15 12	15 16	15 21	18 26	11 06
23 Su	12 34	20 10	04 26	06 30	18 40	15 17	15 18	15 20	18 26	11 06
24 Mo	12 53	21 16	05 12	06 58	18 51	15 21	15 20	15 20	18 26	11 06
25 Tu	13 13	21 25	05 59	07 27	19 02	15 25	15 22	15 19	18 26	11 05
26 We	13 33	20 37	06 46	07 55	19 13	15 29	15 24	15 19	18 26	11 05
27 Th	13 52	18 54	07 34	08 23	19 24	15 34	15 27	15 18	18 25	11 05
28 Fr	14 11	16 18	08 22	08 50	19 35	15 38	15 29	15 18	18 25	11 04
29 Sa	14 29	12 54	09 11	09 18	19 45	15 42	15 31	15 18	18 25	11 04
30 Su	14 48	08 51	10 00	09 45	19 56	15 46	15 33	15 17	18 25	11 04

ASPECTARIAN

Day	Time	Aspect
01	03:01	☽ # ♄
	07:36	☽ # ♂
	08:34	☽ # ♃
	21:47	☽ ✶ ♂
02	00:20	☽ ∥ ♇
	01:55	☽ ✶ ♃
	06:19	☉ △ ♇
	08:12	☽ □ ♇
	10:19	☿ ✶ ♄
	13:29	☽ ✶ ♄
	13:46	☽ ☌ ☿
	22:06	☽ ∥ ☿
03	07:45	☽ ☌ ♀
	07:47	☽ # ☉
	21:16	☽ ∥ ♀
04	02:27	☽ ✶ ♆
	12:06	♂ ∥ ♄
	13:53	☽ △ ♇
	21:46	☽ # ♀
05	02:04	☽ ✶ ♅
	11:08	☿ # ☉
	16:37	☽ # ☿
	17:37	☽ ∥ ☉
06	06:09	☽ □ ♆
	06:41	♂ ☌ ♃
	13:09	☽ ☌ ♃
	13:24	☽ ☌ ♂
	18:07	☽ # ♇
	18:39	♀ ♈
	22:50	☽ ☌ ♄
07	05:04	☽ □ ♅
	08:25	☽ ✶ ☿
	11:00	☽ ∥ ♃
	14:37	☽ ∥ ♄
	18:53	☽ ∥ ♂
	[illegible]	☽ [illegible] ♅
08	00:33	☽ ✶ ♀
	01:22	♂ # ♅
	08:33	☽ △ ♆
	17:43	☽ # ♆
	19:17	☽ ☍ ♇
09	07:00	☽ ✶ ☉
	07:25	☽ △ ♅
	13:09	☉ ✶ ♅
	16:01	☽ □ ☿
10	07:38	☽ □ ♀
	19:34	☽ ✶ ♃
	23:33	☽ ✶ ♂
11	04:32	☽ ✶ ♄
	22:45	♀ ✶ ♇
12	00:46	☽ △ ☿
	14:17	☽ ☍ ♆
	15:45	☽ △ ♀
	21:44	☽ # ♆
	23:49	☽ □ ♃
13	00:17	☿ ♈
	01:09	☽ △ ♇
	05:50	☽ □ ♂
	08:35	☽ □ ♄
	12:02	☽ ∥ ♂
	14:03	☽ ☍ ♅
	15:26	☿ # ♀
	21:10	☽ # ♅
	21:14	☽ △ ☉
14	00:06	☽ ∥ ♄
	02:17	☽ ∥ ♃
	23:10	☽ # ♆
15	05:15	☽ △ ♃
	05:36	☽ □ ♆
	06:05	☽ ∥ ☉
	13:27	☽ △ ♂
	13:45	☽ △ ♄
	20:28	♂ ☌ ♄
16	15:26	☽ ∥ ♀
	23:53	☽ ☍ ☿
17	00:10	♀ △ ♇
	00:15	☽ △ ♆
	02:55	☿ ✶ ♆
	06:52	☽ ∥ ☿
	10:59	☽ # ☿
	11:28	☽ ✶ ♇
	12:39	☽ ☍ ♀
18	01:31	☽ △ ♅
	06:04	☽ # ♀
	18:14	☉ # ♇
19	07:45	☽ □ ♆
	18:41	☉ ♉
	20:11	☽ ∥ ♆
	21:10	☽ ☍ ♃
	22:42	☽ # ☉
20	02:59	♂ □ ♅
	05:21	☽ ☍ ♄
	10:11	☽ □ ♅
	10:37	☽ ☍ ♂
	18:20	☿ △ ♇
	22:43	☽ # ♃
	23:24	☽ # ♄
21	00:20	☽ ∥ ♅
	17:44	☽ ✶ ♆
	20:06	♂ # ♆
22	02:49	☽ ∥ ♆
	03:26	☽ # ♂
	05:41	☽ ☌ ♇
	11:46	☽ △ ☿
	20:15	☽ △ ♀
	21:26	☽ ✶ ♅
23	07:54	♀ ✶ ♅
	13:22	♃ ∥ ♄
	17:07	♃ # ♅
	20:08	♄ # ♅
24	01:15	☽ △ ☉
	22:44	☽ △ ♃
25	04:29	☿ ✶ ♅
	06:17	☽ △ ♄
	11:27	☽ □ ☿
	16:08	☽ □ ♀
	18:13	☽ △ ♂
26	11:36	☉ □ ♆
	18:38	☽ # ♂
	18:52	☽ ☌ ♆
27	05:01	☽ ∥ ♆
	06:38	☽ ✶ ♇
	12:26	☽ □ ♃
	19:13	☽ □ ♄
	22:39	☽ ☌ ♅
28	05:04	☽ # ♃
	06:16	☽ ∥ ♄
	07:41	☽ ∥ ♅
	09:45	☽ □ ♂
	10:34	☽ ✶ ☿
	10:45	☽ ✶ ♀
	13:27	☿ ☌ ♀
	14:21	☽ # ☉
29	08:17	☿ ∥ ♀
	11:05	☽ ✶ ☉
	11:23	☽ ∥ ♇
	16:37	☽ □ ♇
	18:39	☽ # ☿
	19:28	☽ # ♀
	23:14	☽ ✶ ♃
[illegible]	[illegible]	☿ ♉
	05:04	☽ ✶ ♄
	21:13	☽ ✶ ♂

⚷ Chiron

Day	Position
01 Dec.	17 S 55
03	17♐12R
06	17 10
09	17 07
12	17 03
15	16 58
18	16 53
21	16 46
24	16 39
27	16 31
30	16 23

May 2000

Day	S. T.	☉	☽	☿	♀	♂	♃	♄	♅	♆	♇	☊ True
	h m s	° ' "	° ' "	° '	° '	° '	° '	° '	° '	° '	° '	° '
01 Mo	14 36 54	10♉55 27	29♓29 21	01♉40	29♈51	28♉02	16♉10	19♉11	20♒35	06♒33	12♐R21	28♋R04
02 Tu	14 40 51	11 53 42	13♈04 48	03 42	01♉05	28 44	16 24	19 18	20 36	06 34	12 19	27 54
03 We	14 44 47	12 51 55	27 05 22	05 45	02 19	29 26	16 39	19 26	20 37	06 34	12 18	27 44
04 Th	14 48 44	13 50 07	11♉27 22	07 49	03 33	00♊08	16 53	19 34	20 38	06 34	12 17	27 33
05 Fr	14 52 41	14 48 16	26 04 38	09 55	04 47	00 50	17 07	19 41	20 39	06 34	12 15	27 23
06 Sa	14 56 37	15 46 24	10♊49 34	12 02	06 01	01 32	17 22	19 49	20 40	06 34	12 14	27 14
07 Su	15 00 34	16 44 31	25 34 20	14 10	07 14	02 14	17 36	19 57	20 41	06 34	12 12	27 08
08 Mo	15 04 30	17 42 35	10♋12 06	16 19	08 28	02 56	17 50	20 05	20 42	06 34	12 11	27 05
09 Tu	15 08 27	18 40 38	24 37 57	18 29	09 42	03 38	18 04	20 12	20 43	06R 34	12 10	27 03
10 We	15 12 23	19 38 38	08♌49 02	20 39	10 56	04 20	18 19	20 20	20 44	06 34	12 08	27 02
11 Th	15 16 20	20 36 37	22 44 22	22 50	12 10	05 02	18 33	20 28	20 44	06 34	12 07	27 02
12 Fr	15 20 17	21 34 34	06♍24 12	25 01	13 24	05 44	18 47	20 35	20 45	06 34	12 05	27 01
13 Sa	15 24 13	22 32 29	19 49 27	27 11	14 37	06 26	19 02	20 43	20 46	06 34	12 04	26 57
14 Su	15 28 10	23 30 22	03♎01 10	29 21	15 51	07 07	19 16	20 51	20 46	06 34	12 02	26 52
15 Mo	15 32 06	24 28 13	16 00 13	01♊30	17 05	07 49	19 30	20 59	20 47	06 34	12 01	26 44
16 Tu	15 36 03	25 26 03	28 47 14	03 38	18 19	08 30	19 44	21 06	20 47	06 33	11 59	26 33
17 We	15 39 59	26 23 51	11♏22 36	05 45	19 32	09 12	19 58	21 14	20 48	06 33	11 58	26 22
18 Th	15 43 56	27 21 38	23 46 40	07 50	20 46	09 53	20 13	21 22	20 48	06 33	11 56	26 11
19 Fr	15 47 52	28 19 23	06♐00 00	09 53	22 00	10 35	20 27	21 29	20 48	06 33	11 54	26 01
20 Sa	15 51 49	29 17 07	18 03 34	11 54	23 14	11 16	20 41	21 37	20 49	06 32	11 53	25 52
21 Su	15 55 45	00♊14 50	29 59 02	13 53	24 27	11 58	20 55	21 45	20 49	06 32	11 51	25 47
22 Mo	15 59 42	01 12 32	11♑48 49	15 50	25 41	12 39	21 09	21 52	20 49	06 31	11 50	25 44
23 Tu	16 03 39	02 10 12	23 36 06	17 44	26 55	13 20	21 24	22 00	20 49	06 31	11 48	25D 44
24 We	16 07 35	03 07 51	05♒24 52	19 35	28 09	14 02	21 38	22 08	20 49	06 31	11 46	25 46
25 Th	16 11 32	04 05 30	17 19 40	21 24	29 22	14 43	21 52	22 15	20 49	06 30	11 45	25 48
26 Fr	16 15 28	05 03 07	29 25 30	23 09	00♊36	15 24	22 06	22 23	20R 49	06 29	11 43	25 50
27 Sa	16 19 25	06 00 43	11♓47 30	24 52	01 50	16 05	22 20	22 31	20 49	06 29	11 42	25R 51
28 Su	16 23 21	06 58 19	24 30 30	26 32	03 04	16 46	22 34	22 38	20 49	06 28	11 40	25 49
29 Mo	16 27 18	07 55 53	07♈38 35	28 09	04 17	17 27	22 48	22 46	20 49	06 28	11 38	25 46
30 Tu	16 31 14	08 53 27	21 14 19	29 43	05 31	18 08	23 02	22 53	20 49	06 27	11 37	25 41
31 We	16 35 11	09 50 59	05♉18 00	01♋14	06 45	18 49	23 16	23 01	20 49	06 26	11 35	25 35

Data for	05-01-2000
Julian Day	2451665.50
Ayanamsa	23 51 25
SVP	05 ♓ 15 02
☽ ☊ Mean	28 ♋ 40 R

● ◐ PHASES ○ ◑

04	04:12	●	14♉00
10	20:01	◐	20♌27
18	07:35	○	27♏40
26	11:55	◑	05♓32

LAST ASPECT ☽ Day	h m	INGRESS Day	h m	
30	21:13	01	00:55	♈
02	12:59	03	04:54	♉
04	15:07	05	06:23	♊
06	16:01	07	07:14	♋
08	16:31	09	09:01	♌
11	00:12	11	12:41	♍
13	15:58	13	18:28	♎
15	08:55	16	02:17	♏
18	07:35	18	12:10	♐
20	05:31	21	00:01	♑
23	07:31	23	13:00	♒
25	09:56	26	01:07	♓
28	04:17	28	10:08	♈
29	23:15	30	15:02	♉

DECLINATION

Day	☉	☽	☿	♀	♂	♃	♄	♅	♆	♇
01 Mo	15N06	04S16	10N49	10N13	20N06	15N50	15N35	15S17	18S25	11S04
02 Tu	15 24	00N41	11 39	10 40	20 16	15 55	15 37	15 17	18 25	11 03
03 We	15 42	05 45	12 29	11 06	20 25	15 59	15 39	15 16	18 25	11 03
04 Th	15 59	10 37	13 18	11 33	20 35	16 03	15 42	15 16	18 25	11 03
05 Fr	16 17	14 59	14 07	11 59	20 44	16 07	15 44	15 16	18 25	11 03
06 Sa	16 34	18 27	14 56	12 25	20 54	16 11	15 46	15 15	18 25	11 02
07 Su	16 50	20 43	15 45	12 50	21 03	16 15	15 48	15 15	18 25	11 02
08 Mo	17 07	21 35	16 32	13 16	21 11	16 19	15 50	15 15	18 25	11 02
09 Tu	17 23	20 59	17 19	13 41	21 20	16 23	15 52	15 15	18 25	11 02
10 We	17 39	19 03	18 04	14 05	21 28	16 27	15 54	15 14	18 25	11 01
11 Th	17 54	16 01	18 48	14 30	21 37	16 31	15 56	15 14	18 25	11 01
12 Fr	18 09	12 09	19 31	14 54	21 44	16 35	15 58	15 14	18 25	11 01
13 Sa	18 24	07 45	20 12	15 17	21 52	16 39	16 00	15 14	18 25	11 01
14 Su	18 39	03 03	20 51	15 40	22 00	16 43	16 02	15 14	18 25	11 00
15 Mo	18 53	01S42	21 27	16 03	22 07	16 47	16 04	15 14	18 25	11 00
16 Tu	19 07	06 19	22 02	16 26	22 14	16 51	16 06	15 13	18 25	11 00
17 We	19 21	10 35	22 33	16 47	22 21	16 55	16 08	15 13	18 25	11 00
18 Th	19 34	14 22	23 03	17 09	22 28	16 58	16 11	15 13	18 25	11 00
19 Fr	19 47	17 28	23 30	17 30	22 35	17 02	16 13	15 13	18 25	10 59
20 Sa	20 00	19 47	23 54	17 51	22 41	17 06	16 15	15 13	18 26	10 59
21 Su	20 12	21 12	24 16	18 11	22 47	17 10	16 16	15 13	18 26	10 59
22 Mo	20 24	21 41	24 35	18 31	22 53	17 14	16 18	15 13	18 26	10 59
23 Tu	20 36	21 11	24 51	18 50	22 59	17 17	16 20	15 13	18 26	10 59
24 We	20 47	19 45	25 05	19 09	23 04	17 21	16 22	15 13	18 26	10 58
25 Th	20 58	17 27	25 16	19 27	23 10	17 25	16 24	15 13	18 26	10 58
26 Fr	21 08	14 21	25 25	19 45	23 15	17 28	16 26	15 13	18 26	10 58
27 Sa	21 19	10 34	25 32	20 02	23 19	17 32	16 28	15 13	18 26	10 58
28 Su	21 28	06 13	25 36	20 19	23 24	17 36	16 30	15 13	18 27	10 58
29 Mo	21 38	01 28	25 38	20 35	23 29	17 39	16 32	15 13	18 27	10 58
30 Tu	21 47	03N32	25 39	20 50	23 33	17 43	16 34	15 13	18 27	10 58
31 We	21 56	08 30	25 37	21 05	23 37	17 46	16 36	15 14	18 27	10 57

ASPECTARIAN

01 02:49 ♀ ♉
06:59 ☿ # ♇
12:35 ☽ ⚹ ♆
13:59 ☉ # ♅
22:41 ☽ △ ♇
02 12:59 ☽ ⚹ ♅
20:05 ☉ ∥ ♄
21:13 ♀ # ♇
03 09:32 ☿ □ ♆
09:37 ☽ ☌ ♀
15:53 ☽ □ ♆
16:57 ☽ ☌ ☿
19:18 ♂ ♊
04 02:13 ☽ # ♇
05:20 ☽ ∥ ♀
06:14 ☉ ∥ ♃
09:05 ☽ ☌ ♃
13:27 ☽ ☌ ♄
15:07 ☽ □ ♅
17:45 ☽ ∥ ☿
05 01:46 ☽ # ♅
04:43 ☽ ∥ ♄
07:18 ☽ ∥ ♃
08:09 ☽ ☌ ♂
08:55 ☽ ∥ ☉
17:05 ☽ △ ♆
23:47 ☽ # ♆
06 02:16 ☽ ☍ ♇
09:23 ☿ # ♅
10:57 ♀ □ ♆
16:01 ☽ △ ♅
07 01:41 ☿ ∥ ♄
06:19 ☽ ∥ ♂
16:43 ☿ ∥ ♃
20:53 ☽ ⚹ ♀
08 04:10 ☉ ☌ ♃
11:54 ☽ ⚹ ☿
12:31 ♆ SR
12:51 ☽ ⚹ ♃
13:20 ☽ ⚹ ☉
16:31 ☽ ⚹ ♄
17:16 ☽ ∥ ♂
18:53 ☿ ☌ ♃
09 03:10 ☿ ∥ ☉
03:51 ☿ ☌ ☉
15:58 ☽ ⚹ ♂
20:10 ☽ ☍ ♆
20:11 ☿ ☌ ♄
10 00:46 ☿ □ ♅
03:57 ☽ □ ♀
05:40 ☽ △ ♇
05:46 ☽ # ♆
06:53 ☽ ∥ ☿
11:06 ☿ # ♆
11:09 ☽ ∥ ☉
16:36 ☽ □ ♃
19:45 ☉ ☌ ♄
20:00 ☽ □ ♄
20:31 ☽ ☍ ♅
20:34 ☽ ∥ ♃
11 00:12 ☽ □ ☿
00:33 ☽ ∥ ♄
03:16 ☉ □ ♅
05:14 ☽ # ♅
09:03 ☽ ∥ ♀
22:44 ☽ □ ♂
12 06:27 ☽ # ♇
10:05 ☽ □ ♇
13:42 ☽ △ ♀
20:44 ♀ # ♅
22:32 ☽ △ ♃
13 01:10 ☉ # ♆
01:37 ☽ △ ♄
04:55 ♂ △ ♆
05:17 ☽ △ ☉
08:35 ♄ □ ♅
15:58 ☽ △ ☿
14 06:30 ☽ △ ♆
07:10 ☿ ♊
07:57 ☽ △ ♂
16:35 ☽ ⚹ ♇
15 01:33 ♀ ∥ ♄
08:55 ☽ △ ♅
16 12:03 ☿ ∥ ♂
14:45 ☽ □ ♆
17 02:28 ☽ ∥ ♇
09:10 ☿ △ ♆
09:36 ♀ ∥ ♃
10:32 ♀ ☌ ♃
16:55 ☽ ☍ ♃
17:29 ☽ ☍ ♀
18:12 ☽ □ ♅
19:14 ☽ ☍ ♄
18 00:39 ♀ □ ♅
06:10 ☽ ∥ ♅
12:56 ♀ ☌ ♄
13:31 ☽ # ♄
20:17 ☽ # ♃
19 00:25 ☽ # ♀
01:04 ☽ ⚹ ♆
08:57 ☽ ∥ ♆
09:15 ☽ ☍ ☿
09:38 ☽ ☍ ♂
11:41 ☽ ☌ ♇
12:26 ☿ ☌ ♂
23:42 ☿ ☍ ♇
20 03:15 ☽ # ☉
05:31 ☽ ⚹ ♅
13:16 ♃ □ ♅
17:49 ☉ ♊
20:26 ♂ ☍ ♇
21 17:42 ♀ # ♆
22 19:24 ☽ △ ♃
20:42 ☽ △ ♄
23 07:31 ☽ △ ♀
10:20 ☽ # ☉
18:57 ☽ △ ☉
24 02:12 ☽ ☌ ♆
06:16 ☽ # ♀
12:49 ☽ ⚹ ♇
14:43 ☽ ∥ ♆
16:24 ☿ △ ♅
18:26 ☽ △ ♂
25 00:16 ☽ # ♃
07:00 ☽ ☌ ♅
08:26 ♅ SR
08:38 ☽ # ♄
09:13 ☽ □ ♃
09:31 ☽ △ ☿
09:56 ☽ □ ♄
12:15 ♀ ♊
17:47 ☽ ∥ ♅
26 02:35 ☽ □ ♀
21:38 ☽ ∥ ♇
23:50 ☽ □ ♇
27 08:39 ☽ □ ♂
11:39 ☉ △ ♆
20:20 ☽ ⚹ ♃
20:30 ☽ ⚹ ♄
28 04:19 ☽ □ ☿
16:05 ♃ ☌ ♄
17:19 ☽ ⚹ ♀
21:52 ☽ ⚹ ♆
29 00:33 ☽ ⚹ ☉
07:09 ☽ △ ♇
18:18 ☽ ⚹ ♂
23:17 ☽ ⚹ ♅
30 04:27 ☿ ♋
16:19 ☽ ⚹ ☿
18:05 ♀ △ ♆
31 01:54 ☽ □ ♆
12:20 ☽ # ♇

⚷ Chiron

01 Dec.	17 S 32
03	16♐14R
06	16 04
09	15 54
12	15 43
15	15 32
18	15 20
21	15 08
24	14 56
27	14 43
30	14 31

June 2000

Day	S. T.	☉	☽	☿	♀	♂	♃	♄	♅	♆	♇	☊ True
	h m s	° ′ ″	° ′ ″	° ′	° ′	° ′	° ′	° ′	° ′	° ′	° ′	° ′
01 Th	16 39 08	10♊ 48 31	19♉ 47 10	02♋ 42	07♊ 58	19♊ 30	23♉ 30	23♉ 08	20♒R48	06♒R26	11♐R34	25♋R28
02 Fr	16 43 04	11 46 02	04♊ 36 19	04 06	09 12	20 11	23 44	23 16	20 48	06 25	11 32	25 21
03 Sa	16 47 01	12 43 33	19 37 34	05 28	10 26	20 52	23 58	23 23	20 48	06 24	11 30	25 16
04 Su	16 50 57	13 41 02	04♋ 41 44	06 46	11 40	21 33	24 11	23 31	20 47	06 23	11 29	25 12
05 Mo	16 54 54	14 38 30	19 39 50	08 01	12 53	22 13	24 25	23 38	20 47	06 22	11 27	25 10
06 Tu	16 58 50	15 35 57	04♌ 24 26	09 13	14 07	22 54	24 39	23 46	20 46	06 22	11 25	25D 10
07 We	17 02 47	16 33 22	18 50 24	10 22	15 21	23 35	24 53	23 53	20 46	06 21	11 24	25 10
08 Th	17 06 44	17 30 47	02♍ 55 06	11 27	16 35	24 15	25 06	24 00	20 45	06 20	11 22	25 11
09 Fr	17 10 40	18 28 10	16 38 01	12 28	17 48	24 56	25 20	24 08	20 44	06 19	11 21	25R 11
10 Sa	17 14 37	19 25 32	00♎ 00 08	13 26	19 02	25 37	25 34	24 15	20 44	06 18	11 19	25 10
11 Su	17 18 33	20 22 53	13 03 19	14 21	20 16	26 17	25 47	24 22	20 43	06 17	11 17	25 08
12 Mo	17 22 30	21 20 14	25 49 52	15 11	21 29	26 58	26 01	24 29	20 42	06 16	11 16	25 04
13 Tu	17 26 26	22 17 33	08♏ 22 05	15 58	22 43	27 38	26 14	24 36	20 41	06 15	11 14	24 59
14 We	17 30 23	23 14 51	20 42 10	16 41	23 57	28 18	26 28	24 44	20 40	06 14	11 13	24 54
15 Th	17 34 19	24 12 09	02♐ 52 07	17 20	25 11	28 59	26 41	24 51	20 39	06 13	11 11	24 48
16 Fr	17 38 16	25 09 26	14 53 43	17 55	26 24	29 39	26 54	24 58	20 38	06 12	11 09	24 43
17 Sa	17 42 13	26 06 42	26 48 45	18 26	27 38	00♋ 19	27 08	25 05	20 37	06 11	11 08	24 39
18 Su	17 46 09	27 03 58	08♑ 39 05	18 52	28 52	01 00	27 21	25 12	20 36	06 09	11 06	24 37
19 Mo	17 50 06	28 01 13	20 26 53	19 14	00♋ 05	01 40	27 34	25 19	20 35	06 08	11 05	24D 37
20 Tu	17 54 02	28 58 28	02♒ 14 41	19 32	01 19	02 20	27 47	25 26	20 34	06 07	11 03	24 38
21 We	17 57 59	29 55 42	14 05 29	19 45	02 33	03 00	28 01	25 32	20 33	06 06	11 02	24 40
22 Th	18 01 55	00♋ 52 56	26 02 49	19 54	03 47	03 40	28 14	25 39	20 31	06 05	11 00	24 43
23 Fr	18 05 52	01 50 10	08♓ 10 35	19 57	05 00	04 20	28 27	25 46	20 30	06 03	10 59	24 45
24 Sa	18 09 48	02 47 24	20 33 01	19 R 57	06 14	05 00	28 40	25 53	20 29	06 02	10 57	24 47
25 Su	18 13 45	03 44 38	03♈ 14 23	19 52	07 28	05 40	28 52	25 59	20 27	06 01	10 56	24 48
26 Mo	18 17 42	04 41 51	16 18 37	19 42	08 41	06 20	29 05	26 06	20 26	05 59	10 54	24R 48
27 Tu	18 21 38	05 39 05	29 48 43	19 28	09 55	07 00	29 18	26 13	20 25	05 58	10 53	24 47
28 We	18 25 35	06 36 19	13♉ 46 07	19 10	11 09	07 40	29 31	26 19	20 23	05 57	10 51	24 45
29 Th	18 29 31	07 33 32	28 09 56	18 48	12 23	08 20	29 43	26 26	20 22	05 55	10 50	24 43
30 Fr	18 33 28	08 30 46	12♊ 56 34	18 23	13 36	09 00	29 56	26 32	20 20	05 54	10 49	24 40

Data for	06-01-2000
Julian Day	2451696.50
Ayanamsa	23 51 29
SVP	05 ♓ 14 58
☽ ☊ Mean	27 ♋ 01 R

● ◐ PHASES ○ ◑

Day	Time	Phase	Position
02	12:15	●	12♊15
09	03:30	◐	18♍37
16	22:27	○	26♐03
25	01:00	◑	03♈47

LAST ASPECT ☽ INGRESS

Day	h m	Day	h m	Sign
01	06:08	01	16:35	♊
03	02:03	03	16:30	♋
05	07:48	05	16:46	♌
07	10:23	07	18:58	♍
09	15:49	09	23:59	♎
12	02:15	12	07:56	♏
14	11:32	14	18:18	♐
17	01:50	17	06:26	♑
19	14:46	19	19:26	♒
22	04:25	22	07:51	♓
24	15:40	24	17:56	♈
26	07:23	27	00:19	♉
29	02:35	29	03:00	♊

DECLINATION

Day	☉	☽	☿	♀	♂	♃	♄	♅	♆	♇
01 Th	22N04	13N10	25N34	21N20	23N41	17N50	16N38	15S 14	18S 27	10S 57
02 Fr	22 12	17 08	25 28	21 34	23 44	17 53	16 39	15 14	18 27	10 57
03 Sa	22 19	20 02	25 22	21 47	23 48	17 57	16 41	15 14	18 28	10 57
04 Su	22 26	21 33	25 13	21 59	23 51	18 00	16 43	15 14	18 28	10 57
05 Mo	22 33	21 31	25 04	22 11	23 54	18 03	16 45	15 14	18 28	10 57
06 Tu	22 40	19 59	24 53	22 23	23 57	18 07	16 47	15 15	18 28	10 57
07 We	22 46	17 10	24 41	22 34	23 59	18 10	16 48	15 15	18 28	10 57
08 Th	22 51	13 24	24 28	22 44	24 01	18 13	16 50	15 15	18 29	10 57
09 Fr	22 56	09 01	24 14	22 53	24 04	18 17	16 52	15 15	18 29	10 57
10 Sa	23 01	04 19	23 59	23 02	24 05	18 20	16 54	15 16	18 29	10 56
11 Su	23 05	00S 28	23 44	23 10	24 07	18 23	16 55	15 16	18 29	10 56
12 Mo	23 09	05 08	23 28	23 17	24 09	18 26	16 57	15 16	18 30	10 56
13 Tu	23 13	09 30	23 11	23 24	24 10	18 30	16 59	15 16	18 30	10 56
14 We	23 16	13 24	22 54	23 30	24 11	18 33	17 00	15 17	18 30	10 56
15 Th	23 19	16 42	22 37	23 35	24 12	18 36	17 02	15 17	18 30	10 56
16 Fr	23 21	19 15	22 19	23 40	24 12	18 39	17 04	15 17	18 31	10 56
17 Sa	23 23	20 57	22 02	23 44	24 13	18 42	17 05	15 18	18 31	10 56
18 Su	23 24	21 43	21 44	23 47	24 13	18 45	17 07	15 18	18 31	10 56
19 Mo	23 25	21 30	21 26	23 50	24 13	18 48	17 09	15 19	18 32	10 56
20 Tu	23 26	20 20	21 08	23 52	24 13	18 51	17 10	15 19	18 32	10 56
21 We	23 26	18 17	20 51	23 53	24 12	18 54	17 12	15 19	18 32	10 56
22 Th	23 26	15 25	20 34	23 54	24 12	18 57	17 13	15 20	18 33	10 56
23 Fr	23 26	11 52	20 18	23 53	24 11	19 00	17 15	15 20	18 33	10 56
24 Sa	23 25	07 46	20 02	23 52	24 10	19 02	17 16	15 21	18 33	10 56
25 Su	23 23	03 13	19 46	23 51	24 09	19 05	17 18	15 21	18 34	10 56
26 Mo	23 21	01N36	19 32	23 48	24 07	19 08	17 19	15 22	18 34	10 56
27 Tu	23 19	06 29	19 18	23 45	24 06	19 11	17 21	15 22	18 34	10 57
28 We	23 16	11 13	19 05	23 42	24 04	19 13	17 22	15 23	18 35	10 57
29 Th	23 13	15 28	18 53	23 37	24 02	19 16	17 23	15 23	18 35	10 57
30 Fr	23 10	18 52	18 42	23 32	24 00	19 19	17 25	15 24	18 35	10 57

ASPECTARIAN

01 01:40 ☽ □ ♅
05:31 ☽ ☌ ♄
06:08 ☽ ☌ ♃
11:52 ☽ ⋕ ♅
18:18 ☉ ☍ ♇
20:48 ☽ ∥ ♄

02 02:54 ☽ △ ♆
05:30 ☽ ∥ ♃
08:01 ☽ ☌ ♀
09:46 ☽ ⋕ ♆
11:04 ☽ ☍ ♇
21:35 ♂ △ ♅

03 01:51 ☽ △ ♅
02:03 ☽ ☌ ♂
20:31 ♀ ☍ ♇

04 03:36 ☽ ☌ ☿

05 06:28 ☽ ⚹ ♄
07:48 ☽ ⚹ ♃

06 03:12 ☽ ☍ ♆
11:34 ☽ △ ♇
13:56 ☽ ⋕ ♆
16:32 ☽ ∥ ♃
17:35 ☽ ⚹ ♀
19:53 ☽ ⚹ ☉

07 02:30 ☽ ∥ ♄
03:14 ☽ ☍ ♅
08:24 ☽ ⚹ ♂
08:35 ☽ □ ♄
10:23 ☽ □ ♃
12:50 ☽ ⋕ ♅

08 13:47 ☽ ⋕ ♆
14:41 ☽ □ ♇
16:03 ☽ ⚹ ☿

09 02:17 ☽ □ ♀
13:29 ☽ △ ♄
15:33 ☿ ∥ ♂
15:37 ☽ □ ♂
15:49 ☽ △ ♃
19:21 ♀ ∥ ☉

10 11:29 ☽ △ ♆
20:43 ☽ ⚹ ♇

11 02:34 ☽ □ ☿
08:12 ☉ △ ♅
08:44 ♀ △ ♅
10:32 ♀ ☌ ☉
14:18 ☽ △ ♅
14:48 ☽ △ ☉
14:54 ☽ △ ♀

12 02:15 ☽ △ ♂
10:52 ☿ ∥ ♀
19:55 ☽ □ ♆
22:10 ☿ ∥ ☉

13 03:01 ♃ ⋕ ♆
08:29 ☽ ∥ ♇
15:40 ☽ △ ☿
23:56 ☽ □ ♅

14 07:58 ☽ ☍ ♄
11:32 ☽ ☍ ♃
13:05 ☽ ∥ ♅

15 02:48 ☽ ⋕ ♄
06:38 ☽ ⚹ ♆
16:12 ☽ ∥ ♆
16:31 ☽ ☌ ♇
17:26 ☽ ⋕ ♃

16 11:32 ☽ ⚹ ♅
12:31 ♂ ♋

17 01:52 ☽ ☍ ♀
07:31 ☽ ☍ ♂

18 00:53 ☽ ⋕ ☿
21:28 ☽ ☍ ☿
22:15 ♀ ♋

19 03:20 ☽ ⋕ ☿
10:01 ☽ △ ♄
14:46 ☽ △ ♃

20 07:52 ☽ ☌ ♆
17:49 ☽ ⚹ ♇
17:50 ☽ ⋕ ♃
21:25 ☽ ∥ ♆

21 01:48 ☉ ♋
09:45 ☽ ⋕ ♄
12:59 ☽ ☌ ♅
19:27 ♀ ☌ ♂
23:12 ☽ □ ♄

22 00:37 ☽ ∥ ♅
04:25 ☽ □ ♃
10:[illegible]6 ☽ [illegible] ☉
16:00 ☽ △ ♂
17:03 ☽ △ ♀

23 05:28 ☽ □ ♇
05:40 ☽ ∥ ♇
08:33 ☿ SR
22:50 ☽ △ ☿

24 10:14 ☽ ⚹ ♄
15:40 ☽ ⚹ ♃

25 04:45 ☽ □ ♂
05:08 ☽ ⚹ ♆
08:38 ☽ □ ♀
14:11 ☽ △ ♇

26 06:00 ☽ □ ☿
07:23 ☽ ⚹ ♅

27 10:33 ☿ ∥ ♃
10:39 ☽ □ ♆
10:52 ☽ ⚹ ☉
13:05 ☽ ⚹ ♂
19:07 ☽ ⚹ ♀
22:34 ☽ ⋕ ♇

28 08:52 ☽ ⚹ ☿
11:05 ☽ □ ♅
21:07 ☽ [illegible] ♄
23:32 ☽ ⋕ ♅

29 02:35 ☽ ☌ ♃
12:38 ☽ △ ♆
12:46 ☽ ∥ ♄
20:34 ☽ ☍ ♇
21:42 ☽ ⋕ ♆
22:40 ☽ ∥ ☿

30 03:58 ☽ ∥ ♃
07:36 ♃ ♊
11:48 ☽ △ ♅
15:06 ☿ ⋕ ♆

⚷ Chiron

01	Dec.	17 S 08
02	14♐18R	
05	14 05	
08	13 53	
11	13 40	
14	13 28	
17	13 15	
20	13 03	
23	12 52	
26	12 40	
[illegible]	[illegible]	

July 2000

01	19:34	10♋14 ◎	Partial Solar Eclipse (mag 0.476)
16	13:57	24♑19 ✷	Lunar Eclipse: Umbral (mag 1.772)
31	02:14	08♌12 ◎	Partial Solar Eclipse (mag 0.604)

Day	S. T.	☉	☽	☿	♀	♂	♃	♄	♅	♆	♇	☊ True
	h m s	° ' "	° ' "	° '	° '	° '	° '	° '	° '	° '	° '	° '
01 Sa	18 37 24	09♋ 28 00	27♊ 59 38	17♋R54	14♋ 50	09♋ 40	00♊ 09	26♉ 39	20♒R18	05♒R53	10♐ R47	24♋R38
02 Su	18 41 21	10 25 14	13♋ 10 38	17 23	16 04	10 19	00 21	26 45	20 17	05 51	10 46	24 37
03 Mo	18 45 17	11 22 28	28 20 02	16 49	17 18	10 59	00 33	26 51	20 15	05 50	10 45	24 36
04 Tu	18 49 14	12 19 41	13♌ 18 46	16 14	18 31	11 39	00 46	26 57	20 13	05 48	10 43	24 36
05 We	18 53 11	13 16 55	27 59 24	15 37	19 45	12 18	00 58	27 04	20 12	05 47	10 42	24D 36
06 Th	18 57 07	14 14 08	12♍ 16 58	15 00	20 59	12 58	01 10	27 10	20 10	05 45	10 41	24 36
07 Fr	19 01 04	15 11 21	26 09 08	14 23	22 13	13 38	01 22	27 16	20 08	05 44	10 39	24R 36
08 Sa	19 05 00	16 08 33	09♎ 35 53	13 46	23 26	14 17	01 34	27 22	20 06	05 42	10 38	24 36
09 Su	19 08 57	17 05 46	22 38 56	13 11	24 40	14 57	01 46	27 28	20 04	05 41	10 37	24 36
10 Mo	19 12 53	18 02 58	05♏ 21 07	12 38	25 54	15 36	01 58	27 34	20 03	05 39	10 36	24 35
11 Tu	19 16 50	19 00 10	17 45 53	12 07	27 08	16 16	02 10	27 39	20 01	05 38	10 35	24 35
12 We	19 20 46	19 57 22	29 56 51	11 40	28 22	16 55	02 22	27 45	19 59	05 36	10 33	24 34
13 Th	19 24 43	20 54 35	11♐ 57 31	11 16	29 35	17 35	02 33	27 51	19 57	05 35	10 32	24 33
14 Fr	19 28 40	21 51 47	23 51 06	10 56	00♌ 49	18 14	02 45	27 56	19 55	05 33	10 31	24D 33
15 Sa	19 32 36	22 48 59	05♑ 40 30	10 40	02 03	18 53	02 56	28 02	19 53	05 32	10 30	24 34
16 Su	19 36 33	23 46 12	17 28 20	10 30	03 17	19 33	03 08	28 07	19 51	05 30	10 29	24 34
17 Mo	19 40 29	24 43 25	29 16 57	10 24	04 30	20 12	03 19	28 13	19 49	05 28	10 28	24 35
18 Tu	19 44 26	25 40 39	11♒ 08 41	10D 24	05 44	20 51	03 30	28 18	19 46	05 27	10 27	24 36
19 We	19 48 22	26 37 53	23 05 49	10 29	06 58	21 30	03 41	28 23	19 44	05 25	10 26	24 36
20 Th	19 52 19	27 35 08	05♓ 10 45	10 40	08 12	22 09	03 52	28 29	19 42	05 24	10 25	24 37
21 Fr	19 56 15	28 32 23	17 26 02	10 56	09 25	22 49	04 03	28 34	19 40	05 22	10 24	24R 37
22 Sa	20 00 12	29 29 39	29 54 29	11 18	10 39	23 28	04 14	28 39	19 38	05 20	10 23	24 37
23 Su	20 04 09	00♌ 26 55	12♈ 39 01	11 46	11 53	24 07	04 25	28 44	19 36	05 19	10 22	24 37
24 Mo	20 08 05	01 24 13	25 42 30	12 20	13 07	24 46	04 35	28 49	19 33	05 17	10 21	24D 37
25 Tu	20 12 02	02 21 31	09♉ 07 27	13 00	14 21	25 25	04 46	28 54	19 31	05 16	10 20	24 37
26 We	20 15 58	03 18 51	22 55 30	13 45	15 34	26 04	04 56	28 58	19 29	05 14	10 20	24 37
27 Th	20 19 55	04 16 11	07♊ 06 58	14 36	16 48	26 43	05 07	29 03	19 27	05 12	10 19	24 38
28 Fr	20 23 51	05 13 33	21 40 11	15 32	18 02	27 22	05 17	29 08	19 24	05 11	10 18	24 38
29 Sa	20 27 48	06 10 55	06♋ 31 13	16 34	19 16	28 01	05 27	29 12	19 22	05 09	10 17	24 39
30 Su	20 31 44	07 08 18	21 33 49	17 41	20 29	28 40	05 37	29 17	19 20	05 07	10 17	24R 38
31 Mo	20 35 41	08 05 43	06♌ 39 53	18 54	21 43	29 19	05 47	29 21	19 17	05 06	10 16	24 38

Data for	07-01-2000
Julian Day	2451726.50
Ayanamsa	23 51 34
SVP	05 ♓ 14 55
☽ ☊ Mean	25 ♋ 26 R

● ◐ PHASES ○ ◑

01	19:21	◎	10♋14
08	12:53	◐	16♎39
16	13:55	✷	24♑19
24	11:03	◑	01♉51
31	02:25	◎	08♌12

LAST ASPECT ☽ INGRESS

Day	h m	Day	h m	
30	11:48	01	03:10	♋
02	21:37	03	02:38	♌
04	22:26	05	03:19	♍
07	01:57	07	06:47	♎
09	04:10	09	13:48	♏
11	20:29	12	00:06	♐
13	16:02	14	12:27	♑
16	21:48	17	01:27	♒
19	10:37	19	13:44	♓
21	23:08	22	00:10	♈
23	22:11	24	07:44	♉
26	10:21	26	12:02	♊
27	20:18	28	13:30	♋
30	12:18	30	13:24	♌

DECLINATION

Day	☉	☽	☿	♀	♂	♃	♄	♅	♆	♇
01 Sa	23N06	21N03	18N32	23N26	23N57	19N21	17N26	15S 24	18S 36	10S 57
02 Su	23 02	21 45	18 23	23 20	23 54	19 24	17 28	15 25	18 36	10 57
03 Mo	22 57	20 50	18 16	23 12	23 52	19 27	17 29	15 25	18 36	10 57
04 Tu	22 52	18 27	18 09	23 04	23 49	19 29	17 30	15 26	18 37	10 57
05 We	22 47	14 55	18 04	22 56	23 45	19 32	17 32	15 26	18 37	10 57
06 Th	22 41	10 34	18 01	22 46	23 42	19 34	17 33	15 27	18 37	10 57
07 Fr	22 34	05 48	17 58	22 36	23 38	19 37	17 34	15 28	18 38	10 58
08 Sa	22 28	00 54	17 57	22 26	23 35	19 39	17 35	15 28	18 38	10 58
09 Su	22 21	03S 54	17 58	22 15	23 31	19 41	17 37	15 29	18 39	10 58
10 Mo	22 13	08 23	17 59	22 03	23 26	19 44	17 38	15 30	18 39	10 58
11 Tu	22 05	12 26	18 02	21 50	23 22	19 46	17 39	15 30	18 39	10 58
12 We	21 57	15 54	18 06	21 37	23 18	19 48	17 40	15 31	18 40	10 58
13 Th	21 49	18 39	18 11	21 23	23 13	19 50	17 41	15 31	18 40	10 59
14 Fr	21 40	20 35	18 18	21 09	23 08	19 53	17 42	15 32	18 40	10 59
15 Sa	21 30	21 36	18 25	20 54	23 03	19 55	17 43	15 33	18 41	10 59
16 Su	21 21	21 38	18 33	20 38	22 58	19 57	17 45	15 33	18 41	10 59
17 Mo	21 11	20 43	18 42	20 22	22 52	19 59	17 46	15 34	18 42	10 59
18 Tu	21 00	18 53	18 52	20 05	22 47	20 01	17 47	15 35	18 42	11 00
19 We	20 50	16 12	19 02	19 48	22 41	20 03	17 48	15 35	18 42	11 00
20 Th	20 39	12 49	19 13	19 30	22 35	20 05	17 49	15 36	18 43	11 00
21 Fr	20 27	08 52	19 24	19 11	22 29	20 07	17 50	15 37	18 43	11 00
22 Sa	20 15	04 28	19 36	18 52	22 22	20 09	17 51	15 38	18 44	11 01
23 Su	20 03	00N13	19 47	18 33	22 16	20 11	17 52	15 38	18 44	11 01
24 Mo	19 51	05 00	19 58	18 13	22 09	20 13	17 53	15 39	18 44	11 01
25 Tu	19 38	09 40	20 09	17 52	22 02	20 15	17 53	15 40	18 45	11 01
26 We	19 25	13 59	20 19	17 31	21 55	20 17	17 54	15 40	18 45	11 02
27 Th	19 11	17 38	20 28	17 09	21 48	20 19	17 55	15 41	18 46	11 02
28 Fr	18 58	20 18	20 37	16 47	21 41	20 20	17 56	15 42	18 46	11 02
29 Sa	18 44	21 38	20 45	16 25	21 34	20 22	17 57	15 43	18 46	11 02
30 Su	18 29	21 25	20 51	16 02	21 26	20 24	17 58	15 43	18 47	11 03
31 Mo	18 15	19 40	20 57	15 39	21 18	20 26	17 58	15 44	18 47	11 03

ASPECTARIAN

01 15:50 ☉☌♂
19:17 ☽☌♂
02 04:57 ☽☌♀
06:24 ☽☌☿
17:41 ☿☌♀
21:37 ☽⚹♄
03 03:35 ☽⚹♃
11:56 ☽☍♆
15:16 ☽∥♃
19:49 ☽△♇
22:43 ☽#♆
04 02:22 ☽∥☿
07:05 ☽∥♄
11:12 ☽☍♅
20:46 ☽#♅
22:26 ☽□♄
05 05:00 ☽□♃
21:16 ☽□♆
21:58 ☽#♇
06 01:13 ☽⚹♂
03:34 ☽⚹☉
04:26 ☽⚹☿
11:35 ☿☌☉
16:25 ☽⚹♀
07 01:57 ☽△♄
09:21 ☽△♃
12:47 ♀∥☉
14:08 ☿☌♂
16:59 ☽△♆
08 01:52 ☽⚹♇
07:15 ☽□☿
08:59 ☽□♂
19:13 ☽△♅
09 04:10 ☽□♀
10 00:34 ☽□♆
13:26 ☽△☿
14:52 ☽∥♇
20:55 ☽△♂
11 02:37 ☽△☉
04:22 ☽□♅
11:08 ♀⚹♄
19:36 ☽☍♄
20:29 ☽△♀
20:59 ☽∥♅
12 04:52 ☽☍♃
11:14 ☽⚹♆
14:38 ☽#♄
19:11 ☽#☿
21:10 ☽☌♇
13 00:05 ☽∥♆
08:02 ♀♌
13:40 ☽#♃
16:04 ☽⚹♅
14 08:37 ☽#♀
21:14 ☽#☉
15 09:59 ☽☍☿
20:33 ♀⚹♃
16 04:27 ☽☍♂
13:09 ☽#☉
21:48 ☽△♄
22:05 ☿#♆
17 06:56 ☽#♀
08:18 ☽△♃
10:49 ☽#♃
11:48 ☽☍♀
12:30 ☽☌♆
13:21 ☿ SD
18:31 ♀☍♆
22:35 ☽⚹♇
18 00:08 ☽#☿
01:53 ☽∥♆
04:36 ♀∥♃
10:41 ☽#♄
17:17 ☽☌♅
19 04:43 ☽∥♅
10:37 ☽□♄
21:22 ☽□♃
20 10:17 ☽□♇
11:00 ☽△☿
11:22 ☽∥♇
13:28 ☿∥♀
21 00:38 ☉⚹♄
10:58 ☽△♂
18:51 ♀△♇
21:34 ☽⚹♄
23:08 ☽△☉
22 08:19 ☽⚹♃
10:16 ☽⚹♆
10:20 ♀#♆
10:24 ☉∥♃
12:43 ☉♌
19:44 ☽△♇
22:17 ☽□☿
22:24 ☽△♀
23 12:48 ☽⚹♅
16:52 ☿∥☉
22:11 ☽□♂
24 17:09 ☽□♆
22:27 ♀∥♄
25 07:10 ☽⚹☿
07:13 ☽#♇
10:03 ☽□♀
18:05 ☽□♅
18:14 ☿∥♃
26 05:37 ☽⚹♂
10:21 ☽☌♄
10:26 ☽#♅
18:53 ☽⚹☉
20:36 ☽☌♃
20:47 ☽∥♀
20:48 ☽△♆
27 02:05 ☽∥♄
05:18 ☽☍♇
08:55 ☽#♆
11:24 ♃△♆
11:39 ☽∥☉
17:30 ☽⚹♀
20:18 ☽△♅
22:50 ☉☍♆
28 00:29 ☽∥♃
01:41 ☉⚹♃
04:19 ☽∥☿
19:14 ☉#♆
21:47 ☽∥♂
29 02:00 ♀☍♅
17:19 ☽☌☿
23:57 ☽∥♂
30 09:48 ☽∥☿
11:47 ☽☌♂
12:18 ☽⚹♄
16:06 ☽∥♃
18:44 ♀#♅
21:30 ☽☍♆
22:34 ☽⚹♃
31 01:21 ♂⚹♄
05:43 ☽△♇
07:51 ☽#♆
13:14 ☽∥☉
14:10 ☽∥♄
20:06 ☽☍♅

⚷ Chiron

01 Dec.	16 S 51
02	12♐19R
05	12 09
08	12 00
11	11 51
14	11 43
17	11 35
20	11 28
23	11 22
26	11 17
29	11 13

August 2000

Day	S. T.	☉	☽	☿	♀	♂	♃	♄	♅	♆	♇	☊ True
	h m s	° ' "	° ' "	° '	° '	° '	° '	° '	° '	° '	° '	° '
01 Tu	20 39 38	09♌ 03 08	21 40 30	20♋ 12	22♌ 57	29♋ 58	05♊ 57	29♉ 25	19♒R15	05♒R04	10♐R15	24♋R36
02 We	20 43 34	10 00 33	06♍ 27 09	21 34	24 11	00♌ 37	06 06	29 29	19 13	05 03	10 15	24 34
03 Th	20 47 31	10 58 00	20 53 00	23 02	25 25	01 15	06 16	29 33	19 10	05 01	10 14	24 31
04 Fr	20 51 27	11 55 27	04♎ 53 36	24 34	26 38	01 54	06 25	29 37	19 08	04 59	10 14	24 28
05 Sa	20 55 24	12 52 55	18 27 12	26 10	27 52	02 33	06 35	29 41	19 06	04 58	10 13	24 26
06 Su	20 59 20	13 50 23	01♏ 34 25	27 51	29 06	03 12	06 44	29 45	19 03	04 56	10 13	24 24
07 Mo	21 03 17	14 47 53	14 17 48	29 35	00♍ 20	03 51	06 53	29 49	19 01	04 54	10 12	24 23
08 Tu	21 07 13	15 45 23	26 41 05	01♌ 22	01 34	04 29	07 02	29 53	18 58	04 53	10 12	24D 23
09 We	21 11 10	16 42 54	08♐ 48 40	03 13	02 47	05 08	07 11	29 56	18 56	04 51	10 11	24 25
10 Th	21 15 07	17 40 25	20 45 09	05 06	04 01	05 47	07 19	30 00	18 54	04 50	10 11	24 27
11 Fr	21 19 03	18 37 58	02♑ 35 02	07 02	05 15	06 25	07 28	00♊ 03	18 51	04 48	10 11	24 30
12 Sa	21 22 60	19 35 32	14 22 24	09 00	06 29	07 04	07 36	00 06	18 49	04 47	10 10	24 32
13 Su	21 26 56	20 33 06	26 10 53	10 59	07 42	07 42	07 45	00 10	18 47	04 45	10 10	24 33
14 Mo	21 30 53	21 30 42	08♒ 03 33	12 59	08 56	08 21	07 53	00 13	18 44	04 43	10 10	24R 33
15 Tu	21 34 49	22 28 19	20 02 53	15 01	10 10	08 59	08 01	00 16	18 42	04 42	10 10	24 32
16 We	21 38 46	23 25 57	02♓ 10 52	17 02	11 24	09 38	08 09	00 19	18 39	04 40	10 09	24 29
17 Th	21 42 42	24 23 36	14 29 02	19 04	12 37	10 16	08 17	00 21	18 37	04 39	10 09	24 24
18 Fr	21 46 39	25 21 17	26 58 37	21 06	13 51	10 55	08 24	00 24	18 35	04 37	10 09	24 19
19 Sa	21 50 36	26 18 59	09♈ 40 41	23 08	15 05	11 33	08 32	00 27	18 32	04 36	10 09	24 14
20 Su	21 54 32	27 16 43	22 36 18	25 09	16 18	12 12	08 39	00 29	18 30	04 34	10 09	24 10
21 Mo	21 58 29	28 14 28	05♉ 46 34	27 10	17 32	12 50	08 46	00 32	18 28	04 33	10D 09	24 07
22 Tu	22 02 25	29 12 16	19 12 36	29 09	18 46	13 29	08 53	00 34	18 25	04 31	10 09	24 05
23 We	22 06 22	00♍ 10 04	02♊ 55 20	01♍ 08	20 00	14 07	09 00	00 36	18 23	04 30	10 09	24D 05
24 Th	22 10 18	01 07 55	16 55 14	03 06	21 13	14 45	09 07	00 38	18 21	04 29	10 09	24 06
25 Fr	22 14 15	02 05 47	01♋ 11 50	05 03	22 27	15 24	09 13	00 40	18 18	04 27	10 09	24 07
26 Sa	22 18 11	03 03 42	15 43 21	06 58	23 41	16 02	09 20	00 42	18 16	04 26	10 09	24 08
27 Su	22 22 08	04 01 38	00♌ 26 12	08 52	24 54	16 40	09 26	00 44	18 14	04 24	10 10	24R 07
28 Mo	22 26 05	04 59 35	15 14 57	10 46	26 08	17 18	09 32	00 46	18 11	04 23	10 10	24 05
29 Tu	22 30 01	05 57 35	00♍ 02 38	12 37	27 22	17 57	09 38	00 47	18 09	04 22	10 10	24 01
30 We	22 33 58	06 55 35	14 41 35	14 28	28 35	18 35	09 44	00 49	18 07	04 20	10 10	23 55
31 Th	22 37 54	07 53 38	29 04 33	16 17	29 49	19 13	09 50	00 50	18 05	04 19	10 11	23 47

Data for	08-01-2000
Julian Day	2451757.50
Ayanamsa	23 51 39
SVP	05♓14 51
☽ ☊ Mean	23♋47 R

● ◐ PHASES ○ ◑

07	01:02	◐	14♏50
15	05:13	○	22♒41
22	18:51	◑	29♉58
29	10:19	●	06♍23

LAST ASPECT Day	h m	☽ INGRESS Day	h m	
01	12:34	01	13:27	♍
03	14:50	03	15:31	♎
05	18:56	05	21:04	♏
08	06:18	08	06:30	♐
09	20:15	10	18:44	♑
11	06:02	13	07:43	♒
15	05:13	15	19:42	♓
16	19:59	18	05:44	♈
20	09:14	20	13:31	♉
22	18:51	22	18:55	♊
24	07:57	24	21:59	♋
26	14:10	26	23:17	♌
28	04:44	28	23:55	♍
31	01:21	31	01:33	♎

DECLINATION

Day	☉	☽	☿	♀	♂	♃	♄	♅	♆	♇
01 Tu	18N00	16N33	21N00	15N15	21N10	20N27	17N59	15S45	18S48	11S03
02 We	17 44	12 25	21 02	14 51	21 02	20 29	18 00	15 46	18 48	11 04
03 Th	17 29	07 39	21 01	14 26	20 54	20 30	18 01	15 46	18 49	11 04
04 Fr	17 13	02 37	20 59	14 01	20 46	20 32	18 01	15 47	18 49	11 04
05 Sa	16 57	02S23	20 54	13 36	20 37	20 33	18 02	15 48	18 49	11 05
06 Su	16 40	07 07	20 47	13 10	20 28	20 35	18 03	15 49	18 50	11 05
07 Mo	16 24	11 23	20 38	12 44	20 20	20 36	18 03	15 49	18 50	11 05
08 Tu	16 07	15 04	20 25	12 18	20 11	20 38	18 04	15 50	18 50	11 06
09 We	15 50	18 02	20 11	11 51	20 02	20 39	18 04	15 51	18 51	11 06
10 Th	15 32	20 11	19 53	11 24	19 52	20 41	18 05	15 52	18 51	11 07
11 Fr	15 14	21 26	19 33	10 57	19 43	20 42	18 06	15 52	18 52	11 07
12 Sa	14 57	21 44	19 10	10 29	19 33	20 43	18 06	15 53	18 52	11 07
13 Su	14 38	21 03	18 45	10 01	19 24	20 45	18 07	15 54	18 52	11 08
14 Mo	14 20	19 26	18 17	09 33	19 14	20 46	18 07	15 55	18 53	11 08
15 Tu	14 01	16 56	17 47	09 05	19 04	20 47	18 07	15 55	18 53	11 09
16 We	13 42	13 41	17 15	08 36	18 54	20 48	18 08	15 56	18 54	11 09
17 Th	13 23	09 48	16 40	08 08	18 44	20 49	18 08	15 57	18 54	11 09
18 Fr	13 04	05 27	16 04	07 38	18 34	20 51	18 09	15 58	18 54	11 10
19 Sa	12 45	00 48	15 27	07 09	18 23	20 52	18 09	15 58	18 55	11 10
20 Su	12 25	03N58	14 47	06 40	18 13	20 53	18 09	15 59	18 55	11 11
21 Mo	12 05	08 38	14 07	06 10	18 02	20 54	18 10	16 00	18 55	11 11
22 Tu	11 45	12 59	13 25	05 40	17 51	20 55	18 10	16 00	18 56	11 12
23 We	11 25	16 46	12 42	05 10	17 41	20 56	18 10	16 01	18 56	11 12
24 Th	11 04	19 40	11 59	04 40	17 30	20 57	18 11	16 02	18 57	11 13
25 Fr	10 44	21 24	11 14	04 10	17 18	20 58	18 11	16 03	18 57	11 13
26 Sa	10 23	21 45	10 29	03 40	17 07	20 59	18 11	16 03	18 57	11 13
27 Su	10 02	20 37	09 43	03 09	16 56	20 59	18 11	16 04	18 58	11 14
28 Mo	09 41	18 03	08 57	02 39	16 45	21 00	18 11	16 05	18 58	11 14
29 Tu	09 20	14 18	08 11	02 08	16 33	21 01	18 12	16 05	18 58	11 15
30 We	08 58	09 42	07 25	01 37	16 21	21 02	18 12	16 06	18 59	11 15
31 Th	08 37	04 39	06 38	01 06	16 10	21 03	18 12	16 07	18 59	11 16

ASPECTARIAN

01 00:33 ☉ ∥ ♄
01:21 ♂ ♌
02:14 ☽ ☌ ♀
05:06 ☽ # ♅
09:02 ☽ ∥ ♀
12:34 ☽ □ ♄
23:25 ☽ □ ♃
02 01:18 ☿ ∥ ♂
05:54 ☉ △ ♇
06:14 ☽ □ ♇
07:04 ☽ # ♇
03 04:03 ☽ ✶ ☿
14:50 ☽ △ ♄
18:33 ☽ ✶ ♂
04 00:09 ☽ △ ♆
02:41 ☽ △ ♃
09:20 ☽ ✶ ♇
13:16 ☽ ✶ ☉
05 01:10 ☽ △ ♅
08:29 ♂ ∥ ♃
16:03 ☽ □ ☿
18:56 ☽ ✶ ♀
06 03:10 ☽ □ ♂
06:15 ☽ □ ♆
13:25 ♀ □ ♄
17:32 ♀ ♍
22:13 ☽ ∥ ♇
07 02:22 ☿ ∥ ♃
03:20 ☿ ✶ ♄
05:42 ☿ ♌
07:28 ☽ # ♀
09:02 ☽ □ ♅
08 05:38 ☽ ∥ ♅
06:18 ☽ ☍ ♄
07:06 ☽ # ☉
10:40 ☽ □ ♀
10:51 ☽ △ ☿
[illegible]
16:08 ☽ ✶ ♆
16:15 ☽ △ ♂
20:41 ☽ ☍ ♃
22:20 ☉ # ♅
09 00:19 ☽ # ♄
02:44 ☽ ☌ ♇
07:58 ☽ ∥ ♆
17:14 ☽ △ ☉
20:03 ☽ # ♂
20:15 ☽ ✶ ♅
20:28 ☽ # ☿
20:33 ☿ ☍ ♆
10 01:45 ☿ ∥ ♂
02:26 ♄ ♊
07:32 ☽ # ♃
12:37 ☿ ☌ ♂
15:23 ♀ # ♇
11 05:20 ☉ ☍ ♅
05:44 ☿ ✶ ♃
06:02 ☽ △ ♀
12 14:15 ☿ △ ♇
16:56 ☿ # ♆
13 00:50 ♀ □ ♃
01:51 ♂ ✶ ♃
05:51 ☽ # ♃
08:05 ☽ △ ♄
17:17 ☽ ☌ ♆
23:38 ☽ △ ♃
14 00:36 ☽ ☍ ♂
02:28 ☽ # ♂
04:13 ☽ ✶ ♇
06:04 ☽ ∥ ♆
07:54 ☿ ∥ ♄
11:53 ☽ ☍ ☿
13:36 ☽ # ♄
[illegible]
21:18 ☽ ☌ ♅
23:56 ♀ □ ♇
15 08:02 ☽ ∥ ♅
20:17 ☽ □ ♄
23:48 ☽ # ☉
16 01:12 ♂ # ♆
11:48 ☽ □ ♃
15:35 ☽ □ ♇
15:59 ☽ ∥ ♇
18:43 ☿ ☍ ♅
19:35 ♂ △ ♇
19:59 ☽ ☍ ♀
17 10:41 ☽ # ♀
18 04:16 ☿ # ♅
06:31 ☽ ✶ ♄
14:27 ☽ ✶ ♆
21:49 ☽ ✶ ♃
19 00:52 ☽ △ ♇
03:41 ☽ △ ♂
16:26 ☽ ✶ ♅
20 05:31 ☽ △ ☿
07:24 ♂ ∥ ♄
09:14 ☽ △ ☉
12:25 ☽ ∥ ♀
21:46 ☽ □ ♆
22:42 ♆ SD
21 13:18 ☽ □ ♂
13:38 ☽ # ♇
17:24 ☽ ∥ ☉
22:36 ☽ □ ♅
23:07 ☽ △ ♀
22 01:07 ☿ ☌ ☉
02:09 ☽ ∥ ☿
10:11 ☿ ♍
17:23 ☿ □ ♄
[illegible]
19:50 ☉ ♍
19:57 ☽ ☌ ♄
20:22 ☽ □ ☿
23 02:43 ☽ △ ♆
06:22 ☽ ∥ ♂
10:33 ☽ ☌ ♃
10:39 ☽ ∥ ♄
11:16 ☉ □ ♄
12:27 ☽ ☍ ♇
14:39 ☉ # ♇
17:07 ☽ # ♆
20:08 ☽ ✶ ♂
24 02:25 ☽ △ ♅
07:57 ☽ □ ♀
15:46 ☽ ∥ ♃
25 00:38 ☿ ∥ [illegible]
01:37 ☽ ✶ ☉
07:22 ☽ ✶ ☿
26 06:02 ☿ ∥ ☉
14:12 ☽ ✶ ♀
18:51 ☽ ∥ ♃
27 00:30 ☽ ✶ ♄
06:25 ☽ ☍ ♆
07:31 ☿ □ ♃
14:41 ☽ ✶ ♃
15:45 ☽ △ ♇
16:24 ☿ □ ♇
16:47 ☽ # ♆
22:58 ☽ ∥ ♄
28 03:28 ☽ ☌ ♂
04:44 ☽ ☍ ♅
[illegible]
13:25 ☽ # ♅
29 01:14 ☽ □ ♄
07:22 ♂ ☍ ♅
15:46 ☽ □ ♃
16:19 ☽ # ♇
16:34 ☽ □ ♇
23:35 ☽ ☌ ☿
30 03:56 ☽ ∥ ☉
13:08 ☽ ∥ ☿
31 01:21 ☽ ☌ ♀
02:58 ☽ △ ♄
03:35 ♀ ♎
05:40 ♂ # ♅
08:52 ☽ △ ♆
18:18 ☽ ∥ ♀
[illegible]
18:56 ☽ ✶ ♇
20:21 ♀ △ ♄

⚷ Chiron

01	Dec.	16 S 48
01	11♐09R	
04	11 06	
07	11 04	
10	11 03	
13	11 03D	
16	11 03	
19	11 04	
22	11 07	
25	11 10	
28	11 14	
31	11 [illegible]	
12	14:42	11♐03 D

September 2000

Day	S. T.	☉	☽	☿	♀	♂	♃	♄	♅	♆	♇	☊ True
	h m s	° ' "	° ' "	° '	° '	° '	° '	° '	° '	° '	° '	° '
01 Fr	22 41 51	08♍ 51 42	13♎ 05 52	18♍ 05	01♎ 03	19♌ 51	09♊ 55	00♊ 52	18♒R 02	04♒R 18	10♐ 11	23♋R 40
02 Sa	22 45 47	09 49 47	26 42 06	19 52	02 16	20 29	10 00	00 53	18 00	04 17	10 11	23 32
03 Su	22 49 44	10 47 54	09♏ 52 26	21 37	03 30	21 08	10 05	00 54	17 58	04 15	10 12	23 26
04 Mo	22 53 40	11 46 02	22 38 18	23 22	04 43	21 46	10 10	00 55	17 56	04 14	10 12	23 22
05 Tu	22 57 37	12 44 12	05♐ 02 53	25 05	05 57	22 24	10 15	00 56	17 54	04 13	10 13	23 19
06 We	23 01 34	13 42 23	17 10 35	26 47	07 11	23 02	10 20	00 56	17 52	04 12	10 13	23D 19
07 Th	23 05 30	14 40 36	29 06 25	28 27	08 24	23 40	10 24	00 57	17 50	04 11	10 14	23 21
08 Fr	23 09 27	15 38 50	10♑ 55 38	00♎ 07	09 38	24 18	10 28	00 58	17 48	04 10	10 14	23 23
09 Sa	23 13 23	16 37 06	22 43 20	01 45	10 51	24 56	10 32	00 58	17 46	04 08	10 15	23 24
10 Su	23 17 20	17 35 23	04♒ 34 12	03 22	12 05	25 34	10 36	00 58	17 44	04 07	10 16	23R 25
11 Mo	23 21 16	18 33 42	16 32 16	04 58	13 18	26 12	10 40	00 59	17 42	04 06	10 16	23 23
12 Tu	23 25 13	19 32 03	28 40 44	06 33	14 32	26 50	10 44	00 59	17 40	04 05	10 17	23 18
13 We	23 29 09	20 30 25	11♓ 01 48	08 07	15 45	27 28	10 47	00R 59	17 38	04 04	10 18	23 11
14 Th	23 33 06	21 28 49	23 36 43	09 40	16 59	28 06	10 50	00 59	17 36	04 03	10 19	23 02
15 Fr	23 37 02	22 27 15	06♈ 25 43	11 11	18 12	28 44	10 53	00 58	17 34	04 02	10 19	22 52
16 Sa	23 40 59	23 25 43	19 28 17	12 42	19 26	29 22	10 56	00 58	17 32	04 01	10 20	22 41
17 Su	23 44 56	24 24 13	02♉ 43 26	14 11	20 39	29 59	10 58	00 58	17 31	04 01	10 21	22 32
18 Mo	23 48 52	25 22 45	16 10 01	15 39	21 53	00♍ 37	11 01	00 57	17 29	04 00	10 22	22 25
19 Tu	23 52 49	26 21 19	29 47 03	17 07	23 06	01 15	11 03	00 56	17 27	03 59	10 23	22 19
20 We	23 56 45	27 19 55	13♊ 33 53	18 33	24 19	01 53	11 05	00 56	17 26	03 58	10 24	22 17
21 Th	00 00 42	28 18 34	27 30 08	19 58	25 33	02 31	11 07	00 55	17 24	03 57	10 25	22 15
22 Fr	00 04 38	29 17 15	11♋ 35 27	21 21	26 46	03 09	11 08	00 54	17 22	03 56	10 26	22 15
23 Sa	00 08 35	00♎ 15 59	25 49 05	22 44	27 59	03 46	11 10	00 53	17 21	03 56	10 27	22 15
24 Su	00 12 31	01 14 44	10♌ 09 29	24 05	29 13	04 24	11 11	00 52	17 19	03 55	10 28	22 13
25 Mo	00 16 28	02 13 32	24 33 50	25 26	00♏ 26	05 02	11 12	00 50	17 18	03 54	10 29	22 09
26 Tu	00 20 25	03 12 22	08♍ 57 54	26 44	01 39	05 40	11 13	00 49	17 16	03 54	10 30	22 02
27 We	00 24 21	04 11 14	23 16 22	28 02	02 53	06 17	11 13	00 47	17 15	03 53	10 32	21 53
28 Th	00 28 18	05 10 08	07♎ 23 26	29 18	04 06	06 55	11 14	00 46	17 14	03 52	10 33	21 42
29 Fr	00 32 14	06 09 04	21 13 50	00♏ 33	05 19	07 33	11 14	00 44	17 12	03 52	10 34	21 29
30 Sa	00 36 11	07 08 02	04♏ 43 38	01 46	06 33	08 10	11R 14	00 42	17 11	03 51	10 35	21 17

Data for	09-01-2000
Julian Day	2451788.50
Ayanamsa	23 51 43
SVP	05 ♓ 14 46
☽ ☊ Mean	22 ♋ 09 R

● ◐ PHASES ○ ◑

05	16:28	◐	13♐24
13	19:37	○	21♓18
21	01:28	◑	28♊22
27	19:54	●	05♎00

LAST ASPECT ☽ Day	h m	INGRESS Day	h m	
01	12:24	02	05:55	♏
04	01:35	04	14:09	♐
06	22:27	07	01:48	♑
08	10:27	09	14:45	♒
11	20:10	12	02:34	♓
13	19:37	14	12:00	♈
16	18:49	16	19:05	♉
18	17:30	19	00:22	♊
21	01:28	21	04:16	♋
23	03:58	23	07:00	♌
25	01:34	25	09:02	♍
26	03:45	27	11:22	♎
28	16:57	29	15:30	♏

DECLINATION

Day	☉	☽	☿	♀	♂	♃	♄	♅	♆	♇
01 Fr	08N15	00S 31	05N51	00N35	15N58	21N03	18N12	16S 07	18S 59	11S 16
02 Sa	07 53	05 30	05 04	00 05	15 46	21 04	18 12	16 08	18 59	11 17
03 Su	07 31	10 05	04 18	00S 26	15 34	21 05	18 12	16 09	19 00	11 17
04 Mo	07 09	14 05	03 31	00 57	15 22	21 05	18 12	16 09	19 00	11 18
05 Tu	06 47	17 21	02 44	01 28	15 10	21 06	18 12	16 10	19 00	11 18
06 We	06 24	19 47	01 58	01 59	14 57	21 06	18 12	16 11	19 01	11 19
07 Th	06 02	21 19	01 12	02 30	14 45	21 07	18 12	16 11	19 01	11 19
08 Fr	05 40	21 52	00 26	03 01	14 32	21 08	18 12	16 12	19 01	11 20
09 Sa	05 17	21 27	00S 19	03 32	14 20	21 08	18 12	16 12	19 02	11 20
10 Su	04 54	20 05	01 05	04 02	14 07	21 09	18 12	16 13	19 02	11 21
11 Mo	04 31	17 48	01 49	04 33	13 55	21 09	18 11	16 14	19 02	11 22
12 Tu	04 09	14 42	02 34	05 04	13 42	21 09	18 11	16 14	19 02	11 22
13 We	03 46	10 55	03 18	05 34	13 29	21 10	18 11	16 15	19 03	11 23
14 Th	03 23	06 36	04 01	06 04	13 16	21 10	18 11	16 15	19 03	11 23
15 Fr	03 00	01 55	04 44	06 35	13 03	21 11	18 11	16 16	19 03	11 24
16 Sa	02 37	02N56	05 26	07 05	12 50	21 11	18 10	16 16	19 03	11 24
17 Su	02 13	07 44	06 08	07 35	12 37	21 11	18 10	16 17	19 04	11 25
18 Mo	01 50	12 14	06 50	08 05	12 23	21 11	18 10	16 17	19 04	11 25
19 Tu	01 27	16 10	07 30	08 34	12 10	21 12	18 10	16 18	19 04	11 26
20 We	01 04	19 17	08 10	09 04	11 57	21 12	18 09	16 18	19 04	11 26
21 Th	00 40	21 18	08 50	09 33	11 43	21 12	18 09	16 19	19 04	11 27
22 Fr	00 17	22 00	09 29	10 02	11 30	21 12	18 09	16 19	19 05	11 28
23 Sa	00S 06	21 18	10 07	10 31	11 16	21 12	18 08	16 20	19 05	11 28
24 Su	00 30	19 12	10 44	11 00	11 03	21 12	18 08	16 20	19 05	11 29
25 Mo	00 53	15 53	11 20	11 28	10 49	21 12	18 07	16 21	19 05	11 29
26 Tu	01 16	11 37	11 56	11 56	10 35	21 12	18 07	16 21	19 05	11 30
27 We	01 40	06 44	12 31	12 24	10 21	21 12	18 06	16 21	19 05	11 30
28 Th	02 03	01 32	13 05	12 52	10 07	21 12	18 06	16 22	19 06	11 31
29 Fr	02 27	03S 37	13 38	13 19	09 54	21 12	18 05	16 22	19 06	11 31
30 Sa	02 50	08 29	14 10	13 46	09 40	21 12	18 05	16 23	19 06	11 32

ASPECTARIAN

01 00:21 ☽ # ♀
08:36 ☽ △ ♅
12:24 ☽ ✶ ♂
22:10 ☽ # ☿
02 04:46 ☉ □ ♃
09:02 ☉ □ ♇
11:15 ☽ # ☉
13:40 ☽ □ ♆
03 01:50 ☽ ✶ ☉
06:51 ☽ ∥ ♇
14:36 ♀ △ ♆
15:05 ☽ □ ♅
22:14 ☽ □ ♂
04 01:35 ☽ ✶ ☿
08:16 ☽ # ♂
10:54 ♃ ☍ ♇
14:36 ☽ ∥ ♅
15:57 ☽ ☍ ♄
22:22 ☽ ✶ ♆
05 01:57 ☽ ✶ ♀
07:28 ☽ # ♄
10:09 ☽ ☌ ♇
10:18 ☽ ☍ ♃
15:21 ☽ ∥ ♆
23:41 ☿ # ♀
06 01:21 ☽ ✶ ♅
12:23 ☽ △ ♂
19:51 ☽ # ♃
22:27 ☽ □ ☿
07 21:02 ☽ □ ♀
22:22 ☿ ♎
08 10:27 ☽ △ ☉
12:05 ♀ ✶ ♇
12:27 ☿ △ ♄
17:29 ♀ △ ♃
09 07:23 ☽ # ♃
16:43 ☽ △ ♄
21:11 ☽ △ ☿
23:05 ☽ ☌ ♆
10 11:10 ☿ △ ♆
11:27 ☽ ✶ ♇
12:10 ☽ ∥ ♆
12:11 ☽ △ ♃
16:48 ☽ △ ♀
20:27 ☽ # ♄
23:19 ♀ # ☉
11 02:17 ☽ ☌ ♅
12:53 ☽ ∥ ♅
20:10 ☽ ☍ ♂
12 04:29 ☽ □ ♄
07:13 ☽ # ♂
11:35 ♄ SR
21:18 ☽ ∥ ♇
22:34 ☽ □ ♇
23:30 ☽ □ ♃
13 10:06 ☿ # ☉
14 02:29 ☽ ∥ ♀
10:37 ☽ ∥ ☿
11:54 ♀ △ ♅
13:50 ☽ ✶ ♄
18:07 ☽ # ☉
19:03 ☿ △ ♃
19:33 ☽ ✶ ♆
15 07:14 ☽ △ ♇
08:17 ☽ ✶ ♃
09:57 ☽ ☍ ☿
20:28 ☽ ✶ ♅
22:31 ☽ ∥ ☉
23:54 ☽ ☍ ♀
16 14:35 ☽ # ☿
18:51 ☽ △ ♂
23:11 ☽ # ♀
17 00:19 ♂ ♍
02:19 ☽ □ ♆
19:32 ☽ # ♇
18 00:54 ☽ ∥ ♂
02:19 ☽ □ ♅
12:18 ♂ □ ♄
17:32 ☽ △ ☉
19 00:55 ☽ # ♅
02:01 ☽ ☌ ♄
02:41 ☽ □ ♂
05:38 ☿ △ ♅
07:19 ☽ △ ♆
14:30 ☽ ∥ ♄
18:31 ☽ ☍ ♇
19:42 ☽ ☌ ♃
22:10 ☽ # ♆
20 06:39 ☽ △ ♅
09:34 ☽ △ ☿
20:19 ☽ △ ♀
22:34 ☽ ∥ ♃
21 08:57 ☽ ✶ ♂
22 03:49 ♂ # ♇
17:29 ☉ ♎
18:15 ☽ □ ☿
23 01:28 ☽ ∥ ♃
03:58 ☽ □ ♀
08:00 ☽ ✶ ☉
08:28 ☽ ✶ ♄
13:34 ☽ ☍ ♆
14:45 ☉ △ ♄
24 00:30 ☽ △ ♇
01:04 ☽ # ♆
01:32 ♀ # ♂
01:42 ☽ ✶ ♃
08:44 ☽ ∥ ♄
08:51 ☿ # ♂
11:54 ☽ ☍ ♅
15:26 ♀ ♏
21:07 ☽ # ♅
25 00:47 ♀ ∥ ♇
01:34 ☽ ✶ ☿
05:54 ☿ ∥ ♇
10:25 ☽ □ ♄
10:40 ☽ ✶ ♀
18:13 ☽ ☌ ♂
22:30 ☽ # ♀
22:34 ☽ # ☿
26 00:40 ☽ # ♇
00:55 ☿ ∥ ♀
02:34 ☽ □ ♇
03:45 ☽ □ ♃
05:34 ☽ ∥ ♂
16:40 ☉ △ ♆
27 12:41 ☽ △ ♄
17:58 ☽ △ ♆
19:34 ♀ □ ♆
21:49 ☽ # ☉
28 05:25 ☽ ✶ ♇
06:36 ☽ △ ♃
13:28 ☿ ♏
16:57 ☽ △ ♅
17:59 ☽ ∥ ☉
29 12:53 ♃ SR
18:07 ☽ ☌ ☿
22:25 ☽ □ ♆
30 03:36 ☽ ☌ ♀
05:52 ☽ # ♂
06:32 ☽ ✶ ♂
16:30 ☽ ∥ ♇
22:43 ☽ □ ♅

⚷ Chiron

01 Dec.	17 S 00
03	11♐24
06	11 31
09	11 38
12	11 46
15	11 55
18	12 04
21	12 14
24	12 25
27	12 37
30	12 50

October 2000

Day	S. T.	☉	☽	☿	♀	♂	♃	♄	♅	♆	♇	☊ True
	h m s	° ' "	° ' "	° '	° '	° '	° '	° '	° '	° '	° '	° '
01 Su	00 40 07	08♎07 01	17♏50 54	02♏57	07♏46	08♍48	11♊R14	00♊R40	17♒R10	03♒R51	10♐37	21♋R07
02 Mo	00 44 04	09 06 03	00♐35 49	04 06	08 59	09 26	11 13	00 38	17 09	03 50	10 38	20 58
03 Tu	00 48 00	10 05 07	13 00 29	05 14	10 12	10 03	11 13	00 36	17 07	03 50	10 39	20 52
04 We	00 51 57	11 04 12	25 08 33	06 20	11 26	10 41	11 12	00 34	17 06	03 50	10 41	20 50
05 Th	00 55 54	12 03 19	07♑04 38	07 23	12 39	11 18	11 11	00 32	17 05	03 49	10 42	20D 49
06 Fr	00 59 50	13 02 28	18 54 03	08 24	13 52	11 56	11 10	00 29	17 04	03 49	10 43	20 49
07 Sa	01 03 47	14 01 39	00♒42 19	09 23	15 05	12 33	11 08	00 27	17 03	03 49	10 45	20 50
08 Su	01 07 43	15 00 52	12 34 51	10 18	16 18	13 11	11 07	00 24	17 02	03 48	10 46	20R 49
09 Mo	01 11 40	16 00 06	24 36 41	11 11	17 31	13 48	11 05	00 21	17 01	03 48	10 48	20 46
10 Tu	01 15 36	16 59 22	06♓51 59	12 00	18 44	14 26	11 03	00 19	17 01	03 48	10 49	20 40
11 We	01 19 33	17 58 40	19 23 52	12 45	19 57	15 03	11 01	00 16	17 00	03 48	10 51	20 32
12 Th	01 23 29	18 58 00	02♈14 03	13 27	21 10	15 41	10 58	00 13	16 59	03 48	10 53	20 21
13 Fr	01 27 26	19 57 22	15 22 42	14 04	22 23	16 18	10 56	00 10	16 58	03 48	10 54	20 09
14 Sa	01 31 23	20 56 46	28 48 24	14 36	23 36	16 56	10 53	00 07	16 58	03 47	10 56	19 56
15 Su	01 35 19	21 56 12	12♉28 30	15 03	24 49	17 33	10 50	00 03	16 57	03 47	10 58	19 44
16 Mo	01 39 16	22 55 40	26 19 34	15 24	26 02	18 10	10 47	00 00	16 57	03D 47	10 59	19 35
17 Tu	01 43 12	23 55 10	10♊18 03	15 39	27 15	18 48	10 44	29♉57	16 56	03 47	11 01	19 28
18 We	01 47 09	24 54 43	24 20 54	15 47	28 28	19 25	10 40	29 53	16 56	03 48	11 03	19 23
19 Th	01 51 05	25 54 18	08♋25 53	15R 47	29 41	20 02	10 36	29 50	16 55	03 48	11 04	19 21
20 Fr	01 55 02	26 53 55	22 31 33	15 40	00♐54	20 40	10 32	29 46	16 55	03 48	11 06	19 20
21 Sa	01 58 58	27 53 35	06♌37 01	15 24	02 06	21 17	10 28	29 42	16 54	03 48	11 08	19 18
22 Su	02 02 55	28 53 17	20 41 32	15 00	03 19	21 54	10 24	29 39	16 54	03 48	11 10	19 16
23 Mo	02 06 52	29 53 01	04♍43 57	14 26	04 32	22 31	10 19	29 35	16 54	03 48	11 12	19 12
24 Tu	02 10 48	00♏52 47	18 42 29	13 44	05 45	23 09	10 15	29 31	16 54	03 49	11 14	19 05
25 We	02 14 45	01 52 36	02♎34 30	12 53	06 58	23 46	10 10	29 27	16 54	03 49	11 16	18 56
26 Th	02 18 41	02 52 26	16 16 50	11 54	08 10	24 23	10 05	29 23	16 54	03 49	11 17	18 45
27 Fr	02 22 38	03 52 19	29 46 09	10 49	09 23	25 00	10 00	29 19	16D 54	03 50	11 19	18 32
28 Sa	02 26 34	04 52 14	12♏59 33	09 37	10 36	25 37	09 54	29 14	16 54	03 50	11 21	18 20
29 Su	02 30 31	05 52 10	25 55 13	08 22	11 48	26 15	09 49	29 10	16 54	03 50	11 23	18 08
30 Mo	02 34 27	06 52 09	08♐32 42	07 05	13 01	26 52	09 43	29 06	16 54	03 51	11 25	17 59
31 Tu	02 38 24	07 52 09	20 53 07	05 48	14 13	27 29	09 37	29 02	16 54	03 51	11 27	17 53

Data for	10-01-2000
Julian Day	2451818.50
Ayanamsa	23 51 45
SVP	05♓14 40
☽ ☊ Mean	20♋34 R

● ◐ PHASES ○ ◑

Day	h m	Phase	Position
05	10:59	◐	12♑30
13	08:53	○	20♈19
20	08:00	◑	27♋14
27	07:59	●	04♏12

LAST ASPECT ☽ Day	h m	INGRESS Day	h m	Sign
30	22:43	01	22:51	♐
03	08:03	04	09:43	♑
05	12:34	06	22:33	♒
08	08:55	09	10:36	♓
11	01:09	11	19:51	♈
13	08:53	14	02:06	♉
16	06:17	16	06:19	♊
18	01:01	18	09:37	♋
20	12:16	20	12:43	♌
22	15:12	22	15:53	♍
24	18:35	24	19:31	♎
26	01:04	27	00:24	♏
29	06:04	29	07:41	♐
31	13:43	31	18:02	♑

DECLINATION

Day	☉	☽	☿	♀	♂	♃	♄	♅	♆	♇
01 Su	03S13	12S50	14S41	14S13	09N26	21N12	18N04	16S23	19S06	11S33
02 Mo	03 36	16 28	15 11	14 39	09 12	21 12	18 04	16 23	19 06	11 33
03 Tu	04 00	19 17	15 40	15 05	08 57	21 12	18 03	16 24	19 06	11 34
04 We	04 23	21 09	16 08	15 31	08 43	21 12	18 03	16 24	19 06	11 34
05 Th	04 46	22 03	16 34	15 56	08 29	21 12	18 02	16 24	19 06	11 35
06 Fr	05 09	21 56	16 59	16 21	08 15	21 11	18 01	16 25	19 07	11 35
07 Sa	05 32	20 51	17 23	16 46	08 01	21 11	18 01	16 25	19 07	11 36
08 Su	05 55	18 50	17 46	17 10	07 46	21 11	18 00	16 25	19 07	11 37
09 Mo	06 18	15 58	18 06	17 34	07 32	21 11	17 59	16 25	19 07	11 37
10 Tu	06 40	12 22	18 25	17 57	07 18	21 10	17 59	16 26	19 07	11 38
11 We	07 03	08 09	18 42	18 20	07 03	21 10	17 58	16 26	19 07	11 38
12 Th	07 26	03 29	18 58	18 42	06 49	21 09	17 57	16 26	19 07	11 39
13 Fr	07 48	01N27	19 11	19 04	06 35	21 09	17 56	16 26	19 07	11 39
14 Sa	08 10	06 26	19 22	19 26	06 20	21 09	17 56	16 26	19 07	11 40
15 Su	08 33	11 11	19 30	19 46	06 06	21 08	17 55	16 26	19 07	11 41
16 Mo	08 55	15 25	19 36	20 07	05 51	21 08	17 54	16 27	19 07	11 41
17 Tu	09 17	18 51	19 38	20 27	05 37	21 07	17 53	16 27	19 07	11 42
18 We	09 39	21 11	19 38	20 46	05 22	21 07	17 52	16 27	19 07	11 42
19 Th	10 00	22 13	19 35	21 05	05 07	21 06	17 52	16 27	19 07	11 43
20 Fr	10 22	21 50	19 27	21 23	04 53	21 06	17 51	16 27	19 07	11 43
21 Sa	10 43	20 05	19 16	21 41	04 38	21 05	17 50	16 27	19 07	11 44
22 Su	11 05	17 06	19 01	21 58	04 24	21 04	17 49	16 27	19 07	11 44
23 Mo	11 26	13 08	18 41	22 15	04 09	21 04	17 48	16 27	19 07	11 45
24 Tu	11 47	08 28	18 17	22 30	03 54	21 03	17 47	16 27	19 07	11 46
25 We	12 08	03 24	17 48	22 46	03 40	21 02	17 46	16 27	19 07	11 46
26 Th	12 28	01S46	17 16	23 00	03 25	21 01	17 45	16 27	19 07	11 47
27 Fr	12 49	06 46	16 39	23 15	03 10	21 01	17 44	16 27	19 07	11 47
28 Sa	13 09	11 22	15 59	23 28	02 56	21 00	17 43	16 27	19 07	11 48
29 Su	13 29	15 21	15 16	23 41	02 41	20 59	17 43	16 27	19 07	11 48
30 Mo	13 48	18 33	14 32	23 53	02 26	20 58	17 42	16 27	19 07	11 49
31 Tu	14 08	20 49	13 47	24 04	02 12	20 57	17 41	16 27	19 06	11 49

ASPECTARIAN

01	09:42	☽ ∥ ♀
	13:28	☽ ∥ ☿
	18:28	☿ □ ♆
	23:22	☽ ∥ ♅
02	00:04	☽ ☍ ♄
	06:12	☽ ⚹ ♆
	12:39	☽ # ♄
	17:47	☽ ⚹ ☉
	17:50	♀ ⚹ ♂
	17:55	☽ □ ♂
	19:23	☽ ☌ ♇
	20:29	☽ ☍ ♃
	22:17	☽ ∥ ♆
03	08:03	☽ ⚹ ♅
	14:13	☉ ⚹ ♇
	23:56	♂ □ ♇
04	00:50	☽ # ♃
	03:08	☉ △ ♃
	14:37	☿ ∥ ♅
	19:28	♂ □ ♃
05	00:40	☽ ⚹ ☿
	09:02	☽ △ ♂
	12:34	☽ ⚹ ♀
06	03:06	♀ ∥ ♅
	18:19	☽ # ♃
	23:28	☽ △ ♄
07	06:17	☽ ☌ ♆
	19:02	☽ □ ☿
	20:22	☽ ⚹ ♇
	21:04	☽ △ ♃
	21:09	☽ ∥ ♆
08	05:20	☽ △ ☉
	07:41	☽ # ♄
	08:19	☽ □ ♀
	08:41	☽ ∥ ☿
	08:55	☽ ☌ ♅
	11:00	☽ ∥ [illegible]
	14:21	♀ □ ♅
	16:08	☿ # ♄
	20:34	☽ ∥ ♅
09	11:16	☽ □ ♄
10	00:28	☉ △ ♅
	01:39	♀ # ♄
	04:24	☽ ∥ ♇
	07:39	☽ □ ♇
	08:04	☽ □ ♃
	10:34	☽ △ ☿
	15:18	☽ ☍ ♂
11	00:12	☉ # ♂
	01:09	☽ △ ♀
	05:22	☽ ∥ ☉
	06:06	☽ # ♂
	20:16	☽ ⚹ ♄
12	02:52	☽ ⚹ ♆
	15:52	☽ △ ♇
	15:57	☽ ⚹ ♃
	16:50	☿ ∥ ♆
13	02:53	☽ ⚹ ♅
	03:14	♀ ∥ ♆
	08:25	♃ ☍ ♇
	15:27	☿ ∥ ♀
	23:36	☽ ∥ ♂
14	08:47	☽ □ ♆
	09:23	☽ # ☉
15	02:40	☽ # ♇
	04:36	☽ ☍ ☿
	07:47	☽ □ ♅
	09:15	☽ △ ♂
	14:13	♆ SD
	23:26	☽ ☍ ♀
16	00:47	♄ ♉
	06:19	☽ ☌ ♄
	06:32	☽ # ♅
	12:51	☽ △ ♆
	16:36	☽ ∥ ♄
17	00:44	☽ ☌ ♃
	01:13	☽ ☍ ♇
	02:16	☽ # ♆
	06:56	☽ # ☿
	11:19	☽ △ ♅
	15:11	☽ □ ♂
	17:37	☽ # ♀
	22:59	☽ ∥ ♃
18	01:01	☽ △ ☉
	13:42	☿ SR
19	01:31	♀ # ♃
	02:46	♀ ☍ ♄
	06:18	♀ ♐
	12:26	☽ △ ☿
	20:40	☽ ⚹ ♂
20	06:47	☽ # ♀
	12:16	☽ ⚹ ♄
	12:25	☽ ∥ ♃
	15:35	☽ △ ♀
	19:11	☽ ☍ ♆
21	06:31	☽ ⚹ ♃
	07:42	☽ △ ♇
	08:07	☽ # ☿
	08:43	☽ # ♆
	14:34	☽ □ ☿
	14:51	☿ ∥ ♆
	17:31	☽ ☍ ♅
	18:50	☽ ∥ ♄
22	04:13	☽ # ♅
	09:31	♀ ⚹ ♆
	15:03	☽ ⚹ ☉
	15:12	☽ □ ♄
	23:37	☽ □ ♀
23	02:49	☉ ♏
	07:20	☽ # ♇
	08:23	☽ # ☉
	09:31	☽ □ ♃
	11:06	☽ □ ♇
	15:52	☽ ⚹ ☿
	22:32	☉ ∥ ♇
24	08:00	☽ ☌ ♂
	18:35	☽ △ ♄
	21:42	☽ ∥ ♂
25	01:44	☿ # ♄
	02:09	☽ △ ♆
	08:22	☽ ⚹ ♀
	13:09	☽ △ ♃
	15:11	☽ ⚹ ♇
26	01:04	☽ △ ♅
	07:23	☽ # ♂
	15:25	♅ SD
	22:55	☉ □ ♆
27	07:18	☿ ∥ ♅
	07:18	☽ □ ♆
	11:17	♀ ☍ ♃
	18:20	☽ ☌ ☿
28	02:21	☽ ∥ ♇
	07:10	☽ □ ♅
	11:03	☽ ∥ ☉
	15:33	♀ ☌ ♇
	23:30	☽ ∥ ☿
29	00:37	☽ ⚹ ♂
	06:04	☽ ☍ ♄
	07:31	☽ ∥ ♅
	14:59	☽ ⚹ ♆
	16:54	☽ # ♄
30	02:09	☿ ☌ ☉
	02:14	☽ ☍ ♃
	05:03	☽ ∥ ♆
	05:33	☽ ☌ ♇
	09:33	☽ ☌ ♀
	16:07	☿ ∥ ☉
	16:11	☽ ⚹ ♅
31	01:50	☽ # ♃
	13:43	☽ □ ♂

⚷ Chiron

01 Dec. 17 S 21

03	13♐03
06	13 16
09	13 31
12	13 46
15	14 01
18	14 17
21	14 34
24	14 51
27	15 08
30	15 26

November 2000

Day	S. T. h m s	☉ ° ' "	☽ ° ' "	☿ ° '	♀ ° '	♂ ° '	♃ ° '	♄ ° '	♅ ° '	♆ ° '	♇ ° '	☊ True ° '
01 We	02 42 20	08♏ 52 11	02♑ 59 01	04♏R34	15♐ 26	28♍ 06	09♊R31	28♉R57	16♒ 54	03♒ 52	11♐ 29	17♋R49
02 Th	02 46 17	09 52 15	14 54 03	03 26	16 38	28 43	09 25	28 53	16 55	03 53	11 31	17 48
03 Fr	02 50 14	10 52 20	26 42 51	02 25	17 51	29 20	09 19	28 48	16 55	03 53	11 33	17D 49
04 Sa	02 54 10	11 52 27	08♒ 30 35	01 33	19 03	29 57	09 12	28 44	16 55	03 54	11 35	17 50
05 Su	02 58 07	12 52 35	20 22 44	00 52	20 16	00♎ 34	09 06	28 39	16 56	03 54	11 38	17R 50
06 Mo	03 02 03	13 52 45	02♓ 24 47	00 22	21 28	01 11	08 59	28 34	16 56	03 55	11 40	17 49
07 Tu	03 05 60	14 52 56	14 41 46	00 03	22 40	01 48	08 52	28 30	16 57	03 56	11 42	17 46
08 We	03 09 56	15 53 09	27 17 54	29♎ 56	23 53	02 25	08 46	28 25	16 58	03 57	11 44	17 40
09 Th	03 13 53	16 53 24	10♈ 16 06	00♏D01	25 05	03 02	08 38	28 20	16 58	03 57	11 46	17 32
10 Fr	03 17 49	17 53 40	23 37 31	00 16	26 17	03 38	08 31	28 16	16 59	03 58	11 48	17 23
11 Sa	03 21 46	18 53 57	07♉ 21 13	00 41	27 29	04 15	08 24	28 11	17 00	03 59	11 50	17 13
12 Su	03 25 43	19 54 17	21 24 07	01 15	28 41	04 52	08 17	28 06	17 01	04 00	11 53	17 04
13 Mo	03 29 39	20 54 38	05♊ 41 26	01 58	29 53	05 29	08 09	28 01	17 01	04 01	11 55	16 56
14 Tu	03 33 36	21 55 00	20 07 19	02 47	01♑ 05	06 06	08 02	27 56	17 02	04 02	11 57	16 51
15 We	03 37 32	22 55 25	04♋ 35 54	03 43	02 17	06 42	07 54	27 52	17 03	04 03	11 59	16 47
16 Th	03 41 29	23 55 52	19 02 06	04 45	03 29	07 19	07 46	27 47	17 04	04 04	12 01	16 45
17 Fr	03 45 25	24 56 20	03♌ 22 05	05 52	04 41	07 56	07 39	27 42	17 05	04 05	12 04	16 45
18 Sa	03 49 22	25 56 50	17 33 25	07 03	05 53	08 33	07 31	27 37	17 06	04 06	12 06	16 45
19 Su	03 53 19	26 57 22	01♍ 34 50	08 17	07 04	09 09	07 23	27 32	17 08	04 07	12 08	16 44
20 Mo	03 57 15	27 57 56	15 25 41	09 35	08 16	09 46	07 15	27 27	17 09	04 08	12 10	16 42
21 Tu	04 01 12	28 58 32	29 05 39	10 55	09 28	10 23	07 07	27 22	17 10	04 10	12 13	16 38
22 We	04 05 08	29 59 09	12♎ 34 23	12 18	10 39	10 59	06 59	27 17	17 11	04 11	12 15	16 32
23 Th	04 09 05	00♐ 59 48	25 51 21	13 42	11 51	11 36	06 51	27 12	17 13	04 12	12 17	16 25
24 Fr	04 13 01	02 00 28	08♏ 55 51	15 08	13 02	12 12	06 43	27 08	17 14	04 13	12 20	16 17
25 Sa	04 16 58	03 01 10	21 47 10	16 36	14 14	12 49	06 34	27 03	17 16	04 15	12 22	16 09
26 Su	04 20 54	04 01 54	04♐ 24 53	18 04	15 25	13 25	06 26	26 58	17 17	04 16	12 24	16 01
27 Mo	04 24 51	05 02 39	16 49 03	19 33	16 36	14 02	06 18	26 53	17 19	04 17	12 26	15 56
28 Tu	04 28 47	06 03 25	29 00 34	21 04	17 47	14 38	06 10	26 48	17 20	04 19	12 29	15 52
29 We	04 32 44	07 04 12	11♑ 01 09	22 34	18 58	15 15	06 02	26 43	17 22	04 20	12 31	15 51
30 Th	04 36 41	08 05 01	22 53 27	24 05	20 09	15 51	05 53	26 39	17 24	04 21	12 33	15D 51

Data for	11-01-2000
Julian Day	2451849.50
Ayanamsa	23 51 49
SVP	05 ♓ 14 35
☽ ☊ Mean	18 ♋ 55 R

● ◐ PHASES ○ ◑

Day	h m	Phase	Position
04	07:26	◐	12♒11
11	21:15	○	19♉47
18	15:25	◑	26♌36
25	23:11	●	04♐00

LAST ASPECT ☽ INGRESS

Last Aspect Day	h m	Ingress Day	h m	Sign
03	05:36	03	06:40	♒
05	16:25	05	19:12	♓
08	02:04	08	05:02	♈
10	05:08	10	11:12	♉
12	11:13	12	14:28	♊
13	18:52	14	16:21	♋
16	14:31	16	18:19	♌
18	17:04	18	21:16	♍
20	23:45	21	01:35	♎
22	08:18	23	07:33	♏
25	09:52	25	15:33	♐
27	00:57	28	01:57	♑
30	07:34	30	14:26	♒

DECLINATION

Day	☉	☽	☿	♀	♂	♃	♄	♅	♆	♇
01 We	14S 27	22S 06	13S 03	24S 15	01N57	20N57	17N40	16S 27	19S 06	11S 50
02 Th	14 46	22 21	12 21	24 25	01 42	20 56	17 39	16 27	19 06	11 50
03 Fr	15 05	21 35	11 42	24 35	01 28	20 55	17 38	16 26	19 06	11 51
04 Sa	15 24	19 52	11 08	24 43	01 13	20 54	17 37	16 26	19 06	11 51
05 Su	15 42	17 18	10 38	24 51	00 58	20 53	17 36	16 26	19 06	11 52
06 Mo	16 00	13 57	10 14	24 59	00 44	20 52	17 35	16 26	19 06	11 52
07 Tu	16 18	09 58	09 56	25 05	00 29	20 51	17 34	16 26	19 05	11 53
08 We	16 36	05 27	09 43	25 11	00 14	20 50	17 33	16 26	19 05	11 53
09 Th	16 53	00 35	09 36	25 16	00S 00	20 49	17 32	16 25	19 05	11 54
10 Fr	17 10	04N28	09 35	25 21	00 15	20 48	17 30	16 25	19 05	11 54
11 Sa	17 26	09 27	09 38	25 24	00 30	20 47	17 29	16 25	19 05	11 55
12 Su	17 43	14 03	09 46	25 27	00 44	20 45	17 28	16 24	19 05	11 55
13 Mo	17 59	17 56	09 59	25 29	00 59	20 44	17 27	16 24	19 04	11 56
14 Tu	18 15	20 46	10 15	25 31	01 13	20 43	17 26	16 24	19 04	11 56
15 We	18 30	22 16	10 33	25 32	01 28	20 42	17 25	16 24	19 04	11 57
16 Th	18 45	22 17	10 55	25 32	01 42	20 41	17 24	16 23	19 04	11 57
17 Fr	19 00	20 50	11 19	25 31	01 57	20 40	17 23	16 23	19 03	11 58
18 Sa	19 15	18 05	11 45	25 30	02 11	20 39	17 22	16 23	19 03	11 58
19 Su	19 29	14 18	12 12	25 27	02 26	20 37	17 21	16 22	19 03	11 59
20 Mo	19 42	09 48	12 40	25 24	02 40	20 36	17 20	16 22	19 03	11 59
21 Tu	19 56	04 51	13 10	25 21	02 54	20 35	17 19	16 21	19 02	12 00
22 We	20 09	00S 14	13 40	25 16	03 09	20 34	17 18	16 21	19 02	12 00
23 Th	20 21	05 15	14 10	25 11	03 23	20 32	17 17	16 20	19 02	12 00
24 Fr	20 34	09 58	14 41	25 05	03 37	20 31	17 16	16 20	19 02	12 01
25 Sa	20 46	14 09	15 12	24 59	03 51	20 30	17 15	16 20	19 01	12 01
26 Su	20 57	17 38	15 43	24 52	04 06	20 29	17 14	16 19	19 01	12 02
27 Mo	21 08	20 16	16 14	24 44	04 20	20 27	17 13	16 19	19 01	12 02
28 Tu	21 19	21 55	16 44	24 35	04 34	20 26	17 12	16 18	19 00	12 02
29 We	21 29	22 32	17 14	24 26	04 48	20 25	17 11	16 17	19 00	12 03
30 Th	21 39	22 07	17 43	24 15	05 02	20 23	17 10	16 17	19 00	12 03

ASPECTARIAN

01 02:53 ☽ ✶ ☿
12:53 ☽ ✶ ☉
14:29 ☿ □ ♆

02 05:25 ♀ ✶ ♅
05:42 ♂ △ ♄
18:26 ☿ ∥ ♇

03 04:13 ☽ △ ♄
05:38 ☽ △ ♂
10:46 ☽ □ ☿
11:05 ☽ # ♃
14:36 ☽ ☌ ♆

04 01:25 ☽ △ ♃
02:00 ♂ ♎
06:17 ☽ ✶ ♇
08:04 ☽ ∥ ♆
17:04 ☽ ☌ ♅
21:33 ☽ # ♄
23:45 ☽ ✶ ♀

05 06:42 ☽ ∥ ♅
11:04 ☽ ∥ ☉
16:27 ☽ □ ♄
20:04 ☽ △ ☿

06 12:47 ☽ □ ♃
12:55 ☽ ∥ ♇
18:11 ☽ □ ♆

07 00:11 ☽ ∥ ☿
00:24 ☽ △ ☉
07:28 ☿ ♎ R
10:19 ☉ ∥ ♅
16:51 ☽ □ ♀

08 02:04 ☽ ✶ ♄
02:29 ☿ SD
10:00 ☽ ☍ ♂
12:23 ☽ ✶ ♆
21:02 ☽ ✶ ♃
21:42 ☿ ♏

09 01:59 ☉ □ ♅
02:34 ☽ ∥ ♂
02:43 ☽ △ ♇
02:53 ☽ # ♂
12:07 ☽ ✶ ♅

10 05:08 ☽ △ ♀
12:00 ☽ ☍ ☿
13:15 ♂ △ ♆
18:09 ☽ □ ♆

11 00:57 ☽ # ☿
04:02 ☉ # ♄
12:31 ☽ # ♇
16:31 ☽ □ ♅

12 11:13 ☽ ☌ ♄
13:51 ☽ # ♅
20:41 ☽ ∥ ♄
21:11 ☽ △ ♆
23:37 ☽ △ ♂

13 00:20 ☽ # ☉
02:14 ♀ ♑
04:03 ☽ ☌ ♃
08:26 ☽ # ♆
10:23 ☽ ☍ ♇
18:52 ☽ △ ♅
23:25 ☽ ∥ ♃

14 19:48 ☽ ☍ ♀
22:25 ☽ △ ☿

15 03:38 ☽ □ ♂
07:58 ☿ □ ♆

16 08:46 ☽ △ ☉
14:31 ☽ ✶ ♄
14:38 ♂ △ ♃

17 01:11 ☽ ☍ ♆
01:52 ☽ ∥ ♃
04:33 ☽ □ ☿
05:34 ☉ ∥ ♆
07:07 ☽ ✶ ♃
08:01 ☽ ✶ ♂
14:42 ☽ △ ♇
15:43 ☽ # ☉
16:33 ☽ # ♆
23:13 ☽ ☍ ♅

18 04:58 ☽ ∥ ♄
11:30 ☽ # ♅
12:15 ☿ ∥ ♇
17:04 ☽ □ ♄

19 09:54 ☽ □ ♃
10:22 ☽ △ ♀
10:27 ☽ # ☿
12:42 ☽ # ♇
12:43 ☉ ☍ ♄
12:45 ☽ ✶ ☿
18:18 ☽ □ ♇

20 20:57 ☽ △ ♄
23:45 ☽ ✶ ☉

21 08:473 ☽ # ♂
08:58 ☽ △ ♆
14:05 ☽ △ ♃
20:12 ☽ □ ♀
21:00 ☽ ☌ ♂
23:24 ☽ ✶ ♇

22 00:19 ☉ ♐
08:18 ☽ △ ♅
13:43 ♀ □ ♂
14:25 ☽ ∥ ♂

23 15:17 ☽ □ ♆
19:18 ☉ # ♃

24 05:05 ♂ ✶ ♇
08:23 ☽ ✶ ♀
11:17 ☽ ∥ ♇
13:01 ☽ ☌ ☿
15:28 ☽ □ ♅

25 07:44 ☽ ∥ ☿
09:52 ☽ ☍ ♄
11:02 ☿ □ ♅
14:13 ☽ ∥ ♅
20:51 ☽ # ♄
23:43 ☽ ✶ ♆

26 03:51 ☽ ☍ ♃
05:38 ☉ ✶ ♆
11:28 ☽ ∥ ♆
15:28 ☽ ☌ ♇
18:18 ☽ ✶ ♂

27 00:57 ☽ ✶ ♅
02:04 ☽ # ♃
03:49 ☿ ∥ ♅
12:06 ☽ ∥ ☉

28 02:14 ☉ ☍ ♃
21:43 ☿ # ♄

29 08:57 ☽ □ ♂
17:49 ☽ ☌ ♀

30 02:47 ☽ ✶ ☿
07:34 ☽ △ ♄
08:51 ☽ ∥ ☉
23:22 ☽ ☌ ♆

⚷ Chiron

01 Dec.	17 S 46
02	15♐45
05	16 03
08	16 23
11	16 42
14	17 02
17	17 22
20	17 42
23	18 02
26	18 23
29	18 43

December 2000

25 17:36 04♑14 ☉ Partial Solar Eclipse (mag 0.723)

Day	S. T. h m s	☉ ° ' "	☽ ° ' "	☿ ° '	♀ ° '	♂ ° '	♃ ° '	♄ ° '	♅ ° '	♆ ° '	♇ ° '	☊ True ° '
01 Fr	04 40 37	09♐ 05 50	04♒ 40 55	25♏ 37	21♑ 20	16♎ 27	05♊ R45	26♉ R34	17♒ 25	04♒ 23	12♐ 36	15♋ 53
02 Sa	04 44 34	10 06 41	16 27 45	27 09	22 31	17 04	05 37	26 29	17 27	04 24	12 38	15 56
03 Su	04 48 30	11 07 32	28 18 38	28 41	23 42	17 40	05 29	26 25	17 29	04 26	12 40	15 58
04 Mo	04 52 27	12 08 24	10♓ 18 41	00♐ 14	24 53	18 16	05 21	26 20	17 31	04 27	12 43	15 59
05 Tu	04 56 23	13 09 17	22 33 02	01 46	26 03	18 52	05 13	26 15	17 33	04 29	12 45	15R 59
06 We	05 00 20	14 10 11	05♈ 06 30	03 19	27 14	19 29	05 05	26 11	17 35	04 31	12 47	15 58
07 Th	05 04 17	15 11 05	18 03 08	04 52	28 24	20 05	04 57	26 06	17 37	04 32	12 50	15 55
08 Fr	05 08 13	16 12 00	01♉ 25 37	06 25	29 34	20 41	04 49	26 02	17 39	04 34	12 52	15 52
09 Sa	05 12 10	17 12 56	15 14 40	07 58	00♒ 44	21 17	04 41	25 58	17 41	04 36	12 54	15 48
10 Su	05 16 06	18 13 53	29 28 33	09 31	01 54	21 53	04 33	25 53	17 43	04 37	12 57	15 43
11 Mo	05 20 03	19 14 50	14♊ 03 03	11 04	03 04	22 29	04 26	25 49	17 45	04 39	12 59	15 40
12 Tu	05 23 59	20 15 49	28 51 38	12 38	04 14	23 05	04 18	25 45	17 47	04 41	13 01	15 38
13 We	05 27 56	21 16 48	13♋ 46 32	14 11	05 24	23 41	04 11	25 41	17 49	04 42	13 04	15 36
14 Th	05 31 52	22 17 48	28 39 41	15 45	06 33	24 17	04 03	25 36	17 52	04 44	13 06	15 36
15 Fr	05 35 49	23 18 50	13♌ 23 59	17 18	07 43	24 53	03 56	25 32	17 54	04 46	13 08	15D 36
16 Sa	05 39 46	24 19 52	27 54 01	18 52	08 52	25 29	03 49	25 28	17 56	04 48	13 10	15 36
17 Su	05 43 42	25 20 55	12♍ 06 23	20 26	10 01	26 04	03 42	25 24	17 59	04 50	13 13	15R 37
18 Mo	05 47 39	26 21 59	25 59 33	22 00	11 10	26 40	03 35	25 21	18 01	04 52	13 15	15 36
19 Tu	05 51 35	27 23 03	09♎ 33 30	23 34	12 19	27 16	03 28	25 17	18 04	04 53	13 17	15 35
20 We	05 55 32	28 24 09	22 49 12	25 08	13 28	27 52	03 21	25 13	18 06	04 55	13 20	15 33
21 Th	05 59 28	29 25 16	05♏ 48 12	26 42	14 36	28 27	03 14	25 10	18 09	04 57	13 22	15 31
22 Fr	06 03 25	00♑ 26 23	18 32 14	28 17	15 44	29 03	03 08	25 06	18 11	04 59	13 24	15 29
23 Sa	06 07 21	01 27 31	01♐ 03 02	29 52	16 53	29 38	03 02	25 03	18 14	05 01	13 26	15 26
24 Su	06 11 18	02 28 39	13 22 14	01♑ 27	18 01	00♏ 14	02 55	24 59	18 17	05 03	13 29	15 24
25 Mo	06 15 15	03 29 49	25 31 21	03 02	19 09	00 49	02 49	24 56	18 19	05 05	13 31	15 23
26 Tu	06 19 11	04 30 58	07♑ 32 01	04 38	20 16	01 25	02 43	24 53	18 22	05 07	13 33	15D 23
27 We	06 23 08	05 32 08	19 25 56	06 13	21 24	02 00	02 38	24 50	18 25	05 09	13 35	15 24
28 Th	06 27 04	06 33 18	01♒ 15 11	07 49	22 31	02 35	02 32	24 47	18 28	05 11	13 37	15 25
29 Fr	06 31 01	07 34 28	13 02 12	09 26	23 38	03 11	02 27	24 44	18 30	05 13	13 40	15 26
30 Sa	06 34 57	08 35 38	24 49 53	11 02	24 45	03 46	02 21	24 41	18 33	05 15	13 42	15 28
31 Su	06 38 54	09 36 48	06♓ 41 40	12 39	25 52	04 21	02 16	24 38	18 36	05 18	13 44	15 29

Data for	12-01-2000
Julian Day	2451879.50
Ayanamsa	23 51 53
SVP	05 ♓ 14 32
☽ ☊ Mean	17 ♋ 20 R

● ◐ PHASES ○ ◑

Day	h m	Phase	Position
04	03:56	◐	12♓18
11	09:04	○	19♊38
18	00:41	◑	26♍24
25	17:22	☉	04♑14

LAST ASPECT Day	h m	☽ INGRESS Day	h m	Sign
03	00:51	03	03:23	♓
05	07:27	05	14:18	♈
07	20:23	07	21:28	♉
09	18:01	10	00:51	♊
11	14:15	12	01:49	♋
13	19:04	14	02:09	♌
15	19:57	16	03:30	♍
18	00:41	18	07:01	♎
20	11:07	20	13:12	♏
22	12:28	22	21:57	♐
24	10:03	25	08:54	♑
27	10:52	27	21:26	♒
29	23:48	30	10:28	♓

DECLINATION

Day	☉	☽	☿	♀	♂	♃	♄	♅	♆	♇
01 Fr	21S 49	20S 43	18S 12	24S 05	05S 16	20N22	17N09	16S 16	18S 59	12S 04
02 Sa	21 58	18 26	18 41	23 53	05 30	20 21	17 08	16 16	18 59	12 04
03 Su	22 07	15 22	19 08	23 41	05 44	20 20	17 07	16 15	18 59	12 04
04 Mo	22 15	11 38	19 35	23 29	05 57	20 18	17 06	16 15	18 58	12 05
05 Tu	22 23	07 21	20 01	23 15	06 11	20 17	17 05	16 14	18 58	12 05
06 We	22 30	02 41	20 26	23 01	06 25	20 16	17 04	16 13	18 58	12 06
07 Th	22 37	02N15	20 50	22 47	06 39	20 15	17 04	16 13	18 57	12 06
08 Fr	22 43	07 15	21 14	22 32	06 52	20 13	17 03	16 12	18 57	12 06
09 Sa	22 49	12 03	21 36	22 16	07 06	20 12	17 02	16 11	18 57	12 07
10 Su	22 55	16 20	21 57	22 00	07 19	20 11	17 01	16 11	18 56	12 07
11 Mo	23 00	19 45	22 18	21 43	07 33	20 10	17 00	16 10	18 56	12 07
12 Tu	23 05	21 55	22 37	21 25	07 46	20 08	16 59	16 09	18 55	12 08
13 We	23 09	22 33	22 55	21 07	07 59	20 07	16 58	16 09	18 55	12 08
14 Th	23 13	21 35	23 12	20 49	08 13	20 06	16 58	16 08	18 55	12 08
15 Fr	23 16	19 09	23 28	20 29	08 26	20 05	16 57	16 07	18 54	12 08
16 Sa	23 19	15 31	23 42	20 10	08 39	20 04	16 56	16 06	18 54	12 09
17 Su	23 21	11 04	23 56	19 50	08 52	20 03	16 55	16 06	18 53	12 09
18 Mo	23 23	06 08	24 08	19 29	09 05	20 01	16 55	16 05	18 53	12 09
19 Tu	23 25	01 02	24 19	19 08	09 18	20 00	16 54	16 04	18 52	12 10
20 We	23 26	04S 01	24 29	18 47	09 31	19 59	16 53	16 03	18 52	12 10
21 Th	23 26	08 47	24 37	18 25	09 43	19 58	16 53	16 03	18 52	12 10
22 Fr	23 26	13 05	24 44	18 02	09 56	19 57	16 52	16 02	18 51	12 10
23 Sa	23 26	16 45	24 50	17 39	10 09	19 56	16 51	16 01	18 51	12 11
24 Su	23 25	19 37	24 54	17 16	10 21	19 55	16 51	16 00	18 50	12 11
25 Mo	23 24	21 33	24 57	16 52	10 34	19 54	16 50	15 59	18 50	12 11
26 Tu	23 22	22 29	24 59	16 28	10 46	19 53	16 50	15 58	18 49	12 11
27 We	23 19	22 23	24 59	16 04	10 59	19 52	16 49	15 58	18 49	12 11
28 Th	23 17	21 17	24 58	15 39	11 11	19 52	16 49	15 57	18 48	12 12
29 Fr	23 13	19 15	24 55	15 14	11 23	19 51	16 48	15 56	18 48	12 12
30 Sa	23 10	16 25	24 51	14 49	11 35	19 50	16 48	15 55	18 47	12 12
31 Su	23 05	12 53	24 45	14 23	11 47	19 49	16 47	15 54	18 47	12 12

ASPECTARIAN

01 02:09 ☽ △ ♃
04:24 ☽ ⋕ ♃
09:50 ☽ ⚹ ☉
14:07 ☿ ☍ ♄
16:10 ☽ ⚹ ♇
18:53 ☽ ∥ ♆
22:06 ☽ ∥ ☿
02 01:16 ☽ △ ♂
02:00 ☽ ☌ ♅
10:52 ☽ ⋕ ♄
15:50 ☿ ∥ ♆
16:18 ♂ △ ♅
17:33 ☽ ∥ ♅
20:10 ☽ □ ♄
03 00:51 ☽ □ ☿
14:13 ☽ □ ♃
20:26 ☿ ♐
21:17 ☽ ∥ ♇
04 04:45 ☽ □ ♇
14:04 ☉ ☌ ♇
05 03:56 ♀ △ ♄
05:53 ☽ ∥ ♂
07:06 ☽ ⚹ ♄
07:27 ☽ ⚹ ♀
14:29 ☿ ⋕ ♃
20:08 ☽ △ ☿
22:51 ☽ ⚹ ♆
23:56 ☽ ⚹ ♃
06 14:21 ☽ △ ♇
18:17 ☽ △ ☉
18:50 ☿ ⚹ ♆
23:10 ☽ ⚹ ♅
07 01:10 ☿ ☍ ♃
03:51 ☽ ☍ ♂
11:02 ♀ ∥ ☉
20:23 ☽ □ ♀
22:05 ☽ ⋕ ♂

08 05:31 ☽ □ ♆
08:48 ♀ ♒
09 00:19 ☽ ⋕ ♇
04:09 ☽ □ ♅
11:20 ☉ ⚹ ♅
14:16 ♃ △ ♆
18:01 ☽ ☌ ♄
23:00 ☽ ⋕ ♅
10 01:24 ☿ ∥ ♀
04:11 ☽ ∥ ♄
04:22 ☽ △ ♀
08:20 ☽ ☌ ♃
08:32 ☽ △ ♆
17:26 ☽ ⋕ ♆
18:32 ☽ ☍ ☿
22:14 ☽ ☍ ♇
11 03:29 ☽ ∥ ♃
06:01 ☽ △ ♅
14:18 ☽ △ ♂
17:47 ☽ ⋕ ♀
12 01:16 ♀ △ ♃
06:10 ☿ ☌ ♇
09:24 ♀ ☌ ♆
13 16:36 ☽ □ ♂
19:04 ☽ ⚹ ♄
14 01:42 ☿ ∥ ☉
08:39 ☽ ⚹ ♃
09:51 ☽ ☍ ♆
10:46 ☽ ⋕ ♀
13:53 ☽ ☍ ♀
16:16 ☽ ∥ ♃
23:33 ☽ △ ♇
15 01:55 ☽ ⋕ ♆

07:10 ☽ △ ☿
07:24 ☽ ☍ ♅
09:26 ☿ ⚹ ♅
15:25 ☽ ∥ ♄
17:35 ☽ △ ☉
19:46 ☽ ⚹ ♂
19:57 ☽ □ ♄
20:33 ☽ ⋕ ♅
16 07:53 ♀ ⋕ ♃
09:49 ☽ □ ♃
18:31 ☽ ⋕ ♇
17 01:53 ☽ □ ♇
10:30 ☽ ⋕ ♂
16:09 ☽ □ ☿
22:51 ☽ △ ♄
18 13:13 ☽ △ ♃
15:39 ☽ △ ♆
17:10 ☉ ⚹ ♂
19 05:25 ☽ △ ♀
06:42 ☽ ⚹ ♇
15:22 ☽ △ ♅
17:48 ♀ ∥ ♆
21:08 ♀ ⚹ ♇
20 04:48 ☽ ⚹ ☿
09:41 ☽ ☌ ♂
11:07 ☽ ⚹ ☉
22:24 ☽ □ ♆
21 05:15 ☽ ∥ ♂
13:37 ☉ ♑
18:09 ☽ □ ♀
18:37 ☽ ∥ ♇
23:19 ☽ □ ♅
22 12:28 ☽ ☍ ♄

18:31 ☿ ⚹ ♂
18:49 ☽ ∥ ♅
23 00:49 ☽ ⋕ ♄
02:03 ☿ ♑
03:48 ☽ ☍ ♃
06:06 ☽ ∥ ♀
07:42 ☽ ⚹ ♆
14:37 ♂ ♏
16:44 ☽ ∥ ♆
24 00:11 ☽ ☌ ♇
03:05 ☽ ⋕ ♃
05:52 ♀ ☌ ♅
09:40 ☽ ⚹ ♅
10:03 ☽ ⚹ ♀

25 02:01 ♀ ⋕ ♄
11:05 ☽ ⚹ ♂
17:15 ☽ ☌ ☿
19:23 ☿ ☌ ☉
27 06:26 ♀ ∥ ♅
10:52 ☽ △ ♄
28 02:34 ☽ △ ♃
02:51 ☽ □ ♂
08:01 ☽ ☌ ♆
17:57 ☽ ⋕ ♃
29 01:15 ☽ ⚹ ♇
04:22 ☽ ∥ ♆
11:10 ☽ ☌ ♅

21:04 ☽ ⋕ ♄
22:35 ♀ □ ♄
23:41 ☽ □ ♄
23:48 ☽ ☌ ♀
30 03:41 ☽ ∥ ♅
13:01 ☽ ∥ ♀
15:07 ☽ □ ♃
19:01 ☽ △ ♂
31 04:16 ☽ ∥ ♇
06:24 ☽ ⚹ ☉
06:29 ☽ ∥ ♂
13:48 ☽ ⚹ ☿
14:08 ☽ □ ♇

⚷ Chiron

01 Dec. 18 S 05

Day	Position
02	19♐04
05	19 25
08	19 46
11	20 07
14	20 28
17	20 49
20	21 10
23	21 31
26	21 51
29	22 12

09 20:24 20♋21 ✷ Total Lunar Eclipse (mag 1.195)

January 2001

Day	S. T.	☉	☽	☿	♀	♂	♃	♄	♅	♆	♇	☊ True
	h m s	° ' "	° ' "	° '	° '	° '	° '	° '	° '	° '	° '	° '
01 Mo	06 42 50	10♑37 57	18♓41 27	14♑16	26♒58	04♏56	02♊R11	24♉R35	18♒39	05♒20	13♐46	15♋31
02 Tu	06 46 47	11 39 09	00♈53 56	15 54	28 04	05 31	02 07	24 33	18 42	05 22	13 48	15 32
03 We	06 50 44	12 40 19	13 22 40	17 32	29 10	06 06	02 02	24 30	18 45	05 24	13 50	15 32
04 Th	06 54 40	13 41 28	26 12 23	19 10	00♓16	06 41	01 58	24 28	18 48	05 26	13 52	15 33
05 Fr	06 58 37	14 42 37	09♉26 42	20 48	01 21	07 16	01 54	24 26	18 51	05 28	13 54	15 33
06 Sa	07 02 33	15 43 45	23 08 04	22 27	02 26	07 51	01 50	24 24	18 54	05 30	13 56	15 33
07 Su	07 06 30	16 44 54	07♊17 00	24 06	03 31	08 26	01 46	24 22	18 57	05 33	13 58	15R 33
08 Mo	07 10 26	17 46 02	21 51 31	25 45	04 36	09 01	01 42	24 20	19 00	05 35	14 00	15 33
09 Tu	07 14 23	18 47 09	06♋46 49	27 25	05 40	09 35	01 39	24 18	19 03	05 37	14 02	15 32
10 We	07 18 19	19 48 17	21 55 27	29 04	06 44	10 10	01 36	24 16	19 06	05 39	14 04	15 32
11 Th	07 22 16	20 49 24	07♌08 11	00♒44	07 47	10 44	01 32	24 15	19 09	05 41	14 06	15 31
12 Fr	07 26 13	21 50 30	22 15 17	02 24	08 51	11 19	01 30	24 13	19 13	05 44	14 08	15 30
13 Sa	07 30 09	22 51 37	07♍08 00	04 04	09 54	11 53	01 27	24 12	19 16	05 46	14 10	15 28
14 Su	07 34 06	23 52 43	21 39 40	05 43	10 56	12 28	01 25	24 10	19 19	05 48	14 12	15 27
15 Mo	07 38 02	24 53 50	05♎46 22	07 23	11 58	13 02	01 22	24 09	19 22	05 50	14 14	15 25
16 Tu	07 41 59	25 54 56	19 26 51	09 02	13 00	13 36	01 20	24 08	19 25	05 52	14 16	15 24
17 We	07 45 55	26 56 02	02♏42 08	10 40	14 02	14 11	01 18	24 07	19 29	05 55	14 18	15 23
18 Th	07 49 52	27 57 07	15 34 45	12 18	15 03	14 45	01 17	24 06	19 32	05 57	14 20	15D 23
19 Fr	07 53 48	28 58 13	28 08 08	13 55	16 03	15 19	01 15	24 06	19 35	05 59	14 21	15 24
20 Sa	07 57 45	29 59 18	10♐26 00	15 30	17 03	15 53	01 14	24 05	19 39	06 02	14 23	15 25
21 Su	08 01 42	01♒00 23	22 31 57	17 04	18 03	16 27	01 13	24 04	19 42	06 04	14 25	15 27
22 Mo	08 05 38	02 01 27	04♑29 19	18 35	19 02	17 00	01 12	24 04	19 45	06 06	14 27	15 28
23 Tu	08 09 35	03 02 31	16 21 03	20 04	20 01	17 34	01 12	24 04	19 49	06 08	14 28	15 29
24 We	08 13 31	04 03 34	28 09 40	21 30	20 59	18 08	01 11	24 04	19 52	06 11	14 30	15R 29
25 Th	08 17 28	05 04 36	09♒57 25	22 53	21 57	18 41	01 11	24D 04	19 55	06 13	14 32	15 28
26 Fr	08 21 24	06 05 37	21 46 19	24 11	22 54	19 15	01D 11	24 04	19 59	06 15	14 33	15 26
27 Sa	08 25 21	07 06 38	03♓38 20	25 25	23 51	19 48	01 12	24 04	20 02	06 17	14 35	15 24
28 Su	08 29 18	08 07 37	15 35 35	26 32	24 47	20 22	01 12	24 04	20 06	06 20	14 37	15 21
29 Mo	08 33 14	09 08 35	27 40 25	27 34	25 43	20 55	01 13	24 04	20 09	06 22	14 38	15 18
30 Tu	08 37 11	10 09 32	09♈55 37	28 28	26 38	21 28	01 13	24 05	20 13	06 24	14 40	15 15
31 We	08 41 07	11 10 28	22 24 18	29 14	27 32	22 01	01 15	24 06	20 16	06 27	14 41	15 13

Data for	01-01-2001
Julian Day	2451910.50
Ayanamsa	23 51 59
SVP	05 ♓ 14 29
☽ ☊ Mean	15 ♋ 41 R

● ◐ PHASES ○ ◑

02	22:32	◐	12♈37
09	20:24	✷	19♋39
16	12:35	◑	26♎27
24	13:08	●	04♒37

LAST ASPECT Day	h m	☽ INGRESS Day	h m	
01	11:38	01	22:16	♈
03	10:11	04	06:58	♉
06	02:11	06	11:45	♊
07	19:21	08	13:10	♋
10	12:40	10	12:45	♌
12	03:09	12	12:27	♍
14	04:14	14	14:06	♎
16	12:36	16	19:04	♏
19	01:46	19	03:37	♐
20	18:20	21	14:59	♑
23	15:41	24	03:45	♒
26	05:30	26	16:40	♓
28	19:49	29	04:36	♈
31	13:37	31	14:22	♉

DECLINATION

Day	☉	☽	☿	♀	♂	♃	♄	♅	♆	♇
01 Mo	23S01	08S49	24S38	13S57	11S59	19N48	16N47	15S53	18S46	12S12
02 Tu	22 56	04 21	24 30	13 31	12 11	19 48	16 47	15 52	18 46	12 13
03 We	22 50	00N24	24 19	13 04	12 22	19 47	16 46	15 51	18 45	12 13
04 Th	22 44	05 16	24 08	12 37	12 34	19 46	16 46	15 50	18 45	12 13
05 Fr	22 38	10 03	23 54	12 10	12 46	19 46	16 46	15 49	18 44	12 13
06 Sa	22 31	14 29	23 39	11 43	12 57	19 45	16 45	15 48	18 44	12 13
07 Su	22 23	18 17	23 23	11 15	13 08	19 45	16 45	15 47	18 43	12 13
08 Mo	22 16	21 04	23 05	10 47	13 20	19 44	16 45	15 46	18 43	12 13
09 Tu	22 07	22 28	22 46	10 20	13 31	19 44	16 45	15 45	18 42	12 14
10 We	21 59	22 14	22 25	09 51	13 42	19 43	16 45	15 44	18 42	12 14
11 Th	21 50	20 23	22 02	09 23	13 53	19 43	16 45	15 43	18 41	12 14
12 Fr	21 40	17 05	21 38	08 55	14 04	19 43	16 44	15 42	18 41	12 14
13 Sa	21 30	12 45	21 12	08 26	14 14	19 42	16 44	15 41	18 40	12 14
14 Su	21 20	07 45	20 45	07 58	14 25	19 42	16 44	15 40	18 40	12 14
15 Mo	21 09	02 30	20 16	07 29	14 36	19 42	16 44	15 39	18 39	12 14
16 Tu	20 58	02S43	19 46	07 00	14 46	19 42	16 44	15 38	18 38	12 14
17 We	20 46	07 40	19 15	06 31	14 57	19 42	16 44	15 37	18 38	12 14
18 Th	20 34	12 07	18 43	06 02	15 07	19 42	16 44	15 36	18 37	12 14
19 Fr	20 22	15 57	18 09	05 33	15 17	19 42	16 45	15 35	18 37	12 14
20 Sa	20 09	19 00	17 34	05 04	15 27	19 42	16 45	15 34	18 36	12 14
21 Su	19 56	21 10	16 58	04 35	15 37	19 42	16 45	15 33	18 36	12 14
22 Mo	19 43	22 22	16 22	04 06	15 47	19 42	16 45	15 32	18 35	12 14
23 Tu	19 29	22 31	15 45	03 37	15 57	19 42	16 45	15 31	18 35	12 14
24 We	19 14	21 40	15 08	03 08	16 06	19 42	16 45	15 30	18 34	12 14
25 Th	19 00	19 51	14 30	02 39	16 16	19 42	16 46	15 29	18 33	12 14
26 Fr	18 45	17 12	13 53	02 10	16 25	19 42	16 46	15 28	18 33	12 14
27 Sa	18 30	13 49	13 16	01 41	16 35	19 43	16 46	15 27	18 32	12 14
28 Su	18 14	09 53	12 40	01 12	16 44	19 43	16 47	15 26	18 32	12 14
29 Mo	17 58	05 31	12 05	00 43	16 53	19 43	16 47	15 24	18 31	12 14
30 Tu	17 42	00 52	11 32	00 15	17 02	19 44	16 47	15 23	18 31	12 14
31 We	17 25	03N54	11 01	00N14	17 11	19 44	16 48	15 22	18 30	12 14

ASPECTARIAN

01 11:38 ☽ ✶ ♄
17:02 ♂ □ ♆
02 02:21 ☽ ✶ ♃
04:01 ♂ ∥ ♇
08:42 ☽ ✶ ♆
03 00:53 ☽ △ ♇
09:00 ☽ □ ☿
10:11 ☽ ✶ ♅
18:15 ♀ ♓

04 01:56 ♀ ∥ ♂
08:07 ☽ ✶ ♀
16:52 ☽ □ ♆
19:57 ☽ ☍ ♂
21:21 ♀ ∥ ♇
05 10:06 ☽ △ ☉
10:10 ☽ # ♀
11:15 ♀ □ ♃
11:28 ☽ # ♇
15:02 ☽ # ♂
16:39 ☽ □ ♅
22:40 ☽ △ ☿
06 02:11 ☽ ☌ ♄
07:45 ☽ # ♅
13:42 ☽ ∥ ♄
14:47 ☽ ☌ ♃
17:11 ☽ □ ♀
21:06 ☽ △ ♆

07 03:11 ☽ # ♆
03:47 ☿ △ ♄
11:08 ☽ ☍ ♇
11:18 ☽ ∥ ♃
19:21 ☽ △ ♅
08 16:07 ☽ # ☉
22:06 ☽ △ ♀

09 04:40 ☽ △ ♂

10 03:43 ☽ ✶ ♄
05:42 ☽ # ☉
12:40 ☽ ☍ ☿
13:27 ☿ ♒
15:13 ☽ ✶ ♃
21:43 ☽ ☍ ♆
11 05:42 ☽ ∥ ♃
05:56 ☽ □ ♂
11:04 ☽ △ ♇
11:22 ☿ △ ♃
13:32 ☽ # ♆
19:08 ☽ ☍ ♅
20:32 ☿ ∥ ☉
12 02:09 ☽ ∥ ♄
03:09 ☽ □ ♄
08:14 ☽ # ♅
14:49 ☽ □ ♃
16:34 ☽ # ♂

13 02:36 ☽ # ♇
04:52 ☽ ☍ ♀
08:07 ☽ ✶ ♂
11:35 ☽ □ ♇
22:58 ☽ # ♀
14 01:11 ☿ ☌ ♆
04:01 ☽ △ ☉
04:14 ☽ △ ♄
06:50 ☉ △ ♄
16:28 ☽ △ ♃
15 00:08 ☽ △ ♆
03:10 ☽ △ ☿
14:49 ☽ ✶ ♇
23:58 ☽ △ ♅

16 03:42 ☿ # ♃
18:50 ☽ ∥ ♀
17 05:57 ☽ □ ♆
06:35 ♀ □ ♇
07:59 ♀ △ ♂
16:56 ☽ □ ☿
22:22 ☽ ☌ ♂
22:55 ☽ △ ♀
18 00:42 ☽ ∥ ♇
03:53 ☿ ∥ ♆
07:33 ☽ □ ♅
16:14 ☽ ☍ ♄
19:18 ☽ ∥ ♂
21:34 ☽ ∥ ♅

19 01:46 ☽ ✶ ☉
05:41 ☽ # ♄
06:03 ☽ ☍ ♃
06:53 ☿ ✶ ♇
13:53 ☽ ∥ ☿
15:20 ☽ ✶ ♆
20:31 ☽ ∥ ♆
20 00:17 ☉ ♒
06:41 ☽ # ♃
07:50 ☽ ☌ ♇
09:06 ☿ □ ♂
10:29 ☽ ∥ ☉
11:29 ☽ ✶ ☿
14:17 ☽ □ ♀
15:07 ♂ ∥ ♅
18:20 ☽ ✶ ♅
21 05:00 ☉ △ ♃
08:58 ☿ # ♄

22 01:29 ☉ # ♃
17:57 ☿ ∥ ♂
19:36 ☿ ☌ ♅
23 02:37 ☽ ✶ ♂
08:08 ☽ ✶ ♀
09:15 ☿ ∥ ♅
15:41 ☽ △ ♄
24 06:11 ☽ △ ♃
16:23 ☽ ☌ ♆

25 00:25 ♄ SD
01:34 ☽ # ♃
08:39 ♃ SD
09:20 ☽ ✶ ♇
09:27 ☽ ∥ ☉
12:37 ☽ ∥ ♆
18:39 ☽ □ ♂
20:22 ☽ ☌ ♅
21:36 ☿ □ ♄
26 03:20 ☽ # ♄
03:56 ☉ ☌ ♆
04:39 ☽ □ ♄
05:30 ☽ ☌ ☿
05:39 ☽ ∥ ♂
12:58 ☽ ∥ ♅
19:05 ☽ □ ♃
19:39 ☉ ∥ ♆

27 04:17 ☽ ∥ ☿
05:26 ♀ ✶ ♄
09:59 ☽ ∥ ♇
11:11 ♂ □ ♅
22:03 ☽ □ ♇
28 07:28 ♂ # ♄
09:59 ☽ △ ♂
16:54 ☽ ✶ ♄
17:28 ☿ ∥ ♇
19:49 ☽ ☌ ♀
29 06:59 ☽ ✶ ♃
17:08 ☽ ✶ ♆
30 00:30 ☽ ✶ ☉
03:32 ☽ ∥ ♀
05:08 ☽ # ♀
09:12 ☽ △ ♇
19:55 ☽ ✶ ♅
31 13:31 ☉ ∥ ♂
13:37 ☽ ✶ ☿

⚷ Chiron

01 Dec.	18 S 14
01	22♐32
04	22 52
07	23 12
10	23 31
13	23 51
16	24 09
19	24 28
22	24 46
25	25 04
28	25 21
31	25 38

February 2001

Day	S. T.	☉	☽	☿	♀	♂	♃	♄	♅	♆	♇	☊ True
	h m s	° ' "	° ' "	° '	° '	° '	° '	° '	° '	° '	° '	° '
01 Th	08 45 04	12♒ 11 23	05♉ 09 53	29♒ 51	28♓ 25	22♏ 34	01♊ 16	24♉ 06	20♒ 19	06♒ 29	14♐ 43	15♋R12
02 Fr	08 49 00	13 12 16	18 15 43	00♓ 18	29 18	23 07	01 17	24 07	20 23	06 31	14 44	15D 13
03 Sa	08 52 57	14 13 08	01♊ 44 50	00 35	00♈ 10	23 40	01 19	24 08	20 26	06 33	14 46	15 13
04 Su	08 56 53	15 13 59	15 39 20	00 42	01 02	24 12	01 21	24 09	20 30	06 36	14 47	15 15
05 Mo	09 00 50	16 14 48	29 59 38	00R 37	01 52	24 45	01 23	24 10	20 33	06 38	14 48	15 15
06 Tu	09 04 47	17 15 36	14♋ 43 42	00 21	02 42	25 18	01 25	24 12	20 37	06 40	14 50	15R 15
07 We	09 08 43	18 16 23	29 46 34	29♒ 55	03 31	25 50	01 28	24 13	20 40	06 42	14 51	15 13
08 Th	09 12 40	19 17 08	15♌ 00 25	29 18	04 19	26 22	01 30	24 15	20 44	06 45	14 52	15 10
09 Fr	09 16 36	20 17 51	00♍ 15 22	28 32	05 06	26 54	01 33	24 16	20 47	06 47	14 53	15 05
10 Sa	09 20 33	21 18 34	15 21 02	27 38	05 52	27 26	01 36	24 18	20 51	06 49	14 55	15 00
11 Su	09 24 29	22 19 15	00♎ 08 05	26 37	06 37	27 58	01 39	24 20	20 54	06 51	14 56	14 53
12 Mo	09 28 26	23 19 55	14 29 45	25 31	07 21	28 30	01 43	24 22	20 58	06 54	14 57	14 48
13 Tu	09 32 22	24 20 34	28 22 28	24 22	08 04	29 02	01 46	24 24	21 01	06 56	14 58	14 45
14 We	09 36 19	25 21 12	11♏ 45 50	23 12	08 46	29 34	01 50	24 26	21 05	06 58	14 59	14 40
15 Th	09 40 15	26 21 49	24 41 59	22 03	09 27	00♐ 05	01 54	24 28	21 08	07 00	15 00	14 38
16 Fr	09 44 12	27 22 24	07♐ 14 49	20 56	10 06	00 36	01 58	24 31	21 12	07 02	15 01	14D 38
17 Sa	09 48 09	28 22 59	19 29 01	19 53	10 45	01 08	02 03	24 33	21 15	07 04	15 02	14 40
18 Su	09 52 05	29 23 32	01♑ 29 35	18 56	11 22	01 39	02 07	24 36	21 19	07 07	15 03	14 41
19 Mo	09 56 02	00♓ 24 04	13 21 19	18 04	11 57	02 10	02 12	24 38	21 22	07 09	15 04	14 42
20 Tu	09 59 58	01 24 34	25 08 34	17 19	12 32	02 41	02 17	24 41	21 25	07 11	15 05	14R 42
21 We	10 03 55	02 25 03	06♒ 55 02	16 42	13 05	03 11	02 22	24 44	21 29	07 13	15 06	14 40
22 Th	10 07 51	03 25 31	18 43 47	16 12	13 36	03 42	02 27	24 47	21 32	07 15	15 07	14 35
23 Fr	10 11 48	04 25 57	00♓ 37 09	15 49	14 06	04 12	02 33	24 50	21 36	07 17	15 08	14 29
24 Sa	10 15 44	05 26 21	12 36 53	15 34	14 34	04 43	02 38	24 53	21 39	07 19	15 08	14 20
25 Su	10 19 41	06 26 44	24 44 16	15 26	15 00	05 13	02 44	24 57	21 43	07 21	15 09	14 11
26 Mo	10 23 38	07 27 04	07♈ 00 27	15D 25	15 25	05 43	02 50	25 00	21 46	07 23	15 10	14 01
27 Tu	10 27 34	08 27 23	19 26 35	15 31	15 48	06 13	02 56	25 04	21 49	07 25	15 10	13 53
28 We	10 31 31	09 27 40	02♉ 04 03	15 42	16 09	06 42	03 02	25 07	21 53	07 27	15 11	13 46

Data for	02-01-2001
Julian Day	2451941.50
Ayanamsa	23 52 03
SVP	05 ♓ 14 25
☽ ☊ Mean	14 ♋ 03 R

● ◐ PHASES ○ ◑

01	14:02	◐	12♉47
08	07:12	○	19♌35
15	03:25	◑	26♏30
23	08:21	●	04♓47

LAST ASPECT ☽ INGRESS

Day	h m	Day	h m	
02	10:33	02	20:57	♊
04	08:14	05	00:01	♋
06	17:32	07	00:22	♌
08	21:26	08	23:37	♍
10	20:20	10	23:48	♎
12	17:33	13	02:53	♏
15	03:26	15	10:04	♐
17	19:24	17	21:01	♑
19	23:05	20	09:55	♒
22	12:19	22	22:46	♓
25	00:26	25	10:21	♈
27	04:35	27	20:07	♉

DECLINATION

Day	☉	☽	☿	♀	♂	♃	♄	♅	♆	♇
01 Th	17S 08	08N36	10S 33	00N42	17S 20	19N45	16N48	15S 21	18S 30	12S 14
02 Fr	16 51	13 03	10 07	01 11	17 28	19 45	16 49	15 20	18 29	12 14
03 Sa	16 34	16 59	09 45	01 39	17 37	19 46	16 49	15 19	18 28	12 14
04 Su	16 16	20 06	09 27	02 07	17 45	19 47	16 50	15 18	18 28	12 14
05 Mo	15 58	22 04	09 13	02 34	17 54	19 47	16 50	15 17	18 27	12 14
06 Tu	15 40	22 35	09 03	03 02	18 02	19 48	16 51	15 16	18 27	12 14
07 We	15 21	21 29	08 58	03 29	18 10	19 49	16 52	15 15	18 26	12 14
08 Th	15 02	18 49	08 57	03 56	18 18	19 49	16 52	15 13	18 26	12 14
09 Fr	14 43	14 50	09 01	04 23	18 26	19 50	16 53	15 12	18 25	12 14
10 Sa	14 24	09 56	09 09	04 49	18 34	19 51	16 54	15 11	18 25	12 14
11 Su	14 04	04 33	09 21	05 15	18 41	19 52	16 54	15 10	18 24	12 14
12 Mo	13 44	00S 56	09 37	05 41	18 49	19 53	16 55	15 09	18 24	12 13
13 Tu	13 24	06 11	09 56	06 07	18 56	19 54	16 56	15 08	18 23	12 13
14 We	13 04	10 58	10 16	06 32	19 03	19 55	16 57	15 07	18 22	12 13
15 Th	12 44	15 05	10 39	06 56	19 11	19 56	16 58	15 06	18 22	12 13
16 Fr	12 23	18 24	11 03	07 21	19 18	19 57	16 58	15 05	18 21	12 13
17 Sa	12 02	20 49	11 27	07 45	19 25	19 58	16 59	15 03	18 21	12 13
18 Su	11 41	22 15	11 51	08 08	19 31	19 59	17 00	15 02	18 20	12 13
19 Mo	11 20	22 39	12 14	08 31	19 38	20 00	17 01	15 01	18 20	12 12
20 Tu	10 58	22 02	12 37	08 53	19 45	20 01	17 02	15 00	18 19	12 12
21 We	10 37	20 26	12 58	09 15	19 51	20 03	17 03	14 59	18 19	12 12
22 Th	10 15	17 57	13 18	09 37	19 58	20 04	17 04	14 58	18 18	12 12
23 Fr	09 53	14 43	13 37	09 58	20 04	20 05	17 05	14 57	18 18	12 12
24 Sa	09 31	10 51	13 53	10 18	20 10	20 06	17 06	14 56	18 17	12 12
25 Su	09 09	06 31	14 08	10 37	20 16	20 08	17 07	14 55	18 17	12 11
26 Mo	08 46	01 52	14 21	10 56	20 22	20 09	17 08	14 54	18 16	12 11
27 Tu	08 24	02N54	14 33	11 14	20 28	20 11	17 09	14 53	18 16	12 11
28 We	08 01	07 38	14 42	11 32	20 34	20 12	17 10	14 51	18 15	12 11

ASPECTARIAN

01 02:28 ☽ □ ♆
07:14 ☿ ♓
09:22 ☽ # ☿
19:31 ☽ # ♇
02 03:18 ☉ # ♄
03:51 ☽ □ ♅
09:07 ☽ ☍ ♂
10:33 ☽ ☌ ♄
13:26 ☽ # ♅
19:15 ♀ ♈
21:04 ☽ ⚹ ♀
21:29 ☽ # ☉
21:57 ☽ □ ☿
22:59 ☽ ∥ ♄
23:15 ☽ ☌ ♃
03 04:34 ☽ # ♂
08:25 ☽ △ ♆
10:35 ☽ # ♆
13:07 ☉ ⚹ ♇
21:08 ☽ ∥ ♃
21:34 ♂ ☍ ♄
22:32 ☽ ☍ ♇
23:14 ☽ △ ☉
04 01:59 ☿ SR
08:14 ☽ △ ♅
09:27 ♀ ⚹ ♃
05 01:02 ☽ △ ☿
03:17 ☽ □ ♀
06 15:11 ☽ ⚹ ♄
17:32 ☽ △ ♂
19:58 ☿ ♒ R
07 02:41 ☽ ⚹ ♃
06:15 ☽ △ ♀
08:56 ☉ ∥ ♅
10:59 ☽ ☍ ♆
16:22 ☽ ∥ ♃
23:48 ☽ △ ♇
08 02:41 ☽ # ♆
03:26 ☽ # ♂
09:03 ☽ ☍ ♅
12:31 ☽ ∥ ♄
14:34 ☽ □ ♄
18:32 ☽ □ ♂
21:26 ☽ ☍ ☿
21:59 ☽ # ♅
22:19 ♂ ∥ ♆
09 00:38 ☽ # ☉
02:04 ☽ □ ♃
12:20 ☉ ☌ ♅
13:11 ☽ # ♇
23:19 ☽ □ ♇
10 02:59 ☿ □ ♂
03:27 ☽ # ☿
14:30 ☽ △ ♄
20:20 ☽ ⚹ ♂
21:08 ☽ ∥ ♀
11 02:32 ☽ △ ♃
08:09 ♀ ⚹ ♆
11:11 ☽ △ ♆
11:20 ☽ ☍ ♀
12 00:47 ☽ ⚹ ♇
11:08 ☽ △ ♅
16:24 ☽ △ ☉
17:33 ☽ △ ☿
23:26 ☿ □ ♄
23:38 ☽ # ♀
13 00:18 ☿ ☌ ☉
01:22 ☉ □ ♄
15:17 ☽ □ ♆
20:03 ☽ ∥ ☿
14 06:54 ☽ ∥ ♇
10:51 ☽ ∥ ☉
17:16 ☽ □ ♅
19:26 ☽ □ ☿
20:06 ♂ ♐
23:35 ☽ ☍ ♄
15 00:05 ☽ ∥ ♅
10:39 ☽ ☌ ♂
12:51 ☽ # ♄
13:47 ☽ ☍ ♃
18:39 ☿ ☌ ♅
23:35 ☽ ⚹ ♆
23:38 ☽ ∥ ♆
16 05:53 ☽ △ ♀
08:14 ☽ ∥ ♂
11:41 ☉ ∥ ♇
14:22 ☽ # ♃
15:13 ☽ ☌ ♇
17 00:45 ☽ ⚹ ☿
03:32 ☽ ⚹ ♅
18:49 ☿ ∥ ☉
19:24 ☽ ⚹ ☉
18 14:28 ☉ ♓
21:02 ☽ □ ♀
22:08 ☿ ∥ ♇
19 01:58 ♂ ☍ ♃
23:05 ☽ △ ♄
20 14:40 ☽ △ ♃
16:05 ☽ ⚹ ♂
22:40 ☉ □ ♃
21 00:38 ☽ ☌ ♆
04:16 ☽ # ♃
06:02 ☽ ∥ ♂
13:08 ☽ ⚹ ♀
16:40 ☽ ⚹ ♇
19:04 ☽ ☌ ☿
20:59 ☽ ∥ ♆
22 05:44 ☽ ☌ ♅
07:02 ☽ # ♄
12:19 ☽ □ ♄
13:12 ☉ □ ♂
21:30 ♀ # ☉
22:21 ☽ ∥ ♅
23 03:55 ☽ □ ♃
05:39 ♂ # ♃
06:38 ☽ ∥ ☿
07:32 ☽ □ ♂
15:57 ☽ ∥ ♇
24 02:54 ☽ # ♀
05:02 ☽ □ ♇
08:17 ☽ ∥ ☉
25 00:26 ☽ ⚹ ♄
08:31 ♀ △ ♇
15:42 ☿ SD
15:49 ☽ ⚹ ♃
21:23 ☽ △ ♂
23:51 ☿ ⚹ ♀
26 00:45 ☽ ⚹ ♆
15:48 ☽ △ ♇
16:23 ☽ ⚹ ☿
16:48 ☽ ☌ ♀
27 04:35 ☽ ⚹ ♅
28 01:49 ☽ # ☉
10:10 ☽ □ ♆
15:04 ☽ ⚹ ☉
22:03 ☽ ∥ ♀

⚷ Chiron

01 Dec.	18 S 11
03	25 ♐ 54
06	26 09
09	26 25
12	26 39
15	26 53
18	27 06
21	27 19
24	27 31
27	27 42

March 2001

Day	S. T.	☉	☽	☿	♀	♂	♃	♄	♅	♆	♇	☊ True
	h m s	° ' "	° ' "	° '	° '	° '	° '	° '	° '	° '	° '	° '
01 Th	10 35 27	10♓27 55	14♉54 37	16♒00	16♈28	07♐12	03♊09	25♉11	21♒56	07♒29	15♐12	13♋R42
02 Fr	10 39 24	11 28 09	28 00 21	16 23	16 45	07 41	03 15	25 15	21 59	07 31	15 12	13 39
03 Sa	10 43 20	12 28 20	11♊23 33	16 51	17 00	08 10	03 22	25 19	22 03	07 33	15 13	13 38
04 Su	10 47 17	13 28 29	25 06 20	17 24	17 13	08 39	03 29	25 23	22 06	07 35	15 13	13 38
05 Mo	10 51 13	14 28 36	09♋10 09	18 01	17 24	09 08	03 36	25 27	22 09	07 37	15 14	13 38
06 Tu	10 55 10	15 28 40	23 34 58	18 42	17 33	09 36	03 43	25 31	22 13	07 39	15 14	13 36
07 We	10 59 07	16 28 43	08♌18 29	19 27	17 39	10 05	03 50	25 35	22 16	07 41	15 15	13 33
08 Th	11 03 03	17 28 43	23 15 39	20 16	17 42	10 33	03 58	25 39	22 19	07 43	15 15	13 27
09 Fr	11 06 60	18 28 42	08♍18 49	21 08	17 44	11 01	04 06	25 44	22 22	07 45	15 15	13 18
10 Sa	11 10 56	19 28 38	23 18 38	22 03	17R43	11 29	04 13	25 48	22 25	07 46	15 16	13 08
11 Su	11 14 53	20 28 33	08♎05 34	23 01	17 39	11 57	04 21	25 53	22 29	07 48	15 16	12 58
12 Mo	11 18 49	21 28 25	22 31 28	24 02	17 33	12 24	04 29	25 58	22 32	07 50	15 16	12 47
13 Tu	11 22 46	22 28 16	06♏30 50	25 05	17 24	12 51	04 38	26 02	22 35	07 52	15 16	12 38
14 We	11 26 42	23 28 05	20 01 21	26 11	17 13	13 18	04 46	26 07	22 38	07 54	15 16	12 31
15 Th	11 30 39	24 27 53	03♐03 44	27 19	17 00	13 45	04 54	26 12	22 41	07 55	15 16	12 26
16 Fr	11 34 36	25 27 39	15 40 59	28 29	16 44	14 12	05 03	26 17	22 44	07 57	15 17	12 24
17 Sa	11 38 32	26 27 23	27 57 39	29 41	16 25	14 38	05 12	26 22	22 47	07 59	15 17	12D 23
18 Su	11 42 29	27 27 06	09♑59 00	00♓56	16 04	15 04	05 20	26 27	22 50	08 00	15R 17	12 24
19 Mo	11 46 25	28 26 46	21 50 34	02 12	15 41	15 30	05 29	26 33	22 53	08 02	15 17	12R 24
20 Tu	11 50 22	29 26 25	03♒37 39	03 29	15 16	15 55	05 39	26 38	22 56	08 04	15 17	12 23
21 We	11 54 18	00♈26 02	15 25 02	04 49	14 49	16 21	05 48	26 43	22 59	08 05	15 16	12 19
22 Th	11 58 15	01 25 38	27 16 48	06 10	14 20	16 46	05 57	26 49	23 02	08 07	15 16	12 12
23 Fr	12 02 11	02 25 11	09♓16 07	07 33	13 49	17 11	06 07	26 54	23 05	08 08	15 16	12 03
24 Sa	12 06 08	03 24 43	21 25 11	08 57	13 16	17 35	06 16	27 00	23 08	08 10	15 16	11 51
25 Su	12 10 05	04 24 12	03♈45 20	10 23	12 42	17 59	06 26	27 05	23 11	08 11	15 16	11 38
26 Mo	12 14 01	05 23 40	16 17 06	11 51	12 07	18 23	06 36	27 11	23 13	08 13	15 16	11 25
27 Tu	12 17 58	06 23 05	29 00 29	13 20	11 31	18 47	06 46	27 17	23 16	08 14	15 15	11 13
28 We	12 21 54	07 22 28	11♉55 16	14 50	10 54	19 10	06 56	27 23	23 19	08 15	15 15	11 03
29 Th	12 25 51	08 21 49	25 01 16	16 22	10 16	19 33	07 06	27 29	23 22	08 17	15 15	10 55
30 Fr	12 29 47	09 21 08	08♊18 42	17 56	09 38	19 56	07 16	27 35	23 24	08 18	15 14	10 51
31 Sa	12 33 44	10 20 25	21 48 07	19 30	09 01	20 18	07 26	27 41	23 27	08 19	15 14	10 49

Data for	03-01-2001
Julian Day	2451969.50
Ayanamsa	23 52 07
SVP	05 ♓ 14 20
☽ ☊ Mean	12 ♋ 34 R

● ◐ PHASES ○ ◑

03	02:04	◐	12♊34
09	17:24	○	19♍12
16	20:46	◑	26♐19
25	01:21	●	04♈28

LAST ASPECT Day	h m	☽ INGRESS Day	h m	
01	18:58	02	03:37	♊
03	18:47	04	08:26	♋
06	03:12	06	10:32	♌
08	03:52	08	10:46	♍
10	04:03	10	10:49	♎
12	02:46	12	12:44	♏
14	12:19	14	18:18	♐
17	03:49	17	04:03	♑
19	14:42	19	16:37	♒
21	23:04	22	05:29	♓
24	11:00	24	16:45	♈
26	13:12	27	01:52	♉
29	04:31	29	09:03	♊
31	02:56	31	14:25	♋

DECLINATION

Day	☉	☽	☿	♀	♂	♃	♄	♅	♆	♇
01 Th	07S39	12N08	14S49	11N48	20S39	20N13	17N11	14S50	18S15	12S11
02 Fr	07 16	16 09	14 55	12 04	20 45	20 15	17 13	14 49	18 14	12 10
03 Sa	06 53	19 27	14 59	12 19	20 50	20 16	17 14	14 48	18 14	12 10
04 Su	06 30	21 44	15 01	12 33	20 56	20 18	17 15	14 47	18 13	12 10
05 Mo	06 07	22 44	15 01	12 46	21 01	20 19	17 16	14 46	18 13	12 10
06 Tu	05 43	22 15	14 59	12 58	21 06	20 21	17 17	14 45	18 12	12 10
07 We	05 20	20 15	14 56	13 09	21 11	20 23	17 19	14 44	18 12	12 09
08 Th	04 57	16 50	14 51	13 19	21 16	20 24	17 20	14 43	18 11	12 09
09 Fr	04 33	12 18	14 45	13 28	21 21	20 26	17 21	14 42	18 11	12 09
10 Sa	04 10	07 01	14 37	13 35	21 25	20 27	17 22	14 41	18 10	12 09
11 Su	03 46	01 25	14 27	13 41	21 30	20 29	17 24	14 40	18 10	12 08
12 Mo	03 23	04S08	14 16	13 46	21 35	20 31	17 25	14 39	18 09	12 08
13 Tu	02 59	09 19	14 03	13 50	21 39	20 32	17 26	14 38	18 09	12 08
14 We	02 36	13 52	13 49	13 52	21 43	20 34	17 28	14 37	18 08	12 08
15 Th	02 12	17 36	13 33	13 53	21 48	20 36	17 29	14 36	18 08	12 07
16 Fr	01 48	20 23	13 16	13 53	21 52	20 38	17 30	14 35	18 08	12 07
17 Sa	01 25	22 09	12 58	13 50	21 56	20 39	17 32	14 34	18 07	12 07
18 Su	01 01	22 50	12 38	13 47	22 00	20 41	17 33	14 33	18 07	12 07
19 Mo	00 37	22 29	12 17	13 42	22 04	20 43	17 35	14 32	18 06	12 06
20 Tu	00 13	21 07	11 54	13 35	22 08	20 45	17 36	14 31	18 06	12 06
21 We	00N10	18 51	11 30	13 27	22 11	20 46	17 38	14 30	18 06	12 06
22 Th	00 34	15 46	11 05	13 17	22 15	20 48	17 39	14 29	18 05	12 06
23 Fr	00 58	12 00	10 38	13 06	22 19	20 50	17 40	14 28	18 05	12 05
24 Sa	01 21	07 43	10 10	12 54	22 22	20 52	17 42	14 27	18 04	12 05
25 Su	01 45	03 03	09 41	12 40	22 26	20 54	17 43	14 26	18 04	12 05
26 Mo	02 09	01N48	09 11	12 25	22 29	20 56	17 45	14 26	18 04	12 05
27 Tu	02 32	06 40	08 39	12 08	22 32	20 57	17 46	14 25	18 03	12 04
28 We	02 56	11 19	08 07	11 51	22 36	20 59	17 48	14 24	18 03	12 04
29 Th	03 19	15 31	07 33	11 32	22 39	21 01	17 50	14 23	18 03	12 04
30 Fr	03 42	19 01	06 58	11 12	22 42	21 03	17 51	14 22	18 02	12 03
31 Sa	04 06	21 32	06 21	10 52	22 45	21 05	17 53	14 21	18 02	12 03

ASPECTARIAN

01 00:15 ☽ # ♇
02:05 ☽ □ ☿
03:40 ☿ ∥ ♅
13:00 ☽ □ ♅
15:33 ♂ ⚹ ♆
15:40 ☽ # ♅
16:03 ☽ # ☿
18:58 ☽ ☌ ♄

02 07:05 ☽ ∥ ♄
09:34 ☽ ☌ ♃
10:00 ♀ # ♇
14:21 ☽ # ♆
17:11 ☽ △ ♆
18:04 ☽ ☍ ♂
03 06:46 ☽ ☍ ♇
07:29 ☽ ∥ ♃
10:01 ☽ △ ☿
10:05 ☽ ⚹ ♀
12:37 ☿ ⚹ ♀
13:41 ☽ # ♂
18:47 ☽ △ ♅

05 09:35 ☽ △ ☉
13:56 ☽ □ ♀
18:11 ☉ □ ♇
06 03:12 ☽ ⚹ ♄
15:15 ☽ # ♂
16:43 ☽ ⚹ ♃
22:53 ☽ ∥ ♃
23:00 ☽ ☍ ♆
07 02:58 ☽ △ ♂
11:11 ☽ △ ♇
15:06 ☽ △ ♀
15:29 ☽ # ♆
18:57 ☽ ☍ ☿
21:01 ☽ ∥ ♄
22:30 ☽ ☍ ♅

08 03:52 ☽ □ ♄
11:20 ☽ # ☿
11:51 ☽ # ♅
17:14 ☽ □ ♃
18:26 ☽ ∥ ♀
09 00:42 ☽ # ♆
01:08 ♀ SR
04:28 ☽ □ ♂
10:15 ☿ ∥ ♅
11:06 ☽ □ ♇

10 04:03 ☽ △ ♄
09:51 ☿ ☌ ♅
13:12 ☽ # ☉
17:51 ☽ △ ♃
23:32 ☽ △ ♆
11 06:34 ☽ ⚹ ♂
11:51 ☽ ⚹ ♇
15:44 ☽ ☍ ♀
20:52 ☽ ∥ ☉

12 00:01 ☽ △ ♅
02:46 ☽ △ ☿
13 02:23 ☽ □ ♆
14:21 ☽ ∥ ♇
19:15 ☿ # ♀
22:31 ☿ □ ♄
23:42 ☽ ∥ ☿
14 00:00 ☽ # ♀
04:20 ☽ ∥ ♅
04:46 ☽ □ ♅
06:47 ☽ △ ☉
11:12 ☽ ☍ ♄
12:19 ☽ □ ☿
23:06 ☽ # ♄

15 03:31 ☽ ☍ ♃
03:57 ☽ ∥ ♆
09:11 ☽ ⚹ ♆
21:02 ☽ ☌ ♂
23:14 ☽ ☌ ♇
16 01:59 ☽ △ ♀
02:32 ☽ # ♃
13:47 ☽ ⚹ ♅
19:47 ☽ ∥ ♂
21:43 ☉ ⚹ ♄
17 03:49 ☽ ⚹ ☿
06:06 ☿ ♓

18 02:37 ♇ SR
11:46 ♂ ☌ ♇
11:56 ☽ □ ♀
19 05:31 ♀ △ ♂
09:01 ☽ ∥ ♂
09:39 ☽ △ ♄
11:02 ☿ ∥ ♇
14:42 ☽ ⚹ ☉
23:325 ♀ △ ♇
20 04:11 ☽ △ ♃
04:39 ☽ # ♃
09:04 ☽ ☌ ♆
13:32 ☉ ♈
22:50 ☽ ⚹ ♀
23:44 ☽ ⚹ ♇

21 01:58 ☽ ⚹ ♂
06:26 ☽ ∥ ♆
10:08 ☽ # ♄
15:25 ☽ ☌ ♅
19:42 ☿ □ ♃
23:04 ☽ □ ♄

22 08:38 ☽ ∥ ♅
17:00 ☽ # ♀
17:38 ☽ □ ♃
20:09 ☽ ☌ ☿
23:28 ☽ ∥ ♇
23 08:53 ☽ ∥ ☿
11:55 ☽ □ ♇
16:13 ☽ □ ♂
24 11:00 ☽ ⚹ ♄
25 05:14 ☽ ⚹ ♃
06:01 ☽ # ☉
08:34 ☽ ⚹ ♆
16:25 ☽ ☌ ♀
22:04 ☽ △ ♇
26 01:49 ☽ ∥ ☉

04:08 ☽ △ ♂
13:12 ☽ ⚹ ♅
27 05:17 ♀ # ♇
09:03 ☽ # ☿
10:58 ☉ ⚹ ♃
17:14 ☽ □ ♆

28 02:40 ☽ ∥ ♀
04:05 ☽ # ♇
06:06 ☽ ⚹ ☿
06:28 ☿ □ ♇
17:11 ☽ # ♅
20:59 ☽ □ ♅
21:56 ☉ ⚹ ♆
29 04:31 ☽ ☌ ♄

15:15 ☽ ∥ ♄
16:39 ☽ # ♆
22:07 ☽ ☌ ♃
23:59 ☽ △ ♆

30 02:02 ☽ ⚹ ☉
02:17 ☽ ⚹ ♀
04:18 ♀ ☌ ☉
12:23 ☽ ☍ ♇
18:33 ☽ ∥ ♃
19:25 ☽ □ ☿
21:18 ☽ ☍ ♂
31 02:56 ☽ △ ♅
15:37 ☿ □ ♂
21:38 ☽ # ♂

⚷ Chiron

01	Dec.	17 S 59
02	27♐52	
05	28 02	
08	28 11	
11	28 19	
14	28 27	
17	28 33	
20	28 39	
23	28 44	
26	28 48	
29	28 51	

April 2001

Day	S.T. h m s	☉ ° ' "	☽ ° ' "	☿ ° '	♀ ° '	♂ ° '	♃ ° '	♄ ° '	♅ ° '	♆ ° '	♇ ° '	☊ True ° '
01 Su	12 37 40	11♈19 39	05♋30 27	21♓06	08♈R23	20♐40	07♊37	27♉47	23♒30	08♒21	15♐R13	10♋R48
02 Mo	12 41 37	12 18 51	19 26 36	22 44	07 46	21 02	07 47	27 53	23 32	08 22	15 13	10 47
03 Tu	12 45 33	13 18 01	03♌36 49	24 23	07 10	21 23	07 58	27 59	23 35	08 23	15 12	10 45
04 We	12 49 30	14 17 08	18 00 08	26 03	06 34	21 44	08 09	28 06	23 37	08 24	15 12	10 41
05 Th	12 53 27	15 16 13	02♍33 46	27 45	06 00	22 05	08 20	28 12	23 40	08 26	15 11	10 34
06 Fr	12 57 23	16 15 16	17 12 52	29 29	05 27	22 25	08 31	28 19	23 42	08 27	15 11	10 25
07 Sa	13 01 20	17 14 16	01♎50 51	01♈13	04 56	22 45	08 42	28 25	23 45	08 28	15 10	10 14
08 Su	13 05 16	18 13 14	16 20 17	03 00	04 27	23 05	08 53	28 31	23 47	08 29	15 09	10 02
09 Mo	13 09 13	19 12 11	00♏34 05	04 47	03 59	23 24	09 04	28 38	23 49	08 30	15 09	09 50
10 Tu	13 13 09	20 11 05	14 26 39	06 36	03 33	23 42	09 15	28 45	23 52	08 31	15 08	09 39
11 We	13 17 06	21 09 58	27 54 46	08 27	03 10	24 01	09 27	28 51	23 54	08 32	15 07	09 31
12 Th	13 21 02	22 08 48	10♐57 44	10 19	02 49	24 19	09 38	28 58	23 56	08 33	15 06	09 25
13 Fr	13 24 59	23 07 37	23 37 13	12 13	02 30	24 36	09 50	29 05	23 58	08 34	15 06	09 22
14 Sa	13 28 56	24 06 25	05♑56 37	14 08	02 14	24 53	10 01	29 12	24 00	08 35	15 05	09D 21
15 Su	13 32 52	25 05 10	18 00 35	16 05	02 00	25 10	10 13	29 19	24 02	08 36	15 04	09 22
16 Mo	13 36 49	26 03 54	29 54 22	18 03	01 49	25 26	10 25	29 25	24 04	08 36	15 03	09 22
17 Tu	13 40 45	27 02 36	11♒43 27	20 03	01 40	25 41	10 37	29 32	24 06	08 37	15 02	09R 22
18 We	13 44 42	28 01 16	23 33 10	22 04	01 33	25 56	10 48	29 39	24 08	08 38	15 01	09 20
19 Th	13 48 38	28 59 55	05♓28 20	24 06	01 29	26 11	11 00	29 46	24 10	08 39	15 00	09 15
20 Fr	13 52 35	29 58 32	17 33 02	26 10	01 27	26 25	11 13	29 53	24 12	08 39	14 59	09 07
21 Sa	13 56 31	00♉57 07	29 50 23	28 14	01D 28	26 39	11 25	00♊01	24 14	08 40	14 58	08 58
22 Su	14 00 28	01 55 40	12♈22 23	00♉20	01 31	26 52	11 37	00 08	24 16	08 41	14 57	08 46
23 Mo	14 04 25	02 54 12	25 09 49	02 27	01 37	27 04	11 49	00 15	24 18	08 41	14 56	08 35
24 Tu	14 08 21	03 52 42	08♉12 25	04 35	01 44	27 16	12 01	00 22	24 19	08 42	14 55	08 24
25 We	14 12 18	04 51 10	21 28 56	06 43	01 54	27 27	12 14	00 29	24 21	08 42	14 54	08 16
26 Th	14 16 14	05 49 36	04♊57 36	08 52	02 06	27 38	12 26	00 37	24 23	08 43	14 53	08 09
27 Fr	14 20 11	06 48 00	18 36 30	11 00	02 20	27 48	12 39	00 44	24 24	08 43	14 52	08 05
28 Sa	14 24 07	07 46 22	02♋24 01	13 09	02 36	27 58	12 51	00 51	24 26	08 44	14 50	08 04
29 Su	14 28 04	08 44 42	16 18 53	15 17	02 54	28 07	13 04	00 59	24 27	08 44	14 49	08D 03
30 Mo	14 32 00	09 43 00	00♌20 17	17 24	03 13	28 15	13 17	01 06	24 29	08 45	14 48	08R 04

Data for	04-01-2001
Julian Day	2452000.50
Ayanamsa	23 52 10
SVP	05♓14 15
☽ ☊ Mean	10♋55 R

● ◐ PHASES ○ ◑

Day	Time	Phase	Position
01	10:50	◐	11♋46
08	03:22	○	18♎22
15	15:31	◑	25♑43
23	15:27	●	03♉32
30	17:08	◐	10♌25

LAST ASPECT ☽ Day	h m	INGRESS Day	h m	Sign
02	14:28	02	17:56	♌
04	16:48	04	19:48	♍
06	18:20	06	20:59	♎
08	12:32	08	23:03	♏
11	01:44	11	03:48	♐
13	01:57	13	12:22	♑
15	23:02	16	00:12	♒
18	12:28	18	13:02	♓
20	17:42	21	00:19	♈
23	03:36	23	08:58	♉
25	05:10	25	15:13	♊
27	16:14	27	19:51	♋
28	21:55	29	23:26	♌

DECLINATION

Day	☉	☽	☿	♀	♂	♃	♄	♅	♆	♇
01 Su	04N29	22N51	05S44	10N30	22S48	21N07	17N54	14S20	18S02	12S03
02 Mo	04 52	22 47	05 05	10 08	22 51	21 09	17 56	14 20	18 01	12 03
03 Tu	05 15	21 15	04 25	09 46	22 54	21 10	17 57	14 19	18 01	12 02
04 We	05 38	18 20	03 45	09 24	22 57	21 12	17 59	14 18	18 01	12 02
05 Th	06 01	14 15	03 03	09 01	23 00	21 14	18 00	14 17	18 00	12 02
06 Fr	06 24	09 17	02 20	08 38	23 02	21 16	18 02	14 16	18 00	12 02
07 Sa	06 46	03 48	01 36	08 15	23 05	21 18	18 04	14 16	18 00	12 01
08 Su	07 09	01S50	00 51	07 53	23 08	21 20	18 05	14 15	18 00	12 01
09 Mo	07 31	07 16	00 05	07 30	23 11	21 22	18 07	14 14	17 59	12 01
10 Tu	07 53	12 13	00N42	07 09	23 13	21 24	18 08	14 13	17 59	12 00
11 We	08 15	16 25	01 29	06 48	23 16	21 25	18 10	14 13	17 59	12 00
12 Th	08 37	19 42	02 18	06 27	23 19	21 27	18 12	14 12	17 58	12 00
13 Fr	08 59	21 54	03 07	06 07	23 21	21 29	18 13	14 11	17 58	12 00
14 Sa	09 21	23 00	03 57	05 48	23 24	21 31	18 15	14 11	17 58	11 59
15 Su	09 43	22 58	04 48	05 30	23 26	21 33	18 17	14 10	17 58	11 59
16 Mo	10 04	21 54	05 40	05 13	23 29	21 35	18 18	14 09	17 58	11 59
17 Tu	10 25	19 52	06 32	04 57	23 32	21 37	18 20	14 09	17 57	11 58
18 We	10 46	17 00	07 25	04 41	23 34	21 38	18 21	14 08	17 57	11 58
19 Th	11 07	13 24	08 18	04 27	23 37	21 40	18 23	14 07	17 57	11 58
20 Fr	11 28	09 14	09 11	04 14	23 39	21 42	18 25	14 07	17 57	11 58
21 Sa	11 48	04 37	10 04	04 02	23 42	21 44	18 26	14 06	17 57	11 57
22 Su	12 09	00N16	10 58	03 51	23 45	21 46	18 28	14 06	17 57	11 57
23 Mo	12 29	05 14	11 51	03 40	23 47	21 47	18 30	14 05	17 56	11 57
24 Tu	12 49	10 06	12 44	03 32	23 50	21 49	18 31	14 05	17 56	11 57
25 We	13 08	14 34	13 36	03 24	23 52	21 51	18 33	14 04	17 56	11 56
26 Th	13 28	18 23	14 28	03 17	23 55	21 53	18 34	14 04	17 56	11 56
27 Fr	13 47	21 15	15 18	03 11	23 58	21 54	18 36	14 03	17 56	11 56
28 Sa	14 06	22 55	16 08	03 06	24 01	21 56	18 38	14 03	17 56	11 56
29 Su	14 25	23 10	16 56	03 02	24 03	21 58	18 39	14 02	17 56	11 55
30 Mo	14 43	21 58	17 43	02 59	24 06	22 00	18 41	14 02	17 56	11 55

ASPECTARIAN

01 01:31 ♀ ⚹ ♆
04:47 ☽ □ ♀
22:08 ☽ # ♂
23:19 ♀ ⚹ ♃
02 05:06 ☿ # ☉
06:22 ☽ △ ☿
14:28 ☽ ⚹ ♄

03 00:50 ☽ ∥ ♃
05:44 ☽ △ ♀
07:24 ☽ ⚹ ♃
08:01 ☽ ☍ ♆
17:24 ☽ △ ☉
19:22 ☽ △ ♇
04 02:12 ☽ # ♆
02:22 ☽ ∥ ♄
06:21 ☽ △ ♂
09:20 ☽ ☍ ♅
16:48 ☽ □ ♄
22:04 ☉ △ ♇
22:57 ♄ # ♆
23:50 ☽ # ♅

05 06:41 ☿ ⚹ ♄
09:35 ☽ □ ♃
11:12 ☽ # ♇
14:27 ♃ △ ♆
20:41 ☽ □ ♇
06 03:12 ☽ ∥ ♀
07:15 ☿ ♈
08:44 ☽ □ ♂
12:03 ☽ ∥ ☉
18:20 ☽ △ ♄
22:51 ☽ ☍ ☿

07 04:56 ☽ ☍ ♀
10:50 ☽ # ☿
10:57 ☽ △ ♆
11:27 ☽ △ ♃
20:20 ☽ ∥ ☿
22:03 ☽ ⚹ ♇
08 11:35 ☽ ⚹ ♂
12:32 ☽ △ ♅
15:26 ☿ ☌ ♀
23:43 ♀ ∥ ☉

09 01:01 ☽ # ♀
01:13 ☽ # ☉
13:40 ☽ □ ♆
22:53 ☽ ∥ ♇
10 10:47 ☽ ∥ ♅
13:27 ♂ ⚹ ♅
16:45 ☽ □ ♅

11 01:03 ☿ ⚹ ♆
01:44 ☽ ☍ ♄
09:20 ☽ △ ♀
10:28 ☽ ∥ ♆
11:57 ☽ # ♄
14:14 ☿ ⚹ ♃
19:31 ☽ ⚹ ♆
21:30 ☽ ☍ ♃
22:37 ☽ △ ☿
12 07:47 ☽ ☌ ♇
18:16 ☽ # ♃
22:59 ☽ △ ☉

13 00:41 ☽ ⚹ ♅
01:57 ☽ ☌ ♂
16:53 ☽ □ ♀
21:25 ☉ ⚹ ♅
14 11:38 ☿ △ ♇
19:24 ☽ □ ☿

15 02:32 ☉ △ ♂
14:32 ☿ ∥ ♀
23:02 ☽ △ ♄
16 03:49 ☽ ⚹ ♀
04:39 ☽ # ♃
17:42 ☽ ☌ ♆
21:43 ☽ △ ♃

17 06:44 ☽ ⚹ ♇
13:34 ☽ # ♄
16:43 ☽ ∥ ♆
20:22 ☽ ⚹ ☿
18 01:12 ☽ ☌ ♅
04:57 ☽ ⚹ ♂
09:51 ☽ ⚹ ☉
12:28 ☽ □ ♄
19:31 ☽ ∥ ♅

19 00:54 ☿ ⚹ ♅
08:39 ☽ ∥ ♇
11:15 ☽ □ ♃
12:30 ☽ # ☉
18:58 ☽ □ ♇
20 00:16 ☽ # ☿
00:37 ☉ ♉
03:22 ☿ △ ♂
04:36 ♀ SD
17:42 ☽ □ ♂
21:59 ♄ ♊

21 00:21 ☽ ⚹ ♄
03:09 ☽ # ♀
03:10 ☽ ☌ ♀
10:39 ☉ # ♇
16:59 ☽ ⚹ ♆
20:09 ☿ ♉
22:33 ☽ ⚹ ♃
22 04:53 ☽ △ ♇
16:42 ☽ ∥ ♀
22:24 ☽ ⚹ ♅

23 02:48 ☿ # ♇
03:36 ☽ △ ♂
09:25 ☿ ☌ ☉
16:06 ☽ ☌ ☿
24 00:55 ☽ □ ♆
03:43 ☿ ∥ ☉
09:36 ☽ # ♇
15:23 ☽ ∥ ☉
17:12 ☽ ∥ ☿
21:09 ☽ # ♅

25 05:10 ☽ □ ♅
12:55 ☿ # ♅
16:14 ☽ ☌ ♄
18:52 ☽ ⚹ ♀
20:50 ☽ # ♆
22:24 ☿ □ ♆
26 01:20 ☽ ∥ ♄
06:39 ☽ △ ♆
13:24 ☽ ☌ ♃
17:28 ☽ ☍ ♇

27 07:28 ☽ ∥ ♃
10:09 ☽ △ ♅
16:14 ☽ ☍ ♂
19:39 ☉ # ♅

28 00:22 ☽ □ ♀
10:00 ☽ ⚹ ☉
21:55 ☽ ⚹ ☿
23:50 ☉ □ ♆

29 23:46 ☽ ∥ ♃

30 01:20 ☽ ⚹ ♄
05:03 ☽ △ ♀
06:33 ☿ # ♆
14:20 ☽ ☍ ♆
22:21 ☽ ⚹ ♃

⚷ Chiron

01 Dec. 17 S 39

Day	Position
01	28♐53
04	28 55
07	28 55
10	28 55R
13	28 54
16	28 52
19	28 50
22	28 46
25	28 42
28	28 37

07 17:12 28♐55 R

May 2001

Day	S. T.	☉	☽	☿	♀	♂	♃	♄	♅	♆	♇	☊ True
	h m s	° ' "	° ' "	° '	° '	° '	° '	° '	° '	° '	° '	° '
01 Tu	14 35 57	10♉41 16	14♌27 25	19♉31	03♈35	28♐23	13♊29	01♊14	24♒30	08♒45	14♐R47	08♋R03
02 We	14 39 54	11 39 30	28 39 09	21 36	03 58	28 30	13 42	01 21	24 32	08 45	14 46	08 01
03 Th	14 43 50	12 37 41	12♍53 39	23 39	04 23	28 36	13 55	01 29	24 33	08 46	14 44	07 57
04 Fr	14 47 47	13 35 51	27 08 06	25 40	04 50	28 42	14 08	01 36	24 34	08 46	14 43	07 51
05 Sa	14 51 43	14 33 58	11♎18 45	27 40	05 18	28 47	14 21	01 44	24 35	08 46	14 42	07 43
06 Su	14 55 40	15 32 04	25 21 11	29 37	05 47	28 51	14 34	01 51	24 37	08 46	14 40	07 34
07 Mo	14 59 36	16 30 08	09♏10 59	01♊31	06 19	28 55	14 47	01 59	24 38	08 46	14 39	07 24
08 Tu	15 03 33	17 28 11	22 44 19	03 22	06 51	28 58	15 00	02 06	24 39	08 46	14 38	07 16
09 We	15 07 29	18 26 12	05♐58 37	05 10	07 25	29 00	15 13	02 14	24 40	08 47	14 36	07 10
10 Th	15 11 26	19 24 11	18 52 55	06 55	08 00	29 02	15 26	02 22	24 41	08 47	14 35	07 06
11 Fr	15 15 23	20 22 09	01♑27 59	08 37	08 37	29 03	15 40	02 29	24 42	08R 47	14 33	07 04
12 Sa	15 19 19	21 20 05	13 46 01	10 16	09 14	29R 03	15 53	02 37	24 43	08 47	14 32	07D 05
13 Su	15 23 16	22 18 01	25 50 29	11 51	09 53	29 02	16 06	02 45	24 43	08 47	14 30	07 07
14 Mo	15 27 12	23 15 55	07♒45 41	13 22	10 33	29 01	16 19	02 52	24 44	08 46	14 29	07 09
15 Tu	15 31 09	24 13 47	19 36 28	14 50	11 14	28 59	16 33	03 00	24 45	08 46	14 27	07 11
16 We	15 35 05	25 11 39	01♓27 56	16 14	11 56	28 56	16 46	03 08	24 46	08 46	14 26	07 12
17 Th	15 39 02	26 09 29	13 25 09	17 35	12 39	28 52	16 59	03 15	24 46	08 46	14 24	07R 11
18 Fr	15 42 58	27 07 18	25 32 46	18 51	13 23	28 48	17 13	03 23	24 47	08 46	14 23	07 09
19 Sa	15 46 55	28 05 05	07♈54 46	20 04	14 08	28 43	17 26	03 31	24 47	08 46	14 21	07 04
20 Su	15 50 52	29 02 52	20 34 10	21 13	14 54	28 37	17 40	03 39	24 48	08 45	14 20	06 59
21 Mo	15 54 48	00♊00 38	03♉32 42	22 18	15 40	28 30	17 53	03 46	24 48	08 45	14 18	06 52
22 Tu	15 58 45	00 58 22	16 50 45	23 19	16 28	28 22	18 07	03 54	24 49	08 45	14 17	06 47
23 We	16 02 41	01 56 05	00♊27 07	24 16	17 16	28 14	18 21	04 02	24 49	08 44	14 15	06 42
24 Th	16 06 38	02 53 47	14 19 20	25 09	18 05	28 05	18 34	04 10	24 49	08 44	14 14	06 39
25 Fr	16 10 34	03 51 27	28 23 51	25 57	18 54	27 56	18 48	04 18	24 50	08 43	14 12	06 37
26 Sa	16 14 31	04 49 07	12♋36 42	26 42	19 44	27 45	19 02	04 25	24 50	08 43	14 10	06D 37
27 Su	16 18 27	05 46 45	26 53 59	27 22	20 35	27 34	19 15	04 33	24 50	08 43	14 09	06 37
28 Mo	16 22 24	06 44 21	11♌12 12	27 57	21 27	27 22	19 29	04 41	24 50	08 42	14 07	06 38
29 Tu	16 26 21	07 41 56	25 28 26	28 28	22 19	27 10	19 43	04 49	24 50	08 41	14 06	06 39
30 We	16 30 17	08 39 29	09♍40 16	28 55	23 12	26 57	19 56	04 56	24R 50	08 41	14 04	06R 38
31 Th	16 34 14	09 37 01	23 45 43	29 17	24 05	26 43	20 10	05 04	24 50	08 40	14 02	06 37

Data for	05-01-2001
Julian Day	2452030.50
Ayanamsa	23 52 13
SVP	05♓14 10
☽ ☊ Mean	09♋20 R

● ◐ PHASES ○ ◑

07	13:53	○	17♏04
15	10:12	◑	24♒38
23	02:46	●	02♊03
29	22:09	◐	08♍35

LAST ASPECT Day	h m	☽ INGRESS Day	h m	
01	23:45	02	02:17	♍
04	02:40	04	04:51	♎
06	06:05	06	08:02	♏
08	03:26	08	13:06	♐
10	19:21	10	21:11	♑
12	16:19	13	08:22	♒
15	18:55	15	21:03	♓
18	06:20	18	08:43	♈
20	14:50	20	17:31	♉
22	14:08	22	23:14	♊
24	23:14	25	02:44	♋
26	12:45	27	05:13	♌
29	05:15	29	07:39	♍
31	09:41	31	10:43	♎

DECLINATION

Day	☉	☽	☿	♀	♂	♃	♄	♅	♆	♇
01 Tu	15N02	19N24	18N28	02N57	24S09	22N01	18N43	14S01	17S55	11S55
02 We	15 20	15 38	19 11	02 56	24 12	22 03	18 44	14 01	17 55	11 55
03 Th	15 38	10 59	19 52	02 56	24 15	22 05	18 46	14 00	17 55	11 54
04 Fr	15 55	05 44	20 31	02 57	24 18	22 06	18 47	14 00	17 55	11 54
05 Sa	16 12	00 12	21 08	02 59	24 21	22 08	18 49	14 00	17 55	11 54
06 Su	16 29	05S17	21 42	03 01	24 24	22 10	18 51	13 59	17 55	11 54
07 Mo	16 46	10 26	22 14	03 05	24 27	22 11	18 52	13 59	17 55	11 53
08 Tu	17 03	14 59	22 43	03 09	24 30	22 13	18 54	13 59	17 55	11 53
09 We	17 19	18 41	23 09	03 14	24 33	22 14	18 55	13 58	17 55	11 53
10 Th	17 35	21 22	23 33	03 19	24 36	22 16	18 57	13 58	17 55	11 53
11 Fr	17 50	22 55	23 55	03 26	24 39	22 17	18 59	13 58	17 55	11 52
12 Sa	18 06	23 20	24 14	03 33	24 43	22 19	19 00	13 57	17 55	11 52
13 Su	18 21	22 37	24 31	03 41	24 46	22 20	19 02	13 57	17 55	11 52
14 Mo	18 35	20 53	24 45	03 49	24 49	22 22	19 03	13 57	17 55	11 52
15 Tu	18 50	18 15	24 57	03 58	24 53	22 23	19 05	13 57	17 55	11 52
16 We	19 04	14 53	25 07	04 08	24 56	22 25	19 06	13 57	17 55	11 51
17 Th	19 18	10 53	25 15	04 18	25 00	22 26	19 08	13 56	17 55	11 51
18 Fr	19 31	06 25	25 21	04 29	25 03	22 28	19 10	13 56	17 55	11 51
19 Sa	19 44	01 36	25 25	04 40	25 07	22 29	19 11	13 56	17 55	11 51
20 Su	19 57	03N23	25 27	04 52	25 10	22 30	19 13	13 56	17 55	11 51
21 Mo	20 09	08 21	25 27	05 04	25 14	22 32	19 14	13 56	17 56	11 51
22 Tu	20 21	13 04	25 26	05 16	25 17	22 33	19 16	13 56	17 56	11 50
23 We	20 33	17 15	25 23	05 29	25 21	22 34	19 17	13 56	17 56	11 50
24 Th	20 44	20 34	25 18	05 43	25 25	22 36	19 19	13 56	17 56	11 50
25 Fr	20 55	22 41	25 13	05 57	25 28	22 37	19 20	13 56	17 56	11 50
26 Sa	21 06	23 23	25 05	06 11	25 32	22 38	19 22	13 56	17 56	11 50
27 Su	21 16	22 34	24 57	06 26	25 35	22 39	19 23	13 56	17 56	11 50
28 Mo	21 26	20 16	24 48	06 41	25 39	22 41	19 24	13 56	17 56	11 50
29 Tu	21 36	16 44	24 37	06 56	25 43	22 42	19 26	13 56	17 57	11 49
30 We	21 45	12 16	24 25	07 11	25 46	22 43	19 27	13 56	17 57	11 49
31 Th	21 54	07 10	24 13	07 27	25 50	22 44	19 29	13 56	17 57	11 49

ASPECTARIAN

01 00:34 ☽ △ ♇
04:56 ☽ ∥ ♄
05:28 ☽ ∥ ☿
08:16 ☿ ∥ ♄
10:03 ☽ □ ☿
10:16 ☽ # ♆
17:02 ☽ ☍ ♅
23:45 ☽ △ ♂
02 01:38 ☽ ∥ ☉
04:36 ☽ □ ♄
08:53 ☽ # ♅
19:33 ☽ # ♇
23:32 ☽ △ ☉
03 01:46 ☽ □ ♃
03:07 ☽ □ ♇
10:43 ☿ □ ♅
21:09 ☽ △ ☿
04 02:40 ☽ □ ♂
07:38 ☽ △ ♄
12:06 ☽ ∥ ♀
13:27 ☽ ☍ ♀
19:42 ☽ △ ♆
05 05:16 ☽ △ ♃
05:46 ☽ ✶ ♇
13:57 ☽ # ♀
22:44 ☽ △ ♅
06 04:54 ☿ ♊
06:05 ☽ ✶ ♂
10:41 ♃ ☍ ♇
22:00 ☿ ∥ ♃
23:18 ☽ □ ♆
07 06:27 ☿ ☌ ♄
07:19 ☽ ∥ ♇
18:23 ☽ ∥ ♅
08 03:26 ☽ □ ♅
13:40 ☽ # ☉
17:06 ☽ ☍ ♄
18:32 ☽ ∥ ♆
22:18 ☽ ☍ ☿
09 01:53 ☽ # ♄
02:47 ☽ △ ♀
05:10 ☽ ✶ ♆
15:57 ☽ ☌ ♇
17:26 ☽ ☍ ♃
10 11:01 ☽ ✶ ♅
11:54 ☽ # ♃
19:21 ☽ ☌ ♂
23:48 ☿ ✶ ♀
11 01:14 ♆ SR
02:16 ☿ △ ♆
06:26 ♀ ✶ ♆
07:21 ☉ # ♆
14:38 ☽ □ ♀
16:09 ♂ SR
12 16:19 ☽ △ ☉
13 04:40 ☽ # ♃
14:02 ☽ △ ♄
14 02:04 ☽ ☌ ♆
06:00 ☽ ✶ ♀
10:33 ☿ # ♂
12:59 ☽ △ ☿
13:36 ☽ ✶ ♇
17:13 ☽ # ♄
17:41 ☽ △ ♃
17:52 ☿ ☍ ♇
19:40 ☽ # ☉
15 02:36 ☽ ∥ ♆
10:27 ☽ ☌ ♅
13:07 ☉ □ ♅
18:55 ☽ ✶ ♂
16 03:24 ☽ □ ♄
05:17 ☉ ∥ ♄
05:56 ☽ ∥ ♅
11:16 ☿ ☌ ♃
18:27 ☽ ∥ ♇
17 01:59 ☽ □ ♇
07:16 ☽ □ ♃
09:17 ☽ □ ☿
18 03:22 ☽ ✶ ☉
06:20 ☽ □ ♂
09:30 ☽ # ♀
15:27 ☽ ✶ ♄
19 01:38 ☽ ✶ ♆
06:57 ♀ △ ♇
12:17 ☽ △ ♇
12:38 ☽ ☌ ♀
18:28 ☽ ✶ ♃
20 01:21 ☽ ✶ ☿
07:25 ☽ ∥ ♀
07:55 ☽ ✶ ♅
14:50 ☽ △ ♂
23:45 ☉ ♊
21 09:28 ☽ □ ♆
17:36 ☽ # ♇
22 04:42 ☽ # ♅
14:08 ☽ □ ♅
23 04:27 ☽ # ♆
05:49 ☿ # ♂
06:19 ☽ ☌ ♄
14:01 ☽ ∥ ♄
14:24 ☽ △ ♆
14:57 ☿ △ ♅
23:51 ☽ ☍ ♇
24 01:44 ☽ ∥ ☉
06:51 ☽ ✶ ♀
07:25 ☽ ☌ ♃
17:57 ☽ △ ♅
19:38 ☽ ☌ ☿
19:54 ♀ ✶ ♃
22:52 ☽ ∥ ♃
23:14 ☽ ☍ ♂
25 12:34 ☉ ☌ ♄
26 12:45 ☽ □ ♀
22:31 ☽ ∥ ♃
27 06:04 ☿ ☍ ♂
12:58 ☽ ✶ ♄
14:13 ☽ ∥ ☉
15:59 ☽ ✶ ☉
19:49 ☽ ☍ ♆
28 04:54 ☽ △ ♇
06:35 ☽ ∥ ♄
14:09 ☽ ✶ ♃
16:37 ☽ # ♆
18:21 ☽ △ ♀
22:57 ☽ ☍ ♅
29 02:49 ☽ △ ♂
05:15 ☽ ✶ ☿
15:12 ♅ SR
15:33 ☽ # ♅
15:56 ☽ □ ♄
30 00:35 ☉ △ ♆
02:09 ☽ # ♇
07:28 ☽ □ ♇
17:46 ☽ □ ♃
22:44 ☽ ∥ ♀
31 04:59 ☽ □ ♂
09:41 ☽ □ ☿
19:37 ☽ △ ♄
20:13 ♀ ✶ ♅

⚷ Chiron

01 Dec. 17 S 20

01	28♐31R
04	28 24
07	28 17
10	28 09
13	28 01
16	27 52
19	27 42
22	27 32
25	27 22
28	27 11
31	27 00

June 2001

21 11:59 00♋10 ☉ Total Solar Eclipse

Day	S. T. h m s	☉ ° ' "	☽ ° ' "	☿ ° '	♀ ° '	♂ ° '	♃ ° '	♄ ° '	♅ ° '	♆ ° '	♇ ° '	☊ True ° '
01 Fr	16 38 10	10♊ 34 32	07♎ 43 01	29♊ 34	24♈ 59	26♐ R29	20♊ 24	05♊ 12	24♒ R50	08♒ R40	14♐ R01	06♋ R34
02 Sa	16 42 07	11 32 01	21 30 28	29 46	25 53	26 14	20 38	05 20	24 50	08 39	13 59	06 31
03 Su	16 46 03	12 29 29	05♏ 06 29	29 54	26 48	25 58	20 51	05 27	24 50	08 38	13 58	06 26
04 Mo	16 49 60	13 26 56	18 29 34	29 58	27 43	25 42	21 05	05 35	24 50	08 37	13 56	06 22
05 Tu	16 53 56	14 24 22	01♐ 38 28	29R 56	28 39	25 26	21 19	05 43	24 49	08 37	13 54	06 19
06 We	16 57 53	15 21 47	14 32 24	29 51	29 35	25 09	21 33	05 51	24 49	08 36	13 53	06 16
07 Th	17 01 50	16 19 11	27 11 16	29 41	00♉ 32	24 51	21 47	05 58	24 49	08 35	13 51	06 15
08 Fr	17 05 46	17 16 34	09♑ 35 51	29 27	01 29	24 34	22 00	06 06	24 48	08 34	13 50	06D 15
09 Sa	17 09 43	18 13 56	21 47 42	29 09	02 27	24 15	22 14	06 14	24 48	08 33	13 48	06 16
10 Su	17 13 39	19 11 18	03♒ 49 17	28 47	03 25	23 57	22 28	06 21	24 47	08 33	13 46	06 18
11 Mo	17 17 36	20 08 39	15 43 42	28 22	04 23	23 38	22 42	06 29	24 47	08 32	13 45	06 21
12 Tu	17 21 32	21 06 00	27 34 45	27 55	05 22	23 19	22 56	06 37	24 46	08 31	13 43	06 23
13 We	17 25 29	22 03 20	09♓ 26 41	27 25	06 21	23 00	23 10	06 44	24 45	08 30	13 41	06 25
14 Th	17 29 25	23 00 39	21 24 07	26 53	07 21	22 41	23 24	06 52	24 45	08 29	13 40	06 27
15 Fr	17 33 22	23 57 59	03♈ 31 42	26 20	08 21	22 21	23 37	06 59	24 44	08 28	13 38	06 27
16 Sa	17 37 19	24 55 18	15 53 55	25 46	09 21	22 02	23 51	07 07	24 43	08 27	13 37	06R 27
17 Su	17 41 15	25 52 36	28 34 38	25 12	10 21	21 42	24 05	07 14	24 42	08 26	13 35	06 26
18 Mo	17 45 12	26 49 54	11♉ 36 49	24 39	11 22	21 23	24 19	07 22	24 41	08 24	13 34	06 25
19 Tu	17 49 08	27 47 12	25 02 07	24 06	12 23	21 03	24 33	07 29	24 40	08 23	13 32	06 24
20 We	17 53 05	28 44 30	08♊ 50 29	23 35	13 25	20 44	24 47	07 37	24 39	08 22	13 30	06 23
21 Th	17 57 01	29 41 47	22 59 56	23 06	14 26	20 25	25 00	07 44	24 38	08 21	13 29	06 23
22 Fr	18 00 58	00♋ 39 04	07♋ 26 32	22 40	15 28	20 06	25 14	07 52	24 37	08 20	13 27	06 22
23 Sa	18 04 54	01 36 20	22 04 43	22 16	16 30	19 47	25 28	07 59	24 36	08 19	13 26	06 22
24 Su	18 08 51	02 33 36	06♌ 48 00	21 56	17 33	19 29	25 42	08 06	24 35	08 17	13 24	06 22
25 Mo	18 12 48	03 30 52	21 29 47	21 40	18 36	19 11	25 56	08 14	24 34	08 16	13 23	06 21
26 Tu	18 16 44	04 28 06	06♍ 04 12	21 27	19 39	18 53	26 09	08 21	24 33	08 15	13 21	06 21
27 We	18 20 41	05 25 21	20 26 36	21 19	20 42	18 36	26 23	08 28	24 31	08 14	13 20	06 20
28 Th	18 24 37	06 22 34	04♎ 33 54	21 16	21 45	18 19	26 37	08 35	24 30	08 12	13 18	06 19
29 Fr	18 28 34	07 19 47	18 24 27	21D 17	22 49	18 03	26 51	08 43	24 29	08 11	13 17	06 18
30 Sa	18 32 30	08 17 00	01♏ 57 52	21 23	23 53	17 48	27 04	08 50	24 27	08 10	13 16	06 17

Data for	06-01-2001
Julian Day	2452061.50
Ayanamsa	23 52 18
SVP	05 ♓ 14 06
☽ ☊ Mean	07 ♋ 42 R

● ◐ PHASES ○ ◑

06	01:40	○	15♐26
14	03:29	◑	23♓09
21	11:58	☉	00♋10
28	03:20	◐	06♎31

LAST ASPECT ☽ Day	h m	INGRESS Day	h m	
02	14:43	02	14:57	♏
04	11:31	04	21:00	♐
07	04:43	07	05:25	♑
07	06:59	09	16:21	♒
12	00:40	12	04:55	♓
14	10:27	14	17:04	♈
16	18:33	17	02:40	♉
18	23:23	19	08:43	♊
21	03:26	21	11:42	♋
22	14:13	23	12:56	♌
25	07:24	25	13:59	♍
27	10:13	27	16:12	♎
29	15:09	29	20:30	♏

DECLINATION

Day	☉	☽	☿	♀	♂	♃	♄	♅	♆	♇
01 Fr	22N02	01N45	23N59	07N43	25S 53	22N45	19N30	13S 56	17S 57	11S 49
02 Sa	22 10	03S 42	23 45	07 59	25 57	22 46	19 32	13 56	17 57	11 49
03 Su	22 18	08 53	23 30	08 16	26 00	22 47	19 33	13 56	17 57	11 49
04 Mo	22 25	13 35	23 15	08 33	26 03	22 48	19 34	13 56	17 58	11 49
05 Tu	22 32	17 34	22 58	08 49	26 06	22 49	19 36	13 56	17 58	11 49
06 We	22 38	20 36	22 42	09 06	26 10	22 50	19 37	13 56	17 58	11 49
07 Th	22 44	22 35	22 25	09 24	26 13	22 51	19 39	13 56	17 58	11 49
08 Fr	22 50	23 24	22 08	09 41	26 16	22 52	19 40	13 57	17 58	11 48
09 Sa	22 55	23 04	21 51	09 58	26 18	22 53	19 41	13 57	17 59	11 48
10 Su	23 00	21 40	21 33	10 16	26 21	22 54	19 43	13 57	17 59	11 48
11 Mo	23 04	19 20	21 16	10 34	26 24	22 55	19 44	13 57	17 59	11 48
12 Tu	23 08	16 11	20 59	10 51	26 26	22 56	19 45	13 58	17 59	11 48
13 We	23 12	12 23	20 42	11 09	26 29	22 57	19 46	13 58	18 00	11 48
14 Th	23 15	08 06	20 26	11 27	26 31	22 57	19 48	13 58	18 00	11 48
15 Fr	23 18	03 26	20 10	11 45	26 33	22 58	19 49	13 58	18 00	11 48
16 Sa	23 20	01N27	19 56	12 02	26 35	22 59	19 50	13 59	18 01	11 48
17 Su	23 22	06 24	19 42	12 20	26 37	23 00	19 51	13 59	18 01	11 48
18 Mo	23 24	11 12	19 29	12 38	26 39	23 00	19 53	13 59	18 01	11 48
19 Tu	23 25	15 38	19 17	12 56	26 40	23 01	19 54	14 00	18 01	11 48
20 We	23 26	19 22	19 07	13 13	26 42	23 02	19 55	14 00	18 02	11 48
21 Th	23 26	22 02	18 58	13 31	26 43	23 02	19 56	14 00	18 02	11 48
22 Fr	23 26	23 20	18 50	13 48	26 45	23 03	19 57	14 01	18 02	11 48
23 Sa	23 26	23 03	18 44	14 06	26 46	23 03	19 59	14 01	18 03	11 48
24 Su	23 25	21 10	18 39	14 23	26 47	23 04	20 00	14 02	18 03	11 48
25 Mo	23 24	17 53	18 37	14 40	26 48	23 04	20 01	14 02	18 03	11 48
26 Tu	23 22	13 32	18 35	14 57	26 48	23 05	20 02	14 02	18 04	11 48
27 We	23 20	08 28	18 36	15 14	26 49	23 05	20 03	14 03	18 04	11 48
28 Th	23 17	03 03	18 38	15 31	26 50	23 06	20 04	14 03	18 04	11 48
29 Fr	23 14	02S 25	18 41	15 47	26 50	23 06	20 05	14 04	18 05	11 49
30 Sa	23 11	07 40	18 46	16 03	26 51	23 07	20 06	14 04	18 05	11 49

ASPECTARIAN

01 01:39 ☽ △ ♆
05:20 ☽ △ ☉
10:55 ☽ ⚹ ♇
22:27 ☽ △ ♃

02 05:51 ☽ △ ♅
07:04 ♀ △ ♂
08:09 ☽ ⚹ ♂
08:15 ☽ ☍ ♀
14:43 ☽ △ ☿
20:50 ☽ # ♀

03 06:18 ☽ □ ♆
14:34 ☽ ∥ ♆

04 01:56 ☽ ∥ ♅
05:23 ☿ SR
11:31 ☽ □ ♅
11:51 ☉ ☍ ♇

05 02:48 ☽ ∥ ♆
07:37 ☽ ☍ ♄
12:36 ☿ ∥ ♃
12:55 ☽ ⚹ ♆
15:16 ☽ # ♄
22:47 ☽ ☌ ♇

06 03:59 ☿ ∥ ☉
05:42 ☿ ⚹ ♀
10:26 ♀ ♉
13:30 ☽ ☍ ♃
19:28 ☽ ⚹ ♅
19:39 ☽ ☌ ♂
21:52 ☽ # ☿

07 03:14 ☽ # ☉
03:55 ♂ ⚹ ♅
04:12 ☽ [illegible] ☿
05:26 ☽ # ♃
06:59 ☽ △ ♀

08 12:59 ☉ ∥ ♃

09 03:20 ☽ # ☉
04:17 ☽ # ♃
23:08 ☽ □ ♀

10 01:35 ☽ # ☿
05:10 ☽ △ ♄
09:30 ☽ ☌ ♆
20:01 ☽ ⚹ ♇
20:21 ☽ # ♄

11 09:44 ☽ △ ☉
10:49 ☽ ∥ ♆
14:25 ☽ △ ♃
15:37 ☽ ⚹ ♂
18:19 ☽ ☌ ♅

12 00:40 ☽ △ ☿
14:26 ☽ ∥ ♅
17:02 ♂ ☍ ♃
17:13 ☽ ⚹ ♀
18:29 ☽ □ ♄

13 03:23 ☽ ∥ ♆
06:41 ☽ # ♀
08:33 ☽ □ ♇
17:47 ☉ ☍ ♂

14 02:30 ☽ □ ♂
04:04 ☽ □ ♃
10:07 ☽ □ [illegible]
12:39 ☉ ☌ ♃

15 02:46 ♀ □ ♆
04:56 ♀ # ♆
06:51 ☽ ⚹ ♄
09:38 ☽ ⚹ ♆
18:58 ☉ △ ♅
19:38 ☽ △ ♇

16 08:07 ☿ ∥ ♄
11:24 ☽ △ ♂
13:27 ☿ ☌ ☉
15:26 ☽ ⚹ ♃
16:46 ☽ ⚹ ♅
17:58 ☽ ⚹ ☿
18:33 ☽ ⚹ ☉

17 18:11 ☽ □ ♆
22:09 ☿ △ ♅
23:32 ☽ ☌ ♀

18 03:06 ☽ # ♆
07:58 ☽ ∥ ♀
10:13 ☿ ☌ ♃
14:46 ☽ # ♅
23:23 ☽ □ ♅

19 12:26 ♃ △ ♅
14:46 ☽ # ♆
21:53 ☽ ☌ ♄
22:17 ☽ ∥ ☿
23:12 ☽ △ ♆

20 04:17 ☽ ∥ ♄
07:58 ☽ ☍ ♇
19:46 ☽ ☍ ♂

21 00:11 ☽ [illegible] ☿
02:46 ☽ △ ♅
03:26 ☽ ☌ ♃
07:39 ☉ ♋
15:29 ☽ ∥ ♃

22 14:13 ☽ ⚹ ♀
17:33 ♀ # ♅
23:49 ☽ ∥ ♃

24 02:10 ☽ ⚹ ♄
02:26 ☽ ☍ ♆
09:40 ☽ ∥ ♄
10:46 ☽ △ ♇
18:54 ☽ □ ♀
19:24 ☽ ∥ ☿
16:16 ☽ [illegible]
22:58 ☽ # ♆

25 00:17 ☽ ⚹ ☿
05:02 ☽ ☍ ♅
07:18 ♄ △ ♆
07:24 ☽ ⚹ ♃
17:08 ☽ ∥ ♀
21:11 ☽ ⚹ ☉
21:25 ☽ # ♅

26 03:50 ☽ □ ♄
08:36 ☽ # ♆
12:07 ☽ □ ♇
20:58 ☽ □ ♂

27 00:28 ☽ △ ♀
01:15 ☽ □ [illegible]
10:13 ☽ □ ♃

28 05:50 ☿ SD
06:16 ☽ △ ♆
07:00 ☽ △ ♄
15:05 ☽ ⚹ ♇
23:24 ☽ ⚹ ♂

29 05:06 ☽ △ ☿
10:41 ☽ △ ♅
15:09 ☽ △ ♃

30 11:08 ☽ □ ♆
12:15 ☽ △ ☉
12:47 ♀ □ ♅
16:33 ☽ ∥ [illegible]
23:45 ☉ ∥ ♃

⚷ Chiron

01 Dec.	17 S 04
03	26♐48R
06	26 36
09	26 24
12	26 12
15	26 00
18	25 48
21	25 36
24	25 24
27	25 12
30	25 00

05 14:56 13♑39 ✹ Partial Lunar Eclipse (mag 0.499)

July 2001

Day	S. T. h m s	☉ ° ' "	☽ ° ' "	☿ ° '	♀ ° '	♂ ° '	♃ ° '	♄ ° '	♅ ° '	♆ ° '	♇ ° '	☊ True ° '
01 Su	18 36 27	09♋ 14 12	15♏ 14 38	21♊ 34	24♉ 57	17♐R33	27♊ 18	08♊ 57	24♒R26	08♒R08	13♐R14	06♋R17
02 Mo	18 40 24	10 11 24	28 15 45	21 50	26 01	17 18	27 32	09 04	24 24	08 07	13 13	06D 17
03 Tu	18 44 20	11 08 35	11♐ 02 29	22 11	27 05	17 04	27 45	09 11	24 23	08 06	13 11	06 17
04 We	18 48 17	12 05 47	23 36 08	22 36	28 10	16 51	27 59	09 18	24 21	08 04	13 10	06 18
05 Th	18 52 13	13 02 58	05♑ 58 07	23 07	29 15	16 39	28 12	09 25	24 20	08 03	13 09	06 18
06 Fr	18 56 10	14 00 09	18 09 54	23 42	00♊ 20	16 27	28 26	09 32	24 18	08 01	13 07	06 18
07 Sa	19 00 06	14 57 21	00♒ 13 09	24 23	01 25	16 16	28 39	09 39	24 17	08 00	13 06	06R 18
08 Su	19 04 03	15 54 32	12 09 49	25 08	02 30	16 06	28 53	09 45	24 15	07 58	13 05	06 18
09 Mo	19 07 59	16 51 44	24 02 08	25 58	03 36	15 57	29 06	09 52	24 13	07 57	13 03	06 17
10 Tu	19 11 56	17 48 55	05♓ 52 47	26 52	04 42	15 48	29 20	09 59	24 11	07 55	13 02	06 16
11 We	19 15 53	18 46 08	17 44 52	27 51	05 48	15 40	29 33	10 05	24 10	07 54	13 01	06 15
12 Th	19 19 49	19 43 20	29 42 04	28 55	06 54	15 33	29 47	10 12	24 08	07 52	13 00	06 15
13 Fr	19 23 46	20 40 33	11♈ 48 27	00♋ 04	08 00	15 27	30 00	10 19	24 06	07 51	12 59	06D 15
14 Sa	19 27 42	21 37 47	24 08 18	01 16	09 06	15 21	00♋ 13	10 25	24 04	07 49	12 57	06 15
15 Su	19 31 39	22 35 01	06♉ 45 54	02 34	10 13	15 17	00 27	10 32	24 02	07 48	12 56	06 16
16 Mo	19 35 35	23 32 16	19 45 03	03 55	11 19	15 13	00 40	10 38	24 00	07 46	12 55	06 18
17 Tu	19 39 32	24 29 31	03♊ 08 46	05 21	12 26	15 10	00 53	10 44	23 58	07 45	12 54	06 19
18 We	19 43 28	25 26 47	16 58 37	06 51	13 33	15 08	01 06	10 51	23 56	07 43	12 53	06 21
19 Th	19 47 25	26 24 04	01♋ 14 10	08 25	14 40	15 07	01 19	10 57	23 54	07 41	12 52	06R 21
20 Fr	19 51 22	27 21 22	15 52 26	10 02	15 48	15D 06	01 32	11 03	23 52	07 40	12 51	06 20
21 Sa	19 55 18	28 18 39	00♌ 47 44	11 44	16 55	15 07	01 45	11 09	23 50	07 38	12 50	06 19
22 Su	19 59 15	29 15 58	15 52 10	13 29	18 03	15 08	01 58	11 15	23 48	07 37	12 49	06 16
23 Mo	20 03 11	00♌ 13 17	00♍ 56 34	15 17	19 10	15 10	02 11	11 22	23 46	07 35	12 48	06 12
24 Tu	20 07 08	01 10 36	15 52 00	17 09	20 18	15 13	02 24	11 27	23 44	07 33	12 47	06 08
25 We	20 11 04	02 07 55	00♎ 30 56	19 03	21 26	15 17	02 37	11 33	23 42	07 32	12 46	06 04
26 Th	20 15 01	03 05 15	14 48 19	21 00	22 34	15 22	02 50	11 39	23 39	07 30	12 45	06 01
27 Fr	20 18 57	04 02 35	28 41 41	22 59	23 42	15 27	03 03	11 45	23 37	07 29	12 44	05 59
28 Sa	20 22 54	04 59 56	12♏ 11 01	25 00	24 51	15 34	03 15	11 51	23 35	07 27	12 43	05D 59
29 Su	20 26 51	05 57 17	25 18 05	27 03	25 59	15 41	03 28	11 56	23 33	07 25	12 43	05 59
30 Mo	20 30 47	06 54 38	08♐ 05 39	29 07	27 07	15 49	03 41	12 02	23 31	07 24	12 42	06 01
31 Tu	20 34 44	07 52 01	20 36 57	01♌ 11	28 16	15 57	03 53	12 07	23 28	07 22	12 41	06 03

Data for	07-01-2001
Julian Day	2452091.50
Ayanamsa	23 52 23
SVP	05 ♓ 14 03
☽ ☊ Mean	06 ♋ 06 R

● ◐ PHASES ○ ◑

Day	Time	Phase	Position
05	15:04	✹	13♑39
13	18:45	◑	21♈25
20	19:45	●	28♋09
27	10:09	◐	04♏27

LAST ASPECT ☽ Day	h m	INGRESS Day	h m	Sign
01	19:27	02	03:15	♐
04	08:38	04	12:23	♑
05	15:05	06	23:35	♒
09	10:29	09	12:06	♓
12	00:10	12	00:37	♈
13	23:53	14	11:14	♉
16	07:42	16	18:27	♊
18	11:47	18	21:58	♋
20	19:46	20	22:45	♌
22	12:36	22	22:31	♍
24	07:50	24	23:10	♎
26	15:11	27	02:18	♏
29	03:52	29	08:46	♐
31	16:26	31	18:17	♑

DECLINATION

Day	☉	☽	☿	♀	♂	♃	♄	♅	♆	♇
01 Su	23N07	12S27	18N53	16N19	26S51	23N07	20N07	14S05	18S05	11S49
02 Mo	23 03	16 35	19 00	16 35	26 51	23 07	20 09	14 05	18 06	11 49
03 Tu	22 58	19 51	19 09	16 51	26 51	23 08	20 10	14 06	18 06	11 49
04 We	22 53	22 07	19 19	17 06	26 51	23 08	20 11	14 06	18 06	11 49
05 Th	22 48	23 16	19 30	17 21	26 51	23 08	20 12	14 07	18 07	11 49
06 Fr	22 42	23 17	19 41	17 36	26 51	23 09	20 13	14 08	18 07	11 49
07 Sa	22 36	22 12	19 54	17 51	26 51	23 09	20 14	14 08	18 08	11 49
08 Su	22 29	20 07	20 07	18 05	26 51	23 09	20 15	14 09	18 08	11 50
09 Mo	22 23	17 12	20 20	18 19	26 51	23 09	20 15	14 09	18 08	11 50
10 Tu	22 15	13 35	20 34	18 32	26 51	23 09	20 16	14 10	18 09	11 50
11 We	22 07	09 27	20 48	18 45	26 51	23 10	20 17	14 11	18 09	11 50
12 Th	21 59	04 56	21 01	18 58	26 51	23 10	20 18	14 11	18 10	11 50
13 Fr	21 51	00 10	21 15	19 11	26 51	23 10	20 19	14 12	18 10	11 50
14 Sa	21 42	04N42	21 28	19 23	26 51	23 10	20 20	14 12	18 10	11 50
15 Su	21 33	09 29	21 40	19 35	26 51	23 10	20 21	14 13	18 11	11 51
16 Mo	21 23	13 59	21 52	19 46	26 50	23 10	20 22	14 14	18 11	11 51
17 Tu	21 13	17 57	22 03	19 57	26 50	23 10	20 23	14 14	18 12	11 51
18 We	21 03	21 03	22 12	20 07	26 50	23 10	20 23	14 15	18 12	11 51
19 Th	20 52	22 57	22 20	20 17	26 50	23 10	20 24	14 16	18 12	11 51
20 Fr	20 41	23 21	22 27	20 27	26 50	23 10	20 25	14 17	18 13	11 52
21 Sa	20 30	22 06	22 31	20 36	26 50	23 10	20 26	14 17	18 13	11 52
22 Su	20 18	19 16	22 34	20 45	26 50	23 10	20 27	14 18	18 14	11 52
23 Mo	20 06	15 08	22 35	20 53	26 50	23 09	20 27	14 19	18 14	11 52
24 Tu	19 54	10 06	22 33	21 01	26 50	23 09	20 28	14 19	18 15	11 53
25 We	19 41	04 34	22 29	21 08	26 50	23 09	20 29	14 20	18 15	11 53
26 Th	19 28	01S04	22 22	21 15	26 51	23 09	20 29	14 21	18 15	11 53
27 Fr	19 15	06 29	22 13	21 21	26 51	23 09	20 30	14 22	18 16	11 53
28 Sa	19 01	11 27	22 01	21 27	26 51	23 09	20 31	14 22	18 16	11 54
29 Su	18 47	15 45	21 47	21 32	26 51	23 08	20 32	14 23	18 17	11 54
30 Mo	18 33	19 13	21 29	21 37	26 52	23 08	20 32	14 24	18 17	11 54
31 Tu	18 18	21 42	21 10	21 41	26 52	23 08	20 33	14 25	18 18	11 55

ASPECTARIAN

01 08:56 ☽ ∥ ♅
16:52 ☽ □ ♅
19:26 ☽ ☍ ♀
02 00:01 ☽ # ♀
10:17 ☽ ∥ ♆
17:53 ☽ # ☿
18:27 ☽ ⚹ ♆
20:28 ☽ ☍ ♄

03 02:43 ☽ # ♄
04:05 ☽ ☌ ♇
11:18 ☽ ☌ ♂
22:01 ☽ ☍ ☿

04 01:28 ☽ ⚹ ♅
08:38 ☽ ☍ ♃
12:15 ☽ # ☉
19:35 ☽ # ♃

05 16:45 ♀ ♊
06 04:50 ☽ # ♃
16:43 ☽ # ☉
20:43 ☿ △ ♅

07 02:39 ☽ △ ♀
15:35 ☽ ☌ ♆
19:07 ☽ △ ♄
22:48 ☽ # ♄
08 00:01 ☽ # ☿
01:51 ☽ ⚹ ♇
05:49 ♀ # ♆
07:52 ☽ ⚹ ♂
14:49 ☿ ∥ ♄
16:07 ☽ # ♀
16:54 ☽ ∥ ♆

09 00:23 ☽ ☌ ♅
04:14 ☽ △ ☿
10:29 ☽ △ ♃
20:23 ☽ ∥ ♅
21:22 ☽ □ ♀
10 08:24 ☽ □ ♄
10:29 ☽ ∥ ♇
14:29 ☽ □ ♇
19:52 ☽ □ ♂

11 02:15 ☽ △ ☉
22:18 ☽ □ ☿
12 00:10 ☽ □ ♃
15:45 ☽ ⚹ ♀
16:13 ☽ ⚹ ♆
20:50 ♀ △ ♆
21:03 ☽ ⚹ ♄
22:30 ☿ ☌ ♃
22:48 ☿ ♋

13 00:03 ♃ ♋
02:18 ☽ △ ♇
07:06 ☽ △ ♂
23:53 ☽ ⚹ ♅

14 11:52 ☽ ⚹ ♃
15:11 ☽ ⚹ ☿
15:41 ☿ ∥ ☉

15 01:56 ☽ □ ♆
07:33 ♀ ☌ ♄
12:21 ☽ # ♇
16 01:23 ☽ # ♅
07:24 ☽ ⚹ ☉
07:42 ☽ □ ♅

17 01:38 ☽ # ♆
08:04 ☽ △ ♆
09:46 ♀ ☍ ♇
13:23 ☽ ☌ ♄
15:20 ☽ ∥ ♀
17:00 ☽ ☍ ♇
17:38 ☽ ☌ ♀
18:01 ☽ ∥ ♄
20:52 ☽ ☍ ♂
23:53 ☽ ∥ ☉
18 11:47 ☽ △ ♅
13:10 ☽ ∥ ☿

19 00:10 ☽ ☌ ♃
04:33 ☽ ∥ ♃
09:21 ♀ ☍ ♂
13:19 ☽ ☌ ☿
18:25 ♀ ∥ ♄
22:45 ♂ SD
20 07:09 ☽ ∥ ♃
16:40 ♀ ∥ ☉
18:56 ☽ ∥ ☿

21 08:03 ☉ ∥ ♄
10:54 ☽ ☍ ♆
13:33 ☽ ∥ ♀
15:32 ☽ ∥ ♄
16:03 ☽ ∥ ☉
16:38 ☽ ⚹ ♄
19:10 ☽ △ ♇
22:51 ☽ △ ♂

22 03:45 ☽ ⚹ ♀
06:42 ☽ # ♆
12:36 ☽ ☍ ♅
18:27 ☉ ♌

23 02:02 ☽ ⚹ ♃
04:10 ☽ # ♅
15:53 ☽ # ♇
16:50 ☽ □ ♄
19:02 ☽ □ ♇
22:58 ☽ □ ♂
24 02:24 ☽ ⚹ ☿
07:50 ☽ □ ♀

25 02:53 ☽ ⚹ ☉
03:34 ☽ □ ♃
11:42 ☽ △ ♆
18:38 ☽ △ ♄
20:32 ☽ ⚹ ♇
26 00:59 ☽ ⚹ ♂
12:23 ☽ □ ☿
14:32 ☽ △ ♀
15:11 ☽ △ ♅
22:19 ♀ △ ♅

27 07:48 ☽ △ ♃
15:32 ☽ □ ♆
28 02:16 ☽ ∥ ♇
15:48 ☽ ∥ ♅
20:47 ☽ □ ♅

29 03:52 ☽ △ ☿
16:15 ☿ ∥ ♀
16:47 ☽ ∥ ♆
19:05 ☽ # ☉
21:35 ☽ △ ☉
22:41 ☽ ⚹ ♆

30 07:34 ☽ ☍ ♄
08:47 ☽ ☌ ♇
10:19 ☿ ♌
11:27 ☽ # ♄
11:49 ☉ ☍ ♆
14:55 ☽ ☌ ♂
18:38 ☽ # ☿
23:39 ☽ # ♀

31 00:51 ☉ # ♆
05:32 ☽ ⚹ ♅
16:26 ☽ ☍ ♀

⚷ Chiron

01	Dec.	16 S 56
03	24♐48R	
06	24 37	
09	24 26	
12	24 16	
15	24 06	
18	23 56	
21	23 47	
24	23 39	
27	23 31	
30	23 24	

August 2001

Day	S. T. h m s	☉ ° ' "	☽ ° ' "	☿ ° '	♀ ° '	♂ ° '	♃ ° '	♄ ° '	♅ ° '	♆ ° '	♇ ° '	☊ True ° '
01 We	20 38 40	08♌ 49 23	02♑ 55 12	03♌ 17	29♊ 25	16♐ 07	04♋ 06	12♊ 13	23♒R26	07♒R20	12♐R40	06♋ 04
02 Th	20 42 37	09 46 47	15 03 22	05 22	00♋ 34	16 17	04 18	12 18	23 24	07 19	12 40	06R 05
03 Fr	20 46 33	10 44 11	27 04 04	07 28	01 43	16 28	04 31	12 23	23 21	07 17	12 39	06 03
04 Sa	20 50 30	11 41 36	08♒ 59 34	09 33	02 52	16 40	04 43	12 29	23 19	07 16	12 38	06 00
05 Su	20 54 26	12 39 02	20 51 54	11 38	04 01	16 52	04 55	12 34	23 17	07 14	12 38	05 55
06 Mo	20 58 23	13 36 29	02♓ 42 52	13 42	05 10	17 05	05 07	12 39	23 14	07 12	12 37	05 49
07 Tu	21 02 20	14 33 57	14 34 22	15 46	06 19	17 19	05 19	12 44	23 12	07 11	12 37	05 42
08 We	21 06 16	15 31 26	26 28 26	17 48	07 29	17 33	05 31	12 49	23 10	07 09	12 36	05 36
09 Th	21 10 13	16 28 57	08♈ 27 37	19 49	08 38	17 48	05 43	12 54	23 07	07 07	12 36	05 30
10 Fr	21 14 09	17 26 28	20 34 54	21 49	09 48	18 04	05 55	12 58	23 05	07 06	12 35	05 26
11 Sa	21 18 06	18 24 01	02♉ 53 48	23 48	10 58	18 20	06 07	13 03	23 03	07 04	12 35	05 24
12 Su	21 22 02	19 21 35	15 28 12	25 45	12 08	18 37	06 19	13 08	23 00	07 03	12 34	05D 23
13 Mo	21 25 59	20 19 11	28 22 03	27 41	13 18	18 55	06 31	13 12	22 58	07 01	12 34	05 24
14 Tu	21 29 55	21 16 49	11♊ 39 03	29 36	14 28	19 13	06 42	13 17	22 55	07 00	12 34	05 25
15 We	21 33 52	22 14 28	25 22 07	01♍ 29	15 38	19 32	06 54	13 21	22 53	06 58	12 33	05 26
16 Th	21 37 49	23 12 08	09♋ 32 31	03 21	16 48	19 51	07 05	13 25	22 51	06 56	12 33	05R 26
17 Fr	21 41 45	24 09 50	24 09 06	05 11	17 59	20 11	07 17	13 30	22 48	06 55	12 33	05 24
18 Sa	21 45 42	25 07 33	09♌ 07 32	07 00	19 09	20 32	07 28	13 34	22 46	06 53	12 33	05 20
19 Su	21 49 38	26 05 18	24 20 17	08 47	20 20	20 53	07 39	13 38	22 43	06 52	12 33	05 14
20 Mo	21 53 35	27 03 04	09♍ 37 20	10 33	21 31	21 15	07 50	13 42	22 41	06 50	12 32	05 06
21 Tu	21 57 31	28 00 51	24 47 48	12 18	22 41	21 37	08 01	13 46	22 39	06 49	12 32	04 57
22 We	22 01 28	28 58 40	09♎ 41 40	14 01	23 52	22 00	08 12	13 49	22 36	06 47	12 32	04 49
23 Th	22 05 24	29 56 29	24 11 31	15 43	25 03	22 23	08 23	13 53	22 34	06 46	12 32	04 42
24 Fr	22 09 21	00♍ 54 20	08♏ 13 15	17 23	26 14	22 47	08 34	13 56	22 32	06 44	12D 32	04 37
25 Sa	22 13 18	01 52 12	21 46 10	19 02	27 25	23 11	08 45	14 00	22 29	06 43	12 32	04 34
26 Su	22 17 14	02 50 06	04♐ 52 12	20 40	28 36	23 36	08 55	14 03	22 27	06 41	12 32	04D 33
27 Mo	22 21 11	03 48 00	17 35 01	22 16	29 47	24 01	09 06	14 07	22 25	06 40	12 32	04 33
28 Tu	22 25 07	04 45 56	29 59 01	23 51	00♌ 59	24 27	09 16	14 10	22 22	06 39	12 33	04 35
29 We	22 29 04	05 43 53	12♑ 08 47	25 25	02 10	24 53	09 26	14 13	22 20	06 37	12 33	04R 35
30 Th	22 33 00	06 41 52	24 08 40	26 57	03 22	25 20	09 37	14 16	22 18	06 36	12 33	04 34
31 Fr	22 36 57	07 39 52	06♒ 02 28	28 28	04 33	25 47	09 47	14 19	22 15	06 35	12 33	04 31

Data for	08-01-2001
Julian Day	2452122.50
Ayanamsa	23 52 28
SVP	05 ♓ 14 00
☽ ☊ Mean	04 ♋ 28 R

● ◐ PHASES ○ ◑

Day	Time	Phase	Position
04	05:55	○	11♒56
12	07:54	◑	19♉41
19	02:56	●	26♌12
25	19:55	◐	02♐40

LAST ASPECT Day	h m	☽ INGRESS Day	h m	Sign
01	02:22	03	05:54	♒
05	04:53	05	18:31	♓
07	05:40	08	07:06	♈
10	04:55	10	18:25	♉
12	22:33	13	03:00	♊
14	19:44	15	07:57	♋
16	13:05	17	09:27	♌
19	02:57	19	08:54	♍
20	20:23	21	08:20	♎
23	01:36	23	09:51	♏
25	11:17	25	15:00	♐
27	12:51	28	00:03	♑
30	06:29	30	11:49	♒

DECLINATION

Day	☉	☽	☿	♀	♂	♃	♄	♅	♆	♇
01 We	18N03	23S 07	20N47	21N45	26S 52	23N07	20N33	14S 25	18S 18	11S 55
02 Th	17 48	23 23	20 23	21 48	26 53	23 07	20 34	14 26	18 18	11 55
03 Fr	17 33	22 34	19 55	21 50	26 53	23 07	20 35	14 27	18 19	11 55
04 Sa	17 17	20 43	19 26	21 52	26 54	23 06	20 35	14 28	18 19	11 56
05 Su	17 01	18 00	18 55	21 53	26 54	23 06	20 36	14 28	18 20	11 56
06 Mo	16 44	14 32	18 22	21 54	26 54	23 06	20 36	14 29	18 20	11 56
07 Tu	16 28	10 30	17 47	21 54	26 55	23 05	20 37	14 30	18 21	11 57
08 We	16 11	06 04	17 11	21 54	26 55	23 05	20 37	14 31	18 21	11 57
09 Th	15 54	01 23	16 33	21 53	26 56	23 04	20 38	14 32	18 21	11 57
10 Fr	15 36	03N25	15 54	21 52	26 56	23 04	20 38	14 32	18 22	11 58
11 Sa	15 19	08 10	15 14	21 49	26 57	23 04	20 39	14 33	18 22	11 58
12 Su	15 01	12 42	14 32	21 47	26 57	23 03	20 39	14 34	18 23	11 59
13 Mo	14 43	16 46	13 50	21 43	26 58	23 03	20 40	14 35	18 23	11 59
14 Tu	14 24	20 07	13 08	21 40	26 58	23 02	20 40	14 36	18 23	11 59
15 We	14 06	22 27	12 24	21 35	26 59	23 02	20 41	14 36	18 24	12 00
16 Th	13 47	23 27	11 41	21 30	26 59	23 01	20 41	14 37	18 24	12 00
17 Fr	13 28	22 54	10 56	21 24	27 00	23 00	20 42	14 38	18 25	12 00
18 Sa	13 09	20 43	10 12	21 18	27 00	23 00	20 42	14 39	18 25	12 01
19 Su	12 49	17 03	09 27	21 11	27 00	22 59	20 42	14 39	18 25	12 01
20 Mo	12 30	12 13	08 42	21 04	27 01	22 59	20 43	14 40	18 26	12 02
21 Tu	12 10	06 40	07 57	20 56	27 01	22 58	20 43	14 41	18 26	12 02
22 We	11 50	00 50	07 12	20 47	27 01	22 58	20 43	14 42	18 27	12 03
23 Th	11 30	04S 54	06 26	20 38	27 01	22 57	20 44	14 43	18 27	12 03
24 Fr	11 09	10 13	05 41	20 28	27 02	22 56	20 44	14 43	18 27	12 03
25 Sa	10 49	14 51	04 56	20 18	27 02	22 56	20 44	14 44	18 28	12 04
26 Su	10 28	18 37	04 11	20 07	27 02	22 55	20 45	14 45	18 28	12 04
27 Mo	10 07	21 22	03 27	19 55	27 01	22 54	20 45	14 46	18 28	12 05
28 Tu	09 46	23 01	02 42	19 43	27 01	22 54	20 45	14 46	18 29	12 05
29 We	09 25	23 33	01 58	19 30	27 01	22 53	20 45	14 47	18 29	12 06
30 Th	09 03	22 58	01 14	19 17	27 01	22 53	20 46	14 48	18 30	12 06
31 Fr	08 42	21 20	00 31	19 03	27 00	22 52	20 46	14 49	18 30	12 06

ASPECTARIAN

01 00:19 ☽ # ♃
02:22 ☽ ☍ ♃
12:19 ♀ ♋
13:23 ☿ ∥ ♄
02 12:01 ☽ # ♃
21:58 ☿ ☍ ♆
03 10:59 ☽ # ♀
20:31 ☽ ☌ ♆
04 01:24 ☽ ☍ ☿
01:29 ☽ # ♄
07:06 ☽ △ ♄
07:22 ☽ ✶ ♇
15:05 ☽ # ☿
15:46 ☽ ✶ ♂
21:29 ☽ ∥ ♆
21:39 ☉ ✶ ♄
23:31 ☉ △ ♇
05 04:53 ☽ ☌ ♅
08:01 ☽ # ☉
11:12 ☿ ✶ ♄
11:28 ☿ △ ♇
17:04 ♄ ☍ ♇
21:52 ☿ ☌ ☉
22:51 ♀ ☌ ♃
06 00:21 ☽ ∥ ♅
01:10 ☿ # ♆
04:58 ☽ △ ♃
05:31 ☽ △ ♀
15:48 ☽ ∥ ♇
20:03 ☽ □ ♇
20:16 ☽ □ ♄
07 05:40 ☽ □ ♂
20:39 ☿ △ ♂
08 18:28 ☽ □ ♅
21:21 ☽ ✶ ♆
09 00:25 ☽ □ ♀
08:14 ☽ △ ♇
08:53 ☽ ✶ ♄
17:18 ☽ △ ☉
18:57 ☽ △ ♂
10 02:56 ☽ △ ☿
04:55 ☽ ✶ ♅
14:58 ☿ ☍ ♅
18:30 ☿ ∥ ☉
21:48 ☉ △ ♂
11 06:19 ☽ ✶ ♃
08:02 ☽ □ ♆
17:03 ☽ ✶ ♀
20:04 ☽ # ♇
23:09 ☿ # ♅
12 08:59 ☽ ∥ ☿
10:41 ☽ # ♅
12:22 ☽ ∥ ☉
14:04 ☽ □ ♅
22:33 ☽ □ ☿
13 10:07 ☉ # ♅
10:51 ☽ # ♆
15:42 ☽ △ ♆
14 01:38 ☽ ☍ ♇
02:55 ☽ ☌ ♄
04:44 ☽ ∥ ♄
05:05 ☿ ♍
13:40 ☽ ☍ ♂
13:52 ☽ ∥ ♀
18:13 ☽ ✶ ☉
19:44 ☽ △ ♅
15 09:17 ☽ ∥ ♃
12:03 ☽ ✶ ☿
13:30 ☿ # ♇
15:26 ☉ ☍ ♅
19:51 ☽ ☌ ♃
16 13:05 ☽ ☌ ♀
22:14 ☽ ∥ ♃
17 18:54 ☽ ∥ ♀
20:28 ☽ ☍ ♆
18 00:13 ☽ ∥ ♄
05:26 ☽ △ ♇
06:58 ☿ ✶ ♃
07:04 ☽ ✶ ♄
16:02 ☽ # ♆
18:27 ☽ △ ♂
21:29 ☽ ☍ ♅
19 12:29 ☽ # ♅
21:11 ☽ ✶ ♃
22:43 ☽ ∥ ☉
20 00:55 ☽ # ♇
01:40 ☽ ☌ ☿
04:37 ☽ □ ♇
06:27 ☽ □ ♄
17:52 ☽ ∥ ☿
18:50 ☽ □ ♂
20:23 ☽ ✶ ♀
21 03:23 ☿ □ ♇
09:08 ☉ # ♇
19:18 ☽ △ ♆
21:11 ☿ □ ♄
21:34 ☽ □ ♃
22 04:40 ☽ ✶ ♇
06:48 ☽ △ ♄
09:17 ♀ ∥ ♄
20:54 ☽ ✶ ♂
21:18 ☽ △ ♅
23 01:28 ☉ ♍
01:36 ☽ □ ♀
05:54 ☽ # ☿
10:00 ♂ ✶ ♅
10:28 ☽ ✶ ☉
16:07 ♇ SD
21:27 ☽ □ ♆
24 00:37 ☽ △ ♃
04:17 ☽ # ☉
09:07 ☽ ∥ ♇
18:25 ☽ ✶ ☿
23:24 ☽ ∥ ♅
25 01:18 ☽ □ ♅
11:17 ☽ △ ♀
23:00 ☽ ∥ ♆
26 03:00 ☽ ✶ ♆
11:11 ☽ # ♀
14:25 ☽ ☌ ♇
17:22 ☽ ☍ ♄
17:44 ☽ # ♄
27 04:13 ♀ ♌
09:16 ☽ ✶ ♅
10:20 ☽ □ ☿
12:51 ☽ ☌ ♂
21:27 ☽ # ♃
28 10:13 ☽ △ ☉
12:43 ☿ □ ♂
18:34 ☽ ☍ ♃
30 01:45 ☽ # ♃
06:29 ☽ △ ☿
20:40 ☽ ☍ ♀
31 01:06 ☽ ☌ ♆
06:08 ☽ # ♄
13:12 ☽ ✶ ♇
16:50 ☽ △ ♄
23:39 ☽ # ♀

⚷ Chiron

01 Dec.	16 S 58
02	23♐17R
05	23 11
08	23 06
11	23 02
14	22 58
17	22 55
20	22 53
23	22 52
26	22 51
29	22 52D

[illegible]

September 2001

Day	S. T.	☉	☽	☿	♀	♂	♃	♄	♅	♆	♇	☊ True
	h m s	° ' "	° ' "	° '	° '	° '	° '	° '	° '	° '	° '	° '
01 Sa	22 40 53	08♍37 53	17♒53 22	29♍58	05♌45	26♐14	09♋57	14♊22	22♒R13	06♒R33	12♐33	04♋R24
02 Su	22 44 50	09 35 56	29 43 57	01♎26	06 57	26 42	10 07	14 24	22 11	06 32	12 34	04 15
03 Mo	22 48 47	10 34 00	11♓36 08	02 53	08 08	27 10	10 16	14 27	22 09	06 31	12 34	04 04
04 Tu	22 52 43	11 32 07	23 31 27	04 19	09 20	27 39	10 26	14 29	22 06	06 29	12 35	03 52
05 We	22 56 40	12 30 14	05♈31 15	05 43	10 32	28 08	10 36	14 32	22 04	06 28	12 35	03 40
06 Th	23 00 36	13 28 24	17 36 58	07 06	11 44	28 37	10 45	14 34	22 02	06 27	12 35	03 29
07 Fr	23 04 33	14 26 35	29 50 25	08 27	12 56	29 07	10 54	14 36	22 00	06 26	12 36	03 20
08 Sa	23 08 29	15 24 49	12♉13 53	09 47	14 09	29 37	11 04	14 38	21 58	06 25	12 36	03 14
09 Su	23 12 26	16 23 04	24 50 07	11 06	15 21	00♑08	11 13	14 40	21 56	06 23	12 37	03 10
10 Mo	23 16 22	17 21 22	07♊42 23	12 23	16 33	00 39	11 22	14 42	21 53	06 22	12 37	03 09
11 Tu	23 20 19	18 19 42	20 54 05	13 38	17 46	01 10	11 31	14 44	21 51	06 21	12 38	03 09
12 We	23 24 15	19 18 04	04♋28 26	14 52	18 58	01 41	11 39	14 46	21 49	06 20	12 38	03 08
13 Th	23 28 12	20 16 28	18 27 41	16 03	20 11	02 13	11 48	14 47	21 47	06 19	12 39	03 07
14 Fr	23 32 09	21 14 54	02♌52 16	17 13	21 23	02 45	11 57	14 49	21 45	06 18	12 40	03 04
15 Sa	23 36 05	22 13 22	17 39 48	18 21	22 36	03 18	12 05	14 50	21 43	06 17	12 40	02 58
16 Su	23 40 02	23 11 52	02♍44 36	19 27	23 49	03 51	12 13	14 52	21 41	06 16	12 41	02 49
17 Mo	23 43 58	24 10 24	17 57 57	20 31	25 01	04 24	12 21	14 53	21 40	06 15	12 42	02 39
18 Tu	23 47 55	25 08 58	03♎09 12	21 33	26 14	04 57	12 29	14 54	21 38	06 14	12 43	02 27
19 We	23 51 51	26 07 34	18 07 36	22 32	27 27	05 31	12 37	14 55	21 36	06 13	12 44	02 16
20 Th	23 55 48	27 06 12	02♏44 10	23 28	28 40	06 05	12 45	14 55	21 34	06 12	12 45	02 06
21 Fr	23 59 44	28 04 51	16 53 05	24 22	29 53	06 39	12 52	14 56	21 32	06 11	12 45	01 58
22 Sa	00 03 41	29 03 32	00♐32 07	25 12	01♍07	07 13	13 00	14 57	21 30	06 11	12 46	01 52
23 Su	00 07 38	00♎02 15	13 42 18	25 59	02 20	07 48	13 07	14 57	21 29	06 10	12 47	01 49
24 Mo	00 11 34	01 01 00	26 26 55	26 43	03 33	08 23	13 14	14 58	21 27	06 09	12 48	01 48
25 Tu	00 15 31	01 59 46	08♑50 38	27 23	04 46	08 58	13 21	14 58	21 25	06 08	12 49	01D 48
26 We	00 19 27	02 58 35	20 58 36	27 58	06 00	09 34	13 28	14 58	21 24	06 07	12 50	01R 48
27 Th	00 23 24	03 57 24	02♒55 56	28 29	07 13	10 10	13 35	14R 58	21 22	06 07	12 52	01 47
28 Fr	00 27 20	04 56 16	14 47 26	28 55	08 27	10 46	13 41	14 58	21 21	06 06	12 53	01 43
29 Sa	00 31 17	05 55 09	26 37 16	29 15	09 40	11 22	13 48	14 58	21 19	06 06	12 54	01 36
30 Su	00 35 13	06 54 04	08♓28 51	29 30	10 54	11 58	13 54	14 58	21 18	06 05	12 55	01 27

Data for	09-01-2001
Julian Day	2452153.50
Ayanamsa	23 52 32
SVP	05 ♓ 13 55
☽ ☊ Mean	02 ♋ 49 R

● ◐ PHASES ○ ◑

02	21:43	○	10♓29
10	19:00	◑	18♊08
17	10:27	●	24♍36
24	09:31	◐	01♑24

LAST ASPECT ☽ Day	h m	INGRESS Day	h m	
01	17:37	02	00:33	♓
04	08:38	04	12:59	♈
06	22:33	07	00:20	♉
08	18:32	09	09:43	♊
11	01:43	11	16:11	♋
13	03:18	13	19:17	♌
15	08:36	15	19:40	♍
17	10:28	17	19:01	♎
19	16:39	19	19:28	♏
21	21:10	21	23:03	♐
24	00:33	24	06:50	♑
26	14:40	26	18:06	♒
29	05:29	29	06:52	♓

DECLINATION

Day	☉	☽	☿	♀	♂	♃	♄	♅	♆	♇
01 Sa	08N20	18S 47	00S 12	18N49	27S 00	22N51	20N46	14S 49	18S 30	12S 07
02 Su	07 58	15 27	00 55	18 34	26 59	22 51	20 46	14 50	18 31	12 07
03 Mo	07 36	11 30	01 37	18 19	26 59	22 50	20 46	14 51	18 31	12 08
04 Tu	07 14	07 06	02 19	18 03	26 58	22 49	20 46	14 51	18 31	12 08
05 We	06 52	02 25	03 00	17 47	26 57	22 49	20 47	14 52	18 32	12 09
06 Th	06 30	02N25	03 40	17 30	26 56	22 48	20 47	14 53	18 32	12 09
07 Fr	06 07	07 12	04 20	17 13	26 55	22 47	20 47	14 54	18 32	12 10
08 Sa	05 45	11 47	05 00	16 55	26 53	22 46	20 47	14 54	18 33	12 10
09 Su	05 22	15 57	05 38	16 36	26 52	22 46	20 47	14 55	18 33	12 11
10 Mo	05 00	19 28	06 16	16 17	26 51	22 45	20 47	14 56	18 33	12 11
11 Tu	04 37	22 04	06 53	15 58	26 49	22 44	20 47	14 56	18 33	12 12
12 We	04 14	23 29	07 30	15 38	26 47	22 44	20 47	14 57	18 34	12 12
13 Th	03 51	23 30	08 05	15 18	26 45	22 43	20 47	14 58	18 34	12 13
14 Fr	03 28	21 57	08 40	14 57	26 43	22 42	20 47	14 58	18 34	12 13
15 Sa	03 05	18 54	09 13	14 36	26 41	22 42	20 47	14 59	18 35	12 14
16 Su	02 42	14 33	09 46	14 15	26 38	22 41	20 47	14 59	18 35	12 14
17 Mo	02 19	09 13	10 17	13 53	26 36	22 40	20 47	15 00	18 35	12 15
18 Tu	01 56	03 20	10 47	13 31	26 33	22 40	20 47	15 01	18 35	12 15
19 We	01 32	02S 39	11 16	13 08	26 30	22 39	20 47	15 01	18 36	12 16
20 Th	01 09	08 22	11 44	12 45	26 27	22 38	20 47	15 02	18 36	12 16
21 Fr	00 46	13 28	12 11	12 21	26 24	22 38	20 47	15 02	18 36	12 17
22 Sa	00 22	17 42	12 35	12 58	26 21	22 37	20 47	15 03	18 36	12 17
23 Su	00S 01	20 53	12 58	11 34	26 17	22 36	20 47	15 03	18 36	12 18
24 Mo	00 24	22 55	13 20	11 09	26 14	22 36	20 47	15 04	18 37	12 18
25 Tu	00 48	23 46	13 39	10 44	26 10	22 35	20 47	15 04	18 37	12 19
26 We	01 11	23 26	13 57	10 19	26 06	22 35	20 47	15 05	18 37	12 20
27 Th	01 34	22 02	14 12	09 54	26 02	22 34	20 47	15 05	18 37	12 20
28 Fr	01 58	19 41	14 25	09 28	25 57	22 33	20 47	15 06	18 37	12 21
29 Sa	02 21	16 30	14 35	09 02	25 53	22 33	20 47	15 06	18 38	12 21
30 Su	02 44	12 40	14 42	08 36	25 48	22 32	20 46	15 07	18 38	12 22

ASPECTARIAN

01 00:38 ☿ ♎
02:14 ☽ ∥ ♆
08:46 ☽ ☌ ♅
15:55 ♀ ☍ ♆
17:37 ☽ ✶ ♂

02 04:01 ☽ ∥ ♅
05:56 ♀ # ♆
15:17 ☉ ✶ ♃
20:26 ☽ ∥ ♇
21:18 ☽ △ ♃

03 01:58 ☽ □ ♇
05:47 ☽ □ ♄
23:18 ☽ # ☉

04 08:38 ☽ □ ♂
21:29 ☽ ∥ ☿

05 00:27 ☽ ☍ ☿
01:54 ☽ ✶ ♆
01:55 ☉ □ ♇
10:15 ☽ □ ♃
11:06 ☽ △ ♀
12:52 ☿ △ ♆
14:03 ☽ △ ♇
17:58 ☽ ✶ ♄

06 07:19 ☽ # ☿
08:42 ☽ ✶ ♅
17:05 ♀ △ ♇
18:59 ☽ ∥ ☉
22:33 ☽ △ ♂

07 04:13 ☉ □ ♄
12:48 ☽ □ ♆
21:44 ☽ ✶ ♃

08 02:11 ☽ # ♇
04:04 ☽ □ ♀
06:38 ☽ △ ☉
10:15 ♀ ✶ ♄
17:48 ☿ # ☉
17:48 ☽ # ♅
17:52 ♂ ♑
18:32 ☽ □ ♅

09 02:31 ☿ □ ♃
03:53 ☽ ∥ ♀
17:18 ☽ # ♆
21:33 ☽ △ ♆

10 04:40 ☿ ✶ ♇
09:02 ☽ ☍ ♇
09:30 ☽ △ ☿
11:06 ☽ ∥ ♄
12:51 ☽ ☌ ♄
17:48 ☽ ✶ ♀

11 01:43 ☽ △ ♅
08:51 ☽ ∥ ♃
18:57 ☽ ☍ ♂
22:07 ☿ △ ♄

12 12:34 ☽ ☌ ♃
19:34 ☽ □ ☿

13 03:18 ☽ ✶ ☉
15:05 ☽ ∥ ♃
23:15 ♀ # ♅

14 05:37 ☽ ☍ ♆
07:07 ♀ ☍ ♅
10:35 ☽ ∥ ♄
15:59 ☽ △ ♇
19:28 ☽ ✶ ♄

15 01:13 ☽ ✶ ☿
02:05 ☽ # ♆
06:30 ☽ ☍ ♅
08:36 ☽ ☌ ♀
21:49 ☽ # ♅

16 01:34 ☽ ∥ ♀
01:49 ☽ △ ♂
10:49 ☽ # ♇
15:06 ☽ ✶ ♃
15:43 ☽ □ ♇
19:09 ☽ □ ♄
19:50 ☽ # ☿

18 01:52 ☿ △ ♅
02:59 ☽ □ ♂
04:55 ☽ △ ♆
06:03 ☽ ∥ ☉
15:03 ☽ □ ♃
15:18 ☽ ✶ ♇
18:49 ☽ △ ♄
19:48 ☽ # ☉

19 05:39 ☽ △ ♅
07:41 ☽ ☌ ☿
16:39 ☽ ✶ ♀

20 05:49 ☽ □ ♆
05:50 ☽ ✶ ♂
17:00 ☽ ∥ ☿
17:04 ☽ △ ♃
18:06 ☽ ∥ ♇
18:54 ☽ # ♀

21 02:10 ♀ ♍
04:37 ♀ # ♇
05:22 ☿ # ♀
06:10 ☿ ∥ ♇
08:05 ☽ □ ♅
08:22 ☽ ∥ ♅
21:10 ☽ ✶ ☉

22 01:09 ☽ □ ♀
06:06 ☽ ∥ ♆
10:11 ☽ ✶ ♆
22:19 ☽ ☌ ♇
23:06 ☉ ♎
23:10 ☽ # ♄

23 02:20 ☽ ☍ ♄
14:32 ☽ ✶ ♅
19:12 ☽ # ♃

24 00:33 ☽ ✶ ☿
15:12 ☽ △ ♀

25 00:17 ☽ ☌ ♂
08:58 ☽ ☍ ♃

26 14:38 ☽ □ ☿
16:40 ☽ # ♃

27 00:05 ♄ SR
02:16 ☽ △ ☉
06:26 ☽ ☌ ♆
13:53 ☽ # ♄
20:08 ☽ ✶ ♇

28 00:23 ☽ △ ♄
08:36 ☽ ∥ ♆
13:17 ☽ ☌ ♅

29 04:11 ☉ △ ♆
05:29 ☽ △ ☿
09:08 ☽ ∥ ♅
12:03 ☽ ∥ ☿

30 01:44 ☽ ∥ ♇
05:27 ☽ ☍ ♀
07:26 ☽ ✶ ♂
08:58 ☽ □ ♇
11:02 ☽ △ ♃
13:04 ☽ □ ♄

⚷ Chiron

01 Dec.	17 S 10
01	22♐53
04	22 55
07	22 57
10	23 01
13	23 05
16	23 10
19	23 16
22	23 23
25	23 30
28	23 38

October 2001

Day	S. T.	☉	☽	☿	♀	♂	♃	♄	♅	♆	♇	☊ True
	h m s	° ' "	° ' "	° '	° '	° '	° '	° '	° '	° '	° '	° '
01 Mo	00 39 10	07♎ 53 01	20♓ 24 47	29♎ 39	12♍ 07	12♑ 35	14♋ 00	14♊ R57	21♒ R16	06♒ R04	12♐ 56	01♋ R15
02 Tu	00 43 07	08 52 00	02♈ 26 48	29R 41	13 21	13 12	14 06	14 57	21 15	06 04	12 57	01 01
03 We	00 47 03	09 51 01	14 36 06	29 36	14 35	13 49	14 12	14 56	21 14	06 03	12 59	00 48
04 Th	00 50 60	10 50 04	26 53 28	29 24	15 49	14 26	14 17	14 55	21 12	06 03	13 00	00 35
05 Fr	00 54 56	11 49 09	09♉ 19 37	29 04	17 02	15 04	14 23	14 55	21 11	06 02	13 01	00 25
06 Sa	00 58 53	12 48 16	21 55 26	28 36	18 16	15 41	14 28	14 54	21 10	06 02	13 03	00 18
07 Su	01 02 49	13 47 26	04♊ 42 13	28 00	19 30	16 19	14 33	14 53	21 09	06 02	13 04	00 13
08 Mo	01 06 46	14 46 38	17 41 40	27 17	20 44	16 57	14 38	14 52	21 08	06 01	13 05	00 11
09 Tu	01 10 42	15 45 52	00♋ 55 59	26 26	21 58	17 35	14 43	14 50	21 06	06 01	13 07	00 10
10 We	01 14 39	16 45 08	14 27 27	25 28	23 12	18 14	14 47	14 49	21 05	06 01	13 08	00 09
11 Th	01 18 36	17 44 27	28 18 02	24 24	24 27	18 52	14 52	14 47	21 04	06 00	13 10	00 08
12 Fr	01 22 32	18 43 48	12♌ 28 33	23 16	25 41	19 31	14 56	14 46	21 03	06 00	13 11	00 05
13 Sa	01 26 29	19 43 12	26 57 59	22 05	26 55	20 10	15 00	14 44	21 03	06 00	13 13	00 00
14 Su	01 30 25	20 42 37	11♍ 42 49	20 52	28 09	20 49	15 04	14 42	21 02	06 00	13 14	29♊ 53
15 Mo	01 34 22	21 42 05	26 36 51	19 40	29 24	21 28	15 07	14 41	21 01	06 00	13 16	29 44
16 Tu	01 38 18	22 41 35	11♎ 31 53	18 32	00♎ 38	22 08	15 11	14 39	21 00	06 00	13 18	29 33
17 We	01 42 15	23 41 07	26 18 45	17 28	01 53	22 47	15 14	14 36	20 59	06 00	13 19	29 22
18 Th	01 46 11	24 40 42	10♏ 49 01	16 31	03 07	23 27	15 17	14 34	20 59	06D 00	13 21	29 13
19 Fr	01 50 08	25 40 18	24 56 13	15 42	04 21	24 07	15 20	14 32	20 58	06 00	13 23	29 05
20 Sa	01 54 04	26 39 56	08♐ 36 52	15 03	05 36	24 47	15 23	14 30	20 58	06 00	13 24	29 00
21 Su	01 58 01	27 39 36	21 50 24	14 35	06 51	25 27	15 26	14 27	20 57	06 00	13 26	28 57
22 Mo	02 01 58	28 39 18	04♑ 38 49	14 18	08 05	26 07	15 28	14 25	20 57	06 00	13 28	28D 56
23 Tu	02 05 54	29 39 01	17 05 49	14 12	09 20	26 48	15 30	14 22	20 56	06 00	13 30	28 57
24 We	02 09 51	00♏ 38 46	29 16 09	14D 17	10 34	27 28	15 32	14 19	20 56	06 00	13 31	28 58
25 Th	02 13 47	01 38 33	11♒ 14 55	14 34	11 49	28 09	15 34	14 16	20 55	06 00	13 33	28R 58
26 Fr	02 17 44	02 38 21	23 07 18	15 00	13 04	28 50	15 36	14 13	20 55	06 01	13 35	28 57
27 Sa	02 21 40	03 38 11	04♓ 58 11	15 36	14 19	29 30	15 37	14 10	20 55	06 01	13 37	28 53
28 Su	02 25 37	04 38 03	16 51 51	16 20	15 33	00♒ 11	15 38	14 07	20 55	06 01	13 39	28 48
29 Mo	02 29 33	05 37 57	28 51 54	17 12	16 48	00 53	15 39	14 04	20 55	06 02	13 41	28 40
30 Tu	02 33 30	06 37 52	11♈ 00 57	18 11	18 03	01 34	15 40	14 01	20 55	06 02	13 43	28 31
31 We	02 37 27	07 37 49	23 20 46	19 16	19 18	02 15	15 41	13 57	20D 55	06 02	13 45	28 22

Data for	10-01-2001
Julian Day	2452183.50
Ayanamsa	23 52 35
SVP	05 ♓ 13 49
☽ ☊ Mean	01 ♋ 14 R

● ◐ PHASES ○ ◑

Day	h m	Phase	Position
02	13:49	○	09♈26
10	04:19	◑	16♋56
16	19:24	●	23♎30
24	02:59	◐	00♒46

LAST ASPECT Day	h m	☽ INGRESS Day	h m	Sign
30	13:04	01	19:09	♈
04	04:46	04	06:03	♉
05	22:35	06	15:13	♊
08	16:25	08	22:20	♋
10	17:48	11	02:55	♌
12	16:35	13	04:59	♍
15	04:53	15	05:27	♎
16	19:25	17	06:04	♏
18	22:31	19	08:48	♐
21	11:44	21	15:13	♑
23	20:13	24	01:28	♒
25	19:33	26	13:57	♓
27	21:33	29	02:16	♈
30	19:18	31	12:49	♉

DECLINATION

Day	☉	☽	☿	♀	♂	♃	♄	♅	♆	♇
01 Mo	03S 08	08S 19	14S 46	08N09	25S 43	22N32	20N46	15S 07	18S 38	12S 22
02 Tu	03 31	03 36	14 47	07 42	25 38	22 31	20 46	15 08	18 38	12 23
03 We	03 54	01N17	14 44	07 15	25 33	22 31	20 46	15 08	18 38	12 23
04 Th	04 17	06 11	14 37	06 48	25 27	22 30	20 46	15 08	18 38	12 24
05 Fr	04 40	10 54	14 26	06 21	25 21	22 30	20 45	15 09	18 38	12 24
06 Sa	05 03	15 14	14 11	05 53	25 15	22 29	20 45	15 09	18 39	12 25
07 Su	05 26	18 57	13 52	05 25	25 09	22 29	20 45	15 09	18 39	12 25
08 Mo	05 49	21 47	13 28	04 57	25 03	22 28	20 45	15 10	18 39	12 26
09 Tu	06 12	23 31	12 59	04 29	24 57	22 28	20 45	15 10	18 39	12 26
10 We	06 35	23 55	12 26	04 01	24 50	22 28	20 44	15 10	18 39	12 27
11 Th	06 58	22 51	11 49	03 32	24 43	22 27	20 44	15 11	18 39	12 27
12 Fr	07 20	20 20	11 09	03 04	24 36	22 27	20 44	15 11	18 39	12 28
13 Sa	07 43	16 30	10 26	02 35	24 29	22 26	20 43	15 11	18 39	12 29
14 Su	08 05	11 36	09 41	02 06	24 21	22 26	20 43	15 12	18 39	12 29
15 Mo	08 27	05 58	08 55	01 37	24 13	22 26	20 43	15 12	18 39	12 30
16 Tu	08 50	00S 01	08 10	01 08	24 06	22 25	20 42	15 12	18 39	12 30
17 We	09 12	05 57	07 27	00 39	23 57	22 25	20 42	15 12	18 39	12 31
18 Th	09 34	11 28	06 46	00 10	23 49	22 25	20 42	15 12	18 39	12 31
19 Fr	09 55	16 14	06 09	00S 19	23 41	22 25	20 41	15 13	18 39	12 32
20 Sa	10 17	19 59	05 36	00 48	23 32	22 24	20 41	15 13	18 39	12 32
21 Su	10 38	22 33	05 09	01 18	23 23	22 24	20 41	15 13	18 39	12 33
22 Mo	11 00	23 52	04 48	01 47	23 14	22 24	20 40	15 13	18 39	12 33
23 Tu	11 21	23 55	04 32	02 16	23 05	22 24	20 40	15 13	18 39	12 34
24 We	11 42	22 49	04 22	02 45	22 55	22 24	20 40	15 13	18 39	12 34
25 Th	12 03	20 42	04 18	03 15	22 46	22 24	20 39	15 13	18 39	12 35
26 Fr	12 23	17 43	04 19	03 44	22 36	22 24	20 39	15 13	18 39	12 35
27 Sa	12 44	14 02	04 26	04 13	22 26	22 23	20 38	15 13	18 39	12 36
28 Su	13 04	09 47	04 37	04 42	22 15	22 23	20 38	15 13	18 39	12 36
29 Mo	13 24	05 08	04 52	05 11	22 05	22 23	20 37	15 13	18 39	12 37
30 Tu	13 44	00 14	05 11	05 40	21 54	22 23	20 37	15 13	18 39	12 37
31 We	14 03	04N46	05 34	06 09	21 44	22 23	20 37	15 13	18 39	12 38

ASPECTARIAN

01 00:54 ☽ # ♀
16:12 ♀ □ ♇
18:04 ♀ △ ♂
19:24 ☿ SR
02 00:26 ☽ ∥ ☉
07:11 ☽ ✶ ♆
15:50 ♀ ✶ ♃
20:49 ☽ △ ♇
22:23 ☽ □ ♂
23:12 ☽ □ ♃
03 00:40 ☽ ✶ ♄
06:55 ♀ □ ♄
12:58 ☽ ✶ ♅
13:55 ☽ # ☉
17:11 ♂ ☍ ♃
04 02:51 ☽ ∥ ♀
04:46 ☽ ☍ ☿
17:42 ☽ □ ♆
05 08:04 ☽ # ♇
09:45 ☽ ✶ ♃
11:33 ☽ △ ♂
16:21 ☽ △ ♀
18:20 ☽ # ☿
22:35 ☽ □ ♅
23:33 ☽ # ♅
06 06:00 ☉ ✶ ♇
21:52 ☽ # ♆
23:29 ♀ # ☉
07 02:29 ☽ △ ♆
14:13 ☽ ∥ ♄
15:33 ☽ ☍ ♇
18:13 ☽ △ ☉
18:49 ☽ ☌ ♄
20:09 ☉ □ ♃
08 01:3[illegible] ☉ △ ♄
06:09 ☽ □ ♀
06:16 ☽ △ ♅
07:43 ☽ ∥ ♃
16:25 ☽ △ ☿
09 23:18 ☿ ∥ ♇
10 00:36 ☽ ☌ ♃
06:56 ☽ ☍ ♂
16:44 ☽ ✶ ♀
17:48 ☽ □ ☿
11 04:52 ☽ ∥ ♃
13:08 ☽ ☍ ♆
20:57 ☽ ∥ ♄
12 01:13 ☽ △ ♇
03:50 ☽ ✶ ♄
11:12 ☽ ✶ ☉
11:29 ☽ # ♆
14:16 ☽ ☍ ♅
16:35 ☽ ✶ ☿
13 06:53 ☽ # ♅
19:58 ☽ # ♇
20:46 ☿ △ ♅
14 00:39 ☿ □ ♂
01:44 ☿ ☌ ☉
02:29 ☽ □ ♇
04:50 ☽ □ ♄
05:27 ☽ ✶ ♃
07:37 ☉ △ ♅
07:39 ☉ □ ♂
09:47 ☽ # ☿
14:18 ☽ # ☉
15:22 ☽ △ ♂
15 04:53 ☽ ☌ ♀
05:45 ☿ ∥ ☉
11:43 ♀ ♎
15:06 ☽ △ ♆
16 02:52 ☽ ✶ ♇
04:08 ☽ # ☿
05:02 ☽ △ ♄
05:56 ☽ □ ♃
10:33 ☽ ☌ ☿
15:20 ☽ △ ♅
17:59 ☽ □ ♂
17 05:53 ☽ ∥ ☿
14:46 ☽ ∥ ☉
15:58 ☽ □ ♆
18 01:50 ♆ SD
04:57 ☽ ∥ ♇
07:34 ☽ △ ♃
17:12 ☽ □ ♅
18:26 ☽ ∥ ♅
22:31 ☽ ✶ ♂
19 11:49 ☿ □ ♃
14:39 ☽ ∥ ♆
18:07 ☽ ✶ ♀
19:21 ☽ ✶ ♆
20 05:32 ☽ # ♄
07:37 ♀ △ ♆
08:38 ☽ ☌ ♇
10:33 ☽ ☍ ♄
11:10 ☽ ✶ ☿
22:14 ☽ # ♃
22:23 ☽ ✶ ♅
21 10:47 ☿ △ ♄
11:10 ☽ ∥ ♂
11:44 ☽ ✶ ☉
22 07:18 ☽ □ ♀
10:2[illegible] ☽ □ ☿
20:54 ☽ ☍ ♃
23 00:24 ☿ SD
08:27 ☉ ♏
20:13 ☽ ☌ ♂
22:19 ☽ ∥ ♂
24 02:43 ☿ △ ♄
05:49 ☽ # ♃
13:28 ☽ ☌ ♆
25 00:27 ☽ # ♄
01:18 ☽ △ ♀
04:41 ☽ ✶ ♇
06:05 ☽ △ ♄
06:55 ☽ △ ☿
17:06 ☽ ∥ ♆
19:33 ☽ ☍ ♅
26 10:19 ♀ ✶ ♇
14:27 ☉ ∥ ♇
16:38 ☽ ∥ ♅
21:04 ☽ △ ☉
21:25 ♀ △ ♄
27 00:56 ☿ □ ♃
05:26 ♂ # ♃
07:02 ☽ ∥ ☉
08:21 ☽ ∥ ♇
16:19 ☿ ∥ ♀
17:20 ♂ ♒
17:32 ☽ □ ♇
18:31 ☽ □ ♄
21:33 ☽ △ ♃
28 01:37 ☿ □ ♃
23:44 ☽ ∥ ♀
29 01:13 ☽ ∥ ☿
04:15 ☽ ✶ ♂
09:32 ☉ □ ♆
14:12 ☽ ✶ ♆
30 05:18 ☽ △ ♇
05:51 ☽ ✶ ♄
09:07 ☽ □ ♃
15:18 ☽ ☍ ♀
15:21 ☽ ☍ ☿
18:55 ☿ ☌ ♀
19:18 ☽ ✶ ♅
22:56 ♅ SD
31 04:10 ☽ # ☿
07:22 ☽ # ☿
18:07 ☽ □ ♂

⚷ Chiron

01 Dec.	17 S 25
01	23♐47
04	23 57
07	24 07
10	24 18
13	24 30
16	24 42
19	24 55
22	25 08
25	25 22
28	25 36
31	25 51

November 2001

Day	S. T. h m s	☉ ° ' "	☽ ° ' "	☿ ° '	♀ ° '	♂ ° '	♃ ° '	♄ ° '	♅ ° '	♆ ° '	♇ ° '	☊ True ° '
01 Th	02 41 23	08♏ 37 48	05♉ 52 15	20♎ 27	20♎ 33	02♒ 56	15♋ 41	13♊R54	20♒ 55	06♒ 03	13♐ 47	28♊R13
02 Fr	02 45 20	09 37 49	18 35 38	21 42	21 48	03 38	15 41	13 50	20 55	06 03	13 49	28 06
03 Sa	02 49 16	10 37 52	01♊ 30 45	23 01	23 03	04 20	15R 41	13 46	20 55	06 04	13 51	28 01
04 Su	02 53 13	11 37 56	14 37 16	24 23	24 18	05 01	15 41	13 43	20 55	06 04	13 53	27 58
05 Mo	02 57 09	12 38 03	27 54 56	25 48	25 33	05 43	15 41	13 39	20 55	06 05	13 55	27 57
06 Tu	03 01 06	13 38 12	11♋ 23 49	27 15	26 48	06 25	15 40	13 35	20 56	06 06	13 57	27D 57
07 We	03 05 02	14 38 23	25 04 12	28 45	28 03	07 07	15 40	13 31	20 56	06 06	13 59	27 57
08 Th	03 08 59	15 38 36	08♌ 56 24	00♏ 16	29 18	07 49	15 39	13 27	20 56	06 07	14 01	27R 58
09 Fr	03 12 56	16 38 51	23 00 18	01 48	00♏ 33	08 31	15 37	13 23	20 57	06 08	14 03	27 57
10 Sa	03 16 52	17 39 08	07♍ 14 55	03 21	01 48	09 13	15 36	13 19	20 57	06 08	14 05	27 54
11 Su	03 20 49	18 39 27	21 38 00	04 55	03 03	09 56	15 35	13 15	20 58	06 09	14 07	27 50
12 Mo	03 24 45	19 39 48	06♎ 05 48	06 30	04 18	10 38	15 33	13 11	20 58	06 10	14 09	27 45
13 Tu	03 28 42	20 40 11	20 33 20	08 05	05 34	11 20	15 31	13 06	20 59	06 11	14 11	27 39
14 We	03 32 38	21 40 35	04♏ 54 44	09 40	06 49	12 03	15 29	13 02	21 00	06 12	14 14	27 32
15 Th	03 36 35	22 41 02	19 04 08	11 16	08 04	12 46	15 26	12 57	21 00	06 13	14 16	27 27
16 Fr	03 40 31	23 41 30	02♐ 56 28	12 52	09 19	13 28	15 24	12 53	21 01	06 14	14 18	27 22
17 Sa	03 44 28	24 42 00	16 28 17	14 28	10 35	14 11	15 21	12 48	21 02	06 15	14 20	27 20
18 Su	03 48 25	25 42 31	29 38 04	16 03	11 50	14 54	15 18	12 44	21 03	06 16	14 22	27 19
19 Mo	03 52 21	26 43 04	12♑ 26 18	17 39	13 05	15 37	15 15	12 39	21 04	06 17	14 25	27D 19
20 Tu	03 56 18	27 43 38	24 55 08	19 15	14 20	16 20	15 11	12 35	21 05	06 18	14 27	27 21
21 We	04 00 14	28 44 13	07♒ 08 00	20 51	15 36	17 03	15 08	12 30	21 06	06 19	14 29	27 23
22 Th	04 04 11	29 44 49	19 09 06	22 26	16 51	17 46	15 04	12 25	21 07	06 20	14 31	27 25
23 Fr	04 08 07	00♐ 45 27	01♓ 03 10	24 02	18 06	18 29	15 00	12 20	21 08	06 21	14 34	27 27
24 Sa	04 12 04	01 46 05	12 55 06	25 37	19 22	19 12	14 56	12 16	21 09	06 22	14 36	27R 27
25 Su	04 16 00	02 46 45	24 49 44	27 12	20 37	19 55	14 52	12 11	21 11	06 24	14 38	27 26
26 Mo	04 19 57	03 47 26	06♈ 51 32	28 47	21 52	20 38	14 47	12 06	21 12	06 25	14 40	27 24
27 Tu	04 23 54	04 48 08	19 04 20	00♐ 22	23 08	21 22	14 43	12 01	21 13	06 26	14 43	27 22
28 We	04 27 50	05 48 51	01♉ 31 06	01 57	24 23	22 05	14 38	11 56	21 15	06 27	14 45	27 19
29 Th	04 31 47	06 49 35	14 13 49	03 32	25 39	22 48	14 33	11 51	21 16	06 29	14 47	27 16
30 Fr	04 35 43	07 50 21	27 13 19	05 06	26 54	23 32	14 28	11 46	21 18	06 30	14 50	27 14

Data for	11-01-2001
Julian Day	2452214.50
Ayanamsa	23 52 38
SVP	05 ♓ 13 44
☽ ☊ Mean	29 ♊ 35 R

● ◐ PHASES ○ ◑

Day	h m	Phase	Position
01	05:41	○	08♉52
08	12:21	◑	16♌10
15	06:41	●	22♏58
22	23:21	◐	00♓44
30	20:49	○	08♊43

LAST ASPECT ☽ INGRESS

Day	h m	Day	h m	Sign
02	04:21	02	21:13	♊
04	19:46	05	03:45	♋
07	07:12	07	08:35	♌
08	20:31	09	11:50	♍
10	18:42	11	13:54	♎
13	00:43	13	15:46	♏
15	06:42	15	18:53	♐
17	08:16	18	00:41	♑
20	05:59	20	09:57	♒
22	07:39	22	21:53	♓
25	05:30	25	10:22	♈
27	04:45	27	21:07	♉
29	23:22	30	05:05	♊

DECLINATION

Day	☉	☽	☿	♀	♂	♃	♄	♅	♆	♇
01 Th	14S23	09N39	05S59	06S37	21S33	22N23	20N36	15S13	18S39	12S38
02 Fr	14 42	14 13	06 27	07 06	21 21	22 24	20 36	15 13	18 39	12 39
03 Sa	15 01	18 13	06 57	07 34	21 10	22 24	20 35	15 13	18 38	12 39
04 Su	15 19	21 22	07 28	08 02	20 58	22 24	20 35	15 13	18 38	12 40
05 Mo	15 38	23 25	08 01	08 30	20 47	22 24	20 34	15 13	18 38	12 40
06 Tu	15 56	24 09	08 36	08 58	20 35	22 24	20 34	15 13	18 38	12 41
07 We	16 14	23 26	09 11	09 26	20 23	22 24	20 33	15 13	18 38	12 41
08 Th	16 31	21 18	09 46	09 53	20 10	22 25	20 33	15 12	18 38	12 42
09 Fr	16 49	17 51	10 23	10 21	19 58	22 25	20 32	15 12	18 38	12 42
10 Sa	17 06	13 21	10 59	10 48	19 45	22 25	20 32	15 12	18 37	12 43
11 Su	17 23	08 03	11 36	11 15	19 32	22 25	20 31	15 12	18 37	12 43
12 Mo	17 39	02 17	12 12	11 41	19 20	22 26	20 31	15 12	18 37	12 44
13 Tu	17 55	03S36	12 48	12 07	19 06	22 26	20 30	15 11	18 37	12 44
14 We	18 11	09 16	13 25	12 33	18 53	22 26	20 29	15 11	18 37	12 44
15 Th	18 27	14 22	14 00	12 59	18 40	22 27	20 29	15 11	18 36	12 45
16 Fr	18 42	18 37	14 36	13 24	18 26	22 27	20 28	15 11	18 36	12 45
17 Sa	18 57	21 46	15 11	13 49	18 12	22 28	20 28	15 10	18 36	12 46
18 Su	19 11	23 38	15 45	14 14	17 58	22 28	20 27	15 10	18 36	12 46
19 Mo	19 25	24 12	16 18	14 38	17 44	22 28	20 27	15 10	18 36	12 47
20 Tu	19 39	23 31	16 51	15 02	17 30	22 29	20 26	15 09	18 35	12 47
21 We	19 53	21 43	17 24	15 26	17 15	22 30	20 26	15 09	18 35	12 47
22 Th	20 06	18 59	17 55	15 49	17 01	22 30	20 25	15 09	18 35	12 48
23 Fr	20 19	15 28	18 26	16 12	16 46	22 31	20 24	15 08	18 34	12 48
24 Sa	20 31	11 23	18 55	16 34	16 31	22 31	20 24	15 08	18 34	12 49
25 Su	20 43	06 51	19 24	16 56	16 16	22 32	20 23	15 07	18 34	12 49
26 Mo	20 55	02 01	19 52	17 18	16 01	22 32	20 23	15 07	18 34	12 49
27 Tu	21 06	02N58	20 19	17 39	15 46	22 33	20 22	15 06	18 33	12 50
28 We	21 17	07 56	20 45	17 59	15 31	22 34	20 21	15 06	18 33	12 50
29 Th	21 27	12 41	21 10	18 19	15 15	22 34	20 21	15 05	18 33	12 51
30 Fr	21 37	16 58	21 34	18 39	14 59	22 35	20 20	15 05	18 32	12 51

ASPECTARIAN

Day	Time	Aspect
01	00:21	☽ □ ♆
	07:00	♀ △ ♅
	09:05	☿ △ ♅
	15:26	☽ # ♇
	18:34	☽ ⚹ ♃
02	02:52	☽ # ☉
	04:21	☽ □ ♅
	05:37	☽ # ♅
	06:09	♄ ☍ ♇
	15:36	♃ SR
03	02:51	☽ # ♆
	05:29	☽ △ ♂
	07:29	☿ ☌ ♀
	08:24	☽ △ ♆
	15:48	☉ ∥ ♅
	17:12	☽ ∥ ♄
	20:43	☽ # ♂
	22:22	☽ ☌ ♄
	22:40	☽ ☍ ♇
04	10:15	☽ ∥ ♃
	11:25	☽ △ ♅
	19:19	☽ △ ♀
	19:46	☽ △ ☿
05	12:45	♂ ☌ ♆
06	02:14	♂ # ♄
	04:17	☽ △ ☉
	07:33	☽ ☌ ♃
07	05:43	☽ □ ♀
	07:12	☽ □ ☿
	13:30	☽ ∥ ♃
	19:09	☽ ☍ ♆
	19:54	☿ ♏
	21:59	☽ ☍ ♂
08	00:02	☉ △ ♃
	06:04	☽ ∥ ♄
	07:43	☽ ⚹ ♄
	08:44	☽ △ ♇
	09:26	☽ # ♂
	13:29	♀ ♏
	19:16	☽ # ♆
	19:16	☿ ∥ ♀
	20:31	☽ ☍ ♅
09	05:38	☽ # ☉
	13:59	☽ ⚹ ♀
	14:40	☽ # ♅
	16:40	☽ ⚹ ☿
10	03:10	☽ # ♇
	09:53	☽ # ☿
	10:07	☽ □ ♄
	10:57	☽ # ♀
	11:28	☽ □ ♇
	13:57	☽ ⚹ ♃
	18:42	☽ ⚹ ☉
11	19:00	☿ □ ♆
12	00:08	☽ △ ♆
	07:55	☽ △ ♂
	11:42	☽ △ ♄
	13:25	☽ ⚹ ♇
	15:39	☽ □ ♃
	21:03	☿ ∥ ♇
13	00:43	☽ △ ♅
	07:33	☉ □ ♅
	12:03	♀ □ ♆
14	02:11	☽ □ ♆
	03:31	☽ ☌ ♀
	09:03	☽ ☌ ☿
	10:37	♀ ∥ ♇
	12:42	☽ □ ♂
	15:56	☽ ∥ ♇
	16:24	☽ ∥ ♀
	17:50	☽ △ ♃
	21:50	☽ ∥ ☿
15	03:20	☽ □ ♅
	04:07	☽ ∥ ♅
	05:37	♂ ∥ ♆
	06:01	♂ △ ♄
	10:50	☉ ∥ ♂
	15:23	☉ ∥ ♆
	22:49	☽ ∥ ♂
	23:50	☽ ∥ ♆
16	00:28	☽ ∥ ☉
	05:48	☽ ⚹ ♆
	12:55	☽ # ♄
	16:34	☿ □ ♂
	17:29	☽ ☍ ♄
	19:41	☽ ⚹ ♂
	20:10	☽ ☌ ♇
	23:53	☿ ∥ ♅
17	05:27	♂ ⚹ ♇
	07:08	☽ # ♃
	08:16	☽ ⚹ ♅
	12:58	☿ △ ♃
19	01:23	☽ ⚹ ♀
	05:20	☽ ☍ ♃
	11:26	☽ ⚹ ☿
20	05:59	☽ ⚹ ☉
	07:01	♀ ∥ ♅
	15:21	☽ # ♃
	15:32	♀ △ ♃
	19:49	☿ ∥ ♂
	22:23	☽ ☌ ♆
21	03:52	☿ □ ♅
	10:37	☽ △ ♄
	12:20	☽ # ♄
	14:42	☽ ⚹ ♇
	15:44	☽ ∥ ☉
	18:51	☽ □ ♀
	21:03	☽ ☌ ♂
22	03:02	☽ ∥ ♆
	03:58	☽ ☌ ♅
	06:02	☉ ♐
	06:47	☽ ∥ ☿
	07:39	☽ □ ☿
	15:04	☽ ∥ ♂
	19:52	☽ ∥ ♀
23	02:09	☽ ∥ ♅
	07:05	☿ ∥ ♆
	10:45	☉ # ♄
	16:00	☽ ∥ ♇
	16:44	♀ □ ♂
	22:10	♀ ∥ ♂
	22:42	☽ □ ♄
24	03:25	☽ □ ♇
	04:04	☽ △ ♃
	14:33	☽ △ ♀
25	05:30	☽ △ ☿
	10:51	♀ □ ♅
	17:22	☽ △ ☉
	23:08	☽ ⚹ ♆
26	10:18	☽ ⚹ ♄
	15:28	☽ △ ♇
	15:33	☽ □ ♃
	18:25	☿ ♐
	19:08	♂ ☌ ♅
27	02:40	☿ # ♄
	04:12	☽ ⚹ ♅
	04:45	☽ ⚹ ♂
28	09:25	☽ □ ♆
	15:36	☉ ⚹ ♆
29	00:37	☽ ⚹ ♃
	00:50	☽ # ♇
	13:00	☽ # ♅
	13:06	☽ □ ♅
	13:08	☽ # ♂
	15:21	♂ ∥ ♅
	16:11	♀ ∥ ♆
	16:51	☽ □ ♂
	23:22	☽ ☍ ♀
30	05:13	☿ ∥ ☉
	09:49	☽ # ♆
	11:34	☽ # ♀
	16:15	☽ ☍ ☿
	16:53	☽ △ ♆
	21:41	☿ ⚹ ♆
	22:30	☽ ∥ ♄

⚷ Chiron

01 Dec.	17 S 41
03	26♐07
06	26 23
09	26 39
12	26 56
15	27 13
18	27 31
21	27 49
24	28 07
27	28 25
30	28 44

December 2001

14	20:46	22♐56	☉	Annular Solar Eclipse
30	10:30	08♋48	✹	Penumbral Lunar Eclipse (mag 0.919)

Day	S. T. h m s	☉ ° ' "	☽ ° ' "	☿ ° '	♀ ° '	♂ ° '	♃ ° '	♄ ° '	♅ ° '	♆ ° '	♇ ° '	☊ True ° '
01 Sa	04 39 40	08♐51 07	10♊29 21	06♐41	28♏09	24♒15	14♋R23	11♊R42	21♒19	06♒32	14♐52	27♊R12
02 Su	04 43 36	09 51 55	24 00 33	08 15	29 25	24 59	14 17	11 37	21 21	06 33	14 54	27 12
03 Mo	04 47 33	10 52 44	07♋44 51	09 49	00♐40	25 42	14 12	11 32	21 22	06 34	14 56	27D 12
04 Tu	04 51 30	11 53 35	21 39 42	11 24	01 56	26 26	14 06	11 27	21 24	06 36	14 59	27 12
05 We	04 55 26	12 54 27	05♌42 24	12 58	03 11	27 09	14 00	11 22	21 26	06 37	15 01	27 12
06 Th	04 59 23	13 55 20	19 50 25	14 32	04 27	27 53	13 54	11 17	21 27	06 39	15 03	27R 12
07 Fr	05 03 19	14 56 14	04♍01 19	16 06	05 42	28 37	13 48	11 12	21 29	06 40	15 06	27 12
08 Sa	05 07 16	15 57 10	18 12 52	17 40	06 58	29 20	13 41	11 07	21 31	06 42	15 08	27 11
09 Su	05 11 12	16 58 06	02♎22 51	19 15	08 13	00♓04	13 35	11 02	21 33	06 44	15 10	27 10
10 Mo	05 15 09	17 59 04	16 29 02	20 49	09 28	00 48	13 28	10 57	21 35	06 45	15 13	27 08
11 Tu	05 19 05	19 00 04	00♏29 04	22 23	10 44	01 31	13 21	10 52	21 37	06 47	15 15	27 07
12 We	05 23 02	20 01 04	14 20 34	23 58	11 59	02 15	13 14	10 48	21 39	06 49	15 17	27 05
13 Th	05 26 59	21 02 05	28 01 07	25 32	13 15	02 59	13 07	10 43	21 41	06 50	15 20	27 04
14 Fr	05 30 55	22 03 08	11♐28 29	27 07	14 30	03 43	13 00	10 38	21 43	06 52	15 22	27 03
15 Sa	05 34 52	23 04 11	24 40 53	28 41	15 46	04 26	12 53	10 33	21 45	06 54	15 24	27 03
16 Su	05 38 48	24 05 15	07♑37 18	00♑16	17 02	05 10	12 46	10 29	21 47	06 56	15 26	27D 03
17 Mo	05 42 45	25 06 20	20 17 37	01 51	18 17	05 54	12 38	10 24	21 50	06 57	15 29	27 03
18 Tu	05 46 41	26 07 25	02♒42 47	03 26	19 33	06 38	12 31	10 19	21 52	06 59	15 31	27 04
19 We	05 50 38	27 08 30	14 54 42	05 01	20 48	07 22	12 23	10 15	21 54	07 01	15 33	27 05
20 Th	05 54 34	28 09 36	26 56 10	06 36	22 04	08 06	12 16	10 10	21 57	07 03	15 36	27 05
21 Fr	05 58 31	29 10 42	08♓50 44	08 12	23 19	08 50	12 08	10 05	21 59	07 05	15 38	27 06
22 Sa	06 02 28	00♑11 49	20 42 35	09 47	24 35	09 34	12 00	10 01	22 01	07 07	15 40	27 06
23 Su	06 06 24	01 12 55	02♈36 15	11 22	25 50	10 18	11 52	09 57	22 04	07 09	15 42	27 07
24 Mo	06 10 21	02 14 02	14 36 32	12 58	27 06	11 01	11 44	09 52	22 06	07 11	15 45	27 08
25 Tu	06 14 17	03 15 09	26 48 06	14 33	28 21	11 45	11 37	09 48	22 09	07 13	15 47	27 09
26 We	06 18 14	04 16 16	09♉15 12	16 09	29 37	12 29	11 29	09 44	22 11	07 14	15 49	27 11
27 Th	06 22 10	05 17 23	22 01 24	17 44	00♑52	13 13	11 21	09 39	22 14	07 16	15 51	27 12
28 Fr	06 26 07	06 18 30	05♊09 06	19 19	02 08	13 57	11 12	09 35	22 17	07 18	15 53	27 13
29 Sa	06 30 03	07 19 38	18 39 17	20 54	03 23	14 41	11 04	09 31	22 19	07 21	15 56	27 14
30 Su	06 34 00	08 20 45	02♋31 05	22 28	04 39	15 25	10 56	09 27	22 22	07 23	15 58	27R 13
31 Mo	06 37 57	09 21 53	16 41 41	24 02	05 54	16 09	10 48	09 23	22 25	07 25	16 00	27 13

Data for	12-01-2001
Julian Day	2452244.50
Ayanamsa	23 52 42
SVP	05 ♓ 13 40
☽ ☊ Mean	28 ♊ 00 R

● ◐ PHASES ○ ◑

07	19:52	◑	15♍47
14	20:47	☉	22♐56
22	20:56	◐	01♈05
30	10:41	✹	08♋48

LAST ASPECT ☽ INGRESS

Day	h m	Day	h m	
02	01:49	02	10:31	♋
03	11:05	04	14:17	♌
06	14:22	06	17:13	♍
07	22:59	08	19:59	♎
10	08:45	10	23:11	♏
12	12:49	13	03:31	♐
15	08:25	15	09:49	♑
16	09:36	17	18:44	♒
20	02:42	20	06:10	♓
22	08:46	22	18:46	♈
25	03:23	25	06:14	♉
27	00:24	27	14:41	♊
29	06:27	29	19:42	♋
31	13:44	31	22:10	♌

DECLINATION

Day	☉	☽	☿	♀	♂	♃	♄	♅	♆	♇
01 Sa	21S 47	20N31	21S 57	18S 58	14S 44	22N36	20N20	15S 04	18S 32	12S 51
02 Su	21 56	23 00	22 19	19 16	14 28	22 36	20 19	15 04	18 32	12 52
03 Mo	22 05	24 10	22 39	19 34	14 12	22 37	20 19	15 03	18 31	12 52
04 Tu	22 13	23 51	22 59	19 52	13 56	22 38	20 18	15 03	18 31	12 52
05 We	22 21	22 01	23 17	20 09	13 40	22 39	20 17	15 02	18 31	12 53
06 Th	22 28	18 50	23 35	20 25	13 23	22 39	20 17	15 02	18 30	12 53
07 Fr	22 35	14 33	23 51	20 41	13 07	22 40	20 16	15 01	18 30	12 53
08 Sa	22 42	09 27	24 06	20 56	12 50	22 41	20 16	15 00	18 29	12 54
09 Su	22 48	03 52	24 19	21 10	12 34	22 42	20 15	15 00	18 29	12 54
10 Mo	22 54	01S 53	24 31	21 24	12 17	22 43	20 15	14 59	18 29	12 54
11 Tu	22 59	07 30	24 42	21 38	12 00	22 43	20 14	14 58	18 28	12 55
12 We	23 04	12 43	24 52	21 50	11 43	22 44	20 13	14 58	18 28	12 55
13 Th	23 08	17 12	25 00	22 02	11 26	22 45	20 13	14 57	18 27	12 55
14 Fr	23 12	20 44	25 07	22 14	11 09	22 46	20 12	14 56	18 27	12 56
15 Sa	23 15	23 06	25 13	22 24	10 52	22 47	20 12	14 56	18 27	12 56
16 Su	23 18	24 11	25 17	22 34	10 35	22 47	20 11	14 55	18 26	12 56
17 Mo	23 21	23 58	25 20	22 44	10 17	22 48	20 11	14 54	18 26	12 56
18 Tu	23 23	22 33	25 21	22 52	10 00	22 49	20 10	14 53	18 25	12 57
19 We	23 24	20 06	25 21	23 00	09 42	22 50	20 10	14 53	18 25	12 57
20 Th	23 26	16 49	25 20	23 08	09 25	22 51	20 09	14 52	18 24	12 57
21 Fr	23 26	12 54	25 17	23 14	09 07	22 52	20 09	14 51	18 24	12 57
22 Sa	23 26	08 30	25 12	23 20	08 49	22 53	20 08	14 50	18 23	12 58
23 Su	23 26	03 47	25 06	23 25	08 32	22 53	20 08	14 49	18 23	12 58
24 Mo	23 25	01N07	24 58	23 30	08 14	22 54	20 07	14 49	18 22	12 58
25 Tu	23 24	06 03	24 49	23 33	07 56	22 55	20 07	14 48	18 22	12 58
26 We	23 22	10 51	24 39	23 36	07 38	22 56	20 06	14 47	18 22	12 59
27 Th	23 20	15 20	24 26	23 39	07 20	22 57	20 06	14 46	18 21	12 59
28 Fr	23 17	19 12	24 13	23 40	07 02	22 58	20 05	14 45	18 21	12 59
29 Sa	23 14	22 10	23 57	23 41	06 44	22 58	20 05	14 44	18 20	12 59
30 Su	23 11	23 54	23 40	23 41	06 26	22 59	20 05	14 43	18 20	12 59
31 Mo	23 07	24 09	23 22	23 40	06 08	23 00	20 04	14 42	18 19	12 59

ASPECTARIAN

01 02:09 ☽ ☌ ♄
07:51 ☽ ☍ ♇
11:20 ☽ # ☉
14:28 ☽ # ☿
19:06 ☽ ∥ ♃
19:18 ☽ △ ♅
02 01:49 ☽ △ ♂
11:13 ♀ ♐
21:09 ☿ # ♃
03 11:05 ☽ ☌ ♃
14:14 ☉ ☍ ♄
04 00:48 ☿ ☍ ♄
11:52 ☽ # ☿
17:49 ☽ ∥ ♃
19:18 ☽ △ ♀
21:05 ☽ # ☉
21:37 ☿ ☌ ☉
05 01:35 ☽ ☍ ♆
09:35 ☽ ⚹ ♄
12:19 ♀ # ♄
13:12 ☽ △ ☉
13:54 ☽ △ ☿
14:08 ☽ # ♀
14:17 ☽ ∥ ♄
15:53 ☽ △ ♇
06 02:08 ☽ # ♆
02:46 ☽ ☍ ♅
08:13 ☿ ☌ ♇
14:22 ☽ ☍ ♂
21:37 ☽ # ♅
07 03:08 ☽ □ ♀
03:53 ☉ ☌ ♇
07:30 ☽ # ♂
08:09 ☽ # ♇
[illegible]
16:25 ☽ ⚹ ♃
18:47 ☽ □ ♇
19:00 ♀ ⚹ ♆
19:01 ♂ ∥ ♇
20:04 ☉ # ♃
22:59 ☽ □ ☿
08 21:53 ♂ ♓
09 07:25 ☽ △ ♆
10:54 ☽ ⚹ ♀
14:39 ☽ △ ♄
18:55 ☽ □ ♃
21:50 ☽ ⚹ ♇
10 02:47 ☽ ⚹ ☉
08:21 ☽ ⚹ ☿
08:45 ☽ △ ♅
11:57 ☿ ⚹ ♅
11 01:54 ☽ △ ♂
02:34 ♀ ☍ ♄
10:55 ☽ □ ♆
19:29 ☽ ∥ ♂
22:07 ☽ △ ♃
12 01:02 ☽ ∥ ♇
11:27 ☽ ∥ ♅
12:49 ☽ □ ♅
13 07:42 ☽ ∥ ♆
09:19 ☽ □ ♂
15:44 ☽ ⚹ ♆
15:50 ☉ ⚹ ♅
19:51 ☽ # ♄
22:31 ☽ ☍ ♄
14 06:04 ☽ ☌ ♀
07:03 ☽ ☌ ♇
[illegible]
16:51 ♀ ☌ ♇
18:38 ☽ ⚹ ♅
19:44 ☽ # ♃
15 02:17 ☽ ∥ ☉
08:25 ☽ ☌ ☿
19:09 ☽ ⚹ ♂
19:56 ☿ ♑
16 09:36 ☽ ☍ ♃
17 12:32 ☽ ∥ ☉
13:53 ♀ # ♃
20:10 ☽ ∥ ♀
20:37 ☽ # ♃
18 08:24 ☽ ☌ ♆
14:50 ☽ △ ♄
23:33 ☽ # ♄
19 01:18 ☽ ⚹ ♇
13:05 ☽ ∥ ♆
13:06 ☽ ⚹ ♀
13:59 ☽ ☌ ♅
21:42 ♀ ⚹ ♅
20 02:42 ☽ ⚹ ☉
12:29 ☽ ∥ ♅
22:30 ☽ ⚹ ☿
23:42 ☽ ∥ ♇
23:59 ☽ ☌ ♂
21 02:31 ☽ □ ♄
06:36 ☽ △ ♃
13:47 ☽ □ ♇
17:49 ☿ ⚹ ♂
19:23 ☉ ♑
22:14 ☽ ∥ ♂
[illegible]
13:36 ♂ □ ♄
23 03:32 ♀ ∥ ☉
07:00 ☿ ☍ ♃
09:09 ☽ ⚹ ♆
14:38 ☽ ⚹ ♄
18:22 ☽ □ ♃
20:15 ☽ □ ☿
24 02:16 ☽ △ ♇
14:52 ☽ ⚹ ♅
19:55 ♂ △ ♃
25 03:23 ☽ △ ♀
08:46 ☽ # ♂
13:38 ☽ △ ☉
[illegible]
26 04:12 ☽ ⚹ ♃
06:32 ☽ ⚹ ♂
07:26 ♀ ♑
11:05 ☽ # ♇
14:53 ☽ △ ☿
20:52 ☽ # ♅
27 00:24 ☽ □ ♅
18:15 ☽ # ♆
28 03:54 ☽ △ ♆
06:21 ☽ ∥ ♄
07:56 ☽ ☌ ♄
16:38 ☽ □ ♂
19:12 ☽ ☍ ♇
[illegible]
08:57 ☽ ∥ ♃
11:57 ☽ # ☉
19:25 ☽ # ♀
20:08 ☽ # ☿
23:24 ☿ ∥ ♀
30 04:00 ☽ ☍ ♀
14:11 ☽ ☌ ♃
18:50 ♂ □ ♇
23:03 ☽ △ ♂
31 12:04 ☽ # ♀
13:45 ☽ ☍ ☿
20:22 ☽ # ☿
[illegible]
21:21 ☽ ∥ ♃

⚷ Chiron

01 Dec.	17 S 49
03	29♐03
06	29 22
09	29 41
12	00♑00
15	00 20
18	00 39
21	00 59
24	01 18
27	01 38
30	01 57

[illegible]

January 2002

Day	S. T.	☉	☽	☿	♀	♂	♃	♄	♅	♆	♇	☊ True
	h m s	° ' "	° ' "	° '	° '	° '	° '	° '	° '	° '	° '	° '
01 Tu	06 41 53	10♑23 01	01♌06 29	25♑35	07♑10	16♓53	10♋R40	09♊R19	22♒27	07♒27	16♐02	27♊R11
02 We	06 45 50	11 24 09	15 39 42	27 07	08 25	17 37	10 32	09 16	22 30	07 29	16 04	27 08
03 Th	06 49 46	12 25 17	00♍15 04	28 39	09 41	18 21	10 24	09 12	22 33	07 31	16 06	27 06
04 Fr	06 53 43	13 26 26	14 46 43	00♒09	10 56	19 05	10 16	09 08	22 36	07 33	16 08	27 03
05 Sa	06 57 39	14 27 34	29 09 48	01 37	12 12	19 48	10 08	09 05	22 39	07 35	16 11	27 01
06 Su	07 01 36	15 28 43	13♎20 54	03 04	13 27	20 32	09 59	09 01	22 42	07 37	16 13	26 59
07 Mo	07 05 32	16 29 52	27 18 05	04 28	14 43	21 16	09 51	08 58	22 45	07 39	16 15	26 59
08 Tu	07 09 29	17 31 02	11♏00 38	05 49	15 58	22 00	09 43	08 55	22 48	07 42	16 17	26D 59
09 We	07 13 26	18 32 11	24 28 45	07 07	17 14	22 44	09 35	08 51	22 51	07 44	16 19	27 00
10 Th	07 17 22	19 33 21	07♐43 05	08 22	18 29	23 28	09 28	08 48	22 54	07 46	16 21	27 01
11 Fr	07 21 19	20 34 30	20 44 24	09 31	19 45	24 12	09 20	08 45	22 57	07 48	16 23	27 01
12 Sa	07 25 15	21 35 40	03♑33 25	10 36	21 00	24 55	09 12	08 42	23 00	07 50	16 25	27R 01
13 Su	07 29 12	22 36 49	16 10 44	11 34	22 16	25 39	09 04	08 39	23 03	07 52	16 27	27 00
14 Mo	07 33 08	23 37 58	28 36 57	12 25	23 31	26 23	08 57	08 37	23 06	07 55	16 29	26 58
15 Tu	07 37 05	24 39 06	10♒52 50	13 09	24 47	27 07	08 49	08 34	23 09	07 57	16 31	26 54
16 We	07 41 02	25 40 14	22 59 30	13 44	26 02	27 51	08 41	08 31	23 12	07 59	16 33	26 50
17 Th	07 44 58	26 41 21	04♓58 30	14 10	27 17	28 35	08 34	08 29	23 16	08 01	16 34	26 45
18 Fr	07 48 55	27 42 28	16 52 05	14 25	28 33	29 18	08 27	08 27	23 19	08 04	16 36	26 41
19 Sa	07 52 51	28 43 33	28 43 08	14R 29	29 48	00♈02	08 20	08 24	23 22	08 06	16 38	26 37
20 Su	07 56 48	29 44 38	10♈35 18	14 22	01♒04	00 46	08 12	08 22	23 25	08 08	16 40	26 35
21 Mo	08 00 44	00♒45 42	22 32 47	14 03	02 19	01 29	08 05	08 20	23 28	08 10	16 42	26 34
22 Tu	08 04 41	01 46 45	04♉40 18	13 32	03 35	02 13	07 59	08 18	23 32	08 13	16 44	26D 34
23 We	08 08 37	02 47 48	17 02 41	12 51	04 50	02 57	07 52	08 16	23 35	08 15	16 45	26 36
24 Th	08 12 34	03 48 49	29 44 36	11 59	06 05	03 40	07 45	08 15	23 38	08 17	16 47	26 38
25 Fr	08 16 31	04 49 49	12♊50 10	10 59	07 21	04 24	07 39	08 13	23 42	08 20	16 49	26 39
26 Sa	08 20 27	05 50 48	26 22 17	09 51	08 36	05 08	07 32	08 12	23 45	08 22	16 50	26R 39
27 Su	08 24 24	06 51 47	10♋21 53	08 39	09 52	05 51	07 26	08 10	23 48	08 24	16 52	26 37
28 Mo	08 28 20	07 52 44	24 47 14	07 24	11 07	06 35	07 20	08 09	23 52	08 26	16 54	26 34
29 Tu	08 32 17	08 53 40	09♌33 34	06 08	12 22	07 18	07 14	08 08	23 55	08 29	16 55	26 28
30 We	08 36 13	09 54 35	24 33 20	04 54	13 38	08 02	07 08	08 07	23 58	08 31	16 57	26 22
31 Th	08 40 10	10 55 30	09♍37 14	03 44	14 53	08 45	07 03	08 06	24 02	08 33	16 59	26 14

Data for	01-01-2002
Julian Day	2452275.50
Ayanamsa	23 52 48
SVP	05♓13 38
☽ ☊ Mean	26 ♊ 22 R

● ◐ PHASES ○ ◑

06	03:55	◑	15♎39
13	13:28	●	23♑11
21	17:47	◐	01♉31
28	22:51	○	08♌51

LAST ASPECT ☽ INGRESS

Day	h m	Day	h m	
02	11:17	02	23:35	♍
04	07:31	05	01:24	♎
06	16:05	07	04:41	♏
08	21:03	09	09:58	♐
11	06:49	11	17:18	♑
13	19:24	14	02:42	♒
16	00:26	16	14:01	♓
19	02:28	19	02:36	♈
21	01:51	21	14:48	♉
23	12:30	24	00:29	♊
25	19:24	26	06:18	♋
26	19:04	28	08:31	♌
29	23:04	30	08:40	♍

DECLINATION

Day	☉	☽	☿	♀	♂	♃	♄	♅	♆	♇
01 Tu	23S 02	22N47	23S 03	23S 39	05S 50	23N01	20N04	14S 41	18S 18	13S 00
02 We	22 57	19 54	22 41	23 37	05 31	23 02	20 03	14 41	18 18	13 00
03 Th	22 52	15 46	22 19	23 34	05 13	23 02	20 03	14 40	18 17	13 00
04 Fr	22 46	10 44	21 55	23 30	04 55	23 03	20 03	14 39	18 17	13 00
05 Sa	22 39	05 09	21 30	23 26	04 37	23 04	20 02	14 38	18 16	13 00
06 Su	22 32	00S 37	21 04	23 21	04 18	23 05	20 02	14 37	18 16	13 00
07 Mo	22 25	06 16	20 37	23 15	04 00	23 05	20 02	14 36	18 15	13 01
08 Tu	22 18	11 32	20 09	23 08	03 42	23 06	20 01	14 35	18 15	13 01
09 We	22 09	16 09	19 41	23 01	03 23	23 07	20 01	14 34	18 14	13 01
10 Th	22 01	19 53	19 12	22 53	03 05	23 08	20 01	14 33	18 14	13 01
11 Fr	21 52	22 33	18 43	22 45	02 46	23 08	20 01	14 32	18 13	13 01
12 Sa	21 42	23 59	18 14	22 35	02 28	23 09	20 00	14 31	18 13	13 01
13 Su	21 33	24 10	17 45	22 25	02 10	23 10	20 00	14 30	18 12	13 01
14 Mo	21 22	23 07	17 17	22 14	01 51	23 10	20 00	14 29	18 11	13 01
15 Tu	21 12	20 59	16 50	22 03	01 33	23 11	20 00	14 28	18 11	13 01
16 We	21 01	17 56	16 25	21 51	01 14	23 11	20 00	14 27	18 10	13 01
17 Th	20 49	14 11	16 01	21 38	00 56	23 12	20 00	14 25	18 10	13 01
18 Fr	20 37	09 54	15 40	21 25	00 38	23 13	19 59	14 24	18 09	13 02
19 Sa	20 25	05 16	15 22	21 11	00 19	23 13	19 59	14 23	18 09	13 02
20 Su	20 12	00 27	15 06	20 56	00 01	23 14	19 59	14 22	18 08	13 02
21 Mo	19 59	04N26	14 54	20 41	00N17	23 14	19 59	14 21	18 07	13 02
22 Tu	19 46	09 14	14 45	20 25	00 36	23 15	19 59	14 20	18 07	13 02
23 We	19 32	13 45	14 40	20 09	00 54	23 16	19 59	14 19	18 06	13 02
24 Th	19 18	17 48	14 39	19 52	01 12	23 16	19 59	14 18	18 06	13 02
25 Fr	19 03	21 07	14 41	19 34	01 31	23 17	19 59	14 17	18 05	13 02
26 Sa	18 49	23 22	14 47	19 16	01 49	23 17	19 59	14 16	18 05	13 02
27 Su	18 33	24 16	14 55	18 57	02 07	23 18	19 59	14 15	18 04	13 02
28 Mo	18 18	23 35	15 06	18 38	02 25	23 18	19 59	14 13	18 03	13 02
29 Tu	18 02	21 16	15 19	18 18	02 43	23 18	19 59	14 12	18 03	13 02
30 We	17 46	17 28	15 33	17 57	03 01	23 19	19 59	14 11	18 02	13 02
31 Th	17 29	12 31	15 48	17 37	03 19	23 19	19 59	14 10	18 02	13 02

ASPECTARIAN

01 00:49 ☿ ∥ ☉
01:58 ☿ # ♃
05:06 ☉ # ♃
05:53 ☉ ☍ ♃
10:30 ☽ ☍ ♆
13:30 ☽ ⚹ ♄
22:59 ☽ ∥ ♄
02 00:40 ☽ △ ♇
10:09 ☽ # ♆
11:17 ☽ ☍ ♅
03 05:39 ☽ # ♅
12:22 ♀ ☍ ♃
13:37 ☽ # ♇
14:42 ☽ □ ♄
16:35 ☽ ⚹ ♃
17:02 ☽ △ ♀
21:37 ☽ △ ☉
21:38 ☿ ♒
04 02:16 ☽ □ ♆
07:31 ☽ ☍ ♂
05 02:25 ☽ # ♂
04:36 ☽ △ ☿
14:14 ☽ △ ♆
16:40 ☽ △ ♄
18:20 ☽ □ ♃
06 00:12 ☽ □ ♀
04:54 ☽ ⚹ ♇
14:45 ☽ ∥ ♂
16:05 ☽ △ ♅
07 13:52 ☽ □ ☿
18:08 ☽ □ ♆
21:45 ☽ △ ♃
08 06:37 ☿ # ♄
07:09 ♀ # ♃
07:21 ☽ ∥ ♇
09:41 ☽ ⚹ ♀
12:29 ☽ ⚹ ☉
15:23 ☽ ∥ ♅
20:41 ☽ △ ♂
21:03 ☽ □ ♅
09 11:56 ☿ ☌ ♆
12:42 ☽ ∥ ♆
19:45 ☽ ∥ ☿
10 00:05 ☽ ⚹ ♆
01:06 ☽ # ♄
01:17 ☽ ⚹ ☿
01:59 ☽ ☍ ♄
08:37 ☿ △ ♄
15:54 ☽ ☌ ♇
17:19 ☽ ∥ ☉
11 02:22 ☽ ∥ ♀
04:07 ☽ ⚹ ♅
06:49 ☽ □ ♂
07:50 ☽ # ♃
12 00:50 ☿ ∥ ♆
10:34 ☽ ☍ ♃
13 13:00 ☽ ☌ ♀
19:24 ☽ ⚹ ♂
23:12 ☽ # ♃
14 11:32 ♀ ☌ ☉
12:21 ☽ ∥ ♀
18:13 ☽ ☌ ♆
19:28 ☽ △ ♄
21:53 ☽ ∥ ☉
15 04:44 ☽ ☌ ☿
08:30 ☽ # ♄
11:09 ☽ ⚹ ♇
22:20 ☽ ∥ ♆
16 00:26 ☽ ☌ ♅
11:26 ☽ ∥ ☿
22:31 ☽ ∥ ♅
17 06:43 ☽ ∥ ♇
07:02 ☽ □ ♄
07:10 ☽ △ ♃
23:28 ☽ □ ♇
18 20:51 ☿ SR
22:54 ♂ ♈
19 00:01 ☽ ⚹ ☉
02:28 ☽ ⚹ ♀
02:50 ☽ ☌ ♂
03:42 ♀ ♒
10:20 ♀ ⚹ ♂
19:02 ☽ ⚹ ♆
19:14 ☽ □ ♃
19:32 ☽ ⚹ ♄
20 02:09 ☽ # ♂
02:17 ☽ ∥ ♂
06:03 ☉ ♒
07:27 ☽ ⚹ ☿
12:16 ☽ △ ♇
21 00:12 ☉ # ♄
01:51 ☽ ⚹ ♅
21:36 ☽ □ ♀
22 06:24 ☽ ⚹ ♃
06:57 ☽ □ ♆
16:23 ☽ □ ☿
20:04 ☽ # ♇
23 03:09 ☽ # ♅
05:08 ☽ # ☿
08:17 ♄ △ ♆
12:26 ☉ ⚹ ♂
12:30 ☽ □ ♅
13:46 ♀ # ♄
24 01:55 ☽ # ♆
07:43 ☽ ⚹ ♂
08:11 ☽ △ ☉
09:25 ☽ # ☉
12:59 ☽ △ ♀
12:59 ☽ # ♀
15:04 ☽ ∥ ♄
15:38 ☽ ☌ ♄
15:48 ☽ △ ♆
20:53 ☽ △ ☿
25 07:09 ☽ ☍ ♇
16:18 ♀ △ ♄
19:17 ♀ ☌ ♆
19:24 ☽ △ ♅
22:49 ☽ ∥ ♃
26 12:17 ☿ ☌ ♀
15:56 ☽ □ ♂
19:04 ☽ ☌ ♃
27 04:39 ☿ ☌ ♆
09:26 ☿ △ ♄
18:55 ☿ ☌ ☉
28 03:57 ☽ ∥ ♃
06:12 ☉ △ ♄
09:52 ☿ ⚹ ♂
13:47 ☉ ☌ ♆
18:55 ☽ ☍ ☿
20:11 ☽ △ ♂
21:42 ☽ ⚹ ♄
21:59 ♂ □ ♃
22:15 ☽ ☍ ♆
22:49 ☉ ∥ ♆
29 04:56 ☽ ☍ ♀
08:59 ☽ ∥ ♄
11:50 ☽ △ ♇
18:16 ♀ ∥ ♆
20:47 ☽ # ♆
20:59 ☽ # ♀
22:11 ☽ # ☉
23:04 ☽ ☍ ♅
30 02:35 ♂ ⚹ ♄
09:22 ☽ # ☿
16:27 ☽ # ♅
17:00 ♂ ⚹ ♆
19:55 ☽ ⚹ ♃
21:34 ☽ □ ♄
21:42 ☽ # ♇
31 11:47 ☽ □ ♇

⚷ Chiron

01 Dec.	17 S 47
02	02♑16
05	02 35
08	02 55
11	03 13
14	03 32
17	03 51
20	04 09
23	04 27
26	04 44
29	05 01

February 2002

Day	S. T. h m s	☉ ° ' "	☽ ° ' "	☿ ° '	♀ ° '	♂ ° '	♃ ° '	♄ ° '	♅ ° '	♆ ° '	♇ ° '	☊ True ° '
01 Fr	08 44 06	11♒ 56 23	24♍ 35 41	02♒R39	16♒ 08	09♈ 29	06♋R57	08♊R05	24♒ 05	08♒ 35	17♐ 00	26♊R07
02 Sa	08 48 03	12 57 16	09♎ 20 23	01 40	17 23	10 12	06 52	08 04	24 09	08 38	17 02	26 01
03 Su	08 51 60	13 58 07	23 45 26	00 50	18 39	10 55	06 47	08 03	24 12	08 40	17 03	25 56
04 Mo	08 55 56	14 58 58	07♏ 47 49	00 07	19 54	11 39	06 42	08 03	24 16	08 42	17 05	25 53
05 Tu	08 59 53	15 59 49	21 27 07	29♑ 32	21 09	12 22	06 37	08 02	24 19	08 45	17 06	25 52
06 We	09 03 49	17 00 38	04♐ 44 52	29 07	22 25	13 05	06 32	08 02	24 22	08 47	17 07	25D 52
07 Th	09 07 46	18 01 27	17 43 45	28 49	23 40	13 49	06 28	08 02	24 26	08 49	17 09	25 53
08 Fr	09 11 42	19 02 14	00♑ 26 50	28 40	24 55	14 32	06 24	08 02	24 29	08 51	17 10	25R 53
09 Sa	09 15 39	20 03 01	12 57 02	28D 38	26 10	15 15	06 19	08D 02	24 33	08 54	17 11	25 51
10 Su	09 19 35	21 03 46	25 16 54	28 44	27 26	15 58	06 15	08 02	24 36	08 56	17 13	25 48
11 Mo	09 23 32	22 04 30	07♒ 28 27	28 56	28 41	16 42	06 12	08 02	24 40	08 58	17 14	25 41
12 Tu	09 27 29	23 05 13	19 33 19	29 15	29 56	17 25	06 08	08 03	24 43	09 00	17 15	25 33
13 We	09 31 25	24 05 55	01♓ 32 45	29 39	01♓ 11	18 08	06 05	08 03	24 47	09 03	17 16	25 22
14 Th	09 35 22	25 06 35	13 27 55	00♒ 09	02 26	18 51	06 02	08 04	24 50	09 05	17 17	25 11
15 Fr	09 39 18	26 07 13	25 20 12	00 44	03 41	19 34	05 59	08 05	24 54	09 07	17 19	25 00
16 Sa	09 43 15	27 07 50	07♈ 11 20	01 23	04 57	20 17	05 56	08 05	24 57	09 09	17 20	24 50
17 Su	09 47 11	28 08 26	19 03 46	02 06	06 12	21 00	05 53	08 06	25 01	09 11	17 21	24 43
18 Mo	09 51 08	29 08 59	01♉ 00 39	02 54	07 27	21 43	05 51	08 07	25 04	09 13	17 22	24 38
19 Tu	09 55 04	00♓ 09 31	13 05 53	03 45	08 42	22 26	05 49	08 09	25 07	09 16	17 23	24 35
20 We	09 59 01	01 10 01	25 23 54	04 39	09 57	23 09	05 47	08 10	25 11	09 18	17 24	24 34
21 Th	10 02 58	02 10 29	07♊ 59 30	05 36	11 12	23 52	05 45	08 11	25 14	09 20	17 25	24D 34
22 Fr	10 06 54	03 10 56	20 57 30	06 36	12 27	24 34	05 43	08 13	25 18	09 22	17 26	24R 34
23 Sa	10 10 51	04 11 20	04♋ 22 08	07 38	13 42	25 17	05 42	08 15	25 21	09 24	17 27	24 33
24 Su	10 14 47	05 11 43	18 16 13	08 43	14 57	26 00	05 41	08 16	25 25	09 26	17 27	24 30
25 Mo	10 18 44	06 12 04	02♌ 40 05	09 50	16 12	26 43	05 40	08 18	25 28	09 28	17 28	24 24
26 Tu	10 22 40	07 12 23	17 30 34	11 00	17 27	27 25	05 39	08 20	25 32	09 30	17 29	24 16
27 We	10 26 37	08 12 40	02♍ 40 47	12 11	18 42	28 08	05 38	08 22	25 35	09 33	17 30	24 06
28 Th	10 30 33	09 12 55	18 00 40	13 24	19 56	28 51	05 38	08 24	25 38	09 35	17 30	23 54

Data for	02-01-2002
Julian Day	2452306.50
Ayanamsa	23 52 54
SVP	05 ♓ 13 35
☽ ☊ Mean	24 ♊ 43 R

● ◐ PHASES ○ ◑

04	13:33	◑	15♏33
12	07:42	●	23♒25
20	12:02	◐	01♊40
27	09:17	○	08♍36

LAST ASPECT Day	h m	☽ INGRESS Day	h m	
31	11:47	01	08:45	♎
03	00:45	03	10:35	♏
05	14:03	05	15:22	♐
07	12:39	07	23:09	♑
10	06:51	10	09:15	♒
12	10:22	12	20:54	♓
14	07:44	15	09:27	♈
17	19:56	17	21:59	♉
19	23:35	20	08:51	♊
22	07:54	22	16:16	♋
24	13:40	24	19:37	♌
26	16:30	26	19:47	♍
28	03:17	28	18:47	♎

DECLINATION

Day	☉	☽	☿	♀	♂	♃	♄	♅	♆	♇
01 Fr	17S 13	06N52	16S 04	17S 15	03N37	23N20	19N59	14S 09	18S 01	13S 02
02 Sa	16 56	00 54	16 20	16 53	03 55	23 20	20 00	14 08	18 01	13 02
03 Su	16 38	04S 59	16 36	16 31	04 13	23 21	20 00	14 07	18 00	13 02
04 Mo	16 20	10 29	16 52	16 08	04 31	23 21	20 00	14 05	17 59	13 01
05 Tu	16 03	15 18	17 07	15 45	04 49	23 21	20 00	14 04	17 59	13 01
06 We	15 44	19 15	17 22	15 21	05 06	23 22	20 00	14 03	17 58	13 01
07 Th	15 26	22 09	17 35	14 57	05 24	23 22	20 01	14 02	17 58	13 01
08 Fr	15 07	23 51	17 48	14 33	05 42	23 22	20 01	14 01	17 57	13 01
09 Sa	14 48	24 18	17 59	14 08	05 59	23 23	20 01	14 00	17 56	13 01
10 Su	14 29	23 33	18 09	13 42	06 17	23 23	20 01	13 59	17 56	13 01
11 Mo	14 09	21 41	18 19	13 17	06 34	23 23	20 02	13 57	17 55	13 01
12 Tu	13 49	18 51	18 27	12 51	06 51	23 24	20 02	13 56	17 55	13 01
13 We	13 29	15 15	18 33	12 24	07 09	23 24	20 02	13 55	17 54	13 01
14 Th	13 09	11 04	18 39	11 58	07 26	23 24	20 03	13 54	17 54	13 01
15 Fr	12 49	06 30	18 43	11 31	07 43	23 24	20 03	13 53	17 53	13 01
16 Sa	12 28	01 41	18 46	11 03	08 00	23 25	20 03	13 52	17 52	13 00
17 Su	12 07	03N11	18 48	10 36	08 17	23 25	20 04	13 50	17 52	13 00
18 Mo	11 46	08 00	18 48	10 08	08 34	23 25	20 04	13 49	17 51	13 00
19 Tu	11 25	12 34	18 47	09 40	08 51	23 25	20 05	13 48	17 51	13 00
20 We	11 04	16 43	18 45	09 11	09 07	23 25	20 05	13 47	17 50	13 00
21 Th	10 42	20 13	18 41	08 43	09 24	23 26	20 06	13 46	17 50	13 00
22 Fr	10 20	22 49	18 36	08 14	09 40	23 26	20 06	13 45	17 49	13 00
23 Sa	09 58	24 14	18 30	07 45	09 57	23 26	20 07	13 43	17 48	12 59
24 Su	09 36	24 13	18 22	07 15	10 13	23 26	20 07	13 42	17 48	12 59
25 Mo	09 14	22 37	18 13	06 46	10 29	23 26	20 08	13 41	17 47	12 59
26 Tu	08 52	19 26	18 03	06 16	10 45	23 26	20 08	13 40	17 47	12 59
27 We	08 29	14 53	17 51	05 46	11 01	23 27	20 09	13 39	17 46	12 59
28 Th	08 07	09 20	17 38	05 16	11 17	23 27	20 09	13 38	17 46	12 59

ASPECTARIAN

01 12:13 ☽ △ ☿
12:28 ☽ ∥ ♂
13:09 ♀ ∥ ☉
16:54 ♀ ✶ ♇
19:58 ☽ □ ♃
21:55 ☽ △ ♄
22:50 ☽ △ ♆

02 01:29 ☽ ☍ ♂
06:25 ☽ △ ☉
12:45 ☽ ✶ ♇
14:36 ☽ △ ♀
20:37 ☿ ∥ ♀
21:41 ☽ # ♂

03 00:45 ☽ △ ♅
01:18 ☿ ∥ ☉
11:24 ☽ □ ☿
22:06 ☽ △ ♃

04 01:35 ☽ □ ♆
04:19 ☿ ♑R
12:11 ☽ ∥ ♇
17:29 ☽ ∥ ♅
23:25 ☽ □ ♀

05 02:14 ☽ ∥ ♀
03:47 ☽ ∥ ☉
05:08 ☽ □ ♅
10:58 ☽ ∥ ☿
14:03 ☽ ✶ ☿
15:31 ☽ ∥ ♆

06 02:44 ☉ ✶ ♇
05:28 ☽ # ♄
06:02 ☽ ☍ ♄
07:25 ☽ ✶ ♆
10:18 [illegible]
22:54 ☽ ☌ ♇

07 00:36 ☽ ✶ ☉
12:22 ☽ ✶ ♀
12:39 ☽ ✶ ♅
15:23 ♀ ☌ ♅
15:26 ☽ # ♃

08 01:33 ♄ SD
11:18 ☽ ☍ ♃
17:29 ☿ SD
18:34 ☿ ∥ ♆

09 04:44 ☽ □ ♂
08:01 ♀ ∥ ♅

10 02:43 ☽ # ♃
06:51 ☽ ☌ ☿

11 01:07 ☽ △ ♄
02:58 ☽ ☌ ♆
14:38 ♀ ∥ ♇
14:47 ☽ # ♄
15:08 ☉ ∥ ♅
18:33 ♂ △ ♇
19:24 ☽ ✶ ♇
19:27 ☽ ✶ ♂

12 01:18 ♀ ♓
02:48 ☽ ∥ ☿
06:41 ☽ ∥ ♆
10:22 ☽ ☌ ♅
23:11 ☽ ☌ ♀

13 07:57 ☽ ∥ ♅
09:04 ☽ △ ♃
11:19 ☽ ∥ ☉
[illegible] ☽ □ ♄
13:09 ☽ ∥ ♇

17:06 ☉ ☌ ♅
17:20 ☿ ♒
18:29 ☽ ∥ ♀

14 07:44 ☽ □ ♇
09:57 ☉ ∥ ♇
18:05 ☽ # ♂

15 11:32 ☽ ✶ ☿
21:28 ☽ □ ♃

16 01:50 ☽ ✶ ♄
03:59 ☽ ✶ ♆
18:18 ♀ △ ♃
20:32 ☽ △ ♇

17 04:09 ☽ ☌ ♂
12:02 ☽ ✶ ♅
19:56 ☽ ✶ ☉

18 03:03 ☽ ∥ ♂
04:02 ☽ □ ☿
09:37 ☽ ✶ ♃
09:58 ☽ # ♀
13:14 ♀ □ ♄
14:18 ☽ ✶ ♀
16:24 ☽ □ ♆
18:15 ☽ # ☉
20:14 ☉ ♓

19 02:24 ☽ # ♇
06:50 ☽ # ♅
23:35 ☽ □ ♅

20 02:07 ♀ # ♂
07:10 ☽ # ♆
[illegible] ☽ ∥ ☿
19:07 ☽ △ ☿

23:04 ☽ ∥ ♄

21 00:22 ☽ ☌ ♄
02:31 ☽ △ ♆
06:39 ☽ □ ♀
17:32 ☽ ☍ ♇

22 06:56 ☽ ✶ ♂
07:54 ☽ △ ♅
07:57 ☽ ∥ ♃
23:39 ☽ △ ☉

23 01:04 ☉ # ♂
02:20 ☽ ☌ ♃
[illegible]
13:52 ☿ △ ♄

17:47 ☽ △ ♀

24 11:16 ☉ △ ♃
13:40 ☽ □ ♂
14:29 ☽ ∥ ♃
15:59 ☿ ☌ ♆

25 09:12 ☽ ✶ ♄
11:07 ☽ ☍ ♆
12:39 ☽ ☍ ☿
19:29 ☽ ∥ ♄
23:57 ☽ △ ♇

26 00:44 ♀ □ ♇
[illegible]
09:28 ☽ # ♆

12:47 ☽ ☍ ♅
16:30 ☽ △ ♂

27 03:54 ☉ □ ♄
04:38 ☽ ✶ ♃
05:40 ☽ # ♅
08:37 ☽ # ♇
08:56 ☽ □ ♄
09:48 ☿ ∥ ♆
16:16 ☽ ∥ ♂
23:13 ☽ □ ♇

28 03:17 ☽ ☍ ♀
[illegible]
17:31 ☽ # ♀

⚷ Chiron

01 Dec.	17 S 33
01	05♑18
04	05 35
07	05 51
10	06 06
13	06 21
16	06 36
19	06 50
22	07 04
25	07 16
28	07 29

March 2002

Day	S. T.	☉	☽	☿	♀	♂	♃	♄	♅	♆	♇	☊ True
	h m s	° ' "	° ' "	° '	° '	° '	° '	° '	° '	° '	° '	° '
01 Fr	10 34 30	10♓ 13 08	03♎ 18 39	14♒ 39	21♓ 11	29♈ 33	05♋R37	08♊ 27	25♒ 42	09♒ 37	17♐ 31	23♊R43
02 Sa	10 38 26	11 13 20	18 23 35	15 56	22 26	00♉ 16	05D 37	08 29	25 45	09 39	17 32	23 33
03 Su	10 42 23	12 13 30	03♏ 06 44	17 14	23 41	00 58	05 38	08 31	25 49	09 41	17 32	23 26
04 Mo	10 46 20	13 13 39	17 22 52	18 34	24 56	01 41	05 38	08 34	25 52	09 43	17 33	23 20
05 Tu	10 50 16	14 13 46	01♐ 10 22	19 55	26 11	02 23	05 39	08 37	25 55	09 45	17 33	23 17
06 We	10 54 13	15 13 52	14 30 35	21 18	27 25	03 06	05 39	08 39	25 59	09 47	17 34	23 16
07 Th	10 58 09	16 13 56	27 26 48	22 42	28 40	03 48	05 40	08 42	26 02	09 48	17 34	23 16
08 Fr	11 02 06	17 13 59	10♑ 03 12	24 07	29 55	04 30	05 41	08 45	26 05	09 50	17 35	23 15
09 Sa	11 06 02	18 14 00	22 24 09	25 34	01♈ 09	05 13	05 43	08 48	26 08	09 52	17 35	23 12
10 Su	11 09 59	19 13 59	04♒ 33 42	27 02	02 24	05 55	05 44	08 52	26 12	09 54	17 36	23 08
11 Mo	11 13 55	20 13 56	16 35 20	28 31	03 39	06 37	05 46	08 55	26 15	09 56	17 36	23 00
12 Tu	11 17 52	21 13 52	28 31 52	00♓ 02	04 53	07 19	05 48	08 58	26 18	09 58	17 36	22 49
13 We	11 21 49	22 13 46	10♓ 25 29	01 33	06 08	08 01	05 50	09 02	26 21	10 00	17 37	22 36
14 Th	11 25 45	23 13 38	22 17 46	03 06	07 22	08 44	05 52	09 05	26 25	10 01	17 37	22 22
15 Fr	11 29 42	24 13 28	04♈ 09 59	04 41	08 37	09 26	05 55	09 09	26 28	10 03	17 37	22 08
16 Sa	11 33 38	25 13 16	16 03 25	06 16	09 52	10 08	05 58	09 13	26 31	10 05	17 37	21 56
17 Su	11 37 35	26 13 02	27 59 36	07 53	11 06	10 50	06 00	09 17	26 34	10 07	17 37	21 45
18 Mo	11 41 31	27 12 46	10♉ 00 37	09 31	12 20	11 32	06 04	09 21	26 37	10 08	17 37	21 38
19 Tu	11 45 28	28 12 28	22 09 11	11 10	13 35	12 14	06 07	09 25	26 40	10 10	17 38	21 33
20 We	11 49 24	29 12 07	04♊ 28 40	12 50	14 49	12 56	06 10	09 29	26 43	10 12	17 38	21 31
21 Th	11 53 21	00♈ 11 45	17 03 00	14 32	16 04	13 37	06 14	09 33	26 46	10 13	17R 38	21 31
22 Fr	11 57 18	01 11 20	29 56 29	16 15	17 18	14 19	06 18	09 38	26 49	10 15	17 38	21 31
23 Sa	12 01 14	02 10 52	13♋ 13 20	17 59	18 32	15 01	06 22	09 42	26 52	10 17	17 38	21 30
24 Su	12 05 11	03 10 23	26 57 02	19 44	19 47	15 43	06 26	09 46	26 55	10 18	17 37	21 27
25 Mo	12 09 07	04 09 51	11♌ 09 15	21 31	21 01	16 25	06 30	09 51	26 58	10 20	17 37	21 21
26 Tu	12 13 04	05 09 17	25 48 46	23 19	22 15	17 06	06 34	09 56	27 01	10 21	17 37	21 14
27 We	12 17 00	06 08 40	10♍ 50 49	25 08	23 29	17 48	06 39	10 00	27 04	10 23	17 37	21 04
28 Th	12 20 57	07 08 01	26 07 04	26 59	24 43	18 29	06 44	10 05	27 07	10 24	17 37	20 53
29 Fr	12 24 53	08 07 20	11♎ 26 41	28 51	25 57	19 11	06 49	10 10	27 10	10 26	17 36	20 42
30 Sa	12 28 50	09 06 38	26 38 06	00♈ 44	27 11	19 52	06 54	10 15	27 13	10 27	17 36	20 32
31 Su	12 32 47	10 05 53	11♏ 31 11	02 38	28 25	20 34	06 59	10 20	27 15	10 28	17 36	20 24

Data for	03-01-2002
Julian Day	2452334.50
Ayanamsa	23 52 57
SVP	05 ♓ 13 30
☽ ☊ Mean	23 ♊ 14 R

● ◐ PHASES ○ ◑

06	01:25	◑	15♐17
14	02:03	●	23♓19
22	02:28	◐	01♋17
28	18:25	○	07♎54

LAST ASPECT Day	h m	☽ INGRESS Day	h m	
02	11:58	02	18:52	♏
04	14:44	04	21:56	♐
07	02:33	07	04:49	♑
08	15:08	09	14:58	♒
11	19:30	12	02:58	♓
14	02:03	14	15:35	♈
16	21:08	17	04:01	♉
19	12:53	19	15:20	♊
21	18:14	22	00:06	♋
23	10:19	24	05:13	♌
26	01:57	26	06:44	♍
28	01:32	28	06:04	♎
30	00:58	30	05:22	♏

DECLINATION

Day	☉	☽	☿	♀	♂	♃	♄	♅	♆	♇
01 Fr	07S44	03N14	17S24	04S46	11N33	23N27	20N10	13S37	17S45	12S58
02 Sa	07 21	02S59	17 08	04 16	11 49	23 27	20 11	13 36	17 45	12 58
03 Su	06 59	08 54	16 52	03 45	12 04	23 27	20 11	13 34	17 44	12 58
04 Mo	06 35	14 10	16 33	03 15	12 20	23 27	20 12	13 33	17 44	12 58
05 Tu	06 12	18 31	16 14	02 44	12 35	23 27	20 12	13 32	17 43	12 58
06 We	05 49	21 46	15 53	02 14	12 50	23 27	20 13	13 31	17 43	12 58
07 Th	05 26	23 46	15 31	01 43	13 05	23 27	20 14	13 30	17 42	12 57
08 Fr	05 03	24 30	15 08	01 12	13 20	23 27	20 15	13 29	17 42	12 57
09 Sa	04 39	24 00	14 43	00 41	13 35	23 27	20 15	13 28	17 41	12 57
10 Su	04 16	22 21	14 17	00 11	13 50	23 27	20 16	13 27	17 41	12 57
11 Mo	03 52	19 42	13 50	00N20	14 05	23 27	20 17	13 26	17 40	12 57
12 Tu	03 29	16 15	13 22	00 51	14 19	23 27	20 17	13 24	17 40	12 56
13 We	03 05	12 10	12 52	01 22	14 33	23 27	20 18	13 23	17 39	12 56
14 Th	02 41	07 38	12 21	01 53	14 48	23 27	20 19	13 22	17 39	12 56
15 Fr	02 18	02 49	11 49	02 24	15 02	23 27	20 20	13 21	17 38	12 56
16 Sa	01 54	02N06	11 16	02 54	15 16	23 27	20 21	13 20	17 38	12 55
17 Su	01 30	06 59	10 41	03 25	15 29	23 27	20 21	13 19	17 37	12 55
18 Mo	01 06	11 39	10 05	03 56	15 43	23 27	20 22	13 18	17 37	12 55
19 Tu	00 43	15 55	09 29	04 26	15 57	23 27	20 23	13 17	17 36	12 55
20 We	00 19	19 35	08 50	04 57	16 10	23 27	20 24	13 16	17 36	12 54
21 Th	00N05	22 25	08 11	05 27	16 23	23 27	20 25	13 15	17 36	12 54
22 Fr	00 28	24 11	07 30	05 57	16 36	23 27	20 26	13 14	17 35	12 54
23 Sa	00 52	24 38	06 49	06 27	16 49	23 27	20 26	13 13	17 35	12 54
24 Su	01 16	23 38	06 06	06 57	17 02	23 27	20 27	13 12	17 34	12 54
25 Mo	01 39	21 07	05 22	07 27	17 15	23 27	20 28	13 11	17 34	12 53
26 Tu	02 03	17 11	04 37	07 57	17 27	23 26	20 29	13 10	17 34	12 53
27 We	02 26	12 04	03 51	08 26	17 39	23 26	20 30	13 09	17 33	12 53
28 Th	02 50	06 08	03 04	08 55	17 52	23 26	20 31	13 08	17 33	12 53
29 Fr	03 13	00S12	02 15	09 25	18 04	23 26	20 32	13 07	17 32	12 52
30 Sa	03 37	06 28	01 26	09 53	18 15	23 26	20 33	13 06	17 32	12 52
31 Su	04 00	12 16	00 36	10 22	18 27	23 26	20 34	13 05	17 32	12 52

ASPECTARIAN

01 03:39 ☽ □ ♃
08:08 ☽ △ ♄
09:59 ☽ △ ♆
15:05 ♂ ♉
15:16 ♃ SD
19:40 ☽ △ ☿
22:37 ☽ ✱ ♇
02 04:38 ☽ ∥ ♀
11:58 ☽ △ ♅
16:26 ☽ ∥ ☉
20:17 ☽ ☍ ♂
03 04:10 ☽ △ ♃
05:40 ☿ ✱ ♇
10:58 ☽ □ ♆
14:43 ☽ # ♂
16:24 ☽ △ ☉
18:11 ☽ ∥ ♇
20:59 ☽ ∥ ♅
04 02:14 ☽ □ ☿
11:35 ☽ ∥ ☿
14:20 ☽ △ ♀
14:44 ☽ □ ♅
19:05 ☽ ∥ ♆
05 11:24 ☽ # ♄
13:20 ☽ ☍ ♄
15:22 ☽ ✱ ♆
06 05:37 ☽ ☌ ♇
11:28 ♂ # ♇
14:01 ☽ ✱ ☿
18:58 ☽ # ♃
21:20 ☽ ✱ ♅
07 02:33 ☽ □ ♀
12:44 ☽ △ ♂
15:37 ☽ ☍ ♃
08 01:42 ♀ ♈
08:28 ☉ □ ♇
12:38 ♂ # ♅
15:08 ☽ ✱ ☉
09 09:41 ☽ # ♃
09:52 ☿ ☌ ♅
17:46 ♂ ✱ ♃
19:13 ☽ ✱ ♀
10 02:51 ☽ □ ♂
08:36 ☽ △ ♄
10:40 ☽ ☌ ♆
15:50 ☿ # ♂
19:24 ☽ # ♄
11 02:02 ☽ ✱ ♇
14:45 ☽ ∥ ♆
19:30 ☽ ☌ ♅
21:48 ☿ ∥ ♅
23:34 ☿ ♓
12 03:27 ☽ ☌ ☿
11:05 ☽ # ♂
14:42 ☽ △ ♃
17:05 ☽ ∥ ♅
18:06 ♀ □ ♃
18:51 ☽ ✱ ♂
19:33 ☽ ∥ ☿
19:41 ☽ ∥ ♇
20:51 ☿ ∥ ♇
21:10 ☽ □ ♄
13 14:32 ☽ □ ♆
14 21:25 ♀ # ☉
15 01:54 ☽ # ♀
02:49 ☽ ∥ ☉
03:33 ☽ □ ♃
10:02 ☽ ☌ ♀
10:07 ☽ ✱ ♄
10:54 ♀ ✱ ♄
11:58 ☽ ✱ ♆
19:14 ☿ △ ♃
22:25 ♂ □ ♆
23:04 ☽ # ☉
16 03:09 ☽ △ ♇
04:21 ☽ ∥ ♀
04:27 ♀ ✱ ♆
21:08 ☽ ✱ ♅
17 16:06 ☽ ✱ ♃
16:45 ☽ # ☿
21:29 ☿ □ ♄
22:51 ☽ ✱ ☿
18 00:16 ☽ □ ♆
03:12 ☽ ☌ ♂
06:50 ☽ # ♇
08:55 ☽ # ♅
19 00:09 ☽ ∥ ♂
08:53 ☽ □ ♅
10:28 ☽ # ♆
12:53 ☽ ✱ ☉
20 02:12 ☿ ✱ ♂
06:09 ☽ ∥ ♄
09:40 ☽ ☌ ♄
11:02 ☽ △ ♆
14:54 ♇ SR
18:29 ☽ □ ☿
19:16 ☉ ♈
21:56 ☽ ✱ ♀
21 01:05 ☽ ☍ ♇
11:57 ☽ ∥ ♃
18:14 ☽ △ ♅
22 06:20 ♀ △ ♇
11:38 ☽ ☌ ♃
19:11 ☿ □ ♇
23 03:21 ☽ ✱ ♂
07:06 ☿ # ♀
09:38 ☽ △ ☿
10:19 ☽ □ ♀
24 02:36 ☽ ∥ ♃
11:25 ☽ △ ☉
21:49 ☽ ✱ ♄
22:37 ☽ ☍ ♆
25 04:36 ☽ ∥ ♄
09:07 ☽ □ ♂
10:40 ☽ △ ♇
17:42 ☽ △ ♀
22:00 ☽ # ♆
22:38 ☽ ∥ ♂
26 01:57 ☽ ☍ ♅
12:09 ♂ # ♆
17:19 ☽ ✱ ♃
19:16 ☽ # ♅
20:28 ☽ # ♇
22:40 ☽ □ ♄
27 10:40 ☽ □ ♇
11:28 ☽ △ ♂
13:24 ☉ □ ♃
13:50 ☽ ∥ ♀
28 01:32 ☽ ☍ ☿
04:38 ☿ # ☉
11:51 ☽ ∥ ☉
13:24 ☽ # ☿
16:42 ☽ □ ♃
21:59 ☽ △ ♄
22:24 ☽ △ ♆
29 06:52 ☽ ∥ ☿
09:41 ☽ ✱ ♇
12:11 ☽ # ☉
14:45 ☿ ♈
30 00:25 ♀ ✱ ♅
00:55 ☽ △ ♅
00:58 ☽ ☍ ♀
15:02 ☽ # ♀
16:35 ☽ △ ♃
22:17 ☽ □ ♆
31 02:39 ☽ ∥ ♇
03:39 ☽ ∥ ♅
06:22 ☉ ✱ ♄
09:19 ☉ ✱ ♆
15:41 ☽ ☍ ♂

⚷ Chiron

01	Dec.	17 S 14
03	07♑40	
06	07 51	
09	08 02	
12	08 11	
15	08 20	
18	08 29	
21	08 36	
24	08 43	
27	08 49	
30	08 54	

April 2002

Day	S. T.	☉	☽	☿	♀	♂	♃	♄	♅	♆	♇	☊ True
	h m s	° ' "	° ' "	° '	° '	° '	° '	° '	° '	° '	° '	° '
01 Mo	12 36 43	11♈05 06	25♏58 41	04♈34	29♈39	21♉15	07♋05	10♊25	27♒18	10♒30	17♐R35	20♊R18
02 Tu	12 40 40	12 04 18	09♐57 00	06 32	00♉53	21 57	07 11	10 31	27 21	10 31	17 35	20 16
03 We	12 44 36	13 03 28	23 26 00	08 30	02 07	22 38	07 16	10 36	27 24	10 32	17 35	20 15
04 Th	12 48 33	14 02 36	06♑28 03	10 30	03 21	23 20	07 22	10 41	27 26	10 34	17 34	20D 15
05 Fr	12 52 29	15 01 42	19 07 08	12 31	04 35	24 01	07 28	10 47	27 29	10 35	17 34	20R 15
06 Sa	12 56 26	16 00 47	01♒27 53	14 33	05 49	24 42	07 35	10 52	27 31	10 36	17 33	20 15
07 Su	13 00 22	16 59 50	13 35 05	16 36	07 03	25 23	07 41	10 58	27 34	10 37	17 33	20 12
08 Mo	13 04 19	17 58 51	25 33 11	18 40	08 17	26 04	07 48	11 03	27 36	10 38	17 32	20 07
09 Tu	13 08 16	18 57 50	07♓26 09	20 44	09 30	26 46	07 54	11 09	27 39	10 39	17 31	19 59
10 We	13 12 12	19 56 47	19 17 14	22 50	10 44	27 27	08 01	11 15	27 41	10 41	17 31	19 49
11 Th	13 16 09	20 55 42	01♈08 57	24 55	11 58	28 08	08 08	11 21	27 44	10 42	17 30	19 39
12 Fr	13 20 05	21 54 36	13 03 14	27 01	13 11	28 49	08 15	11 27	27 46	10 43	17 29	19 27
13 Sa	13 24 02	22 53 27	25 01 33	29 07	14 25	29 30	08 23	11 33	27 48	10 44	17 29	19 17
14 Su	13 27 58	23 52 17	07♉05 14	01♉12	15 39	00♊11	08 30	11 39	27 51	10 45	17 28	19 09
15 Mo	13 31 55	24 51 04	19 15 40	03 17	16 52	00 52	08 38	11 45	27 53	10 46	17 27	19 03
16 Tu	13 35 51	25 49 50	01♊34 33	05 20	18 06	01 33	08 45	11 51	27 55	10 46	17 26	19 00
17 We	13 39 48	26 48 33	14 03 54	07 23	19 19	02 14	08 53	11 57	27 57	10 47	17 26	18D 59
18 Th	13 43 44	27 47 15	26 46 14	09 23	20 33	02 55	09 01	12 04	28 00	10 48	17 25	19 00
19 Fr	13 47 41	28 45 54	09♋44 24	11 22	21 46	03 35	09 09	12 10	28 02	10 49	17 24	19 01
20 Sa	13 51 38	29 44 31	23 01 22	13 19	22 59	04 16	09 17	12 16	28 04	10 50	17 23	19 02
21 Su	13 55 34	00♉43 05	06♌39 46	15 12	24 13	04 57	09 26	12 23	28 06	10 51	17 22	19R 01
22 Mo	13 59 31	01 41 38	20 41 05	17 03	25 26	05 38	09 34	12 29	28 08	10 51	17 21	18 59
23 Tu	14 03 27	02 40 08	05♍04 59	18 51	26 39	06 18	09 43	12 36	28 10	10 52	17 20	18 54
24 We	14 07 24	03 38 36	19 48 30	20 35	27 52	06 59	09 52	12 43	28 12	10 53	17 19	18 48
25 Th	14 11 20	04 37 02	04♎45 56	22 15	29 05	07 39	10 00	12 49	28 13	10 53	17 18	18 41
26 Fr	14 15 17	05 35 25	19 49 08	23 52	00♊18	08 20	10 09	12 56	28 15	10 54	17 17	18 34
27 Sa	14 19 13	06 33 47	04♏48 45	25 24	01 31	09 00	10 18	13 03	28 17	10 54	17 16	18 27
28 Su	14 23 10	07 32 07	19 35 36	26 53	02 44	09 41	10 27	13 10	28 19	10 55	17 15	18 22
29 Mo	14 27 07	08 30 26	04♐02 15	28 17	03 57	10 21	10 37	13 16	28 20	10 55	17 13	18 19
30 Tu	14 31 03	09 28 43	18 03 51	29 37	05 10	11 02	10 46	13 23	28 22	10 56	17 12	18 17

Data for	04-01-2002
Julian Day	2452365.50
Ayanamsa	23 53 00
SVP	05 ♓ 13 25
☽ ☊ Mean	21 ♊ 36 R

● ◐ PHASES ○ ◑

04	15:30	◑	14♑41
12	19:21	●	22♈42
20	12:49	◐	00♌16
27	03:01	○	06♏41

LAST ASPECT ☽ INGRESS

Day	h m	Day	h m	
01	02:15	01	06:49	♐
03	07:14	03	12:00	♑
05	10:01	05	21:08	♒
08	04:09	08	08:58	♓
10	17:32	10	21:41	♈
13	09:52	13	09:55	♉
15	16:53	15	20:57	♊
18	02:17	18	06:02	♋
19	23:56	20	12:22	♌
22	12:31	22	15:36	♍
24	14:07	24	16:23	♎
26	13:30	26	16:16	♏
28	14:26	28	17:14	♐
30	18:11	30	21:03	♑

DECLINATION

Day	☉	☽	☿	♀	♂	♃	♄	♅	♆	♇
01 Mo	04N23	17S 13	00N15	10N50	18N38	23N25	20N35	13S 04	17S 31	12S 52
02 Tu	04 46	21 02	01 07	11 18	18 50	23 25	20 36	13 03	17 31	12 51
03 We	05 09	23 33	02 00	11 46	19 01	23 25	20 37	13 03	17 31	12 51
04 Th	05 32	24 41	02 54	12 14	19 12	23 25	20 38	13 02	17 30	12 51
05 Fr	05 55	24 30	03 48	12 41	19 23	23 24	20 38	13 01	17 30	12 51
06 Sa	06 18	23 05	04 43	13 08	19 33	23 24	20 39	13 00	17 30	12 50
07 Su	06 41	20 38	05 38	13 35	19 44	23 24	20 40	12 59	17 29	12 50
08 Mo	07 03	17 20	06 34	14 01	19 54	23 24	20 41	12 58	17 29	12 50
09 Tu	07 26	13 21	07 29	14 27	20 04	23 23	20 42	12 57	17 29	12 50
10 We	07 48	08 53	08 25	14 53	20 14	23 23	20 43	12 57	17 28	12 49
11 Th	08 10	04 05	09 21	15 18	20 24	23 23	20 44	12 56	17 28	12 49
12 Fr	08 32	00N53	10 16	15 43	20 34	23 22	20 45	12 55	17 28	12 49
13 Sa	08 54	05 51	11 11	16 07	20 43	23 22	20 46	12 54	17 28	12 49
14 Su	09 16	10 39	12 06	16 31	20 52	23 21	20 47	12 53	17 27	12 48
15 Mo	09 37	15 06	12 59	16 55	21 01	23 21	20 48	12 53	17 27	12 48
16 Tu	09 59	18 58	13 51	17 18	21 10	23 21	20 49	12 52	17 27	12 48
17 We	10 20	22 03	14 42	17 41	21 19	23 20	20 50	12 51	17 27	12 48
18 Th	10 41	24 06	15 31	18 04	21 27	23 20	20 51	12 51	17 26	12 47
19 Fr	11 02	24 54	16 19	18 25	21 36	23 19	20 52	12 50	17 26	12 47
20 Sa	11 23	24 19	17 04	18 47	21 44	23 19	20 53	12 49	17 26	12 47
21 Su	11 43	22 17	17 48	19 08	21 52	23 18	20 54	12 48	17 26	12 47
22 Mo	12 04	18 53	18 29	19 28	22 00	23 18	20 55	12 48	17 26	12 46
23 Tu	12 24	14 17	19 09	19 48	22 07	23 17	20 56	12 47	17 25	12 46
24 We	12 44	08 45	19 45	20 08	22 14	23 17	20 57	12 47	17 25	12 46
25 Th	13 04	02 38	20 19	20 26	22 22	23 16	20 58	12 46	17 25	12 46
26 Fr	13 23	03S 41	20 51	20 45	22 28	23 15	20 59	12 45	17 25	12 45
27 Sa	13 42	09 47	21 20	21 03	22 35	23 15	21 00	12 45	17 25	12 45
28 Su	14 02	15 14	21 47	21 20	22 42	23 14	21 01	12 44	17 25	12 45
29 Mo	14 20	19 42	22 11	21 36	22 48	23 13	21 02	12 44	17 24	12 45
30 Tu	14 39	22 53	22 33	21 52	22 54	23 13	21 03	12 43	17 24	12 44

ASPECTARIAN

01 01:39 ☽ ∥ ♆
02:15 ☽ □ ♅
06:40 ♀ ♉
08:29 ☽ # ♂
17:03 ☽ △ ☿
20:43 ☽ # ♄
02 00:59 ☽ ☍ ♄
01:00 ☽ ⚹ ♆
02:41 ♄ △ ♆
04:01 ☽ △ ☉
08:19 ☿ □ ♃
13:29 ☽ ☌ ♇
22:15 ☽ # ♃
03 07:14 ☽ ⚹ ♅
17:35 ☽ △ ♀
04 00:45 ☿ ⚹ ♆
01:42 ☽ ☍ ♃
02:23 ☿ ⚹ ♄
09:00 ☽ □ ☿
05 08:22 ♀ # ♇
10:01 ☽ △ ♂
16:55 ♀ # ♅
20:01 ☽ # ♃
06 09:32 ☽ □ ♀
18:05 ☽ ☌ ♆
18:44 ☽ △ ♄
23:40 ☽ # ♄
07 06:48 ☽ # ♂
07:16 ☽ ⚹ ☿
07:26 ☽ ⚹ ☉
07:54 ☽ ⚹ ♇
08:56 ☿ ☌ ☉
10:59 ☿ △ ♇
13:15 ☉ △ ♇
13:39 ♀ ⚹ ♃
[illegible] ∥ ♆
08 01:07 ☽ □ ♂
04:09 ☽ ☌ ♅
18:20 ☽ # ♀
21:21 ☿ ∥ ☉
09 00:58 ☽ △ ♃
02:16 ☽ ∥ ♅
02:59 ☽ ∥ ♇
04:40 ☽ ⚹ ♀
07:35 ☽ □ ♄
20:25 ☽ □ ♇
22:51 ♀ □ ♆
10 02:22 ☽ # ☿
05:10 ☽ # ☉
09:04 ♂ □ ♅
17:32 ☽ ⚹ ♂
11 14:15 ☽ □ ♃
19:17 ☽ ⚹ ♆
20:44 ☽ ⚹ ♄
12 08:46 ☿ ⚹ ♅
08:54 ☽ △ ♇
13 05:34 ☽ ⚹ ♅
09:09 ♂ ∥ ♄
09:52 ☽ ☌ ☿
10:11 ☿ ♉
16:21 ☽ ∥ ☉
17:36 ♂ ♊
14 02:50 ☽ ⚹ ♃
07:15 ☽ □ ♆
09:22 ☽ ∥ ☿
11:18 ☽ # ♇
11:44 ☽ # ♅
18:47 ☽ ☌ ♀
19:11 ☿ # ♇
21:15 [illegible] # ♅
15 11:56 ☽ ∥ ♀
14:03 ☽ # ♆
16:53 ☽ □ ♅
23:56 ☽ ☌ ♂
16 08:40 ♀ # ♆
13:35 ☽ ∥ ♄
17:11 ☽ ∥ ♂
17:44 ☽ △ ♆
19:56 ☽ ☌ ♄
17 06:23 ☽ ☍ ♇
13:19 ☽ ∥ ♃
19:12 ☿ ⚹ ♃
18 02:03 ☽ ⚹ ☉
02:17 ☽ △ ♅
05:15 ☉ ⚹ ♅
17:13 ☿ □ ♆
22:55 ☽ ☌ ♃
19 03:30 ☽ ⚹ ☿
23:56 ☽ ⚹ ♀
20 06:21 ☉ ♉
11:37 ☿ # ♆
14:01 ☽ ∥ ♃
20:52 ☽ ⚹ ♂
21 03:27 ☽ ∥ ♂
07:14 ☽ ☍ ♆
09:57 ☽ ⚹ ♄
10:50 ☽ ∥ ♄
16:55 ☽ □ ☿
18:21 ☽ △ ♇
20:44 ☽ ∥ ♀
22 02:00 ☽ ∥ ☿
08:16 ☽ # ♆
08:43 ☽ □ ♀
12:31 ☽ ☍ ♅
19:44 ☽ △ ☉
23 02:06 ☽ □ ♂
06:52 ☽ # ♅
06:56 ☽ # ♇
07:40 ☽ ⚹ ♃
08:05 ☽ ∥ ☉
12:24 ☽ □ ♄
19:58 ☽ □ ♇
24 01:24 ☽ △ ☿
02:26 ☉ # ♇
03:10 ☉ # ♅
06:32 ♀ □ ♅
14:07 ☽ △ ♀
25 04:50 ☽ △ ♂
08:37 ☽ □ ♃
09:46 ☽ △ ♆
12:18 ☿ ∥ ♀
12:57 ☽ △ ♄
17:57 ♀ ♊
18:05 ♅ ∥ ♆
19:57 ☽ ⚹ ♇
26 06:48 ☿ ∥ ♄
13:30 ☽ △ ♅
20:46 ♀ ∥ ♄
27 08:58 ☽ △ ♃
09:51 ☽ □ ♆
12:36 ☽ ∥ ♅
12:39 ☽ ∥ ♇
18:00 ☽ # ☉
28 10:56 ☽ ∥ ♆
13:20 ☽ ☍ ♀
14:26 ☽ □ ♅
23:51 ☽ ☍ ♀
29 01:03 ☿ □ ♅
08:58 ☽ # ♄
11:16 ☽ ☍ ♂
11:42 ☽ ⚹ ♆
14:22 ☽ # ♀
15:51 ☽ ☍ ♄
20:25 ☽ # ☿
20:27 ♂ △ ♆
22:31 ☽ ☌ ♇
30 00:17 ☽ # ♂
03:22 ☽ # ♃
07:16 ☿ ♊
18:11 ☽ ⚹ ♅

⚷ Chiron

01	Dec.	16 S 50
02	08♑59	
05	09 02	
08	09 05	
11	09 07	
14	09 08	
17	09 09	
20	09 09R	
23	09 07	
26	09 06	
29	09 03	
17	04:54	09♑09 R

26 12:04 ✹ 05♐04 Penumbral Lunar Eclipse (mag 0.714)

May 2002

Day	S. T.	☉	☽	☿	♀	♂	♃	♄	♅	♆	♇	☊ True
	h m s	° ' "	° ' "	° '	° '	° '	° '	° '	° '	° '	° '	° '
01 We	14 34 60	10♉ 26 58	01♑ 38 31	00♊ 52	06♊ 23	11♊ 42	10♋ 56	13♊ 30	28♒ 24	10♒ 56	17♐R11	18♊D18
02 Th	14 38 56	11 25 12	14 46 57	02 03	07 36	12 23	11 05	13 37	28 25	10 57	17 10	18 19
03 Fr	14 42 53	12 23 24	27 31 46	03 09	08 49	13 03	11 15	13 44	28 27	10 57	17 09	18 21
04 Sa	14 46 49	13 21 35	09♒ 56 47	04 10	10 02	13 43	11 25	13 51	28 28	10 57	17 07	18 23
05 Su	14 50 46	14 19 44	22 06 32	05 07	11 14	14 23	11 35	13 58	28 30	10 58	17 06	18R 24
06 Mo	14 54 42	15 17 52	04♓ 05 43	05 58	12 27	15 04	11 45	14 06	28 31	10 58	17 05	18 23
07 Tu	14 58 39	16 15 58	15 58 55	06 45	13 40	15 44	11 55	14 13	28 32	10 58	17 04	18 20
08 We	15 02 36	17 14 03	27 50 18	07 27	14 52	16 24	12 05	14 20	28 34	10 58	17 02	18 17
09 Th	15 06 32	18 12 06	09♈ 43 28	08 04	16 05	17 04	12 15	14 27	28 35	10 59	17 01	18 12
10 Fr	15 10 29	19 10 08	21 41 21	08 35	17 17	17 44	12 26	14 35	28 36	10 59	17 00	18 07
11 Sa	15 14 25	20 08 09	03♉ 46 12	09 02	18 30	18 24	12 36	14 42	28 37	10 59	16 58	18 03
12 Su	15 18 22	21 06 08	15 59 50	09 24	19 42	19 04	12 47	14 49	28 38	10 59	16 57	17 59
13 Mo	15 22 18	22 04 05	28 23 34	09 40	20 54	19 44	12 57	14 57	28 39	10 59	16 55	17 57
14 Tu	15 26 15	23 02 01	10♊ 58 29	09 51	22 07	20 24	13 08	15 04	28 40	10R 59	16 54	17 56
15 We	15 30 11	23 59 56	23 45 28	09 58	23 19	21 04	13 19	15 12	28 41	10 59	16 52	17D 57
16 Th	15 34 08	24 57 49	06♋ 45 20	09R 59	24 31	21 44	13 30	15 19	28 42	10 59	16 51	17 58
17 Fr	15 38 05	25 55 40	19 58 58	09 56	25 43	22 24	13 41	15 26	28 43	10 59	16 50	17 59
18 Sa	15 42 01	26 53 30	03♌ 27 13	09 48	26 56	23 04	13 52	15 34	28 44	10 59	16 48	18 00
19 Su	15 45 58	27 51 18	17 10 43	09 35	28 08	23 44	14 03	15 42	28 45	10 58	16 47	18R 01
20 Mo	15 49 54	28 49 04	01♍ 09 33	09 19	29 20	24 23	14 14	15 49	28 45	10 58	16 45	18 00
21 Tu	15 53 51	29 46 48	15 22 49	08 58	00♋ 32	25 03	14 26	15 57	28 46	10 58	16 44	17 59
22 We	15 57 47	00♊ 44 31	29 48 20	08 34	01 44	25 43	14 37	16 04	28 47	10 58	16 42	17 56
23 Th	16 01 44	01 42 12	14♎ 22 22	08 07	02 55	26 23	14 48	16 12	28 47	10 57	16 41	17 53
24 Fr	16 05 40	02 39 51	28 59 51	07 38	04 07	27 02	15 00	16 20	28 48	10 57	16 39	17 51
25 Sa	16 09 37	03 37 30	13♏ 34 44	07 07	05 19	27 42	15 12	16 27	28 48	10 57	16 37	17 48
26 Su	16 13 34	04 35 06	28 00 42	06 34	06 31	28 22	15 23	16 35	28 49	10 56	16 36	17 46
27 Mo	16 17 30	05 32 42	12♐ 12 03	06 00	07 42	29 01	15 35	16 43	28 49	10 56	16 34	17 45
28 Tu	16 21 27	06 30 17	26 04 23	05 26	08 54	29 41	15 47	16 50	28 49	10 55	16 33	17D 45
29 We	16 25 23	07 27 50	09♑ 35 04	04 52	10 05	00♋ 20	15 59	16 58	28 50	10 55	16 31	17 45
30 Th	16 29 20	08 25 23	22 43 26	04 19	11 17	01 00	16 10	17 06	28 50	10 55	16 30	17 47
31 Fr	16 33 16	09 22 54	05♒ 30 29	03 48	12 28	01 39	16 22	17 14	28 50	10 54	16 28	17 48

Data for	05-01-2002
Julian Day	2452395.50
Ayanamsa	23 53 03
SVP	05 ♓ 13 20
☽ ☊ Mean	20 ♊ 00 R

● ◐ PHASES ○ ◑

04	07:16	◑	13♒39
12	10:46	●	21♉32
19	19:43	◐	28♌39
26	11:51	✹	05♐04

LAST ASPECT ☽ Day	h m	INGRESS Day	h m	
01	17:17	03	04:44	♒
05	12:47	05	15:46	♓
07	02:11	08	04:22	♈
10	13:47	10	16:32	♉
13	00:30	13	03:05	♊
15	09:10	15	11:35	♋
17	11:28	17	17:53	♌
19	20:35	19	22:02	♍
21	16:54	22	00:19	♎
23	23:40	24	01:39	♏
26	01:20	26	03:20	♐
28	06:40	28	06:54	♑
29	11:46	30	13:35	♒

DECLINATION

Day	☉	☽	☿	♀	♂	♃	♄	♅	♆	♇
01 We	14N57	24S 38	22N52	22N08	23N00	23N12	21N04	12S 43	17S 24	12S 44
02 Th	15 15	24 55	23 08	22 23	23 06	23 11	21 05	12 42	17 24	12 44
03 Fr	15 33	23 52	23 23	22 37	23 12	23 11	21 06	12 42	17 24	12 44
04 Sa	15 51	21 40	23 35	22 50	23 17	23 10	21 07	12 41	17 24	12 44
05 Su	16 08	18 32	23 44	23 03	23 22	23 09	21 08	12 41	17 24	12 43
06 Mo	16 25	14 41	23 52	23 16	23 27	23 08	21 09	12 40	17 24	12 43
07 Tu	16 42	10 18	23 57	23 27	23 32	23 07	21 10	12 40	17 24	12 43
08 We	16 59	05 33	24 00	23 38	23 36	23 07	21 11	12 39	17 24	12 43
09 Th	17 15	00 35	24 01	23 49	23 41	23 06	21 12	12 39	17 24	12 43
10 Fr	17 31	04N26	24 00	23 58	23 45	23 05	21 13	12 39	17 24	12 42
11 Sa	17 47	09 21	23 57	24 07	23 49	23 04	21 14	12 38	17 24	12 42
12 Su	18 02	13 59	23 52	24 16	23 53	23 03	21 15	12 38	17 24	12 42
13 Mo	18 17	18 06	23 45	24 23	23 56	23 02	21 16	12 38	17 24	12 42
14 Tu	18 32	21 28	23 37	24 30	23 59	23 01	21 17	12 37	17 24	12 42
15 We	18 46	23 50	23 26	24 36	24 03	23 00	21 18	12 37	17 24	12 41
16 Th	19 00	24 58	23 14	24 42	24 06	22 59	21 19	12 37	17 24	12 41
17 Fr	19 14	24 43	23 00	24 47	24 08	22 58	21 20	12 36	17 24	12 41
18 Sa	19 28	23 02	22 44	24 51	24 11	22 57	21 21	12 36	17 24	12 41
19 Su	19 41	20 00	22 27	24 54	24 13	22 56	21 21	12 36	17 24	12 41
20 Mo	19 54	15 46	22 09	24 57	24 15	22 55	21 22	12 36	17 24	12 41
21 Tu	20 06	10 36	21 50	24 59	24 17	22 54	21 23	12 36	17 24	12 41
22 We	20 18	04 48	21 29	25 00	24 19	22 52	21 24	12 35	17 24	12 40
23 Th	20 30	01S 20	21 08	25 01	24 20	22 51	21 25	12 35	17 24	12 40
24 Fr	20 42	07 25	20 46	25 01	24 22	22 50	21 26	12 35	17 24	12 40
25 Sa	20 53	13 05	20 23	25 00	24 23	22 49	21 27	12 35	17 24	12 40
26 Su	21 03	17 59	20 01	24 58	24 24	22 47	21 28	12 35	17 24	12 40
27 Mo	21 14	21 45	19 38	24 56	24 24	22 46	21 28	12 35	17 25	12 40
28 Tu	21 24	24 09	19 16	24 53	24 25	22 45	21 29	12 35	17 25	12 40
29 We	21 33	25 03	18 54	24 49	24 25	22 44	21 30	12 35	17 25	12 39
30 Th	21 43	24 31	18 33	24 45	24 25	22 42	21 31	12 35	17 25	12 39
31 Fr	21 51	22 41	18 13	24 40	24 25	22 41	21 32	12 35	17 25	12 39

ASPECTARIAN

01 12:11 ☉ □ ♆
14:09 ☉ ⚹ ♃
17:05 ☽ ☍ ♃
17:17 ☽ △ ☉
18:37 ☿ ∥ ♂
02 04:28 ☿ ∥ ♃
19:53 ♂ ∥ ♃
03 05:55 ☽ # ☿
08:23 ☽ # ♂
09:01 ☽ # ♃
11:47 ☽ △ ☿
13:44 ☽ # ♀
04 00:10 ☽ △ ♀
01:59 ☽ ☌ ♆
04:45 ☽ # ♄
05:58 ♂ ☌ ♄
07:45 ☽ △ ♄
07:50 ☽ △ ♂
14:05 ☽ ⚹ ♇
18:29 ♀ △ ♆
05 07:35 ☽ ∥ ♆
10:15 ♀ ∥ ♃
12:47 ☽ ☌ ♅
14:24 ☽ # ☉
06 04:04 ☽ □ ☿
11:07 ☽ ∥ ♇
11:24 ☽ ∥ ♅
15:39 ☽ △ ♃
18:46 ☽ □ ♀
20:23 ☽ □ ♄
23:28 ☽ □ ♂
07 00:38 ☽ ⚹ ☉
02:11 ☽ □ ♇
12:11 ♀ ☌ ♄
16:46 ♀ ∥ ♂
08 20:29 ☽ ⚹ ☿
22:13 ♂ ☍ ♇
09 02:31 ☽ ⚹ ♆
05:10 ☽ □ ♃
09:37 ☽ ⚹ ♄
13:00 ☉ # ♆
14:12 ☽ ⚹ ♀
14:37 ☽ △ ♇
15:38 ☽ ⚹ ♂
18:15 ♀ ☍ ♇
10 03:44 ☿ ∥ ♀
13:47 ☽ ⚹ ♅
19:56 ♀ ☌ ♂
11 14:11 ☽ □ ♆
16:49 ☽ # ♅
17:11 ☽ # ♇
17:37 ☽ ⚹ ♃
22:44 ☿ ∥ ♂
12 19:38 ☽ # ♆
13 00:30 ☽ □ ♅
01:18 ☽ ∥ ☉
12:12 ♆ SR
21:51 ☽ ☌ ☿
22:28 ☽ ∥ ♄
14 00:01 ☽ △ ♆
07:48 ☽ ☌ ♄
11:09 ☽ ☍ ♇
14:15 ☽ ∥ ♃
18:43 ☽ ☌ ♂
19:27 ☽ ∥ ☿
23:06 ☽ ☌ ♀
15 03:07 ☽ ∥ ♂
09:10 ☽ △ ♅
14:15 ☽ ∥ ♀
18:51 ☿ SR
16 12:28 ☽ ☌ ♃
22:23 ☽ ∥ ♀
17 03:17 ☿ ∥ ♃
10:25 ☽ ∥ ♂
11:28 ☽ ⚹ ☉
18 00:49 ☽ ∥ ♃
03:09 ☽ ∥ ☿
10:59 ☽ ⚹ ☿
13:12 ☽ ☍ ♆
14:26 ☽ ∥ ♄
21:24 ☽ ⚹ ♄
23:18 ☽ △ ♇
19 01:51 ☽ ∥ ☉
11:52 ☽ ⚹ ♂
12:27 ♀ △ ♅
15:23 ☽ # ♆
19:54 ☽ ☍ ♅
20:35 ☽ ⚹ ♀
22:26 ☉ □ ♅
20 13:27 ♀ ♋
13:29 ☽ □ ☿
14:47 ☽ # ♇
15:10 ☽ # ♅
22:23 ☽ ⚹ ♃
21 00:57 ☽ □ ♄
02:15 ☽ □ ♇
05:29 ☉ ♊
16:54 ☽ □ ♂
22 01:39 ☽ △ ☉
03:27 ☽ □ ♀
05:32 ☿ ∥ ♄
14:02 ☽ △ ☿
18:23 ☽ △ ♆
23 00:43 ☽ □ ♃
03:02 ☽ △ ♄
03:46 ☽ ⚹ ♇
20:38 ☽ △ ♂
23:40 ☽ △ ♅
24 03:01 ☿ ∥ ☉
09:10 ☽ △ ♀
19:39 ☽ □ ♆
21:47 ☽ ∥ ♅
22:09 ☽ ∥ ♇
25 02:42 ☽ △ ♃
20:57 ☽ ∥ ♆
26 01:20 ☽ □ ♅
02:34 ♄ ☍ ♇
10:53 ☽ # ☿
13:51 ☽ ☍ ☿
16:31 ♂ △ ♅
19:59 ☽ # ☉
21:50 ☽ ⚹ ♆
21:59 ☽ # ♄
27 07:10 ☿ ☌ ☉
07:29 ☽ ☌ ♇
07:49 ☽ ☍ ♄
08:33 ☽ # ♃
28 04:19 ☽ # ♂
04:50 ☽ ⚹ ♅
06:40 ☽ ☍ ♂
11:43 ♂ ♋
13:53 ☽ # ♀
15:16 ☉ ∥ ♄
29 01:00 ☽ ☍ ♀
11:46 ☽ ☍ ♃
18:09 ☽ # ♀
30 01:51 ☽ # ♂
20:53 ☽ △ ☿
23:59 ☽ # ♃
31 07:19 ☽ # ☉
08:01 ☽ △ ☉
10:18 ☽ ☌ ♆
10:29 ☽ # ♄
21:01 ☽ ⚹ ♇
22:47 ☽ △ ♄

⚷ Chiron

01 Dec.	16 S 31
02	09♑00R
05	08 55
08	08 51
11	08 45
14	08 39
17	08 32
20	08 25
23	08 16
26	08 08
29	07 59

June 2002

10	23:49	19♊54	Annular Solar Eclipse
24	21:28	03♑11	Penumbral Lunar Eclipse (mag 0.234)

Day	S. T.	☉	☽	☿	♀	♂	♃	♄	♅	♆	♇	☊ True
	h m s	° ' "	° ' "	° '	° '	° '	° '	° '	° '	° '	° '	° '
01 Sa	16 37 13	10♊ 20 25	17♒ 58 39	03♊R19	13♋ 39	02♋ 19	16♋ 34	17♊ 21	28♒ 50	10♒R53	16♐R26	17♊ 49
02 Su	16 41 10	11 17 55	00♓ 11 23	02 52	14 51	02 58	16 47	17 29	28 50	10 53	16 25	17 51
03 Mo	16 45 06	12 15 24	12 12 51	02 28	16 02	03 37	16 59	17 37	28R 50	10 52	16 23	17 52
04 Tu	16 49 03	13 12 52	24 07 35	02 07	17 13	04 17	17 11	17 45	28 50	10 52	16 22	17 52
05 We	16 52 59	14 10 20	06♈ 00 12	01 50	18 24	04 56	17 23	17 52	28 50	10 51	16 20	17 53
06 Th	16 56 56	15 07 47	17 55 06	01 36	19 35	05 35	17 35	18 00	28 50	10 50	16 18	17 53
07 Fr	17 00 52	16 05 13	29 56 16	01 27	20 46	06 15	17 48	18 08	28 50	10 49	16 17	17 53
08 Sa	17 04 49	17 02 39	12♉ 07 07	01 22	21 57	06 54	18 00	18 16	28 50	10 49	16 15	17 54
09 Su	17 08 45	18 00 04	24 30 21	01D 22	23 08	07 33	18 13	18 24	28 49	10 48	16 14	17 54
10 Mo	17 12 42	18 57 28	07♊ 07 56	01 25	24 18	08 13	18 25	18 31	28 49	10 47	16 12	17 55
11 Tu	17 16 39	19 54 51	20 01 00	01 34	25 29	08 52	18 38	18 39	28 49	10 46	16 10	17 55
12 We	17 20 35	20 52 14	03♋ 09 48	01 47	26 40	09 31	18 50	18 47	28 48	10 45	16 09	17R 55
13 Th	17 24 32	21 49 36	16 33 40	02 04	27 50	10 10	19 03	18 55	28 48	10 44	16 07	17 55
14 Fr	17 28 28	22 46 57	00♌ 11 20	02 26	29 01	10 49	19 16	19 03	28 47	10 43	16 06	17 54
15 Sa	17 32 25	23 44 18	14 00 57	02 52	00♌ 11	11 28	19 28	19 10	28 47	10 42	16 04	17 52
16 Su	17 36 21	24 41 37	28 00 26	03 23	01 21	12 07	19 41	19 18	28 46	10 41	16 02	17 51
17 Mo	17 40 18	25 38 56	12♍ 07 32	03 58	02 31	12 47	19 54	19 26	28 45	10 40	16 01	17 50
18 Tu	17 44 14	26 36 13	26 19 52	04 38	03 42	13 26	20 07	19 34	28 45	10 39	15 59	17 49
19 We	17 48 11	27 33 30	10♎ 34 57	05 21	04 52	14 05	20 20	19 41	28 44	10 38	15 58	17 48
20 Th	17 52 08	28 30 46	24 50 10	06 09	06 02	14 44	20 33	19 49	28 43	10 37	15 56	17 47
21 Fr	17 56 04	29 28 01	09♏ 02 46	07 00	07 11	15 23	20 45	19 57	28 42	10 36	15 54	17D 47
22 Sa	18 00 01	00♋ 25 15	23 09 50	07 56	08 21	16 02	20 58	20 05	28 41	10 35	15 53	17 47
23 Su	18 03 57	01 22 29	07♐ 08 21	08 56	09 31	16 40	21 11	20 12	28 41	10 34	15 51	17 48
24 Mo	18 07 54	02 19 43	20 55 22	09 59	10 41	17 19	21 24	20 20	28 40	10 33	15 50	17R 48
25 Tu	18 11 50	03 16 56	04♑ 28 15	11 06	11 50	17 58	21 38	20 28	28 39	10 32	15 48	17 47
26 We	18 15 47	04 14 09	17 45 02	12 17	12 59	18 37	21 51	20 35	28 38	10 30	15 47	17 46
27 Th	18 19 43	05 11 21	00♒ 44 40	13 32	14 09	19 16	22 04	20 43	28 36	10 29	15 45	17 44
28 Fr	18 23 40	06 08 33	13 27 12	14 50	15 18	19 55	22 17	20 51	28 35	10 28	15 44	17 41
29 Sa	18 27 37	07 05 46	25 53 45	16 12	16 27	20 34	22 30	20 58	28 34	10 27	15 42	17 39
30 Su	18 31 33	08 02 58	08♓ 06 29	17 37	17 36	21 12	22 43	21 06	28 33	10 25	15 41	17 37

Data for	06-01-2002
Julian Day	2452426.50
Ayanamsa	23 53 09
SVP	05 ♓ 13 16
☽ ☊ Mean	18 ♊ 22 R

● ◐ PHASES ○ ◑

03	00:05	◑	12♓16
10	23:47	☼	19♊54
18	00:29	◐	26♍37
24	21:43	✸	03♑11

LAST ASPECT Day	h m	☽ INGRESS Day	h m	
01	21:19	01	23:37	♓
03	10:59	04	11:52	♈
06	21:48	07	00:07	♉
09	08:15	09	10:30	♊
11	16:06	11	18:16	♋
13	21:45	13	23:40	♌
16	01:18	16	03:24	♍
18	00:30	18	06:11	♎
20	06:39	20	08:43	♏
22	09:27	22	11:42	♐
24	13:38	24	16:02	♑
26	07:38	26	22:37	♒
29	05:13	29	08:01	♓

DECLINATION

Day	☉	☽	☿	♀	♂	♃	♄	♅	♆	♇
01 Sa	22N00	19S 48	17N54	24N34	24N25	22N39	21N33	12S 35	17S 25	12S 39
02 Su	22 08	16 07	17 36	24 27	24 24	22 38	21 33	12 35	17 25	12 39
03 Mo	22 16	11 50	17 21	24 20	24 23	22 37	21 34	12 35	17 26	12 39
04 Tu	22 23	07 09	17 07	24 12	24 22	22 35	21 35	12 35	17 26	12 39
05 We	22 30	02 14	16 55	24 04	24 21	22 34	21 36	12 35	17 26	12 39
06 Th	22 37	02N48	16 44	23 55	24 20	22 32	21 37	12 35	17 26	12 39
07 Fr	22 43	07 46	16 36	23 45	24 18	22 31	21 37	12 35	17 26	12 39
08 Sa	22 49	12 32	16 30	23 34	24 17	22 29	21 38	12 35	17 27	12 39
09 Su	22 54	16 52	16 26	23 23	24 15	22 27	21 39	12 35	17 27	12 39
10 Mo	22 59	20 32	16 24	23 11	24 13	22 26	21 40	12 35	17 27	12 39
11 Tu	23 03	23 16	16 24	22 59	24 10	22 24	21 40	12 35	17 27	12 39
12 We	23 08	24 48	16 26	22 46	24 08	22 22	21 41	12 36	17 28	12 39
13 Th	23 11	24 56	16 30	22 32	24 05	22 21	21 42	12 36	17 28	12 38
14 Fr	23 15	23 36	16 36	22 18	24 02	22 19	21 43	12 36	17 28	12 38
15 Sa	23 17	20 50	16 44	22 03	23 59	22 17	21 43	12 36	17 28	12 38
16 Su	23 20	16 50	16 53	21 48	23 56	22 16	21 44	12 37	17 29	12 38
17 Mo	23 22	11 52	17 04	21 32	23 52	22 14	21 45	12 37	17 29	12 38
18 Tu	23 24	06 15	17 16	21 15	23 49	22 12	21 45	12 37	17 29	12 38
19 We	23 25	00 18	17 30	20 58	23 45	22 10	21 46	12 37	17 30	12 38
20 Th	23 26	05S 41	17 45	20 41	23 41	22 08	21 47	12 38	17 30	12 38
21 Fr	23 26	11 22	18 01	20 23	23 37	22 06	21 47	12 38	17 30	12 38
22 Sa	23 26	16 26	18 18	20 04	23 32	22 04	21 48	12 38	17 30	12 39
23 Su	23 26	20 32	18 35	19 45	23 28	22 03	21 48	12 39	17 31	12 39
24 Mo	23 25	23 24	18 54	19 25	23 23	22 01	21 49	12 39	17 31	12 39
25 Tu	23 24	24 52	19 13	19 05	23 18	21 59	21 50	12 39	17 31	12 39
26 We	23 22	24 52	19 33	18 45	23 13	21 57	21 50	12 40	17 32	12 39
27 Th	23 20	23 29	19 53	18 23	23 07	21 55	21 51	12 40	17 32	12 39
28 Fr	23 18	20 56	20 13	18 02	23 02	21 53	21 51	12 41	17 32	12 39
29 Sa	23 15	17 28	20 33	17 40	22 56	21 50	21 52	12 41	17 33	12 39
30 Su	23 12	13 19	20 53	17 18	22 51	21 48	21 52	12 42	17 33	12 39

ASPECTARIAN

01 13:38 ☉ △ ♆
14:03 ☽ ∦ ☿
15:57 ☽ ∥ ♆
21:19 ☽ ☌ ♅
02 05:08 ☽ □ ☿
05:50 ☽ △ ♂
12:35 ♀ ∥ ♂
16:13 ☿ ∦ ♆
19:36 ☽ ∥ ♇
20:00 ☽ ∥ ♅

03 00:12 ♅ SR
08:22 ☽ □ ♇
08:31 ☽ △ ♀
09:45 ☽ △ ♃
10:59 ☽ □ ♄
23:12 ♀ ☌ ♃
04 15:45 ☽ ⚹ ☿
21:43 ☽ □ ♂

05 09:46 ☽ ⚹ ♆
10:29 ☉ ∥ ♃
17:54 ☽ ⚹ ☉
20:46 ☽ △ ♇
23:20 ☽ □ ♃
06 00:10 ☽ ⚹ ♄
03:42 ☽ □ ♀
21:48 ☽ ⚹ ♅

07 04:44 ☉ ☍ ♆
13:11 ☽ ⚹ ♂
21:27 ☽ □ ♆
08 00:18 ☽ ∦ ♅
00:37 ☽ ∦ ♇
11:39 ☽ ⚹ ♃
15:12 ☿ SD
[illegible]
21:32 ☽ ∥ ☿

09 03:32 ☽ ∦ ♆
08:15 ☽ □ ♅
11:24 ☉ ☌ ♄
13:08 ☽ ☌ ☿
10 06:51 ☽ △ ♆
08:52 ☽ ∥ ♄
15:25 ☽ ∥ ♃
16:54 ☽ ☍ ♇
17:41 ♀ ∥ ☉
21:15 ☽ ∥ ♀
21:27 ☽ ☌ ♄
21:42 ☽ ∥ ☉

11 11:19 ☽ ∥ ♂
16:06 ☽ △ ♅
12 12:01 ☽ ☌ ♂

13 04:29 ☽ ☌ ♃
18:12 ☽ ∥ ♂
21:45 ☽ ☌ ♀
22:03 ♀ ∥ ♃
14 03:41 ☽ ∥ ☉
04:02 ☽ ⚹ ☿
12:35 ☽ ∥ ♃
13:47 ☽ ∥ ♀
17:22 ☽ ∥ ♄
18:18 ☽ ☍ ♆
20:17 ♀ ♌

15 03:32 ☽ △ ♇
08:58 ☽ ⚹ ♄
17:55 ☽ ⚹ ☉
20:28 ☽ ∦ ♆
[illegible]
16 01:18 ☽ ☍ ♅

05:44 ♀ ∥ ♄
09:33 ☽ □ ☿
20:28 ☽ ∦ ♇
20:36 ☽ ∦ ♅

17 01:09 ☽ ⚹ ♂
06:34 ☽ □ ♇
12:28 ☽ □ ♄
13:21 ☽ ⚹ ♃
18 13:31 ☽ ⚹ ♀
14:43 ☽ △ ☿
23:27 ☿ ∦ ♆

19 00:06 ☽ △ ♆
06:10 ☽ □ ♂
09:02 ☽ ⚹ ♇
15:28 ☽ △ ♄
16:39 ☽ □ ♃
20 05:08 ☉ △ ♅
06:33 ☽ △ ♅
06:39 ☽ △ ☉
08:52 ☿ ⚹ ♀
20:35 ☽ □ ♀

21 02:38 ☽ □ ♆
05:28 ☽ ∥ ♅
05:42 ☽ ∥ ♇
11:15 ☽ △ ♂
13:25 ☉ ♋
20:12 ☽ △ ♃
22 05:45 ☽ ∥ ♆
09:27 ☽ □ ♅
10:54 ☽ ∦ ☿
16:35 ♅ ∥ ♇
19:16 ☽ ∦ ♀

23 03:20 ☽ ☍ ☿
04:29 ☽ △ ♀
05:55 ☽ ⚹ ♆
09:21 ☽ ∦ ♄
09:36 ☉ ∥ ♂
11:06 ☽ ∦ ♃
15:06 ☽ ☌ ♇
21:21 ♀ ☍ ♆
22:58 ☽ ☍ ♄
23:43 ☽ ∦ ♂

24 00:08 ☽ ∦ ☉
12:00 ☿ △ ♆
[illegible]
19:07 ☿ ∥ ♀

26 01:40 ☽ ☍ ♂
07:38 ☽ ☍ ♃

27 01:46 ☽ ∦ ☉
04:18 ☽ ∦ ♂
16:08 ☽ ∦ ♃
16:26 ☽ ∦ ♄
18:19 ☽ ☌ ♆

28 02:57 ☽ △ ☿
03:53 ☽ ☍ ♀
04:21 ☽ ⚹ ♇
05:00 ☽ ∦ ☿
[illegible]
11:07 ♃ ∥ ♄
14:21 ☽ △ ♄
15:37 ☿ ☍ ♇
22:36 ☽ ∦ ♀
23:30 ☽ ∥ ♆

29 05:13 ☽ ☌ ♅
07:46 ♀ ∦ ♆
22:16 ☿ ⚹ ♀
23:52 ☽ △ ☉

30 03:24 ☽ ∥ ♅
03:38 ☽ ∥ ♇
15:02 ☽ □ ♇
21:37 ☽ □ ☿

⚷ Chiron

01	Dec.	16 S 20
01	07♑49R	
04	07 39	
07	07 29	
10	07 18	
13	07 07	
16	06 56	
19	06 44	
22	06 33	
25	06 21	
28	06 10	

July 2002

Day	S. T. h m s	☉ ° ' "	☽ ° ' "	☿ ° '	♀ ° '	♂ ° '	♃ ° '	♄ ° '	♅ ° '	♆ ° '	♇ ° '	☊ True ° '
01 Mo	18 35 30	09♋00 10	20♓08 32	19♊06	18♌45	21♋51	22♋56	21♊13	28♒R32	10♒R24	15♐R40	17♊R36
02 Tu	18 39 26	09 57 22	02♈03 45	20 39	19 54	22 30	23 10	21 21	28 30	10 23	15 38	17D 35
03 We	18 43 23	10 54 35	13 56 35	22 15	21 03	23 09	23 23	21 29	28 29	10 21	15 37	17 36
04 Th	18 47 19	11 51 48	25 51 45	23 54	22 11	23 47	23 36	21 36	28 28	10 20	15 35	17 38
05 Fr	18 51 16	12 49 00	07♉53 58	25 37	23 20	24 26	23 49	21 44	28 26	10 19	15 34	17 40
06 Sa	18 55 12	13 46 14	20 07 44	27 22	24 28	25 05	24 03	21 51	28 25	10 17	15 32	17 43
07 Su	18 59 09	14 43 27	02♊36 56	29 11	25 36	25 43	24 16	21 58	28 23	10 16	15 31	17 45
08 Mo	19 03 06	15 40 41	15 24 40	01♋03	26 44	26 22	24 29	22 06	28 22	10 14	15 30	17 46
09 Tu	19 07 02	16 37 55	28 32 51	02 57	27 52	27 01	24 43	22 13	28 20	10 13	15 28	17R 45
10 We	19 10 59	17 35 09	12♋01 53	04 54	29 00	27 39	24 56	22 21	28 19	10 11	15 27	17 43
11 Th	19 14 55	18 32 24	25 50 25	06 54	00♍08	28 18	25 09	22 28	28 17	10 10	15 26	17 40
12 Fr	19 18 52	19 29 38	09♌55 21	08 56	01 16	28 57	25 23	22 35	28 15	10 08	15 25	17 35
13 Sa	19 22 48	20 26 53	24 12 13	10 59	02 23	29 35	25 36	22 42	28 14	10 07	15 23	17 29
14 Su	19 26 45	21 24 07	08♍35 46	13 04	03 31	00♌14	25 50	22 50	28 12	10 05	15 22	17 24
15 Mo	19 30 42	22 21 22	23 00 45	15 11	04 38	00 52	26 03	22 57	28 10	10 04	15 21	17 19
16 Tu	19 34 38	23 18 37	07♎22 32	17 18	05 45	01 31	26 16	23 04	28 08	10 02	15 20	17 16
17 We	19 38 35	24 15 52	21 37 35	19 26	06 52	02 09	26 30	23 11	28 06	10 01	15 19	17 14
18 Th	19 42 31	25 13 06	05♏43 32	21 35	07 59	02 48	26 43	23 18	28 05	09 59	15 17	17D 14
19 Fr	19 46 28	26 10 21	19 39 04	23 43	09 05	03 26	26 57	23 25	28 03	09 58	15 16	17 14
20 Sa	19 50 24	27 07 37	03♐23 34	25 52	10 12	04 05	27 10	23 32	28 01	09 56	15 15	17 15
21 Su	19 54 21	28 04 52	16 56 49	28 00	11 18	04 43	27 23	23 39	27 59	09 54	15 14	17R 16
22 Mo	19 58 17	29 02 08	00♑18 32	00♌07	12 24	05 22	27 37	23 46	27 57	09 53	15 13	17 15
23 Tu	20 02 14	29 59 24	13 28 24	02 13	13 30	06 00	27 50	23 53	27 55	09 51	15 12	17 12
24 We	20 06 11	00♌56 41	26 25 55	04 19	14 36	06 39	28 04	23 59	27 53	09 50	15 11	17 08
25 Th	20 10 07	01 53 58	09♒10 43	06 23	15 42	07 17	28 17	24 06	27 51	09 48	15 10	17 01
26 Fr	20 14 04	02 51 16	21 42 46	08 26	16 47	07 56	28 30	24 13	27 49	09 46	15 09	16 54
27 Sa	20 18 00	03 48 34	04♓02 34	10 27	17 52	08 34	28 44	24 19	27 47	09 45	15 08	16 45
28 Su	20 21 57	04 45 54	16 11 21	12 28	18 57	09 12	28 57	24 26	27 44	09 43	15 07	16 38
29 Mo	20 25 53	05 43 14	28 11 09	14 26	20 02	09 51	29 10	24 33	27 42	09 42	15 07	16 31
30 Tu	20 29 50	06 40 35	10♈04 53	16 23	21 07	10 29	29 24	24 39	27 40	09 40	15 06	16 26
31 We	20 33 46	07 37 57	21 56 19	18 19	22 11	11 07	29 37	24 46	27 38	09 38	15 05	16 24

Data for	07-01-2002
Julian Day	2452456.50
Ayanamsa	23 53 14
SVP	05 ♓ 13 13
☽ ☊ Mean	16 ♊ 47 R

● ◐ PHASES ○ ◑

Day	Time	Phase	Position
02	17:20	◑	10♈39
10	10:26	●	18♋00
17	04:47	◐	24♎27
24	09:08	○	01♒18

LAST ASPECT ☽ Day	h m	INGRESS Day	h m	Sign
01	05:43	01	19:50	♈
04	05:12	04	08:17	♉
06	15:57	06	19:01	♊
08	23:37	09	02:37	♋
11	04:26	11	07:08	♌
13	06:43	13	09:41	♍
15	05:09	15	11:39	♎
17	10:58	17	14:13	♏
19	14:35	19	18:03	♐
21	19:45	21	23:26	♑
24	03:06	24	06:41	♒
26	11:47	26	16:05	♓
29	02:01	29	03:39	♈
31	15:49	31	16:17	♉

DECLINATION

Day	☉	☽	☿	♀	♂	♃	♄	♅	♆	♇
01 Mo	23N08	08S43	21N12	16N55	22N45	21N46	21N53	12S42	17S34	12S39
02 Tu	23 04	03 51	21 31	16 32	22 38	21 44	21 53	12 42	17 34	12 39
03 We	22 59	01N10	21 50	16 08	22 32	21 42	21 54	12 43	17 34	12 39
04 Th	22 55	06 08	22 07	15 45	22 26	21 40	21 54	12 43	17 35	12 39
05 Fr	22 49	10 57	22 24	15 20	22 19	21 38	21 55	12 44	17 35	12 39
06 Sa	22 44	15 25	22 39	14 56	22 12	21 35	21 55	12 45	17 35	12 40
07 Su	22 38	19 20	22 53	14 31	22 05	21 33	21 56	12 45	17 36	12 40
08 Mo	22 31	22 25	23 05	14 06	21 58	21 31	21 56	12 46	17 36	12 40
09 Tu	22 24	24 25	23 15	13 40	21 51	21 29	21 57	12 46	17 37	12 40
10 We	22 17	25 02	23 23	13 14	21 43	21 26	21 57	12 47	17 37	12 40
11 Th	22 09	24 08	23 29	12 48	21 35	21 24	21 58	12 47	17 37	12 40
12 Fr	22 01	21 43	23 33	12 21	21 28	21 22	21 58	12 48	17 38	12 40
13 Sa	21 53	17 57	23 34	11 55	21 20	21 19	21 59	12 49	17 38	12 41
14 Su	21 44	13 06	23 32	11 28	21 12	21 17	21 59	12 49	17 39	12 41
15 Mo	21 35	07 31	23 28	11 01	21 03	21 14	21 59	12 50	17 39	12 41
16 Tu	21 26	01 33	23 21	10 33	20 55	21 12	22 00	12 51	17 40	12 41
17 We	21 16	04S27	23 11	10 05	20 46	21 09	22 00	12 51	17 40	12 41
18 Th	21 06	10 10	22 59	09 38	20 38	21 07	22 00	12 52	17 40	12 41
19 Fr	20 55	15 19	22 44	09 09	20 29	21 04	22 01	12 53	17 41	12 42
20 Sa	20 44	19 36	22 26	08 41	20 20	21 02	22 01	12 53	17 41	12 42
21 Su	20 33	22 46	22 06	08 13	20 11	20 59	22 02	12 54	17 42	12 42
22 Mo	20 21	24 36	21 44	07 44	20 01	20 57	22 02	12 55	17 42	12 42
23 Tu	20 09	25 01	21 19	07 15	19 52	20 54	22 02	12 55	17 43	12 43
24 We	19 57	24 03	20 52	06 47	19 42	20 52	22 02	12 56	17 43	12 43
25 Th	19 44	21 51	20 23	06 18	19 33	20 49	22 03	12 57	17 43	12 43
26 Fr	19 31	18 38	19 53	05 48	19 23	20 47	22 03	12 58	17 44	12 43
27 Sa	19 18	14 40	19 20	05 19	19 13	20 44	22 03	12 58	17 44	12 44
28 Su	19 04	10 09	18 47	04 50	19 03	20 41	22 04	12 59	17 45	12 44
29 Mo	18 50	05 19	18 12	04 20	18 53	20 39	22 04	13 00	17 45	12 44
30 Tu	18 36	00 19	17 35	03 51	18 42	20 36	22 04	13 01	17 46	12 44
31 We	18 22	04N40	16 58	03 21	18 32	20 33	22 04	13 01	17 46	12 45

ASPECTARIAN

Day	Time	Aspect
01	02:12	☽ □ ♄
	03:38	☽ △ ♂
	05:43	☽ △ ♃
02	11:35	☿ ☌ ♄
	14:44	☿ ∥ ♃
	16:46	☽ ⚹ ♆
03	03:22	☽ △ ♇
	05:46	☿ ∥ ♄
	10:14	♀ ⚹ ♄
	13:26	♂ ☌ ♃
	15:21	☽ ⚹ ♄
	15:50	☽ △ ♀
	19:23	☽ □ ♃
	19:25	☽ ⚹ ☿
	19:36	☽ □ ♂
04	05:12	☽ ⚹ ♅
	18:46	☿ ∥ ♂
05	04:45	☽ □ ♆
	08:49	☽ # ♇
	09:14	☽ # ♅
	10:32	☽ ⚹ ☉
	21:24	☽ ∥ ♀
06	05:28	☿ ∥ ☉
	07:44	☽ ⚹ ♃
	09:15	☽ □ ♀
	10:06	☽ ⚹ ♂
	12:44	☽ # ♆
	13:42	☿ △ ♅
	15:57	☽ □ ♅
07	10:36	☿ ♋
	14:23	☽ △ ♆
	16:08	☽ ∥ ♃
	19:36	☽ ∥ ♄
	20:00	☽ ∥ ♂
08	00:09	☽ ☍ ♇
	00:47	☽ ∥ ☉
	04:33	♂ ∥ ♄
	06:59	☽ ∥ ☿
	12:25	☽ ☌ ♄
	22:40	☽ ⚹ ♀
	23:37	☽ △ ♅
09	09:15	☽ ☌ ☿
	09:38	♀ ☍ ♅
10	21:09	♀ ♍
	22:48	☽ ☌ ♃
11	00:27	♀ # ♅
	04:26	☽ ☌ ♂
	06:56	♀ # ♇
	07:54	☽ ∥ ☿
	21:34	☽ ∥ ☉
	22:05	☽ ∥ ♄
12	00:22	☽ ☍ ♆
	02:03	☽ ∥ ♂
	02:45	☽ ∥ ♃
	08:56	☉ ∥ ♄
	09:14	☽ △ ♇
	21:28	☽ ⚹ ♄
13	01:40	☽ # ♆
	02:17	♂ ∥ ♃
	06:43	☽ ☍ ♅
	14:49	☽ ☌ ♀
	15:24	♂ ♌
14	01:15	☽ # ♅
	01:53	☽ # ♇
	07:56	☽ ∥ ♀
	08:43	☽ ⚹ ☿
	11:15	☽ □ ♇
	22:50	☽ ⚹ ☉
	23:53	☽ □ ♄
15	05:09	☽ ⚹ ♃
	13:44	☽ ⚹ ♂
16	04:28	☽ △ ♆
	13:21	☽ ⚹ ♇
	19:38	☽ □ ☿
17	02:39	☽ △ ♄
	08:23	☽ □ ♃
	10:58	☽ △ ♅
	18:45	☽ □ ♂
	19:37	☉ ∥ ♃
	21:48	☽ # ♀
18	04:12	☽ ⚹ ♀
	07:18	☽ □ ♆
	11:19	☽ ∥ ♇
	12:08	☽ ∥ ♅
19	08:23	☽ △ ☿
	12:12	☽ △ ☉
	12:32	☽ ∥ ♆
	12:54	☽ △ ♃
	14:35	☽ □ ♅
20	01:16	☽ △ ♂
	01:19	☉ ☌ ♃
	04:35	☽ # ♂
	07:11	☽ # ☉
	09:36	☽ # ♃
	11:31	☽ ⚹ ♆
	13:04	☽ □ ♀
	16:24	☿ ☌ ♃
	17:26	☽ # ♄
	18:45	☽ # ☿
	20:57	☽ ☌ ♇
21	01:47	☿ ☌ ☉
	05:05	☿ ∥ ♄
	12:06	☽ ☍ ♄
	19:45	☽ ⚹ ♅
	22:42	☿ ♌
23	00:04	☽ △ ♀
	00:15	☉ ♌
24	00:15	☿ ∥ ♃
	03:06	☽ ☍ ♃
	12:41	♀ □ ♇
	17:40	☽ ☍ ☿
	20:13	☽ ☍ ♂
	22:23	☽ # ♄
25	01:11	☽ ☌ ♆
	08:42	☽ # ♃
	11:24	☽ ⚹ ♇
	14:02	☽ # ☿
	15:20	☿ ☌ ♂
	17:42	☽ # ☉
	18:49	☽ # ♂
26	04:53	☽ △ ♄
	05:54	☽ ∥ ♆
	11:47	☽ ☌ ♅
	15:40	☿ ☍ ♆
27	02:53	☿ ∥ ☉
	07:34	☿ ∥ ♂
	09:18	☽ ∥ ♅
	10:38	☽ ∥ ♇
	21:53	☽ □ ♇
28	06:03	☽ ☍ ♀
	09:26	☉ ∥ ♂
	16:37	☽ □ ♄
	18:33	♂ ☍ ♆
29	02:01	☽ △ ♃
	05:16	☽ # ♀
	08:13	☿ △ ♇
	16:31	☽ △ ☉
	17:15	☿ # ♆
	23:10	☽ ⚹ ♆
30	00:52	☽ △ ♂
	10:08	☽ △ ♇
	15:14	☽ △ ☿
	18:15	☽ ∥ ♀
31	05:45	☽ ⚹ ♄
	11:28	☽ ⚹ ♅
	15:49	☽ □ ♃

⚷ Chiron

01	Dec.	16 S 19
01	05♑58R	
04	05 46	
07	05 35	
10	05 23	
13	05 12	
16	05 01	
19	04 51	
22	04 40	
25	04 31	
28	04 21	
31	04 12	

August 2002

Day	S. T.	☉	☽	☿	♀	♂	♃	♄	♅	♆	♇	☊ True
	h m s	° ' "	° ' "	° '	° '	° '	° '	° '	° '	° '	° '	° '
01 Th	20 37 43	08♌35 20	03♉49 54	20♌12	23♍15	11♌46	29♋50	24♊52	27♒R36	09♒R37	15♐R04	16♊D23
02 Fr	20 41 40	09 32 45	15 50 31	22 05	24 19	12 24	00♌04	24 58	27 34	09 35	15 03	16 25
03 Sa	20 45 36	10 30 10	28 03 17	23 55	25 23	13 02	00 17	25 05	27 31	09 33	15 03	16 26
04 Su	20 49 33	11 27 37	10♊33 07	25 44	26 27	13 41	00 30	25 11	27 29	09 32	15 02	16 28
05 Mo	20 53 29	12 25 05	23 24 20	27 31	27 30	14 19	00 43	25 17	27 27	09 30	15 01	16R 28
06 Tu	20 57 26	13 22 34	06♋40 07	29 17	28 33	14 57	00 57	25 23	27 24	09 29	15 01	16 26
07 We	21 01 22	14 20 04	20 21 54	01♍01	29 36	15 36	01 10	25 29	27 22	09 27	15 00	16 21
08 Th	21 05 19	15 17 36	04♌28 40	02 44	00♎39	16 14	01 23	25 35	27 20	09 25	14 59	16 14
09 Fr	21 09 15	16 15 08	18 56 39	04 25	01 41	16 52	01 36	25 41	27 18	09 24	14 59	16 05
10 Sa	21 13 12	17 12 42	03♍39 34	06 04	02 43	17 31	01 49	25 47	27 15	09 22	14 58	15 56
11 Su	21 17 09	18 10 16	18 29 23	07 42	03 45	18 09	02 03	25 53	27 13	09 20	14 58	15 46
12 Mo	21 21 05	19 07 52	03♎17 44	09 18	04 46	18 47	02 16	25 58	27 10	09 19	14 57	15 37
13 Tu	21 25 02	20 05 28	17 57 09	10 53	05 48	19 25	02 29	26 04	27 08	09 17	14 57	15 31
14 We	21 28 58	21 03 05	02♏22 13	12 26	06 49	20 04	02 42	26 10	27 06	09 16	14 56	15 26
15 Th	21 32 55	22 00 43	16 29 52	13 58	07 49	20 42	02 55	26 15	27 03	09 14	14 56	15 24
16 Fr	21 36 51	22 58 23	00♐19 15	15 28	08 49	21 20	03 08	26 21	27 01	09 12	14 56	15 23
17 Sa	21 40 48	23 56 03	13 51 08	16 56	09 49	21 58	03 21	26 26	26 59	09 11	14 55	15 23
18 Su	21 44 44	24 53 44	27 07 08	18 23	10 49	22 37	03 34	26 31	26 56	09 09	14 55	15 23
19 Mo	21 48 41	25 51 26	10♑09 09	19 48	11 48	23 15	03 46	26 37	26 54	09 08	14 55	15 21
20 Tu	21 52 38	26 49 10	22 58 49	21 12	12 47	23 53	03 59	26 42	26 51	09 06	14 55	15 17
21 We	21 56 34	27 46 54	05♒37 26	22 34	13 45	24 31	04 12	26 47	26 49	09 05	14 54	15 09
22 Th	22 00 31	28 44 40	18 05 55	23 54	14 43	25 09	04 25	26 52	26 47	09 03	14 54	14 59
23 Fr	22 04 27	29 42 27	00♓24 56	25 13	15 41	25 48	04 37	26 57	26 44	09 02	14 54	14 48
24 Sa	22 08 24	00♍40 16	12 35 09	26 29	16 38	26 26	04 50	27 02	26 42	09 00	14 54	14 35
25 Su	22 12 20	01 38 06	24 37 26	27 44	17 35	27 04	05 03	27 06	26 39	08 59	14 54	14 22
26 Mo	22 16 17	02 35 57	06♈33 11	28 57	18 31	27 42	05 15	27 11	26 37	08 57	14 54	14 10
27 Tu	22 20 13	03 33 50	18 24 27	00♎08	19 27	28 20	05 28	27 16	26 35	08 56	14D 54	14 01
28 We	22 24 10	04 31 45	00♉14 07	01 17	20 22	28 59	05 40	27 20	26 32	08 54	14 54	13 55
29 Th	22 28 06	05 29 42	12 05 53	02 24	21 17	29 37	05 53	27 25	26 30	08 53	14 54	13 52
30 Fr	22 32 03	06 27 40	24 04 15	03 29	22 11	00♍15	06 05	27 29	26 28	08 52	14 54	13 51
31 Sa	22 35 60	07 25 41	06♊14 11	04 31	23 05	00 53	06 18	27 33	26 25	08 50	14 54	13D 51

Data for	08-01-2002
Julian Day	2452487.50
Ayanamsa	23 53 19
SVP	05 ♓ 13 10
☽ ☊ Mean	15 ♊ 08 R

● ◐ PHASES ○ ◑

Day	h m	Phase	Position
01	10:22	◑	09♉00
08	19:15	●	16♌04
15	10:13	◐	22♏25
22	22:29	○	29♒39
31	02:31	◑	07♊32

LAST ASPECT ☽ INGRESS

Day	h m	Day	h m	Sign
02	22:58	03	03:47	♊
05	08:42	05	12:02	♋
05	08:42	07	16:27	♌
09	13:37	09	18:03	♍
11	12:02	11	18:39	♎
13	15:11	13	20:01	♏
15	18:14	15	23:26	♐
17	23:40	18	05:16	♑
19	20:14	20	13:17	♒
22	22:30	22	23:11	♓
25	06:58	25	10:48	♈
27	21:18	27	23:31	♉
30	04:44	30	11:45	♊

DECLINATION

Day	☉	☽	☿	♀	♂	♃	♄	♅	♆	♇
01 Th	18N07	09N32	16N19	02N52	18N21	20N31	22N05	13S02	17S47	12S45
02 Fr	17 52	14 05	15 40	02 22	18 11	20 28	22 05	13 03	17 47	12 45
03 Sa	17 36	18 09	15 00	01 52	18 00	20 25	22 05	13 04	17 47	12 45
04 Su	17 21	21 30	14 20	01 23	17 49	20 22	22 05	13 04	17 48	12 46
05 Mo	17 05	23 53	13 39	00 53	17 38	20 20	22 05	13 05	17 48	12 46
06 Tu	16 48	25 02	12 57	00 23	17 27	20 17	22 06	13 06	17 49	12 46
07 We	16 32	24 42	12 15	00S07	17 15	20 14	22 06	13 07	17 49	12 47
08 Th	16 15	22 49	11 33	00 36	17 04	20 11	22 06	13 08	17 50	12 47
09 Fr	15 58	19 25	10 50	01 06	16 52	20 08	22 06	13 08	17 50	12 47
10 Sa	15 41	14 46	10 08	01 36	16 41	20 06	22 06	13 09	17 51	12 48
11 Su	15 23	09 12	09 25	02 05	16 29	20 03	22 07	13 10	17 51	12 48
12 Mo	15 05	03 07	08 42	02 35	16 17	20 00	22 07	13 11	17 51	12 48
13 Tu	14 47	03S05	08 00	03 05	16 05	19 57	22 07	13 12	17 52	12 49
14 We	14 29	09 02	07 17	03 34	15 53	19 54	22 07	13 12	17 52	12 49
15 Th	14 10	14 25	06 35	04 03	15 41	19 51	22 07	13 13	17 53	12 49
16 Fr	13 52	18 55	05 52	04 33	15 29	19 48	22 07	13 14	17 53	12 50
17 Sa	13 33	22 19	05 10	05 02	15 16	19 45	22 07	13 15	17 54	12 50
18 Su	13 13	24 26	04 29	05 31	15 04	19 43	22 07	13 16	17 54	12 50
19 Mo	12 54	25 10	03 47	06 00	14 51	19 40	22 08	13 17	17 54	12 51
20 Tu	12 34	24 31	03 06	06 28	14 39	19 37	22 08	13 18	17 55	12 51
21 We	12 15	22 37	02 26	06 57	14 26	19 34	22 08	13 18	17 55	12 52
22 Th	11 55	19 39	01 46	07 25	14 13	19 31	22 08	13 19	17 56	12 52
23 Fr	11 35	15 51	01 06	07 54	14 00	19 28	22 08	13 20	17 56	12 52
24 Sa	11 14	11 27	00 27	08 22	13 47	19 25	22 08	13 21	17 56	12 53
25 Su	10 54	06 39	00S11	08 50	13 34	19 22	22 08	13 22	17 57	12 53
26 Mo	10 33	01 39	00 49	09 17	13 21	19 19	22 08	13 22	17 57	12 54
27 Tu	10 12	03N23	01 26	09 45	13 08	19 16	22 08	13 23	17 58	12 54
28 We	09 51	08 18	02 02	10 12	12 54	19 13	22 08	13 24	17 58	12 54
29 Th	09 30	12 57	02 37	10 39	12 41	19 10	22 08	13 25	17 58	12 55
30 Fr	09 08	17 08	03 12	11 06	12 27	19 07	22 08	13 26	17 59	12 55
31 Sa	08 47	20 42	03 45	11 32	12 14	19 04	22 08	13 26	17 59	12 56

ASPECTARIAN

01 11:34 ☽ □ ♆
16:47 ☽ # ♇
16:47 ☽ □ ♂
17:21 ♃ ♌
18:22 ☽ # ♅
02 00:57 ☉ ☍ ♆
07:14 ☉ # ♆
07:46 ☽ ∥ ☿
14:30 ☽ □ ☿
16:11 ♀ □ ♄
18:20 ☽ △ ♀
20:48 ☽ ∥ ☉
21:44 ☽ # ♆
22:58 ☽ □ ♅
23:06 ☽ ∥ ♂
03 04:24 ☽ ✶ ♃
15:18 ☽ ∥ ♃
16:11 ☿ ✶ ♄
22:04 ☽ △ ♆
04 01:52 ☽ ✶ ☉
02:00 ♂ # ♆
04:57 ☽ ∥ ♄
06:13 ☽ ✶ ♂
08:26 ☽ ☍ ♇
22:59 ☿ ☍ ♅
05 03:28 ☽ ☌ ♄
07:22 ☽ △ ♅
08:09 ☽ □ ♀
08:42 ☽ ✶ ☿
18:53 ☿ # ♅
06 01:57 ♂ △ ♇
06:05 ☿ # ♇
09:51 ☿ ♍
07 09:09 ♀ ♎
16:29 ☉ △ ♇
17:02 ☽ ✶ ♀
18:43 ☽ ☌ ♃
08 06:02 ☽ ∥ ♄
08:15 ☽ ☍ ♆
17:28 ☽ △ ♇
19:38 ☽ ∥ ♃
20:26 ☽ ☌ ♂
21:41 ♀ ✶ ♃
09 08:52 ☽ # ♆
11:06 ☽ ✶ ♄
13:37 ☽ ☍ ♅
14:27 ☽ ∥ ♂
19:28 ☽ ∥ ☉
10 04:23 ☽ ☌ ☿
07:17 ☽ # ♅
08:53 ☽ # ♇
18:18 ☽ □ ♇
22:17 ☉ ☌ ♂
23:00 ☽ ∥ ☿
11 12:02 ☽ □ ♄
22:17 ☽ ✶ ♃
12 01:55 ☽ # ♀
02:35 ☽ ☌ ♀
09:48 ☽ △ ♆
19:03 ☽ ✶ ♇
13 02:32 ☽ ✶ ♂
03:47 ☽ ✶ ☉
13:32 ☽ △ ♄
15:11 ☽ △ ♅
17:32 ☽ # ☿
14 00:33 ☽ □ ♃
11:37 ☽ □ ♆
16:33 ☽ ∥ ♇
18:22 ☽ ∥ ♅
19:07 ☽ ✶ ☿
22:55 ☽ # ☉
15 06:00 ☽ # ♂
07:35 ☽ □ ☉
15:31 ☿ □ ♇
17:58 ☽ ∥ ♆
18:14 ☽ □ ♅
16 05:01 ☽ △ ♃
05:27 ☽ # ♃
09:01 ♀ △ ♆
15:41 ☽ ✶ ♆
16:13 ☽ ✶ ♀
22:19 ☽ # ♄
17 01:55 ☽ ☌ ♇
02:55 ☿ # ♀
06:13 ☽ □ ☿
15:23 ☽ △ ♂
19:38 ☽ △ ☉
21:05 ☉ # ♅
22:54 ☽ ☍ ♄
23:40 ☽ ✶ ♅
19 03:19 ☽ □ ♀
03:51 ☉ # ♇
20:14 ☽ △ ☿
20:35 ☉ ✶ ♄
20 00:54 ☉ ☍ ♅
21:14 ☽ ☍ ♃
21 04:37 ☽ # ♄
06:36 ☽ ☌ ♆
07:18 ♄ △ ♅
16:55 ☽ △ ♀
17:49 ☽ ✶ ♇
22 01:01 ☽ # ♃
04:34 ♀ ✶ ♇
11:30 ☽ ∥ ♆
14:28 ☽ ☍ ♂
16:49 ☽ ☌ ♅
17:10 ☽ △ ♄
23 07:17 ☉ ♍
11:00 ☽ # ♂
14:04 ☽ ∥ ♅
16:33 ☽ ∥ ♇
24 01:14 ☽ # ☉
04:36 ☽ □ ♇
09:29 ♂ ☍ ♅
10:58 ☿ □ ♄
14:17 ☽ ∥ ♀
25 01:40 ♂ ✶ ♄
05:01 ☽ □ ♄
06:58 ☽ ☍ ☿
21:20 ☽ △ ♃
21:30 ♂ # ♅
26 03:35 ☽ ∥ ☿
04:51 ☽ ✶ ♆
11:00 ♇ SD
13:26 ☽ # ☿
16:53 ☽ △ ♇
21:10 ☿ ♎
27 02:18 ☽ ☍ ♀
13:36 ♀ # ☉
16:32 ☽ ✶ ♅
18:05 ☽ ✶ ♄
21:18 ☽ △ ♂
23:53 ♂ # ♇
28 07:19 ☽ ∥ ☉
09:30 ☽ △ ☉
10:41 ☽ # ♀
11:13 ☽ □ ♃
17:33 ☽ □ ♆
22:43 ☽ ∥ ♂
23:54 ☽ # ♇
29 02:37 ☽ # ♅
14:38 ♂ ♍
30 04:44 ☽ □ ♅
05:19 ☽ # ♆
12:38 ☽ ∥ ♃
12:55 ☽ □ ♂
20:20 ☽ △ ☿
31 00:07 ☽ ✶ ♃
05:03 ☽ △ ♆
11:49 ☽ ∥ ♄
10:47 [illegible]

⚷ Chiron

01 Dec.	16 S 27	
03	04♑04R	
06	03	56
09	03	49
12	03	42
15	03	36
18	03	31
21	03	26
24	03	22
27	03	19
30	03	16

September 2002

Day	S. T.	☉	☽	☿	♀	♂	♃	♄	♅	♆	♇	☊ True
	h m s	° ' "	° ' "	° '	° '	° '	° '	° '	° '	° '	° '	° '
01 Su	22 39 56	08♍23 43	18♊40 57	05♎31	23♎58	01♍31	06♌30	27♊37	26♒R23	08♒R49	14♐54	13♊R52
02 Mo	22 43 53	09 21 48	01♋29 36	06 28	24 51	02 09	06 42	27 42	26 21	08 47	14 55	13 51
03 Tu	22 47 49	10 19 54	14 44 30	07 22	25 43	02 48	06 54	27 46	26 18	08 46	14 55	13 48
04 We	22 51 46	11 18 02	28 28 22	08 14	26 35	03 26	07 06	27 50	26 16	08 45	14 55	13 42
05 Th	22 55 42	12 16 12	12♌41 29	09 02	27 25	04 04	07 19	27 53	26 14	08 44	14 55	13 34
06 Fr	22 59 39	13 14 24	27 20 51	09 47	28 16	04 42	07 31	27 57	26 11	08 42	14 56	13 24
07 Sa	23 03 35	14 12 37	12♍19 58	10 28	29 05	05 20	07 43	28 01	26 09	08 41	14 56	13 13
08 Su	23 07 32	15 10 52	27 29 36	11 05	29 54	05 59	07 54	28 04	26 07	08 40	14 57	13 01
09 Mo	23 11 29	16 09 09	12♎39 08	11 39	00♏42	06 37	08 06	28 08	26 05	08 39	14 57	12 50
10 Tu	23 15 25	17 07 28	27 38 29	12 08	01 29	07 15	08 18	28 11	26 03	08 37	14 57	12 42
11 We	23 19 22	18 05 48	12♏19 45	12 32	02 15	07 53	08 30	28 15	26 00	08 36	14 58	12 36
12 Th	23 23 18	19 04 10	26 38 08	12 51	03 01	08 31	08 41	28 18	25 58	08 35	14 58	12 32
13 Fr	23 27 15	20 02 33	10♐32 03	13 05	03 46	09 09	08 53	28 21	25 56	08 34	14 59	12 31
14 Sa	23 31 11	21 00 58	24 02 23	13 13	04 30	09 48	09 04	28 24	25 54	08 33	15 00	12 30
15 Su	23 35 08	21 59 25	07♑11 37	13R 15	05 12	10 26	09 16	28 27	25 52	08 32	15 00	12 30
16 Mo	23 39 04	22 57 53	20 02 54	13 10	05 54	11 04	09 27	28 29	25 50	08 31	15 01	12 28
17 Tu	23 43 01	23 56 23	02♒39 23	12 59	06 35	11 42	09 38	28 32	25 48	08 30	15 02	12 24
18 We	23 46 58	24 54 54	15 03 55	12 41	07 15	12 20	09 50	28 35	25 46	08 29	15 02	12 17
19 Th	23 50 54	25 53 27	27 18 51	12 17	07 53	12 58	10 01	28 37	25 44	08 28	15 03	12 07
20 Fr	23 54 51	26 52 02	09♓26 04	11 45	08 31	13 37	10 12	28 40	25 42	08 27	15 04	11 56
21 Sa	23 58 47	27 50 38	21 27 02	11 06	09 07	14 15	10 23	28 42	25 40	08 26	15 05	11 43
22 Su	00 02 44	28 49 17	03♈22 59	10 21	09 42	14 53	10 34	28 44	25 38	08 25	15 05	11 30
23 Mo	00 06 40	29 47 57	15 15 12	09 29	10 16	15 31	10 44	28 46	25 36	08 24	15 06	11 18
24 Tu	00 10 37	00♎46 40	27 05 14	08 32	10 48	16 09	10 55	28 48	25 35	08 23	15 07	11 09
25 We	00 14 33	01 45 24	08♉55 14	07 31	11 19	16 47	11 06	28 50	25 33	08 22	15 08	11 02
26 Th	00 18 30	02 44 11	20 47 59	06 26	11 49	17 26	11 16	28 52	25 31	08 21	15 09	10 59
27 Fr	00 22 27	03 43 00	02♊47 02	05 20	12 17	18 04	11 27	28 53	25 29	08 21	15 10	10 57
28 Sa	00 26 23	04 41 51	14 56 35	04 13	12 43	18 42	11 37	28 55	25 28	08 20	15 11	10D 57
29 Su	00 30 20	05 40 45	27 21 18	03 08	13 08	19 20	11 47	28 56	25 26	08 19	15 12	10 58
30 Mo	00 34 16	06 39 40	10♋06 03	02 06	13 31	19 58	11 57	28 58	25 24	08 19	15 13	10R 58

Data for	09-01-2002
Julian Day	2452518.50
Ayanamsa	23 53 23
SVP	05 ♓ 13 05
☽ ☊ Mean	13 ♊ 30 R

● ◐ PHASES ○ ◑

07	03:11	●	14♍20
13	18:08	◐	20♐47
21	13:59	○	28♓25
29	17:03	◑	06♋23

LAST ASPECT Day	h m	☽ INGRESS Day	h m	
01	16:55	01	21:14	♋
03	20:31	04	02:37	♌
06	01:34	06	04:17	♍
08	00:55	08	03:58	♎
10	00:53	10	03:49	♏
11	22:52	12	05:45	♐
14	07:55	14	10:48	♑
16	05:58	16	18:55	♒
19	02:35	19	05:18	♓
21	14:36	21	17:11	♈
24	03:29	24	05:55	♉
26	09:27	26	18:27	♊
29	03:02	29	05:02	♋

DECLINATION

Day	☉	☽	☿	♀	♂	♃	♄	♅	♆	♇
01 Su	08N25	23N23	04S17	11S58	12N00	19N01	22N08	13S27	18S00	12S56
02 Mo	08 04	24 58	04 48	12 24	11 47	18 59	22 08	13 28	18 00	12 57
03 Tu	07 42	25 12	05 18	12 50	11 33	18 56	22 08	13 29	18 00	12 57
04 We	07 20	23 56	05 47	13 16	11 19	18 53	22 08	13 30	18 01	12 57
05 Th	06 57	21 08	06 14	13 41	11 05	18 50	22 08	13 30	18 01	12 58
06 Fr	06 35	16 54	06 39	14 05	10 51	18 47	22 08	13 31	18 01	12 58
07 Sa	06 13	11 33	07 03	14 30	10 37	18 44	22 08	13 32	18 02	12 59
08 Su	05 50	05 26	07 25	14 54	10 23	18 41	22 08	13 33	18 02	12 59
09 Mo	05 28	01S00	07 45	15 18	10 09	18 38	22 08	13 33	18 02	13 00
10 Tu	05 05	07 19	08 03	15 41	09 54	18 35	22 08	13 34	18 03	13 00
11 We	04 42	13 07	08 19	16 04	09 40	18 32	22 08	13 35	18 03	13 01
12 Th	04 20	18 04	08 32	16 27	09 26	18 29	22 08	13 35	18 03	13 01
13 Fr	03 57	21 51	08 42	16 49	09 11	18 26	22 08	13 36	18 04	13 02
14 Sa	03 34	24 19	08 49	17 11	08 57	18 23	22 08	13 37	18 04	13 02
15 Su	03 11	25 21	08 53	17 32	08 42	18 20	22 08	13 38	18 04	13 03
16 Mo	02 48	24 59	08 54	17 53	08 28	18 18	22 08	13 38	18 05	13 03
17 Tu	02 24	23 20	08 51	18 14	08 13	18 15	22 08	13 39	18 05	13 04
18 We	02 01	20 35	08 44	18 34	07 59	18 12	22 08	13 40	18 05	13 04
19 Th	01 38	16 57	08 33	18 53	07 44	18 09	22 08	13 40	18 05	13 04
20 Fr	01 15	12 40	08 18	19 12	07 29	18 06	22 08	13 41	18 06	13 05
21 Sa	00 51	07 55	07 59	19 31	07 14	18 03	22 08	13 41	18 06	13 05
22 Su	00 28	02 56	07 36	19 49	07 00	18 01	22 08	13 42	18 06	13 06
23 Mo	00 05	02N09	07 08	20 06	06 45	17 58	22 08	13 43	18 06	13 06
24 Tu	00S19	07 10	06 36	20 23	06 30	17 55	22 08	13 43	18 07	13 07
25 We	00 42	11 55	06 01	20 39	06 15	17 52	22 08	13 44	18 07	13 07
26 Th	01 05	16 16	05 23	20 55	06 00	17 49	22 08	13 44	18 07	13 08
27 Fr	01 29	20 01	04 42	21 10	05 45	17 47	22 08	13 45	18 07	13 08
28 Sa	01 52	22 57	04 00	21 24	05 30	17 44	22 08	13 46	18 07	13 09
29 Su	02 15	24 52	03 17	21 38	05 15	17 41	22 08	13 46	18 08	13 09
30 Mo	02 39	25 33	02 35	21 50	05 00	17 39	22 08	13 47	18 08	13 10

ASPECTARIAN

01 01:04 ♀ # ♂
10:45 ☽ △ ♀
14:28 ☽ △ ♅
16:55 ☽ ☌ ♄
02 01:17 ☽ ✶ ♂
07:54 ☿ ✶ ♃
09:47 ☽ □ ☿
15:29 ☽ ✶ ☉
03 06:34 ♀ ∥ ♆
15:41 ♀ △ ♅
20:31 ☽ □ ♄
04 13:45 ♀ ∥ ♅
14:53 ☽ ☌ ♃
14:54 ☿ △ ♆
16:42 ☽ ∥ ♄
17:24 ☽ ☍ ♆
17:33 ☽ ✶ ☿
05 03:42 ☽ △ ♆
14:08 ☽ ∥ ♃
14:26 ♀ △ ♄
18:19 ☽ # ♆
21:52 ☿ # ☉
22:08 ☽ ☍ ♅
06 00:59 ☽ ✶ ♄
01:34 ☽ ✶ ♀
12:13 ☽ # ♀
12:21 ☽ ☌ ♂
15:38 ☽ # ♅
18:00 ☽ # ♆
07 04:00 ☽ ∥ ♂
04:08 ☽ □ ♆
16:52 ☽ # ☿
18:05 ☉ □ ♆
22:24 ☽ ∥ ☉
08 00:55 ☽ □ ♄
03:05 ♀ ♏
16:41 ☽ ✶ ♃
17:39 ☽ △ ♆
22:21 ☽ ☌ ☿
09 03:39 ☽ ✶ ♆
15:54 ☽ # ☉
21:26 ☽ △ ♅
10 00:53 ☽ △ ♄
03:04 ☽ ∥ ☿
06:34 ☽ ☌ ♀
09:59 ☽ # ♂
16:20 ☽ ✶ ♂
17:36 ☽ □ ♃
17:52 ☽ □ ♆
23:33 ☽ ∥ ♆
11 02:05 ☽ ∥ ♅
10:18 ☽ ✶ ☉
12:07 ♃ ☍ ♆
14:56 ☽ ∥ ♀
22:52 ☽ □ ♅
12 00:02 ☽ ∥ ♆
02:23 ☽ # ♃
20:34 ☽ ✶ ♆
21:04 ☽ △ ♃
21:28 ☽ □ ♂
13 02:17 ☽ # ♄
04:31 ☽ ✶ ☿
07:50 ☽ ☌ ♆
14 03:21 ☽ ✶ ♅
07:55 ☿ ☍ ♄
09:35 ☿ # ♂
19:40 ☿ SR
20:08 ☽ ✶ ♀
15 06:18 ☽ △ ♂
11:11 ☽ □ ☿
16 05:58 ☽ △ ☉
13:15 ♀ ∥ ♆
17 00:58 ♀ # ♃
07:59 ☽ □ ♀
11:14 ☽ ☌ ♆
11:29 ☽ # ♄
13:40 ☽ ☍ ♃
19:30 ☽ △ ☿
23:57 ☽ ✶ ♆
18 12:49 ☽ ∥ ♀
16:31 ☽ # ♃
17:00 ☽ ∥ ♆
20:54 ☽ ☌ ♅
19 02:35 ☽ △ ♄
18:35 ☽ ∥ ♅
21:19 ♀ □ ♆
21:48 ☽ ∥ ♆
22:06 ☽ △ ♀
20 04:23 ♃ # ♆
08:47 ☽ ☍ ♂
11:14 ☽ □ ♆
23:42 ☽ ∥ ♆
21 03:23 ☽ # ♂
14:36 ☽ □ ♄
18:44 ☿ ✶ ♃
21:47 ☉ □ ♄
22 08:10 ♂ □ ♆
10:09 ☽ ✶ ♆
12:35 ☽ # ☉
13:08 ☽ ☍ ☿
14:43 ☽ △ ♃
14:55 ☽ ∥ ☉
23:42 ☽ △ ♆
23 04:56 ☉ ♎
20:56 ☽ ∥ ♂
20:57 ☽ ✶ ♅
21:35 ☽ # ☿
24 03:29 ☽ ✶ ♄
03:46 ☿ △ ♆
07:53 ♀ □ ♃
08:19 ☿ # ♂
22:53 ☽ □ ♆
25 04:28 ☽ □ ♃
05:04 ☽ ☍ ♀
06:24 ☽ # ♆
09:42 ☽ # ♅
16:49 ☽ △ ♂
26 09:20 ☽ ∥ ♃
09:27 ☽ □ ♅
11:19 ☽ # ♆
27 02:01 ☽ △ ☉
04:38 ☽ △ ☿
09:19 ☽ # ♀
11:01 ☽ △ ♆
16:32 ☽ ∥ ♄
17:23 ☽ ✶ ♃
18:31 ☿ ☌ ☉
28 00:29 ☽ ☍ ♆
07:43 ☽ □ ♂
20:20 ☽ △ ♅
29 03:02 ☽ ☌ ♄
10:08 ☽ □ ☿
22:40 ☿ ∥ ☉
30 06:29 ☽ △ ♀
19:00 ☽ ✶ ♂

⚷ Chiron

01	Dec.	16 S 41
02	03♑15R	
05	03 14	
08	03 13D	
11	03 14	
14	03 15	
17	03 17	
20	03 20	
23	03 23	
26	03 28	
29	03 33	

07 11:02 03♑13 D

October 2002

Day	S. T. h m s	☉ ° ' "	☽ ° ' "	☿ ° '	♀ ° '	♂ ° '	♃ ° '	♄ ° '	♅ ° '	♆ ° '	♇ ° '	☊ True ° '
01 Tu	00 38 13	07♎ 38 39	23♋ 15 25	01♎R09	13♏ 52	20♍ 37	12♌ 07	28♊ 59	25♒R23	08♒R18	15♐ 15	10♊R57
02 We	00 42 09	08 37 39	06♌ 52 57	00 18	14 12	21 15	12 17	29 00	25 21	08 17	15 16	10 53
03 Th	00 46 06	09 36 42	21 00 13	29♍ 35	14 29	21 53	12 27	29 01	25 20	08 17	15 17	10 47
04 Fr	00 50 02	10 35 47	05♍ 35 51	29 01	14 45	22 31	12 37	29 02	25 18	08 16	15 18	10 39
05 Sa	00 53 59	11 34 54	20 34 56	28 37	14 59	23 09	12 47	29 03	25 17	08 16	15 19	10 30
06 Su	00 57 56	12 34 03	05♎ 49 09	28 23	15 11	23 48	12 56	29 03	25 15	08 15	15 21	10 21
07 Mo	01 01 52	13 33 14	21 07 54	28D 19	15 20	24 26	13 06	29 04	25 14	08 15	15 22	10 12
08 Tu	01 05 49	14 32 27	06♏ 19 59	28 27	15 28	25 04	13 15	29 04	25 13	08 14	15 23	10 05
09 We	01 09 45	15 31 43	21 15 34	28 44	15 33	25 42	13 24	29 05	25 11	08 14	15 25	10 00
10 Th	01 13 42	16 31 00	05♐ 47 41	29 12	15 36	26 21	13 33	29 05	25 10	08 14	15 26	09 57
11 Fr	01 17 38	17 30 19	19 52 45	29 49	15R 37	26 59	13 42	29 05	25 09	08 13	15 27	09 56
12 Sa	01 21 35	18 29 39	03♑ 30 27	00♎ 35	15 35	27 37	13 51	29R 05	25 08	08 13	15 29	09D 57
13 Su	01 25 31	19 29 02	16 42 41	01 29	15 31	28 15	14 00	29 05	25 07	08 13	15 30	09 57
14 Mo	01 29 28	20 28 26	29 32 48	02 31	15 24	28 54	14 08	29 05	25 06	08 12	15 32	09R 57
15 Tu	01 33 25	21 27 52	12♒ 04 40	03 39	15 15	29 32	14 17	29 04	25 05	08 12	15 33	09 56
16 We	01 37 21	22 27 20	24 22 10	04 52	15 04	00♎ 10	14 25	29 04	25 04	08 12	15 35	09 53
17 Th	01 41 18	23 26 49	06♓ 28 58	06 11	14 50	00 48	14 33	29 03	25 03	08 12	15 36	09 47
18 Fr	01 45 14	24 26 20	18 28 16	07 33	14 34	01 27	14 42	29 03	25 02	08 12	15 38	09 41
19 Sa	01 49 11	25 25 54	00♈ 22 46	09 00	14 15	02 05	14 50	29 02	25 01	08 12	15 40	09 33
20 Su	01 53 07	26 25 29	12 14 37	10 29	13 54	02 43	14 57	29 01	25 00	08 12	15 41	09 25
21 Mo	01 57 04	27 25 05	24 05 38	12 01	13 31	03 22	15 05	29 00	25 00	08D 12	15 43	09 18
22 Tu	02 01 00	28 24 44	05♉ 57 29	13 36	13 06	04 00	15 13	28 59	24 59	08 12	15 45	09 12
23 We	02 04 57	29 24 25	17 51 54	15 11	12 39	04 38	15 20	28 58	24 58	08 12	15 46	09 09
24 Th	02 08 53	00♏ 24 08	29 50 54	16 49	12 10	05 16	15 27	28 57	24 58	08 12	15 48	09 07
25 Fr	02 12 50	01 23 54	11♊ 56 55	18 27	11 40	05 55	15 35	28 55	24 57	08 12	15 50	09D 07
26 Sa	02 16 47	02 23 41	24 12 50	20 06	11 08	06 33	15 42	28 54	24 57	08 12	15 52	09 08
27 Su	02 20 43	03 23 31	06♋ 41 59	21 45	10 34	07 11	15 49	28 52	24 56	08 12	15 53	09 10
28 Mo	02 24 40	04 23 22	19 27 59	23 25	10 00	07 50	15 55	28 50	24 56	08 13	15 55	09 11
29 Tu	02 28 36	05 23 16	02♌ 34 28	25 05	09 24	08 28	16 02	28 48	24 55	08 13	15 57	09R 11
30 We	02 32 33	06 23 13	16 04 34	26 45	08 48	09 06	16 08	28 46	24 55	08 13	15 59	09 10
31 Th	02 36 29	07 23 11	00♍ 00 14	28 25	08 12	09 45	16 15	28 44	24 55	08 14	16 01	09 08

Data for	10-01-2002
Julian Day	2452548.50
Ayanamsa	23 53 26
SVP	05 ♓ 13 00
☽ ☊ Mean	11 ♊ 54 R

● ◐ PHASES ○ ◑

Day	Time	Phase	Position
06	11:18	●	13♎02
13	05:33	◐	19♑43
21	07:20	○	27♈43
29	05:28	◑	05♌37

LAST ASPECT Day	h m	☽ INGRESS Day	h m	Sign
30	19:00	01	11:59	♌
03	13:17	03	14:53	♍
05	13:22	05	14:52	♎
07	12:30	07	13:58	♏
09	12:38	09	14:21	♐
11	16:08	11	17:45	♑
13	22:42	14	00:52	♒
16	09:16	16	11:07	♓
18	21:17	18	23:14	♈
21	09:55	21	11:57	♉
23	14:15	24	00:18	♊
26	09:02	26	11:11	♋
28	08:22	28	19:20	♌
30	21:51	31	00:00	♍

ASPECTARIAN

01 10:31 ♀ # ♄
13:09 ☽ ⚹ ☿
15:50 ☉ △ ♆
02 02:26 ☽ ☍ ♆
03:00 ☽ # ♀
03:14 ☽ ⚹ ☉
03:59 ☽ ∥ ♄
09:24 ☽ ☌ ♃
09:26 ☿ ♍R
12:48 ☽ □ ♀
14:22 ☽ △ ♇
03 04:48 ☽ # ♆
07:10 ☽ ☍ ♅
08:09 ☽ ∥ ♃
13:17 ☽ ⚹ ♄
16:00 ☉ # ♂
23:08 ☿ □ ♄
04 01:32 ☽ # ♅
04:12 ☽ # ♇
14:57 ☽ ⚹ ♀
15:37 ☽ □ ♇
05 04:15 ☽ ☌ ♂
12:28 ☽ # ☉
12:22 ☽ ☌ ☿
13:23 ☽ □ ♄
18:01 ☽ ∥ ♂
06 03:49 ☽ △ ♆
04:09 ☽ ∥ ☿
09:59 ☽ # ☿
10:38 ☉ ⚹ ♃
11:16 ☽ ⚹ ♃
14:57 ☽ ⚹ ♇
18:51 ☽ # ♂
19:27 ☿ SD
07 02:40 ☽ ∥ ☉
06:26 ☽ △ ♅
12:30 ☽ △ ♄
08 [illegible]
09:26 ☽ ∥ ♆
11:10 ☽ □ ♃
12:03 ☽ ∥ ♅
14:42 ☽ ☌ ♀
21:04 ☉ ⚹ ♇
09 03:44 ☽ # ♃
06:25 ☽ □ ♅
07:36 ☽ ⚹ ♂
08:19 ☽ ∥ ♆
12:38 ☽ ⚹ ☿
18:28 ☿ □ ♄
10 04:05 ☽ ⚹ ♆
08:21 ☽ # ♄
13:16 ☽ △ ♃
16:22 ☽ ☌ ♇
17:31 ☽ ∥ ♀
18:36 ♀ SR
19:35 ☽ ⚹ ☉
11 05:57 ☿ ♎
09:10 ☽ ⚹ ♅
13:02 ☽ □ ♂
13:02 ♄ SR
16:08 ☽ ☍ ♄
18:28 ☽ □ ☿
12 21:48 ☽ ⚹ ♀
13 22:42 ☽ △ ♂
14 06:10 ☽ △ ☿
06:57 ♂ □ ♄
08:31 ☽ ∥ ♀
16:32 ☽ ☌ ♆
19:12 ☽ # ♄
15 04:19 ☽ ☍ ♃
06:04 ☽ □ ♀
06:46 ☽ ⚹ ♇
17:00 [illegible]
19:54 ☽ △ ☉
23:21 ☽ ∥ ♆
16 01:22 ☽ ☌ ♅
06:36 ☽ # ♃
09:16 ☽ △ ♄
23:59 ☽ ∥ ♅
17 00:24 ☿ # ♂
03:08 ☽ ∥ ♇
16:20 ☽ △ ♀
16:34 ♀ □ ♃
18:17 ☽ □ ♇
22:51 ☽ ∥ ☉
18 10:46 ☿ △ ♆
14:09 ☉ △ ♅
21:17 ☽ □ ♄
19 03:38 ☽ ☍ ♂
15:48 ☽ ⚹ ♆
19:42 ☽ ∥ ♂
19:56 ☽ ☍ ☿
20:18 ☽ # ♂
20 05:33 ☽ △ ♃
06:59 ☽ △ ♇
07:34 ☽ # ☿
13:54 ♆ SD
21 01:49 ☽ ⚹ ♅
09:55 ☽ ⚹ ♄
22 00:25 ☽ # ☉
04:31 ☽ □ ♆
13:05 ☽ # ♇
13:32 ☉ △ ♄
13:54 ☽ ☍ ♀
16:04 ☽ # ♅
18:51 ☽ □ ♃
23 02:21 ☿ ⚹ ♃
07:5[illegible] ☽ ∥ [illegible]
08:47 ♀ # ♄
08:50 ☿ ⚹ ♇
14:15 ☽ □ ♅
14:18 ☉ ♏
16:39 ☽ # ♆
24 11:24 ☽ △ ♂
16:36 ☽ △ ♆
17:54 ☽ # ♀
20:46 ☽ ∥ ♄
25 07:12 ☽ ⚹ ♃
07:39 ☽ ☍ ♇
14:44 ☽ △ ☿
26 01:25 ☽ △ ♅
09:02 ☽ ☌ ♄
17:00 [illegible]
27 00:59 ☽ □ ♂
07:01 ☽ △ ♀
23:42 ♃ △ ♇
28 08:22 ☽ □ ☿
14:29 ♂ △ ♆
21:43 ☿ △ ♅
29 06:06 ☉ ∥ ♇
10:07 ☽ ☍ ♆
11:05 ☽ ⚹ ♂
11:43 ☽ □ ♀
13:31 ☽ ∥ ♄
23:50 ☽ △ ♇
30 00:07 ☽ ☌ ♃
03:35 ☽ # ♀
14:31 [illegible]
15:19 ☽ ☍ ♅
20:19 ☉ ∥ ♅
20:57 ☽ ⚹ ☿
21:51 ☽ ⚹ ♄
22:53 ♀ □ ♆
23:00 ☽ ∥ ♃
31 04:28 ☿ △ ♄
10:16 ☽ # ☉
11:07 ☽ # ♅
12:06 ♀ ☌ ☉
13:14 ☽ ⚹ ♀
13:22 ☽ ⚹ ☉
13:23 ☽ # ♇
20:18 [illegible]
22:43 ☿ ♏

DECLINATION

Day	☉	☽	☿	♀	♂	♃	♄	♅	♆	♇
01 Tu	03S 02	24N49	01S 54	22S 03	04N44	17N36	22N08	13S 47	18S 08	13S 10
02 We	03 25	22 38	01 15	22 14	04 29	17 33	22 08	13 48	18 08	13 11
03 Th	03 49	19 02	00 39	22 25	04 14	17 31	22 08	13 48	18 08	13 11
04 Fr	04 12	14 10	00 07	22 34	03 59	17 28	22 07	13 49	18 08	13 12
05 Sa	04 35	08 20	00N20	22 43	03 44	17 26	22 07	13 49	18 09	13 12
06 Su	04 58	01 55	00 43	22 51	03 28	17 23	22 07	13 50	18 09	13 13
07 Mo	05 21	04S 40	01 00	22 58	03 13	17 20	22 07	13 50	18 09	13 13
08 Tu	05 44	10 57	01 12	23 04	02 58	17 18	22 07	13 50	18 09	13 14
09 We	06 07	16 29	01 18	23 09	02 42	17 15	22 07	13 51	18 09	13 14
10 Th	06 30	20 54	01 19	23 13	02 27	17 13	22 07	13 51	18 09	13 15
11 Fr	06 52	23 56	01 15	23 16	02 12	17 11	22 07	13 52	18 09	13 15
12 Sa	07 15	25 27	01 06	23 18	01 57	17 08	22 07	13 52	18 09	13 16
13 Su	07 37	25 27	00 53	23 18	01 41	17 06	22 07	13 52	18 10	13 16
14 Mo	08 00	24 04	00 35	23 18	01 26	17 04	22 07	13 53	18 10	13 17
15 Tu	08 22	21 32	00 14	23 16	01 10	17 01	22 07	13 53	18 10	13 17
16 We	08 44	18 04	00 11	23 13	00 55	16 59	22 07	13 53	18 10	13 18
17 Th	09 06	13 53	00 39	23 08	00 40	16 57	22 07	13 53	18 10	13 18
18 Fr	09 28	09 14	01 09	23 02	00 24	16 55	22 07	13 54	18 10	13 19
19 Sa	09 50	04 15	01 42	22 55	00 09	16 52	22 07	13 54	18 10	13 19
20 Su	10 12	00N51	02 17	22 46	00S 06	16 50	22 07	13 54	18 10	13 20
21 Mo	10 33	05 56	02 53	22 36	00 22	16 48	22 07	13 54	18 10	13 20
22 Tu	10 55	10 50	03 31	22 24	00 37	16 46	22 06	13 55	18 10	13 21
23 We	11 16	15 21	04 10	22 11	00 52	16 44	22 06	13 55	18 10	13 21
24 Th	11 37	19 18	04 49	21 57	01 08	16 42	22 06	13 55	18 10	13 22
25 Fr	11 58	22 28	05 30	21 41	01 23	16 40	22 06	13 55	18 10	13 22
26 Sa	12 18	24 40	06 10	21 25	01 38	16 38	22 06	13 55	18 10	13 23
27 Su	12 39	25 41	06 51	21 06	01 54	16 36	22 06	13 55	18 10	13 23
28 Mo	12 59	25 23	07 33	20 47	02 09	16 35	22 06	13 55	18 10	13 24
29 Tu	13 19	23 40	08 14	20 26	02 24	16 33	22 06	13 55	18 10	13 24
30 We	13 39	20 36	08 55	20 05	02 40	16 31	22 06	13 56	18 10	13 25
31 Th	13 59	16 17	09 36	19 43	02 55	16 29	22 06	13 56	18 10	13 25

⚷ Chiron

01	Dec.	16 S 55
02	03♑39	
05	03 45	
08	03 52	
11	04 00	
14	04 09	
17	04 18	
20	04 28	
23	04 39	
26	04 50	
29	05 01	

20 01:48 27♉33 ✷ Penumbral Lunar Eclipse (mag 0.886)

November 2002

Day	S. T. h m s	☉ ° ′ ″	☽ ° ′ ″	☿ ° ′	♀ ° ′	♂ ° ′	♃ ° ′	♄ ° ′	♅ ° ′	♆ ° ′	♇ ° ′	☊ True ° ′
01 Fr	02 40 26	08♏23 11	14♍21 21	00♏05	07♏R35	10♎23	16♌21	28♊R42	24♒R55	08♒14	16♐03	09♊R04
02 Sa	02 44 22	09 23 14	29 05 10	01 45	06 59	11 02	16 27	28 40	24 55	08 14	16 05	09 00
03 Su	02 48 19	10 23 19	14♎06 00	03 25	06 23	11 40	16 33	28 38	24 55	08 15	16 06	08 55
04 Mo	02 52 16	11 23 25	29 15 34	05 04	05 47	12 18	16 38	28 35	24 54	08 15	16 08	08 50
05 Tu	02 56 12	12 23 34	14♏24 09	06 43	05 13	12 57	16 44	28 33	24D 55	08 16	16 10	08 47
06 We	03 00 09	13 23 45	29 22 03	08 22	04 39	13 35	16 49	28 30	24 55	08 16	16 12	08 45
07 Th	03 04 05	14 23 57	14♐01 05	10 00	04 07	14 13	16 54	28 27	24 55	08 17	16 14	08 44
08 Fr	03 08 02	15 24 11	28 15 41	11 38	03 36	14 52	16 59	28 24	24 55	08 17	16 16	08D 44
09 Sa	03 11 58	16 24 26	12♑03 16	13 16	03 06	15 30	17 04	28 21	24 55	08 18	16 19	08 45
10 Su	03 15 55	17 24 43	25 23 57	14 53	02 39	16 09	17 09	28 18	24 55	08 19	16 21	08 46
11 Mo	03 19 51	18 25 02	08♒19 52	16 30	02 13	16 47	17 13	28 15	24 56	08 20	16 23	08 47
12 Tu	03 23 48	19 25 21	20 54 27	18 07	01 50	17 26	17 18	28 12	24 56	08 20	16 25	08 48
13 We	03 27 45	20 25 43	03♓11 51	19 43	01 28	18 04	17 22	28 09	24 56	08 21	16 27	08R 48
14 Th	03 31 41	21 26 05	15 16 26	21 19	01 09	18 42	17 26	28 05	24 57	08 22	16 29	08 48
15 Fr	03 35 38	22 26 29	27 12 28	22 55	00 52	19 21	17 29	28 02	24 57	08 23	16 31	08 46
16 Sa	03 39 34	23 26 55	09♈03 54	24 30	00 38	19 59	17 33	27 58	24 58	08 24	16 33	08 45
17 Su	03 43 31	24 27 21	20 54 13	26 05	00 26	20 38	17 36	27 55	24 59	08 24	16 35	08 44
18 Mo	03 47 27	25 27 50	02♉46 20	27 40	00 17	21 16	17 40	27 51	24 59	08 25	16 38	08 43
19 Tu	03 51 24	26 28 19	14 42 39	29 15	00 10	21 55	17 43	27 47	25 00	08 26	16 40	08 42
20 We	03 55 20	27 28 50	26 45 11	00♐49	00 05	22 33	17 46	27 43	25 01	08 27	16 42	08D 42
21 Th	03 59 17	28 29 23	08♊55 40	02 23	00 03	23 12	17 48	27 39	25 02	08 28	16 44	08 42
22 Fr	04 03 14	29 29 58	21 15 37	03 57	00D 04	23 50	17 51	27 35	25 03	08 29	16 46	08 43
23 Sa	04 07 10	00♐30 33	03♋46 32	05 31	00 07	24 29	17 53	27 31	25 03	08 31	16 49	08 43
24 Su	04 11 07	01 31 11	16 29 54	07 05	00 12	25 07	17 55	27 27	25 04	08 32	16 51	08 44
25 Mo	04 15 03	02 31 50	29 27 21	08 38	00 20	25 46	17 57	27 23	25 06	08 33	16 53	08 44
26 Tu	04 18 60	03 32 31	12♌40 29	10 11	00 30	26 24	17 59	27 19	25 07	08 34	16 55	08R 44
27 We	04 22 56	04 33 13	26 10 46	11 45	00 42	27 03	18 01	27 14	25 08	08 35	16 57	08 43
28 Th	04 26 53	05 33 57	09♍59 08	13 18	00 56	27 41	18 02	27 10	25 09	08 36	17 00	08 43
29 Fr	04 30 50	06 34 42	24 05 34	14 51	01 12	28 20	18 03	27 06	25 10	08 38	17 02	08 42
30 Sa	04 34 46	07 35 29	08♎28 40	16 23	01 31	28 58	18 04	27 01	25 11	08 39	17 04	08 41

Data for	11-01-2002
Julian Day	2452579.50
Ayanamsa	23 53 29
SVP	05 ♓ 12 55
☽ ☊ Mean	10 ♊ 16 R

● ◐ PHASES ○ ◑

Day	Time	Phase	Position
04	20:34	●	12♏15
11	20:53	◐	19♒18
20	01:34	✷	27♉33
27	15:46	◑	05♍13

LAST ASPECT ☽ INGRESS

Day	h m	Day	h m	Sign
01	23:19	02	01:28	♎
03	22:56	04	1:10	♏
05	16:48	06	01:01	♐
08	00:15	08	02:59	♑
09	08:22	10	08:27	♒
12	14:07	12	17:43	♓
15	01:39	15	05:39	♈
17	14:06	17	18:24	♉
20	01:34	20	06:26	♊
22	12:07	22	16:48	♋
24	16:51	25	01:00	♌
27	01:51	27	06:42	♍
29	05:01	29	09:54	♎

DECLINATION

Day	☉	☽	☿	♀	♂	♃	♄	♅	♆	♇
01 Fr	14S18	10N57	10S16	19S20	03S10	16N28	22N06	13S56	18S09	13S25
02 Sa	14 37	04 52	10 57	18 56	03 25	16 26	22 06	13 56	18 09	13 26
03 Su	14 56	01S37	11 36	18 32	03 41	16 24	22 06	13 56	18 09	13 26
04 Mo	15 15	08 05	12 16	18 07	03 56	16 23	22 06	13 56	18 09	13 27
05 Tu	15 33	14 05	12 55	17 42	04 11	16 21	22 06	13 56	18 09	13 27
06 We	15 52	19 10	13 33	17 17	04 26	16 20	22 06	13 56	18 09	13 28
07 Th	16 10	22 57	14 10	16 53	04 41	16 19	22 06	13 55	18 09	13 28
08 Fr	16 27	25 11	14 47	16 28	04 56	16 17	22 06	13 55	18 09	13 29
09 Sa	16 45	25 46	15 23	16 04	05 11	16 16	22 06	13 55	18 08	13 29
10 Su	17 02	24 49	15 58	15 40	05 26	16 15	22 05	13 55	18 08	13 29
11 Mo	17 19	22 33	16 33	15 17	05 41	16 14	22 05	13 55	18 08	13 30
12 Tu	17 35	19 15	17 07	14 54	05 56	16 13	22 05	13 55	18 08	13 30
13 We	17 51	15 12	17 40	14 32	06 11	16 12	22 05	13 55	18 08	13 31
14 Th	18 07	10 37	18 12	14 11	06 26	16 11	22 05	13 54	18 08	13 31
15 Fr	18 23	05 42	18 43	13 51	06 41	16 10	22 05	13 54	18 07	13 32
16 Sa	18 38	00 36	19 13	13 32	06 55	16 09	22 05	13 54	18 07	13 32
17 Su	18 53	04N32	19 42	13 14	07 10	16 08	22 05	13 54	18 07	13 32
18 Mo	19 08	09 31	20 10	12 57	07 25	16 07	22 05	13 54	18 07	13 33
19 Tu	19 22	14 12	20 38	12 41	07 39	16 06	22 05	13 53	18 06	13 33
20 We	19 36	18 22	21 04	12 27	07 54	16 06	22 05	13 53	18 06	13 34
21 Th	19 49	21 48	21 29	12 13	08 08	16 05	22 05	13 53	18 06	13 34
22 Fr	20 03	24 18	21 53	12 01	08 23	16 04	22 05	13 52	18 06	13 34
23 Sa	20 15	25 38	22 16	11 50	08 37	16 04	22 05	13 52	18 05	13 35
24 Su	20 28	25 38	22 38	11 40	08 51	16 03	22 05	13 52	18 05	13 35
25 Mo	20 40	24 16	22 59	11 31	09 06	16 03	22 05	13 51	18 05	13 35
26 Tu	20 52	21 34	23 19	11 24	09 20	16 03	22 05	13 51	18 05	13 36
27 We	21 03	17 39	23 37	11 17	09 34	16 03	22 05	13 50	18 04	13 36
28 Th	21 14	12 43	23 54	11 12	09 48	16 02	22 05	13 50	18 04	13 37
29 Fr	21 25	07 02	24 10	11 08	10 02	16 02	22 04	13 50	18 04	13 37
30 Sa	21 35	00 51	24 25	11 05	10 16	16 02	22 04	13 49	18 03	13 37

ASPECTARIAN

01 02:30 ☽ # ☿
02:47 ☽ □ ♇
23:19 ☽ □ ♄
02 05:12 ☽ # ♂
14:41 ☽ △ ♆
19:57 ☽ ☌ ♂
03 03:12 ☽ ✶ ♇
03:54 ☽ ✶ ♃
07:51 ☽ ∥ ♂
17:07 ☽ △ ♅
22:00 ♀ ∥ ♆
22:56 ☽ △ ♄
04 06:28 ♅ SD
07:43 ☿ ☌ ♀
09:57 ☽ ☌ ♀
10:19 ☽ ☌ ☿
14:15 ☽ □ ♆
18:26 ☽ ∥ ☿
21:19 ☽ ∥ ♇
23:19 ☽ ∥ ♅
05 03:44 ☽ □ ♃
06:49 ☽ ∥ ☉
10:02 ☽ # ♃
15:14 ☽ ∥ ♀
16:49 ☽ □ ♅
18:42 ☽ ∥ ♆
20:50 ☿ ∥ ♇
22:38 ☿ □ ♆
06 14:31 ☽ ✶ ♆
14:33 ☿ ∥ ♅
17:38 ☽ # ♄
07 00:21 ☽ ✶ ♂
03:42 ☽ ☌ ♇
04:50 ☽ △ ♃
11:30 ☉ # ♃
18:17 ☽ ✶ ♅
08 00:15 ☽ ☍ ♄
00:28 ♀ ∥ ☉
08:52 ☽ ✶ ♀
11:05 ♀ # ♃
09 02:26 ☽ ✶ ☿
06:26 ☽ □ ♂
08:22 ☽ ✶ ☉
16:26 ☿ ∥ ♀
17:06 ☉ □ ♃
10 07:53 ♂ ✶ ♇
11:00 ☿ # ♃
12:55 ☽ □ ♀
11 00:00 ☽ ☌ ♆
03:51 ☽ # ♄
11:09 ☿ □ ♃
15:20 ☽ ✶ ♇
16:56 ☽ △ ♂
17:00 ☽ ☍ ♃
17:49 ☽ □ ☿
18:23 ♂ ✶ ♃
12 07:07 ☽ ∥ ♆
07:49 ☽ ☌ ♅
09:45 ☽ ∥ ☉
11:38 ☽ ∥ ☿
14:07 ☽ △ ♄
18:29 ☽ # ♃
20:41 ☽ △ ♀
13 03:59 ☽ ∥ ♀
07:02 ☽ ∥ ♅
09:06 ☽ ∥ ♇
17:31 ☿ ∥ ☉
20:59 ☿ ∥ ♆
14 00:26 ☉ ∥ ♆
02:26 ☽ □ ♇
04:40 ☿ ☌ ☉
13:30 ☽ △ ☉
14:00 ☽ △ ☿
19:31 ☽ ∥ ♂
20:15 ♀ ∥ ♅
15 01:39 ☽ □ ♄
22:38 ☽ ✶ ♆
16 00:13 ♀ ∥ ♇
07:03 ☿ □ ♅
15:14 ☽ △ ♇
17:17 ☽ △ ♃
23:25 ☽ ☍ ♂
17 08:15 ☽ ✶ ♅
12:34 ☉ □ ♅
13:10 ☽ # ♂
14:06 ☽ ✶ ♄
19:01 ☽ ☍ ♀
18 11:24 ☽ □ ♆
16:26 ☽ # ♀
20:33 ☽ # ♇
22:20 ☽ # ♅
19 06:02 ☽ □ ♃
10:31 ☽ ∥ ♃
11:30 ☿ ♐
20:33 ☽ □ ♅
22:21 ☽ # ♆
20 08:28 ☽ # ☉
09:14 ☽ ☍ ☿
21:00 ☽ # ☿
23:06 ☽ △ ♆
21 02:10 ☽ ∥ ♄
07:14 ♀ SD
15:17 ☽ ☍ ♇
17:22 ☽ ✶ ♃
22 05:14 ☽ △ ♂
07:18 ☽ △ ♅
11:54 ☉ ♐
12:04 ☿ # ♄
12:07 ☽ ☌ ♄
16:59 ☽ △ ♀
23 22:20 ♂ △ ♅
24 16:51 ☽ □ ♂
22:38 ☿ ✶ ♆
25 01:37 ☽ □ ♀
06:06 ☽ △ ☉
11:30 ☽ # ☿
16:35 ☽ ☍ ♆
18:56 ☽ △ ☿
20:14 ☽ ∥ ♄
26 04:38 ☽ # ☉
07:37 ☽ △ ♇
09:31 ☽ ☌ ♃
21:41 ☽ # ♆
22:09 ☽ ☍ ♅
27 01:36 ☽ ✶ ♂
01:51 ☽ ✶ ♄
06:34 ♂ △ ♄
08:02 ☽ ✶ ♀
08:20 ☽ ∥ ♃
18:57 ☽ # ♅
19:58 ☽ # ♇
28 06:23 ☽ □ ☿
06:47 ☽ # ♀
12:01 ☽ □ ♇
12:09 ☽ # ♂
29 05:01 ☽ □ ♄
22:25 ☽ ✶ ☉
30 00:17 ☽ △ ♆
10:51 ☿ ☌ ♇
14:11 ☽ ✶ ♇
14:35 ☽ ✶ ☿
15:48 ☽ ✶ ♃

⚷ Chiron

01 Dec.	17 S 06
01	05♑14
04	05 26
07	05 40
10	05 54
13	06 08
16	06 23
19	06 38
22	06 54
25	07 09
28	07 26

December 2002

04 07:40 11♐58 ☉ Total Solar Eclipse

Day	S. T.	☉	☽	☿	♀	♂	♃	♄	♅	♆	♇	☊ True
	h m s	° ' "	° ' "	° '	° '	° '	° '	° '	° '	° '	° '	° '
01 Su	04 38 43	08♐36 18	23♎05 19	17♐56	01♏51	29♎37	18♌05	26♊R57	25♒13	08♒40	17♐07	08♊R40
02 Mo	04 42 39	09 37 08	07♏50 40	19 29	02 14	00♏15	18 06	26 52	25 14	08 42	17 09	08 39
03 Tu	04 46 36	10 37 59	22 38 21	21 01	02 38	00 54	18 06	26 47	25 16	08 43	17 11	08 39
04 We	04 50 32	11 38 52	07♐21 12	22 34	03 04	01 32	18 06	26 43	25 17	08 44	17 13	08 38
05 Th	04 54 29	12 39 46	21 52 05	24 06	03 32	02 11	18R 06	26 38	25 19	08 46	17 16	08 38
06 Fr	04 58 25	13 40 41	06♑05 00	25 38	04 01	02 50	18 06	26 33	25 20	08 47	17 18	08 37
07 Sa	05 02 22	14 41 37	19 55 41	27 10	04 32	03 28	18 06	26 28	25 22	08 49	17 20	08 36
08 Su	05 06 19	15 42 34	03♒22 07	28 42	05 04	04 07	18 05	26 24	25 23	08 50	17 23	08 36
09 Mo	05 10 15	16 43 31	16 24 24	00♑14	05 38	04 45	18 04	26 19	25 25	08 52	17 25	08 35
10 Tu	05 14 12	17 44 29	29 04 26	01 45	06 14	05 24	18 03	26 14	25 27	08 54	17 27	08 34
11 We	05 18 08	18 45 28	11♓25 31	03 16	06 50	06 03	18 02	26 09	25 29	08 55	17 29	08D 34
12 Th	05 22 05	19 46 27	23 31 48	04 47	07 28	06 41	18 01	26 04	25 30	08 57	17 32	08 35
13 Fr	05 26 01	20 47 27	05♈27 57	06 17	08 08	07 20	17 59	25 59	25 32	08 58	17 34	08 36
14 Sa	05 29 58	21 48 27	17 18 45	07 47	08 48	07 58	17 57	25 54	25 34	09 00	17 36	08 38
15 Su	05 33 54	22 49 28	29 08 52	09 16	09 30	08 37	17 55	25 49	25 36	09 02	17 39	08 40
16 Mo	05 37 51	23 50 29	11♉02 32	10 44	10 13	09 16	17 53	25 44	25 38	09 04	17 41	08 43
17 Tu	05 41 48	24 51 31	23 03 26	12 11	10 56	09 54	17 51	25 40	25 40	09 05	17 43	08 45
18 We	05 45 44	25 52 34	05♊14 37	13 38	11 41	10 33	17 48	25 35	25 42	09 07	17 45	08 46
19 Th	05 49 41	26 53 37	17 38 19	15 02	12 27	11 11	17 46	25 30	25 44	09 09	17 48	08R 46
20 Fr	05 53 37	27 54 41	00♋15 57	16 25	13 14	11 50	17 43	25 25	25 47	09 11	17 50	08 45
21 Sa	05 57 34	28 55 45	13 08 02	17 47	14 02	12 29	17 40	25 20	25 49	09 12	17 52	08 43
22 Su	06 01 30	29 56 50	26 14 24	19 06	14 51	13 07	17 36	25 15	25 51	09 14	17 55	08 40
23 Mo	06 05 27	00♑57 56	09♌34 15	20 22	15 40	13 46	17 33	25 10	25 53	09 16	17 57	08 36
24 Tu	06 09 23	01 59 02	23 06 29	21 35	16 31	14 25	17 29	25 05	25 56	09 18	17 59	08 32
25 We	06 13 20	03 00 08	06♍49 52	22 44	17 22	15 03	17 25	25 00	25 58	09 20	18 01	08 28
26 Th	06 17 17	04 01 16	20 43 10	23 50	18 14	15 42	17 21	24 55	26 01	09 22	18 03	08 26
27 Fr	06 21 13	05 02 24	04♎45 07	24 50	19 07	16 21	17 17	24 50	26 03	09 24	18 06	08 24
28 Sa	06 25 10	06 03 32	18 54 19	25 45	20 00	16 59	17 13	24 46	26 05	09 26	18 08	08 24
29 Su	06 29 06	07 04 41	03♏09 06	26 33	20 54	17 38	17 08	24 41	26 08	09 28	18 10	08D 24
30 Mo	06 33 03	08 05 51	17 27 16	27 14	21 49	18 17	17 03	24 36	26 11	09 30	18 12	08 25
31 Tu	06 36 59	09 07 01	01♐45 56	27 47	22 44	18 55	16 58	24 31	26 13	09 32	18 14	08R 25

Data for 12-01-2002
Julian Day 2452609.50
Ayanamsa 23 53 34
SVP 05 ♓ 12 52
☽ ☊ Mean 08 ♊ 40 R

Last Aspect Day	h m	☽ Ingress Day	h m	Sign
01	11:07	01	11:16	♏
03	04:16	03	11:59	♐
05	07:56	05	13:39	♑
05	20:21	07	17:55	♒
09	18:36	10	01:47	♓
12	05:03	12	12:59	♈
14	17:19	15	01:43	♉
17	05:12	17	13:43	♊
19	19:10	19	23:30	♋
21	09:31	22	06:49	♌
24	04:58	24	12:05	♍
26	07:10	26	15:54	♎
28	12:16	28	18:42	♏
30	17:05	30	21:02	♐

● ◐ PHASES ○ ◑

Day	Time	Phase	Position
04	07:35	☉ (eclipse)	11♐58
11	15:49	◐	19♓26
19	19:10	○	27♊42
27	00:31	◑	05♎04

DECLINATION

Day	☉	☽	☿	♀	♂	♃	♄	♅	♆	♇
01 Su	21S44	05S28	24S39	11S03	10S30	16N02	22N04	13S49	18S03	13S38
02 Mo	21 54	11 33	24 51	11 01	10 44	16 02	22 04	13 48	18 03	13 38
03 Tu	22 02	17 01	25 02	11 01	10 58	16 02	22 04	13 48	18 02	13 38
04 We	22 11	21 25	25 11	11 02	11 11	16 02	22 04	13 47	18 02	13 39
05 Th	22 19	24 24	25 20	11 03	11 25	16 03	22 04	13 47	18 02	13 39
06 Fr	22 27	25 44	25 27	11 06	11 38	16 03	22 04	13 46	18 01	13 39
07 Sa	22 34	25 25	25 32	11 09	11 52	16 03	22 04	13 45	18 01	13 40
08 Su	22 40	23 36	25 36	11 13	12 05	16 04	22 04	13 45	18 00	13 40
09 Mo	22 47	20 34	25 39	11 18	12 19	16 04	22 04	13 44	18 00	13 40
10 Tu	22 52	16 40	25 40	11 23	12 32	16 05	22 04	13 44	18 00	13 40
11 We	22 58	12 09	25 40	11 29	12 45	16 05	22 04	13 43	17 59	13 41
12 Th	23 03	07 15	25 38	11 36	12 58	16 06	22 04	13 42	17 59	13 41
13 Fr	23 07	02 09	25 35	11 43	13 11	16 06	22 04	13 42	17 58	13 41
14 Sa	23 11	02N59	25 30	11 51	13 24	16 07	22 04	13 41	17 58	13 42
15 Su	23 15	08 01	25 24	11 59	13 37	16 08	22 03	13 40	17 57	13 42
16 Mo	23 18	12 48	25 16	12 08	13 49	16 09	22 03	13 40	17 57	13 42
17 Tu	23 20	17 09	25 07	12 18	14 02	16 10	22 03	13 39	17 57	13 42
18 We	23 23	20 52	24 56	12 27	14 14	16 11	22 03	13 38	17 56	13 43
19 Th	23 24	23 41	24 45	12 38	14 27	16 12	22 03	13 37	17 56	13 43
20 Fr	23 25	25 23	24 31	12 48	14 39	16 13	22 03	13 37	17 55	13 43
21 Sa	23 26	25 45	24 17	12 59	14 51	16 14	22 03	13 36	17 55	13 43
22 Su	23 26	24 42	24 01	13 10	15 03	16 15	22 03	13 35	17 54	13 44
23 Mo	23 26	22 16	23 44	13 22	15 15	16 17	22 03	13 34	17 54	13 44
24 Tu	23 26	18 35	23 26	13 34	15 27	16 18	22 03	13 33	17 53	13 44
25 We	23 24	13 51	23 07	13 46	15 39	16 19	22 03	13 33	17 53	13 44
26 Th	23 23	08 22	22 47	13 58	15 50	16 21	22 03	13 32	17 52	13 44
27 Fr	23 21	02 25	22 27	14 11	16 02	16 22	22 03	13 31	17 52	13 45
28 Sa	23 18	03S43	22 06	14 23	16 13	16 24	22 03	13 30	17 51	13 45
29 Su	23 15	09 44	21 45	14 36	16 25	16 25	22 03	13 29	17 51	13 45
30 Mo	23 12	15 15	21 24	14 49	16 36	16 27	22 02	13 28	17 50	13 45
31 Tu	23 08	19 56	21 03	15 02	16 47	16 29	22 02	13 27	17 50	13 45

ASPECTARIAN

Day	Time	Aspect
01	01:38	☉ ⚹ ♆
	02:19	☿ △ ♃
	03:28	☽ △ ♅
	06:15	☽ △ ♄
	11:07	☽ ☌ ♂
	14:26	♂ ♏
	14:38	☽ ☌ ♀
	20:32	☽ ∥ ♂
	21:51	☽ ∥ ♀
02	01:23	☽ □ ♆
	08:44	☽ ∥ ♇
	09:27	☽ ∥ ♅
	16:38	☽ □ ♃
	19:24	☽ # ♃
03	04:16	☽ □ ♅
	04:50	☉ # ♄
	05:02	☽ ∥ ♆
	06:03	♀ ∥ ♂
04	02:17	☽ ⚹ ♆
	04:22	☽ # ♄
	05:22	☽ ∥ ☉
	12:23	♃ SR
	16:19	☽ ☌ ♇
	17:44	☽ △ ♃
05	04:11	☽ ☌ ☿
	05:46	☽ ⚹ ♅
	07:56	☽ ☍ ♄
	14:20	☽ ∥ ☿
	18:11	☽ ⚹ ♂
	19:09	☿ ⚹ ♅
	20:21	☽ ⚹ ♀
06	13:36	☿ ☍ ♄
	21:23	☽ ∥ ☿
08	01:25	☽ □ ♂
	03:14	☽ □ ♀
	08:02	☽ ∥ ☉
	10:00	☽ [illegible] ♀
	13:14	☽ # ♄
	20:21	☿ ♑
09	00:39	☽ ⚹ ☉
	01:53	☽ ⚹ ♇
	03:07	☽ ☍ ♃
	16:22	☽ ∥ ♆
	16:56	☉ ☌ ♇
	17:03	☽ ☌ ♅
	18:36	☽ △ ♄
10	03:20	☽ # ♃
	05:53	☽ ⚹ ☿
	07:16	☉ △ ♃
	12:54	☽ △ ♂
	14:33	☽ △ ♀
	15:58	☽ ∥ ♅
	16:11	☽ ∥ ♇
	21:08	☽ ∥ ♂
11	03:19	☽ ∥ ♀
	12:01	☽ □ ♇
12	05:03	☽ □ ♄
13	01:54	☽ □ ☿
	07:07	☽ ⚹ ♆
	07:48	♅ ∥ ♇
14	00:36	☽ △ ♇
	01:18	☽ △ ♃
	05:25	☿ ⚹ ♂
	07:18	♀ □ ♆
	09:59	☽ △ ☉
	16:48	☽ ⚹ ♅
	17:19	☽ ⚹ ♄
15	06:41	♂ ∥ ♅
	07:09	☿ ⚹ ♀
	10:18	♂ ∥ ♇
	16:08	♂ □ ♇
	20:00	☽ □ ♆
	20:13	☽ ☍ ♂
	20:25	☽ # ♀
	22:13	☽ ☍ ♀
	23:18	☽ △ ☿
16	04:29	☽ # ♅
	04:43	☽ # ♇
	05:37	☽ # ♂
	13:40	☽ □ ♃
	18:14	☽ ∥ ♃
	21:37	♄ △ ♅
17	04:42	☽ # ♆
	05:12	☽ □ ♅
	17:28	☉ ☍ ♄
	19:50	☉ ⚹ ♅
18	07:34	☽ △ ♆
	09:09	☽ ∥ ♄
	13:41	♃ △ ♇
	21:11	☽ # ☉
19	00:14	☽ ⚹ ♃
	00:18	☽ ☍ ♇
	11:25	☽ # ☿
	14:53	☽ ☌ ♄
	15:30	☽ △ ♅
20	22:44	☽ △ ♂
21	01:46	☽ △ ♀
	09:31	☽ ☍ ☿
22	01:15	☉ ♑
	09:34	☽ # ☿
	14:08	☽ # ☉
	23:28	☽ ☍ ♆
23	01:43	☽ ∥ ♄
	07:51	☽ □ ♂
	11:34	☽ □ ♆
	14:07	☽ ☌ ♃
	14:56	☽ △ ♇
	23:18	♀ ∥ ♅
24	00:14	☿ ∥ ☉
	03:27	☽ ⚹ ♄
	03:51	☽ # ♆
	04:58	☽ ☍ ♅
	12:07	☽ ∥ ♃
	15:47	☽ # ♂
	16:48	☽ △ ☉
	20:41	♀ ∥ ♇
25	00:26	☽ # ♀
	00:35	☽ # ♇
	01:27	♀ □ ♃
	01:29	☽ # ♅
	14:56	☽ ⚹ ♂
	19:25	☽ □ ♇
	19:26	☽ ⚹ ♀
26	05:46	☽ △ ☿
	07:10	☽ □ ♄
27	07:55	☽ △ ♆
	21:09	☽ ⚹ ♃
	22:41	☽ ⚹ ♇
28	03:36	☿ # ♄
	07:22	♂ □ ♃
	09:49	☽ △ ♄
	12:09	☽ △ ♅
	12:16	☽ □ ☿
29	02:03	♂ # ♃
	07:06	☽ ⚹ ☉
	10:37	☽ □ ♆
	16:01	☽ ∥ ♅
	17:14	☽ ∥ ♇
	21:57	☽ ∥ ♀
	23:20	☽ □ ♃
30	01:27	☽ ☌ ♂
	05:48	☽ # ♃
	06:44	☽ ∥ ♂
	07:49	☽ ☌ ♀
	12:39	☽ ∥ ♆
	14:40	☽ □ ♅
	17:05	☽ ⚹ ☿
31	06:21	☽ ∥ ☿
	13:05	☽ ⚹ ♆
	13:31	☽ # ♄
	21:11	☽ ∥ ☉

⚷ Chiron

01 Dec.	17 S 07
01	07♑42
04	07 59
07	08 16
10	08 34
13	08 51
16	09 09
19	09 27
22	09 45
25	10 03
28	10 21
31	10 39

January 2003

Day	S. T.	☉	☽	☿	♀	♂	♃	♄	♅	♆	♇	☊ True
	h m s	° ' "	° ' "	° '	° '	° '	° '	° '	° '	° '	° '	° '
01 We	06 40 56	10♑ 08 12	16♐ 01 28	28♑ 11	23♏ 40	19♏ 34	16♌ R53	24♊ R27	26♒ 16	09♒ 34	18♐ 17	08♊ R24
02 Th	06 44 52	11 09 23	00♑ 09 37	28 24	24 37	20 13	16 48	24 22	26 18	09 36	18 19	08 22
03 Fr	06 48 49	12 10 34	14 05 58	28R 27	25 34	20 51	16 43	24 17	26 21	09 38	18 21	08 19
04 Sa	06 52 46	13 11 45	27 46 33	28 19	26 32	21 30	16 37	24 13	26 24	09 40	18 23	08 14
05 Su	06 56 42	14 12 56	11♒ 08 24	27 59	27 30	22 09	16 32	24 08	26 27	09 42	18 25	08 08
06 Mo	07 00 39	15 14 06	24 10 01	27 27	28 28	22 48	16 26	24 04	26 29	09 44	18 27	08 02
07 Tu	07 04 35	16 15 17	06♓ 51 34	26 43	29 27	23 26	16 20	24 00	26 32	09 46	18 29	07 56
08 We	07 08 32	17 16 27	19 14 49	25 49	00♐ 27	24 05	16 14	23 55	26 35	09 48	18 32	07 52
09 Th	07 12 28	18 17 36	01♈ 22 51	24 45	01 27	24 44	16 07	23 51	26 38	09 51	18 34	07 49
10 Fr	07 16 25	19 18 45	13 19 48	23 34	02 27	25 22	16 01	23 47	26 41	09 53	18 36	07D 48
11 Sa	07 20 22	20 19 54	25 10 30	22 17	03 28	26 01	15 54	23 43	26 44	09 55	18 38	07 49
12 Su	07 24 18	21 21 02	07♉ 00 07	20 58	04 30	26 40	15 48	23 38	26 47	09 57	18 40	07 51
13 Mo	07 28 15	22 22 10	18 53 54	19 38	05 31	27 18	15 41	23 34	26 50	09 59	18 42	07 53
14 Tu	07 32 11	23 23 17	00♊ 56 48	18 21	06 33	27 57	15 34	23 30	26 53	10 02	18 44	07 55
15 We	07 36 08	24 24 23	13 13 11	17 07	07 36	28 36	15 27	23 27	26 56	10 04	18 46	07R 55
16 Th	07 40 04	25 25 29	25 46 36	16 01	08 38	29 14	15 20	23 23	26 59	10 06	18 48	07 53
17 Fr	07 44 01	26 26 34	08♋ 39 16	15 01	09 41	29 53	15 13	23 19	27 02	10 08	18 49	07 49
18 Sa	07 47 57	27 27 39	21 51 57	14 11	10 45	00♐ 32	15 06	23 15	27 05	10 10	18 51	07 42
19 Su	07 51 54	28 28 43	05♌ 23 36	13 30	11 48	01 10	14 58	23 12	27 08	10 13	18 53	07 34
20 Mo	07 55 51	29 29 47	19 11 34	12 58	12 52	01 49	14 51	23 08	27 11	10 15	18 55	07 25
21 Tu	07 59 47	00♒ 30 50	03♍ 12 00	12 36	13 57	02 28	14 43	23 05	27 14	10 17	18 57	07 15
22 We	08 03 44	01 31 52	17 20 26	12 22	15 01	03 06	14 36	23 02	27 18	10 19	18 59	07 07
23 Th	08 07 40	02 32 54	01♎ 32 34	12 18	16 06	03 45	14 28	22 58	27 21	10 22	19 01	07 01
24 Fr	08 11 37	03 33 56	15 44 46	12D 21	17 11	04 24	14 20	22 55	27 24	10 24	19 02	06 56
25 Sa	08 15 33	04 34 57	29 54 23	12 33	18 16	05 02	14 13	22 52	27 27	10 26	19 04	06 54
26 Su	08 19 30	05 35 57	13♏ 59 47	12 51	19 22	05 41	14 05	22 49	27 31	10 28	19 06	06 53
27 Mo	08 23 26	06 36 58	28 00 02	13 15	20 28	06 20	13 57	22 46	27 34	10 31	19 08	06 52
28 Tu	08 27 23	07 37 57	11♐ 54 32	13 46	21 34	06 58	13 49	22 44	27 37	10 33	19 09	06 51
29 We	08 31 20	08 38 57	25 42 34	14 22	22 40	07 37	13 41	22 41	27 40	10 35	19 11	06 49
30 Th	08 35 16	09 39 55	09♑ 22 57	15 02	23 47	08 16	13 33	22 38	27 44	10 38	19 13	06 45
31 Fr	08 39 13	10 40 53	22 53 57	15 47	24 53	08 54	13 25	22 36	27 47	10 40	19 14	06 38

Data for	01-01-2003
Julian Day	2452640.50
Ayanamsa	23 53 40
SVP	05 ♓ 12 49
☽ ☊ Mean	07 ♊ 02 R

● ◐ PHASES ○ ◑

02	20:24	●	12♑01
10	13:14	◐	19♈53
18	10:48	○	27♋55
25	08:34	◑	04♏57

LAST ASPECT Day	h m	☽ INGRESS Day	h m	
01	17:25	01	23:44	♑
04	00:56	04	03:57	♒
06	08:45	06	10:57	♓
08	11:55	08	21:15	♈
11	03:10	11	09:48	♉
13	17:44	13	22:08	♊
16	02:17	16	07:56	♋
18	10:48	18	14:30	♌
20	13:47	20	18:32	♍
22	09:35	22	21:24	♎
24	19:49	25	00:10	♏
26	23:15	27	03:26	♐
29	03:27	29	07:30	♑
30	10:35	31	12:45	♒

DECLINATION

Day	☉	☽	☿	♀	♂	♃	♄	♅	♆	♇
01 We	23S 03	23S 24	20S 43	15S 15	16S 58	16N31	22N02	13S 26	17S 49	13S 45
02 Th	22 58	25 22	20 24	15 28	17 09	16 32	22 02	13 25	17 49	13 46
03 Fr	22 53	25 43	20 07	15 41	17 19	16 34	22 02	13 25	17 48	13 46
04 Sa	22 47	24 28	19 50	15 54	17 30	16 36	22 02	13 24	17 48	13 46
05 Su	22 41	21 52	19 36	16 07	17 41	16 38	22 02	13 23	17 47	13 46
06 Mo	22 34	18 12	19 23	16 20	17 51	16 40	22 02	13 22	17 46	13 46
07 Tu	22 27	13 48	19 13	16 33	18 01	16 42	22 02	13 21	17 46	13 46
08 We	22 19	08 55	19 04	16 45	18 11	16 44	22 02	13 20	17 45	13 46
09 Th	22 11	03 48	18 58	16 58	18 21	16 46	22 02	13 19	17 45	13 46
10 Fr	22 03	01N23	18 54	17 11	18 31	16 48	22 02	13 18	17 44	13 47
11 Sa	21 54	06 28	18 52	17 23	18 41	16 50	22 02	13 17	17 44	13 47
12 Su	21 45	11 20	18 52	17 35	18 50	16 52	22 02	13 16	17 43	13 47
13 Mo	21 35	15 49	18 53	17 47	19 00	16 55	22 02	13 15	17 42	13 47
14 Tu	21 25	19 44	18 56	17 59	19 09	16 57	22 02	13 14	17 42	13 47
15 We	21 14	22 52	19 00	18 11	19 18	16 59	22 02	13 13	17 41	13 47
16 Th	21 03	24 58	19 05	18 22	19 27	17 01	22 02	13 11	17 41	13 47
17 Fr	20 52	25 48	19 11	18 33	19 36	17 04	22 02	13 10	17 40	13 47
18 Sa	20 40	25 12	19 18	18 44	19 44	17 06	22 02	13 09	17 40	13 47
19 Su	20 28	23 07	19 26	18 54	19 53	17 08	22 02	13 08	17 39	13 47
20 Mo	20 15	19 40	19 34	19 05	20 01	17 11	22 02	13 07	17 38	13 47
21 Tu	20 02	15 04	19 43	19 15	20 10	17 13	22 02	13 06	17 38	13 47
22 We	19 49	09 37	19 51	19 24	20 18	17 15	22 02	13 05	17 37	13 47
23 Th	19 35	03 39	20 00	19 33	20 26	17 18	22 02	13 04	17 37	13 47
24 Fr	19 21	02S 31	20 09	19 42	20 34	17 20	22 02	13 03	17 36	13 47
25 Sa	19 07	08 33	20 18	19 51	20 41	17 22	22 02	13 02	17 35	13 47
26 Su	18 52	14 08	20 26	19 59	20 49	17 25	22 02	13 00	17 35	13 47
27 Mo	18 37	18 57	20 34	20 06	20 56	17 27	22 02	12 59	17 34	13 47
28 Tu	18 22	22 40	20 42	20 13	21 03	17 30	22 02	12 58	17 34	13 47
29 We	18 06	25 01	20 49	20 20	21 10	17 32	22 02	12 57	17 33	13 47
30 Th	17 50	25 50	20 55	20 27	21 17	17 34	22 02	12 56	17 32	13 47
31 Fr	17 33	25 06	21 01	20 33	21 24	17 37	22 02	12 55	17 32	13 47

ASPECTARIAN

01 01:27 ☽ △ ♃
03:49 ☽ ☌ ♇
14:11 ☽ ☍ ♄
17:25 ☽ ✶ ♅
02 18:20 ☿ SR
03 12:23 ☽ ✶ ♂
20:38 ♀ □ ♅
21:37 ☽ ✶ ♀
04 00:56 ☽ ☌ ☿
17:20 ☽ ∥ ☉
21:23 ☽ ☌ ♆
22:43 ☽ # ♄
05 08:01 ☿ ✶ ♀
09:46 ☽ ☍ ♃
13:22 ☽ ✶ ♇
14:14 ♂ ∥ ♆
16:27 ☽ ∥ ☿
21:18 ☽ □ ♂
23:49 ☽ △ ♄
06 01:59 ☽ ∥ ♂
02:29 ☽ ∥ ♆
04:22 ☽ ☌ ♅
08:41 ☽ # ♃
08:45 ☽ □ ♀
10:06 ☽ ∥ ♀
07 00:07 ☽ ∥ ♇
02:18 ☽ ∥ ♅
13:08 ♀ ♐
19:48 ☽ ✶ ☉
20:35 ♀ # ♃
22:35 ☽ □ ♇
08 09:08 ☽ □ ♄
10:02 ☽ △ ♂
11:55 ☽ ✶ ☿
09 00:09 ☽ △ ♀
00:18 ☿ ✶ ♂
17:01 ☽ ✶ ♆
10 02:40 ☉ # ♄
05:23 ☽ △ ♃
10:41 ☽ △ ♇
18:43 ☽ □ ☿
21:04 ☽ ✶ ♄
11 03:10 ☽ ✶ ♅
20:02 ☿ ☌ ☉
12 04:19 ☿ ∥ ♂
04:47 ♂ □ ♅
05:59 ☽ □ ♆
09:57 ☽ # ♅
12:47 ☽ # ♇
14:57 ♀ ∥ ♆
17:36 ☽ □ ♃
13 01:20 ☽ △ ☿
06:21 ☽ ∥ ♃
07:36 ☽ △ ☉
11:03 ☽ # ♆
12:10 ☽ # ♀
15:54 ☽ □ ♅
17:44 ☽ ☍ ♂
18:38 ☽ # ☿
19:56 ☽ # ♂
14 11:23 ☽ # ☉
12:03 ☽ ☍ ♀
16:53 ☽ ∥ ♄
17:51 ☽ △ ♆
15 04:16 ☽ ✶ ♃
10:41 ☽ ☍ ♇
19:29 ☽ ☌ ♄
16 02:17 ☽ △ ♅
17 04:22 ♂ ♐
10:35 ♀ ✶ ♆
10:55 ☽ ☍ ☿
18 16:12 ☽ △ ♂
19 08:27 ☽ ☍ ♆
08:34 ☽ ∥ ♄
12:08 ☽ △ ♀
16:33 ☽ ☌ ♃
20:14 ☽ # ☉
21:56 ☽ # ♂
23:32 ☽ △ ♇
20 00:35 ☽ # ☿
03:18 ☽ # ♀
06:46 ☽ ✶ ♄
11:16 ☽ # ♆
11:53 ☉ ♒
13:32 ☽ ∥ ♃
13:47 ☽ ☍ ♅
15:53 ☉ ∥ ♂
22:41 ☽ □ ♂
21 05:55 ☽ # ♇
09:03 ☽ # ♅
15:36 ♀ △ ♃
15:41 ☽ △ ☿
19:45 ☽ □ ♀
21:45 ☿ ∥ ☉
22 02:47 ☽ □ ♇
09:35 ☽ □ ♄
23 01:09 ☿ SD
01:50 ☽ △ ☉
02:22 ♀ ∥ ☉
03:54 ☽ ✶ ♂
14:56 ☽ △ ♆
18:13 ☽ □ ☿
21:39 ☽ ✶ ♃
24 02:38 ☽ ✶ ♀
05:35 ☽ ✶ ♇
12:06 ☽ △ ♄
19:49 ☽ △ ♅
25 17:59 ☽ □ ♆
17:59 ♀ ☌ ♇
18:55 ☽ ∥ ♅
21:59 ☽ ✶ ☿
22:26 ☽ ∥ ♇
26 00:09 ☽ □ ♃
05:26 ☉ ✶ ♂
15:59 ☽ # ♃
16:40 ☽ ∥ ♆
22:20 ☽ ∥ ☉
23:15 ☽ □ ♅
27 06:57 ☽ ∥ ♀
09:54 ☽ ∥ ☿
12:17 ☽ ∥ ♂
15:03 ☽ ☌ ♂
16:01 ☽ ✶ ☉
19:21 ☽ # ♄
21:38 ☽ ✶ ♆
28 03:17 ☽ △ ♃
12:36 ☽ ☌ ♇
18:14 ☽ ☌ ♀
18:44 ☽ ☍ ♄
29 00:13 ♀ ☍ ♄
03:27 ☽ ✶ ♅
07:03 ♃ # ♆
30 10:35 ☽ ☌ ☿
19:38 ☉ # ♃
23:35 ☉ ☌ ♆
31 02:50 ☉ ∥ ♆

⚷ Chiron

01 Dec.	16 S 57
03	10♑57
06	11 16
09	11 34
12	11 52
15	12 10
18	12 27
21	12 45
24	13 02
27	13 19
30	13 36

February 2003

Day	S. T.	☉	☽	☿	♀	♂	♃	♄	♅	♆	♇	☊ True
	h m s	° ' "	° ' "	° '	° '	° '	° '	° '	° '	° '	° '	° '
01 Sa	08 43 09	11♒41 49	06♒13 28	16♑37	26♐00	09♐33	13♌R17	22♊R33	27♒50	10♒42	19♐16	06♊R29
02 Su	08 47 06	12 42 45	19 19 26	17 29	27 07	10 12	13 09	22 31	27 54	10 44	19 17	06 18
03 Mo	08 51 02	13 43 39	02♓10 15	18 25	28 15	10 50	13 01	22 29	27 57	10 47	19 19	06 06
04 Tu	08 54 59	14 44 33	14 45 17	19 24	29 22	11 29	12 53	22 27	28 01	10 49	19 20	05 54
05 We	08 58 55	15 45 25	27 05 06	20 26	00♑30	12 07	12 45	22 25	28 04	10 51	19 22	05 44
06 Th	09 02 52	16 46 15	09♈11 29	21 30	01 38	12 46	12 37	22 23	28 07	10 54	19 23	05 37
07 Fr	09 06 49	17 47 05	21 07 30	22 37	02 45	13 25	12 29	22 21	28 11	10 56	19 25	05 32
08 Sa	09 10 45	18 47 52	02♉57 12	23 46	03 54	14 03	12 21	22 20	28 14	10 58	19 26	05 29
09 Su	09 14 42	19 48 39	14 45 29	24 57	05 02	14 42	12 13	22 18	28 18	11 00	19 28	05D 29
10 Mo	09 18 38	20 49 24	26 37 43	26 09	06 10	15 20	12 06	22 17	28 21	11 03	19 29	05 30
11 Tu	09 22 35	21 50 07	08♊39 26	27 23	07 19	15 59	11 58	22 15	28 25	11 05	19 30	05R 30
12 We	09 26 31	22 50 49	20 56 03	28 39	08 27	16 37	11 50	22 14	28 28	11 07	19 31	05 29
13 Th	09 30 28	23 51 29	03♋32 20	29 57	09 36	17 16	11 42	22 13	28 31	11 09	19 33	05 25
14 Fr	09 34 24	24 52 07	16 31 59	01♒16	10 45	17 55	11 35	22 12	28 35	11 12	19 34	05 19
15 Sa	09 38 21	25 52 44	29 56 54	02 36	11 54	18 33	11 27	22 11	28 38	11 14	19 35	05 10
16 Su	09 42 18	26 53 20	13♌46 43	03 57	13 03	19 12	11 20	22 10	28 42	11 16	19 36	04 59
17 Mo	09 46 14	27 53 53	27 58 22	05 20	14 13	19 50	11 12	22 10	28 45	11 18	19 37	04 47
18 Tu	09 50 11	28 54 26	12♍26 23	06 44	15 22	20 29	11 05	22 09	28 49	11 20	19 39	04 34
19 We	09 54 07	29 54 56	27 03 44	08 09	16 31	21 07	10 58	22 09	28 52	11 23	19 40	04 23
20 Th	09 58 04	00♓55 26	11♎42 50	09 35	17 41	21 45	10 51	22 08	28 56	11 25	19 41	04 14
21 Fr	10 02 00	01 55 54	26 17 00	11 02	18 51	22 24	10 44	22 08	28 59	11 27	19 42	04 07
22 Sa	10 05 57	02 56 20	10♏41 13	12 31	20 01	23 02	10 37	22 08	29 03	11 29	19 43	04 03
23 Su	10 09 53	03 56 46	24 52 29	14 00	21 11	23 41	10 30	22 D 08	29 06	11 31	19 44	04 01
24 Mo	10 13 50	04 57 10	08♐49 43	15 31	22 21	24 19	10 23	22 08	29 10	11 33	19 45	04 00
25 Tu	10 17 47	05 57 33	22 33 09	17 02	23 31	24 58	10 16	22 09	29 13	11 35	19 45	03 58
26 We	10 21 43	06 57 54	06♑03 42	18 35	24 41	25 36	10 10	22 09	29 16	11 38	19 46	03 56
27 Th	10 25 40	07 58 14	19 22 20	20 08	25 51	26 14	10 04	22 09	29 20	11 40	19 47	03 51
28 Fr	10 29 36	08 58 33	02♒29 44	21 43	27 02	26 53	09 57	22 10	29 23	11 42	19 48	03 43

Data for	02-01-2003
Julian Day	2452671.50
Ayanamsa	23 53 46
SVP	05 ♓ 12 46
☽ ☊ Mean	05 ♊ 23 R

● ◐ PHASES ○ ◑

01	10:48	●	12♒09
09	11:12	◐	20♉17
16	23:52	○	27♌54
23	16:46	◑	04♐39

LAST ASPECT Day	h m	☽ INGRESS Day	h m	
02	16:02	02	19:55	♓
04	14:53	05	05:44	♈
07	14:22	07	18:00	♉
10	03:29	10	06:46	♊
12	14:30	12	17:20	♋
13	12:23	15	00:05	♌
17	01:19	17	03:23	♍
18	15:57	19	04:49	♎
21	04:30	21	06:10	♏
23	07:15	23	08:46	♐
25	11:51	25	13:11	♑
27	12:58	27	19:25	♒

DECLINATION

Day	☉	☽	☿	♀	♂	♃	♄	♅	♆	♇
01 Sa	17S 17	22S 56	21S 05	20S 38	21S 30	17N39	22N02	12S 53	17S 31	13S 47
02 Su	17 00	19 36	21 09	20 43	21 37	17 42	22 02	12 52	17 31	13 47
03 Mo	16 42	15 24	21 12	20 47	21 43	17 44	22 02	12 51	17 30	13 47
04 Tu	16 25	10 36	21 14	20 51	21 49	17 46	22 03	12 50	17 29	13 47
05 We	16 07	05 29	21 15	20 55	21 55	17 49	22 03	12 49	17 29	13 47
06 Th	15 49	00 14	21 15	20 58	22 01	17 51	22 03	12 48	17 28	13 47
07 Fr	15 30	04N57	21 14	21 00	22 06	17 54	22 03	12 46	17 27	13 47
08 Sa	15 11	09 56	21 12	21 02	22 12	17 56	22 03	12 45	17 27	13 47
09 Su	14 52	14 33	21 08	21 04	22 17	17 58	22 03	12 44	17 26	13 47
10 Mo	14 33	18 38	21 04	21 04	22 22	18 00	22 03	12 43	17 26	13 47
11 Tu	14 14	22 01	20 58	21 05	22 27	18 03	22 03	12 42	17 25	13 47
12 We	13 54	24 29	20 51	21 05	22 32	18 05	22 04	12 40	17 24	13 47
13 Th	13 34	25 47	20 43	21 04	22 36	18 07	22 04	12 39	17 24	13 47
14 Fr	13 14	25 43	20 33	21 02	22 41	18 09	22 04	12 38	17 23	13 47
15 Sa	12 54	24 11	20 23	21 01	22 45	18 12	22 04	12 37	17 23	13 46
16 Su	12 33	21 10	20 11	20 58	22 49	18 14	22 04	12 36	17 22	13 46
17 Mo	12 12	16 51	19 58	20 55	22 53	18 16	22 04	12 34	17 21	13 46
18 Tu	11 51	11 29	19 43	20 51	22 56	18 18	22 05	12 33	17 21	13 46
19 We	11 30	05 25	19 28	20 47	23 00	18 20	22 05	12 32	17 20	13 46
20 Th	11 09	00S 58	19 11	20 42	23 03	18 22	22 05	12 31	17 20	13 46
21 Fr	10 47	07 16	18 52	20 37	23 07	18 24	22 05	12 30	17 19	13 46
22 Sa	10 26	13 07	18 33	20 31	23 10	18 26	22 05	12 28	17 18	13 46
23 Su	10 04	18 12	18 12	20 25	23 12	18 28	22 06	12 27	17 18	13 45
24 Mo	09 42	22 11	17 50	20 18	23 15	18 30	22 06	12 26	17 17	13 45
25 Tu	09 20	24 49	17 26	20 10	23 18	18 31	22 06	12 25	17 17	13 45
26 We	08 57	25 58	17 02	20 02	23 20	18 33	22 06	12 24	17 16	13 45
27 Th	08 35	25 35	16 36	19 53	23 22	18 35	22 06	12 22	17 16	13 45
28 Fr	08 12	23 46	16 08	19 44	23 24	18 37	22 07	12 21	17 15	13 45

ASPECTARIAN

01 06:22 ☽ ✶ ♂
07:15 ☽ # ♄
08:11 ☽ ☌ ♆
10:51 ☽ ∥ ♂
12:45 ☽ ☍ ♃
13:54 ☽ ∥ ☿
16:54 ☽ ∥ ♀
23:56 ☽ ✶ ♇

02 05:54 ☽ △ ♄
09:12 ☉ ☍ ♃
11:17 ☽ # ♃
12:28 ☽ ∥ ♆
15:54 ☽ ✶ ♀
16:02 ☽ ☌ ♅
16:25 ☽ ∥ ☉
17:27 ♀ ✶ ♅
21:40 ♂ ✶ ♆

03 08:16 ☽ ∥ ♇
13:01 ☽ ∥ ♅
17:22 ☽ □ ♂

04 08:53 ☽ □ ♇
09:48 ☽ ✶ ☿
13:27 ♀ ♑
14:53 ☽ □ ♄

05 07:25 ☽ □ ♀
19:26 ♂ △ ♃

06 03:25 ☽ ✶ ♆
06:47 ☽ △ ♃
07:34 ☽ △ ♂
08:45 ♂ # ♄
16:38 ☽ ✶ ☉
20:32 ☽ △ ♇

07 02:29 ☽ ✶ ♄
03:20 ☽ □ ☿
14:22 ☽ ✶ ♅

08 02:07 ☽ △ ♀
14:20 ☽ # ♅
15:30 ☉ ✶ ♇
16:21 ☽ □ ♆
18:55 ☽ □ ♃
19:51 ☽ # ♇

09 01:40 ☽ # ☉
16:31 ☽ # ♆
19:59 ☽ ∥ ♃
20:47 ☿ ∥ ♀
22:56 ☽ △ ☿

10 03:29 ☽ □ ♅
16:05 ☽ # ☿
16:39 ☽ # ♀

11 00:16 ☽ ∥ ♄
03:37 ☽ # ♂
04:48 ☽ △ ♆
06:27 ☽ ✶ ♃
09:46 ☉ △ ♄
15:11 ☽ ☍ ♂
21:16 ☽ ☍ ♇

12 02:30 ☽ ☌ ♄
04:01 ☽ △ ☉
08:52 ☉ ∥ ♇
14:30 ☽ △ ♅

13 01:00 ☿ ♒
12:23 ☽ ☍ ♀

15 05:09 ☽ ☍ ☿
12:28 ☽ # ♂
17:49 ☽ ∥ ♄
19:41 ☽ ☍ ♇
19:50 ☽ ☌ ♃
20:50 ☉ ∥ ♅

16 01:15 ☽ # ♀
06:25 ☽ # ☿
09:12 ♃ ☍ ♇
09:39 ☽ △ ♂
09:56 ☽ △ ♇
14:15 ☽ ✶ ♄
15:58 ♂ ☌ ♇
16:50 ☽ ∥ ♃
21:25 ☽ # ♆

17 01:19 ☽ ☍ ♅
14:11 ☽ # ♇
19:26 ☽ # ♅
21:38 ☉ ☌ ♅
22:18 ☽ # ☉

18 05:14 ☽ △ ♀
11:51 ☽ □ ♇
13:49 ☽ □ ♂
15:57 ☽ □ ♄

19 02:00 ☉ ♓
20:08 ☽ △ ☿
22:35 ☽ ✶ ♃
23:30 ☽ △ ♆

20 10:40 ☽ □ ♀
13:06 ☽ ✶ ♇
14:14 ♂ ☍ ♄
17:09 ☽ △ ♄
17:17 ☽ ✶ ♂
19:13 ☿ ☍ ♃

21 04:30 ☽ △ ♅
06:51 ☿ ☌ ♆
10:05 ☽ △ ☉
13:18 ☽ ∥ ☉
21:12 ☽ ∥ ♅
23:52 ☽ □ ♃

22 01:21 ☽ □ ♆
02:48 ☽ ∥ ♇
03:25 ☽ □ ☿
07:31 ☿ # ♃
07:42 ♄ SD
17:08 ☽ ✶ ♀
19:23 ☽ ∥ ♆
23:59 ☽ ∥ ☿

23 01:23 ☽ # ♃
07:15 ☽ □ ♅
12:03 ☽ ∥ ♀
23:22 ☽ # ♄

24 02:41 ☽ △ ♃
04:45 ☽ ✶ ♆
08:21 ☽ ∥ ♂
13:05 ☽ ✶ ☿
19:05 ☽ ☌ ♇
23:17 ☽ ☍ ♄

25 04:28 ☽ ☌ ♂
09:42 ☿ ∥ ♇
11:51 ☽ ✶ ♅

26 01:45 ☽ ✶ ☉
18:33 ☿ ✶ ♇

27 12:58 ☽ ☌ ♀

28 03:28 ☽ ∥ ♂
06:51 ☿ △ ♄
13:41 ☽ ☍ ♃
14:12 ☽ # ♄
17:05 ☽ ☌ ♆

⚷ Chiron

01	Dec.	16 S 36
02	13♑53	
05	14 09	
08	14 25	
11	14 41	
14	14 56	
17	15 11	
20	15 25	
23	15 39	
26	15 52	

March 2003

Day	S. T. h m s	☉ ° ′ ″	☽ ° ′ ″	☿ ° ′	♀ ° ′	♂ ° ′	♃ ° ′	♄ ° ′	♅ ° ′	♆ ° ′	♇ ° ′	☊ True ° ′
01 Sa	10 33 33	09♓58 49	15♒26 07	23♒19	28♑12	27♐31	09♌R51	22♊11	29♒27	11♒44	19♐49	03♊R33
02 Su	10 37 29	10 59 05	28 11 21	24 55	29 23	28 09	09 45	22 11	29 30	11 46	19 49	03 21
03 Mo	10 41 26	11 59 18	10♓45 07	26 33	00♒33	28 48	09 40	22 12	29 34	11 48	19 50	03 08
04 Tu	10 45 22	12 59 30	23 07 19	28 12	01 44	29 26	09 34	22 13	29 37	11 50	19 51	02 54
05 We	10 49 19	13 59 39	05♈18 17	29 51	02 55	00♑04	09 28	22 15	29 40	11 52	19 51	02 43
06 Th	10 53 16	14 59 47	17 19 12	01♓32	04 06	00 43	09 23	22 16	29 44	11 54	19 52	02 33
07 Fr	10 57 12	15 59 53	29 12 08	03 14	05 17	01 21	09 18	22 17	29 47	11 56	19 53	02 27
08 Sa	11 01 09	16 59 57	11♉00 06	04 57	06 27	01 59	09 13	22 19	29 50	11 58	19 53	02 23
09 Su	11 05 05	17 59 59	22 47 05	06 41	07 38	02 37	09 08	22 20	29 54	12 00	19 54	02 22
10 Mo	11 09 02	18 59 58	04♊37 47	08 26	08 50	03 15	09 03	22 22	29 57	12 02	19 54	02D 23
11 Tu	11 12 58	19 59 56	16 37 26	10 13	10 01	03 53	08 59	22 24	00♓00	12 04	19 54	02 23
12 We	11 16 55	20 59 51	28 51 31	12 00	11 12	04 31	08 54	22 26	00 04	12 06	19 55	02R 23
13 Th	11 20 51	21 59 44	11♋25 22	13 49	12 23	05 10	08 50	22 28	00 07	12 07	19 55	02 21
14 Fr	11 24 48	22 59 35	24 23 39	15 39	13 34	05 48	08 46	22 30	00 10	12 09	19 56	02 16
15 Sa	11 28 45	23 59 24	07♌49 41	17 30	14 46	06 26	08 42	22 32	00 14	12 11	19 56	02 09
16 Su	11 32 41	24 59 10	21 44 32	19 22	15 57	07 04	08 38	22 34	00 17	12 13	19 56	02 00
17 Mo	11 36 38	25 58 54	06♍06 24	21 15	17 09	07 42	08 35	22 37	00 20	12 15	19 56	01 49
18 Tu	11 40 34	26 58 37	20 50 15	23 10	18 20	08 19	08 32	22 39	00 23	12 16	19 57	01 39
19 We	11 44 31	27 58 17	05♎48 17	25 05	19 32	08 57	08 28	22 42	00 26	12 18	19 57	01 29
20 Th	11 48 27	28 57 55	20 51 07	27 02	20 43	09 35	08 25	22 45	00 29	12 20	19 57	01 21
21 Fr	11 52 24	29 57 31	05♏49 24	28 59	21 55	10 13	08 23	22 48	00 33	12 22	19 57	01 15
22 Sa	11 56 20	00♈57 06	20 35 14	00♈58	23 06	10 51	08 20	22 51	00 36	12 23	19 57	01 11
23 Su	12 00 17	01 56 38	05♐03 13	02 58	24 18	11 29	08 18	22 54	00 39	12 25	19 57	01 10
24 Mo	12 04 13	02 56 09	19 10 37	04 58	25 30	12 06	08 16	22 57	00 42	12 27	19R 57	01D 10
25 Tu	12 08 10	03 55 39	02♑57 01	06 59	26 42	12 44	08 13	23 00	00 45	12 28	19 57	01R 10
26 We	12 12 07	04 55 07	16 23 33	09 01	27 53	13 22	08 12	23 04	00 48	12 30	19 57	01 09
27 Th	12 16 03	05 54 32	29 32 11	11 03	29 05	14 00	08 10	23 07	00 51	12 31	19 57	01 07
28 Fr	12 19 60	06 53 57	12♒25 09	13 05	00♓17	14 37	08 09	23 11	00 54	12 33	19 57	01 03
29 Sa	12 23 56	07 53 19	25 04 34	15 07	01 29	15 15	08 07	23 14	00 57	12 34	19 56	00 56
30 Su	12 27 53	08 52 39	07♓32 14	17 09	02 41	15 52	08 06	23 18	01 00	12 36	19 56	00 47
31 Mo	12 31 49	09 51 57	19 49 39	19 10	03 53	16 30	08 05	23 22	01 03	12 37	19 56	00 38

Data for	03-01-2003
Julian Day	2452699.50
Ayanamsa	23 53 49
SVP	05 ♓ 12 42
☽ ☊ Mean	03 ♊ 54 R

● ◐ PHASES ○ ◑

03	02:35	●	12♓06
11	07:16	◐	20♊18
18	10:34	○	27♍25
25	01:51	◑	04♑00

LAST ASPECT Day	h m	☽ INGRESS Day	h m	
02	02:30	02	03:26	♓
04	13:05	04	13:30	♈
07	01:11	07	01:37	♉
09	14:30	09	14:39	♊
11	11:25	12	02:13	♋
13	21:14	14	10:07	♌
16	01:25	16	13:53	♍
18	10:35	18	14:43	♎
20	03:02	20	14:38	♏
22	04:31	22	15:33	♐
24	11:59	24	18:48	♑
25	18:16	27	00:51	♒
28	20:28	29	09:26	♓
31	07:00	31	20:05	♈

DECLINATION

Day	☉	☽	☿	♀	♂	♃	♄	♅	♆	♇
01 Sa	07S 50	20S 44	15S 40	19S 34	23S 26	18N38	22N07	12S 20	17S 14	13S 45
02 Su	07 27	16 46	15 10	19 24	23 27	18 40	22 07	12 19	17 14	13 44
03 Mo	07 04	12 07	14 39	19 13	23 29	18 41	22 07	12 18	17 13	13 44
04 Tu	06 41	07 03	14 06	19 01	23 30	18 43	22 08	12 16	17 13	13 44
05 We	06 18	01 47	13 32	18 49	23 31	18 44	22 08	12 15	17 12	13 44
06 Th	05 55	03N29	12 57	18 36	23 32	18 46	22 08	12 14	17 12	13 44
07 Fr	05 31	08 35	12 21	18 23	23 33	18 47	22 09	12 13	17 11	13 44
08 Sa	05 08	13 22	11 43	18 10	23 34	18 49	22 09	12 12	17 11	13 43
09 Su	04 45	17 39	11 05	17 56	23 34	18 50	22 09	12 11	17 10	13 43
10 Mo	04 21	21 16	10 24	17 41	23 34	18 51	22 09	12 09	17 10	13 43
11 Tu	03 58	24 01	09 43	17 26	23 34	18 52	22 10	12 08	17 09	13 43
12 We	03 34	25 42	09 01	17 10	23 34	18 53	22 10	12 07	17 09	13 43
13 Th	03 11	26 08	08 17	16 54	23 34	18 54	22 10	12 06	17 08	13 42
14 Fr	02 47	25 10	07 32	16 37	23 34	18 56	22 11	12 05	17 08	13 42
15 Sa	02 23	22 46	06 45	16 20	23 33	18 56	22 11	12 04	17 07	13 42
16 Su	02 00	18 58	05 58	16 03	23 32	18 57	22 11	12 03	17 07	13 42
17 Mo	01 36	13 57	05 10	15 45	23 31	18 58	22 12	12 01	17 06	13 42
18 Tu	01 12	08 01	04 20	15 26	23 30	18 59	22 12	12 00	17 06	13 41
19 We	00 48	01 32	03 29	15 07	23 29	19 00	22 12	11 59	17 05	13 41
20 Th	00 25	05S 05	02 38	14 48	23 28	19 01	22 13	11 58	17 05	13 41
21 Fr	00 01	11 24	01 45	14 28	23 26	19 01	22 13	11 57	17 04	13 41
22 Sa	00N23	16 59	00 51	14 08	23 24	19 02	22 14	11 56	17 04	13 41
23 Su	00 46	21 28	00N03	13 47	23 22	19 03	22 14	11 55	17 03	13 40
24 Mo	01 10	24 34	00 58	13 26	23 20	19 03	22 14	11 54	17 03	13 40
25 Tu	01 34	26 05	01 54	13 05	23 18	19 04	22 15	11 53	17 02	13 40
26 We	01 57	26 01	02 50	12 43	23 16	19 04	22 15	11 52	17 02	13 40
27 Th	02 21	24 29	03 47	12 21	23 13	19 04	22 15	11 51	17 01	13 40
28 Fr	02 44	21 42	04 43	11 59	23 11	19 05	22 16	11 50	17 01	13 39
29 Sa	03 08	17 56	05 40	11 36	23 08	19 05	22 16	11 49	17 01	13 39
30 Su	03 31	13 26	06 37	11 13	23 05	19 05	22 16	11 48	17 00	13 39
31 Mo	03 55	08 27	07 33	10 49	23 02	19 05	22 17	11 47	17 00	13 39

⚷ Chiron

01 Dec. 16 S 11

01	16♑05
04	16 18
07	16 29
10	16 41
13	16 51
16	17 01
19	17 11
22	17 19
25	17 27
28	17 35
31	17 42

ASPECTARIAN

01 08:01 ☽ ∥ ♀
08:12 ☽ ⚹ ♇
12:39 ☽ △ ♄
13:15 ☽ # ♃
16:55 ☽ ☌ ☿
21:26 ☽ ∥ ♆
23:56 ☽ ⚹ ♂
02 02:30 ☽ ☌ ♅
09:41 ☽ ∥ ☿
12:40 ♀ ♒
15:56 ☽ ∥ ♇
23:08 ☽ ∥ ♅
03 17:36 ☽ □ ♇
22:15 ☽ □ ♄
04 01:48 ☽ ∥ ☉
07:30 ♂ ⚹ ♅
13:05 ☽ □ ♂
15:49 ☿ ∥ ♇
18:45 ☽ ⚹ ♀
21:17 ☿ ☌ ♅
21:18 ♂ ♑
05 02:04 ☿ ♓
05:01 ☿ ⚹ ♂
08:05 ♀ # ♃
08:14 ☽ △ ♃
13:06 ☽ ⚹ ♆
06 05:08 ☽ △ ♇
09:59 ☽ ⚹ ♄
10:25 ☽ # ☉
07 01:11 ☽ ⚹ ♅
04:36 ☽ △ ♂
05:26 ☿ ∥ ♅
09:35 ☽ ⚹ ☿
13:43 ☽ □ ♀
16:29 ☽ # ☿
17:54 ☽ # ♅
20:23 ☽ □ ♃
08 01:51 ☽ # ♇
01:58 ☽ □ ♆
13:21 ☽ ⚹ ☉
21:06 ☽ # ♆
09 01:33 ☽ # ♀
07:20 ☽ ∥ ♃
14:30 ☽ □ ♅
10 04:18 ♀ ☍ ♃
06:57 ☽ ∥ ♄
08:50 ☽ ⚹ ♃
09:00 ☽ □ ☿
09:22 ☽ △ ♀
14:54 ☽ △ ♆
19:29 ☽ # ♂
20:54 ♅ ♓
21:50 ☉ □ ♇
11 06:30 ☽ ☍ ♇
11:25 ☽ ☌ ♄
12 02:20 ♀ ∥ ♆
02:20 ☽ △ ♅
11:29 ☽ ☍ ♂
18:35 ♀ ☌ ♆
13 05:14 ☽ △ ☿
11:38 ☉ □ ♄
21:14 ☽ △ ☉
14 17:33 ☽ # ♂
15 01:31 ☽ ☌ ♃
04:15 ☽ ∥ ♄
07:37 ☽ ☍ ♆
13:11 ☽ ☍ ♀
20:56 ☽ △ ♇
16 00:01 ☽ ∥ ♃
01:25 ☽ ⚹ ♄
07:18 ☿ □ ♇
09:30 ☽ # ♆
14:24 ☽ ☍ ♅
15:29 ☽ # ♀
17 01:06 ☽ # ♇
02:44 ☽ △ ♂
08:11 ☽ # ♅
17:34 ☿ □ ♄
22:33 ☽ □ ♇
18 02:57 ☽ □ ♄
04:18 ☽ ☍ ☿
15:51 ☽ # ☿
19 02:49 ☽ # ☉
04:15 ☽ ⚹ ♃
05:15 ☽ □ ♂
08:01 ☽ ∥ ☉
08:29 ♀ ⚹ ♇
10:23 ☽ △ ♆
16:06 ☽ ∥ ☿
22:33 ☽ ⚹ ♇
23:46 ☽ △ ♀
20 03:02 ☽ △ ♄
15:29 ☽ △ ♅
21 01:00 ☉ ♈
02:14 ☽ ∥ ♅
04:07 ☽ □ ♃
07:25 ☽ ⚹ ♂
09:21 ☽ ∥ ♇
10:36 ☽ □ ♆
11:59 ☽ ∥ ♀
12:16 ☿ ♈
18:31 ♀ △ ♄
23:35 ☿ ☌ ☉
22 00:22 ☽ ∥ ♆
04:31 ☽ □ ♀
08:51 ☿ # ☉
10:10 ☽ # ♃
16:36 ☽ □ ♅
18:24 ☽ △ ☉
19:56 ☽ △ ☿
23 04:56 ☽ # ♄
05:12 ♇ SR
05:26 ☽ △ ♃
07:53 ♀ ∥ ♇
12:27 ☽ ⚹ ♆
13:08 ☽ ∥ ♂
24 01:20 ☽ ☌ ♇
06:32 ☽ ☍ ♄
09:00 ☿ ∥ ☉
11:59 ☽ ⚹ ♀
20:07 ☽ ⚹ ♅
25 08:24 ☽ □ ☿
14:27 ☿ △ ♃
18:16 ☽ ☌ ♂
27 12:24 ☽ ∥ ♂
12:48 ☽ ⚹ ☉
16:00 ☽ ☍ ♃
17:36 ☿ ⚹ ♆
18:14 ♀ ♓
19:56 ☽ # ♄
28 00:14 ☽ ☌ ♆
01:29 ☽ ⚹ ☿
10:03 ♀ ∥ ♅
12:44 ♀ ☌ ♅
14:13 ☽ ⚹ ♇
17:15 ☽ # ♃
20:28 ☽ △ ♄
29 02:12 ☿ □ ♂
05:13 ☽ ∥ ♆
05:32 ☉ △ ♃
11:18 ☽ ☌ ♅
13:37 ☽ ☌ ♀
22:57 ☽ ∥ ♇
30 08:10 ☽ ∥ ♅
11:53 ☽ ∥ ♀
17:07 ☽ ⚹ ♂
31 00:12 ☽ □ ♇
03:35 ☽ # ☿
07:00 ☽ □ ♄
09:12 ☿ △ ♇
19:28 ☽ # ☉

April 2003

Day	S. T.	☉	☽	☿	♀	♂	♃	♄	♅	♆	♇	☊ True
	h m s	° ' "	° ' "	° '	° '	° '	° '	° '	° '	° '	° '	° '
01 Tu	12 35 46	10♈51 14	01♈58 03	21♈10	05♓05	17♑07	08♌R05	23♊26	01♓06	12♒39	19♐R56	00♊R29
02 We	12 39 42	11 50 28	13 58 38	23 08	06 17	17 45	08 04	23 30	01 08	12 40	19 55	00 20
03 Th	12 43 39	12 49 41	25 52 43	25 05	07 29	18 22	08 04	23 34	01 11	12 41	19 55	00 14
04 Fr	12 47 36	13 48 51	07♉42 04	27 00	08 42	18 59	08 04	23 38	01 14	12 43	19 55	00 10
05 Sa	12 51 32	14 47 59	19 29 00	28 53	09 54	19 36	08D 04	23 42	01 17	12 44	19 54	00 08
06 Su	12 55 29	15 47 05	01♊16 29	00♉42	11 06	20 14	08 04	23 47	01 19	12 45	19 54	00D 08
07 Mo	12 59 25	16 46 09	13 08 10	02 29	12 18	20 51	08 04	23 51	01 22	12 47	19 53	00 10
08 Tu	13 03 22	17 45 10	25 08 18	04 12	13 30	21 28	08 05	23 56	01 25	12 48	19 53	00 12
09 We	13 07 18	18 44 10	07♋21 32	05 50	14 42	22 05	08 06	24 00	01 27	12 49	19 52	00 14
10 Th	13 11 15	19 43 07	19 52 44	07 25	15 55	22 42	08 07	24 05	01 30	12 50	19 52	00R 15
11 Fr	13 15 11	20 42 02	02♌46 32	08 55	17 07	23 19	08 08	24 10	01 32	12 51	19 51	00 14
12 Sa	13 19 08	21 40 54	16 06 49	10 21	18 19	23 55	08 10	24 15	01 35	12 52	19 51	00 11
13 Su	13 23 05	22 39 44	29 55 50	11 41	19 32	24 32	08 11	24 20	01 37	12 53	19 50	00 06
14 Mo	13 27 01	23 38 32	14♍13 30	12 56	20 44	25 09	08 13	24 25	01 40	12 54	19 49	00 00
15 Tu	13 30 58	24 37 17	28 56 37	14 06	21 56	25 46	08 15	24 30	01 42	12 55	19 49	29♉54
16 We	13 34 54	25 36 01	13♎58 54	15 10	23 09	26 22	08 17	24 35	01 45	12 56	19 48	29 48
17 Th	13 38 51	26 34 42	29 11 27	16 09	24 21	26 59	08 19	24 40	01 47	12 57	19 47	29 44
18 Fr	13 42 47	27 33 22	14♏24 14	17 02	25 33	27 35	08 22	24 46	01 49	12 58	19 46	29 41
19 Sa	13 46 44	28 31 59	29 27 30	17 49	26 46	28 12	08 24	24 51	01 52	12 59	19 45	29 39
20 Su	13 50 40	29 30 35	14♐13 22	18 30	27 58	28 48	08 27	24 56	01 54	13 00	19 45	29D 39
21 Mo	13 54 37	00♉29 10	28 36 36	19 05	29 11	29 24	08 30	25 02	01 56	13 01	19 44	29 40
22 Tu	13 58 34	01 27 42	12♑34 47	19 35	00♈23	00♒00	08 33	25 08	01 58	13 02	19 43	29 42
23 We	14 02 30	02 26 13	26 07 54	19 58	01 36	00 36	08 37	25 13	02 00	13 03	19 42	29 43
24 Th	14 06 27	03 24 43	09♒17 37	20 15	02 48	01 12	08 40	25 19	02 02	13 03	19 41	29R 43
25 Fr	14 10 23	04 23 11	22 06 41	20 27	04 01	01 48	08 44	25 25	02 04	13 04	19 40	29 42
26 Sa	14 14 20	05 21 37	04♓38 18	20 33	05 13	02 24	08 48	25 31	02 06	13 05	19 39	29 40
27 Su	14 18 16	06 20 01	16 55 46	20R 33	06 26	03 00	08 52	25 37	02 08	13 05	19 38	29 38
28 Mo	14 22 13	07 18 24	29 02 14	20 27	07 39	03 36	08 56	25 43	02 10	13 06	19 37	29 34
29 Tu	14 26 09	08 16 45	11♈00 28	20 16	08 51	04 11	09 01	25 49	02 12	13 06	19 36	29 31
30 We	14 30 06	09 15 05	22 52 58	20 01	10 04	04 47	09 05	25 55	02 14	13 07	19 35	29 28

Data for	04-01-2003
Julian Day	2452730.50
Ayanamsa	23 53 52
SVP	05 ♓ 12 36
☽ ☊ Mean	02 ♊ 16 R

● ◐ PHASES ○ ◑

Day	Time	Phase	Position
01	19:19	●	11♈39
09	23:40	◐	19♋42
16	19:36	○	26♎24
23	12:19	◑	02♒56

LAST ASPECT ☽ INGRESS

Day	h m	Day	h m	Sign
02	22:06	03	08:21	♉
05	00:16	05	21:25	♊
07	21:35	08	09:36	♋
10	05:34	10	18:54	♌
12	14:19	13	00:07	♍
14	18:38	15	01:42	♎
16	20:22	17	01:16	♏
18	21:53	19	00:52	♐
21	01:03	21	02:21	♑
22	12:41	23	06:59	♒
25	06:20	25	15:03	♓
27	17:19	28	01:55	♈
30	06:12	30	14:27	♉

DECLINATION

Day	☉	☽	☿	♀	♂	♃	♄	♅	♆	♇
01 Tu	04N18	03S 13	08N29	10S 26	22S 58	19N06	22N17	11S 46	16S 59	13S 38
02 We	04 41	02N06	09 24	10 02	22 55	19 06	22 18	11 45	16 59	13 38
03 Th	05 04	07 17	10 18	09 37	22 51	19 06	22 18	11 44	16 59	13 38
04 Fr	05 27	12 13	11 10	09 13	22 48	19 06	22 18	11 43	16 58	13 38
05 Sa	05 50	16 41	12 02	08 48	22 44	19 06	22 19	11 42	16 58	13 38
06 Su	06 13	20 31	12 51	08 23	22 40	19 05	22 19	11 41	16 58	13 37
07 Mo	06 35	23 32	13 39	07 57	22 36	19 05	22 20	11 40	16 57	13 37
08 Tu	06 58	25 33	14 24	07 32	22 32	19 05	22 20	11 39	16 57	13 37
09 We	07 20	26 22	15 08	07 06	22 27	19 05	22 20	11 38	16 57	13 37
10 Th	07 43	25 53	15 48	06 40	22 23	19 04	22 21	11 37	16 56	13 36
11 Fr	08 05	24 01	16 27	06 14	22 18	19 04	22 21	11 36	16 56	13 36
12 Sa	08 27	20 48	17 03	05 47	22 13	19 04	22 21	11 35	16 56	13 36
13 Su	08 49	16 20	17 36	05 21	22 08	19 03	22 22	11 34	16 55	13 36
14 Mo	09 11	10 50	18 06	04 54	22 03	19 03	22 22	11 34	16 55	13 36
15 Tu	09 32	04 35	18 33	04 27	21 58	19 02	22 23	11 33	16 55	13 35
16 We	09 54	02S 05	18 58	04 00	21 53	19 01	22 23	11 32	16 55	13 35
17 Th	10 15	08 43	19 20	03 33	21 48	19 01	22 23	11 31	16 54	13 35
18 Fr	10 36	14 51	19 39	03 05	21 42	19 00	22 24	11 30	16 54	13 35
19 Sa	10 57	20 01	19 55	02 38	21 37	18 59	22 24	11 30	16 54	13 34
20 Su	11 18	23 49	20 08	02 10	21 31	18 58	22 25	11 29	16 54	13 34
21 Mo	11 39	25 58	20 18	01 43	21 25	18 58	22 25	11 28	16 53	13 34
22 Tu	11 59	26 23	20 26	01 15	21 19	18 57	22 25	11 27	16 53	13 34
23 We	12 19	25 12	20 30	00 47	21 13	18 56	22 26	11 27	16 53	13 34
24 Th	12 39	22 39	20 32	00 19	21 07	18 55	22 26	11 26	16 53	13 33
25 Fr	12 59	19 02	20 30	00N08	21 01	18 54	22 26	11 25	16 53	13 33
26 Sa	13 19	14 40	20 26	00 36	20 55	18 53	22 27	11 25	16 52	13 33
27 Su	13 38	09 46	20 19	01 04	20 48	18 52	22 27	11 24	16 52	13 33
28 Mo	13 57	04 34	20 09	01 32	20 42	18 50	22 27	11 23	16 52	13 33
29 Tu	14 16	00N44	19 57	02 00	20 36	18 49	22 28	11 23	16 52	13 32
30 We	14 35	05 59	19 42	02 28	20 29	18 48	22 28	11 22	16 52	13 32

ASPECTARIAN

01 12:10 ☽ △ ♃
21:22 ☽ ✶ ♆
02 04:29 ☿ ✶ ♄
07:59 ☽ □ ♂
11:32 ☿ # ♀
11:58 ☽ △ ♇
12:48 ☽ ∥ ☉
19:17 ☽ ✶ ♄
20:33 ☉ ✶ ♆
22:06 ☽ ☌ ☿
03 10:17 ☽ # ♀
10:48 ☽ ✶ ♅
17:37 ☽ ∥ ☿
21:29 ☽ # ♅
04 00:44 ☽ □ ♃
02:15 ☽ ✶ ♀
03:05 ♃ SD
07:19 ☽ # ♇
10:13 ☽ □ ♆
14:45 ☿ # ♅
05 00:16 ☽ △ ♂
01:40 ☽ # ♆
14:33 ☽ ∥ ♃
14:38 ☿ ♉
06 00:06 ☽ □ ♅
08:27 ☿ ✶ ♅
13:29 ☽ ∥ ♄
13:47 ☽ ✶ ♃
15:55 ☽ # ♂
22:08 ☽ □ ♀
23:13 ☿ # ♇
23:16 ☽ △ ♆
07 07:57 ☽ ✶ ☉
13:32 ☽ ☍ ♇
[illegible]
08 12:26 ☽ △ ♅
16:46 ♀ # ☉
20:36 ☽ ✶ ☿
09 15:40 ☽ △ ♀
10 03:31 ☉ △ ♇
05:34 ☽ ☍ ♂
09:35 ♂ # ♄
11:08 ☿ □ ♃
11 09:46 ☽ ☌ ♃
12:31 ☽ □ ☿
13:40 ☽ ∥ ♄
14:23 ☽ # ♂
18:14 ☽ ☍ ♆
19:18 ☿ # ♆
12 06:33 ☽ △ ♇
10:11 ☽ ∥ ♃
10:31 ☽ △ ☉
14:19 ☽ ✶ ♄
18:30 ☽ ∥ ☿
21:13 ☽ # ♆
13 02:54 ☽ ☍ ♅
06:03 ♀ □ ♇
12:31 ☽ # ♇
21:06 ☽ # ♅
21:40 ☽ △ ☿
23:27 ☿ □ ♆
14 06:16 ☽ ∥ ☉
09:11 ☽ □ ♇
11:38 ☽ ☍ ♀
16:46 ☽ □ ♄
[illegible]
20:38 ☉ ✶ ♄
15 00:33 ☽ # ♀
14:56 ☽ ✶ ♃
22:21 ☽ △ ♆
16 03:21 ☿ ∥ ♃
06:28 ☽ ∥ ♀
09:11 ☽ ✶ ♇
16:50 ☽ △ ♄
20:22 ☽ □ ♂
17 04:06 ☽ △ ♅
06:09 ☽ # ☉
06:52 ♀ □ ♄
10:39 ☽ ∥ ♅
14:26 ☽ □ ♃
18:49 ☽ ∥ ♇
21:44 ☽ □ ♆
18 01:55 ☉ □ ♂
04:24 ☽ ☍ ☿
08:55 ☽ ∥ ♆
18:47 ☽ # ♃
19:18 ☽ △ ♀
21:53 ☽ ✶ ♂
23:28 ☽ # ☿
19 03:53 ☽ □ ♅
08:47 ☽ ∥ ♂
13:56 ☽ # ♄
14:31 ☽ △ ♃
21:59 ☽ ✶ ♆
20 09:07 ☽ ☌ ♇
12:03 ☉ ♉
12:11 ☉ # ♅
17:56 ☽ ☍ ♄
21 01:03 ☽ □ ♀
03:25 ☽ △ ☉
05:39 ☽ ✶ ♅
08:50 ♀ ✶ ♂
16:18 ♀ ♈
23:49 ♂ ♒
22 12:41 ☽ △ ☿
12:57 ☉ ✶ ♅
23 08:28 ☽ ☌ ♂
10:53 ☽ ✶ ♀
22:51 ☽ ☍ ♃
24 01:41 ☽ # ♄
06:59 ☽ ☌ ♆
[illegible]
14:55 ☽ # ☿
19:23 ☽ ✶ ♇
20:49 ☽ □ ☿
25 00:52 ☽ # ♃
06:20 ☽ △ ♄
12:18 ☽ ∥ ♆
19:06 ☽ ☌ ♅
26 01:31 ☽ ✶ ☉
05:38 ☽ ∥ ♇
06:24 ☽ # ☉
12:00 ☿ SR
16:11 ☽ ∥ ♅
17:46 ☉ # ♇
27 05:19 ☽ □ ♇
07:06 ☽ ✶ ☿
17:19 ☽ □ ♄
28 09:35 ☽ ✶ ♂
12:38 ☽ # ♀
19:10 ☽ ☌ ♀
19:57 ☽ △ ♃
29 03:19 ♀ △ ♃
04:14 ☽ ✶ ♆
06:19 ☽ ∥ ♀
17:19 ☽ △ ♇
19:32 ☉ □ ♃
30 00:11 [illegible]
19:02 ☽ ✶ ♅

⚷ Chiron

Day	
01	Dec. 15 S 42
03	17♑47
06	17 53
09	17 57
12	18 01
15	18 04
18	18 06
21	18 08
24	18 09
27	18 09R
30	18 08

23 15:57 18♑09 R

16 03:41 24♏53 ✹ Total Lunar Eclipse (mag 1.133)
31 04:09 09♊20 ⭘ Annular Solar Eclipse

May 2003

Day	S. T.	☉	☽	☿	♀	♂	♃	♄	♅	♆	♇	☊ True
	h m s	° ' "	° ' "	° '	° '	° '	° '	° '	° '	° '	° '	° '
01 Th	14 34 03	10♉ 13 22	04♉ 41 55	19♉ R41	11♈ 16	05♒ 22	09♌ 10	26♊ 01	02♓ 16	13♒ 07	19♐ R34	29♉ R27
02 Fr	14 37 59	11 11 38	16 29 27	19 16	12 29	05 57	09 15	26 07	02 17	13 08	19 32	29 26
03 Sa	14 41 56	12 09 53	28 17 44	18 48	13 42	06 32	09 20	26 14	02 19	13 08	19 31	29D 26
04 Su	14 45 52	13 08 05	10♊ 09 07	18 17	14 54	07 07	09 25	26 20	02 21	13 09	19 30	29 27
05 Mo	14 49 49	14 06 16	22 06 14	17 44	16 07	07 42	09 30	26 26	02 22	13 09	19 29	29 29
06 Tu	14 53 45	15 04 25	04♋ 12 05	17 08	17 20	08 17	09 36	26 33	02 24	13 10	19 28	29 31
07 We	14 57 42	16 02 32	16 30 00	16 32	18 32	08 52	09 41	26 40	02 25	13 10	19 26	29 33
08 Th	15 01 38	17 00 37	29 03 33	15 54	19 45	09 26	09 47	26 46	02 27	13 10	19 25	29 34
09 Fr	15 05 35	17 58 40	11♌ 56 20	15 17	20 58	10 01	09 53	26 53	02 28	13 10	19 24	29R 34
10 Sa	15 09 32	18 56 41	25 11 38	14 41	22 10	10 35	09 59	26 59	02 30	13 11	19 23	29 33
11 Su	15 13 28	19 54 40	08♍ 51 52	14 06	23 23	11 09	10 05	27 06	02 31	13 11	19 21	29 32
12 Mo	15 17 25	20 52 38	22 57 54	13 32	24 36	11 43	10 12	27 13	02 32	13 11	19 20	29 31
13 Tu	15 21 21	21 50 33	07♎ 28 26	13 02	25 49	12 17	10 18	27 20	02 34	13 11	19 18	29 29
14 We	15 25 18	22 48 27	22 19 34	12 34	27 01	12 51	10 25	27 27	02 35	13 11	19 17	29 27
15 Th	15 29 14	23 46 19	07♏ 24 50	12 09	28 14	13 24	10 32	27 34	02 36	13 11	19 16	29 26
16 Fr	15 33 11	24 44 09	22 35 42	11 49	29 27	13 58	10 39	27 41	02 37	13R 11	19 14	29 25
17 Sa	15 37 07	25 41 59	07♐ 42 48	11 32	00♉ 40	14 31	10 46	27 48	02 38	13 11	19 13	29 24
18 Su	15 41 04	26 39 46	22 37 12	11 19	01 52	15 04	10 53	27 55	02 39	13 11	19 11	29 24
19 Mo	15 45 01	27 37 33	07♑ 11 41	11 11	03 05	15 37	11 00	28 02	02 40	13 11	19 10	29D 24
20 Tu	15 48 57	28 35 18	21 21 28	11 08	04 18	16 10	11 08	28 09	02 41	13 11	19 09	29 24
21 We	15 52 54	29 33 02	05♒ 04 28	11 D 08	05 31	16 43	11 15	28 16	02 42	13 11	19 07	29 24
22 Th	15 56 50	00♊ 30 45	18 21 00	11 14	06 44	17 15	11 23	28 23	02 43	13 11	19 06	29 24
23 Fr	16 00 47	01 28 27	01♓ 13 13	11 24	07 56	17 48	11 31	28 30	02 43	13 10	19 04	29 25
24 Sa	16 04 43	02 26 08	13 44 27	11 38	09 09	18 20	11 39	28 38	02 44	13 10	19 03	29 25
25 Su	16 08 40	03 23 48	25 58 46	11 57	10 22	18 52	11 47	28 45	02 45	13 10	19 01	29 26
26 Mo	16 12 36	04 21 27	08♈ 00 24	12 20	11 35	19 23	11 55	28 52	02 46	13 10	19 00	29 27
27 Tu	16 16 33	05 19 04	19 53 31	12 48	12 48	19 55	12 03	29 00	02 46	13 09	18 58	29 28
28 We	16 20 30	06 16 41	01♉ 41 55	13 20	14 01	20 26	12 12	29 07	02 47	13 09	18 56	29 30
29 Th	16 24 26	07 14 17	13 28 58	13 55	15 14	20 57	12 20	29 14	02 47	13 08	18 55	29 31
30 Fr	16 28 23	08 11 52	25 17 38	14 35	16 27	21 28	12 29	29 22	02 48	13 08	18 53	29 32
31 Sa	16 32 19	09 09 26	07♊ 10 27	15 19	17 39	21 59	12 38	29 29	02 48	13 08	18 52	29 33

Data for	05-01-2003
Julian Day	2452760.50
Ayanamsa	23 53 56
SVP	05 ♓ 12 32
☽ ☊ Mean	00 ♊ 41 R

● ◐ PHASES ○ ◑

01	12:15	●	10♉43
09	11:53	◐	18♌27
16	03:37	✹	24♏53
23	00:31	◑	01♓30
31	04:20	⭘	09♊20

LAST ASPECT ☽ INGRESS

Day	h m	Day	h m	
02	05:27	03	03:28	♊
05	08:43	05	15:42	♋
07	04:21	08	01:46	♌
10	03:13	10	08:31	♍
12	07:09	12	11:43	♎
14	08:14	14	12:14	♏
16	03:37	16	11:44	♐
18	08:43	18	12:04	♑
20	13:30	20	15:02	♒
22	18:50	22	21:42	♓
25	05:33	25	07:59	♈
27	18:41	27	20:32	♉
29	15:53	30	09:32	♊

DECLINATION

Day	☉	☽	☿	♀	♂	♃	♄	♅	♆	♇
01 Th	14N53	11N01	19N25	02N56	20S22	18N47	22N28	11S21	16S52	13S32
02 Fr	15 11	15 38	19 05	03 24	20 15	18 45	22 29	11 21	16 51	13 32
03 Sa	15 29	19 41	18 44	03 52	20 09	18 44	22 29	11 20	16 51	13 32
04 Su	15 47	22 57	18 21	04 20	20 02	18 42	22 29	11 20	16 51	13 31
05 Mo	16 04	25 15	17 56	04 47	19 55	18 41	22 30	11 19	16 51	13 31
06 Tu	16 21	26 24	17 30	05 15	19 48	18 39	22 30	11 19	16 51	13 31
07 We	16 38	26 16	17 03	05 43	19 41	18 38	22 30	11 18	16 51	13 31
08 Th	16 55	24 49	16 36	06 10	19 34	18 36	22 31	11 18	16 51	13 31
09 Fr	17 11	22 03	16 08	06 37	19 26	18 34	22 31	11 17	16 51	13 31
10 Sa	17 27	18 05	15 41	07 05	19 19	18 33	22 31	11 17	16 51	13 30
11 Su	17 43	13 05	15 14	07 32	19 12	18 31	22 32	11 16	16 51	13 30
12 Mo	17 58	07 15	14 48	07 59	19 05	18 29	22 32	11 16	16 51	13 30
13 Tu	18 14	00 52	14 24	08 25	18 57	18 27	22 32	11 15	16 51	13 30
14 We	18 28	05S44	14 00	08 52	18 50	18 26	22 32	11 15	16 51	13 30
15 Th	18 43	12 07	13 39	09 18	18 43	18 24	22 33	11 15	16 51	13 30
16 Fr	18 57	17 49	13 19	09 45	18 35	18 22	22 33	11 14	16 51	13 29
17 Sa	19 11	22 22	13 01	10 11	18 28	18 20	22 33	11 14	16 51	13 29
18 Su	19 25	25 20	12 46	10 37	18 20	18 18	22 33	11 14	16 51	13 29
19 Mo	19 38	26 30	12 33	11 02	18 13	18 16	22 34	11 13	16 51	13 29
20 Tu	19 51	25 52	12 22	11 28	18 05	18 14	22 34	11 13	16 51	13 29
21 We	20 03	23 41	12 13	11 53	17 58	18 11	22 34	11 13	16 51	13 29
22 Th	20 16	20 16	12 07	12 18	17 50	18 09	22 34	11 12	16 51	13 29
23 Fr	20 27	15 59	12 04	12 42	17 42	18 07	22 35	11 12	16 51	13 29
24 Sa	20 39	11 08	12 02	13 06	17 35	18 05	22 35	11 12	16 51	13 28
25 Su	20 50	05 57	12 03	13 30	17 27	18 02	22 35	11 12	16 51	13 28
26 Mo	21 01	00 39	12 07	13 54	17 20	18 00	22 35	11 12	16 51	13 28
27 Tu	21 11	04N38	12 12	14 17	17 12	17 58	22 35	11 11	16 51	13 28
28 We	21 21	09 44	12 19	14 41	17 05	17 55	22 35	11 11	16 51	13 28
29 Th	21 31	14 29	12 29	15 03	16 58	17 53	22 36	11 11	16 52	13 28
30 Fr	21 40	18 43	12 40	15 26	16 50	17 50	22 36	11 11	16 52	13 28
31 Sa	21 49	22 12	12 53	15 48	16 43	17 48	22 36	11 11	16 52	13 28

⚷ Chiron

01 Dec.	15 S 21
03	18♑07R
06	18 04
09	18 02
12	17 58
15	17 54
18	17 49
21	17 43
24	17 37
27	17 30
30	17 23

ASPECTARIAN

01 01:26 ☽ □ ♂
01:44 ☽ # ♅
09:09 ☽ □ ♃
12:47 ☽ # ♇
17:10 ☽ □ ♆
21:24 ☽ ∥ ☉
02 05:27 ☽ ☌ ☿
06:52 ☽ # ♆
12:59 ♀ ✶ ♆
18:02 ☽ ∥ ♃
18:31 ☽ ∥ ☿
03 00:15 ☿ ∥ ♃
02:58 ☽ # ♂
08:11 ☽ □ ♅
17:34 ☽ △ ♂
20:14 ☽ ∥ ♄
22:30 ☽ ✶ ♃
04 00:18 ☉ □ ♆
06:02 ☽ △ ♆
10:39 ☽ ✶ ♀
18:46 ☽ ☍ ♇
05 08:43 ☽ ☌ ♄
20:26 ☽ △ ♅
06 23:02 ☽ ✶ ☉
07 00:03 ☽ ✶ ☿
04:21 ☽ □ ♀
07:21 ☿ ☌ ☉
10:55 ☿ # ♆
13:37 ☿ ∥ ☉
17:35 ♀ △ ♇
18:07 ☉ # ♆
08 17:37 ♂ ☍ ♃
20:11 ☽ ☌ ♃
20:17 ☽ ☍ ♂
20:38 ☽ ∥ ♄
09 02:16 ☽ ☍ ♆
05:51 ☽ □ ☿
13:34 ☽ △ ♇
17:00 ☽ # ♂
18:03 ☽ △ ♀
21:30 ☽ ∥ ♃
10 03:07 ☽ ∥ ☉
03:13 ☽ ✶ ♄
06:23 ☽ # ♆
12:56 ☽ ☍ ♅
13:14 ☽ ∥ ☿
22:07 ☽ # ♇
11 07:47 ☽ # ♅
08:38 ☽ △ ☿
17:54 ☽ □ ♆
20:14 ☽ △ ☉
21:21 ☽ ∥ ♀
12 07:09 ☽ □ ♄
16:31 ☿ □ ♆
13 04:39 ☽ ✶ ♃
08:08 ☽ △ ♂
09:17 ☽ △ ♆
17:20 ☿ □ ♂
19:08 ☽ ✶ ♆
19:49 ☉ ∥ ♃
14 08:09 ☽ ☍ ♀
08:14 ☽ △ ♄
09:15 ♀ ✶ ♄
12:30 ☽ # ♀
14:37 ♂ ☌ ♆
16:22 ☽ △ ♅
20:40 ☽ ∥ ♅
23:29 ☉ # ♂
15 04:58 ☽ □ ♃
05:31 ☽ ∥ ♇
05:46 ☽ # ☿
07:20 ☽ ☍ ☿
09:08 ☽ □ ♆
09:50 ☽ □ ♂
10:48 ☿ # ♇
19:36 ☽ ∥ ♆
16 00:48 ♆ SR
02:35 ☽ # ♃
03:34 ☽ ∥ ♂
05:42 ☽ # ☉
10:58 ♀ ♉
15:54 ☽ □ ♅
17 01:18 ☽ # ♄
04:55 ☽ △ ♃
08:46 ☽ ✶ ♆
11:19 ☽ ✶ ♂
18:27 ☽ ☌ ♇
18 08:43 ☽ ☍ ♄
10:30 ♂ # ♃
15:40 ♀ ✶ ♅
16:28 ☽ ✶ ♅
16:33 ☽ △ ♀
19 06:39 ☽ △ ☿
10:18 ♀ # ♅
23:46 ☿ □ ♃
20 07:34 ☿ SD
13:30 ☽ △ ☉
21 00:51 ☽ □ ♀
08:40 ☽ # ♄
10:56 ☽ □ ☿
11:11 ☽ ☍ ♃
11:13 ☉ ♊
14:33 ☽ ☌ ♆
15:51 ☿ ∥ ♀
21:54 ☽ ☌ ♂
22 00:03 ☽ # ☉
01:22 ☽ ✶ ♇
12:26 ☽ # ♃
14:32 ☽ ∥ ♂
18:50 ☽ △ ♄
19:26 ☽ ∥ ♆
23 02:51 ☽ ☌ ♅
12:41 ☽ ∥ ♇
14:11 ☽ ✶ ♀
15:13 ☽ # ♀
19:39 ☽ # ☿
19:50 ☽ ✶ ☿
23:41 ☽ ∥ ♅
24 01:26 ☿ □ ♃
07:37 ☉ □ ♅
10:19 ☽ □ ♇
21:56 ♀ # ♇
25 05:33 ☽ □ ♄
06:56 ♂ ✶ ♇
16:03 ☽ ✶ ☉
26 07:29 ♀ □ ♃
07:58 ☽ △ ♃
10:22 ☽ ✶ ♆
22:08 ☽ △ ♇
27 00:03 ☽ ✶ ♂
00:04 ☿ ☌ ♀
07:01 ♀ □ ♆
16:17 ☿ □ ♆
18:41 ☽ ✶ ♄
28 02:12 ☽ ✶ ♅
07:05 ☽ # ♅
13:13 ☽ ∥ ☿
18:39 ☽ # ♇
21:39 ☽ □ ♃
23:18 ☽ □ ♆
29 00:57 ☽ ☌ ☿
03:18 ☽ ∥ ♀
03:59 ☽ ☌ ♀
13:04 ☽ # ♆
13:14 ☽ # ♂
15:53 ☽ □ ♂
18:47 ♂ ∥ ♆
18:49 ☽ ∥ ♃
30 15:11 ☽ □ ♅
20:57 ☽ ∥ ☉
31 03:09 ☽ ∥ ♄
11:05 ☽ ✶ ♃
11:56 ☽ △ ♆
23:21 ☽ ☍ ♇

June 2003

Day	S. T.	☉	☽	☿	♀	♂	♃	♄	♅	♆	♇	☊ True
	h m s	° ' "	° ' "	° '	° '	° '	° '	° '	° '	° '	° '	° '
01 Su	16 36 16	10♊06 58	19♊09 40	16♉06	18♉52	22♒29	12♌47	29♊37	02♓48	13♒R07	18♐R50	29♉R33
02 Mo	16 40 12	11 04 30	01♋17 18	16 57	20 05	22 59	12 56	29 44	02 49	13 07	18 49	29 32
03 Tu	16 44 09	12 02 01	13 35 10	17 51	21 18	23 29	13 05	29 52	02 49	13 06	18 47	29 31
04 We	16 48 05	12 59 30	26 05 07	18 49	22 31	23 59	13 14	30 00	02 49	13 05	18 45	29 29
05 Th	16 52 02	13 56 58	08♌48 57	19 50	23 44	24 28	13 23	00♋07	02 49	13 05	18 44	29 27
06 Fr	16 55 59	14 54 25	21 48 32	20 55	24 57	24 57	13 33	00 15	02 49	13 04	18 42	29 25
07 Sa	16 59 55	15 51 51	05♍05 41	22 03	26 10	25 26	13 42	00 22	02 49	13 03	18 41	29 23
08 Su	17 03 52	16 49 15	18 41 49	23 14	27 23	25 54	13 52	00 30	02R 49	13 03	18 39	29 22
09 Mo	17 07 48	17 46 39	02♎37 42	24 28	28 36	26 23	14 02	00 38	02 49	13 02	18 37	29 22
10 Tu	17 11 45	18 44 01	16 52 58	25 45	29 49	26 51	14 11	00 45	02 49	13 01	18 36	29D 22
11 We	17 15 41	19 41 22	01♏25 43	27 05	01♊02	27 18	14 21	00 53	02 49	13 01	18 34	29 23
12 Th	17 19 38	20 38 42	16 12 08	28 28	02 15	27 46	14 31	01 01	02 49	13 00	18 33	29R 23
13 Fr	17 23 34	21 36 02	01♐06 33	29 54	03 28	28 13	14 41	01 09	02 48	12 59	18 31	29 22
14 Sa	17 27 31	22 33 20	16 01 44	01♊23	04 41	28 39	14 51	01 16	02 48	12 58	18 29	29 21
15 Su	17 31 28	23 30 38	00♑49 45	02 55	05 54	29 06	15 02	01 24	02 48	12 57	18 28	29 19
16 Mo	17 35 24	24 27 55	15 23 06	04 30	07 08	29 32	15 12	01 32	02 47	12 56	18 26	29 16
17 Tu	17 39 21	25 25 12	29 35 41	06 08	08 21	29 57	15 23	01 40	02 47	12 55	18 25	29 12
18 We	17 43 17	26 22 28	13♒23 36	07 49	09 34	00♓23	15 33	01 47	02 46	12 54	18 23	29 08
19 Th	17 47 14	27 19 44	26 45 25	09 32	10 47	00 48	15 44	01 55	02 46	12 53	18 22	29 05
20 Fr	17 51 10	28 17 00	09♓41 55	11 18	12 00	01 12	15 54	02 03	02 45	12 52	18 20	29 03
21 Sa	17 55 07	29 14 15	22 15 46	13 07	13 13	01 36	16 05	02 11	02 45	12 51	18 18	29 02
22 Su	17 59 04	00♋11 30	04♈30 53	14 59	14 26	02 00	16 16	02 19	02 44	12 50	18 17	29D 03
23 Mo	18 03 00	01 08 45	16 31 54	16 53	15 40	02 24	16 27	02 26	02 43	12 49	18 15	29 05
24 Tu	18 06 57	02 06 00	28 23 47	18 50	16 53	02 46	16 38	02 34	02 43	12 48	18 14	29 07
25 We	18 10 53	03 03 15	10♉11 24	20 49	18 06	03 09	16 49	02 42	02 42	12 47	18 12	29 10
26 Th	18 14 50	04 00 29	21 59 15	22 50	19 19	03 31	17 00	02 50	02 41	12 46	18 11	29 12
27 Fr	18 18 46	04 57 44	03♊51 21	24 53	20 33	03 53	17 11	02 58	02 40	12 44	18 09	29 13
28 Sa	18 22 43	05 54 58	15 51 01	26 59	21 46	04 14	17 22	03 05	02 39	12 43	18 08	29R 13
29 Su	18 26 39	06 52 12	28 00 52	29 05	22 59	04 34	17 34	03 13	02 38	12 42	18 06	29 10
30 Mo	18 30 36	07 49 27	10♋22 47	01♋13	24 13	04 54	17 45	03 21	02 37	12 41	18 05	29 05

Data for	06-01-2003
Julian Day	2452791.50
Ayanamsa	23 54 01
SVP	05 ♓ 12 28
☽ ☊ Mean	29 ♉ 02 R

● ◐ PHASES ○ ◑

07	20:28	◐	16♍41
14	11:16	○	23♐00
21	14:45	◑	29♓49
29	18:40	●	07♋37

LAST ASPECT Day	h m	☽ INGRESS Day	h m	
01	20:55	01	21:28	♋
03	16:28	04	07:26	♌
06	06:19	06	14:52	♍
08	16:28	08	19:31	♎
10	17:01	10	21:40	♏
12	21:51	12	22:13	♐
14	21:05	14	22:39	♑
15	03:13	17	00:42	♒
19	01:08	19	05:57	♓
21	14:45	21	15:06	♈
23	03:28	24	03:15	♉
25	13:42	26	16:14	♊
29	02:32	29	03:53	♋

DECLINATION

Day	☉	☽	☿	♀	♂	♃	♄	♅	♆	♇
01 Su	21N58	24N47	13N08	16N09	16S 35	17N45	22N36	11S11	16S 52	13S 28
02 Mo	22 06	26 14	13 25	16 30	16 28	17 43	22 36	11 11	16 52	13 28
03 Tu	22 14	26 25	13 43	16 51	16 21	17 40	22 36	11 11	16 52	13 27
04 We	22 21	25 16	14 02	17 12	16 14	17 37	22 36	11 11	16 53	13 27
05 Th	22 29	22 50	14 23	17 31	16 07	17 35	22 36	11 11	16 53	13 27
06 Fr	22 35	19 11	14 44	17 51	15 59	17 32	22 37	11 11	16 53	13 27
07 Sa	22 41	14 31	15 07	18 10	15 52	17 29	22 37	11 11	16 53	13 27
08 Su	22 47	09 03	15 31	18 29	15 46	17 26	22 37	11 11	16 53	13 27
09 Mo	22 53	03 00	15 56	18 47	15 39	17 24	22 37	11 11	16 54	13 27
10 Tu	22 58	03S 20	16 21	19 04	15 32	17 21	22 37	11 11	16 54	13 27
11 We	23 02	09 39	16 48	19 22	15 25	17 18	22 37	11 11	16 54	13 27
12 Th	23 07	15 32	17 14	19 38	15 19	17 15	22 37	11 11	16 54	13 27
13 Fr	23 10	20 32	17 41	19 54	15 12	17 12	22 37	11 11	16 54	13 27
14 Sa	23 14	24 12	18 09	20 10	15 06	17 09	22 37	11 11	16 55	13 27
15 Su	23 17	26 11	18 37	20 25	14 59	17 06	22 37	11 11	16 55	13 27
16 Mo	23 19	26 18	19 04	20 40	14 53	17 03	22 37	11 12	16 55	13 27
17 Tu	23 22	24 41	19 32	20 54	14 47	17 00	22 37	11 12	16 56	13 27
18 We	23 23	21 37	19 59	21 07	14 41	16 56	22 37	11 12	16 56	13 27
19 Th	23 25	17 30	20 26	21 20	14 35	16 53	22 37	11 12	16 56	13 27
20 Fr	23 26	12 42	20 52	21 32	14 29	16 50	22 37	11 12	16 56	13 27
21 Sa	23 26	07 30	21 18	21 44	14 24	16 47	22 37	11 13	16 57	13 27
22 Su	23 26	02 08	21 42	21 55	14 18	16 44	22 37	11 13	16 57	13 27
23 Mo	23 26	03N13	22 06	22 05	14 13	16 40	22 37	11 13	16 57	13 27
24 Tu	23 25	08 23	22 28	22 15	14 08	16 37	22 37	11 14	16 58	13 27
25 We	23 24	13 15	22 48	22 24	14 03	16 34	22 37	11 14	16 58	13 27
26 Th	23 23	17 37	23 07	22 33	13 58	16 30	22 37	11 14	16 58	13 27
27 Fr	23 21	21 20	23 24	22 41	13 53	16 27	22 37	11 15	16 59	13 27
28 Sa	23 18	24 10	23 39	22 48	13 49	16 23	22 36	11 15	16 59	13 28
29 Su	23 16	25 56	23 51	22 55	13 44	16 20	22 36	11 15	16 59	13 28
30 Mo	23 13	26 27	24 01	23 01	13 40	16 16	22 36	11 16	17 00	13 28

ASPECTARIAN

01 06:53 ☽ △ ♂
20:55 ☽ ☌ ♄
22:06 ♀ # ♂
02 02:59 ☽ △ ♅
04:02 ☿ # ♇
03 01:26 ♀ # ♆
02:58 ♃ ☍ ♆
08:55 ☽ ✶ ☿
16:28 ☽ ✶ ♀
04 01:29 ♄ ♋
02:27 ☉ △ ♆
07:14 ☉ ✶ ♃
05 01:35 ☽ ∥ ♄
02:30 ☽ ∥ ☉
03:27 ♀ ∥ ♃
07:56 ☽ ☍ ♆
08:37 ☽ ☌ ♃
10:18 ☽ ✶ ☉
18:19 ☽ △ ♇
22:13 ☽ □ ☿
23:53 ♀ □ ♂
06 05:25 ☉ ∥ ♄
05:56 ☽ ☍ ♂
06:19 ☽ □ ♀
06:51 ☽ ∥ ♀
09:07 ☽ ∥ ♃
12:24 ☽ # ♆
15:27 ☽ ✶ ♄
17:19 ☽ # ♂
19:56 ☽ ☍ ♅
21:20 ☽ ∥ ☿
07 04:55 ☽ # ♇
06:59 ♅ SR
15:00 ☽ # ♅
23:55 ☽ □ ♇
08 08:37 ☽ △ ☿
10:59 ☿ # ♂
16:28 ☽ △ ♀
20:33 ☽ □ ♄
09 17:33 ☽ △ ♆
19:27 ☽ ✶ ♃
20:44 ☉ ☍ ♇
10 02:51 ☽ ✶ ♇
03:18 ☽ △ ☉
03:32 ♀ ♊
17:01 ☽ △ ♂
23:06 ☽ △ ♄
11 02:16 ☽ △ ♅
05:49 ☿ □ ♂
05:51 ☿ # ♆
06:01 ☽ ∥ ♅
15:12 ☽ ∥ ♇
18:49 ☽ □ ♆
21:15 ☽ □ ♃
23:04 ☽ ∥ ♂
12 00:25 ☿ ∥ ♃
06:06 ☽ ∥ ♆
07:36 ☽ # ♃
08:22 ☽ # ☿
10:56 ♀ □ ♅
19:12 ☽ □ ♂
20:28 ☽ # ♀
21:51 ☽ ☍ ☿
13 01:34 ☿ ♊
02:44 ☽ □ ♅
04:08 ☽ ☍ ♀
12:22 ☽ # ♄
16:26 ☽ # ☉
19:04 ☽ ✶ ♆
22:05 ☽ △ ♃
14 03:58 ☽ ☌ ♆
21:05 ☽ ✶ ♂
22:04 ☿ □ ♅
15 00:56 ☽ ☍ ♄
03:13 ☽ ✶ ♅
17 02:27 ♂ ♓
11:28 ☽ # ☉
12:50 ☽ △ ☿
16:36 ☽ △ ♀
17:10 ☽ # ♄
23:08 ☽ ☌ ♆
18 03:04 ☽ # ♀
03:52 ☽ ☍ ♃
04:30 ♃ # ♆
08:53 ☽ ✶ ♇
09:08 ☽ # ☿
19 01:08 ☽ △ ☉
02:59 ☽ ∥ ♆
03:16 ☽ # ♃
07:39 ☽ ☌ ♂
09:35 ☽ △ ♄
11:03 ☽ ☌ ♅
15:10 ☽ ∥ ♂
20:26 ☽ ∥ ♇
20 03:32 ☽ □ ☿
04:49 ☽ □ ♀
06:59 ☽ ∥ ♅
16:25 ☽ □ ♇
16:52 ♀ △ ♆
20:33 ☿ △ ♆
21 03:53 ☿ ☌ ♀
19:11 ☉ ♋
19:36 ☽ □ ♄
22 16:33 ☽ ✶ ♆
17:56 ☿ ✶ ♃
22:03 ☽ ✶ ♀
23:02 ☿ ∥ ♀
23:49 ☽ △ ♃
23 00:51 ☽ ✶ ☿
03:28 ☽ △ ♇
04:23 ♂ △ ♄
16:45 ☿ ☍ ♇
18:08 ♀ ✶ ♃
20:03 ♂ ☌ ♅
24 08:11 ☽ ✶ ☉
08:35 ☽ ✶ ♄
08:46 ☽ ✶ ♅
09:12 ☽ ✶ ♂
10:10 ☿ ∥ ♄
13:40 ☉ ☌ ♄
13:44 ☽ # ♅
15:08 ☉ △ ♅
23:34 ♄ △ ♅
25 01:02 ☽ # ♇
01:58 ♀ ☍ ♇
03:54 ☉ △ ♂
04:02 ☽ # ♂
05:16 ☽ □ ♆
13:42 ☽ □ ♃
17:35 ☽ ∥ ♃
20:10 ☽ # ♆
26 11:02 ♀ ∥ ♄
19:41 ☿ ∥ ☉
21:37 ☽ □ ♅
27 00:02 ☽ □ ♂
09:45 ☽ ∥ ♄
10:48 ☽ ∥ ♀
15:50 ☽ ∥ ☉
17:47 ☽ △ ♆
18:18 ☽ ∥ ☿
28 03:04 ☽ ✶ ♃
04:31 ☽ ☍ ♇
13:02 ☽ ☌ ♀
29 02:32 ☽ ☌ ☿
09:00 ☽ △ ♅
10:16 ☽ ☌ ♄
10:17 ☿ ♋
13:08 ☽ △ ♂
30 15:27 ☿ △ ♅

⚷ Chiron		
01	Dec.	15 S 11
02	17♑15R	
05	17	07
08	16	58
11	16	49
14	16	39
17	16	29
20	16	19
23	16	08
26	15	57
29	15	46

July 2003

Day	S. T. h m s	☉ ° ' "	☽ ° ' "	☿ ° '	♀ ° '	♂ ° '	♃ ° '	♄ ° '	♅ ° '	♆ ° '	♇ ° '	☊ True ° '
01 Tu	18 34 33	08♋46 40	22♋57 49	03♋22	25♊26	05♓14	17♌56	03♋29	02♓R36	12♒R40	18♐R03	28♉R58
02 We	18 38 29	09 43 54	05♌46 26	05 32	26 39	05 33	18 08	03 37	02 35	12 38	18 02	28 51
03 Th	18 42 26	10 41 08	18 48 36	07 43	27 53	05 51	18 20	03 44	02 34	12 37	18 00	28 43
04 Fr	18 46 22	11 38 21	02♍04 02	09 53	29 06	06 09	18 31	03 52	02 33	12 36	17 59	28 37
05 Sa	18 50 19	12 35 34	15 32 25	12 04	00♋19	06 27	18 43	04 00	02 31	12 34	17 57	28 32
06 Su	18 54 15	13 32 46	29 13 27	14 14	01 33	06 44	18 55	04 08	02 30	12 33	17 56	28 29
07 Mo	18 58 12	14 29 58	13♎06 49	16 24	02 46	07 00	19 06	04 15	02 29	12 32	17 55	28 28
08 Tu	19 02 08	15 27 10	27 12 03	18 33	04 00	07 15	19 18	04 23	02 27	12 30	17 53	28D 27
09 We	19 06 05	16 24 22	11♏28 09	20 41	05 13	07 31	19 30	04 31	02 26	12 29	17 52	28 28
10 Th	19 10 02	17 21 34	25 53 16	22 47	06 27	07 45	19 42	04 38	02 24	12 27	17 51	28R 28
11 Fr	19 13 58	18 18 46	10♐24 22	24 53	07 40	07 59	19 54	04 46	02 23	12 26	17 49	28 26
12 Sa	19 17 55	19 15 57	24 57 01	26 56	08 54	08 12	20 06	04 54	02 21	12 24	17 48	28 23
13 Su	19 21 51	20 13 09	09♑25 41	28 59	10 07	08 25	20 18	05 01	02 20	12 23	17 47	28 18
14 Mo	19 25 48	21 10 21	23 44 18	00♌59	11 21	08 37	20 31	05 09	02 18	12 21	17 45	28 11
15 Tu	19 29 44	22 07 33	07♒47 09	02 58	12 34	08 48	20 43	05 17	02 17	12 20	17 44	28 03
16 We	19 33 41	23 04 46	21 29 46	04 55	13 48	08 58	20 55	05 24	02 15	12 18	17 43	27 54
17 Th	19 37 37	24 01 59	04♓49 33	06 51	15 01	09 08	21 07	05 32	02 13	12 17	17 42	27 46
18 Fr	19 41 34	24 59 12	17 46 02	08 44	16 15	09 17	21 20	05 39	02 12	12 15	17 41	27 39
19 Sa	19 45 31	25 56 26	00♈20 42	10 36	17 29	09 26	21 32	05 47	02 10	12 14	17 39	27 34
20 Su	19 49 27	26 53 41	12 36 42	12 26	18 42	09 33	21 45	05 54	02 08	12 12	17 38	27 32
21 Mo	19 53 24	27 50 57	24 38 19	14 14	19 56	09 40	21 57	06 02	02 06	12 11	17 37	27D 31
22 Tu	19 57 20	28 48 13	06♉30 34	16 01	21 10	09 47	22 10	06 09	02 04	12 09	17 36	27 33
23 We	20 01 17	29 45 30	18 18 47	17 45	22 24	09 52	22 22	06 17	02 03	12 08	17 35	27 35
24 Th	20 05 13	00♌42 48	00♊08 18	19 28	23 37	09 57	22 35	06 24	02 01	12 06	17 34	27 36
25 Fr	20 09 10	01 40 07	12 04 03	21 09	24 51	10 01	22 47	06 31	01 59	12 04	17 33	27R 36
26 Sa	20 13 06	02 37 27	24 10 20	22 48	26 05	10 04	23 00	06 39	01 57	12 03	17 32	27 33
27 Su	20 17 03	03 34 47	06♋30 35	24 26	27 19	10 06	23 13	06 46	01 55	12 01	17 31	27 27
28 Mo	20 20 60	04 32 08	19 07 02	26 01	28 33	10 07	23 25	06 53	01 53	12 00	17 30	27 19
29 Tu	20 24 56	05 29 31	02♌00 41	27 35	29 46	10 08	23 38	07 00	01 51	11 58	17 29	27 08
30 We	20 28 53	06 26 53	15 11 07	29 07	01♌00	10R 08	23 51	07 07	01 49	11 56	17 28	26 56
31 Th	20 32 49	07 24 17	28 36 48	00♍37	02 14	10 07	24 04	07 15	01 46	11 55	17 27	26 45

Data for	07-01-2003
Julian Day	2452821.50
Ayanamsa	23 54 06
SVP	05 ♓ 12 26
☽ ☊ Mean	27 ♉ 27 R

● ◐ PHASES ○ ◑

Day	h m	Phase	Position
07	02:33	◐	14♎36
13	19:21	○	20♑59
21	07:02	◑	28♈08
29	06:53	●	05♌46

LAST ASPECT Day	h m	☽ INGRESS Day	h m	Sign
29	18:40	01	13:14	♌
03	18:07	03	20:17	♍
05	04:16	06	01:21	♎
07	10:24	08	04:44	♏
09	17:59	10	06:49	♐
11	15:53	12	08:21	♑
13	19:21	14	10:38	♒
15	22:58	16	15:14	♓
18	14:49	18	23:20	♈
21	07:02	21	10:49	♉
23	09:15	23	23:43	♊
25	21:39	26	11:24	♋
28	19:26	28	20:17	♌
30	15:47	31	02:27	♍

DECLINATION

Day	☉	☽	☿	♀	♂	♃	♄	♅	♆	♇
01 Tu	23N09	25N36	24N09	23N06	13S 36	16N13	22N36	11S 16	17S 00	13S 28
02 We	23 05	23 26	24 13	23 11	13 33	16 09	22 36	11 17	17 01	13 28
03 Th	23 01	20 01	24 15	23 15	13 29	16 06	22 36	11 17	17 01	13 28
04 Fr	22 56	15 32	24 15	23 18	13 26	16 02	22 36	11 18	17 01	13 28
05 Sa	22 51	10 15	24 11	23 21	13 23	15 59	22 36	11 18	17 02	13 28
06 Su	22 45	04 23	24 05	23 22	13 20	15 55	22 35	11 19	17 02	13 28
07 Mo	22 39	01S 47	23 55	23 24	13 17	15 51	22 35	11 19	17 02	13 28
08 Tu	22 33	07 58	23 44	23 24	13 14	15 47	22 35	11 20	17 03	13 29
09 We	22 26	13 50	23 29	23 24	13 12	15 44	22 35	11 20	17 03	13 29
10 Th	22 19	19 00	23 13	23 23	13 10	15 40	22 35	11 21	17 04	13 29
11 Fr	22 11	23 03	22 53	23 22	13 08	15 36	22 34	11 21	17 04	13 29
12 Sa	22 03	25 37	22 32	23 20	13 07	15 32	22 34	11 22	17 05	13 29
13 Su	21 55	26 27	22 09	23 17	13 05	15 28	22 34	11 22	17 05	13 29
14 Mo	21 46	25 29	21 44	23 13	13 04	15 25	22 34	11 23	17 05	13 29
15 Tu	21 37	22 56	21 17	23 09	13 03	15 21	22 33	11 24	17 06	13 30
16 We	21 28	19 07	20 48	23 04	13 03	15 17	22 33	11 24	17 06	13 30
17 Th	21 18	14 27	20 18	22 59	13 02	15 13	22 33	11 25	17 07	13 30
18 Fr	21 08	09 14	19 47	22 52	13 02	15 09	22 33	11 25	17 07	13 30
19 Sa	20 58	03 48	19 14	22 45	13 02	15 05	22 32	11 26	17 08	13 30
20 Su	20 47	01N39	18 40	22 38	13 03	15 01	22 32	11 27	17 08	13 31
21 Mo	20 36	06 57	18 06	22 29	13 03	14 57	22 32	11 27	17 08	13 31
22 Tu	20 24	11 57	17 30	22 20	13 04	14 53	22 32	11 28	17 09	13 31
23 We	20 12	16 29	16 54	22 11	13 05	14 49	22 31	11 29	17 09	13 31
24 Th	20 00	20 24	16 17	22 00	13 07	14 45	22 31	11 30	17 10	13 31
25 Fr	19 47	23 30	15 40	21 50	13 09	14 41	22 31	11 30	17 10	13 32
26 Sa	19 34	25 35	15 02	21 38	13 11	14 37	22 30	11 31	17 11	13 32
27 Su	19 21	26 28	14 23	21 26	13 13	14 32	22 30	11 32	17 11	13 32
28 Mo	19 08	26 01	13 45	21 13	13 15	14 28	22 30	11 32	17 12	13 32
29 Tu	18 54	24 09	13 06	21 00	13 18	14 24	22 29	11 33	17 12	13 33
30 We	18 40	20 59	12 27	20 46	13 21	14 20	22 29	11 34	17 13	13 33
31 Th	18 25	16 40	11 47	20 31	13 24	14 16	22 29	11 35	17 13	13 33

⚷ Chiron

Day	Position
01 Dec.	15 S 14
02	15♑35 R
05	15 24
08	15 13
11	15 01
14	14 50
17	14 39
20	14 28
23	14 18
26	14 07
29	13 57

ASPECTARIAN

Day	Time	Aspect
01	01:14	☿ ☌ ♄
	08:10	♀ ∥ ☉
	12:40	♃ △ ♇
	16:55	☽ ∥ ☿
02	00:06	☿ △ ♂
	02:02	☽ ∥ ♀
	02:55	☽ ∥ ☉
	06:38	☽ ∥ ♄
	12:40	☽ ☍ ♆
	22:32	☽ △ ♇
	23:06	☽ ☌ ♃
03	07:29	♂ ∥ ♇
	16:35	☽ # ♆
	18:07	☽ ⚹ ♀
	21:31	☽ ∥ ♃
04	00:51	☽ ☍ ♅
	03:16	☽ ⚹ ♄
	07:29	☽ ☍ ♂
	09:46	☽ # ♇
	10:03	☽ # ♂
	16:40	☽ ⚹ ☿
	17:39	♀ ♋
	18:23	☽ ⚹ ☉
	19:26	☽ # ♅
05	04:16	☽ □ ♇
	10:21	☿ ☌ ☉
06	04:26	☽ □ ♀
	08:35	☽ □ ♄
	18:21	♀ △ ♅
	23:00	☽ △ ♆
07	06:39	☽ □ ☿
	08:12	☽ ⚹ ♇
	10:24	☽ ⚹ ♃
	15:20	☉ ∥ ♄
08	08:31	♀ ☌ ♄
	08:51	☽ △ ♅
	12:14	☽ △ ♄
	12:32	☽ △ ♀
	13:30	☽ ∥ ♅
	17:15	☽ △ ♂
	21:19	☽ ∥ ♂
	22:28	☽ ∥ ♇
09	01:41	☽ □ ♆
	08:01	☿ ∥ ♀
	08:14	☽ # ♃
	08:50	☽ △ ☉
	13:35	☽ □ ♃
	14:28	☽ ∥ ♆
	17:59	☽ △ ☿
10	10:46	☽ □ ♅
	18:18	☽ # ☉
	19:56	☽ □ ♂
	20:38	☽ # ♄
	22:55	☽ # ☿
11	02:11	☽ # ♀
	03:20	☽ ⚹ ♆
	07:28	♀ △ ♂
	12:13	☽ ☌ ♇
	15:53	☽ △ ♃
	21:49	☿ ∥ ♄
12	12:14	☽ ⚹ ♅
	16:37	☽ ☍ ♄
	22:17	☽ ⚹ ♂
13	01:16	☽ ☍ ♀
	12:10	☿ ♌
	20:05	☿ ∥ ☉
14	14:21	☽ ☍ ☿
	22:25	☽ # ♀
15	02:48	☽ # ♄
	07:52	☽ ☌ ♆
	09:31	☽ # ☉
	12:55	☽ # ☿
	17:20	☽ ⚹ ♇
	22:58	☽ ☍ ♃
16	10:49	☽ ∥ ♆
	19:16	☽ ☌ ♅
	20:13	☽ # ♃
17	01:18	☽ △ ♄
	04:31	☽ ∥ ♇
	06:42	☽ ∥ ♂
	08:01	☽ ☌ ♂
	14:10	☽ ∥ ♅
	20:51	☽ △ ♀
	23:50	☽ □ ♇
18	14:49	☽ △ ☉
19	10:40	☽ □ ♄
	20:58	☿ ☍ ♆
	23:12	☽ ⚹ ♆
	23:36	☽ △ ☿
20	09:58	☽ △ ♇
	13:30	☽ □ ♀
	16:32	♀ ∥ ♄
	18:30	☽ △ ♃
21	15:02	☽ ⚹ ♅
	21:35	☽ # ♅
	23:16	☽ ⚹ ♄
22	05:41	☽ # ♂
	06:42	☽ ⚹ ♂
	07:58	☽ # ♇
	11:27	☽ □ ♆
	13:58	☿ # ♆
	14:56	☽ ∥ ♃
	21:38	☿ △ ♇
	22:40	☽ □ ☿
23	02:01	☽ ∥ ☿
	03:48	☽ # ♆
	06:04	☉ ♌
	08:23	☽ □ ♃
	09:15	☽ ⚹ ♀
	21:26	☽ ∥ ☉
24	01:16	☽ ⚹ ☉
	03:46	☽ □ ♅
	10:54	☽ ∥ ♀
	15:30	☽ ∥ ♄
	19:52	☽ □ ♂
25	00:01	☽ △ ♆
	10:54	☽ ☍ ♇
	20:53	☽ ⚹ ☿
	21:39	☽ ⚹ ♃
26	03:17	☿ ☌ ♃
	15:08	☽ △ ♅
	17:35	☿ ∥ ♃
27	00:30	☽ ☌ ♄
	06:54	☽ △ ♂
28	07:32	☿ # ♇
	16:57	☿ # ♂
	19:26	☽ ☌ ♀
29	04:25	♀ ♌
	07:38	♂ SR
	13:49	☽ ∥ ♄
	18:09	☽ ☍ ♆
30	01:33	☽ ∥ ♀
	04:06	☽ △ ♇
	14:06	☿ ♍
	14:25	☽ ∥ ☉
	15:47	☽ ☌ ♃
	21:15	☽ # ♆
31	03:59	☽ ☌ ☿
	05:35	☽ ☍ ♅
	07:37	☿ # ♅
	11:40	☽ ∥ ♃
	14:44	☽ # ♇
	15:15	☽ # ♂
	15:21	☽ ⚹ ♄
	18:24	☿ ☍ ♅
	20:13	☽ ☍ ♂
	23:22	☽ # ♅

August 2003

Day	S. T.	☉	☽	☿	♀	♂	♃	♄	♅	♆	♇	☊ True
	h m s	° ' "	° ' "	° '	° '	° '	° '	° '	° '	° '	° '	° '
01 Fr	20 36 46	08♌ 21 41	12♍ 15 17	02♍ 05	03♌ 28	10♓ R05	24♌ 16	07♋ 22	01♓ R44	11♒ R53	17♐ R26	26♉ R34
02 Sa	20 40 42	09 19 06	26 03 48	03 32	04 42	10 03	24 29	07 29	01 42	11 51	17 25	26 25
03 Su	20 44 39	10 16 31	09♎ 59 44	04 56	05 56	09 59	24 42	07 36	01 40	11 50	17 25	26 19
04 Mo	20 48 35	11 13 57	24 00 58	06 19	07 10	09 55	24 55	07 43	01 38	11 48	17 24	26 16
05 Tu	20 52 32	12 11 24	08♏ 05 57	07 40	08 24	09 50	25 08	07 50	01 36	11 47	17 23	26 14
06 We	20 56 29	13 08 52	22 13 38	08 59	09 38	09 45	25 21	07 57	01 33	11 45	17 22	26 14
07 Th	21 00 25	14 06 20	06♐ 22 59	10 16	10 52	09 38	25 34	08 03	01 31	11 43	17 22	26 13
08 Fr	21 04 22	15 03 49	20 32 41	11 30	12 06	09 31	25 47	08 10	01 29	11 42	17 21	26 11
09 Sa	21 08 18	16 01 18	04♑ 40 41	12 43	13 20	09 23	26 00	08 17	01 27	11 40	17 20	26 06
10 Su	21 12 15	16 58 49	18 44 04	13 53	14 34	09 15	26 13	08 24	01 24	11 38	17 20	25 59
11 Mo	21 16 11	17 56 20	02♒ 39 11	15 01	15 48	09 06	26 26	08 30	01 22	11 37	17 19	25 50
12 Tu	21 20 08	18 53 53	16 22 04	16 07	17 02	08 56	26 39	08 37	01 20	11 35	17 19	25 39
13 We	21 24 05	19 51 27	29 49 09	17 10	18 16	08 46	26 52	08 43	01 17	11 34	17 18	25 27
14 Th	21 28 01	20 49 01	12♓ 57 56	18 11	19 30	08 35	27 05	08 50	01 15	11 32	17 18	25 15
15 Fr	21 31 58	21 46 37	25 47 19	19 09	20 45	08 23	27 18	08 56	01 13	11 30	17 17	25 04
16 Sa	21 35 54	22 44 15	08♈ 17 55	20 04	21 59	08 11	27 31	09 03	01 10	11 29	17 17	24 56
17 Su	21 39 51	23 41 54	20 31 52	20 56	23 13	07 58	27 44	09 09	01 08	11 27	17 16	24 51
18 Mo	21 43 47	24 39 34	02♉ 32 37	21 45	24 27	07 45	27 57	09 15	01 06	11 26	17 16	24 48
19 Tu	21 47 44	25 37 16	14 24 35	22 31	25 41	07 31	28 10	09 22	01 03	11 24	17 16	24D 48
20 We	21 51 40	26 34 59	26 12 55	23 13	26 56	07 17	28 23	09 28	01 01	11 22	17 15	24 49
21 Th	21 55 37	27 32 45	08♊ 03 05	23 52	28 10	07 02	28 36	09 34	00 59	11 21	17 15	24 49
22 Fr	21 59 33	28 30 32	20 00 33	24 26	29 24	06 47	28 49	09 40	00 56	11 19	17 15	24R 49
23 Sa	22 03 30	29 28 20	02♋ 10 27	24 57	00♍ 38	06 32	29 03	09 46	00 54	11 18	17 15	24 46
24 Su	22 07 27	00♍ 26 10	14 37 07	25 23	01 53	06 17	29 16	09 52	00 51	11 16	17 14	24 40
25 Mo	22 11 23	01 24 02	27 23 45	25 44	03 07	06 01	29 29	09 58	00 49	11 15	17 14	24 32
26 Tu	22 15 20	02 21 56	10♌ 31 59	26 01	04 21	05 45	29 42	10 04	00 47	11 13	17 14	24 21
27 We	22 19 16	03 19 51	24 01 31	26 12	05 36	05 29	29 55	10 09	00 44	11 12	17 14	24 09
28 Th	22 23 13	04 17 48	07♍ 50 09	26 18	06 50	05 13	00♍ 08	10 15	00 42	11 10	17 14	23 56
29 Fr	22 27 09	05 15 46	21 53 56	26R 19	08 04	04 57	00 21	10 21	00 39	11 09	17D 14	23 45
30 Sa	22 31 06	06 13 45	06♎ 07 53	26 13	09 19	04 41	00 34	10 26	00 37	11 07	17 14	23 35
31 Su	22 35 02	07 11 46	20 26 49	26 02	10 33	04 25	00 47	10 32	00 35	11 06	17 14	23 29

Data for	08-01-2003
Julian Day	2452852.50
Ayanamsa	23 54 12
SVP	05 ♓ 12 22
☽ ☊ Mean	25 ♉ 48 R

● ◐ PHASES ○ ◑

Day	Time	Phase	Position
05	07:27	◐	12♏29
12	04:49	○	19♒05
20	00:49	◑	26♉37
27	17:26	●	04♍02

LAST ASPECT ☽ Day	h m	INGRESS Day	h m	Sign
01	09:02	02	06:48	♎
04	01:34	04	10:13	♏
06	05:22	06	13:11	♐
08	09:01	08	16:03	♑
09	14:57	10	19:24	♒
12	18:36	13	00:20	♓
14	10:30	15	08:01	♈
17	14:37	17	18:53	♉
20	04:30	20	07:41	♊
22	18:16	22	19:45	♋
24	20:51	25	04:49	♌
26	12:00	27	10:27	♍
29	07:26	29	13:41	♎
30	18:37	31	16:00	♏

DECLINATION

Day	☉	☽	☿	♀	♂	♃	♄	♅	♆	♇
01 Fr	18N10	11N26	11N08	20N16	13S 28	14N11	22N28	11S 36	17S 13	13S 33
02 Sa	17 55	05 35	10 29	20 00	13 31	14 07	22 28	11 36	17 14	13 34
03 Su	17 40	00S 36	09 50	19 44	13 35	14 03	22 28	11 37	17 14	13 34
04 Mo	17 24	06 48	09 11	19 27	13 39	13 59	22 27	11 38	17 15	13 34
05 Tu	17 08	12 43	08 33	19 09	13 43	13 54	22 27	11 39	17 15	13 34
06 We	16 52	17 59	07 54	18 51	13 48	13 50	22 27	11 40	17 16	13 35
07 Th	16 36	22 15	07 17	18 32	13 53	13 46	22 26	11 40	17 16	13 35
08 Fr	16 19	25 11	06 39	18 13	13 57	13 41	22 26	11 41	17 17	13 35
09 Sa	16 02	26 29	06 02	17 54	14 02	13 37	22 26	11 42	17 17	13 36
10 Su	15 45	26 05	05 26	17 34	14 07	13 33	22 25	11 43	17 18	13 36
11 Mo	15 27	24 03	04 50	17 13	14 13	13 28	22 25	11 44	17 18	13 36
12 Tu	15 10	20 38	04 15	16 52	14 18	13 24	22 24	11 44	17 19	13 37
13 We	14 52	16 12	03 40	16 30	14 23	13 19	22 24	11 45	17 19	13 37
14 Th	14 33	11 06	03 07	16 08	14 29	13 15	22 24	11 46	17 19	13 37
15 Fr	14 15	05 38	02 34	15 46	14 34	13 10	22 23	11 47	17 20	13 38
16 Sa	13 56	00 04	02 03	15 23	14 40	13 06	22 23	11 48	17 20	13 38
17 Su	13 37	05N23	01 33	15 00	14 46	13 02	22 22	11 49	17 21	13 38
18 Mo	13 18	10 34	01 04	14 36	14 51	12 57	22 22	11 50	17 21	13 39
19 Tu	12 59	15 18	00 36	14 12	14 57	12 53	22 22	11 50	17 22	13 39
20 We	12 39	19 26	00 09	13 47	15 03	12 48	22 21	11 51	17 22	13 39
21 Th	12 20	22 48	00 15	13 22	15 09	12 44	22 21	11 52	17 23	13 40
22 Fr	12 00	25 13	00S 38	12 57	15 14	12 39	22 20	11 53	17 23	13 40
23 Sa	11 39	26 30	01 00	12 31	15 20	12 35	22 20	11 54	17 23	13 40
24 Su	11 19	26 29	01 19	12 06	15 25	12 30	22 20	11 55	17 24	13 41
25 Mo	10 59	25 04	01 36	11 39	15 31	12 26	22 19	11 55	17 24	13 41
26 Tu	10 38	22 17	01 50	11 13	15 36	12 21	22 19	11 56	17 25	13 42
27 We	10 17	18 14	02 02	10 46	15 41	12 17	22 18	11 57	17 25	13 42
28 Th	09 56	13 08	02 12	10 18	15 46	12 12	22 18	11 58	17 26	13 42
29 Fr	09 35	07 16	02 18	09 51	15 50	12 07	22 18	11 59	17 26	13 43
30 Sa	09 14	00 58	02 22	09 23	15 55	12 03	22 17	12 00	17 26	13 43
31 Su	08 52	05S 27	02 22	08 55	15 59	11 58	22 17	12 01	17 27	13 44

ASPECTARIAN

01 01:29 ☽ ∥ ☿
09:02 ☽ □ ♇
02 16:07 ♂ ∥ ♆
16:21 ☽ ✶ ♀
19:51 ☽ □ ♄
03 00:31 ☽ ✶ ☉
03:09 ☽ △ ♆
12:42 ☽ ✶ ♇
04 01:34 ☽ ✶ ♃
08:33 ☽ ⋕ ☿
12:57 ☽ △ ♅
13:55 ☉ ☍ ♆
14:01 ☉ ⋕ ♆
19:31 ☽ ∥ ♅
23:11 ☽ ✶ ☿
23:32 ☽ △ ♄
05 00:33 ☽ □ ♀
02:56 ☽ △ ♂
03:13 ☿ ✶ ♄
03:42 ☽ ∥ ♇
04:25 ☽ ∥ ♂
05:04 ☽ ⋕ ♃
06:14 ☽ □ ♆
18:54 ☽ ⋕ ☉
20:29 ☽ ∥ ♆
06 04:09 ☽ ⋕ ♀
05:22 ☽ □ ♃
05:30 ♂ ⋕ ♃
13:12 ☿ ☍ ♂
15:47 ☽ □ ♅
07 01:15 ☽ ⋕ ♄
05:28 ☽ □ ♂
07:13 ☽ □ ☿
08:19 ☽ △ ♀
09:02 ☽ ✶ ♆
14:02 ☽ △ ☉
10:20 [illegible]
18:35 ☽ ☌ ♇
08 09:01 ☽ △ ♃
18:31 ☽ ✶ ♅
09 06:11 ☽ ☍ ♄
06:37 ♃ ⋕ ♇
07:57 ☽ ✶ ♂
14:57 ☽ △ ☿
10 08:41 ☉ △ ♇
18:16 ♀ ⋕ ♆
11 12:38 ☽ ⋕ ♄
15:36 ☽ ☌ ♆
12 01:18 ☽ ☍ ♀
01:40 ☽ ✶ ♇
05:19 ♀ △ ♇
18:24 ☽ ∥ ♆
18:36 ☽ ☍ ♃
22:23 ☽ ⋕ ♀
13 02:39 ☽ ☌ ♅
02:56 ♂ △ ♄
03:00 ☿ □ ♇
07:00 ☽ ⋕ ☉
08:40 ☽ ∥ ♂
12:28 ☽ ∥ ♇
14:03 ☽ ⋕ ♃
16:02 ☽ ☌ ♂
16:20 ☽ △ ♄
20:59 ☽ ∥ ♅
14 04:27 ☉ ⋕ ♂
08:02 ☽ □ ♇
10:30 ☽ ☍ ☿
15 14:35 ☽ ⋕ ☿
16 01:28 ☽ □ ♄
06:11 ☽ ✶ ♆
00:28 ☽ ∥ [illegible]
17:34 ☽ △ ♇
22:51 ☉ ⋕ ♇
17 05:56 ☽ △ ♀
06:50 ☽ △ ☉
11:22 ♀ ⋕ ♂
14:37 ☽ △ ♃
21:06 ☽ ✶ ♅
18 06:13 ☽ ⋕ ♅
10:18 ☽ ✶ ♂
11:39 ☽ ∥ ♃
12:45 ☽ ∥ ☉
13:40 ☽ ✶ ♄
15:22 ☽ ⋕ ♇
17:54 ☽ □ ♆
18:05 ♀ ☌ ☉
18:41 ☽ ∥ ♀
22:10 ☽ ⋕ ♂
19 09:56 ☉ ∥ ♃
11:34 ☽ ⋕ ♆
17:32 ☽ △ ☿
20 01:37 ☽ □ ♀
04:30 ☽ □ ♃
07:39 ♀ ⋕ ♇
09:43 ☽ □ ♅
20:25 ☽ ∥ ♄
22:00 ☽ □ ♂
21 06:38 ☽ △ ♆
10:23 ♀ ☌ ♃
18:29 ☽ ☍ ♇
22 07:37 ☉ ⋕ ♅
09:12 ☽ □ ☿
10:08 ☉ ☌ ♃
11:36 ♀ ♍
17:45 ☽ ✶ ♃
18:16 ☽ ✶ ☉
20:26 ♀ ∥ ♃
20:40 ☽ ✶ ♀
21:31 ☽ △ ♅
23 04:49 ♀ ☍ ♅
08:19 ☽ △ ♂
13:08 ☉ ♍
14:50 ☽ ☌ ♄
24 09:38 ♀ ⋕ ♅
10:02 ☉ ☍ ♅
20:51 ☽ ✶ ☿
25 23:47 ☽ ∥ ♄
26 01:14 ☽ ☍ ♆
12:00 ☽ △ ♇
22:17 ☿ ☍ ♂
27 04:11 ☽ ⋕ ♆
09:27 ♃ ♍
10:28 ☽ ☌ ♃
11:42 ☽ ☍ ♅
12:25 ☽ ⋕ ♂
19:35 ☽ ☍ ♂
21:34 ☽ ⋕ ♇
22:08 ☽ ☌ ♀
28 04:08 ☽ ✶ ♄
04:12 ☽ ∥ ♃
05:03 ☽ ⋕ ♅
12:57 ☽ ∥ ♀
13:42 ☿ SR
14:19 ☽ ∥ [illegible]
16:05 ☽ □ ♇
17:59 ☉ ☍ ♂
29 03:34 ♇ SD
07:26 ☽ ☌ ☿
18:51 ☽ ⋕ ☿
30 04:37 ♃ ☍ ♅
07:16 ☽ □ ♄
08:22 ☽ △ ♆
13:52 ♃ ⋕ ♅
18:37 ☽ ✶ ♇
23:24 ♀ ✶ ♄
31 11:00 ♀ ∥ ☉
12:27 ☽ ⋕ ♀
12:28 ☽ ⋕ ☉
16:55 ☽ △ ♅
17:35 ☽ ✶ ♃
23:00 ☽ △ ♂

⚷ Chiron

01	Dec.	15 S 28
01	13♑47R	
04	13 38	
07	13 29	
10	13 20	
13	13 12	
16	13 04	
19	12 57	
22	12 51	
25	12 45	
28	12 40	
31	[illegible]	

September 2003

Day	S. T.	☉	☽	☿	♀	♂	♃	♄	♅	♆	♇	☊ True
	h m s	° ' "	° ' "	° '	° '	° '	° '	° '	° '	° '	° '	° '
01 Mo	22 38 59	08♍09 48	04♏46 06	25♍R44	11♍48	04♓R09	01♍00	10♋37	00♓R32	11♒R05	17♐14	23♉R24
02 Tu	22 42 56	09 07 52	19 02 15	25 21	13 02	03 54	01 13	10 42	00 30	11 03	17 14	23 22
03 We	22 46 52	10 05 57	03♐13 02	24 51	14 17	03 38	01 26	10 47	00 28	11 02	17 14	23 21
04 Th	22 50 49	11 04 04	17 17 11	24 15	15 31	03 23	01 39	10 53	00 25	11 00	17 15	23 21
05 Fr	22 54 45	12 02 12	01♑14 01	23 33	16 45	03 09	01 52	10 58	00 23	10 59	17 15	23 19
06 Sa	22 58 42	13 00 22	15 02 58	22 46	18 00	02 54	02 05	11 03	00 21	10 58	17 15	23 16
07 Su	23 02 38	13 58 32	28 43 12	21 55	19 14	02 40	02 18	11 08	00 18	10 56	17 15	23 10
08 Mo	23 06 35	14 56 45	12♒13 29	20 59	20 29	02 27	02 31	11 12	00 16	10 55	17 16	23 02
09 Tu	23 10 31	15 54 59	25 32 16	20 01	21 43	02 14	02 44	11 17	00 14	10 54	17 16	22 52
10 We	23 14 28	16 53 14	08♓37 54	19 01	22 58	02 01	02 57	11 22	00 11	10 53	17 16	22 41
11 Th	23 18 25	17 51 31	21 29 05	18 01	24 12	01 49	03 10	11 27	00 09	10 51	17 17	22 30
12 Fr	23 22 21	18 49 50	04♈05 10	17 02	25 27	01 37	03 23	11 31	00 07	10 50	17 17	22 20
13 Sa	23 26 18	19 48 11	16 26 31	16 05	26 41	01 26	03 35	11 35	00 05	10 49	17 18	22 13
14 Su	23 30 14	20 46 34	28 34 36	15 11	27 56	01 16	03 48	11 40	00 03	10 48	17 18	22 08
15 Mo	23 34 11	21 44 59	10♉31 58	14 23	29 10	01 07	04 01	11 44	00 00	10 47	17 19	22 05
16 Tu	23 38 07	22 43 26	22 22 09	13 41	00♎25	00 58	04 14	11 48	29♒58	10 46	17 19	22D 05
17 We	23 42 04	23 41 56	04♊09 28	13 06	01 39	00 49	04 26	11 52	29 56	10 44	17 20	22 06
18 Th	23 46 00	24 40 27	15 58 54	12 39	02 54	00 42	04 39	11 56	29 54	10 43	17 20	22 08
19 Fr	23 49 57	25 39 00	27 55 43	12 21	04 09	00 35	04 52	12 00	29 52	10 42	17 21	22 09
20 Sa	23 53 54	26 37 36	10♋05 19	12 13	05 23	00 29	05 04	12 04	29 50	10 41	17 22	22R 08
21 Su	23 57 50	27 36 14	22 32 40	12D 14	06 38	00 23	05 17	12 08	29 48	10 40	17 23	22 05
22 Mo	00 01 47	28 34 54	05♌21 57	12 25	07 52	00 19	05 29	12 12	29 46	10 39	17 23	22 00
23 Tu	00 05 43	29 33 37	18 35 58	12 46	09 07	00 15	05 42	12 15	29 44	10 38	17 24	21 53
24 We	00 09 40	00♎32 21	02♍15 33	13 16	10 22	00 12	05 54	12 19	29 42	10 38	17 25	21 45
25 Th	00 13 36	01 31 07	16 19 08	13 55	11 36	00 09	06 07	12 22	29 40	10 37	17 26	21 36
26 Fr	00 17 33	02 29 56	00♎42 44	14 43	12 51	00 08	06 19	12 25	29 38	10 36	17 27	21 28
27 Sa	00 21 29	03 28 46	15 20 15	15 39	14 05	00 07	06 32	12 28	29 36	10 35	17 28	21 21
28 Su	00 25 26	04 27 39	00♏04 29	16 42	15 20	00D 07	06 44	12 31	29 34	10 34	17 29	21 16
29 Mo	00 29 23	05 26 33	14 48 09	17 52	16 35	00 08	06 56	12 34	29 32	10 33	17 29	21 13
30 Tu	00 33 19	06 25 29	29 24 59	19 08	17 49	00 10	07 08	12 37	29 31	10 33	17 30	21 12

Data for	09-01-2003
Julian Day	2452883.50
Ayanamsa	23 54 15
SVP	05 ♓ 12 18
☽ ☊ Mean	24 ♉ 10 R

● ◐ PHASES ○ ◑

Day	Time	Phase	Position
03	12:35	◐	10♐36
10	16:37	○	17♓34
18	19:02	◑	25♊27
26	03:09	●	02♎38

LAST ASPECT Day	h m	☽ INGRESS Day	h m	
02	10:19	02	18:32	♐
04	11:24	04	21:52	♑
06	12:44	07	02:16	♒
08	09:02	09	08:08	♓
11	05:42	11	16:10	♈
13	01:40	14	02:50	♉
16	15:26	16	15:32	♊
19	03:50	19	04:07	♋
21	10:21	21	14:03	♌
23	19:34	23	20:05	♍
25	01:52	25	22:49	♎
27	23:11	27	23:53	♏
30	00:09	30	00:58	♐

DECLINATION

Day	☉	☽	☿	♀	♂	♃	♄	♅	♆	♇
01 Mo	08N31	11S35	02S18	08N27	16S03	11N54	22N16	12S01	17S27	13S44
02 Tu	08 09	17 07	02 11	07 58	16 07	11 49	22 16	12 02	17 28	13 44
03 We	07 47	21 39	02 01	07 30	16 10	11 45	22 16	12 03	17 28	13 45
04 Th	07 25	24 53	01 46	07 01	16 14	11 40	22 15	12 04	17 28	13 45
05 Fr	07 03	26 33	01 28	06 32	16 17	11 35	22 15	12 05	17 29	13 46
06 Sa	06 41	26 33	01 06	06 02	16 19	11 31	22 14	12 06	17 29	13 46
07 Su	06 18	24 55	00 41	05 33	16 22	11 26	22 14	12 06	17 30	13 46
08 Mo	05 56	21 53	00 13	05 03	16 24	11 22	22 14	12 07	17 30	13 47
09 Tu	05 33	17 45	00N19	04 34	16 25	11 17	22 13	12 08	17 30	13 47
10 We	05 11	12 50	00 53	04 04	16 27	11 13	22 13	12 09	17 31	13 48
11 Th	04 48	07 27	01 28	03 34	16 28	11 08	22 12	12 10	17 31	13 48
12 Fr	04 25	01 51	02 05	03 03	16 28	11 03	22 12	12 10	17 31	13 49
13 Sa	04 02	03N43	02 42	02 33	16 29	10 59	22 12	12 11	17 32	13 49
14 Su	03 39	09 04	03 19	02 03	16 29	10 54	22 11	12 12	17 32	13 49
15 Mo	03 16	14 01	03 55	01 32	16 29	10 50	22 11	12 13	17 32	13 50
16 Tu	02 53	18 24	04 29	01 02	16 28	10 45	22 11	12 13	17 33	13 50
17 We	02 30	22 02	05 00	00 31	16 27	10 41	22 10	12 14	17 33	13 51
18 Th	02 07	24 47	05 29	00 01	16 26	10 36	22 10	12 15	17 33	13 51
19 Fr	01 44	26 26	05 54	00S30	16 24	10 32	22 10	12 16	17 33	13 52
20 Sa	01 20	26 52	06 15	01 00	16 22	10 27	22 09	12 16	17 34	13 52
21 Su	00 57	25 58	06 32	01 31	16 20	10 23	22 09	12 17	17 34	13 53
22 Mo	00 34	23 41	06 44	02 01	16 18	10 18	22 09	12 18	17 34	13 53
23 Tu	00 10	20 07	06 52	02 32	16 15	10 14	22 08	12 18	17 35	13 53
24 We	00S13	15 22	06 55	03 03	16 11	10 09	22 08	12 19	17 35	13 54
25 Th	00 36	09 41	06 53	03 33	16 08	10 05	22 08	12 20	17 35	13 54
26 Fr	01 00	03 22	06 47	04 03	16 04	10 00	22 07	12 20	17 35	13 55
27 Sa	01 23	03S15	06 37	04 34	16 00	09 56	22 07	12 21	17 36	13 55
28 Su	01 46	09 45	06 22	05 04	15 55	09 51	22 07	12 22	17 36	13 56
29 Mo	02 10	15 43	06 03	05 34	15 50	09 47	22 07	12 22	17 36	13 56
30 Tu	02 33	20 45	05 41	06 04	15 45	09 42	22 06	12 23	17 36	13 57

ASPECTARIAN

01 01:16 ☽ # ♃
01:49 ☽ ∥ ♅
06:07 ☽ ⚹ ☉
08:57 ☽ ∥ ♇
09:53 ☽ △ ♄
10:35 ☽ □ ♆
12:56 ☽ ⚹ ♀
19:21 ☽ ∥ ♂
02 01:41 ☽ ∥ ♆
10:19 ☽ ⚹ ☿
19:20 ☽ □ ♅
20:56 ☽ □ ♃
03 00:42 ☽ □ ♂
03:50 ☽ # ♄
13:17 ☽ ⚹ ♆
18:49 ☉ ⚹ ♄
20:41 ☽ □ ♀
23:55 ☽ ☌ ♇
04 11:24 ☽ □ ☿
22:32 ☽ ⚹ ♅
05 01:07 ☽ △ ♃
03:15 ☽ ⚹ ♂
09:31 ♀ □ ♇
16:59 ☽ ☍ ♄
20:10 ☽ △ ☉
06 05:40 ☽ △ ♀
12:44 ☽ △ ☿
07 20:00 ♂ ☍ ♃
21:40 ☽ ☌ ♆
21:43 ☽ # ♄
08 05:35 ☿ ☌ ♀
09:02 ☽ ⚹ ♇
09 01:16 ☽ ∥ ♆
06:45 ☽ ∥ ♂
08:31 ☽ ☌ ♅
12:01 ☽ ☌ ♂
13:21 ☽ ☍ ♃
19:31 ☽ ∥ ♇
10 03:08 ☽ ∥ ♅
05:06 ☽ △ ♄
07:29 ☽ # ♃
09:36 ☉ □ ♇
16:05 ☽ □ ♇
17:57 ☽ ☍ ☿
11 01:57 ☿ ☌ ☉
05:42 ☽ ☍ ♀
12:13 ☽ # ☉
17:45 ☿ □ ♇
18:20 ☽ # ♀
23:06 ☽ # ☿
12 13:02 ☽ ⚹ ♆
14:28 ☽ □ ♄
19:01 ☽ ∥ ☿
19:21 ☽ ∥ ♀
20:51 ☿ ∥ ♀
13 01:19 ☽ ∥ ☉
01:40 ☽ △ ♇
14 02:55 ☽ ⚹ ♅
05:18 ☽ ⚹ ♂
08:18 ☽ ∥ ♃
08:34 ☿ ∥ ☉
10:39 ☽ △ ♃
14:59 ☽ # ♅
23:05 ☽ # ♇
15 00:30 ☽ □ ♆
02:27 ☽ ⚹ ♄
03:48 ♅ ♒R
07:19 ☽ △ ☿
13:05 ☽ # ♂
15:58 ♀ ♎
19:05 ☽ # ♆
16 00:47 ☽ △ ☉
15:26 ☽ □ ♅
17:17 ☽ □ ♂
18:19 ☽ △ ♀
17 00:35 ☽ □ ♃
01:04 ☽ ∥ ♄
13:22 ☽ △ ♆
17:28 ☽ □ ☿
18 02:45 ☽ ☍ ♇
19 03:52 ☽ △ ♅
05:14 ☽ △ ♂
13:44 ☽ □ ♀
13:59 ☽ ⚹ ♃
20 03:53 ☽ ☌ ♄
04:08 ☽ ⚹ ☿
08:53 ☿ SD
09:02 ♀ # ☉
21 10:21 ☽ ⚹ ☉
22 05:05 ☽ ⚹ ♀
09:41 ☽ ☍ ♆
11:23 ☽ ∥ ♄
21:51 ☽ △ ♇
23 10:47 ☉ ♎
13:28 ☽ # ♆
19:34 ☽ ☍ ♅
20:08 ☽ # ♂
20:26 ☽ ☍ ♂
24 05:05 ♀ △ ♆
06:23 ☽ ☌ ♃
06:32 ☽ # ♇
13:16 ☽ # ♅
17:17 ☽ ⚹ ♄
19:44 ☽ ☌ ☿
22:25 ☽ ∥ ♃
25 01:52 ☽ □ ♇
11:00 ☽ ∥ ☿
15:21 ♀ □ ♄
21:38 ☽ # ♀
26 08:11 ☽ # ☉
16:14 ☽ △ ♆
16:51 ☽ ∥ ☉
19:18 ☽ □ ♄
21:47 ☽ ☌ ♀
27 03:28 ☽ ⚹ ♇
05:11 ☽ ∥ ♀
07:53 ♂ SD
11:52 ☽ # ☿
23:11 ☽ △ ♅
28 00:05 ☽ △ ♂
00:24 ☽ # ♃
10:09 ☽ ∥ ♅
10:59 ☽ ⚹ ♃
16:29 ☽ ∥ ♇
16:31 ☿ □ ♇
17:04 ☽ □ ♆
20:21 ☽ △ ♄
29 00:30 ☽ ∥ ♂
05:28 ☽ ⚹ ☿
08:22 ☽ ∥ ♆
13:38 ☿ # ♀
17:51 ♀ ⚹ ♇
30 00:09 ☽ □ ♅
01:15 ☽ □ ♂
07:48 ☽ # ♄
12:28 ☽ ⚹ ☉
12:59 ☽ □ ♃
18:28 ☽ ⚹ ♆

⚷ Chiron

Day	Position
01 Dec.	15 S 46
03	12♑31 R
06	12 28
09	12 25
12	12 24
15	12 22
18	12 22
21	12 22D
24	12 23
27	12 25
30	12 28

18 03:53 12♑22 D

October 2003

Day	S. T.	☉	☽	☿	♀	♂	♃	♄	♅	♆	♇	☊ True
	h m s	° ' "	° ' "	° '	° '	° '	° '	° '	° '	° '	° '	° '
01 We	00 37 16	07♎ 24 27	13♐ 50 25	20♍ 29	19♎ 04	00♓ 13	07♍ 20	12♋ 40	29♒R29	10♒R32	17♐ 32	21♉D12
02 Th	00 41 12	08 23 27	28 01 38	21 55	20 19	00 16	07 33	12 43	29 27	10 31	17 33	21 13
03 Fr	00 45 09	09 22 29	11♑ 57 26	23 25	21 33	00 20	07 45	12 45	29 26	10 31	17 34	21R 13
04 Sa	00 49 05	10 21 32	25 37 42	24 58	22 48	00 25	07 57	12 48	29 24	10 30	17 35	21 12
05 Su	00 53 02	11 20 37	09♒ 02 55	26 35	24 03	00 31	08 09	12 50	29 22	10 29	17 36	21 09
06 Mo	00 56 58	12 19 44	22 13 46	28 13	25 17	00 37	08 20	12 52	29 21	10 29	17 37	21 05
07 Tu	01 00 55	13 18 52	05♓ 11 00	29 54	26 32	00 45	08 32	12 54	29 19	10 28	17 38	21 00
08 We	01 04 52	14 18 02	17 55 13	01♎ 36	27 46	00 53	08 44	12 56	29 18	10 28	17 40	20 54
09 Th	01 08 48	15 17 14	00♈ 27 02	03 19	29 01	01 01	08 56	12 58	29 16	10 27	17 41	20 48
10 Fr	01 12 45	16 16 28	12 47 07	05 03	00♏ 16	01 11	09 07	13 00	29 15	10 27	17 42	20 43
11 Sa	01 16 41	17 15 45	24 56 26	06 47	01 30	01 21	09 19	13 02	29 14	10 26	17 44	20 39
12 Su	01 20 38	18 15 03	06♉ 56 23	08 32	02 45	01 32	09 30	13 03	29 12	10 26	17 45	20 37
13 Mo	01 24 34	19 14 23	18 49 00	10 17	04 00	01 44	09 42	13 05	29 11	10 26	17 46	20D 37
14 Tu	01 28 31	20 13 46	00♊ 36 56	12 02	05 14	01 56	09 53	13 06	29 10	10 25	17 48	20 38
15 We	01 32 27	21 13 11	12 23 29	13 47	06 29	02 09	10 05	13 08	29 09	10 25	17 49	20 40
16 Th	01 36 24	22 12 38	24 12 34	15 32	07 44	02 22	10 16	13 09	29 07	10 25	17 51	20 42
17 Fr	01 40 20	23 12 07	06♋ 08 36	17 16	08 58	02 36	10 27	13 10	29 06	10 25	17 52	20 44
18 Sa	01 44 17	24 11 39	18 16 24	19 00	10 13	02 51	10 38	13 11	29 05	10 24	17 54	20 45
19 Su	01 48 14	25 11 13	00♌ 40 48	20 43	11 27	03 07	10 49	13 12	29 04	10 24	17 55	20R 45
20 Mo	01 52 10	26 10 49	13 26 23	22 26	12 42	03 23	11 00	13 12	29 03	10 24	17 57	20 44
21 Tu	01 56 07	27 10 28	26 36 52	24 09	13 57	03 39	11 11	13 13	29 02	10 24	17 58	20 42
22 We	02 00 03	28 10 08	10♍ 14 34	25 51	15 11	03 57	11 22	13 14	29 01	10 24	18 00	20 39
23 Th	02 03 60	29 09 51	24 19 37	27 32	16 26	04 14	11 32	13 14	29 01	10D 24	18 02	20 35
24 Fr	02 07 56	00♏ 09 36	08♎ 49 36	29 13	17 41	04 33	11 43	13 14	29 00	10 24	18 03	20 32
25 Sa	02 11 53	01 09 23	23 39 19	00♏ 53	18 55	04 52	11 54	13 14	28 59	10 24	18 05	20 29
26 Su	02 15 49	02 09 12	08♏ 41 16	02 32	20 10	05 11	12 04	13 R 14	28 58	10 24	18 07	20 27
27 Mo	02 19 46	03 09 04	23 46 38	04 11	21 25	05 31	12 14	13 14	28 58	10 24	18 08	20 26
28 Tu	02 23 43	04 08 57	08♐ 46 32	05 50	22 39	05 52	12 25	13 14	28 57	10 24	18 10	20 26
29 We	02 27 39	05 08 52	23 33 17	07 28	23 54	06 13	12 35	13 14	28 56	10 25	18 12	20D 26
30 Th	02 31 36	06 08 48	08♑ 01 20	09 05	25 09	06 34	12 45	13 14	28 56	10 25	18 14	20 26
31 Fr	02 35 32	07 08 46	22 07 30	10 42	26 23	06 56	12 55	13 13	28 56	10 25	18 15	20 27

Data for	10-01-2003
Julian Day	2452913.50
Ayanamsa	23 54 19
SVP	05 ♓ 12 13
☽ ☊ Mean	22 ♉ 35 R

● ◐ PHASES ○ ◑

Day	Time	Phase	Position
02	19:10	◐	09♑11
10	07:27	○	16♈35
18	12:32	◑	24♋43
25	12:51	●	01♏41

LAST ASPECT Day	h m	☽ INGRESS Day	h m	Sign
02	02:26	02	03:22	♑
03	22:41	04	07:46	♒
06	13:07	06	14:21	♓
07	23:30	08	23:08	♈
11	08:31	11	10:05	♉
13	21:03	13	22:45	♊
16	09:54	16	11:41	♋
18	12:32	18	22:42	♌
21	04:19	21	06:02	♍
22	13:20	23	09:28	♎
25	08:31	25	10:09	♏
27	08:16	27	09:56	♐
29	08:52	29	10:37	♑
31	08:07	31	13:42	♒

DECLINATION

Day	☉	☽	☿	♀	♂	♃	♄	♅	♆	♇
01 We	02S 56	24S 26	05N16	06S 34	15S 40	09N38	22N06	12S 23	17S 36	13S 57
02 Th	03 20	26 32	04 47	07 04	15 34	09 34	22 06	12 24	17 37	13 58
03 Fr	03 43	26 54	04 16	07 34	15 28	09 29	22 06	12 24	17 37	13 58
04 Sa	04 06	25 36	03 42	08 03	15 22	09 25	22 05	12 25	17 37	13 58
05 Su	04 29	22 53	03 06	08 32	15 16	09 21	22 05	12 26	17 37	13 59
06 Mo	04 52	19 00	02 28	09 02	15 09	09 16	22 05	12 26	17 37	13 59
07 Tu	05 15	14 17	01 49	09 30	15 02	09 12	22 05	12 27	17 38	14 00
08 We	05 38	09 02	01 08	09 59	14 55	09 08	22 05	12 27	17 38	14 00
09 Th	06 01	03 30	00 26	10 28	14 47	09 03	22 05	12 28	17 38	14 01
10 Fr	06 24	02N06	00S 17	10 56	14 40	08 59	22 04	12 28	17 38	14 01
11 Sa	06 47	07 32	01 00	11 24	14 32	08 55	22 04	12 28	17 38	14 02
12 Su	07 09	12 39	01 44	11 52	14 23	08 51	22 04	12 29	17 38	14 02
13 Mo	07 32	17 15	02 28	12 19	14 15	08 47	22 04	12 29	17 38	14 03
14 Tu	07 54	21 10	03 13	12 46	14 06	08 42	22 04	12 30	17 38	14 03
15 We	08 17	24 13	03 57	13 13	13 58	08 38	22 04	12 30	17 38	14 04
16 Th	08 39	26 13	04 42	13 40	13 49	08 34	22 04	12 30	17 39	14 04
17 Fr	09 01	27 03	05 26	14 06	13 39	08 30	22 03	12 31	17 39	14 04
18 Sa	09 23	26 35	06 11	14 32	13 30	08 26	22 03	12 31	17 39	14 05
19 Su	09 45	24 49	06 55	14 57	13 20	08 22	22 03	12 32	17 39	14 05
20 Mo	10 06	21 47	07 38	15 22	13 11	08 18	22 03	12 32	17 39	14 06
21 Tu	10 28	17 33	08 22	15 47	13 01	08 14	22 03	12 32	17 39	14 06
22 We	10 49	12 19	09 04	16 12	12 51	08 10	22 03	12 32	17 39	14 07
23 Th	11 11	06 16	09 47	16 36	12 40	08 06	22 03	12 33	17 39	14 07
24 Fr	11 32	00S 16	10 28	16 59	12 30	08 02	22 03	12 33	17 39	14 08
25 Sa	11 53	06 58	11 10	17 22	12 19	07 58	22 03	12 33	17 39	14 08
26 Su	12 13	13 22	11 50	17 45	12 08	07 54	22 03	12 33	17 39	14 08
27 Mo	12 34	19 00	12 30	18 07	11 57	07 51	22 03	12 34	17 39	14 09
28 Tu	12 54	23 24	13 09	18 29	11 46	07 47	22 03	12 34	17 39	14 09
29 We	13 14	26 10	13 48	18 50	11 35	07 43	22 03	12 34	17 39	14 10
30 Th	13 34	27 05	14 26	19 11	11 23	07 39	22 03	12 34	17 39	14 10
31 Fr	13 54	26 12	15 03	19 31	11 12	07 36	22 03	12 34	17 39	14 11

ASPECTARIAN

01 06:12 ☽ ☌ ♇
09:38 ☽ ⚹ ♀
12:25 ☽ □ ☿
02 02:26 ☽ ⚹ ♅
03:51 ☽ ⚹ ♂
16:35 ☽ △ ♃
03 01:23 ☽ ☍ ♄
13:54 ☿ # ☉
18:29 ☽ □ ♀
22:41 ☽ △ ☿
04 03:22 ☉ △ ♆
05 02:36 ☽ ☌ ♆
04:29 ☽ △ ☉
05:23 ☽ # ♄
15:33 ☽ ⚹ ♇
06 06:13 ☽ △ ♀
07:20 ☽ ∥ ♆
10:35 ♀ # ♃
13:07 ☽ ☌ ♅
13:41 ☉ □ ♄
15:39 ☽ ☌ ♂
20:16 ☽ ∥ ♂
07 01:20 ☽ ∥ ♇
01:28 ☿ ♎
06:23 ☽ ☍ ♃
08:35 ☽ ∥ ♅
14:33 ☽ △ ♄
20:05 ☽ ∥ ♀
23:30 ☽ □ ♇
23:33 ☽ # ♃
08 13:50 ☽ ∥ ☉
09 04:48 ♀ △ ♅
06:26 ☽ ☍ ☿
14:56 ☽ ∥ ☿
15:00 ☽ # ☿
18:56 ♀ ♏
19:25 ☽ ⚹ ♆
10 00:26 ☽ □ ♄
09:41 ☽ △ ♇
20:18 ☽ # ☉
20:32 ♀ △ ♂
11 06:10 ☽ ∥ ♃
08:31 ☽ ⚹ ♅
11:33 ☉ ⚹ ♇
12:59 ☽ ⚹ ♂
14:37 ☽ ☍ ♀
19:44 ☽ # ♀
23:08 ☽ # ♅
12 05:15 ☽ △ ♃
06:53 ☽ # ♇
07:02 ☽ □ ♆
08:26 ☽ # ♂
12:22 ☽ ⚹ ♄
13 01:53 ☿ △ ♆
02:10 ☽ # ♆
09:07 ♀ ∥ ♅
21:03 ☽ □ ♅
14 02:44 ☽ □ ♂
06:21 ☽ ∥ ♄
08:54 ♂ ∥ ♇
14:49 ☿ □ ♄
19:13 ☽ □ ♃
19:59 ☽ △ ♆
15 03:20 ☽ △ ☿
11:04 ☽ ☍ ♇
19:35 ☽ △ ☉
19:37 ☉ # ♃
16 06:04 ♀ ∥ ♂
09:54 ☽ △ ♅
16:47 ☽ [illegible] ♂
22:37 ♀ ∥ ♇
17 06:17 ☽ △ ♀
08:27 ☿ ⚹ ♇
08:42 ☽ ⚹ ♃
13:58 ☽ ☌ ♄
18 01:39 ☽ □ ☿
03:44 ♀ □ ♆
09:33 ♀ ⚹ ♃
19 18:21 ☽ ☍ ♆
22:08 ☽ ∥ ♄
22:29 ☽ □ ♀
20 08:19 ☽ △ ♇
09:49 ♀ △ ♄
18:55 ☽ ⚹ ☿
20:06 ☿ # ♃
23:31 ☽ # ♆
21 01:05 ☽ ⚹ ☉
04:19 ☽ ☍ ♅
07:56 ☽ # ♀
12:46 ☽ ☍ ♂
16:13 ☽ # ♇
21:38 ☽ # ♂
23:00 ☽ # ♅
22 01:58 ☽ ☌ ♃
05:09 ☽ ⚹ ♄
05:48 ☽ # ☉
09:21 ☽ ⚹ ♀
11:48 ☽ # ☿
13:20 ☽ □ ♇
16:54 ☽ ∥ ♃
20:18 ☉ △ ♅
23 01:55 ♆ SD
17:06 ♂ ∥ [illegible]
20:09 ☉ ♏
20:57 ☿ △ ♅
24 02:34 ☽ △ ♆
07:11 ☽ □ ♄
11:20 ☿ ♏
15:01 ☽ ⚹ ♇
25 03:37 ☽ # ♃
08:31 ☽ △ ♅
09:58 ☿ ☌ ☉
13:00 ☽ ☌ ☿
17:19 ☽ ∥ ☿
17:31 ♀ ∥ ♆
18:18 ☽ △ ♂
19:16 ☽ ∥ ☉
19:20 ☽ ∥ ♂
20:12 ☉ ∥ ♂
20:47 ☽ ∥ ♅
23:43 ♄ SR
26 02:44 ☽ □ ♆
03:03 ☽ ∥ ♇
05:26 ☽ ⚹ ♃
07:14 ☽ △ ♄
08:32 ☿ ∥ ♂
17:45 ☽ ∥ ♆
19:34 ☽ ∥ ♀
19:54 ☽ ☌ ♀
23:39 ☉ ∥ ♅
27 02:07 ☿ ∥ ♅
04:44 ☿ ∥ ☉
08:16 ☽ □ ♅
15:42 ☽ # ♄
19:12 ☽ □ ♂
28 00:29 ☿ △ ♂
02:38 ☽ ⚹ ♆
05:56 ☽ □ ♃
15:14 ☽ ☌ ♇
29 08:52 ☽ ⚹ ♅
14:05 ☿ ∥ ♇
20:37 ☽ ⚹ ☉
21:30 ☽ ⚹ ♂
30 02:01 ☽ ⚹ ☿
08:04 ☽ △ ♃
08:46 ☽ ☍ ♄
15:55 ☉ △ ♂
19:42 [illegible] □ ♆
31 08:07 ☽ ⚹ ♀
21:04 ☉ ∥ ♇

⚷ Chiron

01 Dec.	16 S 02
03	12♑31
06	12 35
09	12 39
12	12 45
15	12 51
18	12 57
21	13 05
24	13 13
27	13 21
30	13 31

09 01:20 16♉13 ✹ Total Lunar Eclipse (mag 1.022)
23 22:50 01♐14 ⊛ Total Solar Eclipse

November 2003

Day	S. T. h m s	☉ ° ' "	☽ ° ' "	☿ ° '	♀ ° '	♂ ° '	♃ ° '	♄ ° '	♅ ° '	♆ ° '	♇ ° '	☊ True ° '
01 Sa	02 39 29	08♏ 08 46	05♒ 50 48	12♏ 19	27♏ 38	07♓ 18	13♍ 05	13♋R12	28♒R55	10♒ 25	18♐ 17	20♉ 27
02 Su	02 43 25	09 08 47	19 11 59	13 55	28 53	07 41	13 15	13 12	28 55	10 26	18 19	20R 27
03 Mo	02 47 22	10 08 50	02♓ 12 58	15 30	00♐ 07	08 05	13 24	13 11	28 54	10 26	18 21	20 26
04 Tu	02 51 18	11 08 55	14 56 12	17 05	01 22	08 29	13 34	13 10	28 54	10 26	18 23	20 26
05 We	02 55 15	12 09 01	27 24 24	18 40	02 36	08 53	13 43	13 09	28 54	10 27	18 25	20 25
06 Th	02 59 12	13 09 08	09♈ 40 10	20 14	03 51	09 17	13 53	13 08	28 54	10 27	18 27	20 25
07 Fr	03 03 08	14 09 17	21 45 58	21 48	05 06	09 42	14 02	13 06	28 54	10 28	18 29	20D 25
08 Sa	03 07 05	15 09 28	03♉ 44 01	23 22	06 20	10 08	14 11	13 05	28 54	10 28	18 31	20 26
09 Su	03 11 01	16 09 41	15 36 24	24 55	07 35	10 34	14 20	13 03	28D 54	10 29	18 33	20 26
10 Mo	03 14 58	17 09 56	27 25 13	26 27	08 49	11 00	14 29	13 02	28 54	10 29	18 35	20 27
11 Tu	03 18 54	18 10 12	09♊ 12 38	28 00	10 04	11 26	14 38	13 00	28 54	10 30	18 37	20 28
12 We	03 22 51	19 10 30	21 01 02	29 32	11 18	11 53	14 47	12 58	28 54	10 31	18 39	20 29
13 Th	03 26 47	20 10 50	02♋ 53 08	01♐ 04	12 33	12 20	14 55	12 56	28 54	10 31	18 41	20 29
14 Fr	03 30 44	21 11 12	14 52 01	02 35	13 48	12 48	15 04	12 54	28 54	10 32	18 43	20R 29
15 Sa	03 34 41	22 11 36	27 01 09	04 06	15 02	13 15	15 12	12 52	28 55	10 33	18 45	20 29
16 Su	03 38 37	23 12 01	09♌ 24 23	05 37	16 17	13 44	15 20	12 50	28 55	10 34	18 47	20 28
17 Mo	03 42 34	24 12 29	22 05 39	07 08	17 31	14 12	15 28	12 48	28 56	10 35	18 49	20 28
18 Tu	03 46 30	25 12 58	05♍ 08 42	08 38	18 46	14 41	15 36	12 45	28 56	10 35	18 51	20 28
19 We	03 50 27	26 13 29	18 36 35	10 08	20 00	15 10	15 44	12 43	28 57	10 36	18 53	20 27
20 Th	03 54 23	27 14 02	02♎ 31 05	11 37	21 15	15 39	15 52	12 40	28 57	10 37	18 56	20D 27
21 Fr	03 58 20	28 14 37	16 52 01	13 06	22 29	16 09	15 59	12 37	28 58	10 38	18 58	20 28
22 Sa	04 02 16	29 15 13	01♏ 36 41	14 35	23 44	16 38	16 07	12 35	28 58	10 39	19 00	20R 28
23 Su	04 06 13	00♐ 15 51	16 39 30	16 03	24 58	17 09	16 14	12 32	28 59	10 40	19 02	20 27
24 Mo	04 10 10	01 16 31	01♐ 52 25	17 31	26 13	17 39	16 21	12 29	29 00	10 41	19 04	20 26
25 Tu	04 14 06	02 17 12	17 05 48	18 59	27 27	18 10	16 28	12 25	29 01	10 42	19 06	20 25
26 We	04 18 03	03 17 54	02♑ 09 47	20 25	28 42	18 41	16 35	12 22	29 02	10 43	19 09	20 23
27 Th	04 21 59	04 18 38	16 55 51	21 52	29 57	19 12	16 42	12 19	29 02	10 45	19 11	20 21
28 Fr	04 25 56	05 19 23	01♒ 17 53	23 17	01♑ 11	19 43	16 48	12 16	29 03	10 46	19 13	20 19
29 Sa	04 29 52	06 20 09	15 12 41	24 41	02 25	20 15	16 55	12 12	29 04	10 47	19 15	20 18
30 Su	04 33 49	07 20 55	28 39 52	26 05	03 40	20 47	17 01	12 09	29 05	10 48	19 18	20 17

Data for	11-01-2003
Julian Day	2452944.50
Ayanamsa	23 54 23
SVP	05 ♓ 12 08
☽ ☊ Mean	20 ♉ 56 R

● ◐ PHASES ○ ◑

Day	Time	Phase	Position
01	04:25	◐	08♒20
09	01:13	✹	16♉13
17	04:16	◑	24♌23
23	22:59	⊛	01♐14
30	17:16	◐	08♓05

LAST ASPECT Day	h m	☽ INGRESS Day	h m	Sign
02	19:41	02	19:52	♓
04	06:36	05	05:03	♈
07	14:16	07	16:29	♉
10	03:00	10	05:15	♊
12	15:58	12	18:11	♋
14	13:40	15	05:49	♌
17	12:40	17	14:37	♍
19	14:16	19	19:43	♎
21	19:45	21	21:24	♏
23	19:28	23	21:03	♐
25	18:58	25	20:32	♑
27	03:52	27	21:48	♒
30	00:47	30	02:26	♓

DECLINATION

Day	☉	☽	☿	♀	♂	♃	♄	♅	♆	♇
01 Sa	14S 13	23S 45	15S 39	19S 51	11S 00	07N32	22N03	12S 34	17S 39	14S 11
02 Su	14 33	20 05	16 14	20 10	10 48	07 29	22 04	12 34	17 38	14 12
03 Mo	14 52	15 31	16 49	20 28	10 36	07 25	22 04	12 34	17 38	14 12
04 Tu	15 10	10 23	17 22	20 46	10 24	07 22	22 04	12 34	17 38	14 12
05 We	15 29	04 55	17 55	21 04	10 12	07 18	22 04	12 35	17 38	14 13
06 Th	15 47	00N39	18 27	21 21	09 59	07 15	22 04	12 35	17 38	14 13
07 Fr	16 05	06 07	18 58	21 37	09 47	07 11	22 04	12 35	17 38	14 14
08 Sa	16 23	11 19	19 28	21 53	09 34	07 08	22 04	12 35	17 38	14 14
09 Su	16 40	16 05	19 57	22 08	09 21	07 05	22 05	12 34	17 38	14 14
10 Mo	16 58	20 12	20 26	22 22	09 08	07 01	22 05	12 34	17 38	14 15
11 Tu	17 15	23 31	20 53	22 36	08 55	06 58	22 05	12 34	17 37	14 15
12 We	17 31	25 49	21 19	22 49	08 42	06 55	22 05	12 34	17 37	14 16
13 Th	17 47	26 59	21 44	23 02	08 29	06 52	22 05	12 34	17 37	14 16
14 Fr	18 03	26 54	22 08	23 13	08 16	06 49	22 06	12 34	17 37	14 16
15 Sa	18 19	25 31	22 31	23 25	08 02	06 46	22 06	12 34	17 37	14 17
16 Su	18 34	22 54	22 53	23 35	07 49	06 43	22 06	12 34	17 36	14 17
17 Mo	18 49	19 08	23 13	23 45	07 35	06 40	22 06	12 34	17 36	14 18
18 Tu	19 04	14 23	23 33	23 54	07 21	06 37	22 07	12 33	17 36	14 18
19 We	19 19	08 48	23 51	24 02	07 08	06 34	22 07	12 33	17 36	14 18
20 Th	19 33	02 37	24 08	24 10	06 54	06 31	22 07	12 33	17 36	14 19
21 Fr	19 46	03S 55	24 24	24 17	06 40	06 29	22 07	12 33	17 35	14 19
22 Sa	19 59	10 25	24 39	24 23	06 26	06 26	22 08	12 32	17 35	14 19
23 Su	20 12	16 28	24 52	24 28	06 11	06 23	22 08	12 32	17 35	14 20
24 Mo	20 25	21 33	25 04	24 33	05 57	06 21	22 08	12 32	17 34	14 20
25 Tu	20 37	25 10	25 15	24 37	05 43	06 18	22 09	12 32	17 34	14 20
26 We	20 49	26 56	25 24	24 40	05 28	06 16	22 09	12 31	17 34	14 21
27 Th	21 00	26 44	25 32	24 42	05 14	06 13	22 09	12 31	17 34	14 21
28 Fr	21 11	24 44	25 39	24 44	04 59	06 11	22 10	12 30	17 33	14 21
29 Sa	21 22	21 17	25 44	24 45	04 45	06 09	22 10	12 30	17 33	14 22
30 Su	21 32	16 49	25 48	24 45	04 30	06 07	22 11	12 30	17 33	14 22

ASPECTARIAN

Day	Time	Aspect
01	08:09	☽ ☌ ♆
	11:59	☽ # ♄
	12:48	☿ ⚹ ♃
	13:07	☽ □ ☿
	13:17	☿ △ ♄
	17:13	♃ ⚹ ♄
	22:24	☽ ⚹ ♇
	23:32	☽ ∥ ♀
02	00:44	♀ □ ♅
	13:17	☽ ∥ ♆
	17:51	☽ ☌ ♅
	18:13	☽ ∥ ☿
	19:41	☽ □ ♀
	21:42	♀ ♐
03	02:59	☽ ∥ ☉
	06:20	☽ ∥ ♇
	06:55	☉ □ ♆
	11:21	☽ ☌ ♂
	13:58	☽ ∥ ♅
	16:11	☽ △ ☉
	20:38	☽ △ ♄
	21:21	☽ ☍ ♃
	23:53	☽ ∥ ♂
04	04:42	☽ △ ☿
	06:36	☽ □ ♇
	11:27	☿ ∥ ♆
	13:30	☽ # ♃
05	11:16	☽ △ ♀
	23:24	☉ △ ♄
06	01:33	☽ ⚹ ♆
	06:49	☽ □ ♄
	17:26	☽ △ ♇
	20:34	☉ ⚹ ♃
07	04:46	☽ ∥ ♃
	14:16	☽ ⚹ ♅
	16:02	☽ # ♂
08	06:04	☽ # ♅
	12:45	♅ SD
	13:23	☽ ⚹ ♂
	13:37	☽ □ ♆
	14:22	☽ # ♇
	18:38	♀ # ♄
	18:51	☽ ⚹ ♄
	21:24	☽ △ ♃
09	03:25	☽ # ☉
	08:30	☽ # ♆
	21:45	☽ ☍ ☿
10	01:35	☽ # ☿
	03:00	☽ □ ♅
	12:44	☽ ∥ ♄
	16:02	☽ # ♀
11	01:57	☽ ☍ ♀
	02:38	☽ △ ♆
	04:42	☽ □ ♂
	08:31	♀ ⚹ ♆
	11:10	☽ □ ♃
	14:05	☿ □ ♅
	19:10	☽ ☍ ♇
12	07:19	☿ ♐
	08:48	☉ ∥ ♆
	15:58	☽ △ ♅
	17:27	♀ □ ♂
13	19:42	☽ △ ♂
	20:06	☽ ☌ ♄
	21:36	☿ # ♄
14	00:23	☽ ⚹ ♃
	05:29	♂ △ ♄
	13:40	☽ △ ☉
15	03:35	♀ □ ♃
	15:43	☽ △ ☿
	19:02	☽ # ♀
16	00:13	☽ # ☿
	02:13	☽ ☍ ♆
	05:47	☽ ∥ ♄
	14:30	☽ △ ♀
	17:51	☽ △ ♇
17	01:42	☽ # ☉
	08:20	☽ # ♆
	12:40	☽ ☍ ♅
18	00:25	☽ # ♇
	01:51	♀ ☌ ♇
	07:05	☽ □ ☿
	08:14	☽ # ♅
	13:37	☽ ⚹ ♄
	17:42	☽ ☍ ♂
	18:54	☽ ☌ ♃
19	00:30	☽ □ ♇
	02:41	☽ □ ♀
	06:58	☽ # ♂
	07:44	☿ ⚹ ♆
	08:57	☽ ∥ ♃
	14:16	☽ ⚹ ☉
20	03:43	☿ ∥ ♀
	13:39	☽ △ ♆
	14:01	♂ ☍ ♃
	17:00	☽ □ ♄
	17:04	☽ ⚹ ☿
21	03:27	☽ ⚹ ♇
	09:18	☽ # ♃
	09:42	☽ ∥ ♂
	10:04	☽ ⚹ ♀
	17:14	☉ □ ♅
	19:45	☽ △ ♅
	23:20	♂ # ♃
22	08:06	☽ ∥ ♅
	14:29	☽ □ ♆
	15:09	☽ ∥ ♇
	17:28	☽ △ ♄
	17:44	☉ ♐
	23:19	☽ ⚹ ♃
23	00:48	☽ △ ♂
	03:10	☿ □ ♃
	04:47	☽ ∥ ♆
	17:46	☽ ∥ ☉
	19:28	☽ □ ♅
24	03:14	☿ □ ♂
	03:15	☽ # ♄
	13:54	☽ ⚹ ♆
	19:15	☽ ∥ ♀
	23:00	☽ □ ♃
25	00:40	☽ ∥ ☿
	01:45	☽ □ ♂
	02:12	☿ ☌ ♇
	03:11	☽ ☌ ♇
	03:18	☽ ☌ ☿
	17:57	☽ ☌ ♀
	18:58	☽ ⚹ ♅
26	06:22	♀ ⚹ ♅
	16:28	☽ ☍ ♄
	23:21	♂ □ ♇
	23:37	☽ △ ♃
27	01:08	♀ ♑
	03:52	☽ ⚹ ♂
	15:43	☽ ∥ ☿
28	00:01	☽ ∥ ♀
	07:24	☽ ⚹ ☉
	16:16	☽ ☌ ♆
	18:38	☽ # ♄
	23:35	☽ ∥ ☉
29	07:09	☽ ⚹ ♇
	18:47	☽ ⚹ ☿
	20:26	☽ ∥ ♆
30	00:47	☽ ☌ ♅
	10:05	☽ ⚹ ♀
	11:51	☽ ∥ ♇
	20:32	☽ ∥ ♅

⚷ Chiron

01 Dec.	16 S 11
02	13♑41
05	13 51
08	14 02
11	14 14
14	14 26
17	14 38
20	14 51
23	15 05
26	15 18
29	15 33

December 2003

Day	S. T. h m s	☉ ° ' "	☽ ° ' "	☿ ° '	♀ ° '	♂ ° '	♃ ° '	♄ ° '	♅ ° '	♆ ° '	♇ ° '	☊ True ° '
01 Mo	04 37 46	08♐21 43	11♓41 19	27♐27	04♑54	21♓19	17♍07	12♋R05	29♒07	10♒49	19♐20	20♉D17
02 Tu	04 41 42	09 22 31	24 20 19	28 48	06 09	21 51	17 13	12 01	29 08	10 51	19 22	20 18
03 We	04 45 39	10 23 21	06♈40 58	00♑08	07 23	22 23	17 19	11 58	29 09	10 52	19 24	20 20
04 Th	04 49 35	11 24 11	18 47 32	01 26	08 38	22 56	17 25	11 54	29 10	10 53	19 27	20 22
05 Fr	04 53 32	12 25 02	00♉44 12	02 42	09 52	23 29	17 30	11 50	29 12	10 55	19 29	20 25
06 Sa	04 57 28	13 25 54	12 34 42	03 55	11 06	24 02	17 36	11 46	29 13	10 56	19 31	20 27
07 Su	05 01 25	14 26 47	24 22 22	05 07	12 21	24 35	17 41	11 42	29 14	10 58	19 33	20 28
08 Mo	05 05 21	15 27 41	06♊09 59	06 14	13 35	25 08	17 46	11 38	29 16	10 59	19 36	20R 28
09 Tu	05 09 18	16 28 36	17 59 53	07 19	14 49	25 42	17 51	11 33	29 17	11 01	19 38	20 26
10 We	05 13 15	17 29 32	29 54 03	08 19	16 04	26 15	17 55	11 29	29 19	11 02	19 40	20 22
11 Th	05 17 11	18 30 29	11♋54 15	09 15	17 18	26 49	18 00	11 25	29 20	11 04	19 42	20 18
12 Fr	05 21 08	19 31 27	24 02 11	10 06	18 32	27 23	18 04	11 21	29 22	11 05	19 45	20 12
13 Sa	05 25 04	20 32 26	06♌19 42	10 50	19 47	27 57	18 08	11 16	29 24	11 07	19 47	20 07
14 Su	05 29 01	21 33 26	18 48 58	11 28	21 01	28 32	18 12	11 12	29 26	11 08	19 49	20 02
15 Mo	05 32 57	22 34 27	01♍32 22	11 58	22 15	29 06	18 16	11 07	29 27	11 10	19 52	19 59
16 Tu	05 36 54	23 35 29	14 32 33	12 19	23 29	29 41	18 20	11 03	29 29	11 12	19 54	19 57
17 We	05 40 50	24 36 32	27 52 06	12 31	24 43	00♈15	18 23	10 58	29 31	11 13	19 56	19 56
18 Th	05 44 47	25 37 35	11♎33 17	12R 33	25 57	00 50	18 27	10 53	29 33	11 15	19 58	19D 56
19 Fr	05 48 44	26 38 40	25 37 24	12 24	27 12	01 25	18 30	10 49	29 35	11 17	20 01	19 56
20 Sa	05 52 40	27 39 46	10♏04 06	12 03	28 26	02 00	18 33	10 44	29 37	11 19	20 03	19R 57
21 Su	05 56 37	28 40 53	24 50 47	11 31	29 40	02 35	18 35	10 39	29 39	11 20	20 05	19 56
22 Mo	06 00 33	29 42 00	09♐52 05	10 47	00♒54	03 11	18 38	10 34	29 41	11 22	20 07	19 53
23 Tu	06 04 30	00♑43 08	25 00 09	09 52	02 08	03 46	18 40	10 29	29 43	11 24	20 10	19 49
24 We	06 08 26	01 44 17	10♑05 27	08 48	03 22	04 22	18 43	10 25	29 45	11 26	20 12	19 44
25 Th	06 12 23	02 45 25	24 58 20	07 35	04 36	04 57	18 45	10 20	29 47	11 28	20 14	19 37
26 Fr	06 16 20	03 46 35	09♒30 27	06 17	05 50	05 33	18 46	10 15	29 50	11 30	20 16	19 30
27 Sa	06 20 16	04 47 44	23 36 07	04 55	07 04	06 09	18 48	10 10	29 52	11 31	20 19	19 24
28 Su	06 24 13	05 48 53	07♓12 50	03 33	08 18	06 45	18 50	10 05	29 54	11 33	20 21	19 19
29 Mo	06 28 09	06 50 02	20 21 01	02 13	09 32	07 21	18 51	10 00	29 57	11 35	20 23	19 15
30 Tu	06 32 06	07 51 11	03♈03 29	00 57	10 45	07 57	18 52	09 55	29 59	11 37	20 25	19 14
31 We	06 36 02	08 52 20	15 24 34	29♐49	11 59	08 34	18 53	09 50	00♓01	11 39	20 27	19D 14

Data for	12-01-2003
Julian Day	2452974.50
Ayanamsa	23 54 28
SVP	05 ♓ 12 05
☽ ☊ Mean	19 ♉ 21 R

● ◐ PHASES ○ ◑

Day	Time	Phase	Position
08	20:38	○	16♊20
16	17:42	◑	24♍21
23	09:43	●	01♑08
30	10:04	◐	08♈17

LAST ASPECT ☽ INGRESS

Day	h m	Day	h m	Sign
02	09:40	02	10:56	♈
04	20:53	04	22:31	♉
07	09:56	07	11:27	♊
09	22:49	10	00:12	♋
12	06:54	12	11:41	♌
14	20:06	14	21:07	♍
16	17:49	17	03:47	♎
19	06:39	19	07:20	♏
21	07:43	21	08:16	♐
23	07:30	23	07:56	♑
24	13:53	25	08:14	♒
27	10:58	27	11:10	♓
29	00:04	29	18:09	♈

DECLINATION

Day	☉	☽	☿	♀	♂	♃	♄	♅	♆	♇
01 Mo	21S42	11S43	25S50	24S45	04S15	06N04	22N11	12S29	17S32	14S22
02 Tu	21 51	06 16	25 51	24 43	04 00	06 02	22 11	12 29	17 32	14 23
03 We	22 00	00 42	25 51	24 41	03 46	06 00	22 12	12 28	17 32	14 23
04 Th	22 09	04N48	25 49	24 38	03 31	05 58	22 12	12 28	17 31	14 23
05 Fr	22 17	10 04	25 46	24 35	03 16	05 56	22 12	12 27	17 31	14 24
06 Sa	22 25	14 56	25 41	24 30	03 01	05 54	22 13	12 27	17 31	14 24
07 Su	22 32	19 13	25 35	24 25	02 46	05 53	22 13	12 26	17 30	14 24
08 Mo	22 39	22 45	25 28	24 20	02 30	05 51	22 14	12 26	17 30	14 24
09 Tu	22 45	25 19	25 20	24 13	02 15	05 49	22 14	12 25	17 29	14 25
10 We	22 51	26 46	25 10	24 06	02 00	05 48	22 15	12 25	17 29	14 25
11 Th	22 57	26 59	24 59	23 58	01 45	05 46	22 15	12 24	17 29	14 25
12 Fr	23 02	25 55	24 47	23 49	01 29	05 45	22 15	12 23	17 28	14 26
13 Sa	23 06	23 35	24 34	23 40	01 14	05 43	22 16	12 23	17 28	14 26
14 Su	23 10	20 07	24 20	23 29	00 59	05 42	22 16	12 22	17 27	14 26
15 Mo	23 14	15 41	24 05	23 19	00 43	05 41	22 17	12 22	17 27	14 26
16 Tu	23 17	10 27	23 50	23 07	00 28	05 40	22 17	12 21	17 26	14 26
17 We	23 20	04 37	23 35	22 55	00 12	05 38	22 18	12 20	17 26	14 27
18 Th	23 22	01S36	23 19	22 42	00N03	05 37	22 18	12 19	17 25	14 27
19 Fr	23 24	07 55	23 03	22 28	00 19	05 36	22 19	12 19	17 25	14 27
20 Sa	23 25	13 59	22 46	22 14	00 34	05 36	22 19	12 18	17 25	14 27
21 Su	23 26	19 25	22 30	21 59	00 50	05 35	22 20	12 17	17 24	14 28
22 Mo	23 26	23 41	22 14	21 44	01 05	05 34	22 20	12 17	17 24	14 28
23 Tu	23 26	26 19	21 59	21 28	01 21	05 33	22 21	12 16	17 23	14 28
24 We	23 26	27 01	21 44	21 11	01 36	05 33	22 21	12 15	17 23	14 28
25 Th	23 25	25 45	21 29	20 54	01 52	05 32	22 21	12 14	17 22	14 28
26 Fr	23 23	22 46	21 15	20 36	02 08	05 32	22 22	12 13	17 22	14 29
27 Sa	23 21	18 30	21 02	20 17	02 23	05 31	22 22	12 12	17 21	14 29
28 Su	23 19	13 25	20 50	19 58	02 39	05 31	22 23	12 12	17 21	14 29
29 Mo	23 16	07 53	20 39	19 38	02 54	05 31	22 23	12 11	17 20	14 29
30 Tu	23 12	02 12	20 30	19 18	03 10	05 31	22 24	12 10	17 19	14 29
31 We	23 09	03N25	20 23	18 57	03 26	05 31	22 24	12 09	17 19	14 29

ASPECTARIAN

Day	Time	Aspect
01	00:44	☽ △ ♄
	10:19	☽ ☍ ♃
	14:28	☽ □ ♇
	19:00	☽ ☌ ♂
02	01:01	☽ # ♃
	05:54	☿ ✶ ♅
	09:40	☽ □ ☿
	10:13	☽ ∥ ♂
	21:34	☿ ♑
03	01:32	☽ □ ♀
	07:58	☽ △ ☉
	08:16	☽ ✶ ♆
	10:21	☽ □ ♄
	11:34	☉ ✶ ♆
	18:34	☽ # ♂
04	01:18	☽ △ ♇
	05:12	☽ ∥ ♃
	09:34	☉ # ♄
	20:53	☽ ✶ ♅
05	04:26	☽ △ ☿
	11:30	☽ # ♅
	20:39	☽ □ ♆
	20:40	☽ △ ♀
	21:16	☽ # ♇
	22:21	☽ ✶ ♄
06	10:16	☽ △ ♃
	12:04	♀ ☍ ♄
	13:59	☽ # ♆
07	00:27	☽ ✶ ♂
	09:56	☽ □ ♅
	20:06	☽ ∥ ♄
	23:14	☽ # ☉
08	09:48	☽ △ ♆
	13:00	☽ # ♀
	23:41	☽ □ ♃
09	00:03	☽ # ♆
	03:19	☽ ☍ ♇
	16:19	☽ □ ♂
	22:49	☽ △ ♅
10	10:59	☉ □ ♃
	18:18	☽ ☍ ☿
	23:02	☽ ☌ ♄
11	11:56	☽ ☍ ♀
	12:10	☽ ✶ ♃
	14:23	♀ △ ♃
12	05:28	☉ ☌ ♇
	06:54	☽ △ ♂
	14:30	☽ # ☿
	23:23	☽ # ♀
13	03:47	☽ # ☉
	09:16	☽ ☍ ♆
	10:04	☽ ∥ ♄
	14:13	☿ ☍ ♄
14	01:56	☽ △ ♇
	05:40	☽ △ ☉
	15:03	☽ # ♆
	20:06	☽ ☍ ♅
15	06:03	☽ # ♇
	07:41	♀ ∥ ☉
	15:40	☽ # ♅
	17:38	☽ ✶ ♄
	19:51	☽ △ ☿
16	06:55	☽ ☌ ♃
	09:45	☽ □ ♇
	13:24	♂ ♈
	17:50	☽ △ ♀
	19:56	☽ ∥ ♃
17	04:25	☽ ☍ ♂
	16:02	☿ SR
	17:53	☽ # ♂
	18:00	☽ ∥ ♂
	19:36	☿ ∥ ☉
	22:51	☽ □ ♄
	23:28	☽ △ ♆
18	01:43	☽ □ ☿
	14:29	☽ ✶ ♇
	15:16	☽ # ♃
19	01:51	☽ ✶ ☉
	02:53	☽ □ ♀
	06:39	☽ △ ♅
	16:00	♀ # ♄
	17:11	☽ ∥ ♅
20	01:05	☽ △ ♄
	01:57	☽ ∥ ♇
	02:02	☽ □ ♆
	03:09	☽ ✶ ☿
	13:52	☽ ✶ ♃
	14:39	☽ ∥ ♆
21	06:33	♀ ♒
	07:43	☽ □ ♅
	08:25	☽ ✶ ♀
	12:44	☽ ∥ ♀
	12:55	☽ △ ♂
	15:31	☽ # ♄
	15:31	☽ ∥ ☿
	15:42	☿ # ♄
	22:24	☽ ∥ ☉
	23:34	☉ ✶ ♅
22	02:23	☽ ✶ ♆
	06:37	☿ ☍ ♄
	07:04	☉ ♑
	13:57	☽ □ ♃
	16:19	☽ ☌ ♇
23	07:30	☽ ✶ ♅
	14:29	☽ □ ♂
	22:04	☽ ☌ ☿
24	00:30	☽ ☍ ♄
	13:53	☽ △ ♃
25	13:29	♀ ✶ ♂
	17:07	☽ ✶ ♂
	17:18	☽ ☌ ♀
	19:46	☽ ∥ ☉
26	02:30	☽ # ♄
	03:20	☽ ☌ ♆
	08:54	☿ □ ♂
	09:45	☽ ∥ ☿
	13:54	☽ ∥ ♀
	18:18	☽ ✶ ♇
27	01:11	☿ ☌ ☉
	05:42	☽ ∥ ♆
	10:58	☽ ☌ ♅
	18:03	☽ ✶ ☿
	19:11	☽ ∥ ♇
	21:17	☽ ✶ ☉
28	05:08	☽ △ ♄
	05:25	☽ ∥ ♅
	21:12	☽ ☍ ♃
29	00:04	☽ □ ♇
	10:03	☽ # ♃
	20:06	☽ # ♂
	20:20	☽ □ ☿
30	05:56	☉ □ ♂
	09:15	♅ ♓
	09:56	☽ ☌ ♂
	13:10	☽ □ ♄
	16:33	☽ ✶ ♀
	16:37	☽ ✶ ♆
	17:19	♀ ☌ ♆
	19:30	☿ ✶ ♅
	19:53	☿ ♐R
31	00:05	☽ ∥ ♂
	09:12	☽ ∥ ♄
	10:00	☽ △ ♇
	20:58	☉ ☍ ♄

⚷ Chiron

01	Dec.	16 S 08
02	15♑47	
05	16 02	
08	16 18	
11	16 33	
14	16 49	
17	17 05	
20	17 22	
23	17 38	
26	17 55	
29	18 11	

January 2004

Day	S. T.	☉	☽	☿	♀	♂	♃	♄	♅	♆	♇	☊ True
	h m s	° ' "	° ' "	° '	° '	° '	° '	° '	° '	° '	° '	° '
01 Th	06 39 59	09♑53 29	27♈29 19	28♐R49	13♒13	09♈10	18♍53	09♋R45	00♓04	11♒41	20♐30	19♉16
02 Fr	06 43 55	10 54 38	09♉23 00	27 58	14 27	09 46	18 54	09 40	00 06	11 43	20 32	19 18
03 Sa	06 47 52	11 55 47	21 10 43	27 18	15 40	10 23	18 54	09 35	00 09	11 45	20 34	19 19
04 Su	06 51 49	12 56 55	02♊57 03	26 48	16 54	11 00	18R 54	09 30	00 12	11 47	20 36	19R 19
05 Mo	06 55 45	13 58 04	14 45 57	26 28	18 08	11 36	18 54	09 25	00 14	11 49	20 38	19 16
06 Tu	06 59 42	14 59 12	26 40 33	26 18	19 21	12 13	18 54	09 20	00 17	11 51	20 40	19 10
07 We	07 03 38	16 00 20	08♋43 09	26D 17	20 35	12 50	18 53	09 16	00 20	11 54	20 43	19 01
08 Th	07 07 35	17 01 28	20 55 17	26 25	21 48	13 27	18 53	09 11	00 22	11 56	20 45	18 51
09 Fr	07 11 31	18 02 36	03♌17 47	26 41	23 01	14 04	18 52	09 06	00 25	11 58	20 47	18 40
10 Sa	07 15 28	19 03 44	15 51 08	27 04	24 15	14 41	18 51	09 01	00 28	12 00	20 49	18 28
11 Su	07 19 24	20 04 51	28 35 36	27 34	25 28	15 18	18 49	08 56	00 31	12 02	20 51	18 18
12 Mo	07 23 21	21 05 59	11♍31 35	28 10	26 41	15 55	18 48	08 51	00 33	12 04	20 53	18 10
13 Tu	07 27 18	22 07 06	24 39 49	28 51	27 54	16 32	18 46	08 47	00 36	12 06	20 55	18 04
14 We	07 31 14	23 08 13	08♎01 20	29 37	29 07	17 09	18 45	08 42	00 39	12 08	20 57	18 00
15 Th	07 35 11	24 09 21	21 37 29	00♑28	00♓20	17 47	18 43	08 37	00 42	12 11	20 59	17 58
16 Fr	07 39 07	25 10 28	05♏29 30	01 22	01 33	18 24	18 40	08 32	00 45	12 13	21 01	17 57
17 Sa	07 43 04	26 11 35	19 38 03	02 20	02 46	19 01	18 38	08 28	00 48	12 15	21 03	17 56
18 Su	07 47 00	27 12 42	04♐02 29	03 21	03 59	19 39	18 35	08 23	00 51	12 17	21 05	17 54
19 Mo	07 50 57	28 13 48	18 40 16	04 24	05 12	20 16	18 33	08 19	00 54	12 19	21 07	17 50
20 Tu	07 54 53	29 14 55	03♑26 31	05 31	06 25	20 54	18 30	08 14	00 57	12 22	21 09	17 43
21 We	07 58 50	00♒16 01	18 14 19	06 39	07 38	21 31	18 26	08 10	01 00	12 24	21 10	17 34
22 Th	08 02 47	01 17 06	02♒55 30	07 50	08 50	22 09	18 23	08 06	01 03	12 26	21 12	17 23
23 Fr	08 06 43	02 18 10	17 21 57	09 02	10 03	22 47	18 20	08 01	01 06	12 28	21 14	17 12
24 Sa	08 10 40	03 19 14	01♓27 02	10 16	11 15	23 25	18 16	07 57	01 09	12 31	21 16	17 01
25 Su	08 14 36	04 20 17	15 06 34	11 32	12 28	24 02	18 12	07 53	01 13	12 33	21 18	16 52
26 Mo	08 18 33	05 21 19	28 19 12	12 50	13 40	24 40	18 08	07 49	01 16	12 35	21 20	16 44
27 Tu	08 22 29	06 22 20	11♈06 13	14 08	14 52	25 18	18 04	07 45	01 19	12 37	21 21	16 40
28 We	08 26 26	07 23 19	23 30 59	15 28	16 04	25 56	17 59	07 41	01 22	12 40	21 23	16 38
29 Th	08 30 22	08 24 18	05♉38 09	16 49	17 16	26 34	17 55	07 37	01 25	12 42	21 25	16D 37
30 Fr	08 34 19	09 25 15	17 33 07	18 11	18 28	27 12	17 50	07 33	01 29	12 44	21 26	16 38
31 Sa	08 38 16	10 26 11	29 21 31	19 35	19 40	27 50	17 45	07 29	01 32	12 47	21 28	16R 38

Data for	01-01-2004
Julian Day	2453005.50
Ayanamsa	23 54 33
SVP	05 ♓ 12 02
☽ ☊ Mean	17 ♉ 42 R

● ◐ PHASES ○ ◑

07	15:40	○	16♋40
15	04:46	◑	24♎22
21	21:06	●	01♒10
29	06:03	◐	08♉40

LAST ASPECT ☽ Day	h m	INGRESS Day	h m	
01	02:28	01	05:03	♉
02	19:21	03	17:59	♊
05	23:14	06	06:39	♋
07	20:00	08	17:38	♌
10	22:00	11	02:38	♍
13	08:01	13	09:38	♎
15	04:46	15	14:33	♏
17	11:49	17	17:19	♐
19	03:59	19	18:25	♑
21	05:35	21	19:12	♒
23	09:34	23	21:30	♓
25	11:10	26	03:07	♈
28	05:00	28	12:47	♉
30	02:05	31	01:18	♊

DECLINATION

Day	☉	☽	☿	♀	♂	♃	♄	♅	♆	♇
01 Th	23S04	08N48	20S18	18S36	03N41	05N31	22N25	12S08	17S18	14S30
02 Fr	23 00	13 47	20 15	18 14	03 57	05 31	22 25	12 07	17 18	14 30
03 Sa	22 54	18 13	20 13	17 52	04 13	05 31	22 26	12 06	17 17	14 30
04 Su	22 49	21 57	20 14	17 29	04 28	05 31	22 26	12 05	17 17	14 30
05 Mo	22 42	24 46	20 17	17 06	04 44	05 31	22 27	12 04	17 16	14 30
06 Tu	22 36	26 31	20 21	16 42	04 59	05 32	22 27	12 03	17 16	14 30
07 We	22 29	27 02	20 26	16 18	05 15	05 32	22 28	12 03	17 15	14 30
08 Th	22 21	26 15	20 33	15 54	05 30	05 33	22 28	12 02	17 14	14 30
09 Fr	22 13	24 10	20 41	15 29	05 46	05 33	22 28	12 01	17 14	14 31
10 Sa	22 05	20 55	20 50	15 04	06 01	05 34	22 29	12 00	17 13	14 31
11 Su	21 56	16 38	20 59	14 38	06 17	05 35	22 29	11 59	17 13	14 31
12 Mo	21 47	11 32	21 08	14 12	06 32	05 35	22 30	11 57	17 12	14 31
13 Tu	21 37	05 51	21 18	13 45	06 47	05 36	22 30	11 56	17 12	14 31
14 We	21 27	00S11	21 27	13 18	07 03	05 37	22 31	11 55	17 11	14 31
15 Th	21 17	06 20	21 37	12 51	07 18	05 38	22 31	11 54	17 10	14 31
16 Fr	21 06	12 19	21 46	12 24	07 33	05 39	22 32	11 53	17 10	14 31
17 Sa	20 55	17 47	21 55	11 56	07 49	05 41	22 32	11 52	17 09	14 31
18 Su	20 43	22 21	22 03	11 28	08 04	05 42	22 32	11 51	17 09	14 31
19 Mo	20 31	25 33	22 11	10 59	08 19	05 43	22 33	11 50	17 08	14 31
20 Tu	20 19	27 00	22 18	10 31	08 34	05 45	22 33	11 49	17 07	14 31
21 We	20 06	26 33	22 24	10 02	08 49	05 46	22 34	11 48	17 07	14 31
22 Th	19 52	24 15	22 30	09 32	09 04	05 48	22 34	11 47	17 06	14 31
23 Fr	19 39	20 25	22 34	09 03	09 19	05 49	22 34	11 46	17 05	14 31
24 Sa	19 25	15 31	22 38	08 33	09 34	05 51	22 35	11 44	17 05	14 31
25 Su	19 11	09 58	22 40	08 03	09 49	05 53	22 35	11 43	17 04	14 31
26 Mo	18 56	04 08	22 42	07 33	10 04	05 54	22 36	11 42	17 04	14 31
27 Tu	18 41	01N41	22 42	07 03	10 18	05 56	22 36	11 41	17 03	14 31
28 We	18 25	07 17	22 42	06 33	10 33	05 58	22 36	11 40	17 02	14 31
29 Th	18 10	12 29	22 40	06 02	10 48	06 00	22 37	11 39	17 02	14 31
30 Fr	17 54	17 09	22 37	05 32	11 02	06 02	22 37	11 38	17 01	14 31
31 Sa	17 37	21 06	22 33	05 01	11 17	06 04	22 37	11 36	17 00	14 31

ASPECTARIAN

Day	Time	Aspect
01	02:28	☽ △ ☿
	05:12	☽ ⚹ ♅
	15:48	☽ # ♅
	20:23	♂ □ ♄
02	00:35	☽ ⚹ ♄
	03:24	☽ △ ☉
	03:41	☽ # ♇
	04:45	☽ □ ♆
	11:29	☽ □ ♀
	18:42	☽ # ♆
	19:21	☽ △ ♃
	22:09	☽ # ♀
03	12:18	☽ # ☿
	18:22	☽ □ ♅
	22:58	♃ SR
04	03:42	☽ ∥ ♄
	06:22	☽ # ☉
	13:27	♀ ∥ ♆
	17:15	☽ ⚹ ♂
	18:01	☽ △ ♆
05	07:34	☽ △ ♀
	08:22	☽ □ ♃
	09:06	♂ ⚹ ♆
	11:54	☽ ☍ ♇
	23:14	☽ ☍ ☿
06	07:14	☽ △ ♅
	13:45	☿ SD
07	01:04	☽ ☌ ♄
	02:43	♀ ⚹ ♇
	04:00	☉ # ♄
	08:33	☽ □ ♂
	20:00	☽ ⚹ ♃
08	03:37	♂ ∥ ♃
09	13:27	☽ ∥ ♄
	15:54	☽ # ☉
	16:39	☽ ☍ ♆
	19:00	☉ △ ♃
	21:39	☽ △ ♂
10	00:30	☽ # ☿
	09:25	☽ △ ♇
	17:31	☽ ☍ ♀
	21:00	☽ # ♆
	22:00	☽ △ ☿
11	03:35	☽ ☍ ♅
	06:26	♀ ∥ ♇
	10:23	☽ # ♇
	10:46	☽ # ♀
	19:06	☽ ⚹ ♄
	22:28	☽ # ♅
12	13:19	☽ ☌ ♃
	17:10	☽ □ ♇
	19:00	☽ △ ☉
	20:21	☽ ∥ ♂
13	01:02	☽ ∥ ♃
	08:01	☽ □ ☿
14	00:06	☿ ∥ ☉
	01:12	☽ □ ♄
	07:20	☽ △ ♆
	11:02	☿ ♑
	16:56	☽ ☍ ♂
	17:16	♀ ♓
	21:16	☽ # ♃
	22:52	☽ ⚹ ♇
15	03:58	☽ # ♂
	06:53	☿ ⚹ ♅
	07:23	♀ ☌ ♅
	08:43	☿ ⚹ ♀
	15:49	☽ △ ♅
	16:24	☽ ⚹ ☿
	16:36	☽ △ ♀
	22:14	☽ ∥ ♅
16	00:18	☽ ∥ ♀
	05:11	☽ △ ♄
	09:18	☽ ∥ ♇
	11:30	☽ □ ♆
	21:03	☽ ∥ ♆
	22:19	☽ ⚹ ♃
17	03:09	♀ ∥ ♅
	11:49	☽ ⚹ ☉
	15:08	☽ ∥ ☉
	18:42	☽ □ ♅
	22:12	☽ ∥ ☿
	23:54	☽ □ ♀
18	01:14	☽ # ♄
	13:36	☽ ⚹ ♆
	23:47	☽ □ ♃
19	02:43	☽ △ ♂
	03:59	☽ ☌ ♇
	19:57	☽ ⚹ ♅
20	03:38	☽ ☌ ☿
	05:15	☽ ⚹ ♀
	07:44	☽ ☍ ♄
	10:00	♂ △ ♇
	17:43	☉ ♒
21	00:20	☽ △ ♃
	05:35	☽ □ ♂
	10:04	♀ △ ♄
22	04:59	☿ ☍ ♄
	11:29	☽ # ♄
	11:43	☽ ∥ ☿
	15:20	♀ # ♂
	15:47	☽ ☌ ♆
23	02:11	☿ # ♄
	04:14	☽ ∥ ☉
	06:33	☽ ⚹ ♇
	09:34	☽ ⚹ ♂
	16:42	☽ ∥ ♆
	23:30	☽ ☌ ♅
24	04:23	☽ ∥ ♇
	11:16	☽ △ ♄
	16:31	☽ ∥ ♅
	16:59	☽ ⚹ ☿
	18:49	☽ ☌ ♀
25	00:34	☽ # ♂
	05:31	☽ ☍ ♃
	08:37	☽ ∥ ♀
	11:10	☽ □ ♇
	16:45	☽ # ♃
26	14:15	☽ ⚹ ☉
	17:39	☽ □ ♄
27	02:55	☽ ⚹ ♆
	06:29	☽ □ ☿
	18:12	☽ ∥ ♃
	19:49	☽ △ ♇
	21:02	☽ # ♀
28	05:00	☽ ☌ ♂
	15:32	☽ ∥ ♂
	15:34	☽ ⚹ ♅
	19:59	☽ # ♅
29	01:26	♀ # ♃
	03:56	☽ ⚹ ♄
	10:06	☽ # ♇
	12:01	♀ ☍ ♃
	14:14	☽ □ ♆
	18:10	☿ △ ♃
	22:50	☿ # ♄
	23:18	☽ # ♆
30	00:34	☽ △ ♃
	01:28	☽ △ ☿
	02:05	☽ ⚹ ♀
	03:58	☽ # ☉
31	04:27	☽ □ ♅
	10:02	☽ # ☿
	10:55	☽ ∥ ♄
	11:08	☿ ⚹ ♀

⚷ Chiron

01 Dec.	15 S 53
01	18♑28
04	18 45
07	19 02
10	19 19
13	19 36
16	19 53
19	20 10
22	20 27
25	20 44
28	21 00
31	21 17

February 2004

Day	S. T.	☉	☽	☿	♀	♂	♃	♄	♅	♆	♇	☊ True
	h m s	° ' "	° ' "	° '	° '	° '	° '	° '	° '	° '	° '	° '
01 Su	08 42 12	11♒27 06	11♊08 49	20♑59	20♓52	28♈28	17♍R40	07♋R26	01♓35	12♒49	21♐30	16♉R36
02 Mo	08 46 09	12 28 00	22 59 59	22 24	22 04	29 06	17 35	07 22	01 39	12 51	21 31	16 32
03 Tu	08 50 05	13 28 52	04♋59 14	23 50	23 15	29 44	17 30	07 18	01 42	12 53	21 33	16 25
04 We	08 54 02	14 29 44	17 09 49	25 17	24 27	00♉22	17 24	07 15	01 45	12 56	21 35	16 14
05 Th	08 57 58	15 30 34	29 33 52	26 45	25 38	01 00	17 19	07 12	01 49	12 58	21 36	16 02
06 Fr	09 01 55	16 31 23	12♌12 19	28 14	26 49	01 38	17 13	07 08	01 52	13 00	21 38	15 48
07 Sa	09 05 51	17 32 10	25 05 03	29 44	28 01	02 17	17 07	07 05	01 55	13 03	21 39	15 34
08 Su	09 09 48	18 32 56	08♍11 02	01♒14	29 12	02 55	17 01	07 02	01 59	13 05	21 41	15 21
09 Mo	09 13 45	19 33 42	21 28 47	02 45	00♈23	03 33	16 55	06 59	02 02	13 07	21 42	15 11
10 Tu	09 17 41	20 34 26	04♎56 44	04 18	01 33	04 11	16 48	06 56	02 05	13 09	21 43	15 03
11 We	09 21 38	21 35 08	18 33 42	05 51	02 44	04 49	16 42	06 53	02 09	13 12	21 45	14 58
12 Th	09 25 34	22 35 50	02♏19 00	07 24	03 55	05 28	16 35	06 51	02 12	13 14	21 46	14 55
13 Fr	09 29 31	23 36 31	16 12 27	08 59	05 05	06 06	16 29	06 48	02 16	13 16	21 47	14 54
14 Sa	09 33 27	24 37 11	00♐14 02	10 34	06 15	06 44	16 22	06 46	02 19	13 18	21 49	14 53
15 Su	09 37 24	25 37 50	14 23 27	12 11	07 26	07 22	16 15	06 43	02 23	13 21	21 50	14 50
16 Mo	09 41 20	26 38 28	28 39 30	13 48	08 36	08 01	16 08	06 41	02 26	13 23	21 51	14 46
17 Tu	09 45 17	27 39 04	12♑59 42	15 26	09 46	08 39	16 01	06 39	02 30	13 25	21 52	14 40
18 We	09 49 14	28 39 40	27 20 07	17 05	10 55	09 18	15 54	06 37	02 33	13 27	21 54	14 31
19 Th	09 53 10	29 40 14	11♒35 36	18 44	12 05	09 56	15 47	06 35	02 36	13 30	21 55	14 21
20 Fr	09 57 07	00♓40 46	25 40 30	20 25	13 14	10 34	15 39	06 33	02 40	13 32	21 56	14 09
21 Sa	10 01 03	01 41 17	09♓29 33	22 07	14 24	11 13	15 32	06 31	02 43	13 34	21 57	13 58
22 Su	10 04 60	02 41 46	22 58 45	23 49	15 33	11 51	15 25	06 29	02 47	13 36	21 58	13 48
23 Mo	10 08 56	03 42 13	06♈06 00	25 33	16 42	12 29	15 17	06 28	02 50	13 38	21 59	13 40
24 Tu	10 12 53	04 42 39	18 51 18	27 17	17 51	13 08	15 09	06 26	02 54	13 40	22 00	13 35
25 We	10 16 49	05 43 03	01♉16 36	29 03	19 00	13 46	15 02	06 25	02 57	13 43	22 01	13 33
26 Th	10 20 46	06 43 25	13 25 21	00♓49	20 08	14 25	14 54	06 24	03 01	13 45	22 02	13D 32
27 Fr	10 24 43	07 43 45	25 22 07	02 36	21 16	15 03	14 46	06 22	03 04	13 47	22 03	13 33
28 Sa	10 28 39	08 44 03	07♊12 06	04 25	22 25	15 42	14 39	06 21	03 08	13 49	22 04	13 34
29 Su	10 32 36	09 44 19	19 00 45	06 14	23 33	16 20	14 31	06 20	03 11	13 51	22 05	13R 34

Data for	02-01-2004
Julian Day	2453036.50
Ayanamsa	23 54 39
SVP	05 ♓ 11 59
☽ ☊ Mean	16 ♉ 04 R

● ◐ PHASES ○ ◑

06	08:47	○	16♌54
13	13:40	◑	24♏11
20	09:18	●	01♓04
28	03:24	◐	08♊53

LAST ASPECT ☽ Day	h m	INGRESS Day	h m	
02	12:56	02	14:03	♋
04	17:53	05	00:50	♌
06	17:38	07	09:03	♍
09	00:24	09	15:13	♎
11	05:43	11	19:58	♏
13	13:41	13	23:36	♐
15	20:21	16	02:15	♑
17	05:01	18	04:28	♒
19	17:34	20	07:27	♓
21	22:10	22	12:45	♈
24	18:55	24	21:30	♉
26	02:55	27	09:23	♊
29	10:09	29	22:12	♋

DECLINATION

Day	☉	☽	☿	♀	♂	♃	♄	♅	♆	♇
01 Su	17S21	24N12	22S27	04S30	11N31	06N07	22N38	11S35	17S00	14S31
02 Mo	17 04	26 15	22 21	03 59	11 45	06 09	22 38	11 34	16 59	14 31
03 Tu	16 47	27 07	22 13	03 28	12 00	06 11	22 38	11 33	16 58	14 31
04 We	16 29	26 41	22 04	02 56	12 14	06 14	22 39	11 32	16 58	14 31
05 Th	16 11	24 56	21 54	02 25	12 28	06 16	22 39	11 30	16 57	14 31
06 Fr	15 53	21 55	21 42	01 54	12 42	06 18	22 39	11 29	16 57	14 31
07 Sa	15 35	17 48	21 29	01 22	12 56	06 21	22 40	11 28	16 56	14 31
08 Su	15 16	12 47	21 15	00 51	13 10	06 23	22 40	11 27	16 55	14 31
09 Mo	14 57	07 06	20 59	00 19	13 23	06 26	22 40	11 26	16 55	14 31
10 Tu	14 38	01 01	20 42	00N12	13 37	06 29	22 41	11 24	16 54	14 31
11 We	14 19	05S12	20 24	00 44	13 51	06 31	22 41	11 23	16 53	14 31
12 Th	13 59	11 14	20 05	01 15	14 04	06 34	22 41	11 22	16 53	14 31
13 Fr	13 39	16 47	19 44	01 46	14 17	06 37	22 42	11 21	16 52	14 31
14 Sa	13 19	21 30	19 22	02 18	14 31	06 40	22 42	11 19	16 52	14 31
15 Su	12 59	25 00	18 58	02 49	14 44	06 42	22 42	11 18	16 51	14 31
16 Mo	12 38	26 55	18 33	03 21	14 57	06 45	22 42	11 17	16 50	14 30
17 Tu	12 17	27 04	18 07	03 52	15 10	06 48	22 43	11 16	16 50	14 30
18 We	11 56	25 23	17 39	04 23	15 23	06 51	22 43	11 14	16 49	14 30
19 Th	11 35	22 07	17 10	04 54	15 36	06 54	22 43	11 13	16 48	14 30
20 Fr	11 14	17 35	16 40	05 25	15 48	06 57	22 43	11 12	16 48	14 30
21 Sa	10 52	12 13	16 08	05 56	16 01	07 00	22 44	11 11	16 47	14 30
22 Su	10 31	06 23	15 35	06 26	16 13	07 03	22 44	11 09	16 47	14 30
23 Mo	10 09	00 25	15 00	06 57	16 26	07 06	22 44	11 08	16 46	14 30
24 Tu	09 47	05N25	14 24	07 27	16 38	07 09	22 44	11 07	16 45	14 30
25 We	09 25	10 54	13 47	07 58	16 50	07 12	22 45	11 06	16 45	14 29
26 Th	09 03	15 51	13 08	08 28	17 02	07 15	22 45	11 04	16 44	14 29
27 Fr	08 40	20 07	12 29	08 58	17 14	07 18	22 45	11 03	16 43	14 29
28 Sa	08 18	23 31	11 47	09 27	17 26	07 21	22 45	11 02	16 43	14 29
29 Su	07 55	25 56	11 05	09 57	17 37	07 25	22 45	11 01	16 42	14 29

⚷ Chiron

01 Dec.	15 S 27
03	21♑33
06	21 49
09	22 04
12	22 20
15	22 35
18	22 49
21	23 04
24	23 18
27	23 31

ASPECTARIAN

01 00:41 ☽ △ ☉
03:24 ☽ △ ♆
06:26 ♂ ⋕ ♅
12:56 ♀ □ ♇
13:08 ☽ □ ♃
21:01 ☽ ☍ ♇
21:54 ☽ □ ♀
02 06:53 ☉ ∥ ♆
09:29 ☉ ☌ ♆
12:56 ☽ ✶ ♂
17:25 ☽ △ ♅
03 04:35 ☽ ☌ ♄
10:05 ♂ ♉
04 00:28 ☽ ✶ ♃
15:39 ☽ △ ♀
17:53 ☽ ☍ ☿
05 02:54 ☽ □ ♂
18:55 ☽ ∥ ♄
06 01:30 ☽ ☍ ♆
01:34 ☽ ⋕ ☿
09:23 ♂ ✶ ☿
17:38 ☽ △ ♇
07 04:20 ☿ ♒
04:27 ☽ ⋕ ♆
11:48 ☽ ⋕ ☉
12:38 ☽ ☍ ♅
13:54 ☽ △ ♂
16:05 ☽ ⋕ ♆
21:55 ☽ ✶ ♄
22:21 ☽ ∥ ♂
08 05:51 ☽ ⋕ ♅
15:51 ☽ ☌ ♃
16:21 ♀ ♈
09 00:24 ☽ □ ♇
02:38 ☽ ∥ ♃

17:25 ☽ ☍ ♀
21:09 ☿ □ ♂
22:42 ☽ △ ☿
10 02:53 ☽ ∥ ♀
03:31 ☽ □ ♄
05:07 ☽ ⋕ ♀
08:42 ☉ ∥ ♆
14:33 ☽ △ ♆
11 03:56 ☉ ✶ ♆
05:11 ☽ ⋕ ♃
05:36 ☽ ✶ ♅
05:43 ☽ △ ☉
20:18 ☉ ⋕ ♂
23:48 ☽ △ ♅
12 00:30 ☽ ∥ ♅
05:43 ☽ ☍ ♂
07:50 ☽ △ ♄
09:57 ☽ □ ☿
10:51 ☽ ∥ ☉
12:21 ☽ ⋕ ♂
13:47 ☽ ∥ ♆
18:56 ☽ □ ♆
13 00:20 ☽ ∥ ♆
00:28 ☽ ✶ ♃
13:15 ☽ ∥ ☿
23:56 ♂ ⋕ ♆
14 00:50 ♂ ✶ ♄
03:34 ☽ □ ♅
07:11 ☽ ⋕ ♄
09:58 ♀ □ ♄
11:10 ☽ △ ♀
19:47 ☽ [illegible] ♀
22:14 ☽ ✶ ♆

15 03:07 ☽ □ ♃
12:33 ☽ ☌ ♇
17:43 ☿ ☌ ♆
20:21 ☽ ✶ ☉
16 06:21 ☽ ✶ ♅
13:24 ☽ ☍ ♄
16:24 ☽ △ ♂
18:07 ☽ □ ♀
17 05:01 ☽ △ ♃
18 20:16 ☽ ⋕ ♄
21:03 ☽ □ ♂
19 00:54 ☽ ✶ ♀
03:13 ☽ ☌ ♆
07:50 ☉ ♓
13:46 ☽ ☌ ☿
17:30 ☿ ∥ ♆
17:34 ☽ ✶ ♇
20 02:27 ☉ ∥ ♅
03:46 ☽ ∥ ♆
04:53 ☽ ∥ ☿
06:10 ♀ ✶ ♇
07:59 ☽ ⋕ ♂
12:07 ☽ ☌ ♅
14:07 ☽ ∥ ♆
18:48 ☽ △ ♄
21:42 ☿ ✶ ♇
21 03:10 ☽ ✶ ♂
03:38 ☿ ⋕ ♂
04:22 ☽ ∥ ♅
05:59 ☽ ∥ ☊
10:34 ☽ ☍ ♃
20:18 ☽ ⋕ ♃

22:10 ☽ □ ♇
23:46 ☽ ⋕ ♀
22 02:07 ☉ ☌ ♅
23 00:40 ☽ □ ♄
07:57 ♀ ∥ ♃
14:09 ☽ ✶ ♆
20:25 ☿ ∥ ♆
21:53 ☽ ☌ ♀
24 06:02 ☽ △ ♇
07:27 ☽ ∥ ♃
09:34 ☽ ∥ ♀
13:44 ♂ ∥ ♅
17:43 ☽ ⋕ ☉
18:55 ☽ ✶ ☿

21:32 ♂ □ ♆
25 00:53 ☽ ⋕ ♅
03:18 ☽ ✶ ♅
09:30 ☽ ✶ ☉
10:04 ☽ ✶ ♄
12:03 ☽ ⋕ ☿
12:58 ☿ ♓
16:15 ☉ △ ♄
17:04 ☽ ⋕ ♇
26 00:39 ☽ □ ♆
02:05 ☽ ☌ ♂
[illegible]
04:37 ☽ ⋕ ♆
06:32 ☽ ∥ ♂

15:13 ♂ △ ♃
16:01 ♀ ⋕ ☉
27 06:22 ♀ ☌ ♅
15:41 ☽ □ ♅
16:32 ♀ △ ♇
17:18 ☽ □ ☿
17:54 ☽ ∥ ♄
28 13:29 ☽ △ ♆
14:58 ☽ □ ♃
29 01:23 ☿ △ ♄
02:15 ☿ ∥ ♅
00:15 ☽ [illegible] ♇
10:09 ☽ ✶ ♀
22:16 ☿ ⋕ ♀

March 2004

Day	S. T.	☉	☽	☿	♀	♂	♃	♄	♅	♆	♇	☊ True
	h m s	° ' "	° ' "	° '	° '	° '	° '	° '	° '	° '	° '	° '
01 Mo	10 36 32	10♓44 33	00♋53 31	08♓04	24♈40	16♉59	14♍R23	06♋R20	03♓14	13♒53	22♐05	13♉R32
02 Tu	10 40 29	11 44 45	12 55 22	09 56	25 48	17 37	14 15	06 19	03 18	13 55	22 06	13 28
03 We	10 44 25	12 44 55	25 10 33	11 48	26 55	18 16	14 07	06 18	03 21	13 57	22 07	13 21
04 Th	10 48 22	13 45 03	07♌42 20	13 41	28 02	18 54	13 59	06 18	03 25	13 59	22 08	13 12
05 Fr	10 52 18	14 45 09	20 32 36	15 35	29 09	19 32	13 52	06 18	03 28	14 01	22 08	13 01
06 Sa	10 56 15	15 45 13	03♍41 46	17 30	00♉16	20 11	13 44	06 17	03 31	14 03	22 09	12 50
07 Su	11 00 12	16 45 15	17 08 40	19 26	01 23	20 49	13 36	06 17	03 35	14 05	22 09	12 40
08 Mo	11 04 08	17 45 15	00♎50 52	21 22	02 29	21 28	13 28	06D 17	03 38	14 07	22 10	12 32
09 Tu	11 08 05	18 45 13	14 44 58	23 19	03 35	22 06	13 20	06 17	03 42	14 09	22 11	12 26
10 We	11 12 01	19 45 10	28 47 21	25 17	04 41	22 45	13 13	06 17	03 45	14 11	22 11	12 22
11 Th	11 15 58	20 45 05	12♏54 39	27 14	05 46	23 23	13 05	06 18	03 48	14 13	22 12	12 21
12 Fr	11 19 54	21 44 58	27 04 10	29 12	06 51	24 02	12 57	06 18	03 52	14 15	22 12	12D 20
13 Sa	11 23 51	22 44 50	11♐13 53	01♈10	07 56	24 40	12 49	06 19	03 55	14 17	22 12	12 21
14 Su	11 27 47	23 44 40	25 22 19	03 07	09 01	25 19	12 42	06 19	03 58	14 19	22 13	12R 20
15 Mo	11 31 44	24 44 28	09♑28 07	05 04	10 05	25 57	12 34	06 20	04 02	14 21	22 13	12 19
16 Tu	11 35 41	25 44 15	23 29 50	06 59	11 10	26 35	12 27	06 21	04 05	14 23	22 13	12 16
17 We	11 39 37	26 44 00	07♒25 37	08 54	12 13	27 14	12 19	06 22	04 08	14 24	22 14	12 11
18 Th	11 43 34	27 43 43	21 13 12	10 46	13 17	27 52	12 12	06 23	04 11	14 26	22 14	12 04
19 Fr	11 47 30	28 43 25	04♓50 04	12 36	14 20	28 31	12 05	06 24	04 15	14 28	22 14	11 57
20 Sa	11 51 27	29 43 04	18 13 41	14 24	15 23	29 09	11 57	06 26	04 18	14 30	22 14	11 50
21 Su	11 55 23	00♈42 41	01♈21 58	16 08	16 26	29 48	11 50	06 27	04 21	14 31	22 14	11 44
22 Mo	11 59 20	01 42 17	14 13 40	17 49	17 28	00♊26	11 43	06 29	04 24	14 33	22 15	11 39
23 Tu	12 03 16	02 41 50	26 48 40	19 26	18 30	01 05	11 36	06 30	04 27	14 35	22 15	11 36
24 We	12 07 13	03 41 21	09♉08 08	20 59	19 32	01 43	11 30	06 32	04 30	14 37	22 15	11D 36
25 Th	12 11 10	04 40 50	21 14 20	22 27	20 33	02 21	11 23	06 34	04 34	14 38	22R 15	11 37
26 Fr	12 15 06	05 40 17	03♊10 35	23 49	21 34	03 00	11 16	06 36	04 37	14 40	22 15	11 40
27 Sa	12 19 03	06 39 42	15 01 01	25 06	22 34	03 38	11 10	06 38	04 40	14 41	22 15	11 42
28 Su	12 22 59	07 39 04	26 50 16	26 17	23 34	04 17	11 03	06 40	04 43	14 43	22 14	11 45
29 Mo	12 26 56	08 38 24	08♋43 23	27 21	24 34	04 55	10 57	06 42	04 46	14 44	22 14	11 46
30 Tu	12 30 52	09 37 42	20 45 24	28 20	25 33	05 34	10 51	06 45	04 49	14 46	22 14	11R 45
31 We	12 34 49	10 36 57	03♌01 05	29 11	26 31	06 12	10 45	06 47	04 52	14 47	22 14	11 43

Data for	03-01-2004
Julian Day	2453065.50
Ayanamsa	23 54 43
SVP	05 ♓ 11 55
☽ ☊ Mean	14 ♉ 32 R

● ◐ PHASES ○ ◑

06	23:15	○	16♍43
13	21:01	◑	23♐37
20	22:41	●	00♈39
28	23:49	◐	08♋38

LAST ASPECT Day	h m	☽ INGRESS Day	h m	
03	03:43	03	09:18	♌
05	17:14	05	17:19	♍
07	08:50	07	22:32	♎
09	12:44	10	02:04	♏
12	04:12	12	04:58	♐
13	21:01	14	07:52	♑
16	05:34	16	11:11	♒
18	12:16	18	15:26	♓
20	20:57	20	21:29	♈
22	15:14	23	06:10	♉
24	22:29	25	17:35	♊
27	22:45	28	06:24	♋
30	16:01	30	18:08	♌

DECLINATION

Day	☉	☽	☿	♀	♂	♃	♄	♅	♆	♇
01 Mo	07S 32	27N11	10S 21	10N26	17N49	07N28	22N46	10S 59	16S 42	14S 29
02 Tu	07 09	27 10	09 36	10 55	18 00	07 31	22 46	10 58	16 41	14 29
03 We	06 47	25 50	08 49	11 24	18 12	07 34	22 46	10 57	16 41	14 29
04 Th	06 23	23 13	08 02	11 53	18 23	07 37	22 46	10 56	16 40	14 28
05 Fr	06 00	19 24	07 13	12 21	18 34	07 40	22 46	10 55	16 39	14 28
06 Sa	05 37	14 34	06 23	12 49	18 44	07 43	22 46	10 53	16 39	14 28
07 Su	05 14	08 56	05 32	13 17	18 55	07 46	22 47	10 52	16 38	14 28
08 Mo	04 50	02 46	04 40	13 44	19 06	07 49	22 47	10 51	16 38	14 28
09 Tu	04 27	03S 37	03 47	14 11	19 16	07 52	22 47	10 50	16 37	14 28
10 We	04 03	09 55	02 53	14 38	19 26	07 55	22 47	10 48	16 37	14 27
11 Th	03 40	15 45	01 58	15 05	19 37	07 58	22 47	10 47	16 36	14 27
12 Fr	03 16	20 46	01 03	15 31	19 47	08 01	22 47	10 46	16 35	14 27
13 Sa	02 53	24 35	00 08	15 57	19 57	08 04	22 47	10 45	16 35	14 27
14 Su	02 29	26 51	00N48	16 22	20 06	08 07	22 48	10 44	16 34	14 27
15 Mo	02 05	27 24	01 44	16 47	20 16	08 10	22 48	10 42	16 34	14 27
16 Tu	01 42	26 11	02 40	17 12	20 25	08 13	22 48	10 41	16 33	14 26
17 We	01 18	23 21	03 36	17 36	20 35	08 16	22 48	10 40	16 33	14 26
18 Th	00 54	19 13	04 31	18 00	20 44	08 18	22 48	10 39	16 32	14 26
19 Fr	00 30	14 08	05 25	18 24	20 53	08 21	22 48	10 38	16 32	14 26
20 Sa	00 07	08 28	06 18	18 47	21 01	08 24	22 48	10 37	16 31	14 26
21 Su	00N17	02 32	07 10	19 10	21 10	08 27	22 48	10 35	16 31	14 25
22 Mo	00 41	03N24	08 01	19 32	21 19	08 29	22 48	10 34	16 30	14 25
23 Tu	01 04	09 06	08 49	19 54	21 27	08 32	22 48	10 33	16 30	14 25
24 We	01 28	14 20	09 36	20 16	21 35	08 34	22 48	10 32	16 29	14 25
25 Th	01 52	18 55	10 20	20 37	21 43	08 37	22 49	10 31	16 29	14 25
26 Fr	02 15	22 41	11 01	20 57	21 51	08 39	22 49	10 30	16 29	14 24
27 Sa	02 39	25 27	11 40	21 18	21 59	08 42	22 49	10 29	16 28	14 24
28 Su	03 02	27 06	12 16	21 37	22 06	08 44	22 49	10 28	16 28	14 24
29 Mo	03 26	27 31	12 49	21 56	22 13	08 47	22 49	10 27	16 27	14 24
30 Tu	03 49	26 39	13 19	22 15	22 21	08 49	22 49	10 26	16 27	14 24
31 We	04 12	24 29	13 45	22 33	22 28	08 51	22 49	10 24	16 26	14 24

ASPECTARIAN

Day	Time	Aspect
01	04:22	☉ # ♃
	04:44	☽ △ ♅
	10:53	☽ ☌ ♄
	16:59	☽ △ ☿
	21:28	☽ △ ☉
02	02:28	♀ # ♅
	02:36	☽ ✶ ♃
	09:46	☽ ✶ ♂
03	03:43	☽ □ ♀
04	01:43	☿ ☌ ☉
	03:15	☽ ∥ ♄
	03:36	☿ ☍ ♃
	05:06	☉ ☍ ♃
	11:28	☿ # ♃
	11:51	☽ ☍ ♆
	22:03	☽ □ ♂
05	02:56	☽ △ ♇
	04:22	☽ ∥ ♂
	14:12	☽ # ♆
	17:14	☽ △ ♀
	18:12	♀ ♉
	23:41	☽ ☍ ♅
06	00:28	☽ # ♇
	04:40	☽ ✶ ♄
	07:12	☽ ∥ ♀
	16:04	☽ # ♅
	17:47	☽ ☌ ♃
07	04:39	☽ ∥ ♃
	04:42	☽ ☍ ☿
	06:48	☽ △ ♂
	08:50	☽ □ ♇
	15:07	☿ ∥ ☉
	15:36	☽ # ☉
	15:38	☽ # ☿
	16:52	♄ SD
08	01:42	☿ ✶ ♂
	09:25	☽ □ ♄
	09:51	☿ □ ♇
	22:59	☽ △ ♆
09	00:32	☽ ∥ ☿
	02:38	♀ ✶ ♅
	02:56	☽ ∥ ☉
	12:44	☽ ✶ ♇
	14:36	♀ # ♇
	16:13	☽ # ♃
10	03:30	☽ ∥ ♅
	08:29	☽ △ ♅
	10:51	☽ ☍ ♀
	12:46	☽ △ ♄
	18:25	☽ ∥ ♇
	20:47	☽ # ♀
11	00:17	☽ ✶ ♃
	02:14	☽ □ ♆
	03:43	☽ ∥ ♆
	11:44	♀ ✶ ♄
	14:18	☽ △ ☉
	18:36	☽ ☍ ♂
	18:37	☽ # ♂
12	04:12	☽ △ ☿
	09:45	☿ ♈
	10:57	☉ □ ♇
	11:33	☽ □ ♅
	11:36	☽ # ♄
13	02:41	☽ □ ♃
	05:11	☽ ✶ ♆
	18:38	☽ ☌ ♇
14	11:31	♀ # ♆
	14:41	☽ ✶ ♅
	15:17	☽ □ ☿
	18:39	☽ ☍ ♄
15	01:09	☽ △ ♀
	05:15	☽ △ ♃
	06:19	☿ # ☉
	15:58	☿ □ ♄
16	04:09	☽ ✶ ☉
	05:34	☽ △ ♂
17	02:00	♀ △ ♃
	02:56	☽ ✶ ☿
	03:43	☽ # ♄
	09:00	☽ □ ♀
	12:08	☽ ☌ ♆
	16:17	☽ # ♂
18	01:46	☽ ✶ ♇
	05:41	☽ # ♀
	09:48	☉ ✶ ♂
	12:16	☽ □ ♂
	13:10	☽ ∥ ♆
	22:45	☽ ∥ ♇
	22:57	☽ ☌ ♅
19	02:48	☽ △ ♄
	03:02	♀ □ ♆
	12:49	☽ ☍ ♃
	15:08	☽ ∥ ♅
	18:26	☽ ✶ ♀
20	00:19	☽ # ♃
	01:22	☿ ✶ ♆
	06:49	☉ ♈
	07:17	☽ □ ♇
	07:41	☽ # ☿
	20:57	☽ ✶ ♂
21	07:40	♂ ♊
	08:28	☽ # ☉
	09:27	☽ □ ♄
	12:07	☽ ∥ ☉
22	00:37	☽ ✶ ♆
	07:49	☽ ☌ ☿
	14:53	☿ ∥ ♃
	15:14	☽ △ ♇
	21:32	☽ ∥ ♃
	22:36	☽ ∥ ☿
23	06:26	☽ # ♅
	14:54	☽ ✶ ♅
	18:53	☽ ✶ ♄
24	00:26	☽ # ♇
	04:36	☽ △ ♃
	10:50	☽ □ ♆
	10:51	☽ # ♆
	15:08	♇ SR
	20:41	☿ △ ♇
	22:29	☽ ☌ ♀
25	06:16	☿ # ♅
	11:12	☽ ∥ ♀
	18:00	☽ ∥ ♂
	23:37	☽ ☌ ♂
26	01:01	☽ ∥ ♄
	02:55	☽ □ ♅
	05:30	☽ ✶ ☉
	16:15	☽ □ ♃
	23:15	☉ □ ♄
	23:20	☽ △ ♆
27	14:40	☽ ☍ ♇
	22:45	☽ ✶ ☿
28	16:00	☽ △ ♅
	17:40	♂ □ ♅
	19:56	☽ ☌ ♄
29	04:26	☽ ✶ ♃
30	10:15	☽ ✶ ♀
	11:30	♀ ∥ ♂
	16:01	☽ □ ☿
31	06:28	☽ ✶ ♂
	13:02	☽ ∥ ♄
	13:38	☽ ∥ ♀
	14:58	☽ ∥ ♂
	15:50	☽ △ ☉
	20:38	♀ ∥ ♄
	22:34	☽ ☍ ♆

⚷ Chiron

01	Dec.	14 S 56
01	23♑45	
04	23 57	
07	24 09	
10	24 21	
13	24 32	
16	24 43	
19	24 53	
22	25 03	
25	25 12	
28	25 20	
31	25 28	

April 2004

19 13:35 29♈49 ☉ Solar Eclipse (mag 0.737)

Day	S. T.	☉	☽	☿	♀	♂	♃	♄	♅	♆	♇	☊ True
	h m s	° ' "	° ' "	° '	° '	° '	° '	° '	° '	° '	° '	° '
01 Th	12 38 45	11♈36 10	15♌34 35	29♈56	27♉30	06♊50	10♍R39	06♋50	04♓55	14♒49	22♐R14	11♉R39
02 Fr	12 42 42	12 35 21	28 29 03	00♉33	28 27	07 29	10 33	06 53	04 58	14 50	22 14	11 34
03 Sa	12 46 39	13 34 30	11♍46 13	01 04	29 25	08 07	10 28	06 55	05 00	14 52	22 13	11 29
04 Su	12 50 35	14 33 36	25 26 03	01 27	00♊21	08 45	10 22	06 58	05 03	14 53	22 13	11 25
05 Mo	12 54 32	15 32 40	09♎26 35	01 44	01 17	09 24	10 17	07 01	05 06	14 54	22 13	11 21
06 Tu	12 58 28	16 31 42	23 44 00	01 53	02 13	10 02	10 12	07 04	05 09	14 56	22 12	11 18
07 We	13 02 25	17 30 42	08♏13 03	01R 55	03 08	10 40	10 07	07 08	05 12	14 57	22 12	11 17
08 Th	13 06 21	18 29 40	22 47 52	01 51	04 03	11 19	10 02	07 11	05 15	14 58	22 11	11D 16
09 Fr	13 10 18	19 28 37	07♐22 40	01 40	04 56	11 57	09 57	07 14	05 17	15 00	22 11	11 17
10 Sa	13 14 14	20 27 31	21 52 24	01 24	05 50	12 35	09 52	07 18	05 20	15 01	22 10	11 18
11 Su	13 18 11	21 26 24	06♑13 09	01 01	06 42	13 14	09 48	07 21	05 23	15 02	22 10	11 18
12 Mo	13 22 07	22 25 16	20 22 13	00 34	07 34	13 52	09 44	07 25	05 25	15 03	22 09	11R 19
13 Tu	13 26 04	23 24 05	04♒17 56	00 02	08 25	14 30	09 40	07 29	05 28	15 04	22 09	11 18
14 We	13 30 01	24 22 53	17 59 29	29♈26	09 16	15 08	09 36	07 33	05 30	15 05	22 08	11 17
15 Th	13 33 57	25 21 39	01♓26 37	28 47	10 05	15 47	09 32	07 37	05 33	15 06	22 07	11 15
16 Fr	13 37 54	26 20 23	14 39 27	28 06	10 54	16 25	09 28	07 41	05 35	15 07	22 07	11 13
17 Sa	13 41 50	27 19 06	27 38 15	27 24	11 43	17 03	09 25	07 45	05 38	15 08	22 06	11 11
18 Su	13 45 47	28 17 46	10♈23 25	26 40	12 30	17 41	09 22	07 49	05 40	15 09	22 05	11 10
19 Mo	13 49 43	29 16 25	22 55 32	25 57	13 16	18 20	09 19	07 53	05 43	15 10	22 04	11 09
20 Tu	13 53 40	00♉15 02	05♉15 31	25 15	14 02	18 58	09 16	07 58	05 45	15 11	22 04	11D 09
21 We	13 57 36	01 13 37	17 24 38	24 34	14 47	19 36	09 13	08 02	05 47	15 12	22 03	11 10
22 Th	14 01 33	02 12 10	29 24 44	23 56	15 30	20 14	09 11	08 07	05 50	15 13	22 02	11 11
23 Fr	14 05 30	03 10 41	11♊18 10	23 21	16 13	20 52	09 08	08 11	05 52	15 14	22 01	11 13
24 Sa	14 09 26	04 09 10	23 07 52	22 49	16 54	21 31	09 06	08 16	05 54	15 15	22 00	11 15
25 Su	14 13 23	05 07 37	04♋57 19	22 21	17 35	22 09	09 04	08 21	05 56	15 15	21 59	11 16
26 Mo	14 17 19	06 06 02	16 50 29	21 57	18 14	22 47	09 03	08 26	05 58	15 16	21 58	11 17
27 Tu	14 21 16	07 04 25	28 51 45	21 37	18 53	23 25	09 01	08 31	06 00	15 17	21 57	11 18
28 We	14 25 12	08 02 45	11♌05 39	21 23	19 29	24 03	09 00	08 36	06 02	15 17	21 56	11R 18
29 Th	14 29 09	09 01 04	23 36 36	21 13	20 05	24 41	08 58	08 41	06 04	15 18	21 55	11 17
30 Fr	14 33 05	09 59 20	06♍28 31	21 08	20 40	25 20	08 57	08 46	06 06	15 19	21 54	11 17

Data for	04-01-2004
Julian Day	2453096.50
Ayanamsa	23 54 46
SVP	05 ♓ 11 50
☽ ☊ Mean	12 ♉ 53 R

● ◐ PHASES ○ ◑

05	11:03	○	16♎00
12	03:46	◑	22♑35
19	13:22	☉	29♈49
27	17:33	◐	07♌47

LAST ASPECT Day	h m	☽ INGRESS Day	h m	
01	23:57	02	02:46	♍
03	18:24	04	07:53	♎
05	21:27	06	10:25	♏
07	11:07	08	11:51	♐
10	00:30	10	13:34	♑
12	03:46	12	16:33	♒
14	19:27	14	21:24	♓
16	13:43	17	04:25	♈
19	13:22	19	13:43	♉
20	19:36	22	01:11	♊
23	23:23	24	13:57	♋
26	09:57	27	02:15	♌
29	02:09	29	12:01	♍

DECLINATION

Day	☉	☽	☿	♀	♂	♃	♄	♅	♆	♇
01 Th	04N35	21N07	14N08	22N51	22N35	08N53	22N49	10S23	16S26	14S23
02 Fr	04 58	16 40	14 27	23 08	22 41	08 55	22 49	10 22	16 26	14 23
03 Sa	05 21	11 18	14 43	23 25	22 48	08 57	22 49	10 21	16 25	14 23
04 Su	05 44	05 15	14 55	23 42	22 54	08 59	22 49	10 20	16 25	14 23
05 Mo	06 07	01S12	15 04	23 57	23 00	09 01	22 49	10 19	16 24	14 23
06 Tu	06 30	07 45	15 08	24 12	23 06	09 03	22 49	10 18	16 24	14 22
07 We	06 52	13 59	15 09	24 27	23 12	09 05	22 49	10 17	16 24	14 22
08 Th	07 15	19 29	15 07	24 41	23 18	09 07	22 49	10 16	16 23	14 22
09 Fr	07 37	23 49	15 00	24 55	23 23	09 08	22 49	10 15	16 23	14 22
10 Sa	08 00	26 36	14 50	25 08	23 28	09 10	22 49	10 14	16 23	14 22
11 Su	08 22	27 35	14 37	25 21	23 33	09 11	22 49	10 13	16 22	14 21
12 Mo	08 44	26 45	14 20	25 33	23 38	09 13	22 48	10 12	16 22	14 21
13 Tu	09 05	24 15	14 00	25 45	23 43	09 14	22 48	10 12	16 22	14 21
14 We	09 27	20 24	13 38	25 56	23 47	09 16	22 48	10 11	16 21	14 21
15 Th	09 49	15 34	13 13	26 06	23 52	09 17	22 48	10 10	16 21	14 21
16 Fr	10 10	10 06	12 46	26 16	23 56	09 18	22 48	10 09	16 21	14 20
17 Sa	10 31	04 16	12 18	26 26	24 00	09 19	22 48	10 08	16 20	14 20
18 Su	10 52	01N38	11 48	26 35	24 04	09 20	22 48	10 07	16 20	14 20
19 Mo	11 13	07 23	11 18	26 43	24 07	09 21	22 48	10 06	16 20	14 20
20 Tu	11 34	12 46	10 47	26 51	24 11	09 22	22 48	10 05	16 20	14 20
21 We	11 54	17 35	10 17	26 58	24 14	09 23	22 47	10 05	16 19	14 19
22 Th	12 14	21 39	09 47	27 05	24 17	09 24	22 47	10 04	16 19	14 19
23 Fr	12 34	24 47	09 18	27 11	24 20	09 25	22 47	10 03	16 19	14 19
24 Sa	12 54	26 48	08 50	27 17	24 23	09 25	22 47	10 02	16 19	14 19
25 Su	13 14	27 37	08 25	27 23	24 25	09 26	22 47	10 01	16 18	14 19
26 Mo	13 33	27 10	08 01	27 28	24 28	09 26	22 47	10 01	16 18	14 19
27 Tu	13 53	25 27	07 39	27 32	24 30	09 27	22 46	10 00	16 18	14 18
28 We	14 11	22 31	07 19	27 36	24 32	09 27	22 46	09 59	16 18	14 18
29 Th	14 30	18 31	07 02	27 39	24 33	09 27	22 46	09 59	16 18	14 18
30 Fr	14 49	13 35	06 48	27 42	24 35	09 28	22 46	09 58	16 18	14 18

ASPECTARIAN

01 02:28 ☿ ♉
12:27 ☽ △ ♇
18:38 ☿ # ♇
23:57 ☽ □ ♀
02 01:13 ☽ # ♆
03:58 ☽ △ ☿
09:52 ☽ ∥ ☿
10:41 ☽ # ♇
11:50 ☽ ☍ ♅
15:18 ☽ ⚹ ♄
17:08 ☽ □ ♂
21:41 ☽ ☌ ♃
03 03:59 ☽ # ♅
04:05 ♂ ∥ ♄
09:35 ☽ ∥ ♃
14:57 ♀ ♊
18:24 ☽ □ ♇
21:18 ☽ ∥ ☉
04 08:08 ☉ ⚹ ♆
09:06 ☽ △ ♀
19:52 ☽ □ ♄
23:55 ☽ △ ♂
05 09:14 ☽ △ ♆
19:08 ☽ # ☉
21:27 ☽ ⚹ ♇
06 04:56 ☽ # ♃
05:16 ♂ □ ♃
09:37 ☽ ∥ ♅
13:36 ☽ ☍ ☿
19:00 ☽ △ ♅
20:28 ☿ SR
22:12 ☽ △ ♄
07 01:36 ☽ ∥ ♇
03:06 ☽ ⚹ ♃
04:48 ☽ # ☿
10:01 ☽ ∥ [illegible]
11:07 ☽ □ ♆
08 17:39 ☽ # ♄
19:43 ☽ ☍ ♀
20:33 ☽ □ ♅
21:06 ☽ # ♂
09 04:13 ☽ □ ♃
07:54 ☽ ☍ ♂
08:28 ☽ # ♀
09:51 ♀ □ ♅
12:36 ☽ ⚹ ♆
21:29 ☽ △ ☉
10 00:30 ☽ ☌ ♇
15:31 ☽ △ ☿
22:35 ☽ ⚹ ♅
11 01:55 ☽ ☍ ♄
06:01 ☽ △ ♃
17:34 ☉ △ ♇
22:21 ☿ # ♇
12 12:25 ☽ # ♀
16:53 ☽ □ ☿
13 01:23 ☿ ♈R
03:46 ☽ # ♂
07:39 ☽ △ ♀
09:55 ☽ # ♄
10:24 ☉ ∥ ♃
18:43 ☽ △ ♂
18:52 ☽ ☌ ♆
22:02 ♂ △ ♆
14 07:20 ☽ ⚹ ♇
08:57 ♀ □ ♃
12:14 ☽ ⚹ ☉
19:27 ☽ ⚹ ♀
20:24 ☽ ∥ ♆
15 05:37 ☽ ∥ ♇
07:26 ☽ ☌ ♅
11:12 ☽ △ ♄
11:32 ☽ # ☿
14:35 ☽ ☍ ♃
16:42 ☽ □ ♀
22:48 ☉ # ♅
23:44 ☽ # ☉
23:48 ☽ ∥ ♅
16 03:21 ☽ # ♃
03:24 ☽ □ ♂
13:43 ☽ □ ♇
17 01:05 ☿ ☌ ☉
19:06 ☽ □ ♄
18 04:16 ☽ ⚹ ♀
09:05 ☽ ⚹ ♆
14:40 ☽ ⚹ ♂
22:21 ☽ △ ♇
19 02:11 ☿ ∥ ☉
05:32 ☽ ☌ ☿
08:37 ☽ ∥ ♃
11:53 ☽ # ♅
15:45 ☽ ∥ ☿
17:51 ☉ ♉
18:05 ☽ ∥ ☉
20 00:58 ☽ ⚹ ♅
05:20 ☽ ⚹ ♄
07:28 ☽ # ♇
07:51 ☽ △ ♃
17:25 ☽ # ♆
19:36 ☽ □ ♆
21 09:50 ♀ # ♅
14:17 ♀ △ ♆
22 07:57 ☽ ∥ ♄
12:58 ☽ □ ♅
18:30 ☿ ∥ ♃
19:38 ☽ □ ♃
20:05 ☽ ∥ ♂
23 07:58 ☽ △ ♆
10:35 ☽ ☌ ♀
20:31 ☽ ☌ ♂
21:43 ☽ ☍ ♇
23:23 ☽ ⚹ ☿
24 11:11 ☽ ∥ ♀
18:11 ♂ ☍ ♇
25 00:23 ☽ ⚹ ☉
01:59 ☽ △ ♅
04:30 ☿ ⚹ ♂
06:55 ☽ ☌ ♄
08:19 ☽ ⚹ ♃
16:38 ☽ ∥ ♀
20:40 ☉ ⚹ ♅
22:20 ☿ △ ♇
26 09:57 ☽ □ ☿
27 08:48 ☽ ∥ ♂
22:18 ☽ ∥ ♄
28 08:07 ☽ ☍ ♆
08:36 ☉ # ♇
14:5[illegible] [illegible] ⚹ ♄
16:59 ☽ ⚹ ♀
19:30 ☽ △ ☿
20:48 ☽ △ ♇
22:54 ☉ △ ♃
29 02:09 ☽ ⚹ ♂
11:20 ☽ # ♆
18:43 ☽ ∥ ☉
20:46 ☽ # ♇
23:19 ☽ ☍ ♅
30 04:14 ☽ ⚹ ♄
04:32 ☽ ☌ ♃
06:57 ☽ △ ☉
13:06 ☿ SD
15:04 ☽ # [illegible]
17:36 ☽ ∥ ♃
20:03 ☿ ⚹ ♀

⚷ Chiron

01 Dec.	14 S 24
03	25♑35
06	25 41
09	25 47
12	25 52
15	25 56
18	26 00
21	26 03
24	26 05
27	26 07
30	26 08

04 20:31 14♏42 ✸ Total Lunar Eclipse (mag 1.309)

May 2004

Day	S. T.	☉	☽	☿	♀	♂	♃	♄	♅	♆	♇	☊ True
	h m s	° ' "	° ' "	° '	° '	° '	° '	° '	° '	° '	° '	° '
01 Sa	14 37 02	10♉57 34	19♍44 29	21♈D07	21♊13	25♊58	08♍R56	08♋51	06♓08	15♒19	21♐R53	11♉R16
02 Su	14 40 59	11 55 47	03♎26 05	21 12	21 44	26 36	08 56	08 57	06 10	15 20	21 52	11 15
03 Mo	14 44 55	12 53 57	17 33 01	21 21	22 14	27 14	08 55	09 02	06 12	15 20	21 51	11 15
04 Tu	14 48 52	13 52 06	02♏02 36	21 36	22 43	27 52	08 55	09 08	06 14	15 21	21 50	11 14
05 We	14 52 48	14 50 13	16 49 42	21 54	23 10	28 30	08 55	09 13	06 15	15 21	21 49	11 14
06 Th	14 56 45	15 48 18	01♐47 13	22 17	23 35	29 08	08D 55	09 19	06 17	15 21	21 47	11 13
07 Fr	15 00 41	16 46 21	16 46 54	22 45	23 58	29 46	08 55	09 24	06 19	15 22	21 46	11 12
08 Sa	15 04 38	17 44 24	01♑40 31	23 16	24 20	00♋24	08 56	09 30	06 20	15 22	21 45	11 11
09 Su	15 08 34	18 42 25	16 21 01	23 52	24 40	01 02	08 56	09 36	06 22	15 22	21 44	11 10
10 Mo	15 12 31	19 40 24	00♒43 14	24 31	24 58	01 40	08 57	09 42	06 24	15 23	21 42	11 09
11 Tu	15 16 28	20 38 22	14 44 13	25 15	25 14	02 18	08 58	09 48	06 25	15 23	21 41	11 08
12 We	15 20 24	21 36 19	28 23 07	26 01	25 28	02 56	08 59	09 54	06 26	15 23	21 40	11D 08
13 Th	15 24 21	22 34 14	11♓40 46	26 52	25 41	03 34	09 01	10 00	06 28	15 23	21 39	11 09
14 Fr	15 28 17	23 32 09	24 39 03	27 45	25 51	04 12	09 02	10 06	06 29	15 23	21 37	11 10
15 Sa	15 32 14	24 30 01	07♈20 26	28 42	25 58	04 50	09 04	10 12	06 31	15 24	21 36	11 11
16 Su	15 36 10	25 27 53	19 47 33	29 42	26 04	05 28	09 06	10 18	06 32	15 24	21 34	11 13
17 Mo	15 40 07	26 25 44	02♉02 54	00♉45	26 07	06 06	09 08	10 25	06 33	15 24	21 33	11 14
18 Tu	15 44 03	27 23 33	14 08 51	01 51	26R 08	06 44	09 10	10 31	06 34	15R 24	21 32	11 15
19 We	15 48 00	28 21 20	26 07 34	03 00	26 07	07 22	09 12	10 38	06 35	15 24	21 30	11R 15
20 Th	15 51 57	29 19 07	08♊01 06	04 11	26 03	08 00	09 15	10 44	06 36	15 24	21 29	11 15
21 Fr	15 55 53	00♊16 52	19 51 29	05 26	25 57	08 38	09 18	10 51	06 37	15 23	21 27	11 13
22 Sa	15 59 50	01 14 36	01♋40 50	06 42	25 49	09 16	09 21	10 57	06 38	15 23	21 26	11 10
23 Su	16 03 46	02 12 18	13 31 33	08 02	25 38	09 54	09 24	11 04	06 39	15 23	21 24	11 07
24 Mo	16 07 43	03 09 59	25 26 24	09 24	25 24	10 32	09 27	11 10	06 40	15 23	21 23	11 04
25 Tu	16 11 39	04 07 38	07♌28 37	10 49	25 08	11 10	09 30	11 17	06 41	15 23	21 21	11 01
26 We	16 15 36	05 05 16	19 41 52	12 16	24 50	11 48	09 34	11 24	06 42	15 22	21 20	10 59
27 Th	16 19 33	06 02 52	02♍10 04	13 45	24 30	12 26	09 38	11 31	06 43	15 22	21 18	10 59
28 Fr	16 23 29	07 00 26	14 57 12	15 17	24 07	13 04	09 42	11 38	06 43	15 22	21 17	10D 59
29 Sa	16 27 26	07 58 00	28 06 55	16 51	23 42	13 41	09 46	11 45	06 44	15 21	21 15	10 59
30 Su	16 31 22	08 55 32	11♎42 07	18 28	23 15	14 19	09 50	11 51	06 44	15 21	21 14	11 00
31 Mo	16 35 19	09 53 02	25 44 15	20 08	22 46	14 57	09 54	11 58	06 45	15 21	21 12	11 01

Data for 05-01-2004
Julian Day 2453126.50
Ayanamsa 23 54 50
SVP 05♓11 45
☽ ☊ Mean 11♉18 R

● ◐ PHASES ○ ◑

04	20:34	✸	14♏42
11	11:05	◑	21♒05
19	04:53	●	28♉33
27	07:57	◐	06♍22

LAST ASPECT ☽ Day	h m	INGRESS Day	h m	
01	11:32	01	18:03	♎
03	16:50	03	20:39	♏
04	21:37	05	21:08	♐
07	11:51	07	21:17	♑
09	13:04	09	22:47	♒
11	19:32	12	02:53	♓
14	02:15	14	10:03	♈
16	12:18	16	19:58	♉
19	04:53	19	07:48	♊
21	12:14	21	20:35	♋
22	18:58	24	09:08	♌
26	09:42	26	19:52	♍
28	16:17	29	03:23	♎
30	19:10	31	07:08	♏

DECLINATION

Day	☉	☽	☿	♀	♂	♃	♄	♅	♆	♇
01 Sa	15N07	07N52	06N36	27N44	24N36	09N28	22N46	09S57	16S17	14S18
02 Su	15 25	01 37	06 26	27 46	24 37	09 28	22 45	09 57	16 17	14 18
03 Mo	15 43	04S55	06 19	27 47	24 38	09 28	22 45	09 56	16 17	14 17
04 Tu	16 00	11 24	06 15	27 48	24 39	09 28	22 45	09 55	16 17	14 17
05 We	16 17	17 22	06 13	27 49	24 40	09 28	22 45	09 55	16 17	14 17
06 Th	16 34	22 21	06 14	27 49	24 40	09 27	22 44	09 54	16 17	14 17
07 Fr	16 51	25 51	06 17	27 48	24 41	09 27	22 44	09 53	16 17	14 17
08 Sa	17 07	27 31	06 22	27 47	24 41	09 27	22 44	09 53	16 17	14 17
09 Su	17 23	27 12	06 30	27 46	24 40	09 26	22 44	09 52	16 17	14 17
10 Mo	17 39	25 04	06 39	27 44	24 40	09 26	22 43	09 52	16 16	14 16
11 Tu	17 55	21 28	06 51	27 41	24 40	09 25	22 43	09 51	16 16	14 16
12 We	18 10	16 47	07 05	27 38	24 39	09 25	22 43	09 51	16 16	14 16
13 Th	18 25	11 24	07 20	27 35	24 38	09 24	22 42	09 50	16 16	14 16
14 Fr	18 39	05 40	07 37	27 31	24 37	09 23	22 42	09 50	16 16	14 16
15 Sa	18 54	00N12	07 56	27 26	24 36	09 23	22 42	09 49	16 16	14 16
16 Su	19 08	05 57	08 17	27 21	24 34	09 22	22 41	09 49	16 16	14 16
17 Mo	19 21	11 24	08 39	27 16	24 33	09 21	22 41	09 49	16 16	14 15
18 Tu	19 35	16 21	09 03	27 10	24 31	09 20	22 40	09 48	16 16	14 15
19 We	19 48	20 36	09 27	27 03	24 29	09 19	22 40	09 48	16 16	14 15
20 Th	20 00	24 00	09 54	26 56	24 27	09 17	22 40	09 47	16 16	14 15
21 Fr	20 13	26 20	10 21	26 49	24 24	09 16	22 39	09 47	16 16	14 15
22 Sa	20 25	27 29	10 49	26 40	24 22	09 15	22 39	09 47	16 16	14 15
23 Su	20 36	27 23	11 19	26 31	24 19	09 14	22 38	09 46	16 16	14 15
24 Mo	20 47	26 01	11 49	26 22	24 16	09 12	22 38	09 46	16 17	14 15
25 Tu	20 58	23 28	12 21	26 12	24 13	09 11	22 37	09 46	16 17	14 15
26 We	21 09	19 49	12 53	26 01	24 10	09 09	22 37	09 46	16 17	14 15
27 Th	21 19	15 16	13 25	25 50	24 06	09 08	22 37	09 45	16 17	14 15
28 Fr	21 29	09 56	13 59	25 38	24 02	09 06	22 36	09 45	16 17	14 14
29 Sa	21 38	04 02	14 33	25 25	23 59	09 04	22 36	09 45	16 17	14 14
30 Su	21 47	02S15	15 07	25 12	23 55	09 02	22 35	09 45	16 17	14 14
31 Mo	21 56	08 39	15 42	24 58	23 50	09 01	22 34	09 45	16 17	14 14

ASPECTARIAN

01 02:43 ☽ □ ♀
03:49 ☽ □ ♇
05:12 ☽ ∥ ☿
11:32 ☽ □ ♂
20:31 ♃ ✶ ♄
02 06:04 ♀ ☍ ♇
09:31 ☽ □ ♄
20:17 ☽ △ ♆
03 05:05 ☽ # ☿
06:27 ☽ ☍ ☿
07:10 ☽ ✶ ♇
08:06 ☽ △ ♀
16:45 ☽ # ♃
16:50 ☽ △ ♂
18:29 ☽ ∥ ♅
04 06:51 ☽ △ ♅
11:13 ☽ ✶ ♃
11:18 ☽ ∥ ♇
11:37 ☽ △ ♄
17:47 ☿ △ ♇
19:14 ☽ # ☉
19:26 ☽ ∥ ♆
21:37 ☽ □ ♆
23:20 ☉ # ♆
05 03:08 ♃ SD
12:49 ☉ □ ♆
06 02:16 ☽ # ♄
07:13 ☽ □ ♅
11:24 ☽ □ ♃
14:35 ☽ # ♂
21:43 ☽ ✶ ♆
07 08:00 ☽ ☌ ♇
08:46 ♂ ♋
09:55 ☽ △ ☿
11:51 ☽ ☍ ♀
21:51 ☽ ☍ ♂
08 07:36 ☽ ✶ ♅
11:49 ☽ △ ♃
12:50 ☽ ☍ ♄
09 04:11 ☽ △ ☉
13:04 ☽ □ ☿
10 03:09 ☽ # ♂
16:30 ☽ # ♄
23:47 ☿ ✶ ♀
11 01:07 ☽ ☌ ♆
12:07 ☽ ✶ ♇
17:36 ☽ # ☉
18:44 ☽ △ ♀
19:32 ☽ ✶ ☿
12 02:19 ☽ ∥ ♆
08:33 ☽ △ ♂
11:27 ☽ ∥ ♇
14:30 ☽ ☌ ♅
19:07 ☽ ☍ ♃
20:54 ☽ △ ♄
13 06:37 ☽ ∥ ♅
08:27 ☽ # ♃
16:24 ☽ # ☿
18:21 ☽ □ ♇
21:45 ☽ ✶ ☉
14 02:15 ☽ □ ♀
18:58 ☽ □ ♂
15 05:32 ☽ □ ♄
15:28 ☽ ✶ ♆
16 03:28 ☽ △ ♇
06:55 ☿ ♉
10:48 ☽ ∥ ☿
12:18 ☽ ✶ ♀
14:48 ☽ ∥ ♃
16:52 ☽ # ♅
21:12 ☽ ☌ ☿
17 08:27 ☽ ✶ ♂
08:55 ☽ ✶ ♅
12:14 ♆ SR
13:34 ☽ # ♇
14:03 ☽ △ ♃
16:43 ☽ ✶ ♄
17:28 ♂ △ ♅
22:29 ♀ SR
23:40 ☽ # ♆
18 02:29 ☽ □ ♆
16:02 ☿ ∥ ♃
18:50 ☽ ∥ ☉
19 13:47 ☽ ∥ ♄
18:32 ☿ # ♅
21:08 ☽ □ ♅
20 02:30 ☽ □ ♃
03:51 ☽ ∥ ♂
14:56 ☽ △ ♆
16:59 ☉ ♊
21 03:14 ☽ ☍ ♇
06:41 ☽ ∥ ♀
12:14 ☽ ☌ ♀
22:43 ☿ ✶ ♅
22 03:05 ♂ ✶ ♃
10:04 ☽ △ ♅
11:28 ☽ ✶ ☿
15:36 ☽ ✶ ♃
16:15 ☽ ☌ ♂
18:58 ☽ ☌ ♄
23 18:59 ☽ ∥ ♀
24 00:52 ☿ △ ♃
16:47 ☽ ✶ ☉
17:46 ☽ ∥ ♂
25 05:40 ♂ ☌ ♄
06:08 ☽ ∥ ♄
07:28 ☽ □ ☿
08:39 ☿ ✶ ♄
10:33 ☿ ✶ ♂
15:34 ☽ ☍ ♆
16:20 ☽ ∥ ☉
26 03:10 ☽ △ ♇
09:42 ☽ ✶ ♀
19:01 ☽ # ♆
27 04:51 ☽ # ♇
07:49 ☽ ∥ ☿
08:36 ☽ ☍ ♅
14:09 ☽ ☌ ♃
16:44 ☉ □ ♅
17:46 ☽ ✶ ♄
20:18 ☽ ✶ ♂
28 00:42 ☽ △ ☿
00:47 ☽ # ♅
01:14 ☿ □ ♆
03:33 ☽ ∥ ♃
11:06 ☿ # ♇
11:37 ☽ □ ♇
16:17 ☽ □ ♀
29 18:48 ☽ △ ☉
30 00:16 ☽ □ ♄
04:45 ☽ □ ♂
06:19 ☽ △ ♆
16:21 ☽ ✶ ♇
19:10 ☽ △ ♀
31 00:34 ☉ □ ♃
01:20 ☽ # ♃
04:08 ☽ ∥ ♅
18:20 ☽ △ ♅
21:42 ☽ ∥ ♇
23:37 ☽ ✶ ♃

⚷ Chiron

01 Dec.	14 S 00
03	26♑08R
06	26 07
09	26 06
12	26 04
15	26 02
18	25 58
21	25 55
24	25 50
27	25 45
30	25 39

02 04:12 26♑08 R

June 2004

Day	S. T.	☉	☽	☿	♀	♂	♃	♄	♅	♆	♇	☊ True
	h m s	° ′ ″	° ′ ″	° ′	° ′	° ′	° ′	° ′	° ′	° ′	° ′	° ′
01 Tu	16 39 15	10♊ 50 32	10♏ 12 34	21♉ 49	22♊ R16	15♋ 35	09♍ 59	12♋ 05	06♓ 45	15♒ R20	21♐ R11	11♉ R00
02 We	16 43 12	11 48 00	25 03 34	23 33	21 43	16 13	10 04	12 13	06 46	15 20	21 09	10 59
03 Th	16 47 08	12 45 27	10♐ 10 43	25 20	21 10	16 51	10 08	12 20	06 46	15 19	21 08	10 56
04 Fr	16 51 05	13 42 53	25 25 03	27 09	20 35	17 29	10 13	12 27	06 47	15 19	21 06	10 52
05 Sa	16 55 02	14 40 18	10♑ 36 23	29 00	19 59	18 06	10 19	12 34	06 47	15 18	21 05	10 47
06 Su	16 58 58	15 37 43	25 34 55	00♊ 53	19 22	18 44	10 24	12 41	06 47	15 18	21 03	10 42
07 Mo	17 02 55	16 35 07	10♒ 12 46	02 49	18 45	19 22	10 29	12 48	06 47	15 17	21 01	10 38
08 Tu	17 06 51	17 32 30	24 24 54	04 47	18 07	20 00	10 35	12 56	06 48	15 16	21 00	10 34
09 We	17 10 48	18 29 52	08♓ 09 24	06 47	17 29	20 38	10 41	13 03	06 48	15 16	20 58	10 32
10 Th	17 14 44	19 27 14	21 27 03	08 49	16 52	21 16	10 47	13 10	06 48	15 15	20 57	10D 32
11 Fr	17 18 41	20 24 35	04♈ 20 32	10 54	16 15	21 53	10 53	13 18	06R 48	15 14	20 55	10 33
12 Sa	17 22 37	21 21 56	16 53 40	12 59	15 38	22 31	10 59	13 25	06 48	15 13	20 53	10 36
13 Su	17 26 34	22 19 17	29 10 38	15 07	15 03	23 09	11 05	13 33	06 48	15 13	20 52	10 38
14 Mo	17 30 31	23 16 37	11♉ 15 31	17 16	14 28	23 47	11 12	13 40	06 47	15 12	20 50	10 39
15 Tu	17 34 27	24 13 56	23 12 08	19 25	13 55	24 25	11 18	13 48	06 47	15 11	20 49	10R 39
16 We	17 38 24	25 11 16	05♊ 03 44	21 36	13 23	25 03	11 25	13 55	06 47	15 10	20 47	10 37
17 Th	17 42 20	26 08 34	16 53 04	23 48	12 53	25 40	11 32	14 03	06 47	15 09	20 45	10 33
18 Fr	17 46 17	27 05 52	28 42 23	25 59	12 25	26 18	11 39	14 10	06 46	15 08	20 44	10 26
19 Sa	17 50 13	28 03 10	10♋ 33 33	28 11	11 59	26 56	11 46	14 18	06 46	15 07	20 42	10 17
20 Su	17 54 10	29 00 27	22 28 17	00♋ 23	11 35	27 34	11 53	14 25	06 46	15 06	20 41	10 08
21 Mo	17 58 06	29 57 44	04♌ 28 18	02 34	11 12	28 12	12 01	14 33	06 45	15 05	20 39	09 59
22 Tu	18 02 03	00♋ 55 00	16 35 39	04 44	10 53	28 49	12 08	14 41	06 45	15 04	20 38	09 51
23 We	18 05 60	01 52 15	28 52 45	06 54	10 35	29 27	12 16	14 48	06 44	15 03	20 36	09 44
24 Th	18 09 56	02 49 29	11♍ 22 29	09 02	10 20	00♌ 05	12 24	14 56	06 43	15 02	20 34	09 40
25 Fr	18 13 53	03 46 43	24 08 04	11 09	10 07	00 43	12 32	15 04	06 43	15 01	20 33	09 38
26 Sa	18 17 49	04 43 57	07♎ 12 54	13 15	09 56	01 21	12 40	15 11	06 42	15 00	20 31	09D 38
27 Su	18 21 46	05 41 10	20 40 11	15 18	09 48	01 58	12 48	15 19	06 41	14 59	20 30	09 38
28 Mo	18 25 42	06 38 22	04♏ 32 22	17 21	09 42	02 36	12 56	15 27	06 41	14 58	20 28	09R 38
29 Tu	18 29 39	07 35 34	18 50 21	19 21	09 39	03 14	13 04	15 35	06 40	14 56	20 27	09 37
30 We	18 33 35	08 32 45	03♐ 32 31	21 19	09D 38	03 52	13 13	15 42	06 39	14 55	20 25	09 34

Data for	06-01-2004
Julian Day	2453157.50
Ayanamsa	23 54 55
SVP	05 ♓ 11 41
☽ ☊ Mean	09 ♉ 39 R

●◐ PHASES ○◑

Day	Time	Phase	Position
03	04:20	○	12♐56
09	20:03	◑	19♓18
17	20:27	●	26♊57
25	19:08	◐	04♎32

LAST ASPECT Day	h m	☽ INGRESS Day	h m	Sign
01	21:16	02	7:53	♐
03	17:13	04	07:13	♑
05	12:29	06	07:11	♒
07	18:10	08	09:39	♓
09	23:38	10	15:50	♈
12	11:32	13	01:37	♉
15	02:35	15	13:44	♊
17	20:27	18	02:37	♋
20	10:46	20	15:05	♌
22	07:54	23	02:10	♍
24	17:20	25	10:50	♎
26	23:42	27	16:14	♏
29	00:58	29	18:16	♐

DECLINATION

Day	☉	☽	☿	♀	♂	♃	♄	♅	♆	♇
01 Tu	22N04	14S 48	16N17	24N44	23N46	08N59	22N34	09S 44	16S 17	14S 14
02 We	22 12	20 16	16 52	24 29	23 42	08 57	22 33	09 44	16 18	14 14
03 Th	22 20	24 30	17 27	24 13	23 37	08 55	22 33	09 44	16 18	14 14
04 Fr	22 27	27 01	18 02	23 57	23 32	08 53	22 32	09 44	16 18	14 14
05 Sa	22 34	27 30	18 36	23 41	23 27	08 51	22 32	09 44	16 18	14 14
06 Su	22 40	25 57	19 10	23 24	23 22	08 48	22 31	09 44	16 18	14 14
07 Mo	22 46	22 41	19 44	23 07	23 16	08 46	22 31	09 44	16 18	14 14
08 Tu	22 51	18 08	20 17	22 49	23 11	08 44	22 30	09 44	16 19	14 14
09 We	22 57	12 47	20 48	22 32	23 05	08 42	22 29	09 44	16 19	14 14
10 Th	23 01	07 01	21 19	22 14	22 59	08 39	22 29	09 44	16 19	14 14
11 Fr	23 06	01 07	21 48	21 56	22 53	08 37	22 28	09 44	16 19	14 14
12 Sa	23 10	04N42	22 16	21 39	22 47	08 34	22 28	09 44	16 20	14 14
13 Su	23 13	10 12	22 41	21 21	22 41	08 32	22 27	09 44	16 20	14 14
14 Mo	23 16	15 15	23 05	21 04	22 34	08 29	22 26	09 44	16 20	14 14
15 Tu	23 19	19 40	23 27	20 47	22 27	08 26	22 26	09 44	16 20	14 14
16 We	23 21	23 15	23 46	20 31	22 20	08 24	22 25	09 44	16 21	14 14
17 Th	23 23	25 50	24 03	20 15	22 13	08 21	22 24	09 45	16 21	14 14
18 Fr	23 25	27 15	24 18	20 00	22 06	08 18	22 23	09 45	16 21	14 14
19 Sa	23 26	27 27	24 29	19 45	21 59	08 15	22 23	09 45	16 21	14 14
20 Su	23 26	26 22	24 38	19 31	21 51	08 12	22 22	09 45	16 22	14 14
21 Mo	23 26	24 05	24 44	19 18	21 44	08 09	22 21	09 45	16 22	14 14
22 Tu	23 26	20 42	24 47	19 05	21 36	08 06	22 21	09 46	16 22	14 14
23 We	23 26	16 23	24 47	18 53	21 28	08 03	22 20	09 46	16 23	14 14
24 Th	23 25	11 19	24 45	18 42	21 20	08 00	22 19	09 46	16 23	14 14
25 Fr	23 23	05 41	24 40	18 32	21 12	07 57	22 18	09 46	16 23	14 14
26 Sa	23 21	00S 20	24 32	18 23	21 03	07 54	22 18	09 47	16 24	14 14
27 Su	23 19	06 31	24 22	18 14	20 55	07 50	22 17	09 47	16 24	14 14
28 Mo	23 16	12 36	24 09	18 06	20 46	07 47	22 16	09 47	16 24	14 15
29 Tu	23 13	18 12	23 54	18 00	20 37	07 44	22 15	09 48	16 25	14 15
30 We	23 10	22 53	23 38	17 54	20 28	07 40	22 14	09 48	16 25	14 15

ASPECTARIAN

01 00:26 ☿ # ♆
03:06 ☽ △ ♄
06:10 ☽ ∥ ♆
06:49 ☽ # ☿
08:21 ☽ □ ♆
09:08 ☽ △ ♂
21:16 ☽ ☍ ☿
02 10:18 ☽ # ☉
11:56 ☽ # ♄
18:20 ☽ # ♂
18:37 ☽ □ ♅
22:15 ☽ # ♀
23:56 ☽ □ ♃
03 01:30 ♀ ☍ ♇
08:06 ☽ ⚹ ♆
16:40 ☽ ☍ ♀
17:13 ☽ ☌ ♇
04 17:55 ☉ ∥ ♄
17:56 ☽ ⚹ ♅
23:32 ☽ △ ♃
05 03:09 ☽ ☍ ♄
12:29 ☽ ☍ ♂
12:48 ☿ ♊
15:40 ☉ △ ♆
06 03:59 ♀ ∥ ♂
09:56 ☽ △ ☿
20:15 ☽ # ♂
21:09 ☽ # ♀
23:30 ☽ # ☉
07 01:01 ☽ # ♄
08:28 ☽ ☌ ♆
11:27 ☽ △ ☉
13:43 ☽ △ ♀
14:29 ☽ # ☿
18:10 ☽ ⚹ ♇
21:40 ♀ ∥ ☉
08 00:30 ☽ ∥ ♇
08:43 ♀ ☌ ☉
17:44 ☽ ∥ ♇
21:09 ☽ □ ☿
21:35 ☽ ☌ ♅
09 00:05 ☿ □ ♅
03:09 ♀ ∥ ♄
04:32 ☽ ☍ ♃
08:50 ☽ △ ♄
12:47 ☽ ∥ ♅
16:01 ☽ □ ♀
17:12 ☽ # ♃
19:21 ☉ ∥ ♂
23:04 ☽ □ ♇
23:38 ☽ △ ♂
10 15:48 ♅ SR
23:51 ☿ □ ♃
11 04:25 ☿ ∥ ♀
12:25 ☉ ☍ ♇
14:57 ☽ ⚹ ☿
17:14 ☽ □ ♄
20:46 ☽ ⚹ ♆
21:41 ☽ ⚹ ♀
12 07:44 ☽ △ ♇
09:25 ☽ ⚹ ☉
10:38 ☿ ∥ ♄
11:32 ☽ □ ♂
16:32 ☽ ∥ ♃
17:08 ♀ △ ♆
21:51 ☽ # ♅
23:25 ☿ ☌ ♀
23:26 ☿ ∥ ♂
13 01:05 ☿ △ ♆
15:05 ☽ ⚹ ♅
18:54 ☽ # ♇
23:52 ☽ △ ♃
14 04:52 ☽ ⚹ ♄
05:32 ☽ # ♆
07:53 ☽ □ ♆
13:33 ☿ ∥ ☉
15 02:35 ☽ ⚹ ♂
06:28 ☽ ∥ ♀
07:09 ♂ ∥ ♄
15:08 ☿ ☍ ♇
17:33 ☽ ∥ ♂
17:52 ☽ ∥ ♄
16 00:52 ☽ ∥ ☉
03:30 ☽ □ ♅
04:41 ☽ ∥ ☿
13:01 ☽ □ ♃
16:13 ☽ ☌ ♀
20:29 ☽ △ ♆
17 07:51 ☽ ☍ ♇
17:14 ☽ ☌ ☿
18 16:20 ☽ △ ♅
21:24 ☿ ☌ ☉
19 02:28 ☽ ⚹ ♃
07:38 ☽ ☌ ♄
09:29 ♀ □ ♃
19:50 ☿ ♋
20 10:46 ☽ ☌ ♂
18:25 ☽ ∥ ☿
21 00:57 ☉ ♋
05:08 ☽ ∥ ☉
13:00 ☽ ⚹ ♀
13:10 ☽ ∥ ♄
18:01 ☽ ∥ ♂
21:00 ☽ ☍ ♆
22 07:54 ☽ △ ♇
10:02 ☽ ∥ ♀
22:11 ☿ △ ♅
23 00:03 ☽ # ♆
06:16 ☽ ⚹ ☉
10:35 ☽ # ♇
15:08 ☽ ☍ ♅
18:38 ☽ ⚹ ☿
20:51 ♂ ♌
22:02 ☽ □ ♀
24 01:58 ☽ ☌ ♃
06:49 ☽ ⚹ ♄
06:50 ☽ # ♅
14:31 ☽ ∥ ♃
17:20 ☽ □ ♇
25 12:46 ☽ ⚹ ♂
16:48 ☿ ⚹ ♃
26 04:51 ☽ △ ♀
12:49 ☽ □ ☿
13:57 ☽ △ ♆
14:27 ☽ □ ♄
23:42 ☽ ⚹ ♇
27 00:08 ☿ ☌ ♄
05:04 ☽ # ♃
12:43 ☽ ∥ ♅
20:32 ☽ □ ♂
28 00:55 ☉ △ ♅
03:38 ☽ △ ♅
05:30 ☽ △ ☉
06:44 ☽ ∥ ♆
14:19 ☽ ⚹ ♃
15:58 ☽ ∥ ♆
17:32 ☽ □ ♆
18:32 ☽ △ ♄
23:01 ☽ # ♀
29 00:58 ☽ △ ☿
11:14 ☽ # ♂
20:14 ☽ # ♄
23:16 ♀ SD
30 00:32 ☽ △ ♂
01:36 ☽ # ☉
04:09 ☽ # ☿
04:59 ☽ □ ♅
09:46 ☽ ☍ ♀
15:38 ☽ □ ♃
18:11 ☽ ⚹ ♆

⚷ Chiron

01	Dec.	13 S 50
02	25♑33R	
05	25 26	
08	25 19	
11	25 11	
14	25 02	
17	24 54	
20	24 45	
23	24 35	
26	24 25	
[illegible]	24 1[illegible]	

July 2004

Day	S. T. h m s	☉ ° ' "	☽ ° ' "	☿ ° '	♀ ° '	♂ ° '	♃ ° '	♄ ° '	♅ ° '	♆ ° '	♇ ° '	☊ True ° '
01 Th	18 37 32	09♋29 57	18♐34 13	23♋16	09♊39	04♌29	13♍21	15♋50	06♓R38	14♒R54	20♐R24	09♉R29
02 Fr	18 41 29	10 27 08	03♑47 37	25 10	09 42	05 07	13 30	15 58	06 37	14 53	20 22	09 22
03 Sa	18 45 25	11 24 19	19 02 37	27 02	09 48	05 45	13 39	16 06	06 36	14 51	20 21	09 13
04 Su	18 49 22	12 21 29	04♒08 27	28 53	09 56	06 23	13 48	16 13	06 35	14 50	20 20	09 04
05 Mo	18 53 18	13 18 40	18 55 27	00♌41	10 06	07 01	13 57	16 21	06 34	14 49	20 18	08 55
06 Tu	18 57 15	14 15 51	03♓16 40	02 27	10 18	07 38	14 06	16 29	06 33	14 47	20 17	08 48
07 We	19 01 11	15 13 03	17 08 32	04 11	10 32	08 16	14 15	16 37	06 31	14 46	20 15	08 42
08 Th	19 05 08	16 10 14	00♈30 53	05 54	10 49	08 54	14 24	16 44	06 30	14 45	20 14	08 40
09 Fr	19 09 05	17 07 26	13 26 06	07 34	11 07	09 32	14 34	16 52	06 29	14 43	20 12	08D 39
10 Sa	19 13 01	18 04 39	25 58 16	09 12	11 27	10 10	14 43	17 00	06 28	14 42	20 11	08 40
11 Su	19 16 58	19 01 52	08♉12 17	10 48	11 48	10 47	14 53	17 08	06 26	14 40	20 10	08 41
12 Mo	19 20 54	19 59 05	20 13 13	12 22	12 12	11 25	15 02	17 16	06 25	14 39	20 08	08R 42
13 Tu	19 24 51	20 56 19	02♊05 55	13 54	12 37	12 03	15 12	17 23	06 24	14 38	20 07	08 40
14 We	19 28 47	21 53 33	13 54 42	15 23	13 03	12 41	15 22	17 31	06 22	14 36	20 06	08 36
15 Th	19 32 44	22 50 48	25 43 14	16 51	13 32	13 19	15 32	17 39	06 21	14 35	20 05	08 29
16 Fr	19 36 40	23 48 04	07♋34 22	18 17	14 01	13 56	15 42	17 47	06 19	14 33	20 03	08 19
17 Sa	19 40 37	24 45 19	19 30 14	19 40	14 32	14 34	15 52	17 55	06 17	14 32	20 02	08 07
18 Su	19 44 34	25 42 36	01♌32 21	21 01	15 05	15 12	16 03	18 02	06 16	14 30	20 01	07 53
19 Mo	19 48 30	26 39 52	13 41 51	22 20	15 39	15 50	16 13	18 10	06 14	14 29	20 00	07 40
20 Tu	19 52 27	27 37 09	25 59 47	23 37	16 14	16 28	16 23	18 18	06 13	14 27	19 58	07 27
21 We	19 56 23	28 34 26	08♍27 17	24 51	16 50	17 06	16 34	18 25	06 11	14 26	19 57	07 17
22 Th	20 00 20	29 31 44	21 05 50	26 03	17 27	17 43	16 44	18 33	06 09	14 24	19 56	07 10
23 Fr	20 04 16	00♌29 02	03♎57 21	27 13	18 06	18 21	16 55	18 41	06 07	14 22	19 55	07 06
24 Sa	20 08 13	01 26 20	17 04 08	28 19	18 46	18 59	17 06	18 49	06 05	14 21	19 54	07 04
25 Su	20 12 09	02 23 39	00♏28 42	29 24	19 26	19 37	17 16	18 56	06 04	14 19	19 53	07 03
26 Mo	20 16 06	03 20 58	14 13 20	00♍25	20 08	20 15	17 27	19 04	06 02	14 18	19 52	07 02
27 Tu	20 20 03	04 18 17	28 19 26	01 24	20 51	20 53	17 38	19 11	06 00	14 16	19 51	07 01
28 We	20 23 59	05 15 37	12♐46 41	02 20	21 34	21 31	17 49	19 19	05 58	14 14	19 50	06 57
29 Th	20 27 56	06 12 57	27 32 16	03 13	22 19	22 09	18 00	19 27	05 56	14 13	19 49	06 51
30 Fr	20 31 52	07 10 18	12♑30 31	04 02	23 04	22 46	18 11	19 34	05 54	14 11	19 48	06 43
31 Sa	20 35 49	08 07 40	27 33 15	04 48	23 51	23 24	18 23	19 42	05 52	14 10	19 47	06 33

Data for	07-01-2004
Julian Day	2453187.50
Ayanamsa	23 55 00
SVP	05 ♓ 11 39
☽ ☊ Mean	08 ♉ 04 R

● ◐ PHASES ○ ◑

Day	Time	Phase	Position
02	11:10	○	10♑54
09	07:34	◑	17♈25
17	11:24	●	25♋13
25	03:38	◐	02♏32
31	18:05	○	08♒51

LAST ASPECT ☽ INGRESS

Day	h m	Day	h m	Sign
01	02:53	01	18:02	♑
03	14:26	03	17:23	♒
05	02:16	05	18:27	♓
07	05:30	07	23:04	♈
09	12:52	10	07:51	♉
11	23:29	12	19:45	♊
14	12:33	15	08:41	♋
17	11:24	17	20:57	♌
19	18:51	20	07:45	♍
21	21:49	22	16:40	♎
24	21:55	24	23:09	♏
26	10:49	27	02:49	♐
28	15:07	29	03:58	♑
30	11:22	31	03:54	♒

DECLINATION

Day	☉	☽	☿	♀	♂	♃	♄	♅	♆	♇
01 Th	23N06	26S08	23N19	17N48	20N19	07N37	22N14	09S48	16S26	14S15
02 Fr	23 02	27 30	22 58	17 44	20 10	07 34	22 13	09 49	16 26	14 15
03 Sa	22 57	26 47	22 35	17 40	20 01	07 30	22 12	09 49	16 26	14 15
04 Su	22 52	24 07	22 11	17 37	19 51	07 26	22 11	09 50	16 27	14 15
05 Mo	22 47	19 53	21 46	17 35	19 41	07 23	22 10	09 50	16 27	14 15
06 Tu	22 41	14 36	21 19	17 33	19 32	07 19	22 09	09 50	16 27	14 15
07 We	22 34	08 44	20 51	17 33	19 22	07 16	22 08	09 51	16 28	14 16
08 Th	22 28	02 41	20 22	17 32	19 12	07 12	22 08	09 51	16 28	14 16
09 Fr	22 21	03N18	19 52	17 33	19 01	07 08	22 07	09 52	16 29	14 16
10 Sa	22 13	08 59	19 21	17 33	18 51	07 04	22 06	09 52	16 29	14 16
11 Su	22 05	14 12	18 50	17 35	18 41	07 01	22 05	09 53	16 30	14 16
12 Mo	21 57	18 46	18 18	17 36	18 30	06 57	22 04	09 53	16 30	14 16
13 Tu	21 49	22 33	17 45	17 39	18 20	06 53	22 03	09 54	16 30	14 16
14 We	21 40	25 22	17 11	17 41	18 09	06 49	22 02	09 55	16 31	14 17
15 Th	21 30	27 04	16 38	17 44	17 58	06 45	22 01	09 55	16 31	14 17
16 Fr	21 21	27 32	16 04	17 48	17 47	06 41	22 00	09 56	16 32	14 17
17 Sa	21 11	26 43	15 30	17 51	17 36	06 37	21 59	09 56	16 32	14 17
18 Su	21 00	24 39	14 55	17 55	17 24	06 33	21 58	09 57	16 33	14 17
19 Mo	20 49	21 28	14 21	18 00	17 13	06 29	21 58	09 58	16 33	14 17
20 Tu	20 38	17 19	13 47	18 04	17 01	06 25	21 57	09 58	16 34	14 18
21 We	20 27	12 22	13 13	18 09	16 50	06 20	21 56	09 59	16 34	14 18
22 Th	20 15	06 51	12 39	18 13	16 38	06 16	21 55	10 00	16 35	14 18
23 Fr	20 03	00 57	12 05	18 18	16 26	06 12	21 54	10 00	16 35	14 18
24 Sa	19 50	05S07	11 32	18 23	16 14	06 08	21 53	10 01	16 35	14 18
25 Su	19 38	11 06	10 59	18 28	16 02	06 03	21 52	10 02	16 36	14 19
26 Mo	19 24	16 43	10 27	18 34	15 50	05 59	21 51	10 02	16 36	14 19
27 Tu	19 11	21 34	09 55	18 39	15 38	05 55	21 50	10 03	16 37	14 19
28 We	18 57	25 15	09 25	18 44	15 26	05 50	21 49	10 04	16 37	14 19
29 Th	18 43	27 18	08 55	18 49	15 13	05 46	21 48	10 05	16 38	14 20
30 Fr	18 29	27 24	08 25	18 54	15 01	05 41	21 47	10 05	16 38	14 20
31 Sa	18 14	25 29	07 57	18 59	14 48	05 37	21 46	10 06	16 39	14 20

ASPECTARIAN

01 02:53 ☽ ☌ ♇
18:34 ☿ ∥ ☉

02 04:26 ☽ ✱ ♅
15:25 ☽ △ ♃
19:18 ☽ ☍ ♄

03 14:26 ☽ ☍ ☿
04 00:24 ☿ ∥ ♄
03:45 ☽ ☍ ♂
08:07 ☽ # ☉
09:26 ☽ △ ♀
11:56 ☽ # ♄
13:09 ☽ # ☿
14:52 ☿ ♌
17:16 ☽ ☌ ♆

05 00:59 ☽ # ♂
02:16 ☽ ✱ ♇
10:57 ☽ # ♀
15:57 ☽ ∥ ♆
18:59 ☉ ✱ ♃

06 01:28 ☽ ∥ ♇
05:34 ☽ ☌ ♅
12:15 ☽ □ ♀
18:52 ☽ ☍ ♃
19:34 ☽ ∥ ♅
20:22 ☽ △ ☉
23:03 ☽ △ ♄

07 05:31 ☽ □ ♇
05:57 ☽ # ♃
08 11:22 ☽ △ ☿
16:17 ☽ △ ♂
16:38 ☉ ☌ ♄
19:31 ☽ ✱ ♀

09 02:26 ☽ ✱ ♆
06:35 ☽ □ ♄
12:52 ☽ △ ♇
16:50 ☽ ∥ ♃

10 03:56 ☽ # ♅
20:31 ☽ ✱ ♅
23:50 ☿ ☌ ♂

11 00:21 ☽ # ♇
01:29 ☉ ∥ ♄
05:25 ☽ □ ♂
05:55 ☽ □ ☿
10:03 ☿ ∥ ♂
11:36 ☽ # ♆
12:51 ☽ □ ♆
13:28 ☽ △ ♃
17:28 ☽ ∥ ♀
17:59 ☽ ✱ ♄
20:30 ☿ ✱ ♀
21:35 ☽ ∥ ☿
22:32 ☽ ∥ ♂
23:29 ☽ ✱ ☉
12 19:00 ☽ ∥ ☉
20:28 ☽ ∥ ♄

13 04:02 ☿ ∥ ♀
08:42 ☽ □ ♅
11:32 ☿ ☍ ♇
21:21 ☽ ✱ ♂
22:11 ☽ ☌ ♀

14 01:24 ☽ △ ♆
03:00 ☽ □ ♃
03:26 ☽ ✱ ☿
12:33 ☽ ☍ ♇

15 04:32 ☿ # ♆
21:28 ☽ △ ♅
22:18 ♀ ∥ ♂
16 16:36 ☽ ✱ ♃
16:46 ♀ ✱ ♂
20:46 ☽ ☌ ♄
22:26 ♂ ☍ ♆
23:32 ♀ △ ♆

17 06:26 ☿ △ ♇
18 20:48 ☽ ∥ ♄

19 01:32 ☽ ☍ ♆
02:38 ☿ # ♇
04:00 ☽ ✱ ♀
04:20 ☽ ∥ ☉
04:25 ☽ ☌ ♂
12:18 ☽ △ ♇
18:51 ☽ ☌ ☿
20:05 ☽ ∥ ♀

20 01:36 ☽ ∥ ♂
03:54 ☽ # ♆
08:58 ♀ □ ♃
15:04 ☽ # ♇
19:39 ☽ ☍ ♅
19:40 ☽ ∥ ☿

21 10:39 ☽ # ♅
15:39 ☽ ☌ ♃
16:46 ☽ □ ♀
19:09 ☽ ✱ ♄
21:49 ☽ □ ♇
22 02:28 ☽ ∥ ♃
07:02 ♂ # ♆
11:50 ☉ ♌
17:03 ☽ ✱ ☉

23 19:04 ☽ △ ♆
24 03:11 ☽ □ ♄
03:13 ☽ △ ♀
03:38 ☽ ✱ ♂
03:58 ☽ # ♃
05:06 ☽ ✱ ♇
19:37 ☽ ∥ ♅
21:55 ☽ ✱ ☿
23:35 ☽ # ☿

25 09:48 ☽ △ ♅
09:50 ♂ △ ♇
13:27 ☽ ∥ ♇
13:58 ☿ ♍
14:56 ♀ ☍ ♇
20:13 ☽ # ♂
23:31 ☽ ∥ ♆

26 00:07 ☽ □ ♆
05:38 ☽ ✱ ♃
08:23 ☽ △ ♄
08:44 ☽ # ♀
10:49 ☽ □ ♂
12:10 ☽ # ☉
18:14 ☿ # ♅

27 01:25 ☽ # ♄
05:31 ☽ □ ☿
08:30 ♀ ✱ ♂
10:42 ☽ △ ☉
12:47 ☽ □ ♅
28 02:24 ☽ ✱ ♆
08:21 ☽ □ ♃
11:31 ☽ ☌ ♇
14:53 ☽ △ ♂
15:07 ☽ ☍ ♀
16:57 ♀ ∥ ☉

29 09:40 ☽ △ ☿
13:27 ☽ ✱ ♅
30 09:10 ☽ △ ♃
11:22 ☽ ☍ ♄

⚷ Chiron

01 Dec. 13 S 56

Day	Position
02	24♑05R
05	23 54
08	23 44
11	23 33
14	23 22
17	23 11
20	23 01
23	22 50
26	22 39
29	22 29

August 2004

Day	S. T. h m s	☉ ° ' "	☽ ° ' "	☿ ° '	♀ ° '	♂ ° '	♃ ° '	♄ ° '	♅ ° '	♆ ° '	♇ ° '	☊ True ° '
01 Su	20 39 45	09♌ 05 02	12♒ 30 50	05♍ 31	24♊ 38	24♌ 02	18♍ 34	19♋ 49	05♓R50	14♒R08	19♐R46	06♉R22
02 Mo	20 43 42	10 02 26	27 13 52	06 10	25 26	24 40	18 45	19 57	05 48	14 06	19 45	06 11
03 Tu	20 47 38	10 59 50	11♓ 34 42	06 45	26 14	25 18	18 57	20 04	05 46	14 05	19 44	06 02
04 We	20 51 35	11 57 15	25 28 35	07 16	27 04	25 56	19 08	20 12	05 44	14 03	19 44	05 55
05 Th	20 55 32	12 54 41	08♈ 54 01	07 43	27 54	26 34	19 19	20 19	05 42	14 01	19 43	05 50
06 Fr	20 59 28	13 52 09	21 52 18	08 05	28 45	27 12	19 31	20 26	05 39	14 00	19 42	05 48
07 Sa	21 03 25	14 49 38	04♉ 26 51	08 23	29 36	27 50	19 43	20 34	05 37	13 58	19 41	05D 48
08 Su	21 07 21	15 47 08	16 42 18	08 36	00♋ 28	28 28	19 54	20 41	05 35	13 57	19 41	05 49
09 Mo	21 11 18	16 44 39	28 43 50	08 44	01 21	29 06	20 06	20 48	05 33	13 55	19 40	05R 49
10 Tu	21 15 14	17 42 12	10♊ 36 46	08 46	02 14	29 44	20 18	20 56	05 31	13 53	19 39	05 48
11 We	21 19 11	18 39 46	22 26 07	08R 44	03 08	00♍ 22	20 30	21 03	05 28	13 52	19 39	05 45
12 Th	21 23 07	19 37 22	04♋ 16 22	08 36	04 03	01 00	20 42	21 10	05 26	13 50	19 38	05 39
13 Fr	21 27 04	20 34 59	16 11 14	08 22	04 57	01 38	20 54	21 17	05 24	13 48	19 37	05 30
14 Sa	21 31 01	21 32 37	28 13 31	08 04	05 53	02 16	21 06	21 24	05 21	13 47	19 37	05 18
15 Su	21 34 57	22 30 17	10♌ 25 08	07 39	06 49	02 54	21 18	21 32	05 19	13 45	19 36	05 05
16 Mo	21 38 54	23 27 58	22 47 11	07 09	07 45	03 32	21 30	21 39	05 17	13 44	19 36	04 52
17 Tu	21 42 50	24 25 40	05♍ 20 11	06 35	08 42	04 10	21 42	21 46	05 14	13 42	19 35	04 40
18 We	21 46 47	25 23 24	18 04 17	05 55	09 40	04 48	21 54	21 53	05 12	13 40	19 35	04 31
19 Th	21 50 43	26 21 08	00♎ 59 31	05 12	10 37	05 26	22 06	21 59	05 10	13 39	19 35	04 23
20 Fr	21 54 40	27 18 54	14 06 08	04 24	11 36	06 04	22 19	22 06	05 07	13 37	19 34	04 19
21 Sa	21 58 36	28 16 41	27 24 43	03 34	12 34	06 42	22 31	22 13	05 05	13 36	19 34	04 17
22 Su	22 02 33	29 14 29	10♏ 56 12	02 42	13 33	07 21	22 43	22 20	05 03	13 34	19 34	04 16
23 Mo	22 06 30	00♍ 12 19	24 41 36	01 49	14 33	07 59	22 56	22 27	05 00	13 33	19 33	04 16
24 Tu	22 10 26	01 10 09	08♐ 41 36	00 56	15 33	08 37	23 08	22 33	04 58	13 31	19 33	04 15
25 We	22 14 23	02 08 01	22 55 57	00 03	16 33	09 15	23 21	22 40	04 56	13 29	19 33	04 13
26 Th	22 18 19	03 05 54	07♑ 22 52	29♌ 13	17 33	09 53	23 33	22 47	04 53	13 28	19 33	04 09
27 Fr	22 22 16	04 03 48	21 58 42	28 26	18 34	10 31	23 46	22 53	04 51	13 26	19 33	04 03
28 Sa	22 26 12	05 01 44	06♒ 37 54	27 44	19 35	11 10	23 58	23 00	04 48	13 25	19 33	03 55
29 Su	22 30 09	05 59 40	21 13 35	27 06	20 37	11 48	24 11	23 06	04 46	13 23	19 33	03 46
30 Mo	22 34 05	06 57 39	05♓ 38 29	26 35	21 39	12 26	24 23	23 12	04 44	13 22	19 32	03 37
31 Tu	22 38 02	07 55 39	19 46 10	26 11	22 41	13 04	24 36	23 19	04 41	13 21	19D 32	03 30

Data for	08-01-2004
Julian Day	2453218.50
Ayanamsa	23 55 06
SVP	05 ♓ 11 36
☽ ☊ Mean	06 ♉ 26 R

● ◐ PHASES ○ ◑

Day	Time	Phase	Position
07	22:01	◑	15♉42
16	01:25	●	23♌31
23	10:12	◐	00♐37
30	02:22	○	07♓03

LAST ASPECT ☽ INGRESS

Day	h m	Day	h m	Sign
01	20:52	02	04:35	♓
04	02:58	04	08:00	♈
06	13:59	06	15:26	♉
09	00:47	09	02:33	♊
10	20:00	11	15:21	♋
13	10:18	14	03:31	♌
16	01:25	16	13:50	♍
18	07:16	18	22:10	♎
21	01:40	21	04:37	♏
22	20:54	23	09:09	♐
25	11:13	25	11:47	♑
27	02:58	27	13:08	♒
29	09:23	29	14:34	♓
31	08:29	31	17:46	♈

DECLINATION

Day	☉	☽	☿	♀	♂	♃	♄	♅	♆	♇
01 Su	17N59	21S49	07N30	19N04	14N35	05N33	21N45	10S07	16S39	14S20
02 Mo	17 44	16 49	07 05	19 09	14 22	05 28	21 44	10 08	16 40	14 21
03 Tu	17 28	11 00	06 40	19 13	14 09	05 23	21 43	10 08	16 40	14 21
04 We	17 12	04 47	06 18	19 18	13 56	05 19	21 42	10 09	16 41	14 21
05 Th	16 56	01N26	05 56	19 22	13 43	05 14	21 41	10 10	16 41	14 21
06 Fr	16 40	07 24	05 37	19 26	13 30	05 10	21 40	10 11	16 42	14 22
07 Sa	16 23	12 54	05 20	19 30	13 17	05 05	21 39	10 12	16 42	14 22
08 Su	16 06	17 45	05 04	19 34	13 03	05 00	21 38	10 13	16 43	14 22
09 Mo	15 49	21 48	04 51	19 37	12 50	04 56	21 37	10 13	16 43	14 23
10 Tu	15 32	24 54	04 40	19 40	12 37	04 51	21 36	10 14	16 44	14 23
11 We	15 14	26 54	04 32	19 43	12 23	04 46	21 34	10 15	16 44	14 23
12 Th	14 56	27 40	04 27	19 46	12 09	04 42	21 33	10 16	16 45	14 24
13 Fr	14 38	27 10	04 24	19 48	11 55	04 37	21 32	10 17	16 45	14 24
14 Sa	14 19	25 23	04 24	19 50	11 42	04 32	21 31	10 18	16 46	14 24
15 Su	14 01	22 26	04 28	19 51	11 28	04 27	21 30	10 18	16 46	14 24
16 Mo	13 42	18 25	04 34	19 53	11 14	04 22	21 29	10 19	16 46	14 25
17 Tu	13 23	13 34	04 43	19 54	11 00	04 18	21 28	10 20	16 47	14 25
18 We	13 04	08 04	04 56	19 54	10 46	04 13	21 27	10 21	16 47	14 25
19 Th	12 44	02 09	05 12	19 54	10 31	04 08	21 26	10 22	16 48	14 26
20 Fr	12 24	03S57	05 30	19 54	10 17	04 03	21 25	10 23	16 48	14 26
21 Sa	12 04	09 59	05 51	19 53	10 03	03 58	21 24	10 24	16 49	14 26
22 Su	11 44	15 41	06 14	19 52	09 48	03 53	21 23	10 25	16 49	14 27
23 Mo	11 24	20 40	06 40	19 51	09 34	03 48	21 22	10 25	16 50	14 27
24 Tu	11 04	24 36	07 07	19 49	09 19	03 43	21 21	10 26	16 50	14 27
25 We	10 43	27 04	07 36	19 46	09 05	03 38	21 20	10 27	16 51	14 28
26 Th	10 22	27 46	08 05	19 44	08 50	03 33	21 19	10 28	16 51	14 28
27 Fr	10 01	26 33	08 34	19 40	08 36	03 28	21 18	10 29	16 51	14 29
28 Sa	09 40	23 31	09 03	19 36	08 21	03 23	21 17	10 30	16 52	14 29
29 Su	09 19	19 01	09 31	19 32	08 06	03 18	21 16	10 31	16 52	14 29
30 Mo	08 57	13 27	09 58	19 28	07 51	03 13	21 15	10 32	16 53	14 30
31 Tu	08 36	07 18	10 24	19 22	07 36	03 08	21 14	10 32	16 53	14 30

ASPECTARIAN

01 00:25 ☽ # ♄
02:37 ☽ ☌ ♆
10:50 ☿ ☍ ♅
11:45 ☽ ⚹ ♇
13:40 ☽ # ♀
19:36 ☽ ☍ ♂
19:47 ☽ # ☉
20:52 ☽ △ ♀
02 00:41 ☽ ∥ ♆
03:07 ♂ # ♇
10:31 ☽ ∥ ♇
10:47 ☽ # ♂
14:13 ☽ ☌ ♅
15:30 ☽ ☍ ☿
03 03:22 ☽ ∥ ♅
12:47 ☽ ☍ ♃
13:59 ☽ □ ♇
14:42 ☽ △ ♄
17:53 ☽ # ☿
21:59 ☽ # ♃
04 02:58 ☽ □ ♀
05 07:55 ☽ △ ☉
09:22 ☽ ⚹ ♆
14:59 ☽ ∥ ♃
17:05 ☽ ∥ ☿
19:56 ☽ △ ♇
21:17 ☽ □ ♄
21:33 ☉ # ♆
06 03:08 ☉ ☍ ♆
10:37 ☽ △ ♂
11:56 ☽ # ♅
13:59 ☽ ⚹ ♀
21:23 ♃ □ ♇
07 01:45 ☽ ∥ ♂
02:16 ☽ ⚹ ♅
06:57 ☽ # ♇
07:10 [illegible]
11:02 ♀ ♋
15:58 ☽ ∥ ☉
18:34 ☽ # ♇
18:34 ☽ □ ♆
08 06:27 ☽ △ ♃
07:59 ☽ ⚹ ♄
10:10 ☿ ∥ ♃
10:17 ☽ ∥ ♀
22:46 ☽ ∥ ♄
09 00:47 ☽ □ ♂
13:42 ☽ □ ♅
20:16 ☽ □ ☿
10 00:34 ☿ SR
06:38 ☽ △ ♆
10:14 ♂ ♍
15:40 ☽ ⚹ ☉
18:20 ☽ ☍ ♇
20:00 ☽ □ ♃
11 17:00 ☽ ⚹ ♂
23:30 ☽ ☌ ♀
12 00:18 ☉ △ ♇
02:20 ☽ △ ♅
08:36 ☽ ⚹ ☿
13 09:35 ☽ ⚹ ♃
10:18 ☽ ☌ ♄
10:57 ♀ △ ♅
18:03 ☉ # ♇
14 23:07 ☿ ∥ ♃
15 06:08 ☽ ∥ ♄
06:29 ☽ ☍ ♆
14:10 ☿ ⚹ ♀
15:59 ☽ ∥ ♀
17:51 ☽ △ ♇
16 00:30 [illegible]
20:06 ☽ # ♇
21:40 ☽ ☌ ♂
23:49 ☽ ☍ ♅
17 00:57 ☽ ∥ ☉
02:15 ☽ ☌ ☿
06:54 ☽ ⚹ ♀
12:04 ☽ ∥ ♂
14:23 ☽ # ♅
16:35 ♃ ⚹ ♄
18 02:50 ☽ □ ♇
07:10 ☽ ⚹ ♄
07:16 ☽ ☌ ♃
12:23 ☽ ∥ ☿
14:14 ♂ ☍ ♅
16:02 ☽ ∥ ♃
19:47 ☿ ☌ ♂
19 01:01 ☿ ☍ ♅
15:00 ♂ # ♅
19:04 ☽ □ ♀
23:08 ☽ △ ♆
20 00:24 ☽ # ♃
06:27 ☽ # ☿
09:54 ☽ ⚹ ♇
14:36 ☽ □ ♄
21 00:14 ☽ # ♂
01:40 ☽ ∥ ♅
01:40 ☽ ⚹ ☉
08:07 ☽ # ☉
10:19 ☽ ⚹ ☿
13:37 ☽ △ ♅
17:21 ☽ ⚹ ♂
18:38 ☽ ∥ ♇
22 04:37 ☽ □ ♆
04:38 ☽ △ [illegible]
05:12 ☽ ∥ ♆
18:53 ☉ ♍
19:46 ☽ # ♀
20:04 ☽ △ ♄
20:54 ☽ ⚹ ♃
23 03:47 ☽ # ♄
11:32 ☽ □ ☿
17:40 ☽ □ ♅
20:51 ☿ ☌ ☉
23:52 ☽ □ ♂
24 08:10 ☽ ⚹ ♆
18:20 ☽ ☌ ♇
25 00:42 ☽ □ ♃
01:34 ☿ ♌R
[illegible]
11:13 ☽ △ ☿
16:25 ☽ △ ☉
17:31 ☉ # ♅
19:53 ☽ ⚹ ♅
26 04:19 ☽ △ ♂
17:59 ☽ ☍ ♀
27 00:54 ☿ ∥ ♂
01:30 ☽ ☍ ♄
02:58 ☽ △ ♃
18:41 ☉ ☍ ♅
28 11:07 ☽ ☌ ♆
12:47 ☽ # ♄
17:57 ☿ ∥ ☉
21:13 ☽ ⚹ ♇
21:27 [illegible]
29 09:23 ☽ ☍ ☿
09:39 ☽ ∥ ♆
19:45 ☽ ∥ ♆
22:28 ☽ ☌ ♅
30 11:34 ☽ ∥ ♅
12:01 ☽ ☍ ♂
12:51 ☽ # ☿
18:43 ☽ # ☉
19:38 ♇ SD
22:46 ☽ # ♂
23:36 ☽ □ ♇
31 05:26 ☽ △ ♀
06:10 ☽ △ ♄
08:29 ☽ ☍ ♃
09:19 [illegible]
15:54 ☽ # ♃
16:08 ♀ ☌ ♄

⚷ Chiron

01	Dec.	14 S 14
01	22♑19R	
04	22 09	
07	21 59	
10	21 50	
13	21 41	
16	21 32	
19	21 24	
22	21 16	
25	21 09	
28	21 03	
31	20 57	

September 2004

Day	S. T.	☉	☽	☿	♀	♂	♃	♄	♅	♆	♇	☊ True
	h m s	° ' "	° ' "	° '	° '	° '	° '	° '	° '	° '	° '	° '
01 We	22 41 59	08♍ 53 40	03♈ 31 57	25♌R54	23♋ 43	13♍ 43	24♍ 49	23♋ 25	04♓R39	13♒R19	19♐ 33	03♉R24
02 Th	22 45 55	09 51 43	16 53 34	25 45	24 46	14 21	25 01	23 31	04 36	13 18	19 33	03 20
03 Fr	22 49 52	10 49 49	29 51 09	25D 45	25 49	14 59	25 14	23 37	04 34	13 16	19 33	03 19
04 Sa	22 53 48	11 47 56	12♉ 26 51	25 53	26 53	15 38	25 27	23 43	04 32	13 15	19 33	03D 19
05 Su	22 57 45	12 46 05	24 44 16	26 10	27 56	16 16	25 40	23 49	04 29	13 14	19 33	03 21
06 Mo	23 01 41	13 44 16	06♊ 47 55	26 36	29 00	16 54	25 53	23 55	04 27	13 12	19 33	03 23
07 Tu	23 05 38	14 42 29	18 42 47	27 10	00♌ 05	17 33	26 05	24 01	04 25	13 11	19 33	03 24
08 We	23 09 34	15 40 44	00♋ 33 57	27 52	01 09	18 11	26 18	24 07	04 22	13 10	19 34	03R 23
09 Th	23 13 31	16 39 01	12 26 24	28 42	02 14	18 49	26 31	24 13	04 20	13 08	19 34	03 21
10 Fr	23 17 28	17 37 20	24 24 34	29 40	03 19	19 28	26 44	24 19	04 18	13 07	19 34	03 16
11 Sa	23 21 24	18 35 42	06♌ 32 11	00♍ 45	04 24	20 06	26 57	24 24	04 15	13 06	19 35	03 09
12 Su	23 25 21	19 34 05	18 52 06	01 57	05 30	20 45	27 10	24 30	04 13	13 04	19 35	03 02
13 Mo	23 29 17	20 32 30	01♍ 26 05	03 15	06 35	21 23	27 23	24 35	04 11	13 03	19 35	02 54
14 Tu	23 33 14	21 30 57	14 14 54	04 38	07 41	22 02	27 36	24 41	04 08	13 02	19 36	02 47
15 We	23 37 10	22 29 26	27 18 23	06 06	08 47	22 40	27 48	24 46	04 06	13 01	19 36	02 40
16 Th	23 41 07	23 27 56	10♎ 35 34	07 39	09 54	23 19	28 01	24 51	04 04	13 00	19 37	02 36
17 Fr	23 45 03	24 26 29	24 05 03	09 15	11 00	23 57	28 14	24 57	04 02	12 59	19 37	02 34
18 Sa	23 48 60	25 25 03	07♏ 45 18	10 55	12 07	24 36	28 27	25 02	04 00	12 57	19 38	02 33
19 Su	23 52 56	26 23 39	21 34 53	12 38	13 14	25 14	28 40	25 07	03 57	12 56	19 38	02D 33
20 Mo	23 56 53	27 22 17	05♐ 32 33	14 22	14 21	25 53	28 53	25 12	03 55	12 55	19 39	02 33
21 Tu	00 00 50	28 20 57	19 37 09	16 09	15 28	26 32	29 06	25 17	03 53	12 54	19 40	02 34
22 We	00 04 46	29 19 38	03♑ 47 19	17 57	16 36	27 10	29 19	25 21	03 51	12 53	19 40	02R 33
23 Th	00 08 43	00♎ 18 21	18 01 14	19 46	17 43	27 49	29 32	25 26	03 49	12 52	19 41	02 32
24 Fr	00 12 39	01 17 05	02♒ 16 26	21 35	18 51	28 28	29 45	25 31	03 47	12 51	19 42	02 29
25 Sa	00 16 36	02 15 52	16 29 42	23 25	19 59	29 06	29 58	25 35	03 45	12 50	19 43	02 26
26 Su	00 20 32	03 14 39	00♓ 37 15	25 15	21 08	29 45	00♎ 11	25 40	03 43	12 49	19 44	02 21
27 Mo	00 24 29	04 13 29	14 35 01	27 05	22 16	00♎ 24	00 24	25 44	03 41	12 49	19 44	02 17
28 Tu	00 28 26	05 12 20	28 19 11	28 55	23 25	01 03	00 37	25 49	03 39	12 48	19 45	02 14
29 We	00 32 22	06 11 14	11♈ 46 40	00♎ 45	24 33	01 41	00 50	25 53	03 37	12 47	19 46	02 12
30 Th	00 36 19	07 10 09	24 55 38	02 34	25 42	02 20	01 03	25 57	03 35	12 46	19 47	02 10

Data for	09-01-2004
Julian Day	2453249.50
Ayanamsa	23 55 10
SVP	05 ♓ 11 32
☽ ☊ Mean	04 ♉ 47 R

● ◐ PHASES ○ ◑

06	15:11	◑	14♊21
14	14:29	●	22♍06
21	15:54	◐	29♐00
28	13:10	○	05♈45

LAST ASPECT Day	h m	☽ INGRESS Day	h m	
02	16:18	03	00:17	♉
05	06:57	05	10:25	♊
07	18:09	07	22:51	♋
10	04:42	10	11:07	♌
12	01:23	12	21:17	♍
15	00:56	15	04:54	♎
17	01:32	17	10:25	♏
19	12:24	19	14:30	♐
21	16:20	21	17:36	♑
23	19:41	23	20:10	♒
25	06:26	25	22:56	♓
28	01:13	28	02:58	♈
30	01:54	30	09:25	♉

DECLINATION

Day	☉	☽	☿	♀	♂	♃	♄	♅	♆	♇
01 We	08N14	00S 56	10N47	19N17	07N21	03N03	21N13	10S 33	16S 54	14S 30
02 Th	07 52	05N18	11 07	19 11	07 06	02 58	21 12	10 34	16 54	14 31
03 Fr	07 30	11 08	11 25	19 04	06 51	02 53	21 11	10 35	16 54	14 31
04 Sa	07 08	16 22	11 40	18 57	06 36	02 48	21 10	10 36	16 55	14 32
05 Su	06 46	20 47	11 52	18 49	06 21	02 43	21 09	10 37	16 55	14 32
06 Mo	06 24	24 15	12 00	18 41	06 06	02 38	21 08	10 38	16 56	14 32
07 Tu	06 01	26 37	12 04	18 32	05 51	02 33	21 07	10 38	16 56	14 33
08 We	05 39	27 46	12 05	18 23	05 35	02 28	21 06	10 39	16 56	14 33
09 Th	05 16	27 39	12 02	18 14	05 20	02 22	21 05	10 40	16 57	14 34
10 Fr	04 53	26 14	11 56	18 04	05 05	02 17	21 05	10 41	16 57	14 34
11 Sa	04 31	23 36	11 45	17 53	04 49	02 12	21 04	10 42	16 58	14 34
12 Su	04 08	19 52	11 31	17 42	04 34	02 07	21 03	10 43	16 58	14 35
13 Mo	03 45	15 11	11 14	17 30	04 19	02 02	21 02	10 43	16 58	14 35
14 Tu	03 22	09 46	10 53	17 18	04 03	01 57	21 01	10 44	16 59	14 36
15 We	02 59	03 49	10 29	17 06	03 48	01 52	21 00	10 45	16 59	14 36
16 Th	02 36	02S 25	10 02	16 53	03 32	01 47	20 59	10 46	16 59	14 36
17 Fr	02 12	08 38	09 31	16 39	03 17	01 41	20 58	10 47	17 00	14 37
18 Sa	01 49	14 33	08 59	16 25	03 01	01 36	20 57	10 47	17 00	14 37
19 Su	01 26	19 48	08 24	16 11	02 45	01 31	20 57	10 48	17 00	14 38
20 Mo	01 03	24 01	07 47	15 56	02 30	01 26	20 56	10 49	17 01	14 38
21 Tu	00 39	26 50	07 08	15 40	02 14	01 21	20 55	10 50	17 01	14 38
22 We	00 16	27 57	06 27	15 25	01 59	01 16	20 54	10 50	17 01	14 39
23 Th	00S 07	27 14	05 45	15 08	01 43	01 11	20 53	10 51	17 01	14 39
24 Fr	00 31	24 44	05 02	14 52	01 27	01 05	20 53	10 52	17 02	14 40
25 Sa	00 54	20 43	04 18	14 34	01 12	01 00	20 52	10 53	17 02	14 40
26 Su	01 17	15 33	03 33	14 17	00 56	00 55	20 51	10 53	17 02	14 41
27 Mo	01 41	09 38	02 48	13 59	00 40	00 50	20 50	10 54	17 03	14 41
28 Tu	02 04	03 21	02 02	13 40	00 25	00 45	20 50	10 55	17 03	14 41
29 We	02 27	02N58	01 16	13 21	00 09	00 40	20 49	10 55	17 03	14 42
30 Th	02 51	09 02	00 29	13 02	00S 07	00N35	20 48	10 56	17 03	14 42

ASPECTARIAN

01 15:02 ☽ ∥ ♃
17:29 ☽ ⚹ ♆
02 04:51 ☽ △ ♇
06:56 ☽ ∥ ♂
07:15 ♀ ⚹ ♃
09:45 ☽ ∥ ☉
12:17 ☽ □ ♄
13:10 ☿ SD
15:47 ☽ □ ♀
16:18 ☽ △ ☿
21:39 ☽ # ♅
03 01:20 ☽ ∥ ☿
08:53 ☽ ⚹ ♅
15:13 ☽ # ♇
22:39 ☽ △ ☉
04 01:33 ☽ □ ♆
02:50 ☽ # ♆
06:29 ☽ △ ♂
13:07 ☽ ∥ ♀
22:11 ☽ ⚹ ♄
05 01:51 ☽ △ ♃
02:19 ☽ ∥ ♄
02:55 ☽ □ ☿
06:57 ☽ ⚹ ♀
19:19 ☽ □ ♅
06 12:51 ☽ △ ♆
21:30 ☽ □ ♂
22:16 ♀ ♌
07 01:42 ☽ ☍ ♇
15:12 ☽ □ ♃
18:09 ☽ ⚹ ☿
08 07:41 ☽ △ ♅
11:12 ☉ ∥ ♂
09 09:13 ☽ ⚹ ☉
13:33 ☽ ⚹ ♂
23:48 ☽ ☌ ♄
10 04:04 ♂ □ ♇
04:42 ☽ ⚹ ♃
07:38 ☿ ♍
19:23 ☽ ☌ ♀
11 12:48 ☽ ☍ ♆
17:05 ☽ ∥ ♄
12 00:23 ☉ □ ♇
01:23 ☽ △ ♇
12:09 ☽ ∥ ♀
15:21 ☽ # ♆
13 02:48 ☽ # ♇
03:49 ☽ ☌ ☿
05:10 ☽ ☍ ♅
15:54 ☿ ☍ ♅
18:56 ☽ ∥ ☿
19:53 ☽ # ♅
14 08:37 ☿ # ♅
09:53 ☽ □ ♇
15:05 ☽ ☌ ♂
19:20 ☽ ⚹ ♄
15 00:05 ☽ ∥ ♂
00:56 ☽ ☌ ♃
03:29 ☽ ∥ ☉
07:43 ☽ ∥ ♃
12:24 ♀ # ♆
12:55 ☉ ☌ ♂
21:37 ☽ # ♃
22:38 ☽ ⚹ ♀
16 00:41 ☽ # ☉
04:09 ☽ # ♂
04:18 ☽ △ ♆
16:06 ☽ ⚹ ♇
17 01:32 ☽ □ ♄
03:14 ☽ # ☿
08:32 ☽ ∥ ♅
13:32 ☉ ⚹ ♄
17:27 ☽ △ ♅
18 00:20 ☽ ∥ ♇
06:17 ☽ ⚹ ☿
07:49 ☽ # ♀
08:16 ☽ □ ♀
09:03 ☽ □ ♆
10:46 ☽ ∥ ♆
17:18 ☉ ∥ ♃
17:51 ♀ ☍ ♆
18:33 ♂ ⚹ ♄
19 05:52 ☽ # ♄
06:08 ☽ △ ♄
06:37 ☽ ⚹ ♂
08:55 ☽ ⚹ ☉
12:24 ☽ ⚹ ♃
21:14 ☽ □ ♅
20 12:35 ☽ ⚹ ♆
16:20 ☽ △ ♀
17:15 ☽ □ ☿
21 00:04 ☽ ☌ ♇
12:17 ☽ □ ♂
16:20 ☽ □ ♃
23:48 ☉ ☌ ♃
22 00:06 ☽ ⚹ ♅
16:30 ☉ ♎
23:03 ☿ □ ♇
23 03:21 ☽ △ ☿
12:33 ☽ ☍ ♄
17:16 ☽ △ ♂
19:41 ☽ △ ♃
22:13 ☽ △ ☉
24 16:09 ♀ # ♇
17:50 ☽ ☌ ♆
18:07 ♀ △ ♇
23:10 ☽ # ♄
25 03:25 ♃ ♎
05:13 ☉ # ♃
05:27 ☽ ⚹ ♇
06:26 ☽ ☍ ♀
10:49 ☉ # ♂
17:30 ☽ ∥ ♆
26 01:55 ♂ ∥ ♃
03:40 ☽ ∥ ♇
05:17 ☽ ☌ ♅
05:38 ☿ ⚹ ♄
05:39 ☽ # ♀
09:16 ♂ ♎
19:01 ☽ ∥ ♅
27 00:17 ♂ ☌ ♃
08:58 ☽ □ ♇
19:33 ☽ △ ♄
23:15 ☿ # ☉
28 01:13 ☽ ☍ ☿
04:08 ☽ ☍ ♃
04:33 ☽ ∥ ☉
05:03 ☽ ☍ ♂
05:39 ☽ # ☿
09:57 ☽ # ♃
11:35 ☽ # ♂
14:13 ☿ ♎
15:17 ☽ ∥ ♃
18:09 ☽ ∥ ☿
21:53 ☽ # ☉
29 01:21 ☿ ☌ ♃
01:49 ☽ ⚹ ♆
14:32 ☽ △ ♇
19:21 ☿ ☌ ♂
20:45 ☿ ∥ ♃
30 01:34 ☽ △ ♀
01:54 ☽ □ ♄
07:57 ☽ # ♅
08:30 ☿ # ♂
16:02 ☽ ∥ ♀
16:06 ☽ ⚹ ♅

⚷ Chiron

01 Dec.	14 S 36
03	20♑51R
06	20 46
09	20 42
12	20 38
15	20 35
18	20 33
21	20 31
24	20 30
27	20 30D
30	20 31
26 14:43	20♑30 D

October 2004

14 03:00 21♎06 ☼ Solar Eclipse (mag 0.929)
28 03:05 05♉02 ✸ Total Lunar Eclipse (mag 1.313)

Day	S. T. h m s	☉ ° ' "	☽ ° ' "	☿ ° '	♀ ° '	♂ ° '	♃ ° '	♄ ° '	♅ ° '	♆ ° '	♇ ° '	☊ True ° '
01 Fr	00 40 15	08♎ 09 07	07♉ 45 40	04♎ 22	26♌ 51	02♎ 59	01♎ 16	26♋ 01	03♓R33	12♒R45	19♐ 48	02♉D11
02 Sa	00 44 12	09 08 07	20 17 48	06 10	28 00	03 38	01 29	26 05	03 32	12 45	19 49	02 12
03 Su	00 48 08	10 07 09	02♊ 34 23	07 57	29 10	04 17	01 42	26 09	03 30	12 44	19 50	02 14
04 Mo	00 52 05	11 06 13	14 38 40	09 44	00♍ 19	04 56	01 55	26 13	03 28	12 43	19 51	02 16
05 Tu	00 56 01	12 05 20	26 34 39	11 30	01 29	05 34	02 08	26 16	03 26	12 43	19 52	02 18
06 We	00 59 58	13 04 29	08♋ 26 50	13 15	02 39	06 13	02 20	26 20	03 25	12 42	19 54	02 19
07 Th	01 03 54	14 03 40	20 19 58	14 59	03 49	06 52	02 33	26 23	03 23	12 41	19 55	02R 19
08 Fr	01 07 51	15 02 54	02♌ 18 52	16 43	04 59	07 31	02 46	26 27	03 21	12 41	19 56	02 19
09 Sa	01 11 48	16 02 10	14 28 01	18 26	06 09	08 10	02 59	26 30	03 20	12 40	19 57	02 17
10 Su	01 15 44	17 01 28	26 51 21	20 08	07 19	08 49	03 12	26 33	03 18	12 40	19 58	02 15
11 Mo	01 19 41	18 00 48	09♍ 31 54	21 49	08 30	09 28	03 24	26 36	03 17	12 39	20 00	02 13
12 Tu	01 23 37	19 00 11	22 31 35	23 29	09 41	10 08	03 37	26 39	03 15	12 39	20 01	02 11
13 We	01 27 34	19 59 35	05♎ 50 59	25 09	10 51	10 47	03 50	26 42	03 14	12 39	20 02	02 10
14 Th	01 31 30	20 59 02	19 29 12	26 48	12 02	11 26	04 03	26 45	03 13	12 38	20 04	02 08
15 Fr	01 35 27	21 58 31	03♏ 23 56	28 27	13 13	12 05	04 15	26 48	03 11	12 38	20 05	02 08
16 Sa	01 39 23	22 58 02	17 31 44	00♏ 04	14 24	12 44	04 28	26 50	03 10	12 38	20 06	02 07
17 Su	01 43 20	23 57 35	01♐ 48 27	01 41	15 36	13 23	04 41	26 53	03 09	12 37	20 08	02 07
18 Mo	01 47 17	24 57 09	16 09 41	03 18	16 47	14 03	04 53	26 55	03 07	12 37	20 09	02 07
19 Tu	01 51 13	25 56 46	00♑ 31 18	04 53	17 58	14 42	05 06	26 58	03 06	12 37	20 11	02 07
20 We	01 55 10	26 56 24	14 49 41	06 28	19 10	15 21	05 18	27 00	03 05	12 37	20 12	02 06
21 Th	01 59 06	27 56 04	29 01 54	08 03	20 21	16 00	05 31	27 02	03 04	12 37	20 14	02 06
22 Fr	02 03 03	28 55 45	13♒ 05 42	09 37	21 33	16 40	05 43	27 04	03 03	12 37	20 15	02 06
23 Sa	02 06 59	29 55 29	26 59 26	11 10	22 45	17 19	05 56	27 06	03 02	12 36	20 17	02 05
24 Su	02 10 56	00♏ 55 13	10♓ 41 52	12 43	23 57	17 58	06 08	27 07	03 01	12 36	20 19	02D 05
25 Mo	02 14 52	01 55 00	24 12 04	14 15	25 09	18 38	06 20	27 09	03 00	12D 36	20 20	02 06
26 Tu	02 18 49	02 54 48	07♈ 29 17	15 47	26 21	19 17	06 33	27 11	02 59	12 36	20 22	02 06
27 We	02 22 46	03 54 38	20 32 58	17 18	27 33	19 57	06 45	27 12	02 59	12 37	20 24	02 07
28 Th	02 26 42	04 54 30	03♉ 22 49	18 48	28 46	20 36	06 57	27 13	02 58	12 37	20 25	02 07
29 Fr	02 30 39	05 54 24	15 58 57	20 18	29 58	21 16	07 09	27 15	02 57	12 37	20 27	02 08
30 Sa	02 34 35	06 54 20	28 22 07	21 48	01♎ 11	21 55	07 21	27 16	02 56	12 37	20 29	02R 08
31 Su	02 38 32	07 54 18	10♊ 33 41	23 17	02 23	22 35	07 33	27 17	02 56	12 37	20 31	02 08

Data for	10-01-2004
Julian Day	2453279.50
Ayanamsa	23 55 13
SVP	05 ♓ 11 26
☽ ☊ Mean	03 ♉ 12 R

● ◐ PHASES ○ ◑

06	10:12	◑	13♋30
14	02:48	☼	21♎06
20	21:59	◐	27♑51
28	03:08	✸	05♉02

LAST ASPECT Day	h m	☽ INGRESS Day	h m	
02	16:35	02	18:56	♊
04	10:28	05	06:55	♋
07	12:14	07	19:23	♌
09	10:42	10	06:00	♍
12	07:32	12	13:32	♎
14	14:22	14	18:11	♏
16	15:43	16	20:58	♐
18	15:47	18	23:08	♑
20	22:00	21	01:39	♒
22	12:21	23	05:14	♓
25	05:18	25	10:25	♈
27	12:25	27	17:38	♉
29	21:50	30	03:11	♊

DECLINATION

Day	☉	☽	☿	♀	♂	♃	♄	♅	♆	♇
01 Fr	03S 14	14N35	00S18	12N42	00S 23	00N29	20N48	10S 57	17S 03	14S 43
02 Sa	03 37	19 24	01 04	12 22	00 38	00 24	20 47	10 57	17 04	14 43
03 Su	04 00	23 17	01 51	12 01	00 54	00 19	20 46	10 58	17 04	14 44
04 Mo	04 24	26 05	02 38	11 41	01 10	00 14	20 46	10 59	17 04	14 44
05 Tu	04 47	27 40	03 24	11 19	01 26	00 09	20 45	10 59	17 04	14 44
06 We	05 10	27 58	04 10	10 58	01 41	00 04	20 44	11 00	17 04	14 45
07 Th	05 33	26 59	04 55	10 36	01 57	00S 01	20 44	11 00	17 05	14 45
08 Fr	05 56	24 46	05 41	10 13	02 13	00 06	20 43	11 01	17 05	14 46
09 Sa	06 19	21 25	06 26	09 51	02 28	00 11	20 43	11 01	17 05	14 46
10 Su	06 41	17 04	07 10	09 28	02 44	00 16	20 42	11 02	17 05	14 47
11 Mo	07 04	11 53	07 54	09 04	03 00	00 21	20 42	11 03	17 05	14 47
12 Tu	07 27	06 04	08 37	08 41	03 15	00 26	20 41	11 03	17 05	14 47
13 We	07 49	00S 10	09 20	08 17	03 31	00 31	20 41	11 03	17 06	14 48
14 Th	08 11	06 33	10 02	07 53	03 47	00 36	20 40	11 04	17 06	14 48
15 Fr	08 34	12 45	10 44	07 28	04 02	00 41	20 40	11 04	17 06	14 49
16 Sa	08 56	18 24	11 25	07 04	04 18	00 46	20 39	11 05	17 06	14 49
17 Su	09 18	23 05	12 05	06 39	04 34	00 51	20 39	11 05	17 06	14 49
18 Mo	09 40	26 22	12 44	06 14	04 49	00 56	20 39	11 06	17 06	14 50
19 Tu	10 01	27 56	13 23	05 48	05 05	01 01	20 38	11 06	17 06	14 50
20 We	10 23	27 38	14 01	05 22	05 20	01 06	20 38	11 06	17 06	14 51
21 Th	10 44	25 31	14 39	04 57	05 36	01 11	20 38	11 07	17 06	14 51
22 Fr	11 06	21 51	15 15	04 31	05 51	01 15	20 37	11 07	17 06	14 52
23 Sa	11 27	17 01	15 51	04 04	06 06	01 20	20 37	11 07	17 06	14 52
24 Su	11 48	11 23	16 26	03 38	06 22	01 25	20 37	11 08	17 06	14 52
25 Mo	12 08	05 18	17 00	03 11	06 37	01 30	20 37	11 08	17 06	14 53
26 Tu	12 29	00N56	17 33	02 45	06 52	01 35	20 36	11 08	17 06	14 53
27 We	12 49	07 02	18 06	02 18	07 08	01 39	20 36	11 09	17 06	14 54
28 Th	13 10	12 45	18 37	01 51	07 23	01 44	20 36	11 09	17 06	14 54
29 Fr	13 29	17 50	19 07	01 24	07 38	01 49	20 36	11 09	17 06	14 54
30 Sa	13 49	22 04	19 37	00 56	07 53	01 54	20 36	11 09	17 06	14 55
31 Su	14 09	25 17	20 06	00 29	08 08	01 58	20 36	11 09	17 06	14 55

ASPECTARIAN

01 00:37 ☽ # ♇
03:50 ☿ ∥ ♂
05:28 ☿ # ♃
07:55 ♂ # ♃
09:29 ☽ □ ♆
11:51 ☽ # ♆
12:52 ☿ ∥ ♂
02 07:51 ☽ ∥ ♄
11:19 ☽ ✶ ♄
16:35 ☽ □ ♀
22:14 ☽ △ ♃
03 01:49 ☽ □ ♅
03:33 ☽ △ ♂
12:30 ☽ △ ☿
16:18 ☽ △ ☉
17:20 ♀ ♍
20:09 ☽ △ ♆
04 10:28 ☽ ☍ ♇
05 10:59 ☽ ✶ ♀
11:25 ☽ □ ♃
13:50 ☽ △ ♅
14:59 ☉ △ ♆
16:30 ☿ △ ♆
18:29 ☿ ☌ ☉
19:14 ☽ □ ♂
21:48 ♀ # ♅
06 11:23 ☽ □ ☿
15:23 ♀ ☍ ♅
07 12:14 ☽ ☌ ♄
08 00:55 ☽ ✶ ♃
10:55 ☽ ✶ ♂
16:14 ☿ ∥ ☉
20:29 ☽ ☍ ♆
09 03:20 ☽ ✶ ☉
04:13 ☽ ∥ ♄
08:58 ☽ [illegible] ☿
10:42 ☽ △ ♇
21:48 ☿ ✶ ♇
23:52 ☽ # ♆
10 11:00 ☽ # ♇
12:16 ☽ ☍ ♅
21:52 ☽ ☌ ♀
11 03:35 ☽ # ♅
12:41 ☽ ∥ ♀
14:49 ☽ # ☿
18:48 ☽ # ☉
19:24 ☽ □ ♆
12 01:19 ☿ # ♀
07:32 ☽ ✶ ♄
10:29 ☽ # ♂
20:21 ☽ ☌ ♃
21:25 ☽ # ♃
13 01:09 ☉ ✶ ♇
01:20 ☽ ∥ ♃
09:10 ☽ ☌ ♂
12:01 ☽ △ ♆
13:07 ☽ ∥ ♂
14:27 ♀ # ☉
23:14 ☿ □ ♄
14 01:00 ☽ ✶ ♇
04:44 ☽ # ♀
06:36 ☽ ∥ ☉
12:38 ☽ □ ♄
14:22 ☽ ☌ ☿
14:59 ☽ ∥ ☿
17:20 ☽ ∥ ♅
23:38 ☽ △ ♅
15 08:22 ☽ ∥ ♇
12:13 ☿ ∥ ♅
[illegible] [illegible] □ [illegible]
18:08 ☽ ∥ ♆
18:14 ☽ ✶ ♀
20:03 ♂ △ ♆
22:57 ☿ ♏
16 10:46 ☽ # ♄
15:43 ☽ △ ♄
17 02:14 ☽ □ ♅
04:53 ☽ ✶ ♃
18:05 ☽ ✶ ♆
20:17 ☽ ✶ ♂
21:30 ☿ △ ♅
18 01:08 ☽ □ ♀
06:41 ☽ ☌ ♇
15:47 ☽ ✶ ☉
19 04:19 ☽ ✶ ♅
07:46 ☽ □ ♃
08:13 ☽ ✶ ☿
20 00:55 ☽ □ ♂
01:22 ♀ # ♂
01:26 ☉ □ ♄
07:58 ☽ △ ♀
20:36 ☽ ☍ ♄
21:27 ♀ □ ♇
21 08:16 ☿ ∥ ♇
11:12 ☽ △ ♃
17:17 ☽ □ ☿
23:10 ☽ ☌ ♆
22 01:43 ☉ ∥ ♅
06:26 ☽ △ ♂
06:39 ☽ # ♄
12:21 ☽ ✶ ♇
23:38 ☽ ∥ ♆
23 01:49 ☉ ♏
04:43 ☽ ∥ ☿
05:30 ☽ △ ☉
09:29 ☽ ∥ ♇
10:32 ☽ ☌ ♅
22:20 ☿ □ ♆
22:24 ☽ ∥ ☉
24 01:01 ☽ ∥ ♅
04:01 ☽ △ ☿
11:58 ♆ SD
17:05 ☽ □ ♇
19:03 ☽ ∥ ♂
25 01:52 ☽ ☍ ♀
04:30 ☿ ∥ ♆
05:18 ☽ △ ♄
08:43 ☽ # ♀
14:25 ☽ ∥ ♃
22:15 ☽ ☍ ♃
26 01:49 ☉ △ ♅
02:30 ☽ # ♃
06:31 ☽ ∥ ♀
09:21 ☽ ✶ ♆
16:49 ♀ ✶ ♄
22:49 ☽ ☍ ♂
23:43 ☽ △ ♇
27 00:23 ☽ # ♂
12:25 ☽ □ ♄
17:02 ☽ # ♅
[illegible] [illegible] ∥ [illegible]
23:13 ☽ ✶ ♅
28 01:57 ☽ # ☉
04:57 ♀ # ♃
09:45 ☽ # ♇
17:32 ☽ □ ♆
20:19 ☽ # ♆
29 00:39 ♀ ♎
07:38 ☽ # ☿
09:28 ☽ ☍ ☿
14:59 ☽ ∥ ♄
21:50 ☽ ✶ ♄
30 06:06 ☽ △ ♀
08:57 ☽ □ ♅
17:37 ☽ △ ♃
31 04:05 ☽ △ ♆
19:52 ☽ ☍ ♇

⚷ Chiron

01 Dec.	14 S 54
03	20♑32
06	20 33
09	20 36
12	20 39
15	20 43
18	20 47
21	20 53
24	20 59
27	21 05
30	21 10

November 2004

Day	S. T. h m s	☉ ° ' "	☽ ° ' "	☿ ° '	♀ ° '	♂ ° '	♃ ° '	♄ ° '	♅ ° '	♆ ° '	♇ ° '	☊ True ° '
01 Mo	02 42 28	08♏54 18	22♊35 43	24♏45	03♎36	23♎14	07♎45	27♋18	02♓55	12♒37	20♐32	02♉R07
02 Tu	02 46 25	09 54 20	04♋30 55	26 13	04 49	23 54	07 57	27 18	02 55	12 38	20 34	02 06
03 We	02 50 21	10 54 25	16 22 40	27 41	06 01	24 33	08 09	27 19	02 54	12 38	20 36	02 06
04 Th	02 54 18	11 54 31	28 14 53	29 07	07 14	25 13	08 21	27 20	02 54	12 38	20 38	02 05
05 Fr	02 58 15	12 54 39	10♌12 00	00♐33	08 27	25 53	08 33	27 20	02 54	12 39	20 40	02 05
06 Sa	03 02 11	13 54 50	22 18 40	01 59	09 40	26 32	08 45	27 20	02 53	12 39	20 42	02D 05
07 Su	03 06 08	14 55 02	04♍39 32	03 24	10 54	27 12	08 56	27 21	02 53	12 40	20 44	02 06
08 Mo	03 10 04	15 55 16	17 18 51	04 48	12 07	27 52	09 08	27 21	02 53	12 40	20 46	02 07
09 Tu	03 14 01	16 55 33	00♎20 08	06 11	13 20	28 32	09 19	27R 21	02 53	12 40	20 47	02 08
10 We	03 17 57	17 55 51	13 45 36	07 33	14 33	29 12	09 31	27 20	02 52	12 41	20 49	02 09
11 Th	03 21 54	18 56 11	27 35 48	08 55	15 47	29 51	09 42	27 20	02 52	12 42	20 51	02R 09
12 Fr	03 25 50	19 56 33	11♏49 03	10 15	17 00	00♏31	09 54	27 20	02 D 52	12 42	20 53	02 08
13 Sa	03 29 47	20 56 57	26 21 17	11 34	18 14	01 11	10 05	27 19	02 52	12 43	20 55	02 06
14 Su	03 33 44	21 57 23	11♐06 19	12 52	19 27	01 51	10 16	27 19	02 52	12 44	20 57	02 04
15 Mo	03 37 40	22 57 50	25 56 37	14 08	20 41	02 31	10 27	27 18	02 53	12 44	20 59	02 01
16 Tu	03 41 37	23 58 19	10♑44 22	15 23	21 55	03 11	10 38	27 17	02 53	12 45	21 02	01 57
17 We	03 45 33	24 58 49	25 22 35	16 36	23 09	03 51	10 49	27 16	02 53	12 46	21 04	01 54
18 Th	03 49 30	25 59 20	09♒45 59	17 47	24 22	04 31	11 00	27 15	02 53	12 47	21 06	01 52
19 Fr	03 53 26	26 59 52	23 51 22	18 55	25 36	05 11	11 11	27 14	02 54	12 47	21 08	01 51
20 Sa	03 57 23	28 00 26	07♓37 35	20 01	26 50	05 51	11 22	27 13	02 54	12 48	21 10	01D 51
21 Su	04 01 19	29 01 00	21 05 03	21 04	28 04	06 31	11 32	27 12	02 55	12 49	21 12	01 52
22 Mo	04 05 16	00♐01 36	04♈15 16	22 03	29 18	07 12	11 43	27 10	02 55	12 50	21 14	01 54
23 Tu	04 09 13	01 02 13	17 10 08	22 58	00♏32	07 52	11 54	27 08	02 56	12 51	21 16	01 56
24 We	04 13 09	02 02 51	29 51 39	23 49	01 46	08 32	12 04	27 07	02 56	12 52	21 18	01 57
25 Th	04 17 06	03 03 31	12♉21 36	24 35	03 01	09 12	12 14	27 05	02 57	12 53	21 21	01R 58
26 Fr	04 21 02	04 04 12	24 41 30	25 15	04 15	09 52	12 24	27 03	02 58	12 54	21 23	01 56
27 Sa	04 24 59	05 04 54	06♊52 43	25 49	05 29	10 33	12 35	27 01	02 58	12 55	21 25	01 53
28 Su	04 28 55	06 05 37	18 56 31	26 15	06 43	11 13	12 45	26 59	02 59	12 56	21 27	01 48
29 Mo	04 32 52	07 06 22	00♋54 17	26 34	07 58	11 53	12 55	26 57	03 00	12 58	21 29	01 41
30 Tu	04 36 49	08 07 08	12 47 43	26 43	09 12	12 34	13 04	26 54	03 01	12 59	21 32	01 35

Data for	11-01-2004
Julian Day	2453310.50
Ayanamsa	23 55 17
SVP	05 ♓ 11 22
☽ ☊ Mean	01 ♉ 33 R

● ◐ PHASES ○ ◑

Day	h m	Phase	Position
05	05:53	◑	13♌09
12	14:28	●	20♏33
19	05:51	◐	27♒15
26	20:07	○	04♊55

LAST ASPECT Day	h m	☽ INGRESS Day	h m	Sign
01	01:22	01	14:53	♋
04	02:00	04	03:32	♌
06	08:45	06	15:00	♍
08	18:33	08	23:23	♎
11	04:03	11	04:06	♏
13	01:35	13	05:57	♐
14	15:58	15	06:34	♑
17	03:08	17	07:40	♒
19	05:51	19	10:38	♓
21	15:36	21	16:12	♈
23	18:47	24	00:16	♉
26	04:37	26	10:25	♊
28	15:04	28	22:11	♋

DECLINATION

Day	☉	☽	☿	♀	♂	♃	♄	♅	♆	♇
01 Mo	14S28	27N18	20S33	00N02	08S23	02S03	20N35	11S10	17S06	14S56
02 Tu	14 47	28 02	21 00	00S26	08 38	02 07	20 35	11 10	17 06	14 56
03 We	15 06	27 28	21 26	00 54	08 53	02 12	20 35	11 10	17 06	14 56
04 Th	15 25	25 39	21 50	01 21	09 08	02 17	20 35	11 10	17 06	14 57
05 Fr	15 43	22 42	22 14	01 49	09 23	02 21	20 35	11 10	17 06	14 57
06 Sa	16 01	18 45	22 36	02 17	09 38	02 26	20 35	11 10	17 06	14 57
07 Su	16 19	13 56	22 57	02 44	09 53	02 30	20 35	11 10	17 05	14 58
08 Mo	16 36	08 26	23 17	03 12	10 07	02 35	20 35	11 10	17 05	14 58
09 Tu	16 54	02 25	23 36	03 40	10 22	02 39	20 36	11 10	17 05	14 59
10 We	17 11	03S53	23 54	04 07	10 36	02 43	20 36	11 10	17 05	14 59
11 Th	17 27	10 13	24 10	04 35	10 51	02 48	20 36	11 10	17 05	14 59
12 Fr	17 44	16 13	24 25	05 03	11 05	02 52	20 36	11 10	17 05	15 00
13 Sa	18 00	21 26	24 39	05 30	11 19	02 56	20 36	11 10	17 05	15 00
14 Su	18 15	25 23	24 52	05 58	11 34	03 01	20 36	11 10	17 04	15 00
15 Mo	18 31	27 36	25 03	06 25	11 48	03 05	20 37	11 10	17 04	15 01
16 Tu	18 46	27 52	25 13	06 52	12 02	03 09	20 37	11 10	17 04	15 01
17 We	19 01	26 10	25 21	07 19	12 16	03 13	20 37	11 10	17 04	15 01
18 Th	19 15	22 48	25 28	07 46	12 30	03 17	20 37	11 10	17 04	15 02
19 Fr	19 29	18 09	25 34	08 13	12 44	03 22	20 38	11 10	17 03	15 02
20 Sa	19 43	12 39	25 38	08 40	12 57	03 26	20 38	11 09	17 03	15 02
21 Su	19 56	06 41	25 40	09 07	13 11	03 30	20 38	11 09	17 03	15 03
22 Mo	20 09	00 32	25 41	09 33	13 25	03 34	20 39	11 09	17 03	15 03
23 Tu	20 22	05N31	25 41	09 59	13 38	03 38	20 39	11 09	17 02	15 03
24 We	20 34	11 15	25 39	10 25	13 52	03 42	20 40	11 09	17 02	15 04
25 Th	20 46	16 27	25 36	10 51	14 05	03 45	20 40	11 08	17 02	15 04
26 Fr	20 58	20 54	25 30	11 17	14 18	03 49	20 41	11 08	17 01	15 04
27 Sa	21 09	24 24	25 24	11 42	14 31	03 53	20 41	11 08	17 01	15 05
28 Su	21 20	26 46	25 15	12 07	14 44	03 57	20 41	11 07	17 01	15 05
29 Mo	21 30	27 53	25 05	12 32	14 57	04 00	20 42	11 07	17 00	15 05
30 Tu	21 40	27 41	24 54	12 56	15 10	04 04	20 43	11 07	17 00	15 06

ASPECTARIAN

Day	Time	Aspect
01	01:22	☽ △ ♂
	01:53	☿ # ♄
	20:46	☽ △ ♅
02	00:40	☽ □ ♀
	07:04	☽ □ ♃
	11:28	☉ ∥ ♆
	11:54	☽ △ ☉
	18:02	☿ △ ♄
03	17:31	☽ □ ♂
	22:08	☽ ☌ ♄
04	02:00	☽ △ ☿
	14:40	☿ ♐
	17:35	☉ □ ♆
	20:07	☽ ✶ ♀
	20:39	☽ ✶ ♃
05	02:11	♀ ☌ ♃
	02:59	☽ # ☿
	04:53	☽ ☍ ♆
	13:34	☽ ∥ ♄
	20:49	☽ △ ♇
06	08:44	☽ # ♆
	08:45	☽ ✶ ♂
	09:29	♀ ∥ ♃
	13:18	☽ # ☉
	15:18	☿ □ ♅
	19:12	☽ # ♇
	20:35	☽ ☍ ♅
	21:15	☽ □ ☿
07	05:03	♂ □ ♄
	12:24	☽ # ♅
	17:15	☽ # ♂
	21:10	☽ ✶ ☉
08	06:27	☽ □ ♇
	06:55	♄ SR
	10:58	♀ △ ♆
	18:33	☽ ✶ ♄
	19:37	☽ # ♀
	23:07	☽ # ♃
09	11:45	☽ ✶ ☿
	16:14	☉ ∥ ♆
	16:23	☽ ☌ ♃
	19:33	☽ ∥ ♃
	22:06	☽ △ ♆
10	00:56	☽ # ♀
	01:32	☽ ☌ ♀
	12:22	☽ ✶ ♇
	23:33	☽ □ ♄
11	02:29	☽ ∥ ♂
	03:39	☽ ∥ ♅
	04:03	☽ ☌ ♂
	05:12	♂ ♏
	08:58	☽ △ ♅
	16:32	☿ ✶ ♃
	18:54	☽ ∥ ♇
	19:12	♅ SD
12	01:29	☽ □ ♆
	03:40	☽ ∥ ♆
	06:50	☽ ∥ ☉
	08:44	♂ ∥ ♅
	19:49	☽ # ♄
13	01:35	☽ △ ♄
	10:38	☽ □ ♅
	19:58	☽ ∥ ☿
	21:21	☿ ✶ ♆
	22:38	☽ ✶ ♃
14	02:38	☽ ✶ ♆
	03:08	☽ ☌ ☿
	14:44	☽ ✶ ♀
	15:58	☽ ☌ ♇
15	06:10	♀ ✶ ♇
	11:09	☽ ✶ ♂
	11:14	☽ ✶ ♅
	12:58	♂ △ ♅
	23:50	☽ □ ♃
16	19:59	☽ □ ♀
	23:18	☽ ✶ ☉
17	03:08	☽ ☍ ♄
	06:42	☽ ∥ ☿
	14:46	☽ □ ♂
18	02:07	☽ △ ♃
	05:05	☽ ☌ ♆
	11:54	☽ # ♄
	14:48	☽ ✶ ☿
	17:53	☽ ∥ ☉
	19:18	☽ ✶ ♇
19	03:19	☽ △ ♀
	05:01	☽ ∥ ♆
	05:33	☉ △ ♄
	13:55	☽ ∥ ♇
	15:42	☽ ☌ ♅
	20:43	☽ △ ♂
	22:48	☽ ∥ ♂
20	06:10	☽ ∥ ♅
	07:13	♀ □ ♄
	15:04	☽ ∥ ♀
	23:58	☽ □ ☿
21	00:13	☽ □ ♇
	03:23	☿ ☌ ♇
	11:03	☽ △ ♄
	12:22	☽ ∥ ♃
	15:36	☽ △ ☉
	23:22	☉ ♐
22	13:32	♀ ♏
	14:00	☽ ☍ ♃
	15:56	☽ ✶ ♆
	16:23	☽ # ♃
23	07:44	☽ △ ♇
	11:44	☽ △ ☿
	18:47	☽ □ ♄
	20:10	☽ # ♀
	23:35	☽ # ♅
24	04:03	☽ ☍ ♀
	05:53	☽ ✶ ♅
	10:55	☉ # ♄
	12:15	☽ # ♂
	17:22	☽ # ♇
	17:33	☽ ☍ ♂
	21:19	☉ □ ♅
	22:46	♀ △ ♅
25	01:01	☽ □ ♆
	02:56	☽ # ♆
	15:59	♀ ∥ ♅
	22:42	☽ ∥ ♄
26	00:26	☽ # ☉
	04:37	☽ ✶ ♄
	16:16	☽ □ ♅
27	08:18	☽ # ☿
	11:28	☽ △ ♃
	12:01	☽ △ ♆
28	05:02	☽ ☍ ♇
	15:04	☽ ☍ ☿
29	04:13	☽ △ ♅
	08:29	♃ △ ♆
	15:28	♂ ∥ ♇
	15:53	☽ △ ♀
	23:30	☽ △ ♂
30	00:34	☽ □ ♃
	12:09	☿ SR
	15:27	♂ □ ♆

⚷ Chiron

01 Dec. 15 S 03

Day	Position
02	21♑20
05	21 28
08	21 37
11	21 47
14	21 57
17	22 07
20	22 18
23	22 30
26	22 42
29	22 54

December 2004

Day	S. T. h m s	☉ ° ' "	☽ ° ' "	☿ ° '	♀ ° '	♂ ° '	♃ ° '	♄ ° '	♅ ° '	♆ ° '	♇ ° '	☊ True ° '
01 We	04 40 45	09♐07 56	24♋38 54	26♐R43	10♏26	13♏14	13♎14	26♋R52	03♓02	13♒00	21♐34	01♉R28
02 Th	04 44 42	10 08 45	06♌30 35	26 33	11 41	13 54	13 24	26 50	03 03	13 01	21 36	01 22
03 Fr	04 48 38	11 09 35	18 26 12	26 12	12 55	14 35	13 33	26 47	03 04	13 03	21 38	01 18
04 Sa	04 52 35	12 10 27	00♍29 45	25 40	14 10	15 15	13 43	26 44	03 05	13 04	21 41	01 15
05 Su	04 56 31	13 11 19	12 45 46	24 57	15 24	15 56	13 52	26 41	03 06	13 05	21 43	01 14
06 Mo	05 00 28	14 12 14	25 18 56	24 02	16 39	16 37	14 01	26 39	03 07	13 07	21 45	01D 15
07 Tu	05 04 24	15 13 09	08♎13 49	22 59	17 54	17 17	14 10	26 36	03 09	13 08	21 47	01 16
08 We	05 08 21	16 14 06	21 34 19	21 47	19 08	17 58	14 19	26 32	03 10	13 09	21 50	01 16
09 Th	05 12 18	17 15 04	05♏22 59	20 28	20 23	18 38	14 28	26 29	03 11	13 11	21 52	01R 16
10 Fr	05 16 14	18 16 03	19 40 05	19 06	21 38	19 19	14 37	26 26	03 13	13 12	21 54	01 13
11 Sa	05 20 11	19 17 03	04♐22 50	17 43	22 52	20 00	14 46	26 23	03 14	13 14	21 56	01 09
12 Su	05 24 07	20 18 05	19 25 07	16 22	24 07	20 40	14 54	26 19	03 16	13 15	21 59	01 02
13 Mo	05 28 04	21 19 07	04♑37 51	15 05	25 22	21 21	15 03	26 16	03 17	13 17	22 01	00 55
14 Tu	05 32 00	22 20 10	19 50 24	13 56	26 37	22 02	15 11	26 12	03 19	13 18	22 03	00 47
15 We	05 35 57	23 21 13	04♒52 21	12 54	27 52	22 43	15 19	26 08	03 20	13 20	22 05	00 39
16 Th	05 39 53	24 22 17	19 35 10	12 03	29 06	23 24	15 27	26 05	03 22	13 22	22 08	00 33
17 Fr	05 43 50	25 23 21	03♓53 24	11 23	00♐21	24 05	15 35	26 01	03 24	13 23	22 10	00 28
18 Sa	05 47 47	26 24 26	17 44 57	10 54	01 36	24 45	15 43	25 57	03 25	13 25	22 12	00 26
19 Su	05 51 43	27 25 30	01♈10 34	10 35	02 51	25 26	15 51	25 53	03 27	13 27	22 14	00D 25
20 Mo	05 55 40	28 26 35	14 12 57	10 27	04 06	26 07	15 58	25 49	03 29	13 28	22 17	00 26
21 Tu	05 59 36	29 27 41	26 55 49	10D 30	05 21	26 48	16 06	25 45	03 31	13 30	22 19	00 27
22 We	06 03 33	00♑28 46	09♉23 04	10 41	06 36	27 29	16 13	25 41	03 33	13 32	22 21	00R 27
23 Th	06 07 29	01 29 52	21 38 24	11 01	07 51	28 10	16 20	25 36	03 35	13 34	22 23	00 25
24 Fr	06 11 26	02 30 58	03♊45 00	11 29	09 06	28 51	16 27	25 32	03 37	13 36	22 26	00 21
25 Sa	06 15 22	03 32 05	15 45 27	12 03	10 21	29 32	16 34	25 28	03 39	13 37	22 28	00 14
26 Su	06 19 19	04 33 12	27 41 47	12 44	11 36	00♐14	16 40	25 23	03 41	13 39	22 30	00 04
27 Mo	06 23 16	05 34 19	09♋35 32	13 30	12 51	00 55	16 47	25 19	03 43	13 41	22 32	29♈52
28 Tu	06 27 12	06 35 26	21 28 00	14 21	14 06	01 36	16 53	25 14	03 46	13 43	22 35	29 38
29 We	06 31 09	07 36 33	03♌20 29	15 17	15 21	02 17	16 59	25 10	03 48	13 45	22 37	29 26
30 Th	06 35 05	08 37 41	15 14 36	16 16	16 36	02 58	17 06	25 05	03 50	13 47	22 39	29 14
31 Fr	06 39 02	09 38 49	27 12 31	17 18	17 51	03 40	17 11	25 01	03 52	13 49	22 41	29 04

Data for	12-01-2004
Julian Day	2453340.50
Ayanamsa	23 55 22
SVP	05 ♓ 11 19
☽ ☊ Mean	29 ♈ 58 R

● ◐ PHASES ○ ◑

Day	h m	Phase	Position
05	00:54	◑	13♍14
12	01:30	●	20♐22
18	16:39	◐	27♓07
26	15:07	○	05♋12

LAST ASPECT Day	h m	☽ INGRESS Day	h m	Sign
01	04:29	01	10:50	♌
03	14:53	03	23:01	♍
06	02:29	06	08:47	♎
08	08:42	08	14:45	♏
10	11:04	10	16:55	♐
12	04:03	12	16:42	♑
14	11:44	14	16:10	♒
16	08:33	16	17:24	♓
18	16:40	18	21:52	♈
21	05:16	21	05:52	♉
23	13:41	23	16:33	♊
25	13:31	26	04:39	♋
28	07:35	28	17:15	♌
30	14:55	31	05:34	♍

DECLINATION

Day	☉	☽	☿	♀	♂	♃	♄	♅	♆	♇
01 We	21S49	26N14	24S40	13S21	15S23	04S08	20N43	11S06	17S00	15S06
02 Th	21 58	23 37	24 25	13 44	15 35	04 11	20 44	11 06	16 59	15 06
03 Fr	22 07	19 59	24 09	14 08	15 48	04 15	20 44	11 05	16 59	15 06
04 Sa	22 15	15 30	23 50	14 31	16 00	04 18	20 45	11 05	16 59	15 07
05 Su	22 23	10 20	23 30	14 54	16 12	04 22	20 46	11 05	16 58	15 07
06 Mo	22 30	04 38	23 08	15 17	16 24	04 25	20 46	11 04	16 58	15 07
07 Tu	22 37	01S25	22 45	15 39	16 36	04 29	20 47	11 04	16 58	15 07
08 We	22 44	07 36	22 21	16 01	16 48	04 32	20 48	11 03	16 57	15 08
09 Th	22 50	13 40	21 55	16 22	17 00	04 35	20 48	11 03	16 57	15 08
10 Fr	22 55	19 13	21 30	16 43	17 12	04 38	20 49	11 02	16 56	15 08
11 Sa	23 00	23 47	21 05	17 04	17 23	04 42	20 50	11 01	16 56	15 08
12 Su	23 05	26 50	20 41	17 24	17 35	04 45	20 50	11 01	16 56	15 09
13 Mo	23 09	27 56	20 19	17 44	17 46	04 48	20 51	11 00	16 55	15 09
14 Tu	23 13	26 56	19 59	18 03	17 57	04 51	20 52	11 00	16 55	15 09
15 We	23 16	23 59	19 42	18 21	18 08	04 54	20 53	10 59	16 54	15 09
16 Th	23 19	19 31	19 29	18 40	18 19	04 57	20 54	10 58	16 54	15 09
17 Fr	23 22	14 03	19 18	18 57	18 30	04 59	20 54	10 58	16 53	15 10
18 Sa	23 24	08 02	19 11	19 15	18 40	05 02	20 55	10 57	16 53	15 10
19 Su	23 25	01 50	19 08	19 31	18 50	05 05	20 56	10 56	16 52	15 10
20 Mo	23 26	04N17	19 07	19 47	19 01	05 08	20 57	10 56	16 52	15 10
21 Tu	23 26	10 05	19 10	20 03	19 11	05 10	20 58	10 55	16 51	15 10
22 We	23 26	15 23	19 15	20 18	19 21	05 13	20 59	10 54	16 51	15 11
23 Th	23 26	19 58	19 22	20 33	19 31	05 15	21 00	10 53	16 50	15 11
24 Fr	23 25	23 39	19 31	20 46	19 40	05 18	21 00	10 53	16 50	15 11
25 Sa	23 24	26 16	19 42	21 00	19 50	05 20	21 01	10 52	16 49	15 11
26 Su	23 22	27 41	19 54	21 12	19 59	05 23	21 02	10 51	16 49	15 11
27 Mo	23 19	27 48	20 07	21 25	20 08	05 25	21 03	10 50	16 48	15 11
28 Tu	23 17	26 38	20 20	21 36	20 17	05 27	21 04	10 50	16 48	15 12
29 We	23 13	24 17	20 34	21 47	20 26	05 29	21 05	10 49	16 47	15 12
30 Th	23 10	20 53	20 49	21 57	20 35	05 31	21 06	10 48	16 47	15 12
31 Fr	23 05	16 36	21 03	22 07	20 43	05 34	21 07	10 47	16 46	15 12

ASPECTARIAN

01 04:29 ☽ ☌ ♄
16:58 ☽ # ☿
02 08:01 ☽ △ ☉
11:13 ☽ # ☉
11:39 ☽ □ ♀
13:09 ☽ ☍ ♆
14:04 ☽ ⚹ ♃
15:48 ☽ □ ♂
19:32 ☽ ∥ ♄
03 02:22 ♀ □ ♆
06:25 ☽ △ ♇
14:53 ☽ △ ☿
16:30 ☽ # ♆
21:40 ☽ # ♂
04 01:58 ☽ # ♇
04:29 ☽ # ♀
05:06 ☽ ☍ ♅
20:45 ☽ # ♅
21:32 ☉ ⚹ ♆
05 05:40 ☽ ⚹ ♀
06:28 ☽ ⚹ ♂
13:33 ♀ ∥ ♇
17:14 ☽ □ ♇
18:58 ☉ ⚹ ♃
21:46 ☽ □ ☿
22:14 ♀ ☌ ♂
06 00:55 ☽ # ♃
02:29 ☽ ⚹ ♄
07 06:03 ☿ ∥ ☉
08:56 ☽ △ ♆
10:55 ☽ ☌ ♃
12:01 ☽ ∥ ♃
13:43 ☽ ⚹ ☉
23:05 ☿ ☌ ♇
08 00:20 ☽ ⚹ ☿
00:27 ☽ ⚹ ♇
0[illegible]:41 ☽ □ ♄
13:29 ☽ ∥ ♅
17:28 ♂ ∥ ♆
20:14 ☽ △ ♅
09 06:03 ☽ ∥ ♇
12:03 ☽ ∥ ♀
13:13 ☽ □ ♆
13:46 ☽ ∥ ♆
14:32 ☽ ∥ ♂
23:23 ☽ ☌ ♂
10 03:32 ☽ ☌ ♀
07:43 ☽ # ♄
08:22 ☿ ☌ ☉
10:15 ☽ ∥ ☿
11:04 ☽ △ ♄
15:02 ♀ ∥ ♆
19:19 ☽ ∥ ☉
22:09 ☽ □ ♅
11 14:12 ☽ ⚹ ♆
14:46 ☿ # ♄
16:46 ☽ ⚹ ♃
19:33 ☽ ☌ ☿
12 04:03 ☽ ☌ ♇
21:53 ☽ ⚹ ♅
13 00:47 ☿ ⚹ ♃
06:50 ♀ ∥ ♂
16:27 ♀ △ ♄
16:34 ☽ □ ♃
17:04 ☉ ☌ ♇
14 03:39 ☽ ⚹ ♂
10:04 ☽ ☍ ♄
11:44 ☽ ⚹ ♀
13:46 ☿ ⚹ ♆
15 04:17 ☽ ∥ ☉
12:17 ☽ ⚹ ☿
13:45 ☽ ☌ ♆
17:08 ☽ △ ♃
17:18 ☽ # ♄
16 00:13 ☽ ∥ ☿
03:48 ☽ ∥ ♀
04:13 ☽ ⚹ ♇
05:26 ☽ ∥ ♂
06:38 ☽ □ ♂
08:33 ☽ ⚹ ☉
11:58 ☽ ∥ ♆
17:10 ♀ ♐
17:25 ☽ □ ♀
19:24 ☽ ∥ ♇
23:09 ☽ ☌ ♅
17 12:24 ☽ □ ☿
12:32 ☽ ∥ ♅
20:34 ☿ ∥ ♀
18 07:54 ☽ □ ♇
11:32 ☽ ∥ ♃
13:06 ☽ △ ♂
14:30 ☽ △ ♄
19 03:22 ☽ △ ♀
11:54 ♀ □ ♅
14:11 ♂ △ ♄
17:03 ☽ △ ☿
22:37 ☽ ⚹ ♆
20 03:18 ☽ ☍ ♃
03:26 ☽ # ♃
06:30 ☿ SD
15:12 ☽ △ ♇
20:40 ☿ ∥ ♂
21:45 ☽ □ ♄
21 03:37 ☽ # ♅
05:16 ☽ △ [illegible]
12:40 ☽ ⚹ ♅
12:42 ☉ ♑
23:06 ☽ # ♇
22 07:19 ☽ # ♆
08:06 ☽ □ ♆
20:36 ☽ # ☿
21:24 ☽ # ♂
23 03:42 ☽ # ♀
06:13 ☽ ∥ ♄
07:47 ☽ ⚹ ♄
13:41 ☽ ☍ ♂
22:21 ☽ # ☉
23:44 ☽ □ ♅
24 11:54 ☽ ☍ ♀
16:11 ☽ ☍ ☿
19:43 ☽ △ ♆
25 01:38 ☽ △ ♃
02:51 ☉ ⚹ ♅
03:16 ♀ # ♄
13:31 ☽ ☍ ♆
16:04 ♂ ♐
26 12:07 ☽ △ ♅
27 05:37 ☿ ⚹ ♆
09:01 ☿ ∥ ♂
14:39 ☽ □ ♃
16:32 ♀ ⚹ ♆
28 07:35 ☽ ☌ ♄
18:23 ☿ ☌ ♀
21:44 ☽ △ ♂
29 08:34 ☽ ∥ [illegible]
17:26 ☽ # ♀
21:03 ☽ ☍ ♆
22:36 ☽ ∥ ♄
30 00:24 ☽ # ☿
01:47 ☽ # ♂
02:15 ☽ △ ☿
03:03 ☽ △ ♀
03:45 ☽ ⚹ ♃
10:19 ♀ ⚹ ♃
14:55 ☽ △ ♇
21:08 ☿ ⚹ ♃
23:10 ☽ # ♆
31 06:59 ☿ # ♄
07:07 ☽ # ♇
[illegible]
13:19 ☽ ☍ ♅
13:38 ☽ □ ♂

⚷ Chiron

01 Dec. 14 S 59

Day	°	'
02	23♑	07
05	23	20
08	23	34
11	23	48
14	24	02
17	24	17
20	24	31
23	24	47
26	25	02
29	25	17

January 2005

Day	S. T.	☉	☽	☿	♀	♂	♃	♄	♅	♆	♇	☊ True
	h m s	° ' "	° ' "	° '	° '	° '	° '	° '	° '	° '	° '	° '
01 Sa	06 42 58	10♑39 58	09♍17 06	18♐24	19♐06	04♐21	17♎17	24♋R56	03♓55	13♒51	22♐43	28♈R58
02 Su	06 46 55	11 41 06	21 31 51	19 32	20 21	05 02	17 23	24 51	03 57	13 53	22 45	28 54
03 Mo	06 50 51	12 42 15	04♎00 55	20 43	21 36	05 44	17 28	24 46	04 00	13 55	22 48	28 52
04 Tu	06 54 48	13 43 25	16 48 44	21 56	22 51	06 25	17 34	24 42	04 02	13 57	22 50	28 51
05 We	06 58 45	14 44 34	29 59 47	23 10	24 06	07 07	17 39	24 37	04 04	13 59	22 52	28 50
06 Th	07 02 41	15 45 44	13♏37 51	24 27	25 22	07 48	17 44	24 32	04 07	14 01	22 54	28 48
07 Fr	07 06 38	16 46 54	27 45 06	25 45	26 37	08 29	17 49	24 27	04 10	14 03	22 56	28 44
08 Sa	07 10 34	17 48 05	12♐20 57	27 04	27 52	09 11	17 53	24 22	04 12	14 05	22 58	28 38
09 Su	07 14 31	18 49 15	27 21 14	28 24	29 07	09 53	17 58	24 17	04 15	14 07	23 00	28 29
10 Mo	07 18 27	19 50 25	12♑37 58	29 46	00♑22	10 34	18 02	24 12	04 17	14 09	23 02	28 19
11 Tu	07 22 24	20 51 35	28 00 19	01♑08	01 37	11 16	18 06	24 07	04 20	14 11	23 05	28 08
12 We	07 26 21	21 52 45	13♒16 20	02 32	02 52	11 57	18 10	24 02	04 23	14 14	23 07	27 57
13 Th	07 30 17	22 53 54	28 15 07	03 56	04 08	12 39	18 14	23 58	04 26	14 16	23 09	27 47
14 Fr	07 34 14	23 55 02	12♓48 36	05 22	05 23	13 21	18 18	23 53	04 28	14 18	23 11	27 40
15 Sa	07 38 10	24 56 10	26 52 32	06 48	06 38	14 02	18 21	23 48	04 31	14 20	23 13	27 35
16 Su	07 42 07	25 57 17	10♈26 24	08 14	07 53	14 44	18 24	23 43	04 34	14 22	23 15	27 33
17 Mo	07 46 03	26 58 23	23 32 28	09 42	09 08	15 26	18 27	23 38	04 37	14 24	23 17	27 32
18 Tu	07 49 60	27 59 29	06♉14 50	11 10	10 23	16 08	18 30	23 33	04 40	14 27	23 19	27 32
19 We	07 53 56	29 00 33	18 38 18	12 39	11 39	16 50	18 33	23 28	04 43	14 29	23 20	27 31
20 Th	07 57 53	00♒01 37	00♊47 43	14 08	12 54	17 31	18 36	23 23	04 46	14 31	23 22	27 29
21 Fr	08 01 50	01 02 40	12 47 37	15 39	14 09	18 13	18 38	23 18	04 49	14 33	23 24	27 24
22 Sa	08 05 46	02 03 42	24 41 52	17 09	15 24	18 55	18 40	23 13	04 52	14 35	23 26	27 16
23 Su	08 09 43	03 04 44	06♋33 41	18 41	16 39	19 37	18 42	23 08	04 55	14 38	23 28	27 05
24 Mo	08 13 39	04 05 44	18 25 27	20 13	17 54	20 19	18 44	23 04	04 58	14 40	23 30	26 52
25 Tu	08 17 36	05 06 44	00♌18 53	21 45	19 10	21 01	18 46	22 59	05 01	14 42	23 32	26 37
26 We	08 21 32	06 07 42	12 15 16	23 18	20 25	21 43	18 47	22 54	05 04	14 44	23 33	26 23
27 Th	08 25 29	07 08 40	24 15 41	24 52	21 40	22 25	18 48	22 49	05 07	14 47	23 35	26 09
28 Fr	08 29 25	08 09 37	06♍21 20	26 27	22 55	23 07	18 49	22 45	05 10	14 49	23 37	25 58
29 Sa	08 33 22	09 10 34	18 33 53	28 02	24 10	23 49	18 50	22 40	05 14	14 51	23 39	25 50
30 Su	08 37 19	10 11 29	00♎55 26	29 37	25 25	24 31	18 51	22 35	05 17	14 54	23 40	25 45
31 Mo	08 41 15	11 12 24	13 28 47	01♒14	26 40	25 14	18 51	22 31	05 20	14 56	23 42	25 43

Data for	01-01-2005
Julian Day	2453371.50
Ayanamsa	23 55 28
SVP	05 ♓ 11 16
☽ ☊ Mean	28 ♈ 19 R

● ◐ PHASES ○ ◑

03	17:46	◑	13♎28
10	12:03	●	20♑21
17	06:58	◐	27♈16
25	10:33	○	05♌34

LAST ASPECT ☽ INGRESS

Day	h m	Day	h m	
02	06:24	02	16:20	♎
04	14:21	05	00:00	♏
06	18:30	07	03:45	♐
09	03:02	09	04:11	♑
10	17:58	11	03:07	♒
12	15:44	13	02:51	♓
14	20:22	15	05:27	♈
17	06:58	17	12:07	♉
19	22:20	19	22:25	♊
21	21:27	22	10:43	♋
24	09:18	24	23:22	♌
26	22:39	27	11:25	♍
29	21:07	29	22:13	♎

DECLINATION

Day	☉	☽	☿	♀	♂	♃	♄	♅	♆	♇
01 Sa	23S01	11N39	21S17	22S15	20S52	05S36	21N08	10S46	16S46	15S12
02 Su	22 56	06 10	21 32	22 24	21 00	05 37	21 09	10 45	16 45	15 12
03 Mo	22 50	00 21	21 45	22 31	21 08	05 39	21 10	10 44	16 44	15 12
04 Tu	22 44	05S38	21 59	22 38	21 15	05 41	21 11	10 43	16 44	15 13
05 We	22 38	11 34	22 12	22 44	21 23	05 43	21 12	10 42	16 43	15 13
06 Th	22 31	17 10	22 24	22 50	21 31	05 45	21 13	10 41	16 43	15 13
07 Fr	22 23	22 02	22 35	22 55	21 38	05 46	21 14	10 41	16 42	15 13
08 Sa	22 15	25 43	22 46	22 59	21 45	05 48	21 15	10 40	16 42	15 13
09 Su	22 07	27 43	22 55	23 02	21 52	05 49	21 16	10 39	16 41	15 13
10 Mo	21 58	27 39	23 04	23 05	21 58	05 51	21 17	10 38	16 40	15 13
11 Tu	21 49	25 28	23 12	23 07	22 05	05 52	21 18	10 37	16 40	15 13
12 We	21 40	21 28	23 19	23 08	22 11	05 53	21 19	10 36	16 39	15 13
13 Th	21 30	16 08	23 25	23 09	22 17	05 54	21 19	10 34	16 38	15 13
14 Fr	21 19	10 02	23 29	23 09	22 23	05 56	21 20	10 33	16 38	15 13
15 Sa	21 09	03 36	23 33	23 08	22 29	05 57	21 21	10 32	16 37	15 13
16 Su	20 57	02N46	23 35	23 06	22 35	05 58	21 22	10 31	16 37	15 14
17 Mo	20 46	08 49	23 36	23 04	22 40	05 59	21 23	10 30	16 36	15 14
18 Tu	20 34	14 19	23 36	23 01	22 45	05 59	21 24	10 29	16 35	15 14
19 We	20 22	19 07	23 35	22 58	22 50	06 00	21 25	10 28	16 35	15 14
20 Th	20 09	23 01	23 33	22 53	22 55	06 01	21 26	10 27	16 34	15 14
21 Fr	19 56	25 53	23 29	22 48	22 59	06 02	21 27	10 26	16 33	15 14
22 Sa	19 42	27 33	23 24	22 42	23 04	06 02	21 28	10 25	16 33	15 14
23 Su	19 28	27 57	23 17	22 36	23 08	06 03	21 29	10 24	16 32	15 14
24 Mo	19 14	27 04	23 10	22 29	23 12	06 03	21 30	10 22	16 32	15 14
25 Tu	18 59	24 57	23 01	22 21	23 15	06 04	21 31	10 21	16 31	15 14
26 We	18 45	21 44	22 50	22 12	23 19	06 04	21 32	10 20	16 30	15 14
27 Th	18 29	17 36	22 39	22 03	23 22	06 04	21 33	10 19	16 30	15 14
28 Fr	18 14	12 44	22 25	21 53	23 25	06 04	21 33	10 18	16 29	15 14
29 Sa	17 58	07 20	22 11	21 43	23 28	06 04	21 34	10 17	16 28	15 14
30 Su	17 41	01 35	21 55	21 32	23 31	06 04	21 35	10 15	16 28	15 14
31 Mo	17 25	04S20	21 38	21 20	23 33	06 04	21 36	10 14	16 27	15 14

ASPECTARIAN

Day	Time	Aspect
01	02:58	☽ △ ☉
	04:00	☽ # ♅
	19:43	☽ □ ☿
	21:27	☽ □ ♀
02	02:18	☽ # ♃
	02:23	☽ □ ♇
	06:24	☽ ⚹ ♄
03	03:26	☽ ⚹ ♂
	07:25	♂ # ♄
	18:40	☽ △ ♆
	23:31	♀ ☌ ♇
04	00:13	☽ ∥ ♃
	01:24	☽ ☌ ♃
	10:23	☽ ⚹ ☿
	11:05	☽ ⚹ ♇
	11:08	♀ ∥ ☉
	12:16	☽ ⚹ ♀
	14:21	☽ □ ♄
	17:58	☿ ☌ ♇
	20:28	☽ ∥ ♅
05	07:17	☽ △ ♅
	15:22	☽ ∥ ♇
	21:58	☽ ∥ ♆
06	00:40	☽ □ ♆
	03:58	☽ ⚹ ☉
	08:51	☿ ∥ ☉
	18:30	☽ △ ♄
	19:37	☽ # ♄
	21:44	☽ ∥ ♂
07	01:54	☽ ∥ ☉
	03:12	☽ ∥ ☿
	05:03	☽ ∥ ♀
	10:40	☽ □ ♅
	18:36	☽ ☌ ♂
08	02:12	☉ □ ♃
	02:49	☽ ⚹ ♆
	08:58	☽ ⚹ ♃
	17:05	☽ ☌ ♇
09	01:49	☽ ☌ ☿
	03:02	☽ ☌ ♀
	10:54	☽ ⚹ ♅
	16:56	♀ ♑
	23:56	☉ ∥ ♂
10	02:57	☿ ∥ ♀
	04:09	☿ ♑
	08:28	☽ □ ♃
	17:58	☽ ☍ ♄
11	14:17	☽ ∥ ☿
	15:05	☽ ∥ ♀
	20:19	☽ ∥ ♂
	21:49	☽ ⚹ ♂
	22:56	☽ ∥ ☉
12	00:45	☽ # ♄
	01:31	☽ ☌ ♆
	07:49	☽ △ ♃
	15:44	☽ ⚹ ♇
	21:53	☽ ∥ ♆
13	03:44	☽ ∥ ♇
	05:59	♀ ⚹ ♅
	08:33	☿ ⚹ ♅
	10:07	☽ ☌ ♅
	10:17	☽ ⚹ ☿
	10:30	☽ ⚹ ♀
	21:58	☉ # ♄
	21:59	☽ ∥ ♅
	23:06	☉ ☍ ♄
14	00:57	☽ □ ♂
	02:36	☿ ☌ ♀
	15:19	☽ ∥ ♃
	17:39	☽ □ ♇
	18:42	☽ △ ♄
	20:22	☽ ⚹ ☉
15	10:40	♂ ⚹ ♆
	18:56	☽ □ ♀
	19:34	☽ □ ☿
16	07:08	☽ ⚹ ♆
	08:13	☽ △ ♂
	12:30	☽ # ♃
	14:33	☽ ☍ ♃
	23:30	☽ △ ♇
17	00:10	☽ □ ♄
	07:04	☽ # ♅
	20:58	☽ ⚹ ♅
18	04:15	☽ # ♇
	08:51	☽ △ ♀
	10:45	☽ △ ☿
	10:50	☽ # ♆
	15:52	☽ □ ♆
19	06:42	☽ # ☉
	09:25	☽ ⚹ ♄
	13:29	☽ ∥ ♄
	19:54	♀ ∥ ♂
	22:20	☽ △ ☉
	23:05	☽ # ♀
	23:14	☽ # ♂
	23:22	☉ ♒
20	03:43	☽ # ☿
	07:57	☽ □ ♅
21	03:33	☽ △ ♆
	11:36	☽ ☍ ♂
	11:47	☽ △ ♃
	14:54	♂ ⚹ ♃
	21:27	☽ ☍ ♇
22	20:39	☽ △ ♅
23	00:25	☿ □ ♃
	20:14	☿ ∥ ♂
	22:50	☽ ☍ ♀
24	00:37	☽ □ ♃
	04:09	☽ ☍ ☿
	09:18	☽ ☌ ♄
	16:10	♀ □ ♃
25	13:17	☽ # ♂
	16:06	☽ # ☿
	18:04	☿ ☍ ♄
	20:43	☽ # ♀
26	01:18	☽ ∥ ♄
	05:00	☽ ☍ ♆
	13:06	☽ ⚹ ♃
	18:52	☽ # ☉
	20:06	☽ △ ♂
	22:39	☽ △ ♇
27	05:45	☽ # ♆
	12:02	☽ # ♇
	20:53	♀ ☍ ♄
	21:39	☽ ☍ ♅
28	11:05	☽ # ♅
	17:45	♂ ☌ ♇
29	05:19	☽ # ♃
	07:57	☽ ⚹ ♄
	09:56	☽ □ ♇
	10:52	☽ □ ♂
	12:09	☽ △ ♀
	17:11	♀ # ♄
	21:07	☽ △ ☿
30	05:37	☿ ♒
	19:19	☽ △ ☉
31	02:13	☿ # ♄
	02:45	☽ △ ♆
	07:03	☽ ∥ ♃
	10:09	☽ ☌ ♃
	16:53	☽ □ ♄
	19:14	☽ ⚹ ♇
	23:18	☽ ⚹ ♂

⚷ Chiron

01 Dec.	14 S 41
01	25♑33
04	25 49
07	26 04
10	26 20
13	26 36
16	26 52
19	27 08
22	27 24
25	27 40
28	27 56
31	28 12

February 2005

Day	S. T. h m s	☉ ° ' "	☽ ° ' "	☿ ° '	♀ ° '	♂ ° '	♃ ° '	♄ ° '	♅ ° '	♆ ° '	♇ ° '	☊ True ° '
01 Tu	08 45 12	12♒13 18	26♎17 11	02♒51	27♑56	25♐56	18♎52	22♋R26	05♓23	14♒58	23♐44	25♈R42
02 We	08 49 08	13 14 11	09♏24 18	04 29	29 11	26 38	18 52	22 22	05 26	15 00	23 45	25 41
03 Th	08 53 05	14 15 03	22 53 36	06 07	00♒26	27 20	18R 52	22 17	05 30	15 03	23 47	25 40
04 Fr	08 57 01	15 15 55	06♐47 43	07 47	01 41	28 03	18 51	22 13	05 33	15 05	23 49	25 37
05 Sa	09 00 58	16 16 46	21 07 25	09 27	02 56	28 45	18 51	22 09	05 36	15 07	23 50	25 32
06 Su	09 04 54	17 17 36	05♑50 41	11 08	04 11	29 27	18 50	22 05	05 40	15 09	23 52	25 25
07 Mo	09 08 51	18 18 25	20 52 07	12 49	05 27	00♑10	18 49	22 00	05 43	15 12	23 53	25 16
08 Tu	09 12 48	19 19 13	06♒03 15	14 31	06 42	00 52	18 48	21 56	05 46	15 14	23 55	25 06
09 We	09 16 44	20 20 00	21 13 37	16 15	07 57	01 34	18 47	21 52	05 50	15 16	23 56	24 56
10 Th	09 20 41	21 20 46	06♓12 34	17 58	09 12	02 17	18 46	21 48	05 53	15 19	23 58	24 47
11 Fr	09 24 37	22 21 30	20 51 03	19 43	10 27	02 59	18 44	21 44	05 56	15 21	23 59	24 41
12 Sa	09 28 34	23 22 12	05♈03 01	21 29	11 42	03 42	18 42	21 40	06 00	15 23	24 00	24 36
13 Su	09 32 30	24 22 53	18 45 56	23 15	12 57	04 24	18 40	21 37	06 03	15 25	24 02	24 35
14 Mo	09 36 27	25 23 32	02♉00 21	25 02	14 12	05 07	18 38	21 33	06 07	15 28	24 03	24D 35
15 Tu	09 40 23	26 24 10	14 49 11	26 50	15 27	05 50	18 36	21 29	06 10	15 30	24 04	24 36
16 We	09 44 20	27 24 45	27 16 44	28 39	16 43	06 32	18 33	21 26	06 13	15 32	24 06	24 36
17 Th	09 48 17	28 25 20	09♊27 55	00♓29	17 58	07 15	18 31	21 22	06 17	15 34	24 07	24R 36
18 Fr	09 52 13	29 25 52	21 27 45	02 19	19 13	07 57	18 28	21 19	06 20	15 37	24 08	24 34
19 Sa	09 56 10	00♓26 23	03♋21 00	04 10	20 28	08 40	18 25	21 16	06 24	15 39	24 09	24 30
20 Su	10 00 06	01 26 51	15 11 55	06 01	21 43	09 23	18 21	21 13	06 27	15 41	24 10	24 23
21 Mo	10 04 03	02 27 18	27 03 59	07 53	22 58	10 05	18 18	21 10	06 31	15 43	24 12	24 15
22 Tu	10 07 59	03 27 44	08♌59 57	09 46	24 13	10 48	18 14	21 07	06 34	15 45	24 13	24 04
23 We	10 11 56	04 28 07	21 01 49	11 38	25 28	11 31	18 11	21 04	06 37	15 48	24 14	23 54
24 Th	10 15 52	05 28 29	03♍10 59	13 31	26 43	12 14	18 07	21 01	06 41	15 50	24 15	23 44
25 Fr	10 19 49	06 28 49	15 28 29	15 24	27 58	12 57	18 03	20 58	06 44	15 52	24 16	23 36
26 Sa	10 23 46	07 29 07	27 55 11	17 17	29 13	13 39	17 58	20 55	06 48	15 54	24 17	23 31
27 Su	10 27 42	08 29 24	10♎31 59	19 09	00 28	14 22	17 54	20 53	06 51	15 56	24 18	23 28
28 Mo	10 31 39	09 29 39	23 20 02	21 01	01♓43	15 05	17 49	20 51	06 55	15 58	24 19	23 27

Data for	02-01-2005
Julian Day	2453402.50
Ayanamsa	23 55 34
SVP	05♓11 13
☽ ☊ Mean	26♈41 R

● ◐ PHASES ○ ◑

Day	Time	Phase	Position
02	07:27	◑	13♏33
08	22:28	●	20♒16
16	00:17	◐	27♉25
24	04:54	○	05♍41

LAST ASPECT ☽ INGRESS

Day	h m	Day	h m	Sign
01	03:22	01	06:51	♏
02	22:57	03	12:22	♐
05	13:08	05	14:32	♑
07	01:48	07	14:27	♒
09	04:19	09	14:00	♓
11	05:14	11	15:22	♈
13	10:55	13	20:19	♉
16	03:09	16	05:19	♊
18	05:23	18	17:13	♋
20	12:07	21	05:55	♌
23	09:48	23	17:44	♍
25	17:00	26	03:59	♎
28	01:49	28	12:21	♏

DECLINATION

Day	☉	☽	☿	♀	♂	♃	♄	♅	♆	♇
01 Tu	17S 08	10S 11	21S 19	21S 07	23S 35	06S 04	21N37	10S 13	16S 26	15S 14
02 We	16 51	15 46	20 59	20 54	23 37	06 04	21 38	10 12	16 26	15 14
03 Th	16 33	20 44	20 37	20 41	23 39	06 04	21 38	10 11	16 25	15 14
04 Fr	16 16	24 43	20 14	20 26	23 40	06 03	21 39	10 09	16 24	15 14
05 Sa	15 57	27 18	19 50	20 11	23 42	06 03	21 40	10 08	16 24	15 14
06 Su	15 39	28 03	19 24	19 56	23 43	06 03	21 41	10 07	16 23	15 13
07 Mo	15 21	26 47	18 57	19 40	23 44	06 02	21 42	10 06	16 22	15 13
08 Tu	15 02	23 33	18 28	19 23	23 44	06 01	21 42	10 05	16 22	15 13
09 We	14 43	18 43	17 58	19 06	23 45	06 01	21 43	10 03	16 21	15 13
10 Th	14 23	12 46	17 26	18 48	23 45	06 00	21 44	10 02	16 20	15 13
11 Fr	14 04	06 14	16 53	18 29	23 45	05 59	21 45	10 01	16 20	15 13
12 Sa	13 44	00N26	16 18	18 11	23 45	05 58	21 45	10 00	16 19	15 13
13 Su	13 24	06 52	15 42	17 51	23 44	05 57	21 46	09 58	16 18	15 13
14 Mo	13 03	12 47	15 05	17 31	23 44	05 56	21 47	09 57	16 18	15 13
15 Tu	12 43	17 58	14 26	17 11	23 43	05 55	21 47	09 56	16 17	15 13
16 We	12 22	22 14	13 46	16 50	23 41	05 54	21 48	09 54	16 16	15 13
17 Th	12 01	25 25	13 04	16 28	23 40	05 53	21 49	09 53	16 16	15 13
18 Fr	11 40	27 25	12 21	16 07	23 38	05 51	21 49	09 52	16 15	15 13
19 Sa	11 19	28 09	11 37	15 44	23 37	05 50	21 50	09 51	16 15	15 13
20 Su	10 58	27 34	10 51	15 22	23 35	05 48	21 51	09 49	16 14	15 12
21 Mo	10 36	25 44	10 04	14 59	23 32	05 47	21 51	09 48	16 13	15 12
22 Tu	10 14	22 45	09 16	14 35	23 30	05 45	21 52	09 47	1 6 13	15 12
23 We	09 52	18 47	08 27	14 11	23 27	05 44	21 52	09 46	16 12	15 12
24 Th	09 30	14 01	07 37	13 47	23 24	05 42	21 53	09 44	16 11	15 12
25 Fr	09 08	08 39	06 46	13 22	23 21	05 40	21 53	09 43	16 11	15 12
26 Sa	08 46	02 52	05 55	12 57	23 18	05 38	21 54	09 42	16 10	15 12
27 Su	08 23	03S 06	05 02	12 31	23 14	05 37	21 54	09 40	16 09	15 12
28 Mo	08 01	09 03	04 09	12 06	23 10	05 35	21 55	09 39	16 09	15 11

ASPECTARIAN

01 00:06 ☽ ∥ ♅
03:22 ☽ □ ♀
13:48 ☽ □ ☿
16:47 ☽ △ ♅
21:36 ☽ ∥ ♇

02 02:27 ♃ SR
03:01 ☽ ∥ ♆
04:41 ☽ ∥ ☉
10:05 ☽ □ ♆
14:27 ☿ ∥ ♀
15:43 ♀ ♒
22:57 ☽ △ ♄
23:27 ☽ ∥ ☿
23:43 ☽ ∥ ♀

03 04:55 ☽ # ♄
11:44 ☉ ∥ ♆
14:24 ☽ ⚹ ♀
16:51 ☽ ∥ ♂
19:29 ☉ ☌ ♆
21:52 ☽ □ ♅

04 01:54 ☽ ⚹ ☿
14:00 ☽ ⚹ ♆
15:21 ☽ ⚹ ☉
20:14 ☽ ⚹ ♃

05 04:29 ☽ ☌ ♇
13:08 ☽ ☌ ♂
23:42 ☽ ⚹ ♅
06 18:32 ♂ ♑
20:46 ☽ □ ♃

07 01:48 ☽ ☍ ♄
09:07 ☉ ∥ ♇
12:04 ☉ △ ♃
[illegible]
08 01:06 ☽ ☌ ♀
09:58 ☽ # ♄
10:10 ☿ ☌ ♆
14:32 ☽ ☌ ♆
15:05 ☽ ☌ ☿
20:08 ☽ △ ♃
22:12 ☽ ∥ ♀

09 03:33 ☽ ∥ ☿
04:19 ☽ ⚹ ♇
09:58 ☽ ∥ ♆
14:29 ☽ ∥ ♇
17:21 ☽ ⚹ ♂
17:25 ☽ ∥ ☉
23:28 ☽ ☌ ♅
10 10:12 ☽ ∥ ♅
10:43 ☿ △ ♃

11 00:52 ☽ ∥ ♃
01:28 ☽ △ ♄
05:14 ☽ □ ♇
21:33 ☽ □ ♂
23:26 ☿ ∥ ♆

12 12:42 ☽ ⚹ ♀
15:30 ☉ ⚹ ♇
18:03 ☽ ⚹ ♆
20:28 ☽ # ♃
23:50 ☽ ☍ ♃

13 05:04 ☽ □ ♄
09:17 ☽ ⚹ ☿
09:28 ☽ △ ♇
10:37 ☿ ⚹ ♇
10:53 ☽ ⚹ ☉
12:12 ☽ # ♅
18:49 ☿ ∥ ♇
14 01:05 ☽ # ☉
06:05 ☽ △ ♂
07:38 ☽ ⚹ ♅
09:02 ☽ # ☿
10:46 ☽ # ♇
10:50 ☿ ☌ ☉
15:45 ☽ # ♆
20:20 ☽ # ♀

15 00:49 ♀ ☌ ♆
01:18 ☽ □ ♆
01:21 ☽ □ ♀
12:32 ♂ ⚹ ♅
12:43 ☽ ⚹ ♄
21:15 ☽ ∥ ♄
16 03:09 ☽ □ ☿
09:49 ☽ # ♂
17:39 ☽ □ ♅
17:46 ☿ ♓

17 10:10 ♀ △ ♃
12:13 ☽ △ ♆
14:26 ♀ ∥ ♆
18:00 ☽ △ ♃
18:57 ☽ △ ♀
18 05:23 ☽ ☍ ♇
13:32 ☉ ♓
17:34 ☽ △ ☉

19 01:57 ☽ △ ☿
06:12 ☽ △ ♅
11:27 ☽ ☍ ♂
17:42 ☿ ∥ ☉
20 05:44 ☿ ☌ ♅
06:22 ☽ □ ♃
09:37 ♀ ∥ ♇
12:07 ☽ ☌ ♄
21 08:26 ☿ ∥ ♅
18:39 ☽ # ♂
23:58 ♀ ⚹ ♇

22 05:53 ☽ ∥ ♄
13:33 ☽ ☍ ♆
18:21 ☽ ⚹ ♃
21:24 ☿ ⚹ ♂

23 06:21 ☽ △ ♇
07:52 ☉ ∥ ♅
09:48 ☽ ☍ ♀
13:29 ☽ # ♆
18:21 ☽ # ♇

24 01:14 ☽ # ♀
06:53 ☽ ☍ ♅
18:47 ☽ △ ♂
19:22 ☽ # ♅
21:45 ☽ # ☉
23:50 ☽ ☍ ☿

25 06:33 ☉ ☌ ♅
09:19 ☽ # ☿
10:36 ☽ ⚹ ♄
12:35 ☽ # ♃
17:00 ☽ □ ♆
26 07:41 ☿ ∥ ♃
15:07 ♀ ♓
27 06:44 ☽ ∥ ☿
07:40 ☽ □ ♂
09:59 ☽ ∥ ♃
10:12 ☽ △ ♆
13:46 ☽ ☌ ♃
19:22 ☽ □ ♄
20:00 ☽ ∥ ☉
21:51 ☿ △ ♄

28 01:49 ☽ ⚹ ♇
03:27 ☽ ∥ ♅
11:45 ☽ ∥ ♀
17:08 ☽ △ ♀

⚷ Chiron

Day	Position
01 Dec.	14 S 10
03	28♑27
06	28 43
09	28 58
12	29 13
15	29 28
18	29 42
21	29 57
24	00♒10
27	00 24

21 18:34 ♒

March 2005

Day	S. T.	☉	☽	☿	♀	♂	♃	♄	♅	♆	♇	☊ True
	h m s	° ' "	° ' "	° '	° '	° '	° '	° '	° '	° '	° '	° '
01 Tu	10 35 35	10♓ 29 53	06♏ 20 47	22♓ 51	02♓ 58	15♑ 48	17♎R44	20♋R48	06♓ 58	16♒ 01	24♐ 19	23♈D27
02 We	10 39 32	11 30 05	19 35 59	24 40	04 13	16 31	17 40	20 46	07 02	16 03	24 20	23 28
03 Th	10 43 28	12 30 16	03♐ 07 27	26 27	05 28	17 14	17 34	20 44	07 05	16 05	24 21	23 29
04 Fr	10 47 25	13 30 25	16 56 37	28 12	06 42	17 57	17 29	20 42	07 08	16 07	24 22	23R 29
05 Sa	10 51 21	14 30 33	01♑ 03 54	29 53	07 57	18 40	17 24	20 40	07 12	16 09	24 23	23 27
06 Su	10 55 18	15 30 39	15 28 06	01♈ 32	09 12	19 23	17 18	20 38	07 15	16 11	24 23	23 23
07 Mo	10 59 15	16 30 44	00♒ 05 59	03 07	10 27	20 06	17 13	20 36	07 19	16 13	24 24	23 19
08 Tu	11 03 11	17 30 47	14 52 07	04 37	11 42	20 49	17 07	20 35	07 22	16 15	24 25	23 13
09 We	11 07 08	18 30 48	29 39 22	06 02	12 57	21 32	17 01	20 33	07 26	16 17	24 25	23 08
10 Th	11 11 04	19 30 48	14♓ 19 53	07 22	14 12	22 15	16 55	20 32	07 29	16 19	24 26	23 03
11 Fr	11 15 01	20 30 45	28 46 11	08 36	15 27	22 58	16 49	20 31	07 32	16 21	24 26	23 00
12 Sa	11 18 57	21 30 41	12♈ 52 24	09 43	16 41	23 42	16 43	20 29	07 36	16 23	24 27	22 58
13 Su	11 22 54	22 30 34	26 34 55	10 44	17 56	24 25	16 36	20 28	07 39	16 25	24 27	22D 58
14 Mo	11 26 50	23 30 26	09♉ 52 36	11 37	19 11	25 08	16 30	20 27	07 42	16 27	24 28	22 59
15 Tu	11 30 47	24 30 15	22 46 30	12 22	20 26	25 51	16 23	20 26	07 46	16 29	24 28	23 02
16 We	11 34 44	25 30 02	05♊ 19 21	12 59	21 41	26 34	16 16	20 26	07 49	16 31	24 29	23 04
17 Th	11 38 40	26 29 47	17 35 02	13 29	22 55	27 18	16 09	20 25	07 52	16 32	24 29	23 06
18 Fr	11 42 37	27 29 30	29 38 05	13 49	24 10	28 01	16 02	20 25	07 56	16 34	24 29	23 07
19 Sa	11 46 33	28 29 10	11♋ 33 18	14 02	25 25	28 44	15 55	20 24	07 59	16 36	24 30	23R 06
20 Su	11 50 30	29 28 48	23 25 25	14 06	26 40	29 27	15 48	20 24	08 02	16 38	24 30	23 05
21 Mo	11 54 26	00♈ 28 24	05♌ 18 50	14R 02	27 54	00♒ 11	15 41	20 24	08 06	16 40	24 30	23 02
22 Tu	11 58 23	01 27 58	17 17 22	13 50	29 09	00 54	15 34	20 24	08 09	16 41	24 30	22 58
23 We	12 02 19	02 27 29	29 24 12	13 31	00♈ 24	01 37	15 27	20D 24	08 12	16 43	24 30	22 54
24 Th	12 06 16	03 26 58	11♍ 41 44	13 05	01 38	02 21	15 19	20 24	08 15	16 45	24 31	22 51
25 Fr	12 10 13	04 26 25	24 11 40	12 33	02 53	03 04	15 12	20 24	08 18	16 47	24 31	22 48
26 Sa	12 14 09	05 25 50	06♎ 54 58	11 55	04 07	03 48	15 04	20 24	08 22	16 48	24 31	22 46
27 Su	12 18 06	06 25 13	19 51 58	11 12	05 22	04 31	14 57	20 25	08 25	16 50	24 31	22 45
28 Mo	12 22 02	07 24 34	03♏ 02 26	10 26	06 37	05 14	14 49	20 26	08 28	16 52	24R 31	22D 45
29 Tu	12 25 59	08 23 53	16 25 51	09 38	07 51	05 58	14 42	20 26	08 31	16 53	24 31	22 46
30 We	12 29 55	09 23 10	00♐ 01 29	08 47	09 06	06 41	14 34	20 27	08 34	16 55	24 31	22 47
31 Th	12 33 52	10 22 26	13 48 31	07 57	10 20	07 25	14 26	20 28	08 37	16 56	24 31	22 48

Data for	03-01-2005
Julian Day	2453430.50
Ayanamsa	23 55 37
SVP	05 ♓ 11 09
☽ ☊ Mean	25 ♈ 12 R

● ◐ PHASES ○ ◑

Day	h m	Phase	Position
03	17:37	◑	13♐14
10	09:11	●	19♓54
17	19:19	◐	27♊18
25	20:59	○	05♎18

LAST ASPECT ☽ INGRESS

Day	h m	Day	h m	Sign
02	10:26	02	18:30	♐
04	21:46	04	22:12	♑
06	08:30	06	23:50	♒
08	15:29	09	00:34	♓
10	16:45	11	02:04	♈
12	20:14	13	06:06	♉
15	06:11	15	13:45	♊
17	19:19	18	00:44	♋
20	12:59	20	13:17	♌
22	14:20	23	01:10	♍
25	00:36	25	11:00	♎
27	08:31	27	18:30	♏
29	07:07	29	23:57	♐

DECLINATION

Day	☉	☽	☿	♀	♂	♃	♄	♅	♆	♇
01 Tu	07S 38	14S 43	03S 16	11S 39	23S 06	05S 33	21N55	09S 38	16S 08	15S 11
02 We	07 15	19 49	02 23	11 13	23 02	05 31	21 56	09 37	16 08	15 11
03 Th	06 52	24 01	01 30	10 46	22 57	05 28	21 56	09 35	16 07	15 11
04 Fr	06 29	26 56	00 37	10 19	22 53	05 26	21 56	09 34	16 06	15 11
05 Sa	06 06	28 13	00N15	09 52	22 48	05 24	21 57	09 33	16 06	15 11
06 Su	05 43	27 37	01 06	09 25	22 43	05 22	21 57	09 32	16 05	15 11
07 Mo	05 19	25 08	01 56	08 57	22 37	05 20	21 58	09 30	16 05	15 11
08 Tu	04 56	20 58	02 44	08 29	22 32	05 17	21 58	09 29	16 04	15 10
09 We	04 33	15 29	03 30	08 00	22 26	05 15	21 58	09 28	16 03	15 10
10 Th	04 09	09 10	04 14	07 32	22 20	05 12	21 58	09 26	16 03	15 10
11 Fr	03 46	02 27	04 56	07 03	22 14	05 10	21 59	09 25	16 02	15 10
12 Sa	03 22	04N15	05 35	06 34	22 07	05 07	21 59	09 24	16 02	15 10
13 Su	02 58	10 34	06 11	06 05	22 01	05 05	21 59	09 23	16 01	15 10
14 Mo	02 35	16 13	06 43	05 36	21 54	05 02	22 00	09 21	16 00	15 10
15 Tu	02 11	20 59	07 12	05 07	21 47	04 59	22 00	09 20	16 00	15 09
16 We	01 47	24 39	07 37	04 37	21 40	04 57	22 00	09 19	15 59	15 09
17 Th	01 24	27 06	07 57	04 08	21 32	04 54	22 00	09 18	15 59	15 09
18 Fr	01 00	28 14	08 14	03 38	21 25	04 51	22 00	09 17	15 58	15 09
19 Sa	00 36	28 02	08 26	03 08	21 17	04 48	22 00	09 15	15 58	15 09
20 Su	00 12	26 33	08 34	02 38	21 09	04 46	22 01	09 14	15 57	15 09
21 Mo	00N11	23 53	08 38	02 08	21 00	04 43	22 01	09 13	15 57	15 08
22 Tu	00 35	20 11	08 37	01 38	20 52	04 40	22 01	09 12	15 56	15 08
23 We	00 59	15 37	08 32	01 08	20 43	04 37	22 01	09 11	15 56	15 08
24 Th	01 22	10 22	08 22	00 38	20 35	04 34	22 01	09 09	15 55	15 08
25 Fr	01 46	04 36	08 08	00 08	20 26	04 31	22 01	09 08	15 55	15 08
26 Sa	02 09	01S 26	07 51	00N22	20 17	04 28	22 01	09 07	15 54	15 08
27 Su	02 33	07 32	07 30	00 53	20 07	04 25	22 01	09 06	15 54	15 07
28 Mo	02 56	13 25	07 06	01 23	19 58	04 22	22 01	09 05	15 53	15 07
29 Tu	03 20	18 47	06 39	01 53	19 48	04 19	22 01	09 04	15 53	15 07
30 We	03 43	23 17	06 10	02 23	19 38	04 16	22 01	09 02	15 52	15 07
31 Th	04 06	26 32	05 39	02 53	19 28	04 13	22 01	09 01	15 52	15 07

ASPECTARIAN

01 01:09 ☽ △ ♅
02:05 ☽ ∥ ♇
06:18 ☽ ∥ ♆
08:11 ☽ △ ☉
17:35 ☽ □ ♆
18:09 ☽ ⚹ ♂
19:39 ☿ □ ♇
02 02:05 ☽ △ ♄
10:26 ☽ △ ☿
11:14 ☽ # ♄
17:22 ☽ ∥ ♂
03 04:30 ☽ □ ♀
06:58 ☽ □ ♅
10:15 ♂ □ ♃
22:34 ☽ ⚹ ♆
04 00:56 ☽ ⚹ ♃
08:44 ♀ ☌ ♅
12:41 ☽ ☌ ♇
21:46 ☽ □ ☿
05 01:34 ☿ ♈
10:19 ☽ ⚹ ♅
12:38 ☽ ⚹ ♀
17:39 ♀ ∥ ♅
06 00:05 ☽ ⚹ ☉
03:01 ☽ □ ♃
06:47 ☽ ☌ ♂
08:30 ☽ ☍ ♄
23:53 ☉ ∥ ♃
07 05:29 ☽ ⚹ ☿
15:50 ☽ ∥ ♂
16:16 ♂ ☍ ♄
19:00 ☽ # ♄
08 02:15 ☽ ☌ ♆
03:37 ☽ △ ♃
15:29 ☽ ⚹ ♇
21:43 ☽ ∥ ♆
09 01:17 ☽ ∥ ♇
12:43 ☽ ☌ ♅
22:04 ☿ # ☉
23:00 ☽ ∥ ♅
23:45 ☽ ☌ ♀
10 06:21 ☽ ∥ ♀
10:14 ☽ △ ♄
13:48 ☽ ⚹ ♂
14:19 ☽ ∥ ♃
15:59 ☽ # ☿
16:45 ☽ □ ♇
19:04 ☽ ∥ ☉
23:54 ☉ △ ♄
11 07:49 ☿ # ♃
18:08 ☽ ☌ ☿
21:00 ☽ # ☉
12 03:13 ☽ # ♃
05:27 ☽ ∥ ☿
06:05 ☽ ⚹ ♆
06:35 ☽ ☍ ♃
08:02 ☽ # ♀
13:14 ☽ □ ♄
19:22 ☽ # ♅
19:56 ☽ □ ♂
20:14 ☽ △ ♇
22:02 ☿ # ♀
13 05:00 ♂ # ♄
19:15 ☽ # ♇
20:01 ☽ ⚹ ♅
23:03 ☽ # ♆
14 07:46 ♃ △ ♆
12:10 ☽ □ ♆
19:07 ☽ ⚹ ♀
19:37 ☽ ⚹ ♄
23:15 ☉ □ ♇
15 00:11 ♀ △ ♄
03:33 ☽ ⚹ ☉
04:31 ☽ # ♂
05:56 ☽ ∥ ♄
06:11 ☽ △ ♂
06:40 ♀ ∥ ♃
16 04:52 ☽ □ ♅
15:36 ☽ ⚹ ☿
21:12 ☽ △ ♃
21:56 ☽ △ ♆
17 11:49 ☽ □ ♀
13:42 ☽ ☍ ♇
18 06:14 ♀ □ ♇
16:46 ☽ △ ♅
19 05:03 ☽ □ ☿
08:45 ☽ □ ♃
17:53 ☽ ☌ ♄
22:02 ☉ ⚹ ♂
20 00:15 ☿ SR
07:18 ☽ △ ♀
12:34 ☉ ♈
12:59 ☽ ☍ ♂
13:21 ☽ △ ☉
18:02 ♂ ♒
21 12:59 ☽ ∥ ♄
17:15 ☽ △ ☿
19:55 ☽ # ♂
20:36 ☽ ⚹ ♃
22:48 ☽ ☍ ♆
22 02:55 ♄ SD
14:20 ☽ △ ♇
16:25 ♀ ♈
22:32 ☽ # ♆
23 02:22 ☽ # ♇
04:13 ♀ # ☉
17:17 ☽ ☍ ♅
24 05:14 ☽ # ♅
08:53 ☽ ∥ ☿
16:46 ☽ ⚹ ♄
25 00:24 ☽ # ♃
00:36 ☽ □ ♇
08:42 ♀ ⚹ ♂
10:44 ☽ ∥ ☉
17:28 ☽ ∥ ♀
17:48 ☽ △ ♂
18:12 ☽ ☍ ♀
19:30 ☽ # ♀
26 03:05 ☽ # ☉
08:51 ☽ ☍ ☿
11:52 ☽ ∥ ♃
15:01 ☽ ☌ ♃
18:24 ☽ △ ♆
23:54 ☽ # ☿
27 01:01 ☽ □ ♄
02:28 ♇ SR
06:16 ☽ ∥ ♅
08:31 ☽ ⚹ ♇
28 04:12 ☽ □ ♂
07:18 ☽ ∥ ♇
09:48 ☽ △ ♅
10:40 ☽ ∥ ♆
29 00:49 ☽ □ ♆
04:49 ☽ ∥ ♂
07:07 ☽ △ ♄
16:11 ☿ ☌ ☉
16:39 ☽ # ♄
20:30 ☿ ☌ ♀
30 12:17 ☽ ⚹ ♂
14:25 ☽ △ ☿
14:58 ☽ □ ♅
17:23 ☽ △ ♀
17:35 ☽ △ ☉
31 01:05 ☽ ⚹ ♃
03:30 ♀ ☌ ☉
05:25 ☽ ⚹ ♆
06:25 ☉ # ♃
08:07 ☿ ⚹ ♂
18:25 ☽ ☌ ♇

⚷ Chiron

Day		
01	Dec.	13 S 37
02	00♒37	
05	00 50	
08	01 02	
11	01 14	
14	01 26	
17	01 37	
20	01 48	
23	01 58	
26	02 07	
29	02 16	

April 2005

08 20:17 19♈06 ☉ Total Solar Eclipse
24 09:56 04♏20 ✷ Penumbral Lunar Eclipse (mag 0.890)

Day	S. T.	☉	☽	☿	♀	♂	♃	♄	♅	♆	♇	☊ True
	h m s	° ' "	° ' "	° '	° '	° '	° '	° '	° '	° '	° '	° '
01 Fr	12 37 48	11♈21 39	27♐45 59	07♈R06	11♈35	08♒08	14♎R19	20♋29	08♓40	16♒58	24♐R30	22♈48
02 Sa	12 41 45	12 20 51	11♑52 35	06 17	12 49	08 52	14 11	20 30	08 43	16 59	24 30	22R48
03 Su	12 45 42	13 20 02	26 06 33	05 31	14 04	09 35	14 03	20 31	08 46	17 01	24 30	22 47
04 Mo	12 49 38	14 19 10	10♒25 24	04 47	15 18	10 19	13 56	20 33	08 49	17 02	24 30	22 46
05 Tu	12 53 35	15 18 17	24 45 55	04 07	16 33	11 02	13 48	20 34	08 52	17 04	24 29	22 45
06 We	12 57 31	16 17 22	09♓04 16	03 32	17 47	11 46	13 40	20 36	08 55	17 05	24 29	22 44
07 Th	13 01 28	17 16 25	23 16 10	03 01	19 01	12 29	13 32	20 38	08 58	17 06	24 29	22 43
08 Fr	13 05 24	18 15 26	07♈17 20	02 35	20 16	13 13	13 25	20 39	09 01	17 08	24 28	22 43
09 Sa	13 09 21	19 14 25	21 03 59	02 14	21 30	13 56	13 17	20 41	09 04	17 09	24 28	22D43
10 Su	13 13 17	20 13 22	04♉33 13	01 59	22 45	14 40	13 09	20 43	09 06	17 10	24 28	22 43
11 Mo	13 17 14	21 12 17	17 43 27	01 49	23 59	15 24	13 02	20 45	09 09	17 12	24 27	22 44
12 Tu	13 21 10	22 11 10	00♊34 32	01 45	25 13	16 07	12 54	20 48	09 12	17 13	24 27	22 45
13 We	13 25 07	23 10 01	13 07 38	01D46	26 28	16 51	12 47	20 50	09 15	17 14	24 26	22 46
14 Th	13 29 04	24 08 50	25 25 06	01 52	27 42	17 34	12 39	20 52	09 17	17 15	24 26	22 46
15 Fr	13 33 00	25 07 36	07♋30 15	02 03	28 56	18 18	12 32	20 55	09 20	17 16	24 25	22 46
16 Sa	13 36 57	26 06 20	19 27 09	02 19	00♉10	19 01	12 24	20 57	09 23	17 17	24 24	22 47
17 Su	13 40 53	27 05 02	01♌20 17	02 40	01 25	19 45	12 17	21 00	09 25	17 18	24 24	22 47
18 Mo	13 44 50	28 03 42	13 14 19	03 06	02 39	20 28	12 10	21 03	09 28	17 19	24 23	22 47
19 Tu	13 48 46	29 02 20	25 13 52	03 35	03 53	21 12	12 03	21 06	09 30	17 21	24 22	22 48
20 We	13 52 43	00♉00 55	07♍23 07	04 09	05 07	21 56	11 56	21 09	09 33	17 21	24 22	22 48
21 Th	13 56 39	00 59 28	19 45 47	04 47	06 21	22 39	11 49	21 12	09 35	17 22	24 21	22 49
22 Fr	14 00 36	01 57 59	02♎24 45	05 29	07 36	23 23	11 42	21 15	09 38	17 23	24 20	22 49
23 Sa	14 04 33	02 56 28	15 21 58	06 14	08 50	24 06	11 35	21 19	09 40	17 24	24 19	22R49
24 Su	14 08 29	03 54 55	28 38 06	07 03	10 04	24 50	11 28	21 22	09 42	17 25	24 19	22 49
25 Mo	14 12 26	04 53 20	12♏12 28	07 55	11 18	25 33	11 22	21 25	09 45	17 26	24 18	22 48
26 Tu	14 16 22	05 51 44	26 02 59	08 50	12 32	26 17	11 15	21 29	09 47	17 27	24 17	22 46
27 We	14 20 19	06 50 05	10♐06 29	09 48	13 46	27 00	11 09	21 33	09 49	17 28	24 16	22 44
28 Th	14 24 15	07 48 25	24 19 02	10 49	15 00	27 44	11 02	21 37	09 51	17 28	24 15	22 42
29 Fr	14 28 12	08 46 44	08♑36 29	11 53	16 14	28 28	10 56	21 40	09 54	17 29	24 14	22 40
30 Sa	14 32 08	09 45 01	22 54 56	13 00	17 28	29 11	10 50	21 44	09 56	17 30	24 13	22 38

Data for	04-01-2005
Julian Day	2453461.50
Ayanamsa	23 55 41
SVP	05 ♓ 11 04
☽ ☊ Mean	23 ♈ 33 R

● ◐ PHASES ○ ◑

02	00:51	◑	12♑23
08	20:32	☉	19♈06
16	14:38	◐	26♋42
24	10:07	✷	04♏20

LAST ASPECT ☽ Day	h m	INGRESS Day	h m	
31	18:25	01	3:49	♑
02	14:35	03	06:32	♒
04	23:32	05	08:46	♓
07	02:04	07	11:29	♈
09	06:00	09	15:50	♉
11	05:37	11	22:55	♊
14	05:01	14	09:03	♋
16	14:38	16	21:18	♌
19	08:13	19	09:28	♍
21	08:45	21	19:28	♎
23	16:47	24	02:26	♏
26	00:25	26	06:47	♐
28	06:03	28	09:33	♑
29	22:01	30	11:54	♒

DECLINATION

Day	☉	☽	☿	♀	♂	♃	♄	♅	♆	♇
01 Fr	04N30	28S12	05N08	03N23	19S18	04S10	22N01	09S00	15S51	15S07
02 Sa	04 53	28 04	04 36	03 53	19 07	04 07	22 01	08 59	15 51	15 06
03 Su	05 16	26 06	04 04	04 23	18 57	04 04	22 01	08 58	15 51	15 06
04 Mo	05 39	22 27	03 33	04 53	18 46	04 01	22 00	08 57	15 50	15 06
05 Tu	06 02	17 29	03 03	05 23	18 35	03 59	22 00	08 56	15 50	15 06
06 We	06 24	11 33	02 34	05 52	18 24	03 56	22 00	08 55	15 49	15 06
07 Th	06 47	05 03	02 07	06 22	18 13	03 53	22 00	08 54	15 49	15 05
08 Fr	07 10	01N36	01 41	06 51	18 01	03 50	22 00	08 53	15 48	15 05
09 Sa	07 32	08 05	01 19	07 20	17 50	03 47	22 00	08 51	15 48	15 05
10 Su	07 54	14 04	00 58	07 49	17 38	03 44	21 59	08 50	15 48	15 05
11 Mo	08 16	19 16	00 40	08 18	17 26	03 41	21 59	08 49	15 47	15 05
12 Tu	08 38	23 27	00 25	08 47	17 14	03 38	21 59	08 48	15 47	15 05
13 We	09 00	26 25	00 12	09 15	17 02	03 35	21 59	08 47	15 47	15 04
14 Th	09 22	28 03	00 02	09 44	16 49	03 32	21 58	08 46	15 46	15 04
15 Fr	09 43	28 19	00S06	10 12	16 37	03 29	21 58	08 45	15 46	15 04
16 Sa	10 05	27 15	00 11	10 39	16 24	03 27	21 58	08 44	15 46	15 04
17 Su	10 26	24 57	00 14	11 07	16 12	03 24	21 57	08 44	15 45	15 04
18 Mo	10 47	21 35	00 14	11 34	15 59	03 21	21 57	08 43	15 45	15 04
19 Tu	11 08	17 18	00 11	12 01	15 46	03 18	21 57	08 42	15 45	15 04
20 We	11 29	12 17	00 07	12 28	15 33	03 16	21 56	08 41	15 44	15 03
21 Th	11 49	06 42	00 00	12 54	15 20	03 13	21 56	08 40	15 44	15 03
22 Fr	12 09	00 44	00N09	13 20	15 06	03 11	21 55	08 39	15 44	15 03
23 Sa	12 30	05S25	00 20	13 46	14 53	03 08	21 55	08 38	15 44	15 03
24 Su	12 49	11 30	00 33	14 12	14 39	03 06	21 54	08 37	15 43	15 03
25 Mo	13 09	17 11	00 48	14 37	14 25	03 03	21 54	08 36	15 43	15 03
26 Tu	13 29	22 05	01 05	15 01	14 12	03 01	21 54	08 36	15 43	15 02
27 We	13 48	25 48	01 23	15 26	13 58	02 58	21 53	08 35	15 43	15 02
28 Th	14 07	27 57	01 43	15 50	13 44	02 56	21 53	08 34	15 43	15 02
29 Fr	14 26	28 15	02 05	16 13	13 29	02 54	21 52	08 33	15 42	15 02
30 Sa	14 44	26 41	02 29	16 37	13 15	02 52	21 51	08 32	15 42	15 02

ASPECTARIAN

01 15:02 ☽ □ ☿
16:42 ☿ ∥ ☉
18:38 ☽ ✶ ♅
02 01:45 ☽ □ ♀
03:52 ☽ □ ♃
10:18 ♀ ⋕ ♃
14:35 ☽ ☍ ♄
16:34 ☿ ∥ ♀
23:45 ☿ ⋕ ♃
23:52 ♀ ☍ ♃
08:04 ☽ ⋕ ♃
10:31 ☽ ☍ ♃
10:50 ☽ ✶ ♂
17:07 ☽ ✶ ♆
20:59 ☽ ∥ ♀
21:48 ☽ ∥ ☉
23:20 ☽ □ ♄
03 15:00 ☽ ✶ ☿
15:30 ☉ ☍ ♃
23:48 ☽ ☌ ♂
04 02:26 ☽ ⋕ ♄
05:48 ☽ △ ♃
07:00 ☽ ✶ ☉
08:56 ☽ ✶ ♀
11:05 ☽ ☌ ♆
18:58 ☽ ∥ ♂
23:32 ☽ ✶ ♇
05 07:02 ☽ ∥ ♆
10:02 ☽ ∥ ♇
10:11 ♀ ✶ ♆
23:44 ☽ ☌ ♅
06 09:57 ☽ ∥ ♅
18:05 ☽ ⋕ ☉
19:30 ☽ △ ♄
19:36 ☽ ⋕ ♀
19:49 ☉ ✶ ♆
07 02:04 ☽ □ ♇
04:19 ☽ ∥ ♃
11:22 ☽ ⋕ ☿
16:08 ☽ ☌ ☿
08 00:20 ☽ ∥ ☿
05:32 ♂ [illegible] ♅
07:46 ♀ □ ♄

09 00:51 ☽ ☌ ♀
03:00 ☽ ⋕ ♅
06:00 ☽ △ ♇
10 04:26 ☽ ⋕ ♇
07:34 ☽ ⋕ ♆
08:16 ☽ ✶ ♅
12:36 ☉ □ ♄
15:25 ☽ ⋕ ♂
17:04 ♀ ∥ ☉
19:27 ☽ □ ♂
23:01 ☽ □ ♆
11 05:37 ☽ ✶ ♄
09:05 ♀ △ ♇
14:52 ☽ ∥ ♄
12 01:11 ♀ ⋕ ♅
02:13 ☽ ✶ ☿
07:47 ☿ SD
10:33 ☉ ⋕ ♅
16:29 ☽ □ ♅
23:20 ☽ △ ♃
13 07:40 ☽ △ ♂
07:59 ☽ △ ♆
13:12 ♂ ☌ ♆
21:17 ☽ [illegible] ☉
22:03 ☽ ☍ ♇

14 05:01 ☽ ✶ ♀
06:50 ☉ △ ♇
12:57 ☽ □ ☿
15 03:40 ☽ △ ♅
09:58 ☽ □ ♃
20:37 ♀ ♉
16 03:03 ☽ ☌ ♄
17 00:10 ☽ □ ♀
02:47 ☽ △ ☿
21:40 ☽ ∥ ♄
21:52 ☽ ✶ ♃
18 08:13 ☽ ☍ ♆
15:27 ☽ ☍ ♂
22:18 ☽ △ ♇
19 02:13 ♂ ∥ ♆
07:49 ☽ ⋕ ♆
08:05 ☽ ⋕ ♂
08:13 ☽ △ ☉
11:09 ☽ ⋕ ♇
19:04 ☽ △ ♀
23:16 ☽ ∥ ♀
23:37 ☉ ♉
20 03:25 ☽ ∥ ☉
04:14 ☽ ☍ ♅
15:47 ☽ ⋕ ♅
21 02:46 ☽ ✶ ♄
08:45 ☽ □ ♇
14:15 ☽ ⋕ ♃
22 02:14 ☽ ∥ ☿
03:33 ☽ ⋕ ☿
05:35 ♂ ∥ ♇

06:05 ☽ ☍ ☿
15:10 ☽ ∥ ♃
17:07 ☽ ☌ ♃
23 03:44 ☽ △ ♆
07:08 ♂ ✶ ♇
10:52 ☽ □ ♄
12:34 ☽ ∥ ♅
16:15 ☽ ✶ ♇
16:47 ☽ △ ♂
16:52 ♀ ✶ ♅
24 05:46 ☽ ⋕ ☉
12:01 ☽ ⋕ ♀
12:33 ☽ ∥ ♂
14:44 ☽ ∥ ♇

16:57 ♀ ⋕ ♂
17:38 ☽ ∥ ♆
19:40 ☽ △ ♅
22:15 ☽ ☍ ♀
25 09:07 ☽ □ ♆
16:06 ☽ △ ♄
23:00 ☽ ⋕ ♄
26 00:25 ☽ □ ♂
00:59 ♀ ⋕ ♇
23:26 ☽ △ ☿
23:31 ☽ □ ♅
27 01:45 ☽ ✶ ♃
07:06 ☉ ⋕ ♂

12:27 ☽ ✶ ♆
16:42 ♀ ⋕ ♆
23:53 ☽ ☌ ♇
28 04:37 ☿ ☍ ♃
06:03 ☽ ✶ ♂
29 00:18 ☽ △ ☉
02:10 ☽ ✶ ♅
03:53 ☽ □ ♃
05:57 ☽ □ ☿
14:00 ☽ △ ♀
22:01 ☽ ☍ ♄
30 00:35 ♀ □ ♆
04:33 ☉ [illegible] ♅
20:29 ☿ ⋕ ♃
23:06 ☉ ⋕ ♇

⚷ Chiron

01 Dec. 13 S 01

01	02♒25
04	02 32
07	02 40
10	02 46
13	02 52
16	02 58
19	03 02
22	03 07
25	03 10
28	03 13

May 2005

Day	S. T.	☉	☽	☿	♀	♂	♃	♄	♅	♆	♇	☊ True
	h m s	° ' "	° ' "	° '	° '	° '	° '	° '	° '	° '	° '	° '
01 Su	14 36 05	10♉ 43 16	07♒ 10 56	14♈ 09	18♉ 42	29♒ 55	10♎ R44	21♋ 48	09♓ 58	17♒ 30	24♐ R12	22♈ R37
02 Mo	14 40 02	11 41 30	21 21 46	15 21	19 56	00♓ 38	10 39	21 53	10 00	17 31	24 11	22D 37
03 Tu	14 43 58	12 39 43	05♓ 25 18	16 35	21 10	01 22	10 33	21 57	10 02	17 31	24 10	22 38
04 We	14 47 55	13 37 54	19 19 59	17 51	22 24	02 05	10 27	22 01	10 04	17 32	24 09	22 39
05 Th	14 51 51	14 36 03	03♈ 04 36	19 10	23 38	02 49	10 22	22 05	10 06	17 33	24 08	22 40
06 Fr	14 55 48	15 34 11	16 38 02	20 31	24 52	03 32	10 17	22 10	10 08	17 33	24 07	22 41
07 Sa	14 59 44	16 32 18	29 59 14	21 55	26 05	04 15	10 12	22 14	10 10	17 33	24 05	22R 41
08 Su	15 03 41	17 30 23	13♉ 07 17	23 20	27 19	04 59	10 07	22 19	10 11	17 34	24 04	22 40
09 Mo	15 07 37	18 28 26	26 01 28	24 48	28 33	05 42	10 02	22 24	10 13	17 34	24 03	22 38
10 Tu	15 11 34	19 26 28	08♊ 41 35	26 18	29 47	06 26	09 57	22 29	10 15	17 35	24 02	22 35
11 We	15 15 31	20 24 28	21 08 05	27 49	01♊ 01	07 09	09 53	22 34	10 16	17 35	24 01	22 31
12 Th	15 19 27	21 22 26	03♋ 22 12	29 23	02 15	07 52	09 48	22 38	10 18	17 35	23 59	22 26
13 Fr	15 23 24	22 20 23	15 25 59	00♉ 59	03 28	08 36	09 44	22 43	10 20	17 35	23 58	22 22
14 Sa	15 27 20	23 18 18	27 22 21	02 37	04 42	09 19	09 40	22 49	10 21	17 36	23 57	22 19
15 Su	15 31 17	24 16 11	09♌ 14 57	04 17	05 56	10 02	09 36	22 54	10 23	17 36	23 56	22 17
16 Mo	15 35 13	25 14 03	21 08 07	06 00	07 10	10 45	09 33	22 59	10 24	17 36	23 54	22 16
17 Tu	15 39 10	26 11 52	03♍ 06 32	07 44	08 23	11 28	09 29	23 04	10 26	17 36	23 53	22D 16
18 We	15 43 06	27 09 40	15 15 03	09 30	09 37	12 12	09 26	23 10	10 27	17 36	23 52	22 18
19 Th	15 47 03	28 07 26	27 38 18	11 19	10 51	12 55	09 23	23 15	10 28	17 36	23 50	22 19
20 Fr	15 50 60	29 05 11	10♎ 20 28	13 09	12 04	13 38	09 20	23 21	10 30	17R 36	23 49	22 20
21 Sa	15 54 56	00♊ 02 54	23 24 50	15 01	13 18	14 21	09 17	23 26	10 31	17 36	23 47	22R 20
22 Su	15 58 53	01 00 36	06♏ 53 16	16 56	14 31	15 04	09 14	23 32	10 32	17 36	23 46	22 18
23 Mo	16 02 49	01 58 16	20 45 40	18 52	15 45	15 47	09 12	23 38	10 33	17 36	23 45	22 15
24 Tu	16 06 46	02 55 54	04♐ 59 33	20 51	16 59	16 30	09 10	23 44	10 34	17 36	23 43	22 10
25 We	16 10 42	03 53 32	19 30 07	22 51	18 12	17 13	09 07	23 49	10 35	17 36	23 42	22 04
26 Th	16 14 39	04 51 09	04♑ 10 46	24 54	19 26	17 55	09 05	23 55	10 36	17 36	23 40	21 57
27 Fr	16 18 36	05 48 44	18 54 04	26 58	20 39	18 38	09 04	24 01	10 37	17 35	23 39	21 51
28 Sa	16 22 32	06 46 19	03♒ 32 52	29 03	21 53	19 21	09 02	24 07	10 38	17 35	23 37	21 46
29 Su	16 26 29	07 43 52	18 01 23	01♊ 10	23 06	20 04	09 01	24 14	10 39	17 35	23 36	21 42
30 Mo	16 30 25	08 41 25	02♓ 15 41	03 19	24 20	20 46	08 59	24 20	10 40	17 35	23 34	21 40
31 Tu	16 34 22	09 38 57	16 13 49	05 29	25 33	21 29	08 58	24 26	10 40	17 34	23 33	21D 40

Data for	05-01-2005
Julian Day	2453491.50
Ayanamsa	23 55 45
SVP	05 ♓ 11 00
☽ ☊ Mean	21 ♈ 58 R

● ◐ PHASES ○ ◑

Day	Time	Phase	Position
01	06:24	◑	10♒59
08	08:45	●	17♉52
16	08:58	◐	25♌36
23	20:18	○	02♐47
30	11:47	◑	09♓10

LAST ASPECT Day	h m	☽ INGRESS Day	h m	Sign
02	04:47	02	14:43	♓
04	08:22	04	18:36	♈
06	13:22	07	00:01	♉
09	05:16	09	07:29	♊
11	14:59	11	17:21	♋
13	15:05	14	05:18	♌
16	08:58	16	17:48	♍
19	01:00	19	04:31	♎
21	00:41	21	11:49	♏
23	04:55	23	15:39	♐
25	06:52	25	17:11	♑
27	15:22	27	18:10	♒
29	09:19	29	20:09	♓

DECLINATION

Day	☉	☽	☿	♀	♂	♃	♄	♅	♆	♇
01 Su	15N02	23S24	02N53	16N59	13S01	02S49	21N51	08S32	15S42	15S02
02 Mo	15 21	18 45	03 20	17 22	12 46	02 47	21 50	08 31	15 42	15 02
03 Tu	15 38	13 07	03 47	17 43	12 32	02 45	21 50	08 30	15 42	15 01
04 We	15 56	06 53	04 16	18 05	12 17	02 43	21 49	08 29	15 41	15 01
05 Th	16 13	00 24	04 46	18 26	12 02	02 41	21 48	08 29	15 41	15 01
06 Fr	16 30	06N01	05 18	18 46	11 48	02 39	21 48	08 28	15 41	15 01
07 Sa	16 47	12 05	05 50	19 06	11 33	02 38	21 47	08 27	15 41	15 01
08 Su	17 03	17 31	06 24	19 26	11 18	02 36	21 46	08 27	15 41	15 01
09 Mo	17 20	22 03	06 58	19 44	11 03	02 34	21 46	08 26	15 41	15 01
10 Tu	17 35	25 28	07 34	20 03	10 48	02 32	21 45	08 26	15 41	15 01
11 We	17 51	27 35	08 10	20 21	10 33	02 31	21 44	08 25	15 41	15 01
12 Th	18 06	28 19	08 48	20 38	10 17	02 29	21 44	08 24	15 41	15 00
13 Fr	18 21	27 41	09 26	20 55	10 02	02 28	21 43	08 24	15 41	15 00
14 Sa	18 36	25 46	10 04	21 11	09 47	02 27	21 42	08 23	15 40	15 00
15 Su	18 50	22 44	10 44	21 26	09 31	02 25	21 41	08 23	15 40	15 00
16 Mo	19 04	18 46	11 24	21 41	09 16	02 24	21 40	08 22	15 40	15 00
17 Tu	19 18	14 02	12 04	21 56	09 00	02 23	21 40	08 22	15 40	15 00
18 We	19 32	08 43	12 45	22 09	08 45	02 22	21 39	08 21	15 40	15 00
19 Th	19 45	02 57	13 26	22 23	08 29	02 21	21 38	08 21	15 40	15 00
20 Fr	19 57	03S06	14 08	22 35	08 13	02 20	21 37	08 20	15 40	15 00
21 Sa	20 10	09 11	14 49	22 47	07 58	02 19	21 36	08 20	15 40	15 00
22 Su	20 22	15 03	15 31	22 58	07 42	02 18	21 35	08 19	15 40	15 00
23 Mo	20 33	20 20	16 12	23 09	07 26	02 17	21 34	08 19	15 40	14 59
24 Tu	20 45	24 35	16 53	23 19	07 10	02 17	21 33	08 19	15 41	14 59
25 We	20 56	27 21	17 34	23 28	06 55	02 16	21 33	08 18	15 41	14 59
26 Th	21 06	28 16	18 14	23 36	06 39	02 16	21 32	08 18	15 41	14 59
27 Fr	21 17	27 12	18 53	23 44	06 23	02 15	21 31	08 18	15 41	14 59
28 Sa	21 27	24 16	19 32	23 51	06 07	02 15	21 30	08 17	15 41	14 59
29 Su	21 36	19 49	20 09	23 58	05 51	02 14	21 29	08 17	15 41	14 59
30 Mo	21 45	14 20	20 45	24 04	05 35	02 14	21 28	08 17	15 41	14 59
31 Tu	21 54	08 12	21 19	24 09	05 19	02 14	21 27	08 17	15 41	14 59

ASPECTARIAN

Day	Time	Aspect
01	02:58	♂ ♓
	05:58	☽ △ ♃
	08:44	☽ # ♄
	12:51	☽ ✶ ☿
	17:28	☽ ☌ ♆
	21:20	☽ □ ♀
02	04:47	☽ ✶ ♇
	05:52	☽ # ♀
	13:26	☽ ∥ ♆
	14:11	☽ # ☉
	16:14	☽ ∥ ♇
	16:40	☽ ☌ ♂
03	02:26	☽ ∥ ♂
	04:23	☉ # ♆
	07:56	☽ ☌ ♅
	13:23	☽ ✶ ☉
	16:11	♀ ✶ ♄
	17:55	☿ ✶ ♆
	17:57	☽ ∥ ♅
04	04:41	☽ △ ♄
	05:51	☽ ✶ ♀
	08:22	☽ □ ♇
	09:02	☽ # ☿
	15:32	☽ ∥ ♃
05	11:24	☽ # ♃
	12:47	☽ ☍ ♃
	21:00	☽ ∥ ☿
06	01:38	☽ ✶ ♆
	07:45	☽ ☌ ☿
	09:26	☽ # ♅
	09:57	☽ □ ♄
	13:22	☽ △ ♆
	21:52	☽ # ♂
07	05:55	☿ □ ♄
	08:13	☽ ✶ ♂
	12:32	☽ # ♇
	15:30	☽ # ♆
	18:36	☽ ✶ ♅
	21:41	☽ ∥ ☉
08	01:28	☉ □ ♆
	08:13	☽ □ ♆
	10:08	☽ ∥ ♀
	11:59	☿ △ ♇
	17:10	☽ ✶ ♄
	22:16	☽ ∥ ♄
09	05:16	☽ ☌ ♀
	19:25	☽ □ ♂
10	02:24	☽ △ ♃
	02:59	☽ □ ♅
	04:15	♀ ♊
	17:06	☽ △ ♆
11	05:36	☽ ☍ ♇
	09:17	☿ # ♅
	14:59	☽ ✶ ☿
12	09:14	☿ ♉
	09:29	☽ △ ♂
	12:42	☽ □ ♃
	13:47	☽ △ ♅
13	10:30	☉ ✶ ♄
	14:44	☽ ☌ ♄
	15:05	☽ ✶ ☉
	16:09	☿ # ♂
14	12:19	☽ □ ☿
	16:31	☽ ✶ ♀
15	00:43	☽ ✶ ♃
	06:54	☽ ∥ ♄
	07:54	☽ ∥ ♀
	11:53	♂ ☌ ♅
	16:52	☽ ☍ ♆
	22:25	☽ ∥ ☉
	22:33	♀ ∥ ♄
16	05:34	☽ △ ♇
	16:04	☽ # ♆
	19:22	☽ # ♇
17	08:05	☽ ∥ ☿
	10:45	☽ △ ☿
	11:40	☽ □ ♀
	14:32	☽ ☍ ♅
	17:37	☽ ☍ ♂
	20:34	♀ △ ♃
	23:49	☽ # ♂
18	01:31	☽ # ♅
	12:47	☿ ✶ ♅
	15:30	☽ ✶ ♄
	16:36	♀ □ ♅
	16:42	☽ □ ♇
19	01:00	☽ △ ☉
	02:23	☽ # ♃
	13:09	♂ ∥ ♅
	20:59	☽ ∥ ♃
	22:07	☽ ☌ ♃
	23:37	♆ SR
20	03:33	☽ △ ♀
	10:02	☿ ✶ ♂
	13:25	☽ △ ♆
	19:21	☽ ∥ ♂
	20:37	☽ ∥ ♅
	22:48	☉ ♊
21	00:03	☽ □ ♄
	00:41	☽ ✶ ♇
	06:04	☿ # ♇
	23:46	☽ ∥ ♇
22	02:16	☽ # ☿
	02:41	☽ ∥ ♆
	05:40	☿ # ♆
	06:23	☽ △ ♅
	08:20	☿ □ ♆
	15:00	☽ △ ♂
	18:36	☽ □ ♆
	20:15	☽ ☍ ☿
23	01:12	☽ # ☉
	01:21	♀ □ ♂
	04:55	☽ △ ♄
	06:21	☽ # ♄
	15:44	☽ # ♀
24	06:55	☽ ✶ ♃
	09:17	☽ □ ♅
	12:13	♀ △ ♆
	20:02	☽ □ ♂
	20:52	☽ ✶ ♆
	21:40	☽ ☍ ♀
25	06:52	☽ ☌ ♇
	12:03	☿ ✶ ♄
26	07:59	☽ □ ♃
	10:29	☽ ✶ ♅
	23:33	☽ ✶ ♂
27	08:26	☽ ☍ ♄
	15:22	☽ △ ☿
28	02:26	☽ # ♀
	05:42	☽ △ ☉
	07:01	☉ ∥ ♄
	09:03	☽ △ ♃
	10:44	☿ ♊
	15:28	☽ # ☉
	15:47	☽ # ♄
	22:37	☽ # ☿
	23:16	☽ ☌ ♆
29	09:18	☽ △ ♀
	09:19	☽ ✶ ♇
	09:31	♀ ☍ ♇
	18:22	☽ ∥ ♆
	21:17	☽ ∥ ♇
30	02:07	☽ □ ☿
	07:23	☉ △ ♃
	14:23	☽ ☌ ♅
	23:41	☽ ∥ ♅
31	05:14	☿ ∥ ♄
	09:39	☽ ☌ ♂
	11:18	☽ ∥ ♂
	12:44	☽ □ ♇
	14:26	☽ △ ♄
	17:53	☽ □ ♀
	22:21	☽ ∥ ♃

⚷ Chiron

01	Dec.	12 S 36
	01	03 ♒ 15
	04	03 16
	07	03 17
	10	03 17R
	13	03 17
	16	03 16
	19	03 14
	22	03 11
	25	03 08
	28	03 04
	31	03 00
08	23:47	03♒17 R

June 2005

Day	S. T.	☉	☽	☿	♀	♂	♃	♄	♅	♆	♇	☊ True
	h m s	° ' "	° ' "	° '	° '	° '	° '	° '	° '	° '	° '	° '
01 We	16 38 18	10♊ 36 28	29♓ 55 29	07♊ 40	26♊ 46	22♓ 12	08♎R58	24♋ 32	10♓ 41	17♒R34	23♐R31	21♈ 41
02 Th	16 42 15	11 33 58	13♈ 21 30	09 51	28 00	22 54	08 57	24 39	10 42	17 34	23 30	21 43
03 Fr	16 46 11	12 31 27	26 33 09	12 03	29 13	23 37	08 56	24 45	10 42	17 33	23 28	21R 43
04 Sa	16 50 08	13 28 56	09♉ 31 50	14 15	00♋ 27	24 19	08 56	24 51	10 43	17 33	23 27	21 42
05 Su	16 54 05	14 26 24	22 18 41	16 27	01 40	25 01	08 56	24 58	10 43	17 32	23 25	21 39
06 Mo	16 58 01	15 23 51	04♊ 54 32	18 39	02 53	25 44	08D 56	25 04	10 44	17 32	23 23	21 33
07 Tu	17 01 58	16 21 18	17 20 03	20 50	04 07	26 26	08 56	25 11	10 44	17 31	23 22	21 25
08 We	17 05 54	17 18 43	29 35 54	23 00	05 20	27 08	08 57	25 18	10 45	17 30	23 20	21 15
09 Th	17 09 51	18 16 08	11♋ 42 56	25 09	06 33	27 50	08 57	25 24	10 45	17 30	23 19	21 04
10 Fr	17 13 47	19 13 31	23 42 27	27 17	07 47	28 32	08 58	25 31	10 45	17 29	23 17	20 54
11 Sa	17 17 44	20 10 54	05♌ 36 22	29 23	09 00	29 14	08 59	25 38	10 45	17 29	23 16	20 45
12 Su	17 21 40	21 08 16	17 27 23	01♋ 28	10 13	29 56	09 00	25 45	10 46	17 28	23 14	20 38
13 Mo	17 25 37	22 05 36	29 18 59	03 31	11 26	00♈ 37	09 01	25 52	10 46	17 27	23 12	20 34
14 Tu	17 29 34	23 02 56	11♍ 15 24	05 31	12 40	01 19	09 03	25 59	10 46	17 26	23 11	20 32
15 We	17 33 30	24 00 15	23 21 24	07 30	13 53	02 01	09 04	26 05	10R 46	17 26	23 09	20D 31
16 Th	17 37 27	24 57 33	05♎ 42 01	09 26	15 06	02 42	09 06	26 12	10 46	17 25	23 08	20 32
17 Fr	17 41 23	25 54 50	18 22 12	11 21	16 19	03 24	09 08	26 20	10 46	17 24	23 06	20R 32
18 Sa	17 45 20	26 52 06	01♏ 26 23	13 13	17 32	04 05	09 10	26 27	10 46	17 23	23 04	20 31
19 Su	17 49 16	27 49 22	14 57 47	15 03	18 45	04 46	09 13	26 34	10 45	17 22	23 03	20 28
20 Mo	17 53 13	28 46 37	28 57 37	16 50	19 58	05 27	09 15	26 41	10 45	17 21	23 01	20 22
21 Tu	17 57 09	29 43 51	13♐ 24 11	18 35	21 11	06 08	09 18	26 48	10 45	17 20	23 00	20 15
22 We	18 01 06	00♋ 41 05	28 12 36	20 18	22 24	06 49	09 21	26 55	10 45	17 19	22 58	20 05
23 Th	18 05 03	01 38 18	13♑ 14 55	21 59	23 37	07 30	09 24	27 02	10 44	17 18	22 57	19 55
24 Fr	18 08 59	02 35 31	28 21 23	23 37	24 50	08 11	09 27	27 10	10 44	17 17	22 55	19 45
25 Sa	18 12 56	03 32 44	13♒ 22 01	25 13	26 03	08 52	09 30	27 17	10 43	17 16	22 54	19 36
26 Su	18 16 52	04 29 56	28 08 18	26 46	27 16	09 32	09 34	27 24	10 43	17 15	22 52	19 30
27 Mo	18 20 49	05 27 09	12♓ 34 23	28 17	28 29	10 13	09 37	27 32	10 42	17 14	22 50	19 26
28 Tu	18 24 45	06 24 22	26 37 24	29 45	29 42	10 53	09 41	27 39	10 42	17 13	22 49	19 24
29 We	18 28 42	07 21 34	10♈ 17 14	01♌ 11	00♌ 55	11 33	09 45	27 47	10 41	17 12	22 47	19D 24
30 Th	18 32 38	08 18 47	23 35 38	02 35	02 08	12 13	09 49	27 54	10 41	17 11	22 46	19 25

Data for	06-01-2005
Julian Day	2453522.50
Ayanamsa	23 55 50
SVP	05 ♓ 10 56
☽ ☊ Mean	20 ♈ 20 R

LAST ASPECT Day	h m	☽ INGRESS Day	h m	
31	17:53	01	00:08	♈
03	05:25	03	06:20	♉
05	05:26	05	14:36	♊
07	18:51	08	00:47	♋
10	10:19	10	12:40	♌
12	11:40	13	01:23	♍
15	05:25	15	12:59	♎
17	15:03	17	21:24	♏
19	20:07	20	01:45	♐
21	15:35	22	02:52	♑
23	22:05	24	02:37	♒
25	15:24	26	03:04	♓
28	05:52	28	05:52	♈
30	07:58	30	11:46	♉

● ◐ PHASES ○ ◑

06	21:56	●	16♊16
15	01:22	◐	24♍04
22	04:14	○	00♑51
28	18:24	◑	07♈08

DECLINATION

Day	☉	☽	☿	♀	♂	♃	♄	♅	♆	♇
01 We	22N02	01S48	21N52	24N13	05S03	02S14	21N26	08S16	15S41	14S59
02 Th	22 10	04N35	22 22	24 17	04 47	02 14	21 24	08 16	15 41	14 59
03 Fr	22 18	10 38	22 51	24 20	04 31	02 14	21 23	08 16	15 42	14 59
04 Sa	22 25	16 09	23 17	24 22	04 15	02 14	21 22	08 16	15 42	14 59
05 Su	22 32	20 52	23 41	24 23	03 59	02 14	21 21	08 16	15 42	14 59
06 Mo	22 38	24 33	24 02	24 24	03 43	02 14	21 20	08 15	15 42	14 59
07 Tu	22 44	27 01	24 20	24 24	03 27	02 15	21 19	08 15	15 42	14 59
08 We	22 50	28 09	24 36	24 24	03 11	02 15	21 18	08 15	15 42	14 59
09 Th	22 55	27 54	24 49	24 22	02 55	02 16	21 17	08 15	15 43	14 59
10 Fr	23 00	26 21	24 59	24 20	02 40	02 16	21 16	08 15	15 43	14 59
11 Sa	23 05	23 37	25 07	24 17	02 24	02 17	21 14	08 15	15 43	14 59
12 Su	23 09	19 55	25 11	24 14	02 08	02 17	21 13	08 15	15 43	14 59
13 Mo	23 12	15 25	25 13	24 10	01 52	02 18	21 12	08 15	15 43	14 59
14 Tu	23 15	10 19	25 13	24 05	01 36	02 19	21 11	08 15	15 44	14 59
15 We	23 18	04 47	25 09	23 59	01 20	02 20	21 10	08 15	15 44	14 59
16 Th	23 21	01S04	25 04	23 53	01 05	02 21	21 08	08 15	15 44	14 59
17 Fr	23 23	07 01	24 56	23 46	00 49	02 22	21 07	08 15	15 45	14 59
18 Sa	23 24	12 52	24 46	23 38	00 33	02 23	21 06	08 15	15 45	14 59
19 Su	23 25	18 19	24 34	23 30	00 18	02 24	21 04	08 15	15 45	14 59
20 Mo	23 26	23 00	24 21	23 21	00 02	02 25	21 03	08 15	15 45	14 59
21 Tu	23 26	26 24	24 05	23 11	00N13	02 27	21 02	08 16	15 46	14 59
22 We	23 26	28 06	23 48	23 00	00 29	02 28	21 01	08 16	15 46	14 59
23 Th	23 26	27 45	23 29	22 49	00 44	02 29	20 59	08 16	15 46	14 59
24 Fr	23 25	25 21	23 09	22 37	01 00	02 31	20 58	08 16	15 47	14 59
25 Sa	23 24	21 13	22 48	22 25	01 15	02 32	20 57	08 16	15 47	14 59
26 Su	23 22	15 49	22 25	22 12	01 30	02 34	20 55	08 17	15 47	15 00
27 Mo	23 20	09 39	22 02	21 58	01 45	02 36	20 54	08 17	15 48	15 00
28 Tu	23 17	03 09	21 38	21 44	02 00	02 37	20 52	08 17	15 48	15 00
29 We	23 14	03N20	21 12	21 29	02 15	02 39	20 51	08 17	15 48	15 00
30 Th	23 11	09 30	20 46	21 14	02 30	02 41	20 50	08 18	15 49	15 00

ASPECTARIAN

01 01:58 ☉ □ ♅
11:04 ☿ ∥ ☉
14:11 ☿ △ ♃
15:04 ☽ ⋕ ♃
16:05 ☽ ☍ ♃
16:27 ☽ ✶ ☿
20:31 ☽ ✶ ☉
02 00:46 ☽ ⋕ ♂
07:35 ☽ ✶ ♆
09:18 ☿ □ ♅
14:23 ☽ ⋕ ♅
18:22 ☽ △ ♇
19:26 ♂ □ ♇
20:40 ☽ □ ♄
03 05:25 ☽ ✶ ♀
09:12 ☿ ☌ ☉
15:18 ♀ ♋
18:39 ☽ ⋕ ♇
21:54 ☽ ⋕ ♆
04 02:13 ☽ ✶ ♅
15:00 ☽ □ ♆
21:44 ♂ △ ♄
05 02:49 ☽ ∥ ♄
05:04 ☽ ✶ ♄
05:26 ☽ ✶ ♂
07:22 ♃ SD
10:16 ☽ ∥ ☉
11:49 ☿ △ ♆
19:32 ☽ ∥ ☿
22:51 ☽ ∥ ♀
06 07:44 ☽ △ ♃
11:13 ☽ □ ♅
07 00:21 ☽ △ ♆
05:29 ☿ ∥ ♀
08:17 ☽ ☌ ☿
11:44 ☽ ☍ ♇
18:31 ☽ □ [illegible]
08 03:42 ☿ ☍ ♇
04:54 ☉ △ ♆
12:36 ☽ ☌ ♀
18:30 ☽ □ ♃
22:05 ☽ △ ♅
10 03:41 ☽ ☌ ♄
10:19 ☽ △ ♂
12:31 ☽ ∥ ☿
18:47 ☽ ∥ ♀
21:18 ☿ □ ♂
23:39 ♀ □ ♃
11 03:50 ☽ ∥ ☉
06:50 ☽ ✶ ♃
07:03 ☿ ♋
10:07 ♂ ∥ ♃
16:08 ☽ ∥ ♄
12 00:01 ☽ ☍ ♆
02:30 ♂ ♈
08:07 ☽ ✶ ☉
10:42 ♀ △ ♅
11:40 ☽ △ ♇
22:30 ☽ ⋕ ♆
13 02:12 ☽ ⋕ ♇
10:11 ☽ ✶ ☿
23:01 ☽ ☍ ♅
14 03:07 ☽ ✶ ♀
03:14 ☉ ☍ ♇
09:09 ☽ ⋕ ♅
22:39 ♅ SR
23:36 ☽ □ ♇
15 05:25 ☽ ✶ ♄
10:08 ☽ ⋕ ♃
14:53 ☽ ⋕ ♂
17:33 ☽ ☌ [illegible]
19:42 ☿ □ ♃
16 00:03 ☽ ∥ ♂
05:11 ☽ ∥ ♃
06:32 ☽ ☌ ♃
08:27 ☽ □ ☿
16:36 ☿ △ ♅
19:45 ☽ □ ♀
22:11 ☽ △ ♆
17 04:59 ☽ ∥ ♅
08:46 ☽ ✶ ♇
14:50 ☽ □ ♄
15:03 ☽ △ ☉
18 08:59 ☽ ∥ ♇
12:20 ☽ ∥ ♆
16:38 ☽ △ ♅
19 00:10 ☽ △ ☿
04:11 ☽ □ ♆
07:13 ☽ △ ♀
10:24 ♀ ∥ ☉
13:25 ☽ ⋕ ♄
20:07 ☽ △ ♄
20 01:58 ☽ ⋕ ♀
02:36 ☽ ⋕ ☉
07:47 ☽ ⋕ ☿
11:25 ☽ △ ♂
17:13 ☽ ✶ ♃
19:39 ☽ □ ♅
21 06:25 ☽ ✶ ♆
06:46 ☉ ♋
15:35 ☽ ☌ ♇
22 14:25 ☽ □ ♂
17:51 ☽ □ ♃
20:01 ☽ ✶ ♅
23 04:23 ☿ ∥ ☉
15:33 ☽ ☍ ☿
17:55 ☽ ☍ ♀
22:05 ☽ ☍ ♄
24 12:26 ☽ ⋕ ☉
15:06 ☽ ⋕ ☿
16:25 ☽ ✶ ♂
17:33 ☽ ⋕ ♀
17:46 ☽ △ ♃
25 01:21 ☽ ⋕ ♄
06:18 ☽ ☌ ♆
15:24 ☽ ✶ ♇
26 00:07 ☽ ∥ ♆
01:04 [illegible] ☍ ♃
02:59 ♀ ☌ ♄
03:20 ☽ ∥ ♇
10:59 ☿ ☌ ♄
11:14 ☽ △ ☉
20:52 ☽ ☌ ♅
27 05:05 ☽ ∥ ♅
08:43 ☿ ∥ ♀
17:26 ☽ □ ♇
18:39 ☿ ☌ ♀
28 01:48 ☽ △ ♄
01:54 ☽ ∥ ♃
04:01 ☽ ⋕ ♂
04:02 ☿ ♌
05:52 ☽ △ ♀
05:53 ♀ ♌
06:05 ☽ △ ☿
19:48 ☽ ∥ ♂
21:26 ☽ ⋕ ♃
23:03 ☽ ☍ ♃
29 02:22 ☽ ☌ ♂
12:22 ☽ ✶ ♆
19:09 ☽ ⋕ ♅
20:56 ☿ ∥ ♄
22:29 ☽ △ ♇
30 07:58 ☽ □ ♄
17:20 ☽ □ ♀
18:29 ☽ □ ☿
19:36 ♂ ⋕ ♃
23:27 ☽ ⋕ ♇

⚷ Chiron

01 Dec.	12 S 24
03	02♒55R
06	02 50
09	02 44
12	02 37
15	02 30
18	02 23
21	02 15
24	02 06
27	01 57
30	01 48

July 2005

Day	S. T. h m s	☉ ° ' "	☽ ° ' "	☿ ° '	♀ ° '	♂ ° '	♃ ° '	♄ ° '	♅ ° '	♆ ° '	♇ ° '	☊ True ° '
01 Fr	18 36 35	09♋ 16 00	06♉ 35 21	03♌ 56	03♌ 21	12♈ 53	09♎ 54	28♋ 02	10♓R40	17♒R10	22♐R44	19♈R25
02 Sa	18 40 32	10 13 13	19 19 25	05 15	04 34	13 33	09 58	28 09	10 39	17 08	22 43	19 23
03 Su	18 44 28	11 10 26	01♊ 50 38	06 31	05 46	14 13	10 03	28 17	10 38	17 07	22 41	19 18
04 Mo	18 48 25	12 07 39	14 11 26	07 44	06 59	14 52	10 08	28 24	10 37	17 06	22 40	19 10
05 Tu	18 52 21	13 04 53	26 23 41	08 55	08 12	15 32	10 13	28 32	10 36	17 05	22 38	19 00
06 We	18 56 18	14 02 06	08♋ 28 54	10 03	09 25	16 11	10 18	28 39	10 35	17 03	22 37	18 47
07 Th	19 00 14	14 59 20	20 28 15	11 08	10 37	16 50	10 23	28 47	10 34	17 02	22 36	18 33
08 Fr	19 04 11	15 56 33	02♌ 22 56	12 10	11 50	17 29	10 28	28 55	10 33	17 01	22 34	18 19
09 Sa	19 08 08	16 53 47	14 14 22	13 09	13 03	18 08	10 34	29 02	10 32	16 59	22 33	18 06
10 Su	19 12 04	17 51 01	26 04 29	14 05	14 15	18 47	10 40	29 10	10 31	16 58	22 31	17 56
11 Mo	19 16 01	18 48 14	07♍ 55 54	14 58	15 28	19 25	10 45	29 17	10 30	16 57	22 30	17 49
12 Tu	19 19 57	19 45 28	19 52 02	15 47	16 41	20 04	10 51	29 25	10 29	16 55	22 29	17 45
13 We	19 23 54	20 42 41	01♎ 56 58	16 33	17 53	20 42	10 58	29 33	10 28	16 54	22 27	17 43
14 Th	19 27 50	21 39 55	14 15 23	17 15	19 06	21 20	11 04	29 41	10 26	16 52	22 26	17D 43
15 Fr	19 31 47	22 37 08	26 52 12	17 54	20 18	21 58	11 10	29 48	10 25	16 51	22 25	17R 43
16 Sa	19 35 43	23 34 22	09♏ 52 14	18 28	21 31	22 35	11 17	29 56	10 24	16 50	22 23	17 42
17 Su	19 39 40	24 31 36	23 19 34	18 59	22 43	23 13	11 23	00♌ 04	10 22	16 48	22 22	17 39
18 Mo	19 43 37	25 28 50	07♐ 16 34	19 25	23 55	23 50	11 30	00 11	10 21	16 47	22 21	17 34
19 Tu	19 47 33	26 26 04	21 42 58	19 47	25 08	24 27	11 37	00 19	10 19	16 45	22 20	17 26
20 We	19 51 30	27 23 18	06♑ 35 00	20 04	26 20	25 04	11 44	00 27	10 18	16 44	22 18	17 17
21 Th	19 55 26	28 20 33	21 45 19	20 17	27 32	25 41	11 51	00 35	10 16	16 42	22 17	17 06
22 Fr	19 59 23	29 17 49	07♒ 03 45	20 25	28 45	26 18	11 59	00 42	10 15	16 41	22 16	16 56
23 Sa	20 03 19	00♌ 15 04	22 18 58	20 28	29 57	26 54	12 06	00 50	10 13	16 39	22 15	16 47
24 Su	20 07 16	01 12 21	07♓ 20 27	20R 26	01♍ 09	27 30	12 14	00 58	10 11	16 37	22 14	16 40
25 Mo	20 11 12	02 09 38	22 00 16	20 19	02 21	28 06	12 21	01 06	10 10	16 36	22 13	16 35
26 Tu	20 15 09	03 06 56	06♈ 13 51	20 08	03 33	28 42	12 29	01 13	10 08	16 34	22 11	16 33
27 We	20 19 06	04 04 15	20 00 03	19 51	04 46	29 17	12 37	01 21	10 06	16 33	22 10	16D 32
28 Th	20 23 02	05 01 35	03♉ 20 20	19 29	05 58	29 52	12 45	01 29	10 04	16 31	22 09	16 33
29 Fr	20 26 59	05 58 56	16 17 46	19 03	07 10	00♉ 27	12 53	01 37	10 02	16 30	22 08	16R 33
30 Sa	20 30 55	06 56 18	28 56 10	18 32	08 22	01 02	13 02	01 44	10 01	16 28	22 07	16 32
31 Su	20 34 52	07 53 42	11♊ 19 24	17 58	09 34	01 37	13 10	01 52	09 59	16 26	22 06	16 29

Data for	07-01-2005
Julian Day	2453552.50
Ayanamsa	23 55 56
SVP	05 ♓ 10 54
☽ ☊ Mean	18 ♈ 44 R

● ◐ PHASES ○ ◑

Day	Time	Phase	Position
06	12:03	●	14♋31
14	15:20	◐	22♎16
21	11:01	○	28♑47
28	03:20	◑	05♉10

LAST ASPECT ☽ INGRESS

Day	h m	Day	h m	Sign
02	17:03	02	20:27	♊
04	16:36	05	07:08	♋
07	16:55	07	19:11	♌
09	16:49	10	07:57	♍
12	19:12	12	20:09	♎
15	05:32	15	05:51	♏
17	02:15	17	11:35	♐
19	06:04	19	13:27	♑
21	11:01	21	12:56	♒
23	07:34	23	12:13	♓
25	00:20	25	13:24	♈
27	17:24	27	17:55	♉
29	04:59	30	02:03	♊

DECLINATION

Day	☉	☽	☿	♀	♂	♃	♄	♅	♆	♇
01 Fr	23N07	15N07	20N20	20N58	02N45	02S43	20N48	08S18	15S49	15S00
02 Sa	23 03	19 58	19 53	20 41	03 00	02 45	20 47	08 18	15 49	15 00
03 Su	22 58	23 51	19 26	20 24	03 15	02 47	20 45	08 19	15 50	15 00
04 Mo	22 53	26 34	18 58	20 06	03 30	02 49	20 44	08 19	15 50	15 00
05 Tu	22 48	28 00	18 30	19 48	03 44	02 51	20 42	08 19	15 51	15 00
06 We	22 42	28 03	18 02	19 29	03 59	02 53	20 41	08 20	15 51	15 01
07 Th	22 36	26 48	17 34	19 09	04 13	02 56	20 40	08 20	15 51	15 01
08 Fr	22 29	24 20	17 06	18 49	04 27	02 58	20 38	08 21	15 52	15 01
09 Sa	22 22	20 51	16 38	18 29	04 41	03 00	20 37	08 21	15 52	15 01
10 Su	22 15	16 32	16 11	18 08	04 56	03 03	20 35	08 21	15 53	15 01
11 Mo	22 07	11 35	15 43	17 47	05 10	03 05	20 34	08 22	15 53	15 01
12 Tu	21 59	06 11	15 17	17 25	05 23	03 08	20 32	08 22	15 54	15 01
13 We	21 51	00 29	14 51	17 02	05 37	03 11	20 31	08 23	15 54	15 02
14 Th	21 42	05S21	14 26	16 40	05 51	03 13	20 29	08 23	15 54	15 02
15 Fr	21 33	11 07	14 01	16 16	06 05	03 16	20 27	08 24	15 55	15 02
16 Sa	21 23	16 35	13 37	15 53	06 18	03 19	20 26	08 25	15 55	15 02
17 Su	21 13	21 27	13 15	15 29	06 31	03 22	20 24	08 25	15 56	15 02
18 Mo	21 03	25 18	12 54	15 04	06 45	03 24	20 23	08 26	15 56	15 02
19 Tu	20 52	27 42	12 33	14 40	06 58	03 27	20 21	08 26	15 57	15 03
20 We	20 41	28 12	12 15	14 14	07 11	03 30	20 20	08 27	15 57	15 03
21 Th	20 30	26 37	11 58	13 49	07 24	03 33	20 18	08 27	15 58	15 03
22 Fr	20 18	23 03	11 42	13 23	07 36	03 36	20 16	08 28	15 58	15 03
23 Sa	20 06	17 56	11 29	12 57	07 49	03 39	20 15	08 29	15 59	15 03
24 Su	19 54	11 46	11 17	12 30	08 02	03 43	20 13	08 29	15 59	15 04
25 Mo	19 41	05 06	11 07	12 03	08 14	03 46	20 12	08 30	16 00	15 04
26 Tu	19 28	01N39	10 59	11 36	08 26	03 49	20 10	08 31	16 00	15 04
27 We	19 14	08 06	10 54	11 08	08 38	03 52	20 08	08 31	16 01	15 04
28 Th	19 01	14 00	10 51	10 41	08 50	03 56	20 07	08 32	16 01	15 04
29 Fr	18 47	19 07	10 50	10 13	09 02	03 59	20 05	08 33	16 02	15 05
30 Sa	18 32	23 15	10 52	09 44	09 14	04 02	20 03	08 34	16 02	15 05
31 Su	18 18	26 14	10 56	09 16	09 26	04 06	20 02	08 34	16 03	15 05

⚷ Chiron

01 Dec.	12 S 30
03	01♒39R
06	01 29
09	01 19
12	01 09
15	00 59
18	00 49
21	00 38
24	00 28
27	00 18
30	00 07

ASPECTARIAN

01 03:13 ☽ # ♆
05:25 ☽ ✶ ☉
07:37 ☽ ✶ ♅
14:52 ♀ ∥ ♄
17:11 ☉ □ ♃
19:52 ☽ □ ♆
23:33 ☽ ∥ ☿
02 03:41 ☽ ∥ ♀
04:28 ☽ ∥ ♄
10:40 ☉ △ ♅
17:03 ☽ ✶ ♄
18:01 ☽ ∥ ☉
03 08:26 ☽ ✶ ♀
10:03 ☽ ✶ ☿
16:01 ☽ △ ♃
17:02 ☽ □ ♅
04 01:25 ☽ ✶ ♂
05:41 ☽ △ ♆
16:36 ☽ ☍ ♇
06 03:39 ☽ □ ♃
04:12 ☽ △ ♅
05:52 ☿ ✶ ♃
16:17 ☽ □ ♂
18:52 ♀ ✶ ♃
07 07:00 ♂ ✶ ♆
16:55 ☽ ☌ ♄
08 14:00 ☽ ∥ ☉
16:30 ☽ ✶ ♃
21:18 ☽ ☌ ♀
21:36 ☽ ☌ ☿
09 01:30 ☽ ∥ ♄
05:34 ☽ ☍ ♆
08:21 ☽ △ ♂
09:43 ☿ ☌ ♀
14:51 ☽ ∥ ♀
16:49 ☽ △ ♇
10 02:01 ☽ ∥ ☿
03:20 ☽ # ♆
07:38 ☽ # ♇
15:34 ☿ # ♆
11 05:10 ☽ ☍ ♅
14:29 ☽ # ♅
23:46 ☽ ✶ ☉
12 03:16 ☽ ∥ ♂
04:46 ♀ ☍ ♆
05:12 ☽ □ ♇
12:52 ☽ # ♃
14:06 ☿ # ♇
19:12 ☽ ✶ ♄
23:00 ☉ □ ♂
13 11:14 ☿ ☍ ♆
15:13 ☽ ∥ ♃
17:46 ☽ ☌ ♃
14 02:11 ☽ # ♂
05:01 ☽ △ ♆
06:05 ☽ ✶ ☿
10:16 ☽ ✶ ♀
12:38 ☽ ∥ ♅
14:15 ☽ ☍ ♂
15:36 ☽ ✶ ♇
15 05:32 ☽ □ ♄
11:41 ☽ # ☿
16:35 ♂ △ ♇
17:01 ☽ ∥ ♇
21:00 ☽ ∥ ♆
21:02 ☽ # ♀
21:29 ♀ # ♆
16 00:57 ☽ △ ♅
12:29 ☽ □ ♆
12:31 ♄ ♌
16:03 ☽ □ ☿
17:12 ♀ △ ♇
18:32 ☽ # ♄
22:49 ☽ # ☉
22:50 ☽ □ ♀
17 02:15 ☽ △ ☉
11:48 ☽ △ ♄
20:27 ♀ △ ♂
18 01:51 ♀ # ♇
05:09 ☽ □ ♅
07:10 ☽ ✶ ♃
15:51 ☽ ✶ ♆
20:45 ☽ △ ☿
19 01:00 ☽ ☌ ♇
04:40 ☽ △ ♂
06:04 ☽ △ ♀
20 05:54 ☽ ✶ ♅
08:16 ☽ □ ♃
21 06:25 ☽ □ ♂
13:57 ☽ ☍ ♄
22 03:36 ☉ ∥ ♄
07:46 ☽ △ ♃
13:53 ☽ # ♄
14:14 ☽ # ☉
15:04 ☽ ☌ ♆
17:41 ☉ ♌
21:05 ☽ ☍ ☿
23:53 ☽ ✶ ♇
23 01:01 ♀ ♍
03:01 ☿ SR
07:34 ☽ ✶ ♂
07:59 ☽ ∥ ♆
11:36 ☽ ∥ ♇
13:11 ☽ ☍ ♀
17:01 ☉ ☌ ♄
21:08 ☽ # ♀
24 01:53 ☽ # ☿
04:36 ☽ ☌ ♅
11:56 ☽ ∥ ♅
13:12 ☽ # ♂
25 00:20 ☽ □ ♇
04:43 ☽ ∥ ♃
15:23 ☽ △ ♄
18:17 ☽ △ ☉
26 07:59 ☽ # ♃
09:18 ♂ # ♅
10:55 ☽ ☍ ♃
17:55 ☽ ✶ ♆
23:44 ☽ △ ☿
27 01:39 ☽ # ♅
02:10 ☽ ∥ ♂
03:51 ☽ △ ♇
10:56 ☽ ∥ ☿
11:09 ☽ ∥ ♀
14:23 ☿ ∥ ♀
17:24 ☽ ☌ ♂
20:35 ☽ □ ♄
28 04:43 ☽ # ♇
05:12 ♂ ♉
05:17 ☽ △ ♀
08:58 ☽ # ♆
12:21 ☽ ✶ ♅
22:20 ☽ ∥ ☉
29 00:22 ☽ □ ♆
04:59 ☽ □ ☿
05:05 ☽ ∥ ♄
30 05:27 ☽ ✶ ♄
16:45 ☽ ✶ ☉
18:09 ♀ ∥ ♂
20:11 ☽ □ ♀
21:23 ☽ □ ♅
31 03:39 ☽ △ ♃
08:09 ♀ ☍ ♅
10:00 ☽ △ ♆
12:24 ☽ ✶ ☿
14:00 ♂ □ ♄
21:10 ☽ ☍ ♇

August 2005

Day	S. T. h m s	☉ ° ′ ″	☽ ° ′ ″	☿ ° ′	♀ ° ′	♂ ° ′	♃ ° ′	♄ ° ′	♅ ° ′	♆ ° ′	♇ ° ′	☊ True ° ′
01 Mo	20 38 48	08♌51 06	23♊31 05	17♌R20	10♍45	02♉11	13♎19	02♌00	09♓R57	16♒R25	22♐R05	16♈R23
02 Tu	20 42 45	09 48 32	05♋34 21	16 38	11 57	02 45	13 27	02 08	09 55	16 23	22 04	16 14
03 We	20 46 41	10 45 58	17 31 48	15 55	13 09	03 19	13 36	02 15	09 53	16 21	22 03	16 03
04 Th	20 50 38	11 43 25	29 25 29	15 09	14 21	03 52	13 45	02 23	09 51	16 20	22 03	15 51
05 Fr	20 54 35	12 40 54	11♌17 03	14 23	15 33	04 25	13 54	02 31	09 49	16 18	22 02	15 39
06 Sa	20 58 31	13 38 23	23 08 00	13 37	16 45	04 58	14 03	02 38	09 47	16 17	22 01	15 28
07 Su	21 02 28	14 35 53	04♍59 55	12 51	17 56	05 31	14 12	02 46	09 45	16 15	22 00	15 19
08 Mo	21 06 24	15 33 25	16 54 48	12 07	19 08	06 03	14 21	02 54	09 43	16 13	21 59	15 12
09 Tu	21 10 21	16 30 57	28 55 08	11 25	20 20	06 35	14 31	03 01	09 40	16 12	21 59	15 09
10 We	21 14 17	17 28 30	11♎04 00	10 47	21 31	07 06	14 40	03 09	09 38	16 10	21 58	15 07
11 Th	21 18 14	18 26 03	23 25 02	10 13	22 43	07 38	14 50	03 16	09 36	16 08	21 57	15D 08
12 Fr	21 22 10	19 23 38	06♏02 20	09 43	23 54	08 09	14 59	03 24	09 34	16 07	21 56	15 09
13 Sa	21 26 07	20 21 14	19 00 04	09 19	25 06	08 39	15 09	03 32	09 32	16 05	21 56	15R 09
14 Su	21 30 04	21 18 51	02♐22 06	09 01	26 17	09 10	15 19	03 39	09 29	16 04	21 55	15 08
15 Mo	21 34 00	22 16 28	16 11 10	08 50	27 28	09 40	15 29	03 47	09 27	16 02	21 55	15 05
16 Tu	21 37 57	23 14 07	00♑27 59	08 45	28 40	10 09	15 39	03 54	09 25	16 00	21 54	15 01
17 We	21 41 53	24 11 47	15 10 25	08D 48	29 51	10 38	15 49	04 02	09 23	15 59	21 54	14 54
18 Th	21 45 50	25 09 27	00♒12 57	08 57	01♎02	11 07	15 59	04 09	09 20	15 57	21 53	14 47
19 Fr	21 49 46	26 07 09	15 27 05	09 15	02 13	11 36	16 10	04 16	09 18	15 55	21 53	14 40
20 Sa	21 53 43	27 04 52	00♓42 22	09 39	03 24	12 04	16 20	04 24	09 16	15 54	21 52	14 34
21 Su	21 57 39	28 02 37	15 48 11	10 12	04 35	12 31	16 30	04 31	09 13	15 52	21 52	14 29
22 Mo	22 01 36	29 00 22	00♈35 28	10 51	05 46	12 59	16 41	04 38	09 11	15 51	21 51	14 26
23 Tu	22 05 33	29 58 10	14 57 59	11 38	06 57	13 26	16 51	04 46	09 09	15 49	21 51	14 25
24 We	22 09 29	00♍55 59	28 52 48	12 32	08 08	13 52	17 02	04 53	09 06	15 48	21 51	14D 25
25 Th	22 13 26	01 53 50	12♉19 58	13 33	09 19	14 18	17 13	05 00	09 04	15 46	21 50	14 26
26 Fr	22 17 22	02 51 43	25 21 44	14 41	10 29	14 44	17 24	05 07	09 02	15 44	21 50	14 28
27 Sa	22 21 19	03 49 37	08♊01 37	15 54	11 40	15 09	17 35	05 15	08 59	15 43	21 50	14 29
28 Su	22 25 15	04 47 34	20 23 44	17 14	12 51	15 33	17 46	05 22	08 57	15 41	21 50	14R 28
29 Mo	22 29 12	05 45 32	02♋32 23	18 39	14 01	15 57	17 57	05 29	08 54	15 40	21 50	14 26
30 Tu	22 33 08	06 43 32	14 31 38	20 09	15 12	16 21	18 08	05 36	08 52	15 38	21 50	14 22
31 We	22 37 05	07 41 34	26 25 13	21 44	16 22	16 44	18 19	05 43	08 50	15 37	21 49	14 17

Data for	08-01-2005
Julian Day	2453583.50
Ayanamsa	23 56 01
SVP	05 ♓ 10 51
☽ ☊ Mean	17 ♈ 06 R

● ◐ PHASES ○ ◑

Day	Time	Phase	Position
05	03:04	●	12♌48
13	02:39	◐	20♏28
19	17:54	○	26♒50
26	15:18	◑	03♊29

LAST ASPECT Day	h m	☽ INGRESS Day	h m	Sign
31	21:10	01	12:53	♋
02	16:00	04	01:10	♌
05	21:44	06	13:54	♍
08	10:10	09	02:09	♎
10	21:11	11	12:35	♏
13	12:07	13	19:48	♐
15	20:44	15	23:14	♑
17	01:03	17	23:39	♒
19	17:54	19	22:53	♓
21	09:46	21	23:02	♈
23	11:46	24	01:58	♉
25	06:14	26	08:43	♊
28	02:49	28	18:57	♋
30	07:23	31	07:15	♌

DECLINATION

Day	☉	☽	☿	♀	♂	♃	♄	♅	♆	♇
01 Mo	18N03	27N56	11N02	08N47	09N37	04S09	20N00	08S35	16S03	15S05
02 Tu	17 48	28 16	11 11	08 18	09 49	04 13	19 59	08 36	16 03	15 06
03 We	17 32	27 17	11 21	07 49	10 00	04 16	19 57	08 37	16 04	15 06
04 Th	17 16	25 03	11 34	07 20	10 11	04 20	19 55	08 37	16 04	15 06
05 Fr	17 00	21 45	11 49	06 50	10 22	04 24	19 54	08 38	16 05	15 06
06 Sa	16 44	17 35	12 05	06 20	10 32	04 27	19 52	08 39	16 05	15 07
07 Su	16 27	12 44	12 22	05 50	10 43	04 31	19 50	08 40	16 06	15 07
08 Mo	16 10	07 23	12 41	05 20	10 53	04 35	19 49	08 41	16 06	15 07
09 Tu	15 53	01 44	13 01	04 50	11 04	04 39	19 47	08 41	16 07	15 08
10 We	15 36	04S03	13 21	04 20	11 14	04 42	19 45	08 42	16 07	15 08
11 Th	15 18	09 48	13 41	03 49	11 24	04 46	19 44	08 43	16 08	15 08
12 Fr	15 00	15 16	14 02	03 19	11 34	04 50	19 42	08 44	16 08	15 08
13 Sa	14 42	20 14	14 22	02 48	11 44	04 54	19 40	08 45	16 09	15 09
14 Su	14 24	24 20	14 41	02 18	11 53	04 58	19 39	08 46	16 09	15 09
15 Mo	14 05	27 11	14 59	01 47	12 02	05 02	19 37	08 46	16 10	15 09
16 Tu	13 46	28 24	15 17	01 16	12 12	05 06	19 35	08 47	16 10	15 10
17 We	13 27	27 40	15 33	00 45	12 21	05 10	19 34	08 48	16 11	15 10
18 Th	13 08	24 55	15 47	00 14	12 30	05 14	19 32	08 49	16 11	15 10
19 Fr	12 49	20 24	15 59	00S17	12 39	05 18	19 30	08 50	16 12	15 10
20 Sa	12 29	14 33	16 10	00 48	12 47	05 22	19 29	08 51	16 12	15 11
21 Su	12 09	07 52	16 18	01 19	12 56	05 26	19 27	08 52	16 13	15 11
22 Mo	11 49	00 54	16 24	01 50	13 04	05 31	19 25	08 53	16 13	15 11
23 Tu	11 29	05N57	16 27	02 21	13 12	05 35	19 24	08 53	16 14	15 12
24 We	11 09	12 18	16 28	02 52	13 20	05 39	19 22	08 54	16 14	15 12
25 Th	10 48	17 51	16 26	03 22	13 28	05 43	19 20	08 55	16 15	15 12
26 Fr	10 27	22 24	16 20	03 53	13 36	05 48	19 19	08 56	16 15	15 13
27 Sa	10 06	25 46	16 12	04 24	13 43	05 52	19 17	08 57	16 16	15 13
28 Su	09 45	27 48	16 01	04 55	13 51	05 56	19 15	08 58	16 16	15 13
29 Mo	09 24	28 29	15 47	05 25	13 58	06 00	19 14	08 59	16 17	15 14
30 Tu	09 03	27 48	15 29	05 56	14 05	06 05	19 12	09 00	16 17	15 14
31 We	08 41	25 50	15 09	06 26	14 12	06 09	19 11	09 01	16 18	15 14

ASPECTARIAN

01 09:48 ♀ # ♅
18:05 ☽ ✶ ♂
02 08:40 ☽ △ ♅
08:52 ☿ ☍ ♆
14:13 ☽ ✶ ♀
16:00 ☽ □ ♃
04 06:03 ☽ ☌ ♄
09:27 ☽ □ ♂
05 05:22 ☽ ✶ ♃
05:54 ☽ ☌ ☿
10:09 ☽ ☍ ♆
11:24 ☽ ∥ ♄
12:45 ☿ ✶ ♃
21:44 ☽ △ ♇
23:36 ☿ ☌ ☉
06 04:45 ☽ ∥ ☉
07:43 ☽ # ♆
12:08 ☉ ✶ ♃
12:36 ☽ # ♇
07 01:05 ☽ △ ♂
01:35 ☽ ∥ ☿
08:59 ☽ ∥ ♂
09:33 ☽ ☍ ♅
18:24 ☽ # ♅
08 04:57 ☽ ☌ ♀
05:22 ☉ # ♆
09:42 ☽ ∥ ♀
10:10 ☽ □ ♇
11:56 ☽ # ♃
16:11 ☉ ☍ ♆
09 08:13 ♀ # ♃
08:14 ☽ ✶ ♄
23:28 ☽ ✶ ☿
10 01:0[illegible] [illegible] ∥ [illegible]
02:46 ☽ ∥ ♃
07:08 ☽ ☌ ♃
08:54 ♀ □ ♇
09:57 ☽ △ ♆
13:34 ☽ ✶ ☉
19:29 ☽ ∥ ♅
21:11 ☽ ✶ ♇
11 07:08 ☽ # ♂
13:27 ☉ # ♇
18:03 ☽ # ☿
18:59 ☽ □ ♄
22:54 ☽ # ☉
23:27 ☽ ∥ ♇
12 04:03 ☽ ∥ ♆
04:07 ☽ ☍ ♂
06:35 ☽ △ ♅
06:40 ☽ □ ☿
18:40 ☽ □ ♆
21:11 ☽ # ♄
13 12:07 ☽ ✶ ♀
13:10 ☿ ∥ ☉
19:41 ☿ □ ♂
14 02:17 ☽ △ ♄
11:29 ☽ △ ☿
12:26 ☽ □ ♅
14:45 ♂ ✶ ♅
15:02 ☉ △ ♇
22:47 ☽ ✶ ♃
23:44 ☽ ✶ ♆
15 09:42 ☽ ☌ ♇
11:04 ☽ △ ☉
10:[illegible] [illegible] ∥ [illegible]
20:44 ☽ □ ♀
16 03:52 ☿ SD
14:38 ☽ ✶ ♅
16:25 ☽ △ ♂
17 01:03 ☽ □ ♃
03:05 ♀ ♎
19:38 ♃ △ ♆
18 01:24 ☽ △ ♀
06:16 ☽ ☍ ♄
14:02 ☽ ☍ ☿
17:45 ☽ □ ♂
19 00:44 ☽ ☌ ♆
01:07 ☽ △ ♃
04:03 ☽ # ♄
08:38 ☉ ∥ ♂
10:05 ☽ ✶ ♇
17:38 ☽ ∥ ♆
17:57 ☽ # ☿
21:36 ☽ ∥ ♇
20 06:25 ☽ # ♂
07:21 ☿ # ♆
08:03 ☽ # ☉
13:31 ☽ ☌ ♅
18:35 ☽ ✶ ♂
20:34 ☽ ∥ ♅
22:28 ♀ ✶ ♄
21 08:20 ☽ ∥ ♃
09:46 ☽ □ ♇
21:02 ☽ ∥ ♀
22 06:45 ☽ △ ♄
09:19 ☽ ☍ ♀
18:1[illegible] [illegible] ∥ [illegible]
18:02 ☽ △ ☿
22:41 ☽ # ♃
23 00:46 ☉ ♍
01:27 ☽ ✶ ♆
03:15 ☽ ☍ ♃
10:54 ☽ # ♅
11:46 ☽ △ ♇
19:44 ☽ ∥ ☉
24 03:53 ☽ △ ☉
04:23 ☽ ∥ ♂
10:42 ☽ □ ♄
12:08 ☽ # ♇
16:41 ☽ # ♆
17:?4 ☽ ∥ ♆
18:07 ☽ ✶ ♅
25 02:25 ☽ □ ☿
03:42 ☽ ☌ ♂
06:14 ☽ □ ♆
07:15 ☽ ∥ ♄
26 01:30 ☿ □ ♂
15:16 ☿ # ♆
18:37 ☽ ✶ ♄
20:26 ☿ ☍ ♆
27 01:50 ☽ □ ♅
07:45 ☽ △ ♀
14:50 ☽ △ ♆
17:03 ☽ ✶ ♆
18:46 ☽ △ ♃
28 02:49 ☽ ☍ ♇
07:32 ♂ □ ♆
10:29 ☿ ✶ ♃
29 06:59 ☽ ✶ ☉
12:41 ☽ △ ♅
30 01:30 ☽ □ ♀
03:06 ☉ # ♅
03:48 ☽ ✶ ♂
07:23 ☽ □ ♃
08:19 ♀ ∥ ♃
08:50 ♀ △ ♆
18:10 ☿ # ♇
31 01:25 ☿ △ ♇
19:00 ☽ ☌ ♄

⚷ Chiron

01 Dec. 12 S 50

Day	Position
02	29♑57 R
05	29 47
08	29 37
11	29 27
14	29 18
17	29 09
20	29 00
23	28 52
26	28 44
29	28 36

01 03:08 ♑ R

September 2005

Day	S. T.	☉	☽	☿	♀	♂	♃	♄	♅	♆	♇	☊ True
	h m s	° ' "	° ' "	° '	° '	° '	° '	° '	° '	° '	° '	° '
01 Th	22 41 02	08♍ 39 37	08♌ 16 16	23♌ 23	17♎ 33	17♉ 07	18♎ 30	05♌ 50	08♓R47	15♒R35	21♐R49	14♈R10
02 Fr	22 44 58	09 37 43	20 07 23	25 05	18 43	17 29	18 42	05 57	08 45	15 34	21 49	14 04
03 Sa	22 48 55	10 35 50	02♍ 00 42	26 50	19 53	17 50	18 53	06 04	08 42	15 32	21D 49	13 58
04 Su	22 52 51	11 33 58	13 58 00	28 38	21 03	18 11	19 05	06 10	08 40	15 31	21 49	13 53
05 Mo	22 56 48	12 32 09	26 01 02	00♍ 28	22 14	18 32	19 16	06 17	08 38	15 30	21 49	13 50
06 Tu	23 00 44	13 30 21	08♎ 11 34	02 20	23 24	18 51	19 28	06 24	08 35	15 28	21 50	13 49
07 We	23 04 41	14 28 34	20 31 33	04 13	24 34	19 11	19 40	06 31	08 33	15 27	21 50	13D 48
08 Th	23 08 37	15 26 50	03♏ 03 15	06 07	25 44	19 29	19 51	06 37	08 31	15 25	21 50	13 49
09 Fr	23 12 34	16 25 07	15 49 09	08 02	26 53	19 47	20 03	06 44	08 28	15 24	21 50	13 51
10 Sa	23 16 31	17 23 25	28 51 54	09 57	28 03	20 05	20 15	06 50	08 26	15 23	21 50	13 52
11 Su	23 20 27	18 21 45	12♐ 14 00	11 52	29 13	20 21	20 27	06 57	08 23	15 21	21 51	13R 52
12 Mo	23 24 24	19 20 07	25 57 22	13 47	00♏ 23	20 37	20 39	07 03	08 21	15 20	21 51	13 52
13 Tu	23 28 20	20 18 30	10♑ 02 41	15 42	01 32	20 53	20 51	07 10	08 19	15 19	21 51	13 50
14 We	23 32 17	21 16 55	24 28 49	17 36	02 41	21 07	21 03	07 16	08 16	15 18	21 52	13 48
15 Th	23 36 13	22 15 21	09♒ 12 20	19 29	03 51	21 21	21 15	07 22	08 14	15 16	21 52	13 45
16 Fr	23 40 10	23 13 49	24 07 27	21 22	05 00	21 35	21 27	07 29	08 12	15 15	21 52	13 43
17 Sa	23 44 06	24 12 18	09♓ 06 33	23 14	06 09	21 47	21 39	07 35	08 10	15 14	21 53	13 40
18 Su	23 48 03	25 10 49	24 01 02	25 05	07 18	21 59	21 51	07 41	08 07	15 13	21 53	13 39
19 Mo	23 51 60	26 09 23	08♈ 42 43	26 55	08 27	22 10	22 04	07 47	08 05	15 12	21 54	13 38
20 Tu	23 55 56	27 07 58	23 04 53	28 45	09 36	22 21	22 16	07 53	08 03	15 11	21 54	13D 38
21 We	23 59 53	28 06 35	07♉ 03 11	00♎ 33	10 45	22 31	22 28	07 59	08 01	15 10	21 55	13 39
22 Th	00 03 49	29 05 15	20 35 48	02 20	11 54	22 39	22 41	08 05	07 58	15 08	21 56	13 40
23 Fr	00 07 46	00♎ 03 56	03♊ 43 16	04 07	13 02	22 48	22 53	08 10	07 56	15 07	21 56	13 41
24 Sa	00 11 42	01 02 40	16 27 50	05 52	14 11	22 55	23 06	08 16	07 54	15 06	21 57	13 42
25 Su	00 15 39	02 01 26	28 52 59	07 37	15 19	23 01	23 18	08 22	07 52	15 05	21 58	13 43
26 Mo	00 19 35	03 00 15	11♋ 02 52	09 20	16 27	23 07	23 31	08 27	07 50	15 04	21 58	13R 43
27 Tu	00 23 32	03 59 06	23 01 59	11 03	17 35	23 12	23 43	08 33	07 48	15 03	21 59	13 43
28 We	00 27 29	04 57 59	04♌ 54 48	12 45	18 44	23 16	23 56	08 38	07 46	15 03	22 00	13 42
29 Th	00 31 25	05 56 54	16 45 33	14 25	19 51	23 19	24 09	08 44	07 44	15 02	22 01	13 42
30 Fr	00 35 22	06 55 51	28 37 58	16 05	20 59	23 21	24 21	08 49	07 42	15 01	22 02	13 41

Data for	09-01-2005
Julian Day	2453614.50
Ayanamsa	23 56 06
SVP	05 ♓ 10 47
☽ ☊ Mean	15 ♈ 27 R

● ◐ PHASES ○ ◑

03	18:46	●	11♍21
11	11:37	◐	18♐50
18	02:01	○	25♓16
25	06:41	◑	02♋18

LAST ASPECT ☽ Day	h m	INGRESS Day	h m	
02	11:45	02	19:57	♍
04	15:40	05	07:53	♎
07	08:34	07	18:11	♏
09	07:32	10	02:04	♐
11	16:53	12	06:57	♑
13	18:22	14	09:03	♒
15	20:23	16	09:25	♓
18	02:01	18	09:43	♈
19	22:36	20	11:48	♉
22	16:42	22	17:07	♊
24	12:58	25	02:11	♋
27	01:25	27	14:03	♌
29	15:13	30	02:45	♍

DECLINATION

Day	☉	☽	☿	♀	♂	♃	♄	♅	♆	♇
01 Th	08N19	22N46	14N46	06S 56	14N19	06S 14	19N09	09S 02	16S 18	15S 15
02 Fr	07 58	18 45	14 20	07 27	14 25	06 18	19 07	09 02	16 18	15 15
03 Sa	07 36	14 00	13 51	07 57	14 32	06 22	19 06	09 03	16 19	15 16
04 Su	07 14	08 42	13 20	08 27	14 38	06 27	19 04	09 04	16 19	15 16
05 Mo	06 51	03 03	12 46	08 56	14 44	06 31	19 03	09 05	16 20	15 16
06 Tu	06 29	02S 47	12 11	09 26	14 50	06 36	19 01	09 06	16 20	15 17
07 We	06 07	08 35	11 33	09 55	14 56	06 40	18 59	09 07	16 21	15 17
08 Th	05 44	14 09	10 53	10 24	15 01	06 45	18 58	09 08	16 21	15 17
09 Fr	05 22	19 15	10 12	10 53	15 07	06 50	18 56	09 09	16 21	15 18
10 Sa	04 59	23 32	09 30	11 22	15 12	06 54	18 55	09 10	16 22	15 18
11 Su	04 36	26 42	08 47	11 51	15 17	06 59	18 53	09 10	16 22	15 18
12 Mo	04 13	28 23	08 02	12 19	15 22	07 03	18 52	09 11	16 23	15 19
13 Tu	03 50	28 17	07 17	12 47	15 27	07 08	18 50	09 12	16 23	15 19
14 We	03 27	26 18	06 31	13 15	15 32	07 12	18 49	09 13	16 23	15 20
15 Th	03 04	22 31	05 44	13 42	15 36	07 17	18 47	09 14	16 24	15 20
16 Fr	02 41	17 15	04 57	14 09	15 41	07 22	18 46	09 15	16 24	15 20
17 Sa	02 18	10 56	04 10	14 36	15 45	07 26	18 44	09 16	16 24	15 21
18 Su	01 55	04 01	03 23	15 03	15 49	07 31	18 43	09 16	16 25	15 21
19 Mo	01 32	03N02	02 35	15 29	15 53	07 36	18 41	09 17	16 25	15 22
20 Tu	01 08	09 46	01 47	15 55	15 57	07 40	18 40	09 18	16 25	15 22
21 We	00 45	15 51	01 00	16 20	16 00	07 45	18 39	09 19	16 26	15 22
22 Th	00 22	20 58	00 12	16 46	16 04	07 50	18 37	09 20	16 26	15 23
23 Fr	00S 02	24 52	00S 35	17 11	16 07	07 54	18 36	09 20	16 26	15 23
24 Sa	00 25	27 25	01 22	17 35	16 10	07 59	18 34	09 21	16 27	15 23
25 Su	00 48	28 32	02 08	17 59	16 13	08 04	18 33	09 22	16 27	15 24
26 Mo	01 12	28 14	02 55	18 23	16 16	08 08	18 32	09 23	16 27	15 24
27 Tu	01 35	26 36	03 41	18 46	16 19	08 13	18 30	09 24	16 28	15 25
28 We	01 58	23 49	04 26	19 09	16 21	08 18	18 29	09 24	16 28	15 25
29 Th	02 22	20 02	05 12	19 32	16 24	08 23	18 28	09 25	16 28	15 25
30 Fr	02 45	15 28	05 56	19 54	16 26	08 27	18 27	09 26	16 28	15 26

ASPECTARIAN

01 03:03 ☉ ☍ ♅
14:48 ☽ ☍ ♆
18:29 ☽ □ ♂
20:07 ☿ ∥ ♂
20:51 ☽ ✶ ♀
21:04 ☽ ✶ ♃
21:56 ☽ ∥ ♄
23:31 ♀ ☌ ♃
02 03:26 ☽ △ ♇
10:51 ♇ SD
11:45 ☽ ☌ ☿
12:43 ☽ # ♆
14:17 ♀ # ☉
17:56 ☽ # ♇
21:31 ☽ ∥ ♂
03 00:46 ☽ ∥ ☿
13:25 ☽ ☍ ♅
22:24 ☽ # ♅
04 01:02 ☽ # ♀
06:50 ☽ ∥ ☉
08:41 ☽ △ ♂
09:34 ☽ # ♃
15:40 ☽ □ ♇
15:45 ♀ ✶ ♇
17:53 ☿ ♍
05 07:26 ♀ ∥ ♅
17:54 ☉ # ♃
20:27 ☽ ✶ ♄
06 14:11 ☽ △ ♆
14:19 ☽ # ☉
15:55 ☽ ∥ ♃
22:18 ☽ ☌ ♃
07 02:13 ☽ ∥ ♅
02:31 ☽ ✶ ♇
06:08 ☽ ∥ ♀
08:34 ☽ ☌ ♀
11:15 ☽ # ☿
08 03:58 ☽ # ♂
05:07 ☽ ∥ ♇
06:49 ☽ □ ♄
06:50 ☽ ✶ ☿
10:00 ☽ ∥ ♆
10:02 ☿ # ♀
10:17 ☽ △ ♅
22:31 ☽ # ♄
23:13 ☽ □ ♆
09 01:12 ☽ ✶ ☉
05:22 ☿ ☍ ♅
07:32 ☽ ☍ ♂
10 11:10 ☿ # ♅
14:31 ☽ △ ♄
17:11 ☽ □ ♅
23:15 ☽ □ ☿
11 05:31 ☽ ✶ ♆
05:38 ♂ # ♇
14:39 ☽ ✶ ♃
16:14 ♀ ♏
16:53 ☽ ☌ ♇
12 08:17 ☽ ✶ ♀
21:05 ☽ ✶ ♅
13 04:15 ☿ # ♃
10:54 ☽ △ ☿
18:16 ☽ □ ♃
18:22 ☽ △ ☉
18:23 ☽ △ ♂
18:49 ☉ △ ♂
14 14:20 ☉ □ ♇
14:34 ☽ □ ♀
21:01 ☽ ☍ ♄
15 09:47 ☽ ☌ ♆
17:39 ☽ # ♄
19:39 ☽ △ ♃
19:52 ☽ □ ♂
20:23 ☽ ✶ ♇
16 03:06 ☿ △ ♂
03:26 ☽ ∥ ♆
06:12 ☽ # ♂
06:31 ☿ □ ♇
07:36 ☽ ∥ ♇
11:20 ☽ ∥ ♀
18:52 ☽ △ ♀
22:29 ☽ ☌ ♅
17 05:53 ☽ ∥ ♅
12:06 ☽ ∥ ♃
20:33 ☽ □ ♇
20:41 ☽ ✶ ♂
18 01:59 ☽ ☍ ☿
02:25 ☽ # ☿
02:38 ☿ ☌ ☉
03:48 ♃ ✶ ♇
07:31 ☽ # ☉
08:34 ♀ □ ♄
16:31 ♀ △ ♅
17:07 ♀ ∥ ♇
19:06 ☽ ∥ ☉
22:27 ☽ △ ♄
22:36 ☽ ∥ ☿
19 10:44 ☽ ✶ ♆
16:13 ☽ # ♃
22:01 ☽ △ ♇
22:15 ☽ # ♅
22:36 ☽ ☍ ♃
20 02:07 ♀ # ♂
16:40 ☿ ♎
21:59 ☽ # ♇
21 00:41 ☽ ∥ ♂
01:38 ☽ □ ♄
01:40 ☽ ✶ ♅
02:17 ☽ # ♀
02:30 ☽ # ♆
05:10 ♀ ∥ ♆
07:04 ☽ ☍ ♀
12:25 ☽ ∥ ♄
14:15 ☽ □ ♆
14:42 ☿ ∥ ☉
22 03:46 ☽ ☌ ♂
11:39 ☿ # ☉
16:42 ☽ △ ☉
22:23 ☉ ♎
23 00:51 ☽ △ ☿
07:51 ☽ □ ♅
08:22 ☽ ✶ ♄
21:25 ☽ △ ♆
24 10:32 ☽ ☍ ♇
12:58 ☽ △ ♃
19:16 ♀ □ ♆
25 10:59 ☿ ✶ ♄
17:38 ☽ △ ♅
20:02 ☽ □ ☿
26 08:39 ♀ # ♄
11:55 ☽ △ ♀
27 00:19 ☽ ✶ ♂
01:25 ☽ □ ♃
28 00:07 ☽ ✶ ☉
07:36 ☽ ☌ ♄
18:29 ☽ ✶ ☿
20:30 ☽ ☍ ♆
29 02:37 ☽ # ♀
06:56 ☽ □ ♀
08:40 ☿ △ ♆
08:40 ☽ ∥ ♄
10:39 ☽ △ ♇
13:18 ☽ □ ♂
15:13 ☽ ✶ ♃
18:57 ☽ # ♆
19:12 ☽ ∥ ♂
30 00:07 ☽ # ♇
18:10 ☽ ☍ ♅

⚷ Chiron

01 Dec.	13 S 16
01	28♑29R
04	28 22
07	28 16
10	28 11
13	28 06
16	28 01
19	27 58
22	27 54
25	27 52
28	27 50

October 2005

03 10:12 10♎19 ☉ Annular Solar Eclipse
17 12:04 24♈13 ☽ Total Lunar Eclipse (mag 0.068)

Day	S. T.	☉	☽	☿	♀	♂	♃	♄	♅	♆	♇	☊ True
	h m s	° ′ ″	° ′ ″	° ′	° ′	° ′	° ′	° ′	° ′	° ′	° ′	° ′
01 Sa	00 39 18	07♎54 51	10♍35 18	17♎44	22♏07	23♉22	24♎34	08♌54	07♓R40	15♒R00	22♐02	13♈R41
02 Su	00 43 15	08 53 52	22 40 07	19 22	23 15	23R 22	24 47	08 59	07 38	14 59	22 03	13 40
03 Mo	00 47 11	09 52 56	04♎54 32	20 59	24 22	23 22	25 00	09 04	07 36	14 58	22 04	13 40
04 Tu	00 51 08	10 52 02	17 20 04	22 36	25 29	23 20	25 13	09 09	07 34	14 58	22 05	13 40
05 We	00 55 04	11 51 10	29 57 51	24 11	26 37	23 18	25 25	09 14	07 32	14 57	22 06	13 40
06 Th	00 59 01	12 50 19	12♏48 39	25 46	27 44	23 15	25 38	09 19	07 30	14 56	22 07	13 39
07 Fr	01 02 57	13 49 31	25 52 57	27 20	28 50	23 11	25 51	09 24	07 29	14 56	22 08	13 39
08 Sa	01 06 54	14 48 45	09♐11 08	28 53	29 57	23 06	26 04	09 28	07 27	14 55	22 10	13 38
09 Su	01 10 51	15 48 00	22 43 25	00♏26	01♐04	23 00	26 17	09 33	07 25	14 54	22 11	13 38
10 Mo	01 14 47	16 47 18	06♑29 48	01 58	02 10	22 53	26 30	09 38	07 23	14 54	22 12	13 37
11 Tu	01 18 44	17 46 37	20 29 47	03 29	03 17	22 45	26 43	09 42	07 22	14 53	22 13	13 36
12 We	01 22 40	18 45 57	04♒42 10	04 59	04 23	22 37	26 56	09 46	07 20	14 53	22 14	13D 36
13 Th	01 26 37	19 45 20	19 04 49	06 28	05 29	22 27	27 09	09 51	07 18	14 52	22 16	13 37
14 Fr	01 30 33	20 44 44	03♓34 29	07 57	06 34	22 17	27 22	09 55	07 17	14 52	22 17	13 37
15 Sa	01 34 30	21 44 10	18 06 52	09 25	07 40	22 06	27 35	09 59	07 15	14 51	22 18	13 38
16 Su	01 38 26	22 43 37	02♈36 48	10 52	08 45	21 54	27 48	10 03	07 14	14 51	22 19	13 39
17 Mo	01 42 23	23 43 07	16 58 43	12 18	09 50	21 42	28 01	10 07	07 13	14 51	22 21	13 39
18 Tu	01 46 20	24 42 38	01♉07 19	13 44	10 55	21 28	28 14	10 10	07 11	14 50	22 22	13R 39
19 We	01 50 16	25 42 12	14 58 12	15 09	12 00	21 14	28 27	10 14	07 10	14 50	22 24	13 38
20 Th	01 54 13	26 41 48	28 28 29	16 33	13 04	20 59	28 40	10 18	07 09	14 50	22 25	13 37
21 Fr	01 58 09	27 41 26	11♊36 58	17 56	14 09	20 43	28 53	10 21	07 07	14 50	22 27	13 35
22 Sa	02 02 06	28 41 06	24 24 14	19 18	15 13	20 27	29 06	10 25	07 06	14 49	22 28	13 33
23 Su	02 06 02	29 40 49	06♋52 21	20 39	16 16	20 10	29 19	10 28	07 05	14 49	22 30	13 31
24 Mo	02 09 59	00♏40 33	19 04 33	21 59	17 20	19 53	29 32	10 31	07 04	14 49	22 31	13 30
25 Tu	02 13 56	01 40 20	01♌04 57	23 18	18 23	19 35	29 45	10 34	07 03	14 49	22 33	13D 29
26 We	02 17 52	02 40 10	12 58 10	24 36	19 26	19 16	29 58	10 37	07 02	14 49	22 34	13 30
27 Th	02 21 49	03 40 01	24 49 03	25 53	20 29	18 57	00♏12	10 40	07 01	14D 49	22 36	13 32
28 Fr	02 25 45	04 39 54	06♍42 22	27 08	21 31	18 37	00 25	10 43	07 00	14 49	22 37	13 33
29 Sa	02 29 42	05 39 50	18 42 34	28 21	22 34	18 17	00 38	10 46	06 59	14 49	22 39	13 35
30 Su	02 33 38	06 39 48	00♎53 31	29 33	23 35	17 57	00 51	10 48	06 58	14 49	22 41	13 37
31 Mo	02 37 35	07 39 47	13 18 20	00♐43	24 37	17 36	01 04	10 51	06 57	14 49	22 43	13R 38

Data for	10-01-2005
Julian Day	2453644.50
Ayanamsa	23 56 09
SVP	05 ♓ 10 42
☽ ☊ Mean	13 ♈ 52 R

PHASES

03	10:29	●	10♎19
10	19:01	◐	17♑34
17	12:14	○	24♈13
25	01:17	◑	01♌44

LAST ASPECT ☽ INGRESS

Day	h m	Day	h m	
02	01:23	02	14:25	♎
04	15:16	05	00:04	♏
07	05:52	07	07:28	♐
09	06:20	09	12:44	♑
11	10:42	11	16:05	♒
13	13:35	13	18:06	♓
15	06:56	15	19:40	♈
17	18:59	17	22:05	♉
19	10:51	20	02:45	♊
22	09:08	22	10:42	♋
24	21:17	24	21:49	♌
27	02:24	27	10:29	♍
29	21:07	29	22:15	♎

DECLINATION

Day	☉	☽	☿	♀	♂	♃	♄	♅	♆	♇
01 Sa	03S 08	10N16	06S 40	20S 15	16N28	08S 32	18N25	09S 27	16S 29	15S 26
02 Su	03 32	04 39	07 24	20 36	16 30	08 37	18 24	09 27	16 29	15 27
03 Mo	03 55	01S 12	08 07	20 57	16 31	08 41	18 23	09 28	16 29	15 27
04 Tu	04 18	07 07	08 50	21 17	16 33	08 46	18 22	09 29	16 29	15 27
05 We	04 41	12 51	09 31	21 37	16 34	08 51	18 21	09 29	16 30	15 28
06 Th	05 04	18 08	10 13	21 56	16 35	08 56	18 19	09 30	16 30	15 28
07 Fr	05 27	22 41	10 53	22 15	16 36	09 00	18 18	09 31	16 30	15 29
08 Sa	05 50	26 08	11 33	22 33	16 37	09 05	18 17	09 31	16 30	15 29
09 Su	06 13	28 10	12 13	22 50	16 38	09 10	18 16	09 32	16 30	15 29
10 Mo	06 36	28 31	12 51	23 08	16 38	09 15	18 15	09 32	16 31	15 30
11 Tu	06 59	27 04	13 29	23 24	16 39	09 19	18 14	09 33	16 31	15 30
12 We	07 21	23 52	14 06	23 40	16 39	09 24	18 13	09 34	16 31	15 30
13 Th	07 44	19 12	14 42	23 56	16 39	09 29	18 12	09 34	16 31	15 31
14 Fr	08 06	13 22	15 18	24 10	16 39	09 33	18 11	09 35	16 31	15 31
15 Sa	08 28	06 48	15 53	24 25	16 39	09 38	18 10	09 35	16 31	15 32
16 Su	08 50	00N06	16 27	24 38	16 38	09 43	18 09	09 36	16 31	15 32
17 Mo	09 12	06 57	17 00	24 52	16 38	09 47	18 08	09 36	16 32	15 32
18 Tu	09 34	13 21	17 32	25 04	16 37	09 52	18 07	09 37	16 32	15 33
19 We	09 56	18 57	18 03	25 16	16 36	09 57	18 06	09 37	16 32	15 33
20 Th	10 18	23 26	18 33	25 27	16 35	10 01	18 05	09 38	16 32	15 33
21 Fr	10 39	26 36	19 03	25 38	16 34	10 06	18 05	09 38	16 32	15 34
22 Sa	11 00	28 17	19 31	25 48	16 32	10 11	18 04	09 39	16 32	15 34
23 Su	11 22	28 28	19 58	25 58	16 31	10 15	18 03	09 39	16 32	15 35
24 Mo	11 43	27 15	20 25	26 07	16 29	10 20	18 02	09 39	16 32	15 35
25 Tu	12 03	24 48	20 50	26 15	16 27	10 25	18 02	09 40	16 32	15 35
26 We	12 24	21 19	21 14	26 22	16 26	10 29	18 01	09 40	16 32	15 36
27 Th	12 44	16 59	21 37	26 29	16 24	10 34	18 00	09 40	16 32	15 36
28 Fr	13 05	11 59	21 59	26 36	16 21	10 38	18 00	09 41	16 32	15 37
29 Sa	13 25	06 30	22 19	26 42	16 19	10 43	17 59	09 41	16 32	15 37
30 Su	13 44	00 43	22 39	26 47	16 17	10 48	17 58	09 41	16 32	15 37
31 Mo	14 04	05S 14	22 57	26 51	16 14	10 52	17 58	09 42	16 32	15 38

ASPECTARIAN

01 03:37 ☽ # ♅
07:27 ☽ # ♃
13:45 ☽ # ☿
14:24 ♂ # ♆
22:05 ♂ SR
22:47 ☽ □ ♆
02 01:15 ☽ ⚹ ♀
01:23 ☽ △ ♂
02:24 ☉ ⚹ ♄
02:46 ☿ ☍ ♂
04:21 ☽ # ☉
03 08:08 ☽ ⚹ ♄
11:42 ☽ ∥ ☉
16:20 ☿ ⚹ ♆
19:27 ☿ △ ♆
21:51 ☿ ∥ ♃
04 06:52 ☽ ∥ ♃
07:58 ☽ ∥ ☿
09:06 ☽ ⚹ ♆
09:43 ☽ ∥ ♅
11:30 ☽ ☌ ☿
15:16 ☽ ☌ ♃
22:43 ☿ ∥ ♅
05 11:32 ☽ ∥ ♆
14:10 ☽ △ ♅
16:13 ☽ ∥ ♆
16:37 ☽ # ♂
17:28 ☽ □ ♄
21:39 ☿ ☌ ♃
06 00:53 ☽ # ♄
03:56 ☽ □ ♆
19:06 ☽ ☍ ♂
21:15 ☽ ∥ ♀
07 05:52 ☽ ☌ ♀
20:54 ☽ □ ♅
08 00:31 ☽ △ ♄
01:00 ☿ [illegible]
02:29 ☉ △ ♆
10:12 ☽ ⚹ ♆
10:49 ☽ ⚹ ☉
17:15 ☿ ♏
23:02 ☽ ☌ ♆
09 06:20 ☽ ⚹ ♃
15:10 ☽ ⚹ ☿
10 01:32 ☽ ⚹ ♅
11 03:48 ☽ △ ♂
10:42 ☽ □ ♃
23:24 ☽ ⚹ ☿
12 00:31 ☽ □ ☿
01:09 ☽ ∥ ♀
08:32 ☽ ☍ ♄
17:00 ☽ ☌ ♆
13 01:12 ☽ △ ☉
04:25 ☽ # ♄
05:17 ☽ ⚹ ♆
05:32 ☽ □ ♂
10:58 ☽ # ♂
11:32 ☽ ∥ ♆
13:20 ♀ △ ♅
13:35 ☽ △ ♃
15:36 ☽ ∥ ♆
17:07 ☽ ∥ ♅
14 05:21 ☽ □ ♀
06:07 ☽ ☌ ♅
07:41 ♃ ∥ ♅
08:02 ☽ △ ☿
09:13 ☿ ∥ ♆
14:00 ☽ ∥ ♃
14:04 ☽ ∥ ♅
15:17 ♀ □ ♅
18:23 ☽ ∥ ☉
15 06:30 ☽ ⚹ ♂
06:56 ☽ □ ♆
09:44 ☿ □ ♄
14:04 ☉ ⚹ ♆
16 03:28 ☿ ∥ ♆
08:16 ☿ # ♂
11:03 ☽ △ ♀
12:26 ☽ △ ♄
20:25 ☽ ⚹ ♆
17 06:25 ♀ △ ♄
08:41 ☽ # ☉
09:04 ☽ △ ♆
09:41 ☽ # ♅
10:29 ☽ # ♃
18:59 ☽ ☍ ♃
18 02:43 ☉ ∥ ♅
08:57 ☽ # ♆
10:25 ☽ ⚹ ♅
13:06 ☽ # ♆
13:27 ☽ ∥ ♂
15:41 ☽ □ ♄
18:41 ☿ □ ♆
19:23 ☽ # ♆
20:06 ☽ ∥ ♄
23:46 ☽ □ ♆
19 00:21 ☽ ☍ ☿
01:00 ☉ ∥ ♃
02:29 ☿ # ♄
10:51 ☽ ☌ ♂
20 14:49 ☽ # ♀
15:44 ☽ □ ♅
21:39 ☽ ⚹ ♄
21 05:07 ☽ ☍ ♀
05:58 ☽ △ ♆
15:19 ♀ ⚹ ♆
20:19 ☽ ☍ ♆
22 08:22 ♂ # ♆
08:53 ☽ △ ☉
09:08 ☽ △ ♃
12:54 ☉ ☌ ♃
16:56 ☿ ☍ ♂
23 00:24 ☽ △ ♅
07:43 ☉ ♏
24 01:33 ☽ ⚹ ♂
06:30 ☽ △ ☿
12:02 ☽ # ♀
21:17 ☽ □ ♃
25 19:14 ☽ ☌ ♄
26 00:25 ☽ # ☿
02:52 ♃ ♏
03:44 ☽ ☍ ♆
12:25 ☽ □ ♂
14:22 ☽ △ ♀
18:38 ☽ ∥ ♄
19:30 ☽ △ ♆
23:25 ♆ SD
27 02:15 ☽ # ♆
02:24 ☽ □ ☿
02:59 ☽ ∥ ♂
06:53 ☽ # ♆
11:04 ☽ ⚹ ♃
19:16 ☽ # ☉
19:31 ☽ ⚹ ☉
28 00:35 ☽ ☍ ♅
05:59 ☽ # ♃
10:18 ☽ # ♆
23:11 ☽ △ ♂
29 02:14 ♀ ☌ ♆
07:50 ☽ □ ♆
08:20 ☽ □ ♀
21:07 ☽ ⚹ ☿
30 07:07 ☉ △ ♅
09:02 ☿ ♐
19:16 ☽ ⚹ ♄
31 02:54 ☽ △ ♆
17:53 ☽ ⚹ ♆
18:13 ☽ ∥ ♅
23:17 ☽ ⚹ ♀
23:24 ☽ ∥ ♃

⚷ Chiron

01 Dec.	13 S 37
01	27♑49R
04	27 48
07	27 48D
10	27 49
13	27 50
16	27 52
19	27 55
22	27 58
25	28 02
28	28 07
31	28 12

05 03:10 27♑48 D

November 2005

Day	S. T.	☉	☽	☿	♀	♂	♃	♄	♅	♆	♇	☊ True
	h m s	° ′ ″	° ′ ″	° ′	° ′	° ′	° ′	° ′	° ′	° ′	° ′	° ′
01 Tu	02 41 31	08♏ 39 49	25♎ 59 11	01♐ 52	25♐ 38	17♉R15	01♏ 17	10♌ 53	06♓R56	14♒ 49	22♐ 44	13♈R37
02 We	02 45 28	09 39 53	08♏ 57 04	02 57	26 39	16 54	01 30	10 56	06 56	14 50	22 46	13 35
03 Th	02 49 24	10 39 59	22 11 47	04 01	27 40	16 33	01 43	10 58	06 55	14 50	22 48	13 31
04 Fr	02 53 21	11 40 06	05♐ 42 00	05 01	28 40	16 12	01 56	11 00	06 54	14 50	22 50	13 26
05 Sa	02 57 18	12 40 15	19 25 27	05 59	29 40	15 50	02 09	11 02	06 54	14 50	22 51	13 22
06 Su	03 01 14	13 40 26	03♑ 19 25	06 52	00♑ 39	15 29	02 22	11 04	06 53	14 51	22 53	13 17
07 Mo	03 05 11	14 40 39	17 20 59	07 42	01 38	15 08	02 35	11 05	06 53	14 51	22 55	13 13
08 Tu	03 09 07	15 40 53	01♒ 27 29	08 28	02 37	14 47	02 48	11 07	06 52	14 51	22 57	13 10
09 We	03 13 04	16 41 08	15 36 34	09 08	03 35	14 26	03 01	11 09	06 52	14 52	22 59	13 09
10 Th	03 17 00	17 41 25	29 46 17	09 44	04 32	14 05	03 13	11 10	06 51	14 52	23 01	13D 10
11 Fr	03 20 57	18 41 43	13♓ 54 56	10 13	05 30	13 44	03 26	11 11	06 51	14 53	23 03	13 11
12 Sa	03 24 54	19 42 03	28 00 48	10 35	06 26	13 24	03 39	11 13	06 51	14 53	23 05	13 12
13 Su	03 28 50	20 42 24	12♈ 02 01	10 49	07 22	13 04	03 52	11 14	06 51	14 54	23 07	13 13
14 Mo	03 32 47	21 42 46	25 56 18	10 56	08 18	12 45	04 05	11 15	06 51	14 55	23 09	13R 12
15 Tu	03 36 43	22 43 10	09♉ 40 58	10R 54	09 13	12 26	04 17	11 16	06 51	14 55	23 11	13 10
16 We	03 40 40	23 43 36	23 13 15	10 42	10 08	12 08	04 30	11 16	06D 51	14 56	23 13	13 06
17 Th	03 44 36	24 44 03	06♊ 30 39	10 20	11 01	11 50	04 43	11 17	06 51	14 57	23 15	13 00
18 Fr	03 48 33	25 44 32	19 31 24	09 47	11 55	11 32	04 56	11 18	06 51	14 57	23 17	12 52
19 Sa	03 52 29	26 45 02	02♋ 14 53	09 04	12 47	11 16	05 08	11 18	06 51	14 58	23 19	12 44
20 Su	03 56 26	27 45 34	14 41 43	08 11	13 39	10 59	05 21	11 18	06 51	14 59	23 21	12 37
21 Mo	04 00 23	28 46 08	26 53 49	07 09	14 30	10 44	05 33	11 19	06 51	15 00	23 23	12 30
22 Tu	04 04 19	29 46 44	08♌ 54 15	05 59	15 21	10 29	05 46	11 19	06 51	15 01	23 25	12 26
23 We	04 08 16	00♐ 47 21	20 47 03	04 42	16 10	10 15	05 58	11R 19	06 52	15 01	23 27	12 23
24 Th	04 12 12	01 48 00	02♍ 36 58	03 21	16 59	10 01	06 11	11 19	06 52	15 02	23 29	12D 23
25 Fr	04 16 09	02 48 41	14 29 11	01 59	17 47	09 49	06 23	11 18	06 53	15 03	23 31	12 23
26 Sa	04 20 05	03 49 23	26 29 00	00 39	18 34	09 37	06 36	11 18	06 53	15 04	23 33	12 25
27 Su	04 24 02	04 50 06	08♎ 41 29	29♏ 22	19 20	09 25	06 48	11 17	06 54	15 05	23 35	12 25
28 Mo	04 27 58	05 50 52	21 11 08	28 12	20 06	09 15	07 00	11 17	06 54	15 06	23 38	12R 25
29 Tu	04 31 55	06 51 38	04♏ 01 27	27 10	20 50	09 05	07 12	11 16	06 55	15 07	23 40	12 22
30 We	04 35 52	07 52 27	17 14 25	26 18	21 33	08 57	07 24	11 15	06 56	15 09	23 42	12 16

Data for	11-01-2005
Julian Day	2453675.50
Ayanamsa	23 56 13
SVP	05 ♓ 10 37
☽ ☊ Mean	12 ♈ 14 R

● ◐ PHASES ○ ◑

02	01:24	●	09♏43
09	01:57	◐	16♒46
16	00:58	○	23♉46
23	22:11	◑	01♍43

LAST ASPECT ☽ Day	h m	INGRESS Day	h m	
31	23:17	01	07:29	♏
02	14:06	03	13:55	♐
05	05:58	05	18:17	♑
06	20:18	07	21:31	♒
09	12:31	10	00:23	♓
11	15:34	12	03:23	♈
13	19:09	14	07:04	♉
16	00:59	16	12:11	♊
18	07:02	18	19:43	♋
21	04:04	21	06:10	♌
23	05:25	23	18:42	♍
25	18:09	26	06:58	♎
28	04:38	28	16:33	♏
30	15:17	30	22:33	♐

DECLINATION

Day	☉	☽	☿	♀	♂	♃	♄	♅	♆	♇
01 Tu	14S 23	11S 05	23S 13	26S 55	16N12	10S 57	17N57	09S 42	16S 32	15S 38
02 We	14 43	16 36	23 29	26 58	16 09	11 01	17 57	09 42	16 32	15 38
03 Th	15 01	21 28	23 42	27 01	16 06	11 06	17 56	09 42	16 32	15 39
04 Fr	15 20	25 19	23 55	27 03	16 04	11 10	17 56	09 43	16 32	15 39
05 Sa	15 39	27 45	24 05	27 04	16 01	11 15	17 56	09 43	16 32	15 39
06 Su	15 57	28 31	24 14	27 05	15 58	11 19	17 55	09 43	16 32	15 40
07 Mo	16 15	27 27	24 22	27 05	15 55	11 23	17 55	09 43	16 32	15 40
08 Tu	16 32	24 39	24 27	27 05	15 52	11 28	17 55	09 43	16 31	15 40
09 We	16 49	20 21	24 31	27 04	15 49	11 32	17 54	09 43	16 31	15 41
10 Th	17 07	14 54	24 32	27 02	15 46	11 37	17 54	09 43	16 31	15 41
11 Fr	17 23	08 40	24 32	27 00	15 44	11 41	17 54	09 43	16 31	15 41
12 Sa	17 40	02 02	24 29	26 57	15 41	11 45	17 54	09 43	16 31	15 42
13 Su	17 56	04N40	24 23	26 54	15 38	11 50	17 54	09 43	16 31	15 42
14 Mo	18 12	11 05	24 16	26 50	15 35	11 54	17 53	09 43	16 31	15 42
15 Tu	18 27	16 54	24 05	26 46	15 33	11 58	17 53	09 43	16 30	15 43
16 We	18 42	21 48	23 51	26 41	15 30	12 02	17 53	09 43	16 30	15 43
17 Th	18 57	25 28	23 35	26 35	15 28	12 07	17 53	09 43	16 30	15 43
18 Fr	19 12	27 44	23 15	26 29	15 25	12 11	17 53	09 43	16 30	15 44
19 Sa	19 26	28 28	22 52	26 23	15 23	12 15	17 53	09 43	16 30	15 44
20 Su	19 40	27 44	22 26	26 16	15 21	12 19	17 53	09 43	16 29	15 44
21 Mo	19 53	25 39	21 57	26 08	15 19	12 23	17 53	09 43	16 29	15 45
22 Tu	20 06	22 28	21 26	26 01	15 17	12 27	17 53	09 43	16 29	15 45
23 We	20 19	18 23	20 52	25 52	15 16	12 31	17 54	09 43	16 29	15 45
24 Th	20 31	13 36	20 17	25 43	15 14	12 36	17 54	09 43	16 28	15 45
25 Fr	20 43	08 20	19 41	25 34	15 13	12 40	17 54	09 42	16 28	15 46
26 Sa	20 55	02 41	19 06	25 25	15 12	12 44	17 54	09 42	16 28	15 46
27 Su	21 06	03S 09	18 33	25 15	15 11	12 48	17 55	09 42	16 27	15 46
28 Mo	21 17	09 00	18 02	25 04	15 10	12 52	17 55	09 42	16 27	15 46
29 Tu	21 27	14 39	17 34	24 54	15 09	12 55	17 55	09 41	16 27	15 47
30 We	21 37	19 48	17 11	24 43	15 09	12 59	17 56	09 41	16 27	15 47

ASPECTARIAN

01 10:02 ☽ ☌ ♃
14:59 ☽ ∥ ☉
19:37 ☽ ∥ ♇
20:17 ☽ △ ♅
21:56 ☽ # ♂
23:39 ☽ ∥ ♆
02 03:37 ☽ □ ♄
06:14 ☽ # ♄
10:43 ☽ □ ♆
14:06 ☽ ☍ ♂
03 07:21 ☉ □ ♄
13:48 ☽ ∥ ☿
22:42 ☽ ☌ ☿
04 02:07 ☽ □ ♅
09:20 ☽ △ ♄
15:23 ☽ ∥ ♀
16:01 ☽ ⚹ ♆
05 01:08 ☉ ∥ ♇
05:58 ☽ ☌ ♇
08:11 ♀ ♑
19:03 ☽ ☌ ♀
22:19 ☽ ⚹ ♃
06 00:20 ☿ □ ♅
01:20 ☉ # ♂
06:06 ☽ ⚹ ♅
19:08 ☽ ⚹ ☉
20:18 ☽ △ ♂
07 04:13 ☉ □ ♆
04:14 ☽ ∥ ♀
07:58 ☉ ☍ ♂
18:26 ♂ □ ♆
22:59 ☉ ∥ ♆
08 01:18 ☽ ∥ ☿
02:18 ☽ □ ♃
05:46 ♀ ⚹ ♃
12:30 ☽ ⚹ ☿
16:25 ☽ ☍ ♄
22:02 ☽ □ ♂
22:44 ☽ ☌ ♆
09 11:19 ☽ # ♄
12:31 ☽ ⚹ ♇
15:14 ☽ ∥ ☉
17:18 ☽ ∥ ♆
20:24 ☽ # ♂
20:48 ☽ ∥ ♇
10 05:57 ☽ △ ♃
08:40 ☽ ⚹ ♀
12:01 ☽ ☌ ♅
12:50 ☽ ∥ ♃
17:30 ☽ □ ☿
20:07 ☽ ∥ ♅
23:42 ☽ ⚹ ♂
11 08:45 ☽ △ ☉
15:34 ☽ □ ♇
16:21 ♂ # ♇
12 10:28 ♀ ⚹ ♅
15:26 ☽ □ ♀
20:32 ☉ # ♄
21:54 ☽ △ ☿
22:37 ☽ △ ♄
13 04:56 ☽ ⚹ ♆
18:46 ☽ # ♅
19:09 ☽ △ ♇
14 03:13 ☽ # ♃
05:44 ☿ SR
14:24 ☽ ☍ ♃
18:08 ☽ ∥ ♂
18:48 ☽ # ♇
19:01 ☽ ⚹ ♅
22:14 ☽ # ♆
23:08 ☽ △ ♀
15 02:47 ☽ □ ♄
04:26 ☽ ∥ ♄
04:44 ☽ ☌ ♂
07:26 ☽ # ☉
09:15 ☽ □ ♆
16 00:09 ♅ SD
11:33 ☽ # ☿
17 00:36 ☽ □ ♅
06:44 ☽ ☍ ☿
08:45 ☽ ⚹ ♄
09:31 ☽ # ♀
15:30 ☽ △ ♆
16:23 ♀ △ ♂
18 07:02 ☽ ☍ ♇
20:28 ♂ □ ♄
19 05:37 ☽ △ ♃
08:49 ☽ △ ♅
16:57 ☽ ⚹ ♂
21:49 ☽ ☍ ♀
20 19:15 ☽ # ♀
21 04:04 ☽ △ ☉
06:18 ☿ □ ♅
17:35 ☽ □ ♃
18:40 ☽ △ ☿
22 03:07 ☽ □ ♂
04:51 ☽ ☌ ♄
05:15 ☉ ♐
07:41 ☽ # ☿
09:02 ♄ SR
12:20 ☽ ☍ ♆
13:45 ☽ # ☉
23 02:38 ☽ ∥ ♄
05:25 ☽ △ ♇
10:00 ☽ # ♆
13:37 ☽ # ♇
16:08 ☽ ∥ ♂
16:37 ☿ ∥ ☉
24 01:21 ☽ □ ☿
04:45 ☽ # ♃
07:21 ☽ ⚹ ♃
08:37 ☽ ☍ ♅
14:44 ☽ △ ♂
15:43 ☿ ☌ ☉
17:57 ☽ # ♅
25 07:06 ☽ △ ♀
18:11 ☽ □ ♇
26 07:26 ☽ ⚹ ☿
11:54 ☿ ♏R
15:48 ☽ ⚹ ☉
27 05:02 ☽ ⚹ ♄
11:57 ♃ △ ♅
12:23 ☽ △ ♆
21:48 ☽ □ ♀
28 02:56 ☽ ∥ ♅
04:38 ☽ ⚹ ♇
05:36 ☿ # ♄
16:29 ☽ ∥ ♃
29 01:18 ☉ □ ♅
02:17 ☽ # ♂
05:06 ☽ ∥ ♇
05:19 ☽ △ ♅
05:56 ☽ ☌ ♃
08:06 ☽ ∥ ♆
09:11 ☽ ☍ ♂
12:20 ☽ ∥ ☿
13:14 ☽ □ ♄
14:58 ☽ # ♄
20:14 ☽ □ ♆
30 08:07 ☽ ⚹ ♀
10:00 ☽ ∥ ☉
15:17 ☽ ☌ ☿

⚷ Chiron

01 Dec.	13 S 48
03	28♑18
06	28 25
09	28 32
12	28 40
15	28 48
18	28 57
21	29 06
24	29 16
27	29 26
30	29 37

December 2005

Day	S. T.	☉	☽	☿	♀	♂	♃	♄	♅	♆	♇	☊ True
	h m s	° ′ ″	° ′ ″	° ′	° ′	° ′	° ′	° ′	° ′	° ′	° ′	° ′
01 Th	04 39 48	08♐53 16	00♐50 05	25♏R38	22♑16	08♉R49	07♏37	11♌R14	06♓56	15♒10	23♐44	12♈R09
02 Fr	04 43 45	09 54 07	14 46 17	25 08	22 57	08 41	07 49	11 13	06 57	15 11	23 46	12 00
03 Sa	04 47 41	10 54 59	28 58 47	24 51	23 37	08 35	08 01	11 12	06 58	15 12	23 49	11 50
04 Su	04 51 38	11 55 52	13♑21 52	24 44	24 16	08 30	08 13	11 11	06 59	15 13	23 51	11 40
05 Mo	04 55 34	12 56 46	27 49 17	24D48	24 53	08 25	08 25	11 10	07 00	15 15	23 53	11 32
06 Tu	04 59 31	13 57 41	12♒15 12	25 02	25 29	08 21	08 36	11 08	07 01	15 16	23 55	11 25
07 We	05 03 27	14 58 36	26 35 02	25 25	26 04	08 18	08 48	11 07	07 02	15 17	23 57	11 21
08 Th	05 07 24	15 59 32	10♓45 46	25 57	26 38	08 16	09 00	11 05	07 03	15 19	24 00	11 19
09 Fr	05 11 21	17 00 29	24 46 00	26 35	27 10	08 15	09 11	11 03	07 04	15 20	24 02	11D19
10 Sa	05 15 17	18 01 26	08♈35 26	27 20	27 40	08 14	09 23	11 01	07 05	15 21	24 04	11R19
11 Su	05 19 14	19 02 24	22 14 24	28 11	28 09	08 D 14	09 35	10 59	07 07	15 23	24 06	11 19
12 Mo	05 23 10	20 03 23	05♉43 18	29 07	28 36	08 15	09 46	10 57	07 08	15 24	24 09	11 16
13 Tu	05 27 07	21 04 22	19 02 11	00♐07	29 01	08 17	09 57	10 55	07 09	15 26	24 11	11 11
14 We	05 31 03	22 05 22	02♊10 39	01 11	29 25	08 20	10 09	10 53	07 11	15 27	24 13	11 04
15 Th	05 34 60	23 06 23	15 07 53	02 18	29 47	08 23	10 20	10 50	07 12	15 29	24 15	10 53
16 Fr	05 38 56	24 07 24	27 53 00	03 29	00♒06	08 27	10 31	10 48	07 13	15 30	24 18	10 41
17 Sa	05 42 53	25 08 26	10♋25 22	04 42	00 24	08 32	10 42	10 45	07 15	15 32	24 20	10 28
18 Su	05 46 50	26 09 29	22 45 01	05 57	00 40	08 38	10 53	10 42	07 17	15 34	24 22	10 15
19 Mo	05 50 46	27 10 32	04♌52 50	07 14	00 54	08 44	11 04	10 40	07 18	15 35	24 24	10 03
20 Tu	05 54 43	28 11 36	16 50 45	08 33	01 05	08 51	11 15	10 37	07 20	15 37	24 27	09 54
21 We	05 58 39	29 12 41	28 41 45	09 53	01 14	08 59	11 26	10 34	07 22	15 39	24 29	09 48
22 Th	06 02 36	00♑13 47	10♍29 49	11 15	01 21	09 07	11 36	10 31	07 23	15 40	24 31	09 44
23 Fr	06 06 32	01 14 53	22 19 42	12 38	01 26	09 16	11 47	10 28	07 25	15 42	24 33	09 42
24 Sa	06 10 29	02 16 00	04♎16 44	14 02	01 28	09 26	11 57	10 24	07 27	15 44	24 36	09 42
25 Su	06 14 25	03 17 08	16 26 28	15 27	01R 28	09 36	12 08	10 21	07 29	15 46	24 38	09 41
26 Mo	06 18 22	04 18 16	28 54 22	16 52	01 25	09 47	12 18	10 18	07 31	15 47	24 40	09 39
27 Tu	06 22 19	05 19 25	11♏45 13	18 19	01 20	09 58	12 28	10 14	07 33	15 49	24 42	09 36
28 We	06 26 15	06 20 34	25 02 35	19 46	01 12	10 10	12 39	10 10	07 35	15 51	24 44	09 29
29 Th	06 30 12	07 21 44	08♐47 54	21 13	01 02	10 23	12 49	10 07	07 37	15 53	24 47	09 20
30 Fr	06 34 08	08 22 55	22 59 50	22 42	00 49	10 36	12 59	10 03	07 39	15 55	24 49	09 09
31 Sa	06 38 05	09 24 05	07♑33 57	24 10	00 34	10 50	13 08	09 59	07 41	15 57	24 51	08 57

Data for	12-01-2005
Julian Day	2453705.50
Ayanamsa	23 56 18
SVP	05 ♓ 10 33
☽ ☊ Mean	10 ♈ 38 R

● ◐ PHASES ○ ◑

01	15:01	●	09♐31
08	09:37	◐	16♓24
15	16:15	○	23♊48
23	19:36	◑	02♎05
31	03:12	●	09♑32

LAST ASPECT ☽ Day	h m	INGRESS Day	h m	
02	15:17	03	01:43	♑
04	18:57	05	03:37	♒
06	21:59	07	05:45	♓
09	04:18	09	09:03	♈
11	10:51	11	13:47	♉
13	18:47	13	20:00	♊
15	17:11	16	04:01	♋
17	00:33	18	14:18	♌
21	01:09	21	02:39	♍
23	04:31	23	15:27	♎
25	15:53	26	02:04	♏
27	07:27	28	08:45	♐
30	03:02	30	11:36	♑

DECLINATION

Day	☉	☽	☿	♀	♂	♃	♄	♅	♆	♇
01 Th	21S47	24S04	16S52	24S31	15N09	13S03	17N56	09S41	16S26	15S47
02 Fr	21 56	27 03	16 38	24 20	15 08	13 07	17 56	09 40	16 26	15 48
03 Sa	22 05	28 23	16 28	24 08	15 09	13 11	17 57	09 40	16 25	15 48
04 Su	22 13	27 48	16 23	23 55	15 09	13 15	17 57	09 40	16 25	15 48
05 Mo	22 21	25 21	16 22	23 43	15 10	13 18	17 58	09 39	16 25	15 48
06 Tu	22 29	21 18	16 26	23 30	15 10	13 22	17 58	09 39	16 24	15 48
07 We	22 36	16 01	16 32	23 17	15 11	13 26	17 59	09 39	16 24	15 49
08 Th	22 42	09 55	16 41	23 04	15 13	13 30	18 00	09 38	16 24	15 49
09 Fr	22 48	03 24	16 53	22 51	15 14	13 33	18 00	09 38	16 23	15 49
10 Sa	22 54	03N11	17 08	22 38	15 16	13 37	18 01	09 37	16 23	15 49
11 Su	22 59	09 34	17 24	22 24	15 17	13 40	18 02	09 37	16 22	15 50
12 Mo	23 04	15 25	17 41	22 11	15 19	13 44	18 02	09 36	16 22	15 50
13 Tu	23 08	20 28	18 00	21 57	15 22	13 47	18 03	09 36	16 22	15 50
14 We	23 12	24 26	18 19	21 43	15 24	13 51	18 04	09 35	16 21	15 50
15 Th	23 16	27 06	18 39	21 29	15 27	13 54	18 05	09 34	16 21	15 50
16 Fr	23 19	28 19	19 00	21 16	15 29	13 58	18 06	09 34	16 20	15 51
17 Sa	23 21	28 02	19 20	21 02	15 32	14 01	18 06	09 33	16 20	15 51
18 Su	23 23	26 21	19 41	20 48	15 36	14 04	18 07	09 33	16 19	15 51
19 Mo	23 25	23 28	20 02	20 34	15 39	14 08	18 08	09 32	16 19	15 51
20 Tu	23 26	19 37	20 22	20 21	15 43	14 11	18 09	09 31	16 18	15 51
21 We	23 26	15 02	20 42	20 07	15 46	14 14	18 10	09 31	16 18	15 51
22 Th	23 26	09 55	21 01	19 54	15 50	14 17	18 11	09 30	16 17	15 52
23 Fr	23 26	04 26	21 20	19 40	15 54	14 21	18 12	09 29	16 17	15 52
24 Sa	23 25	01S16	21 38	19 27	15 58	14 24	18 13	09 29	16 16	15 52
25 Su	23 24	07 01	21 56	19 14	16 03	14 27	18 14	09 28	16 16	15 52
26 Mo	23 22	12 39	22 12	19 01	16 07	14 30	18 15	09 27	16 15	15 52
27 Tu	23 20	17 55	22 28	18 49	16 12	14 33	18 16	09 26	16 15	15 52
28 We	23 17	22 31	22 43	18 36	16 17	14 36	18 17	09 26	16 14	15 52
29 Th	23 14	26 03	22 57	18 24	16 22	14 39	18 19	09 25	16 14	15 53
30 Fr	23 11	28 04	23 09	18 12	16 27	14 42	18 20	09 24	16 13	15 53
31 Sa	23 06	28 13	23 21	18 01	16 32	14 44	18 21	09 23	16 12	15 53

ASPECTARIAN

Day	Time	Aspect
01	02:54	☽ ∥ ♀
	10:35	☽ □ ♅
	17:57	☽ △ ♄
02	00:42	☽ ⚹ ♆
	15:17	☽ ☌ ♇
03	06:42	☉ △ ♄
	11:44	☿ ∥ ♆
	13:22	☽ ⚹ ♅
	15:18	☽ ⚹ ♃
	15:57	☽ △ ♂
04	02:24	☿ SD
	18:55	☽ ☌ ♀
	18:57	☽ ⚹ ☿
	20:07	☿ ⚹ ♀
05	00:45	♂ ☍ ♃
	11:22	☽ ∥ ♀
	17:32	☽ □ ♂
	17:50	☽ □ ♃
	18:01	☽ ∥ ☉
	18:24	☿ ∥ ♆
	22:09	☽ ☍ ♄
06	03:04	☽ ⚹ ☉
	05:02	☽ ☌ ♆
	15:35	☽ # ♄
	19:34	☽ ⚹ ♇
	21:53	☽ ∥ ☿
	21:59	☽ □ ☿
	22:24	☽ ∥ ♆
07	00:51	☽ ∥ ♇
	03:24	☽ # ♂
	07:30	☉ ⚹ ♆
	10:23	☽ ∥ ♃
	17:41	☽ ☌ ♅
	[illegible]	[illegible] ∥ [illegible]
	20:57	☽ △ ♃
08	01:06	☽ ∥ ♅
	22:44	☽ □ ♇
09	03:16	♀ ∥ ☉
	03:19	☽ △ ☿
	04:18	☽ ⚹ ♀
10	04:05	♂ SD
	04:15	☽ △ ♄
	11:53	☽ ⚹ ♆
	17:53	☽ △ ☉
	22:00	☿ ⚹ ♀
11	00:14	☽ # ♅
	03:19	☽ △ ♇
	10:51	☽ □ ♀
	16:47	☽ # ♃
	23:39	☽ ∥ ♂
12	01:53	☽ # ♇
	02:32	☽ ⚹ ♅
	04:15	☽ # ♆
	04:33	☽ ☌ ♂
	07:22	☽ ☍ ♃
	09:22	☽ □ ♄
	10:57	☽ # ☿
	12:00	☽ ∥ ♄
	17:27	☽ □ ♆
	21:19	☿ ♐
13	04:23	☿ # ♄
	07:46	☽ # ♀
	15:37	☽ # ☉
	18:47	☽ △ ♀
	22:00	☽ ☍ ☿
[illegible]	[illegible]	[illegible] □ [illegible]
	16:02	☽ ⚹ ♄
15	00:39	☽ △ ♆
	15:58	♀ ♒
	17:11	☽ ☍ ♇
16	04:11	☉ ☌ ♇
	17:52	☽ △ ♅
	20:20	☽ ⚹ ♂
17	00:33	☽ △ ♃
	05:17	♃ □ ♄
18	15:56	☽ ☍ ♀
19	00:23	☽ # ☉
	01:14	☿ □ ♅
	05:17	☽ △ ☿
	07:46	☽ □ ♂
	11:31	☽ ☌ ♄
	12:34	☽ □ ♃
	19:37	☽ # ♀
	20:03	☽ # ☿
	21:31	☽ ☍ ♆
	22:59	☿ ∥ ♀
20	08:02	☽ ∥ ♄
	15:25	☽ △ ♇
	17:42	☽ # ♆
	19:55	☽ # ♇
	20:24	☽ ∥ ♂
21	01:09	☽ △ ☉
	03:51	☽ # ♃
	11:28	☿ △ ♄
	17:40	☽ ☍ ♅
	18:35	☉ ♑
	21:09	☽ △ ♂
[illegible]	01:11	[illegible] □ [illegible]
	01:52	☽ # ♅
	02:17	☽ ⚹ ♃
	09:23	♂ # ♇
23	04:31	☽ □ ♇
	18:22	☽ △ ♀
24	09:38	♀ SR
	12:06	☽ ⚹ ♄
	21:48	☽ ⚹ ☿
	22:40	☽ △ ♆
25	05:25	☿ ⚹ ♆
	10:19	☽ ∥ ♅
	15:53	☽ ⚹ ♇
26	04:43	☽ □ ♀
	08:17	☽ ∥ ♃
	11:03	☽ ⚹ ☉
	14:26	☽ ∥ ♇
	15:49	☽ # ♂
	16:09	☽ ∥ ♆
	16:12	☽ △ ♅
	20:40	☽ ☍ ♂
27	01:20	☽ ☌ ♃
	01:45	☽ # ♄
	04:12	☽ ∥ ♀
	07:27	☽ □ ♆
	12:41	♂ # ♆
28	00:13	♂ □ ♄
	01:18	☽ ∥ ☿
	04:37	☽ ∥ ☉
	10:43	☽ ⚹ ♀
	21:57	☽ □ ♅
29	02:15	☽ △ ♄
	06:07	☉ ⚹ ♅
	10:12	♀ # ♄
	12:05	☽ ⚹ ♆
	23:26	☽ ☌ ☿
30	01:37	☿ ∥ ☉
	03:02	☽ ☌ ♇
31	00:12	☽ ⚹ ♅
	05:25	☽ △ ♂
	09:10	☽ ⚹ ♃
	11:16	☿ ☌ ♆

⚷ Chiron

01 Dec.	13 S 44
03	29♑48
06	30 00
09	00♒12
12	00 24
15	00 37
18	00 50
21	01 04
24	01 17
27	01 31
30	01 46

06 01:18 ♒

January 2006

Day	S. T.	☉	☽	☿	♀	♂	♃	♄	♅	♆	♇	☊ True
	h m s	° ' "	° ' "	° '	° '	° '	° '	° '	° '	° '	° '	° '
01 Su	06 42 01	10♑ 25 16	22♑ 23 01	25♐ 39	00♒R17	11♉ 05	13♏ 18	09♌ R55	07♓ 43	15♒ 59	24♐ 53	08♈R45
02 Mo	06 45 58	11 26 27	07♒ 18 10	27 09	29♑ 57	11 19	13 28	09 51	07 45	16 01	24 55	08 35
03 Tu	06 49 55	12 27 37	22 10 21	28 39	29 35	11 35	13 37	09 47	07 48	16 03	24 58	08 27
04 We	06 53 51	13 28 48	06♓ 51 53	00♑ 10	29 10	11 51	13 47	09 43	07 50	16 05	25 00	08 21
05 Th	06 57 48	14 29 58	21 17 25	01 41	28 44	12 07	13 56	09 39	07 52	16 07	25 02	08 18
06 Fr	07 01 44	15 31 07	05♈ 24 16	03 12	28 15	12 24	14 05	09 35	07 55	16 09	25 04	08 16
07 Sa	07 05 41	16 32 16	19 12 06	04 44	27 45	12 42	14 14	09 31	07 57	16 11	25 06	08 16
08 Su	07 09 37	17 33 25	02♉ 42 04	06 16	27 13	13 00	14 23	09 26	08 00	16 13	25 08	08 15
09 Mo	07 13 34	18 34 34	15 56 07	07 48	26 40	13 18	14 32	09 22	08 02	16 15	25 10	08 13
10 Tu	07 17 30	19 35 42	28 56 18	09 21	26 05	13 37	14 41	09 17	08 05	16 17	25 12	08 08
11 We	07 21 27	20 36 49	11♊ 44 20	10 55	25 30	13 56	14 49	09 13	08 07	16 19	25 15	08 01
12 Th	07 25 24	21 37 56	24 21 30	12 29	24 54	14 16	14 58	09 08	08 10	16 21	25 17	07 50
13 Fr	07 29 20	22 39 03	06♋ 48 38	14 03	24 17	14 36	15 06	09 04	08 12	16 23	25 19	07 38
14 Sa	07 33 17	23 40 09	19 06 18	15 38	23 40	14 57	15 14	08 59	08 15	16 25	25 21	07 24
15 Su	07 37 13	24 41 15	01♌ 15 01	17 13	23 03	15 17	15 23	08 55	08 18	16 27	25 23	07 11
16 Mo	07 41 10	25 42 20	13 15 36	18 49	22 27	15 39	15 31	08 50	08 20	16 30	25 25	06 59
17 Tu	07 45 06	26 43 25	25 09 21	20 25	21 50	16 00	15 38	08 45	08 23	16 32	25 27	06 49
18 We	07 49 03	27 44 29	06♍ 58 21	22 02	21 15	16 22	15 46	08 40	08 26	16 34	25 29	06 42
19 Th	07 52 59	28 45 33	18 45 31	23 39	20 41	16 45	15 54	08 36	08 29	16 36	25 31	06 38
20 Fr	07 56 56	29 46 37	00♎ 34 37	25 17	20 08	17 07	16 01	08 31	08 32	16 38	25 33	06 36
21 Sa	08 00 53	00♒ 47 40	12 30 09	26 55	19 37	17 30	16 09	08 26	08 35	16 40	25 35	06D 36
22 Su	08 04 49	01 48 43	24 37 10	28 34	19 07	17 54	16 16	08 21	08 37	16 43	25 36	06 36
23 Mo	08 08 46	02 49 46	07♏ 01 00	00♒ 14	18 39	18 17	16 23	08 16	08 40	16 45	25 38	06R 36
24 Tu	08 12 42	03 50 48	19 46 51	01 54	18 13	18 41	16 30	08 11	08 43	16 47	25 40	06 34
25 We	08 16 39	04 51 50	02♐ 59 15	03 35	17 49	19 05	16 36	08 06	08 46	16 49	25 42	06 30
26 Th	08 20 35	05 52 51	16 41 08	05 16	17 28	19 30	16 43	08 01	08 49	16 52	25 44	06 24
27 Fr	08 24 32	06 53 51	00♑ 52 57	06 58	17 08	19 55	16 50	07 57	08 52	16 54	25 46	06 15
28 Sa	08 28 28	07 54 51	15 31 52	08 41	16 51	20 20	16 56	07 52	08 55	16 56	25 48	06 06
29 Su	08 32 25	08 55 51	00♒ 31 36	10 24	16 37	20 45	17 02	07 47	08 58	16 58	25 49	05 57
30 Mo	08 36 22	09 56 49	15 43 00	12 08	16 25	21 11	17 08	07 42	09 02	17 01	25 51	05 49
31 Tu	08 40 18	10 57 46	00♓ 55 30	13 52	16 15	21 37	17 14	07 37	09 05	17 03	25 53	05 43

Data for	01-01-2006
Julian Day	2453736.50
Ayanamsa	23 56 24
SVP	05 ♓ 10 31
☽ ☊ Mean	09 ♈ 00 R

●◐ PHASES ○◑

06	18:57	◐	16♈19
14	09:48	○	24♋05
22	15:15	◑	02♏28
29	14:15	●	09♒32

LAST ASPECT ☽ Day	h m	INGRESS Day	h m	
31	09:10	01	12:15	♒
03	11:45	03	12:44	♓
05	12:11	05	14:45	♈
07	14:34	07	19:09	♉
09	18:56	10	01:59	♊
12	01:46	12	10:50	♋
14	09:48	14	21:31	♌
17	00:35	17	09:49	♍
19	22:14	19	22:50	♎
22	08:55	22	10:30	♏
23	21:54	24	18:39	♐
26	15:25	26	22:32	♑
28	07:57	28	23:10	♒
30	16:01	30	22:32	♓

DECLINATION

Day	☉	☽	☿	♀	♂	♃	♄	♅	♆	♇
01 Su	23S 02	26S 23	23S 32	17S 49	16N37	14S 47	18N22	09S 22	16S 12	15S 53
02 Mo	22 57	22 42	23 42	17 38	16 43	14 50	18 23	09 21	16 11	15 53
03 Tu	22 51	17 33	23 50	17 28	16 48	14 53	18 24	09 21	16 11	15 53
04 We	22 45	11 26	23 57	17 17	16 54	14 55	18 26	09 20	16 10	15 53
05 Th	22 39	04 49	24 03	17 07	17 00	14 58	18 27	09 19	16 10	15 53
06 Fr	22 32	01N55	24 08	16 58	17 06	15 01	18 28	09 18	16 09	15 53
07 Sa	22 25	08 24	24 12	16 48	17 12	15 03	18 29	09 17	16 08	15 54
08 Su	22 17	14 22	24 14	16 40	17 18	15 06	18 31	09 16	16 08	15 54
09 Mo	22 09	19 33	24 15	16 31	17 24	15 08	18 32	09 15	16 07	15 54
10 Tu	22 01	23 43	24 14	16 23	17 30	15 11	18 33	09 14	16 07	15 54
11 We	21 52	26 39	24 13	16 16	17 37	15 13	18 35	09 13	16 06	15 54
12 Th	21 42	28 11	24 10	16 09	17 43	15 15	18 36	09 12	16 05	15 54
13 Fr	21 32	28 16	24 05	16 02	17 49	15 18	18 37	09 11	16 05	15 54
14 Sa	21 22	26 56	24 00	15 56	17 56	15 20	18 39	09 10	16 04	15 54
15 Su	21 11	24 21	23 52	15 50	18 03	15 22	18 40	09 09	16 03	15 54
16 Mo	21 00	20 44	23 44	15 45	18 09	15 24	18 41	09 08	16 03	15 54
17 Tu	20 49	16 18	23 34	15 41	18 16	15 27	18 43	09 07	16 02	15 54
18 We	20 37	11 18	23 22	15 36	18 23	15 29	18 44	09 06	16 01	15 54
19 Th	20 25	05 53	23 09	15 33	18 30	15 31	18 45	09 05	16 01	15 54
20 Fr	20 12	00 16	22 55	15 29	18 36	15 33	18 47	09 04	16 00	15 54
21 Sa	19 59	05S 25	22 39	15 27	18 43	15 35	18 48	09 03	16 00	15 54
22 Su	19 45	11 00	22 22	15 24	18 50	15 37	18 50	09 01	15 59	15 54
23 Mo	19 32	16 18	22 03	15 22	18 57	15 38	18 51	09 00	15 58	15 54
24 Tu	19 18	21 02	21 43	15 21	19 04	15 40	18 52	08 59	15 58	15 54
25 We	19 03	24 55	21 21	15 20	19 11	15 42	18 54	08 58	15 57	15 54
26 Th	18 48	27 32	20 57	15 20	19 18	15 44	18 55	08 57	15 56	15 54
27 Fr	18 33	28 30	20 33	15 19	19 25	15 45	18 56	08 56	15 56	15 54
28 Sa	18 17	27 31	20 06	15 20	19 32	15 47	18 58	08 55	15 55	15 54
29 Su	18 02	24 33	19 38	15 20	19 39	15 49	18 59	08 53	15 54	15 54
30 Mo	17 45	19 52	19 09	15 21	19 46	15 50	19 01	08 52	15 54	15 54
31 Tu	17 29	13 52	18 38	15 23	19 53	15 52	19 02	08 51	15 53	15 54

ASPECTARIAN

01 12:26 ☽ ☌ ♀
18:40 ☽ ∥ ☿
20:18 ♀ ♑ R
20:22 ☉ △ ♂
22:37 ☽ ∥ ☉
02 04:05 ☽ ☍ ♄
06:35 ☽ □ ♂
10:02 ☽ □ ♃
14:04 ☽ ☌ ♆
20:23 ☽ # ♄
03 00:24 ☽ ∥ ♀
03:05 ☽ # ♂
04:32 ☽ ⚹ ♇
05:40 ☽ ∥ ♆
06:50 ☽ ∥ ♇
10:46 ☽ ∥ ♃
11:45 ☽ ⚹ ☿
21:26 ☿ ♑
04 01:36 ☽ ☌ ♅
07:47 ☽ ∥ ♅
08:16 ☉ ⚹ ♃
08:24 ☽ ⚹ ♂
11:34 ☽ △ ♃
11:47 ☽ ⚹ ☉
05 06:19 ☽ □ ♇
11:38 ♀ # ♂
12:11 ☽ ⚹ ♀
19:45 ☽ □ ☿
06 07:10 ☽ △ ♄
18:41 ☽ ⚹ ♆
07 03:24 ☽ # ♅
10:27 ☽ △ ♇
14:34 ☽ □ ♀
08 03:10 ☽ # ♃
06:39 ☽ # ♇
07:16 ☽ △ ☿
07:41 ☽ # ♆
09:34 ☽ ⚹ ♅
09:50 ☽ # ♀
12:06 ☽ □ ♄
13:18 ☽ ∥ ♂
18:53 ☽ ∥ ♄
19:05 ☽ ☌ ♂
21:25 ☽ ☍ ♃
09 00:34 ☽ □ ♆
03:37 ☿ ⚹ ♅
05:15 ☽ △ ☉
13:41 ☽ # ☉
18:56 ☽ △ ♀
10 03:33 ☽ # ☿
17:09 ☽ □ ♅
19:16 ☽ ⚹ ♄
11 08:41 ☽ △ ♆
12 01:46 ☽ ☍ ♇
13:39 ♀ ∥ ♆
13 02:43 ☽ △ ♅
10:40 ☿ △ ♂
15:36 ☽ ⚹ ♂
16:11 ☽ ☍ ☿
16:20 ☽ △ ♃
17:32 ☿ ⚹ ♃
23:59 ♀ ☌ ☉
14 08:12 ♀ ∥ ♇
08:33 ☽ ☍ ♀
15 03:42 ☽ # ☿
09:29 ♂ ☍ ♃
15:11 ☽ ☌ ♄
22:15 ☽ # ☉
16 04:34 ☽ □ ♃
04:57 ☽ □ ♂
06:31 ☽ ☍ ♆
11:26 ☽ ∥ ♄
14:01 ☽ ∥ ♂
17 00:35 ☽ △ ♇
01:18 ☽ # ♆
01:58 ☽ # ♇
03:09 ☽ # ♀
04:13 ☽ # ♃
15:32 ☿ ☌ ♀
18 02:59 ☽ ☍ ♅
09:54 ☽ # ♅
13:55 ♂ □ ♆
18:07 ☽ ⚹ ♃
19:46 ☽ △ ♂
19 03:44 ☽ △ ♀
08:16 ♀ ∥ ♃
11:32 ☽ △ ☿
13:46 ☽ □ ♇
22:14 ☽ △ ☉
20 05:16 ☉ ♒
15:54 ☽ ⚹ ♄
21 08:20 ☽ △ ♆
13:34 ☽ □ ♀
15:24 ☽ ∥ ♅
21:32 ♂ ∥ ♄
22 01:56 ☽ ⚹ ♇
08:55 ☽ □ ☿
19:42 ☽ ∥ ♀
20:42 ☿ ♒
20:55 ☽ ∥ ♃
22:10 ☽ ∥ ♇
22:29 ☽ ∥ ♆
23 02:22 ☽ □ ♄
03:10 ☽ △ ♅
10:22 ♀ △ ♂
12:33 ☽ # ♄
13:20 ☽ # ♂
15:12 ☽ ∥ ☉
17:50 ☽ ☌ ♃
18:25 ☽ □ ♆
21:11 ☽ ⚹ ♀
21:54 ☽ ☍ ♂
24 03:28 ☽ ∥ ☿
15:09 ☉ # ♂
25 01:12 ☽ ⚹ ☿
03:37 ☽ ⚹ ☉
09:01 ☽ △ ♄
10:17 ☽ □ ♅
13:51 ☉ # ♄
26 00:18 ☽ ⚹ ♆
15:25 ☽ ☌ ♇
21:34 ☿ ☌ ☉
27 13:04 ☿ ☍ ♄
13:13 ☽ ⚹ ♅
19:07 ♀ ⚹ ♃
22:49 ☉ ☍ ♄
28 01:30 ♃ □ ♆
02:06 ☽ ☌ ♀
02:17 ☽ ⚹ ♃
07:57 ☽ △ ♂
21:26 ♆ ∥ ♇
23:29 ☿ # ♂
29 11:25 ☽ ☍ ♄
17:37 ☽ ☌ ☿
30 00:24 ☽ # ♂
02:03 ☽ ☌ ♆
02:15 ☽ □ ♃
03:24 ☽ ∥ ☿
03:41 ☽ # ♄
06:15 ☿ # ♄
08:52 ☽ □ ♂
09:22 ☽ ∥ ☉
16:01 ☽ ⚹ ♇
16:18 ☽ ∥ ♇
16:22 ☽ ∥ ♆
16:29 ☽ ∥ ♃
18:22 ☽ ∥ ♀
31 12:38 ♃ ∥ ♆
13:00 ☽ ☌ ♅
18:03 ☽ ∥ ♅

⚷ Chiron

01 Dec.	13 S 24
02	02♒00
05	02 15
08	02 29
11	02 44
14	02 59
17	03 14
20	03 29
23	03 44
26	04 00
29	04 15

February 2006

Day	S. T.	☉	☽	☿	♀	♂	♃	♄	♅	♆	♇	☊ True
	h m s	° ' "	° ' "	° '	° '	° '	° '	° '	° '	° '	° '	° '
01 We	08 44 15	11♒ 58 42	15♓ 58 53	15♒ 37	16♑R08	22♉ 03	17♏ 20	07♌R32	09♓ 08	17♒ 05	25♐ 54	05♈R38
02 Th	08 48 11	12 59 37	00♈ 45 01	17 23	16 04	22 30	17 25	07 27	09 11	17 07	25 56	05 36
03 Fr	08 52 08	14 00 30	15 08 39	19 09	16 01	22 56	17 31	07 22	09 14	17 10	25 58	05D 36
04 Sa	08 56 04	15 01 22	29 07 40	20 56	16D 02	23 23	17 36	07 17	09 17	17 12	25 59	05 37
05 Su	09 00 01	16 02 13	12♉ 42 26	22 43	16 04	23 50	17 41	07 12	09 21	17 14	26 01	05 38
06 Mo	09 03 57	17 03 02	25 55 00	24 31	16 09	24 18	17 46	07 08	09 24	17 17	26 03	05R 38
07 Tu	09 07 54	18 03 50	08♊ 48 15	26 18	16 17	24 45	17 50	07 03	09 27	17 19	26 04	05 36
08 We	09 11 51	19 04 36	21 25 18	28 06	16 26	25 13	17 55	06 58	09 30	17 21	26 06	05 32
09 Th	09 15 47	20 05 21	03♋ 49 08	29 54	16 38	25 41	17 59	06 53	09 34	17 23	26 07	05 26
10 Fr	09 19 44	21 06 05	16 02 23	01♓ 42	16 52	26 09	18 04	06 49	09 37	17 26	26 09	05 18
11 Sa	09 23 40	22 06 47	28 07 18	03 29	17 08	26 38	18 08	06 44	09 40	17 28	26 10	05 10
12 Su	09 27 37	23 07 27	10♌ 05 42	05 15	17 26	27 06	18 12	06 39	09 44	17 30	26 12	05 01
13 Mo	09 31 33	24 08 06	21 59 10	07 01	17 46	27 35	18 15	06 35	09 47	17 33	26 13	04 54
14 Tu	09 35 30	25 08 44	03♍ 49 18	08 46	18 08	28 04	18 19	06 30	09 50	17 35	26 14	04 48
15 We	09 39 27	26 09 20	15 37 52	10 28	18 31	28 33	18 22	06 26	09 54	17 37	26 16	04 44
16 Th	09 43 23	27 09 55	27 27 02	12 09	18 57	29 02	18 25	06 21	09 57	17 39	26 17	04 42
17 Fr	09 47 20	28 10 28	09♎ 19 30	13 48	19 24	29 32	18 28	06 17	10 00	17 42	26 18	04D 42
18 Sa	09 51 16	29 11 00	21 18 33	15 23	19 53	00♊ 02	18 31	06 13	10 04	17 44	26 20	04 43
19 Su	09 55 13	00♓ 11 31	03♏ 27 57	16 55	20 23	00 31	18 34	06 08	10 07	17 46	26 21	04 45
20 Mo	09 59 09	01 12 01	15 51 54	18 22	20 55	01 01	18 36	06 04	10 11	17 48	26 22	04 47
21 Tu	10 03 06	02 12 29	28 34 49	19 45	21 28	01 31	18 39	06 00	10 14	17 51	26 23	04 47
22 We	10 07 02	03 12 56	11♐ 40 51	21 03	22 03	02 02	18 41	05 56	10 17	17 53	26 25	04R 47
23 Th	10 10 59	04 13 22	25 13 21	22 14	22 39	02 32	18 43	05 52	10 21	17 55	26 26	04 45
24 Fr	10 14 56	05 13 46	09♑ 13 59	23 18	23 17	03 03	18 44	05 48	10 24	17 57	26 27	04 41
25 Sa	10 18 52	06 14 10	23 41 59	24 15	23 55	03 33	18 46	05 44	10 28	17 59	26 28	04 38
26 Su	10 22 49	07 14 31	08♒ 33 28	25 04	24 35	04 04	18 47	05 40	10 31	18 02	26 29	04 33
27 Mo	10 26 45	08 14 51	23 41 28	25 45	25 16	04 35	18 49	05 36	10 35	18 04	26 30	04 30
28 Tu	10 30 42	09 15 09	08♓ 56 42	26 17	25 58	05 06	18 50	05 33	10 38	18 06	26 31	04 27

Data for	02-01-2006
Julian Day	2453767.50
Ayanamsa	23 56 30
SVP	05 ♓ 10 28
☽ ☊ Mean	07 ♈ 21 R

● ◐ PHASES ○ ◑

05	06:29	◐	16♉19
13	04:45	○	24♌20
21	07:17	◑	02♐31
28	00:31	●	09♓16

LAST ASPECT Day	h m	☽ INGRESS Day	h m	
01	16:06	01	22:46	♈
03	18:33	04	01:31	♉
05	21:00	06	07:33	♊
08	15:04	08	16:34	♋
10	20:54	11	03:45	♌
13	11:49	13	16:14	♍
16	03:22	16	05:10	♎
18	17:00	18	17:12	♏
20	10:03	21	02:38	♐
23	02:06	23	08:16	♑
25	00:58	25	10:15	♒
27	04:26	27	09:56	♓

DECLINATION

Day	☉	☽	☿	♀	♂	♃	♄	♅	♆	♇
01 We	17S 12	07S 07	18S 06	15S 24	20N00	15S 53	19N03	08S 50	15S 52	15S 54
02 Th	16 55	00 06	17 32	15 26	20 07	15 55	19 05	08 49	15 51	15 54
03 Fr	16 38	06N45	16 56	15 28	20 14	15 56	19 06	08 47	15 51	15 54
04 Sa	16 20	13 04	16 19	15 30	20 21	15 57	19 07	08 46	15 50	15 54
05 Su	16 02	18 36	15 41	15 33	20 28	15 58	19 09	08 45	15 49	15 54
06 Mo	15 44	23 04	15 01	15 36	20 35	16 00	19 10	08 44	15 49	15 54
07 Tu	15 25	26 17	14 20	15 38	20 42	16 01	19 11	08 43	15 48	15 54
08 We	15 06	28 08	13 38	15 41	20 49	16 02	19 13	08 41	15 47	15 54
09 Th	14 47	28 31	12 55	15 44	20 56	16 03	19 14	08 40	15 47	15 54
10 Fr	14 28	27 29	12 10	15 47	21 02	16 04	19 15	08 39	15 46	15 54
11 Sa	14 08	25 10	11 25	15 50	21 09	16 05	19 16	08 38	15 45	15 54
12 Su	13 49	21 46	10 38	15 53	21 16	16 06	19 18	08 36	15 45	15 54
13 Mo	13 29	17 30	09 51	15 56	21 22	16 07	19 19	08 35	15 44	15 54
14 Tu	13 08	12 36	09 03	15 59	21 29	16 08	19 20	08 34	15 43	15 53
15 We	12 48	07 14	08 15	16 01	21 36	16 08	19 21	08 32	15 43	15 53
16 Th	12 27	01 37	07 26	16 04	21 42	16 09	19 22	08 31	15 42	15 53
17 Fr	12 07	04S 05	06 38	16 07	21 49	16 10	19 24	08 30	15 41	15 53
18 Sa	11 46	09 41	05 50	16 09	21 55	16 10	19 25	08 29	15 41	15 53
19 Su	11 24	15 02	05 02	16 11	22 01	16 11	19 26	08 27	15 40	15 53
20 Mo	11 03	19 52	04 16	16 13	22 08	16 11	19 27	08 26	15 39	15 53
21 Tu	10 41	23 57	03 30	16 14	22 14	16 12	19 28	08 25	15 39	15 53
22 We	10 20	26 56	02 47	16 16	22 20	16 12	19 29	08 23	15 38	15 53
23 Th	09 58	28 30	02 05	16 17	22 26	16 13	19 30	08 22	15 37	15 53
24 Fr	09 36	28 18	01 26	16 18	22 32	16 13	19 31	08 21	15 37	15 53
25 Sa	09 13	26 13	00 50	16 18	22 38	16 13	19 32	08 19	15 36	15 52
26 Su	08 51	22 18	00 17	16 19	22 43	16 13	19 33	08 18	15 35	15 52
27 Mo	08 29	16 51	00N13	16 18	22 49	16 14	19 34	08 17	15 35	15 52
28 Tu	08 06	10 18	00 39	16 18	22 55	16 14	19 35	08 16	15 34	15 52

ASPECTARIAN

01 00:15 ☽ ✱ ♀
02:11 ☽ △ ♃
10:06 ☽ ✱ ♂
16:06 ☽ □ ♆
17:24 ♃ ∥ ♆
20:22 ☿ ☌ ♆
02 00:29 ☿ □ ♃
11:02 ☽ △ ♄
21:56 ☽ ✱ ☉
03 01:29 ☽ □ ♀
03:26 ☽ ✱ ♆
07:28 ☽ # ♅
07:48 ☽ ✱ ☿
09:20 ♀ SD
18:33 ☽ △ ♆
23:25 ☿ ∥ ☉
04 10:08 ☽ # ♀
11:27 ☽ # ♆
11:45 ☽ # ♇
12:02 ☽ # ♃
12:11 ☽ # ☿
12:55 ☽ # ☉
13:33 ☿ ∥ ♃
14:15 ☽ □ ♄
15:57 ☿ ∥ ♆
17:58 ☽ ✱ ♅
18:44 ☿ ∥ ♆
05 02:37 ☽ ∥ ♄
04:11 ☉ ∥ ♃
04:42 ☿ ∥ ♀
06:05 ☽ △ ♀
08:11 ☽ □ ♆
09:01 ☽ ☍ ♃
09:30 ☽ ∥ ♂
10:10 ☉ ∥ ♀
16:58 ☉ ∥ ♆
20:09 ☿ □ ♂
20:55 ☽ ☌ ♂
21:00 ☽ □ ☿
06 05:35 ☉ ☌ ♆
09:06 ♀ ∥ ☉
18:17 ☉ □ ♃
20:43 ☽ ✱ ♄
20:52 ☿ ✱ ♆
07 01:14 ☽ □ ♅
16:11 ☽ △ ♆
19:07 ☽ △ ☉
08 09:01 ☽ ☍ ♆
15:04 ☽ △ ☿
09 01:22 ☿ ♓
11:17 ☽ △ ♅
17:25 ♀ ∥ ♆
10 01:40 ☽ ☍ ♀
04:01 ☽ △ ♃
20:54 ☽ ✱ ♂
11 17:08 ☽ ☌ ♄
12 03:00 ☽ ∥ ♆
05:52 ♀ ∥ ♆
14:22 ☽ ∥ ♄
14:59 ☽ ☍ ♆
16:25 ☽ □ ♃
13 07:04 ☽ # ♃
07:54 ☽ # ♀
08:11 ☽ # ♆
08:35 ☽ △ ♆
08:59 ☽ # ♀
11:49 ☽ □ ♂
21:15 ☽ # ☉
14 11:45 ☽ ☍ ♆
12:17 ☽ ☍ ♅
13:26 ♀ ✱ ♃
14:54 ☿ ∥ ♅
15:34 ☿ ☌ ♅
18:14 ☽ # ♅
18:48 ☽ # ☿
15 02:40 ☉ ✱ ♆
05:36 ☽ ✱ ♃
06:05 ☽ △ ♀
21:38 ☽ □ ♆
16 03:22 ☽ △ ♂
17:54 ☽ ✱ ♄
17 09:27 ☽ ∥ ☿
16:50 ☽ △ ♆
18:43 ☽ ∥ ♅
21:02 ☽ □ ♀
22:45 ♂ ♊
18 08:16 ☽ ∥ ☉
09:59 ☽ ✱ ♆
17:00 ☽ △ ☉
19:26 ☉ ♓
23:21 ♀ ∥ ♃
19 02:59 ☽ ∥ ♆
04:02 ☽ ∥ ♆
05:11 ☽ □ ♄
05:27 ☽ ∥ ♃
05:29 ☽ ∥ ♀
13:00 ☽ △ ☿
15:33 ☉ □ ♂
21:45 ☽ # ♄
20 03:43 ☽ □ ♆
04:03 ☿ △ ♃
05:15 ☽ ☌ ♃
05:24 ☽ △ ☿
10:03 ☽ ✱ ♀
12:53 ☽ # ♂
21 05:41 ☽ ☍ ♂
13:37 ☽ △ ♄
21:29 ☽ □ ♅
22 11:07 ☽ ✱ ♆
18:17 ☽ □ ☿
23 02:06 ☽ ☌ ♆
16:42 ☽ ✱ ☉
22:16 ☿ ✱ ♀
24 01:59 ☽ ✱ ♅
15:53 ☽ ✱ ♃
25 00:23 ☽ ☌ ♀
00:58 ☽ ✱ ☿
16:33 ☽ △ ♂
19:24 ☽ ☍ ♄
21:55 ☽ # ♂
26 12:49 ☽ # ♄
15:05 ☽ ☌ ♆
16:17 ☽ □ ♃
27 02:10 ☽ ∥ ♀
02:29 ☽ ∥ ♃
03:50 ☽ ∥ ♆
04:26 ☽ ✱ ♆
04:59 ☽ ∥ ♆
13:20 ☉ ∥ ♅
17:44 ☽ □ ♂
28 02:40 ☽ ☌ ♅
07:04 ☽ ∥ ♅
08:01 ☽ ∥ ☉
14:40 ☿ □ ♆
15:35 ☽ ∧ ♃
18:21 ♂ ✱ ♄
22:07 ☿ ✱ ♀

⚷ Chiron

01 Dec.	12 S 52
01	04♒30
04	04 45
07	05 00
10	05 14
13	05 29
16	05 43
19	05 57
22	06 11
25	06 25
28	06 39

14 23:49 24♍15 ✷ Penumbral Lunar Eclipse (mag 1.056)
29 10:34 08♈35 ❂ Total Solar Eclipse

March 2006

Day	S. T. h m s	☉ ° ' "	☽ ° ' "	☿ ° '	♀ ° '	♂ ° '	♃ ° '	♄ ° '	♅ ° '	♆ ° '	♇ ° '	☊ True ° '
01 We	10 34 38	10♓ 15 26	24♓ 08 52	26♓ 39	26♑ 41	05♊ 37	18♏ 50	05♌ R29	10♓ 42	18♒ 08	26♐ 32	04♈ R26
02 Th	10 38 35	11 15 40	09♈ 08 18	26 52	27 25	06 09	18 51	05 26	10 45	18 10	26 33	04D 25
03 Fr	10 42 31	12 15 53	23 47 25	26R 55	28 10	06 40	18 51	05 22	10 48	18 12	26 34	04 26
04 Sa	10 46 28	13 16 04	08♉ 01 26	26 49	28 56	07 12	18 52	05 19	10 52	18 14	26 34	04 28
05 Su	10 50 24	14 16 13	21 48 31	26 34	29 43	07 43	18R 52	05 15	10 55	18 17	26 35	04 29
06 Mo	10 54 21	15 16 19	05♊ 09 17	26 10	00♒ 31	08 15	18 52	05 12	10 59	18 19	26 36	04 31
07 Tu	10 58 18	16 16 24	18 06 04	25 38	01 19	08 47	18 51	05 09	11 02	18 21	26 37	04 31
08 We	11 02 14	17 16 26	00♋ 42 14	24 58	02 08	09 19	18 51	05 06	11 06	18 23	26 38	04R 31
09 Th	11 06 11	18 16 26	13 01 42	24 13	02 58	09 51	18 50	05 03	11 09	18 25	26 38	04 30
10 Fr	11 10 07	19 16 25	25 08 22	23 22	03 49	10 23	18 49	05 00	11 12	18 27	26 39	04 28
11 Sa	11 14 04	20 16 21	07♌ 06 00	22 27	04 41	10 56	18 48	04 58	11 16	18 29	26 39	04 25
12 Su	11 18 00	21 16 14	18 57 56	21 30	05 33	11 28	18 47	04 55	11 19	18 31	26 40	04 23
13 Mo	11 21 57	22 16 06	00♍ 47 03	20 31	06 25	12 01	18 45	04 53	11 23	18 33	26 41	04 21
14 Tu	11 25 53	23 15 56	12 35 51	19 33	07 19	12 33	18 44	04 50	11 26	18 35	26 41	04 20
15 We	11 29 50	24 15 43	24 26 34	18 35	08 13	13 06	18 42	04 48	11 29	18 37	26 42	04 19
16 Th	11 33 47	25 15 29	06♎ 21 16	17 41	09 07	13 38	18 40	04 45	11 33	18 39	26 42	04D 19
17 Fr	11 37 43	26 15 13	18 22 02	16 50	10 03	14 11	18 38	04 43	11 36	18 41	26 43	04 20
18 Sa	11 41 40	27 14 55	00♏ 30 59	16 03	10 58	14 44	18 35	04 41	11 40	18 42	26 43	04 20
19 Su	11 45 36	28 14 35	12 50 27	15 21	11 55	15 17	18 33	04 39	11 43	18 44	26 43	04 21
20 Mo	11 49 33	29 14 13	25 22 55	14 44	12 51	15 50	18 30	04 37	11 46	18 46	26 44	04 22
21 Tu	11 53 29	00♈ 13 50	08♐ 11 06	14 14	13 48	16 23	18 27	04 36	11 50	18 48	26 44	04 22
22 We	11 57 26	01 13 25	21 17 39	13 49	14 46	16 57	18 24	04 34	11 53	18 50	26 44	04R 23
23 Th	12 01 22	02 12 58	04♑ 44 55	13 30	15 44	17 30	18 21	04 33	11 56	18 52	26 44	04 22
24 Fr	12 05 19	03 12 30	18 34 27	13 18	16 43	18 03	18 17	04 31	11 59	18 53	26 45	04 22
25 Sa	12 09 16	04 11 59	02♒ 46 25	13 12	17 42	18 37	18 14	04 30	12 03	18 55	26 45	04 22
26 Su	12 13 12	05 11 27	17 19 03	13D 11	18 41	19 10	18 10	04 29	12 06	18 57	26 45	04 21
27 Mo	12 17 09	06 10 53	02♓ 08 20	13 17	19 41	19 44	18 06	04 28	12 09	18 59	26 45	04 21
28 Tu	12 21 05	07 10 18	17 08 03	13 27	20 41	20 18	18 02	04 27	12 12	19 00	26 45	04D 21
29 We	12 25 02	08 09 40	02♈ 10 14	13 43	21 41	20 51	17 58	04 26	12 16	19 02	26 45	04 21
30 Th	12 28 58	09 09 00	17 06 18	14 05	22 42	21 25	17 53	04 25	12 19	19 04	26R 45	04R 21
31 Fr	12 32 55	10 08 18	01♉ 48 07	14 30	23 43	21 59	17 49	04 24	12 22	19 05	26 45	04 21

Data for 03-01-2006
Julian Day 2453795.50
Ayanamsa 23 56 34
SVP 05 ♓ 10 25
☽ ☊ Mean 05 ♈ 52 R

● ◐ PHASES ○ ◑

06	20:16	◐	16♊07
14	23:36	✷	24♍15
22	19:10	◑	02♑01
29	10:16	❂	08♈35

LAST ASPECT ☽ INGRESS

Day	h m	Day	h m	
01	04:14	01	09:19	♈
03	07:43	03	10:23	♉
05	08:15	05	14:38	♊
07	16:10	07	22:39	♋
09	20:42	10	09:43	♌
12	15:39	12	22:24	♍
15	04:33	15	11:13	♎
17	16:31	17	22:59	♏
20	07:54	20	08:43	♐
22	09:47	22	15:36	♑
23	23:31	24	19:22	♒
26	15:19	26	20:33	♓
28	15:21	28	20:32	♈
30	15:42	30	21:02	♉

DECLINATION

Day	☉	☽	☿	♀	♂	♃	♄	♅	♆	♇
01 We	07S 43	03S 11	01N00	16S 17	23N00	16S 14	19N36	08S 14	15S 33	15S 52
02 Th	07 21	04N01	01 17	16 16	23 06	16 14	19 37	08 13	15 33	15 52
03 Fr	06 58	10 52	01 29	16 14	23 11	16 14	19 38	08 12	15 32	15 52
04 Sa	06 35	16 56	01 37	16 12	23 16	16 14	19 39	08 10	15 31	15 52
05 Su	06 11	21 57	01 40	16 10	23 22	16 13	19 40	08 09	15 31	15 52
06 Mo	05 48	25 40	01 37	16 07	23 27	16 13	19 40	08 08	15 30	15 51
07 Tu	05 25	27 56	01 30	16 04	23 31	16 13	19 41	08 06	15 29	15 51
08 We	05 02	28 41	01 19	16 00	23 36	16 13	19 42	08 05	15 29	15 51
09 Th	04 38	27 59	01 03	15 56	23 41	16 12	19 43	08 04	15 28	15 51
10 Fr	04 15	25 57	00 43	15 52	23 46	16 12	19 44	08 03	15 28	15 51
11 Sa	03 51	22 47	00 20	15 47	23 50	16 11	19 44	08 01	15 27	15 51
12 Su	03 28	18 42	00S 06	15 41	23 55	16 11	19 45	08 00	15 26	15 51
13 Mo	03 04	13 55	00 34	15 36	23 59	16 10	19 46	07 59	15 26	15 51
14 Tu	02 40	08 38	01 04	15 29	24 03	16 10	19 46	07 57	15 25	15 50
15 We	02 17	03 02	01 35	15 22	24 07	16 09	19 47	07 56	15 25	15 50
16 Th	01 53	02S 42	02 06	15 15	24 11	16 08	19 47	07 55	15 24	15 50
17 Fr	01 29	08 24	02 37	15 07	24 15	16 08	19 48	07 54	15 23	15 50
18 Sa	01 06	13 51	03 07	14 59	24 18	16 07	19 48	07 52	15 23	15 50
19 Su	00 42	18 50	03 36	14 50	24 22	16 06	19 49	07 51	15 22	15 50
20 Mo	00 18	23 07	04 03	14 41	24 25	16 05	19 49	07 50	15 22	15 50
21 Tu	00N06	26 22	04 29	14 32	24 29	16 04	19 50	07 48	15 21	15 49
22 We	00 29	28 18	04 52	14 22	24 32	16 03	19 50	07 47	15 21	15 49
23 Th	00 53	28 39	05 13	14 11	24 35	16 02	19 51	07 46	15 20	15 49
24 Fr	01 17	27 14	05 32	14 00	24 38	16 01	19 51	07 45	15 20	15 49
25 Sa	01 40	24 04	05 48	13 49	24 41	16 00	19 51	07 44	15 19	15 49
26 Su	02 04	19 20	06 02	13 37	24 43	15 59	19 52	07 42	15 19	15 49
27 Mo	02 27	13 21	06 13	13 24	24 46	15 58	19 52	07 41	15 18	15 49
28 Tu	02 51	06 31	06 22	13 11	24 48	15 56	19 52	07 40	15 18	15 48
29 We	03 14	00N41	06 29	12 58	24 50	15 55	19 52	07 39	15 17	15 48
30 Th	03 38	07 49	06 33	12 44	24 52	15 54	19 53	07 37	15 17	15 48
31 Fr	04 01	14 24	06 34	12 30	24 54	15 52	19 53	07 36	15 16	15 48

ASPECTARIAN

Day	Time	Aspect
01	03:47	☽ □ ♇
	04:03	☽ ☌ ☿
	04:14	☽ ✶ ♀
	06:57	☽ # ☿
	11:02	☉ ☌ ♅
	14:29	☽ ∥ ☿
	18:02	☽ △ ♄
	18:59	☽ ✶ ♂
02	10:49	☽ # ☉
	14:27	☽ # ♅
	14:46	☽ ✶ ♆
	20:30	☿ SR
03	04:37	☽ △ ♇
	07:43	☽ □ ♀
	09:11	♀ ∥ ♃
	18:04	☽ # ♆
	19:23	☽ □ ♄
	19:27	☽ # ♇
	20:55	☽ # ♀
	20:58	☽ # ♃
04	04:54	☽ ✶ ♅
	09:45	☽ ✶ ☉
	12:17	☽ ∥ ♄
	17:45	☽ □ ♆
	18:03	♃ SR
	18:48	☽ ☍ ♃
	22:29	☿ □ ♇
05	08:13	☽ ✶ ☿
	08:15	☽ ∥ ♂
	08:39	♀ ♒
	15:01	☽ △ ♀
06	00:05	☽ ✶ ♄
	05:55	☽ ☌ ♂
	10:46	☽ □ ♅
07	00:28	☽ △ ♆
	13:35	☽ □ ☿
	16:10	☽ ☍ ♇
08	20:18	☽ △ ♅
09	11:17	☽ △ ☉
	11:27	☽ △ ♃
	13:15	☉ △ ♃
	20:42	☽ △ ☿
10	03:56	♀ ∥ ♇
	17:00	☽ ∥ ♂
	18:44	☽ ☍ ♀
	19:42	☽ ☌ ♄
11	07:32	♀ ☍ ♄
	08:06	☽ ✶ ♂
	16:49	♂ □ ♅
	18:19	☽ ∥ ♄
	23:05	☽ ☍ ♆
	23:37	☽ □ ♃
12	02:45	☿ ☌ ☉
	13:05	☽ # ♃
	14:44	☽ # ♇
	15:39	☽ △ ♇
	15:48	☽ # ♀
	16:45	☽ # ♆
13	21:38	☽ ☍ ♅
	23:54	☽ □ ♂
14	03:01	☽ # ♅
	12:24	☽ ✶ ♃
	13:02	☽ ☍ ☿
	15:31	♀ ∥ ♆
	21:10	☿ △ ♃
15	03:27	☽ # ☉
	04:33	☽ □ ♇
	05:39	☽ # ☿
	18:30	☿ ∥ ☉
	20:48	☽ ✶ ♄
	20:50	☽ ∥ ☉
	21:15	☽ ∥ ☿
16	06:01	☽ △ ♀
	07:02	♃ □ ♆
	15:17	☽ △ ♂
	21:53	☽ ∥ ♅
17	00:37	☽ △ ♆
	11:06	☉ □ ♇
	16:31	☽ ✶ ♇
18	05:08	☽ ∥ ♀
	07:05	☽ ∥ ♆
	08:09	☽ □ ♄
	09:14	☽ ∥ ♇
	10:32	☽ ∥ ♃
	21:49	☽ △ ♅
	22:03	☽ □ ♀
19	01:09	☿ □ ♂
	04:35	☽ △ ☿
	05:08	☽ # ♄
	10:56	☽ ☌ ♃
	11:22	☽ □ ♆
20	07:54	☽ △ ☉
	08:53	☽ # ♂
	17:22	☽ △ ♄
	18:26	☉ ♈
21	06:45	☽ □ ♅
	10:48	☽ □ ☿
	11:11	☽ ✶ ♀
	15:45	☽ ☍ ♂
	19:32	☽ ✶ ♆
22	09:47	☽ ☌ ♇
23	12:37	☽ ✶ ♅
	15:02	☽ ✶ ☿
	23:31	☽ ✶ ♃
24	20:19	☽ # ♂
25	02:33	☽ ✶ ☉
	02:52	☽ ☍ ♄
	07:04	☉ △ ♄
	12:10	♀ □ ♃
	13:43	☿ SD
	13:54	♂ △ ♆
	21:40	☽ # ♄
26	01:23	☽ □ ♃
	02:24	☽ ☌ ♀
	02:40	☽ ☌ ♆
	03:09	☽ △ ♂
	06:32	♀ ☌ ♆
	14:01	☽ ∥ ♃
	14:39	☽ ∥ ♇
	15:19	☽ ✶ ♇
	16:38	☽ ∥ ♆
	23:49	☽ ∥ ♀
27	02:41	♀ △ ♂
	16:06	☽ ☌ ♅
	18:03	☽ ☌ ☿
	20:08	☽ ∥ ♅
28	00:32	☽ ∥ ☿
	01:26	☽ △ ♃
	05:14	☽ □ ♂
	11:42	☽ # ☉
	15:21	☽ □ ♇
29	03:36	☽ △ ♄
	09:00	☽ ∥ ☉
	12:39	♇ SR
	19:40	☽ # ☿
	23:23	☽ # ♅
30	03:10	☽ ✶ ♆
	07:16	☽ ✶ ♂
	09:46	☽ ✶ ♀
	15:42	☽ △ ♇
	17:05	☽ # ♀
31	03:25	☽ # ♆
	04:18	☽ □ ♄
	05:33	☽ # ♇
	05:50	☽ # ♃
	17:40	☽ ✶ ♅
	21:59	☽ ✶ ☿
	23:12	☽ ∥ ♄

⚷ Chiron

01 Dec.	12 S 16
03	06♒52
06	07 05
09	07 17
12	07 29
15	07 41
18	07 52
21	08 03
24	08 13
27	08 23
30	08 33

April 2006

Day	S. T.	☉	☽	☿	♀	♂	♃	♄	♅	♆	♇	☊ True
	h m s	° ' "	° ' "	° '	° '	° '	° '	° '	° '	° '	° '	° '
01 Sa	12 36 51	11♈07 34	16♉09 14	15♓01	24♒45	22♊33	17♏R44	04♌R24	12♓25	19♒07	26♐R45	04♈R21
02 Su	12 40 48	12 06 48	00♊05 32	15 35	25 47	23 07	17 39	04 23	12 28	19 08	26 45	04 20
03 Mo	12 44 45	13 06 00	13 35 23	16 14	26 49	23 41	17 34	04 23	12 31	19 10	26 45	04 19
04 Tu	12 48 41	14 05 10	26 39 29	16 56	27 51	24 15	17 29	04 23	12 34	19 11	26 45	04 19
05 We	12 52 38	15 04 17	09♋20 20	17 42	28 54	24 49	17 23	04 23	12 37	19 13	26 45	04 19
06 Th	12 56 34	16 03 21	21 41 37	18 31	29 56	25 23	17 18	04D 23	12 40	19 14	26 44	04D 19
07 Fr	13 00 31	17 02 24	03♌47 46	19 24	01♓00	25 57	17 12	04 23	12 43	19 16	26 44	04 19
08 Sa	13 04 27	18 01 24	15 43 26	20 20	02 03	26 32	17 06	04 23	12 46	19 17	26 44	04 20
09 Su	13 08 24	19 00 22	27 33 08	21 18	03 07	27 06	17 01	04 23	12 49	19 18	26 43	04 22
10 Mo	13 12 20	19 59 18	09♍21 05	22 20	04 10	27 41	16 55	04 24	12 52	19 20	26 43	04 23
11 Tu	13 16 17	20 58 11	21 10 59	23 24	05 15	28 15	16 48	04 24	12 55	19 21	26 43	04 24
12 We	13 20 14	21 57 02	03♎06 03	24 31	06 19	28 49	16 42	04 25	12 58	19 22	26 42	04 25
13 Th	13 24 10	22 55 51	15 08 53	25 40	07 23	29 24	16 36	04 26	13 01	19 24	26 42	04R 24
14 Fr	13 28 07	23 54 39	27 21 33	26 51	08 28	29 59	16 29	04 26	13 04	19 25	26 41	04 23
15 Sa	13 32 03	24 53 24	09♏45 36	28 04	09 33	00♋33	16 23	04 27	13 06	19 26	26 41	04 20
16 Su	13 35 60	25 52 07	22 22 05	29 20	10 38	01 08	16 16	04 28	13 09	19 27	26 40	04 17
17 Mo	13 39 56	26 50 49	05♐11 46	00♈38	11 43	01 43	16 09	04 30	13 12	19 28	26 40	04 14
18 Tu	13 43 53	27 49 29	18 15 11	01 58	12 49	02 17	16 03	04 31	13 15	19 30	26 39	04 10
19 We	13 47 49	28 48 07	01♑32 52	03 20	13 55	02 52	15 56	04 32	13 17	19 31	26 39	04 07
20 Th	13 51 46	29 46 43	15 05 10	04 44	15 01	03 27	15 49	04 34	13 20	19 32	26 38	04 06
21 Fr	13 55 43	00♉45 18	28 52 19	06 10	16 07	04 02	15 41	04 36	13 22	19 33	26 37	04 05
22 Sa	13 59 39	01 43 51	12♒54 05	07 37	17 13	04 37	15 34	04 37	13 25	19 34	26 37	04D 05
23 Su	14 03 36	02 42 23	27 09 33	09 07	18 19	05 12	15 27	04 39	13 28	19 35	26 36	04 06
24 Mo	14 07 32	03 40 53	11♓36 43	10 38	19 26	05 47	15 20	04 41	13 30	19 36	26 35	04 07
25 Tu	14 11 29	04 39 21	26 12 18	12 11	20 32	06 22	15 12	04 43	13 32	19 37	26 34	04 08
26 We	14 15 25	05 37 48	10♈51 34	13 46	21 39	06 57	15 05	04 45	13 35	19 37	26 33	04R 07
27 Th	14 19 22	06 36 12	25 28 40	15 23	22 46	07 32	14 57	04 47	13 37	19 38	26 33	04 06
28 Fr	14 23 18	07 34 36	09♉57 09	17 01	23 53	08 07	14 50	04 50	13 40	19 39	26 32	04 03
29 Sa	14 27 15	08 32 57	24 10 50	18 41	25 01	08 42	14 42	04 52	13 42	19 40	26 31	03 59
30 Su	14 31 12	09 31 17	08♊04 45	20 23	26 08	09 17	14 35	04 55	13 44	19 41	26 30	03 53

Data for	04-01-2006
Julian Day	2453826.50
Ayanamsa	23 56 37
SVP	05 ♓ 10 19
☽ ☊ Mean	04 ♈ 14 R

● ◐ PHASES ○ ◑

Day	Time	Phase	Position
05	12:01	◐	15♋34
13	16:40	○	23♎37
21	03:29	◑	00♒54
27	19:45	●	07♉24

LAST ASPECT ☽ INGRESS

Day	h m	Day	h m	Sign
01	15:53	01	23:50	♊
04	02:25	04	06:15	♋
05	17:20	06	16:26	♌
08	23:02	09	04:59	♍
11	14:59	11	17:47	♎
13	22:42	14	05:09	♏
15	18:29	16	14:20	♐
18	18:41	18	21:14	♑
20	01:16	21	01:57	♒
22	23:04	23	04:44	♓
25	00:36	25	06:13	♈
27	01:45	27	07:28	♉
29	01:32	29	09:58	♊

DECLINATION

Day	☉	☽	☿	♀	♂	♃	♄	♅	♆	♇
01 Sa	04N24	20N04	06S 34	12S 16	24N56	15S 51	19N53	07S 35	15S 16	15S 48
02 Su	04 47	24 27	06 31	12 01	24 58	15 50	19 53	07 34	15 15	15 48
03 Mo	05 10	27 21	06 26	11 45	24 59	15 48	19 53	07 33	15 15	15 47
04 Tu	05 33	28 38	06 19	11 29	25 01	15 46	19 53	07 32	15 14	15 47
05 We	05 56	28 22	06 10	11 13	25 02	15 45	19 53	07 30	15 14	15 47
06 Th	06 19	26 40	05 59	10 56	25 03	15 43	19 53	07 29	15 13	15 47
07 Fr	06 42	23 47	05 46	10 39	25 04	15 42	19 53	07 28	15 13	15 47
08 Sa	07 04	19 55	05 32	10 22	25 04	15 40	19 53	07 27	15 12	15 47
09 Su	07 27	15 18	05 15	10 04	25 05	15 38	19 53	07 26	15 12	15 47
10 Mo	07 49	10 08	04 57	09 46	25 05	15 36	19 53	07 25	15 12	15 46
11 Tu	08 11	04 36	04 37	09 27	25 06	15 35	19 53	07 24	15 11	15 46
12 We	08 33	01S 08	04 16	09 08	25 06	15 33	19 53	07 23	15 11	15 46
13 Th	08 55	06 53	03 53	08 49	25 06	15 31	19 53	07 22	15 10	15 46
14 Fr	09 17	12 28	03 29	08 30	25 06	15 29	19 53	07 21	15 10	15 46
15 Sa	09 38	17 38	03 03	08 10	25 05	15 27	19 52	07 20	15 10	15 46
16 Su	10 00	22 09	02 36	07 49	25 05	15 25	19 52	07 18	15 09	15 46
17 Mo	10 21	25 41	02 08	07 29	25 04	15 23	19 52	07 17	15 09	15 45
18 Tu	10 42	27 57	01 38	07 08	25 03	15 21	19 51	07 16	15 09	15 45
19 We	11 03	28 40	01 07	06 47	25 03	15 19	19 51	07 15	15 08	15 45
20 Th	11 24	27 42	00 35	06 25	25 01	15 17	19 51	07 14	15 08	15 45
21 Fr	11 44	25 03	00 02	06 04	25 00	15 15	19 50	07 14	15 08	15 45
22 Sa	12 05	20 51	00N33	05 42	24 59	15 13	19 50	07 13	15 07	15 45
23 Su	12 25	15 25	01 09	05 20	24 57	15 11	19 50	07 12	15 07	15 45
24 Mo	12 45	09 04	01 45	04 57	24 55	15 09	19 49	07 11	15 07	15 45
25 Tu	13 04	02 10	02 23	04 34	24 53	15 07	19 49	07 10	15 07	15 44
26 We	13 24	04N52	03 02	04 11	24 51	15 05	19 48	07 09	15 06	15 44
27 Th	13 43	11 37	03 42	03 48	24 49	15 02	19 48	07 08	15 06	15 44
28 Fr	14 02	17 42	04 22	03 25	24 47	15 00	19 47	07 07	15 06	15 44
29 Sa	14 21	22 41	05 04	03 01	24 44	14 58	19 46	07 06	15 06	15 44
30 Su	14 40	26 16	05 46	02 38	24 42	14 56	19 46	07 05	15 05	15 44

ASPECTARIAN

01 02:40 ☽ ☍ ♃
05:02 ☽ □ ♆
15:53 ☽ □ ♀
02 03:30 ☽ ∥ ♂
07:33 ☽ ⚹ ♄
22:04 ☽ □ ♅
22:34 ♀ ⚹ ♇
23:03 ☽ ⚹ ☉
03 05:03 ☽ □ ☿
08:31 ♃ ∥ ♇
10:10 ☽ △ ♆
19:19 ☽ ☌ ♂
04 00:10 ☽ ☍ ♇
02:25 ☽ △ ♀
15:25 ☿ △ ♃
05 06:21 ☽ △ ♅
09:55 ☿ # ☉
12:56 ♄ SD
15:28 ☽ △ ♃
17:20 ☽ △ ☿
06 01:21 ♀ ♓
14:29 ☽ ∥ ♂
07 01:10 ☽ ☌ ♄
08 00:09 ☽ ∥ ♄
02:47 ☽ □ ♃
05:05 ☽ △ ☉
07:14 ☽ ☍ ♆
08:20 ♂ ☍ ♇
21:39 ☽ # ♇
22:19 ☽ △ ♇
22:20 ☽ # ♃
23:02 ☽ ⚹ ♂
23:19 ☉ # ♅
[illegible]
07:33 ☉ ⚹ ♆
12:25 ☽ ☍ ♀
10 01:46 ☽ # ♀
07:11 ☽ ☍ ♅
09:36 ☽ ∥ ☉
12:01 ☽ # ♅
15:13 ☽ ⚹ ♃
23:56 ☽ # ☿
11 04:56 ☽ ☍ ☿
11:09 ☽ □ ♇
14:59 ☽ □ ♂
12 02:38 ☽ ⚹ ♄
12:19 ☽ ∥ ☿
20:37 ♀ # ☉
13 02:03 ☽ ∥ ♅
07:48 ☽ ∥ ♀
08:24 ☽ △ ♆
09:15 ☽ # ☉
20:54 ☿ □ ♇
22:42 ☽ ⚹ ♇
14 00:59 ♂ ♋
05:21 ☽ △ ♂
12:17 ☽ ∥ ♆
13:41 ☽ ∥ ♃
13:46 ☽ □ ♄
15:04 ☽ ∥ ♇
23:33 ☽ △ ♀
15 06:26 ☽ △ ♅
11:24 ☽ # ♄
12:33 ☽ ☌ ♃
18:29 ☽ □ ♆
16 12:20 ☿ ♈
14:33 ☽ △ ☿
19:18 ☽ # ♂
19:34 ☉ △ ♇
22:42 ☽ △ ♄
17 13:09 ☽ □ ♀
13:51 ♀ ∥ ♅
14:48 ☽ □ ♅
18 02:16 ☽ ⚹ ♆
09:44 ♀ ☌ ♅
09:56 ☿ □ ♂
15:12 ☽ ☌ ♇
18:41 ☽ △ ☉
19 02:28 ☽ ☍ ♂
03:33 ☽ □ ☿
20:54 ☽ ⚹ ♅
21:09 ☿ △ ♄
23:51 ☽ ⚹ ♀
20 01:16 ☽ ⚹ ♃
05:26 ☉ ♉
15:46 ♀ △ ♃
21 00:17 ☽ # ♂
09:51 ☽ ☍ ♄
13:58 ☽ ⚹ ☿
22 04:29 ☽ □ ♃
04:55 ☽ # ♄
11:16 ☽ ☌ ♆
22:40 ☽ ∥ ♇
23:04 ☽ ⚹ ♇
23 00:56 ☽ ∥ ♃
01:12 ☽ ∥ ♆
09:55 ☽ ⚹ ☉
11:07 ☽ # ☉
13:56 ☽ △ ♂
24 03:07 ☽ ☌ ♅
06:05 ☽ △ ♃
06:43 ☽ ∥ ♅
13:56 ☽ ☌ ♀
15:17 ☽ ∥ ♀
23:20 ☽ # ☿
25 00:36 ☽ □ ♇
01:33 ☉ □ ♄
[illegible]
13:59 ☽ △ ♄
17:04 ☽ ∥ ☿
17:19 ☽ □ ♂
21:49 ☽ # ♀
26 05:21 ☽ ☌ ☿
07:55 ☽ # ♅
14:23 ☽ ⚹ ♆
27 01:45 ☽ △ ♇
02:34 ☿ # ♀
08:20 ☽ ∥ ☉
12:59 ☽ # ♃
13:18 ☽ # ♆
15:27 ☽ □ ♄
15:51 ☽ # ♇
20:48 ☽ ⚹ ♂
28 06:14 ☽ ⚹ ♅
08:06 ☽ ☍ ♃
09:18 ☽ ∥ ♄
16:19 ☽ □ ♆
29 01:32 ☽ ⚹ ♀
09:33 ☉ ⚹ ♂
12:19 ☽ ∥ ♂
13:56 ☿ ⚹ ♆
18:27 ☽ ⚹ ♄
30 07:45 ♀ □ ♇
09:59 ☽ □ ♅
18:48 ☉ # ♃
20:34 ☽ △ ♆

⚷ Chiron

01 Dec.	11 S 38
02	08♒42
05	08 50
08	08 58
11	09 05
14	09 12
17	09 18
20	09 24
23	09 29
26	09 33
29	09 37

May 2006

Day	S. T. h m s	☉ ° ' "	☽ ° ' "	☿ ° '	♀ ° '	♂ ° '	♃ ° '	♄ ° '	♅ ° '	♆ ° '	♇ ° '	☊ True ° '
01 Mo	14 35 08	10♉ 29 34	21♊ 35 45	22♈ 07	27♓ 15	09♋ 53	14♏R27	04♌ 57	13♓ 47	19♒ 41	26♐R29	03♈R48
02 Tu	14 39 05	11 27 50	04♋ 42 49	23 53	28 23	10 28	14 20	05 00	13 49	19 42	26 28	03 43
03 We	14 43 01	12 26 04	17 26 56	25 40	29 31	11 03	14 12	05 03	13 51	19 43	26 27	03 39
04 Th	14 46 58	13 24 15	29 50 51	27 29	00♈ 38	11 39	14 04	05 06	13 53	19 43	26 26	03 36
05 Fr	14 50 54	14 22 25	11♌ 58 30	29 20	01 46	12 14	13 57	05 09	13 55	19 44	26 25	03 35
06 Sa	14 54 51	15 20 33	23 54 40	01♉ 13	02 54	12 49	13 49	05 12	13 57	19 45	26 24	03D 36
07 Su	14 58 47	16 18 38	05♍ 44 27	03 08	04 02	13 25	13 41	05 15	13 59	19 45	26 23	03 38
08 Mo	15 02 44	17 16 42	17 32 57	05 04	05 11	14 00	13 34	05 19	14 01	19 46	26 22	03 39
09 Tu	15 06 41	18 14 44	29 25 01	07 02	06 19	14 36	13 26	05 22	14 03	19 46	26 21	03 40
10 We	15 10 37	19 12 44	11♎ 24 54	09 02	07 27	15 11	13 19	05 26	14 05	19 46	26 19	03R 40
11 Th	15 14 34	20 10 42	23 36 10	11 04	08 36	15 47	13 11	05 29	14 07	19 47	26 18	03 38
12 Fr	15 18 30	21 08 39	06♏ 01 30	13 07	09 44	16 22	13 03	05 33	14 09	19 47	26 17	03 33
13 Sa	15 22 27	22 06 34	18 42 29	15 12	10 53	16 58	12 56	05 37	14 10	19 48	26 16	03 26
14 Su	15 26 23	23 04 28	01♐ 39 33	17 18	12 02	17 34	12 48	05 40	14 12	19 48	26 15	03 18
15 Mo	15 30 20	24 02 20	14 52 04	19 26	13 11	18 09	12 41	05 44	14 14	19 48	26 13	03 09
16 Tu	15 34 16	25 00 11	28 18 31	21 34	14 20	18 45	12 34	05 48	14 15	19 48	26 12	03 01
17 We	15 38 13	25 58 00	11♑ 56 55	23 44	15 29	19 21	12 26	05 53	14 17	19 49	26 11	02 53
18 Th	15 42 10	26 55 48	25 45 09	25 55	16 38	19 56	12 19	05 57	14 19	19 49	26 09	02 48
19 Fr	15 46 06	27 53 35	09♒ 41 21	28 06	17 47	20 32	12 12	06 01	14 20	19 49	26 08	02 45
20 Sa	15 50 03	28 51 21	23 43 55	00♊ 17	18 56	21 08	12 05	06 05	14 22	19 49	26 07	02 43
21 Su	15 53 59	29 49 06	07♓ 51 35	02 29	20 06	21 44	11 58	06 10	14 23	19 49	26 05	02D 43
22 Mo	15 57 56	00♊ 46 50	22 03 09	04 40	21 15	22 20	11 51	06 14	14 24	19 49	26 04	02 44
23 Tu	16 01 52	01 44 32	06♈ 17 05	06 51	22 25	22 55	11 44	06 19	14 26	19 R 49	26 03	02R 44
24 We	16 05 49	02 42 14	20 31 15	09 01	23 34	23 31	11 37	06 24	14 27	19 49	26 01	02 42
25 Th	16 09 45	03 39 55	04♉ 42 40	11 10	24 44	24 07	11 31	06 29	14 28	19 49	26 00	02 39
26 Fr	16 13 42	04 37 34	18 47 29	13 17	25 54	24 43	11 24	06 33	14 29	19 49	25 58	02 33
27 Sa	16 17 39	05 35 13	02♊ 41 28	15 23	27 03	25 19	11 18	06 38	14 31	19 49	25 57	02 24
28 Su	16 21 35	06 32 50	16 20 33	17 28	28 13	25 55	11 11	06 43	14 32	19 49	25 56	02 14
29 Mo	16 25 32	07 30 26	29 41 30	19 30	29 23	26 31	11 05	06 48	14 33	19 48	25 54	02 03
30 Tu	16 29 28	08 28 01	12♋ 42 36	21 31	00♉ 33	27 07	10 59	06 54	14 34	19 48	25 53	01 53
31 We	16 33 25	09 25 35	25 23 42	23 29	01 43	27 43	10 53	06 59	14 35	19 48	25 51	01 44

Data for	05-01-2006
Julian Day	2453856.50
Ayanamsa	23 56 41
SVP	05 ♓ 10 15
☽ ☊ Mean	02 ♈ 38 R

● ◐ PHASES ○ ◑

05	05:13	◐	14♌35
13	06:52	○	22♏23
20	09:21	◑	29♒14
27	05:25	●	05♊48

LAST ASPECT Day	h m	☽ INGRESS Day	h m	
01	11:14	01	15:18	♋
03	18:35	04	00:18	♌
06	05:02	06	12:20	♍
08	17:49	09	01:10	♎
11	05:15	11	12:25	♏
13	06:52	13	20:57	♐
15	20:16	16	03:00	♑
18	02:11	18	07:20	♒
20	09:21	20	10:40	♓
22	06:46	22	13:24	♈
24	09:16	24	16:01	♉
26	10:39	26	19:19	♊
28	23:23	29	00:34	♋
31	04:42	31	08:52	♌

DECLINATION

Day	☉	☽	☿	♀	♂	♃	♄	♅	♆	♇
01 Mo	14N58	28N14	06N29	02S 14	24N39	14S 54	19N45	07S 05	15S 05	15S 44
02 Tu	15 16	28 33	07 13	01 50	24 36	14 52	19 45	07 04	15 05	15 44
03 We	15 34	27 18	07 57	01 26	24 33	14 49	19 44	07 03	15 05	15 44
04 Th	15 52	24 44	08 42	01 01	24 29	14 47	19 43	07 02	15 05	15 43
05 Fr	16 09	21 07	09 28	00 37	24 26	14 45	19 42	07 01	15 04	15 43
06 Sa	16 26	16 41	10 14	00 12	24 22	14 43	19 42	07 01	15 04	15 43
07 Su	16 43	11 40	11 00	00N12	24 18	14 41	19 41	07 00	15 04	15 43
08 Mo	17 00	06 15	11 47	00 37	24 14	14 38	19 40	06 59	15 04	15 43
09 Tu	17 16	00 35	12 34	01 02	24 10	14 36	19 39	06 58	15 04	15 43
10 We	17 32	05S 09	13 21	01 27	24 06	14 34	19 38	06 58	15 04	15 43
11 Th	17 47	10 48	14 08	01 52	24 01	14 32	19 38	06 57	15 04	15 43
12 Fr	18 03	16 08	14 55	02 17	23 57	14 30	19 37	06 56	15 04	15 43
13 Sa	18 18	20 54	15 41	02 42	23 52	14 28	19 36	06 56	15 03	15 43
14 Su	18 33	24 46	16 27	03 07	23 47	14 25	19 35	06 55	15 03	15 43
15 Mo	18 47	27 24	17 12	03 33	23 42	14 23	19 34	06 54	15 03	15 42
16 Tu	19 01	28 31	17 56	03 58	23 37	14 21	19 33	06 54	15 03	15 42
17 We	19 15	27 56	18 39	04 23	23 32	14 19	19 32	06 53	15 03	15 42
18 Th	19 28	25 37	19 21	04 48	23 26	14 17	19 31	06 53	15 03	15 42
19 Fr	19 42	21 46	20 02	05 13	23 20	14 15	19 30	06 52	15 03	15 42
20 Sa	19 54	16 39	20 40	05 39	23 15	14 13	19 29	06 52	15 03	15 42
21 Su	20 07	10 37	21 17	06 04	23 09	14 11	19 28	06 51	15 03	15 42
22 Mo	20 19	04 02	21 52	06 29	23 03	14 09	19 27	06 51	15 03	15 42
23 Tu	20 31	02N47	22 24	06 54	22 56	14 07	19 25	06 50	15 03	15 42
24 We	20 42	09 27	22 54	07 19	22 50	14 05	19 24	06 50	15 03	15 42
25 Th	20 53	15 38	23 22	07 43	22 43	14 03	19 23	06 49	15 03	15 42
26 Fr	21 04	20 56	23 47	08 08	22 36	14 01	19 22	06 49	15 03	15 42
27 Sa	21 14	25 00	24 09	08 33	22 30	14 00	19 21	06 48	15 03	15 42
28 Su	21 24	27 34	24 29	08 57	22 23	13 58	19 19	06 48	15 03	15 42
29 Mo	21 34	28 29	24 46	09 22	22 15	13 56	19 18	06 48	15 03	15 42
30 Tu	21 43	27 47	25 00	09 46	22 08	13 54	19 17	06 47	15 04	15 42
31 We	21 52	25 37	25 11	10 10	22 01	13 53	19 16	06 47	15 04	15 42

ASPECTARIAN

01 01:05 ☽ ✶ ☿
08:51 ☽ ☍ ♇
09:04 ☉ # ♆
11:14 ☽ □ ♀
19:07 ☿ # ♅
02 11:16 ☽ ☌ ♂
13:41 ☽ ✶ ☉
17:08 ☽ △ ♅
17:52 ☽ △ ♃
03 10:15 ☿ △ ♇
10:25 ♀ ♈
12:41 ☉ # ♇
18:35 ☽ □ ☿
04 01:43 ☽ △ ♀
01:53 ☽ ∥ ♂
10:23 ☽ ☌ ♄
12:19 ☉ ✶ ♅
14:37 ☉ ☍ ♃
05 03:50 ♃ △ ♅
03:54 ☽ □ ♃
08:02 ☽ ∥ ♄
08:28 ☿ ♉
15:35 ☽ ☍ ♆
06 01:10 ☽ ∥ ☉
04:47 ☽ # ♇
05:02 ☽ △ ♇
07:58 ☽ # ♆
09:47 ☽ # ♃
17:40 ☽ △ ☿
07 02:37 ☽ ∥ ☿
09:14 ♂ △ ♃
15:59 ☽ ✶ ♃
16:25 ☽ ✶ ♂
16:49 ☽ ☍ ♅
20:48 ☽ # ♅
23:24 ☽ △ ☉
08 00:38 ♂ △ ♅
02:57 ♀ △ ♄
03:02 ☿ □ ♄
17:49 ☽ □ ♇
09 07:18 ☽ # ♀
12:00 ☽ ✶ ♄
15:17 ☽ ☍ ♀
10 07:34 ☽ ∥ ♅
07:51 ☽ □ ♂
14:04 ☉ □ ♆
16:31 ☽ △ ♆
11 05:15 ☽ ✶ ♇
11:44 ☿ # ♃
16:27 ☽ ∥ ♃
17:18 ☽ # ☿
18:58 ☽ ∥ ♆
22:00 ☽ ∥ ♇
23:05 ☽ □ ♄
23:20 ☿ ☍ ♃
12 04:35 ☿ # ♆
09:42 ☽ # ☉
12:03 ☿ ✶ ♅
13:14 ☽ ☌ ♃
15:27 ☽ △ ♅
16:07 ☽ ☍ ☿
17:05 ☽ # ♄
20:35 ☽ △ ♂
13 00:54 ☿ # ♇
02:02 ☽ □ ♆
17:27 ☽ # ♂
14 04:05 ☿ ✶ ♂
07:23 ☽ △ ♄
20:40 ☽ △ ♀
22:51 ☽ □ ♅
15 04:12 ☿ □ ♆
08:51 ☽ ✶ ♆
20:16 ☽ ☌ ♇
17 00:51 ☽ ✶ ♃
04:05 ☽ ✶ ♅
06:43 ☽ □ ♀
13:28 ☽ ☍ ♂
18 00:19 ☽ △ ☿
02:11 ☽ △ ☉
04:11 ☉ ∥ ♄
05:30 ☿ ∥ ♄
06:12 ☿ ∥ ☉
14:59 ☽ # ♂
17:40 ☽ ☍ ♄
20:02 ☿ ☌ ☉
19 04:16 ☽ □ ♃
07:45 ☽ # ☿
09:57 ☽ # ☉
11:19 ☽ # ♄
15:06 ☽ ✶ ♀
17:19 ☽ ☌ ♆
20:52 ☿ ♊
20 03:59 ☽ ∥ ♇
04:03 ☽ ✶ ♇
06:40 ☽ ∥ ♆
10:07 ☽ ∥ ♃
13:12 ☽ □ ☿
18:13 ♀ ✶ ♆
21 04:32 ☉ ♊
06:54 ☽ △ ♃
11:04 ☽ ☌ ♅
13:54 ☽ ∥ ♅
15:45 ☽ # ♀
22 00:29 ☽ △ ♂
06:46 ☽ □ ♇
13:07 ♆ SR
15:47 ☽ ✶ ☉
17:58 ☿ ✶ ♄
20:36 ♀ # ♅
23 00:03 ☽ △ ♄
01:07 ☽ ✶ ☿
14:24 ☽ # ♅
15:37 ☽ ∥ ♀
20:55 ☿ ∥ ♂
22:00 ♀ □ ♂
22:49 ☽ ✶ ♆
24 05:17 ☽ □ ♂
05:36 ☽ ☌ ♀
09:16 ☽ △ ♇
17:37 ☽ # ♃
21:38 ☽ # ♆
25 00:16 ☽ # ♇
03:01 ☽ □ ♄
11:28 ☽ ☍ ♃
16:23 ☽ ∥ ♄
16:38 ☽ ✶ ♅
26 00:41 ☽ ∥ ☉
01:42 ♀ △ ♇
01:45 ☽ □ ♆
08:41 ☽ ∥ ♂
10:39 ☽ ✶ ♂
13:48 ☿ □ ♅
17:35 ☽ ∥ ☿
27 06:56 ☽ ✶ ♄
020:47 ☽ □ ♅
28 02:21 ☽ ☌ ☿
04:49 ☉ ✶ ♄
06:11 ☽ △ ♆
17:08 ☽ ☍ ♇
23:23 ☽ ✶ ♀
29 03:34 ☿ △ ♆
12:41 ♀ ♉
20:48 ☽ △ ♃
30 03:28 ☽ △ ♅
31 03:22 ☽ ∥ ☿
04:42 ☽ ☌ ♂
13:00 ☉ ∥ ♂
13:27 ☽ □ ♀
22:37 ☽ ☌ ♄

⚷ Chiron

01 Dec.	11 S 10
02	09♒40
05	09 42
08	09 44
11	09 45
14	09 46
17	09 46R
20	09 45
23	09 44
26	09 42
29	09 40

15 05:49 09♒46 R

June 2006

Day	S. T.	☉	☽	☿	♀	♂	♃	♄	♅	♆	♇	☊ True
	h m s	° ' "	° ' "	° '	° '	° '	° '	° '	° '	° '	° '	° '
01 Th	16 37 21	10♊ 23 07	07♌ 46 25	25♊ 24	02♉ 53	28♋ 20	10♏R47	07♌ 04	14♓ 36	19♒R48	25♐R50	01♈R37
02 Fr	16 41 18	11 20 38	19 53 45	27 18	04 03	28 56	10 41	07 10	14 36	19 47	25 48	01 33
03 Sa	16 45 14	12 18 07	01♍ 49 51	29 09	05 13	29 32	10 36	07 15	14 37	19 47	25 47	01 31
04 Su	16 49 11	13 15 36	13 39 40	00♋ 57	06 24	00♌ 08	10 30	07 20	14 38	19 47	25 45	01D 31
05 Mo	16 53 08	14 13 03	25 28 34	02 42	07 34	00 44	10 25	07 26	14 39	19 46	25 44	01 32
06 Tu	16 57 04	15 10 29	07♎ 21 58	04 25	08 44	01 20	10 20	07 32	14 39	19 46	25 42	01R 32
07 We	17 01 01	16 07 53	19 25 03	06 05	09 54	01 57	10 15	07 37	14 40	19 45	25 41	01 31
08 Th	17 04 57	17 05 17	01♏ 42 20	07 43	11 05	02 33	10 10	07 43	14 41	19 45	25 39	01 27
09 Fr	17 08 54	18 02 40	14 17 25	09 17	12 15	03 09	10 05	07 49	14 41	19 44	25 37	01 20
10 Sa	17 12 50	19 00 01	27 12 36	10 49	13 26	03 45	10 00	07 55	14 42	19 44	25 36	01 11
11 Su	17 16 47	19 57 22	10♐ 28 27	12 18	14 36	04 22	09 56	08 01	14 42	19 43	25 34	01 00
12 Mo	17 20 43	20 54 42	24 03 48	13 44	15 47	04 58	09 51	08 07	14 42	19 42	25 33	00 48
13 Tu	17 24 40	21 52 02	07♑ 55 40	15 08	16 58	05 34	09 47	08 13	14 43	19 42	25 31	00 36
14 We	17 28 37	22 49 20	21 59 53	16 28	18 08	06 11	09 43	08 19	14 43	19 41	25 30	00 26
15 Th	17 32 33	23 46 38	06♒ 11 44	17 46	19 19	06 47	09 39	08 25	14 43	19 40	25 28	00 18
16 Fr	17 36 30	24 43 56	20 26 43	19 00	20 30	07 24	09 36	08 32	14 43	19 40	25 26	00 13
17 Sa	17 40 26	25 41 13	04♓ 41 11	20 12	21 41	08 00	09 32	08 38	14 44	19 39	25 25	00 10
18 Su	17 44 23	26 38 30	18 52 37	21 20	22 52	08 37	09 29	08 44	14 44	19 38	25 23	00 09
19 Mo	17 48 19	27 35 47	02♈ 59 28	22 25	24 03	09 13	09 26	08 51	14 44	19 37	25 22	00 09
20 Tu	17 52 16	28 33 03	17 00 51	23 27	25 14	09 50	09 23	08 57	14 R 44	19 36	25 20	00 08
21 We	17 56 13	29 30 20	00♉ 56 07	24 26	26 25	10 26	09 20	09 04	14 44	19 36	25 19	00 06
22 Th	18 00 09	00♋ 27 36	14 44 22	25 21	27 36	11 03	09 17	09 10	14 44	19 35	25 17	00 02
23 Fr	18 04 06	01 24 52	28 24 13	26 13	28 47	11 39	09 15	09 17	14 43	19 34	25 15	29♓ 55
24 Sa	18 08 02	02 22 07	11♊ 53 47	27 01	29 58	12 16	09 12	09 23	14 43	19 33	25 14	29 45
25 Su	18 11 59	03 19 23	25 11 00	27 46	01♊ 10	12 53	09 10	09 30	14 43	19 32	25 12	29 33
26 Mo	18 15 55	04 16 38	08♋ 13 53	28 27	02 21	13 29	09 08	09 37	14 43	19 31	25 11	29 21
27 Tu	18 19 52	05 13 53	21 01 11	29 03	03 32	14 06	09 07	09 44	14 42	19 30	25 09	29 08
28 We	18 23 48	06 11 08	03♌ 32 34	29 36	04 44	14 43	09 05	09 50	14 42	19 29	25 08	28 57
29 Th	18 27 45	07 08 22	15 49 00	00♌ 05	05 55	15 19	09 04	09 57	14 42	19 28	25 06	28 48
30 Fr	18 31 42	08 05 36	27 52 35	00 29	07 07	15 56	09 02	10 04	14 41	19 26	25 05	28 42

Data for	06-01-2006
Julian Day	2453887.50
Ayanamsa	23 56 46
SVP	05 ♓ 10 12
☽ ☊ Mean	01 ♈ 00 R

● ◐ PHASES ○ ◑

Day	h m	Phase	Position
03	23:06	◐	13♍13
11	18:04	○	20♐41
18	14:08	◑	27♓12
25	16:05	●	03♋58

LAST ASPECT ☽ INGRESS

Day	h m	Day	h m	Sign
02	17:35	02	20:18	♍
05	00:30	05	09:10	♎
07	12:16	07	20:42	♏
09	10:11	10	05:06	♐
12	02:35	12	10:20	♑
13	16:51	14	13:33	♒
16	08:24	16	16:06	♓
18	14:09	18	18:54	♈
20	21:21	20	22:23	♉
23	00:45	23	02:49	♊
25	00:03	25	08:48	♋
27	16:03	27	17:10	♌
29	18:25	30	04:16	♍

DECLINATION

Day	☉	☽	☿	♀	♂	♃	♄	♅	♆	♇
01 Th	22N00	22N17	25N20	10N34	21N53	13S 51	19N14	06S 47	15S 04	15S 42
02 Fr	22 09	18 03	25 27	10 57	21 45	13 49	19 13	06 46	15 04	15 42
03 Sa	22 16	13 11	25 31	11 21	21 37	13 48	19 11	06 46	15 04	15 42
04 Su	22 24	07 52	25 33	11 44	21 29	13 46	19 10	06 46	15 04	15 42
05 Mo	22 30	02 18	25 32	12 07	21 21	13 45	19 09	06 46	15 04	15 42
06 Tu	22 37	03S 23	25 29	12 30	21 13	13 43	19 07	06 45	15 04	15 42
07 We	22 43	09 02	25 25	12 53	21 05	13 42	19 06	06 45	15 05	15 42
08 Th	22 49	14 27	25 18	13 15	20 56	13 41	19 04	06 45	15 05	15 42
09 Fr	22 54	19 24	25 10	13 37	20 47	13 40	19 03	06 45	15 05	15 42
10 Sa	22 59	23 35	25 00	13 59	20 38	13 38	19 01	06 45	15 05	15 42
11 Su	23 04	26 40	24 49	14 21	20 29	13 37	19 00	06 44	15 05	15 42
12 Mo	23 08	28 16	24 36	14 42	20 20	13 36	18 58	06 44	15 06	15 42
13 Tu	23 11	28 10	24 22	15 03	20 11	13 35	18 57	06 44	15 06	15 42
14 We	23 15	26 15	24 06	15 24	20 02	13 34	18 55	06 44	15 06	15 42
15 Th	23 18	22 40	23 50	15 44	19 52	13 33	18 54	06 44	15 06	15 42
16 Fr	23 20	17 44	23 33	16 04	19 43	13 32	18 52	06 44	15 06	15 42
17 Sa	23 22	11 49	23 15	16 24	19 33	13 31	18 50	06 44	15 07	15 42
18 Su	23 24	05 19	22 56	16 43	19 23	13 30	18 49	06 44	15 07	15 42
19 Mo	23 25	01N25	22 36	17 02	19 13	13 29	18 47	06 44	15 07	15 42
20 Tu	23 26	08 02	22 16	17 20	19 03	13 29	18 45	06 44	15 08	15 42
21 We	23 26	14 13	21 56	17 39	18 53	13 28	18 44	06 44	15 08	15 42
22 Th	23 26	19 38	21 35	17 56	18 42	13 28	18 42	06 44	15 08	15 42
23 Fr	23 26	23 58	21 14	18 14	18 32	13 27	18 40	06 44	15 08	15 42
24 Sa	23 25	26 56	20 53	18 30	18 21	13 26	18 39	06 44	15 09	15 43
25 Su	23 24	28 20	20 31	18 47	18 10	13 26	18 37	06 45	15 09	15 43
26 Mo	23 22	28 07	20 10	19 03	18 00	13 26	18 35	06 45	15 09	15 43
27 Tu	23 20	26 23	19 50	19 18	17 49	13 25	18 33	06 45	15 10	15 43
28 We	23 18	23 23	19 29	19 33	17 37	13 25	18 31	06 45	15 10	15 43
29 Th	23 15	19 22	19 09	19 48	17 26	13 25	18 30	06 45	15 10	15 43
30 Fr	23 12	14 38	18 49	20 02	17 15	13 25	18 28	06 46	15 11	15 43

ASPECTARIAN

01 01:39 ☽ ∥ ☉
02:33 ☽ ∥ ♂
05:14 ☿ ☍ ♇
05:34 ☽ ⚹ ☉
05:52 ☽ □ ♃
17:46 ☽ ∥ ♄
23:47 ☽ ☍ ♆
02 11:49 ☽ △ ♇
11:56 ☽ # ♇
15:01 ☽ # ♆
17:35 ☽ ⚹ ☿
21:03 ☽ # ♃
03 07:37 ☽ △ ♀
07:51 ☽ ∥ ♀
11:21 ☿ ♋
17:38 ☽ ⚹ ♃
18:44 ♂ ♌
04 01:59 ☽ ☍ ♅
04:49 ☽ # ♅
21:09 ♀ □ ♄
05 00:30 ☽ □ ♇
10:52 ☉ □ ♅
11:13 ☽ ⚹ ♂
17:05 ☽ □ ☿
06 00:20 ☽ ⚹ ♄
14:11 ☽ ∥ ♅
16:56 ☽ △ ☉
07 00:40 ☽ △ ♆
06:24 ♀ ☍ ♃
12:16 ☽ ⚹ ♇
18:07 ☽ # ♀
20:29 ☽ ∥ ♃
08 01:42 ☽ □ ♂
02:52 ☽ ∥ ♆
05:46 ☽ ∥ ♇
[illegible]
13:11 ☽ △ ☿
16:05 ☽ ☌ ♃
19:46 ☽ ☍ ♀
22:10 ☽ # ♄
09 00:45 ☽ △ ♅
02:13 ♀ # ♃
07:08 ☽ # ♂
10:11 ☽ □ ♆
11:46 ☿ △ ♃
20:04 ☽ # ☉
10 09:17 ☽ # ☿
12:30 ☽ △ ♂
18:03 ☉ △ ♆
19:34 ☽ △ ♄
11 01:55 ♀ ⚹ ♅
07:32 ☽ □ ♅
16:22 ☽ ⚹ ♆
12 02:35 ☽ ☌ ♇
16:44 ☿ △ ♅
13 03:08 ♀ # ♆
03:10 ☽ ⚹ ♃
11:37 ☽ ⚹ ♅
13:37 ☽ ☍ ☿
16:51 ☽ △ ♀
14 16:42 ☽ # ☿
20:29 ☽ # ☉
21:31 ♀ # ♇
15 01:02 ☽ ☍ ♂
03:47 ☽ ☍ ♄
05:49 ☽ □ ♃
07:05 ♀ □ ♆
14:44 ☽ # ♂
[illegible]
22:41 ☽ ☌ ♆
16 00:06 ☽ □ ♀
06:42 ☽ # ♀
07:44 ☽ △ ☉
08:24 ☽ ⚹ ♇
08:37 ☽ ∥ ♇
11:02 ☽ ∥ ♆
15:02 ☿ ∥ ☉
17:23 ☉ ☍ ♇
17:23 ☽ ∥ ♃
17 08:10 ☽ △ ♃
16:58 ☽ ☌ ♅
18:52 ☽ ∥ ♅
18 04:31 ☽ △ ☿
06:05 ♂ ☌ ♄
07:23 ☽ ⚹ ♀
11:02 ☽ □ ♇
19 07:41 ♅ SR
07:42 ♂ □ ♃
10:04 ☽ △ ♄
11:07 ☽ △ ♂
19:12 ☽ # ♅
20 04:27 ☽ ⚹ ♆
11:56 ☽ □ ☿
14:18 ☽ △ ♇
20:57 ☽ # ♃
21:21 ☽ ⚹ ☉
21 03:48 ☽ # ♆
06:13 ☽ # ♇
12:26 ☉ ♋
14:13 ☽ □ ♄
14:31 ☽ ☍ ♃
[illegible]
17:15 ☽ □ ♂
19:35 ☽ ∥ ♄
19:43 ☽ ∥ ♂
23:59 ☽ ⚹ ♅
22 00:30 ♂ ∥ ♄
08:27 ☽ □ ♆
09:14 ☽ ∥ ☿
18:44 ♃ □ ♄
19:53 ☽ ⚹ ☿
20:36 ☽ ∥ ☉
23 00:45 ☽ ☌ ♀
15:48 ♀ ∥ ♂
19:29 ☽ ⚹ ♄
24 [illegible]
00:42 ☽ ⚹ ♂
05:04 ☽ □ ♅
10:39 ♀ ∥ ♄
13:45 ☽ △ ♆
25 00:03 ☽ ☍ ♇
26 01:41 ☽ △ ♃
12:06 ☽ △ ♅
27 16:03 ☽ ☌ ☿
21:03 ☿ ∥ ♀
28 00:34 ☽ ∥ ☉
02:33 ☽ ⚹ ♀
10:45 ☽ □ ♃
[illegible]
19:57 ☿ ♌
21:46 ☽ ∥ ♀
22:58 ☽ ☌ ♂
29 01:18 ☽ ∥ ☿
04:43 ☽ ∥ ♄
07:12 ☽ ☍ ♆
10:36 ☽ ∥ ♂
18:25 ☽ △ ♇
18:46 ☽ # ♇
21:24 ☽ # ♆
30 05:50 ☽ # ♃
20:41 ☽ □ ♀
22:24 ☽ ⚹ ☉
[illegible]
23:22 ☉ △ ♃

⚷ Chiron

01	Dec.	10 S 56
01	09♒36R	
04	09 33	
07	09 28	
10	09 24	
13	09 18	
16	09 12	
19	09 06	
22	08 59	
25	08 52	
28	08 44	

July 2006

Day	S. T. h m s	☉ ° ' "	☽ ° ' "	☿ ° '	♀ ° '	♂ ° '	♃ ° '	♄ ° '	♅ ° '	♆ ° '	♇ ° '	☊ True ° '
01 Sa	18 35 38	09♋02 49	09♍46 35	00♌49	08♊18	16♌33	09♏R01	10♌11	14♓R40	19♒R25	25♐R03	28♓R39
02 Su	18 39 35	10 00 02	21 35 14	01 04	09 30	17 10	09 00	10 18	14 40	19 24	25 02	28 38
03 Mo	18 43 31	10 57 14	03♎23 29	01 15	10 41	17 47	09 00	10 25	14 39	19 23	25 00	28D 38
04 Tu	18 47 28	11 54 27	15 16 40	01 21	11 53	18 23	08 59	10 32	14 39	19 22	24 59	28 39
05 We	18 51 24	12 51 39	27 20 15	01R 22	13 04	19 00	08 59	10 39	14 38	19 21	24 57	28R 38
06 Th	18 55 21	13 48 50	09♏39 27	01 19	14 16	19 37	08 59	10 47	14 37	19 19	24 56	28 35
07 Fr	18 59 17	14 46 02	22 18 44	01 11	15 28	20 14	08D 59	10 54	14 36	19 18	24 54	28 30
08 Sa	19 03 14	15 43 13	05♐21 22	00 59	16 40	20 51	08 59	11 01	14 35	19 17	24 53	28 22
09 Su	19 07 11	16 40 25	18 48 51	00 42	17 51	21 28	08 59	11 08	14 35	19 16	24 51	28 13
10 Mo	19 11 07	17 37 36	02♑40 26	00 21	19 03	22 05	09 00	11 15	14 34	19 14	24 50	28 02
11 Tu	19 15 04	18 34 47	16 52 55	29♋56	20 15	22 42	09 01	11 23	14 33	19 13	24 49	27 51
12 We	19 19 00	19 31 59	01♒20 59	29 27	21 27	23 19	09 02	11 30	14 32	19 12	24 47	27 42
13 Th	19 22 57	20 29 11	15 57 55	28 55	22 39	23 56	09 03	11 37	14 31	19 10	24 46	27 34
14 Fr	19 26 53	21 26 23	00♓36 42	28 20	23 51	24 33	09 04	11 45	14 29	19 09	24 44	27 29
15 Sa	19 30 50	22 23 35	15 11 07	27 43	25 03	25 10	09 06	11 52	14 28	19 07	24 43	27 26
16 Su	19 34 46	23 20 49	29 36 30	27 04	26 15	25 47	09 07	12 00	14 27	19 06	24 42	27D 26
17 Mo	19 38 43	24 18 02	13♈49 57	26 24	27 27	26 25	09 09	12 07	14 26	19 05	24 40	27 26
18 Tu	19 42 40	25 15 17	27 50 08	25 44	28 40	27 02	09 11	12 15	14 25	19 03	24 39	27 26
19 We	19 46 36	26 12 32	11♉36 47	25 04	29 52	27 39	09 13	12 22	14 23	19 02	24 38	27R 26
20 Th	19 50 33	27 09 48	25 10 09	24 25	01♋04	28 16	09 15	12 30	14 22	19 00	24 37	27 23
21 Fr	19 54 29	28 07 05	08♊30 40	23 48	02 17	28 53	09 18	12 37	14 20	18 59	24 35	27 18
22 Sa	19 58 26	29 04 22	21 38 33	23 13	03 29	29 31	09 21	12 45	14 19	18 57	24 34	27 11
23 Su	20 02 22	00♌01 40	04♋33 54	22 41	04 41	00♍08	09 24	12 52	14 18	18 56	24 33	27 02
24 Mo	20 06 19	00 58 59	17 16 38	22 13	05 54	00 45	09 27	13 00	14 16	18 54	24 32	26 52
25 Tu	20 10 16	01 56 19	29 46 50	21 49	07 06	01 23	09 30	13 07	14 15	18 53	24 31	26 42
26 We	20 14 12	02 53 39	12♌04 51	21 30	08 19	02 00	09 33	13 15	14 13	18 51	24 29	26 33
27 Th	20 18 09	03 51 00	24 11 42	21 16	09 31	02 37	09 37	13 23	14 11	18 49	24 28	26 26
28 Fr	20 22 05	04 48 21	06♍09 06	21 07	10 44	03 15	09 40	13 30	14 10	18 48	24 27	26 21
29 Sa	20 26 02	05 45 43	17 59 33	21 04	11 57	03 52	09 44	13 38	14 08	18 46	24 26	26 19
30 Su	20 29 58	06 43 05	29 46 24	21D 07	13 09	04 30	09 48	13 46	14 06	18 45	24 25	26D 19
31 Mo	20 33 55	07 40 28	11♎33 42	21 16	14 22	05 07	09 52	13 53	14 04	18 43	24 24	26 20

Data for	07-01-2006
Julian Day	2453917.50
Ayanamsa	23 56 52
SVP	05 ♓ 10 09
☽ ☊ Mean	29 ♓ 25 R

● ◐ PHASES ○ ◑

03	16:38	◐	11♎37
11	03:02	○	18♑42
17	19:13	◑	25♈04
25	04:32	●	02♌07

LAST ASPECT ☽ INGRESS

Day	h m	Day	h m	
02	06:59	02	17:07	♎
04	19:18	05	05:14	♏
06	19:55	07	14:14	♐
09	10:32	09	19:25	♑
11	20:59	11	21:46	♒
13	14:23	13	23:00	♓
15	19:57	16	00:39	♈
18	01:34	18	03:45	♉
20	05:49	20	08:39	♊
22	15:18	22	15:29	♋
24	09:08	25	00:25	♌
27	00:33	27	11:37	♍
29	13:06	30	00:28	♎

DECLINATION

Day	☉	☽	☿	♀	♂	♃	♄	♅	♆	♇
01 Sa	23N08	09N25	18N30	20N16	17N04	13S 25	18N26	06S 46	15S 11	15S 43
02 Su	23 04	03 54	18 12	20 29	16 52	13 25	18 24	06 46	15 12	15 43
03 Mo	22 59	01S 44	17 55	20 41	16 40	13 25	18 22	06 46	15 12	15 43
04 Tu	22 54	07 22	17 38	20 53	16 29	13 25	18 20	06 47	15 12	15 44
05 We	22 49	12 48	17 23	21 05	16 17	13 25	18 19	06 47	15 13	15 44
06 Th	22 43	17 52	17 08	21 15	16 05	13 25	18 17	06 47	15 13	15 44
07 Fr	22 37	22 17	16 55	21 26	15 53	13 25	18 15	06 48	15 14	15 44
08 Sa	22 31	25 46	16 44	21 36	15 41	13 26	18 13	06 48	15 14	15 44
09 Su	22 24	27 55	16 33	21 45	15 29	13 26	18 11	06 48	15 14	15 44
10 Mo	22 17	28 26	16 24	21 53	15 16	13 27	18 09	06 49	15 15	15 44
11 Tu	22 09	27 06	16 17	22 01	15 04	13 27	18 07	06 49	15 15	15 45
12 We	22 01	23 57	16 11	22 09	14 51	13 28	18 05	06 50	15 16	15 45
13 Th	21 53	19 15	16 07	22 16	14 39	13 28	18 03	06 50	15 16	15 45
14 Fr	21 44	13 24	16 04	22 22	14 26	13 29	18 01	06 50	15 17	15 45
15 Sa	21 35	06 50	16 03	22 27	14 13	13 30	17 59	06 51	15 17	15 45
16 Su	21 25	00N01	16 04	22 32	14 00	13 30	17 57	06 51	15 17	15 45
17 Mo	21 15	06 47	16 06	22 37	13 47	13 31	17 55	06 52	15 18	15 46
18 Tu	21 05	13 07	16 10	22 40	13 34	13 32	17 53	06 52	15 18	15 46
19 We	20 55	18 42	16 15	22 43	13 21	13 33	17 51	06 53	15 19	15 46
20 Th	20 44	23 14	16 21	22 46	13 08	13 34	17 49	06 54	15 19	15 46
21 Fr	20 32	26 29	16 29	22 48	12 54	13 35	17 47	06 54	15 20	15 46
22 Sa	20 21	28 13	16 38	22 49	12 41	13 36	17 45	06 55	15 20	15 46
23 Su	20 09	28 24	16 48	22 49	12 27	13 37	17 42	06 55	15 21	15 47
24 Mo	19 56	27 03	16 58	22 49	12 14	13 38	17 40	06 56	15 21	15 47
25 Tu	19 44	24 22	17 10	22 48	12 00	13 40	17 38	06 57	15 22	15 47
26 We	19 31	20 36	17 22	22 47	11 46	13 41	17 36	06 57	15 22	15 47
27 Th	19 17	16 02	17 34	22 45	11 32	13 42	17 34	06 58	15 23	15 47
28 Fr	19 04	10 54	17 47	22 42	11 19	13 44	17 32	06 58	15 23	15 48
29 Sa	18 50	05 26	18 00	22 38	11 05	13 45	17 30	06 59	15 24	15 48
30 Su	18 36	00S 12	18 13	22 34	10 50	13 47	17 28	07 00	15 24	15 48
31 Mo	18 21	05 50	18 25	22 30	10 36	13 48	17 25	07 01	15 25	15 48

ASPECTARIAN

01 06:12 ☿ ∥ ♄
09:56 ☽ ☍ ♅
11:41 ☽ # ♅
02 06:59 ☽ □ ♆
18:05 ♀ ⚹ ♄
19:35 ☽ ⚹ ☿
03 14:22 ☽ ⚹ ♄
16:24 ☽ △ ♀
21:30 ☽ ∥ ♅
04 06:34 ☽ ⚹ ♂
08:10 ☽ △ ♆
19:18 ☽ ⚹ ♇
19:34 ☿ SR
05 02:49 ☽ ∥ ♃
07:54 ☽ □ ☿
11:10 ☽ ∥ ♆
12:48 ♂ ☍ ♆
13:36 ☽ ∥ ♇
15:38 ☽ # ♂
20:36 ☽ # ☿
22:42 ☽ ☌ ♃
06 02:05 ☽ # ♄
02:10 ☽ □ ♄
06:57 ♀ □ ♅
07:20 ♃ SD
08:37 ☽ △ ☉
09:29 ☽ △ ♅
18:22 ☽ □ ♆
18:47 ☽ # ♀
19:55 ☽ □ ♂
19:59 ☉ △ ♅
07 01:59 ☽ # ☉
16:10 ☽ △ ☿
17:33 ♂ # ♇
08 10:16 ☽ △ ♄
16:32 ☽ □ ♅
22:09 ☽ ☍ ♀
09 00:47 ☽ ⚹ ♆
04:52 ☽ △ ♂
10:32 ☽ ☌ ♇
10 02:36 ♂ # ♆
03:36 ♀ △ ♆
10:46 ☽ ⚹ ♃
20:05 ☽ ⚹ ♅
20:18 ☿ ♋R
11 12:04 ♀ ∥ ☉
20:59 ☽ ☍ ☿
12 09:54 ☽ # ♀
11:08 ☽ # ☉
12:38 ☽ □ ♃
16:49 ☽ ☍ ♄
13 05:14 ☽ ☌ ♆
05:22 ☽ # ♄
11:56 ☽ △ ♀
13:30 ☽ # ☿
13:38 ☽ ☍ ♂
14:23 ☽ ⚹ ♇
14:52 ☽ ∥ ♇
16:45 ☽ ∥ ♆
19:55 ☽ # ♂
23:44 ☽ ∥ ♃
14 07:06 ♂ △ ♇
13:55 ☽ △ ♃
17:28 ♀ ☍ ♇
22:49 ☽ ☌ ♅
23:58 ☽ ∥ ♅
15 04:50 ♀ ⚹ ♂
12:48 ☽ △ ☉
15:48 ☽ □ ♇
17:53 ☽ □ ♀
19:57 ☽ △ ☿
16 21:04 ☽ △ ♄
17 00:19 ☽ # ♅
08:56 ☽ ⚹ ♆
18:31 ☽ △ ♇
20:33 ☽ □ ☿
22:32 ☽ △ ♂
18 01:34 ☽ ⚹ ♀
01:42 ☽ # ♃
01:46 ☽ ∥ ♂
03:253 ♂ # ♃
07:08 ☿ ☌ ☉
09:01 ☽ # ♆
10:56 ☽ # ♇
12:50 ☽ ∥ ☿
19:47 ☽ ☍ ♃
20:06 ☽ ∥ ♄
19 01:20 ☽ □ ♄
02:41 ♀ ♋
04:52 ☽ ⚹ ♅
10:31 ☽ ∥ ☉
13:03 ☽ □ ♆
21:10 ☽ ∥ ♀
22:44 ☽ ⚹ ☿
20 03:50 ☽ ⚹ ☉
05:49 ☽ □ ♂
21 07:32 ☽ ⚹ ♄
10:35 ☽ □ ♅
19:04 ☽ △ ♆
22 05:23 ☽ ☍ ♇
15:18 ☽ ⚹ ♂
18:53 ♂ ♍
23:18 ☉ ♌
23 00:15 ☽ ☌ ♀
09:06 ☽ △ ♃
18:18 ☽ △ ♅
24 09:08 ☽ ☌ ☿
25 10:48 ☽ ∥ ♀
19:01 ☽ □ ♃
26 02:20 ☽ ☌ ♄
06:23 ☽ ∥ ☉
13:20 ☽ ☍ ♆
16:16 ☽ ∥ ♄
16:38 ☽ ∥ ☿
23:40 ☿ ∥ ♄
27 00:33 ☽ △ ♇
01:11 ☽ # ♇
01:50 ♀ △ ♃
03:11 ☽ # ♆
11:06 ☽ # ♃
17:49 ☽ ☌ ♂
22:06 ☽ ∥ ♂
28 07:09 ☽ ⚹ ♃
10:19 ☽ ⚹ ♀
16:11 ☽ ☍ ♅
17:19 ☽ # ♅
29 00:41 ☿ SD
06:15 ☽ ⚹ ☿
13:06 ☽ □ ♇
30 15:24 ☽ ⚹ ☉
18:20 ♀ △ ♅
20:22 ☿ ∥ ☉
31 04:46 ☽ ⚹ ♄
05:09 ☽ ∥ ♅
06:21 ☽ □ ♀
14:28 ☽ △ ♆
20:03 ☽ # ♂
20:042 ☽ □ ☿

⚷ Chiron

01	Dec.	11 S 01
01	08♒36R	
04	08 28	
07	08 19	
10	08 10	
13	08 00	
16	07 51	
19	07 41	
22	07 31	
25	07 21	
28	07 11	
31	07 01	

August 2006

Day	S. T.	☉	☽	☿	♀	♂	♃	♄	♅	♆	♇	☊ True
	h m s	° ' "	° ' "	° '	° '	° '	° '	° '	° '	° '	° '	° '
01 Tu	20 37 51	08♌ 37 51	23♎ 26 07	21♋ 31	15♋ 35	05♍ 45	09♏ 57	14♌ 01	14♓ R03	18♒ R42	24♐ R23	26♓ 22
02 We	20 41 48	09 35 15	05♏ 28 43	21 52	16 48	06 23	10 01	14 09	14 01	18 40	24 22	26 23
03 Th	20 45 44	10 32 40	17 46 39	22 20	18 01	07 00	10 06	14 16	13 59	18 38	24 21	26R 23
04 Fr	20 49 41	11 30 06	00♐ 24 48	22 54	19 13	07 38	10 11	14 24	13 57	18 37	24 20	26 21
05 Sa	20 53 38	12 27 32	13 27 13	23 35	20 26	08 15	10 16	14 32	13 55	18 35	24 19	26 17
06 Su	20 57 34	13 24 59	26 56 32	24 22	21 39	08 53	10 21	14 39	13 53	18 33	24 18	26 12
07 Mo	21 01 31	14 22 26	10♑ 53 14	25 15	22 52	09 31	10 26	14 47	13 51	18 32	24 17	26 06
08 Tu	21 05 27	15 19 55	25 15 13	26 14	24 05	10 08	10 32	14 55	13 49	18 30	24 17	26 00
09 We	21 09 24	16 17 24	09♒ 57 33	27 19	25 18	10 46	10 37	15 03	13 47	18 29	24 16	25 54
10 Th	21 13 20	17 14 55	24 53 04	28 30	26 32	11 24	10 43	15 10	13 45	18 27	24 15	25 50
11 Fr	21 17 17	18 12 26	09♓ 53 13	29 46	27 45	12 02	10 49	15 18	13 43	18 25	24 14	25 47
12 Sa	21 21 14	19 09 59	24 49 29	01♌ 08	28 58	12 40	10 55	15 26	13 41	18 24	24 13	25 46
13 Su	21 25 10	20 07 33	09♈ 34 35	02 35	00♌ 11	13 18	11 01	15 33	13 39	18 22	24 13	25D 46
14 Mo	21 29 07	21 05 08	24 03 15	04 06	01 24	13 55	11 08	15 41	13 37	18 20	24 12	25 47
15 Tu	21 33 03	22 02 45	08♉ 12 29	05 42	02 38	14 33	11 14	15 49	13 35	18 19	24 11	25 48
16 We	21 36 60	23 00 24	22 01 17	07 23	03 51	15 11	11 21	15 56	13 32	18 17	24 11	25 49
17 Th	21 40 56	23 58 04	05♊ 30 06	09 06	05 05	15 49	11 27	16 04	13 30	18 16	24 10	25R 49
18 Fr	21 44 53	24 55 45	18 40 19	10 54	06 18	16 27	11 34	16 12	13 28	18 14	24 10	25 47
19 Sa	21 48 49	25 53 29	01♋ 33 46	12 44	07 32	17 05	11 41	16 19	13 26	18 12	24 09	25 44
20 Su	21 52 46	26 51 14	14 12 23	14 37	08 45	17 43	11 48	16 27	13 24	18 11	24 09	25 40
21 Mo	21 56 43	27 49 00	26 38 04	16 31	09 59	18 22	11 56	16 35	13 21	18 09	24 08	25 36
22 Tu	22 00 39	28 46 48	08♌ 52 34	18 28	11 12	19 00	12 03	16 42	13 19	18 07	24 08	25 31
23 We	22 04 36	29 44 37	20 57 32	20 26	12 26	19 38	12 11	16 50	13 17	18 06	24 07	25 27
24 Th	22 08 32	00♍ 42 28	02♍ 54 41	22 24	13 40	20 16	12 18	16 57	13 14	18 04	24 07	25 24
25 Fr	22 12 29	01 40 20	14 45 53	24 24	14 53	20 54	12 26	17 05	13 12	18 03	24 06	25 22
26 Sa	22 16 25	02 38 14	26 33 18	26 23	16 07	21 33	12 34	17 12	13 10	18 01	24 06	25D 22
27 Su	22 20 22	03 36 09	08♎ 19 34	28 23	17 21	22 11	12 42	17 20	13 07	18 00	24 06	25 23
28 Mo	22 24 18	04 34 05	20 07 45	00♍ 22	18 35	22 49	12 50	17 27	13 05	17 58	24 06	25 24
29 Tu	22 28 15	05 32 03	02♏ 01 25	02 21	19 49	23 28	12 59	17 35	13 03	17 56	24 05	25 26
30 We	22 32 11	06 30 02	14 04 36	04 20	21 02	24 06	13 07	17 42	13 00	17 55	24 05	25 27
31 Th	22 36 08	07 28 03	26 21 41	06 17	22 16	24 44	13 15	17 50	12 58	17 53	24 05	25 28

Data for	08-01-2006
Julian Day	2453948.50
Ayanamsa	23 56 57
SVP	05 ♓ 10 06
☽ ☊ Mean	27 ♓ 46 R

● ◐ PHASES ○ ◑

Day	h m	Phase	Position
02	08:46	◐	09♏56
09	10:54	○	16♒44
16	01:52	◑	23♉05
23	19:10	●	00♍31
31	22:57	◐	08♐24

LAST ASPECT Day	h m	☽ INGRESS Day	h m	Sign
01	01:54	01	13:08	♏
03	09:08	03	23:14	♐
05	19:22	06	05:20	♑
08	01:44	08	07:48	♒
09	22:59	10	08:11	♓
12	07:18	12	08:23	♈
14	00:15	14	10:01	♉
16	01:52	16	14:08	♊
18	12:31	18	21:04	♋
20	07:07	21	06:34	♌
23	06:19	23	18:08	♍
25	19:00	26	07:01	♎
28	08:01	28	19:56	♏
30	20:42	31	07:00	♐

DECLINATION

Day	☉	☽	☿	♀	♂	♃	♄	♅	♆	♇
01 Tu	18N06	11S 18	18N38	22N24	10N22	13S 50	17N23	07S 01	15S 25	15S 49
02 We	17 51	16 26	18 49	22 18	10 08	13 52	17 21	07 02	15 26	15 49
03 Th	17 36	21 01	19 00	22 11	09 54	13 53	17 19	07 03	15 26	15 49
04 Fr	17 20	24 47	19 10	22 04	09 39	13 55	17 17	07 03	15 27	15 49
05 Sa	17 04	27 25	19 19	21 56	09 25	13 57	17 15	07 04	15 27	15 50
06 Su	16 48	28 34	19 27	21 48	09 10	13 59	17 13	07 05	15 28	15 50
07 Mo	16 31	27 58	19 34	21 38	08 56	14 01	17 10	07 06	15 28	15 50
08 Tu	16 14	25 31	19 38	21 28	08 41	14 02	17 08	07 07	15 29	15 50
09 We	15 57	21 20	19 41	21 18	08 26	14 04	17 06	07 07	15 29	15 51
10 Th	15 40	15 44	19 42	21 07	08 12	14 06	17 04	07 08	15 30	15 51
11 Fr	15 22	09 11	19 41	20 55	07 57	14 09	17 02	07 09	15 30	15 51
12 Sa	15 05	02 08	19 37	20 43	07 42	14 11	16 59	07 10	15 31	15 51
13 Su	14 47	04N57	19 32	20 30	07 27	14 13	16 57	07 11	15 31	15 52
14 Mo	14 28	11 38	19 23	20 16	07 12	14 15	16 55	07 11	15 32	15 52
15 Tu	14 10	17 35	19 12	20 02	06 57	14 17	16 53	07 12	15 32	15 52
16 We	13 51	22 28	18 58	19 47	06 42	14 19	16 51	07 13	15 33	15 53
17 Th	13 32	26 02	18 42	19 32	06 27	14 22	16 48	07 14	15 33	15 53
18 Fr	13 13	28 07	18 22	19 16	06 12	14 24	16 46	07 15	15 34	15 53
19 Sa	12 53	28 38	18 00	19 00	05 56	14 26	16 44	07 16	15 34	15 53
20 Su	12 34	27 37	17 36	18 43	05 41	14 29	16 42	07 17	15 35	15 54
21 Mo	12 14	25 14	17 09	18 26	05 26	14 31	16 40	07 18	15 35	15 54
22 Tu	11 54	21 43	16 39	18 08	05 10	14 34	16 37	07 18	15 36	15 54
23 We	11 34	17 20	16 07	17 49	04 55	14 36	16 35	07 19	15 36	15 55
24 Th	11 13	12 19	15 33	17 30	04 40	14 39	16 33	07 20	15 37	15 55
25 Fr	10 53	06 54	14 57	17 11	04 24	14 41	16 31	07 21	15 37	15 55
26 Sa	10 32	01 16	14 20	16 51	04 09	14 44	16 29	07 22	15 38	15 56
27 Su	10 11	04S 23	13 40	16 30	03 53	14 47	16 26	07 23	15 38	15 56
28 Mo	09 50	09 55	12 59	16 09	03 38	14 49	16 24	07 24	15 39	15 56
29 Tu	09 29	15 08	12 17	15 48	03 22	14 52	16 22	07 25	15 39	15 57
30 We	09 08	19 51	11 34	15 26	03 06	14 55	16 20	07 26	15 40	15 57
31 Th	08 46	23 50	10 50	15 04	02 51	14 57	16 18	07 27	15 40	15 57

ASPECTARIAN

01 01:54 ☽ ⚹ ♇
11:42 ☽ ∥ ♃
19:11 ☽ ∥ ♆
21:03 ☽ ∥ ♇
02 01:52 ☽ ⚹ ♂
04:35 ☽ ⋕ ♄
06:46 ☽ ⋕ ☉
08:59 ☽ ☌ ♃
11:52 ☉ □ ♃
12:38 ☽ ⋕ ☿
16:41 ☽ △ ♅
17:09 ☽ □ ♄
03 00:30 ☽ △ ♀
01:39 ☽ □ ♆
06:43 ☽ ⋕ ♀
09:08 ☽ △ ☿
04 05:31 ☉ ∥ ♄
14:03 ☽ □ ♂
22:03 ☽ △ ☉
05 00:51 ☽ □ ♅
01:58 ☽ △ ♄
09:13 ☽ ⚹ ♆
19:22 ☽ ☌ ♇
06 21:34 ☽ △ ♂
23:14 ☽ ⚹ ♃
07 05:00 ☽ ⚹ ♅
11:55 ☉ ☌ ♄
21:54 ☽ ☍ ♀
08 01:44 ☽ ☍ ☿
17:21 ♂ ⚹ ♃
09 00:11 ☽ ⋕ ♀
01:05 ☽ □ ♃
07:38 ☽ ⋕ ☿
08:16 ☽ ☍ ♄
09:10 ☉ ⋕ ♇
13:42 ☽ ☌ ♆
19:44 ☽ ∥ ♄
22:59 ☽ ⚹ ♇

23:36 ☽ ∥ ♇
10 00:19 ☽ ⋕ ☉
00:57 ☽ ∥ ♆
06:13 ☽ ∥ ♃
13:29 ☉ ⋕ ♆
11 01:30 ☽ △ ♃
03:35 ☽ ☍ ♂
04:10 ☿ ♌
04:27 ☽ ⋕ ♂
05:14 ☉ ☍ ♆
06:08 ☽ ☌ ♅
07:01 ☽ ∥ ♅
23:02 ☽ □ ♇
12 07:18 ☽ △ ♀
11:18 ☽ △ ☿
20:21 ♀ ♌
13 07:50 ☽ ⋕ ♅
08:28 ☽ ∥ ♂
09:56 ☽ △ ♄
12:51 ♂ ☍ ♅
14:29 ☽ ⚹ ♆
18:41 ☽ △ ☉
14 00:15 ☽ △ ♇
00:47 ♂ ⋕ ♅
10:12 ☽ ⋕ ♃
10:29 ☽ ∥ ☉
13:34 ☽ □ ♀
15:21 ☽ ⋕ ♆
15:28 ☉ ⋕ ♃
16:43 ☽ ⋕ ♇
19:09 ☽ □ ☿
21:00 ☽ ∥ ♄
15 05:15 ☽ ☍ ♃
07:04 ☽ ∥ ☿

09:14 ☽ ⚹ ♅
10:49 ☽ ∥ ♀
11:29 ☽ △ ♂
13:16 ☽ □ ♄
17:28 ☽ □ ♆
16 23:09 ☽ ⚹ ♀
17 05:01 ☉ △ ♇
07:31 ☽ ⚹ ☿
14:29 ☽ □ ♅
19:24 ☽ ⚹ ♄
19:43 ☽ □ ♂
23:11 ☽ △ ♆
18 09:30 ☿ □ ♃
10:09 ☽ ☍ ♇
12:31 ☽ ⚹ ☉
19 19:22 ☽ △ ♃
22:27 ☽ △ ♅
20 07:07 ☽ ⚹ ♂
21 00:43 ☿ ☌ ♄
19:53 ☿ ☍ ♆
22 01:29 ☿ ∥ ♄
05:07 ☽ ☌ ♀
06:21 ☽ □ ♃
15:41 ☽ ☌ ♄
18:18 ☽ ☍ ♆
18:27 ♀ □ ♃
21:18 ☽ ∥ ♀
22:44 ☽ ☌ ☿
23 03:44 ☽ ∥ ♄
06:19 ☽ △ ♇
06:23 ☉ ♍
06:46 ☽ ∥ ☿
07:02 ☽ ⋕ ♇
08:30 ☽ ⋕ ♆

09:01 ☿ ⋕ ♇
13:13 ☽ ⋕ ♃
21:29 ☿ ⋕ ♆
24 05:17 ☽ ∥ ☉
19:13 ☽ ⚹ ♃
20:34 ☿ △ ♇
20:50 ☽ ☍ ♅
22:02 ☽ ⋕ ♅
25 09:37 ☿ ⋕ ♃
11:13 ☽ ∥ ♂
13:12 ☽ ☌ ♂
19:00 ☽ □ ♇
26 21:58 ☽ ⋕ ♂
23:39 ♀ ☌ ♄
27 04:55 ♀ ∥ ♄

12:19 ♀ ☍ ♆
12:56 ☽ ∥ ♅
18:31 ☽ ⚹ ♄
19:31 ☿ ♍
19:39 ☽ △ ♆
20:31 ☽ ⚹ ♀
23:42 ☽ ⋕ ☉
28 08:01 ☽ ⚹ ♇
12:18 ☽ ⋕ ☿
14:30 ♀ ⋕ ♇
22:44 ☽ ∥ ♃
29 00:48 ☽ ⚹ ☿
02:34 ☽ ∥ ♆
03:01 ☽ ⋕ ☿

03:57 ☽ ∥ ♇
05:59 ☽ ⋕ ♄
07:39 ☽ ⚹ ☉
09:07 ♀ ⋕ ♆
09:12 ♃ △ ♅
21:53 ☽ △ ♅
22:05 ☽ ☌ ♃
23:27 ♂ □ ♇
30 07:13 ☽ □ ♄
07:32 ☽ □ ♆
15:12 ☽ □ ♀
20:42 ☽ ⚹ ♂
31 06:03 ♀ ⋕ ♃
09:56 ♄ ☍ ♆
22:25 ☽ □ ☿

⚷ Chiron

01	Dec.	11 S 22
03	06♒51R	
06	06 41	
09	06 31	
12	06 21	
15	06 12	
18	06 02	
21	05 53	
24	05 44	
27	05 36	
30	05 28	

07 18:52 15♓00 ✹ Total Lunar Eclipse (mag 0.189)
22 12:08 29♍20 ✪ Annular Solar Eclipse

September 2006

Day	S. T. h m s	☉ ° ' "	☽ ° ' "	☿ ° '	♀ ° '	♂ ° '	♃ ° '	♄ ° '	♅ ° '	♆ ° '	♇ ° '	☊ True ° '
01 Fr	22 40 05	08♍26 05	08♐57 04	08♍14	23♌30	25♍23	13♏24	17♌57	12♓R56	17♒R52	24♐R05	25♓R28
02 Sa	22 44 01	09 24 08	21 54 53	10 10	24 44	26 01	13 33	18 05	12 53	17 50	24 05	25 28
03 Su	22 47 58	10 22 13	05♑18 26	12 05	25 58	26 40	13 42	18 12	12 51	17 49	24 05	25 27
04 Mo	22 51 54	11 20 19	19 09 29	13 59	27 12	27 19	13 51	18 19	12 48	17 47	24 05	25 25
05 Tu	22 55 51	12 18 26	03♒27 39	15 52	28 27	27 57	14 00	18 26	12 46	17 46	24D05	25 24
06 We	22 59 47	13 16 35	18 09 53	17 43	29 41	28 36	14 09	18 34	12 44	17 45	24 05	25 22
07 Th	23 03 44	14 14 46	03♓10 21	19 34	00♍55	29 14	14 18	18 41	12 41	17 43	24 05	25 21
08 Fr	23 07 41	15 12 58	18 20 53	21 23	02 09	29 53	14 28	18 48	12 39	17 42	24 05	25 21
09 Sa	23 11 37	16 11 12	03♈32 07	23 11	03 23	00♎32	14 37	18 55	12 36	17 40	24 05	25D21
10 Su	23 15 34	17 09 27	18 34 42	24 58	04 37	01 11	14 47	19 02	12 34	17 39	24 05	25 21
11 Mo	23 19 30	18 07 45	03♉20 44	26 44	05 52	01 49	14 57	19 09	12 32	17 38	24 05	25 21
12 Tu	23 23 27	19 06 05	17 44 41	28 29	07 06	02 28	15 06	19 17	12 29	17 36	24 05	25 22
13 We	23 27 23	20 04 27	01♊43 34	00♎12	08 20	03 07	15 16	19 24	12 27	17 35	24 06	25 22
14 Th	23 31 20	21 02 51	15 16 56	01 55	09 35	03 46	15 26	19 31	12 24	17 34	24 06	25 22
15 Fr	23 35 16	22 01 18	28 26 07	03 36	10 49	04 25	15 36	19 37	12 22	17 32	24 06	25R22
16 Sa	23 39 13	22 59 46	11♋13 46	05 16	12 04	05 04	15 47	19 44	12 20	17 31	24 06	25 22
17 Su	23 43 10	23 58 17	23 43 11	06 56	13 18	05 43	15 57	19 51	12 17	17 30	24 07	25D22
18 Mo	23 47 06	24 56 50	05♌57 52	08 34	14 33	06 22	16 07	19 58	12 15	17 29	24 07	25 22
19 Tu	23 51 03	25 55 24	18 01 15	10 11	15 47	07 01	16 18	20 05	12 13	17 27	24 08	25 23
20 We	23 54 59	26 54 01	29 56 32	11 47	17 02	07 40	16 28	20 11	12 10	17 26	24 08	25 23
21 Th	23 58 56	27 52 40	11♍46 33	13 22	18 16	08 19	16 39	20 18	12 08	17 25	24 09	25 24
22 Fr	00 02 52	28 51 21	23 33 52	14 56	19 31	08 58	16 50	20 25	12 06	17 24	24 09	25 24
23 Sa	00 06 49	29 50 04	05♎20 51	16 29	20 46	09 38	17 01	20 31	12 04	17 23	24 10	25R24
24 Su	00 10 45	00♎48 49	17 09 46	18 02	22 00	10 17	17 11	20 38	12 01	17 22	24 10	25 24
25 Mo	00 14 42	01 47 35	29 02 52	19 33	23 15	10 56	17 22	20 44	11 59	17 21	24 11	25 23
26 Tu	00 18 38	02 46 24	11♏02 34	21 03	24 30	11 35	17 34	20 51	11 57	17 20	24 12	25 22
27 We	00 22 35	03 45 14	23 11 32	22 32	25 44	12 15	17 45	20 57	11 55	17 19	24 12	25 21
28 Th	00 26 32	04 44 07	05♐32 40	24 01	26 59	12 54	17 56	21 03	11 53	17 18	24 13	25 19
29 Fr	00 30 28	05 43 01	18 09 09	25 28	28 14	13 34	18 07	21 10	11 51	17 17	24 14	25 18
30 Sa	00 34 25	06 41 57	01♑04 12	26 54	29 29	14 13	18 19	21 16	11 48	17 16	24 15	25 18

Data for	09-01-2006
Julian Day	2453979.50
Ayanamsa	23 57 01
SVP	05 ♓ 10 02
☽ ☊ Mean	26 ♓ 08 R

● ◐ PHASES ○ ◑

07	18:43	✹	15♓00
14	11:16	◑	21♊30
22	11:45	✪	29♍20
30	11:04	◐	07♑09

LAST ASPECT ☽ Day	h m	INGRESS Day	h m	
02	07:50	02	14:35	♑
04	14:25	04	18:15	♒
06	09:30	06	18:57	♓
08	09:03	08	18:24	♈
10	08:53	10	18:31	♉
12	20:59	12	21:00	♊
14	16:01	15	02:54	♋
17	00:32	17	12:15	♌
19	12:16	20	00:07	♍
22	11:45	22	13:07	♎
24	14:11	25	01:55	♏
27	05:33	27	13:17	♐
29	20:46	29	22:02	♑

DECLINATION

Day	☉	☽	☿	♀	♂	♃	♄	♅	♆	♇
01 Fr	08N24	26S48	10N05	14N41	02N35	15S00	16N15	07S27	15S41	15S58
02 Sa	08 03	28 27	09 20	14 18	02 19	15 03	16 13	07 28	15 41	15 58
03 Su	07 41	28 32	08 33	13 55	02 04	15 06	16 11	07 29	15 42	15 58
04 Mo	07 19	26 52	07 47	13 31	01 48	15 09	16 09	07 30	15 42	15 59
05 Tu	06 57	23 26	07 00	13 06	01 32	15 12	16 07	07 31	15 43	15 59
06 We	06 34	18 26	06 13	12 42	01 16	15 14	16 05	07 32	15 43	15 59
07 Th	06 12	12 13	05 26	12 17	01 01	15 17	16 03	07 33	15 44	16 00
08 Fr	05 50	05 13	04 38	11 51	00 45	15 20	16 00	07 34	15 44	16 00
09 Sa	05 27	02N06	03 51	11 26	00 29	15 23	15 58	07 35	15 44	16 00
10 Su	05 04	09 13	03 04	11 00	00 13	15 26	15 56	07 36	15 45	16 01
11 Mo	04 42	15 42	02 16	10 34	00S03	15 29	15 54	07 37	15 45	16 01
12 Tu	04 19	21 09	01 29	10 07	00 19	15 32	15 52	07 38	15 46	16 01
13 We	03 56	25 15	00 43	09 40	00 34	15 35	15 50	07 38	15 46	16 02
14 Th	03 33	27 48	00S04	09 13	00 50	15 39	15 48	07 39	15 47	16 02
15 Fr	03 10	28 43	00 50	08 46	01 06	15 42	15 46	07 40	15 47	16 02
16 Sa	02 47	28 03	01 36	08 18	01 22	15 45	15 44	07 41	15 47	16 03
17 Su	02 24	25 58	02 22	07 50	01 38	15 48	15 42	07 42	15 48	16 03
18 Mo	02 00	22 41	03 07	07 22	01 54	15 51	15 40	07 43	15 48	16 03
19 Tu	01 37	18 30	03 52	06 54	02 10	15 54	15 38	07 44	15 48	16 04
20 We	01 14	13 38	04 36	06 25	02 26	15 57	15 36	07 45	15 49	16 04
21 Th	00 51	08 18	05 20	05 56	02 42	16 01	15 34	07 46	15 49	16 04
22 Fr	00 27	02 43	06 03	05 27	02 57	16 04	15 32	07 46	15 50	16 05
23 Sa	00 04	02S58	06 46	04 58	03 13	16 07	15 30	07 47	15 50	16 05
24 Su	00S19	08 33	07 28	04 29	03 29	16 10	15 28	07 48	15 50	16 05
25 Mo	00 43	13 53	08 09	04 00	03 45	16 13	15 26	07 49	15 51	16 06
26 Tu	01 06	18 44	08 50	03 30	04 01	16 17	15 24	07 50	15 51	16 06
27 We	01 30	22 54	09 31	03 01	04 17	16 20	15 22	07 51	15 51	16 07
28 Th	01 53	26 07	10 10	02 31	04 32	16 23	15 20	07 51	15 51	16 07
29 Fr	02 16	28 08	10 49	02 01	04 48	16 26	15 18	07 52	15 52	16 07
30 Sa	02 40	28 41	11 27	01 31	05 04	16 30	15 16	07 53	15 52	16 08

ASPECTARIAN

01 04:50 ☿ ☌ ☉
07:25 ☽ □ ♅
11:10 ♀ △ ♇
16:33 ☽ ⚹ ♆
16:54 ☽ △ ♄
02 03:56 ☽ ☌ ♇
05:39 ☽ △ ♀
07:50 ☽ □ ♂
03 09:22 ☿ ☍ ♅
09:32 ☽ △ ☉
12:01 ☉ # ♅
13:07 ☽ ⚹ ♅
13:44 ☽ △ ☿
14:47 ☽ ⚹ ♃
22:05 ☿ ⚹ ♃
04 08:22 ☿ # ♅
14:25 ☽ △ ♂
23:21 ♇ SD
05 03:14 ☿ ‖ ☉
10:54 ☉ ☍ ♅
17:26 ☽ □ ♃
23:19 ☽ ☌ ♆
06 00:39 ☽ ☍ ♄
06:15 ♀ ♍
09:30 ☽ ⚹ ♇
09:36 ☽ # ♄
09:54 ☽ ‖ ♇
10:56 ☽ ‖ ♆
12:41 ☽ ‖ ♃
20:05 ☽ ☍ ♀
23:43 ☽ # ♀
07 01:47 ☉ ⚹ ♃
15:01 ☽ ☌ ♅
16:07 ☽ ‖ ♅
17:48 ☽ △ ♃
20:49 ☽ # ☉
08 02:06 ☽ # ☿
04:19 ♂ ♎
05:27 ☽ ☍ ☿
06:00 ♄ # ♇
09:03 ☽ □ ♇
15:12 ☽ # ♂
18:52 ☽ ‖ ♂
19:02 ☽ ☍ ♂
09 05:13 ☽ ‖ ☿
10:34 ☽ ‖ ☉
11:59 ☿ □ ♇
18:22 ☽ # ♅
22:31 ☽ ⚹ ♆
10 00:45 ☽ △ ♄
05:55 ☽ ‖ ♀
08:53 ☽ △ ♇
23:11 ☽ # ♃
11 00:15 ☽ # ♆
00:50 ☽ ‖ ♄
01:17 ☽ # ♇
04:32 ☽ △ ♀
15:11 ☽ ⚹ ♅
19:30 ☽ ☍ ♃
23:46 ☽ □ ♆
12 02:28 ☽ △ ☉
02:37 ☽ □ ♄
20:59 ☽ △ ☿
21:08 ☿ ♎
13 02:33 ☽ △ ♂
03:08 ☿ # ♂
12:47 ☽ □ ♀
18:52 ☽ □ ♅
14 04:06 ☽ △ ♆
07:42 ☽ ⚹ ♄
13:00 ♄ # ♆
16:01 ☽ ☍ ♇
15 11:04 ☽ □ ☿
11:44 ☽ □ ♂
12:44 ☿ ‖ ♂
19:01 ☿ ☌ ♂
19:33 ♃ # ♄
16 01:45 ☽ ⚹ ♀
02:05 ☽ △ ♅
05:01 ♀ ☍ ♅
08:48 ☽ △ ♃
22:51 ♃ ‖ ♆
17 00:32 ☽ ⚹ ☉
00:43 ☿ # ☉
03:34 ☉ □ ♇
06:38 ♀ # ♅
18 00:50 ☽ ⚹ ♂
04:04 ☉ # ♂
05:57 ☽ ⚹ ☿
20:30 ☽ □ ♃
22:52 ☽ ☍ ♆
19 04:10 ☽ ☌ ♄
11:28 ♀ ⚹ ♃
12:16 ☽ △ ♇
12:21 ☽ # ♇
12:59 ☽ # ♃
13:36 ☽ # ♆
14:35 ☽ ‖ ♄
21 00:44 ☽ ☍ ♅
02:22 ☽ # ♅
10:04 ☽ ⚹ ♃
11:13 ☽ ‖ ♀
11:24 ☽ # ☿
12:13 ☿ # ♀
14:47 ☽ ☌ ♀
23:01 ☽ # ♂
22 01:12 ☽ □ ♇
08:32 ♃ ‖ ♇
10:15 ☽ ‖ ☉
12:32 ☽ # ☉
23 01:08 ☽ ‖ ♂
04:04 ☉ ♎
07:51 ☽ # ♀
09:13 ☽ ☌ ♂
13:45 ☿ △ ♆
18:32 ☽ ‖ ☿
20:42 ☽ ‖ ♅
24 00:24 ☽ △ ♆
02:00 ☽ ☌ ☿
07:05 ☽ ⚹ ♄
11:59 ☿ ‖ ♅
14:11 ☽ ⚹ ♇
20:34 ♃ □ ♆
25 07:20 ☽ # ♄
07:52 ♀ # ♂
09:23 ☽ ‖ ♆
10:38 ☽ ‖ ♇
11:22 ☽ ‖ ♃
18:08 ♀ □ ♇
20:28 ☿ ⚹ ♄
26 01:48 ☽ △ ♅
12:27 ☽ □ ♆
13:07 ☽ ☌ ♃
19:34 ☽ □ ♄
27 05:33 ☽ ⚹ ♀
22:18 ☽ ⚹ ☉
28 03:27 ☿ ⚹ ♇
12:05 ☽ □ ♅
14:51 ☽ ⚹ ♂
17:13 ♀ # ☉
22:21 ☽ ⚹ ♆
29 05:41 ☽ △ ♄
11:23 ☽ ☌ ♇
15:22 ☽ ⚹ ☿
20:46 ☽ □ ♀
30 10:02 ♀ ♎
19:25 ☽ ⚹ ♅

⚷ Chiron

01 Dec. 11 S 50

02	05♒20R
05	05 13
08	05 06
11	04 59
14	04 53
17	04 48
20	04 43
23	04 39
26	04 35
29	04 32

October 2006

Day	S. T. h m s	☉ ° ' "	☽ ° ' "	☿ ° '	♀ ° '	♂ ° '	♃ ° '	♄ ° '	♅ ° '	♆ ° '	♇ ° '	☊ True ° '
01 Su	00 38 21	07♎ 40 54	14♑ 20 47	28♎ 20	00♎ 44	14♎ 53	18♏ 30	21♌ 22	11♓R46	17♒R15	24♐ 15	25♓D18
02 Mo	00 42 18	08 39 54	28 01 11	29 44	01 58	15 32	18 41	21 28	11 44	17 14	24 16	25 18
03 Tu	00 46 14	09 38 55	12♒ 06 24	01♏ 07	03 13	16 12	18 53	21 34	11 42	17 13	24 17	25 19
04 We	00 50 11	10 37 57	26 35 32	02 29	04 28	16 51	19 05	21 40	11 40	17 12	24 18	25 21
05 Th	00 54 07	11 37 02	11♓ 25 17	03 50	05 43	17 31	19 16	21 46	11 38	17 12	24 19	25 22
06 Fr	00 58 04	12 36 08	26 29 46	05 10	06 58	18 11	19 28	21 52	11 36	17 11	24 20	25R 22
07 Sa	01 02 01	13 35 16	11♈ 40 51	06 29	08 13	18 51	19 40	21 57	11 34	17 10	24 21	25 22
08 Su	01 05 57	14 34 27	26 49 06	07 46	09 28	19 30	19 52	22 03	11 32	17 09	24 22	25 20
09 Mo	01 09 54	15 33 39	11♉ 45 08	09 02	10 43	20 10	20 04	22 09	11 31	17 09	24 23	25 18
10 Tu	01 13 50	16 32 54	26 21 04	10 17	11 58	20 50	20 16	22 14	11 29	17 08	24 24	25 15
11 We	01 17 47	17 32 11	10♊ 31 28	11 30	13 13	21 30	20 28	22 20	11 27	17 07	24 25	25 12
12 Th	01 21 43	18 31 30	24 13 51	12 41	14 28	22 10	20 40	22 25	11 25	17 07	24 26	25 09
13 Fr	01 25 40	19 30 52	07♋ 28 28	13 51	15 43	22 50	20 52	22 31	11 23	17 06	24 27	25 07
14 Sa	01 29 36	20 30 16	20 17 41	14 58	16 58	23 30	21 05	22 36	11 22	17 06	24 29	25 07
15 Su	01 33 33	21 29 42	02♌ 45 17	16 04	18 13	24 10	21 17	22 41	11 20	17 05	24 30	25D 07
16 Mo	01 37 30	22 29 11	14 55 46	17 07	19 28	24 50	21 29	22 46	11 18	17 05	24 31	25 09
17 Tu	01 41 26	23 28 42	26 53 51	18 08	20 43	25 30	21 42	22 51	11 17	17 04	24 32	25 11
18 We	01 45 23	24 28 15	08♍ 44 07	19 07	21 58	26 10	21 54	22 56	11 15	17 04	24 34	25 13
19 Th	01 49 19	25 27 50	20 30 44	20 02	23 13	26 51	22 07	23 01	11 14	17 04	24 35	25 14
20 Fr	01 53 16	26 27 27	02♎ 17 22	20 54	24 28	27 31	22 19	23 06	11 12	17 03	24 36	25R 15
21 Sa	01 57 12	27 27 06	14 07 04	21 42	25 43	28 11	22 32	23 11	11 11	17 03	24 38	25 13
22 Su	02 01 09	28 26 48	26 02 17	22 26	26 58	28 52	22 44	23 15	11 09	17 03	24 39	25 10
23 Mo	02 05 05	29 26 32	08♏ 04 58	23 06	28 14	29 32	22 57	23 20	11 08	17 03	24 41	25 04
24 Tu	02 09 02	00♏ 26 17	20 16 35	23 41	29 29	00♏ 12	23 10	23 24	11 07	17 02	24 42	24 58
25 We	02 12 59	01 26 05	02♐ 38 23	24 11	00♏ 44	00 53	23 22	23 29	11 06	17 02	24 44	24 51
26 Th	02 16 55	02 25 54	15 11 35	24 35	01 59	01 33	23 35	23 33	11 04	17 02	24 45	24 44
27 Fr	02 20 52	03 25 45	27 57 29	24 52	03 14	02 14	23 48	23 37	11 03	17 02	24 47	24 39
28 Sa	02 24 48	04 25 38	10♑ 57 36	25 02	04 30	02 55	24 01	23 41	11 02	17 02	24 48	24 35
29 Su	02 28 45	05 25 33	24 13 34	25R 05	05 45	03 35	24 14	23 45	11 01	17 02	24 50	24 33
30 Mo	02 32 41	06 25 29	07♒ 47 01	24 59	07 00	04 16	24 27	23 49	11 00	17 D 02	24 51	24D 32
31 Tu	02 36 38	07 25 26	21 39 12	24 44	08 15	04 57	24 40	23 53	10 59	17 02	24 53	24 33

Data for	10-01-2006
Julian Day	2454009.50
Ayanamsa	23 57 05
SVP	05 ♓ 09 57
☽ ☊ Mean	24 ♓ 32 R

LAST ASPECT Day	h m	☽ INGRESS Day	h m	
02	03:17	02	03:25	♒
03	20:14	04	05:34	♓
05	20:34	06	05:33	♈
07	20:05	08	05:05	♉
09	17:08	10	06:06	♊
12	00:22	12	10:21	♋
14	06:27	14	18:38	♌
16	21:01	17	06:16	♍
19	08:19	19	19:20	♎
22	05:59	22	07:55	♏
24	06:57	24	18:54	♐
26	18:03	27	03:48	♑
29	01:31	29	10:17	♒
31	05:31	31	14:11	♓

● ◐ PHASES ○ ◑

Day	Time	Phase	Position
07	03:13	○	13♈43
14	00:26	◑	20♋31
22	05:15	●	28♎40
29	21:25	◐	06♒19

DECLINATION

Day	☉	☽	☿	♀	♂	♃	♄	♅	♆	♇
01 Su	03S 03	27S 38	12S 05	01N01	05S 20	16S 33	15N15	07S 54	15S 52	16S 08
02 Mo	03 26	24 55	12 42	00 31	05 35	16 36	15 13	07 54	15 53	16 08
03 Tu	03 49	20 38	13 18	00 01	05 51	16 40	15 11	07 55	15 53	16 09
04 We	04 13	15 03	13 53	00S 29	06 07	16 43	15 09	07 56	15 53	16 09
05 Th	04 36	08 28	14 27	00 59	06 22	16 46	15 07	07 57	15 53	16 09
06 Fr	04 59	01 18	15 01	01 29	06 38	16 50	15 06	07 57	15 54	16 10
07 Sa	05 22	05N59	15 33	01 59	06 54	16 53	15 04	07 58	15 54	16 10
08 Su	05 45	12 54	16 05	02 29	07 09	16 56	15 02	07 59	15 54	16 10
09 Mo	06 08	18 59	16 36	02 59	07 25	17 00	15 01	08 00	15 54	16 11
10 Tu	06 30	23 48	17 06	03 29	07 40	17 03	14 59	08 00	15 54	16 11
11 We	06 53	27 03	17 34	03 59	07 55	17 06	14 57	08 01	15 55	16 12
12 Th	07 16	28 33	18 02	04 29	08 11	17 10	14 56	08 02	15 55	16 12
13 Fr	07 38	28 21	18 28	04 59	08 26	17 13	14 54	08 02	15 55	16 12
14 Sa	08 01	26 36	18 53	05 28	08 41	17 16	14 53	08 03	15 55	16 13
15 Su	08 23	23 36	19 17	05 58	08 57	17 20	14 51	08 03	15 55	16 13
16 Mo	08 45	19 36	19 40	06 27	09 12	17 23	14 50	08 04	15 55	16 13
17 Tu	09 07	14 53	20 01	06 57	09 27	17 26	14 48	08 05	15 56	16 14
18 We	09 29	09 41	20 21	07 26	09 42	17 30	14 47	08 05	15 56	16 14
19 Th	09 51	04 09	20 39	07 55	09 57	17 33	14 45	08 06	15 56	16 14
20 Fr	10 13	01S 30	20 55	08 24	10 12	17 36	14 44	08 06	15 56	16 15
21 Sa	10 34	07 08	21 10	08 52	10 27	17 40	14 42	08 07	15 56	16 15
22 Su	10 55	12 33	21 23	09 21	10 42	17 43	14 41	08 07	15 56	16 15
23 Mo	11 17	17 33	21 33	09 49	10 57	17 46	14 40	08 08	15 56	16 16
24 Tu	11 38	21 55	21 42	10 17	11 11	17 49	14 38	08 08	15 56	16 16
25 We	11 58	25 23	21 48	10 45	11 26	17 53	14 37	08 09	15 56	16 16
26 Th	12 19	27 41	21 52	11 13	11 41	17 56	14 36	08 09	15 56	16 17
27 Fr	12 40	28 35	21 53	11 40	11 55	17 59	14 35	08 10	15 56	16 17
28 Sa	13 00	27 56	21 51	12 07	12 09	18 03	14 33	08 10	15 56	16 17
29 Su	13 20	25 41	21 46	12 34	12 24	18 06	14 32	08 10	15 56	16 18
30 Mo	13 40	21 57	21 38	13 00	12 38	18 09	14 31	08 11	15 56	16 18
31 Tu	13 59	16 56	21 26	13 26	12 52	18 12	14 30	08 11	15 56	16 18

ASPECTARIAN

Day	Time	Aspect
01	01:00	☽ □ ♂
	07:28	☽ ✶ ♃
02	03:17	☽ □ ☿
	04:38	☿ ♏
	07:28	☽ △ ♀
	19:33	☽ △ ☉
03	07:10	☽ △ ♂
	08:32	☽ ☌ ♆
	11:28	☽ □ ♃
	15:51	☽ ☍ ♄
	17:24	☽ ∥ ♃
	19:37	☽ ∥ ♇
	20:14	☽ ✶ ♇
	20:41	☽ ∥ ♆
	23:34	☽ # ♄
04	04:25	☽ ∥ ☿
	10:34	☽ △ ☿
	12:23	♂ △ ♆
05	00:21	☽ ☌ ♅
	01:46	☽ ∥ ♅
	06:53	☽ ∥ ♂
	12:26	☽ ∥ ☉
	12:42	☽ △ ♃
	20:34	☽ □ ♇
	23:26	☽ ∥ ♀
06	03:21	☿ # ♄
	09:47	☽ # ♀
	18:02	☽ ☍ ♀
	21:49	☽ # ☉
07	03:10	☽ # ♂
	06:42	☽ # ♅
	08:40	☽ ✶ ♆
	11:51	☽ ☍ ♂
	15:29	☿ ∥ ♆
	16:22	☽ △ ♄
	20:05	☽ △ ♇
08	04:10	☿ ∥ ♇
	07:58	☽ ∥ ♄
	11:19	☽ # ♆
	12:24	☽ # ♇
	13:09	☽ # ☿
	15:34	☽ # ♃
	19:11	☽ ☍ ☿
	23:36	☽ ✶ ♅
09	08:47	☽ □ ♆
	13:47	☽ ☍ ♃
	17:08	☽ □ ♄
	21:36	☿ ∥ ♃
10	14:05	☉ △ ♆
	23:07	☿ △ ♅
11	01:35	☽ □ ♅
	05:06	☽ △ ♀
	08:56	♂ ∥ ♅
	11:27	☽ △ ♆
	13:07	☽ △ ☉
	20:08	☽ △ ♂
	20:46	☽ ✶ ♄
12	00:22	☽ ☍ ♇
	10:39	♂ ✶ ♄
13	07:14	☽ △ ♅
	12:59	☽ △ ☿
	17:00	☽ □ ♀
14	01:31	☽ △ ♃
	02:28	☉ ∥ ♅
	02:37	♀ △ ♆
	06:27	☽ □ ♂
15	12:19	♂ ✶ ♇
	23:00	☿ □ ♆
	23:42	☽ # ☿
16	04:17	☽ ☍ ♆
	04:47	☽ □ ☿
	07:29	☉ ✶ ♄
	10:06	☽ ∥ ♀
	11:33	☽ # ♃
	13:20	☽ □ ♃
	15:48	☽ ☌ ♄
	16:29	☽ ✶ ☉
	17:29	☽ # ♇
	18:58	☽ # ♆
	19:14	☽ △ ♇
	21:01	☽ ✶ ♂
17	00:26	☽ ∥ ♄
	23:54	☽ # ♂
18	00:48	☽ # ☉
	02:16	☉ ✶ ♇
	05:07	☽ ☍ ♅
	07:00	☽ # ♅
	09:05	☽ # ♀
	19:54	♀ ✶ ♄
	22:56	☽ ✶ ☿
19	03:19	☽ ✶ ♃
	08:19	☽ □ ♇
	09:09	♀ ∥ ♅
	22:15	☉ ∥ ♂
20	02:42	♀ ✶ ♇
21	04:16	☽ ∥ ♅
	05:55	☽ △ ♆
	08:18	☽ ∥ ♀
	15:14	☽ ∥ ♂
	16:08	☽ ∥ ☉
	18:23	☽ ✶ ♄
	21:13	☽ ✶ ♇
22	02:06	☽ ☌ ♀
	05:59	☽ ☌ ♂
	09:54	☽ # ♄
	15:20	☿ ☌ ♃
	15:58	☽ ∥ ♆
	17:33	☽ ∥ ♇
23	01:05	☽ ∥ ♃
	06:02	☽ △ ♅
	06:46	☉ ☌ ♂
	10:07	☿ □ ♄
	13:27	☉ ♏
	16:38	♂ ♏
	17:40	☽ □ ♆
	22:37	☽ ∥ ☿
24	05:44	☽ ☌ ♃
	06:09	☽ □ ♄
	06:57	☽ ☌ ☿
	09:58	♀ ♏
25	06:11	♀ ☌ ♂
	16:10	☽ □ ♅
	17:27	☽ □ ♄
26	03:29	☽ ✶ ♆
	15:51	☽ △ ♄
	18:03	☽ ☌ ♇
27	08:23	☽ ✶ ♂
	10:51	☽ ✶ ♀
	11:00	☽ ✶ ☉
	17:50	♀ ☌ ☉
28	00:08	☽ ✶ ♅
	04:53	♀ ∥ ♂
	19:17	☿ SR
29	00:01	☽ ✶ ♃
	01:31	☽ ✶ ☿
	07:58	♆ SD
	17:30	☽ □ ♂
	22:29	☽ □ ♀
30	01:55	☽ ∥ ♀
	16:04	☽ ☌ ♆
	18:33	☽ ∥ ♃
31	02:45	☽ ∥ ♇
	03:02	☿ ☌ ♃
	03:50	☽ ☍ ♄
	04:17	☽ ∥ ♆
	05:08	☽ □ ☿
	05:13	☽ □ ♃
	05:31	☽ ✶ ♇
	10:13	☽ # ♄
	11:35	☽ ∥ ☉
	13:28	☽ ∥ ♀
	15:58	☽ ∥ ♂
	23:37	☽ △ ♂

⚷ Chiron

01	Dec. 12 S 15
02	04♒29R
05	04 27
08	04 26
11	04 25
14	04 25D
17	04 26
20	04 27
23	04 29
26	04 31
29	04 34

12 19:42 04♒25 D

November 2006

Day	S. T. h m s	☉ ° ' "	☽ ° ' "	☿ ° '	♀ ° '	♂ ° '	♃ ° '	♄ ° '	♅ ° '	♆ ° '	♇ ° '	☊ True ° '
01 We	02 40 34	08♏ 25 26	05♓ 50 32	24♏R20	09♏ 30	05♏ 37	24♏ 53	23♌ 57	10♓R58	17♒ 02	24♐ 55	24♓ 34
02 Th	02 44 31	09 25 26	20 19 59	23 47	10 46	06 18	25 06	24 01	10 57	17 02	24 56	24 35
03 Fr	02 48 28	10 25 29	05♈ 04 29	23 04	12 01	06 59	25 19	24 04	10 56	17 02	24 58	24R 34
04 Sa	02 52 24	11 25 33	19 58 44	22 12	13 16	07 40	25 32	24 08	10 55	17 02	25 00	24 32
05 Su	02 56 21	12 25 39	04♉ 55 20	21 11	14 31	08 21	25 45	24 11	10 54	17 03	25 02	24 27
06 Mo	03 00 17	13 25 46	19 45 45	20 03	15 47	09 02	25 58	24 14	10 54	17 03	25 03	24 21
07 Tu	03 04 14	14 25 56	04♊ 21 33	18 49	17 02	09 43	26 11	24 17	10 53	17 03	25 05	24 13
08 We	03 08 10	15 26 08	18 35 46	17 31	18 17	10 24	26 25	24 21	10 52	17 03	25 07	24 04
09 Th	03 12 07	16 26 21	02♋ 23 54	16 12	19 32	11 05	26 38	24 24	10 52	17 04	25 09	23 56
10 Fr	03 16 03	17 26 36	15 44 22	14 53	20 48	11 46	26 51	24 26	10 51	17 04	25 11	23 50
11 Sa	03 20 00	18 26 54	28 38 15	13 38	22 03	12 27	27 04	24 29	10 51	17 05	25 12	23 45
12 Su	03 23 57	19 27 13	11♌ 08 40	12 29	23 18	13 09	27 17	24 32	10 50	17 05	25 14	23 43
13 Mo	03 27 53	20 27 34	23 20 08	11 28	24 34	13 50	27 31	24 35	10 50	17 05	25 16	23D 42
14 Tu	03 31 50	21 27 58	05♍ 17 51	10 37	25 49	14 31	27 44	24 37	10 50	17 06	25 18	23 43
15 We	03 35 46	22 28 23	17 07 15	09 56	27 04	15 13	27 57	24 39	10 49	17 07	25 20	23 44
16 Th	03 39 43	23 28 50	28 53 34	09 27	28 19	15 54	28 11	24 42	10 49	17 07	25 22	23R 44
17 Fr	03 43 39	24 29 18	10♎ 41 39	09 09	29 35	16 35	28 24	24 44	10 49	17 08	25 24	23 42
18 Sa	03 47 36	25 29 49	22 35 35	09D 04	00♐ 50	17 17	28 37	24 46	10 49	17 08	25 26	23 38
19 Su	03 51 32	26 30 21	04♏ 38 39	09 09	02 05	17 58	28 51	24 48	10 49	17 09	25 28	23 31
20 Mo	03 55 29	27 30 55	16 53 06	09 25	03 21	18 40	29 04	24 50	10 49	17 10	25 30	23 22
21 Tu	03 59 26	28 31 30	29 20 13	09 50	04 36	19 22	29 17	24 51	10D 49	17 11	25 32	23 10
22 We	04 03 22	29 32 07	12♐ 00 20	10 24	05 51	20 03	29 31	24 53	10 49	17 11	25 34	22 58
23 Th	04 07 19	00♐ 32 46	24 53 08	11 05	07 07	20 45	29 44	24 55	10 49	17 12	25 36	22 46
24 Fr	04 11 15	01 33 26	07♑ 57 55	11 54	08 22	21 27	29 57	24 56	10 49	17 13	25 38	22 36
25 Sa	04 15 12	02 34 07	21 13 57	12 49	09 37	22 09	00♐ 11	24 57	10 49	17 14	25 40	22 27
26 Su	04 19 08	03 34 49	04♒ 40 46	13 49	10 53	22 50	00 24	24 59	10 49	17 15	25 42	22 22
27 Mo	04 23 05	04 35 32	18 18 17	14 54	12 08	23 32	00 37	25 00	10 50	17 16	25 44	22 19
28 Tu	04 27 01	05 36 16	02♓ 06 48	16 03	13 23	24 14	00 51	25 01	10 50	17 17	25 47	22 17
29 We	04 30 58	06 37 01	16 06 37	17 15	14 39	24 56	01 04	25 01	10 51	17 18	25 49	22 17
30 Th	04 34 55	07 37 47	00♈ 17 39	18 30	15 54	25 38	01 17	25 02	10 51	17 19	25 51	22 16

Data for	11-01-2006
Julian Day	2454040.50
Ayanamsa	23 57 09
SVP	05 ♓ 09 52
☽ ☊ Mean	22 ♓ 54 R

● ◐ PHASES ○ ◑

Day	Time	Phase	Position
05	12:58	○	12♉58
12	17:46	◑	20♌12
20	22:18	●	28♏27
28	06:29	◐	05♓53

LAST ASPECT Day	h m	☽ INGRESS Day	h m	Sign
02	07:55	02	15:47	♈
04	08:04	04	16:05	♉
06	10:18	06	16:47	♊
08	11:16	08	19:46	♋
10	20:59	11	02:35	♌
13	08:30	13	13:20	♍
15	22:42	16	02:15	♎
18	05:42	18	14:48	♏
20	23:54	21	01:16	♐
23	01:20	23	09:26	♑
25	01:44	25	15:41	♒
27	13:00	27	20:21	♓
29	16:29	29	23:30	♈

DECLINATION

Day	☉	☽	☿	♀	♂	♃	♄	♅	♆	♇
01 We	14S19	10S54	21S09	13S52	13S06	18S16	14N29	08S11	15S56	16S19
02 Th	14 38	04 11	20 49	14 17	13 20	18 19	14 28	08 12	15 56	16 19
03 Fr	14 57	02N53	20 25	14 42	13 34	18 22	14 27	08 12	15 56	16 19
04 Sa	15 16	09 52	19 57	15 07	13 48	18 25	14 26	08 12	15 56	16 20
05 Su	15 34	16 18	19 24	15 31	14 02	18 28	14 25	08 13	15 56	16 20
06 Mo	15 52	21 44	18 49	15 55	14 16	18 32	14 24	08 13	15 56	16 20
07 Tu	16 10	25 45	18 10	16 19	14 29	18 35	14 23	08 13	15 56	16 21
08 We	16 28	28 01	17 28	16 42	14 43	18 38	14 22	08 13	15 56	16 21
09 Th	16 45	28 28	16 46	17 05	14 56	18 41	14 21	08 13	15 56	16 21
10 Fr	17 02	27 13	16 03	17 27	15 09	18 44	14 21	08 14	15 56	16 21
11 Sa	17 19	24 32	15 22	17 49	15 22	18 47	14 20	08 14	15 56	16 22
12 Su	17 36	20 46	14 43	18 10	15 35	18 50	14 19	08 14	15 56	16 22
13 Mo	17 52	16 11	14 08	18 31	15 48	18 53	14 19	08 14	15 55	16 22
14 Tu	18 08	11 04	13 37	18 51	16 01	18 57	14 18	08 14	15 55	16 23
15 We	18 23	05 38	13 11	19 11	16 14	19 00	14 17	08 14	15 55	16 23
16 Th	18 39	00 01	12 50	19 30	16 26	19 03	14 17	08 14	15 55	16 23
17 Fr	18 54	05S37	12 35	19 49	16 39	19 06	14 16	08 14	15 55	16 23
18 Sa	19 08	11 05	12 26	20 07	16 51	19 09	14 16	08 14	15 55	16 24
19 Su	19 22	16 13	12 22	20 24	17 03	19 12	14 15	08 14	15 54	16 24
20 Mo	19 36	20 47	12 22	20 41	17 15	19 15	14 15	08 14	15 54	16 24
21 Tu	19 50	24 31	12 27	20 58	17 27	19 18	14 14	08 14	15 54	16 25
22 We	20 03	27 08	12 37	21 14	17 39	19 20	14 14	08 14	15 54	16 25
23 Th	20 16	28 22	12 49	21 29	17 51	19 23	14 14	08 14	15 53	16 25
24 Fr	20 28	28 03	13 05	21 44	18 02	19 26	14 13	08 14	15 53	16 25
25 Sa	20 41	26 08	13 24	21 57	18 14	19 29	14 13	08 14	15 53	16 26
26 Su	20 52	22 43	13 44	22 11	18 25	19 32	14 13	08 14	15 53	16 26
27 Mo	21 04	18 02	14 07	22 23	18 36	19 35	14 13	08 14	15 52	16 26
28 Tu	21 14	12 21	14 31	22 35	18 47	19 38	14 13	08 13	15 52	16 26
29 We	21 25	05 58	14 56	22 47	18 58	19 40	14 13	08 13	15 52	16 27
30 Th	21 35	00N46	15 22	22 57	19 08	19 43	14 13	08 13	15 51	16 27

ASPECTARIAN

Day	Time	Aspect
01	04:38	☽ △ ☉
	06:42	☽ △ ♀
	08:32	☽ ☌ ♅
	09:57	☽ ∥ ♅
	11:56	☉ # ♄
	15:43	☿ □ ♄
02	03:32	♀ △ ♅
	05:24	☽ △ ☿
	07:33	☽ □ ♇
	07:55	☽ △ ♃
	09:35	♀ # ♄
03	12:02	☉ △ ♅
	18:11	☽ # ♅
	19:17	☽ ⚹ ♆
04	06:41	☽ △ ♄
	08:04	☽ △ ♇
	14:52	☽ # ♂
	16:40	☽ ∥ ♄
	20:43	☽ # ♀
	20:57	☽ # ☉
	22:33	☽ # ♆
05	00:08	☽ # ♇
	05:47	☽ ☍ ♂
	09:03	☽ # ♃
	09:38	☽ ⚹ ♅
	10:45	♀ ∥ ☉
	11:46	☽ # ☿
	16:55	☽ ☍ ♀
	19:35	☽ □ ♆
06	00:26	☽ ☍ ☿
	00:44	♀ ∥ ♆
	05:10	☉ ∥ ♆
	07:20	☽ □ ♄
	09:50	☿ ∥ ♃
	10:18	☽ ☍ ♃
	14:04	♂ # ♄
07	00:24	♀ □ ♆
	01:43	♀ ∥ ♇
	10:54	☽ □ ♅
	14:18	☉ ∥ ♇
	16:51	☿ ☌ ♀
	21:22	☽ △ ♆
08	08:24	☿ □ ♆
	09:56	☽ ⚹ ♄
	11:16	☽ ☍ ♇
	16:23	♂ △ ♅
	17:04	☿ ∥ ♀
	21:31	☿ ☌ ☉
09	00:15	☿ ∥ ☉
	13:49	☿ ∥ ♇
	15:00	☉ □ ♆
	15:07	☽ △ ♅
	16:23	☽ △ ♂
	22:35	☽ △ ☿
10	03:23	☽ △ ☉
	04:23	☿ ∥ ♆
	10:19	☽ △ ♀
	20:59	☽ △ ♃
	23:57	☿ ∥ ♂
11	15:16	☿ ☌ ♂
12	02:24	☽ □ ☿
	04:08	☽ □ ♂
	10:25	☽ # ♃
	11:38	☽ ☍ ♆
	13:04	☽ # ♀
	16:03	☽ # ☉
	16:20	☿ # ♄
	23:06	☽ # ♇
13	00:20	♀ □ ♄
	01:19	☽ # ♆
	01:49	☽ # ♂
	02:29	☽ ☌ ♄
	02:43	☽ □ ♀
	03:52	☽ △ ♇
	08:30	☽ □ ♃
	09:04	☽ ∥ ♄
	11:05	☽ # ☿
	13:16	♂ ∥ ♆
	17:27	☿ △ ♅
14	07:51	♀ ∥ ♃
	10:08	☽ ⚹ ☿
	11:12	☽ ☍ ♅
	12:39	☽ # ♅
	19:52	☽ ⚹ ♂
15	11:56	☽ ⚹ ☉
	16:47	☽ □ ♇
	18:02	♂ ∥ ♇
	20:36	♀ ☌ ♃
	22:31	☽ ⚹ ♃
	22:42	☽ ⚹ ♀
17	05:59	☉ □ ♄
	08:02	♀ ♐
	11:25	☽ ∥ ♅
	13:01	☽ △ ♆
	19:02	♂ □ ♆
18	00:26	☿ SD
	00:52	☉ ∥ ♃
	04:22	☽ ⚹ ♄
	05:42	☽ ⚹ ♇
	06:01	☽ ∥ ☿
	14:37	☽ # ♄
	22:31	☽ ∥ ♆
19	00:56	☽ ∥ ♇
	04:22	☽ ∥ ♂
	09:02	☽ ☌ ☿
	12:08	☽ △ ♅
	15:26	☽ ∥ ♃
	17:08	☽ ∥ ☉
	23:29	☽ ∥ ♀
20	00:32	☽ □ ♆
	03:40	☽ ☌ ♂
	06:10	♅ SD
	15:24	☽ □ ♄
	23:54	☽ ☌ ♃
21	11:08	☽ ☌ ♀
	21:45	☽ □ ♅
	23:16	☉ ☌ ♃
22	09:43	☽ ⚹ ♆
	11:02	☉ ♐
	14:51	☿ △ ♅
23	00:03	☽ △ ♄
	01:20	☽ ☌ ♇
24	04:44	♃ ♐
	05:11	☽ ⚹ ♅
	07:40	☽ ⚹ ☿
25	01:44	☽ ⚹ ♂
	16:16	☽ ⚹ ♃
	21:54	☽ ⚹ ☉
	22:56	♀ □ ♅
26	02:58	☽ ∥ ♀
	09:48	☽ ∥ ☉
	12:04	☽ ⚹ ♀
	16:46	☽ ∥ ♃
	17:30	☽ □ ☿
	21:29	☽ ∥ ♂
	22:10	☽ ☌ ♆
27	06:10	☿ # ♄
	07:09	☽ ∥ ♇
	09:35	☽ ∥ ♆
	09:37	☽ □ ♂
	11:41	☽ ☍ ♄
	13:00	☽ ⚹ ♇
	15:52	☽ ∥ ☿
	16:32	☽ # ♄
	21:47	☽ □ ♃
28	15:00	☽ ☌ ♅
	15:48	☽ ∥ ♅
	21:16	☽ □ ♀
29	00:54	☿ □ ♆
	02:07	☽ △ ☿
	03:01	♂ □ ♄
	15:45	☽ △ ♂
	16:29	☽ □ ♇
30	01:42	☽ △ ♃
	13:14	☽ △ ☉

⚷ Chiron

01 Dec. 12 S 28

Day	Position
01	04♒38
04	04 42
07	04 47
10	04 53
13	04 59
16	05 05
19	05 13
22	05 20
25	05 29
28	05 38

December 2006

Day	S. T.	☉	☽	☿	♀	♂	♃	♄	♅	♆	♇	☊ True
	h m s	° ' "	° ' "	° '	° '	° '	° '	° '	° '	° '	° '	° '
01 Fr	04 38 51	08♐38 34	14♈38 47	19♏48	17♐09	26♏20	01♐31	25♌03	10♓52	17♒20	25♐53	22♓R14
02 Sa	04 42 48	09 39 22	29 07 24	21 08	18 25	27 02	01 44	25 03	10 52	17 21	25 55	22 10
03 Su	04 46 44	10 40 11	13♉39 07	22 30	19 40	27 45	01 57	25 04	10 53	17 22	25 57	22 03
04 Mo	04 50 41	11 41 01	28 08 04	23 53	20 55	28 27	02 11	25 04	10 53	17 23	25 59	21 54
05 Tu	04 54 37	12 41 52	12♊27 35	25 18	22 11	29 09	02 24	25 04	10 54	17 24	26 02	21 42
06 We	04 58 34	13 42 45	26 31 17	26 44	23 26	29 51	02 37	25 04	10 55	17 26	26 04	21 30
07 Th	05 02 31	14 43 38	10♋14 10	28 11	24 41	00♐34	02 50	25R04	10 56	17 27	26 06	21 18
08 Fr	05 06 27	15 44 32	23 33 21	29 38	25 57	01 16	03 04	25 04	10 57	17 28	26 08	21 07
09 Sa	05 10 24	16 45 28	06♌28 22	01♐07	27 12	01 58	03 17	25 04	10 58	17 29	26 10	20 59
10 Su	05 14 20	17 46 24	19 00 58	02 36	28 27	02 41	03 30	25 03	10 59	17 31	26 13	20 53
11 Mo	05 18 17	18 47 22	01♍14 39	04 06	29 43	03 23	03 43	25 03	11 00	17 32	26 15	20 50
12 Tu	05 22 13	19 48 21	13 14 06	05 36	00♑58	04 06	03 56	25 02	11 01	17 34	26 17	20 49
13 We	05 26 10	20 49 21	25 04 41	07 06	02 13	04 48	04 09	25 02	11 02	17 35	26 19	20 49
14 Th	05 30 06	21 50 22	06♎52 01	08 37	03 28	05 31	04 22	25 01	11 03	17 36	26 22	20 48
15 Fr	05 34 03	22 51 24	18 41 39	10 08	04 44	06 14	04 35	25 00	11 04	17 38	26 24	20 46
16 Sa	05 37 60	23 52 27	00♏38 41	11 39	05 59	06 56	04 48	24 59	11 05	17 39	26 26	20 41
17 Su	05 41 56	24 53 31	12 47 27	13 11	07 14	07 39	05 01	24 58	11 07	17 41	26 28	20 34
18 Mo	05 45 53	25 54 36	25 11 13	14 42	08 30	08 22	05 14	24 56	11 08	17 43	26 30	20 24
19 Tu	05 49 49	26 55 42	07♐51 59	16 14	09 45	09 05	05 27	24 55	11 10	17 44	26 33	20 12
20 We	05 53 46	27 56 48	20 50 14	17 47	11 00	09 48	05 40	24 54	11 11	17 46	26 35	19 59
21 Th	05 57 42	28 57 55	04♑05 01	19 19	12 16	10 31	05 53	24 52	11 13	17 47	26 37	19 46
22 Fr	06 01 39	29 59 03	17 34 14	20 52	13 31	11 14	06 06	24 50	11 14	17 49	26 39	19 34
23 Sa	06 05 35	01♑00 11	01♒14 59	22 25	14 46	11 57	06 19	24 48	11 16	17 51	26 42	19 24
24 Su	06 09 32	02 01 19	15 04 19	23 58	16 02	12 40	06 31	24 47	11 17	17 52	26 44	19 18
25 Mo	06 13 29	03 02 27	28 59 36	25 31	17 17	13 23	06 44	24 45	11 19	17 54	26 46	19 14
26 Tu	06 17 25	04 03 36	12♓59 00	27 04	18 32	14 06	06 57	24 42	11 21	17 56	26 48	19 12
27 We	06 21 22	05 04 44	27 01 21	28 38	19 47	14 49	07 09	24 40	11 23	17 58	26 50	19 12
28 Th	06 25 18	06 05 53	11♈05 59	00♑12	21 03	15 32	07 22	24 38	11 25	17 59	26 53	19 11
29 Fr	06 29 15	07 07 01	25 12 11	01 46	22 18	16 16	07 34	24 35	11 26	18 01	26 55	19 09
30 Sa	06 33 11	08 08 09	09♉18 50	03 21	23 33	16 59	07 47	24 33	11 28	18 03	26 57	19 06
31 Su	06 37 08	09 09 18	23 23 55	04 56	24 48	17 42	07 59	24 30	11 30	18 05	26 59	19 00

Data for	12-01-2006
Julian Day	2454070.50
Ayanamsa	23 57 14
SVP	05 ♓ 09 49
☽ ☊ Mean	21 ♓ 19 R

LAST ASPECT ☽ Day	h m	INGRESS Day	h m	
01	18:41	02	01:27	♉
04	00:33	04	03:06	♊
05	23:12	06	06:02	♋
07	01:14	08	11:53	♌
10	20:36	10	21:32	♍
13	02:32	13	10:01	♎
15	15:33	15	22:43	♏
17	23:32	18	09:10	♐
20	14:01	20	16:39	♑
21	16:05	22	21:49	♒
24	20:10	25	01:44	♓
27	03:06	27	05:05	♈
29	02:55	29	08:09	♉
31	02:38	31	11:17	♊

● ◐ PHASES ○ ◑

05	00:26	○	12♊43
12	14:32	◑	20♍25
20	14:01	●	28♐33
27	14:48	◐	05♈42

DECLINATION

Day	☉	☽	☿	♀	♂	♃	♄	♅	♆	♇
01 Fr	21S45	07N33	15S48	23S07	19S19	19S46	14N13	08S13	15S51	16S27
02 Sa	21 54	14 00	16 15	23 17	19 29	19 49	14 13	08 12	15 51	16 27
03 Su	22 03	19 42	16 42	23 25	19 39	19 51	14 13	08 12	15 50	16 27
04 Mo	22 11	24 13	17 10	23 33	19 49	19 54	14 13	08 12	15 50	16 28
05 Tu	22 19	27 12	17 37	23 40	19 59	19 57	14 13	08 12	15 50	16 28
06 We	22 27	28 23	18 04	23 47	20 09	19 59	14 13	08 11	15 49	16 28
07 Th	22 34	27 47	18 30	23 52	20 18	20 02	14 13	08 11	15 49	16 28
08 Fr	22 41	25 34	18 56	23 57	20 28	20 05	14 13	08 10	15 49	16 28
09 Sa	22 47	22 04	19 22	24 02	20 37	20 07	14 14	08 10	15 48	16 29
10 Su	22 53	17 39	19 47	24 05	20 46	20 10	14 14	08 10	15 48	16 29
11 Mo	22 58	12 37	20 11	24 08	20 55	20 12	14 14	08 09	15 47	16 29
12 Tu	23 03	07 12	20 35	24 10	21 03	20 15	14 15	08 09	15 47	16 29
13 We	23 07	01 36	20 57	24 11	21 12	20 17	14 15	08 08	15 46	16 29
14 Th	23 11	04S01	21 19	24 12	21 20	20 20	14 16	08 08	15 46	16 30
15 Fr	23 15	09 31	21 40	24 11	21 28	20 22	14 16	08 07	15 46	16 30
16 Sa	23 18	14 44	22 00	24 10	21 36	20 24	14 17	08 07	15 45	16 30
17 Su	23 21	19 28	22 19	24 09	21 44	20 27	14 17	08 06	15 45	16 30
18 Mo	23 23	23 28	22 37	24 06	21 51	20 29	14 18	08 06	15 44	16 30
19 Tu	23 24	26 27	22 54	24 03	21 59	20 31	14 19	08 05	15 44	16 30
20 We	23 25	28 07	23 10	23 59	22 06	20 34	14 19	08 05	15 43	16 31
21 Th	23 26	28 14	23 24	23 54	22 12	20 36	14 20	08 04	15 43	16 31
22 Fr	23 26	26 41	23 38	23 49	22 19	20 38	14 21	08 03	15 42	16 31
23 Sa	23 26	23 32	23 50	23 43	22 26	20 40	14 22	08 03	15 42	16 31
24 Su	23 26	19 02	24 02	23 36	22 32	20 43	14 22	08 02	15 41	16 31
25 Mo	23 24	13 28	24 11	23 28	22 38	20 45	14 23	08 01	15 41	16 31
26 Tu	23 23	07 12	24 20	23 20	22 44	20 47	14 24	08 01	15 40	16 31
27 We	23 21	00 33	24 27	23 11	22 49	20 49	14 25	08 00	15 40	16 31
28 Th	23 18	06N08	24 34	23 01	22 55	20 51	14 26	07 59	15 39	16 32
29 Fr	23 15	12 31	24 38	22 50	23 00	20 53	14 27	07 58	15 39	16 32
30 Sa	23 11	18 17	24 42	22 39	23 05	20 55	14 28	07 58	15 38	16 32
31 Su	23 07	23 02	24 44	22 27	23 10	20 57	14 29	07 57	15 37	16 32

ASPECTARIAN

01 02:24 ☽ # ♅
02:31 ☿ ∥ ♆
03:22 ♀ ✶ ♆
04:28 ☽ ✶ ♆
04:34 ☽ △ ♀
17:16 ☽ △ ♄
18:41 ☽ △ ♇
02 00:50 ☽ ∥ ♄
07:22 ☽ # ♆
09:48 ☽ # ☿
09:53 ☽ # ♇
10:44 ☿ ∥ ♇
19:25 ☽ ✶ ♅
23:49 ☽ # ♂
03 00:45 ☽ # ♃
05:00 ☉ □ ♅
06:09 ☽ □ ♆
11:56 ☽ # ☉
16:11 ☽ ☍ ☿
18:54 ☽ □ ♄
19:48 ☽ # ♀
04 00:33 ☽ ☍ ♂
06:51 ☽ ☍ ♃
15:17 ♂ ∥ ♃
20:08 ☿ □ ♄
21:22 ☽ □ ♅
05 08:24 ☽ △ ♆
18:09 ☽ ☍ ♀
21:30 ☽ ✶ ♄
23:12 ☽ ☍ ♇
06 04:08 ♄ SR
04:58 ♂ ♐
07 01:14 ☽ △ ♅
07:18 ♀ △ ♄
08 03:51 ♀ ☌ ♇
05:52 ☿ ♐
11:44 ☽ # ♀
12:33 ☽ □ ♆
15:04 ☽ △ ♂
17:54 ☽ △ ♃
19:47 ☽ # ☉
09 08:07 ☽ # ♂
11:02 ☽ # ♃
13:54 ☽ # ☿
17:45 ☉ ✶ ♆
21:05 ☽ ☍ ♆
21:23 ☽ △ ☉
10 02:28 ☿ ☌ ♂
05:48 ☽ # ♇
09:08 ☽ # ♆
11:47 ☽ ☌ ♄
14:06 ☽ △ ♇
16:30 ☽ ∥ ♄
16:59 ☿ ☌ ♃
20:36 ☽ △ ♀
11 01:09 ☿ ∥ ♃
04:32 ☽ □ ♂
05:01 ☽ □ ♃
05:33 ♀ ♑
06:28 ☽ □ ☿
16:12 ♂ ☌ ♃
19:31 ☽ ☍ ♅
19:54 ☽ # ♅
13 02:32 ☽ □ ♇
16:16 ☽ □ ♀
18:50 ☽ ✶ ♃
21:05 ☽ ✶ ♂
14 01:39 ☿ ∥ ♂
04:04 ☽ ✶ ☿
17:52 ☽ ∥ ♅
21:51 ☽ △ ♆
15 09:11 ☽ ✶ ☉
10:41 ☽ ✶ ♄
15:03 ☿ □ ♅
15:33 ☽ ✶ ♇
21:53 ☽ # ♄
16 04:58 ☽ ∥ ♆
08:40 ☽ ∥ ♇
11:50 ☽ ✶ ♀
20:42 ☽ △ ♅
17 01:36 ☉ △ ♄
05:33 ☽ ∥ ♃
09:33 ☽ □ ♆
13:24 ☽ ∥ ♂
18:02 ☽ ∥ ☿
23:27 ☽ ∥ ☉
23:32 ☽ □ ♄
18 04:26 ☽ ∥ ♀
14:39 ☉ ☌ ♇
19:24 ☽ ☌ ♃
19 02:24 ☽ ☌ ♂
06:09 ☽ □ ♅
17:38 ☽ ☌ ☿
18:21 ☽ ✶ ♆
23:44 ☿ ✶ ♆
20 03:30 ♀ ✶ ♅
07:23 ☽ △ ♄
10:30 ☽ ☌ ♇
21 03:02 ☿ ∥ ☉
12:45 ☽ ✶ ♅
16:05 ☽ ☌ ♀
22 00:18 ♂ □ ♅
00:22 ☉ ♑
14:01 ☿ ∥ ♀
22:11 ☽ ∥ ☿
22:52 ☽ ∥ ♀
23 00:36 ☽ ∥ ☉
06:01 ☽ ∥ ♂
08:57 ☽ ✶ ♃
15:49 ☽ ∥ ♃
19:36 ☽ ✶ ♂
24 04:51 ☽ ☌ ♆
11:18 ☽ ∥ ♇
12:19 ☿ △ ♄
14:53 ☽ ∥ ♆
16:42 ☽ ☍ ♄
17:16 ☽ ✶ ☿
20:10 ☽ ✶ ♇
20:15 ☽ # ♄
25 07:30 ☽ ✶ ☉
13:30 ☽ □ ♃
13:42 ♀ ∥ ☉
19:47 ☿ ☌ ♇
20:58 ☽ ∥ ♅
21:12 ☽ ☌ ♅
26 02:01 ☽ □ ♂
10:26 ☽ ✶ ♀
23:41 ☽ □ ♇
27 03:06 ☽ □ ☿
17:33 ☽ △ ♃
20:55 ☿ ♑
28 06:47 ☽ # ♅
07:58 ☽ △ ♂
09:25 ♀ ∥ ♂
11:45 ☽ ✶ ♆
18:34 ☽ □ ♀
22:58 ☽ △ ♄
29 02:55 ☽ △ ♇
07:42 ☽ ∥ ♄
12:34 ☽ # ♆
12:34 ☽ △ ☿
16:20 ☽ # ♇
21:50 ☽ △ ☉
30 03:41 ☽ ✶ ♅
12:42 ☽ # ♃
14:55 ☽ □ ♆
17:49 ☉ ∥ ♂
20:55 ☽ # ♀
31 00:34 ☽ # ☉
00:49 ☽ # ♂
01:53 ☽ □ ♄
02:38 ☽ △ ♀
10:48 ☽ # ☿
13:07 ♂ ✶ ♆

⚷ Chiron

01 Dec.	12 S 25
01	05♒47
04	05 57
07	06 07
10	06 18
13	06 29
16	06 40
19	06 52
22	07 04
25	07 17
28	07 29
31	07 42

January 2007

Day	S. T. h m s	☉ ° ' "	☽ ° ' "	☿ ° '	♀ ° '	♂ ° '	♃ ° '	♄ ° '	♅ ° '	♆ ° '	♇ ° '	☊ True ° '
01 Mo	06 41 04	10♑10 26	07♊24 34	06♑31	26♑03	18♐26	08♐12	24♌R28	11♓32	18♒07	27♐01	18♓R51
02 Tu	06 45 01	11 11 34	21 17 06	08 06	27 19	19 09	08 24	24 25	11 34	18 09	27 04	18 41
03 We	06 48 58	12 12 42	04♋57 35	09 42	28 34	19 53	08 36	24 22	11 36	18 11	27 06	18 30
04 Th	06 52 54	13 13 50	18 22 28	11 18	29 49	20 36	08 48	24 19	11 39	18 13	27 08	18 18
05 Fr	06 56 51	14 14 59	01♌29 13	12 54	01♒04	21 20	09 01	24 16	11 41	18 15	27 10	18 08
06 Sa	07 00 47	15 16 07	14 16 48	14 31	02 19	22 03	09 13	24 13	11 43	18 17	27 12	18 00
07 Su	07 04 44	16 17 15	26 45 52	16 08	03 34	22 47	09 25	24 09	11 45	18 19	27 14	17 55
08 Mo	07 08 40	17 18 23	08♍58 33	17 46	04 50	23 31	09 37	24 06	11 48	18 21	27 17	17 52
09 Tu	07 12 37	18 19 31	20 58 21	19 24	06 05	24 14	09 48	24 03	11 50	18 23	27 19	17D 51
10 We	07 16 34	19 20 40	02♎49 43	21 02	07 20	24 58	10 00	23 59	11 52	18 25	27 21	17 52
11 Th	07 20 30	20 21 48	14 37 44	22 41	08 35	25 42	10 12	23 56	11 55	18 27	27 23	17 53
12 Fr	07 24 27	21 22 56	26 27 48	24 20	09 50	26 26	10 24	23 52	11 57	18 29	27 25	17R 53
13 Sa	07 28 23	22 24 04	08♏25 21	26 00	11 05	27 10	10 35	23 48	12 00	18 31	27 27	17 51
14 Su	07 32 20	23 25 13	20 35 28	27 40	12 20	27 54	10 47	23 44	12 02	18 33	27 29	17 47
15 Mo	07 36 16	24 26 21	03♐02 31	29 20	13 35	28 38	10 58	23 41	12 05	18 35	27 31	17 41
16 Tu	07 40 13	25 27 29	15 49 44	01♒01	14 50	29 22	11 10	23 37	12 07	18 37	27 33	17 33
17 We	07 44 09	26 28 36	28 58 50	02 43	16 05	00♑06	11 21	23 33	12 10	18 39	27 35	17 24
18 Th	07 48 06	27 29 44	12♑29 40	04 24	17 20	00 50	11 32	23 29	12 12	18 41	27 37	17 16
19 Fr	07 52 03	28 30 51	26 20 07	06 07	18 35	01 34	11 43	23 24	12 15	18 44	27 39	17 08
20 Sa	07 55 59	29 31 57	10♒26 21	07 49	19 50	02 18	11 54	23 20	12 18	18 46	27 41	17 01
21 Su	07 59 56	00♒33 02	24 43 25	09 31	21 05	03 02	12 05	23 16	12 21	18 48	27 43	16 57
22 Mo	08 03 52	01 34 07	09♓06 00	11 14	22 20	03 47	12 16	23 12	12 23	18 50	27 45	16 55
23 Tu	08 07 49	02 35 11	23 29 15	12 57	23 35	04 31	12 27	23 07	12 26	18 52	27 47	16D 54
24 We	08 11 45	03 36 14	07♈49 15	14 39	24 50	05 15	12 38	23 03	12 29	18 55	27 49	16 55
25 Th	08 15 42	04 37 16	22 03 15	16 22	26 05	06 00	12 48	22 58	12 32	18 57	27 51	16 56
26 Fr	08 19 38	05 38 16	06♉09 27	18 04	27 19	06 44	12 59	22 54	12 35	18 59	27 53	16R 56
27 Sa	08 23 35	06 39 16	20 06 45	19 45	28 34	07 28	13 09	22 49	12 38	19 01	27 55	16 55
28 Su	08 27 32	07 40 15	03♊54 20	21 26	29 49	08 13	13 20	22 45	12 41	19 03	27 56	16 53
29 Mo	08 31 28	08 41 12	17 31 28	23 05	01♓04	08 57	13 30	22 40	12 44	19 06	27 58	16 48
30 Tu	08 35 25	09 42 08	00♋57 18	24 44	02 18	09 42	13 40	22 35	12 47	19 08	28 00	16 43
31 We	08 39 21	10 43 03	14 10 52	26 20	03 33	10 26	13 50	22 31	12 50	19 10	28 02	16 36

Data for	01-01-2007
Julian Day	2454101.50
Ayanamsa	23 57 20
SVP	05 ♓ 09 47
☽ ☊ Mean	19 ♓ 40 R

● ◐ PHASES ○ ◑

Day	Time	Phase	Position
03	13:58	○	12♋48
11	12:44	◑	20♎54
19	04:01	●	28♑41
25	23:02	◐	05♉36

LAST ASPECT Day	h m	☽ INGRESS Day	h m	Sign
02	10:07	02	15:15	♋
03	13:58	04	21:15	♌
07	00:56	07	06:19	♍
09	12:51	09	18:15	♎
12	01:56	12	07:08	♏
14	15:50	14	18:12	♐
16	21:29	17	01:50	♑
19	04:02	19	06:16	♒
21	05:01	21	08:49	♓
23	07:12	23	10:53	♈
25	09:51	25	13:29	♉
27	16:09	27	17:10	♊
29	18:41	29	22:17	♋

DECLINATION

Day	☉	☽	☿	♀	♂	♃	♄	♅	♆	♇
01 Mo	23S 03	26N25	24S 45	22S 15	23S 14	20S 59	14N30	07S 56	15S 37	16S 32
02 Tu	22 58	28 11	24 44	22 02	23 19	21 01	14 31	07 55	15 36	16 32
03 We	22 53	28 11	24 42	21 48	23 23	21 03	14 33	07 54	15 36	16 32
04 Th	22 47	26 31	24 38	21 33	23 26	21 05	14 34	07 54	15 35	16 32
05 Fr	22 41	23 26	24 33	21 18	23 30	21 07	14 35	07 53	15 34	16 32
06 Sa	22 34	19 15	24 27	21 02	23 33	21 09	14 36	07 52	15 34	16 32
07 Su	22 27	14 20	24 19	20 46	23 37	21 11	14 37	07 51	15 33	16 32
08 Mo	22 19	08 57	24 09	20 29	23 40	21 12	14 39	07 50	15 33	16 32
09 Tu	22 11	03 20	23 59	20 11	23 42	21 14	14 40	07 49	15 32	16 33
10 We	22 03	02S 20	23 46	19 53	23 45	21 16	14 41	07 48	15 31	16 33
11 Th	21 54	07 53	23 32	19 35	23 47	21 18	14 43	07 47	15 31	16 33
12 Fr	21 44	13 11	23 17	19 15	23 49	21 19	14 44	07 46	15 30	16 33
13 Sa	21 35	18 03	23 00	18 55	23 51	21 21	14 45	07 45	15 30	16 33
14 Su	21 25	22 17	22 41	18 35	23 52	21 23	14 47	07 44	15 29	16 33
15 Mo	21 14	25 37	22 21	18 14	23 53	21 24	14 48	07 43	15 28	16 33
16 Tu	21 03	27 46	21 59	17 53	23 54	21 26	14 50	07 42	15 28	16 33
17 We	20 52	28 27	21 36	17 31	23 55	21 27	14 51	07 41	15 27	16 33
18 Th	20 40	27 28	21 11	17 08	23 56	21 29	14 53	07 40	15 26	16 33
19 Fr	20 28	24 47	20 45	16 45	23 56	21 30	14 54	07 39	15 26	16 33
20 Sa	20 15	20 34	20 17	16 22	23 56	21 32	14 56	07 38	15 25	16 33
21 Su	20 02	15 07	19 47	15 58	23 56	21 33	14 57	07 37	15 24	16 33
22 Mo	19 49	08 49	19 17	15 34	23 55	21 35	14 59	07 36	15 24	16 33
23 Tu	19 35	02 03	18 44	15 09	23 54	21 36	15 01	07 35	15 23	16 33
24 We	19 21	04N47	18 11	14 44	23 53	21 37	15 02	07 34	15 22	16 33
25 Th	19 07	11 21	17 35	14 18	23 52	21 39	15 04	07 33	15 22	16 33
26 Fr	18 52	17 16	16 59	13 53	23 51	21 40	15 05	07 32	15 21	16 33
27 Sa	18 37	22 13	16 22	13 26	23 49	21 41	15 07	07 30	15 20	16 33
28 Su	18 21	25 53	15 43	13 00	23 47	21 43	15 09	07 29	15 19	16 33
29 Mo	18 05	28 00	15 03	12 33	23 45	21 44	15 10	07 28	15 19	16 33
30 Tu	17 49	28 27	14 23	12 06	23 42	21 45	15 12	07 27	15 18	16 33
31 We	17 33	27 15	13 42	11 38	23 40	21 46	15 14	07 26	15 17	16 33

ASPECTARIAN

01 01:22 ☽ ☍ ♃
07:08 ☽ □ ♅
18:32 ☽ △ ♆
20:05 ☽ ☍ ♂
02 05:26 ☽ ⚹ ♄
09:16 ☉ ⚹ ♅
10:07 ☽ ☍ ♇
03 09:33 ☽ ☍ ☿
11:52 ☽ △ ♅
04 03:32 ♀ ♒
05:20 ☿ ⚹ ♅
16:01 ☽ # ☿
23:09 ☽ ☍ ♀
23:28 ☽ # ♂
05 04:49 ☽ # ☉
13:39 ☽ # ♀
13:47 ☽ # ♃
14:16 ☽ △ ♃
15:32 ♀ ∥ ♃
06 07:38 ☽ ☍ ♆
13:32 ☽ # ♇
15:48 ☽ △ ♂
18:14 ☽ # ♆
18:58 ☽ ☌ ♄
22:37 ☽ ∥ ♄
07 00:56 ☽ △ ♇
06:05 ☿ ☌ ☉
08 01:17 ☽ □ ♃
04:47 ☽ # ♅
05:37 ☽ ☍ ♅
18:06 ♂ △ ♄
18:11 ☽ △ ☉
20:19 ☽ △ ☿
09 07:01 ☽ □ ♂
12:51 ☽ □ ♇
10 02:09 ☿ ∥ ♂
10:14 ☽ △ ♀
14:50 ☽ ⚹ ♃
23:33 ☽ ∥ ♅
11 07:47 ☽ △ ♆
18:47 ☽ ⚹ ♄
19:00 ☽ □ ☿
23:56 ☽ ⚹ ♂
12 01:56 ☽ ⚹ ♇
07:24 ☽ # ♄
11:05 ☽ ∥ ♆
12:45 ♀ ⚹ ♃
16:17 ☽ ∥ ♇
13 04:18 ☽ ∥ ♀
05:54 ☽ □ ♀
07:07 ☽ △ ♅
10:01 ♂ ☌ ♇
18:26 ☽ ∥ ♃
18:53 ☽ ∥ ☉
20:00 ☽ □ ♆
14 02:21 ☽ ∥ ☿
04:07 ☉ ∥ ♃
06:00 ☽ ⚹ ☉
06:06 ☽ □ ♄
10:34 ☽ ∥ ♂
15:50 ☽ ⚹ ☿
15 09:25 ☿ ♒
15:11 ☽ ☌ ♃
17:05 ☽ □ ♅
21:58 ☽ ⚹ ♀
16 05:10 ☽ ⚹ ♆
14:13 ☽ △ ♄
20:54 ♂ ♑
21:29 ☽ ☌ ♇
17 02:07 ☽ ☌ ♂
08:01 ☿ ∥ ♃
23:30 ☽ ⚹ ♅
19 02:48 ♀ ☌ ♆
05:36 ☽ ∥ ♂
12:41 ♀ ∥ ♇
18:57 ☽ ☌ ☿
19:12 ☽ ∥ ♃
20 01:34 ☽ ∥ ☿
01:38 ☽ ∥ ☉
02:30 ☽ ⚹ ♃
02:45 ☿ ∥ ☉
11:01 ☉ ♒
14:03 ☽ ☌ ♆
17:19 ☽ ☌ ♀
18:08 ☽ ∥ ♇
20:17 ☽ ∥ ♀
21:34 ☽ ☍ ♄
22:50 ☽ ∥ ♆
21 00:39 ☽ # ♄
05:01 ☽ ⚹ ♇
14:38 ☽ ⚹ ♂
22 04:22 ☽ ∥ ♅
05:21 ☽ □ ♃
05:30 ☽ ☌ ♅
10:13 ♀ ∥ ♆
15:39 ♀ ☍ ♄
16:13 ☿ ⚹ ♃
21:43 ♃ □ ♅
23 07:12 ☽ □ ♇
07:44 ♀ # ♄
16:23 ☽ ⚹ ☉
19:27 ☽ □ ♂
24 08:11 ☽ △ ♃
09:53 ☽ # ♅
13:04 ☽ ⚹ ☿
18:44 ☽ ⚹ ♆
25 01:33 ☽ △ ♄
07:29 ☽ ⚹ ♀
09:51 ☽ △ ♇
10:47 ☽ # ♀
14:42 ☽ ∥ ♄
15:50 ☽ # ♆
20:52 ☽ # ♇
22:52 ☽ # ☿
26 01:02 ☽ △ ♂
06:49 ☽ # ☉
10:59 ♀ ⚹ ♇
11:03 ☽ ⚹ ♅
13:20 ☿ ☌ ♆
16:46 ☿ ∥ ♇
21:07 ☽ # ♃
22:06 ☽ □ ♆
23:18 ☽ □ ☿
27 04:40 ☽ □ ♄
09:17 ☽ # ♂
16:09 ☽ □ ♀
28 03:32 ♀ ♓
07:08 ☽ △ ☉
14:28 ☿ ∥ ♆
15:29 ☽ □ ♅
16:47 ☽ ☍ ♃
18:07 ☿ ☍ ♄
19:57 ☿ # ♄
29 02:48 ☽ △ ♆
09:06 ☽ ⚹ ♄
11:17 ☽ △ ☿
18:41 ☽ ☍ ♇
30 02:41 ☽ △ ♀
16:46 ☽ ☍ ♂
21:31 ☽ △ ♅

⚷ Chiron

01	Dec.	12 S 06
03	07♒56	
06	08 09	
09	08 23	
12	08 37	
15	08 51	
18	09 05	
21	09 19	
24	09 34	
27	09 48	
30	10 02	

February 2007

Day	S. T. h m s	☉ ° ' "	☽ ° ' "	☿ ° '	♀ ° '	♂ ° '	♃ ° '	♄ ° '	♅ ° '	♆ ° '	♇ ° '	☊ True ° '
01 Th	08 43 18	11♒43 57	27♋11 13	27♒54	04♓48	11♑11	14♐00	22♌R26	12♓53	19♒13	28♐03	16♓R30
02 Fr	08 47 14	12 44 50	09♌57 41	29 25	06 02	11 56	14 10	22 21	12 56	19 15	28 05	16 24
03 Sa	08 51 11	13 45 42	22 30 03	00♓53	07 17	12 40	14 19	22 16	12 59	19 17	28 07	16 20
04 Su	08 55 07	14 46 33	04♍48 55	02 18	08 32	13 25	14 29	22 12	13 02	19 19	28 09	16 18
05 Mo	08 59 04	15 47 22	16 55 45	03 38	09 46	14 10	14 39	22 07	13 05	19 22	28 10	16D 17
06 Tu	09 03 01	16 48 11	28 52 55	04 52	11 01	14 55	14 48	22 02	13 08	19 24	28 12	16 18
07 We	09 06 57	17 48 58	10♎43 36	06 01	12 15	15 39	14 57	21 57	13 11	19 26	28 13	16 20
08 Th	09 10 54	18 49 45	22 31 41	07 03	13 30	16 24	15 06	21 52	13 15	19 28	28 15	16 22
09 Fr	09 14 50	19 50 30	04♏21 40	07 57	14 44	17 09	15 15	21 47	13 18	19 31	28 17	16 24
10 Sa	09 18 47	20 51 15	16 18 23	08 43	15 58	17 54	15 24	21 42	13 21	19 33	28 18	16 25
11 Su	09 22 43	21 51 59	28 26 50	09 21	17 13	18 39	15 33	21 38	13 24	19 35	28 20	16R 25
12 Mo	09 26 40	22 52 41	10♐51 47	09 48	18 27	19 24	15 42	21 33	13 28	19 38	28 21	16 23
13 Tu	09 30 36	23 53 23	23 37 28	10 06	19 41	20 09	15 50	21 28	13 31	19 40	28 23	16 21
14 We	09 34 33	24 54 03	06♑47 00	10 14	20 56	20 54	15 59	21 23	13 34	19 42	28 24	16 18
15 Th	09 38 30	25 54 42	20 21 56	10R 10	22 10	21 39	16 07	21 18	13 37	19 44	28 25	16 15
16 Fr	09 42 26	26 55 20	04♒21 43	09 57	23 24	22 24	16 15	21 13	13 41	19 47	28 27	16 12
17 Sa	09 46 23	27 55 57	18 43 26	09 33	24 38	23 09	16 23	21 08	13 44	19 49	28 28	16 10
18 Su	09 50 19	28 56 32	03♓21 57	08 59	25 52	23 55	16 31	21 03	13 47	19 51	28 30	16 09
19 Mo	09 54 16	29 57 05	18 10 27	08 16	27 06	24 40	16 39	20 59	13 51	19 53	28 31	16D 09
20 Tu	09 58 12	00♓57 37	03♈01 20	07 26	28 20	25 25	16 47	20 54	13 54	19 56	28 32	16 09
21 We	10 02 09	01 58 07	17 47 25	06 30	29 34	26 10	16 54	20 49	13 58	19 58	28 33	16 10
22 Th	10 06 05	02 58 35	02♉22 38	05 29	00♈48	26 56	17 02	20 44	14 01	20 00	28 35	16 11
23 Fr	10 10 02	03 59 02	16 42 39	04 24	02 02	27 41	17 09	20 40	14 04	20 02	28 36	16 12
24 Sa	10 13 59	04 59 26	00♊44 59	03 18	03 16	28 26	17 16	20 35	14 08	20 05	28 37	16 12
25 Su	10 17 55	05 59 49	14 28 43	02 12	04 30	29 12	17 23	20 30	14 11	20 07	28 38	16R 12
26 Mo	10 21 52	07 00 09	27 54 09	01 08	05 43	29 57	17 30	20 26	14 15	20 09	28 39	16 11
27 Tu	10 25 48	08 00 28	11♋02 23	00 07	06 57	00♒42	17 36	20 21	14 18	20 11	28 40	16 10
28 We	10 29 45	09 00 45	23 54 58	29♒11	08 11	01 28	17 43	20 17	14 21	20 13	28 41	16 09

Data for	02-01-2007
Julian Day	2454132.50
Ayanamsa	23 57 26
SVP	05 ♓ 09 44
☽ ☊ Mean	18 ♓ 02 R

● ◐ PHASES ○ ◑

Day	Time	Phase	Position
02	05:45	○	12♌59
10	09:52	◑	21♏16
17	16:15	●	28♒37
24	07:56	◐	05♊19

LAST ASPECT Day	h m	☽ INGRESS Day	h m	Sign
30	21:31	01	05:15	♌
03	10:55	03	14:34	♍
05	22:37	06	02:15	♎
08	11:39	08	15:10	♏
10	10:40	11	03:02	♐
13	08:46	13	11:43	♑
15	03:26	15	16:36	♒
17	16:15	17	18:31	♓
19	16:44	19	19:07	♈
21	17:42	21	20:04	♉
23	19:47	23	22:42	♊
26	01:21	26	03:48	♋
27	06:04	28	11:30	♌

DECLINATION

Day	☉	☽	☿	♀	♂	♃	♄	♅	♆	♇
01 Th	17S 16	24N35	13S 00	11S 10	23S 37	21S 47	15N15	07S 25	15S 17	16S 33
02 Fr	16 59	20 44	12 18	10 42	23 33	21 48	15 17	07 23	15 16	16 33
03 Sa	16 42	16 01	11 36	10 13	23 30	21 50	15 19	07 22	15 15	16 33
04 Su	16 24	10 43	10 54	09 45	23 26	21 51	15 20	07 21	15 15	16 33
05 Mo	16 06	05 07	10 13	09 16	23 22	21 52	15 22	07 20	15 14	16 33
06 Tu	15 48	00S 36	09 33	08 47	23 18	21 53	15 24	07 19	15 13	16 33
07 We	15 30	06 14	08 55	08 17	23 13	21 54	15 25	07 17	15 13	16 33
08 Th	15 11	11 39	08 18	07 47	23 09	21 55	15 27	07 16	15 12	16 33
09 Fr	14 52	16 39	07 43	07 18	23 04	21 56	15 29	07 15	15 11	16 32
10 Sa	14 33	21 03	07 11	06 47	22 59	21 57	15 30	07 14	15 10	16 32
11 Su	14 13	24 40	06 42	06 17	22 53	21 58	15 32	07 12	15 10	16 32
12 Mo	13 53	27 14	06 16	05 47	22 48	21 59	15 34	07 11	15 09	16 32
13 Tu	13 34	28 29	05 54	05 16	22 42	21 59	15 35	07 10	15 08	16 32
14 We	13 13	28 10	05 37	04 46	22 35	22 00	15 37	07 09	15 08	16 32
15 Th	12 53	26 12	05 24	04 15	22 29	22 01	15 39	07 07	15 07	16 32
16 Fr	12 32	22 35	05 16	03 44	22 22	22 02	15 40	07 06	15 06	16 32
17 Sa	12 12	17 31	05 12	03 13	22 15	22 03	15 42	07 05	15 05	16 32
18 Su	11 51	11 21	05 14	02 42	22 08	22 03	15 44	07 03	15 05	16 32
19 Mo	11 29	04 30	05 20	02 11	22 01	22 04	15 45	07 02	15 04	16 32
20 Tu	11 08	02N37	05 31	01 39	21 53	22 05	15 47	07 01	15 03	16 32
21 We	10 46	09 32	05 46	01 08	21 46	22 06	15 48	06 59	15 03	16 32
22 Th	10 25	15 52	06 04	00 37	21 38	22 06	15 50	06 58	15 02	16 31
23 Fr	10 03	21 14	06 26	00 05	21 29	22 07	15 51	06 57	15 01	16 31
24 Sa	09 41	25 17	06 50	00N26	21 21	22 08	15 53	06 55	15 01	16 31
25 Su	09 19	27 47	07 16	00 57	21 12	22 08	15 55	06 54	15 00	16 31
26 Mo	08 56	28 36	07 43	01 29	21 03	22 09	15 56	06 53	14 59	16 31
27 Tu	08 34	27 46	08 10	02 00	20 54	22 09	15 58	06 52	14 59	16 31
28 We	08 11	25 26	08 38	02 31	20 45	22 10	15 59	06 50	14 58	16 31

ASPECTARIAN

01 02:33 ☿ ✶ ♇
06:50 ☽ # ♂
15:00 ♄ # ♆
17:55 ☽ # ♃

02 08:06 ☽ △ ♃
09:20 ☿ ♓
17:47 ☽ ☍ ♆
20:31 ☽ # ☉
21:27 ☽ # ♇
23:34 ☽ ☌ ♄

03 03:18 ☽ ∥ ♄
03:35 ☽ # ♆
10:41 ♂ ✶ ♅
10:55 ☽ △ ♇
12:20 ☉ ∥ ♇
15:48 ☉ ✶ ♃
18:26 ☽ ☍ ☿
23:03 ☽ # ☿

04 04:39 ☽ # ♀
08:09 ☽ ☍ ♀
14:34 ☽ # ♅
16:19 ☽ ☍ ♅
18:07 ☽ △ ♂
19:23 ☽ □ ♃

05 22:37 ☽ □ ♇

07 04:33 ☽ ∥ ♅
05:07 ☉ # ♄
08:12 ☽ ∥ ♀
08:42 ☽ ✶ ♃
10:29 ☽ ∥ ☿
10:42 ☽ □ ♂
15:46 ☽ △ ☉
17:46 ☽ △ ♆
18:58 ♀ ☌ ♅
22:40 ☽ ✶ ♄
22:48 ☉ ∥ ♆

08 11:39 ☽ ✶ ♇
15:42 ☽ ∥ ☉
15:53 ☉ ☌ ♆
16:45 ☽ ∥ ♆
18:09 ☽ # ♄
23:29 ☽ ∥ ♇

09 02:14 ♀ ∥ ♅
07:47 ☽ △ ☿
11:32 ♀ □ ♃
18:04 ☽ △ ♅
21:45 ☿ ∥ ♅
23:15 ☽ △ ♀

10 03:23 ☽ ✶ ♂
05:21 ☽ ∥ ♃
06:29 ☽ □ ♆
10:40 ☽ □ ♄
11:38 ☽ ∥ ♂
18:42 ☉ ☍ ♄

11 21:55 ☽ □ ☿

12 04:58 ☽ □ ♅
09:17 ☽ ☌ ♃
15:53 ☽ □ ♀
16:36 ☽ ✶ ♆
20:01 ☽ △ ♄

13 00:32 ☽ ✶ ☉
00:40 ☽ ☌ [illegible]
22:47 ♀ ✶ ♂

14 04:38 ☿ SR
06:10 ☽ ✶ ☿
12:08 ☽ ✶ ♅

15 02:22 ☽ ☌ ♂
03:26 ☽ ✶ ♀

16 01:08 ☽ ∥ ♂
02:56 ☽ ∥ ♃
20:06 ☽ ✶ ♃

17 01:48 ☽ ☌ ♆
03:58 ☽ ☍ ♄
04:08 ☽ ∥ ♇
07:28 ☽ # ♄
09:56 ☽ ∥ ♆
13:06 ☉ ✶ ♇
16:02 ☽ ✶ ♇
22:09 ☽ ∥ ☉

18 08:44 ☽ ☌ ☿
14:37 ♂ ∥ ♃
15:17 ☽ ∥ ♅
16:59 ☽ ☌ ♅
21:14 ☽ ∥ ☿
21:31 ☽ □ ♃

19 01:09 ☉ ♓
08:30 ☽ ∥ ♀
11:03 ☽ ✶ ♂
15:44 ☽ ☌ ♀
16:44 ☽ □ ♇
21:00 ☽ # ♀

20 03:56 ♀ □ ♆
10:13 ☽ # ☿
15:02 ☽ # ♅
22:33 ☽ △ ♃

21 03:34 ☽ ✶ ♆
04:13 ☽ # ☉
04:56 ☽ △ ♄
08:22 ♀ ♈
14:30 ☽ □ ♂
17:42 ☽ △ ♇
20:38 ☽ # ♆
23:50 ☽ ∥ ♄

22 01:04 ☽ ✶ ☉
02:40 ☽ # ♇
04:48 ☽ ✶ ☿
19:32 ☽ ✶ ♅

23 01:15 ☽ # ♂
04:36 ☽ # ♃
04:46 ☿ ☌ ☉
05:40 ☽ □ ♆
06:40 ☽ □ ♄
19:47 ☽ △ ♂

24 04:06 ☽ □ ☿
04:47 ☽ ✶ ♀
05:17 ☿ ∥ ♅
23:29 ☽ □ ♅

25 05:12 ☽ ☍ ♃
10:02 ☽ △ ♆
10:39 ☽ ✶ ♄

26 01:21 ☽ ☍ ♇
01:33 ♂ ♒
05:26 ☽ △ ☿
15:42 ☽ □ ♀
17:57 ☽ △ ☉

27 03:01 ☿ ♒R
06:04 ☽ △ ♅
11:29 ☿ ∥ ☉

28 12:01 ♄ ☍ ♆
13:13 ☿ ✶ ♇
15:12 ☽ ☍ ♂
22:22 ☽ # ♃

⚷ Chiron

01 Dec.	11 S 32
02	10♒17
05	10 31
08	10 45
11	11 00
14	11 14
17	11 28
20	11 42
23	11 55
26	12 09

03	23:22	13♍00	✷	Total Lunar Eclipse (mag 1.237)
19	02:33	28♓07	⦻	Solar Eclipse (mag 0.876)

March 2007

Day	S. T.	☉	☽	☿	♀	♂	♃	♄	♅	♆	♇	☊ True
	h m s	° ' "	° ' "	° '	° '	° '	° '	° '	° '	° '	° '	° '
01 Th	10 33 41	10♓00 59	06♌33 30	28♒R19	09♈24	02♒13	17♐49	20♌R12	14♓25	20♒16	28♐42	16♓R08
02 Fr	10 37 38	11 01 12	18 59 37	27 34	10 38	02 59	17 56	20 08	14 28	20 18	28 43	16 07
03 Sa	10 41 34	12 01 23	01♍14 53	26 55	11 51	03 44	18 02	20 04	14 32	20 20	28 44	16 06
04 Su	10 45 31	13 01 32	13 20 53	26 23	13 05	04 30	18 08	19 59	14 35	20 22	28 45	16D 06
05 Mo	10 49 28	14 01 39	25 19 18	25 58	14 18	05 16	18 13	19 55	14 39	20 24	28 46	16 06
06 Tu	10 53 24	15 01 44	07♎12 06	25 40	15 31	06 01	18 19	19 51	14 42	20 26	28 47	16 07
07 We	10 57 21	16 01 48	19 01 33	25 29	16 44	06 47	18 24	19 47	14 45	20 28	28 48	16 07
08 Th	11 01 17	17 01 50	00♏50 21	25 25	17 58	07 32	18 30	19 43	14 49	20 31	28 48	16 08
09 Fr	11 05 14	18 01 50	12 41 38	25D 27	19 11	08 18	18 35	19 39	14 52	20 33	28 49	16 08
10 Sa	11 09 10	19 01 49	24 39 03	25 35	20 24	09 04	18 40	19 35	14 56	20 35	28 50	16R 08
11 Su	11 13 07	20 01 46	06♐46 38	25 48	21 37	09 49	18 44	19 31	14 59	20 37	28 51	16 08
12 Mo	11 17 03	21 01 41	19 08 39	26 08	22 50	10 35	18 49	19 27	15 03	20 39	28 51	16D 08
13 Tu	11 20 60	22 01 35	01♑49 22	26 32	24 03	11 21	18 53	19 23	15 06	20 41	28 52	16 08
14 We	11 24 57	23 01 27	14 52 35	27 01	25 15	12 07	18 58	19 20	15 10	20 43	28 53	16 08
15 Th	11 28 53	24 01 17	28 21 16	27 35	26 28	12 52	19 02	19 16	15 13	20 45	28 53	16 09
16 Fr	11 32 50	25 01 06	12♒16 51	28 13	27 41	13 38	19 06	19 13	15 16	20 47	28 54	16 10
17 Sa	11 36 46	26 00 53	26 38 38	28 55	28 53	14 24	19 10	19 09	15 20	20 49	28 54	16 10
18 Su	11 40 43	27 00 38	11♓23 12	29 41	00♉06	15 10	19 13	19 06	15 23	20 51	28 55	16 10
19 Mo	11 44 39	28 00 21	26 24 23	00♓30	01 18	15 56	19 17	19 03	15 27	20 53	28 55	16R 10
20 Tu	11 48 36	29 00 02	11♈33 48	01 22	02 31	16 42	19 20	18 59	15 30	20 55	28 56	16 10
21 We	11 52 32	29 59 41	26 41 59	02 18	03 43	17 28	19 23	18 56	15 33	20 56	28 56	16 08
22 Th	11 56 29	00♈59 18	11♉39 43	03 16	04 56	18 13	19 26	18 53	15 37	20 58	28 56	16 07
23 Fr	12 00 25	01 58 52	26 19 25	04 17	06 08	18 59	19 28	18 50	15 40	21 00	28 57	16 05
24 Sa	12 04 22	02 58 25	10♊36 02	05 21	07 20	19 45	19 31	18 48	15 43	21 02	28 57	16 03
25 Su	12 08 19	03 57 55	24 27 12	06 27	08 32	20 31	19 33	18 45	15 47	21 04	28 57	16 02
26 Mo	12 12 15	04 57 23	07♋53 02	07 35	09 44	21 17	19 35	18 42	15 50	21 06	28 57	16 02
27 Tu	12 16 12	05 56 49	20 55 29	08 46	10 56	22 03	19 37	18 40	15 53	21 07	28 57	16D 02
28 We	12 20 08	06 56 12	03♌37 36	09 59	12 08	22 49	19 39	18 37	15 56	21 09	28 58	16 03
29 Th	12 24 05	07 55 33	16 02 57	11 13	13 19	23 35	19 41	18 35	16 00	21 11	28 58	16 04
30 Fr	12 28 01	08 54 51	28 15 08	12 30	14 31	24 21	19 42	18 33	16 03	21 12	28 58	16 06
31 Sa	12 31 58	09 54 07	10♍17 30	13 48	15 42	25 07	19 43	18 31	16 06	21 14	28 58	16 07

Data for	03-01-2007
Julian Day	2454160.50
Ayanamsa	23 57 30
SVP	05 ♓ 09 40
☽ ☊ Mean	16 ♓ 33 R

● ◐ PHASES ○ ◑

03	23:18	✷	13♍00
12	03:55	◑	21♐11
19	02:43	⦻	28♓07
25	18:16	◐	04♋43

LAST ASPECT ☽ Day	h m	INGRESS Day	h m	
02	19:03	02	21:32	♍
05	06:57	05	09:26	♎
07	19:52	07	22:18	♏
10	01:53	10	10:38	♐
12	18:27	12	20:35	♑
14	20:22	15	02:53	♒
17	04:02	17	05:31	♓
19	03:59	19	05:42	♈
21	03:33	21	05:16	♉
22	15:12	23	06:07	♊
25	07:57	25	09:49	♋
26	14:36	27	17:05	♌
30	01:25	30	03:28	♍

DECLINATION

Day	☉	☽	☿	♀	♂	♃	♄	♅	♆	♇
01 Th	07S 49	21N54	09S 04	03N03	20S 35	22S 11	16N00	06S 49	14S 57	16S 31
02 Fr	07 26	17 25	09 30	03 34	20 25	22 11	16 02	06 48	14 57	16 31
03 Sa	07 03	12 18	09 55	04 05	20 15	22 12	16 03	06 46	14 56	16 31
04 Su	06 40	06 47	10 18	04 36	20 05	22 12	16 05	06 45	14 55	16 30
05 Mo	06 17	01 05	10 39	05 07	19 55	22 13	16 06	06 44	14 55	16 30
06 Tu	05 54	04S 37	10 58	05 38	19 44	22 13	16 07	06 42	14 54	16 30
07 We	05 31	10 07	11 15	06 08	19 33	22 13	16 09	06 41	14 53	16 30
08 Th	05 07	15 16	11 29	06 39	19 22	22 14	16 10	06 40	14 53	16 30
09 Fr	04 44	19 52	11 42	07 09	19 11	22 14	16 11	06 38	14 52	16 30
10 Sa	04 20	23 42	11 52	07 40	18 59	22 15	16 13	06 37	14 51	16 30
11 Su	03 57	26 35	12 01	08 10	18 48	22 15	16 14	06 36	14 51	16 30
12 Mo	03 33	28 15	12 07	08 40	18 36	22 15	16 15	06 34	14 50	16 30
13 Tu	03 10	28 31	12 11	09 09	18 24	22 16	16 16	06 33	14 49	16 29
14 We	02 46	27 13	12 13	09 39	18 12	22 16	16 17	06 32	14 49	16 29
15 Th	02 22	24 20	12 13	10 08	17 59	22 16	16 18	06 30	14 48	16 29
16 Fr	01 59	19 57	12 11	10 37	17 47	22 17	16 20	06 29	14 47	16 29
17 Sa	01 35	14 18	12 07	11 06	17 34	22 17	16 21	06 28	14 47	16 29
18 Su	01 11	07 42	12 02	11 35	17 21	22 17	16 22	06 26	14 46	16 29
19 Mo	00 48	00 33	11 54	12 03	17 08	22 17	16 23	06 25	14 46	16 29
20 Tu	00 24	06N40	11 45	12 31	16 55	22 18	16 24	06 24	14 45	16 29
21 We	00 00	13 30	11 34	12 59	16 41	22 18	16 25	06 22	14 44	16 28
22 Th	00N24	19 28	11 22	13 26	16 28	22 18	16 26	06 21	14 44	16 28
23 Fr	00 47	24 09	11 08	13 53	16 14	22 18	16 26	06 20	14 43	16 28
24 Sa	01 11	27 14	10 52	14 20	16 00	22 18	16 27	06 19	14 43	16 28
25 Su	01 35	28 32	10 35	14 47	15 46	22 18	16 28	06 17	14 42	16 28
26 Mo	01 58	28 06	10 16	15 13	15 32	22 19	16 29	06 16	14 42	16 28
27 Tu	02 22	26 05	09 56	15 39	15 17	22 19	16 30	06 15	14 41	16 28
28 We	02 45	22 48	09 34	16 04	15 03	22 19	16 30	06 14	14 41	16 28
29 Th	03 09	18 33	09 11	16 29	14 48	22 19	16 31	06 12	14 40	16 27
30 Fr	03 32	13 36	08 46	16 54	14 33	22 19	16 32	06 11	14 40	16 27
31 Sa	03 55	08 12	08 21	17 18	14 19	22 19	16 32	06 10	14 39	16 27

ASPECTARIAN

01 06:03 ☽ △ ♀
07:50 ☽ # ♂
21:54 ☽ △ ♃
02 02:12 ☽ ☌ ♄
02:33 ☽ ☍ ♆
04:29 ☽ # ♇
06:46 ☽ ∥ ♄
11:59 ☽ # ♆
15:54 ☽ ☍ ☿
19:03 ☽ △ ♇
03 09:52 ☽ # ☿
18:46 ☉ ∥ ♅
04 00:09 ☽ # ♅
00:31 ☽ # ☉
02:29 ☽ ☍ ♅
08:29 ☽ ∥ ♀
09:38 ☽ □ ♃
05 06:57 ☽ □ ♇
15:40 ☉ ☌ ♅
21:26 ☽ △ ♂
06 04:47 ☽ # ♀
05:09 ☽ ∥ ☉
07:13 ♀ # ☉
08:57 ☽ ∥ ♅
18:49 ☽ ☍ ♀
22:44 ☽ ✶ ♃
07 01:31 ☽ ✶ ♄
02:57 ☽ △ ♆
05:22 ☽ ∥ ☿
13:01 ☽ △ ☿
19:52 ☽ ✶ ♇
22:07 ☽ ∥ ♆
08 00:29 ♀ # ♅
04:31 ☽ # ♄
04:46 ☿ SD
06:10 ☽ ∥ ♇
11:16 ♀ △ ♃
14:31 ☽ □ ♂
20:24 ☽ ∥ ♂
09 04:25 ☽ △ ♅
08:41 ♀ △ ♄
11:44 ☽ △ ☉
13:54 ☽ □ ♄
14:12 ☽ ∥ ♃
14:19 ☉ □ ♃
15:50 ☽ □ ♆
10 01:53 ☽ □ ☿
03:42 ♀ ✶ ♆
11 06:21 ☽ ✶ ♂
16:04 ☽ □ ♅
23:22 ☽ ☌ ♃
12 00:35 ☽ △ ♄
02:53 ☽ ✶ ♆
07:47 ☽ △ ♀
13:43 ☽ ✶ ☿
18:27 ☽ ☌ ♇
14 00:31 ☽ ✶ ♅
15:46 ☽ ✶ ☉
20:22 ☽ □ ♀
15 12:15 ☽ ∥ ♃
16 02:26 ☽ ☌ ♂
10:12 ☽ ∥ ♂
11:32 ☽ ✶ ♃
11:37 ☽ ☍ ♄
14:19 ☽ ☌ ♆
15:15 ☽ ∥ ♇
15:51 ☽ # ♄
22:07 ☽ ∥ ♆
22:44 ♃ △ ♄
23:33 ☿ ✶ ♇
17 00:17 ♀ △ ♇
01:23 ☿ ✶ ♀
03:43 ☽ ✶ ♇
03:56 ☽ ☌ ☿
04:02 ☽ ✶ ♀
08:21 ☽ ∥ ☿
11:10 ☽ # ♀
22:01 ♀ ♉
18 04:22 ☽ ∥ ♅
06:27 ☽ ☌ ♅
09:35 ☿ ♓
12:36 ☽ □ ♃
18:12 ☿ # ♀
23:13 ☽ ∥ ☉
19 03:59 ☽ □ ♇
04:16 ☽ # ☉
22:13 ☉ □ ♇
23:07 ☽ # ♅
20 08:33 ☽ ✶ ♂
11:43 ☽ △ ♄
12:20 ☽ △ ♃
14:50 ☽ ✶ ♆
17:12 ☽ # ☿
21:58 ☽ ∥ ♀
21 00:08 ☉ ♈
03:33 ☽ △ ♇
04:40 ☽ # ♆
09:33 ☽ ✶ ☿
11:12 ☽ ∥ ♄
11:26 ☽ # ♇
11:51 ☽ # ♂
12:12 ☽ ☌ ♀
22:55 ♂ ∥ ♆
22 03:34 ♂ # ♄
06:26 ☽ ✶ ♅
11:16 ☽ □ ♂
11:43 ☽ □ ♄
13:34 ☽ # ♃
15:12 ☽ □ ♆
19:34 ♂ ☍ ♄
23 10:08 ☽ ✶ ☉
14:22 ☽ □ ☿
16:02 ♂ ✶ ♃
24 08:49 ☽ □ ♅
14:03 ☽ ✶ ♄
15:24 ☽ ☍ ♃
16:42 ☽ △ ♂
18:03 ☽ △ ♆
19:04 ♄ # ♇
19:59 ♀ # ♆
25 07:57 ☽ ☍ ♇
17:39 ♂ ☌ ♆
23:25 ☽ △ ☿
26 03:42 ☽ ✶ ♀
11:20 ♀ # ♂
14:36 ☽ △ ♅
28 03:03 ☽ # ♃
06:54 ☽ △ ☉
18:07 ☽ □ ♀
22:32 ♀ # ♇
29 02:10 ♀ ∥ ♄
04:56 ☽ ☌ ♄
07:07 ☽ △ ♃
09:32 ☽ ∥ ♀
10:04 ☽ ☍ ♆
10:08 ☽ ∥ ♄
10:28 ☽ # ♇
13:53 ♂ ∥ ♆
15:46 ☽ ☍ ♂
19:04 ☽ # ♆
19:19 ☽ # ♂
30 01:25 ☽ △ ♇
23:20 ☽ # ☿
31 07:56 ☽ ☍ ☿
08:22 ♀ ✶ ♅
08:48 ☽ # ♅
11:43 ☽ ☍ ♅
12:05 ☽ △ ♀
17:08 ☽ ∥ ☉
18:59 ☽ □ ♃
22:44 ♇ SR

⚷ Chiron

01 Dec.	10 S 55
01	12♒22
04	12 35
07	12 48
10	13 00
13	13 12
16	13 24
19	13 35
22	13 46
25	13 57
28	14 07
31	14 17

April 2007

Day	S. T.	☉	☽	☿	♀	♂	♃	♄	♅	♆	♇	☊ True
	h m s	° ′ ″	° ′ ″	° ′	° ′	° ′	° ′	° ′	° ′	° ′	° ′	° ′
01 Su	12 35 54	10♈ 53 21	22♍ 13 03	15♓ 09	16♉ 54	25♒ 53	19♐ 44	18♌ R28	16♓ 09	21♒ 16	28♐ R58	16♓ R07
02 Mo	12 39 51	11 52 33	04♎ 04 25	16 31	18 05	26 39	19 45	18 27	16 13	21 17	28 58	16 05
03 Tu	12 43 48	12 51 43	15 53 53	17 55	19 16	27 25	19 46	18 25	16 16	21 19	28 58	16 03
04 We	12 47 44	13 50 51	27 43 27	19 20	20 28	28 11	19 46	18 23	16 19	21 20	28 58	15 59
05 Th	12 51 41	14 49 57	09♏ 35 04	20 47	21 39	28 57	19 47	18 21	16 22	21 22	28 58	15 54
06 Fr	12 55 37	15 49 01	21 30 45	22 16	22 50	29 43	19 47	18 20	16 25	21 24	28 57	15 49
07 Sa	12 59 34	16 48 03	03♐ 32 48	23 46	24 00	00♓ 29	19 R 47	18 18	16 28	21 25	28 57	15 44
08 Su	13 03 30	17 47 04	15 43 54	25 18	25 11	01 15	19 46	18 17	16 31	21 27	28 57	15 40
09 Mo	13 07 27	18 46 02	28 07 05	26 52	26 22	02 01	19 46	18 16	16 34	21 28	28 57	15 37
10 Tu	13 11 23	19 44 59	10♑ 45 46	28 27	27 33	02 47	19 45	18 15	16 37	21 29	28 57	15 36
11 We	13 15 20	20 43 54	23 43 24	00♈ 04	28 43	03 33	19 44	18 14	16 40	21 31	28 56	15D 36
12 Th	13 19 17	21 42 48	07♒ 03 17	01 42	29 53	04 19	19 43	18 13	16 43	21 32	28 56	15 37
13 Fr	13 23 13	22 41 39	20 48 01	03 21	01♊ 04	05 05	19 42	18 12	16 46	21 33	28 56	15 38
14 Sa	13 27 10	23 40 29	04♓ 58 47	05 03	02 14	05 51	19 41	18 11	16 49	21 35	28 55	15 39
15 Su	13 31 06	24 39 17	19 34 33	06 46	03 24	06 37	19 39	18 11	16 52	21 36	28 55	15R 39
16 Mo	13 35 03	25 38 04	04♈ 31 21	08 30	04 34	07 24	19 37	18 10	16 55	21 37	28 54	15 37
17 Tu	13 38 59	26 36 48	19 42 10	10 16	05 44	08 10	19 36	18 10	16 58	21 39	28 54	15 33
18 We	13 42 56	27 35 31	04♉ 57 33	12 03	06 53	08 56	19 33	18 10	17 01	21 40	28 53	15 28
19 Th	13 46 52	28 34 11	20 06 59	13 53	08 03	09 42	19 31	18 09	17 03	21 41	28 53	15 22
20 Fr	13 50 49	29 32 50	05♊ 00 37	15 43	09 13	10 28	19 29	18 D 09	17 06	21 42	28 52	15 16
21 Sa	13 54 46	00♉ 31 27	19 30 46	17 36	10 22	11 14	19 26	18 09	17 09	21 43	28 52	15 10
22 Su	13 58 42	01 30 01	03♋ 32 56	19 30	11 31	12 00	19 23	18 10	17 12	21 44	28 51	15 05
23 Mo	14 02 39	02 28 33	17 05 53	21 25	12 40	12 46	19 20	18 10	17 14	21 45	28 50	15 02
24 Tu	14 06 35	03 27 03	00♌ 11 09	23 22	13 49	13 32	19 17	18 10	17 17	21 46	28 50	15 00
25 We	14 10 32	04 25 31	12 52 11	25 21	14 58	14 18	19 13	18 11	17 19	21 47	28 49	15D 01
26 Th	14 14 28	05 23 57	25 13 26	27 21	16 07	15 03	19 10	18 11	17 22	21 48	28 48	15 02
27 Fr	14 18 25	06 22 20	07♍ 19 47	29 23	17 15	15 49	19 06	18 12	17 25	21 49	28 47	15 03
28 Sa	14 22 21	07 20 42	19 15 55	01♉ 26	18 24	16 35	19 02	18 13	17 27	21 50	28 47	15R 03
29 Su	14 26 18	08 19 01	01♎ 06 09	03 31	19 32	17 21	18 58	18 14	17 30	21 51	28 46	15 02
30 Mo	14 30 15	09 17 18	12 54 13	05 36	20 40	18 07	18 54	18 15	17 32	21 52	28 45	14 58

Data for	04-01-2007
Julian Day	2454191.50
Ayanamsa	23 57 33
SVP	05 ♓ 09 35
☽ ☊ Mean	14 ♓ 54 R

● ◐ PHASES ○ ◑

02	17:16	○	12♎35
10	18:04	◑	20♑29
17	11:36	●	27♈05
24	06:37	◐	03♌43

LAST ASPECT ☽ Day	h m	INGRESS Day	h m	
01	13:39	01	15:45	♎
04	02:31	04	04:37	♏
06	02:55	06	16:57	♐
09	01:35	09	03:36	♑
11	09:57	11	11:23	♒
13	13:50	13	15:39	♓
15	15:02	15	16:47	♈
17	14:27	17	16:12	♉
19	02:30	19	15:52	♊
21	15:53	21	17:51	♋
23	09:12	23	23:39	♌
26	07:03	26	09:25	♍
28	19:15	28	21:46	♎

DECLINATION

Day	☉	☽	☿	♀	♂	♃	♄	♅	♆	♇
01 Su	04N19	02N34	07S 54	17N42	14S 03	22S 19	16N33	06S 09	14S 38	16S 27
02 Mo	04 42	03S 07	07 25	18 05	13 48	22 19	16 34	06 07	14 38	16 27
03 Tu	05 05	08 40	06 55	18 28	13 33	22 19	16 34	06 06	14 37	16 27
04 We	05 28	13 55	06 25	18 50	13 17	22 19	16 35	06 05	14 37	16 27
05 Th	05 51	18 40	05 52	19 12	13 02	22 19	16 35	06 04	14 37	16 27
06 Fr	06 13	22 44	05 19	19 34	12 46	22 19	16 36	06 03	14 36	16 26
07 Sa	06 36	25 52	04 45	19 55	12 30	22 19	16 36	06 01	14 36	16 26
08 Su	06 59	27 51	04 09	20 15	12 14	22 19	16 36	06 00	14 35	16 26
09 Mo	07 21	28 31	03 32	20 35	11 58	22 19	16 37	05 59	14 35	16 26
10 Tu	07 44	27 42	02 54	20 55	11 42	22 19	16 37	05 58	14 34	16 26
11 We	08 06	25 24	02 15	21 14	11 26	22 19	16 37	05 57	14 34	16 26
12 Th	08 28	21 39	01 35	21 32	11 10	22 19	16 38	05 56	14 33	16 26
13 Fr	08 50	16 38	00 54	21 50	10 53	22 19	16 38	05 54	14 33	16 26
14 Sa	09 11	10 35	00 12	22 07	10 37	22 18	16 38	05 53	14 33	16 25
15 Su	09 33	03 48	00N31	22 24	10 20	22 18	16 38	05 52	14 32	16 25
16 Mo	09 55	03N21	01 15	22 40	10 03	22 18	16 38	05 51	14 32	16 25
17 Tu	10 16	10 24	02 00	22 56	09 46	22 18	16 38	05 50	14 31	16 25
18 We	10 37	16 53	02 46	23 11	09 29	22 18	16 38	05 49	14 31	16 25
19 Th	10 58	22 16	03 33	23 25	09 12	22 18	16 38	05 48	14 31	16 25
20 Fr	11 19	26 07	04 20	23 39	08 55	22 17	16 38	05 47	14 30	16 25
21 Sa	11 39	28 09	05 08	23 52	08 38	22 17	16 38	05 46	14 30	16 25
22 Su	12 00	28 17	05 57	24 05	08 21	22 17	16 38	05 45	14 30	16 25
23 Mo	12 20	26 41	06 47	24 17	08 04	22 17	16 38	05 44	14 29	16 24
24 Tu	12 40	23 40	07 37	24 28	07 46	22 16	16 38	05 43	14 29	16 24
25 We	13 00	19 36	08 27	24 39	07 29	22 16	16 37	05 42	14 29	16 24
26 Th	13 19	14 46	09 18	24 49	07 12	22 16	16 37	05 41	14 28	16 24
27 Fr	13 39	09 28	10 09	24 58	06 54	22 16	16 37	05 40	14 28	16 24
28 Sa	13 58	03 54	11 00	25 07	06 37	22 15	16 37	05 39	14 28	16 24
29 Su	14 17	01S 45	11 52	25 15	06 19	22 15	16 36	05 38	14 28	16 24
30 Mo	14 35	07 18	12 43	25 22	06 01	22 15	16 36	05 37	14 27	16 24

ASPECTARIAN

01	13:39	☽ □ ♇
	18:30	☿ ☌ ♅
02	07:01	♀ □ ♄
	07:17	☽ # ☉
	12:52	☽ ∥ ♅
	17:02	☽ ∥ ☿
03	05:05	☽ ✶ ♄
	07:51	☽ ✶ ♃
	11:01	☽ △ ♆
	21:13	☽ ∥ ♂
04	01:00	☽ △ ♂
	02:31	☽ ✶ ♇
	03:23	☽ ∥ ♆
	07:18	☿ □ ♃
	12:25	☽ ∥ ♇
	13:07	☽ # ♄
	15:22	☿ ∥ ♅
	18:16	♀ □ ♆
05	00:20	♂ ✶ ♇
	00:44	☿ # ☉
	03:12	☽ # ♀
	13:01	☉ # ♅
	13:44	☽ △ ♅
	17:38	☽ □ ♄
	21:23	☽ ∥ ♃
	23:46	☽ □ ♆
06	01:24	♃ SR
	01:44	☽ △ ☿
	02:55	☽ ☍ ♀
	08:51	♂ ♓
	17:30	☽ □ ♂
07	16:10	☿ ✶ ♀
08	01:33	☽ □ ♅
	04:21	☽ △ ☉
	04:58	☽ △ ♄
	[illegible]	
	11:08	☽ ✶ ♆
	11:56	☉ △ ♄
	21:15	☽ □ ☿
09	01:35	☽ ☌ ♇
	07:57	☽ ✶ ♂
10	00:07	☉ △ ♃
	07:23	☿ □ ♇
	10:58	☽ ✶ ♅
	23:07	☿ ♈
11	09:57	☽ △ ♀
	13:06	☽ ✶ ☿
	19:35	☉ ✶ ♆
	20:20	☽ ∥ ♃
12	00:36	☽ # ♀
	02:15	♀ ♊
	19:31	☽ ☍ ♄
	22:07	☽ ✶ ♃
13	00:02	☽ # ♄
	00:53	☽ ∥ ♇
	01:18	☽ ☌ ♆
	03:30	☽ ✶ ☉
	08:42	☽ ∥ ♆
	13:50	☽ ✶ ♇
	19:00	☽ □ ♀
	23:54	☽ ∥ ♂
14	01:33	☽ ☌ ♂
	04:50	☽ # ☉
	15:40	♀ # ♃
	16:51	☽ ∥ ♅
	19:35	☽ ☌ ♅
15	00:08	☽ □ ♃
	10:04	☽ # ☿
	[illegible]	☽ □
	16:14	☽ ∥ ☿
16	00:04	☽ ✶ ♀
	05:24	☉ # ♂
	07:08	☽ ☌ ☿
	08:24	☽ # ♅
	21:35	☽ △ ♄
	21:56	☽ # ♂
	23:31	☽ ∥ ☉
	23:50	☽ △ ♃
17	03:03	☽ ✶ ♆
	14:27	☽ △ ♇
	14:55	☽ # ♆
	22:11	☽ # ♇
	23:03	☽ ∥ ♄
18	06:35	☽ ✶ ♂
	19:07	☽ ✶ ♅
	20:53	☽ □ ♄
19	00:10	☽ # ♃
	02:30	☽ □ ♆
	06:35	☽ ∥ ♀
	07:34	☉ △ ♇
	21:25	♄ SD
20	07:28	☽ ☌ ♀
	09:26	☽ □ ♂
	11:07	☉ ♉
	20:01	☽ □ ♅
	20:18	☽ ✶ ☿
	21:43	☽ ✶ ♄
	23:52	☽ ☍ ♃
21	03:43	☽ △ ♆
	07:10	☿ △ ♄
	15:53	☽ ☍ ♇
	18:03	☿ ∥ ♅
	20:11	☽ ✶ ☉
	22:40	☿ △ ♃
22	15:45	☽ △ ♂
23	00:15	☽ △ ♅
	04:12	☿ ✶ ♆
	05:30	♀ □ ♂
	09:12	☽ □ ☿
	18:45	☽ ∥ ♀
24	03:28	☿ # ♂
	08:48	☽ # ♃
25	04:27	☽ ✶ ♀
	10:15	☽ ☌ ♄
	12:13	☽ △ ♃
	15:04	☽ ∥ ♄
	16:08	☽ # ♇
	17:17	☽ ☍ ♆
26	01:20	☽ # ♆
	05:01	☽ △ ☿
	06:17	☽ ∥ ☉
	07:03	☽ △ ♇
	17:05	☿ △ ♇
	21:22	☽ ∥ ☿
	21:55	☽ △ ☉
27	03:21	♀ □ ♅
	07:16	☿ ♉
	11:45	☽ # ♂
	16:31	☽ # ♅
	18:13	☽ ☍ ♂
	20:07	☿ ✶ ♄
	20:19	☽ ☍ ♅
	22:04	☽ □ ♀
	23:32	☽ □ ♃
28	12:44	♀ ☍ ♃
	19:15	☽ □ ♇
29	04:37	♂ ☌ ♅
	13:47	☉ # ♆
	16:39	☽ ∥ ♅
	18:41	☽ ∥ ♂
30	10:53	☽ ✶ ♄
	12:07	☽ ✶ ♃
	17:27	☽ △ ♀
	18:13	☽ △ ♆
	22:22	☉ □ ♃

⚷ Chiron

01 Dec.	10 S 15
03	14♒26
06	14 35
09	14 43
12	14 51
15	14 58
18	15 05
21	15 11
24	15 17
27	15 22
30	15 26

May 2007

Day	S. T. h m s	☉ ° ' "	☽ ° ' "	☿ ° '	♀ ° '	♂ ° '	♃ ° '	♄ ° '	♅ ° '	♆ ° '	♇ ° '	☊ True ° '
01 Tu	14 34 11	10♉15 34	24♎43 10	07♉43	21♊48	18♓53	18♐R50	18♌16	17♓34	21♒53	28♐R44	14♓R51
02 We	14 38 08	11 13 48	06♏35 23	09 51	22 56	19 39	18 45	18 17	17 37	21 53	28 43	14 42
03 Th	14 42 04	12 12 00	18 32 40	12 00	24 04	20 25	18 40	18 19	17 39	21 54	28 42	14 32
04 Fr	14 46 01	13 10 10	00♐36 25	14 09	25 11	21 10	18 35	18 20	17 41	21 55	28 41	14 21
05 Sa	14 49 57	14 08 19	12 47 56	16 19	26 18	21 56	18 30	18 22	17 44	21 55	28 40	14 10
06 Su	14 53 54	15 06 26	25 08 31	18 29	27 26	22 42	18 25	18 23	17 46	21 56	28 39	14 00
07 Mo	14 57 50	16 04 31	07♑39 48	20 39	28 32	23 28	18 20	18 25	17 48	21 57	28 38	13 53
08 Tu	15 01 47	17 02 36	20 23 46	22 48	29 39	24 14	18 15	18 27	17 50	21 57	28 37	13 49
09 We	15 05 44	18 00 38	03♒22 43	24 57	00♋46	24 59	18 09	18 29	17 52	21 58	28 36	13 46
10 Th	15 09 40	18 58 40	16 39 15	27 05	01 52	25 45	18 03	18 31	17 54	21 58	28 35	13D46
11 Fr	15 13 37	19 56 40	00♓15 51	29 11	02 58	26 31	17 58	18 33	17 56	21 59	28 34	13 46
12 Sa	15 17 33	20 54 39	14 14 22	01♊16	04 05	27 16	17 52	18 35	17 58	21 59	28 33	13R46
13 Su	15 21 30	21 52 36	28 35 13	03 20	05 10	28 02	17 46	18 38	18 00	22 00	28 32	13 45
14 Mo	15 25 26	22 50 32	13♈16 33	05 21	06 16	28 48	17 39	18 40	18 02	22 00	28 30	13 42
15 Tu	15 29 23	23 48 27	28 13 40	07 20	07 21	29 33	17 33	18 43	18 04	22 00	28 29	13 36
16 We	15 33 19	24 46 21	13♉19 03	09 16	08 27	00♈19	17 27	18 45	18 06	22 01	28 28	13 28
17 Th	15 37 16	25 44 13	28 23 13	11 10	09 32	01 04	17 20	18 48	18 08	22 01	28 27	13 18
18 Fr	15 41 13	26 42 04	13♊16 13	13 01	10 37	01 50	17 13	18 51	18 09	22 01	28 26	13 08
19 Sa	15 45 09	27 39 53	27 49 22	14 49	11 41	02 35	17 07	18 54	18 11	22 01	28 24	12 58
20 Su	15 49 06	28 37 41	11♋56 35	16 35	12 45	03 21	17 00	18 57	18 13	22 02	28 23	12 49
21 Mo	15 53 02	29 35 27	25 35 00	18 17	13 50	04 06	16 53	19 00	18 14	22 02	28 22	12 43
22 Tu	15 56 59	00♊33 11	08♌44 55	19 56	14 53	04 51	16 46	19 03	18 16	22 02	28 20	12 39
23 We	16 00 55	01 30 54	21 29 03	21 32	15 57	05 37	16 39	19 07	18 17	22 02	28 19	12 37
24 Th	16 04 52	02 28 35	03♍51 43	23 04	17 00	06 22	16 32	19 10	18 19	22 02	28 18	12D36
25 Fr	16 08 49	03 26 15	15 58 01	24 33	18 03	07 07	16 25	19 14	18 20	22 02	28 16	12 37
26 Sa	16 12 45	04 23 53	27 53 19	25 59	19 06	07 52	16 17	19 17	18 22	22R02	28 15	12R36
27 Su	16 16 42	05 21 30	09♎42 47	27 22	20 08	08 37	16 10	19 21	18 23	22 02	28 14	12 34
28 Mo	16 20 38	06 19 05	21 31 05	28 41	21 11	09 23	16 02	19 25	18 24	22 02	28 12	12 29
29 Tu	16 24 35	07 16 39	03♏22 11	29 57	22 12	10 08	15 55	19 29	18 26	22 02	28 11	12 21
30 We	16 28 31	08 14 12	15 19 09	01♋09	23 14	10 53	15 48	19 33	18 27	22 02	28 09	12 10
31 Th	16 32 28	09 11 43	27 24 14	02 18	24 15	11 38	15 40	19 37	18 28	22 01	28 08	11 57

Data for	05-01-2007
Julian Day	2454221.50
Ayanamsa	23 57 36
SVP	05 ♓ 09 30
☽ ☊ Mean	13 ♓ 19 R

● ◐ PHASES ○ ◑

Day	Time	Phase	Position
02	10:09	○	11♏38
10	04:27	◑	19♒09
16	19:28	●	25♉33
23	21:03	◐	02♍22

LAST ASPECT Day	h m	☽ INGRESS Day	h m	Sign
01	08:07	01	10:42	♏
03	06:42	03	22:48	♐
06	06:46	06	09:21	♑
08	07:35	08	17:48	♒
10	21:48	10	23:32	♓
12	23:54	13	02:20	♈
15	00:25	15	02:50	♉
16	19:28	17	02:35	♊
19	00:58	19	03:39	♋
21	07:47	21	07:57	♌
23	13:09	23	16:27	♍
26	00:44	26	04:17	♎
28	16:17	28	17:12	♏
30	17:12	31	05:07	♐

DECLINATION

Day	☉	☽	☿	♀	♂	♃	♄	♅	♆	♇
01 Tu	14N54	12S37	13N34	25N29	05S44	22S14	16N36	05S36	14S27	16S24
02 We	15 12	17 29	14 24	25 35	05 26	22 14	16 35	05 35	14 27	16 24
03 Th	15 30	21 43	15 14	25 41	05 08	22 13	16 35	05 34	14 27	16 24
04 Fr	15 48	25 05	16 03	25 45	04 50	22 13	16 34	05 33	14 26	16 24
05 Sa	16 05	27 21	16 51	25 49	04 32	22 13	16 34	05 33	14 26	16 23
06 Su	16 22	28 19	17 38	25 53	04 15	22 12	16 33	05 32	14 26	16 23
07 Mo	16 39	27 51	18 24	25 56	03 57	22 12	16 32	05 31	14 26	16 23
08 Tu	16 56	25 56	19 07	25 58	03 39	22 11	16 32	05 30	14 26	16 23
09 We	17 12	22 37	19 49	25 59	03 21	22 11	16 31	05 29	14 26	16 23
10 Th	17 28	18 04	20 29	26 00	03 03	22 10	16 30	05 29	14 25	16 23
11 Fr	17 44	12 30	21 07	26 00	02 45	22 10	16 30	05 28	14 25	16 23
12 Sa	17 59	06 10	21 42	26 00	02 27	22 09	16 29	05 27	14 25	16 23
13 Su	18 14	00N37	22 15	25 58	02 09	22 09	16 28	05 26	14 25	16 23
14 Mo	18 29	07 32	22 46	25 56	01 51	22 08	16 27	05 26	14 25	16 23
15 Tu	18 44	14 09	23 14	25 54	01 33	22 08	16 26	05 25	14 25	16 23
16 We	18 58	19 59	23 39	25 51	01 15	22 07	16 26	05 24	14 25	16 23
17 Th	19 12	24 32	24 02	25 47	00 57	22 07	16 25	05 24	14 25	16 23
18 Fr	19 25	27 23	24 22	25 43	00 39	22 06	16 24	05 23	14 25	16 23
19 Sa	19 38	28 17	24 40	25 38	00 21	22 06	16 23	05 22	14 25	16 23
20 Su	19 51	27 17	24 55	25 32	00 03	22 05	16 22	05 22	14 25	16 23
21 Mo	20 04	24 40	25 08	25 26	00N14	22 04	16 21	05 21	14 25	16 23
22 Tu	20 16	20 48	25 19	25 19	00 32	22 04	16 20	05 20	14 24	16 23
23 We	20 28	16 04	25 27	25 12	00 50	22 03	16 19	05 20	14 24	16 23
24 Th	20 39	10 49	25 33	25 04	01 08	22 03	16 17	05 19	14 24	16 23
25 Fr	20 51	05 16	25 37	24 55	01 26	22 02	16 16	05 19	14 24	16 23
26 Sa	21 01	00S23	25 39	24 46	01 44	22 01	16 15	05 18	14 25	16 23
27 Su	21 12	05 58	25 39	24 36	02 01	22 01	16 14	05 18	14 25	16 22
28 Mo	21 22	11 20	25 37	24 26	02 19	22 00	16 13	05 17	14 25	16 22
29 Tu	21 32	16 18	25 34	24 15	02 37	21 59	16 11	05 17	14 25	16 22
30 We	21 41	20 42	25 29	24 04	02 54	21 59	16 10	05 16	14 25	16 22
31 Th	21 50	24 17	25 23	23 52	03 12	21 58	16 09	05 16	14 25	16 22

ASPECTARIAN

01	01:34	♀ △ ♆
	05:23	☽ # ☿
	08:07	☽ ⚹ ♇
	08:44	☽ ∥ ♆
	10:46	♂ ∥ ♅
	11:37	☽ # ☉
	18:22	☽ ∥ ♇
	19:20	☽ # ♄
02	01:09	☿ # ♆
	08:00	☽ ☍ ☿
	22:12	☽ △ ♅
	23:32	☽ □ ♄
03	03:10	☽ ∥ ♃
	03:59	☽ △ ♂
	04:06	☿ ☌ ☉
	06:42	☽ □ ♆
	11:50	☿ ∥ ☉
04	06:04	☽ # ♀
	10:00	☿ # ♇
	15:06	☿ ∥ ♄
05	09:39	☽ □ ♅
	10:53	☽ △ ♄
	11:04	☽ ☌ ♃
	15:53	☿ ⚹ ♅
	17:47	☽ ⚹ ♆
	18:58	☽ □ ♂
	22:55	☿ □ ♄
06	01:51	☉ # ♇
	04:50	☽ ☍ ♀
	06:46	☽ ☌ ♇
	07:12	♃ △ ♄
	14:57	☉ ∥ ♄
07	02:05	♀ ☍ ♇
	14:29	☿ □ ♆
	17:12	☽ △ ☉
	19:12	☽ ⚹ ♅
	23:39	☽ # ♀
08	05:23	☽ △ ☿
	07:28	♀ ♋
	07:35	☽ ⚹ ♂
	20:25	☉ ⚹ ♅
09	00:39	☿ ⚹ ♂
	02:31	☽ ∥ ♃
	12:05	☉ □ ♄
	13:27	☽ # ☿
10	02:29	☽ ⚹ ♃
	02:38	☽ # ☉
	03:20	☽ ☍ ♄
	07:06	☽ # ♄
	07:39	☽ ∥ ♇
	09:27	☽ ☌ ♆
	16:05	☽ ∥ ♆
	21:03	☽ ⚹ ♇
	21:48	☽ □ ☿
11	03:34	♃ □ ♅
	05:07	☽ △ ♀
	09:17	☿ ♊
12	02:34	☽ ∥ ♅
	06:04	☽ □ ♃
	06:19	☽ ☌ ♅
	12:03	☽ ⚹ ☉
	13:50	☽ ∥ ♂
	19:15	☿ # ♃
	23:02	☽ ☌ ♂
	23:54	☽ □ ♇
13	02:56	☉ □ ♆
	05:02	☽ # ♂
	09:03	☽ ⚹ ☿
	11:42	☽ □ ♀
	15:14	♂ □ ♇
	16:37	☽ # ♅
14	07:01	☽ △ ♃
	08:43	☽ △ ♄
	14:03	☽ ⚹ ♆
15	00:25	☽ △ ♇
	00:58	☽ # ♆
	08:40	☽ # ♇
	08:53	☽ ∥ ♄
	14:07	♂ ♈
	15:40	☽ ⚹ ♀
	19:13	☽ ∥ ☉
16	07:37	☽ ⚹ ♅
	08:40	☽ □ ♄
	10:15	☽ # ♃
	13:50	☽ □ ♆
	20:35	☽ ∥ ☿
17	04:32	☽ ⚹ ♂
	08:32	☽ ∥ ♀
	23:32	☽ ☌ ☿
18	06:25	☽ ☍ ♃
	08:00	☽ □ ♅
	09:09	☽ ⚹ ♄
	14:21	☽ △ ♆
19	00:24	♄ # ♇
	00:58	☽ ☍ ♇
	08:28	☽ □ ♂
20	01:32	☽ ☌ ♀
	05:31	☿ ☍ ♃
	10:57	☽ △ ♅
	17:57	☽ ∥ ♀
	20:41	☽ ∥ ☿
	23:26	☿ □ ♅
21	07:47	☽ ⚹ ☉
	10:12	☉ ♊
	10:48	☿ ⚹ ♄
	16:23	☽ △ ♂
	16:46	☽ # ♃
22	00:35	☿ ∥ ♀
	02:45	☽ ∥ ☉
	14:53	☽ △ ♃
	19:27	☽ ☌ ♄
	22:32	☽ # ♇
	22:53	☽ ∥ ♄
23	00:05	☽ ⚹ ☿
	01:03	☽ ☍ ♆
	07:48	☽ # ♆
	07:49	☿ △ ♆
	13:09	☽ △ ♇
24	22:47	☽ # ♅
25	00:52	☽ □ ♃
	01:10	♆ SR
	04:35	☽ ⚹ ♀
	04:46	☽ ☍ ♅
	06:42	♀ △ ♅
	15:29	☽ ∥ ♂
	19:39	☽ □ ☿
26	00:44	☽ □ ♇
	06:02	☽ # ♂
	14:22	☽ △ ☉
	21:06	☽ ∥ ♅
	21:38	☽ ☍ ♂
27	12:59	☽ ⚹ ♃
	15:13	☿ ☍ ♇
	19:42	☽ ⚹ ♄
	23:14	☽ □ ♀
28	01:03	☽ △ ♆
	13:32	☽ ⚹ ♇
	14:36	☽ ∥ ♆
	16:17	☽ △ ☿
	23:26	☽ # ♄
29	00:23	☽ ∥ ♆
	00:56	☿ ♋
30	06:16	☽ # ☉
	06:16	☽ △ ♅
	07:52	☽ ∥ ♃
	08:28	☽ □ ♄
	13:21	☽ □ ♆
	17:12	☽ △ ♀
	21:02	☽ # ♀
31	08:42	☽ # ☿
	20:53	☉ # ♃

⚷ Chiron

01 Dec.	09 S 44
03	15♒30
06	15 34
09	15 36
12	15 39
15	15 40
18	15 41
21	15 41R
24	15 41
27	15 40
30	15 39

20 23:11 15♒41 R

June 2007

Day	S. T. h m s	☉ ° ' "	☽ ° ' "	☿ ° '	♀ ° '	♂ ° '	♃ ° '	♄ ° '	♅ ° '	♆ ° '	♇ ° '	☊ True ° '
01 Fr	16 36 24	10♊ 09 13	09♐ 38 51	03♋ 23	25♋ 16	12♈ 22	15♐ R32	19♌ 41	18♓ 29	22♒ R01	28♐ R06	11♓ R44
02 Sa	16 40 21	11 06 43	22 03 46	04 25	26 16	13 07	15 25	19 45	18 30	22 01	28 05	11 30
03 Su	16 44 18	12 04 11	04♑ 39 23	05 22	27 17	13 52	15 17	19 49	18 31	22 01	28 03	11 18
04 Mo	16 48 14	13 01 39	17 25 57	06 16	28 16	14 37	15 10	19 54	18 32	22 00	28 02	11 09
05 Tu	16 52 11	13 59 05	00♒ 23 49	07 06	29 16	15 22	15 02	19 58	18 33	22 00	28 00	11 02
06 We	16 56 07	14 56 31	13 33 43	07 52	00♌ 15	16 06	14 54	20 03	18 34	22 00	27 59	10 58
07 Th	17 00 04	15 53 56	26 56 42	08 34	01 13	16 51	14 47	20 07	18 35	21 59	27 57	10 57
08 Fr	17 04 00	16 51 21	10♓ 34 08	09 12	02 12	17 36	14 39	20 12	18 36	21 59	27 56	10D 57
09 Sa	17 07 57	17 48 44	24 27 13	09 45	03 09	18 20	14 31	20 17	18 36	21 59	27 54	10R 57
10 Su	17 11 53	18 46 08	08♈ 36 29	10 15	04 07	19 05	14 24	20 22	18 37	21 58	27 53	10 56
11 Mo	17 15 50	19 43 31	23 01 01	10 39	05 03	19 49	14 16	20 26	18 38	21 57	27 51	10 53
12 Tu	17 19 47	20 40 53	07♉ 37 58	11 00	06 00	20 34	14 09	20 31	18 38	21 57	27 50	10 47
13 We	17 23 43	21 38 15	22 22 13	11 16	06 56	21 18	14 01	20 37	18 39	21 56	27 48	10 39
14 Th	17 27 40	22 35 36	07♊ 06 46	11 27	07 51	22 02	13 54	20 42	18 39	21 56	27 47	10 29
15 Fr	17 31 36	23 32 57	21 43 40	11 34	08 46	22 46	13 46	20 47	18 40	21 55	27 45	10 18
16 Sa	17 35 33	24 30 17	06♋ 05 16	11 R 36	09 41	23 30	13 39	20 52	18 40	21 54	27 44	10 07
17 Su	17 39 29	25 27 37	20 05 34	11 34	10 34	24 15	13 31	20 57	18 40	21 54	27 42	09 58
18 Mo	17 43 26	26 24 56	03♌ 40 57	11 27	11 28	24 59	13 24	21 03	18 41	21 53	27 40	09 50
19 Tu	17 47 22	27 22 14	16 50 35	11 16	12 20	25 42	13 17	21 08	18 41	21 52	27 39	09 45
20 We	17 51 19	28 19 31	29 35 58	11 01	13 12	26 26	13 10	21 14	18 41	21 52	27 37	09 43
21 Th	17 55 16	29 16 47	12♍ 00 28	10 42	14 04	27 10	13 02	21 20	18 41	21 51	27 36	09D 43
22 Fr	17 59 12	00♋ 14 03	24 08 34	10 19	14 55	27 54	12 55	21 25	18 41	21 50	27 34	09 43
23 Sa	18 03 09	01 11 18	06♎ 05 25	09 53	15 45	28 38	12 48	21 31	18 42	21 49	27 33	09R 44
24 Su	18 07 05	02 08 32	17 56 21	09 24	16 34	29 21	12 42	21 37	18R 42	21 48	27 31	09 43
25 Mo	18 11 02	03 05 46	29 46 32	08 53	17 23	00♉ 05	12 35	21 43	18 41	21 47	27 29	09 40
26 Tu	18 14 58	04 02 59	11♏ 40 38	08 20	18 10	00 48	12 28	21 49	18 41	21 46	27 28	09 34
27 We	18 18 55	05 00 11	23 42 36	07 45	18 57	01 31	12 21	21 55	18 41	21 45	27 26	09 26
28 Th	18 22 51	05 57 24	05♐ 55 25	07 10	19 44	02 15	12 15	22 01	18 41	21 44	27 25	09 16
29 Fr	18 26 48	06 54 36	18 21 06	06 34	20 29	02 58	12 09	22 07	18 41	21 43	27 23	09 05
30 Sa	18 30 45	07 51 47	01♑ 00 38	05 59	21 13	03 41	12 02	22 13	18 41	21 42	27 22	08 54

Data for	06-01-2007
Julian Day	2454252.50
Ayanamsa	23 57 42
SVP	05 ♓ 09 27
☽ ☊ Mean	11 ♓ 40 R

● ◐ PHASES ○ ◑

Day	h m	Phase	Position
01	01:04	○	10♐12
08	11:44	◑	17♓19
15	03:13	●	23♊41
22	13:15	◐	00♎46
30	13:50	○	08♑25

LAST ASPECT Day	h m	☽ INGRESS Day	h m	Sign
02	11:29	02	15:10	♑
04	21:45	04	23:16	♒
07	01:48	07	05:25	♓
09	05:53	09	09:27	♈
11	07:58	11	11:30	♉
12	23:18	13	12:24	♊
15	09:59	15	13:46	♋
17	07:40	17	17:25	♌
19	21:22	20	00:46	♍
22	06:50	22	11:44	♎
24	19:23	25	00:27	♏
26	20:24	27	12:25	♐
29	17:09	29	22:06	♑

DECLINATION

Day	☉	☽	☿	♀	♂	♃	♄	♅	♆	♇
01 Fr	21N58	26S 49	25N16	23N40	03N29	21S 57	16N07	05S 16	14S 25	16S 22
02 Sa	22 07	28 06	25 07	23 27	03 47	21 56	16 06	05 15	14 25	16 23
03 Su	22 14	27 57	24 57	23 14	04 04	21 56	16 05	05 15	14 25	16 23
04 Mo	22 22	26 19	24 46	23 00	04 21	21 55	16 03	05 15	14 25	16 23
05 Tu	22 29	23 16	24 34	22 46	04 39	21 54	16 02	05 14	14 25	16 23
06 We	22 35	18 59	24 21	22 32	04 56	21 54	16 00	05 14	14 25	16 23
07 Th	22 42	13 41	24 08	22 17	05 13	21 53	15 59	05 14	14 26	16 23
08 Fr	22 47	07 38	23 53	22 01	05 30	21 52	15 57	05 13	14 26	16 23
09 Sa	22 53	01 07	23 39	21 45	05 47	21 51	15 56	05 13	14 26	16 23
10 Su	22 58	05N34	23 23	21 29	06 04	21 51	15 54	05 13	14 26	16 23
11 Mo	23 03	12 05	23 07	21 13	06 21	21 50	15 53	05 13	14 26	16 23
12 Tu	23 07	18 02	22 51	20 56	06 38	21 49	15 51	05 12	14 26	16 23
13 We	23 11	22 57	22 35	20 39	06 54	21 48	15 49	05 12	14 27	16 23
14 Th	23 14	26 25	22 19	20 21	07 11	21 48	15 48	05 12	14 27	16 23
15 Fr	23 17	28 05	22 02	20 03	07 27	21 47	15 46	05 12	14 27	16 23
16 Sa	23 20	27 48	21 46	19 45	07 44	21 46	15 44	05 12	14 27	16 23
17 Su	23 22	25 44	21 30	19 27	08 00	21 45	15 42	05 12	14 28	16 23
18 Mo	23 24	22 13	21 14	19 08	08 16	21 45	15 41	05 12	14 28	16 23
19 Tu	23 25	17 40	20 58	18 49	08 32	21 44	15 39	05 12	14 28	16 23
20 We	23 26	12 26	20 42	18 30	08 48	21 43	15 37	05 12	14 28	16 23
21 Th	23 26	06 52	20 28	18 11	09 04	21 42	15 35	05 12	14 29	16 23
22 Fr	23 26	01 10	20 13	17 51	09 20	21 42	15 33	05 12	14 29	16 23
23 Sa	23 26	04S 30	20 00	17 31	09 36	21 41	15 31	05 12	14 29	16 24
24 Su	23 25	09 56	19 47	17 11	09 51	21 40	15 30	05 12	14 30	16 24
25 Mo	23 24	15 02	19 35	16 51	10 07	21 40	15 28	05 12	14 30	16 24
26 Tu	23 23	19 35	19 24	16 31	10 22	21 39	15 26	05 12	14 30	16 24
27 We	23 21	23 24	19 14	16 11	10 38	21 38	15 24	05 12	14 31	16 24
28 Th	23 18	26 15	19 05	15 50	10 53	21 38	15 22	05 12	14 31	16 24
29 Fr	23 16	27 53	18 57	15 30	11 08	21 37	15 20	05 12	14 31	16 24
30 Sa	23 12	28 07	18 50	15 09	11 23	21 36	15 18	05 12	14 32	16 24

ASPECTARIAN

01 05:39 ☽ △ ♂
11:19 ☽ ☌ ♃
17:09 ☽ □ ♅
19:32 ☽ △ ♄
23:55 ☽ ✶ ♆
02 11:29 ☽ ☌ ♆
03 01:28 ☽ ☍ ☿
18:25 ☽ □ ♂
04 02:04 ☽ ✶ ♅
14:21 ☽ # ☿
14:58 ♂ △ ♃
21:45 ☽ ☍ ♀
05 03:20 ☽ # ♀
04:47 ☽ # ☉
08:21 ☽ ∥ ♃
17:59 ♀ ☊
19:43 ♀ ∥ ☉
23:13 ☉ ☍ ♃
06 02:24 ☽ ✶ ♃
02:41 ☽ △ ☉
04:52 ☽ ✶ ♂
11:45 ☽ ☍ ♄
12:17 ☽ ∥ ♆
14:01 ☽ # ♄
15:10 ☽ ☌ ♆
20:50 ☽ ∥ ♆
07 00:58 ♂ # ♅
01:48 ☽ ✶ ♆
21:30 ☽ △ ☿
08 07:02 ☽ □ ♃
07:39 ☽ # ♂
09:02 ☽ ∥ ♅
13:56 ☽ ☌ ♅
14:43 ♀ # ♃
09 05:53 ☽ □ ♆
15:53 ☽ △ ♀
20:08 ☉ □ ♅
23:43 ☽ # ♅
10 01:52 ☽ ∥ ♂
02:50 ☽ □ ☿
09:36 ☽ △ ♃
18:10 ☽ ✶ ☉
18:25 ☽ ☌ ♂
19:43 ☽ △ ♄
22:15 ☽ ✶ ♆
11 05:48 ☿ ∥ ☉
07:58 ☽ △ ♆
09:08 ☽ # ♆
10:27 ☉ ✶ ♂
14:50 ☽ ∥ ♄
16:59 ☽ # ♆
19:40 ☉ ✶ ♄
21:09 ☽ □ ♀
22:44 ♂ △ ♄
12 05:36 ☽ ✶ ☿
12:37 ☽ ∥ ♀
17:46 ☽ # ♃
17:56 ☽ ✶ ♅
21:07 ☽ □ ♄
22:01 ☽ ∥ ☿
23:18 ☽ □ ♆
13 01:13 ☽ ∥ ☉
07:30 ☉ △ ♆
20:30 ☉ ✶ ♆
14 01:18 ☽ ✶ ♀
11:00 ☽ ☍ ♃
18:56 ☽ □ ♅
22:25 ☽ ✶ ♄
15 00:19 ☽ △ ♆
01:49 ☽ ✶ ♂
09:59 ☽ ☍ ♆
23:41 ☿ SR
23:47 ☿ # ♃
16 09:21 ☽ ☌ ☿
21:32 ☽ △ ♅
17 07:40 ☽ □ ♂
16:58 ☽ ∥ ☉
18 02:48 ☽ # ♃
06:03 ☽ ∥ ☿
15:06 ☽ ☌ ♀
17:29 ☽ △ ♃
17:56 ☽ ∥ ♀
19 06:07 ☽ # ♆
06:48 ☉ ☍ ♆
08:04 ☽ ☌ ♄
09:22 ☽ ☍ ♆
09:38 ☽ ∥ ♄
14:58 ☽ # ♆
17:37 ☽ △ ♂
20:14 ☽ △ ♆
21:22 ☽ ✶ ☉
22:48 ♀ △ ♃
20 15:04 ☽ ∥ ♂
21:30 ☽ ✶ ☿
21 02:00 ☽ □ ♃
[illegible] ☽ ∥ ♅
13:10 ☽ ☍ ♅
13:35 ♂ △ ♆
18:07 ☉ ♋
22 06:50 ☽ □ ♆
23 03:02 ☽ ∥ ♅
07:23 ☽ □ ☿
13:28 ☽ ✶ ♃
14:44 ♅ SR
21:01 ☽ ✶ ♀
23:37 ☽ # ♂
24 07:31 ☽ ✶ ♄
07:50 ☽ △ ♆
19:23 ☽ ✶ ♆
21:00 ☽ ∥ ♆
21:27 ♂ ♉
25 00:39 ☽ ☍ ♂
02:07 ☽ # ♄
06:51 ☽ ∥ ♆
07:18 ☽ △ ☉
08:35 ☽ # ♀
15:57 ♄ ☍ ♆
17:34 ☽ △ ☿
22:57 ☽ # ☿
26 08:26 ♀ # ♆
12:14 ☽ ∥ ♃
13:54 ☽ □ ♀
14:01 ☽ △ ♅
20:08 ☽ □ ♆
20:24 ☽ □ ♄
23:32 ☽ # ☉
28 12:10 ☽ ☌ ♃
18:40 ☿ ☌ ☉
29 00:38 ☽ □ ♅
04:20 ☽ △ ♀
06:25 ☽ ✶ ♆
07:14 ☽ △ ♄
13:03 ♀ ∥ ♄
17:09 ☽ ☌ ♆
30 05:19 ☽ △ ♂
08:54 ☽ ☍ ☿
15:31 ♀ ☍ ♆

⚷ Chiron

01	Dec.	09 S 29
02	15♒37R	
05	15 34	
08	15 31	
11	15 27	
14	15 23	
17	15 18	
20	15 12	
23	15 07	
26	15 00	
29	14 53	

July 2007

Day	S. T.	☉	☽	☿	♀	♂	♃	♄	♅	♆	♇	☊ True
	h m s	° ' "	° ' "	° '	° '	° '	° '	° '	° '	° '	° '	° '
01 Su	18 34 41	08♋ 48 59	13♑ 54 03	05♋R25	21♌ 57	04♉ 24	11♐R56	22♌ 19	18♓R40	21♒R41	27♐R20	08♓R44
02 Mo	18 38 38	09 46 10	27 00 43	04 52	22 39	05 07	11 50	22 25	18 40	21 40	27 19	08 36
03 Tu	18 42 34	10 43 21	10♒ 19 32	04 22	23 21	05 50	11 44	22 32	18 39	21 39	27 17	08 31
04 We	18 46 31	11 40 32	23 49 23	03 54	24 01	06 33	11 39	22 38	18 39	21 38	27 16	08 28
05 Th	18 50 27	12 37 44	07♓ 29 16	03 29	24 41	07 16	11 33	22 44	18 38	21 37	27 14	08D 27
06 Fr	18 54 24	13 34 55	21 18 35	03 09	25 19	07 58	11 27	22 51	18 38	21 36	27 13	08 28
07 Sa	18 58 21	14 32 07	05♈ 16 51	02 52	25 56	08 41	11 22	22 57	18 37	21 34	27 11	08 29
08 Su	19 02 17	15 29 19	19 23 30	02 39	26 32	09 23	11 17	23 04	18 37	21 33	27 10	08R 29
09 Mo	19 06 14	16 26 32	03♉ 37 23	02 31	27 07	10 06	11 12	23 11	18 36	21 32	27 08	08 28
10 Tu	19 10 10	17 23 45	17 56 26	02 28	27 40	10 48	11 07	23 17	18 35	21 31	27 07	08 25
11 We	19 14 07	18 20 58	02♊ 17 25	02D 30	28 12	11 30	11 02	23 24	18 34	21 29	27 05	08 20
12 Th	19 18 03	19 18 12	16 36 01	02 38	28 43	12 13	10 57	23 31	18 34	21 28	27 04	08 13
13 Fr	19 21 60	20 15 26	00♋ 47 12	02 50	29 12	12 55	10 53	23 37	18 33	21 27	27 03	08 05
14 Sa	19 25 56	21 12 41	14 45 54	03 08	29 40	13 37	10 48	23 44	18 32	21 25	27 01	07 57
15 Su	19 29 53	22 09 56	28 27 48	03 31	00♍ 06	14 18	10 44	23 51	18 31	21 24	27 00	07 51
16 Mo	19 33 50	23 07 11	11♌ 49 56	04 00	00 31	15 00	10 40	23 58	18 30	21 23	26 58	07 45
17 Tu	19 37 46	24 04 27	24 51 11	04 34	00 53	15 42	10 36	24 05	18 29	21 21	26 57	07 42
18 We	19 41 43	25 01 42	07♍ 32 09	05 13	01 15	16 23	10 33	24 12	18 28	21 20	26 56	07 41
19 Th	19 45 39	25 58 58	19 55 05	05 58	01 34	17 05	10 29	24 19	18 26	21 18	26 54	07D 42
20 Fr	19 49 36	26 56 14	02♎ 03 22	06 47	01 51	17 46	10 26	24 26	18 25	21 17	26 53	07 43
21 Sa	19 53 32	27 53 30	14 01 16	07 42	02 07	18 27	10 22	24 33	18 24	21 15	26 52	07 45
22 Su	19 57 29	28 50 46	25 53 32	08 42	02 21	19 08	10 19	24 40	18 23	21 14	26 51	07 46
23 Mo	20 01 25	29 48 03	07♏ 45 11	09 48	02 32	19 49	10 17	24 47	18 21	21 13	26 49	07R 46
24 Tu	20 05 22	00♌ 45 20	19 41 08	10 58	02 42	20 30	10 14	24 55	18 20	21 11	26 48	07 45
25 We	20 09 19	01 42 37	01♐ 45 58	12 13	02 49	21 11	10 12	25 02	18 19	21 09	26 47	07 41
26 Th	20 13 15	02 39 55	14 03 35	13 32	02 54	21 52	10 09	25 09	18 17	21 08	26 46	07 37
27 Fr	20 17 12	03 37 13	26 36 57	14 57	02 57	22 32	10 07	25 16	18 16	21 06	26 45	07 32
28 Sa	20 21 08	04 34 32	09♑ 27 56	16 26	02R 57	23 13	10 05	25 24	18 14	21 05	26 43	07 26
29 Su	20 25 05	05 31 51	22 37 04	17 59	02 55	23 53	10 03	25 31	18 13	21 03	26 42	07 21
30 Mo	20 29 01	06 29 11	06♒ 03 41	19 36	02 51	24 33	10 02	25 38	18 11	21 02	26 41	07 17
31 Tu	20 32 58	07 26 32	19 45 53	21 18	02 45	25 13	10 00	25 46	18 10	21 00	26 40	07 15

Data for	07-01-2007
Julian Day	2454282.50
Ayanamsa	23 57 47
SVP	05 ♓ 09 24
☽ ☊ Mean	10 ♓ 05 R

● ◐ PHASES ○ ◑

07	16:54	◑	15♈12
14	12:04	●	21♋41
22	06:30	◐	29♎06
30	00:48	○	06♒31

LAST ASPECT Day	h m	☽ INGRESS Day	h m	
01	08:46	02	05:25	♒
04	06:03	04	10:53	♓
06	10:09	06	14:57	♈
08	13:06	08	17:54	♉
10	16:54	10	20:10	♊
12	21:12	12	22:40	♋
14	12:04	15	02:44	♌
17	03:55	17	09:40	♍
19	13:45	19	19:54	♎
22	06:30	22	08:19	♏
24	10:32	24	20:31	♐
27	00:14	27	06:23	♑
29	02:24	29	13:14	♒
31	11:56	31	17:41	♓

DECLINATION

Day	☉	☽	☿	♀	♂	♃	♄	♅	♆	♇
01 Su	23N09	26S 49	18N45	14N49	11N37	21S 36	15N16	05S 12	14S 32	16S 24
02 Mo	23 05	24 03	18 41	14 28	11 52	21 35	15 14	05 13	14 32	16 25
03 Tu	23 00	19 58	18 38	14 08	12 07	21 34	15 12	05 13	14 33	16 25
04 We	22 56	14 48	18 37	13 47	12 21	21 34	15 09	05 13	14 33	16 25
05 Th	22 50	08 50	18 37	13 26	12 35	21 33	15 07	05 13	14 33	16 25
06 Fr	22 45	02 24	18 39	13 06	12 49	21 33	15 05	05 13	14 34	16 25
07 Sa	22 39	04N14	18 42	12 45	13 03	21 32	15 03	05 14	14 34	16 25
08 Su	22 32	10 42	18 46	12 25	13 17	21 32	15 01	05 14	14 35	16 25
09 Mo	22 26	16 41	18 51	12 05	13 31	21 31	14 59	05 14	14 35	16 25
10 Tu	22 19	21 47	18 58	11 45	13 45	21 31	14 57	05 15	14 35	16 26
11 We	22 11	25 36	19 05	11 25	13 58	21 30	14 54	05 15	14 36	16 26
12 Th	22 03	27 48	19 13	11 05	14 11	21 30	14 52	05 15	14 36	16 26
13 Fr	21 55	28 10	19 23	10 46	14 24	21 29	14 50	05 16	14 37	16 26
14 Sa	21 46	26 42	19 33	10 27	14 37	21 29	14 48	05 16	14 37	16 26
15 Su	21 37	23 40	19 43	10 08	14 50	21 28	14 45	05 16	14 38	16 26
16 Mo	21 28	19 24	19 55	09 49	15 03	21 28	14 43	05 17	14 38	16 27
17 Tu	21 18	14 19	20 06	09 31	15 15	21 28	14 41	05 17	14 39	16 27
18 We	21 08	08 45	20 17	09 13	15 28	21 27	14 38	05 18	14 39	16 27
19 Th	20 57	02 58	20 29	08 55	15 40	21 27	14 36	05 18	14 40	16 27
20 Fr	20 46	02S 48	20 41	08 38	15 52	21 27	14 34	05 19	14 40	16 27
21 Sa	20 35	08 22	20 52	08 21	16 04	21 26	14 31	05 19	14 40	16 28
22 Su	20 24	13 37	21 02	08 05	16 16	21 26	14 29	05 20	14 41	16 28
23 Mo	20 12	18 21	21 12	07 50	16 27	21 26	14 27	05 20	14 41	16 28
24 Tu	19 59	22 24	21 21	07 34	16 39	21 26	14 24	05 21	14 42	16 28
25 We	19 47	25 34	21 29	07 20	16 50	21 26	14 22	05 21	14 42	16 28
26 Th	19 34	27 36	21 36	07 06	17 01	21 26	14 19	05 22	14 43	16 29
27 Fr	19 21	28 18	21 41	06 53	17 12	21 25	14 17	05 23	14 43	16 29
28 Sa	19 07	27 30	21 44	06 40	17 23	21 25	14 15	05 23	14 44	16 29
29 Su	18 53	25 09	21 46	06 28	17 33	21 25	14 12	05 24	14 44	16 29
30 Mo	18 39	21 22	21 46	06 17	17 44	21 25	14 10	05 25	14 45	16 29
31 Tu	18 25	16 22	21 43	06 07	17 54	21 25	14 07	05 25	14 45	16 30

ASPECTARIAN

01 08:46 ☽ ⚹ ♅
14:39 ☿ ☌ ♄
19:08 ☿ ⚹ ♂
19:15 ♀ # ♆
02 06:37 ☽ # ☉
15:24 ☽ ∥ ♃
15:29 ☽ □ ♂
03 02:30 ☽ ⚹ ♃
06:38 ☽ # ☿
16:56 ☽ ∥ ♇
20:08 ☽ ☌ ♆
21:53 ☽ ☍ ♄
22:27 ☽ # ♄
04 00:22 ☽ ☍ ♀
01:04 ☽ ∥ ♆
04:34 ☽ # ♀
06:03 ☽ ⚹ ♇
09:47 ☽ # ♂
17:12 ☽ △ ☿
23:35 ☽ ⚹ ♂
05 07:02 ☽ □ ♃
09:37 ☽ △ ☉
13:38 ☽ ∥ ♅
19:22 ☽ ☌ ♅
06 10:09 ☽ □ ♇
11:30 ♀ ∥ ♂
19:56 ☽ □ ☿
07 03:40 ☽ # ♅
10:19 ☽ △ ♃
08 03:39 ☽ ⚹ ♆
06:16 ☽ △ ♄
06:19 ☽ ∥ ♀
10:29 ☽ ∥ ♂
12:35 ☽ △ ♀
13:06 ☽ △ ♇
15:16 ☽ # ♆
16:57 ☽ ∥ ♄
22:10 ☽ ⚹ ☿
22:56 ☽ # ♇
09 01:07 ♀ △ ♇
09:50 ☽ ∥ ☿
11:26 ☽ ☌ ♂
22:36 ☽ # ♃
23:01 ☽ ⚹ ☉
10 01:05 ☽ ⚹ ♅
02:16 ☿ SD
02:48 ☽ ∥ ☉
05:58 ☽ □ ♆
09:01 ☽ □ ♄
16:54 ☽ □ ♀
11 05:33 ☉ △ ♅
14:34 ☽ ☍ ♃
12 03:18 ☽ □ ♅
08:11 ☽ △ ♆
11:45 ☽ ⚹ ♄
17:39 ☽ ☍ ♇
21:12 ☽ ⚹ ♀
13 03:34 ☽ ☌ ☿
21:53 ☽ ⚹ ♂
23:36 ♂ # ♆
14 06:32 ☽ △ ♅
16:14 ♂ ∥ ♄
18:24 ♀ ♍
15 12:44 ☽ ∥ ☉
13:04 ☽ # ♃
21:31 ☽ ∥ ☿
21:54 ☽ △ ♃
22:58 ☉ # ♃
16 06:06 ☽ □ ♂
14:21 ☽ # ♇
17:30 ☽ ☍ ♆
19:58 ☽ ∥ ♂
22:24 ☽ ∥ ♄
22:33 ☽ ☌ ♄
22:35 ☽ # ♆
17 03:55 ☽ △ ♇
11:41 ☽ ☌ ♀
18:53 ♄ # ♆
19:18 ☽ ⚹ ☿
21:58 ☽ ∥ ♀
18 05:45 ☽ □ ♃
14:24 ☽ # ♅
18:07 ☽ △ ♂
21:07 ☽ ☍ ♅
19 12:58 ☽ ⚹ ☉
13:45 ☽ □ ♇
20 06:15 ☿ ∥ ☉
10:13 ☽ □ ☿
10:46 ☽ ∥ ♅
16:41 ☽ ⚹ ♃
22:13 ♂ ⚹ ♅
23:59 ☽ # ♀
21 14:35 ☽ △ ♆
21:30 ☽ ⚹ ♄
22 01:55 ☽ ⚹ ♇
04:13 ☽ # ♄
05:14 ☽ ∥ ♆
13:17 ☽ ⚹ ♀
13:38 ☽ # ♂
14:06 ☽ ∥ ♇
23 01:16 ♂ # ♇
04:33 ☽ △ ☿
05:00 ☉ ♌
09:54 ☽ # ☉
17:02 ☽ # ☿
17:48 ☽ ∥ ♃
21:18 ☽ △ ♅
24 01:44 ☽ ☍ ♂
02:59 ☽ □ ♆
10:32 ☽ □ ♄
13:52 ☿ # ♃
23:10 ♂ □ ♆
23:53 ☽ △ ☉
25 02:05 ☽ □ ♀
16:27 ☽ ☌ ♃
26 08:08 ☽ □ ♅
13:33 ☽ ⚹ ♆
21:26 ☽ △ ♄
27 00:14 ☽ ☌ ♇
11:55 ☽ △ ♀
17:29 ♀ SR
28 14:28 ☽ ☍ ☿
16:02 ☽ ⚹ ♅
29 02:24 ☽ △ ♂
03:26 ☿ △ ♅
21:52 ☽ # ☿
23:45 ☽ ∥ ♃
30 06:59 ☽ ⚹ ♃
14:24 ☽ # ☉
17:25 ☽ # ♂
23:30 ☽ ∥ ♇
31 02:09 ☽ ☌ ♆
06:53 ☽ ∥ ♆
09:36 ☽ # ♄
09:55 ☽ □ ♂
10:28 ☽ ☍ ♄
11:56 ☽ ⚹ ♇
22:10 ☽ ☍ ♀

⚷ Chiron

01	Dec.	09 S 32
02	14♒46R	
05	14 39	
08	14 31	
11	14 23	
14	14 14	
17	14 05	
20	13 56	
23	13 47	
26	13 37	
29	13 28	

August 2007

28 10:39 04♓46 ✸ Total Lunar Eclipse (mag 1.481)

Day	S. T.	☉	☽	☿	♀	♂	♃	♄	♅	♆	♇	☊ True
	h m s	° ′ ″	° ′ ″	° ′	° ′	° ′	° ′	° ′	° ′	° ′	° ′	° ′
01 We	20 36 54	08♌ 23 54	03♓ 40 55	23♋ 03	02♍R36	25♉ 53	09♐R59	25♌ 53	18♓R08	20♒R59	26♐R39	07♓R14
02 Th	20 40 51	09 21 16	17 45 35	24 51	02 24	26 33	09 58	26 01	18 06	20 57	26 38	07D 14
03 Fr	20 44 48	10 18 40	01♈ 56 37	26 42	02 11	27 13	09 57	26 08	18 05	20 55	26 37	07 16
04 Sa	20 48 44	11 16 04	16 10 59	28 36	01 55	27 52	09 57	26 15	18 03	20 54	26 36	07 17
05 Su	20 52 41	12 13 30	00♉ 26 00	00♌ 33	01 36	28 32	09 56	26 23	18 01	20 52	26 35	07 18
06 Mo	20 56 37	13 10 57	14 39 14	02 32	01 16	29 11	09 56	26 30	17 59	20 51	26 34	07R 18
07 Tu	21 00 34	14 08 26	28 48 29	04 32	00 53	29 50	09 56	26 38	17 57	20 49	26 33	07 17
08 We	21 04 30	15 05 56	12♊ 51 34	06 34	00 28	00♊ 29	09D 56	26 45	17 56	20 47	26 32	07 15
09 Th	21 08 27	16 03 27	26 46 20	08 36	00 01	01 08	09 56	26 53	17 54	20 46	26 31	07 13
10 Fr	21 12 23	17 00 59	10♋ 30 37	10 40	29♌ 33	01 47	09 57	27 01	17 52	20 44	26 30	07 10
11 Sa	21 16 20	17 58 33	24 02 23	12 43	29 02	02 26	09 57	27 08	17 50	20 43	26 30	07 07
12 Su	21 20 17	18 56 08	07♌ 19 52	14 47	28 30	03 04	09 58	27 16	17 48	20 41	26 29	07 04
13 Mo	21 24 13	19 53 44	20 21 54	16 50	27 57	03 42	09 59	27 23	17 46	20 39	26 28	07 02
14 Tu	21 28 10	20 51 21	03♍ 08 02	18 53	27 22	04 20	10 00	27 31	17 44	20 38	26 27	07 01
15 We	21 32 06	21 49 00	15 38 48	20 56	26 47	04 58	10 02	27 38	17 42	20 36	26 27	07D 01
16 Th	21 36 03	22 46 39	27 55 41	22 57	26 11	05 36	10 03	27 46	17 40	20 34	26 26	07 02
17 Fr	21 39 59	23 44 20	10♎ 01 01	24 58	25 34	06 14	10 05	27 54	17 37	20 33	26 25	07 03
18 Sa	21 43 56	24 42 01	21 57 55	26 58	24 57	06 52	10 07	28 01	17 35	20 31	26 25	07 04
19 Su	21 47 52	25 39 44	03♏ 50 04	28 56	24 19	07 29	10 09	28 09	17 33	20 29	26 24	07 06
20 Mo	21 51 49	26 37 28	15 41 43	00♍ 54	23 42	08 06	10 11	28 17	17 31	20 28	26 23	07 06
21 Tu	21 55 46	27 35 13	27 37 26	02 50	23 05	08 43	10 14	28 24	17 29	20 26	26 23	07 07
22 We	21 59 42	28 32 59	09♐ 41 55	04 44	22 29	09 20	10 16	28 32	17 27	20 25	26 22	07R 07
23 Th	22 03 39	29 30 46	21 59 39	06 38	21 54	09 57	10 19	28 40	17 24	20 23	26 22	07 06
24 Fr	22 07 35	00♍ 28 34	04♑ 34 38	08 30	21 20	10 33	10 22	28 47	17 22	20 21	26 21	07 06
25 Sa	22 11 32	01 26 24	17 29 58	10 21	20 47	11 09	10 25	28 55	17 20	20 20	26 21	07 06
26 Su	22 15 28	02 24 15	00♒ 47 30	12 10	20 15	11 46	10 29	29 02	17 18	20 18	26 20	07 05
27 Mo	22 19 25	03 22 07	14 27 33	13 58	19 46	12 22	10 32	29 10	17 15	20 17	26 20	07D 05
28 Tu	22 23 21	04 20 01	28 28 37	15 45	19 18	12 57	10 36	29 18	17 13	20 15	26 20	07 05
29 We	22 27 18	05 17 56	12♓ 47 20	17 30	18 52	13 33	10 40	29 25	17 11	20 13	26 19	07 06
30 Th	22 31 15	06 15 53	27 18 44	19 14	18 28	14 08	10 44	29 33	17 08	20 12	26 19	07R 06
31 Fr	22 35 11	07 13 51	11♈ 56 52	20 57	18 06	14 44	10 48	29 40	17 06	20 10	26 19	07 06

Data for	08-01-2007
Julian Day	2454313.50
Ayanamsa	23 57 53
SVP	05 ♓ 09 21
☽ ☊ Mean	08 ♓ 26 R

● ◐ PHASES ○ ◑

05	21:20	◑	13♉05
12	23:03	●	19♌51
20	23:55	◐	27♏35
28	10:35	✸	04♓46

LAST ASPECT ☽ INGRESS

Day	h m	Day	h m	Sign
02	15:37	02	20:43	♈
04	17:31	04	23:16	♉
07	01:50	07	02:02	♊
09	05:28	09	05:37	♋
10	12:58	11	10:42	♌
13	13:35	13	18:04	♍
15	21:03	16	04:05	♎
18	12:22	18	16:14	♏
21	01:35	21	04:45	♐
23	12:55	23	15:20	♑
24	23:41	25	22:35	♒
28	01:24	28	02:35	♓
29	22:22	30	04:25	♈

DECLINATION

Day	☉	☽	☿	♀	♂	♃	♄	♅	♆	♇
01 We	18N10	10S 27	21N39	05N57	18N04	21S 25	14N05	05S 26	14S 46	16S 30
02 Th	17 55	03 57	21 31	05 49	18 14	21 25	14 02	05 27	14 47	16 30
03 Fr	17 39	02N47	21 22	05 41	18 24	21 25	14 00	05 27	14 47	16 30
04 Sa	17 24	09 24	21 09	05 34	18 34	21 26	13 57	05 28	14 48	16 31
05 Su	17 08	15 33	20 54	05 29	18 43	21 26	13 55	05 29	14 48	16 31
06 Mo	16 52	20 52	20 36	05 24	18 52	21 26	13 52	05 29	14 49	16 31
07 Tu	16 35	24 57	20 16	05 20	19 01	21 26	13 50	05 30	14 49	16 31
08 We	16 18	27 31	19 53	05 17	19 10	21 26	13 47	05 31	14 50	16 32
09 Th	16 01	28 21	19 28	05 15	19 19	21 27	13 45	05 32	14 50	16 32
10 Fr	15 44	27 24	19 00	05 14	19 28	21 27	13 42	05 32	14 51	16 32
11 Sa	15 27	24 50	18 30	05 14	19 36	21 27	13 40	05 33	14 51	16 32
12 Su	15 09	20 58	17 58	05 15	19 44	21 27	13 37	05 34	14 52	16 33
13 Mo	14 51	16 08	17 24	05 18	19 52	21 28	13 34	05 35	14 52	16 33
14 Tu	14 33	10 42	16 48	05 21	20 00	21 28	13 32	05 36	14 53	16 33
15 We	14 14	04 56	16 11	05 24	20 08	21 29	13 29	05 36	14 53	16 33
16 Th	13 55	00S 54	15 32	05 29	20 15	21 29	13 27	05 37	14 54	16 34
17 Fr	13 37	06 37	14 52	05 35	20 23	21 29	13 24	05 38	14 54	16 34
18 Sa	13 17	12 01	14 10	05 41	20 30	21 30	13 22	05 39	14 55	16 34
19 Su	12 58	16 57	13 28	05 48	20 37	21 30	13 19	05 40	14 56	16 35
20 Mo	12 38	21 15	12 45	05 56	20 44	21 31	13 16	05 41	14 56	16 35
21 Tu	12 19	24 42	12 01	06 05	20 51	21 31	13 14	05 42	14 57	16 35
22 We	11 59	27 07	11 17	06 14	20 57	21 32	13 11	05 42	14 57	16 35
23 Th	11 39	28 18	10 32	06 23	21 04	21 33	13 09	05 43	14 58	16 36
24 Fr	11 18	28 03	09 46	06 33	21 10	21 33	13 06	05 44	14 58	16 36
25 Sa	10 58	26 18	09 01	06 44	21 16	21 34	13 03	05 45	14 59	16 36
26 Su	10 37	23 03	08 15	06 54	21 22	21 35	13 01	05 46	14 59	16 37
27 Mo	10 16	18 27	07 28	07 05	21 28	21 35	12 58	05 47	15 00	16 37
28 Tu	09 55	12 45	06 42	07 16	21 33	21 36	12 56	05 48	15 00	16 37
29 We	09 34	06 17	05 56	07 27	21 39	21 37	12 53	05 49	15 01	16 37
30 Th	09 13	00N36	05 10	07 38	21 44	21 37	12 50	05 50	15 01	16 38
31 Fr	08 51	07 30	04 24	07 49	21 49	21 38	12 48	05 51	15 02	16 38

⚷ Chiron

01 Dec. 09 S 53

01	13♒18R
04	13 08
07	12 58
10	12 49
13	12 39
16	12 30
19	12 20
22	12 11
25	12 02
28	11 53
31	11 44

ASPECTARIAN

01 00:01 ♂ □ ♄
05:26 ☉ ∥ ♂
10:46 ☽ □ ♃
17:09 ☽ # ♀
18:39 ☽ ∥ ♅
02 00:35 ☽ ☌ ♅
13:48 ☽ △ ☿
15:01 ☽ □ ♇
15:13 ☉ △ ♃
15:17 ☿ # ♃
15:37 ☽ ⚹ ♂
03 09:35 ☽ # ♅
09:53 ☿ ⚹ ♂
10:14 ☽ ∥ ♀
13:30 ☽ △ ♃
15:07 ☽ △ ☉
04 07:55 ☽ ⚹ ♆
17:07 ☽ △ ♄
17:15 ☿ ♌
17:23 ☽ ∥ ♄
17:31 ☽ △ ♇
20:56 ☽ # ♆
05 00:09 ♀ # ♅
00:14 ☽ □ ☿
01:56 ☽ △ ♀
04:04 ☽ # ♇
06:24 ☽ ∥ ☉
14:12 ☽ ∥ ♂
22:51 ☽ ∥ ☿
06 02:59 ☽ # ♃
05:38 ☽ ⚹ ♅
10:20 ♄ △ ♇
10:27 ☽ □ ♆
20:16 ☽ □ ♄
07 01:50 ☽ ☌ ♂
02:06 ♃ SD
03:26 ☽ □ ♀
05:35 ☉ # ♇
06:01 ♂ ♊
[illegible]
18:59 ☽ ☍ ♃

23:37 ♀ □ ♂
08 04:08 ☽ ⚹ ☉
08:41 ☽ □ ♅
13:37 ☽ △ ♆
23:34 ☽ ☍ ♇
09 00:12 ☽ ⚹ ♄
01:10 ♀ ♌R
05:28 ☽ ⚹ ♀
05:48 ☿ ∥ ♂
15:37 ☿ △ ♃
10 12:58 ☽ △ ♅
11 15:51 ☽ ⚹ ♂
21:19 ☽ # ♃
12 04:49 ☽ △ ♃
06:18 ☽ ∥ ♂
16:14 ☽ ☌ ☿
17:20 ☽ ∥ ☿
22:06 ☽ # ♇
22:07 ☉ # ♆
13 00:32 ☽ ☍ ♆
05:46 ☽ # ♆
06:14 ☽ ∥ ☉
11:24 ☽ △ ♇
11:39 ☽ ∥ ♄
13:16 ☽ ☌ ♄
13:35 ☽ ☌ ♀
18:26 ☉ ☍ ♆
19:16 ♀ ☌ ♄
14 02:25 ☽ □ ♂
09:33 ☿ # ♇
13:08 ☽ □ ♃
20:09 ☿ ☍ ♆
21:13 ☽ # ♅
22:04 ☽ ∥ ♀
15 03:58 ☽ ☍ ♅
[illegible]
19:57 ☿ ☌ ☉

21:03 ☽ □ ♇
16 16:02 ☽ △ ♂
19:31 ☽ # ♀
19:49 ☽ ∥ ♅
22:17 ☿ # ♆
17 00:08 ☽ ⚹ ♃
05:27 ☿ ☌ ♀
14:28 ♀ # ♅
17:23 ☿ △ ♇
18:02 ☉ ∥ ♄
21:05 ☽ △ ♆
18 03:41 ♀ ☌ ☉
05:37 ☽ # ☉
05:43 ☽ ⚹ ♀
06:00 ☽ ⚹ ☉
06:14 ☽ # ♄
08:56 ☽ # ☿
08:58 ☽ ⚹ ♇
12:07 ☽ ⚹ ☿
12:22 ☽ ⚹ ♄
13:45 ☿ ☌ ♄
13:49 ☽ ∥ ♆
22:05 ☽ ∥ ♇
19 05:26 ☿ ∥ ♄
13:01 ☿ ♍
18:13 ☉ △ ♇
20:51 ☽ # ♂
20 01:43 ☽ ∥ ♃
03:40 ☽ △ ♅
06:28 ☿ ∥ ☉
09:36 ☽ □ ♆
15:21 ☽ □ ♀
21 01:35 ☽ □ ♄
12:21 ☽ □ ☿
23:14 ☽ ☍ ♂
23:29 ☉ ☌ ♄
22 01:08 ☽ ☌ ♃

15:08 ☽ □ ♅
20:53 ☽ ⚹ ♆
23:50 ☽ △ ♀
23 08:24 ☽ ☌ ♇
12:08 ☉ ♍
12:55 ☽ △ ♄
15:36 ☽ △ ☉
16:03 ♂ ☍ ♃
24 08:35 ☽ △ ☿
23:41 ☽ ⚹ ♅
25 01:04 ☿ □ ♃
15:59 ☿ □ ♂
21:49 ♀ ☍ ♆
26 08:22 ☽ ∥ ♃
09:20 ☽ # ♂
17:09 ☽ ⚹ ♃

20:11 ☽ △ ♂
27 08:11 ☽ ∥ ♇
08:51 ☽ ☍ ♀
09:57 ☿ ∥ ♀
10:01 ☽ ☌ ♆
15:00 ☽ ∥ ♆
20:22 ☽ ⚹ ♇
23:18 ☽ # ♄
28 01:24 ☽ ☍ ♄
11:23 ☽ # ☉
14:14 ♂ # ♃
19:39 ☿ ☍ ♅
19:56 ☽ # ♀
20:26 ☽ □ ♃
29 01:19 ☽ □ ♂
01:22 ☽ # ☿

01:39 ☽ ∥ ♅
03:42 ☿ # ♅
07:16 ☽ ☌ ♅
08:53 ☽ ☍ ☿
22:22 ☽ □ ♇
30 14:11 ☽ ∥ ☿
18:07 ☽ # ♅
22:06 ☽ △ ♃
31 01:07 ☽ ∥ ♀
04:34 ☽ ∥ ☉
04:44 ☽ ⚹ ♂
09:50 ☽ △ ♀
13:27 ☽ ⚹ ♆
19:13 ☽ ∥ ♄
23:32 ☽ △ ♇

11 12:32 18♍25 ☉ Solar Eclipse (mag 0.751)

September 2007

Day	S. T.	☉	☽	☿	♀	♂	♃	♄	♅	♆	♇	☊ True
	h m s	° ' "	° ' "	° '	° '	° '	° '	° '	° '	° '	° '	° '
01 Sa	22 39 08	08♍11 51	26♈35 30	22♍38	17♌R46	15♊19	10♐52	29♌48	17♓R04	20♒R09	26♐R19	07♓R05
02 Su	22 43 04	09 09 53	11♉08 53	24 19	17 29	15 53	10 57	29 56	17 01	20 07	26 18	07 05
03 Mo	22 47 01	10 07 57	25 32 20	25 58	17 14	16 28	11 01	00♍03	16 59	20 06	26 18	07 04
04 Tu	22 50 57	11 06 03	09♊42 33	27 35	17 02	17 02	11 06	00 11	16 56	20 04	26 18	07 04
05 We	22 54 54	12 04 11	23 37 35	29 12	16 52	17 37	11 11	00 18	16 54	20 03	26 18	07 03
06 Th	22 58 50	13 02 22	07♋16 42	00♎47	16 44	18 10	11 16	00 26	16 52	20 01	26 18	07D 04
07 Fr	23 02 47	14 00 34	20 40 02	02 22	16 39	18 44	11 22	00 33	16 49	20 00	26 18	07 04
08 Sa	23 06 44	14 58 48	03♌48 15	03 54	16 36	19 18	11 27	00 41	16 47	19 58	26D 18	07 05
09 Su	23 10 40	15 57 04	16 42 13	05 26	16D 36	19 51	11 33	00 48	16 45	19 57	26 18	07 05
10 Mo	23 14 37	16 55 21	29 22 58	06 57	16 37	20 24	11 39	00 56	16 42	19 55	26 18	07 06
11 Tu	23 18 33	17 53 41	11♍51 30	08 26	16 42	20 57	11 45	01 03	16 40	19 54	26 18	07R 06
12 We	23 22 30	18 52 03	24 09 01	09 54	16 48	21 29	11 51	01 11	16 37	19 53	26 18	07 06
13 Th	23 26 26	19 50 26	06♎16 55	11 21	16 57	22 02	11 57	01 18	16 35	19 51	26 18	07 04
14 Fr	23 30 23	20 48 51	18 16 55	12 47	17 08	22 34	12 03	01 25	16 33	19 50	26 19	07 02
15 Sa	23 34 19	21 47 18	00♏11 09	14 12	17 21	23 05	12 10	01 33	16 30	19 49	26 19	07 00
16 Su	23 38 16	22 45 46	12 02 10	15 35	17 35	23 37	12 17	01 40	16 28	19 47	26 19	06 57
17 Mo	23 42 13	23 44 16	23 53 05	16 57	17 52	24 08	12 24	01 47	16 25	19 46	26 19	06 55
18 Tu	23 46 09	24 42 48	05♐47 34	18 18	18 11	24 39	12 31	01 55	16 23	19 45	26 20	06 53
19 We	23 50 06	25 41 22	17 49 47	19 37	18 32	25 10	12 38	02 02	16 21	19 43	26 20	06 52
20 Th	23 54 02	26 39 57	00♑04 12	20 55	18 54	25 40	12 45	02 09	16 18	19 42	26 20	06D 52
21 Fr	23 57 59	27 38 34	12 35 19	22 11	19 19	26 10	12 52	02 16	16 16	19 41	26 21	06 54
22 Sa	00 01 55	28 37 13	25 27 18	23 26	19 44	26 40	13 00	02 23	16 14	19 40	26 21	06 55
23 Su	00 05 52	29 35 53	08♒43 30	24 39	20 12	27 10	13 08	02 30	16 11	19 39	26 22	06 57
24 Mo	00 09 48	00♎34 35	22 25 58	25 51	20 41	27 39	13 15	02 38	16 09	19 38	26 22	06 59
25 Tu	00 13 45	01 33 19	06♓34 46	27 00	21 12	28 08	13 23	02 45	16 07	19 36	26 23	07 00
26 We	00 17 42	02 32 05	21 07 26	28 08	21 44	28 36	13 31	02 52	16 04	19 35	26 23	06R 59
27 Th	00 21 38	03 30 52	05♈58 42	29 14	22 17	29 05	13 40	02 58	16 02	19 34	26 24	06 58
28 Fr	00 25 35	04 29 42	21 00 57	00♏18	22 52	29 32	13 48	03 05	16 00	19 33	26 25	06 55
29 Sa	00 29 31	05 28 33	06♉05 05	01 19	23 28	00♋00	13 56	03 12	15 58	19 32	26 25	06 51
30 Su	00 33 28	06 27 27	21 01 59	02 18	24 06	00 27	14 05	03 19	15 55	19 31	26 26	06 46

Data for	09-01-2007
Julian Day	2454344.50
Ayanamsa	23 57 57
SVP	05 ♓ 09 17
☽ ☊ Mean	06 ♓ 48 R

● ◐ PHASES ○ ◑

04	02:33	◑	11♊12
11	12:45	☉	18♍25
19	16:48	◐	26♐22
26	19:46	○	03♈21

LAST ASPECT ☽ Day	h m	INGRESS Day	h m	
01	05:19	01	05:36	♉
03	00:48	03	07:31	♊
05	11:02	05	11:09	♋
06	17:05	07	17:00	♌
09	18:08	10	01:11	♍
12	04:14	12	11:32	♎
14	16:10	14	23:37	♏
16	23:41	17	12:21	♐
19	16:48	19	23:52	♑
22	06:16	22	08:18	♒
24	09:15	24	12:56	♓
26	12:32	26	14:23	♈
28	14:00	28	14:18	♉
30	05:11	30	14:35	♊

DECLINATION

Day	☉	☽	☿	♀	♂	♃	♄	♅	♆	♇
01 Sa	08N30	14N00	03N38	07N59	21N54	21S39	12N45	05S51	15S02	16S38
02 Su	08 08	19 41	02 52	08 10	21 59	21 40	12 43	05 52	15 03	16 39
03 Mo	07 46	24 10	02 06	08 21	22 03	21 41	12 40	05 53	15 03	16 39
04 Tu	07 24	27 08	01 21	08 31	22 08	21 41	12 37	05 54	15 04	16 39
05 We	07 02	28 21	00 36	08 41	22 12	21 42	12 35	05 55	15 04	16 40
06 Th	06 40	27 49	00S09	08 50	22 16	21 43	12 32	05 56	15 05	16 40
07 Fr	06 17	25 39	00 53	08 59	22 20	21 44	12 30	05 57	15 05	16 40
08 Sa	05 55	22 08	01 37	09 08	22 24	21 45	12 27	05 58	15 06	16 41
09 Su	05 32	17 36	02 21	09 16	22 28	21 46	12 25	05 59	15 06	16 41
10 Mo	05 10	12 22	03 04	09 24	22 32	21 47	12 22	06 00	15 06	16 41
11 Tu	04 47	06 43	03 46	09 32	22 35	21 48	12 19	06 01	15 07	16 41
12 We	04 24	00 54	04 28	09 39	22 39	21 49	12 17	06 02	15 07	16 42
13 Th	04 01	04S51	05 10	09 45	22 42	21 50	12 14	06 03	15 08	16 42
14 Fr	03 38	10 22	05 51	09 51	22 45	21 51	12 12	06 04	15 08	16 42
15 Sa	03 15	15 28	06 31	09 57	22 48	21 52	12 09	06 05	15 09	16 43
16 Su	02 52	19 59	07 11	10 01	22 51	21 53	12 07	06 05	15 09	16 43
17 Mo	02 29	23 42	07 50	10 06	22 54	21 54	12 04	06 06	15 09	16 43
18 Tu	02 06	26 26	08 28	10 10	22 57	21 55	12 02	06 07	15 10	16 44
19 We	01 43	28 01	09 05	10 13	22 59	21 56	11 59	06 08	15 10	16 44
20 Th	01 20	28 16	09 42	10 15	23 02	21 57	11 57	06 09	15 11	16 44
21 Fr	00 56	27 06	10 18	10 17	23 04	21 58	11 54	06 10	15 11	16 45
22 Sa	00 33	24 29	10 53	10 19	23 06	22 00	11 52	06 11	15 11	16 45
23 Su	00 10	20 29	11 27	10 20	23 08	22 01	11 50	06 12	15 12	16 45
24 Mo	00S14	15 18	12 01	10 20	23 10	22 02	11 47	06 13	15 12	16 46
25 Tu	00 37	09 09	12 33	10 19	23 12	22 03	11 45	06 14	15 13	16 46
26 We	01 00	02 21	13 04	10 19	23 14	22 04	11 42	06 14	15 13	16 46
27 Th	01 24	04N42	13 34	10 17	23 16	22 05	11 40	06 15	15 13	16 47
28 Fr	01 47	11 34	14 03	10 15	23 18	22 06	11 38	06 16	15 14	16 47
29 Sa	02 11	17 46	14 30	10 12	23 19	22 08	11 35	06 17	15 14	16 47
30 Su	02 34	22 50	14 57	10 09	23 21	22 09	11 33	06 18	15 14	16 48

ASPECTARIAN

01	04:05	☽ # ♆
	05:19	☽ △ ♄
	10:37	☽ # ♇
	20:29	☽ △ ☉
	22:30	♀ ∥ ☉
02	09:44	☽ # ♃
	09:44	☽ ✱ ♅
	10:20	☽ □ ♀
	11:33	☽ ∥ ♂
	13:49	♄ ♍
	14:54	☽ □ ♆
03	00:48	☽ △ ☿
	05:02	☿ □ ♇
	07:40	☽ □ ♄
	20:09	♂ □ ♅
	23:37	♀ ✱ ♂
04	00:05	☉ □ ♃
	02:24	☽ ☍ ♃
	12:23	☽ □ ♅
	12:24	☽ ✱ ♀
	13:07	☽ ☌ ♂
	17:48	☽ △ ♆
05	04:40	☽ ☍ ♇
	11:02	☽ □ ☿
	11:47	☽ ✱ ♄
	12:03	☿ ♎
06	11:04	☽ ✱ ☉
	17:05	☽ △ ♅
07	14:54	♇ SD
	20:59	☉ # ♅
	22:21	☽ ∥ ♂
08	00:13	☽ ✱ ☿
	02:11	☽ # ♃
	14:17	☽ △ ♃
	16:15	♀ SD
	23:47	☽ ☌ ♀
09	04:06	♂ △ ♆
	04:23	☽ # ♇
	06:06	☽ ☍ ♆
	06:11	☽ ✱ ♂
	11:44	☽ # ♆
	18:08	☽ △ ♇
	18:46	☉ ☍ ♅
	23:57	☽ ∥ ♄
10	02:59	☽ ☌ ♄
	12:25	☽ ∥ ♀
	23:47	☽ □ ♃
11	02:54	☽ # ♅
	08:33	☽ ∥ ☉
	09:18	☽ ☍ ♅
	10:52	☽ # ☿
	18:32	☽ □ ♂
	22:31	☿ # ☉
12	04:14	☽ □ ♇
	20:42	☽ # ☉
13	01:28	☽ ∥ ☿
	05:04	☽ ∥ ♅
	10:44	☿ ✱ ♃
	11:24	☽ ✱ ♃
	11:30	☽ ☌ ☿
	21:37	☽ # ♀
	21:39	☽ ✱ ♀
14	03:07	☽ △ ♆
	07:51	☿ ∥ ♅
	08:14	☽ # ♄
	09:01	☽ △ ♂
	16:10	☽ ✱ ♇
	22:22	☽ ∥ ♆
15	02:47	☽ ✱ ♄
	06:16	☽ ∥ ♇
16	08:56	☽ △ ♅
	11:31	☽ □ ♀
	11:39	☽ ∥ ♃
	15:41	☽ □ ♆
	18:15	☽ # ♂
	23:41	☽ ✱ ☉
17	16:07	☽ □ ♄
	20:50	☉ □ ♂
	21:28	☿ ✱ ♀
18	13:34	☽ ☌ ♃
	21:04	☽ □ ♅
19	01:26	☽ △ ♀
	02:01	☿ △ ♆
	03:44	☽ ✱ ♆
	03:57	☽ ✱ ☿
	15:04	☽ ☍ ♂
	15:58	☉ □ ♇
	16:44	☽ ☌ ♇
20	04:04	☽ △ ♄
	23:28	☿ # ♀
21	06:55	☽ ✱ ♅
	08:37	♂ ☍ ♇
	19:53	☽ □ ☿
	20:05	♀ ☍ ♆
22	06:16	☽ △ ☉
	09:07	☽ # ♂
	15:43	☽ ∥ ♃
23	07:52	☽ ✱ ♃
	09:52	☉ ♎
	14:55	☿ # ♄
	17:42	☽ ∥ ♇
	19:09	☽ ☌ ♆
	20:52	☽ ☍ ♀
24	00:21	☽ ∥ ♆
	06:23	☽ △ ☿
	06:46	☽ ✱ ♇
	09:15	☽ △ ♂
	10:55	☿ ✱ ♇
	12:07	☽ ∥ ☿
	14:08	☽ # ♄
	17:30	☽ ☍ ♄
	19:35	☽ # ♀
25	10:27	☽ ∥ ♅
	11:25	☽ □ ♃
	15:45	☽ ☌ ♅
26	04:20	☽ ∥ ☉
	08:34	☽ □ ♇
	12:06	☽ # ☉
	12:32	☽ □ ♂
	17:53	☿ △ ♂
27	05:19	☽ # ♅
	12:24	☽ △ ♃
	17:18	☿ ♏
	19:16	☽ ∥ ♀
	21:41	☽ ✱ ♆
28	00:12	☽ ∥ ♄
	03:04	☽ △ ♀
	08:35	☽ △ ♇
	09:54	☽ # ☿
	13:43	☽ # ♆
	14:00	☽ ✱ ♂
	15:51	☽ ☍ ☿
	19:22	☽ △ ♄
	19:55	☽ # ♇
	23:56	♂ ♋
29	15:47	☽ ✱ ♅
	20:13	☽ # ♃
	21:33	☽ □ ♆
30	02:48	☽ ∥ ♂
	05:11	☽ □ ♀
	16:55	☿ ∥ ♆
	20:11	☽ □ ♄

⚷ Chiron

01 Dec.	10 S 23
03	11♒36R
06	11 28
09	11 21
12	11 14
15	11 07
18	11 01
21	10 55
24	10 50
27	10 45
30	10 41

October 2007

Day	S. T.	☉	☽	☿	♀	♂	♃	♄	♅	♆	♇	☊ True
	h m s	° ' "	° ' "	° '	° '	° '	° '	° '	° '	° '	° '	° '
01 Mo	00 37 24	07♎26 23	05♊43 49	03♏14	24♌45	00♋54	14♐14	03♍26	15♓R53	19♒R30	26♐27	06♓R42
02 Tu	00 41 21	08 25 22	20 05 09	04 07	25 25	01 21	14 23	03 33	15 51	19 29	26 27	06 39
03 We	00 45 17	09 24 23	04♋03 11	04 56	26 06	01 47	14 32	03 39	15 49	19 28	26 28	06 38
04 Th	00 49 14	10 23 26	17 37 39	05 43	26 48	02 13	14 41	03 46	15 47	19 27	26 29	06D 37
05 Fr	00 53 11	11 22 32	00♌50 06	06 25	27 31	02 38	14 50	03 53	15 45	19 27	26 30	06 38
06 Sa	00 57 07	12 21 39	13 43 12	07 04	28 15	03 03	14 59	03 59	15 43	19 26	26 31	06 40
07 Su	01 01 04	13 20 49	26 20 02	07 38	29 00	03 28	15 09	04 06	15 41	19 25	26 32	06 41
08 Mo	01 05 00	14 20 02	08♍43 37	08 07	29 47	03 52	15 18	04 12	15 39	19 24	26 33	06R 42
09 Tu	01 08 57	15 19 16	20 56 44	08 31	00♍34	04 16	15 28	04 19	15 37	19 23	26 34	06 41
10 We	01 12 53	16 18 33	03♎01 43	08 48	01 22	04 39	15 37	04 25	15 35	19 23	26 35	06 38
11 Th	01 16 50	17 17 51	15 00 37	09 00	02 10	05 02	15 47	04 31	15 33	19 22	26 36	06 33
12 Fr	01 20 46	18 17 12	26 55 08	09 05	03 00	05 25	15 57	04 38	15 31	19 21	26 37	06 26
13 Sa	01 24 43	19 16 35	08♏46 54	09R 02	03 50	05 47	16 07	04 44	15 29	19 21	26 38	06 18
14 Su	01 28 39	20 16 00	20 37 37	08 52	04 41	06 08	16 18	04 50	15 27	19 20	26 39	06 09
15 Mo	01 32 36	21 15 26	02♐29 15	08 34	05 33	06 29	16 28	04 56	15 25	19 20	26 40	06 00
16 Tu	01 36 33	22 14 55	14 24 20	08 07	06 26	06 50	16 38	05 02	15 24	19 19	26 41	05 53
17 We	01 40 29	23 14 25	26 25 57	07 32	07 19	07 10	16 49	05 08	15 22	19 19	26 42	05 48
18 Th	01 44 26	24 13 58	08♑37 51	06 48	08 13	07 30	16 59	05 14	15 20	19 18	26 44	05 45
19 Fr	01 48 22	25 13 32	21 04 13	05 56	09 07	07 49	17 10	05 20	15 18	19 18	26 45	05D 44
20 Sa	01 52 19	26 13 08	03♒49 25	04 57	10 02	08 07	17 21	05 25	15 17	19 17	26 46	05 45
21 Su	01 56 15	27 12 45	16 57 43	03 52	10 58	08 25	17 31	05 31	15 15	19 17	26 47	05 46
22 Mo	02 00 12	28 12 24	00♓32 36	02 41	11 54	08 43	17 42	05 37	15 14	19 17	26 49	05 47
23 Tu	02 04 08	29 12 05	14 36 06	01 27	12 51	09 00	17 53	05 42	15 12	19 16	26 50	05R 47
24 We	02 08 05	00♏11 48	29 07 42	00 11	13 49	09 16	18 05	05 48	15 11	19 16	26 52	05 44
25 Th	02 12 02	01 11 32	14♈03 38	28♎57	14 46	09 32	18 16	05 53	15 09	19 16	26 53	05 40
26 Fr	02 15 58	02 11 18	29 16 39	27 45	15 45	09 47	18 27	05 59	15 08	19 15	26 54	05 33
27 Sa	02 19 55	03 11 07	14♉36 39	26 39	16 44	10 02	18 38	06 04	15 06	19 15	26 56	05 25
28 Su	02 23 51	04 10 57	29 52 16	25 41	17 43	10 15	18 50	06 09	15 05	19 15	26 57	05 16
29 Mo	02 27 48	05 10 49	14♊52 49	24 51	18 43	10 29	19 01	06 14	15 04	19 15	26 59	05 07
30 Tu	02 31 44	06 10 44	29 30 03	24 12	19 43	10 42	19 13	06 19	15 02	19 15	27 00	05 00
31 We	02 35 41	07 10 41	13♋39 13	23 43	20 44	10 54	19 25	06 24	15 01	19 15	27 02	04 55

Data for	10-01-2007
Julian Day	2454374.50
Ayanamsa	23 58 00
SVP	05 ♓ 09 12
☽ ☊ Mean	05 ♓ 13 R

● ◐ PHASES ○ ◑

03	10:06	◑	09♋49
11	05:01	●	17♎30
19	08:33	◐	25♑35
26	04:52	○	02♉23

LAST ASPECT Day	h m	☽ INGRESS Day	h m	
02	10:52	02	16:58	♋
03	20:42	04	22:28	♌
07	05:29	07	07:04	♍
09	11:08	09	17:58	♎
11	23:23	12	06:14	♏
13	21:23	14	18:58	♐
17	00:33	17	07:04	♑
19	08:34	19	16:52	♒
21	19:37	21	23:03	♓
23	20:18	24	01:25	♈
25	21:47	26	01:08	♉
27	07:17	28	00:12	♊
29	19:51	30	00:50	♋

DECLINATION

Day	☉	☽	☿	♀	♂	♃	♄	♅	♆	♇
01 Mo	02S 57	26N23	15S 22	10N05	23N22	22S 10	11N31	06S 19	15S 14	16S 48
02 Tu	03 20	28 07	15 45	10 00	23 24	22 11	11 28	06 19	15 15	16 48
03 We	03 44	28 00	16 07	09 55	23 25	22 12	11 26	06 20	15 15	16 49
04 Th	04 07	26 11	16 27	09 50	23 26	22 13	11 24	06 21	15 15	16 49
05 Fr	04 30	22 57	16 45	09 43	23 27	22 15	11 21	06 22	15 16	16 49
06 Sa	04 53	18 39	17 01	09 36	23 29	22 16	11 19	06 23	15 16	16 50
07 Su	05 16	13 37	17 15	09 29	23 30	22 17	11 17	06 23	15 16	16 50
08 Mo	05 39	08 07	17 27	09 21	23 31	22 18	11 15	06 24	15 16	16 50
09 Tu	06 02	02 25	17 36	09 13	23 32	22 19	11 12	06 25	15 17	16 51
10 We	06 25	03S 19	17 42	09 03	23 33	22 21	11 10	06 26	15 17	16 51
11 Th	06 48	08 52	17 45	08 54	23 34	22 22	11 08	06 26	15 17	16 51
12 Fr	07 10	14 04	17 45	08 44	23 35	22 23	11 06	06 27	15 17	16 52
13 Sa	07 33	18 44	17 41	08 33	23 36	22 24	11 04	06 28	15 17	16 52
14 Su	07 55	22 40	17 34	08 22	23 37	22 25	11 02	06 29	15 18	16 52
15 Mo	08 18	25 40	17 23	08 10	23 38	22 27	11 00	06 29	15 18	16 53
16 Tu	08 40	27 33	17 07	07 58	23 39	22 28	10 58	06 30	15 18	16 53
17 We	09 02	28 11	16 47	07 45	23 40	22 29	10 56	06 31	15 18	16 53
18 Th	09 24	27 27	16 22	07 32	23 41	22 30	10 54	06 31	15 18	16 53
19 Fr	09 46	25 21	15 53	07 18	23 42	22 31	10 52	06 32	15 18	16 54
20 Sa	10 07	21 56	15 19	07 04	23 43	22 32	10 50	06 32	15 19	16 54
21 Su	10 29	17 20	14 42	06 50	23 44	22 34	10 48	06 33	15 19	16 54
22 Mo	10 50	11 43	14 01	06 35	23 45	22 35	10 46	06 34	15 19	16 55
23 Tu	11 11	05 20	13 17	06 19	23 46	22 36	10 44	06 34	15 19	16 55
24 We	11 33	01N31	12 32	06 03	23 47	22 37	10 42	06 35	15 19	16 55
25 Th	11 53	08 28	11 47	05 47	23 48	22 38	10 40	06 35	15 19	16 56
26 Fr	12 14	15 03	11 02	05 30	23 49	22 39	10 39	06 36	15 19	16 56
27 Sa	12 35	20 45	10 19	05 13	23 50	22 40	10 37	06 36	15 19	16 56
28 Su	12 55	25 03	09 39	04 55	23 51	22 41	10 35	06 37	15 19	16 56
29 Mo	13 15	27 33	09 04	04 37	23 53	22 42	10 33	06 37	15 19	16 57
30 Tu	13 35	28 03	08 33	04 19	23 54	22 43	10 32	06 38	15 19	16 57
31 We	13 55	26 41	08 08	04 00	23 55	22 45	10 30	06 38	15 19	16 57

ASPECTARIAN

01 03:02 ☽ △ ☉
06:16 ☿ ✶ ♄
14:17 ☽ ☍ ♃
16:52 ☽ □ ♅
22:59 ☽ △ ♆
02 09:32 ☽ ✶ ♀
10:52 ☽ ☍ ♇
19:56 ☽ ☌ ♂
23:18 ☽ ✶ ♄
03 01:39 ☽ △ ☿
13:14 ♀ △ ♇
20:42 ☽ △ ♅
04 20:46 ☽ ∥ ♂
05 04:18 ☽ # ♃
06:26 ☿ ∥ ♇
10:54 ☽ □ ☿
21:14 ☽ ✶ ☉
06 02:25 ☽ △ ♃
07:46 ☽ # ☿
09:02 ☽ # ♇
10:48 ☽ ☍ ♆
16:26 ☽ # ♆
07 00:22 ☽ △ ♇
05:29 ☽ ☌ ♀
10:29 ☽ ∥ ♄
14:14 ☽ ✶ ♂
15:07 ☽ ☌ ♄
18:39 ☽ ∥ ♀
22:46 ☽ ✶ ☿
08 06:53 ♀ ♍
07:18 ☽ # ♅
09:49 ☽ # ☉
13:03 ☽ □ ♃
13:31 ☽ ☍ ♅
09 03:45 ♂ ✶ ♄
04:05 ☉ ✶ ♃
11:08 ☽ □ ♇
18:24 ♃ □ ♅
10 00:58 ☉ ∥ ♅
03:21 ☽ □ ♂
13:21 ☽ ∥ ♅
14:14 ☽ ∥ ☉
11 00:08 ☽ # ♀
01:35 ☽ ✶ ♃
08:46 ☽ △ ♆
10:08 ☽ # ♄
23:23 ☽ ✶ ♇
12 04:01 ☿ SR
05:58 ☽ ∥ ♆
13:13 ☽ ✶ ♀
13:58 ☽ ∥ ♇
15:43 ☽ ✶ ♄
17:44 ☽ △ ♂
18:25 ☽ ∥ ☿
13 00:31 ☽ ☌ ☿
01:40 ☉ △ ♆
13:32 ☽ △ ♅
21:23 ☽ □ ♆
22:20 ☽ ∥ ♃
14 04:29 ♀ ☌ ♄
06:48 ☽ # ♂
18:49 ♀ # ☉
15 04:59 ☽ □ ♄
06:41 ☽ □ ♀
16 01:58 ☽ □ ♅
04:32 ☽ ☌ ♃
09:50 ☽ ✶ ♆
17:05 ☽ ✶ ☉
17:07 ☿ ∥ ♇
17:37 ☿ ✶ ☊
17 00:33 ☽ ☌ ♇
03:18 ☿ ✶ ♀
08:37 ☿ △ ♂
17:18 ☽ △ ♄
20:39 ☽ ✶ ☿
21:43 ☽ ☍ ♂
23:07 ☽ △ ♀
18 12:58 ☽ ✶ ♅
19 12:44 ☽ # ♂
13:54 ☿ ✶ ♄
20:22 ☽ ∥ ♃
20 00:32 ☿ ∥ ♆
01:56 ☽ □ ☿
13:38 ☉ ✶ ♇
21 01:02 ☽ ✶ ♃
01:57 ☽ ∥ ♇
04:09 ☽ ☌ ♆
09:05 ☽ ∥ ♆
13:19 ☽ ∥ ☿
17:29 ☽ ✶ ♇
19:30 ☉ # ♄
19:37 ☽ △ ☉
22 01:32 ♀ # ♅
03:17 ☽ ∥ ☉
03:24 ☽ △ ☿
03:46 ☽ # ♄
08:48 ☽ ☍ ♄
14:20 ☽ △ ♂
19:32 ☽ ∥ ♅
20:18 ☽ # ♀
20:51 ☽ ☍ ♀
23 01:00 ☽ ☌ ♅
05:34 ☽ □ ♃
20:18 ☽ □ ♇
23:55 ☿ ☌ ☉
24 03:37 ☿ ♎R
15:02 ☽ ∥ ♀
16:39 ☽ □ ♂
17:27 ☽ # ♅
21:36 ☿ ∥ ☉
25 06:45 ☽ △ ♃
07:47 ☽ ∥ ♄
08:14 ☽ ✶ ♆
09:08 ♀ ☍ ♅
10:36 ☽ # ☿
12:53 ☽ # ☉
21:47 ☽ ☍ ☿
26 01:02 ☽ # ♆
07:25 ☽ # ♇
10:33 ☽ △ ♄
13:29 ☿ # ♄
16:43 ☽ ✶ ♂
17:55 ☿ ✶ ♇
27 00:46 ☽ ✶ ♅
03:33 ☽ △ ♀
07:17 ☽ □ ♆
09:40 ☽ # ♃
16:17 ☽ ∥ ♂
28 10:02 ☽ □ ♄
29 00:18 ☽ □ ♅
06:49 ☽ ☍ ♃
07:06 ☽ △ ♆
09:09 ♀ □ ♃
15:33 ☽ △ ☿
19:51 ☽ ☍ ♇
30 03:43 ☉ ✶ ♄
04:01 ♃ ✶ ♆
11:32 ☽ ✶ ♄
12:04 ☽ △ ☉
19:11 ☽ ☌ ♂
31 02:22 ☽ △ ♅
13:19 ☽ ✶ ♀
17:13 ☽ □ ☿
22:37 ☽ ∥ ♂

⚷ Chiron

01	Dec.	10 S 50
03	10♒37R	
06	10 34	
09	10 32	
12	10 30	
15	10 28	
18	10 28	
21	10 28D	
24	10 28	
27	10 29	
30	10 31	

November 2007

Day	S. T.	☉	☽	☿	♀	♂	♃	♄	♅	♆	♇	☊ True
	h m s	° ' "	° ' "	° '	° '	° '	° '	° '	° '	° '	° '	° '
01 Th	02 39 37	08♏ 10 40	27♋ 19 09	23♎R27	21♍ 45	11♋ 05	19♐ 36	06♍ 29	15♓R00	19♒D15	27♐ 04	04♓R51
02 Fr	02 43 34	09 10 41	10♌ 31 34	23D 22	22 47	11 16	19 48	06 34	14 59	19 15	27 05	04 50
03 Sa	02 47 31	10 10 45	23 20 01	23 28	23 49	11 26	20 00	06 39	14 58	19 15	27 07	04D 50
04 Su	02 51 27	11 10 50	05♍ 48 57	23 44	24 51	11 35	20 12	06 43	14 57	19 15	27 09	04R 50
05 Mo	02 55 24	12 10 58	18 02 57	24 11	25 54	11 44	20 24	06 48	14 56	19 15	27 10	04 50
06 Tu	02 59 20	13 11 07	00♎ 06 16	24 47	26 57	11 51	20 36	06 52	14 55	19 15	27 12	04 47
07 We	03 03 17	14 11 19	12 02 39	25 31	28 00	11 59	20 49	06 57	14 54	19 16	27 14	04 41
08 Th	03 07 13	15 11 32	23 55 09	26 23	29 04	12 05	21 01	07 01	14 53	19 16	27 15	04 33
09 Fr	03 11 10	16 11 47	05♏ 46 10	27 21	00♎ 08	12 10	21 13	07 05	14 52	19 16	27 17	04 22
10 Sa	03 15 06	17 12 05	17 37 30	28 25	01 12	12 15	21 25	07 10	14 52	19 16	27 19	04 08
11 Su	03 19 03	18 12 24	29 30 29	29 34	02 17	12 19	21 38	07 14	14 51	19 17	27 21	03 54
12 Mo	03 22 60	19 12 44	11♐ 26 25	00♏ 47	03 22	12 22	21 50	07 18	14 50	19 17	27 23	03 41
13 Tu	03 26 56	20 13 07	23 26 41	02 04	04 27	12 25	22 03	07 21	14 50	19 17	27 24	03 29
14 We	03 30 53	21 13 31	05♑ 33 11	03 24	05 33	12 26	22 16	07 25	14 49	19 18	27 26	03 19
15 Th	03 34 49	22 13 56	17 48 23	04 47	06 38	12 27	22 28	07 29	14 49	19 18	27 28	03 12
16 Fr	03 38 46	23 14 23	00♒ 15 20	06 12	07 44	12R 27	22 41	07 33	14 48	19 19	27 30	03 09
17 Sa	03 42 42	24 14 51	12 57 40	07 38	08 51	12 26	22 54	07 36	14 48	19 19	27 32	03 07
18 Su	03 46 39	25 15 20	25 59 19	09 07	09 57	12 24	23 07	07 40	14 48	19 20	27 34	03 07
19 Mo	03 50 35	26 15 51	09♓ 24 07	10 36	11 04	12 21	23 19	07 43	14 47	19 21	27 36	03 07
20 Tu	03 54 32	27 16 23	23 15 07	12 06	12 11	12 18	23 32	07 46	14 47	19 21	27 38	03 05
21 We	03 58 29	28 16 56	07♈ 33 33	13 38	13 18	12 14	23 45	07 49	14 47	19 22	27 40	03 02
22 Th	04 02 25	29 17 30	22 17 47	15 10	14 26	12 08	23 58	07 52	14 47	19 23	27 42	02 56
23 Fr	04 06 22	00♐ 18 05	07♉ 22 43	16 42	15 33	12 02	24 11	07 55	14 46	19 23	27 44	02 47
24 Sa	04 10 18	01 18 42	22 39 46	18 15	16 41	11 55	24 25	07 58	14 46	19 24	27 46	02 37
25 Su	04 14 15	02 19 21	07♊ 57 57	19 49	17 49	11 47	24 38	08 01	14D 46	19 25	27 48	02 25
26 Mo	04 18 11	03 20 01	23 05 46	21 22	18 58	11 39	24 51	08 03	14 47	19 26	27 50	02 14
27 Tu	04 22 08	04 20 42	07♋ 53 10	22 56	20 06	11 29	25 04	08 06	14 47	19 27	27 52	02 04
28 We	04 26 05	05 21 25	22 13 16	24 30	21 15	11 19	25 17	08 08	14 47	19 27	27 54	01 57
29 Th	04 30 01	06 22 09	06♌ 02 56	26 04	22 24	11 07	25 31	08 11	14 47	19 28	27 56	01 51
30 Fr	04 33 58	07 22 55	19 22 33	27 38	23 33	10 55	25 44	08 13	14 47	19 29	27 58	01 49

Data for	11-01-2007
Julian Day	2454405.50
Ayanamsa	23 58 05
SVP	05 ♓ 09 08
☽ ☊ Mean	03 ♓ 34 R

● ◐ PHASES ○ ◑

Day	Time	Phase	Position
01	21:18	◑	09♌04
09	23:03	●	17♏10
17	22:33	◐	25♒12
24	14:30	○	01♊55

LAST ASPECT Day	h m	☽ INGRESS Day	h m	Sign
31	17:13	01	04:48	♌
03	07:14	03	12:45	♍
05	18:10	05	23:47	♎
08	06:46	08	12:19	♏
10	03:20	11	00:59	♐
13	07:54	13	13:01	♑
15	09:20	15	23:31	♒
18	02:52	18	07:15	♓
20	07:27	20	11:25	♈
22	08:40	22	12:19	♉
23	18:53	24	11:29	♊
26	07:38	26	11:07	♋
28	04:22	28	13:23	♌
30	17:25	30	19:45	♍

DECLINATION

Day	☉	☽	☿	♀	♂	♃	♄	♅	♆	♇
01 Th	14S14	23N44	07S49	03N41	23N57	22S46	10N28	06S39	15S19	16S58
02 Fr	14 33	19 37	07 36	03 22	23 58	22 47	10 27	06 39	15 19	16 58
03 Sa	14 52	14 42	07 28	03 02	24 00	22 48	10 25	06 39	15 19	16 58
04 Su	15 11	09 18	07 26	02 42	24 02	22 49	10 24	06 40	15 19	16 59
05 Mo	15 30	03 39	07 29	02 22	24 04	22 50	10 22	06 40	15 19	16 59
06 Tu	15 48	02S02	07 37	02 01	24 05	22 51	10 21	06 40	15 19	16 59
07 We	16 06	07 35	07 50	01 40	24 07	22 52	10 19	06 41	15 19	16 59
08 Th	16 24	12 50	08 06	01 19	24 10	22 53	10 18	06 41	15 19	17 00
09 Fr	16 41	17 36	08 25	00 58	24 12	22 54	10 16	06 41	15 19	17 00
10 Sa	16 58	21 41	08 48	00 36	24 14	22 54	10 15	06 41	15 19	17 00
11 Su	17 15	24 55	09 13	00 14	24 16	22 55	10 14	06 42	15 19	17 00
12 Mo	17 32	27 04	09 40	00S08	24 19	22 56	10 12	06 42	15 19	17 01
13 Tu	17 48	27 59	10 09	00 30	24 21	22 57	10 11	06 42	15 19	17 01
14 We	18 04	27 34	10 39	00 53	24 24	22 58	10 10	06 42	15 18	17 01
15 Th	18 20	25 48	11 10	01 16	24 27	22 59	10 09	06 42	15 18	17 01
16 Fr	18 35	22 46	11 42	01 38	24 30	23 00	10 08	06 43	15 18	17 02
17 Sa	18 50	18 35	12 15	02 02	24 33	23 00	10 07	06 43	15 18	17 02
18 Su	19 05	13 26	12 48	02 25	24 36	23 01	10 06	06 43	15 18	17 02
19 Mo	19 19	07 31	13 21	02 48	24 39	23 02	10 05	06 43	15 18	17 02
20 Tu	19 33	01 05	13 55	03 11	24 42	23 03	10 04	06 43	15 17	17 03
21 We	19 47	05N37	14 28	03 35	24 45	23 03	10 03	06 43	15 17	17 03
22 Th	20 00	12 13	15 01	03 59	24 49	23 04	10 02	06 43	15 17	17 03
23 Fr	20 13	18 15	15 34	04 22	24 52	23 05	10 01	06 43	15 17	17 03
24 Sa	20 25	23 12	16 06	04 46	24 56	23 06	10 00	06 43	15 17	17 04
25 Su	20 38	26 33	16 38	05 10	25 00	23 06	09 59	06 43	15 16	17 04
26 Mo	20 49	27 56	17 09	05 34	25 03	23 07	09 58	06 43	15 16	17 04
27 Tu	21 01	27 16	17 40	05 58	25 07	23 07	09 58	06 43	15 16	17 04
28 We	21 12	24 47	18 10	06 22	25 11	23 08	09 57	06 43	15 15	17 04
29 Th	21 22	20 53	18 39	06 46	25 15	23 09	09 56	06 43	15 15	17 05
30 Fr	21 33	16 02	19 08	07 10	25 19	23 09	09 56	06 42	15 15	17 05

ASPECTARIAN

01 06:15 ☽ # ♃
23:00 ☿ SD

02 13:20 ☽ # ♇
16:16 ☽ ☍ ♆
17:36 ☽ △ ♃
21:07 ☽ # ♆
23:16 ☽ # ☉

03 00:15 ☽ ⚹ ☿
07:14 ☽ △ ♇
19:14 ☽ ∥ ♄
04 01:46 ☽ ☌ ♄
07:59 ☽ # ☿
10:38 ☉ ∥ ♆
11:18 ☽ # ♅
11:19 ☉ △ ♂
11:24 ☽ ⚹ ♂
11:24 ☽ ⚹ ☉
17:51 ☽ ☍ ♅

05 04:45 ☽ □ ♃
05:49 ☽ ∥ ♀
17:04 ☽ ☌ ♀
18:10 ☽ □ ♇
23:59 ☽ # ♀
06 06:00 ♀ □ ♇
20:03 ☽ ∥ ♅
23:52 ☽ □ ♂

07 01:10 ☽ ∥ ☿
12:16 ☽ # ♄
14:35 ☽ △ ♆
16:49 ☉ △ ♅
18:01 ☽ ⚹ ♃
08 05:24 ☽ ☌ ☿
06:46 ☽ ⚹ ♇
12:11 ☽ ∥ ♆
18:50 ☽ ∥ ☉
20:48 ☽ ∥ ♇
21:05 ♀ ♎
22:26 ☿ ⚹ ♇

09 02:41 ☽ ⚹ ♄
13:04 ☽ △ ♂
18:25 ☽ △ ♅
10 02:43 ☉ ∥ ♇
03:20 ☽ □ ♆
08:18 ☽ ∥ ♃
18:31 ☽ # ♂

11 06:09 ☽ ⚹ ♀
08:41 ☿ ♏
15:38 ☽ □ ♄
12 01:42 ☉ □ ♆
06:48 ☽ □ ♅
15:42 ☽ ⚹ ♆
21:10 ☽ ☌ ♃

13 01:58 ☿ # ♄
07:54 ☽ ☌ ♇
19:13 ☽ ⚹ ☿
23:59 ☽ □ ♀
14 03:42 ☽ △ ♄
13:33 ☽ ☍ ♂
18:10 ☽ ⚹ ♅

15 08:26 ♂ SR
09:20 ☽ ⚹ ☉
11:44 ☽ # ♂
22:30 ☽ ∥ ♃
16 12:43 ☽ □ ☿
15:33 ☽ △ ♀
22:47 ☽ ∥ ☉
23:22 ☿ ⚹ ♄
17 07:43 ☽ ∥ ♇
11:49 ☽ ☌ ♆
15:48 ☽ ∥ ♆
18:40 ☽ ⚹ ♃

18 02:30 ☽ ∥ ☿
02:52 ☽ ⚹ ♇
13:58 ☽ # ♄
21:00 ☽ ☍ ♄

19 02:22 ☽ △ ☿
03:08 ☽ ∥ ♅
05:10 ☽ △ ♂
09:25 ☽ ☌ ♅
16:45 ☽ ∥ ♀

20 00:30 ☽ □ ♃
02:19 ♀ □ ♂
02:53 ☿ △ ♂
07:21 ☽ △ ☉
07:27 ☽ □ ♇
16:17 ☽ # ♀

21 03:56 ☽ # ♅
07:38 ☽ □ ♂
10:13 ☽ ☍ ♀
15:56 ☽ ∥ ♄
17:58 ☿ △ ♅
19:17 ☽ ⚹ ♆

22 02:44 ☽ △ ♃
08:40 ☽ △ ♇
11:34 ☿ ∥ ♆
11:49 ☽ # ♆
11:51 ☽ # ☿
16:50 ☉ ♐
18:59 ☽ # ♇

23 00:51 ☽ △ ♄
07:17 ☽ ⚹ ♂
09:11 ☽ # ☉
11:38 ☽ ⚹ ♅
16:19 ☽ ☍ ☿
18:53 ☽ □ ♆
23:26 ☽ # ♃

24 10:16 ♅ SD
11:03 ☽ ∥ ♂
17:50 ☿ □ ♆

25 00:04 ☽ □ ♄
10:45 ☽ □ ♅
16:52 ☽ △ ♀
18:08 ☽ △ ♆
19:44 ☿ ∥ ♇

26 02:51 ☽ ☍ ♃
07:38 ☽ ☍ ♇
09:55 ♀ △ ♆

27 00:21 ☽ ⚹ ♄
05:53 ☽ ☌ ♂
11:26 ☽ △ ♅
20:59 ☽ ∥ ♂
22:12 ☽ □ ♀

28 04:22 ☽ △ ☿
11:01 ☽ # ♃
20:58 ♀ ∥ ♅
21:27 ☽ # ☉
29 00:37 ☽ △ ☉
10:29 ☽ # ☿
19:10 ☽ # ♇

30 00:12 ☽ ☍ ♆
03:40 ☽ # ♆
08:27 ☽ ⚹ ♀
11:58 ☽ △ ♃
15:58 ☽ △ ♇
17:25 ☽ □ ☿
20:23 ☉ □ ♄

⚷ Chiron

01 Dec.	11 S 06
02	10♒34
05	10 37
08	10 40
11	10 44
14	10 49
17	10 54
20	11 00
23	11 07
26	11 14
29	11 21

December 2007

Day	S. T.	☉	☽	☿	♀	♂	♃	♄	♅	♆	♇	☊ True
	h m s	° ' "	° ' "	° '	° '	° '	° '	° '	° '	° '	° '	° '
01 Sa	04 37 54	08♐23 42	02♍15 09	29♏12	24♎42	10♋R42	25♐57	08♍15	14♓48	19♒30	28♐00	01♓R47
02 Su	04 41 51	09 24 31	14 45 14	00♐46	25 51	10 29	26 11	08 17	14 48	19 31	28 02	01 47
03 Mo	04 45 47	10 25 21	26 57 59	02 20	27 01	10 14	26 24	08 19	14 48	19 32	28 05	01 46
04 Tu	04 49 44	11 26 13	08♎58 32	03 54	28 11	09 59	26 38	08 21	14 49	19 33	28 07	01 43
05 We	04 53 40	12 27 05	20 51 42	05 28	29 21	09 43	26 51	08 22	14 49	19 35	28 09	01 38
06 Th	04 57 37	13 28 00	02♏41 43	07 02	00♏31	09 26	27 05	08 24	14 50	19 36	28 11	01 30
07 Fr	05 01 34	14 28 55	14 31 59	08 36	01 41	09 08	27 18	08 25	14 50	19 37	28 13	01 19
08 Sa	05 05 30	15 29 52	26 25 07	10 10	02 51	08 50	27 32	08 27	14 51	19 38	28 15	01 06
09 Su	05 09 27	16 30 50	08♐22 55	11 44	04 02	08 31	27 45	08 28	14 52	19 39	28 17	00 52
10 Mo	05 13 23	17 31 49	20 26 37	13 18	05 12	08 12	27 59	08 29	14 53	19 41	28 20	00 39
11 Tu	05 17 20	18 32 49	02♑37 06	14 52	06 23	07 52	28 13	08 30	14 53	19 42	28 22	00 26
12 We	05 21 16	19 33 49	14 55 15	16 26	07 34	07 31	28 26	08 31	14 54	19 43	28 24	00 17
13 Th	05 25 13	20 34 50	27 22 08	18 00	08 45	07 10	28 40	08 32	14 55	19 45	28 26	00 10
14 Fr	05 29 09	21 35 52	09♒59 18	19 34	09 56	06 48	28 54	08 32	14 56	19 46	28 28	00 06
15 Sa	05 33 06	22 36 55	22 48 47	21 09	11 07	06 26	29 07	08 33	14 57	19 47	28 31	00 04
16 Su	05 37 03	23 37 58	05♓53 08	22 43	12 18	06 04	29 21	08 33	14 58	19 49	28 33	00D 04
17 Mo	05 40 59	24 39 01	19 15 06	24 17	13 29	05 41	29 35	08 34	14 59	19 50	28 35	00 04
18 Tu	05 44 56	25 40 05	02♈57 13	25 52	14 41	05 18	29 48	08 34	15 01	19 52	28 37	00R 04
19 We	05 48 52	26 41 09	17 00 59	27 27	15 52	04 55	00♑02	08 34	15 02	19 53	28 39	00 02
20 Th	05 52 49	27 42 13	01♉26 06	29 02	17 04	04 31	00 16	08R 34	15 03	19 55	28 42	29♒57
21 Fr	05 56 45	28 43 18	16 09 41	00♑37	18 16	04 08	00 30	08 34	15 04	19 56	28 44	29 51
22 Sa	06 00 42	29 44 23	01♊05 59	02 12	19 28	03 44	00 43	08 34	15 06	19 58	28 46	29 43
23 Su	06 04 38	00♑45 29	16 06 50	03 48	20 40	03 20	00 57	08 33	15 07	20 00	28 48	29 33
24 Mo	06 08 35	01 46 34	01♋02 42	05 23	21 52	02 56	01 11	08 33	15 09	20 01	28 51	29 24
25 Tu	06 12 32	02 47 41	15 44 21	06 59	23 04	02 33	01 25	08 32	15 10	20 03	28 53	29 16
26 We	06 16 28	03 48 47	00♌04 19	08 35	24 16	02 09	01 38	08 32	15 12	20 05	28 55	29 09
27 Th	06 20 25	04 49 54	13 57 59	10 11	25 28	01 46	01 52	08 31	15 13	20 06	28 57	29 05
28 Fr	06 24 21	05 51 02	27 23 49	11 48	26 41	01 23	02 06	08 30	15 15	20 08	28 59	29 03
29 Sa	06 28 18	06 52 10	10♍23 06	13 25	27 53	01 00	02 20	08 29	15 17	20 10	29 02	29D 03
30 Su	06 32 14	07 53 19	22 59 03	15 02	29 05	00 37	02 33	08 28	15 18	20 12	29 04	29 03
31 Mo	06 36 11	08 54 27	05♎16 09	16 39	00♐18	00 15	02 47	08 27	15 20	20 14	29 06	29 04

Data for	12-01-2007
Julian Day	2454435.50
Ayanamsa	23 58 10
SVP	05 ♓ 09 04
☽ ☊ Mean	01 ♓ 59 R

● ◐ PHASES ○ ◑

01	12:44	◑	08♍56
09	17:41	●	17♐16
17	10:17	◐	25♓05
24	01:16	○	01♋50
31	07:52	◑	09♎15

LAST ASPECT ☽ Day	h m	INGRESS Day	h m	
03	02:12	03	06:02	♎
05	14:49	05	18:32	♏
07	10:18	08	07:12	♐
10	15:37	10	18:51	♑
11	23:58	13	05:02	♒
15	11:51	15	13:15	♓
17	18:27	17	18:53	♈
19	19:33	19	21:38	♉
21	06:06	21	22:14	♊
23	20:26	23	22:19	♋
25	13:18	25	23:53	♌
28	02:54	28	04:45	♍
30	13:10	30	13:38	♎

DECLINATION

Day	☉	☽	☿	♀	♂	♃	♄	♅	♆	♇
01 Sa	21S42	10N38	19S35	07S34	25N23	23S10	09N55	06S42	15S15	17S05
02 Su	21 52	04 57	20 02	07 57	25 27	23 10	09 55	06 42	15 14	17 05
03 Mo	22 01	00S46	20 28	08 21	25 31	23 11	09 54	06 42	15 14	17 05
04 Tu	22 09	06 22	20 53	08 45	25 35	23 11	09 54	06 42	15 14	17 06
05 We	22 17	11 41	21 17	09 09	25 40	23 11	09 53	06 41	15 13	17 06
06 Th	22 25	16 32	21 40	09 32	25 44	23 12	09 53	06 41	15 13	17 06
07 Fr	22 32	20 47	22 02	09 56	25 48	23 12	09 53	06 41	15 12	17 06
08 Sa	22 39	24 12	22 22	10 19	25 52	23 13	09 52	06 41	15 12	17 06
09 Su	22 45	26 36	22 42	10 42	25 56	23 13	09 52	06 40	15 12	17 06
10 Mo	22 51	27 48	23 01	11 05	26 00	23 13	09 52	06 40	15 11	17 07
11 Tu	22 57	27 40	23 18	11 28	26 04	23 14	09 52	06 40	15 11	17 07
12 We	23 02	26 10	23 34	11 51	26 07	23 14	09 52	06 39	15 11	17 07
13 Th	23 06	23 22	23 49	12 14	26 11	23 14	09 52	06 39	15 10	17 07
14 Fr	23 10	19 26	24 03	12 36	26 15	23 14	09 52	06 38	15 10	17 07
15 Sa	23 14	14 31	24 16	12 58	26 18	23 15	09 52	06 38	15 09	17 07
16 Su	23 17	08 52	24 27	13 20	26 22	23 15	09 52	06 38	15 09	17 07
17 Mo	23 20	02 43	24 37	13 42	26 25	23 15	09 52	06 37	15 08	17 08
18 Tu	23 22	03N43	24 46	14 03	26 28	23 15	09 52	06 37	15 08	17 08
19 We	23 24	10 07	24 53	14 24	26 31	23 15	09 52	06 36	15 07	17 08
20 Th	23 25	16 09	24 59	14 45	26 34	23 15	09 52	06 36	15 07	17 08
21 Fr	23 26	21 23	25 04	15 06	26 37	23 15	09 53	06 35	15 06	17 08
22 Sa	23 26	25 20	25 07	15 26	26 40	23 15	09 53	06 34	15 06	17 08
23 Su	23 26	27 33	25 09	15 46	26 42	23 15	09 53	06 34	15 05	17 08
24 Mo	23 26	27 46	25 09	16 05	26 44	23 15	09 54	06 33	15 05	17 08
25 Tu	23 25	26 00	25 08	16 25	26 46	23 15	09 54	06 33	15 04	17 08
26 We	23 23	22 34	25 05	16 43	26 48	23 15	09 55	06 32	15 04	17 09
27 Th	23 21	17 55	25 01	17 02	26 50	23 15	09 55	06 31	15 03	17 09
28 Fr	23 19	12 31	24 56	17 20	26 52	23 15	09 56	06 31	15 03	17 09
29 Sa	23 16	06 44	24 49	17 38	26 53	23 15	09 56	06 30	15 02	17 09
30 Su	23 12	00 52	24 40	17 55	26 54	23 15	09 57	06 29	15 02	17 09
31 Mo	23 08	04S53	24 30	18 12	26 56	23 14	09 58	06 29	15 01	17 09

ASPECTARIAN

01 03:05 ☽ ∥ ♄
11:28 ☽ ☌ ♄
12:14 ☽ ‡ ♀
12:21 ☿ ♐
15:52 ☽ ✱ ♂
16:42 ☽ ‡ ♅
02 00:05 ☽ ☍ ♅
08:15 ♀ ✱ ♃
22:52 ☽ □ ♃
03 02:12 ☽ □ ♇
12:17 ☽ ✱ ☿
22:34 ♀ ✱ ♇
04 01:29 ☽ ∥ ♅
01:59 ☽ □ ♂
05:25 ☽ ✱ ☉
11:27 ☽ ∥ ♀
15:47 ☽ ‡ ♄
21:24 ☽ △ ♆
05 12:23 ☽ ✱ ♃
13:29 ♀ ♏
14:49 ☽ ✱ ♇
17:14 ☽ ∥ ♆
19:05 ☽ ☌ ♀
06 03:00 ☽ ∥ ♇
11:35 ☽ ✱ ♄
13:20 ☽ △ ♂
20:54 ♀ ‡ ♄
21:17 ☿ □ ♄
07 00:37 ☽ △ ♅
08:35 ☉ □ ♅
08:58 ☽ ∥ ☿
10:18 ☽ □ ♆
11:58 ☽ ∥ ☉
16:22 ☽ ∥ ♃
08 15:49 ☽ ‡ ♂
09 00:10 ☽ □ ♄
04:00 ♂ ✱ ♄
06:18 ☿ ∥ ☉
07:41 ☽ ☌ ☿
12:56 ☽ □ ♅
22:29 ☽ ✱ ♆
10 15:10 ☽ ☌ ♃
15:37 ☽ ☌ ♇
17:41 ☿ ∥ ♃
11 00:25 ☿ □ ♅
08:09 ☽ ✱ ♀
09:59 ☽ ☍ ♂
11:31 ☽ △ ♄
19:26 ♃ ☌ ♇
23:16 ♀ △ ♂
23:58 ☽ ✱ ♅
12 00:24 ☽ ‡ ♂
03:48 ☉ ✱ ♆
19:33 ♀ ✱ ♄
20:56 ☽ ∥ ☿
13 00:55 ☽ ∥ ♃
01:46 ☽ ∥ ☉
23:52 ☽ □ ♀
14 03:02 ☿ ✱ ♆
11:46 ☽ ∥ ♇
18:22 ☽ ☌ ♆
20:28 ☽ ✱ ☿
21:05 ☽ ∥ ♆
23:36 ☽ ✱ ☉
15 03:28 ☉ ∥ ♃
06:26 ☽ ∥ ♀
10:33 ☽ ✱ ♇
11:51 ☽ ✱ ♃
19:58 ☽ ‡ ♄
16 00:19 ☽ △ ♂
04:51 ☽ ☍ ♄
08:57 ☽ ∥ ♅
11:43 ☽ △ [illegible]
16:24 ☽ ☌ ♅
17 10:03 ☽ □ ☿
15:27 ☿ ☌ ☉
16:27 ☽ □ ♇
18:27 ☽ □ ♃
18 03:56 ☽ □ ♂
06:43 ♀ △ ♅
10:47 ☽ ‡ ♅
20:11 ♃ ♑
23:04 ☽ ∥ ♄
19 04:50 ☽ ✱ ♆
14:11 ♄ SR
17:23 ☽ △ ☉
17:52 ☽ ‡ ♀
18:49 ☿ ☌ ♇
19:28 ☽ △ ♇
19:33 ☽ △ ☿
19:42 ☽ ‡ ♆
22:03 ☽ △ ♃
20 04:12 ☽ ‡ ♇
04:56 ☽ ✱ ♂
11:41 ☽ △ ♄
14:43 ☿ ♑
21:54 ☿ ☌ ♃
22:14 ☽ ✱ ♅
21 00:16 ☉ ☌ ♇
00:59 ♀ ∥ ♆
03:43 ☽ ☍ ♀
06:06 ☽ □ ♆
10:16 ☽ ‡ ♃
11:21 ☽ ‡ ☉
22:22 ☽ ‡ ☿
22 06:08 ☉ ♑
10:22 ♀ □ ♆
11:00 [illegible] ♄
12:04 ☽ ∥ ♂
18:29 ☿ ☍ ♂
22:24 ☽ □ ♅
23 05:56 ☉ ☌ ♃
06:14 ☽ △ ♆
20:26 ☽ ☍ ♇
24 00:14 ☽ ☍ ♃
02:59 ☽ ☌ ♂
07:54 ☽ ☍ ☿
12:11 ☽ ✱ ♄
16:13 ☽ ∥ ♂
19:47 ☉ ☍ ♂
23:03 ☽ △ ♅
25 07:09 ☽ ‡ ☿
13:18 ☽ △ ♀
18:59 ☽ ‡ ☉
19:51 ☽ ‡ ♃
23:10 ☿ △ ♄
26 19:52 ♂ ☍ ♃
27 03:36 ☽ ‡ ♇
03:54 ☽ ‡ ♀
09:06 ♀ ∥ ♇
10:54 ☽ ☍ ♆
13:03 ☽ ‡ ♆
22:34 ☽ □ ♀
28 02:54 ☽ △ ♇
07:04 ☽ ✱ ♂
08:45 ☽ △ ♃
10:51 ☽ ∥ ♄
16:52 ☽ △ ☉
20:27 ☽ ☌ ♄
29 00:58 ☽ ‡ ♅
06:32 ☽ △ ☿
07:28 ☉ ∥ ♃
09:16 ☽ ☍ ♅
30 04:10 ☿ ✱ ♅
11:51 ☽ □ ♇
13:10 ☽ ✱ ♀
13:24 ☉ △ ♄
14:25 ☽ □ ♂
18:02 ♀ ♐
19:01 ☽ □ ♃
31 06:50 ☽ ∥ ♅
16:01 ♂ ∥R
22:18 ☽ ‡ ♄

⚷ Chiron

01 Dec.	11 S 05
02	11♒29
05	11 38
08	11 47
11	11 56
14	12 06
17	12 16
20	12 27
23	12 38
26	12 49
29	13 01

January 2008

Day	S. T. h m s	☉ ° ' "	☽ ° ' "	☿ ° '	♀ ° '	♂ ° '	♃ ° '	♄ ° '	♅ ° '	♆ ° '	♇ ° '	☊ True ° '
01 Tu	06 40 08	09♑ 55 37	17♎ 19 25	18♑ 16	01♐ 31	29♊ R53	03♑ 01	08♍ R26	15♓ 22	20♒ 15	29♐ 08	29♒ R04
02 We	06 44 04	10 56 47	29 13 58	19 54	02 43	29 31	03 15	08 24	15 24	20 17	29 10	29 02
03 Th	06 48 01	11 57 57	11♏ 04 45	21 32	03 56	29 10	03 28	08 23	15 26	20 19	29 12	28 58
04 Fr	06 51 57	12 59 07	22 56 10	23 10	05 09	28 50	03 42	08 21	15 28	20 21	29 15	28 52
05 Sa	06 55 54	14 00 18	04♐ 51 55	24 47	06 22	28 29	03 56	08 19	15 30	20 23	29 17	28 45
06 Su	06 59 50	15 01 29	16 54 52	26 25	07 35	28 10	04 09	08 17	15 32	20 25	29 19	28 36
07 Mo	07 03 47	16 02 40	29 07 04	28 03	08 48	27 51	04 23	08 15	15 34	20 27	29 21	28 28
08 Tu	07 07 43	17 03 51	11♑ 29 45	29 41	10 01	27 33	04 37	08 13	15 36	20 29	29 23	28 21
09 We	07 11 40	18 05 01	24 03 35	01♒ 18	11 14	27 15	04 50	08 11	15 38	20 31	29 25	28 15
10 Th	07 15 37	19 06 12	06♒ 48 49	02 55	12 27	26 58	05 04	08 09	15 40	20 33	29 28	28 11
11 Fr	07 19 33	20 07 22	19 45 30	04 31	13 40	26 42	05 18	08 07	15 43	20 35	29 30	28 10
12 Sa	07 23 30	21 08 32	02♓ 53 45	06 06	14 54	26 27	05 31	08 04	15 45	20 37	29 32	28D 10
13 Su	07 27 26	22 09 41	16 13 55	07 40	16 07	26 12	05 45	08 02	15 47	20 39	29 34	28 11
14 Mo	07 31 23	23 10 50	29 46 32	09 13	17 20	25 58	05 58	07 59	15 50	20 41	29 36	28 12
15 Tu	07 35 19	24 11 57	13♈ 32 12	10 45	18 33	25 45	06 12	07 56	15 52	20 43	29 38	28 14
16 We	07 39 16	25 13 05	27 31 05	12 14	19 47	25 33	06 25	07 54	15 54	20 45	29 40	28R 14
17 Th	07 43 12	26 14 11	11♉ 42 35	13 41	21 00	25 21	06 38	07 51	15 57	20 47	29 42	28 12
18 Fr	07 47 09	27 15 17	26 04 46	15 05	22 14	25 11	06 52	07 48	15 59	20 49	29 44	28 10
19 Sa	07 51 06	28 16 22	10♊ 34 17	16 25	23 27	25 01	07 05	07 44	16 02	20 51	29 46	28 06
20 Su	07 55 02	29 17 26	25 06 16	17 41	24 41	24 52	07 18	07 41	16 05	20 54	29 48	28 02
21 Mo	07 58 59	00♒ 18 29	09♋ 34 52	18 52	25 54	24 43	07 32	07 38	16 07	20 56	29 50	27 57
22 Tu	08 02 55	01 19 32	23 53 52	19 58	27 08	24 36	07 45	07 35	16 10	20 58	29 52	27 53
23 We	08 06 52	02 20 34	07♌ 57 38	20 57	28 21	24 29	07 58	07 31	16 12	21 00	29 54	27 50
24 Th	08 10 48	03 21 35	21 41 56	21 49	29 35	24 23	08 11	07 28	16 15	21 02	29 56	27 48
25 Fr	08 14 45	04 22 36	05♍ 04 21	22 33	00♑ 49	24 18	08 24	07 24	16 18	21 04	29 58	27D 48
26 Sa	08 18 41	05 23 35	18 04 30	23 08	02 03	24 14	08 37	07 21	16 21	21 07	30 00	27 49
27 Su	08 22 38	06 24 34	00♎ 43 50	23 34	03 16	24 11	08 50	07 17	16 23	21 09	00♑ 02	27 50
28 Mo	08 26 35	07 25 33	13 05 10	23 49	04 30	24 08	09 03	07 13	16 26	21 11	00 03	27 52
29 Tu	08 30 31	08 26 31	25 12 23	23R 53	05 44	24 06	09 16	07 09	16 29	21 13	00 05	27 53
30 We	08 34 28	09 27 28	07♏ 09 57	23 45	06 58	24 05	09 29	07 05	16 32	21 16	00 07	27 54
31 Th	08 38 24	10 28 25	19 02 41	23 27	08 11	24 D 05	09 42	07 01	16 35	21 18	00 09	27R 53

Data for	01-01-2008
Julian Day	2454466.50
Ayanamsa	23 58 15
SVP	05 ♓ 09 02
☽ ☊ Mean	00 ♓ 20 R

● ◐ PHASES ○ ◑

08	11:37	●	17♑33
15	19:46	◐	25♈02
22	13:35	○	01♌54
30	05:03	◑	09♏40

LAST ASPECT ☽ Day	h m	INGRESS Day	h m	
02	00:34	02	01:33	♏
04	00:31	04	14:14	♐
07	00:28	07	01:43	♑
08	11:37	09	11:13	♒
11	17:52	11	18:44	♓
13	23:41	14	00:24	♈
16	03:40	16	04:13	♉
18	02:06	18	06:30	♊
20	07:47	20	08:06	♋
21	10:57	22	10:21	♌
24	14:44	24	14:49	♍
26	11:33	26	22:36	♎
28	21:48	29	09:35	♏
31	08:35	31	22:08	♐

DECLINATION

Day	☉	☽	☿	♀	♂	♃	♄	♅	♆	♇
01 Tu	23S 04	10S 20	24S 19	18S 28	26N56	23S 14	09N58	06S 28	15S 00	17S 09
02 We	22 59	15 22	24 05	18 44	26 57	23 14	09 59	06 27	15 00	17 09
03 Th	22 54	19 47	23 51	19 00	26 58	23 14	10 00	06 26	14 59	17 09
04 Fr	22 48	23 26	23 34	19 15	26 58	23 13	10 01	06 26	14 59	17 09
05 Sa	22 42	26 07	23 17	19 29	26 59	23 13	10 02	06 25	14 58	17 09
06 Su	22 36	27 39	22 57	19 43	26 59	23 13	10 02	06 24	14 57	17 09
07 Mo	22 29	27 52	22 36	19 57	26 59	23 12	10 03	06 23	14 57	17 09
08 Tu	22 21	26 41	22 14	20 10	26 59	23 12	10 04	06 22	14 56	17 10
09 We	22 13	24 09	21 50	20 22	26 59	23 12	10 05	06 21	14 56	17 10
10 Th	22 05	20 24	21 25	20 34	26 59	23 11	10 06	06 20	14 55	17 10
11 Fr	21 56	15 37	20 58	20 45	26 58	23 11	10 08	06 20	14 54	17 10
12 Sa	21 47	10 03	20 30	20 56	26 58	23 10	10 09	06 19	14 54	17 10
13 Su	21 37	03 57	20 01	21 06	26 57	23 10	10 10	06 18	14 53	17 10
14 Mo	21 27	02N25	19 30	21 16	26 57	23 09	10 11	06 17	14 52	17 10
15 Tu	21 16	08 45	18 59	21 25	26 56	23 09	10 12	06 16	14 52	17 10
16 We	21 06	14 46	18 26	21 34	26 55	23 08	10 13	06 15	14 51	17 10
17 Th	20 54	20 05	17 53	21 41	26 54	23 08	10 15	06 14	14 50	17 10
18 Fr	20 43	24 19	17 19	21 49	26 54	23 07	10 16	06 13	14 50	17 10
19 Sa	20 30	27 03	16 45	21 55	26 53	23 06	10 17	06 12	14 49	17 10
20 Su	20 18	27 59	16 11	22 01	26 52	23 06	10 19	06 11	14 48	17 10
21 Mo	20 05	26 59	15 38	22 06	26 51	23 05	10 20	06 10	14 48	17 10
22 Tu	19 52	24 12	15 04	22 11	26 50	23 05	10 22	06 09	14 47	17 10
23 We	19 38	20 00	14 32	22 15	26 49	23 04	10 23	06 08	14 46	17 10
24 Th	19 24	14 48	14 01	22 19	26 48	23 03	10 25	06 07	14 46	17 10
25 Fr	19 10	09 02	13 32	22 21	26 47	23 02	10 26	06 06	14 45	17 10
26 Sa	18 55	03 02	13 05	22 24	26 46	23 02	10 28	06 04	14 44	17 10
27 Su	18 40	02S 55	12 40	22 25	26 45	23 01	10 29	06 03	14 44	17 10
28 Mo	18 25	08 36	12 19	22 26	26 44	23 00	10 31	06 02	14 43	17 10
29 Tu	18 09	13 52	12 01	22 26	26 43	22 59	10 32	06 01	14 42	17 10
30 We	17 53	18 32	11 46	22 25	26 42	22 58	10 34	06 00	14 41	17 10
31 Th	17 37	22 28	11 36	22 24	26 41	22 58	10 36	05 59	14 41	17 10

ASPECTARIAN

01 02:12 ☽ □ ☿
05:54 ☽ △ ♆
22:11 ☽ ∥ ♆
23:53 ☽ ⚹ ♇
02 00:34 ☽ △ ♂
08:17 ☽ ⚹ ♃
09:18 ☽ ∥ ♇
18:32 ☽ ⚹ ♄
19:07 ☽ ∥ ♀
21:29 ♂ ☍ ♇
03 01:58 ☽ ⚹ ☉
08:50 ☽ △ ♅
18:46 ☽ □ ♆
19:32 ☽ ∥ ☉
22:26 ☽ ∥ ♃
04 00:31 ☽ ⚹ ☿
00:57 ☽ ∥ ☿
05 03:20 ☽ ☌ ♀
04:27 ☿ ∥ ♃
06:54 ☽ □ ♄
11:09 ☽ # ♂
21:15 ☽ □ ♅
06 06:56 ☽ ⚹ ♆
12:19 ☉ ⚹ ♅
13:39 ♀ □ ♄
21:35 ☽ ☍ ♂
07 00:28 ☽ ☌ ♇
10:27 ☽ ☌ ♃
12:34 ☿ ∥ ☉
17:42 ☽ △ ♄
19:56 ☽ # ♂
08 04:46 ☿ ♒
07:54 ☽ ⚹ ♅
09 06:55 ☽ ∥ ♃
13:45 ☽ ∥ ☉
15:39 ☽ ☌ ☿
17:29 ☽ ∥ ☿
23:05 ☽ ∥ ♀
10 11:35 ☽ ⚹ ♀
16:42 ☽ ∥ ♇
11 01:31 ☽ ☌ ♆
03:13 ☽ ∥ ♆
07:46 ☿ ∥ ♀
12:29 ☽ △ ♂
17:52 ☽ ⚹ ♇
23:34 ☽ # ♄
12 04:50 ☽ ⚹ ♃
09:20 ☽ ☍ ♄
14:54 ☽ ∥ ♅
17:23 ♀ □ ♅
23:12 ☽ ☌ ♅
23:46 ☽ □ ♀
13 11:25 ☽ ⚹ ☉
17:24 ☽ □ ♂
23:41 ☽ □ ♇
14 11:02 ☽ □ ♃
13:26 ♀ ∥ ☉
14:31 ☽ # ♅
18:34 ☽ ⚹ ☿
15 05:38 ☽ ∥ ♄
09:30 ☽ △ ♀
12:24 ☽ ⚹ ♆
20:41 ☽ ⚹ ♂
16 00:22 ☽ # ♆
03:40 ☽ △ ♇
10:20 ☽ # ♇
14:34 ☽ # ☿
15:20 ☽ △ ♃
17:31 ☽ △ ♄
19:36 ♀ ⚹ ♆
17 03:40 ☽ □ ☿
03:58 ☽ # ☉
07:08 ☽ ⚹ ♅
08:30 ☽ # ♀
15:14 ☽ □ ♆
16:23 ☽ # ♃
18 02:06 ☽ △ ☉
06:46 ☿ ∥ ♇
19:20 ☽ □ ♄
21:48 ☽ ∥ ♂
19 09:03 ☽ □ ♅
10:35 ☽ △ ☿
17:02 ☽ △ ♆
23:14 ☽ ☍ ♀
23:36 ☽ ☌ ♂
20 03:11 ♀ ☍ ♂
07:47 ☽ ☍ ♇
16:44 ☉ ♒
20:32 ☽ ☍ ♃
20:46 ☽ ⚹ ♄
21 01:35 ☽ ∥ ♂
09:16 ♃ △ ♄
10:57 ☽ △ ♅
22 07:12 ☽ # ♃
12:10 ☽ # ♀
13:00 ☿ ∥ ♆
23 01:17 ☿ ☌ ♆
01:53 ☽ # ☉
13:32 ☽ # ♇
22:50 ☽ ☍ ♆
24 00:10 ☽ # ♆
00:14 ☽ ☍ ☿
03:42 ☽ # ☿
04:45 ☽ ⚹ ♂
06:56 ♀ ☌ ♇
08:06 ♀ ♑
14:44 ☽ △ ♇
15:30 ☽ △ ♀
18:18 ☽ ∥ ♄
25 04:14 ☽ ☌ ♄
06:12 ☽ △ ♃
11:51 ☽ # ♅
20:45 ☽ ☍ ♅
26 03:45 ♇ ♑
11:33 ☽ □ ♂
22:39 ☽ □ ♇
27 05:26 ☽ □ ♀
11:57 ☽ △ ☉
13:01 ☽ ∥ ♅
15:58 ☽ □ ♃
28 08:29 ☽ # ♄
15:42 ☽ ∥ ☿
16:03 ☽ △ ♆
20:31 ☿ SR
21:21 ☽ △ ☿
21:48 ☽ △ ♂
29 04:02 ☽ ∥ ♆
09:47 ☽ ⚹ ♇
16:31 ☽ ∥ ♇
20:36 ☽ ∥ ☉
23:32 ☽ ⚹ ♀
23:50 ☽ ⚹ ♄
30 02:19 ♀ △ ♄
04:46 ☽ ⚹ ♃
19:00 ☽ △ ♅
22:34 ♂ SD
23:35 ☽ ∥ ♀
31 03:24 ☽ ∥ ♃
04:34 ☽ □ ♆
08:35 ☽ □ ☿

⚷ Chiron

01 Dec. 10 S 47

01	13♒13
04	13 25
07	13 38
10	13 51
13	14 04
16	14 17
19	14 30
22	14 43
25	14 57
28	15 11
31	15 24

February 2008

07	03:56	17♒44	☉	Annular Solar Eclipse
21	03:27	01♍53	✹	Total Lunar Eclipse (mag 1.111)

Day	S. T.	☉	☽	☿	♀	♂	♃	♄	♅	♆	♇	☊ True
	h m s	° ' "	° ' "	° '	° '	° '	° '	° '	° '	° '	° '	° '
01 Fr	08 42 21	11♒29 21	00♐55 22	22♒R57	09♑25	24♊05	09♑55	06♍R57	16♓38	21♒20	00♑11	27♒R53
02 Sa	08 46 17	12 30 16	12 52 35	22 17	10 39	24 06	10 08	06 53	16 41	21 22	00 12	27 51
03 Su	08 50 14	13 31 10	24 58 22	21 27	11 53	24 08	10 20	06 49	16 44	21 25	00 14	27 49
04 Mo	08 54 10	14 32 04	07♑16 01	20 29	13 07	24 11	10 33	06 45	16 47	21 27	00 16	27 47
05 Tu	08 58 07	15 32 56	19 48 03	19 24	14 21	24 14	10 45	06 40	16 50	21 29	00 18	27 46
06 We	09 02 04	16 33 48	02♒35 56	18 15	15 35	24 18	10 58	06 36	16 53	21 31	00 19	27 44
07 Th	09 06 00	17 34 38	15 40 11	17 03	16 49	24 23	11 10	06 31	16 56	21 34	00 21	27 44
08 Fr	09 09 57	18 35 27	29 00 12	15 51	18 03	24 29	11 23	06 27	16 59	21 36	00 23	27D 44
09 Sa	09 13 53	19 36 15	12♓34 33	14 40	19 17	24 35	11 35	06 23	17 02	21 38	00 24	27 45
10 Su	09 17 50	20 37 02	26 21 10	13 32	20 31	24 41	11 47	06 18	17 05	21 41	00 26	27 46
11 Mo	09 21 46	21 37 47	10♈17 40	12 28	21 45	24 49	11 59	06 13	17 09	21 43	00 27	27 46
12 Tu	09 25 43	22 38 30	24 21 33	11 31	22 59	24 57	12 12	06 09	17 12	21 45	00 29	27 47
13 We	09 29 39	23 39 12	08♉30 25	10 41	24 13	25 06	12 24	06 04	17 15	21 47	00 30	27 47
14 Th	09 33 36	24 39 52	22 41 54	09 58	25 27	25 15	12 36	05 59	17 18	21 50	00 32	27R 47
15 Fr	09 37 33	25 40 30	06♊53 41	09 22	26 41	25 25	12 47	05 55	17 21	21 52	00 33	27 47
16 Sa	09 41 29	26 41 07	21 03 26	08 55	27 55	25 36	12 59	05 50	17 25	21 54	00 35	27 46
17 Su	09 45 26	27 41 42	05♋08 47	08 36	29 09	25 47	13 11	05 45	17 28	21 56	00 36	27 45
18 Mo	09 49 22	28 42 15	19 07 12	08 24	00♒24	25 58	13 23	05 41	17 31	21 59	00 37	27 45
19 Tu	09 53 19	29 42 47	02♌56 10	08 19	01 38	26 11	13 34	05 36	17 35	22 01	00 39	27 44
20 We	09 57 15	00♓43 16	16 33 12	08 D 22	02 52	26 23	13 46	05 31	17 38	22 03	00 40	27 44
21 Th	10 01 12	01 43 44	29 56 10	08 31	04 06	26 37	13 57	05 26	17 41	22 06	00 41	27 44
22 Fr	10 05 08	02 44 11	13♍03 31	08 46	05 20	26 50	14 08	05 21	17 45	22 08	00 43	27 43
23 Sa	10 09 05	03 44 36	25 54 34	09 08	06 34	27 05	14 20	05 16	17 48	22 10	00 44	27 43
24 Su	10 13 02	04 44 59	08♎29 38	09 34	07 48	27 19	14 31	05 12	17 51	22 12	00 45	27 42
25 Mo	10 16 58	05 45 21	20 50 02	10 06	09 02	27 35	14 42	05 07	17 55	22 15	00 46	27 41
26 Tu	10 20 55	06 45 41	02♏58 07	10 42	10 17	27 50	14 53	05 02	17 58	22 17	00 48	27 40
27 We	10 24 51	07 46 00	14 57 03	11 22	11 31	28 06	15 04	04 57	18 01	22 19	00 49	27 39
28 Th	10 28 48	08 46 17	26 50 43	12 07	12 45	28 23	15 15	04 52	18 05	22 21	00 50	27 39
29 Fr	10 32 44	09 46 33	08♐43 33	12 55	13 59	28 40	15 25	04 47	18 08	22 23	00 51	27D 39

Data for	02-01-2008
Julian Day	2454497.50
Ayanamsa	23 58 21
SVP	05♓08 59
☽ ☊ Mean	28♒42 R

● ◐ PHASES ○ ◑

07	03:45	☉	17♒44
14	03:34	◐	24♉49
21	03:31	✹	01♍53
29	02:18	◑	09♐52

LAST ASPECT Day	h m	☽ INGRESS Day	h m	
02	22:21	03	09:52	♑
04	18:20	05	19:10	♒
07	15:51	08	01:47	♓
09	21:06	10	06:18	♈
12	01:01	12	09:35	♉
14	05:06	14	12:20	♊
16	10:18	16	15:13	♋
17	21:14	18	18:52	♌
20	17:53	21	00:07	♍
23	02:15	23	07:45	♎
25	13:35	25	18:06	♏
27	14:54	28	06:23	♐

DECLINATION

Day	☉	☽	☿	♀	♂	♃	♄	♅	♆	♇
01 Fr	17S20	25S28	11S30	22S22	26N40	22S57	10N37	05S58	14S40	17S10
02 Sa	17 03	27 22	11 28	22 20	26 39	22 56	10 39	05 56	14 39	17 10
03 Su	16 46	28 01	11 31	22 16	26 38	22 55	10 41	05 55	14 39	17 09
04 Mo	16 28	27 18	11 37	22 13	26 37	22 54	10 43	05 54	14 38	17 09
05 Tu	16 11	25 11	11 48	22 08	26 37	22 53	10 44	05 53	14 37	17 09
06 We	15 52	21 45	12 01	22 03	26 36	22 52	10 46	05 52	14 36	17 09
07 Th	15 34	17 11	12 17	21 57	26 35	22 51	10 48	05 50	14 36	17 09
08 Fr	15 15	11 42	12 35	21 50	26 34	22 50	10 50	05 49	14 35	17 09
09 Sa	14 56	05 35	12 55	21 43	26 33	22 49	10 51	05 48	14 34	17 09
10 Su	14 37	00N52	13 15	21 35	26 32	22 48	10 53	05 47	14 34	17 09
11 Mo	14 18	07 22	13 36	21 27	26 31	22 47	10 55	05 45	14 33	17 09
12 Tu	13 58	13 32	13 57	21 18	26 31	22 46	10 57	05 44	14 32	17 09
13 We	13 38	19 03	14 18	21 08	26 30	22 45	10 59	05 43	14 31	17 09
14 Th	13 18	23 32	14 38	20 58	26 29	22 44	11 01	05 42	14 31	17 09
15 Fr	12 58	26 36	14 56	20 47	26 28	22 43	11 02	05 40	14 30	17 09
16 Sa	12 37	27 58	15 14	20 35	26 27	22 42	11 04	05 39	14 29	17 09
17 Su	12 16	27 32	15 30	20 23	26 27	22 41	11 06	05 38	14 28	17 09
18 Mo	11 56	25 20	15 45	20 10	26 26	22 40	11 08	05 36	14 28	17 09
19 Tu	11 34	21 39	15 58	19 57	26 25	22 39	11 10	05 35	14 27	17 08
20 We	11 13	16 51	16 10	19 43	26 24	22 38	11 12	05 34	14 26	17 08
21 Th	10 52	11 18	16 20	19 28	26 24	22 37	11 14	05 33	14 26	17 08
22 Fr	10 30	05 22	16 28	19 13	26 23	22 36	11 16	05 31	14 25	17 08
23 Sa	10 08	00S40	16 35	18 57	26 22	22 35	11 17	05 30	14 24	17 08
24 Su	09 46	06 33	16 40	18 41	26 21	22 34	11 19	05 29	14 23	17 08
25 Mo	09 24	12 03	16 43	18 24	26 20	22 32	11 21	05 27	14 23	17 08
26 Tu	09 02	17 00	16 45	18 07	26 19	22 31	11 23	05 26	14 22	17 08
27 We	08 39	21 14	16 46	17 49	26 19	22 30	11 25	05 25	14 21	17 08
28 Th	08 17	24 34	16 45	17 31	26 18	22 29	11 27	05 23	14 21	17 08
29 Fr	07 54	26 51	16 42	17 12	26 17	22 28	11 28	05 22	14 20	17 08

ASPECTARIAN

01	11:34	♀ ☌ ♃
	12:04	☽ □ ♄
	13:06	☽ # ♂
	15:04	☉ ∥ ♇
	23:11	☽ ✶ ☉
02	07:37	☽ □ ♅
	16:57	☽ ✶ ♆
	17:31	☽ ✶ ☿
	22:21	☽ ☍ ♂
03	00:56	☿ ☌ ♆
	10:21	☽ ☌ ♇
	23:01	☽ △ ♄
04	06:26	☽ ☌ ♃
	09:37	☽ # ♂
	12:30	☽ ☌ ♀
	18:20	☽ ✶ ♅
05	17:06	☽ ∥ ♃
	22:15	☽ ∥ ♀
06	18:19	☿ ☌ ☉
07	00:10	☽ ∥ ♇
	02:18	☽ ☌ ☿
	02:23	♀ ✶ ♅
	07:57	☽ ∥ ☉
	10:41	☽ ☌ ♆
	11:49	☽ ∥ ♆
	15:51	☽ △ ♂
	20:44	☽ ∥ ☿
08	02:27	☽ ✶ ♇
	03:34	☽ # ♄
	13:09	☽ ☍ ♄
	22:14	☽ ✶ ♃
	23:12	☽ ∥ ♅
09	07:50	☽ ☌ ♅
	12:53	☽ ✶ ♀
	21:06	☽ □ ♂
10	04:45	☉ ∥ ♆
	07:03	☽ □ ♇
	18:02	☽ # ♅
11	02:04	☉ ☌ ♆
	02:57	☽ □ ♃
	03:29	☽ ✶ ☿
	13:38	☽ ∥ ♄
	19:33	☽ ✶ ♆
	20:51	☽ ✶ ☉
	21:26	☽ □ ♀
12	00:27	☿ ∥ ☉
	01:01	☽ ✶ ♂
	01:38	☽ # ☉
	01:46	☽ # ☿
	04:03	☽ # ♆
	10:25	☽ △ ♇
	15:14	☽ # ♇
	19:54	☽ △ ♄
13	03:29	☽ □ ☿
	06:40	☽ △ ♃
	09:59	☽ # ♀
	14:51	☽ ✶ ♅
	15:42	☿ ∥ ♆
	19:15	☽ # ♃
	22:31	☽ □ ♆
14	05:06	☽ △ ♆
	16:39	☉ △ ♂
	22:21	☽ □ ♄
	22:41	☽ ∥ ♂
15	04:03	☽ △ ☿
	17:47	☽ □ ♅
16	01:26	☽ △ ♆
	07:49	☽ ☌ ♂
	10:18	☽ △ ☉
	16:14	☽ ☍ ♇
17	01:02	☽ ✶ ♄
	13:58	☽ ☍ ♃
	14:12	☽ ∥ ♂
	16:22	♀ ♒
	21:14	☽ △ ♅
18	18:18	☽ # ♃
	21:29	☽ ☍ ♀
19	02:58	☿ SD
	06:50	☉ ♓
	09:27	☽ ☍ ☿
	09:36	☽ # ♀
	22:42	☽ # ♇
	22:47	☉ ✶ ♇
20	01:20	☉ # ♄
	03:04	☽ # ☿
	09:50	☽ ☍ ♆
	10:47	☽ # ♆
	17:53	☽ ✶ ♂
21	00:20	☽ ∥ ♄
	01:22	☽ △ ♇
	01:58	☽ # ☉
	09:56	☽ ☌ ♄
	23:23	☽ # ♅
22	02:02	☽ △ ♃
	08:44	☽ ☍ ♅
23	02:15	☽ □ ♂
	09:09	☽ □ ♇
	19:36	☽ ∥ ♅
	22:32	☽ △ ♀
24	02:10	☽ △ ☿
	[illegible]	[illegible]
	11:50	☽ □ ♃
	12:59	☽ ∥ ☉
	20:52	☽ # ♄
25	02:46	☽ △ ♆
	10:55	☽ ∥ ♆
	13:35	☽ △ ♂
	19:40	☽ ✶ ♇
	22:45	☽ ∥ ☿
26	00:42	☽ ∥ ♇
	04:05	☽ ✶ ♄
	05:35	☽ ∥ ♀
	08:16	☽ △ ☉
	[illegible]	[illegible]
	16:22	☽ □ ☿
	17:52	☿ ☌ ♀
27	00:14	☽ ✶ ♃
	06:13	☽ △ ♅
	08:19	☽ ∥ ♃
	14:54	☽ □ ♆
28	16:07	☽ □ ♄
	16:37	☽ # ♂
29	05:45	♀ ∥ ♇
	09:05	☽ ✶ ☿
	11:49	☽ ✶ ♀
	10:01	☽ □ ♅

⚷ Chiron

01 Dec.	10 S 13
03	15♒38
06	15 52
09	16 06
12	16 19
15	16 33
18	16 46
21	17 00
24	17 13
27	17 26

March 2008

Day	S. T.	☉	☽	☿	♀	♂	♃	♄	♅	♆	♇	☊ True
	h m s	° ' "	° ' "	° '	° '	° '	° '	° '	° '	° '	° '	° '
01 Sa	10 36 41	10♓ 46 48	20♐ 40 14	13♒ 47	15♒ 13	28♊ 57	15♑ 36	04♍R43	18♓ 12	22♒ 26	00♑ 52	27♒ 39
02 Su	10 40 37	11 47 01	02♑ 45 28	14 41	16 27	29 15	15 46	04 38	18 15	22 28	00 53	27 40
03 Mo	10 44 34	12 47 13	15 03 44	15 39	17 42	29 33	15 57	04 33	18 18	22 30	00 54	27 42
04 Tu	10 48 31	13 47 22	27 38 54	16 39	18 56	29 52	16 07	04 28	18 22	22 32	00 55	27 43
05 We	10 52 27	14 47 31	10♒ 34 01	17 42	20 10	00♋ 11	16 17	04 24	18 25	22 34	00 56	27 45
06 Th	10 56 24	15 47 37	23 50 53	18 47	21 24	00 31	16 27	04 19	18 29	22 36	00 57	27 45
07 Fr	11 00 20	16 47 42	07♓ 29 37	19 55	22 38	00 50	16 37	04 14	18 32	22 39	00 58	27R 45
08 Sa	11 04 17	17 47 45	21 28 27	21 05	23 53	01 10	16 47	04 10	18 36	22 41	00 58	27 44
09 Su	11 08 13	18 47 46	05♈ 43 42	22 16	25 07	01 31	16 57	04 05	18 39	22 43	00 59	27 42
10 Mo	11 12 10	19 47 45	20 10 12	23 30	26 21	01 52	17 07	04 00	18 43	22 45	01 00	27 39
11 Tu	11 16 06	20 47 42	04♉ 41 59	24 46	27 35	02 13	17 16	03 56	18 46	22 47	01 01	27 36
12 We	11 20 03	21 47 36	19 13 06	26 03	28 49	02 35	17 26	03 51	18 49	22 49	01 01	27 34
13 Th	11 23 60	22 47 29	03♊ 38 25	27 22	00♓ 04	02 56	17 35	03 47	18 53	22 51	01 02	27 31
14 Fr	11 27 56	23 47 20	17 54 01	28 42	01 18	03 19	17 44	03 43	18 56	22 53	01 03	27 30
15 Sa	11 31 53	24 47 08	01♋ 57 25	00♓ 04	02 32	03 41	17 53	03 38	19 00	22 55	01 03	27D 29
16 Su	11 35 49	25 46 54	15 47 28	01 28	03 46	04 04	18 02	03 34	19 03	22 57	01 04	27 30
17 Mo	11 39 46	26 46 37	29 23 58	02 53	05 00	04 27	18 11	03 30	19 07	22 59	01 04	27 31
18 Tu	11 43 42	27 46 19	12♌ 47 17	04 19	06 14	04 50	18 20	03 25	19 10	23 01	01 05	27 31
19 We	11 47 39	28 45 58	25 57 57	05 47	07 29	05 14	18 28	03 21	19 13	23 03	01 05	27R 32
20 Th	11 51 35	29 45 34	08♍ 56 25	07 16	08 43	05 38	18 37	03 17	19 17	23 05	01 06	27 31
21 Fr	11 55 32	00♈ 45 09	21 43 03	08 47	09 57	06 02	18 45	03 13	19 20	23 07	01 06	27 28
22 Sa	11 59 29	01 44 42	04♎ 18 09	10 19	11 11	06 26	18 53	03 09	19 23	23 09	01 07	27 24
23 Su	12 03 25	02 44 12	16 42 09	11 52	12 25	06 51	19 01	03 05	19 27	23 11	01 07	27 18
24 Mo	12 07 22	03 43 41	28 55 49	13 27	13 39	07 16	19 09	03 02	19 30	23 12	01 07	27 12
25 Tu	12 11 18	04 43 08	11♏ 00 25	15 03	14 53	07 41	19 17	02 58	19 34	23 14	01 08	27 05
26 We	12 15 15	05 42 33	22 57 54	16 40	16 08	08 06	19 25	02 54	19 37	23 16	01 08	26 58
27 Th	12 19 11	06 41 56	04♐ 50 56	18 19	17 22	08 32	19 32	02 51	19 40	23 18	01 08	26 53
28 Fr	12 23 08	07 41 17	16 42 58	19 59	18 36	08 58	19 40	02 47	19 44	23 20	01 08	26 50
29 Sa	12 27 04	08 40 37	28 38 06	21 41	19 50	09 24	19 47	02 44	19 47	23 21	01 08	26 48
30 Su	12 31 01	09 39 54	10♑ 40 59	23 23	21 04	09 50	19 54	02 40	19 50	23 23	01 09	26D 48
31 Mo	12 34 58	10 39 10	22 56 26	25 08	22 18	10 16	20 01	02 37	19 53	23 25	01 09	26 50

Data for	03-01-2008
Julian Day	2454526.50
Ayanamsa	23 58 25
SVP	05 ♓ 08 55
☽ ☊ Mean	27 ♒ 10 R

● ◐ PHASES ○ ◑

07	17:15	●	17♓31
14	10:46	◐	24♊14
21	18:40	○	01♎31
29	21:48	◑	09♑35

LAST ASPECT ☽ Day	h m	INGRESS Day	h m	
01	16:54	01	18:33	♑
03	06:16	04	04:25	♒
05	21:47	06	10:54	♓
07	19:05	08	14:24	♈
10	11:10	10	16:15	♉
12	17:27	12	17:55	♊
14	20:24	14	20:38	♋
16	18:58	17	01:04	♌
18	18:39	19	07:25	♍
20	19:29	21	15:45	♎
23	12:42	24	02:07	♏
26	00:37	26	14:12	♐
28	13:22	29	02:44	♑
31	04:55	31	13:35	♒

DECLINATION

Day	☉	☽	☿	♀	♂	♃	♄	♅	♆	♇
01 Sa	07S 31	27S 57	16S 38	16S 53	26N16	22S 27	11N30	05S 21	14S 19	17S 07
02 Su	07 09	27 43	16 32	16 33	26 15	22 26	11 32	05 19	14 18	17 07
03 Mo	06 46	26 08	16 25	16 13	26 14	22 25	11 34	05 18	14 18	17 07
04 Tu	06 23	23 13	16 16	15 52	26 13	22 23	11 36	05 17	14 17	17 07
05 We	05 59	19 05	16 06	15 31	26 12	22 22	11 37	05 15	14 16	17 07
06 Th	05 36	13 54	15 54	15 09	26 10	22 21	11 39	05 14	14 16	17 07
07 Fr	05 13	07 55	15 41	14 47	26 09	22 20	11 41	05 13	14 15	17 07
08 Sa	04 49	01 26	15 27	14 25	26 08	22 19	11 43	05 11	14 14	17 07
09 Su	04 26	05N15	15 11	14 02	26 07	22 18	11 44	05 10	14 14	17 07
10 Mo	04 02	11 44	14 54	13 39	26 05	22 17	11 46	05 09	14 13	17 07
11 Tu	03 39	17 38	14 36	13 15	26 04	22 16	11 48	05 07	14 12	17 06
12 We	03 15	22 31	14 16	12 51	26 02	22 14	11 49	05 06	14 12	17 06
13 Th	02 52	25 59	13 55	12 27	26 01	22 13	11 51	05 04	14 11	17 06
14 Fr	02 28	27 47	13 32	12 02	25 59	22 12	11 53	05 03	14 10	17 06
15 Sa	02 04	27 45	13 09	11 37	25 58	22 11	11 54	05 02	14 10	17 06
16 Su	01 41	25 58	12 43	11 12	25 56	22 10	11 56	05 00	14 09	17 06
17 Mo	01 17	22 41	12 17	10 46	25 54	22 09	11 57	04 59	14 08	17 06
18 Tu	00 53	18 15	11 50	10 21	25 52	22 08	11 59	04 58	14 08	17 06
19 We	00 29	13 00	11 21	09 54	25 50	22 07	12 00	04 56	14 07	17 06
20 Th	00 06	07 15	10 51	09 28	25 48	22 06	12 02	04 55	14 07	17 05
21 Fr	00N18	01 19	10 19	09 01	25 46	22 05	12 03	04 54	14 06	17 05
22 Sa	00 42	04S 34	09 47	08 34	25 44	22 04	12 05	04 52	14 05	17 05
23 Su	01 05	10 12	09 13	08 07	25 42	22 03	12 06	04 51	14 05	17 05
24 Mo	01 29	15 21	08 38	07 40	25 39	22 02	12 07	04 50	14 04	17 05
25 Tu	01 53	19 51	08 02	07 12	25 37	22 01	12 09	04 49	14 04	17 05
26 We	02 16	23 30	07 25	06 44	25 34	22 00	12 10	04 47	14 03	17 05
27 Th	02 40	26 08	06 46	06 17	25 32	21 59	12 11	04 46	14 02	17 05
28 Fr	03 03	27 37	06 07	05 48	25 29	21 58	12 12	04 45	14 02	17 05
29 Sa	03 26	27 49	05 26	05 20	25 26	21 57	12 14	04 43	14 01	17 04
30 Su	03 50	26 44	04 44	04 52	25 23	21 56	12 15	04 42	14 01	17 04
31 Mo	04 13	24 20	04 01	04 23	25 20	21 55	12 16	04 41	14 00	17 04

ASPECTARIAN

01 03:31 ☽ ⚹ ♆
16:54 ☽ ☍ ♂
20:18 ☽ ☌ ♇
02 02:00 ☿ ∥ ♀
03:40 ☽ △ ♄
19:13 ☽ ⚹ ☉
23:03 ☽ # ♂
03 01:44 ☽ ☌ ♃
06:16 ☽ ⚹ ♅
04 05:27 ☽ ∥ ♃
10:01 ♂ ♋
05 09:38 ☽ ∥ ♇
14:07 ☽ ☌ ☿
14:54 ☽ ∥ ☿
18:11 ☽ ∥ ♀
19:11 ☽ ☌ ♀
21:47 ☽ ☌ ♆
22:29 ☽ ∥ ♆
06 09:20 ☽ # ♄
12:05 ☽ △ ♂
12:34 ☽ ⚹ ♇
18:22 ☽ ☍ ♄
19:05 ☉ ⚹ ♃
07 00:03 ♀ ☌ ♆
00:17 ☉ ∥ ♅
09:06 ♂ ☍ ♇
10:15 ☽ ∥ ♅
10:51 ☽ ∥ ☉
15:55 ☽ ⚹ ♃
19:05 ☽ ☌ ♅
08 11:28 ♀ ∥ ♆
16:03 ☽ □ ♇
16:46 ☽ □ ♂
20:19 ☉ ☌ ♅
21:14 ☽ # ☉
23:43 ☽ # ♅
09 08:53 ☿ ☌ ♆
18:53 ☽ □ ♃
10 00:08 ☽ ∥ ♄
04:16 ☽ ⚹ ♆
06:02 ☽ ⚹ ☿
07:00 ☽ # ♀
09:43 ☽ # ♆
11:10 ☽ ⚹ ♀
11:54 ☽ # ☿
17:55 ☽ △ ♇
19:48 ☽ ⚹ ♂
21:44 ☽ # ♇
22:44 ☽ △ ♄
11 21:00 ☽ △ ♃
22:32 ☽ # ♃
23:21 ☽ ⚹ ♅
12 04:35 ☽ ⚹ ☉
05:04 ☿ ∥ ♆
05:59 ☽ □ ♆
12:28 ☽ □ ☿
17:27 ☽ □ ♀
22:51 ♀ ♓
13 00:14 ☽ □ ♄
00:16 ☽ ∥ ♂
19:06 ♀ ⚹ ♇
14 01:46 ☽ □ ♅
08:29 ☽ △ ♆
08:50 ♀ # ♄
20:24 ☽ △ ☿
21:25 ♂ ⚹ ♄
22:27 ☽ ☍ ♇
22:46 ☿ ♓
15 01:05 ☽ △ ♀
02:53 ☽ ⚹ ♄
03:04 ☽ ☌ ♂
17:08 ☿ ⚹ ♇
20:17 ♀ ☍ ♄
16 00:19 ☽ ∥ ♂
03:59 ☽ ☍ ♃
05:44 ☽ △ ♅
08:21 ♀ △ ♂
18:58 ☽ △ ☉
17 03:16 ☽ # ♃
09:47 ☿ ☍ ♄
16:29 ☿ # ♄
18 05:34 ☽ # ♇
11:38 ☿ △ ♂
18:39 ☽ ☍ ♆
19:06 ☽ # ♆
19 04:16 ☽ ∥ ♄
07:46 ☽ # ☿
09:26 ☽ △ ♇
13:33 ☽ ☌ ♄
14:12 ☽ # ♀
17:39 ☽ ⚹ ♂
20:29 ☽ ☍ ☿
23:32 ☽ ☍ ♀
20 05:49 ☉ ♈
09:33 ☽ # ♅
18:20 ☽ △ ♃
19:29 ☽ ☍ ♅
21 03:53 ☽ ∥ ☉
07:02 ☽ # ☉
08:37 ☉ □ ♇
17:53 ☽ □ ♇
22 01:17 ☽ ∥ ♅
04:15 ☽ □ ♂
15:40 ☽ ∥ ♀
20:08 ☽ ∥ ☿
23 04:35 ☽ □ ♃
08:38 ☽ # ♄
12:42 ☽ △ ♆
17:49 ☽ ∥ ♆
24 04:20 ☽ ⚹ ♇
08:04 ☽ ⚹ ♄
08:50 ☽ ∥ ♇
13:29 ☿ ☌ ♀
17:08 ☽ △ ♂
25 08:40 ☽ △ ♀
09:22 ☽ △ ☿
13:28 ☽ ∥ ♃
16:46 ☽ ⚹ ♃
17:14 ☽ △ ♅
26 00:37 ☽ □ ♆
17:42 ☽ # ♂
19:58 ☽ □ ♄
27 04:05 ☽ △ ☉
18:57 ☿ ⚹ ♃
20:08 ☿ ☌ ♅
28 04:14 ☽ □ ♀
06:06 ☽ □ ♅
07:41 ☽ □ ☿
13:22 ☽ ⚹ ♆
22:54 ♀ ⚹ ♃
22:58 ♀ ☌ ♅
29 00:15 ♃ ⚹ ♅
05:01 ☽ ☌ ♇
08:09 ☽ △ ♄
11:06 ☿ ∥ ♀
22:15 ☽ ☍ ♂
30 01:19 ☿ ∥ ♅
07:21 ☉ □ ♂
08:23 ♀ ∥ ♅
15:11 ☽ # ♂
18:03 ☽ ⚹ ♅
18:16 ☽ ☌ ♃
19:50 ☿ # ☉
22:38 ☽ ⚹ ♀
31 04:38 ♀ # ☉
04:55 ☽ ⚹ ☿
16:56 ☽ ∥ ♃

⚷ Chiron

01 Dec.	09 S 34
01	17♒39
04	17 52
07	18 05
10	18 17
13	18 29
16	18 41
19	18 52
22	19 03
25	19 14
28	19 24
31	19 34

April 2008

Day	S. T.	☉	☽	☿	♀	♂	♃	♄	♅	♆	♇	☊ True
	h m s	° ' "	° ' "	° '	° '	° '	° '	° '	° '	° '	° '	° '
01 Tu	12 38 54	11♈38 25	05♒29 16	26♓53	23♓32	10♋43	20♑08	02♍R34	19♓57	23♒26	01♑09	26♒51
02 We	12 42 51	12 37 37	18 23 51	28 40	24 46	11 10	20 14	02 31	20 00	23 28	01 09	26 52
03 Th	12 46 47	13 36 47	01♓43 37	00♈28	26 00	11 37	20 21	02 27	20 03	23 30	01R 09	26R 52
04 Fr	12 50 44	14 35 56	15 30 18	02 18	27 15	12 04	20 27	02 25	20 06	23 31	01 09	26 51
05 Sa	12 54 40	15 35 03	29 43 10	04 09	28 29	12 32	20 33	02 22	20 10	23 33	01 09	26 47
06 Su	12 58 37	16 34 08	14♈18 35	06 02	29 43	12 59	20 39	02 19	20 13	23 34	01 09	26 41
07 Mo	13 02 33	17 33 10	29 10 01	07 56	00♈57	13 27	20 45	02 16	20 16	23 36	01 08	26 34
08 Tu	13 06 30	18 32 11	14♉08 53	09 51	02 11	13 55	20 51	02 14	20 19	23 38	01 08	26 26
09 We	13 10 26	19 31 10	29 05 54	11 48	03 25	14 23	20 56	02 11	20 22	23 39	01 08	26 18
10 Th	13 14 23	20 30 06	13♊52 35	13 46	04 39	14 52	21 02	02 09	20 25	23 40	01 08	26 12
11 Fr	13 18 20	21 29 00	28 22 32	15 46	05 53	15 20	21 07	02 06	20 28	23 42	01 08	26 08
12 Sa	13 22 16	22 27 52	12♋32 02	17 47	07 07	15 49	21 12	02 04	20 31	23 43	01 07	26 05
13 Su	13 26 13	23 26 42	26 19 57	19 49	08 21	16 18	21 17	02 02	20 34	23 45	01 07	26 04
14 Mo	13 30 09	24 25 29	09♌47 10	21 52	09 35	16 47	21 22	02 00	20 37	23 46	01 07	26D 04
15 Tu	13 34 06	25 24 14	22 55 50	23 57	10 49	17 16	21 26	01 58	20 40	23 47	01 06	26R 04
16 We	13 38 02	26 22 57	05♍48 34	26 02	12 03	17 45	21 31	01 56	20 43	23 49	01 06	26 03
17 Th	13 41 59	27 21 37	18 27 59	28 08	13 17	18 14	21 35	01 54	20 46	23 50	01 05	26 01
18 Fr	13 45 55	28 20 15	00♎56 21	00♉15	14 31	18 44	21 39	01 53	20 49	23 51	01 05	25 56
19 Sa	13 49 52	29 18 51	13 15 29	02 23	15 45	19 14	21 43	01 51	20 52	23 52	01 04	25 48
20 Su	13 53 49	00♉17 26	25 26 47	04 30	16 59	19 43	21 46	01 50	20 55	23 54	01 04	25 37
21 Mo	13 57 45	01 15 58	07♏31 22	06 38	18 13	20 13	21 50	01 49	20 58	23 55	01 03	25 25
22 Tu	14 01 42	02 14 28	19 30 20	08 45	19 27	20 44	21 53	01 47	21 00	23 56	01 03	25 12
23 We	14 05 38	03 12 57	01♐24 57	10 51	20 41	21 14	21 56	01 46	21 03	23 57	01 02	25 00
24 Th	14 09 35	04 11 24	13 16 58	12 57	21 55	21 44	21 59	01 45	21 06	23 58	01 02	24 50
25 Fr	14 13 31	05 09 49	25 08 47	15 01	23 09	22 15	22 02	01 44	21 09	23 59	01 01	24 42
26 Sa	14 17 28	06 08 13	07♑03 33	17 04	24 23	22 45	22 05	01 43	21 11	24 00	01 00	24 36
27 Su	14 21 24	07 06 35	19 05 11	19 05	25 36	23 16	22 07	01 43	21 14	24 01	00 59	24 34
28 Mo	14 25 21	08 04 55	01♒18 10	21 04	26 50	23 47	22 10	01 42	21 17	24 02	00 59	24D 33
29 Tu	14 29 18	09 03 14	13 47 18	23 01	28 04	24 18	22 12	01 42	21 19	24 03	00 58	24 34
30 We	14 33 14	10 01 31	26 37 28	24 54	29 18	24 49	22 14	01 41	21 22	24 04	00 57	24R 34

Data for	04-01-2008
Julian Day	2454557.50
Ayanamsa	23 58 29
SVP	05 ♓ 08 50
☽ ☊ Mean	25 ♒ 31 R

● ◐ PHASES ○ ◑

06	03:56	●	16♈44
12	18:32	◐	23♋13
20	10:26	○	00♏43
28	14:12	◑	08♒39

LAST ASPECT ☽ INGRESS

Day	h m	Day	h m	
02	09:15	02	20:56	♓
04	21:44	05	00:28	♈
06	15:02	07	01:20	♉
08	15:13	09	01:27	♊
10	16:11	11	02:43	♋
12	18:32	13	06:29	♌
15	04:57	15	13:07	♍
17	06:00	17	22:11	♎
19	20:55	20	09:02	♏
22	08:55	22	21:08	♐
24	21:39	25	09:48	♑
27	14:19	27	21:28	♒
30	05:25	30	06:11	♓

DECLINATION

Day	☉	☽	☿	♀	♂	♃	♄	♅	♆	♇
01 Tu	04N36	20S 44	03S 17	03S 54	25N17	21S 54	12N17	04S 40	14S 00	17S 04
02 We	04 59	16 03	02 32	03 25	25 14	21 53	12 18	04 38	13 59	17 04
03 Th	05 22	10 27	01 46	02 56	25 11	21 52	12 19	04 37	13 59	17 04
04 Fr	05 45	04 12	00 59	02 27	25 08	21 51	12 20	04 36	13 58	17 04
05 Sa	06 08	02N27	00 11	01 58	25 04	21 51	12 21	04 35	13 58	17 04
06 Su	06 31	09 08	00N37	01 29	25 00	21 50	12 22	04 33	13 57	17 04
07 Mo	06 53	15 25	01 27	01 00	24 57	21 49	12 23	04 32	13 57	17 04
08 Tu	07 16	20 51	02 18	00 31	24 53	21 48	12 24	04 31	13 56	17 03
09 We	07 38	24 56	03 09	00 01	24 49	21 48	12 25	04 30	13 56	17 03
10 Th	08 01	27 18	04 01	00N28	24 45	21 47	12 25	04 28	13 55	17 03
11 Fr	08 23	27 45	04 54	00 57	24 41	21 46	12 26	04 27	13 55	17 03
12 Sa	08 45	26 22	05 48	01 27	24 37	21 46	12 27	04 26	13 54	17 03
13 Su	09 06	23 23	06 41	01 56	24 32	21 45	12 28	04 25	13 54	17 03
14 Mo	09 28	19 11	07 36	02 25	24 28	21 44	12 28	04 24	13 53	17 03
15 Tu	09 50	14 09	08 30	02 55	24 23	21 44	12 29	04 23	13 53	17 03
16 We	10 11	08 35	09 25	03 24	24 19	21 43	12 29	04 21	13 53	17 03
17 Th	10 32	02 46	10 19	03 53	24 14	21 43	12 30	04 20	13 52	17 03
18 Fr	10 53	03S 03	11 13	04 22	24 09	21 42	12 30	04 19	13 52	17 03
19 Sa	11 14	08 40	12 07	04 51	24 04	21 42	12 31	04 18	13 51	17 02
20 Su	11 34	13 53	13 01	05 20	23 59	21 41	12 31	04 17	13 51	17 02
21 Mo	11 55	18 32	13 53	05 49	23 53	21 41	12 32	04 16	13 51	17 02
22 Tu	12 15	22 25	14 45	06 17	23 48	21 40	12 32	04 15	13 50	17 02
23 We	12 35	25 20	15 35	06 46	23 42	21 40	12 32	04 14	13 50	17 02
24 Th	12 55	27 08	16 24	07 14	23 37	21 39	12 33	04 13	13 50	17 02
25 Fr	13 15	27 42	17 11	07 43	23 31	21 39	12 33	04 12	13 49	17 02
26 Sa	13 34	26 59	17 57	08 11	23 25	21 39	12 33	04 11	13 49	17 02
27 Su	13 53	24 59	18 41	08 39	23 19	21 39	12 33	04 10	13 49	17 02
28 Mo	14 12	21 49	19 22	09 06	23 13	21 38	12 33	04 09	13 48	17 02
29 Tu	14 31	17 36	20 01	09 34	23 07	21 38	12 33	04 08	13 48	17 02
30 We	14 49	12 29	20 38	10 01	23 00	21 38	12 33	04 07	13 48	17 02

ASPECTARIAN

01 03:16 ☉ # ♅
12:29 ☽ ✱ ☉
19:07 ☽ ∥ ♇
02 09:14 ☽ ∥ ♆
09:15 ☽ ☌ ♆
09:22 ♇ SR
16:22 ☽ # ♄
17:45 ☿ ♈
22:58 ☽ ✱ ♇
03 01:17 ☽ ☍ ♄
08:53 ☿ □ ♇
17:53 ☽ △ ♂
18:32 ☽ # ☉
22:31 ☽ ∥ ♅
04 06:52 ☽ ∥ ♀
07:53 ☽ ☌ ♅
08:30 ☽ ✱ ♃
13:16 ☽ ∥ ☿
16:48 ☽ # ☿
21:44 ☽ ☌ ♀
22:25 ☽ # ♀
05 02:22 ☽ □ ♇
07:34 ☽ # ♅
08:26 ☽ ☌ ☿
14:58 ☽ ∥ ☉
21:47 ☽ □ ♂
06 05:36 ♀ ♈
10:22 ☽ □ ♃
12:07 ☽ ∥ ♄
15:02 ☽ ✱ ♆
15:43 ☿ # ♀
18:10 ☽ # ♆
07 03:10 ☽ △ ♇
03:48 ♀ □ ♇
04:58 ☽ △ ♄
06:48 ☽ # ♇
[illegible]
08 04:57 ☽ # ♃
09:55 ☽ ✱ ♅
10:48 ☽ △ ♃
15:13 ☽ □ ♆
23:16 ☽ ∥ ♂
09 04:58 ☽ □ ♄
07:36 ☽ ✱ ♀
23:48 ☽ ✱ ☿
10 10:48 ☽ □ ♅
11:42 ☽ ✱ ☉
12:01 ☿ # ♅
14:11 ☉ □ ♃
16:11 ☽ △ ♆
17:14 ☿ □ ♂
11 04:37 ☽ ☍ ♇
06:15 ☽ ✱ ♄
13:52 ☽ □ ♀
12 05:51 ☽ ☌ ♂
10:37 ☽ □ ☿
13:52 ☽ △ ♅
15:06 ☽ ☍ ♃
15:43 ☽ ∥ ♂
13 07:31 ☉ ✱ ♆
10:06 ☽ # ♃
17:49 ☿ □ ♃
23:36 ☽ △ ♀
14 10:33 ☽ # ♇
22:10 ☿ ✱ ♆
15 01:09 ☽ # ♆
01:35 ☽ ☍ ♆
02:14 ☽ △ ☿
04:57 ☽ △ ☉
07:20 ☽ ∥ ♄
10:10 [illegible]
16:45 ☽ ☌ ♄
17:39 ☽ ∥ ☉
20:59 ☽ ∥ ☿
16 07:24 ☿ ☌ ☉
17:34 ☽ # ♅
19:48 ☽ ∥ ♀
23:33 ☽ ✱ ♂
17 04:25 ☽ ☍ ♅
06:00 ☽ △ ♃
09:16 ☿ ∥ ☉
21:07 ☿ ♉
21:41 ♀ # ♅
18 00:17 ☽ □ ♇
05:20 ☽ ∥ ♅
06:04 ☽ # ♀
09:21 ☿ △ ♇
18:11 ☿ △ ♄
19 05:26 ☽ ☍ ♀
10:39 ☿ ∥ ♄
12:13 ☽ □ ♂
12:18 ☽ # ☉
16:42 ☽ □ ♃
16:51 ☉ ♉
17:27 ☽ # ♄
18:53 ☽ # ☿
20:55 ☽ △ ♆
23:48 ☽ ∥ ♆
20 11:08 ☽ ✱ ♇
12:38 ☽ ✱ ♄
15:51 ☽ ∥ ♇
18:55 ☉ △ ♇
21:50 ☽ ☍ ☿
22:51 ☿ # ♆
21 13:05 [illegible]
18:55 ☽ ∥ ♃
22 02:34 ☽ △ ♂
03:02 ☽ △ ♅
04:49 ☽ ✱ ♃
08:55 ☽ □ ♆
10:00 ☽ # ♂
14:48 ♂ △ ♅
20:37 ☉ ∥ ♄
23 00:43 ☽ □ ♄
18:07 ♀ □ ♂
24 01:36 ♀ □ ♃
13:24 ♂ ☍ ♃
15:53 ☽ □ ♅
19:10 ☿ # ♇
19:29 ☽ △ ♀
21:39 ☽ ✱ ♆
25 11:50 ☽ ☌ ♇
13:17 ☽ △ ♄
16:34 ♀ ✱ ♆
21:59 ☽ △ ☉
26 18:29 ☉ # ♆
27 00:00 ☽ △ ☿
04:16 ☽ ✱ ♅
06:01 ☽ ☌ ♃
08:37 ☽ ☍ ♂
14:07 ☽ # ♂
14:19 ☽ □ ♀
28 01:10 ☽ ∥ ♃
02:36 ☿ ✱ ♅
12:38 ☽ # ☿
13:41 ☿ △ ♃
29 02:51 ☽ ∥ ♇
13:10 ☿ □ ♆
14:02 ☽ # ☉
18:07 ☽ ∥ ♆
19:16 ☽ ☌ ♆
20:17 ☽ □ ☿
22:20 ☿ ✱ ♂
23:37 ☽ # ♄
30 05:25 ☽ ✱ ♀
07:55 ☽ ✱ ♇
09:16 ☽ ☍ ♄
09:38 ☽ # ♀
13:34 ♀ ♉

⚷ Chiron

01 Dec.	08 S 52
03	19♒43
06	19 53
09	20 01
12	20 09
15	20 17
18	20 24
21	20 31
24	20 37
27	20 43
30	20 48

May 2008

Day	S. T. h m s	☉ ° ′ ″	☽ ° ′ ″	☿ ° ′	♀ ° ′	♂ ° ′	♃ ° ′	♄ ° ′	♅ ° ′	♆ ° ′	♇ ° ′	☊ True ° ′
01 Th	14 37 11	10♉59 47	09♓53 06	26♉45	00♉32	25♋20	22♑15	01♍R41	21♓24	24♒05	00♑R56	24♒R33
02 Fr	14 41 07	11 58 01	23 37 20	28 33	01 46	25 51	22 17	01 41	21 27	24 05	00 55	24 30
03 Sa	14 45 04	12 56 14	07♈51 05	00♊17	03 00	26 22	22 18	01 41	21 29	24 06	00 55	24 25
04 Su	14 49 00	13 54 25	22 32 03	01 58	04 14	26 54	22 19	01D 41	21 32	24 07	00 54	24 17
05 Mo	14 52 57	14 52 35	07♉34 15	03 35	05 28	27 25	22 20	01 41	21 34	24 08	00 53	24 07
06 Tu	14 56 53	15 50 43	22 48 31	05 09	06 41	27 57	22 21	01 41	21 36	24 08	00 52	23 56
07 We	15 00 50	16 48 49	08♊03 50	06 39	07 55	28 29	22 22	01 42	21 39	24 09	00 51	23 46
08 Th	15 04 47	17 46 54	23 09 18	08 05	09 09	29 01	22 22	01 42	21 41	24 10	00 50	23 37
09 Fr	15 08 43	18 44 57	07♋55 56	09 27	10 23	29 33	22 22	01 43	21 43	24 10	00 49	23 30
10 Sa	15 12 40	19 42 58	22 18 00	10 45	11 37	00♌05	22R 22	01 43	21 45	24 11	00 48	23 25
11 Su	15 16 36	20 40 57	06♌13 13	11 58	12 51	00 37	22 22	01 44	21 47	24 11	00 47	23 23
12 Mo	15 20 33	21 38 54	19 42 12	13 08	14 04	01 09	22 22	01 45	21 49	24 12	00 45	23 22
13 Tu	15 24 29	22 36 49	02♍47 39	14 13	15 18	01 42	22 21	01 46	21 52	24 12	00 44	23 21
14 We	15 28 26	23 34 43	15 33 12	15 15	16 32	02 14	22 20	01 47	21 54	24 13	00 43	23 20
15 Th	15 32 23	24 32 34	28 02 46	16 11	17 46	02 47	22 19	01 48	21 56	24 13	00 42	23 17
16 Fr	15 36 19	25 30 24	10♎20 01	17 04	19 00	03 19	22 18	01 50	21 58	24 13	00 41	23 12
17 Sa	15 40 16	26 28 13	22 28 09	17 51	20 13	03 52	22 17	01 51	21 59	24 14	00 40	23 03
18 Su	15 44 12	27 25 59	04♏29 46	18 35	21 27	04 25	22 15	01 53	22 01	24 14	00 38	22 52
19 Mo	15 48 09	28 23 45	16 26 56	19 13	22 41	04 58	22 14	01 54	22 03	24 14	00 37	22 38
20 Tu	15 52 05	29 21 29	28 21 11	19 47	23 55	05 31	22 12	01 56	22 05	24 15	00 36	22 24
21 We	15 56 02	00♊19 11	10♐13 51	20 17	25 08	06 04	22 10	01 58	22 07	24 15	00 35	22 10
22 Th	15 59 58	01 16 53	22 06 20	20 41	26 22	06 37	22 07	02 00	22 08	24 15	00 33	21 57
23 Fr	16 03 55	02 14 33	04♑00 19	21 01	27 36	07 10	22 05	02 02	22 10	24 15	00 32	21 48
24 Sa	16 07 52	03 12 12	15 58 04	21 16	28 50	07 43	22 02	02 04	22 12	24 15	00 31	21 41
25 Su	16 11 48	04 09 49	28 02 30	21 26	00♊03	08 16	22 00	02 06	22 13	24 15	00 29	21 38
26 Mo	16 15 45	05 07 26	10♒17 11	21 31	01 17	08 50	21 57	02 08	22 15	24 15	00 28	21 37
27 Tu	16 19 41	06 05 02	22 46 11	21R 32	02 31	09 23	21 54	02 11	22 16	24R 15	00 27	21D 37
28 We	16 23 38	07 02 37	05♓33 56	21 28	03 45	09 57	21 50	02 13	22 18	24 15	00 25	21 38
29 Th	16 27 34	08 00 11	18 44 45	21 20	04 58	10 30	21 47	02 16	22 19	24 15	00 24	21R 38
30 Fr	16 31 31	08 57 44	02♈22 13	21 08	06 12	11 04	21 43	02 18	22 20	24 15	00 23	21 36
31 Sa	16 35 27	09 55 16	16 28 15	20 51	07 26	11 38	21 39	02 21	22 22	24 15	00 21	21 32

Data for	05-01-2008
Julian Day	2454587.50
Ayanamsa	23 58 32
SVP	05♓08 45
☽☊ Mean	23♒56 R

●◐ PHASES ○◑

Day	Time	Phase	Position
05	12:18	●	15♉22
12	03:48	◐	21♌48
20	02:12	○	29♏27
28	02:56	◑	07♓10

LAST ASPECT ☽ INGRESS

Day	h m	Day	h m	Sign
02	09:35	02	10:51	♈
04	07:16	04	11:58	♉
06	08:22	06	11:18	♊
08	01:37	08	11:03	♋
10	00:07	10	13:11	♌
12	08:10	12	18:49	♍
14	16:39	15	03:48	♎
17	03:30	17	15:00	♏
20	02:12	20	03:20	♐
22	04:20	22	15:56	♑
24	12:26	25	03:52	♒
27	02:49	27	13:39	♓
29	06:23	29	19:53	♈
31	12:54	31	22:19	♉

DECLINATION

Day	☉	☽	☿	♀	♂	♃	♄	♅	♆	♇
01 Th	15N08	06S38	21N13	10N28	22N54	21S38	12N33	04S06	13S48	17S02
02 Fr	15 26	00 17	21 45	10 55	22 47	21 38	12 33	04 05	13 47	17 02
03 Sa	15 43	06N17	22 14	11 22	22 41	21 38	12 33	04 04	13 47	17 01
04 Su	16 01	12 44	22 41	11 48	22 34	21 37	12 33	04 03	13 47	17 01
05 Mo	16 18	18 34	23 06	12 14	22 27	21 37	12 33	04 02	13 47	17 01
06 Tu	16 35	23 19	23 28	12 40	22 20	21 37	12 33	04 01	13 46	17 01
07 We	16 52	26 26	23 47	13 05	22 12	21 38	12 33	04 00	13 46	17 01
08 Th	17 08	27 37	24 04	13 30	22 05	21 38	12 32	03 59	13 46	17 01
09 Fr	17 24	26 47	24 19	13 55	21 58	21 38	12 32	03 58	13 46	17 01
10 Sa	17 40	24 10	24 32	14 20	21 50	21 38	12 32	03 58	13 46	17 01
11 Su	17 55	20 10	24 42	14 44	21 42	21 38	12 31	03 57	13 46	17 01
12 Mo	18 11	15 13	24 51	15 07	21 35	21 38	12 31	03 56	13 45	17 01
13 Tu	18 26	09 43	24 57	15 31	21 27	21 38	12 30	03 55	13 45	17 01
14 We	18 40	03 57	25 01	15 54	21 19	21 39	12 30	03 54	13 45	17 01
15 Th	18 54	01S51	25 04	16 16	21 10	21 39	12 29	03 54	13 45	17 01
16 Fr	19 08	07 28	25 04	16 39	21 02	21 39	12 29	03 53	13 45	17 01
17 Sa	19 22	12 44	25 03	17 00	20 54	21 40	12 28	03 52	13 45	17 01
18 Su	19 35	17 28	25 01	17 22	20 45	21 40	12 27	03 51	13 45	17 01
19 Mo	19 48	21 30	24 56	17 43	20 36	21 40	12 27	03 51	13 45	17 01
20 Tu	20 01	24 37	24 50	18 03	20 28	21 41	12 26	03 50	13 45	17 01
21 We	20 13	26 40	24 43	18 23	20 19	21 41	12 25	03 49	13 45	17 01
22 Th	20 25	27 31	24 34	18 43	20 10	21 42	12 24	03 49	13 45	17 01
23 Fr	20 37	27 06	24 24	19 02	20 01	21 42	12 24	03 48	13 45	17 01
24 Sa	20 48	25 24	24 12	19 20	19 51	21 43	12 23	03 48	13 45	17 01
25 Su	20 59	22 32	24 00	19 38	19 42	21 44	12 22	03 47	13 45	17 01
26 Mo	21 09	18 37	23 46	19 55	19 33	21 44	12 21	03 46	13 45	17 01
27 Tu	21 19	13 50	23 31	20 12	19 23	21 45	12 20	03 46	13 45	17 01
28 We	21 29	08 20	23 15	20 29	19 13	21 45	12 19	03 45	13 45	17 01
29 Th	21 39	02 20	22 59	20 45	19 03	21 46	12 18	03 45	13 45	17 01
30 Fr	21 48	03N58	22 41	21 00	18 54	21 47	12 17	03 44	13 45	17 01
31 Sa	21 56	10 17	22 23	21 14	18 43	21 48	12 16	03 44	13 45	17 01

ASPECTARIAN

01 02:08 ☽ ⚹ ☉
07:46 ♀ △ ♇
09:46 ☽ ∥ ♅
18:36 ☿ # ♃
20:14 ☽ ☌ ♅
21:41 ☽ ⚹ ♃
22:21 ♀ △ ♄
02 03:58 ☽ △ ♂
09:35 ☽ ⚹ ☿
12:24 ☽ □ ♇
15:54 ☽ # ♅
20:00 ☿ ♊
03 03:08 ♄ SD
18:40 ☿ ∥ ♂
19:51 ☿ □ ♄
20:11 ☽ ∥ ♀
23:20 ☽ ∥ ♄
23:39 ☽ □ ♃
04 02:33 ☽ ⚹ ♆
04:07 ☽ # ♆
07:16 ☽ □ ♂
13:23 ☽ △ ♇
13:45 ☽ ∥ ☉
14:39 ☽ △ ♄
17:17 ☽ # ♇
20:22 ☽ ☌ ♀
05 14:37 ☽ # ♃
17:36 ♀ ∥ ♄
18:31 ☽ ∥ ♂
22:06 ☽ ⚹ ♅
23:17 ☽ △ ♃
06 00:59 ☽ ∥ ☿
02:05 ☽ □ ♆
08:22 ☽ ⚹ ♂
13:57 ☽ □ ♄
21:32 ☽ ☌ ☿
07 14:09 ☉ # ♇
21:38 ☽ □ ♅

08 01:37 ☽ △ ♆
12:22 ☽ ☍ ♇
13:49 ☽ ⚹ ♄
15:12 ♀ # ♆
09 04:25 ☽ ⚹ ♀
11:25 ♃ SR
19:19 ☽ ⚹ ☉
20:20 ♂ ♌
21:29 ☽ ∥ ☿
23:04 ☽ △ ♅
10 00:07 ☽ ☍ ♃
13:51 ☽ ☌ ♂
15:15 ☽ ∥ ♂
15:56 ☽ # ♃
11 10:46 ☽ ∥ ☉
11:07 ☽ ⚹ ☿
12:52 ☽ □ ♀
13:19 ♂ # ♃
15:39 ☽ # ♇
12 00:23 ☽ ∥ ♀
04:33 ☉ ⚹ ♅
06:32 ☽ # ♆
08:10 ☽ ☍ ♆
12:00 ☽ ∥ ♄
17:30 ☉ △ ♃
20:12 ☽ △ ♇
22:05 ☽ ☌ ♄
13 23:22 ☽ □ ☿
14 00:08 ☽ # ♅
02:04 ☽ △ ♀
12:09 ☽ ☍ ♅
12:58 ☽ △ ♃
15:51 ☉ □ ♆
16:39 ☽ △ ☉

15 05:09 ☽ □ ♇
08:34 ☽ ∥ ♅
09:38 ☽ ⚹ ♂
16 14:13 ☽ △ ☿
22:43 ☽ # ♄
23:37 ☽ □ ♃
17 00:47 ♀ # ♇
03:30 ☽ △ ♆
04:51 ☽ ∥ ♆
16:18 ☽ ⚹ ♇
18:45 ☽ ⚹ ♄
21:30 ☽ ∥ ♇
23:18 ☽ # ♀
23:50 ☽ □ ♂
18 11:23 ♀ ⚹ ♅
12:39 ☽ # ☉
15:18 ♀ △ ♃
18:25 ☽ # ♂
19 01:08 ☽ ∥ ♃
11:19 ☽ △ ♅
11:37 ☽ ⚹ ♃
14:00 ☽ ☍ ♀
15:42 ☽ □ ♆
20 01:50 ☽ # ☿
06:28 ♀ □ ♆
07:15 ☽ □ ♄
15:10 ☽ △ ♂
16:01 ☉ ♊
21 06:30 ☉ ∥ ♂
18:09 ♃ ⚹ ♅
21:02 ☽ ☍ ☿
22 00:04 ☽ □ ♅
04:20 ☽ ⚹ ♆

17:01 ☽ ☌ ♇
18:26 ☉ □ ♄
20:00 ☽ △ ♄
24 12:03 ☽ # ☿
12:04 ☽ ☌ ♃
12:26 ☽ ⚹ ♅
22:52 ♀ ♊
25 03:35 ♀ ∥ ♂
04:25 ☽ △ ♀
05:29 ☽ ∥ ♃
09:49 ☽ # ☉
13:04 ☽ △ ☉
17:07 ☽ # ♀
18:38 ☽ # ♂

21:02 ☽ ☍ ♂
26 08:29 ☽ ∥ ♇
15:49 ☿ SR
16:16 ♆ SR
17:11 ♀ □ ♄
21:39 ☽ △ ☿
27 00:24 ☽ ∥ ♆
02:49 ☽ ☌ ♆
06:51 ☽ # ♄
14:27 ☽ ⚹ ♇
17:46 ☽ ☍ ♄
20:16 ☽ □ ♀
28 18:28 ☽ ∥ ♅
29 04:34 ☽ □ ☿

05:23 ☽ ⚹ ♃
06:23 ☽ ☌ ♅
20:33 ☽ □ ♇
21:42 ☉ # ♃
23:07 ☽ # ♅
30 07:14 ☽ ⚹ ♀
12:09 ☽ ⚹ ☉
15:31 ☽ △ ♂
31 07:09 ☽ ⚹ ☿
07:40 ☽ ∥ ♄
08:35 ☽ □ ♃
12:54 ☽ ⚹ ♆
13:37 ☽ # ♆
22:51 ☽ △ ♇

⚷ Chiron

01 Dec.	08 S 20
03	20♒53
06	20 57
09	21 00
12	21 03
15	21 05
18	21 07
21	21 08
24	21 09
27	21 09R
30	21 08

25 09:25 21♒09 R

June 2008

Day	S. T.	☉	☽	☿	♀	♂	♃	♄	♅	♆	♇	☊ True
	h m s	° ' "	° ' "	° '	° '	° '	° '	° '	° '	° '	° '	° '
01 Su	16 39 24	10♊52 48	01♉02 01	20♊R31	08♊40	12♌12	21♑R35	02♍24	22♓23	24♒R15	00♑R20	21♒R25
02 Mo	16 43 21	11 50 18	15 59 16	20 08	09 53	12 46	21 31	02 27	22 24	24 15	00 18	21 16
03 Tu	16 47 17	12 47 48	01♊12 10	19 41	11 07	13 20	21 27	02 30	22 25	24 14	00 17	21 07
04 We	16 51 14	13 45 17	16 30 22	19 12	12 21	13 54	21 22	02 33	22 27	24 14	00 15	20 57
05 Th	16 55 10	14 42 46	01♋42 35	18 42	13 35	14 28	21 17	02 37	22 28	24 14	00 14	20 48
06 Fr	16 59 07	15 40 13	16 38 37	18 09	14 48	15 02	21 13	02 40	22 29	24 14	00 12	20 42
07 Sa	17 03 03	16 37 39	01♌10 58	17 36	16 02	15 36	21 08	02 43	22 30	24 13	00 11	20 37
08 Su	17 06 60	17 35 04	15 15 37	17 03	17 16	16 10	21 03	02 47	22 31	24 13	00 09	20 35
09 Mo	17 10 56	18 32 27	28 51 53	16 29	18 30	16 45	20 57	02 51	22 32	24 12	00 08	20D35
10 Tu	17 14 53	19 29 50	12♍01 40	15 57	19 43	17 19	20 52	02 54	22 32	24 12	00 06	20 35
11 We	17 18 50	20 27 11	24 48 26	15 26	20 57	17 54	20 46	02 58	22 33	24 12	00 05	20R35
12 Th	17 22 46	21 24 32	07♎16 24	14 56	22 11	18 28	20 41	03 02	22 34	24 11	00 03	20 34
13 Fr	17 26 43	22 21 51	19 29 57	14 29	23 24	19 03	20 35	03 06	22 35	24 10	00 02	20 31
14 Sa	17 30 39	23 19 10	01♏33 12	14 05	24 38	19 38	20 29	03 10	22 35	24 10	00 00	20 26
15 Su	17 34 36	24 16 28	13 29 53	13 44	25 52	20 12	20 23	03 14	22 36	24 09	29♐59	20 18
16 Mo	17 38 32	25 13 45	25 23 05	13 27	27 06	20 47	20 17	03 18	22 36	24 09	29 57	20 08
17 Tu	17 42 29	26 11 01	07♐15 14	13 14	28 19	21 22	20 10	03 23	22 37	24 08	29 56	19 57
18 We	17 46 25	27 08 17	19 08 19	13 05	29 33	21 57	20 04	03 27	22 37	24 07	29 54	19 47
19 Th	17 50 22	28 05 32	01♑03 54	13 00	00♋47	22 32	19 57	03 32	22 38	24 07	29 52	19 38
20 Fr	17 54 19	29 02 47	13 03 32	12D59	02 00	23 07	19 51	03 36	22 38	24 06	29 51	19 31
21 Sa	17 58 15	00♋00 01	25 08 55	13 03	03 14	23 42	19 44	03 41	22 38	24 05	29 49	19 26
22 Su	18 02 12	00 57 15	07♒22 00	13 12	04 28	24 17	19 37	03 45	22 39	24 04	29 48	19 24
23 Mo	18 06 08	01 54 28	19 45 14	13 25	05 41	24 52	19 30	03 50	22 39	24 04	29 46	19D24
24 Tu	18 10 05	02 51 42	02♓21 23	13 44	06 55	25 27	19 23	03 55	22 39	24 03	29 45	19 26
25 We	18 14 01	03 48 55	15 13 37	14 07	08 09	26 03	19 16	04 00	22 39	24 02	29 43	19 28
26 Th	18 17 58	04 46 08	28 25 09	14 34	09 23	26 38	19 09	04 05	22 39	24 01	29 42	19 29
27 Fr	18 21 54	05 43 21	11♈58 47	15 06	10 36	27 14	19 02	04 10	22R39	24 00	29 40	19R30
28 Sa	18 25 51	06 40 35	25 56 16	15 43	11 50	27 49	18 54	04 15	22 39	23 59	29 38	19 28
29 Su	18 29 48	07 37 48	10♉17 27	16 25	13 04	28 24	18 47	04 20	22 39	23 58	29 37	19 25
30 Mo	18 33 44	08 35 02	24 59 36	17 10	14 18	29 00	18 39	04 25	22 39	23 57	29 35	19 20

Data for	06-01-2008
Julian Day	2454618.50
Ayanamsa	23 58 37
SVP	05♓08 41
☽ ☊ Mean	22♒17 R

● ◐ PHASES ○ ◑

Day	Time	Phase	Position
03	19:23	●	13♊34
10	15:04	◐	20♍06
18	17:30	○	27♐50
26	12:10	◑	05♈15

LAST ASPECT Day	h m	☽ INGRESS Day	h m	Sign
02	13:03	02	22:07	♊
04	12:10	04	21:17	♋
06	09:33	06	22:01	♌
08	15:42	09	02:02	♍
10	19:43	11	09:56	♎
13	09:16	13	20:54	♏
15	21:30	16	09:20	♐
18	21:37	18	21:52	♑
20	19:02	21	09:34	♒
23	19:04	23	19:33	♓
26	02:17	26	02:50	♈
28	06:15	28	06:51	♉
30	06:44	30	08:04	♊

DECLINATION

Day	☉	☽	☿	♀	♂	♃	♄	♅	♆	♇
01 Su	22N05	16N15	22N05	21N29	18N33	21S48	12N14	03S43	13S45	17S01
02 Mo	22 12	21 25	21 45	21 42	18 23	21 49	12 13	03 43	13 45	17 01
03 Tu	22 20	25 15	21 26	21 55	18 13	21 50	12 12	03 42	13 45	17 01
04 We	22 27	27 16	21 07	22 07	18 02	21 51	12 11	03 42	13 45	17 01
05 Th	22 34	27 14	20 47	22 19	17 52	21 52	12 09	03 42	13 45	17 01
06 Fr	22 40	25 12	20 28	22 30	17 41	21 53	12 08	03 41	13 45	17 01
07 Sa	22 46	21 32	20 10	22 40	17 30	21 54	12 07	03 41	13 46	17 01
08 Su	22 52	16 42	19 52	22 50	17 19	21 55	12 05	03 41	13 46	17 02
09 Mo	22 57	11 10	19 34	22 59	17 08	21 55	12 04	03 40	13 46	17 02
10 Tu	23 01	05 19	19 18	23 07	16 57	21 56	12 03	03 40	13 46	17 02
11 We	23 06	00S34	19 03	23 15	16 46	21 57	12 01	03 40	13 46	17 02
12 Th	23 10	06 17	18 49	23 22	16 34	21 58	12 00	03 39	13 46	17 02
13 Fr	23 13	11 40	18 37	23 28	16 23	21 59	11 58	03 39	13 47	17 02
14 Sa	23 16	16 31	18 26	23 34	16 11	22 01	11 56	03 39	13 47	17 02
15 Su	23 19	20 42	18 17	23 39	16 00	22 02	11 55	03 39	13 47	17 02
16 Mo	23 21	24 00	18 10	23 43	15 48	22 03	11 53	03 39	13 47	17 02
17 Tu	23 23	26 18	18 04	23 47	15 36	22 04	11 52	03 38	13 47	17 02
18 We	23 25	27 24	18 01	23 50	15 24	22 05	11 50	03 38	13 48	17 02
19 Th	23 26	27 15	17 59	23 52	15 12	22 06	11 48	03 38	13 48	17 02
20 Fr	23 26	25 49	17 59	23 54	15 00	22 07	11 46	03 38	13 48	17 02
21 Sa	23 26	23 11	18 00	23 54	14 48	22 08	11 45	03 38	13 48	17 03
22 Su	23 26	19 28	18 04	23 55	14 36	22 09	11 43	03 38	13 49	17 03
23 Mo	23 26	14 51	18 09	23 54	14 23	22 11	11 41	03 38	13 49	17 03
24 Tu	23 25	09 33	18 15	23 53	14 11	22 12	11 39	03 38	13 49	17 03
25 We	23 23	03 44	18 24	23 50	13 58	22 13	11 37	03 38	13 50	17 03
26 Th	23 21	02N22	18 33	23 48	13 46	22 14	11 36	03 38	13 50	17 03
27 Fr	23 19	08 31	18 44	23 44	13 33	22 15	11 34	03 38	13 50	17 03
28 Sa	23 16	14 26	18 56	23 40	13 20	22 16	11 32	03 38	13 51	17 03
29 Su	23 13	19 44	19 08	23 35	13 07	22 17	11 30	03 38	13 51	17 03
30 Mo	23 10	23 59	19 22	23 30	12 54	22 19	11 28	03 38	13 51	17 04

ASPECTARIAN

01 00:00 ☿ ∥ ☉
02:14 ☽ △ ♄
03:17 ☽ # ♇
09:48 ☽ ∥ ♂
18:40 ☽ □ ♂
19:35 ☿ # ♃
02 01:35 ☽ ∥ ♀
01:42 ☽ ∥ ☿
02:08 ☽ # ♃
02:35 ☿ ∥ ♀
04:22 ☽ ∥ ☉
08:43 ☽ △ ♃
10:10 ☽ ⚹ ♅
13:03 ☽ □ ♆
14:09 ♀ # ♃
03 02:03 ☽ □ ♄
16:54 ☽ ☌ ♀
19:45 ☽ ⚹ ♂
04 04:07 ☽ ☌ ☿
08:35 ☉ ⚹ ♂
09:21 ☽ □ ♅
12:10 ☽ △ ♆
21:39 ☽ ☍ ♇
05 01:26 ☽ ⚹ ♄
06 07:25 ☽ ☍ ♃
08:16 ♀ ⚹ ♂
09:33 ☽ △ ♅
17:07 ☽ ∥ ☉
17:48 ☽ ∥ ♀
22:04 ☽ # ♃
07 07:53 ☽ ∥ ☿
15:27 ☿ ☌ ☉
21:03 ☽ ∥ ♂
21:06 ☿ ☌ ♀
22:33 ☽ # ♇
08 01:39 ☽ ☌ ♂
02:59 ☽ ⚹ ☿
03:49 ☽ ⚹ ♀
04:21 ☽ ⚹ ☉
10:21 ♀ ∥ ☉
13:03 ☽ # ♇
15:42 ☽ ☍ ♆
18:29 ☿ ⚹ ♂
20:16 ☽ ∥ ♄
09 02:16 ☽ △ ♇
04:23 ♀ ☌ ☉
07:13 ☽ ☌ ♄
14:05 ♂ # ♇
10 06:44 ☽ # ♅
07:00 ☽ □ ☿
15:54 ☽ □ ♀
16:24 ☽ △ ♃
19:43 ☽ ☍ ♅
11 10:04 ☽ □ ♇
12:49 ☽ ∥ ♅
12 07:38 ♀ □ ♅
14:28 ☽ △ ☿
23:04 ☽ ⚹ ♂
13 01:25 ☽ # ♄
02:07 ☽ □ ♃
05:24 ☉ □ ♅
06:10 ☽ △ ☉
08:38 ☽ △ ♀
09:16 ☽ △ ♆
10:06 ☽ ∥ ♆
14:54 ♀ △ ♆
20:54 ☽ ⚹ ♇
22:21 ☽ # ♂
14 02:44 ☽ ∥ ♇
03:15 ☽ ⚹ ♄
03:47 ♇ ♐R
10:07 ☽ # ☿
21:04 ☉ △ ♆
15 08:53 ☽ ∥ ♃
13:46 ☽ ⚹ ♃
14:14 ☽ □ ♂
18:23 ☽ △ ♅
18:34 ☽ # ☉
21:30 ☽ □ ♆
21:32 ☽ # ♀
16 16:07 ☽ □ ♄
17 11:54 ☽ ☍ ☿
18 05:57 ☽ △ ♂
06:45 ♀ ☍ ♇
07:02 ☽ □ ♅
08:49 ♀ ♋
10:02 ☽ ⚹ ♆
21:37 ☽ ☌ ♇
23:21 ☽ ☍ ♀
19 04:58 ☽ △ ♄
14:32 ☿ SD
20 13:22 ☽ ☌ ♃
18:22 ☽ # ♀
19:02 ☽ ⚹ ♅
19:41 ☉ ☍ ♇
22:03 ☽ # ☉
21 00:00 ☉ ♋
07:25 ☽ ∥ ♃
09:15 ♀ ⚹ ♄
15:34 ♂ ☍ ♆
22 07:38 ☽ [illegible]
11:32 ☽ △ ☿
13:07 ☽ ∥ ♇
23 02:22 ☽ # ♂
04:57 ☽ ∥ ♆
08:14 ☽ ☌ ♆
10:17 ☽ ☍ ♂
14:47 ☽ # ♄
19:04 ☽ ⚹ ♇
24 01:02 ☽ △ ☉
02:57 ☽ ☍ ♄
09:29 ☽ △ ♀
21:52 ☽ □ ☿
25 00:28 ☽ ∥ ♅
05:01 ☽ ⚹ ♄
07:21 ☽ ⚹ ♃
13:36 ☽ ☌ ♅
15:58 ♂ # ♆
26 02:17 ☽ □ ♇
04:57 ☽ # ♅
21:22 ☽ □ ♀
27 00:02 ♅ SR
05:41 ☽ ⚹ ☿
12:06 ☽ □ ♃
12:06 ☽ ∥ ♄
19:35 ☽ ∥ ♂
20:41 ☽ ⚹ ♆
21:32 ☽ # ♆
28 03:10 ☽ △ ♂
06:15 ☽ △ ♇
11:25 ☽ # ♇
14:04 ☽ △ ♄
19:17 ☽ ⚹ ☉
20:58 ☽ ∥ ☿
29 04:59 ☽ ⚹ ♀
13:34 ☽ # ♃
13:48 ☽ △ ♃
18:45 ☽ ∥ ☉
20:13 ☽ ⚹ ♅
20:50 ☽ ∥ ♀
22:19 ☽ □ ♆
30 06:44 ☽ □ ♂
15:15 ☽ □ ♄
22:54 ♂ △ ♇

⚷ Chiron

01 Dec.	08 S 03
02	21♒07R
05	21 05
08	21 03
11	21 00
14	20 57
17	20 53
20	20 49
23	20 44
26	20 38
29	20 32

July 2008

Day	S. T.	☉	☽	☿	♀	♂	♃	♄	♅	♆	♇	☊ True
	h m s	° ' "	° ' "	° '	° '	° '	° '	° '	° '	° '	° '	° '
01 Tu	18 37 41	09♋ 32 15	09♊ 57 12	18♊ 01	15♋ 31	29♌ 36	18♑R32	04♍ 31	22♓R39	23♒R56	29♐R34	19♒R14
02 We	18 41 37	10 29 29	25 02 12	18 55	16 45	00♍ 11	18 24	04 36	22 39	23 55	29 32	19 08
03 Th	18 45 34	11 26 43	10♋ 05 12	19 54	17 59	00 47	18 17	04 42	22 39	23 54	29 31	19 03
04 Fr	18 49 30	12 23 56	24 56 52	20 58	19 13	01 23	18 09	04 47	22 38	23 53	29 29	18 59
05 Sa	18 53 27	13 21 10	09♌ 29 23	22 05	20 26	01 59	18 02	04 53	22 38	23 52	29 28	18 56
06 Su	18 57 24	14 18 23	23 37 34	23 17	21 40	02 35	17 54	04 58	22 37	23 51	29 26	18 55
07 Mo	19 01 20	15 15 37	07♍ 19 09	24 32	22 54	03 11	17 46	05 04	22 37	23 49	29 25	18D 56
08 Tu	19 05 17	16 12 50	20 34 28	25 52	24 08	03 47	17 38	05 10	22 36	23 48	29 23	18 57
09 We	19 09 13	17 10 02	03♎ 25 49	27 15	25 21	04 23	17 31	05 16	22 36	23 47	29 22	18 58
10 Th	19 13 10	18 07 15	15 56 43	28 43	26 35	04 59	17 23	05 22	22 35	23 46	29 20	18 59
11 Fr	19 17 06	19 04 28	28 11 25	00♋ 14	27 49	05 35	17 15	05 28	22 35	23 44	29 19	18R 59
12 Sa	19 21 03	20 01 40	10♏ 14 22	01 49	29 03	06 11	17 08	05 34	22 34	23 43	29 18	18 57
13 Su	19 24 59	20 58 53	22 09 56	03 28	00♌ 16	06 47	17 00	05 40	22 33	23 42	29 16	18 55
14 Mo	19 28 56	21 56 05	04♐ 02 10	05 11	01 30	07 24	16 52	05 46	22 32	23 41	29 15	18 51
15 Tu	19 32 53	22 53 18	15 54 35	06 56	02 44	08 00	16 45	05 52	22 32	23 39	29 13	18 47
16 We	19 36 49	23 50 31	27 50 06	08 45	03 58	08 36	16 37	05 58	22 31	23 38	29 12	18 43
17 Th	19 40 46	24 47 44	09♑ 51 04	10 37	05 11	09 13	16 29	06 05	22 30	23 37	29 11	18 39
18 Fr	19 44 42	25 44 58	21 59 24	12 32	06 25	09 49	16 22	06 11	22 29	23 35	29 09	18 36
19 Sa	19 48 39	26 42 12	04♒ 16 40	14 29	07 39	10 26	16 14	06 17	22 28	23 34	29 08	18 35
20 Su	19 52 35	27 39 26	16 44 16	16 29	08 53	11 03	16 07	06 24	22 27	23 32	29 07	18D 35
21 Mo	19 56 32	28 36 41	29 23 27	18 30	10 06	11 39	16 00	06 30	22 26	23 31	29 05	18 36
22 Tu	20 00 28	29 33 57	12♓ 15 29	20 34	11 20	12 16	15 52	06 37	22 25	23 29	29 04	18 37
23 We	20 04 25	00♌ 31 13	25 21 38	22 38	12 34	12 53	15 45	06 43	22 24	23 28	29 03	18 39
24 Th	20 08 22	01 28 30	08♈ 43 10	24 44	13 48	13 29	15 38	06 50	22 22	23 27	29 02	18 41
25 Fr	20 12 18	02 25 48	22 21 07	26 51	15 01	14 06	15 31	06 57	22 21	23 25	29 00	18 41
26 Sa	20 16 15	03 23 07	06♉ 15 54	28 57	16 15	14 43	15 24	07 03	22 20	23 24	28 59	18R 41
27 Su	20 20 11	04 20 27	20 26 58	01♌ 05	17 29	15 20	15 17	07 10	22 18	23 22	28 58	18 41
28 Mo	20 24 08	05 17 48	04♊ 52 19	03 11	18 43	15 57	15 10	07 17	22 17	23 21	28 57	18 39
29 Tu	20 28 04	06 15 10	19 28 22	05 18	19 57	16 34	15 03	07 24	22 16	23 19	28 56	18 38
30 We	20 32 01	07 12 34	04♋ 09 57	07 24	21 10	17 11	14 56	07 30	22 14	23 17	28 54	18 36
31 Th	20 35 57	08 09 58	18 50 45	09 29	22 24	17 49	14 50	07 37	22 13	23 16	28 53	18 34

Data for	07-01-2008
Julian Day	2454648.50
Ayanamsa	23 58 43
SVP	05 ♓ 08 39
☽ ☊ Mean	20 ♒ 42 R

● ◐ PHASES ○ ◑

Day	Time	Phase	Position
03	02:19	●	11♋32
10	04:35	◐	18♎18
18	07:59	○	26♑04
25	18:42	◑	03♉11

LAST ASPECT ☽ Day	h m	INGRESS Day	h m	
02	07:09	02	07:54	♋
03	20:14	04	08:16	♌
06	10:04	06	11:04	♍
08	16:21	08	17:32	♎
11	02:14	11	03:35	♏
13	03:05	13	15:50	♐
16	02:44	16	04:20	♑
18	08:00	18	15:41	♒
20	23:26	21	01:09	♓
23	06:40	23	08:23	♈
25	11:31	25	13:15	♉
27	04:53	27	15:56	♊
29	15:26	29	17:12	♋
31	05:31	31	18:22	♌

DECLINATION

Day	☉	☽	☿	♀	♂	♃	♄	♅	♆	♇
01 Tu	23N06	26N43	19N37	23N23	12N41	22S20	11N26	03S38	13S52	17S04
02 We	23 02	27 32	19 52	23 16	12 28	22 21	11 24	03 38	13 52	17 04
03 Th	22 57	26 18	20 08	23 09	12 14	22 22	11 22	03 38	13 52	17 04
04 Fr	22 52	23 13	20 23	23 00	12 01	22 23	11 19	03 38	13 53	17 04
05 Sa	22 46	18 41	20 40	22 51	11 48	22 25	11 17	03 39	13 53	17 04
06 Su	22 40	13 14	20 56	22 42	11 34	22 26	11 15	03 39	13 54	17 04
07 Mo	22 34	07 17	21 12	22 31	11 21	22 27	11 13	03 39	13 54	17 05
08 Tu	22 27	01 13	21 27	22 20	11 07	22 28	11 11	03 39	13 54	17 05
09 We	22 20	04S43	21 42	22 09	10 53	22 29	11 09	03 40	13 55	17 05
10 Th	22 13	10 19	21 57	21 56	10 39	22 30	11 06	03 40	13 55	17 05
11 Fr	22 05	15 23	22 10	21 44	10 26	22 31	11 04	03 40	13 56	17 05
12 Sa	21 57	19 47	22 22	21 30	10 12	22 33	11 02	03 40	13 56	17 05
13 Su	21 48	23 20	22 33	21 16	09 58	22 34	11 00	03 41	13 57	17 06
14 Mo	21 39	25 54	22 43	21 01	09 44	22 35	10 57	03 41	13 57	17 06
15 Tu	21 30	27 19	22 51	20 46	09 29	22 36	10 55	03 41	13 57	17 06
16 We	21 20	27 29	22 57	20 30	09 15	22 37	10 53	03 42	13 58	17 06
17 Th	21 10	26 21	23 01	20 13	09 01	22 38	10 50	03 42	13 58	17 06
18 Fr	21 00	23 58	23 02	19 56	08 46	22 39	10 48	03 43	13 59	17 06
19 Sa	20 49	20 27	23 02	19 38	08 32	22 40	10 45	03 43	13 59	17 07
20 Su	20 38	15 59	22 58	19 20	08 18	22 41	10 43	03 43	14 00	17 07
21 Mo	20 26	10 45	22 52	19 01	08 03	22 42	10 41	03 44	14 00	17 07
22 Tu	20 15	05 00	22 44	18 42	07 48	22 43	10 38	03 44	14 01	17 07
23 We	20 02	01N04	22 33	18 22	07 34	22 44	10 36	03 45	14 01	17 07
24 Th	19 50	07 11	22 19	18 02	07 19	22 45	10 33	03 45	14 02	17 08
25 Fr	19 37	13 05	22 02	17 41	07 04	22 46	10 31	03 46	14 02	17 08
26 Sa	19 24	18 28	21 43	17 20	06 50	22 47	10 28	03 46	14 03	17 08
27 Su	19 10	22 56	21 22	16 58	06 35	22 48	10 26	03 47	14 03	17 08
28 Mo	18 57	26 05	20 58	16 36	06 20	22 49	10 23	03 48	14 04	17 09
29 Tu	18 43	27 32	20 31	16 13	06 05	22 50	10 20	03 48	14 04	17 09
30 We	18 28	27 04	20 03	15 50	05 50	22 50	10 18	03 49	14 05	17 09
31 Th	18 13	24 43	19 32	15 26	05 35	22 51	10 15	03 49	14 05	17 09

ASPECTARIAN

01 13:39 ☽ ☌ ☿
16:21 ♂ ♍
20:12 ☽ □ ♅
22:13 ☽ △ ♆
02 07:09 ☽ ☍ ♇
08:32 ☽ ⚹ ♂
15:20 ☽ ⚹ ♄
03 05:19 ♀ ☍ ♃
13:04 ☽ ☍ ♃
13:50 ☽ ☌ ♀
20:14 ☽ △ ♅
04 01:21 ☽ ∥ ♀
02:13 ☽ ∥ ☉
04:56 ☽ # ♃
14:53 ☽ ∥ ☿
05 07:30 ☽ # ♇
11:07 ☿ □ ♅
21:17 ☽ # ♆
23:20 ☽ ⚹ ☿
06 00:22 ☽ ☍ ♆
07:09 ☽ ∥ ♂
08:13 ☽ ∥ ♄
08:40 ♀ ∥ ☉
10:04 ☽ △ ♇
10:48 ☿ △ ♆
16:19 ☽ ☌ ♂
18:32 ♀ △ ♅
19:58 ☽ ☌ ♄
07 09:10 ♀ # ♃
14:22 ☽ # ♅
15:24 ☽ ⚹ ☉
16:01 ♂ ∥ ♄
18:41 ☽ △ ♃
22:03 ☉ # ♃
08 03:45 ☽ ☍ ♅
07:15 ☽ ⚹ ♀
10:58 ☽ □ ☿
16:21 ☽ □ ♇
19:38 ☽ ∥ ♅
09 07:40 ☉ ☍ ♃
23:56 ☿ ∥ ♀
10 01:31 ☽ # ♂
02:46 ☽ □ ♃
03:36 ☽ # ♄
09:50 ☿ ☍ ♇
15:15 ☽ △ ♆
16:50 ☽ ∥ ♆
18:12 ♂ ☌ ♄
18:17 ☿ ∥ ☉
20:17 ☿ ♋
23:10 ☽ □ ♀
11 02:14 ☽ ⚹ ♇
04:40 ☽ △ ☿
08:51 ☽ ∥ ♇
14:34 ☽ ⚹ ♄
15:28 ☽ ⚹ ♂
12 10:13 ☽ # ♀
13:21 ☽ # ☉
13:42 ☽ ⚹ ♃
17:52 ☽ # ☿
18:11 ☽ ∥ ♃
18:39 ♀ ♌
21:24 ☽ △ ☉
13 00:46 ☿ # ♃
00:47 ☽ △ ♅
03:05 ☽ □ ♆
18:17 ☽ △ ♀
14 03:32 ☽ □ ♄
07:10 ☽ □ ♂
08:36 ☿ ⚹ ♄
15:03 ☉ △ ♅
15 13:19 ☽ □ ♅
15:34 ☽ ⚹ ♆
21:11 ☿ ⚹ ♂
16 02:44 ☽ ☌ ♇
16:25 ☽ △ ♄
22:40 ☽ △ ♂
17 01:48 ☽ ☍ ☿
13:01 ☽ ☌ ♃
18 00:58 ☽ ⚹ ♅
07:03 ☽ # ☿
09:47 ☽ ∥ ♃
21:40 ☽ # ☉
19 05:05 ☽ # ♀
07:15 ☽ ☍ ♀
18:19 ☽ ∥ ♇
19:54 ☿ ☍ ♃
20 09:27 ☽ ∥ ♆
12:55 ☽ ☌ ♆
23:26 ☽ ⚹ ♇
21 00:22 ☽ # ♄
12:01 ☽ # ♂
13:26 ☽ ☍ ♄
22 00:01 ☽ ☍ ♂
02:06 ☿ # ♃
05:05 ☽ ∥ ♅
06:36 ☽ ⚹ ♃
10:55 ☉ ♌
18:07 ☽ △ ☿
18:37 ☽ ☌ ♅
21:11 ☿ △ ♅
23 06:40 ☽ □ ♇
10:03 ☽ △ ☉
10:03 ☽ # ♅
24 00:34 ☽ ∥ ♂
09:53 ☽ △ ♀
12:07 ☽ □ ♃
13:27 ☽ ∥ ♄
25 01:51 ☽ ⚹ ♆
04:04 ☽ # ♆
09:12 ☽ □ ☿
11:31 ☽ △ ♇
17:47 ☽ # ♇
19:01 ☽ ∥ ♀
26 01:21 ☽ △ ♄
04:25 ☽ ∥ ☉
11:49 ☿ ♌
12:36 ♀ # ♇
15:01 ☽ △ ♂
15:22 ☽ △ ♃
15:36 ☽ ∥ ☿
18:33 ☽ □ ♀
22:04 ♂ △ ♃
23:12 ☽ # ♃
27 03:06 ☽ ⚹ ♅
04:53 ☽ □ ♆
20:45 ☽ ⚹ ☿
28 00:45 ☽ ⚹ ☉
04:00 ☽ □ ♄
19:02 ☽ □ ♂
29 00:50 ☽ ⚹ ♀
04:33 ☽ □ ♅
06:16 ☽ △ ♆
15:26 ☽ ☍ ♇
20:05 ☿ ☌ ☉
30 05:30 ☽ ⚹ ♄
17:28 ☽ ☍ ♃
22:13 ☽ ⚹ ♂
31 05:31 ☽ △ ♅
12:14 ☽ # ♃
16:30 ♀ ☍ ♆

⚷ Chiron

01 Dec.	08 S 05
02	20♒26R
05	20 19
08	20 12
11	20 05
14	19 57
17	19 49
20	19 40
23	19 31
26	19 23
29	19 13

August 2008

01	09:48	09♌32	☉	Total Solar Eclipse
16	21:11	24♒21	✹	Total Lunar Eclipse (mag 0.813)

Day	S. T. h m s	☉ ° ' "	☽ ° ' "	☿ ° '	♀ ° '	♂ ° '	♃ ° '	♄ ° '	♅ ° '	♆ ° '	♇ ° '	☊ True ° '
01 Fr	20 39 54	09♌ 07 23	03♌ 24 04	11♌ 33	23♌ 38	18♍ 26	14♑R43	07♍ 44	22♓R11	23♒R14	28♐R52	18♒R33
02 Sa	20 43 51	10 04 49	17 43 36	13 36	24 52	19 03	14 37	07 51	22 10	23 13	28 51	18 32
03 Su	20 47 47	11 02 15	01♍ 44 24	15 38	26 05	19 40	14 31	07 58	22 08	23 11	28 50	18 32
04 Mo	20 51 44	11 59 42	15 23 20	17 38	27 19	20 18	14 25	08 05	22 06	23 10	28 49	18D 32
05 Tu	20 55 40	12 57 11	28 39 17	19 37	28 33	20 55	14 19	08 12	22 05	23 08	28 48	18 32
06 We	20 59 37	13 54 40	11♎ 33 00	21 35	29 47	21 33	14 13	08 19	22 03	23 06	28 47	18 32
07 Th	21 03 33	14 52 09	24 06 42	23 31	01♍ 00	22 10	14 07	08 27	22 01	23 05	28 46	18 32
08 Fr	21 07 30	15 49 40	06♏ 23 42	25 25	02 14	22 48	14 02	08 34	22 00	23 03	28 45	18 32
09 Sa	21 11 26	16 47 11	18 28 04	27 18	03 28	23 25	13 56	08 41	21 58	23 01	28 44	18 32
10 Su	21 15 23	17 44 43	00♐ 24 15	29 10	04 42	24 03	13 51	08 48	21 56	23 00	28 43	18 32
11 Mo	21 19 20	18 42 16	12 16 49	01♍ 00	05 55	24 41	13 46	08 55	21 54	22 58	28 42	18 33
12 Tu	21 23 16	19 39 50	24 10 11	02 49	07 09	25 18	13 41	09 03	21 52	22 57	28 42	18 33
13 We	21 27 13	20 37 25	06♑ 08 23	04 36	08 23	25 56	13 36	09 10	21 50	22 55	28 41	18 34
14 Th	21 31 09	21 35 01	18 14 51	06 21	09 37	26 34	13 31	09 17	21 48	22 53	28 40	18 35
15 Fr	21 35 06	22 32 38	00♒ 32 26	08 05	10 50	27 12	13 27	09 25	21 46	22 52	28 39	18 36
16 Sa	21 39 02	23 30 16	13 03 11	09 48	12 04	27 50	13 23	09 32	21 44	22 50	28 39	18 37
17 Su	21 42 59	24 27 55	25 48 27	11 29	13 18	28 28	13 18	09 39	21 42	22 48	28 38	18 37
18 Mo	21 46 55	25 25 36	08♓ 48 45	13 09	14 31	29 06	13 14	09 47	21 40	22 47	28 37	18R 37
19 Tu	21 50 52	26 23 18	22 03 48	14 47	15 45	29 44	13 11	09 54	21 38	22 45	28 37	18 37
20 We	21 54 49	27 21 01	05♈ 32 46	16 23	16 59	00♎ 22	13 07	10 02	21 36	22 44	28 36	18 36
21 Th	21 58 45	28 18 46	19 14 19	17 59	18 12	01 00	13 03	10 09	21 34	22 42	28 35	18 35
22 Fr	22 02 42	29 16 33	03♉ 06 52	19 33	19 26	01 39	13 00	10 17	21 32	22 40	28 35	18 34
23 Sa	22 06 38	00♍ 14 21	17 08 40	21 05	20 40	02 17	12 57	10 24	21 30	22 39	28 34	18 33
24 Su	22 10 35	01 12 11	01♊ 17 45	22 36	21 53	02 55	12 54	10 32	21 27	22 37	28 34	18 32
25 Mo	22 14 31	02 10 03	15 31 58	24 06	23 07	03 34	12 51	10 39	21 25	22 35	28 33	18D 32
26 Tu	22 18 28	03 07 57	29 48 51	25 34	24 20	04 12	12 49	10 47	21 23	22 34	28 33	18 32
27 We	22 22 24	04 05 53	14♋ 05 36	27 01	25 34	04 51	12 46	10 54	21 21	22 32	28 32	18 33
28 Th	22 26 21	05 03 50	28 19 05	28 26	26 48	05 29	12 44	11 02	21 18	22 31	28 32	18 34
29 Fr	22 30 18	06 01 49	12♌ 25 50	29 50	28 01	06 08	12 42	11 09	21 16	22 29	28 32	18R 34
30 Sa	22 34 14	06 59 50	26 22 16	01♎ 13	29 15	06 46	12 40	11 17	21 14	22 27	28 31	18 34
31 Su	22 38 11	07 57 52	10♍ 05 03	02 33	00♎ 29	07 25	12 38	11 24	21 12	22 26	28 31	18 32

Data for	08-01-2008
Julian Day	2454679.50
Ayanamsa	23 58 48
SVP	05 ♓ 08 36
☽ ☊ Mean	19 ♒ 04 R

● ◐ PHASES ○ ◑

01	10:12	☉	09♌32
08	20:20	◐	16♏38
16	21:17	✹	24♒21
23	23:49	◑	01♊12
30	19:58	●	07♍48

LAST ASPECT ☽ Day	h m	INGRESS Day	h m	
02	18:59	02	20:59	♍
05	00:16	05	02:28	♎
07	09:02	07	11:27	♏
09	21:03	09	23:11	♐
12	09:05	12	11:43	♑
14	17:10	14	22:57	♒
17	05:15	17	07:47	♓
19	11:41	19	14:11	♈
21	16:54	21	18:38	♉
23	09:20	23	21:49	♊
25	21:52	26	00:19	♋
28	00:14	28	02:51	♌
30	03:44	30	06:19	♍

DECLINATION

Day	☉	☽	☿	♀	♂	♃	♄	♅	♆	♇
01 Fr	17N58	20N44	19N00	15N02	05N20	22S52	10N13	03S50	14S06	17S09
02 Sa	17 43	15 36	18 26	14 38	05 04	22 53	10 10	03 51	14 06	17 10
03 Su	17 28	09 44	17 51	14 13	04 49	22 54	10 07	03 51	14 07	17 10
04 Mo	17 12	03 33	17 14	13 48	04 34	22 54	10 05	03 52	14 08	17 10
05 Tu	16 56	02S36	16 36	13 22	04 19	22 55	10 02	03 53	14 08	17 10
06 We	16 39	08 27	15 57	12 56	04 03	22 56	09 59	03 53	14 09	17 11
07 Th	16 23	13 49	15 17	12 30	03 48	22 57	09 57	03 54	14 09	17 11
08 Fr	16 06	18 31	14 36	12 03	03 33	22 57	09 54	03 55	14 10	17 11
09 Sa	15 48	22 24	13 54	11 37	03 17	22 58	09 51	03 55	14 10	17 11
10 Su	15 31	25 18	13 12	11 09	03 02	22 59	09 49	03 56	14 11	17 12
11 Mo	15 13	27 04	12 29	10 42	02 46	22 59	09 46	03 57	14 11	17 12
12 Tu	14 55	27 37	11 46	10 14	02 31	23 00	09 43	03 58	14 12	17 12
13 We	14 37	26 52	11 02	09 46	02 15	23 01	09 40	03 59	14 12	17 12
14 Th	14 19	24 51	10 18	09 18	01 59	23 01	09 38	03 59	14 13	17 13
15 Fr	14 00	21 39	09 34	08 49	01 44	23 02	09 35	04 00	14 14	17 13
16 Sa	13 41	17 24	08 50	08 21	01 28	23 02	09 32	04 01	14 14	17 13
17 Su	13 22	12 18	08 06	07 52	01 12	23 03	09 29	04 02	14 15	17 13
18 Mo	13 03	06 34	07 21	07 22	00 57	23 03	09 27	04 03	14 15	17 14
19 Tu	12 43	00 28	06 37	06 53	00 41	23 04	09 24	04 03	14 16	17 14
20 We	12 24	05N45	05 53	06 23	00 25	23 04	09 21	04 04	14 16	17 14
21 Th	12 04	11 47	05 09	05 54	00 10	23 05	09 18	04 05	14 17	17 14
22 Fr	11 44	17 19	04 25	05 24	00S06	23 05	09 15	04 06	14 17	17 15
23 Sa	11 23	22 00	03 42	04 54	00 22	23 05	09 13	04 07	14 18	17 15
24 Su	11 03	25 27	02 58	04 24	00 38	23 06	09 10	04 08	14 18	17 15
25 Mo	10 42	27 20	02 16	03 53	00 54	23 06	09 07	04 09	14 19	17 16
26 Tu	10 21	27 24	01 33	03 23	01 09	23 06	09 04	04 09	14 19	17 16
27 We	10 00	25 40	00 51	02 52	01 25	23 07	09 01	04 10	14 20	17 16
28 Th	09 39	22 17	00 09	02 22	01 41	23 07	08 58	04 11	14 21	17 16
29 Fr	09 18	17 37	00S32	01 51	01 57	23 07	08 56	04 12	14 21	17 17
30 Sa	08 57	12 04	01 13	01 20	02 13	23 08	08 53	04 13	14 22	17 17
31 Su	08 35	06 00	01 53	00 49	02 29	23 08	08 50	04 14	14 22	17 17

ASPECTARIAN

01	09:42	☽ ∥ ☿
	14:07	☽ ∥ ☉
	15:52	☽ ☌ ☿
	17:08	☽ # ♇
02	04:26	☽ ∥ ♀
	06:17	☽ # ♆
	09:18	☽ ☍ ♆
	13:19	☽ ☌ ♀
	18:59	☽ △ ♇
	22:26	☽ ∥ ♄
03	05:30	♀ # ♆
	10:58	☽ ☌ ♄
	19:56	☽ ∥ ♂
	22:16	☽ △ ♃
	22:49	☽ # ♅
04	02:23	☉ # ♇
	02:24	☿ # ♇
	02:24	☿ ∥ ☉
	09:14	☽ ☌ ♂
	12:03	☽ ☍ ♅
05	00:16	☽ □ ♇
	04:51	♀ △ ♇
	05:09	☽ ∥ ♅
	06:36	☽ # ♂
06	04:20	♀ ♍
	04:50	☽ ✶ ☉
	05:00	☽ □ ♃
	06:35	☽ # ♄
	14:55	♂ # ♅
	18:19	☽ # ♀
	18:38	♂ ☍ ♅
	18:41	☿ ☍ ♆
	22:01	☽ △ ♆
	22:38	☽ ✶ ☿
07	01:35	☽ ∥ ♆
	06:11	☽ # ☿
	09:02	☽ ✶ ♇
	11:51	☽ # ☉
	14:55	☽ ✶ ♀
	16:46	☽ ∥ ♇
08	04:20	☽ ✶ ♄
	14:47	☿ # ♆
	15:01	☽ ✶ ♃
09	04:10	☽ ∥ ♃
	06:59	☽ △ ♅
	09:07	☽ □ ♆
	10:29	☽ ✶ ♂
	18:16	☿ △ ♇
	21:03	☽ □ ☿
10	09:40	☽ □ ♀
	10:51	☿ ♍
	17:09	☽ □ ♄
11	14:07	☽ △ ☉
	19:23	☽ □ ♅
	21:32	☽ ✶ ♆
12	02:25	☽ □ ♂
	09:05	☽ ☌ ♇
	20:23	☽ △ ☿
13	04:58	☽ △ ♀
	05:29	♀ ∥ ♄
	06:05	☽ △ ♄
	14:44	☽ ☌ ♃
	17:03	♀ ☌ ♄
14	06:58	☽ ✶ ♅
	07:06	☉ # ♆
	14:37	☽ ∥ ♃
	17:10	☽ △ ♂
	23:33	☿ ∥ ♄
15	07:43	☉ ☍ ♆
	20:00	☿ ☌ ♄
16	00:51	☽ ∥ ♇
	15:14	☽ ∥ ♆
	[illegible]	☽ ☌ ♆
	18:53	☽ # ☉
17	00:16	♀ △ ♃
	05:15	☽ ✶ ♇
	06:13	♂ □ ♇
	12:07	☽ # ♄
	20:22	☽ # ☿
	20:27	☽ # ♀
	22:16	☿ ∥ ♀
18	01:23	☿ △ ♃
	01:47	☽ ☍ ♄
	08:02	☽ ✶ ♃
	09:01	☽ ☍ ☿
	10:03	☽ ∥ ♅
	11:27	☽ ☍ ♀
	23:08	☽ # ♂
	23:14	☽ ☌ ♅
19	04:17	☽ ∥ ♂
	10:03	♂ ♎
	11:41	☽ □ ♇
	14:23	☽ ☍ ♂
	17:30	☽ # ♅
20	00:29	☽ ∥ ☿
	02:20	☽ ∥ ♀
	13:15	☽ □ ♃
	14:03	☽ ∥ ♄
21	01:05	☽ ∥ ☉
	06:00	☽ ✶ ♆
	06:50	☉ △ ♇
	10:29	☽ # ♆
	15:45	☿ ☌ ♀
	16:11	☽ △ ♇
	16:54	☽ △ ☉
	23:40	☽ # ♇
22	10:25	☿ # ♅
	[illegible]	☽ △ ♄
	16:53	☽ △ ♃
	18:03	☉ ♍
23	06:13	☿ ☍ ♅
	06:33	☽ △ ♀
	06:38	☽ # ♃
	07:22	☽ ✶ ♅
	07:31	☽ △ ☿
	09:20	☽ □ ♆
	15:49	♀ ☍ ♅
24	02:52	☽ △ ♂
	12:16	♀ # ♅
	15:43	☽ □ ♄
25	09:52	☽ □ ♅
	11:51	☽ △ ♆
	13:57	☽ □ ♀
	[illegible]	☽ □ [illegible]
	21:52	☽ ☍ ♇
26	05:59	☽ ✶ ☉
	07:43	☽ □ ♂
	09:42	☿ # ♂
	18:35	☽ ✶ ♄
	21:47	☽ ☍ ♃
27	12:11	☽ △ ♅
	18:58	☽ # ♃
	21:11	☽ ✶ ♀
28	00:14	☽ ✶ ☿
	01:37	☿ □ ♇
	12:44	☽ ✶ ♂
	20:55	♀ # ♂
29	01:34	☽ # ♇
	[illegible]	☿ ♎
	09:51	♀ □ ♇
	14:26	☽ # ♆
	17:14	☽ ☍ ♆
30	02:27	☿ # ♀
	03:44	☽ △ ♇
	04:55	☉ ∥ ♄
	12:50	☽ ∥ ♄
	13:16	☽ ∥ ☉
	14:42	♀ ♎
31	02:21	☽ ☌ ♄
	04:31	☽ △ ♃
	06:50	☽ # ♅
	13:04	☽ # ♂
	14:23	☽ # ☿
	19:44	☽ ☍ ♅
	21:52	☽ ∥ ♀

⚷ Chiron

01 Dec.	08 S 26
01	19♒04R
04	18 55
07	18 45
10	18 36
13	18 27
16	18 17
19	18 08
22	17 58
25	17 49
28	17 40
31	17 32

September 2008

Day	S. T. h m s	☉ ° ' "	☽ ° ' "	☿ ° '	♀ ° '	♂ ° '	♃ ° '	♄ ° '	♅ ° '	♆ ° '	♇ ° '	☊ True ° '
01 Mo	22 42 07	08♍ 55 56	23♍ 31 32	03♎ 53	01♎ 42	08♎ 04	12♑R37	11♍ 32	21♓R09	22♒R24	28♐R31	18♒R30
02 Tu	22 46 04	09 54 01	06♎ 40 01	05 10	02 56	08 43	12 36	11 39	21 07	22 23	28 30	18 27
03 We	22 50 00	10 52 08	19 30 07	06 26	04 09	09 22	12 35	11 47	21 05	22 21	28 30	18 23
04 Th	22 53 57	11 50 17	02♏ 02 41	07 40	05 23	10 01	12 34	11 54	21 02	22 20	28 30	18 19
05 Fr	22 57 53	12 48 27	14 19 47	08 53	06 36	10 40	12 33	12 02	21 00	22 18	28 30	18 15
06 Sa	23 01 50	13 46 38	26 24 30	10 03	07 50	11 19	12 32	12 10	20 57	22 17	28 30	18 13
07 Su	23 05 47	14 44 51	08♐ 20 45	11 12	09 03	11 58	12 32	12 17	20 55	22 15	28 30	18 12
08 Mo	23 09 43	15 43 06	20 13 02	12 18	10 17	12 37	12 32	12 25	20 53	22 14	28 30	18D 12
09 Tu	23 13 40	16 41 22	02♑ 06 11	13 22	11 30	13 16	12D 32	12 32	20 50	22 12	28D 30	18 13
10 We	23 17 36	17 39 40	14 05 03	14 24	12 44	13 55	12 32	12 40	20 48	22 11	28 30	18 16
11 Th	23 21 33	18 37 59	26 14 15	15 24	13 57	14 34	12 33	12 47	20 45	22 09	28 30	18 19
12 Fr	23 25 29	19 36 20	08♒ 37 50	16 20	15 11	15 14	12 33	12 55	20 43	22 08	28 30	18 21
13 Sa	23 29 26	20 34 42	21 19 00	17 14	16 24	15 53	12 34	13 02	20 41	22 07	28 30	18 22
14 Su	23 33 22	21 33 06	04♓ 19 50	18 05	17 38	16 33	12 35	13 10	20 38	22 05	28 30	18R 21
15 Mo	23 37 19	22 31 32	17 40 52	18 53	18 51	17 12	12 36	13 17	20 36	22 04	28 30	18 19
16 Tu	23 41 16	23 29 59	01♈ 21 01	19 37	20 04	17 52	12 38	13 25	20 33	22 03	28 30	18 15
17 We	23 45 12	24 28 29	15 17 33	20 18	21 18	18 31	12 39	13 32	20 31	22 01	28 31	18 09
18 Th	23 49 09	25 27 01	29 26 27	20 54	22 31	19 11	12 41	13 40	20 29	22 00	28 31	18 03
19 Fr	23 53 05	26 25 34	13♉ 42 59	21 26	23 44	19 50	12 43	13 47	20 26	21 59	28 31	17 58
20 Sa	23 57 02	27 24 10	28 02 22	21 54	24 58	20 30	12 45	13 55	20 24	21 57	28 31	17 53
21 Su	00 00 58	28 22 48	12♊ 20 28	22 16	26 11	21 10	12 48	14 02	20 22	21 56	28 32	17 50
22 Mo	00 04 55	29 21 29	26 34 01	22 33	27 24	21 50	12 50	14 10	20 19	21 55	28 32	17 49
23 Tu	00 08 51	00♎ 20 12	10♋ 40 50	22 45	28 38	22 30	12 53	14 17	20 17	21 54	28 33	17D 49
24 We	00 12 48	01 18 57	24 39 34	22 50	29 51	23 10	12 56	14 24	20 14	21 53	28 33	17 49
25 Th	00 16 45	02 17 44	08♌ 29 30	22R 49	01♏ 04	23 50	12 59	14 32	20 12	21 51	28 34	17 50
26 Fr	00 20 41	03 16 33	22 10 01	22 40	02 17	24 30	13 02	14 39	20 10	21 50	28 34	17R 50
27 Sa	00 24 38	04 15 25	05♍ 40 24	22 25	03 31	25 10	13 06	14 46	20 08	21 49	28 35	17 48
28 Su	00 28 34	05 14 19	18 59 42	22 02	04 44	25 50	13 09	14 54	20 05	21 48	28 35	17 44
29 Mo	00 32 31	06 13 14	02♎ 06 45	21 32	05 57	26 30	13 13	15 01	20 03	21 47	28 36	17 38
30 Tu	00 36 27	07 12 12	15 00 32	20 54	07 10	27 11	13 17	15 08	20 01	21 46	28 36	17 30

Data for	09-01-2008
Julian Day	2454710.50
Ayanamsa	23 58 53
SVP	05 ♓ 08 32
☽ ☊ Mean	17 ♒ 25 R

● ◐ PHASES ○ ◑

Day	h m	Phase	Position
07	14:05	◐	15♐19
15	09:13	○	22♓54
22	05:04	◑	29♊34
29	08:13	●	06♎33

LAST ASPECT Day	h m	☽ INGRESS Day	h m	Sign
01	09:02	01	11:45	♎
03	17:10	03	20:03	♏
05	15:46	06	07:12	♐
08	16:43	08	19:46	♑
10	13:16	11	07:20	♒
13	13:19	13	16:05	♓
15	19:03	15	21:39	♈
17	22:26	18	00:57	♉
19	22:51	20	03:17	♊
22	05:05	22	05:49	♋
23	21:17	24	09:14	♌
26	11:21	26	13:53	♍
28	17:31	28	20:06	♎

DECLINATION

Day	☉	☽	☿	♀	♂	♃	♄	♅	♆	♇
01 Mo	08N13	00S 11	02S 33	00N19	02S 45	23S 08	08N47	04S 15	14S 23	17S 18
02 Tu	07 51	06 14	03 12	00S 12	03 00	23 08	08 44	04 16	14 23	17 18
03 We	07 29	11 52	03 50	00 43	03 16	23 08	08 41	04 17	14 24	17 18
04 Th	07 07	16 53	04 27	01 14	03 32	23 09	08 39	04 18	14 24	17 18
05 Fr	06 45	21 07	05 04	01 45	03 48	23 09	08 36	04 19	14 25	17 19
06 Sa	06 23	24 23	05 40	02 16	04 04	23 09	08 33	04 20	14 25	17 19
07 Su	06 00	26 33	06 15	02 47	04 20	23 09	08 30	04 21	14 26	17 19
08 Mo	05 38	27 30	06 49	03 18	04 35	23 09	08 27	04 21	14 26	17 20
09 Tu	05 15	27 12	07 22	03 49	04 51	23 09	08 24	04 22	14 27	17 20
10 We	04 53	25 38	07 54	04 19	05 07	23 09	08 22	04 23	14 27	17 20
11 Th	04 30	22 51	08 24	04 50	05 23	23 09	08 19	04 24	14 28	17 20
12 Fr	04 07	18 58	08 54	05 20	05 39	23 09	08 16	04 25	14 28	17 21
13 Sa	03 44	14 09	09 22	05 51	05 54	23 09	08 13	04 26	14 28	17 21
14 Su	03 21	08 35	09 49	06 21	06 10	23 09	08 10	04 27	14 29	17 21
15 Mo	02 58	02 31	10 14	06 51	06 26	23 09	08 07	04 28	14 29	17 22
16 Tu	02 35	03N47	10 37	07 22	06 41	23 09	08 05	04 29	14 30	17 22
17 We	02 12	10 01	10 59	07 51	06 57	23 09	08 02	04 30	14 30	17 22
18 Th	01 48	15 51	11 18	08 21	07 13	23 09	07 59	04 31	14 31	17 23
19 Fr	01 25	20 52	11 36	08 51	07 28	23 09	07 56	04 32	14 31	17 23
20 Sa	01 02	24 40	11 51	09 20	07 44	23 09	07 53	04 33	14 31	17 23
21 Su	00 39	26 56	12 04	09 49	07 59	23 08	07 51	04 34	14 32	17 23
22 Mo	00 15	27 25	12 14	10 18	08 15	23 08	07 48	04 35	14 32	17 24
23 Tu	00S 08	26 07	12 21	10 47	08 30	23 08	07 45	04 36	14 33	17 24
24 We	00 31	23 11	12 24	11 16	08 45	23 08	07 42	04 36	14 33	17 24
25 Th	00 55	18 57	12 25	11 44	09 01	23 08	07 40	04 37	14 33	17 25
26 Fr	01 18	13 45	12 22	12 12	09 16	23 07	07 37	04 38	14 34	17 25
27 Sa	01 42	07 58	12 15	12 40	09 31	23 07	07 34	04 39	14 34	17 25
28 Su	02 05	01 54	12 03	13 07	09 47	23 07	07 32	04 40	14 34	17 26
29 Mo	02 28	04S 07	11 48	13 34	10 02	23 06	07 29	04 41	14 35	17 26
30 Tu	02 52	09 53	11 28	14 01	10 17	23 06	07 26	04 42	14 35	17 26

ASPECTARIAN

01 00:25 ☽ # ♀
09:02 ☽ □ ♇
10:19 ☽ ∥ ☿
10:26 ☽ ∥ ♂
12:15 ☿ ∥ ♂
16:01 ☽ ∥ ♅
16:24 ☽ ☌ ♀
20:56 ☽ ☌ ☿
02 03:59 ☽ ☌ ♂
06:18 ☽ # ☉
10:18 ☽ # ♄
11:00 ☽ □ ♃
03 05:24 ☽ △ ♆
11:39 ☽ ∥ ♆
17:10 ☽ ⚹ ♇
17:37 ☿ ∥ ♅
04 02:00 ☉ ☌ ♄
02:11 ☽ ∥ ♇
17:42 ☉ △ ♃
19:26 ☽ ⚹ ♄
20:30 ☽ ⚹ ♃
20:45 ☽ ⚹ ☉
05 13:10 ☽ △ ♅
14:00 ☽ ∥ ♃
15:46 ☽ □ ♆
06 17:55 ☿ # ☉
07 01:27 ♂ ∥ ♅
01:36 ☽ ⚹ ♀
06:21 ☽ ⚹ ☿
07:43 ☽ ⚹ ♂
08:03 ☽ □ ♄
21:06 ♂ □ ♃
08 01:20 ☽ □ ♅
04:04 ☽ ⚹ ♆
04:17 ♃ SD
05:09 ☿ □ ♃
16:43 ☽ ☌ ♇
17:45 ☿ ☌ ♂
23:18 ♃ △ ♄
09 03:13 ♇ SD
14:59 ☉ # ♂
20:13 ♀ □ ♃
20:55 ☽ ☌ ♃
21:00 ☽ □ ♀
21:08 ☽ △ ♄
23:39 ☽ □ ♂
10 00:41 ☽ □ ☿
03:20 ♀ ∥ ♅
07:44 ☽ △ ☉
13:16 ☽ ⚹ ♅
14:59 ♀ # ☉
19:51 ☿ # ♄
21:42 ☽ ∥ ♃
11 05:32 ☉ # ♅
12 02:06 ♀ ☌ ♂
08:30 ☽ ∥ ♇
13:15 ☽ △ ♂
13:48 ☽ △ ♀
15:47 ☽ △ ☿
22:28 ☽ ∥ ♆
13 01:29 ☽ ☌ ♆
02:22 ☉ ☍ ♅
05:34 ♀ ∥ ♂
13:19 ☽ ⚹ ♇
19:17 ☽ ∥ ☿
14 01:43 ☽ # ♄
08:19 ☽ ∥ ♀
09:21 ☽ ∥ ♂
14:56 ☽ ⚹ ♃
16:06 ☽ ☍ ♄
16:28 ☽ ∥ ♅
22:10 ☽ # ☉
15 01:38 ☿ ☌ ♀
05:09 ☽ ☌ ♅
19:03 ☽ □ ♇
19:42 ☽ ∥ ☉
16 02:39 ☽ # ♅
11:32 ☽ # ♂
14:48 ☽ # ♀
16:16 ☽ ∥ ♄
19:29 ☽ □ ♃
17 03:59 ☽ # ☿
05:46 ☽ ☍ ♂
07:42 ♀ # ♄
08:55 ☽ ☍ ☿
11:11 ☽ ☍ ♀
11:26 ☽ ⚹ ♆
13:59 ♀ △ ♆
18:15 ☽ # ♆
22:26 ☽ △ ♇
18 06:51 ☽ # ♇
22:19 ☽ △ ♃
19 00:07 ☽ △ ♄
11:14 ☽ ⚹ ♅
13:18 ☽ # ♃
13:49 ☽ □ ♆
22:51 ☽ △ ☉
20 03:23 ☿ △ ♆
12:55 ♂ # ♄
21 02:53 ☽ □ ♄
03:45 ☉ □ ♇
13:28 ☽ □ ♅
15:36 ☽ △ ♂
16:09 ☽ △ ♆
17:06 ☽ △ ☿
22 01:33 ☽ △ ♀
02:59 ♂ △ ♆
03:20 ☽ ☍ ♇
15:45 ☉ ♎
22:24 ♀ ⚹ ♇
23 03:47 ☽ ☍ ♃
06:13 ☽ ⚹ ♄
10:51 ☿ ☌ ♂
16:25 ☽ △ ♅
20:50 ☽ □ ☿
21:17 ☽ □ ♂
24 00:26 ☽ # ♃
02:59 ♀ ♏
07:18 ☿ SR
09:50 ☽ □ ♀
12:24 ☽ ⚹ ☉
25 07:30 ☽ # ♇
20:29 ☽ # ♆
23:25 ☽ ☍ ♆
26 00:53 ☽ ⚹ ☿
04:20 ☽ ⚹ ♂
06:03 ☽ # ☿
06:07 ☽ # ♀
06:56 ☿ ∥ ♀
11:21 ☽ △ ♇
17:57 ☽ # ♂
19:45 ☽ ⚹ ♀
27 01:35 ☽ ∥ ♄
13:08 ☽ # ♅
13:23 ☽ △ ♃
16:30 ☽ ☌ ♄
23:20 ☽ # ☉
28 01:59 ☽ ☍ ♅
12:25 ☿ △ ♆
16:52 ☽ ∥ ☉
17:31 ☽ □ ♇
29 02:14 ☽ ∥ ♅
13:39 ☽ # ♄
20:45 ☽ □ ♃
30 01:48 ☽ ∥ ♂
06:29 ☽ ∥ ☿
10:31 ☽ ☌ ☿
12:44 ☽ △ ♆
20:30 ☽ ∥ ♀
21:25 ☽ ∥ ♆

⚷ Chiron

01 Dec.	08 S 57
03	17♒23R
06	17 15
09	17 07
12	16 59
15	16 52
18	16 45
21	16 39
24	16 33
27	16 28
30	16 23

October 2008

Day	S. T.	☉	☽	☿	♀	♂	♃	♄	♅	♆	♇	☊ True
	h m s	° ′ ″	° ′ ″	° ′	° ′	° ′	° ′	° ′	° ′	° ′	° ′	° ′
01 We	00 40 24	08♎ 11 12	27♎ 40 25	20♎ R09	08♏ 23	27♎ 51	13♑ 21	15♍ 15	19♓ R58	21♒ R45	28♐ 37	17♒ R20
02 Th	00 44 20	09 10 14	10♏ 06 29	19 17	09 36	28 31	13 26	15 22	19 56	21 44	28 38	17 10
03 Fr	00 48 17	10 09 18	22 19 44	18 19	10 50	29 12	13 30	15 29	19 54	21 43	28 39	17 01
04 Sa	00 52 13	11 08 23	04♐ 22 09	17 16	12 03	29 52	13 35	15 37	19 52	21 42	28 39	16 53
05 Su	00 56 10	12 07 31	16 16 39	16 09	13 16	00♏ 33	13 40	15 44	19 50	21 41	28 40	16 47
06 Mo	01 00 07	13 06 40	28 07 08	14 59	14 29	01 13	13 45	15 51	19 47	21 40	28 41	16 44
07 Tu	01 04 03	14 05 51	09♑ 58 10	13 49	15 42	01 54	13 50	15 58	19 45	21 39	28 42	16D 43
08 We	01 07 60	15 05 04	21 54 53	12 40	16 55	02 35	13 55	16 05	19 43	21 39	28 43	16 44
09 Th	01 11 56	16 04 18	04♒ 02 34	11 34	18 08	03 16	14 01	16 12	19 41	21 38	28 44	16 45
10 Fr	01 15 53	17 03 35	16 26 22	10 34	19 21	03 56	14 06	16 18	19 39	21 37	28 45	16 46
11 Sa	01 19 49	18 02 53	29 10 49	09 40	20 34	04 37	14 12	16 25	19 37	21 36	28 46	16R 46
12 Su	01 23 46	19 02 13	12♓ 19 20	08 54	21 47	05 18	14 18	16 32	19 35	21 36	28 47	16 43
13 Mo	01 27 42	20 01 34	25 53 34	08 18	23 00	05 59	14 25	16 39	19 33	21 35	28 48	16 38
14 Tu	01 31 39	21 00 58	09♈ 52 50	07 53	24 13	06 40	14 31	16 46	19 31	21 34	28 49	16 31
15 We	01 35 36	22 00 24	24 13 41	07 38	25 25	07 21	14 37	16 52	19 29	21 34	28 50	16 21
16 Th	01 39 32	22 59 51	08♉ 50 15	07D 34	26 38	08 02	14 44	16 59	19 27	21 33	28 51	16 11
17 Fr	01 43 29	23 59 21	23 34 53	07 41	27 51	08 44	14 51	17 05	19 25	21 33	28 52	16 01
18 Sa	01 47 25	24 58 53	08♊ 19 34	07 59	29 04	09 25	14 58	17 12	19 24	21 32	28 53	15 52
19 Su	01 51 22	25 58 28	22 57 03	08 27	00♐ 17	10 06	15 05	17 18	19 22	21 32	28 54	15 46
20 Mo	01 55 18	26 58 05	07♋ 22 01	09 05	01 29	10 47	15 12	17 25	19 20	21 31	28 56	15 41
21 Tu	01 59 15	27 57 44	21 31 24	09 51	02 42	11 29	15 20	17 31	19 18	21 31	28 57	15 39
22 We	02 03 11	28 57 25	05♌ 24 14	10 45	03 55	12 10	15 27	17 38	19 17	21 30	28 58	15 38
23 Th	02 07 08	29 57 08	19 01 04	11 45	05 07	12 52	15 35	17 44	19 15	21 30	29 00	15 37
24 Fr	02 11 05	00♏ 56 54	02♍ 23 17	12 52	06 20	13 33	15 43	17 50	19 13	21 30	29 01	15 36
25 Sa	02 15 01	01 56 42	15 32 27	14 05	07 33	14 15	15 51	17 56	19 12	21 29	29 02	15 33
26 Su	02 18 58	02 56 32	28 29 49	15 22	08 45	14 57	15 59	18 02	19 10	21 29	29 04	15 27
27 Mo	02 22 54	03 56 24	11♎ 16 16	16 43	09 58	15 39	16 07	18 08	19 09	21 29	29 05	15 18
28 Tu	02 26 51	04 56 19	23 52 11	18 08	11 10	16 20	16 15	18 14	19 07	21 29	29 06	15 06
29 We	02 30 47	05 56 15	06♏ 17 47	19 35	12 23	17 02	16 24	18 20	19 06	21 28	29 08	14 53
30 Th	02 34 44	06 56 13	18 33 19	21 05	13 35	17 44	16 33	18 26	19 05	21 28	29 09	14 38
31 Fr	02 38 40	07 56 14	00♐ 39 24	22 36	14 48	18 26	16 42	18 32	19 03	21 28	29 11	14 24

Data for	10-01-2008
Julian Day	2454740.50
Ayanamsa	23 58 56
SVP	05 ♓ 08 26
☽ ☊ Mean	15 ♒ 50 R

● ◐ PHASES ○ ◑

Day	Time	Phase	Position
07	09:04	◐	14♑28
14	20:02	○	21♈51
21	11:55	◑	28♋27
28	23:14	●	05♏54

LAST ASPECT ☽ INGRESS

Day	h m	Day	h m	Sign
01	01:49	01	04:27	♏
02	22:47	03	15:15	♐
06	01:09	06	03:49	♑
07	19:38	08	16:03	♒
10	23:13	11	01:31	♓
13	05:02	13	07:07	♈
15	07:36	15	09:31	♉
17	07:34	17	10:26	♊
19	09:53	19	11:41	♋
21	11:56	21	14:36	♌
23	17:54	23	19:41	♍
26	01:03	26	02:48	♎
28	10:06	28	11:48	♏
30	05:45	30	22:41	♐

DECLINATION

Day	☉	☽	☿	♀	♂	♃	♄	♅	♆	♇
01 We	03S 15	15S 07	11S 03	14S 28	10S 32	23S 06	07N23	04S 43	14S 36	17S 26
02 Th	03 38	19 38	10 34	14 54	10 47	23 05	07 21	04 44	14 36	17 27
03 Fr	04 01	23 15	10 01	15 20	11 02	23 05	07 18	04 44	14 36	17 27
04 Sa	04 24	25 47	09 25	15 45	11 17	23 04	07 16	04 45	14 36	17 27
05 Su	04 48	27 09	08 45	16 10	11 31	23 04	07 13	04 46	14 37	17 28
06 Mo	05 11	27 15	08 02	16 35	11 46	23 04	07 10	04 47	14 37	17 28
07 Tu	05 34	26 07	07 18	17 00	12 01	23 03	07 08	04 48	14 37	17 28
08 We	05 57	23 46	06 33	17 23	12 15	23 03	07 05	04 49	14 38	17 29
09 Th	06 19	20 20	05 49	17 47	12 30	23 02	07 03	04 49	14 38	17 29
10 Fr	06 42	15 56	05 06	18 10	12 44	23 01	07 00	04 50	14 38	17 29
11 Sa	07 05	10 44	04 26	18 33	12 59	23 01	06 57	04 51	14 38	17 29
12 Su	07 27	04 54	03 50	18 55	13 13	23 00	06 55	04 52	14 39	17 30
13 Mo	07 50	01N19	03 18	19 16	13 27	23 00	06 52	04 52	14 39	17 30
14 Tu	08 12	07 39	02 51	19 38	13 41	22 59	06 50	04 53	14 39	17 30
15 We	08 34	13 44	02 29	19 58	13 55	22 58	06 48	04 54	14 39	17 31
16 Th	08 56	19 10	02 13	20 18	14 09	22 58	06 45	04 55	14 39	17 31
17 Fr	09 18	23 30	02 03	20 38	14 23	22 57	06 43	04 55	14 40	17 31
18 Sa	09 40	26 17	01 58	20 57	14 37	22 56	06 40	04 56	14 40	17 31
19 Su	10 02	27 16	01 59	21 16	14 50	22 56	06 38	04 57	14 40	17 32
20 Mo	10 24	26 22	02 05	21 34	15 04	22 55	06 36	04 57	14 40	17 32
21 Tu	10 45	23 45	02 16	21 51	15 17	22 54	06 33	04 58	14 40	17 32
22 We	11 06	19 47	02 31	22 08	15 31	22 53	06 31	04 59	14 40	17 32
23 Th	11 27	14 50	02 50	22 24	15 44	22 52	06 29	04 59	14 40	17 33
24 Fr	11 48	09 15	03 13	22 40	15 57	22 51	06 26	05 00	14 40	17 33
25 Sa	12 09	03 22	03 38	22 55	16 10	22 51	06 24	05 00	14 41	17 33
26 Su	12 30	02S 33	04 07	23 09	16 23	22 50	06 22	05 01	14 41	17 34
27 Mo	12 50	08 17	04 38	23 23	16 35	22 49	06 20	05 02	14 41	17 34
28 Tu	13 10	13 36	05 10	23 36	16 48	22 48	06 17	05 02	14 41	17 34
29 We	13 30	18 16	05 44	23 49	17 01	22 47	06 15	05 03	14 41	17 34
30 Th	13 50	22 08	06 20	24 00	17 13	22 46	06 13	05 03	14 41	17 35
31 Fr	14 09	24 59	06 56	24 11	17 25	22 45	06 11	05 04	14 41	17 35

ASPECTARIAN

01 00:21 ☽ ☌ ♂
01:49 ☽ ⚹ ♇
07:06 ♀ ∥ ♆
11:47 ☽ ∥ ♇
17:31 ☿ ∥ ♂
22:55 ☽ ☌ ♀
02 03:59 ♂ ⚹ ♇
06:31 ☽ ⚹ ♃
10:23 ☽ ⚹ ♄
19:13 ☽ △ ♅
22:42 ☽ ∥ ♃
22:47 ☽ □ ♆
04 04:34 ♂ ♏
14:51 ☽ ⚹ ☉
22:24 ☉ ∥ ♅
22:53 ☽ □ ♄
23:45 ☽ ⚹ ☿
05 07:10 ☽ □ ♅
08:23 ♀ ⚹ ♃
10:56 ☽ ⚹ ♆
06 01:09 ☽ ☌ ♇
06:41 ☽ ⚹ ♂
16:53 ☉ □ ♃
20:53 ☿ ☌ ☉
23:46 ☿ □ ♃
07 05:43 ♀ ⚹ ♄
05:49 ☿ # ♄
07:04 ☽ □ ☿
07:51 ☽ ☌ ♃
12:11 ☽ △ ♄
12:51 ☽ ⚹ ♀
19:38 ☽ ⚹ ♅
08 05:11 ♀ ∥ ♇
05:50 ☽ ∥ ♃
13:02 ☿ ∥ ☉
22:22 ☽ □ ♂
09 13:22 ☽ ∥ ♀
13:32 ☽ △ ☿
16:06 ☽ ∥ ♇
10 01:17 ☽ △ ☉
05:48 ♀ △ ♅
06:08 ☽ □ ♀
06:20 ☽ ∥ ♆
09:11 ☿ ∥ ♅
09:50 ☽ ☌ ♆
14:29 ☽ ∥ ♂
17:04 ☉ # ♄
23:13 ☽ ⚹ ♇
11 10:35 ☽ △ ♂
14:23 ☽ ∥ ☉
15:54 ☽ # ♄
20:22 ♀ □ ♆
12 00:11 ☽ ∥ ♅
03:35 ☽ ⚹ ♃
04:42 ☽ ∥ ☿
07:36 ☽ ☍ ♄
12:54 ☽ ☌ ♅
18:27 ☽ △ ♀
13 05:02 ☽ □ ♇
06:59 ☽ # ☿
13:31 ☽ # ♅
20:41 ☽ ☍ ☿
20:57 ☽ ∥ ♄
14 02:17 ☽ # ☉
07:52 ☽ □ ♃
13:19 ☉ △ ♆
19:35 ☽ ⚹ ♆
15 00:48 ☽ # ♂
03:49 ☽ # ♇
07:36 ☽ △ ♇
16:14 ☽ # ♇
20:07 ☿ SD
22:38 ☽ ☍ ♂
16 06:05 ☽ # ♀
09:41 ☽ △ ♃
13:22 ☽ △ ♄
17:15 ☽ ⚹ ♅
20:29 ☽ # ♃
20:41 ☽ □ ♆
17 07:34 ☽ ☍ ♀
23:26 ☽ △ ☿
18 05:29 ♂ ∥ ♆
14:38 ☽ □ ♄
18:06 ☽ □ ♅
18:31 ♀ ♐
21:39 ☽ △ ♆
19 05:22 ☽ △ ☉
09:53 ☽ ☍ ♇
20 03:01 ☽ □ ☿
06:03 ☽ △ ♂
13:21 ☽ ☍ ♃
17:07 ☽ ⚹ ♄
20:13 ☽ △ ♅
21 05:52 ☽ # ♃
11:38 ☽ # ♀
21:09 ☽ △ ♀
22 00:22 ☉ ⚹ ♇
10:05 ☽ ⚹ ☿
11:23 ☽ # ♇
12:31 ☽ □ ♂
20:06 ☽ # ♂
23 00:45 ☽ # ♆
01:09 ☉ ♏
04:25 ☽ ☍ ♆
13:57 ☽ # ☉
17:54 ☽ △ ♇
21:11 ☽ ⚹ ☉
24 07:53 ☽ □ ♀
11:41 ☽ ∥ ♄
17:25 ☽ # ♅
17:37 ♀ ∥ ♃
21:30 ☽ ⚹ ♂
23:01 ☽ # ☿
25 00:34 ☽ △ ♃
04:27 ☽ ☌ ♄
06:43 ☽ ☍ ♅
26 01:03 ☽ □ ♇
07:03 ☽ ∥ ☿
10:12 ☽ ∥ ♅
12:15 ☿ □ ♃
15:44 ☽ # ♄
21:16 ☽ ⚹ ♀
27 09:18 ☽ □ ♃
11:38 ☽ ☌ ☿
18:08 ☿ ∥ ♅
19:25 ☽ △ ♆
20:30 ♂ ⚹ ♃
21:51 ☽ ∥ ☉
28 05:16 ☽ ∥ ♆
10:06 ☽ ⚹ ♇
16:48 ☽ ∥ ♂
20:08 ☽ ∥ ♇
29 19:48 ☿ # ♄
20:00 ☽ ⚹ ♃
22:17 ☽ ☌ ♂
23:46 ☽ ⚹ ♄
30 01:01 ☽ △ ♅
04:36 ☽ ∥ ♃
05:45 ☽ □ ♆
06:12 ☿ △ ♆
15:47 ☽ ∥ ♀
31 03:52 ☽ ⚹ ♄
19:17 ♂ ∥ ♇
20:33 ♂ △ ♅

⚷ Chiron

01 Dec.	09 S 26
03	16♒18R
06	16 14
09	16 11
12	16 08
15	16 06
18	16 04
21	16 03
24	16 02
27	16 02D
30	16 03
25 05:16	16♒02 D

November 2008

Day	S. T.	☉	☽	☿	♀	♂	♃	♄	♅	♆	♇	☊ True
	h m s	° ' "	° ' "	° '	° '	° '	° '	° '	° '	° '	° '	° '
01 Sa	02 42 37	08♏ 56 16	12♐ 37 12	24♎ 10	16♐ 00	19♏ 08	16♑ 50	18♍ 38	19♓R02	21♒R28	29♐ 13	14♒R12
02 Su	02 46 34	09 56 19	24 28 43	25 44	17 12	19 50	17 00	18 43	19 01	21 28	29 14	14 03
03 Mo	02 50 30	10 56 25	06♑ 16 54	27 19	18 25	20 32	17 09	18 49	18 59	21D 28	29 16	13 56
04 Tu	02 54 27	11 56 32	18 05 32	28 56	19 37	21 15	17 18	18 55	18 58	21 28	29 17	13 53
05 We	02 58 23	12 56 40	29 59 15	00♏ 32	20 49	21 57	17 27	19 00	18 57	21 28	29 19	13 52
06 Th	03 02 20	13 56 51	12♒ 03 12	02 09	22 02	22 39	17 37	19 05	18 56	21 28	29 21	13D 52
07 Fr	03 06 16	14 57 02	24 22 54	03 47	23 14	23 21	17 47	19 11	18 55	21 28	29 22	13R 52
08 Sa	03 10 13	15 57 15	07♓ 03 40	05 24	24 26	24 04	17 56	19 16	18 54	21 29	29 24	13 51
09 Su	03 14 09	16 57 30	20 10 10	07 02	25 38	24 46	18 06	19 21	18 53	21 29	29 26	13 48
10 Mo	03 18 06	17 57 46	03♈ 45 36	08 39	26 50	25 29	18 16	19 26	18 52	21 29	29 27	13 43
11 Tu	03 22 03	18 58 04	17 50 44	10 17	28 02	26 11	18 27	19 31	18 51	21 29	29 29	13 35
12 We	03 25 59	19 58 23	02♉ 23 10	11 54	29 14	26 54	18 37	19 36	18 51	21 30	29 31	13 25
13 Th	03 29 56	20 58 44	17 17 03	13 31	00♑ 26	27 37	18 47	19 41	18 50	21 30	29 33	13 13
14 Fr	03 33 52	21 59 06	02♊ 23 34	15 08	01 38	28 19	18 58	19 46	18 49	21 30	29 35	13 02
15 Sa	03 37 49	22 59 30	17 32 23	16 45	02 49	29 02	19 08	19 51	18 48	21 31	29 36	12 52
16 Su	03 41 45	23 59 57	02♋ 33 21	18 22	04 01	29 45	19 19	19 55	18 48	21 31	29 38	12 44
17 Mo	03 45 42	25 00 24	17 18 15	19 58	05 13	00♐ 28	19 30	20 00	18 47	21 32	29 40	12 39
18 Tu	03 49 38	26 00 54	01♌ 41 49	21 34	06 24	01 11	19 40	20 04	18 47	21 32	29 42	12 36
19 We	03 53 35	27 01 25	15 41 55	23 10	07 36	01 54	19 51	20 09	18 46	21 33	29 44	12 34
20 Th	03 57 32	28 01 59	29 19 01	24 46	08 47	02 37	20 03	20 13	18 46	21 33	29 46	12 33
21 Fr	04 01 28	29 02 34	12♍ 35 13	26 21	09 59	03 20	20 14	20 17	18 45	21 34	29 48	12 32
22 Sa	04 05 25	00♐ 03 10	25 33 21	27 57	11 10	04 03	20 25	20 21	18 45	21 35	29 50	12 29
23 Su	04 09 21	01 03 49	08♎ 16 22	29 32	12 21	04 46	20 36	20 26	18 45	21 35	29 52	12 24
24 Mo	04 13 18	02 04 29	20 46 54	01♐ 07	13 33	05 30	20 48	20 29	18 45	21 36	29 54	12 16
25 Tu	04 17 14	03 05 11	03♏ 07 07	02 41	14 44	06 13	20 59	20 33	18 44	21 37	29 56	12 06
26 We	04 21 11	04 05 54	15 18 43	04 16	15 55	06 56	21 11	20 37	18 44	21 38	29 58	11 54
27 Th	04 25 07	05 06 39	27 22 57	05 50	17 06	07 40	21 23	20 41	18 44	21 38	30 00	11 41
28 Fr	04 29 04	06 07 25	09♐ 20 56	07 25	18 17	08 23	21 34	20 44	18D 44	21 39	00♑ 02	11 28
29 Sa	04 33 01	07 08 12	21 13 48	08 59	19 28	09 07	21 46	20 48	18 44	21 40	00 04	11 17
30 Su	04 36 57	08 09 01	03♑ 03 04	10 33	20 38	09 50	21 58	20 51	18 44	21 41	00 06	11 08

Data for	11-01-2008
Julian Day	2454771.50
Ayanamsa	23 58 59
SVP	05 ♓ 08 22
☽ ☊ Mean	14 ♒ 11 R

LAST ASPECT ☽ Day	h m	INGRESS Day	h m	
02	09:41	02	11:13	♑
04	06:47	05	00:02	♒
07	09:33	07	10:44	♓
09	16:29	09	17:27	♈
11	19:18	11	20:06	♉
13	17:14	13	20:13	♊
15	19:18	15	19:53	♋
17	13:44	17	21:08	♌
20	00:48	20	01:13	♍
22	08:02	22	08:20	♎
24	17:45	24	17:54	♏
26	12:33	27	05:14	♐
29	00:54	29	17:48	♑

● ◐ PHASES ○ ◑

06	04:03	◐	14♒07
13	06:18	○	21♉15
19	21:31	◑	27♌56
27	16:54	●	05♐49

DECLINATION

Day	☉	☽	☿	♀	♂	♃	♄	♅	♆	♇
01 Sa	14S 29	26S 41	07S 34	24S 22	17S 37	22S 44	06N09	05S 04	14S 41	17S 35
02 Su	14 48	27 09	08 12	24 32	17 49	22 43	06 07	05 05	14 41	17 35
03 Mo	15 07	26 22	08 50	24 41	18 01	22 41	06 05	05 05	14 41	17 36
04 Tu	15 25	24 23	09 29	24 49	18 13	22 40	06 03	05 05	14 41	17 36
05 We	15 43	21 20	10 07	24 57	18 25	22 39	06 01	05 06	14 41	17 36
06 Th	16 02	17 20	10 46	25 04	18 36	22 38	05 59	05 06	14 41	17 36
07 Fr	16 19	12 31	11 24	25 10	18 47	22 37	05 57	05 07	14 41	17 37
08 Sa	16 37	07 04	12 02	25 16	18 58	22 36	05 55	05 07	14 41	17 37
09 Su	16 54	01 08	12 40	25 20	19 09	22 34	05 53	05 07	14 41	17 37
10 Mo	17 11	05N02	13 18	25 24	19 20	22 33	05 52	05 08	14 41	17 37
11 Tu	17 28	11 11	13 55	25 28	19 31	22 32	05 50	05 08	14 41	17 37
12 We	17 44	16 55	14 31	25 30	19 41	22 30	05 48	05 08	14 40	17 38
13 Th	18 00	21 47	15 07	25 32	19 52	22 29	05 46	05 09	14 40	17 38
14 Fr	18 16	25 17	15 42	25 33	20 02	22 27	05 45	05 09	14 40	17 38
15 Sa	18 31	26 58	16 16	25 34	20 12	22 26	05 43	05 09	14 40	17 38
16 Su	18 46	26 40	16 50	25 33	20 22	22 25	05 41	05 09	14 40	17 39
17 Mo	19 01	24 28	17 22	25 32	20 31	22 23	05 40	05 09	14 40	17 39
18 Tu	19 16	20 43	17 54	25 31	20 41	22 22	05 38	05 10	14 40	17 39
19 We	19 30	15 52	18 25	25 28	20 50	22 20	05 37	05 10	14 39	17 39
20 Th	19 43	10 20	18 56	25 25	20 59	22 18	05 35	05 10	14 39	17 39
21 Fr	19 57	04 30	19 25	25 21	21 08	22 17	05 34	05 10	14 39	17 40
22 Sa	20 10	01S 24	19 53	25 16	21 17	22 15	05 32	05 10	14 39	17 40
23 Su	20 22	07 07	20 21	25 11	21 25	22 14	05 31	05 10	14 39	17 40
24 Mo	20 35	12 27	20 47	25 04	21 34	22 12	05 30	05 10	14 38	17 40
25 Tu	20 46	17 13	21 13	24 58	21 42	22 10	05 28	05 10	14 38	17 40
26 We	20 58	21 14	21 37	24 50	21 50	22 08	05 27	05 10	14 38	17 40
27 Th	21 09	24 18	22 00	24 42	21 58	22 07	05 26	05 10	14 38	17 41
28 Fr	21 20	26 16	22 22	24 33	22 05	22 05	05 25	05 10	14 37	17 41
29 Sa	21 30	27 02	22 43	24 23	22 12	22 03	05 24	05 10	14 37	17 41
30 Su	21 40	26 33	23 03	24 13	22 20	22 01	05 23	05 10	14 37	17 41

ASPECTARIAN

01	07:36	☽ ☌ ♀
	12:14	☽ □ ♄
	12:56	☽ □ ♅
	15:26	☉ ∥ ♆
	17:54	☽ ✶ ♆
02	02:57	☽ ✶ ☿
	06:40	♆ SD
	09:41	☽ ☌ ♇
03	08:42	♀ □ ♄
	10:21	☽ ✶ ☉
	11:19	♀ □ ♅
	20:03	☽ ∥ ♀
	22:22	☽ ☌ ♃
04	01:40	☽ △ ♄
	01:47	☽ ✶ ♅
	05:30	☿ ✶ ♇
	06:47	☽ ✶ ♂
	07:45	♂ □ ♆
	13:34	♄ ☍ ♅
	14:35	☽ ∥ ♃
	16:00	☿ ♏
05	01:17	☽ □ ☿
	12:58	♀ ✶ ♆
	17:16	☽ ∥ ♂
	22:32	☽ ∥ ♇
06	06:30	☽ ∥ ☉
	13:42	☽ ∥ ♆
	18:24	☽ ☌ ♆
	21:33	☽ ✶ ♀
	21:55	☽ □ ♂
07	04:36	☽ ∥ ☿
	09:33	☽ ✶ ♇
	20:27	☽ △ ☿
08	04:49	☽ # ♄
	08:05	☽ ∥ ♅
	17:43	☽ △ ☉
	20:14	☽ ✶ ♃
	21:41	☽ ☌ ♅
	22:31	☽ ☍ ♄
09	08:41	☽ △ ♂
	10:42	☽ □ ♀
	16:29	☽ □ ♇
10	00:23	☽ # ♅
	03:12	☽ ∥ ♄
	08:55	☉ ✶ ♃
	21:21	☉ △ ♅
11	01:01	☽ □ ♃
	06:05	☽ ✶ ♆
	12:26	☽ # ☿
	14:19	☽ # ♆
	14:25	☉ ✶ ♄
	14:26	☉ ∥ ♇
	18:23	☽ △ ♀
	19:18	☽ △ ♇
12	03:14	☽ # ♇
	03:55	☽ # ☉
	05:54	♀ ☌ ♇
	06:21	☿ ∥ ♆
	13:29	☽ # ♂
	15:25	♀ ♑
	17:15	☽ ☍ ☿
13	02:25	☽ △ ♃
	02:28	☽ ✶ ♅
	03:51	☽ △ ♄
	04:03	☽ # ♃
	05:40	♃ ✶ ♅
	06:43	☽ □ ♆
	12:32	☉ □ ♆
	17:14	☽ ☍ ♂
14	02:42	☽ # ♀
15	02:01	☽ □ ♅
	03:41	☽ □ ♄
	06:19	☽ △ ♆
	19:18	☽ ☍ ♇
16	02:34	☽ ☍ ♀
	06:27	☿ △ ♅
	08:27	♂ ♐
	14:25	☽ # ♀
	16:00	☿ ✶ ♃
17	00:28	☿ ✶ ♄
	02:27	☽ △ ♅
	03:39	☽ ☍ ♃
	04:28	☽ ✶ ♄
	04:56	☽ △ ☿
	12:16	☿ ∥ ♇
	13:44	☽ △ ☉
	14:26	☽ # ♃
	23:05	☽ △ ♂
	23:31	☿ □ ♆
18	00:11	☽ # ♂
	07:19	☽ # ☉
	13:02	☽ # ☿
	15:40	☽ # ♇
19	05:27	☽ # ♆
	10:14	☽ ☍ ♆
	14:50	☽ □ ☿
20	00:48	☽ △ ♇
	06:14	☽ □ ♂
	18:46	☽ △ ♀
	19:39	☽ ∥ ♄
	21:18	☽ # ♅
21	11:21	☽ ☍ ♅
	12:15	♃ △ ♄
	14:16	☽ ☌ ♄
	14:17	☽ △ ♃
	22:45	☉ ♐
22	05:06	☽ ✶ ☿
	08:02	☽ □ ♇
	09:10	☽ ✶ ☉
	15:42	☽ ∥ ♅
	16:57	☽ ✶ ♂
	17:13	☽ # ♄
23	02:41	☿ ∥ ☉
	07:09	☿ ♐
	08:36	☽ □ ♀
24	00:02	☽ □ ♃
	01:35	☽ △ ♆
	10:37	☽ ∥ ♆
	17:45	☽ ✶ ♇
25	02:31	☽ ∥ ♇
	16:52	☿ ☌ ☉
	22:13	☽ ∥ ☉
26	01:19	☽ ✶ ♀
	03:03	☽ ∥ ☿
	04:23	☽ ∥ ♂
	06:22	☽ ∥ ♃
	06:47	☽ △ ♅
	10:35	☽ ✶ ♄
	11:50	☽ ✶ ♃
	12:33	☽ □ ♆
	20:02	☿ ∥ ♂
27	02:06	♇ ♑
	03:48	☽ ∥ ♀
	06:26	☿ ∥ ♃
	16:09	♅ SD
	19:31	☽ ☌ ☿
	21:56	☽ ☌ ♂
28	09:21	♀ ✶ ♅
	18:57	☽ □ ♅
	23:07	☽ □ ♄
29	00:54	☽ ✶ ♆
	03:40	☿ ☌ ♂
	17:59	☽ ☌ ♇
30	04:37	♀ △ ♄

⚷ Chiron

01 Dec.	09 S 45
02	16♒04
05	16 06
08	16 08
11	16 11
14	16 15
17	16 19
20	16 24
23	16 29
26	16 35
29	16 41

December 2008

Day	S. T.	☉	☽	☿	♀	♂	♃	♄	♅	♆	♇	☊ True
	h m s	° ' "	° ' "	° '	° '	° '	° '	° '	° '	° '	° '	° '
01 Mo	04 40 54	09♐ 09 50	14♑ 50 48	12♐ 07	21♑ 49	10♐ 34	22♑ 10	20♍ 55	18♓ 45	21♒ 42	00♑ 08	11♒R02
02 Tu	04 44 50	10 10 41	26 39 48	13 41	23 00	11 18	22 23	20 58	18 45	21 43	00 10	10 59
03 We	04 48 47	11 11 33	08♒ 33 39	15 15	24 10	12 01	22 35	21 01	18 45	21 44	00 12	10D 59
04 Th	04 52 43	12 12 25	20 36 37	16 49	25 21	12 45	22 47	21 04	18 45	21 45	00 14	10 59
05 Fr	04 56 40	13 13 18	02♓ 53 29	18 23	26 31	13 29	22 59	21 07	18 46	21 46	00 16	11 01
06 Sa	05 00 37	14 14 12	15 29 17	19 57	27 41	14 13	23 12	21 10	18 46	21 47	00 19	11R 01
07 Su	05 04 33	15 15 07	28 28 55	21 31	28 51	14 57	23 24	21 12	18 47	21 48	00 21	11 01
08 Mo	05 08 30	16 16 03	11♈ 56 25	23 04	00♒ 01	15 41	23 37	21 15	18 47	21 50	00 23	10 58
09 Tu	05 12 26	17 16 59	25 54 00	24 38	01 11	16 25	23 49	21 18	18 48	21 51	00 25	10 53
10 We	05 16 23	18 17 56	10♉ 21 13	26 12	02 21	17 09	24 02	21 20	18 48	21 52	00 27	10 46
11 Th	05 20 19	19 18 54	25 14 06	27 46	03 30	17 53	24 15	21 22	18 49	21 53	00 29	10 38
12 Fr	05 24 16	20 19 52	10♊ 25 12	29 20	04 40	18 37	24 28	21 24	18 50	21 55	00 31	10 30
13 Sa	05 28 12	21 20 51	25 44 24	00♑ 54	05 49	19 21	24 41	21 27	18 50	21 56	00 34	10 23
14 Su	05 32 09	22 21 51	11♋ 00 32	02 28	06 58	20 06	24 54	21 29	18 51	21 57	00 36	10 17
15 Mo	05 36 06	23 22 52	26 03 16	04 02	08 07	20 50	25 07	21 30	18 52	21 59	00 38	10 13
16 Tu	05 40 02	24 23 54	10♌ 44 42	05 36	09 16	21 34	25 20	21 32	18 53	22 00	00 40	10 11
17 We	05 43 59	25 24 57	25 00 14	07 10	10 25	22 19	25 33	21 34	18 54	22 01	00 42	10D 11
18 Th	05 47 55	26 26 00	08♍ 48 28	08 43	11 34	23 03	25 46	21 36	18 55	22 03	00 45	10 11
19 Fr	05 51 52	27 27 05	22 10 37	10 17	12 42	23 48	25 59	21 37	18 56	22 04	00 47	10 12
20 Sa	05 55 48	28 28 10	05♎ 09 27	11 50	13 50	24 32	26 13	21 38	18 57	22 06	00 49	10R 11
21 Su	05 59 45	29 29 16	17 48 34	13 23	14 58	25 17	26 26	21 40	18 58	22 07	00 51	10 09
22 Mo	06 03 41	00♑ 30 23	00♏ 11 46	14 56	16 06	26 02	26 39	21 41	18 59	22 09	00 53	10 06
23 Tu	06 07 38	01 31 31	12 22 41	16 28	17 14	26 47	26 53	21 42	19 01	22 11	00 56	10 01
24 We	06 11 35	02 32 39	24 24 35	18 00	18 22	27 31	27 06	21 43	19 02	22 12	00 58	09 54
25 Th	06 15 31	03 33 48	06♐ 20 14	19 31	19 29	28 16	27 20	21 43	19 03	22 14	01 00	09 47
26 Fr	06 19 28	04 34 58	18 11 56	21 01	20 36	29 01	27 33	21 44	19 05	22 15	01 02	09 41
27 Sa	06 23 24	05 36 07	00♑ 01 36	22 30	21 43	29 46	27 47	21 45	19 06	22 17	01 04	09 35
28 Su	06 27 21	06 37 18	11 50 57	23 58	22 50	00♑ 31	28 00	21 45	19 08	22 19	01 07	09 30
29 Mo	06 31 17	07 38 28	23 41 48	25 24	23 57	01 16	28 14	21 46	19 09	22 21	01 09	09 28
30 Tu	06 35 14	08 39 38	05♒ 36 11	26 48	25 03	02 01	28 28	21 46	19 11	22 22	01 11	09D 27
31 We	06 39 10	09 40 49	17 36 31	28 10	26 09	02 46	28 42	21 46	19 13	22 24	01 13	09 28

Data for	12-01-2008
Julian Day	2454801.50
Ayanamsa	23 59 04
SVP	05♓ 08 18
☽ ☊ Mean	12♒ 36 R

● ◐ PHASES ○ ◑

Day	Time	Phase	Position
05	21:26	◐	14♓08
12	16:38	○	21♊02
19	10:29	◑	27♍54
27	12:23	●	06♑08

LAST ASPECT Day	h m	☽ INGRESS Day	h m	
01	15:44	02	06:45	♒
04	02:15	04	18:24	♓
07	00:44	07	02:45	♈
08	21:36	09	06:53	♉
10	22:24	11	07:34	♊
12	18:02	13	06:41	♋
14	22:27	15	06:23	♌
17	00:46	17	08:36	♍
19	10:30	19	14:23	♎
21	16:58	21	23:37	♏
24	05:30	24	11:14	♐
26	23:26	26	23:57	♑
29	09:21	29	12:44	♒

DECLINATION

Day	☉	☽	☿	♀	♂	♃	♄	♅	♆	♇
01 Mo	21S49	24S51	23S22	24S02	22S27	21S59	05N21	05S10	14S36	17S41
02 Tu	21 59	22 04	23 39	23 50	22 33	21 57	05 20	05 10	14 36	17 41
03 We	22 07	18 20	23 56	23 38	22 40	21 55	05 19	05 10	14 36	17 42
04 Th	22 15	13 48	24 11	23 25	22 46	21 53	05 18	05 09	14 35	17 42
05 Fr	22 23	08 39	24 25	23 11	22 52	21 51	05 18	05 09	14 35	17 42
06 Sa	22 31	03 02	24 37	22 57	22 58	21 49	05 17	05 09	14 35	17 42
07 Su	22 37	02N52	24 49	22 42	23 04	21 47	05 16	05 09	14 34	17 42
08 Mo	22 44	08 51	24 59	22 27	23 09	21 45	05 15	05 09	14 34	17 42
09 Tu	22 50	14 36	25 07	22 11	23 14	21 43	05 14	05 08	14 34	17 42
10 We	22 55	19 47	25 15	21 54	23 19	21 41	05 14	05 08	14 33	17 43
11 Th	23 01	23 53	25 20	21 37	23 24	21 39	05 13	05 08	14 33	17 43
12 Fr	23 05	26 24	25 25	21 20	23 28	21 37	05 12	05 08	14 32	17 43
13 Sa	23 09	26 59	25 28	21 01	23 32	21 34	05 12	05 07	14 32	17 43
14 Su	23 13	25 30	25 30	20 43	23 36	21 32	05 11	05 07	14 31	17 43
15 Mo	23 16	22 11	25 30	20 23	23 40	21 30	05 11	05 06	14 31	17 43
16 Tu	23 19	17 30	25 28	20 04	23 44	21 27	05 10	05 06	14 31	17 43
17 We	23 22	11 57	25 26	19 43	23 47	21 25	05 10	05 06	14 30	17 43
18 Th	23 23	05 58	25 21	19 23	23 50	21 23	05 09	05 05	14 30	17 43
19 Fr	23 25	00S04	25 15	19 01	23 53	21 20	05 09	05 05	14 29	17 44
20 Sa	23 26	05 56	25 08	18 40	23 55	21 18	05 09	05 04	14 29	17 44
21 Su	23 26	11 24	24 59	18 18	23 57	21 15	05 09	05 04	14 28	17 44
22 Mo	23 26	16 18	24 49	17 55	23 59	21 13	05 08	05 03	14 28	17 44
23 Tu	23 26	20 28	24 37	17 32	24 01	21 10	05 08	05 03	14 27	17 44
24 We	23 25	23 43	24 24	17 09	24 03	21 08	05 08	05 02	14 27	17 44
25 Th	23 23	25 56	24 09	16 45	24 04	21 05	05 08	05 02	14 26	17 44
26 Fr	23 22	26 58	23 53	16 21	24 05	21 03	05 08	05 01	14 26	17 44
27 Sa	23 19	26 45	23 35	15 56	24 06	21 00	05 08	05 00	14 25	17 44
28 Su	23 16	25 20	23 16	15 32	24 06	20 57	05 08	05 00	14 24	17 44
29 Mo	23 13	22 46	22 56	15 06	24 06	20 55	05 08	04 59	14 24	17 44
30 Tu	23 09	19 13	22 34	14 41	24 06	20 52	05 08	04 59	14 23	17 44
31 We	23 05	14 51	22 12	14 15	24 06	20 49	05 09	04 58	14 23	17 44

ASPECTARIAN

01 07:56 ☽ ⚹ ♅
08:38 ☽ ∥ ♀
08:45 ♀ ☌ ♃
12:23 ☽ △ ♄
12:32 ☽ ∥ ☿
15:09 ☽ ☌ ♃
15:44 ☽ ☌ ♀
20:27 ☽ ∥ ♂
21:17 ☉ ∥ ♃
02 00:36 ☽ ∥ ☉
00:47 ☽ ∥ ♃
08:56 ☿ ∥ ♀
03 03:37 ☽ ∥ ♇
05:46 ☽ ⚹ ☉
07:23 ☽ ⚹ ♂
15:22 ☽ ⚹ ☿
20:05 ☽ ∥ ♆
04 02:15 ☽ ☌ ♆
18:55 ☽ ⚹ ♇
05 05:52 ☿ □ ♅
14:36 ☽ # ♄
15:10 ☽ ∥ ♅
21:27 ☽ □ ♂
22:04 ☉ ☌ ♂
23:07 ♀ ∥ ♂
06 06:08 ☽ ☌ ♅
09:28 ☽ □ ☿
10:37 ☽ ☍ ♄
14:34 ☽ ⚹ ♃
19:14 ☿ □ ♄
07 00:44 ☽ ⚹ ♀
03:23 ☽ □ ♇
04:38 ☿ ⚹ ♆
05:30 ♀ ∥ ☉
09:09 ☽ # ♅
09:36 ☽ ∥ ♄
23:37 ♀ ♒
08 06:53 ☽ △ ♂
08:08 ☽ △ ☉
17:07 ☽ ⚹ ♆
20:26 ☽ □ ♃
21:36 ☽ △ ☿
23:48 ☽ # ♆
09 07:36 ☽ △ ♇
09:39 ☽ □ ♀
13:56 ☽ # ♇
10 10:10 ☽ # ♃
10:49 ☽ # ♀
12:03 ☉ □ ♅
13:43 ☽ ⚹ ♅
17:49 ☽ △ ♄
18:06 ☽ # ☉
18:39 ☽ □ ♆
20:41 ☽ # ♂
21:42 ♀ ∥ ♃
22:24 ☽ △ ♃
11 12:13 ☽ # ☿
14:12 ☽ △ ♀
12 06:51 ♂ □ ♅
10:13 ☿ ♑
13:11 ☽ □ ♅
13:30 ☽ ☍ ♂
17:15 ☽ □ ♄
18:02 ☽ △ ♆
18:41 ☿ ☌ ♇
13 02:20 ☉ □ ♄
07:34 ☽ ☍ ♇
09:00 ☽ ☍ ☿
14:06 ☉ ⚹ ♆
23:57 ☽ # ☿
14 12:28 ☽ △ ♅
14:44 ☽ # ♂
16:41 ☽ ⚹ ♄
17:20 ☽ # ☉
22:27 ☽ ☍ ♃
15 03:57 ☽ # ♃
10:38 ☽ # ♀
21:21 ☽ ☍ ♀
22:47 ♂ □ ♄
22:57 ☽ # ♇
16 13:15 ☽ # ♆
14:18 ♂ ⚹ ♆
18:55 ☽ ☍ ♆
19:10 ☽ △ ♂
17 00:46 ☽ △ ☉
09:51 ☽ △ ♇
23:50 ☽ △ ☿
18 03:13 ☽ ∥ ♄
03:30 ☽ # ♅
18:06 ☽ ☍ ♅
22:59 ☽ ☌ ♄
19 03:08 ☽ □ ♂
07:06 ☽ △ ♃
15:52 ☽ □ ♇
20:23 ☽ ∥ ♅
20:42 ☽ # ♄
20 14:22 ☽ □ ☿
18:02 ☽ △ ♀
21 08:19 ☽ △ ♆
12:04 ☉ ♑
14:37 ☽ ∥ ♆
15:21 ☽ ⚹ ♂
16:58 ☽ □ ♃
22 00:40 ☽ ⚹ ☉
01:21 ☽ ⚹ ♇
07:44 ☽ ∥ ♇
08:04 ☽ ∥ ♀
09:22 ☉ ☌ ♇
11:42 ♀ ∥ ♇
23 04:37 ☽ ∥ ♃
09:20 ☽ ⚹ ☿
10:39 ☽ □ ♀
13:14 ☽ △ ♅
18:35 ☽ ⚹ ♄
19:34 ☽ □ ♆
21:25 ☽ ∥ ☉
24 02:54 ☽ ∥ ♂
05:30 ☽ ⚹ ♃
05:36 ☽ ∥ ♆
16:33 ☿ ⚹ ♅
25 06:48 ☿ ∥ ♂
26 01:47 ☽ □ ♅
05:23 ☽ ⚹ ♀
07:11 ☽ □ ♄
08:15 ☽ ⚹ ♆
11:38 ☿ △ ♄
23:26 ☽ ☌ ♂
27 02:08 ☽ ☌ ♇
07:31 ♂ ♑
12:29 ♀ ☌ ♆
23:10 ☿ ∥ ☉
28 12:37 ☽ ∥ ♂
14:47 ☽ ⚹ ♅
20:00 ♂ ☌ ♇
20:05 ☽ △ ♄
20:15 ☽ ∥ ☉
22:30 ☽ ∥ ☿
29 03:54 ☽ ☌ ☿
09:21 ☽ ☌ ♃
13:24 ☽ ∥ ♃
30 08:32 ☽ ∥ ♇
16:49 ♀ ∥ ♆
31 02:22 ☽ ∥ ♆
03:17 ☽ ∥ ♀
09:32 ☽ ☌ ♆
11:41 ☿ ☌ ♃
18:09 ♄ SR
18:35 ☽ ☌ ♀

⚷ Chiron

01 Dec.	09 S 45
02	16♒48
05	16 55
08	17 03
11	17 12
14	17 20
17	17 29
20	17 39
23	17 49
26	17 59
29	18 10

26 07:47 06♒30 ☉ Annular Solar Eclipse

January 2009

Day	S. T.	☉	☽	☿	♀	♂	♃	♄	♅	♆	♇	☊ True
	h m s	° ' "	° ' "	° '	° '	° '	° '	° '	° '	° '	° '	° '
01 Th	06 43 07	10♑41 59	29♒45 40	29♑28	27♒15	03♑31	28♑55	21♍R46	19♓14	22♒26	01♑15	09♒30
02 Fr	06 47 04	11 43 09	12♓06 58	00♒44	28 20	04 16	29 09	21 46	19 16	22 28	01 17	09 33
03 Sa	06 51 00	12 44 19	24 44 05	01 56	29 26	05 02	29 23	21 46	19 18	22 30	01 20	09 35
04 Su	06 54 57	13 45 29	07♈40 49	03 03	00♓31	05 47	29 37	21 45	19 20	22 31	01 22	09 36
05 Mo	06 58 53	14 46 39	21 00 35	04 05	01 36	06 32	29 51	21 45	19 21	22 33	01 24	09R 36
06 Tu	07 02 50	15 47 48	04♉45 55	05 02	02 40	07 18	00♒05	21 45	19 23	22 35	01 26	09 35
07 We	07 06 46	16 48 57	18 57 36	05 51	03 44	08 03	00 19	21 44	19 25	22 37	01 28	09 32
08 Th	07 10 43	17 50 05	03♊33 54	06 33	04 48	08 48	00 33	21 43	19 27	22 39	01 30	09 30
09 Fr	07 14 40	18 51 13	18 30 13	07 06	05 52	09 34	00 47	21 42	19 29	22 41	01 32	09 26
10 Sa	07 18 36	19 52 21	03♋39 10	07 29	06 55	10 19	01 01	21 41	19 31	22 43	01 35	09 24
11 Su	07 22 33	20 53 28	18 51 27	07 43	07 58	11 05	01 15	21 40	19 33	22 45	01 37	09 21
12 Mo	07 26 29	21 54 35	03♌57 06	07R 45	09 00	11 50	01 29	21 39	19 36	22 47	01 39	09 20
13 Tu	07 30 26	22 55 42	18 47 08	07 36	10 02	12 36	01 43	21 38	19 38	22 49	01 41	09 19
14 We	07 34 22	23 56 48	03♍14 46	07 14	11 04	13 22	01 57	21 36	19 40	22 51	01 43	09D 19
15 Th	07 38 19	24 57 54	17 16 01	06 42	12 05	14 07	02 11	21 35	19 42	22 53	01 45	09 20
16 Fr	07 42 15	25 59 00	00♎49 44	05 58	13 06	14 53	02 25	21 33	19 45	22 55	01 47	09 21
17 Sa	07 46 12	27 00 06	13 57 10	05 03	14 07	15 39	02 40	21 32	19 47	22 57	01 49	09 21
18 Su	07 50 09	28 01 12	26 41 09	04 00	15 07	16 25	02 54	21 30	19 49	22 59	01 51	09R 21
19 Mo	07 54 05	29 02 17	09♏05 34	02 50	16 06	17 11	03 08	21 28	19 52	23 02	01 53	09 21
20 Tu	07 58 02	00♒03 22	21 14 45	01 35	17 05	17 57	03 22	21 26	19 54	23 04	01 55	09 20
21 We	08 01 58	01 04 27	03♐13 05	00 18	18 04	18 43	03 36	21 24	19 57	23 06	01 57	09 19
22 Th	08 05 55	02 05 31	15 04 44	29♑01	19 02	19 29	03 50	21 22	19 59	23 08	01 59	09 18
23 Fr	08 09 51	03 06 35	26 53 24	27 46	20 00	20 15	04 05	21 19	20 02	23 10	02 01	09 17
24 Sa	08 13 48	04 07 38	08♑42 19	26 35	20 57	21 01	04 19	21 17	20 04	23 12	02 03	09 17
25 Su	08 17 44	05 08 40	20 34 10	25 30	21 53	21 47	04 33	21 14	20 07	23 15	02 05	09D 17
26 Mo	08 21 41	06 09 42	02♒31 14	24 32	22 49	22 33	04 47	21 12	20 10	23 17	02 07	09 18
27 Tu	08 25 38	07 10 43	14 35 27	23 42	23 44	23 19	05 01	21 09	20 12	23 19	02 09	09 18
28 We	08 29 34	08 11 43	26 48 32	23 01	24 39	24 05	05 16	21 06	20 15	23 21	02 11	09 19
29 Th	08 33 31	09 12 42	09♓12 02	22 29	25 33	24 51	05 30	21 04	20 18	23 23	02 12	09 20
30 Fr	08 37 27	10 13 39	21 47 32	22 06	26 26	25 38	05 44	21 01	20 21	23 26	02 14	09 21
31 Sa	08 41 24	11 14 36	04♈36 40	21 51	27 19	26 24	05 58	20 58	20 24	23 28	02 16	09 21

Data for	01-01-2009
Julian Day	2454832.50
Ayanamsa	23 59 10
SVP	05 ♓ 08 16
☽ ☊ Mean	10 ♒ 57 R

● ◐ PHASES ○ ◑			
04	11:57	◐	14♈16
11	03:27	○	21♋02
18	02:46	◑	28♎08
26	07:56	☉	06♒30

LAST ASPECT Day	h m	☽ INGRESS Day	h m	
31	18:35	01	00:28	♓
03	08:51	03	09:50	♈
05	02:45	05	15:46	♉
07	06:05	07	18:12	♊
09	06:40	09	18:14	♋
11	04:27	11	17:41	♌
13	06:38	13	18:33	♍
15	14:38	15	22:31	♎
18	02:47	18	06:21	♏
20	03:38	20	17:31	♐
22	16:25	23	06:19	♑
25	09:09	25	18:57	♒
27	17:13	28	06:13	♓
30	09:24	30	15:25	♈

DECLINATION

Day	☉	☽	☿	♀	♂	♃	♄	♅	♆	♇
01 Th	23S 01	09S 51	21S 48	13S 49	24S 06	20S 47	05N09	04S 57	14S 22	17S 44
02 Fr	22 55	04 24	21 24	13 23	24 05	20 44	05 09	04 56	14 22	17 45
03 Sa	22 50	01N20	20 59	12 56	24 04	20 41	05 09	04 56	14 21	17 45
04 Su	22 44	07 09	20 33	12 29	24 02	20 38	05 10	04 55	14 20	17 45
05 Mo	22 37	12 48	20 08	12 02	24 01	20 35	05 10	04 54	14 20	17 45
06 Tu	22 30	18 02	19 42	11 35	23 59	20 32	05 11	04 53	14 19	17 45
07 We	22 23	22 26	19 17	11 08	23 57	20 29	05 11	04 53	14 19	17 45
08 Th	22 15	25 34	18 53	10 40	23 54	20 27	05 12	04 52	14 18	17 45
09 Fr	22 07	27 00	18 29	10 12	23 52	20 24	05 12	04 51	14 17	17 45
10 Sa	21 58	26 28	18 07	09 44	23 49	20 21	05 13	04 50	14 17	17 45
11 Su	21 49	23 56	17 47	09 16	23 45	20 18	05 14	04 49	14 16	17 45
12 Mo	21 39	19 45	17 29	08 48	23 42	20 15	05 14	04 48	14 15	17 45
13 Tu	21 29	14 22	17 14	08 19	23 38	20 12	05 15	04 47	14 15	17 45
14 We	21 19	08 19	17 01	07 51	23 34	20 08	05 16	04 47	14 14	17 45
15 Th	21 08	02 02	16 51	07 22	23 30	20 05	05 17	04 46	14 13	17 45
16 Fr	20 57	04S 07	16 44	06 53	23 26	20 02	05 18	04 45	14 13	17 45
17 Sa	20 45	09 53	16 40	06 25	23 21	19 59	05 18	04 44	14 12	17 45
18 Su	20 33	15 05	16 39	05 56	23 16	19 56	05 19	04 43	14 11	17 45
19 Mo	20 21	19 31	16 41	05 27	23 11	19 53	05 20	04 42	14 11	17 45
20 Tu	20 08	23 03	16 46	04 58	23 05	19 50	05 21	04 41	14 10	17 45
21 We	19 55	25 32	16 53	04 29	22 59	19 46	05 23	04 40	14 09	17 45
22 Th	19 42	26 52	17 01	04 00	22 53	19 43	05 24	04 39	14 09	17 45
23 Fr	19 28	26 58	17 11	03 31	22 47	19 40	05 25	04 38	14 08	17 45
24 Sa	19 14	25 50	17 22	03 03	22 40	19 37	05 26	04 37	14 07	17 45
25 Su	18 59	23 33	17 34	02 34	22 34	19 33	05 27	04 36	14 06	17 45
26 Mo	18 44	20 12	17 46	02 05	22 26	19 30	05 28	04 35	14 06	17 45
27 Tu	18 29	15 59	17 58	01 37	22 19	19 27	05 30	04 33	14 05	17 45
28 We	18 13	11 05	18 11	01 08	22 12	19 23	05 31	04 32	14 04	17 45
29 Th	17 57	05 40	18 23	00 40	22 04	19 20	05 32	04 31	14 04	17 45
30 Fr	17 41	00N02	18 35	00 11	21 56	19 16	05 34	04 30	14 03	17 44
31 Sa	17 24	05 50	18 46	00N17	21 47	19 13	05 35	04 29	14 02	17 44

ASPECTARIAN

01 02:56 ☽ ⚹ ♇
07:50 ☽ ⚹ ♂
09:52 ☿ ♒
20:46 ☽ # ♄
20:41 ☽ ∥ ♅
23:10 ☽ ⚹ ☉
02 13:42 ☽ ☌ ♅
18:24 ☽ ☍ ♄
03 08:51 ☽ ⚹ ♃
12:20 ☽ □ ♇
12:36 ♀ ♓
14:43 ☽ ⚹ ☿
14:48 ☽ # ♅
15:48 ☽ ∥ ♄
19:00 ☿ ∥ ♃
20:18 ☽ □ ♂
04 19:31 ♀ ⚹ ♇
20:55 ☽ # ♀
05 02:45 ☽ ⚹ ♆
06:43 ☽ # ♆
15:42 ♃ ♒
15:46 ☽ □ ♃
18:15 ☽ △ ♇
20:05 ☽ ⚹ ♀
22:37 ☽ # ♇
06 00:29 ☽ □ ☿
04:34 ☽ △ ♂
07:48 ☽ # ☿
12:51 ☽ # ♃
20:09 ☽ △ ☉
23:43 ☽ # ☉
07 00:46 ☽ ⚹ ♅
04:36 ☽ △ ♄
06:05 ☽ □ ♆
10:05 ☽ # ♂
19:01 ☽ △ ♃
08 02:10 ☽ □ ♀
05:02 ☽ △ ☿
09 01:34 ☽ □ ♅
05:05 ☽ □ ♄
06:40 ☽ △ ♆
15:28 ☉ ⚹ ♅
20:43 ☽ ☍ ♇
10 05:32 ☽ △ ♀
11:05 ☽ ☍ ♂
11 01:07 ☽ △ ♅
01:16 ☽ # ♂
03:01 ☿ ∥ ♇
04:27 ☽ ⚹ ♄
13:40 ☽ # ☉
16:45 ☿ SR
18:02 ☉ △ ♄
19:59 ☽ ☍ ♃
21:29 ☽ # ♃
12 06:04 ☽ ☍ ☿
09:25 ☽ # ♇
11:10 ☽ # ☿
13 00:31 ☽ # ♆
06:38 ☽ ☍ ♆
21:25 ☽ △ ♇
14 01:58 ☽ # ♀
11:39 ☽ ∥ ♄
13:34 ☽ # ♅
14:20 ☽ ☍ ♀
18:14 ☽ △ ♂
15 04:16 ☽ ☍ ♅
07:32 ☽ ☌ ♄
14:38 ☽ △ ☉
16 01:43 ☽ □ ♇
02:29 ☽ ∥ ♅
02:56 ☽ △ ♃
04:43 ☽ # ♄
08:44 ☽ △ ☿
10:23 ☽ ∥ ♀
17 03:22 ☽ □ ♂
16:57 ☽ △ ♆
19:36 ☽ ∥ ♆
18 08:01 ☽ ∥ ☿
09:57 ☽ ⚹ ♇
12:10 ☽ □ ♃
12:54 ☽ □ ☿
13:49 ☽ ∥ ♇
19:07 ☿ ☌ ♃
19 02:06 ☽ ∥ ♃
04:48 ☽ ∥ ☉
05:18 ♀ # ♄
15:01 ☽ △ ♀
17:00 ☽ ⚹ ♂
21:19 ☽ △ ♅
22:41 ☉ ♒
20 00:12 ☽ ∥ ♂
00:22 ☽ ⚹ ♄
03:38 ☽ □ ♆
14:54 ♀ ∥ ♅
16:00 ☿ ☌ ☉
18:42 ☽ ⚹ ☿
19:17 ☽ ⚹ ☉
21 00:48 ☽ ⚹ ♃
05:37 ☿ ♑R
20:46 ☉ ∥ ♃
22 08:44 ☽ □ ♀
10:00 ☽ □ ♅
12:43 ☽ □ ♄
16:25 ☽ ⚹ ♆
17:00 ♂ ⚹ ♅
23 01:01 ♀ ☌ ♅
10:27 ☽ ☌ ♇
24 05:45 ☉ ☌ ♃
08:05 ♂ △ ♄
08:16 ♀ ☍ ♄
09:05 ♀ ⚹ ♂
23:05 ☽ ⚹ ♅
25 01:21 ☽ △ ♄
02:36 ☽ ☌ ♂
02:53 ☽ ⚹ ♀
08:09 ☽ ∥ ♂
09:09 ☽ ☌ ☿
21:35 ☿ ∥ ♇
26 04:25 ☽ ∥ ♃
04:37 ☽ ☌ ♃
09:29 ☽ ∥ ☉
13:45 ☽ ∥ ☿
14:32 ☽ ∥ ♇
23:31 ☿ ⚹ ♀
27 06:12 ☿ ☌ ♂
09:41 ☽ ∥ ♆
17:13 ☽ ☌ ♆
28 01:59 ☿ ∥ ☉
10:28 ☽ ⚹ ♇
29 00:33 ☽ # ♄
04:55 ☽ ∥ ♅
18:33 ☉ ∥ ♇
21:15 ☽ ☌ ♅
22:32 ☽ ☍ ♄
30 00:34 ☽ ⚹ ☿
00:35 ☽ # ♀
07:41 ☽ ⚹ ♂
09:24 ☽ ☌ ♀
18:26 ☽ # ♅
19:38 ☽ □ ♇
22:59 ☽ ∥ ♄
31 02:34 ☽ ⚹ ♃
13:16 ☽ ⚹ ☉

⚷ Chiron

01	Dec.	09 S 27
01	18♒21	
04	18 32	
07	18 44	
10	18 56	
13	19 08	
16	19 20	
19	19 33	
22	19 46	
25	19 58	
28	20 11	
31	20 24	

February 2009

09 14:39 21♌00 ✸ Penumbral Lunar Eclipse (mag 0.924)

Day	S. T.	☉	☽	☿	♀	♂	♃	♄	♅	♆	♇	☊ True
	h m s	° ′ ″	° ′ ″	° ′	° ′	° ′	° ′	° ′	° ′	° ′	° ′	° ′
01 Su	08 45 20	12♒15 31	17♈41 07	21♑R45	28♓11	27♑10	06♒12	20♍R54	20♓26	23♒30	02♑18	09♒R21
02 Mo	08 49 17	13 16 26	01♉02 32	21D 47	29 02	27 56	06 27	20 51	20 29	23 32	02 20	09 21
03 Tu	08 53 13	14 17 18	14 42 11	21 55	29 52	28 43	06 41	20 48	20 32	23 35	02 21	09 21
04 We	08 57 10	15 18 10	28 40 36	22 11	00♈42	29 29	06 55	20 45	20 35	23 37	02 23	09 21
05 Th	09 01 07	16 19 00	12♊57 10	22 33	01 30	00♒16	07 09	20 41	20 38	23 39	02 25	09D 21
06 Fr	09 05 03	17 19 49	27 29 37	23 01	02 18	01 02	07 23	20 38	20 41	23 41	02 26	09 21
07 Sa	09 08 60	18 20 36	12♋13 57	23 34	03 05	01 49	07 37	20 34	20 44	23 44	02 28	09R 21
08 Su	09 12 56	19 21 21	27 04 20	24 12	03 51	02 35	07 51	20 30	20 47	23 46	02 30	09 21
09 Mo	09 16 53	20 22 06	11♌53 43	24 54	04 36	03 21	08 05	20 27	20 50	23 48	02 31	09 20
10 Tu	09 20 49	21 22 48	26 34 35	25 40	05 19	04 08	08 19	20 23	20 53	23 50	02 33	09 19
11 We	09 24 46	22 23 30	11♍00 00	26 30	06 02	04 55	08 33	20 19	20 56	23 53	02 35	09 17
12 Th	09 28 42	23 24 10	25 04 32	27 24	06 44	05 41	08 47	20 15	20 59	23 55	02 36	09 15
13 Fr	09 32 39	24 24 49	08♎44 51	28 20	07 24	06 28	09 01	20 11	21 03	23 57	02 38	09 13
14 Sa	09 36 36	25 25 27	21 59 58	29 20	08 03	07 14	09 15	20 07	21 06	24 00	02 39	09 11
15 Su	09 40 32	26 26 04	04♏50 56	00♒22	08 41	08 01	09 29	20 03	21 09	24 02	02 41	09 09
16 Mo	09 44 29	27 26 39	17 20 32	01 26	09 18	08 48	09 43	19 58	21 12	24 04	02 42	09 07
17 Tu	09 48 25	28 27 14	29 32 41	02 33	09 53	09 34	09 57	19 54	21 15	24 06	02 44	09D 07
18 We	09 52 22	29 27 47	11♐32 02	03 42	10 27	10 21	10 11	19 50	21 19	24 09	02 45	09 08
19 Th	09 56 18	00♓28 19	23 23 28	04 53	10 59	11 08	10 25	19 46	21 22	24 11	02 46	09 09
20 Fr	10 00 15	01 28 49	05♑11 49	06 06	11 30	11 55	10 38	19 41	21 25	24 13	02 48	09 11
21 Sa	10 04 11	02 29 18	17 01 37	07 21	11 59	12 41	10 52	19 37	21 28	24 16	02 49	09 14
22 Su	10 08 08	03 29 46	28 56 50	08 37	12 27	13 28	11 06	19 32	21 32	24 18	02 50	09 15
23 Mo	10 12 05	04 30 12	11♒00 51	09 55	12 53	14 15	11 19	19 28	21 35	24 20	02 52	09 16
24 Tu	10 16 01	05 30 37	23 16 16	11 14	13 17	15 02	11 33	19 23	21 38	24 22	02 53	09R 16
25 We	10 19 58	06 31 00	05♓44 55	12 35	13 39	15 49	11 47	19 19	21 42	24 25	02 54	09 14
26 Th	10 23 54	07 31 21	18 27 43	13 57	13 59	16 36	12 00	19 14	21 45	24 27	02 55	09 11
27 Fr	10 27 51	08 31 41	01♈24 48	15 20	14 18	17 22	12 14	19 09	21 48	24 29	02 56	09 07
28 Sa	10 31 47	09 31 59	14 35 43	16 45	14 34	18 09	12 27	19 05	21 52	24 31	02 58	09 02

Data for	02-01-2009
Julian Day	2454863.50
Ayanamsa	23 59 16
SVP	05 ♓ 08 12
☽ ☊ Mean	09 ♒ 19 R

● ◐ PHASES ○ ◑

02	23:13	◐	14♉15
09	14:50	✸	21♌00
16	21:38	◑	28♏21
25	01:35	●	06♓35

LAST ASPECT Day	h m	☽ INGRESS Day	h m	
01	18:08	01	22:09	♉
04	01:27	04	02:15	♊
05	17:44	06	04:06	♋
07	19:08	08	04:44	♌
09	19:29	10	05:39	♍
12	04:18	12	08:34	♎
14	14:47	14	14:52	♏
16	21:38	17	00:54	♐
19	01:37	19	13:26	♑
21	09:01	22	02:06	♒
24	02:08	24	13:00	♓
26	06:10	26	21:24	♈

DECLINATION

Day	☉	☽	☿	♀	♂	♃	♄	♅	♆	♇
01 Su	17S 07	11N29	18S 57	00N45	21S 39	19S 10	05N37	04S 28	14S 01	17S 44
02 Mo	16 50	16 44	19 07	01 13	21 30	19 06	05 38	04 27	14 01	17 44
03 Tu	16 33	21 16	19 17	01 41	21 21	19 03	05 39	04 26	14 00	17 44
04 We	16 15	24 43	19 25	02 08	21 12	18 59	05 41	04 24	13 59	17 44
05 Th	15 57	26 43	19 33	02 35	21 02	18 56	05 43	04 23	13 58	17 44
06 Fr	15 38	26 56	19 40	03 02	20 53	18 52	05 44	04 22	13 58	17 44
07 Sa	15 20	25 16	19 46	03 29	20 43	18 49	05 46	04 21	13 57	17 44
08 Su	15 01	21 50	19 50	03 55	20 33	18 45	05 47	04 20	13 56	17 44
09 Mo	14 42	16 59	19 54	04 22	20 22	18 42	05 49	04 18	13 55	17 44
10 Tu	14 23	11 10	19 57	04 47	20 12	18 38	05 51	04 17	13 55	17 44
11 We	14 03	04 52	19 58	05 13	20 01	18 35	05 52	04 16	13 54	17 44
12 Th	13 43	01S 32	19 59	05 38	19 50	18 31	05 54	04 15	13 53	17 44
13 Fr	13 23	07 39	19 58	06 03	19 38	18 27	05 56	04 13	13 52	17 44
14 Sa	13 03	13 14	19 56	06 27	19 27	18 24	05 58	04 12	13 52	17 44
15 Su	12 42	18 05	19 52	06 51	19 15	18 20	05 59	04 11	13 51	17 44
16 Mo	12 22	22 00	19 48	07 15	19 03	18 17	06 01	04 10	13 50	17 43
17 Tu	12 01	24 52	19 42	07 38	18 51	18 13	06 03	04 08	13 49	17 43
18 We	11 40	26 34	19 35	08 00	18 39	18 09	06 05	04 07	13 49	17 43
19 Th	11 18	27 03	19 27	08 22	18 27	18 06	06 07	04 06	13 48	17 43
20 Fr	10 57	26 17	19 17	08 44	18 14	18 02	06 09	04 04	13 47	17 43
21 Sa	10 35	24 20	19 07	09 04	18 01	17 59	06 10	04 03	13 46	17 43
22 Su	10 13	21 17	18 55	09 25	17 48	17 55	06 12	04 02	13 46	17 43
23 Mo	09 52	17 18	18 41	09 44	17 35	17 51	06 14	04 00	13 45	17 43
24 Tu	09 29	12 33	18 27	10 03	17 21	17 48	06 16	03 59	13 44	17 43
25 We	09 07	07 13	18 11	10 21	17 08	17 44	06 18	03 58	13 44	17 43
26 Th	08 45	01 30	17 53	10 39	16 54	17 40	06 20	03 57	13 43	17 43
27 Fr	08 22	04N21	17 35	10 55	16 40	17 37	06 22	03 55	13 42	17 42
28 Sa	08 00	10 08	17 15	11 11	16 26	17 33	06 24	03 54	13 41	17 42

ASPECTARIAN

01 07:11 ☿ SD
07:21 ☽ □ ☿
10:33 ☽ ✶ ♆
11:16 ☽ # ♆
18:08 ☽ □ ♂
22:03 ☿ ∥ ♃

02 00:27 ☽ # ☉
02:17 ☽ △ ♇
04:56 ☽ # ♇
09:43 ☽ □ ♃
11:47 ☽ # ♃
12:28 ☽ # ☿

03 00:27 ☽ # ♂
03:41 ♀ ♈
10:07 ☽ ✶ ♅
10:30 ☽ △ ♄
12:41 ☽ △ ☿
15:21 ☽ □ ♆

04 01:27 ☽ △ ♂
03:38 ☽ ✶ ♀
14:08 ☽ △ ♃
15:56 ♂ ♒

05 06:00 ☽ △ ☉
11:00 ♄ ☍ ♅
12:46 ☽ □ ♄
12:47 ☽ □ ♅
17:44 ☽ △ ♆

06 04:23 ♀ □ ♇
08:06 ☽ ☍ ♇
08:18 ☽ □ ♀

07 13:26 ☽ ✶ ♄
13:48 ☽ △ ♅
19:08 ☽ ☍ ☿

08 07:22 ☽ # ♂
09:24 ☽ ☍ ♂
10:31 ☽ # ☿
11:32 ☽ △ ♀
16:08 ☽ # ♃
17:43 ☽ ☍ ♃
20:41 ☽ # ♇
21:13 ♀ # ♅

09 10:26 ☽ # ☉
13:04 ☽ # ♆
19:29 ☽ ☍ ♆

10 09:54 ☽ △ ♇
20:15 ☽ ∥ ♄
22:47 ☽ ∥ ♀

11 02:16 ☽ # ♅
05:05 ☿ ∥ ♂
11:23 ☉ ∥ ♆
15:44 ☽ ☌ ♄
16:56 ☽ ☍ ♅

12 04:18 ☽ △ ☿
10:27 ☽ ∥ ♅
12:41 ☉ ☌ ♆
13:09 ☽ □ ♇
16:46 ♀ ∥ ♄
17:05 ☽ # ♄
17:06 ☽ # ♀
19:41 ☽ △ ♂
21:29 ☽ ☍ ♀

13 00:30 ☽ △ ♃
23:11 ☽ ∥ ☉

14 02:53 ☽ ∥ ♆
03:41 ☽ △ ♆
06:52 ☽ △ ☉
14:47 ☽ □ ☿
15:39 ☿ ♒
19:53 ☽ ✶ ♇
22:06 ☽ ∥ ♇

15 01:23 ☽ ∥ ♃
06:17 ☽ ∥ ♂
06:25 ☽ □ ♂
09:00 ☽ □ ♃
10:03 ☽ ∥ ☿

16 05:06 ☽ ✶ ♄
07:34 ☽ △ ♅
13:13 ☽ □ ♆

17 04:48 ♀ ✶ ♃
06:37 ☽ ✶ ☿
16:28 ♂ ☌ ♃
21:13 ☽ ✶ ♃
21:27 ☽ ✶ ♂
21:42 ☽ △ ♀

18 09:31 ♀ ✶ ♂
12:46 ☉ ♓
16:41 ☽ □ ♄
19:52 ☽ □ ♅

19 01:37 ☽ ✶ ♆
15:44 ☽ ✶ ☉
19:06 ☽ ☌ ♇

20 13:21 ☽ □ ♀

21 05:12 ☽ △ ♄
06:18 ♂ ∥ ♃
08:02 ☉ ✶ ♇
09:01 ☽ ✶ ♅

22 09:04 ♂ ∥ ♇
15:49 ☽ ∥ ☿
20:58 ☽ ∥ ♃
21:34 ☽ ☌ ☿
21:47 ☽ ∥ ♇
22:28 ☽ ∥ ♂

23 00:37 ☽ ☌ ♃
03:48 ☽ ✶ ♀
04:17 ☿ # ☉
06:49 ☽ ☌ ♂
18:21 ☽ ∥ ♆

24 02:08 ☽ ☌ ♆
06:53 ☿ ☌ ♃
10:56 ☽ # ♀
15:07 ☽ ∥ ☉
18:33 ☽ ✶ ♇

25 03:58 ☽ # ♄
08:19 ♃ ∥ ♇
13:54 ☽ ∥ ♅

26 01:00 ☿ ✶ ♀
01:26 ☽ ☍ ♄
06:10 ☽ ☌ ♅
14:22 ☿ ∥ ♇
21:27 ☿ ∥ ♃
22:15 ☽ # ♅

27 02:48 ☽ □ ♇
08:20 ☽ ∥ ♄
15:36 ☽ # ☉
20:04 ☽ ✶ ♃
23:57 ☽ ☌ ♀

28 04:21 ☽ ✶ ☿
04:48 ☽ ∥ ♀
06:49 ☽ ✶ ♂
15:32 ☽ # ♇
17:52 ☽ ✶ ♆

⚷ Chiron

01	Dec.	08 S 54
03	20♒38	
06	20 51	
09	21 04	
12	21 17	
15	21 30	
18	21 43	
21	21 56	
24	22 09	
27	22 22	

March 2009

Day	S. T. h m s	☉ ° ' "	☽ ° ' "	☿ ° '	♀ ° '	♂ ° '	♃ ° '	♄ ° '	♅ ° '	♆ ° '	♇ ° '	☊ True ° '
01 Su	10 35 44	10♓ 32 14	27♈ 59 35	18♒ 11	14♈ 49	18♒ 56	12♒ 41	19♍R00	21♓ 55	24♒ 33	02♑ 59	08♒R58
02 Mo	10 39 40	11 32 28	11♉ 35 27	19 38	15 01	19 43	12 54	18 55	21 59	24 36	03 00	08 54
03 Tu	10 43 37	12 32 40	25 22 16	21 06	15 11	20 30	13 07	18 51	22 02	24 38	03 01	08 52
04 We	10 47 34	13 32 50	09♊ 19 07	22 36	15 18	21 17	13 21	18 46	22 05	24 40	03 02	08 51
05 Th	10 51 30	14 32 58	23 24 56	24 06	15 24	22 04	13 34	18 41	22 09	24 42	03 03	08D 50
06 Fr	10 55 27	15 33 03	07♋ 38 24	25 38	15 27	22 51	13 47	18 36	22 12	24 44	03 04	08 51
07 Sa	10 59 23	16 33 06	21 57 43	27 11	15R 27	23 38	14 00	18 32	22 16	24 47	03 05	08 51
08 Su	11 03 20	17 33 08	06♌ 20 12	28 45	15 25	24 25	14 13	18 27	22 19	24 49	03 06	08R 51
09 Mo	11 07 16	18 33 07	20 42 18	00♓ 20	15 21	25 12	14 26	18 22	22 22	24 51	03 06	08 49
10 Tu	11 11 13	19 33 04	04♍ 59 29	01 56	15 14	25 59	14 39	18 17	22 26	24 53	03 07	08 46
11 We	11 15 09	20 32 58	19 06 51	03 34	15 05	26 46	14 52	18 12	22 29	24 55	03 08	08 41
12 Th	11 19 06	21 32 51	02♎ 59 38	05 13	14 53	27 33	15 05	18 08	22 33	24 57	03 09	08 34
13 Fr	11 23 03	22 32 42	16 33 57	06 52	14 38	28 20	15 17	18 03	22 36	24 59	03 10	08 26
14 Sa	11 26 59	23 32 32	29 47 28	08 33	14 21	29 06	15 30	17 58	22 40	25 01	03 10	08 19
15 Su	11 30 56	24 32 19	12♏ 39 36	10 15	14 02	29 53	15 43	17 53	22 43	25 03	03 11	08 11
16 Mo	11 34 52	25 32 05	25 11 32	11 59	13 40	00♓ 40	15 55	17 49	22 46	25 05	03 12	08 05
17 Tu	11 38 49	26 31 49	07♐ 26 02	13 43	13 16	01 27	16 08	17 44	22 50	25 07	03 12	08 02
18 We	11 42 45	27 31 31	19 27 04	15 29	12 50	02 14	16 20	17 39	22 53	25 09	03 13	08 00
19 Th	11 46 42	28 31 12	01♑ 19 24	17 16	12 22	03 01	16 33	17 35	22 57	25 11	03 13	08D 00
20 Fr	11 50 38	29 30 51	13 08 14	19 04	11 52	03 48	16 45	17 30	23 00	25 13	03 14	08 01
21 Sa	11 54 35	00♈ 30 28	24 58 53	20 54	11 21	04 35	16 57	17 25	23 04	25 15	03 14	08 03
22 Su	11 58 32	01 30 03	06♒ 56 25	22 44	10 47	05 22	17 09	17 21	23 07	25 17	03 15	08 04
23 Mo	12 02 28	02 29 37	19 05 24	24 36	10 13	06 09	17 21	17 16	23 10	25 19	03 15	08R 04
24 Tu	12 06 25	03 29 09	01♓ 29 36	26 30	09 37	06 56	17 33	17 12	23 14	25 21	03 16	08 01
25 We	12 10 21	04 28 38	14 11 41	28 24	09 01	07 43	17 45	17 07	23 17	25 23	03 16	07 56
26 Th	12 14 18	05 28 06	27 12 54	00♈ 20	08 24	08 30	17 57	17 03	23 21	25 25	03 16	07 49
27 Fr	12 18 14	06 27 32	10♈ 32 53	02 17	07 46	09 17	18 09	16 59	23 24	25 27	03 17	07 40
28 Sa	12 22 11	07 26 55	24 09 42	04 15	07 08	10 04	18 20	16 54	23 27	25 28	03 17	07 30
29 Su	12 26 07	08 26 17	08♉ 00 10	06 14	06 31	10 51	18 32	16 50	23 31	25 30	03 17	07 20
30 Mo	12 30 04	09 25 37	22 00 27	08 14	05 53	11 38	18 43	16 46	23 34	25 32	03 17	07 12
31 Tu	12 34 00	10 24 54	06♊ 06 40	10 16	05 17	12 25	18 55	16 41	23 37	25 34	03 18	07 06

Data for	03-01-2009
Julian Day	2454891.50
Ayanamsa	23 59 19
SVP	05 ♓ 08 08
☽ ☊ Mean	07 ♒ 50 R

● ◐ PHASES ○ ◑

04	07:46	◐	13♊52
11	02:39	○	20♍40
18	17:47	◑	28♐16
26	16:06	●	06♈08

LAST ASPECT ☽ Day	h m	INGRESS Day	h m	
28	17:52	01	03:34	♉
02	22:43	03	08:00	♊
05	02:11	05	11:08	♋
07	00:30	07	13:26	♌
09	07:57	09	15:35	♍
11	05:49	11	18:47	♎
13	22:40	14	00:23	♏
16	00:43	16	09:22	♐
18	17:48	18	21:19	♑
20	20:06	21	10:07	♒
23	12:09	23	21:08	♓
25	16:54	26	05:03	♈
28	02:18	28	10:10	♉
30	06:02	30	13:37	♊

DECLINATION

Day	☉	☽	☿	♀	♂	♃	♄	♅	♆	♇
01 Su	07S 37	15N32	16S 54	11N26	16S 11	17S 29	06N26	03S 53	13S 41	17S 42
02 Mo	07 14	20 15	16 32	11 40	15 57	17 26	06 28	03 51	13 40	17 42
03 Tu	06 51	23 57	16 08	11 53	15 42	17 22	06 30	03 50	13 39	17 42
04 We	06 28	26 17	15 43	12 05	15 27	17 18	06 32	03 48	13 38	17 42
05 Th	06 05	26 58	15 17	12 16	15 12	17 15	06 34	03 47	13 38	17 42
06 Fr	05 42	25 54	14 50	12 26	14 57	17 11	06 35	03 46	13 37	17 42
07 Sa	05 18	23 08	14 21	12 35	14 42	17 07	06 37	03 44	13 36	17 42
08 Su	04 55	18 54	13 51	12 42	14 27	17 04	06 39	03 43	13 36	17 42
09 Mo	04 32	13 34	13 20	12 49	14 11	17 00	06 41	03 42	13 35	17 42
10 Tu	04 08	07 33	12 48	12 54	13 55	16 56	06 43	03 40	13 34	17 41
11 We	03 45	01 13	12 14	12 57	13 40	16 53	06 45	03 39	13 34	17 41
12 Th	03 21	05S 02	11 39	12 59	13 24	16 49	06 47	03 38	13 33	17 41
13 Fr	02 57	10 55	11 03	13 00	13 07	16 46	06 49	03 36	13 32	17 41
14 Sa	02 34	16 09	10 26	13 00	12 51	16 42	06 51	03 35	13 32	17 41
15 Su	02 10	20 31	09 47	12 57	12 35	16 38	06 53	03 34	13 31	17 41
16 Mo	01 46	23 50	09 07	12 54	12 18	16 35	06 55	03 32	13 30	17 41
17 Tu	01 23	25 59	08 26	12 48	12 02	16 31	06 56	03 31	13 29	17 41
18 We	00 59	26 52	07 44	12 42	11 45	16 28	06 58	03 29	13 29	17 41
19 Th	00 35	26 31	07 01	12 33	11 28	16 24	07 00	03 28	13 28	17 41
20 Fr	00 12	24 57	06 16	12 24	11 11	16 20	07 02	03 27	13 28	17 40
21 Sa	00N12	22 17	05 30	12 13	10 54	16 17	07 04	03 25	13 27	17 40
22 Su	00 36	18 38	04 44	12 00	10 37	16 13	07 06	03 24	13 26	17 40
23 Mo	01 00	14 10	03 56	11 46	10 20	16 10	07 07	03 23	13 26	17 40
24 Tu	01 23	09 02	03 07	11 31	10 03	16 06	07 09	03 21	13 25	17 40
25 We	01 47	03 25	02 17	11 14	09 45	16 03	07 11	03 20	13 24	17 40
26 Th	02 10	02N27	01 26	10 57	09 28	15 59	07 13	03 19	13 24	17 40
27 Fr	02 34	08 20	00 34	10 38	09 10	15 56	07 14	03 17	13 23	17 40
28 Sa	02 57	13 57	00N19	10 18	08 52	15 52	07 16	03 16	13 23	17 40
29 Su	03 21	18 58	01 12	09 58	08 35	15 49	07 18	03 15	13 22	17 40
30 Mo	03 44	23 00	02 07	09 37	08 17	15 46	07 19	03 13	13 21	17 40
31 Tu	04 07	25 42	03 02	09 16	07 59	15 42	07 21	03 12	13 21	17 39

ASPECTARIAN

01 03:00 ☽ # ♂
06:08 ☽ # ☿
08:51 ☽ △ ♇
09:21 ☽ # ♃
10:33 ☽ # ♇
23:54 ☽ ✶ ☉
02 02:20 ☽ □ ♃
03:06 ☿ ☌ ♂
12:44 ☽ △ ♄
15:03 ☽ □ ♂
15:43 ☽ □ ☿
18:11 ☽ ✶ ♅
22:43 ☽ □ ♆
03 20:41 ☉ # ♄
04 06:59 ☽ △ ♃
10:18 ☽ ✶ ♀
16:01 ☽ □ ♄
21:35 ☽ △ ♂
21:50 ☽ □ ♅
05 01:18 ☽ △ ☿
02:11 ☽ △ ♆
09:40 ☿ ☌ ♆
09:51 ☿ ∥ ♂
16:17 ☽ ☍ ♇
06 13:07 ☽ □ ♀
14:16 ☽ △ ☉
17:18 ♀ SR
18:17 ☽ ✶ ♄
07 00:30 ☽ △ ♅
08 05:51 ☽ # ♇
08:53 ☽ # ♃
12:32 ☿ ∥ ♆
12:51 ♂ ☌ ♆
13:21 ☽ ☍ ♃
15:06 ☽ △ ♀
18:56 ☿ ♓
19:54 ☉ ☍ ♄
21:19 ☽ # ♂
23:58 ☽ # ♆
09 01:05 ☽ # ☿
03:07 ☽ ∥ ♀
06:57 ☽ ☍ ♆
07:57 ☽ ☍ ♂
18:12 ☽ ☍ ☿
20:15 ☿ # ♀
20:51 ☽ △ ♇
10 03:10 ☽ ∥ ♄
13:51 ☽ # ☉
14:48 ☽ # ♅
17:38 ☿ ✶ ♇
22:27 ☽ ☌ ♄
11 05:49 ☽ ☍ ♅
06:13 ☉ ∥ ♅
09:28 ♂ ∥ ♆
12:36 ♀ ✶ ♃
17:50 ☽ ∥ ☉
18:32 ☽ ∥ ♅
12 00:16 ☽ □ ♇
06:58 ☽ # ♄
20:37 ☽ ☍ ♀
21:41 ☽ △ ♃
13 00:31 ☽ ∥ ☿
01:28 ☉ ☌ ♅
09:09 ☽ # ♀
09:14 ☽ ∥ ♂
11:08 ♀ # ♂
11:33 ☽ ∥ ♆
15:15 ☽ △ ♆
22:40 ☽ △ ♂
14 02:43 ☽ ∥ ♃
06:15 ☽ ✶ ♇
07:52 ☽ ∥ ♇
18:46 ☽ △ ☿
15 03:20 ♂ ♓
05:53 ☽ □ ♃
09:53 ☽ ✶ ♄
19:18 ☽ △ ♅
23:48 ☽ □ ♆
16 00:43 ☽ △ ☉
11:25 ☽ □ ♂
17 11:13 ☽ △ ♀
14:39 ☽ □ ☿
17:38 ☽ ✶ ♃
20:25 ☽ □ ♄
18 06:58 ☽ □ ♅
11:32 ☽ ✶ ♆
19 00:18 ☿ # ♄
03:42 ☽ ✶ ♂
03:52 ☽ ☌ ♇
04:00 ☿ ☍ ♄
06:13 ♂ ✶ ♇
21:32 ☽ □ ♀
20 08:48 ☽ △ ♄
11:44 ☉ ♈
14:13 ☽ ✶ ☿
20:06 ☽ ✶ ♅
21 12:08 ☽ ✶ ☉
22 05:02 ☿ ☌ ♅
05:33 ☽ ∥ ♇
07:20 ☽ ✶ ♀
13:37 ☽ ∥ ♃
20:33 ☽ ☌ ♃
23 03:39 ☽ ∥ ♆
12:09 ☽ ☌ ♆
12:10 ☽ # ♀
16:44 ☿ ∥ ♅
18:34 ☉ □ ♇
19:14 ☽ ∥ ♂
24 03:23 ☽ ✶ ♇
08:13 ☽ # ♄
11:03 ☽ ☌ ♂
25 00:25 ☽ ∥ ♅
05:25 ☽ ☍ ♄
05:34 ☽ ∥ ☿
06:24 ☽ # ☉
09:44 ☿ # ☉
16:54 ☽ ☌ ♅
19:56 ☿ ♈
20:26 ☽ # ☿
22:51 ☽ ∥ ☉
26 03:32 ☽ # ♅
06:37 ☽ ☌ ☿
10:59 ☽ □ ♇
19:16 ☽ ☌ ♀
19:29 ☽ ∥ ♄
27 03:19 ☽ # ♂
09:07 ☽ ∥ ♀
12:15 ☿ □ ♇
13:39 ☽ ✶ ♃
19:24 ♀ ☌ ☉
21:26 ☽ # ♆
28 02:18 ☽ ✶ ♆
08:41 ☽ # ♃
15:51 ☽ △ ♇
17:23 ☽ # ♇
18:09 ☉ # ♅
29 02:32 ☿ ☌ ♀
05:12 ☽ ✶ ♂
15:05 ☽ △ ♄
18:19 ☽ □ ♃
30 02:40 ☽ ✶ ♅
06:02 ☽ □ ♆
22:38 ☽ ✶ ♀
31 03:30 ☿ ☌ ☉
04:22 ☿ # ♅
07:51 ☽ ✶ ☉
08:14 ☽ ✶ ☿
11:20 ☽ □ ♂
17:52 ☽ □ ♄
22:01 ☽ △ ♃

⚷ Chiron

01 Dec.	08 S 16
02	22♒35
05	22 47
08	23 00
11	23 12
14	23 24
17	23 35
20	23 47
23	23 58
26	24 09
29	24 19

April 2009

Day	S. T.	☉	☽	☿	♀	♂	♃	♄	♅	♆	♇	☊ True
	h m s	° ' "	° ' "	° '	° '	° '	° '	° '	° '	° '	° '	° '
01 We	12 37 57	11♈24 09	20♊15 29	12♈18	04♈R41	13♓12	19♒06	16♍R37	23♓41	25♒36	03♑18	07♒R02
02 Th	12 41 54	12 23 22	04♋24 20	14 21	04 06	13 59	19 17	16 33	23 44	25 37	03 18	07 00
03 Fr	12 45 50	13 22 32	18 31 34	16 25	03 32	14 46	19 28	16 29	23 47	25 39	03 18	06 59
04 Sa	12 49 47	14 21 40	02♌36 06	18 29	03 00	15 33	19 39	16 25	23 50	25 41	03 18	06 58
05 Su	12 53 43	15 20 45	16 37 03	20 33	02 29	16 19	19 50	16 21	23 54	25 42	03R 18	06 57
06 Mo	12 57 40	16 19 48	00♍33 24	22 37	02 01	17 06	20 01	16 18	23 57	25 44	03 18	06 54
07 Tu	13 01 36	17 18 49	14 23 30	24 41	01 34	17 53	20 12	16 14	24 00	25 45	03 18	06 48
08 We	13 05 33	18 17 48	28 05 09	26 45	01 09	18 40	20 22	16 10	24 03	25 47	03 18	06 40
09 Th	13 09 29	19 16 44	11♎35 42	28 47	00 47	19 27	20 33	16 07	24 07	25 49	03 18	06 30
10 Fr	13 13 26	20 15 39	24 52 30	00♉49	00 26	20 14	20 43	16 03	24 10	25 50	03 17	06 18
11 Sa	13 17 23	21 14 32	07♏53 28	02 48	00 09	21 00	20 54	16 00	24 13	25 52	03 17	06 05
12 Su	13 21 19	22 13 22	20 37 32	04 46	29♓53	21 47	21 04	15 56	24 16	25 53	03 17	05 53
13 Mo	13 25 16	23 12 11	03♐04 55	06 42	29 40	22 34	21 14	15 53	24 19	25 55	03 17	05 42
14 Tu	13 29 12	24 10 58	15 17 13	08 35	29 30	23 20	21 24	15 50	24 22	25 56	03 17	05 34
15 We	13 33 09	25 09 43	27 17 21	10 25	29 21	24 07	21 34	15 46	24 25	25 57	03 16	05 29
16 Th	13 37 05	26 08 27	09♑09 16	12 11	29 16	24 54	21 44	15 43	24 28	25 59	03 16	05 27
17 Fr	13 41 02	27 07 09	20 57 49	13 55	29 13	25 40	21 53	15 40	24 31	26 00	03 16	05D 27
18 Sa	13 44 58	28 05 49	02♒48 18	15 34	29D 12	26 27	22 03	15 38	24 34	26 01	03 15	05 27
19 Su	13 48 55	29 04 27	14 46 14	17 09	29 14	27 14	22 12	15 35	24 37	26 03	03 15	05R 28
20 Mo	13 52 52	00♉03 04	26 56 55	18 40	29 18	28 00	22 22	15 32	24 40	26 04	03 14	05 27
21 Tu	13 56 48	01 01 39	09♓25 11	20 07	29 24	28 47	22 31	15 29	24 43	26 05	03 14	05 23
22 We	14 00 45	02 00 12	22 14 48	21 29	29 32	29 33	22 40	15 27	24 46	26 06	03 13	05 17
23 Th	14 04 41	02 58 43	05♈28 01	22 46	29 43	00♈20	22 49	15 24	24 49	26 07	03 13	05 09
24 Fr	14 08 38	03 57 13	19 05 04	23 59	29 56	01 06	22 57	15 22	24 52	26 09	03 12	04 58
25 Sa	14 12 34	04 55 41	03♉03 47	25 06	00♈10	01 53	23 06	15 20	24 55	26 10	03 12	04 46
26 Su	14 16 31	05 54 08	17 19 46	26 09	00 27	02 39	23 15	15 17	24 58	26 11	03 11	04 34
27 Mo	14 20 27	06 52 32	01♊46 58	27 06	00 46	03 26	23 23	15 15	25 00	26 12	03 10	04 24
28 Tu	14 24 24	07 50 54	16 18 46	27 58	01 06	04 12	23 31	15 13	25 03	26 13	03 10	04 16
29 We	14 28 21	08 49 15	00♋48 58	28 45	01 29	04 58	23 39	15 11	25 06	26 14	03 09	04 11
30 Th	14 32 17	09 47 33	15 12 42	29 26	01 53	05 45	23 47	15 10	25 08	26 15	03 08	04 07

Data for	04-01-2009
Julian Day	2454922.50
Ayanamsa	23 59 23
SVP	05 ♓ 08 03
☽ ☊ Mean	06 ♒ 12 R

● ◐ PHASES ○ ◑

Day	Time	Phase	Position
02	14:34	◐	12♋59
09	14:56	○	19♎53
17	13:37	◑	27♑40
25	03:23	●	05♉04

LAST ASPECT Day	h m	☽ INGRESS Day	h m	Sign
01	09:04	01	16:31	♋
03	09:00	03	19:33	♌
05	15:39	05	23:02	♍
07	16:53	08	03:23	♎
10	01:45	10	09:23	♏
12	17:29	12	18:01	♐
15	04:08	15	05:28	♑
17	16:43	17	18:20	♒
19	22:16	20	05:56	♓
22	13:30	22	14:10	♈
24	12:12	24	18:47	♉
26	15:43	26	21:03	♊
28	16:23	28	22:39	♋

DECLINATION

Day	☉	☽	☿	♀	♂	♃	♄	♅	♆	♇
01 We	04N31	26N47	03N57	08N54	07S 41	15S 39	07N22	03S 11	13S 20	17S 39
02 Th	04 54	26 06	04 53	08 31	07 23	15 36	07 24	03 10	13 20	17 39
03 Fr	05 17	23 45	05 50	08 09	07 05	15 32	07 25	03 08	13 19	17 39
04 Sa	05 40	19 57	06 46	07 46	06 47	15 29	07 27	03 07	13 19	17 39
05 Su	06 03	15 02	07 42	07 24	06 29	15 26	07 28	03 06	13 18	17 39
06 Mo	06 25	09 22	08 39	07 02	06 10	15 22	07 30	03 04	13 18	17 39
07 Tu	06 48	03 18	09 34	06 40	05 52	15 19	07 31	03 03	13 17	17 39
08 We	07 10	02S 51	10 30	06 19	05 34	15 16	07 32	03 02	13 17	17 39
09 Th	07 33	08 47	11 24	05 58	05 15	15 13	07 34	03 01	13 16	17 39
10 Fr	07 55	14 12	12 17	05 38	04 57	15 10	07 35	02 59	13 16	17 39
11 Sa	08 17	18 53	13 09	05 19	04 38	15 07	07 36	02 58	13 15	17 39
12 Su	08 39	22 36	14 00	05 00	04 20	15 03	07 38	02 57	13 15	17 38
13 Mo	09 01	25 10	14 49	04 43	04 01	15 00	07 39	02 56	13 14	17 38
14 Tu	09 23	26 29	15 36	04 26	03 43	14 57	07 40	02 54	13 14	17 38
15 We	09 44	26 33	16 21	04 10	03 24	14 54	07 41	02 53	13 13	17 38
16 Th	10 06	25 22	17 04	03 55	03 06	14 51	07 42	02 52	13 13	17 38
17 Fr	10 27	23 04	17 45	03 41	02 47	14 48	07 43	02 51	13 12	17 38
18 Sa	10 48	19 46	18 23	03 29	02 29	14 46	07 44	02 50	13 12	17 38
19 Su	11 09	15 38	18 59	03 17	02 10	14 43	07 45	02 49	13 11	17 38
20 Mo	11 29	10 48	19 32	03 06	01 51	14 40	07 46	02 47	13 11	17 38
21 Tu	11 50	05 25	20 02	02 57	01 33	14 37	07 47	02 46	13 11	17 38
22 We	12 10	00N18	20 30	02 48	01 14	14 34	07 48	02 45	13 10	17 38
23 Th	12 30	06 10	20 56	02 40	00 55	14 32	07 49	02 44	13 10	17 38
24 Fr	12 50	11 56	21 18	02 34	00 37	14 29	07 50	02 43	13 10	17 38
25 Sa	13 10	17 14	21 39	02 28	00 18	14 26	07 50	02 42	13 09	17 38
26 Su	13 29	21 41	21 56	02 24	00N00	14 24	07 51	02 41	13 09	17 38
27 Mo	13 49	24 52	22 11	02 20	00 19	14 21	07 52	02 40	13 09	17 38
28 Tu	14 08	26 27	22 24	02 18	00 38	14 18	07 52	02 39	13 08	17 38
29 We	14 26	26 12	22 34	02 16	00 56	14 16	07 53	02 37	13 08	17 38
30 Th	14 45	24 12	22 41	02 16	01 15	14 14	07 54	02 36	13 08	17 37

ASPECTARIAN

01 05:49 ☽ □ ♅
09:04 ☽ △ ♆
22:07 ☽ ☍ ♇
23:00 ♂ # ♄
23:30 ☽ □ ♀
02 00:17 ☿ ∥ ☉
17:13 ☽ △ ♂
19:47 ☽ □ ☿
20:33 ☽ ✶ ♄
03 09:00 ☽ △ ♅
10:28 ♀ □ ♆
04 00:14 ☿ # ♂
00:39 ☽ △ ♀
11:50 ☽ # ♇
14:59 ☿ ✶ ♃
17:34 ♇ SR
17:47 ☿ ∥ ♄
18:25 ☿ ∥ ♀
19:51 ♀ ∥ ♄
21:39 ☽ △ ☉
22:14 ☽ # ♃
05 00:56 ♂ ☍ ♄
05:36 ☽ ☍ ♃
07:39 ☽ # ♆
07:56 ☽ △ ☿
15:15 ☉ # ♂
15:39 ☽ ☍ ♆
06 02:32 ☽ ∥ ☿
04:44 ☽ △ ♇
07:30 ☽ ∥ ♄
09:57 ☽ ∥ ♀
11:05 ☽ ∥ ☉
13:26 ☽ # ♂
19:54 ♀ ∥ ☉
07 00:58 ☽ # ♅
03:12 ☽ ☌ ♄
[illegible]
12:38 ☿ ✶ ♆
16:53 ☽ ☍ ♅
08 00:42 ☽ ∥ ♅
05:16 ☽ ☍ ♀
09:13 ☽ □ ♇
10:16 ☽ ∥ ♂
13:05 ☽ # ♀
18:30 ☽ # ☉
18:55 ☽ # ♄
09 01:08 ☉ ∥ ♄
13:28 ☽ # ☿
14:22 ☿ ♉
16:21 ☽ △ ♃
19:36 ☽ ∥ ♆
10 01:45 ☽ △ ♆
04:31 ☽ ∥ ♃
12:51 ☽ ☍ ☿
13:44 ☉ ✶ ♃
15:27 ☽ ✶ ♇
17:12 ☽ ∥ ♇
11 02:34 ☿ # ♆
05:51 ☿ △ ♇
12:47 ♀ ♓R
15:08 ☽ ✶ ♄
12 00:51 ☽ □ ♃
02:21 ☽ △ ♂
07:00 ☽ △ ♅
10:05 ☽ □ ♆
17:29 ☽ △ ♀
13 05:19 ☿ # ♃
14 01:04 ☽ □ ♄
12:21 ☽ ✶ ♃
17:11 ☽ □ ♂
[illegible]
19:20 ☽ △ ☉
21:19 ☽ ✶ ♆
15 04:08 ☽ □ ♀
10:00 ♂ ☌ ♅
12:04 ☽ ☌ ♇
19:55 ☉ ✶ ♆
16 07:14 ☽ △ ☿
13:17 ☽ △ ♄
18:55 ♂ ∥ ♅
20:02 ☿ # ♇
17 07:16 ☽ ✶ ♅
10:14 ☽ ✶ ♂
16:43 ☽ ✶ ♀
19:26 ♀ SD
18 00:52 ☿ △ ♄
07:23 ☽ # ☿
12:52 ☽ ∥ ♇
19 04:48 ☽ ∥ ♃
05:26 ☽ □ ☿
12:27 ☽ ∥ ♆
14:54 ☽ ☌ ♃
20:53 ☽ # ☉
22:16 ☽ ☌ ♆
22:45 ☉ ♉
20 06:33 ☽ ✶ ☉
12:10 ☽ ✶ ♇
13:42 ☽ # ♄
21 10:48 ☽ # ♀
11:17 ☽ ∥ ♅
11:24 ☽ ☍ ♄
17:17 ☽ ∥ ♂
22:26 ☽ ✶ ☿
23:19 ♀ ☌ ♂
22 [illegible]
04:39 ☽ ☌ ♅
10:00 ☽ # ♅
10:00 ☽ ∥ ♀
10:31 ♀ # ♅
13:30 ☽ ☌ ♀
13:45 ♂ ♈
14:11 ☽ ☌ ♂
19:58 ☽ □ ♇
23 00:53 ☿ □ ♃
05:43 ☉ △ ♇
06:45 ☽ ∥ ♄
24 04:11 ☽ ∥ ☉
05:20 ☽ # ♆
[illegible]
07:19 ♀ ♈
11:06 ☽ # ♃
12:12 ☽ ✶ ♆
19:38 ☿ ✶ ♅
23:11 ☉ # ♆
25 00:13 ☽ △ ♇
01:57 ☽ # ♇
20:36 ☽ △ ♄
26 00:55 ☿ □ ♆
01:42 ☽ ∥ ☿
09:57 ☽ □ ♃
12:44 ☽ ✶ ♅
14:44 ☽ □ ♆
15:43 ☽ ☌ ☿
[illegible]
22:17 ☽ ✶ ♀
27 02:52 ☽ ✶ ♂
22:12 ☽ □ ♄
28 12:02 ☽ △ ♃
12:14 ☉ # ♃
14:29 ☽ □ ♅
16:23 ☽ △ ♆
29 01:08 ☽ □ ♀
03:52 ☽ ☍ ♇
07:18 ☽ □ ♂
14:17 ☽ ✶ ☉
23:55 ☽ ✶ ♄
30 11:09 ☽ ∥ ☿
[illegible]
22:29 ☿ ♊

⚷ Chiron

01 Dec.	07 S 33
01	24♒29
04	24 39
07	24 48
10	24 57
13	25 06
16	25 14
19	25 21
22	25 28
25	25 35
28	25 41

May 2009

Day	S. T.	☉	☽	☿	♀	♂	♃	♄	♅	♆	♇	☊ True
	h m s	° ′ ″	° ′ ″	° ′	° ′	° ′	° ′	° ′	° ′	° ′	° ′	° ′
01 Fr	14 36 14	10♉ 45 50	29♋ 26 51	00♊ 02	02♈ 18	06♈ 31	23♒ 55	15♍R08	25♓ 11	26♒ 16	03♑R07	04♒R06
02 Sa	14 40 10	11 44 04	13♌ 29 51	00 33	02 46	07 17	24 03	15 06	25 14	26 17	03 07	04 05
03 Su	14 44 07	12 42 16	27 21 19	00 58	03 14	08 03	24 11	15 05	25 16	26 17	03 06	04 04
04 Mo	14 48 03	13 40 26	11♍ 01 30	01 18	03 45	08 49	24 18	15 04	25 19	26 18	03 05	04 01
05 Tu	14 51 60	14 38 34	24 30 41	01 32	04 16	09 35	24 25	15 02	25 21	26 19	03 04	03 56
06 We	14 55 56	15 36 40	07♎ 48 57	01 41	04 50	10 22	24 32	15 01	25 24	26 20	03 03	03 48
07 Th	14 59 53	16 34 44	20 55 55	01 44	05 24	11 08	24 39	15 00	25 26	26 21	03 02	03 37
08 Fr	15 03 50	17 32 47	03♏ 51 01	01R 43	06 00	11 54	24 46	14 59	25 29	26 21	03 01	03 25
09 Sa	15 07 46	18 30 48	16 33 35	01 37	06 37	12 40	24 53	14 58	25 31	26 22	03 00	03 12
10 Su	15 11 43	19 28 47	29 03 17	01 25	07 15	13 25	24 59	14 57	25 33	26 23	02 59	02 59
11 Mo	15 15 39	20 26 45	11♐ 20 23	01 10	07 54	14 11	25 06	14 57	25 36	26 23	02 58	02 48
12 Tu	15 19 36	21 24 41	23 26 02	00 50	08 34	14 57	25 12	14 56	25 38	26 24	02 57	02 40
13 We	15 23 32	22 22 36	05♑ 22 17	00 27	09 16	15 43	25 18	14 56	25 40	26 24	02 56	02 34
14 Th	15 27 29	23 20 29	17 12 12	30♉ 00	09 58	16 29	25 24	14 55	25 42	26 25	02 55	02 32
15 Fr	15 31 25	24 18 22	28 59 44	29 30	10 42	17 15	25 30	14 55	25 45	26 25	02 54	02D 31
16 Sa	15 35 22	25 16 12	10♒ 49 34	28 59	11 26	18 00	25 35	14 55	25 47	26 26	02 53	02 33
17 Su	15 39 19	26 14 02	22 46 51	28 25	12 11	18 46	25 41	14 55	25 49	26 26	02 52	02 34
18 Mo	15 43 15	27 11 51	04♓ 56 58	27 51	12 57	19 31	25 46	14D 55	25 51	26 27	02 51	02R 35
19 Tu	15 47 12	28 09 38	17 25 05	27 16	13 44	20 17	25 51	14 55	25 53	26 27	02 49	02 33
20 We	15 51 08	29 07 24	00♈ 15 47	26 41	14 32	21 03	25 56	14 55	25 55	26 27	02 48	02 30
21 Th	15 55 05	00♊ 05 09	13 32 23	26 07	15 20	21 48	26 01	14 55	25 57	26 28	02 47	02 24
22 Fr	15 59 01	01 02 53	27 16 15	25 34	16 09	22 33	26 05	14 56	25 59	26 28	02 46	02 15
23 Sa	16 02 58	02 00 36	11♉ 26 13	25 03	16 59	23 19	26 10	14 57	26 01	26 28	02 44	02 06
24 Su	16 06 54	02 58 18	25 58 11	24 35	17 50	24 04	26 14	14 57	26 02	26 28	02 43	01 57
25 Mo	16 10 51	03 55 58	10♊ 45 33	24 09	18 41	24 49	26 18	14 58	26 04	26 28	02 42	01 49
26 Tu	16 14 48	04 53 38	25 40 04	23 46	19 33	25 35	26 22	14 59	26 06	26 29	02 41	01 42
27 We	16 18 44	05 51 16	10♋ 33 11	23 28	20 25	26 20	26 26	15 00	26 07	26 29	02 39	01 38
28 Th	16 22 41	06 48 53	25 17 28	23 12	21 18	27 05	26 29	15 01	26 09	26 29	02 38	01 36
29 Fr	16 26 37	07 46 28	09♌ 47 25	23 01	22 12	27 50	26 33	15 02	26 11	26R 29	02 37	01 35
30 Sa	16 30 34	08 44 02	23 59 54	22 55	23 06	28 35	26 36	15 03	26 12	26 29	02 35	01D 35
31 Su	16 34 30	09 41 34	07♍ 53 45	22 52	24 01	29 20	26 39	15 05	26 14	26 29	02 34	01R 35

Data for	05-01-2009
Julian Day	2454952.50
Ayanamsa	23 59 27
SVP	05 ♓ 07 59
☽ ☊ Mean	04 ♒ 36 R

● ◐ PHASES ○ ◑

Day	Time	Phase	Position
01	20:44	◐	11♌36
09	04:02	○	18♏41
17	07:27	◑	26♒32
24	12:11	●	03♊28
31	03:22	◐	09♍50

LAST ASPECT ☽ INGRESS

Day	h m	Day	h m	Sign
30	16:46	01	00:56	♌
02	22:09	03	04:37	♍
05	01:31	05	09:52	♎
07	10:01	07	16:48	♏
09	18:49	10	01:50	♐
12	05:56	12	13:10	♑
15	01:00	15	02:03	♒
17	10:41	17	14:18	♓
19	21:44	19	23:31	♈
21	22:37	22	04:41	♉
24	00:49	24	06:34	♊
26	01:18	26	06:58	♋
28	03:06	28	07:45	♌
30	08:19	30	10:18	♍

DECLINATION

Day	☉	☽	☿	♀	♂	♃	♄	♅	♆	♇
01 Fr	15N03	20N40	22N46	02N16	01N33	14S11	07N54	02S35	13S07	17S37
02 Sa	15 21	15 59	22 49	02 17	01 52	14 09	07 55	02 34	13 07	17 37
03 Su	15 39	10 31	22 49	02 19	02 10	14 06	07 55	02 33	13 07	17 37
04 Mo	15 57	04 38	22 47	02 22	02 28	14 04	07 55	02 32	13 06	17 37
05 Tu	16 14	01S23	22 42	02 25	02 47	14 02	07 56	02 31	13 06	17 37
06 We	16 31	07 14	22 36	02 29	03 05	14 00	07 56	02 30	13 06	17 37
07 Th	16 48	12 42	22 27	02 35	03 23	13 58	07 56	02 30	13 06	17 37
08 Fr	17 04	17 30	22 16	02 40	03 41	13 56	07 57	02 29	13 06	17 37
09 Sa	17 20	21 27	22 02	02 47	04 00	13 54	07 57	02 28	13 05	17 37
10 Su	17 36	24 21	21 47	02 54	04 18	13 52	07 57	02 27	13 05	17 37
11 Mo	17 52	26 03	21 30	03 02	04 36	13 50	07 57	02 26	13 05	17 37
12 Tu	18 07	26 28	21 11	03 11	04 54	13 48	07 57	02 25	13 05	17 37
13 We	18 22	25 39	20 51	03 20	05 12	13 46	07 57	02 24	13 05	17 37
14 Th	18 37	23 40	20 29	03 29	05 29	13 44	07 57	02 23	13 04	17 37
15 Fr	18 51	20 40	20 06	03 40	05 47	13 43	07 57	02 22	13 04	17 37
16 Sa	19 05	16 49	19 43	03 51	06 05	13 41	07 57	02 22	13 04	17 37
17 Su	19 19	12 15	19 18	04 02	06 23	13 39	07 57	02 21	13 04	17 37
18 Mo	19 32	07 09	18 53	04 14	06 40	13 38	07 57	02 20	13 04	17 37
19 Tu	19 45	01 40	18 28	04 26	06 58	13 36	07 56	02 19	13 04	17 37
20 We	19 58	04N03	18 04	04 39	07 15	13 35	07 56	02 19	13 04	17 37
21 Th	20 10	09 46	17 40	04 52	07 32	13 33	07 56	02 18	13 04	17 37
22 Fr	20 22	15 12	17 16	05 06	07 50	13 32	07 56	02 17	13 04	17 37
23 Sa	20 34	20 01	16 54	05 20	08 07	13 31	07 55	02 16	13 04	17 37
24 Su	20 45	23 45	16 33	05 34	08 24	13 30	07 55	02 16	13 04	17 37
25 Mo	20 56	25 58	16 13	05 49	08 41	13 28	07 54	02 15	13 04	17 37
26 Tu	21 07	26 21	15 56	06 04	08 58	13 27	07 54	02 14	13 03	17 37
27 We	21 17	24 51	15 40	06 20	09 14	13 26	07 53	02 14	13 03	17 38
28 Th	21 27	21 38	15 26	06 35	09 31	13 25	07 53	02 13	13 03	17 38
29 Fr	21 36	17 05	15 14	06 51	09 47	13 24	07 52	02 12	13 03	17 38
30 Sa	21 46	11 40	15 04	07 08	10 04	13 24	07 51	02 12	13 04	17 38
31 Su	21 54	05 47	14 57	07 24	10 20	13 23	07 51	02 11	13 04	17 38

ASPECTARIAN

01 01:02 ☽ ⚹ ☿
05:01 ☽ △ ♀
12:44 ☽ △ ♂
16:08 ☽ # ♇
02 02:45 ☽ ∥ ☉
08:26 ☽ # ♃
12:55 ☽ # ♆
17:07 ♀ □ ♇
18:25 ☽ ☍ ♃
22:09 ☽ ☍ ♆
03 06:28 ☽ □ ☿
10:02 ☽ △ ♇
10:42 ☽ ∥ ♄
13:26 ♀ ∥ ♂
04 05:03 ☽ △ ☉
05:05 ♂ # ♅
07:08 ☽ ☌ ♄
08:10 ☽ ∥ ♂
08:20 ☽ # ♅
08:57 ☽ ∥ ♀
05 01:31 ☽ ☍ ♅
04:12 ☽ # ♀
04:34 ☽ ∥ ♅
05:54 ☽ # ♂
09:34 ☉ △ ♄
12:47 ☽ △ ☿
15:23 ☽ □ ♇
18:20 ☽ ☍ ♀
06 02:55 ☽ # ♄
04:21 ♀ # ♅
04:55 ☽ ☍ ♂
07 01:52 ☽ ∥ ♆
05:01 ☿ SR
05:56 ☽ ∥ ♃
06:56 ☽ △ ♃
10:01 ☽ △ ♆
21:27 ☽ # ☉
22:27 ☽ ⚹ ♇
08 00:36 ☽ ∥ ♇
20:58 ☽ ⚹ ♄
09 03:51 ☽ # ☿
16:05 ☽ □ ♃
17:13 ☽ △ ♅
18:49 ☽ □ ♆
10 01:56 ☉ # ♇
04:31 ☽ ☍ ☿
16:51 ☽ △ ♀
11 06:00 ☽ △ ♂
07:06 ☽ □ ♄
12 03:34 ☽ ⚹ ♃
04:25 ☽ □ ♅
05:56 ☽ ⚹ ♆
19:05 ☽ ☌ ♇
13 08:22 ☽ □ ♀
19:22 ☽ △ ♄
22:26 ☽ □ ♂
23:53 ☿ ♉R
14 13:36 ☽ △ ☉
17:22 ☽ ⚹ ♅
15 01:00 ☽ △ ☿
04:16 ☽ # ☿
11:14 ☽ # ☉
19:19 ☽ ∥ ♇
16 01:18 ☽ ⚹ ♀
08:46 ☉ □ ♃
13:09 ☉ ⚹ ♅
15:26 ☽ ⚹ ♂
16:55 ☽ ∥ ♃
19:58 ☽ ∥ ♆
23:45 ☿ ∥ ☉
17 02:08 ♄ SD
05:06 ☉ □ ♆
05:48 ☽ ☌ ♃
07:16 ☽ ☌ ♆
10:41 ☽ □ ☿
19:53 ☽ ⚹ ♇
20:24 ☽ # ♄
18 02:04 ☽ # ♂
10:02 ☿ ☌ ☉
12:28 ☽ # ♀
19:14 ☽ ☍ ♄
21:10 ☽ ∥ ♅
19 15:56 ☽ ☌ ♅
16:44 ☽ # ♅
17:40 ☽ ⚹ ☿
21:44 ☽ ⚹ ☉
20 02:36 ☽ ∥ ♀
04:39 ☽ □ ♇
09:22 ☿ □ ♆
14:04 ☽ ∥ ♂
16:14 ☽ ∥ ♄
21:51 ☉ ♊
21 02:16 ☿ # ♇
03:23 ☽ ☌ ♀
03:45 ☿ □ ♃
06:47 ☿ ⚹ ♅
14:16 ☽ # ♆
15:23 ☽ ☌ ♂
16:25 ☽ # ♃
21:57 ☽ ⚹ ♃
22:37 ☽ ⚹ ♆
22 08:11 ♂ ∥ ♄
09:04 ☽ ∥ ☿
09:22 ☽ △ ♇
11:32 ☽ # ♇
23 03:12 ☽ ∥ ☉
05:51 ☽ △ ♄
21:47 ☽ ☌ ☿
24 00:07 ☽ ⚹ ♅
00:26 ☽ □ ♃
00:49 ☽ □ ♆
25 06:47 ☽ □ ♄
13:33 ☽ ⚹ ♀
23:51 ☽ ⚹ ♂
26 00:41 ☽ □ ♅
01:08 ☽ △ ♃
01:18 ☽ △ ♆
11:16 ☽ ☍ ♇
27 03:20 ♂ ⚹ ♃
04:38 ♂ ⚹ ♆
07:13 ☽ ⚹ ♄
17:03 ☽ □ ♀
20:13 ♃ ☌ ♆
20:38 ☽ ⚹ ☿
28 01:04 ☽ ∥ ☉
01:25 ☽ △ ♅
03:06 ☽ □ ♂
20:24 ☽ ⚹ ☉
21:25 ☽ # ♇
29 04:31 ♆ SR
08:51 ☽ ∥ ☿
16:39 ☽ # ♃
18:07 ☽ # ♆
22:09 ☽ □ ☿
22:22 ☽ △ ♀
30 04:15 ☽ ☍ ♆
04:28 ☽ ☍ ♃
06:23 ☽ ∥ ♂
08:19 ☽ △ ♂
14:45 ☽ △ ♇
15:43 ☽ ∥ ♄
17:49 ☽ ∥ ♀
31 01:23 ☿ SD
12:38 ☽ ☌ ♄
14:20 ☽ # ♅
21:19 ♂ ♉

⚷ Chiron

01 Dec.	06 S 59
01	25 ♒ 47
04	25 52
07	25 57
10	26 01
13	26 04
16	26 07
19	26 10
22	26 12
25	26 13
28	26 14
31	26 14R

30 09:14 26♒14 R

June 2009

Day	S. T.	☉	☽	☿	♀	♂	♃	♄	♅	♆	♇	☊ True
	h m s	° ′ ″	° ′ ″	° ′	° ′	° ′	° ′	° ′	° ′	° ′	° ′	° ′
01 Mo	16 38 27	10♊ 39 05	21♍ 29 21	22♉D54	24♈ 56	00♉ 05	26♒ 41	15♍ 06	26♓ 15	26♒R29	02♑R32	01♒R34
02 Tu	16 42 23	11 36 34	04♎ 47 54	23 01	25 51	00 50	26 44	15 08	26 17	26 28	02 31	01 32
03 We	16 46 20	12 34 02	17 50 55	23 11	26 47	01 35	26 47	15 10	26 18	26 28	02 30	01 27
04 Th	16 50 17	13 31 29	00♏ 39 57	23 27	27 44	02 19	26 49	15 11	26 19	26 28	02 28	01 21
05 Fr	16 54 13	14 28 55	13 16 22	23 47	28 41	03 04	26 51	15 13	26 21	26 28	02 27	01 12
06 Sa	16 58 10	15 26 20	25 41 20	24 11	29 38	03 49	26 53	15 15	26 22	26 28	02 25	01 04
07 Su	17 02 06	16 23 44	07♐ 55 56	24 40	00♉ 36	04 33	26 54	15 17	26 23	26 27	02 24	00 55
08 Mo	17 06 03	17 21 07	20 01 16	25 12	01 34	05 18	26 56	15 19	26 24	26 27	02 22	00 48
09 Tu	17 09 59	18 18 30	01♑ 58 45	25 49	02 33	06 02	26 57	15 22	26 25	26 27	02 21	00 42
10 We	17 13 56	19 15 51	13 50 14	26 30	03 32	06 47	26 58	15 24	26 26	26 26	02 19	00 39
11 Th	17 17 53	20 13 12	25 38 07	27 15	04 31	07 31	26 59	15 27	26 27	26 26	02 18	00D 39
12 Fr	17 21 49	21 10 32	07♒ 25 28	28 04	05 30	08 15	27 00	15 29	26 28	26 26	02 16	00 40
13 Sa	17 25 46	22 07 51	19 16 00	28 57	06 30	09 00	27 01	15 32	26 29	26 25	02 15	00 42
14 Su	17 29 42	23 05 11	01♓ 13 56	29 53	07 31	09 44	27 01	15 35	26 30	26 25	02 13	00 45
15 Mo	17 33 39	24 02 29	13 23 55	00♊ 53	08 31	10 28	27 01	15 37	26 31	26 24	02 12	00 47
16 Tu	17 37 35	24 59 47	25 50 46	01 57	09 32	11 12	27R 01	15 40	26 32	26 24	02 10	00 49
17 We	17 41 32	25 57 05	08♈ 39 03	03 04	10 34	11 56	27 01	15 43	26 32	26 23	02 09	00R 48
18 Th	17 45 28	26 54 23	21 52 35	04 15	11 35	12 40	27 00	15 47	26 33	26 22	02 07	00 46
19 Fr	17 49 25	27 51 40	05♉ 33 44	05 29	12 37	13 24	27 00	15 50	26 34	26 22	02 06	00 43
20 Sa	17 53 22	28 48 58	19 42 44	06 47	13 39	14 08	26 59	15 53	26 34	26 21	02 04	00 39
21 Su	17 57 18	29 46 15	04♊ 17 02	08 07	14 41	14 52	26 58	15 56	26 35	26 20	02 03	00 34
22 Mo	18 01 15	00♋ 43 31	19 11 08	09 32	15 44	15 35	26 57	16 00	26 35	26 20	02 01	00 30
23 Tu	18 05 11	01 40 48	04♋ 17 07	10 59	16 47	16 19	26 55	16 04	26 36	26 19	02 00	00 26
24 We	18 09 08	02 38 04	19 25 41	12 30	17 50	17 03	26 54	16 07	26 36	26 18	01 58	00 24
25 Th	18 13 04	03 35 19	04♌ 27 34	14 04	18 53	17 46	26 52	16 11	26 36	26 17	01 56	00 23
26 Fr	18 17 01	04 32 34	19 14 56	15 41	19 57	18 29	26 50	16 15	26 37	26 17	01 55	00D 23
27 Sa	18 20 57	05 29 49	03♍ 42 14	17 21	21 00	19 13	26 48	16 19	26 37	26 16	01 53	00 24
28 Su	18 24 54	06 27 02	17 46 27	19 04	22 04	19 56	26 46	16 23	26 37	26 15	01 52	00 25
29 Mo	18 28 51	07 24 16	01♎ 26 55	20 51	23 08	20 39	26 43	16 27	26 37	26 14	01 50	00R 25
30 Tu	18 32 47	08 21 28	14 44 39	22 40	24 13	21 23	26 40	16 31	26 37	26 13	01 49	00 25

Data for	06-01-2009
Julian Day	2454983.50
Ayanamsa	23 59 31
SVP	05 ♓ 07 55
☽ ☊ Mean	02 ♒ 58 R

● ◐ PHASES ○ ◑

Day	Time	Phase	Position
07	18:13	○	17♐07
15	22:15	◑	24♓56
22	19:35	●	01♋30
29	11:29	◐	07♎52

LAST ASPECT ☽ INGRESS

Day	h m	Day	h m	Sign
01	08:33	01	15:17	♎
03	18:01	03	22:45	♏
06	02:19	06	08:25	♐
08	13:52	08	20:01	♑
11	03:32	11	08:53	♒
13	21:05	13	21:33	♓
16	01:18	16	07:52	♈
18	09:35	18	14:21	♉
20	12:02	20	17:00	♊
22	12:20	22	17:12	♋
24	11:25	24	16:51	♌
26	12:29	26	17:47	♍
28	15:27	28	21:25	♎

DECLINATION

Day	☉	☽	☿	♀	♂	♃	♄	♅	♆	♇
01 Mo	22N03	00S 14	14N52	07N41	10N36	13S 22	07N50	02S 11	13S 04	17S 38
02 Tu	22 11	06 07	14 49	07 58	10 52	13 21	07 49	02 10	13 04	17 38
03 We	22 18	11 36	14 48	08 15	11 08	13 21	07 48	02 10	13 04	17 38
04 Th	22 25	16 30	14 49	08 32	11 24	13 20	07 47	02 09	13 04	17 38
05 Fr	22 32	20 36	14 53	08 50	11 40	13 20	07 47	02 09	13 04	17 38
06 Sa	22 39	23 42	14 58	09 07	11 56	13 20	07 46	02 08	13 04	17 38
07 Su	22 45	25 41	15 06	09 25	12 11	13 19	07 45	02 08	13 04	17 38
08 Mo	22 50	26 25	15 15	09 43	12 26	13 19	07 44	02 07	13 04	17 38
09 Tu	22 55	25 55	15 26	10 01	12 42	13 19	07 43	02 07	13 04	17 38
10 We	23 00	24 13	15 38	10 19	12 57	13 19	07 42	02 07	13 05	17 38
11 Th	23 05	21 28	15 52	10 37	13 12	13 19	07 40	02 06	13 05	17 38
12 Fr	23 09	17 49	16 08	10 55	13 26	13 19	07 39	02 06	13 05	17 39
13 Sa	23 12	13 27	16 25	11 13	13 41	13 19	07 38	02 06	13 05	17 39
14 Su	23 16	08 33	16 43	11 31	13 55	13 19	07 37	02 05	13 05	17 39
15 Mo	23 18	03 15	17 02	11 49	14 10	13 19	07 36	02 05	13 05	17 39
16 Tu	23 21	02N17	17 22	12 07	14 24	13 19	07 34	02 05	13 06	17 39
17 We	23 23	07 52	17 43	12 25	14 38	13 19	07 33	02 05	13 06	17 39
18 Th	23 24	13 17	18 04	12 43	14 52	13 20	07 32	02 04	13 06	17 39
19 Fr	23 25	18 16	18 27	13 01	15 06	13 20	07 30	02 04	13 06	17 39
20 Sa	23 26	22 24	18 49	13 19	15 19	13 21	07 29	02 04	13 07	17 39
21 Su	23 26	25 15	19 12	13 37	15 33	13 21	07 27	02 04	13 07	17 40
22 Mo	23 26	26 26	19 35	13 55	15 46	13 22	07 26	02 04	13 07	17 40
23 Tu	23 26	25 40	19 59	14 12	15 59	13 23	07 24	02 03	13 07	17 40
24 We	23 25	23 00	20 22	14 30	16 12	13 24	07 23	02 03	13 08	17 40
25 Th	23 23	18 46	20 44	14 47	16 25	13 24	07 21	02 03	13 08	17 40
26 Fr	23 22	13 25	21 07	15 04	16 38	13 25	07 19	02 03	13 08	17 40
27 Sa	23 20	07 27	21 29	15 21	16 50	13 26	07 18	02 03	13 09	17 40
28 Su	23 17	01 16	21 50	15 38	17 02	13 27	07 16	02 03	13 09	17 40
29 Mo	23 14	04S 48	22 09	15 54	17 14	13 29	07 14	02 03	13 09	17 41
30 Tu	23 11	10 28	22 28	16 11	17 26	13 30	07 13	02 03	13 10	17 41

ASPECTARIAN

01 02:32 ☽ △ ☿
07:49 ☽ ∥ ♅
08:33 ☽ ☍ ♅
12:21 ♀ ∥ ♄
19:52 ☽ □ ♇
02 07:13 ☽ # ♄
08:17 ☽ # ♀
13:27 ☽ △ ☉
15:55 ♀ ✶ ♆
21:46 ☽ # ♂
23:40 ♀ ✶ ♃
03 06:48 ☽ ∥ ♆
08:09 ☽ ∥ ♃
15:18 ☽ # ☿
16:06 ☽ △ ♆
16:43 ☽ △ ♃
18:01 ☽ ☍ ♀
04 03:20 ☽ ☍ ♂
03:24 ☽ ✶ ♇
04:35 ♂ △ ♇
06:09 ☽ ∥ ♇
05 03:45 ☽ ✶ ♄
14:26 ☽ # ☉
19:11 ☉ □ ♄
20:58 ☽ ☍ ☿
06 01:19 ☽ △ ♅
01:30 ☽ □ ♆
02:19 ☽ □ ♃
09:07 ♀ ♉
07 14:37 ☽ □ ♄
08 12:48 ☽ □ ♅
12:52 ☽ ✶ ♆
13:52 ☽ ✶ ♃
19:21 ♀ △ ♇
09 00:44 ☽ ☌ ♇
01:14 ☽ △ ♀
08:45 ☽ △ ♂
21:46 ☿ ✶ ♅
21:54 ☿ □ ♆
10 03:11 ☽ △ ♄
11:19 ☽ # ☉
12:49 ♂ # ♆
15:34 ☿ □ ♃
11 01:40 ☽ ✶ ♅
03:32 ☽ △ ☿
11:15 ♂ # ♃
19:45 ☽ □ ♀
12 01:04 ☽ ∥ ♇
01:48 ☽ □ ♂
09:10 ☽ # ☿
22:57 ☽ # ♂
13 00:50 ☽ ∥ ♃
01:58 ☽ ∥ ♆
06:17 ☽ △ ☉
10:41 ☽ # ♀
14:23 ☽ ☌ ♆
15:35 ☽ ☌ ♃
21:05 ☽ □ ☿
14 01:58 ☽ ✶ ♇
02:48 ☿ ♊
04:25 ☽ # ♄
13:34 ☽ ✶ ♀
17:54 ☽ ✶ ♂
15 04:21 ☽ ☍ ♄
05:10 ☽ ∥ ♅
07:51 ♃ SR
23:10 ☽ # ♅
16 01:18 ☽ ☌ ♅
11:55 ☽ □ ♇
12:37 ☽ ✶ ☿
19:53 ☿ # ♇
22:39 ☽ ∥ ♄
17 10:47 ☉ △ ♆
14:57 ☉ □ ♅
21:15 ☽ ∥ ♀
23:08 ☽ # ♆
18 00:11 ☽ # ♃
02:29 ☉ △ ♃
07:39 ☽ ∥ ♂
07:58 ☽ ✶ ♆
09:05 ☽ ✶ ♃
09:35 ☽ ✶ ☉
18:00 ☽ △ ♇
20:53 ☽ # ♇
19 01:01 ☽ ∥ ☿
06:53 ♀ # ♆
13:00 ☽ ☌ ♀
14:07 ☽ ☌ ♂
17:33 ☽ △ ♄
20 02:25 ♀ # ♃
07:24 ☽ ∥ ☉
11:00 ☽ □ ♆
11:23 ☽ ✶ ♅
12:02 ☽ □ ♃
21 05:46 ☉ ♋
06:52 ☽ ☌ ☿
13:10 ♀ ☌ ♂
18:53 ☽ □ ♄
22 06:31 ♀ △ ♄
11:22 ☽ △ ♆
11:48 ☽ □ ♅
12:20 ☽ △ ♃
14:45 ♂ △ ♄
20:22 ☽ ☍ ♇
23 07:40 ☉ ☍ ♇
18:44 ☽ ✶ ♄
20:01 ☽ ✶ ♂
21:13 ☽ ∥ ☉
21:16 ☽ ✶ ♀
24 11:25 ☽ △ ♅
14:39 ☽ ∥ ☿
25 05:21 ☽ # ♇
10:41 ☽ ∥ ♂
17:24 ☽ ∥ ♀
17:26 ☽ ✶ ☿
22:42 ☽ □ ♂
26 00:03 ☽ # ♃
01:14 ☽ □ ♀
01:14 ☽ # ♆
08:33 ☿ □ ♄
11:35 ☽ ☍ ♆
12:29 ☽ ☍ ♃
20:57 ☽ △ ♇
27 00:37 ☽ ∥ ♄
03:14 ☽ ✶ ☉
20:58 ☽ # ♅
21:35 ☽ ☌ ♄
28 02:34 ☽ □ ☿
03:57 ☽ △ ♂
08:06 ☽ △ ♀
13:00 ☽ ∥ ♅
15:27 ☽ ☍ ♅
29 00:41 ☽ □ ♇
10:02 ☽ # ♄
30 12:18 ☽ ∥ ♆
13:57 ☽ ∥ ♃
17:04 [illegible]
21:12 ☽ △ ♆
22:00 ☽ △ ♃

⚷ Chiron

01 Dec.	06 S 40
03	26♒13R
06	26 12
09	26 11
12	26 09
15	26 06
18	26 03
21	26 00
24	25 56
27	25 51
30	25 46

07 09:40 15♑24 ✹ Penumbral Lunar Eclipse (mag 0.182)
22 02:34 29♋27 ✺ Total Solar Eclipse

July 2009

Day	S. T.	☉	☽	☿	♀	♂	♃	♄	♅	♆	♇	☊ True
	h m s	° ' "	° ' "	° '	° '	° '	° '	° '	° '	° '	° '	° '
01 We	18 36 44	09♋ 18 41	27♎ 41 48	24♊ 32	25♉ 17	22♉ 06	26♒R37	16♍ 35	26♓ 37	26♒R12	01♑R47	00♒R23
02 Th	18 40 40	10 15 53	10♏ 21 07	26 27	26 22	22 49	26 34	16 39	26R 37	26 11	01 46	00 21
03 Fr	18 44 37	11 13 05	22 45 31	28 24	27 27	23 32	26 31	16 44	26 37	26 10	01 44	00 18
04 Sa	18 48 33	12 10 16	04♐ 57 50	00♋ 23	28 32	24 14	26 28	16 48	26 37	26 09	01 43	00 15
05 Su	18 52 30	13 07 28	17 00 43	02 25	29 37	24 57	26 24	16 53	26 37	26 08	01 41	00 13
06 Mo	18 56 26	14 04 39	28 56 33	04 29	00♊ 43	25 40	26 20	16 58	26 37	26 07	01 40	00 11
07 Tu	19 00 23	15 01 50	10♑ 47 33	06 34	01 48	26 23	26 16	17 02	26 37	26 06	01 38	00 09
08 We	19 04 20	15 59 01	22 35 52	08 41	02 54	27 05	26 12	17 07	26 36	26 05	01 37	00D 09
09 Th	19 08 16	16 56 13	04♒ 23 46	10 49	04 00	27 48	26 08	17 12	26 36	26 03	01 35	00 10
10 Fr	19 12 13	17 53 24	16 13 41	12 58	05 06	28 30	26 04	17 17	26 36	26 02	01 34	00 11
11 Sa	19 16 09	18 50 36	28 08 21	15 07	06 12	29 13	25 59	17 22	26 35	26 01	01 32	00 13
12 Su	19 20 06	19 47 48	10♓ 10 51	17 17	07 19	29 55	25 54	17 27	26 35	26 00	01 31	00 15
13 Mo	19 24 02	20 45 00	22 24 34	19 26	08 25	00♊ 37	25 49	17 32	26 34	25 59	01 29	00 17
14 Tu	19 27 59	21 42 13	04♈ 53 13	21 35	09 32	01 19	25 44	17 37	26 34	25 57	01 28	00 18
15 We	19 31 55	22 39 27	17 40 31	23 44	10 39	02 01	25 39	17 42	26 33	25 56	01 26	00 19
16 Th	19 35 52	23 36 41	00♉ 49 52	25 52	11 46	02 43	25 33	17 48	26 32	25 55	01 25	00R 19
17 Fr	19 39 49	24 33 56	14 23 55	27 59	12 53	03 25	25 28	17 53	26 31	25 53	01 24	00 18
18 Sa	19 43 45	25 31 12	28 23 51	00♌ 05	14 01	04 07	25 22	17 59	26 31	25 52	01 22	00 18
19 Su	19 47 42	26 28 28	12♊ 48 52	02 09	15 08	04 49	25 16	18 04	26 30	25 51	01 21	00 17
20 Mo	19 51 38	27 25 45	27 35 35	04 12	16 16	05 31	25 10	18 10	26 29	25 49	01 20	00 17
21 Tu	19 55 35	28 23 03	12♋ 38 03	06 14	17 23	06 12	25 04	18 15	26 28	25 48	01 18	00 16
22 We	19 59 31	29 20 21	27 48 09	08 14	18 31	06 54	24 58	18 21	26 27	25 47	01 17	00 15
23 Th	20 03 28	00♌ 17 40	12♌ 56 39	10 12	19 39	07 35	24 52	18 27	26 26	25 45	01 16	00 15
24 Fr	20 07 25	01 14 59	27 54 25	12 09	20 47	08 16	24 45	18 33	26 25	25 44	01 14	00 14
25 Sa	20 11 21	02 12 19	12♍ 33 50	14 04	21 55	08 58	24 39	18 39	26 24	25 42	01 13	00 13
26 Su	20 15 18	03 09 39	26 49 36	15 58	23 04	09 39	24 32	18 44	26 23	25 41	01 12	00 12
27 Mo	20 19 14	04 07 00	10♎ 39 06	17 49	24 12	10 20	24 25	18 50	26 22	25 39	01 11	00 11
28 Tu	20 23 11	05 04 20	24 02 11	19 39	25 21	11 01	24 18	18 57	26 21	25 38	01 09	00 10
29 We	20 27 07	06 01 42	07♏ 00 39	21 28	26 29	11 42	24 11	19 03	26 19	25 36	01 08	00 10
30 Th	20 31 04	06 59 04	19 37 35	23 14	27 38	12 23	24 04	19 09	26 18	25 35	01 07	00D 10
31 Fr	20 35 00	07 56 26	01♐ 56 49	24 59	28 47	13 04	23 57	19 15	26 17	25 33	01 06	00 10

Data for	07-01-2009
Julian Day	2455013.50
Ayanamsa	23 59 37
SVP	05 ♓ 07 52
☽ ☊ Mean	01 ♒ 22 R

● ◐ PHASES ○ ◑

07	09:21	✹	15♑24
15	09:53	◑	23♈03
22	02:35	✺	29♋27
28	22:00	◐	05♏57

LAST ASPECT Day	h m	☽ INGRESS Day	h m	
30	22:00	01	04:20	♏
03	10:04	03	14:12	♐
05	19:18	06	02:08	♑
08	09:43	08	15:04	♒
11	02:17	11	03:44	♓
13	08:03	13	14:40	♈
15	15:07	15	22:30	♉
17	20:49	18	02:42	♊
19	22:13	20	03:52	♋
22	02:36	22	03:28	♌
23	20:29	24	03:24	♍
25	23:15	26	05:26	♎
28	02:54	28	10:57	♏
30	12:55	30	20:10	♐

DECLINATION

Day	☉	☽	☿	♀	♂	♃	♄	♅	♆	♇
01 We	23N07	15S 33	22N46	16N27	17N38	13S 31	07N11	02S 03	13S 10	17S 41
02 Th	23 03	19 50	23 01	16 43	17 50	13 32	07 09	02 03	13 10	17 41
03 Fr	22 58	23 10	23 15	16 58	18 01	13 34	07 07	02 03	13 11	17 41
04 Sa	22 53	25 23	23 27	17 14	18 12	13 35	07 05	02 03	13 11	17 41
05 Su	22 48	26 24	23 37	17 29	18 23	13 36	07 03	02 03	13 11	17 41
06 Mo	22 42	26 10	23 45	17 44	18 34	13 38	07 01	02 03	13 12	17 42
07 Tu	22 36	24 45	23 50	17 58	18 45	13 40	07 00	02 04	13 12	17 42
08 We	22 29	22 14	23 52	18 12	18 55	13 41	06 58	02 04	13 13	17 42
09 Th	22 22	18 47	23 52	18 26	19 06	13 43	06 56	02 04	13 13	17 42
10 Fr	22 15	14 34	23 49	18 40	19 16	13 45	06 53	02 04	13 13	17 42
11 Sa	22 07	09 47	23 44	18 53	19 26	13 46	06 51	02 04	13 14	17 42
12 Su	21 59	04 35	23 35	19 06	19 36	13 48	06 49	02 04	13 14	17 43
13 Mo	21 50	00N51	23 24	19 18	19 45	13 50	06 47	02 05	13 15	17 43
14 Tu	21 41	06 21	23 10	19 30	19 55	13 52	06 45	02 05	13 15	17 43
15 We	21 32	11 43	22 54	19 42	20 04	13 54	06 43	02 05	13 16	17 43
16 Th	21 23	16 43	22 35	19 53	20 13	13 56	06 41	02 06	13 16	17 43
17 Fr	21 13	21 03	22 14	20 04	20 21	13 58	06 38	02 06	13 16	17 44
18 Sa	21 02	24 21	21 50	20 14	20 30	14 00	06 36	02 06	13 17	17 44
19 Su	20 52	26 12	21 24	20 24	20 38	14 03	06 34	02 07	13 17	17 44
20 Mo	20 40	26 17	20 57	20 33	20 47	14 05	06 32	02 07	13 18	17 44
21 Tu	20 29	24 26	20 28	20 42	20 55	14 07	06 29	02 07	13 18	17 44
22 We	20 17	20 49	19 57	20 51	21 02	14 09	06 27	02 08	13 19	17 45
23 Th	20 05	15 48	19 25	20 59	21 10	14 12	06 25	02 08	13 19	17 45
24 Fr	19 53	09 53	18 51	21 07	21 18	14 14	06 22	02 09	13 20	17 45
25 Sa	19 40	03 32	18 16	21 14	21 25	14 16	06 20	02 09	13 20	17 45
26 Su	19 27	02S 48	17 40	21 20	21 32	14 19	06 17	02 09	13 21	17 45
27 Mo	19 14	08 48	17 03	21 26	21 39	14 21	06 15	02 10	13 21	17 46
28 Tu	19 00	14 13	16 25	21 32	21 45	14 24	06 13	02 10	13 22	17 46
29 We	18 46	18 49	15 47	21 37	21 52	14 26	06 10	02 11	13 22	17 46
30 Th	18 32	22 27	15 08	21 41	21 58	14 29	06 08	02 12	13 23	17 46
31 Fr	18 17	24 58	14 28	21 45	22 04	14 31	06 05	02 12	13 24	17 47

ASPECTARIAN

01 04:58 ☽ # ♀
05:40 ♂ # ♇
07:38 ♅ SR
07:41 ☽ ⚹ ♇
11:21 ☽ ∥ ♇
11:37 ☽ # ♂
20:01 ♀ □ ♆
20:48 ☿ △ ♆
23:49 ☽ △ ☉
02 01:27 ☿ ∥ ☉
01:35 ☿ △ ♃
02:13 ☿ □ ♅
04:25 ♀ □ ♃
05:42 ♀ ⚹ ♅
12:13 ☽ ⚹ ♄
22:25 ☽ # ☉
03 00:55 ☽ # ☿
01:35 ☽ ☍ ♂
06:39 ☽ □ ♆
07:19 ☽ □ ♃
07:33 ☽ △ ♅
10:04 ☽ ☍ ♀
19:20 ☿ ♋
04 15:29 ☿ ☍ ♇
23:44 ☽ □ ♄
05 08:23 ♀ ♊
18:18 ☽ ⚹ ♆
18:46 ☽ ⚹ ♃
19:18 ☽ □ ♅
20:38 ♀ # ♇
06 05:29 ☽ ☌ ♇
13:36 ☽ ☍ ☿
14:43 ♂ □ ♆
20:48 ♂ □ ♃
07 07:48 ♂ ⚹ ♅
09:43 ☽ # ☿
12:46 ☽ △ ♄
21:54 ☽ # ☉
08 08:09 ☽ ⚹ ♅
09:43 ☽ △ ♂
22:07 ☽ # ♂
23:07 ☽ △ ♀
09 02:01 ☽ # ♀
06:32 ☽ ∥ ♇
07:10 ☉ ⚹ ♄
10 04:19 ☽ ∥ ♃
07:03 ☽ ∥ ♆
09:08 ♃ ☌ ♆
19:42 ☽ ☌ ♃
19:45 ☽ ☌ ♆
11 02:17 ☽ □ ♂
06:49 ☽ ⚹ ♇
13:50 ☽ # ♄
17:46 ☽ □ ♀
12 01:55 ☿ ⚹ ♄
02:56 ♂ ♊
11:15 ☽ ∥ ♅
14:25 ☽ ☍ ♄
17:00 ☽ △ ☿
20:30 ☽ △ ☉
13 05:27 ☽ # ♅
08:03 ☽ ☌ ♅
16:47 ☽ ⚹ ♂
17:29 ☽ □ ♇
14 01:50 ☽ ∥ ♄
02:16 ☿ ☌ ☉
09:38 ☽ ⚹ ♀
15 07:13 ☽ # ♆
10:18 ☽ # ♃
13:18 ☽ □ ☿
14:32 ☽ ⚹ ♃
15:07 ☽ ⚹ ♆
16 01:03 ☽ △ ♇
05:11 ☽ # ♇
07:33 ☿ △ ♅
17:48 ☽ ∥ ♀
19:37 ☽ ∥ ♂
17 00:54 ☽ ∥ ☉
06:05 ☽ △ ♄
06:49 ☽ ∥ ☿
18:46 ☽ ⚹ ☉
18:54 ☽ □ ♃
19:43 ☽ □ ♆
20:49 ☽ ⚹ ♅
23:08 ☿ ♌
18 03:19 ☽ ⚹ ☿
10:05 ☽ ☌ ♂
19 00:36 ☉ △ ♅
04:07 ☽ ☌ ♀
08:39 ☽ □ ♄
16:18 ☉ ∥ ♂
20:08 ☽ △ ♃
21:09 ☽ △ ♆
22:13 ☽ □ ♅
20 05:59 ☽ ☍ ♇
06:47 ☿ ∥ ♂
08:15 ♀ ∥ ☉
14:54 ☿ ∥ ♀
22:20 ☿ ∥ ☉
23:32 ☿ ⚹ ♂
21 08:58 ☽ ⚹ ♄
20:06 ♀ □ ♄
21:52 ☽ △ ♅
22:50 ☽ ∥ ♂
23:52 ☽ ∥ ♀
22 03:00 ☽ ∥ ☉
05:14 ☽ ∥ ☿
15:05 ☽ ⚹ ♂
15:21 ☽ # ♇
16:36 ☉ ♌
18:59 ☽ ☌ ☿
23 06:46 ☽ # ♃
10:23 ☽ # ♆
11:35 ☽ ⚹ ♀
18:57 ☽ ☍ ♃
20:29 ☽ ☍ ♆
24 05:24 ☽ △ ♇
13:28 ☽ ∥ ♄
17:45 ☽ □ ♂
25 05:13 ☽ # ♅
10:13 ☽ ☌ ♄
17:02 ☽ □ ♀
20:27 ☿ # ♇
21:32 ☽ ∥ ♅
23:15 ☽ ☍ ♅
26 07:29 ☽ □ ♇
11:42 ☽ ⚹ ☉
13:38 ☽ # ♄
23:25 ☽ △ ♂
27 04:08 ♀ △ ♃
14:48 ☽ ⚹ ☿
20:01 ☽ ∥ ♆
28 00:29 ☽ △ ♃
00:53 ☽ ∥ ♃
02:37 ☽ △ ♀
02:54 ☽ △ ♆
05:52 ♀ △ ♆
09:37 ☽ # ☿
13:04 ☽ ⚹ ♇
18:05 ☽ ∥ ♇
20:38 ♀ □ ♅
23:43 ☽ # ☉
29 18:14 ☽ # ♀
20:16 ☽ # ♂
23:04 ☽ ⚹ ♄
30 08:08 ☽ □ ☿
08:30 ☽ □ ♃
10:38 ☿ ☍ ♃
11:30 ☽ □ ♆
12:55 ☽ △ ♅
21:56 ☿ # ♃
31 07:47 ☿ ☍ ♆
12:52 ☽ △ ☉
23:21 ☽ ☍ ♂

⚷ Chiron

01 Dec.	06 S 40
03	25♒40R
06	25 34
09	25 28
12	25 21
15	25 14
18	25 07
21	24 59
24	24 51
27	24 42
30	24 34

August 2009

06 00:40 13♒43 ✷ Penumbral Lunar Eclipse (mag 0.428)

Day	S. T.	☉	☽	☿	♀	♂	♃	♄	♅	♆	♇	☊ True
	h m s	° ' "	° ' "	° '	° '	° '	° '	° '	° '	° '	° '	° '
01 Sa	20 38 57	08♌ 53 49	14♐ 02 28	26♌ 42	29♊ 56	13♊ 44	23♒R50	19♍ 21	26♓R15	25♒R32	01♑R05	00♒ 11
02 Su	20 42 54	09 51 13	25 58 32	28 24	01♋ 05	14 25	23 42	19 28	26 14	25 30	01 04	00 13
03 Mo	20 46 50	10 48 37	07♑ 48 47	00♍ 04	02 14	15 05	23 35	19 34	26 13	25 29	01 02	00 14
04 Tu	20 50 47	11 46 03	19 36 32	01 42	03 23	15 46	23 27	19 40	26 11	25 27	01 01	00 16
05 We	20 54 43	12 43 29	01♒ 24 42	03 18	04 33	16 26	23 20	19 47	26 10	25 25	01 00	00 17
06 Th	20 58 40	13 40 55	13 15 51	04 53	05 42	17 06	23 12	19 53	26 08	25 24	00 59	00R 17
07 Fr	21 02 36	14 38 23	25 12 13	06 26	06 52	17 46	23 05	20 00	26 07	25 22	00 58	00 17
08 Sa	21 06 33	15 35 52	07♓ 15 49	07 58	08 01	18 26	22 57	20 06	26 05	25 21	00 57	00 16
09 Su	21 10 29	16 33 22	19 28 36	09 28	09 11	19 06	22 49	20 13	26 03	25 19	00 56	00 14
10 Mo	21 14 26	17 30 53	01♈ 52 32	10 56	10 21	19 46	22 41	20 20	26 02	25 17	00 55	00 12
11 Tu	21 18 22	18 28 26	14 29 38	12 23	11 31	20 26	22 34	20 26	26 00	25 16	00 54	00 09
12 We	21 22 19	19 26 00	27 22 06	13 48	12 41	21 06	22 26	20 33	25 58	25 14	00 53	00 08
13 Th	21 26 16	20 23 35	10♉ 32 04	15 11	13 51	21 45	22 18	20 40	25 56	25 12	00 53	00 07
14 Fr	21 30 12	21 21 12	24 01 25	16 32	15 02	22 25	22 10	20 47	25 54	25 11	00 52	00D 07
15 Sa	21 34 09	22 18 50	07♊ 51 28	17 51	16 12	23 04	22 02	20 54	25 53	25 09	00 51	00 07
16 Su	21 38 05	23 16 30	22 02 24	19 09	17 22	23 43	21 54	21 01	25 51	25 08	00 50	00 08
17 Mo	21 42 02	24 14 12	06♋ 32 48	20 25	18 33	24 23	21 46	21 07	25 49	25 06	00 49	00 09
18 Tu	21 45 58	25 11 55	21 19 18	21 39	19 44	25 02	21 39	21 14	25 47	25 04	00 49	00 10
19 We	21 49 55	26 09 39	06♌ 16 17	22 50	20 54	25 41	21 31	21 21	25 45	25 03	00 48	00R 09
20 Th	21 53 52	27 07 25	21 16 21	24 00	22 05	26 20	21 23	21 28	25 43	25 01	00 47	00 08
21 Fr	21 57 48	28 05 12	06♍ 11 03	25 08	23 16	26 58	21 15	21 36	25 41	24 59	00 47	00 04
22 Sa	22 01 45	29 03 01	20 52 08	26 13	24 27	27 37	21 08	21 43	25 39	24 58	00 46	00 00
23 Su	22 05 41	00♍ 00 51	05♎ 12 46	27 16	25 38	28 16	21 00	21 50	25 37	24 56	00 45	29♑ 55
24 Mo	22 09 38	00 58 42	19 08 22	28 16	26 49	28 54	20 52	21 57	25 35	24 54	00 45	29 50
25 Tu	22 13 34	01 56 34	02♏ 37 03	29 13	28 01	29 32	20 45	22 04	25 32	24 53	00 44	29 46
26 We	22 17 31	02 54 28	15 39 27	00♎ 08	29 12	00♋ 11	20 37	22 11	25 30	24 51	00 44	29 42
27 Th	22 21 27	03 52 23	28 18 08	01 00	00♌ 23	00 49	20 30	22 19	25 28	24 50	00 43	29 41
28 Fr	22 25 24	04 50 19	10♐ 37 02	01 49	01 35	01 27	20 22	22 26	25 26	24 48	00 43	29D 41
29 Sa	22 29 20	05 48 17	22 40 48	02 34	02 46	02 05	20 15	22 33	25 24	24 46	00 42	29 42
30 Su	22 33 17	06 46 15	04♑ 34 23	03 16	03 58	02 43	20 08	22 40	25 22	24 45	00 42	29 45
31 Mo	22 37 14	07 44 16	16 22 39	03 54	05 10	03 20	20 01	22 48	25 19	24 43	00 42	29 47

Data for	08-01-2009
Julian Day	2455044.50
Ayanamsa	23 59 42
SVP	05 ♓ 07 49
☽ ☊ Mean	29 ♑ 44 R

LAST ASPECT Day	h m	☽ INGRESS Day	h m	
02	05:42	02	08:09	♑
04	13:21	04	21:08	♒
07	00:20	07	09:35	♓
09	12:45	09	20:24	♈
11	20:04	12	04:51	♉
14	03:18	14	10:27	♊
16	06:20	16	13:14	♋
18	07:10	18	13:57	♌
20	10:02	20	14:01	♍
22	11:44	22	15:12	♎
24	18:10	24	19:17	♏
26	18:35	27	03:16	♐
29	05:26	29	14:44	♑

● ◐ PHASES ○ ◑

06	00:55	✷	13♒43
13	18:56	◑	21♉09
20	10:02	●	27♌32
27	11:42	◐	04♐21

DECLINATION

Day	☉	☽	☿	♀	♂	♃	♄	♅	♆	♇
01 Sa	18N02	26S 18	13N48	21N48	22N10	14S 34	06N03	02S 13	13S 24	17S 47
02 Su	17 47	26 22	13 07	21 51	22 15	14 37	06 00	02 13	13 25	17 47
03 Mo	17 31	25 14	12 26	21 53	22 21	14 39	05 57	02 14	13 25	17 47
04 Tu	17 16	22 59	11 45	21 55	22 26	14 42	05 55	02 14	13 26	17 47
05 We	16 59	19 45	11 04	21 56	22 31	14 45	05 52	02 15	13 26	17 48
06 Th	16 43	15 42	10 22	21 56	22 36	14 47	05 50	02 16	13 27	17 48
07 Fr	16 26	11 01	09 41	21 56	22 40	14 50	05 47	02 16	13 27	17 48
08 Sa	16 10	05 54	09 00	21 56	22 45	14 52	05 44	02 17	13 28	17 48
09 Su	15 52	00 29	08 19	21 54	22 49	14 55	05 42	02 18	13 28	17 49
10 Mo	15 35	05N00	07 37	21 53	22 53	14 58	05 39	02 18	13 29	17 49
11 Tu	15 17	10 23	06 57	21 50	22 57	15 01	05 36	02 19	13 30	17 49
12 We	15 00	15 26	06 16	21 47	23 01	15 03	05 34	02 20	13 30	17 49
13 Th	14 41	19 53	05 36	21 44	23 04	15 06	05 31	02 21	13 31	17 50
14 Fr	14 23	23 26	04 56	21 39	23 07	15 09	05 28	02 21	13 31	17 50
15 Sa	14 04	25 43	04 16	21 34	23 10	15 11	05 25	02 22	13 32	17 50
16 Su	13 46	26 26	03 37	21 29	23 13	15 14	05 23	02 23	13 32	17 50
17 Mo	13 27	25 23	02 59	21 23	23 16	15 16	05 20	02 24	13 33	17 51
18 Tu	13 07	22 33	02 21	21 16	23 18	15 19	05 17	02 24	13 33	17 51
19 We	12 48	18 10	01 43	21 09	23 21	15 22	05 14	02 25	13 34	17 51
20 Th	12 28	12 37	01 07	21 01	23 23	15 24	05 11	02 26	13 35	17 51
21 Fr	12 08	06 23	00 31	20 53	23 25	15 27	05 09	02 27	13 35	17 52
22 Sa	11 48	00S 06	00S 04	20 44	23 27	15 29	05 06	02 28	13 36	17 52
23 Su	11 28	06 25	00 38	20 35	23 28	15 32	05 03	02 28	13 36	17 52
24 Mo	11 08	12 14	01 11	20 24	23 30	15 35	05 00	02 29	13 37	17 52
25 Tu	10 47	17 16	01 43	20 14	23 31	15 37	04 57	02 30	13 37	17 53
26 We	10 26	21 19	02 14	20 02	23 32	15 39	04 54	02 31	13 38	17 53
27 Th	10 05	24 15	02 44	19 51	23 33	15 42	04 52	02 32	13 38	17 53
28 Fr	09 44	25 57	03 12	19 38	23 33	15 44	04 49	02 33	13 39	17 53
29 Sa	09 23	26 23	03 39	19 25	23 34	15 47	04 46	02 34	13 40	17 54
30 Su	09 02	25 35	04 04	19 12	23 34	15 49	04 43	02 35	13 40	17 54
31 Mo	08 40	23 39	04 27	18 58	23 34	15 51	04 40	02 35	13 41	17 54

ASPECTARIAN

01 01:28 ♀ ♋
10:45 ☽ □ ♄
13:45 ☿ # ♆
19:28 ☽ ✶ ♃
23:03 ☽ ✶ ♆
23:35 ♀ ☍ ♇
23:47 ☉ # ♇
02 00:31 ☽ □ ♅
05:42 ☽ △ ☿
10:16 ☽ ☌ ♇
11:27 ☽ ☍ ♀
23:07 ☿ ♍
03 14:10 ☿ △ ♇
04 00:08 ☽ △ ♄
04:27 ☽ # ♂
08:36 ☽ # ♀
13:21 ☽ ✶ ♅
05 12:07 ☽ ∥ ♇
17:57 ☽ # ☉
06 04:53 ☽ ∥ ♃
08:12 ☽ △ ♂
11:54 ☽ ∥ ♆
19:47 ☽ ☌ ♃
07 00:20 ☽ ☌ ♆
07:26 ☽ # ☿
11:30 ☽ ✶ ♇
08 00:42 ☽ # ♄
01:35 ☽ ☍ ☿
01:40 ☽ △ ♀
03:53 ☿ ✶ ♀
16:05 ☽ ∥ ♅
23:14 ☽ □ ♂
09 01:27 ☽ ☍ ♄
12:12 ☽ # ♅
12:45 ☽ ☌ ♅
22:10 ☽ □ ♇
10 02:51 ☽ ∥ ♄
10:18 ☽ ∥ ☿
17:49 ☽ □ ♀
11 00:16 ♂ □ ♄
08:05 ☽ △ ☉
11:44 ☽ ✶ ♂
14:33 ☽ # ♆
14:57 ☽ ✶ ♃
19:47 ☉ # ♃
20:04 ☽ ✶ ♆
21:57 ☽ ∥ ☉
22:05 ☽ # ♃
12 06:28 ☽ △ ♇
12:22 ☽ # ♇
13 03:03 ☿ ∥ ♄
06:32 ☽ ✶ ♀
09:16 ☽ △ ☿
11:19 ☽ ∥ ♀
16:30 ♂ △ ♃
18:14 ☽ △ ♄
20:46 ☽ □ ♃
21:30 ☽ ∥ ♂
14 02:02 ☽ □ ♆
03:18 ☽ ✶ ♅
17:53 ☉ ☍ ♃
15 18:41 ☽ □ ☿
22:16 ☽ □ ♄
23:47 ☽ △ ♃
16 02:13 ☽ ✶ ☉
02:57 ☽ ☌ ♂
05:08 ☽ △ ♆
06:20 ☽ □ ♅
14:36 ☽ ☍ ♇
16:18 ☉ # ♆
17 10:58 ☉ ✶ ♂
15:15 ☿ ☌ ♄
18:48 ☽ ∥ ♂
20:55 ☉ ☍ ♆
21:12 ☽ ☌ ♀
21:38 ☿ # ♅
23:52 ☽ ✶ ♄
18 00:34 ☽ ✶ ☿
01:30 ♂ △ ♆
07:10 ☽ △ ♅
08:01 ☽ ∥ ♀
19 01:27 ☽ # ♇
02:24 ♂ □ ♅
10:11 ♀ ✶ ♄
12:31 ☽ # ♃
20:06 ☽ # ♆
20 00:11 ☽ ☍ ♃
00:37 ☽ ∥ ☉
06:00 ☽ ☍ ♆
08:29 ☽ ✶ ♂
15:16 ☽ △ ♇
21 04:36 ☽ ∥ ♄
11:47 ☿ ☍ ♅
14:31 ☽ # ♅
23:24 ☽ # ☿
23:51 ☽ ∥ ☿
22 01:24 ☽ ☌ ♄
06:28 ☽ ✶ ♀
07:54 ☽ ☍ ♅
08:49 ☽ ∥ ♅
09:34 ☽ ☌ ☿
11:44 ☽ □ ♂
16:29 ☽ □ ♇
18:42 ☽ # ♄
23:32 ♀ △ ♅
23:39 ☉ ♍
23 18:19 ☉ △ ♇
19:30 ☽ # ☉
24 03:00 ☽ △ ♃
06:13 ☽ ∥ ♆
10:09 ☽ △ ♆
14:54 ☽ □ ♀
15:36 ☽ ∥ ♃
18:10 ☽ △ ♂
20:36 ☽ ✶ ♇
22:41 ☽ ✶ ☉
25 03:17 ☽ ∥ ♇
16:07 ☽ # ♀
17:16 ♂ ♋
20:18 ☿ ♎
26 03:58 ☿ □ ♂
09:14 ☽ □ ♃
12:26 ☽ ✶ ♄
13:57 ☿ ∥ ♅
16:08 ☿ □ ♇
16:12 ♀ ♌
17:14 ☽ # ♂
17:21 ☽ □ ♆
18:35 ☽ △ ♅
20:32 ♂ ☍ ♇
27 04:27 ☽ △ ♆
05:35 ☽ ✶ ☿
28 13:26 ☿ ✶ ♀
19:11 ☽ ✶ ♃
23:44 ☽ □ ♄
29 04:12 ☽ ✶ ♆
05:26 ☽ □ ♅
16:09 ☽ ☌ ♇
20:01 ☽ ☍ ♂
21:12 ☽ □ ☿
30 04:51 ☽ △ ☉
31 00:37 ☽ # ♂
12:06 ☿ # ♄
13:12 ☽ △ ♄
18:09 ☽ ✶ ♅

⚷ Chiron

01	Dec.	07 S 00
02	24♒25R	
05	24 16	
08	24 07	
11	23 58	
14	23 49	
17	23 39	
20	23 30	
23	23 21	
26	23 12	
29	23 03	

September 2009

Day	S. T. h m s	☉ ° ′ ″	☽ ° ′ ″	☿ ° ′	♀ ° ′	♂ ° ′	♃ ° ′	♄ ° ′	♅ ° ′	♆ ° ′	♇ ° ′	☊ True ° ′
01 Tu	22 41 10	08♍ 42 17	28♑ 10 06	04♎ 28	06♌ 21	03♋ 58	19♒R53	22♍ 55	25♓R17	24♒R42	00♑R41	29♑ 49
02 We	22 45 07	09 40 20	10♒ 00 42	04 58	07 33	04 35	19 46	23 02	25 15	24 40	00 41	29R 50
03 Th	22 49 03	10 38 25	21 57 43	05 23	08 45	05 13	19 40	23 10	25 12	24 39	00 41	29 48
04 Fr	22 52 60	11 36 31	04♓ 03 41	05 44	09 57	05 50	19 33	23 17	25 10	24 37	00 40	29 44
05 Sa	22 56 56	12 34 38	16 20 25	05 59	11 09	06 27	19 26	23 24	25 08	24 35	00 40	29 38
06 Su	23 00 53	13 32 48	28 49 01	06 09	12 21	07 04	19 20	23 32	25 06	24 34	00 40	29 31
07 Mo	23 04 49	14 30 59	11♈ 30 03	06 13	13 33	07 41	19 13	23 39	25 03	24 32	00 40	29 23
08 Tu	23 08 46	15 29 12	24 23 45	06R 11	14 46	08 17	19 07	23 47	25 01	24 31	00 40	29 16
09 We	23 12 43	16 27 27	07♉ 30 14	06 03	15 58	08 54	19 01	23 54	24 58	24 29	00 40	29 09
10 Th	23 16 39	17 25 44	20 49 42	05 49	17 11	09 30	18 55	24 02	24 56	24 28	00 39	29 05
11 Fr	23 20 36	18 24 03	04♊ 22 28	05 28	18 23	10 07	18 49	24 09	24 54	24 26	00 39	29 02
12 Sa	23 24 32	19 22 24	18 08 54	05 00	19 36	10 43	18 43	24 16	24 51	24 25	00D 39	29 01
13 Su	23 28 29	20 20 48	02♋ 09 10	04 25	20 48	11 19	18 38	24 24	24 49	24 24	00 39	29D 01
14 Mo	23 32 25	21 19 14	16 22 53	03 45	22 01	11 55	18 32	24 31	24 47	24 22	00 39	29 02
15 Tu	23 36 22	22 17 41	00♌ 48 34	02 58	23 14	12 31	18 27	24 39	24 44	24 21	00 40	29R 01
16 We	23 40 18	23 16 11	15 23 17	02 06	24 27	13 06	18 22	24 46	24 42	24 19	00 40	28 59
17 Th	23 44 15	24 14 43	00♍ 02 22	01 09	25 40	13 42	18 17	24 54	24 39	24 18	00 40	28 55
18 Fr	23 48 12	25 13 17	14 39 41	00 09	26 53	14 17	18 12	25 01	24 37	24 17	00 40	28 49
19 Sa	23 52 08	26 11 52	29 08 17	29♍ 06	28 06	14 53	18 07	25 09	24 35	24 15	00 40	28 41
20 Su	23 56 05	27 10 30	13♎ 21 28	28 02	29 19	15 28	18 03	25 16	24 32	24 14	00 40	28 31
21 Mo	00 00 01	28 09 10	27 13 53	26 58	00♍ 32	16 03	17 58	25 24	24 30	24 13	00 41	28 20
22 Tu	00 03 58	29 07 51	10♏ 42 17	25 57	01 45	16 37	17 54	25 31	24 27	24 11	00 41	28 11
23 We	00 07 54	00♎ 06 34	23 45 51	24 59	02 58	17 12	17 50	25 39	24 25	24 10	00 41	28 03
24 Th	00 11 51	01 05 19	06♐ 26 01	24 05	04 12	17 46	17 46	25 46	24 23	24 09	00 42	27 57
25 Fr	00 15 47	02 04 06	18 45 58	23 19	05 25	18 21	17 43	25 54	24 20	24 08	00 42	27 54
26 Sa	00 19 44	03 02 54	00♑ 50 09	22 40	06 39	18 55	17 39	26 01	24 18	24 06	00 43	27D 53
27 Su	00 23 41	04 01 44	12 43 40	22 10	07 52	19 29	17 36	26 09	24 15	24 05	00 43	27 54
28 Mo	00 27 37	05 00 36	24 31 58	21 49	09 06	20 03	17 33	26 16	24 13	24 04	00 44	27 55
29 Tu	00 31 34	05 59 30	06♒ 20 25	21 38	10 19	20 36	17 30	26 23	24 11	24 03	00 44	27R 56
30 We	00 35 30	06 58 25	18 13 57	21D 38	11 33	21 10	17 27	26 31	24 08	24 02	00 45	27 55

Data for 09-01-2009
Julian Day 2455075.50
Ayanamsa 23 59 46
SVP 05♓07 45
☽ ☊ Mean 28♑05 R

LAST ASPECT Day	h m	☽ INGRESS Day	h m	
31	18:09	01	03:43	♒
03	05:20	03	15:59	♓
05	16:54	06	02:15	♈
08	00:13	08	10:19	♉
10	07:18	10	16:18	♊
12	11:31	12	20:20	♋
14	13:58	14	22:40	♌
16	16:11	16	23:56	♍
18	23:56	19	01:26	♎
20	18:43	21	04:52	♏
23	03:33	23	11:44	♐
25	14:16	25	22:19	♑
28	03:34	28	11:07	♒
30	11:35	30	23:27	♓

● ◐ PHASES ○ ◑

04	16:03	○	12♓15
12	02:16	◑	19♊28
18	18:44	●	25♍59
26	04:50	◐	03♑15

DECLINATION

Day	☉	☽	☿	♀	♂	♃	♄	♅	♆	♇
01 Tu	08N18	20S41	04S49	18N43	23N34	15S54	04N37	02S36	13S41	17S55
02 We	07 57	16 51	05 09	18 28	23 34	15 56	04 34	02 37	13 42	17 55
03 Th	07 35	12 20	05 26	18 12	23 34	15 58	04 31	02 38	13 42	17 55
04 Fr	07 13	07 18	05 41	17 56	23 33	16 00	04 28	02 39	13 43	17 55
05 Sa	06 50	01 56	05 53	17 39	23 33	16 02	04 25	02 40	13 43	17 56
06 Su	06 28	03N35	06 03	17 22	23 32	16 04	04 22	02 41	13 44	17 56
07 Mo	06 06	09 02	06 09	17 04	23 31	16 06	04 20	02 42	13 44	17 56
08 Tu	05 43	14 12	06 13	16 46	23 30	16 08	04 17	02 43	13 45	17 56
09 We	05 21	18 48	06 13	16 28	23 28	16 10	04 14	02 44	13 45	17 57
10 Th	04 58	22 33	06 09	16 08	23 27	16 12	04 11	02 45	13 46	17 57
11 Fr	04 35	25 08	06 02	15 49	23 25	16 14	04 08	02 46	13 46	17 57
12 Sa	04 12	26 15	05 51	15 29	23 24	16 16	04 05	02 47	13 47	17 58
13 Su	03 49	25 44	05 35	15 08	23 22	16 17	04 02	02 48	13 47	17 58
14 Mo	03 26	23 33	05 16	14 47	23 20	16 19	03 59	02 49	13 48	17 58
15 Tu	03 03	19 49	04 53	14 26	23 18	16 21	03 56	02 50	13 48	17 58
16 We	02 40	14 50	04 25	14 04	23 15	16 22	03 53	02 50	13 49	17 59
17 Th	02 17	08 59	03 54	13 42	23 13	16 24	03 50	02 51	13 49	17 59
18 Fr	01 54	02 40	03 20	13 19	23 10	16 25	03 47	02 52	13 50	17 59
19 Sa	01 31	03S43	02 43	12 57	23 08	16 27	03 44	02 53	13 50	17 59
20 Su	01 07	09 47	02 05	12 33	23 05	16 28	03 41	02 54	13 51	18 00
21 Mo	00 44	15 13	01 24	12 10	23 02	16 29	03 38	02 55	13 51	18 00
22 Tu	00 21	19 44	00 44	11 46	22 59	16 30	03 36	02 56	13 51	18 00
23 We	00S03	23 09	00 03	11 21	22 56	16 32	03 33	02 57	13 52	18 01
24 Th	00 26	25 18	00N36	10 56	22 52	16 33	03 30	02 58	13 52	18 01
25 Fr	00 49	26 10	01 12	10 31	22 49	16 34	03 27	02 59	13 53	18 01
26 Sa	01 13	25 45	01 46	10 06	22 45	16 35	03 24	03 00	13 53	18 01
27 Su	01 36	24 10	02 17	09 40	22 42	16 36	03 21	03 01	13 53	18 02
28 Mo	01 59	21 31	02 43	09 15	22 38	16 37	03 18	03 02	13 54	18 02
29 Tu	02 23	17 58	03 05	08 48	22 34	16 37	03 15	03 03	13 54	18 02
30 We	02 46	13 41	03 21	08 22	22 30	16 38	03 12	03 04	13 55	18 02

ASPECTARIAN

01 13:22 ☽ △ ☿
13:48 ☽ # ♀
17:49 ☽ ∥ ♇
18:28 ☽ ☍ ♀
02 05:09 ☽ ∥ ♃
17:04 ☽ ∥ ♆
19:27 ☽ ☌ ♃
03 05:20 ☽ ☌ ♆
16:20 ☿ □ ♂
17:19 ☽ ⚹ ♇
04 00:26 ☽ # ☉
00:58 ♀ # ♇
03:40 ☽ △ ♂
07:04 ☽ ∥ ☿
12:53 ☽ # ♄
20:46 ☽ ∥ ♅
05 13:47 ☽ ☍ ♄
16:54 ☽ ☌ ♅
20:03 ☽ # ♅
06 03:23 ☽ ∥ ♄
03:31 ☽ □ ♇
10:58 ☽ # ☿
11:46 ☽ ∥ ☉
14:01 ☽ ☍ ☿
16:26 ☽ □ ♂
20:52 ☿ # ☉
07 04:15 ☽ △ ♀
04:46 ☿ SR
14:18 ☽ ⚹ ♃
21:46 ☽ # ♆
08 00:13 ☽ ⚹ ♆
09:43 ☽ # ♃
11:31 ☽ △ ♇
12:07 ☽ ∥ ♀
19:11 ☽ # ♇
09 02:39 ☽ ⚹ ♂
16:49 ☽ □ ♀
17:26 ☽ △ ☉
19:57 ♀ # ♃
20:36 ☽ □ ♃
10 05:45 ☽ △ ♄
06:28 ☽ □ ♆
07:00 ☽ ∥ ♂
07:18 ☽ ⚹ ♅
11 01:51 ☽ △ ☿
07:55 ♀ ☍ ♃
16:56 ♇ SD
12 00:59 ☽ △ ♃
02:44 ☽ ⚹ ♀
09:04 ☉ ∥ ♄
10:38 ☽ □ ♄
10:46 ☽ △ ♆
11:31 ☽ □ ♅
21:27 ☽ ☍ ♇
13 03:41 ☽ □ ☿
16:11 ☽ ☌ ♂
14 01:44 ☽ ∥ ♂
08:51 ☽ ⚹ ☉
13:42 ☽ ⚹ ♄
13:58 ☽ △ ♅
15 03:22 ☽ ⚹ ☿
09:30 ☽ # ♇
12:50 ♄ ☍ ♅
13:54 ☉ # ♅
17:10 ☽ # ♃
21:37 ♀ ☍ ♆
16 03:33 ☽ ∥ ♀
04:25 ☽ # ♆
04:51 ☽ ☍ ♃
14:37 ☽ ☍ ♆
16:11 ☽ ☌ ♀
16:30 ♀ # ♆
17 01:01 ☽ △ ♇
03:26 ☿ # ♄
09:41 ☉ ☍ ♅
11:50 ☿ □ ♇
18:23 ☉ ☌ ♄
19:46 ☽ ∥ ♄
21:13 ☽ # ☿
23:13 ☽ # ♅
23:22 ☽ ⚹ ♂
18 03:03 ☽ ∥ ☉
03:26 ☿ ♍R
16:07 ☽ # ☉
16:25 ☽ ☍ ♅
17:17 ☽ ☌ ♄
17:55 ☿ ∥ ♅
20:34 ☽ ∥ ☿
22:51 ☽ ∥ ♅
23:56 ☽ ☌ ☿
19 00:05 ☽ # ♄
02:34 ☽ □ ♇
20 03:45 ☽ □ ♂
07:59 ☽ △ ♃
10:05 ☿ ☌ ☉
11:01 ☽ # ♀
13:32 ♀ ♍
17:36 ☽ ∥ ♆
18:43 ☽ △ ♆
21 02:57 ♀ △ ♇
06:04 ☽ ⚹ ♇
06:16 ☽ ∥ ♃
06:23 ☽ ⚹ ♀
14:09 ☽ ∥ ♇
22 09:08 ☿ ☌ ♄
11:17 ☽ △ ♂
13:04 ☽ □ ♃
21:19 ☉ ♎
22:11 ☽ # ♂
23 00:18 ☿ ∥ ☉
00:45 ☽ □ ♆
01:13 ☽ △ ♅
02:07 ☽ ⚹ ☿
03:33 ☽ ⚹ ♄
08:56 ☿ # ☉
12:56 ☽ ⚹ ☉
14:21 ☉ □ ♇
15:36 ☿ ☍ ♅
19:15 ☽ □ ♀
24 21:56 ☽ ⚹ ♃
25 08:30 ☽ □ ☿
10:35 ☽ ⚹ ♆
10:59 ☽ □ ♅
14:16 ☽ □ ♄
23:45 ☽ ☌ ♇
26 13:02 ☽ △ ♀
27 14:25 ☽ ☍ ♂
14:38 ☽ # ♂
18:36 ☽ △ ☿
23:22 ☽ ⚹ ♅
28 03:34 ☽ △ ♄
21:34 ☿ # ♅
23:14 ☽ △ ☉
23:33 ☽ ∥ ♇
29 07:54 ☽ ∥ ♃
12:19 ☿ ∥ ♄
13:15 ☿ SD
22:27 ☽ ☌ ♃
22:49 ☽ ∥ ♆
30 11:35 ☽ ☌ ♆
18:13 ☉ ∥ ♅

⚷ Chiron

01 Dec. 07 S 31

01	22♒54R
04	22 45
07	22 37
10	22 29
13	22 21
16	22 13
19	22 06
22	21 59
25	21 53
28	21 47

October 2009

Day	S. T.	☉	☽	☿	♀	♂	♃	♄	♅	♆	♇	☊ True
	h m s	° ' "	° ' "	° '	° '	° '	° '	° '	° '	° '	° '	° '
01 Th	00 39 27	07♎ 57 22	00♓ 16 52	21♍ 47	12♍ 47	21♋ 43	17♒R25	26♍ 38	24♓R06	24♒R01	00♑ 45	27♑R51
02 Fr	00 43 23	08 56 21	12 32 31	22 07	14 00	22 16	17 22	26 46	24 04	24 00	00 46	27 44
03 Sa	00 47 20	09 55 22	25 03 07	22 37	15 14	22 49	17 20	26 53	24 01	23 59	00 47	27 35
04 Su	00 51 16	10 54 24	07♈ 49 39	23 16	16 28	23 22	17 18	27 00	23 59	23 58	00 47	27 24
05 Mo	00 55 13	11 53 29	20 51 51	24 03	17 42	23 55	17 17	27 08	23 57	23 57	00 48	27 12
06 Tu	00 59 10	12 52 36	04♉ 08 22	24 59	18 56	24 27	17 15	27 15	23 55	23 56	00 49	27 00
07 We	01 03 06	13 51 45	17 37 13	26 02	20 10	24 59	17 14	27 22	23 53	23 55	00 49	26 49
08 Th	01 07 03	14 50 56	01♊ 16 13	27 11	21 24	25 31	17 13	27 29	23 50	23 54	00 50	26 41
09 Fr	01 10 59	15 50 10	15 03 23	28 27	22 38	26 03	17 12	27 37	23 48	23 53	00 51	26 35
10 Sa	01 14 56	16 49 26	28 57 15	29 47	23 52	26 35	17 11	27 44	23 46	23 52	00 52	26 32
11 Su	01 18 52	17 48 44	12♋ 56 55	01♎ 12	25 06	27 06	17 10	27 51	23 44	23 52	00 53	26 31
12 Mo	01 22 49	18 48 05	27 01 47	02 40	26 21	27 38	17 10	27 58	23 42	23 51	00 54	26 30
13 Tu	01 26 45	19 47 28	11♌ 11 12	04 12	27 35	28 09	17 10	28 06	23 40	23 50	00 55	26 28
14 We	01 30 42	20 46 53	25 23 55	05 46	28 49	28 39	17D 10	28 13	23 38	23 49	00 56	26 25
15 Th	01 34 39	21 46 20	09♍ 37 43	07 22	00♎ 04	29 10	17 10	28 20	23 36	23 49	00 57	26 20
16 Fr	01 38 35	22 45 50	23 49 14	09 00	01 18	29 40	17 11	28 27	23 34	23 48	00 58	26 12
17 Sa	01 42 32	23 45 22	07♎ 54 07	10 40	02 33	00♌ 11	17 11	28 34	23 32	23 47	00 59	26 01
18 Su	01 46 28	24 44 56	21 47 35	12 20	03 47	00 41	17 12	28 41	23 30	23 47	01 00	25 49
19 Mo	01 50 25	25 44 32	05♏ 25 11	14 01	05 02	01 10	17 13	28 48	23 28	23 46	01 01	25 36
20 Tu	01 54 21	26 44 10	18 43 38	15 43	06 16	01 40	17 15	28 55	23 26	23 46	01 02	25 23
21 We	01 58 18	27 43 49	01♐ 41 20	17 25	07 31	02 09	17 16	29 02	23 24	23 45	01 03	25 12
22 Th	02 02 14	28 43 31	14 18 34	19 07	08 46	02 38	17 18	29 08	23 22	23 45	01 05	25 04
23 Fr	02 06 11	29 43 15	26 37 24	20 50	10 00	03 07	17 20	29 15	23 20	23 44	01 06	24 58
24 Sa	02 10 07	00♏ 43 00	08♑ 41 19	22 32	11 15	03 35	17 22	29 22	23 18	23 44	01 07	24 56
25 Su	02 14 04	01 42 47	20 34 52	24 14	12 30	04 03	17 24	29 29	23 17	23 43	01 08	24D 55
26 Mo	02 18 01	02 42 36	02♒ 23 16	25 55	13 44	04 31	17 26	29 36	23 15	23 43	01 10	24 56
27 Tu	02 21 57	03 42 26	14 12 02	27 36	14 59	04 59	17 29	29 42	23 13	23 43	01 11	24R 56
28 We	02 25 54	04 42 18	26 06 43	29 17	16 14	05 26	17 32	29 49	23 12	23 42	01 12	24 55
29 Th	02 29 50	05 42 12	08♓ 12 27	00♏ 58	17 29	05 53	17 35	29 55	23 10	23 42	01 14	24 52
30 Fr	02 33 47	06 42 07	20 33 39	02 38	18 44	06 20	17 38	00♎ 02	23 09	23 42	01 15	24 46
31 Sa	02 37 43	07 42 04	03♈ 13 35	04 18	19 59	06 46	17 42	00 08	23 07	23 42	01 17	24 37

Data for	10-01-2009
Julian Day	2455105.50
Ayanamsa	23 59 49
SVP	05 ♓ 07 40
☽ ☊ Mean	26 ♑ 30 R

● ◐ PHASES ○ ◑

04	06:10	○	11♈10
11	08:56	◑	18♋11
18	05:34	●	24♎59
26	00:43	◐	02♒44

LAST ASPECT ☽ INGRESS

Day	h m	Day	h m	
03	03:30	03	09:21	♈
05	05:47	05	16:34	♉
07	17:19	07	21:47	♊
10	01:35	10	01:48	♋
12	01:37	12	05:03	♌
13	21:21	14	07:46	♍
16	10:19	16	10:30	♎
18	05:34	18	14:23	♏
20	18:58	20	20:50	♐
23	06:40	23	06:40	♑
25	18:15	25	19:08	♒
28	07:23	28	07:46	♓
30	04:56	30	17:57	♈

DECLINATION

Day	☉	☽	☿	♀	♂	♃	♄	♅	♆	♇
01 Th	03S 09	08S 49	03N33	07N55	22N26	16S 39	03N09	03S 04	13S 55	18S 03
02 Fr	03 33	03 33	03 39	07 28	22 22	16 39	03 07	03 05	13 55	18 03
03 Sa	03 56	01N56	03 41	07 01	22 18	16 40	03 04	03 06	13 56	18 03
04 Su	04 19	07 27	03 37	06 34	22 14	16 40	03 01	03 07	13 56	18 04
05 Mo	04 42	12 45	03 29	06 06	22 10	16 41	02 58	03 08	13 56	18 04
06 Tu	05 05	17 34	03 16	05 38	22 05	16 41	02 55	03 09	13 57	18 04
07 We	05 28	21 35	02 59	05 10	22 01	16 42	02 52	03 10	13 57	18 04
08 Th	05 51	24 27	02 38	04 42	21 56	16 42	02 50	03 11	13 57	18 05
09 Fr	06 14	25 55	02 14	04 14	21 52	16 42	02 47	03 11	13 57	18 05
10 Sa	06 37	25 46	01 47	03 45	21 47	16 42	02 44	03 12	13 58	18 05
11 Su	06 59	23 59	01 16	03 17	21 42	16 42	02 41	03 13	13 58	18 05
12 Mo	07 22	20 42	00 44	02 48	21 37	16 42	02 39	03 14	13 58	18 06
13 Tu	07 44	16 10	00 09	02 19	21 32	16 42	02 36	03 15	13 58	18 06
14 We	08 07	10 45	00S 28	01 50	21 28	16 42	02 33	03 16	13 59	18 06
15 Th	08 29	04 45	01 06	01 21	21 23	16 42	02 30	03 16	13 59	18 06
16 Fr	08 51	01S 26	01 46	00 52	21 18	16 42	02 28	03 17	13 59	18 07
17 Sa	09 13	07 30	02 26	00 23	21 13	16 41	02 25	03 18	13 59	18 07
18 Su	09 35	13 07	03 08	00S 06	21 08	16 41	02 22	03 19	14 00	18 07
19 Mo	09 57	17 59	03 50	00 35	21 03	16 41	02 20	03 19	14 00	18 07
20 Tu	10 19	21 49	04 32	01 05	20 57	16 40	02 17	03 20	14 00	18 08
21 We	10 40	24 27	05 15	01 34	20 52	16 39	02 14	03 21	14 00	18 08
22 Th	11 01	25 47	05 57	02 03	20 47	16 39	02 12	03 22	14 00	18 08
23 Fr	11 22	25 48	06 40	02 32	20 42	16 38	02 09	03 22	14 00	18 08
24 Sa	11 43	24 34	07 23	03 02	20 37	16 37	02 07	03 23	14 00	18 09
25 Su	12 04	22 14	08 05	03 31	20 32	16 37	02 04	03 24	14 01	18 09
26 Mo	12 25	18 59	08 48	04 00	20 27	16 36	02 02	03 24	14 01	18 09
27 Tu	12 45	14 58	09 29	04 29	20 21	16 35	01 59	03 25	14 01	18 09
28 We	13 05	10 20	10 11	04 58	20 16	16 34	01 57	03 25	14 01	18 09
29 Th	13 25	05 16	10 52	05 27	20 11	16 33	01 54	03 26	14 01	18 10
30 Fr	13 45	00N06	11 32	05 56	20 06	16 32	01 52	03 27	14 01	18 10
31 Sa	14 05	05 35	12 12	06 25	20 01	16 31	01 49	03 27	14 01	18 10

ASPECTARIAN

01 00:08 ☉ # ♄
00:56 ☽ ⚹ ♇
04:39 ☽ # ♀
06:00 ☿ ⚹ ♂
23:35 ☽ # ☿
02 00:05 ☽ ∥ ☉
02:01 ☽ # ♄
02:05 ☽ ∥ ♅
03:09 ☽ ☍ ♀
07:56 ♄ # ♅
08:13 ☿ # ☉
19:09 ☽ ☍ ☿
19:34 ☽ △ ♂
22:03 ☽ ☌ ♅
03 03:30 ☽ ☍ ♄
04:51 ☽ ∥ ♄
05:05 ☽ # ♅
07:32 ☽ ∥ ☿
09:18 ☽ # ☉
10:49 ☽ □ ♇
20:24 ☽ ∥ ♀
04 11:37 ☿ ⚹ ♂
17:27 ☽ ⚹ ♃
21:06 ☿ ☍ ♅
05 01:38 ♂ △ ♅
05:35 ☽ # ♆
05:37 ☽ ⚹ ♆
05:47 ☽ □ ♂
18:01 ☽ △ ♇
19:19 ☽ # ♃
06 02:40 ☽ # ♇
10:40 ☿ # ♅
15:40 ♀ # ☉
23:19 ☽ □ ♃
07 02:55 ☽ ∥ ♂
04:57 ☽ △ ♀
09:48 ☿ ∥ ♄
11:00 ☽ ⚹ ♅
11:05 ☽ □ ♆
13:31 ☽ ⚹ ♂
16:10 ☽ △ ☿
17:19 ☽ △ ♄
08 06:35 ☿ ☌ ♄
09 01:27 ☽ △ ☉
03:42 ☽ △ ♃
14:23 ☽ □ ♀
15:05 ☽ □ ♅
15:15 ☽ △ ♆
21:53 ☽ □ ♄
22:01 ♀ ☍ ♅
10 01:35 ☽ □ ☿
03:17 ☽ ☍ ♇
03:46 ☿ ♎
08:39 ☉ △ ♃
18:44 ☿ □ ♇
11 03:04 ♀ # ♅
18:04 ☽ ∥ ♂
18:21 ☽ △ ♅
22:44 ☽ ⚹ ♀
12 01:03 ☽ ☌ ♂
01:37 ☽ ⚹ ♄
08:53 ♀ ∥ ♄
10:43 ☽ ⚹ ☿
14:32 ☽ # ♇
20:53 ♂ ⚹ ♄
21:30 ☽ # ♃
13 04:36 ♃ SD
10:06 ☽ ☍ ♃
10:10 ☽ # ♆
10:52 ♀ ☌ ♄
15:38 ☽ ⚹ ☉
18:31 ♀ ⚹ ♂
21:21 ☽ ☍ ♆
14 09:20 ☽ △ ♇
10:00 [illegible] ∥ [illegible]
22:47 ♀ ♎
15 05:20 ☿ # ♀
05:47 ☽ # ♅
08:50 ☽ ∥ ♄
12:49 ☽ # ☿
14:19 ☽ ∥ ♀
17:21 ♀ □ ♇
21:58 ☽ # ♀
23:33 ☽ ☍ ♅
16 01:26 ☽ ∥ ☿
03:57 ☽ # ♄
07:13 ☽ ∥ ♅
07:55 ☽ ☌ ♄
10:19 ☽ ⚹ ♂
12:09 ☽ □ ♇
13:57 ☽ ☌ ♀
15:33 ♂ ♌
22:15 ☿ # ♄
17 00:47 ☉ △ ♆
05:23 ☽ ☌ ☿
07:34 ☽ ∥ ☉
16:01 ☽ △ ♃
18 03:28 ☽ △ ♆
04:01 ☽ ∥ ♆
06:25 ☿ ∥ ♅
16:10 ☽ □ ♂
16:11 ☽ ⚹ ♇
17:09 ☽ ∥ ♃
19 00:49 ☽ ∥ ♇
18:07 ☽ # ♂
21:17 ☽ □ ♃
20 08:37 ☽ △ ♅
09:14 ☽ □ ♆
18:58 ☽ ⚹ ♄
21:50 ☿ △ ♃
21 00:54 ☽ △ ♂
12:12 ☽ ⚹ ♀
22 05:47 ☽ ⚹ ♃
06:35 ♀ # ♄
10:48 ☽ ⚹ ☿
17:33 ☽ □ ♅
18:20 ☽ ⚹ ♆
23 05:15 ☽ □ ♄
06:40 ☽ ⚹ ☉
06:44 ☉ ♏
08:52 ☽ ☌ ♇
24 05:45 ☽ □ ♀
09:56 ☉ ⚹ ♇
16:56 ☿ △ ♆
17:56 ♀ ∥ ♅
25 05:28 ☽ ⚹ ♅
08:38 ☽ □ ☿
13:46 ☽ # ♂
18:15 ☽ △ ♄
26 04:31 ☽ ☍ ♂
05:22 ☽ ∥ ♇
14:49 ☽ ∥ ♃
27 01:47 ☽ △ ♀
05:12 ☽ ∥ ♆
06:40 ☽ ☌ ♃
11:03 ☽ ∥ ☉
19:11 ☽ ☌ ♆
28 00:44 ☽ ∥ ☿
07:23 ☽ △ ☿
10:09 ☿ ♏
10:11 ☽ ⚹ ♇
18:38 ☽ △ ☉
23:16 ☽ ∥ ♀
29 02:02 ♀ △ ♃
03:53 ☿ ⚹ ♇
07:56 ☉ □ ♂
08:20 ☽ ∥ ♅
15:19 ☽ # ♄
17:09 ♄ ♎
30 04:56 ☽ ☌ ♅
07:42 ☽ ∥ ♄
14:42 ☽ # ♅
18:09 ☽ ☍ ♄
19:33 ☉ ∥ ♆
20:21 ☽ □ ♇
31 04:00 ☽ # ☿
06:50 ☽ △ ♂

⚷ Chiron

01	Dec.	08 S 02
	01	21♒41R
	04	21 36
	07	21 31
	10	21 27
	13	21 23
	16	21 20
	19	21 18
	22	21 16
	25	21 14
	28	21 13
	31	21 13R
31	05:55	21♒13 D

November 2009

Day	S. T.	☉	☽	☿	♀	♂	♃	♄	♅	♆	♇	☊ True
	h m s	° ' "	° ' "	° '	° '	° '	° '	° '	° '	° '	° '	° '
01 Su	02 41 40	08♏ 42 03	16♈ 13 59	05♏ 57	21♎ 14	07♌ 13	17♒ 45	00♎ 15	23♓R06	23♒R42	01♑ 18	24♑R27
02 Mo	02 45 36	09 42 04	29 34 50	07 36	22 29	07 39	17 49	00 21	23 04	23 42	01 20	24 15
03 Tu	02 49 33	10 42 06	13♉ 14 18	09 14	23 44	08 04	17 53	00 27	23 03	23 42	01 21	24 03
04 We	02 53 30	11 42 10	27 08 55	10 52	24 59	08 29	17 57	00 34	23 01	23 41	01 23	23 53
05 Th	02 57 26	12 42 17	11♊ 14 19	12 30	26 14	08 54	18 02	00 40	23 00	23 D 41	01 24	23 44
06 Fr	03 01 23	13 42 25	25 25 50	14 07	27 29	09 19	18 06	00 46	22 59	23 41	01 26	23 38
07 Sa	03 05 19	14 42 35	09♋ 39 24	15 44	28 44	09 43	18 11	00 52	22 58	23 42	01 28	23 34
08 Su	03 09 16	15 42 48	23 51 51	17 20	29 59	10 07	18 16	00 58	22 56	23 42	01 29	23 33
09 Mo	03 13 12	16 43 02	08♌ 01 11	18 56	01♏ 14	10 31	18 21	01 04	22 55	23 42	01 31	23 32
10 Tu	03 17 09	17 43 18	22 06 11	20 31	02 29	10 54	18 26	01 10	22 54	23 42	01 32	23 31
11 We	03 21 06	18 43 37	06♍ 06 05	22 07	03 44	11 17	18 31	01 16	22 53	23 42	01 34	23 29
12 Th	03 25 02	19 43 57	20 00 06	23 42	04 59	11 40	18 37	01 22	22 52	23 42	01 36	23 25
13 Fr	03 28 59	20 44 19	03♎ 47 03	25 16	06 15	12 02	18 43	01 28	22 51	23 43	01 38	23 18
14 Sa	03 32 55	21 44 43	17 25 20	26 50	07 30	12 23	18 49	01 33	22 50	23 43	01 39	23 10
15 Su	03 36 52	22 45 09	00♏ 52 52	28 24	08 45	12 45	18 55	01 39	22 49	23 43	01 41	23 00
16 Mo	03 40 48	23 45 37	14 07 29	29 58	10 00	13 06	19 01	01 44	22 49	23 44	01 43	22 49
17 Tu	03 44 45	24 46 07	27 07 21	01♐ 32	11 16	13 26	19 08	01 50	22 48	23 44	01 45	22 38
18 We	03 48 41	25 46 38	09♐ 51 19	03 05	12 31	13 46	19 14	01 55	22 47	23 44	01 47	22 29
19 Th	03 52 38	26 47 10	22 19 23	04 38	13 46	14 06	19 21	02 01	22 46	23 45	01 48	22 22
20 Fr	03 56 35	27 47 44	04♑ 32 44	06 11	15 02	14 25	19 28	02 06	22 46	23 45	01 50	22 18
21 Sa	04 00 31	28 48 19	16 33 46	07 43	16 17	14 44	19 35	02 11	22 45	23 46	01 52	22 16
22 Su	04 04 28	29 48 56	28 25 56	09 16	17 32	15 02	19 42	02 16	22 45	23 47	01 54	22D 16
23 Mo	04 08 24	00♐ 49 34	10♒ 13 32	10 48	18 48	15 20	19 50	02 21	22 44	23 47	01 56	22 18
24 Tu	04 12 21	01 50 13	22 01 29	12 20	20 03	15 37	19 57	02 26	22 44	23 48	01 58	22 20
25 We	04 16 17	02 50 53	03♓ 55 05	13 52	21 18	15 54	20 05	02 31	22 43	23 48	02 00	22 21
26 Th	04 20 14	03 51 34	15 59 42	15 23	22 34	16 10	20 13	02 36	22 43	23 49	02 02	22R 21
27 Fr	04 24 10	04 52 16	28 20 24	16 55	23 49	16 26	20 21	02 41	22 43	23 50	02 04	22 18
28 Sa	04 28 07	05 52 59	11♈ 01 34	18 26	25 04	16 41	20 29	02 46	22 43	23 51	02 06	22 14
29 Su	04 32 04	06 53 43	24 06 16	19 57	26 20	16 56	20 38	02 50	22 42	23 51	02 08	22 08
30 Mo	04 36 00	07 54 29	07♉ 35 51	21 27	27 35	17 10	20 46	02 55	22 42	23 52	02 10	22 01

Data for	11-01-2009
Julian Day	2455136.50
Ayanamsa	23 59 53
SVP	05 ♓ 07 35
☽ ☊ Mean	24 ♑ 52 R

● ◐ PHASES ○ ◑

Day	Time	Phase	Position
02	19:13	○	10♉30
09	15:56	◑	17♌23
16	19:14	●	24♏34
24	21:39	◐	02♓45

LAST ASPECT ☽ Day	h m	INGRESS Day	h m	
01	13:29	02	00:45	♉
03	18:04	04	04:53	♊
06	03:47	06	07:43	♋
07	22:26	08	10:23	♌
10	02:44	10	13:31	♍
12	07:14	12	17:23	♎
14	11:11	14	22:25	♏
16	19:15	17	05:23	♐
19	02:47	19	15:01	♑
22	03:04	22	03:11	♒
24	03:36	24	16:08	♓
26	14:18	27	03:11	♈
28	23:33	29	10:35	♉

DECLINATION

Day	☉	☽	☿	♀	♂	♃	♄	♅	♆	♇
01 Su	14S 24	10N59	12S 51	06S 53	19N56	16S 29	01N47	03S 28	14S 01	18S 10
02 Mo	14 43	16 00	13 30	07 22	19 51	16 28	01 44	03 28	14 01	18 11
03 Tu	15 02	20 20	14 08	07 50	19 46	16 27	01 42	03 29	14 01	18 11
04 We	15 21	23 37	14 45	08 18	19 41	16 25	01 40	03 29	14 01	18 11
05 Th	15 39	25 30	15 21	08 46	19 36	16 24	01 37	03 30	14 01	18 11
06 Fr	15 57	25 45	15 57	09 14	19 31	16 22	01 35	03 30	14 01	18 11
07 Sa	16 15	24 18	16 32	09 42	19 26	16 21	01 33	03 31	14 01	18 12
08 Su	16 33	21 18	17 06	10 09	19 21	16 19	01 31	03 31	14 01	18 12
09 Mo	16 50	17 02	17 39	10 36	19 17	16 17	01 28	03 32	14 01	18 12
10 Tu	17 07	11 50	18 11	11 03	19 12	16 16	01 26	03 32	14 01	18 12
11 We	17 24	06 04	18 42	11 30	19 08	16 14	01 24	03 32	14 01	18 12
12 Th	17 40	00 03	19 13	11 56	19 03	16 12	01 22	03 33	14 01	18 13
13 Fr	17 56	05S 54	19 42	12 22	18 59	16 10	01 20	03 33	14 01	18 13
14 Sa	18 12	11 30	20 11	12 48	18 54	16 08	01 18	03 33	14 01	18 13
15 Su	18 28	16 29	20 38	13 13	18 50	16 06	01 16	03 34	14 01	18 13
16 Mo	18 43	20 36	21 05	13 39	18 46	16 04	01 14	03 34	14 00	18 13
17 Tu	18 58	23 36	21 30	14 04	18 42	16 02	01 12	03 34	14 00	18 13
18 We	19 12	25 21	21 55	14 28	18 38	16 00	01 10	03 35	14 00	18 14
19 Th	19 26	25 47	22 18	14 52	18 34	15 58	01 08	03 35	14 00	18 14
20 Fr	19 40	24 56	22 40	15 16	18 30	15 55	01 06	03 35	14 00	18 14
21 Sa	19 54	22 56	23 01	15 39	18 27	15 53	01 04	03 35	14 00	18 14
22 Su	20 07	19 56	23 21	16 02	18 23	15 51	01 02	03 35	13 59	18 14
23 Mo	20 19	16 09	23 40	16 25	18 20	15 48	01 00	03 35	13 59	18 14
24 Tu	20 32	11 45	23 57	16 47	18 17	15 46	00 59	03 36	13 59	18 15
25 We	20 44	06 53	24 13	17 09	18 14	15 43	00 57	03 36	13 59	18 15
26 Th	20 55	01 42	24 28	17 30	18 11	15 41	00 55	03 36	13 59	18 15
27 Fr	21 06	03N40	24 42	17 51	18 08	15 38	00 54	03 36	13 58	18 15
28 Sa	21 17	09 01	24 55	18 11	18 06	15 35	00 52	03 36	13 58	18 15
29 Su	21 28	14 08	25 06	18 31	18 03	15 32	00 50	03 36	13 58	18 15
30 Mo	21 38	18 44	25 16	18 50	18 01	15 30	00 49	03 36	13 58	18 15

ASPECTARIAN

01	02:47	☽ ⚹ ♃
	10:00	☽ ☍ ♀
	10:00	☽ # ☿
	13:29	☽ ⚹ ♆
	14:16	☽ # ♆
	17:14	☽ # ☉
02	00:51	☿ □ ♂
	02:24	☽ # ♃
	03:06	☽ △ ♇
	11:28	☽ # ♇
	14:41	☽ □ ♂
	16:05	☽ ☍ ☿
	19:48	☿ ∥ ♆
	20:38	☽ ∥ ♂
	23:23	♀ △ ♆
03	08:06	☽ □ ♃
	16:56	☽ ⚹ ♅
	18:04	☽ □ ♆
04	05:53	☽ △ ♄
	18:11	♆ SD
	19:57	☽ ⚹ ♂
05	08:03	☿ ☌ ☉
	11:33	☽ △ ♃
	19:52	☽ □ ♅
	21:04	☽ △ ♆
06	00:16	☿ ∥ ☉
	03:47	☽ △ ♀
	09:04	☽ □ ♄
	10:09	☽ ☍ ♇
	16:40	☿ ∥ ♃
07	07:09	☉ ∥ ♃
	09:11	☽ △ ☉
	11:33	☽ △ ☿
	22:26	☽ △ ♅
08	00:24	♀ ♏
	11:22	☽ □ ♀
	11:55	☽ ∥ ♂
	12:07	☽ ⚹ ♄
	14:43	☿ □ ♃
	18:00	☽ # ♇
	21:11	☽ # ☿
09	00:56	☽ # ☉
	03:40	☽ # ♃
	04:22	☽ ☌ ♂
	05:34	♀ ⚹ ♇
	14:19	☽ # ♆
	17:42	☽ ☍ ♃
	20:57	☽ □ ☿
10	00:49	☿ ∥ ♇
	02:44	☽ ☍ ♆
	03:09	☽ # ♀
	16:12	☽ △ ♇
	18:42	☉ □ ♃
	19:32	☽ ⚹ ♀
11	10:09	☽ # ♅
	11:38	☿ △ ♅
	17:10	☿ # ♂
	18:47	☽ ∥ ♄
	23:30	☽ ⚹ ☉
12	00:12	☿ □ ♆
	04:58	☽ ☍ ♅
	05:40	☽ # ♄
	07:14	☽ ⚹ ☿
	14:28	☽ ∥ ♅
	19:55	☽ ☌ ♄
	20:13	☽ □ ♇
13	14:52	☽ ⚹ ♂
14	01:22	☉ ∥ ♇
	02:29	☽ △ ♃
	06:30	☽ ∥ ♀
	11:11	☽ △ ♆
	11:40	☽ ∥ ♆
	22:04	☽ ∥ ♃
15	01:27	☽ ⚹ ♇
	01:41	☉ △ ♅
	09:29	☽ ∥ ♇
	11:34	☽ ∥ ☉
	12:50	☽ # ♂
	14:43	♄ □ ♇
	15:42	☽ ☌ ♀
	22:04	☽ □ ♂
	23:11	☉ □ ♆
16	00:28	☿ ♐
	03:52	☽ ∥ ☿
	03:59	☉ # ♂
	09:03	☽ □ ♃
	15:58	☽ △ ♅
	17:41	☽ □ ♆
	20:55	♀ ∥ ♆
17	05:01	☿ ⚹ ♄
	08:53	☽ ⚹ ♄
	09:23	☽ ☌ ☿
18	07:41	☽ △ ♂
	18:11	☽ ⚹ ♃
19	00:53	☽ □ ♅
	02:47	☽ ⚹ ♆
	08:23	♀ □ ♂
	18:38	☽ ☌ ♇
	19:08	☽ □ ♄
20	23:13	☽ ∥ ☿
	23:22	☽ ⚹ ♀
21	12:29	☽ ⚹ ♅
	12:55	♀ ∥ ♃
	22:51	☽ ∥ ☉
22	03:04	☽ ⚹ ☉
	04:23	☉ ♐
	07:52	☽ △ ♄
	10:30	☽ # ♂
	11:19	☽ ∥ ♇
	22:35	☽ ∥ ♀
23	01:20	☽ ⚹ ☿
	02:02	☽ ∥ ♃
	10:39	☽ ☍ ♂
	12:10	☽ ∥ ♆
	19:31	☽ □ ♀
	19:45	☽ ☌ ♃
	22:02	♀ □ ♃
24	03:36	☽ ☌ ♆
	15:35	☉ ⚹ ♄
	18:03	♂ # ♇
	20:08	☽ ⚹ ♇
25	15:20	☽ ∥ ♅
	22:38	☽ □ ☿
26	03:00	♀ △ ♅
	03:32	☽ # ♄
	11:43	☽ # ♄
	13:08	☽ ☌ ♅
	14:18	☽ △ ♀
	14:54	☿ △ ♂
	23:44	☽ # ♅
27	00:16	♀ □ ♆
	07:08	☽ □ ♇
	08:21	☽ ☍ ♄
	13:31	☽ △ ☉
	18:33	♀ # ♂
28	05:07	♀ ∥ ♇
	10:41	☽ △ ♂
	15:28	☽ △ ☿
	17:37	☽ ⚹ ♃
	23:13	☽ # ♆
	23:33	☽ ⚹ ♆
29	06:58	☽ # ♃
	11:54	☿ ⚹ ♃
	14:24	☽ △ ♇
	20:05	☽ ∥ ♂
	21:21	☽ # ♇
30	00:39	☽ # ♀
	16:53	☽ □ ♂
	18:52	☽ # ☉
	19:53	☿ □ ♅
	23:00	☽ □ ♃

⚷ Chiron

01 Dec.	08 S 23
03	21♒13
06	21 14
09	21 15
12	21 17
15	21 20
18	21 23
21	21 27
24	21 31
27	21 36
30	21 41

December 2009

31 19:24 10♋15 ✹ Total Lunar Eclipse (mag 0.082)

Day	S. T.	☉	☽	☿	♀	♂	♃	♄	♅	♆	♇	☊ True
	h m s	° ' "	° ' "	° '	° '	° '	° '	° '	° '	° '	° '	° '
01 Tu	04 39 57	08♐ 55 15	21♉ 29 27	22♐ 58	28♏ 51	17♌ 24	20♒ 55	02♎ 59	22♓ R42	23♒ 53	02♑ 12	21♑ R54
02 We	04 43 53	09 56 03	05♊ 43 52	24 28	00♐ 06	17 37	21 03	03 03	22D 42	23 54	02 14	21 48
03 Th	04 47 50	10 56 51	20 13 48	25 58	01 22	17 49	21 12	03 08	22 42	23 55	02 16	21 43
04 Fr	04 51 46	11 57 41	04♋ 52 41	27 27	02 37	18 01	21 21	03 12	22 42	23 56	02 18	21 39
05 Sa	04 55 43	12 58 33	19 33 35	28 56	03 52	18 12	21 30	03 16	22 43	23 57	02 20	21 37
06 Su	04 59 39	13 59 25	04♌ 10 16	00♑ 24	05 08	18 23	21 40	03 20	22 43	23 58	02 22	21 37
07 Mo	05 03 36	15 00 19	18 37 52	01 52	06 23	18 33	21 49	03 24	22 43	23 59	02 24	21D 37
08 Tu	05 07 33	16 01 14	02♍ 53 10	03 19	07 39	18 43	21 59	03 28	22 43	24 00	02 26	21 37
09 We	05 11 29	17 02 10	16 54 24	04 45	08 54	18 51	22 08	03 31	22 44	24 01	02 29	21R 37
10 Th	05 15 26	18 03 07	00♎ 40 57	06 10	10 10	19 00	22 18	03 35	22 44	24 02	02 31	21 35
11 Fr	05 19 22	19 04 06	14 12 56	07 34	11 25	19 07	22 28	03 39	22 44	24 04	02 33	21 33
12 Sa	05 23 19	20 05 06	27 30 46	08 56	12 41	19 14	22 38	03 42	22 45	24 05	02 35	21 29
13 Su	05 27 15	21 06 07	10♏ 34 57	10 17	13 56	19 20	22 48	03 45	22 45	24 06	02 37	21 24
14 Mo	05 31 12	22 07 09	23 25 56	11 35	15 12	19 25	22 58	03 49	22 46	24 07	02 39	21 19
15 Tu	05 35 08	23 08 12	06♐ 04 11	12 52	16 27	19 30	23 09	03 52	22 47	24 09	02 41	21 14
16 We	05 39 05	24 09 15	18 30 10	14 06	17 43	19 34	23 19	03 55	22 47	24 10	02 44	21 09
17 Th	05 43 02	25 10 20	00♑ 44 40	15 16	18 58	19 37	23 30	03 58	22 48	24 11	02 46	21 07
18 Fr	05 46 58	26 11 25	12 48 53	16 23	20 14	19 39	23 41	04 01	22 49	24 13	02 48	21 05
19 Sa	05 50 55	27 12 30	24 44 38	17 26	21 29	19 41	23 51	04 04	22 50	24 14	02 50	21D 06
20 Su	05 54 51	28 13 36	06♒ 34 20	18 24	22 45	19 42	24 02	04 06	22 51	24 16	02 52	21 07
21 Mo	05 58 48	29 14 43	18 21 07	19 16	24 00	19R 42	24 14	04 09	22 52	24 17	02 54	21 09
22 Tu	06 02 44	00♑ 15 49	00♓ 08 42	20 02	25 16	19 41	24 25	04 11	22 53	24 19	02 57	21 12
23 We	06 06 41	01 16 56	12 01 21	20 41	26 31	19 39	24 36	04 14	22 54	24 20	02 59	21 14
24 Th	06 10 37	02 18 03	24 03 45	21 12	27 47	19 37	24 47	04 16	22 55	24 22	03 01	21 16
25 Fr	06 14 34	03 19 11	06♈ 20 46	21 34	29 02	19 34	24 59	04 18	22 56	24 23	03 03	21 17
26 Sa	06 18 31	04 20 18	18 57 05	21 46	00♑ 18	19 30	25 10	04 20	22 57	24 25	03 05	21R 17
27 Su	06 22 27	05 21 25	01♉ 56 47	21R 47	01 33	19 25	25 22	04 22	22 58	24 26	03 08	21 16
28 Mo	06 26 24	06 22 33	15 22 46	21 37	02 49	19 20	25 34	04 24	23 00	24 28	03 10	21 14
29 Tu	06 30 20	07 23 40	29 16 06	21 15	04 04	19 13	25 46	04 26	23 01	24 30	03 12	21 12
30 We	06 34 17	08 24 48	13♊ 35 29	20 41	05 20	19 06	25 57	04 27	23 02	24 31	03 14	21 11
31 Th	06 38 13	09 25 56	28 16 56	19 56	06 35	18 58	26 09	04 29	23 04	24 33	03 16	21 09

Data for	12-01-2009
Julian Day	2455166.50
Ayanamsa	23 59 57
SVP	05 ♓ 07 31
☽ ☊ Mean	23 ♑ 16 R

● ◐ PHASES ○ ◑

02	07:31	○	10♊15
09	00:14	◑	17♍03
16	12:02	●	24♐40
24	17:36	◐	03♈03
31	19:14	✹	10♋15

LAST ASPECT Day	h m	☽ INGRESS Day	h m	
01	13:40	01	14:24	♊
03	10:28	03	16:01	♋
05	05:09	05	17:08	♌
07	08:59	07	19:07	♍
09	10:05	09	22:48	♎
11	17:45	12	04:32	♏
14	01:18	14	12:25	♐
16	12:02	16	22:32	♑
18	20:08	19	10:39	♒
21	12:54	21	23:42	♓
24	08:10	24	11:40	♈
26	11:46	26	20:27	♉
28	17:55	29	01:15	♊
30	20:31	31	02:46	♋

DECLINATION

Day	☉	☽	☿	♀	♂	♃	♄	♅	♆	♇
01 Tu	21S 47	22N28	25S 24	19S 09	17N59	15S 27	00N47	03S 36	13S 57	18S 16
02 We	21 56	24 55	25 31	19 27	17 57	15 24	00 46	03 36	13 57	18 16
03 Th	22 05	25 47	25 37	19 45	17 56	15 21	00 44	03 36	13 57	18 16
04 Fr	22 13	24 51	25 41	20 02	17 54	15 18	00 43	03 36	13 56	18 16
05 Sa	22 21	22 12	25 44	20 19	17 53	15 15	00 41	03 36	13 56	18 16
06 Su	22 29	18 07	25 46	20 35	17 52	15 12	00 40	03 36	13 56	18 16
07 Mo	22 36	13 00	25 46	20 50	17 51	15 09	00 39	03 35	13 55	18 16
08 Tu	22 42	07 14	25 44	21 05	17 50	15 06	00 37	03 35	13 55	18 16
09 We	22 48	01 14	25 42	21 19	17 50	15 02	00 36	03 35	13 55	18 17
10 Th	22 54	04S 44	25 37	21 33	17 49	14 59	00 35	03 35	13 54	18 17
11 Fr	22 59	10 22	25 31	21 45	17 49	14 56	00 34	03 35	13 54	18 17
12 Sa	23 04	15 25	25 24	21 58	17 49	14 52	00 33	03 35	13 53	18 17
13 Su	23 08	19 40	25 16	22 09	17 50	14 49	00 32	03 34	13 53	18 17
14 Mo	23 12	22 55	25 06	22 20	17 51	14 46	00 30	03 34	13 53	18 17
15 Tu	23 16	24 59	24 55	22 31	17 51	14 42	00 29	03 34	13 52	18 17
16 We	23 19	25 46	24 42	22 40	17 53	14 39	00 29	03 33	13 52	18 17
17 Th	23 21	25 17	24 29	22 49	17 54	14 35	00 28	03 33	13 51	18 17
18 Fr	23 23	23 35	24 14	22 58	17 56	14 31	00 27	03 33	13 51	18 17
19 Sa	23 25	20 52	23 58	23 05	17 58	14 28	00 26	03 32	13 50	18 17
20 Su	23 26	17 17	23 42	23 12	18 00	14 24	00 25	03 32	13 50	18 17
21 Mo	23 26	13 02	23 24	23 18	18 02	14 20	00 24	03 32	13 49	18 17
22 Tu	23 26	08 19	23 07	23 24	18 05	14 17	00 24	03 31	13 49	18 18
23 We	23 26	03 16	22 49	23 29	18 08	14 13	00 23	03 31	13 48	18 18
24 Th	23 25	01N57	22 31	23 33	18 11	14 09	00 22	03 30	13 48	18 18
25 Fr	23 24	07 12	22 13	23 36	18 14	14 05	00 22	03 30	13 47	18 18
26 Sa	23 22	12 17	21 55	23 39	18 18	14 01	00 21	03 29	13 47	18 18
27 Su	23 20	17 00	21 38	23 40	18 22	13 57	00 21	03 29	13 46	18 18
28 Mo	23 17	21 03	21 22	23 41	18 26	13 53	00 20	03 28	13 46	18 18
29 Tu	23 14	24 03	21 06	23 42	18 30	13 49	00 20	03 28	13 45	18 18
30 We	23 10	25 38	20 52	23 41	18 35	13 45	00 19	03 27	13 44	18 18
31 Th	23 06	25 29	20 40	23 40	18 40	13 41	00 19	03 26	13 44	18 18

ASPECTARIAN

01 02:04 ☽ ⚹ ♅
04:05 ☽ □ ♆
13:40 ☽ ☍ ♀
14:53 ☿ ⚹ ♆
19:31 ☽ △ ♄
20:28 ♅ SD
22:04 ♀ ♐
02 12:20 ☽ # ☿
19:59 ☽ ⚹ ♂
03 01:37 ☽ △ ♃
04:04 ☽ □ ♅
06:04 ☽ △ ♆
08:34 ☽ # ☿
10:28 ☽ ☍ ☿
19:47 ☽ ☍ ♇
21:14 ☽ □ ♄
04 11:45 ♀ ⚹ ♄
22:54 ☽ # ☉
05 05:09 ☽ △ ♅
11:16 ☽ # ♀
17:25 ☿ ♑
22:37 ☽ ⚹ ♄
23:11 ☽ # ♇
06 01:17 ☽ ∥ ♂
01:44 ☽ △ ♀
14:15 ☽ # ♃
17:30 ☽ △ ☉
19:53 ☽ # ♆
23:52 ☽ ☌ ♂
07 05:23 ☽ ☍ ♃
08:59 ☽ ☍ ♆
09:12 ☿ ☌ ♇
23:15 ☽ △ ♇
08 00:48 ☽ △ ☿
02:35 ☿ □ ♄
[illegible]
14:35 ☽ # ♅

09 02:28 ☽ ∥ ♄
07:15 ☽ # ♄
10:05 ☽ ☍ ♅
19:17 ☽ ∥ ♅
10 03:14 ☽ □ ♇
05:08 ☽ ☌ ♄
10:47 ☽ □ ☿
18:30 ☽ ⚹ ♀
11 01:18 ☉ △ ♂
08:53 ☽ ⚹ ♂
09:26 ☽ ⚹ ☉
15:02 ☽ △ ♃
16:24 ☽ ∥ ♆
17:45 ☽ △ ♆
21:16 ☽ ∥ ♃
12 09:17 ☽ ⚹ ♇
12:59 ☽ # ♂
15:34 ☽ ∥ ♇
23:22 ☽ ⚹ ☿
13 16:25 ☽ □ ♂
18:49 ☽ ∥ ♀
22:45 ☽ △ ♅
23:07 ☽ □ ♃
14 01:18 ☽ □ ♆
02:48 ☽ ∥ ☉
15:27 ☉ □ ♅
19:46 ☽ ⚹ ♄
23:08 ☽ ∥ ☿
15 00:17 ☉ ⚹ ♃
22:17 ☽ ☌ ♀
16 00:17 ☉ ⚹ ♆
[illegible]
08:22 ☽ □ ♅

09:33 ☽ ⚹ ♃
11:05 ☽ ⚹ ♆
17 04:00 ☽ ☌ ♇
06:24 ☽ □ ♄
12:39 ♀ △ ♂
15:19 ☽ ∥ ☿
18 02:02 ☽ ∥ ☉
05:57 ☽ ∥ ♀
07:52 ☽ ☌ ☿
20:08 ☽ ⚹ ♅
19 17:39 ☽ ∥ ♇
18:58 ☽ △ ♄
19:34 ☽ # ♂
20 01:53 ♀ □ ♅
13:27 ♂ SR
16:51 ☽ ∥ ♃
19:45 ☽ ∥ ♆
21:36 ☿ ∥ ☉
21 02:44 ☽ ☍ ♂
04:53 ♀ ⚹ ♃
05:24 ♀ ⚹ ♆
06:13 ☿ ∥ ♀
08:54 ♃ ☌ ♆
12:07 ☽ ☌ ♆
12:09 ☽ ☌ ♃
12:54 ☽ ⚹ ♀
17:47 ☉ ♑
22 00:16 ☽ ⚹ ☉
05:42 ☽ ⚹ ♇
11:24 ♀ ∥ ☉
22:52 ☽ ∥ ♅
23 [illegible]
16:47 ☽ ∥ ♄

18:07 ☽ ⚹ ☿
21:43 ☽ ☌ ♅
24 07:03 ☽ # ♅
08:10 ☽ □ ♀
17:31 ☉ ☌ ♇
17:36 ☽ □ ♇
20:02 ☽ ☍ ♄
25 18:17 ♀ ♑
23:00 ♂ # ♇
23:58 ☉ □ ♄
26 01:01 ☽ △ ♂
05:18 ☽ □ ☿
[illegible]
08:21 ☽ # ♃

10:12 ☽ ⚹ ♆
11:46 ☽ ⚹ ♃
14:39 ☿ SR
23:13 ☽ △ ♀
27 02:09 ☽ △ ♇
06:41 ☽ △ ☉
07:08 ☽ # ♇
07:39 ☽ ∥ ♂
28 01:59 ☽ # ☿
06:47 ♀ ☌ ♇
06:51 ☽ □ ♂
10:38 ☽ △ ☿
[illegible]
15:49 ☽ □ ♆

16:32 ☽ # ☉
17:55 ☽ □ ♃
20:33 ☽ # ♀
29 06:56 ♀ □ ♄
08:45 ☽ △ ♄
30 02:25 ♃ ∥ ♆
08:59 ☽ ⚹ ♂
15:31 ☽ □ ♅
17:57 ☽ △ ♆
20:31 ☽ △ ♃
31 08:04 ☽ ☍ ♇
10:01 ☽ □ ♄
[illegible]
22:48 ☽ # ♀

⚷ Chiron

01 Dec.	08 S 26
03	21♒47
06	21 53
09	22 00
12	22 07
15	22 15
18	22 23
21	22 32
24	22 41
27	22 50
30	23 00

15 07:21 25♑01 ☼ Annular Solar Eclipse

January 2010

Day	S. T. h m s	☉ ° ' "	☽ ° ' "	☿ ° '	♀ ° '	♂ ° '	♃ ° '	♄ ° '	♅ ° '	♆ ° '	♇ ° '	☊ True ° '
01 Fr	06 42 10	10♑27 03	13♋13 56	19♑R00	07♑51	18♌R49	26♒22	04♎30	23♓05	24♒35	03♑18	21♑R08
02 Sa	06 46 07	11 28 11	28 18 12	17 54	09 06	18 39	26 34	04 32	23 07	24 37	03 21	21 08
03 Su	06 50 03	12 29 19	13♌20 49	16 42	10 22	18 29	26 46	04 33	23 09	24 38	03 23	21 07
04 Mo	06 53 60	13 30 27	28 13 33	15 23	11 37	18 18	26 58	04 34	23 10	24 40	03 25	21 07
05 Tu	06 57 56	14 31 35	12♍49 49	14 03	12 53	18 06	27 11	04 35	23 12	24 42	03 27	21 07
06 We	07 01 53	15 32 44	27 05 22	12 42	14 08	17 53	27 23	04 36	23 14	24 44	03 29	21 06
07 Th	07 05 49	16 33 52	10♎58 15	11 23	15 24	17 39	27 36	04 37	23 15	24 46	03 31	21 06
08 Fr	07 09 46	17 35 01	24 28 29	10 10	16 39	17 24	27 48	04 37	23 17	24 48	03 33	21 05
09 Sa	07 13 42	18 36 10	07♏37 33	09 03	17 55	17 09	28 01	04 38	23 19	24 50	03 36	21 04
10 Su	07 17 39	19 37 20	20 27 45	08 05	19 10	16 53	28 14	04 38	23 21	24 52	03 38	21 04
11 Mo	07 21 36	20 38 29	03♐01 48	07 15	20 26	16 37	28 27	04 39	23 23	24 54	03 40	21 04
12 Tu	07 25 32	21 39 38	15 22 25	06 36	21 41	16 19	28 40	04 39	23 25	24 56	03 42	21D 04
13 We	07 29 29	22 40 47	27 32 08	06 06	22 57	16 01	28 53	04 39	23 27	24 58	03 44	21 04
14 Th	07 33 25	23 41 56	09♑33 17	05 46	24 12	15 42	29 06	04R 39	23 29	25 00	03 46	21 04
15 Fr	07 37 22	24 43 04	21 27 59	05 35	25 28	15 23	29 19	04 39	23 31	25 02	03 48	21 05
16 Sa	07 41 18	25 44 12	03♒18 16	05D 34	26 43	15 03	29 32	04 39	23 33	25 04	03 50	21 06
17 Su	07 45 15	26 45 19	15 06 12	05 40	27 59	14 43	29 45	04 39	23 35	25 06	03 52	21 06
18 Mo	07 49 11	27 46 26	26 53 58	05 54	29 14	14 22	29 59	04 38	23 38	25 08	03 54	21R 06
19 Tu	07 53 08	28 47 32	08♓44 03	06 16	00♒30	14 00	00♓12	04 38	23 40	25 10	03 56	21 06
20 We	07 57 05	29 48 38	20 39 19	06 43	01 45	13 38	00 26	04 37	23 42	25 12	03 58	21 06
21 Th	08 01 01	00♒49 42	02♈43 05	07 17	03 00	13 16	00 39	04 36	23 44	25 14	04 00	21 05
22 Fr	08 04 58	01 50 46	14 59 03	07 55	04 16	12 53	00 53	04 35	23 47	25 16	04 02	21 05
23 Sa	08 08 54	02 51 49	27 31 12	08 39	05 31	12 30	01 06	04 34	23 49	25 18	04 04	21D 05
24 Su	08 12 51	03 52 50	10♉23 29	09 27	06 47	12 07	01 20	04 33	23 52	25 20	04 06	21 06
25 Mo	08 16 47	04 53 51	23 39 25	10 19	08 02	11 43	01 34	04 32	23 54	25 23	04 08	21 06
26 Tu	08 20 44	05 54 51	07♊21 36	11 14	09 17	11 19	01 47	04 31	23 57	25 25	04 10	21 07
27 We	08 24 40	06 55 50	21 31 01	12 12	10 33	10 56	02 01	04 29	23 59	25 27	04 12	21 08
28 Th	08 28 37	07 56 47	06♋06 14	13 13	11 48	10 32	02 15	04 28	24 02	25 29	04 14	21 08
29 Fr	08 32 34	08 57 44	21 03 01	14 17	13 03	10 08	02 29	04 26	24 05	25 31	04 16	21R 08
30 Sa	08 36 30	09 58 39	06♌14 12	15 24	14 19	09 44	02 43	04 25	24 07	25 33	04 18	21 07
31 Su	08 40 27	10 59 34	21 30 28	16 32	15 34	09 20	02 57	04 23	24 10	25 36	04 19	21 05

Data for	01-01-2010
Julian Day	2455197.50
Ayanamsa	24 00 04
SVP	05 ♓ 07 29
☽ ☊ Mean	21 ♑ 38 R

● ◐ PHASES ○ ◑

07	10:39	◑	17♎01
15	07:11	☼	25♑01
23	10:54	◐	03♉20
30	06:18	○	10♌15

LAST ASPECT Day	h m	☽ INGRESS Day	h m	
01	15:44	02	02:42	♌
03	21:56	04	02:53	♍
05	17:25	06	04:59	♎
08	06:07	08	10:01	♏
10	15:02	10	18:10	♐
13	02:43	13	04:54	♑
15	09:03	15	17:17	♒
17	20:23	18	06:18	♓
20	06:07	20	18:37	♈
22	19:47	23	04:41	♉
25	03:04	25	11:12	♊
27	06:33	27	14:02	♋
29	04:49	29	14:10	♌
31	06:27	31	13:23	♍

DECLINATION

Day	☉	☽	☿	♀	♂	♃	♄	♅	♆	♇
01 Fr	23S02	23N30	20S28	23S39	18N45	13S37	00N19	03S26	13S43	18S18
02 Sa	22 57	19 51	20 19	23 36	18 51	13 32	00 18	03 25	13 43	18 18
03 Su	22 51	14 54	20 10	23 33	18 56	13 28	00 18	03 24	13 42	18 18
04 Mo	22 45	09 05	20 03	23 29	19 02	13 24	00 18	03 24	13 42	18 18
05 Tu	22 39	02 54	19 58	23 24	19 08	13 20	00 18	03 23	13 41	18 18
06 We	22 32	03S17	19 54	23 18	19 14	13 15	00 18	03 22	13 40	18 18
07 Th	22 25	09 08	19 51	23 12	19 21	13 11	00 18	03 22	13 40	18 18
08 Fr	22 17	14 24	19 50	23 05	19 27	13 06	00 18	03 21	13 39	18 18
09 Sa	22 09	18 51	19 51	22 58	19 34	13 02	00 18	03 20	13 38	18 18
10 Su	22 00	22 19	19 52	22 49	19 41	12 58	00 18	03 19	13 38	18 18
11 Mo	21 51	24 38	19 55	22 40	19 48	12 53	00 18	03 18	13 37	18 18
12 Tu	21 42	25 43	19 59	22 31	19 55	12 48	00 18	03 18	13 36	18 18
13 We	21 32	25 32	20 04	22 20	20 03	12 44	00 18	03 17	13 36	18 18
14 Th	21 22	24 09	20 11	22 09	20 10	12 39	00 19	03 16	13 35	18 18
15 Fr	21 11	21 41	20 17	21 57	20 18	12 35	00 19	03 15	13 34	18 18
16 Sa	21 00	18 20	20 25	21 45	20 25	12 30	00 19	03 14	13 34	18 18
17 Su	20 48	14 15	20 33	21 32	20 33	12 25	00 20	03 13	13 33	18 18
18 Mo	20 36	09 38	20 41	21 18	20 41	12 21	00 20	03 12	13 32	18 18
19 Tu	20 24	04 40	20 49	21 04	20 48	12 16	00 20	03 11	13 32	18 18
20 We	20 11	00N30	20 58	20 49	20 56	12 11	00 21	03 11	13 31	18 18
21 Th	19 58	05 41	21 06	20 33	21 04	12 06	00 21	03 10	13 30	18 18
22 Fr	19 45	10 44	21 14	20 17	21 12	12 02	00 22	03 09	13 30	18 18
23 Sa	19 31	15 29	21 22	20 00	21 19	11 57	00 23	03 08	13 29	18 18
24 Su	19 17	19 39	21 29	19 43	21 27	11 52	00 23	03 07	13 28	18 18
25 Mo	19 02	22 59	21 35	19 25	21 34	11 47	00 24	03 06	13 27	18 18
26 Tu	18 48	25 08	21 41	19 06	21 42	11 42	00 25	03 05	13 27	18 18
27 We	18 32	25 46	21 46	18 47	21 49	11 37	00 26	03 04	13 26	18 18
28 Th	18 17	24 39	21 50	18 27	21 56	11 32	00 27	03 02	13 25	18 18
29 Fr	18 01	21 47	21 53	18 07	22 03	11 27	00 27	03 01	13 25	18 18
30 Sa	17 45	17 22	21 55	17 47	22 10	11 22	00 28	03 00	13 24	18 17
31 Su	17 28	11 48	21 57	17 25	22 17	11 17	00 29	02 59	13 23	18 17

ASPECTARIAN

01 03:55 ☽ # ☉
08:36 ☽ ☍ ☿
15:44 ☽ △ ♅
21:21 ☽ # ☿
02 05:15 ☽ ∥ ♂
08:06 ☽ # ♇
09:56 ☽ ⚹ ♄
03 05:08 ☽ # ♆
06:12 ☽ # ♃
08:09 ☽ ☌ ♂
18:13 ☽ ☍ ♆
21:56 ☽ ☍ ♃
04 08:29 ☽ △ ♇
19:06 ☿ ☌ ☉
22:07 ☽ # ♅
05 00:06 ☽ △ ♀
01:51 ☽ △ ☿
03:02 ☽ △ ☉
10:01 ☽ ∥ ♄
10:40 ☿ ☌ ♀
12:18 ☽ # ♄
17:26 ☽ ☍ ♅
06 00:21 ☽ ∥ ♅
11:00 ☽ □ ♇
12:55 ☽ ☌ ♄
07 00:40 ☽ □ ☿
08:36 ☽ □ ♀
11:35 ☽ ⚹ ♂
17:54 ☽ ∥ ♃
20:23 ☽ ∥ ♆
08 00:35 ☽ △ ♆
06:07 ☽ △ ♃
16:33 ☽ ⚹ ♇
20:45 ☽ ∥ ♇
09 02:27 ☽ ⚹ ☿
04:33 ☽ # ♂
06:13 ☽ ∥ ☿
17:24 ☽ □ ♂
21:18 ☽ ⚹ ♀
21:37 ☽ ∥ ☉
22:16 ☽ ⚹ ☉
10 04:08 ☽ ∥ ♀
05:29 ☽ △ ♅
08:22 ☽ □ ♆
15:02 ☽ □ ♃
11 03:07 ☽ ⚹ ♄
21:06 ♀ ☌ ☉
12 01:49 ☽ △ ♂
15:52 ☽ □ ♅
18:52 ☽ ⚹ ♆
13 02:43 ☽ ⚹ ♃
09:48 ♀ ⚹ ♅
12:23 ☽ ☌ ♇
14:11 ☽ □ ♄
15:58 ♄ SR
16:35 ☽ ☌ ☿
18:42 ☉ ⚹ ♅
14 14:44 ☿ # ♂
21:42 ☽ ∥ ♀
15 04:10 ☽ ⚹ ♅
04:20 ☽ ∥ ☉
09:03 ☽ ☌ ♀
10:20 ☽ # ♂
10:22 ☽ ∥ ☿
16:53 ☿ SD
16 00:10 ☽ ∥ ♇
02:44 ☽ △ ♄
23:13 ☽ ☍ ♂
17 01:50 ☿ # ♂
03:46 ☽ ∥ ♆
09:55 ☽ ∥ ♃
18:28 ☿ ∥ ☉
18:54 ☉ # ♂
20:23 ☽ ☌ ♆
18 02:11 ♃ ♓
06:23 ☽ ☌ ♃
14:16 ☽ ⚹ ♇
14:35 ♀ ♒
18:49 ☽ ⚹ ☿
19 06:56 ☽ ∥ ♅
14:37 ☿ ∥ ♀
16:13 ♀ # ♂
20:07 ☽ # ♄
23:20 ☽ ∥ ♄
20 04:28 ☉ ♒
06:07 ☽ ☌ ♅
12:20 ☽ # ♅
19:55 ☽ ⚹ ☉
21 00:38 ☽ ⚹ ♀
02:33 ☽ □ ♇
03:43 ☽ ☍ ♄
09:27 ☽ □ ☿
20:03 ☽ △ ♂
22 06:08 ♀ △ ♄
06:10 ☽ # ♃
13:36 ☽ # ♆
19:47 ☽ ⚹ ♆
23 06:52 ☽ ⚹ ♃
12:20 ☽ △ ♇
15:40 ☽ # ♇
16:37 ☽ □ ♀
21:44 ☽ # ☉
22:09 ☽ △ ☿
24 00:15 ☽ # ♀
03:04 ☽ □ ♂
12:23 ☽ ∥ ♂
12:31 ☽ # ☿
15:38 ☉ △ ♄
25 00:26 ☽ ⚹ ♅
03:04 ☽ □ ♆
09:04 ☿ # ♂
14:10 ☽ □ ♃
19:05 ☽ △ ♄
21:18 ☽ △ ☉
26 03:39 ☽ △ ♀
06:37 ☽ ⚹ ♂
27 04:08 ☽ □ ♅
05:30 ♀ ☍ ♂
06:33 ☽ △ ♆
17:37 ☽ △ ♃
20:57 ☽ ☍ ♇
21:21 ☽ □ ♄
22:52 ☉ ∥ ♇
28 11:46 ♀ ∥ ♇
12:22 ☽ ☍ ☿
22:20 ☽ ∥ ♂
23:26 ☽ # ☿
29 04:49 ☽ △ ♅
19:35 ☽ # ♇
19:43 ☉ ☍ ♂
21:08 ☽ ⚹ ♄
21:57 ☽ # ♀
22:08 ☽ # ☉
30 05:21 ☽ ☌ ♂
09:03 ♀ ∥ ☉
13:50 ☽ ☍ ♀
17:34 ☽ # ♆
31 02:06 ☽ # ♃
06:27 ☽ ☍ ♆
18:20 ☽ ☍ ♃
20:17 ☽ △ ♇
22:07 ♄ □ ♇

⚷ Chiron

01 Dec.	08 S 10
02	23♒10
05	23 21
08	23 32
11	23 43
14	23 54
17	24 06
20	24 17
23	24 29
26	24 41
29	24 54

February 2010

Day	S. T.	☉	☽	☿	♀	♂	♃	♄	♅	♆	♇	☊ True
	h m s	° ' "	° ' "	° '	° '	° '	° '	° '	° '	° '	° '	° '
01 Mo	08 44 23	12♒ 00 27	06♍ 41 33	17♑ 43	16♒ 49	08♌ R56	03♓ 11	04♎ R21	24♓ 13	25♒ 38	04♑ 21	21♑ R02
02 Tu	08 48 20	13 01 20	21 37 52	18 55	18 05	08 32	03 25	04 19	24 15	25 40	04 23	20 58
03 We	08 52 16	14 02 11	06♎ 11 59	20 09	19 20	08 08	03 39	04 17	24 18	25 42	04 25	20 54
04 Th	08 56 13	15 03 02	20 19 18	21 25	20 35	07 45	03 53	04 15	24 21	25 45	04 27	20 51
05 Fr	09 00 09	16 03 52	03♏ 58 19	22 42	21 51	07 22	04 07	04 13	24 24	25 47	04 28	20 49
06 Sa	09 04 06	17 04 41	17 10 09	24 01	23 06	06 59	04 21	04 10	24 27	25 49	04 30	20 48
07 Su	09 08 03	18 05 29	29 57 44	25 21	24 21	06 37	04 35	04 08	24 30	25 51	04 32	20D 48
08 Mo	09 11 59	19 06 17	12♐ 25 05	26 42	25 36	06 14	04 49	04 05	24 32	25 54	04 34	20 49
09 Tu	09 15 56	20 07 03	24 36 34	28 05	26 51	05 53	05 04	04 02	24 35	25 56	04 35	20 51
10 We	09 19 52	21 07 49	06♑ 36 30	29 28	28 07	05 32	05 18	04 00	24 38	25 58	04 37	20 53
11 Th	09 23 49	22 08 33	18 28 53	00♒ 53	29 22	05 11	05 32	03 57	24 41	26 01	04 38	20 54
12 Fr	09 27 45	23 09 16	00♒ 17 10	02 18	00♓ 37	04 51	05 47	03 54	24 44	26 03	04 40	20R 54
13 Sa	09 31 42	24 09 58	12 04 18	03 45	01 52	04 31	06 01	03 51	24 47	26 05	04 42	20 53
14 Su	09 35 38	25 10 38	23 52 43	05 13	03 07	04 12	06 15	03 48	24 51	26 07	04 43	20 50
15 Mo	09 39 35	26 11 17	05♓ 44 23	06 41	04 23	03 54	06 30	03 45	24 54	26 10	04 45	20 45
16 Tu	09 43 32	27 11 55	17 41 00	08 11	05 38	03 36	06 44	03 41	24 57	26 12	04 46	20 39
17 We	09 47 28	28 12 30	29 44 13	09 42	06 53	03 19	06 58	03 38	25 00	26 14	04 48	20 33
18 Th	09 51 25	29 13 05	11♈ 55 50	11 13	08 08	03 03	07 13	03 35	25 03	26 16	04 49	20 27
19 Fr	09 55 21	00♓ 13 37	24 17 56	12 46	09 23	02 47	07 27	03 31	25 06	26 19	04 51	20 21
20 Sa	09 59 18	01 14 08	06♉ 52 58	14 19	10 38	02 32	07 42	03 28	25 09	26 21	04 52	20 18
21 Su	10 03 14	02 14 37	19 43 43	15 53	11 53	02 18	07 56	03 24	25 13	26 23	04 53	20 16
22 Mo	10 07 11	03 15 04	02♊ 53 05	17 28	13 08	02 05	08 11	03 20	25 16	26 26	04 55	20D 15
23 Tu	10 11 07	04 15 30	16 23 50	19 05	14 23	01 52	08 25	03 16	25 19	26 28	04 56	20 16
24 We	10 15 04	05 15 53	00♋ 18 04	20 42	15 38	01 41	08 40	03 13	25 22	26 30	04 57	20 16
25 Th	10 19 01	06 16 15	14 36 28	22 20	16 53	01 30	08 54	03 09	25 26	26 32	04 59	20R 16
26 Fr	10 22 57	07 16 34	29 17 27	23 59	18 08	01 19	09 09	03 05	25 29	26 35	05 00	20 15
27 Sa	10 26 54	08 16 51	14♌ 16 37	25 39	19 23	01 10	09 23	03 01	25 32	26 37	05 01	20 11
28 Su	10 30 50	09 17 07	29 26 39	27 20	20 37	01 01	09 38	02 57	25 36	26 39	05 02	20 06

Data for	02-01-2010
Julian Day	2455228.50
Ayanamsa	24 00 09
SVP	05 ♓ 07 26
☽ ☊ Mean	19 ♑ 59 R

● ◐ PHASES ○ ◑

Day	h m	Phase	Position
05	23:49	◑	17♏04
14	02:52	●	25♒18
22	00:42	◐	03♊17
28	16:38	○	09♍59

LAST ASPECT ☽ INGRESS

Day	h m	Day	h m	Sign
02	04:17	02	13:42	♎
04	09:28	04	16:56	♏
06	16:12	07	00:04	♐
09	04:59	09	10:45	♑
11	12:40	11	23:25	♒
14	04:34	14	12:24	♓
16	14:33	17	00:31	♈
19	03:53	19	10:56	♉
21	12:16	21	18:47	♊
23	17:30	23	23:29	♋
25	17:48	26	01:09	♌
27	20:15	28	00:53	♍

DECLINATION

Day	☉	☽	☿	♀	♂	♃	♄	♅	♆	♇
01 Mo	17S 11	05N33	21S 57	17S 04	22N24	11S 12	00N30	02S 58	13S 22	18S 17
02 Tu	16 54	00S 54	21 56	16 42	22 30	11 07	00 31	02 57	13 22	18 17
03 We	16 37	07 08	21 54	16 19	22 36	11 02	00 32	02 56	13 21	18 17
04 Th	16 19	12 48	21 51	15 56	22 42	10 57	00 33	02 55	13 20	18 17
05 Fr	16 01	17 39	21 47	15 32	22 48	10 52	00 35	02 54	13 19	18 17
06 Sa	15 43	21 28	21 42	15 08	22 53	10 46	00 36	02 52	13 19	18 17
07 Su	15 24	24 07	21 35	14 44	22 58	10 41	00 37	02 51	13 18	18 17
08 Mo	15 06	25 31	21 28	14 19	23 03	10 36	00 38	02 50	13 17	18 17
09 Tu	14 47	25 39	21 19	13 54	23 08	10 31	00 39	02 49	13 16	18 17
10 We	14 27	24 34	21 09	13 29	23 13	10 26	00 41	02 48	13 15	18 17
11 Th	14 08	22 22	20 57	13 03	23 17	10 20	00 42	02 47	13 15	18 17
12 Fr	13 48	19 15	20 44	12 37	23 21	10 15	00 43	02 45	13 14	18 17
13 Sa	13 28	15 21	20 30	12 10	23 25	10 10	00 45	02 44	13 13	18 17
14 Su	13 08	10 52	20 15	11 43	23 28	10 05	00 46	02 43	13 12	18 16
15 Mo	12 47	05 59	19 59	11 16	23 31	09 59	00 48	02 42	13 12	18 16
16 Tu	12 27	00 51	19 41	10 48	23 34	09 54	00 49	02 40	13 11	18 16
17 We	12 06	04N20	19 22	10 21	23 37	09 49	00 51	02 39	13 10	18 16
18 Th	11 45	09 25	19 01	09 53	23 39	09 43	00 52	02 38	13 09	18 16
19 Fr	11 24	14 13	18 39	09 24	23 42	09 38	00 54	02 37	13 09	18 16
20 Sa	11 02	18 29	18 16	08 56	23 43	09 33	00 55	02 35	13 08	18 16
21 Su	10 41	22 00	17 52	08 27	23 45	09 27	00 57	02 34	13 07	18 16
22 Mo	10 19	24 28	17 26	07 58	23 47	09 22	00 59	02 33	13 06	18 16
23 Tu	09 57	25 36	16 59	07 29	23 48	09 16	01 00	02 31	13 06	18 16
24 We	09 35	25 12	16 31	06 59	23 49	09 11	01 02	02 30	13 05	18 16
25 Th	09 13	23 08	16 01	06 29	23 49	09 06	01 04	02 29	13 04	18 16
26 Fr	08 50	19 30	15 30	06 00	23 50	09 00	01 06	02 27	13 03	18 15
27 Sa	08 28	14 31	14 58	05 30	23 50	08 55	01 07	02 26	13 03	18 15
28 Su	08 05	08 36	14 24	05 00	23 50	08 49	01 09	02 25	13 02	18 15

ASPECTARIAN

01 09:39 ☽ # ♅
18:44 ☽ ∥ ♄
19:13 ☽ △ ☿
22:36 ☽ # ♄

02 04:17 ☽ ☍ ♅
07:45 ☽ ∥ ♅
20:49 ☽ ☌ ♄
21:01 ☽ □ ♆

03 03:10 ☽ ✶ ♂
14:15 ☽ △ ☉
15:57 ☽ ∥ ♃

04 00:30 ☽ △ ♀
02:05 ☽ □ ☿
02:27 ☽ ∥ ♆
09:28 ☽ △ ♆
13:50 ☽ ∥ ♀
15:55 ☽ ∥ ☉

05 00:16 ☽ △ ♃
00:54 ☽ ✶ ♇
03:34 ☽ ∥ ♆
05:55 ☽ □ ♂

06 01:39 ☽ ∥ ☿
08:02 ☿ ✶ ♅
11:49 ☽ # ♂
12:14 ☽ □ ♀
13:37 ☽ △ ♅
14:14 ☽ ✶ ☿
16:12 ☽ □ ♆
17:41 ♃ ✶ ♇

07 07:56 ☽ ✶ ♄
09:01 ☽ □ ♃
11:11 ☽ △ [illegible]

08 05:44 ♀ ☌ ♆
14:18 ☽ ✶ ☉
23:58 ☽ □ ♅

09 02:38 ☽ ✶ ♆
04:59 ☽ ✶ ♀
18:46 ☽ □ ♄
19:59 ☽ ☌ ♇
21:19 ☽ ✶ ♃

10 09:06 ☿ ♒
12:40 ♀ ∥ ♆
15:41 ☽ # ♂

11 12:10 ♀ ♓
12:39 ☽ ∥ ☿
12:40 ☽ ✶ ♅

12 04:41 ☽ ☌ ☿
06:27 ☽ ∥ ♆
07:20 ☽ △ ♄
09:02 ☽ ☍ ♂

13 01:35 ☿ △ ♄
10:20 ☿ ☍ ♂
11:20 ☽ ∥ ☉
11:50 ☽ ∥ ♆
18:13 ☉ ∥ ♆
19:15 ☽ ∥ ♀

14 04:08 ☽ ∥ ♃
04:34 ☽ ☌ ♆
20:55 ☽ ☌ ♀
21:39 ☽ ✶ ♆
23:19 ☉ ☌ ♆

15 01:33 ☽ ☌ ♃
07:16 ♀ ✶ ♇
15:20 ♂ ✶ ♄
15:34 ☽ ∥ ♅

16 00:11 ☽ # ♄
07:48 ☽ ∥ ♄
14:33 ☽ ☌ ♅
16:15 ☽ # ♅

17 02:15 ♀ ☌ ♃
06:56 ☽ △ ♂
07:40 ☽ ☍ ♄
10:01 ☽ □ ♇
22:24 ☽ ✶ ☿

18 01:26 ☽ # ♃
02:01 ☽ # ♀
09:37 ♀ ∥ ♃
10:38 ☽ # ☉
18:30 ☽ # ♆
18:36 ☉ ♓

19 03:53 ☽ ✶ ♆
12:21 ☽ ✶ ☉
15:56 ☽ □ ♂
20:11 ☽ △ ♇
22:40 ☽ # ♇
22:49 ☽ # ☿

20 00:14 ☿ ∥ ♇
01:34 ☽ ✶ ♃
07:49 ☽ ✶ ♀
15:53 ☽ □ ☿

21 10:07 ☽ ✶ ♅
12:16 ☽ □ ♆
15:57 ☽ ∥ ♂
22:35 ☽ ✶ ♂

22 00:49 ☽ △ ♄
09:39 ☽ □ ♃
20:06 ☽ □ ♀

23 05:18 ☽ △ ☿
15:32 ☽ □ ♅
16:30 ☉ ✶ ♇
17:30 ☽ △ ♆

24 04:55 ☽ □ ♄
07:54 ☽ ☍ ♇
09:02 ☽ △ ☉
14:21 ☽ △ ♃
17:54 ☽ ∥ ♂

25 04:06 ☽ △ ♀
09:53 ☉ ∥ ♃
17:48 ☽ △ ♅

26 03:15 ☽ ☌ ♂
06:05 ☽ ✶ ♄
06:35 ☽ # ♇
21:54 ☽ # ☿

27 06:22 ☽ # ♆
14:03 ☿ ☌ ♆
19:35 ☽ ☍ ♆
20:15 ☽ ☍ ☿
23:11 ☽ # ♃

28 02:09 ☽ # ☉
08:50 ☽ △ ♇
10:45 ☉ ☌ ♃
14:48 ☽ # ♀
16:20 ☽ ☍ ♃
23:17 ☽ # ♅

⚷ Chiron

01 Dec.	07 S 37
01	25♒06
04	25 19
07	25 31
10	25 44
13	25 57
16	26 10
19	26 22
22	26 35
25	26 47
28	27 00

March 2010

Day	S. T.	☉	☽	☿	♀	♂	♃	♄	♅	♆	♇	☊ True
	h m s	° ' "	° ' "	° '	° '	° '	° '	° '	° '	° '	° '	° '
01 Mo	10 34 47	10♓ 17 21	14♍ 37 58	29♒ 02	21♓ 52	00♌ R54	09♓ 52	02♎ R53	25♓ 39	26♒ 41	05♑ 03	19♑ R59
02 Tu	10 38 43	11 17 33	29 40 17	00♓ 45	23 07	00 46	10 07	02 48	25 42	26 44	05 05	19 50
03 We	10 42 40	12 17 43	14♎ 24 11	02 30	24 22	00 40	10 21	02 44	25 46	26 46	05 06	19 42
04 Th	10 46 36	13 17 52	28 42 43	04 15	25 37	00 35	10 36	02 40	25 49	26 48	05 07	19 34
05 Fr	10 50 33	14 17 59	12♏ 32 09	06 01	26 51	00 30	10 50	02 36	25 52	26 50	05 08	19 27
06 Sa	10 54 30	15 18 05	25 51 59	07 48	28 06	00 26	11 05	02 31	25 56	26 52	05 09	19 23
07 Su	10 58 26	16 18 09	08♐ 44 26	09 37	29 21	00 23	11 19	02 27	25 59	26 55	05 10	19 20
08 Mo	11 02 23	17 18 12	21 13 30	11 26	00♈ 36	00 20	11 33	02 22	26 03	26 57	05 11	19D 20
09 Tu	11 06 19	18 18 13	03♑ 24 09	13 17	01 50	00 19	11 48	02 18	26 06	26 59	05 12	19 21
10 We	11 10 16	19 18 12	15 21 42	15 09	03 05	00 18	12 02	02 13	26 09	27 01	05 13	19 22
11 Th	11 14 12	20 18 09	27 11 19	17 02	04 20	00D 18	12 17	02 09	26 13	27 03	05 13	19R 22
12 Fr	11 18 09	21 18 05	08♒ 57 43	18 56	05 34	00 18	12 31	02 04	26 16	27 05	05 14	19 21
13 Sa	11 22 05	22 18 00	20 44 58	20 50	06 49	00 20	12 46	02 00	26 20	27 07	05 15	19 16
14 Su	11 26 02	23 17 52	02♓ 36 21	22 46	08 03	00 22	13 00	01 55	26 23	27 10	05 16	19 09
15 Mo	11 29 59	24 17 42	14 34 18	24 43	09 18	00 24	13 14	01 50	26 26	27 12	05 17	19 00
16 Tu	11 33 55	25 17 31	26 40 30	26 41	10 32	00 28	13 29	01 46	26 30	27 14	05 17	18 49
17 We	11 37 52	26 17 17	08♈ 56 01	28 40	11 47	00 32	13 43	01 41	26 33	27 16	05 18	18 37
18 Th	11 41 48	27 17 02	21 21 32	00♈ 39	13 01	00 37	13 57	01 36	26 37	27 18	05 19	18 25
19 Fr	11 45 45	28 16 44	03♉ 57 35	02 39	14 16	00 42	14 12	01 31	26 40	27 20	05 19	18 15
20 Sa	11 49 41	29 16 24	16 44 53	04 39	15 30	00 48	14 26	01 27	26 44	27 22	05 20	18 07
21 Su	11 53 38	00♈ 16 03	29 44 24	06 39	16 44	00 55	14 40	01 22	26 47	27 24	05 20	18 02
22 Mo	11 57 34	01 15 38	12♊ 57 33	08 39	17 59	01 02	14 54	01 17	26 50	27 26	05 21	17 59
23 Tu	12 01 31	02 15 12	26 26 00	10 39	19 13	01 10	15 09	01 13	26 54	27 28	05 21	17 57
24 We	12 05 27	03 14 43	10♋ 11 30	12 39	20 27	01 19	15 23	01 08	26 57	27 30	05 22	17 57
25 Th	12 09 24	04 14 12	24 15 10	14 38	21 41	01 28	15 37	01 03	27 01	27 32	05 22	17 56
26 Fr	12 13 21	05 13 39	08♌ 36 56	16 35	22 55	01 38	15 51	00 58	27 04	27 34	05 23	17 53
27 Sa	12 17 17	06 13 03	23 14 41	18 31	24 10	01 48	16 05	00 54	27 08	27 35	05 23	17 48
28 Su	12 21 14	07 12 25	08♍ 03 51	20 24	25 24	01 59	16 19	00 49	27 11	27 37	05 23	17 41
29 Mo	12 25 10	08 11 45	22 57 28	22 16	26 38	02 11	16 33	00 44	27 14	27 39	05 24	17 31
30 Tu	12 29 07	09 11 02	07♎ 47 02	24 04	27 52	02 23	16 47	00 40	27 18	27 41	05 24	17 19
31 We	12 33 03	10 10 18	22 23 47	25 49	29 06	02 35	17 01	00 35	27 21	27 43	05 24	17 07

Data for	03-01-2010
Julian Day	2455256.50
Ayanamsa	24 00 13
SVP	05 ♓ 07 21
☽ ☊ Mean	18 ♑ 30 R

● ◐ PHASES ○ ◑

Day	Time	Phase	Position
07	15:43	◑	16♐58
15	21:01	●	25♓10
23	11:00	◐	02♋43
30	02:26	○	09♎17

LAST ASPECT Day	h m	☽ INGRESS Day	h m	
01	17:36	02	00:32	♎
03	20:44	04	02:12	♏
06	04:33	06	07:37	♐
08	11:15	08	17:15	♑
10	22:00	11	05:44	♒
13	12:58	13	18:44	♓
16	00:01	16	06:32	♈
18	11:23	18	16:30	♉
20	19:41	21	00:29	♊
23	01:49	23	06:16	♋
25	04:40	25	09:40	♌
27	07:05	27	10:58	♍
29	06:56	29	11:22	♎
31	12:14	31	12:42	♏

ASPECTARIAN

01 03:43 ☽ ∥ ♄
12:30 ☽ # ♄
12:33 ☽ ☍ ♀
13:28 ☿ ♓
16:53 ☽ ∥ ♅
17:36 ☽ ☍ ♅
23:00 ☽ ∥ ♀
02 01:46 ☽ ⚹ ♂
05:02 ☽ ☌ ♄
08:24 ☿ ∥ ♆
08:45 ☽ □ ♇
11:05 ☽ ∥ ☉
16:43 ☽ ∥ ♃
03 08:30 ☽ ∥ ☿
11:15 ☽ ∥ ♆
20:44 ☽ △ ♆
04 03:10 ☽ □ ♂
04:07 ♀ ☌ ♅
10:54 ☽ △ ☿
11:01 ☽ ⚹ ♇
11:53 ☿ ⚹ ♇
13:08 ☽ ∥ ♇
20:57 ☽ △ ♃
05 03:23 ☽ △ ☉
07:43 ♀ ∥ ♅
06 00:07 ☽ △ ♅
01:51 ☽ □ ♆
04:33 ☽ △ ♀
05:30 ☽ # ♂
08:23 ☽ △ ♂
12:14 ☽ ⚹ ♄
07 01:56 ☽ □ ☿
03:05 ♀ # ♄
05:00 ☽ □ ♃
12:34 ♀ ♈
19:16 ♀ △ ♂
08 01:46 ☿ ☌ ♃
09:28 ☽ □ ♅
11:15 ☽ ⚹ ♆
20:32 ☽ □ ♀
21:49 ☽ □ ♄
09 03:35 ☽ ☌ ♇
08:22 ♀ ☍ ♄
13:22 ☿ ∥ ♃
15:54 ☽ # ♂
17:10 ☽ ⚹ ♃
23:29 ☽ ⚹ ☿
10 08:43 ☽ ⚹ ☉
17:10 ♂ SD
22:00 ☽ ⚹ ♅
11 06:20 ☽ ☍ ♂
10:02 ☽ △ ♄
12:21 ☽ ∥ ♇
16:16 ☽ ⚹ ♀
17:35 ♀ □ ♇
12 19:08 ☿ ∥ ♄
19:40 ☽ ∥ ♆
13 12:58 ☽ ☌ ♆
20:55 ♀ # ♅
22:42 ☽ ∥ ♃
14 05:22 ☽ ⚹ ♇
13:05 ♀ # ☉
13:17 ☿ ☌ ☉
17:05 ☽ ∥ ☿
21:17 ☽ ☌ ♃
22:05 ☽ # ♀
23:48 ☽ ∥ ☉
15 00:43 ☽ ∥ ♅
02:49 ☽ # ♄
12:22 ☉ ∥ ♅
13:25 ☿ # ♀
17:55 ☽ ∥ ♄
19:19 ☽ # ☉
19:49 ☽ # ♅
21:40 ☿ ☌ ♅
22:28 ☽ # ☿
23:39 ☽ ☌ ♅
16 00:01 ☽ ☌ ☿
00:11 ☽ # ☿
07:29 ☽ △ ♂
09:56 ☽ ☍ ♄
12:23 ☉ # ♄
14:23 ☿ ∥ ♅
16:55 ☽ □ ♇
20:12 ☽ # ♃
23:51 ☽ # ♄
17 06:08 ☽ ☌ ♀
06:50 ☉ ☌ ♅
09:27 ☿ ∥ ☉
16:12 ☿ ♈
23:08 ☽ # ♆
23:33 ☿ △ ♂
18 11:05 ☿ ☍ ♄
11:23 ☽ ⚹ ♆
17:47 ☽ □ ♂
19 02:34 ☽ △ ♇
05:00 ☽ # ♇
09:35 ☽ # ☉
19:36 ☽ ⚹ ♃
20 08:15 ☿ □ ♇
17:33 ☉ ♈
17:34 ☽ ∥ ♂
18:20 ☿ ∥ ♄
18:33 ☽ ⚹ ♅
19:41 ☽ □ ♆
21:48 ☿ # ♅
21 01:03 ☽ ⚹ ☉
02:10 ☽ ⚹ ♂
02:58 ☽ △ ♄
14:51 ☽ ⚹ ☿
17:54 ☉ △ ♂
22 00:38 ☉ ☍ ♄
03:34 ☽ □ ♃
09:54 ☽ ⚹ ♀
23 00:03 ♀ # ♃
00:49 ☽ □ ♅
01:49 ☽ △ ♆
03:55 ♂ ⚹ ♄
07:36 ♄ # ♅
08:21 ☽ □ ♄
15:38 ☽ ☍ ♇
24 04:56 ☽ □ ☿
07:30 ☽ ∥ ♂
09:04 ☽ △ ♃
19:15 ☽ □ ♀
25 04:40 ☽ △ ♅
09:27 ☉ # ♅
11:22 ☽ ⚹ ♄
12:18 ☽ ☌ ♂
14:36 ☽ # ♇
16:56 ☉ ∥ ♄
17:55 ☿ # ♃
17:58 ☽ △ ☉
26 03:44 ☉ □ ♇
15:07 ☽ △ ☿
16:45 ☽ # ♆
27 01:38 ☽ △ ♀
07:05 ☽ ☍ ♆
08:37 ☽ ∥ ♀
11:48 ☽ ∥ ☿
18:45 ☽ # ♃
19:41 ☽ △ ♇
28 07:29 ☽ ∥ ☉
11:01 ☽ ∥ ♄
12:04 ☽ # ♅
13:31 ☽ ☍ ♃
29 01:21 ☽ ∥ ♅
02:32 ☽ # ♄
06:56 ☽ ☍ ♅
07:29 ☽ # ☉
12:30 ☽ ☌ ♄
15:06 ☽ ⚹ ♂
16:06 ☿ ∥ ♀
18:07 ☽ ∥ ♃
20:07 ☽ □ ♇
20:25 ♀ ⚹ ♆
30 11:12 ☽ # ♀
12:38 ☽ # ☿
21:00 ☽ ∥ ♆
31 06:29 ☽ ☍ ☿
08:53 ☽ △ ♆
12:14 ☽ ☍ ♀
17:20 ☽ □ ♂
17:35 ♀ ♉
21:51 ☽ ⚹ ♇

DECLINATION

Day	☉	☽	☿	♀	♂	♃	♄	♅	♆	♇
01 Mo	07S43	02N11	13S49	04S29	23N50	08S44	01N11	02S23	13S01	18S15
02 Tu	07 20	04S17	13 13	03 59	23 49	08 39	01 13	02 22	13 00	18 15
03 We	06 57	10 23	12 35	03 29	23 49	08 33	01 14	02 21	13 00	18 15
04 Th	06 34	15 44	11 57	02 58	23 48	08 28	01 16	02 19	12 59	18 15
05 Fr	06 11	20 05	11 16	02 27	23 47	08 22	01 18	02 18	12 58	18 15
06 Sa	05 48	23 14	10 35	01 57	23 46	08 17	01 20	02 17	12 57	18 15
07 Su	05 24	25 04	09 52	01 26	23 45	08 11	01 22	02 15	12 57	18 15
08 Mo	05 01	25 34	09 08	00 55	23 43	08 06	01 24	02 14	12 56	18 15
09 Tu	04 38	24 49	08 23	00 24	23 41	08 00	01 26	02 13	12 55	18 14
10 We	04 14	22 55	07 37	00N07	23 39	07 55	01 28	02 11	12 54	18 14
11 Th	03 51	20 02	06 49	00 37	23 37	07 50	01 29	02 10	12 54	18 14
12 Fr	03 27	16 21	06 00	01 08	23 35	07 44	01 31	02 09	12 53	18 14
13 Sa	03 03	12 03	05 10	01 39	23 33	07 39	01 33	02 07	12 52	18 14
14 Su	02 40	07 17	04 19	02 10	23 30	07 33	01 35	02 06	12 52	18 14
15 Mo	02 16	02 13	03 27	02 41	23 28	07 28	01 37	02 04	12 51	18 14
16 Tu	01 52	02N58	02 34	03 12	23 25	07 22	01 39	02 03	12 50	18 14
17 We	01 29	08 06	01 41	03 42	23 22	07 17	01 41	02 02	12 49	18 14
18 Th	01 05	12 59	00 46	04 13	23 19	07 11	01 43	02 00	12 49	18 14
19 Fr	00 41	17 24	00N09	04 43	23 16	07 06	01 45	01 59	12 48	18 14
20 Sa	00 17	21 05	01 05	05 14	23 12	07 01	01 47	01 58	12 47	18 13
21 Su	00N06	23 47	02 02	05 44	23 09	06 55	01 49	01 56	12 47	18 13
22 Mo	00 30	25 14	02 58	06 14	23 05	06 50	01 51	01 55	12 46	18 13
23 Tu	00 54	25 15	03 55	06 44	23 01	06 44	01 53	01 54	12 45	18 13
24 We	01 17	23 44	04 51	07 14	22 57	06 39	01 54	01 52	12 45	18 13
25 Th	01 41	20 43	05 48	07 44	22 53	06 34	01 56	01 51	12 44	18 13
26 Fr	02 05	16 23	06 44	08 13	22 49	06 28	01 58	01 49	12 44	18 13
27 Sa	02 28	11 00	07 39	08 43	22 45	06 23	02 00	01 48	12 43	18 13
28 Su	02 52	04 57	08 33	09 12	22 40	06 17	02 02	01 47	12 42	18 13
29 Mo	03 15	01S24	09 26	09 41	22 36	06 12	02 04	01 45	12 42	18 13
30 Tu	03 38	07 38	10 17	10 10	22 31	06 07	02 06	01 44	12 41	18 13
31 We	04 02	13 21	11 07	10 38	22 26	06 01	02 08	01 43	12 40	18 13

⚷ Chiron

01	Dec.	06 S 59
03	27♒12	
06	27 25	
09	27 37	
12	27 49	
15	28 01	
18	28 12	
21	28 23	
24	28 34	
27	28 45	
30	28 56	

April 2010

Day	S. T.	☉	☽	☿	♀	♂	♃	♄	♅	♆	♇	☊ True
	h m s	° ' "	° ' "	° '	° '	° '	° '	° '	° '	° '	° '	° '
01 Th	12 36 60	11♈ 09 31	06♏ 40 14	27♈ 31	00♉ 20	02♌ 48	17♓ 15	00♎R31	27♓ 24	27♒ 45	05♑ 24	16♑R56
02 Fr	12 40 56	12 08 43	20 31 21	29 09	01 34	03 02	17 29	00 26	27 28	27 46	05 25	16 46
03 Sa	12 44 53	13 07 53	03♐ 55 01	00♉ 42	02 48	03 16	17 43	00 21	27 31	27 48	05 25	16 39
04 Su	12 48 50	14 07 01	16 51 57	02 10	04 01	03 30	17 56	00 17	27 34	27 50	05 25	16 34
05 Mo	12 52 46	15 06 07	29 25 04	03 34	05 15	03 45	18 10	00 12	27 38	27 52	05 25	16 32
06 Tu	12 56 43	16 05 12	11♑ 38 47	04 52	06 29	04 01	18 24	00 08	27 41	27 53	05 25	16D 32
07 We	13 00 39	17 04 15	23 38 22	06 05	07 43	04 17	18 37	00 03	27 44	27 55	05R 25	16 32
08 Th	13 04 36	18 03 16	05♒ 29 21	07 11	08 57	04 33	18 51	29♍ 59	27 48	27 57	05 25	16R 32
09 Fr	13 08 32	19 02 15	17 17 07	08 12	10 10	04 50	19 04	29 55	27 51	27 58	05 25	16 30
10 Sa	13 12 29	20 01 12	29 06 37	09 07	11 24	05 07	19 18	29 50	27 54	28 00	05 25	16 26
11 Su	13 16 25	21 00 08	11♓ 02 05	09 56	12 38	05 24	19 31	29 46	27 57	28 01	05 25	16 19
12 Mo	13 20 22	21 59 01	23 06 49	10 38	13 51	05 42	19 45	29 42	28 01	28 03	05 25	16 09
13 Tu	13 24 19	22 57 53	05♈ 23 06	11 14	15 05	06 01	19 58	29 38	28 04	28 04	05 25	15 57
14 We	13 28 15	23 56 43	17 52 14	11 44	16 18	06 19	20 11	29 34	28 07	28 06	05 24	15 44
15 Th	13 32 12	24 55 31	00♉ 34 33	12 07	17 32	06 39	20 25	29 30	28 10	28 07	05 24	15 31
16 Fr	13 36 08	25 54 17	13 29 38	12 23	18 45	06 58	20 38	29 26	28 13	28 09	05 24	15 19
17 Sa	13 40 05	26 53 01	26 36 41	12 34	19 59	07 18	20 51	29 22	28 16	28 10	05 24	15 10
18 Su	13 44 01	27 51 43	09♊ 54 47	12 38	21 12	07 38	21 04	29 18	28 19	28 11	05 23	15 04
19 Mo	13 47 58	28 50 23	23 23 19	12R 36	22 25	07 59	21 17	29 14	28 23	28 13	05 23	15 00
20 Tu	13 51 54	29 49 00	07♋ 02 05	12 28	23 39	08 20	21 30	29 10	28 26	28 14	05 22	14 59
21 We	13 55 51	00♉ 47 36	20 51 14	12 15	24 52	08 41	21 43	29 07	28 29	28 15	05 22	14 58
22 Th	13 59 48	01 46 09	04♌ 50 59	11 56	26 05	09 03	21 55	29 03	28 32	28 17	05 22	14 57
23 Fr	14 03 44	02 44 40	19 01 06	11 33	27 18	09 24	22 08	29 00	28 35	28 18	05 21	14 55
24 Sa	14 07 41	03 43 08	03♍ 20 21	11 05	28 31	09 47	22 21	28 56	28 38	28 19	05 21	14 52
25 Su	14 11 37	04 41 35	17 46 02	10 34	29 45	10 09	22 33	28 53	28 41	28 20	05 20	14 45
26 Mo	14 15 34	05 39 59	02♎ 13 54	10 00	00♊ 58	10 32	22 46	28 49	28 43	28 21	05 20	14 36
27 Tu	14 19 30	06 38 22	16 38 23	09 23	02 11	10 55	22 58	28 46	28 46	28 23	05 19	14 26
28 We	14 23 27	07 36 42	00♏ 53 16	08 45	03 23	11 18	23 10	28 43	28 49	28 24	05 18	14 15
29 Th	14 27 23	08 35 01	14 52 44	08 05	04 36	11 42	23 23	28 40	28 52	28 25	05 18	14 04
30 Fr	14 31 20	09 33 17	28 32 17	07 26	05 49	12 06	23 35	28 37	28 55	28 26	05 17	13 55

Data for	04-01-2010
Julian Day	2455287.50
Ayanamsa	24 00 16
SVP	05 ♓ 07 16
☽ ☊ Mean	16 ♑ 52 R

● ◐ PHASES ○ ◑

06	09:37	◑	16♑29
14	12:29	●	24♈27
21	18:21	◐	01♌32
28	12:19	○	08♏07

LAST ASPECT Day	h m	☽ INGRESS Day	h m	
02	12:55	02	16:54	♐
04	20:59	05	01:08	♑
07	08:19	07	12:51	♒
09	21:44	10	01:48	♓
12	12:51	12	13:31	♈
14	19:23	14	22:55	♉
17	04:58	17	06:09	♊
19	10:22	19	11:40	♋
21	14:08	21	15:43	♌
23	15:36	23	18:25	♍
25	18:22	25	20:18	♎
27	19:46	27	22:30	♏
30	00:40	30	02:36	♐

DECLINATION

Day	☉	☽	☿	♀	♂	♃	♄	♅	♆	♇
01 Th	04N25	18S 12	11N54	11N06	22N21	05S 56	02N09	01S 41	12S 40	18S 12
02 Fr	04 48	21 55	12 40	11 34	22 16	05 51	02 11	01 40	12 39	18 12
03 Sa	05 11	24 18	13 24	12 02	22 11	05 46	02 13	01 39	12 39	18 12
04 Su	05 34	25 17	14 04	12 29	22 06	05 40	02 15	01 37	12 38	18 12
05 Mo	05 57	24 56	14 43	12 57	22 01	05 35	02 16	01 36	12 38	18 12
06 Tu	06 20	23 22	15 18	13 23	21 55	05 30	02 18	01 35	12 37	18 12
07 We	06 42	20 46	15 51	13 50	21 50	05 25	02 20	01 34	12 36	18 12
08 Th	07 05	17 19	16 21	14 16	21 44	05 19	02 22	01 32	12 36	18 12
09 Fr	07 27	13 12	16 48	14 42	21 38	05 14	02 23	01 31	12 35	18 12
10 Sa	07 50	08 35	17 11	15 07	21 32	05 09	02 25	01 30	12 35	18 12
11 Su	08 12	03 38	17 32	15 32	21 26	05 04	02 27	01 28	12 34	18 12
12 Mo	08 34	01N30	17 49	15 57	21 20	04 59	02 28	01 27	12 34	18 12
13 Tu	08 56	06 39	18 03	16 21	21 14	04 54	02 30	01 26	12 33	18 12
14 We	09 17	11 37	18 14	16 45	21 07	04 48	02 31	01 25	12 33	18 12
15 Th	09 39	16 11	18 22	17 09	21 01	04 43	02 33	01 23	12 32	18 12
16 Fr	10 00	20 06	18 27	17 32	20 54	04 38	02 34	01 22	12 32	18 12
17 Sa	10 22	23 03	18 28	17 54	20 48	04 33	02 36	01 21	12 31	18 12
18 Su	10 43	24 48	18 26	18 16	20 41	04 28	02 37	01 20	12 31	18 12
19 Mo	11 04	25 08	18 21	18 38	20 34	04 23	02 39	01 19	12 30	18 11
20 Tu	11 24	23 57	18 12	18 59	20 27	04 18	02 40	01 17	12 30	18 11
21 We	11 45	21 18	18 01	19 20	20 20	04 13	02 41	01 16	12 30	18 11
22 Th	12 05	17 23	17 46	19 40	20 12	04 09	02 43	01 15	12 29	18 11
23 Fr	12 25	12 25	17 29	20 00	20 05	04 04	02 44	01 14	12 29	18 11
24 Sa	12 45	06 45	17 10	20 19	19 57	03 59	02 45	01 13	12 28	18 11
25 Su	13 05	00 42	16 48	20 37	19 50	03 54	02 47	01 11	12 28	18 11
26 Mo	13 25	05S 24	16 24	20 55	19 42	03 49	02 48	01 10	12 28	18 11
27 Tu	13 44	11 11	15 58	21 13	19 34	03 44	02 49	01 09	12 27	18 11
28 We	14 03	16 18	15 31	21 30	19 26	03 40	02 50	01 08	12 27	18 11
29 Th	14 22	20 27	15 02	21 46	19 18	03 35	02 51	01 07	12 27	18 11
30 Fr	14 40	23 21	14 34	22 02	19 10	03 30	02 52	01 06	12 26	18 11

ASPECTARIAN

01 00:04 ☽ ∥ ♇
03:21 ☿ ✱ ♆
18:34 ☽ △ ♃
23:32 ☿ ⋕ ♆

02 02:53 ☽ ⋕ ♂
12:23 ☽ △ ♅
12:55 ☽ □ ♆
13:06 ☿ ♉
17:34 ☽ ✱ ♄
22:47 ☽ △ ♂
03 11:25 ♀ □ ♂
18:25 ☽ △ ☉

04 02:04 ☽ □ ♃
05:17 ☉ ⋕ ♃
07:26 ♀ ⋕ ♆
20:31 ☽ □ ♅
20:59 ☽ ✱ ♆
05 01:31 ☽ □ ♄
03:11 ♀ △ ♇
04:17 ☿ □ ♂
09:03 ☽ △ ☿
11:42 ☽ ☌ ♇
12:40 ☽ △ ♀

06 10:46 ☿ △ ♇
13:43 ☽ ✱ ♃
14:48 ☽ ⋕ ♂
07 02:33 ♇ SR
08:19 ☽ ✱ ♅
12:54 ☽ △ ♄
18:14 ☽ ∥ ♇
18:55 ♄ ♍R
22:03 ☽ ☍ ♂

08 03:48 ☽ □ ☿
[illegible]
07:50 ☽ □ ♀
16:23 ☽ ⋕ ♀
09 03:17 ☽ ∥ ♆
03:53 ☽ ✱ ☉
21:44 ☽ ☌ ♆

10 03:30 ☽ ⋕ ☉
12:43 ☽ ✱ ♇
17:04 ☽ ∥ ♃
21:38 ☽ ✱ ☿
11 03:33 ☽ ✱ ♀
05:35 ☽ ⋕ ♄
10:12 ☽ ∥ ♅
17:13 ☽ ☌ ♃
23:47 ☽ ⋕ ♅

12 04:31 ☽ ∥ ♄
09:40 ☽ ☌ ♅
12:51 ☽ ☍ ♄
15:54 ☽ ⋕ ♃
13 00:03 ☽ □ ♇
01:15 ☽ △ ♂
11:40 ☽ ∥ ☉
17:36 ☿ ⋕ ♇

14 04:39 ☽ ⋕ ♆
19:23 ☽ ✱ ♆
15 06:01 ☽ ∥ ♀
09:00 ☽ △ ♇
11:37 ☽ □ ♂
11:42 ☽ ⋕ ♇
13:04 ☽ ∥ ☿
21:56 ☽ ☌ ☿
[illegible]
10:40 ☽ ☌ ♀
13:19 ☽ ✱ ♃
17 02:50 ☽ □ ♆
03:02 ☽ ✱ ♅
04:58 ☽ △ ♄
18:40 ♀ ⋕ ♇
19:49 ☽ ✱ ♂
20:44 ♀ ✱ ♃

18 04:07 ☿ SR
08:16 ☉ ✱ ♆
08:37 ☿ ∥ ♀
20:12 ☽ □ ♃
19 08:32 ☽ △ ♆
08:50 ☽ □ ♅
10:16 ☽ □ ♄
10:22 ☽ ✱ ☉
21:06 ☽ ☍ ♇

20 01:43 ☿ ⋕ ♇
04:30 ☉ ♉
09:20 ☽ ✱ ☿
21 01:30 ☽ △ ♃
06:54 ☽ ∥ ♂
07:34 ☽ ✱ ♀
11:53 ☽ ∥ ♀
13:09 ☽ △ ♅
14:08 ☽ ✱ ♄
19:31 ☽ ⋕ ♇
21:41 ☽ ∥ ☿

22 07:19 ☽ ☌ ♂
11:44 ☽ □ ☿
23:43 ☽ ⋕ ♆
[illegible] ☽ ∥ ☉
23 03:55 ☉ ⋕ ♆
04:38 ♀ ∥ ♂
15:12 ☽ □ ♀
15:36 ☽ ☍ ♆
19:54 ♀ □ ♆

24 00:41 ☽ △ ☉
02:06 ♀ ✱ ♅
03:20 ☽ △ ♇
07:44 ♀ △ ♄
11:13 ☽ ⋕ ♃
12:28 ☽ △ ☿
15:51 ☽ ∥ ♄
22:02 ☽ ⋕ ♅
[illegible]
07:20 ☽ ∥ ♅
08:03 ☽ ☍ ♃
10:42 ☿ □ ♂
13:37 ☽ ⋕ ♄
15:43 ☉ △ ♇
17:46 ☽ ∥ ♃
18:09 ☽ ☍ ♅
18:22 ☽ ☌ ♄
21:42 ☽ △ ♀

26 05:08 ☽ □ ♆
14:11 ☽ ✱ ♂
23:27 ♄ ☍ ♅
27 05:36 ☽ ∥ ♆
[illegible]
19:46 ☽ △ ♆
20:20 ☽ ⋕ ☿
28 07:31 ☽ ✱ ♇
10:08 ☽ ∥ ♇
12:49 ☽ ☍ ☿
16:44 ☿ ☌ ☉
16:57 ☽ ⋕ ♂
18:21 ☽ □ ♂

29 10:33 ☽ ⋕ ♀
15:05 ☽ △ ♃
20:33 ☿ ∥ ☉
23:48 ☽ □ ♆
30 00:08 ☽ ✱ ♄
[illegible]
14:23 ☽ ☍ ♀

⚷ Chiron

01	Dec.	06 S 16
02	29♒06	
05	29 16	
08	29 25	
11	29 34	
14	29 43	
17	29 51	
20	29 59	
23	00♓07	
26	00 14	
29	00 20	

20 07:70 ♓

May 2010

Day	S. T. h m s	☉ ° ' "	☽ ° ' "	☿ ° '	♀ ° '	♂ ° '	♃ ° '	♄ ° '	♅ ° '	♆ ° '	♇ ° '	☊ True ° '
01 Sa	14 35 17	10♉31 33	11♐49 24	06♉R46	07♊02	12♌30	23♓47	28♍R34	28♓58	28♒27	05♑R16	13♑R48
02 Su	14 39 13	11 29 46	24 43 42	06 08	08 15	12 55	23 59	28 31	29 00	28 28	05 16	13 44
03 Mo	14 43 10	12 27 58	07♑16 50	05 32	09 27	13 19	24 11	28 28	29 03	28 29	05 15	13 42
04 Tu	14 47 06	13 26 09	19 32 04	04 58	10 40	13 44	24 23	28 26	29 06	28 30	05 14	13D 42
05 We	14 51 03	14 24 18	01♒33 43	04 27	11 53	14 09	24 35	28 23	29 08	28 30	05 13	13 44
06 Th	14 54 59	15 22 26	13 26 49	03 59	13 05	14 35	24 46	28 21	29 11	28 31	05 12	13 46
07 Fr	14 58 56	16 20 32	25 16 40	03 35	14 18	15 01	24 58	28 18	29 14	28 32	05 12	13R 46
08 Sa	15 02 52	17 18 36	07♓08 29	03 15	15 30	15 26	25 09	28 16	29 16	28 33	05 11	13 45
09 Su	15 06 49	18 16 40	19 07 10	03 00	16 43	15 53	25 21	28 14	29 19	28 34	05 10	13 41
10 Mo	15 10 46	19 14 41	01♈16 50	02 48	17 55	16 19	25 32	28 12	29 21	28 34	05 09	13 35
11 Tu	15 14 42	20 12 42	13 40 45	02 42	19 08	16 45	25 43	28 10	29 24	28 35	05 08	13 27
12 We	15 18 39	21 10 41	26 21 00	02D 40	20 20	17 12	25 55	28 08	29 26	28 36	05 07	13 18
13 Th	15 22 35	22 08 39	09♉18 25	02 42	21 32	17 39	26 06	28 06	29 28	28 36	05 06	13 09
14 Fr	15 26 32	23 06 35	22 32 36	02 49	22 44	18 06	26 17	28 04	29 31	28 37	05 05	13 01
15 Sa	15 30 28	24 04 30	06♊01 59	03 01	23 57	18 34	26 27	28 03	29 33	28 38	05 04	12 55
16 Su	15 34 25	25 02 23	19 44 11	03 18	25 09	19 01	26 38	28 01	29 35	28 38	05 03	12 51
17 Mo	15 38 21	26 00 15	03♋36 31	03 39	26 21	19 29	26 49	28 00	29 38	28 39	05 02	12 48
18 Tu	15 42 18	26 58 05	17 36 22	04 04	27 33	19 57	26 59	27 58	29 40	28 39	05 01	12D 48
19 We	15 46 15	27 55 53	01♌41 29	04 33	28 45	20 25	27 10	27 57	29 42	28 39	04 59	12 48
20 Th	15 50 11	28 53 40	15 50 03	05 06	29 57	20 54	27 20	27 56	29 44	28 40	04 58	12 49
21 Fr	15 54 08	29 51 25	00♍00 27	05 44	01♋09	21 22	27 30	27 55	29 46	28 40	04 57	12R 49
22 Sa	15 58 04	00♊49 08	14 11 01	06 25	02 20	21 51	27 41	27 54	29 48	28 41	04 56	12 47
23 Su	16 02 01	01 46 49	28 19 49	07 10	03 32	22 20	27 51	27 53	29 50	28 41	04 55	12 44
24 Mo	16 05 57	02 44 29	12♎24 25	07 58	04 44	22 49	28 00	27 52	29 52	28 41	04 54	12 39
25 Tu	16 09 54	03 42 08	26 21 56	08 50	05 55	23 18	28 10	27 52	29 54	28 41	04 52	12 33
26 We	16 13 50	04 39 45	10♏09 14	09 46	07 07	23 48	28 20	27 51	29 56	28 42	04 51	12 26
27 Th	16 17 47	05 37 21	23 43 17	10 44	08 18	24 17	28 29	27 51	29 58	28 42	04 50	12 20
28 Fr	16 21 44	06 34 55	07♐01 33	11 46	09 30	24 47	28 39	27 50	30 00	28 42	04 48	12 14
29 Sa	16 25 40	07 32 29	20 02 30	12 51	10 41	25 17	28 48	27 50	00♈02	28 42	04 47	12 10
30 Su	16 29 37	08 30 01	02♑45 51	13 59	11 53	25 47	28 57	27 50	00 03	28 42	04 46	12 08
31 Mo	16 33 33	09 27 32	15 12 34	15 11	13 04	26 17	29 06	27D 50	00 05	28 42	04 44	12D 08

Data for	05-01-2010
Julian Day	2455317.50
Ayanamsa	24 00 19
SVP	05 ♓ 07 11
☽ ☊ Mean	15 ♑ 17 R

● ◐ PHASES ○ ◑

06	04:15	◑	15♒33
14	01:05	●	23♉09
20	23:43	◐	29♌51
27	23:07	○	06♐33

LAST ASPECT Day	h m	☽ INGRESS Day	h m	
02	08:08	02	10:00	♑
04	19:07	04	20:52	♒
07	06:37	07	09:34	♓
09	20:13	09	21:30	♈
12	04:12	12	06:49	♉
14	12:29	14	13:19	♊
16	17:07	16	17:47	♋
18	20:36	18	21:07	♌
20	23:44	20	23:59	♍
23	02:34	23	02:50	♎
25	04:01	25	06:18	♏
27	11:14	27	11:16	♐
29	16:40	29	18:44	♑

DECLINATION

Day	☉	☽	☿	♀	♂	♃	♄	♅	♆	♇
01 Sa	14N59	24S52	14N05	22N17	19N02	03S26	02N53	01S05	12S26	18S11
02 Su	15 17	25 00	13 36	22 31	18 54	03 21	02 54	01 04	12 26	18 11
03 Mo	15 35	23 49	13 07	22 45	18 45	03 17	02 55	01 03	12 25	18 11
04 Tu	15 52	21 30	12 40	22 59	18 37	03 12	02 56	01 02	12 25	18 11
05 We	16 10	18 17	12 14	23 11	18 28	03 07	02 57	01 01	12 25	18 11
06 Th	16 27	14 21	11 49	23 23	18 19	03 03	02 58	01 00	12 24	18 11
07 Fr	16 43	09 54	11 26	23 34	18 10	02 59	02 59	00 59	12 24	18 11
08 Sa	17 00	05 05	11 05	23 45	18 01	02 54	02 59	00 58	12 24	18 11
09 Su	17 16	00 02	10 47	23 55	17 52	02 50	03 00	00 57	12 24	18 11
10 Mo	17 32	05N05	10 30	24 04	17 43	02 45	03 01	00 56	12 23	18 11
11 Tu	17 48	10 05	10 16	24 13	17 34	02 41	03 01	00 55	12 23	18 11
12 We	18 03	14 47	10 05	24 21	17 24	02 37	03 02	00 54	12 23	18 11
13 Th	18 18	18 55	09 56	24 28	17 15	02 33	03 02	00 53	12 23	18 11
14 Fr	18 33	22 12	09 49	24 35	17 05	02 29	03 03	00 52	12 23	18 11
15 Sa	18 47	24 20	09 45	24 41	16 55	02 24	03 03	00 51	12 22	18 11
16 Su	19 02	25 03	09 44	24 46	16 46	02 20	03 04	00 50	12 22	18 11
17 Mo	19 15	24 12	09 45	24 50	16 36	02 16	03 04	00 49	12 22	18 11
18 Tu	19 29	21 51	09 48	24 54	16 26	02 12	03 05	00 48	12 22	18 11
19 We	19 42	18 09	09 53	24 57	16 15	02 08	03 05	00 47	12 22	18 11
20 Th	19 55	13 24	10 00	24 59	16 05	02 04	03 05	00 47	12 22	18 12
21 Fr	20 07	07 55	10 10	25 01	15 55	02 01	03 05	00 46	12 22	18 12
22 Sa	20 19	02 02	10 21	25 02	15 44	01 57	03 06	00 45	12 22	18 12
23 Su	20 31	03S56	10 35	25 02	15 34	01 53	03 06	00 44	12 22	18 12
24 Mo	20 42	09 39	10 50	25 02	15 23	01 49	03 06	00 44	12 21	18 12
25 Tu	20 53	14 50	11 07	25 00	15 12	01 46	03 06	00 43	12 21	18 12
26 We	21 04	19 12	11 25	24 58	15 02	01 42	03 06	00 42	12 21	18 12
27 Th	21 15	22 28	11 45	24 56	14 51	01 38	03 06	00 41	12 21	18 12
28 Fr	21 24	24 26	12 06	24 52	14 40	01 35	03 06	00 41	12 21	18 12
29 Sa	21 34	25 01	12 29	24 48	14 28	01 31	03 06	00 40	12 21	18 12
30 Su	21 43	24 15	12 53	24 43	14 17	01 28	03 06	00 39	12 21	18 12
31 Mo	21 52	22 17	13 18	24 38	14 06	01 24	03 05	00 39	12 21	18 12

ASPECTARIAN

01 01:17 ☽ △ ♂
22:34 ☽ □ ♃
02 07:05 ☽ ✶ ♆
07:09 ☽ □ ♄
08:08 ☽ □ ♅
20:04 ☽ ☌ ♇
20:46 ☽ △ ☿
03 10:58 ☽ △ ☉
11:15 ☽ # ♀
12:07 ☿ △ ♇
04 09:47 ☽ ✶ ♃
13:09 ☉ □ ♂
13:39 ☿ # ♇
17:39 ☽ △ ♄
19:07 ☽ ✶ ♅
22:43 ☽ # ♂
05 00:38 ☽ ∥ ♇
05:35 ☽ □ ☿
12:31 ☽ # ☉
23:12 ☽ △ ♀
06 02:23 ☽ ☍ ♂
10:48 ☽ ∥ ♆
15:16 ☽ # ☿
21:31 ♂ # ♇
07 00:00 ♃ # ♄
06:37 ☽ ☌ ♆
16:21 ☽ ✶ ☿
20:03 ☽ ✶ ♇
21:55 ♀ ✶ ♂
08 10:00 ☽ # ♄
10:35 ☽ ∥ ♃
18:40 ☽ □ ♀
19:42 ☽ ∥ ♅
22:11 ☽ ✶ ☉
09 04:34 ☽ # ♅
12:32 ☽ ☌ ♃
13:15 ☽ # ♃
14:16 ☽ ∥ ♄
17:58 ☽ ☍ ♄
20:13 ☽ ☌ ♅
10 07:32 ☽ □ ♇
10:23 ☉ ∥ ♂
11 00:52 ☽ ∥ ☿
06:06 ☽ △ ♂
11:26 ☽ # ♆
11:29 ☽ ✶ ♀
22:28 ☿ SD
12 04:12 ☽ ✶ ♆
11:47 ☽ ☌ ☿
12:53 ☉ # ♇
14:06 ☽ ∥ ♂
16:16 ☽ △ ♇
19:21 ☽ # ♇
19:47 ☽ ∥ ☉
13 15:43 ☽ □ ♂
14 06:47 ☽ ✶ ♃
09:52 ☽ △ ♄
10:52 ☽ □ ♆
12:29 ☽ ✶ ♅
15 07:02 ☽ ∥ ♀
22:43 ☽ ✶ ♂
16 10:17 ☽ ☌ ♀
12:08 ☽ □ ♃
12:34 ☽ ∥ ♀
14:20 ☽ □ ♄
15:26 ☽ △ ♆
17:07 ☽ □ ♅
17 00:04 ☽ ✶ ☿
02:26 ☽ ☍ ♇
10:57 ♀ □ ♃
18 00:42 ☉ ✶ ♃
08:19 ♀ □ ♄
15:20 ☽ ∥ ☉
16:12 ☽ △ ♃
17:08 ☽ ✶ ☉
17:39 ☽ ✶ ♄
20:36 ☽ △ ♅
22:13 ♀ △ ♆
23:45 ☽ # ♇
19 00:27 ☉ △ ♄
05:03 ☽ □ ☿
10:28 ☽ ∥ ♂
18:15 ☉ □ ♆
18:44 ☿ △ ♇
19:41 ♀ □ ♅
20 01:05 ♀ ♋
04:43 ☽ # ♆
08:52 ☽ ☌ ♂
14:43 ☽ ∥ ☿
21:44 ☽ ☍ ♆
21:48 ☉ ✶ ♅
21 02:06 ☽ ✶ ♀
03:34 ☉ ♊
08:21 ☽ △ ♇
10:10 ☽ △ ☿
19:44 ☽ ∥ ♄
22 00:22 ☽ # ♃
05:10 ☽ # ♅
11:09 ☽ ∥ ♅
15:49 ☽ ∥ ♃
20:36 ☽ # ♄
23:10 ☽ ☍ ♃
23:14 ☽ ☌ ♄
23 02:34 ☽ ☍ ♅
05:39 ♃ ☍ ♄
06:18 ☽ △ ☉
09:41 ☽ □ ♀
11:11 ☽ □ ♇
24 03:14 ♀ ☍ ♇
05:27 ☽ # ☿
12:07 ☽ ∥ ♆
18:31 ☽ ✶ ♂
25 01:46 ☽ # ♂
04:01 ☽ △ ♆
14:44 ☽ ✶ ♇
17:58 ☽ ∥ ♇
18:11 ☽ △ ♀
23:16 ☽ ☍ ☿
26 13:22 ☽ # ☉
27 01:03 ☽ □ ♂
07:23 ☽ ✶ ♄
08:39 ☽ △ ♃
08:55 ☽ □ ♆
11:14 ☽ △ ♅
28 01:50 ♅ ♈
09:50 ☽ # ♀
16:04 ☿ # ♆
29 10:13 ☽ △ ♂
13:22 ☽ # ♀
14:38 ☽ □ ♄
16:17 ☽ ✶ ♆
16:40 ☽ □ ♃
18:50 ☽ □ ♅
30 03:49 ☽ ☌ ♇
18:10 ♄ SD
19:23 ☽ ☍ ♀
23:56 ☽ △ ☿
31 03:48 ☽ # ☉
18:50 ♆ SR

⚷ Chiron

01 Dec.	05 S 41
02	00♓26
05	00 32
08	00 37
11	00 42
14	00 46
17	00 49
20	00 52
23	00 55
26	00 57
29	00 58

June 2010

26 11:39 04♑46 ✹ Total Lunar Eclipse (mag 0.541)

Day	S. T.	☉	☽	☿	♀	♂	♃	♄	♅	♆	♇	☊ True
	h m s	° ' "	° ' "	° '	° '	° '	° '	° '	° '	° '	° '	° '
01 Tu	16 37 30	10♊ 25 03	27♑ 24 49	16♉ 25	14♋ 15	26♌ 47	29♓ 15	27♍ 50	00♈ 07	28♒R42	04♑R43	12♑ 10
02 We	16 41 26	11 22 32	09♒ 25 48	17 41	15 26	27 18	29 24	27 50	00 08	28 42	04 42	12 13
03 Th	16 45 23	12 20 01	21 19 28	19 01	16 37	27 48	29 33	27 50	00 10	28 42	04 40	12 16
04 Fr	16 49 19	13 17 29	03♓ 10 19	20 24	17 48	28 19	29 41	27 51	00 11	28 42	04 39	12 18
05 Sa	16 53 16	14 14 56	15 03 12	21 49	18 59	28 50	29 50	27 51	00 13	28 42	04 38	12 19
06 Su	16 57 13	15 12 23	27 03 01	23 17	20 10	29 21	29 58	27 52	00 14	28 42	04 36	12R 20
07 Mo	17 01 09	16 09 48	09♈ 14 23	24 48	21 21	29 52	00♈ 06	27 53	00 16	28 42	04 35	12 18
08 Tu	17 05 06	17 07 13	21 41 23	26 21	22 31	00♍ 23	00 14	27 53	00 17	28 41	04 33	12 16
09 We	17 09 02	18 04 38	04♉ 27 09	27 57	23 42	00 55	00 22	27 54	00 18	28 41	04 32	12 13
10 Th	17 12 59	19 02 02	17 33 41	29 36	24 53	01 26	00 29	27 55	00 20	28 41	04 30	12 10
11 Fr	17 16 55	19 59 25	01♊ 01 25	01♊ 18	26 03	01 58	00 37	27 56	00 21	28 40	04 29	12 06
12 Sa	17 20 52	20 56 48	14 49 10	03 02	27 14	02 30	00 44	27 58	00 22	28 40	04 27	12 04
13 Su	17 24 48	21 54 10	28 54 09	04 48	28 24	03 02	00 52	27 59	00 23	28 40	04 26	12 02
14 Mo	17 28 45	22 51 32	13♋ 12 12	06 38	29 34	03 34	00 59	28 00	00 24	28 39	04 24	12 02
15 Tu	17 32 42	23 48 52	27 38 24	08 30	00♌ 44	04 06	01 06	28 02	00 25	28 39	04 23	12D 02
16 We	17 36 38	24 46 12	12♌ 07 38	10 24	01 54	04 38	01 13	28 03	00 26	28 38	04 21	12 02
17 Th	17 40 35	25 43 31	26 35 10	12 20	03 05	05 11	01 19	28 05	00 27	28 38	04 20	12 03
18 Fr	17 44 31	26 40 49	10♍ 57 01	14 19	04 14	05 43	01 26	28 07	00 28	28 37	04 18	12R 03
19 Sa	17 48 28	27 38 06	25 10 09	16 20	05 24	06 16	01 32	28 09	00 29	28 37	04 17	12 02
20 Su	17 52 24	28 35 22	09♎ 12 19	18 24	06 34	06 49	01 38	28 11	00 30	28 36	04 15	12 01
21 Mo	17 56 21	29 32 37	23 02 06	20 29	07 44	07 22	01 44	28 13	00 30	28 36	04 14	12 00
22 Tu	18 00 17	00♋ 29 52	06♏ 38 37	22 35	08 53	07 55	01 50	28 15	00 31	28 35	04 12	11 58
23 We	18 04 14	01 27 06	20 01 21	24 43	10 03	08 28	01 56	28 17	00 32	28 34	04 11	11 56
24 Th	18 08 11	02 24 19	03♐ 10 04	26 52	11 12	09 01	02 01	28 20	00 32	28 34	04 09	11 54
25 Fr	18 12 07	03 21 32	16 04 45	29 03	12 21	09 34	02 07	28 22	00 33	28 33	04 08	11 53
26 Sa	18 16 04	04 18 45	28 45 40	01♋ 13	13 31	10 08	02 12	28 25	00 33	28 32	04 06	11 52
27 Su	18 20 00	05 15 57	11♑ 13 28	03 25	14 40	10 41	02 17	28 27	00 34	28 31	04 05	11D 52
28 Mo	18 23 57	06 13 10	23 29 20	05 36	15 49	11 15	02 22	28 30	00 34	28 30	04 03	11 53
29 Tu	18 27 53	07 10 21	05♒ 34 57	07 47	16 58	11 48	02 27	28 33	00 34	28 30	04 02	11 54
30 We	18 31 50	08 07 33	17 32 36	09 57	18 06	12 22	02 31	28 36	00 35	28 29	04 00	11 55

Data for	06-01-2010
Julian Day	2455348.50
Ayanamsa	24 00 24
SVP	05 ♓ 07 08
☽ ☊ Mean	13 ♑ 38 R

● ◐ PHASES ○ ◑

04	22:14	◑	14♓11
12	11:15	●	21♊24
19	04:29	◐	27♍49
26	11:31	✹	04♑46

LAST ASPECT Day	h m	☽ INGRESS Day	h m	
01	03:42	01	05:08	♒
03	14:56	03	17:34	♓
06	05:50	06	05:51	♈
08	13:14	08	15:42	♉
10	19:51	10	22:12	♊
12	23:36	13	01:51	♋
15	00:39	15	03:55	♌
17	03:24	17	05:41	♍
19	05:04	19	08:13	♎
21	09:45	21	12:14	♏
23	15:33	23	18:11	♐
25	23:34	26	02:22	♑
28	09:57	28	12:53	♒

DECLINATION

Day	☉	☽	☿	♀	♂	♃	♄	♅	♆	♇
01 Tu	22N01	19S 19	13N44	24N32	13N55	01S 21	03N05	00S 38	12S 21	18S 12
02 We	22 09	15 34	14 11	24 25	13 43	01 18	03 05	00 37	12 21	18 12
03 Th	22 16	11 15	14 39	24 17	13 31	01 15	03 05	00 37	12 21	18 12
04 Fr	22 24	06 32	15 07	24 09	13 20	01 11	03 04	00 36	12 21	18 12
05 Sa	22 31	01 35	15 37	24 00	13 08	01 08	03 04	00 36	12 21	18 13
06 Su	22 37	03N28	16 06	23 51	12 56	01 05	03 03	00 35	12 22	18 13
07 Mo	22 43	08 28	16 37	23 41	12 44	01 02	03 03	00 35	12 22	18 13
08 Tu	22 49	13 14	17 07	23 30	12 32	00 59	03 02	00 34	12 22	18 13
09 We	22 54	17 33	17 38	23 18	12 20	00 56	03 02	00 34	12 22	18 13
10 Th	22 59	21 10	18 09	23 06	12 08	00 54	03 01	00 33	12 22	18 13
11 Fr	23 04	23 44	18 40	22 54	11 55	00 51	03 01	00 33	12 22	18 13
12 Sa	23 08	24 57	19 10	22 40	11 43	00 48	03 00	00 32	12 22	18 13
13 Su	23 11	24 37	19 40	22 26	11 30	00 46	02 59	00 32	12 22	18 13
14 Mo	23 15	22 40	20 10	22 12	11 18	00 43	02 59	00 32	12 23	18 14
15 Tu	23 18	19 14	20 39	21 57	11 05	00 40	02 58	00 31	12 23	18 14
16 We	23 20	14 37	21 07	21 41	10 52	00 38	02 57	00 31	12 23	18 14
17 Th	23 22	09 10	21 34	21 25	10 40	00 36	02 56	00 30	12 23	18 14
18 Fr	23 24	03 16	22 00	21 08	10 27	00 33	02 55	00 30	12 23	18 14
19 Sa	23 25	02S 43	22 25	20 51	10 14	00 31	02 54	00 30	12 24	18 14
20 Su	23 26	08 30	22 47	20 33	10 01	00 29	02 53	00 30	12 24	18 14
21 Mo	23 26	13 46	23 08	20 15	09 48	00 27	02 52	00 29	12 24	18 14
22 Tu	23 26	18 16	23 27	19 56	09 34	00 25	02 51	00 29	12 24	18 15
23 We	23 26	21 45	23 44	19 36	09 21	00 22	02 50	00 29	12 25	18 15
24 Th	23 25	24 02	23 58	19 17	09 08	00 21	02 49	00 29	12 25	18 15
25 Fr	23 24	25 00	24 10	18 56	08 54	00 19	02 48	00 28	12 25	18 15
26 Sa	23 22	24 38	24 19	18 36	08 41	00 17	02 46	00 28	12 25	18 15
27 Su	23 20	23 02	24 25	18 14	08 27	00 15	02 45	00 28	12 26	18 15
28 Mo	23 18	20 21	24 29	17 53	08 13	00 13	02 44	00 28	12 26	18 15
29 Tu	23 15	16 48	24 30	17 31	08 00	00 12	02 43	00 28	12 26	18 16
30 We	23 11	12 38	24 27	17 08	07 46	00 10	02 41	00 28	12 27	18 16

ASPECTARIAN

01 00:50 ☽ △ ♄
03:42 ☽ ⚹ ♃
05:22 ☽ ⚹ ♅
06:42 ☿ ∥ ♂
07:42 ☽ ∥ ♇
02 04:15 ☽ △ ☉
07:18 ☽ # ☿
11:12 ☽ # ♂
18:10 ☽ ∥ ♆
18:44 ☽ □ ☿
03 13:43 ☽ ☍ ♂
14:56 ☽ ☌ ♆
04 02:59 ☽ ⚹ ♇
16:56 ☽ # ♄
17:53 ♂ ☍ ♆
05 02:11 ☽ ∥ ♃
04:46 ☽ ∥ ♅
08:46 ☽ △ ♀
10:23 ☽ # ♅
12:51 ☽ # ♃
15:27 ☽ ⚹ ☿
22:05 ☽ ∥ ♄
06 01:37 ☽ ☍ ♄
05:50 ☽ ☌ ♃
06:20 ☽ ☌ ♅
06:28 ♃ ♈
14:54 ☽ □ ♇
07 06:11 ♂ ♍
14:32 ☽ ⚹ ☉
19:30 ☽ # ♆
20:32 ☽ ∥ ♂
08 01:45 ☽ □ ♀
11:26 ♃ ☌ ♅
13:14 ☽ ⚹ ♆
17:07 ☽ △ ♂
[illegible]
23:14 ☿ △ ♄
09 00:09 ☽ △ ♇
00:31 ☽ ∥ ☿
04:01 ☽ # ♇
10:39 ☿ □ ♆
10 03:25 ☿ # ♇
05:41 ☿ ♊
10:06 ♀ ∥ ☉
10:28 ☿ ⚹ ♅
13:42 ☿ ⚹ ♃
14:22 ☽ ⚹ ♀
15:35 ☽ ∥ ♀
16:16 ☽ ∥ ☉
18:33 ☽ △ ♄
19:51 ☽ □ ♆
22:48 ☽ ⚹ ♅
23:17 ☽ ⚹ ♃
11 00:33 ☽ ☌ ☿
01:43 ☽ □ ♂
13:28 ☿ □ ♂
12 15:19 ♀ ⚹ ♄
22:26 ☽ □ ♄
23:36 ☽ △ ♆
13 02:30 ☽ □ ♅
03:20 ☽ □ ♃
07:13 ☽ ⚹ ♂
09:18 ☽ ☍ ♇
18:43 ☽ ∥ ☉
14 04:19 ☽ ∥ ♀
08:50 ♀ ♌
16:14 ☽ ∥ ☿
17:20 ♀ △ ♅
15 00:39 ☽ ⚹ ♄
04:37 ☽ △ ♅
05:35 ☽ ☌ ♀
05:44 ☽ # ♇
05:46 ☽ △ ♃
08:08 ♀ △ ♃
12:09 ♂ △ ♇
20:42 ☽ ⚹ ☿
16 10:11 ☽ # ♆
17:26 ☽ ∥ ♂
18:48 ☿ ∥ ♀
22:28 ☽ ⚹ ☉
17 03:24 ☽ ☍ ♆
12:54 ☽ △ ♇
14:54 ☽ ☌ ♂
18 01:27 ☽ ∥ ♄
06:36 ☽ □ ☿
10:58 ☽ # ♃
11:06 ☽ # ♅
15:06 ☽ ∥ ♅
15:14 ☽ ∥ ♃
19 00:47 ☽ # ♄
05:04 ☽ ☌ ♄
09:03 ☽ ☍ ♅
10:55 ☽ ☍ ♃
13:18 ☉ □ ♄
13:20 ♃ ∥ ♅
15:31 ☽ □ ♇
19:03 ☽ ⚹ ♀
20 00:22 ☉ △ ♆
06:26 ☽ # ♂
17:33 ☽ ∥ ♆
18:44 ☽ △ ☿
21 09:45 ☽ △ ♆
11:29 ☉ ♋
12:17 ☽ △ ☉
19:41 ☽ ⚹ ♇
22:56 ☿ ∥ ☉
23:57 ☽ ∥ ♇
22 00:27 ☉ □ ♅
02:21 ☽ ⚹ ♂
04:23 ☽ □ ♀
09:43 ☽ # ♀
23 13:21 ☉ □ ♃
15:05 ☽ ⚹ ♄
15:33 ☽ □ ♆
16:05 ☽ # ☉
19:09 ☽ △ ♅
21:53 ☽ △ ♃
22:52 ☽ # ☿
24 11:18 ☽ □ ♂
16:21 ☽ △ ♀
16:24 ☿ □ ♄
18:33 ☿ △ ♆
25 10:32 ☿ ♋
16:36 ☿ □ ♅
18:51 ☉ ☍ ♇
23:20 ☽ □ ♄
23:34 ☽ ⚹ ♆
26 03:26 ☽ □ ♅
05:42 ☽ ☍ ☿
06:02 ☽ # ☿
06:37 ☽ □ ♃
10:13 ☽ ☌ ♇
11:09 ☿ □ ♃
20:27 ☽ # ☉
22:54 ☽ △ ♂
22:57 ♀ # ♇
27 07:14 ☿ ☍ ♇
28 09:57 ☽ △ ♄
12:07 ☿ ☌ ☉
14:01 ☽ ⚹ ♅
14:45 ☽ ∥ ♇
17:42 ☽ ⚹ ♃
19:09 ☽ # ♀
30 01:03 ☽ ∥ ♆
01:15 ☽ ☍ ♀
22:04 ☽ ☌ ♆

⚷ Chiron

01 Dec. 05 S 19

01	00♓59
04	00 59
07	00 59R
10	00 58
13	00 57
16	00 55
19	00 53
22	00 50
25	00 47
28	00 43

04 05:19 00♓59 R

11 19:52 19♋24 ● Total Solar Eclipse

July 2010

Day	S. T.	☉	☽	☿	♀	♂	♃	♄	♅	♆	♇	☊ True
	h m s	° ' "	° ' "	° '	° '	° '	° '	° '	° '	° '	° '	° '
01 Th	18 35 47	09♋ 04 45	29♒ 25 07	12♋ 07	19♌ 15	12♍ 56	02♈ 35	28♍ 39	00♈ 35	28♒R28	03♑R58	11♑ 56
02 Fr	18 39 43	10 01 57	11♓ 15 55	14 16	20 24	13 30	02 40	28 42	00 35	28 27	03 57	11 57
03 Sa	18 43 40	10 59 08	23 08 51	16 24	21 32	14 04	02 44	28 45	00 35	28 26	03 55	11 58
04 Su	18 47 36	11 56 21	05♈ 08 15	18 31	22 40	14 39	02 47	28 48	00 35	28 25	03 54	11 59
05 Mo	18 51 33	12 53 33	17 18 37	20 36	23 48	15 13	02 51	28 52	00 35	28 24	03 52	12 00
06 Tu	18 55 29	13 50 45	29 44 21	22 39	24 57	15 47	02 54	28 55	00R 35	28 23	03 51	12 01
07 We	18 59 26	14 47 58	12♉ 29 26	24 41	26 04	16 22	02 58	28 59	00 35	28 22	03 49	12 01
08 Th	19 03 22	15 45 11	25 37 04	26 41	27 12	16 56	03 01	29 02	00 35	28 21	03 48	12 02
09 Fr	19 07 19	16 42 25	09♊ 09 08	28 40	28 20	17 31	03 04	29 06	00 35	28 20	03 46	12 02
10 Sa	19 11 16	17 39 39	23 05 49	00♌ 36	29 28	18 06	03 06	29 10	00 35	28 19	03 45	12 03
11 Su	19 15 12	18 36 53	07♋ 25 09	02 31	00♍ 35	18 41	03 09	29 14	00 35	28 17	03 43	12R 03
12 Mo	19 19 09	19 34 08	22 02 52	04 24	01 42	19 16	03 11	29 18	00 35	28 16	03 42	12 02
13 Tu	19 23 05	20 31 22	06♌ 52 41	06 15	02 50	19 51	03 13	29 22	00 34	28 15	03 41	12 01
14 We	19 27 02	21 28 37	21 47 00	08 04	03 57	20 26	03 15	29 26	00 34	28 14	03 39	11 59
15 Th	19 30 58	22 25 52	06♍ 37 58	09 51	05 03	21 01	03 17	29 30	00 33	28 13	03 38	11 57
16 Fr	19 34 55	23 23 07	21 18 34	11 36	06 10	21 37	03 19	29 34	00 33	28 11	03 36	11 55
17 Sa	19 38 51	24 20 22	05♎ 43 23	13 19	07 17	22 12	03 20	29 39	00 32	28 10	03 35	11 53
18 Su	19 42 48	25 17 37	19 49 05	15 00	08 23	22 48	03 21	29 43	00 32	28 09	03 33	11 52
19 Mo	19 46 45	26 14 52	03♏ 34 18	16 40	09 29	23 23	03 22	29 48	00 31	28 08	03 32	11 51
20 Tu	19 50 41	27 12 07	16 59 21	18 17	10 35	23 59	03 23	29 52	00 31	28 06	03 31	11D 51
21 We	19 54 38	28 09 23	00♐ 05 40	19 53	11 41	24 35	03 24	29 57	00 30	28 05	03 29	11 52
22 Th	19 58 34	29 06 39	12 55 18	21 26	12 47	25 11	03 24	00♎ 02	00 29	28 04	03 28	11 53
23 Fr	20 02 31	00♌ 03 55	25 30 30	22 58	13 53	25 47	03 24	00 07	00 28	28 02	03 27	11 55
24 Sa	20 06 27	01 01 12	07♑ 53 27	24 28	14 58	26 23	03R 24	00 11	00 28	28 01	03 25	11 56
25 Su	20 10 24	01 58 29	20 06 14	25 56	16 03	26 59	03 24	00 16	00 27	27 59	03 24	11R 56
26 Mo	20 14 20	02 55 46	02♒ 10 45	27 22	17 08	27 35	03 24	00 21	00 26	27 58	03 23	11 56
27 Tu	20 18 17	03 53 05	14 08 53	28 46	18 13	28 11	03 23	00 27	00 25	27 56	03 21	11 54
28 We	20 22 14	04 50 23	26 02 32	00♍ 08	19 17	28 47	03 22	00 32	00 24	27 55	03 20	11 51
29 Th	20 26 10	05 47 43	07♓ 53 44	01 28	20 22	29 24	03 21	00 37	00 23	27 53	03 19	11 48
30 Fr	20 30 07	06 45 04	19 44 47	02 45	21 26	00♎ 00	03 20	00 42	00 22	27 52	03 18	11 44
31 Sa	20 34 03	07 42 25	01♈ 38 28	04 01	22 30	00 37	03 19	00 48	00 20	27 50	03 16	11 41

Data for	07-01-2010
Julian Day	2455378.50
Ayanamsa	24 00 29
SVP	05 ♓ 07 05
☽ ☊ Mean	12 ♑ 03 R

● ◐ PHASES ○ ◑

04	14:36	◑	12♈31
11	19:40	●	19♋24
18	10:11	◐	25♎42
26	01:37	○	03♒00

LAST ASPECT Day	h m	☽ INGRESS Day	h m	
30	22:04	01	01:11	♓
03	11:18	03	13:45	♈
05	21:25	06	00:30	♉
08	06:10	08	07:51	♊
10	10:17	10	11:38	♋
12	11:49	12	12:54	♌
14	10:23	14	13:15	♍
16	13:46	16	14:25	♎
18	14:26	18	17:43	♏
20	23:44	20	23:49	♐
23	04:52	23	08:40	♑
25	14:21	25	19:39	♒
28	03:47	28	08:00	♓
30	03:44	30	20:42	♈

DECLINATION

Day	☉	☽	☿	♀	♂	♃	♄	♅	♆	♇
01 Th	23N08	08S 00	24N23	16N45	07N32	00S 09	02N40	00S 28	12S 27	18S 16
02 Fr	23 04	03 07	24 15	16 22	07 18	00 07	02 38	00 28	12 27	18 16
03 Sa	22 59	01N53	24 05	15 58	07 04	00 06	02 37	00 28	12 28	18 16
04 Su	22 54	06 52	23 52	15 34	06 50	00 05	02 36	00 28	12 28	18 16
05 Mo	22 49	11 39	23 37	15 10	06 36	00 04	02 34	00 28	12 28	18 17
06 Tu	22 43	16 05	23 20	14 45	06 22	00 03	02 32	00 28	12 29	18 17
07 We	22 37	19 55	23 00	14 20	06 08	00 02	02 31	00 28	12 29	18 17
08 Th	22 31	22 53	22 38	13 55	05 53	00 01	02 29	00 28	12 30	18 17
09 Fr	22 24	24 40	22 15	13 29	05 39	00N00	02 28	00 28	12 30	18 17
10 Sa	22 16	24 59	21 50	13 03	05 25	00 01	02 26	00 28	12 30	18 17
11 Su	22 09	23 39	21 23	12 37	05 10	00 02	02 24	00 28	12 31	18 18
12 Mo	22 01	20 43	20 55	12 10	04 56	00 02	02 22	00 28	12 31	18 18
13 Tu	21 52	16 24	20 25	11 44	04 41	00 03	02 21	00 28	12 32	18 18
14 We	21 43	11 02	19 54	11 17	04 26	00 03	02 19	00 29	12 32	18 18
15 Th	21 34	05 04	19 22	10 49	04 12	00 04	02 17	00 29	12 33	18 18
16 Fr	21 25	01S 06	18 49	10 22	03 57	00 04	02 15	00 29	12 33	18 19
17 Sa	21 15	07 05	18 16	09 54	03 42	00 04	02 13	00 29	12 33	18 19
18 Su	21 05	12 35	17 41	09 26	03 28	00 05	02 11	00 30	12 34	18 19
19 Mo	20 54	17 19	17 06	08 58	03 13	00 05	02 09	00 30	12 34	18 19
20 Tu	20 43	21 03	16 31	08 30	02 58	00 05	02 07	00 30	12 35	18 19
21 We	20 32	23 37	15 54	08 01	02 43	00 05	02 05	00 30	12 35	18 20
22 Th	20 20	24 54	15 18	07 33	02 28	00 05	02 03	00 31	12 36	18 20
23 Fr	20 08	24 52	14 41	07 04	02 13	00 04	02 01	00 31	12 36	18 20
24 Sa	19 56	23 35	14 04	06 35	01 58	00 04	01 59	00 31	12 37	18 20
25 Su	19 43	21 12	13 27	06 06	01 43	00 04	01 57	00 32	12 37	18 20
26 Mo	19 30	17 54	12 49	05 37	01 27	00 03	01 55	00 32	12 38	18 21
27 Tu	19 17	13 54	12 12	05 08	01 12	00 03	01 53	00 33	12 38	18 21
28 We	19 03	09 24	11 35	04 39	00 57	00 02	01 51	00 33	12 39	18 21
29 Th	18 49	04 35	10 58	04 09	00 42	00 02	01 48	00 34	12 39	18 21
30 Fr	18 35	00N24	10 21	03 40	00 27	00 01	01 46	00 34	12 40	18 22
31 Sa	18 21	05 22	09 45	03 10	00 11	00 00	01 44	00 34	12 40	18 22

⚷ Chiron

01 Dec.	05 S 17
01	00♓39R
04	00 34
07	00 28
10	00 23
13	00 17
16	00 10
19	00 03
22	29♒56
25	29 49
28	29 41
31	29 33
20	08:38 ♒ R

ASPECTARIAN

01	02:32	☽ # ♂
	09:13	☽ ⚹ ♇
	12:19	☿ ⚹ ♂
	21:17	☽ △ ☉
02	02:21	☽ # ♄
	04:46	☽ ☍ ♂
	07:25	☽ △ ☿
	12:48	☽ ∥ ♅
	14:29	☽ ∥ ♃
	15:32	☽ # ♃
	17:13	☽ # ♅
03	03:31	☽ ∥ ♄
	11:18	☽ ☍ ♄
	14:56	☽ ☌ ♅
	19:18	☽ ☌ ♃
	21:32	☽ □ ♇
	23:54	☽ ∥ ♂
05	04:15	☽ # ♆
	07:41	☽ □ ☿
	13:53	☽ △ ♀
	16:50	♅ SR
	17:09	☽ ∥ ♀
	21:25	☽ ⚹ ♆
06	07:47	☽ △ ♇
	13:09	☽ # ♇
07	04:37	☽ ⚹ ☉
	07:29	☽ △ ♂
	20:35	☽ ∥ ☉
	21:58	☽ ∥ ☿
08	02:16	☽ ⚹ ☿
	03:07	☽ □ ♀
	04:54	☽ □ ♆
	06:10	☽ △ ♄
	08:54	☽ ⚹ ♅
	11:32	☿ ∥ ☉
	13:15	☽ ⚹ ♃
	23:55	♀ ☍ ♆
09	05:35	☿ ⚹ ♄
	15:06	☽ □ ♂
	16:30	☿ ♌
	23:45	☿ △ ♅
10	08:48	☽ △ ♆
	10:17	☽ □ ♄
	11:32	♀ ♍
	11:39	☽ ⚹ ♀
	12:37	☽ □ ♅
	16:53	☽ □ ♃
	17:52	☽ ☍ ♇
11	04:13	☉ ⚹ ♂
	05:31	♀ # ♆
	08:13	☿ △ ♃
	14:30	☽ ∥ ☉
	19:17	☽ ⚹ ♂
	22:35	☽ ∥ ☿
12	11:49	☽ ⚹ ♄
	13:49	☽ △ ♅
	14:15	☽ # ♇
	18:05	☽ △ ♃
	22:50	☽ ☌ ☿
13	17:39	☽ # ♆
	17:54	♀ △ ♇
	22:55	☽ ∥ ♀
14	10:23	☽ ☍ ♆
	19:08	☽ △ ♇
	21:14	☽ ☌ ♀
15	03:34	☽ ∥ ♂
	10:55	☽ ∥ ♄
	17:51	☽ # ♅
	19:28	☽ ∥ ♃
	20:00	☽ # ♃
	21:37	☽ ∥ ♅
16	00:31	☽ ☌ ♂
	03:40	☽ ⚹ ☉
	04:32	☽ # ♄
	10:50	☽ # ♂
	13:46	☽ ☌ ♄
	15:19	☽ ☍ ♅
	19:59	☽ ☍ ♃
	20:24	☽ □ ♇
	21:51	☿ # ♇
17	11:02	☽ # ♀
	14:37	☽ ⚹ ☿
	23:56	☽ ∥ ♆
18	14:26	☽ △ ♆
	22:59	☽ # ☿
	23:56	☽ ⚹ ♇
19	05:51	☽ ∥ ♆
	11:27	☽ ⚹ ♀
	21:42	☽ # ☉
20	02:41	☽ □ ☿
	13:21	☽ ⚹ ♂
	20:08	☽ △ ☉
	20:17	☽ □ ♆
	23:44	☽ ⚹ ♄
21	00:45	☽ △ ♅
	06:08	☽ △ ♃
	15:09	♄ ♎
	23:43	☽ □ ♀
22	18:27	☽ △ ☿
	22:22	☉ ♌
23	00:32	☽ □ ♂
	01:13	☉ ⚹ ♄
	04:52	☽ ⚹ ♆
	08:56	☽ □ ♄
	09:34	☽ □ ♅
	10:06	☉ △ ♅
	12:04	♃ SR
	15:15	☽ □ ♃
	15:19	☽ ☌ ♇
	21:14	♂ ∥ ♄
24	15:13	☽ △ ♀
25	01:51	♃ □ ♇
	12:19	☽ # ☉
	14:21	☽ △ ♂
	20:20	☽ △ ♄
	20:31	☽ ⚹ ♅
	21:06	☽ ∥ ♇
26	02:25	☽ ⚹ ♃
	07:25	☿ # ♆
	10:03	☿ ☍ ♆
	11:31	☉ △ ♃
	17:05	♄ ☍ ♅
27	06:58	☽ ∥ ♆
	10:48	☽ # ☿
	21:43	☿ ♍
28	03:47	☽ ☌ ♆
	09:19	☽ ☍ ☿
	14:44	☽ ⚹ ♇
29	02:17	☽ # ♀
	12:47	♂ # ♅
	13:31	☽ # ♄
	19:23	☽ ∥ ♅
	19:45	☽ # ♂
	22:01	☽ # ♃
	22:10	☽ ∥ ♃
	23:47	♂ ♎
30	00:13	☽ ∥ ♂
	00:49	☽ # ♅
	03:44	☽ ☍ ♀
	06:33	☽ ∥ ♄
	10:03	☿ △ ♇
	13:31	♂ ☍ ♅
	14:19	☽ ∥ ♀
	21:23	☽ ☌ ♅
	21:50	☽ ☍ ♂
	21:59	☉ # ♇
	22:17	☽ ☍ ♄
31	03:17	☽ □ ♇
	03:21	☽ ☌ ♃
	08:08	♂ ☌ ♄
	13:14	☽ △ ☉
	16:56	♂ # ♃
	18:43	♂ ∥ ♃
	19:17	☽ ∥ ☿

August 2010

Day	S. T.	☉	☽	☿	♀	♂	♃	♄	♅	♆	♇	☊ True
	h m s	° ′ ″	° ′ ″	° ′	° ′	° ′	° ′	° ′	° ′	° ′	° ′	° ′
01 Su	20 37 60	08♌39 48	13♈38 02	05♍14	23♍33	01♎14	03♈R17	00♎53	00♈R19	27♒R49	03♑R15	11♑R38
02 Mo	20 41 56	09 37 12	25 47 15	06 25	24 37	01 50	03 15	00 58	00 18	27 47	03 14	11 37
03 Tu	20 45 53	10 34 37	08♉10 13	07 34	25 40	02 27	03 13	01 04	00 17	27 46	03 13	11D 37
04 We	20 49 49	11 32 03	20 51 06	08 41	26 43	03 04	03 11	01 10	00 15	27 44	03 12	11 38
05 Th	20 53 46	12 29 30	03♊53 50	09 44	27 46	03 41	03 09	01 15	00 14	27 43	03 11	11 40
06 Fr	20 57 43	13 26 59	17 21 38	10 46	28 48	04 18	03 06	01 21	00 13	27 41	03 10	11 41
07 Sa	21 01 39	14 24 28	01♋16 19	11 44	29 50	04 55	03 04	01 27	00 11	27 40	03 09	11 42
08 Su	21 05 36	15 21 59	15 37 41	12 40	00♎52	05 32	03 01	01 33	00 10	27 38	03 07	11R 42
09 Mo	21 09 32	16 19 32	00♌22 43	13 32	01 54	06 10	02 58	01 38	00 08	27 36	03 06	11 40
10 Tu	21 13 29	17 17 05	15 25 23	14 22	02 55	06 47	02 54	01 44	00 07	27 35	03 05	11 37
11 We	21 17 25	18 14 40	00♍37 03	15 08	03 56	07 25	02 51	01 50	00 05	27 33	03 05	11 32
12 Th	21 21 22	19 12 15	15 47 40	15 50	04 57	08 02	02 47	01 56	00 04	27 32	03 04	11 26
13 Fr	21 25 18	20 09 51	00♎47 21	16 29	05 57	08 40	02 43	02 03	00 02	27 30	03 03	11 20
14 Sa	21 29 15	21 07 29	15 27 55	17 05	06 57	09 17	02 39	02 09	00 00	27 28	03 02	11 14
15 Su	21 33 12	22 05 07	29 44 02	17 36	07 57	09 55	02 35	02 15	29♓59	27 27	03 01	11 09
16 Mo	21 37 08	23 02 46	13♏33 27	18 02	08 56	10 33	02 31	02 21	29 57	27 25	03 00	11 07
17 Tu	21 41 05	24 00 27	26 56 42	18 24	09 55	11 11	02 26	02 27	29 55	27 24	02 59	11 06
18 We	21 45 01	24 58 08	09♐56 13	18 42	10 54	11 49	02 21	02 34	29 53	27 22	02 58	11D 07
19 Th	21 48 58	25 55 51	22 35 33	18 54	11 52	12 27	02 16	02 40	29 51	27 20	02 58	11 08
20 Fr	21 52 54	26 53 34	04♑58 40	19 02	12 50	13 05	02 11	02 47	29 49	27 19	02 57	11 10
21 Sa	21 56 51	27 51 19	17 09 23	19R 03	13 48	13 43	02 06	02 53	29 48	27 17	02 56	11 11
22 Su	22 00 47	28 49 05	29 11 12	19 00	14 45	14 22	02 01	03 00	29 46	27 15	02 55	11R 10
23 Mo	22 04 44	29 46 52	11♒07 08	18 50	15 41	15 00	01 55	03 06	29 44	27 14	02 55	11 07
24 Tu	22 08 41	00♍44 40	22 59 40	18 35	16 37	15 39	01 50	03 13	29 42	27 12	02 54	11 01
25 We	22 12 37	01 42 30	04♓50 52	18 14	17 33	16 17	01 44	03 20	29 40	27 10	02 53	10 53
26 Th	22 16 34	02 40 21	16 42 27	17 47	18 28	16 56	01 38	03 26	29 38	27 09	02 53	10 44
27 Fr	22 20 30	03 38 14	28 36 00	17 14	19 22	17 34	01 32	03 33	29 36	27 07	02 52	10 34
28 Sa	22 24 27	04 36 08	10♈33 17	16 36	20 16	18 13	01 25	03 40	29 33	27 06	02 52	10 24
29 Su	22 28 23	05 34 05	22 36 23	15 52	21 10	18 52	01 19	03 47	29 31	27 04	02 51	10 16
30 Mo	22 32 20	06 32 02	04♉47 57	15 05	22 03	19 31	01 13	03 53	29 29	27 02	02 51	10 10
31 Tu	22 36 16	07 30 02	17 11 07	14 13	22 55	20 09	01 06	04 00	29 27	27 01	02 50	10 07

Data for	08-01-2010
Julian Day	2455409.50
Ayanamsa	24 00 34
SVP	05 ♓ 07 01
☽ ☊ Mean	10 ♑ 24 R

● ◐ PHASES ○ ◑

03	04:58	◑	10♉47
10	03:08	●	17♌25
16	18:15	◐	23♏47
24	17:04	○	01♓26

LAST ASPECT Day	h m	☽ INGRESS Day	h m	
02	03:54	02	08:13	♉
04	12:44	04	16:54	♊
06	21:22	06	21:50	♋
07	18:46	08	23:23	♌
10	19:11	10	23:02	♍
12	00:05	12	22:44	♎
14	20:07	15	00:27	♏
17	05:25	17	05:35	♐
19	13:59	19	14:18	♑
22	01:09	22	01:38	♒
24	08:30	24	14:11	♓
27	01:59	27	02:49	♈
29	08:48	29	14:36	♉

DECLINATION

Day	☉	☽	☿	♀	♂	♃	♄	♅	♆	♇
01 Su	18N06	10N11	09N08	02N41	00S 04	00S 01	01N42	00S 35	12S 41	18S 22
02 Mo	17 51	14 40	08 33	02 11	00 19	00 02	01 39	00 35	12 42	18 22
03 Tu	17 35	18 39	07 57	01 42	00 35	00 03	01 37	00 36	12 42	18 22
04 We	17 19	21 52	07 23	01 12	00 50	00 04	01 35	00 37	12 43	18 23
05 Th	17 03	24 04	06 49	00 42	01 06	00 05	01 32	00 37	12 43	18 23
06 Fr	16 47	24 59	06 15	00 13	01 21	00 06	01 30	00 38	12 44	18 23
07 Sa	16 30	24 23	05 43	00S 17	01 36	00 08	01 27	00 38	12 44	18 23
08 Su	16 14	22 10	05 11	00 47	01 52	00 09	01 25	00 39	12 45	18 24
09 Mo	15 57	18 27	04 41	01 16	02 07	00 11	01 23	00 40	12 45	18 24
10 Tu	15 39	13 27	04 11	01 46	02 23	00 12	01 20	00 40	12 46	18 24
11 We	15 22	07 35	03 43	02 15	02 38	00 14	01 18	00 41	12 47	18 24
12 Th	15 04	01 18	03 16	02 45	02 54	00 16	01 15	00 41	12 47	18 25
13 Fr	14 46	04S 58	02 51	03 14	03 09	00 17	01 12	00 42	12 48	18 25
14 Sa	14 27	10 50	02 27	03 43	03 25	00 19	01 10	00 43	12 48	18 25
15 Su	14 09	15 58	02 05	04 12	03 40	00 21	01 07	00 44	12 49	18 25
16 Mo	13 50	20 04	01 44	04 41	03 56	00 23	01 05	00 44	12 49	18 26
17 Tu	13 31	22 59	01 26	05 10	04 12	00 25	01 02	00 45	12 50	18 26
18 We	13 12	24 36	01 10	05 39	04 27	00 27	01 00	00 46	12 51	18 26
19 Th	12 52	24 53	00 56	06 08	04 43	00 29	00 57	00 46	12 51	18 26
20 Fr	12 33	23 54	00 45	06 36	04 58	00 32	00 54	00 47	12 52	18 27
21 Sa	12 13	21 48	00 36	07 05	05 14	00 34	00 52	00 48	12 52	18 27
22 Su	11 53	18 46	00 31	07 33	05 29	00 36	00 49	00 49	12 53	18 27
23 Mo	11 33	14 58	00 28	08 01	05 45	00 39	00 46	00 50	12 53	18 27
24 Tu	11 13	10 37	00 28	08 29	06 00	00 41	00 44	00 50	12 54	18 28
25 We	10 52	05 54	00 32	08 56	06 16	00 44	00 41	00 51	12 55	18 28
26 Th	10 31	00 58	00 40	09 24	06 31	00 46	00 38	00 52	12 55	18 28
27 Fr	10 10	04N00	00 50	09 51	06 47	00 49	00 35	00 53	12 56	18 28
28 Sa	09 49	08 51	01 05	10 18	07 02	00 51	00 33	00 54	12 56	18 29
29 Su	09 28	13 24	01 22	10 45	07 17	00 54	00 30	00 55	12 57	18 29
30 Mo	09 07	17 28	01 43	11 11	07 33	00 57	00 27	00 55	12 57	18 29
31 Tu	08 45	20 51	02 08	11 37	07 48	01 00	00 24	00 56	12 58	18 29

ASPECTARIAN

01 13:07 ☽ # ♆
02 03:54 ☽ ⚹ ♆
14:28 ☽ △ ♇
17:37 ☽ ∥ ☉
22:14 ☽ # ♇
22:44 ☽ △ ☿
03 02:09 ♂ ∥ ♅
04:05 ♀ ∥ ♄
08:03 ♃ □ ♇
04 04:20 ♂ ☍ ♃
04:56 ♂ □ ♇
11:40 ♀ # ♂
11:49 ☽ △ ♀
12:44 ☽ □ ♆
17:21 ☽ ⚹ ♅
19:10 ☽ △ ♄
22:39 ☽ ⚹ ♃
23:36 ☽ △ ♂
05 04:08 ♀ ∥ ♅
11:23 ☽ □ ☿
16:35 ☽ ⚹ ☉
06 04:52 ♀ # ♃
11:59 ♂ # ♄
16:10 ♀ ∥ ♃
17:51 ☽ △ ♆
21:22 ☽ □ ♀
22:10 ☽ □ ♅
07 00:18 ☽ □ ♄
03:01 ☽ □ ♃
03:10 ☽ ☍ ♇
03:48 ♀ ♎
06:27 ☽ □ ♂
07:59 ♀ ☍ ♅
17:38 ♀ # ♅
18:46 ☽ ⚹ ☿
08 17:24 ♀ ☌ ♄
[illegible]
09 00:12 ☽ # ♇
02:02 ☽ ⚹ ♄
02:37 ☽ ⚹ ♀
04:08 ☽ △ ♃
04:46 ♀ # ♄
09:40 ☽ ⚹ ♂
13:24 ☽ ∥ ☉
23:41 ♀ ☍ ♃
10 02:56 ☽ # ♆
04:04 ♀ □ ♇
19:11 ☽ ☍ ♆
11 03:52 ☽ △ ♇
16:00 ☽ ∥ ☿
18:11 ☽ # ♂
18:56 ☽ # ♀
12 00:05 ☽ ☌ ☿
00:11 ☽ ∥ ♄
02:17 ☽ # ♅
03:54 ☽ # ♃
05:55 ☽ ∥ ♃
07:32 ☽ ∥ ♅
09:35 ☽ ∥ ♄
12:52 ☿ # ♂
13:40 ☿ # ♀
16:03 ♀ ∥ ♂
16:15 ☽ # ☿
16:38 ☽ ∥ ♂
16:39 ☽ ∥ ♀
22:47 ☽ ☍ ♅
13 02:02 ☽ ☌ ♄
03:07 ☽ ☍ ♃
03:39 ☽ □ ♇
08:59 ☽ ☌ ♀
[illegible]
14 03:33 ♅ ♓R
08:45 ☽ ∥ ♆
10:07 ☽ ⚹ ☉
15:32 ☽ # ☉
20:07 ☽ △ ♆
15 05:37 ☽ ⚹ ♇
13:35 ☽ ∥ ♇
16 08:11 ☽ ⚹ ☿
20:45 ♃ ☍ ♄
17 00:49 ☽ □ ♆
05:25 ☽ △ ♅
10:00 ☽ △ ♃
10:11 ☽ ⚹ ♄
18 01:57 ☽ ⚹ ♀
03:43 ☽ ⚹ ♂
16:51 ☽ □ ☿
22:00 ☿ ∥ ♄
19 01:38 ☉ # ♆
06:58 ☽ △ ☉
09:07 ☽ ⚹ ♆
13:59 ☽ □ ♅
18:36 ☽ □ ♃
18:48 ☿ # ♅
19:40 ☽ □ ♄
20:02 ☽ ☌ ♇
20 10:07 ☉ ☍ ♆
16:46 ☽ □ ♀
16:49 ☽ □ ♂
18:49 ♀ ☌ ♂
20:00 ☿ SR
21 03:46 ☽ △ ☿
06:36 ☿ # ♃
09:56 ♄ □ ♇
[illegible]
01:09 ☽ ⚹ ♅
02:08 ☽ ∥ ♇
05:38 ☽ ⚹ ♃
07:43 ☽ △ ♄
23 05:27 ☉ ♍
08:17 ☽ △ ♂
10:00 ☽ △ ♀
11:46 ☽ ∥ ♆
20:36 ☽ # ☉
24 08:30 ☽ ☌ ♆
10:03 ☽ ∥ ♀
11:40 ♃ # ♄
20:03 ☽ ⚹ ♇
22:15 ☽ ∥ ♂
25 20:46 ☿ ∥ ♄
26 00:27 ☽ ∥ ♅
00:55 ☽ ∥ ♃
01:24 ☽ # ☿
01:35 ☽ # ♄
02:04 ☽ ☍ ☿
05:11 ☉ △ ♇
07:37 ☽ ∥ ♄
08:05 ☽ ∥ ☿
08:24 ☽ # ♃
08:50 ☽ # ♅
19:42 ☿ # ♃
27 02:01 ☽ ☌ ♅
04:57 ☿ # ♅
05:51 ☽ ☌ ♃
08:37 ☽ □ ♆
09:45 ♀ # ☉
10:03 ☽ ☍ ♄
14:22 ☽ # ♂
28 04:37 ☽ ∥ ☉
08:13 ☽ # ♀
16:09 ☽ ☍ ♂
20:55 ☽ ☍ ♀
21:31 ☽ # ♆
29 08:45 ♃ ∥ ♅
08:48 ☽ ⚹ ♆
20:11 ☽ △ ♇
30 03:41 ☽ △ ☉
06:37 ☽ # ♇
18:40 ☽ △ ☿
31 18:40 ☽ □ ♆
23:14 ☽ ⚹ ♅

⚷ Chiron

01 Dec.	05 S 35
03	29♒24R
06	29 16
09	29 07
12	28 58
15	28 50
18	28 41
21	28 32
24	28 23
27	28 14
30	28 05

September 2010

Day	S. T.	☉	☽	☿	♀	♂	♃	♄	♅	♆	♇	☊ True
	h m s	° ′ ″	° ′ ″	° ′	° ′	° ′	° ′	° ′	° ′	° ′	° ′	° ′
01 We	22 40 13	08♍ 28 04	29♉ 49 29	13♍R19	23♎ 47	20♎ 48	00♈R59	04♎ 07	29♓R25	26♒R59	02♑R50	10♑R06
02 Th	22 44 10	09 26 08	12♊ 46 51	12 22	24 38	21 28	00 52	04 14	29 23	26 57	02 50	10D 06
03 Fr	22 48 06	10 24 13	26 06 54	11 25	25 28	22 07	00 45	04 21	29 20	26 56	02 49	10 06
04 Sa	22 52 03	11 22 21	09♋ 52 36	10 28	26 18	22 46	00 38	04 28	29 18	26 54	02 49	10R 06
05 Su	22 55 59	12 20 30	24 05 21	09 32	27 07	23 25	00 31	04 35	29 16	26 53	02 49	10 05
06 Mo	22 59 56	13 18 42	08♌ 44 04	08 40	27 55	24 05	00 24	04 42	29 14	26 51	02 48	10 01
07 Tu	23 03 52	14 16 55	23 44 21	07 51	28 43	24 44	00 17	04 49	29 11	26 50	02 48	09 55
08 We	23 07 49	15 15 10	08♍ 58 30	07 08	29 30	25 24	00 09	04 56	29 09	26 48	02 48	09 47
09 Th	23 11 45	16 13 27	24 16 13	06 31	00♏ 16	26 03	00 02	05 03	29 07	26 47	02 48	09 37
10 Fr	23 15 42	17 11 46	09♎ 26 18	06 01	01 01	26 43	29♓ 54	05 11	29 04	26 45	02 48	09 26
11 Sa	23 19 39	18 10 06	24 18 33	05 39	01 45	27 23	29 46	05 18	29 02	26 44	02 47	09 17
12 Su	23 23 35	19 08 28	08♏ 45 28	05 26	02 28	28 02	29 39	05 25	29 00	26 42	02 47	09 08
13 Mo	23 27 32	20 06 52	22 43 08	05D 22	03 11	28 42	29 31	05 32	28 57	26 41	02 47	09 02
14 Tu	23 31 28	21 05 17	06♐ 11 08	05 27	03 52	29 22	29 23	05 39	28 55	26 39	02 47	08 59
15 We	23 35 25	22 03 44	19 11 46	05 41	04 32	00♏ 02	29 15	05 47	28 53	26 38	02D 47	08 57
16 Th	23 39 21	23 02 12	01♑ 49 04	06 05	05 12	00 42	29 07	05 54	28 50	26 36	02 47	08D 57
17 Fr	23 43 18	24 00 42	14 07 52	06 37	05 50	01 23	28 59	06 01	28 48	26 35	02 47	08 58
18 Sa	23 47 14	24 59 13	26 13 05	07 18	06 27	02 03	28 51	06 09	28 45	26 33	02 47	08R 57
19 Su	23 51 11	25 57 47	08♒ 09 23	08 08	07 02	02 43	28 43	06 16	28 43	26 32	02 48	08 55
20 Mo	23 55 08	26 56 22	20 00 52	09 05	07 37	03 23	28 35	06 23	28 41	26 31	02 48	08 50
21 Tu	23 59 04	27 54 58	01♓ 50 59	10 10	08 10	04 04	28 27	06 31	28 38	26 29	02 48	08 42
22 We	00 03 01	28 53 37	13 42 25	11 21	08 41	04 44	28 19	06 38	28 36	26 28	02 48	08 31
23 Th	00 06 57	29 52 17	25 37 06	12 39	09 12	05 25	28 11	06 45	28 33	26 27	02 48	08 19
24 Fr	00 10 54	00♎ 50 59	07♈ 36 28	14 02	09 40	06 06	28 03	06 53	28 31	26 25	02 49	08 05
25 Sa	00 14 50	01 49 43	19 41 38	15 29	10 07	06 46	27 55	07 00	28 29	26 24	02 49	07 51
26 Su	00 18 47	02 48 30	01♉ 53 42	17 01	10 33	07 27	27 47	07 07	28 26	26 23	02 49	07 40
27 Mo	00 22 43	03 47 18	14 14 02	18 36	10 57	08 08	27 39	07 15	28 24	26 22	02 50	07 30
28 Tu	00 26 40	04 46 09	26 44 28	20 14	11 19	08 49	27 31	07 22	28 21	26 20	02 50	07 24
29 We	00 30 36	05 45 02	09♊ 27 17	21 55	11 40	09 30	27 24	07 30	28 19	26 19	02 51	07 20
30 Th	00 34 33	06 43 57	22 25 18	23 37	11 58	10 11	27 16	07 37	28 17	26 18	02 51	07 19

Data for	09-01-2010
Julian Day	2455440.50
Ayanamsa	24 00 38
SVP	05 ♓ 06 56
☽ ☊ Mean	08 ♑ 46 R

● ◐ PHASES ○ ◑

Day	Time	Phase	Position
01	17:22	◑	09♊10
08	10:30	●	15♍41
15	05:50	◐	22♐18
23	09:17	○	00♈15

LAST ASPECT Day	h m	☽ INGRESS Day	h m	
31	23:14	01	00:20	♊
03	05:41	03	06:51	♋
05	08:32	05	09:46	♌
07	08:18	07	09:54	♍
09	09:00	09	09:02	♎
11	05:16	11	09:22	♏
13	11:53	13	12:52	♐
15	18:52	15	20:30	♑
18	05:13	18	07:35	♒
20	13:09	20	20:15	♓
23	05:53	23	08:47	♈
25	13:12	25	20:17	♉
28	03:04	28	06:12	♊
30	10:38	30	13:47	♋

DECLINATION

Day	☉	☽	☿	♀	♂	♃	♄	♅	♆	♇
01 We	08N24	23N19	02N34	12S03	08S04	01S02	00N21	00S57	12S59	18S30
02 Th	08 02	24 39	03 04	12 29	08 19	01 05	00 19	00 58	12 59	18 30
03 Fr	07 40	24 36	03 35	12 54	08 34	01 08	00 16	00 59	13 00	18 30
04 Sa	07 18	23 05	04 08	13 19	08 49	01 11	00 13	01 00	13 00	18 30
05 Su	06 56	20 05	04 42	13 44	09 05	01 14	00 10	01 01	13 01	18 31
06 Mo	06 34	15 43	05 16	14 08	09 20	01 17	00 07	01 02	13 01	18 31
07 Tu	06 11	10 18	05 50	14 32	09 35	01 20	00 05	01 03	13 02	18 31
08 We	05 49	04 11	06 22	14 56	09 50	01 23	00 02	01 04	13 02	18 31
09 Th	05 26	02S11	06 53	15 19	10 05	01 26	00S01	01 04	13 03	18 32
10 Fr	05 03	08 22	07 22	15 42	10 20	01 30	00 04	01 05	13 03	18 32
11 Sa	04 41	13 57	07 48	16 05	10 35	01 33	00 07	01 06	13 04	18 32
12 Su	04 18	18 34	08 11	16 27	10 50	01 36	00 10	01 07	13 04	18 32
13 Mo	03 55	21 59	08 31	16 49	11 05	01 39	00 13	01 08	13 05	18 33
14 Tu	03 32	24 02	08 46	17 10	11 20	01 42	00 16	01 09	13 05	18 33
15 We	03 09	24 42	08 58	17 31	11 34	01 46	00 18	01 10	13 06	18 33
16 Th	02 46	24 03	09 05	17 51	11 49	01 49	00 21	01 11	13 06	18 33
17 Fr	02 23	22 14	09 09	18 11	12 04	01 52	00 24	01 12	13 07	18 34
18 Sa	02 00	19 26	09 08	18 30	12 18	01 55	00 27	01 13	13 07	18 34
19 Su	01 36	15 50	09 02	18 49	12 33	01 58	00 30	01 14	13 08	18 34
20 Mo	01 13	11 39	08 53	19 07	12 47	02 02	00 33	01 15	13 08	18 34
21 Tu	00 50	07 04	08 40	19 25	13 01	02 05	00 36	01 16	13 09	18 35
22 We	00 26	02 13	08 22	19 42	13 16	02 08	00 39	01 17	13 09	18 35
23 Th	00 03	02N44	08 02	19 59	13 30	02 11	00 42	01 18	13 10	18 35
24 Fr	00S20	07 36	07 37	20 14	13 44	02 14	00 45	01 19	13 10	18 35
25 Sa	00 44	12 13	07 10	20 30	13 58	02 18	00 48	01 20	13 11	18 36
26 Su	01 07	16 24	06 40	20 44	14 12	02 21	00 50	01 21	13 11	18 36
27 Mo	01 30	19 56	06 07	20 58	14 26	02 24	00 53	01 22	13 11	18 36
28 Tu	01 54	22 36	05 32	21 11	14 40	02 27	00 56	01 23	13 12	18 36
29 We	02 17	24 11	04 54	21 24	14 53	02 30	00 59	01 23	13 12	18 37
30 Th	02 40	24 30	04 15	21 35	15 07	02 33	01 02	01 24	13 13	18 37

ASPECTARIAN

01 02:10 ☽ ✶ ♃
08:06 ☽ △ ♄
13:05 ☉ # ♂
23:18 ☽ □ ☿
02 16:30 ☽ △ ♂
22:47 ☽ △ ♀
03 01:27 ☽ △ ♆
05:22 ♀ ∥ ♆
05:41 ☽ □ ♅
08:07 ☽ □ ♃
11:47 ☽ ☍ ♇
12:35 ☿ ☌ ☉
14:35 ☽ □ ♄
04 00:56 ☽ ✶ ☿
02:45 ☽ ✶ ☉
17:13 ♀ △ ♆
22:50 ☽ □ ♂
05 05:19 ☽ □ ♀
08:32 ☽ △ ♅
09:25 ☽ # ♇
10:32 ☽ △ ♃
17:24 ☽ ✶ ♄
06 06:53 ☽ # ♀
12:24 ☽ # ♆
07 01:39 ☽ ✶ ♂
02:47 ☽ # ♂
04:53 ☽ ☍ ♆
08:18 ☽ ✶ ♀
09:19 ☿ ∥ ☉
14:18 ☽ △ ♇
16:16 ☽ ∥ ☿
17:21 ☽ ∥ ☉
21:14 ☽ ☌ ☿
08 10:26 ☽ # ♃
11:44 ☽ # ♅
15:44 ☽ # ♄
15:45 ☽ ∥ ♄
15:46 ♀ ♏
19:46 ☽ ∥ ♅
21:08 ☽ ∥ ♃
09 04:51 ♃ ♓R
07:36 ☽ ☍ ♅
09:00 ☽ ☍ ♃
11:41 ☽ # ☉
13:26 ☽ □ ♇
17:10 ☽ ☌ ♄
19:38 ☽ # ☿
10 01:18 ♂ △ ♆
08:26 ☽ ∥ ♂
19:57 ☽ ∥ ♆
11 03:57 ☽ △ ♆
05:16 ☽ ☌ ♂
11:14 ☽ ∥ ♀
12:55 ☽ ☌ ♀
13:59 ☽ ✶ ♇
18:29 ☽ ✶ ☿
23:48 ☽ ∥ ♇
12 10:41 ♀ ✶ ♇
19:07 ☽ ✶ ☉
23:10 ☿ SD
13 06:57 ☽ □ ♆
10:58 ☽ △ ♅
11:53 ☽ △ ♃
22:39 ☽ □ ☿
23:02 ☽ ✶ ♄
14 04:35 ♇ SD
22:38 ♂ ♏
15 14:01 ☽ ✶ ♆
18:17 ☽ □ ♅
18:52 ☽ □ ♃
21:44 ☽ ✶ ♂
16 01:52 ☽ ☌ ♇
06:53 ☽ ✶ ♀
07:59 ☽ □ ♄
08:35 ☽ △ ☿
17 21:19 ☽ △ ☉
18 03:58 ☉ # ♃
04:57 ♀ ∥ ♇
05:04 ☽ ✶ ♅
05:13 ☽ ✶ ♃
06:09 ☽ ∥ ♀
06:16 ☽ ∥ ♇
12:23 ☽ □ ♂
20:09 ☽ △ ♄
21:39 ☽ □ ♀
19 01:07 ♃ ☌ ♅
02:46 ♂ ✶ ♇
14:10 ☉ # ♅
15:53 ☽ ∥ ♆
18:11 ☽ ∥ ♂
22:08 ☉ # ♅
20 13:09 ☽ ☌ ♆
15:24 ☽ # ☿
21 01:57 ☽ ✶ ♇
04:45 ☽ △ ♂
11:36 ☉ ☍ ♃
12:41 ☉ # ♄
12:51 ♂ ∥ ♆
13:23 ☽ △ ♀
16:58 ☉ ☍ ♅
18:42 ☽ ☍ ☿
22 00:23 ☽ ∥ ♃
04:31 ☽ ∥ ♅
07:32 ☽ ∥ ♄
09:21 ☽ # ☉
11:57 ☽ ∥ ☉
14:01 ☽ # ♄
17:01 ☽ # ♅
23:20 ☽ # ♃
23 03:09 ☉ ♎
05:06 ☽ ☌ ♃
05:53 ☽ ☌ ♅
14:25 ☽ □ ♇
22:32 ☽ ☍ ♄
24 00:06 ☽ ∥ ☿
25 04:35 ☉ ∥ ♄
05:13 ☽ # ♆
10:11 ☽ # ♂
13:12 ☽ ✶ ♆
26 00:24 ☉ □ ♇
01:49 ☽ △ ♇
11:28 ☽ ☍ ♂
14:17 ☽ # ♇
14:38 ☉ ∥ ♅
17:26 ☽ ☍ ♀
27 08:58 ☽ # ♀
09:40 ☽ △ ☿
23:14 ☽ □ ♆
28 01:28 ☽ ✶ ♃
03:04 ☽ ✶ ♅
16:28 ☽ △ ☉
20:17 ☽ △ ♄
29 15:34 ☉ ∥ ♃
30 02:32 ☽ □ ☿
07:04 ☽ △ ♆
08:44 ☽ □ ♃
10:38 ☽ □ ♅
18:55 ☽ ☍ ♇

⚷ Chiron

01 Dec. 06 S 07

Day	⚷
02	27♒56R
05	27 47
08	27 39
11	27 30
14	27 22
17	27 14
20	27 07
23	26 59
26	26 53
29	26 46

October 2010

Day	S. T.	☉	☽	☿	♀	♂	♃	♄	♅	♆	♇	☊ True
	h m s	° ' "	° ' "	° '	° '	° '	° '	° '	° '	° '	° '	° '
01 Fr	00 38 30	07♎ 42 55	05♋ 41 34	25♍ 21	12♏ 15	10♏ 52	27♓R08	07♎ 44	28♓R14	26♒R17	02♑ 52	07♑R18
02 Sa	00 42 26	08 41 55	19 18 59	27 07	12 30	11 33	27 00	07 52	28 12	26 16	02 52	07 17
03 Su	00 46 23	09 40 57	03♌ 19 33	28 53	12 42	12 14	26 53	07 59	28 09	26 15	02 53	07 15
04 Mo	00 50 19	10 40 01	17 43 31	00♎ 40	12 53	12 56	26 45	08 07	28 07	26 14	02 53	07 11
05 Tu	00 54 16	11 39 08	02♍ 28 28	02 27	13 02	13 37	26 38	08 14	28 05	26 13	02 54	07 04
06 We	00 58 12	12 38 17	17 28 53	04 14	13 08	14 19	26 30	08 21	28 03	26 12	02 55	06 55
07 Th	01 02 09	13 37 28	02♎ 36 23	06 01	13 12	15 00	26 23	08 29	28 00	26 11	02 55	06 44
08 Fr	01 06 05	14 36 41	17 40 51	07 48	13 14	15 42	26 16	08 36	27 58	26 10	02 56	06 32
09 Sa	01 10 02	15 35 56	02♏ 32 05	09 35	13R 13	16 23	26 08	08 44	27 56	26 09	02 57	06 21
10 Su	01 13 59	16 35 13	17 01 38	11 21	13 11	17 05	26 01	08 51	27 53	26 08	02 58	06 11
11 Mo	01 17 55	17 34 32	01♐ 04 02	13 07	13 05	17 47	25 54	08 58	27 51	26 07	02 58	06 04
12 Tu	01 21 52	18 33 53	14 37 18	14 52	12 58	18 29	25 48	09 06	27 49	26 06	02 59	05 59
13 We	01 25 48	19 33 16	27 42 30	16 37	12 48	19 11	25 41	09 13	27 47	26 05	03 00	05 56
14 Th	01 29 45	20 32 41	10♑ 22 57	18 21	12 35	19 53	25 34	09 20	27 45	26 05	03 01	05D 56
15 Fr	01 33 41	21 32 07	22 43 16	20 04	12 20	20 35	25 28	09 28	27 42	26 04	03 02	05 56
16 Sa	01 37 38	22 31 35	04♒ 48 38	21 47	12 03	21 17	25 22	09 35	27 40	26 03	03 03	05R 56
17 Su	01 41 34	23 31 05	16 44 16	23 29	11 43	21 59	25 15	09 42	27 38	26 02	03 04	05 55
18 Mo	01 45 31	24 30 36	28 35 07	25 11	11 22	22 42	25 09	09 49	27 36	26 02	03 05	05 51
19 Tu	01 49 28	25 30 09	10♓ 25 32	26 51	10 58	23 24	25 03	09 57	27 34	26 01	03 06	05 45
20 We	01 53 24	26 29 44	22 19 09	28 32	10 32	24 06	24 58	10 04	27 32	26 00	03 07	05 36
21 Th	01 57 21	27 29 21	04♈ 18 43	00♏ 11	10 04	24 49	24 52	10 11	27 30	26 00	03 08	05 25
22 Fr	02 01 17	28 29 00	16 26 06	01 50	09 34	25 31	24 47	10 18	27 28	25 59	03 09	05 13
23 Sa	02 05 14	29 28 41	28 42 28	03 28	09 03	26 14	24 41	10 25	27 26	25 59	03 11	05 01
24 Su	02 09 10	00♏ 28 24	11♉ 08 26	05 06	08 30	26 57	24 36	10 33	27 24	25 58	03 12	04 50
25 Mo	02 13 07	01 28 08	23 44 22	06 43	07 56	27 39	24 31	10 40	27 22	25 58	03 13	04 42
26 Tu	02 17 03	02 27 55	06♊ 30 39	08 20	07 21	28 22	24 27	10 47	27 21	25 57	03 14	04 36
27 We	02 20 60	03 27 44	19 27 55	09 56	06 45	29 05	24 22	10 54	27 19	25 57	03 15	04 33
28 Th	02 24 57	04 27 36	02♋ 37 12	11 31	06 09	29 48	24 18	11 01	27 17	25 57	03 17	04 31
29 Fr	02 28 53	05 27 29	15 59 55	13 06	05 32	00♐ 31	24 13	11 08	27 15	25 56	03 18	04D 31
30 Sa	02 32 50	06 27 25	29 37 37	14 41	04 56	01 14	24 09	11 15	27 14	25 56	03 19	04R 31
31 Su	02 36 46	07 27 23	13♌ 31 31	16 15	04 19	01 57	24 06	11 22	27 12	25 56	03 21	04 30

Data for	10-01-2010
Julian Day	2455470.50
Ayanamsa	24 00 41
SVP	05 ♓ 06 51
☽ ☊ Mean	07 ♑ 10 R

LAST ASPECT ☽ Day	h m	INGRESS Day	h m	
02	15:23	02	18:22	♌
04	13:53	04	20:00	♍
06	16:43	06	19:52	♎
08	13:38	08	19:52	♏
10	18:27	10	22:09	♐
13	00:08	13	04:17	♑
15	09:49	15	14:24	♒
17	18:49	18	02:52	♓
20	10:26	20	15:24	♈
23	01:38	23	02:31	♉
25	07:50	25	11:48	♊
27	14:20	27	19:15	♋
29	19:49	30	00:39	♌

● ◐ PHASES ○ ◑

01	03:53	◑	07♋52
07	18:44	●	14♎24
14	21:27	◐	21♑26
23	01:37	○	29♈33
30	12:46	◑	06♌59

DECLINATION

Day	☉	☽	☿	♀	♂	♃	♄	♅	♆	♇
01 Fr	03S 04	23N27	03N35	21S 46	15S 20	02S 36	01S 05	01S 25	13S 13	18S 37
02 Sa	03 27	21 00	02 53	21 56	15 34	02 39	01 08	01 26	13 13	18 37
03 Su	03 50	17 15	02 10	22 05	15 47	02 42	01 11	01 27	13 14	18 38
04 Mo	04 13	12 24	01 26	22 13	16 00	02 45	01 14	01 28	13 14	18 38
05 Tu	04 36	06 44	00 42	22 20	16 13	02 48	01 16	01 29	13 14	18 38
06 We	05 00	00 35	00S 03	22 26	16 26	02 51	01 19	01 30	13 15	18 38
07 Th	05 23	05S 37	00 48	22 31	16 39	02 54	01 22	01 31	13 15	18 39
08 Fr	05 46	11 29	01 34	22 35	16 52	02 57	01 25	01 32	13 15	18 39
09 Sa	06 08	16 35	02 19	22 38	17 04	02 59	01 28	01 33	13 16	18 39
10 Su	06 31	20 34	03 05	22 40	17 17	03 02	01 31	01 33	13 16	18 39
11 Mo	06 54	23 12	03 51	22 40	17 29	03 05	01 34	01 34	13 16	18 40
12 Tu	07 17	24 23	04 36	22 40	17 41	03 07	01 36	01 35	13 17	18 40
13 We	07 39	24 08	05 21	22 38	17 54	03 10	01 39	01 36	13 17	18 40
14 Th	08 01	22 38	06 06	22 34	18 06	03 12	01 42	01 37	13 17	18 40
15 Fr	08 24	20 05	06 50	22 30	18 17	03 15	01 45	01 38	13 17	18 40
16 Sa	08 46	16 41	07 34	22 23	18 29	03 17	01 48	01 39	13 18	18 41
17 Su	09 08	12 40	08 18	22 16	18 41	03 20	01 50	01 39	13 18	18 41
18 Mo	09 30	08 11	09 01	22 07	18 52	03 22	01 53	01 40	13 18	18 41
19 Tu	09 52	03 26	09 43	21 57	19 03	03 24	01 56	01 41	13 18	18 41
20 We	10 13	01N28	10 25	21 45	19 14	03 26	01 59	01 42	13 19	18 41
21 Th	10 35	06 20	11 07	21 32	19 25	03 28	02 02	01 43	13 19	18 42
22 Fr	10 56	11 01	11 47	21 17	19 36	03 30	02 04	01 43	13 19	18 42
23 Sa	11 17	15 19	12 27	21 01	19 47	03 32	02 07	01 44	13 19	18 42
24 Su	11 38	19 02	13 07	20 44	19 57	03 34	02 10	01 45	13 19	18 42
25 Mo	11 59	21 55	13 45	20 25	20 08	03 36	02 12	01 46	13 19	18 43
26 Tu	12 20	23 45	14 23	20 05	20 18	03 37	02 15	01 46	13 20	18 43
27 We	12 40	24 20	15 00	19 45	20 28	03 39	02 18	01 47	13 20	18 43
28 Th	13 00	23 34	15 36	19 23	20 38	03 41	02 20	01 48	13 20	18 43
29 Fr	13 20	21 28	16 12	19 00	20 47	03 42	02 23	01 48	13 20	18 43
30 Sa	13 40	18 06	16 46	18 37	20 57	03 43	02 26	01 49	13 20	18 43
31 Su	14 00	13 40	17 20	18 13	21 06	03 45	02 28	01 50	13 20	18 44

ASPECTARIAN

01 00:42 ☉ ☌ ♄
03:41 ☽ □ ♄
09:41 ☽ △ ♂
11:34 ☿ # ☉
11:51 ☽ △ ♀
16:46 ☽ # ♀
22:36 ☿ ☍ ♃
02 07:15 ☿ # ♃
13:08 ☽ △ ♃
14:24 ☿ ☍ ♅
15:15 ☽ △ ♅
15:23 ☽ ⚹ ☿
16:03 ☽ # ♇
03 07:26 ☽ # ♂
07:54 ☽ ⚹ ♄
11:27 ☽ ⚹ ☉
15:04 ☿ ♎
15:41 ☽ □ ♂
15:54 ☽ □ ♀
20:12 ☽ # ♆
21:58 ♀ ☌ ♂
23:07 ☿ # ♅
04 06:33 ☿ # ♄
13:53 ☽ ☍ ♆
05 00:41 ☽ △ ♇
06:08 ☿ □ ♇
07:56 ☽ # ☉
15:22 ☽ # ♃
17:02 ☽ ⚹ ♀
18:42 ☽ ⚹ ♂
20:32 ☽ # ♅
21:13 ☽ # ♄
06 01:52 ☽ # ☿
02:49 ☽ ∥ ☿
07:25 ☽ ∥ ♄
08:04 ☽ ∥ ♅
13:21 ☽ ∥ ♃
[illegible] ♃
16:43 ☽ ☍ ♅
23:00 ☽ ∥ ☉
07 00:30 ☽ □ ♇
06:09 ☽ ☌ ☿
09:24 ☽ ☌ ♄
19:08 ☿ ∥ ♄
22:56 ☿ ∥ ♅
08 07:06 ♀ SR
07:56 ☽ ∥ ♆
11:36 ☿ ☌ ♄
13:38 ☽ △ ♆
09 00:40 ☽ ⚹ ♇
02:47 ☽ ∥ ♂
11:35 ☽ ∥ ♇
17:35 ☽ ☌ ♀
22:24 ☿ ∥ ♃
10 00:06 ☽ ☌ ♂
15:09 ☽ △ ♃
15:27 ☽ □ ♆
17:58 ☽ ∥ ♀
18:27 ☽ △ ♅
11 09:00 ♄ ∥ ♅
14:01 ☽ ⚹ ♄
12 00:31 ☽ ⚹ ☿
07:44 ☽ ⚹ ☉
20:16 ☽ □ ♃
20:59 ☽ ⚹ ♆
13 00:08 ☽ □ ♅
09:57 ☽ ☌ ♇
21:59 ☽ □ ♄
14 00:53 ☽ ∥ ♀
04:10 ☽ ⚹ ♀
17:57 ☽ □ ☿
19:33 ☽ ⚹ ♂
15 03:22 ☽ [illegible] ♃
09:49 ☽ ⚹ ♅
10:37 ☽ ∥ ♇
12:36 ☽ ∥ ♂
16 09:40 ☽ △ ♄
14:10 ☽ □ ♀
20:26 ☽ ∥ ♆
17 00:20 ♂ ∥ ♇
01:05 ☿ ☌ ☉
11:18 ☽ □ ♂
14:59 ☽ △ ☉
15:56 ☽ △ ☿
17:42 ☽ ∥ ☉
18:49 ☽ ☌ ♆
20:19 ☽ ∥ ☿
18 09:08 ☽ ⚹ ♇
12:04 ☿ △ ♆
19 00:12 ☽ ∥ ♃
01:03 ☽ △ ♀
07:21 ☽ ∥ ♄
08:38 ☽ ∥ ♅
09:36 ☿ ∥ ☉
12:19 ☉ △ ♆
20 01:11 ☽ # ♅
02:36 ☽ # ♄
03:49 ☽ △ ♂
05:16 ☽ ☌ ♃
09:46 ☽ # ♃
10:26 ☽ ☌ ♅
21:19 ☿ ♏
21:39 ☽ □ ♇
21 01:43 ♂ △ ♃
11:47 ☽ ☍ ♄
23:32 ☽ # ☉
22 04:52 ☽ # ☿
12:30 ☽ # ♆
15:36 ♂ □ ♆
18:42 ☽ ⚹ ♆
19:38 ☿ ⚹ ♇
23 08:40 ☽ △ ♇
10:38 ☽ ☍ ☿
12:35 ☉ ♏
19:08 ☽ ☍ ♀
21:41 ☽ # ♇
24 07:19 ☽ # ♂
07:50 ☿ ∥ ♆
11:53 ☽ # ♀
14:57 ♂ △ ♅
25 01:28 ☽ ⚹ ♃
04:12 ☽ □ ♆
00:31 ☽ [illegible] ♅
07:50 ☽ ☍ ♂
13:18 ☿ ☌ ♀
14:02 ♀ ∥ ♂
26 08:01 ☽ △ ♄
18:59 ☉ ⚹ ♇
27 08:56 ☽ □ ♃
11:52 ☽ △ ♆
14:20 ☽ □ ♅
28 01:12 ☽ ☍ ♇
03:36 ☽ △ ☉
06:05 ☽ △ ♀
06:48 ♂ ♐
15:14 ☽ □ ♄
18:09 ☽ △ ☿
23:18 ☉ ∥ ♆
29 01:11 ♀ ☌ ☉
05:16 ☽ # ♂
14:28 ☽ △ ♃
17:31 ♀ ∥ ♇
19:49 ☽ △ ♅
20:05 ☽ # ♇
20:25 ☽ # ♀
30 02:57 ☽ △ ♂
06:52 ☽ # ☿
08:49 ☽ □ ♀
20:16 ☽ ⚹ ♄
22:30 ☽ # ☉
31 01:40 ☽ # ♆
05:13 ☽ □ ☿
21:01 ☽ ☍ ♆
22:15 ☿ ∥ ♀

⚷ Chiron

01 Dec.	06 S 39
02	26♒40R
05	26 34
08	26 29
11	26 24
14	26 20
17	26 16
20	26 13
23	26 10
26	26 08
29	26 06

November 2010

Day	S. T. h m s	☉ ° ' "	☽ ° ' "	☿ ° '	♀ ° '	♂ ° '	♃ ° '	♄ ° '	♅ ° '	♆ ° '	♇ ° '	☊ True ° '
01 Mo	02 40 43	08♏27 22	27♌41 53	17♏48	03♏R43	02♐40	24♓R02	11♎29	27♓R10	25♒R56	03♑22	04♑R27
02 Tu	02 44 39	09 27 25	12♍07 24	19 21	03 08	03 23	23 58	11 35	27 09	25 55	03 24	04 23
03 We	02 48 36	10 27 29	26 44 40	20 54	02 34	04 07	23 55	11 42	27 07	25 55	03 25	04 16
04 Th	02 52 32	11 27 35	11♎28 10	22 26	02 01	04 50	23 52	11 49	27 06	25 55	03 27	04 08
05 Fr	02 56 29	12 27 43	26 10 43	23 58	01 29	05 34	23 49	11 56	27 04	25 55	03 28	04 00
06 Sa	03 00 26	13 27 54	10♏44 32	25 29	00 59	06 17	23 46	12 02	27 03	25 55	03 30	03 51
07 Su	03 04 22	14 28 06	25 02 21	27 00	00 30	07 01	23 44	12 09	27 01	25 55	03 31	03 44
08 Mo	03 08 19	15 28 20	08♐58 41	28 31	00 03	07 44	23 42	12 16	27 00	25D55	03 33	03 38
09 Tu	03 12 15	16 28 35	22 30 28	00♐01	29♎39	08 28	23 39	12 22	26 59	25 55	03 34	03 35
10 We	03 16 12	17 28 52	05♑37 15	01 31	29 16	09 12	23 38	12 29	26 57	25 55	03 36	03 33
11 Th	03 20 08	18 29 11	18 20 48	03 00	28 56	09 56	23 36	12 35	26 56	25 55	03 38	03D34
12 Fr	03 24 05	19 29 31	00♒44 30	04 29	28 38	10 39	23 34	12 42	26 55	25 55	03 39	03 36
13 Sa	03 28 01	20 29 53	12 52 45	05 58	28 22	11 23	23 33	12 48	26 54	25 55	03 41	03 37
14 Su	03 31 58	21 30 15	24 50 29	07 26	28 09	12 07	23 32	12 55	26 53	25 56	03 43	03 38
15 Mo	03 35 55	22 30 40	06♓42 45	08 53	27 58	12 51	23 31	13 01	26 51	25 56	03 44	03R38
16 Tu	03 39 51	23 31 05	18 34 29	10 20	27 50	13 36	23 30	13 07	26 50	25 56	03 46	03 36
17 We	03 43 48	24 31 32	00♈30 10	11 46	27 44	14 20	23 30	13 13	26 50	25 56	03 48	03 32
18 Th	03 47 44	25 32 00	12 33 37	13 12	27 40	15 04	23 30	13 19	26 49	25 57	03 50	03 27
19 Fr	03 51 41	26 32 30	24 47 46	14 37	27D39	15 48	23D30	13 25	26 48	25 57	03 51	03 21
20 Sa	03 55 37	27 33 01	07♉14 40	16 02	27 41	16 33	23 30	13 32	26 47	25 58	03 53	03 15
21 Su	03 59 34	28 33 34	19 55 19	17 25	27 45	17 17	23 30	13 37	26 46	25 58	03 55	03 10
22 Mo	04 03 30	29 34 08	02♊49 57	18 48	27 51	18 01	23 31	13 43	26 45	25 59	03 57	03 06
23 Tu	04 07 27	00♐34 43	15 58 00	20 09	28 00	18 46	23 32	13 49	26 45	25 59	03 59	03 03
24 We	04 11 24	01 35 20	29 18 29	21 29	28 10	19 30	23 33	13 55	26 44	26 00	04 01	03 01
25 Th	04 15 20	02 35 59	12♋50 16	22 48	28 23	20 15	23 34	14 01	26 43	26 00	04 02	03D01
26 Fr	04 19 17	03 36 39	26 32 16	24 06	28 39	21 00	23 35	14 06	26 43	26 01	04 04	03 02
27 Sa	04 23 13	04 37 21	10♌23 32	25 21	28 56	21 45	23 37	14 12	26 42	26 02	04 06	03 03
28 Su	04 27 10	05 38 04	24 23 12	26 35	29 15	22 29	23 39	14 18	26 42	26 02	04 08	03R03
29 Mo	04 31 06	06 38 49	08♍30 11	27 46	29 36	23 14	23 41	14 23	26 42	26 03	04 10	03 02
30 Tu	04 35 03	07 39 35	22 42 54	28 54	29 59	23 59	23 43	14 28	26 41	26 04	04 12	03 00

Data for 11-01-2010
Julian Day 2455501.50
Ayanamsa 24 00 45
SVP 05 ♓ 06 46
☽ ☊ Mean 05 ♑ 32 R

● ◐ PHASES ○ ◑

Day	h m	Phase	Position
06	04:52	●	13♏40
13	16:39	◐	21♒12
21	17:27	○	29♉18
28	20:36	◑	06♍30

LAST ASPECT ☽ INGRESS

Day	h m	Day	h m	Sign
31	21:01	01	03:51	♍
03	00:36	03	05:19	♎
04	23:34	05	06:16	♏
07	03:45	07	08:28	♐
09	12:36	09	13:37	♑
11	19:58	11	22:33	♒
14	06:34	14	10:25	♓
16	16:38	16	23:00	♈
19	05:34	19	10:05	♉
21	17:28	21	18:46	♊
23	21:57	24	01:14	♋
26	03:44	26	06:01	♌
28	08:30	28	09:34	♍
30	11:17	30	12:16	♎

DECLINATION

Day	☉	☽	☿	♀	♂	♃	♄	♅	♆	♇
01 Mo	14S19	08N25	17S53	17S49	21S15	03S46	02S31	01S50	13S20	18S44
02 Tu	14 39	02 38	18 25	17 24	21 24	03 47	02 33	01 51	13 20	18 44
03 We	14 57	03S21	18 56	16 59	21 33	03 48	02 36	01 51	13 20	18 44
04 Th	15 16	09 12	19 26	16 35	21 41	03 49	02 38	01 52	13 20	18 44
05 Fr	15 35	14 31	19 55	16 10	21 50	03 50	02 41	01 53	13 20	18 45
06 Sa	15 53	18 56	20 24	15 45	21 58	03 51	02 43	01 53	13 20	18 45
07 Su	16 11	22 09	20 51	15 21	22 06	03 52	02 46	01 54	13 20	18 45
08 Mo	16 28	23 55	21 17	14 58	22 13	03 53	02 48	01 54	13 20	18 45
09 Tu	16 46	24 13	21 42	14 35	22 21	03 53	02 51	01 55	13 20	18 45
10 We	17 03	23 08	22 06	14 12	22 28	03 54	02 53	01 55	13 20	18 45
11 Th	17 20	20 52	22 29	13 51	22 35	03 54	02 55	01 56	13 20	18 46
12 Fr	17 36	17 41	22 51	13 30	22 42	03 55	02 58	01 56	13 20	18 46
13 Sa	17 52	13 48	23 12	13 10	22 49	03 55	03 00	01 56	13 20	18 46
14 Su	18 08	09 26	23 32	12 52	22 55	03 55	03 03	01 57	13 20	18 46
15 Mo	18 24	04 46	23 50	12 34	23 01	03 55	03 05	01 57	13 20	18 46
16 Tu	18 39	00N05	24 07	12 17	23 07	03 55	03 07	01 58	13 20	18 46
17 We	18 54	04 56	24 23	12 02	23 13	03 55	03 09	01 58	13 20	18 46
18 Th	19 09	09 40	24 38	11 48	23 19	03 55	03 12	01 58	13 20	18 47
19 Fr	19 23	14 05	24 51	11 35	23 24	03 55	03 14	01 59	13 19	18 47
20 Sa	19 37	17 59	25 03	11 23	23 29	03 54	03 16	01 59	13 19	18 47
21 Su	19 50	21 08	25 14	11 12	23 34	03 54	03 18	01 59	13 19	18 47
22 Mo	20 03	23 18	25 24	11 02	23 38	03 54	03 20	01 59	13 19	18 47
23 Tu	20 16	24 13	25 32	10 54	23 43	03 53	03 22	02 00	13 19	18 47
24 We	20 29	23 47	25 39	10 47	23 47	03 52	03 24	02 00	13 19	18 47
25 Th	20 41	21 56	25 44	10 41	23 51	03 52	03 26	02 00	13 18	18 47
26 Fr	20 52	18 48	25 48	10 36	23 54	03 51	03 28	02 00	13 18	18 48
27 Sa	21 04	14 34	25 50	10 32	23 58	03 50	03 30	02 00	13 18	18 48
28 Su	21 15	09 32	25 51	10 29	24 01	03 49	03 32	02 01	13 18	18 48
29 Mo	21 25	03 58	25 51	10 27	24 04	03 48	03 34	02 01	13 17	18 48
30 Tu	21 35	01S50	25 49	10 26	24 06	03 47	03 36	02 01	13 17	18 48

ASPECTARIAN

01 08:45 ☽ □ ♂
09:30 ☽ △ ♆
09:40 ☽ ✶ ♀
13:43 ♀ ✶ ♇
19:16 ☽ ✶ ☉
19:24 ☽ # ♃
02 00:23 ☽ # ♄
03:14 ☽ # ♅
13:18 ☽ ✶ ☿
14:42 ☿ ∥ ♇
18:03 ☽ ∥ ♅
19:24 ☽ ☍ ♃
21:00 ☽ ∥ ♄
03 00:36 ☽ ☍ ♅
01:53 ☽ ∥ ♃
10:54 ☽ □ ♇
12:38 ☽ ✶ ♂
04 00:34 ☽ ☌ ♄
18:27 ☿ ∥ ♆
21:46 ☿ △ ♃
23:34 ☽ △ ♆
05 05:43 ☽ ∥ ☉
07:41 ☽ ∥ ♀
08:24 ☽ ☌ ♀
11:59 ☽ ✶ ♇
19:46 ♀ ∥ ☉
22:54 ☽ ∥ ☿
06 06:45 ☿ □ ♆
11:22 ☽ ∥ ☿
21:47 ☽ △ ♃
23:37 ☽ ∥ ♂
07 00:15 ☿ △ ♅
01:29 ☽ □ ♆
03:22 ☽ △ ♅
03:45 ☽ ☌ ☿
06:05 ♆ SD
21:43 ☽ ☌ ♂
08 03:06 ♀ ♎R
05:48 ☽ ✶ ♄
23:43 ☿ ♐
09 02:04 ☽ □ ♃
06:10 ☽ ✶ ♆
08:05 ☽ □ ♅
12:36 ☽ ✶ ♀
20:14 ☽ ☌ ♇
10 07:55 ☽ ∥ ♂
10:31 ☽ ∥ ☿
12:58 ☽ □ ♄
11 00:17 ☽ ✶ ☉
09:18 ☿ ∥ ♂
10:04 ☽ ✶ ♃
16:32 ☽ ✶ ♅
16:33 ☽ □ ♆
19:58 ☽ □ ♀
12 00:31 ☽ ∥ ☉
08:22 ☽ ✶ ☿
11:49 ♀ ∥ ♆
20:50 ☽ ✶ ♂
23:51 ☽ △ ♄
13 02:42 ☽ ∥ ♆
03:55 ☽ ∥ ♀
14 02:11 ☽ ☌ ♆
06:34 ☽ △ ♀
17:58 ☽ ✶ ♇
15 04:13 ☽ ∥ ♃
05:01 ☽ □ ☿
05:58 ♂ ✶ ♄
08:20 ☽ ∥ ♄
13:16 ☽ □ ♂
13:58 ☽ ∥ ♅
23:45 ☉ △ ♃
16 09:18 ☽ # ♅
09:56 ☽ ☌ ♃
10:53 ☽ △ ☉
11:39 ☉ ∥ ♆
15:07 ☽ # ♄
16:38 ☽ ☌ ♅
18:58 ☽ # ♃
17 06:36 ☽ □ ♆
18 01:27 ☽ △ ☿
01:31 ☽ ☍ ♄
02:09 ☿ ✶ ♄
05:16 ☽ △ ♂
09:54 ☉ □ ♆
10:48 ☽ # ♀
16:55 ♃ SD
19:46 ☽ ∥ ♆
21:19 ♀ SD
19 02:15 ☽ ✶ ♆
05:34 ☽ ☍ ♀
05:57 ☉ △ ♅
17:33 ☽ △ ♇
20 05:31 ☽ # ♆
12:29 ☽ # ☉
18:53 ☿ ☌ ♂
21 06:42 ☽ ✶ ♃
11:18 ☽ □ ♆
12:46 ☽ ✶ ♅
22 06:23 ☽ # ♂
10:15 ☉ ♐
20:05 ☽ △ ♄
23 05:22 ☽ ☍ ♂
08:25 ☽ ☍ ☿
13:40 ☽ □ ♃
18:04 ☽ △ ♆
19:24 ☽ □ ♅
21:57 ☽ △ ♀
23:42 ☽ # ♂
24 08:24 ☽ ☍ ♇
25 02:05 ☽ □ ♄
10:02 ☽ # ☉
14:19 ☿ □ ♃
18:51 ☽ △ ♃
23:59 ☽ # ♆
26 00:19 ☽ △ ♅
03:44 ☽ □ ♀
13:15 ☽ △ ☉
27 06:23 ☽ # ♆
06:36 ☽ ✶ ♄
13:15 ☿ ✶ ♆
19:41 ☽ # ♀
20:35 ☽ △ ♂
28 02:27 ☿ □ ♅
02:49 ☽ ☍ ♆
04:05 ☽ △ ☿
08:30 ☽ ✶ ♀
16:38 ☽ △ ♆
29 00:41 ☽ # ♃
01:38 ☽ # ♄
08:07 ☽ # ♅
14:51 ♂ □ ♃
30 00:33 ♀ ♏
00:43 ☽ ∥ ♅
01:41 ☽ ☍ ♃
02:15 ☽ □ ♂
06:41 ☽ □ ☉
07:21 ☽ ∥ ♄
08:01 ☽ ∥ ♃
11:17 ☽ □ ☿
19:22 ☽ □ ♆

⚷ Chiron

01 Dec. 07 S 03

Day	Position
01	26♒05R
04	26 04
07	26 04D
10	26 05
13	26 06
16	26 07
19	26 10
22	26 12
25	26 16
28	26 19

05 16:23 26♒04 D

December 2010

21 08:18 29♊21 ✷ Total Lunar Eclipse (mag 1.261)

Day	S. T. h m s	☉ ° ' "	☽ ° ' "	☿ ° '	♀ ° '	♂ ° '	♃ ° '	♄ ° '	♅ ° '	♆ ° '	♇ ° '	☊ True ° '
01 We	04 38 59	08♐ 40 23	06♎ 59 03	30♐ 00	00♏ 24	24♐ 44	23♓ 45	14♎ 34	26♓R41	26♒ 04	04♑ 14	02♑R58
02 Th	04 42 56	09 41 13	21 15 32	01♑ 01	00 51	25 29	23 48	14 39	26 41	26 05	04 16	02 54
03 Fr	04 46 53	10 42 04	05♏ 28 30	01 59	01 19	26 14	23 51	14 44	26 41	26 06	04 18	02 50
04 Sa	04 50 49	11 42 56	19 33 41	02 52	01 50	26 59	23 54	14 49	26 40	26 07	04 20	02 47
05 Su	04 54 46	12 43 49	03♐ 26 52	03 40	02 21	27 44	23 57	14 54	26 40	26 08	04 22	02 44
06 Mo	04 58 42	13 44 44	17 04 23	04 22	02 54	28 30	24 00	14 59	26 40	26 09	04 24	02 42
07 Tu	05 02 39	14 45 40	00♑ 23 43	04 58	03 29	29 15	24 04	15 04	26D 40	26 10	04 26	02 41
08 We	05 06 35	15 46 37	13 23 45	05 25	04 05	00♑ 00	24 08	15 09	26 40	26 11	04 28	02D 41
09 Th	05 10 32	16 47 34	26 04 54	05 45	04 42	00 46	24 12	15 14	26 41	26 12	04 30	02 42
10 Fr	05 14 28	17 48 32	08♒ 29 00	05 55	05 21	01 31	24 16	15 19	26 41	26 13	04 33	02 44
11 Sa	05 18 25	18 49 31	20 38 57	05R 55	06 00	02 17	24 20	15 23	26 41	26 14	04 35	02 46
12 Su	05 22 22	19 50 31	02♓ 38 32	05 44	06 41	03 02	24 25	15 28	26 41	26 15	04 37	02 48
13 Mo	05 26 18	20 51 31	14 32 05	05 22	07 24	03 48	24 30	15 32	26 42	26 16	04 39	02 49
14 Tu	05 30 15	21 52 31	26 24 21	04 48	08 07	04 33	24 35	15 37	26 42	26 18	04 41	02 50
15 We	05 34 11	22 53 32	08♈ 20 06	04 03	08 51	05 19	24 40	15 41	26 42	26 19	04 43	02 50
16 Th	05 38 08	23 54 34	20 23 59	03 08	09 36	06 05	24 45	15 45	26 43	26 20	04 45	02R 50
17 Fr	05 42 04	24 55 36	02♉ 40 09	02 02	10 23	06 51	24 50	15 49	26 43	26 21	04 47	02 50
18 Sa	05 46 01	25 56 39	15 11 59	00 49	11 10	07 36	24 56	15 53	26 44	26 23	04 50	02 50
19 Su	05 49 57	26 57 42	28 01 55	29♐ 29	11 58	08 22	25 02	15 57	26 45	26 24	04 52	02 49
20 Mo	05 53 54	27 58 45	11♊ 11 09	28 07	12 47	09 08	25 08	16 01	26 45	26 25	04 54	02 49
21 Tu	05 57 51	28 59 49	24 39 27	26 44	13 37	09 54	25 14	16 05	26 46	26 27	04 56	02D 49
22 We	06 01 47	00♑ 00 54	08♋ 25 12	25 24	14 28	10 40	25 21	16 08	26 47	26 28	04 58	02 49
23 Th	06 05 44	01 01 59	22 25 28	24 09	15 19	11 26	25 27	16 12	26 48	26 30	05 00	02 49
24 Fr	06 09 40	02 03 05	06♌ 36 26	23 00	16 12	12 12	25 34	16 15	26 49	26 31	05 02	02R 48
25 Sa	06 13 37	03 04 11	20 53 52	22 01	17 05	12 58	25 41	16 19	26 50	26 33	05 05	02 47
26 Su	06 17 33	04 05 18	05♍ 13 34	21 11	17 58	13 44	25 48	16 22	26 51	26 34	05 07	02 46
27 Mo	06 21 30	05 06 25	19 31 47	20 32	18 53	14 31	25 55	16 25	26 52	26 36	05 09	02 45
28 Tu	06 25 26	06 07 33	03♎ 45 24	20 04	19 48	15 17	26 02	16 28	26 53	26 37	05 11	02D 44
29 We	06 29 23	07 08 41	17 51 59	19 46	20 43	16 03	26 10	16 31	26 54	26 39	05 13	02 43
30 Th	06 33 20	08 09 50	01♏ 49 44	19 38	21 40	16 50	26 17	16 34	26 55	26 41	05 15	02 43
31 Fr	06 37 16	09 11 00	15 37 22	19D 40	22 36	17 36	26 25	16 37	26 56	26 42	05 18	02 42

Data for	12-01-2010
Julian Day	2455531.50
Ayanamsa	24 00 49
SVP	05 ♓ 06 42
☽ ☊ Mean	03 ♑ 57 R

● ◐ PHASES ○ ◑

05	17:36	●	13♐29
13	13:59	◐	21♓27
21	08:13	✷	29♊21
28	04:19	◑	06♎19

LAST ASPECT ☽ Day	h m	INGRESS Day	h m	
02	08:09	02	14:44	♏
04	12:14	04	18:00	♐
06	21:47	06	23:17	♑
09	01:08	09	07:32	♒
11	11:10	11	18:41	♓
14	00:36	14	07:15	♈
16	11:41	16	18:49	♉
18	21:37	19	03:38	♊
21	08:14	21	09:22	♋
23	07:26	23	12:51	♌
25	09:29	25	15:15	♍
27	12:22	27	17:39	♎
29	15:06	29	20:50	♏

DECLINATION

Day	☉	☽	☿	♀	♂	♃	♄	♅	♆	♇
01 We	21S 45	07S 34	25S 46	10S 26	24S 09	03S 46	03S 38	02S 01	13S 17	18S 48
02 Th	21 54	12 54	25 42	10 27	24 11	03 44	03 40	02 01	13 17	18 48
03 Fr	22 03	17 30	25 36	10 29	24 13	03 43	03 42	02 01	13 16	18 48
04 Sa	22 11	21 05	25 29	10 32	24 14	03 42	03 43	02 01	13 16	18 48
05 Su	22 19	23 22	25 20	10 35	24 15	03 40	03 45	02 01	13 16	18 48
06 Mo	22 27	24 14	25 10	10 39	24 17	03 38	03 47	02 01	13 15	18 49
07 Tu	22 34	23 39	25 00	10 44	24 17	03 37	03 49	02 01	13 15	18 49
08 We	22 41	21 48	24 48	10 50	24 18	03 35	03 50	02 01	13 15	18 49
09 Th	22 47	18 53	24 35	10 56	24 18	03 33	03 52	02 01	13 14	18 49
10 Fr	22 53	15 10	24 20	11 03	24 18	03 31	03 53	02 01	13 14	18 49
11 Sa	22 58	10 54	24 05	11 11	24 18	03 29	03 55	02 00	13 14	18 49
12 Su	23 03	06 17	23 50	11 19	24 17	03 27	03 57	02 00	13 13	18 49
13 Mo	23 07	01 28	23 33	11 28	24 16	03 25	03 58	02 00	13 13	18 49
14 Tu	23 11	03N22	23 16	11 37	24 15	03 23	03 59	02 00	13 12	18 49
15 We	23 15	08 07	22 58	11 47	24 14	03 21	04 01	02 00	13 12	18 49
16 Th	23 18	12 37	22 39	11 57	24 12	03 18	04 02	02 00	13 12	18 49
17 Fr	23 20	16 42	22 20	12 07	24 10	03 16	04 04	01 59	13 11	18 49
18 Sa	23 23	20 07	22 01	12 18	24 08	03 14	04 05	01 59	13 11	18 49
19 Su	23 24	22 39	21 43	12 30	24 06	03 11	04 06	01 59	13 10	18 49
20 Mo	23 25	24 03	21 24	12 41	24 03	03 08	04 07	01 58	13 10	18 49
21 Tu	23 26	24 05	21 07	12 53	24 00	03 06	04 09	01 58	13 09	18 49
22 We	23 26	22 39	20 51	13 05	23 57	03 03	04 10	01 58	13 09	18 49
23 Th	23 26	19 49	20 36	13 18	23 53	03 00	04 11	01 57	13 08	18 49
24 Fr	23 25	15 46	20 24	13 30	23 50	02 57	04 12	01 57	13 08	18 49
25 Sa	23 24	10 48	20 14	13 43	23 46	02 54	04 13	01 57	13 07	18 50
26 Su	23 22	05 14	20 07	13 56	23 41	02 51	04 14	01 56	13 07	18 50
27 Mo	23 20	00S 35	20 02	14 09	23 37	02 48	04 15	01 56	13 06	18 50
28 Tu	23 18	06 20	20 00	14 23	23 32	02 45	04 16	01 55	13 06	18 50
29 We	23 15	11 43	20 00	14 36	23 27	02 42	04 17	01 55	13 05	18 50
30 Th	23 11	16 26	20 02	14 50	23 21	02 39	04 18	01 54	13 04	18 50
31 Fr	23 07	20 13	20 07	15 03	23 15	02 35	04 19	01 54	13 04	18 50

ASPECTARIAN

01 00:11 ☿ ♑
03:03 ☽ ⚹ ☉
12:39 ☽ ∥ ♀
12:49 ☽ ☌ ♄
16:35 ☿ ⚹ ♀
02 01:50 ☽ ∥ ♆
07:31 ☽ ⚹ ♂
08:09 ☽ △ ♆
16:44 ☽ ☌ ♀
17:41 ☽ ⚹ ☿
19:41 ♂ ⚹ ♆
22:01 ☽ ⚹ ♇
03 07:52 ☽ ∥ ♇
10:11 ♃ ∥ ♄
14:00 ♂ □ ♅
04 07:29 ☽ △ ♃
10:28 ☽ ∥ ☉
11:18 ☽ □ ♆
12:14 ☽ △ ♅
05 20:17 ☽ ⚹ ♄
06 01:18 ☿ ☌ ♇
01:51 ♅ SD
12:28 ☽ □ ♃
16:18 ☽ ⚹ ♆
17:14 ☽ □ ♅
21:47 ☽ ☌ ♂
07 05:54 ☽ ⚹ ♀
07:25 ☽ ☌ ♇
08:00 ☉ ⚹ ♄
08:42 ☽ ☌ ☿
14:53 ☽ ∥ ☉
23:49 ♂ ♑
08 03:19 ☽ □ ♄
16:17 ♀ ⚹ ♇
20:23 ☽ ⚹ ♃
09 [illegible]
01:08 ☽ ⚹ ♅
17:32 ☽ □ ♀
10 03:52 ☿ ∥ ♂
11:10 ☽ ∥ ♆
12:06 ☿ SR
13:30 ☽ △ ♄
20:03 ☽ ⚹ ☉
21:01 ☿ ⚹ ♀
22:29 ☽ ∥ ♀
11 11:10 ☽ ☌ ♆
12 00:51 ☽ ⚹ ♂
03:59 ☽ ⚹ ♇
06:05 ☽ ⚹ ☿
08:40 ☽ △ ♀
11:37 ☽ ∥ ♄
14:13 ☽ ∥ ♃
21:19 ☽ ∥ ♅
13 17:08 ☽ # ♅
20:17 ☽ ☌ ♃
14 00:00 ☽ # ♃
00:36 ☽ ☌ ♅
03:03 ☽ # ♄
04:09 ☿ ☌ ♇
04:10 ☿ ☌ ♂
04:13 ♂ ☌ ♇
04:53 ☿ ∥ ☉
15:58 ☽ □ ☿
16:43 ☽ □ ♇
17:32 ☽ □ ♂
15 14:44 ☽ ☍ ♄
20:07 ☽ # ♀
16 03:11 ☽ # ♆
[illegible]
11:41 ☽ ⚹ ♆
21:47 ☉ □ ♃
22:52 ☽ △ ☿
17 04:07 ☽ △ ♇
08:35 ☽ △ ♂
14:14 ☽ # ♇
15:49 ☽ ☍ ♀
18 10:30 ☉ ⚹ ♆
14:54 ☿ ♐R
14:59 ☽ # ☿
18:24 ☽ ⚹ ♃
18:48 ☉ □ ♅
20:59 ☽ □ ♆
21:37 ☽ ⚹ ♅
19 10:25 ☽ # ☉
20 00:17 ☽ # ♂
01:23 ☿ ☌ ☉
08:42 ☽ △ ♄
23:31 ☿ □ ♅
21 01:02 ☽ □ ♃
02:24 ☽ # ♂
03:09 ☽ △ ♆
03:21 ☽ ☍ ☿
03:43 ☽ □ ♅
05:03 ☿ ⚹ ♆
13:44 ☽ # ☉
18:01 ☽ ☍ ♇
23:39 ☉ ♑
22 00:59 ☿ □ ♃
04:06 ☽ ☍ ♂
06:23 ♀ ∥ ♆
11:05 ☽ △ ♀
13:20 ☽ □ ♄
[illegible]
23 05:11 ☽ △ ♃
06:30 ☽ # ♇
07:26 ☽ △ ♅
24 10:53 ☽ # ♀
13:11 ☽ # ♆
16:17 ☽ ⚹ ♄
17:10 ☽ □ ♀
25 01:45 ☽ △ ☿
09:29 ☽ ☍ ♆
21:57 ☽ △ ☉
23:49 ☽ △ ♇
26 04:07 ☽ # ♄
[illegible]
13:38 ☽ # ♅
15:05 ☽ △ ♂
22:50 ☽ ⚹ ♀
27 01:03 ☉ ☌ ♇
01:37 ☽ □ ☿
05:30 ☽ ∥ ♅
09:03 ☽ ∥ ♃
10:50 ☽ ☍ ♃
12:22 ☽ ☍ ♅
15:12 ☽ ∥ ♄
28 02:26 ☽ □ ♇
20:43 ☽ □ ♂
21:42 ☽ ☌ ♄
29 03:12 ☽ [illegible]
06:31 ☽ ∥ ♆
14:52 ☽ ∥ ♀
15:06 ☽ △ ♆
15:29 ♂ □ ♄
30 05:57 ☽ ⚹ ♇
07:22 ☿ SD
11:52 ☽ ⚹ ☉
14:19 ☽ ∥ ♇
23:10 ☽ ∥ ☿
31 03:40 ☽ ⚹ ♂
13:12 ☽ ☌ ♀
19:12 ☽ △ ♃
19:34 ☽ □ ♆
19:58 ☽ △ ♅

⚷ Chiron

01 Dec.	07 S 08
01	26♒24
04	26 29
07	26 34
10	26 40
13	26 46
16	26 53
19	27 01
22	27 08
25	27 16
28	27 25
31	27 34

04 08:52 13♑39 ⊙ Solar Eclipse (mag 0.858)

January 2011

Day	S. T.	☉	☽	☿	♀	♂	♃	♄	♅	♆	♇	☊ True
	h m s	° ′ ″	° ′ ″	° ′	° ′	° ′	° ′	° ′	° ′	° ′	° ′	° ′
01 Sa	06 41 13	10♑ 12 10	29♏ 13 53	19♐ 50	23♏ 34	18♑ 22	26♓ 33	16♎ 40	26♓ 58	26♒ 44	05♑ 20	02♑ 42
02 Su	06 45 09	11 13 20	12 38 24	20 09	24 32	19 09	26 41	16 42	26 59	26 46	05 22	02 43
03 Mo	06 49 06	12 14 31	25 50 11	20 35	25 30	19 55	26 49	16 45	27 00	26 48	05 24	02 43
04 Tu	06 53 02	13 15 42	08♑ 48 38	21 08	26 29	20 42	26 58	16 47	27 02	26 49	05 26	02R 43
05 We	06 56 59	14 16 53	21 33 25	21 46	27 28	21 28	27 06	16 49	27 03	26 51	05 28	02 43
06 Th	07 00 56	15 18 03	04♒ 04 45	22 30	28 28	22 15	27 15	16 52	27 05	26 53	05 30	02 42
07 Fr	07 04 52	16 19 14	16 23 28	23 19	29 28	23 02	27 24	16 54	27 06	26 55	05 33	02 41
08 Sa	07 08 49	17 20 24	28 31 10	24 12	00♐ 29	23 48	27 33	16 56	27 08	26 57	05 35	02 39
09 Su	07 12 45	18 21 34	10♓ 30 11	25 09	01 30	24 35	27 42	16 58	27 10	26 58	05 37	02 38
10 Mo	07 16 42	19 22 43	22 23 34	26 09	02 32	25 22	27 51	16 59	27 11	27 00	05 39	02 36
11 Tu	07 20 38	20 23 52	04♈ 15 06	27 12	03 33	26 08	28 01	17 01	27 13	27 02	05 41	02 36
12 We	07 24 35	21 25 00	16 09 06	28 18	04 36	26 55	28 10	17 03	27 15	27 04	05 43	02D 36
13 Th	07 28 31	22 26 08	28 10 16	29 27	05 38	27 42	28 20	17 04	27 17	27 06	05 45	02 36
14 Fr	07 32 28	23 27 15	10♉ 23 22	00♑ 37	06 41	28 29	28 30	17 06	27 19	27 08	05 47	02 38
15 Sa	07 36 25	24 28 22	22 53 01	01 50	07 44	29 16	28 40	17 07	27 20	27 10	05 49	02 40
16 Su	07 40 21	25 29 28	05♊ 43 12	03 04	08 48	00♒ 03	28 50	17 08	27 22	27 12	05 52	02 42
17 Mo	07 44 18	26 30 33	18 56 54	04 20	09 52	00 49	29 00	17 09	27 24	27 14	05 54	02 43
18 Tu	07 48 14	27 31 38	02♋ 35 33	05 37	10 56	01 36	29 10	17 10	27 27	27 16	05 56	02R 43
19 We	07 52 11	28 32 42	16 38 27	06 56	12 00	02 23	29 20	17 11	27 29	27 18	05 58	02 42
20 Th	07 56 07	29 33 46	01♌ 02 28	08 16	13 05	03 10	29 31	17 11	27 31	27 20	06 00	02 39
21 Fr	08 00 04	00♒ 34 48	15 42 05	09 37	14 10	03 57	29 42	17 12	27 33	27 22	06 02	02 35
22 Sa	08 04 00	01 35 50	00♍ 30 05	10 59	15 15	04 44	29 52	17 13	27 35	27 24	06 04	02 31
23 Su	08 07 57	02 36 52	15 18 31	12 22	16 20	05 31	00♈ 03	17 13	27 37	27 26	06 06	02 26
24 Mo	08 11 54	03 37 53	29 59 58	13 47	17 26	06 19	00 14	17 13	27 40	27 29	06 08	02 21
25 Tu	08 15 50	04 38 54	14♎ 28 30	15 12	18 32	07 06	00 25	17 14	27 42	27 31	06 10	02 17
26 We	08 19 47	05 39 54	28 40 15	16 38	19 38	07 53	00 36	17 14	27 44	27 33	06 12	02 15
27 Th	08 23 43	06 40 54	12♏ 33 29	18 04	20 45	08 40	00 47	17R 14	27 47	27 35	06 13	02 14
28 Fr	08 27 40	07 41 53	26 08 14	19 32	21 51	09 27	00 59	17 13	27 49	27 37	06 15	02D 14
29 Sa	08 31 36	08 42 52	09♐ 25 45	21 00	22 58	10 14	01 10	17 13	27 52	27 39	06 17	02 15
30 Su	08 35 33	09 43 50	22 27 52	22 30	24 05	11 02	01 22	17 13	27 54	27 41	06 19	02 16
31 Mo	08 39 29	10 44 47	05♑ 16 30	23 59	25 12	11 49	01 33	17 12	27 57	27 44	06 21	02R 16

Data for	01-01-2011
Julian Day	2455562.50
Ayanamsa	24 00 55
SVP	05 ♓ 06 40
☽ ☊ Mean	02 ♑ 18 R

● ◐ PHASES ○ ◑

04	09:03	⊙	13♑39
12	11:31	◐	21♈54
19	21:22	○	29♋27
26	12:58	◑	06♏13

LAST ASPECT Day	h m	☽ INGRESS Day	h m	
31	19:58	01	01:22	♐
03	02:09	03	07:39	♑
05	12:16	05	16:08	♒
07	20:51	08	02:57	♓
10	11:12	10	15:24	♈
13	02:47	13	03:37	♉
15	12:47	15	13:23	♊
17	17:58	17	19:30	♋
19	21:27	19	22:17	♌
21	18:59	21	23:11	♍
23	20:09	24	00:00	♎
25	22:05	26	02:16	♏
28	03:01	28	06:55	♐
30	10:11	30	14:04	♑

DECLINATION

Day	☉	☽	☿	♀	♂	♃	♄	♅	♆	♇
01 Sa	23S 03	22S 50	20S 13	15S 17	23S 10	02S 32	04S 19	01S 53	13S 03	18S 50
02 Su	22 58	24 07	20 21	15 30	23 03	02 29	04 20	01 53	13 03	18 50
03 Mo	22 52	24 00	20 30	15 44	22 57	02 25	04 21	01 52	13 02	18 50
04 Tu	22 47	22 35	20 39	15 57	22 50	02 22	04 21	01 51	13 02	18 50
05 We	22 40	20 02	20 50	16 10	22 43	02 18	04 22	01 51	13 01	18 50
06 Th	22 34	16 34	21 01	16 24	22 36	02 14	04 23	01 50	13 00	18 50
07 Fr	22 26	12 27	21 12	16 37	22 28	02 11	04 23	01 49	13 00	18 50
08 Sa	22 19	07 54	21 24	16 50	22 21	02 07	04 24	01 49	12 59	18 50
09 Su	22 11	03 07	21 35	17 03	22 13	02 03	04 24	01 48	12 58	18 49
10 Mo	22 02	01N44	21 46	17 16	22 04	01 59	04 25	01 47	12 58	18 49
11 Tu	21 53	06 30	21 57	17 28	21 56	01 55	04 25	01 47	12 57	18 49
12 We	21 44	11 03	22 07	17 41	21 47	01 51	04 25	01 46	12 56	18 49
13 Th	21 34	15 14	22 17	17 53	21 38	01 47	04 26	01 45	12 56	18 49
14 Fr	21 24	18 52	22 26	18 05	21 29	01 43	04 26	01 44	12 55	18 49
15 Sa	21 13	21 44	22 34	18 17	21 19	01 39	04 26	01 43	12 54	18 49
16 Su	21 02	23 36	22 42	18 28	21 10	01 35	04 26	01 43	12 54	18 49
17 Mo	20 51	24 13	22 48	18 39	21 00	01 31	04 26	01 42	12 53	18 49
18 Tu	20 39	23 25	22 54	18 50	20 49	01 26	04 27	01 41	12 52	18 49
19 We	20 27	21 08	22 59	19 01	20 39	01 22	04 27	01 40	12 52	18 49
20 Th	20 14	17 29	23 03	19 11	20 28	01 18	04 27	01 39	12 51	18 49
21 Fr	20 02	12 43	23 05	19 21	20 18	01 13	04 27	01 38	12 50	18 49
22 Sa	19 48	07 09	23 07	19 30	20 06	01 09	04 27	01 37	12 50	18 49
23 Su	19 34	01 12	23 07	19 40	19 55	01 04	04 26	01 37	12 49	18 49
24 Mo	19 20	04S 46	23 07	19 48	19 43	01 00	04 26	01 36	12 48	18 49
25 Tu	19 06	10 24	23 05	19 57	19 32	00 55	04 26	01 35	12 47	18 49
26 We	18 51	15 21	23 02	20 05	19 20	00 51	04 26	01 34	12 47	18 49
27 Th	18 36	19 23	22 58	20 12	19 08	00 46	04 26	01 33	12 46	18 49
28 Fr	18 21	22 16	22 52	20 20	18 55	00 41	04 25	01 32	12 45	18 49
29 Sa	18 05	23 52	22 45	20 26	18 42	00 37	04 25	01 31	12 44	18 49
30 Su	17 49	24 06	22 37	20 33	18 30	00 32	04 25	01 30	12 44	18 49
31 Mo	17 32	23 03	22 28	20 38	18 17	00 27	04 24	01 29	12 43	18 48

ASPECTARIAN

01 02:31 ☽ ∥ ☉
03:56 ☽ ∥ ♂
02 07:22 ☽ ⚹ ♄
14:03 ☽ ☌ ☿
03 01:45 ☽ ⚹ ♆
01:50 ☽ □ ♃
02:09 ☽ □ ♅
17:42 ☽ ☌ ♇
20:48 ☽ ∥ ♂
21:37 ☽ ∥ ☉
04 08:28 ♀ □ ♆
12:52 ♃ ☌ ♅
13:34 ♀ △ ♅
13:40 ♀ △ ♃
15:01 ☽ □ ♄
17:49 ☽ ∥ ☿
23:50 ☽ ☌ ♂
05 08:59 ☽ ∥ ♇
10:30 ☽ ⚹ ♅
10:43 ☽ ⚹ ♃
12:16 ☽ ⚹ ♀
06 01:01 ☽ ∥ ♀
20:59 ☽ ∥ ♆
07 01:00 ☽ △ ♄
12:31 ♀ ♐
14:01 ☉ □ ♄
14:44 ☽ ⚹ ☿
20:51 ☽ ☌ ♆
08 04:17 ☽ □ ♀
14:09 ☽ ⚹ ♇
17:37 ☽ ∥ ♄
09 05:22 ☽ ∥ ♃
06:32 ☽ ∥ ♅
17:20 ☽ ⚹ ☉
10 00:16 ☽ # ♅
01:14 ☽ # ♃
06:26 ☽ ⚹ ♂
08:20 ☽ □ ☿
09:44 ☽ ☌ ♅
11:12 ☽ ☌ ♃
13:23 ☽ # ♄
20:01 ☿ ∥ ☉
20:04 ☿ ⚹ ♆
22:28 ☽ △ ♀
23:06 ☿ ∥ ♂
11 00:13 ☿ □ ♅
02:54 ☽ □ ♇
20:37 ☿ □ ♃
12 01:48 ☽ ☍ ♄
10:26 ☽ # ♆
10:30 ♂ ⚹ ♅
21:52 ☽ ⚹ ♆
23:00 ☽ □ ♂
13 02:47 ☽ △ ☿
11:25 ☿ ♑
15:00 ☽ △ ♇
15:21 ♃ ∥ ♅
18:00 ☽ # ♀
23:36 ☽ # ♇
14 00:36 ♂ ⚹ ♃
19:21 ☽ # ☉
20:11 ☽ # ♂
15 03:16 ☽ △ ☉
08:06 ☽ □ ♆
08:26 ☽ ⚹ ♅
09:35 ☽ # ☿
11:02 ☽ ⚹ ♃
12:47 ☽ △ ♂
22:42 ♂ ♒
16 06:08 ☽ ☍ ♀
20:47 ☽ △ ♄
17 14:42 ☽ △ ♆
15:00 ☽ □ ♅
17:58 ☽ □ ♃
21:57 ☉ ⚹ ♅
21:58 ♀ ∥ ♇
18 05:46 ☽ ☍ ☿
05:46 ☽ ☍ ♇
05:50 ☿ ☌ ♇
06:46 ☽ # ☿
19 00:54 ☽ □ ♄
04:02 ☽ # ♂
05:39 ☽ # ☉
14:21 ☽ # ♀
16:12 ☽ # ♇
18:09 ☽ △ ♅
21:27 ☽ △ ♃
22:39 ☉ ⚹ ♃
20 03:43 ☽ ☍ ♂
10:19 ☉ ♒
21:18 ☽ △ ♀
23:26 ☽ # ♆
21 02:26 ☽ ⚹ ♄
18:59 ☽ ☍ ♆
22 09:01 ☽ △ ♇
11:02 ☽ # ♄
17:11 ♃ ♈
18:34 ♀ ∥ ☉
18:44 ☽ △ ☿
21:21 ☽ # ♅
23 00:30 ☽ # ♃
01:49 ☽ □ ♀
08:57 ☽ ∥ ♃
11:10 ☽ ∥ ♅
18:08 ♀ ∥ ♂
19:19 ♀ ⚹ ♄
20:09 ☽ ☍ ♅
23:38 ☽ ∥ ♄
24 00:23 ☽ ☍ ♃
06:26 ☽ △ ☉
10:08 ☽ □ ♇
11:00 ☽ △ ♂
25 01:20 ☽ □ ☿
04:37 ☽ ☌ ♄
07:23 ☽ ⚹ ♀
11:05 ☽ ∥ ♆
22:05 ☽ △ ♆
26 03:53 ☉ ∥ ♇
06:11 ♄ SR
09:58 ☿ □ ♄
12:58 ☽ ⚹ ♇
16:49 ☽ □ ♂
19:06 ☽ ∥ ☉
20:08 ☽ ∥ ♇
22:18 ☽ ∥ ♂
27 06:06 ☽ ∥ ♀
10:51 ☽ ⚹ ☿
28 02:39 ☽ □ ♆
03:01 ☽ △ ♅
06:31 ☽ ∥ ☿
08:49 ☽ △ ♃
12:18 ♂ ∥ ♇
22:35 ☽ ⚹ ☉
29 01:34 ☽ ⚹ ♂
14:17 ☽ ⚹ ♄
30 03:18 ☽ ☌ ♀
09:46 ☽ ⚹ ♆
10:11 ☽ □ ♅
16:53 ☽ □ ♃
31 02:02 ☽ ☌ ♇
08:37 ☽ ∥ ☿
22:40 ☽ □ ♄

⚷ Chiron

01	Dec.	06 S 54
03	27♒43	
06	27 53	
09	28 03	
12	28 13	
15	28 24	
18	28 35	
21	28 46	
24	28 57	
27	29 09	
30	29 21	

February 2011

Day	S. T.	☉	☽	☿	♀	♂	♃	♄	♅	♆	♇	☊ True
	h m s	° ' "	° ' "	° '	° '	° '	° '	° '	° '	° '	° '	° '
01 Tu	08 43 26	11♒45 43	17♑53 23	25♑30	26♐20	12♒36	01♈45	17♎R12	27♓59	27♒46	06♑23	02♑R15
02 We	08 47 23	12 46 39	00♒19 54	27 02	27 27	13 23	01 57	17 11	28 02	27 48	06 25	02 12
03 Th	08 51 19	13 47 33	12 37 11	28 34	28 35	14 11	02 09	17 10	28 05	27 50	06 27	02 07
04 Fr	08 55 16	14 48 27	24 46 15	00♒07	29 43	14 58	02 21	17 09	28 07	27 53	06 28	02 01
05 Sa	08 59 12	15 49 19	06♓48 11	01 40	00♑51	15 45	02 33	17 09	28 10	27 55	06 30	01 53
06 Su	09 03 09	16 50 10	18 44 20	03 14	01 59	16 32	02 45	17 07	28 13	27 57	06 32	01 45
07 Mo	09 07 05	17 50 59	00♈36 33	04 50	03 08	17 20	02 58	17 06	28 16	27 59	06 34	01 37
08 Tu	09 11 02	18 51 48	12 27 21	06 25	04 16	18 07	03 10	17 05	28 18	28 02	06 35	01 31
09 We	09 14 58	19 52 34	24 20 01	08 02	05 25	18 55	03 22	17 04	28 21	28 04	06 37	01 27
10 Th	09 18 55	20 53 20	06♉18 31	09 39	06 34	19 42	03 35	17 02	28 24	28 06	06 39	01 24
11 Fr	09 22 52	21 54 03	18 27 23	11 18	07 43	20 29	03 47	17 00	28 27	28 08	06 40	01D 24
12 Sa	09 26 48	22 54 46	00♊51 27	12 57	08 52	21 17	04 00	16 59	28 30	28 11	06 42	01 25
13 Su	09 30 45	23 55 26	13 35 37	14 36	10 01	22 04	04 13	16 57	28 33	28 13	06 44	01 26
14 Mo	09 34 41	24 56 05	26 44 17	16 17	11 10	22 51	04 26	16 55	28 36	28 15	06 45	01R 26
15 Tu	09 38 38	25 56 43	10♋20 47	17 59	12 19	23 39	04 39	16 53	28 39	28 17	06 47	01 25
16 We	09 42 34	26 57 18	24 26 24	19 41	13 29	24 26	04 51	16 51	28 42	28 20	06 48	01 21
17 Th	09 46 31	27 57 52	08♌59 31	21 24	14 38	25 13	05 04	16 49	28 45	28 22	06 50	01 16
18 Fr	09 50 27	28 58 24	23 55 01	23 08	15 48	26 01	05 18	16 46	28 48	28 24	06 51	01 08
19 Sa	09 54 24	29 58 55	09♍04 36	24 53	16 58	26 48	05 31	16 44	28 51	28 27	06 53	00 59
20 Su	09 58 21	00♓59 24	24 17 54	26 39	18 08	27 36	05 44	16 42	28 54	28 29	06 54	00 49
21 Mo	10 02 17	01 59 52	09♎24 10	28 26	19 18	28 23	05 57	16 39	28 57	28 31	06 56	00 40
22 Tu	10 06 14	03 00 19	24 14 10	00♓14	20 28	29 10	06 10	16 36	29 00	28 33	06 57	00 32
23 We	10 10 10	04 00 44	08♏41 28	02 03	21 38	29 58	06 24	16 34	29 04	28 36	06 58	00 26
24 Th	10 14 07	05 01 07	22 43 00	03 52	22 49	00♓45	06 37	16 31	29 07	28 38	07 00	00 23
25 Fr	10 18 03	06 01 30	06♐18 41	05 43	23 59	01 33	06 51	16 28	29 10	28 40	07 01	00 21
26 Sa	10 21 60	07 01 51	19 30 39	07 34	25 09	02 20	07 04	16 25	29 13	28 42	07 02	00D 21
27 Su	10 25 56	08 02 11	02♑22 14	09 27	26 20	03 07	07 18	16 22	29 17	28 45	07 04	00R 21
28 Mo	10 29 53	09 02 29	14 57 14	11 20	27 31	03 55	07 32	16 19	29 20	28 47	07 05	00 20

Data for	02-01-2011
Julian Day	2455593.50
Ayanamsa	24 01 01
SVP	05 ♓ 06 37
☽ ☊ Mean	00 ♑ 40 R

● ◐ PHASES ○ ◑

03	02:30	●	13♒54
11	07:19	◐	22♉13
18	08:36	○	29♌20
24	23:26	◑	06♐00

LAST ASPECT Day	h m	☽ INGRESS Day	h m	
01	19:32	01	23:21	♒
04	06:11	04	10:24	♓
06	19:14	06	22:46	♈
09	07:32	09	11:23	♉
11	19:28	11	22:22	♊
14	03:21	14	05:50	♋
16	07:07	16	09:15	♌
18	08:36	18	09:40	♍
20	07:19	20	09:01	♎
22	08:35	22	09:29	♏
24	11:15	24	12:46	♐
26	18:09	26	19:32	♑

DECLINATION

Day	☉	☽	☿	♀	♂	♃	♄	♅	♆	♇
01 Tu	17S 16	20S 51	22S 17	20S 44	18S 03	00S 22	04S 24	01S 28	12S 42	18S 48
02 We	16 58	17 42	22 05	20 48	17 50	00 18	04 23	01 26	12 41	18 48
03 Th	16 41	13 48	21 52	20 53	17 36	00 13	04 23	01 25	12 41	18 48
04 Fr	16 24	09 24	21 37	20 56	17 23	00 08	04 22	01 24	12 40	18 48
05 Sa	16 06	04 41	21 21	21 00	17 09	00 03	04 21	01 23	12 39	18 48
06 Su	15 47	00N09	21 04	21 03	16 54	00N02	04 21	01 22	12 38	18 48
07 Mo	15 29	04 57	20 45	21 05	16 40	00 07	04 20	01 21	12 38	18 48
08 Tu	15 10	09 33	20 25	21 06	16 26	00 12	04 19	01 20	12 37	18 48
09 We	14 51	13 49	20 03	21 08	16 11	00 17	04 18	01 19	12 36	18 48
10 Th	14 32	17 35	19 41	21 08	15 56	00 22	04 18	01 17	12 35	18 48
11 Fr	14 12	20 40	19 16	21 08	15 41	00 27	04 17	01 16	12 34	18 48
12 Sa	13 53	22 53	18 51	21 08	15 26	00 33	04 16	01 15	12 34	18 48
13 Su	13 33	23 59	18 24	21 07	15 11	00 38	04 15	01 14	12 33	18 47
14 Mo	13 13	23 49	17 55	21 05	14 55	00 43	04 14	01 13	12 32	18 47
15 Tu	12 52	22 14	17 25	21 03	14 39	00 48	04 13	01 11	12 31	18 47
16 We	12 32	19 15	16 54	21 00	14 24	00 53	04 12	01 10	12 31	18 47
17 Th	12 11	14 59	16 21	20 57	14 08	00 59	04 11	01 09	12 30	18 47
18 Fr	11 50	09 43	15 47	20 53	13 52	01 04	04 10	01 08	12 29	18 47
19 Sa	11 29	03 48	15 12	20 48	13 35	01 09	04 09	01 07	12 28	18 47
20 Su	11 07	02S 22	14 35	20 43	13 19	01 15	04 08	01 05	12 27	18 47
21 Mo	10 46	08 20	13 57	20 38	13 03	01 20	04 06	01 04	12 27	18 47
22 Tu	10 24	13 43	13 17	20 32	12 46	01 25	04 05	01 03	12 26	18 47
23 We	10 02	18 11	12 36	20 25	12 29	01 31	04 04	01 01	12 25	18 47
24 Th	09 40	21 28	11 54	20 17	12 12	01 36	04 03	01 00	12 24	18 47
25 Fr	09 18	23 25	11 11	20 10	11 55	01 42	04 01	00 59	12 24	18 46
26 Sa	08 56	23 59	10 26	20 01	11 38	01 47	04 00	00 58	12 23	18 46
27 Su	08 33	23 15	09 39	19 52	11 21	01 52	03 59	00 56	12 22	18 46
28 Mo	08 11	21 20	08 52	19 42	11 04	01 58	03 57	00 55	12 21	18 46

ASPECTARIAN

01 01:11 ☽ ∥ ♀
16:19 ☽ ∥ ♇
16:42 ☽ ☌ ☿
19:32 ☽ ✶ ♅
23:04 ☽ ∥ ♂

02 03:12 ☽ ✶ ♃
05:13 ☽ ∥ ☉
07:35 ♀ ✶ ♆
12:47 ♀ □ ♅
16:16 ☿ ✶ ♅

03 03:16 ☽ ☌ ♂
06:25 ☽ ∥ ♇
08:57 ☽ △ ♄
22:19 ☿ ♒

04 05:59 ♀ ♑
06:11 ☽ ☌ ♆
10:52 ☽ ✶ ♀
16:40 ☉ ☌ ♂
23:24 ☽ ✶ ♇

05 01:42 ☽ ∥ ♄
15:30 ☿ ✶ ♃
16:31 ☽ ∥ ♅
23:09 ☽ # ♃
23:28 ☽ ∥ ♃

06 01:24 ☿ ∥ ♀
06:04 ☽ # ♅
06:41 ☉ △ ♄
17:16 ♂ △ ♄
19:14 ☽ ☌ ♅
19:40 ♀ □ ♃
20:56 ☽ # ♄

07 04:51 ☽ ☌ ♃
[illegible] ☽ □ ♆
09:52 ☽ ✶ ☿
12:05 ☽ □ ♇

08 09:21 ☽ ☍ ♄
12:16 ☽ ✶ ♂
14:10 ☽ ✶ ☉
16:54 ☽ # ♆

09 05:45 ☽ # ☉
07:32 ☽ ✶ ♆
13:41 ☽ # ♂

10 00:33 ☽ △ ♀
00:40 ☽ △ ♇
01:51 ♀ ☌ ♇
07:42 ☽ □ ☿
08:42 ☽ # ♇
13:42 ☽ # ☿

11 04:14 ☽ □ ♂
04:16 ☽ # ♀
18:51 ☽ □ ♆
19:28 ☽ ✶ ♅

12 02:49 ☿ ∥ ♇
06:06 ☽ ✶ ♃

13 02:09 ☽ △ ☿
06:11 ☽ △ ♄
16:33 ☽ △ ♂
20:29 ☽ △ ☉

14 02:43 ☽ △ ♆
03:21 ☽ □ ♅
08:49 ☿ △ ♄
13:53 ☽ □ ♃
17:47 ☽ ☍ ♇

15 03:44 ☽ ☍ ♀
11:05 ☽ # ♀
11:12 ☽ □ ♄

16 01:13 ☉ ∥ ♆
03:00 ☽ # ♇
07:07 ☽ △ ♅
16:01 ☽ # ☿
17:31 ☽ △ ♃

17 04:29 ☽ # ♂
09:58 ☉ ☌ ♆
11:55 ☽ # ♆
12:36 ☽ ✶ ♄
14:13 ☽ # ☉
22:36 ☽ ☍ ☿

18 03:31 ☽ ☍ ♂
07:09 ☽ ☍ ♆
14:10 ♃ # ♅
19:22 ♀ □ ♄
20:32 ☽ △ ♇
22:39 ☽ # ♄

19 00:26 ☉ ♓
10:12 ☽ ∥ ♃
10:33 ☽ # ♅
13:28 ☽ △ ♀
19:04 ☽ ∥ ♅
19:36 ☽ # ♃

20 06:56 ☽ ∥ ♄
07:19 ☽ ☍ ♅
18:24 ☽ ☍ ♃
20:02 ☽ □ ♇
22:44 ☿ ☌ ♂

21 01:07 ☿ ☌ ♆
04:18 ♂ ☌ ♆
09:48 ☽ ∥ ☉
11:37 ☽ ☌ ♄
17:19 ☽ □ ♀
17:59 ☽ ∥ ♆
19:45 ☽ ∥ ♂
20:54 ☿ ♓
22:14 ☽ ∥ ☿

22 07:07 ☽ △ ♆
08:35 ☽ △ ♂
11:17 ☽ △ ☿
15:34 ☽ △ ☉
21:06 ☽ ✶ ♇

23 01:06 ♂ ♓
03:49 ☽ ∥ ♇
06:08 ♂ ∥ ♆
06:37 ☿ ∥ ♆
06:57 ☿ ∥ ♂
14:45 ☽ ∥ ♀

24 00:11 ☽ ✶ ♀
10:23 ☽ □ ♆
11:15 ☽ △ ♅
14:58 ☽ □ ♂
22:46 ☽ □ ☿

25 00:58 ☽ △ ♃
08:48 ☿ ☌ ☉
17:05 ☿ ✶ ♇
18:20 ☽ ✶ ♄
20:29 ♃ □ ♇

26 00:16 ☉ ✶ ♇
17:09 ☽ ✶ ♆
18:09 ☽ □ ♅

27 01:31 ☽ ✶ ♂
08:54 ☽ ☌ ♇
09:31 ☽ □ ♃
11:41 ☽ ✶ ☉
15:48 ☽ ✶ ☿

28 02:36 ☽ □ ♄
15:15 ☽ ∥ ♀
21:37 ☽ ∥ ♇

⚷ Chiron

01 Dec.	06 S 23
02	29♒32
05	29 44
08	29 57
11	00♓09
14	00 21
17	00 33
20	00 46
23	00 58
26	01 10

08 20:16 ♓

March 2011

Day	S. T. h m s	☉ ° ' "	☽ ° ' "	☿ ° '	♀ ° '	♂ ° '	♃ ° '	♄ ° '	♅ ° '	♆ ° '	♇ ° '	☊ True ° '
01 Tu	10 33 50	10♓02 46	27♑19 13	13♓13	28♑41	04♓42	07♈45	16♎R15	29♓23	28♒49	07♑06	00♑R17
02 We	10 37 46	11 03 01	09♒31 20	15 08	29 52	05 30	07 59	16 12	29 26	28 51	07 07	00 12
03 Th	10 41 43	12 03 15	21 36 08	17 03	01♒03	06 17	08 13	16 09	29 30	28 54	07 09	00 03
04 Fr	10 45 39	13 03 26	03♓35 37	18 58	02 14	07 04	08 27	16 05	29 33	28 56	07 10	29♐52
05 Sa	10 49 36	14 03 36	15 31 16	20 54	03 25	07 52	08 40	16 02	29 36	28 58	07 11	29 40
06 Su	10 53 32	15 03 45	27 24 20	22 50	04 36	08 39	08 54	15 58	29 40	29 00	07 12	29 26
07 Mo	10 57 29	16 03 51	09♈16 03	24 45	05 47	09 26	09 08	15 54	29 43	29 03	07 13	29 13
08 Tu	11 01 25	17 03 55	21 08 00	26 41	06 58	10 14	09 22	15 51	29 46	29 05	07 14	29 02
09 We	11 05 22	18 03 57	03♉02 21	28 36	08 09	11 01	09 36	15 47	29 50	29 07	07 15	28 53
10 Th	11 09 18	19 03 58	15 01 56	00♈29	09 21	11 48	09 50	15 43	29 53	29 09	07 16	28 48
11 Fr	11 13 15	20 03 56	27 10 21	02 22	10 32	12 35	10 04	15 39	29 56	29 11	07 17	28 44
12 Sa	11 17 12	21 03 52	09♊31 44	04 13	11 43	13 23	10 19	15 35	30 00	29 13	07 18	28 43
13 Su	11 21 08	22 03 46	22 10 44	06 01	12 55	14 10	10 33	15 31	00♈03	29 16	07 19	28 43
14 Mo	11 25 05	23 03 37	05♋12 03	07 48	14 06	14 57	10 47	15 27	00 07	29 18	07 19	28 42
15 Tu	11 29 01	24 03 27	18 39 53	09 31	15 18	15 44	11 01	15 23	00 10	29 20	07 20	28 41
16 We	11 32 58	25 03 14	02♌37 03	11 10	16 29	16 32	11 15	15 19	00 14	29 22	07 21	28 36
17 Th	11 36 54	26 02 59	17 03 51	12 46	17 41	17 19	11 30	15 14	00 17	29 24	07 22	28 30
18 Fr	11 40 51	27 02 41	01♍57 06	14 17	18 53	18 06	11 44	15 10	00 20	29 26	07 22	28 20
19 Sa	11 44 47	28 02 22	17 09 47	15 43	20 04	18 53	11 58	15 06	00 24	29 28	07 23	28 09
20 Su	11 48 44	29 02 00	02♎31 39	17 04	21 16	19 40	12 13	15 01	00 27	29 30	07 24	27 57
21 Mo	11 52 41	00♈01 37	17 50 55	18 19	22 28	20 27	12 27	14 57	00 31	29 32	07 24	27 46
22 Tu	11 56 37	01 01 11	02♏56 17	19 27	23 40	21 14	12 41	14 53	00 34	29 34	07 25	27 36
23 We	12 00 34	02 00 44	17 38 55	20 29	24 52	22 01	12 56	14 48	00 37	29 36	07 26	27 29
24 Th	12 04 30	03 00 15	01♐53 34	21 24	26 03	22 48	13 10	14 44	00 41	29 38	07 26	27 24
25 Fr	12 08 27	03 59 44	15 38 40	22 12	27 15	23 35	13 25	14 39	00 44	29 40	07 27	27 21
26 Sa	12 12 23	04 59 12	28 55 43	22 52	28 27	24 22	13 39	14 35	00 48	29 42	07 27	27 21
27 Su	12 16 20	05 58 38	11♑48 08	23 25	29 39	25 09	13 53	14 30	00 51	29 44	07 27	27 20
28 Mo	12 20 16	06 58 02	24 20 19	23 51	00♓51	25 56	14 08	14 25	00 55	29 46	07 28	27 20
29 Tu	12 24 13	07 57 24	06♒36 53	24 08	02 03	26 43	14 22	14 21	00 58	29 48	07 28	27 17
30 We	12 28 10	08 56 44	18 42 11	24 19	03 16	27 30	14 37	14 16	01 01	29 50	07 29	27 12
31 Th	12 32 06	09 56 03	00♓40 02	24R 21	04 28	28 17	14 51	14 11	01 05	29 52	07 29	27 05

Data for	03-01-2011
Julian Day	2455621.50
Ayanamsa	24 01 04
SVP	05 ♓ 06 32
☽ ☊ Mean	29 ♐ 11 R

● ◐ PHASES ○ ◑

04	20:47	●	13♓56
12	23:45	◐	22♊03
19	18:10	○	28♍48
26	12:08	◑	05♑29

LAST ASPECT ☽ Day	h m	INGRESS Day	h m	
01	04:03	01	05:15	♒
03	14:37	03	16:48	♓
06	04:35	06	05:15	♈
08	16:05	08	17:53	♉
11	05:27	11	05:32	♊
13	13:11	13	14:30	♋
15	10:06	15	19:34	♌
17	19:59	17	20:53	♍
19	18:10	19	20:03	♎
21	18:35	21	19:17	♏
23	20:08	23	20:46	♐
26	01:26	26	01:58	♑
28	03:19	28	11:01	♒
30	22:23	30	22:39	♓

DECLINATION

Day	☉	☽	☿	♀	♂	♃	♄	♅	♆	♇
01 Tu	07S 48	18S 26	08S 03	19S 32	10S 46	02N03	03S 56	00S 54	12S 20	18S 46
02 We	07 25	14 47	07 14	19 21	10 29	02 09	03 54	00 52	12 20	18 46
03 Th	07 02	10 34	06 23	19 10	10 11	02 14	03 53	00 51	12 19	18 46
04 Fr	06 39	06 00	05 31	18 58	09 54	02 20	03 51	00 50	12 18	18 46
05 Sa	06 16	01 14	04 38	18 46	09 36	02 25	03 50	00 48	12 17	18 46
06 Su	05 53	03N33	03 45	18 33	09 18	02 31	03 48	00 47	12 17	18 46
07 Mo	05 30	08 12	02 51	18 20	09 00	02 37	03 47	00 46	12 16	18 46
08 Tu	05 06	12 33	01 56	18 06	08 42	02 42	03 45	00 44	12 15	18 45
09 We	04 43	16 26	01 01	17 51	08 24	02 48	03 43	00 43	12 14	18 45
10 Th	04 20	19 40	00 06	17 36	08 06	02 53	03 42	00 42	12 14	18 45
11 Fr	03 56	22 06	00N49	17 21	07 47	02 59	03 40	00 40	12 13	18 45
12 Sa	03 33	23 31	01 44	17 05	07 29	03 04	03 38	00 39	12 12	18 45
13 Su	03 09	23 47	02 38	16 48	07 11	03 10	03 37	00 38	12 11	18 45
14 Mo	02 45	22 46	03 31	16 31	06 52	03 16	03 35	00 36	12 11	18 45
15 Tu	02 22	20 26	04 24	16 14	06 34	03 21	03 33	00 35	12 10	18 45
16 We	01 58	16 50	05 15	15 56	06 15	03 27	03 32	00 33	12 09	18 45
17 Th	01 34	12 08	06 04	15 38	05 56	03 33	03 30	00 32	12 09	18 45
18 Fr	01 10	06 35	06 51	15 19	05 38	03 38	03 28	00 31	12 08	18 45
19 Sa	00 47	00 32	07 36	15 00	05 19	03 44	03 26	00 29	12 07	18 45
20 Su	00 23	05S 35	08 19	14 40	05 00	03 49	03 25	00 28	12 06	18 44
21 Mo	00N01	11 21	08 59	14 20	04 42	03 55	03 23	00 27	12 06	18 44
22 Tu	00 24	16 20	09 36	13 59	04 23	04 01	03 21	00 25	12 05	18 44
23 We	00 48	20 11	10 10	13 39	04 04	04 06	03 19	00 24	12 04	18 44
24 Th	01 12	22 40	10 41	13 17	03 45	04 12	03 17	00 23	12 04	18 44
25 Fr	01 35	23 41	11 08	12 56	03 26	04 18	03 16	00 21	12 03	18 44
26 Sa	01 59	23 18	11 31	12 34	03 07	04 23	03 14	00 20	12 02	18 44
27 Su	02 22	21 40	11 51	12 11	02 49	04 29	03 12	00 18	12 02	18 44
28 Mo	02 46	19 00	12 07	11 49	02 30	04 35	03 10	00 17	12 01	18 44
29 Tu	03 09	15 32	12 19	11 26	02 11	04 40	03 08	00 16	12 00	18 44
30 We	03 33	11 29	12 27	11 02	01 52	04 46	03 06	00 14	12 00	18 44
31 Th	03 56	07 02	12 31	10 39	01 33	04 51	03 05	00 13	11 59	18 44

ASPECTARIAN

01 02:58 ☽ ☌ ♀
04:03 ☽ ✶ ♅
13:38 ☿ ∥ ☉
14:49 ♀ ✶ ♅
20:54 ☽ ✶ ♃
02 02:39 ♀ ♒
13:11 ☽ △ ♄
14:22 ☽ ∥ ♆
03 02:15 ☽ ∥ ♂
14:37 ☽ ☌ ♆
20:19 ☽ ∥ ☉
04 02:50 ♂ ✶ ♇
03:00 ☽ ∥ ☿
07:11 ☽ ✶ ♇
07:29 ☽ ☌ ♂
10:54 ☽ ∥ ♄
18:09 ☽ # ♃
05 00:16 ♀ ∥ ♇
02:08 ☽ ∥ ♅
10:08 ☽ # ♅
12:57 ☽ ☌ ☿
18:39 ☽ ∥ ♃
22:24 ☿ ∥ ♄
06 00:49 ☽ # ☿
01:55 ☽ # ♄
04:35 ☽ ☌ ♅
10:59 ☽ # ☉
16:10 ☽ ✶ ♀
19:50 ☽ □ ♇
23:44 ☽ ☌ ♃
07 04:00 ☽ # ♂
05:38 ☿ # ♃
13:22 ☽ ☍ ♄
22:20 ☽ # ♆
08 16:05 ☽ ✶ ♆
09 08:03 ☿ ∥ ♅
08:28 ☽ △ ♇
09:15 ☽ # ♀
11:24 ☽ □ ♀
16:06 ☿ ☌ ♅
16:42 ☽ # ♇
17:07 ☽ ✶ ♂
17:47 ☿ ♈
10 08:44 ☽ ✶ ☉
12:26 ♀ ✶ ♃
20:16 ☿ # ♅
11 03:58 ☽ □ ♆
05:27 ☽ ✶ ♅
11:56 ☽ ✶ ☿
17:29 ☉ ∥ ♄
12 00:53 ♅ ♈
01:32 ☽ ✶ ♃
04:39 ☽ △ ♀
07:52 ☽ □ ♂
11:30 ☽ △ ♄
23:04 ☉ # ♃
13 09:37 ☿ # ☉
13:11 ☽ △ ♆
14:40 ☽ □ ♅
16:06 ☿ ∥ ♃
17:30 ☿ □ ♇
14 01:40 ☿ # ♄
03:51 ☽ ☍ ♇
05:23 ☽ □ ☿
10:14 ☽ □ ♃
18:15 ☽ □ ♄
18:33 ☽ △ ♂
15 01:35 ♀ △ ♄
10:06 ☽ △ ☉
12:11 ☽ # ♇
19:56 ☽ △ ♅
16 01:27 ☿ ☌ ♃
05:23 ☽ # ♀
14:41 ☽ △ ♃
15:27 ♃ # ♄
16:05 ☽ △ ☿
21:02 ☽ ✶ ♄
21:23 ☿ # ♂
23:56 ☽ # ♆
17 01:06 ☽ ☍ ♀
19:59 ☽ ☍ ♆
23:02 ☽ ∥ ☿
18 04:07 ☽ # ♂
08:36 ☽ △ ♆
11:42 ☽ ∥ ♃
12:36 ☽ # ♄
13:55 ☿ ☍ ♄
23:02 ☽ # ☉
19 00:14 ☽ # ♅
02:50 ☽ ☍ ♂
04:02 ☽ ∥ ♅
04:52 ☽ ∥ ☉
15:29 ☽ ∥ ♄
16:58 ☽ # ♃
18:42 ☉ ∥ ♅
20:45 ☽ ☍ ♅
21:52 ☽ ∥ ♂
20 07:37 ☽ □ ♇
12:36 ☽ # ☿
15:23 ☽ ☍ ♃
19:28 ☽ ☌ ♄
23:21 ☉ ♈
21 00:47 ☽ ☍ ☿
03:21 ☽ ∥ ♆
07:55 ☽ △ ♀
12:24 ☉ ☌ ♅
12:56 ☽ ∥ ♀
18:35 ☽ △ ♆
22 00:57 ☉ # ♅
07:14 ☽ ✶ ♇
14:02 ☽ ∥ ♇
21:49 ♂ # ♃
23 07:42 ☽ △ ♂
13:09 ☽ □ ♀
20:08 ☽ □ ♆
21:55 ☽ △ ♅
24 02:03 ☽ △ ☉
19:58 ☽ △ ♃
22:15 ☽ ✶ ♄
25 12:24 ☽ △ ☿
15:09 ☽ □ ♂
15:13 ♂ ∥ ♄
23:03 ☽ ✶ ♀
26 01:26 ☽ ✶ ♆
03:27 ☽ □ ♅
15:49 ☽ ☌ ♇
27 01:40 ♀ ☌ ♆
04:02 ☽ □ ♃
05:05 ☽ □ ♄
06:53 ♀ ♓
10:43 ♀ ∥ ♆
12:20 ☿ # ♀
14:29 ☿ # ♆
14:49 ☉ # ♂
23:01 ☽ □ ☿
28 02:00 ☽ ∥ ♇
03:19 ☽ ✶ ♂
12:11 ☉ □ ♇
12:51 ☽ ✶ ♅
21:56 ♃ ☍ ♄
22:57 ☉ # ♄
29 02:53 ☽ ✶ ☉
15:13 ☽ △ ♄
15:41 ☽ ✶ ♃
18:37 ☽ # ☿
21:04 ☽ ∥ ♆
30 02:40 ☽ ∥ ♀
11:18 ☽ ✶ ☿
20:49 ☿ SR
22:23 ☽ ☌ ♆
31 08:30 ☽ ☌ ♀
11:02 ☽ # ♃
13:45 ☽ ✶ ♇
14:45 ☽ # ☉
20:29 ☽ ∥ ♄

⚷ Chiron

01 Dec.	05 S 45
01	01♓22
04	01 35
07	01 47
10	01 59
13	02 10
16	02 22
19	02 33
22	02 45
25	02 56
28	03 06
31	03 17

April 2011

Day	S. T.	☉	☽	☿	♀	♂	♃	♄	♅	♆	♇	☊ True
	h m s	° ' "	° ' "	° '	° '	° '	° '	° '	° '	° '	° '	° '
01 Fr	12 36 03	10♈55 19	12♓33 39	24♈R17	05♓40	29♓04	15♈06	14♎R07	01♈08	29♒54	07♑29	26♐R54
02 Sa	12 39 59	11 54 34	24 25 27	24 05	06 52	29 51	15 20	14 02	01 12	29 55	07 29	26 42
03 Su	12 43 56	12 53 47	06♈17 17	23 47	08 04	00♈37	15 35	13 57	01 15	29 57	07 30	26 28
04 Mo	12 47 52	13 52 57	18 10 34	23 23	09 17	01 24	15 49	13 53	01 18	29 59	07 30	26 15
05 Tu	12 51 49	14 52 06	00♉06 36	22 53	10 29	02 11	16 04	13 48	01 22	00♓01	07 30	26 04
06 We	12 55 45	15 51 13	12 06 51	22 19	11 41	02 57	16 18	13 43	01 25	00 02	07 30	25 55
07 Th	12 59 42	16 50 17	24 13 09	21 40	12 53	03 44	16 33	13 39	01 28	00 04	07 30	25 49
08 Fr	13 03 39	17 49 20	06♊27 54	20 58	14 06	04 31	16 48	13 34	01 32	00 06	07 30	25 45
09 Sa	13 07 35	18 48 20	18 54 02	20 14	15 18	05 17	17 02	13 30	01 35	00 08	07 30	25 44
10 Su	13 11 32	19 47 18	01♋35 02	19 28	16 30	06 04	17 17	13 25	01 38	00 09	07R 30	25D 45
11 Mo	13 15 28	20 46 13	14 34 42	18 42	17 43	06 50	17 31	13 20	01 42	00 11	07 30	25 45
12 Tu	13 19 25	21 45 06	27 56 44	17 56	18 55	07 37	17 46	13 16	01 45	00 12	07 30	25R 44
13 We	13 23 21	22 43 57	11♌44 02	17 11	20 08	08 23	18 00	13 11	01 48	00 14	07 30	25 42
14 Th	13 27 18	23 42 46	25 57 48	16 27	21 20	09 10	18 14	13 07	01 51	00 16	07 30	25 37
15 Fr	13 31 14	24 41 32	10♍36 35	15 47	22 33	09 56	18 29	13 02	01 55	00 17	07 30	25 30
16 Sa	13 35 11	25 40 16	25 35 35	15 10	23 45	10 42	18 43	12 58	01 58	00 19	07 30	25 22
17 Su	13 39 08	26 38 58	10♎46 51	14 36	24 57	11 29	18 58	12 53	02 01	00 20	07 29	25 12
18 Mo	13 43 04	27 37 38	26 00 12	14 07	26 10	12 15	19 12	12 49	02 04	00 21	07 29	25 03
19 Tu	13 47 01	28 36 16	11♏04 54	13 43	27 22	13 01	19 27	12 44	02 07	00 23	07 29	24 55
20 We	13 50 57	29 34 52	25 51 29	13 23	28 35	13 47	19 41	12 40	02 11	00 24	07 29	24 49
21 Th	13 54 54	00♉33 26	10♐13 15	13 08	29 48	14 33	19 55	12 35	02 14	00 26	07 28	24 45
22 Fr	13 58 50	01 31 59	24 06 47	12 58	01♈00	15 19	20 10	12 31	02 17	00 27	07 28	24 44
23 Sa	14 02 47	02 30 30	07♑31 54	12 54	02 13	16 05	20 24	12 27	02 20	00 28	07 27	24D 44
24 Su	14 06 43	03 29 00	20 30 48	12D 54	03 25	16 51	20 38	12 23	02 23	00 30	07 27	24 45
25 Mo	14 10 40	04 27 27	03♒07 12	12 59	04 38	17 37	20 53	12 19	02 26	00 31	07 27	24 46
26 Tu	14 14 37	05 25 53	15 25 40	13 10	05 51	18 23	21 07	12 14	02 29	00 32	07 26	24R 47
27 We	14 18 33	06 24 18	27 30 54	13 25	07 03	19 09	21 21	12 10	02 32	00 33	07 26	24 45
28 Th	14 22 30	07 22 41	09♓27 29	13 45	08 16	19 55	21 36	12 06	02 35	00 34	07 25	24 41
29 Fr	14 26 26	08 21 02	21 19 35	14 09	09 29	20 41	21 50	12 02	02 38	00 36	07 24	24 36
30 Sa	14 30 23	09 19 22	03♈10 42	14 37	10 41	21 27	22 04	11 59	02 41	00 37	07 24	24 28

Data for	04-01-2011
Julian Day	2455652.50
Ayanamsa	24 01 07
SVP	05 ♓ 06 27
☽ ☊ Mean	27 ♐ 32 R

● ◐ PHASES ○ ◑

Day	Time	Phase	Position
03	14:33	●	13♈30
11	12:05	◐	21♋16
18	02:44	○	27♎44
25	02:48	◑	04♒34

LAST ASPECT ☽ INGRESS

Day	h m	Day	h m	Sign
31	13:45	02	11:17	♈
04	10:05	04	23:47	♉
05	23:03	07	11:22	♊
09	02:24	09	21:02	♋
11	12:06	12	03:37	♌
13	19:59	14	06:41	♍
15	20:49	16	06:59	♎
18	02:45	18	06:20	♏
20	04:54	20	06:51	♐
21	16:58	22	10:26	♑
24	00:15	24	18:00	♒
26	11:29	27	04:58	♓
27	19:53	29	17:34	♈

DECLINATION

Day	☉	☽	☿	♀	♂	♃	♄	♅	♆	♇
01 Fr	04N19	02S 22	12N30	10S 15	01S 14	04N57	03S 03	00S 12	11S 58	18S 44
02 Sa	04 42	02N22	12 26	09 51	00 55	05 03	03 01	00 10	11 58	18 44
03 Su	05 06	07 01	12 18	09 26	00 36	05 08	02 59	00 09	11 57	18 44
04 Mo	05 29	11 25	12 06	09 01	00 17	05 14	02 57	00 08	11 57	18 43
05 Tu	05 51	15 24	11 50	08 36	00N02	05 19	02 55	00 06	11 56	18 43
06 We	06 14	18 47	11 31	08 11	00 21	05 25	02 54	00 05	11 55	18 43
07 Th	06 37	21 24	11 09	07 45	00 39	05 31	02 52	00 04	11 55	18 43
08 Fr	07 00	23 02	10 44	07 19	00 58	05 36	02 50	00 02	11 54	18 43
09 Sa	07 22	23 35	10 17	06 53	01 17	05 42	02 48	00 01	11 54	18 43
10 Su	07 44	22 55	09 48	06 27	01 36	05 47	02 46	00N00	11 53	18 43
11 Mo	08 07	21 00	09 18	06 01	01 55	05 53	02 45	00 02	11 53	18 43
12 Tu	08 29	17 55	08 47	05 34	02 13	05 58	02 43	00 03	11 52	18 43
13 We	08 51	13 45	08 16	05 08	02 32	06 04	02 41	00 04	11 52	18 43
14 Th	09 12	08 43	07 44	04 41	02 51	06 09	02 39	00 05	11 51	18 43
15 Fr	09 34	03 03	07 14	04 14	03 09	06 15	02 38	00 07	11 50	18 43
16 Sa	09 55	02S 54	06 44	03 46	03 28	06 20	02 36	00 08	11 50	18 43
17 Su	10 17	08 45	06 15	03 19	03 46	06 26	02 34	00 09	11 49	18 43
18 Mo	10 38	14 06	05 49	02 52	04 05	06 31	02 33	00 10	11 49	18 43
19 Tu	10 59	18 30	05 24	02 24	04 23	06 37	02 31	00 12	11 49	18 43
20 We	11 19	21 37	05 02	01 56	04 41	06 42	02 29	00 13	11 48	18 43
21 Th	11 40	23 15	04 42	01 29	05 00	06 47	02 28	00 14	11 48	18 43
22 Fr	12 00	23 21	04 24	01 01	05 18	06 53	02 26	00 15	11 47	18 43
23 Sa	12 21	22 05	04 09	00 33	05 36	06 58	02 25	00 17	11 47	18 43
24 Su	12 41	19 40	03 57	00 05	05 54	07 04	02 23	00 18	11 46	18 43
25 Mo	13 00	16 22	03 47	00N23	06 12	07 09	02 22	00 19	11 46	18 43
26 Tu	13 20	12 25	03 40	00 51	06 30	07 14	02 20	00 20	11 45	18 43
27 We	13 39	08 03	03 35	01 19	06 48	07 20	02 19	00 21	11 45	18 43
28 Th	13 58	03 26	03 33	01 47	07 06	07 25	02 17	00 23	11 45	18 43
29 Fr	14 17	01N16	03 33	02 15	07 23	07 30	02 16	00 24	11 44	18 43
30 Sa	14 36	05 54	03 36	02 43	07 41	07 35	02 14	00 25	11 44	18 43

ASPECTARIAN

01 06:07 ☽ ∥ ♂
11:00 ☽ ∥ ♅
12:52 ☽ # ♅
17:03 ☽ # ♂
02 03:13 ☽ # ♄
04:51 ♂ ♈
11:44 ☽ ☌ ♂
12:28 ♀ ⚹ ♇
13:00 ☽ ∥ ☉
13:46 ☽ ☌ ♅
13:56 ☽ ∥ ♃
03 02:26 ☽ □ ♇
03:38 ☉ ∥ ♃
11:49 ☽ # ♀
15:24 ☽ ☍ ♄
19:09 ☽ ☌ ♃
20:52 ♂ ☌ ♅
23:57 ☉ ☍ ♄
04 03:00 ☽ # ♆
03:41 ☽ ∥ ☿
10:05 ☽ ☌ ☿
12:58 ♂ ∥ ♅
13:37 ♆ ♓
15:20 ☿ # ♆
23:48 ☽ ⚹ ♆
05 05:30 ♂ # ♅
14:48 ☽ △ ♇
23:03 ☽ ⚹ ♀
23:31 ☽ # ♇
06 14:41 ☉ ☌ ♃
07 11:32 ☽ □ ♆
14:19 ☽ ⚹ ♅
19:56 ☽ ⚹ ♂
08 09:52 ♀ # ☉
13:41 ☽ △ ♄
16:22 ☽ □ ♀
[illegible]
23:48 ☽ ⚹ ☉
09 02:24 ☽ ⚹ ☿
08:49 ♇ SR
19:37 ☿ ☌ ☉
21:19 ☽ △ ♆
10 00:06 ☽ □ ♅
08:53 ☽ □ ♂
11:01 ☽ ☍ ♇
21:45 ☽ □ ♄
11 05:26 ☽ □ ♃
06:03 ♀ # ♃
06:16 ☽ △ ♀
07:03 ☽ □ ☿
18:28 ☽ # ♇
20:39 ♂ □ ♇
12 03:59 ☿ ☌ ♃
06:43 ☽ △ ♅
08:15 ☿ ∥ ☉
17:54 ☽ △ ♂
13 02:28 ☽ ⚹ ♄
08:49 ☽ △ ☿
09:26 ☽ # ♆
10:44 ♂ # ♄
10:51 ☽ △ ♃
19:59 ☽ △ ☉
21:55 ☽ ∥ ☉
14 04:43 ☽ ∥ ☿
07:07 ☽ ☍ ♆
10:54 ☽ ∥ ♃
18:46 ☽ # ♀
18:57 ☽ △ ♇
23:36 ☽ ∥ ♂
15 01:44 ☽ # ♄
11:5[illegible] ☽ ∥ [illegible]
12:51 ☽ # ♅
20:49 ☽ ☍ ♀
22:48 ☽ ∥ ♄
16 02:24 ☽ # ♂
03:16 ☽ ∥ ♀
09:45 ♀ # ♂
10:08 ☽ ☍ ♅
14:09 ☽ # ♃
14:22 ☽ # ☿
16:35 ☿ ∥ ♃
18:49 ☽ □ ♇
17 01:09 ☽ ☍ ♂
03:18 ☽ ☌ ♄
05:50 ☽ ☍ ☿
06:58 ☽ # ☉
13:06 ☽ ☍ ♃
13:22 ☽ ∥ ♆
18 06:54 ☽ △ ♆
16:03 ♂ ☍ ♄
17:37 ♀ ∥ ♄
18:14 ☽ ⚹ ♇
19 01:22 ☽ ∥ ♇
14:59 ☿ ☌ ♂
20 04:54 ☽ △ ♀
07:32 ☽ □ ♆
10:18 ☉ ♉
10:30 ☽ △ ♅
12:31 ☿ ∥ ♂
20:43 ☉ ⚹ ♆
21 04:01 ☽ ⚹ ♄
04:07 ♀ ♈
04:54 ☽ △ ☿
07:50 ☽ △ ♂
08:42 [illegible]
16:58 ☽ △ ♃
22 11:15 ☽ ⚹ ♆
13:26 ☽ □ ♀
14:12 ☽ △ ☉
14:34 ☽ □ ♅
23:52 ☽ ☌ ♇
23 02:29 ♀ ☌ ♅
08:57 ☽ □ ♄
09:48 ☽ □ ☿
10:05 ☿ SD
13:34 ♀ # ♅
16:44 ☽ □ ♂
24 00:15 ☽ □ ♃
07:29 ☽ ∥ ♇
20:33 [illegible]
22:40 ☽ ⚹ ♅
25 03:14 ☽ ⚹ ♀
17:46 ☽ △ ♄
19:03 ☽ # ☉
19:28 ☽ ⚹ ☿
26 03:43 ☽ ∥ ♆
06:14 ☽ ⚹ ♂
11:29 ☽ ⚹ ♃
27 03:44 ☽ # ♃
06:06 ☽ ☌ ♆
06:11 ☽ # ♂
07:19 ♀ □ ♇
19:26 ☽ ⚹ ☉
19:53 ☽ ⚹ ♇
23:25 ☽ # [illegible]
28 00:56 ☉ △ ♇
05:55 ☽ ∥ ♄
07:42 ☽ # ♀
15:34 ☽ # ♅
19:33 ☽ ∥ ♅
29 00:40 ♀ # ♄
05:04 ☽ # ♄
05:35 ☽ ∥ ♀
11:50 ☽ ∥ ☿
13:02 ♂ ∥ ♃
22:59 ☽ ☌ ♅
30 08:31 ☽ □ ♇
09:05 ☽ ∥ ♃
10:04 ☽ ∥ ♂
16:54 ☽ ☌ ♀
17:41 ☽ ☍ ♄

⚷ Chiron

01	Dec.	05 S 01
03	03♓27	
06	03 37	
09	03 47	
12	03 56	
15	04 05	
18	04 13	
21	04 21	
24	04 29	
27	04 37	
30	04 43	

May 2011

Day	S. T. h m s	☉ ° ' "	☽ ° ' "	☿ ° '	♀ ° '	♂ ° '	♃ ° '	♄ ° '	♅ ° '	♆ ° '	♇ ° '	☊ True ° '
01 Su	14 34 19	10♉17 40	15♈03 39	15♈10	11♈54	22♈12	22♈18	11♎R55	02♈44	00♓38	07♑R23	24♐R20
02 Mo	14 38 16	11 15 56	27 00 38	15 47	13 07	22 58	22 32	11 51	02 47	00 39	07 23	24 12
03 Tu	14 42 12	12 14 11	09♉03 19	16 27	14 19	23 43	22 46	11 47	02 49	00 40	07 22	24 05
04 We	14 46 09	13 12 24	21 13 09	17 11	15 32	24 29	23 00	11 44	02 52	00 41	07 21	24 00
05 Th	14 50 05	14 10 35	03♊31 25	17 59	16 45	25 15	23 14	11 40	02 55	00 42	07 20	23 57
06 Fr	14 54 02	15 08 44	15 59 31	18 50	17 58	26 00	23 28	11 37	02 58	00 43	07 20	23 55
07 Sa	14 57 59	16 06 52	28 39 03	19 44	19 10	26 45	23 42	11 33	03 01	00 44	07 19	23D 56
08 Su	15 01 55	17 04 58	11♋31 50	20 42	20 23	27 31	23 56	11 30	03 03	00 44	07 18	23 57
09 Mo	15 05 52	18 03 02	24 39 58	21 42	21 36	28 16	24 10	11 26	03 06	00 45	07 17	23 58
10 Tu	15 09 48	19 01 04	08♌05 29	22 46	22 49	29 01	24 24	11 23	03 09	00 46	07 16	23 59
11 We	15 13 45	19 59 04	21 50 04	23 52	24 01	29 47	24 38	11 20	03 11	00 47	07 15	23R 59
12 Th	15 17 41	20 57 02	05♍54 21	25 01	25 14	00♉32	24 51	11 17	03 14	00 48	07 14	23 57
13 Fr	15 21 38	21 54 58	20 17 21	26 12	26 27	01 17	25 05	11 14	03 16	00 48	07 14	23 54
14 Sa	15 25 34	22 52 52	04♎55 58	27 27	27 40	02 02	25 19	11 11	03 19	00 49	07 13	23 50
15 Su	15 29 31	23 50 45	19 44 55	28 43	28 53	02 47	25 32	11 08	03 21	00 50	07 12	23 45
16 Mo	15 33 28	24 48 36	04♏37 06	00♉02	00♉05	03 32	25 46	11 06	03 24	00 50	07 11	23 40
17 Tu	15 37 24	25 46 25	19 24 27	01 24	01 18	04 17	25 59	11 03	03 26	00 51	07 10	23 36
18 We	15 41 21	26 44 13	03♐59 11	02 48	02 31	05 02	26 13	11 00	03 28	00 51	07 08	23 32
19 Th	15 45 17	27 42 00	18 14 57	04 14	03 44	05 47	26 26	10 58	03 31	00 52	07 07	23 31
20 Fr	15 49 14	28 39 45	02♑07 33	05 43	04 57	06 31	26 40	10 56	03 33	00 52	07 06	23D 31
21 Sa	15 53 10	29 37 29	15 35 15	07 13	06 10	07 16	26 53	10 53	03 35	00 53	07 05	23 32
22 Su	15 57 07	00♊35 12	28 38 35	08 47	07 23	08 01	27 06	10 51	03 38	00 53	07 04	23 34
23 Mo	16 01 04	01 32 54	11♒19 48	10 22	08 35	08 45	27 19	10 49	03 40	00 54	07 03	23 36
24 Tu	16 05 00	02 30 35	23 42 23	12 00	09 48	09 30	27 32	10 47	03 42	00 54	07 02	23 38
25 We	16 08 57	03 28 15	05♓50 32	13 40	11 01	10 14	27 46	10 45	03 44	00 54	07 01	23 39
26 Th	16 12 53	04 25 53	17 48 52	15 22	12 14	10 59	27 59	10 43	03 46	00 55	06 59	23R 39
27 Fr	16 16 50	05 23 31	29 41 57	17 07	13 27	11 43	28 12	10 41	03 48	00 55	06 58	23 39
28 Sa	16 20 46	06 21 08	11♈34 08	18 53	14 40	12 28	28 24	10 40	03 50	00 55	06 57	23 37
29 Su	16 24 43	07 18 43	23 29 17	20 42	15 53	13 12	28 37	10 38	03 52	00 55	06 56	23 35
30 Mo	16 28 39	08 16 18	05♉30 37	22 33	17 06	13 56	28 50	10 37	03 54	00 55	06 54	23 33
31 Tu	16 32 36	09 13 52	17 40 46	24 27	18 19	14 40	29 03	10 35	03 56	00 55	06 53	23 32

Data for	05-01-2011
Julian Day	2455682.50
Ayanamsa	24 01 10
SVP	05 ♓ 06 21
☽ ☊ Mean	25 ♐ 57 R

● ◐ PHASES ○ ◑

Day	Time	Phase	Position
03	06:50	●	12♉31
10	20:33	◐	19♌51
17	11:09	○	26♏13
24	18:52	◑	03♓16

LAST ASPECT ☽ INGRESS

Day	h m	Day	h m	Sign
01	15:21	02	05:59	♉
03	06:51	04	17:09	♊
06	20:13	07	02:32	♋
09	06:53	09	09:36	♌
11	04:53	11	14:00	♍
13	02:53	13	15:57	♎
15	16:02	15	16:33	♏
17	11:10	17	17:23	♐
19	14:18	19	20:17	♑
21	21:05	22	02:32	♒
24	07:41	24	12:24	♓
25	18:15	27	00:36	♈
29	10:28	29	13:02	♉
31	15:37	31	23:57	♊

DECLINATION

Day	☉	☽	☿	♀	♂	♃	♄	♅	♆	♇
01 Su	14N54	10N21	03N41	03N11	07N58	07N41	02S13	00N26	11S44	18S43
02 Mo	15 13	14 25	03 48	03 39	08 16	07 46	02 12	00 27	11 43	18 43
03 Tu	15 30	17 57	03 58	04 07	08 33	07 51	02 10	00 28	11 43	18 43
04 We	15 48	20 45	04 09	04 34	08 50	07 56	02 09	00 29	11 43	18 43
05 Th	16 06	22 37	04 23	05 02	09 08	08 01	02 08	00 30	11 42	18 43
06 Fr	16 23	23 24	04 38	05 30	09 25	08 06	02 06	00 31	11 42	18 43
07 Sa	16 40	23 00	04 55	05 57	09 41	08 11	02 05	00 33	11 42	18 43
08 Su	16 56	21 22	05 14	06 25	09 58	08 17	02 04	00 34	11 41	18 43
09 Mo	17 12	18 34	05 35	06 52	10 15	08 22	02 03	00 35	11 41	18 43
10 Tu	17 28	14 44	05 57	07 19	10 32	08 27	02 02	00 36	11 41	18 43
11 We	17 44	10 03	06 20	07 46	10 48	08 32	02 01	00 37	11 41	18 43
12 Th	18 00	04 44	06 45	08 13	11 04	08 37	02 00	00 38	11 40	18 43
13 Fr	18 15	00S56	07 12	08 40	11 21	08 41	01 59	00 39	11 40	18 43
14 Sa	18 29	06 38	07 39	09 07	11 37	08 46	01 58	00 40	11 40	18 43
15 Su	18 44	12 03	08 08	09 33	11 53	08 51	01 57	00 41	11 40	18 43
16 Mo	18 58	16 45	08 38	09 59	12 08	08 56	01 56	00 42	11 40	18 43
17 Tu	19 12	20 24	09 09	10 25	12 24	09 01	01 55	00 42	11 39	18 43
18 We	19 26	22 39	09 41	10 51	12 40	09 06	01 54	00 43	11 39	18 43
19 Th	19 39	23 23	10 14	11 16	12 55	09 11	01 53	00 44	11 39	18 44
20 Fr	19 52	22 38	10 48	11 42	13 11	09 15	01 53	00 45	11 39	18 44
21 Sa	20 04	20 33	11 23	12 07	13 26	09 20	01 52	00 46	11 39	18 44
22 Su	20 16	17 28	11 58	12 31	13 41	09 25	01 51	00 47	11 39	18 44
23 Mo	20 28	13 38	12 34	12 56	13 56	09 29	01 51	00 48	11 38	18 44
24 Tu	20 40	09 19	13 11	13 20	14 10	09 34	01 50	00 49	11 38	18 44
25 We	20 51	04 43	13 48	13 44	14 25	09 39	01 50	00 49	11 38	18 44
26 Th	21 02	00 00	14 25	14 07	14 39	09 43	01 49	00 50	11 38	18 44
27 Fr	21 12	04N40	15 03	14 31	14 54	09 48	01 49	00 51	11 38	18 44
28 Sa	21 22	09 10	15 41	14 54	15 08	09 52	01 48	00 52	11 38	18 44
29 Su	21 32	13 21	16 19	15 16	15 22	09 57	01 48	00 52	11 38	18 44
30 Mo	21 41	17 02	16 56	15 38	15 36	10 01	01 47	00 53	11 38	18 44
31 Tu	21 50	20 03	17 34	16 00	15 49	10 05	01 47	00 54	11 38	18 45

ASPECTARIAN

01 00:13 ☽ ☌ ☿
00:15 ♀ ☍ ♄
04:26 ♂ ☌ ♃
07:48 ☽ # ♆
14:51 ☽ ☌ ♃
15:21 ☽ ☌ ♂
02 05:24 ☽ ∥ ☉
07:17 ☽ ✶ ♆
12:06 ☿ ∥ ♀
20:39 ☽ △ ♇
03 05:51 ☽ # ♇
04 18:30 ☽ □ ♆
22:49 ☽ ✶ ♅
05 15:38 ☽ △ ♄
06 04:09 ☽ ✶ ♀
05:50 ☽ ✶ ☿
14:30 ☽ ✶ ♃
20:13 ☽ ✶ ♂
07 03:54 ☽ △ ♆
08:12 ☽ □ ♅
16:10 ☽ ☍ ♇
23:56 ☽ □ ♄
08 11:01 ☽ ✶ ☉
17:53 ☽ □ ♀
18:11 ☽ □ ☿
22:53 ☽ # ♇
23:05 ☽ □ ♃
09 06:53 ☽ □ ♂
08:36 ☽ ∥ ☉
15:11 ☽ △ ♅
15:45 ☿ ☌ ♀
10 05:47 ☽ ✶ ♄
16:03 ☽ # ♆
20:34 ☽ ∥ ♂
11 03:48 ☽ △ ☿
04:08 ☽ △ ♀
04:53 ☽ △ ♃
07:01 ☽ ∥ ♃
07:05 ♂ ♉
09:44 ☽ ∥ ♀
14:23 ☽ △ ♂
14:43 ♀ ☌ ♃
15:20 ☽ ☍ ♆
15:47 ☽ ∥ ☿
19:57 ☿ ☌ ♃
12 02:15 ☽ △ ♇
08:30 ♂ ✶ ♆
11:46 ☽ # ♄
17:26 ☽ ∥ ♅
22:48 ☽ # ♅
13 01:37 ♀ ∥ ♃
02:53 ☽ △ ☉
04:22 ☽ ∥ ♄
21:21 ☽ ☍ ♅
14 03:42 ☽ □ ♇
04:45 ☽ # ☿
04:46 ♂ # ♆
09:20 ☽ # ♃
10:07 ☽ ☌ ♄
11:37 ☽ # ♀
22:11 ☽ ∥ ♆
22:59 ☉ # ♇
23:08 ☽ # ♂
15 09:29 ☽ ☍ ♃
15:53 ☽ ☍ ☿
16:02 ☽ ☍ ♀
17:53 ☽ △ ♆
22:09 ☽ ☍ ♂
22:12 ♀ ♉
23:19 ☿ ♉
16 04:08 ☽ ✶ ♇
09:25 ☿ ☌ ♀
11:54 ☽ ∥ ♇
14:17 ☿ ✶ ♆
14:28 ☽ # ☉
14:52 ♀ ✶ ♆
16:27 ☿ ∥ ♃
17 18:48 ☽ □ ♆
23:09 ☽ △ ♅
18 11:42 ☽ ✶ ♄
19 14:18 ☽ △ ♃
21:23 ♀ # ♆
21:48 ☽ ✶ ♆
20 02:31 ☽ □ ♅
05:27 ☽ △ ♀
07:06 ☽ △ ☿
08:13 ☽ △ ♂
08:47 ☽ ☌ ♇
15:34 ☽ □ ♄
18:20 ♂ △ ♇
21:54 ☿ △ ♇
21 01:20 ☿ ☌ ♂
04:07 ☽ # ☉
09:22 ☉ ♊
10:42 ☿ # ♆
15:04 ☽ ∥ ♇
18:03 ♀ △ ♇
21:05 ☽ □ ♃
22 03:56 ☽ △ ☉
07:32 ☉ □ ♆
09:22 ☽ ✶ ♅
18:12 ☽ □ ♀
18:46 ☽ □ ♂
21:53 ☽ □ ☿
22:23 ☽ # ♂
23:01 ☽ △ ♄
23 03:44 ☽ # ♀
05:22 ☽ # ☿
08:26 ♀ ☌ ♂
11:20 ☽ ∥ ♆
22:41 ☽ # ♃
24 07:41 ☽ ✶ ♃
14:11 ☽ ☌ ♆
16:44 ☿ ∥ ♀
25 02:21 ☽ ✶ ♇
06:47 ☉ ✶ ♅
09:22 ☽ ✶ ♂
11:31 ☽ ✶ ♀
14:47 ☽ ∥ ♄
18:15 ☽ ✶ ☿
19:48 ☽ # ♅
26 04:19 ☽ ∥ ♅
09:18 ☽ # ♄
14:34 ☿ ∥ ♂
27 08:19 ☽ ☌ ♅
12:32 ☽ ✶ ☉
14:41 ☽ □ ♇
22:10 ☽ ☍ ♄
28 03:57 ☽ ∥ ♃
13:54 ☽ # ♆
29 10:28 ☽ ☌ ♃
13:22 ☽ ∥ ♀
13:30 ☽ ∥ ♂
14:52 ☽ ✶ ♆
16:40 ♀ ∥ ♂
23:09 ☽ ∥ ☿
30 02:46 ☽ △ ♇
12:47 ☽ # ♇
17:43 ☽ ☌ ♂
31 01:23 ☽ ☌ ♀
015:37 ☽ ☌ ☿
20:35 ☽ ∥ ☉

⚷ Chiron

01 Dec. 04 S 25

Day	° '
03	04♓50
06	04 56
09	05 01
12	05 06
15	05 11
18	05 15
21	05 19
24	05 22
27	05 24
30	05 26

June 2011

01 21:17 11♊02 ⊙ Solar Eclipse (mag 0.601)
15 20:12 24♐23 ✷ Total Lunar Eclipse (mag 1.704)

Day	S. T.	☉	☽	☿	♀	♂	♃	♄	♅	♆	♇	☊ True
	h m s	° ′ ″	° ′ ″	° ′	° ′	° ′	° ′	° ′	° ′	° ′	° ′	° ′
01 We	16 36 33	10♊ 11 25	00♊ 01 45	26♉ 22	19♉ 32	15♉ 25	29♈ 15	10♎R34	03♈ 58	00♓ 56	06♑R52	23♐R30
02 Th	16 40 29	11 08 57	12 34 57	28 20	20 45	16 09	29 28	10 33	03 59	00 56	06 50	23 30
03 Fr	16 44 26	12 06 28	25 21 18	00♊ 20	21 58	16 53	29 40	10 32	04 01	00 56	06 49	23D 30
04 Sa	16 48 22	13 03 57	08♋ 21 12	02 22	23 11	17 37	29 53	10 31	04 03	00R 56	06 48	23 30
05 Su	16 52 19	14 01 26	21 34 44	04 25	24 24	18 20	00♉ 05	10 30	04 05	00 56	06 46	23 31
06 Mo	16 56 15	14 58 54	05♌ 01 43	06 31	25 37	19 04	00 17	10 29	04 06	00 56	06 45	23 32
07 Tu	17 00 12	15 56 20	18 41 51	08 38	26 50	19 48	00 30	10 29	04 08	00 55	06 44	23R 32
08 We	17 04 08	16 53 45	02♍ 34 30	10 46	28 03	20 32	00 42	10 28	04 09	00 55	06 42	23 31
09 Th	17 08 05	17 51 09	16 38 39	12 56	29 16	21 16	00 54	10 28	04 11	00 55	06 41	23 31
10 Fr	17 12 02	18 48 32	00♎ 52 40	15 06	00♊ 29	21 59	01 06	10 27	04 12	00 55	06 39	23 29
11 Sa	17 15 58	19 45 53	15 14 03	17 17	01 42	22 43	01 18	10 27	04 14	00 55	06 38	23 28
12 Su	17 19 55	20 43 14	29 39 30	19 29	02 55	23 26	01 29	10 27	04 15	00 54	06 36	23 27
13 Mo	17 23 51	21 40 34	14♏ 04 52	21 41	04 08	24 10	01 41	10 27	04 16	00 54	06 35	23 26
14 Tu	17 27 48	22 37 53	28 25 31	23 53	05 22	24 53	01 53	10D 27	04 18	00 54	06 33	23 25
15 We	17 31 44	23 35 11	12♐ 36 37	26 05	06 35	25 36	02 04	10 27	04 19	00 53	06 32	23 24
16 Th	17 35 41	24 32 28	26 33 50	28 16	07 48	26 20	02 16	10 27	04 20	00 53	06 31	23 24
17 Fr	17 39 37	25 29 45	10♑ 13 44	00♋ 26	09 01	27 03	02 27	10 27	04 21	00 53	06 29	23 24
18 Sa	17 43 34	26 27 01	23 34 12	02 35	10 14	27 46	02 38	10 28	04 22	00 52	06 28	23 24
19 Su	17 47 31	27 24 17	06♒ 34 37	04 43	11 27	28 29	02 49	10 28	04 23	00 52	06 26	23 24
20 Mo	17 51 27	28 21 32	19 15 48	06 50	12 41	29 12	03 00	10 29	04 24	00 51	06 25	23 24
21 Tu	17 55 24	29 18 47	01♓ 39 49	08 54	13 54	29 55	03 11	10 30	04 25	00 51	06 23	23D 24
22 We	17 59 20	00♋ 16 02	13 49 49	10 57	15 07	00♊ 38	03 22	10 31	04 26	00 50	06 21	23 24
23 Th	18 03 17	01 13 16	25 49 38	12 58	16 20	01 21	03 33	10 32	04 27	00 49	06 20	23 24
24 Fr	18 07 13	02 10 30	07♈ 43 43	14 58	17 33	02 03	03 44	10 33	04 28	00 49	06 18	23 25
25 Sa	18 11 10	03 07 45	19 36 42	16 55	18 47	02 46	03 54	10 34	04 28	00 48	06 17	23 27
26 Su	18 15 06	04 04 59	01♉ 33 11	18 50	20 00	03 29	04 05	10 35	04 29	00 47	06 15	23 28
27 Mo	18 19 03	05 02 13	13 37 26	20 43	21 13	04 11	04 15	10 36	04 30	00 47	06 14	23 30
28 Tu	18 22 60	05 59 27	25 53 14	22 34	22 27	04 54	04 25	10 38	04 30	00 46	06 12	23 32
29 We	18 26 56	06 56 41	08♊ 23 39	24 23	23 40	05 37	04 36	10 39	04 31	00 45	06 11	23 33
30 Th	18 30 53	07 53 55	21 10 53	26 10	24 53	06 19	04 46	10 41	04 31	00 44	06 09	23 34

Data for	06-01-2011
Julian Day	2455713.50
Ayanamsa	24 01 15
SVP	05♓06 18
☽ ☊ Mean	24♐18 R

● ◐ PHASES ○ ◑

01	21:03	⊙	11♊02
09	02:11	◐	17♍56
15	20:13	✷	24♐23
23	11:48	◑	01♈41

LAST ASPECT Day	h m	☽ INGRESS Day	h m	
03	08:09	03	08:37	♋
05	05:34	05	15:04	♌
07	15:28	07	19:34	♍
09	08:14	09	22:32	♎
11	08:05	12	00:34	♏
13	17:44	14	02:39	♐
16	03:31	16	05:59	♑
18	08:07	18	11:47	♒
20	20:22	20	20:45	♓
22	02:51	23	08:24	♈
24	22:08	25	20:53	♉
27	16:25	28	07:57	♊
30	07:34	30	16:14	♋

DECLINATION

Day	☉	☽	☿	♀	♂	♃	♄	♅	♆	♇
01 We	21N59	22N12	18N11	16N21	16N03	10N10	01S47	00N55	11S38	18S45
02 Th	22 07	23 17	18 48	16 42	16 16	10 14	01 47	00 55	11 38	18 45
03 Fr	22 14	23 10	19 24	17 03	16 29	10 18	01 46	00 56	11 38	18 45
04 Sa	22 22	21 49	19 59	17 23	16 42	10 23	01 46	00 57	11 38	18 45
05 Su	22 29	19 14	20 33	17 43	16 55	10 27	01 46	00 57	11 38	18 45
06 Mo	22 36	15 36	21 06	18 02	17 08	10 31	01 46	00 58	11 38	18 45
07 Tu	22 42	11 05	21 37	18 21	17 20	10 35	01 46	00 58	11 38	18 45
08 We	22 48	05 56	22 07	18 39	17 33	10 39	01 46	00 59	11 38	18 45
09 Th	22 53	00 26	22 35	18 57	17 45	10 43	01 46	01 00	11 38	18 46
10 Fr	22 58	05S10	23 01	19 15	17 57	10 48	01 46	01 00	11 38	18 46
11 Sa	23 03	10 31	23 24	19 32	18 08	10 52	01 46	01 01	11 39	18 46
12 Su	23 07	15 20	23 46	19 48	18 20	10 55	01 46	01 01	11 39	18 46
13 Mo	23 11	19 16	24 04	20 04	18 31	10 59	01 46	01 02	11 39	18 46
14 Tu	23 14	21 59	24 20	20 19	18 43	11 03	01 47	01 02	11 39	18 46
15 We	23 17	23 18	24 34	20 34	18 54	11 07	01 47	01 03	11 39	18 46
16 Th	23 19	23 06	24 44	20 48	19 05	11 11	01 47	01 03	11 39	18 46
17 Fr	23 22	21 31	24 52	21 02	19 15	11 15	01 48	01 03	11 39	18 47
18 Sa	23 23	18 45	24 57	21 15	19 26	11 18	01 48	01 04	11 40	18 47
19 Su	23 25	15 07	24 59	21 27	19 36	11 22	01 49	01 04	11 40	18 47
20 Mo	23 26	10 53	24 58	21 39	19 46	11 26	01 49	01 05	11 40	18 47
21 Tu	23 26	06 17	24 55	21 51	19 56	11 29	01 50	01 05	11 40	18 47
22 We	23 26	01 33	24 49	22 01	20 06	11 33	01 50	01 05	11 41	18 47
23 Th	23 26	03N11	24 40	22 12	20 15	11 36	01 51	01 06	11 41	18 48
24 Fr	23 25	07 46	24 30	22 21	20 24	11 40	01 51	01 06	11 41	18 48
25 Sa	23 24	12 04	24 17	22 30	20 33	11 43	01 52	01 06	11 41	18 48
26 Su	23 22	15 55	24 02	22 38	20 42	11 47	01 53	01 06	11 42	18 48
27 Mo	23 20	19 10	23 45	22 46	20 51	11 50	01 53	01 07	11 42	18 48
28 Tu	23 18	21 37	23 26	22 53	20 59	11 53	01 54	01 07	11 42	18 48
29 We	23 15	23 04	23 06	22 59	21 08	11 57	01 55	01 07	11 42	18 49
30 Th	23 12	23 21	22 44	23 04	21 16	12 00	01 56	01 07	11 43	18 49

ASPECTARIAN

01 01:44 ☽ □ ♆
07:35 ☽ ✶ ♅
09:16 ☉ △ ♄
20:09 ☽ △ ♄
21:56 ☿ # ♇
02 20:03 ☿ ☌ ♊
03 07:06 ☿ □ ♆
07:29 ♆ SR
08:09 ☽ ✶ ♃
10:20 ☽ △ ♆
16:05 ☽ □ ♅
17:08 ☽ ∥ ☉
21:09 ☽ ☍ ♇

04 03:57 ☽ □ ♄
13:58 ♃ ♉
15:00 ☽ ∥ ☿
17:49 ☽ ✶ ♂
19:57 ☿ ✶ ♅

05 03:40 ☽ # ♇
05:34 ☽ ✶ ♀
09:58 ☽ ∥ ♀
15:10 ☽ ∥ ♂
15:28 ☽ □ ♃
22:21 ☽ △ ♅
06 03:06 ☽ ✶ ☿
09:37 ☽ ✶ ♄
18:50 ☽ ✶ ☉
21:16 ☽ # ♆

07 02:02 ☽ □ ♂
02:25 ☽ ∥ ♃
15:28 ☽ □ ♀
20:41 ☿ △ ♄
20:43 ☽ △ ♃
21:10 ☽ ☍ ♆
08 07:04 ☽ △ [illegible]
08:05 ♀ # ♇
16:32 ☽ □ ☿
18:18 ☽ # ♄
21:37 ☽ ∥ ♅
09 02:42 ♃ ✶ ♆
06:08 ☽ # ♅
08:14 ☽ △ ♂
09:26 ☽ ∥ ♄
14:24 ♀ ♊
20:36 ☿ ∥ ☉
23:17 ☽ △ ♀

10 05:35 ☽ ☍ ♅
08:24 ♀ □ ♆
09:40 ☽ □ ♇
16:01 ☽ ☌ ♄
11 01:36 ☽ # ♃
04:03 ☽ △ ☿
05:18 ☽ ∥ ♆
08:05 ☽ △ ☉

12 02:04 ☽ △ ♆
03:05 ☽ ☍ ♃
11:32 ☽ ✶ ♇
18:40 ☽ # ♂
20:34 ☽ ∥ ♇
23:45 ☿ ☌ ☉

13 02:36 ♀ ✶ ♅
03:52 ♄ SD
06:32 ☽ # ♀
17:44 ☽ ☍ ♂
14 04:09 ☽ □ ♆
07:44 ♂ # ♇
09:54 ☽ △ ♅
12:47 ☽ ☍ ♀
20:19 ☽ ✶ ♄
23:29 ☽ # ☉
15 16:55 ☽ ∥ ☉

16 03:31 ☽ ☍ ☿
07:31 ☽ ✶ ♆
10:05 ☽ △ ♃
13:36 ☽ □ ♅
17:23 ☽ ☌ ♇
19:09 ☿ ♋
17 00:24 ☽ □ ♄
04:37 ☽ # ♀
04:52 ☿ △ ♆
19:14 ☽ # ♂
23:54 ☽ ∥ ♇
18 00:36 ☿ ✶ ♃
04:31 ♀ △ ♄
08:07 ☽ △ ♂
16:55 ☽ □ ♃
19:55 ☽ ✶ ♅
20:12 ☿ □ ♅
19 07:19 ☽ △ ♄
10:08 ☽ △ ♀
19:17 ☿ ☍ ♇
19:47 ☽ ∥ ♆
21:07 ☽ # ♃

20 19:02 ☽ △ ☉
20:24 ☽ □ ♂
22:24 ☽ ☌ ♆
21 02:50 ♂ ♊
03:02 ☽ ✶ ♃
09:15 ☽ ✶ ♇
17:10 ☽ △ [illegible]
17:17 ☉ ♋
18:43 ☿ □ ♄
22:36 ☽ ∥ ♄
22 02:21 ☽ # ♅
02:51 ☽ □ ♀
06:45 ♂ □ ♆
13:21 ☽ ∥ ♅
14:07 ☉ △ ♆
17:10 ☽ # ♄

23 11:49 ☽ ✶ ♂
17:23 ☽ ☌ ♅
21:08 ☽ □ ♇
24 05:42 ☽ ☍ ♄
08:05 ♃ # ♆
17:30 ☽ □ ☿
21:52 ☽ # ♆
22:03 ☽ ∥ ♃
22:08 ☽ ✶ ♀
25 22:29 ☽ ✶ ♆
23:55 ☉ ✶ ♃

26 05:07 ☽ ☌ ♃
05:30 ☽ ✶ ☉
09:22 ☽ △ ♇
10:14 ☉ □ ♅
21:09 ☽ # ♇
27 10:26 ♂ ✶ ♅
16:25 ☽ ✶ ☿
16:39 ☽ ∥ ♂
28 05:18 ☉ ☍ ♇
09:25 ☽ □ ♆
11:39 ☿ ∥ ☉
16:36 ☽ ✶ ♅
18:23 ☽ ☌ ♂
21:48 ☽ ∥ ♀
29 00:50 ☽ ∥ ☿
04:17 ☽ △ ♄
05:48 ☽ ∥ ☉
06:24 ☿ ∥ ♀

30 07:34 ☽ ☌ ♀
07:41 ☽ ∥ ☉
09:53 ☽ ∥ [illegible]
17:34 ☽ △ ♆
21:24 ♀ ∥ ☉

⚷ Chiron

01	Dec.	04 S 01
02	05♓28	
05	05 29	
08	05 29	
11	05 29R	
14	05 28	
17	05 27	
20	05 25	
23	05 23	
26	05 21	
29	05 17	
08	17:27	05♓29 R

01 08:39 09♋12 ⊛ Solar Eclipse (mag 0.097)

Day	S. T.	☉	☽	☿	♀	♂	♃	♄	♅	♆	♇	☊ True
	h m s	° ' "	° ' "	° '	° '	° '	° '	° '	° '	° '	° '	° '
01 Fr	18 34 49	08♋51 09	04♋16 00	27♋54	26♊07	07♊01	04♉56	10♎43	04♈32	00♓R44	06♑R08	23♐R34
02 Sa	18 38 46	09 48 23	17 38 53	29 36	27 20	07 44	05 05	10 44	04 32	00 43	06 06	23 32
03 Su	18 42 42	10 45 37	01♌18 09	01♌16	28 33	08 26	05 15	10 46	04 33	00 42	06 05	23 30
04 Mo	18 46 39	11 42 50	15 11 19	02 54	29 47	09 08	05 25	10 48	04 33	00 41	06 03	23 27
05 Tu	18 50 35	12 40 03	29 15 10	04 30	01♋00	09 50	05 34	10 50	04 33	00 40	06 02	23 24
06 We	18 54 32	13 37 16	13♍26 10	06 04	02 14	10 32	05 43	10 53	04 33	00 39	06 00	23 21
07 Th	18 58 29	14 34 29	27 40 49	07 35	03 27	11 14	05 53	10 55	04 34	00 38	05 59	23 18
08 Fr	19 02 25	15 31 41	11♎55 56	09 04	04 41	11 56	06 02	10 57	04 34	00 37	05 57	23 17
09 Sa	19 06 22	16 28 53	26 08 48	10 31	05 54	12 38	06 11	11 00	04 34	00 36	05 56	23 16
10 Su	19 10 18	17 26 05	10♏17 10	11 56	07 08	13 20	06 19	11 02	04R 34	00 35	05 54	23D 16
11 Mo	19 14 15	18 23 17	24 19 06	13 18	08 21	14 02	06 28	11 05	04 34	00 34	05 53	23 17
12 Tu	19 18 11	19 20 29	08♐12 54	14 38	09 35	14 43	06 37	11 08	04 34	00 33	05 51	23 18
13 We	19 22 08	20 17 41	21 56 52	15 56	10 48	15 25	06 45	11 11	04 34	00 32	05 50	23R 18
14 Th	19 26 05	21 14 53	05♑29 17	17 11	12 02	16 06	06 53	11 14	04 33	00 30	05 48	23 18
15 Fr	19 30 01	22 12 05	18 48 34	18 23	13 15	16 48	07 02	11 17	04 33	00 29	05 47	23 16
16 Sa	19 33 58	23 09 17	01♒53 25	19 33	14 29	17 29	07 10	11 20	04 33	00 28	05 45	23 13
17 Su	19 37 54	24 06 30	14 43 07	20 41	15 43	18 11	07 17	11 23	04 33	00 27	05 44	23 08
18 Mo	19 41 51	25 03 43	27 17 45	21 46	16 56	18 52	07 25	11 26	04 32	00 26	05 42	23 03
19 Tu	19 45 47	26 00 57	09♓38 19	22 48	18 10	19 33	07 33	11 30	04 32	00 24	05 41	22 58
20 We	19 49 44	26 58 11	21 46 46	23 47	19 24	20 14	07 40	11 33	04 31	00 23	05 40	22 54
21 Th	19 53 40	27 55 26	03♈45 55	24 43	20 37	20 56	07 48	11 37	04 31	00 22	05 38	22 51
22 Fr	19 57 37	28 52 42	15 39 28	25 35	21 51	21 37	07 55	11 40	04 30	00 20	05 37	22 49
23 Sa	20 01 33	29 49 58	27 31 46	26 25	23 05	22 18	08 02	11 44	04 30	00 19	05 36	22D 49
24 Su	20 05 30	00♌47 15	09♉27 36	27 11	24 19	22 59	08 09	11 48	04 29	00 18	05 34	22 50
25 Mo	20 09 27	01 44 34	21 31 54	27 54	25 32	23 39	08 15	11 52	04 29	00 16	05 33	22 53
26 Tu	20 13 23	02 41 53	03♊49 27	28 33	26 46	24 20	08 22	11 56	04 28	00 15	05 32	22 55
27 We	20 17 20	03 39 13	16 24 30	29 08	28 00	25 01	08 28	12 00	04 27	00 14	05 30	22 57
28 Th	20 21 16	04 36 34	29 20 28	29 39	29 14	25 42	08 35	12 04	04 26	00 12	05 29	22R 57
29 Fr	20 25 13	05 33 55	12♋39 24	00♍06	00♌28	26 22	08 41	12 08	04 25	00 11	05 28	22 55
30 Sa	20 29 09	06 31 18	26 21 30	00 29	01 42	27 03	08 47	12 12	04 24	00 09	05 26	22 51
31 Su	20 33 06	07 28 41	10♌24 49	00 47	02 56	27 43	08 52	12 17	04 23	00 08	05 25	22 45

Data for	07-01-2011
Julian Day	2455743.50
Ayanamsa	24 01 20
SVP	05 ♓ 06 15
☽ ☊ Mean	22 ♐ 43 R

● ◐ PHASES ○ ◑

01	08:55	⊛	09♋12
08	06:29	◐	15♎47
15	06:39	○	22♑28
23	05:03	◑	00♉02
30	18:40	●	07♌16

LAST ASPECT ☽ Day	h m	INGRESS Day	h m	
01	11:39	02	21:44	♌
03	16:27	05	01:16	♍
06	00:20	07	03:54	♎
08	06:30	09	06:32	♏
10	13:05	11	09:47	♐
12	12:21	13	14:14	♑
15	06:40	15	20:30	♒
17	12:24	18	05:13	♓
20	11:15	20	16:26	♈
22	21:36	23	04:59	♉
25	13:13	25	16:35	♊
28	00:36	28	01:12	♋
28	23:04	30	06:16	♌

DECLINATION

Day	☉	☽	☿	♀	♂	♃	♄	♅	♆	♇
01 Fr	23N09	22N23	22N21	23N09	21N24	12N03	01S57	01N07	11S43	18S49
02 Sa	23 05	20 08	21 56	23 14	21 31	12 06	01 58	01 07	11 43	18 49
03 Su	23 00	16 43	21 31	23 17	21 39	12 09	01 59	01 08	11 44	18 49
04 Mo	22 55	12 20	21 04	23 20	21 46	12 12	02 00	01 08	11 44	18 49
05 Tu	22 50	07 14	20 36	23 22	21 53	12 15	02 01	01 08	11 44	18 50
06 We	22 44	01 44	20 07	23 24	22 00	12 18	02 02	01 08	11 45	18 50
07 Th	22 38	03S53	19 38	23 25	22 06	12 21	02 03	01 08	11 45	18 50
08 Fr	22 32	09 18	19 08	23 25	22 13	12 24	02 04	01 08	11 46	18 50
09 Sa	22 25	14 12	18 37	23 25	22 19	12 27	02 05	01 08	11 46	18 50
10 Su	22 18	18 18	18 07	23 23	22 25	12 29	02 07	01 08	11 46	18 51
11 Mo	22 11	21 19	17 35	23 21	22 31	12 32	02 08	01 08	11 47	18 51
12 Tu	22 03	23 02	17 04	23 19	22 36	12 35	02 09	01 08	11 47	18 51
13 We	21 54	23 19	16 32	23 16	22 42	12 37	02 10	01 08	11 48	18 51
14 Th	21 46	22 14	16 00	23 12	22 47	12 40	02 12	01 08	11 48	18 51
15 Fr	21 36	19 54	15 28	23 07	22 52	12 42	02 13	01 07	11 48	18 52
16 Sa	21 27	16 34	14 57	23 02	22 56	12 45	02 15	01 07	11 49	18 52
17 Su	21 17	12 31	14 25	22 56	23 01	12 47	02 16	01 07	11 49	18 52
18 Mo	21 07	08 01	13 54	22 49	23 05	12 49	02 18	01 07	11 50	18 52
19 Tu	20 57	03 16	13 23	22 42	23 09	12 52	02 19	01 07	11 50	18 52
20 We	20 46	01N30	12 53	22 34	23 13	12 54	02 21	01 07	11 51	18 53
21 Th	20 35	06 10	12 23	22 25	23 17	12 56	02 22	01 06	11 51	18 53
22 Fr	20 23	10 34	11 54	22 16	23 20	12 58	02 24	01 06	11 52	18 53
23 Sa	20 11	14 35	11 25	22 06	23 24	13 00	02 25	01 06	11 52	18 53
24 Su	19 59	18 02	10 58	21 55	23 27	13 02	02 27	01 06	11 53	18 53
25 Mo	19 46	20 45	10 31	21 44	23 29	13 04	02 29	01 05	11 53	18 54
26 Tu	19 33	22 35	10 05	21 32	23 32	13 06	02 31	01 05	11 54	18 54
27 We	19 20	23 20	09 41	21 19	23 34	13 08	02 32	01 05	11 54	18 54
28 Th	19 07	22 52	09 18	21 06	23 37	13 10	02 34	01 04	11 55	18 54
29 Fr	18 53	21 06	08 57	20 52	23 39	13 12	02 36	01 04	11 55	18 55
30 Sa	18 38	18 05	08 37	20 38	23 40	13 13	02 38	01 03	11 56	18 55
31 Su	18 24	13 58	08 18	20 23	23 42	13 15	02 40	01 03	11 56	18 55

ASPECTARIAN

01 00:29 ☽ □ ♅
00:37 ☽ ∥ ☿
01:13 ☽ ⚹ ♃
03:22 ☽ ☍ ♇
11:31 ☽ ∥ ♂
11:39 ☽ □ ♄
02 05:38 ☿ ♌
10:09 ☽ # ♆
18:08 ☿ ∥ ♂
23:57 ☽ ☌ ☿
03 00:17 ☉ □ ♄
05:38 ☽ △ ♅
06:56 ☽ □ ♃
13:02 ☽ ⚹ ♂
16:27 ☽ ⚹ ♄
04 00:38 ☽ ∥ ♃
02:58 ☽ # ♆
04:17 ♀ ♋
17:27 ♀ △ ♆
05 00:47 ☿ △ ♅
02:24 ☽ ☍ ♆
03:15 ☽ ⚹ ♀
10:49 ☽ △ ♃
11:28 ☽ △ ♇
18:09 ☿ □ ♃
18:51 ☽ □ ♂
22:45 ☽ # ♄
06 00:20 ☽ ⚹ ☉
02:36 ☽ ∥ ♅
12:14 ☽ # ♅
12:17 ♂ △ ♄
16:09 ☽ ∥ ♄
07 10:38 ☽ □ ♀
11:35 ☽ ☍ ♅
13:53 ♃ △ ♇
13:57 ☽ □ ♇
18:37 ☽ ⚹ ☿
21:44 ♀ □ ♅
22:21 ☽ ☌ ♄
08 00:00 ☽ △ ♂
11:42 ☽ ∥ ♆
13:57 ☿ # ♇
14:58 ☽ # ♃
09 00:30 ♀ ☍ ♇
06:06 ♀ ⚹ ♃
07:32 ☽ △ ♆
08:19 ☿ ⚹ ♄
11:19 ☉ ∥ ♂
16:33 ☽ ⚹ ♇
17:11 ☽ ☍ ♃
18:07 ☽ △ ♀
22:54 ☽ # ☿
10 00:36 ♅ SR
03:06 ☽ □ ☿
03:44 ☽ ∥ ♇
13:05 ☽ △ ☉
11 09:07 ☽ # ☉
10:44 ☽ □ ♆
15:26 ☽ # ♂
17:40 ☽ △ ♅
12 03:29 ☿ ⚹ ♂
05:05 ☽ ⚹ ♄
08:39 ☽ # ♀
11:56 ☽ ☍ ♂
12:21 ☽ △ ☿
13 03:12 ☽ # ♀
07:36 ♀ □ ♄
15:08 ☽ ⚹ ♆
15:43 ☽ # ♂
22:20 ☽ □ ♅
14 00:34 ☽ ☌ ♇
02:32 ☽ △ ♃
06:19 ☽ # ☉
10:20 ☽ □ ♄
12:56 ☽ ☍ ♀
15 08:08 ☽ ∥ ♇
16 04:56 ☽ ⚹ ♅
09:54 ☽ □ ♃
11:33 ☽ # ☿
12:03 ♀ ∥ ♂
17:41 ☽ △ ♄
22:31 ☽ # ♃
17 03:50 ☽ ∥ ♆
06:56 ☽ △ ♂
12:24 ☽ ☍ ☿
18 06:02 ☽ ☌ ♆
16:17 ☽ ⚹ ♇
19:52 ☽ ⚹ ♃
19 04:46 ☽ ∥ ♄
10:50 ☽ # ♅
18:43 ☽ △ ♀
20:45 ☽ □ ♂
22:00 ☽ ∥ ♅
22:58 ☿ ∥ ♃
20 04:16 ☽ # ♄
11:15 ☽ △ ☉
21 01:31 ☽ ☌ ♅
03:46 ☽ □ ♇
15:54 ☽ ☍ ♄
22 01:35 ☿ # ♆
06:49 ☽ ∥ ☿
07:26 ☽ # ♆
12:46 ☽ ⚹ ♂
13:58 ☽ □ ♀
14:09 ☽ ∥ ♃
21:36 ☽ △ ☿
23 04:12 ☉ ♌
05:37 ☽ ⚹ ♆
16:13 ☽ △ ♇
21:20 ☽ ☌ ♃
24 06:56 ☽ # ♇
15:14 ☽ ∥ ☉
25 08:46 ☽ ⚹ ♀
10:11 ☽ ∥ ♀
13:13 ☽ □ ☿
17:05 ☽ □ ♆
21:38 ☽ ⚹ ☉
26 01:14 ☽ ⚹ ♅
15:36 ☽ △ ♄
27 16:56 ☽ ☌ ♂
19:43 ☉ △ ♅
28 00:36 ☽ ⚹ ☿
01:34 ☽ △ ♆
09:15 ☽ □ ♅
11:08 ☽ ☍ ♇
14:59 ♀ ♌
16:51 ☽ ⚹ ♃
17:59 ☿ ♍
20:44 ☉ # ♇
23:04 ☽ □ ♄
29 02:22 ☽ ∥ ♀
04:03 ☿ ☍ ♆
18:13 ☽ # ♇
19:52 ☽ ∥ ☉
30 10:04 ☽ ☌ ♀
13:48 ☽ △ ♅
21:23 ☽ □ ♃
31 03:10 ☽ ⚹ ♄
03:40 ☽ ∥ ♃
10:11 ☽ # ♆

⚷ Chiron

01 Dec.	03 S 58
02	05♓14R
05	05 10
08	05 05
11	05 00
14	04 54
17	04 49
20	04 42
23	04 36
26	04 29
29	04 21

August 2011

Day	S. T.	☉	☽	☿	♀	♂	♃	♄	♅	♆	♇	☊ True
	h m s	° ′ ″	° ′ ″	° ′	° ′	° ′	° ′	° ′	° ′	° ′	° ′	° ′
01 Mo	20 37 03	08♌ 26 06	24♌ 45 09	01♍ 00	04♌ 09	28♊ 24	08♉ 58	12♎ 21	04♈R22	00♓R06	05♑R24	22♐R37
02 Tu	20 40 59	09 23 30	09♍ 16 39	01 09	05 23	29 04	09 03	12 26	04 21	00 05	05 23	22 30
03 We	20 44 56	10 20 56	23 52 34	01 12	06 37	29 44	09 09	12 30	04 20	00 03	05 22	22 22
04 Th	20 48 52	11 18 22	08♎ 26 23	01R 10	07 51	00♋ 25	09 14	12 35	04 19	00 02	05 20	22 16
05 Fr	20 52 49	12 15 49	22 52 39	01 03	09 05	01 05	09 19	12 40	04 18	00 00	05 19	22 12
06 Sa	20 56 45	13 13 17	07♏ 07 33	00 51	10 19	01 45	09 23	12 44	04 17	29♒ 59	05 18	22 09
07 Su	21 00 42	14 10 45	21 09 05	00 33	11 33	02 25	09 28	12 49	04 16	29 57	05 17	22D 09
08 Mo	21 04 38	15 08 14	04♐ 56 37	00 11	12 47	03 05	09 32	12 54	04 14	29 56	05 16	22 09
09 Tu	21 08 35	16 05 44	18 30 32	29♌ 43	14 01	03 44	09 36	12 59	04 13	29 54	05 15	22 10
10 We	21 12 32	17 03 14	01♑ 51 35	29 10	15 16	04 24	09 40	13 04	04 12	29 52	05 14	22R 09
11 Th	21 16 28	18 00 46	15 00 32	28 33	16 30	05 04	09 44	13 09	04 10	29 51	05 13	22 07
12 Fr	21 20 25	18 58 18	27 57 51	27 52	17 44	05 44	09 48	13 15	04 09	29 49	05 12	22 03
13 Sa	21 24 21	19 55 52	10♒ 43 47	27 08	18 58	06 23	09 51	13 20	04 07	29 48	05 11	21 56
14 Su	21 28 18	20 53 26	23 18 25	26 21	20 12	07 03	09 55	13 25	04 06	29 46	05 10	21 47
15 Mo	21 32 14	21 51 02	05♓ 41 58	25 32	21 26	07 42	09 58	13 31	04 04	29 44	05 09	21 36
16 Tu	21 36 11	22 48 39	17 55 01	24 42	22 40	08 22	10 01	13 36	04 02	29 43	05 08	21 25
17 We	21 40 07	23 46 17	29 58 44	23 51	23 54	09 01	10 03	13 42	04 01	29 41	05 07	21 15
18 Th	21 44 04	24 43 57	11♈ 55 00	23 01	25 09	09 40	10 06	13 47	03 59	29 39	05 06	21 06
19 Fr	21 48 00	25 41 38	23 46 35	22 14	26 23	10 19	10 08	13 53	03 57	29 38	05 05	21 00
20 Sa	21 51 57	26 39 21	05♉ 37 06	21 29	27 37	10 58	10 11	13 58	03 56	29 36	05 05	20 57
21 Su	21 55 54	27 37 05	17 30 54	20 47	28 51	11 38	10 12	14 04	03 54	29 35	05 04	20 55
22 Mo	21 59 50	28 34 52	29 32 57	20 11	00♍ 06	12 17	10 14	14 10	03 52	29 33	05 03	20D 56
23 Tu	22 03 47	29 32 39	11♊ 48 27	19 39	01 20	12 55	10 16	14 16	03 50	29 31	05 02	20 57
24 We	22 07 43	00♍ 30 29	24 22 32	19 14	02 34	13 34	10 17	14 22	03 48	29 30	05 02	20R 57
25 Th	22 11 40	01 28 20	07♋ 19 44	18 56	03 49	14 13	10 18	14 28	03 47	29 28	05 01	20 56
26 Fr	22 15 36	02 26 13	20 43 28	18 45	05 03	14 52	10 19	14 34	03 45	29 26	05 00	20 52
27 Sa	22 19 33	03 24 08	04♌ 35 05	18 D 42	06 17	15 30	10 20	14 40	03 43	29 25	05 00	20 46
28 Su	22 23 29	04 22 04	18 53 17	18 46	07 32	16 09	10 21	14 46	03 41	29 23	04 59	20 37
29 Mo	22 27 26	05 20 02	03♍ 33 33	18 59	08 46	16 47	10 21	14 52	03 39	29 21	04 58	20 27
30 Tu	22 31 23	06 18 01	18 28 33	19 20	10 00	17 26	10 21	14 58	03 37	29 20	04 58	20 16
31 We	22 35 19	07 16 02	03♎ 29 05	19 49	11 15	18 04	10R 21	15 05	03 35	29 18	04 57	20 05

Data for	08-01-2011
Julian Day	2455774.50
Ayanamsa	24 01 25
SVP	05 ♓ 06 11
☽ ☊ Mean	21 ♐ 05 R

● ◐ PHASES ○ ◑

Day	Time	Phase	Position
06	11:08	◐	13♏40
13	18:58	○	20♒41
21	21:55	◑	28♉30
29	03:04	●	05♍27

LAST ASPECT Day	h m	☽ INGRESS Day	h m	Sign
01	06:20	01	08:42	♍
01	23:38	03	10:04	♎
05	11:56	05	11:57	♏
07	15:14	07	15:21	♐
09	20:25	09	20:38	♑
10	20:34	12	03:48	♒
14	12:26	14	12:55	♓
15	08:22	17	00:03	♈
19	11:50	19	12:37	♉
22	00:00	22	00:53	♊
24	09:33	24	10:31	♋
25	13:04	26	16:09	♌
28	17:11	28	18:13	♍
29	22:15	30	18:26	♎

DECLINATION

Day	☉	☽	☿	♀	♂	♃	♄	♅	♆	♇
01 Mo	18N09	09N00	08N02	20N07	23N43	13N17	02S42	01N03	11S57	18S55
02 Tu	17 54	03 28	07 48	19 51	23 44	13 18	02 44	01 02	11 57	18 56
03 We	17 39	02S16	07 35	19 35	23 45	13 20	02 46	01 02	11 58	18 56
04 Th	17 23	07 52	07 25	19 17	23 46	13 21	02 47	01 01	11 59	18 56
05 Fr	17 07	12 59	07 18	18 59	23 47	13 22	02 50	01 01	11 59	18 56
06 Sa	16 51	17 19	07 13	18 41	23 47	13 24	02 52	01 00	12 00	18 56
07 Su	16 34	20 36	07 11	18 22	23 47	13 25	02 54	01 00	12 00	18 57
08 Mo	16 18	22 37	07 11	18 03	23 47	13 26	02 56	00 59	12 01	18 57
09 Tu	16 01	23 16	07 14	17 43	23 47	13 27	02 58	00 59	12 01	18 57
10 We	15 43	22 34	07 20	17 22	23 46	13 28	03 00	00 58	12 02	18 57
11 Th	15 26	20 37	07 29	17 02	23 46	13 29	03 02	00 57	12 02	18 58
12 Fr	15 08	17 39	07 40	16 40	23 45	13 30	03 04	00 57	12 03	18 58
13 Sa	14 50	13 52	07 54	16 18	23 44	13 31	03 06	00 56	12 04	18 58
14 Su	14 32	09 33	08 11	15 56	23 43	13 32	03 09	00 56	12 04	18 58
15 Mo	14 13	04 54	08 30	15 33	23 41	13 33	03 11	00 55	12 05	18 59
16 Tu	13 55	00 08	08 51	15 10	23 40	13 33	03 13	00 54	12 05	18 59
17 We	13 36	04N34	09 13	14 47	23 38	13 34	03 15	00 54	12 06	18 59
18 Th	13 17	09 03	09 37	14 23	23 36	13 35	03 18	00 53	12 07	18 59
19 Fr	12 57	13 11	10 02	13 58	23 34	13 35	03 20	00 52	12 07	19 00
20 Sa	12 38	16 47	10 28	13 34	23 31	13 36	03 22	00 52	12 08	19 00
21 Su	12 18	19 44	10 53	13 08	23 29	13 36	03 25	00 51	12 08	19 00
22 Mo	11 58	21 51	11 19	12 43	23 26	13 36	03 27	00 50	12 09	19 00
23 Tu	11 38	23 00	11 43	12 17	23 23	13 37	03 30	00 49	12 09	19 01
24 We	11 18	23 01	12 07	11 51	23 20	13 37	03 32	00 49	12 10	19 01
25 Th	10 57	21 48	12 29	11 24	23 17	13 37	03 34	00 48	12 11	19 01
26 Fr	10 36	19 21	12 49	10 58	23 13	13 37	03 37	00 47	12 11	19 01
27 Sa	10 15	15 43	13 08	10 31	23 10	13 37	03 39	00 46	12 12	19 01
28 Su	09 54	11 05	13 24	10 03	23 06	13 37	03 42	00 45	12 12	19 02
29 Mo	09 33	05 43	13 37	09 36	23 02	13 37	03 44	00 45	12 13	19 02
30 Tu	09 12	00S 03	13 48	09 08	22 58	13 37	03 47	00 44	12 14	19 02
31 We	08 51	05 51	13 56	08 39	22 54	13 37	03 49	00 43	12 14	19 02

ASPECTARIAN

01 04:09 ♀ △ ♅
04:29 ☽ ∥ ☿
06:20 ☽ ⚹ ♂
08:51 ☽ ☍ ♆
10:28 ☽ ☌ ☿
14:42 ☉ □ ♃
17:35 ☽ △ ♇
23:38 ☽ △ ♃
02 03:06 ☽ # ♄
10:12 ☽ ∥ ♅
18:49 ☽ # ♅
03 02:04 ☽ ∥ ♄
03:51 ☿ SR
09:23 ♂ ♋
10:07 ☽ □ ♂
10:54 ♂ △ ♆
17:12 ☽ ☍ ♅
18:53 ☽ □ ♆
22:07 ☽ # ☿
22:57 ☽ ⚹ ♀
04 05:05 ☽ ⚹ ☉
06:54 ☽ ☌ ♄
19:02 ☽ ∥ ♆
23:20 ☿ ⚹ ♂
05 01:57 ☽ # ♃
03:12 ♆ ♒R
04:13 ♀ # ♇
04:35 ♀ □ ♃
10:48 ☉ ⚹ ♄
11:56 ☽ △ ♆
13:33 ☽ ⚹ ☿
14:27 ☽ △ ♂
20:55 ☽ ⚹ ♇
21:17 ☽ # ☉
06 03:52 ☽ ☍ ♃
05:57 ☽ □ ♀
08:14 ☽ # ♀
10:00 [illegible] ∥ [illegible]
07 15:14 ☽ □ ♆
15:54 ☽ □ ☿
22:46 ☽ △ ♅
08 02:20 ♀ ⚹ ♄
09:46 ☿ ♌R
14:07 ☽ ⚹ ♄
14:28 ☿ ☍ ♆
15:13 ☽ △ ♀
19:22 ☽ △ ☉
09 16:33 ♂ □ ♅
19:21 ☽ △ ☿
20:25 ☽ ⚹ ♆
10 04:13 ☽ □ ♅
04:52 ☽ ☍ ♂
06:06 ☽ ☌ ♇
14:17 ☽ △ ♃
20:34 ☽ □ ♄
11 05:12 ♂ ☍ ♇
14:14 ☽ ∥ ♇
12 07:20 ☽ # ♀
11:33 ☽ ⚹ ♅
17:44 ☽ # ☉
22:20 ☽ □ ♃
13 02:00 ☽ # ♃
04:58 ☽ △ ♄
10:17 ☽ ∥ ♆
17:23 ☽ ☍ ♀
14 05:30 ☽ ☍ ☿
06:40 ☽ # ☿
12:26 ☽ ☌ ♆
22:56 ☽ ⚹ ♇
15 04:08 ☽ △ ♂
08:22 ☽ ⚹ ♃
08:34 ☽ ∥ ♄
16:03 ☽ # [illegible]
16 05:12 ☽ ∥ ♅
12:08 ♀ ☌ ☉
17:08 ☽ # ♄
23:21 ☿ ☌ ♀
17 01:04 ☿ ☌ ☉
02:02 ☉ ∥ ♃
08:04 ☽ ☌ ♅
10:18 ☽ □ ♇
19:12 ☽ □ ♂
18 03:27 ☽ ∥ ☿
03:48 ☽ ☍ ♄
16:50 ♂ ⚹ ♃
17:32 ☽ # ♆
21:03 ☽ △ ☿
22:43 ☽ ∥ ☉
19 02:32 ☽ ∥ ♃
04:14 ☽ △ ☉
04:28 ☽ ∥ ♀
05:54 ☽ △ ♀
11:50 ☽ ⚹ ♆
22:01 ♀ ∥ ♃
22:54 ☽ △ ♇
20 09:14 ☽ ☌ ♃
11:27 ☽ ⚹ ♂
17:22 ☽ # ♇
21 06:14 ☽ □ ☿
11:20 ☉ # ♆
13:41 ♀ ☍ ♆
22:11 ♀ ♍
22 00:00 ☽ □ ♆
01:12 ☽ □ ♀
08:30 ☽ ⚹ ♅
21:08 ☿ ∥ ☉
23:28 ☉ ☍ ♆
23 04:47 ☽ △ ♄
06:55 ♀ # ♆
11:21 ☉ ♍
14:34 ☽ ⚹ ☿
16:19 ☿ ∥ ♀
24 03:31 ☿ # ♆
09:33 ☽ △ ♆
12:23 ☽ ⚹ ☉
16:53 ☽ ⚹ ♀
17:31 ☽ □ ♅
19:46 ☽ ☍ ♇
25 05:25 ☽ ⚹ ♃
10:47 ♂ □ ♄
12:59 ☽ □ ♄
13:04 ☽ ☌ ♂
23:12 ♀ △ ♇
26 02:32 ☽ # ♇
22:04 ☿ SD
22:31 ☽ △ ♅
27 09:45 ☽ □ ♃
11:28 ☽ ∥ ♃
13:12 ☽ ∥ ☿
17:06 ☽ ⚹ ♄
18:37 ☽ # ♆
23:49 ☽ ☌ ☿
28 05:20 ☽ ∥ ♀
05:56 ☽ ∥ ☉
15:11 ☉ △ ♇
17:11 ☽ ☍ ♆
23:33 ☿ ∥ ♃
29 02:17 ☽ △ ♇
08:17 ☽ # ♄
08:19 ♀ ∥ ☉
09:10 ☽ ☌ ♀
10:58 ☽ △ ♃
20:46 ☽ ∥ ♅
22:15 ☽ ⚹ ♂
30 02:45 ☽ # ♅
06:47 ♀ △ ♃
09:18 ♃ SR
15:25 ☽ ∥ ♄
31 00:09 ☽ ☍ ♅
02:21 ☽ □ ♇
11:04 ☽ # ♀
12:02 ☽ # ☉
18:44 ☽ ☌ ♄

⚷ Chiron

01	Dec.	04 S 14
01	04♓14R	
04	04 06	
07	03 58	
10	03 50	
13	03 41	
16	03 33	
19	03 24	
22	03 15	
25	03 06	
28	02 57	
[illegible]	[illegible]	

September 2011

Day	S. T.	☉	☽	☿	♀	♂	♃	♄	♅	♆	♇	☊ True
	h m s	° ' "	° ' "	° '	° '	° '	° '	° '	° '	° '	° '	° '
01 Th	22 39 16	08♍ 14 05	18♎ 25 38	20♌ 26	12♍ 29	18♋ 42	10♉R21	15♎ 11	03♈R32	29♒R17	04♑R57	19♐R55
02 Fr	22 43 12	09 12 09	03♏ 10 02	21 11	13 44	19 20	10 21	15 17	03 30	29 15	04 56	19 48
03 Sa	22 47 09	10 10 14	17 36 29	22 04	14 58	19 59	10 20	15 24	03 28	29 13	04 56	19 43
04 Su	22 51 05	11 08 21	01♐ 42 08	23 04	16 12	20 37	10 19	15 30	03 26	29 12	04 56	19 41
05 Mo	22 55 02	12 06 29	15 26 36	24 10	17 27	21 15	10 18	15 37	03 24	29 10	04 55	19 40
06 Tu	22 58 58	13 04 38	28 51 18	25 24	18 41	21 52	10 17	15 43	03 22	29 09	04 55	19 40
07 We	23 02 55	14 02 49	11♑ 58 33	26 43	19 56	22 30	10 16	15 50	03 19	29 07	04 55	19 39
08 Th	23 06 52	15 01 02	24 50 55	28 07	21 10	23 08	10 14	15 56	03 17	29 05	04 54	19 36
09 Fr	23 10 48	15 59 16	07♒ 30 41	29 37	22 25	23 45	10 12	16 03	03 15	29 04	04 54	19 30
10 Sa	23 14 45	16 57 31	19 59 44	01♍ 11	23 39	24 23	10 10	16 10	03 13	29 02	04 54	19 22
11 Su	23 18 41	17 55 48	02♓ 19 27	02 49	24 54	25 00	10 08	16 16	03 10	29 01	04 54	19 11
12 Mo	23 22 38	18 54 07	14 30 56	04 30	26 08	25 38	10 05	16 23	03 08	28 59	04 53	18 58
13 Tu	23 26 34	19 52 28	26 35 05	06 14	27 23	26 15	10 03	16 30	03 06	28 58	04 53	18 44
14 We	23 30 31	20 50 50	08♈ 32 52	08 01	28 37	26 52	10 00	16 37	03 03	28 56	04 53	18 31
15 Th	23 34 27	21 49 15	20 25 41	09 49	29 52	27 29	09 57	16 44	03 01	28 55	04 53	18 19
16 Fr	23 38 24	22 47 41	02♉ 15 29	11 39	01♎ 06	28 06	09 54	16 51	02 59	28 53	04 53	18 10
17 Sa	23 42 21	23 46 10	14 05 01	13 30	02 21	28 43	09 50	16 58	02 56	28 52	04D 53	18 04
18 Su	23 46 17	24 44 40	25 57 49	15 21	03 35	29 20	09 47	17 04	02 54	28 50	04 53	18 01
19 Mo	23 50 14	25 43 13	07♊ 58 11	17 13	04 50	29 57	09 43	17 11	02 52	28 49	04 53	18 00
20 Tu	23 54 10	26 41 48	20 11 00	19 05	06 04	00♌ 34	09 39	17 18	02 49	28 48	04 53	18D 00
21 We	23 58 07	27 40 25	02♋ 41 24	20 58	07 19	01 11	09 35	17 25	02 47	28 46	04 53	18R 00
22 Th	00 02 03	28 39 05	15 34 27	22 50	08 34	01 47	09 31	17 33	02 44	28 45	04 54	17 59
23 Fr	00 05 60	29 37 47	28 54 27	24 41	09 48	02 24	09 26	17 40	02 42	28 43	04 54	17 56
24 Sa	00 09 56	00♎ 36 30	12♌ 44 08	26 33	11 03	03 00	09 22	17 47	02 40	28 42	04 54	17 50
25 Su	00 13 53	01 35 17	27 03 36	28 23	12 17	03 36	09 17	17 54	02 37	28 41	04 54	17 42
26 Mo	00 17 50	02 34 05	11♍ 49 34	00♎ 13	13 32	04 12	09 12	18 01	02 35	28 39	04 54	17 32
27 Tu	00 21 46	03 32 55	26 55 09	02 02	14 47	04 48	09 07	18 08	02 32	28 38	04 55	17 21
28 We	00 25 43	04 31 47	12♎ 10 36	03 51	16 01	05 24	09 01	18 15	02 30	28 37	04 55	17 10
29 Th	00 29 39	05 30 42	27 24 48	05 38	17 16	06 00	08 56	18 23	02 27	28 36	04 55	17 01
30 Fr	00 33 36	06 29 38	12♏ 27 16	07 25	18 30	06 36	08 50	18 30	02 25	28 34	04 56	16 53

Data for	09-01-2011
Julian Day	2455805.50
Ayanamsa	24 01 29
SVP	05 ♓ 06 06
☽ ☊ Mean	19 ♐ 26 R

● ◐ PHASES ○ ◑

Day	h m	Phase	Position
04	17:40	◐	11♐51
12	09:27	○	19♓17
20	13:38	◑	27♊15
27	11:09	●	04♎00

LAST ASPECT ☽ Day	h m	INGRESS Day	h m	Sign
01	17:35	01	18:48	♏
03	19:42	03	21:04	♐
06	00:31	06	02:04	♑
07	20:36	08	09:43	♒
10	17:32	10	19:27	♓
13	01:46	13	06:50	♈
15	17:10	15	19:25	♉
18	07:09	18	08:06	♊
20	16:34	20	18:54	♋
23	01:22	23	01:56	♌
25	02:40	25	04:50	♍
25	19:48	27	04:51	♎
29	01:52	29	04:06	♏

DECLINATION

Day	☉	☽	☿	♀	♂	♃	♄	♅	♆	♇
01 Th	08N29	11S 18	14N01	08N11	22N49	13N37	03S 52	00N42	12S 15	19S 03
02 Fr	08 07	16 00	14 02	07 42	22 45	13 36	03 54	00 41	12 15	19 03
03 Sa	07 45	19 39	14 00	07 14	22 40	13 36	03 57	00 40	12 16	19 03
04 Su	07 23	22 01	13 55	06 45	22 35	13 35	04 00	00 40	12 16	19 03
05 Mo	07 01	23 01	13 46	06 15	22 30	13 35	04 02	00 39	12 17	19 04
06 Tu	06 39	22 37	13 34	05 46	22 25	13 34	04 05	00 38	12 18	19 04
07 We	06 17	20 59	13 19	05 16	22 20	13 34	04 07	00 37	12 18	19 04
08 Th	05 54	18 18	13 00	04 47	22 14	13 33	04 10	00 36	12 19	19 04
09 Fr	05 32	14 47	12 38	04 17	22 09	13 32	04 13	00 35	12 19	19 05
10 Sa	05 09	10 40	12 13	03 47	22 03	13 31	04 15	00 34	12 20	19 05
11 Su	04 46	06 11	11 45	03 17	21 57	13 31	04 18	00 33	12 20	19 05
12 Mo	04 23	01 30	11 15	02 46	21 51	13 30	04 21	00 32	12 21	19 05
13 Tu	04 01	03N10	10 42	02 16	21 45	13 29	04 23	00 31	12 21	19 06
14 We	03 38	07 42	10 07	01 46	21 38	13 28	04 26	00 30	12 22	19 06
15 Th	03 15	11 54	09 29	01 15	21 32	13 26	04 29	00 30	12 22	19 06
16 Fr	02 52	15 38	08 50	00 45	21 25	13 25	04 31	00 29	12 23	19 06
17 Sa	02 28	18 45	08 09	00 14	21 19	13 24	04 34	00 28	12 23	19 07
18 Su	02 05	21 05	07 27	00S 16	21 12	13 23	04 37	00 27	12 24	19 07
19 Mo	01 42	22 30	06 44	00 47	21 05	13 21	04 40	00 26	12 24	19 07
20 Tu	01 19	22 52	06 00	01 17	20 58	13 20	04 42	00 25	12 25	19 07
21 We	00 56	22 07	05 14	01 48	20 51	13 19	04 45	00 24	12 25	19 07
22 Th	00 32	20 11	04 29	02 19	20 44	13 17	04 48	00 23	12 26	19 08
23 Fr	00 09	17 08	03 42	02 49	20 36	13 16	04 51	00 22	12 26	19 08
24 Sa	00S 15	13 01	02 55	03 20	20 29	13 14	04 53	00 21	12 27	19 08
25 Su	00 38	08 03	02 08	03 50	20 21	13 12	04 56	00 20	12 27	19 08
26 Mo	01 01	02 30	01 21	04 21	20 13	13 11	04 59	00 19	12 28	19 09
27 Tu	01 25	03S 20	00 34	04 51	20 06	13 09	05 02	00 18	12 28	19 09
28 We	01 48	09 01	00S 13	05 21	19 58	13 07	05 04	00 17	12 29	19 09
29 Th	02 11	14 08	01 00	05 51	19 50	13 05	05 07	00 16	12 29	19 09
30 Fr	02 35	18 17	01 47	06 21	19 41	13 03	05 10	00 15	12 29	19 09

ASPECTARIAN

01 00:28 ☽ □ ♂
03:25 ☽ ⚹ ☿
04:29 ☽ ∥ ♆
11:11 ☽ # ♃
13:21 ☽ # ☿
17:35 ☽ △ ♆
02 02:55 ☽ ⚹ ♇
10:41 ☽ ⚹ ☉
11:51 ☽ ☍ ♃
19:09 ☽ ⚹ ♀
19:25 ☽ ∥ ♇
03 04:00 ☉ △ ♃
04:11 ☽ △ ♂
08:04 ☽ □ ☿
19:42 ☽ □ ♆
04 02:59 ☽ △ ♅
08:49 ☽ # ♂
05 00:18 ☽ ⚹ ♄
03:55 ☽ □ ♀
17:06 ☽ △ ☿
23:57 ☿ ∥ ♃
06 00:31 ☽ ⚹ ♆
04:46 ☽ # ♂
08:10 ☽ □ ♅
11:01 ☽ ☌ ♇
20:50 ☽ △ ♃
07 04:09 ☽ △ ☉
07:12 ☽ □ ♄
16:22 ☽ △ ♀
17:55 ☽ ∥ ♇
20:36 ☽ ☍ ♂
08 15:26 ☿ ☍ ♆
15:54 ☽ ⚹ ♅
09 03:01 ♀ # ♄
05:08 ☽ □ ♃
05:59 ☿ ♍
07:36 ☽ # ♃
14:16 ☽ # ☿
14:40 ☽ ∥ ♆
16:32 ☽ △ ♄
18:09 ☿ # ♆
10 17:32 ☽ ☌ ♆
11 01:07 ☽ ☍ ☿
04:22 ♀ ⚹ ♂
05:02 ☽ ⚹ ♇
07:55 ☽ # ☉
09:36 ☽ ∥ ♄
15:17 ☽ ⚹ ♃
16:43 ☽ # ♀
12 02:37 ☉ # ♄
04:55 ☽ # ♅
05:27 ☿ △ ♇
10:23 ☽ ∥ ♅
19:45 ☽ ∥ ♀
23:18 ☽ △ ♂
13 01:46 ☽ ☍ ♀
03:59 ☽ ∥ ☉
06:22 ☽ # ♄
13:00 ☽ ☌ ♅
16:38 ☽ □ ♇
14 11:45 ☽ ∥ ☿
16:26 ☽ ☍ ♄
15 01:44 ☿ △ ♃
02:40 ♀ ♎
02:51 ☽ # ♆
09:25 ☽ ∥ ♃
15:07 ☽ □ ♂
17:10 ☽ ⚹ ♆
16 05:20 ☽ △ ♇
13:14 ♀ ∥ ♅
15:26 ☽ ☌ ♃
18:23 ♆ SD
22:35 ☽ △ ☿
17 03:15 ☽ # ♇
11:06 ♀ ☍ ♅
21:20 ☽ △ ☉
18 01:26 ☽ ∥ ♂
05:46 ☽ □ ♆
07:09 ☽ ⚹ ♂
07:59 ♀ # ♅
13:53 ☽ ⚹ ♅
17:02 ☽ △ ♀
19 01:07 ♀ □ ♇
01:51 ♂ ♌
18:20 ☽ △ ♄
21:30 ☽ □ ☿
20 00:37 ♀ # ☉
16:34 ☽ △ ♆
21 00:10 ☽ □ ♅
04:09 ☽ ☍ ♇
09:38 ☽ □ ♀
12:52 ☽ ⚹ ♃
14:30 ☿ # ♄
18:22 ☽ ∥ ♂
22 03:38 ☽ □ ♄
09:24 ☽ # ♇
09:58 ☉ ∥ ♅
15:18 ☽ ⚹ ☿
23 01:22 ☽ ⚹ ☉
06:25 ☽ ☌ ♂
06:39 ☽ △ ♅
09:05 ☉ ♎
11:24 ♂ △ ♅
16:29 ☿ # ♀
18:15 ☽ □ ♃
20:50 ☽ ⚹ ♀
22:55 ☽ ∥ ♃
24 03:01 ☽ # ♆
06:23 ☉ # ♅
08:37 ☽ ⚹ ♄
25 02:40 ☽ ☍ ♆
12:50 ☽ △ ♇
13:37 ☽ # ♄
16:54 ☽ # ♀
19:48 ☽ △ ♃
21:09 ☿ ♎
26 00:15 ☉ ☍ ♅
05:31 ☽ ∥ ☿
05:45 ☽ # ☉
06:48 ☿ # ☉
09:04 ☽ ∥ ♅
11:37 ☽ # ♅
14:01 ☽ # ☿
15:35 ☽ ∥ ☉
27 06:30 ☿ ☍ ♅
06:55 ☽ ∥ ♀
07:07 ☽ ∥ ♄
08:15 ☿ ∥ ♅
08:50 ☽ ☍ ♅
09:09 ☽ ☌ ☿
09:24 ♀ ∥ ♄
12:36 ☽ □ ♇
12:56 ☽ ⚹ ♂
28 02:00 ☿ # ♅
06:34 ☽ ☌ ♀
09:35 ☉ □ ♇
09:38 ☽ ☌ ♄
14:25 ☿ □ ♇
15:51 ☽ ∥ ♆
18:48 ☽ # ♃
20:16 ☿ ☌ ☉
29 01:52 ☽ △ ♆
07:28 ☿ ⚹ ♂
11:56 ☽ ⚹ ♇
14:13 ☽ □ ♂
18:13 ☽ ☍ ♃
23:48 ♀ ☌ ♄
30 06:05 ☽ ∥ ♇
06:44 ☉ ⚹ ♂
09:45 ☽ # ♂

⚷ Chiron

01 Dec.	04 S 45
03	02♓40R
06	02 31
09	02 23
12	02 14
15	02 06
18	01 58
21	01 50
24	01 43
27	01 35
30	01 29

October 2011

Day	S. T.	☉	☽	☿	♀	♂	♃	♄	♅	♆	♇	☊ True
	h m s	° ' "	° ' "	° '	° '	° '	° '	° '	° '	° '	° '	° '
01 Sa	00 37 32	07♎28 36	27♏09 50	09♎11	19♎45	07♌12	08♉R44	18♎37	02♈R23	28♒R33	04♑56	16♐R48
02 Su	00 41 29	08 27 36	11♐27 41	10 56	21 00	07 47	08 38	18 44	02 20	28 32	04 57	16 45
03 Mo	00 45 25	09 26 38	25 19 16	12 40	22 14	08 23	08 32	18 52	02 18	28 31	04 57	16 45
04 Tu	00 49 22	10 25 41	08♑45 44	14 23	23 29	08 58	08 26	18 59	02 15	28 30	04 58	16 44
05 We	00 53 19	11 24 46	21 49 49	16 06	24 43	09 34	08 20	19 06	02 13	28 29	04 58	16 44
06 Th	00 57 15	12 23 53	04♒35 01	17 48	25 58	10 09	08 13	19 13	02 11	28 28	04 59	16 43
07 Fr	01 01 12	13 23 02	17 04 51	19 28	27 13	10 44	08 06	19 21	02 08	28 27	05 00	16 39
08 Sa	01 05 08	14 22 12	29 22 36	21 08	28 27	11 19	08 00	19 28	02 06	28 26	05 00	16 33
09 Su	01 09 05	15 21 24	11♓31 08	22 48	29 42	11 54	07 53	19 35	02 04	28 25	05 01	16 24
10 Mo	01 13 01	16 20 38	23 32 45	24 26	00♏56	12 29	07 46	19 43	02 01	28 24	05 02	16 14
11 Tu	01 16 58	17 19 54	05♈29 17	26 04	02 11	13 03	07 39	19 50	01 59	28 23	05 02	16 03
12 We	01 20 54	18 19 13	17 22 17	27 40	03 26	13 38	07 31	19 57	01 57	28 22	05 03	15 53
13 Th	01 24 51	19 18 33	29 13 10	29 17	04 40	14 12	07 24	20 05	01 55	28 21	05 04	15 43
14 Fr	01 28 47	20 17 55	11♉03 33	00♏52	05 55	14 47	07 17	20 12	01 52	28 20	05 05	15 36
15 Sa	01 32 44	21 17 19	22 55 26	02 27	07 09	15 21	07 09	20 19	01 50	28 19	05 06	15 31
16 Su	01 36 41	22 16 46	04♊51 26	04 01	08 24	15 55	07 02	20 27	01 48	28 18	05 06	15 29
17 Mo	01 40 37	23 16 15	16 54 46	05 35	09 39	16 29	06 54	20 34	01 46	28 18	05 07	15D29
18 Tu	01 44 34	24 15 46	29 09 13	07 07	10 53	17 03	06 46	20 41	01 43	28 17	05 08	15 30
19 We	01 48 30	25 15 19	11♋39 03	08 40	12 08	17 37	06 38	20 49	01 41	28 16	05 09	15 31
20 Th	01 52 27	26 14 55	24 28 44	10 11	13 23	18 10	06 30	20 56	01 39	28 15	05 10	15R31
21 Fr	01 56 23	27 14 33	07♌42 25	11 42	14 37	18 44	06 23	21 03	01 37	28 15	05 11	15 30
22 Sa	02 00 20	28 14 13	21 23 19	13 12	15 52	19 17	06 15	21 10	01 35	28 14	05 12	15 27
23 Su	02 04 16	29 13 55	05♍32 43	14 42	17 06	19 50	06 07	21 18	01 33	28 14	05 13	15 23
24 Mo	02 08 13	00♏13 40	20 09 11	16 11	18 21	20 23	05 58	21 25	01 31	28 13	05 15	15 16
25 Tu	02 12 10	01 13 27	05♎07 52	17 39	19 36	20 56	05 50	21 32	01 29	28 12	05 16	15 09
26 We	02 16 06	02 13 16	20 20 48	19 07	20 50	21 29	05 42	21 40	01 27	28 12	05 17	15 02
27 Th	02 20 03	03 13 07	05♏37 45	20 34	22 05	22 02	05 34	21 47	01 25	28 11	05 18	14 56
28 Fr	02 23 59	04 13 00	20 47 54	22 01	23 19	22 35	05 26	21 54	01 23	28 11	05 19	14 51
29 Sa	02 27 56	05 12 55	05♐41 39	23 27	24 34	23 07	05 18	22 01	01 21	28 11	05 21	14 48
30 Su	02 31 52	06 12 52	20 12 02	24 52	25 49	23 39	05 10	22 08	01 19	28 10	05 22	14 47
31 Mo	02 35 49	07 12 50	04♑15 20	26 16	27 03	24 11	05 01	22 16	01 17	28 10	05 23	14D47

Data for	10-01-2011
Julian Day	2455835.50
Ayanamsa	24 01 31
SVP	05 ♓ 06 01
☽ ☊ Mean	17 ♐ 51 R

● ◐ PHASES ○ ◑

04	03:16	◐	10♑34
12	02:05	○	18♈24
20	03:31	◑	26♋24
26	19:56	●	03♏03

LAST ASPECT ☽ Day	h m	INGRESS Day	h m	
01	02:18	01	04:42	♐
03	05:38	03	08:16	♑
05	05:59	05	15:19	♒
07	22:08	08	01:13	♓
08	16:51	10	12:57	♈
13	00:08	13	01:35	♉
15	10:52	15	14:15	♊
17	22:18	18	01:39	♋
20	03:31	20	10:07	♌
22	12:36	22	14:42	♍
23	20:48	24	15:50	♎
26	12:19	26	15:09	♏
28	11:49	28	14:46	♐
30	13:30	30	16:39	♑

DECLINATION

Day	☉	☽	☿	♀	♂	♃	♄	♅	♆	♇
01 Sa	02S 58	21S 11	02S 34	06S 51	19N33	13N01	05S 13	00N14	12S 30	19S 10
02 Su	03 21	22 37	03 20	07 21	19 25	12 59	05 15	00 13	12 30	19 10
03 Mo	03 45	22 35	04 06	07 50	19 17	12 57	05 18	00 12	12 31	19 10
04 Tu	04 08	21 15	04 52	08 20	19 08	12 55	05 21	00 11	12 31	19 10
05 We	04 31	18 47	05 37	08 49	19 00	12 53	05 24	00 11	12 31	19 11
06 Th	04 54	15 27	06 22	09 18	18 51	12 51	05 26	00 10	12 32	19 11
07 Fr	05 17	11 30	07 06	09 47	18 42	12 49	05 29	00 09	12 32	19 11
08 Sa	05 40	07 09	07 50	10 15	18 34	12 46	05 32	00 08	12 33	19 11
09 Su	06 03	02 34	08 33	10 44	18 25	12 44	05 35	00 07	12 33	19 11
10 Mo	06 26	02N03	09 16	11 12	18 16	12 42	05 38	00 06	12 33	19 12
11 Tu	06 48	06 34	09 58	11 40	18 07	12 39	05 40	00 05	12 34	19 12
12 We	07 11	10 49	10 39	12 07	17 58	12 37	05 43	00 04	12 34	19 12
13 Th	07 33	14 39	11 20	12 35	17 49	12 35	05 46	00 03	12 34	19 12
14 Fr	07 56	17 54	12 00	13 02	17 40	12 32	05 49	00 02	12 34	19 12
15 Sa	08 18	20 24	12 39	13 28	17 30	12 30	05 51	00 02	12 35	19 13
16 Su	08 40	22 02	13 18	13 55	17 21	12 27	05 54	00 01	12 35	19 13
17 Mo	09 02	22 40	13 56	14 21	17 12	12 25	05 57	00S 00	12 35	19 13
18 Tu	09 24	22 12	14 33	14 46	17 02	12 22	05 59	00 01	12 36	19 13
19 We	09 46	20 39	15 09	15 12	16 53	12 20	06 02	00 02	12 36	19 13
20 Th	10 08	18 01	15 45	15 37	16 44	12 17	06 05	00 03	12 36	19 13
21 Fr	10 29	14 23	16 20	16 01	16 34	12 14	06 08	00 04	12 36	19 14
22 Sa	10 51	09 54	16 54	16 25	16 25	12 12	06 10	00 04	12 36	19 14
23 Su	11 12	04 45	17 27	16 49	16 15	12 09	06 13	00 05	12 37	19 14
24 Mo	11 33	00S 49	17 59	17 12	16 05	12 07	06 16	00 06	12 37	19 14
25 Tu	11 54	06 29	18 30	17 35	15 56	12 04	06 18	00 07	12 37	19 14
26 We	12 15	11 51	19 01	17 58	15 46	12 01	06 21	00 07	12 37	19 15
27 Th	12 35	16 29	19 30	18 20	15 37	11 59	06 24	00 08	12 37	19 15
28 Fr	12 55	20 00	19 59	18 41	15 27	11 56	06 26	00 09	12 37	19 15
29 Sa	13 16	22 04	20 26	19 02	15 17	11 53	06 29	00 10	12 38	19 15
30 Su	13 35	22 35	20 53	19 23	15 07	11 51	06 31	00 10	12 38	19 15
31 Mo	13 55	21 38	21 18	19 42	14 58	11 48	06 34	00 11	12 38	19 15

ASPECTARIAN

01 02:18 ☽ □ ♆
08:39 ☽ △ ♅
17:30 ☽ △ ♂
18:31 ☽ ✶ ☉
22:58 ☽ ✶ ☿
02 01:02 ☿ ∥ ☉
12:37 ☽ ✶ ♄
18:03 ☽ ✶ ♀
03 05:22 ♂ □ ♃
05:38 ☽ ✶ ♆
12:20 ☽ □ ♅
17:08 ☽ ☌ ♇
18:06 ♂ # ♇
23:24 ☽ △ ♃
04 11:48 ☽ □ ☿
16:19 ☿ ∥ ♄
18:53 ☽ □ ♄
20:45 ☽ ∥ ♇
22:11 ☽ # ♂
05 05:59 ☽ □ ♀
19:27 ☽ ✶ ♅
06 06:52 ☽ □ ♃
11:09 ☽ ☍ ♂
16:14 ☽ △ ☉
16:22 ☽ # ♃
18:03 ☽ ∥ ♆
22:02 ☿ ☌ ♄
07 04:26 ☽ △ ♄
05:22 ☽ △ ☿
08:47 ☽ ∥ ♀
14:37 ☉ ∥ ♄
20:52 ☽ ∥ ☿
21:59 ☽ △ ♀
22:08 ☽ ☌ ♆
23:28 ♀ △ ♆
08 07:17 ☽ ∥ ☉
[illegible] ∥ [illegible]
11:06 ☽ ✶ ♇
16:51 ☽ ✶ ♃
09 05:50 ♀ ♏
12:49 ☽ # ♅
13:55 ☽ ∥ ♅
10 16:58 ☽ ☌ ♅
19:08 ☽ # ♄
23:06 ☽ □ ♇
11 01:27 ☽ # ☉
16:03 ☽ △ ♂
22:53 ☽ # ☿
12 05:17 ☽ ☍ ♄
08:55 ☽ # ♀
10:10 ☿ △ ♆
10:37 ☽ # ♆
10:49 ☽ ∥ ♃
22:14 ☽ ✶ ♆
23:36 ♀ ∥ ♆
13 00:00 ♀ # ♃
00:08 ☽ ☍ ☿
03:48 ♃ # ♇
07:42 ♀ ✶ ♇
10:52 ☿ ♏
11:52 ☽ △ ♇
12:21 ☽ ☍ ♀
16:25 ☽ ☌ ♃
21:14 ☉ ☌ ♄
22:14 ☽ ∥ ♂
14 07:54 ☽ □ ♂
11:38 ☽ # ♇
18:22 ☿ # ♃
21:06 ☿ ∥ ♆
23:54 ♀ ☍ ♃
15 [illegible] ☽ □ [illegible]
17:53 ☽ ✶ ♅
16 16:56 ☿ ✶ ♇
23:06 ☽ ✶ ♂
17 07:17 ☽ △ ♄
13:37 ☽ △ ☉
18:56 ☿ ☍ ♃
22:18 ☽ △ ♆
18 04:58 ☽ □ ♅
11:35 ☽ ☍ ♇
14:32 ☽ ✶ ♃
17:31 ☽ △ ☿
19 01:01 ☽ △ ♀
04:56 ☿ ∥ ♀
14:13 ☽ # ♇
17:22 ☽ □ ♄
20 09:46 ☽ ∥ ♂
13:05 ☽ △ ♅
13:37 ☽ # ☿
15:00 ☽ # ♀
21:39 ☽ □ ♃
21 07:51 ☿ # ♂
07:58 ☽ □ ☿
10:06 ☽ # ♆
12:11 ☽ ∥ ♃
13:27 ☽ □ ♀
19:39 ☽ # ☉
20:13 ☽ ☌ ♂
23:28 ♀ # ♂
23:38 ☽ ✶ ♄
23:59 ☉ △ ♆
22 11:42 ☽ ☍ ♆
12:36 ☽ ✶ ☉
17:30 ☽ # ♄
23:28 ☽ △ ♆
23 00:56 ☽ △ ♃
16:49 ☽ ✶ ☿
18:31 ☉ ♏
20:08 ☽ # ♅
20:48 ☽ ✶ ♀
20:58 ☽ ∥ ♅
24 18:12 ☽ ☍ ♅
23:15 ☽ ∥ ♄
25 00:12 ☽ □ ♇
10:16 ☉ # ♃
26 00:50 ☽ # ♃
01:52 ☽ ✶ ♂
02:03 ☽ ∥ ☉
02:05 ☽ ☌ ♄
03:43 ☽ ∥ ♆
09:39 ♂ ✶ ♄
11:14 ☿ ∥ ♇
12:19 ☽ △ ♆
19:15 ☽ # ♂
22:24 ♀ □ ♂
23:29 ☽ ✶ ♇
23:54 ☽ ☍ ♃
27 02:36 ☉ ∥ ♆
12:50 ☽ ∥ ♀
18:05 ☽ ∥ ♇
23:54 ☽ ∥ ☿
28 02:09 ☽ ☌ ☿
02:57 ☽ □ ♂
04:34 ☽ ☌ [illegible]
11:49 ☽ □ ♆
15:08 ☿ □ ♂
16:46 ♃ △ ♇
16:58 ☽ △ ♅
29 01:42 ☉ ☍ ♃
03:09 ☉ ✶ ♇
15:14 ♀ ∥ ♇
30 03:18 ☽ ✶ ♄
06:03 ☽ △ ♂
13:30 ☽ ✶ ♆
18:53 ☽ □ ♅
31 01:19 ☽ △ ♃
01:58 ☽ ☌ ♇
03:49 ☽ ∥ ☿
[illegible] ∥ [illegible]
19:10 ☽ ∥ ♀
21:20 ♀ □ ♆

⚷ Chiron

01 Dec. 05 S 18

03	01♓22R
06	01 16
09	01 10
12	01 05
15	01 00
18	00 56
21	00 52
24	00 48
27	00 45
30	00 43

25 06:21 02♐37 ⊛ Solar Eclipse (mag 0.905)

November 2011

Day	S. T. h m s	☉ ° ' "	☽ ° ' "	☿ ° '	♀ ° '	♂ ° '	♃ ° '	♄ ° '	♅ ° '	♆ ° '	♇ ° '	☊ True ° '
01 Tu	02 39 45	08♏ 12 50	17♑ 50 56	27♏ 40	28♏ 18	24♌ 43	04♉R53	22♎ 23	01♈R16	28♒R10	05♑ 24	14♐ 48
02 We	02 43 42	09 12 52	01♒ 00 35	29 02	29 32	25 15	04 45	22 30	01 14	28 09	05 26	14 49
03 Th	02 47 39	10 12 55	13 47 29	00♐ 24	00♐ 47	25 47	04 37	22 37	01 12	28 09	05 27	14 50
04 Fr	02 51 35	11 13 00	26 15 35	01 45	02 02	26 18	04 29	22 44	01 10	28 09	05 29	14R 49
05 Sa	02 55 32	12 13 06	08♓ 29 01	03 05	03 16	26 50	04 21	22 51	01 09	28 09	05 30	14 47
06 Su	02 59 28	13 13 14	20 31 46	04 23	04 31	27 21	04 13	22 58	01 07	28 08	05 31	14 44
07 Mo	03 03 25	14 13 23	02♈ 27 31	05 40	05 45	27 52	04 05	23 05	01 06	28 08	05 33	14 40
08 Tu	03 07 21	15 13 34	14 19 23	06 56	07 00	28 23	03 57	23 12	01 04	28 08	05 34	14 36
09 We	03 11 18	16 13 47	26 10 00	08 10	08 14	28 54	03 50	23 19	01 03	28 08	05 36	14 31
10 Th	03 15 14	17 14 02	08♉ 01 32	09 23	09 29	29 24	03 42	23 26	01 01	28D 08	05 38	14 28
11 Fr	03 19 11	18 14 18	19 55 51	10 33	10 43	29 55	03 34	23 33	01 00	28 08	05 39	14 25
12 Sa	03 23 08	19 14 36	01♊ 54 45	11 42	11 58	00♍ 25	03 27	23 40	00 58	28 08	05 41	14 23
13 Su	03 27 04	20 14 55	14 00 02	12 48	13 13	00 55	03 19	23 47	00 57	28 08	05 42	14D 23
14 Mo	03 31 01	21 15 17	26 13 47	13 51	14 27	01 25	03 12	23 54	00 56	28 09	05 44	14 24
15 Tu	03 34 57	22 15 40	08♋ 38 17	14 51	15 42	01 54	03 05	24 00	00 54	28 09	05 46	14 25
16 We	03 38 54	23 16 05	21 16 09	15 47	16 56	02 24	02 57	24 07	00 53	28 09	05 47	14 27
17 Th	03 42 50	24 16 32	04♌ 10 12	16 40	18 11	02 53	02 50	24 14	00 52	28 09	05 49	14 28
18 Fr	03 46 47	25 17 01	17 23 12	17 28	19 25	03 22	02 43	24 20	00 51	28 09	05 51	14R 28
19 Sa	03 50 43	26 17 32	00♍ 57 31	18 11	20 40	03 51	02 36	24 27	00 50	28 10	05 52	14 27
20 Su	03 54 40	27 18 04	14 54 26	18 48	21 54	04 20	02 30	24 34	00 49	28 10	05 54	14 26
21 Mo	03 58 37	28 18 38	29 13 37	19 19	23 09	04 48	02 23	24 40	00 48	28 10	05 56	14 24
22 Tu	04 02 33	29 19 14	13♎ 52 28	19 43	24 23	05 16	02 17	24 47	00 47	28 11	05 58	14 21
23 We	04 06 30	00♐ 19 52	28 45 59	19 59	25 37	05 44	02 10	24 53	00 46	28 11	06 00	14 19
24 Th	04 10 26	01 20 32	13♏ 46 57	20 07	26 52	06 12	02 04	24 59	00 45	28 12	06 01	14 16
25 Fr	04 14 23	02 21 13	28 46 51	20R 05	28 06	06 40	01 58	25 06	00 44	28 12	06 03	14 15
26 Sa	04 18 19	03 21 55	13♐ 36 57	19 53	29 21	07 07	01 52	25 12	00 44	28 13	06 05	14 14
27 Su	04 22 16	04 22 39	28 09 40	19 30	00♑ 35	07 34	01 47	25 18	00 43	28 13	06 07	14D 14
28 Mo	04 26 12	05 23 24	12♑ 19 30	18 57	01 50	08 01	01 41	25 25	00 42	28 14	06 09	14 14
29 Tu	04 30 09	06 24 10	26 03 34	18 13	03 04	08 28	01 36	25 31	00 42	28 15	06 11	14 14
30 We	04 34 06	07 24 57	09♒ 21 31	17 18	04 19	08 54	01 31	25 37	00 41	28 15	06 13	14 15

Data for	11-01-2011
Julian Day	2455866.50
Ayanamsa	24 01 35
SVP	05 ♓ 05 56
☽ ☊ Mean	16 ♐ 12 R

● ◐ PHASES ○ ◑

02	16:38	◐	09♒55
10	20:16	○	18♉05
18	15:10	◑	25♌55
25	06:09	⊛	02♐37

LAST ASPECT ☽ INGRESS

Day	h m	Day	h m	
01	21:00	01	22:08	♒
04	03:41	04	07:18	♓
05	08:05	06	19:02	♈
09	05:47	09	07:46	♉
11	16:28	11	20:11	♊
14	03:43	14	07:20	♋
16	05:24	16	16:18	♌
18	19:06	18	22:20	♍
20	22:22	21	01:17	♎
22	23:04	23	01:59	♏
24	23:04	25	01:58	♐
27	00:06	27	03:05	♑
28	23:01	29	07:02	♒

DECLINATION

Day	☉	☽	☿	♀	♂	♃	♄	♅	♆	♇
01 Tu	14S 15	19S 26	21S 42	20S 02	14N48	11N45	06S 37	00S 12	12S 38	19S 15
02 We	14 34	16 16	22 06	20 21	14 38	11 43	06 39	00 12	12 38	19 16
03 Th	14 53	12 25	22 28	20 39	14 28	11 40	06 42	00 13	12 38	19 16
04 Fr	15 12	08 07	22 49	20 57	14 19	11 38	06 44	00 14	12 38	19 16
05 Sa	15 30	03 35	23 08	21 14	14 09	11 35	06 47	00 14	12 38	19 16
06 Su	15 48	01N01	23 27	21 30	13 59	11 33	06 49	00 15	12 38	19 16
07 Mo	16 06	05 32	23 44	21 46	13 49	11 30	06 52	00 16	12 38	19 16
08 Tu	16 24	09 49	24 00	22 02	13 40	11 28	06 54	00 16	12 38	19 16
09 We	16 41	13 44	24 15	22 17	13 30	11 25	06 57	00 17	12 38	19 17
10 Th	16 59	17 07	24 28	22 31	13 20	11 23	06 59	00 17	12 38	19 17
11 Fr	17 16	19 49	24 40	22 44	13 11	11 20	07 02	00 18	12 38	19 17
12 Sa	17 32	21 39	24 51	22 57	13 01	11 18	07 04	00 18	12 38	19 17
13 Su	17 48	22 31	25 00	23 09	12 51	11 15	07 07	00 19	12 38	19 17
14 Mo	18 04	22 18	25 08	23 20	12 42	11 13	07 09	00 19	12 38	19 17
15 Tu	18 20	21 00	25 14	23 31	12 32	11 11	07 11	00 20	12 38	19 17
16 We	18 35	18 38	25 18	23 41	12 23	11 08	07 14	00 20	12 38	19 17
17 Th	18 50	15 18	25 21	23 50	12 13	11 06	07 16	00 21	12 38	19 18
18 Fr	19 05	11 08	25 22	23 59	12 04	11 04	07 18	00 21	12 38	19 18
19 Sa	19 19	06 20	25 22	24 07	11 54	11 02	07 21	00 22	12 38	19 18
20 Su	19 33	01 05	25 19	24 14	11 45	11 00	07 23	00 22	12 38	19 18
21 Mo	19 47	04S 22	25 15	24 21	11 36	10 58	07 25	00 22	12 37	19 18
22 Tu	20 00	09 42	25 09	24 27	11 26	10 56	07 27	00 23	12 37	19 18
23 We	20 13	14 33	25 01	24 32	11 17	10 54	07 30	00 23	12 37	19 18
24 Th	20 26	18 32	24 51	24 36	11 08	10 52	07 32	00 23	12 37	19 18
25 Fr	20 38	21 16	24 38	24 39	10 59	10 50	07 34	00 23	12 37	19 18
26 Sa	20 50	22 30	24 24	24 42	10 50	10 48	07 36	00 24	12 37	19 18
27 Su	21 01	22 09	24 07	24 44	10 41	10 47	07 38	00 24	12 36	19 18
28 Mo	21 12	20 23	23 47	24 46	10 32	10 45	07 40	00 24	12 36	19 19
29 Tu	21 23	17 28	23 26	24 46	10 23	10 43	07 43	00 24	12 36	19 19
30 We	21 33	13 44	23 02	24 46	10 14	10 42	07 45	00 25	12 36	19 19

ASPECTARIAN

01 01:31 ☽ ∥ ♇
08:15 ☽ □ ♄
08:39 ☿ □ ♆
19:56 ☽ ✶ ☿
21:00 ☽ ✶ ♀

02 00:25 ☽ ✶ ♅
03:38 ☉ # ♂
06:53 ☽ □ ♃
08:52 ♀ ♐
10:12 ☽ ∥ ☉
11:03 ☽ # ♂
16:55 ☿ ♐
22:42 ☽ ∥ ♆

03 04:21 ☽ # ♃
07:54 ♀ △ ♅
13:58 ☿ △ ♅
17:06 ☽ △ ♄

04 00:06 ☽ ☍ ♂
03:41 ☽ ☌ ♆
07:20 ☽ ∥ ♄
12:02 ☽ □ ☿
12:33 ☽ □ ♀
15:56 ☽ ✶ ♃
18:06 ☽ ✶ ♇

05 08:05 ☽ △ ☉
17:27 ☽ ∥ ♅
20:02 ☽ # ♅

06 21:15 ☽ ☌ ♅

07 06:15 ☽ □ ♇
07:16 ☽ △ ☿
07:24 ☽ # ♄
07:26 ☽ △ ♀
12:36 ♂ ☍ ♆

08 09:39 ☽ ∥ ♃
17:01 ☽ # ♆
18:10 ☽ ☍ ♄
22:34 ☽ ∥ ♂

09 03:59 ☽ ✶ ♆
05:47 ☽ △ ♂
15:21 ☽ ☌ ♃
18:55 ♆ SD
19:08 ☽ △ ♇
22:49 ☽ # ☉

10 18:42 ☽ # ♇

11 04:16 ♂ ♍
16:28 ☽ □ ♆
20:53 ☽ □ ♂
22:08 ☽ ✶ ♅

12 21:23 ☽ ☍ ☿
22:15 ☽ ☍ ♀

13 19:24 ☽ △ ♄

14 03:43 ☽ △ ♆
08:59 ♂ # ♆
09:07 ☽ □ ♅
10:29 ☽ ✶ ♂
13:24 ☽ ✶ ♃
18:27 ☽ ☍ ♇

15 18:11 ☽ # ♇

16 00:18 ☽ # ☉
04:04 ☽ △ ☉
05:24 ☽ □ ♄
17:55 ☽ △ ♅
21:34 ☽ □ ♃
22:09 ♂ △ ♃

17 15:52 ☽ # ♆
18:51 ☽ ∥ ♂

18 00:09 ☽ △ ☿
00:24 ☽ ∥ ♃
04:00 ☽ △ ♀
12:29 ☽ ✶ ♄
19:06 ☽ ☍ ♆
19:12 ☽ # ♄
21:24 ☉ ∥ ♇

19 02:51 ☽ △ ♃
05:13 ☽ ☌ ♂
08:33 ☽ △ ♇

20 03:11 ☽ # ♅
06:25 ☽ ∥ ♅
06:52 ☽ □ ☿
12:55 ☽ □ ♀
20:44 ☉ □ ♆
22:22 ☽ ✶ ☉

21 02:36 ☽ ☍ ♅
11:04 ☽ □ ♇
13:41 ☽ ∥ ♄

22 05:46 ☽ # ♃
07:59 ☽ # ♂
08:20 ♀ ✶ ♄
09:40 ☽ ✶ ☿
14:02 ☽ ∥ ♆
16:08 ☉ ♐
17:44 ☽ ☌ ♄
18:30 ☽ ✶ ♀
23:04 ☽ △ ♆

23 05:25 ☽ ☍ ♃
10:14 ☉ △ ♅
11:31 ☽ ✶ ♂
11:35 ☽ ✶ ♇
14:02 ♂ △ ♇

24 05:42 ☽ ∥ ♇
07:21 ☿ SR
16:34 ☽ ∥ ☉
22:27 ☿ ∥ ♀
23:04 ☽ □ ♆

25 01:52 ♀ ✶ ♆
03:09 ☽ △ ♅
13:06 ☽ □ ♂

26 04:42 ♂ ∥ ♃
10:03 ☽ ☌ ☿
12:36 ♀ ♑
19:13 ☽ ✶ ♄

27 00:06 ☽ ✶ ♆
02:27 ♀ □ ♅
04:17 ☽ □ ♅
04:27 ☽ ☌ ♀
06:01 ☽ △ ♃
13:25 ☽ ☌ ♇
15:47 ☽ ∥ ☉
16:23 ☽ △ ♂
21:24 ♀ △ ♃

28 09:51 ☽ ∥ ♇
23:01 ☽ □ ♄

29 08:16 ☽ ✶ ♅
09:50 ☽ □ ♃
20:09 ☽ ✶ ☉

30 06:33 ☽ ∥ ♆
13:36 ☽ ✶ ☿
17:18 ☽ # ♃
20:23 ☽ # ♂

⚷ Chiron

01	Dec.	05 S 44
	02	00♓41R
	05	00 40
	08	00 39
	11	00 38D
	14	00 39
	17	00 40
	20	00 41
	23	00 43
	26	00 45
	29	00 48

10 21:01 00♓38 D

December 2011

10 14:33 18♊11 ✸ Total Lunar Eclipse (mag 1.111)

Day	S. T. h m s	☉ ° ' "	☽ ° ' "	☿ ° '	♀ ° '	♂ ° '	♃ ° '	♄ ° '	♅ ° '	♆ ° '	♇ ° '	☊ True ° '
01 Th	04 38 02	08♐25 45	22♒15 00	16♐R14	05♑33	09♍20	01♉R25	25♎43	00♈R41	28♒16	06♑15	14♐16
02 Fr	04 41 59	09 26 34	04♓47 10	15 02	06 47	09 46	01 21	25 49	00 40	28 17	06 17	14 17
03 Sa	04 45 55	10 27 23	17 02 06	13 43	08 02	10 11	01 16	25 55	00 40	28 17	06 19	14 17
04 Su	04 49 52	11 28 14	29 04 15	12 21	09 16	10 37	01 11	26 01	00 40	28 18	06 21	14 18
05 Mo	04 53 48	12 29 05	10♈58 13	10 59	10 30	11 02	01 07	26 07	00 39	28 19	06 23	14 19
06 Tu	04 57 45	13 29 58	22 48 22	09 38	11 45	11 26	01 03	26 12	00 39	28 20	06 25	14 19
07 We	05 01 41	14 30 51	04♉38 38	08 21	12 59	11 51	00 59	26 18	00 39	28 21	06 27	14 20
08 Th	05 05 38	15 31 45	16 32 27	07 12	14 13	12 15	00 55	26 24	00 39	28 22	06 29	14 21
09 Fr	05 09 35	16 32 40	28 32 40	06 12	15 28	12 39	00 52	26 29	00 39	28 23	06 31	14 22
10 Sa	05 13 31	17 33 36	10♊41 31	05 21	16 42	13 02	00 48	26 35	00 39	28 24	06 33	14 23
11 Su	05 17 28	18 34 33	23 00 45	04 42	17 56	13 26	00 45	26 40	00D 39	28 25	06 35	14 23
12 Mo	05 21 24	19 35 30	05♋31 36	04 14	19 10	13 48	00 42	26 45	00 39	28 26	06 37	14R 23
13 Tu	05 25 21	20 36 29	18 14 57	03 57	20 24	14 11	00 39	26 51	00 39	28 27	06 39	14 23
14 We	05 29 17	21 37 28	01♌11 22	03 51	21 39	14 33	00 37	26 56	00 39	28 28	06 41	14 21
15 Th	05 33 14	22 38 29	14 21 20	03D 56	22 53	14 55	00 34	27 01	00 39	28 29	06 43	14 20
16 Fr	05 37 10	23 39 30	27 45 14	04 09	24 07	15 17	00 32	27 06	00 39	28 30	06 45	14 18
17 Sa	05 41 07	24 40 33	11♍23 22	04 32	25 21	15 38	00 30	27 11	00 40	28 32	06 47	14 17
18 Su	05 45 04	25 41 36	25 15 41	05 02	26 35	15 59	00 28	27 16	00 40	28 33	06 49	14 16
19 Mo	05 49 00	26 42 40	09♎21 34	05 39	27 49	16 19	00 27	27 21	00 40	28 34	06 51	14 15
20 Tu	05 52 57	27 43 45	23 39 36	06 22	29 03	16 39	00 26	27 26	00 41	28 35	06 54	14D 15
21 We	05 56 53	28 44 52	08♏07 13	07 11	00♒17	16 59	00 24	27 31	00 41	28 37	06 56	14 15
22 Th	06 00 50	29 45 59	22 40 41	08 05	01 31	17 18	00 23	27 35	00 42	28 38	06 58	14 15
23 Fr	06 04 46	00♑47 06	07♐15 04	09 03	02 45	17 37	00 23	27 40	00 43	28 40	07 00	14R 15
24 Sa	06 08 43	01 48 15	21 44 40	10 04	03 59	17 55	00 22	27 44	00 43	28 41	07 02	14 15
25 Su	06 12 40	02 49 23	06♑03 33	11 09	05 13	18 14	00 22	27 49	00 44	28 42	07 04	14 14
26 Mo	06 16 36	03 50 32	20 06 23	12 17	06 27	18 31	00D 22	27 53	00 45	28 44	07 06	14 12
27 Tu	06 20 33	04 51 42	03♒49 08	13 28	07 41	18 48	00 22	27 57	00 46	28 45	07 09	14 09
28 We	06 24 29	05 52 51	17 09 32	14 41	08 54	19 05	00 22	28 02	00 47	28 47	07 11	14 06
29 Th	06 28 26	06 54 01	00♓07 17	15 56	10 08	19 21	00 23	28 06	00 48	28 49	07 13	14 03
30 Fr	06 32 22	07 55 10	12 43 52	17 12	11 22	19 37	00 24	28 10	00 48	28 50	07 15	14 01
31 Sa	06 36 19	08 56 20	25 02 12	18 30	12 36	19 52	00 25	28 14	00 50	28 52	07 17	13 59

Data for	12-01-2011
Julian Day	2455896.50
Ayanamsa	24 01 39
SVP	05 ♓ 05 52
☽ ☊ Mean	14 ♐ 37 R

● ◐ PHASES ○ ◑

02	09:52	◐	09♓52
10	14:37	✸	18♊11
18	00:47	◑	25♍44
24	18:07	●	02♑34

LAST ASPECT Day	h m	☽ INGRESS Day	h m	
01	11:27	01	14:46	♓
02	18:06	04	01:52	♈
06	11:14	06	14:36	♉
08	23:40	09	02:54	♊
11	10:25	11	13:27	♋
13	16:06	13	21:49	♌
16	01:20	16	03:59	♍
18	02:29	18	08:06	♎
20	09:49	20	10:33	♏
22	09:49	22	12:03	♐
24	11:36	24	13:48	♑
26	13:37	26	17:15	♒
28	21:32	28	23:46	♓
30	13:38	31	09:49	♈

DECLINATION

Day	☉	☽	☿	♀	♂	♃	♄	♅	♆	♇
01 Th	21S43	09S27	22S35	24S45	10N06	10N40	07S47	00S25	12S35	19S19
02 Fr	21 52	04 54	22 07	24 43	09 57	10 39	07 49	00 25	12 35	19 19
03 Sa	22 01	00 15	21 38	24 41	09 49	10 37	07 51	00 25	12 35	19 19
04 Su	22 09	04N19	21 08	24 38	09 40	10 36	07 53	00 25	12 35	19 19
05 Mo	22 17	08 41	20 38	24 34	09 32	10 35	07 55	00 25	12 34	19 19
06 Tu	22 25	12 43	20 10	24 29	09 24	10 34	07 56	00 25	12 34	19 19
07 We	22 32	16 15	19 43	24 23	09 16	10 33	07 58	00 25	12 34	19 19
08 Th	22 39	19 09	19 18	24 17	09 08	10 32	08 00	00 25	12 33	19 19
09 Fr	22 45	21 14	18 57	24 10	09 00	10 31	08 02	00 25	12 33	19 19
10 Sa	22 51	22 23	18 39	24 03	08 52	10 30	08 04	00 25	12 33	19 19
11 Su	22 57	22 27	18 26	23 54	08 44	10 29	08 06	00 25	12 32	19 19
12 Mo	23 02	21 24	18 16	23 45	08 37	10 28	08 07	00 25	12 32	19 19
13 Tu	23 06	19 16	18 11	23 35	08 29	10 27	08 09	00 25	12 32	19 19
14 We	23 10	16 07	18 09	23 25	08 22	10 27	08 11	00 25	12 31	19 19
15 Th	23 14	12 08	18 10	23 14	08 15	10 26	08 13	00 25	12 31	19 19
16 Fr	23 17	07 30	18 15	23 02	08 08	10 26	08 14	00 25	12 30	19 19
17 Sa	23 20	02 25	18 22	22 49	08 01	10 25	08 16	00 25	12 30	19 19
18 Su	23 22	02S52	18 31	22 36	07 54	10 25	08 17	00 24	12 29	19 19
19 Mo	23 24	08 06	18 43	22 22	07 48	10 25	08 19	00 24	12 29	19 19
20 Tu	23 25	12 58	18 56	22 07	07 41	10 25	08 20	00 24	12 29	19 19
21 We	23 26	17 09	19 10	21 52	07 35	10 24	08 22	00 24	12 28	19 19
22 Th	23 26	20 19	19 26	21 36	07 29	10 24	08 23	00 23	12 28	19 19
23 Fr	23 26	22 09	19 42	21 20	07 23	10 24	08 25	00 23	12 27	19 19
24 Sa	23 25	22 30	19 58	21 03	07 17	10 25	08 26	00 23	12 27	19 19
25 Su	23 24	21 21	20 15	20 45	07 11	10 25	08 28	00 23	12 26	19 19
26 Mo	23 23	18 53	20 32	20 27	07 06	10 25	08 29	00 22	12 26	19 19
27 Tu	23 21	15 24	20 49	20 08	07 01	10 25	08 30	00 22	12 25	19 19
28 We	23 18	11 14	21 06	19 49	06 56	10 26	08 31	00 21	12 25	19 19
29 Th	23 15	06 40	21 22	19 29	06 51	10 26	08 33	00 21	12 24	19 19
30 Fr	23 12	01 56	21 38	19 08	06 46	10 27	08 34	00 21	12 23	19 19
31 Sa	23 08	02N45	21 53	18 47	06 41	10 27	08 35	00 20	12 23	19 19

ASPECTARIAN

01 06:37 ☽ △ ♄
08:51 ☽ ∥ ♄
11:27 ☽ ☌ ♆
13:50 ♀ ☌ ♇
17:23 ☽ ⚹ ♃

02 02:54 ☽ ⚹ ♇
04:20 ☽ ⚹ ♀
09:48 ☿ ∥ ☉
10:03 ☽ ☍ ♂
13:08 ☉ □ ♂
18:06 ☽ □ ☿
23:08 ☽ ∥ ♅

03 03:27 ☽ # ♅
04 03:11 ☽ ☌ ♅
08:53 ☿ ☌ ☉
14:41 ☽ □ ♇
19:33 ☽ # ♄
22:57 ☽ □ ♀
23:19 ☿ □ ♂

05 00:01 ☽ △ ☿
03:21 ☽ △ ☉
04:40 ☽ ∥ ♂
10:55 ☽ ∥ ♃
15:07 ♀ △ ♂
23:04 ☽ # ♆

06 06:57 ☽ ☍ ♄
11:14 ☽ ⚹ ♆
16:38 ☽ ☌ ♃

07 03:39 ☽ △ ♇
15:04 ☽ △ ♂
18:48 ☽ △ ♀
22:54 ☿ ∥ ♇
08 01:17 ☽ ∥ [illegible]
01:39 ☽ # ♇
23:40 ☽ □ ♆

09 04:10 ☽ ⚹ ♅
14:08 ☽ ☍ ☿
10 04:45 ☽ □ ♂
07:05 ♅ SD

11 07:06 ☽ △ ♄
10:25 ☽ △ ♆
14:41 ☽ □ ♅
14:50 ☽ ⚹ ♃
12 02:04 ☽ ☍ ♇
16:09 ☽ ⚹ ♂
23:27 ☽ # ♇

13 04:28 ☽ ☍ ♀
09:17 ☽ # ☿
16:06 ☽ □ ♄
22:57 ☽ □ ♃
23:00 ☽ △ ♅

14 01:43 ☿ SD
04:54 ☽ △ ☿
21:53 ☽ # ♆
23:27 ♀ ∥ ☉

15 06:27 ♂ # ♄
09:08 ☽ ∥ ♃
16:07 ☽ △ ☉
20:21 ☽ # ♄
20:46 ☽ ∥ ♂
22:50 ☽ ⚹ ♄
16 01:20 ☽ ☍ ♆
[illegible]
11:36 ☽ □ ☿
15:55 ☽ △ ♇

17 07:34 ☽ ☌ ♂
09:08 ☽ # ♅
12:51 ☽ ∥ ♅
18 02:29 ☽ △ ♀
09:15 ☽ ☍ ♅
14:19 ♀ □ ♄
17:24 ☽ ⚹ ☿
19:45 ☽ □ ♇
22:35 ☽ # ♂

19 01:00 ☽ ∥ ♄
11:04 ☽ # ♃
16:25 ☉ ⚹ ♄
21:25 ☽ ∥ ♆

20 06:19 ☽ ☌ ♄
07:17 ☽ ⚹ ☉
08:13 ☽ △ ♆
09:49 ☽ □ ♀
11:15 ☽ ☍ ♃
18:26 ♀ ♒
20:48 ☉ ⚹ ♆
22:01 ☽ ⚹ ♇

21 02:20 ♀ □ ♃
07:57 ♀ ⚹ ♅
14:36 ☿ ∥ ♇
14:57 ☽ ⚹ ♂
15:23 ☽ ∥ ♇
15:28 ☽ ∥ ☿
22 05:30 ☉ ♑
[illegible]
12:37 ☽ ∥ ♀
13:13 ☽ △ ♅
14:33 ☉ △ ♃
15:54 ☽ ⚹ ♀
22:14 ☉ □ ♅
23 03:11 ☽ ☌ ☿
17:31 ☽ □ ♂

24 10:04 ☽ ⚹ ♄
11:36 ☽ ⚹ ♆
14:25 ☽ △ ♃
15:01 ☽ □ ♅

25 01:43 ☽ ☌ ♇
08:00 ☽ ∥ ♀
10:48 ☽ ∥ ☿
20:23 ☽ ∥ ♇
20:32 ☿ ∥ ♀
21:12 ☽ △ ♂
22:09 ♃ SD
26 13:37 ☽ □ ♄
17:54 ☽ □ ♃
18:35 ☽ ⚹ ♅

27 07:34 ☽ ☌ ♀
17:29 ☽ ∥ ♆
19:02 ☽ ⚹ ☿

28 04:18 [illegible]
14:18 ☽ ∥ ♄
20:11 ☽ △ ♄
21:32 ☽ ☌ ♆
23:01 ☽ # ♂

29 00:29 ☽ ⚹ ♃
07:41 ☉ ☌ ♇
11:07 ♀ ∥ ♇
13:28 ☽ ⚹ ♇
13:57 ☽ ⚹ ☉

30 08:05 ☽ ∥ ♅
09:41 ☽ □ ☿
11:34 ☽ # ♅
[illegible]
31 11:29 ☽ ☌ ♅
20:36 ☽ ∥ ♂

⚷ Chiron

01	Dec.	05 S 52
02	00♓52	
05	00 56	
08	01 01	
11	01 06	
14	01 11	
17	01 17	
20	01 24	
23	01 31	
26	01 38	
29	01 46	

January 2012

Day	S. T.	☉	☽	☿	♀	♂	♃	♄	♅	♆	♇	☊ True
	h m s	° ' "	° ' "	° '	° '	° '	° '	° '	° '	° '	° '	° '
01 Su	06 40 15	09♑57 29	07♈06 18	19♐50	13♒49	20♍07	00♉26	28♎17	00♈51	28♒53	07♑19	13♐D59
02 Mo	06 44 12	10 58 38	19 00 50	21 11	15 03	20 21	00 27	28 21	00 52	28 55	07 21	14 00
03 Tu	06 48 09	11 59 47	00♉50 49	22 33	16 17	20 35	00 29	28 25	00 53	28 57	07 24	14 02
04 We	06 52 05	13 00 56	12 41 13	23 56	17 30	20 48	00 30	28 28	00 54	28 58	07 26	14 05
05 Th	06 56 02	14 02 04	24 36 45	25 20	18 44	21 01	00 32	28 32	00 55	29 00	07 28	14 07
06 Fr	06 59 58	15 03 13	06♊41 35	26 44	19 57	21 13	00 35	28 35	00 57	29 02	07 30	14 09
07 Sa	07 03 55	16 04 21	18 59 10	28 10	21 10	21 24	00 37	28 38	00 58	29 04	07 32	14 09
08 Su	07 07 51	17 05 29	01♋32 02	29 36	22 24	21 35	00 40	28 42	01 00	29 05	07 34	14R 08
09 Mo	07 11 48	18 06 37	14 21 32	01♑03	23 37	21 46	00 42	28 45	01 01	29 07	07 36	14 05
10 Tu	07 15 44	19 07 44	27 27 49	02 31	24 50	21 56	00 45	28 48	01 03	29 09	07 38	14 00
11 We	07 19 41	20 08 51	10♌49 51	03 59	26 03	22 05	00 49	28 51	01 04	29 11	07 41	13 54
12 Th	07 23 38	21 09 58	24 25 36	05 28	27 16	22 14	00 52	28 53	01 06	29 13	07 43	13 47
13 Fr	07 27 34	22 11 05	08♍12 34	06 57	28 29	22 22	00 56	28 56	01 07	29 15	07 45	13 41
14 Sa	07 31 31	23 12 12	22 08 02	08 27	29 42	22 29	00 59	28 59	01 09	29 17	07 47	13 35
15 Su	07 35 27	24 13 19	06♎09 31	09 58	00♓55	22 36	01 03	29 01	01 11	29 19	07 49	13 31
16 Mo	07 39 24	25 14 25	20 14 56	11 29	02 08	22 42	01 07	29 04	01 13	29 20	07 51	13 29
17 Tu	07 43 20	26 15 32	04♏22 37	13 00	03 21	22 47	01 12	29 06	01 14	29 22	07 53	13 28
18 We	07 47 17	27 16 38	18 31 08	14 32	04 34	22 52	01 16	29 08	01 16	29 24	07 55	13D 28
19 Th	07 51 13	28 17 44	02♐39 04	16 05	05 46	22 56	01 21	29 10	01 18	29 26	07 57	13 28
20 Fr	07 55 10	29 18 50	16 44 40	17 38	06 59	23 00	01 26	29 12	01 20	29 28	07 59	13R 28
21 Sa	07 59 07	00♒19 56	00♑45 36	19 12	08 12	23 02	01 31	29 14	01 22	29 31	08 01	13 26
22 Su	08 03 03	01 21 01	14 38 57	20 46	09 24	23 04	01 36	29 16	01 24	29 33	08 03	13 22
23 Mo	08 06 60	02 22 05	28 21 24	22 21	10 36	23 05	01 42	29 18	01 26	29 35	08 05	13 16
24 Tu	08 10 56	03 23 09	11♒49 42	23 56	11 49	23 06	01 47	29 19	01 29	29 37	08 07	13 08
25 We	08 14 53	04 24 12	25 01 18	25 32	13 01	23R 05	01 53	29 21	01 31	29 39	08 09	12 59
26 Th	08 18 49	05 25 14	07♓54 44	27 09	14 13	23 04	01 59	29 22	01 33	29 41	08 11	12 50
27 Fr	08 22 46	06 26 15	20 30 01	28 46	15 25	23 02	02 05	29 23	01 35	29 43	08 13	12 42
28 Sa	08 26 42	07 27 15	02♈48 36	00♒24	16 37	23 00	02 12	29 25	01 37	29 45	08 15	12 35
29 Su	08 30 39	08 28 13	14 53 18	02 02	17 49	22 56	02 18	29 26	01 40	29 47	08 17	12 31
30 Mo	08 34 36	09 29 11	26 48 03	03 41	19 01	22 52	02 25	29 27	01 42	29 49	08 19	12 29
31 Tu	08 38 32	10 30 08	08♉37 37	05 21	20 13	22 47	02 32	29 27	01 45	29 52	08 21	12D 28

Data for	01-01-2012
Julian Day	2455927.50
Ayanamsa	24 01 45
SVP	05 ♓ 05 49
☽ ☊ Mean	12 ♐ 58 R

● ◐ PHASES ○ ◑

Day	Time	Phase	Position
01	06:15	◐	10♈13
09	07:30	○	18♋26
16	09:08	◑	25♎38
23	07:40	●	02♒42
31	04:10	◐	10♉41

LAST ASPECT Day	h m	☽ INGRESS Day	h m	Sign
02	20:08	02	22:17	♉
05	08:47	05	10:45	♊
07	19:52	07	21:06	♋
10	02:25	10	04:35	♌
12	08:23	12	09:44	♍
14	01:59	14	13:29	♎
16	15:29	16	16:34	♏
18	18:32	18	19:30	♐
20	21:51	20	22:42	♑
23	01:40	23	02:54	♒
25	08:34	25	09:12	♓
27	04:53	27	18:29	♈
30	06:09	30	06:29	♉

ASPECTARIAN

01 00:26 ☽ □ ♇
06:06 ☿ □ ♂
07:35 ☽ # ♄
15:04 ☽ ⚹ ♀
18:23 ☽ ∥ ♃
02 04:57 ☽ △ ☿
05:47 ☽ # ♆
19:02 ☽ ☍ ♄
20:08 ☽ ⚹ ♆
23:15 ☽ ☌ ♃
03 13:19 ☽ △ ♇
17:05 ☽ # ♀
04 00:44 ☽ △ ☉
00:51 ☿ ∥ ☉
09:56 ☽ # ♇
10:50 ☽ □ ♀
16:39 ☽ △ ♂
05 08:47 ☽ □ ♆
12:36 ☽ ⚹ ♅
06 16:09 ☽ # ☉
07 04:41 ☽ △ ♀
04:44 ☽ □ ♂
08:14 ☿ ⚹ ♄
08:34 ☽ # ☉
15:15 ☿ ⚹ ♆
18:35 ☽ △ ♄
19:21 ☽ △ ♆
19:52 ☽ ☍ ☿
22:20 ☽ ⚹ ♃
22:58 ☽ □ ♅
08 06:34 ☿ ♑
11:24 ☽ ☍ ♇
18:05 ☿ △ ♃
23:24 ☿ □ ♅
09 06:57 ☽ # ♇
13:48 ☽ ⚹ ♂
10 02:25 ☽ □ ♄
05:59 ☽ □ ♃
06:29 ☽ △ ♅
16:54 ☽ # ♀
11 05:55 ☽ # ♆
14:25 ☽ ∥ ♃
12 00:00 ☽ # ♄
05:28 ☽ ☍ ♀
07:50 ☽ ⚹ ♄
08:23 ☽ ☍ ♆
11:18 ☽ △ ♃
13:06 ☽ ∥ ♂
21:34 ☽ △ ☿
23:12 ☽ △ ♇
13 04:46 ☉ △ ♂
09:06 ♀ △ ♄
13:01 ☿ ☌ ♇
15:17 ♀ ☌ ♆
15:58 ☽ # ♅
17:47 ☽ ∥ ♅
14 00:36 ☽ ☌ ♂
01:59 ☽ △ ☉
05:48 ♀ ♓
15:29 ☽ ☍ ♅
20:11 ☽ # ♂
15 02:48 ♀ ⚹ ♃
02:50 ☽ □ ♇
07:16 ☽ □ ☿
09:24 ☽ ∥ ♄
19:04 ☽ # ♃
21:49 ♀ ∥ ♆
16 02:00 ☽ ∥ ♀
02:25 ☽ ∥ ♆
15:01 ☽ ☌ ♄
15:29 ☽ △ ♆
18:34 ☽ ☍ ♃
22:06 ☽ △ ♀
17 05:58 ☽ ⚹ ♇
16:25 ☽ ⚹ ☿
22:57 ☽ ∥ ♇
18 07:25 ☽ ⚹ ♂
11:02 ☽ ∥ ☉
16:01 ☽ ⚹ ☉
18:32 ☽ □ ♆
18:33 ♀ # ♃
21:42 ☽ △ ♅
19 05:48 ☽ □ ♀
21:19 ☉ □ ♄
20 10:43 ☽ □ ♂
16:10 ☉ ♒
20:31 ♀ ⚹ ♇
21:22 ☽ ⚹ ♄
21:51 ☽ ⚹ ♆
21 01:03 ☽ □ ♅
01:19 ☽ △ ♃
12:33 ☽ ☌ ♇
14:02 ☽ ⚹ ♀
22 01:17 ☽ ∥ ☉
01:21 ☉ ⚹ ♅
06:04 ☽ ∥ ♇
06:35 ☉ □ ♃
12:03 ☽ ☌ ☿
14:43 ☽ △ ♂
21:22 ♀ ∥ ♄
23 01:40 ☽ □ ♄
05:28 ☽ ⚹ ♅
05:57 ☽ □ ♃
11:16 ☿ △ ♂
24 00:55 ♂ SR
05:12 ☽ ∥ ♆
09:23 ☉ ∥ ♇
10:57 ☽ # ♃
22:33 ☽ ∥ ♄
25 04:31 ☽ ∥ ♀
08:00 ☽ △ ♄
08:34 ☽ ☌ ♆
12:48 ☽ ⚹ ♃
12:49 ☽ # ♂
26 00:31 ☽ ⚹ ♇
13:13 ☽ ☌ ♀
19:25 ☽ ∥ ♅
19:37 ☽ # ♅
27 04:53 ☽ ☍ ♂
09:20 ☿ □ ♄
18:12 ☿ ♒
18:31 ☽ ⚹ ☿
21:40 ☽ ☌ ♅
28 03:42 ☽ ∥ ♂
03:49 ☽ # ♀
04:42 ♀ # ♂
10:01 ☽ ⚹ ☉
10:47 ☽ □ ♇
18:26 ☿ ⚹ ♅
18:31 ☽ # ♄
29 04:08 ☿ □ ♃
08:23 ☽ ∥ ♃
13:12 ☽ # ♆
30 05:22 ☽ ☍ ♄
06:09 ☽ ⚹ ♆
11:29 ☽ ☌ ♃
16:15 ☽ □ ☿
23:26 ☽ △ ♇
31 03:57 ☽ # ☉
19:42 ☽ # ♇

DECLINATION

Day	☉	☽	☿	♀	♂	♃	♄	♅	♆	♇
01 Su	23S 04	07N15	22S 08	18S 26	06N37	10N28	08S 36	00S 20	12S 22	19S 19
02 Mo	22 59	11 25	22 22	18 04	06 33	10 29	08 37	00 19	12 22	19 19
03 Tu	22 54	15 08	22 35	17 41	06 29	10 30	08 38	00 19	12 21	19 19
04 We	22 48	18 15	22 47	17 18	06 26	10 31	08 39	00 18	12 21	19 19
05 Th	22 42	20 37	22 59	16 55	06 22	10 32	08 40	00 18	12 20	19 19
06 Fr	22 35	22 05	23 09	16 31	06 19	10 33	08 41	00 17	12 19	19 19
07 Sa	22 28	22 32	23 19	16 07	06 16	10 34	08 42	00 16	12 19	19 19
08 Su	22 21	21 51	23 27	15 42	06 13	10 35	08 43	00 16	12 18	19 19
09 Mo	22 13	20 02	23 34	15 17	06 10	10 36	08 44	00 15	12 17	19 19
10 Tu	22 04	17 09	23 40	14 51	06 08	10 38	08 45	00 15	12 17	19 19
11 We	21 56	13 20	23 46	14 25	06 06	10 39	08 46	00 14	12 16	19 19
12 Th	21 46	08 46	23 49	13 59	06 04	10 40	08 46	00 13	12 15	19 19
13 Fr	21 37	03 43	23 52	13 32	06 03	10 42	08 47	00 13	12 15	19 19
14 Sa	21 27	01S 35	23 53	13 05	06 01	10 44	08 48	00 12	12 14	19 19
15 Su	21 16	06 50	23 54	12 38	06 00	10 45	08 48	00 11	12 13	19 19
16 Mo	21 05	11 45	23 52	12 10	06 00	10 47	08 49	00 10	12 13	19 19
17 Tu	20 54	16 03	23 50	11 42	05 59	10 49	08 50	00 10	12 12	19 19
18 We	20 42	19 26	23 46	11 14	05 59	10 51	08 50	00 09	12 11	19 19
19 Th	20 30	21 38	23 41	10 46	05 59	10 52	08 51	00 08	12 11	19 19
20 Fr	20 18	22 29	23 35	10 17	05 59	10 54	08 51	00 07	12 10	19 19
21 Sa	20 05	21 54	23 27	09 48	06 00	10 56	08 51	00 06	12 09	19 18
22 Su	19 51	19 59	23 18	09 18	06 00	10 59	08 52	00 06	12 09	19 18
23 Mo	19 38	16 56	23 07	08 49	06 02	11 01	08 52	00 05	12 08	19 18
24 Tu	19 24	13 02	22 55	08 19	06 03	11 03	08 52	00 04	12 07	19 18
25 We	19 09	08 36	22 42	07 49	06 05	11 05	08 53	00 03	12 06	19 18
26 Th	18 55	03 52	22 27	07 19	06 07	11 08	08 53	00 02	12 06	19 18
27 Fr	18 40	00N53	22 11	06 49	06 09	11 10	08 53	00 01	12 05	19 18
28 Sa	18 24	05 31	21 54	06 18	06 12	11 12	08 53	00 00	12 04	19 18
29 Su	18 09	09 51	21 34	05 48	06 15	11 15	08 53	00N01	12 03	19 18
30 Mo	17 53	13 44	21 14	05 17	06 18	11 17	08 54	00 02	12 03	19 18
31 Tu	17 36	17 04	20 52	04 46	06 21	11 20	08 54	00 03	12 02	19 18

⚷ Chiron

01 Dec.	05 S 40
01	01♓54
04	02 03
07	02 12
10	02 21
13	02 31
16	02 41
19	02 51
22	03 01
25	03 12
28	03 23
31	03 34

February 2012

Day	S. T.	☉	☽	☿	♀	♂	♃	♄	♅	♆	♇	☊ True
	h m s	° ' "	° ' "	° '	° '	° '	° '	° '	° '	° '	° '	° '
01 We	08 42 29	11♒31 03	20♉27 14	07♒02	21♓24	22♍R41	02♉38	29♎28	01♈47	29♒54	08♑23	12♐29
02 Th	08 46 25	12 31 57	02♊22 21	08 43	22 36	22 35	02 46	29 29	01 49	29 56	08 24	12 31
03 Fr	08 50 22	13 32 50	14 28 12	10 25	23 47	22 27	02 53	29 29	01 52	29 58	08 26	12R 31
04 Sa	08 54 18	14 33 41	26 49 35	12 07	24 58	22 19	03 00	29 30	01 55	00♓00	08 28	12 30
05 Su	08 58 15	15 34 31	09♋30 22	13 50	26 09	22 10	03 08	29 30	01 57	00 03	08 30	12 26
06 Mo	09 02 11	16 35 20	22 33 01	15 35	27 21	22 00	03 16	29 30	02 00	00 05	08 32	12 19
07 Tu	09 06 08	17 36 08	05♌58 12	17 19	28 31	21 50	03 24	29 30	02 02	00 07	08 33	12 10
08 We	09 10 05	18 36 54	19 44 28	19 05	29 42	21 39	03 32	29R 30	02 05	00 09	08 35	12 00
09 Th	09 14 01	19 37 39	03♍48 13	20 51	00♈53	21 27	03 40	29 30	02 08	00 12	08 37	11 48
10 Fr	09 17 58	20 38 22	18 04 20	22 38	02 03	21 14	03 48	29 30	02 11	00 14	08 39	11 37
11 Sa	09 21 54	21 39 05	02♎26 56	24 26	03 14	21 00	03 57	29 30	02 13	00 16	08 40	11 27
12 Su	09 25 51	22 39 46	16 50 21	26 14	04 24	20 46	04 05	29 29	02 16	00 18	08 42	11 20
13 Mo	09 29 47	23 40 27	01♏09 57	28 03	05 34	20 31	04 14	29 29	02 19	00 21	08 44	11 15
14 Tu	09 33 44	24 41 06	15 22 34	29 53	06 44	20 15	04 23	29 28	02 22	00 23	08 45	11 12
15 We	09 37 40	25 41 44	29 26 31	01♓42	07 54	19 59	04 32	29 28	02 25	00 25	08 47	11 11
16 Th	09 41 37	26 42 22	13♐21 13	03 33	09 04	19 42	04 41	29 27	02 28	00 27	08 48	11 10
17 Fr	09 45 34	27 42 58	27 06 44	05 24	10 14	19 24	04 50	29 26	02 31	00 30	08 50	11 08
18 Sa	09 49 30	28 43 33	10♑43 12	07 14	11 23	19 06	05 00	29 25	02 33	00 32	08 52	11 05
19 Su	09 53 27	29 44 06	24 10 24	09 05	12 33	18 47	05 09	29 24	02 36	00 34	08 53	10 59
20 Mo	09 57 23	00♓44 39	07♒27 37	10 56	13 42	18 27	05 19	29 22	02 39	00 37	08 55	10 50
21 Tu	10 01 20	01 45 09	20 33 41	12 46	14 51	18 07	05 29	29 21	02 43	00 39	08 56	10 39
22 We	10 05 16	02 45 38	03♓27 19	14 35	16 00	17 46	05 39	29 20	02 46	00 41	08 58	10 27
23 Th	10 09 13	03 46 06	16 07 28	16 24	17 08	17 25	05 49	29 18	02 49	00 43	08 59	10 13
24 Fr	10 13 09	04 46 32	28 33 49	18 10	18 17	17 03	05 59	29 16	02 52	00 46	09 00	10 01
25 Sa	10 17 06	05 46 56	10♈46 56	19 56	19 25	16 41	06 09	29 15	02 55	00 48	09 02	09 50
26 Su	10 21 03	06 47 18	22 48 33	21 39	20 33	16 19	06 19	29 13	02 58	00 50	09 03	09 42
27 Mo	10 24 59	07 47 38	04♉41 29	23 19	21 41	15 56	06 30	29 11	03 01	00 53	09 04	09 37
28 Tu	10 28 56	08 47 57	16 29 37	24 56	22 49	15 33	06 40	29 09	03 04	00 55	09 06	09 34
29 We	10 32 52	09 48 13	28 17 39	26 29	23 57	15 10	06 51	29 07	03 08	00 57	09 07	09D 34

Data for	02-01-2012
Julian Day	2455958.50
Ayanamsa	24 01 50
SVP	05 ♓ 05 45
☽ ☊ Mean	11 ♐ 20 R

● ◐ PHASES ○ ◑

07	21:54	○	18♌32
14	17:05	◑	25♏24
21	22:35	●	02♓42

LAST ASPECT Day	h m	☽ INGRESS Day	h m	
01	19:06	01	19:15	♊
04	05:07	04	06:04	♋
06	12:31	06	13:24	♌
08	16:43	08	17:33	♍
10	05:12	10	19:55	♎
12	21:10	12	22:02	♏
14	17:05	15	00:57	♐
17	04:04	17	05:04	♑
19	09:23	19	10:29	♒
21	16:17	21	17:32	♓
23	02:24	24	02:48	♈
26	12:52	26	14:30	♉
28	19:46	29	03:28	♊

DECLINATION

Day	☉	☽	☿	♀	♂	♃	♄	♅	♆	♇
01 We	17S20	19N43	20S29	04S15	06N25	11N23	08S54	00N04	12S01	19S18
02 Th	17 03	21 31	20 04	03 44	06 29	11 25	08 54	00 05	12 00	19 18
03 Fr	16 45	22 21	19 37	03 13	06 33	11 28	08 54	00 06	12 00	19 17
04 Sa	16 28	22 08	19 09	02 42	06 38	11 31	08 53	00 07	11 59	19 17
05 Su	16 10	20 46	18 40	02 10	06 43	11 34	08 53	00 08	11 58	19 17
06 Mo	15 52	18 18	18 09	01 39	06 48	11 36	08 53	00 09	11 57	19 17
07 Tu	15 33	14 48	17 37	01 07	06 54	11 39	08 53	00 10	11 56	19 17
08 We	15 15	10 27	17 03	00 36	06 59	11 42	08 53	00 11	11 56	19 17
09 Th	14 56	05 27	16 28	00 05	07 05	11 45	08 52	00 12	11 55	19 17
10 Fr	14 37	00 06	15 51	00N27	07 11	11 48	08 52	00 13	11 54	19 17
11 Sa	14 17	05S18	15 13	00 58	07 18	11 51	08 52	00 14	11 53	19 17
12 Su	13 58	10 25	14 33	01 30	07 25	11 55	08 51	00 16	11 53	19 17
13 Mo	13 38	14 56	13 52	02 01	07 31	11 58	08 51	00 17	11 52	19 17
14 Tu	13 18	18 34	13 10	02 33	07 39	12 01	08 50	00 18	11 51	19 16
15 We	12 57	21 02	12 27	03 04	07 46	12 04	08 50	00 19	11 50	19 16
16 Th	12 37	22 13	11 42	03 35	07 54	12 08	08 49	00 20	11 49	19 16
17 Fr	12 16	22 01	10 56	04 06	08 01	12 11	08 49	00 21	11 49	19 16
18 Sa	11 55	20 31	10 08	04 37	08 09	12 14	08 48	00 23	11 48	19 16
19 Su	11 34	17 53	09 20	05 08	08 17	12 18	08 48	00 24	11 47	19 16
20 Mo	11 13	14 21	08 31	05 39	08 26	12 21	08 47	00 25	11 46	19 16
21 Tu	10 51	10 10	07 41	06 10	08 34	12 24	08 46	00 26	11 45	19 16
22 We	10 29	05 36	06 50	06 41	08 42	12 28	08 46	00 28	11 45	19 16
23 Th	10 08	00 53	05 59	07 11	08 51	12 32	08 45	00 29	11 44	19 16
24 Fr	09 46	03N48	05 08	07 41	09 00	12 35	08 44	00 30	11 43	19 16
25 Sa	09 23	08 14	04 17	08 11	09 08	12 39	08 43	00 31	11 42	19 15
26 Su	09 01	12 17	03 25	08 41	09 17	12 42	08 42	00 33	11 41	19 15
27 Mo	08 39	15 48	02 34	09 11	09 26	12 46	08 41	00 34	11 41	19 15
28 Tu	08 16	18 40	01 44	09 41	09 35	12 50	08 40	00 35	11 40	19 15
29 We	07 54	20 45	00 55	10 10	09 43	12 53	08 39	00 36	11 39	19 15

ASPECTARIAN

Day	Time	Aspect
01	02:08	☽ ✶ ♀
	04:29	☽ △ ♂
	07:18	☽ # ☿
	19:06	☽ □ ♆
	22:54	☽ ✶ ♅
	23:42	♀ ☍ ♂
02	14:41	☽ △ ☿
	22:01	☽ △ ☉
03	15:24	☽ □ ♂
	17:12	☿ ∥ ♆
	18:53	♆ ♓
	20:03	☽ □ ♀
04	05:07	☽ △ ♄
	06:06	☽ △ ♆
	09:44	☽ □ ♅
	11:54	☽ ✶ ♃
	22:07	☽ ☍ ♇
05	15:45	☽ # ♇
	23:02	☽ ✶ ♂
06	01:35	☽ # ☿
	09:29	☽ △ ♀
	12:31	☽ □ ♄
	17:01	☽ △ ♅
	19:03	☽ # ☉
	19:24	☽ □ ♃
07	09:03	☿ ☌ ☉
	14:04	♄ SR
	16:20	☽ # ♆
	17:37	☽ ∥ ♃
	22:42	☽ ☍ ☿
08	06:01	♀ ♈
	07:52	☽ # ♄
	16:35	☽ ∥ ♂
	16:43	☽ ✶ ♄
	17:52	☽ ☍ ♆
	18:20	♀ # ♅
	23:46	☽ △ ♃
09	08:08	☽ △ ♇
	13:15	♀ ∥ ♅
	22:37	☽ ∥ ♀
	23:29	☽ ∥ ♅
10	01:27	☽ # ♅
	02:29	♀ ☌ ♅
	02:43	☽ # ♀
	05:12	☽ ☌ ♂
	23:37	☽ ☍ ♅
11	01:25	☽ ☍ ♀
	09:22	☽ # ♂
	10:24	☽ □ ♇
	11:15	♃ # ♆
	16:29	☽ ∥ ♄
12	07:21	☽ ∥ ♆
	07:38	☽ # ♃
	10:29	☽ △ ☉
	17:10	☽ ∥ ☉
	18:00	☽ △ ☿
	18:49	☽ ∥ ☿
	21:10	☽ ☌ ♄
	22:37	☽ △ ♆
13	05:13	☽ ☍ ♃
	12:46	☽ ✶ ♇
	16:11	☿ ∥ ☉
	18:44	☿ △ ♄
14	01:38	☿ ♓
	05:51	☽ ∥ ♇
	06:47	☿ ☌ ♆
	08:08	☽ ✶ ♂
15	01:41	☽ □ ♆
	04:29	☽ □ ☿
	05:07	☽ △ ♅
	11:13	☿ # ♃
	15:54	☽ △ ♀
	18:31	♀ □ ♇
	19:54	☿ ∥ ♆
16	10:48	☽ □ ♂
	16:07	☿ ✶ ♃
17	01:09	☽ ✶ ☉
	04:04	☽ ✶ ♄
	05:02	☉ # ♃
	05:58	☽ ✶ ♆
	09:31	☽ □ ♅
	13:45	☽ △ ♃
	16:52	☽ ✶ ☿
	20:42	☽ ☌ ♇
18	01:18	☽ □ ♀
	08:33	☉ ∥ ♆
	12:35	☽ ∥ ♇
	14:34	☽ △ ♂
	16:03	☉ △ ♄
	21:22	☿ ✶ ♇
19	06:18	☉ ♓
	09:23	☽ □ ♄
	15:15	☽ ✶ ♅
	16:11	☿ ∥ ♄
	20:03	☽ □ ♃
	20:41	☉ ☌ ♆
20	02:18	☿ # ♂
	11:42	☽ # ♃
	12:28	☽ ✶ ♀
	15:13	☽ ∥ ♆
	19:58	☽ ∥ ☉
21	07:33	☽ ∥ ♄
	08:22	☽ # ♂
	16:13	☽ ∥ ☿
	16:18	☽ △ ♄
	18:48	☽ ☌ ♆
	19:03	☽ # ♀
22	02:56	☿ # ♀
	04:10	☽ ✶ ♃
	08:01	♂ # ♄
	10:23	☽ ✶ ♇
23	00:36	☽ ☌ ☿
	02:04	☽ # ♅
	02:25	☽ ☍ ♂
	06:59	☽ ∥ ♅
	11:26	☿ ☍ ♂
24	06:00	☽ # ☿
	08:26	☽ ☌ ♅
	20:31	☽ □ ♇
	23:49	☽ ∥ ♀
25	02:48	☽ # ♄
	05:24	☽ ∥ ♂
	06:07	☽ # ☉
	10:33	☉ ✶ ♃
	11:40	☉ # ♂
	19:00	☽ ☌ ♀
	20:23	☽ # ♆
26	00:37	♀ # ♄
	02:47	☽ ∥ ♃
	09:04	♀ # ☉
	12:52	☽ ☍ ♄
	16:15	☽ ✶ ♆
	21:08	☉ ∥ ♄
27	03:43	☽ ☌ ♃
	06:53	☽ ✶ ☉
	08:55	☽ △ ♇
	16:53	♀ ∥ ♂
	22:09	☽ △ ♂
28	05:58	☽ # ♇
	07:14	☉ ✶ ♇
	19:46	☽ ✶ ☿
29	05:24	☽ □ ♆
	09:04	☿ # ♅
	09:50	☽ ✶ ♅

⚷ Chiron

01	Dec.	05 S 10
03	03♓45	
06	03 57	
09	04 09	
12	04 20	
15	04 32	
18	04 44	
21	04 56	
24	05 08	
27	05 20	

March 2012

Day	S. T. h m s	☉ ° ' "	☽ ° ' "	☿ ° '	♀ ° '	♂ ° '	♃ ° '	♄ ° '	♅ ° '	♆ ° '	♇ ° '	☊ True ° '
01 Th	10 36 49	10♓48 28	10♊10 53	27♓58	25♈04	14♍R47	07♉02	29♎R05	03♈11	00♓59	09♑08	09♐34
02 Fr	10 40 45	11 48 40	22 14 52	29 21	26 11	14 23	07 13	29 02	03 14	01 02	09 09	09R 34
03 Sa	10 44 42	12 48 51	04♋35 04	00♈39	27 18	13 59	07 24	29 00	03 17	01 04	09 11	09 32
04 Su	10 48 38	13 48 59	17 16 28	01 51	28 25	13 36	07 35	28 57	03 21	01 06	09 12	09 28
05 Mo	10 52 35	14 49 05	00♌22 55	02 57	29 31	13 12	07 46	28 55	03 24	01 08	09 13	09 21
06 Tu	10 56 32	15 49 10	13 56 28	03 55	00♉37	12 48	07 57	28 52	03 27	01 11	09 14	09 12
07 We	11 00 28	16 49 12	27 56 38	04 45	01 43	12 25	08 09	28 49	03 30	01 13	09 15	09 01
08 Th	11 04 25	17 49 12	12♍20 00	05 27	02 49	12 01	08 20	28 46	03 34	01 15	09 16	08 48
09 Fr	11 08 21	18 49 10	27 00 26	06 01	03 55	11 38	08 32	28 44	03 37	01 17	09 17	08 36
10 Sa	11 12 18	19 49 07	11♎49 55	06 26	05 00	11 14	08 43	28 41	03 40	01 19	09 18	08 26
11 Su	11 16 14	20 49 01	26 39 54	06 42	06 05	10 52	08 55	28 37	03 44	01 22	09 19	08 17
12 Mo	11 20 11	21 48 54	11♏22 43	06 49	07 10	10 29	09 07	28 34	03 47	01 24	09 20	08 12
13 Tu	11 24 07	22 48 45	25 52 39	06R 47	08 14	10 07	09 19	28 31	03 51	01 26	09 21	08 09
14 We	11 28 04	23 48 35	10♐06 21	06 37	09 18	09 45	09 31	28 28	03 54	01 28	09 22	08 07
15 Th	11 32 00	24 48 23	24 02 41	06 19	10 22	09 23	09 43	28 24	03 57	01 30	09 23	08 07
16 Fr	11 35 57	25 48 10	07♑42 03	05 53	11 25	09 02	09 55	28 21	04 01	01 32	09 23	08 06
17 Sa	11 39 54	26 47 54	21 05 45	05 21	12 29	08 41	10 07	28 17	04 04	01 34	09 24	08 03
18 Su	11 43 50	27 47 37	04♒15 18	04 42	13 31	08 21	10 20	28 14	04 08	01 36	09 25	07 59
19 Mo	11 47 47	28 47 18	17 12 06	03 57	14 34	08 02	10 32	28 10	04 11	01 39	09 26	07 51
20 Tu	11 51 43	29 46 58	29 57 09	03 09	15 36	07 42	10 44	28 06	04 14	01 41	09 26	07 42
21 We	11 55 40	00♈46 35	12♓31 08	02 18	16 38	07 24	10 57	28 03	04 18	01 43	09 27	07 31
22 Th	11 59 36	01 46 11	24 54 29	01 24	17 40	07 06	11 09	27 59	04 21	01 45	09 28	07 19
23 Fr	12 03 33	02 45 44	07♈07 41	00 30	18 41	06 49	11 22	27 55	04 25	01 47	09 28	07 07
24 Sa	12 07 29	03 45 15	19 11 31	29♓36	19 41	06 33	11 35	27 51	04 28	01 49	09 29	06 57
25 Su	12 11 26	04 44 45	01♉07 17	28 44	20 42	06 17	11 48	27 47	04 32	01 51	09 29	06 50
26 Mo	12 15 23	05 44 12	12 57 05	27 54	21 42	06 02	12 00	27 43	04 35	01 53	09 30	06 45
27 Tu	12 19 19	06 43 37	24 43 48	27 08	22 41	05 47	12 13	27 39	04 38	01 55	09 30	06 43
28 We	12 23 16	07 43 00	06♊31 10	26 25	23 40	05 34	12 26	27 35	04 42	01 57	09 31	06D 43
29 Th	12 27 12	08 42 21	18 23 36	25 47	24 39	05 21	12 39	27 30	04 45	01 59	09 31	06 44
30 Fr	12 31 09	09 41 39	00♋26 05	25 14	25 37	05 09	12 52	27 26	04 49	02 00	09 31	06 46
31 Sa	12 35 05	10 40 55	12 43 51	24 46	26 35	04 57	13 06	27 22	04 52	02 02	09 32	06R 46

Data for 03-01-2012
Julian Day 2455987.50
Ayanamsa 24 01 53
SVP 05 ♓ 05 41
☽ ☊ Mean 09 ♐ 48 R

● ◐ **PHASES** ○ ◑

01	01:22	◐	10♊52
08	09:40	○	18♍13
15	01:25	◑	24♐52
22	14:37	●	02♈22
30	19:41	◐	10♋30

LAST ASPECT Day	h m	☽ INGRESS Day	h m	
02	13:14	02	15:09	♋
04	22:18	04	23:19	♌
07	01:29	07	03:28	♍
08	09:41	09	04:51	♎
11	03:10	11	05:25	♏
12	18:31	13	06:54	♐
15	07:35	15	10:24	♑
17	13:01	17	16:12	♒
19	20:31	20	00:05	♓
21	08:39	22	09:58	♈
24	17:18	24	21:44	♉
27	04:36	27	10:44	♊
29	18:06	29	23:08	♋

DECLINATION

Day	☉	☽	☿	♀	♂	♃	♄	♅	♆	♇
01 Th	07S31	21N55	00S07	10N39	09N52	12N57	08S38	00N38	11S38	19S15
02 Fr	07 08	22 06	00N39	11 08	10 01	13 01	08 37	00 39	11 37	19 15
03 Sa	06 45	21 13	01 23	11 37	10 10	13 04	08 36	00 40	11 37	19 15
04 Su	06 22	19 15	02 04	12 05	10 18	13 08	08 35	00 42	11 36	19 15
05 Mo	05 59	16 14	02 43	12 33	10 27	13 12	08 34	00 43	11 35	19 15
06 Tu	05 35	12 17	03 19	13 01	10 35	13 16	08 33	00 44	11 34	19 15
07 We	05 12	07 33	03 52	13 29	10 43	13 20	08 32	00 46	11 34	19 14
08 Th	04 49	02 18	04 20	13 56	10 51	13 24	08 31	00 47	11 33	19 14
09 Fr	04 25	03S11	04 45	14 23	10 59	13 27	08 29	00 48	11 32	19 14
10 Sa	04 02	08 32	05 06	14 50	11 07	13 31	08 28	00 50	11 31	19 14
11 Su	03 38	13 24	05 22	15 16	11 14	13 35	08 27	00 51	11 30	19 14
12 Mo	03 15	17 25	05 34	15 42	11 22	13 39	08 26	00 52	11 30	19 14
13 Tu	02 51	20 17	05 41	16 08	11 29	13 43	08 24	00 54	11 29	19 14
14 We	02 27	21 48	05 44	16 33	11 35	13 47	08 23	00 55	11 28	19 14
15 Th	02 04	21 56	05 41	16 58	11 42	13 51	08 22	00 56	11 27	19 14
16 Fr	01 40	20 44	05 35	17 23	11 48	13 55	08 20	00 58	11 27	19 14
17 Sa	01 16	18 23	05 23	17 47	11 55	13 59	08 19	00 59	11 26	19 14
18 Su	00 53	15 07	05 08	18 11	12 00	14 03	08 18	01 00	11 25	19 14
19 Mo	00 29	11 11	04 49	18 34	12 06	14 07	08 16	01 02	11 24	19 13
20 Tu	00 05	06 49	04 26	18 57	12 11	14 11	08 15	01 03	11 24	19 13
21 We	00N19	02 14	04 00	19 20	12 16	14 15	08 13	01 04	11 23	19 13
22 Th	00 42	02N23	03 32	19 42	12 21	14 19	08 12	01 06	11 22	19 13
23 Fr	01 06	06 51	03 03	20 04	12 25	14 23	08 10	01 07	11 22	19 13
24 Sa	01 30	10 59	02 31	20 25	12 29	14 27	08 09	01 09	11 21	19 13
25 Su	01 53	14 39	02 00	20 46	12 33	14 31	08 07	01 10	11 20	19 13
26 Mo	02 17	17 41	01 28	21 06	12 36	14 36	08 05	01 11	11 20	19 13
27 Tu	02 40	19 59	00 56	21 26	12 39	14 40	08 04	01 13	11 19	19 13
28 We	03 04	21 25	00 26	21 46	12 42	14 44	08 02	01 14	11 18	19 13
29 Th	03 27	21 54	00S03	22 05	12 45	14 48	08 01	01 15	11 18	19 13
30 Fr	03 50	21 22	00 31	22 23	12 47	14 52	07 59	01 17	11 17	19 13
31 Sa	04 14	19 50	00 56	22 41	12 49	14 56	07 58	01 18	11 16	19 13

ASPECTARIAN

01 08:54 ☽ □ ♂
02 00:09 ☿ ∥ ♅
08:30 ☽ ⚹ ♀
11:41 ☿ ♈
13:14 ☽ △ ♄
15:34 ☽ □ ☿
17:12 ☽ △ ♆
21:30 ☽ □ ♅
23:45 ♀ # ♆
03 05:28 ☽ ⚹ ♃
08:47 ☽ ☍ ♇
16:59 ☽ △ ☉
17:19 ☽ ⚹ ♂
20:11 ☉ ☍ ♂
04 00:04 ☽ # ♇
11:19 ♀ ☍ ♄
21:21 ☽ □ ♄
22:18 ☽ □ ♀
05 04:58 ☽ △ ☿
05:26 ☽ △ ♅
10:25 ♀ ♉
11:35 ☿ ☌ ♅
13:21 ☽ □ ♃
18:36 ☽ ∥ ♃
20:18 ☽ ∥ ♀
06 03:53 ☽ # ♆
08:49 ☽ ∥ ♂
12:26 ♀ ⚹ ♆
14:38 ♀ ∥ ♃
19:19 ☽ # ♄
07 01:29 ☽ ⚹ ♄
05:31 ☽ ☍ ♆
06:53 ☽ △ ♀
11:52 ☽ # ☉
15:39 ☽ ∥ ☿
17:18 ☽ △ ♃
18:56 ☽ △ ♇
23:30 ☽ ☌ ♂
08 06:41 ☽ ∥ ♅
13:36 ☽ # ♅
13:57 ☿ # ☉
09 05:06 ☽ ∥ ☉
07:25 ☽ # ☿
10:46 ☽ ☍ ♅
15:02 ☽ ☍ ☿
19:54 ☽ □ ♇
23:41 ☽ ∥ ♄
10 12:36 ☽ # ♂
14:16 ☽ ∥ ♆
11 01:01 ☽ # ♃
03:10 ☽ ☌ ♄
07:39 ☽ △ ♆
11:41 ☽ # ♀
16:32 ☽ ☍ ♀
20:14 ☽ ☍ ♃
20:38 ☽ ⚹ ♇
22:34 ☽ ⚹ ♂
12 07:50 ☿ SR
14:00 ☽ ∥ ♇
18:31 ☽ △ ☉
13 01:03 ♂ # ♆
04:31 ♃ △ ♇
09:20 ☽ □ ♆
13:26 ☽ △ ♅
18:10 ☽ △ ☿
23:24 ☽ □ ♂
14 01:26 ♀ △ ♇
05:55 ♀ ☌ ♃
07:28 ♀ △ ♂
09:51 ♂ △ ♃
15 00:32 ♂ △ ♇
07:35 ☽ ⚹ ♄
13:05 ☽ ⚹ ♆
17:27 ☽ □ ♅
20:54 ☽ □ ☿
16 02:18 ☽ △ ♂
03:00 ☽ ☌ ♇
04:00 ☽ △ ♃
07:11 ☽ △ ♀
16:24 ☽ ∥ ♇
17 04:16 ☽ # ♀
11:12 ☽ ⚹ ☉
13:01 ☽ □ ♄
16:35 ☉ # ♅
23:46 ☽ ⚹ ♅
18 00:46 ☽ ⚹ ☿
06:44 ☽ # ♃
11:23 ☽ □ ♃
17:23 ☿ ☌ ♅
18:39 ☽ □ ♀
18:47 ☽ # ♂
22:41 ☽ ∥ ♆
19 16:16 ☽ ∥ ♄
20:31 ☽ △ ♄
20 03:17 ☽ ☌ ♆
05:15 ☉ ♈
13:45 ☽ # ☿
14:25 ☽ ☍ ♂
17:14 ♀ # ♇
18:06 ☽ ⚹ ♇
20:56 ☽ ⚹ ♃
21 05:55 ☽ # ♅
08:39 ☽ ⚹ ♀
09:08 ☽ # ☉
14:22 ☽ ∥ ☉
17:12 ☽ ∥ ♅
19:21 ☿ ☌ ☉
22 05:30 ☽ ∥ ☿
11:51 ☽ ☌ ☿
18:37 ☽ ☌ ♅
23 01:25 ☉ ∥ ♅
04:38 ☽ □ ♇
07:24 ☽ # ♄
13:23 ☿ ♓R
24 02:16 ☽ # ♆
09:37 ☽ ∥ ♂
17:18 ☽ ☍ ♄
18:20 ☉ ☌ ♅
23:09 ☽ ∥ ♃
25 01:28 ☽ ⚹ ♆
02:51 ☿ ∥ ☉
10:13 ☽ △ ♂
16:58 ☽ △ ♇
22:03 ☽ ☌ ♃
26 12:05 ☿ ∥ ♅
15:04 ☽ # ♇
19:27 ☽ ☌ ♀
27 04:36 ☽ ⚹ ☿
14:40 ☽ □ ♆
20:17 ☽ ⚹ ♅
22:05 ☽ □ ♂
28 02:39 ☽ ⚹ ☉
29 14:06 ☽ □ ☿
17:43 ☿ ⚹ ♀
18:06 ☽ △ ♄
19:54 ☉ □ ♇
30 03:07 ☽ △ ♆
08:39 ☽ □ ♅
09:07 ☽ ⚹ ♂
17:48 ☽ ☍ ♇
31 00:43 ☽ ⚹ ♃
06:38 ☽ # ♇
22:14 ☽ △ ☿
23:57 ☿ # ♅

⚷ Chiron

01 Dec. 04 S 31

01	05♓32
04	05 44
07	05 55
10	06 07
13	06 19
16	06 30
19	06 42
22	06 53
25	07 04
28	07 14
31	07 25

April 2012

Day	S. T.	☉	☽	☿	♀	♂	♃	♄	♅	♆	♇	☊ True
	h m s	° ' "	° ' "	° '	° '	° '	° '	° '	° '	° '	° '	° '
01 Su	12 39 02	11♈ 40 09	25♋ 22 03	24♓ R24	27♉ 32	04♍ R47	13♉ 19	27♎ R18	04♈ 55	02♓ 04	09♑ 32	06♐ R45
02 Mo	12 42 58	12 39 20	08♌ 25 13	24 07	28 28	04 37	13 32	27 13	04 59	02 06	09 32	06 41
03 Tu	12 46 55	13 38 29	21 56 36	23 57	29 25	04 28	13 45	27 09	05 02	02 08	09 33	06 35
04 We	12 50 52	14 37 36	05♍ 57 14	23 51	00♊ 20	04 20	13 59	27 04	05 06	02 10	09 33	06 28
05 Th	12 54 48	15 36 40	20 25 15	23D 52	01 15	04 13	14 12	27 00	05 09	02 12	09 33	06 19
06 Fr	12 58 45	16 35 42	05♎ 15 33	23 58	02 09	04 06	14 25	26 55	05 12	02 13	09 33	06 11
07 Sa	13 02 41	17 34 43	20 20 10	24 09	03 03	04 00	14 39	26 51	05 16	02 15	09 33	06 03
08 Su	13 06 38	18 33 41	05♏ 29 25	24 25	03 56	03 55	14 52	26 46	05 19	02 17	09 33	05 57
09 Mo	13 10 34	19 32 37	20 33 31	24 46	04 48	03 51	15 06	26 42	05 23	02 18	09 34	05 53
10 Tu	13 14 31	20 31 32	05♐ 24 06	25 11	05 40	03 47	15 20	26 37	05 26	02 20	09 34	05 52
11 We	13 18 27	21 30 25	19 55 16	25 41	06 31	03 45	15 33	26 33	05 29	02 22	09R 34	05D 52
12 Th	13 22 24	22 29 16	04♑ 03 54	26 16	07 21	03 43	15 47	26 28	05 33	02 23	09 34	05 53
13 Fr	13 26 21	23 28 05	17 49 20	26 54	08 11	03 41	16 01	26 24	05 36	02 25	09 34	05 53
14 Sa	13 30 17	24 26 53	01♒ 12 41	27 36	09 00	03 41	16 14	26 19	05 39	02 27	09 33	05R 53
15 Su	13 34 14	25 25 39	14 16 06	28 21	09 48	03D 41	16 28	26 14	05 42	02 28	09 33	05 52
16 Mo	13 38 10	26 24 23	27 02 11	29 11	10 35	03 42	16 42	26 10	05 46	02 30	09 33	05 49
17 Tu	13 42 07	27 23 05	09♓ 33 38	00♈ 03	11 21	03 44	16 56	26 05	05 49	02 31	09 33	05 45
18 We	13 46 03	28 21 46	21 52 54	00 58	12 06	03 46	17 10	26 01	05 52	02 33	09 33	05 39
19 Th	13 49 60	29 20 25	04♈ 02 12	01 57	12 51	03 49	17 24	25 56	05 55	02 34	09 33	05 32
20 Fr	13 53 56	00♉ 19 02	16 03 25	02 58	13 34	03 53	17 38	25 51	05 59	02 36	09 32	05 26
21 Sa	13 57 53	01 17 37	27 58 16	04 02	14 16	03 58	17 51	25 47	06 02	02 37	09 32	05 21
22 Su	14 01 49	02 16 10	09♉ 48 30	05 09	14 58	04 03	18 05	25 42	06 05	02 38	09 32	05 17
23 Mo	14 05 46	03 14 42	21 36 03	06 18	15 38	04 09	18 19	25 38	06 08	02 40	09 31	05 15
24 Tu	14 09 43	04 13 11	03♊ 23 15	07 30	16 17	04 15	18 34	25 33	06 11	02 41	09 31	05D 15
25 We	14 13 39	05 11 39	15 12 53	08 43	16 55	04 22	18 48	25 29	06 14	02 42	09 30	05 17
26 Th	14 17 36	06 10 04	27 08 17	10 00	17 31	04 30	19 02	25 24	06 17	02 44	09 30	05 19
27 Fr	14 21 32	07 08 28	09♋ 13 14	11 18	18 07	04 39	19 16	25 20	06 21	02 45	09 30	05 22
28 Sa	14 25 29	08 06 49	21 31 55	12 39	18 41	04 48	19 30	25 15	06 24	02 46	09 29	05 23
29 Su	14 29 25	09 05 08	04♌ 08 42	14 01	19 13	04 57	19 44	25 11	06 27	02 47	09 28	05 24
30 Mo	14 33 22	10 03 26	17 07 44	15 26	19 45	05 08	19 58	25 06	06 30	02 48	09 28	05R 24

Data for	04-01-2012
Julian Day	2456018.50
Ayanamsa	24 01 56
SVP	05 ♓ 05 35
☽ ☊ Mean	08 ♐ 09 R

● ◐ PHASES ○ ◑

Day	Time	Phase	Position
06	19:19	○	17♎23
13	10:50	◑	23♑55
21	07:19	●	01♉36
29	09:58	◐	09♌29

LAST ASPECT Day	h m	☽ INGRESS Day	h m	Sign
01	04:21	01	08:37	♌
03	13:48	03	13:54	♍
05	05:38	05	15:33	♎
07	10:16	07	15:18	♏
09	06:56	09	15:13	♐
11	11:06	11	17:02	♑
13	17:05	13	21:48	♒
15	22:42	16	05:38	♓
17	14:35	18	16:00	♈
20	19:36	21	04:06	♉
22	17:12	23	17:06	♊
25	20:32	26	05:43	♋
28	07:06	28	16:11	♌
30	14:18	30	23:03	♍

DECLINATION

Day	☉	☽	☿	♀	♂	♃	♄	♅	♆	♇
01 Su	04N37	17N17	01S19	22N59	12N50	15N00	07S56	01N19	11S16	19S13
02 Mo	05 00	13 48	01 40	23 16	12 52	15 04	07 54	01 21	11 15	19 13
03 Tu	05 23	09 30	01 59	23 32	12 52	15 08	07 53	01 22	11 14	19 13
04 We	05 46	04 34	02 15	23 48	12 53	15 12	07 51	01 23	11 14	19 13
05 Th	06 09	00S46	02 28	24 04	12 53	15 16	07 49	01 25	11 13	19 13
06 Fr	06 31	06 12	02 39	24 19	12 53	15 21	07 48	01 26	11 12	19 13
07 Sa	06 54	11 20	02 47	24 33	12 53	15 25	07 46	01 27	11 12	19 12
08 Su	07 16	15 48	02 53	24 47	12 53	15 29	07 44	01 29	11 11	19 12
09 Mo	07 39	19 12	02 56	25 01	12 52	15 33	07 43	01 30	11 11	19 12
10 Tu	08 01	21 15	02 57	25 14	12 51	15 37	07 41	01 31	11 10	19 12
11 We	08 23	21 48	02 55	25 26	12 49	15 41	07 39	01 33	11 09	19 12
12 Th	08 45	20 56	02 51	25 38	12 48	15 45	07 38	01 34	11 09	19 12
13 Fr	09 07	18 50	02 45	25 50	12 46	15 49	07 36	01 35	11 08	19 12
14 Sa	09 28	15 44	02 37	26 00	12 43	15 53	07 34	01 37	11 08	19 12
15 Su	09 50	11 57	02 26	26 11	12 41	15 57	07 33	01 38	11 07	19 12
16 Mo	10 11	07 42	02 14	26 21	12 38	16 01	07 31	01 39	11 07	19 12
17 Tu	10 32	03 12	02 00	26 30	12 35	16 05	07 29	01 41	11 06	19 12
18 We	10 53	01N21	01 44	26 39	12 32	16 09	07 28	01 42	11 06	19 12
19 Th	11 14	05 47	01 26	26 47	12 29	16 13	07 26	01 43	11 05	19 12
20 Fr	11 35	09 57	01 07	26 55	12 25	16 17	07 24	01 44	11 05	19 12
21 Sa	11 55	13 42	00 46	27 02	12 21	16 21	07 23	01 46	11 04	19 12
22 Su	12 16	16 53	00 23	27 09	12 17	16 25	07 21	01 47	11 04	19 12
23 Mo	12 36	19 21	00 01	27 15	12 13	16 29	07 20	01 48	11 03	19 12
24 Tu	12 55	21 00	00 27	27 20	12 08	16 33	07 18	01 49	11 03	19 12
25 We	13 15	21 43	00 54	27 26	12 03	16 37	07 16	01 51	11 02	19 12
26 Th	13 35	21 27	01 22	27 30	11 58	16 41	07 15	01 52	11 02	19 12
27 Fr	13 54	20 11	01 52	27 34	11 53	16 45	07 13	01 53	11 02	19 12
28 Sa	14 13	17 58	02 23	27 38	11 47	16 49	07 12	01 54	11 01	19 12
29 Su	14 31	14 51	02 55	27 41	11 42	16 53	07 10	01 55	11 01	19 12
30 Mo	14 50	10 57	03 28	27 44	11 36	16 57	07 09	01 56	11 00	19 12

ASPECTARIAN

01 03:34 ☽ □ ♄
04:21 ☽ ✶ ♀
16:01 ☽ ∥ ♃
17:44 ☽ △ ♅

02 05:34 ☽ ∥ ♂
08:13 ☽ △ ☉
09:20 ☽ □ ♃
14:44 ☽ # ♆

03 08:14 ☽ # ♄
08:58 ☽ ✶ ♄
13:48 ☽ □ ♀
15:18 ♀ ♊
17:34 ☽ ☍ ♆
18:47 ☽ ∥ ☉
21:17 ☽ ☌ ♂

04 06:02 ☽ △ ♇
10:07 ☽ # ☿
10:12 ☿ SD
13:36 ☽ △ ♃
14:21 ☽ ∥ ♅

05 02:51 ☽ # ♅
05:38 ☽ ☍ ☿
07:45 ☽ ∥ ☿
18:42 ☽ △ ♀
23:55 ☽ ☍ ♅

06 01:34 ☽ # ☉
01:51 ♀ □ ♆
06:52 ☽ □ ♇
07:10 ☽ ∥ ♄
23:15 ☽ ∥ ♆

07 07:45 ☽ # ♂
10:16 ☽ ☌ ♄
18:54 ☽ △ ♆
21:31 ☽ ✶ ♂
21:59 ☽ # ♃
23:37 ♀ □ ♂

08 06:27 ☽ ✶ ♇
15:08 ☽ ☍ ♃

09 00:01 ☽ ∥ ♇
03:55 ☉ # ♄
06:56 ☽ △ ☿
16:52 ♀ ✶ ♅
19:00 ☽ □ ♆
21:23 ☽ □ ♂

10 00:03 ☽ △ ♅
00:28 ☽ ☍ ♀
16:22 ♇ SR

11 02:51 ☽ △ ☉
10:06 ☽ □ ☿
11:06 ☽ ✶ ♄
21:07 ☽ ✶ ♆
23:24 ☽ △ ♂

12 02:33 ☽ □ ♅
09:30 ☽ ☌ ♇
20:24 ☽ ∥ ♇
20:45 ☽ △ ♃

13 15:12 ☽ □ ♄
17:05 ☽ ✶ ☿
22:58 ☽ # ♃

14 03:54 ♂ SD
08:08 ☽ ✶ ♅
15:10 ☽ △ ♀
19:30 ☽ # ♂

15 04:11 ☽ □ ♃
04:48 ☽ ∥ ♆
11:12 ☽ # ☉
18:27 ☉ ☍ ♄
22:21 ☽ △ ♄
22:42 ☽ ✶ ☉

16 00:57 ☽ ∥ ♄
10:26 ☽ ☌ ♆
12:45 ☽ ☍ ♂
22:42 ☿ ♈
23:59 ☽ ✶ ♇

17 03:41 ☽ □ ♀
06:42 ☽ ∥ ☿
07:59 ☽ # ♅
14:35 ☽ ✶ ♃

18 01:52 ☽ ∥ ♅
01:56 ☽ # ☿
02:49 ☿ # ♅
13:46 ☉ # ♆
19:29 ☽ ☌ ☿

19 03:46 ☽ ☌ ♅
09:13 ☽ # ♄
10:57 ☽ □ ♇
16:12 ☉ ♉
18:41 ☽ ✶ ♀

20 06:53 ☽ # ♆
11:03 ☽ ∥ ☉
15:09 ☽ ∥ ♂
19:36 ☽ ☍ ♄

21 09:25 ☽ ✶ ♆
12:13 ☽ △ ♂
20:07 ☽ ∥ ♃
23:26 ☽ △ ♇

22 01:20 ☉ ∥ ♂
09:19 ☉ ✶ ♆
17:12 ☽ ☌ ♃
20:29 ☿ ☌ ♅
22:21 ☽ # ♇

23 22:34 ☽ □ ♆

24 01:00 ☉ △ ♂
01:47 ☽ □ ♂
05:43 ☽ ✶ ♅
09:18 ☽ ✶ ☿

25 03:37 ☽ ☌ ♀
14:50 ☿ □ ♇
20:32 ☽ △ ♄

26 11:10 ☽ △ ♆
14:51 ☽ ✶ ♂
18:18 ☽ □ ♅
19:32 ☽ ✶ ☉

27 00:32 ☽ ☍ ♇
00:54 ☿ ∥ ♅
04:35 ☽ □ ☿
11:51 ☽ # ♇
20:00 ☽ ✶ ♃

28 07:06 ☽ □ ♄
09:27 ☽ ∥ ♃

29 02:00 ☽ ∥ ☉
04:19 ☽ △ ♅
09:33 ☉ △ ♇
20:10 ☽ ∥ ♂
20:32 ☽ △ ☿
23:38 ☽ # ♆

30 04:56 ☽ ✶ ♀
05:15 ☽ □ ♃
14:18 ☽ ✶ ♄
20:17 ☽ # ♄

⚷ Chiron

01	Dec.	03 S 47
03	07♓35	
06	07 45	
09	07 55	
12	08 04	
15	08 13	
18	08 22	
21	08 30	
24	08 38	
27	08 46	
30	08 53	

21 00:00 00♊21 ⊙ Annular Solar Eclipse

May 2012

Day	S. T. h m s	☉ ° ' "	☽ ° ' "	☿ ° '	♀ ° '	♂ ° '	♃ ° '	♄ ° '	♅ ° '	♆ ° '	♇ ° '	☊ True ° '
01 Tu	14 37 18	11♉ 01 41	00♍ 32 26	16♈ 53	20♊ 14	05♍ 18	20♉ 12	25♎R02	06♈ 33	02♓ 49	09♑R27	05♐R22
02 We	14 41 15	11 59 54	14 24 49	18 21	20 42	05 30	20 26	24 58	06 35	02 51	09 27	05 19
03 Th	14 45 12	12 58 05	28 44 39	19 52	21 09	05 42	20 41	24 53	06 38	02 52	09 26	05 15
04 Fr	14 49 08	13 56 14	13♎ 28 54	21 25	21 34	05 54	20 55	24 49	06 41	02 53	09 25	05 11
05 Sa	14 53 05	14 54 21	28 31 34	22 59	21 57	06 07	21 09	24 45	06 44	02 54	09 25	05 08
06 Su	14 57 01	15 52 26	13♏ 44 14	24 36	22 18	06 21	21 23	24 41	06 47	02 55	09 24	05 06
07 Mo	15 00 58	16 50 30	28 57 07	26 15	22 37	06 35	21 37	24 37	06 50	02 56	09 23	05 04
08 Tu	15 04 54	17 48 33	14♐ 00 37	27 55	22 55	06 49	21 52	24 33	06 53	02 57	09 22	05D 04
09 We	15 08 51	18 46 33	28 46 39	29 37	23 10	07 04	22 06	24 29	06 55	02 57	09 22	05 04
10 Th	15 12 47	19 44 33	13♑ 09 41	01♉ 22	23 24	07 20	22 20	24 25	06 58	02 58	09 21	05 05
11 Fr	15 16 44	20 42 31	27 06 51	03 08	23 35	07 36	22 34	24 21	07 01	02 59	09 20	05 06
12 Sa	15 20 41	21 40 28	10♒ 37 51	04 56	23 45	07 52	22 49	24 17	07 04	03 00	09 19	05 07
13 Su	15 24 37	22 38 23	23 44 15	06 46	23 52	08 09	23 03	24 13	07 06	03 01	09 18	05 08
14 Mo	15 28 34	23 36 17	06♓ 28 53	08 39	23 57	08 26	23 17	24 09	07 09	03 01	09 17	05R 08
15 Tu	15 32 30	24 34 10	18 55 16	10 33	23 59	08 44	23 31	24 06	07 11	03 02	09 16	05 07
16 We	15 36 27	25 32 02	01♈ 07 10	12 29	23R 59	09 02	23 45	24 02	07 14	03 03	09 15	05 07
17 Th	15 40 23	26 29 52	13 08 16	14 26	23 57	09 21	24 00	23 59	07 16	03 03	09 14	05 06
18 Fr	15 44 20	27 27 41	25 01 55	16 26	23 53	09 40	24 14	23 55	07 19	03 04	09 13	05 05
19 Sa	15 48 16	28 25 29	06♉ 51 07	18 28	23 46	09 59	24 28	23 52	07 21	03 04	09 12	05 05
20 Su	15 52 13	29 23 16	18 38 31	20 31	23 37	10 19	24 42	23 48	07 24	03 05	09 11	05D 05
21 Mo	15 56 10	00♊ 21 01	00♊ 26 32	22 36	23 25	10 39	24 56	23 45	07 26	03 06	09 10	05 06
22 Tu	16 00 06	01 18 45	12 17 27	24 42	23 11	11 00	25 11	23 42	07 28	03 06	09 09	05 07
23 We	16 04 03	02 16 28	24 13 30	26 50	22 54	11 20	25 25	23 39	07 31	03 06	09 08	05 07
24 Th	16 07 59	03 14 09	06♋ 17 01	28 59	22 35	11 42	25 39	23 36	07 33	03 07	09 07	05 08
25 Fr	16 11 56	04 11 49	18 30 29	01♊ 10	22 14	12 03	25 53	23 33	07 35	03 07	09 06	05 09
26 Sa	16 15 52	05 09 28	00♌ 56 37	03 20	21 51	12 25	26 07	23 30	07 38	03 08	09 04	05R 09
27 Su	16 19 49	06 07 05	13 38 14	05 32	21 25	12 48	26 21	23 27	07 40	03 08	09 03	05 09
28 Mo	16 23 45	07 04 40	26 38 12	07 44	20 58	13 10	26 35	23 25	07 42	03 08	09 02	05 08
29 Tu	16 27 42	08 02 14	09♍ 59 06	09 56	20 28	13 33	26 49	23 22	07 44	03 08	09 01	05 07
30 We	16 31 39	08 59 47	23 42 44	12 08	19 57	13 57	27 03	23 19	07 46	03 09	08 59	05 07
31 Th	16 35 35	09 57 18	07♎ 49 41	14 19	19 24	14 20	27 18	23 17	07 48	03 09	08 58	05 06

Data for	05-01-2012
Julian Day	2456048.50
Ayanamsa	24 01 59
SVP	05 ♓ 05 30
☽ ☊ Mean	06 ♐ 34 R

● ◐ PHASES ○ ◑

Day	h m	Phase	Position
06	03:35	○	16♏01
12	21:47	◑	22♒33
20	23:47	⊙	00♊21
28	20:16	◐	07♍53

LAST ASPECT Day	h m	☽ INGRESS Day	h m	Sign
02	10:58	03	02:04	♎
04	18:02	05	02:20	♏
06	12:14	07	01:40	♐
09	01:35	09	02:01	♑
10	19:12	11	05:04	♒
13	00:54	13	11:43	♓
15	12:01	15	21:47	♈
17	21:46	18	10:04	♉
20	12:35	20	23:06	♊
22	22:51	23	11:32	♋
25	14:34	25	22:12	♌
27	23:55	28	06:07	♍
30	05:51	30	10:46	♎

DECLINATION

Day	☉	☽	☿	♀	♂	♃	♄	♅	♆	♇
01 Tu	15N08	06N23	04N02	27N46	11N30	17N00	07S 07	01N58	11S 00	19S 13
02 We	15 26	01 21	04 38	27 48	11 24	17 04	07 06	01 59	11 00	19 13
03 Th	15 44	03S 55	05 14	27 49	11 17	17 08	07 04	02 00	10 59	19 13
04 Fr	16 01	09 07	05 52	27 49	11 10	17 12	07 03	02 01	10 59	19 13
05 Sa	16 18	13 52	06 30	27 49	11 04	17 16	07 01	02 02	10 59	19 13
06 Su	16 35	17 47	07 09	27 49	10 57	17 20	07 00	02 03	10 58	19 13
07 Mo	16 52	20 28	07 49	27 48	10 50	17 23	06 59	02 04	10 58	19 13
08 Tu	17 08	21 39	08 29	27 47	10 42	17 27	06 57	02 05	10 58	19 13
09 We	17 24	21 18	09 11	27 45	10 35	17 31	06 56	02 07	10 57	19 13
10 Th	17 40	19 32	09 53	27 42	10 27	17 35	06 55	02 08	10 57	19 13
11 Fr	17 56	16 39	10 35	27 40	10 19	17 38	06 53	02 09	10 57	19 13
12 Sa	18 11	12 56	11 18	27 36	10 11	17 42	06 52	02 10	10 57	19 13
13 Su	18 26	08 43	12 01	27 32	10 03	17 46	06 51	02 11	10 56	19 13
14 Mo	18 40	04 13	12 45	27 27	09 55	17 49	06 49	02 12	10 56	19 13
15 Tu	18 55	00N21	13 29	27 22	09 46	17 53	06 48	02 13	10 56	19 13
16 We	19 09	04 48	14 13	27 17	09 38	17 56	06 47	02 14	10 56	19 13
17 Th	19 22	09 01	14 57	27 10	09 29	18 00	06 46	02 15	10 55	19 13
18 Fr	19 35	12 51	15 40	27 03	09 20	18 03	06 45	02 16	10 55	19 14
19 Sa	19 48	16 10	16 24	26 56	09 11	18 07	06 44	02 17	10 55	19 14
20 Su	20 01	18 48	17 07	26 48	09 02	18 10	06 43	02 18	10 55	19 14
21 Mo	20 13	20 39	17 49	26 39	08 53	18 14	06 42	02 18	10 55	19 14
22 Tu	20 25	21 36	18 31	26 29	08 43	18 17	06 41	02 19	10 55	19 14
23 We	20 37	21 34	19 11	26 19	08 34	18 21	06 40	02 20	10 54	19 14
24 Th	20 48	20 32	19 50	26 08	08 24	18 24	06 39	02 21	10 54	19 14
25 Fr	20 59	18 32	20 28	25 57	08 14	18 28	06 38	02 22	10 54	19 14
26 Sa	21 09	15 39	21 04	25 45	08 04	18 31	06 37	02 23	10 54	19 14
27 Su	21 20	11 59	21 39	25 32	07 54	18 34	06 36	02 24	10 54	19 14
28 Mo	21 29	07 42	22 11	25 18	07 44	18 38	06 35	02 24	10 54	19 15
29 Tu	21 39	02 56	22 41	25 04	07 33	18 41	06 35	02 25	10 54	19 15
30 We	21 48	02S 07	23 09	24 49	07 23	18 44	06 34	02 26	10 54	19 15
31 Th	21 56	07 12	23 35	24 34	07 12	18 47	06 33	02 27	10 54	19 15

ASPECTARIAN

01 04:01 ☽ ☍ ♆
08:27 ☽ ☌ ♂
10:12 ☽ ∥ ☿
15:30 ☽ △ ♇
19:34 ☽ △ ☉
21:05 ☽ ∥ ♅
02 10:21 ☽ △ ♃
10:58 ☽ □ ♀
15:16 ☽ # ♅
03 06:47 ☽ # ☿
12:58 ☽ ☍ ♅
14:20 ☽ ∥ ♄
17:27 ☽ □ ♇
04 03:01 ☿ ⚹ ♀
09:00 ☽ ∥ ♆
09:45 ☽ # ♂
13:17 ☽ △ ♀
14:11 ☽ ☍ ☿
18:02 ☽ ☌ ♄
05 06:55 ☽ △ ♆
12:11 ☽ ⚹ ♂
15:13 ☽ # ☉
17:10 ☽ ⚹ ♇
18:31 ♂ # ♆
18:43 ☿ # ♄
20:41 ☽ # ♃
06 01:08 ☿ ☍ ♄
11:10 ☽ ∥ ♇
12:14 ☽ ☍ ♃
07 06:18 ☽ □ ♆
12:18 ☽ □ ♂
12:33 ☽ △ ♅
08 14:40 ☽ ☍ ♀
16:59 ☽ ⚹ ♄
09 01:35 ☽ △ ☿
05:15 ☿ ♉
06:55 ☽ ⚹ ♆
12:39 ☉ ∥ ♃
13:33 ☽ □ ♅
14:00 ☽ △ ♂
17:34 ☽ ☌ ♇
10 03:10 ☽ ∥ ♇
12:04 ☽ △ ☉
15:04 ☽ # ☉
15:58 ☽ △ ♃
16:28 ☿ ∥ ♂
16:45 ☽ # ♃
19:12 ☽ □ ♄
21:59 ☿ ⚹ ♆
11 12:03 ☿ # ♆
12:13 ☽ □ ☿
17:34 ☽ ⚹ ♅
12 08:09 ☽ # ☿
11:37 ☽ ∥ ♆
16:22 ☽ # ♂
22:41 ☽ □ ♃
13 00:14 ☽ △ ♀
00:54 ☽ △ ♄
10:07 ☽ ∥ ♄
13:23 ☉ ☌ ♃
17:25 ☽ ☌ ♆
20:54 ☿ △ ♂
14 03:50 ☽ ☍ ♂
04:52 ☽ ⚹ ☿
05:22 ☽ ⚹ ♇
08:09 ☿ △ ♇
10:36 ☽ # ♅
15 09:11 ☽ ⚹ ♃
09:55 ☽ □ ♀
09:59 ☽ ∥ ♅
12:01 ☽ ⚹ ☉
14:33 ♀ SR
16 08:34 ☉ # ♇
11:00 ☽ # ♄
12:13 ☽ ☌ ♅
16:11 ☽ □ ♇
16:20 ♂ △ ♇
17 02:39 ☽ ∥ ♂
11:32 ☽ # ♆
21:41 ☽ ⚹ ♀
21:46 ☽ ☍ ♄
18 16:19 ☽ ⚹ ♆
19 02:28 ☽ ∥ ☿
04:47 ☽ △ ♇
06:33 ☽ △ ♂
17:25 ☽ ∥ ♃
20 04:36 ☽ # ♇
04:38 ☽ ☌ ☿
12:35 ☽ ☌ ♃
15:16 ☉ ♊
16:16 ☽ ∥ ☉
21 05:23 ☽ □ ♆
14:14 ☽ ⚹ ♅
15:39 ☿ ∥ ♃
21:18 ☽ □ ♂
22 05:58 ☿ ☌ ♃
21:25 ☽ ☌ ♀
22:51 ☽ △ ♄
23 01:50 ☿ # ♇
17:43 ☽ △ ♆
20:09 ☽ ∥ ☉
20:56 ☉ □ ♆
24 02:31 ☽ □ ♅
05:35 ☽ ☍ ♇
07:18 ☽ ∥ ☿
11:00 ☽ ⚹ ♂
11:12 ☿ ♊
16:50 ☽ # ♇
25 00:47 ☽ ∥ ♃
09:45 ☽ □ ♄
14:34 ☽ ⚹ ♃
21:37 ☿ □ ♆
26 04:51 ☿ ∥ ☉
05:32 ☽ ⚹ ☿
08:41 ☽ ⚹ ☉
12:44 ☽ △ ♅
27 06:27 ☽ # ♆
11:20 ☿ ☌ ☉
13:57 ☽ ⚹ ♀
18:07 ☽ ⚹ ♄
23:36 ☿ ⚹ ♅
23:52 ☽ ∥ ♂
23:55 ☽ □ ♃
28 05:48 ☽ # ♄
11:46 ☽ ☍ ♆
16:04 ☉ ⚹ ♅
22:17 ☽ △ ♇
23:53 ☽ □ ☿
29 02:29 ☽ ∥ ♅
06:30 ☽ ☌ ♂
17:44 ☽ □ ♀
30 01:31 ☽ # ♅
05:51 ☽ △ ♃
20:56 ☽ ∥ ♄
23:57 ☽ ☍ ♅
31 00:02 ☽ # ♂
00:20 ☿ □ ♂
01:55 ☽ □ ♇
03:49 ☽ △ ☉
12:44 ☽ △ ☿
18:12 ☽ ∥ ♆
18:31 ☽ △ ♀

⚷ Chiron

01 Dec. 03 S 10

Day	Position
03	09♓00
06	09 06
09	09 12
12	09 18
15	09 23
18	09 27
21	09 31
24	09 35
27	09 38
30	09 40

June 2012

04 11:04 14♐14 ✹ Total Lunar Eclipse (mag 0.376)

Day	S. T. h m s	☉ ° ' "	☽ ° ' "	☿ ° '	♀ ° '	♂ ° '	♃ ° '	♄ ° '	♅ ° '	♆ ° '	♇ ° '	☊ True ° '
01 Fr	16 39 32	10♊ 54 48	22♎ 18 41	16♊ 29	18♊R50	14♍ 44	27♉ 32	23♎R15	07♈ 50	03♓ 09	08♑R57	05♐R06
02 Sa	16 43 28	11 52 16	07♏ 06 15	18 39	18 15	15 09	27 45	23 12	07 52	03 09	08 56	05 05
03 Su	16 47 25	12 49 44	22 06 33	20 47	17 39	15 33	27 59	23 10	07 54	03 09	08 54	05 05
04 Mo	16 51 21	13 47 10	07♐ 11 53	22 54	17 02	15 58	28 13	23 08	07 56	03 09	08 53	05 04
05 Tu	16 55 18	14 44 36	22 13 33	24 59	16 24	16 23	28 27	23 06	07 57	03R 09	08 52	05 04
06 We	16 59 14	15 42 00	07♑ 02 59	27 02	15 47	16 48	28 41	23 04	07 59	03 09	08 50	05 02
07 Th	17 03 11	16 39 24	21 33 03	29 04	15 09	17 14	28 55	23 02	08 01	03 09	08 49	05 01
08 Fr	17 07 08	17 36 47	05♒ 38 51	01♋ 03	14 31	17 40	29 09	23 01	08 03	03 09	08 48	05 00
09 Sa	17 11 04	18 34 10	19 18 02	03 00	13 54	18 06	29 23	22 59	08 04	03 09	08 46	04 58
10 Su	17 15 01	19 31 31	02♓ 30 41	04 55	13 18	18 33	29 36	22 57	08 06	03 09	08 45	04 57
11 Mo	17 18 57	20 28 53	15 18 53	06 47	12 43	18 59	29 50	22 56	08 07	03 08	08 43	04D 57
12 Tu	17 22 54	21 26 13	27 46 03	08 37	12 09	19 26	00♊ 04	22 55	08 09	03 08	08 42	04 58
13 We	17 26 50	22 23 34	09♈ 56 29	10 25	11 36	19 54	00 17	22 53	08 10	03 08	08 40	04 59
14 Th	17 30 47	23 20 54	21 54 44	12 10	11 05	20 21	00 31	22 52	08 12	03 08	08 39	05 01
15 Fr	17 34 43	24 18 13	03♉ 45 22	13 52	10 35	20 49	00 45	22 51	08 13	03 07	08 38	05 04
16 Sa	17 38 40	25 15 32	15 32 37	15 33	10 08	21 17	00 58	22 50	08 14	03 07	08 36	05 06
17 Su	17 42 37	26 12 51	27 20 16	17 10	09 42	21 45	01 12	22 49	08 16	03 07	08 35	05 08
18 Mo	17 46 33	27 10 09	09♊ 11 35	18 45	09 18	22 13	01 25	22 48	08 17	03 06	08 33	05 09
19 Tu	17 50 30	28 07 27	21 09 17	20 18	08 57	22 42	01 39	22 48	08 18	03 06	08 32	05R 09
20 We	17 54 26	29 04 44	03♋ 15 35	21 48	08 38	23 10	01 52	22 47	08 19	03 05	08 30	05 07
21 Th	17 58 23	00♋ 02 01	15 32 12	23 15	08 21	23 39	02 05	22 47	08 20	03 05	08 29	05 04
22 Fr	18 02 19	00 59 18	28 00 29	24 40	08 07	24 09	02 19	22 46	08 21	03 04	08 27	04 59
23 Sa	18 06 16	01 56 33	10♌ 41 31	26 02	07 54	24 38	02 32	22 46	08 22	03 04	08 26	04 55
24 Su	18 10 13	02 53 49	23 36 17	27 22	07 45	25 08	02 45	22 46	08 23	03 03	08 24	04 50
25 Mo	18 14 09	03 51 03	06♍ 45 43	28 39	07 37	25 38	02 58	22 46	08 24	03 03	08 23	04 47
26 Tu	18 18 06	04 48 17	20 10 41	29 53	07 32	26 08	03 11	22D 46	08 25	03 02	08 21	04 44
27 We	18 22 02	05 45 30	03♎ 51 55	01♌ 04	07 30	26 38	03 24	22 46	08 26	03 01	08 20	04 43
28 Th	18 25 59	06 42 43	17 49 40	02 12	07D 29	27 09	03 37	22 46	08 27	03 01	08 18	04D 43
29 Fr	18 29 55	07 39 56	02♏ 03 26	03 18	07 31	27 39	03 50	22 46	08 27	03 00	08 17	04 43
30 Sa	18 33 52	08 37 08	16 31 33	04 20	07 36	28 10	04 03	22 47	08 28	02 59	08 15	04 44

Data for	06-01-2012
Julian Day	2456079.50
Ayanamsa	24 02 04
SVP	05 ♓ 05 26
☽ ☊ Mean	04 ♐ 55 R

● ◐ PHASES ○ ◑

Day	Time	Phase	Position
04	11:12	✹	14♐14
11	10:42	◑	20♓54
19	15:02	●	28♊43
27	03:31	◐	05♎54

LAST ASPECT ☽ INGRESS

Day	h m	Day	h m	Sign
01	01:31	01	12:32	♏
03	09:30	03	12:33	♐
05	05:09	05	12:32	♑
07	12:39	07	14:18	♒
09	18:34	09	19:23	♓
11	10:42	12	04:22	♈
14	03:09	14	16:22	♉
16	12:09	17	05:24	♊
19	15:02	19	17:34	♋
21	16:49	22	03:48	♌
23	22:27	24	11:43	♍
26	10:54	26	17:16	♎
28	08:23	28	20:33	♏
30	19:47	30	22:05	♐

DECLINATION

Day	☉	☽	☿	♀	♂	♃	♄	♅	♆	♇
01 Fr	22N05	12S01	23N58	24N17	07N02	18N51	06S33	02N28	10S54	19S15
02 Sa	22 13	16 14	24 18	24 01	06 51	18 54	06 32	02 28	10 54	19 15
03 Su	22 20	19 27	24 35	23 44	06 40	18 57	06 31	02 29	10 54	19 15
04 Mo	22 27	21 19	24 50	23 26	06 29	19 00	06 31	02 30	10 54	19 15
05 Tu	22 34	21 39	25 01	23 09	06 18	19 03	06 30	02 30	10 54	19 16
06 We	22 40	20 27	25 11	22 51	06 06	19 06	06 30	02 31	10 54	19 16
07 Th	22 46	17 55	25 17	22 32	05 55	19 09	06 29	02 32	10 54	19 16
08 Fr	22 52	14 23	25 21	22 14	05 43	19 12	06 29	02 32	10 54	19 16
09 Sa	22 57	10 12	25 22	21 56	05 32	19 15	06 29	02 33	10 54	19 16
10 Su	23 01	05 39	25 21	21 38	05 20	19 18	06 28	02 34	10 54	19 16
11 Mo	23 06	00 59	25 17	21 20	05 08	19 21	06 28	02 34	10 54	19 16
12 Tu	23 10	03N35	25 12	21 02	04 57	19 24	06 28	02 35	10 54	19 17
13 We	23 13	07 56	25 04	20 44	04 45	19 27	06 27	02 35	10 55	19 17
14 Th	23 16	11 54	24 54	20 27	04 33	19 30	06 27	02 36	10 55	19 17
15 Fr	23 19	15 21	24 43	20 11	04 20	19 33	06 27	02 36	10 55	19 17
16 Sa	23 21	18 11	24 30	19 55	04 08	19 36	06 27	02 37	10 55	19 17
17 Su	23 23	20 15	24 15	19 40	03 56	19 38	06 27	02 37	10 55	19 17
18 Mo	23 24	21 27	23 59	19 25	03 43	19 41	06 27	02 38	10 55	19 18
19 Tu	23 25	21 41	23 41	19 12	03 31	19 44	06 27	02 38	10 55	19 18
20 We	23 26	20 55	23 22	18 59	03 18	19 47	06 27	02 39	10 56	19 18
21 Th	23 26	19 09	23 02	18 47	03 06	19 49	06 27	02 39	10 56	19 18
22 Fr	23 26	16 27	22 41	18 35	02 53	19 52	06 27	02 39	10 56	19 18
23 Sa	23 25	12 57	22 19	18 25	02 40	19 55	06 27	02 40	10 56	19 18
24 Su	23 24	08 48	21 57	18 15	02 27	19 57	06 27	02 40	10 57	19 19
25 Mo	23 23	04 10	21 34	18 06	02 14	20 00	06 27	02 40	10 57	19 19
26 Tu	23 21	00S45	21 10	17 58	02 01	20 02	06 28	02 41	10 57	19 19
27 We	23 19	05 44	20 45	17 51	01 48	20 05	06 28	02 41	10 57	19 19
28 Th	23 16	10 31	20 21	17 45	01 35	20 07	06 28	02 41	10 58	19 19
29 Fr	23 13	14 50	19 56	17 40	01 21	20 10	06 29	02 42	10 58	19 20
30 Sa	23 09	18 20	19 31	17 35	01 08	20 12	06 29	02 42	10 58	19 20

ASPECTARIAN

01 01:31 ☽ ☌ ♄
12:51 ☿ ∥ ♀
17:37 ☽ △ ♆
20:31 ☿ ☌ ♀
02 02:56 ☽ ✱ ♇
13:15 ☽ ✱ ♂
19:34 ☽ # ♃
22:18 ☽ ∥ ♇

03 09:30 ☽ ☍ ♃
17:34 ☽ □ ♆
19:24 ♂ # ♄
04 01:10 ☽ △ ♅
02:39 ☿ △ ♄
14:23 ☽ □ ♂
15:03 ☽ ☍ ♀
20:06 ♆ SR

05 00:28 ♀ □ ♂
01:24 ☽ ✱ ♄
05:09 ☽ ☍ ☿
17:39 ☽ ✱ ♆

06 01:09 ♀ ☌ ☉
01:32 ☽ □ ♅
02:55 ☽ ☌ ♇
10:22 ♀ ∥ ☉
12:38 ☽ ∥ ♇
13:50 ☽ # ♃
16:34 ☽ △ ♂

07 02:30 ☽ □ ♄
11:16 ☿ ♋
12:39 ☽ △ ♃
08 02:30 ☉ □ ♂
04:10 ☽ ✱ ♅
14:50 ☽ △ ♀
20:08 ☽ ∥ ♆
22:36 ☽ △ ☉

09 01:50 ☿ △ ♆
06:12 ♃ # ♇
06:36 ☽ △ ♄
18:34 ☽ □ ♃
19:44 ☽ ∥ ♄

10 01:10 ☽ ☌ ♆
01:41 ☽ # ♂
05:12 ☽ △ ☿
11:34 ☽ ✱ ♇
15:53 ☽ # ♅
19:17 ☽ □ ♀

11 07:17 ☽ ☍ ♂
17:23 ♃ ♊
17:42 ☿ □ ♅
18:39 ☽ ∥ ♅

12 01:03 ☿ ☍ ♇
04:34 ☽ ✱ ♃
07:03 ☽ ∥ ♂
15:43 ☽ # ♄
20:28 ☽ ☌ ♅
21:29 ☽ □ ♇

13 01:06 ☽ □ ☿
03:10 ☽ ✱ ♀
12:13 ☉ △ ♄
17:51 ☽ # ♆
14 01:56 ☽ ☍ ♄
03:09 ☽ ✱ ☉
22:43 ☽ ✱ ♆

15 09:53 ☽ [illegible] ♅
16 00:01 ☽ ✱ ☿
11:38 ☽ # ♇
12:10 ☽ △ ♂
15:38 ☽ ∥ ♃
17:03 ☽ ∥ ♀

17 02:03 ♀ ∥ ♃
07:59 ☽ ☌ ♃
11:42 ☽ □ ♆
22:10 ☽ ✱ ♅
18 00:13 ☽ ☌ ♀
13:19 ♀ # ♇

19 03:12 ☽ □ ♂
03:16 ☽ △ ♄
19:22 ☿ ∥ ☉
23:40 ☽ △ ♆

20 09:57 ☽ □ ♅
10:16 ☽ ☍ ♇
16:06 ☿ □ ♄
16:25 ☽ ∥ ♃
22:17 ☽ # ♇
23:09 ☉ ♋

21 00:57 ♀ ✱ ♅
04:11 ☽ ∥ ♀
10:15 ☿ ✱ ♂
13:59 ☽ □ ♄
16:18 ☽ ✱ ♂
16:49 ☽ ☌ ☿
22 08:20 ☽ ✱ ♃
18:50 ☽ ✱ ♀
19:38 ☽ △ ♅

23 00:31 ♂ ∥ ♅
12:05 ☽ # ♆
22:27 ☽ ✱ ♄
24 03:54 ☉ △ ♆
08:15 ♅ □ ♇
12:26 ☽ # ♄
17:01 ☽ □ ♃
17:16 ☽ ☍ ♆
18:19 ☽ ✱ ☉

25 01:32 ☽ □ ♀
02:55 ☽ △ ♇
07:25 ☽ ∥ ♅
07:59 ♃ □ ♆
08:02 ♄ SD
10:02 ☽ ∥ ♂

26 02:24 ☿ ♌
05:53 ☽ # ♂
09:19 ☽ # ♅
10:54 ☽ ☌ ♂
18:41 ☽ ✱ ☿
23:11 ☽ △ ♃

27 03:38 ☽ ∥ ♄
06:17 ☽ △ ♀
07:42 ☽ □ ♇
07:55 ☽ ☍ ♅
15:08 ♀ SD

28 02:19 ☽ ∥ ♆
08:23 ☽ ☌ ♄
11:59 ☿ ∥ ♃
29 01:34 ☽ △ ♆
02:14 ☽ □ ☿
10:00 ☽ △ ☉
10:20 ☽ ✱ ♇
15:00 ☉ ☍ ♇
15:38 ☿ ✱ ♃
18:19 ☽ # ♀
20:10 ☉ □ ♅

30 08:34 ☽ ∥ ♇
08:48 ☽ # ☿
10:22 ☿ # ♇
17:52 ☽ # ♃
19:47 ☽ ✱ ♂

⚷ Chiron

01 Dec. 02 S 46

Day	Position
02	09♓42
05	09 44
08	09 45
11	09 45
14	09 45R
17	09 45
20	09 43
23	09 42
26	09 40
[illegible]	09 37

12 03:30 09♓45 R

July 2012

Day	S. T. h m s	☉ ° ' "	☽ ° ' "	☿ ° '	♀ ° '	♂ ° '	♃ ° '	♄ ° '	♅ ° '	♆ ° '	♇ ° '	☊ True ° '
01 Su	18 37 48	09♋34 19	01♐10 49	05♌19	07♊42	28♍41	04♊16	22♎47	08♈29	02♓R58	08♑R14	04♐R44
02 Mo	18 41 45	10 31 31	15 56 21	06 15	07 51	29 12	04 28	22 48	08 29	02 57	08 12	04 43
03 Tu	18 45 42	11 28 42	00♑41 51	07 07	08 01	29 44	04 41	22 49	08 30	02 57	08 10	04 40
04 We	18 49 38	12 25 53	15 20 11	07 56	08 14	00♎15	04 54	22 50	08 30	02 56	08 09	04 36
05 Th	18 53 35	13 23 04	29 44 22	08 42	08 29	00 47	05 06	22 50	08 31	02 55	08 07	04 31
06 Fr	18 57 31	14 20 15	13♒48 40	09 24	08 46	01 19	05 19	22 51	08 31	02 54	08 06	04 25
07 Sa	19 01 28	15 17 26	27 29 16	10 02	09 04	01 51	05 31	22 53	08 31	02 53	08 04	04 19
08 Su	19 05 24	16 14 37	10♓44 41	10 36	09 25	02 23	05 44	22 54	08 32	02 52	08 03	04 13
09 Mo	19 09 21	17 11 49	23 35 43	11 06	09 47	02 55	05 56	22 55	08 32	02 51	08 01	04 09
10 Tu	19 13 17	18 09 01	06♈04 57	11 31	10 11	03 28	06 08	22 56	08 32	02 50	08 00	04 08
11 We	19 17 14	19 06 13	18 16 20	11 53	10 37	04 00	06 20	22 58	08 32	02 49	07 58	04D 07
12 Th	19 21 11	20 03 26	00♉14 37	12 10	11 04	04 33	06 32	23 00	08 32	02 48	07 57	04 09
13 Fr	19 25 07	21 00 40	12 04 59	12 22	11 33	05 06	06 44	23 01	08 32	02 47	07 55	04 11
14 Sa	19 29 04	21 57 54	23 52 32	12 30	12 03	05 39	06 56	23 03	08R 32	02 46	07 54	04 14
15 Su	19 33 00	22 55 08	05♊42 06	12 33	12 34	06 13	07 08	23 05	08 32	02 44	07 53	04 15
16 Mo	19 36 57	23 52 23	17 38 00	12R 31	13 07	06 46	07 20	23 07	08 32	02 43	07 51	04R 15
17 Tu	19 40 53	24 49 39	29 43 46	12 24	13 41	07 20	07 32	23 09	08 32	02 42	07 50	04 12
18 We	19 44 50	25 46 55	12♋02 03	12 12	14 17	07 54	07 43	23 11	08 32	02 41	07 48	04 07
19 Th	19 48 46	26 44 12	24 34 32	11 56	14 54	08 28	07 55	23 13	08 32	02 40	07 47	03 59
20 Fr	19 52 43	27 41 29	07♌21 52	11 35	15 31	09 02	08 06	23 16	08 31	02 38	07 45	03 50
21 Sa	19 56 40	28 38 47	20 23 47	11 10	16 10	09 36	08 17	23 18	08 31	02 37	07 44	03 40
22 Su	20 00 36	29 36 04	03♍39 21	10 40	16 51	10 10	08 29	23 21	08 31	02 36	07 43	03 30
23 Mo	20 04 33	00♌33 23	17 07 13	10 07	17 32	10 45	08 40	23 23	08 30	02 35	07 41	03 22
24 Tu	20 08 29	01 30 41	00♎46 03	09 31	18 14	11 19	08 51	23 26	08 30	02 33	07 40	03 16
25 We	20 12 26	02 28 00	14 34 41	08 52	18 57	11 54	09 02	23 29	08 29	02 32	07 39	03 12
26 Th	20 16 22	03 25 19	28 32 11	08 11	19 41	12 29	09 13	23 32	08 29	02 30	07 37	03 10
27 Fr	20 20 19	04 22 39	12♏37 48	07 28	20 25	13 04	09 24	23 35	08 28	02 29	07 36	03 10
28 Sa	20 24 15	05 19 59	26 50 35	06 44	21 11	13 39	09 34	23 38	08 27	02 28	07 35	03 10
29 Su	20 28 12	06 17 20	11♐09 02	06 00	21 58	14 14	09 45	23 41	08 27	02 26	07 33	03 09
30 Mo	20 32 09	07 14 41	25 30 42	05 17	22 45	14 50	09 55	23 44	08 26	02 25	07 32	03 07
31 Tu	20 36 05	08 12 03	09♑51 58	04 36	23 33	15 25	10 06	23 48	08 25	02 23	07 31	03 02

Data for	07-01-2012
Julian Day	2456109.50
Ayanamsa	24 02 09
SVP	05 ♓ 05 23
☽ ☊ Mean	03 ♐ 20 R

● ◐ PHASES ○ ◑

Day	Time	Phase	Position
03	18:52	○	12♑14
11	01:47	◑	19♈11
19	04:25	●	26♋55
26	08:57	◐	03♏47

LAST ASPECT Day	h m	☽ INGRESS Day	h m	Sign
02	22:21	02	22:52	♑
04	12:26	05	00:26	♒
06	15:49	07	04:29	♓
08	11:00	09	12:14	♈
11	09:23	11	23:30	♉
13	19:46	14	12:27	♊
16	10:57	17	00:32	♋
19	04:25	19	10:14	♌
21	05:19	21	17:25	♍
23	00:45	23	22:39	♎
25	15:23	26	02:30	♏
26	15:39	28	05:18	♐
29	21:02	30	07:30	♑

DECLINATION

Day	☉	☽	☿	♀	♂	♃	♄	♅	♆	♇
01 Su	23N05	20S43	19N06	17N31	00N55	20N14	06S30	02N42	10S58	19S20
02 Mo	23 01	21 42	18 40	17 28	00 41	20 17	06 30	02 42	10 59	19 20
03 Tu	22 56	21 10	18 16	17 26	00 28	20 19	06 30	02 42	10 59	19 20
04 We	22 51	19 12	17 51	17 24	00 14	20 21	06 31	02 43	10 59	19 21
05 Th	22 46	16 03	17 27	17 23	00 00	20 24	06 32	02 43	11 00	19 21
06 Fr	22 40	12 02	17 03	17 23	00S13	20 26	06 32	02 43	11 00	19 21
07 Sa	22 34	07 31	16 39	17 23	00 27	20 28	06 33	02 43	11 01	19 21
08 Su	22 27	02 47	16 17	17 24	00 41	20 30	06 34	02 43	11 01	19 21
09 Mo	22 20	01N56	15 55	17 26	00 55	20 32	06 34	02 43	11 01	19 22
10 Tu	22 12	06 27	15 34	17 27	01 09	20 34	06 35	02 43	11 02	19 22
11 We	22 05	10 36	15 14	17 30	01 23	20 36	06 36	02 43	11 02	19 22
12 Th	21 56	14 16	14 55	17 33	01 37	20 38	06 37	02 43	11 03	19 22
13 Fr	21 48	17 19	14 38	17 36	01 51	20 40	06 38	02 43	11 03	19 22
14 Sa	21 39	19 39	14 22	17 39	02 05	20 42	06 38	02 43	11 03	19 23
15 Su	21 29	21 08	14 07	17 43	02 19	20 44	06 39	02 43	11 04	19 23
16 Mo	21 20	21 40	13 54	17 47	02 33	20 46	06 40	02 43	11 04	19 23
17 Tu	21 10	21 13	13 43	17 52	02 47	20 48	06 41	02 43	11 05	19 23
18 We	20 59	19 45	13 33	17 56	03 02	20 50	06 42	02 43	11 05	19 24
19 Th	20 48	17 19	13 26	18 01	03 16	20 52	06 43	02 43	11 06	19 24
20 Fr	20 37	14 01	13 20	18 06	03 30	20 54	06 45	02 42	11 06	19 24
21 Sa	20 26	09 59	13 16	18 12	03 45	20 56	06 46	02 42	11 07	19 24
22 Su	20 14	05 25	13 15	18 17	03 59	20 57	06 47	02 42	11 07	19 24
23 Mo	20 02	00 32	13 15	18 22	04 13	20 59	06 48	02 42	11 08	19 25
24 Tu	19 49	04S27	13 17	18 28	04 28	21 01	06 49	02 42	11 08	19 25
25 We	19 36	09 16	13 22	18 33	04 42	21 02	06 51	02 41	11 09	19 25
26 Th	19 23	13 40	13 28	18 39	04 57	21 04	06 52	02 41	11 09	19 25
27 Fr	19 10	17 20	13 36	18 44	05 11	21 06	06 53	02 41	11 10	19 26
28 Sa	18 56	20 00	13 46	18 49	05 26	21 07	06 55	02 41	11 10	19 26
29 Su	18 42	21 24	13 58	18 55	05 40	21 09	06 56	02 40	11 11	19 26
30 Mo	18 28	21 26	14 11	19 00	05 55	21 10	06 57	02 40	11 11	19 26
31 Tu	18 13	20 03	14 25	19 05	06 10	21 12	06 59	02 40	11 12	19 27

⚷ Chiron

01 Dec. 02 S 41

Day	Position
02	09♓34R
05	09 31
08	09 27
11	09 22
14	09 17
17	09 12
20	09 06
23	09 00
26	08 54
29	08 47

ASPECTARIAN

01 02:55 ☽ □ ♆
05:06 ☽ ☍ ♃
07:12 ☽ △ ☿
10:43 ☽ ☍ ♀
11:53 ☽ △ ♅
02 11:09 ☽ ⚹ ♄
22:21 ☽ □ ♂
03 03:40 ☽ ⚹ ♆
12:02 ☽ # ♃
12:12 ☽ ☌ ♇
12:33 ♂ ♎
12:45 ☽ □ ♅
22:35 ☽ ∥ ♇
04 12:26 ☽ □ ♄
12:52 ☽ # ☿
13:26 ☿ ⚹ ♀
14:37 ☽ # ♀
17:53 ☿ △ ♅
05 01:49 ☽ △ ♂
02:35 ♀ ⚹ ♅
03:14 ☿ ∥ ♀
09:13 ☽ △ ♃
14:54 ☽ ⚹ ♅
15:08 ☽ △ ♀
16:01 ☽ ☍ ☿
06 05:40 ☽ ∥ ♆
15:49 ☽ △ ♄
07 04:57 ☽ ∥ ♄
09:40 ☽ ☌ ♆
14:41 ☽ □ ♃
19:04 ☽ ⚹ ♇
21:29 ☽ □ ♀
08 00:20 ☽ # ♅
10:07 ☽ ∥ ♂
11:00 ☽ △ ☉
18:27 ☽ # ♂
09 04:04 ☽ ∥ ♅
18:41 ☽ ☍ ♂
10 00:06 ☽ ⚹ ♃
00:46 ☽ # ♄
03:44 ☽ □ ♇
04:49 ☽ ☌ ♅
08:18 ☽ ⚹ ♀
10:59 ☽ △ ☿
11 02:43 ☽ # ♆
09:23 ☽ ☍ ♄
12 04:25 ☽ ∥ ☿
05:09 ☽ ⚹ ♆
15:34 ☽ △ ♇
13 00:35 ☽ □ ☿
02:38 ☽ ∥ ♀
09:50 ♅ SR
19:46 ☽ ⚹ ☉
20:53 ☽ # ♇
14 16:07 ☽ ∥ ♃
18:01 ☽ □ ♆
22:42 ☿ ⚹ ♀
15 01:05 ☽ △ ♂
02:17 ☿ SR
02:57 ☽ ☌ ♃
04:15 ☉ □ ♄
05:44 ☽ ⚹ ♅
08:47 ☽ ∥ ☉
13:47 ☽ ⚹ ☿
14:31 ☽ ☌ ♀
16 10:57 ☽ △ ♄
16:29 ♂ # ♅
17 01:34 ☽ ∥ ☉
05:50 ☽ △ ♆
08:24 ☽ ∥ ♃
12:39 ♂ △ ♃
15:35 ☽ □ ♂
15:49 ☽ ☍ ♇
17:13 ☽ □ ♅
20:23 ♂ □ ♇
18 04:09 ☽ # ♇
17:16 ☉ ∥ ♃
18:05 ☽ ∥ ♀
21:26 ☽ □ ♄
19 02:55 ♂ ☍ ♅
20 01:23 ☽ ⚹ ♃
02:09 ☽ △ ♅
03:14 ☽ ⚹ ♂
04:23 ☽ ∥ ☿
07:35 ☽ ☌ ☿
15:52 ☽ ⚹ ♀
17:38 ☽ # ♆
21 05:19 ☽ ⚹ ♄
17:04 ☽ # ♄
22:06 ☽ ☍ ♆
22 04:04 ♃ ⚹ ♅
06:49 ☽ # ♂
07:15 ☽ △ ♇
08:45 ☽ □ ♃
10:01 ☉ ♌
10:55 ☿ ⚹ ♂
13:27 ☽ ∥ ♅
23 00:45 ☽ □ ♀
15:31 ☽ # ♅
24 00:04 ☽ ∥ ♂
01:24 ☽ ⚹ ☉
11:40 ☽ ∥ ♄
12:00 ☽ □ ♇
13:27 ☽ ☍ ♅
14:16 ☽ △ ♃
14:33 ☽ ⚹ ☿
19:10 ☽ ☌ ♂
19:15 ☿ ⚹ ♃
25 07:57 ☽ △ ♀
09:52 ☽ ∥ ♆
13:27 ☿ △ ♅
15:23 ☽ ☌ ♄
20:16 ☉ # ♇
22:51 ☽ # ☿
26 06:46 ☽ △ ♆
15:28 ☽ ⚹ ♇
15:39 ☽ □ ☿
27 11:46 ☽ # ♀
14:04 ☽ # ☉
17:56 ☽ ∥ ♇
28 08:09 ♀ ∥ ☉
09:25 ☽ □ ♆
15:17 ☽ △ ☉
15:48 ☽ △ ☿
17:03 ☽ # ♃
19:28 ☽ △ ♅
19:58 ☿ ☌ ☉
21:38 ☽ ☍ ♃
29 05:23 ☽ ⚹ ♂
19:07 ☽ ☍ ♀
21:02 ☽ ⚹ ♄
30 06:34 ☽ # ♃
11:31 ☽ ⚹ ♆
20:04 ☽ ☌ ♇
21:35 ☽ □ ♅
31 05:22 ☉ △ ♅
06:25 ☽ ∥ ♇
07:39 ♀ △ ♄
09:33 ☽ # ♀
09:43 ☽ □ ♂
19:12 ☽ # ☉
23:31 ☽ □ ♄

August 2012

Day	S. T.	☉	☽	☿	♀	♂	♃	♄	♅	♆	♇	☊ True
	h m s	° ' "	° ' "	° '	° '	° '	° '	° '	° '	° '	° '	° '
01 We	20 40 02	09♌ 09 25	24♑ 08 13	03♌R57	24♊ 22	16♎ 01	10♊ 16	23♎ 51	08♈R24	02♓R22	07♑R30	02♐R55
02 Th	20 43 58	10 06 48	08♒ 14 14	03 21	25 12	16 37	10 26	23 55	08 23	02 20	07 28	02 46
03 Fr	20 47 55	11 04 12	22 05 02	02 49	26 02	17 13	10 36	23 58	08 22	02 19	07 27	02 35
04 Sa	20 51 51	12 01 37	05♓ 36 43	02 21	26 53	17 49	10 46	24 02	08 21	02 17	07 26	02 25
05 Su	20 55 48	12 59 03	18 46 59	01 59	27 44	18 25	10 56	24 05	08 20	02 16	07 25	02 14
06 Mo	20 59 44	13 56 30	01♈ 35 32	01 42	28 37	19 01	11 06	24 09	08 19	02 14	07 24	02 06
07 Tu	21 03 41	14 53 58	14 03 54	01 31	29 30	19 38	11 16	24 13	08 18	02 13	07 22	02 00
08 We	21 07 38	15 51 28	26 15 10	01 26	00♋ 23	20 14	11 25	24 17	08 17	02 11	07 21	01 57
09 Th	21 11 34	16 48 59	08♉ 13 35	01 D 28	01 17	20 51	11 35	24 21	08 16	02 10	07 20	01D 56
10 Fr	21 15 31	17 46 31	20 04 10	01 36	02 12	21 27	11 44	24 25	08 15	02 08	07 19	01 57
11 Sa	21 19 27	18 44 05	01♊ 52 21	01 52	03 07	22 04	11 53	24 30	08 13	02 07	07 18	01 58
12 Su	21 23 24	19 41 40	13 43 35	02 14	04 02	22 41	12 02	24 34	08 12	02 05	07 17	01R 58
13 Mo	21 27 20	20 39 16	25 43 01	02 44	04 58	23 18	12 11	24 38	08 11	02 03	07 16	01 57
14 Tu	21 31 17	21 36 54	07♋ 55 09	03 20	05 55	23 55	12 20	24 43	08 09	02 02	07 15	01 53
15 We	21 35 13	22 34 33	20 23 32	04 03	06 52	24 33	12 29	24 47	08 08	02 00	07 14	01 45
16 Th	21 39 10	23 32 14	03♌ 10 23	04 54	07 49	25 10	12 37	24 52	08 06	01 59	07 13	01 36
17 Fr	21 43 07	24 29 56	16 16 26	05 50	08 47	25 48	12 46	24 57	08 05	01 57	07 12	01 24
18 Sa	21 47 03	25 27 39	29 40 44	06 54	09 46	26 25	12 54	25 01	08 03	01 55	07 11	01 11
19 Su	21 50 60	26 25 24	13♍ 20 55	08 03	10 44	27 03	13 02	25 06	08 02	01 54	07 10	00 58
20 Mo	21 54 56	27 23 10	27 13 31	09 19	11 44	27 41	13 10	25 11	08 00	01 52	07 10	00 47
21 Tu	21 58 53	28 20 57	11♎ 14 41	10 40	12 43	28 19	13 18	25 16	07 58	01 50	07 09	00 39
22 We	22 02 49	29 18 45	25 20 48	12 06	13 43	28 57	13 26	25 21	07 57	01 49	07 08	00 33
23 Th	22 06 46	00♍ 16 35	09♏ 28 59	13 37	14 43	29 35	13 33	25 26	07 55	01 47	07 07	00 29
24 Fr	22 10 42	01 14 25	23 37 15	15 13	15 44	00♏ 14	13 41	25 31	07 53	01 45	07 07	00 28
25 Sa	22 14 39	02 12 17	07♐ 44 18	16 53	16 45	00 52	13 48	25 36	07 51	01 44	07 06	00 27
26 Su	22 18 36	03 10 10	21 49 15	18 36	17 46	01 31	13 55	25 42	07 49	01 42	07 05	00 26
27 Mo	22 22 32	04 08 04	05♑ 51 07	20 23	18 48	02 09	14 02	25 47	07 48	01 41	07 04	00 23
28 Tu	22 26 29	05 06 00	19 48 27	22 12	19 50	02 48	14 09	25 52	07 46	01 39	07 04	00 18
29 We	22 30 25	06 03 57	03♒ 39 06	24 04	20 52	03 27	14 16	25 58	07 44	01 37	07 03	00 11
30 Th	22 34 22	07 01 55	17 20 22	25 57	21 54	04 06	14 23	26 03	07 42	01 36	07 03	00 01
31 Fr	22 38 18	07 59 55	00♓ 49 16	27 52	22 57	04 45	14 29	26 09	07 40	01 34	07 02	29♏ 49

Data for	08-01-2012
Julian Day	2456140.50
Ayanamsa	24 02 14
SVP	05 ♓ 05 19
☽ ☊ Mean	01 ♐ 42 R

LAST ASPECT ☽ Day	h m	INGRESS Day	h m	
31	23:31	01	09:56	♒
03	07:24	03	13:58	♓
05	17:56	05	20:59	♈
07	20:04	08	07:28	♉
09	18:56	10	20:12	♊
12	21:51	13	08:29	♋
15	08:22	15	18:06	♌
17	17:56	18	00:34	♍
18	23:27	20	04:46	♎
22	07:14	22	07:54	♏
23	09:34	24	10:50	♐
26	06:40	26	13:59	♑
28	10:34	28	17:39	♒
30	17:49	30	22:32	♓

● ◐ PHASES ○ ◑

02	03:27	○	10♒15
09	18:55	◑	17♉34
17	15:55	●	25♌08
24	13:53	◐	01♐48
31	13:58	○	08♓34

DECLINATION

Day	☉	☽	☿	♀	♂	♃	♄	♅	♆	♇
01 We	17N58	17S24	14N40	19N10	06S24	21N13	07S00	02N39	11S12	19S27
02 Th	17 43	13 46	14 55	19 15	06 39	21 15	07 02	02 39	11 13	19 27
03 Fr	17 27	09 27	15 12	19 20	06 53	21 16	07 03	02 39	11 13	19 27
04 Sa	17 11	04 46	15 28	19 25	07 08	21 17	07 05	02 38	11 14	19 27
05 Su	16 55	00N01	15 45	19 29	07 22	21 19	07 06	02 38	11 14	19 28
06 Mo	16 38	04 40	16 02	19 33	07 37	21 20	07 08	02 37	11 15	19 28
07 Tu	16 22	09 00	16 18	19 37	07 52	21 21	07 10	02 37	11 16	19 28
08 We	16 05	12 53	16 34	19 41	08 06	21 23	07 11	02 36	11 16	19 28
09 Th	15 48	16 11	16 49	19 44	08 21	21 24	07 13	02 36	11 17	19 29
10 Fr	15 30	18 46	17 03	19 47	08 35	21 25	07 15	02 35	11 17	19 29
11 Sa	15 12	20 32	17 16	19 50	08 50	21 26	07 17	02 35	11 18	19 29
12 Su	14 55	21 25	17 27	19 53	09 04	21 27	07 18	02 34	11 19	19 29
13 Mo	14 36	21 19	17 37	19 55	09 19	21 29	07 20	02 34	11 19	19 30
14 Tu	14 18	20 14	17 45	19 57	09 33	21 30	07 22	02 33	11 20	19 30
15 We	13 59	18 09	17 52	19 58	09 48	21 31	07 24	02 32	11 20	19 30
16 Th	13 40	15 09	17 56	19 59	10 02	21 32	07 26	02 32	11 21	19 30
17 Fr	13 21	11 20	17 58	20 00	10 17	21 33	07 27	02 31	11 21	19 31
18 Sa	13 02	06 54	17 57	20 00	10 31	21 34	07 29	02 31	11 22	19 31
19 Su	12 42	02 02	17 54	20 00	10 46	21 35	07 31	02 30	11 23	19 31
20 Mo	12 23	03S01	17 49	20 00	11 00	21 36	07 33	02 29	11 23	19 31
21 Tu	12 03	07 57	17 40	19 59	11 14	21 37	07 35	02 29	11 24	19 32
22 We	11 43	12 30	17 29	19 58	11 28	21 38	07 37	02 28	11 24	19 32
23 Th	11 22	16 22	17 15	19 56	11 43	21 39	07 39	02 27	11 25	19 32
24 Fr	11 02	19 15	16 58	19 54	11 57	21 39	07 41	02 27	11 26	19 32
25 Sa	10 41	20 58	16 38	19 51	12 11	21 40	07 43	02 26	11 26	19 32
26 Su	10 21	21 20	16 15	19 48	12 25	21 41	07 45	02 25	11 27	19 33
27 Mo	10 00	20 23	15 50	19 44	12 39	21 42	07 48	02 24	11 27	19 33
28 Tu	09 38	18 11	15 22	19 40	12 53	21 43	07 50	02 24	11 28	19 33
29 We	09 17	14 58	14 51	19 36	13 07	21 43	07 52	02 23	11 29	19 33
30 Th	08 56	10 58	14 18	19 31	13 21	21 44	07 54	02 22	11 29	19 34
31 Fr	08 34	06 30	13 43	19 25	13 35	21 45	07 56	02 21	11 30	19 34

ASPECTARIAN

01 15:58 ☽ ☍ ☿
17:22 ☽ ⋕ ☿
02 00:16 ☽ ✶ ♅
03:50 ☽ △ ♃
09:56 ☉ ✶ ♃
14:30 ☽ ∥ ♆
15:07 ☽ △ ♂
03 03:19 ☽ △ ♄
07:24 ☽ △ ♀
12:21 ☽ ∥ ♄
12:37 ☽ ∥ ♂
18:04 ☽ ☌ ♆
18:30 ♂ ∥ ♄
04 03:16 ☽ ✶ ♇
09:27 ☽ □ ♃
10:42 ☽ ⋕ ♅
15:53 ♀ ⋕ ♇
05 13:19 ☽ ∥ ♅
17:56 ☽ □ ♀
06 00:12 ☽ △ ☿
11:04 ☽ □ ☿
12:51 ☽ ☌ ♅
13:27 ☽ ⋕ ♄
17:02 ☽ ⋕ ♂
18:29 ☽ ✶ ♃
07 01:46 ☽ △ ☉
02:41 ☿ ∥ ☉
11:27 ☽ ☍ ♂
13:34 ☽ ⋕ ♆
13:43 ♀ ♋
20:04 ☽ ☍ ♄
08 05:41 ☿ SD
08:54 ☽ ✶ ♀
10:20 ☽ □ ☿
11:50 ☽ ✶ ♆
21:13 ☽ ∥ ☉
22:12 ☽ △ ♆
09 05:50 ☽ ∥ ☿
22:36 ♀ △ ♆
10 08:26 ☽ ⋕ ♆
12:48 ☽ ∥ ♀
23:58 ☽ ✶ ☿
11 00:29 ☽ □ ♆
12:51 ☽ ✶ ♅
20:33 ☽ ☌ ♃
12 03:28 ☽ ⋕ ♃
13:02 ☽ ✶ ☉
15:13 ☽ ∥ ♃
18:56 ☽ △ ♂
21:51 ☽ △ ♄
13 12:30 ☽ △ ♆
19:46 ☽ ☌ ♀
22:42 ☽ ☍ ♇
14 00:27 ☽ □ ♅
03:59 ☽ ∥ ♀
09:45 ☽ ⋕ ♆
15 02:33 ☽ ∥ ☿
08:16 ☽ □ ♂
08:22 ☽ □ ♄
09:09 ♀ ☍ ♇
10:36 ♂ ☌ ♄
16 03:25 ☽ ☌ ☿
06:48 ♀ □ ♅
09:05 ☽ △ ♅
10:44 ☽ ∥ ☉
17:33 ☽ [illegible]
23:51 ☽ ⋕ ♆
17 05:39 ☽ ⋕ ♂
12:04 ☉ ✶ ♄
15:40 ☽ ✶ ♄
17:56 ☽ ✶ ♂
20:55 ☽ ⋕ ♄
18 03:58 ☽ ☍ ♆
13:13 ☽ △ ♇
19:06 ☽ ✶ ♀
21:42 ☽ ∥ ♅
23:27 ☽ □ ♃
23:29 ☿ △ ♅
19 21:29 ☽ ⋕ ♅
20 17:01 ☽ □ ♇
18:25 ☽ ☍ ♅
21:46 ☉ ✶ ♂
22:08 ☽ ∥ ♄
22:54 ☽ ✶ ☿
21 02:42 ☽ □ ♀
03:32 ☽ △ ♃
16:54 ♂ ∥ ♆
17:55 ☽ ∥ ♆
17:58 ☽ ∥ ♂
19:54 ☽ ⋕ ☉
22 00:00 ☽ ☌ ♄
06:25 ☽ ☌ ♂
07:14 ☽ ✶ ☉
09:53 ☉ ⋕ ♂
10:58 ☽ △ ♆
17:07 ☉ ♍
20:00 ☽ ✶ ♇
[illegible]
22:56 ☿ ✶ ♃
23 05:57 ☽ ⋕ ☿
07:54 ☽ □ ☿
09:34 ☽ △ ♀
15:24 ♂ ♏
24 02:58 ☽ ∥ ♆
06:57 ☽ ⋕ ♀
12:32 ☉ ☍ ♆
13:48 ☽ □ ♆
25 00:12 ☽ △ ♅
10:25 ☽ ☍ ♃
17:44 ☽ △ ☿
26 06:40 ☽ ✶ ♄
[illegible]
16:52 ☽ ✶ ♆
17:22 ☽ ✶ ♂
20:50 ☽ △ ☉
27 02:06 ☽ ☌ ♆
03:19 ☽ □ ♅
08:47 ☽ ⋕ ♀
10:42 ☽ ∥ ♇
28 00:02 ☽ ☍ ♀
10:34 ☽ □ ♄
23:38 ☽ □ ♂
29 00:51 ☽ ⋕ ☿
07:06 ☽ ✶ ♅
10:56 ☽ ∥ ♂
11:56 ♀ ⋕ ♇
18:44 ☽ △ ♃
21:06 ☽ ∥ ♆
30 00:20 ☉ △ ♆
01:25 ☿ ✶ ♄
12:09 ☽ ⋕ ☉
15:35 ☽ △ ♄
16:31 ☽ ∥ ♄
17:49 ☽ ☍ ☿
31 01:20 ☽ ☌ ♆
04:05 ☿ ⋕ ♂
[illegible]
11:12 ☽ ✶ ♇
21:12 ☽ ⋕ ♅

⚷ Chiron

01	Dec.	02 S 56
01	08♓40R	
04	08 33	
07	08 25	
10	08 17	
13	08 09	
16	08 01	
19	07 52	
22	07 44	
25	07 35	
28	07 26	
31	07 18	

September 2012

Day	S. T.	☉	☽	☿	♀	♂	♃	♄	♅	♆	♇	☊ True
	h m s	° ' "	° ' "	° '	° '	° '	° '	° '	° '	° '	° '	° '
01 Sa	22 42 15	08♍ 57 56	14♓ 03 08	29♌ 48	24♋ 00	05♏ 24	14♊ 35	26♎ 15	07♈R38	01♓R32	07♑R02	29♏R37
02 Su	22 46 11	09 55 59	27 00 05	01♍ 44	25 04	06 03	14 41	26 20	07 36	01 31	07 01	29 25
03 Mo	22 50 08	10 54 04	09♈ 39 29	03 41	26 07	06 42	14 47	26 26	07 34	01 29	07 01	29 15
04 Tu	22 54 04	11 52 10	22 02 07	05 38	27 11	07 22	14 53	26 32	07 32	01 28	07 00	29 08
05 We	22 58 01	12 50 18	04♉ 10 09	07 35	28 16	08 01	14 59	26 38	07 29	01 26	07 00	29 03
06 Th	23 01 58	13 48 29	16 06 55	09 32	29 20	08 41	15 04	26 44	07 27	01 24	06 59	29 01
07 Fr	23 05 54	14 46 41	27 56 45	11 28	00♌ 25	09 21	15 09	26 50	07 25	01 23	06 59	29D 01
08 Sa	23 09 51	15 44 55	09♊ 44 38	13 24	01 30	10 00	15 14	26 56	07 23	01 21	06 59	29 02
09 Su	23 13 47	16 43 12	21 35 59	15 18	02 35	10 40	15 19	27 02	07 21	01 20	06 58	29 03
10 Mo	23 17 44	17 41 30	03♋ 36 16	17 12	03 41	11 20	15 24	27 08	07 19	01 18	06 58	29R 02
11 Tu	23 21 40	18 39 50	15 50 41	19 05	04 46	12 00	15 29	27 14	07 16	01 16	06 58	28 59
12 We	23 25 37	19 38 13	28 23 42	20 58	05 52	12 40	15 33	27 20	07 14	01 15	06 58	28 53
13 Th	23 29 33	20 36 38	11♌ 18 36	22 49	06 58	13 21	15 37	27 27	07 12	01 13	06 58	28 45
14 Fr	23 33 30	21 35 04	24 36 59	24 39	08 05	14 01	15 41	27 33	07 10	01 12	06 57	28 35
15 Sa	23 37 27	22 33 33	08♍ 18 22	26 28	09 11	14 42	15 45	27 39	07 07	01 10	06 57	28 24
16 Su	23 41 23	23 32 03	22 20 03	28 16	10 18	15 22	15 49	27 46	07 05	01 09	06 57	28 13
17 Mo	23 45 20	24 30 36	06♎ 37 23	00 03	11 25	16 03	15 53	27 52	07 03	01 07	06 57	28 04
18 Tu	23 49 16	25 29 10	21 04 28	01♎ 49	12 32	16 44	15 56	27 59	07 00	01 06	06D 57	27 56
19 We	23 53 13	26 27 46	05♏ 35 06	03 34	13 39	17 24	15 59	28 05	06 58	01 04	06 57	27 51
20 Th	23 57 09	27 26 24	20 03 44	05 17	14 47	18 05	16 02	28 12	06 55	01 03	06 57	27 48
21 Fr	00 01 06	28 25 03	04♐ 26 07	07 00	15 55	18 46	16 05	28 18	06 53	01 02	06 57	27 46
22 Sa	00 05 02	29 23 44	18 39 31	08 42	17 02	19 27	16 07	28 25	06 51	01 00	06 57	27 46
23 Su	00 08 59	00♎ 22 27	02♑ 42 28	10 23	18 10	20 09	16 10	28 32	06 48	00 59	06 57	27 46
24 Mo	00 12 56	01 21 12	16 34 24	12 03	19 19	20 50	16 12	28 38	06 46	00 57	06 58	27 45
25 Tu	00 16 52	02 19 58	00♒ 15 05	13 41	20 27	21 31	16 14	28 45	06 44	00 56	06 58	27 42
26 We	00 20 49	03 18 46	13 44 22	15 19	21 36	22 13	16 16	28 52	06 41	00 55	06 58	27 37
27 Th	00 24 45	04 17 35	27 01 53	16 56	22 44	22 54	16 17	28 59	06 39	00 53	06 58	27 30
28 Fr	00 28 42	05 16 27	10♓ 07 04	18 32	23 53	23 36	16 19	29 05	06 36	00 52	06 59	27 22
29 Sa	00 32 38	06 15 20	22 59 18	20 08	25 02	24 18	16 20	29 12	06 34	00 51	06 59	27 13
30 Su	00 36 35	07 14 15	05♈ 38 10	21 42	26 11	24 59	16 21	29 19	06 31	00 50	06 59	27 04

Data for	09-01-2012
Julian Day	2456171.50
Ayanamsa	24 02 17
SVP	05 ♓ 05 14
☽ ☊ Mean	00 ♐ 03 R

LAST ASPECT ☽ Day	h m	INGRESS Day	h m	
01	20:03	02	05:38	♈
04	11:07	04	15:42	♉
05	18:56	07	04:11	♊
09	11:00	09	16:50	♋
11	21:59	12	03:01	♌
14	05:14	14	09:31	♍
16	11:26	16	12:55	♎
18	11:30	18	14:46	♏
20	13:11	20	16:34	♐
22	16:46	22	19:21	♑
24	21:20	24	23:33	♒
27	03:34	27	05:25	♓
29	02:36	29	13:15	♈

● ◐ PHASES ○ ◑

08	13:16	◑	16♊17
16	02:10	●	23♍37
22	19:41	◐	00♑12
30	03:19	○	07♈22

DECLINATION

Day	☉	☽	☿	♀	♂	♃	♄	♅	♆	♇
01 Sa	08N12	01S 47	13N06	19N20	13S 49	21N45	07S 58	02N20	11S 30	19S 34
02 Su	07 51	02N54	12 28	19 13	14 02	21 46	08 01	02 20	11 31	19 34
03 Mo	07 29	07 21	11 47	19 06	14 16	21 47	08 03	02 19	11 32	19 35
04 Tu	07 07	11 24	11 06	18 59	14 30	21 47	08 05	02 18	11 32	19 35
05 We	06 44	14 54	10 23	18 51	14 43	21 48	08 07	02 17	11 33	19 35
06 Th	06 22	17 44	09 39	18 42	14 56	21 48	08 10	02 16	11 33	19 35
07 Fr	06 00	19 47	08 54	18 33	15 10	21 49	08 12	02 15	11 34	19 35
08 Sa	05 37	20 58	08 08	18 24	15 23	21 49	08 14	02 14	11 34	19 36
09 Su	05 15	21 13	07 22	18 14	15 36	21 50	08 16	02 14	11 35	19 36
10 Mo	04 52	20 30	06 35	18 04	15 49	21 50	08 19	02 13	11 36	19 36
11 Tu	04 29	18 48	05 48	17 53	16 02	21 51	08 21	02 12	11 36	19 36
12 We	04 06	16 12	05 01	17 41	16 15	21 51	08 23	02 11	11 37	19 37
13 Th	03 43	12 44	04 13	17 29	16 28	21 52	08 26	02 10	11 37	19 37
14 Fr	03 20	08 33	03 26	17 17	16 41	21 52	08 28	02 09	11 38	19 37
15 Sa	02 57	03 50	02 38	17 04	16 53	21 52	08 31	02 08	11 38	19 37
16 Su	02 34	01S 12	01 50	16 51	17 06	21 53	08 33	02 07	11 39	19 37
17 Mo	02 11	06 16	01 03	16 37	17 18	21 53	08 35	02 06	11 39	19 38
18 Tu	01 48	11 02	00 16	16 23	17 30	21 53	08 38	02 05	11 40	19 38
19 We	01 24	15 12	00S 31	16 08	17 43	21 54	08 40	02 04	11 40	19 38
20 Th	01 01	18 25	01 18	15 52	17 55	21 54	08 43	02 04	11 41	19 38
21 Fr	00 38	20 26	02 04	15 37	18 07	21 54	08 45	02 03	11 41	19 39
22 Sa	00 14	21 08	02 50	15 20	18 18	21 54	08 48	02 02	11 42	19 39
23 Su	00S 09	20 28	03 36	15 04	18 30	21 54	08 50	02 01	11 42	19 39
24 Mo	00 32	18 35	04 21	14 47	18 42	21 55	08 52	02 00	11 43	19 39
25 Tu	00 56	15 39	05 06	14 29	18 53	21 55	08 55	01 59	11 43	19 39
26 We	01 19	11 56	05 50	14 11	19 04	21 55	08 57	01 58	11 44	19 40
27 Th	01 42	07 41	06 34	13 53	19 16	21 55	09 00	01 57	11 44	19 40
28 Fr	02 06	03 09	07 17	13 34	19 27	21 55	09 02	01 56	11 45	19 40
29 Sa	02 29	01N27	08 00	13 15	19 38	21 55	09 05	01 55	11 45	19 40
30 Su	02 52	05 55	08 42	12 55	19 48	21 55	09 07	01 54	11 46	19 40

ASPECTARIAN

01 00:59 ☽ □ ♃
02:33 ☿ ♍
14:05 ☉ ♯ ♄
20:03 ☽ △ ♀
21:03 ☽ ∥ ♅
21:16 ☿ ☍ ♆
02 18:56 ☽ □ ♇
20:00 ☽ ☌ ♅
03 00:39 ☽ ∥ ☉
03:58 ☽ ♯ ♄
07:44 ♀ □ ♄
09:05 ☿ ♯ ♆
09:58 ☽ ⚹ ♃
11:02 ♂ ⚹ ♇
22:22 ☽ ∥ ☿
04 00:50 ☽ ♯ ♆
08:55 ☽ ☍ ♄
11:07 ☽ □ ♀
16:45 ☿ △ ♇
18:34 ☽ ⚹ ♆
22:26 ☽ ♯ ♂
05 05:39 ☽ △ ♇
08:06 ☿ ⚹ ♂
08:09 ☽ ☍ ♂
08:09 ☽ △ ☿
18:56 ☽ △ ☉
06 09:27 ☽ ∥ ♀
14:48 ♀ ♌
21:08 ☽ ♯ ♆
07 05:32 ☽ ⚹ ♀
06:58 ☽ □ ♆
10:15 ☉ □ ♃
19:13 ☽ ⚹ ♅
21:04 ☿ ♯ ♄
08 08:50 ☽ □ ☿
11:14 ☽ ☌ ♃
09 00:14 ☿ □ ♃
11:00 ☽ △ ♄
19:26 ☽ △ ♆
10 06:39 ☽ ☍ ♇
07:18 ☽ □ ♅
12:44 ☿ ☌ ☉
14:23 ☽ ♯ ♇
16:06 ☽ △ ♂
11 05:55 ☽ ⚹ ☉
07:23 ☽ ⚹ ☿
10:21 ☽ ∥ ♀
21:59 ☽ □ ♄
23:37 ☽ ♯ ♂
12 15:16 ☽ ☌ ♀
16:27 ☽ △ ♅
13 03:55 ☽ □ ♂
04:45 ♀ △ ♅
06:47 ☽ ♯ ♆
07:54 ☽ ⚹ ♃
14 00:28 ☽ ♯ ♄
05:14 ☽ ⚹ ♄
05:19 ☿ ∥ ☉
11:36 ☽ ☍ ♆
21:39 ☽ △ ♇
15 04:38 ☽ ∥ ☉
06:54 ☽ ∥ ☿
08:13 ☽ ∥ ♅
10:00 ♀ ♯ ♂
11:33 ☽ ⚹ ♂
12:52 ☽ □ ♃
15:23 ☿ ∥ ♅
16 02:37 ☽ ♯ ☿
04:19 ☽ ♯ ♅
05:59 ☽ ♯ ☉
11:26 ☽ ☌ ☿
23:22 ☿ ♎
17 00:34 ☽ □ ♇
00:43 ☽ ☍ ♅
04:53 ☉ ∥ ♅
08:39 ☽ ⚹ ♀
11:29 ☽ ∥ ♄
15:27 ☽ △ ♃
18 03:20 ☽ ∥ ♆
05:06 ♇ SD
11:30 ☽ ☌ ♄
16:33 ☽ △ ♆
19 02:16 ☽ ⚹ ♇
05:46 ☽ ♯ ♀
06:55 ♅ □ ♇
14:29 ☽ □ ♀
18:14 ☿ ♯ ☉
19:18 ☽ ∥ ♂
20:33 ☽ ☌ ♂
20 12:35 ☽ ∥ ♇
13:11 ☽ ⚹ ☉
18:18 ☽ □ ♆
22:23 ☿ ☍ ♅
23:07 ☿ ♯ ♅
23:19 ☿ □ ♇
21 03:45 ♀ ⚹ ♃
04:06 ☽ △ ♅
04:54 ☽ ⚹ ☿
19:41 ☽ ☍ ♃
21:01 ☽ △ ♀
22 14:49 ☉ ♎
16:46 ☽ ⚹ ♄
21:02 ☽ ⚹ ♆
23 07:02 ☽ □ ♅
07:19 ☽ ☌ ♇
12:20 ☽ ∥ ♇
15:03 ☽ □ ☿
23:01 ☽ ∥ ♂
24 07:50 ☽ ⚹ ♂
21:20 ☽ □ ♄
25 03:58 ☽ △ ☉
08:48 ☽ ♯ ♀
11:27 ☽ ⚹ ♅
26 01:16 ☽ ∥ ♆
03:14 ☽ △ ☿
04:32 ☽ △ ♃
14:09 ☿ △ ♃
15:28 ☽ ☍ ♀
16:06 ☽ □ ♂
16:57 ☽ ∥ ♄
27 03:34 ☽ △ ♄
05:14 ☽ ∥ ☿
07:02 ☽ ☌ ♆
08:49 ♀ □ ♂
14:21 ☉ ♯ ♅
18:12 ☽ ⚹ ♇
28 05:06 ☽ ∥ ☉
06:24 ☽ ♯ ♅
11:31 ☽ □ ♃
29 02:27 ☽ ∥ ♅
02:36 ☽ △ ♂
05:58 ☽ ♯ ☉
05:58 ♂ ∥ ♇
07:15 ☉ ☍ ♅
17:54 ☉ □ ♇
30 01:42 ☽ ☌ ♅
02:35 ☽ □ ♇
15:36 ☿ ∥ ♄
18:31 ☽ ♯ ♄
19:05 ☽ ♯ ☿
20:41 ☽ ⚹ ♃

⚷ Chiron

01 Dec.	03 S 26
03	07♓09R
06	07 00
09	06 52
12	06 43
15	06 35
18	06 27
21	06 19
24	06 11
27	06 04
30	05 57

October 2012

Day	S. T.	☉	☽	☿	♀	♂	♃	♄	♅	♆	♇	☊ True
	h m s	° ' "	° ' "	° '	° '	° '	° '	° '	° '	° '	° '	° '
01 Mo	00 40 31	08♎ 13 13	18♈ 03 39	23♎ 15	27♌ 21	25♏ 41	16♊ 22	29♎ 26	06♈R29	00♓R48	07♑ 00	26♏R57
02 Tu	00 44 28	09 12 12	00♉ 16 32	24 48	28 30	26 23	16 22	29 33	06 27	00 47	07 00	26 52
03 We	00 48 25	10 11 13	12 18 23	26 20	29 40	27 05	16 23	29 40	06 24	00 46	07 01	26 49
04 Th	00 52 21	11 10 17	24 11 43	27 50	00♍ 49	27 47	16 23	29 47	06 22	00 45	07 01	26D 48
05 Fr	00 56 18	12 09 23	05♊ 59 50	29 21	01 59	28 29	16R 23	29 54	06 19	00 44	07 01	26 49
06 Sa	01 00 14	13 08 31	17 46 49	00♏ 50	03 09	29 12	16 23	00♏ 01	06 17	00 43	07 02	26 51
07 Su	01 04 11	14 07 42	29 37 20	02 18	04 20	29 54	16 22	00 08	06 15	00 42	07 03	26 53
08 Mo	01 08 07	15 06 54	11♋ 36 26	03 46	05 30	00♐ 37	16 22	00 15	06 12	00 40	07 03	26 54
09 Tu	01 12 04	16 06 09	23 49 17	05 12	06 40	01 19	16 21	00 22	06 10	00 39	07 04	26R 54
10 We	01 16 00	17 05 27	06♌ 20 51	06 38	07 51	02 02	16 20	00 29	06 07	00 38	07 04	26 52
11 Th	01 19 57	18 04 47	19 15 19	08 03	09 02	02 44	16 19	00 37	06 05	00 37	07 05	26 49
12 Fr	01 23 53	19 04 08	02♍ 35 33	09 27	10 13	03 27	16 17	00 44	06 03	00 36	07 06	26 44
13 Sa	01 27 50	20 03 33	16 22 31	10 50	11 24	04 10	16 16	00 51	06 00	00 36	07 07	26 38
14 Su	01 31 47	21 02 59	00♎ 34 41	12 12	12 35	04 53	16 14	00 58	05 58	00 35	07 07	26 32
15 Mo	01 35 43	22 02 28	15 07 55	13 33	13 46	05 36	16 12	01 05	05 56	00 34	07 08	26 26
16 Tu	01 39 40	23 01 58	29 55 47	14 53	14 57	06 19	16 10	01 12	05 54	00 33	07 09	26 22
17 We	01 43 36	24 01 31	14♏ 50 25	16 12	16 09	07 02	16 07	01 20	05 51	00 32	07 10	26 19
18 Th	01 47 33	25 01 05	29 43 41	17 30	17 20	07 45	16 05	01 27	05 49	00 31	07 11	26 17
19 Fr	01 51 29	26 00 42	14♐ 28 24	18 46	18 32	08 29	16 02	01 34	05 47	00 30	07 12	26D 17
20 Sa	01 55 26	27 00 20	28 59 04	20 01	19 43	09 12	15 59	01 41	05 45	00 30	07 13	26 18
21 Su	01 59 22	28 00 00	13♑ 12 19	21 14	20 55	09 56	15 56	01 49	05 42	00 29	07 14	26 18
22 Mo	02 03 19	28 59 42	27 06 39	22 26	22 07	10 39	15 52	01 56	05 40	00 28	07 15	26 19
23 Tu	02 07 16	29 59 25	10♒ 42 04	23 36	23 19	11 23	15 49	02 03	05 38	00 28	07 16	26R 18
24 We	02 11 12	00♏ 59 10	23 59 31	24 44	24 31	12 06	15 45	02 10	05 36	00 27	07 17	26 17
25 Th	02 15 09	01 58 57	07♓ 00 26	25 49	25 43	12 50	15 41	02 18	05 34	00 27	07 18	26 15
26 Fr	02 19 05	02 58 45	19 46 28	26 53	26 56	13 34	15 37	02 25	05 32	00 26	07 19	26 12
27 Sa	02 23 02	03 58 35	02♈ 19 14	27 54	28 08	14 18	15 32	02 32	05 30	00 25	07 20	26 09
28 Su	02 26 58	04 58 27	14 40 13	28 52	29 20	15 02	15 28	02 39	05 28	00 25	07 21	26 06
29 Mo	02 30 55	05 58 21	26 50 53	29 46	00♎ 33	15 46	15 23	02 47	05 26	00 24	07 23	26 04
30 Tu	02 34 51	06 58 17	08♉ 52 43	00♐ 37	01 46	16 30	15 18	02 54	05 24	00 24	07 24	26 03
31 We	02 38 48	07 58 14	20 47 28	01 24	02 58	17 14	15 13	03 01	05 22	00 24	07 25	26D 03

Data for	10-01-2012
Julian Day	2456201.50
Ayanamsa	24 02 20
SVP	05 ♓ 05 08
☽ ☊ Mean	28 ♏ 28 R

PHASES

Day	Time	Phase	Position
08	07:33	◑	15♋26
15	12:02	●	22♎32
22	03:33	◐	29♑09
29	19:49	○	06♉48

LAST ASPECT ☽ INGRESS

Day	h m	Day	h m	Sign
01	22:33	01	23:27	♉
04	07:45	04	11:47	♊
05	21:09	07	00:46	♋
08	07:34	09	11:55	♌
10	21:40	11	19:24	♍
12	23:48	13	23:02	♎
15	12:03	16	00:07	♏
17	02:24	18	00:26	♐
19	20:27	20	01:42	♑
22	03:33	22	05:03	♒
24	01:28	24	11:01	♓
26	15:05	26	19:32	♈
28	01:33	29	06:16	♉
29	21:01	31	18:40	♊

DECLINATION

Day	☉	☽	☿	♀	♂	♃	♄	♅	♆	♇
01 Mo	03S 16	10N04	09S 23	12N35	19S 59	21N55	09S 10	01N53	11S 46	19S 41
02 Tu	03 39	13 44	10 04	12 15	20 09	21 55	09 12	01 52	11 47	19 41
03 We	04 02	16 46	10 44	11 54	20 20	21 55	09 15	01 51	11 47	19 41
04 Th	04 25	19 03	11 24	11 33	20 30	21 55	09 17	01 50	11 47	19 41
05 Fr	04 48	20 29	12 03	11 11	20 40	21 55	09 20	01 49	11 48	19 41
06 Sa	05 11	21 01	12 41	10 49	20 49	21 55	09 22	01 48	11 48	19 41
07 Su	05 34	20 37	13 18	10 27	20 59	21 55	09 25	01 47	11 49	19 42
08 Mo	05 57	19 17	13 55	10 04	21 08	21 55	09 28	01 46	11 49	19 42
09 Tu	06 20	17 02	14 31	09 42	21 18	21 55	09 30	01 46	11 49	19 42
10 We	06 43	13 58	15 06	09 18	21 27	21 55	09 33	01 45	11 50	19 42
11 Th	07 05	10 09	15 40	08 55	21 36	21 55	09 35	01 44	11 50	19 42
12 Fr	07 28	05 43	16 13	08 31	21 44	21 54	09 38	01 43	11 50	19 43
13 Sa	07 50	00 52	16 46	08 07	21 53	21 54	09 40	01 42	11 51	19 43
14 Su	08 13	04S 11	17 18	07 42	22 01	21 54	09 43	01 41	11 51	19 43
15 Mo	08 35	09 08	17 48	07 18	22 09	21 54	09 45	01 40	11 51	19 43
16 Tu	08 57	13 37	18 18	06 53	22 17	21 54	09 48	01 39	11 52	19 43
17 We	09 19	17 17	18 47	06 28	22 25	21 53	09 50	01 38	11 52	19 43
18 Th	09 41	19 47	19 14	06 02	22 33	21 53	09 53	01 37	11 52	19 44
19 Fr	10 03	20 55	19 41	05 37	22 40	21 53	09 55	01 37	11 52	19 44
20 Sa	10 24	20 36	20 06	05 11	22 47	21 52	09 58	01 36	11 53	19 44
21 Su	10 46	18 58	20 31	04 45	22 54	21 52	10 00	01 35	11 53	19 44
22 Mo	11 07	16 15	20 54	04 19	23 01	21 52	10 03	01 34	11 53	19 44
23 Tu	11 28	12 41	21 16	03 52	23 07	21 51	10 05	01 33	11 53	19 44
24 We	11 49	08 34	21 36	03 26	23 14	21 51	10 08	01 32	11 54	19 44
25 Th	12 10	04 09	21 56	02 59	23 20	21 50	10 10	01 32	11 54	19 45
26 Fr	12 30	00N23	22 13	02 32	23 26	21 50	10 13	01 31	11 54	19 45
27 Sa	12 51	04 49	22 30	02 05	23 31	21 49	10 15	01 30	11 54	19 45
28 Su	13 11	09 00	22 45	01 38	23 37	21 49	10 18	01 29	11 54	19 45
29 Mo	13 31	12 45	22 58	01 10	23 42	21 49	10 20	01 28	11 54	19 45
30 Tu	13 50	15 56	23 09	00 43	23 47	21 48	10 23	01 28	11 55	19 45
31 We	14 10	18 25	23 19	00 16	23 51	21 47	10 25	01 27	11 55	19 45

ASPECTARIAN

Day	Time	Aspect
01	10:44	☽ # ♆
	11:38	☽ ☍ ☿
	14:43	☽ ∥ ♀
	20:07	☽ △ ♀
	22:33	☽ ☍ ♄
02	01:01	☽ ⚹ ♆
	13:23	☽ △ ♇
03	00:06	♀ ⚹ ♄
	06:59	♀ ♍
	07:48	♀ # ♆
	22:28	♀ ☍ ♆
04	03:28	☿ # ♀
	07:45	☽ ☍ ♂
	08:58	☽ # ♇
	13:17	☽ □ ♆
	13:19	♃ SR
	14:31	☿ ∥ ♆
	14:57	☽ □ ♀
05	00:40	☽ ⚹ ♅
	05:42	☽ # ♂
	09:44	☿ ☌ ♄
	10:36	☿ ♏
	13:42	☽ △ ☉
	20:34	♄ ♏
	21:09	☽ ☌ ♃
	22:06	☿ △ ♆
06	13:29	☽ # ♂
07	01:03	☽ △ ♄
	02:09	☽ △ ♆
	03:21	♂ ♐
	06:09	☽ △ ☿
	10:30	☽ ⚹ ♀
	13:16	☽ □ ♅
	14:55	☽ ☍ ♇
	18:07	☽ # ♇
08	02:12	♂ □ ♆
09	05:53	☉ △ ♃
	08:03	♀ △ ♇
	10:46	♀ # ♄
	12:45	☽ □ ♄
	15:18	☽ △ ♂
	17:07	☽ # ☿
	23:35	☽ △ ♅
10	00:37	☽ □ ☿
	07:31	☿ ⚹ ♇
	13:59	☽ # ♆
	18:36	☽ ⚹ ♃
	21:40	☽ ⚹ ☉
11	02:42	♄ △ ♆
	03:14	☽ # ♄
	07:44	☽ ∥ ♀
	15:37	☽ # ☉
	20:29	☽ ☍ ♆
	20:40	☽ ⚹ ♄
12	01:36	☽ □ ♂
	07:56	☽ △ ♇
	13:23	☽ ⚹ ☿
	14:36	☽ ☌ ♀
	20:01	☽ ∥ ♅
	23:48	☽ □ ♃
13	03:37	♂ # ♃
	08:26	♀ # ☉
	12:13	☽ # ♅
14	07:31	☽ ⚹ ♂
	08:56	☽ ☍ ♅
	10:52	☽ □ ♇
	15:39	☽ # ♀
	21:04	☽ ∥ ☉
15	01:44	☽ △ ♃
	03:12	☽ ∥ ♄
	10:33	♂ △ ♅
	14:10	☽ ∥ ♆
16	01:00	☽ △ ♆
	02:05	☽ ☌ ♄
	11:39	☽ ⚹ ♇
	12:31	☿ ⚹ ♀
	23:35	♀ □ ♃
17	02:17	☽ ⚹ ♀
	02:24	☽ ☌ ☿
	15:53	☽ ∥ ☿
	23:15	☽ ∥ ♇
18	01:17	☽ □ ♆
	09:51	☽ △ ♅
	13:42	☽ ☌ ♂
	14:42	☉ ∥ ♄
19	02:33	☽ ☍ ♃
	02:39	☿ ∥ ♇
	07:15	☽ □ ♀
	20:27	☽ ⚹ ☉
20	02:31	☽ ⚹ ♆
	04:34	☽ ⚹ ♄
	07:23	☽ ∥ ☿
	11:19	☽ □ ♅
	13:50	☽ ☌ ♇
	14:54	☽ ∥ ♇
21	14:30	☽ △ ♀
	15:06	☽ ⚹ ☿
22	08:32	☽ □ ♄
	15:01	☽ ⚹ ♅
23	00:14	☉ ♏
	01:17	☽ ⚹ ♂
	04:52	☽ ∥ ♆
	06:48	☽ ∥ ☉
	09:07	☽ △ ♃
	11:15	☉ △ ♆
	15:16	☽ ∥ ♄
24	01:28	☽ □ ☿
	05:21	☉ ∥ ♆
	11:51	☽ ☌ ♆
	13:54	☽ △ ☉
	15:10	☽ △ ♄
	17:39	☿ # ♃
25	00:33	☽ ⚹ ♇
	06:53	☽ # ♀
	08:33	☉ ☌ ♄
	11:34	☽ □ ♂
	13:57	☽ # ♅
	16:10	☽ □ ♃
	17:10	☿ ⚹ ♀
26	06:02	☽ ∥ ♅
	10:29	☽ ∥ ♀
	14:45	☽ △ ☿
	15:05	☽ ☍ ♀
27	06:07	☽ ☌ ♅
	09:43	☽ □ ♇
28	00:45	☽ △ ♂
	01:33	☽ ⚹ ♃
	07:45	♀ ∥ ♅
	08:05	☽ # ♄
	12:55	♂ ☍ ♃
	13:04	♀ ♎
	18:21	☽ # ♆
29	05:57	☽ # ☉
	06:18	☿ ♐
	07:04	☽ ⚹ ♆
	11:55	☽ ☍ ♄
	17:40	☿ □ ♆
	21:01	☽ △ ♇
30	10:28	☉ ⚹ ♇
31	18:09	☽ # ♇
	19:28	☽ □ ♆
	22:55	☽ ☍ ☿

⚷ Chiron

01 Dec.	04 S 00
03	05♓50R
06	05 43
09	05 37
12	05 31
15	05 26
18	05 21
21	05 17
24	05 13
27	05 09
30	05 06

13	22:19	21♏57	✪ Total Solar Eclipse
28	14:34	06♊47	✷ Penumbral Lunar Eclipse (mag 0.942)

November 2012

Day	S. T. h m s	☉ ° ' "	☽ ° ' "	☿ ° '	♀ ° '	♂ ° '	♃ ° '	♄ ° '	♅ ° '	♆ ° '	♇ ° '	☊ True ° '
01 Th	02 42 45	08♏ 58 14	02♊ 37 08	02♐ 07	04♎ 11	17♐ 58	15♊R08	03♏ 08	05♈R20	00♓R23	07♑ 26	26♏ 03
02 Fr	02 46 41	09 58 15	14 24 13	02 45	05 24	18 43	15 03	03 16	05 18	00 23	07 28	26 05
03 Sa	02 50 38	10 58 19	26 11 41	03 17	06 37	19 27	14 57	03 23	05 16	00 23	07 29	26 07
04 Su	02 54 34	11 58 25	08♋ 03 00	03 43	07 50	20 11	14 52	03 30	05 14	00 22	07 30	26 08
05 Mo	02 58 31	12 58 32	20 02 07	04 02	09 03	20 56	14 46	03 37	05 13	00 22	07 32	26 10
06 Tu	03 02 27	13 58 42	02♌ 13 21	04 14	10 16	21 41	14 40	03 44	05 11	00 22	07 33	26 10
07 We	03 06 24	14 58 54	14 41 14	04R 18	11 30	22 25	14 34	03 52	05 09	00 22	07 35	26R 10
08 Th	03 10 20	15 59 08	27 30 04	04 13	12 43	23 10	14 28	03 59	05 08	00 22	07 36	26 10
09 Fr	03 14 17	16 59 24	10♍ 43 32	03 59	13 56	23 55	14 21	04 06	05 06	00 22	07 38	26 09
10 Sa	03 18 14	17 59 41	24 24 01	03 35	15 10	24 39	14 15	04 13	05 04	00 22	07 39	26 07
11 Su	03 22 10	19 00 01	08♎ 31 59	03 02	16 23	25 24	14 08	04 20	05 03	00 22	07 41	26 06
12 Mo	03 26 07	20 00 23	23 05 24	02 18	17 37	26 09	14 01	04 27	05 01	00D 22	07 42	26 05
13 Tu	03 30 03	21 00 47	07♏ 59 25	01 24	18 51	26 54	13 54	04 34	05 00	00 22	07 44	26 04
14 We	03 33 60	22 01 12	23 06 37	00 22	20 04	27 39	13 47	04 41	04 58	00 22	07 45	26 03
15 Th	03 37 56	23 01 39	08♐ 17 54	29♏ 12	21 18	28 25	13 40	04 49	04 57	00 22	07 47	26 02
16 Fr	03 41 53	24 02 08	23 23 44	27 56	22 32	29 10	13 33	04 56	04 56	00 22	07 49	26 02
17 Sa	03 45 49	25 02 38	08♑ 15 34	26 36	23 46	29 55	13 25	05 03	04 54	00 22	07 50	26 01
18 Su	03 49 46	26 03 09	22 46 57	25 15	25 00	00♑ 40	13 18	05 09	04 53	00 22	07 52	26 01
19 Mo	03 53 43	27 03 42	06♒ 54 01	23 55	26 14	01 26	13 10	05 16	04 52	00 23	07 54	26 01
20 Tu	03 57 39	28 04 16	20 35 35	22 39	27 28	02 11	13 03	05 23	04 51	00 23	07 55	26 01
21 We	04 01 36	29 04 51	03♓ 52 30	21 30	28 42	02 57	12 55	05 30	04 50	00 23	07 57	26D 01
22 Th	04 05 32	00♐ 05 27	16 47 08	20 30	29 56	03 42	12 47	05 37	04 49	00 24	07 59	26 01
23 Fr	04 09 29	01 06 04	29 22 41	19 39	01♏ 10	04 28	12 39	05 44	04 47	00 24	08 01	26 02
24 Sa	04 13 25	02 06 43	11♈ 42 38	18 59	02 24	05 14	12 31	05 51	04 46	00 24	08 02	26 03
25 Su	04 17 22	03 07 22	23 50 27	18 31	03 38	05 59	12 23	05 57	04 46	00 25	08 04	26 05
26 Mo	04 21 18	04 08 03	05♉ 49 18	18 15	04 53	06 45	12 15	06 04	04 45	00 25	08 06	26 07
27 Tu	04 25 15	05 08 45	17 42 01	18D 10	06 07	07 31	12 07	06 11	04 44	00 26	08 08	26 08
28 We	04 29 12	06 09 29	29 31 10	18 16	07 21	08 17	11 59	06 17	04 43	00 26	08 10	26 09
29 Th	04 33 08	07 10 13	11♊ 19 00	18 32	08 36	09 03	11 51	06 24	04 42	00 27	08 12	26R 10
30 Fr	04 37 05	08 10 59	23 07 38	18 57	09 50	09 48	11 43	06 31	04 42	00 28	08 13	26 09

Data for	11-01-2012
Julian Day	2456232.50
Ayanamsa	24 02 23
SVP	05 ♓ 05 03
☽ ☊ Mean	26 ♏ 49 R

● ◐ PHASES ○ ◑

07	00:36	◑	15♌00
13	22:09	✪	21♏57
20	14:31	◐	28♒41
28	14:46	✷	06♊47

LAST ASPECT ☽ INGRESS

Day	h m	Day	h m	
02	09:22	03	07:43	♋
04	08:37	05	19:40	♌
07	15:28	08	04:36	♍
10	00:28	10	09:36	♎
12	05:15	12	11:11	♏
14	10:40	14	10:53	♐
16	09:45	16	10:36	♑
18	05:55	18	12:11	♒
20	14:32	20	16:55	♓
22	06:32	23	01:12	♈
24	01:35	25	12:18	♉
27	00:57	28	00:59	♊
29	01:05	30	13:56	♋

DECLINATION

Day	☉	☽	☿	♀	♂	♃	♄	♅	♆	♇
01 Th	14S 29	20N06	23S 27	00S 12	23S 56	21N47	10S 28	01N26	11S 55	19S 45
02 Fr	14 48	20 52	23 32	00 40	24 00	21 46	10 30	01 25	11 55	19 46
03 Sa	15 07	20 44	23 36	01 07	24 04	21 46	10 32	01 25	11 55	19 46
04 Su	15 26	19 39	23 37	01 35	24 08	21 45	10 35	01 24	11 55	19 46
05 Mo	15 44	17 42	23 36	02 03	24 11	21 44	10 37	01 23	11 55	19 46
06 Tu	16 02	14 57	23 32	02 30	24 14	21 44	10 40	01 23	11 55	19 46
07 We	16 20	11 28	23 25	02 58	24 17	21 43	10 42	01 22	11 55	19 46
08 Th	16 37	07 22	23 15	03 26	24 20	21 42	10 44	01 21	11 55	19 46
09 Fr	16 54	02 48	23 02	03 54	24 23	21 42	10 47	01 21	11 55	19 46
10 Sa	17 11	02S 04	22 45	04 21	24 25	21 41	10 49	01 20	11 55	19 46
11 Su	17 28	06 59	22 25	04 49	24 27	21 40	10 51	01 20	11 55	19 46
12 Mo	17 44	11 40	22 01	05 17	24 28	21 40	10 54	01 19	11 55	19 47
13 Tu	18 00	15 44	21 33	05 44	24 30	21 39	10 56	01 19	11 55	19 47
14 We	18 16	18 49	21 02	06 12	24 31	21 38	10 58	01 18	11 55	19 47
15 Th	18 32	20 35	20 28	06 39	24 32	21 37	11 01	01 17	11 55	19 47
16 Fr	18 47	20 51	19 51	07 06	24 33	21 36	11 03	01 17	11 55	19 47
17 Sa	19 01	19 38	19 13	07 33	24 33	21 36	11 05	01 16	11 55	19 47
18 Su	19 16	17 09	18 34	08 00	24 33	21 35	11 07	01 16	11 55	19 47
19 Mo	19 30	13 42	17 55	08 27	24 33	21 34	11 10	01 16	11 55	19 47
20 Tu	19 44	09 37	17 18	08 54	24 33	21 33	11 12	01 15	11 55	19 47
21 We	19 57	05 12	16 44	09 21	24 32	21 32	11 14	01 15	11 55	19 47
22 Th	20 10	00 39	16 13	09 47	24 31	21 31	11 16	01 14	11 55	19 47
23 Fr	20 23	03N49	15 47	10 13	24 30	21 30	11 18	01 14	11 54	19 47
24 Sa	20 35	08 03	15 25	10 39	24 28	21 29	11 21	01 14	11 54	19 47
25 Su	20 47	11 53	15 09	11 05	24 26	21 29	11 23	01 13	11 54	19 47
26 Mo	20 58	15 12	14 57	11 30	24 24	21 28	11 25	01 13	11 54	19 48
27 Tu	21 09	17 51	14 51	11 55	24 22	21 27	11 27	01 13	11 54	19 48
28 We	21 20	19 44	14 49	12 20	24 19	21 26	11 29	01 12	11 53	19 48
29 Th	21 30	20 46	14 52	12 45	24 16	21 25	11 31	01 12	11 53	19 48
30 Fr	21 40	20 52	14 58	13 10	24 13	21 24	11 33	01 12	11 53	19 48

ASPECTARIAN

01 03:33 ☽ △ ♀
05:30 ☽ ✶ ♅
22:05 ♀ ☍ ♅

02 01:18 ☽ ☌ ♃
09:22 ☽ ☍ ♂

03 08:29 ☽ △ ♆
14:43 ☽ △ ♄
14:51 ♀ # ♅
17:26 ♀ □ ♇
18:21 ☽ □ ♅
22:15 ☽ # ♇
22:54 ☽ ☍ ♇
23:31 ☽ □ ♀

04 08:37 ☽ △ ☉

05 16:07 ☽ # ☉

06 02:59 ☽ □ ♄
03:57 ☽ △ ☿
05:44 ☽ △ ♅
17:15 ☽ ✶ ♀
21:03 ☽ # ♆
23:05 ☿ SR
23:46 ☽ ✶ ♃

07 04:39 ☽ # ♄
15:28 ☽ △ ♂

08 05:15 ☽ ☍ ♆
11:58 ☽ ✶ ♄
12:07 ☽ □ ☿
18:26 ☽ △ ♇
18:55 ☽ # ♀

09 06:24 ☽ □ ♃
07:17 ☽ ∥ ♅
07:27 ♀ △ ♃
11:59 ☽ ✶ ☉
20:29 ☽ # ♅

10 00:28 ☽ □ ♂
12:19 ☽ ∥ ♀
15:07 ☽ ✶ ☿
18:10 ☽ ☍ ♅
22:34 ☽ □ ♇

11 07:54 ♆ SD
09:14 ☽ △ ♃
14:14 ☽ ☌ ♀
19:54 ☽ ∥ ♄

12 01:27 ☽ ∥ ♆
05:15 ☽ ✶ ♂
11:46 ☽ △ ♆
18:30 ☽ ☌ ♄
19:01 ☿ # ♃
23:35 ☽ ✶ ♇

13 18:29 ☽ ∥ ☉

14 00:03 ☿ □ ♆
07:43 ☿ ♏R
10:40 ☽ ☌ ☿
10:45 ☽ ∥ ♇
11:27 ☽ □ ♆
18:43 ☽ △ ♅
22:25 ☽ ∥ ☿

15 08:26 ☽ ☍ ♃
22:30 ☽ ✶ ♀

16 02:44 ☿ ∥ ♇
09:45 ☽ ☌ ♂
11:12 ☽ ✶ ♆
18:33 ☽ □ ♅
18:43 ☽ ✶ ♄
21:51 ☽ ∥ ♇
23:18 ☽ ☌ ♇

17 02:37 ♂ ♑
05:08 ☿ ∥ ☉
06:12 ☽ ∥ ☉
06:48 ☽ ∥ ☿
14:22 ♂ ✶ ♆
15:47 ☿ ☌ ☉

18 03:47 ☽ ✶ ☿
04:04 ☽ □ ♀
05:55 ☽ ✶ ☉
20:30 ☽ ✶ ♅
21:10 ☽ □ ♄

19 10:48 ☽ △ ♃
10:50 ☽ ∥ ♆
15:05 ☽ ∥ ♄

20 03:23 ☽ □ ☿
03:36 ☽ ∥ ♀
06:18 ☉ ∥ ♇
13:35 ☽ △ ♀
17:37 ☽ ☌ ♆
22:12 ☽ ✶ ♂

21 03:01 ☽ △ ♄
07:31 ☽ ✶ ♇
16:34 ☽ □ ♃
20:51 ☽ # ♅
21:51 ☉ ♐

22 01:20 ♀ ♏
06:32 ☽ △ ☿
07:12 ☉ □ ♆
08:59 ♀ △ ♆
10:00 ☽ ∥ ♅

23 03:37 ☽ △ ☉
10:05 ♂ □ ♅
10:28 ☽ ☌ ♅
10:29 ☽ □ ♂
16:47 ☽ □ ♇

24 01:35 ☽ ✶ ♃
18:02 ☽ # ♀
20:38 ☽ # ♄
22:52 ♂ ✶ ♄

25 00:07 ☽ # ♆
13:09 ☽ ✶ ♆
18:28 ♀ ∥ ♄
21:53 ☽ ☍ ♀
22:13 ☽ # ☿

26 00:30 ☽ ☍ ♄
02:00 ☽ △ ♂
04:36 ☽ △ ♇
14:15 ☉ △ ♅
22:17 ♀ ∥ ♆
22:49 ☿ SD

27 00:57 ☽ ☍ ☿
01:20 ♀ ☌ ♄
20:14 ♂ ☌ ♇

28 00:57 ☽ # ♇
01:52 ☽ □ ♆
10:34 ☽ ✶ ♅
12:03 ☉ # ♃
16:01 ♀ ✶ ♇

29 01:05 ☽ ☌ ♃
22:33 ♀ ✶ ♂

30 14:52 ☽ △ ♆
23:23 ☽ □ ♅

⚷ Chiron

01 Dec.	04 S 27
02	05♓04R
05	05 02
08	05 00
11	04 59
14	04 59
17	04 59D
20	05 00
23	05 01
26	05 02
29	05 05

14 18:56 04♓59 D

11:13 21 December (Winter Solstice): Mayan calendar ends

December 2012

Day	S. T.	☉	☽	☿	♀	♂	♃	♄	♅	♆	♇	☊ True
	h m s	° ' "	° ' "	° '	° '	° '	° '	° '	° '	° '	° '	° '
01 Sa	04 41 01	09♐ 11 47	04♋ 59 10	19♏ 30	11♏ 05	10♑ 34	11♊ R35	06♏ 37	04♈ R41	00♓ 28	08♑ 15	26♏ R08
02 Su	04 44 58	10 12 35	16 55 52	20 11	12 19	11 21	11 27	06 44	04 40	00 29	08 17	26 05
03 Mo	04 48 54	11 13 25	29 00 11	20 58	13 34	12 07	11 18	06 50	04 40	00 30	08 19	26 03
04 Tu	04 52 51	12 14 16	11♌ 14 59	21 52	14 48	12 53	11 10	06 56	04 39	00 30	08 21	26 01
05 We	04 56 47	13 15 08	23 43 26	22 50	16 03	13 39	11 02	07 03	04 39	00 31	08 23	25 59
06 Th	05 00 44	14 16 02	06♍ 28 54	23 52	17 18	14 25	10 54	07 09	04 38	00 32	08 25	25 57
07 Fr	05 04 41	15 16 57	19 34 43	24 59	18 32	15 11	10 46	07 15	04 38	00 33	08 27	25 57
08 Sa	05 08 37	16 17 53	03♎ 03 47	26 09	19 47	15 58	10 38	07 21	04 38	00 34	08 29	25D 57
09 Su	05 12 34	17 18 51	16 57 58	27 22	21 02	16 44	10 30	07 28	04 37	00 35	08 31	25 58
10 Mo	05 16 30	18 19 50	01♏ 17 32	28 37	22 16	17 31	10 22	07 34	04 37	00 36	08 33	25 58
11 Tu	05 20 27	19 20 50	16 00 20	29 55	23 31	18 17	10 14	07 40	04 37	00 37	08 35	25R 59
12 We	05 24 23	20 21 51	01♐ 01 30	01♐ 14	24 46	19 04	10 06	07 46	04 37	00 38	08 37	25 58
13 Th	05 28 20	21 22 53	16 13 25	02 35	26 01	19 50	09 58	07 52	04 37	00 39	08 39	25 56
14 Fr	05 32 16	22 23 56	01♑ 26 34	03 57	27 16	20 37	09 50	07 58	04D 37	00 40	08 41	25 53
15 Sa	05 36 13	23 24 59	16 30 52	05 21	28 31	21 23	09 42	08 03	04 37	00 41	08 43	25 50
16 Su	05 40 10	24 26 03	01♒ 17 14	06 46	29 46	22 10	09 34	08 09	04 37	00 42	08 45	25 45
17 Mo	05 44 06	25 27 08	15 38 59	08 11	01♐ 00	22 57	09 27	08 15	04 37	00 43	08 47	25 41
18 Tu	05 48 03	26 28 13	29 32 28	09 37	02 15	23 43	09 19	08 21	04 37	00 44	08 50	25 38
19 We	05 51 59	27 29 18	12♓ 57 06	11 05	03 30	24 30	09 12	08 26	04 38	00 46	08 52	25 35
20 Th	05 55 56	28 30 24	25 54 49	12 32	04 45	25 17	09 05	08 32	04 38	00 47	08 54	25D 35
21 Fr	05 59 52	29 31 29	08♈ 29 13	14 01	06 00	26 04	08 57	08 37	04 38	00 48	08 56	25 35
22 Sa	06 03 49	00♑ 32 35	20 44 53	15 29	07 15	26 51	08 50	08 43	04 39	00 50	08 58	25 37
23 Su	06 07 45	01 33 41	02♉ 46 38	16 59	08 30	27 38	08 43	08 48	04 39	00 51	09 00	25 40
24 Mo	06 11 42	02 34 48	14 39 12	18 28	09 45	28 24	08 36	08 53	04 40	00 52	09 02	25 42
25 Tu	06 15 39	03 35 54	26 26 55	19 58	11 00	29 11	08 30	08 58	04 40	00 54	09 04	25 44
26 We	06 19 35	04 37 01	08♊ 13 33	21 29	12 15	29 58	08 23	09 03	04 41	00 55	09 06	25R 43
27 Th	06 23 32	05 38 08	20 02 13	23 00	13 30	00♒ 45	08 17	09 08	04 42	00 57	09 09	25 41
28 Fr	06 27 28	06 39 16	01♋ 55 25	24 31	14 45	01 32	08 10	09 13	04 42	00 58	09 11	25 36
29 Sa	06 31 25	07 40 23	13 55 00	26 02	16 00	02 19	08 04	09 18	04 43	01 00	09 13	25 30
30 Su	06 35 21	08 41 31	26 02 25	27 34	17 16	03 07	07 58	09 23	04 44	01 01	09 15	25 21
31 Mo	06 39 18	09 42 39	08♌ 18 49	29 06	18 31	03 54	07 52	09 28	04 45	01 03	09 17	25 13

Data for	12-01-2012
Julian Day	2456262.50
Ayanamsa	24 02 27
SVP	05 ♓ 04 59
☽ ☊ Mean	25 ♏ 14 R

● ◐ PHASES ○ ◑

06	15:32	◑	14♍56
13	08:42	●	21♐45
20	05:19	◐	28♓44
28	10:22	○	07♋06

LAST ASPECT Day	h m	☽ INGRESS Day	h m	
02	06:56	03	01:58	♌
04	22:09	05	11:53	♍
07	10:37	07	18:37	♎
09	00:38	09	21:52	♏
11	13:09	11	22:22	♐
13	08:42	13	21:43	♑
15	21:15	15	21:53	♒
17	18:11	18	00:48	♓
20	05:19	20	07:44	♈
22	12:57	22	18:26	♉
25	05:59	25	07:14	♊
27	06:52	27	20:08	♋
28	14:44	30	07:47	♌

DECLINATION

Day	☉	☽	☿	♀	♂	♃	♄	♅	♆	♇
01 Sa	21S 50	20N02	15S 08	13S 34	24S 10	21N23	11S 35	01N12	11S 53	19S 48
02 Su	21 59	18 18	15 21	13 57	24 06	21 22	11 37	01 11	11 52	19 48
03 Mo	22 07	15 45	15 36	14 21	24 02	21 21	11 39	01 11	11 52	19 48
04 Tu	22 15	12 30	15 53	14 44	23 58	21 20	11 41	01 11	11 52	19 48
05 We	22 23	08 38	16 12	15 07	23 53	21 19	11 43	01 11	11 52	19 48
06 Th	22 31	04 19	16 33	15 29	23 49	21 18	11 45	01 11	11 51	19 48
07 Fr	22 37	00S 18	16 55	15 51	23 43	21 17	11 47	01 11	11 51	19 48
08 Sa	22 44	05 04	17 17	16 13	23 38	21 16	11 49	01 11	11 51	19 48
09 Su	22 50	09 43	17 40	16 34	23 33	21 15	11 51	01 11	11 50	19 48
10 Mo	22 55	13 59	18 03	16 55	23 27	21 14	11 52	01 10	11 50	19 48
11 Tu	23 01	17 30	18 27	17 15	23 20	21 13	11 54	01 10	11 50	19 48
12 We	23 05	19 54	18 50	17 35	23 14	21 12	11 56	01 10	11 49	19 48
13 Th	23 09	20 55	19 13	17 54	23 07	21 11	11 58	01 10	11 49	19 48
14 Fr	23 13	20 23	19 36	18 13	23 00	21 10	12 00	01 10	11 49	19 48
15 Sa	23 16	18 24	19 59	18 32	22 53	21 09	12 01	01 11	11 48	19 48
16 Su	23 19	15 13	20 21	18 50	22 46	21 08	12 03	01 11	11 48	19 48
17 Mo	23 21	11 13	20 42	19 07	22 38	21 07	12 05	01 11	11 47	19 48
18 Tu	23 23	06 45	21 03	19 24	22 30	21 06	12 06	01 11	11 47	19 48
19 We	23 25	02 05	21 23	19 41	22 22	21 05	12 08	01 11	11 46	19 48
20 Th	23 26	02N31	21 43	19 57	22 13	21 04	12 10	01 11	11 46	19 48
21 Fr	23 26	06 53	22 01	20 12	22 05	21 03	12 11	01 11	11 45	19 48
22 Sa	23 26	10 52	22 19	20 27	21 56	21 02	12 13	01 12	11 45	19 48
23 Su	23 26	14 20	22 35	20 41	21 46	21 01	12 14	01 12	11 45	19 48
24 Mo	23 25	17 11	22 51	20 54	21 37	21 01	12 16	01 12	11 44	19 48
25 Tu	23 23	19 17	23 05	21 07	21 27	21 00	12 17	01 12	11 44	19 48
26 We	23 21	20 33	23 19	21 20	21 17	20 59	12 19	01 13	11 43	19 48
27 Th	23 19	20 55	23 31	21 31	21 07	20 58	12 20	01 13	11 42	19 48
28 Fr	23 16	20 21	23 42	21 43	20 57	20 57	12 22	01 13	11 42	19 48
29 Sa	23 13	18 52	23 52	21 53	20 46	20 57	12 23	01 13	11 41	19 47
30 Su	23 09	16 31	24 01	22 03	20 35	20 56	12 24	01 14	11 41	19 47
31 Mo	23 05	13 26	24 09	22 12	20 24	20 55	12 26	01 14	11 40	19 47

ASPECTARIAN

01 03:19 ☽ △ ♄
04:00 ☽ # ♇
06:36 ☽ ☍ ♇
12:02 ☽ ☍ ♂
13:41 ☽ △ ♀
11:29 ☽ ∥ ♆
11:37 ☽ ∥ ♄
22:51 ☽ △ ♆

02 06:56 ☽ △ ☿

03 01:08 ☽ # ☿
01:46 ☉ ☍ ♃
09:46 ☽ # ♀
11:08 ☽ △ ♅
15:32 ☽ □ ♄
23:51 ☽ ✱ ♃

04 02:05 ☽ △ ☉
04:08 ☽ # ♇
05:15 ☽ # ♄
07:39 ☽ □ ♀
22:09 ☽ □ ☿

05 12:52 ☽ ☍ ♆

06 01:15 ☽ ✱ ♄
03:36 ☽ △ ♇
08:05 ☽ □ ♃
15:32 ☽ △ ♂
16:21 ☽ ∥ ♅
21:55 ☽ ✱ ♀

07 04:21 ☽ # ♅
10:37 ☽ ✱ ☿

08 02:44 ☽ ☍ ♅
09:28 ☽ □ ♇
13:01 ☽ △ ♃
21:12 ♄ ∥ ♆
23:35 ☽ □ ♂

09 00:38 ☽ ✱ ☉

10 10:22 ☽ ☌ ♄
11:56 ☽ ✱ ♇
21:50 ☽ ∥ ♀

11 01:40 ☿ ♐
03:52 ☽ ✱ ♂
09:25 ☽ ∥ ☿
12:57 ☿ □ ♆
13:09 ☽ ☌ ♀
22:31 ☽ ∥ ♇
23:22 ☽ □ ♆

12 00:22 ☽ ☌ ☿
05:41 ☽ △ ♅
14:13 ☽ ☍ ♃
19:40 ☉ ∥ ♂

13 12:03 ♅ SD
22:46 ☽ ✱ ♆

14 05:01 ☽ □ ♅
08:58 ☽ ∥ ♇
09:32 ☽ ∥ ☿
10:24 ☽ ✱ ♄
11:24 ☿ △ ♅
11:31 ☽ ☌ ♇
12:21 ☿ ∥ ♇
22:53 ☽ ∥ ♀

15 08:17 ☽ ☌ ♂
21:15 ☽ ✱ ♀

16 04:39 ♀ ♐
05:30 ☽ ✱ ♅
10:03 ☽ ✱ ☿
11:27 ☽ □ ♄
13:38 ☽ △ ♃
18:24 ♀ □ ♆
19:12 ☽ ∥ ♄
20:49 ☽ ∥ ♆

17 18:13 ☽ ✱ ☉
19:23 ☿ ☍ ♃

18 02:07 ☽ ☌ ♆
03:12 ☿ # ♃
05:17 ☽ □ ♀
15:46 ☽ △ ♄
16:34 ☽ ✱ ♇
17:16 ☽ □ ♃
20:10 ☽ □ ☿

19 04:39 ☽ # ♅
10:38 ♀ ∥ ♇
16:57 ☽ ∥ ♅
21:38 ♀ △ ♅
22:44 ☽ ✱ ♂

20 16:35 ☽ ☌ ♅
18:40 ☽ △ ♀

21 00:52 ☽ □ ♇
00:54 ☽ ✱ ♃
03:17 ☿ ∥ ♂
11:12 ☉ ♑
12:13 ☽ △ ☿

22 05:44 ☽ # [illegible]
06:48 ☉ ✱ ♆
08:54 ☽ # ♄
12:57 ☽ □ ♂
20:07 ☽ ✱ ♆
21:20 ☽ △ ☉

23 03:50 ♀ ☍ ♃
12:14 ☽ ☍ ♄
12:35 ☽ △ ♇

24 10:41 ♀ # ♃

25 05:59 ☽ △ ♂
07:44 ☽ # ♇
[illegible] ☽ □ ♆
16:46 ☽ ✱ ♅
21:17 ♀ ∥ ♂

26 00:19 ☽ ☌ ♃
00:49 ♂ ♒
01:31 ☉ □ ♅
04:25 ☿ ∥ ☉
09:10 ☽ ☍ ♀

27 00:48 ♄ ✱ ♇
06:52 ☽ ☍ ☿
22:04 ☽ △ ♆
22:13 ♂ # ♃

28 05:35 ☽ □ ♅
10:51 ☽ # ♇
14:35 ☽ ☍ ♇
14:44 ☽ △ ♄

30 13:37 ☉ ☌ ♇
14:48 ☽ ☍ ♂
17:03 ☽ △ ♅
17:46 ☉ ✱ ♄
23:09 ☽ ✱ ♃

31 02:15 ☽ □ ♄
06:48 ☽ # ♄
11:49 ☽ # ♆
14:03 ☿ ♑
21:53 ☽ △ ♀

⚷ Chiron

01 Dec.	04 S 37
02	05♓07
05	05 11
08	05 15
11	05 19
14	05 24
17	05 29
20	05 35
23	05 41
26	05 48
29	05 55

January 2013

Day	S. T. h m s	☉ ° ' "	☽ ° ' "	☿ ° '	♀ ° '	♂ ° '	♃ ° '	♄ ° '	♅ ° '	♆ ° '	♇ ° '	☊ True ° '
01 Tu	06 43 15	10♑ 43 47	20♌ 45 21	00♑ 38	19♐ 46	04♒ 41	07♊ R46	09♏ 33	04♈ 46	01♓ 04	09♑ 19	25♏ R04
02 We	06 47 11	11 44 55	03♍ 23 21	02 11	21 01	05 28	07 41	09 37	04 47	01 06	09 21	24 57
03 Th	06 51 08	12 46 04	16 14 27	03 44	22 16	06 15	07 35	09 42	04 48	01 08	09 23	24 52
04 Fr	06 55 04	13 47 13	29 20 31	05 18	23 31	07 02	07 30	09 46	04 49	01 09	09 26	24 49
05 Sa	06 59 01	14 48 22	12♎ 43 38	06 51	24 46	07 49	07 25	09 51	04 50	01 11	09 28	24 47
06 Su	07 02 57	15 49 31	26 25 45	08 26	26 01	08 37	07 20	09 55	04 51	01 13	09 30	24 47
07 Mo	07 06 54	16 50 41	10♏ 28 08	10 00	27 17	09 24	07 15	09 59	04 52	01 14	09 32	24 46
08 Tu	07 10 50	17 51 51	24 50 45	11 35	28 32	10 11	07 11	10 03	04 53	01 16	09 34	24 46
09 We	07 14 47	18 53 01	09♐ 31 33	13 10	29 47	10 59	07 06	10 07	04 55	01 18	09 36	24 44
10 Th	07 18 44	19 54 10	24 25 51	14 46	01♑ 02	11 46	07 02	10 11	04 56	01 20	09 38	24 40
11 Fr	07 22 40	20 55 20	09♑ 26 32	16 22	02 17	12 33	06 58	10 15	04 58	01 22	09 40	24 33
12 Sa	07 26 37	21 56 30	24 24 42	17 59	03 32	13 21	06 54	10 19	04 59	01 23	09 42	24 25
13 Su	07 30 33	22 57 39	09♒ 11 00	19 36	04 48	14 08	06 51	10 23	05 01	01 25	09 45	24 16
14 Mo	07 34 30	23 58 48	23 37 18	21 14	06 03	14 55	06 47	10 26	05 02	01 27	09 47	24 07
15 Tu	07 38 26	24 59 56	07♓ 37 53	22 52	07 18	15 43	06 44	10 30	05 04	01 29	09 49	23 58
16 We	07 42 23	26 01 03	21 10 03	24 30	08 33	16 30	06 41	10 33	05 05	01 31	09 51	23 51
17 Th	07 46 19	27 02 10	04♈ 14 10	26 09	09 48	17 17	06 38	10 37	05 07	01 33	09 53	23 47
18 Fr	07 50 16	28 03 16	16 52 57	27 49	11 04	18 05	06 36	10 40	05 09	01 35	09 55	23 44
19 Sa	07 54 13	29 04 21	29 10 46	29 29	12 19	18 52	06 33	10 43	05 11	01 37	09 57	23D 44
20 Su	07 58 09	00♒ 05 25	11♉ 12 49	01♒ 10	13 34	19 40	06 31	10 46	05 12	01 39	09 59	23 45
21 Mo	08 02 06	01 06 29	23 04 37	02 51	14 49	20 27	06 29	10 49	05 14	01 41	10 01	23 46
22 Tu	08 06 02	02 07 31	04♊ 51 34	04 33	16 04	21 15	06 27	10 52	05 16	01 43	10 03	23R 46
23 We	08 09 59	03 08 33	16 38 36	06 15	17 20	22 02	06 25	10 55	05 18	01 45	10 05	23 43
24 Th	08 13 55	04 09 33	28 29 59	07 58	18 35	22 49	06 24	10 57	05 20	01 47	10 07	23 38
25 Fr	08 17 52	05 10 33	10♋ 29 06	09 41	19 50	23 37	06 23	11 00	05 22	01 49	10 09	23 30
26 Sa	08 21 48	06 11 32	22 38 22	11 25	21 05	24 24	06 22	11 02	05 24	01 51	10 11	23 19
27 Su	08 25 45	07 12 30	04♌ 59 13	13 09	22 20	25 12	06 21	11 05	05 26	01 53	10 13	23 07
28 Mo	08 29 42	08 13 27	17 32 12	14 54	23 35	25 59	06 20	11 07	05 29	01 55	10 15	22 53
29 Tu	08 33 38	09 14 23	00♍ 17 13	16 39	24 51	26 47	06 20	11 09	05 31	01 58	10 17	22 40
30 We	08 37 35	10 15 18	13 13 48	18 24	26 06	27 34	06 20	11 11	05 33	02 00	10 19	22 28
31 Th	08 41 31	11 16 12	26 21 29	20 10	27 21	28 21	06D 20	11 13	05 35	02 02	10 20	22 19

Data for	01-01-2013
Julian Day	2456293.50
Ayanamsa	24 02 33
SVP	05 ♓ 04 56
☽ ☊ Mean	23 ♏ 36 R

● ◐ PHASES ○ ◑

Day	h m	Phase	Position
05	03:58	◑	14♎59
11	19:43	●	21♑46
18	23:46	◐	29♈04
27	04:39	○	07♌24

LAST ASPECT Day	h m	☽ INGRESS Day	h m	Sign
31	21:53	01	17:36	♍
03	12:16	04	01:12	♎
05	23:14	06	06:10	♏
07	11:32	08	08:29	♐
09	02:29	10	08:55	♑
11	19:44	12	09:02	♒
13	08:37	14	10:50	♓
16	09:33	16	16:08	♈
19	00:42	19	01:37	♉
20	18:17	21	14:06	♊
23	11:43	24	03:01	♋
25	20:36	26	14:21	♌
28	17:00	28	23:28	♍
31	01:59	31	06:36	♎

DECLINATION

Day	☉	☽	☿	♀	♂	♃	♄	♅	♆	♇
01 Tu	23S 00	09N43	24S 15	22S 21	20S 12	20N54	12S 27	01N15	11S 40	19S 47
02 We	22 55	05 31	24 20	22 29	20 01	20 54	12 28	01 15	11 39	19 47
03 Th	22 49	01 01	24 24	22 36	19 49	20 53	12 29	01 15	11 39	19 47
04 Fr	22 43	03S 37	24 26	22 42	19 37	20 52	12 31	01 16	11 38	19 47
05 Sa	22 37	08 11	24 28	22 48	19 25	20 52	12 32	01 16	11 37	19 47
06 Su	22 30	12 28	24 27	22 53	19 12	20 51	12 33	01 17	11 37	19 47
07 Mo	22 22	16 09	24 26	22 58	19 00	20 51	12 34	01 17	11 36	19 47
08 Tu	22 15	18 58	24 23	23 01	18 47	20 50	12 35	01 18	11 35	19 47
09 We	22 06	20 35	24 19	23 05	18 34	20 50	12 36	01 19	11 35	19 47
10 Th	21 58	20 47	24 13	23 07	18 20	20 49	12 37	01 19	11 34	19 47
11 Fr	21 48	19 30	24 06	23 08	18 07	20 49	12 38	01 20	11 34	19 47
12 Sa	21 39	16 53	23 57	23 09	17 53	20 48	12 39	01 20	11 33	19 47
13 Su	21 29	13 13	23 47	23 10	17 39	20 48	12 40	01 21	11 32	19 47
14 Mo	21 19	08 50	23 36	23 09	17 25	20 48	12 41	01 22	11 32	19 46
15 Tu	21 08	04 07	23 23	23 08	17 11	20 47	12 42	01 22	11 31	19 46
16 We	20 57	00N40	23 08	23 06	16 57	20 47	12 43	01 23	11 30	19 46
17 Th	20 45	05 15	22 52	23 03	16 42	20 47	12 44	01 24	11 29	19 46
18 Fr	20 33	09 27	22 34	23 00	16 27	20 46	12 44	01 24	11 29	19 46
19 Sa	20 20	13 10	22 15	22 56	16 13	20 46	12 45	01 25	11 28	19 46
20 Su	20 08	16 15	21 55	22 51	15 57	20 46	12 46	01 26	11 27	19 46
21 Mo	19 55	18 36	21 33	22 46	15 42	20 46	12 47	01 27	11 27	19 46
22 Tu	19 41	20 08	21 09	22 40	15 27	20 46	12 47	01 27	11 26	19 46
23 We	19 27	20 47	20 44	22 33	15 11	20 46	12 48	01 28	11 25	19 46
24 Th	19 13	20 31	20 17	22 25	14 55	20 46	12 49	01 29	11 24	19 46
25 Fr	18 58	19 20	19 49	22 17	14 40	20 46	12 49	01 30	11 24	19 46
26 Sa	18 43	17 15	19 19	22 08	14 24	20 46	12 50	01 31	11 23	19 45
27 Su	18 28	14 23	18 48	21 59	14 07	20 46	12 50	01 32	11 22	19 45
28 Mo	18 12	10 49	18 15	21 48	13 51	20 46	12 51	01 33	11 21	19 45
29 Tu	17 56	06 43	17 41	21 37	13 35	20 46	12 51	01 33	11 21	19 45
30 We	17 40	02 15	17 05	21 26	13 18	20 47	12 52	01 34	11 20	19 45
31 Th	17 24	02S 23	16 28	21 14	13 01	20 47	12 52	01 35	11 19	19 45

ASPECTARIAN

01 02:32 ♂ ⚹ ♅
06:50 ☿ ⚹ ♆
19:40 ☽ ☍ ♆
21:25 ☽ △ ☿
02 08:00 ☽ □ ♃
11:14 ☽ △ ♇
11:46 ☽ ⚹ ♄
17:00 ☽ △ ☉
22:44 ☽ ∥ ♅
03 03:49 ♂ ∥ ♇
11:49 ☽ # ♅
12:16 ☽ □ ♀
16:30 ☿ □ ♅
04 02:05 ♀ ∥ ☉
09:53 ☽ ☍ ♅
12:09 ☽ □ ☿
12:46 ♂ △ ♃
14:36 ☽ △ ♃
14:44 ☽ △ ♂
18:11 ☽ □ ♇
05 18:58 ☽ ∥ ♆
23:14 ☽ ⚹ ♀
06 00:28 ☽ ∥ ♄
08:15 ☽ △ ♆
16:43 ☿ ☌ ♇
22:05 ☽ □ ♂
22:25 ☽ ⚹ ♇
23:07 ☽ ⚹ ☿
23:11 ☽ ☌ ♄
23:45 ☿ ⚹ ♄
07 11:32 ☽ ⚹ ☉
19:33 ♂ □ ♄
22:13 ☽ ∥ ♂
08 09:45 ☽ ∥ ♇
10:35 ☽ □ ♆
16:30 ☽ △ ♅
20:06 ☽ ☍ ♃
09 02:29 ☽ ⚹ ♂
04:11 ♀ ♑
08:41 ☽ ∥ ♃
22:09 ☽ # ♃
10 05:46 ♀ ⚹ ♆
11:03 ☽ ⚹ ♆
11:32 ☽ ☌ ♀
16:50 ☽ □ ♅
20:34 ☽ ∥ ♇
11 00:22 ☽ ☌ ♇
01:18 ☽ ⚹ ♄
12:25 ☽ ☌ ☿
15:24 ☽ ∥ ♂
12 17:10 ☽ ⚹ ♅
20:11 ☽ △ ♃
13 01:58 ☽ □ ♄
03:11 ☽ ∥ ♄
04:13 ♀ □ ♅
08:37 ☽ ☌ ♂
09:37 ☽ ∥ ♆
14 13:21 ☽ ☌ ♆
22:27 ☽ □ ♃
23:22 ☽ ⚹ ♀
15 03:49 ☽ ⚹ ♇
05:02 ☽ △ ♄
13:44 ☽ # ♅
16 03:44 ☽ ∥ ♅
03:45 ☿ ∥ ♀
06:55 ☽ ⚹ ☿
09:33 ☽ ⚹ ☉
20:18 ☉ # ♃
17 01:28 ♀ ☌ ♇
01:39 ☽ ☌ ♅
04:29 ☽ ⚹ ♃
10:39 ☽ □ ♇
11:38 ☽ □ ♀
16:03 ♀ ⚹ ♄
18 02:28 ☽ ⚹ ♂
08:57 ☿ ☌ ☉
12:36 ☽ # ♆
21:08 ☽ # ♄
19 00:42 ☽ □ ☿
04:50 ☽ ⚹ ♆
07:26 ☿ ♒
21:31 ☽ △ ♇
21:44 ☽ # ♂
21:52 ☉ ♒
23:06 ☽ ☍ ♄
20 05:18 ☽ △ ♀
18:17 ☽ □ ♂
21 15:31 ☉ ∥ ♇
16:38 ☽ # ☉
16:49 ☽ # ♇
17:34 ☽ □ ♆
17:54 ☽ △ ☉
23:15 ☽ △ ☿
22 00:50 ☽ ⚹ ♅
03:14 ☽ ☌ ♃
10:29 ☿ ⚹ ♅
21:13 ☽ ∥ ♃
21:48 ☽ # ☿
22:12 ☿ # ♃
23 02:26 ☿ △ ♃
11:43 ☽ △ ♂
13:14 ☽ # ♃
24 06:37 ☽ △ ♆
11:51 ☽ # ☿
13:46 ☽ □ ♅
17:24 ☽ # ♇
23:20 ☽ ☍ ♇
25 01:01 ☽ △ ♄
02:52 ☿ ∥ ♇
04:46 ☉ ⚹ ♅
05:54 ☽ # ☉
18:43 ☿ □ ♄
20:36 ☽ ☍ ♀
26 03:56 ☉ △ ♃
27 00:53 ☽ △ ♅
02:05 ☽ # ♂
02:37 ☽ ⚹ ♃
10:54 ☽ # ♄
11:44 ☽ □ ♄
18:10 ☽ ☍ ☿
20:37 ☽ # ♆
28 03:43 ☿ ∥ ☉
17:00 ☽ ☍ ♂
29 03:08 ☽ ☍ ♆
11:15 ☽ □ ♃
18:36 ☽ △ ♇
20:14 ☽ ⚹ ♄
30 03:33 ☽ ∥ ♅
11:38 ♃ SD
19:56 ☽ # ♅
22:49 ☉ □ ♄
31 01:59 ☽ △ ♀
13:03 ♂ ∥ ♄
16:44 ☽ ☍ ♅
18:01 ☽ △ ♃

⚷ Chiron	
01 Dec.	04 S 26
01	06♓02
04	06 10
07	06 19
10	06 27
13	06 36
16	06 46
19	06 55
22	07 05
25	07 15
28	07 26
31	07 36

February 2013

Day	S. T.	☉	☽	☿	♀	♂	♃	♄	♅	♆	♇	☊ True
	h m s	° ' "	° ' "	° '	° '	° '	° '	° '	° '	° '	° '	° '
01 Fr	08 45 28	12♒ 17 06	09♎ 40 02	21♒ 55	28♑ 36	29♒ 09	06♊ 20	11♏ 15	05♈ 38	02♓ 04	10♑ 22	22♏R13
02 Sa	08 49 24	13 17 59	23 09 43	23 41	29 51	29 56	06 20	11 17	05 40	02 06	10 24	22 09
03 Su	08 53 21	14 18 51	06♏ 51 10	25 26	01♒ 06	00♓ 44	06 21	11 19	05 42	02 08	10 26	22 07
04 Mo	08 57 17	15 19 42	20 45 09	27 11	02 22	01 31	06 22	11 20	05 45	02 11	10 28	22 06
05 Tu	09 01 14	16 20 32	04♐ 52 04	28 55	03 37	02 18	06 23	11 22	05 47	02 13	10 30	22 04
06 We	09 05 11	17 21 22	19 11 20	00♓ 39	04 52	03 06	06 24	11 23	05 50	02 15	10 31	22 01
07 Th	09 09 07	18 22 11	03♑ 40 42	02 21	06 07	03 53	06 25	11 24	05 52	02 17	10 33	21 56
08 Fr	09 13 04	19 22 59	18 16 02	04 02	07 22	04 41	06 27	11 26	05 55	02 19	10 35	21 48
09 Sa	09 17 00	20 23 45	02♒ 51 21	05 41	08 37	05 28	06 29	11 27	05 58	02 22	10 37	21 38
10 Su	09 20 57	21 24 31	17 19 33	07 17	09 53	06 15	06 31	11 28	06 00	02 24	10 38	21 27
11 Mo	09 24 53	22 25 15	01♓ 33 34	08 51	11 08	07 03	06 33	11 28	06 03	02 26	10 40	21 15
12 Tu	09 28 50	23 25 57	15 27 32	10 21	12 23	07 50	06 36	11 29	06 06	02 28	10 42	21 03
13 We	09 32 46	24 26 38	28 57 43	11 46	13 38	08 37	06 38	11 30	06 08	02 31	10 44	20 53
14 Th	09 36 43	25 27 18	12♈ 02 57	13 07	14 53	09 25	06 41	11 30	06 11	02 33	10 45	20 46
15 Fr	09 40 40	26 27 56	24 44 24	14 23	16 08	10 12	06 44	11 31	06 14	02 35	10 47	20 42
16 Sa	09 44 36	27 28 32	07♉ 05 14	15 32	17 23	10 59	06 47	11 31	06 17	02 37	10 48	20 40
17 Su	09 48 33	28 29 06	19 09 55	16 35	18 38	11 47	06 51	11 31	06 20	02 40	10 50	20D 40
18 Mo	09 52 29	29 29 39	01♊ 03 42	17 30	19 53	12 34	06 54	11 32	06 22	02 42	10 51	20 41
19 Tu	09 56 26	00♓ 30 10	12 52 12	18 17	21 08	13 21	06 58	11R 32	06 25	02 44	10 53	20R 40
20 We	10 00 22	01 30 39	24 40 54	18 55	22 24	14 08	07 02	11 32	06 28	02 47	10 55	20 39
21 Th	10 04 19	02 31 07	06♋ 34 55	19 24	23 39	14 56	07 06	11 31	06 31	02 49	10 56	20 34
22 Fr	10 08 15	03 31 32	18 38 40	19 43	24 54	15 43	07 10	11 31	06 34	02 51	10 57	20 27
23 Sa	10 12 12	04 31 56	00♌ 55 34	19 52	26 09	16 30	07 15	11 31	06 37	02 53	10 59	20 17
24 Su	10 16 09	05 32 18	13 27 53	19R 51	27 24	17 17	07 19	11 30	06 40	02 56	11 00	20 04
25 Mo	10 20 05	06 32 38	26 16 35	19 40	28 39	18 04	07 24	11 30	06 43	02 58	11 02	19 51
26 Tu	10 24 02	07 32 56	09♍ 21 19	19 19	29 54	18 51	07 29	11 29	06 46	03 00	11 03	19 39
27 We	10 27 58	08 33 13	22 40 35	18 50	01♓ 09	19 38	07 35	11 28	06 49	03 03	11 04	19 27
28 Th	10 31 55	09 33 28	06♎ 12 12	18 12	02 24	20 25	07 40	11 27	06 53	03 05	11 06	19 18

Data for	02-01-2013
Julian Day	2456324.50
Ayanamsa	24 02 37
SVP	05 ♓ 04 52
☽ ☊ Mean	21 ♏ 57 R

● ◐ PHASES ○ ◑

03	13:56	◑	14♏54
10	07:21	●	21♒43
17	20:31	◐	29♉21
25	20:26	○	07♍24

LAST ASPECT Day	h m	☽ INGRESS Day	h m	
02	01:03	02	12:02	♏
04	12:32	04	15:46	♐
05	20:43	06	17:56	♑
07	12:45	08	19:18	♒
10	07:21	10	21:21	♓
11	17:04	13	01:53	♈
15	03:37	15	10:09	♉
17	20:32	17	21:51	♊
19	18:48	20	10:45	♋
22	02:08	22	22:12	♌
25	04:51	25	06:53	♍
26	18:13	27	13:02	♎

DECLINATION

Day	☉	☽	☿	♀	♂	♃	♄	♅	♆	♇
01 Fr	17S 07	06S 58	15S 50	21S 01	12S 44	20N47	12S 52	01N36	11S 18	19S 45
02 Sa	16 50	11 16	15 10	20 47	12 27	20 47	12 53	01 37	11 18	19 45
03 Su	16 32	15 03	14 29	20 33	12 10	20 48	12 53	01 38	11 17	19 45
04 Mo	16 14	18 03	13 47	20 19	11 53	20 48	12 53	01 39	11 16	19 45
05 Tu	15 56	20 00	13 04	20 03	11 36	20 48	12 53	01 40	11 15	19 44
06 We	15 38	20 42	12 20	19 48	11 18	20 49	12 54	01 41	11 14	19 44
07 Th	15 19	20 02	11 35	19 31	11 01	20 49	12 54	01 42	11 14	19 44
08 Fr	15 00	18 03	10 50	19 14	10 43	20 50	12 54	01 43	11 13	19 44
09 Sa	14 41	14 54	10 04	18 56	10 26	20 50	12 54	01 44	11 12	19 44
10 Su	14 22	10 52	09 18	18 38	10 08	20 51	12 54	01 45	11 11	19 44
11 Mo	14 02	06 18	08 32	18 20	09 50	20 52	12 54	01 47	11 10	19 44
12 Tu	13 42	01 31	07 47	18 00	09 32	20 52	12 54	01 48	11 10	19 44
13 We	13 22	03N14	07 02	17 41	09 14	20 53	12 54	01 49	11 09	19 44
14 Th	13 02	07 40	06 18	17 21	08 56	20 53	12 54	01 50	11 08	19 44
15 Fr	12 42	11 39	05 36	17 00	08 37	20 54	12 54	01 51	11 07	19 43
16 Sa	12 21	15 00	04 56	16 39	08 19	20 55	12 54	01 52	11 06	19 43
17 Su	12 00	17 39	04 17	16 17	08 01	20 56	12 54	01 53	11 06	19 43
18 Mo	11 39	19 29	03 42	15 55	07 42	20 57	12 53	01 54	11 05	19 43
19 Tu	11 18	20 27	03 09	15 32	07 24	20 57	12 53	01 56	11 04	19 43
20 We	10 56	20 30	02 40	15 09	07 05	20 58	12 53	01 57	11 03	19 43
21 Th	10 35	19 39	02 15	14 46	06 47	20 59	12 53	01 58	11 02	19 43
22 Fr	10 13	17 54	01 54	14 22	06 28	21 00	12 52	01 59	11 02	19 43
23 Sa	09 51	15 19	01 37	13 58	06 09	21 01	12 52	02 00	11 01	19 43
24 Su	09 29	11 59	01 25	13 33	05 50	21 02	12 52	02 02	11 00	19 43
25 Mo	09 07	08 03	01 18	13 08	05 32	21 03	12 51	02 03	10 59	19 42
26 Tu	08 44	03 41	01 16	12 43	05 13	21 04	12 51	02 04	10 58	19 42
27 We	08 22	00S 58	01 18	12 18	04 54	21 05	12 50	02 05	10 57	19 42
28 Th	07 59	05 38	01 26	11 52	04 35	21 06	12 50	02 07	10 57	19 42

ASPECTARIAN

01 01:16 ☽ □ ♇
05:04 ☽ △ ☉

02 00:09 ☽ ∥ ♆
00:12 ♀ ∦ ♃
01:03 ☽ △ ☿
01:54 ♂ ♓
02:47 ♀ ♒
06:39 ☽ ∥ ♂
09:44 ☽ ∥ ♄
12:39 ☽ △ ♂
12:58 ☽ □ ♀
15:45 ☽ △ ♆
20:43 ☽ ∥ ☿

03 06:14 ☽ ✶ ♇
07:45 ☽ ☌ ♄
09:58 ☽ ∥ ☉

04 12:32 ☽ □ ☿
19:25 ☽ □ ♂
19:30 ☽ □ ♆
19:44 ☽ ∥ ♇
20:58 ♂ ☌ ♆
21:41 ☽ ✶ ♀

05 00:48 ☽ ∥ ♀
01:33 ☽ △ ♅
02:33 ☽ ☍ ♃
05:43 ☿ ∥ ♄
14:56 ☿ ♓
20:43 ☽ ✶ ☉

06 04:40 ♀ ∥ ♇
05:47 ♂ ∥ ♆
19:08 ♀ ✶ ♅
21:32 ☽ ✶ ☿
21:42 ☽ ✶ [illegible]
23:00 ♀ ☌ ♆

07 00:22 ☽ ✶ ♂
03:38 ☽ □ ♅
04:53 ☽ ∥ ♇
05:58 ♀ △ ♃
09:26 ☽ ∥ ♀
11:21 ☽ ☌ ♇
11:33 ☿ ∥ ♆
12:45 ☽ ✶ ♄

08 05:26 ☿ ∥ ♂
17:57 ☿ ☌ ♂

09 01:31 ☽ ∥ ☉
05:08 ☽ ✶ ♅
06:00 ☽ △ ♃
10:26 ☽ ☌ ♀
12:08 ☿ □ ♃
12:27 ☽ ∥ ♄
14:13 ☽ □ ♄
22:17 ☽ ∥ ♆

10 04:23 ☽ ∥ ♂
08:12 ♂ □ ♃
10:16 ☽ ∥ ☿

11 01:30 ☽ ☌ ♆
06:43 ♀ □ ♄
08:34 ☽ □ ♃
09:58 ☽ ☌ ♂
14:01 ☽ ☌ ☿
15:41 ☽ ✶ ♇
17:04 ☽ △ ♄
22:38 ☽ ∦ ♅

12 05:58 ☿ ✶ ♇
16:1[illegible] ☽ ∥ ♆
19:16 ☿ △ ♄

13 13:07 ☽ ☌ ♅
14:02 ☽ ✶ ♃
17:30 ☽ ∦ ☿
21:35 ☽ □ ♆

14 05:52 ☽ ✶ ♀
06:47 ☽ ∦ ♂
09:32 ☉ ∥ ♄
20:41 ☽ ∦ ♆

15 03:37 ☽ ✶ ☉
06:25 ☽ ∦ ☉
08:27 ☽ ∦ ♄
15:14 ☽ ✶ ♆
18:15 ♂ ✶ ♇

16 07:21 ☽ △ ♇
08:14 ☽ ✶ ♂
08:46 ☽ ☍ ♄
12:24 ☽ ∦ ♀
16:15 ♂ △ ♄
18:22 ☽ ✶ ☿
22:49 ☽ □ ♀

18 03:20 ☽ □ ♆
04:29 ☽ ∦ ♇
10:50 ☽ ✶ ♅
11:55 ☽ ☌ ♃
12:02 ☉ ♓
17:04 ♄ SR

19 01:03 ☽ □ ♂
11:40 ☽ □ ☿
13:32 [illegible] ∥ ♆
18:48 ☽ △ ♀

20 15:05 ☽ △ ☉
16:24 ☽ △ ♆
22:39 ☽ ∦ ♇
23:53 ☽ □ ♅

21 07:19 ☉ ☌ ♆
08:43 ☽ ☍ ♇
09:52 ☽ △ ♄
17:34 ☿ ∦ ♅
17:48 ☽ △ ♂

22 02:08 ☽ △ ☿

23 09:42 ☿ SR
11:00 ☽ △ [illegible]
11:48 ☽ ∦ ♀
12:14 ☽ ✶ ♃
18:11 ☽ ∦ ♄
20:17 ☽ □ ♄

24 06:24 ☽ ∦ ♆
17:17 ☽ ∦ ☉

25 04:51 ☽ ☍ ♀
12:22 ☽ ☍ ♆
15:10 ☽ ∦ ♂
16:40 ♀ ∥ ♄
20:35 ☽ □ ♃
22:27 ☉ □ ♃

26 [illegible]
03:05 ☽ △ ♆
03:51 ☽ ✶ ♄
08:23 ☽ ∥ ♅
09:10 ☿ ☌ ♂
12:31 ☽ ∦ ☿
17:23 ☽ ☍ ☿
18:13 ☽ ☍ ♂

27 01:48 ☽ ∥ ☿
05:47 ☽ ∦ ♅
18:55 ☽ ∥ ♂

28 01:11 ☽ ☍ ♅
02:36 ☽ △ ♃
08:37 ☽ □ [illegible]
11:31 ☽ ∥ ☉
13:37 ♀ ☌ ♆

⚷ Chiron

01 Dec.	03 S 58
03	07♓47
06	07 58
09	08 09
12	08 21
15	08 32
18	08 44
21	08 55
24	09 07
27	09 18

March 2013

Day	S. T. h m s	☉ ° ' "	☽ ° ' "	☿ ° '	♀ ° '	♂ ° '	♃ ° '	♄ ° '	♅ ° '	♆ ° '	♇ ° '	☊ True ° '
01 Fr	10 35 51	10♓33 41	19♎53 42	17♓R27	03♓38	21♓12	07♊45	11♏R26	06♈56	03♓07	11♑07	19♏R12
02 Sa	10 39 48	11 33 53	03♏42 56	16 35	04 53	21 59	07 51	11 25	06 59	03 09	11 08	19 08
03 Su	10 43 44	12 34 03	17♏38 14	15 39	06 08	22 46	07 57	11 24	07 02	03 12	11 09	19 06
04 Mo	10 47 41	13 34 12	01 38 33	14 40	07 23	23 33	08 03	11 22	07 05	03 14	11 11	19 06
05 Tu	10 51 37	14 34 19	15♐43 11	13 38	08 38	24 20	08 09	11 21	07 08	03 16	11 12	19 05
06 We	10 55 34	15 34 25	29 51 20	12 36	09 53	25 07	08 16	11 20	07 12	03 18	11 13	19 03
07 Th	10 59 31	16 34 29	14♑01 43	11 36	11 08	25 54	08 22	11 18	07 15	03 21	11 14	19 00
08 Fr	11 03 27	17 34 32	28 12 11	10 37	12 23	26 40	08 29	11 16	07 18	03 23	11 15	18 54
09 Sa	11 07 24	18 34 33	12♒19 34	09 42	13 38	27 27	08 36	11 14	07 22	03 25	11 16	18 47
10 Su	11 11 20	19 34 32	26 19 57	08 51	14 53	28 14	08 43	11 13	07 25	03 27	11 17	18 37
11 Mo	11 15 17	20 34 30	10♓09 02	08 05	16 07	29 01	08 50	11 11	07 28	03 29	11 18	18 28
12 Tu	11 19 13	21 34 25	23 42 54	07 25	17 22	29 47	08 57	11 09	07 31	03 32	11 19	18 18
13 We	11 23 10	22 34 19	06♈58 35	06 51	18 37	00♈34	09 04	11 06	07 35	03 34	11 20	18 10
14 Th	11 27 06	23 34 10	19 54 41	06 24	19 52	01 21	09 12	11 04	07 38	03 36	11 21	18 05
15 Fr	11 31 03	24 34 00	02♉31 27	06 03	21 07	02 07	09 20	11 02	07 41	03 38	11 22	18 01
16 Sa	11 34 60	25 33 47	14 50 45	05 48	22 22	02 54	09 28	10 59	07 45	03 40	11 23	18D 01
17 Su	11 38 56	26 33 32	26 55 46	05 40	23 36	03 40	09 36	10 57	07 48	03 42	11 24	18 02
18 Mo	11 42 53	27 33 15	08♊50 43	05D 38	24 51	04 27	09 44	10 54	07 52	03 45	11 25	18 04
19 Tu	11 46 49	28 32 56	20 40 28	05 42	26 06	05 13	09 52	10 51	07 55	03 47	11 25	18 05
20 We	11 50 46	29 32 35	02♋30 17	05 52	27 20	06 00	10 00	10 49	07 58	03 49	11 26	18R 06
21 Th	11 54 42	00♈32 11	14 25 24	06 07	28 35	06 46	10 09	10 46	08 02	03 51	11 27	18 05
22 Fr	11 58 39	01 31 45	26 30 50	06 27	29 50	07 33	10 18	10 43	08 05	03 53	11 28	18 02
23 Sa	12 02 35	02 31 17	08♌50 53	06 52	01♈05	08 19	10 26	10 40	08 09	03 55	11 28	17 57
24 Su	12 06 32	03 30 46	21 28 59	07 22	02 19	09 05	10 35	10 37	08 12	03 57	11 29	17 50
25 Mo	12 10 29	04 30 13	04♍27 13	07 56	03 34	09 51	10 44	10 34	08 15	03 59	11 30	17 42
26 Tu	12 14 25	05 29 38	17 46 08	08 35	04 48	10 37	10 54	10 30	08 19	04 01	11 30	17 34
27 We	12 18 22	06 29 01	01♎24 33	09 17	06 03	11 24	11 03	10 27	08 22	04 03	11 31	17 27
28 Th	12 22 18	07 28 22	15 19 43	10 03	07 18	12 10	11 12	10 24	08 26	04 05	11 31	17 21
29 Fr	12 26 15	08 27 41	29 27 41	10 52	08 32	12 56	11 22	10 20	08 29	04 07	11 32	17 17
30 Sa	12 30 11	09 26 58	13♏43 53	11 44	09 47	13 42	11 31	10 17	08 33	04 09	11 32	17 16
31 Su	12 34 08	10 26 13	28 03 53	12 40	11 01	14 28	11 41	10 13	08 36	04 11	11 32	17D 15

Data for	03-01-2013
Julian Day	2456352.50
Ayanamsa	24 02 40
SVP	05 ♓ 04 47
☽ ☊ Mean	20 ♏ 28 R

● ◐ PHASES ○ ◑

04	21:53	◑	14♐29
11	19:52	●	21♓24
19	17:27	◐	29♊16
27	09:28	○	06♎52

LAST ASPECT ☽ Day	h m	INGRESS Day	h m	
28	08:37	01	17:34	♏
03	09:20	03	21:11	♐
05	15:29	06	00:15	♑
07	21:16	08	03:03	♒
08	22:09	10	06:20	♓
11	19:52	12	11:18	♈
13	08:02	14	19:09	♉
16	23:11	17	06:10	♊
19	17:27	19	18:56	♋
20	18:02	22	06:50	♌
23	03:28	24	15:50	♍
25	12:46	26	21:33	♎
27	18:16	29	00:55	♏
29	20:26	31	03:14	♐

DECLINATION

Day	☉	☽	☿	♀	♂	♃	♄	♅	♆	♇
01 Fr	07S 36	10S 05	01S 38	11S 25	04S 16	21N07	12S 49	02N08	10S 56	19S 42
02 Sa	07 14	14 02	01 55	10 59	03 57	21 08	12 49	02 09	10 55	19 42
03 Su	06 51	17 13	02 15	10 32	03 38	21 10	12 48	02 10	10 54	19 42
04 Mo	06 28	19 25	02 38	10 05	03 19	21 11	12 48	02 12	10 53	19 42
05 Tu	06 04	20 25	03 04	09 37	03 00	21 12	12 47	02 13	10 53	19 42
06 We	05 41	20 07	03 32	09 09	02 41	21 13	12 46	02 14	10 52	19 42
07 Th	05 18	18 34	04 01	08 41	02 22	21 14	12 46	02 15	10 51	19 41
08 Fr	04 55	15 52	04 31	08 13	02 03	21 16	12 45	02 17	10 50	19 41
09 Sa	04 31	12 16	05 01	07 45	01 44	21 17	12 44	02 18	10 49	19 41
10 Su	04 08	08 01	05 30	07 16	01 25	21 18	12 43	02 19	10 49	19 41
11 Mo	03 44	03 23	05 59	06 48	01 06	21 19	12 43	02 21	10 48	19 41
12 Tu	03 20	01N19	06 26	06 19	00 47	21 21	12 42	02 22	10 47	19 41
13 We	02 57	05 52	06 51	05 49	00 28	21 22	12 41	02 23	10 46	19 41
14 Th	02 33	10 02	07 15	05 20	00 09	21 23	12 40	02 25	10 46	19 41
15 Fr	02 10	13 38	07 36	04 51	00N10	21 25	12 39	02 26	10 45	19 41
16 Sa	01 46	16 34	07 55	04 21	00 29	21 26	12 38	02 27	10 44	19 41
17 Su	01 22	18 41	08 12	03 51	00 48	21 27	12 37	02 29	10 43	19 41
18 Mo	00 58	19 58	08 26	03 22	01 07	21 29	12 36	02 30	10 43	19 41
19 Tu	00 35	20 20	08 38	02 52	01 26	21 30	12 35	02 31	10 42	19 41
20 We	00 11	19 48	08 48	02 22	01 45	21 32	12 34	02 33	10 41	19 41
21 Th	00N13	18 24	08 55	01 52	02 04	21 33	12 33	02 34	10 40	19 40
22 Fr	00 36	16 09	09 00	01 22	02 23	21 34	12 32	02 35	10 40	19 40
23 Sa	01 00	13 09	09 02	00 51	02 42	21 36	12 31	02 37	10 39	19 40
24 Su	01 24	09 29	09 03	00 21	03 00	21 37	12 30	02 38	10 38	19 40
25 Mo	01 47	05 17	09 01	00N09	03 19	21 39	12 29	02 40	10 37	19 40
26 Tu	02 11	00 44	08 58	00 39	03 38	21 40	12 28	02 41	10 37	19 40
27 We	02 34	03S 58	08 52	01 09	03 56	21 42	12 26	02 42	10 36	19 40
28 Th	02 58	08 34	08 44	01 40	04 15	21 43	12 25	02 44	10 35	19 40
29 Fr	03 21	12 46	08 35	02 10	04 33	21 45	12 24	02 45	10 35	19 40
30 Sa	03 45	16 16	08 23	02 40	04 52	21 46	12 23	02 46	10 34	19 40
31 Su	04 08	18 47	08 10	03 10	05 10	21 48	12 22	02 48	10 33	19 40

ASPECTARIAN

01 04:55 ☽ ∥ ♆
07:02 ☽ ∥ ♀
13:36 ☉ ⚹ ♇
16:15 ☽ ∥ ♄
20:32 ☉ △ ♄
23:02 ☽ △ ♆
02 02:14 ☽ △ ♀
03:18 ♀ ∥ ♆
12:50 ☽ ⚹ ♇
13:17 ☽ ☌ ♄
14:36 ☽ △ ☉
18:51 ☿ # ♅
20:49 ☽ △ ☿
03 09:20 ☽ △ ♂
04 02:43 ☽ □ ♆
04:42 ☽ ∥ ♇
09:20 ☽ △ ♅
10:46 ☽ □ ♀
11:01 ☽ ☍ ♃
12:58 ☿ ☌ ☉
13:53 ♀ □ ♃
20:42 ☽ □ ☿
21:58 ☿ ∥ ♂
05 15:29 ☽ □ ♂
06 05:52 ☽ ⚹ ♆
08:39 ☽ ∥ ♇
12:29 ☽ □ ♅
18:38 ☽ ⚹ ♀
19:16 ☽ ☌ ♇
19:23 ☽ ⚹ ♄
20:09 ☽ ⚹ ☿
07 02:02 ♀ ⚹ ♇
03:08 ♀ △ ♄
04:38 ☽ ⚹ ☉
04:55 ☿ ☌ ♀
07:17 ☿ △ ♄
07:36 ♂ # ♅
08:28 ☿ ⚹ ♇
21:16 ☽ ⚹ ♂
08 08:04 ♄ ⚹ ♇
10:40 ☿ ∥ ☉
15:31 ☽ ⚹ ♅
17:35 ☽ △ ♃
21:05 ☽ ∥ ♄
22:09 ☽ □ ♄
09 08:27 ☽ ∥ ♆
10 03:38 ☿ □ ♃
04:21 ☽ ∥ ♀
11:55 ☽ ∥ ☿
12:21 ☽ ☌ ♆
20:34 ☽ ☌ ☿
21:40 ☽ □ ♃
22:04 ☽ ∥ ☉
11 01:48 ☽ △ ♄
02:02 ☽ ⚹ ♇
05:17 ☽ # ♅
11:34 ☽ ☌ ♀
12:31 ☽ ∥ ♂
20:57 ☿ ∥ ♀
21:25 ☽ # ♂
12 05:28 ☽ ∥ ♅
06:26 ♂ ♈
09:42 ☽ # ☉
11:36 ☽ ☌ ♂
23:51 ☽ # ♀
13 01:07 ☽ ☌ ♅
03:53 ☽ ⚹ ♃
06:04 ☽ # ☿
08:02 ☽ □ ♇
14 04:35 ☽ # ♆
08:08 ☉ # ♅
17:03 ☽ # ♄
15 02:09 ☽ ⚹ ♆
06:40 ☽ ⚹ ☿
16:27 ☽ ☍ ♄
17:12 ☽ △ ♇
16 16:35 ☽ ⚹ ♀
23:11 ☽ ⚹ ☉
17 13:40 ☽ □ ♆
14:30 ☽ ⚹ ♂
16:56 ☽ # ♇
17:30 ☽ □ ☿
18:57 ☉ # ♂
20:04 ☿ SD
22:00 ☽ ⚹ ♅
18 01:49 ☽ ☌ ♃
19 12:18 ☽ □ ♀
15:32 ♀ # ♅
20 02:39 ☽ △ ♆
02:53 ☽ # ♇
06:55 ☽ △ ☿
07:33 ☽ □ ♂
11:02 ☉ ♈
11:06 ☽ □ ♅
16:41 ☽ △ ♄
17:57 ♀ # ♂
18:02 ☽ ☍ ♇
22 03:16 ♀ ♈
07:14 ☽ △ ♀
10:41 ☽ △ ☉
17:29 ♂ ∥ ♅
18:17 ♂ ☌ ♅
20:06 ♀ # ☉
22:38 ☽ △ ♅
22:54 ☽ △ ♂
23 03:06 ☽ ⚹ ♃
03:28 ☽ □ ♄
04:23 ☽ # ♄
16:46 ☽ # ♆
24 02:33 ☽ # ☿
23:09 ☽ ☍ ♆
25 06:39 ☽ ☍ ☿
09:52 ☽ ∥ ♂
11:02 ☽ ⚹ ♄
11:32 ☽ □ ♃
12:46 ☽ △ ♇
13:55 ☽ ∥ ♅
17:04 ☽ ∥ ☉
26 00:21 ☽ ∥ ♀
07:56 ☽ # ♀
10:27 ♂ ⚹ ♃
16:13 ☽ # ☉
17:30 ☽ # ♅
23:48 ☽ # ♂
27 03:45 ♂ □ ♇
08:28 ☉ ∥ ♅
08:50 ☽ ☍ ♀
12:06 ☽ ☍ ♅
16:51 ☽ △ ♃
17:28 ☽ □ ♇
18:16 ☽ ☍ ♂
28 00:53 ☽ ∥ ☿
09:49 ☿ △ ♄
11:07 ☽ ∥ ♆
17:05 ♀ ☌ ☉
21:46 ☽ ∥ ♄
23:01 ♀ ☌ ♅
29 00:38 ☉ ☌ ♅
07:52 ☽ △ ♆
16:56 ☿ □ ♃
18:14 ☽ ☌ ♄
18:32 ☿ ⚹ ♇
20:19 ☽ ⚹ ♇
20:26 ☽ △ ☿
30 05:16 ♀ ∥ ♅
31 10:11 ♀ □ ♇
10:16 ☽ □ ♆
13:27 ☽ ∥ ♇
14:49 ♀ ⚹ ♃
17:42 ☽ △ ♅
22:15 ☽ △ ☉
23:04 ☽ ☍ ♃
23:45 ☽ △ ♀

⚷ Chiron

01 Dec.	03 S 21
02	09♓30
05	09 42
08	09 53
11	10 05
14	10 16
17	10 28
20	10 39
23	10 50
26	11 01
29	11 12

April 2013

25 20:09 05♏46 ✸ Total Lunar Eclipse (mag 0.020)

Day	S. T. h m s	☉ ° ' "	☽ ° ' "	☿ ° '	♀ ° '	♂ ° '	♃ ° '	♄ ° '	♅ ° '	♆ ° '	♇ ° '	☊ True ° '
01 Mo	12 38 04	11♈25 26	12♐23 48	13♓38	12♈16	15♈14	11♊51	10♏R09	08♈39	04♓13	11♑33	17♏16
02 Tu	12 42 01	12 24 38	26 40 34	14 40	13 30	16 00	12 01	10 06	08 43	04 15	11 33	17 17
03 We	12 45 58	13 23 48	10♑51 51	15 43	14 45	16 45	12 11	10 02	08 46	04 17	11 34	17R 17
04 Th	12 49 54	14 22 56	24 55 55	16 50	15 59	17 31	12 21	09 58	08 50	04 19	11 34	17 17
05 Fr	12 53 51	15 22 03	08♒51 23	17 58	17 14	18 17	12 31	09 54	08 53	04 20	11 34	17 15
06 Sa	12 57 47	16 21 07	22 36 56	19 10	18 28	19 03	12 42	09 50	08 57	04 22	11 34	17 11
07 Su	13 01 44	17 20 10	06♓11 19	20 23	19 42	19 48	12 52	09 46	09 00	04 24	11 35	17 07
08 Mo	13 05 40	18 19 11	19 33 15	21 38	20 57	20 34	13 03	09 42	09 03	04 26	11 35	17 02
09 Tu	13 09 37	19 18 10	02♈41 36	22 55	22 11	21 20	13 13	09 38	09 07	04 28	11 35	16 58
10 We	13 13 33	20 17 07	15 35 30	24 15	23 25	22 05	13 24	09 34	09 10	04 29	11 35	16 55
11 Th	13 17 30	21 16 02	28 14 39	25 36	24 40	22 51	13 35	09 30	09 13	04 31	11 35	16 53
12 Fr	13 21 26	22 14 55	10♉39 34	26 59	25 54	23 36	13 46	09 26	09 17	04 33	11 35	16D 52
13 Sa	13 25 23	23 13 47	22 51 34	28 24	27 08	24 21	13 57	09 21	09 20	04 34	11R 35	16 53
14 Su	13 29 20	24 12 36	04♊52 53	29 50	28 23	25 07	14 08	09 17	09 24	04 36	11 35	16 55
15 Mo	13 33 16	25 11 22	16 46 31	01♈19	29 37	25 52	14 19	09 13	09 27	04 38	11 35	16 57
16 Tu	13 37 13	26 10 07	28 36 11	02 49	00♉51	26 37	14 31	09 08	09 30	04 39	11 35	17 00
17 We	13 41 09	27 08 50	10♋26 10	04 21	02 06	27 23	14 42	09 04	09 34	04 41	11 35	17 02
18 Th	13 45 06	28 07 30	22 21 06	05 54	03 20	28 08	14 54	09 00	09 37	04 42	11 35	17 03
19 Fr	13 49 02	29 06 08	04♌25 49	07 30	04 34	28 53	15 05	08 55	09 40	04 44	11 34	17R 03
20 Sa	13 52 59	00♉04 44	16 44 59	09 06	05 48	29 38	15 17	08 51	09 43	04 45	11 34	17 02
21 Su	13 56 55	01 03 17	29 22 46	10 45	07 02	00♉23	15 28	08 46	09 47	04 47	11 34	17 00
22 Mo	14 00 52	02 01 49	12♍22 29	12 25	08 16	01 08	15 40	08 42	09 50	04 48	11 34	16 58
23 Tu	14 04 49	03 00 18	25 46 07	14 07	09 30	01 53	15 52	08 37	09 53	04 50	11 33	16 55
24 We	14 08 45	03 58 45	09♎33 55	15 51	10 45	02 38	16 04	08 33	09 56	04 51	11 33	16 53
25 Th	14 12 42	04 57 11	23 44 02	17 36	11 59	03 22	16 16	08 28	09 59	04 52	11 33	16 51
26 Fr	14 16 38	05 55 34	08♏12 31	19 23	13 13	04 07	16 28	08 24	10 03	04 54	11 32	16 50
27 Sa	14 20 35	06 53 56	22 53 37	21 12	14 27	04 52	16 40	08 19	10 06	04 55	11 32	16 49
28 Su	14 24 31	07 52 16	07♐40 30	23 02	15 41	05 36	16 52	08 14	10 09	04 56	11 32	16D 49
29 Mo	14 28 28	08 50 34	22 26 12	24 54	16 55	06 21	17 05	08 10	10 12	04 58	11 31	16 50
30 Tu	14 32 24	09 48 51	07♑04 26	26 48	18 09	07 06	17 17	08 05	10 15	04 59	11 31	16 50

Data for	04-01-2013
Julian Day	2456383.50
Ayanamsa	24 02 43
SVP	05 ♓ 04 42
☽ ☊ Mean	18 ♏ 50 R

● ◐ PHASES ○ ◑

Day	Time	Phase	Position
03	04:37	◑	13♑35
10	09:35	●	20♈41
18	12:32	◐	28♋38
25	19:58	✸	05♏46

LAST ASPECT Day	h m	☽ INGRESS Day	h m	Sign
01	05:01	02	05:36	♑
03	10:36	04	08:42	♒
05	17:22	06	13:01	♓
08	04:10	08	19:03	♈
10	16:25	11	03:22	♉
13	12:31	13	14:13	♊
15	19:42	16	02:50	♋
18	12:32	18	15:15	♌
19	21:07	21	01:10	♍
22	06:04	23	07:26	♎
24	12:13	25	10:27	♏
26	08:57	27	11:33	♐
29	04:38	29	12:22	♑

DECLINATION

Day	☉	☽	☿	♀	♂	♃	♄	♅	♆	♇
01 Mo	04N31	20S 05	07S 55	03N40	05N28	21N49	12S 20	02N49	10S 32	19S 40
02 Tu	04 54	20 06	07 39	04 10	05 46	21 51	12 19	02 50	10 32	19 40
03 We	05 17	18 49	07 21	04 40	06 04	21 52	12 18	02 52	10 31	19 40
04 Th	05 40	16 24	07 01	05 10	06 23	21 54	12 16	02 53	10 31	19 40
05 Fr	06 03	13 04	06 40	05 39	06 41	21 55	12 15	02 54	10 30	19 40
06 Sa	06 26	09 04	06 17	06 09	06 58	21 56	12 14	02 56	10 29	19 40
07 Su	06 48	04 40	05 53	06 38	07 16	21 58	12 12	02 57	10 29	19 40
08 Mo	07 11	00 05	05 27	07 08	07 34	21 59	12 11	02 58	10 28	19 40
09 Tu	07 33	04N25	05 00	07 37	07 52	22 01	12 10	03 00	10 27	19 40
10 We	07 56	08 39	04 32	08 06	08 09	22 02	12 08	03 01	10 27	19 40
11 Th	08 18	12 25	04 02	08 34	08 27	22 04	12 07	03 02	10 26	19 40
12 Fr	08 40	15 33	03 31	09 03	08 44	22 05	12 06	03 04	10 26	19 40
13 Sa	09 02	17 57	02 59	09 31	09 01	22 07	12 04	03 05	10 25	19 40
14 Su	09 23	19 30	02 26	10 00	09 18	22 08	12 03	03 06	10 24	19 40
15 Mo	09 45	20 09	01 52	10 27	09 36	22 10	12 01	03 08	10 24	19 40
16 Tu	10 06	19 55	01 16	10 55	09 52	22 11	12 00	03 09	10 23	19 40
17 We	10 27	18 48	00 39	11 22	10 09	22 13	11 58	03 10	10 23	19 40
18 Th	10 48	16 51	00 02	11 50	10 26	22 14	11 57	03 12	10 22	19 40
19 Fr	11 09	14 09	00N37	12 16	10 43	22 15	11 56	03 13	10 22	19 40
20 Sa	11 30	10 47	01 17	12 43	10 59	22 17	11 54	03 14	10 21	19 40
21 Su	11 50	06 52	01 58	13 09	11 16	22 18	11 53	03 15	10 21	19 40
22 Mo	12 11	02 31	02 39	13 35	11 32	22 20	11 51	03 17	10 20	19 40
23 Tu	12 31	02S 05	03 22	14 01	11 48	22 21	11 50	03 18	10 20	19 40
24 We	12 51	06 44	04 05	14 26	12 04	22 22	11 48	03 19	10 19	19 40
25 Th	13 10	11 08	04 50	14 51	12 20	22 24	11 47	03 20	10 19	19 40
26 Fr	13 30	14 59	05 35	15 15	12 36	22 25	11 45	03 22	10 18	19 40
27 Sa	13 49	17 57	06 20	15 40	12 51	22 27	11 44	03 23	10 18	19 40
28 Su	14 08	19 43	07 07	16 03	13 07	22 28	11 43	03 24	10 17	19 40
29 Mo	14 27	20 08	07 54	16 27	13 22	22 29	11 41	03 25	10 17	19 40
30 Tu	14 45	19 11	08 41	16 50	13 37	22 30	11 40	03 26	10 16	19 40

ASPECTARIAN

01 02:14 ☽ □ ☿
03:04 ☉ □ ♇
05:01 ☽ △ ♂
12:25 ☉ ⚹ ♃
02 10:48 ☽ ∥ ♇
12:49 ☽ ⚹ ♆
20:26 ☽ □ ♅
22:36 ☽ ⚹ ♄

03 01:11 ☽ ☌ ♇
07:14 ☽ □ ♀
08:58 ☽ ⚹ ☿
10:36 ☽ □ ♂
04 23:26 ☿ # ♂

05 00:03 ☽ ⚹ ♅
01:48 ☽ □ ♄
05:13 ☽ ∥ ♄
06:27 ☽ △ ♃
12:11 ☽ ⚹ ☉
15:50 ☽ ∥ ♆
16:00 ☽ ⚹ ♀
17:22 ☽ ⚹ ♂
19:22 ☿ # ☉

06 03:34 ☿ # ♀
10:51 ☽ # ♂
13:24 ☽ # ☉
14:27 ☽ # ♀
16:50 ☽ ∥ ☿
20:49 ☽ ☌ ♆

07 04:58 ♀ ☌ ♂
06:22 ☽ △ ♄
08:57 ☽ # ♅
09:38 ☽ ⚹ ♇
12:07 ☽ □ ♃
08 04:10 ☽ ☌ ☿
11:34 ☿ ∥ ☉
16:16 ☽ ∥ ♅
09 02:53 ☽ # ☿
11:56 ☽ ☌ ♅
16:29 ☽ □ ♇
19:19 ☽ ∥ ☉
19:50 ☽ ⚹ ♃
20:18 ☽ ∥ ♀
20:51 ☽ ∥ ♂

10 07:25 ♀ ∥ ♂
10:59 ☽ # ♆
13:02 ☽ ☌ ♂
16:25 ☽ ☌ ♀
21:58 ☽ # ♄

11 12:06 ☽ ⚹ ♆
21:37 ☽ ☍ ♄
12 01:48 ☽ △ ♇
19:33 ♇ SR
19:59 ☿ # ♅
22:57 ☉ ∥ ♂

13 12:31 ☽ ⚹ ☿
23:26 ☽ □ ♆
14 02:37 ☿ ♈
03:45 ☽ # ♇
09:07 ☽ ⚹ ♅
18:58 ☽ ☌ ♃
20:56 ♀ # ♆

15 07:25 ♀ ♉
18:37 ☽ ⚹ ☉
19:42 ☽ ⚹ ♂
16 05:06 ☽ ⚹ [illegible]
07:35 ☽ # ♇
09:49 ☽ □ ☿
12:18 ☽ △ ♆
18:52 ☉ # ♆
21:15 ☽ △ ♄
22:13 ☽ □ ♅

17 02:19 ☽ ☍ ♇
18:32 ♂ # ♆
18 00:20 ☉ ☌ ♂
06:14 ♀ # ♄
12:18 ☽ □ ♂

19 00:18 ☽ □ ♀
03:16 ♀ ⚹ ♆
06:55 ☽ △ ☿
08:45 ☽ □ ♄
10:19 ☽ △ ♅
12:18 ☽ ∥ ♀
16:26 ☽ # ♄
19:40 ☽ ∥ ☉
21:07 ☽ ⚹ ♃
22:04 ☉ ♉
22:45 ☽ ∥ ♂

20 02:48 ☽ # ♆
09:21 ☿ ☌ ♅
11:49 ♂ ♉

21 01:59 ☽ △ ♂
02:33 ☉ # ♄
03:23 ☽ △ ☉
10:05 ☽ ☍ ♆
11:46 ☿ □ ♇
15:4[illegible] ☽ [illegible]
17:19 ☽ ⚹ ♄
19:55 ☽ ∥ ♅
22:31 ☽ △ ♇
23:20 ☽ ∥ ☿
22 06:04 ☽ □ ♃
07:43 ♀ ☍ ♄
21:42 ☿ ∥ ♅

23 02:28 ♂ # ♄
06:14 ☽ # ♅
07:44 ☽ # ☿
24 00:39 ☽ ☍ ♅
03:24 ☽ □ ♇
03:25 ☿ ⚹ ♃
11:1[illegible] ☽ △ ♃
12:13 ☽ ☍ ☿
15:42 ♀ △ ♇
19:18 ☽ ∥ ♆
21:59 ☉ ⚹ ♆
25 03:41 ☽ ∥ ♄
07:27 ☽ # ♂
13:11 ☽ # ☉
16:54 ☽ ☍ ♂
18:32 ☽ △ ♆

26 00:18 ☽ ☌ ♄
02:06 ☽ # ♀
05:28 ☽ ⚹ ♇
08:57 ☽ ☍ ♀
27 01:48 ☽ ⚹ ♆
19:33 ☽ □ ♆
22:46 ☽ ∥ ♇
28 04:02 ☽ △ ♅
08:27 ☉ ☍ ♄
15:09 ☽ ☍ ♃
29 04:38 ☽ △ ☿
15:36 ☽ ∥ ♇
20:33 ☽ ⚹ ♆

30 00:02 ☽ △ ♂
01:40 ☽ ⚹ ♄
04:51 ☽ △ ☉
05:16 ☽ □ ♅
07:20 ☽ ☌ ♇
20:06 ☽ △ ♀
22:21 ☽ # ♀

⚷ Chiron

01 Dec.	02 S 37
01	11♓22
04	11 32
07	11 42
10	11 52
13	12 02
16	12 11
19	12 20
22	12 28
25	12 36
28	12 44

May 2013

10	00:21	19♉31	☉	Annular Solar Eclipse
25	04:11	04♐08	✷	Penumbral Lunar Eclipse (mag 0.040)

Day	S. T. h m s	☉ ° ' "	☽ ° ' "	☿ ° '	♀ ° '	♂ ° '	♃ ° '	♄ ° '	♅ ° '	♆ ° '	♇ ° '	☊ True ° '
01 We	14 36 21	10♉47 06	21♑30 15	28♈44	19♉23	07♉50	17♊29	08♏R01	10♈18	05♓00	11♑R30	16♏50
02 Th	14 40 18	11 45 20	05♒40 18	00♉41	20 37	08 35	17 42	07 56	10 21	05 01	11 29	16R 50
03 Fr	14 44 14	12 43 32	19 32 53	02 40	21 50	09 19	17 54	07 52	10 24	05 02	11 29	16 50
04 Sa	14 48 11	13 41 43	03♓07 32	04 41	23 04	10 03	18 07	07 47	10 27	05 03	11 28	16 50
05 Su	14 52 07	14 39 52	16 24 49	06 43	24 18	10 48	18 19	07 43	10 30	05 04	11 28	16 49
06 Mo	14 56 04	15 38 00	29 25 53	08 47	25 32	11 32	18 32	07 38	10 33	05 06	11 27	16D 49
07 Tu	15 00 00	16 36 07	12♈12 06	10 52	26 46	12 16	18 45	07 34	10 36	05 07	11 26	16 49
08 We	15 03 57	17 34 11	24 44 59	12 58	28 00	13 00	18 57	07 29	10 39	05 08	11 26	16 50
09 Th	15 07 53	18 32 15	07♉06 01	15 06	29 14	13 44	19 10	07 25	10 42	05 08	11 25	16 50
10 Fr	15 11 50	19 30 17	19 16 45	17 15	00♊27	14 28	19 23	07 20	10 45	05 09	11 24	16 51
11 Sa	15 15 47	20 28 17	01♊18 55	19 24	01 41	15 12	19 36	07 16	10 48	05 10	11 23	16 52
12 Su	15 19 43	21 26 16	13 14 24	21 35	02 55	15 56	19 49	07 12	10 50	05 11	11 22	16 52
13 Mo	15 23 40	22 24 13	25 05 27	23 46	04 09	16 40	20 02	07 07	10 53	05 12	11 22	16R 52
14 Tu	15 27 36	23 22 09	06♋54 40	25 57	05 23	17 24	20 15	07 03	10 56	05 13	11 21	16 52
15 We	15 31 33	24 20 02	18 45 08	28 07	06 36	18 08	20 28	06 59	10 59	05 14	11 20	16 52
16 Th	15 35 29	25 17 54	00♌40 24	00♊18	07 50	18 52	20 41	06 54	11 01	05 14	11 19	16 51
17 Fr	15 39 26	26 15 45	12 44 27	02 28	09 04	19 35	20 54	06 50	11 04	05 15	11 18	16 51
18 Sa	15 43 22	27 13 33	25 01 33	04 37	10 17	20 19	21 08	06 46	11 07	05 16	11 17	16 50
19 Su	15 47 19	28 11 20	07♍35 56	06 44	11 31	21 02	21 21	06 42	11 09	05 16	11 16	16D 50
20 Mo	15 51 16	29 09 05	20 31 32	08 51	12 45	21 46	21 34	06 38	11 12	05 17	11 15	16 50
21 Tu	15 55 12	00♊06 49	03♎51 33	10 55	13 58	22 29	21 47	06 34	11 14	05 18	11 14	16 51
22 We	15 59 09	01 04 31	17 37 51	12 58	15 12	23 13	22 01	06 30	11 17	05 18	11 13	16 51
23 Th	16 03 05	02 02 12	01♏50 25	14 58	16 25	23 56	22 14	06 26	11 19	05 19	11 12	16R 51
24 Fr	16 07 02	02 59 51	16 26 45	16 57	17 39	24 39	22 28	06 22	11 22	05 19	11 11	16 51
25 Sa	16 10 58	03 57 28	01♐21 36	18 53	18 52	25 23	22 41	06 19	11 24	05 20	11 10	16 50
26 Su	16 14 55	04 55 05	16 27 23	20 46	20 06	26 06	22 54	06 15	11 26	05 20	11 09	16 48
27 Mo	16 18 51	05 52 41	01♑34 58	22 36	21 19	26 49	23 08	06 11	11 29	05 20	11 07	16 46
28 Tu	16 22 48	06 50 15	16 35 07	24 24	22 33	27 32	23 21	06 08	11 31	05 21	11 06	16 43
29 We	16 26 45	07 47 48	01♒19 49	26 09	23 46	28 15	23 35	06 04	11 33	05 21	11 05	16 40
30 Th	16 30 41	08 45 21	15 43 17	27 52	25 00	28 58	23 49	06 01	11 36	05 21	11 04	16 38
31 Fr	16 34 38	09 42 53	29 42 24	29 31	26 13	29 41	24 02	05 57	11 38	05 22	11 03	16 37

Data for	05-01-2013
Julian Day	2456413.50
Ayanamsa	24 02 47
SVP	05♓04 37
☽ ☊ Mean	17♏14 R

● ◐ PHASES ○ ◑

02	11:14	◑	12♒13
10	00:29	☉	19♉31
18	04:35	◐	27♌25
25	04:25	✷	04♐08
31	18:58	◑	10♓28

LAST ASPECT Day	h m	☽ INGRESS Day	h m	
01	14:08	01	14:20	♒
03	04:25	03	18:26	♓
05	16:01	06	01:04	♈
07	12:41	08	10:09	♉
10	00:29	10	21:22	♊
12	13:33	13	09:58	♋
15	12:15	15	22:39	♌
18	04:36	18	09:34	♍
20	16:49	20	17:08	♎
22	07:36	22	20:56	♏
24	13:56	24	21:50	♐
26	10:23	26	21:29	♑
28	18:41	28	21:49	♒
30	23:57	31	00:31	♓

DECLINATION

Day	☉	☽	☿	♀	♂	♃	♄	♅	♆	♇
01 We	15N03	16S59	09N29	17N12	13N52	22N32	11S38	03N28	10S16	19S40
02 Th	15 22	13 49	10 18	17 34	14 07	22 33	11 37	03 29	10 16	19 40
03 Fr	15 39	09 55	11 07	17 56	14 22	22 34	11 35	03 30	10 15	19 40
04 Sa	15 57	05 36	11 55	18 17	14 37	22 36	11 34	03 31	10 15	19 40
05 Su	16 14	01 07	12 44	18 38	14 51	22 37	11 33	03 32	10 14	19 40
06 Mo	16 31	03N21	13 33	18 58	15 05	22 38	11 31	03 33	10 14	19 40
07 Tu	16 48	07 36	14 22	19 17	15 20	22 39	11 30	03 35	10 14	19 40
08 We	17 04	11 26	15 10	19 37	15 34	22 40	11 29	03 36	10 13	19 41
09 Th	17 20	14 44	15 58	19 55	15 48	22 42	11 27	03 37	10 13	19 41
10 Fr	17 36	17 19	16 45	20 13	16 01	22 43	11 26	03 38	10 13	19 41
11 Sa	17 52	19 07	17 31	20 31	16 15	22 44	11 24	03 39	10 13	19 41
12 Su	18 07	20 02	18 15	20 48	16 28	22 45	11 23	03 40	10 12	19 41
13 Mo	18 22	20 03	18 59	21 04	16 41	22 46	11 22	03 41	10 12	19 41
14 Tu	18 37	19 11	19 41	21 20	16 54	22 47	11 21	03 42	10 12	19 41
15 We	18 51	17 29	20 21	21 35	17 07	22 48	11 19	03 43	10 11	19 41
16 Th	19 05	15 02	20 59	21 50	17 20	22 49	11 18	03 44	10 11	19 41
17 Fr	19 19	11 55	21 35	22 04	17 32	22 50	11 17	03 45	10 11	19 41
18 Sa	19 32	08 15	22 09	22 17	17 45	22 51	11 16	03 46	10 11	19 41
19 Su	19 45	04 08	22 40	22 30	17 57	22 52	11 14	03 47	10 11	19 42
20 Mo	19 58	00S17	23 09	22 42	18 09	22 53	11 13	03 48	10 10	19 42
21 Tu	20 10	04 51	23 36	22 54	18 21	22 54	11 12	03 49	10 10	19 42
22 We	20 22	09 18	23 59	23 05	18 32	22 55	11 11	03 50	10 10	19 42
23 Th	20 34	13 24	24 20	23 15	18 44	22 56	11 10	03 51	10 10	19 42
24 Fr	20 45	16 47	24 39	23 24	18 55	22 57	11 09	03 52	10 10	19 42
25 Sa	20 56	19 08	24 55	23 33	19 06	22 58	11 08	03 53	10 10	19 42
26 Su	21 07	20 09	25 08	23 41	19 17	22 59	11 06	03 54	10 09	19 42
27 Mo	21 17	19 43	25 18	23 49	19 28	22 59	11 05	03 55	10 09	19 43
28 Tu	21 27	17 53	25 27	23 56	19 38	23 00	11 04	03 56	10 09	19 43
29 We	21 36	14 55	25 32	24 02	19 48	23 01	11 03	03 57	10 09	19 43
30 Th	21 45	11 06	25 36	24 07	19 58	23 02	11 02	03 57	10 09	19 43
31 Fr	21 54	06 46	25 38	24 12	20 08	23 02	11 01	03 58	10 09	19 43

ASPECTARIAN

01 05:12 ♂ ☍ ♄
14:03 ☽ ⋕ ☉
14:08 ☽ □ ☿
15:37 ☿ ♉
17:34 ☉ △ ♇
22:04 ☽ ⋕ ♂
22:53 ☿ ⋕ ♆
02 03:52 ☽ □ ♄
05:16 ☽ □ ♂
08:04 ☽ ✶ ♅
14:06 ☽ ∥ ♄
18:14 ☽ ⋕ ☿
21:05 ☽ △ ♃
22:07 ☽ ∥ ♆
03 04:25 ☽ □ ♀
13:48 ☿ ⋕ ♄
04 03:16 ☽ ✶ ☿
03:28 ☽ ☌ ♆
04:31 ☿ ✶ ♆
08:19 ☽ △ ♄
11:11 ☽ ⋕ ♅
13:11 ☽ ✶ ♂
15:00 ☽ ✶ ♇
20:34 ☽ ✶ ☉
05 03:33 ☽ □ ♃
11:13 ☿ ☍ ♄
16:01 ☽ ✶ ♀
21:26 ♂ △ ♇
06 01:08 ☽ ∥ ♅
20:58 ☽ ☌ ♅
22:33 ☽ □ ♇
07 06:34 ☿ △ ♇
12:41 ☽ ✶ ♃
16:05 ☽ ⋕ ♆
08 00:15 ☽ ⋕ ♄
00:34 ☿ ☌ ♂
05:10 ♀ ⋕ ♇
16:47 ☿ ∥ ♂
20:10 ☽ ✶ ♆
09 00:36 ☽ ☍ ♄
08:27 ☽ △ ♇
09:50 ☽ ∥ ♂
13:53 ☽ ☌ ♂
15:04 ♀ ♊
15:07 ☽ ∥ ☿
19:06 ☽ ☌ ☿
10 03:31 ☽ ∥ ☉
11 00:50 ☽ ☌ ♀
07:45 ☽ □ ♆
11:34 ☽ ⋕ ♇
17:15 ☿ ∥ ☉
19:08 ☽ ✶ ♅
21:10 ☿ ☌ ☉
12 13:33 ☽ ☌ ♃
13 13:35 ☽ ⋕ ♇
17:45 ☽ ∥ ☿
20:33 ☽ △ ♆
20:49 ♀ □ ♆
14 00:15 ☿ ⋕ ♇
00:17 ☽ △ ♄
08:12 ☽ □ ♅
08:25 ☽ ∥ ☉
08:59 ☽ ☍ ♇
22:40 ☽ ✶ ♂
15 03:51 ☽ ∥ ♂
12:15 ☽ ✶ ☉
20:42 ☿ ♊
23:05 ☽ ✶ ☿
16 12:22 ☽ □ ♄
15:54 ☽ ✶ ♀
20:41 ☽ △ ♅
17 04:27 ☽ ⋕ ♄
11:46 ☽ ⋕ ♆
14:17 ☽ □ ♂
16:18 ☽ ✶ ♃
18 07:20 ☿ □ ♆
10:32 ☿ ∥ ♀
16:40 ♀ ✶ ♅
17:17 ☉ ⋕ ♇
19:36 ☽ ☍ ♆
22:04 ☽ □ ☿
22:19 ☽ ✶ ♄
19 01:53 ☽ ∥ ♅
06:53 ☽ △ ♇
08:07 ☽ □ ♀
09:54 ☿ ∥ ♃
20 01:56 ☽ □ ♃
02:24 ☽ △ ♂
16:49 ☽ △ ☉
18:37 ☽ ⋕ ♅
21:10 ☉ ♊
22:22 ♅ □ ♇
21 00:47 ♀ ∥ ♃
03:47 ☿ ✶ ♅
12:56 ☽ □ ♇
13:00 ☽ ☍ ♅
14:34 ☽ △ ☿
19:24 ☽ △ ♀
22 04:50 ☽ ∥ ♆
07:36 ☽ △ ♃
10:35 ☽ ∥ ♄
23 05:46 ☽ △ ♆
07:35 ☽ ☌ ♄
15:26 ☽ ✶ ♇
24 13:56 ☽ ☍ ♂
23:30 ☽ ⋕ ♂
23:54 ☿ ☌ ♀
25 06:20 ☽ □ ♆
09:30 ☽ ∥ ♇
16:01 ☽ △ ♅
26 06:17 ☽ ☍ ♀
07:47 ☽ ☍ ☿
10:23 ☽ ☍ ♃
10:28 ☉ □ ♆
27 00:09 ☽ ∥ ♇
04:15 ☽ ⋕ ♂
05:59 ☽ ✶ ♆
07:18 ☽ ✶ ♄
07:56 ☿ ☌ ♃
15:12 ☽ ☌ ♇
15:50 ☽ □ ♅
28 10:59 ♂ ⋕ ♇
18:41 ☽ △ ♂
19:29 ♀ ☌ ♃
29 07:48 ☽ □ ♄
11:28 ☽ △ ☉
17:02 ☽ ✶ ♅
30 00:21 ☽ ∥ ♄
05:28 ☽ ∥ ♆
14:01 ☽ △ ♃
17:22 ☽ △ ♀
23:38 ☽ △ ☿
23:57 ☽ □ ♂
31 07:07 ☿ ♋
09:55 ☽ ☌ ♆
10:39 ♂ ♊
10:55 ☽ △ ♄
14:49 ☽ ⋕ ♅
19:58 ☽ ✶ ♇

⚷ Chiron

01 Dec. 01 S 59

01	12♓52
04	12 59
07	13 05
10	13 12
13	13 17
16	13 23
19	13 28
22	13 32
25	13 36
28	13 39
31	13 42

June 2013

Day	S. T.	☉	☽	☿	♀	♂	♃	♄	♅	♆	♇	☊ True
	h m s	° ' "	° ' "	° '	° '	° '	° '	° '	° '	° '	° '	° '
01 Sa	16 38 34	10♊ 40 24	13♓ 16 35	01♋ 08	27♊ 26	00♊ 24	24♊ 16	05♏R54	11♈ 40	05♓ 22	11♑R01	16♏D36
02 Su	16 42 31	11 37 54	26 27 15	02 41	28 40	01 07	24 29	05 51	11 42	05 22	11 00	16 37
03 Mo	16 46 27	12 35 23	09♈ 17 04	04 12	29 53	01 49	24 43	05 47	11 44	05 22	10 59	16 39
04 Tu	16 50 24	13 32 52	21 49 20	05 39	01♋ 07	02 32	24 57	05 44	11 46	05 22	10 58	16 41
05 We	16 54 20	14 30 19	04♉ 07 29	07 04	02 20	03 15	25 10	05 41	11 48	05 22	10 56	16 43
06 Th	16 58 17	15 27 47	16 14 44	08 25	03 33	03 57	25 24	05 38	11 50	05 22	10 55	16 45
07 Fr	17 02 14	16 25 13	28 14 01	09 43	04 47	04 40	25 38	05 35	11 52	05 22	10 54	16R 45
08 Sa	17 06 10	17 22 39	10♊ 07 53	10 59	06 00	05 23	25 51	05 33	11 54	05R 22	10 52	16 43
09 Su	17 10 07	18 20 04	21 58 35	12 11	07 13	06 05	26 05	05 30	11 56	05 22	10 51	16 40
10 Mo	17 14 03	19 17 28	03♋ 48 07	13 19	08 26	06 47	26 19	05 27	11 57	05 22	10 50	16 35
11 Tu	17 17 60	20 14 51	15 38 25	14 25	09 40	07 30	26 33	05 25	11 59	05 22	10 48	16 30
12 We	17 21 56	21 12 13	27 31 33	15 27	10 53	08 12	26 46	05 22	12 01	05 22	10 47	16 23
13 Th	17 25 53	22 09 35	09♌ 29 51	16 25	12 06	08 54	27 00	05 20	12 02	05 22	10 45	16 18
14 Fr	17 29 49	23 06 55	21 36 10	17 20	13 19	09 36	27 14	05 17	12 04	05 22	10 44	16 13
15 Sa	17 33 46	24 04 15	03♍ 53 43	18 11	14 32	10 19	27 28	05 15	12 06	05 22	10 43	16 09
16 Su	17 37 43	25 01 33	16 26 07	18 59	15 45	11 01	27 41	05 13	12 07	05 21	10 41	16 08
17 Mo	17 41 39	25 58 51	29 17 07	19 42	16 58	11 43	27 55	05 11	12 09	05 21	10 40	16D 07
18 Tu	17 45 36	26 56 08	12♎ 30 18	20 22	18 12	12 25	28 09	05 09	12 10	05 21	10 38	16 08
19 We	17 49 32	27 53 24	26 08 37	20 58	19 25	13 07	28 23	05 07	12 11	05 20	10 37	16 09
20 Th	17 53 29	28 50 39	10♏ 13 40	21 29	20 38	13 48	28 37	05 05	12 13	05 20	10 35	16R 09
21 Fr	17 57 25	29 47 54	24 44 54	21 57	21 51	14 30	28 50	05 04	12 14	05 19	10 34	16 08
22 Sa	18 01 22	00♋ 45 08	09♐ 38 52	22 19	23 04	15 12	29 04	05 02	12 15	05 19	10 32	16 05
23 Su	18 05 18	01 42 22	24 48 54	22 38	24 17	15 54	29 18	05 00	12 17	05 19	10 31	16 01
24 Mo	18 09 15	02 39 35	10♑ 05 44	22 52	25 30	16 35	29 32	04 59	12 18	05 18	10 29	15 55
25 Tu	18 13 12	03 36 48	25 18 43	23 01	26 42	17 17	29 45	04 58	12 19	05 17	10 28	15 48
26 We	18 17 08	04 34 01	10♒ 17 34	23 06	27 55	17 59	29 59	04 56	12 20	05 17	10 26	15 40
27 Th	18 21 05	05 31 13	24 54 07	23R 06	29 08	18 40	00♋ 13	04 55	12 21	05 16	10 25	15 34
28 Fr	18 25 01	06 28 25	09♓ 03 17	23 02	00♌ 21	19 21	00 26	04 54	12 22	05 16	10 23	15 29
29 Sa	18 28 58	07 25 38	22 43 22	22 53	01 34	20 03	00 40	04 53	12 23	05 15	10 22	15 27
30 Su	18 32 54	08 22 50	05♈ 55 30	22 40	02 47	20 44	00 54	04 52	12 24	05 14	10 20	15D 26

Data for	06-01-2013
Julian Day	2456444.50
Ayanamsa	24 02 51
SVP	05 ♓ 04 33
☽ ☊ Mean	15 ♏ 36 R

● ◐ PHASES ○ ◑

08	15:57	●	18♊01
16	17:24	◐	25♍43
23	11:32	○	02♑10
30	04:54	◑	08♈35

LAST ASPECT ☽ Day	h m	INGRESS Day	h m	
02	04:31	02	06:34	♈
04	06:10	04	15:55	♉
05	13:26	07	03:33	♊
09	08:30	09	16:17	♋
10	21:16	12	04:59	♌
14	11:15	14	16:26	♍
16	21:26	17	01:19	♎
19	03:56	19	06:39	♏
20	19:17	21	08:31	♐
23	07:09	23	08:09	♑
25	02:25	25	07:27	♒
26	13:09	27	08:33	♓
29	00:17	29	13:08	♈

DECLINATION

Day	☉	☽	☿	♀	♂	♃	♄	♅	♆	♇
01 Sa	22N03	02S 13	25N37	24N16	20N18	23N03	11S 01	03N59	10S 09	19S 43
02 Su	22 11	02N19	25 34	24 19	20 28	23 04	11 00	04 00	10 09	19 43
03 Mo	22 18	06 38	25 30	24 22	20 37	23 04	10 59	04 01	10 09	19 44
04 Tu	22 25	10 34	25 24	24 24	20 46	23 05	10 58	04 01	10 09	19 44
05 We	22 32	13 58	25 17	24 25	20 55	23 06	10 57	04 02	10 09	19 44
06 Th	22 39	16 44	25 07	24 25	21 04	23 06	10 56	04 03	10 09	19 44
07 Fr	22 45	18 45	24 57	24 25	21 12	23 07	10 56	04 04	10 09	19 44
08 Sa	22 50	19 55	24 45	24 24	21 20	23 07	10 55	04 04	10 09	19 44
09 Su	22 55	20 11	24 32	24 22	21 28	23 08	10 54	04 05	10 09	19 45
10 Mo	23 00	19 34	24 19	24 19	21 36	23 08	10 53	04 06	10 09	19 45
11 Tu	23 05	18 05	24 04	24 16	21 44	23 09	10 53	04 06	10 09	19 45
12 We	23 09	15 50	23 48	24 12	21 51	23 09	10 52	04 07	10 09	19 45
13 Th	23 12	12 54	23 32	24 08	21 59	23 10	10 52	04 08	10 09	19 45
14 Fr	23 15	09 24	23 15	24 02	22 06	23 10	10 51	04 08	10 09	19 45
15 Sa	23 18	05 28	22 57	23 56	22 13	23 11	10 51	04 09	10 09	19 46
16 Su	23 21	01 13	22 39	23 50	22 19	23 11	10 50	04 09	10 10	19 46
17 Mo	23 22	03S 11	22 21	23 42	22 26	23 11	10 50	04 10	10 10	19 46
18 Tu	23 24	07 35	22 03	23 34	22 32	23 12	10 49	04 10	10 10	19 46
19 We	23 25	11 45	21 44	23 25	22 38	23 12	10 49	04 11	10 10	19 46
20 Th	23 26	15 24	21 25	23 16	22 44	23 12	10 48	04 11	10 10	19 47
21 Fr	23 26	18 13	21 07	23 06	22 49	23 12	10 48	04 12	10 10	19 47
22 Sa	23 26	19 52	20 49	22 55	22 54	23 12	10 48	04 12	10 11	19 47
23 Su	23 25	20 07	20 31	22 43	23 00	23 13	10 48	04 13	10 11	19 47
24 Mo	23 25	18 54	20 13	22 31	23 04	23 13	10 47	04 13	10 11	19 47
25 Tu	23 23	16 21	19 56	22 19	23 09	23 13	10 47	04 14	10 11	19 48
26 We	23 21	12 44	19 39	22 05	23 14	23 13	10 47	04 14	10 11	19 48
27 Th	23 19	08 27	19 23	21 51	23 18	23 13	10 47	04 14	10 12	19 48
28 Fr	23 17	03 48	19 08	21 37	23 22	23 13	10 47	04 15	10 12	19 48
29 Sa	23 14	00N52	18 54	21 21	23 26	23 13	10 47	04 15	10 12	19 48
30 Su	23 10	05 22	18 41	21 06	23 29	23 13	10 47	04 15	10 13	19 49

ASPECTARIAN

01 20:19 ☽ □ ♃

02 01:46 ☉ ⚹ ♅
04:31 ☽ □ ♀
09:09 ☽ ⚹ ♂
09:17 ☽ ∥ ♅
13:08 ☽ □ ☿
22:08 ☿ ∥ ♀

03 02:13 ♀ ♋
03:13 ☽ □ ♆
04:40 ☽ ☌ ♅
06:48 ☽ ⚹ ☉
19:18 ☿ △ ♆
21:23 ☽ # ♆

04 01:23 ☿ △ ♄
02:41 ☽ # ♄
06:10 ☽ ⚹ ♃
20:05 ☽ ⚹ ♀

05 02:27 ☽ ⚹ ♆
03:04 ☽ ☍ ♄
06:31 ☽ ⚹ ☿
13:26 ☽ △ ♇

07 08:27 ♆ SR
11:46 ♀ △ ♆
13:47 ☽ ☌ ♂
14:23 ☽ □ ♆
15:24 ♀ △ ♄
19:00 ☽ # ♇
22:02 ☿ ☍ ♇
23:58 ♂ □ ♆

08 03:35 ☽ ⚹ ♅
18:47 ☿ □ ♅

09 08:30 ☽ ☌ ♃
19:31 ☽ # ♇

10 03:11 ☽ △ ♆
03:20 ☽ △ ♄
10:29 ☽ ☌ ♀
14:13 ☽ ☍ ♇
16:35 ☽ □ ♅
21:16 ☽ ☌ ☿

11 22:07 ♀ ☍ ♇
23:19 ♄ △ ♆

12 05:22 ☉ ∥ ♃
15:42 ☽ □ ♄
22:45 ☽ ⚹ ♂
22:49 ♀ □ ♅

13 05:05 ☽ △ ♅
14:30 ☽ # ♄
19:11 ☽ # ♆
23:12 ☿ ∥ ☉

14 03:14 ☽ ⚹ ☉
06:08 ☿ ∥ ♃
11:15 ☽ ⚹ ♃

15 02:37 ☽ ⚹ ♄
02:50 ☽ ☍ ♆
07:39 ☽ ∥ ♅
13:04 ☽ □ ♂
13:05 ☽ △ ♆
22:35 ☽ ⚹ ♀

16 05:06 ☽ ⚹ ♅
19:33 ☿ ∥ ♂
21:26 ☽ □ ♃

17 05:19 ☽ # ♅
15:20 ♂ ⚹ ♅
20:40 ☽ □ ♆
23:24 ☽ ☍ ♅
23:49 ☽ △ ♂

18 11:06 ☽ □ ♀
14:35 ☽ □ ☿
14:39 ☽ ∥ ♆
18:27 ☽ ∥ ♄

19 00:28 ♀ ∥ ☉
03:14 ☽ △ ☉
03:56 ☽ △ ♃
15:21 ☽ ☌ ♄
15:45 ☽ △ ♆
16:12 ☉ ☌ ♃

20 00:36 ☽ ⚹ ♇
09:00 ♀ ∥ ♃
18:49 ☽ △ ♀
19:17 ☽ △ ☿

21 02:57 ☿ ☌ ♀
05:04 ☉ ♋
17:05 ☽ □ ♆
22:23 ☽ ∥ ♇

22 00:40 ♀ ∥ ♂
04:09 ☽ △ ♅
09:15 ☽ ☍ ♂

23 07:09 ☽ ☍ ♃
09:49 ☽ ∥ ♇
15:59 ☽ ⚹ ♄
16:28 ☽ ⚹ ♆

24 00:37 ☽ ☌ ♇
03:28 ☽ □ ♅
20:20 ☽ ☍ ☿

25 02:25 ☽ ☍ ♀
11:46 ☿ # ♇
15:23 ☽ □ ♄
20:37 ♂ ∥ ♃

26 01:40 ♃ ♋
03:19 ☽ ⚹ ♅
09:13 ☉ △ ♄
11:18 ☽ ∥ ♄
13:09 ☿ SR
13:09 ☽ △ ♂
14:34 ☽ ∥ ♆
17:48 ☉ △ ♆

27 04:48 ☉ ∥ ♂
09:03 ☽ △ ♃
16:53 ☽ △ ♄
17:04 ♀ ♌
17:29 ☽ ☌ ♆
19:14 ☽ △ ☉
[illegible] ☽ ∥ ♆

28 02:18 ☽ ⚹ ♇
18:59 ☽ □ ♂

29 00:17 ☽ △ ☿
03:03 ☉ ∥ ♃
14:36 ☽ □ ♃
17:37 ☽ △ ♀
17:56 ☽ ∥ ♅

30 08:11 ☽ □ ♇
12:04 ☽ ☌ ♅

⚷ Chiron

01 Dec.	01 S 33
03	13♓45
06	13 47
09	13 48
12	13 49
15	13 50
18	13 50R
21	13 49
24	13 48
27	13 47
30	13 45

16 07:20 13♓50 R

July 2013

Day	S. T.	☉	☽	☿	♀	♂	♃	♄	♅	♆	♇	☊ True
	h m s	° ' "	° ' "	° '	° '	° '	° '	° '	° '	° '	° '	° '
01 Mo	18 36 51	09♋20 03	18♈42 54	22♋R23	03♌59	21♊26	01♋08	04♏R52	12♈25	05♓R14	10♑R19	15♏27
02 Tu	18 40 47	10 17 16	01♉09 49	22 01	05 12	22 07	01 21	04 51	12 25	05 13	10 17	15 28
03 We	18 44 44	11 14 29	13 20 58	21 36	06 25	22 48	01 35	04 50	12 26	05 12	10 16	15 30
04 Th	18 48 41	12 11 42	25 20 53	21 08	07 38	23 29	01 49	04 50	12 27	05 11	10 14	15R 31
05 Fr	18 52 37	13 08 55	07♊13 45	20 36	08 50	24 10	02 02	04 50	12 27	05 10	10 13	15 29
06 Sa	18 56 34	14 06 09	19 03 09	20 03	10 03	24 51	02 16	04 49	12 28	05 10	10 11	15 25
07 Su	19 00 30	15 03 23	00♋52 01	19 27	11 16	25 32	02 29	04 49	12 29	05 09	10 10	15 18
08 Mo	19 04 27	16 00 36	12 42 39	18 50	12 28	26 13	02 43	04 49	12 29	05 08	10 08	15 08
09 Tu	19 08 23	16 57 50	24 36 50	18 12	13 41	26 54	02 56	04D 49	12 29	05 07	10 07	14 57
10 We	19 12 20	17 55 04	06♌36 00	17 34	14 53	27 35	03 10	04 49	12 30	05 06	10 05	14 45
11 Th	19 16 16	18 52 18	18 41 36	16 57	16 06	28 16	03 24	04 49	12 30	05 05	10 04	14 33
12 Fr	19 20 13	19 49 32	00♍55 12	16 20	17 19	28 57	03 37	04 50	12 31	05 04	10 02	14 23
13 Sa	19 24 10	20 46 47	13 18 49	15 46	18 31	29 37	03 50	04 50	12 31	05 03	10 01	14 16
14 Su	19 28 06	21 44 01	25 54 50	15 14	19 43	00♋18	04 04	04 51	12 31	05 02	09 59	14 11
15 Mo	19 32 03	22 41 15	08♎46 06	14 45	20 56	00 59	04 17	04 51	12 31	05 01	09 58	14 08
16 Tu	19 35 59	23 38 29	21 55 40	14 20	22 08	01 39	04 31	04 52	12 31	05 00	09 56	14 07
17 We	19 39 56	24 35 43	05♏26 34	13 59	23 21	02 20	04 44	04 53	12 31	04 58	09 55	14 07
18 Th	19 43 52	25 32 58	19 21 03	13 42	24 33	03 00	04 57	04 54	12R 31	04 57	09 54	14 06
19 Fr	19 47 49	26 30 12	03♐39 56	13 30	25 45	03 40	05 10	04 55	12 31	04 56	09 52	14 04
20 Sa	19 51 45	27 27 27	18 21 36	13 23	26 57	04 21	05 24	04 56	12 31	04 55	09 51	14 00
21 Su	19 55 42	28 24 42	03♑21 22	13D 22	28 10	05 01	05 37	04 57	12 31	04 54	09 49	13 54
22 Mo	19 59 39	29 21 58	18 31 32	13 26	29 22	05 41	05 50	04 58	12 31	04 52	09 48	13 45
23 Tu	20 03 35	00♌19 14	03♒42 09	13 36	00♍34	06 22	06 03	05 00	12 31	04 51	09 46	13 35
24 We	20 07 32	01 16 30	18 42 42	13 51	01 46	07 02	06 16	05 01	12 30	04 50	09 45	13 24
25 Th	20 11 28	02 13 48	03♓23 49	14 13	02 58	07 42	06 29	05 03	12 30	04 49	09 44	13 15
26 Fr	20 15 25	03 11 05	17 38 52	14 40	04 10	08 22	06 42	05 05	12 30	04 47	09 42	13 07
27 Sa	20 19 21	04 08 24	01♈24 38	15 14	05 22	09 02	06 55	05 06	12 29	04 46	09 41	13 01
28 Su	20 23 18	05 05 44	14 41 12	15 53	06 34	09 42	07 08	05 08	12 29	04 45	09 40	12 58
29 Mo	20 27 14	06 03 05	27 31 09	16 38	07 46	10 22	07 21	05 10	12 28	04 43	09 38	12D 57
30 Tu	20 31 11	07 00 26	09♉58 46	17 29	08 58	11 02	07 33	05 12	12 28	04 42	09 37	12 57
31 We	20 35 08	07 57 49	22 09 04	18 26	10 10	11 41	07 46	05 14	12 27	04 40	09 36	12 58

Data for	07-01-2013
Julian Day	2456474.50
Ayanamsa	24 02 56
SVP	05 ♓ 04 29
☽ ☊ Mean	14 ♏ 00 R

● ◐ PHASES ○ ◑

08	07:14	●	16♋18
16	03:18	◐	23♎46
22	18:16	○	00♒06
29	17:44	◑	06♉45

LAST ASPECT ☽ Day	h m	INGRESS Day	h m	
01	06:49	01	21:44	♉
03	15:52	04	09:23	♊
06	12:31	06	22:14	♋
08	11:44	09	10:48	♌
11	19:55	11	22:12	♍
13	15:26	14	07:41	♎
16	03:19	16	14:25	♏
18	11:13	18	17:55	♐
20	15:01	20	18:40	♑
21	15:54	22	18:08	♒
23	14:02	24	18:23	♓
25	18:44	26	21:30	♈
28	02:20	29	04:44	♉
30	15:58	31	15:42	♊

DECLINATION

Day	☉	☽	☿	♀	♂	♃	♄	♅	♆	♇
01 Mo	23N06	09N29	18N28	20N49	23N33	23N13	10S47	04N16	10S13	19S49
02 Tu	23 02	13 05	18 17	20 32	23 36	23 13	10 47	04 16	10 13	19 49
03 We	22 58	16 02	18 07	20 15	23 39	23 13	10 47	04 16	10 13	19 49
04 Th	22 53	18 15	17 58	19 57	23 42	23 13	10 47	04 16	10 14	19 49
05 Fr	22 47	19 39	17 50	19 38	23 44	23 13	10 47	04 17	10 14	19 50
06 Sa	22 41	20 10	17 44	19 19	23 47	23 13	10 47	04 17	10 14	19 50
07 Su	22 35	19 48	17 39	18 59	23 49	23 13	10 47	04 17	10 15	19 50
08 Mo	22 29	18 35	17 36	18 39	23 51	23 12	10 47	04 17	10 15	19 50
09 Tu	22 22	16 32	17 33	18 18	23 52	23 12	10 48	04 17	10 15	19 51
10 We	22 14	13 47	17 33	17 57	23 54	23 12	10 48	04 17	10 16	19 51
11 Th	22 06	10 26	17 33	17 35	23 55	23 12	10 48	04 18	10 16	19 51
12 Fr	21 58	06 37	17 35	17 13	23 56	23 12	10 49	04 18	10 17	19 51
13 Sa	21 50	02 28	17 38	16 51	23 57	23 11	10 49	04 18	10 17	19 51
14 Su	21 41	01S51	17 42	16 28	23 57	23 11	10 50	04 18	10 17	19 52
15 Mo	21 32	06 10	17 48	16 04	23 58	23 11	10 50	04 18	10 18	19 52
16 Tu	21 22	10 19	17 54	15 40	23 58	23 10	10 50	04 18	10 18	19 52
17 We	21 12	14 03	18 02	15 16	23 58	23 10	10 51	04 18	10 19	19 52
18 Th	21 02	17 07	18 10	14 52	23 58	23 09	10 52	04 18	10 19	19 53
19 Fr	20 51	19 13	18 20	14 27	23 57	23 09	10 52	04 18	10 20	19 53
20 Sa	20 40	20 06	18 30	14 01	23 57	23 09	10 53	04 18	10 20	19 53
21 Su	20 29	19 34	18 40	13 35	23 56	23 08	10 53	04 18	10 21	19 53
22 Mo	20 17	17 39	18 51	13 09	23 55	23 08	10 54	04 18	10 21	19 54
23 Tu	20 05	14 31	19 02	12 43	23 54	23 07	10 55	04 17	10 22	19 54
24 We	19 52	10 28	19 14	12 16	23 52	23 07	10 56	04 17	10 22	19 54
25 Th	19 40	05 52	19 25	11 49	23 51	23 06	10 56	04 17	10 23	19 54
26 Fr	19 27	01 04	19 36	11 22	23 49	23 06	10 57	04 17	10 23	19 55
27 Sa	19 13	03N38	19 47	10 54	23 47	23 05	10 58	04 17	10 24	19 55
28 Su	18 59	07 59	19 58	10 26	23 45	23 04	10 59	04 17	10 24	19 55
29 Mo	18 45	11 51	20 07	09 58	23 42	23 04	11 00	04 16	10 25	19 55
30 Tu	18 31	15 03	20 16	09 30	23 39	23 03	11 01	04 16	10 25	19 56
31 We	18 16	17 31	20 24	09 01	23 37	23 02	11 02	04 16	10 26	19 56

ASPECTARIAN

01 04:39 ☽ # ♆
05:29 ☽ ✶ ♂
06:49 ☽ □ ☿
08:16 ☽ # ♄
17:04 ♀ □ ♄

02 00:04 ☉ ☍ ♆
00:23 ☽ ✶ ♃
07:12 ☽ ☍ ♄
07:55 ☽ ✶ ♆
08:47 ☽ □ ♀
17:54 ☽ △ ♇
19:28 ☽ ✶ ☉

03 15:52 ☽ ✶ ☿
20:42 ☽ ∥ ☿

04 06:23 ☉ □ ♅
09:12 ♀ # ♇
19:50 ☽ □ ♆
23:50 ☽ ∥ ♀

05 03:38 ☽ ✶ ♀
05:08 ☽ # ♇
10:37 ☽ ✶ ♅

06 12:31 ☽ ☌ ♂
22:58 ☽ # ♇

07 03:21 ☽ ☌ ♃
08:01 ☽ △ ♄
08:40 ☽ △ ♆
18:48 ☽ ☍ ♇
22:29 ☽ ∥ ♀
23:32 ☽ □ ♅

08 00:14 ♀ △ ♅
05:13 ♄ SD
11:44 ☽ ☌ ☿
12:52 ☽ ∥ ☿

09 18:42 ☿ ☌ ☉
20:27 ☽ □ ♄

10 11:44 ☽ △ ♅
18:18 ☽ ☌ ♀
21:28 ☽ # ♄

11 01:02 ☽ # ♆
02:32 ☿ ∥ ♀
19:55 ☽ ✶ ♂

12 05:21 ☽ ✶ ♃
07:37 ☽ ✶ ♄
08:03 ☽ ☍ ♆
13:27 ☽ ∥ ♅
17:39 ☽ △ ♇

13 04:31 ☽ ✶ ☿
13:23 ♂ ♋
15:26 ☽ ✶ ☉

14 08:42 ☽ □ ♂
13:33 ☽ # ♅
15:33 ☽ □ ♃

15 02:12 ☽ □ ♇
06:54 ☽ ☍ ♅
10:38 ☽ □ ☿
23:56 ☽ ∥ ♆

16 00:25 ☽ ✶ ♀
03:11 ☽ ∥ ♄
18:14 ☽ △ ♂
22:44 ☽ △ ♃
23:01 ☽ ☌ ♄
23:11 ☽ △ ♆

17 07:47 ☽ ✶ ♇
07:49 ☽ # ♀
14:31 ☽ △ ☿
17:21 ♅ SR
17:33 ♃ △ ♄

18 00:17 ♃ △ ♆
09:36 ☽ □ ♀
11:13 ☽ △ ☉
11:23 ☽ # ☿

19 02:06 ☽ □ ♆
13:29 ♄ △ ♆
14:01 ☽ ∥ ♇
14:32 ☽ △ ♅

20 15:01 ☽ △ ♀
16:14 ☽ ∥ ♇
18:23 ☿ SD
19:43 ♂ △ ♆
21:35 ♂ △ ♄

21 02:26 ☽ ✶ ♆
02:32 ☽ ✶ ♄
02:46 ☽ ☍ ♂
03:38 ☽ ☍ ♃
10:14 ☽ ☌ ♇
12:08 ☽ # ☿
14:30 ☽ □ ♅
15:54 ☽ ☍ ☿

22 07:35 ♂ ☌ ♃
12:41 ♀ ♍
15:56 ☉ ♌

23 02:04 ☽ □ ♄
12:31 ☽ # ♀
14:02 ☽ ✶ ♅
20:43 ☉ # ♇
21:25 ☽ ∥ ♄

24 00:28 ☽ ∥ ♆
23:14 ☽ ☍ ♀

25 02:20 ☽ ☌ ♆
02:45 ☽ △ ♄
05:13 ☽ △ ♃
07:30 ☽ △ ♂
07:53 ☽ # ♅
10:33 ☽ ✶ ♇
14:14 ☿ ∥ ☉
18:44 ☽ △ ☿

26 12:08 ♀ ☍ ♆
18:36 ♀ ✶ ♄
20:51 ♀ # ♄

27 03:24 ☽ ∥ ♅
05:14 ☽ △ ☉
10:01 ☽ □ ♃
14:24 ☽ □ ♂
14:50 ☽ □ ♇
17:36 ☿ # ♇
19:57 ☽ ☌ ♅
22:48 ♂ ☍ ♇

28 01:06 ☉ □ ♄
01:55 ♀ # ♆
02:20 ☽ □ ☿
13:11 ☽ ∥ ♀
13:42 ♀ ✶ ♃
14:36 ☽ # ♆
18:21 ☽ # ♄

29 13:46 ☽ ✶ ♆
14:42 ☽ ☍ ♄
19:12 ☽ ✶ ♃
21:49 ☽ △ ♀
23:18 ☽ △ ♇

30 02:10 ☽ ✶ ♂
12:51 ♀ △ ♇
15:58 ☽ ✶ ☿

31 08:18 ☽ ∥ ☉

⚷ Chiron

01 Dec.	01 S 26
03	13♓42R
06	13 39
09	13 36
12	13 32
15	13 28
18	13 23
21	13 18
24	13 12
27	13 06
30	13 00

August 2013

Day	S. T. h m s	☉ ° ' "	☽ ° ' "	☿ ° '	♀ ° '	♂ ° '	♃ ° '	♄ ° '	♅ ° '	♆ ° '	♇ ° '	☊ True ° '
01 Th	20 39 04	08♌55 13	04♊07 19	19♋28	11♍22	12♋21	07♋59	05♏17	12♈R26	04♓R39	09♑R34	12♏R58
02 Fr	20 43 01	09 52 38	15 58 30	20 36	12 33	13 01	08 12	05 19	12 26	04 37	09 33	12 56
03 Sa	20 46 57	10 50 05	27 47 10	21 49	13 45	13 41	08 24	05 21	12 25	04 36	09 32	12 51
04 Su	20 50 54	11 47 32	09♋37 06	23 07	14 57	14 20	08 37	05 24	12 24	04 35	09 31	12 42
05 Mo	20 54 50	12 45 00	21 31 18	24 31	16 08	15 00	08 49	05 27	12 23	04 33	09 30	12 31
06 Tu	20 58 47	13 42 29	03♌31 54	25 59	17 20	15 39	09 02	05 29	12 22	04 32	09 28	12 18
07 We	21 02 43	14 40 00	15 40 18	27 32	18 32	16 19	09 14	05 32	12 22	04 30	09 27	12 04
08 Th	21 06 40	15 37 31	27 57 28	29 09	19 43	16 58	09 26	05 35	12 21	04 29	09 26	11 50
09 Fr	21 10 37	16 35 03	10♍24 02	00♌50	20 55	17 38	09 39	05 38	12 20	04 27	09 25	11 38
10 Sa	21 14 33	17 32 37	23 00 43	02 35	22 06	18 17	09 51	05 41	12 18	04 25	09 24	11 28
11 Su	21 18 30	18 30 11	05♎48 29	04 23	23 18	18 56	10 03	05 44	12 17	04 24	09 23	11 22
12 Mo	21 22 26	19 27 46	18 48 39	06 14	24 29	19 35	10 15	05 47	12 16	04 22	09 22	11 18
13 Tu	21 26 23	20 25 22	02♏02 59	08 08	25 40	20 15	10 27	05 51	12 15	04 21	09 21	11 16
14 We	21 30 19	21 22 59	15 33 28	10 04	26 51	20 54	10 39	05 54	12 14	04 19	09 19	11 15
15 Th	21 34 16	22 20 37	29 21 53	12 02	28 03	21 33	10 51	05 58	12 13	04 18	09 18	11 15
16 Fr	21 38 12	23 18 16	13♐29 11	14 01	29 14	22 12	11 02	06 01	12 11	04 16	09 17	11 13
17 Sa	21 42 09	24 15 56	27 54 44	16 01	00♎25	22 51	11 14	06 05	12 10	04 14	09 17	11 09
18 Su	21 46 06	25 13 37	12♑35 38	18 02	01 36	23 30	11 26	06 09	12 08	04 13	09 16	11 03
19 Mo	21 50 02	26 11 19	27 26 29	20 03	02 47	24 09	11 37	06 12	12 07	04 11	09 15	10 55
20 Tu	21 53 59	27 09 02	12♒19 42	22 05	03 58	24 48	11 49	06 16	12 06	04 09	09 14	10 45
21 We	21 57 55	28 06 47	27 06 34	24 06	05 09	25 26	12 00	06 20	12 04	04 08	09 13	10 35
22 Th	22 01 52	29 04 32	11♓38 35	26 07	06 19	26 05	12 11	06 24	12 02	04 06	09 12	10 25
23 Fr	22 05 48	00♍02 19	25 49 00	28 07	07 30	26 44	12 23	06 28	12 01	04 05	09 11	10 16
24 Sa	22 09 45	01 00 08	09♈33 41	00♍07	08 41	27 22	12 34	06 33	11 59	04 03	09 10	10 10
25 Su	22 13 41	01 57 59	22 51 27	02 06	09 51	28 01	12 45	06 37	11 58	04 01	09 10	10 06
26 Mo	22 17 38	02 55 51	05♉43 42	04 04	11 02	28 40	12 56	06 41	11 56	04 00	09 09	10 05
27 Tu	22 21 35	03 53 45	18 13 45	06 00	12 12	29 18	13 07	06 46	11 54	03 58	09 08	10D06
28 We	22 25 31	04 51 40	00♊26 07	07 56	13 23	29 57	13 18	06 50	11 52	03 56	09 07	10 07
29 Th	22 29 28	05 49 38	12 25 52	09 51	14 33	00♌35	13 28	06 55	11 51	03 55	09 07	10 08
30 Fr	22 33 24	06 47 37	24 18 16	11 44	15 44	01 13	13 39	06 59	11 49	03 53	09 06	10R07
31 Sa	22 37 21	07 45 39	06♋08 21	13 36	16 54	01 52	13 50	07 04	11 47	03 51	09 06	10 04

Data for 08-01-2013
Julian Day 2456505.50
Ayanamsa 24 03 00
SVP 05 ♓ 04 25
☽ ☊ Mean 12 ♏ 22 R

● ◐ PHASES ○ ◑

Day	Time	Phase	Position
06	21:51	●	14♌35
14	10:57	◐	21♏49
21	01:45	○	28♒11
28	09:35	◑	05♊15

LAST ASPECT ☽ INGRESS

Day	h m	Day	h m	Sign
01	16:49	03	04:30	♋
05	06:49	05	16:58	♌
06	21:51	08	03:58	♍
09	22:06	10	13:09	♎
12	01:30	12	20:19	♏
14	21:31	15	01:05	♐
16	17:33	17	03:26	♑
18	18:27	19	04:07	♒
21	01:45	21	04:44	♓
23	01:39	23	07:13	♈
25	10:02	25	13:14	♉
27	22:58	27	23:08	♊
29	04:45	30	11:33	♋

DECLINATION

Day	☉	☽	☿	♀	♂	♃	♄	♅	♆	♇
01 Th	18N02	19N10	20N31	08N32	23N34	23N02	11S03	04N16	10S26	19S56
02 Fr	17 46	19 58	20 36	08 03	23 30	23 01	11 04	04 15	10 27	19 56
03 Sa	17 31	19 52	20 39	07 34	23 27	23 00	11 05	04 15	10 27	19 56
04 Su	17 15	18 54	20 41	07 05	23 23	23 00	11 06	04 15	10 28	19 57
05 Mo	16 59	17 06	20 41	06 35	23 20	22 59	11 07	04 14	10 28	19 57
06 Tu	16 43	14 34	20 39	06 05	23 16	22 58	11 08	04 14	10 29	19 57
07 We	16 26	11 23	20 34	05 35	23 12	22 57	11 09	04 13	10 30	19 57
08 Th	16 09	07 41	20 27	05 05	23 07	22 57	11 10	04 13	10 30	19 58
09 Fr	15 52	03 37	20 18	04 35	23 03	22 56	11 12	04 13	10 31	19 58
10 Sa	15 34	00S40	20 06	04 04	22 58	22 55	11 13	04 12	10 31	19 58
11 Su	15 17	04 59	19 51	03 34	22 53	22 54	11 14	04 12	10 32	19 58
12 Mo	14 59	09 09	19 33	03 03	22 48	22 53	11 15	04 11	10 32	19 59
13 Tu	14 41	12 56	19 13	02 33	22 43	22 53	11 17	04 11	10 33	19 59
14 We	14 22	16 07	18 51	02 02	22 38	22 52	11 18	04 10	10 34	19 59
15 Th	14 04	18 28	18 25	01 31	22 32	22 51	11 19	04 10	10 34	19 59
16 Fr	13 45	19 44	17 57	01 00	22 27	22 50	11 21	04 09	10 35	20 00
17 Sa	13 26	19 45	17 27	00 29	22 21	22 49	11 22	04 09	10 35	20 00
18 Su	13 07	18 26	16 55	00S01	22 15	22 48	11 24	04 08	10 36	20 00
19 Mo	12 47	15 52	16 21	00 32	22 09	22 47	11 25	04 08	10 37	20 00
20 Tu	12 28	12 16	15 45	01 03	22 02	22 46	11 27	04 07	10 37	20 01
21 We	12 08	07 56	15 07	01 34	21 56	22 45	11 28	04 06	10 38	20 01
22 Th	11 48	03 12	14 27	02 05	21 49	22 45	11 30	04 06	10 38	20 01
23 Fr	11 27	01N35	13 47	02 36	21 42	22 44	11 31	04 05	10 39	20 01
24 Sa	11 07	06 09	13 05	03 07	21 35	22 43	11 33	04 04	10 40	20 01
25 Su	10 46	10 17	12 22	03 38	21 28	22 42	11 34	04 04	10 40	20 02
26 Mo	10 26	13 47	11 38	04 09	21 21	22 41	11 36	04 03	10 41	20 02
27 Tu	10 05	16 33	10 53	04 39	21 13	22 40	11 38	04 02	10 42	20 02
28 We	09 44	18 30	10 08	05 10	21 06	22 39	11 39	04 02	10 42	20 02
29 Th	09 22	19 34	09 22	05 40	20 58	22 38	11 41	04 01	10 43	20 03
30 Fr	09 01	19 46	08 36	06 11	20 50	22 37	11 43	04 00	10 43	20 03
31 Sa	08 39	19 04	07 50	06 41	20 42	22 36	11 45	04 00	10 44	20 03

ASPECTARIAN

01 01:04 ☽ □ ♆
03:08 ♂ □ ♅
10:33 ☽ ✶ ☉
16:17 ☽ □ ♀
16:49 ☽ ✶ ♅
22:34 ☽ ∥ ♇

02 19:57 ☽ ∥ ♇
20:36 ♀ ✶ ♂

03 13:48 ☽ △ ♆
15:25 ☽ △ ♄
21:56 ☽ ☌ ♃
23:47 ☽ ☍ ♇

04 05:37 ☽ □ ♅
10:06 ☽ ☌ ♂
11:58 ☽ ✶ ♀
15:05 ☉ △ ♅
05 01:27 ☽ ∥ ☉
06:49 ☽ ☌ ☿

06 03:54 ☽ □ ♄
17:29 ☽ △ ♅
07 01:29 ☽ ∥ ♄
05:57 ☽ ∥ ♆
23:35 ♃ ☍ ♇

08 12:13 ☿ ♌
12:35 ☽ ☍ ♆
14:48 ☽ ✶ ♄
17:35 ☽ ∥ ♀
20:30 ☽ ∥ ♅
22:07 ☽ △ ♇
22:31 ☽ ✶ ♃

09 14:33 ☽ ✶ ♂
17:45 ♀ ∥ ♅
22:06 ☽ ☌ ♀

10 12:28 ☿ ∥ ♇
16:52 ☽ ∥ ♀
18:30 ♂ ∥ ♃
19:33 ☽ ∥ ♅
20:55 ☽ ✶ ☿

11 06:37 ☽ □ ♇
08:00 ☽ □ ♃
12:00 ☽ ☍ ♅
18:05 ☿ □ ♄

12 01:17 ☽ ✶ ☉
01:30 ☽ □ ♂
08:28 ☽ ∥ ♆
13:02 ☽ ∥ ♄

13 04:06 ☽ △ ♆
06:49 ☽ ☌ ♄
11:19 ☽ ∥ ☉
12:41 ☽ □ ☿
13:00 ☽ ✶ ♇
15:12 ☽ △ ♃

14 09:48 ☽ △ ♂
21:31 ☽ ✶ ♀
23:27 ☽ ∥ ☿
15 02:11 ☿ △ ♅
08:25 ☽ □ ♆
21:49 ☽ △ ♅

16 01:02 ☽ △ ☿
15:37 ♀ ♎
17:33 ☽ △ ☉
17 04:29 ☽ □ ♀

10:22 ☽ ✶ ♆
13:27 ☽ ✶ ♄
18:35 ☽ ☌ ♇
22:05 ☽ ☍ ♃
23:16 ☽ □ ♅

18 18:27 ☽ ☍ ♂
19:18 ☽ ∥ ☿

19 09:21 ☽ △ ♀
14:11 ☽ □ ♄
22:47 ☽ ∥ ☉
23:37 ☽ ✶ ♅

20 04:49 ☽ ∥ ♄
09:30 ☽ ∥ ♆
18:18 ☽ ☍ ☿

21 07:15 ♃ □ ♅
11:31 ☽ ☌ ♆
15:15 ☽ △ ♄
19:34 ☽ ∥ ♅
19:56 ☽ ✶ ♇

22 00:56 ☽ △ ♃
05:03 ☽ ∥ ♀
19:45 ☉ ∥ ♄
23:02 ☉ ♍

23 01:39 ☽ △ ♂
05:51 ☽ ∥ ♀
12:52 ☽ ∥ ♅
22:17 ☽ ☍ ♀
22:37 ☿ ♍
23:19 ☽ □ ♇

24 04:19 ☽ ☌ ♅
05:25 ☽ □ ♃
10:00 ♀ □ ♇
20:56 ☿ ☌ ☉

25 02:27 ☽ ∥ ♆
02:50 ☽ ∥ ☉
06:52 ☉ ∥ ♆
08:22 ☽ ∥ ♄
10:02 ☽ □ ♂
11:18 ☽ ∥ ☿
18:17 ☽ △ ☉
19:47 ♀ ∥ ♅
20:16 ☽ △ ☿
20:44 ☽ ✶ ♆

23:13 ☿ ☍ ♆
26 00:59 ☿ ∥ ♄
01:50 ☽ ☍ ♄
06:30 ☽ △ ♇
13:57 ☽ ✶ ♃
17:56 ♀ ☍ ♅

27 01:43 ☉ ☍ ♆
06:13 ☿ ∥ ♆
09:45 ☿ ✶ ♄
21:54 ♀ □ ♃
22:58 ☽ ✶ ♂
28 02:05 ♂ ♌
06:30 ☽ □ [illegible]
14:51 ☿ △ ♇

17:49 ☽ □ ☿
22:49 ☽ ✶ ♅
23:56 ☿ ∥ ☉
29 04:45 ☽ △ ♀
30 05:18 ☉ ✶ ♄
19:23 ☽ △ ♆

31 01:54 ☽ △ ♄
03:11 ☿ ✶ ♃
03:35 ☽ ✶ ☉
05:59 ☽ ☍ ♇
11:24 ☽ □ ♅
15:47 ☽ ☌ ♃
17:53 ☽ ✶ ♀
21:19 ☿ ∥ ♀

⚷ Chiron

01 Dec. 01 S 39

Day	Position
02	12♓54R
05	12 47
08	12 39
11	12 32
14	12 24
17	12 16
20	12 08
23	12 00
26	11 51
29	11 43

September 2013

Day	S. T. h m s	☉ ° ' "	☽ ° ' "	☿ ° '	♀ ° '	♂ ° '	♃ ° '	♄ ° '	♅ ° '	♆ ° '	♇ ° '	☊ True ° '
01 Su	22 41 17	08♍43 42	18♋00 42	15♍27	18♎04	02♌30	14♋00	07♏09	11♈R45	03♓R50	09♑R05	09♏R58
02 Mo	22 45 14	09 41 47	29 59 10	17 17	19 14	03 08	14 10	07 14	11 43	03 48	09 04	09 50
03 Tu	22 49 10	10 39 53	12♌06 44	19 05	20 24	03 47	14 21	07 19	11 41	03 47	09 04	09 40
04 We	22 53 07	11 38 02	24 25 22	20 52	21 34	04 25	14 31	07 24	11 39	03 45	09 03	09 30
05 Th	22 57 04	12 36 12	06♍56 10	22 38	22 44	05 03	14 41	07 29	11 37	03 43	09 03	09 19
06 Fr	23 01 00	13 34 24	19 39 29	24 23	23 54	05 41	14 51	07 34	11 35	03 42	09 02	09 09
07 Sa	23 04 57	14 32 38	02♎35 05	26 07	25 04	06 19	15 00	07 39	11 33	03 40	09 02	09 02
08 Su	23 08 53	15 30 53	15 42 28	27 49	26 13	06 57	15 10	07 44	11 31	03 38	09 02	08 56
09 Mo	23 12 50	16 29 10	29 01 12	29 30	27 23	07 35	15 20	07 50	11 29	03 37	09 01	08 53
10 Tu	23 16 46	17 27 29	12♏31 01	01♎10	28 32	08 12	15 29	07 55	11 27	03 35	09 01	08 52
11 We	23 20 43	18 25 49	26 11 58	02 49	29 42	08 50	15 39	08 00	11 25	03 34	09 01	08D 52
12 Th	23 24 39	19 24 10	10♐04 14	04 27	00♏51	09 28	15 48	08 06	11 22	03 32	09 00	08 53
13 Fr	23 28 36	20 22 34	24 07 39	06 04	02 01	10 06	15 57	08 12	11 20	03 31	09 00	08R 52
14 Sa	23 32 33	21 20 59	08♑21 24	07 39	03 10	10 43	16 06	08 17	11 18	03 29	09 00	08 51
15 Su	23 36 29	22 19 25	22 43 26	09 14	04 19	11 21	16 15	08 23	11 16	03 27	09 00	08 47
16 Mo	23 40 26	23 17 53	07♒10 19	10 47	05 28	11 58	16 24	08 29	11 14	03 26	09 00	08 42
17 Tu	23 44 22	24 16 22	21 37 19	12 19	06 37	12 36	16 33	08 34	11 11	03 24	08 59	08 36
18 We	23 48 19	25 14 54	05♓58 45	13 51	07 45	13 13	16 41	08 40	11 09	03 23	08 59	08 30
19 Th	23 52 15	26 13 26	20 08 46	15 21	08 54	13 51	16 50	08 46	11 07	03 21	08 59	08 23
20 Fr	23 56 12	27 12 01	04♈02 13	16 50	10 03	14 28	16 58	08 52	11 04	03 20	08 59	08 18
21 Sa	00 00 08	28 10 38	17 35 28	18 18	11 11	15 05	17 06	08 58	11 02	03 18	08D 59	08 14
22 Su	00 04 05	29 09 17	00♉46 46	19 45	12 20	15 42	17 14	09 04	11 00	03 17	08 59	08 12
23 Mo	00 08 01	00♎07 58	13 36 26	21 11	13 28	16 20	17 22	09 10	10 57	03 16	08 59	08D 12
24 Tu	00 11 58	01 06 41	26 06 27	22 36	14 36	16 57	17 30	09 16	10 55	03 14	08 59	08 14
25 We	00 15 55	02 05 27	08♊20 08	23 59	15 44	17 34	17 37	09 22	10 53	03 13	09 00	08 16
26 Th	00 19 51	03 04 14	20 21 41	25 22	16 52	18 11	17 45	09 29	10 50	03 11	09 00	08 18
27 Fr	00 23 48	04 03 04	02♋15 47	26 43	18 00	18 48	17 52	09 35	10 48	03 10	09 00	08 19
28 Sa	00 27 44	05 01 57	14 07 25	28 03	19 07	19 25	17 59	09 41	10 45	03 09	09 00	08R 19
29 Su	00 31 41	06 00 51	26 01 26	29 22	20 15	20 02	18 06	09 48	10 43	03 07	09 00	08 18
30 Mo	00 35 37	06 59 48	08♌02 27	00♏40	21 22	20 38	18 13	09 54	10 41	03 06	09 01	08 15

Data for	09-01-2013
Julian Day	2456536.50
Ayanamsa	24 03 04
SVP	05 ♓ 04 20
☽ ☊ Mean	10 ♏ 43 R

● ◐ PHASES ○ ◑

Day	Time	Phase	Position
05	11:37	●	13♍04
12	17:09	◐	20♐06
19	11:13	○	26♓41
27	03:56	◑	04♋13

LAST ASPECT Day	h m	☽ INGRESS Day	h m	Sign
01	00:07	02	00:02	♌
03	17:54	04	10:45	♍
06	10:11	06	19:14	♎
08	20:47	09	01:45	♏
10	09:22	11	06:36	♐
12	17:09	13	09:56	♑
14	23:17	15	12:06	♒
16	08:20	17	13:59	♓
19	11:13	19	16:58	♈
21	01:26	21	22:34	♉
23	07:14	24	07:35	♊
26	11:22	26	19:25	♋
29	07:32	29	07:58	♌

DECLINATION

Day	☉	☽	☿	♀	♂	♃	♄	♅	♆	♇
01 Su	08N18	17N33	07N03	07S 11	20N34	22N35	11S 46	03N59	10S 45	20S 03
02 Mo	07 56	15 15	06 16	07 42	20 26	22 34	11 48	03 58	10 45	20 04
03 Tu	07 34	12 17	05 29	08 11	20 18	22 33	11 50	03 57	10 46	20 04
04 We	07 12	08 45	04 42	08 41	20 09	22 32	11 52	03 56	10 46	20 04
05 Th	06 50	04 47	03 55	09 11	20 00	22 31	11 53	03 56	10 47	20 04
06 Fr	06 27	00 32	03 08	09 40	19 52	22 30	11 55	03 55	10 48	20 04
07 Sa	06 05	03S 48	02 21	10 10	19 43	22 29	11 57	03 54	10 48	20 05
08 Su	05 43	08 02	01 35	10 39	19 34	22 28	11 59	03 53	10 49	20 05
09 Mo	05 20	11 57	00 49	11 07	19 25	22 27	12 01	03 52	10 49	20 05
10 Tu	04 57	15 17	00 03	11 36	19 15	22 26	12 03	03 51	10 50	20 05
11 We	04 35	17 48	00S 43	12 04	19 06	22 25	12 05	03 51	10 50	20 05
12 Th	04 12	19 19	01 28	12 33	18 56	22 24	12 07	03 50	10 51	20 06
13 Fr	03 49	19 39	02 13	13 00	18 47	22 23	12 09	03 49	10 52	20 06
14 Sa	03 26	18 44	02 58	13 28	18 37	22 22	12 11	03 48	10 52	20 06
15 Su	03 03	16 38	03 42	13 55	18 27	22 21	12 13	03 47	10 53	20 06
16 Mo	02 40	13 30	04 25	14 22	18 17	22 20	12 15	03 46	10 53	20 07
17 Tu	02 16	09 33	05 08	14 49	18 07	22 19	12 17	03 45	10 54	20 07
18 We	01 53	05 04	05 51	15 16	17 57	22 18	12 19	03 44	10 54	20 07
19 Th	01 30	00 21	06 33	15 42	17 47	22 17	12 21	03 44	10 55	20 07
20 Fr	01 07	04N17	07 14	16 07	17 37	22 16	12 23	03 43	10 55	20 07
21 Sa	00 43	08 36	07 55	16 33	17 26	22 15	12 25	03 42	10 56	20 07
22 Su	00 20	12 22	08 35	16 58	17 16	22 14	12 27	03 41	10 57	20 08
23 Mo	00S 03	15 27	09 15	17 23	17 05	22 13	12 29	03 40	10 57	20 08
24 Tu	00 27	17 43	09 53	17 47	16 54	22 12	12 31	03 39	10 58	20 08
25 We	00 50	19 06	10 32	18 11	16 44	22 12	12 33	03 38	10 58	20 08
26 Th	01 13	19 36	11 09	18 34	16 33	22 11	12 35	03 37	10 59	20 08
27 Fr	01 37	19 12	11 45	18 57	16 22	22 10	12 37	03 36	10 59	20 09
28 Sa	02 00	17 57	12 21	19 20	16 11	22 09	12 39	03 35	11 00	20 09
29 Su	02 23	15 56	12 56	19 42	16 00	22 08	12 41	03 34	11 00	20 09
30 Mo	02 47	13 13	13 30	20 04	15 48	22 07	12 43	03 33	11 00	20 09

⚷ Chiron

01 Dec. 02 S 09

Day	Position
01	11♓34R
04	11 26
07	11 17
10	11 09
13	11 00
16	10 52
19	10 44
22	10 36
25	10 28
28	10 20

ASPECTARIAN

01 00:07 ☽ □ ♀
08:44 ☉ △ ♇
02 06:37 ☽ ☌ ♂
06:41 ♀ # ☉
14:28 ☽ □ ♄
23:10 ☽ △ ♅
03 03:18 ☽ # ♄
10:46 ☽ # ♆
17:54 ☽ ⚹ ♀
04 00:22 ☽ # ♀
10:39 ☽ ∥ ☉
13:45 ♂ # ♇
17:53 ☽ ☍ ♆
23:39 ☿ ∥ ♅
05 01:02 ☽ ⚹ ♄
04:01 ☽ △ ♇
04:56 ☽ ∥ ♅
06:07 ☽ ∥ ☿
14:51 ☽ ⚹ ♃
06 10:11 ☽ ☌ ☿
17:11 ☽ # ☿
07 00:30 ☽ # ♅
07:12 ☽ ⚹ ♂
11:43 ☽ # ☉
11:50 ☽ □ ♇
13:47 ☉ ⚹ ♃
16:24 ☽ ☍ ♅
23:01 ☽ □ ♃
08 08:37 ♀ ∥ ♆
16:47 ☽ ∥ ♆
17:57 ☽ ∥ ♀
20:47 ☽ ☌ ♀
09 00:29 ☽ ∥ ♄
07:07 ☿ ♎
08:11 ☽ △ ♆
11:07 ♂ □ ♄
15:48 ☽ ☌ ♄
16:00 ☽ □ ♂
17:48 ☽ ⚹ ♇
10 05:18 ☽ △ ♃
09:22 ☽ ⚹ ☉
11 00:17 ♀ ∥ ♄
06:16 ♀ ♏
12:45 ☽ □ ♆
13:02 ☽ ⚹ ☿
16:56 ☽ # ♂
22:55 ☽ △ ♂
12 02:14 ☽ △ ♅
23:49 ☉ ∥ ♅
13 14:30 ☽ ⚹ ♀
15:49 ☽ ⚹ ♆
22:40 ☽ □ ☿
23:53 ☽ ⚹ ♄
14 01:05 ☽ ☌ ♇
02:12 ☽ # ♂
04:55 ☽ □ ♅
06:33 ♀ △ ♆
10:04 ☿ # ☉
13:06 ☽ ☍ ♃
20:29 ☿ □ ♇
20:57 ♂ △ ♅
23:17 ☽ △ ☉
15 02:58 ☿ # ♅
18:43 ☽ ∥ ♀
20:55 ☽ □ ♀
16 02:11 ☽ □ ♄
06:41 ☿ ☍ ♅
06:42 ☽ ⚹ ♅
06:43 ☽ △ ☿
08:04 ☽ ∥ ♄
08:20 ☽ ☍ ♂
16:17 ☽ ∥ ♆
17 07:18 ☿ ⚹ ♂
19:39 ☽ ☌ ♆
20:30 ☽ ∥ ☿
18 03:15 ☽ △ ♀
04:34 ☽ △ ♄
05:04 ☽ ⚹ ♇
06:51 ☽ # ♅
17:43 ☽ # ☉
18:17 ☽ △ ♃
20:54 ♀ ☌ ♄
19 01:50 ♀ ⚹ ♇
08:49 ☽ ∥ ☉
21:02 ☽ ∥ ♅
20 02:21 ☿ □ ♃
08:42 ☽ □ ♇
12:21 ☽ ☌ ♅
15:28 ♇ SD
19:19 ☽ △ ♂
19:19 ☽ # ☿
23:07 ☽ □ ♃
21 01:26 ☽ ☍ ☿
05:19 ♄ ⚹ ♇
14:25 ☽ # ♆
22 00:30 ☽ # ♄
04:38 ☽ ⚹ ♆
12:06 ♀ # ♂
15:17 ☽ △ ♇
15:33 ☽ ☍ ♄
20:45 ☉ ♎
23:42 ☽ ☍ ♀
23 05:26 ☽ □ ♂
07:14 ☽ ⚹ ♃
15:02 ☽ ∥ ♂
24 01:02 ☽ # ♀
10:37 ☽ △ ☉
13:54 ☽ □ ♆
25 05:01 ☽ ⚹ ♅
17:13 ☿ ∥ ♆
19:23 ☽ ⚹ ♂
26 11:22 ☽ △ ☿
21:01 ♀ △ ♃
27 01:50 ☽ △ ♆
04:30 ☽ # ♀
13:38 ☽ ☍ ♇
14:57 ☽ △ ♄
17:13 ☽ □ ♅
28 07:54 ☽ ☌ ♃
11:09 ☽ △ ♀
13:01 ☿ ∥ ♄
13:33 ♀ □ ♂
23:23 ☽ ∥ ♂
29 07:32 ☽ □ ☿
11:39 ☿ ♏
21:45 ☽ ⚹ ☉
22:04 ☽ # ☿
30 03:43 ☽ □ ♄
03:48 ☽ # ♄
05:12 ☽ △ ♅
05:28 ♀ ∥ ♇
16:22 ☽ # ♆

October 2013

18 23:51 25♈45 ✷ Penumbral Lunar Eclipse (mag 0.790)

Day	S. T. h m s	☉ ° ' "	☽ ° ' "	☿ ° '	♀ ° '	♂ ° '	♃ ° '	♄ ° '	♅ ° '	♆ ° '	♇ ° '	☊ True ° '
01 Tu	00 39 34	07♎58 47	20♌14 23	01♏56	22♏30	21♌15	18♋20	10♏00	10♈R38	03♓R05	09♑01	08♏R11
02 We	00 43 30	08 57 48	02♍40 20	03 11	23 37	21 52	18 27	10 07	10 36	03 04	09 01	08 06
03 Th	00 47 27	09 56 51	15 22 19	04 24	24 44	22 28	18 33	10 13	10 33	03 02	09 02	08 01
04 Fr	00 51 24	10 55 56	28 21 13	05 35	25 51	23 05	18 40	10 20	10 31	03 01	09 02	07 57
05 Sa	00 55 20	11 55 04	11♎36 48	06 45	26 57	23 42	18 46	10 27	10 28	03 00	09 02	07 53
06 Su	00 59 17	12 54 13	25 07 43	07 53	28 04	24 18	18 52	10 33	10 26	02 59	09 03	07 51
07 Mo	01 03 13	13 53 25	08♏51 51	08 58	29 11	24 54	18 58	10 40	10 24	02 58	09 03	07 50
08 Tu	01 07 10	14 52 38	22 46 40	10 02	00♐17	25 31	19 03	10 47	10 21	02 56	09 04	07D 50
09 We	01 11 06	15 51 54	06♐49 28	11 03	01 23	26 07	19 09	10 53	10 19	02 55	09 04	07 50
10 Th	01 15 03	16 51 11	20 57 43	12 02	02 29	26 43	19 14	11 00	10 16	02 54	09 05	07 51
11 Fr	01 18 59	17 50 30	05♑09 06	12 57	03 35	27 19	19 19	11 07	10 14	02 53	09 06	07 51
12 Sa	01 22 56	18 49 51	19 21 22	13 50	04 40	27 56	19 24	11 14	10 12	02 52	09 06	07R 51
13 Su	01 26 53	19 49 13	03♒32 20	14 39	05 46	28 32	19 29	11 21	10 09	02 51	09 07	07 50
14 Mo	01 30 49	20 48 37	17 39 43	15 25	06 51	29 08	19 34	11 27	10 07	02 50	09 08	07 49
15 Tu	01 34 46	21 48 03	01♓41 04	16 06	07 56	29 43	19 38	11 34	10 04	02 49	09 08	07 47
16 We	01 38 42	22 47 31	15 33 52	16 43	09 01	00♍19	19 43	11 41	10 02	02 48	09 09	07 45
17 Th	01 42 39	23 47 00	29 15 33	17 15	10 05	00 55	19 47	11 48	10 00	02 47	09 10	07 44
18 Fr	01 46 35	24 46 31	12♈43 47	17 41	11 09	01 31	19 51	11 55	09 57	02 47	09 11	07 42
19 Sa	01 50 32	25 46 05	25 56 45	18 02	12 13	02 06	19 54	12 02	09 55	02 46	09 12	07 42
20 Su	01 54 28	26 45 40	08♉53 25	18 16	13 17	02 42	19 58	12 09	09 53	02 45	09 13	07D 42
21 Mo	01 58 25	27 45 18	21 33 45	18 23	14 21	03 18	20 02	12 16	09 51	02 44	09 13	07 43
22 Tu	02 02 21	28 44 57	03♊58 47	18R 22	15 24	03 53	20 05	12 23	09 48	02 43	09 14	07 44
23 We	02 06 18	29 44 39	16 10 34	18 14	16 27	04 28	20 08	12 30	09 46	02 43	09 15	07 45
24 Th	02 10 15	00♏44 23	28 11 56	17 57	17 30	05 04	20 11	12 38	09 44	02 42	09 16	07 46
25 Fr	02 14 11	01 44 09	10♋06 30	17 31	18 33	05 39	20 13	12 45	09 42	02 41	09 17	07 47
26 Sa	02 18 08	02 43 57	21 58 24	16 55	19 35	06 14	20 16	12 52	09 39	02 41	09 18	07 48
27 Su	02 22 04	03 43 48	03♌52 11	16 11	20 37	06 49	20 18	12 59	09 37	02 40	09 19	07 48
28 Mo	02 26 01	04 43 40	15 52 34	15 19	21 38	07 24	20 20	13 06	09 35	02 39	09 21	07R 48
29 Tu	02 29 57	05 43 35	28 04 10	14 18	22 40	07 59	20 22	13 13	09 33	02 39	09 22	07 48
30 We	02 33 54	06 43 32	10♍31 13	13 10	23 41	08 34	20 24	13 20	09 31	02 38	09 23	07 48
31 Th	02 37 50	07 43 31	23 17 07	11 57	24 41	09 09	20 25	13 28	09 29	02 38	09 24	07 48

Data for	10-01-2013
Julian Day	2456566.50
Ayanamsa	24 03 06
SVP	05 ♓ 04 15
☽ ☊ Mean	09 ♏ 08 R

● ◐ PHASES ○ ◑

Day	h m	Phase	Position
05	00:35	●	11♎57
11	23:02	◐	18♑48
18	23:38	✷	25♈45
26	23:41	◑	03♌43

LAST ASPECT Day	h m	☽ INGRESS Day	h m	Sign
01	04:50	01	18:53	♍
03	18:59	04	03:00	♎
05	22:28	06	08:33	♏
08	04:54	08	12:22	♐
10	10:11	10	15:18	♑
12	00:05	12	18:00	♒
14	20:29	14	21:06	♓
16	07:16	17	01:19	♈
18	23:39	19	07:28	♉
20	21:03	21	16:15	♊
23	00:36	24	03:37	♋
25	20:32	26	16:13	♌
28	12:27	29	03:45	♍
31	02:49	31	12:22	♎

DECLINATION

Day	☉	☽	☿	♀	♂	♃	♄	♅	♆	♇
01 Tu	03S 10	09N54	14S 04	20S 26	15N37	22N06	12S 46	03N32	11S 01	20S 09
02 We	03 33	06 05	14 36	20 47	15 26	22 06	12 48	03 31	11 01	20 09
03 Th	03 56	01 56	15 07	21 07	15 14	22 05	12 50	03 30	11 02	20 10
04 Fr	04 20	02S 24	15 37	21 27	15 03	22 04	12 52	03 30	11 02	20 10
05 Sa	04 43	06 44	16 07	21 46	14 51	22 03	12 54	03 29	11 03	20 10
06 Su	05 06	10 48	16 35	22 05	14 40	22 03	12 56	03 28	11 03	20 10
07 Mo	05 29	14 22	17 01	22 24	14 28	22 02	12 59	03 27	11 04	20 10
08 Tu	05 52	17 09	17 27	22 42	14 16	22 01	13 01	03 26	11 04	20 10
09 We	06 15	18 56	17 51	22 59	14 05	22 01	13 03	03 25	11 04	20 11
10 Th	06 37	19 32	18 14	23 16	13 53	22 00	13 05	03 24	11 05	20 11
11 Fr	07 00	18 55	18 35	23 33	13 41	21 59	13 07	03 23	11 05	20 11
12 Sa	07 23	17 06	18 55	23 48	13 29	21 59	13 09	03 22	11 05	20 11
13 Su	07 45	14 15	19 13	24 04	13 17	21 58	13 12	03 21	11 06	20 11
14 Mo	08 07	10 35	19 29	24 18	13 05	21 58	13 14	03 20	11 06	20 11
15 Tu	08 30	06 22	19 43	24 32	12 53	21 57	13 16	03 19	11 06	20 11
16 We	08 52	01 51	19 56	24 46	12 41	21 57	13 18	03 18	11 07	20 12
17 Th	09 14	02N43	20 05	24 59	12 28	21 56	13 20	03 17	11 07	20 12
18 Fr	09 36	07 05	20 13	25 11	12 16	21 56	13 23	03 17	11 07	20 12
19 Sa	09 57	11 02	20 17	25 23	12 04	21 55	13 25	03 16	11 08	20 12
20 Su	10 19	14 22	20 19	25 34	11 52	21 55	13 27	03 15	11 08	20 12
21 Mo	10 40	16 56	20 18	25 45	11 39	21 54	13 29	03 14	11 08	20 12
22 Tu	11 02	18 39	20 13	25 54	11 27	21 54	13 31	03 13	11 09	20 12
23 We	11 23	19 27	20 05	26 04	11 14	21 54	13 34	03 12	11 09	20 12
24 Th	11 44	19 21	19 53	26 12	11 02	21 53	13 36	03 11	11 09	20 13
25 Fr	12 05	18 23	19 36	26 21	10 49	21 53	13 38	03 10	11 09	20 13
26 Sa	12 25	16 37	19 16	26 28	10 37	21 53	13 40	03 10	11 09	20 13
27 Su	12 45	14 09	18 51	26 35	10 24	21 52	13 42	03 09	11 10	20 13
28 Mo	13 06	11 05	18 21	26 41	10 12	21 52	13 45	03 08	11 10	20 13
29 Tu	13 26	07 29	17 48	26 47	09 59	21 52	13 47	03 07	11 10	20 13
30 We	13 45	03 30	17 11	26 52	09 47	21 52	13 49	03 06	11 10	20 13
31 Th	14 05	00S 45	16 30	26 56	09 34	21 52	13 51	03 06	11 10	20 13

ASPECTARIAN

01 02:04 ☽ ☌ ♂
04:50 ☽ □ ♀
21:46 ☿ △ ♆
22:17 ☉ # ♅

02 00:44 ☽ ☍ ♆
01:04 ☽ ⚹ ☿
01:26 ☉ □ ♇
12:04 ☽ △ ♇
13:36 ☽ # ☉
14:15 ☽ ⚹ ♄
15:05 ☽ ∥ ♅

03 04:09 ☿ # ♂
05:59 ☽ ⚹ ♃
14:12 ☉ ☍ ♅
18:59 ☽ ⚹ ♀

04 05:59 ☽ # ♅
11:39 ☽ ∥ ☉
19:23 ☽ □ ♆
21:58 ☽ ☍ ♅

05 12:51 ☽ □ ♃
20:31 ♀ # ♃
22:28 ☽ ⚹ ♂

06 01:37 ☽ ∥ ♆
13:44 ☽ △ ♆
14:04 ☽ ∥ ♄

07 00:12 ☽ ☌ ☿
00:20 ☽ ⚹ ♇
00:47 ☽ # ♂
01:56 ☿ ⚹ ♇
03:09 ☽ ☌ ♄
17:33 ☽ △ ♃
17:54 ♀ ♐

08 03:59 ☽ ∥ ☿
04:54 ☽ □ ♂
13:56 ☽ ☌ ♀
17:22 ☽ □ ♆
19:40 ☿ ☌ ♄

09 05:55 ☽ △ ♅
16:31 ☽ ⚹ ☉

10 09:06 ♀ □ ♆
10:11 ☽ △ ♂
20:11 ☽ ⚹ ♆

11 04:35 ☽ ∥ ☿
06:40 ☽ ☌ ♇
08:34 ☽ □ ♅
10:09 ☽ ⚹ ♄
14:04 ☽ ⚹ ☿

12 00:05 ☽ ☍ ♃
15:09 ☉ □ ♃

13 04:05 ☽ ⚹ ♀
07:18 ☽ # ♂
07:23 ☽ ∥ ♄
08:40 ♂ # ♄
11:11 ☽ ⚹ ♄
13:21 ☽ □ ☿
19:57 ☽ □ ☉
20:56 ☽ ∥ ♄

14 05:47 ☽ △ ☉
13:12 ☽ ∥ ☉
20:29 ☽ ☍ ♂

15 01:57 ☽ ☌ ♆
11:06 ♂ ♍
11:40 ☽ □ ♀
12:52 ☽ ⚹ ♇
16:20 ☽ # ♅
17:13 ☽ △ ♄

16 02:05 ☽ △ ☿
07:16 ☽ △ ♃
22:04 ♀ △ ♅

17 03:03 ☽ ∥ ♅
17:38 ☽ □ ♇
19:02 ☽ ☌ ♅
20:32 ☿ ∥ ♆
20:56 ☽ △ ♀

18 12:55 ☽ □ ♃
16:27 ☽ # ☉

19 00:41 ☽ # ♆
06:35 ☽ ∥ ♂
11:54 ☽ △ ♂
12:34 ☽ ⚹ ♆
16:52 ☽ # ♄

20 00:36 ☽ △ ♆
01:55 ♂ ☍ ♆
06:11 ☽ ☍ ♄
17:54 ☽ ☍ ☿
21:03 ☽ ⚹ ♃

21 10:30 ☿ SR
[illegible] ☽ □ ♆
23:48 ☽ □ ♂

22 03:09 ☿ ∥ ♆
07:51 ☉ ∥ ♆
11:23 ☽ ⚹ ♅
17:59 ☉ # ♂

23 00:36 ☽ ☍ ♀
06:10 ☉ ♏
10:38 ♂ # ♆

24 05:34 ☽ △ ☉
09:02 ☽ △ ♆
14:32 ☽ ⚹ ♂
22:20 ☽ ☍ ♇
[illegible] [illegible] □ ☿

25 05:23 ☽ △ ♄
14:18 ☽ △ ☿
20:32 ☽ ☌ ♃
22:41 ☉ △ ♆

27 03:42 ☽ # ♄
10:15 ☽ # ☉
11:30 ☽ △ ♅
18:25 ☽ □ ♄
22:57 ☽ □ ☿
23:20 ☽ # ♆

28 06:31 ☽ ∥ ♂
12:27 ☽ △ ☿

29 08:53 ☽ ☍ ♆
16:07 ☽ ⚹ ☉
20:06 ☽ ☌ ♂
20:48 ☿ ☌ ♄
21:49 ☽ △ ♇

30 02:17 ☽ ∥ ♅
04:36 ☽ ⚹ ☿
04:46 ☉ ∥ ♄
05:24 ☽ ⚹ ♄
18:40 ☽ ⚹ ♃

31 02:49 ☽ □ ♀
10:46 ♂ △ ♇
13:59 ☽ # ♅

⚷ Chiron

01 Dec. 02 S 43

Day	°	'
01	10♓13R	
04	10	06
07	09	59
10	09	52
13	09	46
16	09	41
19	09	35
22	09	30
25	09	26
28	09	22
01	[illegible]	[illegible]

November 2013

03 12:40 11♏16 ☉ Total Solar Eclipse

Day	S. T. h m s	☉ ° ' "	☽ ° ' "	☿ ° '	♀ ° '	♂ ° '	♃ ° '	♄ ° '	♅ ° '	♆ ° '	♇ ° '	☊ True ° '
01 Fr	02 41 47	08♏ 43 33	06♎ 24 13	10♏R40	25♐ 42	09♍ 44	20♋ 27	13♏ 35	09♈R27	02♓R37	09♑ 25	07♏R48
02 Sa	02 45 44	09 43 36	19 53 24	09 23	26 41	10 18	20 28	13 42	09 25	02 37	09 27	07 48
03 Su	02 49 40	10 43 41	03♏ 43 49	08 06	27 41	10 53	20 29	13 49	09 23	02 37	09 28	07 47
04 Mo	02 53 37	11 43 48	17 52 46	06 53	28 40	11 28	20 30	13 56	09 21	02 36	09 29	07 47
05 Tu	02 57 33	12 43 57	02♐ 15 56	05 45	29 39	12 02	20 30	14 03	09 19	02 36	09 30	07 46
06 We	03 01 30	13 44 08	16 47 45	04 46	00♑ 37	12 36	20 31	14 11	09 17	02 36	09 32	07 45
07 Th	03 05 26	14 44 21	01♑ 22 17	03 56	01 35	13 11	20 31	14 18	09 15	02 36	09 33	07 43
08 Fr	03 09 23	15 44 35	15 53 48	03 17	02 32	13 45	20R 31	14 25	09 13	02 35	09 34	07 42
09 Sa	03 13 19	16 44 50	00♒ 17 33	02 50	03 29	14 19	20 30	14 32	09 12	02 35	09 36	07 41
10 Su	03 17 16	17 45 07	14 30 02	02 34	04 26	14 53	20 30	14 39	09 10	02 35	09 37	07 40
11 Mo	03 21 13	18 45 25	28 29 06	02D 30	05 21	15 27	20 29	14 47	09 08	02 35	09 39	07D 40
12 Tu	03 25 09	19 45 45	12♓ 13 46	02 36	06 17	16 01	20 28	14 54	09 07	02 35	09 40	07 41
13 We	03 29 06	20 46 06	25 43 56	02 54	07 11	16 34	20 27	15 01	09 05	02 35	09 42	07 42
14 Th	03 33 02	21 46 28	09♈ 00 01	03 21	08 05	17 08	20 26	15 08	09 03	02D 35	09 43	07 43
15 Fr	03 36 59	22 46 52	22 02 37	03 56	08 59	17 42	20 25	15 15	09 02	02 35	09 45	07 45
16 Sa	03 40 55	23 47 18	04♉ 52 24	04 40	09 52	18 15	20 23	15 23	09 00	02 35	09 46	07 46
17 Su	03 44 52	24 47 45	17 29 58	05 31	10 44	18 48	20 21	15 30	08 59	02 35	09 48	07R 47
18 Mo	03 48 48	25 48 13	29 56 04	06 28	11 35	19 22	20 19	15 37	08 57	02 35	09 50	07 46
19 Tu	03 52 45	26 48 43	12♊ 11 37	07 30	12 26	19 55	20 17	15 44	08 56	02 35	09 51	07 44
20 We	03 56 42	27 49 15	24 17 54	08 37	13 16	20 28	20 14	15 51	08 55	02 35	09 53	07 42
21 Th	04 00 38	28 49 48	06♋ 16 41	09 48	14 05	21 01	20 12	15 58	08 53	02 36	09 55	07 38
22 Fr	04 04 35	29 50 23	18 10 14	11 03	14 54	21 34	20 09	16 05	08 52	02 36	09 56	07 34
23 Sa	04 08 31	00♐ 51 00	00♌ 01 29	12 21	15 41	22 07	20 06	16 12	08 51	02 36	09 58	07 31
24 Su	04 12 28	01 51 38	11 53 59	13 41	16 28	22 39	20 03	16 19	08 50	02 37	10 00	07 29
25 Mo	04 16 24	02 52 18	23 51 55	15 03	17 14	23 12	19 59	16 26	08 48	02 37	10 02	07 27
26 Tu	04 20 21	03 52 59	05♍ 59 50	16 27	17 58	23 44	19 56	16 33	08 47	02 37	10 03	07D 27
27 We	04 24 17	04 53 42	18 22 26	17 53	18 42	24 17	19 52	16 40	08 46	02 38	10 05	07 28
28 Th	04 28 14	05 54 27	01♎ 04 15	19 20	19 25	24 49	19 48	16 47	08 45	02 38	10 07	07 29
29 Fr	04 32 11	06 55 13	14 09 14	20 48	20 06	25 21	19 44	16 54	08 44	02 39	10 09	07 30
30 Sa	04 36 07	07 56 00	27 40 08	22 16	20 47	25 53	19 39	17 01	08 43	02 39	10 11	07 31

Data for	11-01-2013
Julian Day	2456597.50
Ayanamsa	24 03 09
SVP	05 ♓ 04 09
☽ ☊ Mean	07 ♏ 30 R

● ◐ PHASES ○ ◑

03	12:50	☉	11♏16
10	05:58	◐	18♒00
17	15:16	○	25♉26
25	19:28	◑	03♍42

LAST ASPECT Day	h m	☽ INGRESS Day	h m	
02	12:47	02	17:35	♏
04	04:23	04	20:14	♐
05	16:49	06	21:44	♑
08	07:40	08	23:31	♒
10	05:58	11	02:37	♓
12	14:35	13	07:40	♈
14	20:59	15	14:50	♉
17	15:17	18	00:08	♊
19	16:00	20	11:24	♋
22	07:12	22	23:57	♌
24	08:59	25	12:11	♍
27	11:44	27	22:00	♎
29	11:14	30	04:04	♏

DECLINATION

Day	☉	☽	☿	♀	♂	♃	♄	♅	♆	♇
01 Fr	14S 24	05S 05	15S 47	27S 00	09N22	21N52	13S 53	03N05	11S 11	20S 13
02 Sa	14 44	09 17	15 03	27 03	09 09	21 52	13 55	03 04	11 11	20 13
03 Su	15 02	13 07	14 19	27 05	08 56	21 52	13 58	03 03	11 11	20 14
04 Mo	15 21	16 16	13 36	27 07	08 44	21 52	14 00	03 03	11 11	20 14
05 Tu	15 39	18 28	12 55	27 09	08 31	21 52	14 02	03 02	11 11	20 14
06 We	15 58	19 28	12 18	27 09	08 19	21 52	14 04	03 01	11 11	20 14
07 Th	16 15	19 11	11 45	27 10	08 06	21 52	14 06	03 00	11 11	20 14
08 Fr	16 33	17 38	11 18	27 09	07 53	21 52	14 08	03 00	11 11	20 14
09 Sa	16 50	14 58	10 56	27 08	07 41	21 52	14 10	02 59	11 11	20 14
10 Su	17 07	11 28	10 39	27 06	07 28	21 52	14 13	02 58	11 11	20 14
11 Mo	17 24	07 22	10 29	27 04	07 15	21 53	14 15	02 58	11 11	20 14
12 Tu	17 40	02 57	10 24	27 02	07 03	21 53	14 17	02 57	11 11	20 14
13 We	17 57	01N33	10 24	26 58	06 50	21 53	14 19	02 56	11 11	20 14
14 Th	18 12	05 54	10 29	26 54	06 38	21 53	14 21	02 56	11 11	20 14
15 Fr	18 28	09 55	10 38	26 50	06 25	21 54	14 23	02 55	11 11	20 14
16 Sa	18 43	13 24	10 51	26 45	06 12	21 54	14 25	02 55	11 11	20 14
17 Su	18 58	16 12	11 07	26 40	06 00	21 55	14 27	02 54	11 11	20 14
18 Mo	19 12	18 12	11 27	26 34	05 47	21 55	14 29	02 54	11 11	20 15
19 Tu	19 26	19 18	11 49	26 28	05 35	21 56	14 31	02 53	11 11	20 15
20 We	19 40	19 30	12 13	26 21	05 22	21 56	14 33	02 53	11 11	20 15
21 Th	19 54	18 49	12 38	26 13	05 10	21 57	14 35	02 52	11 11	20 15
22 Fr	20 07	17 19	13 05	26 06	04 57	21 57	14 37	02 52	11 11	20 15
23 Sa	20 19	15 04	13 33	25 58	04 45	21 58	14 39	02 51	11 11	20 15
24 Su	20 32	12 13	14 02	25 49	04 33	21 58	14 41	02 51	11 11	20 15
25 Mo	20 44	08 50	14 31	25 40	04 20	21 59	14 43	02 50	11 10	20 15
26 Tu	20 55	05 03	15 01	25 31	04 08	22 00	14 45	02 50	11 10	20 15
27 We	21 07	00 58	15 31	25 21	03 56	22 00	14 47	02 50	11 10	20 15
28 Th	21 17	03S 16	16 01	25 11	03 44	22 01	14 49	02 49	11 10	20 15
29 Fr	21 28	07 30	16 30	25 00	03 31	22 02	14 51	02 49	11 10	20 15
30 Sa	21 38	11 29	17 00	24 50	03 19	22 03	14 53	02 49	11 10	20 15

ASPECTARIAN

01 05:26 ☽ □ ♇
05:28 ☽ ☍ ♅
11:14 ♅ □ ♇
12:03 ☿ ✶ ♂
17:03 ☉ ✶ ♇
20:20 ☿ ☌ ☉
22:47 ☿ ✶ ♇
23:16 ☽ # ♂

02 01:01 ☽ □ ♃
07:29 ☿ ∥ ☉
11:30 ☽ ∥ ♆
12:47 ☽ ✶ ♀
22:05 ☽ △ ♆

03 05:59 ☽ ∥ ♄
06:51 ☽ ☌ ☿
06:58 ☽ ∥ ☿
08:49 ☉ ✶ ♂
09:48 ☽ ✶ ♇
11:16 ☿ ∥ ♄
12:43 ☽ ✶ ♂
15:30 ☽ ∥ ☉
17:18 ☽ ☌ ♄
04 04:23 ☽ △ ♃

05 00:33 ☽ □ ♆
08:43 ♀ ♑
11:38 ☽ △ ♅
16:49 ☽ □ ♂
06 12:01 ☉ ☌ ♄

07 00:22 ☽ ☌ ♀
02:01 ☽ ✶ ♆
04:02 ☽ ✶ ☿
05:05 ♃ SR
12:59 ☽ □ ♅
13:31 ☽ ☌ ♇
20:18 ☽ △ ♂
21:32 ☽ ✶ ♄
23:44 ☽ ✶ ☉

08 01:16 ♀ ✶ ♆
06:23 ☿ ∥ ♆
07:40 ☽ ☍ ♃
09:42 ☽ ∥ ☉
12:24 ☿ ✶ ♀

09 04:10 ☽ □ ☿
05:49 ☽ ∥ ♄
11:58 ♂ ✶ ♄
14:58 ☽ ✶ ♅
21:33 ☿ △ ♆

10 00:16 ☽ □ ♄
01:40 ☽ ∥ ♆
05:13 ☽ ∥ ☿
21:13 ☿ SD

11 00:37 ☽ # ♂
06:59 ☽ △ ☿
07:06 ☽ ☌ ♆
12:48 ☽ ✶ ♀
19:30 ☽ ✶ ♇
20:49 ☿ △ ♆
23:57 ☽ # ♅

12 04:45 ☽ △ ♄
06:58 ☽ ☍ ♂
14:25 ☽ △ ☉
14:35 ☽ △ ♃
16:40 ☉ △ ♃

13 07:30 ☽ ∥ ♅
18:44 ♆ SD
22:13 ☽ □ ♀
14 00:06 ☽ ☌ ♅
01:19 ☽ □ ♇
03:56 ☽ ∥ ♂
20:59 ☽ □ ♃

15 01:13 ♀ □ ♅
04:52 ☽ # ☿
08:20 ☽ # ♆
19:41 ☽ ✶ ♆
21:31 ♀ ☌ ♇
23:35 ☽ ☍ ☿

16 08:09 ☽ # ♄
09:17 ☽ △ ♇
10:08 ☽ △ ♀
20:08 ☽ ☍ ♄

17 02:37 ☽ △ ♂
05:02 ☿ ∥ ♆
05:27 ☽ ✶ ♃
18 05:10 ☽ □ ♆
17:36 ☽ ✶ ♅

19 14:49 ♂ ✶ ♃
16:00 ☽ □ ♂
20 16:36 ☽ △ ♆

21 02:08 ☿ ✶ ♇
05:15 ☽ □ ♅
07:20 ☽ ☍ ♇
07:55 ☽ △ ☿
16:54 ☽ ☍ ♀
19:45 ☽ △ ♄

22 03:49 ☉ ♐
03:59 ☽ ☌ ♃
07:12 ☽ ✶ ♂
14:46 ☉ ∥ ♇

23 01:50 ☽ △ ☉
03:44 ☽ # ♄
11:24 ☽ # ☿
17:48 ☽ △ ♅
18:46 ♀ ✶ ♄

24 04:03 ☽ □ ☿
07:38 ☽ # ♆
08:59 ☽ □ ♄
17:53 ☉ □ ♆

25 10:26 ☿ ∥ ♄
17:22 ☽ ☍ ♆
26 01:55 ☿ ☌ ♄
05:41 ☽ ∥ ♂
07:57 ☽ △ ♇
13:07 ☽ ∥ ♅
20:42 ☽ ✶ ♄
22:56 ☽ ✶ ☿

27 00:40 ☽ △ ♀
02:50 ☽ ✶ ♃
11:44 ☽ ☌ ♂
22:25 ☽ # ♅

28 02:25 ☽ # ♂
02:46 ☿ ✶ ♀
07:24 ☿ △ ♃
09:43 ☽ ✶ ☉
12:01 ♀ ☍ ♃
14:10 ☽ ☍ ♅
16:43 ☽ □ ♇

29 09:57 ☽ □ ♃
11:14 ☽ □ ♀
21:59 ☽ ∥ ♆

30 08:40 ☽ △ ♆
18:25 ☉ △ ♅
21:35 ☽ ✶ ♇
23:35 ☽ ∥ ♄

⚷ Chiron

01 Dec.	03 S 12
03	09♓15R
06	09 13
09	09 10
12	09 09
15	09 08
18	09 07
21	09 07D
24	09 08
27	09 09
30	09 10

19 11:21 09♓07D

December 2013

Day	S. T.	☉	☽	☿	♀	♂	♃	♄	♅	♆	♇	☊ True
	h m s	° ' "	° ' "	° '	° '	° '	° '	° '	° '	° '	° '	° '
01 Su	04 40 04	08♐ 56 50	11♏ 37 55	23♏ 46	21♑ 26	26♍ 25	19♋R35	17♏ 08	08♈R42	02♓ 40	10♑ 12	07♏R30
02 Mo	04 44 00	09 57 40	26 01 00	25 16	22 04	26 57	19 30	17 15	08 42	02 40	10 14	07 29
03 Tu	04 47 57	10 58 32	10♐ 44 58	26 47	22 41	27 28	19 25	17 22	08 41	02 41	10 16	07 25
04 We	04 51 53	11 59 25	25 42 43	28 18	23 17	28 00	19 20	17 28	08 40	02 42	10 18	07 20
05 Th	04 55 50	13 00 19	10♑ 45 19	29 50	23 51	28 31	19 15	17 35	08 40	02 42	10 20	07 15
06 Fr	04 59 46	14 01 14	25 43 25	01♐ 21	24 24	29 02	19 10	17 42	08 39	02 43	10 22	07 09
07 Sa	05 03 43	15 02 10	10♒ 28 45	02 53	24 55	29 33	19 04	17 49	08 38	02 44	10 24	07 04
08 Su	05 07 40	16 03 06	24 55 13	04 26	25 25	00♎ 04	18 59	17 55	08 38	02 45	10 26	07 00
09 Mo	05 11 36	17 04 03	08♓ 59 28	05 58	25 53	00 35	18 53	18 02	08 37	02 46	10 28	06 58
10 Tu	05 15 33	18 05 01	22 40 45	07 31	26 19	01 06	18 47	18 08	08 37	02 47	10 30	06D 58
11 We	05 19 29	19 05 59	06♈ 00 21	09 03	26 44	01 36	18 41	18 15	08 37	02 47	10 32	06 59
12 Th	05 23 26	20 06 58	19 00 43	10 36	27 06	02 07	18 34	18 22	08 36	02 48	10 34	07 01
13 Fr	05 27 22	21 07 57	01♉ 44 49	12 09	27 27	02 37	18 28	18 28	08 36	02 49	10 36	07 02
14 Sa	05 31 19	22 08 57	14 15 35	13 42	27 46	03 07	18 21	18 34	08 36	02 50	10 38	07R 03
15 Su	05 35 15	23 09 58	26 35 36	15 15	28 03	03 37	18 15	18 41	08 36	02 51	10 40	07 01
16 Mo	05 39 12	24 10 59	08♊ 47 03	16 49	28 18	04 07	18 08	18 47	08 35	02 52	10 42	06 57
17 Tu	05 43 09	25 12 01	20 51 40	18 22	28 30	04 36	18 01	18 53	08 35	02 54	10 44	06 51
18 We	05 47 05	26 13 03	02♋ 50 52	19 55	28 41	05 06	17 54	19 00	08D 35	02 55	10 46	06 43
19 Th	05 51 02	27 14 06	14 45 58	21 29	28 49	05 35	17 47	19 06	08 35	02 56	10 48	06 33
20 Fr	05 54 58	28 15 10	26 38 24	23 03	28 55	06 04	17 40	19 12	08 35	02 57	10 50	06 22
21 Sa	05 58 55	29 16 15	08♌ 29 53	24 37	28 58	06 33	17 32	19 18	08 36	02 58	10 52	06 12
22 Su	06 02 51	00♑ 17 20	20 22 48	26 11	28R 59	07 02	17 25	19 24	08 36	03 00	10 54	06 04
23 Mo	06 06 48	01 18 26	02♍ 20 09	27 45	28 57	07 30	17 18	19 30	08 36	03 01	10 56	05 58
24 Tu	06 10 45	02 19 32	14 25 39	29 20	28 54	07 59	17 10	19 36	08 36	03 02	10 58	05 54
25 We	06 14 41	03 20 39	26 43 36	00♑ 54	28 47	08 27	17 02	19 42	08 37	03 03	11 00	05 52
26 Th	06 18 38	04 21 47	09♎ 18 39	02 29	28 38	08 55	16 55	19 48	08 37	03 05	11 03	05 52
27 Fr	06 22 34	05 22 55	22 15 29	04 05	28 27	09 23	16 47	19 54	08 38	03 06	11 05	05 52
28 Sa	06 26 31	06 24 04	05♏ 38 16	05 40	28 13	09 50	16 39	19 59	08 38	03 08	11 07	05 51
29 Su	06 30 27	07 25 14	19 29 55	07 16	27 57	10 18	16 31	20 05	08 39	03 09	11 09	05 49
30 Mo	06 34 24	08 26 24	03♐ 51 00	08 52	27 38	10 45	16 23	20 10	08 39	03 11	11 11	05 45
31 Tu	06 38 20	09 27 34	18 38 51	10 28	27 17	11 12	16 15	20 16	08 40	03 12	11 13	05 39

Data for	12-01-2013
Julian Day	2456627.50
Ayanamsa	24 03 13
SVP	05 ♓ 04 05
☽ ☊ Mean	05 ♏ 54 R

● ◐ PHASES ○ ◑

Day	Time	Phase	Position
03	00:23	●	11♐00
09	15:12	◐	17♓43
17	09:28	○	25♊36
25	13:48	◑	03♎56

LAST ASPECT Day	h m	☽ INGRESS Day	h m	
02	01:35	02	06:32	♐
04	03:46	04	06:51	♑
06	05:33	06	06:55	♒
07	12:12	08	08:35	♓
10	06:42	10	13:06	♈
12	15:38	12	20:41	♉
15	02:55	15	06:41	♊
17	09:28	17	18:17	♋
20	04:37	20	06:48	♌
22	13:26	22	19:20	♍
25	03:56	25	06:18	♎
27	11:01	27	13:59	♏
29	13:56	29	17:38	♐
30	11:37	31	18:02	♑

DECLINATION

Day	☉	☽	☿	♀	♂	♃	♄	♅	♆	♇
01 Su	21S 47	14S 58	17S 29	24S 39	03N07	22N03	14S 55	02N48	11S 09	20S 15
02 Mo	21 56	17 39	17 58	24 27	02 55	22 04	14 57	02 48	11 09	20 15
03 Tu	22 05	19 13	18 26	24 16	02 43	22 05	14 58	02 48	11 09	20 15
04 We	22 13	19 29	18 53	24 04	02 31	22 06	15 00	02 47	11 09	20 15
05 Th	22 21	18 23	19 20	23 52	02 19	22 07	15 02	02 47	11 08	20 15
06 Fr	22 29	16 01	19 46	23 40	02 07	22 08	15 04	02 47	11 08	20 15
07 Sa	22 36	12 38	20 12	23 27	01 55	22 08	15 06	02 47	11 08	20 15
08 Su	22 42	08 34	20 36	23 15	01 44	22 09	15 07	02 47	11 07	20 15
09 Mo	22 48	04 07	21 00	23 02	01 32	22 10	15 09	02 46	11 07	20 15
10 Tu	22 54	00N25	21 22	22 49	01 20	22 11	15 11	02 46	11 07	20 15
11 We	22 59	04 50	21 44	22 36	01 09	22 12	15 13	02 46	11 06	20 15
12 Th	23 04	08 55	22 04	22 23	00 57	22 13	15 14	02 46	11 06	20 15
13 Fr	23 08	12 31	22 24	22 10	00 46	22 14	15 16	02 46	11 06	20 15
14 Sa	23 12	15 29	22 43	21 57	00 35	22 15	15 18	02 46	11 05	20 15
15 Su	23 16	17 42	23 00	21 43	00 23	22 16	15 19	02 46	11 05	20 14
16 Mo	23 18	19 05	23 16	21 30	00 12	22 17	15 21	02 46	11 05	20 14
17 Tu	23 21	19 34	23 31	21 17	00 01	22 19	15 23	02 46	11 04	20 14
18 We	23 23	19 10	23 45	21 04	00S 10	22 20	15 24	02 46	11 04	20 14
19 Th	23 24	17 55	23 58	20 51	00 21	22 21	15 26	02 46	11 03	20 14
20 Fr	23 25	15 54	24 10	20 38	00 32	22 22	15 27	02 46	11 03	20 14
21 Sa	23 26	13 13	24 20	20 25	00 43	22 23	15 29	02 46	11 02	20 14
22 Su	23 26	10 00	24 29	20 12	00 54	22 24	15 30	02 46	11 02	20 14
23 Mo	23 26	06 23	24 37	19 59	01 04	22 25	15 32	02 46	11 02	20 14
24 Tu	23 25	02 27	24 43	19 47	01 15	22 26	15 33	02 47	11 01	20 14
25 We	23 24	01S 39	24 48	19 34	01 25	22 27	15 35	02 47	11 01	20 14
26 Th	23 22	05 47	24 52	19 22	01 36	22 28	15 36	02 47	11 00	20 14
27 Fr	23 20	09 46	24 54	19 10	01 46	22 30	15 38	02 47	11 00	20 14
28 Sa	23 17	13 25	24 55	18 58	01 56	22 31	15 39	02 47	10 59	20 14
29 Su	23 14	16 26	24 54	18 46	02 06	22 32	15 40	02 48	10 58	20 14
30 Mo	23 10	18 34	24 52	18 35	02 17	22 33	15 42	02 48	10 58	20 14
31 Tu	23 06	19 30	24 49	18 24	02 26	22 34	15 43	02 48	10 57	20 14

ASPECTARIAN

01 09:20 ☽ ☌ ♄
13:16 ☽ △ ♃
17:12 ☽ ✶ ♀
22:38 ☽ ☌ ☿

02 01:35 ☽ ✶ ♂
04:47 ☽ ∥ ☿
10:55 ☽ □ ♆
19:29 ♂ ∥ ♅
20:40 ☽ △ ♅
23:25 ☉ # ♃

03 16:35 ☿ ✶ ♂

04 03:46 ☽ □ ♂
11:09 ☽ ✶ ♆
11:47 ☽ ∥ ☿
20:39 ☽ □ ♅
23:19 ☽ ☌ ♇

05 02:43 ☿ ♐
11:00 ☽ ✶ ♄
13:31 ☽ ☍ ♃
21:47 ☽ ☌ ♀

06 05:33 ☽ △ ♂
07:17 ☽ ∥ ♄
10:10 ☽ ✶ ☿
20:59 ☽ ✶ ♅
21:31 ☿ □ ♆

07 03:04 ☿ ∥ ♇
08:05 ☽ ✶ ☉
09:14 ☽ ∥ ♆
12:12 ☽ □ ♄
20:42 ♂ ♎

08 13:17 ☽ ☌ ♆
18:09 ☽ □ ♇

09 02:33 ☽ ✶ ♇
07:06 ☽ # ♅
14:14 ☽ # ♂
15:54 ☽ △ ♄
17:09 ☽ △ ♃
17:30 ♀ ∥ ☉

10 04:41 ☽ ∥ ♂
06:42 ☽ ✶ ♀
12:36 ☽ ∥ ♅
15:41 ☽ ☍ ♂
17:05 ☿ △ ♅

11 04:45 ☽ ☌ ♅
06:20 ☽ △ ☿
08:18 ☽ □ ♇
23:11 ☽ □ ♃

12 02:14 ☽ △ ☉
11:19 ☿ # ♃
13:31 ☿ ∥ ♀
14:01 ☽ # ♆
15:38 ☽ □ ♀
16:27 ♀ # ♃

13 00:01 ♃ △ ♄
02:03 ☽ ✶ ♆
16:59 ☽ △ ♇
22:09 ☽ # ♄

14 07:52 ☽ ✶ ♃
08:26 ☽ ☍ ♄

15 02:55 ☽ △ ♀
12:19 ☽ △ ♆

14:23 ☽ △ ♂
23:37 ☽ ✶ ♅

16 03:58 ☿ ∥ ☉
18:17 ☽ ☍ ☿

17 17:40 ♅ SD

18 00:08 ☽ △ ♆
04:42 ☽ □ ♂
11:33 ☽ □ ♅
15:59 ☽ ☍ ♇

19 06:02 ☽ ☌ ♃
08:49 ☽ △ ♄

20 04:19 ☽ # ♄
04:37 ☽ ☍ ♀
19:53 ☽ ✶ ♂

21 00:12 ☽ △ ♅
16:42 ☽ # ♆
17:11 ☉ ♑
20:01 ♀ ∥ ♇
21:53 ♀ SR
22:01 ☽ □ ♄

22 13:26 ☽ △ ☿
21:45 ☽ △ ☉

23 01:21 ☽ ☍ ♆
17:10 ☽ △ ♇
22:01 ☽ ∥ ♅

24 05:20 ☽ ✶ ♃
06:46 ☽ # ♂
10:13 ☿ ♑
10:14 ☽ ✶ ♄
17:06 ☉ ✶ ♆
22:35 ☽ ∥ ♂

25 03:56 ☽ △ ♀
06:31 ☽ # ♅
08:33 ♂ ☍ ♅
09:12 ☽ □ ☿
22:42 ☽ ☍ ♅
23:13 ☽ ☌ ♂

26 03:16 ☽ □ [illegible]
09:04 ☿ ✶ ♆

14:02 ☽ □ ♃

27 07:39 ☽ ∥ ♆
11:01 ☽ □ ♀
19:33 ☽ △ ♆

28 00:04 ☽ ✶ ☿
01:27 ☽ ✶ ☉
09:36 ☽ ✶ ♇
17:13 ☽ ∥ ♄
18:57 ☽ △ ♃

29 01:00 ☽ ☌ ♄
[illegible]
13:56 ☽ ✶ ♀

20:49 ☿ □ ♅
22:53 ☽ □ ♆

30 00:14 ☽ ∥ ♀
05:05 ☉ □ ♅
07:52 ☽ △ ♅
11:37 ☽ ✶ ♂

31 01:17 ♂ □ ♇
11:24 ☿ ☌ ♇
14:59 ☿ □ ♂
23:07 ☽ ✶ ♆

⚷ Chiron

01 Dec.	03 S 24
03	09♓12
06	09 15
09	09 18
12	09 22
15	09 26
18	09 31
21	09 36
24	09 41
27	09 47
30	09 54

January 2014

Day	S.T. h m s	☉ ° ′ ″	☽ ° ′ ″	☿ ° ′	♀ ° ′	♂ ° ′	♃ ° ′	♄ ° ′	♅ ° ′	♆ ° ′	♇ ° ′	☊ True ° ′
01 We	06 42 17	10♑28 45	03♑47 06	12♑05	26♑R54	11♎39	16♋R07	20♏21	08♈41	03♓14	11♑15	05♏R30
02 Th	06 46 14	11 29 56	19 06 06	13 42	26 29	12 05	15 59	20 27	08 41	03 15	11 17	05 21
03 Fr	06 50 10	12 31 07	04♒24 28	15 19	26 01	12 31	15 51	20 32	08 42	03 17	11 19	05 11
04 Sa	06 54 07	13 32 17	19 30 54	16 57	25 32	12 58	15 43	20 37	08 43	03 18	11 22	05 01
05 Su	06 58 03	14 33 28	04♓16 16	18 35	25 01	13 23	15 35	20 43	08 44	03 20	11 24	04 53
06 Mo	07 01 60	15 34 38	18 34 49	20 13	24 29	13 49	15 27	20 48	08 45	03 22	11 26	04 48
07 Tu	07 05 56	16 35 48	02♈24 28	21 52	23 55	14 14	15 18	20 53	08 46	03 23	11 28	04 45
08 We	07 09 53	17 36 57	15 46 17	23 31	23 20	14 39	15 10	20 58	08 47	03 25	11 30	04 44
09 Th	07 13 49	18 38 06	28 43 27	25 10	22 44	15 04	15 02	21 03	08 48	03 27	11 32	04D 44
10 Fr	07 17 46	19 39 14	11♉20 13	26 50	22 08	15 29	14 54	21 08	08 49	03 29	11 34	04R 44
11 Sa	07 21 43	20 40 22	23 41 03	28 30	21 31	15 53	14 46	21 12	08 51	03 30	11 36	04 43
12 Su	07 25 39	21 41 29	05♊50 10	00♒10	20 54	16 17	14 38	21 17	08 52	03 32	11 38	04 40
13 Mo	07 29 36	22 42 36	17 51 16	01 50	20 17	16 41	14 30	21 22	08 53	03 34	11 40	04 34
14 Tu	07 33 32	23 43 43	29 47 22	03 31	19 41	17 04	14 22	21 26	08 54	03 36	11 43	04 24
15 We	07 37 29	24 44 49	11♋40 48	05 11	19 05	17 27	14 14	21 31	08 56	03 38	11 45	04 12
16 Th	07 41 25	25 45 54	23 33 17	06 52	18 31	17 50	14 07	21 35	08 57	03 39	11 47	03 58
17 Fr	07 45 22	26 46 59	05♌26 06	08 32	17 57	18 13	13 59	21 40	08 59	03 41	11 49	03 44
18 Sa	07 49 18	27 48 03	17 20 27	10 13	17 25	18 35	13 51	21 44	09 00	03 43	11 51	03 29
19 Su	07 53 15	28 49 07	29 17 42	11 52	16 54	18 57	13 43	21 48	09 02	03 45	11 53	03 17
20 Mo	07 57 12	29 50 11	11♍19 40	13 31	16 25	19 18	13 36	21 52	09 04	03 47	11 55	03 07
21 Tu	08 01 08	00♒51 13	23 28 49	15 10	15 58	19 40	13 29	21 56	09 05	03 49	11 57	03 00
22 We	08 05 05	01 52 16	05♎48 15	16 47	15 33	20 01	13 21	22 00	09 07	03 51	11 59	02 56
23 Th	08 09 01	02 53 18	18 21 42	18 22	15 10	20 21	13 14	22 04	09 09	03 53	12 01	02 54
24 Fr	08 12 58	03 54 20	01♏13 20	19 56	14 50	20 41	13 07	22 07	09 11	03 55	12 03	02 53
25 Sa	08 16 54	04 55 21	14 27 22	21 28	14 32	21 01	13 00	22 11	09 12	03 57	12 05	02 52
26 Su	08 20 51	05 56 22	28 07 25	22 57	14 16	21 21	12 53	22 15	09 14	03 59	12 07	02 49
27 Mo	08 24 47	06 57 22	12♐15 36	24 23	14 03	21 40	12 46	22 18	09 16	04 01	12 09	02 45
28 Tu	08 28 44	07 58 22	26 51 20	25 46	13 52	21 59	12 39	22 22	09 18	04 03	12 11	02 38
29 We	08 32 41	08 59 21	11♑50 31	27 03	13 43	22 17	12 33	22 25	09 20	04 06	12 13	02 29
30 Th	08 36 37	10 00 20	27 05 23	28 16	13 38	22 35	12 26	22 28	09 22	04 08	12 15	02 18
31 Fr	08 40 34	11 01 17	12♒25 19	29 23	13 34	22 52	12 20	22 31	09 25	04 10	12 17	02 07

Data for	01-01-2014
Julian Day	2456658.50
Ayanamsa	24 03 19
SVP	05 ♓ 04 02
☽ ☊ Mean	04 ♏ 16 R

● ◐ PHASES ○ ◑

Day	h m	Phase	Position
01	11:15	●	10♑57
08	03:39	◐	17♈46
16	04:53	○	25♋58
24	05:20	◑	04♏08
30	21:38	●	10♒55

LAST ASPECT Day	h m	☽ INGRESS Day	h m	Sign
02	11:13	02	17:04	♒
04	01:48	04	16:59	♓
06	09:45	06	19:46	♈
08	16:22	09	02:24	♉
11	10:59	11	12:26	♊
12	21:34	14	00:25	♋
16	04:53	16	13:01	♌
18	08:53	19	01:25	♍
20	20:57	21	12:45	♎
23	03:52	23	21:45	♏
25	13:56	26	03:14	♐
27	22:03	28	05:05	♑
29	16:47	30	04:33	♒

DECLINATION

Day	☉	☽	☿	♀	♂	♃	♄	♅	♆	♇
01 We	23S01	19S05	24S44	18S13	02S36	22N35	15S44	02N49	10S57	20S13
02 Th	22 56	17 16	24 38	18 02	02 46	22 36	15 46	02 49	10 56	20 13
03 Fr	22 51	14 15	24 30	17 52	02 56	22 37	15 47	02 49	10 56	20 13
04 Sa	22 45	10 19	24 21	17 42	03 05	22 38	15 48	02 50	10 55	20 13
05 Su	22 38	05 50	24 10	17 32	03 14	22 39	15 50	02 50	10 54	20 13
06 Mo	22 32	01 09	23 57	17 23	03 24	22 41	15 51	02 50	10 54	20 13
07 Tu	22 24	03N27	23 43	17 14	03 33	22 42	15 52	02 51	10 53	20 13
08 We	22 17	07 44	23 28	17 05	03 42	22 43	15 53	02 51	10 53	20 13
09 Th	22 08	11 31	23 11	16 57	03 51	22 44	15 54	02 52	10 52	20 13
10 Fr	22 00	14 40	22 52	16 49	04 00	22 45	15 55	02 52	10 51	20 13
11 Sa	21 51	17 06	22 32	16 42	04 08	22 46	15 56	02 53	10 51	20 13
12 Su	21 41	18 42	22 10	16 35	04 17	22 47	15 57	02 53	10 50	20 13
13 Mo	21 31	19 26	21 47	16 28	04 25	22 48	15 59	02 54	10 49	20 12
14 Tu	21 21	19 18	21 22	16 22	04 33	22 49	16 00	02 54	10 49	20 12
15 We	21 10	18 19	20 56	16 17	04 42	22 50	16 01	02 55	10 48	20 12
16 Th	20 59	16 32	20 28	16 12	04 50	22 51	16 01	02 56	10 47	20 12
17 Fr	20 48	14 04	19 59	16 07	04 58	22 52	16 02	02 56	10 47	20 12
18 Sa	20 36	11 00	19 28	16 03	05 05	22 53	16 03	02 57	10 46	20 12
19 Su	20 24	07 30	18 56	15 59	05 13	22 53	16 04	02 58	10 45	20 12
20 Mo	20 11	03 40	18 23	15 56	05 20	22 54	16 05	02 58	10 45	20 12
21 Tu	19 58	00S21	17 48	15 53	05 28	22 55	16 06	02 59	10 44	20 12
22 We	19 44	04 25	17 13	15 51	05 35	22 56	16 07	03 00	10 43	20 11
23 Th	19 31	08 22	16 36	15 49	05 42	22 57	16 08	03 00	10 42	20 11
24 Fr	19 16	12 02	15 59	15 48	05 49	22 58	16 08	03 01	10 42	20 11
25 Sa	19 02	15 12	15 21	15 47	05 56	22 59	16 09	03 02	10 41	20 11
26 Su	18 47	17 38	14 42	15 46	06 02	22 59	16 10	03 03	10 40	20 11
27 Mo	18 32	19 05	14 03	15 46	06 09	23 00	16 11	03 04	10 39	20 11
28 Tu	18 16	19 19	13 25	15 47	06 15	23 01	16 11	03 04	10 39	20 11
29 We	18 00	18 14	12 46	15 47	06 21	23 02	16 12	03 05	10 38	20 11
30 Th	17 44	15 50	12 09	15 48	06 27	23 02	16 13	03 06	10 37	20 11
31 Fr	17 28	12 20	11 32	15 49	06 33	23 03	16 13	03 07	10 36	20 10

ASPECTARIAN

01 07:41 ☽ □ ♅
11:45 ☽ ☌ ♇
12:42 ☽ □ ♂
14:33 ☽ ☌ ☿
15:06 ☽ ∥ ♀
18:55 ☉ ☌ ♇
19:10 ☽ ☍ ♃

02 02:07 ☽ ✶ ♄
07:42 ♂ # ♅
11:13 ☽ ☌ ♀
13:00 ☽ ∥ ♄

03 00:15 ☉ □ ♂
06:47 ☽ ✶ ♅
07:11 ☿ ☍ ♃
13:13 ☽ △ ♂
20:37 ☽ ∥ ♆

04 01:48 ☽ □ ♄
20:37 ☉ # ♃
22:27 ☽ ☌ ♆

05 11:53 ☽ ✶ ♇
12:55 ☽ ∥ ♂
15:23 ☽ # ♅
18:30 ☽ ✶ ☉
18:43 ☽ △ ♃
21:12 ☉ ☍ ♃

06 03:11 ☽ ✶ ☿
03:49 ☽ △ ♄
08:52 ☿ ✶ ♄
09:45 ☽ ✶ ♀
20:49 ☽ ∥ ♅

07 00:34 ☽ # ♂
11:20 ☽ ☌ ♅
16:13 ☽ □ ♇
21:54 ☽ ☍ ♂
22:02 ☿ ☌ ♀
22:55 ☽ □ ♃

08 13:19 ☽ □ ♀
16:22 ☽ □ ☿
19:37 ☽ # ♆
22:37 ♂ □ ♃

09 08:56 ☽ ✶ ♆

10 00:27 ☽ △ ♇
06:49 ☽ ✶ ♃
08:34 ☿ # ♃
11:29 ☽ # ♄
17:34 ☽ △ ☉
19:07 ☽ ☍ ♄
19:40 ☽ # ♀
19:58 ☽ △ ♀

11 10:44 ♀ ✶ ♄
10:59 ☽ △ ☿
12:25 ♀ ☌ ☉
13:39 ☉ ✶ ♄
19:25 ☽ □ ♆
21:35 ☿ ♒

12 06:02 ☽ ✶ ♅
21:34 ☽ △ ♂

14 01:40 ☿ ∥ ☉
07:42 ☽ △ ♆
18:26 ☽ □ ♅

15 00:08 ☽ ☍ ♇
05:07 ☽ ☌ ♃
12:04 ☽ □ ♂
14:17 ☽ ☍ ♀
20:00 ☽ △ ♄

16 03:56 ☽ # ♀
05:33 ☽ # ♄
13:13 ☿ ∥ ♇
17:13 ♀ □ ♂

17 06:25 ☿ ✶ ♅
07:10 ☽ △ ♅
07:17 ☽ ☍ ☿
21:04 ♀ ∥ ♄

18 01:47 ☽ # ♆
02:35 ☽ ✶ ♂
08:53 ☽ □ ♄

19 08:56 ☽ ☍ ♆
14:03 ☽ # ♂
22:28 ☉ ∥ ♇

20 01:10 ☽ △ ♇
03:52 ☉ ♒
04:14 ☽ ∥ ♅
04:28 ☽ ✶ ♃
09:43 ☽ △ ♀
20:57 ☽ ✶ ♄

21 15:36 ☽ # ♅
15:42 ☽ △ ☉

22 06:24 ☽ ☍ ♅
07:13 ☽ ∥ ♂
11:54 ☽ □ ♇
14:21 ☽ □ ♃
18:07 ☽ □ ♀

23 00:02 ☽ △ ☿
03:52 ☽ ☌ ♂
14:59 ☽ ∥ ♆
17:53 ☿ ∥ ♄

24 04:58 ☽ △ ♆
07:04 ☿ ∥ ♀
14:57 ☿ △ ♂
19:44 ☽ ✶ ♇
21:25 ☽ △ ♃

25 00:07 ☽ ✶ ♀
00:58 ☽ ∥ ☿
04:59 ☽ ∥ ♀
08:26 ☽ ∥ ♄
11:56 ☿ □ ♄
13:44 ☽ ☌ ♄
13:56 ☽ □ ☿

26 10:05 ☽ □ ♆
14:17 ☽ ∥ ☉
14:24 ☽ ✶ ☉
18:59 ☽ △ ♅

27 15:53 ☽ ✶ ♂
22:03 ☽ ✶ ☿

28 11:38 ☽ ✶ ♆
20:01 ☽ □ ♅

29 00:35 ☽ ☌ ♇
01:06 ☽ ☍ ♃
02:57 ☽ ☌ ♀
03:30 ☽ ∥ ☉
08:33 ☉ ✶ ♅
16:44 ☽ ✶ ♄
16:47 ☽ □ ♂
21:02 ☽ ∥ ♄

30 00:21 ☽ ∥ ♀
19:16 ☽ ✶ ♅

31 05:37 ☽ ∥ ☿
09:37 ♃ ☍ ♇
10:08 ☽ ∥ ♆
14:29 ☿ ♓
15:56 ☽ □ ♄
16:45 ☽ △ ♂
20:50 ♀ SD

⚷ Chiron

01	Dec.	03 S 15
02	10♓01	
05	10 08	
08	10 16	
11	10 24	
14	10 32	
17	10 41	
20	10 50	
23	10 59	
26	11 09	
29	11 19	

February 2014

Day	S. T. h m s	☉ ° ' "	☽ ° ' "	☿ ° '	♀ ° '	♂ ° '	♃ ° '	♄ ° '	♅ ° '	♆ ° '	♇ ° '	☊ True ° '
01 Sa	08 44 30	12♒02 13	27♒38 37	00♓23	13♑D33	23♎09	12♋R14	22♏34	09♈27	04♓12	12♑18	01♏R57
02 Su	08 48 27	13 03 08	12♓34 35	01 16	13 35	23 26	12 08	22 37	09 29	04 14	12 20	01 48
03 Mo	08 52 23	14 04 02	27 05 16	02 00	13 39	23 42	12 02	22 40	09 31	04 16	12 22	01 41
04 Tu	08 56 20	15 04 55	11♈06 28	02 35	13 45	23 58	11 56	22 43	09 33	04 18	12 24	01 37
05 We	09 00 16	16 05 46	24 37 34	03 01	13 54	24 13	11 50	22 45	09 36	04 21	12 26	01 35
06 Th	09 04 13	17 06 36	07♉40 51	03 16	14 04	24 28	11 45	22 48	09 38	04 23	12 28	01D35
07 Fr	09 08 10	18 07 24	20 20 19	03R20	14 17	24 42	11 40	22 50	09 40	04 25	12 29	01 35
08 Sa	09 12 06	19 08 11	02♊40 51	03 14	14 32	24 56	11 35	22 53	09 43	04 27	12 31	01R35
09 Su	09 16 03	20 08 56	14 47 26	02 56	14 49	25 09	11 30	22 55	09 45	04 29	12 33	01 33
10 Mo	09 19 59	21 09 40	26 44 50	02 28	15 08	25 22	11 25	22 57	09 48	04 32	12 35	01 28
11 Tu	09 23 56	22 10 23	08♋37 16	01 50	15 29	25 34	11 20	22 59	09 50	04 34	12 36	01 20
12 We	09 27 52	23 11 03	20 28 11	01 03	15 52	25 46	11 16	23 01	09 53	04 36	12 38	01 10
13 Th	09 31 49	24 11 43	02♌20 19	00 09	16 16	25 57	11 12	23 03	09 56	04 38	12 40	00 58
14 Fr	09 35 45	25 12 21	14 15 36	29♒08	16 43	26 07	11 08	23 05	09 58	04 41	12 42	00 45
15 Sa	09 39 42	26 12 57	26 15 29	28 02	17 10	26 17	11 04	23 06	10 01	04 43	12 43	00 32
16 Su	09 43 39	27 13 32	08♍21 02	26 54	17 40	26 27	11 00	23 08	10 04	04 45	12 45	00 20
17 Mo	09 47 35	28 14 05	20 33 21	25 45	18 11	26 36	10 57	23 09	10 06	04 47	12 46	00 11
18 Tu	09 51 32	29 14 37	02♎53 44	24 37	18 44	26 44	10 53	23 11	10 09	04 50	12 48	00 05
19 We	09 55 28	00♓15 08	15 23 50	23 31	19 18	26 52	10 50	23 12	10 12	04 52	12 50	00 02
20 Th	09 59 25	01 15 37	28 05 52	22 30	19 53	26 59	10 47	23 13	10 15	04 54	12 51	00 01
21 Fr	10 03 21	02 16 05	11♏02 26	21 33	20 30	27 05	10 44	23 14	10 18	04 56	12 53	00D01
22 Sa	10 07 18	03 16 32	24 16 27	20 43	21 08	27 11	10 42	23 15	10 20	04 59	12 54	00 01
23 Su	10 11 14	04 16 57	07♐50 42	20 00	21 47	27 16	10 40	23 16	10 23	05 01	12 56	00 00
24 Mo	10 15 11	05 17 21	21 47 07	19 23	22 28	27 20	10 37	23 17	10 26	05 03	12 57	29♎58
25 Tu	10 19 08	06 17 44	06♑05 57	18 54	23 09	27 24	10 35	23 17	10 29	05 06	12 59	29 54
26 We	10 23 04	07 18 05	20 45 00	18 33	23 52	27 27	10 34	23 18	10 32	05 08	13 00	29 48
27 Th	10 27 01	08 18 25	05♒39 15	18 18	24 35	27 29	10 32	23 18	10 35	05 10	13 01	29 41
28 Fr	10 30 57	09 18 44	20 41 00	18 11	25 20	27 31	10 31	23 19	10 38	05 12	13 03	29 33

Data for 02-01-2014
Julian Day 2456689.50
Ayanamsa 24 03 24
SVP 05 ♓ 03 58
☽ ☊ Mean 02 ♏ 37 R

● ◐ PHASES ○ ◑

06	19:22	◐	17♉56
14	23:54	○	26♌13
22	17:15	◑	04♐00

LAST ASPECT ☽ Day	h m	INGRESS Day	h m	
31	16:45	01	03:45	♓
02	16:35	03	04:55	♈
04	23:15	05	09:47	♉
07	04:50	07	18:45	♊
09	21:09	10	06:34	♋
12	10:53	12	19:17	♌
15	03:14	15	07:27	♍
17	05:06	17	18:24	♎
19	21:53	20	03:33	♏
21	22:10	22	10:12	♐
24	09:26	24	13:51	♑
26	10:52	26	14:56	♒
28	10:55	28	14:53	♓

DECLINATION

Day	☉	☽	☿	♀	♂	♃	♄	♅	♆	♇
01 Sa	17S11	08S02	10S57	15S51	06S38	23N04	16S14	03N08	10S36	20S10
02 Su	16 54	03 19	10 24	15 53	06 44	23 04	16 14	03 09	10 35	20 10
03 Mo	16 36	01N28	09 53	15 55	06 49	23 05	16 15	03 10	10 34	20 10
04 Tu	16 19	06 02	09 25	15 57	06 54	23 06	16 15	03 10	10 33	20 10
05 We	16 01	10 07	09 00	15 59	06 59	23 06	16 16	03 11	10 32	20 10
06 Th	15 42	13 34	08 39	16 02	07 04	23 07	16 16	03 12	10 32	20 10
07 Fr	15 24	16 16	08 21	16 04	07 08	23 07	16 17	03 13	10 31	20 10
08 Sa	15 05	18 08	08 08	16 07	07 13	23 08	16 17	03 14	10 30	20 10
09 Su	14 46	19 08	08 00	16 10	07 17	23 09	16 17	03 15	10 29	20 09
10 Mo	14 27	19 15	07 56	16 13	07 21	23 09	16 18	03 16	10 28	20 09
11 Tu	14 07	18 31	07 57	16 15	07 25	23 10	16 18	03 17	10 28	20 09
12 We	13 47	16 59	08 03	16 18	07 28	23 10	16 18	03 18	10 27	20 09
13 Th	13 27	14 44	08 12	16 21	07 32	23 11	16 19	03 19	10 26	20 09
14 Fr	13 07	11 52	08 26	16 23	07 35	23 11	16 19	03 20	10 25	20 09
15 Sa	12 47	08 29	08 43	16 26	07 38	23 11	16 19	03 22	10 24	20 09
16 Su	12 26	04 45	09 04	16 28	07 41	23 12	16 19	03 23	10 23	20 09
17 Mo	12 05	00 46	09 26	16 30	07 43	23 12	16 19	03 24	10 23	20 09
18 Tu	11 44	03S17	09 50	16 32	07 46	23 13	16 20	03 25	10 22	20 08
19 We	11 23	07 15	10 15	16 34	07 48	23 13	16 20	03 26	10 21	20 08
20 Th	11 01	10 58	10 40	16 35	07 50	23 13	16 20	03 27	10 20	20 08
21 Fr	10 40	14 13	11 05	16 36	07 52	23 14	16 20	03 28	10 19	20 08
22 Sa	10 18	16 48	11 29	16 37	07 53	23 14	16 20	03 29	10 19	20 08
23 Su	09 56	18 31	11 52	16 38	07 55	23 14	16 20	03 31	10 18	20 08
24 Mo	09 34	19 10	12 14	16 39	07 56	23 14	16 20	03 32	10 17	20 08
25 Tu	09 12	18 38	12 35	16 39	07 57	23 15	16 20	03 33	10 16	20 08
26 We	08 50	16 51	12 54	16 38	07 57	23 15	16 20	03 34	10 15	20 08
27 Th	08 27	13 56	13 11	16 38	07 58	23 15	16 20	03 35	10 14	20 07
28 Fr	08 05	10 04	13 26	16 37	07 58	23 15	16 20	03 36	10 14	20 07

ASPECTARIAN

01 04:39 ☽ ☌ ☿
07:08 ☽ ∥ ♂
10:29 ☽ ☌ ♆
15:38 ☿ ∥ ♆
23:16 ☽ △ ♃
23:37 ☽ ⚹ ♇

02 00:54 ☽ # ♅
01:39 ☽ ⚹ ♀
16:35 ☽ △ ♄

03 08:44 ☽ ∥ ♅
21:18 ☽ ☌ ♅

04 01:26 ☽ □ ♃
02:16 ☽ □ ♇
04:18 ☉ ∥ ♄
04:40 ☽ □ ♀
05:02 ☽ # ♂
07:32 ☽ ⚹ ☉
17:45 ☽ # ☿
23:15 ☽ ☍ ♂

05 01:30 ♀ ∥ ☉
02:45 ☽ # ♆
15:40 ☽ ⚹ ☿
17:50 ☽ ⚹ ♆

06 07:35 ☽ ⚹ ♃
09:00 ☽ △ ♇
12:14 ☽ △ ♀
16:26 ☽ # ☉
21:43 ☿ SR
22:04 ☽ # ♀

07 00:12 ☽ # ♄
04:50 ☽ ☍ ♄

08 01:03 ☽ □ ☿
03:30 ☽ □ ♆
13:57 ☽ ⚹ ♅

09 11:43 ☽ △ ☉
21:09 ☽ △ ♂

10 11:00 ☽ △ ☿
15:46 ☽ △ ♆

11 02:29 ☽ □ ♅
05:28 ☽ ☌ ♃
08:06 ☽ ☍ ♇
14:21 ☽ ☍ ♀
19:57 ☉ □ ♄

12 03:58 ♀ ∥ ♄
05:11 ☽ △ ♄
07:59 ☽ # ♀
08:03 ☽ # ♄
10:53 ☽ □ ♂

13 03:30 ☿ ♒R
12:46 ☽ # ☉
15:21 ☽ △ ♅

14 10:43 ☽ # ♆
17:42 ☽ □ ♄
22:32 ☽ # ☿

15 00:04 ☽ ⚹ ♂
02:06 ☉ △ ♂
03:14 ☽ ☍ ☿
05:35 ☽ # ♂
[illegible]
20:22 ☿ ☌ ☉

16 05:12 ☽ ⚹ ♃
08:20 ☽ ∥ ♅
08:20 ☿ △ ♂
08:42 ☽ △ ♇
19:09 ☽ △ ♀

17 05:06 ☽ ⚹ ♄

18 00:48 ☽ # ♅
14:01 ☽ ☍ ♅
15:19 ☽ □ ♃
18:00 ☉ ♓
19:05 ☽ □ ♇

19 03:29 ☽ ∥ ♂
06:00 ☿ ∥ ♆
07:11 ☿ □ ♄
07:46 ☽ □ ♀
14:14 ☽ △ ☿
19:52 ☽ ∥ ♆
21:46 ☽ ∥ ☿
21:53 ☽ ☌ ♂

20 00:28 ☽ ∥ ☉
06:25 ☽ △ ☉
11:12 ☿ ∥ ☉
12:43 ☽ △ ♆
23:27 ☽ △ ♃

21 03:22 ☽ ⚹ ♇
17:58 ☽ □ ☿
18:04 ☽ ⚹ ♀
19:08 ☽ ∥ ♄
22:09 ☽ ∥ ♀
22:10 ☽ ☌ ♄
23:34 ☉ ∥ ♆

22 19:02 ☽ □ ♆

23 04:27 ☽ △ ♅
18:11 ☉ ☌ ♆
20:04 ☽ ⚹ ☿

24 09:26 ☽ ⚹ ♂
22:20 ☽ ⚹ ♆

25 00:21 ☽ ⚹ ☉
04:50 [illegible] ♄
07:16 ☽ □ ♅
07:24 ☽ ☍ ♃
11:21 ☽ ☌ ♇

26 02:07 ☽ ∥ ♀
04:08 ☽ ⚹ ♄
05:01 ☽ ∥ ♄
05:18 ☽ ☌ ♀
07:34 ♃ □ ♅
10:52 ☽ □ ♂

27 04:47 ☽ ∥ ☿
07:55 ☽ ⚹ ♅
20:02 ☽ ☌ ☿
[illegible] ☽ ∥ ♆

28 04:12 ☽ □ ♄
07:20 ☉ ∥ ♂
10:55 ☽ △ ♂
11:37 ☽ ∥ ♂
12:00 ☽ ∥ ☉
14:01 ☿ SD
23:18 ☽ ☌ ♆

⚷ Chiron

01 Dec. 02 S 48

01	11♓29
04	11 39
07	11 50
10	12 01
13	12 12
16	12 23
19	12 34
22	12 45
25	12 56
28	13 08

March 2014

Day	S. T.	☉	☽	☿	♀	♂	♃	♄	♅	♆	♇	☊ True
	h m s	° ' "	° ' "	° '	° '	° '	° '	° '	° '	° '	° '	° '
01 Sa	10 34 54	10♓19 00	05♓40 56	18♒D10	26♑06	27♎32	10♋R29	23♏19	10♈41	05♓15	13♑04	29♎R26
02 Su	10 38 50	11 19 15	20 29 35	18 16	26 52	27R 32	10 28	23 19	10 44	05 17	13 05	29 19
03 Mo	10 42 47	12 19 28	04♈58 53	18 28	27 39	27 31	10 28	23R 19	10 47	05 19	13 07	29 15
04 Tu	10 46 43	13 19 39	19 03 21	18 46	28 27	27 30	10 27	23 19	10 51	05 21	13 08	29 12
05 We	10 50 40	14 19 48	02♉40 35	19 09	29 16	27 28	10 27	23 19	10 54	05 24	13 09	29D 12
06 Th	10 54 36	15 19 55	15 50 56	19 37	00♒06	27 25	10 27	23 19	10 57	05 26	13 10	29 13
07 Fr	10 58 33	16 20 00	28 36 57	20 10	00 57	27 21	10D 27	23 18	11 00	05 28	13 11	29 15
08 Sa	11 02 30	17 20 03	11♊02 34	20 47	01 48	27 17	10 27	23 18	11 03	05 30	13 13	29 17
09 Su	11 06 26	18 20 04	23 12 26	21 28	02 40	27 12	10 27	23 17	11 06	05 33	13 14	29R 17
10 Mo	11 10 23	19 20 03	05♋11 28	22 14	03 32	27 06	10 28	23 16	11 10	05 35	13 15	29 16
11 Tu	11 14 19	20 19 59	17 04 29	23 02	04 25	26 59	10 29	23 16	11 13	05 37	13 16	29 14
12 We	11 18 16	21 19 53	28 55 53	23 54	05 19	26 52	10 30	23 15	11 16	05 39	13 17	29 09
13 Th	11 22 12	22 19 45	10♌49 26	24 49	06 13	26 44	10 31	23 14	11 19	05 42	13 18	29 02
14 Fr	11 26 09	23 19 35	22 48 12	25 47	07 08	26 35	10 32	23 13	11 23	05 44	13 19	28 55
15 Sa	11 30 05	24 19 23	04♍54 30	26 48	08 04	26 25	10 34	23 11	11 26	05 46	13 20	28 49
16 Su	11 34 02	25 19 09	17 10 00	27 51	09 00	26 14	10 35	23 10	11 29	05 48	13 21	28 42
17 Mo	11 37 59	26 18 53	29 35 50	28 57	09 56	26 03	10 37	23 09	11 33	05 50	13 22	28 38
18 Tu	11 41 55	27 18 35	12♎12 46	00♓05	10 53	25 51	10 40	23 07	11 36	05 53	13 23	28 35
19 We	11 45 52	28 18 15	25 01 19	01 15	11 51	25 38	10 42	23 06	11 39	05 55	13 23	28 34
20 Th	11 49 48	29 17 53	08♏01 59	02 27	12 49	25 25	10 44	23 04	11 43	05 57	13 24	28D 34
21 Fr	11 53 45	00♈17 29	21 15 21	03 41	13 47	25 10	10 47	23 02	11 46	05 59	13 25	28 35
22 Sa	11 57 41	01 17 04	04♐42 07	04 57	14 46	24 56	10 50	23 00	11 49	06 01	13 26	28 36
23 Su	12 01 38	02 16 36	18 22 56	06 15	15 45	24 40	10 53	22 58	11 53	06 03	13 26	28 37
24 Mo	12 05 34	03 16 08	02♑18 07	07 35	16 45	24 24	10 56	22 56	11 56	06 05	13 27	28R 37
25 Tu	12 09 31	04 15 37	16 27 08	08 56	17 45	24 07	11 00	22 54	12 00	06 07	13 28	28 36
26 We	12 13 28	05 15 05	00♒48 19	10 19	18 45	23 49	11 03	22 52	12 03	06 09	13 28	28 34
27 Th	12 17 24	06 14 31	15 18 29	11 44	19 46	23 31	11 07	22 50	12 06	06 11	13 29	28 32
28 Fr	12 21 21	07 13 55	29 52 59	13 10	20 47	23 12	11 11	22 47	12 10	06 13	13 30	28 28
29 Sa	12 25 17	08 13 17	14♓26 03	14 38	21 48	22 53	11 15	22 45	12 13	06 15	13 30	28 25
30 Su	12 29 14	09 12 37	28 51 25	16 07	22 50	22 33	11 19	22 42	12 17	06 17	13 31	28 23
31 Mo	12 33 10	10 11 55	13♈03 14	17 38	23 52	22 13	11 24	22 40	12 20	06 19	13 31	28 22

Data for	03-01-2014
Julian Day	2456717.50
Ayanamsa	24 03 27
SVP	05 ♓ 03 53
☽ ☊ Mean	01 ♏ 08 R

● ◐ PHASES ○ ◑

01	08:00	●	10♓39
08	13:27	◐	17♊54
16	17:08	○	26♍02
24	01:46	◑	03♑21
30	18:46	●	09♈59

LAST ASPECT ☽ Day	h m	INGRESS Day	h m	
02	11:05	02	15:41	♈
04	17:32	04	19:13	♉
06	13:56	07	02:39	♊
09	07:54	09	13:34	♋
11	19:52	12	02:10	♌
14	07:25	14	14:18	♍
16	17:09	17	00:46	♎
19	01:07	19	09:14	♏
21	03:12	21	15:39	♐
23	10:41	23	20:04	♑
25	12:35	25	22:40	♒
27	13:14	28	00:12	♓
29	13:45	30	01:55	♈

DECLINATION

Day	☉	☽	☿	♀	♂	♃	♄	♅	♆	♇
01 Sa	07S 42	05S 35	13S 39	16S 36	07S 58	23N15	16S 19	03N38	10S 13	20S 07
02 Su	07 19	00 48	13 50	16 34	07 57	23 16	16 19	03 39	10 12	20 07
03 Mo	06 56	03N55	13 59	16 32	07 57	23 16	16 19	03 40	10 11	20 07
04 Tu	06 33	08 18	14 06	16 29	07 56	23 16	16 19	03 41	10 10	20 07
05 We	06 10	12 06	14 11	16 26	07 55	23 16	16 19	03 43	10 09	20 07
06 Th	05 47	15 10	14 15	16 23	07 54	23 16	16 18	03 44	10 09	20 07
07 Fr	05 24	17 22	14 16	16 19	07 52	23 16	16 18	03 45	10 08	20 07
08 Sa	05 00	18 41	14 16	16 15	07 50	23 16	16 18	03 46	10 07	20 07
09 Su	04 37	19 06	14 14	16 11	07 48	23 16	16 17	03 48	10 06	20 06
10 Mo	04 13	18 37	14 10	16 06	07 46	23 16	16 17	03 49	10 05	20 06
11 Tu	03 50	17 20	14 05	16 00	07 44	23 16	16 17	03 50	10 05	20 06
12 We	03 26	15 19	13 58	15 54	07 41	23 16	16 16	03 52	10 04	20 06
13 Th	03 03	12 39	13 49	15 48	07 38	23 16	16 16	03 53	10 03	20 06
14 Fr	02 39	09 27	13 38	15 41	07 35	23 16	16 15	03 54	10 02	20 06
15 Sa	02 15	05 49	13 26	15 34	07 32	23 16	16 15	03 55	10 01	20 06
16 Su	01 52	01 54	13 13	15 26	07 28	23 16	16 14	03 57	10 01	20 06
17 Mo	01 28	02S 09	12 58	15 18	07 24	23 16	16 14	03 58	10 00	20 06
18 Tu	01 04	06 11	12 42	15 09	07 20	23 16	16 13	03 59	09 59	20 06
19 We	00 40	09 59	12 24	15 00	07 16	23 16	16 13	04 01	09 58	20 06
20 Th	00 17	13 22	12 04	14 50	07 11	23 16	16 12	04 02	09 58	20 06
21 Fr	00N07	16 07	11 44	14 40	07 07	23 15	16 12	04 03	09 57	20 06
22 Sa	00 31	18 02	11 22	14 29	07 02	23 15	16 11	04 05	09 56	20 06
23 Su	00 54	18 56	10 58	14 18	06 56	23 15	16 10	04 06	09 55	20 06
24 Mo	01 18	18 43	10 33	14 07	06 51	23 15	16 10	04 07	09 55	20 05
25 Tu	01 42	17 20	10 07	13 55	06 46	23 15	16 09	04 09	09 54	20 05
26 We	02 05	14 52	09 40	13 42	06 40	23 14	16 08	04 10	09 53	20 05
27 Th	02 29	11 27	09 11	13 29	06 34	23 14	16 07	04 11	09 52	20 05
28 Fr	02 52	07 19	08 41	13 16	06 28	23 14	16 07	04 13	09 52	20 05
29 Sa	03 16	02 46	08 10	13 02	06 22	23 14	16 06	04 14	09 51	20 05
30 Su	03 39	01N55	07 38	12 48	06 15	23 13	16 05	04 15	09 50	20 05
31 Mo	04 02	06 26	07 04	12 33	06 09	23 13	16 04	04 17	09 49	20 05

ASPECTARIAN

01 04:06 ☉ △ ♃
07:44 ☽ △ ♃
09:52 ☽ # ♅
11:56 ☽ ⚹ ♇
16:25 ♂ SR

02 04:38 ☽ △ ♄
11:05 ☽ ⚹ ♀
16:20 ♄ SR
20:04 ♀ □ ♂
22:41 ☽ ∥ ♅

03 09:15 ☽ □ ♃
09:51 ☽ ☌ ♅
13:47 ☽ □ ♇
14:54 ☽ # ☉
19:14 ☉ ⚹ ♇
21:52 ☽ # ♂
23:29 ☽ ⚹ ☿

04 11:14 ☽ # ♆
14:45 ☽ ☍ ♂
17:32 ☽ □ ♀

05 04:54 ☽ ⚹ ♆
14:03 ☽ ⚹ ♃
16:00 ☽ # ☿
19:03 ☽ △ ♇
21:04 ♀ ♒
22:58 ☽ ⚹ ☉

06 07:19 ☽ □ ☿
10:43 ♃ SD
11:12 ☽ # ♄
11:47 ☽ # ♀
13:56 ☽ ☍ ♄

07 04:46 ☽ △ ♀
09:09 ♀ ∥ ♄
13:12 ☽ □ ♆

08 00:01 ☽ ⚹ ♅
20:20 ☽ △ ☿

09 07:54 ☽ △ ♂

10 00:47 ☽ △ ♆
10:38 ☽ ☌ ♃
12:06 ☽ □ ♅
16:17 ☽ ☍ ♇
23:33 ☉ # ♅

11 06:15 ☿ □ ♄
07:12 ☽ △ ☉
12:30 ☽ △ ♄
13:33 ☽ # ♄
17:24 ☽ # ♀
19:52 ☽ □ ♂

12 13:30 ☽ # ☿
13:58 ☽ ☍ ♀

13 01:01 ☽ △ ♅
19:45 ☽ # ♆
21:16 ☉ △ ♄

14 00:49 ☽ □ ♄
06:28 ☽ ☍ ☿
07:25 ☽ ⚹ ♂
12:46 ☽ # ♂
16:17 ☿ △ ♂

15 01:42 ☽ ☍ ♆
11:08 ☽ ⚹ ♃
11:40 ☽ ∥ ♅
16:33 ☽ △ ♇

16 00:15 ☽ # ☉
11:37 ☽ ⚹ ♄
20:18 ☽ ∥ ☉

17 10:47 ☽ # ♅
21:03 ☽ □ ♃
21:17 ☽ △ ♀
22:24 ☿ ♓
22:50 ☽ ☍ ♅

18 02:12 ☽ □ ♇
06:59 ☽ ∥ ♂
19:00 ♀ ⚹ ♅
23:55 ☽ ∥ ♆

19 01:07 ☽ ☌ ♂
12:42 ☽ △ ☿
15:15 ☽ ∥ ☿
20:10 ☽ △ ♆

20 04:58 ☽ △ ♃
09:24 ☽ □ ♀
09:48 ☽ ⚹ ♇
11:20 ☽ ∥ ♀
16:58 ☉ ♈

21 00:46 ☽ ∥ ♄
03:12 ☽ ☌ ♄
17:27 ☽ △ ☉

22 00:29 ☽ □ ☿
02:20 ☽ □ ♆
12:36 ☽ △ ♅
19:03 ☽ ⚹ ♀
20:16 ☿ ☌ ♆

23 10:41 ☽ ⚹ ♂

24 06:28 ☽ ⚹ ♆
09:56 ☽ ⚹ ☿
14:45 ☽ ☍ ♃
16:27 ☽ □ ♅
18:58 ☽ ☌ ♇

25 10:48 ☽ ⚹ ♄
12:14 ☿ ∥ ♆
12:35 ☽ □ ♂
12:51 ☽ ∥ ♄

26 07:55 ☽ ⚹ ☉
09:25 ☽ ∥ ♀
13:11 ☿ △ ♃
18:42 ☽ ⚹ ♅

27 07:53 ☽ ☌ ♀
09:35 ☽ ∥ ♆
12:21 ☽ □ ♄
13:14 ☽ △ ♂
15:21 ☽ ∥ ☿

28 04:43 ☽ ∥ ♂
05:31 ☿ ⚹ ♇
10:28 ☽ ☌ ♆
16:24 ☽ # ♅
18:43 ☽ △ ♃
21:37 ☽ # ☉
22:28 ☽ ⚹ ♇

29 00:21 ☽ ☌ ☿
13:45 ☽ △ ♄
19:05 ♀ △ ♂
21:15 ♀ □ ♄

30 09:09 ☽ ∥ ☉
12:15 ☽ ∥ ♅
21:10 ☽ □ ♃
22:27 ☽ # ♂
22:46 ☽ ☌ ♅

31 00:48 ☽ □ ♇
03:07 ☽ # ☿
15:22 ☽ ☍ ♂
15:48 ☉ ∥ ♅
19:40 ☽ # ♆
20:08 ☽ ⚹ ♀

⚷ Chiron

01	Dec.	02 S 13
03	13♓19	
06	13 31	
09	13 42	
12	13 53	
15	14 05	
18	14 16	
21	14 27	
24	14 38	
27	14 49	
30	14 59	

April 2014

15	07:47	25♎16	✹	Total Lunar Eclipse (mag 1.295)
29	06:05	08♉52	✪	Annular Solar Eclipse

Day	S. T.	☉	☽	☿	♀	♂	♃	♄	♅	♆	♇	☊ True
	h m s	° ' "	° ' "	° '	° '	° '	° '	° '	° '	° '	° '	° '
01 Tu	12 37 07	11♈ 11 12	26♈ 56 45	19♓ 10	24♒ 54	21♎R52	11♋ 29	22♏R37	12♈ 24	06♓ 21	13♑ 32	28♎D22
02 We	12 41 03	12 10 26	10♉ 29 02	20 43	25 56	21 31	11 33	22 34	12 27	06 23	13 32	28 22
03 Th	12 44 60	13 09 38	23 39 02	22 19	26 59	21 09	11 38	22 31	12 30	06 25	13 32	28 24
04 Fr	12 48 56	14 08 48	06♊ 27 34	23 55	28 02	20 47	11 44	22 28	12 34	06 27	13 33	28 26
05 Sa	12 52 53	15 07 55	18 56 57	25 33	29 06	20 25	11 49	22 25	12 37	06 29	13 33	28 28
06 Su	12 56 50	16 07 00	01♋ 10 34	27 13	00♓ 09	20 02	11 54	22 22	12 41	06 31	13 33	28 29
07 Mo	13 00 46	17 06 03	13 12 35	28 54	01 13	19 40	12 00	22 19	12 44	06 33	13 34	28 30
08 Tu	13 04 43	18 05 04	25 07 37	00♈ 36	02 17	19 17	12 06	22 16	12 48	06 35	13 34	28R 30
09 We	13 08 39	19 04 02	07♌ 00 20	02 20	03 21	18 54	12 12	22 12	12 51	06 36	13 34	28 30
10 Th	13 12 36	20 02 58	18 55 16	04 05	04 26	18 31	12 18	22 09	12 54	06 38	13 34	28 29
11 Fr	13 16 32	21 01 52	00♍ 56 31	05 52	05 30	18 08	12 24	22 05	12 58	06 40	13 34	28 28
12 Sa	13 20 29	22 00 44	13 07 36	07 41	06 35	17 45	12 30	22 02	13 01	06 42	13 35	28 26
13 Su	13 24 25	22 59 33	25 31 18	09 30	07 40	17 22	12 37	21 58	13 05	06 43	13 35	28 25
14 Mo	13 28 22	23 58 21	08♎ 09 39	11 22	08 46	16 59	12 44	21 54	13 08	06 45	13 35	28 25
15 Tu	13 32 19	24 57 06	21 03 43	13 15	09 51	16 36	12 51	21 51	13 11	06 47	13R 35	28 24
16 We	13 36 15	25 55 49	04♏ 13 42	15 09	10 57	16 14	12 58	21 47	13 15	06 48	13 35	28 24
17 Th	13 40 12	26 54 30	17 38 49	17 05	12 03	15 52	13 05	21 43	13 18	06 50	13 35	28 24
18 Fr	13 44 08	27 53 10	01♐ 17 36	19 03	13 09	15 30	13 12	21 39	13 21	06 52	13 35	28 24
19 Sa	13 48 05	28 51 48	15 08 03	21 02	14 15	15 08	13 19	21 35	13 25	06 53	13 34	28 23
20 Su	13 52 01	29 50 24	29 07 56	23 02	15 21	14 47	13 27	21 31	13 28	06 55	13 34	28 23
21 Mo	13 55 58	00♉ 48 58	13♑ 14 53	25 04	16 28	14 26	13 34	21 27	13 31	06 56	13 34	28 22
22 Tu	13 59 54	01 47 31	27 26 29	27 07	17 34	14 05	13 42	21 23	13 35	06 58	13 34	28 22
23 We	14 03 51	02 46 03	11♒ 40 17	29 12	18 41	13 45	13 50	21 19	13 38	06 59	13 34	28 22
24 Th	14 07 48	03 44 32	25 53 46	01♉ 17	19 48	13 25	13 58	21 15	13 41	07 01	13 33	28D 21
25 Fr	14 11 44	04 43 00	10♓ 04 19	03 24	20 55	13 06	14 06	21 11	13 45	07 02	13 33	28 22
26 Sa	14 15 41	05 41 26	24 09 15	05 31	22 02	12 48	14 15	21 07	13 48	07 04	13 33	28 22
27 Su	14 19 37	06 39 51	08♈ 05 46	07 40	23 10	12 30	14 23	21 02	13 51	07 05	13 32	28 23
28 Mo	14 23 34	07 38 14	21 51 07	09 48	24 17	12 13	14 32	20 58	13 54	07 06	13 32	28 23
29 Tu	14 27 30	08 36 36	05♉ 22 51	11 57	25 25	11 56	14 40	20 54	13 57	07 08	13 32	28 23
30 We	14 31 27	09 34 55	18 39 03	14 06	26 32	11 40	14 49	20 49	14 01	07 09	13 31	28R 23

Data for	04-01-2014
Julian Day	2456748.50
Ayanamsa	24 03 29
SVP	05 ♓ 03 47
☽ ☊ Mean	29 ♎ 30 R

● ◐ PHASES ○ ◑

07	08:31	◐	17♋27
15	07:42	✹	25♎16
22	07:52	◑	02♒07
29	06:15	✪	08♉52

LAST ASPECT Day	h m	☽ INGRESS Day	h m	
31	20:08	01	05:21	♉
03	06:44	03	11:49	♊
05	14:56	05	21:40	♋
07	18:14	08	09:51	♌
10	06:26	10	22:08	♍
12	17:12	13	08:34	♎
15	07:43	15	16:21	♏
17	07:10	17	21:44	♐
20	01:18	20	01:29	♑
21	23:22	22	04:19	♒
23	16:12	24	06:56	♓
25	20:04	26	10:02	♈
27	11:03	28	14:24	♉
30	15:54	30	20:56	♊

DECLINATION

Day	☉	☽	☿	♀	♂	♃	♄	♅	♆	♇
01 Tu	04N26	10N30	06S 29	12S 18	06S 02	23N13	16S 03	04N18	09S 49	20S 05
02 We	04 49	13 54	05 53	12 03	05 56	23 12	16 03	04 19	09 48	20 05
03 Th	05 12	16 29	05 16	11 47	05 49	23 12	16 02	04 21	09 47	20 05
04 Fr	05 35	18 10	04 38	11 31	05 42	23 12	16 01	04 22	09 47	20 05
05 Sa	05 58	18 55	03 59	11 14	05 35	23 11	16 00	04 23	09 46	20 05
06 Su	06 20	18 44	03 19	10 57	05 28	23 11	15 59	04 25	09 45	20 05
07 Mo	06 43	17 43	02 37	10 39	05 21	23 10	15 58	04 26	09 45	20 05
08 Tu	07 06	15 56	01 55	10 21	05 14	23 10	15 57	04 27	09 44	20 05
09 We	07 28	13 28	01 11	10 03	05 07	23 09	15 56	04 29	09 43	20 05
10 Th	07 50	10 27	00 27	09 45	05 00	23 09	15 55	04 30	09 43	20 05
11 Fr	08 12	06 58	00N18	09 26	04 53	23 08	15 54	04 31	09 42	20 05
12 Sa	08 34	03 10	01 05	09 06	04 46	23 08	15 53	04 33	09 41	20 05
13 Su	08 56	00S 51	01 52	08 47	04 39	23 07	15 52	04 34	09 41	20 05
14 Mo	09 18	04 55	02 40	08 27	04 33	23 07	15 51	04 35	09 40	20 05
15 Tu	09 40	08 50	03 29	08 06	04 26	23 06	15 50	04 37	09 40	20 05
16 We	10 01	12 25	04 19	07 46	04 19	23 05	15 49	04 38	09 39	20 05
17 Th	10 22	15 24	05 09	07 25	04 13	23 05	15 48	04 39	09 38	20 05
18 Fr	10 43	17 36	06 00	07 04	04 07	23 04	15 47	04 41	09 38	20 05
19 Sa	11 04	18 46	06 51	06 42	04 01	23 03	15 46	04 42	09 37	20 05
20 Su	11 25	18 49	07 43	06 21	03 55	23 03	15 45	04 43	09 37	20 05
21 Mo	11 45	17 41	08 36	05 59	03 49	23 02	15 44	04 44	09 36	20 05
22 Tu	12 06	15 29	09 29	05 36	03 43	23 01	15 43	04 46	09 36	20 05
23 We	12 26	12 20	10 21	05 14	03 38	23 01	15 42	04 47	09 35	20 05
24 Th	12 46	08 28	11 14	04 51	03 32	23 00	15 40	04 48	09 35	20 05
25 Fr	13 05	04 08	12 07	04 28	03 27	22 59	15 39	04 49	09 34	20 05
26 Sa	13 25	00N25	12 59	04 05	03 22	22 58	15 38	04 51	09 34	20 05
27 Su	13 44	04 54	13 51	03 41	03 18	22 57	15 37	04 52	09 33	20 05
28 Mo	14 03	09 04	14 42	03 18	03 14	22 56	15 36	04 53	09 33	20 06
29 Tu	14 22	12 42	15 32	02 54	03 09	22 56	15 35	04 54	09 32	20 06
30 We	14 41	15 36	16 22	02 30	03 06	22 55	15 34	04 56	09 32	20 06

⚷ Chiron

01 Dec.	01 S 29
02	15♓10
05	15 20
08	15 30
11	15 40
14	15 50
17	15 59
20	16 08
23	16 17
26	16 25
29	16 33

ASPECTARIAN

Day	Time	Aspect
01	07:40	☉ □ ♃
	11:11	☽ ⋕ ♀
	16:39	☽ ⚹ ♆
	22:15	☿ ∥ ♂
02	01:57	☽ ⚹ ♃
	05:30	☽ △ ♇
	07:09	☉ ☌ ♅
	18:59	☽ ⋕ ♄
	21:11	☽ ⚹ ☿
	21:55	☽ ☍ ♄
03	01:52	☿ ⋕ ☉
	03:03	☿ △ ♄
	06:44	☽ □ ♀
	09:20	☉ □ ♇
	23:59	☽ □ ♆
04	05:56	☉ ⋕ ♂
	09:40	☿ ⋕ ♅
	11:43	☽ ⚹ ♅
	15:58	☽ ⚹ ☉
05	02:46	☽ △ ♂
	14:56	☽ □ ☿
	20:31	♀ ♓
	21:47	☽ △ ♀
06	10:38	☽ △ ♆
	21:33	☽ ☌ ♃
	23:03	☽ □ ♅
07	00:42	☽ ☍ ♇
	12:35	☽ □ ♂
	15:35	☿ ♈
	18:14	☽ △ ♄
	23:40	☽ ⋕ ♄
08	12:56	☽ △ ☿
	21:04	☉ ☍ ♂
09	11:51	☽ △ ♅
	23:13	☽ ⚹ ♂
10	02:28	☽ △ ☉
	02:37	♀ ∥ ♆
	05:21	☽ ⋕ ♆
	05:38	☽ ⋕ ♀
	06:26	☽ □ ♄
	16:34	☽ ∥ ☉
11	09:55	☽ ☍ ♀
	11:21	☽ ☍ ♆
	13:45	☽ ⋕ ♂
	15:32	☽ ∥ ♅
	22:47	☽ ⚹ ♃
12	00:53	☽ △ ♇
	02:24	♀ ☌ ♆
	10:30	☽ ∥ ☿
	17:12	☽ ⚹ ♄
	18:33	♀ ⋕ ☉
13	07:25	☽ ⋕ ☿
	16:12	♂ ⋕ ♅
	21:51	☽ ∥ ♂
	22:03	☽ ⋕ ♅
14	07:02	☽ ☍ ☿
	08:38	☽ □ ♃
	09:21	☽ ☍ ♅
	10:08	☽ □ ♇
	16:00	☽ ☌ ♂
	18:33	☿ □ ♃
	19:46	☽ ∥ ♀
	23:16	☿ ☌ ♅
	23:45	♇ SR
15	00:07	☉ ⋕ ♆
	04:12	☿ □ ♇
	05:15	☽ ∥ ♆
	05:47	☽ ⋕ ☉
16	00:21	☿ ⋕ ♂
	04:39	☽ [illegible] ♅
	09:29	☿ ∥ ♅
	11:16	☿ ☍ ♂
	13:09	☽ △ ♀
	15:48	☽ △ ♃
	16:46	☽ ⚹ ♇
17	03:36	☽ ∥ ♄
	07:10	☽ ☌ ♄
18	01:20	♀ △ ♃
	09:25	♀ ⚹ ♇
	09:43	☽ □ ♆
	21:00	☿ ⋕ ♀
	21:01	☽ △ ♅
	22:20	☽ □ ♀
19	00:00	☽ ⚹ ♂
	11:49	☽ △ ☿
20	01:18	☽ △ ☉
	03:56	☉ ♉
	07:26	♃ □ ♅
	13:16	☽ ⚹ ♆
	23:06	♃ ☍ ♇
21	00:28	☽ □ ♅
	00:33	☽ ☌ ♇
	00:34	☽ ☍ ♃
	01:57	☽ □ ♂
	05:54	☽ ⚹ ♀
	13:49	☽ ⚹ ♄
	18:39	♅ □ ♇
	21:58	☽ ∥ ♄
	23:22	☽ □ ☿
22	03:14	☿ ⋕ ♆
	19:29	♂ □ ♃
	23:25	☽ ⋕ ☉
23	03:19	☽ ⚹ ♅
	[illegible]	☽ [illegible]
	07:11	♂ ☍ ♅
	09:16	☿ ♉
	10:22	☽ ⋕ ☿
	13:48	♂ □ ♇
	16:12	☽ □ ♄
	17:26	☽ ∥ ♆
24	02:39	♀ ⋕ ♅
	10:42	☽ ⚹ ☿
	14:15	☽ ⚹ ☉
	18:50	☽ ☌ ♆
	20:17	☽ ⋕ ♅
	22:02	☽ ∥ ♀
25	03:40	☽ ∥ ♂
	05:10	☿ △ ♄
	05:55	☽ ⚹ ♇
	06:55	☽ △ ♃
	18:49	☽ △ ♄
	20:04	☽ ☌ ♀
26	03:28	☿ ☌ ☉
	15:27	☽ ⋕ ♂
	17:27	☿ ⚹ ♆
	17:55	☽ ⋕ ♀
	19:03	☿ ∥ ☉
	23:49	☽ ∥ ♅
27	07:29	☽ ☍ ♂
	09:27	☽ □ ♇
	10:02	☽ ☌ ♅
	10:34	☉ ⚹ ♆
	11:03	☽ □ ♃
28	02:53	☽ ⋕ ♆
	05:02	♀ ∥ ♂
29	01:04	☿ ⋕ ♄
	03:08	☽ ⚹ ♆
	14:07	☽ ☌ ☿
	14:29	☽ ∥ ☉
	14:40	☽ △ ♇
	16:56	☽ ⚹ ♃
	17:29	☿ △ ♇
	23:29	☽ ⋕ ♄
30	03:57	☽ ☍ ♄
	[illegible]	☿ ∥ ♃
	12:19	☽ ∥ ☿
	15:54	☽ ⚹ ♀

May 2014

Day	S. T. h m s	☉ ° ' "	☽ ° ' "	☿ ° '	♀ ° '	♂ ° '	♃ ° '	♄ ° '	♅ ° '	♆ ° '	♇ ° '	☊ True ° '
01 Th	14 35 23	10♉33 13	01♊38 38	16♉15	27♓40	11♎R25	14♋58	20♏R45	14♈04	07♓10	13♑R31	28♎R22
02 Fr	14 39 20	11 31 29	14 21 31	18 23	28 48	11 10	15 07	20 40	14 07	07 11	13 30	28 21
03 Sa	14 43 17	12 29 43	26 48 42	20 31	29 56	10 56	15 16	20 36	14 10	07 13	13 30	28 20
04 Su	14 47 13	13 27 55	09♋02 10	22 37	01♈04	10 43	15 25	20 32	14 13	07 14	13 29	28 18
05 Mo	14 51 10	14 26 05	21 04 51	24 42	02 12	10 31	15 35	20 27	14 16	07 15	13 29	28 17
06 Tu	14 55 06	15 24 13	03♌00 29	26 45	03 21	10 19	15 44	20 23	14 19	07 16	13 28	28 16
07 We	14 59 03	16 22 20	14 53 21	28 46	04 29	10 09	15 54	20 18	14 22	07 17	13 27	28D 16
08 Th	15 02 59	17 20 24	26 48 06	00♊44	05 38	09 59	16 04	20 14	14 25	07 18	13 27	28 16
09 Fr	15 06 56	18 18 26	08♍49 27	02 41	06 46	09 50	16 13	20 09	14 28	07 19	13 26	28 17
10 Sa	15 10 52	19 16 27	21 01 55	04 34	07 55	09 41	16 23	20 05	14 31	07 20	13 25	28 19
11 Su	15 14 49	20 14 26	03♎29 31	06 25	09 04	09 34	16 33	20 00	14 34	07 21	13 25	28 20
12 Mo	15 18 46	21 12 23	16 15 35	08 13	10 13	09 27	16 43	19 56	14 37	07 22	13 24	28 21
13 Tu	15 22 42	22 10 18	29 22 17	09 57	11 22	09 21	16 53	19 51	14 40	07 23	13 23	28R 21
14 We	15 26 39	23 08 11	12♏50 25	11 38	12 31	09 16	17 04	19 47	14 43	07 24	13 22	28 19
15 Th	15 30 35	24 06 03	26 38 54	13 16	13 40	09 11	17 14	19 42	14 46	07 25	13 22	28 17
16 Fr	15 34 32	25 03 54	10♐44 55	14 51	14 49	09 08	17 24	19 38	14 49	07 26	13 21	28 13
17 Sa	15 38 28	26 01 43	25 03 59	16 22	15 58	09 05	17 35	19 33	14 51	07 26	13 20	28 09
18 Su	15 42 25	26 59 31	09♑30 41	17 50	17 07	09 03	17 45	19 29	14 54	07 27	13 19	28 05
19 Mo	15 46 21	27 57 18	23 59 22	19 13	18 17	09 02	17 56	19 24	14 57	07 28	13 18	28 01
20 Tu	15 50 18	28 55 04	08♒24 55	20 34	19 26	09 02	18 07	19 20	15 00	07 29	13 17	27 58
21 We	15 54 15	29 52 48	22 43 14	21 50	20 36	09D 02	18 18	19 15	15 02	07 29	13 16	27 57
22 Th	15 58 11	00♊50 32	06♓51 29	23 03	21 46	09 03	18 29	19 11	15 05	07 30	13 15	27D 57
23 Fr	16 02 08	01 48 14	20 48 06	24 12	22 55	09 05	18 40	19 07	15 07	07 31	13 14	27 58
24 Sa	16 06 04	02 45 55	04♈32 24	25 18	24 05	09 07	18 51	19 02	15 10	07 31	13 13	27 59
25 Su	16 10 01	03 43 36	18 04 20	26 19	25 15	09 11	19 02	18 58	15 13	07 32	13 12	28 00
26 Mo	16 13 57	04 41 15	01♉24 00	27 16	26 25	09 15	19 13	18 54	15 15	07 32	13 11	28R 01
27 Tu	16 17 54	05 38 53	14 31 28	28 10	27 35	09 20	19 25	18 49	15 18	07 33	13 10	28 00
28 We	16 21 50	06 36 31	27 26 41	28 59	28 45	09 25	19 36	18 45	15 20	07 33	13 09	27 57
29 Th	16 25 47	07 34 07	10♊09 36	29 44	29 55	09 31	19 47	18 41	15 22	07 33	13 08	27 53
30 Fr	16 29 44	08 31 42	22 40 23	00♋25	01♉05	09 38	19 59	18 37	15 25	07 34	13 06	27 46
31 Sa	16 33 40	09 29 16	04♋59 34	01 01	02 15	09 46	20 11	18 33	15 27	07 34	13 05	27 39

Data for 05-01-2014
Julian Day 2456778.50
Ayanamsa 24 03 32
SVP 05♓03 42
☽☊ Mean 27♎55 R

●◐ PHASES ○◑

07	03:15	◐	16♌30
14	19:17	○	23♏55
21	12:59	◑	00♓24
28	18:40	●	07♊21

LAST ASPECT ☽ Day	h m	INGRESS Day	h m	
01	23:32	03	06:13	♋
05	08:46	05	17:56	♌
07	10:51	08	06:25	♍
09	22:09	10	17:20	♎
12	00:52	13	01:08	♏
14	19:17	15	05:45	♐
16	07:44	17	08:13	♑
19	07:03	19	09:59	♒
20	22:22	21	12:19	♓
23	06:26	23	16:02	♈
25	15:58	25	21:28	♉
27	09:10	28	04:48	♊
29	09:59	30	14:14	♋

DECLINATION

Day	☉	☽	☿	♀	♂	♃	♄	♅	♆	♇
01 Th	14N59	17N39	17N10	02S06	03S02	22N54	15S32	04N57	09S31	20S06
02 Fr	15 17	18 45	17 56	01 41	02 59	22 53	15 31	04 58	09 31	20 06
03 Sa	15 35	18 54	18 41	01 17	02 56	22 52	15 30	04 59	09 30	20 06
04 Su	15 53	18 10	19 23	00 52	02 53	22 51	15 29	05 00	09 30	20 06
05 Mo	16 10	16 37	20 04	00 28	02 51	22 50	15 28	05 02	09 30	20 06
06 Tu	16 27	14 22	20 43	00 03	02 49	22 49	15 27	05 03	09 29	20 06
07 We	16 44	11 32	21 19	00N22	02 47	22 48	15 25	05 04	09 29	20 06
08 Th	17 00	08 14	21 53	00 47	02 46	22 47	15 24	05 05	09 29	20 06
09 Fr	17 17	04 33	22 24	01 12	02 44	22 45	15 23	05 06	09 28	20 06
10 Sa	17 33	00 37	22 53	01 37	02 44	22 44	15 22	05 07	09 28	20 06
11 Su	17 48	03S25	23 19	02 02	02 43	22 43	15 21	05 09	09 28	20 07
12 Mo	18 03	07 25	23 43	02 27	02 43	22 42	15 20	05 10	09 27	20 07
13 Tu	18 19	11 09	24 04	02 53	02 43	22 41	15 18	05 11	09 27	20 07
14 We	18 33	14 26	24 23	03 18	02 43	22 39	15 17	05 12	09 27	20 07
15 Th	18 48	16 59	24 39	03 43	02 44	22 38	15 16	05 13	09 26	20 07
16 Fr	19 02	18 33	24 53	04 09	02 45	22 37	15 15	05 14	09 26	20 07
17 Sa	19 15	18 58	25 05	04 34	02 46	22 36	15 14	05 15	09 26	20 07
18 Su	19 29	18 10	25 14	04 59	02 47	22 34	15 13	05 16	09 26	20 07
19 Mo	19 42	16 11	25 22	05 25	02 49	22 33	15 12	05 17	09 25	20 08
20 Tu	19 55	13 12	25 27	05 50	02 51	22 32	15 10	05 18	09 25	20 08
21 We	20 07	09 27	25 31	06 15	02 53	22 30	15 09	05 19	09 25	20 08
22 Th	20 19	05 12	25 32	06 40	02 56	22 29	15 08	05 20	09 25	20 08
23 Fr	20 31	00 44	25 32	07 05	02 59	22 27	15 07	05 21	09 24	20 08
24 Sa	20 43	03N43	25 30	07 30	03 02	22 26	15 06	05 22	09 24	20 08
25 Su	20 54	07 55	25 27	07 55	03 05	22 24	15 05	05 23	09 24	20 08
26 Mo	21 04	11 40	25 22	08 20	03 09	22 23	15 04	05 24	09 24	20 09
27 Tu	21 15	14 47	25 16	08 44	03 13	22 21	15 03	05 25	09 24	20 09
28 We	21 25	17 06	25 09	09 09	03 17	22 20	15 02	05 26	09 24	20 09
29 Th	21 34	18 32	25 00	09 33	03 21	22 18	15 01	05 27	09 24	20 09
30 Fr	21 43	19 01	24 50	09 58	03 26	22 16	15 00	05 28	09 23	20 09
31 Sa	21 52	18 36	24 39	10 22	03 31	22 15	14 59	05 29	09 23	20 09

ASPECTARIAN

01 10:23 ☽ □ ♆
18:02 ☽ △ ♂
23:32 ☽ ✶ ♅
02 17:36 ☉ # ♄
03 00:57 ☿ ☍ ♄
01:22 ♀ ♈
04:59 ☽ ∥ ☿
06:43 ☽ □ ♀
20:25 ☽ △ ♆
04 00:35 ☉ △ ♇
03:17 ☽ □ ♂
08:50 ☽ ☍ ♇
09:33 ☽ ✶ ☉
10:20 ☽ □ ♅
12:51 ☽ ☌ ♃
22:45 ☽ △ ♄
05 01:10 ☿ # ♇
04:51 ☽ ∥ ☉
08:46 ☽ ✶ ☿
13:24 ☽ # ♄
06 00:45 ☽ △ ♀
09:54 ☉ ✶ ♃
14:33 ☽ ✶ ♂
22:57 ☽ △ ♅
07 10:51 ☽ □ ♄
14:57 ☿ ♊
15:17 ☽ # ♆
08 09:26 ☽ □ ☿
20:31 ☽ ∥ ♅
21:01 ☽ ☍ ♆
09 09:06 ☽ △ ♇
11:15 ☽ # ♂
14:48 ☽ ✶ ♃
16:51 ☿ ∥ ♃
18:36 ☽ ∥ ♀
20:17 ☽ △ ☉
22:09 ☽ ✶ ♄
10 14:54 ☽ # ♀
18:29 ☉ ☍ ♄
19:52 ☽ ∥ ♂
11 06:29 ☽ △ ☿
09:27 ♀ ☍ ♂
10:20 ☽ # ♅
11:23 ☽ ☌ ♂
11:35 ☽ ☍ ♀
12:35 ☿ □ ♆
18:41 ☽ □ ♇
20:56 ☽ ☍ ♅
12 00:52 ☽ □ ♃
12:47 ☽ ∥ ♆
14:28 ♀ # ♂
16:02 ☿ △ ♂
13 14:22 ☽ △ ♆
14 00:56 ☽ ✶ ♇
07:08 ☽ ∥ ♄
07:29 ☽ △ ♃
12:03 ☽ ☌ ♄
17:47 ♀ □ ♇
15 18:23 ☽ □ ♆
21:17 ☽ ✶ ♂
21:55 ☿ ✶ ♀
23:23 ☿ ✶ ♅
23:54 ♀ ☌ ♅
16 06:52 ☽ △ ♅
07:27 ☽ △ ♀
07:44 ☽ ☍ ☿
17 20:35 ☽ ✶ ♆
23:14 ☽ □ ♂
18 06:18 ☽ ☌ ♇
08:58 ☽ □ ♅
13:43 ☽ □ ♀
13:50 ☽ ☍ ♃
15:31 ♀ □ ♃
16:26 ☽ ✶ ♄
16:41 ♀ ∥ ♅
19 07:03 ☽ △ ☉
08:54 ☽ ∥ ♄
20 01:01 ☽ △ ♂
01:32 ♂ SD
11:02 ☽ ✶ ♅
18:12 ☽ □ ♄
20:06 ☽ ✶ ♀
22:22 ☽ △ ☿
21 00:16 ☽ ∥ ♆
01:04 ☉ # ♇
02:59 ☉ ♊
16:44 ☽ # ♀
22:21 ☽ # ♅
22 01:06 ☽ ☌ ♆
10:57 ☽ ✶ ♇
12:13 ☽ ∥ ♂
20:15 ☽ △ ♃
21:05 ☽ △ ♄
23 06:26 ☽ □ ☿
20:15 ☽ # ♂
20:39 ☽ ✶ ☉
24 08:07 ☽ ☍ ♂
09:16 ☽ ∥ ♅
15:20 ☽ □ ♇
17:48 ♃ △ ♄
18:53 ☽ ☌ ♅
23:59 ☽ ∥ ♀
25 01:45 ☽ □ ♃
09:04 ☽ # ♆
14:07 ☽ ☌ ♀
15:58 ☽ ✶ ☿
26 11:11 ☽ ✶ ♆
21:30 ☽ △ ♇
27 02:22 ☽ # ♄
07:54 ☽ ☍ ♄
09:10 ☽ ✶ ♃
28 13:58 ☿ ✶ ♀
14:26 ♀ # ♆
19:03 ☽ □ ♆
22:46 ☽ △ ♂
23:45 ☉ □ ♆
29 01:46 ♀ ♉
09:12 ☿ ♋
09:59 ☽ ✶ ♅
30 15:50 ☽ ☌ ☿
18:04 ☽ ✶ ♀
31 05:04 ☽ △ ♆
08:01 ☉ △ ♂
09:30 ☽ □ ♂
15:56 ☽ ☍ ♇
20:43 ☽ □ ♅

⚷ Chiron

01 Dec. 00 S 51

02	16♓41
05	16 48
08	16 55
11	17 01
14	17 08
17	17 13
20	17 19
23	17 23
26	17 28
29	17 32

June 2014

Day	S. T.	☉	☽	☿	♀	♂	♃	♄	♅	♆	♇	☊ True
	h m s	° ' "	° ' "	° '	° '	° '	° '	° '	° '	° '	° '	° '
01 Su	16 37 37	10♊ 26 48	17♋ 08 17	01♋ 33	03♉ 25	09♎ 54	20♋ 22	18♏R29	15♈ 29	07♓ 34	13♑R04	27♎R32
02 Mo	16 41 33	11 24 20	29 08 21	02 01	04 36	10 03	20 34	18 25	15 32	07 35	13 03	27 25
03 Tu	16 45 30	12 21 50	11♌ 02 25	02 24	05 46	10 13	20 46	18 21	15 34	07 35	13 02	27 19
04 We	16 49 26	13 19 19	22 53 56	02 42	06 56	10 23	20 58	18 17	15 36	07 35	13 00	27 15
05 Th	16 53 23	14 16 47	04♍ 47 05	02 56	08 07	10 34	21 10	18 13	15 38	07 35	12 59	27 13
06 Fr	16 57 19	15 14 13	16 46 33	03 05	09 17	10 45	21 22	18 10	15 40	07 36	12 58	27D 13
07 Sa	17 01 16	16 11 38	28 57 17	03 09	10 28	10 57	21 34	18 06	15 42	07 36	12 56	27 14
08 Su	17 05 13	17 09 03	11♎ 24 11	03R 09	11 38	11 10	21 46	18 02	15 44	07 36	12 55	27 15
09 Mo	17 09 09	18 06 26	24 11 42	03 05	12 49	11 23	21 58	17 59	15 46	07 36	12 54	27R 16
10 Tu	17 13 06	19 03 48	07♏ 23 25	02 56	14 00	11 37	22 10	17 55	15 48	07R 36	12 53	27 15
11 We	17 17 02	20 01 09	21 01 20	02 43	15 10	11 51	22 22	17 52	15 50	07 36	12 51	27 12
12 Th	17 20 59	20 58 29	05♐ 05 12	02 26	16 21	12 06	22 35	17 48	15 52	07 36	12 50	27 06
13 Fr	17 24 55	21 55 48	19 32 00	02 06	17 32	12 22	22 47	17 45	15 54	07 36	12 48	26 59
14 Sa	17 28 52	22 53 07	04♑ 15 57	01 42	18 43	12 38	22 59	17 42	15 56	07 35	12 47	26 51
15 Su	17 32 48	23 50 25	19 09 10	01 15	19 54	12 54	23 12	17 38	15 58	07 35	12 46	26 42
16 Mo	17 36 45	24 47 43	04♒ 02 52	00 45	21 05	13 11	23 24	17 35	15 59	07 35	12 44	26 34
17 Tu	17 40 42	25 45 00	18 48 51	00♊ 14	22 16	13 29	23 37	17 32	16 01	07 35	12 43	26 28
18 We	17 44 38	26 42 16	03♓ 20 40	29 41	23 27	13 47	23 49	17 29	16 03	07 35	12 41	26 24
19 Th	17 48 35	27 39 33	17 34 19	29 07	24 38	14 05	24 02	17 26	16 04	07 34	12 40	26 21
20 Fr	17 52 31	28 36 49	01♈ 28 14	28 32	25 49	14 24	24 15	17 23	16 06	07 34	12 39	26D 21
21 Sa	17 56 28	29 34 05	15 02 49	27 58	27 00	14 44	24 27	17 21	16 07	07 34	12 37	26 22
22 Su	18 00 24	00♋ 31 21	28 19 42	27 25	28 11	15 03	24 40	17 18	16 09	07 33	12 36	26 23
23 Mo	18 04 21	01 28 37	11♉ 21 01	26 53	29 23	15 24	24 53	17 15	16 10	07 33	12 34	26R 22
24 Tu	18 08 17	02 25 52	24 08 54	26 23	00♊ 34	15 45	25 06	17 13	16 11	07 33	12 33	26 20
25 We	18 12 14	03 23 08	06♊ 45 12	25 55	01 45	16 06	25 18	17 10	16 13	07 32	12 31	26 15
26 Th	18 16 11	04 20 23	19 11 20	25 31	02 57	16 27	25 31	17 08	16 14	07 32	12 30	26 06
27 Fr	18 20 07	05 17 38	01♋ 28 23	25 09	04 08	16 49	25 44	17 06	16 15	07 31	12 28	25 56
28 Sa	18 24 04	06 14 53	13 37 17	24 51	05 20	17 12	25 57	17 04	16 16	07 30	12 27	25 43
29 Su	18 28 00	07 12 07	25 38 59	24 38	06 31	17 35	26 10	17 01	16 18	07 30	12 25	25 30
30 Mo	18 31 57	08 09 21	07♌ 34 47	24 28	07 43	17 58	26 23	16 59	16 19	07 29	12 24	25 18

Data for	06-01-2014
Julian Day	2456809.50
Ayanamsa	24 03 36
SVP	05 ♓ 03 38
☽ ☊ Mean	26 ♎ 16 R

● ◐ PHASES ○ ◑

Day	Time	Phase	Position
05	20:40	◐	15♍06
13	04:12	○	22♐06
19	18:38	◑	28♓24
27	08:09	●	05♋37

LAST ASPECT ☽ INGRESS

Day	h m	Day	h m	Sign
01	06:33	02	01:44	♌
03	14:43	04	14:21	♍
06	09:14	07	02:02	♎
08	19:48	09	10:39	♏
11	02:22	11	15:24	♐
13	04:12	13	17:05	♑
15	06:36	15	17:28	♒
17	18:08	17	18:26	♓
19	19:06	19	21:26	♈
21	22:24	22	03:03	♉
24	01:49	24	11:06	♊
26	11:57	26	21:06	♋
29	01:04	29	08:44	♌

DECLINATION

Day	☉	☽	☿	♀	♂	♃	♄	♅	♆	♇
01 Su	22N01	17N19	24N28	10N45	03S 36	22N13	14S 58	05N29	09S 23	20S 09
02 Mo	22 09	15 17	24 15	11 09	03 42	22 11	14 57	05 30	09 23	20 10
03 Tu	22 16	12 39	24 02	11 33	03 47	22 10	14 56	05 31	09 23	20 10
04 We	22 24	09 29	23 47	11 56	03 53	22 08	14 55	05 32	09 23	20 10
05 Th	22 31	05 57	23 33	12 19	03 59	22 06	14 54	05 33	09 23	20 10
06 Fr	22 37	02 08	23 17	12 42	04 05	22 04	14 53	05 34	09 23	20 10
07 Sa	22 43	01S 50	23 02	13 04	04 12	22 03	14 53	05 34	09 23	20 11
08 Su	22 49	05 49	22 45	13 27	04 19	22 01	14 52	05 35	09 23	20 11
09 Mo	22 54	09 39	22 29	13 49	04 25	21 59	14 51	05 36	09 23	20 11
10 Tu	22 59	13 08	22 12	14 11	04 33	21 57	14 50	05 37	09 23	20 11
11 We	23 04	16 01	21 56	14 32	04 40	21 55	14 49	05 37	09 23	20 11
12 Th	23 08	18 03	21 39	14 53	04 47	21 53	14 49	05 38	09 23	20 12
13 Fr	23 11	18 59	21 22	15 14	04 55	21 51	14 48	05 39	09 23	20 12
14 Sa	23 15	18 40	21 06	15 35	05 03	21 49	14 47	05 39	09 23	20 12
15 Su	23 17	17 05	20 49	15 55	05 11	21 47	14 46	05 40	09 23	20 12
16 Mo	23 20	14 20	20 34	16 15	05 19	21 45	14 46	05 41	09 23	20 12
17 Tu	23 22	10 42	20 19	16 34	05 27	21 43	14 45	05 41	09 24	20 13
18 We	23 24	06 28	20 04	16 53	05 36	21 41	14 44	05 42	09 24	20 13
19 Th	23 25	01 58	19 50	17 12	05 44	21 39	14 44	05 42	09 24	20 13
20 Fr	23 26	02N34	19 38	17 30	05 53	21 36	14 43	05 43	09 24	20 13
21 Sa	23 26	06 51	19 26	17 48	06 02	21 34	14 43	05 44	09 24	20 13
22 Su	23 26	10 43	19 16	18 06	06 11	21 32	14 42	05 44	09 24	20 14
23 Mo	23 26	13 59	19 06	18 23	06 20	21 30	14 42	05 45	09 25	20 14
24 Tu	23 25	16 31	18 59	18 40	06 30	21 28	14 41	05 45	09 25	20 14
25 We	23 23	18 12	18 52	18 56	06 39	21 25	14 41	05 46	09 25	20 14
26 Th	23 22	18 59	18 47	19 12	06 49	21 23	14 40	05 46	09 25	20 14
27 Fr	23 20	18 52	18 44	19 27	06 58	21 21	14 40	05 46	09 25	20 15
28 Sa	23 17	17 52	18 42	19 42	07 08	21 18	14 39	05 47	09 26	20 15
29 Su	23 14	16 05	18 42	19 56	07 18	21 16	14 39	05 47	09 26	20 15
30 Mo	23 11	13 38	18 43	20 10	07 28	21 14	14 39	05 48	09 26	20 15

ASPECTARIAN

01 02:40 ☽ △ ♄
06:33 ☽ ☌ ♃

02 03:29 ☽ # ♄
06:52 ☉ ∥ ♃
12:11 ☽ ☐ ♀
22:18 ☽ ⚹ ♂

03 02:55 ☽ ⚹ ☉
07:48 ☽ ∥ ♀
09:11 ☽ △ ♅
14:43 ☽ ☐ ♄

04 00:46 ☽ # ♆
13:17 ♀ ⚹ ♆
20:12 ☽ ⚹ ☿

05 02:36 ☽ ∥ ♅
05:38 ☽ ☍ ♆
07:25 ☽ △ ♀
12:12 ☽ # ♂
16:25 ☽ △ ♇

06 02:44 ☽ ⚹ ♄
09:14 ☽ ⚹ ♃
11:21 ☉ ⚹ ♅

07 08:11 ☽ ☐ ☿
11:58 ☿ SR
14:40 ☽ ∥ ♂
20:17 ☿ ∥ ☉
22:40 ☽ # ♅
23:32 ☽ ☌ ♂

08 02:53 ☽ ☐ ♇
08:14 ☽ ☍ ♅
11:45 ☽ △ ☉
19:48 ☽ ☐ ♃
22:22 ☽ ∥ ♆

09 01:39 ♀ △ ♇
16:05 ☽ △ ☿
19:52 ♆ SR

10 00:22 ☽ △ ♆
09:12 ☽ # ♀
09:44 ☽ ⚹ ♇
12:50 ☽ ☍ ♀
13:25 ☽ ∥ ♄
18:31 ☽ ☌ ♄

11 01:05 ☿ ∥ ♃
02:22 ☽ △ ♃
18:53 ♀ # ♄

12 04:12 ☽ ☐ ♆
11:57 ☽ ⚹ ♂
18:00 ☽ △ ♅

13 04:09 ♀ ☍ ♄
19:57 ☽ ☍ ☿

14 05:22 ☽ ⚹ ♆
12:34 ♂ ☐ ♇
13:44 ☽ ☌ ♇
13:46 ☽ ☐ ♂
18:51 ☽ ☐ ♅
21:34 ☽ ⚹ ♄

15 01:18 ☽ △ ♀
06:36 ☽ ☍ ♃
10:07 ☽ # ♀
20:49 ☽ ∥ ♄

16 15:07 ☽ △ ♂
19:25 ☽ ⚹ ♅
21:55 ☽ ☐ ♄

17 06:09 ☽ ☐ ♀
07:44 ☽ ∥ ♆
09:45 ☿ # ♇
10:05 ☿ ♊ R
12:12 ☽ △ ☉
18:08 ☽ △ ☿

18 04:13 ☽ # ♅
04:39 ☽ ∥ ♂
07:05 ☽ ☌ ♆
09:16 ♀ ⚹ ♃
15:41 ☽ ⚹ ♇
18:52 ♂ # ♅
23:46 ☽ △ ♄

19 11:15 ☽ △ ♃
13:15 ☽ ⚹ ♀
19:06 ☽ ☐ ☿
22:51 ☿ ☌ ☉

20 17:31 ☽ ∥ ♅
19:06 ☽ # ♂
19:40 ☽ ☐ ♇
23:25 ☽ ☍ ♂

21 01:58 ☽ ☌ ♅
10:52 ☉ ♋
15:29 ☽ # ♆
17:14 ☽ ☐ ♃
22:24 ☽ ⚹ ☿

22 04:20 ☽ ⚹ ☉
16:57 ☽ ⚹ ♆

23 02:16 ☽ △ ♇
06:01 ☽ # ♄
10:59 ☽ ☍ ♄
12:34 ♀ ♊

24 01:49 ☽ ⚹ ♃
13:27 ☽ ☌ ♀
19:55 ☿ ∥ ♀

25 01:30 ☽ ☐ ♆
08:26 ♂ ☍ ♅
15:51 ☽ ∥ ☿
18:16 ☽ ⚹ ♅
18:32 ☽ △ ♂

26 11:57 ☽ ☌ ☿

27 04:32 ☽ ∥ ☿
11:54 ☽ △ ♆
21:40 ☽ ☍ ♇

28 05:17 ☽ ☐ ♅
06:49 ☽ △ ♄
07:20 ☽ ☐ ♂

29 01:04 ☽ ☌ ♃
07:23 ☉ △ ♆
14:40 ☽ # ♄
19:31 ♀ ☐ ♆

30 00:18 ☽ ⚹ ♀
09:55 ♀ # ♇
17:41 ☽ △ ♅
18:59 ☽ ☐ ♄
21:44 ☽ ⚹ ♂

⚷ Chiron

01 Dec. 00 S 23

Day	Position
01	17♓35
04	17 38
07	17 40
10	17 42
13	17 44
16	17 45
19	17 45
22	17 45R
25	17 45
28	17 44

20 10:46 17♓46 R

July 2014

Day	S. T.	☉	☽	☿	♀	♂	♃	♄	♅	♆	♇	☊ True
	h m s	° ' "	° ' "	° '	° '	° '	° '	° '	° '	° '	° '	° '
01 Tu	18 35 53	09♋ 06 35	19♌ 26 27	24♊ R24	08♊ 54	18♎ 21	26♋ 36	16♏ R58	16♈ 20	07♓ R29	12♑ R22	25♎ R07
02 We	18 39 50	10 03 49	01♍ 16 33	24D 23	10 06	18 45	26 49	16 56	16 21	07 28	12 21	24 59
03 Th	18 43 46	11 01 02	13 08 25	24 28	11 18	19 10	27 02	16 54	16 22	07 27	12 19	24 54
04 Fr	18 47 43	11 58 15	25 06 10	24 38	12 29	19 34	27 15	16 52	16 23	07 27	12 18	24 52
05 Sa	18 51 40	12 55 27	07♎ 14 29	24 52	13 41	19 59	27 29	16 51	16 23	07 26	12 16	24 51
06 Su	18 55 36	13 52 40	19 38 24	25 12	14 53	20 25	27 42	16 49	16 24	07 25	12 15	24D 51
07 Mo	18 59 33	14 49 52	02♏ 22 58	25 36	16 05	20 51	27 55	16 48	16 25	07 24	12 13	24R 51
08 Tu	19 03 29	15 47 04	15 32 42	26 06	17 17	21 17	28 08	16 47	16 26	07 23	12 12	24 49
09 We	19 07 26	16 44 15	29 10 51	26 41	18 29	21 43	28 21	16 45	16 26	07 22	12 10	24 45
10 Th	19 11 22	17 41 27	13♐ 18 30	27 20	19 40	22 10	28 35	16 44	16 27	07 22	12 09	24 39
11 Fr	19 15 19	18 38 39	27 53 39	28 05	20 52	22 37	28 48	16 43	16 27	07 21	12 07	24 30
12 Sa	19 19 15	19 35 50	12♑ 50 47	28 54	22 04	23 04	29 01	16 42	16 28	07 20	12 06	24 20
13 Su	19 23 12	20 33 02	28 01 14	29 49	23 17	23 32	29 14	16 42	16 28	07 19	12 04	24 09
14 Mo	19 27 09	21 30 14	13♒ 14 25	00♋ 48	24 29	24 00	29 28	16 41	16 29	07 18	12 03	23 58
15 Tu	19 31 05	22 27 27	28 19 41	01 51	25 41	24 28	29 41	16 40	16 29	07 17	12 01	23 50
16 We	19 35 02	23 24 40	13♓ 08 08	03 00	26 53	24 57	29 54	16 40	16 30	07 16	12 00	23 44
17 Th	19 38 58	24 21 54	27 33 50	04 13	28 05	25 25	00♌ 07	16 39	16 30	07 15	11 58	23 40
18 Fr	19 42 55	25 19 08	11♈ 34 12	05 30	29 17	25 55	00 21	16 39	16 30	07 13	11 57	23 38
19 Sa	19 46 51	26 16 23	25 09 29	06 52	00♋ 30	26 24	00 34	16 39	16 30	07 12	11 56	23D 38
20 Su	19 50 48	27 13 38	08♉ 21 54	08 18	01 42	26 54	00 47	16 39	16 30	07 11	11 54	23 39
21 Mo	19 54 44	28 10 55	21 14 40	09 49	02 55	27 23	01 01	16D 39	16 30	07 10	11 53	23R 38
22 Tu	19 58 41	29 08 12	03♊ 51 17	11 23	04 07	27 54	01 14	16 39	16 R 30	07 09	11 51	23 35
23 We	20 02 38	00♌ 05 30	16 15 01	13 02	05 20	28 24	01 27	16 39	16 30	07 08	11 50	23 30
24 Th	20 06 34	01 02 49	28 28 40	14 44	06 32	28 55	01 41	16 39	16 30	07 06	11 48	23 21
25 Fr	20 10 31	02 00 08	10♋ 34 34	16 30	07 45	29 26	01 54	16 39	16 30	07 05	11 47	23 10
26 Sa	20 14 27	02 57 28	22 34 32	18 19	08 57	29 57	02 08	16 40	16 30	07 04	11 46	22 57
27 Su	20 18 24	03 54 49	04♌ 29 59	20 11	10 10	00♏ 28	02 21	16 40	16 30	07 03	11 44	22 43
28 Mo	20 22 20	04 52 11	16 22 14	22 06	11 23	01 00	02 34	16 41	16 30	07 01	11 43	22 30
29 Tu	20 26 17	05 49 33	28 12 45	24 03	12 35	01 32	02 48	16 42	16 29	07 00	11 42	22 18
30 We	20 30 13	06 46 56	10♍ 03 23	26 02	13 48	02 04	03 01	16 43	16 29	06 59	11 40	22 09
31 Th	20 34 10	07 44 19	21 56 40	28 04	15 01	02 36	03 14	16 44	16 29	06 57	11 39	22 03

Data for	07-01-2014
Julian Day	2456839.50
Ayanamsa	24 03 41
SVP	05 ♓ 03 34
☽ ☊ Mean	24 ♎ 41 R

● ◐ PHASES ○ ◑

05	11:59	◐	13♎24
12	11:25	○	20♑03
19	02:09	◑	26♈22
26	22:42	●	03♌52

LAST ASPECT Day	h m	☽ INGRESS Day	h m	
01	10:01	01	21:25	♍
04	04:22	04	09:44	♎
06	15:31	06	19:34	♏
08	22:33	09	01:25	♐
11	00:19	11	03:25	♑
13	01:57	13	03:07	♒
14	19:23	15	02:41	♓
17	00:58	17	04:07	♈
19	02:19	19	08:43	♉
21	14:13	21	16:37	♊
24	00:54	24	03:00	♋
25	13:55	26	14:56	♌
28	00:38	29	03:37	♍
31	14:48	31	16:10	♎

DECLINATION

Day	☉	☽	☿	♀	♂	♃	♄	♅	♆	♇
01 Tu	23N07	10N38	18N46	20N23	07S 39	21N11	14S 38	05N48	09S 26	20S 16
02 We	23 03	07 13	18 50	20 36	07 49	21 09	14 38	05 48	09 27	20 16
03 Th	22 59	03 31	18 55	20 48	07 59	21 06	14 38	05 49	09 27	20 16
04 Fr	22 54	00S 21	19 02	21 00	08 10	21 04	14 38	05 49	09 27	20 16
05 Sa	22 48	04 17	19 10	21 11	08 21	21 01	14 37	05 49	09 28	20 17
06 Su	22 43	08 06	19 19	21 22	08 31	20 59	14 37	05 50	09 28	20 17
07 Mo	22 37	11 41	19 29	21 32	08 42	20 56	14 37	05 50	09 28	20 17
08 Tu	22 30	14 47	19 40	21 41	08 53	20 54	14 37	05 50	09 29	20 17
09 We	22 23	17 12	19 51	21 50	09 04	20 51	14 37	05 50	09 29	20 18
10 Th	22 16	18 39	20 03	21 59	09 15	20 49	14 37	05 51	09 29	20 18
11 Fr	22 08	18 56	20 16	22 06	09 26	20 46	14 37	05 51	09 30	20 18
12 Sa	22 00	17 56	20 29	22 13	09 37	20 43	14 37	05 51	09 30	20 18
13 Su	21 52	15 40	20 42	22 20	09 49	20 41	14 37	05 51	09 30	20 19
14 Mo	21 43	12 18	20 55	22 26	10 00	20 38	14 37	05 51	09 31	20 19
15 Tu	21 34	08 10	21 08	22 31	10 11	20 35	14 37	05 51	09 31	20 19
16 We	21 24	03 36	21 20	22 36	10 23	20 33	14 37	05 51	09 32	20 19
17 Th	21 14	01N04	21 32	22 40	10 34	20 30	14 37	05 52	09 32	20 19
18 Fr	21 04	05 33	21 43	22 43	10 46	20 27	14 37	05 52	09 32	20 20
19 Sa	20 54	09 37	21 53	22 46	10 57	20 24	14 37	05 52	09 33	20 20
20 Su	20 43	13 05	22 02	22 48	11 09	20 21	14 38	05 52	09 33	20 20
21 Mo	20 31	15 49	22 09	22 49	11 21	20 19	14 38	05 52	09 34	20 20
22 Tu	20 20	17 43	22 15	22 50	11 33	20 16	14 38	05 52	09 34	20 21
23 We	20 08	18 45	22 19	22 50	11 44	20 13	14 38	05 52	09 35	20 21
24 Th	19 55	18 53	22 22	22 50	11 56	20 10	14 39	05 52	09 35	20 21
25 Fr	19 43	18 09	22 22	22 48	12 08	20 07	14 39	05 52	09 36	20 21
26 Sa	19 30	16 37	22 20	22 47	12 20	20 04	14 39	05 51	09 36	20 22
27 Su	19 16	14 23	22 15	22 44	12 32	20 01	14 40	05 51	09 37	20 22
28 Mo	19 03	11 34	22 08	22 41	12 44	19 58	14 40	05 51	09 37	20 22
29 Tu	18 49	08 17	21 59	22 37	12 56	19 55	14 41	05 51	09 38	20 22
30 We	18 34	04 40	21 47	22 33	13 07	19 52	14 41	05 51	09 38	20 23
31 Th	18 20	00 52	21 32	22 28	13 19	19 49	14 42	05 51	09 39	20 23

ASPECTARIAN

01 08:40 ☽ # ♆
10:01 ☽ ⚹ ☿
12:41 ☿ SD
20:09 ☽ # ♂

02 09:19 ☽ ∥ ♅
12:32 ☽ ☍ ♆
19:20 ☽ ⚹ ☉
19:52 ☽ □ ♀
22:21 ☽ △ ♇

03 07:33 ☽ ⚹ ♄
23:02 ☽ □ ☿

04 04:22 ☽ ⚹ ♃
06:43 ♀ ∥ ♃
08:01 ☉ ☍ ♇

05 09:34 ☽ # ♅
09:47 ☽ □ ♇
13:53 ☽ △ ♀
17:47 ☽ ☍ ♅

06 01:32 ☽ ☌ ♂
02:49 ☽ ∥ ♂
08:53 ☽ ∥ ♆
10:53 ☽ △ ☿
15:31 ☽ □ ♃

07 06:49 ♀ ⚹ ♅
09:14 ☽ △ ♆
17:59 ☽ ⚹ ♇
22:36 ☽ ∥ ♄

08 00:28 ☽ △ ☉
02:12 ☽ ☌ ♄
16:23 ☉ □ ♅
22:33 ☽ △ ♃

09 00:29 ☉ △ ♄
14:00 ☽ □ ♆

10 05:14 ☽ △ ♅
11:30 ☽ ☍ ♀
15:07 ☽ ⚹ ♂

11 00:19 ☽ ☍ ☿
03:13 ♀ ∥ ☉
03:50 ☿ # ♆
07:49 ♂ ∥ ♆
15:12 ☽ ⚹ ♆
22:49 ☽ ☌ ♇

12 05:45 ☽ □ ♅
06:07 ☽ ⚹ ♄
16:42 ☽ □ ♂
22:07 ☿ ∥ ♃

13 01:57 ☽ ☍ ♃
04:45 ☿ ♋
08:15 ☽ ∥ ♄
08:22 ♀ △ ♂

14 05:08 ☽ ⚹ ♅
05:27 ☽ □ ♄
13:12 ☽ ∥ ♂
16:32 ☽ ∥ ♆
17:38 ☽ △ ♂
19:23 ☽ △ ♀

15 06:08 ☽ △ ☿
12:18 ☽ # ♅
14:25 ☽ ☌ ♆
22:08 ☽ ⚹ ♇

16 04:57 ☿ ∥ ☉
05:48 ☽ △ ♄
10:31 ♃ ♌
18:14 ☽ △ ☉

17 00:58 ☽ □ ♀
04:24 ☽ △ ♃
12:26 ☽ □ ☿

18 00:40 ☽ □ ♇
01:44 ☽ ∥ ♅
08:38 ☽ ☌ ♅
14:07 ♀ ♋
23:36 ☽ # ♆

19 02:19 ☽ ☍ ♂
05:37 ☿ △ ♆
06:32 ☉ □ ♂
09:18 ☽ # ♂
09:55 ☽ □ ♃
10:35 ☽ ⚹ ♀
21:50 ☽ ⚹ ♆
23:53 ☽ ⚹ ☿

20 06:31 ☽ △ ♇
09:10 ♃ # ♇
12:46 ☽ # ♄
15:22 ☽ ☍ ♄
20:37 ♄ SD

21 14:13 ☽ ⚹ ☉
18:53 ☽ ⚹ ♃
21:49 ☉ # ♇

22 02:54 ♅ SR
06:20 ☽ □ ♆
06:47 ☿ ☍ ♇
10:19 ☉ ∥ ♃
21:42 ☉ ♌

23 00:30 ☽ ⚹ ♅

24 00:54 ☽ △ ♂
11:10 ♀ △ ♆
17:04 ☽ △ ♆
17:44 ☽ ☌ ♀
20:44 ☉ ☌ ♃

25 00:08 ☿ □ ♅
02:11 ☿ △ ♄
02:24 ☽ ☍ ♇
11:50 ☽ □ ♅
12:09 ☽ △ ♄
13:55 ☽ ☌ ☿

26 02:25 ♂ ♏
15:30 ☽ □ ♂
19:35 ☽ ☌ ♃
21:13 ☽ # ♄

27 15:08 ☽ # ♂

28 00:15 ☽ △ ♅
00:38 ☽ □ ♄
06:38 ♀ ☍ ♇
14:26 ☽ # ♆

29 07:03 ☽ ⚹ ♂
16:20 ☽ ∥ ♅
17:46 ☽ ☍ ♆

30 03:16 ☽ △ ♇
08:26 ☽ ⚹ ♀
13:28 ☽ ⚹ ♄

31 14:48 ☽ ⚹ ☿
22:46 ☿ ♌
23:03 ☽ ⚹ ♃

⚷ Chiron

01	Dec.	00 S 14
01	17♓42 R	
04	17 41	
07	17 38	
10	17 35	
13	17 32	
16	17 28	
19	17 24	
22	17 19	
25	17 14	
28	17 09	
31	17 03	

August 2014

Day	S. T. h m s	☉ ° ' "	☽ ° ' "	☿ ° '	♀ ° '	♂ ° '	♃ ° '	♄ ° '	♅ ° '	♆ ° '	♇ ° '	☊ True ° '
01 Fr	20 38 07	08♌41 43	03♎55 46	00♌06	16♋14	03♏09	03♌27	16♏45	16♈R28	06♓R56	11♑R38	21♎R59
02 Sa	20 42 03	09 39 08	16 04 34	02 10	17 27	03 42	03 41	16 46	16 28	06 54	11 36	21 58
03 Su	20 45 60	10 36 33	28 27 29	04 15	18 39	04 15	03 54	16 47	16 27	06 53	11 35	21D 58
04 Mo	20 49 56	11 33 59	11♏09 13	06 20	19 52	04 48	04 07	16 48	16 26	06 52	11 34	21R 58
05 Tu	20 53 53	12 31 26	24 14 21	08 25	21 05	05 22	04 21	16 50	16 26	06 50	11 33	21 58
06 We	20 57 49	13 28 53	07♐46 43	10 30	22 18	05 55	04 34	16 51	16 25	06 49	11 31	21 55
07 Th	21 01 46	14 26 21	21 48 24	12 35	23 32	06 29	04 47	16 53	16 24	06 47	11 30	21 50
08 Fr	21 05 42	15 23 50	06♑18 51	14 39	24 45	07 03	05 00	16 55	16 24	06 46	11 29	21 43
09 Sa	21 09 39	16 21 20	21 14 01	16 42	25 58	07 37	05 14	16 57	16 23	06 44	11 28	21 34
10 Su	21 13 36	17 18 50	06♒26 21	18 45	27 11	08 12	05 27	16 58	16 22	06 43	11 27	21 25
11 Mo	21 17 32	18 16 22	21 45 37	20 46	28 24	08 46	05 40	17 00	16 21	06 41	11 25	21 16
12 Tu	21 21 29	19 13 54	07♓00 32	22 46	29 37	09 21	05 53	17 02	16 20	06 40	11 24	21 09
13 We	21 25 25	20 11 28	22 00 43	24 45	00♌51	09 56	06 06	17 05	16 19	06 38	11 23	21 03
14 Th	21 29 22	21 09 03	06♈38 22	26 43	02 04	10 31	06 19	17 07	16 18	06 36	11 22	21 00
15 Fr	21 33 18	22 06 40	20 49 05	28 40	03 17	11 07	06 32	17 09	16 17	06 35	11 21	20 59
16 Sa	21 37 15	23 04 18	04♉31 50	00♍35	04 31	11 42	06 45	17 12	16 16	06 33	11 20	20D 59
17 Su	21 41 11	24 01 58	17 48 09	02 28	05 44	12 18	06 58	17 14	16 15	06 32	11 19	21 00
18 Mo	21 45 08	24 59 39	00♊41 14	04 21	06 58	12 54	07 11	17 17	16 13	06 30	11 18	21 01
19 Tu	21 49 05	25 57 22	13 15 00	06 11	08 11	13 30	07 24	17 20	16 12	06 28	11 17	21R 00
20 We	21 53 01	26 55 07	25 33 30	08 01	09 25	14 06	07 37	17 22	16 11	06 27	11 16	20 57
21 Th	21 56 58	27 52 53	07♋40 39	09 49	10 39	14 43	07 50	17 25	16 09	06 25	11 15	20 52
22 Fr	22 00 54	28 50 40	19 39 54	11 35	11 52	15 19	08 03	17 28	16 08	06 24	11 14	20 45
23 Sa	22 04 51	29 48 30	01♌34 13	13 20	13 06	15 56	08 16	17 31	16 07	06 22	11 14	20 36
24 Su	22 08 47	00♍46 20	13 25 56	15 04	14 20	16 33	08 29	17 35	16 05	06 20	11 13	20 26
25 Mo	22 12 44	01 44 13	25 16 59	16 47	15 33	17 10	08 41	17 38	16 04	06 19	11 12	20 16
26 Tu	22 16 40	02 42 06	07♍09 01	18 28	16 47	17 48	08 54	17 41	16 02	06 17	11 11	20 08
27 We	22 20 37	03 40 02	19 03 42	20 08	18 01	18 25	09 07	17 45	16 01	06 15	11 10	20 02
28 Th	22 24 34	04 37 58	01♎02 56	21 46	19 15	19 03	09 20	17 48	15 59	06 14	11 09	19 58
29 Fr	22 28 30	05 35 56	13 08 58	23 23	20 29	19 40	09 32	17 52	15 57	06 12	11 09	19 56
30 Sa	22 32 27	06 33 56	25 24 30	24 59	21 43	20 18	09 45	17 55	15 56	06 10	11 08	19D 55
31 Su	22 36 23	07 31 57	07♏52 45	26 33	22 57	20 56	09 57	17 59	15 54	06 09	11 07	19 56

Data for	08-01-2014
Julian Day	2456870.50
Ayanamsa	24 03 46
SVP	05♓03 30
☽ ☊ Mean	23♎02 R

LAST ASPECT ☽ INGRESS

Day	h m	Day	h m	
02	02:58	03	02:57	♏
04	17:43	05	10:19	♐
06	14:52	07	13:39	♑
09	08:09	09	13:53	♒
10	22:13	11	12:56	♓
12	16:02	13	13:01	♈
15	15:51	15	15:59	♉
17	12:27	17	22:42	♊
20	02:54	20	08:46	♋
21	19:35	22	20:50	♌
24	08:26	25	09:33	♍
27	02:29	27	21:54	♎
29	16:00	30	08:53	♏

● ◐ PHASES ○ ◑

04	00:49	◐	11♏36
10	18:10	○	18♒02
17	12:27	◑	24♉32
25	14:12	●	02♍19

DECLINATION

Day	☉	☽	☿	♀	♂	♃	♄	♅	♆	♇
01 Fr	18N05	03S00	21N14	22N22	13S31	19N46	14S42	05N50	09S39	20S23
02 Sa	17 50	06 48	20 54	22 15	13 43	19 43	14 43	05 50	09 40	20 23
03 Su	17 34	10 23	20 32	22 08	13 55	19 40	14 44	05 50	09 40	20 24
04 Mo	17 19	13 35	20 07	22 01	14 07	19 37	14 44	05 50	09 41	20 24
05 Tu	17 03	16 11	19 39	21 52	14 19	19 34	14 45	05 50	09 42	20 24
06 We	16 46	18 00	19 10	21 43	14 31	19 31	14 45	05 49	09 42	20 24
07 Th	16 30	18 48	18 38	21 34	14 43	19 28	14 46	05 49	09 43	20 25
08 Fr	16 13	18 25	18 05	21 23	14 55	19 25	14 47	05 49	09 43	20 25
09 Sa	15 56	16 46	17 29	21 13	15 07	19 22	14 48	05 48	09 44	20 25
10 Su	15 39	13 56	16 53	21 01	15 19	19 19	14 48	05 48	09 44	20 25
11 Mo	15 21	10 07	16 14	20 49	15 31	19 16	14 49	05 48	09 45	20 26
12 Tu	15 03	05 39	15 35	20 36	15 42	19 13	14 50	05 47	09 46	20 26
13 We	14 45	00 54	14 54	20 23	15 54	19 09	14 51	05 47	09 46	20 26
14 Th	14 27	03N47	14 13	20 09	16 06	19 06	14 52	05 46	09 47	20 26
15 Fr	14 08	08 08	13 31	19 55	16 18	19 03	14 53	05 46	09 47	20 27
16 Sa	13 49	11 53	12 48	19 40	16 29	19 00	14 54	05 45	09 48	20 27
17 Su	13 30	14 55	12 04	19 24	16 41	18 57	14 55	05 45	09 49	20 27
18 Mo	13 11	17 05	11 20	19 08	16 52	18 54	14 56	05 44	09 49	20 27
19 Tu	12 52	18 22	10 35	18 51	17 04	18 50	14 57	05 44	09 50	20 28
20 We	12 32	18 45	09 50	18 34	17 15	18 47	14 58	05 43	09 50	20 28
21 Th	12 12	18 15	09 05	18 16	17 27	18 44	14 59	05 43	09 51	20 28
22 Fr	11 52	16 56	08 20	17 58	17 38	18 41	15 00	05 42	09 52	20 28
23 Sa	11 32	14 55	07 34	17 39	17 49	18 37	15 01	05 42	09 52	20 29
24 Su	11 12	12 17	06 49	17 20	18 01	18 34	15 02	05 41	09 53	20 29
25 Mo	10 51	09 09	06 03	17 00	18 12	18 31	15 03	05 41	09 54	20 29
26 Tu	10 31	05 38	05 18	16 40	18 23	18 28	15 04	05 40	09 54	20 29
27 We	10 10	01 54	04 32	16 19	18 34	18 25	15 05	05 39	09 55	20 29
28 Th	09 49	01S57	03 47	15 58	18 45	18 21	15 07	05 39	09 55	20 30
29 Fr	09 27	05 45	03 02	15 36	18 55	18 18	15 08	05 38	09 56	20 30
30 Sa	09 06	09 22	02 17	15 14	19 06	18 15	15 09	05 37	09 57	20 30
31 Su	08 45	12 37	01 33	14 52	19 17	18 12	15 10	05 37	09 57	20 30

ASPECTARIAN

Day	Time	Aspect
01	04:43	♀ □ ♅
	10:16	☽ ✶ ☉
	10:23	♀ △ ♄
	15:14	☽ □ ♇
	17:50	☽ # ♅
	22:47	♂ □ ♃
02	00:45	☽ ☍ ♅
	02:58	☽ □ ♀
	19:04	☽ ∥ ♆
	19:34	☿ ☌ ♃
03	00:02	☿ □ ♂
	07:47	☿ # ♇
	10:33	☽ □ ♃
	11:32	☽ ☌ ♂
	13:11	☽ □ ☿
	15:58	☽ △ ♆
04	00:46	☽ ✶ ♇
	04:51	☽ ∥ ♂
	09:53	☽ ∥ ♄
	10:29	☽ ☌ ♄
	17:43	☽ △ ♀
05	04:37	☿ ∥ ♃
	08:37	☽ # ☉
	18:17	☽ △ ♃
	22:19	☽ □ ♆
06	05:33	☽ △ ☿
	10:35	☽ △ ☉
	14:52	☽ △ ♅
	19:22	☽ # ☿
07	06:33	♂ ∥ ♄
	12:15	♂ △ ♆
08	00:44	☽ ✶ ♆
	01:15	☽ ✶ ♂
	08:22	☽ ☌ ♇
	11:03	☽ # ☿
	16:16	☽ □ ♅
	16:22	☿ ☌ ☉
	17:08	☽ ✶ ♄
	20:14	☿ △ ♅
09	00:36	☉ △ ♅
	02:50	☿ □ ♄
	08:09	☽ ☍ ♀
	09:04	☽ # ☉
	14:05	☽ ∥ ♂
	15:11	☉ □ ♄
	17:29	☽ ∥ ♄
	22:25	☽ ☍ ♃
10	02:52	☽ □ ♂
	15:32	☽ ✶ ♅
	16:10	☉ # ♂
	16:32	☽ □ ♄
	22:13	☽ ☍ ☿
11	02:04	☽ ∥ ♆
	20:32	☿ # ♂
	23:18	☽ # ♅
	23:27	☽ ☌ ♆
12	03:52	☽ △ ♂
	06:58	☽ ✶ ♇
	07:24	♀ ♌
	16:02	☽ △ ♄
	16:38	☉ # ♄
	18:32	♀ # ♇
13	02:01	☿ # ♄
	09:48	☿ ∥ ☉
	15:44	☽ △ ♀
	23:28	☽ △ ♃
14	07:55	☽ □ ♇
	10:34	☽ ∥ ♅
	16:15	☽ ☌ ♅
15	02:24	☽ △ ☉
	09:26	♂ ✶ ♇
	10:02	☽ # ♆
	15:51	☽ △ ☿
	16:44	☿ ♍
	23:58	☽ □ ♀
16	03:36	☽ ✶ ♆
	04:02	☽ □ ♃
	05:17	☽ ∥ ☿
	12:12	☽ △ ♇
	13:08	☽ ∥ ☉
	13:29	☽ ☍ ♂
	22:58	☽ ☍ ♄
	23:55	☽ # ♄
17	20:44	☽ # ♂
18	05:22	♀ ☌ ♃
	08:07	☽ □ ☿
	11:01	☽ □ ♆
	12:34	☽ ✶ ♃
	13:13	☽ ✶ ♀
19	01:43	♀ ∥ ♃
	03:41	☿ ☍ ♆
	05:42	☽ ✶ ♅
	13:34	☽ ∥ ♀
	23:51	☿ # ♆
20	02:54	☽ ✶ ☉
	21:30	☽ △ ♆
	23:11	☽ ∥ ♀
21	05:00	☽ ✶ ☿
	07:08	☽ ☍ ♇
	14:23	☽ # ♂
	14:49	☽ △ ♂
	16:55	☽ □ ♅
	19:20	☿ △ ♇
	19:35	☽ △ ♄
22	15:53	♀ # ♂
	23:03	☽ # ♄
23	04:46	☉ ♍
	13:47	☽ ☌ ♃
24	02:01	☽ ☌ ♀
	05:22	☽ △ ♅
	06:40	☽ □ ♂
	08:26	☽ □ ♄
	09:44	☽ ∥ ☉
	18:35	☽ # ♆
25	08:47	☿ ✶ ♂
	09:38	♀ △ ♅
	12:03	☿ ∥ ♅
	[illegible]	☿ ✶ ♄
	19:30	♂ ☌ ♄
	22:15	☽ ☍ ♆
	23:49	☽ ∥ ♅
26	02:50	☽ ∥ ☿
	08:08	☽ △ ♇
	08:22	♂ # ♃
	18:22	♀ □ ♄
	21:20	☽ ✶ ♄
	22:38	☽ ✶ ♂
27	02:29	☽ ☌ ☿
	15:46	♀ □ ♂
	16:33	☉ # ♆
28	09:37	☽ # ☿
	[illegible]	[illegible]
	20:03	☽ □ ♇
	23:16	☽ # ♅
29	05:31	☽ ☍ ♅
	14:33	☉ ☍ ♆
	16:00	☽ ✶ ♀
	22:23	☽ # ☉
30	04:06	☽ ∥ ♆
	05:17	♀ # ♄
	20:42	☽ △ ♆
	23:17	☽ ✶ ☉
31	04:01	☽ □ ♃
	06:09	☽ ✶ ♇
	16:49	☽ # ♀
	[illegible]	[illegible]
	22:28	☽ ∥ ♄

⚷ Chiron

01 Dec.	00 S 25
03	16♓57R
06	16 50
09	16 43
12	16 36
15	16 29
18	16 21
21	16 13
24	16 05
27	15 57
30	15 49

September 2014

Day	S. T. h m s	☉ ° ' "	☽ ° ' "	☿ ° '	♀ ° '	♂ ° '	♃ ° '	♄ ° '	♅ ° '	♆ ° '	♇ ° '	☊ True ° '
01 Mo	22 40 20	08♍ 29 59	20♏ 37 11	28♍ 07	24♌ 11	21♏ 34	10♌ 10	18♏ 03	15♈R52	06♓R07	11♑R07	19♎ 58
02 Tu	22 44 16	09 28 03	03♐ 41 27	29 39	25 25	22 13	10 22	18 07	15 50	06 06	11 06	19 58
03 We	22 48 13	10 26 08	17 08 44	01♎ 09	26 39	22 51	10 34	18 11	15 49	06 04	11 06	19R 58
04 Th	22 52 09	11 24 14	01♑ 01 13	02 39	27 53	23 30	10 47	18 15	15 47	06 02	11 05	19 56
05 Fr	22 56 06	12 22 22	15 19 09	04 07	29 07	24 09	10 59	18 19	15 45	06 01	11 04	19 53
06 Sa	23 00 03	13 20 31	00♒ 00 12	05 33	00♍ 21	24 47	11 11	18 23	15 43	05 59	11 04	19 48
07 Su	23 03 59	14 18 42	14 59 04	06 59	01 35	25 27	11 23	18 27	15 41	05 57	11 03	19 43
08 Mo	23 07 56	15 16 54	00♓ 07 48	08 23	02 50	26 06	11 36	18 32	15 39	05 56	11 03	19 38
09 Tu	23 11 52	16 15 08	15 16 45	09 45	04 04	26 45	11 48	18 36	15 37	05 54	11 03	19 34
10 We	23 15 49	17 13 24	00♈ 16 05	11 07	05 18	27 24	12 00	18 41	15 35	05 52	11 02	19 31
11 Th	23 19 45	18 11 41	14 57 22	12 27	06 32	28 04	12 12	18 45	15 33	05 51	11 02	19 30
12 Fr	23 23 42	19 10 00	29 14 41	13 45	07 47	28 44	12 23	18 50	15 31	05 49	11 01	19D 30
13 Sa	23 27 38	20 08 22	13♉ 05 09	15 02	09 01	29 23	12 35	18 55	15 29	05 48	11 01	19 31
14 Su	23 31 35	21 06 46	26 28 36	16 17	10 16	00♐ 03	12 47	19 00	15 27	05 46	11 01	19 33
15 Mo	23 35 31	22 05 11	09♊ 27 02	17 31	11 30	00 43	12 59	19 04	15 24	05 45	11 01	19 34
16 Tu	23 39 28	23 03 39	22 03 47	18 42	12 44	01 24	13 10	19 09	15 22	05 43	11 00	19 35
17 We	23 43 25	24 02 09	04♋ 22 56	19 52	13 59	02 04	13 22	19 14	15 20	05 41	11 00	19R 35
18 Th	23 47 21	25 00 41	16 28 49	21 01	15 14	02 44	13 33	19 19	15 18	05 40	11 00	19 34
19 Fr	23 51 18	25 59 16	28 25 46	22 07	16 28	03 25	13 45	19 25	15 16	05 38	11 00	19 32
20 Sa	23 55 14	26 57 52	10♌ 17 46	23 11	17 43	04 06	13 56	19 30	15 13	05 37	11 00	19 29
21 Su	23 59 11	27 56 31	22 08 21	24 12	18 57	04 46	14 07	19 35	15 11	05 35	11 00	19 26
22 Mo	00 03 07	28 55 11	04♍ 00 30	25 11	20 12	05 27	14 19	19 40	15 09	05 34	11 00	19 22
23 Tu	00 07 04	29 53 54	15 56 39	26 08	21 27	06 08	14 30	19 46	15 06	05 32	11D 00	19 19
24 We	00 11 00	00♎ 52 39	27 58 49	27 01	22 41	06 50	14 41	19 51	15 04	05 31	11 00	19 17
25 Th	00 14 57	01 51 25	10♎ 08 44	27 52	23 56	07 31	14 52	19 57	15 02	05 30	11 00	19 16
26 Fr	00 18 54	02 50 14	22 28 00	28 39	25 11	08 12	15 03	20 02	14 59	05 28	11 00	19D 16
27 Sa	00 22 50	03 49 04	04♏ 58 07	29 22	26 25	08 54	15 13	20 08	14 57	05 27	11 00	19 16
28 Su	00 26 47	04 47 57	17 40 40	00♏ 02	27 40	09 35	15 24	20 13	14 55	05 25	11 00	19 17
29 Mo	00 30 43	05 46 51	00♐ 37 18	00 37	28 55	10 17	15 35	20 19	14 52	05 24	11 00	19 18
30 Tu	00 34 40	06 45 47	13 49 43	01 08	00♎ 10	10 59	15 45	20 25	14 50	05 23	11 00	19 18

Data for 09-01-2014
Julian Day 2456901.50
Ayanamsa 24 03 49
SVP 05 ♓ 03 25
☽ ☊ Mean 21 ♎ 24 R

● ◐ PHASES ○ ◑

Day	Time	Phase	Position
02	11:11	◐	09♐55
09	01:39	○	16♓19
16	02:05	◑	23♊09
24	06:14	●	01♎08

LAST ASPECT Day	h m	☽ INGRESS Day	h m	Sign
01	15:40	01	17:17	♐
03	18:07	03	22:16	♑
05	15:09	06	00:00	♒
07	17:20	07	23:48	♓
09	19:10	09	23:34	♈
11	00:59	12	01:17	♉
13	13:32	14	06:27	♊
16	02:05	16	15:24	♋
18	18:38	19	03:10	♌
21	04:34	21	15:54	♍
23	12:16	24	04:00	♎
26	12:40	26	14:30	♏
28	20:32	28	22:51	♐

DECLINATION

Day	☉	☽	☿	♀	♂	♃	♄	♅	♆	♇
01 Mo	08N23	15S 20	00N49	14N29	19S 27	18N08	15S 12	05N36	09S 58	20S 31
02 Tu	08 01	17 21	00 05	14 05	19 38	18 05	15 13	05 35	09 58	20 31
03 We	07 39	18 28	00S 38	13 42	19 48	18 02	15 14	05 35	09 59	20 31
04 Th	07 17	18 31	01 21	13 18	19 58	17 59	15 15	05 34	10 00	20 31
05 Fr	06 55	17 25	02 03	12 53	20 08	17 55	15 17	05 33	10 00	20 31
06 Sa	06 33	15 09	02 45	12 28	20 18	17 52	15 18	05 32	10 01	20 32
07 Su	06 10	11 50	03 27	12 03	20 28	17 49	15 20	05 32	10 02	20 32
08 Mo	05 48	07 43	04 08	11 38	20 38	17 46	15 21	05 31	10 02	20 32
09 Tu	05 25	03 05	04 48	11 12	20 47	17 42	15 22	05 30	10 03	20 32
10 We	05 03	01N41	05 27	10 46	20 57	17 39	15 24	05 29	10 03	20 32
11 Th	04 40	06 16	06 06	10 19	21 06	17 36	15 25	05 28	10 04	20 33
12 Fr	04 17	10 22	06 45	09 52	21 16	17 33	15 27	05 28	10 04	20 33
13 Sa	03 54	13 46	07 22	09 25	21 25	17 30	15 28	05 27	10 05	20 33
14 Su	03 31	16 18	07 59	08 58	21 34	17 26	15 30	05 26	10 06	20 33
15 Mo	03 08	17 54	08 35	08 30	21 42	17 23	15 31	05 25	10 06	20 33
16 Tu	02 45	18 34	09 10	08 03	21 51	17 20	15 33	05 24	10 07	20 34
17 We	02 22	18 19	09 44	07 35	21 59	17 17	15 34	05 23	10 07	20 34
18 Th	01 59	17 13	10 17	07 06	22 08	17 14	15 36	05 23	10 08	20 34
19 Fr	01 36	15 24	10 49	06 38	22 16	17 11	15 37	05 22	10 09	20 34
20 Sa	01 12	12 56	11 20	06 09	22 24	17 07	15 39	05 21	10 09	20 34
21 Su	00 49	09 56	11 49	05 40	22 32	17 04	15 41	05 20	10 10	20 35
22 Mo	00 26	06 33	12 18	05 11	22 39	17 01	15 42	05 19	10 10	20 35
23 Tu	00 02	02 52	12 45	04 42	22 47	16 58	15 44	05 18	10 11	20 35
24 We	00S 21	00S 57	13 10	04 13	22 54	16 55	15 45	05 17	10 11	20 35
25 Th	00 44	04 47	13 35	03 44	23 01	16 52	15 47	05 16	10 12	20 35
26 Fr	01 08	08 28	13 57	03 14	23 08	16 49	15 49	05 15	10 12	20 35
27 Sa	01 31	11 50	14 18	02 44	23 15	16 46	15 50	05 15	10 13	20 36
28 Su	01 54	14 41	14 37	02 15	23 22	16 43	15 52	05 14	10 13	20 36
29 Mo	02 18	16 51	14 53	01 45	23 28	16 40	15 54	05 13	10 14	20 36
30 Tu	02 41	18 10	15 08	01 15	23 34	16 37	15 55	05 12	10 14	20 36

ASPECTARIAN

01 01:52 ☽ ☌ ♂
07:18 ☽ □ ♀
15:40 ☽ ⚹ ☿

02 04:20 ☽ □ ♆
05:38 ☿ ♎
12:11 ☽ △ ♃
12:25 ☽ # ♃
21:39 ☽ △ ♅

03 16:09 ☉ △ ♇
18:07 ☽ △ ♀

04 03:05 ☽ □ ☿
08:29 ☽ ⚹ ♆
15:34 ☽ # ♃
16:57 ☽ ☌ ♇
18:45 ☽ △ ☉

05 00:42 ☽ □ ♅
04:58 ☽ ⚹ ♄
15:09 ☽ ⚹ ♂
17:07 ♀ ♍
22:44 ☽ ∥ ♄

06 09:53 ☽ △ ☿
18:12 ☽ ☍ ♃
22:31 ☽ # ♀

07 01:06 ☽ ⚹ ♅
05:33 ☽ □ ♄
09:18 ♂ ∥ ♇
11:03 ☽ ∥ ♆
17:20 ☽ □ ♂

08 04:39 ☽ ☍ ♀
09:09 ☽ ☌ ♆
11:05 ☽ # ☉
11:42 ☽ # ♅
16:26 ☽ ∥ ☿
17:17 ☽ ⚹ ♇
18:53 ☉ ∥ ♅

09 05:19 ☽ △ ♄
14:27 ☿ # ☉
19:10 ☽ △ ♂
22:39 ☿ □ ♇

10 01:05 ☿ # ♅
10:51 ♀ ☍ ♆
16:03 ☽ ∥ ☉
17:32 ☽ □ ♇
18:38 ☿ ⚹ ♃
19:22 ☽ △ ♃
19:26 ☽ ☍ ☿
19:41 ☽ ∥ ♅
22:56 ☽ # ☿

11 00:59 ☽ ☌ ♅
13:21 ♀ # ♆
15:04 ☉ ⚹ ♄
21:11 ☽ ∥ ♀
22:08 ☽ # ♆

12 11:17 ☽ ⚹ ♆
16:10 ☽ △ ♀
20:22 ☽ △ ♇
23:06 ☽ □ ♃

13 08:17 ☿ ☍ ♅
10:24 ☽ ☍ ♄
13:32 ☽ △ ☉
15:19 ☽ # ♄
21:57 ♂ ♐

14 06:54 ☽ ☍ ♂
14:34 ♀ △ ♇
14:59 ☽ ∥ ♃
17:05 ☽ □ ♆
22:26 ☿ # ♀

15 04:16 ☽ □ ♀
06:45 ☽ ⚹ ♃
11:13 ☽ ⚹ ♅
16:53 ☽ △ ☿

17 02:34 ☽ △ ♆
13:05 ☽ ☍ ♇
17:31 ☿ ∥ ♆
21:12 ☽ ⚹ ♀
21:39 ☽ □ ♅
23:59 ☽ ∥ ♃

18 05:44 ☽ △ ♄
10:00 ☽ □ ☿
18:38 ☽ ⚹ ☉
21:33 ☽ # ♄

19 10:41 ☽ △ ♂

20 07:30 ☽ ☌ ♃
09:57 ☽ △ ♅
11:31 ☽ # ☿
18:47 ☽ □ ♄
22:26 ☽ # ♆

21 04:34 ☽ ⚹ ☿
13:04 ♀ ⚹ ♄
17:30 ♀ ∥ ♅

22 03:06 ☽ □ ♂
03:08 ☽ ☍ ♆
03:41 ♂ □ ♆
08:17 ☽ ∥ ♅
10:28 ☽ ∥ ♀
14:04 ☽ △ ♇

23 00:35 ♇ SD
02:30 ☉ ♎
07:42 ☽ ⚹ ♄
12:16 ☽ ☌ ♀
16:39 ☽ # ☉
19:50 ☽ ∥ ☉

24 18:05 ☽ # ♀
18:32 ☽ ⚹ ♂

25 01:40 ☽ □ ♇
03:06 ☽ # ♅
09:22 ☽ ⚹ ♃
09:31 ☽ ☍ ♅
18:18 ♃ △ ♅

26 12:06 ☽ ∥ ♆
12:40 ☽ ☌ ☿

27 00:54 ☽ △ ♆
11:26 ☽ ⚹ ♇
19:40 ☽ □ ♃
22:40 ☿ ♏
23:12 ☽ ∥ ☿

28 04:48 ☽ ☌ ♄
09:06 ♀ # ☉
12:09 ☽ ∥ ♄
20:32 ☽ ⚹ ♀
21:33 ☽ # ♃

29 08:43 ☽ □ ♆
10:12 ☽ ⚹ ☉
18:35 ☽ ☌ ♂
20:53 ♀ ♎

30 01:48 ☽ △ ♅
03:30 ☽ △ ♃

⚷ Chiron

01 Dec.	00 S 54
02	15♓41R
05	15 32
08	15 24
11	15 15
14	15 07
17	14 59
20	14 50
23	14 42
26	14 34
29	14 26

October 2014

08	10:56	15♈05	✷	Total Lunar Eclipse (mag 1.171)
23	21:46	00♏25	✪	Solar Eclipse (mag 0.812)

Day	S. T. h m s	☉ ° ' "	☽ ° ' "	☿ ° '	♀ ° '	♂ ° '	♃ ° '	♄ ° '	♅ ° '	♆ ° '	♇ ° '	☊ True ° '
01 We	00 38 36	07♎44 45	27♐19 24	01♏34	01♎25	11♐41	15♌56	20♏31	14♈R47	05♓R21	11♑01	19♎R18
02 Th	00 42 33	08 43 45	11♑07 22	01 54	02 39	12 23	16 06	20 37	14 45	05 20	11 01	19 18
03 Fr	00 46 29	09 42 46	25 13 35	02 09	03 54	13 05	16 16	20 43	14 43	05 19	11 01	19 17
04 Sa	00 50 26	10 41 49	09♒36 39	02 17	05 09	13 47	16 26	20 49	14 40	05 17	11 01	19 17
05 Su	00 54 23	11 40 53	24 13 24	02R 18	06 24	14 30	16 36	20 55	14 38	05 16	11 02	19 16
06 Mo	00 58 19	12 40 00	08♓58 51	02 13	07 39	15 12	16 46	21 01	14 35	05 15	11 02	19 15
07 Tu	01 02 16	13 39 08	23 46 37	02 00	08 54	15 55	16 56	21 07	14 33	05 14	11 03	19 14
08 We	01 06 12	14 38 18	08♈29 28	01 38	10 09	16 37	17 06	21 13	14 31	05 12	11 03	19 14
09 Th	01 10 09	15 37 30	23 00 20	01 09	11 24	17 20	17 16	21 19	14 28	05 11	11 04	19D 14
10 Fr	01 14 05	16 36 44	07♉13 18	00 32	12 39	18 03	17 25	21 26	14 26	05 10	11 04	19 15
11 Sa	01 18 02	17 36 01	21 04 13	29♎47	13 54	18 46	17 35	21 32	14 23	05 09	11 05	19 15
12 Su	01 21 58	18 35 19	04♊31 08	28 54	15 09	19 29	17 44	21 38	14 21	05 08	11 05	19 15
13 Mo	01 25 55	19 34 40	17 34 14	27 55	16 24	20 12	17 53	21 45	14 18	05 07	11 06	19 15
14 Tu	01 29 51	20 34 04	00♋15 22	26 50	17 39	20 55	18 02	21 51	14 16	05 06	11 06	19R 15
15 We	01 33 48	21 33 29	12 37 43	25 40	18 54	21 38	18 11	21 57	14 14	05 05	11 07	19D 15
16 Th	01 37 45	22 32 57	24 45 20	24 27	20 09	22 22	18 20	22 04	14 11	05 04	11 08	19 15
17 Fr	01 41 41	23 32 27	06♌42 42	23 14	21 24	23 05	18 29	22 11	14 09	05 03	11 08	19 16
18 Sa	01 45 38	24 31 59	18 34 30	22 02	22 39	23 49	18 38	22 17	14 06	05 02	11 09	19 16
19 Su	01 49 34	25 31 34	00♍25 16	20 53	23 54	24 32	18 46	22 24	14 04	05 01	11 10	19 17
20 Mo	01 53 31	26 31 11	12 19 13	19 50	25 09	25 16	18 54	22 30	14 02	05 00	11 11	19 18
21 Tu	01 57 27	27 30 50	24 19 58	18 54	26 24	26 00	19 03	22 37	13 59	04 59	11 12	19 19
22 We	02 01 24	28 30 31	06♎30 32	18 07	27 39	26 44	19 11	22 44	13 57	04 58	11 12	19 20
23 Th	02 05 20	29 30 14	18 53 12	17 30	28 55	27 28	19 19	22 51	13 55	04 58	11 13	19R 21
24 Fr	02 09 17	00♏29 59	01♏29 30	17 04	00♏10	28 12	19 27	22 57	13 52	04 57	11 14	19 20
25 Sa	02 13 14	01 29 46	14 20 09	16 49	01 25	28 56	19 34	23 04	13 50	04 56	11 15	19 19
26 Su	02 17 10	02 29 36	27 25 11	16D 46	02 40	29 40	19 42	23 11	13 48	04 55	11 16	19 17
27 Mo	02 21 07	03 29 27	10♐44 03	16 54	03 55	00♑25	19 50	23 18	13 46	04 55	11 17	19 15
28 Tu	02 25 03	04 29 20	24 15 50	17 12	05 11	01 09	19 57	23 25	13 43	04 54	11 18	19 12
29 We	02 28 60	05 29 14	07♑59 25	17 40	06 26	01 53	20 04	23 32	13 41	04 53	11 19	19 10
30 Th	02 32 56	06 29 10	21 53 32	18 18	07 41	02 38	20 11	23 39	13 39	04 53	11 20	19 09
31 Fr	02 36 53	07 29 08	05♒56 48	19 04	08 56	03 22	20 18	23 45	13 37	04 52	11 22	19 08

Data for	10-01-2014
Julian Day	2456931.50
Ayanamsa	24 03 51
SVP	05 ♓ 03 19
☽ ☊ Mean	19 ♎ 48 R

● ◐ PHASES ○ ◑

01	19:33	◐	08♑33
08	10:50	✷	15♈05
15	19:12	◑	22♋21
23	21:58	✪	00♏25
31	02:48	◐	07♒36

LAST ASPECT Day	h m	☽ INGRESS Day	h m	
30	03:30	01	04:42	♑
02	16:19	03	08:01	♒
04	18:33	05	09:25	♓
06	19:39	07	10:07	♈
08	14:20	09	11:44	♉
11	00:49	11	15:51	♊
13	17:59	13	23:31	♋
15	23:27	16	10:30	♌
18	13:11	18	23:09	♍
21	03:31	21	11:13	♎
23	17:23	23	21:11	♏
25	16:12	26	04:41	♐
27	16:19	28	10:04	♑
30	03:02	30	13:52	♒

DECLINATION

Day	☉	☽	☿	♀	♂	♃	♄	♅	♆	♇
01 We	03S 04	18S 30	15S 20	00N45	23S 40	16N34	15S 57	05N11	10S 15	20S 36
02 Th	03 28	17 45	15 29	00 15	23 46	16 31	15 59	05 10	10 15	20 36
03 Fr	03 51	15 55	15 35	00S 15	23 51	16 28	16 00	05 09	10 16	20 37
04 Sa	04 14	13 04	15 39	00 45	23 57	16 25	16 02	05 08	10 16	20 37
05 Su	04 37	09 21	15 38	01 15	24 02	16 22	16 04	05 07	10 17	20 37
06 Mo	05 00	05 01	15 35	01 46	24 07	16 19	16 06	05 06	10 17	20 37
07 Tu	05 23	00 22	15 27	02 16	24 12	16 17	16 07	05 05	10 18	20 37
08 We	05 46	04N17	15 15	02 46	24 16	16 14	16 09	05 04	10 18	20 37
09 Th	06 09	08 38	14 59	03 16	24 20	16 11	16 11	05 03	10 18	20 37
10 Fr	06 32	12 23	14 38	03 46	24 25	16 08	16 13	05 02	10 19	20 37
11 Sa	06 54	15 20	14 13	04 16	24 28	16 06	16 14	05 02	10 19	20 38
12 Su	07 17	17 21	13 44	04 45	24 32	16 03	16 16	05 01	10 20	20 38
13 Mo	07 39	18 22	13 10	05 15	24 35	16 00	16 18	05 00	10 20	20 38
14 Tu	08 02	18 26	12 32	05 45	24 38	15 58	16 20	04 59	10 20	20 38
15 We	08 24	17 37	11 51	06 14	24 41	15 55	16 21	04 58	10 21	20 38
16 Th	08 46	16 00	11 07	06 44	24 44	15 52	16 23	04 57	10 21	20 38
17 Fr	09 08	13 42	10 22	07 13	24 46	15 50	16 25	04 56	10 21	20 38
18 Sa	09 30	10 52	09 36	07 42	24 49	15 47	16 27	04 55	10 22	20 38
19 Su	09 52	07 35	08 51	08 11	24 51	15 45	16 29	04 54	10 22	20 39
20 Mo	10 14	03 59	08 08	08 40	24 52	15 43	16 30	04 53	10 22	20 39
21 Tu	10 35	00 12	07 28	09 08	24 54	15 40	16 32	04 52	10 23	20 39
22 We	10 57	03S 40	06 52	09 37	24 55	15 38	16 34	04 51	10 23	20 39
23 Th	11 18	07 26	06 21	10 05	24 56	15 35	16 36	04 51	10 23	20 39
24 Fr	11 39	10 57	05 56	10 33	24 56	15 33	16 38	04 50	10 24	20 39
25 Sa	12 00	14 00	05 36	11 00	24 57	15 31	16 39	04 49	10 24	20 39
26 Su	12 20	16 24	05 22	11 28	24 57	15 29	16 41	04 48	10 24	20 39
27 Mo	12 41	17 58	05 14	11 55	24 57	15 26	16 43	04 47	10 24	20 39
28 Tu	13 01	18 33	05 11	12 22	24 57	15 24	16 45	04 46	10 25	20 39
29 We	13 21	18 03	05 14	12 49	24 56	15 22	16 47	04 45	10 25	20 39
30 Th	13 41	16 28	05 22	13 15	24 55	15 20	16 48	04 45	10 25	20 39
31 Fr	14 00	13 54	05 34	13 41	24 54	15 18	16 50	04 44	10 25	20 40

ASPECTARIAN

01 07:39 ☽ ✱ ☿
07:53 ☽ □ ♀
14:01 ☽ ✱ ♆
23:49 ☽ ☌ ♇
02 06:12 ☽ □ ♅
16:19 ☽ ✱ ♄
17:57 ☽ # ♃
23:05 ☽ ∥ ♄

03 03:09 ☽ ∥ ☿
11:44 ☽ □ ☿
15:54 ☽ △ ♀
04 01:56 ☽ △ ☉
07:15 ☽ ✱ ♂
08:04 ☉ □ ♇
08:19 ☽ ✱ ♅
11:23 ☽ ☍ ♃
17:03 ☿ SR
18:27 ☽ ∥ ♆
18:33 ☽ □ ♄

05 04:18 ♂ △ ♅
13:06 ☽ △ ☿
17:57 ☽ ☌ ♆
22:36 ☽ # ♅
06 00:07 ☽ ∥ ☉
03:20 ☽ ✱ ♇
06:01 ☉ # ♅
10:36 ☽ □ ♂
15:19 ☽ ∥ ♀
19:39 ☽ △ ♄

07 15:09 ☽ # ♀
20:58 ☉ ☍ ♅
08 02:58 ☽ ☍ ♀
04:12 ☽ □ ♇
04:14 ☽ ∥ ♅
08:45 ☽ # ☉
09:53 ☽ ☌ ♅

14:05 ☽ △ ♂
14:20 ☽ △ ♃
17:32 ♀ □ ♇
20:44 ♂ △ ♃

09 01:31 ♃ # ♄
10:17 ☽ # ♆
13:08 ☽ ☍ ☿
20:30 ☽ ✱ ♆
10 06:36 ☽ △ ♇
15:30 ☽ # ☿
17:27 ☿ ♎R
17:49 ☽ □ ♃
23:18 ☉ ✱ ♃

11 00:49 ☽ ☍ ♄
07:46 ☽ ∥ ♃
09:10 ♀ ☍ ♅
09:40 ☽ # ♄
12 01:07 ☽ □ ♆
11:53 ♀ # ♅
17:57 ☽ ✱ ♅
21:34 ☽ △ ♀

13 00:36 ☽ ✱ ♃
04:04 ☽ △ ☉
05:13 ☽ ☍ ♂
17:59 ☽ △ ☿
14 08:33 ♀ ✱ ♃
09:19 ☽ △ ♆
21:02 ☽ ☍ ♇

15 03:08 ☽ □ ♅
[illegible]
13:46 ☽ □ ♀

18:36 ☽ △ ♄
19:02 ☽ # ♄
23:27 ☽ □ ☿
16 01:23 ☽ ∥ ♃
20:40 ☿ ☌ ☉

17 00:07 ☿ ∥ ♆
01:49 ☿ ✱ ♂
14:59 ☽ △ ♅
17:56 ☿ ☌ ♀

18 00:06 ☽ ☌ ♃
02:01 ☿ ∥ ☉
03:51 ☽ # ♆
06:23 ☽ ✱ ☿
07:35 ☽ □ ♄
09:13 ☽ # ☉
09:14 ☽ ✱ ♀
11:19 ☽ △ ♂
12:28 ☽ # ☿
13:11 ☽ ✱ ☉
20:20 ☽ # ♀

19 09:17 ☽ ☍ ♆
13:18 ☿ ∥ ♀
18:09 ☽ ∥ ♅
21:42 ☽ △ ♇
20 05:19 ♀ ✱ ♂
09:50 ☉ ∥ ♆
20:34 ☽ ✱ ♄
20:38 ☿ ✱ ♃

21 03:31 ☽ □ ♂
[illegible]
09:10 ☽ □ ♇

14:26 ☽ ☍ ♅
17:50 ☽ ∥ ☿
21:27 ☽ ☌ ☿

23 00:50 ☽ ✱ ♃
11:57 ☉ ♏
15:58 ♀ ∥ ♆
17:23 ☽ ✱ ♂
20:04 ☽ ∥ ♆
20:41 ☽ ∥ ♀
20:53 ♀ ♏
21:13 ☽ ☌ ♀
24 05:46 ☽ ∥ ☉
[illegible]
18:16 ☽ ✱ ♇

25 07:31 ♀ ☌ ☉
09:45 ☽ □ ♃
14:00 ☽ # ♃
16:12 ☽ ☌ ♄
19:18 ☿ SD
26 03:33 ☽ ∥ ♄
10:43 ♂ ♑
13:34 ☽ □ ♆

27 05:23 ☽ △ ♅
11:11 ☽ ✱ ☿
16:19 ☽ △ ♃
18:48 ♀ △ ♆
[illegible]
12:46 ☽ ☌ ♂

18:37 ☽ ✱ ♆
19:18 ☽ ✱ ☉
21:01 ☽ ✱ ♀
29 05:47 ☽ ☌ ♇
09:50 ☽ □ ♅
17:30 ☽ □ ☿
19:59 ☽ ∥ ♄

30 03:02 ☽ ✱ ♄
11:46 ☽ # ♃
23:13 ☽ ∥ ☉
31 01:31 ☽ ∥ ♀
05:34 ☽ □ ♀
12:58 ☽ ✱ ♅
23:41 ☽ △ ☿

⚷ Chiron

01 Dec.	01 S 28
02	14♓19R
05	14 12
08	14 05
11	13 58
14	13 51
17	13 45
20	13 40
23	13 34
26	13 29
29	13 25

November 2014

Day	S. T.	☉	☽	☿	♀	♂	♃	♄	♅	♆	♇	☊ True
	h m s	° ' "	° ' "	° '	° '	° '	° '	° '	° '	° '	° '	° '
01 Sa	02 40 49	08♏ 29 07	20♒ 07 35	19♎ 57	10♏ 11	04♑ 07	20♌ 25	23♏ 52	13♈R35	04♓R52	11♑ 23	19♎D08
02 Su	02 44 46	09 29 08	04♓ 23 55	20 57	11 27	04 52	20 31	23 59	13 32	04 51	11 24	19 09
03 Mo	02 48 43	10 29 11	18 43 17	22 03	12 42	05 37	20 38	24 06	13 30	04 51	11 25	19 10
04 Tu	02 52 39	11 29 15	03♈ 02 35	23 14	13 57	06 22	20 44	24 13	13 28	04 50	11 26	19 11
05 We	02 56 36	12 29 20	17 18 06	24 30	15 12	07 06	20 50	24 21	13 26	04 50	11 28	19 12
06 Th	03 00 32	13 29 28	01♉ 25 40	25 49	16 28	07 51	20 56	24 28	13 24	04 50	11 29	19R 12
07 Fr	03 04 29	14 29 37	15 21 08	27 11	17 43	08 36	21 02	24 35	13 22	04 49	11 30	19 10
08 Sa	03 08 25	15 29 48	29 00 46	28 36	18 58	09 22	21 08	24 42	13 20	04 49	11 31	19 08
09 Su	03 12 22	16 30 01	12♊ 21 55	00♏ 03	20 13	10 07	21 14	24 49	13 18	04 49	11 33	19 04
10 Mo	03 16 18	17 30 16	25 23 16	01 32	21 29	10 52	21 19	24 56	13 16	04 49	11 34	18 59
11 Tu	03 20 15	18 30 32	08♋ 05 05	03 03	22 44	11 37	21 24	25 03	13 15	04 48	11 36	18 54
12 We	03 24 12	19 30 51	20 29 04	04 34	23 59	12 23	21 29	25 10	13 13	04 48	11 37	18 50
13 Th	03 28 08	20 31 11	02♌ 38 11	06 07	25 14	13 08	21 34	25 17	13 11	04 48	11 38	18 47
14 Fr	03 32 05	21 31 34	14 36 21	07 40	26 30	13 54	21 39	25 24	13 09	04 48	11 40	18 46
15 Sa	03 36 01	22 31 58	26 28 12	09 14	27 45	14 39	21 43	25 32	13 08	04 48	11 41	18D 46
16 Su	03 39 58	23 32 24	08♍ 18 44	10 49	29 00	15 25	21 48	25 39	13 06	04 48	11 43	18 47
17 Mo	03 43 54	24 32 52	20 13 02	12 24	00♐ 15	16 10	21 52	25 46	13 04	04D 48	11 44	18 49
18 Tu	03 47 51	25 33 22	02♎ 15 56	13 59	01 31	16 56	21 56	25 53	13 03	04 48	11 46	18 50
19 We	03 51 47	26 33 53	14 31 47	15 34	02 46	17 42	22 00	26 00	13 01	04 48	11 48	18 51
20 Th	03 55 44	27 34 27	27 04 06	17 09	04 01	18 28	22 03	26 07	12 59	04 48	11 49	18R 51
21 Fr	03 59 41	28 35 02	09♏ 55 16	18 45	05 17	19 13	22 07	26 14	12 58	04 48	11 51	18 48
22 Sa	04 03 37	29 35 38	23 06 11	20 20	06 32	19 59	22 10	26 22	12 56	04 48	11 52	18 44
23 Su	04 07 34	00♐ 36 16	06♐ 36 10	21 55	07 47	20 45	22 13	26 29	12 55	04 49	11 54	18 38
24 Mo	04 11 30	01 36 56	20 22 51	23 30	09 03	21 31	22 16	26 36	12 54	04 49	11 56	18 31
25 Tu	04 15 27	02 37 37	04♑ 22 37	25 06	10 18	22 17	22 19	26 43	12 52	04 49	11 57	18 23
26 We	04 19 23	03 38 19	18 31 08	26 41	11 33	23 03	22 22	26 50	12 51	04 50	11 59	18 16
27 Th	04 23 20	04 39 02	02♒ 44 03	28 16	12 49	23 50	22 24	26 57	12 50	04 50	12 01	18 11
28 Fr	04 27 16	05 39 47	16 57 33	29 50	14 04	24 36	22 26	27 04	12 49	04 50	12 03	18 07
29 Sa	04 31 13	06 40 32	01♓ 08 39	01♐ 25	15 19	25 22	22 28	27 12	12 47	04 51	12 04	18 05
30 Su	04 35 10	07 41 18	15 15 25	03 00	16 35	26 08	22 30	27 19	12 46	04 51	12 06	18D 05

Data for	11-01-2014
Julian Day	2456962.50
Ayanamsa	24 03 55
SVP	05 ♓ 03 14
☽ ☊ Mean	18 ♎ 10 R

● ◐ PHASES ○ ◑

06	22:23	○	14♉26
14	15:16	◑	22♌10
22	12:32	●	00♐07
29	10:06	◐	07♓06

LAST ASPECT Day	h m	☽ INGRESS Day	h m	
01	06:22	01	16:37	♓
03	09:06	03	18:54	♈
05	13:26	05	21:34	♉
07	16:17	08	01:45	♊
09	16:22	10	08:39	♋
12	09:18	12	18:45	♌
15	02:54	15	07:09	♍
17	11:12	17	19:31	♎
19	14:26	20	05:32	♏
22	05:54	22	12:20	♐
24	03:16	24	16:32	♑
26	15:30	26	19:23	♒
28	17:15	28	22:04	♓

DECLINATION

Day	☉	☽	☿	♀	♂	♃	♄	♅	♆	♇
01 Sa	14S 20	10S 29	05S 51	14S 06	24S 53	15N16	16S 52	04N43	10S 25	20S 40
02 Su	14 39	06 27	06 11	14 32	24 51	15 14	16 54	04 42	10 25	20 40
03 Mo	14 58	02 01	06 34	14 56	24 49	15 12	16 55	04 41	10 26	20 40
04 Tu	15 17	02N33	06 59	15 21	24 47	15 11	16 57	04 41	10 26	20 40
05 We	15 35	06 57	07 28	15 45	24 44	15 09	16 59	04 40	10 26	20 40
06 Th	15 53	10 56	07 58	16 09	24 42	15 07	17 01	04 39	10 26	20 40
07 Fr	16 11	14 14	08 29	16 32	24 39	15 06	17 03	04 38	10 26	20 40
08 Sa	16 29	16 40	09 02	16 55	24 35	15 04	17 04	04 38	10 26	20 40
09 Su	16 46	18 07	09 36	17 17	24 32	15 02	17 06	04 37	10 26	20 40
10 Mo	17 03	18 35	10 10	17 39	24 28	15 01	17 08	04 36	10 26	20 40
11 Tu	17 20	18 05	10 46	18 01	24 24	14 59	17 10	04 35	10 26	20 40
12 We	17 36	16 44	11 21	18 22	24 20	14 58	17 11	04 35	10 26	20 40
13 Th	17 53	14 39	11 57	18 42	24 15	14 57	17 13	04 34	10 27	20 40
14 Fr	18 08	11 58	12 32	19 02	24 10	14 55	17 15	04 33	10 27	20 40
15 Sa	18 24	08 49	13 08	19 22	24 05	14 54	17 17	04 33	10 27	20 40
16 Su	18 39	05 19	13 43	19 41	24 00	14 53	17 18	04 32	10 27	20 40
17 Mo	18 54	01 35	14 18	19 59	23 54	14 52	17 20	04 32	10 27	20 40
18 Tu	19 09	02S 16	14 53	20 17	23 48	14 50	17 22	04 31	10 27	20 40
19 We	19 23	06 05	15 27	20 34	23 42	14 49	17 23	04 30	10 26	20 40
20 Th	19 37	09 44	16 01	20 51	23 35	14 48	17 25	04 30	10 26	20 40
21 Fr	19 50	13 00	16 34	21 07	23 29	14 48	17 27	04 29	10 26	20 40
22 Sa	20 04	15 43	17 06	21 23	23 22	14 47	17 29	04 29	10 26	20 40
23 Su	20 16	17 38	17 38	21 37	23 14	14 46	17 30	04 28	10 26	20 40
24 Mo	20 29	18 34	18 09	21 52	23 07	14 45	17 32	04 28	10 26	20 40
25 Tu	20 41	18 23	18 39	22 05	22 59	14 44	17 34	04 27	10 26	20 40
26 We	20 53	17 04	19 08	22 18	22 51	14 44	17 35	04 27	10 26	20 40
27 Th	21 04	14 42	19 36	22 31	22 43	14 43	17 37	04 26	10 26	20 40
28 Fr	21 15	11 27	20 03	22 42	22 35	14 43	17 38	04 26	10 26	20 40
29 Sa	21 25	07 33	20 30	22 53	22 26	14 42	17 40	04 25	10 25	20 40
30 Su	21 35	03 14	20 55	23 03	22 17	14 42	17 42	04 25	10 25	20 40

ASPECTARIAN

01 00:24 ☽ ∥ ♆
00:29 ☽ ☍ ♃
06:22 ☽ □ ♄
12:45 ☿ ⚹ ♃
23:08 ♀ ⚹ ♇
23:44 ♂ ⚹ ♆

02 00:46 ☽ ☌ ♆
00:49 ☽ ⚹ ♂
01:22 ☽ ∥ ☿
09:10 ☽ △ ☉
09:36 ☽ # ♅
11:45 ☽ ⚹ ♇
12:57 ☽ △ ♀

03 05:53 ♀ ∥ ☉
09:06 ☽ △ ♄
14:37 ♀ # ♃
17:06 ☉ # ♃
22:49 ☉ ⚹ ♇

04 05:53 ☽ □ ♂
11:21 ☽ ∥ ♅
14:08 ☽ □ ♇
17:30 ☽ ☌ ♅

05 03:16 ☽ # ☿
06:02 ☽ △ ♃
13:26 ☽ ☍ ☿
20:49 ☽ # ♆

06 05:49 ☽ ⚹ ♆
11:40 ☽ △ ♂
17:19 ☽ △ ♇

07 04:31 ☽ ☍ ♀
07:31 ☽ ∥ ♃
10:00 ☽ □ ♃
16:17 ☽ ☍ ♄
21:31 ☽ # ☉

08 04:09 ☽ # ♀
05:31 ☽ # ♄
10:22 ☽ □ ♆
10:57 ♀ ∥ ♄
23:09 ☿ ♏

09 01:43 ☽ ⚹ ♅
16:22 ☽ ⚹ ♃
20:42 ♀ □ ♃

10 07:33 ☉ ∥ ♄
10:57 ☿ ∥ ♆
13:06 ☽ △ ☿
17:45 ☽ △ ♆
23:04 ♂ ☌ ♇

11 01:08 ☽ # ♀
06:45 ☽ ☍ ♇
07:14 ☽ ☍ ♂
09:54 ☽ □ ♅
12:23 ☽ # ☉
17:14 ☽ # ♄
21:56 ☽ △ ☉

12 03:39 ☿ △ ♆
07:39 ☽ △ ♀
09:18 ☽ △ ♄
20:50 ☽ ∥ ♃

13 01:03 ♀ ☌ ♄
01:29 ♂ □ ♅
07:59 ☽ □ ☿
19:57 ☽ # ☿
21:05 ☽ △ ♅

14 03:06 ☉ □ ♃
11:53 ☽ # ♆
14:19 ☽ ☌ ♃
22:04 ☽ □ ♄

15 02:54 ☽ □ ♀
16:53 ☽ ☍ ♆

16 05:05 ☽ ∥ ♅
05:50 ☽ ⚹ ☿
06:54 ☽ △ ♇
07:07 ♆ SD
13:55 ☿ ⚹ ♇
15:19 ☽ △ ♂
19:04 ♀ ♐

17 09:28 ☽ ⚹ ☉
11:12 ☽ ⚹ ♄
22:14 ☿ # ♃
22:20 ☽ ⚹ ♀

18 08:51 ☉ ☌ ♄
14:01 ☽ # ♅
18:40 ☽ □ ♇
21:04 ☽ ☍ ♅

19 06:31 ☽ □ ♂
08:26 ♀ ∥ ♇
14:26 ☽ ⚹ ♃

20 04:55 ☽ ∥ ♆
14:31 ☽ △ ♆
14:56 ♀ □ ♆

21 03:33 ☽ ⚹ ♇
14:01 ☿ ⚹ ♂
14:59 ☽ # ♃
18:02 ☽ ⚹ ♂
18:19 ☽ ☌ ☿
22:19 ☽ □ ♃

22 05:54 ☽ ☌ ♄
09:39 ☉ ♐
17:52 ☿ ∥ ♄
20:51 ☽ □ ♆
21:54 ☽ ∥ ♄
23:57 ☽ ∥ ☿

23 02:18 ☽ ☌ ♀
04:44 ☿ □ ♃
11:02 ☽ △ ♅

24 03:16 ☽ △ ♃
18:18 ☽ ∥ ☿
22:33 ☉ ∥ ♇

25 00:45 ☽ ⚹ ♆
12:55 ☽ ☌ ♇
14:25 ☽ □ ♅
16:46 ☽ ∥ ♄

26 02:37 ☿ ☌ ♄
08:07 ☽ ☌ ♂
14:10 ☽ ⚹ ♄
15:30 ☽ ⚹ ☿
23:49 ☽ # ♃

27 00:23 ♀ △ ♅
03:29 ☽ ⚹ ☉
04:20 ☉ □ ♆
14:46 ♀ ∥ ♂
17:00 ☽ ⚹ ♅
18:39 ☽ ⚹ ♀

28 02:26 ☿ ♐
06:40 ☽ ∥ ♆
09:17 ☽ ☍ ♃
17:15 ☽ □ ♄

29 00:31 ☽ □ ☿
06:17 ☽ ☌ ♆
09:43 ☿ ∥ ♇
17:33 ☽ # ♅
18:37 ☽ ⚹ ♇

30 02:28 ☽ □ ♀
19:42 ☽ ⚹ ♂
20:48 ☽ △ ♄

⚷ Chiron

01 Dec.	01 S 58
01	13♓21R
04	13 17
07	13 14
10	13 11
13	13 09
16	13 08
19	13 06
22	13 06
25	13 06D
28	13 06

23 21:55 13♓06 D

December 2014

Day	S. T.	☉	☽	☿	♀	♂	♃	♄	♅	♆	♇	☊ True
	h m s	° ' "	° ' "	° '	° '	° '	° '	° '	° '	° '	° '	° '
01 Mo	04 39 06	08♐ 42 05	29♓ 16 38	04♐ 34	17♐ 50	26♑ 55	22♌ 32	27♏ 26	12♈R45	04♓ 52	12♑ 08	18♎ 05
02 Tu	04 43 03	09 42 53	13♈ 11 32	06 09	19 05	27 41	22 33	27 33	12 44	04 52	12 10	18 06
03 We	04 46 59	10 43 42	26 59 24	07 43	20 21	28 27	22 34	27 40	12 43	04 53	12 12	18R 06
04 Th	04 50 56	11 44 32	10♉ 39 10	09 17	21 36	29 14	22 35	27 47	12 42	04 53	12 14	18 04
05 Fr	04 54 52	12 45 23	24 09 20	10 52	22 51	00♒ 00	22 36	27 54	12 42	04 54	12 15	17 59
06 Sa	04 58 49	13 46 14	07♊ 28 05	12 26	24 06	00 47	22 37	28 01	12 41	04 55	12 17	17 52
07 Su	05 02 45	14 47 07	20 33 34	14 00	25 22	01 33	22 37	28 08	12 40	04 55	12 19	17 43
08 Mo	05 06 42	15 48 01	03♋ 24 22	15 34	26 37	02 20	22 38	28 15	12 39	04 56	12 21	17 32
09 Tu	05 10 39	16 48 56	15 59 54	17 09	27 52	03 06	22R 38	28 22	12 39	04 57	12 23	17 21
10 We	05 14 35	17 49 52	28 20 38	18 43	29 08	03 53	22 38	28 29	12 38	04 58	12 25	17 11
11 Th	05 18 32	18 50 49	10♌ 28 14	20 17	00♑ 23	04 39	22 37	28 35	12 37	04 58	12 27	17 02
12 Fr	05 22 28	19 51 48	22 25 33	21 51	01 38	05 26	22 37	28 42	12 37	04 59	12 29	16 56
13 Sa	05 26 25	20 52 47	04♍ 16 23	23 26	02 53	06 13	22 36	28 49	12 36	05 00	12 31	16 52
14 Su	05 30 21	21 53 47	16 05 26	25 00	04 09	06 59	22 35	28 56	12 36	05 01	12 33	16 50
15 Mo	05 34 18	22 54 49	27 57 55	26 35	05 24	07 46	22 34	29 03	12 35	05 02	12 35	16D 50
16 Tu	05 38 14	23 55 51	09♎ 59 18	28 10	06 39	08 33	22 33	29 09	12 35	05 03	12 37	16 50
17 We	05 42 11	24 56 54	22 14 57	29 45	07 55	09 20	22 31	29 16	12 35	05 04	12 39	16R 50
18 Th	05 46 08	25 57 59	04♏ 49 44	01♑ 20	09 10	10 06	22 29	29 23	12 35	05 05	12 41	16 48
19 Fr	05 50 04	26 59 04	17 47 29	02 55	10 25	10 53	22 27	29 29	12 34	05 06	12 43	16 43
20 Sa	05 54 01	28 00 10	01♐ 10 25	04 30	11 40	11 40	22 25	29 36	12 34	05 07	12 45	16 36
21 Su	05 57 57	29 01 17	14 58 26	06 05	12 56	12 27	22 23	29 43	12 34	05 09	12 47	16 27
22 Mo	06 01 54	00♑ 02 24	29 08 53	07 41	14 11	13 14	22 21	29 49	12D 34	05 10	12 49	16 15
23 Tu	06 05 50	01 03 32	13♑ 36 31	09 17	15 26	14 01	22 18	29 56	12 34	05 11	12 51	16 04
24 We	06 09 47	02 04 40	28 14 19	10 52	16 42	14 47	22 15	00♐ 02	12 34	05 12	12 53	15 53
25 Th	06 13 43	03 05 49	12♒ 54 34	12 28	17 57	15 34	22 12	00 08	12 34	05 13	12 55	15 44
26 Fr	06 17 40	04 06 57	27 30 13	14 04	19 12	16 21	22 09	00 15	12 35	05 15	12 57	15 37
27 Sa	06 21 37	05 08 06	11♓ 55 48	15 40	20 27	17 08	22 05	00 21	12 35	05 16	12 59	15 32
28 Su	06 25 33	06 09 15	26 08 03	17 16	21 42	17 55	22 02	00 27	12 35	05 17	13 01	15 30
29 Mo	06 29 30	07 10 23	10♈ 05 39	18 52	22 58	18 42	21 58	00 34	12 35	05 19	13 03	15 29
30 Tu	06 33 26	08 11 32	23 48 51	20 28	24 13	19 29	21 54	00 40	12 36	05 20	13 06	15 29
31 We	06 37 23	09 12 40	07♉ 18 41	22 04	25 28	20 16	21 50	00 46	12 36	05 22	13 08	15 28

Data for	12-01-2014
Julian Day	2456992.50
Ayanamsa	24 03 58
SVP	05 ♓ 03 09
☽ ☊ Mean	16 ♎ 35 R

● ◐ PHASES ○ ◑

Day	Time	Phase	Position
06	12:27	○	14♊18
14	12:51	◑	22♍27
22	01:36	●	00♑07
28	18:32	◐	06♈57

LAST ASPECT Day	h m	☽ INGRESS Day	h m	Sign
30	20:48	01	01:14	♈
03	02:43	03	05:16	♉
05	06:46	05	10:29	♊
07	09:53	07	17:35	♋
10	00:16	10	03:15	♌
12	12:49	12	15:20	♍
15	02:11	15	04:05	♎
17	05:40	17	14:52	♏
19	21:12	19	21:56	♐
21	12:35	22	01:26	♑
23	03:18	24	02:53	♒
25	15:12	26	04:08	♓
27	15:45	28	06:36	♈
30	00:47	30	10:57	♉

DECLINATION

Day	☉	☽	☿	♀	♂	♃	♄	♅	♆	♇
01 Mo	21S 45	01N14	21S 19	23S 13	22S 08	14N42	17S 43	04N25	10S 25	20S 40
02 Tu	21 54	05 36	21 43	23 22	21 58	14 41	17 45	04 24	10 25	20 40
03 We	22 03	09 39	22 05	23 30	21 48	14 41	17 46	04 24	10 25	20 40
04 Th	22 11	13 08	22 26	23 38	21 38	14 41	17 48	04 24	10 24	20 40
05 Fr	22 19	15 53	22 46	23 44	21 28	14 41	17 50	04 23	10 24	20 40
06 Sa	22 27	17 44	23 05	23 50	21 18	14 41	17 51	04 23	10 24	20 40
07 Su	22 34	18 35	23 23	23 56	21 07	14 41	17 53	04 23	10 24	20 40
08 Mo	22 41	18 28	23 40	24 00	20 56	14 41	17 54	04 22	10 23	20 40
09 Tu	22 47	17 27	23 55	24 04	20 45	14 42	17 56	04 22	10 23	20 40
10 We	22 53	15 37	24 09	24 07	20 34	14 42	17 57	04 22	10 23	20 40
11 Th	22 58	13 07	24 22	24 10	20 22	14 42	17 59	04 22	10 22	20 40
12 Fr	23 03	10 07	24 34	24 11	20 10	14 43	18 00	04 22	10 22	20 40
13 Sa	23 07	06 44	24 44	24 12	19 58	14 43	18 01	04 21	10 22	20 39
14 Su	23 11	03 05	24 53	24 12	19 46	14 44	18 03	04 21	10 21	20 39
15 Mo	23 15	00S 42	25 00	24 12	19 34	14 44	18 04	04 21	10 21	20 39
16 Tu	23 18	04 31	25 07	24 10	19 21	14 45	18 06	04 21	10 21	20 39
17 We	23 20	08 13	25 12	24 08	19 08	14 45	18 07	04 21	10 20	20 39
18 Th	23 22	11 39	25 15	24 05	18 55	14 46	18 09	04 21	10 20	20 39
19 Fr	23 24	14 37	25 17	24 02	18 42	14 47	18 10	04 21	10 19	20 39
20 Sa	23 25	16 55	25 18	23 57	18 29	14 48	18 11	04 21	10 19	20 39
21 Su	23 26	18 19	25 17	23 52	18 15	14 49	18 13	04 21	10 19	20 39
22 Mo	23 26	18 37	25 15	23 46	18 01	14 50	18 14	04 21	10 18	20 39
23 Tu	23 26	17 45	25 11	23 40	17 47	14 51	18 15	04 21	10 18	20 39
24 We	23 25	15 42	25 06	23 32	17 33	14 52	18 17	04 21	10 17	20 39
25 Th	23 24	12 38	24 59	23 24	17 19	14 53	18 18	04 21	10 17	20 39
26 Fr	23 22	08 49	24 50	23 16	17 04	14 55	18 19	04 21	10 16	20 38
27 Sa	23 20	04 30	24 40	23 06	16 49	14 56	18 20	04 21	10 16	20 38
28 Su	23 17	00N00	24 29	22 56	16 34	14 57	18 22	04 22	10 15	20 38
29 Mo	23 14	04 26	24 16	22 45	16 19	14 59	18 23	04 22	10 15	20 38
30 Tu	23 11	08 33	24 02	22 33	16 04	15 00	18 24	04 22	10 14	20 38
31 We	23 07	12 10	23 45	22 21	15 49	15 02	18 25	04 22	10 14	20 38

ASPECTARIAN

01 04:27 ☿ □ ♆
10:16 ☽ △ ☿
17:15 ☽ ∥ ♅
17:31 ☽ △ ☉
19:01 ♂ ✶ ♄
22:13 ☽ □ ♇
23:13 ☽ ☌ ♅

02 04:59 ☉ ∥ ♂
11:15 ☽ △ ♀
11:23 ☿ ∥ ♂
16:17 ☽ △ ♃
20:16 ☿ ∥ ☉

03 02:43 ☽ □ ♂
04:48 ☽ # ♆
13:50 ☽ ✶ ♆

04 02:47 ☽ △ ♇
12:34 ☽ ∥ ♃
19:11 ♀ △ ♃
21:13 ☽ □ ♃
22:30 ☉ △ ♅
23:57 ♂ ♒

05 06:46 ☽ ☍ ♄
11:08 ☽ △ ♂
19:21 ☽ □ ♆

06 02:19 ☽ # ♄
03:44 ☿ △ ♅
09:29 ☽ ✶ ♅
10:17 ☽ ☍ ☿

07 03:49 ☽ ✶ ♃
09:53 ☽ ☍ ♀

08 [illegible] ☽ △ ♆
09:52 ☿ ☌ ☉

15:17 ☽ # ♄
17:03 ☽ ☍ ♇
17:34 ☽ □ ♅
20:42 ♃ SR

09 11:29 ♂ ∥ ♆
19:59 ☿ ∥ ♀

10 00:16 ☽ △ ♄
09:30 ☽ ∥ ♃
11:39 ☽ ☍ ♂
16:42 ♀ ♑

11 04:18 ☽ △ ♅
18:21 ☽ △ ☉
22:09 ☽ # ♆
22:41 ☽ △ ☿

12 00:22 ☽ ☌ ♃
11:25 ☿ △ ♃
12:49 ☽ □ ♄
20:52 ☽ △ ♀

13 01:29 ☽ ☍ ♆
15:46 ☽ ∥ ♅
16:47 ☽ △ ♇

14 15:57 ☉ △ ♃
16:56 ♀ ✶ ♆
20:48 ☽ □ ☿

15 02:11 ☽ ✶ ♄
06:15 ♅ □ ♇
16:37 ☽ □ ♀
20:57 ☽ △ [illegible]

22:59 ☽ # ♅

16 05:08 ☽ ☍ ♅
05:12 ☽ □ ♇

17 00:31 ☽ ✶ ♃
03:53 ☿ ♑
05:40 ☽ ✶ ☉
14:31 ☽ ∥ ♆
16:26 ☽ ✶ ☿

18 00:29 ☽ △ ♆
08:59 ☽ ✶ ♀
10:30 ☽ □ ♂
14:40 ☽ ✶ ♇

19 01:28 ☽ # ♃
08:26 ☽ □ ♃
21:12 ☽ ☌ ♄

20 06:57 ☽ □ ♆
09:31 ☿ ✶ ♆
17:09 ♀ □ ♅
19:24 ☽ ✶ ♂
19:52 ☽ △ ♅
21:04 ☽ ∥ ♄
21:08 ♀ ☌ ♇
22:28 ☽ ∥ ♂

21 03:47 ♂ ∥ ♄
03:47 ♂ ✶ ♅
12:35 ☽ △ ♃
22:46 ♅ SD
23:03 ☉ ♑

22 10:02 ☽ ✶ ♆
14:27 ☽ ∥ ♄
15:58 ☽ ☌ ☿
22:17 ☽ □ ♅
22:45 ☽ ☌ ♇
23:18 ☽ ∥ ♂

23 03:18 ☽ ☌ ♀
16:34 ♄ ♐

24 02:57 ☽ ✶ ♄
07:13 ☽ # ♃
23:27 ☽ ✶ ♅

25 01:14 ♀ ∥ ☉
01:32 ☿ □ ♅
04:37 ☽ ☌ ♂
06:53 ☿ ☌ ♇
15:12 ☽ ☍ ♃
15:17 ☽ ∥ ♆

26 04:34 ☽ □ ♄
11:47 ☽ ✶ ☉
12:51 ☽ ☌ ♆

27 00:46 ☽ # ♅
01:47 ☽ ✶ ♇
03:11 ☉ ✶ ♆
07:05 ☽ ✶ ☿
15:45 ☽ ✶ ♀

28 07:26 ☽ △ ♄
23:35 ☽ ∥ ♅

29 04:20 ☽ ☌ ♅
05:10 ☽ □ ♇
15:55 ☽ ✶ ♂
17:20 ☽ □ ☿
20:39 ☽ △ ♃

30 00:47 ☽ □ ♀
10:35 ☽ # ♆
20:30 ☽ ✶ ♆

31 03:41 ☽ △ ☉
10:29 ☽ △ ♇
23:35 ☽ ∥ ♃

⚷ Chiron

01 Dec.	02 S 12
01	13♓07
04	13 09
07	13 11
10	13 13
13	13 16
16	13 20
19	13 24
22	13 28
25	13 33
28	13 38
31	13 44

January 2015

Day	S. T. h m s	☉ ° ' "	☽ ° ' "	☿ ° '	♀ ° '	♂ ° '	♃ ° '	♄ ° '	♅ ° '	♆ ° '	♇ ° '	☊ True ° '
01 Th	06 41 19	10♑13 48	20♉36 24	23♑39	26♑43	21♒03	21♌R46	00♐52	12♈37	05♓23	13♑10	15♎R24
02 Fr	06 45 16	11 14 57	03♊42 56	25 14	27 58	21 50	21 41	00 58	12 37	05 25	13 12	15 18
03 Sa	06 49 12	12 16 05	16 38 47	26 49	29 14	22 37	21 36	01 04	12 38	05 26	13 14	15 09
04 Su	06 53 09	13 17 13	29 23 57	28 23	00♒29	23 24	21 32	01 10	12 39	05 28	13 16	14 57
05 Mo	06 57 06	14 18 21	11♋58 12	29 56	01 44	24 11	21 27	01 16	12 39	05 29	13 18	14 43
06 Tu	07 01 02	15 19 29	24 21 24	01♒27	02 59	24 58	21 21	01 22	12 40	05 31	13 20	14 29
07 We	07 04 59	16 20 37	06♌33 45	02 58	04 14	25 45	21 16	01 27	12 41	05 32	13 22	14 15
08 Th	07 08 55	17 21 45	18 36 12	04 27	05 29	26 32	21 11	01 33	12 42	05 34	13 24	14 03
09 Fr	07 12 52	18 22 53	00♍30 31	05 53	06 44	27 19	21 05	01 39	12 43	05 36	13 26	13 54
10 Sa	07 16 48	19 24 00	12 19 31	07 18	07 59	28 06	20 59	01 44	12 44	05 37	13 29	13 48
11 Su	07 20 45	20 25 08	24 06 55	08 39	09 14	28 53	20 53	01 50	12 45	05 39	13 31	13 44
12 Mo	07 24 41	21 26 16	05♎57 16	09 57	10 29	29 40	20 47	01 55	12 46	05 41	13 33	13 43
13 Tu	07 28 38	22 27 24	17 55 46	11 11	11 45	00♓27	20 41	02 01	12 47	05 43	13 35	13 43
14 We	07 32 35	23 28 31	00♏07 54	12 20	13 00	01 14	20 35	02 06	12 48	05 44	13 37	13 42
15 Th	07 36 31	24 29 39	12 39 09	13 24	14 15	02 01	20 28	02 11	12 49	05 46	13 39	13 40
16 Fr	07 40 28	25 30 46	25 34 29	14 21	15 30	02 48	20 22	02 17	12 50	05 48	13 41	13 36
17 Sa	07 44 24	26 31 54	08♐57 34	15 11	16 45	03 35	20 15	02 22	12 52	05 50	13 43	13 30
18 Su	07 48 21	27 33 01	22 49 59	15 53	17 59	04 21	20 08	02 27	12 53	05 52	13 45	13 21
19 Mo	07 52 17	28 34 08	07♑10 21	16 26	19 14	05 08	20 01	02 32	12 54	05 54	13 47	13 10
20 Tu	07 56 14	29 35 14	21 53 56	16 50	20 29	05 55	19 54	02 37	12 56	05 56	13 49	12 59
21 We	08 00 10	00♒36 20	06♒53 00	17 03	21 44	06 42	19 47	02 42	12 57	05 58	13 51	12 49
22 Th	08 04 07	01 37 25	21 57 52	17R 05	22 59	07 29	19 40	02 46	12 59	05 59	13 53	12 39
23 Fr	08 08 04	02 38 29	06♓58 39	16 55	24 14	08 16	19 33	02 51	13 00	06 01	13 55	12 32
24 Sa	08 12 00	03 39 32	21 46 52	16 34	25 29	09 03	19 25	02 56	13 02	06 03	13 57	12 28
25 Su	08 15 57	04 40 34	06♈16 38	16 01	26 44	09 50	19 18	03 00	13 04	06 05	13 59	12 26
26 Mo	08 19 53	05 41 35	20 24 57	15 18	27 59	10 37	19 10	03 05	13 06	06 07	14 01	12 25
27 Tu	08 23 50	06 42 35	04♉11 25	14 25	29 13	11 23	19 03	03 09	13 07	06 09	14 03	12D 25
28 We	08 27 46	07 43 34	17 37 24	13 23	00♓28	12 10	18 55	03 14	13 09	06 12	14 05	12R 25
29 Th	08 31 43	08 44 32	00♊45 12	12 15	01 43	12 57	18 47	03 18	13 11	06 14	14 07	12 23
30 Fr	08 35 39	09 45 29	13 37 21	11 03	02 57	13 44	18 39	03 22	13 13	06 16	14 09	12 19
31 Sa	08 39 36	10 46 24	26 16 14	09 48	04 12	14 30	18 32	03 26	13 15	06 18	14 11	12 12

Data for	01-01-2015
Julian Day	2457023.50
Ayanamsa	24 04 04
SVP	05 ♓ 03 06
☽ ☊ Mean	14 ♎ 56 R

● ◐ PHASES ○ ◑

05	04:53	○	14♋31
13	09:47	◑	22♎52
20	13:14	●	00♒09
27	04:49	◐	06♉55

LAST ASPECT Day	h m	☽ INGRESS Day	h m	
01	12:20	01	17:10	♊
03	11:56	04	01:08	♋
05	04:54	06	11:03	♌
08	17:05	08	22:58	♍
10	15:46	11	11:57	♎
13	09:47	13	23:45	♏
15	23:53	16	08:02	♐
17	19:26	18	12:05	♑
19	10:52	20	13:01	♒
22	01:46	22	12:49	♓
23	11:14	24	13:32	♈
26	14:24	26	16:38	♉
28	02:19	28	22:37	♊
30	09:25	31	07:09	♋

DECLINATION

Day	☉	☽	☿	♀	♂	♃	♄	♅	♆	♇
01 Th	23S 02	15N06	23S 28	22S 08	15S 33	15N04	18S 26	04N22	10S 13	20S 38
02 Fr	22 57	17 12	23 09	21 55	15 18	15 05	18 27	04 23	10 12	20 38
03 Sa	22 52	18 23	22 48	21 41	15 02	15 07	18 29	04 23	10 12	20 38
04 Su	22 46	18 37	22 26	21 26	14 46	15 09	18 30	04 23	10 11	20 38
05 Mo	22 40	17 55	22 03	21 10	14 30	15 11	18 31	04 23	10 11	20 37
06 Tu	22 33	16 23	21 38	20 54	14 13	15 12	18 32	04 24	10 10	20 37
07 We	22 26	14 08	21 13	20 37	13 57	15 14	18 33	04 24	10 10	20 37
08 Th	22 18	11 18	20 46	20 20	13 40	15 16	18 34	04 25	10 09	20 37
09 Fr	22 10	08 03	20 18	20 02	13 24	15 18	18 35	04 25	10 08	20 37
10 Sa	22 02	04 30	19 49	19 44	13 07	15 20	18 36	04 25	10 08	20 37
11 Su	21 53	00 47	19 19	19 25	12 50	15 22	18 37	04 26	10 07	20 37
12 Mo	21 44	03S 00	18 49	19 05	12 33	15 25	18 38	04 26	10 06	20 37
13 Tu	21 34	06 41	18 18	18 45	12 16	15 27	18 39	04 27	10 06	20 37
14 We	21 24	10 10	17 48	18 24	11 59	15 29	18 40	04 27	10 05	20 36
15 Th	21 13	13 18	17 18	18 03	11 41	15 31	18 41	04 28	10 04	20 36
16 Fr	21 02	15 52	16 48	17 41	11 24	15 33	18 42	04 28	10 04	20 36
17 Sa	20 51	17 41	16 20	17 19	11 06	15 36	18 43	04 29	10 03	20 36
18 Su	20 39	18 32	15 52	16 56	10 49	15 38	18 44	04 29	10 02	20 36
19 Mo	20 27	18 14	15 27	16 33	10 31	15 40	18 45	04 30	10 02	20 36
20 Tu	20 14	16 44	15 03	16 10	10 13	15 43	18 45	04 30	10 01	20 36
21 We	20 01	14 05	14 43	15 45	09 55	15 45	18 46	04 31	10 00	20 35
22 Th	19 48	10 29	14 25	15 21	09 37	15 48	18 47	04 32	10 00	20 35
23 Fr	19 34	06 13	14 10	14 56	09 19	15 50	18 48	04 32	09 59	20 35
24 Sa	19 20	01 36	13 59	14 31	09 01	15 53	18 49	04 33	09 58	20 35
25 Su	19 05	03N00	13 52	14 05	08 42	15 55	18 49	04 34	09 57	20 35
26 Mo	18 51	07 20	13 49	13 39	08 24	15 58	18 50	04 34	09 57	20 35
27 Tu	18 36	11 09	13 50	13 13	08 06	16 00	18 51	04 35	09 56	20 35
28 We	18 20	14 17	13 54	12 46	07 47	16 03	18 51	04 36	09 55	20 35
29 Th	18 04	16 36	14 01	12 19	07 29	16 05	18 52	04 37	09 54	20 34
30 Fr	17 48	18 01	14 12	11 51	07 10	16 08	18 53	04 37	09 54	20 34
31 Sa	17 32	18 31	14 24	11 23	06 52	16 10	18 53	04 38	09 53	20 34

ASPECTARIAN

01 00:52 ☽ □ ♂
02:05 ☽ □ ♃
03:59 ☽ # ♂
06:19 ☽ △ ☿
12:20 ☽ △ ♀
18:54 ☽ ☍ ♄
19:50 ♂ ☍ ♃
02 03:08 ☽ □ ♆
16:31 ☽ ⚹ ♅
16:53 ♂ # ♃
18:10 ☿ ∥ ☉
03 03:31 ☽ # ♄
08:40 ☉ □ ♅
09:14 ☽ ⚹ ♃
11:56 ☽ △ ♂
14:49 ♀ ♒
23:33 ☉ ☌ ♇
04 07:23 ☽ # ♄
11:33 ☽ △ ♆
14:16 ♀ ⚹ ♄
05 01:08 ☿ ♒
01:19 ☽ □ ♅
02:34 ☽ ☍ ♇
22:22 ☿ ⚹ ♄
06 13:21 ☽ ∥ ♃
13:50 ☽ △ ♄
15:54 ☽ ☍ ☿
18:52 ☽ ☍ ♀
07 00:20 ♀ ∥ ♇
02:02 ☽ # ♂
12:10 ☽ △ ♅
08 05:08 ☽ ☌ ♃
07:22 ☿ ∥ ♇
08:57 ☽ # ♆
17:05 ☽ ☍ ♂
09 02:19 ☽ □ ♄
10:20 ☽ ☍ ♆
10 00:32 ☽ ∥ ♅
02:21 ☽ △ ♇
11:10 ☿ ∥ ♀
15:46 ☽ △ ☉
11 15:47 ☽ ⚹ ♄
12 08:15 ☿ ∥ ♄
09:00 ☽ △ ☿
09:19 ☽ # ♅
10:12 ☽ △ ♀
10:20 ♂ ♓
13:42 ☽ ☍ ♅
15:18 ☽ □ ♇
13 05:25 ☽ ⚹ ♃
06:30 ♀ ∥ ♄
20:12 ♀ ⚹ ♅
23:22 ☽ ∥ ♆
14 02:16 ☽ △ ♂
10:25 ☿ ⚹ ♅
10:52 ☽ △ ♆
12:18 ☽ ∥ ♂
15 01:31 ☽ □ ☿
01:53 ☽ ⚹ ♇
03:19 ☽ □ ♀
06:11 ♂ □ ♄
14:30 ☽ □ ♃
20:41 ☽ # ♃
23:53 ☽ ⚹ ☉
16 08:43 ☽ ∥ ☿
12:12 ☽ ☌ ♄
13:52 ☽ □ ♂
18:27 ☽ □ ♆
19:11 ☽ ∥ ♀
17 06:51 ☽ △ ♅
11:29 ☽ ⚹ ☿
14:54 ☽ ⚹ ♀
19:26 ☽ △ ♃
18 05:52 ☉ ∥ ♇
12:00 ☿ # ♃
20:27 ☽ ⚹ ♂
21:53 ☽ ⚹ ♆
19 09:26 ☽ □ ♅
10:52 ☽ ☌ ♇
13:45 ♀ ☍ ♃
20 00:09 ♂ ☌ ♆
07:31 ☽ ∥ ♀
09:44 ☉ ♒
10:24 ☽ # ♃
16:45 ♂ ∥ ♆
17:17 ☽ ⚹ ♄
18:37 ☽ ∥ ☿
21 00:14 ♀ # ♃
09:41 ☽ ⚹ ♅
15:55 ☿ SR
16:14 ☽ ☌ ☿
20:22 ☽ ☍ ♃
22 01:46 ☽ ☌ ♀
02:57 ☽ ∥ ♆
05:32 ☽ ∥ ♂
17:21 ☽ □ ♄
22:28 ☽ ☌ ♆
23 02:11 ☽ ☌ ♂
05:24 ☉ ⚹ ♄
08:49 ☽ # ♅
11:14 ☽ ⚹ ♇
24 18:30 ☽ △ ♄
21:07 ☽ ⚹ ☉
25 08:28 ☽ ∥ ♅
11:28 ☽ ☌ ♅
13:02 ☽ □ ♆
13:55 ☿ ∥ ♀
15:42 ☽ ⚹ ☿
21:53 ☽ △ ♃
26 00:59 ☉ ∥ ♄
05:58 ☽ # ♂
14:24 ☽ ⚹ ♀
16:00 ☽ # ♆
27 03:29 ☽ ⚹ ♆
13:16 ☽ # ♀
13:35 ☽ ⚹ ♂
15:00 ♀ ♓
16:56 ☽ □ ☿
17:36 ☽ △ ♇
20:42 ☽ # ☿
28 02:19 ☽ □ ♃
05:03 ☿ ⚹ ♅
17:52 ☽ ∥ ♃
29 01:58 ☽ □ ♀
04:44 ☽ ☍ ♄
10:11 ☽ □ ♆
16:07 ☽ △ ☉
19:36 ☽ △ ☿
20:11 ☽ # ☉
23:14 ☽ ⚹ ♅
30 00:13 ☽ □ ♂
08:24 ♀ □ ♄
09:25 ☽ ⚹ ♃
13:37 ♂ ⚹ ♇
13:46 ☿ ☌ ☉
31 16:56 ☽ △ ♀
19:20 ☽ △ ♆

⚷ Chiron

01 Dec.	02 S 06
03	13♓51
06	13 57
09	14 04
12	14 12
15	14 20
18	14 28
21	14 37
24	14 45
27	14 55
30	15 04

February 2015

Day	S. T. h m s	☉ ° ' "	☽ ° ' "	☿ ° '	♀ ° '	♂ ° '	♃ ° '	♄ ° '	♅ ° '	♆ ° '	♇ ° '	☊ True ° '
01 Su	08 43 33	11♒47 19	08♋43 49	08♒R34	05♓27	15♓17	18♌R24	03♐30	13♈17	06♓20	14♑13	12♎R03
02 Mo	08 47 29	12 48 12	21 01 43	07 21	06 41	16 04	18 16	03 34	13 19	06 22	14 15	11 52
03 Tu	08 51 26	13 49 04	03♌11 06	06 12	07 56	16 50	18 08	03 38	13 21	06 24	14 17	11 40
04 We	08 55 22	14 49 54	15 13 03	05 09	09 10	17 37	18 00	03 42	13 23	06 26	14 19	11 29
05 Th	08 59 19	15 50 44	27 08 40	04 12	10 25	18 24	17 52	03 45	13 25	06 28	14 20	11 19
06 Fr	09 03 15	16 51 32	08♍59 26	03 23	11 39	19 10	17 44	03 49	13 27	06 31	14 22	11 11
07 Sa	09 07 12	17 52 20	20 47 22	02 42	12 54	19 57	17 36	03 53	13 30	06 33	14 24	11 07
08 Su	09 11 08	18 53 06	02♎35 12	02 09	14 08	20 43	17 28	03 56	13 32	06 35	14 26	11 04
09 Mo	09 15 05	19 53 51	14 26 23	01 44	15 23	21 30	17 20	03 59	13 34	06 37	14 28	11D 04
10 Tu	09 19 02	20 54 36	26 25 03	01 28	16 37	22 16	17 12	04 03	13 36	06 39	14 29	11 05
11 We	09 22 58	21 55 19	08♏35 50	01 19	17 51	23 03	17 04	04 06	13 39	06 42	14 31	11 06
12 Th	09 26 55	22 56 01	21 03 43	01D 19	19 05	23 49	16 56	04 09	13 41	06 44	14 33	11R 07
13 Fr	09 30 51	23 56 42	03♐53 35	01 25	20 20	24 36	16 49	04 12	13 44	06 46	14 35	11 06
14 Sa	09 34 48	24 57 22	17 09 43	01 38	21 34	25 22	16 41	04 15	13 46	06 48	14 36	11 03
15 Su	09 38 44	25 58 01	00♑54 55	01 57	22 48	26 08	16 33	04 17	13 49	06 50	14 38	10 58
16 Mo	09 42 41	26 58 39	15 09 38	02 21	24 02	26 55	16 25	04 20	13 51	06 53	14 40	10 52
17 Tu	09 46 37	27 59 16	29 51 11	02 51	25 16	27 41	16 17	04 23	13 54	06 55	14 41	10 46
18 We	09 50 34	28 59 51	14♒53 26	03 26	26 30	28 27	16 10	04 25	13 57	06 57	14 43	10 39
19 Th	09 54 31	00♓00 24	00♓07 27	04 06	27 44	29 13	16 02	04 28	13 59	07 00	14 45	10 34
20 Fr	09 58 27	01 00 57	15 22 44	04 49	28 58	30 00	15 55	04 30	14 02	07 02	14 46	10 30
21 Sa	10 02 24	02 01 27	00♈28 57	05 37	00♈12	00♈46	15 47	04 32	14 05	07 04	14 48	10 27
22 Su	10 06 20	03 01 56	15 17 39	06 27	01 26	01 32	15 40	04 34	14 07	07 06	14 49	10D 27
23 Mo	10 10 17	04 02 23	29 43 08	07 22	02 40	02 18	15 33	04 36	14 10	07 09	14 51	10 28
24 Tu	10 14 13	05 02 48	13♉42 51	08 19	03 53	03 04	15 25	04 38	14 13	07 11	14 52	10 29
25 We	10 18 10	06 03 11	27 16 49	09 19	05 07	03 50	15 18	04 40	14 16	07 13	14 54	10 31
26 Th	10 22 06	07 03 32	10♊26 55	10 21	06 21	04 36	15 11	04 42	14 19	07 15	14 55	10R 31
27 Fr	10 26 03	08 03 51	23 16 08	11 26	07 34	05 22	15 05	04 43	14 22	07 18	14 57	10 30
28 Sa	10 29 60	09 04 08	05♋47 52	12 33	08 48	06 08	14 58	04 45	14 25	07 20	14 58	10 28

Data for	02-01-2015
Julian Day	2457054.50
Ayanamsa	24 04 08
SVP	05 ♓ 03 02
☽ ☊ Mean	13 ♎ 18 R

● ◐ PHASES ○ ◑

Day	Time	Phase	Position
03	23:09	○	14♌48
12	03:51	◑	23♏06
18	23:48	●	00♓00
25	17:14	◐	06♊47

LAST ASPECT ☽ INGRESS

Day	h m	Day	h m	Sign
01	13:37	02	17:41	♌
04	05:31	05	05:46	♍
06	22:10	07	18:45	♎
09	11:59	10	07:06	♏
12	05:34	12	16:48	♐
14	15:17	14	22:26	♑
16	20:18	17	00:14	♒
18	23:48	18	23:48	♓
19	23:02	20	23:14	♈
22	00:36	23	00:28	♉
24	02:58	25	04:54	♊
26	08:44	27	12:50	♋

DECLINATION

Day	☉	☽	☿	♀	♂	♃	♄	♅	♆	♇
01 Su	17S 15	18N06	14S 39	10S 55	06S 33	16N13	18S 54	04N39	09S 52	20S 34
02 Mo	16 58	16 51	14 55	10 27	06 14	16 15	18 55	04 40	09 51	20 34
03 Tu	16 41	14 51	15 12	09 58	05 55	16 18	18 55	04 41	09 50	20 34
04 We	16 23	12 13	15 30	09 30	05 37	16 21	18 56	04 41	09 50	20 34
05 Th	16 05	09 07	15 47	09 00	05 18	16 23	18 56	04 42	09 49	20 34
06 Fr	15 47	05 40	16 04	08 31	04 59	16 26	18 57	04 43	09 48	20 33
07 Sa	15 28	02 00	16 21	08 02	04 40	16 28	18 57	04 44	09 47	20 33
08 Su	15 10	01S 44	16 37	07 32	04 21	16 31	18 58	04 45	09 46	20 33
09 Mo	14 51	05 25	16 52	07 02	04 02	16 33	18 58	04 46	09 46	20 33
10 Tu	14 31	08 55	17 06	06 32	03 43	16 36	18 59	04 47	09 45	20 33
11 We	14 12	12 06	17 19	06 01	03 24	16 38	18 59	04 48	09 44	20 33
12 Th	13 52	14 49	17 31	05 31	03 05	16 41	19 00	04 49	09 43	20 33
13 Fr	13 32	16 53	17 41	05 00	02 46	16 43	19 00	04 50	09 42	20 32
14 Sa	13 12	18 07	17 50	04 29	02 27	16 46	19 00	04 51	09 41	20 32
15 Su	12 52	18 22	17 58	03 59	02 08	16 48	19 01	04 52	09 41	20 32
16 Mo	12 31	17 28	18 04	03 28	01 49	16 51	19 01	04 53	09 40	20 32
17 Tu	12 10	15 25	18 09	02 57	01 30	16 53	19 01	04 54	09 39	20 32
18 We	11 49	12 17	18 13	02 25	01 11	16 55	19 02	04 55	09 38	20 32
19 Th	11 28	08 17	18 15	01 54	00 52	16 58	19 02	04 56	09 37	20 32
20 Fr	11 07	03 44	18 16	01 23	00 33	17 00	19 02	04 57	09 36	20 32
21 Sa	10 45	01N01	18 16	00 52	00 14	17 02	19 02	04 58	09 36	20 31
22 Su	10 23	05 37	18 14	00 20	00N05	17 04	19 03	04 59	09 35	20 31
23 Mo	10 02	09 46	18 11	00N11	00 24	17 07	19 03	05 00	09 34	20 31
24 Tu	09 40	13 14	18 06	00 43	00 43	17 09	19 03	05 01	09 33	20 31
25 We	09 17	15 51	18 00	01 14	01 02	17 11	19 03	05 02	09 32	20 31
26 Th	08 55	17 33	17 53	01 45	01 21	17 13	19 03	05 04	09 31	20 31
27 Fr	08 33	18 18	17 44	02 17	01 40	17 15	19 04	05 05	09 31	20 31
28 Sa	08 10	18 08	17 34	02 48	01 59	17 17	19 04	05 06	09 30	20 31

ASPECTARIAN

01 08:52 ☽ □ ♅
10:42 ☽ ☍ ♇
13:37 ☽ △ ♂
17:36 ♀ ☌ ♆
21:53 ☽ ⋕ ☉
19:43 ☽ ⋕ ♅
22:14 ☽ ☍ ♅

02 07:49 ☽ ∥ ♃
12:30 ☉ ⚹ ♅
20:42 ☽ ⋕ ☿

03 00:54 ☽ △ ♄
05:30 ☽ ☍ ☿
06:53 ♀ ∥ ♆
20:19 ☽ △ ♅

04 02:45 ☉ ⋕ ♃
05:31 ☽ ☌ ♃
18:53 ☽ ⋕ ♆

05 00:58 ☽ ⋕ ♀
11:39 ☿ ⚹ ♄
12:06 ☿ ∥ ☉
13:27 ☽ □ ♄
18:57 ☽ ☍ ♆

06 05:03 ☽ ⋕ ♂
06:04 ☽ ☍ ♀
06:19 ☽ ∥ ♅
10:58 ☽ △ ♇
18:21 ☉ ☍ ♃
18:57 ♂ ⋕ ♅
22:10 ☽ ☍ ♂

07 12:37 ☿ ⋕ ♃
23:08 ☽ △ ☿

08 02:45 ☽ ⚹ ♄
05:51 ♀ ⚹ ♇
15:40 ☽ ∥ ♂

09 00:02 ☽ □ ♇
05:46 ☽ ⚹ ♃
09:31 ☽ ∥ ♀
11:59 ☽ △ ☉

10 05:57 ☽ ∥ ♆
09:51 ☽ □ ☿
20:16 ☽ △ ♆

11 11:30 ☽ ⚹ ♇
14:58 ☿ SD
16:02 ☽ ∥ ☉
16:13 ☽ □ ♃
19:50 ☽ △ ♀

12 05:34 ☽ △ ♂
19:22 ☽ ⚹ ☿
21:43 ☽ ⋕ ♃

13 00:34 ☽ ☌ ♄
05:17 ☽ □ ♆
07:55 ♀ ⋕ ♅
15:27 ☽ ∥ ☿
17:56 ☽ △ ♅
23:09 ☽ △ ♃

14 08:33 ☽ □ ♀
14:47 ☽ ⚹ [illegible]
[illegible] ☽ □ ♂

15 10:07 ☽ ⚹ ♆
12:47 ☽ ∥ ☿
21:50 ☽ □ ♅
23:10 ☽ ☌ ♇

16 08:37 ☽ ⋕ ♃
15:54 ☽ ⚹ ♀
20:18 ☽ ⚹ ♂

17 05:01 ☽ ☌ ☿
07:18 ☽ ⚹ ♄
22:30 ☽ ⚹ ♅

18 02:00 ☽ ☍ ♃
03:19 ☽ ∥ ☉
16:23 ☽ ∥ ♆
23:50 ☉ ♓

19 06:50 ☽ □ ♄
10:49 ☽ ☌ ♆
13:06 ☿ ⚹ ♄
17:46 ☽ ⋕ ♅
23:02 ☽ ⚹ ♇

20 00:12 ♂ ♈
13:20 ☽ ∥ ♀
17:14 ☽ ∥ ♂
20:06 ♀ ♈
20:16 ☽ ⋕ ♂
23:17 ☽ ⋕ ♀
23:30 ☽ ☌ ♀

21 00:28 ☽ ☌ ♂
06:32 ☽ △ ♄
08:44 ☽ ⚹ ☿
20:36 ☽ ∥ ♅
22:04 ☽ ☌ ♅
23:13 ☽ □ ♇

22 00:36 ☽ △ ♃
05:13 ♀ ☌ ♂
07:12 ♀ ⋕ ♂
22:48 ☽ ⋕ ♆

23 01:32 ☽ ⋕ [illegible]
07:[illegible]1 ☽ ⚹ ☉
12:40 ☽ ⚹ ♆
13:57 ☉ □ ♄
13:58 ☽ □ ☿

24 01:03 ♀ ∥ ♂
02:01 ☽ △ ♇
02:58 ☽ □ ♃
07:16 ☉ ∥ ♆
14:58 ♀ △ ♄

25 12:36 ☽ ⚹ ♂
13:25 ☽ ☍ ♄
15:40 ☽ ∥ ♆
18:03 ☽ ∥ ♃
18:06 ☽ □ ♆
23:48 ☽ △ ☿

26 03:09 ♂ △ ♄
04:55 ☉ ☌ ♆
06:44 ☽ ⋕ ☿
07:12 ☽ ⚹ ♅
08:44 ☽ ⚹ ♃

28 00:41 ☽ □ ♂
02:59 ☽ △ ♆
06:28 ☽ □ [illegible]
[illegible] ☽ △ ☉
16:50 ☽ □ ♅
17:47 ☽ ⋕ ☿
17:53 ☽ ☍ ♇
20:08 ☽ ∥ ♃

⚷ Chiron

01 Dec.	01 S 41
02	15♓14
05	15 24
08	15 34
11	15 44
14	15 55
17	16 06
20	16 16
23	16 27
26	16 38

20 10:18 29♓27 ☀ Total Solar Eclipse

March 2015

Day	S. T. h m s	☉ ° ' "	☽ ° ' "	☿ ° '	♀ ° '	♂ ° '	♃ ° '	♄ ° '	♅ ° '	♆ ° '	♇ ° '	☊ True ° '
01 Su	10 33 56	10♓ 04 23	18♋ 05 36	13♒ 42	10♈ 01	06♈ 54	14♌ R51	04♐ 46	14♈ 27	07♓ 22	14♑ 59	10♎ R24
02 Mo	10 37 53	11 04 37	00♌ 12 34	14 54	11 15	07 40	14 45	04 48	14 30	07 25	15 01	10 19
03 Tu	10 41 49	12 04 48	12 11 38	16 07	12 28	08 25	14 38	04 49	14 33	07 27	15 02	10 14
04 We	10 45 46	13 04 57	24 05 13	17 22	13 42	09 11	14 32	04 50	14 36	07 29	15 03	10 09
05 Th	10 49 42	14 05 04	05♍ 55 28	18 38	14 55	09 57	14 26	04 51	14 40	07 31	15 05	10 04
06 Fr	10 53 39	15 05 10	17 44 17	19 57	16 08	10 42	14 20	04 52	14 43	07 34	15 06	10 01
07 Sa	10 57 35	16 05 14	29 33 38	21 17	17 21	11 28	14 14	04 53	14 46	07 36	15 07	10 00
08 Su	11 01 32	17 05 15	11♎ 25 40	22 38	18 34	12 13	14 08	04 54	14 49	07 38	15 08	09D 59
09 Mo	11 05 28	18 05 16	23 22 49	24 01	19 47	12 59	14 03	04 54	14 52	07 40	15 09	10 00
10 Tu	11 09 25	19 05 14	05♏ 27 55	25 25	21 00	13 44	13 57	04 55	14 55	07 43	15 11	10 02
11 We	11 13 22	20 05 11	17 44 10	26 50	22 13	14 30	13 52	04 55	14 58	07 45	15 12	10 04
12 Th	11 17 18	21 05 06	00♐ 15 06	28 17	23 26	15 15	13 47	04 55	15 01	07 47	15 13	10 05
13 Fr	11 21 15	22 04 59	13 04 25	29 46	24 39	16 01	13 42	04 56	15 05	07 49	15 14	10 06
14 Sa	11 25 11	23 04 51	26 15 34	01♓ 15	25 51	16 46	13 37	04 56	15 08	07 52	15 15	10R 06
15 Su	11 29 08	24 04 41	09♑ 51 20	02 46	27 04	17 31	13 33	04 R 56	15 11	07 54	15 16	10 05
16 Mo	11 33 04	25 04 30	23 53 02	04 18	28 17	18 17	13 28	04 56	15 14	07 56	15 17	10 04
17 Tu	11 37 01	26 04 16	08♒ 19 50	05 51	29 29	19 02	13 24	04 56	15 18	07 58	15 18	10 02
18 We	11 40 57	27 04 01	23 08 15	07 26	00♉ 41	19 47	13 20	04 55	15 21	08 00	15 19	10 00
19 Th	11 44 54	28 03 44	08♓ 12 03	09 02	01 54	20 32	13 16	04 55	15 24	08 03	15 20	09 58
20 Fr	11 48 51	29 03 26	23 22 47	10 39	03 06	21 17	13 12	04 54	15 28	08 05	15 20	09 57
21 Sa	11 52 47	00♈ 03 05	08♈ 30 53	12 17	04 18	22 02	13 08	04 54	15 31	08 07	15 21	09D 57
22 Su	11 56 44	01 02 42	23 27 08	13 57	05 30	22 47	13 05	04 53	15 34	08 09	15 22	09 57
23 Mo	12 00 40	02 02 17	08♉ 03 57	15 38	06 43	23 32	13 02	04 52	15 38	08 11	15 23	09 58
24 Tu	12 04 37	03 01 50	22 16 15	17 20	07 55	24 17	12 59	04 51	15 41	08 13	15 24	09 59
25 We	12 08 33	04 01 21	06♊ 01 45	19 04	09 06	25 02	12 56	04 50	15 44	08 15	15 24	10 00
26 Th	12 12 30	05 00 49	19 20 35	20 48	10 18	25 46	12 53	04 49	15 48	08 17	15 25	10 01
27 Fr	12 16 26	06 00 15	02♋ 14 50	22 34	11 30	26 31	12 51	04 48	15 51	08 20	15 26	10 01
28 Sa	12 20 23	06 59 39	14 47 51	24 22	12 42	27 16	12 48	04 47	15 54	08 22	15 26	10R 01
29 Su	12 24 20	07 59 00	27 03 42	26 10	13 53	28 00	12 46	04 45	15 58	08 24	15 27	10 01
30 Mo	12 28 16	08 58 19	09♌ 06 41	28 00	15 05	28 45	12 44	04 44	16 01	08 26	15 28	10 01
31 Tu	12 32 13	09 57 36	21 00 58	29 52	16 16	29 30	12 43	04 42	16 05	08 28	15 28	10 01

Data for	03-01-2015
Julian Day	2457082.50
Ayanamsa	24 04 11
SVP	05 ♓ 02 57
☽ ☊ Mean	11 ♎ 49 R

● ◐ PHASES ○ ◑

Day	Time	Phase	Position
05	18:06	○	14♍50
13	17:48	◑	22♐49
20	09:36	☀	29♓27
27	07:43	◐	06♋19

LAST ASPECT ☽ INGRESS

Day	h m	Day	h m	Sign
28	17:53	01	23:35	♌
03	08:49	04	11:59	♍
05	18:37	07	00:53	♎
09	01:25	09	13:11	♏
11	19:47	11	23:31	♐
13	23:12	14	06:41	♑
16	08:03	16	10:14	♒
17	18:19	18	10:58	♓
20	09:37	20	10:28	♈
21	22:51	22	10:41	♉
23	14:25	24	13:23	♊
26	12:36	26	19:46	♋
29	01:59	29	05:49	♌
30	13:58	31	18:14	♍

DECLINATION

Day	☉	☽	☿	♀	♂	♃	♄	♅	♆	♇
01 Su	07S 47	17N06	17S 23	03N19	02N17	17N19	19S 04	05N07	09S 29	20S 31
02 Mo	07 25	15 19	17 10	03 50	02 36	17 21	19 04	05 08	09 28	20 30
03 Tu	07 02	12 53	16 56	04 21	02 55	17 23	19 04	05 09	09 27	20 30
04 We	06 39	09 56	16 41	04 52	03 13	17 25	19 04	05 11	09 26	20 30
05 Th	06 16	06 36	16 24	05 23	03 32	17 26	19 04	05 12	09 26	20 30
06 Fr	05 52	03 01	16 06	05 54	03 51	17 28	19 04	05 13	09 25	20 30
07 Sa	05 29	00S 42	15 47	06 25	04 09	17 30	19 04	05 14	09 24	20 30
08 Su	05 06	04 24	15 26	06 55	04 27	17 32	19 04	05 15	09 23	20 30
09 Mo	04 43	07 57	15 04	07 25	04 46	17 33	19 04	05 17	09 22	20 30
10 Tu	04 19	11 12	14 41	07 56	05 04	17 35	19 04	05 18	09 21	20 30
11 We	03 56	14 01	14 17	08 26	05 22	17 36	19 04	05 19	09 21	20 30
12 Th	03 32	16 14	13 51	08 55	05 41	17 38	19 04	05 20	09 20	20 30
13 Fr	03 08	17 42	13 24	09 25	05 59	17 39	19 03	05 22	09 19	20 29
14 Sa	02 45	18 16	12 56	09 54	06 17	17 40	19 03	05 23	09 18	20 29
15 Su	02 21	17 49	12 26	10 24	06 35	17 42	19 03	05 24	09 17	20 29
16 Mo	01 57	16 17	11 55	10 53	06 53	17 43	19 03	05 25	09 17	20 29
17 Tu	01 34	13 41	11 23	11 21	07 10	17 44	19 03	05 27	09 16	20 29
18 We	01 10	10 10	10 50	11 50	07 28	17 45	19 03	05 28	09 15	20 29
19 Th	00 46	05 55	10 16	12 18	07 46	17 46	19 02	05 29	09 14	20 29
20 Fr	00 23	01 15	09 40	12 46	08 03	17 47	19 02	05 31	09 13	20 29
21 Sa	00N01	03 29	09 03	13 13	08 21	17 48	19 02	05 32	09 13	20 29
22 Su	00 25	07 57	08 25	13 41	08 38	17 49	19 02	05 33	09 12	20 29
23 Mo	00 49	11 49	07 46	14 08	08 55	17 50	19 01	05 34	09 11	20 29
24 Tu	01 12	14 53	07 06	14 34	09 13	17 51	19 01	05 36	09 10	20 29
25 We	01 36	16 59	06 25	15 01	09 30	17 52	19 01	05 37	09 09	20 29
26 Th	02 00	18 04	05 42	15 27	09 47	17 52	19 00	05 38	09 09	20 29
27 Fr	02 23	18 10	04 58	15 52	10 03	17 53	19 00	05 40	09 08	20 29
28 Sa	02 47	17 22	04 13	16 17	10 20	17 54	19 00	05 41	09 07	20 29
29 Su	03 10	15 46	03 28	16 42	10 37	17 54	18 59	05 42	09 06	20 29
30 Mo	03 33	13 30	02 41	17 07	10 53	17 55	18 59	05 44	09 06	20 28
31 Tu	03 57	10 42	01 53	17 31	11 10	17 55	18 58	05 45	09 05	20 28

ASPECTARIAN

01 06:47 ☿ # ♃
15:56 ☿ ⚹ ♅
21:16 ☿ ☍ ♃
02 09:10 ☽ △ ♄
15:54 ☽ △ ♂
03 00:37 ☽ △ ♀
04:47 ☽ △ ♅
04:53 ☽ ☌ ♃
08:49 ☽ ☍ ☿
12:28 ♃ △ ♅
04 03:43 ☽ # ♆
14:50 ♀ ∥ ♅
15:15 ♀ △ ♃
18:46 ♀ ☌ ♅
21:49 ☽ □ ♄
05 02:36 ☽ # ☉
03:14 ♀ □ ♇
03:15 ☽ ☍ ♆
07:13 ☽ ∥ ♀
09:29 ☽ ∥ ♅
18:37 ☽ △ ♇
18:58 ☽ ∥ ♂
23:22 ♀ # ☉
06 00:17 ☉ ⚹ ♇
07 10:47 ☽ ⚹ ♄
14:41 ☉ # ♅
08 00:25 ☽ # ♂
01:43 ☽ ☍ ♂
04:11 ☽ ∥ ☉
05:25 ☽ ⚹ ♃
05:43 ☽ # ♅
06:51 ☽ ☍ ♅
07:29 ☽ □ ♇
16:00 ☽ ☍ ♀
19:43 ☽ # ♀
22:06 ☉ # ♂
09 01:25 ☽ △ ☿
10:08 ☽ ∥ ♆
10 04:26 ☽ △ ♆
06:05 ♂ △ ♃
16:32 ☽ □ ♃
19:03 ☽ ⚹ ♇
19:26 ♂ ∥ ♅
11 02:10 ☽ ∥ ☿
04:57 ☽ △ ☉
16:06 ♂ ☌ ♅
19:47 ☽ □ ☿
22:38 ♂ □ ♇
12 08:49 ☽ ☌ ♄
14:13 ☽ □ ♆
19:13 ♀ # ♆
22:59 ☽ # ♃
13 01:09 ☽ △ ♃
03:42 ☽ △ ♅
03:52 ☿ ♓
05:44 ☽ △ ♂
23:12 ☽ △ ♀
14 10:01 ☽ ⚹ ☿
15:03 ♄ SR
20:35 ☽ ⚹ ♆
15 02:38 ☽ # ♃
09:15 ☽ □ ♅
09:21 ☽ ☌ ♇
13:57 ☽ □ ♂
16 02:09 ☽ ⚹ ☉
08:03 ☽ □ ♀
09:41 ☿ □ ♄
18:24 ☽ ⚹ ♄
17 00:50 ☿ # ♀
01:51 ♅ □ ♇
08:14 ☽ ☍ ♃
10:15 ♀ ♉
11:24 ☽ ⚹ ♅
14:33 ☽ # ♀
18:19 ☽ ⚹ ♂
19:01 ☽ ∥ ☿
18 05:26 ☽ ∥ ♆
08:49 ☿ ☌ ♆
13:07 ☽ ⚹ ♀
14:33 ☽ # ♂
18:47 ☽ □ ♄
23:45 ☽ ☌ ♆
19 01:28 ☽ ☌ ☿
02:12 ☽ # ♅
11:17 ☽ ⚹ ♇
20 04:45 ☽ ∥ ☉
07:30 ☽ # ☉
17:58 ☿ ∥ ♆
18:15 ☽ △ ♄
22:46 ☉ ♈
21 07:22 ☽ △ ♃
10:43 ☽ ∥ ♅
10:57 ☽ □ ♇
11:14 ☽ ☌ ♅
18:33 ☿ # ♂
22:51 ☽ ☌ ♂
22 02:20 ☽ # ☿
04:13 ☽ ∥ ♂
07:13 ☽ # ♆
20:25 ☿ ⚹ ♇
21:32 ☽ ☌ ♀
23 00:12 ☽ ⚹ ♆
08:16 ☽ □ ♃
12:17 ☽ △ ♇
14:25 ☽ ⚹ ☿
20:40 ☽ ∥ ♀
20:55 ♂ # ♆
24 06:26 ♀ ⚹ ♆
20:10 ☽ ⚹ ☉
21:54 ☽ ☍ ♄
25 03:58 ☽ □ ♆
12:18 ☽ ⚹ ♃
17:30 ☽ ⚹ ♅
17:34 ☽ ∥ ♃
19:25 ☉ △ ♄
26 01:55 ☿ # ♅
03:06 ☽ □ ☿
12:36 ☽ ⚹ ♂
27 11:29 ☽ ∥ ♃
11:35 ☽ △ ♆
19:30 ☽ ⚹ ♀
28 01:15 ☽ ☍ ♇
02:10 ☽ □ ♅
02:11 ♀ □ ♃
13:54 ☽ ∥ ♀
21:56 ☽ △ ☿
29 01:59 ☽ □ ♂
06:03 ☿ # ☉
15:15 ☽ △ ♄
23:42 ☽ △ ☉
30 07:16 ☽ ☌ ♃
07:46 ♀ △ ♇
13:20 ☽ □ ♀
13:58 ☽ △ ♅
20:35 ☽ ∥ ♂
31 01:44 ☿ ♈
12:21 ☽ # ♆
16:27 ♂ ♉
18:21 ☽ △ ♂

⚷ Chiron

01 Dec.	01 S 07
01	16♓49
04	17 01
07	17 12
10	17 23
13	17 34
16	17 45
19	17 56
22	18 07
25	18 18
28	18 29
31	18 40

April 2015

04 12:01 14♎24 ✸ Total Lunar Eclipse (mag 1.006)

Day	S. T.	☉	☽	☿	♀	♂	♃	♄	♅	♆	♇	☊ True
	h m s	° ' "	° ' "	° '	° '	° '	° '	° '	° '	° '	° '	° '
01 We	12 36 09	10♈ 56 51	02♍ 50 26	01♈ 45	17♉ 27	00♉ 14	12♌R41	04♐R41	16♈ 08	08♓ 30	15♑ 29	10♎D01
02 Th	12 40 06	11 56 03	14 38 30	03 39	18 39	00 58	12 40	04 39	16 12	08 32	15 29	10 01
03 Fr	12 44 02	12 55 13	26 28 05	05 34	19 50	01 43	12 39	04 37	16 15	08 34	15 30	10 01
04 Sa	12 47 59	13 54 21	08♎ 21 44	07 31	21 01	02 27	12 38	04 35	16 18	08 36	15 30	10 01
05 Su	12 51 55	14 53 27	20 21 40	09 29	22 12	03 11	12 37	04 33	16 22	08 38	15 30	10R 01
06 Mo	12 55 52	15 52 31	02♏ 29 48	11 28	23 22	03 56	12 36	04 31	16 25	08 39	15 31	10 01
07 Tu	12 59 48	16 51 33	14 47 53	13 29	24 33	04 40	12 36	04 29	16 29	08 41	15 31	10 00
08 We	13 03 45	17 50 33	27 17 37	15 31	25 44	05 24	12 35	04 27	16 32	08 43	15 31	09 59
09 Th	13 07 42	18 49 32	10♐ 00 40	17 34	26 54	06 08	12D 35	04 24	16 36	08 45	15 32	09 58
10 Fr	13 11 38	19 48 28	22 58 48	19 38	28 04	06 52	12 36	04 22	16 39	08 47	15 32	09 57
11 Sa	13 15 35	20 47 23	06♑ 13 41	21 42	29 15	07 36	12 36	04 19	16 42	08 49	15 32	09 57
12 Su	13 19 31	21 46 16	19 46 48	23 48	00♊ 25	08 20	12 36	04 17	16 46	08 51	15 32	09 56
13 Mo	13 23 28	22 45 08	03♒ 38 57	25 53	01 35	09 04	12 37	04 14	16 49	08 52	15 32	09D 56
14 Tu	13 27 24	23 43 57	17 49 58	28 00	02 45	09 48	12 38	04 11	16 53	08 54	15 32	09 57
15 We	13 31 21	24 42 45	02♓ 18 15	00♉ 06	03 55	10 32	12 39	04 08	16 56	08 56	15 33	09 57
16 Th	13 35 17	25 41 31	17 00 23	02 12	05 04	11 15	12 40	04 05	16 59	08 57	15 33	09 58
17 Fr	13 39 14	26 40 16	01♈ 51 07	04 18	06 14	11 59	12 42	04 02	17 03	08 59	15 33	09 58
18 Sa	13 43 11	27 38 58	16 43 40	06 22	07 24	12 43	12 44	03 59	17 06	09 01	15R 33	09R 58
19 Su	13 47 07	28 37 39	01♉ 30 24	08 26	08 33	13 26	12 45	03 56	17 10	09 02	15 33	09 57
20 Mo	13 51 04	29 36 18	16 03 57	10 28	09 42	14 10	12 47	03 53	17 13	09 04	15 32	09 56
21 Tu	13 55 00	00♉ 34 54	00♊ 18 05	12 29	10 51	14 53	12 50	03 49	17 16	09 06	15 32	09 53
22 We	13 58 57	01 33 29	14 08 37	14 27	12 00	15 37	12 52	03 46	17 20	09 07	15 32	09 51
23 Th	14 02 53	02 32 02	27 33 42	16 23	13 09	16 20	12 54	03 43	17 23	09 09	15 32	09 48
24 Fr	14 06 50	03 30 32	10♋ 33 44	18 16	14 18	17 04	12 57	03 39	17 26	09 10	15 32	09 46
25 Sa	14 10 46	04 29 01	23 11 00	20 06	15 26	17 47	13 00	03 35	17 30	09 12	15 32	09 44
26 Su	14 14 43	05 27 27	05♌ 29 07	21 53	16 35	18 30	13 03	03 32	17 33	09 13	15 31	09D 44
27 Mo	14 18 40	06 25 51	17 32 36	23 37	17 43	19 13	13 06	03 28	17 36	09 15	15 31	09 44
28 Tu	14 22 36	07 24 13	29 26 18	25 17	18 51	19 56	13 10	03 24	17 40	09 16	15 31	09 46
29 We	14 26 33	08 22 33	11♍ 15 05	26 53	19 59	20 40	13 13	03 21	17 43	09 18	15 31	09 48
30 Th	14 30 29	09 20 50	23 03 32	28 25	21 07	21 23	13 17	03 17	17 46	09 19	15 30	09 50

Data for	04-01-2015
Julian Day	2457113.50
Ayanamsa	24 04 14
SVP	05 ♓ 02 51
☽ ☊ Mean	10 ♎ 10 R

● ◐ PHASES ○ ◑

04	12:06	✸	14♎24
12	03:44	◑	21♑55
18	18:58	●	28♈25
25	23:56	◐	05♌27

LAST ASPECT ☽ INGRESS

Day	h m	Day	h m	
02	09:02	03	07:09	♎
04	16:00	05	19:05	♏
07	20:42	08	05:09	♐
09	17:42	10	12:47	♑
12	08:15	12	17:44	♒
14	19:45	14	20:12	♓
15	21:37	16	21:01	♈
18	18:58	18	21:32	♉
19	23:08	20	23:29	♊
22	05:39	23	04:27	♋
24	17:05	25	13:14	♌
27	14:14	28	01:08	♍
30	12:24	30	14:03	♎

DECLINATION

Day	☉	☽	☿	♀	♂	♃	♄	♅	♆	♇
01 We	04N20	07N28	01S 04	17N54	11N26	17N55	18S 58	05N46	09S 04	20S 28
02 Th	04 43	03 57	00 14	18 17	11 42	17 56	18 58	05 48	09 04	20 28
03 Fr	05 06	00 16	00N37	18 40	11 58	17 56	18 57	05 49	09 03	20 28
04 Sa	05 29	03S 27	01 29	19 02	12 14	17 56	18 57	05 50	09 02	20 28
05 Su	05 52	07 04	02 21	19 24	12 30	17 56	18 56	05 52	09 01	20 28
06 Mo	06 15	10 25	03 14	19 45	12 46	17 57	18 56	05 53	09 01	20 28
07 Tu	06 37	13 22	04 08	20 06	13 01	17 57	18 55	05 54	09 00	20 28
08 We	07 00	15 45	05 02	20 26	13 16	17 57	18 55	05 56	08 59	20 28
09 Th	07 22	17 24	05 57	20 46	13 32	17 56	18 54	05 57	08 59	20 28
10 Fr	07 45	18 11	06 52	21 05	13 47	17 56	18 53	05 58	08 58	20 28
11 Sa	08 07	18 01	07 48	21 24	14 02	17 56	18 53	06 00	08 57	20 28
12 Su	08 29	16 49	08 43	21 42	14 17	17 56	18 52	06 01	08 57	20 28
13 Mo	08 51	14 37	09 39	22 00	14 31	17 56	18 52	06 02	08 56	20 28
14 Tu	09 13	11 31	10 34	22 17	14 46	17 55	18 51	06 03	08 55	20 28
15 We	09 34	07 38	11 29	22 33	15 00	17 55	18 50	06 05	08 55	20 28
16 Th	09 56	03 14	12 23	22 49	15 15	17 55	18 50	06 06	08 54	20 28
17 Fr	10 17	01N25	13 16	23 04	15 29	17 54	18 49	06 07	08 54	20 29
18 Sa	10 38	06 00	14 08	23 19	15 43	17 53	18 48	06 09	08 53	20 29
19 Su	10 59	10 10	14 59	23 33	15 56	17 53	18 48	06 10	08 52	20 29
20 Mo	11 20	13 39	15 48	23 47	16 10	17 52	18 47	06 11	08 52	20 29
21 Tu	11 40	16 14	16 35	24 00	16 24	17 52	18 46	06 13	08 51	20 29
22 We	12 01	17 47	17 21	24 12	16 37	17 51	18 46	06 14	08 51	20 29
23 Th	12 21	18 16	18 05	24 24	16 50	17 50	18 45	06 15	08 50	20 29
24 Fr	12 41	17 47	18 46	24 35	17 03	17 49	18 44	06 16	08 50	20 29
25 Sa	13 01	16 24	19 25	24 45	17 16	17 48	18 44	06 18	08 49	20 29
26 Su	13 20	14 18	20 02	24 55	17 28	17 47	18 43	06 19	08 48	20 29
27 Mo	13 40	11 36	20 36	25 04	17 41	17 46	18 42	06 20	08 48	20 29
28 Tu	13 59	08 28	21 08	25 12	17 53	17 45	18 41	06 21	08 47	20 29
29 We	14 18	05 01	21 37	25 20	18 05	17 44	18 41	06 23	08 47	20 29
30 Th	14 36	01 22	22 04	25 27	18 17	17 43	18 40	06 24	08 46	20 29

ASPECTARIAN

01 01:26 ♀ ∥ ♃
03:44 ☽ □ ♄
11:32 ☽ ☍ ♆
11:40 ☽ ∥ ♅
19:22 ☽ ∥ ☉
02 01:43 ☽ △ ♃
09:02 ☽ △ ♀
12:21 ☿ △ ♄
17:21 ☉ △ ♃
22:09 ☽ ∥ ☿
03 07:20 ☽ # ☿
16:25 ☽ ⚹ ♄
18:02 ♀ # ♄
21:58 ☽ ☍ ☿
04 08:33 ☽ ⚹ ♃
14:19 ☽ □ ♆
14:54 ☽ # ☉
15:46 ☽ # ♅
16:00 ☽ ☍ ♅
23:36 ☉ ∥ ♅
05 13:37 ☽ ∥ ♆
15:06 ☉ □ ♆
06 02:59 ☽ ☍ ♂
12:06 ☽ △ ♆
13:26 ☿ △ ♃
14:08 ☉ ☌ ♅
19:44 ☽ □ ♃
20:30 ☽ # ♂
07 01:23 ☽ ⚹ ♆
20:42 ☽ ☍ ♀
08 00:05 ☿ □ ♆
02:36 ♀ # ♆
12:20 ☿ ☌ ♅
13:30 ☽ [illegible] ♄
16:58 ♃ SD
21:38 ☽ □ ♆
23:49 ☿ ∥ ♅
09 04:49 ☽ △ ♃
12:18 ☽ △ ♅
12:40 ☽ # ♃
16:41 ☽ △ ☿
17:42 ☽ △ ☉
10 04:01 ☿ ☌ ☉
11 02:36 ☽ △ ♂
02:41 ☽ # ♃
04:38 ☽ ⚹ ♆
13:38 ☿ ∥ ☉
15:29 ♀ ♊
16:33 ☽ ☌ ♆
18:41 ☽ □ ☿
12 05:41 ☽ # ♆
08:15 ☽ □ ☿
17:23 ♂ ⚹ ♆
20:08 ☽ △ ♀
13 00:50 ☽ # ♂
00:59 ☽ ⚹ ♄
05:32 ☉ # ♆
09:44 ☽ □ ♂
15:15 ☽ ☍ ♃
22:24 ☽ ⚹ ♅
14 05:00 ☽ # ☿
10:33 ☽ ⚹ ☉
13:28 ☽ # ☉
16:29 ☽ ∥ ♆
19:43 ☽ ⚹ ♆
22:52 ☿ ♉
15 02:52 ☽ □ ♀
03:00 ☽ □ ♄
04:26 ♀ ☍ ♄
08:41 ☽ # ♅
10:53 ☽ ☌ ♆
14:10 ☽ ⚹ ♂
21:37 ☽ ⚹ ♆
17 03:31 ☽ △ ♄
03:52 ♆ SR
07:40 ☽ ⚹ ♀
17:31 ☽ △ ♃
22:05 ☽ □ ♆
18 00:30 ♂ □ ♃
00:37 ☽ ☌ ♅
00:47 ☽ ∥ ♅
16:10 ☽ # ♅
19 05:37 ☽ ∥ ☉
07:13 ☿ ⚹ ♆
10:30 ♀ □ ♆
12:23 ☽ ⚹ ♆
13:13 ☽ ☌ ☿
18:33 ☽ □ ♃
20:40 ☽ ☌ ♂
23:08 ☽ △ ♆
20 09:42 ☉ ♉
15:31 ☿ ∥ ♂
21 01:59 ☽ ∥ ♂
04:16 ☿ □ ♃
06:01 ☽ ☍ ♄
07:16 ☽ ∥ ☿
15:11 ☽ □ ♆
19:54 ☽ ☌ ♀
21:32 ♂ △ ♆
21:45 ☽ ⚹ ♃
22 01:24 ☽ ∥ ♃
05:39 ☽ ⚹ ♅
13:25 ☿ △ ♆
15:56 ☿ ∥ ♃
18:41 ♀ ⚹ ♃
23:04 ☿ ☌ ♂
23 05:45 ☽ ∥ ☿
09:50 ☽ ⚹ ☉
21:24 ☽ △ ♆
22:57 ♀ # ♄
22:59 ☽ ∥ ♃
24 09:22 ☽ ☍ ♆
12:33 ☽ ∥ ♂
13:01 ☽ ⚹ ♂
13:03 ☽ □ ♅
17:05 ☽ ⚹ ☿
25 20:10 ☽ △ ♄
26 08:08 ☽ ∥ ☉
15:05 ☽ ☌ ♃
18:46 ☿ # ♆
21:31 ♀ ⚹ ♅
27 00:08 ☽ △ ♅
00:23 ☽ ⚹ ♀
03:35 ☽ □ ♂
09:37 ♂ ∥ ♃
14:14 ☽ □ ☿
21:43 ☽ # ♆
28 08:01 ☽ □ ♄
14:51 ☽ ∥ ♅
17:38 ☽ △ ☉
20:00 ☽ ☍ ♆
29 08:39 ☽ △ ♆
19:38 ☽ □ ♀
20:22 ☽ △ ♂
23:11 ☉ ⚹ ♆
30 12:24 ☽ △ ☿
20:34 ☽ ⚹ ♄

⚷ Chiron

01 Dec.	00 S 23
03	18♓50
06	19 00
09	19 10
12	19 20
15	19 30
18	19 39
21	19 49
24	19 57
27	20 06
30	20 14

19 N Declination

May 2015

Day	S. T. h m s	☉ ° ' "	☽ ° ' "	☿ ° '	♀ ° '	♂ ° '	♃ ° '	♄ ° '	♅ ° '	♆ ° '	♇ ° '	☊ True ° '
01 Fr	14 34 26	10♉ 19 06	04♎ 55 47	29♉ 53	22♊ 15	22♉ 06	13♌ 21	03♐R 13	17♈ 49	09♓ 20	15♑R 30	09♎ 51
02 Sa	14 38 22	11 17 20	16 55 21	01♊ 17	23 22	22 49	13 25	03 09	17 53	09 22	15 29	09R 50
03 Su	14 42 19	12 15 32	29 05 02	02 36	24 29	23 31	13 29	03 05	17 56	09 23	15 29	09 48
04 Mo	14 46 15	13 13 42	11♏ 26 54	03 51	25 36	24 14	13 34	03 01	17 59	09 24	15 28	09 45
05 Tu	14 50 12	14 11 51	24 02 13	05 02	26 43	24 57	13 38	02 57	18 02	09 25	15 28	09 40
06 We	14 54 09	15 09 58	06♐ 51 29	06 08	27 50	25 40	13 43	02 53	18 05	09 27	15 27	09 34
07 Th	14 58 05	16 08 03	19 54 34	07 09	28 57	26 23	13 48	02 49	18 08	09 28	15 27	09 28
08 Fr	15 02 02	17 06 07	03♑ 11 00	08 06	00♋ 03	27 05	13 53	02 44	18 12	09 29	15 26	09 22
09 Sa	15 05 58	18 04 10	16 40 10	08 58	01 09	27 48	13 58	02 40	18 15	09 30	15 26	09 18
10 Su	15 09 55	19 02 11	00♒ 21 22	09 46	02 15	28 30	14 04	02 36	18 18	09 31	15 25	09 15
11 Mo	15 13 51	20 00 10	14 13 58	10 28	03 21	29 13	14 09	02 32	18 21	09 32	15 24	09 13
12 Tu	15 17 48	20 58 09	28 17 15	11 06	04 27	29 55	14 15	02 27	18 24	09 33	15 24	09D 14
13 We	15 21 44	21 56 06	12♓ 30 09	11 38	05 32	00♊ 38	14 21	02 23	18 27	09 34	15 23	09 14
14 Th	15 25 41	22 54 02	26 51 00	12 06	06 37	01 20	14 27	02 19	18 30	09 35	15 22	09 15
15 Fr	15 29 38	23 51 56	11♈ 17 14	12 28	07 42	02 02	14 33	02 14	18 33	09 36	15 21	09R 15
16 Sa	15 33 34	24 49 50	25 45 04	12 46	08 47	02 45	14 39	02 10	18 36	09 37	15 21	09 14
17 Su	15 37 31	25 47 42	10♉ 09 44	12 58	09 51	03 27	14 46	02 05	18 39	09 38	15 20	09 11
18 Mo	15 41 27	26 45 32	24 25 45	13 06	10 56	04 09	14 52	02 01	18 41	09 39	15 19	09 06
19 Tu	15 45 24	27 43 22	08♊ 27 47	13 09	12 00	04 51	14 59	01 56	18 44	09 39	15 18	08 59
20 We	15 49 20	28 41 10	22 11 18	13R 07	13 03	05 33	15 06	01 52	18 47	09 40	15 17	08 50
21 Th	15 53 17	29 38 56	05♋ 33 26	13 00	14 07	06 15	15 13	01 47	18 50	09 41	15 16	08 42
22 Fr	15 57 13	00♊ 36 41	18 33 09	12 49	15 10	06 57	15 20	01 43	18 53	09 42	15 15	08 35
23 Sa	16 01 10	01 34 24	01♌ 11 24	12 34	16 13	07 39	15 27	01 39	18 55	09 42	15 14	08 29
24 Su	16 05 07	02 32 06	13 30 47	12 15	17 16	08 21	15 34	01 34	18 58	09 43	15 13	08 25
25 Mo	16 09 03	03 29 46	25 35 10	11 52	18 18	09 03	15 42	01 30	19 01	09 44	15 12	08 23
26 Tu	16 12 60	04 27 25	07♍ 29 19	11 27	19 20	09 45	15 49	01 25	19 03	09 44	15 11	08D 23
27 We	16 16 56	05 25 02	19 18 27	10 58	20 22	10 26	15 57	01 21	19 06	09 45	15 10	08 24
28 Th	16 20 53	06 22 37	01♎ 07 52	10 28	21 24	11 08	16 05	01 16	19 09	09 45	15 09	08 25
29 Fr	16 24 49	07 20 12	13 02 39	09 55	22 25	11 50	16 13	01 12	19 11	09 46	15 08	08R 26
30 Sa	16 28 46	08 17 45	25 07 22	09 22	23 25	12 31	16 21	01 07	19 14	09 46	15 07	08 24
31 Su	16 32 42	09 15 16	07♏ 25 45	08 48	24 26	13 13	16 29	01 03	19 16	09 47	15 06	08 20

Data for	05-01-2015
Julian Day	2457143.50
Ayanamsa	24 04 17
SVP	05 ♓ 02 46
☽ ☊ Mean	08 ♎ 35 R

● ◐ PHASES ○ ◑

Day	Time	Phase	Position
04	03:42	○	13♏23
11	10:37	◑	20♒26
18	04:14	●	26♉56
25	17:19	◐	04♍11

LAST ASPECT Day	h m	☽ INGRESS Day	h m	Sign
02	14:04	03	01:48	♏
05	01:50	05	11:13	♐
07	17:52	07	18:17	♑
09	20:36	09	23:23	♒
11	10:37	12	02:54	♓
13	16:56	14	05:15	♈
15	12:05	16	07:04	♉
18	04:15	18	09:28	♊
19	17:58	20	13:57	♋
22	00:37	22	21:43	♌
24	10:50	25	08:52	♍
27	02:22	27	21:43	♎
29	20:21	30	09:34	♏

DECLINATION

Day	☉	☽	☿	♀	♂	♃	♄	♅	♆	♇
01 Fr	14N55	02S 22	22N29	25N34	18N29	17N42	18S 39	06N25	08S 46	20S 29
02 Sa	15 13	06 03	22 50	25 40	18 41	17 40	18 38	06 26	08 46	20 29
03 Su	15 31	09 32	23 10	25 45	18 52	17 39	18 37	06 27	08 45	20 30
04 Mo	15 48	12 39	23 27	25 49	19 03	17 38	18 37	06 29	08 45	20 30
05 Tu	16 06	15 15	23 41	25 53	19 14	17 36	18 36	06 30	08 44	20 30
06 We	16 23	17 08	23 54	25 56	19 25	17 35	18 35	06 31	08 44	20 30
07 Th	16 40	18 10	24 04	25 59	19 36	17 33	18 34	06 32	08 43	20 30
08 Fr	16 56	18 14	24 12	26 01	19 46	17 32	18 33	06 33	08 43	20 30
09 Sa	17 13	17 17	24 17	26 02	19 56	17 30	18 32	06 35	08 43	20 30
10 Su	17 29	15 20	24 21	26 02	20 06	17 29	18 32	06 36	08 42	20 30
11 Mo	17 44	12 29	24 23	26 02	20 16	17 27	18 31	06 37	08 42	20 30
12 Tu	18 00	08 52	24 22	26 01	20 26	17 25	18 30	06 38	08 42	20 31
13 We	18 15	04 42	24 20	26 00	20 36	17 23	18 29	06 39	08 41	20 31
14 Th	18 30	00 14	24 16	25 58	20 45	17 22	18 28	06 40	08 41	20 31
15 Fr	18 44	04N18	24 10	25 55	20 54	17 20	18 27	06 41	08 41	20 31
16 Sa	18 58	08 35	24 02	25 52	21 03	17 18	18 27	06 43	08 40	20 31
17 Su	19 12	12 21	23 53	25 48	21 12	17 16	18 26	06 44	08 40	20 31
18 Mo	19 26	15 20	23 42	25 43	21 20	17 14	18 25	06 45	08 40	20 31
19 Tu	19 39	17 21	23 30	25 38	21 28	17 12	18 24	06 46	08 39	20 31
20 We	19 52	18 18	23 15	25 32	21 37	17 10	18 23	06 47	08 39	20 32
21 Th	20 04	18 12	23 00	25 26	21 44	17 08	18 22	06 48	08 39	20 32
22 Fr	20 16	17 09	22 43	25 19	21 52	17 06	18 22	06 49	08 39	20 32
23 Sa	20 28	15 16	22 25	25 11	22 00	17 03	18 21	06 50	08 38	20 32
24 Su	20 40	12 44	22 06	25 03	22 07	17 01	18 20	06 51	08 38	20 32
25 Mo	20 51	09 42	21 46	24 55	22 14	16 59	18 19	06 52	08 38	20 32
26 Tu	21 02	06 18	21 26	24 45	22 21	16 56	18 18	06 53	08 38	20 33
27 We	21 12	02 41	21 04	24 35	22 28	16 54	18 17	06 54	08 38	20 33
28 Th	21 22	01S 03	20 43	24 25	22 34	16 52	18 17	06 55	08 37	20 33
29 Fr	21 32	04 47	20 21	24 14	22 40	16 49	18 16	06 56	08 37	20 33
30 Sa	21 41	08 21	19 59	24 03	22 46	16 47	18 15	06 57	08 37	20 33
31 Su	21 50	11 39	19 38	23 51	22 52	16 44	18 14	06 58	08 37	20 33

ASPECTARIAN

01 02:00 ☿ ♊
16:59 ☽ ✶ ♃
19:26 ♂ # ♄
21:09 ☽ □ ♇

02 01:54 ☽ ☍ ♅
02:37 ☽ # ♅
14:04 ☽ △ ♀
18:28 ☽ ∥ ♆

03 08:36 ☿ ☍ ♄
20:03 ☽ △ ♆

04 04:05 ☽ □ ♃
07:43 ☽ ✶ ♇
09:03 ☉ □ ♃

05 01:50 ☽ ☍ ♂
11:12 ☽ # ☉
16:38 ☽ ☌ ♄
22:32 ☽ ☍ ☿

06 04:48 ☽ □ ♆
07:09 ☉ △ ♇
07:40 ☽ # ♃
12:45 ☽ △ ♃
20:46 ☽ △ ♅

07 17:52 ☽ ☍ ♀
22:52 ♀ ☌ ♋

08 11:16 ☽ ✶ ♆
20:19 ☽ # ♃
21:48 ☽ ☌ ♇

09 01:11 ☽ # ☉
02:40 ☽ △ ☉
02:47 ☽ □ ♅
16:09 ☿ □ ♆
20:36 ☽ △ ♂

10 00:03 ☉ ∥ ♃
03:53 ☽ ✶ ♄
17:11 ☽ △ ☿
23:52 ☽ ☍ ♃

11 07:05 ☽ ✶ ♅

12 01:07 ☽ ∥ ♆
02:41 ♂ ♊
02:55 ☽ □ ♂
07:02 ☽ □ ♄
11:17 ♂ # ♇
11:18 ☽ △ ♀
13:10 ☽ # ♅
19:04 ☽ ☌ ♆
22:29 ☽ □ ☿

13 04:50 ☽ ✶ ♇
16:56 ☽ ✶ ☉
21:58 ☉ # ♄

14 07:51 ☽ ✶ ♂
09:02 ☽ △ ♄
17:34 ☽ □ ♀

15 02:01 ☽ ✶ ☿
05:27 ☽ △ ♃
06:03 ♂ ☍ ♄
06:45 ☽ □ ♇
12:05 ☽ ☌ ♅
13:12 ☽ ∥ ♅

16 00:31 ☽ # ♆
18:49 ♀ △ ♆
23:06 ☽ ✶ ♆
23:27 ☽ ✶ ♀

17 07:45 ☽ □ ♃
08:39 ☽ △ ♇

18 12:51 ☽ ☍ ♄
17:27 ☽ ☌ ♂
21:47 ☽ ∥ ♃

19 01:50 ☿ SR
02:04 ☽ □ ♆
08:07 ☽ ☌ ☿
11:25 ☽ ✶ ♃
17:58 ☽ ✶ ♅

20 07:11 ☽ # ♄
12:55 ☽ ∥ ♄

21 07:33 ☽ △ ♆
08:45 ☉ ♊
17:08 ☽ ☌ ♀
17:51 ☽ ☍ ♇

22 00:37 ☽ □ ♅
00:59 ☽ ∥ ♃
01:55 ♀ ☍ ♇

23 00:48 ☽ ✶ ☉
00:52 ☽ △ ♄
01:36 ☉ ☍ ♄
07:58 ☉ # ♇
13:16 ☽ ✶ ♂
21:35 ☽ ✶ ☿
23:21 ☿ ∥ ♂

24 04:06 ☽ ☌ ♃
10:50 ☽ △ ♅

25 07:48 ☽ # ♆
11:48 ☽ □ ♄
17:10 ♀ □ ♅
20:06 ☽ ∥ ♅
23:39 ♂ □ ♆

26 04:33 ☽ ☍ ♆
04:52 ☽ □ ♂
07:43 ☽ □ ☿
15:36 ☽ △ ♇
18:05 ☿ ∥ ☉

27 02:22 ☽ ✶ ♀
10:37 ☿ ☌ ♂

28 00:17 ☽ ✶ ♄
10:33 ☿ # ♇
11:32 ☽ △ ☉
18:01 ☽ △ ☿
21:25 ☽ △ ♂

29 04:11 ☽ □ ♇
06:24 ☽ ✶ ♃
07:01 ☿ □ ♆
12:18 ☽ ☍ ♅
14:23 ☽ # ♅
20:21 ☽ □ ♀

30 01:53 ☽ ∥ ♆
16:56 ☿ ☌ ☉

31 04:31 ☽ △ ♆
13:08 ☉ □ ♆
14:40 ☽ ✶ ♇
17:32 ☽ □ ♃

⚷ Chiron

01 Dec. 00 N 16

Day	° '
03	20♓22
06	20 30
09	20 37
12	20 44
15	20 50
18	20 56
21	21 02
24	21 07
27	21 12
30	21 16

June 2015

Day	S. T. h m s	☉ ° ' "	☽ ° ' "	☿ ° '	♀ ° '	♂ ° '	♃ ° '	♄ ° '	♅ ° '	♆ ° '	♇ ° '	☊ True ° '
01 Mo	16 36 39	10♊ 12 47	20♏ 00 34	08♊R15	25♋ 26	13♊ 54	16♌ 37	00♐R59	19♈ 19	09♓ 47	15♑R05	08♎R13
02 Tu	16 40 36	11 10 16	02♐ 53 16	07 42	26 26	14 36	16 46	00 54	19 21	09 47	15 04	08 04
03 We	16 44 32	12 07 44	16 03 51	07 10	27 25	15 17	16 54	00 50	19 23	09 48	15 02	07 54
04 Th	16 48 29	13 05 11	29 30 55	06 40	28 24	15 59	17 03	00 45	19 26	09 48	15 01	07 43
05 Fr	16 52 25	14 02 38	13♑ 11 58	06 12	29 22	16 40	17 12	00 41	19 28	09 48	15 00	07 33
06 Sa	16 56 22	15 00 03	27 03 51	05 47	00♌ 20	17 21	17 21	00 37	19 30	09 48	14 59	07 24
07 Su	17 00 18	15 57 28	11♒ 03 21	05 25	01 18	18 03	17 30	00 33	19 33	09 49	14 58	07 18
08 Mo	17 04 15	16 54 52	25 07 39	05 07	02 15	18 44	17 39	00 28	19 35	09 49	14 56	07 14
09 Tu	17 08 11	17 52 16	09♓ 14 34	04 53	03 12	19 25	17 48	00 24	19 37	09 49	14 55	07 13
10 We	17 12 08	18 49 39	23 22 33	04 42	04 08	20 06	17 57	00 20	19 39	09 49	14 54	07D13
11 Th	17 16 05	19 47 01	07♈ 30 29	04 36	05 04	20 47	18 07	00 16	19 41	09 49	14 52	07R13
12 Fr	17 20 01	20 44 23	21 37 14	04D34	05 59	21 28	18 16	00 12	19 43	09 49	14 51	07 12
13 Sa	17 23 58	21 41 45	05♉ 41 18	04 36	06 54	22 09	18 26	00 08	19 45	09 R 49	14 50	07 09
14 Su	17 27 54	22 39 06	19 40 28	04 43	07 48	22 50	18 36	00 04	19 47	09 49	14 48	07 04
15 Mo	17 31 51	23 36 26	03♊ 31 52	04 55	08 42	23 31	18 45	00 00	19 49	09 49	14 47	06 56
16 Tu	17 35 47	24 33 46	17 12 10	05 11	09 35	24 12	18 55	29♏ 56	19 51	09 49	14 46	06 46
17 We	17 39 44	25 31 06	00♋ 38 08	05 32	10 27	24 53	19 05	29 52	19 53	09 49	14 44	06 34
18 Th	17 43 40	26 28 25	13 47 15	05 57	11 19	25 33	19 15	29 49	19 55	09 49	14 43	06 22
19 Fr	17 47 37	27 25 43	26 38 08	06 27	12 10	26 14	19 26	29 45	19 57	09 48	14 41	06 10
20 Sa	17 51 34	28 23 01	09♌ 10 56	07 01	13 01	26 55	19 36	29 41	19 58	09 48	14 40	06 00
21 Su	17 55 30	29 20 18	21 27 14	07 39	13 51	27 35	19 46	29 38	20 00	09 48	14 39	05 53
22 Mo	17 59 27	00♋ 17 34	03♍ 29 59	08 22	14 40	28 16	19 57	29 34	20 02	09 48	14 37	05 48
23 Tu	18 03 23	01 14 49	15 23 16	09 09	15 28	28 57	20 07	29 31	20 03	09 47	14 36	05 46
24 We	18 07 20	02 12 04	27 11 56	10 00	16 16	29 37	20 18	29 27	20 05	09 47	14 34	05D46
25 Th	18 11 16	03 09 18	09♎ 01 20	10 55	17 03	00♋ 18	20 28	29 24	20 06	09 46	14 33	05 47
26 Fr	18 15 13	04 06 32	20 56 59	11 54	17 48	00 58	20 39	29 21	20 08	09 46	14 31	05R46
27 Sa	18 19 09	05 03 45	03♏ 04 10	12 57	18 34	01 38	20 50	29 17	20 09	09 46	14 30	05 44
28 Su	18 23 06	06 00 57	15 27 34	14 04	19 18	02 19	21 01	29 14	20 11	09 45	14 28	05 40
29 Mo	18 27 03	06 58 09	28 10 53	15 15	20 01	02 59	21 12	29 11	20 12	09 45	14 27	05 33
30 Tu	18 30 59	07 55 21	11♐ 16 22	16 29	20 43	03 39	21 23	29 08	20 13	09 44	14 25	05 23

Data for	06-01-2015
Julian Day	2457174.50
Ayanamsa	24 04 21
SVP	05 ♓ 02 41
☽ ☊ Mean	06 ♎ 56 R

● ◐ PHASES ○ ◑		
02	16:19 ○	11♐49
09	15:42 ◑	18♓30
16	14:05 ●	25♊07
24	11:03 ◐	02♎38

LAST ASPECT Day	h m	☽ INGRESS Day	h m	
01	11:02	01	18:40	♐
03	06:00	04	00:51	♑
05	10:55	06	05:03	♒
07	14:32	08	08:17	♓
09	18:09	10	11:15	♈
11	23:44	12	14:17	♉
13	22:07	14	17:51	♊
16	14:06	16	22:51	♋
19	05:52	19	06:23	♌
21	16:10	21	16:59	♍
24	05:13	24	05:42	♎
25	23:24	26	17:58	♏
29	01:52	29	03:23	♐

DECLINATION

Day	☉	☽	☿	♀	♂	♃	♄	♅	♆	♇
01 Mo	21N59	14S 28	19N17	23N39	22N58	16N42	18S 13	06N59	08S 37	20S 34
02 Tu	22 07	16 40	18 56	23 26	23 03	16 39	18 13	06 59	08 37	20 34
03 We	22 14	18 02	18 37	23 13	23 08	16 36	18 12	07 00	08 37	20 34
04 Th	22 22	18 26	18 19	22 59	23 13	16 34	18 11	07 01	08 37	20 34
05 Fr	22 29	17 47	18 03	22 45	23 18	16 31	18 10	07 02	08 37	20 34
06 Sa	22 35	16 04	17 48	22 30	23 23	16 28	18 09	07 03	08 37	20 35
07 Su	22 42	13 24	17 34	22 15	23 27	16 25	18 09	07 04	08 37	20 35
08 Mo	22 47	09 56	17 23	22 00	23 31	16 23	18 08	07 05	08 36	20 35
09 Tu	22 53	05 53	17 13	21 44	23 35	16 20	18 07	07 05	08 36	20 35
10 We	22 58	01 31	17 06	21 28	23 39	16 17	18 06	07 06	08 36	20 36
11 Th	23 02	02N57	17 00	21 12	23 42	16 14	18 06	07 07	08 36	20 36
12 Fr	23 07	07 16	16 56	20 55	23 45	16 11	18 05	07 08	08 37	20 36
13 Sa	23 10	11 08	16 55	20 38	23 48	16 08	18 04	07 08	08 37	20 36
14 Su	23 14	14 22	16 55	20 21	23 51	16 05	18 04	07 09	08 37	20 36
15 Mo	23 17	16 44	16 57	20 03	23 54	16 02	18 03	07 10	08 37	20 37
16 Tu	23 19	18 07	17 01	19 45	23 56	15 58	18 02	07 11	08 37	20 37
17 We	23 22	18 27	17 07	19 27	23 58	15 55	18 02	07 11	08 37	20 37
18 Th	23 23	17 47	17 15	19 08	24 00	15 52	18 01	07 12	08 37	20 37
19 Fr	23 25	16 12	17 24	18 50	24 02	15 49	18 00	07 13	08 37	20 38
20 Sa	23 25	13 53	17 35	18 31	24 04	15 46	18 00	07 13	08 37	20 38
21 Su	23 26	11 00	17 47	18 11	24 05	15 42	17 59	07 14	08 37	20 38
22 Mo	23 26	07 42	18 00	17 52	24 06	15 39	17 59	07 14	08 37	20 38
23 Tu	23 26	04 08	18 14	17 33	24 07	15 36	17 58	07 15	08 38	20 38
24 We	23 25	00 25	18 30	17 13	24 08	15 32	17 57	07 15	08 38	20 39
25 Th	23 24	03S 19	18 46	16 53	24 08	15 29	17 57	07 16	08 38	20 39
26 Fr	23 22	06 57	19 04	16 34	24 08	15 25	17 56	07 17	08 38	20 39
27 Sa	23 20	10 21	19 21	16 14	24 09	15 22	17 56	07 17	08 38	20 39
28 Su	23 18	13 23	19 40	15 54	24 08	15 18	17 55	07 18	08 39	20 40
29 Mo	23 15	15 51	19 58	15 34	24 08	15 15	17 55	07 18	08 39	20 40
30 Tu	23 12	17 35	20 17	15 14	24 08	15 11	17 54	07 19	08 39	20 40

ASPECTARIAN

01 11:02 ☽ △ ♀
20:21 ☽ ☌ ♄
23:47 ☽ ∥ ♃

02 08:28 ☽ ☍ ☿
12:38 ☽ □ ♆
22:32 ☽ ☍ ♂

03 01:32 ☽ △ ♃
04:43 ☽ ∥ ♄
05:49 ♀ ∥ ♂
06:00 ☽ △ ♅
16:19 ☽ ∥ ☿

04 12:34 ☿ ∥ ♄
14:08 ☽ ∥ ♄
14:57 ☽ ∥ ☿
18:04 ☽ ✶ ♆

05 03:08 ☽ ☌ ♇
10:55 ☽ □ ♅
15:33 ♀ ♌
18:18 ♀ ∥ ☉
19:20 ☽ ∥ ♃
23:39 ♂ ✶ ♃

06 06:03 ☽ ☍ ♀
06:05 ☽ ✶ ♄
06:24 ♀ △ ♄
14:36 ☽ △ ☿

07 08:59 ☽ △ ☉
11:07 ☽ ☍ ♃
12:32 ☽ △ ♂
14:32 ☽ ✶ ♅

08 08:12 ☽ ∥ ♆
09:03 ☽ □ ♄
10:42 ☽ □ ☿
17:12 ☽ ∥ ♅
21:56 ☉ ✶ ♃

09 00:58 ☽ ☌ ♆
07:30 ♂ ✶ ♅
09:37 ☽ ✶ ♇
18:09 ☽ □ ♂

10 11:46 ☽ △ ♄
12:52 ☿ ✶ ♀
19:05 ☽ ✶ ☿
19:34 ☽ △ ♀
21:31 ☉ ✶ ♅

11 12:30 ☽ □ ♇
18:14 ☽ △ ♃
20:46 ☽ ☌ ♅
22:24 ☽ ✶ ☉
22:34 ☿ SD
23:16 ☽ ∥ ♅
23:44 ☽ ✶ ♂

12 08:01 ☽ ∥ ♆
09:10 ♆ SR

13 02:13 ☽ □ ♀
02:35 ♀ ∥ ♇
07:04 ☽ ✶ ♆
15:38 ☽ △ ♇
22:07 ☽ □ ♃

14 15:57 ☉ ☌ ♂
16:01 ☽ ∥ ♃
17:[illegible] ☽ ☌ ♄

15 00:39 ♄ ♏R
02:27 ☽ ☌ ☿
03:04 ☽ ∥ ☿
09:39 ☽ ✶ ♀
10:59 ☽ □ ♆
22:12 ☽ ∥ ♄

16 03:05 ☽ ✶ ♃
04:43 ☽ ✶ ♅
13:06 ☽ ☌ ♂

17 16:41 ☽ △ ♆
18:14 ☽ ∥ ♄

18 01:43 ☽ ☍ ♇
08:41 ☽ ∥ ☿
11:24 ☽ □ ♅

19 04:41 ☽ ∥ ♃
05:52 ☽ △ ♄
19:36 ☽ ✶ ☿

20 07:59 ☽ ☌ ♀
20:38 ☽ ☌ ♃
21:07 ☽ △ ♅

21 12:54 ☽ ✶ ♂
15:52 ♀ ∥ ♄
16:10 ☽ □ ♄
16:38 ☉ ♋
17:01 ☽ ✶ ☉
17:32 ☽ ∥ ♆
[illegible] ☿ ∥ [illegible]
21:37 ☿ ∥ ♄

22 03:11 ☽ ∥ ♅
10:28 ☽ □ ☿
12:40 ☽ ☍ ♆
13:45 ♃ △ ♅
22:24 ☽ △ ♇

23 18:09 ☿ □ ♆

24 04:34 ☽ ✶ ♄
05:13 ☽ □ ♂
13:34 ♂ ♋

25 04:10 ☽ △ ☿
11:08 ☽ □ ♇
17:17 ☽ ✶ ♀
22:21 ☽ ☍ ♅
23:24 ☽ ✶ ♃

26 02:15 ☽ ∥ ♅
11:39 ☽ ∥ ♆
21:02 ☽ △ ♂

27 04:14 ☽ △ ☉
13:02 ☽ △ ♆
22:07 ☽ ✶ ♇

28 07:45 ☽ □ ♀
10:43 ☽ □ ♃
17:40 ☽ ∥ ♃
21:14 ☽ ∥ ♀

29 01:52 ☽ ☌ ♄
06:26 ♀ △ ♅
21:13 ☽ □ ♆

30 03:20 ♀ ∥ ♃
06:32 ☽ ∥ ♄
10:22 ☽ ☍ ☿
16:03 ☽ △ ♅
17:48 ☽ △ ♀
18:19 ☽ △ [illegible]

⚷ Chiron

01	Dec.	00 N 44
02	21♓20	
05	21 23	
08	21 26	
11	21 28	
14	21 30	
17	21 32	
20	21 33	
23	21 33	
26	21 33R	
29	21 33	
24	09:39	21♓33 R

July 2015

Day	S. T. h m s	☉ ° ' "	☽ ° ' "	☿ ° '	♀ ° '	♂ ° '	♃ ° '	♄ ° '	♅ ° '	♆ ° '	♇ ° '	☊ True ° '
01 We	18 34 56	08♋52 32	24♐44 23	17♊48	21♌24	04♋20	21♌34	29♏R05	20♈14	09♓R44	14♑R24	05♎R12
02 Th	18 38 52	09 49 44	08♑33 12	19 10	22 04	05 00	21 45	29 02	20 16	09 43	14 23	05 00
03 Fr	18 42 49	10 46 55	22 39 08	20 35	22 43	05 40	21 56	29 00	20 17	09 42	14 21	04 48
04 Sa	18 46 45	11 44 06	06♒56 59	22 05	23 21	06 20	22 08	28 57	20 18	09 42	14 20	04 38
05 Su	18 50 42	12 41 17	21 21 01	23 37	23 58	07 00	22 19	28 54	20 19	09 41	14 18	04 31
06 Mo	18 54 38	13 38 28	05♓45 46	25 14	24 33	07 40	22 30	28 52	20 20	09 40	14 17	04 26
07 Tu	18 58 35	14 35 40	20 06 52	26 54	25 08	08 20	22 42	28 49	20 21	09 39	14 15	04 24
08 We	19 02 32	15 32 52	04♈21 23	28 37	25 41	09 00	22 53	28 47	20 22	09 39	14 14	04 24
09 Th	19 06 28	16 30 04	18 27 36	00♋23	26 12	09 40	23 05	28 45	20 23	09 38	14 12	04D 24
10 Fr	19 10 25	17 27 16	02♉24 48	02 12	26 42	10 20	23 17	28 43	20 24	09 37	14 11	04R 23
11 Sa	19 14 21	18 24 30	16 12 39	04 05	27 11	11 00	23 28	28 40	20 24	09 36	14 09	04 21
12 Su	19 18 18	19 21 43	29 50 47	06 00	27 38	11 40	23 40	28 38	20 25	09 35	14 08	04 16
13 Mo	19 22 14	20 18 57	13♊18 34	07 58	28 04	12 19	23 52	28 36	20 26	09 34	14 06	04 09
14 Tu	19 26 11	21 16 12	26 34 59	09 58	28 28	12 59	24 04	28 35	20 26	09 33	14 05	04 00
15 We	19 30 07	22 13 27	09♋38 52	12 00	28 50	13 39	24 16	28 33	20 27	09 33	14 03	03 48
16 Th	19 34 04	23 10 42	22 29 06	14 04	29 10	14 19	24 28	28 31	20 28	09 32	14 02	03 36
17 Fr	19 38 01	24 07 57	05♌05 05	16 09	29 29	14 58	24 40	28 30	20 28	09 31	14 00	03 24
18 Sa	19 41 57	25 05 13	17 27 00	18 15	29 46	15 38	24 52	28 28	20 28	09 29	13 59	03 14
19 Su	19 45 54	26 02 29	29 35 59	20 23	00♍01	16 17	25 04	28 27	20 29	09 28	13 57	03 07
20 Mo	19 49 50	26 59 46	11♍34 15	22 31	00 14	16 57	25 17	28 25	20 29	09 27	13 56	03 02
21 Tu	19 53 47	27 57 02	23 24 57	24 39	00 25	17 37	25 29	28 24	20 29	09 26	13 54	03 00
22 We	19 57 43	28 54 19	05♎12 11	26 47	00 33	18 16	25 41	28 23	20 30	09 25	13 53	03D 00
23 Th	20 01 40	29 51 36	17 00 41	28 55	00 40	18 55	25 53	28 22	20 30	09 24	13 52	03 01
24 Fr	20 05 36	00♌48 54	28 55 41	01♌02	00 44	19 35	26 06	28 21	20 30	09 23	13 50	03 01
25 Sa	20 09 33	01 46 12	11♏02 34	03 09	00 46	20 14	26 18	28 20	20 30	09 22	13 49	03R 01
26 Su	20 13 30	02 43 30	23 26 30	05 14	00R 46	20 54	26 31	28 19	20 30	09 20	13 47	02 59
27 Mo	20 17 26	03 40 49	06♐12 00	07 19	00 43	21 33	26 43	28 19	20R 30	09 19	13 46	02 55
28 Tu	20 21 23	04 38 08	19 22 18	09 22	00 38	22 12	26 56	28 18	20 30	09 18	13 45	02 48
29 We	20 25 19	05 35 27	02♑58 51	11 24	00 31	22 51	27 08	28 18	20 30	09 17	13 43	02 40
30 Th	20 29 16	06 32 48	17 00 41	13 25	00 21	23 31	27 21	28 17	20 30	09 15	13 42	02 31
31 Fr	20 33 12	07 30 09	01♒24 14	15 24	00 09	24 10	27 33	28 17	20 30	09 14	13 41	02 23

Data for	07-01-2015
Julian Day	2457204.50
Ayanamsa	24 04 25
SVP	05♓02 38
☽ ☊ Mean	05♎21 R

● ◐ PHASES ○ ◑

Day	Time	Phase	Position
02	02:20	○	09♑55
08	20:24	◑	16♈22
16	01:24	●	23♋14
24	04:05	◐	00♏59
31	10:43	○	07♒56

LAST ASPECT ☽ INGRESS

Day	h m	Day	h m	Sign
30	18:19	01	09:12	♑
03	10:39	03	12:22	♒
05	12:32	05	14:24	♓
07	14:37	07	16:38	♈
09	13:47	09	19:50	♉
11	21:52	12	00:16	♊
14	03:32	14	06:14	♋
16	11:25	16	14:16	♌
18	21:42	19	00:48	♍
21	10:08	21	13:24	♎
23	18:13	24	02:09	♏
26	09:16	26	12:26	♐
28	13:38	28	18:48	♑
30	18:51	30	21:41	♒

DECLINATION

Day	☉	☽	☿	♀	♂	♃	♄	♅	♆	♇
01 We	23N08	18S24	20N36	14N53	24N07	15N08	17S54	07N19	08S39	20S40
02 Th	23 04	18 10	20 55	14 33	24 06	15 04	17 54	07 19	08 39	20 41
03 Fr	23 00	16 50	21 13	14 13	24 05	15 00	17 53	07 20	08 40	20 41
04 Sa	22 55	14 26	21 31	13 53	24 03	14 57	17 53	07 20	08 40	20 41
05 Su	22 50	11 07	21 49	13 33	24 02	14 53	17 52	07 21	08 40	20 41
06 Mo	22 44	07 09	22 05	13 14	24 00	14 49	17 52	07 21	08 41	20 42
07 Tu	22 38	02 46	22 21	12 54	23 58	14 45	17 52	07 21	08 41	20 42
08 We	22 32	01N44	22 35	12 34	23 56	14 42	17 52	07 22	08 41	20 42
09 Th	22 25	06 06	22 48	12 15	23 53	14 38	17 51	07 22	08 42	20 42
10 Fr	22 18	10 04	22 59	11 56	23 51	14 34	17 51	07 22	08 42	20 43
11 Sa	22 10	13 27	23 09	11 37	23 48	14 30	17 51	07 22	08 42	20 43
12 Su	22 02	16 03	23 16	11 18	23 45	14 26	17 50	07 23	08 43	20 43
13 Mo	21 54	17 43	23 22	10 59	23 42	14 22	17 50	07 23	08 43	20 43
14 Tu	21 45	18 24	23 25	10 41	23 38	14 18	17 50	07 23	08 43	20 44
15 We	21 36	18 06	23 25	10 23	23 35	14 14	17 50	07 23	08 44	20 44
16 Th	21 27	16 52	23 23	10 05	23 31	14 10	17 50	07 23	08 44	20 44
17 Fr	21 17	14 50	23 18	09 48	23 27	14 06	17 50	07 24	08 45	20 45
18 Sa	21 07	12 09	23 11	09 31	23 23	14 02	17 50	07 24	08 45	20 45
19 Su	20 56	08 59	23 00	09 15	23 19	13 58	17 50	07 24	08 45	20 45
20 Mo	20 45	05 30	22 48	08 59	23 14	13 54	17 49	07 24	08 46	20 45
21 Tu	20 34	01 50	22 32	08 44	23 10	13 50	17 49	07 24	08 46	20 46
22 We	20 22	01S53	22 14	08 29	23 05	13 46	17 49	07 24	08 47	20 46
23 Th	20 11	05 32	21 53	08 14	23 00	13 42	17 49	07 24	08 47	20 46
24 Fr	19 58	08 59	21 30	08 01	22 54	13 37	17 49	07 24	08 48	20 46
25 Sa	19 46	12 08	21 05	07 47	22 49	13 33	17 49	07 24	08 48	20 47
26 Su	19 33	14 48	20 37	07 35	22 44	13 29	17 50	07 24	08 49	20 47
27 Mo	19 20	16 51	20 08	07 23	22 38	13 25	17 50	07 24	08 49	20 47
28 Tu	19 06	18 04	19 37	07 12	22 32	13 20	17 50	07 24	08 50	20 47
29 We	18 52	18 18	19 04	07 02	22 26	13 16	17 50	07 24	08 50	20 48
30 Th	18 38	17 27	18 30	06 53	22 20	13 12	17 50	07 24	08 51	20 48
31 Fr	18 23	15 29	17 54	06 44	22 13	13 07	17 50	07 24	08 51	20 48

ASPECTARIAN

01 05:35 ☿ # ♇
07:52 ♀ ☌ ♃
17:34 ☽ ☍ ♂
21:10 ☉ △ ♆
22:43 ♂ △ ♆

02 02:00 ☽ ⚹ ♆
06:52 ☽ ∥ ♄
09:57 ☽ ☌ ♇
18:50 ☿ ⚹ ♅
19:59 ☽ □ ♅

03 10:39 ☽ ⚹ ♄
19:21 ☽ # ♃

04 00:54 ☿ ⚹ ♃
04:47 ☽ # ♀
22:17 ☽ ⚹ ♅

05 01:38 ☽ ☍ ♃
04:15 ☽ △ ☿
04:33 ☽ ☍ ♀
08:19 ☿ ⚹ ♀
12:32 ☽ □ ♄
15:06 ☽ ∥ ♆
22:49 ☽ # ♅

06 03:20 ☽ △ ♂
06:31 ☽ ☌ ♆
14:06 ☽ △ ☉
14:12 ☽ ⚹ ♇
15:37 ☉ ☍ ♇

07 12:57 ☽ □ ☿
14:37 ☽ △ ♄
19:50 ☿ ∥ ☉

08 08:16 ☽ □ ♂
16:44 ☽ □ ♇
18:52 ☿ ♋

09 03:17 ☽ ☌ ♅
07:21 ☽ ∥ ♅
08:02 ☽ △ ♃
13:47 ☽ △ ♀
15:21 ☽ # ♆
23:35 ☽ ⚹ ☿

10 11:28 ☽ ∥ ♀
12:29 ☽ ⚹ ♆
14:26 ☽ ⚹ ♂
20:24 ☽ △ ♇

11 04:08 ☽ ⚹ ☉
08:35 ☽ ∥ ♃
12:56 ☽ □ ♃
19:57 ☽ □ ♀
21:52 ☽ ☍ ♄

12 17:19 ☽ □ ♆

13 02:32 ☽ # ♄
02:53 ☉ □ ♅
12:50 ☽ ⚹ ♅
19:14 ☿ △ ♆
19:21 ☽ ⚹ ♃

14 03:32 ☽ ⚹ ♀
06:49 ♀ □ ♄
23:48 ☽ △ ♆

15 05:11 ☽ ☌ ☿
06:45 ☽ # ♄
07:50 ☽ ☌ ♂
08:10 ☽ ☍ ♇
14:09 ♂ ☍ ♇
20:11 ☽ □ ♅
23:38 ☿ ☍ ♇

16 04:15 ☿ ☌ ♂
11:25 ☽ △ ♄

17 07:08 ☽ ∥ ♃

18 05:56 ☽ △ ♅
14:51 ☽ ☌ ♃
21:42 ☽ □ ♄
21:55 ☽ ∥ ♀
22:39 ♀ ♍

19 00:50 ☽ ☌ ♀
01:08 ☿ □ ♅
01:36 ☽ # ♆
11:07 ☽ ∥ ♅
19:45 ☽ ☍ ♆
23:54 ☉ # ♇

20 04:45 ☽ △ ♇
11:31 ☽ ⚹ ♂
19:44 ♀ # ♆

21 03:04 ☽ ⚹ ☿
10:02 ☽ ⚹ ☉
10:08 ☽ ⚹ ♄
11:08 ☉ △ ♄

22 17:37 ☽ □ ♇
17:51 ☿ △ ♄

23 03:31 ☉ ♌
04:06 ☽ □ ♂
07:03 ☽ ☍ ♅
12:15 ☿ ♌
12:42 ☽ # ♅
17:23 ☽ # ♀
18:13 ☽ ⚹ ♃
19:24 ☿ ☌ ☉
22:31 ☽ ∥ ♆

24 03:38 ☽ ⚹ ♀
05:07 ☽ □ ☿
20:42 ☽ △ ♆

25 05:24 ☽ ⚹ ♇
09:30 ♀ SR
09:43 ♂ □ ♅
11:45 ☽ # ♃
15:57 ☿ # ♆
18:51 ☽ △ ♂

26 05:56 ☽ □ ♃
09:16 ☽ ☌ ♄
10:39 ♅ SR
13:50 ☽ □ ♀
18:57 ☽ △ ☉
22:32 ♀ ∥ ♅

27 02:27 ☽ △ ☿
05:45 ☽ □ ♆
17:34 ☽ ∥ ♄

28 02:01 ☽ △ ♅
13:38 ☽ △ ♃
19:45 ☽ △ ♀

29 10:50 ☽ ⚹ ♆
14:27 ☿ ∥ ☉
16:40 ☽ ∥ ♄
18:24 ☽ ☌ ♇

30 05:52 ☽ □ ♅
11:25 ☽ ☍ ♂
18:51 ☽ ⚹ ♄

31 02:37 ☿ # ♄
15:28 ♀ ♌R
19:55 ☽ # ♃

⚷ Chiron

01	Dec.	00 N 55
02	21♓32R	
05	21 30	
08	21 28	
11	21 26	
14	21 23	
17	21 20	
20	21 16	
23	21 12	
26	21 07	
29	21 02	

August 2015

Day	S. T.	☉	☽	☿	♀	♂	♃	♄	♅	♆	♇	☊ True
	h m s	° ' "	° ' "	° '	° '	° '	° '	° '	° '	° '	° '	° '
01 Sa	20 37 09	08♌ 27 30	16♒ 03 37	17♌ 22	29♌R55	24♋ 49	27♌ 46	28♏R17	20♈R29	09♓R13	13♑R39	02♎R15
02 Su	20 41 05	09 24 53	00♓ 51 23	19 18	29 38	25 28	27 59	28 17	20 29	09 11	13 38	02 10
03 Mo	20 45 02	10 22 16	15 39 44	21 12	29 18	26 07	28 11	28D 17	20 29	09 10	13 37	02 06
04 Tu	20 48 59	11 19 40	00♈ 21 41	23 05	28 57	26 46	28 24	28 17	20 28	09 08	13 35	02 05
05 We	20 52 55	12 17 06	14 51 57	24 57	28 33	27 25	28 37	28 17	20 28	09 07	13 34	02D 05
06 Th	20 56 52	13 14 33	29 07 12	26 46	28 08	28 04	28 50	28 18	20 27	09 06	13 33	02 06
07 Fr	21 00 48	14 12 01	13♉ 05 57	28 34	27 40	28 43	29 02	28 18	20 27	09 04	13 32	02 07
08 Sa	21 04 45	15 09 30	26 48 02	00♍ 21	27 11	29 22	29 15	28 19	20 26	09 03	13 30	02R 06
09 Su	21 08 41	16 07 01	10♊ 14 03	02 06	26 40	00♌ 01	29 28	28 19	20 26	09 01	13 29	02 04
10 Mo	21 12 38	17 04 34	23 24 59	03 49	26 07	00 40	29 41	28 20	20 25	09 00	13 28	02 01
11 Tu	21 16 34	18 02 07	06♋ 21 49	05 31	25 33	01 18	29 54	28 21	20 24	08 58	13 27	01 55
12 We	21 20 31	18 59 42	19 05 29	07 11	24 59	01 57	00♍ 07	28 22	20 24	08 57	13 26	01 48
13 Th	21 24 28	19 57 18	01♌ 36 46	08 50	24 23	02 36	00 20	28 23	20 23	08 55	13 24	01 40
14 Fr	21 28 24	20 54 55	13 56 30	10 27	23 46	03 15	00 33	28 24	20 22	08 54	13 23	01 33
15 Sa	21 32 21	21 52 34	26 05 40	12 03	23 09	03 53	00 46	28 25	20 21	08 52	13 22	01 26
16 Su	21 36 17	22 50 14	08♍ 05 38	13 37	22 32	04 32	00 59	28 26	20 20	08 51	13 21	01 22
17 Mo	21 40 14	23 47 54	19 58 17	15 10	21 55	05 11	01 12	28 28	20 19	08 49	13 20	01 19
18 Tu	21 44 10	24 45 36	01♎ 46 07	16 41	21 18	05 49	01 25	28 29	20 18	08 47	13 19	01D 19
19 We	21 48 07	25 43 20	13 32 18	18 11	20 41	06 28	01 38	28 31	20 17	08 46	13 18	01 19
20 Th	21 52 03	26 41 04	25 20 36	19 39	20 05	07 07	01 51	28 32	20 16	08 44	13 17	01 21
21 Fr	21 55 60	27 38 49	07♏ 15 22	21 05	19 31	07 45	02 04	28 34	20 15	08 43	13 16	01 23
22 Sa	21 59 57	28 36 36	19 21 19	22 30	18 57	08 24	02 17	28 36	20 13	08 41	13 15	01 25
23 Su	22 03 53	29 34 24	01♐ 43 22	23 54	18 24	09 02	02 30	28 38	20 12	08 39	13 14	01R 25
24 Mo	22 07 50	00♍ 32 13	14 26 10	25 16	17 54	09 41	02 43	28 40	20 11	08 38	13 13	01 24
25 Tu	22 11 46	01 30 03	27 33 40	26 36	17 24	10 19	02 56	28 42	20 10	08 36	13 12	01 22
26 We	22 15 43	02 27 54	11♑ 08 25	27 54	16 57	10 57	03 09	28 44	20 08	08 34	13 12	01 19
27 Th	22 19 39	03 25 47	25 10 56	29 11	16 32	11 36	03 22	28 47	20 07	08 33	13 11	01 15
28 Fr	22 23 36	04 23 41	09♒ 39 06	00♎ 26	16 08	12 14	03 35	28 49	20 05	08 31	13 10	01 11
29 Sa	22 27 32	05 21 36	24 28 01	01 38	15 47	12 52	03 48	28 51	20 04	08 30	13 09	01 08
30 Su	22 31 29	06 19 33	09♓ 30 20	02 49	15 28	13 31	04 01	28 54	20 02	08 28	13 08	01 06
31 Mo	22 35 26	07 17 31	24 37 09	03 58	15 12	14 09	04 14	28 57	20 01	08 26	13 08	01 04

Data for	08-01-2015
Julian Day	2457235.50
Ayanamsa	24 04 30
SVP	05 ♓ 02 34
☽ ☊ Mean	03 ♎ 43 R

● ◐ PHASES ○ ◑

07	02:03	◑	14♉17
14	14:54	●	21♌31
22	19:31	◐	29♏24
29	18:35	○	06♓07

LAST ASPECT ☽ Day	h m	INGRESS Day	h m	
01	22:03	01	22:37	♓
03	20:36	03	23:24	♈
05	23:30	06	01:30	♉
08	04:46	08	05:40	♊
10	11:46	10	12:09	♋
12	17:45	12	20:53	♌
15	04:38	15	07:47	♍
17	17:17	17	20:24	♎
20	02:57	20	09:25	♏
22	19:32	22	20:42	♐
24	22:04	25	04:22	♑
27	07:20	27	08:04	♒
29	07:03	29	08:52	♓
31	06:54	31	08:34	♈

DECLINATION

Day	☉	☽	☿	♀	♂	♃	♄	♅	♆	♇
01 Sa	18N09	12S30	17N17	06N37	22N06	13N03	17S50	07N24	08S52	20S48
02 Su	17 54	08 41	16 39	06 30	22 00	12 59	17 51	07 24	08 52	20 49
03 Mo	17 38	04 19	16 00	06 24	21 53	12 54	17 51	07 24	08 53	20 49
04 Tu	17 23	00N17	15 21	06 19	21 46	12 50	17 51	07 23	08 53	20 49
05 We	17 07	04 48	14 40	06 16	21 38	12 46	17 51	07 23	08 54	20 49
06 Th	16 50	08 58	13 59	06 13	21 31	12 41	17 52	07 23	08 54	20 50
07 Fr	16 34	12 32	13 18	06 10	21 24	12 37	17 52	07 23	08 55	20 50
08 Sa	16 17	15 20	12 35	06 09	21 16	12 32	17 52	07 22	08 56	20 50
09 Su	16 00	17 14	11 53	06 09	21 08	12 28	17 53	07 22	08 56	20 51
10 Mo	15 43	18 11	11 10	06 10	21 00	12 23	17 53	07 22	08 57	20 51
11 Tu	15 25	18 09	10 27	06 12	20 52	12 19	17 54	07 22	08 57	20 51
12 We	15 08	17 12	09 44	06 15	20 43	12 14	17 54	07 21	08 58	20 51
13 Th	14 50	15 26	09 01	06 18	20 35	12 10	17 55	07 21	08 59	20 52
14 Fr	14 31	12 59	08 18	06 22	20 26	12 05	17 55	07 21	08 59	20 52
15 Sa	14 13	10 00	07 35	06 28	20 18	12 01	17 56	07 20	09 00	20 52
16 Su	13 54	06 37	06 52	06 33	20 09	11 56	17 56	07 20	09 00	20 52
17 Mo	13 35	03 02	06 09	06 40	20 00	11 51	17 57	07 20	09 01	20 52
18 Tu	13 16	00S40	05 26	06 47	19 51	11 47	17 57	07 19	09 02	20 53
19 We	12 57	04 19	04 44	06 55	19 41	11 42	17 58	07 19	09 02	20 53
20 Th	12 37	07 49	04 01	07 03	19 32	11 38	17 58	07 18	09 03	20 53
21 Fr	12 17	11 01	03 20	07 12	19 22	11 33	17 59	07 18	09 03	20 53
22 Sa	11 57	13 48	02 38	07 21	19 13	11 28	18 00	07 17	09 04	20 54
23 Su	11 37	16 02	01 57	07 31	19 03	11 24	18 00	07 17	09 05	20 54
24 Mo	11 17	17 33	01 16	07 41	18 53	11 19	18 01	07 16	09 05	20 54
25 Tu	10 56	18 11	00 36	07 51	18 43	11 14	18 02	07 16	09 06	20 54
26 We	10 36	17 50	00S03	08 01	18 33	11 10	18 02	07 15	09 06	20 55
27 Th	10 15	16 23	00 42	08 11	18 22	11 05	18 03	07 15	09 07	20 55
28 Fr	09 54	13 52	01 20	08 21	18 12	11 00	18 04	07 14	09 08	20 55
29 Sa	09 33	10 24	01 58	08 32	18 01	10 56	18 05	07 14	09 08	20 55
30 Su	09 11	06 12	02 35	08 42	17 51	10 51	18 05	07 13	09 09	20 55
31 Mo	08 50	01 35	03 11	08 52	17 40	10 46	18 06	07 12	09 10	20 56

ASPECTARIAN

01 02:27 ☽ ☍ ☿
07:12 ☽ ⚹ ♅
19:16 ☽ ☍ ♃
19:50 ☽ □ ♄
22:03 ☽ ☍ ♀
22:51 ☽ ∥ ♆
02 04:39 ☉ ⋕ ♄
05:54 ♄ SD
07:16 ☽ ⋕ ♅
12:29 ☽ ⋕ ♀
13:28 ☽ ☌ ♆
14:51 ☿ △ ♅
20:40 ☽ ⚹ ♇
03 10:38 ♃ □ ♄
17:50 ☽ △ ♂
20:36 ☽ △ ♄
04 19:24 ☽ △ ☉
21:47 ♀ ☌ ♃
21:50 ☽ □ ♇
05 07:59 ☽ ∥ ♀
09:22 ☽ ☌ ♅
14:33 ☽ ∥ ♅
15:08 ♀ □ ♄
19:26 ☽ △ ☿
22:07 ☽ □ ♂
22:22 ☽ △ ♀
23:30 ☽ △ ♃
23:39 ☽ ⋕ ♆
06 08:29 ♂ △ ♄
14:25 ☿ ☌ ♀
17:03 ☽ ⚹ ♆
20:20 ☿ □ ♄
07 00:33 ☽ ∥ ♃
00:44 ☽ △ ♇
04:44 ☽ ∥ ☿
07:09 ☿ ☌ ♃
19:15 ☿ ♍
08 00:39 ☽ □ ♀
02:04 ☿ ∥ ♃
02:40 ☽ ☍ ♄
04:25 ☽ □ ♃
04:46 ☽ ⚹ ♂
07:14 ☽ □ ☿
09:10 ☽ ∥ ☉
21:49 ☽ □ ♆
23:33 ♂ ♌
09 11:29 ☽ ⚹ ☉
13:14 ☽ ⋕ ♄
18:30 ☽ ⚹ ♅
10 04:47 ☽ ⚹ ♀
11:46 ☽ ⚹ ♃
22:11 ☽ ⚹ ☿
11 02:09 ♂ ⋕ ♇
04:52 ☽ △ ♆
09:05 ☽ ⋕ ♄
11:11 ♃ ♍
13:17 ☽ ☍ ♇
12 02:28 ☽ □ ♅
17:45 ☽ △ ♄
13 01:12 ☿ ☍ ♆
01:24 ☿ ⋕ ♆
02:01 ☽ ☌ ♂
07:30 ☽ ∥ ☉
10:27 ☉ △ ♅
14 07:48 ☽ ∥ ♃
13:30 [illegible]
18:27 ☽ ☌ ♀
15 04:38 ☽ □ ♄
07:19 ☽ ⋕ ♆
08:07 ☿ ∥ ♅
09:28 ☽ ☌ ♃
19:07 ☽ ∥ ♅
19:22 ♀ ☌ ☉
19:55 ☿ △ ♇
22:57 ☽ ∥ ☿
16 00:27 ☽ ∥ ♀
01:30 ☽ ☍ ♆
08:53 ☿ ∥ ♀
10:35 ☽ △ ♇
12:49 ☽ ☌ ☿
17 17:17 ☽ ⚹ ♄
18 08:45 ☽ ⚹ ♂
23:31 ☽ □ ♇
19 02:15 ☽ ⋕ ☿
13:42 ☽ ☍ ♅
13:51 ☽ ⚹ ♀
16:45 ♀ △ ♅
18:24 ☽ ⋕ ♀
20:22 ☽ ⋕ ♅
20 02:57 ☽ ⚹ ☉
08:54 ☽ ∥ ♆
13:23 ☽ ⚹ ♃
21 01:03 ☽ □ ♂
02:54 ☽ △ ♆
04:05 ☽ ⋕ ♃
09:15 ☽ ⋕ ☉
11:58 ☽ ⚹ ♇
13:48 ☿ ∥ ♅
23:14 ☽ □ ♀
23:41 ☉ □ ♄
22 06:58 ☽ ⚹ ☿
18:02 ☽ ☌ ♄
23 01:31 ☽ □ ♃
10:38 ☉ ♍
13:09 ☽ □ ♆
14:38 ☽ △ ♂
20:46 ☉ ∥ ♃
24 06:10 ☽ △ ♀
10:35 ☽ △ ♅
13:09 ☽ ∥ ♄
22:04 ☽ □ ☿
25 07:35 ☽ △ ☉
09:45 ☽ △ ♃
17:24 ☽ ∥ ♄
19:32 ☽ ⚹ ♆
26 03:33 ☽ ☌ ♇
15:26 ☽ □ ♅
16:08 ☿ ⚹ ♄
22:03 ☉ ☌ ♃
27 06:02 ☽ ⚹ ♄
07:20 ☽ △ ☿
15:45 ☿ ♎
28 04:25 ☽ ☍ ♂
10:10 [illegible]
16:56 ☽ ⚹ ♅
17:12 ♂ ⋕ ♄
20:39 ☽ ⋕ ♃
29 05:44 ☽ ⋕ ☉
07:03 ☽ □ ♄
07:36 ☽ ∥ ♇
10:44 ☽ ⋕ ♀
15:09 ☽ ☍ ♃
18:31 ☽ ⋕ ♅
22:21 ☽ ☌ ♆
30 02:35 ☉ ⋕ ♆
05:46 ☽ ⚹ ♇
16:50 ☽ ∥ ☿
22:41 ☿ ∥ ☉
31 06:54 ☽ △ ♄
16:06 ☽ ☍ ☿

⚷ Chiron

01	Dec.	00 N 46
01	20♓57R	
04	20 51	
07	20 45	
10	20 39	
13	20 32	
16	20 25	
19	20 18	
22	20 10	
25	20 02	
28	[illegible]	
31	19 47	

13 06:55 20♍10 ☉ Solar Eclipse (mag 0.788)
28 02:49 04♈40 ✸ Total Lunar Eclipse (mag 1.282)

September 2015

Day	S. T. h m s	☉ ° ' "	☽ ° ' "	☿ ° '	♀ ° '	♂ ° '	♃ ° '	♄ ° '	♅ ° '	♆ ° '	♇ ° '	☊ True ° '
01 Tu	22 39 22	08♍15 31	09♈39 26	05♎05	14♌R58	14♌47	04♍27	28♏59	19♈R59	08♓R25	13♑R07	01♎D04
02 We	22 43 19	09 13 33	24 29 14	06 10	14 46	15 25	04 40	29 02	19 57	08 23	13 06	01 05
03 Th	22 47 15	10 11 37	09♉00 40	07 12	14 37	16 03	04 53	29 05	19 56	08 21	13 06	01 06
04 Fr	22 51 12	11 09 42	23 10 15	08 12	14 30	16 42	05 06	29 08	19 54	08 20	13 05	01 07
05 Sa	22 55 08	12 07 50	06♊56 48	09 09	14 25	17 20	05 19	29 11	19 52	08 18	13 04	01 08
06 Su	22 59 05	13 06 00	20 20 54	10 04	14 23	17 58	05 32	29 15	19 51	08 16	13 04	01R 08
07 Mo	23 03 01	14 04 12	03♋24 19	10 55	14D 24	18 36	05 45	29 18	19 49	08 15	13 03	01 07
08 Tu	23 06 58	15 02 25	16 09 28	11 43	14 26	19 14	05 58	29 21	19 47	08 13	13 03	01 05
09 We	23 10 55	16 00 41	28 39 04	12 28	14 31	19 52	06 11	29 25	19 45	08 11	13 02	01 04
10 Th	23 14 51	16 58 59	10♌55 44	13 10	14 39	20 30	06 24	29 28	19 43	08 10	13 02	01 01
11 Fr	23 18 48	17 57 19	23 02 00	13 47	14 48	21 08	06 37	29 32	19 41	08 08	13 01	01 00
12 Sa	23 22 44	18 55 40	05♍00 09	14 20	15 00	21 46	06 50	29 36	19 39	08 07	13 01	00 58
13 Su	23 26 41	19 54 03	16 52 19	14 49	15 13	22 24	07 03	29 39	19 37	08 05	13 01	00 57
14 Mo	23 30 37	20 52 29	28 40 38	15 13	15 29	23 02	07 16	29 43	19 35	08 03	13 00	00D 57
15 Tu	23 34 34	21 50 56	10♎27 20	15 32	15 46	23 39	07 29	29 47	19 33	08 02	13 00	00 57
16 We	23 38 30	22 49 24	22 14 53	15 46	16 06	24 17	07 42	29 51	19 31	08 00	13 00	00 58
17 Th	23 42 27	23 47 55	04♏06 00	15 53	16 27	24 55	07 54	29 55	19 29	07 59	12 59	00 59
18 Fr	23 46 23	24 46 27	16 03 49	15R 55	16 50	25 33	08 07	30 00	19 27	07 57	12 59	00 59
19 Sa	23 50 20	25 45 01	28 11 52	15 50	17 15	26 11	08 20	00♐04	19 25	07 55	12 59	01 00
20 Su	23 54 17	26 43 37	10♐33 57	15 39	17 41	26 48	08 33	00 08	19 22	07 54	12 59	01 00
21 Mo	23 58 13	27 42 15	23 14 02	15 20	18 09	27 26	08 45	00 12	19 20	07 52	12 59	01R 00
22 Tu	00 02 10	28 40 54	06♑15 51	14 54	18 39	28 04	08 58	00 17	19 18	07 51	12 58	01 00
23 We	00 06 06	29 39 35	19 42 27	14 22	19 10	28 41	09 11	00 22	19 16	07 49	12 58	01D 00
24 Th	00 10 03	00♎38 17	03♒35 36	13 42	19 43	29 19	09 23	00 26	19 14	07 48	12 58	01 00
25 Fr	00 13 59	01 37 01	17 55 05	12 55	20 17	29 56	09 36	00 31	19 11	07 46	12 58	01 00
26 Sa	00 17 56	02 35 47	02♓38 15	12 03	20 52	00♍34	09 48	00 36	19 09	07 45	12D 58	01 01
27 Su	00 21 52	03 34 35	17 39 38	11 04	21 28	01 11	10 01	00 40	19 07	07 43	12 58	01 01
28 Mo	00 25 49	04 33 24	02♈51 25	10 02	22 06	01 49	10 13	00 45	19 04	07 42	12 58	01 01
29 Tu	00 29 46	05 32 16	18 04 13	08 56	22 45	02 26	10 26	00 50	19 02	07 41	12 59	01R 01
30 We	00 33 42	06 31 09	03♉08 23	07 48	23 25	03 04	10 38	00 55	19 00	07 39	12 59	01 00

Data for	09-01-2015
Julian Day	2457266.50
Ayanamsa	24 04 33
SVP	05♓02 29
☽ ☊ Mean	02♎04 R

● ◐ PHASES ○ ◑

05	09:55	◑	12♊32
13	06:41	☉	20♍10
21	08:59	◐	28♐04
28	02:51	✸	04♈40

LAST ASPECT ☽ Day	h m	INGRESS Day	h m	
01	16:38	02	09:03	♉
04	10:21	04	11:49	♊
05	23:05	06	17:41	♋
09	01:29	09	02:37	♌
11	13:04	11	13:56	♍
14	02:08	14	02:42	♎
16	04:22	16	15:43	♏
18	19:49	19	03:32	♐
21	08:59	21	12:33	♑
22	23:13	23	17:52	♒
25	04:03	25	19:44	♓
26	16:33	27	19:30	♈
29	07:46	29	18:58	♉

DECLINATION

Day	☉	☽	☿	♀	♂	♃	♄	♅	♆	♇
01 Tu	08N28	03N06	03S46	09N01	17N29	10N42	18S07	07N12	09S10	20S56
02 We	08 07	07 32	04 20	09 11	17 18	10 37	18 08	07 11	09 11	20 56
03 Th	07 45	11 25	04 53	09 20	17 07	10 32	18 09	07 10	09 12	20 56
04 Fr	07 23	14 32	05 25	09 29	16 56	10 27	18 10	07 10	09 12	20 57
05 Sa	07 01	16 43	05 56	09 38	16 44	10 23	18 10	07 09	09 13	20 57
06 Su	06 38	17 55	06 25	09 46	16 33	10 18	18 11	07 08	09 13	20 57
07 Mo	06 16	18 07	06 53	09 54	16 22	10 13	18 12	07 08	09 14	20 57
08 Tu	05 54	17 23	07 20	10 01	16 10	10 09	18 13	07 07	09 15	20 57
09 We	05 31	15 50	07 45	10 08	15 58	10 04	18 14	07 06	09 15	20 58
10 Th	05 08	13 35	08 08	10 15	15 46	09 59	18 15	07 06	09 16	20 58
11 Fr	04 46	10 45	08 30	10 21	15 35	09 54	18 16	07 05	09 16	20 58
12 Sa	04 23	07 31	08 49	10 27	15 23	09 50	18 17	07 04	09 17	20 58
13 Su	04 00	03 59	09 07	10 32	15 11	09 45	18 18	07 03	09 18	20 58
14 Mo	03 37	00 20	09 22	10 36	14 58	09 40	18 19	07 03	09 18	20 58
15 Tu	03 14	03S20	09 34	10 40	14 46	09 36	18 20	07 02	09 19	20 59
16 We	02 51	06 53	09 44	10 44	14 34	09 31	18 21	07 01	09 19	20 59
17 Th	02 28	10 09	09 50	10 47	14 21	09 26	18 22	07 00	09 20	20 59
18 Fr	02 05	13 02	09 54	10 49	14 09	09 21	18 23	06 59	09 21	20 59
19 Sa	01 41	15 24	09 54	10 51	13 56	09 17	18 24	06 58	09 21	20 59
20 Su	01 18	17 06	09 50	10 52	13 44	09 12	18 26	06 58	09 22	21 00
21 Mo	00 55	18 01	09 42	10 53	13 31	09 07	18 27	06 57	09 22	21 00
22 Tu	00 31	18 01	09 31	10 53	13 18	09 03	18 28	06 56	09 23	21 00
23 We	00 08	17 01	09 15	10 53	13 05	08 58	18 29	06 55	09 24	21 00
24 Th	00S15	15 00	08 54	10 52	12 53	08 53	18 30	06 54	09 24	21 00
25 Fr	00 39	12 00	08 30	10 50	12 40	08 49	18 31	06 53	09 25	21 00
26 Sa	01 02	08 10	08 01	10 48	12 27	08 44	18 32	06 53	09 25	21 01
27 Su	01 25	03 44	07 28	10 46	12 13	08 40	18 34	06 52	09 26	21 01
28 Mo	01 49	00N59	06 52	10 42	12 00	08 35	18 35	06 51	09 26	21 01
29 Tu	02 12	05 39	06 13	10 38	11 47	08 30	18 36	06 50	09 27	21 01
30 We	02 35	09 55	05 31	10 34	11 34	08 26	18 37	06 49	09 27	21 01

ASPECTARIAN

Day	Time	Aspect
01	03:40	☉ ☍ ♆
	03:56	☽ # ☿
	05:04	♀ ☌ ♂
	05:33	☽ □ ♇
	08:25	☽ △ ♀
	08:37	☽ △ ♂
	16:38	☽ ☌ ♅
	22:00	☽ ∥ ♅
	23:48	♀ # ♆
02	03:02	☽ ∥ ☉
	09:41	☽ # ♆
	10:04	☽ ∥ ♀
	17:02	☽ △ ♃
	18:14	☽ ∥ ♃
	22:54	☽ ✱ ♆
03	02:07	☽ △ ☉
	06:51	☽ △ ♇
	09:20	☽ □ ♀
	12:25	☽ □ ♂
04	10:21	☽ ☍ ♄
	14:27	☉ ∥ ♅
	21:06	☽ □ ♃
05	00:17	☽ ∥ ♂
	02:23	☽ □ ♆
	04:12	☽ △ ☿
	13:16	☽ ✱ ♀
	19:28	☽ ✱ ♂
	23:05	☽ ✱ ♅
	23:09	☉ △ ♇
06	06:15	☿ # ☉
	08:30	♀ SD
07	04:28	☽ ✱ ♃
	09:02	☽ △ ♆
	12:32	☿ # ♅
	15:02	☽ □ ☿
	18:06	☽ ☍ ♆
	21:42	☽ ✱ ☉
08	06:54	☽ □ ♅
	14:37	♀ ∥ ♃
	19:53	♂ △ ♅
	22:13	☽ ∥ ♂
09	01:29	☽ △ ♄
	19:24	☿ □ ♇
10	04:39	☽ ✱ ☿
	07:25	☽ ☌ ♀
	17:21	☽ △ ♅
	20:00	☽ ☌ ♂
11	03:06	☽ ∥ ♀
	06:45	☽ ∥ ♃
	11:17	☽ # ♆
	13:04	☽ □ ♄
	15:26	☽ # ☿
12	03:09	☽ ∥ ♅
	03:46	☽ ☌ ♃
	06:15	☽ ☍ ♆
	16:11	☽ △ ♇
	23:58	☽ ∥ ☉
13	17:50	☿ ∥ ♆
14	02:08	☽ ✱ ♄
	23:25	☽ # ☉
15	02:14	☿ # ♃
	05:11	☽ □ ♇
	10:34	☽ ☌ ☿
	11:08	☽ ✱ ♀
	18:28	☽ ☍ ♅
16	01:01	☽ # ♅
	04:22	☽ ✱ ♂
	17:47	☽ ∥ ♆
	18:42	☽ # ♃
	21:34	☽ ∥ ☿
17	05:01	☽ # ♀
	06:56	♃ ☍ ♆
	07:48	☽ △ ♆
	07:49	☽ ✱ ♃
	17:51	☽ ✱ ♇
	18:11	☿ SR
18	01:35	☽ □ ♀
	02:47	♄ ♐
	03:51	♃ # ♆
	09:44	☽ # ♂
	18:46	☽ ✱ ☉
	19:49	☽ □ ♂
19	03:40	☽ ☌ ♄
	18:53	☽ □ ♆
	20:03	☽ □ ♃
20	09:29	☽ ✱ ☿
	14:05	☽ △ ♀
	16:42	☽ △ ♅
21	08:12	☽ △ ♂
22	02:52	☽ ✱ ♆
	04:58	☽ △ ♃
	11:59	☿ ∥ ♆
	12:04	☽ ☌ ♇
	14:56	☽ □ ☿
	23:13	☽ □ ♅
23	04:00	♀ △ ♅
	08:21	☉ ♎
	18:35	☽ △ ☉
	18:35	☽ ✱ ♄
	18:37	☉ ✱ ♄
24	01:11	☿ # ♃
	16:09	☽ △ ☿
	18:58	☽ # ♂
	22:32	☿ □ ♇
25	02:05	☽ ✱ ♅
	02:19	♂ ♍
	04:03	☽ ☍ ♀
	06:56	♇ SD
	07:53	☽ # ♀
	16:42	☽ ∥ ♆
	20:31	☽ ☍ ♂
	20:40	☽ # ♃
	20:41	☽ □ ♄
26	01:03	☽ ∥ ☿
	01:13	♂ □ ♄
	07:20	☽ # ♅
	08:12	☽ ☌ ♆
	11:40	☽ ☍ ♃
	16:33	☽ ✱ ♇
27	10:59	☽ ∥ ☉
	20:40	☽ △ ♄
28	00:42	☿ # ♅
	04:35	☽ # ☉
	10:33	☽ ☍ ☿
	15:57	☽ □ ♇
29	01:31	☽ ☌ ♅
	02:37	☽ # ☿
	06:22	☽ ∥ ♅
	07:46	☽ △ ♀
	15:29	☽ ∥ ♃
	21:18	☽ # ♆
	23:52	☽ △ ♂
30	04:02	☽ ∥ ♀
	07:15	☽ ✱ ♆
	09:57	☽ ∥ ♂
	12:16	☽ △ ♃
	14:39	☿ ☌ ☉
	15:54	☽ △ ♇

⚷ Chiron

01 Dec.	00 N 19
03	19♓38R
06	19 30
09	19 22
12	19 13
15	19 05
18	18 57
21	18 49
24	18 40
27	18 32
30	18 24

October 2015

Day	S. T.	☉	☽	☿	♀	♂	♃	♄	♅	♆	♇	☊ True
	h m s	° ' "	° ' "	° '	° '	° '	° '	° '	° '	° '	° '	° '
01 Th	00 37 39	07♎ 30 05	17♉ 55 33	06♎R41	24♌ 07	03♍ 41	10♍ 50	01♐ 00	18♈R57	07♓R38	12♑ 59	01♎R00
02 Fr	00 41 35	08 29 03	02♊ 19 33	05 35	24 49	04 19	11 03	01 06	18 55	07 36	12 59	00 59
03 Sa	00 45 32	09 28 04	16 17 03	04 33	25 33	04 56	11 15	01 11	18 52	07 35	12 59	00 58
04 Su	00 49 28	10 27 06	29 47 22	03 36	26 17	05 33	11 27	01 16	18 50	07 34	12 59	00 57
05 Mo	00 53 25	11 26 11	12♋ 52 03	02 46	27 03	06 11	11 39	01 21	18 48	07 33	13 00	00 56
06 Tu	00 57 21	12 25 19	25 34 04	02 04	27 49	06 48	11 51	01 27	18 45	07 31	13 00	00D 57
07 We	01 01 18	13 24 28	07♌ 57 15	01 31	28 36	07 25	12 03	01 32	18 43	07 30	13 00	00 58
08 Th	01 05 15	14 23 40	20 05 44	01 08	29 24	08 03	12 15	01 38	18 40	07 29	13 01	00 59
09 Fr	01 09 11	15 22 54	02♍ 03 34	00 56	00♍ 13	08 40	12 27	01 43	18 38	07 27	13 01	01 01
10 Sa	01 13 08	16 22 11	13 54 31	00D 55	01 03	09 17	12 39	01 49	18 36	07 26	13 02	01 02
11 Su	01 17 04	17 21 29	25 41 57	01 04	01 54	09 54	12 51	01 55	18 33	07 25	13 02	01 04
12 Mo	01 21 01	18 20 50	07♎ 28 44	01 24	02 45	10 31	13 03	02 00	18 31	07 24	13 03	01R 04
13 Tu	01 24 57	19 20 12	19 17 23	01 54	03 37	11 08	13 15	02 06	18 28	07 23	13 03	01 03
14 We	01 28 54	20 19 37	01♏ 10 05	02 33	04 30	11 45	13 26	02 12	18 26	07 22	13 04	01 01
15 Th	01 32 50	21 19 04	13 08 48	03 20	05 23	12 23	13 38	02 18	18 23	07 21	13 04	00 58
16 Fr	01 36 47	22 18 33	25 15 26	04 16	06 17	13 00	13 49	02 24	18 21	07 20	13 05	00 54
17 Sa	01 40 43	23 18 03	07♐ 32 00	05 18	07 12	13 37	14 01	02 30	18 19	07 19	13 05	00 50
18 Su	01 44 40	24 17 36	20 00 42	06 27	08 07	14 13	14 12	02 36	18 16	07 18	13 06	00 47
19 Mo	01 48 37	25 17 10	02♑ 43 58	07 42	09 03	14 50	14 23	02 42	18 14	07 17	13 07	00 44
20 Tu	01 52 33	26 16 46	15 44 22	09 01	09 59	15 27	14 35	02 48	18 11	07 16	13 08	00 43
21 We	01 56 30	27 16 24	29 04 21	10 24	10 56	16 04	14 46	02 54	18 09	07 15	13 08	00D 42
22 Th	02 00 26	28 16 04	12♒ 45 55	11 50	11 54	16 41	14 57	03 01	18 06	07 14	13 09	00 43
23 Fr	02 04 23	29 15 45	26 50 06	13 20	12 52	17 18	15 08	03 07	18 04	07 13	13 10	00 45
24 Sa	02 08 19	00♏ 15 28	11♓ 16 21	14 52	13 50	17 55	15 19	03 13	18 02	07 12	13 11	00 46
25 Su	02 12 16	01 15 12	26 01 58	16 26	14 49	18 31	15 30	03 20	17 59	07 11	13 12	00 47
26 Mo	02 16 12	02 14 59	11♈ 01 46	18 01	15 49	19 08	15 41	03 26	17 57	07 10	13 13	00R 47
27 Tu	02 20 09	03 14 47	26 08 13	19 38	16 48	19 45	15 51	03 33	17 55	07 10	13 14	00 45
28 We	02 24 06	04 14 37	11♉ 12 15	21 15	17 49	20 21	16 02	03 39	17 52	07 09	13 15	00 41
29 Th	02 28 02	05 14 29	26 04 35	22 54	18 49	20 58	16 12	03 46	17 50	07 08	13 16	00 37
30 Fr	02 31 59	06 14 23	10♊ 37 08	24 33	19 51	21 35	16 23	03 52	17 48	07 08	13 17	00 31
31 Sa	02 35 55	07 14 19	24 44 11	26 12	20 52	22 11	16 33	03 59	17 46	07 07	13 18	00 26

Data for	10-01-2015
Julian Day	2457296.50
Ayanamsa	24 04 36
SVP	05 ♓ 02 23
☽ ☊ Mean	00 ♎ 29 R

● ◐ PHASES ○ ◑

04	21:07	◑	11♋19
13	00:06	●	19♎20
20	20:32	◐	27♑08
27	12:06	○	03♉45

LAST ASPECT Day	h m	☽ INGRESS Day	h m	
01	10:45	01	20:04	♊
03	17:19	04	00:23	♋
05	11:05	06	08:31	♌
07	21:11	08	19:51	♍
09	22:12	11	08:46	♎
13	00:06	13	21:39	♏
15	00:59	16	09:19	♐
18	08:49	18	18:53	♑
20	20:33	21	01:39	♒
23	04:23	23	05:19	♓
24	11:19	25	06:23	♈
26	12:26	27	06:08	♉
28	15:21	29	06:25	♊
31	02:52	31	09:09	♋

DECLINATION

Day	☉	☽	☿	♀	♂	♃	♄	♅	♆	♇
01 Th	02S 59	13N27	04S 48	10N29	11N20	08N21	18S 38	06N48	09S 28	21S 01
02 Fr	03 22	16 04	04 05	10 24	11 07	08 17	18 40	06 47	09 28	21 01
03 Sa	03 45	17 38	03 22	10 17	10 54	08 12	18 41	06 46	09 29	21 01
04 Su	04 08	18 09	02 40	10 11	10 40	08 08	18 42	06 45	09 29	21 02
05 Mo	04 31	17 39	02 02	10 03	10 27	08 03	18 43	06 44	09 30	21 02
06 Tu	04 54	16 17	01 26	09 56	10 13	07 59	18 45	06 44	09 30	21 02
07 We	05 17	14 11	00 55	09 47	09 59	07 54	18 46	06 43	09 31	21 02
08 Th	05 40	11 29	00 29	09 38	09 46	07 50	18 47	06 42	09 31	21 02
09 Fr	06 03	08 20	00 08	09 29	09 32	07 45	18 48	06 41	09 32	21 02
10 Sa	06 26	04 53	00N08	09 19	09 18	07 41	18 50	06 40	09 32	21 02
11 Su	06 49	01 15	00 19	09 09	09 04	07 36	18 51	06 39	09 32	21 02
12 Mo	07 11	02S 26	00 24	08 58	08 50	07 32	18 52	06 38	09 33	21 03
13 Tu	07 34	06 01	00 23	08 46	08 36	07 28	18 54	06 37	09 33	21 03
14 We	07 56	09 24	00 18	08 34	08 23	07 23	18 55	06 36	09 34	21 03
15 Th	08 19	12 25	00 07	08 22	08 09	07 19	18 56	06 35	09 34	21 03
16 Fr	08 41	14 56	00S 08	08 09	07 55	07 15	18 57	06 34	09 34	21 03
17 Sa	09 03	16 48	00 26	07 55	07 41	07 10	18 59	06 33	09 35	21 03
18 Su	09 25	17 55	00 49	07 41	07 26	07 06	19 00	06 33	09 35	21 03
19 Mo	09 47	18 10	01 15	07 27	07 12	07 02	19 01	06 32	09 36	21 03
20 Tu	10 09	17 29	01 43	07 12	06 58	06 58	19 03	06 31	09 36	21 03
21 We	10 30	15 50	02 14	06 57	06 44	06 54	19 04	06 30	09 36	21 03
22 Th	10 51	13 14	02 48	06 41	06 30	06 49	19 05	06 29	09 37	21 03
23 Fr	11 13	09 49	03 23	06 25	06 16	06 45	19 07	06 28	09 37	21 03
24 Sa	11 34	05 43	03 59	06 08	06 02	06 41	19 08	06 27	09 37	21 04
25 Su	11 54	01 11	04 37	05 51	05 48	06 37	19 09	06 26	09 37	21 04
26 Mo	12 15	03N31	05 15	05 34	05 33	06 33	19 11	06 25	09 38	21 04
27 Tu	12 36	08 00	05 55	05 16	05 19	06 29	19 12	06 24	09 38	21 04
28 We	12 56	11 58	06 35	04 58	05 05	06 25	19 13	06 24	09 38	21 04
29 Th	13 16	15 06	07 15	04 39	04 51	06 21	19 15	06 23	09 39	21 04
30 Fr	13 36	17 12	07 55	04 20	04 36	06 17	19 16	06 22	09 39	21 04
31 Sa	13 56	18 10	08 36	04 01	04 22	06 13	19 17	06 21	09 39	21 04

ASPECTARIAN

01 10:45 ☽ □ ♀
21:54 ☽ ☍ ♄
02 03:31 ☽ □ ♂
05:09 ☽ △ ☿
08:59 ☽ □ ♆
11:17 ☽ △ ☉
15:07 ☽ □ ♃
15:29 ☿ ∥ ☉

03 04:32 ☽ ⚹ ♅
17:19 ☽ ⚹ ♀
04 06:28 ☽ □ ☿
11:01 ☽ ⚹ ♂
14:09 ☽ △ ♆
21:42 ☽ ⚹ ♃

05 00:14 ☽ ☍ ♇
11:05 ☽ □ ♅
06 11:24 ☽ △ ♄
11:57 ☽ ⚹ ☿
14:13 ☉ □ ♇
22:58 ☿ ⚹ ♄
07 02:53 ♂ ☍ ♆
11:41 ☽ ⚹ ☉
21:11 ☽ △ ♅
08 14:35 ☽ ∥ ♂
15:10 ☽ ∥ ♀
15:18 ☽ # ♆
17:30 ♀ ♍
17:46 ♀ # ♆
20:01 ☽ ☌ ♀
23:19 ☽ □ ♄

09 00:20 ♂ # ♆
04:19 ☽ ∥ ♃
10:54 ☽ ☍ ♆
11:49 ☽ ∥ ♅
[illegible]
14:30 ☽ # ☉
14:58 ☿ SD
17:24 ♀ ∥ ♂
21:24 ☽ ☌ ♃
22:12 ☽ △ ♇
10 13:56 ☉ # ♅

11 00:32 ♀ □ ♄
05:59 ☽ ∥ ☿
10:32 ☽ # ☿
11:13 ☽ ☌ ☿
12:46 ☽ ⚹ ♄
23:38 ♃ △ ♇
12 03:50 ☉ ☍ ♅
11:20 ☽ □ ♇
18:19 ☉ # ♃
22:21 ☽ ☍ ♅

13 04:08 ☽ # ♅
09:49 ☽ # ♃
09:55 ☿ ⚹ ♄
12:07 ☽ ∥ ☉
17:02 ☽ # ♂
18:18 ☽ # ♀
14 01:17 ☽ ∥ ♆
07:13 ☽ ⚹ ☿
12:25 ☽ △ ♆
17:15 ☉ # ♂
22:23 ☽ ⚹ ♂
23:51 ☽ ⚹ ♇

15 00:59 ☽ ⚹ ♃
01:55 ♀ # ☉
16 03:32 ♂ △ ♇
[illegible]
19:15 ☽ ⚹ ☿
23:18 ☽ □ ♀
23:34 ☽ □ ♆

17 02:55 ♀ ☍ ♆
12:21 ☽ □ ♂
12:42 ☽ □ ♃
20:41 ☽ △ ♅
22:41 ♂ ☌ ♃
18 08:49 ☽ ⚹ ☉
11:25 ☉ ∥ ♆
19 08:26 ☽ ⚹ ♆
10:15 ☽ □ ☿
12:38 ☽ △ ♀
19:13 ☽ ☌ ♇
21:51 ☽ △ ♃
23:27 ☽ △ ♂
20 01:25 ♂ ∥ ♃
04:26 ☽ □ ♅

21 06:28 ♀ ∥ ♃
06:50 ☽ ⚹ ♄
22:13 ☽ △ ☿
22 02:05 ♂ ∥ ♅
09:10 ☽ ⚹ ♅
15:39 ☽ ∥ ☉
18:56 ♀ ∥ ♅
21:20 ☿ □ ♇

23 01:14 ☽ ∥ ♆
04:23 ☽ △ ☉
07:41 ♀ △ ♇
10:35 ☽ □ ♄
17:18 ☽ ☌ ♆
[illegible]
18:30 ☽ # ♃
19:54 ☽ # ♅
21:31 ☽ # ♀
22:10 ☽ # ♂
24 03:08 ☽ ⚹ ♇
04:30 ☽ ☍ ♀
06:42 ☽ ☍ ♃
08:14 ☽ ∥ ☿
11:19 ☽ ☍ ♂

25 11:48 ☽ △ ♄
20:04 ♀ ☌ ♃
23:00 ☿ ☍ ♅
26 03:29 ☽ □ ♇
[illegible]
07:49 ☿ # ♀
08:06 ☿ # ♂
10:06 ☽ ∥ ♀
10:11 ☽ ∥ ♂
10:39 ☽ # ☿
10:58 ☽ ☌ ♅
12:26 ☽ ☍ ☿
15:18 ☽ ∥ ♅
15:49 ☽ ∥ ♃

27 09:25 ☽ # ♆
17:32 ☽ ⚹ ♆
17:35 ☿ # ♅
18:53 ☿ # ♃
[illegible]
07:27 ☽ # ☉
07:50 ☽ △ ♃
11:23 ☽ △ ♀
12:03 ♃ ∥ ♅
15:21 ☽ △ ♂
29 12:42 ☽ ☍ ♄
18:11 ☽ □ ♆

30 09:49 ☽ □ ♃
12:04 ☽ ⚹ ♅
16:49 ☽ □ ♀
19:24 ☽ □ ♂
21:06 ☉ △ ♆
31 02:52 ☽ △ ☿
[illegible]
23:43 ☽ △ ☉

⚷ Chiron

01 Dec.	00 N 15
03	18♓17R
06	18 09
09	18 02
12	17 55
15	17 48
18	17 42
21	17 36
24	17 30
27	17 25
30	17 20

November 2015

Day	S. T.	☉	☽	☿	♀	♂	♃	♄	♅	♆	♇	☊ True
	h m s	° ' "	° ' "	° '	° '	° '	° '	° '	° '	° '	° '	° '
01 Su	02 39 52	08♏ 14 18	08♋ 23 00	27♎ 51	21♍ 54	22♍ 48	16♍ 43	04♐ 05	17♈R43	07♓R06	13♑ 19	00♎R21
02 Mo	02 43 48	09 14 18	21 33 43	29 31	22 57	23 24	16 54	04 12	17 41	07 06	13 20	00 17
03 Tu	02 47 45	10 14 21	04♌ 18 47	01♏ 10	23 59	24 01	17 04	04 19	17 39	07 05	13 21	00 15
04 We	02 51 41	11 14 26	16 42 09	02 49	25 02	24 37	17 14	04 25	17 37	07 05	13 22	00D 15
05 Th	02 55 38	12 14 33	28 48 40	04 28	26 06	25 14	17 24	04 32	17 35	07 04	13 23	00 16
06 Fr	02 59 35	13 14 42	10♍ 43 27	06 07	27 10	25 50	17 33	04 39	17 32	07 04	13 24	00 18
07 Sa	03 03 31	14 14 53	22 31 31	07 46	28 14	26 26	17 43	04 46	17 30	07 03	13 26	00 20
08 Su	03 07 28	15 15 06	04♎ 17 30	09 25	29 18	27 03	17 53	04 52	17 28	07 03	13 27	00R 20
09 Mo	03 11 24	16 15 21	16 05 25	11 03	00♎ 23	27 39	18 02	04 59	17 26	07 03	13 28	00 19
10 Tu	03 15 21	17 15 37	27 58 36	12 41	01 28	28 15	18 12	05 06	17 24	07 02	13 29	00 15
11 We	03 19 17	18 15 56	09♏ 59 35	14 18	02 33	28 51	18 21	05 13	17 22	07 02	13 31	00 09
12 Th	03 23 14	19 16 17	22 10 08	15 55	03 39	29 27	18 30	05 20	17 20	07 02	13 32	00 01
13 Fr	03 27 10	20 16 39	04♐ 31 21	17 32	04 45	00♎ 03	18 39	05 27	17 18	07 02	13 34	29♍ 51
14 Sa	03 31 07	21 17 03	17 03 50	19 09	05 51	00 40	18 48	05 34	17 16	07 02	13 35	29 41
15 Su	03 35 04	22 17 28	29 48 01	20 45	06 57	01 16	18 57	05 41	17 14	07 01	13 36	29 32
16 Mo	03 39 00	23 17 55	12♑ 44 16	22 21	08 04	01 52	19 06	05 48	17 13	07 01	13 38	29 25
17 Tu	03 42 57	24 18 23	25 53 12	23 57	09 10	02 28	19 14	05 55	17 11	07 01	13 39	29 19
18 We	03 46 53	25 18 53	09♒ 15 41	25 32	10 18	03 03	19 23	06 02	17 09	07 01	13 41	29 16
19 Th	03 50 50	26 19 23	22 52 46	27 07	11 25	03 39	19 31	06 09	17 07	07 D 01	13 42	29 15
20 Fr	03 54 46	27 19 55	06♓ 45 24	28 42	12 32	04 15	19 39	06 16	17 06	07 01	13 44	29D 15
21 Sa	03 58 43	28 20 29	20 53 59	00♐ 17	13 40	04 51	19 47	06 23	17 04	07 01	13 45	29 15
22 Su	04 02 39	29 21 03	05♈ 17 48	01 51	14 48	05 27	19 55	06 30	17 02	07 01	13 47	29R 15
23 Mo	04 06 36	00♐ 21 38	19 54 24	03 26	15 56	06 02	20 03	06 37	17 01	07 01	13 49	29 13
24 Tu	04 10 33	01 22 15	04♉ 39 12	05 00	17 04	06 38	20 11	06 44	16 59	07 02	13 50	29 08
25 We	04 14 29	02 22 53	19 25 42	06 34	18 13	07 14	20 18	06 51	16 58	07 02	13 52	29 01
26 Th	04 18 26	03 23 33	04♊ 06 08	08 07	19 22	07 49	20 26	06 59	16 56	07 02	13 54	28 52
27 Fr	04 22 22	04 24 14	18 32 38	09 41	20 30	08 25	20 33	07 06	16 55	07 02	13 55	28 42
28 Sa	04 26 19	05 24 56	02♋ 38 36	11 14	21 39	09 00	20 40	07 13	16 53	07 03	13 57	28 31
29 Su	04 30 15	06 25 40	16 19 39	12 48	22 49	09 36	20 47	07 20	16 52	07 03	13 59	28 21
30 Mo	04 34 12	07 26 25	29 34 10	14 21	23 58	10 11	20 54	07 27	16 51	07 03	14 00	28 13

Data for	11-01-2015
Julian Day	2457327.50
Ayanamsa	24 04 39
SVP	05 ♓ 02 17
☽ ☊ Mean	28 ♍ 50 R

● ◐ PHASES ○ ◑

03	12:24	◑	10♌45
11	17:48	●	19♏01
19	06:28	◐	26♒36
25	22:44	○	03♊20

LAST ASPECT Day	h m	☽ INGRESS Day	h m	
02	03:35	02	15:48	♌
04	01:47	05	02:23	♍
07	12:48	07	15:15	♎
09	02:43	10	04:04	♏
12	14:56	12	15:15	♐
14	03:20	15	00:22	♑
16	20:54	17	07:26	♒
19	08:20	19	12:22	♓
21	13:24	21	15:13	♈
22	19:17	23	16:26	♉
25	01:26	25	17:16	♊
27	03:36	27	19:27	♋
29	12:46	30	00:48	♌

DECLINATION

Day	☉	☽	☿	♀	♂	♃	♄	♅	♆	♇
01 Su	14S 15	18N03	09S 16	03N42	04N08	06N10	19S 19	06N20	09S 39	21S 04
02 Mo	14 34	16 56	09 57	03 22	03 54	06 06	19 20	06 19	09 39	21 04
03 Tu	14 53	15 00	10 37	03 02	03 39	06 02	19 21	06 19	09 40	21 04
04 We	15 12	12 25	11 16	02 41	03 25	05 58	19 23	06 18	09 40	21 04
05 Th	15 30	09 20	11 56	02 20	03 11	05 55	19 24	06 17	09 40	21 04
06 Fr	15 49	05 56	12 34	01 59	02 56	05 51	19 25	06 16	09 40	21 04
07 Sa	16 07	02 19	13 13	01 38	02 42	05 47	19 27	06 15	09 40	21 04
08 Su	16 24	01S 22	13 50	01 16	02 28	05 44	19 28	06 15	09 40	21 04
09 Mo	16 42	05 01	14 27	00 54	02 14	05 40	19 29	06 14	09 40	21 04
10 Tu	16 59	08 30	15 04	00 32	01 59	05 37	19 31	06 13	09 41	21 04
11 We	17 16	11 41	15 39	00 10	01 45	05 33	19 32	06 12	09 41	21 04
12 Th	17 32	14 24	16 14	00S 13	01 31	05 30	19 33	06 12	09 41	21 04
13 Fr	17 49	16 30	16 48	00 35	01 17	05 27	19 35	06 11	09 41	21 04
14 Sa	18 05	17 51	17 21	00 58	01 02	05 23	19 36	06 10	09 41	21 04
15 Su	18 20	18 20	17 54	01 21	00 48	05 20	19 37	06 10	09 41	21 04
16 Mo	18 36	17 53	18 25	01 44	00 34	05 17	19 39	06 09	09 41	21 04
17 Tu	18 51	16 29	18 56	02 08	00 20	05 14	19 40	06 08	09 41	21 04
18 We	19 05	14 10	19 25	02 31	00 06	05 10	19 41	06 08	09 41	21 04
19 Th	19 20	11 02	19 54	02 55	00S 08	05 07	19 42	06 07	09 41	21 04
20 Fr	19 34	07 13	20 22	03 19	00 22	05 04	19 44	06 06	09 41	21 04
21 Sa	19 47	02 56	20 49	03 42	00 36	05 01	19 45	06 06	09 41	21 04
22 Su	20 00	01N36	21 14	04 06	00 50	04 58	19 46	06 05	09 41	21 04
23 Mo	20 13	06 06	21 39	04 30	01 04	04 56	19 47	06 04	09 41	21 04
24 Tu	20 26	10 17	22 02	04 54	01 18	04 53	19 49	06 04	09 41	21 04
25 We	20 38	13 49	22 25	05 18	01 32	04 50	19 50	06 03	09 41	21 04
26 Th	20 50	16 27	22 46	05 42	01 46	04 47	19 51	06 03	09 40	21 04
27 Fr	21 01	17 59	23 06	06 07	02 00	04 45	19 52	06 02	09 40	21 04
28 Sa	21 12	18 22	23 25	06 31	02 14	04 42	19 54	06 02	09 40	21 04
29 Su	21 23	17 39	23 43	06 55	02 28	04 39	19 55	06 01	09 40	21 04
30 Mo	21 33	16 00	24 00	07 19	02 41	04 37	19 56	06 01	09 40	21 04

ASPECTARIAN

01 08:53 ☽ ☍ ♇
13:43 ☿ ∥ ♆
15:17 ☽ ⚹ ♃
16:52 ☽ □ ♅

02 02:47 ☽ ⚹ ♀
03:35 ☽ ⚹ ♂
07:07 ☿ ♏
17:06 ☽ □ ☿

03 00:00 ☽ △ ♄
00:57 ☽ # ☉
01:11 ♀ ☌ ♂

04 01:47 ☽ △ ♅
07:36 ☽ # ☿
21:35 ☽ # ♆

05 11:36 ☽ □ ♄
13:12 ☽ ⚹ ☿
16:36 ☽ ☍ ♆
21:41 ☽ ∥ ♅

06 00:34 ☽ ∥ ♃
03:59 ☉ ⚹ ♇
05:27 ☽ △ ♇
05:35 ☽ ⚹ ☉
13:39 ☿ △ ♆
14:04 ☽ ☌ ♃
21:19 ☽ ∥ ♂

07 04:58 ☽ ∥ ♀
08:25 ☽ ☌ ♂
12:48 ☽ ☌ ♀
22:23 ☽ # ♀

08 01:12 ☽ ⚹ ♄
06:40 ☽ # ♂
15:31 ♀ ♎
18:40 ☽ □ ♇

09 02:43 ☽ ☍ ♅
04:15 ☽ # ♃
08:04 ☽ # ♅

10 08:29 ☽ ∥ ♆
12:12 ☿ ⚹ ♇
18:07 ☽ △ ♆

11 02:18 ☉ ⚹ ♃
06:59 ☽ ⚹ ♇
09:51 ☽ ☌ ☿
16:43 ☽ ⚹ ♃

12 14:56 ☽ ⚹ ♂
21:41 ♂ ♎

13 00:28 ☽ ⚹ ♀
01:48 ☽ ☌ ♄
04:49 ☽ □ ♆
06:47 ☽ ∥ ☿
17:12 ♀ ⚹ ♄
18:18 ☿ ⚹ ♃

14 00:24 ☽ △ ♅
02:49 ♀ # ♂
03:20 ☽ □ ♃
12:40 ☽ # ☉
23:14 ☽ ∥ ☉

15 02:52 ☽ □ ♂
13:27 ☽ ⚹ ♆
13:34 ☽ ∥ ☿
14:33 ☽ □ ♀

16 01:39 ☽ ☌ ♇
08:11 ☽ □ ♅
11:47 ☽ △ ♃
16:01 ☿ ∥ ☉
20:00 ☽ ⚹ ☿
20:54 ☽ ⚹ ☉

17 12:24 ☽ △ ♂
14:53 ☿ ☌ ☉
18:12 ☽ ⚹ ♄

18 02:00 ☽ △ ♀
13:38 ☿ ∥ ♄
13:56 ☽ ⚹ ♅
16:33 ♆ SD

19 08:20 ☽ □ ☿
08:52 ☽ ∥ ♆
23:09 ☽ □ ♄

20 00:27 ☽ ☌ ♆
06:26 ☽ # ♅
11:55 ☽ ⚹ ♇
12:23 ☽ # ♃
19:31 ☉ ∥ ♄
19:44 ☿ ♐
20:07 ☽ ∥ ♀
22:07 ☽ ☍ ♃

21 01:59 ♀ □ ♇
11:45 ☽ ∥ ♂
13:24 ☽ △ ☉
14:20 ☿ ∥ ♇
17:36 ☽ △ ☿
19:45 ☽ # ♂

22 00:15 ☽ ☍ ♂
02:01 ☽ △ ♄
14:00 ☽ □ ♇
14:29 ☽ # ♀
15:26 ☉ ♐
16:57 ☽ ☍ ♀
17:39 ☽ ∥ ♃
19:17 ☽ ☌ ♅
23:48 ☽ ∥ ♅

23 20:19 ☽ # ♆
22:13 ♀ ☍ ♅
22:41 ♀ # ♃

24 03:51 ☽ ⚹ ♆
05:19 ♂ ⚹ ♄
14:56 ☽ △ ♇

25 01:26 ☽ △ ♃
04:58 ☿ ☌ ♄
07:15 ☿ □ ♆
16:32 ☿ ⚹ ♂

26 04:47 ☽ ☍ ♄
04:50 ☽ □ ♆
06:24 ☽ △ ♂
07:26 ☽ ☍ ☿
12:19 ♄ □ ♆
19:53 ♀ # ♅
21:16 ☽ ⚹ ♅

27 03:24 ☽ □ ♃
03:36 ☽ △ ♀
05:40 ☉ ∥ ♇

28 07:38 ☽ △ ♆
11:34 ☽ □ ♂
19:49 ☽ ☍ ♇

29 00:58 ☽ □ ♅
08:04 ☽ ⚹ ♃
12:46 ☽ □ ♀
14:50 ☉ □ ♆

30 00:16 ☉ ☌ ♄
14:48 ☽ △ ♄
15:55 ☽ △ ☉
20:47 ☽ ⚹ ♂

⚷ Chiron

01 Dec.	00 N 46
02	17♓15R
05	17 11
08	17 08
11	17 05
14	17 02
17	17 00
20	16 58
23	16 57
26	16 56
29	16 56D

28 04:57 16♓56D

December 2015

Day	S. T.	☉	☽	☿	♀	♂	♃	♄	♅	♆	♇	☊ True
	h m s	° ' "	° ' "	° '	° '	° '	° '	° '	° '	° '	° '	° '
01 Tu	04 38 08	08♐ 27 11	12♌ 23 06	15♐ 54	25♎ 08	10♎ 46	21♍ 01	07♐ 34	16♈R49	07♓ 04	14♑ 02	28♍R07
02 We	04 42 05	09 27 59	24 49 31	17 28	26 17	11 22	21 07	07 41	16 48	07 04	14 04	28 03
03 Th	04 46 02	10 28 49	06♍ 57 53	19 01	27 27	11 57	21 14	07 48	16 47	07 05	14 06	28 02
04 Fr	04 49 58	11 29 40	18 53 28	20 34	28 37	12 32	21 20	07 55	16 46	07 05	14 08	28D 01
05 Sa	04 53 55	12 30 32	00♎ 41 52	22 07	29 48	13 07	21 26	08 03	16 45	07 06	14 09	28R 01
06 Su	04 57 51	13 31 25	12 28 34	23 39	00♏ 58	13 42	21 32	08 10	16 44	07 06	14 11	28 00
07 Mo	05 01 48	14 32 20	24 18 41	25 12	02 08	14 17	21 38	08 17	16 43	07 07	14 13	27 57
08 Tu	05 05 44	15 33 16	06♏ 16 38	26 45	03 19	14 52	21 44	08 24	16 42	07 08	14 15	27 52
09 We	05 09 41	16 34 14	18 25 52	28 18	04 30	15 27	21 49	08 31	16 41	07 08	14 17	27 43
10 Th	05 13 37	17 35 12	00♐ 48 46	29 50	05 40	16 02	21 55	08 38	16 40	07 09	14 19	27 32
11 Fr	05 17 34	18 36 11	13 26 26	01♑ 22	06 51	16 37	22 00	08 45	16 39	07 10	14 21	27 19
12 Sa	05 21 31	19 37 12	26 18 53	02 55	08 03	17 12	22 05	08 52	16 39	07 10	14 22	27 05
13 Su	05 25 27	20 38 13	09♑ 25 06	04 26	09 14	17 47	22 10	08 59	16 38	07 11	14 24	26 53
14 Mo	05 29 24	21 39 15	22 43 28	05 58	10 25	18 21	22 15	09 06	16 37	07 12	14 26	26 41
15 Tu	05 33 20	22 40 17	06♒ 12 16	07 29	11 36	18 56	22 19	09 13	16 37	07 13	14 28	26 33
16 We	05 37 17	23 41 20	19 50 00	09 00	12 48	19 30	22 24	09 20	16 36	07 14	14 30	26 27
17 Th	05 41 13	24 42 24	03♓ 35 43	10 30	14 00	20 05	22 28	09 27	16 36	07 15	14 32	26 24
18 Fr	05 45 10	25 43 28	17 29 02	12 00	15 11	20 39	22 32	09 34	16 35	07 16	14 34	26 23
19 Sa	05 49 06	26 44 32	01♈ 29 51	13 29	16 23	21 14	22 36	09 41	16 35	07 17	14 36	26 22
20 Su	05 53 03	27 45 36	15 37 53	14 56	17 35	21 48	22 39	09 48	16 34	07 18	14 38	26 21
21 Mo	05 56 60	28 46 41	29 52 11	16 23	18 47	22 22	22 43	09 55	16 34	07 19	14 40	26 18
22 Tu	06 00 56	29 47 46	14♉ 10 37	17 48	19 59	22 56	22 46	10 02	16 34	07 20	14 42	26 13
23 We	06 04 53	00♑ 48 51	28 29 40	19 12	21 11	23 30	22 49	10 08	16 34	07 21	14 44	26 05
24 Th	06 08 49	01 49 57	12♊ 44 32	20 33	22 23	24 04	22 52	10 15	16 34	07 22	14 46	25 55
25 Fr	06 12 46	02 51 03	26 49 42	21 52	23 35	24 38	22 55	10 22	16 34	07 24	14 48	25 43
26 Sa	06 16 42	03 52 09	10♋ 39 51	23 09	24 48	25 12	22 58	10 29	16 33	07 25	14 50	25 30
27 Su	06 20 39	04 53 16	24 10 48	24 22	26 00	25 46	23 00	10 35	16D 34	07 26	14 52	25 18
28 Mo	06 24 35	05 54 23	07♌ 20 05	25 31	27 13	26 19	23 02	10 42	16 34	07 27	14 54	25 08
29 Tu	06 28 32	06 55 30	20 07 23	26 36	28 25	26 53	23 04	10 49	16 34	07 29	14 57	25 00
30 We	06 32 29	07 56 38	02♍ 34 19	27 36	29 38	27 27	23 06	10 55	16 34	07 30	14 59	24 55
31 Th	06 36 25	08 57 46	14 44 09	28 30	00♐ 51	28 00	23 08	11 02	16 34	07 31	15 01	24 52

Data for	12-01-2015
Julian Day	2457357.50
Ayanamsa	24 04 43
SVP	05 ♓ 02 13
☽ ☊ Mean	27 ♍ 15 R

● ◐ PHASES ○ ◑

Day	Time	Phase	Position
03	07:41	◑	10♍48
11	10:30	●	19♐03
18	15:14	◐	26♓22
25	11:12	○	03♋20

LAST ASPECT ☽ INGRESS

Day	h m	Day	h m	Sign
02	03:10	02	10:10	♍
04	05:00	04	22:35	♎
07	02:04	07	11:27	♏
09	06:40	09	22:26	♐
11	16:07	12	06:47	♑
13	23:08	14	12:59	♒
16	07:17	16	17:45	♓
18	15:15	18	21:27	♈
20	22:01	21	00:13	♉
22	14:27	23	02:32	♊
24	20:05	25	05:28	♋
27	03:37	27	10:32	♌
29	17:39	29	19:00	♍

DECLINATION

Day	☉	☽	☿	♀	♂	♃	♄	♅	♆	♇
01 Tu	21S 43	13N36	24S 15	07S 43	02S 55	04N35	19S 57	06N00	09S 40	21S 04
02 We	21 52	10 37	24 29	08 07	03 09	04 32	19 58	06 00	09 40	21 03
03 Th	22 01	07 15	24 42	08 31	03 23	04 30	20 00	06 00	09 39	21 03
04 Fr	22 09	03 39	24 54	08 55	03 36	04 28	20 01	05 59	09 39	21 03
05 Sa	22 17	00S 03	25 04	09 19	03 50	04 25	20 02	05 59	09 39	21 03
06 Su	22 25	03 45	25 13	09 42	04 03	04 23	20 03	05 58	09 39	21 03
07 Mo	22 32	07 19	25 20	10 06	04 17	04 21	20 04	05 58	09 38	21 03
08 Tu	22 39	10 38	25 27	10 29	04 30	04 19	20 05	05 58	09 38	21 03
09 We	22 45	13 33	25 31	10 52	04 43	04 17	20 07	05 57	09 38	21 03
10 Th	22 51	15 55	25 35	11 16	04 57	04 15	20 08	05 57	09 38	21 03
11 Fr	22 57	17 35	25 37	11 39	05 10	04 14	20 09	05 57	09 37	21 03
12 Sa	23 02	18 23	25 37	12 01	05 23	04 12	20 10	05 57	09 37	21 03
13 Su	23 06	18 13	25 36	12 24	05 36	04 10	20 11	05 56	09 37	21 03
14 Mo	23 10	17 05	25 33	12 46	05 49	04 08	20 12	05 56	09 36	21 03
15 Tu	23 14	14 58	25 29	13 08	06 02	04 07	20 13	05 56	09 36	21 02
16 We	23 17	12 01	25 24	13 30	06 15	04 05	20 14	05 56	09 36	21 02
17 Th	23 20	08 22	25 17	13 52	06 28	04 04	20 15	05 56	09 35	21 02
18 Fr	23 22	04 14	25 09	14 13	06 41	04 03	20 16	05 55	09 35	21 02
19 Sa	23 24	00N11	24 59	14 34	06 54	04 01	20 17	05 55	09 35	21 02
20 Su	23 25	04 36	24 47	14 55	07 07	04 00	20 18	05 55	09 34	21 02
21 Mo	23 26	08 49	24 34	15 16	07 19	03 59	20 19	05 55	09 34	21 02
22 Tu	23 26	12 31	24 20	15 36	07 32	03 58	20 20	05 55	09 33	21 02
23 We	23 26	15 29	24 05	15 56	07 44	03 57	20 21	05 55	09 33	21 02
24 Th	23 25	17 29	23 48	16 15	07 57	03 56	20 22	05 55	09 32	21 01
25 Fr	23 24	18 23	23 30	16 34	08 09	03 55	20 23	05 55	09 32	21 01
26 Sa	23 23	18 10	23 11	16 53	08 21	03 54	20 24	05 55	09 31	21 01
27 Su	23 21	16 56	22 51	17 11	08 34	03 54	20 25	05 55	09 31	21 01
28 Mo	23 18	14 49	22 30	17 29	08 46	03 53	20 26	05 55	09 31	21 01
29 Tu	23 15	12 01	22 09	17 47	08 58	03 53	20 27	05 55	09 30	21 01
30 We	23 12	08 45	21 47	18 04	09 10	03 52	20 28	05 55	09 30	21 01
31 Th	23 08	05 11	21 25	18 21	09 22	03 52	20 29	05 56	09 29	21 01

ASPECTARIAN

01 07:41 ☽ △ ☿
08:28 ☽ △ ♅
14:00 ☿ △ ♅

02 03:10 ☽ ✶ ♀
07:05 ☽ # ♆
16:10 ☽ # ♀

03 00:14 ☽ ☍ ♆
01:42 ☽ □ ♄
08:33 ☽ ∥ ♅
14:21 ☽ △ ♇
18:39 ☽ ∥ ♃

04 00:18 ☽ # ♂
03:54 ☽ □ ☿
05:00 ☽ ☌ ♃
12:53 ☿ □ ♃

05 04:15 ♀ ♏
15:07 ☽ ✶ ♄
20:27 ♀ ∥ ♆

06 02:02 ☽ ∥ ♂
02:20 ☽ ✶ ☉
02:38 ☽ ☌ ♂
03:29 ☽ □ ♇
04:04 ☽ # ♃
08:38 ☽ ☍ ♅
10:14 ☉ ✶ ♂
14:40 ☽ # ♅
20:48 ♂ □ ♇

07 02:04 ☽ ✶ ☿
07:01 ♂ # ♃
16:24 ☽ ∥ ♆
17:27 ☽ ☌ ♀
22:34 ☽ ∥ ♀

08 01:41 ☽ △ ♆
15:50 ☽ ✶ ♇

09 02:38 ☉ △ ♅
06:40 ☽ ✶ ♃

10 02:35 ☿ ♑
12:07 ☽ □ ♆
15:04 ☽ ☌ ♄

11 01:31 ♂ ☍ ♅
06:02 ☽ △ ♅
06:14 ♀ △ ♆
06:15 ☽ ✶ ♂
16:07 ☽ □ ♃

12 13:44 ☽ ☌ ☿
19:56 ☽ ✶ ♆
23:37 ☽ ✶ ♀

13 09:04 ☽ ☌ ♇
13:03 ☽ □ ♅
15:48 ☽ □ ♂
23:08 ☽ △ ♃

14 12:14 ♂ # ♅
15:03 ☉ □ ♃
19:40 ☿ ✶ ♆

15 05:23 ☽ ✶ ♄
10:27 ☽ □ ♀
13:53 ☽ ∥ ♀
18:20 ☽ ✶ ♅
23:34 ☽ △ ♂

16 07:17 ☽ ✶ ☉
16:21 ☽ ∥ ♆

17 06:20 ☽ ☌ ♆
10:14 ☽ □ ♄
10:43 ☽ ∥ ♂
11:18 ♀ ✶ ♇
13:25 ☽ ✶ ☿
14:25 ☽ # ♅
18:58 ☽ ✶ ♇
19:40 ☽ △ ♀

18 01:02 ☽ # ♃
08:42 ☽ ☍ ♃

19 14:02 ☽ △ ♄
18:54 ☿ ☌ ♇
20:41 ☽ ∥ ♃
22:19 ☽ □ ♇
22:42 ☽ □ ☿

20 01:36 ☽ ☌ ♅
07:17 ☽ ∥ ♅
10:51 ☽ ☍ ♂
14:46 ☽ # ♂
22:01 ☽ △ ☉

21 03:08 ☿ □ ♅
04:33 ☽ # ♆
12:31 ☽ ✶ ♆

22 00:53 ☽ △ ♇
04:48 ☉ ♑
06:44 ☽ △ ♅
10:37 ☽ ☍ ♀
14:27 ☽ △ ♃

23 05:11 ☽ # ♀
14:55 ☽ □ ♆
19:46 ☽ ☍ ♄

24 06:28 ☽ ✶ ♅
10:03 ♀ ✶ ♃
17:17 ☽ □ ♃
20:05 ☽ △ ♂

25 08:14 ☿ ∥ ☉
18:18 ☽ △ ♆
20:19 ☿ △ ♃

26 03:54 ♅ SD
07:22 ☽ ☍ ♇
10:23 ☽ □ ♅
20:59 ☽ # ♀
21:52 ☽ ✶ ♃

27 00:22 ☽ ☍ ☿
02:58 ☽ □ ♂
03:37 ☽ △ ♀

28 06:19 ☽ △ ♄
17:15 ☽ △ ♅

29 13:18 ☉ ✶ ♆
13:34 ☽ ✶ ♂
14:53 ☿ □ ♂
17:39 ☽ □ ♀
18:49 ☽ # ♆
21:19 ☽ # ♂

30 07:17 ♀ ♐
09:41 ☽ ☍ ♆
11:30 ☽ △ ☉
16:35 ☽ □ ♄
19:08 ☽ ∥ ♅

31 00:33 ☽ △ ♇
07:38 ☽ ∥ ♃
14:08 ♂ ∥ ♆
16:51 ☽ ☌ ♃

⚷ Chiron

01 Dec.	01 N 02
02	16♓57
05	16 58
08	16 59
11	17 01
14	17 03
17	17 06
20	17 10
23	17 14
26	17 18
29	17 23

January 2016

Day	S. T. h m s	☉ ° ' "	☽ ° ' "	☿ ° '	♀ ° '	♂ ° '	♃ ° '	♄ ° '	♅ ° '	♆ ° '	♇ ° '	☊ True ° '
01 Fr	06 40 22	09♑58 55	26♍41 21	29♑17	02♐03	28♎33	23♍09	11♐09	16♈34	07♓33	15♑03	24♍R52
02 Sa	06 44 18	11 00 04	08♎31 06	29 57	03 16	29 07	23 11	11 15	16 35	07 34	15 05	24 52
03 Su	06 48 15	12 01 13	20 18 57	00♒28	04 29	29 40	23 12	11 22	16 35	07 36	15 07	24 51
04 Mo	06 52 11	13 02 23	02♏10 28	00 49	05 42	00♏13	23 13	11 28	16 35	07 37	15 09	24 49
05 Tu	06 56 08	14 03 33	14 10 50	01 01	06 55	00 46	23 13	11 34	16 36	07 39	15 11	24 45
06 We	07 00 04	15 04 43	26 24 37	01R 01	08 08	01 19	23 14	11 41	16 37	07 40	15 13	24 38
07 Th	07 04 01	16 05 54	08♐55 16	00 50	09 21	01 52	23 14	11 47	16 37	07 42	15 15	24 29
08 Fr	07 07 58	17 07 04	21 44 57	00 27	10 35	02 25	23 14	11 53	16 38	07 43	15 17	24 18
09 Sa	07 11 54	18 08 14	04♑54 10	29♑53	11 48	02 57	23R 14	12 00	16 38	07 45	15 19	24 07
10 Su	07 15 51	19 09 25	18 21 44	29 07	13 01	03 30	23 14	12 06	16 39	07 47	15 21	23 55
11 Mo	07 19 47	20 10 35	02♒05 01	28 11	14 14	04 02	23 14	12 12	16 40	07 48	15 24	23 46
12 Tu	07 23 44	21 11 45	16 00 19	27 05	15 28	04 35	23 13	12 18	16 41	07 50	15 26	23 38
13 We	07 27 40	22 12 54	00♓03 38	25 53	16 41	05 07	23 12	12 24	16 42	07 52	15 28	23 33
14 Th	07 31 37	23 14 03	14 11 16	24 36	17 55	05 39	23 11	12 30	16 43	07 53	15 30	23 31
15 Fr	07 35 33	24 15 11	28 20 19	23 17	19 08	06 11	23 10	12 36	16 44	07 55	15 32	23 30
16 Sa	07 39 30	25 16 18	12♈28 44	21 58	20 22	06 43	23 08	12 42	16 45	07 57	15 34	23D 30
17 Su	07 43 27	26 17 25	26 35 13	20 42	21 35	07 15	23 07	12 48	16 46	07 59	15 36	23R 30
18 Mo	07 47 23	27 18 31	10♉38 47	19 31	22 49	07 47	23 05	12 53	16 47	08 00	15 38	23 29
19 Tu	07 51 20	28 19 36	24 38 23	18 26	24 02	08 18	23 03	12 59	16 48	08 02	15 40	23 26
20 We	07 55 16	29 20 40	08♊32 34	17 28	25 16	08 50	23 01	13 05	16 49	08 04	15 42	23 21
21 Th	07 59 13	00♒21 43	22 19 24	16 40	26 30	09 21	22 58	13 10	16 51	08 06	15 44	23 14
22 Fr	08 03 09	01 22 46	05♋56 27	16 01	27 43	09 52	22 56	13 16	16 52	08 08	15 46	23 05
23 Sa	08 07 06	02 23 48	19 21 10	15 31	28 57	10 24	22 53	13 21	16 53	08 10	15 48	22 56
24 Su	08 11 03	03 24 49	02♌31 14	15 10	00♑11	10 55	22 50	13 27	16 55	08 12	15 50	22 47
25 Mo	08 14 59	04 25 49	15 25 05	14 58	01 24	11 25	22 47	13 32	16 56	08 14	15 52	22 39
26 Tu	08 18 56	05 26 49	28 02 15	14 D 55	02 38	11 56	22 44	13 37	16 58	08 16	15 54	22 34
27 We	08 22 52	06 27 47	10♍23 31	14 59	03 52	12 27	22 40	13 43	16 59	08 18	15 56	22 30
28 Th	08 26 49	07 28 46	22 30 57	15 11	05 06	12 57	22 37	13 48	17 01	08 20	15 58	22 29
29 Fr	08 30 45	08 29 43	04♎27 44	15 30	06 20	13 28	22 33	13 53	17 03	08 22	16 00	22D 30
30 Sa	08 34 42	09 30 40	16 17 55	15 55	07 34	13 58	22 29	13 58	17 04	08 24	16 02	22 31
31 Su	08 38 38	10 31 36	28 06 13	16 26	08 48	14 28	22 25	14 03	17 06	08 26	16 04	22 33

Data for	01-01-2016
Julian Day	2457388.50
Ayanamsa	24 04 48
SVP	05 ♓ 02 10
☽ ☊ Mean	25 ♍ 36 R

● ◐ PHASES ○ ◑

Day	h m	Phase	Position
02	05:31	◑	11♎14
10	01:30	●	19♑13
16	23:27	◐	26♈16
24	01:47	○	03♌29

LAST ASPECT ☽ Day	h m	INGRESS Day	h m	Sign
01	05:35	01	06:42	♎
02	16:25	03	19:37	♏
05	17:48	06	06:57	♐
08	02:45	08	15:07	♑
10	17:40	10	20:23	♒
12	01:09	12	23:54	♓
14	16:32	15	02:49	♈
16	23:27	17	05:49	♉
19	06:51	19	09:14	♊
21	08:02	21	13:29	♋
23	06:22	23	19:22	♌
25	02:52	26	03:47	♍
28	00:11	28	15:00	♎
30	01:35	31	03:51	♏

DECLINATION

Day	☉	☽	☿	♀	♂	♃	♄	♅	♆	♇
01 Fr	23S04	01N28	21S03	18S37	09S34	03N51	20S30	05N56	09S28	21S00
02 Sa	22 59	02S17	20 41	18 53	09 45	03 51	20 30	05 56	09 28	21 00
03 Su	22 53	05 55	20 20	19 08	09 57	03 51	20 31	05 56	09 27	21 00
04 Mo	22 48	09 20	19 59	19 23	10 09	03 51	20 32	05 56	09 27	21 00
05 Tu	22 42	12 25	19 40	19 37	10 20	03 51	20 33	05 56	09 26	21 00
06 We	22 35	15 00	19 23	19 51	10 31	03 51	20 34	05 57	09 26	21 00
07 Th	22 28	16 58	19 07	20 04	10 43	03 51	20 35	05 57	09 25	21 00
08 Fr	22 20	18 08	18 53	20 17	10 54	03 51	20 35	05 57	09 24	21 00
09 Sa	22 12	18 23	18 42	20 29	11 05	03 51	20 36	05 57	09 24	20 59
10 Su	22 04	17 37	18 33	20 41	11 16	03 52	20 37	05 58	09 23	20 59
11 Mo	21 55	15 49	18 26	20 52	11 27	03 52	20 38	05 58	09 23	20 59
12 Tu	21 46	13 04	18 22	21 03	11 38	03 53	20 38	05 59	09 22	20 59
13 We	21 36	09 32	18 20	21 12	11 49	03 53	20 39	05 59	09 21	20 59
14 Th	21 26	05 27	18 20	21 22	12 00	03 54	20 40	05 59	09 21	20 59
15 Fr	21 16	01 03	18 22	21 31	12 10	03 55	20 41	06 00	09 20	20 59
16 Sa	21 05	03N24	18 26	21 39	12 21	03 56	20 41	06 00	09 19	20 58
17 Su	20 53	07 39	18 31	21 46	12 31	03 56	20 42	06 01	09 19	20 58
18 Mo	20 42	11 27	18 37	21 53	12 41	03 57	20 43	06 01	09 18	20 58
19 Tu	20 30	14 35	18 44	21 59	12 52	03 58	20 43	06 02	09 17	20 58
20 We	20 17	16 51	18 52	22 05	13 02	04 00	20 44	06 02	09 17	20 58
21 Th	20 04	18 07	19 01	22 10	13 12	04 01	20 45	06 03	09 16	20 58
22 Fr	19 51	18 20	19 10	22 15	13 22	04 02	20 45	06 03	09 15	20 57
23 Sa	19 37	17 30	19 19	22 18	13 31	04 03	20 46	06 04	09 14	20 57
24 Su	19 23	15 45	19 29	22 21	13 41	04 05	20 46	06 04	09 14	20 57
25 Mo	19 09	13 14	19 38	22 24	13 51	04 06	20 47	06 05	09 13	20 57
26 Tu	18 54	10 08	19 48	22 26	14 00	04 08	20 48	06 06	09 12	20 57
27 We	18 39	06 39	19 57	22 27	14 10	04 09	20 48	06 06	09 11	20 57
28 Th	18 24	02 58	20 06	22 27	14 19	04 11	20 49	06 07	09 11	20 57
29 Fr	18 08	00S47	20 14	22 27	14 28	04 13	20 49	06 07	09 10	20 56
30 Sa	17 52	04 29	20 22	22 26	14 37	04 15	20 50	06 08	09 09	20 56
31 Su	17 36	07 59	20 29	22 24	14 46	04 16	20 50	06 09	09 08	20 56

ASPECTARIAN

01 02:29 ☿ ∥ ♇
05:35 ☽ △ ☿
12:07 ☽ ✶ ♀

02 02:20 ☿ ♒
05:37 ☽ ✶ ♄
10:16 ☽ # ♃
11:15 ☿ ∥ ♄
13:24 ☽ □ ♇
16:25 ☽ ☍ ♅

03 00:08 ☽ # ♅
14:33 ♂ ♏
19:52 ☽ ☌ ♂
21:13 ☽ □ ☿

04 00:50 ☽ ∥ ♆
06:24 ☽ ∥ ♂
10:57 ☽ △ ♆
23:44 ☽ ✶ ☉

05 02:00 ☽ ✶ ♇
02:16 ☿ ∥ ♀
12:01 ☿ □ ♂
13:07 ☿ SR
14:34 ♀ □ ♆
17:48 ☽ ✶ ♃

06 03:27 ☉ ☌ ♇
08:50 ☽ ✶ ☿
21:40 ☽ □ ♆

07 00:55 ☽ ☌ ♀
05:27 ☽ ☌ ♄
12:22 ☉ □ ♅
14:29 ☽ △ ♅

08 02:45 ☽ □ ♃
04:41 ♃ SR
19:37 ☿ ♑R
20:20 ☽ ✶ ♂

09 04:11 ♀ ☌ ♄
05:08 ☽ ✶ ♆
15:17 ♀ ∥ ♄
18:40 ☽ ☌ ♇
20:59 ☽ □ ♅

10 08:34 ☽ △ ♃
17:40 ☽ ☌ ☿

11 03:32 ☽ □ ♂
15:56 ♀ ∥ ♇
17:35 ☽ ✶ ♄
22:59 ☽ ✶ ♀

12 01:09 ☽ ✶ ♅
09:52 ☽ ∥ ♂

13 00:11 ♀ △ ♅
01:12 ☽ ∥ ♆
08:56 ☽ △ ♂
13:17 ☽ ☌ ♆
21:02 ☽ # ♅
21:07 ☽ □ ♄
22:50 ☉ △ ♃

14 02:13 ☽ ✶ ♇
05:20 ♀ ∥ ☉
06:55 ☽ □ ♀
08:36 ☽ # ♃
14:05 ☿ ☌ ☉
15:14 ☽ ☍ ♃
16:10 ☽ ✶ ☿
16:32 ☽ ✶ ☉

15 02:16 ☿ △ ♃

16 00:22 ☽ △ ♄
02:55 ☽ ∥ ♃
05:15 ☽ □ ♇
07:16 ☽ ☌ ♅
13:39 ☉ ∥ ♇
14:31 ☽ ∥ ♅
14:40 ☽ △ ♀
14:48 ☽ □ ☿

17 10:03 ☽ # ♆
18:54 ☽ ☍ ♂
19:29 ☽ ✶ ♆
22:05 ☉ ∥ ♄

18 05:11 ♀ □ ♃
08:33 ☽ △ ♇
09:20 ☽ # ♂
11:02 ♂ △ ♆
14:04 ☽ △ ☿
21:16 ☽ △ ♃

19 06:51 ☽ △ ☉
23:11 ☽ □ ♆

20 07:55 ☽ ☍ ♄
14:25 ☽ ✶ ♅
15:28 ☉ ♒
18:27 ☿ □ ♅

21 01:08 ☽ □ ♃
08:02 ☽ ☍ ♀

22 03:54 ☽ △ ♆
07:17 ☽ △ ♂
10:05 ☿ ☌ ♇
17:18 ☽ ☍ ☿
17:35 ☽ ☍ ♇
19:33 ☽ □ ♅

23 06:22 ☽ ✶ ♃
18:24 ☿ ∥ ☉
20:32 ♀ ♑

24 16:12 ☽ □ ♂
19:12 ☽ # ♂
20:26 ☽ △ ♄

25 02:52 ☽ △ ♅
21:51 ☿ SD

26 06:44 ☽ # ♆
09:51 ☽ △ ♀
19:53 ☽ ☍ ♆

27 03:42 ☽ ∥ ♅
04:13 ☽ ✶ ♂
06:34 ☽ □ ♄
09:11 ☽ △ ☿
10:57 ☽ △ ♇
16:16 ☽ ∥ ♃

28 00:11 ☽ ☌ ♃

29 04:13 ☽ □ ♀
08:56 ☽ △ ☉
19:13 ☽ ✶ ♄
22:27 ☽ # ♃
23:12 ☽ □ ☿
23:27 ☽ □ ♇

30 01:35 ☽ ☍ ♅
05:53 ☿ ☌ ♇
11:14 ☽ # ♅
16:45 ♀ ✶ ♆

31 08:26 ☽ ∥ ♆
20:58 ☽ △ ♆

⚷ Chiron

01 Dec. 00 N 59

Day	Position
01	17♓28
04	17 34
07	17 40
10	17 46
13	17 53
16	18 01
19	18 09
22	18 17
25	18 25
28	18 34
31	18 43

February 2016

Day	S. T.	☉	☽	☿	♀	♂	♃	♄	♅	♆	♇	☊ True
	h m s	° ' "	° ' "	° '	° '	° '	° '	° '	° '	° '	° '	° '
01 Mo	08 42 35	11♒ 32 31	09♏ 57 46	17♑ 02	10♑ 02	14♏ 58	22♍R20	14♐ 08	17♈ 08	08♓ 28	16♑ 06	22♍ 34
02 Tu	08 46 31	12 33 25	21 57 47	17 43	11 15	15 28	22 16	14 12	17 10	08 30	16 08	22R 33
03 We	08 50 28	13 34 19	04♐ 11 20	18 28	12 29	15 57	22 11	14 17	17 12	08 32	16 10	22 31
04 Th	08 54 25	14 35 12	16 42 50	19 18	13 43	16 27	22 06	14 22	17 14	08 34	16 11	22 27
05 Fr	08 58 21	15 36 04	29 35 45	20 10	14 58	16 56	22 01	14 26	17 16	08 36	16 13	22 22
06 Sa	09 02 18	16 36 56	12♑ 52 02	21 06	16 12	17 25	21 56	14 31	17 18	08 38	16 15	22 16
07 Su	09 06 14	17 37 46	26 31 47	22 05	17 26	17 54	21 51	14 35	17 20	08 41	16 17	22 10
08 Mo	09 10 11	18 38 35	10♒ 32 57	23 07	18 40	18 23	21 46	14 40	17 22	08 43	16 19	22 05
09 Tu	09 14 07	19 39 23	24 51 27	24 11	19 54	18 51	21 40	14 44	17 24	08 45	16 21	22 02
10 We	09 18 04	20 40 10	09♓ 21 44	25 18	21 08	19 20	21 34	14 48	17 26	08 47	16 22	22 00
11 Th	09 22 00	21 40 55	23 57 33	26 27	22 22	19 48	21 28	14 52	17 28	08 49	16 24	21D 59
12 Fr	09 25 57	22 41 38	08♈ 32 50	27 37	23 36	20 16	21 22	14 56	17 31	08 52	16 26	22 00
13 Sa	09 29 54	23 42 20	23 02 27	28 50	24 50	20 44	21 16	15 00	17 33	08 54	16 28	22 01
14 Su	09 33 50	24 43 01	07♉ 22 35	00♒ 04	26 04	21 11	21 10	15 04	17 35	08 56	16 29	22 02
15 Mo	09 37 47	25 43 40	21 30 43	01 20	27 18	21 39	21 04	15 08	17 38	08 58	16 31	22 03
16 Tu	09 41 43	26 44 17	05♊ 25 28	02 37	28 33	22 06	20 57	15 12	17 40	09 00	16 33	22R 03
17 We	09 45 40	27 44 52	19 06 19	03 56	29 47	22 33	20 50	15 15	17 43	09 03	16 34	22 01
18 Th	09 49 36	28 45 25	02♋ 33 12	05 16	01♒ 01	23 00	20 44	15 19	17 45	09 05	16 36	21 58
19 Fr	09 53 33	29 45 57	15 46 18	06 38	02 15	23 26	20 37	15 22	17 48	09 07	16 38	21 55
20 Sa	09 57 29	00♓ 46 27	28 45 54	08 01	03 29	23 53	20 30	15 26	17 50	09 09	16 39	21 51
21 Su	10 01 26	01 46 56	11♌ 32 18	09 25	04 43	24 19	20 23	15 29	17 53	09 12	16 41	21 48
22 Mo	10 05 23	02 47 22	24 05 55	10 50	05 58	24 45	20 16	15 32	17 55	09 14	16 43	21 45
23 Tu	10 09 19	03 47 47	06♍ 27 25	12 16	07 12	25 10	20 09	15 35	17 58	09 16	16 44	21 43
24 We	10 13 16	04 48 10	18 37 52	13 44	08 26	25 36	20 01	15 38	18 01	09 19	16 46	21 42
25 Th	10 17 12	05 48 32	00♎ 38 53	15 12	09 40	26 01	19 54	15 41	18 03	09 21	16 47	21D 43
26 Fr	10 21 09	06 48 52	12 32 43	16 42	10 54	26 26	19 47	15 44	18 06	09 23	16 49	21 44
27 Sa	10 25 05	07 49 11	24 22 12	18 13	12 09	26 51	19 39	15 47	18 09	09 25	16 50	21 46
28 Su	10 29 02	08 49 28	06♏ 10 46	19 44	13 23	27 15	19 32	15 50	18 12	09 28	16 52	21 47
29 Mo	10 32 58	09 49 43	18 02 26	21 17	14 37	27 39	19 24	15 52	18 15	09 30	16 53	21 49

Data for	02-01-2016
Julian Day	2457419.50
Ayanamsa	24 04 52
SVP	05 ♓ 02 06
☽ ☊ Mean	23 ♍ 58 R

● ◐ PHASES ○ ◑

01	03:28	◑	11♏41
08	14:39	●	19♒16
15	07:47	◐	26♉03
22	18:20	○	03♍34

LAST ASPECT ☽ INGRESS

Day	h m	Day	h m	
02	00:36	02	15:50	♐
04	10:04	05	00:44	♑
06	15:54	07	06:00	♒
08	14:40	09	08:32	♓
11	04:26	11	09:56	♈
13	10:34	13	11:37	♉
15	10:55	15	14:36	♊
17	16:38	17	19:25	♋
19	14:36	20	02:18	♌
22	01:17	22	11:25	♍
24	14:23	24	22:42	♎
26	11:19	27	11:27	♏
29	19:56	29	23:57	♐

DECLINATION

Day	☉	☽	☿	♀	♂	♃	♄	♅	♆	♇
01 Mo	17S 19	11S 10	20S 35	22S 22	14S 55	04N18	20S 51	06N10	09S 08	20S 56
02 Tu	17 02	13 56	20 41	22 19	15 04	04 20	20 51	06 10	09 07	20 56
03 We	16 45	16 09	20 46	22 16	15 13	04 22	20 52	06 11	09 06	20 56
04 Th	16 27	17 38	20 50	22 12	15 21	04 25	20 52	06 12	09 05	20 55
05 Fr	16 09	18 17	20 53	22 07	15 30	04 27	20 53	06 13	09 04	20 55
06 Sa	15 51	17 59	20 55	22 01	15 38	04 29	20 53	06 13	09 04	20 55
07 Su	15 33	16 38	20 56	21 55	15 46	04 31	20 53	06 14	09 03	20 55
08 Mo	15 14	14 16	20 56	21 48	15 54	04 34	20 54	06 15	09 02	20 55
09 Tu	14 55	10 59	20 54	21 40	16 02	04 36	20 54	06 16	09 01	20 55
10 We	14 36	07 00	20 52	21 32	16 10	04 38	20 55	06 17	09 00	20 55
11 Th	14 17	02 34	20 48	21 23	16 18	04 41	20 55	06 18	09 00	20 54
12 Fr	13 57	02N01	20 44	21 14	16 26	04 44	20 55	06 19	08 59	20 54
13 Sa	13 37	06 27	20 38	21 04	16 33	04 46	20 56	06 19	08 58	20 54
14 Su	13 17	10 27	20 31	20 53	16 41	04 49	20 56	06 20	08 57	20 54
15 Mo	12 56	13 47	20 22	20 42	16 48	04 51	20 56	06 21	08 56	20 54
16 Tu	12 36	16 16	20 13	20 30	16 56	04 54	20 57	06 22	08 55	20 54
17 We	12 15	17 46	20 02	20 17	17 03	04 57	20 57	06 23	08 55	20 54
18 Th	11 54	18 16	19 50	20 04	17 10	05 00	20 57	06 24	08 54	20 53
19 Fr	11 33	17 44	19 37	19 51	17 17	05 03	20 57	06 25	08 53	20 53
20 Sa	11 12	16 18	19 22	19 36	17 24	05 05	20 58	06 26	08 52	20 53
21 Su	10 50	14 03	19 06	19 21	17 30	05 08	20 58	06 27	08 51	20 53
22 Mo	10 29	11 11	18 49	19 06	17 37	05 11	20 58	06 28	08 50	20 53
23 Tu	10 07	07 51	18 31	18 50	17 43	05 14	20 58	06 29	08 49	20 53
24 We	09 45	04 14	18 11	18 33	17 50	05 17	20 59	06 30	08 49	20 53
25 Th	09 23	00 30	17 50	18 16	17 56	05 20	20 59	06 31	08 48	20 53
26 Fr	09 01	03S 13	17 28	17 59	18 02	05 23	20 59	06 32	08 47	20 52
27 Sa	08 38	06 47	17 04	17 41	18 09	05 26	20 59	06 34	08 46	20 52
28 Su	08 16	10 05	16 39	17 22	18 15	05 29	20 59	06 35	08 45	20 52
29 Mo	07 53	12 58	16 13	17 03	18 20	05 33	20 59	06 36	08 44	20 52

ASPECTARIAN

01 00:08 ☽ ✶ ♀
03:42 ☿ □ ♅
10:29 ☽ ☌ ♂
12:21 ☽ ✶ ♇
15:02 ☽ ✶ ☿

02 00:36 ☽ ✶ ♃
12:18 ☽ ∥ ♂

03 07:08 ☽ ∥ ☉
08:25 ☽ □ ♆
10:48 ♂ ✶ ♇
18:18 ☉ ✶ ♄
19:31 ☽ ☌ ♄
19:37 ☽ ✶ ☉

04 00:58 ☽ △ ♅
10:04 ☽ □ ♃
21:12 ☿ ∥ ♄

05 16:25 ☽ ✶ ♆

06 01:12 ♀ ☌ ♇
06:01 ☽ ☌ ♇
06:30 ☽ ☌ ♀
07:43 ☿ ∥ ♇
07:52 ☽ □ ♅
08:22 ☽ ✶ ♂
11:55 ☉ ∥ ♂
15:40 ☽ ☌ ☿
15:54 ☽ △ ♃
16:39 ☉ ✶ ♅
18:46 ☿ △ ♃
22:03 ♀ □ ♅

07 09:22 ☽ ∥ ♂
12:08 ☉ □ ♂
14:01 ☽ ∥ [illegible]
15:01 ♀ ✶ ♂

08 06:59 ☽ ✶ ♄
11:31 ☽ ✶ ♅
13:38 ☽ □ ♂
16:34 ☿ ∥ ♇

09 01:33 ☿ ∥ ♄
12:19 ☽ ∥ ♆
23:03 ☽ ☌ ♆

10 03:33 ♄ ∥ ♇
04:00 ☽ ⧣ ♅
07:55 ♀ △ ♃
09:00 ☽ □ ♄
11:34 ☽ ✶ ♇
12:50 ☽ ⧣ ♃
16:56 ☽ △ ♂
19:57 ☽ ☍ ♃
21:08 ☽ ✶ ♀

11 04:26 ☽ ✶ ☿

12 10:37 ☽ △ ♄
13:03 ☽ □ ♇
14:39 ☽ ∥ ♃
14:51 ☽ ☌ ♅
23:21 ☽ ∥ ♅

13 01:11 ☽ ✶ ☉
03:16 ☽ □ ♀
10:34 ☽ □ ☿
14:42 ☽ ⧣ ♆
18:20 ♀ ∥ ♄
22:17 ♀ ∥ ♇
22:43 ☿ ♒
23:00 ♂ ✶ ♃

14 02:38 ☽ ✶ ♆
15:28 ☽ △ ♇
18:07 ☽ ⧣ ☉
23:14 ☽ △ ♃

15 00:14 ☽ ☍ ♂
10:55 ☽ △ ♀
18:38 ☽ △ ☿

16 06:16 ☽ □ ♆
09:30 ☽ ⧣ ♂
17:11 ☽ ☍ ♄
21:32 ☽ ✶ ♅

17 03:03 ☽ □ ♃
04:17 ♀ ♒
16:38 ☽ △ ☉

18 11:50 ☽ △ ♆

19 01:35 ☽ ☍ ♇
03:43 ☽ □ ♅
05:34 ☉ ♓
08:32 ☽ ⧣ ♂
08:49 ☽ ✶ ♃
14:36 ☽ △ ♂

20 09:46 ☽ ☍ ♀
19:28 ☽ ☍ ☿

21 07:32 ☽ △ ♄
12:06 ☽ △ ♅

22 01:17 [illegible]
05:57 ☽ ⧣ ☉

17:11 ☽ ⧣ ♆

23 05:32 ☽ ☍ ♆
09:09 ☽ ∥ ♅
17:13 ☽ ∥ ♃
18:03 ☽ □ ♄
20:17 ☽ △ ♇

24 02:44 ☽ ☌ ♃
14:23 ☽ ✶ ♂
18:34 ☿ ∥ ♂

25 08:06 ☿ ✶ ♄
20:10 [illegible]
20:20 ♀ ∥ ♂

26 06:29 ☽ ✶ ♄
08:40 ☽ □ ♇
09:39 ☽ △ ☿
11:19 ☽ ☍ ♅
14:36 ☽ ⧣ ♃
15:07 ☉ ∥ ♆
22:24 ☽ ⧣ ♅
23:02 ☿ ✶ ♅

27 11:49 ☽ ∥ ☉
14:05 ☽ ∥ ♆

28 05:52 ☽ △ ☉
06:11 [illegible]
15:49 ☉ ☌ ♆

16:17 ☽ □ ♀
21:40 ☽ ✶ ♇

29 02:42 ☽ ✶ ♃
07:30 ☽ □ ☿
19:56 ☽ ☌ ♂

⚷ Chiron

01 Dec. 00 N 35

03	18♓52
06	19 02
09	19 11
12	19 21
15	19 32
18	19 42
21	19 52
24	20 03
27	20 14

09	02:07	18♓56	☼ Total Solar Eclipse
23	11:48	03♎17	✷ Penumbral Lunar Eclipse (mag 0.801)

March 2016

Day	S. T. h m s	☉ ° ' "	☽ ° ' "	☿ ° '	♀ ° '	♂ ° '	♃ ° '	♄ ° '	♅ ° '	♆ ° '	♇ ° '	☊ True ° '
01 Tu	10 36 55	10♓49 57	00♐01 36	22♒51	15♒51	28♏03	19♍R16	15♐55	18♈17	09♓32	16♑54	21♍50
02 We	10 40 52	11 50 10	12 12 56	24 26	17 05	28 26	19 09	15 57	18 20	09 34	16 56	21 50
03 Th	10 44 48	12 50 21	24 41 01	26 02	18 20	28 50	19 01	15 59	18 23	09 37	16 57	21R 50
04 Fr	10 48 45	13 50 31	07♑30 01	27 39	19 34	29 13	18 53	16 01	18 26	09 39	16 58	21 49
05 Sa	10 52 41	14 50 39	20 43 17	29 17	20 48	29 35	18 45	16 03	18 29	09 41	17 00	21 49
06 Su	10 56 38	15 50 45	04♒22 39	00♓56	22 02	29 58	18 38	16 05	18 32	09 44	17 01	21 48
07 Mo	11 00 34	16 50 50	18 27 59	02 37	23 17	00♐20	18 30	16 07	18 35	09 46	17 02	21 47
08 Tu	11 04 31	17 50 53	02♓56 44	04 18	24 31	00 41	18 22	16 09	18 38	09 48	17 03	21 47
09 We	11 08 27	18 50 54	17 43 47	06 01	25 45	01 03	18 14	16 11	18 41	09 50	17 05	21D 47
10 Th	11 12 24	19 50 53	02♈41 55	07 44	26 59	01 24	18 06	16 12	18 44	09 53	17 06	21 47
11 Fr	11 16 21	20 50 51	17 42 46	09 29	28 14	01 44	17 58	16 14	18 48	09 55	17 07	21 47
12 Sa	11 20 17	21 50 46	02♉37 58	11 15	29 28	02 05	17 51	16 15	18 51	09 57	17 08	21 47
13 Su	11 24 14	22 50 39	17 20 16	13 02	00♓42	02 24	17 43	16 17	18 54	09 59	17 09	21R 47
14 Mo	11 28 10	23 50 30	01♊44 23	14 50	01 56	02 44	17 35	16 18	18 57	10 02	17 10	21 47
15 Tu	11 32 07	24 50 18	15 47 15	16 40	03 10	03 03	17 27	16 19	19 00	10 04	17 11	21 47
16 We	11 36 03	25 50 05	29 27 56	18 30	04 25	03 22	17 20	16 20	19 03	10 06	17 12	21 46
17 Th	11 39 60	26 49 49	12♋47 10	20 22	05 39	03 40	17 12	16 21	19 07	10 08	17 13	21 46
18 Fr	11 43 56	27 49 31	25 46 49	22 15	06 53	03 58	17 04	16 22	19 10	10 10	17 14	21D 46
19 Sa	11 47 53	28 49 10	08♌29 20	24 09	08 07	04 16	16 57	16 22	19 13	10 13	17 15	21 46
20 Su	11 51 49	29 48 48	20 57 20	26 04	09 21	04 33	16 49	16 23	19 16	10 15	17 16	21 47
21 Mo	11 55 46	00♈48 23	03♍13 24	28 01	10 36	04 49	16 42	16 23	19 20	10 17	17 17	21 48
22 Tu	11 59 43	01 47 56	15 19 54	29 58	11 50	05 06	16 35	16 24	19 23	10 19	17 18	21 48
23 We	12 03 39	02 47 27	27 18 59	01♈57	13 04	05 21	16 27	16 24	19 26	10 21	17 18	21R 48
24 Th	12 07 36	03 46 55	09♎12 44	03 57	14 18	05 37	16 20	16 24	19 30	10 23	17 19	21 47
25 Fr	12 11 32	04 46 22	21 03 09	05 57	15 32	05 51	16 13	16 24	19 33	10 26	17 20	21 46
26 Sa	12 15 29	05 45 47	02♏52 21	07 58	16 47	06 06	16 06	16R 24	19 36	10 28	17 21	21 45
27 Su	12 19 25	06 45 10	14 42 42	10 00	18 01	06 20	15 59	16 24	19 40	10 30	17 21	21 43
28 Mo	12 23 22	07 44 31	26 36 50	12 02	19 15	06 33	15 52	16 24	19 43	10 32	17 22	21 41
29 Tu	12 27 18	08 43 50	08♐37 55	14 05	20 29	06 46	15 45	16 24	19 46	10 34	17 23	21 39
30 We	12 31 15	09 43 08	20 49 28	16 08	21 43	06 58	15 39	16 23	19 50	10 36	17 23	21 37
31 Th	12 35 12	10 42 23	03♑15 19	18 10	22 57	07 10	15 32	16 23	19 53	10 38	17 24	21 37

Data for	03-01-2016
Julian Day	2457448.50
Ayanamsa	24 04 56
SVP	05 ♓ 02 01
☽ ☊ Mean	22 ♍ 26 R

● ◐ PHASES ○ ◑

01	23:11	◑	11♐48
09	01:55	☼	18♓56
15	17:03	◐	25♊33
23	12:01	✷	03♎17
31	15:18	◑	11♑20

LAST ASPECT Day	h m	☽ INGRESS Day	h m	
03	02:56	03	10:02	♑
05	16:07	05	16:23	♒
07	08:48	07	19:10	♓
09	01:56	09	19:41	♈
11	18:25	11	19:45	♉
13	09:47	13	21:04	♊
15	17:03	16	00:57	♋
18	04:09	18	07:55	♌
19	20:43	20	17:40	♍
22	03:55	23	05:24	♎
24	20:56	25	18:10	♏
27	07:27	28	06:47	♐
30	01:56	30	17:46	♑

DECLINATION

Day	☉	☽	☿	♀	♂	♃	♄	♅	♆	♇
01 Tu	07S30	15S21	15S46	16S43	18S26	05N36	21S00	06N37	08S44	20S52
02 We	07 07	17 05	15 17	16 23	18 32	05 39	21 00	06 38	08 43	20 52
03 Th	06 44	18 03	14 47	16 03	18 38	05 42	21 00	06 39	08 42	20 52
04 Fr	06 21	18 09	14 16	15 42	18 43	05 45	21 00	06 40	08 41	20 52
05 Sa	05 58	17 16	13 43	15 20	18 48	05 48	21 00	06 41	08 40	20 52
06 Su	05 35	15 22	13 10	14 58	18 54	05 51	21 00	06 43	08 39	20 51
07 Mo	05 12	12 31	12 34	14 36	18 59	05 54	21 00	06 44	08 38	20 51
08 Tu	04 48	08 50	11 58	14 14	19 04	05 57	21 00	06 45	08 38	20 51
09 We	04 25	04 31	11 21	13 50	19 09	06 01	21 00	06 46	08 37	20 51
10 Th	04 01	00N09	10 42	13 27	19 14	06 04	21 00	06 47	08 36	20 51
11 Fr	03 38	04 48	10 02	13 03	19 19	06 07	21 00	06 48	08 35	20 51
12 Sa	03 14	09 07	09 20	12 39	19 23	06 10	21 00	06 50	08 34	20 51
13 Su	02 50	12 48	08 38	12 14	19 28	06 13	21 00	06 51	08 33	20 51
14 Mo	02 27	15 38	07 54	11 50	19 33	06 16	21 00	06 52	08 33	20 51
15 Tu	02 03	17 26	07 09	11 24	19 37	06 19	21 00	06 53	08 32	20 51
16 We	01 39	18 11	06 23	10 59	19 41	06 22	21 00	06 55	08 31	20 51
17 Th	01 16	17 54	05 36	10 33	19 46	06 25	21 00	06 56	08 30	20 50
18 Fr	00 52	16 40	04 47	10 07	19 50	06 28	21 00	06 57	08 29	20 50
19 Sa	00 28	14 37	03 58	09 41	19 54	06 31	21 00	06 58	08 29	20 50
20 Su	00 04	11 56	03 08	09 14	19 58	06 34	21 00	07 00	08 28	20 50
21 Mo	00N19	08 45	02 16	08 47	20 02	06 37	21 00	07 01	08 27	20 50
22 Tu	00 43	05 14	01 24	08 20	20 06	06 40	21 00	07 02	08 26	20 50
23 We	01 07	01 32	00 30	07 53	20 09	06 42	21 00	07 03	08 25	20 50
24 Th	01 30	02S11	00N24	07 25	20 13	06 45	21 00	07 05	08 25	20 50
25 Fr	01 54	05 49	01 19	06 58	20 17	06 48	21 00	07 06	08 24	20 50
26 Sa	02 17	09 12	02 14	06 30	20 20	06 51	21 00	07 07	08 23	20 50
27 Su	02 41	12 13	03 11	06 02	20 24	06 53	21 00	07 08	08 22	20 50
28 Mo	03 04	14 44	04 07	05 33	20 27	06 56	20 59	07 10	08 21	20 50
29 Tu	03 28	16 40	05 04	05 05	20 30	06 58	20 59	07 11	08 21	20 50
30 We	03 51	17 52	06 00	04 36	20 34	07 01	20 59	07 12	08 20	20 50
31 Th	04 14	18 15	06 57	04 08	20 37	07 03	20 59	07 14	08 19	20 50

ASPECTARIAN

01 01:08 ♀ ⚹ ♄
04:02 ☽ ∥ ☿
15:02 ☽ ∥ ♀
18:49 ☽ □ ♆
02 07:16 ☽ ☌ ♄
10:30 ☽ ⚹ ♀
11:55 ☽ △ ♅
13:16 ☽ □ ♃
03 01:13 ♀ ⚹ ♅
02:56 ☽ ⚹ ☿
05:13 ☉ # ♅
04 03:58 ☽ ⚹ ♆
12:34 ☽ ⚹ ☉
17:18 ☽ ☌ ♇
19:59 ☽ □ ♅
20:31 ☽ △ ♃
05 05:44 ☿ □ ♂
09:07 ☉ # ♃
10:24 ☿ ♓
16:07 ☽ ⚹ ♂
06 02:31 ♂ ♐
04:29 ☽ ∥ ♀
06:04 ☉ □ ♄
20:03 ☽ ⚹ ♄
23:28 ☽ ∥ ☿
07 00:12 ☽ ⚹ ♅
04:39 ☉ ⚹ ♇
08:48 ☽ ☌ ♀
20:12 ☽ □ ♂
08 01:09 ☽ ∥ ♆
02:30 ☽ ☌ ☿
10:58 ☉ ☍ ♃
11:13 ☽ ☌ ♆
11:48 ☽ # ♅
16:01 ☽ # ♃
21:30 ☽ □ ♄
22:57 ☽ ⚹ ♇
09 00:31 ☽ ∥ ☉
00:48 ☽ ☍ ♃
21:52 ☽ △ ♂
10 18:19 ☽ # ☉
21:38 ☽ △ ♄
23:02 ☽ □ ♇
11 01:44 ☽ ☌ ♅
06:01 ☿ ☌ ♆
07:06 ☽ ∥ ♃
10:51 ☽ ∥ ♅
18:25 ☽ ⚹ ♀
20:47 ☽ # ♆
12 01:07 ☽ # ☿
10:24 ♀ ♓
11:55 ☽ ⚹ ♆
15:56 ☽ ⚹ ☿
20:24 ☽ # ♀
23:42 ☽ △ ♇
13 00:37 ☽ △ ♃
02:30 ☿ ∥ ♆
09:47 ☽ ⚹ ☉
14 00:22 ☽ □ ♀
01:43 ☽ ☍ ♂
14:07 ☽ □ ♆
19:26 ☿ □ ♄
20:48 ♀ □ ♂
15 00:55 ☽ ☍ ♄
01:45 ☽ □ ☿
02:52 ☽ □ ♃
05:36 ☽ ⚹ ♅
06:57 ☿ ⚹ ♇
08:09 ☿ # ♅
09:42 ☿ ☍ ♃
16 00:32 ☿ # ♃
09:44 ☽ △ ♀
19:10 ☽ △ ♆
20:27 ♃ △ ♇
17 08:00 ☽ ⚹ ♃
08:08 ☽ ☍ ♇
11:39 ☽ □ ♅
16:18 ☽ △ ☿
18 04:09 ☽ △ ☉
15:46 ☽ △ ♂
19 15:08 ☽ △ ♄
20:43 ☽ △ ♅
20 04:31 ☉ ♈
17:47 ♀ ☌ ♆
23:37 ☽ # ♀
21 02:06 ☽ # ♆
03:14 ☽ □ ♂
11:57 ☽ ∥ ♅
14:00 ☽ ☍ ♆
14:33 ☽ ∥ ♃
16:14 ☽ ☍ ♀
18:44 ♀ ∥ ♆
22 00:20 ☿ ♈
02:07 ☽ □ ♄
02:27 ☽ ☌ ♃
03:55 ☽ △ ♇
12:42 ☿ # ☉
23 02:27 ☽ ∥ ☉
08:42 ☽ # ☿
10:16 ♃ □ ♄
10:32 ☽ ∥ ☿
11:12 ☽ ☍ ☿
16:34 ☽ ⚹ ♂
18:59 ☽ # ☉
20:11 ☿ ☌ ☉
24 14:34 ☽ ⚹ ♄
16:27 ☽ □ ♇
17:19 ♀ # ♅
20:56 ☽ ☍ ♅
22:45 ☿ △ ♂
25 06:50 ☽ # ♃
06:57 ☽ ∥ ♀
07:48 ♀ # ♃
08:53 ☽ # ♅
10:02 ♄ SR
11:58 ♀ ☍ ♃
16:49 ♀ □ ♄
18:00 ☽ ∥ ♆
26 02:08 ☿ ∥ ☉
10:33 ☉ △ ♂
11:11 ♀ ⚹ ♇
15:26 ☽ △ ♆
27 02:33 ☽ ⚹ ♃
05:21 ☽ ⚹ ♇
07:27 ☽ △ ♀
28 20:13 ☽ ☌ ♂
29 00:13 ☽ △ ☉
00:25 ☿ # ♀
03:50 ☽ □ ♆
12:57 ☽ △ ☿
13:57 ☽ □ ♃
15:19 ☽ ☌ ♄
22:03 ☽ △ ♅
30 01:56 ☽ □ ♀
03:02 ☿ △ ♄
14:52 ☿ □ ♇
21:02 ♀ # ☉
31 02:49 ☿ ∥ ♃
07:14 ☿ ∥ ♅
14:01 ☽ ⚹ ♆
20:50 ☿ ☌ ♅
22:58 ☽ △ ♃

⚷ Chiron

01 Dec.	00 N 01
01	20♓25
04	20 36
07	20 47
10	20 58
13	21 09
16	21 20
19	21 31
22	21 41
25	21 52
28	22 03
31	22 14

April 2016

Day	S. T.	☉	☽	☿	♀	♂	♃	♄	♅	♆	♇	☊ True
	h m s	° ' "	° ' "	° '	° '	° '	° '	° '	° '	° '	° '	° '
01 Fr	12 39 08	11♈41 37	15♑59 25	20♈12	24♓11	07♐21	15♍R26	16♐R22	19♈57	10♓40	17♑24	21♍D37
02 Sa	12 43 05	12 40 50	29 05 28	22 14	25 26	07 31	15 19	16 21	20 00	10 42	17 25	21 38
03 Su	12 47 01	13 40 00	12♒36 30	24 14	26 40	07 41	15 13	16 21	20 03	10 44	17 25	21 39
04 Mo	12 50 58	14 39 09	26 34 14	26 12	27 54	07 51	15 07	16 20	20 07	10 46	17 26	21 40
05 Tu	12 54 54	15 38 15	10♓58 20	28 09	29 08	08 00	15 01	16 19	20 10	10 48	17 26	21 41
06 We	12 58 51	16 37 20	25 45 49	00♉04	00♈22	08 08	14 55	16 18	20 14	10 50	17 27	21R 41
07 Th	13 02 47	17 36 23	10♈50 40	01 56	01 36	08 15	14 50	16 16	20 17	10 52	17 27	21 40
08 Fr	13 06 44	18 35 24	26 04 18	03 45	02 50	08 22	14 44	16 15	20 21	10 54	17 27	21 38
09 Sa	13 10 41	19 34 23	11♉16 43	05 31	04 04	08 28	14 39	16 14	20 24	10 56	17 28	21 34
10 Su	13 14 37	20 33 20	26 17 58	07 13	05 18	08 34	14 34	16 12	20 27	10 57	17 28	21 30
11 Mo	13 18 34	21 32 14	10♊59 45	08 51	06 33	08 39	14 29	16 11	20 31	10 59	17 28	21 27
12 Tu	13 22 30	22 31 07	25 16 28	10 25	07 47	08 43	14 24	16 09	20 34	11 01	17 28	21 23
13 We	13 26 27	23 29 57	09♋05 35	11 55	09 01	08 47	14 19	16 07	20 38	11 03	17 28	21 21
14 Th	13 30 23	24 28 45	22 27 25	13 19	10 15	08 50	14 14	16 06	20 41	11 05	17 29	21 19
15 Fr	13 34 20	25 27 30	05♌24 18	14 39	11 29	08 52	14 10	16 04	20 45	11 06	17 29	21D 20
16 Sa	13 38 16	26 26 14	17 59 55	15 54	12 43	08 53	14 06	16 02	20 48	11 08	17 29	21 21
17 Su	13 42 13	27 24 55	00♍18 27	17 03	13 57	08 54	14 01	15 59	20 51	11 10	17 29	21 22
18 Mo	13 46 09	28 23 33	12 24 09	18 07	15 11	08R 54	13 58	15 57	20 55	11 12	17 29	21 24
19 Tu	13 50 06	29 22 10	24 20 59	19 06	16 25	08 53	13 54	15 55	20 58	11 13	17R 29	21R 24
20 We	13 54 03	00♉20 45	06♎12 26	19 59	17 39	08 52	13 50	15 53	21 02	11 15	17 29	21 23
21 Th	13 57 59	01 19 17	18 01 29	20 46	18 53	08 50	13 47	15 50	21 05	11 17	17 29	21 19
22 Fr	14 01 56	02 17 48	29 50 36	21 28	20 07	08 47	13 43	15 48	21 08	11 18	17 29	21 14
23 Sa	14 05 52	03 16 16	11♏41 47	22 03	21 21	08 43	13 40	15 45	21 12	11 20	17 29	21 07
24 Su	14 09 49	04 14 43	23 36 47	22 33	22 35	08 38	13 37	15 42	21 15	11 21	17 28	20 59
25 Mo	14 13 45	05 13 08	05♐37 17	22 57	23 49	08 33	13 35	15 40	21 19	11 23	17 28	20 51
26 Tu	14 17 42	06 11 32	17 45 05	23 15	25 02	08 27	13 32	15 37	21 22	11 24	17 28	20 43
27 We	14 21 38	07 09 54	00♑02 22	23 28	26 16	08 20	13 30	15 34	21 25	11 26	17 28	20 37
28 Th	14 25 35	08 08 14	12 31 39	23 35	27 30	08 13	13 28	15 31	21 29	11 27	17 28	20 33
29 Fr	14 29 32	09 06 32	25 15 48	23R 36	28 44	08 05	13 26	15 28	21 32	11 29	17 27	20 31
30 Sa	14 33 28	10 04 49	08♒17 56	23 32	29 58	07 56	13 24	15 25	21 35	11 30	17 27	20D 30

Data for 04-01-2016
Julian Day 2457479.50
Ayanamsa 24 04 58
SVP 05 ♓ 01 55
☽ ☊ Mean 20 ♍ 47 R

● ◐ PHASES ○ ◑

07	11:24	●	18♈04
14	03:59	◐	24♋39
22	05:25	○	02♏31
30	03:29	◑	10♒13

LAST ASPECT Day	h m	☽ INGRESS Day	h m	
01	16:40	02	01:38	♒
03	23:17	04	05:47	♓
05	10:34	06	06:47	♈
07	14:57	08	06:11	♉
09	09:50	10	05:59	♊
11	18:57	12	08:07	♋
14	04:00	14	13:54	♌
16	17:49	16	23:24	♍
18	12:31	19	11:25	♎
21	06:15	22	00:19	♏
23	21:47	24	12:47	♐
26	15:52	26	23:55	♑
29	07:08	29	08:47	♒

DECLINATION

Day	☉	☽	☿	♀	♂	♃	♄	♅	♆	♇
01 Fr	04N37	17S43	07N53	03S39	20S40	07N06	20S59	07N15	08S18	20S50
02 Sa	05 00	16 15	08 49	03 10	20 43	07 08	20 59	07 16	08 18	20 50
03 Su	05 24	13 50	09 44	02 41	20 46	07 10	20 58	07 17	08 17	20 50
04 Mo	05 46	10 33	10 38	02 12	20 49	07 13	20 58	07 19	08 16	20 50
05 Tu	06 09	06 33	11 31	01 43	20 52	07 15	20 58	07 20	08 15	20 50
06 We	06 32	02 02	12 23	01 13	20 54	07 17	20 58	07 21	08 15	20 50
07 Th	06 55	02N42	13 12	00 44	20 57	07 19	20 58	07 23	08 14	20 50
08 Fr	07 17	07 18	14 01	00 15	21 00	07 21	20 57	07 24	08 13	20 50
09 Sa	07 39	11 24	14 47	00N15	21 02	07 23	20 57	07 25	08 13	20 50
10 Su	08 02	14 43	15 30	00 44	21 05	07 25	20 57	07 26	08 12	20 50
11 Mo	08 24	17 00	16 12	01 13	21 07	07 27	20 57	07 28	08 11	20 50
12 Tu	08 46	18 08	16 51	01 43	21 09	07 29	20 56	07 29	08 11	20 50
13 We	09 07	18 09	17 28	02 12	21 12	07 30	20 56	07 30	08 10	20 50
14 Th	09 29	17 08	18 02	02 41	21 14	07 32	20 56	07 32	08 09	20 50
15 Fr	09 51	15 16	18 33	03 11	21 16	07 33	20 56	07 33	08 09	20 50
16 Sa	10 12	12 41	19 01	03 40	21 18	07 35	20 55	07 34	08 08	20 50
17 Su	10 33	09 36	19 27	04 09	21 20	07 36	20 55	07 36	08 07	20 50
18 Mo	10 54	06 09	19 50	04 38	21 22	07 38	20 55	07 37	08 07	20 50
19 Tu	11 15	02 30	20 10	05 07	21 24	07 39	20 54	07 38	08 06	20 50
20 We	11 35	01S14	20 28	05 36	21 26	07 40	20 54	07 39	08 06	20 50
21 Th	11 56	04 55	20 42	06 05	21 28	07 42	20 54	07 41	08 05	20 50
22 Fr	12 16	08 23	20 54	06 33	21 30	07 43	20 53	07 42	08 04	20 50
23 Sa	12 36	11 33	21 03	07 02	21 31	07 44	20 53	07 43	08 04	20 50
24 Su	12 56	14 14	21 09	07 30	21 33	07 45	20 53	07 44	08 03	20 50
25 Mo	13 16	16 21	21 13	07 58	21 34	07 45	20 52	07 46	08 03	20 51
26 Tu	13 35	17 45	21 13	08 26	21 36	07 46	20 52	07 47	08 02	20 51
27 We	13 54	18 22	21 11	08 54	21 37	07 47	20 52	07 48	08 02	20 51
28 Th	14 13	18 06	21 07	09 22	21 38	07 48	20 51	07 49	08 01	20 51
29 Fr	14 32	16 55	20 59	09 49	21 39	07 48	20 51	07 51	08 01	20 51
30 Sa	14 50	14 50	20 49	10 16	21 41	07 49	20 50	07 52	08 00	20 51

⚷ Chiron

01	Dec.	00 N 42
03	22♓24	
06	22 34	
09	22 44	
12	22 54	
15	23 04	
18	23 14	
21	23 23	
24	23 32	
27	23 41	
30	23 49	

ASPECTARIAN

01 02:38 ☽ ☌ ♇
07:21 ☽ □ ♅
09:14 ☽ □ ☿
10:38 ☿ ⋕ ♆
16:40 ☽ ✶ ♀

02 15:15 ☽ ✶ ♂

03 01:59 ☽ ✶ ☉
06:30 ☽ ✶ ♄
12:57 ☽ ✶ ♅
23:17 ☽ ✶ ☿
23:33 ☽ ⋕ ☿

04 08:54 ♂ ∥ ♇
14:10 ☽ ∥ ♆
19:03 ☽ □ ♂
19:34 ☽ ⋕ ♅
20:04 ☽ ⋕ ♃
23:43 ☽ ☌ ♆

05 01:59 ☽ ⋕ ☉
06:35 ☽ ☍ ♃
08:43 ☽ □ ♄
10:34 ☽ ✶ ♇
16:10 ☉ △ ♄
16:51 ♀ ♈
23:10 ☿ ♉

06 04:34 ☽ ∥ ♀
08:02 ☽ ☌ ♀
14:59 ☽ ⋕ ♀
19:52 ☽ △ ♂
20:11 ☉ □ ♇

07 05:29 ♂ ∥ ♄
08:34 ☽ △ ♄
10:26 ☽ □ ♇
14:57 ☽ ☌ ♅
23:54 ☽ ∥ ☉

08 00:17 ☽ ∥ ♃
00:32 ☽ ∥ ♅
04:46 ☉ ∥ ♃
05:04 ☽ ⋕ ♆
07:50 ☉ ∥ ♅
13:42 ☽ ☌ ☿
23:26 ☽ ✶ ♆

09 05:19 ☽ △ ♃
09:50 ☽ △ ♇
21:27 ☉ ☌ ♅

10 09:57 ☽ ∥ ☿
10:54 ☉ ⋕ ♆
15:59 ☽ ✶ ♀
20:06 ☽ ☍ ♂

11 00:00 ☽ □ ♆
05:45 ☽ □ ♃
08:37 ☽ ☍ ♄
15:59 ☽ ✶ ♅
18:57 ☽ ✶ ☉

12 09:40 ☿ ✶ ♆
19:17 ♀ △ ♂
23:50 ☽ □ ♀

13 03:28 ☽ △ ♆
05:36 ☽ ✶ ☿
07:49 ♃ ∥ ♅
09:14 ☽ ✶ ♃
12:22 ☽ ∥ ☿
14:58 ☽ ☍ ♇
20:46 ☽ □ ♅

14 15:28 ☿ △ ♃

15 06:33 ☽ △ ♂
12:45 ☽ △ ♀
19:31 ☽ □ ☿
20:13 ☽ △ ♄

16 05:27 ☽ △ ♅
17:40 ☽ ∥ ☉
17:49 ☽ △ ☉

17 09:22 ☿ △ ♇
10:33 ☽ ⋕ ♆
12:15 ♂ SR
14:00 ☽ ∥ ♃
14:08 ☽ ∥ ♅
17:01 ☽ □ ♂
21:35 ☽ ☍ ♆

18 03:06 ☽ ☌ ♃
07:05 ☽ □ ♄
07:24 ♇ SR
08:53 ☽ ∥ ♀
10:10 ☽ △ ♇
12:31 ☽ △ ☿
14:38 ♀ △ ♄

19 15:30 ☉ ♉
20:50 ♀ □ ♇

20 05:22 ☽ ✶ ♂
19:34 ☽ ✶ ♄
22:54 ☽ □ ♇

21 01:58 ☽ ☍ ♀
06:15 ☽ ☍ ♅
09:01 ☽ ⋕ ♀
15:52 ☿ ⋕ ♇
19:00 ☽ ⋕ ♅
19:06 ☽ ⋕ ♃
21:42 ☽ ∥ ♆
22:37 ☿ ⋕ ♄

22 21:00 ♀ ☌ ♅
23:15 ☽ △ ♆

23 03:58 ☽ ✶ ♃
10:03 ☽ ⋕ ☉
11:39 ☽ ✶ ♇
21:47 ☽ ☍ ☿

24 12:54 ♆ ∥ ♅
12:54 ♀ ∥ ♃
13:15 ♃ ∥ ♅

25 03:50 ♀ ⋕ ♆
05:47 ☽ ☌ ♂
11:27 ☽ □ ♆
15:43 ☽ □ ♃
19:48 ☽ ☌ ♄

26 07:08 ☽ △ ♅
15:52 ☽ △ ♀

27 14:54 ☽ △ ☉
21:57 ☽ ✶ ♆

28 01:46 ☽ △ ♃
09:21 ☽ ☌ ♇
16:59 ☽ □ ♅
17:21 ☿ SR
20:54 ☽ △ ☿
21:05 ♄ ∥ ♇

29 07:08 ☽ □ ♀
20:38 ☿ ⋕ ♇
21:43 ☿ ⋕ ♄
23:20 ☽ ✶ ♂

30 00:04 ☽ ⋕ ☉
00:36 ♀ ♉
12:47 ☽ ✶ ♄
23:55 ☽ ✶ ♅

May 2016

Day	S. T. h m s	☉ ° ' "	☽ ° ' "	☿ ° '	♀ ° '	♂ ° '	♃ ° '	♄ ° '	♅ ° '	♆ ° '	♇ ° '	☊ True ° '
01 Su	14 37 25	11♉ 03 05	21♒ 41 05	23♉R22	01♉ 12	07♐R46	13♍R22	15♐R21	21♈ 38	11♓ 32	17♑R27	20♍ 31
02 Mo	14 41 21	12 01 19	05♓ 27 45	23 08	02 26	07 35	13 21	15 18	21 42	11 33	17 26	20 32
03 Tu	14 45 18	12 59 31	19 39 13	22 50	03 40	07 24	13 19	15 15	21 45	11 34	17 26	20R 32
04 We	14 49 14	13 57 42	04♈ 14 35	22 27	04 54	07 12	13 18	15 11	21 48	11 36	17 25	20 31
05 Th	14 53 11	14 55 52	19 10 09	22 00	06 08	06 59	13 17	15 08	21 51	11 37	17 25	20 28
06 Fr	14 57 07	15 54 00	04♉ 19 05	21 30	07 22	06 46	13 16	15 04	21 55	11 38	17 24	20 22
07 Sa	15 01 04	16 52 06	19 32 04	20 58	08 35	06 32	13 16	15 01	21 58	11 39	17 24	20 15
08 Su	15 05 01	17 50 11	04♊ 38 37	20 24	09 49	06 17	13 16	14 57	22 01	11 40	17 23	20 07
09 Mo	15 08 57	18 48 14	19 28 50	19 48	11 03	06 01	13 15	14 53	22 04	11 42	17 23	19 58
10 Tu	15 12 54	19 46 16	03♋ 55 03	19 12	12 17	05 45	13D 15	14 50	22 07	11 43	17 22	19 50
11 We	15 16 50	20 44 15	17 52 50	18 35	13 31	05 29	13 16	14 46	22 11	11 44	17 21	19 44
12 Th	15 20 47	21 42 13	01♌ 21 04	17 59	14 45	05 12	13 16	14 42	22 14	11 45	17 21	19 39
13 Fr	15 24 43	22 40 09	14 21 28	17 24	15 58	04 54	13 16	14 38	22 17	11 46	17 20	19 37
14 Sa	15 28 40	23 38 03	26 57 43	16 51	17 12	04 36	13 17	14 34	22 20	11 47	17 19	19D 37
15 Su	15 32 36	24 35 55	09♍ 14 34	16 20	18 26	04 17	13 18	14 30	22 23	11 48	17 19	19 37
16 Mo	15 36 33	25 33 46	21 17 10	15 52	19 40	03 58	13 19	14 26	22 26	11 49	17 18	19 38
17 Tu	15 40 30	26 31 34	03♎ 10 33	15 27	20 54	03 39	13 20	14 22	22 29	11 50	17 17	19R 37
18 We	15 44 26	27 29 22	14 59 19	15 06	22 07	03 19	13 22	14 18	22 32	11 51	17 16	19 34
19 Th	15 48 23	28 27 07	26 47 30	14 48	23 21	02 59	13 24	14 14	22 35	11 51	17 15	19 28
20 Fr	15 52 19	29 24 51	08♏ 38 16	14 35	24 35	02 39	13 25	14 09	22 38	11 52	17 15	19 20
21 Sa	15 56 16	00♊ 22 34	20 34 06	14 26	25 49	02 18	13 27	14 05	22 40	11 53	17 14	19 09
22 Su	16 00 12	01 20 15	02♐ 36 44	14 21	27 03	01 57	13 29	14 01	22 43	11 54	17 13	18 56
23 Mo	16 04 09	02 17 55	14 47 22	14D 21	28 16	01 36	13 32	13 57	22 46	11 55	17 12	18 43
24 Tu	16 08 05	03 15 34	27 06 54	14 25	29 30	01 15	13 34	13 52	22 49	11 55	17 11	18 31
25 We	16 12 02	04 13 11	09♑ 36 13	14 34	00♊ 44	00 54	13 37	13 48	22 52	11 56	17 10	18 20
26 Th	16 15 59	05 10 48	22 16 23	14 47	01 58	00 33	13 40	13 44	22 55	11 57	17 09	18 13
27 Fr	16 19 55	06 08 23	05♒ 08 46	15 05	03 11	00 12	13 43	13 39	22 57	11 57	17 08	18 08
28 Sa	16 23 52	07 05 58	18 15 09	15 27	04 25	29♏ 51	13 46	13 35	23 00	11 58	17 07	18 05
29 Su	16 27 48	08 03 31	01♓ 37 38	15 53	05 39	29 30	13 49	13 30	23 03	11 58	17 06	18 05
30 Mo	16 31 45	09 01 04	15 18 16	16 24	06 53	29 10	13 53	13 26	23 05	11 59	17 05	18D 05
31 Tu	16 35 41	09 58 36	29 18 35	16 59	08 06	28 49	13 57	13 22	23 08	11 59	17 04	18R 04

Data for 05-01-2016
Julian Day 2457509.50
Ayanamsa 24 05 01
SVP 05 ♓ 01 50
☽ ☊ Mean 19 ♍ 12 R

● ◐ PHASES ○ ◑

06	19:30	●	16♉41
13	17:03	◐	23♌21
21	21:15	○	01♐14
29	12:12	◑	08♓33

LAST ASPECT Day	h m	☽ INGRESS Day	h m	
01	02:56	01	14:34	♓
03	05:09	03	17:05	♈
05	04:18	05	17:11	♉
07	02:11	07	16:35	♊
09	04:16	09	17:25	♋
11	07:35	11	21:33	♌
13	17:04	14	05:53	♍
16	09:22	16	17:34	♎
18	15:24	19	06:31	♏
21	11:40	21	18:49	♐
23	15:38	24	05:34	♑
26	01:12	26	14:27	♒
28	20:19	28	21:06	♓
30	23:11	31	01:10	♈

DECLINATION

Day	☉	☽	☿	♀	♂	♃	♄	♅	♆	♇
01 Su	15N08	11S 55	20N37	10N43	21S 42	07N49	20S 50	07N53	08S 00	20S 51
02 Mo	15 26	08 15	20 22	11 10	21 42	07 50	20 50	07 54	07 59	20 51
03 Tu	15 44	04 01	20 05	11 36	21 43	07 50	20 49	07 56	07 59	20 51
04 We	16 02	00N34	19 46	12 02	21 44	07 50	20 49	07 57	07 58	20 51
05 Th	16 19	05 14	19 25	12 28	21 45	07 50	20 48	07 58	07 58	20 52
06 Fr	16 36	09 37	19 02	12 54	21 45	07 51	20 48	07 59	07 57	20 52
07 Sa	16 52	13 23	18 38	13 19	21 46	07 51	20 48	08 00	07 57	20 52
08 Su	17 09	16 14	18 13	13 44	21 46	07 51	20 47	08 02	07 56	20 52
09 Mo	17 25	17 57	17 47	14 09	21 46	07 50	20 47	08 03	07 56	20 52
10 Tu	17 41	18 26	17 20	14 33	21 46	07 50	20 46	08 04	07 56	20 52
11 We	17 56	17 47	16 53	14 57	21 46	07 50	20 46	08 05	07 55	20 52
12 Th	18 11	16 08	16 26	15 21	21 46	07 50	20 45	08 06	07 55	20 52
13 Fr	18 26	13 41	16 00	15 44	21 46	07 49	20 45	08 07	07 54	20 53
14 Sa	18 41	10 40	15 35	16 07	21 46	07 49	20 44	08 08	07 54	20 53
15 Su	18 55	07 14	15 10	16 29	21 45	07 48	20 44	08 10	07 54	20 53
16 Mo	19 09	03 35	14 48	16 51	21 45	07 48	20 44	08 11	07 53	20 53
17 Tu	19 22	00S 10	14 26	17 13	21 44	07 47	20 43	08 12	07 53	20 53
18 We	19 36	03 54	14 07	17 34	21 43	07 46	20 43	08 13	07 53	20 53
19 Th	19 49	07 28	13 50	17 54	21 43	07 45	20 42	08 14	07 53	20 54
20 Fr	20 01	10 45	13 34	18 15	21 42	07 45	20 42	08 15	07 52	20 54
21 Sa	20 14	13 38	13 21	18 34	21 41	07 44	20 41	08 16	07 52	20 54
22 Su	20 25	15 57	13 11	18 53	21 40	07 43	20 41	08 17	07 52	20 54
23 Mo	20 37	17 36	13 02	19 12	21 38	07 42	20 40	08 18	07 52	20 54
24 Tu	20 48	18 27	12 56	19 30	21 37	07 40	20 40	08 19	07 51	20 54
25 We	20 59	18 25	12 53	19 48	21 36	07 39	20 39	08 20	07 51	20 55
26 Th	21 10	17 28	12 51	20 05	21 34	07 38	20 39	08 21	07 51	20 55
27 Fr	21 20	15 38	12 52	20 22	21 33	07 37	20 38	08 22	07 51	20 55
28 Sa	21 29	12 58	12 55	20 38	21 31	07 35	20 38	08 23	07 50	20 55
29 Su	21 39	09 33	13 00	20 54	21 30	07 34	20 37	08 24	07 50	20 55
30 Mo	21 48	05 34	13 08	21 08	21 28	07 32	20 37	08 25	07 50	20 56
31 Tu	21 56	01 10	13 17	21 23	21 26	07 30	20 36	08 26	07 50	20 56

ASPECTARIAN

01 02:56 ☽ ☐ ☿
07:28 ☽ # ♀
12:01 ☉ ⚹ ♆
18:16 ☽ ⚹ ♀
02 01:40 ☽ ∥ ♆
02:08 ☽ # ♅
02:36 ☽ # ♃
03:36 ☽ ☐ ♂
10:24 ☽ ☌ ♆
12:00 ☽ ⚹ ☉
13:23 ☽ ☍ ♃
16:39 ☽ ☐ ♄
20:17 ☽ ⚹ ♇
03 05:09 ☽ ⚹ ☿
07:59 ☉ △ ♃
04 04:43 ☽ △ ♂
17:35 ☽ △ ♄
20:19 ♅ # ♆
21:12 ☽ ☐ ♇
05 04:18 ☽ ☌ ♅
14:01 ☽ ∥ ♃
14:38 ☽ # ♆
14:45 ☽ ∥ ♅
06 05:13 ☽ ☌ ♀
11:33 ☽ ⚹ ♆
14:07 ☽ △ ♃
20:38 ☽ △ ♇
23:26 ☽ ∥ ♀
07 02:11 ☽ ☌ ☿
13:01 ☉ △ ♇
08 02:35 ☽ ☍ ♂
11:19 ☽ ☐ ♆
12:37 ☽ ∥ ☉
13:52 ☽ ☐ ♃
16:33 ☽ ☍ ♄
21:26 ☽ ∥ ☿
09 04:16 ☽ ⚹ ♅
12:15 ♃ SD
12:17 ♀ ∥ ☉
12:41 ♀ ⚹ ♆
15:13 ☿ ☌ ☉
10 13:19 ☽ △ ♆
15:40 ☽ ⚹ ♀
15:58 ☽ ⚹ ♃
19:01 ♀ △ ♃
21:28 ☽ ∥ ☉
23:05 ☽ ☍ ♇
11 01:11 ☽ ⚹ ☿
05:24 ☽ ⚹ ☉
07:35 ☽ ☐ ♅
19:22 ☽ ∥ ☿
12 06:51 ☽ △ ♂
07:21 ☽ ∥ ♀
13 00:31 ☽ △ ♄
02:53 ☿ △ ♇
03:22 ☽ ☐ ♀
05:29 ☽ ☐ ☿
08:01 ☿ ∥ ♀
15:04 ☽ △ ♅
19:11 ☿ ☌ ♀
14 02:19 ♀ △ ♇
14:29 ☽ ☐ ♂
17:51 ☽ ∥ ♅
19:36 ☽ # ♆
20:13 ☽ ∥ ♃
15 05:04 ☽ ☍ ♆
08:03 ☽ ☌ ♃
10:22 ☽ ☐ ♄
13:33 ☽ △ ☿
16:01 ☽ △ ♇
20:23 ☽ △ ♀
16 09:22 ☽ △ ☉
17 00:56 ☽ ⚹ ♂
22:36 ☽ ⚹ ♄
18 04:38 ☽ ☐ ♇
15:24 ☽ ☍ ♅
19 02:02 ☽ # ♃
02:53 ☽ ∥ ♆
05:25 ☽ # ♅
20 06:32 ☽ △ ♆
09:40 ☽ ⚹ ♃
11:49 ☽ ☍ ☿
14:37 ☉ ♊
17:19 ☽ ⚹ ♇
21:43 ☽ # ☿
21 11:40 ☽ ☍ ♀
22:44 ☽ ☌ ♂
22 11:17 ☉ ☍ ♂
13:21 ☿ SD
18:21 ☽ ☐ ♆
21:31 ☽ ☐ ♃
22:21 ☽ ☌ ♄
23 06:33 ☉ # ♄
15:38 ☽ △ ♅
24 09:45 ♀ ♊
13:58 ☉ # ♇
25 02:39 ♀ ☍ ♂
04:26 ☽ ⚹ ♆
07:40 ☽ △ ♃
09:35 ☽ △ ☿
14:21 ☽ ☌ ♇
26 01:12 ☽ ☐ ♅
12:29 ♃ ☐ ♄
15:04 ☽ ⚹ ♂
20:00 ☽ △ ♀
27 01:59 ☽ △ ☉
13:52 ♂ ♏R
15:33 ☽ ⚹ ♄
18:45 ☽ ☐ ☿
23:44 ♀ # ♄
28 00:20 ☽ # ☿
04:06 ☉ # ♂
08:36 ☽ ⚹ ♅
20:19 ☽ ☐ ♂
29 03:08 ♀ # ♇
07:15 ☽ # ♅
07:49 ☽ ☐ ♀
10:43 ☽ ∥ ♆
12:28 ☽ # ♃
18:13 ☽ ☌ ♆
20:46 ☽ ☐ ♄
21:31 ☽ ☍ ♃
30 01:59 ☽ ⚹ ☿
03:04 ☽ ⚹ ♇
23:11 ☽ △ ♂
31 03:06 ☿ △ ♇
05:33 ♀ # ♂
16:10 ☽ ⚹ ♀
19:11 ☽ ⚹ ☉
23:24 ☽ △ ♄

⚷ Chiron

01 Dec. 01 N 21

03	23♓57
06	24 05
09	24 12
12	24 19
15	24 26
18	24 32
21	24 38
24	24 44
27	24 49
30	24 54

June 2016

Day	S. T. h m s	☉ ° ' "	☽ ° ' "	☿ ° '	♀ ° '	♂ ° '	♃ ° '	♄ ° '	♅ ° '	♆ ° '	♇ ° '	☊ True ° '
01 We	16 39 38	10♊56 07	13♈38 44	17♉38	09♊20	28♏R29	14♍00	13♐R17	23♈10	12♓00	17♑R03	18♍R02
02 Th	16 43 34	11 53 38	28 16 51	18 20	10 34	28 09	14 04	13 13	23 13	12 00	17 01	17 58
03 Fr	16 47 31	12 51 07	13♉08 21	19 07	11 47	27 49	14 09	13 08	23 15	12 00	17 00	17 51
04 Sa	16 51 28	13 48 36	28 06 10	19 57	13 01	27 30	14 13	13 04	23 18	12 01	16 59	17 42
05 Su	16 55 24	14 46 04	13♊01 26	20 51	14 15	27 12	14 17	12 59	23 20	12 01	16 58	17 31
06 Mo	16 59 21	15 43 32	27 44 56	21 48	15 29	26 53	14 22	12 55	23 23	12 01	16 57	17 20
07 Tu	17 03 17	16 40 58	12♋08 40	22 49	16 42	26 35	14 27	12 50	23 25	12 02	16 56	17 09
08 We	17 07 14	17 38 23	26 07 09	23 53	17 56	26 18	14 32	12 46	23 27	12 02	16 54	17 00
09 Th	17 11 10	18 35 47	09♌37 55	25 00	19 10	26 02	14 37	12 42	23 30	12 02	16 53	16 54
10 Fr	17 15 07	19 33 10	22 41 25	26 11	20 24	25 46	14 42	12 37	23 32	12 02	16 52	16 50
11 Sa	17 19 03	20 30 32	05♍20 27	27 25	21 37	25 30	14 47	12 33	23 34	12 02	16 51	16 48
12 Su	17 23 00	21 27 53	17 39 17	28 42	22 51	25 16	14 53	12 28	23 36	12 02	16 49	16D48
13 Mo	17 26 57	22 25 13	29 43 02	00♊02	24 05	25 02	14 59	12 24	23 39	12 02	16 48	16R48
14 Tu	17 30 53	23 22 32	11♎37 02	01 25	25 19	24 48	15 05	12 20	23 41	12 R 02	16 47	16 47
15 We	17 34 50	24 19 50	23 26 34	02 52	26 32	24 36	15 11	12 15	23 43	12 02	16 45	16 45
16 Th	17 38 46	25 17 07	05♏16 22	04 21	27 46	24 24	15 17	12 11	23 45	12 02	16 44	16 40
17 Fr	17 42 43	26 14 24	17 10 33	05 53	29 00	24 13	15 23	12 07	23 47	12 02	16 43	16 32
18 Sa	17 46 39	27 11 40	29 12 18	07 28	00♋13	24 03	15 29	12 03	23 49	12 02	16 41	16 21
19 Su	17 50 36	28 08 55	11♐23 53	09 06	01 27	23 53	15 36	11 59	23 51	12 02	16 40	16 08
20 Mo	17 54 32	29 06 09	23 46 38	10 47	02 41	23 45	15 43	11 54	23 53	12 02	16 38	15 55
21 Tu	17 58 29	00♋03 23	06♑21 10	12 31	03 54	23 37	15 49	11 50	23 54	12 02	16 37	15 43
22 We	18 02 26	01 00 37	19 07 30	14 18	05 08	23 30	15 56	11 46	23 56	12 01	16 36	15 32
23 Th	18 06 22	01 57 50	02♒05 24	16 08	06 22	23 24	16 03	11 42	23 58	12 01	16 34	15 23
24 Fr	18 10 19	02 55 04	15 14 36	18 00	07 36	23 18	16 11	11 38	24 00	12 01	16 33	15 18
25 Sa	18 14 15	03 52 16	28 35 06	19 55	08 49	23 14	16 18	11 34	24 01	12 00	16 31	15 16
26 Su	18 18 12	04 49 29	12♓07 19	21 52	10 03	23 10	16 25	11 30	24 03	12 00	16 30	15D15
27 Mo	18 22 08	05 46 42	25 51 51	23 52	11 17	23 07	16 33	11 26	24 04	12 00	16 28	15 15
28 Tu	18 26 05	06 43 55	09♈49 16	25 54	12 30	23 05	16 41	11 23	24 06	11 59	16 27	15R16
29 We	18 30 01	07 41 08	23 59 30	27 58	13 44	23 04	16 48	11 19	24 07	11 59	16 25	15 15
30 Th	18 33 58	08 38 21	08♉21 16	00♋03	14 58	23D03	16 56	11 15	24 09	11 58	16 24	15 12

Data for	06-01-2016
Julian Day	2457540.50
Ayanamsa	24 05 05
SVP	05 ♓ 01 45
☽ ☊ Mean	17 ♍ 34 R

● ◐ PHASES ○ ◑

Day	Time	Phase	Position
05	03:00	●	14♊53
12	08:10	◐	21♍47
20	11:02	○	29♐33
27	18:19	◑	06♈30

LAST ASPECT ☽ INGRESS

Day	h m	Day	h m	Sign
01	15:43	02	02:48	♉
03	23:04	04	03:03	♊
05	16:49	06	03:43	♋
08	00:19	08	06:48	♌
10	07:15	10	13:47	♍
12	14:48	13	00:34	♎
15	07:01	15	13:19	♏
17	13:53	18	01:34	♐
20	11:03	20	11:55	♑
22	08:58	22	20:09	♒
24	15:49	25	02:31	♓
26	19:56	27	07:09	♈
29	07:47	29	10:05	♉

DECLINATION

Day	☉	☽	☿	♀	♂	♃	♄	♅	♆	♇
01 We	22N05	03N23	13N28	21N37	21S25	07N29	20S36	08N27	07S50	20S56
02 Th	22 13	07 49	13 41	21 50	21 23	07 27	20 35	08 28	07 50	20 56
03 Fr	22 20	11 51	13 55	22 02	21 21	07 25	20 35	08 29	07 50	20 56
04 Sa	22 27	15 08	14 12	22 14	21 19	07 23	20 34	08 30	07 50	20 57
05 Su	22 34	17 24	14 29	22 25	21 18	07 22	20 34	08 31	07 49	20 57
06 Mo	22 40	18 29	14 48	22 36	21 16	07 20	20 33	08 31	07 49	20 57
07 Tu	22 46	18 21	15 08	22 46	21 14	07 18	20 33	08 32	07 49	20 57
08 We	22 52	17 05	15 29	22 55	21 13	07 15	20 32	08 33	07 49	20 58
09 Th	22 57	14 54	15 52	23 04	21 11	07 13	20 32	08 34	07 49	20 58
10 Fr	23 01	12 00	16 15	23 12	21 10	07 11	20 32	08 35	07 49	20 58
11 Sa	23 06	08 37	16 39	23 20	21 08	07 09	20 31	08 36	07 49	20 58
12 Su	23 10	04 57	17 04	23 26	21 07	07 07	20 31	08 36	07 49	20 59
13 Mo	23 13	01 09	17 29	23 32	21 06	07 04	20 30	08 37	07 49	20 59
14 Tu	23 16	02S 39	17 55	23 38	21 04	07 02	20 30	08 38	07 49	20 59
15 We	23 19	06 18	18 21	23 42	21 03	06 59	20 29	08 39	07 49	20 59
16 Th	23 21	09 43	18 47	23 46	21 02	06 57	20 29	08 39	07 49	21 00
17 Fr	23 23	12 45	19 14	23 49	21 02	06 54	20 28	08 40	07 49	21 00
18 Sa	23 24	15 18	19 40	23 52	21 01	06 51	20 28	08 41	07 50	21 00
19 Su	23 25	17 12	20 06	23 54	21 00	06 49	20 27	08 41	07 50	21 00
20 Mo	23 26	18 20	20 32	23 55	21 00	06 46	20 27	08 42	07 50	21 01
21 Tu	23 26	18 36	20 57	23 55	20 59	06 43	20 27	08 43	07 50	21 01
22 We	23 26	17 56	21 22	23 55	20 59	06 40	20 26	08 43	07 50	21 01
23 Th	23 25	16 19	21 45	23 54	20 59	06 37	20 26	08 44	07 50	21 01
24 Fr	23 24	13 50	22 07	23 52	20 59	06 34	20 25	08 45	07 50	21 02
25 Sa	23 23	10 36	22 29	23 50	20 59	06 31	20 25	08 45	07 50	21 02
26 Su	23 21	06 45	22 48	23 46	21 00	06 28	20 25	08 46	07 51	21 02
27 Mo	23 19	02 30	23 06	23 43	21 00	06 25	20 24	08 46	07 51	21 02
28 Tu	23 16	01N57	23 22	23 38	21 01	06 22	20 24	08 47	07 51	21 03
29 We	23 13	06 21	23 36	23 33	21 02	06 19	20 24	08 47	07 51	21 03
30 Th	23 09	10 27	23 48	23 27	21 03	06 16	20 23	08 48	07 51	21 03

ASPECTARIAN

01 05:36 ☽ □ ♇
15:43 ☽ ☌ ♅
22:00 ☽ ∥ ♃

02 00:05 ☽ # ♆
02:43 ☉ □ ♆
03:41 ☽ ∥ ♅
22:11 ☽ ✶ ♆

03 01:37 ☽ △ ♃
04:15 ♀ □ ♆
06:12 ☽ △ ♇
06:38 ☉ ☍ ♄
10:09 ☽ ☌ ☿
15:45 ☽ ∥ ☿
23:04 ☽ ☍ ♂

04 00:47 ♀ ☍ ♄
10:58 ☉ □ ♃
22:22 ☽ □ ♆
23:57 ☽ ☍ ♄

05 00:49 ♀ □ ♃
02:03 ☽ □ ♃
02:10 ☽ ☌ ♀
16:49 ☽ ✶ ♅

06 21:40 ♀ ☌ ☉
23:48 ☽ △ ♆
23:48 ♀ ∥ ☉

07 03:55 ☽ ✶ ♃
08:07 ☽ ☍ ♇
19:21 ☽ □ ♅
19:46 ☽ ✶ ☿

08 00:19 ☽ △ ♂
15:48 ☽ ∥ ☿

09 05:31 ☽ △ ♄
17:01 ☿ ☍ ♂
17:42 ☽ ✶ ☉
19:16 ☽ ✶ ♀

10 01:35 ☽ △ ♅
05:39 ☽ □ ♂
07:15 ☽ □ ☿

11 00:08 ☽ ∥ ♅
05:19 ☽ # ♆
09:52 ☽ ∥ ♃
12:59 ☽ ☍ ♆
13:54 ☽ □ ♄
18:30 ☽ ☌ ♃
22:21 ☽ △ ♇

12 11:27 ☽ □ ♀
14:48 ☽ ✶ ♂
15:13 ♀ ✶ ♅
23:23 ☿ ♊

13 00:43 ☽ △ ☿
20:44 ♆ SR

14 01:26 ☽ ✶ ♄
07:54 ☉ ✶ ♅
10:26 ☽ □ ♇

15 00:33 ☽ ☍ ♅
01:58 ☽ △ ☉
04:38 ☽ # ♃
07:01 ☽ △ ♀
10:27 ☽ ∥ ♅
16:19 ☽ # ♅

16 13:40 ☽ △ ♆
20:22 ☽ ✶ ♃
23:04 ☽ ✶ ♇

17 13:53 ☽ ☌ ♂
19:39 ♀ ♋

18 03:24 ♄ □ ♆
18:49 ☽ ☍ ☿
19:02 ♂ ∥ ♇

19 01:07 ☽ ☌ ♄
01:14 ☽ □ ♆
08:16 ☽ □ ♃
19:31 ☿ # ♄

20 00:11 ☽ △ ♅
14:56 ☿ ☍ ♄
17:12 ☿ □ ♆
18:52 ☽ ☍ ♀
22:35 ☉ ♋

21 02:09 ☿ # ♂
03:38 ☿ # ♇
10:42 ☽ ✶ ♆
18:00 ☽ △ ♃
19:17 ☽ ☌ ♇

22 08:04 ☽ ✶ ♂
08:58 ☽ □ ♅
22:59 ☿ □ ♃

23 17:29 ☽ ✶ ♄

24 05:49 ☽ △ ☿
14:27 ☽ □ ♂
15:49 ☽ ✶ ♅

25 10:08 ☽ △ ☉
11:54 ☽ # ♅
17:32 ☽ ∥ ♆
19:59 ☽ △ ♀
22:55 ☽ □ ♄
23:47 ☽ ☌ ♆

26 01:41 ☽ # ♃
07:37 ☽ ☍ ♃
07:40 ☽ ✶ ♇
13:11 ☽ △ ♆
19:15 ☽ △ ♂
19:56 ☽ □ ☿

27 02:32 ☿ ✶ ♅
13:53 ♀ △ ♆
15:40 ☿ ∥ ☉

28 02:38 ☽ △ ♄
05:01 ☽ □ ♀
11:15 ☽ □ ♇
19:42 ☿ ∥ ♀
23:49 ☽ ∥ ♃

29 00:13 ☽ ☌ ♅
07:47 ☽ ✶ ☿
08:32 ☽ # ♆
14:01 ☽ ∥ ♅
23:25 ☿ ♋
23:39 ♂ SD

30 00:30 ☽ ✶ ☉
03:21 ♂ ∥ ♇
06:00 ☽ ✶ ♆
11:58 ☽ ✶ ♀
13:19 ☽ △ ♇
14:21 ☽ △ ♃

⚷ Chiron

01 Dec. 01 N 51

Day	Position
02	24♓58
05	25 02
08	25 05
11	25 08
14	25 10
17	25 12
20	25 13
23	25 14
26	25 15
29	25 15R

27 09:04 25♓15 R

July 2016

Day	S. T. h m s	☉ ° ' "	☽ ° ' "	☿ ° '	♀ ° '	♂ ° '	♃ ° '	♄ ° '	♅ ° '	♆ ° '	♇ ° '	☊ True ° '
01 Fr	18 37 55	09♋ 35 34	22♉ 51 42	02♋ 10	16♋ 12	23♏ 04	17♍ 04	11♐R12	24♈ 10	11♓R58	16♑R23	15♍R06
02 Sa	18 41 51	10 32 48	07♊ 26 04	04 18	17 25	23 05	17 13	11 08	24 12	11 57	16 21	14 59
03 Su	18 45 48	11 30 01	21 58 16	06 28	18 39	23 07	17 21	11 04	24 13	11 57	16 20	14 49
04 Mo	18 49 44	12 27 15	06♋ 21 29	08 38	19 53	23 10	17 29	11 01	24 14	11 56	16 18	14 40
05 Tu	18 53 41	13 24 28	20 29 20	10 48	21 07	23 14	17 38	10 58	24 15	11 55	16 17	14 30
06 We	18 57 37	14 21 42	04♌ 16 50	12 58	22 20	23 18	17 46	10 54	24 17	11 55	16 15	14 22
07 Th	19 01 34	15 18 55	17 41 10	15 09	23 34	23 24	17 55	10 51	24 18	11 54	16 14	14 16
08 Fr	19 05 31	16 16 08	00♍ 41 52	17 18	24 48	23 30	18 04	10 48	24 19	11 53	16 12	14 13
09 Sa	19 09 27	17 13 22	13 20 32	19 27	26 02	23 37	18 13	10 45	24 20	11 52	16 11	14 11
10 Su	19 13 24	18 10 35	25 40 21	21 36	27 15	23 44	18 22	10 41	24 21	11 52	16 09	14D 12
11 Mo	19 17 20	19 07 48	07♎ 45 36	23 43	28 29	23 53	18 31	10 39	24 22	11 51	16 08	14 13
12 Tu	19 21 17	20 05 01	19 41 14	25 49	29 43	24 02	18 40	10 36	24 23	11 50	16 06	14 14
13 We	19 25 13	21 02 14	01♏ 32 22	27 53	00♌ 57	24 12	18 50	10 33	24 24	11 49	16 05	14R 14
14 Th	19 29 10	21 59 27	13 24 09	29 56	02 10	24 22	18 59	10 30	24 24	11 48	16 03	14 12
15 Fr	19 33 06	22 56 40	25 21 16	01♌ 57	03 24	24 33	19 09	10 27	24 25	11 47	16 02	14 07
16 Sa	19 37 03	23 53 53	07♐ 27 45	03 57	04 38	24 45	19 18	10 25	24 26	11 46	16 00	14 01
17 Su	19 40 59	24 51 06	19 46 46	05 55	05 52	24 58	19 28	10 22	24 26	11 45	15 59	13 53
18 Mo	19 44 56	25 48 20	02♑ 20 23	07 52	07 05	25 11	19 38	10 20	24 27	11 44	15 57	13 45
19 Tu	19 48 53	26 45 34	15 09 38	09 46	08 19	25 25	19 48	10 17	24 28	11 43	15 56	13 36
20 We	19 52 49	27 42 48	28 14 24	11 39	09 33	25 40	19 58	10 15	24 28	11 42	15 55	13 29
21 Th	19 56 46	28 40 03	11♒ 33 41	13 30	10 47	25 55	20 08	10 13	24 29	11 41	15 53	13 24
22 Fr	20 00 42	29 37 18	25 05 52	15 19	12 00	26 11	20 18	10 11	24 29	11 40	15 52	13 21
23 Sa	20 04 39	00♌ 34 34	08♓ 48 59	17 06	13 14	26 28	20 28	10 09	24 29	11 39	15 50	13 19
24 Su	20 08 35	01 31 51	22 41 11	18 52	14 28	26 45	20 39	10 07	24 30	11 38	15 49	13D 20
25 Mo	20 12 32	02 29 09	06♈ 40 47	20 36	15 41	27 02	20 49	10 05	24 30	11 37	15 47	13 21
26 Tu	20 16 28	03 26 27	20 46 22	22 18	16 55	27 20	21 00	10 03	24 30	11 36	15 46	13 22
27 We	20 20 25	04 23 47	04♉ 56 30	23 58	18 09	27 39	21 10	10 01	24 30	11 34	15 45	13R 23
28 Th	20 24 22	05 21 07	19 09 27	25 37	19 23	27 59	21 21	10 00	24 30	11 33	15 43	13 22
29 Fr	20 28 18	06 18 29	03♊ 23 00	27 14	20 36	28 18	21 32	09 58	24 30	11 32	15 42	13 19
30 Sa	20 32 15	07 15 52	17 34 18	28 49	21 50	28 39	21 42	09 57	24R 30	11 31	15 40	13 16
31 Su	20 36 11	08 13 15	01♋ 39 55	00♍ 22	23 04	29 00	21 53	09 55	24 30	11 29	15 39	13 10

Data for	07-01-2016
Julian Day	2457570.50
Ayanamsa	24 05 10
SVP	05 ♓ 01 42
☽ ☊ Mean	15 ♍ 58 R

● ◐ PHASES ○ ◑

Day	Time	Phase	Position
04	11:02	●	12♋54
12	00:52	◐	20♎07
19	22:58	○	27♑40
26	23:00	◑	04♉21

LAST ASPECT ☽ INGRESS

Day	h m	Day	h m	Sign
01	00:20	01	11:46	♊
03	03:44	03	13:21	♋
05	06:30	05	16:29	♌
07	12:07	07	22:42	♍
10	03:28	10	08:33	♎
12	15:02	12	20:53	♏
14	22:23	15	09:15	♐
17	08:58	17	19:34	♑
19	22:58	20	03:12	♒
22	01:57	22	08:37	♓
24	07:08	24	12:34	♈
26	06:20	26	15:38	♉
28	15:14	28	18:17	♊
30	11:47	30	21:09	♋

DECLINATION

Day	☉	☽	☿	♀	♂	♃	♄	♅	♆	♇
01 Fr	23N05	13N57	23N57	23N20	21S04	06N12	20S23	08N48	07S52	21S03
02 Sa	23 01	16 36	24 04	23 13	21 06	06 09	20 23	08 49	07 52	21 04
03 Su	22 56	18 11	24 08	23 05	21 07	06 06	20 22	08 49	07 52	21 04
04 Mo	22 51	18 35	24 09	22 56	21 09	06 02	20 22	08 50	07 52	21 04
05 Tu	22 46	17 49	24 08	22 47	21 11	05 59	20 22	08 50	07 53	21 05
06 We	22 40	16 00	24 03	22 37	21 13	05 55	20 21	08 51	07 53	21 05
07 Th	22 33	13 22	23 56	22 26	21 15	05 52	20 21	08 51	07 53	21 05
08 Fr	22 27	10 07	23 46	22 15	21 17	05 48	20 21	08 51	07 54	21 05
09 Sa	22 20	06 29	23 34	22 02	21 20	05 45	20 20	08 52	07 54	21 06
10 Su	22 12	02 40	23 19	21 50	21 22	05 41	20 20	08 52	07 54	21 06
11 Mo	22 04	01S 11	23 01	21 37	21 25	05 37	20 20	08 52	07 55	21 06
12 Tu	21 56	04 56	22 42	21 23	21 28	05 33	20 20	08 53	07 55	21 06
13 We	21 47	08 28	22 20	21 08	21 31	05 30	20 19	08 53	07 55	21 07
14 Th	21 38	11 39	21 56	20 53	21 34	05 26	20 19	08 53	07 56	21 07
15 Fr	21 29	14 23	21 30	20 37	21 37	05 22	20 19	08 54	07 56	21 07
16 Sa	21 19	16 32	21 02	20 21	21 41	05 18	20 19	08 54	07 56	21 08
17 Su	21 09	17 58	20 33	20 04	21 44	05 14	20 19	08 54	07 57	21 08
18 Mo	20 59	18 34	20 02	19 47	21 48	05 10	20 19	08 54	07 57	21 08
19 Tu	20 48	18 14	19 30	19 29	21 52	05 06	20 18	08 54	07 58	21 08
20 We	20 37	16 56	18 57	19 10	21 56	05 02	20 18	08 55	07 58	21 09
21 Th	20 25	14 42	18 22	18 51	22 00	04 58	20 18	08 55	07 59	21 09
22 Fr	20 14	11 38	17 47	18 31	22 04	04 54	20 18	08 55	07 59	21 09
23 Sa	20 01	07 53	17 11	18 11	22 08	04 50	20 18	08 55	07 59	21 10
24 Su	19 49	03 41	16 34	17 51	22 12	04 46	20 18	08 55	08 00	21 10
25 Mo	19 36	00N46	15 56	17 30	22 16	04 41	20 18	08 55	08 00	21 10
26 Tu	19 23	05 11	15 18	17 08	22 21	04 37	20 18	08 55	08 01	21 10
27 We	19 09	09 19	14 40	16 46	22 25	04 33	20 18	08 55	08 01	21 11
28 Th	18 56	12 57	14 00	16 23	22 30	04 29	20 18	08 55	08 02	21 11
29 Fr	18 42	15 48	13 21	16 01	22 34	04 24	20 18	08 55	08 02	21 11
30 Sa	18 27	17 42	12 42	15 37	22 39	04 20	20 18	08 55	08 03	21 11
31 Su	18 12	18 30	12 02	15 13	22 44	04 16	20 18	08 55	08 03	21 12

ASPECTARIAN

Day	Time	Aspect
01	00:20	☽ ☍ ♂
	03:29	♀ ☍ ♇
	19:19	♀ ⚹ ♃
02	06:04	☽ ☍ ♄
	07:26	☽ □ ♆
	16:17	☽ □ ♃
03	03:44	☽ ⚹ ♅
	11:02	☉ △ ♆
04	04:31	☽ ☌ ☿
	09:24	☽ △ ♆
	16:48	☽ ☍ ♇
	19:03	☽ ⚹ ♃
05	01:10	☽ ☌ ♀
	04:45	☽ △ ♂
	06:24	♀ ∥ ☉
	06:30	☽ □ ♅
	12:20	☿ △ ♆
06	11:43	☽ △ ♄
	20:17	♀ △ ♂
07	03:24	☿ ☌ ☉
	10:31	☽ □ ♂
	11:55	☿ ☍ ♇
	12:07	☽ △ ♅
	14:25	♀ □ ♅
	22:25	☉ ☍ ♇
08	08:27	☽ ∥ ♅
	09:06	☿ ⚹ ♃
	14:48	☽ # ♆
	19:02	☽ □ ♄
	21:11	☽ ☍ ♆
09	04:44	☽ ∥ ♃
	05:28	☽ △ ♇
	08:07	☽ ⚹ ☉
	09:32	☽ ☌ ♃
	14:19	☽ ⚹ ☿
	20:09	☽ ⚹ ♂
10	03:28	☽ ⚹ ♀
	05:42	☉ ⚹ ♃
11	02:01	☿ △ ♂
	05:45	☽ ⚹ ♄
	07:29	☿ □ ♅
	16:48	☽ □ ♇
	16:49	♀ # ♂
12	04:01	☽ # ♃
	05:35	♀ ♌
	09:30	☽ ☍ ♅
	15:02	☽ □ ☿
	20:11	☽ ∥ ♆
	22:39	☽ □ ♀
13	02:12	♀ # ♇
	03:01	☽ # ♅
	20:47	☽ △ ♆
14	00:47	☿ ♌
	05:20	☽ ⚹ ♇
	08:30	☉ # ♂
	11:24	☽ ⚹ ♃
	18:01	☿ # ♂
	18:46	☽ △ ☉
	22:23	☽ ☌ ♂
15	01:03	☿ ∥ ☉
	15:44	☽ △ ☿
	17:48	☽ △ ♀
	19:24	☿ # ♇
16	03:05	♀ # ♄
	05:46	☽ ☌ ♄
	08:26	☽ □ ♆
	13:32	☉ □ ♅
	22:00	☿ ☌ ♀
	23:23	☽ □ ♃
17	03:08	☉ # ♇
	03:45	☉ △ ♂
	08:58	☽ △ ♅
	11:13	☿ # ♄
18	17:37	☽ ⚹ ♆
19	01:26	☽ ☌ ♇
	02:11	☿ ∥ ♀
	06:28	☿ △ ♄
	08:40	☽ △ ♃
	17:07	☽ □ ♅
	19:13	☽ ⚹ ♂
20	13:20	♀ △ ♄
	21:36	☽ ⚹ ♄
	22:27	☽ ☍ ♀
21	04:00	☽ ☍ ☿
	14:54	☉ # ♄
	22:55	☽ ⚹ ♅
22	01:57	☽ □ ♂
	09:31	☉ ♌
	17:44	☽ # ♅
	23:22	☽ ∥ ♆
23	02:18	☽ □ ♄
	04:55	☽ ☌ ♆
	12:09	☽ ⚹ ♇
	17:55	☽ # ♃
	20:26	☽ ☍ ♃
24	07:08	☽ △ ♂
	16:18	☽ △ ☉
25	05:47	☽ △ ♄
	15:30	☽ □ ♇
	16:50	☽ △ ♀
	20:56	☽ ∥ ♃
26	02:56	☽ △ ☿
	06:20	☽ ☌ ♅
	16:09	☽ # ♆
	21:31	☽ ∥ ♅
27	07:45	☿ △ ♅
	11:11	☽ ⚹ ♆
	18:13	☽ △ ♇
28	00:24	☽ □ ♀
	03:44	☽ △ ♃
	06:37	☽ ∥ ☿
	12:17	☽ □ ☿
	15:14	☽ ☍ ♂
29	01:43	☽ ∥ ♀
	05:18	☽ ⚹ ☉
	11:06	☽ ☍ ♄
	13:45	☽ □ ♆
	20:49	☿ □ ♂
	21:07	♅ SR
30	07:07	☽ □ ♃
	07:56	☽ ⚹ ♀
	11:47	☽ ⚹ ♅
	13:59	☽ ∥ ☉
	18:19	☿ ♍
	21:30	☽ ⚹ ☿
31	16:51	☽ △ ♆

⚷ Chiron

01 Dec. 02 N 03

Day	Position
02	25♓14R
05	25 13
08	25 12
11	25 10
14	25 08
17	25 05
20	25 02
23	24 58
26	24 54
29	24 49

August 2016

18 09:44 25♒52 ✹ Penumbral Lunar Eclipse (mag 0.016)

Day	S. T. h m s	☉ ° ' "	☽ ° ' "	☿ ° '	♀ ° '	♂ ° '	♃ ° '	♄ ° '	♅ ° '	♆ ° '	♇ ° '	☊ True ° '
01 Mo	20 40 08	09♌ 10 40	15♋ 36 04	01♍ 53	24♌ 18	29♏ 21	22♍ 04	09♐R54	24♈R30	11♓R28	15♑R38	13♍R05
02 Tu	20 44 04	10 08 06	29 19 07	03 23	25 31	29 43	22 15	09 53	24 30	11 27	15 36	13 00
03 We	20 48 01	11 05 33	12♌ 46 04	04 51	26 45	00♐ 06	22 26	09 52	24 30	11 25	15 35	12 55
04 Th	20 51 58	12 03 00	25 54 58	06 17	27 59	00 29	22 38	09 51	24 30	11 24	15 34	12 52
05 Fr	20 55 54	13 00 29	08♍ 45 16	07 42	29 13	00 52	22 49	09 50	24 30	11 23	15 33	12 50
06 Sa	20 59 51	13 57 58	21 17 50	09 04	00♍ 26	01 16	23 00	09 49	24 29	11 21	15 31	12 50
07 Su	21 03 47	14 55 28	03♎ 34 48	10 25	01 40	01 41	23 11	09 49	24 29	11 20	15 30	12D 51
08 Mo	21 07 44	15 52 59	15 39 18	11 43	02 54	02 06	23 23	09 48	24 28	11 18	15 29	12 53
09 Tu	21 11 40	16 50 30	27 35 14	13 00	04 07	02 31	23 34	09 48	24 28	11 17	15 28	12 55
10 We	21 15 37	17 48 03	09♏ 27 03	14 15	05 21	02 57	23 46	09 47	24 27	11 15	15 26	12 56
11 Th	21 19 33	18 45 37	21 19 32	15 27	06 35	03 23	23 58	09 47	24 27	11 14	15 25	12R 57
12 Fr	21 23 30	19 43 11	03♐ 17 31	16 37	07 48	03 49	24 09	09 47	24 26	11 12	15 24	12 56
13 Sa	21 27 26	20 40 46	15 25 35	17 46	09 02	04 16	24 21	09 47	24 26	11 11	15 23	12 55
14 Su	21 31 23	21 38 22	27 47 48	18 51	10 16	04 44	24 33	09D 47	24 25	11 09	15 22	12 53
15 Mo	21 35 20	22 36 00	10♑ 27 23	19 55	11 30	05 12	24 45	09 47	24 24	11 08	15 21	12 50
16 Tu	21 39 16	23 33 38	23 26 23	20 55	12 43	05 40	24 56	09 47	24 23	11 06	15 20	12 47
17 We	21 43 13	24 31 18	06♒ 45 33	21 53	13 57	06 08	25 08	09 47	24 22	11 05	15 18	12 45
18 Th	21 47 09	25 28 58	20 24 05	22 49	15 10	06 37	25 20	09 48	24 22	11 03	15 17	12 44
19 Fr	21 51 06	26 26 40	04♓ 19 47	23 41	16 24	07 06	25 32	09 48	24 21	11 02	15 16	12 43
20 Sa	21 55 02	27 24 23	18 29 12	24 30	17 38	07 36	25 44	09 49	24 20	11 00	15 15	12D 43
21 Su	21 58 59	28 22 08	02♈ 48 02	25 16	18 51	08 06	25 57	09 50	24 19	10 58	15 14	12 44
22 Mo	22 02 55	29 19 54	17 11 46	25 58	20 05	08 36	26 09	09 50	24 18	10 57	15 13	12 44
23 Tu	22 06 52	00♍ 17 42	01♉ 36 04	26 37	21 19	09 07	26 21	09 51	24 16	10 55	15 12	12 45
24 We	22 10 49	01 15 32	15 57 05	27 12	22 32	09 38	26 33	09 52	24 15	10 54	15 12	12 46
25 Th	22 14 45	02 13 24	00♊ 11 42	27 42	23 46	10 09	26 45	09 53	24 14	10 52	15 11	12R 46
26 Fr	22 18 42	03 11 17	14 17 29	28 08	24 59	10 41	26 58	09 54	24 13	10 50	15 10	12 45
27 Sa	22 22 38	04 09 12	28 12 40	28 30	26 13	11 12	27 10	09 56	24 12	10 49	15 09	12 44
28 Su	22 26 35	05 07 09	11♋ 55 54	28 46	27 26	11 45	27 23	09 57	24 10	10 47	15 08	12 43
29 Mo	22 30 31	06 05 08	25 26 15	28 58	28 40	12 17	27 35	09 58	24 09	10 45	15 07	12 42
30 Tu	22 34 28	07 03 09	08♌ 43 03	29 04	29 54	12 50	27 47	10 00	24 08	10 44	15 06	12 41
31 We	22 38 24	08 01 11	21 45 49	29R 04	01♎ 07	13 23	28 00	10 02	24 06	10 42	15 06	12 40

Data for	08-01-2016
Julian Day	2457601.50
Ayanamsa	24 05 15
SVP	05 ♓ 01 38
☽ ☊ Mean	14 ♍ 20 R

LAST ASPECT ☽ Day	h m	INGRESS Day	h m	
02	00:44	02	01:12	♌
04	04:13	04	07:34	♍
06	03:21	06	16:57	♎
08	17:42	09	04:52	♏
11	05:23	11	17:25	♐
13	17:38	14	04:13	♑
16	02:46	16	11:54	♒
18	09:28	18	16:35	♓
20	12:22	20	19:19	♈
22	11:48	22	21:20	♉
24	19:38	24	23:40	♊
27	00:31	27	03:07	♋
29	06:24	29	08:12	♌
31	04:20	31	15:23	♍

● ◐ PHASES ○ ◑

02	20:44	●	10♌58
10	18:22	◐	18♏32
18	09:27	✹	25♒52
25	03:41	◑	02♊22

DECLINATION

Day	☉	☽	☿	♀	♂	♃	♄	♅	♆	♇
01 Mo	17N57	18N10	11N22	14N49	22S48	04N11	20S18	08N55	08S04	21S12
02 Tu	17 42	16 46	10 42	14 24	22 53	04 07	20 18	08 55	08 04	21 12
03 We	17 26	14 27	10 03	13 59	22 58	04 02	20 18	08 55	08 05	21 13
04 Th	17 11	11 26	09 23	13 34	23 03	03 58	20 18	08 55	08 06	21 13
05 Fr	16 54	07 56	08 43	13 08	23 08	03 53	20 18	08 55	08 06	21 13
06 Sa	16 38	04 10	08 04	12 42	23 13	03 49	20 18	08 55	08 07	21 13
07 Su	16 21	00 17	07 25	12 16	23 18	03 44	20 18	08 54	08 07	21 14
08 Mo	16 04	03S32	06 46	11 49	23 23	03 39	20 18	08 54	08 08	21 14
09 Tu	15 47	07 09	06 08	11 22	23 28	03 35	20 19	08 54	08 08	21 14
10 We	15 30	10 29	05 31	10 55	23 33	03 30	20 19	08 54	08 09	21 14
11 Th	15 12	13 22	04 53	10 27	23 38	03 26	20 19	08 54	08 10	21 15
12 Fr	14 54	15 44	04 17	09 59	23 43	03 21	20 19	08 53	08 10	21 15
13 Sa	14 36	17 25	03 41	09 31	23 47	03 16	20 19	08 53	08 11	21 15
14 Su	14 17	18 20	03 06	09 03	23 52	03 11	20 19	08 53	08 11	21 15
15 Mo	13 59	18 23	02 31	08 34	23 57	03 07	20 20	08 52	08 12	21 16
16 Tu	13 40	17 28	01 58	08 05	24 02	03 02	20 20	08 52	08 12	21 16
17 We	13 21	15 35	01 25	07 36	24 07	02 57	20 20	08 52	08 13	21 16
18 Th	13 01	12 47	00 54	07 07	24 12	02 52	20 20	08 51	08 14	21 16
19 Fr	12 42	09 12	00 23	06 37	24 16	02 47	20 21	08 51	08 14	21 17
20 Sa	12 22	05 02	00S06	06 08	24 21	02 43	20 21	08 51	08 15	21 17
21 Su	12 02	00 33	00 33	05 38	24 26	02 38	20 21	08 50	08 16	21 17
22 Mo	11 42	03N59	00 59	05 08	24 30	02 33	20 22	08 50	08 16	21 17
23 Tu	11 22	08 18	01 24	04 38	24 34	02 28	20 22	08 50	08 17	21 18
24 We	11 02	12 05	01 47	04 07	24 39	02 23	20 22	08 49	08 17	21 18
25 Th	10 41	15 08	02 08	03 37	24 43	02 18	20 23	08 49	08 18	21 18
26 Fr	10 20	17 15	02 27	03 06	24 47	02 13	20 23	08 48	08 19	21 18
27 Sa	09 59	18 19	02 43	02 36	24 52	02 08	20 24	08 48	08 19	21 18
28 Su	09 38	18 18	02 57	02 05	24 56	02 03	20 24	08 47	08 20	21 19
29 Mo	09 17	17 13	03 09	01 34	25 00	01 58	20 24	08 47	08 21	21 19
30 Tu	08 55	15 13	03 18	01 04	25 03	01 53	20 25	08 46	08 21	21 19
31 We	08 34	12 27	03 24	00 33	25 07	01 48	20 25	08 46	08 22	21 19

ASPECTARIAN

01 00:03 ☽ ☍ ♇
04:09 ♀ △ ♅
06:40 ☽ ∥ ☉
11:25 ☽ ✶ ♃
15:32 ☽ □ ♅
17:50 ☉ △ ♄

02 00:44 ☽ △ ♂
17:50 ♂ ♐
18:47 ☽ △ ♄

03 04:48 ☽ ∥ ♀
21:23 ☽ △ ♅

04 04:13 ☽ ☌ ♀
08:44 ☽ □ ♂
17:07 ☿ ∥ ♅
17:35 ☽ ∥ ♅
17:41 ☽ ∥ ☿
21:45 ☽ ☌ ☿
22:56 ☽ # ♆

05 02:03 ☽ □ ♄
04:58 ☽ ☍ ♆
12:54 ☽ △ ♇
15:27 ♀ ♍
22:30 ☿ # ♆

06 02:11 ☽ ∥ ♃
03:21 ☽ ☌ ♃
13:18 ☿ □ ♄
20:07 ☽ ✶ ♂

07 00:20 ♀ □ ♂
12:20 ☽ ✶ ♄
16:25 ☿ ☍ ♆
23:39 ☽ □ ♇

08 00:30 ☽ ✶ ☉
00:45 ☽ # ♃
17:42 ☽ ☍ ♅
18:04 ☽ # ☿

09 06:48 ☽ ∥ ♆
12:12 ☽ # ♅
14:45 ☽ ✶ ♀

10 02:54 ☽ # ♀
03:39 ☽ △ ♆
10:48 ☽ ✶ ☿
12:06 ☽ ✶ ♇
23:24 ☿ △ ♇

11 05:23 ☽ ✶ ♃
15:52 ☽ # ☉

12 01:06 ☽ ☌ ♂
10:00 ☽ □ ♀
12:53 ☽ ☌ ♄
15:40 ☽ □ ♆

13 05:01 ☽ □ ☿
09:51 ♄ SD
11:07 ☽ △ ☉
14:31 ♀ □ ♄
17:30 ☽ △ ♅
17:38 ☽ □ ♃
19:24 ☿ ∥ ♃

14 08:21 ♀ ∥ ♅
17:04 ♀ ☍ ♆

15 01:15 ☽ ✶ ♆
02:08 ☽ △ ♀
09:06 ☽ ☌ ♇
17:59 ♀ # ♆
19:01 ☽ △ ☿

16 01:44 ☽ □ ♅
02:46 ☽ △ ♃
20:23 ☉ △ ♅
22:51 ☽ ✶ ♂

17 05:23 ☽ ✶ ♄
21:58 ☽ # ☉

18 02:16 ♀ △ ♇
06:52 ☽ ✶ ♅

19 02:09 ☽ # ♅
04:54 ☽ □ ♂
05:48 ☽ ∥ ♆
09:20 ☽ □ ♄
11:22 ☽ ☌ ♆
17:12 ☽ # ♀
18:33 ☽ ✶ ♇
22:25 ☽ ☍ ♀

20 10:41 ☽ ☍ ☿
12:22 ☽ ☍ ♃
12:50 ☽ # ♃

21 00:00 ☽ ∥ ☿
06:27 ☽ # ☿
09:10 ☽ △ ♂
11:44 ☽ △ ♄
16:28 ☽ ∥ ♃
20:43 ☽ □ ♇

22 05:33 ☽ ∥ ♀
09:05 ☿ ☌ ♃
11:48 ☽ ☌ ♅
16:39 ☉ ♍
21:40 ☽ △ ☉
23:56 ☽ # ♆

23 03:09 ☽ ∥ ♅
15:32 ☽ ✶ ♆
17:26 ☽ ∥ ☉
22:44 ☽ △ ♇

24 11:28 ♂ ☌ ♄
12:07 ☽ △ ♀
18:06 ☽ △ ♃
19:38 ☽ △ ☿

25 10:13 ☿ # ♃
16:30 ☽ ☍ ♄
17:35 ☽ ☍ ♂
18:06 ☽ □ ♆

26 06:57 ♂ □ ♆
17:03 ☽ ✶ ♅
20:12 ☽ □ ♀
20:13 ☿ # ♀
22:10 ☽ □ ♃

27 00:31 ☽ □ ☿
11:08 ☽ ✶ ☉
21:59 ☽ △ ♆
22:29 ♀ ☌ ♃

28 01:55 ♀ ∥ ♃
05:39 ☽ ☍ ♆
21:42 ☽ □ ♅

29 03:55 ☽ ✶ ♃
06:23 ☽ ✶ ♀
06:24 ☽ ✶ ☿
06:33 ☿ ☌ ♀

30 02:07 ♀ ♎
02:21 ☽ △ ♄
07:51 ☽ △ ♂
10:25 ☉ ∥ ♅
13:05 ☿ SR

31 04:20 ☽ △ ♅
12:40 ☉ # ♆

⚷ Chiron

01 Dec. 01 N 55

01	24♓44R
04	24 39
07	24 33
10	24 27
13	24 21
16	24 14
19	24 07
22	24 00
25	23 53
28	23 45
31	23 37

01 09:19 09♍21 ☉ Annular Solar Eclipse
16 18:55 24♓20 ✷ Penumbral Lunar Eclipse (mag 0.933)

September 2016

Day	S. T.	☉	☽	☿	♀	♂	♃	♄	♅	♆	♇	☊ True
	h m s	° ' "	° ' "	° '	° '	° '	° '	° '	° '	° '	° '	° '
01 Th	22 42 21	08♍ 59 15	04♍ 34 26	28♍R58	02♎ 21	13♐ 56	28♍ 13	10♐ 03	24♈R05	10♓R40	15♑R05	12♍R40
02 Fr	22 46 18	09 57 21	17 09 11	28 47	03 34	14 30	28 25	10 05	24 03	10 39	15 04	12 40
03 Sa	22 50 14	10 55 28	29 30 58	28 29	04 48	15 04	28 38	10 07	24 02	10 37	15 04	12 40
04 Su	22 54 11	11 53 37	11♎ 41 17	28 04	06 01	15 38	28 50	10 09	24 00	10 35	15 03	12 40
05 Mo	22 58 07	12 51 47	23 42 18	27 34	07 15	16 13	29 03	10 11	23 58	10 34	15 02	12 40
06 Tu	23 02 04	13 49 59	05♏ 36 45	26 57	08 28	16 47	29 16	10 14	23 57	10 32	15 02	12 40
07 We	23 06 00	14 48 12	17 27 58	26 15	09 42	17 22	29 28	10 16	23 55	10 31	15 01	12 39
08 Th	23 09 57	15 46 27	29 19 50	25 27	10 55	17 57	29 41	10 18	23 53	10 29	15 01	12 39
09 Fr	23 13 53	16 44 44	11♐ 16 39	24 34	12 08	18 33	29 54	10 21	23 51	10 27	15 00	12D 39
10 Sa	23 17 50	17 43 02	23 23 02	23 38	13 22	19 08	00♎ 07	10 23	23 50	10 26	15 00	12 39
11 Su	23 21 47	18 41 22	05♑ 43 31	22 39	14 35	19 44	00 20	10 26	23 48	10 24	14 59	12 40
12 Mo	23 25 43	19 39 43	18 22 16	21 38	15 49	20 21	00 32	10 29	23 46	10 22	14 59	12 41
13 Tu	23 29 40	20 38 06	01♒ 22 44	20 36	17 02	20 57	00 45	10 32	23 44	10 21	14 58	12 42
14 We	23 33 36	21 36 30	14 47 09	19 36	18 15	21 33	00 58	10 34	23 42	10 19	14 58	12 44
15 Th	23 37 33	22 34 56	28 36 06	18 38	19 29	22 10	01 11	10 37	23 40	10 18	14 57	12 45
16 Fr	23 41 29	23 33 24	12♓ 48 06	17 45	20 42	22 47	01 24	10 41	23 38	10 16	14 57	12 45
17 Sa	23 45 26	24 31 54	27 19 24	16 56	21 55	23 24	01 37	10 44	23 36	10 14	14 57	12R 45
18 Su	23 49 22	25 30 25	12♈ 04 11	16 14	23 09	24 02	01 50	10 47	23 34	10 13	14 57	12 44
19 Mo	23 53 19	26 28 59	26 55 11	15 40	24 22	24 39	02 03	10 50	23 32	10 11	14 56	12 43
20 Tu	23 57 15	27 27 35	11♉ 44 42	15 14	25 35	25 17	02 16	10 54	23 30	10 10	14 56	12 41
21 We	00 01 12	28 26 12	26 25 40	14 57	26 48	25 55	02 28	10 57	23 28	10 08	14 56	12 38
22 Th	00 05 09	29 24 53	10♊ 52 30	14 50	28 02	26 33	02 41	11 01	23 26	10 07	14 56	12 37
23 Fr	00 09 05	00♎ 23 35	25 01 35	14 D 52	29 15	27 12	02 54	11 05	23 23	10 05	14 56	12 35
24 Sa	00 13 02	01 22 20	08♋ 51 18	15 05	00♏ 28	27 50	03 07	11 08	23 21	10 03	14 56	12D 35
25 Su	00 16 58	02 21 07	22 21 40	15 27	01 41	28 29	03 20	11 12	23 19	10 02	14 55	12 36
26 Mo	00 20 55	03 19 56	05♌ 33 54	15 59	02 54	29 08	03 33	11 16	23 17	10 00	14 55	12 37
27 Tu	00 24 51	04 18 47	18 29 49	16 40	04 07	29 47	03 46	11 20	23 14	09 59	14D 55	12 38
28 We	00 28 48	05 17 41	01♍ 11 24	17 29	05 21	00♑ 26	03 59	11 24	23 12	09 58	14 55	12 40
29 Th	00 32 44	06 16 37	13 40 34	18 26	06 34	01 05	04 12	11 28	23 10	09 56	14 56	12 40
30 Fr	00 36 41	07 15 35	25 59 07	19 31	07 47	01 45	04 25	11 33	23 08	09 55	14 56	12R 40

Data for	09-01-2016
Julian Day	2457632.50
Ayanamsa	24 05 18
SVP	05 ♓ 01 33
☽ ☊ Mean	12 ♍ 41 R

● ◐ PHASES ○ ◑

01	09:04	☉	09♍21
09	11:50	◐	17♐14
16	19:05	✷	24♓20
23	09:56	◑	00♋48

LAST ASPECT Day	h m	☽ INGRESS Day	h m	
02	22:14	03	00:57	♎
05	00:32	05	12:40	♏
08	00:44	08	01:21	♐
10	00:52	10	12:56	♑
12	10:01	12	21:29	♒
14	15:32	15	02:23	♓
16	19:06	17	04:23	♈
18	20:11	19	04:59	♉
21	03:33	21	05:54	♊
23	07:58	23	08:34	♋
25	01:43	25	13:49	♌
27	08:54	27	21:44	♍
29	10:06	30	07:54	♎

DECLINATION

Day	☉	☽	☿	♀	♂	♃	♄	♅	♆	♇
01 Th	08N12	09N08	03S 27	00N02	25S 11	01N43	20S 26	08N45	08S 23	21S 20
02 Fr	07 50	05 27	03 26	00S 29	25 14	01 38	20 26	08 44	08 23	21 20
03 Sa	07 28	01 36	03 22	01 00	25 18	01 33	20 27	08 44	08 24	21 20
04 Su	07 06	02S 16	03 14	01 31	25 21	01 28	20 27	08 43	08 24	21 20
05 Mo	06 44	05 58	03 03	02 02	25 24	01 23	20 28	08 43	08 25	21 20
06 Tu	06 21	09 24	02 47	02 33	25 27	01 18	20 28	08 42	08 26	21 21
07 We	05 59	12 27	02 28	03 03	25 30	01 13	20 29	08 41	08 26	21 21
08 Th	05 37	14 59	02 05	03 34	25 33	01 08	20 29	08 41	08 27	21 21
09 Fr	05 14	16 54	01 38	04 05	25 35	01 03	20 30	08 40	08 28	21 21
10 Sa	04 51	18 05	01 08	04 36	25 38	00 57	20 31	08 39	08 28	21 21
11 Su	04 28	18 27	00 36	05 06	25 40	00 52	20 31	08 39	08 29	21 21
12 Mo	04 06	17 55	00 01	05 37	25 42	00 47	20 32	08 38	08 29	21 22
13 Tu	03 43	16 26	00N36	06 07	25 44	00 42	20 32	08 37	08 30	21 22
14 We	03 20	14 01	01 14	06 37	25 46	00 37	20 33	08 37	08 31	21 22
15 Th	02 57	10 44	01 52	07 08	25 48	00 32	20 34	08 36	08 31	21 22
16 Fr	02 33	06 45	02 30	07 38	25 49	00 27	20 34	08 35	08 32	21 22
17 Sa	02 10	02 17	03 06	08 07	25 51	00 22	20 35	08 34	08 33	21 23
18 Su	01 47	02N24	03 41	08 37	25 52	00 16	20 36	08 34	08 33	21 23
19 Mo	01 24	06 57	04 12	09 07	25 53	00 11	20 36	08 33	08 34	21 23
20 Tu	01 01	11 04	04 41	09 36	25 53	00 06	20 37	08 32	08 34	21 23
21 We	00 37	14 26	05 05	10 05	25 54	00 01	20 38	08 31	08 35	21 23
22 Th	00 14	16 52	05 26	10 34	25 54	00S 04	20 38	08 30	08 35	21 23
23 Fr	00S 09	18 12	05 42	11 02	25 55	00 09	20 39	08 30	08 36	21 23
24 Sa	00 33	18 25	05 53	11 31	25 55	00 14	20 40	08 29	08 37	21 24
25 Su	00 56	17 35	05 59	11 59	25 54	00 19	20 40	08 28	08 37	21 24
26 Mo	01 19	15 47	06 01	12 27	25 54	00 25	20 41	08 27	08 38	21 24
27 Tu	01 43	13 14	05 58	12 55	25 53	00 30	20 42	08 26	08 38	21 24
28 We	02 06	10 04	05 51	13 22	25 53	00 35	20 43	08 25	08 39	21 24
29 Th	02 30	06 30	05 39	13 49	25 52	00 40	20 43	08 25	08 39	21 24
30 Fr	02 53	02 42	05 23	14 16	25 50	00 45	20 44	08 24	08 40	21 24

ASPECTARIAN

Day	Time	Aspect
01	02:36	☽ ∥ ♅
	05:05	☽ # ♆
	06:58	☽ ∥ ☉
	10:26	☽ □ ♄
	11:34	☽ ☍ ♆
	18:40	☽ □ ♂
	20:00	☽ △ ♇
02	03:23	☉ □ ♄
	12:49	☽ # ☿
	16:38	☉ ☍ ♆
	17:18	☿ ☌ ♃
	22:01	☽ ☌ ☿
	22:14	☽ ☌ ♃
03	00:18	☽ ∥ ♃
	03:16	☽ # ♀
	11:31	☽ ☌ ♀
	18:35	☽ ∥ ♀
	19:08	☽ # ♃
	20:57	☽ ⚹ ♄
	22:09	♀ # ♃
04	05:55	☽ ∥ ☿
	06:41	☽ □ ♇
	08:15	☽ ⚹ ♂
05	00:32	☽ ☍ ♅
	04:38	☽ # ☉
	16:54	☽ ∥ ♆
	18:52	☽ # ♅
06	07:08	☿ ∥ ♀
	09:56	☽ △ ♆
	18:07	☽ ⚹ ☉
	19:03	☽ ⚹ ♇
07	05:18	☉ △ ♇
	11:35	♀ ⚹ ♄
	16:40	☽ ⚹ ☿
08	00:44	☽ ⚹ ♃
	22:08	☽ ☌ ♄
	22:21	☽ □ ♆
09	01:55	☽ ⚹ ♀
	11:19	♃ ♎
	15:12	☽ ☌ ♂
10	00:27	☽ □ ☿
	00:52	☽ △ ♅
	06:59	♀ # ☉
	09:46	☿ # ♃
	13:11	♄ □ ♆
	13:23	☽ □ ♃
11	07:46	♀ □ ♇
	08:56	☽ ⚹ ♆
	17:37	☽ ☌ ♇
	18:41	☽ □ ♀
12	02:37	☽ △ ☉
	05:38	☽ △ ☿
	10:01	☽ □ ♅
	18:59	☿ □ ♂
	22:51	☽ △ ♃
	23:41	☿ ☌ ☉
13	03:22	☿ ∥ ♃
	16:31	☽ ⚹ ♄
	20:39	☉ □ ♂
14	06:41	☽ △ ♀
	12:24	☽ ⚹ ♂
	15:32	☽ ⚹ ♅
15	13:26	☽ # ♅
	13:48	☽ ∥ ♆
	19:36	☽ ∥ ♀
	19:46	☽ ☌ ♆
	20:26	☽ □ ♄
16	01:27	☿ ∥ ☉
	03:35	☽ ⚹ ♇
	07:46	☽ ☍ ☿
	17:17	☽ □ ♂
	20:14	☽ # ☿
17	00:39	☽ # ☉
	07:05	♂ △ ♅
	07:07	☽ ☍ ♃
	10:07	☽ # ♃
	13:23	☽ ∥ ♃
	20:42	♀ ∥ ♆
	21:09	☽ ∥ ☉
	21:11	♀ # ♅
	21:55	☽ △ ♄
18	04:39	☽ □ ♇
	07:07	♅ # ♆
	07:34	☽ ∥ ☿
	08:06	♀ ☍ ♅
	18:32	☽ ☌ ♅
	19:30	☽ ☍ ♀
	20:11	☽ △ ♂
19	08:58	☽ ∥ ♅
	09:06	☽ # ♆
	11:53	♀ ⚹ ♂
	13:54	☽ # ♀
	21:26	☽ ⚹ ♆
20	05:11	☽ △ ♇
	05:32	☽ △ ☿
21	01:42	☿ △ ♇
	03:33	☽ △ ☉
	10:09	☽ △ ♃
	22:43	☽ □ ♆
22	00:14	☽ ☍ ♄
	05:32	☿ SD
	06:39	☽ □ ☿
	08:21	☉ # ♃
	14:22	☉ ♎
	21:12	☽ ⚹ ♅
	23:47	☉ ∥ ♃
23	03:54	☽ ☍ ♂
	07:58	☽ △ ♀
	08:37	☿ △ ♇
	13:49	☽ □ ♃
	14:51	♀ ♏
24	02:07	☽ △ ♆
	10:43	☽ ☍ ♇
	11:16	☽ ⚹ ☿
25	01:43	☽ □ ♅
	18:37	☽ □ ♀
	19:34	☽ ⚹ ☉
	20:15	☽ ⚹ ♃
26	07:00	☉ ☌ ♃
	10:35	☽ △ ♄
	15:00	♇ SD
27	02:15	☽ # ♀
	08:07	♂ ♑
	08:54	☽ △ ♅
	22:29	☽ △ ♂
28	08:48	☽ ⚹ ♀
	09:47	☽ # ♆
	11:22	☽ ∥ ♅
	16:47	☽ ☍ ♆
	19:43	☽ □ ♄
29	02:25	☽ △ ♇
	05:53	☽ ∥ ☿
	10:06	☽ ☌ ☿
	23:00	☽ # ☉
30	11:51	☽ # ♃
	12:00	☽ □ ♂
	16:55	☽ ☌ ♃
	21:57	☽ ∥ ♃

⚷ Chiron

01 Dec. 01 N 29

03	23♓29R
06	23 21
09	23 13
12	23 05
15	22 56
18	22 48
21	22 40
24	22 32
27	22 24
30	22 16

October 2016

Day	S. T.	☉	☽	☿	♀	♂	♃	♄	♅	♆	♇	☊ True
	h m s	° ' "	° ' "	° '	° '	° '	° '	° '	° '	° '	° '	° '
01 Sa	00 40 38	08♎ 14 35	08♎ 08 38	20♍ 42	09♏ 00	02♑ 25	04♎ 38	11♐ 37	23♈R05	09♓R53	14♑ 56	12♍R38
02 Su	00 44 34	09 13 36	20 10 42	21 59	10 13	03 04	04 51	11 41	23 03	09 52	14 56	12 34
03 Mo	00 48 31	10 12 40	02♏ 06 58	23 21	11 26	03 45	05 04	11 46	23 01	09 50	14 56	12 29
04 Tu	00 52 27	11 11 46	13 59 17	24 47	12 39	04 25	05 17	11 50	22 58	09 49	14 56	12 24
05 We	00 56 24	12 10 54	25 49 50	26 18	13 52	05 05	05 30	11 55	22 56	09 48	14 57	12 18
06 Th	01 00 20	13 10 04	07♐ 41 20	27 51	15 05	05 46	05 43	12 00	22 53	09 46	14 57	12 14
07 Fr	01 04 17	14 09 15	19 37 06	29 28	16 18	06 26	05 56	12 04	22 51	09 45	14 57	12 10
08 Sa	01 08 13	15 08 29	01♑ 41 01	01♎ 06	17 31	07 07	06 09	12 09	22 49	09 44	14 57	12 08
09 Su	01 12 10	16 07 44	13 57 23	02 46	18 44	07 48	06 22	12 14	22 46	09 43	14 58	12D 07
10 Mo	01 16 07	17 07 01	26 30 44	04 28	19 57	08 29	06 35	12 19	22 44	09 41	14 58	12 08
11 Tu	01 20 03	18 06 20	09♒ 25 22	06 11	21 09	09 10	06 47	12 24	22 41	09 40	14 59	12 10
12 We	01 23 60	19 05 40	22 45 01	07 54	22 22	09 51	07 00	12 29	22 39	09 39	14 59	12 12
13 Th	01 27 56	20 05 02	06♓ 32 08	09 38	23 35	10 33	07 13	12 34	22 36	09 38	15 00	12 13
14 Fr	01 31 53	21 04 26	20 47 05	11 22	24 48	11 14	07 26	12 40	22 34	09 37	15 00	12R 13
15 Sa	01 35 49	22 03 52	05♈ 27 27	13 06	26 01	11 56	07 39	12 45	22 32	09 36	15 01	12 11
16 Su	01 39 46	23 03 20	20 27 38	14 51	27 13	12 38	07 51	12 50	22 29	09 34	15 01	12 07
17 Mo	01 43 42	24 02 49	05♉ 39 09	16 35	28 26	13 19	08 04	12 56	22 27	09 33	15 02	12 02
18 Tu	01 47 39	25 02 21	20 51 47	18 18	29 39	14 01	08 17	13 01	22 24	09 32	15 02	11 56
19 We	01 51 35	26 01 55	05♊ 55 15	20 02	00♐ 51	14 44	08 30	13 07	22 22	09 31	15 03	11 49
20 Th	01 55 32	27 01 32	20 40 48	21 45	02 04	15 26	08 42	13 12	22 19	09 30	15 04	11 43
21 Fr	01 59 29	28 01 11	05♋ 02 33	23 28	03 17	16 08	08 55	13 18	22 17	09 29	15 04	11 38
22 Sa	02 03 25	29 00 51	18 57 49	25 10	04 29	16 50	09 07	13 23	22 14	09 28	15 05	11 36
23 Su	02 07 22	00♏ 00 35	02♌ 26 50	26 52	05 42	17 33	09 20	13 29	22 12	09 28	15 06	11 35
24 Mo	02 11 18	01 00 20	15 31 54	28 33	06 54	18 16	09 33	13 35	22 10	09 27	15 07	11D 35
25 Tu	02 15 15	02 00 08	28 16 32	00♏ 13	08 07	18 58	09 45	13 41	22 07	09 26	15 08	11 36
26 We	02 19 11	02 59 58	10♍ 44 36	01 54	09 19	19 41	09 57	13 47	22 05	09 25	15 08	11 37
27 Th	02 23 08	03 59 50	22 59 53	03 33	10 32	20 24	10 10	13 53	22 02	09 24	15 09	11R 36
28 Fr	02 27 04	04 59 44	05♎ 05 43	05 12	11 44	21 07	10 22	13 59	22 00	09 23	15 10	11 33
29 Sa	02 31 01	05 59 40	17 04 56	06 51	12 56	21 50	10 35	14 05	21 58	09 23	15 11	11 28
30 Su	02 34 58	06 59 38	28 59 50	08 29	14 09	22 33	10 47	14 11	21 55	09 22	15 12	11 20
31 Mo	02 38 54	07 59 39	10♏ 52 12	10 06	15 21	23 16	10 59	14 17	21 53	09 21	15 13	11 10

Data for 10-01-2016
Julian Day 2457662.50
Ayanamsa 24 05 20
SVP 05 ♓ 01 27
☽ ☊ Mean 11 ♍ 06 R

● ◐ PHASES ○ ◑

01	00:12	●	08♎15
09	04:33	◐	16♑19
16	04:24	○	23♈14
22	19:15	◑	29♋49
30	17:38	●	07♏44

LAST ASPECT Day	h m	☽ INGRESS Day	h m	
02	05:44	02	19:44	♏
05	01:05	05	08:27	♐
07	06:26	07	20:40	♑
09	16:51	10	06:33	♒
11	23:49	12	12:43	♓
14	07:14	14	15:09	♈
16	04:24	16	15:05	♉
17	14:47	18	14:31	♊
20	11:18	20	15:29	♋
22	19:15	22	19:35	♌
24	12:22	25	03:17	♍
26	18:34	27	13:52	♎
29	10:10	30	02:01	♏

DECLINATION

Day	☉	☽	☿	♀	♂	♃	♄	♅	♆	♇
01 Sa	03S 16	01S 10	05N03	14S 42	25S 49	00S 50	20S 45	08N23	08S 40	21S 24
02 Su	03 39	04 56	04 39	15 08	25 47	00 55	20 46	08 22	08 41	21 25
03 Mo	04 03	08 29	04 12	15 34	25 45	01 01	20 46	08 21	08 41	21 25
04 Tu	04 26	11 40	03 43	15 59	25 43	01 06	20 47	08 20	08 42	21 25
05 We	04 49	14 22	03 10	16 24	25 41	01 11	20 48	08 19	08 42	21 25
06 Th	05 12	16 29	02 36	16 48	25 39	01 16	20 49	08 18	08 43	21 25
07 Fr	05 35	17 54	01 59	17 13	25 36	01 21	20 50	08 18	08 43	21 25
08 Sa	05 58	18 32	01 21	17 36	25 33	01 26	20 50	08 17	08 44	21 25
09 Su	06 21	18 19	00 41	18 00	25 30	01 31	20 51	08 16	08 44	21 25
10 Mo	06 43	17 12	00S 00	18 22	25 26	01 36	20 52	08 15	08 45	21 25
11 Tu	07 06	15 10	00 42	18 45	25 23	01 41	20 53	08 14	08 45	21 25
12 We	07 29	12 17	01 25	19 07	25 19	01 46	20 54	08 13	08 46	21 26
13 Th	07 51	08 38	02 08	19 28	25 15	01 51	20 54	08 12	08 46	21 26
14 Fr	08 13	04 21	02 52	19 49	25 10	01 56	20 55	08 11	08 47	21 26
15 Sa	08 36	00N18	03 36	20 09	25 06	02 01	20 56	08 10	08 47	21 26
16 Su	08 58	05 02	04 20	20 29	25 01	02 06	20 57	08 09	08 47	21 26
17 Mo	09 20	09 29	05 04	20 49	24 56	02 11	20 58	08 09	08 48	21 26
18 Tu	09 41	13 20	05 48	21 07	24 51	02 16	20 59	08 08	08 48	21 26
19 We	10 03	16 15	06 32	21 26	24 45	02 21	20 59	08 07	08 49	21 26
20 Th	10 25	18 03	07 16	21 43	24 40	02 26	21 00	08 06	08 49	21 26
21 Fr	10 46	18 37	07 59	22 01	24 34	02 31	21 01	08 05	08 49	21 26
22 Sa	11 07	18 03	08 42	22 17	24 27	02 36	21 02	08 04	08 50	21 26
23 Su	11 28	16 27	09 24	22 33	24 21	02 41	21 03	08 03	08 50	21 26
24 Mo	11 49	14 01	10 06	22 48	24 14	02 45	21 04	08 02	08 50	21 26
25 Tu	12 10	10 58	10 48	23 03	24 07	02 50	21 04	08 01	08 51	21 26
26 We	12 31	07 28	11 28	23 17	24 00	02 55	21 05	08 01	08 51	21 26
27 Th	12 51	03 43	12 09	23 31	23 53	03 00	21 06	08 00	08 51	21 26
28 Fr	13 11	00S08	12 48	23 44	23 45	03 05	21 07	07 59	08 51	21 26
29 Sa	13 31	03 58	13 27	23 56	23 38	03 09	21 08	07 58	08 52	21 26
30 Su	13 51	07 36	14 05	24 07	23 30	03 14	21 09	07 57	08 52	21 26
31 Mo	14 10	10 55	14 43	24 18	23 21	03 19	21 09	07 56	08 52	21 26

ASPECTARIAN

01 06:57 ☽ ⚹ ♄
13:30 ☽ □ ♆
14:49 ☽ ∥ ☉
17:14 ♀ △ ♆
22:22 ☽ # ☿
02 05:44 ☽ ☍ ♅
23:08 ☽ # ♅
03 01:33 ☽ ∥ ♆
03:29 ☽ ⚹ ♂
04:31 ☿ # ☉
15:35 ☽ △ ♆
20:59 ☽ ☌ ♀
04 01:56 ☽ ⚹ ♇
17:00 ☉ ⚹ ♄
05 01:05 ☽ ⚹ ☿
19:56 ☽ ⚹ ♃
21:21 ♀ ⚹ ♆
21:45 ♂ □ ♃
06 04:12 ☽ □ ♆
06:21 ☽ ∥ ♀
08:45 ☽ ☌ ♄
12:03 ☽ ⚹ ☉
07 06:26 ☽ △ ♅
07:56 ☿ ♎
19:32 ☉ □ ♇
21:04 ☿ # ♃
22:40 ☽ □ ☿
08 08:56 ☽ □ ♃
11:19 ☽ ☌ ♂
15:46 ☽ ⚹ ♆
09 01:57 ☽ ☌ ♇
06:22 ☽ ∥ ♀
10:11 ☽ ⚹ ♀
16:51 ☽ □ ♅
10 17:08 ☽ △ ☿
19:05 ☽ △ ♃
11 05:28 ☽ ⚹ ♄
09:46 ☿ ☌ ♃
16:59 ☽ △ ☉
17:04 ♂ ⚹ ♆
23:16 ☽ □ ♀
23:49 ☽ ⚹ ♅
12 13:21 ☿ ∥ ♃
23:11 ☽ ∥ ♆
13 02:35 ☽ # ♅
04:15 ☽ ∥ ☉
05:16 ☽ ☌ ♆
07:11 ☽ ⚹ ♂
10:19 ☽ □ ♄
14:21 ☽ ⚹ ♇
21:00 ☿ □ ♂
21:52 ☉ # ♅
14 06:48 ☽ ∥ ☿
07:14 ☽ △ ♀
12:28 ☽ ∥ ♃
18:47 ☿ ⚹ ♄
15 03:35 ☽ ☍ ♃
08:54 ☽ # ♃
10:43 ☉ ☍ ♅
10:55 ☽ □ ♂
11:47 ☽ △ ♄
12:42 ☉ ∥ ♆
13:54 ☽ ☍ ☿
15:20 ☽ □ ♇
19:50 ☽ # ☿
16 02:29 ☿ □ ♇
03:12 ☽ ☌ ♅
16:32 ☽ ∥ ♅
20:07 ☽ # ♆
23:01 ☽ # ☉
17 06:09 ☽ ⚹ ♆
11:56 ♀ ∥ ♄
12:40 ☽ △ ♂
14:47 ☽ △ ♇
18 07:01 ♀ ♐
15:10 ☽ ☍ ♀
19 00:23 ♀ ∥ ♆
04:12 ☽ △ ♃
05:48 ☽ □ ♆
11:19 ♂ ☌ ♇
11:41 ☽ ☍ ♄
20 02:00 ☽ △ ☿
02:42 ☽ ⚹ ♅
07:48 ☿ ☍ ♅
11:18 ☽ △ ☉
21 03:16 ☿ # ♅
06:42 ☽ □ ♃
07:34 ☽ △ ♆
17:14 ☽ ☍ ♇
20:05 ☽ ☍ ♂
22 04:24 ☿ ∥ ♆
05:45 ☽ □ ♅
12:31 ☽ □ ☿
23:46 ☉ ♏
23 06:29 ☽ △ ♀
12:45 ☽ ⚹ ♃
20:21 ☽ △ ♄
24 12:22 ☽ △ ♅
15:50 ☽ # ☉
20:47 ☿ ♏
25 00:57 ☽ # ☿
04:18 ☽ ⚹ ☿
07:44 ☽ ⚹ ☉
14:45 ☽ # ♆
20:22 ☽ ∥ ♅
20:56 ☽ □ ♀
21:26 ☽ ☍ ♆
26 01:55 ♀ □ ♆
05:57 ☽ □ ♄
08:35 ☽ △ ♇
15:20 ♀ ⚹ ♃
18:34 ☽ △ ♂
27 04:21 ☽ # ♃
16:17 ☿ ☌ ☉
28 02:19 ♀ ∥ ♂
10:43 ☽ ☌ ♃
14:45 ☽ ⚹ ♀
17:55 ☽ ⚹ ♄
18:45 ☽ ∥ ♃
20:11 ☽ □ ♇
29 04:05 ♂ □ ♅
05:07 ☿ ∥ ☉
09:47 ☽ ☍ ♅
10:10 ☽ □ ♂
30 00:46 ♀ ☌ ♄
02:25 ☽ # ♅
08:52 ☽ ∥ ♆
12:58 ☿ △ ♆
20:56 ☽ △ ♆
22:13 ☽ ☌ ☿
31 08:49 ☽ ⚹ ♇

⚷ Chiron

01 Dec.	00 N 55
03	22♓08R
06	22 00
09	21 53
12	21 46
15	21 39
18	21 32
21	21 26
24	21 20
27	21 14
30	21 09

November 2016

Day	S. T. h m s	☉ ° ' "	☽ ° ' "	☿ ° '	♀ ° '	♂ ° '	♃ ° '	♄ ° '	♅ ° '	♆ ° '	♇ ° '	☊ True ° '
01 Tu	02 42 51	08♏59 41	22♏43 32	11♏43	16♐33	24♑00	11♎11	14♐23	21♈R51	09♓R21	15♑14	10♍R58
02 We	02 46 47	09 59 45	04♐35 17	13 20	17 45	24 43	11 24	14 29	21 48	09 20	15 15	10 46
03 Th	02 50 44	10 59 50	16 29 06	14 56	18 58	25 27	11 36	14 36	21 46	09 19	15 16	10 35
04 Fr	02 54 40	11 59 58	28 27 05	16 32	20 10	26 10	11 48	14 42	21 44	09 19	15 18	10 26
05 Sa	02 58 37	13 00 07	10♑31 58	18 07	21 22	26 54	12 00	14 48	21 42	09 18	15 19	10 19
06 Su	03 02 33	14 00 17	22 47 06	19 42	22 34	27 38	12 12	14 55	21 39	09 18	15 20	10 15
07 Mo	03 06 30	15 00 30	05♒16 25	21 16	23 46	28 21	12 24	15 01	21 37	09 17	15 21	10 14
08 Tu	03 10 27	16 00 43	18 04 10	22 51	24 58	29 05	12 36	15 08	21 35	09 17	15 22	10D 14
09 We	03 14 23	17 00 58	01♓14 37	24 24	26 10	29 49	12 47	15 14	21 33	09 17	15 23	10 14
10 Th	03 18 20	18 01 15	14 51 25	25 58	27 22	00♒33	12 59	15 21	21 31	09 16	15 25	10R 14
11 Fr	03 22 16	19 01 33	28 56 48	27 31	28 34	01 17	13 11	15 27	21 29	09 16	15 26	10 12
12 Sa	03 26 13	20 01 52	13♈30 25	29 03	29 45	02 01	13 22	15 34	21 27	09 16	15 27	10 08
13 Su	03 30 09	21 02 13	28 28 35	00♐36	00♑57	02 46	13 34	15 40	21 25	09 15	15 29	10 02
14 Mo	03 34 06	22 02 36	13♉43 54	02 08	02 09	03 30	13 45	15 47	21 23	09 15	15 30	09 53
15 Tu	03 38 02	23 03 00	29 06 00	03 40	03 20	04 14	13 57	15 54	21 21	09 15	15 31	09 43
16 We	03 41 59	24 03 25	14♊23 04	05 11	04 32	04 58	14 08	16 01	21 19	09 15	15 33	09 32
17 Th	03 45 56	25 03 53	29 24 05	06 43	05 43	05 43	14 19	16 07	21 17	09 15	15 34	09 22
18 Fr	03 49 52	26 04 22	14♋00 31	08 14	06 55	06 27	14 31	16 14	21 15	09 15	15 36	09 14
19 Sa	03 53 49	27 04 53	28 07 36	09 44	08 06	07 12	14 42	16 21	21 13	09 14	15 37	09 07
20 Su	03 57 45	28 05 26	11♌44 16	11 15	09 17	07 56	14 53	16 28	21 11	09 14	15 39	09 04
21 Mo	04 01 42	29 06 01	24 52 30	12 45	10 28	08 41	15 04	16 35	21 09	09D 14	15 40	09 02
22 Tu	04 05 38	00♐06 37	07♍36 14	14 15	11 40	09 25	15 15	16 42	21 08	09 15	15 42	09 01
23 We	04 09 35	01 07 15	20 00 15	15 44	12 51	10 10	15 26	16 48	21 06	09 15	15 43	09 00
24 Th	04 13 31	02 07 54	02♎09 32	17 13	14 02	10 55	15 36	16 55	21 04	09 15	15 45	08 59
25 Fr	04 17 28	03 08 36	14 08 45	18 42	15 13	11 40	15 47	17 02	21 03	09 15	15 47	08 55
26 Sa	04 21 25	04 09 19	26 02 00	20 10	16 23	12 24	15 58	17 09	21 01	09 15	15 48	08 48
27 Su	04 25 21	05 10 03	07♏52 39	21 38	17 34	13 09	16 08	17 16	21 00	09 15	15 50	08 38
28 Mo	04 29 18	06 10 49	19 43 20	23 05	18 45	13 54	16 19	17 23	20 58	09 16	15 52	08 25
29 Tu	04 33 14	07 11 36	01♐35 56	24 32	19 56	14 39	16 29	17 30	20 56	09 16	15 53	08 11
30 We	04 37 11	08 12 24	13 31 48	25 57	21 06	15 24	16 39	17 37	20 55	09 16	15 55	07 56

Data for	11-01-2016
Julian Day	2457693.50
Ayanamsa	24 05 23
SVP	05 ♓ 01 21
☽ ☊ Mean	09 ♍ 27 R

● ◐ PHASES ○ ◑

Day	h m	Phase	Position
07	19:51	◐	15♒50
14	13:53	○	22♉38
21	08:33	◑	29♌28
29	12:18	●	07♐43

LAST ASPECT ☽ INGRESS

Day	h m	Day	h m	Sign
01	02:44	01	14:43	♐
03	10:35	04	03:05	♑
06	09:57	06	13:56	♒
08	13:55	08	21:46	♓
10	23:17	11	01:46	♈
12	12:46	13	02:25	♉
14	13:53	15	01:24	♊
16	10:59	17	00:58	♋
18	22:03	19	03:15	♌
21	08:34	21	09:35	♍
22	17:42	23	19:42	♎
25	13:53	26	08:02	♏
27	21:49	28	20:46	♐

ASPECTARIAN

01 02:44 ☽ ⚹ ♂
07:37 ☽ ∥ ☉
08:17 ☉ △ ♆
20:52 ☽ ∥ ☿

02 09:35 ☽ □ ♆
13:59 ☽ ⚹ ♃
20:09 ☽ ☌ ♄

03 05:08 ☿ ⚹ ♇
05:32 ☽ ☌ ♀
10:35 ☽ △ ♅

04 21:35 ☽ ⚹ ♆

05 02:56 ☽ □ ♃
05:18 ☽ ⚹ ☉
06:24 ♀ △ ♅
09:25 ☽ ☌ ♇
17:07 ☽ ⚹ ☿
19:10 ☽ ∥ ☿
21:49 ☽ □ ♅

06 09:57 ☽ ☌ ♂
21:50 ☽ ∥ ☉

07 08:23 ☉ ⚹ ♇
13:39 ☽ △ ♃
18:29 ☽ ⚹ ♄

08 06:28 ☽ ⚹ ♅
09:58 ☽ □ ☿
13:55 ☽ ⚹ ♀

09 05:52 ♂ ♒
08:52 ☽ ∥ ♆
14:15 ☽ ☌ ♆
15:29 ☽ # ♅

10 00:51 ☽ □ ♄
00:58 ☽ ⚹ ♇
05:53 ☽ △ ☉
12:19 ☽ ∥ ♃
21:18 ☽ △ ☿
23:17 ☽ □ ♀

11 04:07 ☽ ⚹ ♂
20:47 ♂ ∥ ♇
23:47 ☽ ☍ ♃

12 03:10 ☽ □ ♇
03:21 ☽ △ ♄
04:55 ♀ ♑
07:34 ☽ # ♃
11:00 ♂ ∥ ♄
12:46 ☽ ☌ ♅
13:18 ☿ ∥ ♂
14:25 ☿ ∥ ♄
14:40 ☿ ♐
20:27 ☿ ∥ ♇

13 01:56 ☽ ∥ ♅
04:15 ☽ △ ♀
07:07 ☽ □ ♂
08:08 ☽ # ♆
16:59 ☽ ⚹ ♆

14 02:46 ☽ △ ♇

15 07:56 ☽ ☍ ☿
08:26 ☽ △ ♂
15:54 ☽ □ ♆
17:23 ☿ ⚹ ♂
23:36 ☽ △ ♃

16 02:35 ☽ ☍ ♄
10:59 ☽ ⚹ ♅

17 11:12 ☽ ☍ ♀
16:05 ☽ △ ♆

18 00:51 ☽ □ ♃
02:40 ☽ ☍ ♇
12:10 ☽ □ ♅
16:04 ☿ □ ♆
22:03 ☽ △ ☉

19 16:49 ☽ ☍ ♂
23:01 ☽ △ ☿
23:04 ♀ ⚹ ♆

20 04:40 ♆ SD
04:42 ♄ ∥ ♇
05:45 ☽ ⚹ ♃
06:03 ☉ ∥ ♂
08:37 ☽ △ ♄
17:09 ☽ △ ♅

21 21:23 ☉ ♐
21:57 ☽ # ♆

22 03:08 ☽ ☍ ♆
06:07 ☽ ∥ ♅
08:36 ☽ △ ♀
14:32 ☽ □ ☿
15:38 ☽ △ ♇
17:42 ☽ □ ♄
18:20 ☿ ⚹ ♃
23:02 ☽ # ♃

23 18:44 ☿ ☌ ♄
23:56 ☽ ⚹ ☉

24 08:18 ☿ ∥ ♀
18:40 ☽ △ ♂
22:43 ♃ □ ♇

25 02:23 ☽ □ ♀
03:17 ☽ □ ♇
03:21 ☽ ☌ ♃
05:53 ☽ ⚹ ♄
10:29 ☽ ⚹ ☿
11:49 ♀ ☌ ♇
13:46 ♀ □ ♃
13:53 ☽ ☍ ♅
14:29 ☽ ∥ ♃
20:20 ♀ □ ♅

26 06:44 ☽ # ♅
13:38 ☿ △ ♅
15:38 ☽ ∥ ♆

27 02:48 ☽ △ ♆
11:25 ☽ □ ♂
16:09 ☽ ⚹ ♇
21:49 ☽ ⚹ ♀

28 12:14 ☉ ∥ ♇

29 05:47 ☉ ∥ ♄
15:26 ☽ □ ♆

30 01:23 ☽ ∥ ♂
04:00 ☽ ⚹ ♂
06:21 ☽ ⚹ ♃
08:17 ☽ ☌ ♄
14:46 ☽ △ ♅

DECLINATION

Day	☉	☽	☿	♀	♂	♃	♄	♅	♆	♇
01 Tu	14S 30	13S 48	15S 19	24S 28	23S 13	03S 24	21S 10	07N55	08S 52	21S 26
02 We	14 49	16 07	15 55	24 38	23 04	03 28	21 11	07 55	08 53	21 26
03 Th	15 07	17 45	16 30	24 46	22 55	03 33	21 12	07 54	08 53	21 26
04 Fr	15 26	18 37	17 04	24 54	22 46	03 38	21 13	07 53	08 53	21 26
05 Sa	15 44	18 39	17 37	25 02	22 37	03 42	21 14	07 52	08 53	21 26
06 Su	16 02	17 50	18 10	25 08	22 27	03 47	21 15	07 51	08 53	21 26
07 Mo	16 20	16 08	18 41	25 14	22 17	03 51	21 15	07 50	08 54	21 26
08 Tu	16 38	13 36	19 12	25 19	22 07	03 56	21 16	07 50	08 54	21 26
09 We	16 55	10 18	19 42	25 24	21 57	04 00	21 17	07 49	08 54	21 26
10 Th	17 12	06 21	20 10	25 28	21 46	04 05	21 18	07 48	08 54	21 26
11 Fr	17 28	01 55	20 38	25 31	21 36	04 09	21 19	07 47	08 54	21 26
12 Sa	17 45	02N46	21 04	25 33	21 25	04 14	21 19	07 47	08 54	21 26
13 Su	18 01	07 24	21 30	25 34	21 14	04 18	21 20	07 46	08 54	21 26
14 Mo	18 16	11 38	21 55	25 35	21 02	04 22	21 21	07 45	08 54	21 26
15 Tu	18 32	15 07	22 18	25 35	20 51	04 27	21 22	07 44	08 54	21 26
16 We	18 47	17 33	22 40	25 35	20 39	04 31	21 23	07 44	08 54	21 26
17 Th	19 02	18 42	23 02	25 33	20 27	04 35	21 23	07 43	08 54	21 26
18 Fr	19 16	18 35	23 22	25 31	20 15	04 39	21 24	07 42	08 54	21 26
19 Sa	19 30	17 17	23 41	25 28	20 03	04 44	21 25	07 42	08 54	21 26
20 Su	19 44	15 02	23 58	25 25	19 50	04 48	21 26	07 41	08 54	21 26
21 Mo	19 57	12 04	24 15	25 20	19 38	04 52	21 27	07 40	08 54	21 26
22 Tu	20 10	08 36	24 30	25 15	19 25	04 56	21 27	07 40	08 54	21 26
23 We	20 23	04 51	24 44	25 10	19 12	05 00	21 28	07 39	08 54	21 26
24 Th	20 35	00 58	24 57	25 03	18 59	05 04	21 29	07 38	08 54	21 26
25 Fr	20 47	02S 54	25 08	24 56	18 45	05 08	21 30	07 38	08 54	21 26
26 Sa	20 58	06 37	25 18	24 48	18 32	05 12	21 30	07 37	08 54	21 26
27 Su	21 09	10 03	25 27	24 40	18 18	05 16	21 31	07 37	08 54	21 26
28 Mo	21 20	13 06	25 34	24 30	18 04	05 20	21 32	07 36	08 54	21 25
29 Tu	21 30	15 37	25 40	24 21	17 50	05 23	21 33	07 36	08 54	21 25
30 We	21 40	17 29	25 45	24 10	17 35	05 27	21 33	07 35	08 54	21 25

⚷ Chiron

01	Dec.	00 N 23
02	21♓04R	
05	20 59	
08	20 55	
11	20 52	
14	20 49	
17	20 46	
20	20 44	
23	20 42	
26	20 41	
29	20 40	

December 2016

Day	S. T.	☉	☽	☿	♀	♂	♃	♄	♅	♆	♇	☊ True
	h m s	° ' "	° ' "	° '	° '	° '	° '	° '	° '	° '	° '	° '
01 Th	04 41 07	09♐13 14	25♐32 01	27♐22	22♑16	16♒09	16♎49	17♐44	20♈R54	09♓16	15♑57	07♍R42
02 Fr	04 45 04	10 14 05	07♑37 43	28 46	23 27	16 54	16 59	17 51	20 52	09 17	15 58	07 30
03 Sa	04 49 00	11 14 57	19 50 23	00♑09	24 37	17 39	17 09	17 58	20 51	09 17	16 00	07 21
04 Su	04 52 57	12 15 50	02♒11 59	01 31	25 47	18 24	17 19	18 05	20 50	09 18	16 02	07 15
05 Mo	04 56 54	13 16 43	14 45 08	02 51	26 57	19 09	17 29	18 13	20 48	09 18	16 04	07 11
06 Tu	05 00 50	14 17 38	27 32 57	04 09	28 07	19 54	17 39	18 20	20 47	09 19	16 05	07 10
07 We	05 04 47	15 18 33	10♓38 59	05 25	29 17	20 39	17 48	18 27	20 46	09 19	16 07	07 10
08 Th	05 08 43	16 19 29	24 06 42	06 39	00♒27	21 25	17 58	18 34	20 45	09 20	16 09	07 09
09 Fr	05 12 40	17 20 25	07♈58 51	07 50	01 36	22 10	18 07	18 41	20 44	09 21	16 11	07 08
10 Sa	05 16 36	18 21 22	22 16 29	08 58	02 45	22 55	18 16	18 48	20 43	09 21	16 13	07 04
11 Su	05 20 33	19 22 20	06♉58 00	10 03	03 55	23 40	18 25	18 55	20 42	09 22	16 15	06 57
12 Mo	05 24 29	20 23 18	21 58 29	11 03	05 04	24 25	18 34	19 02	20 41	09 23	16 17	06 49
13 Tu	05 28 26	21 24 17	07♊09 48	11 58	06 13	25 11	18 43	19 09	20 40	09 23	16 18	06 38
14 We	05 32 23	22 25 17	22 21 34	12 48	07 22	25 56	18 52	19 16	20 40	09 24	16 20	06 27
15 Th	05 36 19	23 26 18	07♋22 56	13 32	08 30	26 41	19 01	19 23	20 39	09 25	16 22	06 17
16 Fr	05 40 16	24 27 19	22 04 21	14 08	09 39	27 26	19 09	19 31	20 38	09 26	16 24	06 08
17 Sa	05 44 12	25 28 22	06♌19 16	14 37	10 47	28 12	19 18	19 38	20 37	09 27	16 26	06 01
18 Su	05 48 09	26 29 25	20 04 36	14 57	11 55	28 57	19 26	19 45	20 37	09 28	16 28	05 57
19 Mo	05 52 05	27 30 29	03♍20 41	15 07	13 03	29 42	19 34	19 52	20 36	09 29	16 30	05 54
20 Tu	05 56 02	28 31 33	16 10 21	15R 06	14 11	00♓28	19 42	19 59	20 36	09 30	16 32	05 54
21 We	05 59 58	29 32 39	28 37 58	14 55	15 19	01 13	19 50	20 06	20 35	09 31	16 34	05 54
22 Th	06 03 55	00♑33 45	10♎48 38	14 31	16 26	01 58	19 58	20 13	20 35	09 32	16 36	05 53
23 Fr	06 07 52	01 34 52	22 47 35	13 56	17 34	02 44	20 06	20 20	20 34	09 33	16 38	05 51
24 Sa	06 11 48	02 36 00	04♏39 46	13 10	18 41	03 29	20 13	20 27	20 34	09 34	16 40	05 46
25 Su	06 15 45	03 37 08	16 29 38	12 12	19 48	04 14	20 21	20 34	20 34	09 35	16 42	05 39
26 Mo	06 19 41	04 38 17	28 20 55	11 06	20 54	04 59	20 28	20 41	20 34	09 36	16 44	05 30
27 Tu	06 23 38	05 39 26	10♐16 27	09 52	22 01	05 45	20 35	20 48	20 34	09 38	16 46	05 19
28 We	06 27 34	06 40 36	22 18 17	08 32	23 07	06 30	20 42	20 55	20 33	09 39	16 48	05 07
29 Th	06 31 31	07 41 46	04♑27 46	07 11	24 13	07 15	20 49	21 02	20 33	09 40	16 50	04 56
30 Fr	06 35 27	08 42 57	16 45 46	05 49	25 19	08 01	20 56	21 09	20D 33	09 41	16 52	04 46
31 Sa	06 39 24	09 44 07	29 12 56	04 30	26 24	08 46	21 02	21 15	20 33	09 43	16 54	04 39

Data for	12-01-2016
Julian Day	2457723.50
Ayanamsa	24 05 27
SVP	05 ♓ 01 17
☽ ☊ Mean	07 ♍ 52 R

● ◐ PHASES ○ ◑

Day	Time	Phase	Position
07	09:04	◐	15♓42
14	00:06	○	22♊26
21	01:56	◑	29♍38
29	06:54	●	07♑59

LAST ASPECT Day	h m	☽ INGRESS Day	h m	Sign
01	04:09	01	08:53	♑
03	10:17	03	19:45	♒
05	11:24	06	04:32	♓
07	14:06	08	10:17	♈
10	1:07	10	12:42	♉
12	04:05	12	12:42	♊
14	05:58	14	12:09	♋
15	21:37	16	13:15	♌
18	16:55	18	17:52	♍
21	01:56	21	02:40	♎
22	19:32	23	14:33	♏
25	07:23	26	03:20	♐
28	01:46	28	15:13	♑
30	08:08	31	01:30	♒

DECLINATION

Day	☉	☽	☿	♀	♂	♃	♄	♅	♆	♇
01 Th	21S50	18S36	25S48	23S59	17S21	05S31	21S34	07N35	08S53	21S25
02 Fr	21 59	18 54	25 50	23 47	17 06	05 35	21 35	07 34	08 53	21 25
03 Sa	22 07	18 18	25 50	23 34	16 52	05 38	21 35	07 34	08 53	21 25
04 Su	22 16	16 51	25 49	23 21	16 37	05 42	21 36	07 33	08 53	21 25
05 Mo	22 23	14 34	25 47	23 07	16 22	05 46	21 37	07 33	08 53	21 25
06 Tu	22 31	11 32	25 43	22 53	16 06	05 49	21 38	07 32	08 52	21 25
07 We	22 38	07 52	25 37	22 38	15 51	05 53	21 38	07 32	08 52	21 25
08 Th	22 44	03 42	25 31	22 22	15 36	05 56	21 39	07 32	08 52	21 24
09 Fr	22 50	00N47	25 23	22 06	15 20	05 59	21 40	07 31	08 52	21 24
10 Sa	22 55	05 21	25 13	21 49	15 04	06 03	21 40	07 31	08 51	21 24
11 Su	23 01	09 43	25 03	21 32	14 48	06 06	21 41	07 31	08 51	21 24
12 Mo	23 05	13 33	24 51	21 14	14 32	06 09	21 41	07 30	08 51	21 24
13 Tu	23 09	16 33	24 39	20 56	14 16	06 12	21 42	07 30	08 51	21 24
14 We	23 13	18 23	24 25	20 37	14 00	06 16	21 43	07 30	08 50	21 24
15 Th	23 16	18 55	24 11	20 17	13 43	06 19	21 43	07 30	08 50	21 24
16 Fr	23 19	18 09	23 55	19 57	13 27	06 22	21 44	07 29	08 50	21 24
17 Sa	23 21	16 15	23 40	19 37	13 10	06 25	21 45	07 29	08 49	21 23
18 Su	23 23	13 28	23 24	19 16	12 53	06 28	21 45	07 29	08 49	21 23
19 Mo	23 25	10 04	23 08	18 55	12 36	06 31	21 46	07 29	08 49	21 23
20 Tu	23 26	06 17	22 51	18 33	12 19	06 34	21 46	07 29	08 48	21 23
21 We	23 26	02 21	22 35	18 11	12 02	06 36	21 47	07 28	08 48	21 23
22 Th	23 26	01S36	22 19	17 48	11 45	06 39	21 47	07 28	08 47	21 23
23 Fr	23 26	05 25	22 04	17 25	11 28	06 42	21 48	07 28	08 47	21 23
24 Sa	23 25	08 59	21 49	17 02	11 11	06 44	21 48	07 28	08 46	21 22
25 Su	23 23	12 11	21 34	16 38	10 53	06 47	21 49	07 28	08 46	21 22
26 Mo	23 21	14 53	21 21	16 14	10 36	06 50	21 50	07 28	08 46	21 22
27 Tu	23 19	17 00	21 08	15 49	10 18	06 52	21 50	07 28	08 45	21 22
28 We	23 16	18 23	20 56	15 24	10 00	06 54	21 51	07 28	08 45	21 22
29 Th	23 13	18 57	20 45	14 59	09 42	06 57	21 51	07 28	08 44	21 22
30 Fr	23 09	18 38	20 36	14 34	09 25	06 59	21 51	07 28	08 44	21 21
31 Sa	23 05	17 25	20 28	14 08	09 07	07 01	21 52	07 28	08 43	21 21

ASPECTARIAN

01 01:17 ☉ □ ♆
04:09 ☽ ☌ ☿

02 03:16 ☽ ⚹ ♆
03:46 ♂ △ ♃
16:28 ☽ ☌ ♇
18:41 ☽ □ ♃
21:18 ☿ ♑

03 01:58 ☽ □ ♅
10:17 ☽ ☌ ♀
12:16 ♂ ⚹ ♄

04 03:25 ☽ ∥ ♂
20:58 ☽ ⚹ ☉

05 05:14 ☽ △ ♃
06:36 ☽ ⚹ ♄
08:50 ☽ ☌ ♂
11:24 ☽ ⚹ ♅

06 13:29 ☽ ⚹ ☿
17:45 ☽ ∥ ♆
21:36 ☽ ☌ ♆

07 00:22 ♀ ∥ ☉
01:58 ☽ # ♅
03:30 ♂ ⚹ ♅
09:52 ☽ ⚹ ♇
11:34 ☽ ∥ ♃
14:06 ☽ □ ♄
14:52 ♀ ♒

08 12:03 ☽ ⚹ ♀
23:44 ☽ □ ☿

09 13:53 ☽ □ ♇
17:00 ☽ △ ☉
17:17 ☽ ☍ ♃
18:11 ☽ △ ♄
21:25 ☽ ☌ ♅
21:40 ☉ ⚹ ♃

10 01:07 ☽ ⚹ ♂
03:46 ☽ # ♃
08:26 ☿ ⚹ ♆
11:39 ☽ ∥ ♅
11:52 ☉ ☌ ♄
12:12 ♀ ∥ ♄
18:38 ☽ □ ♀
19:05 ☽ # ♆

11 03:52 ☽ ⚹ ♆
05:20 ☽ △ ☿
10:38 ♀ ∥ ♇
14:55 ☽ △ ♇

12 04:05 ☽ □ ♂
06:26 ☽ # ♂
06:56 ☉ △ ♅
22:23 ☽ △ ♀

13 03:31 ☽ □ ♆
18:25 ☽ △ ♃
19:04 ☽ ☍ ♄
21:18 ☽ ⚹ ♅

14 05:58 ☽ △ ♂

15 03:18 ☽ △ ♆
10:25 ☽ ☍ ☿
14:38 ☽ ☍ ♇
19:08 ☽ □ ♃
21:37 ☽ □ ♅

17 08:23 ☽ ☍ ♀
22:51 ☽ ⚹ ♃
23:24 ☽ △ ♄

18 00:42 ☿ ∥ ☉
00:57 ☽ △ ♅
04:47 ☽ # ♂
12:27 ☽ △ ☉
16:55 ☽ ☍ ♂

19 08:12 ☽ # ♆
09:23 ♂ ♓
10:56 ☿ SR
11:24 ☽ ☍ ♆
16:38 ☽ ∥ ♅
21:59 ☽ △ ☿
22:21 ☽ # ♃

20 00:41 ☽ △ ♇
07:20 ☽ □ ♄

21 10:45 ☉ ♑

22 07:05 ☽ □ ☿
11:36 ☽ □ ♇
12:24 ☽ △ ♀
18:31 ☽ ☌ ♃
19:00 ☽ ⚹ ♄
19:32 ☽ ☍ ♅

23 08:29 ☽ ∥ ♃
13:33 ☽ # ♅
19:26 ☽ ⚹ ☉
21:26 ☽ △ ♂
22:33 ☽ ∥ ♆

24 00:26 ☿ ∥ ♄
09:58 ☽ △ ♆
14:44 ☽ ∥ ♂
15:59 ☽ ⚹ ☿

25 00:18 ♄ △ ♅
00:25 ☽ ⚹ ♇
07:23 ☽ □ ♀
13:23 ♀ △ ♃
16:36 ♀ ⚹ ♅
18:34 ♀ ⚹ ♄
21:29 ☿ ∥ ♇

26 11:58 ☽ ∥ ♀
14:18 ☽ □ [illegible]
18:33 ♃ ☍ ♅
22:42 ☽ □ ♆

27 04:19 ☿ ⚹ ♆
08:08 ☉ ⚹ ♂
20:32 ☽ △ ♅
20:47 ☽ ⚹ ♃
21:12 ☽ ☌ ♄

28 01:46 ☽ ⚹ ♀
18:48 ☿ ☌ ☉
23:06 ☿ ⚹ ♂

29 04:47 ☽ ☌ ☿
[illegible] ☽ ⚹ [illegible]
09:30 ♅ SD
10:12 ☽ ⚹ ♆

30 00:13 ☽ ☌ ♇
07:21 ☽ □ ♅
08:08 ☽ □ ♃
23:27 ☉ ⚹ ♆

⚷ Chiron

01 Dec.	00 N 06
02	20♓40D
05	20 41
08	20 41
11	20 43
14	20 45
17	20 47
20	20 50
23	20 53
26	20 57
29	21 01

01 07:53 20♓40 D

January 2017

Day	S. T. h m s	☉ ° ′ ″	☽ ° ′ ″	☿ ° ′	♀ ° ′	♂ ° ′	♃ ° ′	♄ ° ′	♅ ° ′	♆ ° ′	♇ ° ′	☊ True ° ′
01 Su	06 43 21	10♑45 18	11♒49 54	03♑R17	27♒29	09♓31	21♎09	21♐22	20♈34	09♓44	16♑56	04♍R34
02 Mo	06 47 17	11 46 28	24 37 36	02 10	28 34	10 17	21 15	21 29	20 34	09 45	16 58	04 32
03 Tu	06 51 14	12 47 38	07♓37 22	01 12	29 39	11 02	21 21	21 36	20 34	09 47	17 00	04D 32
04 We	06 55 10	13 48 48	20 50 53	00 24	00♓43	11 47	21 27	21 43	20 34	09 48	17 03	04 33
05 Th	06 59 07	14 49 58	04♈20 03	29♐46	01 48	12 33	21 33	21 49	20 35	09 50	17 05	04 34
06 Fr	07 03 03	15 51 07	18 06 33	29 18	02 51	13 18	21 39	21 56	20 35	09 51	17 07	04R 34
07 Sa	07 06 60	16 52 16	02♉11 08	29 00	03 55	14 03	21 44	22 03	20 35	09 53	17 09	04 33
08 Su	07 10 56	17 53 24	16 33 06	28 52	04 58	14 48	21 49	22 09	20 36	09 54	17 11	04 29
09 Mo	07 14 53	18 54 33	01♊09 34	28D 52	06 01	15 34	21 54	22 16	20 36	09 56	17 13	04 24
10 Tu	07 18 50	19 55 40	15 55 25	29 02	07 03	16 19	22 00	22 23	20 37	09 57	17 15	04 18
11 We	07 22 46	20 56 47	00♋43 38	29 19	08 05	17 04	22 04	22 29	20 38	09 59	17 17	04 11
12 Th	07 26 43	21 57 54	15 26 07	29 43	09 06	17 49	22 09	22 36	20 38	10 01	17 19	04 04
13 Fr	07 30 39	22 59 01	29 55 01	00♑13	10 08	18 34	22 14	22 42	20 39	10 02	17 21	03 58
14 Sa	07 34 36	24 00 07	14♌03 59	00 50	11 08	19 20	22 18	22 48	20 40	10 04	17 23	03 53
15 Su	07 38 32	25 01 12	27 48 54	01 31	12 09	20 05	22 22	22 55	20 41	10 06	17 25	03 51
16 Mo	07 42 29	26 02 18	11♍08 22	02 18	13 09	20 50	22 26	23 01	20 41	10 08	17 27	03D 50
17 Tu	07 46 26	27 03 23	24 03 19	03 08	14 08	21 35	22 30	23 08	20 42	10 09	17 29	03 51
18 We	07 50 22	28 04 28	06♎36 32	04 02	15 07	22 20	22 34	23 14	20 43	10 11	17 31	03 53
19 Th	07 54 19	29 05 32	18 52 03	05 00	16 05	23 05	22 37	23 20	20 44	10 13	17 33	03 54
20 Fr	07 58 15	00 06 37	00♏54 34	06 01	17 03	23 50	22 40	23 26	20 46	10 15	17 35	03 54
21 Sa	08 02 12	01♒07 40	12 49 03	07 05	18 01	24 35	22 43	23 32	20 47	10 17	17 37	03R 54
22 Su	08 06 08	02 08 44	24 40 25	08 11	18 58	25 20	22 46	23 38	20 48	10 19	17 39	03 52
23 Mo	08 10 05	03 09 47	06♐33 15	09 19	19 54	26 05	22 49	23 44	20 49	10 20	17 41	03 48
24 Tu	08 14 01	04 10 50	18 31 31	10 30	20 49	26 50	22 52	23 50	20 50	10 22	17 43	03 44
25 We	08 17 58	05 11 52	00♑38 27	11 42	21 45	27 35	22 54	23 56	20 52	10 24	17 45	03 39
26 Th	08 21 55	06 12 53	12 56 27	12 56	22 39	28 20	22 56	24 02	20 53	10 26	17 47	03 34
27 Fr	08 25 51	07 13 54	25 27 03	14 12	23 33	29 05	22 58	24 08	20 55	10 28	17 49	03 30
28 Sa	08 29 48	08 14 54	08♒11 01	15 29	24 26	29 49	23 00	24 14	20 56	10 30	17 51	03 27
29 Su	08 33 44	09 15 52	21 08 26	16 47	25 18	00♈34	23 02	24 20	20 58	10 32	17 53	03 26
30 Mo	08 37 41	10 16 50	04♓18 50	18 07	26 10	01 19	23 03	24 25	20 59	10 34	17 55	03D 26
31 Tu	08 41 37	11 17 47	17 41 30	19 28	27 01	02 04	23 05	24 31	21 01	10 36	17 57	03 27

Data for	01-01-2017
Julian Day	2457754.50
Ayanamsa	24 05 33
SVP	05 ♓ 01 14
☽ ☊ Mean	06 ♍ 14 R

● ◐ PHASES ○ ◑

Day	Time	Phase	Position
05	19:47	◐	15♈40
12	11:34	○	22♋27
19	22:14	◑	00♏02
28	00:07	●	08♒15

LAST ASPECT Day	h m	☽ INGRESS Day	h m	Sign
02	08:00	02	09:58	♓
04	16:15	04	16:20	♈
06	18:42	06	20:18	♉
08	02:23	08	22:06	♊
10	21:39	10	22:49	♋
12	11:35	13	00:08	♌
14	15:18	15	03:53	♍
17	06:11	17	11:17	♎
19	08:56	19	22:11	♏
22	01:26	22	10:46	♐
24	17:34	24	22:44	♑
27	07:18	27	08:37	♒
29	05:53	29	16:11	♓
31	17:36	31	21:47	♈

DECLINATION

Day	☉	☽	☿	♀	♂	♃	♄	♅	♆	♇
01 Su	23S00	15S20	20S21	13S42	08S49	07S04	21S52	07N28	08S43	21S21
02 Mo	22 55	12 28	20 16	13 16	08 31	07 06	21 53	07 28	08 42	21 21
03 Tu	22 49	08 58	20 14	12 49	08 12	07 08	21 53	07 28	08 41	21 21
04 We	22 43	04 57	20 12	12 22	07 54	07 10	21 54	07 29	08 41	21 21
05 Th	22 37	00 37	20 13	11 55	07 36	07 12	21 54	07 29	08 40	21 21
06 Fr	22 30	03N49	20 15	11 28	07 18	07 14	21 55	07 29	08 40	21 20
07 Sa	22 22	08 09	20 19	11 01	06 59	07 15	21 55	07 29	08 39	21 20
08 Su	22 14	12 05	20 24	10 33	06 41	07 17	21 55	07 29	08 39	21 20
09 Mo	22 06	15 22	20 31	10 05	06 22	07 19	21 56	07 29	08 38	21 20
10 Tu	21 57	17 41	20 38	09 37	06 04	07 21	21 56	07 30	08 37	21 20
11 We	21 48	18 50	20 46	09 09	05 46	07 22	21 57	07 30	08 37	21 20
12 Th	21 39	18 42	20 54	08 41	05 27	07 24	21 57	07 30	08 36	21 19
13 Fr	21 29	17 20	21 03	08 13	05 08	07 25	21 57	07 31	08 35	21 19
14 Sa	21 18	14 55	21 12	07 44	04 50	07 27	21 58	07 31	08 35	21 19
15 Su	21 07	11 43	21 22	07 16	04 31	07 28	21 58	07 31	08 34	21 19
16 Mo	20 56	08 00	21 31	06 47	04 13	07 29	21 58	07 32	08 33	21 19
17 Tu	20 45	04 01	21 39	06 19	03 54	07 30	21 59	07 32	08 33	21 18
18 We	20 33	00S02	21 48	05 50	03 35	07 31	21 59	07 32	08 32	21 18
19 Th	20 20	03 59	21 55	05 22	03 17	07 33	21 59	07 33	08 31	21 18
20 Fr	20 07	07 42	22 03	04 53	02 58	07 34	22 00	07 33	08 31	21 18
21 Sa	19 54	11 03	22 09	04 24	02 39	07 34	22 00	07 34	08 30	21 18
22 Su	19 41	13 57	22 15	03 56	02 20	07 35	22 00	07 34	08 29	21 18
23 Mo	19 27	16 17	22 20	03 27	02 02	07 36	22 00	07 35	08 29	21 17
24 Tu	19 13	17 56	22 24	02 59	01 43	07 37	22 01	07 35	08 28	21 17
25 We	18 58	18 48	22 27	02 30	01 24	07 37	22 01	07 36	08 27	21 17
26 Th	18 43	18 48	22 29	02 02	01 06	07 38	22 01	07 36	08 26	21 17
27 Fr	18 28	17 53	22 30	01 34	00 47	07 39	22 01	07 37	08 26	21 17
28 Sa	18 12	16 03	22 30	01 05	00 28	07 39	22 02	07 37	08 25	21 17
29 Su	17 56	13 23	22 29	00 37	00 09	07 39	22 02	07 38	08 24	21 16
30 Mo	17 40	10 00	22 27	00 09	00N09	07 40	22 02	07 39	08 23	21 16
31 Tu	17 23	06 03	22 24	00N18	00 28	07 40	22 02	07 39	08 23	21 16

⚷ Chiron

01 Dec. 00 N 08

Day	Position
01	21♓06
04	21 11
07	21 17
10	21 23
13	21 30
16	21 36
19	21 44
22	21 51
25	21 59
28	22 08
31	22 16

ASPECTARIAN

01 06:53 ♂ ☌ ♆
08:21 ♂ ∥ ♆
16:25 ☽ ⚹ ♅
17:00 ☽ ∥ ♀
17:39 ☽ △ ♃
18:05 ☽ ⚹ ♄
02 08:00 ☽ ☌ ♀
12:59 ☽ ⚹ ☿
03 01:46 ☽ ∥ ♆
03:57 ☽ ☌ ♆
05:10 ☽ ∥ ♂
06:37 ☽ ☌ ♂
07:47 ♀ ♓
09:15 ☽ # ♅
10:14 ☽ ⚹ ☉
11:13 ☽ ∥ ♃
17:07 ☽ ⚹ ♇
19:41 ☿ ⚹ ♀
04 01:34 ☽ □ ♄
14:18 ☿ ♐R
16:15 ☽ □ ☿
05 09:32 ♂ # ♅
22:16 ☽ □ ♇
06 04:15 ☽ ☌ ♅
04:47 ♂ ∥ ♃
06:07 ☽ ☍ ♃
06:37 ☽ △ ♄
17:53 ☽ # ♂
18:42 ☽ △ ☿
18:56 ☽ # ♃
20:14 ☽ ∥ ♅
07 02:56 ☽ # ♆
03:08 ☽ ⚹ ♀
06:43 ☉ ☌ ♇
12:56 ☽ ⚹ ♆
15:17 ☽ # ♀
20:57 ☽ ⚹ ♂
08 01:02 ☽ △ ♇
02:23 ☽ △ ☉
09:44 ☿ SD
09 08:30 ☽ □ ♀
14:18 ☽ □ ♆
10 00:40 ☽ □ ♂
02:59 ☉ ∥ ♄
07:37 ☽ ⚹ ♅
09:53 ☽ △ ♃
10:32 ☽ ☍ ♄
16:22 ☉ □ ♅
21:39 ☽ ☍ ☿
11 07:12 ♂ ⚹ ♇
12:52 ☽ △ ♀
15:06 ☽ △ ♆
12 03:06 ☽ ☍ ♇
04:08 ☽ △ ♂
04:21 ♀ ∥ ♆
04:43 ☉ □ ♃
08:34 ☽ □ ♅
11:08 ☽ □ ♃
14:04 ☿ ♑
21:54 ♀ ☌ ♆
13 22:12 ☉ ∥ ♇
14 07:03 ☿ ∥ ☉
11:16 ♀ # ♅
11:26 ☽ △ ♅
14:21 ☽ ⚹ ♃
14:23 ♀ ∥ ♃
15:18 ☽ △ ♄
17:03 ☿ ∥ ♇
15 06:59 ☽ △ ☿
20:31 ☽ # ♆
22:09 ☽ ☍ ♆
16 02:53 ☽ ∥ ♅
03:08 ☽ # ♃
03:59 ☽ ☍ ♀
08:23 ☽ # ♀
11:40 ☽ △ ♇
19:03 ☽ ☍ ♂
22:14 ☽ □ ♄
17 00:43 ☽ # ♂
06:11 ☽ △ ☉
18:38 ☽ □ ☿
18 19:57 ☽ ∥ ♂
21:24 ☽ □ ♇
19 03:43 ☽ ☍ ♅
07:28 ☽ ☌ ♃
07:44 ☽ ∥ ♀
08:56 ☽ ⚹ ♄
09:16 ♂ □ ♄
12:49 ♃ # ♅
13:14 ☿ ∥ ♄
21:24 ☉ ♒
23:05 ☽ # ♅
23:06 ☽ ∥ ♃
20 05:36 ☽ ∥ ♆
11:16 ☽ ⚹ ☿
13:55 ♀ ⚹ ♇
18:51 ☽ △ ♆
21 09:45 ☽ ⚹ ♇
11:25 ☽ △ ♀
22 01:26 ☽ △ ♂
16:31 ☽ ⚹ ☉
23 07:38 ☽ □ ♆
21:29 ☿ ⚹ ♆
24 04:37 ☽ △ ♅
04:57 ☽ □ ♀
08:40 ☽ ⚹ ♃
10:39 ☽ ☌ ♄
17:34 ☽ □ ♂
25 07:38 ☽ ∥ ☉
19:08 ☽ ⚹ ♆
23:59 ☽ ☌ ☿
26 05:43 ☽ ∥ ☉
09:23 ☽ ☌ ♇
15:19 ☽ □ ♅
19:16 ☽ □ ♃
20:06 ☽ ⚹ ♀
27 07:18 ☽ ⚹ ♂
17:51 ♀ □ ♄
28 05:39 ♂ ♈
23:40 ☽ ⚹ ♅
29 03:29 ☽ △ ♃
05:53 ☽ ⚹ ♄
20:20 ☿ ☌ ♇
30 00:04 ♀ # ♂
10:13 ☽ ∥ ♆
11:18 ☽ ☌ ♆
14:34 ☽ ∥ ♃
14:39 ☽ # ♅
31 00:28 ☽ ⚹ ♇
03:31 ☽ ⚹ ☿
12:11 ☽ □ ♄
17:36 ☽ ☌ ♀

February 2017

11	00:45	22♌28 ✷	Penumbral Lunar Eclipse (mag 1.014)
26	14:40	08♓12 ☼	Annular Solar Eclipse

Day	S. T. h m s	☉ ° ' "	☽ ° ' "	☿ ° '	♀ ° '	♂ ° '	♃ ° '	♄ ° '	♅ ° '	♆ ° '	♇ ° '	☊ True ° '
01 We	08 45 34	12♒ 18 43	01♈ 15 40	20♑ 50	27♓ 51	02♈ 49	23♎ 06	24♐ 36	21♈ 02	10♓ 38	17♑ 59	03♍ 28
02 Th	08 49 30	13 19 37	15 00 36	22 14	28 40	03 33	23 07	24 42	21 04	10 40	18 01	03 30
03 Fr	08 53 27	14 20 30	28 55 31	23 38	29 29	04 18	23 07	24 47	21 06	10 42	18 03	03 31
04 Sa	08 57 24	15 21 21	12♉ 59 27	25 03	00♈ 16	05 03	23 08	24 53	21 08	10 44	18 05	03R 31
05 Su	09 01 20	16 22 12	27 10 59	26 29	01 03	05 47	23 08	24 58	21 09	10 47	18 07	03 30
06 Mo	09 05 17	17 23 00	11♊ 27 55	27 57	01 48	06 32	23 08	25 03	21 11	10 49	18 08	03 29
07 Tu	09 09 13	18 23 47	25 47 19	29 25	02 33	07 16	23R 08	25 08	21 13	10 51	18 10	03 27
08 We	09 13 10	19 24 33	10♋ 05 25	00♒ 53	03 16	08 01	23 08	25 13	21 15	10 53	18 12	03 24
09 Th	09 17 06	20 25 18	24 17 54	02 23	03 58	08 45	23 08	25 18	21 17	10 55	18 14	03 22
10 Fr	09 21 03	21 26 00	08♌ 20 19	03 54	04 40	09 30	23 07	25 23	21 19	10 57	18 16	03 20
11 Sa	09 24 59	22 26 42	22 08 36	05 25	05 20	10 14	23 06	25 28	21 21	11 00	18 18	03 19
12 Su	09 28 56	23 27 22	05♍ 39 37	06 58	05 58	10 58	23 05	25 33	21 24	11 02	18 19	03 18
13 Mo	09 32 53	24 28 01	18 51 32	08 31	06 36	11 43	23 04	25 38	21 26	11 04	18 21	03D 19
14 Tu	09 36 49	25 28 38	01♎ 44 00	10 05	07 12	12 27	23 03	25 43	21 28	11 06	18 23	03 19
15 We	09 40 46	26 29 14	14 18 10	11 40	07 47	13 11	23 01	25 47	21 30	11 08	18 24	03 20
16 Th	09 44 42	27 29 49	26 36 20	13 16	08 20	13 56	23 00	25 52	21 33	11 11	18 26	03 21
17 Fr	09 48 39	28 30 23	08♏ 41 53	14 52	08 52	14 40	22 58	25 56	21 35	11 13	18 28	03 21
18 Sa	09 52 35	29 30 55	20 38 51	16 30	09 22	15 24	22 55	26 01	21 37	11 15	18 30	03 22
19 Su	09 56 32	00♓ 31 26	02♐ 31 48	18 08	09 51	16 08	22 53	26 05	21 40	11 17	18 31	03 22
20 Mo	10 00 28	01 31 56	14 25 27	19 48	10 18	16 52	22 51	26 09	21 42	11 19	18 33	03 22
21 Tu	10 04 25	02 32 25	26 24 28	21 28	10 43	17 36	22 48	26 13	21 45	11 22	18 34	03 22
22 We	10 08 21	03 32 52	08♑ 33 10	23 09	11 06	18 20	22 45	26 17	21 47	11 24	18 36	03 23
23 Th	10 12 18	04 33 18	20 55 16	24 51	11 28	19 04	22 42	26 21	21 50	11 26	18 38	03 23
24 Fr	10 16 15	05 33 43	03♒ 33 44	26 35	11 48	19 48	22 39	26 25	21 52	11 29	18 39	03 23
25 Sa	10 20 11	06 34 05	16 30 27	28 19	12 05	20 32	22 36	26 29	21 55	11 31	18 41	03 24
26 Su	10 24 08	07 34 27	29 46 05	00♓ 04	12 21	21 16	22 32	26 33	21 57	11 33	18 42	03 24
27 Mo	10 28 04	08 34 46	13♓ 19 56	01 50	12 34	22 00	22 28	26 37	22 00	11 35	18 44	03 25
28 Tu	10 32 01	09 35 04	27 09 52	03 37	12 46	22 43	22 24	26 40	22 03	11 38	18 45	03R 25

Data for	02-01-2017
Julian Day	2457785.50
Ayanamsa	24 05 37
SVP	05 ♓ 01 10
☽ ☊ Mean	04 ♍ 35 R

● ◐ PHASES ○ ◑

04	04:19	◐	15♉32
11	00:34	✷	22♌28
18	19:34	◑	00♐20
26	14:58	☼	08♓12

LAST ASPECT Day	h m	☽ INGRESS Day	h m	
02	16:51	03	01:51	♉
04	22:42	05	04:45	♊
06	22:54	07	07:03	♋
08	22:01	09	09:42	♌
11	05:54	11	13:53	♍
13	12:38	13	20:44	♎
16	01:55	16	06:42	♏
17	19:39	18	18:53	♐
20	23:38	21	07:08	♑
23	03:24	23	17:18	♒
25	18:12	26	00:25	♓
27	23:09	28	04:52	♈

DECLINATION

Day	☉	☽	☿	♀	♂	♃	♄	♅	♆	♇
01 We	17S 06	01S 45	22S 19	00N46	00N46	07S 40	22S 02	07N40	08S 22	21S 16
02 Th	16 49	02N42	22 13	01 13	01 05	07 40	22 03	07 41	08 21	21 16
03 Fr	16 31	07 02	22 06	01 41	01 24	07 40	22 03	07 41	08 20	21 16
04 Sa	16 14	11 02	21 58	02 08	01 42	07 40	22 03	07 42	08 19	21 15
05 Su	15 56	14 26	21 49	02 34	02 01	07 40	22 03	07 43	08 19	21 15
06 Mo	15 37	17 00	21 38	03 01	02 19	07 40	22 03	07 44	08 18	21 15
07 Tu	15 19	18 31	21 26	03 27	02 38	07 40	22 03	07 44	08 17	21 15
08 We	15 00	18 51	21 13	03 53	02 56	07 39	22 04	07 45	08 16	21 15
09 Th	14 41	17 59	20 58	04 18	03 14	07 39	22 04	07 46	08 15	21 15
10 Fr	14 21	16 01	20 42	04 43	03 33	07 39	22 04	07 47	08 14	21 14
11 Sa	14 02	13 10	20 25	05 08	03 51	07 38	22 04	07 47	08 14	21 14
12 Su	13 42	09 38	20 06	05 32	04 09	07 37	22 04	07 48	08 13	21 14
13 Mo	13 22	05 43	19 46	05 56	04 27	07 37	22 04	07 49	08 12	21 14
14 Tu	13 02	01 37	19 25	06 20	04 45	07 36	22 04	07 50	08 11	21 14
15 We	12 41	02S 26	19 03	06 43	05 03	07 35	22 04	07 51	08 10	21 14
16 Th	12 20	06 18	18 39	07 06	05 21	07 34	22 05	07 52	08 09	21 13
17 Fr	11 59	09 51	18 14	07 28	05 39	07 33	22 05	07 53	08 09	21 13
18 Sa	11 38	12 56	17 47	07 49	05 57	07 32	22 05	07 54	08 08	21 13
19 Su	11 17	15 29	17 19	08 10	06 15	07 31	22 05	07 54	08 07	21 13
20 Mo	10 56	17 23	16 50	08 30	06 32	07 30	22 05	07 55	08 06	21 13
21 Tu	10 34	18 32	16 19	08 50	06 50	07 29	22 05	07 56	08 05	21 13
22 We	10 12	18 51	15 47	09 09	07 07	07 28	22 05	07 57	08 04	21 13
23 Th	09 50	18 17	15 14	09 27	07 25	07 27	22 05	07 58	08 03	21 12
24 Fr	09 28	16 48	14 39	09 44	07 42	07 25	22 05	07 59	08 03	21 12
25 Sa	09 06	14 25	14 03	10 01	07 59	07 24	22 05	08 00	08 02	21 12
26 Su	08 44	11 15	13 26	10 17	08 17	07 22	22 05	08 01	08 01	21 12
27 Mo	08 21	07 24	12 47	10 32	08 34	07 21	22 05	08 02	08 00	21 12
28 Tu	07 58	03 05	12 07	10 46	08 51	07 19	22 05	08 03	07 59	21 12

ASPECTARIAN

01 01:16 ♀ ∥ ♂
02:52 ☽ ☌ ♂
03:32 ☿ □ ♅
04:52 ☽ # ♀
04:59 ☽ # ♂
07:48 ♃ # ♅
14:43 ☽ ∥ ♂
15:12 ☽ ∥ ♀
20:51 ☽ ⚹ ☉

02 05:13 ☽ □ ♇
10:30 ☽ ☌ ♅
13:53 ☽ □ ☿
14:01 ☽ ☍ ♃
15:16 ☿ □ ♃
16:51 ☽ △ ♄

03 03:41 ☽ # ♃
03:47 ☽ ∥ ♅
07:31 ☽ # ♆
10:18 ☿ ∥ ♄
15:52 ♀ ♈
20:10 ☽ ⚹ ♆

04 08:39 ☽ △ ♇
22:42 ☽ △ ☿

05 06:52 ☽ ⚹ ♀
11:34 ☽ # ☉
15:16 ☽ ⚹ ♂
22:54 ☽ □ ♆

06 06:54 ♃ SR
10:40 ☽ △ ☉
16:20 ☽ ⚹ ♅
19:34 ☽ △ ♃
22:54 ☽ ☍ ♄

07 09:36 ☿ ♒
11:56 ☽ □ ♀
20:12 ☿ ∥ ♇
20:19 ☽ □ ♂

08 01:20 ☽ △ ♆
13:42 ☽ ☍ ♇
18:53 ☽ □ ♅
22:01 ☽ □ ♃

09 15:27 ☽ ☍ ☿
17:21 ☽ △ ♀
21:16 ☉ ⚹ ♅

10 02:06 ☽ △ ♂
16:30 ☽ # ☉
21:19 ☿ ⚹ ♀
22:37 ☽ △ ♅

11 01:41 ☽ ⚹ ♃
05:54 ☽ △ ♄
15:26 ☉ △ ♃

12 08:55 ☽ # ♆
09:43 ☽ ☍ ♆
11:22 ☽ ∥ ♅
12:32 ☽ # ♃
22:45 ☽ ∥ ♀
23:04 ☽ △ ♇

13 06:55 ☽ ∥ ♂
12:38 ☽ □ ♄

14 05:59 ☉ ⚹ ♄
10:52 ☽ ☍ ♀
18:11 ☽ △ ☿
21:43 ☽ ☍ ♂

15 07:58 ☽ □ ♇
14:02 ☽ ☍ ♅
16:55 ☽ ☌ ♃
17:24 ☽ # ♂
22:32 ☽ ⚹ ♄

16 01:55 ☽ △ ☉
05:42 ☽ # ♀
08:17 ☽ ∥ ♃
10:17 ☽ # ♅
12:12 ☽ ∥ ♆
18:15 ☿ ⚹ ♂

17 05:03 ☽ △ ♆
06:18 ♀ # ♃
14:19 ☽ □ ☿
14:34 ☽ ∥ ☉
19:39 ☽ ⚹ ♇

18 05:22 ♀ ∥ ♅
11:32 ☉ ♓
20:41 ♀ # ♆

19 15:21 ☽ △ ♀
17:44 ☽ □ ♆
17:55 ☽ ∥ ☿

20 05:14 ☽ △ ♂
12:32 ☽ ⚹ ☿
14:39 ☽ △ ♅
16:50 ☽ ⚹ ♃
23:38 ☽ ☌ ♄

21 04:02 ☿ ⚹ ♅
13:16 ☽ ⚹ ☉
18:28 ☿ △ ♃

22 05:09 ☽ □ ♀
05:35 ☽ ⚹ ♆
09:00 ♂ □ ♇
19:34 ☽ ☌ ♇
20:13 ☽ □ ♂

23 01:45 ☽ □ ♅
02:10 ♂ # ♃
03:24 ☽ □ ♃
14:09 ♀ # ☉
21:46 ☿ ⚹ ♄

24 15:41 ☽ ⚹ ♀

25 01:09 ♂ ∥ ♅
03:03 ♂ # ♆
03:54 ☽ ∥ ☿
07:46 ☽ ⚹ ♂
09:53 ☽ ⚹ ♅
11:02 ☽ △ ♃
18:12 ☽ ⚹ ♄
19:27 ♅ # ♆
23:08 ☿ ♓

26 00:36 ☽ ☌ ☿
05:56 ☽ # ♀
10:19 [illegible] # [illegible]
17:32 ☽ # ♂
17:45 ☽ ∥ ☉
20:14 ☽ # ♅
20:25 ☽ ∥ ♆
20:56 ☽ ☌ ♆

27 00:19 ♂ ☌ ♅
00:20 ☽ ∥ ♃
09:26 ☽ ⚹ ♇
14:25 ♂ ☍ ♃
19:08 ☉ # ♅
23:09 ☽ □ ♄
23:13 ☉ ∥ ♆

⚷ Chiron

01 Dec.	00 N 30
03	22♓25
06	22 34
09	22 44
12	22 53
15	23 03
18	23 13
21	23 23
24	23 34
27	23 44

March 2017

Day	S. T. h m s	☉ ° ' "	☽ ° ' "	☿ ° '	♀ ° '	♂ ° '	♃ ° '	♄ ° '	♅ ° '	♆ ° '	♇ ° '	☊ True ° '
01 We	10 35 57	10♓35 20	11♈12 43	05♓25	12♈55	23♈27	22♎R20	26♐44	22♈06	11♓40	18♑46	03♍R24
02 Th	10 39 54	11 35 34	25 24 35	07 14	13 02	24 11	22 16	26 47	22 08	11 42	18 48	03 23
03 Fr	10 43 50	12 35 46	09♉41 22	09 05	13 06	24 55	22 12	26 50	22 11	11 44	18 49	03 22
04 Sa	10 47 47	13 35 56	23 59 11	10 56	13 09	25 38	22 07	26 54	22 14	11 47	18 51	03 21
05 Su	10 51 44	14 36 04	08♊14 38	12 48	13R 08	26 22	22 02	26 57	22 17	11 49	18 52	03 21
06 Mo	10 55 40	15 36 10	22 24 58	14 41	13 06	27 05	21 57	27 00	22 20	11 51	18 53	03 20
07 Tu	10 59 37	16 36 14	06♋28 07	16 35	13 00	27 49	21 52	27 03	22 23	11 54	18 55	03D 20
08 We	11 03 33	17 36 15	20 22 29	18 30	12 53	28 32	21 47	27 06	22 26	11 56	18 56	03 20
09 Th	11 07 30	18 36 15	04♌06 53	20 26	12 42	29 16	21 42	27 08	22 29	11 58	18 57	03 21
10 Fr	11 11 26	19 36 12	17 40 15	22 23	12 29	29 59	21 36	27 11	22 32	12 00	18 58	03 21
11 Sa	11 15 23	20 36 07	01♍01 36	24 20	12 14	00♉42	21 31	27 14	22 35	12 03	18 59	03R 21
12 Su	11 19 19	21 36 00	14 10 02	26 18	11 56	01 26	21 25	27 16	22 38	12 05	19 01	03 20
13 Mo	11 23 16	22 35 51	27 04 53	28 16	11 36	02 09	21 19	27 19	22 41	12 07	19 02	03 19
14 Tu	11 27 13	23 35 40	09♎45 52	00♈14	11 13	02 52	21 13	27 21	22 44	12 09	19 03	03 16
15 We	11 31 09	24 35 27	22 13 25	02 13	10 49	03 35	21 07	27 23	22 47	12 12	19 04	03 13
16 Th	11 35 06	25 35 13	04♏28 37	04 11	10 22	04 18	21 01	27 25	22 50	12 14	19 05	03 10
17 Fr	11 39 02	26 34 57	16 33 25	06 09	09 53	05 01	20 55	27 27	22 53	12 16	19 06	03 06
18 Sa	11 42 59	27 34 38	28 30 34	08 06	09 22	05 44	20 48	27 29	22 56	12 18	19 07	03 03
19 Su	11 46 55	28 34 19	10♐23 34	10 01	08 50	06 27	20 41	27 31	23 00	12 21	19 08	03 01
20 Mo	11 50 52	29 33 57	22 16 37	11 56	08 16	07 10	20 35	27 33	23 03	12 23	19 09	03 00
21 Tu	11 54 48	00♈33 34	04♑14 16	13 48	07 41	07 53	20 28	27 35	23 06	12 25	19 10	03D 01
22 We	11 58 45	01 33 09	16 21 19	15 38	07 05	08 36	20 21	27 36	23 09	12 27	19 11	03 02
23 Th	12 02 41	02 32 42	28 42 27	17 25	06 28	09 19	20 14	27 38	23 12	12 29	19 12	03 04
24 Fr	12 06 38	03 32 13	11♒21 57	19 09	05 51	10 02	20 07	27 39	23 16	12 31	19 12	03 06
25 Sa	12 10 35	04 31 43	24 23 18	20 50	05 13	10 44	20 00	27 40	23 19	12 34	19 13	03 07
26 Su	12 14 31	05 31 11	07♓48 41	22 26	04 36	11 27	19 53	27 41	23 22	12 36	19 14	03R 07
27 Mo	12 18 28	06 30 36	21 38 24	23 58	03 58	12 10	19 45	27 43	23 26	12 38	19 15	03 06
28 Tu	12 22 24	07 30 00	05♈50 22	25 25	03 21	12 52	19 38	27 43	23 29	12 40	19 15	03 03
29 We	12 26 21	08 29 22	20 20 08	26 46	02 45	13 35	19 31	27 44	23 32	12 42	19 16	02 59
30 Th	12 30 17	09 28 42	05♉01 15	28 03	02 09	14 17	19 23	27 45	23 36	12 44	19 17	02 53
31 Fr	12 34 14	10 27 59	19 46 11	29 13	01 35	15 00	19 16	27 46	23 39	12 46	19 17	02 48

Data for	03-01-2017
Julian Day	2457813.50
Ayanamsa	24 05 40
SVP	05 ♓ 01 05
☽ ☊ Mean	03 ♍ 06 R

● ◐ PHASES ○ ◑

Day	h m	Phase	Position
05	11:33	◐	15♊05
12	14:54	○	22♍13
20	15:58	◑	00♑14
28	02:58	●	07♈37

LAST ASPECT Day	h m	☽ INGRESS Day	h m	Sign
02	02:19	02	07:43	♉
03	15:21	04	10:07	♊
06	08:23	06	12:55	♋
08	15:01	08	16:47	♌
10	17:07	10	22:08	♍
13	02:37	13	05:29	♎
15	10:06	15	15:11	♏
17	21:57	18	03:00	♐
20	10:38	20	15:31	♑
22	13:20	23	02:29	♒
25	05:57	25	10:07	♓
27	10:20	27	14:12	♈
29	12:08	29	15:49	♉
30	23:13	31	16:42	♊

DECLINATION

Day	☉	☽	☿	♀	♂	♃	♄	♅	♆	♇
01 We	07S 36	01N27	11S 26	10N59	09N08	07S 17	22S 05	08N04	07S 58	21S 12
02 Th	07 13	05 57	10 43	11 11	09 25	07 15	22 05	08 05	07 57	21 12
03 Fr	06 50	10 07	09 59	11 22	09 41	07 14	22 05	08 06	07 57	21 11
04 Sa	06 27	13 43	09 14	11 32	09 58	07 12	22 05	08 08	07 56	21 11
05 Su	06 04	16 29	08 27	11 40	10 14	07 10	22 05	08 09	07 55	21 11
06 Mo	05 40	18 14	07 40	11 48	10 31	07 08	22 05	08 10	07 54	21 11
07 Tu	05 17	18 52	06 51	11 54	10 47	07 06	22 05	08 11	07 53	21 11
08 We	04 54	18 20	06 01	11 59	11 03	07 04	22 05	08 12	07 52	21 11
09 Th	04 30	16 43	05 10	12 03	11 20	07 02	22 05	08 13	07 51	21 11
10 Fr	04 07	14 11	04 18	12 05	11 36	06 59	22 05	08 14	07 51	21 11
11 Sa	03 43	10 55	03 25	12 05	11 51	06 57	22 05	08 15	07 50	21 11
12 Su	03 20	07 09	02 31	12 05	12 07	06 55	22 05	08 17	07 49	21 10
13 Mo	02 56	03 07	01 36	12 02	12 23	06 53	22 05	08 18	07 48	21 10
14 Tu	02 33	00S 59	00 41	11 58	12 38	06 50	22 05	08 19	07 47	21 10
15 We	02 09	04 58	00N15	11 53	12 54	06 48	22 05	08 20	07 46	21 10
16 Th	01 45	08 40	01 11	11 46	13 09	06 45	22 05	08 21	07 46	21 10
17 Fr	01 22	11 58	02 07	11 38	13 24	06 43	22 05	08 22	07 45	21 10
18 Sa	00 58	14 44	03 03	11 28	13 39	06 40	22 05	08 24	07 44	21 10
19 Su	00 34	16 52	03 59	11 17	13 54	06 38	22 05	08 25	07 43	21 10
20 Mo	00 10	18 17	04 54	11 04	14 09	06 35	22 05	08 26	07 42	21 10
21 Tu	00N13	18 54	05 49	10 50	14 23	06 32	22 05	08 27	07 41	21 10
22 We	00 37	18 39	06 43	10 35	14 38	06 30	22 05	08 28	07 41	21 10
23 Th	01 01	17 31	07 36	10 19	14 52	06 27	22 05	08 30	07 40	21 10
24 Fr	01 24	15 30	08 27	10 01	15 06	06 24	22 05	08 31	07 39	21 10
25 Sa	01 48	12 38	09 16	09 43	15 20	06 22	22 05	08 32	07 38	21 10
26 Su	02 12	09 02	10 03	09 23	15 34	06 19	22 05	08 33	07 37	21 10
27 Mo	02 35	04 51	10 49	09 03	15 48	06 16	22 05	08 35	07 37	21 09
28 Tu	02 59	00 17	11 32	08 42	16 02	06 13	22 05	08 36	07 36	21 09
29 We	03 22	04N24	12 12	08 21	16 15	06 10	22 05	08 37	07 35	21 09
30 Th	03 45	08 52	12 49	08 00	16 28	06 07	22 05	08 38	07 34	21 09
31 Fr	04 09	12 49	13 24	07 38	16 41	06 05	22 05	08 39	07 33	21 09

ASPECTARIAN

01 02:55 ☽ ☌ ♀
11:50 ☿ # ♀
12:50 ☽ □ ♇
18:28 ☽ ☌ ♅
18:44 ☽ ☍ ♃
21:03 ☉ ∥ ♃
21:49 ☽ ☌ ♂
02 02:19 ☽ △ ♄
02:46 ☉ ☌ ♆
06:28 ☽ # ☉
07:14 ☽ # ♃
11:13 ☽ # ♆
12:03 ☽ ∥ ♅
21:11 ☽ ∥ ♂
22:49 ☽ ⚹ ☿
23:19 ☽ # ☿
03 01:20 ♃ ☍ ♅
03:27 ☽ ⚹ ♆
05:15 ☽ ⚹ ☉
06:59 ☿ # ♂
08:11 ☽ ∥ ♀
15:21 ☽ △ ♇
04 09:10 ♀ SR
11:11 ☿ ☌ ♆
05 06:03 ☽ □ ♆
08:15 ☽ ⚹ ♀
08:52 ☽ □ ☿
09:21 ☿ # ♅
16:47 ☿ ∥ ♆
20:47 ♂ △ ♄
23:13 ☽ △ ♃
23:51 ☽ ⚹ ♅
06 07:50 ☽ ☍ ♄
08:23 ☽ ⚹ ♂
16:27 ☿ ∥ ♃
07 00:29 ☿ ☌ ☉
09:21 ☽ △ ♆
11:09 ☽ □ ♀
18:49 ☽ △ ☉
20:14 ☽ △ ☿
21:29 ☽ ☍ ♇
08 02:26 ☽ □ ♃
03:35 ☽ □ ♅
05:25 ☿ ⚹ ♇
15:01 ☽ □ ♂
09 08:31 ☉ ⚹ ♇
14:56 ☽ △ ♀
10 00:34 ♂ ♉
06:59 ☽ ⚹ ♃
08:43 ☽ △ ♅
09:00 ☿ ∥ ☉
15:48 ☽ ∥ ♀
17:07 ☽ △ ♄
17:54 ☽ ∥ ♂
23:23 ☽ △ ♂
11 17:02 ☽ ∥ ♅
19:50 ☽ # ♆
20:09 ☽ ☍ ♆
20:21 ♀ ∥ ♂
12 01:22 ☽ # ♃
08:58 ☽ △ ♇
12:11 ☿ □ ♄
13 00:26 ☽ □ ♄
01:04 ☽ # ☉
02:37 ☽ ☍ ☿
11:15 ☽ # ☿
21:08 ☿ ♈
22:30 ☽ ∥ ☿
14 02:42 ☽ ☍ ♀
08:22 ☽ ∥ ☉
17:52 ☽ □ ♇
21:52 ☽ ☌ ♃
15 01:05 ☽ ☍ ♅
10:06 ☽ ⚹ ♄
11:23 ☽ ∥ ♃
17:50 ☽ ∥ ♆
21:46 ☽ # ♅
23:38 ☽ ☍ ♂
16 12:24 ☿ # ☉
15:25 ☽ △ ♆
21:30 ☽ # ♀
17 05:06 ☽ ⚹ ♇
12:58 ☽ # ♂
21:48 ☉ □ ♄
21:57 ☽ △ ☉
18 12:27 ☿ ☌ ♀
20:59 ☽ △ ♀
23:06 ☽ △ ☿
19 03:57 ☽ □ ♆
20:37 ☽ ⚹ ♃
20 01:33 ☽ △ ♅
10:29 ☉ ♈
10:38 ☽ ☌ ♄
21 06:33 ☽ □ ♀
07:43 ☽ △ ♂
16:17 ☽ ⚹ ♆
18:21 ☿ # ♃
22:20 ☽ □ ☿
22 05:32 ☽ ☌ ♇
07:45 ☽ □ ♃
13:20 ☽ □ ♅
23 01:55 ☿ # ♆
07:58 ☽ ⚹ ☉
14:07 ☽ ⚹ ♀
21:21 ☽ □ ♂
24 00:43 ☿ □ ♆
02:02 ☿ ∥ ♅
03:28 ☽ # ♂
12:46 ☿ ☍ ♃
16:04 ☽ △ ♃
16:35 ☽ ⚹ ☿
22:03 ☽ ⚹ ♅
25 05:57 ☽ ⚹ ♄
09:32 ☿ ∥ ♀
10:18 ♀ ☌ ☉
18:43 ☽ # ☿
21:37 ☽ # ♀
26 02:53 ☽ # ♅
06:44 ☽ ⚹ ♂
08:24 ☽ ☌ ♆
08:26 ☽ ∥ ♆
15:06 ☿ ☌ ♅
16:02 ☽ ∥ ♃
19:53 ☽ ⚹ ♇
27 10:20 ☽ □ ♄
11:03 ☽ # ☉
16:46 ♂ ⚹ ♆
20:00 ☽ ☌ ♀
28 07:08 ♀ ∥ ♅
18:07 ☽ ∥ ☉
22:15 ☽ □ ♇
22:39 ☽ ☍ ♃
29 05:16 ☽ ☌ ♅
09:09 ☽ # ♃
11:34 ☽ ☌ ☿
12:08 ☽ △ ♄
16:47 ☽ # ♆
18:17 ☿ △ ♄
19:28 ☽ ∥ ♀
22:39 ☽ ∥ ♅
30 12:35 ☽ ⚹ ♆
15:50 ☽ ☌ ♂
18:42 ♃ □ ♇
23:13 ☽ △ ♇
31 04:42 ☽ ∥ ☿
04:44 ♀ # ♆
17:31 ☿ ♉
18:35 ☽ ⚹ ♀

⚷ Chiron

01 Dec. 01 N 02

Day	Position
02	23♓55
05	24 06
08	24 16
11	24 27
14	24 38
17	24 49
20	25 00
23	25 11
26	25 21
29	25 32

April 2017

Day	S. T.	☉	☽	☿	♀	♂	♃	♄	♅	♆	♇	☊ True
	h m s	° ' "	° ' "	° '	° '	° '	° '	° '	° '	° '	° '	° '
01 Sa	12 38 10	11♈ 27 15	04♊ 27 31	00♉ 17	01♈R02	15♉ 42	19♎R08	27♐ 46	23♈ 42	12♓ 48	19♑ 18	02♍R43
02 Su	12 42 07	12 26 28	18 58 58	01 14	00 30	16 25	19 01	27 47	23 46	12 50	19 19	02 39
03 Mo	12 46 04	13 25 39	03♋ 16 08	02 05	00 01	17 07	18 53	27 47	23 49	12 52	19 19	02 37
04 Tu	12 50 00	14 24 47	17 16 37	02 50	29♓ 33	17 49	18 45	27 47	23 53	12 54	19 20	02 36
05 We	12 53 57	15 23 53	00♌ 59 53	03 27	29 07	18 32	18 38	27 48	23 56	12 56	19 20	02D 36
06 Th	12 57 53	16 22 57	14 26 39	03 58	28 43	19 14	18 30	27 48	23 59	12 58	19 21	02 37
07 Fr	13 01 50	17 21 58	27 38 21	04 21	28 21	19 56	18 22	27R 48	24 03	13 00	19 21	02R 37
08 Sa	13 05 46	18 20 57	10♍ 36 36	04 38	28 02	20 38	18 14	27 48	24 06	13 02	19 21	02 36
09 Su	13 09 43	19 19 54	23 22 49	04 48	27 45	21 20	18 07	27 47	24 10	13 04	19 22	02 34
10 Mo	13 13 39	20 18 49	05♎ 58 09	04 R 51	27 30	22 02	17 59	27 47	24 13	13 06	19 22	02 29
11 Tu	13 17 36	21 17 42	18 23 28	04 47	27 18	22 44	17 51	27 47	24 17	13 08	19 22	02 22
12 We	13 21 33	22 16 32	00♏ 39 35	04 38	27 09	23 26	17 44	27 46	24 20	13 10	19 23	02 14
13 Th	13 25 29	23 15 21	12 47 19	04 22	27 02	24 08	17 36	27 45	24 23	13 12	19 23	02 04
14 Fr	13 29 26	24 14 08	24 47 52	04 01	26 57	24 50	17 28	27 45	24 27	13 14	19 23	01 54
15 Sa	13 33 22	25 12 53	06♐ 42 55	03 35	26 55	25 32	17 21	27 44	24 30	13 15	19 23	01 45
16 Su	13 37 19	26 11 36	18 34 53	03 04	26D 55	26 13	17 13	27 43	24 34	13 17	19 23	01 38
17 Mo	13 41 15	27 10 18	00♑ 26 55	02 30	26 57	26 55	17 06	27 42	24 37	13 19	19 24	01 33
18 Tu	13 45 12	28 08 58	12 22 57	01 52	27 02	27 37	16 58	27 41	24 41	13 21	19 24	01 30
19 We	13 49 08	29 07 36	24 27 27	01 12	27 09	28 19	16 51	27 40	24 44	13 23	19 24	01D 29
20 Th	13 53 05	00♉ 06 12	06♒ 45 18	00 31	27 19	29 00	16 43	27 38	24 47	13 24	19 24	01 30
21 Fr	13 57 02	01 04 47	19 21 26	29♈ 49	27 30	29 42	16 36	27 37	24 51	13 26	19R 24	01 31
22 Sa	14 00 58	02 03 20	02♓ 20 26	29 07	27 44	00♊ 23	16 29	27 36	24 54	13 28	19 24	01R 32
23 Su	14 04 55	03 01 51	15 45 59	28 25	28 00	01 05	16 22	27 34	24 58	13 29	19 24	01 31
24 Mo	14 08 51	04 00 21	29 39 59	27 45	28 17	01 46	16 15	27 32	25 01	13 31	19 24	01 27
25 Tu	14 12 48	04 58 49	14♈ 01 39	27 07	28 37	02 28	16 08	27 31	25 04	13 32	19 23	01 22
26 We	14 16 44	05 57 16	28 46 57	26 32	28 58	03 09	16 01	27 29	25 08	13 34	19 23	01 14
27 Th	14 20 41	06 55 40	13♉ 48 34	26 00	29 21	03 50	15 54	27 27	25 11	13 36	19 23	01 04
28 Fr	14 24 37	07 54 03	28 56 54	25 32	29 46	04 32	15 47	27 25	25 15	13 37	19 23	00 55
29 Sa	14 28 34	08 52 24	14♊ 01 35	25 08	00♈ 12	05 13	15 41	27 23	25 18	13 39	19 23	00 46
30 Su	14 32 30	09 50 43	28 53 21	24 48	00 40	05 54	15 34	27 21	25 21	13 40	19 22	00 38

Data for	04-01-2017
Julian Day	2457844.50
Ayanamsa	24 05 43
SVP	05 ♓ 00 59
☽ ☊ Mean	01 ♍ 28 R

● ◐ PHASES ○ ◑

Day	h m	Phase	Position
03	18:40	◐	14♋12
11	06:08	○	21♎33
19	09:57	◑	29♑32
26	12:17	●	06♉27

LAST ASPECT ☽ INGRESS

Day	h m	Day	h m	
02	14:44	02	18:28	♋
04	20:46	04	22:14	♌
07	00:17	07	04:20	♍
09	08:22	09	12:35	♎
11	18:19	11	22:42	♏
14	04:18	14	10:27	♐
16	18:27	16	23:06	♑
19	09:58	19	10:53	♒
21	18:24	21	19:44	♓
23	21:36	24	00:34	♈
25	21:54	26	01:57	♉
28	01:20	28	01:40	♊
29	21:29	30	01:49	♋

DECLINATION

Day	☉	☽	☿	♀	♂	♃	♄	♅	♆	♇
01 Sa	04N32	15N56	13N55	07N16	16N54	06S 02	22S 05	08N41	07S 33	21S 09
02 Su	04 55	18 01	14 24	06 54	17 07	05 59	22 05	08 42	07 32	21 09
03 Mo	05 18	18 56	14 49	06 32	17 20	05 56	22 05	08 43	07 31	21 09
04 Tu	05 41	18 39	15 10	06 10	17 32	05 53	22 04	08 45	07 30	21 09
05 We	06 04	17 16	15 29	05 49	17 45	05 50	22 04	08 46	07 30	21 09
06 Th	06 26	14 57	15 43	05 28	17 57	05 47	22 04	08 47	07 29	21 09
07 Fr	06 49	11 52	15 55	05 08	18 09	05 44	22 04	08 48	07 28	21 09
08 Sa	07 12	08 15	16 02	04 48	18 21	05 41	22 04	08 50	07 27	21 09
09 Su	07 34	04 18	16 06	04 30	18 32	05 38	22 04	08 51	07 27	21 09
10 Mo	07 56	00 14	16 06	04 12	18 44	05 35	22 04	08 52	07 26	21 09
11 Tu	08 18	03S 48	16 03	03 54	18 55	05 33	22 04	08 53	07 25	21 09
12 We	08 40	07 37	15 56	03 38	19 06	05 30	22 04	08 55	07 25	21 09
13 Th	09 02	11 05	15 46	03 23	19 17	05 27	22 04	08 56	07 24	21 09
14 Fr	09 24	14 03	15 33	03 08	19 28	05 24	22 04	08 57	07 23	21 10
15 Sa	09 45	16 25	15 16	02 55	19 39	05 21	22 04	08 58	07 23	21 10
16 Su	10 07	18 05	14 56	02 43	19 49	05 18	22 04	09 00	07 22	21 10
17 Mo	10 28	18 58	14 34	02 31	19 59	05 15	22 04	09 01	07 21	21 10
18 Tu	10 49	19 00	14 10	02 21	20 09	05 13	22 04	09 02	07 21	21 10
19 We	11 10	18 11	13 43	02 12	20 19	05 10	22 03	09 03	07 20	21 10
20 Th	11 30	16 30	13 15	02 04	20 29	05 07	22 03	09 05	07 19	21 10
21 Fr	11 51	13 59	12 46	01 57	20 39	05 04	22 03	09 06	07 19	21 10
22 Sa	12 11	10 42	12 16	01 50	20 48	05 02	22 03	09 07	07 18	21 10
23 Su	12 31	06 46	11 45	01 45	20 57	04 59	22 03	09 08	07 17	21 10
24 Mo	12 51	02 20	11 15	01 41	21 06	04 57	22 03	09 10	07 17	21 10
25 Tu	13 11	02N22	10 46	01 38	21 15	04 54	22 03	09 11	07 16	21 10
26 We	13 30	07 03	10 17	01 36	21 23	04 51	22 03	09 12	07 16	21 10
27 Th	13 49	11 22	09 50	01 34	21 32	04 49	22 03	09 13	07 15	21 10
28 Fr	14 08	15 00	09 24	01 34	21 40	04 46	22 03	09 15	07 15	21 11
29 Sa	14 27	17 36	09 01	01 34	21 48	04 44	22 03	09 16	07 14	21 11
30 Su	14 46	18 59	08 39	01 36	21 56	04 42	22 03	09 17	07 14	21 11

ASPECTARIAN

01 10:33 ☽ ∥ ♂
12:21 ☽ ⚹ ☉
13:47 ☽ □ ♆

02 00:03 ☽ △ ♃
08:01 ☽ ⚹ ♅
14:44 ☽ ☍ ♄
18:40 ☽ □ ♀
21:53 ☽ ⚹ ☿

03 00:26 ♀ ♓R
16:26 ☽ △ ♆

04 01:00 ☽ ⚹ ♂
02:32 ☽ □ ♃
03:33 ☽ ☍ ♇
11:16 ☉ ⋕ ♃
11:32 ☽ □ ♅
15:57 ♀ ∥ ☉
18:22 ☽ ∥ ♂
20:46 ☽ △ ♀
22:34 ♀ ⋕ ♃

05 04:32 ☽ □ ☿
17:28 ☽ ∥ ☿

06 03:47 ☽ △ ☉
03:55 ♂ △ ♇
05:07 ♄ SR
07:15 ☽ ⚹ ♃
09:08 ☽ □ ♂
17:24 ☽ △ ♅

07 00:17 ☽ △ ♄
12:40 ☽ △ ☿
20:22 ☽ ∥ ♅
21:40 ☉ ☍ ♃

08 04:33 ☽ ☍ ♆
04:58 ☽ ⋕ ♆
06:00 ☽ ∥ ☉
15:56 ☽ ⋕ ♃
16:24 ☽ △ ♇
16:29 ☉ ⋕ ♆
19:54 ☽ △ ♂
20:28 ♀ □ ♄
22:47 ☽ ∥ ♀

09 00:47 ☉ □ ♇
08:08 ☽ ☍ ♀
08:22 ☽ □ ♄
23:15 ☿ SR

10 22:58 ☽ ☌ ♃

11 00:36 ☽ ⋕ ♀
01:55 ☽ □ ♇
10:35 ☽ ∥ ♃
11:32 ☽ ☍ ♅
18:19 ☽ ⚹ ♄
22:37 ☽ ∥ ♆

12 07:41 ☽ ☍ ☿
07:46 ☽ ⋕ ☉
08:39 ☽ ⋕ ♅
16:46 ☉ ∥ ♅

13 00:49 ☽ △ ♆
13:09 ☽ ⚹ ♇

14 00:04 ☽ ☍ ♂
04:18 ☽ △ ♀
05:30 ☉ ☌ ♅
12:50 ☽ ⋕ ☿

15 10:19 ♀ SD
13:16 ☽ □ ♆
21:16 ☽ ⚹ ♃

16 12:10 ☽ △ ♅
16:47 ☽ △ ☉
16:55 ☽ □ ♀
18:27 ☽ ☌ ♄

17 01:27 ♀ ⚹ ♂
03:56 ☽ △ ☿
12:44 ☉ △ ♄

18 01:56 ☽ ⚹ ♆
09:04 ☽ □ ♃
13:59 ☽ ☌ ♇

19 00:33 ☽ □ ♅
05:22 ☽ ⚹ ♀
08:02 ☽ △ ♂
12:32 ☽ □ ☿
21:28 ☉ ♉

20 05:55 ☿ ☌ ☉
12:47 ♇ SR
17:38 ☿ ♈R
18:52 ☽ △ ♃

21 10:17 ☽ ⚹ ♅
10:32 ♂ ♊
11:10 ♀ □ ♄
11:21 ☽ ⋕ ☿
14:45 ☽ ⋕ ☉
15:20 ☽ ⚹ ♄
18:24 ☽ ⚹ ☿
20:15 ☽ □ ♂
23:26 ☽ ⚹ ☉

22 02:11 ☿ ∥ ☉
10:03 ☽ ⋕ ♅
19:59 ☽ ☌ ♆
20:59 ☽ ∥ ♆

23 06:21 ☽ ⚹ ♇
10:01 ☽ ∥ ♃
20:23 ☽ □ ♄
21:36 ☽ ☌ ♀

24 03:25 ☽ ⋕ ♀
03:44 ☽ ⚹ ♂
08:16 ☿ △ ♄
11:25 ♂ ⋕ ♇
20:19 ☽ ∥ ♀

25 03:25 ☽ ☍ ♃
08:47 ☽ □ ♇
12:45 ☽ ⋕ ♃
18:05 ☽ ☌ ♅
20:31 ☽ ☌ ☿
21:54 ☽ △ ♄

26 01:07 ☽ ⋕ ♆
11:38 ☽ ∥ ♅
15:56 ☽ ∥ ☿
23:39 ☽ ⚹ ♆

27 08:51 ☽ △ ♇
17:07 ☽ ∥ ☉

28 01:20 ☽ ⚹ ♀
09:01 ☿ ∥ ♅
09:17 ☽ ☌ ♂
13:14 ♀ ♈
14:50 ☿ ☌ ♅
23:23 ☽ □ ♆

29 02:37 ☽ △ ♃
17:29 ☽ ⚹ ☿
18:14 ☽ ⚹ ♅
21:29 ☽ ☍ ♄

30 03:00 ☽ □ ♀
19:20 ☽ ⚹ ☉
21:45 ♂ ⋕ ♄

⚷ Chiron

01	Dec.	01 N 45
01		25♓43
04		25 53
07		26 03
10		26 14
13		26 24
16		26 33
19		26 43
22		26 52
25		27 01
28		27 10

May 2017

Day	S. T. h m s	☉ ° ' "	☽ ° ' "	☿ ° '	♀ ° '	♂ ° '	♃ ° '	♄ ° '	♅ ° '	♆ ° '	♇ ° '	☊ True ° '
01 Mo	14 36 27	10♉49 00	13♋25 19	24♈R33	01♈09	06♊35	15♎R28	27♐R18	25♈25	13♓41	19♑R22	00♍R33
02 Tu	14 40 24	11 47 14	27 33 47	24 23	01 40	07 17	15 21	27 16	25 28	13 43	19 22	00 30
03 We	14 44 20	12 45 27	11♌17 57	24 17	02 13	07 58	15 15	27 13	25 31	13 44	19 21	00 29
04 Th	14 48 17	13 43 38	24 39 16	24D16	02 46	08 39	15 09	27 11	25 35	13 46	19 21	00 28
05 Fr	14 52 13	14 41 46	07♍40 31	24 20	03 21	09 20	15 03	27 08	25 38	13 47	19 21	00 28
06 Sa	14 56 10	15 39 53	20 24 59	24 29	03 58	10 01	14 57	27 06	25 41	13 48	19 20	00 26
07 Su	15 00 06	16 37 58	02♎55 53	24 42	04 35	10 42	14 52	27 03	25 45	13 50	19 20	00 22
08 Mo	15 04 03	17 36 01	15 16 00	25 00	05 14	11 22	14 46	27 00	25 48	13 51	19 19	00 15
09 Tu	15 07 59	18 34 02	27 27 40	25 22	05 53	12 03	14 41	26 57	25 51	13 52	19 19	00 06
10 We	15 11 56	19 32 01	09♏32 39	25 49	06 34	12 44	14 35	26 54	25 54	13 53	19 18	29♌54
11 Th	15 15 53	20 29 59	21 32 21	26 19	07 16	13 25	14 30	26 51	25 57	13 54	19 18	29 40
12 Fr	15 19 49	21 27 55	03♐27 58	26 54	07 59	14 06	14 25	26 48	26 01	13 56	19 17	29 26
13 Sa	15 23 46	22 25 50	15 20 48	27 33	08 43	14 46	14 20	26 45	26 04	13 57	19 16	29 12
14 Su	15 27 42	23 23 43	27 12 27	28 15	09 27	15 27	14 16	26 41	26 07	13 58	19 16	29 01
15 Mo	15 31 39	24 21 35	09♑05 12	29 01	10 13	16 08	14 11	26 38	26 10	13 59	19 15	28 53
16 Tu	15 35 35	25 19 26	21 02 00	29 51	10 59	16 48	14 07	26 35	26 13	14 00	19 14	28 47
17 We	15 39 32	26 17 16	03♒06 31	00♉44	11 46	17 29	14 03	26 31	26 16	14 01	19 14	28 45
18 Th	15 43 28	27 15 04	15 23 03	01 40	12 34	18 09	13 59	26 28	26 19	14 02	19 13	28D45
19 Fr	15 47 25	28 12 51	27 56 16	02 40	13 23	18 50	13 55	26 24	26 22	14 03	19 12	28 45
20 Sa	15 51 22	29 10 37	10♓50 58	03 42	14 13	19 30	13 51	26 21	26 25	14 04	19 11	28R45
21 Su	15 55 18	00♊08 22	24 11 26	04 48	15 03	20 11	13 47	26 17	26 28	14 04	19 11	28 44
22 Mo	15 59 15	01 06 05	08♈00 44	05 56	15 54	20 51	13 44	26 13	26 31	14 05	19 10	28 41
23 Tu	16 03 11	02 03 48	22 19 34	07 08	16 45	21 31	13 41	26 09	26 34	14 06	19 09	28 34
24 We	16 07 08	03 01 30	07♉05 20	08 22	17 37	22 12	13 38	26 05	26 37	14 07	19 08	28 26
25 Th	16 11 04	03 59 10	22 11 46	09 38	18 30	22 52	13 35	26 02	26 40	14 08	19 07	28 16
26 Fr	16 15 01	04 56 50	07♊29 21	10 58	19 24	23 32	13 32	25 58	26 43	14 08	19 06	28 05
27 Sa	16 18 57	05 54 28	22 46 46	12 20	20 17	24 12	13 30	25 54	26 46	14 09	19 05	27 55
28 Su	16 22 54	06 52 05	07♋52 57	13 44	21 12	24 53	13 27	25 50	26 49	14 10	19 04	27 46
29 Mo	16 26 51	07 49 41	22 38 55	15 12	22 07	25 33	13 25	25 46	26 52	14 10	19 03	27 40
30 Tu	16 30 47	08 47 15	06♌59 01	16 41	23 02	26 13	13 23	25 41	26 54	14 11	19 02	27 35
31 We	16 34 44	09 44 48	20 51 08	18 13	23 58	26 53	13 21	25 37	26 57	14 11	19 01	27 34

Data for	05-01-2017
Julian Day	2457874.50
Ayanamsa	24 05 46
SVP	05 ♓ 00 54
☽ ☊ Mean	29 ♌ 52 R

● ◐ PHASES ○ ◑

Day	Time	Phase	Position
03	02:47	◐	12♌52
10	21:43	○	20♏25
19	00:33	◑	28♒14
25	19:45	●	04♊47

LAST ASPECT Day	h m	☽ INGRESS Day	h m	Sign
01	20:23	02	04:12	♌
04	04:36	04	09:47	♍
06	12:43	06	18:21	♎
08	23:00	09	05:02	♏
10	21:44	11	17:01	♐
14	02:16	14	05:39	♑
16	10:24	16	17:51	♒
19	00:34	19	03:53	♓
21	03:40	21	10:12	♈
23	07:00	23	12:33	♉
24	19:09	25	12:16	♊
27	06:19	27	11:25	♋
29	07:00	29	12:13	♌
31	11:15	31	16:17	♍

DECLINATION

Day	☉	☽	☿	♀	♂	♃	♄	♅	♆	♇
01 Mo	15N04	19N04	08N19	01N38	22N03	04S39	22S03	09N18	07S13	21S11
02 Tu	15 22	17 56	08 02	01 41	22 11	04 37	22 02	09 20	07 12	21 11
03 We	15 40	15 46	07 48	01 45	22 18	04 35	22 02	09 21	07 12	21 11
04 Th	15 57	12 48	07 36	01 49	22 25	04 33	22 02	09 22	07 12	21 11
05 Fr	16 15	09 16	07 26	01 54	22 32	04 31	22 02	09 23	07 11	21 11
06 Sa	16 32	05 23	07 19	02 00	22 38	04 28	22 02	09 24	07 11	21 11
07 Su	16 48	01 19	07 15	02 07	22 45	04 26	22 02	09 26	07 10	21 12
08 Mo	17 05	02S44	07 13	02 15	22 51	04 24	22 02	09 27	07 10	21 12
09 Tu	17 21	06 38	07 14	02 23	22 57	04 23	22 02	09 28	07 09	21 12
10 We	17 37	10 13	07 16	02 31	23 02	04 21	22 02	09 29	07 09	21 12
11 Th	17 52	13 22	07 22	02 41	23 08	04 19	22 02	09 30	07 08	21 12
12 Fr	18 08	15 57	07 29	02 50	23 13	04 17	22 02	09 31	07 08	21 12
13 Sa	18 22	17 51	07 38	03 01	23 19	04 15	22 01	09 33	07 08	21 13
14 Su	18 37	18 59	07 50	03 12	23 23	04 14	22 01	09 34	07 07	21 13
15 Mo	18 51	19 17	08 03	03 23	23 28	04 12	22 01	09 35	07 07	21 13
16 Tu	19 05	18 44	08 19	03 35	23 33	04 11	22 01	09 36	07 06	21 13
17 We	19 19	17 20	08 36	03 48	23 37	04 09	22 01	09 37	07 06	21 13
18 Th	19 32	15 06	08 54	04 01	23 41	04 08	22 01	09 38	07 06	21 13
19 Fr	19 46	12 07	09 15	04 14	23 45	04 07	22 01	09 39	07 05	21 14
20 Sa	19 58	08 28	09 37	04 28	23 49	04 06	22 01	09 40	07 05	21 14
21 Su	20 11	04 17	10 00	04 42	23 52	04 04	22 01	09 41	07 05	21 14
22 Mo	20 23	00N15	10 24	04 57	23 56	04 03	22 01	09 42	07 05	21 14
23 Tu	20 34	04 56	10 50	05 12	23 59	04 02	22 01	09 43	07 04	21 14
24 We	20 46	09 28	11 17	05 27	24 02	04 01	22 00	09 45	07 04	21 15
25 Th	20 56	13 30	11 45	05 43	24 04	04 00	22 00	09 46	07 04	21 15
26 Fr	21 07	16 41	12 14	05 59	24 07	04 00	22 00	09 47	07 04	21 15
27 Sa	21 17	18 42	12 44	06 15	24 09	03 59	22 00	09 48	07 03	21 15
28 Su	21 27	19 22	13 15	06 31	24 11	03 58	22 00	09 49	07 03	21 15
29 Mo	21 37	18 41	13 46	06 48	24 13	03 57	22 00	09 50	07 03	21 16
30 Tu	21 46	16 48	14 18	07 05	24 14	03 57	22 00	09 51	07 03	21 16
31 We	21 54	13 59	14 51	07 22	24 16	03 56	22 00	09 52	07 02	21 16

ASPECTARIAN

01 00:27 ☽ △ ♆
03:23 ☽ □ ♃
10:00 ☽ ☍ ♇
18:35 ☽ □ ☿
20:23 ☽ □ ♅

02 07:24 ☽ △ ♀
17:47 ☽ ⚹ ♂

03 00:58 ☽ ∥ ☉
06:59 ☽ ⚹ ♃
16:34 ☿ SD
23:18 ☽ △ ☿

04 00:51 ☉ ⚹ ♆
01:41 ☽ △ ♅
04:36 ☽ △ ♄
23:18 ☽ ∥ ♅

05 03:16 ☽ □ ♂
11:28 ☽ ☍ ♆
11:55 ☽ ∥ ☿
13:07 ☽ # ♆
14:15 ☽ △ ☉
21:57 ☽ △ ♇

06 05:28 ☽ # ♃
12:43 ☽ □ ♄
19:28 ☽ ∥ ♀

07 03:22 ☽ ☍ ♀
15:57 ☽ △ ♂
20:59 ☽ # ♀
23:02 ☽ ☌ ♃

08 07:57 ☽ □ ♇
10:05 ☽ ∥ ♃
19:43 ☽ ☍ ☿
20:48 ☽ ☍ ♅
23:00 ☽ ⚹ ♄

09 03:23 ☽ ∥ ♆
03:53 ☽ # ☿
18:22 ☉ △ ♇
18:52 ☽ # ♅

10 05:20 ☿ ☌ ♅
08:41 ☽ △ ♆
19:30 ☽ ⚹ ♇

11 17:53 ♂ □ ♆
20:15 ☿ △ ♄

12 09:42 ☽ △ ♀
10:20 ♂ △ ♃
21:10 ☽ □ ♆
21:59 ☽ ⚹ ♃
22:46 ☽ ☍ ♂

13 11:23 ☽ # ☉
21:47 ☽ △ ♅
22:58 ☽ ☌ ♄

14 02:16 ☽ △ ☿

15 02:26 ☽ □ ♀
09:52 ☽ ⚹ ♆
10:12 ☽ □ ♃
16:12 ☽ # ☉
20:25 ☽ ☌ ♇

16 04:07 ☿ ♉
09:18 ☽ △ ☉
10:24 ☽ □ ♅
18:56 ☽ □ ☿

17 18:10 ☽ ⚹ ♀
21:17 ☽ △ ♃

18 05:39 ☽ △ ♂
11:57 ♀ # ♃
21:02 ☽ ⚹ ♅
21:07 ☽ ⚹ ♄

19 06:20 ♄ △ ♅
09:39 ☽ ⚹ ☿
14:12 ♀ ☍ ♃
16:36 ☽ # ♅
17:35 ☽ # ☿

20 04:13 ☿ ∥ ♅
05:51 ☽ ☌ ♆
08:19 ☽ ∥ ♆
15:05 ☽ ⚹ ♇
16:29 ☽ □ ♂
20:31 ☉ ♊
21:54 ☽ # ♀

21 01:15 ☽ ∥ ♃
03:40 ☽ □ ♄
11:13 ☽ ⚹ ☉

22 09:39 ☽ ☍ ♃
14:10 ☽ ☌ ♀
18:45 ☽ □ ♇
19:27 ☽ # ♃
22:37 ☽ ⚹ ♂

23 01:28 ☽ ∥ ♀
06:16 ☽ △ ♄
07:00 ☽ ☌ ♅
11:09 ☽ # ♆

24 01:35 ☽ ∥ ♅
02:14 ☽ ☌ ☿
11:14 ☽ ⚹ ♆
11:42 ☽ ∥ ☿
19:09 ☽ △ ♇

25 16:20 ♀ □ ♆

26 09:27 ☽ △ ♃
10:26 ☽ □ ♆
18:54 ☉ # ♇
19:50 ☽ ⚹ ♀

27 02:21 ☽ ☌ ♂
04:54 ☽ ☍ ♄
06:19 ☽ ⚹ ♅

28 07:04 ☿ ⚹ ♆
08:58 ☽ □ ♃
10:08 ☽ △ ♆
10:28 ☽ ⚹ ☿
18:06 ☽ ☍ ♇
23:03 ☽ □ ♀

29 06:56 ♂ ☍ ♄
07:00 ☽ □ ♅
21:09 ♀ # ♆

30 03:18 ☽ ⚹ ☉
10:57 ☽ ⚹ ♃
18:18 ☽ ∥ ☿
18:48 ☽ □ ☿

31 02:40 ♂ ⚹ ♅
05:54 ☽ △ ♀
08:24 ☽ △ ♄
10:51 ☽ △ ♅
11:15 ☽ ⚹ ♂
12:02 ☿ △ ♇
14:55 ☉ # ♄

⚷ Chiron

01 Dec.	02 N 24
01	27♓19
04	27 27
07	27 35
10	27 43
13	27 50
16	27 57
19	28 04
22	28 10
25	28 16
28	28 21
31	28 26

June 2017

Day	S. T.	☉	☽	☿	♀	♂	♃	♄	♅	♆	♇	☊ True
	h m s	° ' "	° ' "	° '	° '	° '	° '	° '	° '	° '	° '	° '
01 Th	16 38 40	10♊ 42 19	04♍ 16 10	19♉ 48	24♈ 54	27♊ 33	13♎ R20	25♐ R33	27♈ 00	14♓ 12	19♑ R00	27♌ R33
02 Fr	16 42 37	11 39 49	17 17 00	21 25	25 51	28 13	13 18	25 29	27 02	14 12	18 59	27 33
03 Sa	16 46 33	12 37 18	29 57 29	23 05	26 48	28 53	13 17	25 25	27 05	14 13	18 58	27 32
04 Su	16 50 30	13 34 46	12♎ 21 47	24 47	27 45	29 33	13 16	25 20	27 08	14 13	18 57	27 29
05 Mo	16 54 26	14 32 12	24 33 53	26 32	28 43	00♋ 13	13 15	25 16	27 10	14 14	18 55	27 23
06 Tu	16 58 23	15 29 37	06♏ 37 19	28 19	29 42	00 53	13 14	25 12	27 13	14 14	18 54	27 15
07 We	17 02 20	16 27 01	18 35 00	00♊ 08	00♉ 41	01 33	13 14	25 07	27 15	14 14	18 53	27 04
08 Th	17 06 16	17 24 24	00♐ 29 17	02 00	01 40	02 12	13 13	25 03	27 18	14 15	18 52	26 51
09 Fr	17 10 13	18 21 47	12 21 57	03 54	02 39	02 52	13 13	24 59	27 20	14 15	18 51	26 38
10 Sa	17 14 09	19 19 08	24 14 31	05 50	03 39	03 32	13D 13	24 54	27 23	14 15	18 50	26 26
11 Su	17 18 06	20 16 29	06♑ 08 24	07 49	04 39	04 12	13 13	24 50	27 25	14 15	18 48	26 16
12 Mo	17 22 02	21 13 49	18 05 17	09 50	05 39	04 51	13 14	24 46	27 28	14 16	18 47	26 08
13 Tu	17 25 59	22 11 08	00♒ 07 17	11 53	06 40	05 31	13 14	24 41	27 30	14 16	18 46	26 03
14 We	17 29 56	23 08 27	12 17 04	13 57	07 41	06 11	13 15	24 37	27 32	14 16	18 44	26 01
15 Th	17 33 52	24 05 45	24 37 50	16 04	08 43	06 50	13 16	24 32	27 34	14 16	18 43	26D 01
16 Fr	17 37 49	25 03 03	07♓ 13 18	18 12	09 44	07 30	13 17	24 28	27 37	14 16	18 42	26 02
17 Sa	17 41 45	26 00 21	20 07 27	20 21	10 46	08 09	13 18	24 23	27 39	14R 16	18 41	26 03
18 Su	17 45 42	26 57 38	03♈ 24 08	22 31	11 48	08 49	13 19	24 19	27 41	14 16	18 39	26R 04
19 Mo	17 49 38	27 54 55	17 06 25	24 42	12 51	09 28	13 21	24 15	27 43	14 16	18 38	26 02
20 Tu	17 53 35	28 52 12	01♉ 15 43	26 54	13 54	10 08	13 23	24 10	27 45	14 16	18 37	25 58
21 We	17 57 31	29 49 28	15 50 48	29 05	14 56	10 47	13 25	24 06	27 47	14 16	18 35	25 53
22 Th	18 01 28	00♋ 46 45	00♊ 47 14	01♋ 17	16 00	11 27	13 27	24 01	27 49	14 15	18 34	25 45
23 Fr	18 05 25	01 44 01	15 57 23	03 28	17 03	12 06	13 29	23 57	27 51	14 15	18 32	25 37
24 Sa	18 09 21	02 41 17	01♋ 11 17	05 39	18 07	12 45	13 31	23 53	27 53	14 15	18 31	25 29
25 Su	18 13 18	03 38 33	16 18 16	07 49	19 11	13 25	13 34	23 48	27 55	14 15	18 30	25 23
26 Mo	18 17 14	04 35 48	01♌ 08 47	09 58	20 15	14 04	13 37	23 44	27 56	14 14	18 28	25 18
27 Tu	18 21 11	05 33 03	15 35 48	12 06	21 19	14 43	13 40	23 40	27 58	14 14	18 27	25 15
28 We	18 25 07	06 30 17	29 35 34	14 12	22 23	15 23	13 43	23 35	28 00	14 14	18 25	25 14
29 Th	18 29 04	07 27 31	13♍ 07 30	16 16	23 28	16 02	13 46	23 31	28 02	14 13	18 24	25D 14
30 Fr	18 33 00	08 24 44	26 13 22	18 19	24 33	16 41	13 50	23 27	28 03	14 13	18 22	25 15

Data for	06-01-2017
Julian Day	2457905.50
Ayanamsa	24 05 50
SVP	05 ♓ 00 50
☽ ☊ Mean	28 ♌ 14 R

● ◐ PHASES ○ ◑

01	12:43	◐	11♍13
09	13:10	○	18♐53
17	11:33	◑	26♓28
24	02:31	●	02♋47

LAST ASPECT Day	h m	☽ INGRESS Day	h m	
02	21:50	03	00:05	♎
05	08:58	05	10:47	♏
07	00:36	07	23:01	♐
10	06:21	10	11:37	♑
12	18:46	12	23:46	♒
15	05:40	15	10:18	♓
17	11:33	17	17:55	♈
19	19:43	19	21:53	♉
21	04:26	21	22:45	♊
23	18:46	23	22:07	♋
25	18:45	25	22:08	♌
27	21:13	28	00:43	♍
29	20:36	30	07:04	♎

DECLINATION

Day	☉	☽	☿	♀	♂	♃	♄	♅	♆	♇
01 Th	22N03	10N29	15N24	07N39	24N17	03S 56	22S 00	09N53	07S 02	21S 16
02 Fr	22 11	06 35	15 58	07 57	24 18	03 56	22 00	09 53	07 02	21 17
03 Sa	22 18	02 30	16 32	08 14	24 19	03 56	21 59	09 54	07 02	21 17
04 Su	22 26	01S 37	17 05	08 32	24 19	03 55	21 59	09 55	07 02	21 17
05 Mo	22 32	05 35	17 39	08 50	24 20	03 55	21 59	09 56	07 02	21 17
06 Tu	22 39	09 17	18 13	09 08	24 20	03 55	21 59	09 57	07 02	21 17
07 We	22 45	12 35	18 46	09 26	24 20	03 55	21 59	09 58	07 02	21 18
08 Th	22 50	15 21	19 19	09 45	24 19	03 55	21 59	09 59	07 02	21 18
09 Fr	22 56	17 29	19 52	10 03	24 19	03 55	21 59	10 00	07 01	21 18
10 Sa	23 00	18 52	20 23	10 21	24 18	03 56	21 59	10 01	07 01	21 18
11 Su	23 05	19 26	20 54	10 40	24 18	03 56	21 59	10 01	07 01	21 19
12 Mo	23 09	19 07	21 24	10 58	24 16	03 56	21 58	10 02	07 01	21 19
13 Tu	23 12	17 57	21 52	11 17	24 15	03 57	21 58	10 03	07 01	21 19
14 We	23 15	15 57	22 19	11 35	24 14	03 57	21 58	10 04	07 01	21 19
15 Th	23 18	13 12	22 44	11 54	24 12	03 58	21 58	10 05	07 01	21 20
16 Fr	23 21	09 47	23 07	12 12	24 10	03 59	21 58	10 05	07 01	21 20
17 Sa	23 22	05 50	23 27	12 30	24 08	03 59	21 58	10 06	07 01	21 20
18 Su	23 24	01 30	23 46	12 49	24 06	04 00	21 58	10 07	07 01	21 21
19 Mo	23 25	03N03	24 02	13 07	24 03	04 01	21 58	10 08	07 02	21 21
20 Tu	23 26	07 34	24 16	13 25	24 01	04 02	21 58	10 08	07 02	21 21
21 We	23 26	11 47	24 26	13 43	23 58	04 03	21 58	10 09	07 02	21 21
22 Th	23 26	15 21	24 35	14 01	23 55	04 04	21 57	10 10	07 02	21 22
23 Fr	23 25	17 57	24 40	14 19	23 52	04 05	21 57	10 11	07 02	21 22
24 Sa	23 24	19 17	24 42	14 36	23 48	04 06	21 57	10 11	07 02	21 22
25 Su	23 23	19 15	24 42	14 54	23 45	04 08	21 57	10 12	07 02	21 23
26 Mo	23 21	17 52	24 39	15 11	23 41	04 09	21 57	10 12	07 02	21 23
27 Tu	23 19	15 21	24 33	15 28	23 37	04 10	21 57	10 13	07 02	21 23
28 We	23 17	11 59	24 24	15 45	23 33	04 12	21 57	10 14	07 03	21 23
29 Th	23 13	08 06	24 13	16 02	23 28	04 13	21 57	10 14	07 03	21 24
30 Fr	23 10	03 57	24 00	16 18	23 24	04 15	21 57	10 15	07 03	21 24

ASPECTARIAN

01 03:55 ☽ ∥ ♅
15:24 ♀ △ ♄
16:27 ☽ ∥ ♀
18:16 ☽ ☍ ♆
21:20 ☽ # ♆

02 03:11 ☽ △ ♇
08:56 ☽ △ ☿
15:22 ☽ □ ♄
15:42 ☽ # ♃
21:50 ☽ □ ♂

03 07:32 ♀ ☌ ♅
16:13 ☉ △ ♃

04 01:45 ☽ ☌ ♃
02:35 ☽ △ ☉
12:53 ☽ □ ♇
13:50 ☽ ∥ ♃
16:14 ☉ □ ♆
16:17 ♂ ♋

05 01:23 ☽ ✶ ♄
05:11 ☽ ☍ ♅
08:58 ☽ ☍ ♀
09:10 ☽ ∥ ♆
11:52 ☽ △ ♂
22:57 ☽ # ♀

06 04:42 ☽ # ♅
07:27 ♀ ♉
15:16 ☽ △ ♆
22:16 ☿ ♊

07 00:36 ☽ ✶ ♇

08 03:37 ☽ ☍ ☿
19:42 ♀ ∥ ♅

09 01:43 ☽ ✶ ♃
03:48 ☽ □ ♆
14:04 ♃ SD
15:41 ♀ ✶ ♂

10 01:20 ☽ ☌ ♄
06:21 ☽ △ ♅
19:51 ☽ ☍ ♂
20:43 ☽ △ ♀

11 14:15 ☽ □ ♃
16:19 ☽ ✶ ♆
20:05 ☿ # ♇

12 01:23 ☽ ☌ ♇
18:46 ☽ □ ♅

13 05:39 ☿ # ♄
14:08 ☽ □ ♀
15:45 ☿ △ ♃

14 01:53 ☽ △ ♃
03:30 ☿ □ ♆
03:57 ☽ △ ☿
22:53 ☽ △ ☉
23:49 ☽ ✶ ♄

15 05:40 ☽ ✶ ♅
08:52 ☽ # ♀
10:19 ☉ ☍ ♄
21:59 ☽ # ♅

16 00:33 ☽ △ ♂
05:09 ☽ ✶ ♀
11:11 ♆ SR

13:11 ☽ ☌ ♆
17:03 ☽ ∥ ♆
17:30 ☿ ∥ ☉
21:21 ☽ ✶ ♇

17 00:30 ☽ □ ☿
07:45 ☽ □ ♄
10:23 ☽ ∥ ♃

18 10:03 ☽ □ ♂
17:29 ☽ ☍ ♃
18:47 ☉ ✶ ♅
19:08 ☿ ☍ ♄

19 01:54 ☿ ∥ ♂
02:37 ☽ □ ♇
05:07 ☽ # ♃
12:08 ☽ △ ♄
15:20 ☽ ✶ ☿
18:06 ☽ ☌ ♅
19:43 ☽ ✶ ☉
21:06 ☽ # ♆

20 08:26 ♀ ✶ ♆
09:31 ☿ ✶ ♅
14:25 ☽ ∥ ♅
15:22 ☽ ✶ ♂
21:25 ☽ ✶ ♆
22:25 ☽ ☌ ♀

21 04:25 ☉ ♋
04:26 ☽ △ ♇
09:58 ☿ ♋
13:21 ☽ ∥ ☊
14:15 ☿ ☌ ☉

22 20:05 ☽ △ ♃
21:19 ☽ □ ♆

23 12:32 ☽ ☍ ♄
18:46 ☽ ✶ ♅

24 08:14 ☽ ☌ ☿
08:56 ♀ △ ♇
19:10 ☽ ☌ ♂
19:37 ☽ □ ♃
20:43 ☽ △ ♆

25 03:30 ☽ ☍ ♇
04:57 ☽ ✶ ♀
06:08 ♂ □ ♃

18:45 ☽ □ ♅

26 06:19 ♂ △ ♆
20:44 ☽ ✶ ♃
23:01 ☽ ∥ ♀

27 10:31 ☽ □ ♀
13:40 ☽ △ ♄
18:21 ☿ □ ♃
21:13 ☽ △ ♅

28 00:23 ☿ △ ♆
11:05 ☽ ∥ ♅
13:05 ☽ ✶ ☉
19:51 ☿ ☌ ♂

29 01:59 ☽ ☍ ♆
05:32 ☽ ✶ ♂
06:07 ☽ # ♆
06:45 ☽ ✶ ☿
09:33 ☽ △ ♇
18:53 ☽ □ ♄
20:36 ☽ △ ♀
22:14 ☽ # ♃

30 00:36 ☿ ☍ ♇

⚷ Chiron

01 Dec.	02 N 55
03	28♓31
06	28 35
09	28 39
12	28 42
15	28 45
18	28 47
21	28 49
24	28 50
27	28 51
30	28 52

July 2017

Day	S. T. h m s	☉ ° ' "	☽ ° ' "	☿ ° '	♀ ° '	♂ ° '	♃ ° '	♄ ° '	♅ ° '	♆ ° '	♇ ° '	☊ True ° '
01 Sa	18 36 57	09♋21 57	08♎56 32	20♋21	25♉38	17♋20	13♎53	23♐R23	28♈05	14♓R12	18♑R21	25♌16
02 Su	18 40 54	10 19 09	21 21 10	22 20	26 43	17 59	13 57	23 19	28 06	14 12	18 19	25R 15
03 Mo	18 44 50	11 16 21	03♏31 40	24 18	27 48	18 38	14 01	23 15	28 08	14 12	18 18	25 13
04 Tu	18 48 47	12 13 33	15 32 19	26 13	28 54	19 18	14 05	23 10	28 09	14 11	18 17	25 09
05 We	18 52 43	13 10 44	27 27 03	28 07	29 59	19 57	14 10	23 06	28 11	14 10	18 15	25 03
06 Th	18 56 40	14 07 56	09♐19 14	29 58	01♊05	20 36	14 14	23 02	28 12	14 10	18 14	24 56
07 Fr	19 00 36	15 05 07	21 11 37	01♌48	02 11	21 15	14 19	22 59	28 14	14 09	18 12	24 49
08 Sa	19 04 33	16 02 18	03♑06 25	03 36	03 17	21 54	14 23	22 55	28 15	14 09	18 11	24 41
09 Su	19 08 29	16 59 29	15 05 25	05 21	04 24	22 33	14 28	22 51	28 16	14 08	18 09	24 35
10 Mo	19 12 26	17 56 41	27 10 16	07 05	05 30	23 12	14 33	22 47	28 17	14 07	18 08	24 31
11 Tu	19 16 23	18 53 52	09♒22 32	08 46	06 37	23 51	14 38	22 43	28 18	14 06	18 06	24 29
12 We	19 20 19	19 51 04	21 43 57	10 26	07 44	24 29	14 44	22 40	28 19	14 06	18 05	24D 29
13 Th	19 24 16	20 48 16	04♓16 30	12 03	08 51	25 08	14 49	22 36	28 21	14 05	18 03	24 30
14 Fr	19 28 12	21 45 29	17 02 22	13 39	09 58	25 47	14 55	22 32	28 22	14 04	18 02	24 32
15 Sa	19 32 09	22 42 42	00♈04 00	15 12	11 05	26 26	15 01	22 29	28 23	14 03	18 00	24 34
16 Su	19 36 05	23 39 56	13 23 46	16 44	12 12	27 05	15 07	22 25	28 23	14 02	17 59	24 35
17 Mo	19 40 02	24 37 10	27 03 37	18 13	13 20	27 44	15 13	22 22	28 24	14 01	17 57	24R 36
18 Tu	19 43 58	25 34 25	11♉04 30	19 41	14 27	28 23	15 19	22 19	28 25	14 00	17 56	24 35
19 We	19 47 55	26 31 41	25 25 40	21 06	15 35	29 01	15 25	22 15	28 26	13 59	17 55	24 32
20 Th	19 51 52	27 28 57	10♊04 14	22 29	16 43	29 40	15 32	22 12	28 27	13 58	17 53	24 29
21 Fr	19 55 48	28 26 15	24 55 01	23 50	17 51	00♌19	15 38	22 09	28 27	13 57	17 52	24 25
22 Sa	19 59 45	29 23 33	09♋50 49	25 09	18 59	00 58	15 45	22 06	28 28	13 56	17 50	24 22
23 Su	20 03 41	00♌20 52	24 43 22	26 25	20 07	01 36	15 52	22 03	28 28	13 55	17 49	24 18
24 Mo	20 07 38	01 18 11	09♌24 31	27 39	21 16	02 15	15 59	22 00	28 29	13 54	17 47	24 16
25 Tu	20 11 34	02 15 31	23 47 26	28 51	22 24	02 54	16 06	21 57	28 29	13 53	17 46	24 15
26 We	20 15 31	03 12 51	07♍47 28	00♍01	23 33	03 32	16 14	21 54	28 30	13 52	17 44	24D 14
27 Th	20 19 27	04 10 12	21 22 29	01 08	24 41	04 11	16 21	21 52	28 30	13 51	17 43	24 15
28 Fr	20 23 24	05 07 33	04♎32 41	02 12	25 50	04 50	16 29	21 49	28 31	13 50	17 42	24 16
29 Sa	20 27 21	06 04 54	17 20 03	03 14	26 59	05 28	16 36	21 46	28 31	13 48	17 40	24 17
30 Su	20 31 17	07 02 17	29 47 52	04 13	28 08	06 07	16 44	21 44	28 31	13 47	17 39	24 18
31 Mo	20 35 14	07 59 39	12♏00 10	05 09	29 17	06 45	16 52	21 42	28 31	13 46	17 38	24R 18

Data for	07-01-2017
Julian Day	2457935.50
Ayanamsa	24 05 55
SVP	05♓00 46
☽ ☊ Mean	26♌39 R

● ◐ PHASES ○ ◑

Day	Time	Phase	Position
01	00:52	◐	09♎24
09	04:06	○	17♑09
16	19:26	◑	24♈26
23	09:46	●	00♌44
30	15:23	◐	07♏39

LAST ASPECT Day	h m	☽ INGRESS Day	h m	Sign
02	13:18	02	17:00	♏
05	01:35	05	05:09	♐
07	14:12	07	17:45	♑
10	02:13	10	05:35	♒
12	12:41	12	15:52	♓
14	17:01	14	23:53	♈
17	02:20	17	05:05	♉
19	06:12	19	07:32	♊
21	05:42	21	08:11	♋
23	06:06	23	08:35	♌
25	09:23	25	10:33	♍
27	06:32	27	15:38	♎
29	21:31	30	00:24	♏

DECLINATION

Day	☉	☽	☿	♀	♂	♃	♄	♅	♆	♇
01 Sa	23N06	00S16	23N44	16N34	23N19	04S16	21S57	10N15	07S03	21S24
02 Su	23 02	04 21	23 27	16 50	23 14	04 18	21 56	10 16	07 03	21 25
03 Mo	22 57	08 11	23 07	17 06	23 09	04 20	21 56	10 16	07 04	21 25
04 Tu	22 52	11 38	22 45	17 21	23 04	04 22	21 56	10 17	07 04	21 25
05 We	22 47	14 34	22 22	17 36	22 58	04 24	21 56	10 17	07 04	21 25
06 Th	22 41	16 55	21 57	17 51	22 53	04 26	21 56	10 18	07 04	21 26
07 Fr	22 35	18 32	21 31	18 05	22 47	04 28	21 56	10 18	07 05	21 26
08 Sa	22 28	19 21	21 03	18 20	22 41	04 30	21 56	10 19	07 05	21 26
09 Su	22 21	19 19	20 34	18 33	22 35	04 32	21 56	10 19	07 05	21 27
10 Mo	22 14	18 23	20 05	18 47	22 29	04 34	21 56	10 20	07 06	21 27
11 Tu	22 06	16 36	19 34	19 00	22 22	04 36	21 56	10 20	07 06	21 27
12 We	21 58	14 01	19 02	19 13	22 15	04 38	21 56	10 20	07 06	21 27
13 Th	21 50	10 46	18 30	19 25	22 09	04 41	21 56	10 21	07 07	21 28
14 Fr	21 41	06 58	17 56	19 37	22 02	04 43	21 56	10 21	07 07	21 28
15 Sa	21 31	02 46	17 23	19 48	21 55	04 46	21 56	10 21	07 07	21 28
16 Su	21 22	01N39	16 49	19 59	21 47	04 48	21 56	10 22	07 08	21 29
17 Mo	21 12	06 05	16 14	20 10	21 40	04 51	21 55	10 22	07 08	21 29
18 Tu	21 01	10 18	15 39	20 20	21 32	04 53	21 55	10 22	07 08	21 29
19 We	20 51	14 02	15 04	20 30	21 24	04 56	21 55	10 22	07 09	21 29
20 Th	20 40	16 58	14 29	20 39	21 16	04 59	21 55	10 23	07 09	21 30
21 Fr	20 28	18 50	13 54	20 48	21 08	05 01	21 55	10 23	07 10	21 30
22 Sa	20 16	19 24	13 19	20 57	21 00	05 04	21 55	10 23	07 10	21 30
23 Su	20 04	18 38	12 44	21 04	20 52	05 07	21 55	10 23	07 11	21 31
24 Mo	19 52	16 36	12 09	21 12	20 43	05 10	21 55	10 23	07 11	21 31
25 Tu	19 39	13 33	11 35	21 19	20 34	05 13	21 55	10 24	07 11	21 31
26 We	19 26	09 48	11 00	21 25	20 26	05 16	21 55	10 24	07 12	21 32
27 Th	19 13	05 38	10 27	21 31	20 17	05 19	21 55	10 24	07 12	21 32
28 Fr	18 59	01 20	09 53	21 36	20 08	05 22	21 55	10 24	07 13	21 32
29 Sa	18 45	02S55	09 21	21 41	19 58	05 25	21 55	10 24	07 13	21 32
30 Su	18 31	06 55	08 49	21 45	19 49	05 28	21 55	10 24	07 14	21 33
31 Mo	18 16	10 32	08 17	21 48	19 39	05 32	21 55	10 24	07 14	21 33

ASPECTARIAN

Day	Time	Aspect
01	09:33	☽ ☌ ♃
	17:05	☽ □ ♂
	18:06	☽ □ ♇
	23:40	☽ ∥ ♃
02	02:17	☽ □ ☿
	03:49	☽ ⚹ ♄
	11:59	♂ ☍ ♇
	13:18	☽ ☍ ♅
	16:44	☽ ∥ ♆
	20:22	☿ ∥ ♂
03	13:52	☿ ∥ ☉
	14:13	☽ # ♅
	16:47	☽ △ ☉
	21:17	☽ △ ♆
04	05:29	☽ ⚹ ♇
	07:59	☽ △ ♂
05	00:12	♀ ♊
	00:52	☿ □ ♅
	01:35	☽ △ ☿
	05:39	☽ ☍ ♀
06	00:20	☿ ♌
	00:47	☉ △ ♆
	00:53	☿ # ♄
	02:45	☉ □ ♃
	09:47	☽ □ ♆
	10:00	☽ ⚹ ♃
	14:47	☽ # ♀
07	03:35	☽ ☌ ♄
	04:13	☿ # ♇
	13:19	☿ ⚹ ♀
	14:12	☽ △ ♅
08	22:05	☽ ⚹ ♆
	22:45	☽ □ ♃
09	06:06	☽ ☌ ♇
	15:41	☽ ☍ ♂
	17:26	☽ # ♀
10	02:13	☽ □ ♅
	04:33	☉ ☍ ♇
	18:03	☽ △ ♀
	22:38	☽ ☍ ☿
11	10:20	☽ △ ♃
	18:16	☿ ∥ ♀
12	01:47	☽ ⚹ ♄
	06:40	☉ # ♄
	12:41	☽ ⚹ ♅
13	02:50	☽ # ♅
	09:28	☽ □ ♀
	18:27	☽ ☌ ♆
	23:06	☽ ∥ ♆
14	01:50	☽ ⚹ ♇
	09:27	☽ △ ☉
	10:09	☽ □ ♄
	12:55	☽ ∥ ♃
	17:01	☽ △ ♂
	20:33	♂ # ♄
	20:46	☿ ⚹ ♃
15	07:22	☉ # ♇
	21:41	☽ ⚹ ♀
16	03:04	☽ ☍ ♃
	06:39	☽ △ ☿
	08:06	☽ □ ♇
	15:51	☽ △ ♄
	17:07	☽ # ♃
17	01:13	☽ □ ♂
	02:20	☽ ☌ ♅
	05:45	☽ # ♆
	14:34	♀ □ ♆
18	00:20	☽ ∥ ♅
	01:37	♂ □ ♅
	04:56	☽ ⚹ ♆
	08:43	♂ # ♇
	11:31	☽ △ ♇
	16:02	☽ □ ☿
	20:09	♀ △ ♃
19	01:57	☽ ⚹ ☉
	06:12	☽ ⚹ ♂
	06:26	☽ ∥ ☿
	19:17	☿ △ ♄
20	00:17	♀ ∥ ☉
	06:20	☽ □ ♆
	08:55	☽ △ ♃
	11:40	☽ ☌ ♀
	12:20	♂ ♌
	19:33	☽ ☍ ♄
	22:05	☽ ⚹ ☿
21	00:26	☉ □ ♅
	05:42	☽ ⚹ ♅
22	05:14	♀ ∥ ♂
	06:35	☽ △ ♆
	09:35	☽ □ ♃
	12:51	☽ ☍ ♇
	15:16	☉ ♌
23	06:06	☽ □ ♅
	11:43	☽ ☌ ♂
24	11:00	☽ ⚹ ♃
	14:54	♀ ☍ ♄
	16:33	☿ △ ♅
	20:54	☽ △ ♄
	21:27	☽ ⚹ ♀
25	07:59	☽ △ ♅
	09:23	☽ ☌ ☿
	15:18	☽ ∥ ☿
	20:20	☽ ∥ ♅
	23:42	☿ ♍
26	10:38	☽ ☍ ♆
	15:04	☽ # ♆
	17:29	☽ △ ♇
27	00:52	☽ □ ♄
	00:58	☉ ☌ ♂
	01:42	☽ # ♃
	02:01	☿ ∥ ♅
	05:16	♀ # ♇
	06:32	☽ □ ♀
28	00:33	☽ ⚹ ♂
	01:10	☽ ⚹ ☉
	22:36	☽ ☌ ♃
29	00:38	☽ □ ♇
	08:27	☽ ⚹ ♄
	15:02	☽ ∥ ♃
	20:26	☽ △ ♀
	21:31	☽ ☍ ♅
30	02:02	☽ ∥ ♆
	08:04	♀ ⚹ ♅
	09:21	☽ ⚹ ☿
	10:42	☽ # ☿
	13:03	☽ □ ♂
	23:05	☽ # ♅
31	03:30	☽ △ ♆
	11:10	☽ ⚹ ♇
	14:54	♀ ♋

⚷ Chiron

01	Dec.	03 N 08
03		28♓52R
06		28 51
09		28 50
12		28 49
15		28 47
18		28 45
21		28 42
24		28 39
27		28 35
30		28 31
01	05:06	28♓52 R

August 2017

07	18:22	15♒25	✹	Total Lunar Eclipse (mag 0.251)
21	18:14	28♌53	✪	Total Solar Eclipse

Day	S. T.	☉	☽	☿	♀	♂	♃	♄	♅	♆	♇	☊ True
	h m s	° ' "	° ' "	° '	° '	° '	° '	° '	° '	° '	° '	° '
01 Tu	20 39 10	08♌ 57 03	24♏ 01 20	06♍ 02	00♋ 26	07♌ 24	17♎ 00	21♐ R39	28♈ 31	13♓ R45	17♑ R36	24♌ R18
02 We	20 43 07	09 54 26	05♐ 55 52	06 51	01 36	08 02	17 08	21 37	28 32	13 43	17 35	24 16
03 Th	20 47 03	10 51 51	17 48 00	07 38	02 45	08 41	17 16	21 35	28 32	13 42	17 34	24 15
04 Fr	20 50 60	11 49 16	29 41 34	08 21	03 54	09 19	17 25	21 33	28R 32	13 41	17 32	24 14
05 Sa	20 54 56	12 46 42	11♑ 39 51	09 00	05 04	09 58	17 33	21 31	28 31	13 39	17 31	24 12
06 Su	20 58 53	13 44 09	23 45 30	09 35	06 14	10 36	17 42	21 29	28 31	13 38	17 30	24 12
07 Mo	21 02 50	14 41 37	06♒ 00 38	10 07	07 23	11 15	17 51	21 27	28 31	13 37	17 28	24 11
08 Tu	21 06 46	15 39 06	18 26 54	10 34	08 33	11 53	17 59	21 26	28 31	13 35	17 27	24D 11
09 We	21 10 43	16 36 36	01♓ 05 25	10 56	09 43	12 32	18 08	21 24	28 31	13 34	17 26	24 12
10 Th	21 14 39	17 34 06	13 56 59	11 14	10 53	13 10	18 17	21 22	28 30	13 32	17 25	24 13
11 Fr	21 18 36	18 31 38	27 02 04	11 27	12 03	13 48	18 26	21 21	28 30	13 31	17 23	24 14
12 Sa	21 22 32	19 29 12	10♈ 20 58	11 36	13 14	14 27	18 36	21 20	28 30	13 30	17 22	24 14
13 Su	21 26 29	20 26 47	23 53 48	11 38	14 24	15 05	18 45	21 18	28 29	13 28	17 21	24 15
14 Mo	21 30 25	21 24 23	07♉ 40 25	11R 36	15 35	15 44	18 54	21 17	28 29	13 27	17 20	24R 15
15 Tu	21 34 22	22 22 01	21 40 13	11 28	16 45	16 22	19 04	21 16	28 28	13 25	17 19	24 14
16 We	21 38 19	23 19 40	05♊ 51 57	11 14	17 56	17 00	19 14	21 15	28 28	13 24	17 18	24 14
17 Th	21 42 15	24 17 21	20 13 25	10 55	19 06	17 39	19 23	21 14	28 27	13 22	17 16	24 14
18 Fr	21 46 12	25 15 04	04♋ 41 22	10 31	20 17	18 17	19 33	21 14	28 26	13 21	17 15	24 13
19 Sa	21 50 08	26 12 48	19 11 36	10 01	21 28	18 55	19 43	21 13	28 26	13 19	17 14	24 13
20 Su	21 54 05	27 10 34	03♌ 39 07	09 25	22 39	19 33	19 53	21 12	28 25	13 17	17 13	24 13
21 Mo	21 58 01	28 08 21	17 58 38	08 45	23 50	20 12	20 03	21 12	28 24	13 16	17 12	24 12
22 Tu	22 01 58	29 06 09	02♍ 05 07	08 01	25 01	20 50	20 13	21 12	28 23	13 14	17 11	24 12
23 We	22 05 54	00♍ 03 59	15 54 27	07 13	26 12	21 28	20 24	21 11	28 22	13 13	17 10	24 11
24 Th	22 09 51	01 01 50	29 23 54	06 22	27 24	22 07	20 34	21 11	28 21	13 11	17 09	24 10
25 Fr	22 13 48	01 59 43	12♎ 32 22	05 29	28 35	22 45	20 45	21 11	28 20	13 09	17 08	24 08
26 Sa	22 17 44	02 57 37	25 20 21	04 34	29 47	23 23	20 55	21D 11	28 19	13 08	17 07	24 07
27 Su	22 21 41	03 55 32	07♏ 49 45	03 40	00♌ 58	24 01	21 06	21 11	28 18	13 06	17 06	24 06
28 Mo	22 25 37	04 53 28	20 03 35	02 47	02 10	24 39	21 16	21 11	28 17	13 05	17 06	24 04
29 Tu	22 29 34	05 51 26	02♐ 05 44	01 56	03 21	25 18	21 27	21 12	28 16	13 03	17 05	24 04
30 We	22 33 30	06 49 25	14 00 35	01 08	04 33	25 56	21 38	21 12	28 14	13 01	17 04	24D 04
31 Th	22 37 27	07 47 26	25 52 49	00 25	05 45	26 34	21 49	21 12	28 13	13 00	17 03	24 06

Data for	08-01-2017
Julian Day	2457966.50
Ayanamsa	24 05 59
SVP	05 ♓ 00 42
☽ ☊ Mean	25 ♌ 00 R

● ◐ PHASES ○ ◑

07	18:11	✹	15♒25
15	01:16	◑	22♉25
21	18:30	✪	28♌53
29	08:13	◐	06♐11

LAST ASPECT Day	h m	☽ INGRESS Day	h m	
31	11:10	01	12:02	♐
03	21:39	04	00:37	♑
06	09:22	06	12:16	♒
08	19:08	08	21:57	♓
10	13:39	11	05:23	♈
13	08:02	13	10:41	♉
15	01:16	15	14:07	♊
17	13:39	17	16:14	♋
19	15:18	19	17:55	♌
21	18:31	21	20:25	♍
23	20:03	24	01:05	♎
26	05:40	26	08:53	♏
28	09:38	28	19:48	♐
31	04:43	31	08:19	♑

DECLINATION

Day	☉	☽	☿	♀	♂	♃	♄	♅	♆	♇
01 Tu	18N01	13S 40	07N47	21N52	19N30	05S 35	21S 55	10N24	07S 15	21S 33
02 We	17 46	16 12	07 18	21 54	19 20	05 38	21 55	10 24	07 15	21 34
03 Th	17 30	18 03	06 49	21 56	19 10	05 42	21 55	10 24	07 16	21 34
04 Fr	17 14	19 08	06 22	21 57	19 00	05 45	21 55	10 24	07 16	21 34
05 Sa	16 58	19 22	05 57	21 58	18 50	05 48	21 56	10 24	07 17	21 34
06 Su	16 42	18 42	05 32	21 58	18 39	05 52	21 56	10 24	07 18	21 35
07 Mo	16 25	17 10	05 09	21 58	18 29	05 55	21 56	10 24	07 18	21 35
08 Tu	16 08	14 47	04 48	21 57	18 18	05 59	21 56	10 24	07 19	21 35
09 We	15 51	11 41	04 29	21 55	18 08	06 02	21 56	10 24	07 19	21 35
10 Th	15 34	07 58	04 12	21 53	17 57	06 06	21 56	10 24	07 20	21 36
11 Fr	15 16	03 49	03 57	21 51	17 46	06 10	21 56	10 23	07 20	21 36
12 Sa	14 58	00N35	03 44	21 47	17 35	06 13	21 56	10 23	07 21	21 36
13 Su	14 40	05 01	03 34	21 43	17 24	06 17	21 56	10 23	07 22	21 36
14 Mo	14 22	09 15	03 26	21 39	17 12	06 21	21 56	10 23	07 22	21 37
15 Tu	14 03	13 04	03 21	21 34	17 01	06 25	21 56	10 23	07 23	21 37
16 We	13 44	16 10	03 19	21 28	16 50	06 28	21 57	10 22	07 23	21 37
17 Th	13 25	18 20	03 20	21 22	16 38	06 32	21 57	10 22	07 24	21 37
18 Fr	13 06	19 19	03 24	21 15	16 26	06 36	21 57	10 22	07 25	21 38
19 Sa	12 47	19 03	03 31	21 07	16 14	06 40	21 57	10 22	07 25	21 38
20 Su	12 27	17 31	03 42	20 59	16 03	06 44	21 57	10 21	07 26	21 38
21 Mo	12 07	14 53	03 55	20 50	15 51	06 48	21 57	10 21	07 26	21 38
22 Tu	11 47	11 24	04 12	20 41	15 38	06 52	21 57	10 21	07 27	21 39
23 We	11 27	07 21	04 32	20 31	15 26	06 56	21 58	10 20	07 28	21 39
24 Th	11 06	03 02	04 54	20 21	15 14	07 00	21 58	10 20	07 28	21 39
25 Fr	10 46	01S 19	05 19	20 10	15 01	07 04	21 58	10 20	07 29	21 39
26 Sa	10 25	05 30	05 46	19 58	14 49	07 08	21 58	10 19	07 30	21 40
27 Su	10 04	09 19	06 15	19 46	14 36	07 12	21 58	10 19	07 30	21 40
28 Mo	09 43	12 41	06 45	19 33	14 24	07 16	21 58	10 18	07 31	21 40
29 Tu	09 22	15 27	07 15	19 20	14 11	07 20	21 59	10 18	07 31	21 40
30 We	09 00	17 33	07 46	19 06	13 58	07 25	21 59	10 18	07 32	21 41
31 Th	08 39	18 53	08 16	18 52	13 45	07 29	21 59	10 17	07 33	21 41

ASPECTARIAN

02 01:58 ☿ # ♆
02:00 ☽ □ ☿
04:30 ☽ △ ♂
08:45 ☽ △ ☉
15:31 ♀ # ♄
15:44 ☽ □ ♆
16:45 ☽ # ☉
22:55 ☽ ✶ ♃

03 05:32 ♅ SR
07:37 ☽ ☌ ♄
20:13 ☽ # ♂
21:39 ☽ △ ♅

04 09:23 ☽ ☍ ♀
18:23 ☽ △ ☿
18:30 ♃ □ ♇

05 03:58 ☽ ✶ ♆
07:03 ☿ # ♃
11:37 ☽ ☌ ♇
11:52 ☽ □ ♃

06 01:19 ☽ # ♂
09:22 ☽ □ ♅
07 09:38 ☽ # ☉
10:42 ☽ ☍ ♂
23:07 ☽ △ ♃

08 05:41 ☽ ✶ ♄
19:08 ☽ ✶ ♅
20:07 ♀ # ♄

09 08:41 ☽ # ♅
17:46 ☽ △ ♀
18:52 ☽ ☍ ☿
23:15 ☽ ☌ ♆

10 03:48 ☽ ∥ ♆
06:23 ☽ ✶ ♇
09:03 ☿ ✶ ♀
10:53 ☽ ∥ ♃
13:39 ☽ □ ♄
21:25 ☉ ✶ ♃
23:13 ☽ # ☿

12 05:18 ♀ △ ♆
05:37 ☽ □ ♀
07:40 ☽ △ ♂
12:28 ☽ □ ♇
14:50 ☽ ☍ ♃
16:19 ☽ ∥ ☿
17:28 ☽ △ ☉
19:27 ☽ △ ♄

13 01:02 ☿ SR
07:02 ☽ # ♃
08:02 ☽ ☌ ♅
13:01 ☽ # ♆
21:07 ☉ △ ♄

14 06:39 ☽ ∥ ♅
06:43 ☽ △ ☿
09:13 ♀ # ♇
09:55 ☽ ✶ ♆
14:31 ☽ □ ♂
14:50 ☽ ✶ ♀
16:34 ☽ △ ♇
15 06:20 ☽ ∥ ☉
11:16 ♀ ☍ ♇

16 05:39 ☽ ∥ ♂
08:50 ☽ □ ☿
12:35 ☽ □ ♆
19:30 ☽ ✶ ♂
22:36 ☽ △ ♃

17 01:41 ☽ ☍ ♄
06:40 ♀ □ ♃
07:14 ☽ ✶ ☉
13:39 ☽ ✶ ♅

18 09:20 ☽ ✶ ☿
14:17 ☽ △ ♆
20:46 ☽ ☍ ♇

19 00:53 ☽ □ ♃
04:06 ☽ ☌ ♀
15:18 ☽ □ ♅

20 15:43 ☽ ∥ ♂
16:43 ♂ ✶ ♃

21 03:33 ☽ ✶ ♃
03:55 ☽ ☌ ♂
05:26 ☽ △ ♄
06:22 ☉ △ ♅
17:40 ☽ △ ♅
21:23 ☽ ∥ ☉

22 06:34 ☽ ∥ ♅
09:41 ☽ ☌ ☿
13:22 ♂ △ ♄
19:17 ☽ ☍ ♆
22:21 ☉ ♍
23:26 ☽ # ♆

23 02:13 ☽ △ ♇
02:24 ☽ # ♃
09:19 ☽ □ ♄
14:34 ☽ ∥ ☿
20:03 ☽ ✶ ♀

24 19:02 ♀ □ ♅
25 08:32 ☽ □ ♇
12:10 ♄ SD
15:31 ☽ ☌ ♃
16:08 ☽ ✶ ♄
20:06 ☽ ✶ ♂

26 01:55 ☽ # ☿
04:30 ♀ ♌
05:40 ☽ ☍ ♅
06:46 ☉ ∥ ♅
09:21 ☽ □ ♀
10:10 ☽ ∥ ♃
12:16 ☽ ∥ ♆
15:48 ☽ ✶ ☉
16:29 ☽ ✶ ☿
20:43 ☿ ☌ ☉

27 04:35 ☽ # ☉
06:43 ☽ # ♅
10:16 ☽ △ ♆
12:16 ♃ ✶ ♄
18:09 ☽ ✶ ♇

28 09:38 ☽ □ ♂
13:10 ☽ # ♂
23:42 ☽ □ ☿
29 02:49 ☽ △ ♀
04:45 ☿ # ♃
13:02 ☿ # ♆
22:01 ☽ □ ♆

30 14:33 ☽ ☌ ♄
15:39 ☽ ✶ ♃
23:31 ☽ # ♀

31 01:28 ☽ △ ♂
04:43 ☽ △ ♅
08:41 ☽ △ ☿
10:41 ☿ ∥ ☉
15:28 ☿ ♌ R

⚷ Chiron

01 Dec.	03 N 02
02	28♓26R
05	28 21
08	28 16
11	28 10
14	28 04
17	27 58
20	27 51
23	27 45
26	27 37
29	27 30

September 2017

Day	S. T. h m s	☉ ° ' "	☽ ° ' "	☿ ° '	♀ ° '	♂ ° '	♃ ° '	♄ ° '	♅ ° '	♆ ° '	♇ ° '	☊ True ° '
01 Fr	22 41 23	08♍45 28	07♑47 06	29♌R47	06♌57	27♌12	22♎00	21♐13	28♈R12	12♓R58	17♑R02	24♌07
02 Sa	22 45 20	09 43 31	19 47 50	29 16	08 09	27 50	22 11	21 14	28 10	12 56	17 01	24 10
03 Su	22 49 16	10 41 36	01♒58 51	28 51	09 21	28 29	22 22	21 14	28 09	12 55	17 01	24 12
04 Mo	22 53 13	11 39 42	14 23 20	28 35	10 33	29 07	22 33	21 15	28 08	12 53	17 00	24 14
05 Tu	22 57 10	12 37 50	27 03 36	28 26	11 45	29 45	22 45	21 16	28 06	12 51	16 59	24 15
06 We	23 01 06	13 35 59	10♓00 54	28D27	12 57	00♍23	22 56	21 17	28 05	12 50	16 59	24R 14
07 Th	23 05 03	14 34 10	23 15 19	28 36	14 10	01 01	23 07	21 19	28 03	12 48	16 58	24 13
08 Fr	23 08 59	15 32 23	06♈45 48	28 53	15 22	01 39	23 19	21 20	28 02	12 46	16 57	24 11
09 Sa	23 12 56	16 30 37	20 30 17	29 20	16 35	02 17	23 30	21 21	28 00	12 45	16 57	24 07
10 Su	23 16 52	17 28 54	04♉26 03	29 55	17 47	02 55	23 42	21 23	27 58	12 43	16 56	24 04
11 Mo	23 20 49	18 27 13	18 30 06	00♍39	19 00	03 33	23 54	21 24	27 57	12 42	16 56	24 01
12 Tu	23 24 45	19 25 33	02♊39 26	01 30	20 12	04 11	24 05	21 26	27 55	12 40	16 55	23 59
13 We	23 28 42	20 23 57	16 51 19	02 30	21 25	04 50	24 17	21 28	27 53	12 38	16 55	23 58
14 Th	23 32 39	21 22 22	01♋03 20	03 36	22 38	05 28	24 29	21 29	27 51	12 37	16 54	23D 57
15 Fr	23 36 35	22 20 49	15 13 23	04 49	23 51	06 06	24 41	21 31	27 49	12 35	16 54	23 58
16 Sa	23 40 32	23 19 19	29 19 31	06 08	25 04	06 44	24 53	21 33	27 48	12 33	16 54	23 59
17 Su	23 44 28	24 17 50	13♌19 48	07 32	26 17	07 22	25 05	21 35	27 46	12 32	16 53	24 00
18 Mo	23 48 25	25 16 24	27 12 05	09 01	27 30	08 00	25 17	21 38	27 44	12 30	16 53	24R 00
19 Tu	23 52 21	26 15 00	10♍54 04	10 34	28 43	08 38	25 29	21 40	27 42	12 29	16 53	23 59
20 We	23 56 18	27 13 37	24 23 26	12 11	29 56	09 16	25 41	21 42	27 40	12 27	16 52	23 55
21 Th	00 00 14	28 12 17	07♎38 09	13 51	01♍09	09 54	25 53	21 45	27 38	12 25	16 52	23 51
22 Fr	00 04 11	29 10 58	20 36 50	15 33	02 23	10 32	26 06	21 47	27 36	12 24	16 52	23 45
23 Sa	00 08 08	00♎09 42	03♏19 04	17 18	03 36	11 10	26 18	21 50	27 34	12 22	16 52	23 38
24 Su	00 12 04	01 08 27	15 45 31	19 04	04 49	11 48	26 30	21 53	27 32	12 21	16 51	23 31
25 Mo	00 16 01	02 07 14	27 58 01	20 52	06 03	12 26	26 43	21 56	27 29	12 19	16 51	23 25
26 Tu	00 19 57	03 06 03	09♐59 24	22 40	07 16	13 04	26 55	21 59	27 27	12 18	16 51	23 21
27 We	00 23 54	04 04 53	21 53 24	24 29	08 30	13 42	27 08	22 02	27 25	12 16	16 51	23 18
28 Th	00 27 50	05 03 46	03♑44 31	26 18	09 44	14 20	27 20	22 05	27 23	12 15	16 51	23D 18
29 Fr	00 31 47	06 02 40	15 37 40	28 07	10 57	14 58	27 33	22 08	27 21	12 13	16D 51	23 19
30 Sa	00 35 43	07 01 35	27 37 58	29 57	12 11	15 36	27 45	22 11	27 18	12 12	16 51	23 21

Data for	09-01-2017
Julian Day	2457997.50
Ayanamsa	24 06 03
SVP	05 ♓ 00 37
☽ ☊ Mean	23 ♌ 22 R

●◐ PHASES ○◑

Day	h m	Phase	Position
06	07:03	○	13♓53
13	06:25	◑	20♊40
20	05:30	●	27♍27
28	02:54	◐	05♑11

LAST ASPECT ☽ Day	h m	INGRESS Day	h m	Sign
02	16:31	02	20:08	♒
05	05:17	05	05:30	♓
06	20:30	07	12:02	♈
09	15:53	09	16:23	♉
11	00:55	11	19:30	♊
13	18:36	13	22:13	♋
15	21:23	16	01:09	♌
18	00:55	18	04:53	♍
20	05:30	20	10:06	♎
22	13:05	22	17:41	♏
24	07:34	25	04:02	♐
27	11:09	27	16:25	♑
30	00:15	30	04:41	♒

DECLINATION

Day	☉	☽	☿	♀	♂	♃	♄	♅	♆	♇
01 Fr	08N17	19S23	08N45	18N37	13N32	07S33	21S59	10N17	07S33	21S41
02 Sa	07 55	19 01	09 13	18 21	13 19	07 37	22 00	10 16	07 34	21 41
03 Su	07 33	17 45	09 39	18 05	13 06	07 42	22 00	10 16	07 35	21 41
04 Mo	07 11	15 38	10 02	17 49	12 52	07 46	22 00	10 15	07 35	21 42
05 Tu	06 49	12 44	10 23	17 32	12 39	07 50	22 00	10 14	07 36	21 42
06 We	06 27	09 08	10 41	17 14	12 26	07 54	22 01	10 14	07 37	21 42
07 Th	06 04	05 01	10 55	16 56	12 12	07 59	22 01	10 13	07 37	21 42
08 Fr	05 42	00 34	11 06	16 38	11 59	08 03	22 01	10 13	07 38	21 42
09 Sa	05 19	03N58	11 14	16 19	11 45	08 08	22 01	10 12	07 39	21 43
10 Su	04 57	08 21	11 17	15 59	11 31	08 12	22 02	10 12	07 39	21 43
11 Mo	04 34	12 19	11 17	15 39	11 18	08 16	22 02	10 11	07 40	21 43
12 Tu	04 11	15 37	11 12	15 19	11 04	08 21	22 02	10 10	07 40	21 43
13 We	03 48	17 59	11 04	14 58	10 50	08 25	22 02	10 10	07 41	21 43
14 Th	03 25	19 15	10 52	14 37	10 36	08 30	22 03	10 09	07 42	21 43
15 Fr	03 02	19 19	10 37	14 16	10 22	08 34	22 03	10 08	07 42	21 44
16 Sa	02 39	18 09	10 17	13 54	10 08	08 39	22 03	10 08	07 43	21 44
17 Su	02 16	15 53	09 55	13 31	09 54	08 43	22 04	10 07	07 44	21 44
18 Mo	01 53	12 43	09 29	13 08	09 39	08 48	22 04	10 06	07 44	21 44
19 Tu	01 29	08 52	09 00	12 45	09 25	08 52	22 04	10 06	07 45	21 44
20 We	01 06	04 38	08 29	12 22	09 11	08 57	22 05	10 05	07 45	21 44
21 Th	00 43	00 16	07 55	11 58	08 57	09 01	22 05	10 04	07 46	21 45
22 Fr	00 19	04S02	07 19	11 33	08 42	09 06	22 05	10 03	07 47	21 45
23 Sa	00S04	08 04	06 41	11 09	08 28	09 10	22 06	10 03	07 47	21 45
24 Su	00 27	11 39	06 01	10 44	08 13	09 15	22 06	10 02	07 48	21 45
25 Mo	00 51	14 41	05 20	10 19	07 59	09 19	22 06	10 01	07 48	21 45
26 Tu	01 14	17 03	04 37	09 53	07 44	09 24	22 07	10 00	07 49	21 45
27 We	01 37	18 39	03 54	09 27	07 30	09 28	22 07	10 00	07 50	21 45
28 Th	02 01	19 27	03 09	09 01	07 15	09 33	22 07	09 59	07 50	21 45
29 Fr	02 24	19 22	02 24	08 35	07 00	09 37	22 08	09 58	07 51	21 46
30 Sa	02 47	18 25	01 38	08 08	06 45	09 42	22 08	09 57	07 51	21 46

ASPECTARIAN

01 02:07 ☽ △ ☉
02:24 ♃ ∥ ♆
10:22 ☽ ⚹ ♆
18:29 ☽ ☌ ♇
02 04:48 ☽ □ ♃
12:14 ♂ △ ♅
16:31 ☽ □ ♅
16:31 ☉ # ♃
18:12 ☽ # ♀
22:35 ☉ # ♆
03 09:38 ☿ ☌ ♂
15:50 ☽ ☍ ♀
04 13:05 ☽ ⚹ ♄
13:53 ☿ ∥ ♅
15:46 ☽ △ ♃
05 00:38 ☽ # ♂
01:57 ☽ ⚹ ♅
02:34 ☽ ☍ ☿
05:17 ☽ ☍ ♂
05:28 ☉ ☍ ♆
09:35 ♂ ♍
11:31 ☿ SD
14:53 ☽ # ☿
17:04 ☽ # ♅
06 05:08 ☽ ☌ ♆
07:19 ☽ ∥ ♃
09:11 ☽ ∥ ♆
12:40 ☽ ⚹ ♇
17:31 ☽ # ☉
20:30 ☽ □ ♄
08 16:31 ☽ △ ♀
17:50 ☽ □ ♇
09 01:28 ☽ △ ♄
05:16 ☽ ☍ ♃
06:44 ☽ ∥ ☉
10:43 ☉ △ ♇
12:55 ☽ ☌ ♅
15:53 ☽ △ ☿
20:04 ☽ # ♆
21:17 ☽ △ ♂
23:08 ☽ # ♃
10 02:53 ☿ ♍
10:43 ☽ ∥ ♅
14:08 ☽ ⚹ ♆
17:25 ☽ ∥ ☿
17:49 ☽ ∥ ♂
21:20 ☽ △ ♇
23:55 ☽ △ ☉
11 00:55 ☽ □ ♀
02:06 ☿ ∥ ♂
21:51 ☽ ∥ ♀
21:55 ☽ □ ☿
12 02:43 ☽ □ ♂
16:53 ☽ □ ♆
13 00:50 ♀ △ ♄
07:48 ☽ ☍ ♄
08:26 ☽ ⚹ ♀
12:44 ☽ △ ♃
18:36 ☽ ⚹ ♅
14 02:59 ☉ □ ♄
04:41 ☽ ⚹ ☿
07:48 ☽ ⚹ ♂
19:32 ☽ △ ♆
15 02:51 ☽ ☍ ♇
13:00 ☽ ⚹ ☉
16:19 ☽ □ ♃
19:45 ♀ ⚹ ♃
21:23 ☽ □ ♅
16 00:12 ♂ ∥ ♅
11:03 ☿ ∥ ♅
19:02 ☿ ☌ ♂
17 02:38 ☿ ∥ ♂
14:18 ☽ △ ♄
20:37 ☽ ⚹ ♃
20:43 ☽ ∥ ♀
18 00:34 ☽ ☌ ♀
00:55 ☽ △ ♅
04:27 ♀ △ ♅
16:44 ☽ ∥ ♅
19:48 ☽ ☌ ♂
20:36 ☽ ∥ ♂
23:10 ☽ ∥ ☿
23:20 ☽ ☌ ☿
19 00:05 ☽ # ♃
02:47 ☽ ☍ ♆
05:42 ☿ # ♃
06:35 ☽ # ♆
10:34 ☽ △ ♇
19:11 ☽ □ ♄
20 01:16 ♀ ♍
03:50 ☿ ☍ ♆
18:19 ♂ # ♃
21:20 ☽ ∥ ☉
21 04:59 ☽ # ☉
06:01 ☿ # ♆
17:01 ☽ □ ♇
22 02:12 ☽ ⚹ ♄
10:28 ☽ ☌ ♃
13:05 ☽ ☍ ♅
16:45 ☽ # ☿
18:01 ☿ △ ♇
20:02 ☉ ♎
22:21 ☽ ∥ ♆
23 00:36 ☽ ⚹ ♀
02:26 ☽ # ♂
07:16 ☽ ∥ ♃
12:50 ☽ # ♅
15:54 ☽ ⚹ ♂
17:23 ☽ △ ♆
18:15 ☽ # ♀
24 02:08 ☽ ⚹ ♇
07:34 ☽ ⚹ ☿
19:50 ♂ ☍ ♆
25 08:59 ☽ ⚹ ☉
14:36 ☿ □ ♄
16:21 ♂ # ♆
17:07 ♀ ∥ ♅
17:56 ☽ □ ♀
26 04:37 ☽ □ ♆
06:32 ☽ □ ♂
23:15 ♀ # ♃
27 00:17 ☽ ☌ ♄
06:12 ☽ □ ☿
10:48 ☽ ⚹ ♃
11:09 ☽ △ ♅
28 04:24 ♃ ☍ ♅
13:30 ☽ △ ♀
17:09 ☽ ⚹ ♆
19:34 ♇ SD
22:36 ☽ △ ♂
29 00:04 ☿ # ☉
02:28 ☽ ☌ ♇
23:21 ☽ □ ♅
30 00:12 ♀ ☍ ♆
00:15 ☽ □ ♃
00:43 ☿ ♎
05:24 ☽ △ ☿
14:58 ♀ # ♆
20:07 ☽ △ ☉

⚷ Chiron

01 Dec. 02 N 38

Day	Position
01	27♓22R
04	27 15
07	27 07
10	26 59
13	26 51
16	26 42
19	26 34
22	26 26
25	26 18
28	26 10

October 2017

Day	S. T.	☉	☽	☿	♀	♂	♃	♄	♅	♆	♇	☊ True
	h m s	° ' "	° ' "	° '	° '	° '	° '	° '	° '	° '	° '	° '
01 Su	00 39 40	08♎00 33	09♒50 23	01♎46	13♍25	16♍14	27♎58	22♐15	27♈R16	12♓R10	16♑51	23♌23
02 Mo	00 43 36	08 59 32	22 19 22	03 35	14 39	16 52	28 11	22 18	27 14	12 09	16 51	23 25
03 Tu	00 47 33	09 58 33	05♓08 30	05 23	15 53	17 30	28 23	22 22	27 12	12 07	16 51	23R 25
04 We	00 51 30	10 57 36	18 20 00	07 10	17 07	18 08	28 36	22 25	27 09	12 06	16 51	23 22
05 Th	00 55 26	11 56 41	01♈54 18	08 58	18 20	18 46	28 49	22 29	27 07	12 05	16 52	23 18
06 Fr	00 59 23	12 55 48	15 49 39	10 44	19 35	19 24	29 01	22 33	27 05	12 03	16 52	23 11
07 Sa	01 03 19	13 54 57	00♉02 17	12 30	20 49	20 02	29 14	22 37	27 02	12 02	16 52	23 03
08 Su	01 07 16	14 54 08	14 26 44	14 15	22 03	20 40	29 27	22 40	27 00	12 01	16 52	22 55
09 Mo	01 11 12	15 53 21	28 56 41	15 59	23 17	21 18	29 40	22 44	26 57	11 59	16 53	22 47
10 Tu	01 15 09	16 52 37	13♊26 01	17 43	24 31	21 56	29 53	22 49	26 55	11 58	16 53	22 41
11 We	01 19 05	17 51 55	27 49 37	19 26	25 45	22 34	00♏06	22 53	26 53	11 57	16 53	22 36
12 Th	01 23 02	18 51 15	12♋03 55	21 08	27 00	23 12	00 19	22 57	26 50	11 55	16 54	22 34
13 Fr	01 26 59	19 50 38	26 06 55	22 49	28 14	23 49	00 32	23 01	26 48	11 54	16 54	22 33
14 Sa	01 30 55	20 50 03	09♌58 00	24 30	29 28	24 27	00 45	23 06	26 45	11 53	16 55	22D 33
15 Su	01 34 52	21 49 30	23 37 20	26 10	00♎43	25 05	00 58	23 10	26 43	11 52	16 55	22R 33
16 Mo	01 38 48	22 49 00	07♍05 24	27 49	01 57	25 43	01 11	23 15	26 41	11 51	16 55	22 32
17 Tu	01 42 45	23 48 31	20 22 31	29 27	03 12	26 21	01 24	23 19	26 38	11 50	16 56	22 28
18 We	01 46 41	24 48 05	03♎28 40	01♏05	04 26	26 59	01 37	23 24	26 36	11 48	16 57	22 22
19 Th	01 50 38	25 47 41	16 23 27	02 42	05 41	27 37	01 50	23 29	26 33	11 47	16 57	22 13
20 Fr	01 54 34	26 47 19	29 06 23	04 19	06 56	28 15	02 03	23 34	26 31	11 46	16 58	22 02
21 Sa	01 58 31	27 46 59	11♏37 07	05 55	08 10	28 53	02 16	23 38	26 28	11 45	16 58	21 50
22 Su	02 02 28	28 46 41	23 55 51	07 30	09 25	29 31	02 29	23 43	26 26	11 44	16 59	21 37
23 Mo	02 06 24	29 46 25	06♐03 27	09 05	10 40	00♎09	02 42	23 48	26 23	11 43	17 00	21 26
24 Tu	02 10 21	00♏46 11	18 01 45	10 39	11 54	00 47	02 55	23 54	26 21	11 42	17 01	21 16
25 We	02 14 17	01 45 59	29 53 28	12 13	13 09	01 25	03 08	23 59	26 18	11 41	17 01	21 09
26 Th	02 18 14	02 45 48	11♑42 21	13 46	14 24	02 02	03 21	24 04	26 16	11 41	17 02	21 05
27 Fr	02 22 10	03 45 39	23 32 55	15 19	15 39	02 40	03 34	24 09	26 14	11 40	17 03	21 03
28 Sa	02 26 07	04 45 32	05♒30 16	16 51	16 54	03 18	03 47	24 15	26 11	11 39	17 04	21D 04
29 Su	02 30 03	05 45 26	17 39 51	18 23	18 09	03 56	04 00	24 20	26 09	11 38	17 05	21 04
30 Mo	02 33 60	06 45 22	00♓07 02	19 54	19 23	04 34	04 14	24 25	26 06	11 37	17 06	21R 04
31 Tu	02 37 57	07 45 20	12 56 39	21 24	20 38	05 12	04 27	24 31	26 04	11 36	17 07	21 03

Data for	10-01-2017
Julian Day	2458027.50
Ayanamsa	24 06 06
SVP	05 ♓ 00 31
☽ ☊ Mean	21 ♌ 46 R

● ◐ PHASES ○ ◑

05	18:40	○	12♈43
12	12:25	◑	19♋22
19	19:13	●	26♎35
27	22:22	◐	04♒42

LAST ASPECT ☽ Day	h m	INGRESS Day	h m	
02	11:14	02	14:27	♓
04	07:20	04	20:40	♈
06	22:38	06	23:56	♉
08	13:46	09	01:45	♊
10	22:25	11	03:39	♋
13	04:00	13	06:42	♌
15	05:28	15	11:19	♍
17	11:28	17	17:36	♎
19	19:13	20	1:42	♏
22	11:37	22	11:58	♐
24	16:46	25	00:13	♑
27	05:23	27	12:59	♒
29	16:22	29	23:47	♓

DECLINATION

Day	☉	☽	☿	♀	♂	♃	♄	♅	♆	♇
01 Su	03S 11	16S 36	00N52	07N41	06N31	09S 47	22S 09	09N56	07S 52	21S 46
02 Mo	03 34	13 58	00 06	07 14	06 16	09 51	22 09	09 56	07 52	21 46
03 Tu	03 57	10 36	00S 40	06 47	06 01	09 56	22 09	09 55	07 53	21 46
04 We	04 20	06 37	01 27	06 19	05 46	10 00	22 10	09 54	07 53	21 46
05 Th	04 43	02 11	02 13	05 52	05 31	10 05	22 10	09 53	07 54	21 46
06 Fr	05 06	02N28	02 59	05 24	05 16	10 10	22 10	09 52	07 54	21 46
07 Sa	05 29	07 05	03 45	04 56	05 01	10 14	22 11	09 51	07 55	21 46
08 Su	05 52	11 21	04 31	04 28	04 46	10 19	22 11	09 50	07 55	21 46
09 Mo	06 15	14 59	05 17	03 59	04 31	10 23	22 11	09 50	07 56	21 46
10 Tu	06 38	17 41	06 02	03 31	04 16	10 28	22 12	09 49	07 56	21 47
11 We	07 00	19 15	06 46	03 02	04 01	10 32	22 12	09 48	07 57	21 47
12 Th	07 23	19 35	07 31	02 33	03 46	10 37	22 13	09 47	07 57	21 47
13 Fr	07 46	18 41	08 14	02 04	03 31	10 42	22 13	09 46	07 58	21 47
14 Sa	08 08	16 41	08 57	01 35	03 16	10 46	22 13	09 45	07 58	21 47
15 Su	08 30	13 44	09 40	01 06	03 01	10 51	22 14	09 44	07 59	21 47
16 Mo	08 52	10 06	10 22	00 37	02 46	10 55	22 14	09 44	07 59	21 47
17 Tu	09 14	06 00	11 03	00 08	02 31	11 00	22 15	09 43	08 00	21 47
18 We	09 36	01 41	11 44	00S 22	02 16	11 04	22 15	09 42	08 00	21 47
19 Th	09 58	02S 40	12 24	00 51	02 01	11 09	22 15	09 41	08 00	21 47
20 Fr	10 20	06 49	13 03	01 20	01 45	11 13	22 16	09 40	08 01	21 47
21 Sa	10 41	10 36	13 41	01 49	01 30	11 18	22 16	09 39	08 01	21 47
22 Su	11 02	13 53	14 19	02 19	01 15	11 23	22 16	09 38	08 02	21 47
23 Mo	11 23	16 31	14 56	02 48	01 00	11 27	22 17	09 37	08 02	21 47
24 Tu	11 44	18 25	15 33	03 17	00 45	11 32	22 17	09 36	08 02	21 47
25 We	12 05	19 30	16 08	03 47	00 30	11 36	22 18	09 36	08 03	21 47
26 Th	12 26	19 43	16 43	04 16	00 15	11 41	22 18	09 35	08 03	21 47
27 Fr	12 46	19 05	17 16	04 45	00S 01	11 45	22 18	09 34	08 03	21 47
28 Sa	13 06	17 34	17 49	05 14	00 16	11 49	22 19	09 33	08 03	21 47
29 Su	13 26	15 15	18 21	05 43	00 31	11 54	22 19	09 32	08 04	21 47
30 Mo	13 46	12 11	18 52	06 11	00 46	11 58	22 19	09 31	08 04	21 47
31 Tu	14 06	08 27	19 22	06 40	01 01	12 03	22 20	09 30	08 04	21 47

ASPECTARIAN

01 23:33 ♂ △ ♆
23:57 ☽ ✱ ♄
02 09:14 ☽ ✱ ♅
11:14 ☽ △ ♃
19:01 ♃ # ♅
03 04:12 ☽ ∥ ♃
04:25 ☽ # ♅
12:46 ☽ ☌ ♆
16:41 ☽ ∥ ♆
19:08 ♀ △ ♇
21:21 ☽ ✱ ♇
21:35 ☽ ☍ ♀
23:37 ☽ ☍ ♂
04 01:49 ☽ # ♀
05:00 ☽ # ♂
07:20 ☽ □ ♄
11:33 ☽ ∥ ☉
23:50 ☽ ∥ ♀
05 14:01 ☽ ☍ ☿
16:54 ♀ ☌ ♂
06 01:46 ☽ □ ♇
03:12 ☽ # ☿
06:21 ☉ # ☿
08:14 ♀ # ☉
11:27 ☽ △ ♄
13:43 ♀ ∥ ♂
13:44 ☽ ∥ ♀
13:44 ☽ ∥ ♂
14:51 ☽ # ☉
18:59 ☽ ☌ ♅
22:38 ☽ ☍ ♃
07 04:32 ☽ # ♆
15:15 ☽ ∥ ♅
17:49 ☽ # ♃
19:58 ☽ ✱ ♆
22:48 ☿ ∥ ♀
08 04:01 ☽ △ ♇
06:03 ☿ # ♂
10:46 ☽ △ ♂
12:55 ♀ □ ♄
13:46 ☽ △ ♀
20:54 ☿ ☌ ☉
09 12:26 ☿ □ ♇
21:34 ☽ □ ♆
10 00:10 ☉ □ ♇
06:09 ☽ △ ☉
08:04 ☽ △ ☿
13:21 ♃ ♏
14:47 ☽ □ ♂
15:41 ☽ ☍ ♄
20:12 ☽ □ ♀
22:25 ☽ ✱ ♅
11 03:52 ☽ △ ♃
13:38 ♂ □ ♄
15:40 ☿ ∥ ☉
23:46 ☽ △ ♆
12 08:13 ☽ ☍ ♇
14:53 ☿ ∥ ♆
17:34 ☽ □ ☿
19:53 ☽ ✱ ♂
13 01:10 ☽ □ ♅
03:03 ☿ ✱ ♄
04:00 ☽ ✱ ♀
07:44 ☽ □ ♃
13:25 ☉ ∥ ♆
14 10:12 ♀ ♎
20:34 ☽ ✱ ☉
23:12 ☽ △ ♄
15 02:33 ☿ # ♅
05:07 ☽ ✱ ☿
05:28 ☽ △ ♅
07:52 ☿ ☍ ♅
13:15 ☽ ✱ ♃
18:57 ☽ # ♃
22:36 ☽ # ☿
16 02:16 ☽ ∥ ♅
06:46 ☽ # ☉
08:32 ☽ ☍ ♆
11:14 ☉ ✱ ♄
12:35 ☽ # ♆
17:45 ☽ △ ♇
21:54 ☿ ∥ ♃
17 05:24 ☿ □ ♄
07:59 ☿ ♏
11:28 ☽ ☌ ♂
21:34 ☽ ∥ ♂
18 01:58 ☽ ☌ ♀
05:54 ☉ # ♅
06:31 ☽ # ♀
08:55 ☿ ☌ ♃
12:37 ☽ ∥ ♀
21:34 ☽ # ♂
19 01:03 ☽ □ ♇
13:25 ☽ ✱ ♄
17:35 ☉ ☍ ♅
19:05 ☽ ☍ ♅
20 05:42 ☽ ☌ ♃
07:21 ☽ ∥ ♆
11:25 ☽ ☌ ☿
13:41 ♀ # ♂
17:46 ☽ # ♅
21 00:16 ☽ △ ♆
00:39 ☽ ∥ ☉
04:56 ☽ ∥ ♃
10:24 ☽ ✱ ♇
22 04:40 ☽ ∥ ☿
11:37 ☽ ✱ ♂
18:30 ♂ ♎
23 05:17 ☉ ∥ ♃
05:27 ☉ ♏
10:16 ☽ ✱ ♀
11:18 ☽ □ ♆
24 11:56 ☽ ☌ ♄
15:56 ☿ △ ♆
16:46 ☽ △ ♅
25 03:15 ☽ □ ♂
04:09 ☽ ✱ ☉
06:43 ☽ ✱ ♃
23:56 ☽ ✱ ♆
26 04:50 ☽ ✱ ☿
06:07 ☽ □ ♀
10:50 ☽ ☌ ♇
18:10 ☉ ☌ ♃
27 05:23 ☽ □ ♅
19:22 ☽ △ ♂
20:31 ☽ □ ♃
21:31 ☽ ∥ ☿
28 03:20 ♀ □ ♇
03:34 ☿ ✱ ♇
29 01:02 ☽ △ ♀
01:35 ☽ □ ☿
13:02 ☽ ✱ ♄
13:22 ☽ ∥ ☉
16:22 ☽ ✱ ♆
30 01:24 ☽ ∥ ♃
07:54 ☽ △ ♃
13:34 ☽ △ ☉
17:30 ☽ # ♅
21:32 ☽ ☌ ♆
31 02:11 ☽ ∥ ♆
07:38 ☽ ✱ ♇
09:15 ☽ ∥ ♀
17:01 ☽ △ ☿
21:08 ☽ □ ♄

⚷ Chiron

01 Dec.	02 N 05
01	26♓02R
04	25 54
07	25 46
10	25 38
13	25 31
16	25 24
19	25 17
22	25 10
25	25 04
28	24 58
31	24 52

November 2017

Day	S. T. h m s	☉ ° ' "	☽ ° ' "	☿ ° '	♀ ° '	♂ ° '	♃ ° '	♄ ° '	♅ ° '	♆ ° '	♇ ° '	☊ True ° '
01 We	02 41 53	08♏45 19	26♓12 24	22♏54	21♎53	05♎50	04♏40	24♐37	26♈R02	11♓R36	17♑07	20♌R59
02 Th	02 45 50	09 45 20	09♈56 03	24 24	23 08	06 28	04 53	24 42	25 59	11 35	17 08	20 52
03 Fr	02 49 46	10 45 23	24 06 41	25 53	24 23	07 05	05 06	24 48	25 57	11 34	17 09	20 43
04 Sa	02 53 43	11 45 27	08♉40 18	27 21	25 38	07 43	05 19	24 54	25 55	11 34	17 11	20 32
05 Su	02 57 39	12 45 33	23 30 01	28 50	26 53	08 21	05 32	24 59	25 52	11 33	17 12	20 21
06 Mo	03 01 36	13 45 41	08♊27 00	00♐17	28 08	08 59	05 45	25 05	25 50	11 33	17 13	20 10
07 Tu	03 05 32	14 45 51	23 21 57	01 44	29 24	09 37	05 58	25 11	25 48	11 32	17 14	20 00
08 We	03 09 29	15 46 04	08♋06 39	03 10	00♏39	10 15	06 11	25 17	25 45	11 32	17 15	19 53
09 Th	03 13 26	16 46 18	22 35 14	04 36	01 54	10 53	06 24	25 23	25 43	11 31	17 16	19 49
10 Fr	03 17 22	17 46 34	06♌44 31	06 01	03 09	11 30	06 37	25 29	25 41	11 31	17 17	19 46
11 Sa	03 21 19	18 46 52	20 33 52	07 25	04 24	12 08	06 50	25 35	25 39	11 30	17 19	19 45
12 Su	03 25 15	19 47 12	04♍04 23	08 49	05 39	12 46	07 03	25 42	25 37	11 30	17 20	19 44
13 Mo	03 29 12	20 47 34	17 18 09	10 12	06 55	13 24	07 16	25 48	25 34	11 30	17 21	19 42
14 Tu	03 33 08	21 47 58	00♎17 24	11 33	08 10	14 02	07 29	25 54	25 32	11 29	17 22	19 37
15 We	03 37 05	22 48 24	13 04 10	12 54	09 25	14 40	07 42	26 00	25 30	11 29	17 24	19 30
16 Th	03 41 01	23 48 52	25 40 00	14 13	10 40	15 17	07 55	26 07	25 28	11 29	17 25	19 21
17 Fr	03 44 58	24 49 21	08♏05 59	15 32	11 56	15 55	08 08	26 13	25 26	11 28	17 26	19 08
18 Sa	03 48 55	25 49 53	20 22 49	16 48	13 11	16 33	08 20	26 19	25 24	11 28	17 28	18 54
19 Su	03 52 51	26 50 25	02♐31 08	18 03	14 26	17 11	08 33	26 26	25 22	11 28	17 29	18 40
20 Mo	03 56 48	27 50 59	14 31 40	19 17	15 42	17 49	08 46	26 32	25 20	11 28	17 31	18 26
21 Tu	04 00 44	28 51 35	26 25 39	20 28	16 57	18 27	08 59	26 39	25 18	11 28	17 32	18 15
22 We	04 04 41	29 52 12	08♑14 58	21 36	18 12	19 04	09 11	26 45	25 16	11 28	17 34	18 06
23 Th	04 08 37	00♐52 51	20 02 20	22 42	19 28	19 42	09 24	26 52	25 14	11D28	17 35	18 00
24 Fr	04 12 34	01 53 30	01♒51 22	23 45	20 43	20 20	09 37	26 59	25 13	11 28	17 37	17 58
25 Sa	04 16 30	02 54 11	13 46 26	24 44	21 58	20 58	09 49	27 05	25 11	11 28	17 38	17D57
26 Su	04 20 27	03 54 53	25 52 32	25 39	23 14	21 35	10 02	27 12	25 09	11 28	17 40	17 58
27 Mo	04 24 24	04 55 35	08♓15 00	26 30	24 29	22 13	10 14	27 19	25 07	11 28	17 41	17R58
28 Tu	04 28 20	05 56 19	20 59 10	27 15	25 45	22 51	10 27	27 25	25 06	11 28	17 43	17 57
29 We	04 32 17	06 57 04	04♈09 44	27 55	27 00	23 29	10 39	27 32	25 04	11 29	17 45	17 54
30 Th	04 36 13	07 57 50	17 49 55	28 27	28 16	24 06	10 52	27 39	25 02	11 29	17 46	17 49

Data for	11-01-2017
Julian Day	2458058.50
Ayanamsa	24 06 08
SVP	05 ♓ 00 26
☽ ☊ Mean	20 ♌ 08 R

● ◐ PHASES ○ ◑

Day	h m	Phase	Position
04	05:23	○	11♉59
10	20:37	◑	18♌38
18	11:42	●	26♏19
26	17:02	◐	04♓38

LAST ASPECT Day	h m	☽ INGRESS Day	h m	Sign
31	21:08	01	06:43	♈
03	03:03	03	09:46	♉
05	09:29	05	10:27	♊
07	10:40	07	10:45	♋
09	05:15	09	12:30	♌
11	08:56	11	16:42	♍
13	15:46	13	23:28	♎
16	00:51	16	08:20	♏
18	11:43	18	19:00	♐
21	00:27	21	07:14	♑
23	10:33	23	20:14	♒
26	02:37	26	08:04	♓
28	12:09	28	16:31	♈
30	18:38	30	20:39	♉

DECLINATION

Day	☉	☽	☿	♀	♂	♃	♄	♅	♆	♇
01 We	14S25	04S11	19S52	07S09	01S16	12S07	22S20	09N30	08S05	21S47
02 Th	14 44	00N26	20 20	07 37	01 31	12 12	22 21	09 29	08 05	21 47
03 Fr	15 03	05 10	20 47	08 05	01 46	12 16	22 21	09 28	08 05	21 47
04 Sa	15 22	09 44	21 13	08 33	02 01	12 20	22 21	09 27	08 05	21 47
05 Su	15 40	13 48	21 38	09 01	02 16	12 25	22 22	09 26	08 06	21 47
06 Mo	15 58	17 01	22 02	09 29	02 31	12 29	22 22	09 25	08 06	21 47
07 Tu	16 16	19 05	22 25	09 57	02 46	12 33	22 22	09 25	08 06	21 47
08 We	16 33	19 51	22 47	10 24	03 01	12 38	22 23	09 24	08 06	21 47
09 Th	16 51	19 16	23 08	10 51	03 16	12 42	22 23	09 23	08 06	21 47
10 Fr	17 08	17 29	23 28	11 18	03 31	12 46	22 23	09 22	08 06	21 47
11 Sa	17 24	14 42	23 46	11 44	03 46	12 50	22 24	09 21	08 07	21 47
12 Su	17 41	11 11	24 03	12 10	04 01	12 55	22 24	09 21	08 07	21 47
13 Mo	17 57	07 10	24 19	12 36	04 16	12 59	22 24	09 20	08 07	21 47
14 Tu	18 13	02 54	24 34	13 02	04 31	13 03	22 25	09 19	08 07	21 47
15 We	18 28	01S26	24 47	13 27	04 45	13 07	22 25	09 18	08 07	21 47
16 Th	18 43	05 39	24 59	13 53	05 00	13 11	22 25	09 18	08 07	21 47
17 Fr	18 58	09 33	25 10	14 17	05 15	13 15	22 25	09 17	08 07	21 46
18 Sa	19 13	13 01	25 19	14 42	05 30	13 20	22 26	09 16	08 07	21 46
19 Su	19 27	15 54	25 27	15 05	05 44	13 24	22 26	09 16	08 07	21 46
20 Mo	19 41	18 04	25 34	15 29	05 59	13 28	22 26	09 15	08 07	21 46
21 Tu	19 54	19 26	25 39	15 52	06 13	13 32	22 27	09 14	08 07	21 46
22 We	20 07	19 57	25 43	16 15	06 28	13 36	22 27	09 13	08 07	21 46
23 Th	20 20	19 35	25 45	16 37	06 42	13 40	22 27	09 13	08 07	21 46
24 Fr	20 32	18 22	25 46	16 59	06 57	13 44	22 27	09 12	08 07	21 46
25 Sa	20 44	16 20	25 45	17 21	07 11	13 48	22 28	09 12	08 07	21 46
26 Su	20 56	13 33	25 43	17 42	07 25	13 52	22 28	09 11	08 07	21 46
27 Mo	21 07	10 07	25 39	18 02	07 40	13 55	22 28	09 10	08 07	21 46
28 Tu	21 18	06 07	25 34	18 22	07 54	13 59	22 28	09 10	08 07	21 45
29 We	21 28	01 43	25 27	18 42	08 08	14 03	22 29	09 09	08 07	21 45
30 Th	21 38	02N56	25 19	19 01	08 22	14 07	22 29	09 09	08 07	21 45

ASPECTARIAN

01 14:29 ☽ ∥ ♂
17:43 ☽ ☍ ♂

02 05:48 ☽ # ♂
12:18 ☽ □ ♆
23:45 ♀ ∥ ♆

03 00:30 ☽ ☍ ♀
01:09 ☽ △ ♄
03:03 ☽ ☌ ♅
08:32 ♀ ⚹ ♄
15:06 ☽ # ♆
16:52 ☽ # ♀
18:26 ☽ ☍ ♃
19:23 ☉ △ ♆
22:26 ☽ ∥ ♅

04 04:42 ☽ ⚹ ♆
05:03 ♀ ☍ ♅
13:49 ☽ △ ♇
15:08 ☽ # ♃

05 08:34 ☿ ∥ ♇
09:29 ☽ ☍ ☿
14:24 ☽ # ☉
19:19 ☿ ♐
20:56 ♀ # ♅

06 00:54 ☽ △ ♂
04:58 ☽ □ ♆
20:33 ☿ ∥ ♄

07 02:58 ☽ ☍ ♄
03:55 ☽ ⚹ ♅
10:40 ☽ △ ♀
11:39 ♀ ♏
20:48 ☽ △ ♃

08 03:40 ☽ □ ♂
05:37 ☽ △ ♆
13:34 ☽ △ ☉
15:07 ☽ ☍ ♇

09 05:15 ☽ □ ♅
12:10 ☉ ⚹ ♇
17:15 ☽ □ ♀
22:37 ☽ △ ☿
23:47 ☽ □ ♃

10 03:09 ☽ # ☉
08:36 ☽ ⚹ ♂

11 08:56 ☽ △ ♄
08:56 ☽ △ ♅
09:42 ♄ △ ♅
12:54 ☽ # ♃
18:15 ☽ # ♀

12 03:09 ☽ ⚹ ♀
05:27 ☽ ⚹ ♃
09:33 ☽ □ ☿
11:14 ☽ ∥ ♅
13:24 ☽ ☍ ♆
18:29 ☽ # ♆

13 00:05 ☽ △ ♇
06:57 ☽ ⚹ ☉
08:16 ♀ ☌ ♃
15:27 ☽ # ♂
15:46 ☽ □ ♄
22:48 ☿ □ ♆

14 00:57 ♀ ∥ ♃
23:38 ☽ ⚹ ☿

15 03:10 ☽ ☌ ♂
08:13 ☽ □ ♇
19:59 ☽ ∥ ♂
23:37 ☽ ☍ ♅

16 00:51 ☽ ⚹ ♄
14:53 ☽ ∥ ♆
15:21 ♀ △ ♆
22:13 ☽ # ♅

17 00:03 ☽ ☌ ♃
06:34 ☽ △ ♆
08:18 ☽ ☌ ♀
14:25 ☿ ⚹ ♂
18:16 ☽ ⚹ ♇

18 02:21 ☽ ∥ ♃
15:28 ☽ ∥ ♀

19 12:12 ♂ □ ♇
17:52 ☽ □ ♆

20 06:58 ☽ ⚹ ♂
10:37 ☽ ☌ ☿
21:44 ☽ △ ♅

21 00:27 ☽ ☌ ♄
11:28 ♀ ⚹ ♇

22 01:57 ☽ ⚹ ♃
03:05 ☉ ♐
06:33 ☽ ⚹ ♆
14:22 ♆ SD
19:00 ☽ ☌ ♇
22:41 ☽ ⚹ ♀
23:16 ☽ □ ♂

23 10:33 ☽ □ ♅

24 00:05 ☽ ⚹ ☉
14:46 ☽ ∥ ♀
15:56 ☽ □ ♃

25 10:56 ☿ △ ♅
15:06 ☽ △ ♂
18:12 ☽ □ ♀
21:40 ☽ ∥ ♃
22:35 ☽ ⚹ ♅
23:32 ☽ ⚹ ☿

26 02:37 ☽ ⚹ ♄

27 03:52 ☽ △ ♃
05:56 ☽ # ♅
06:09 ☽ ☌ ♆
12:21 ☽ ∥ ♆
14:14 ☽ ∥ ♂
17:54 ☽ ⚹ ♇

28 06:59 ☿ ☌ ♄
09:42 ☽ △ ♀
11:56 ☽ □ ♄
12:09 ☽ □ ☿
22:12 ♂ ∥ ♆

29 05:22 ☽ △ ☉
23:54 ☽ □ ♇

30 11:14 ☽ ☍ ♂
12:17 ☽ ☌ ♅
16:50 ☽ △ ♄
17:58 ☉ ∥ ♇
18:38 ☽ △ ☿

⚷ Chiron

01 Dec. 01 N 32

Day	Position
03	24♓47R
06	24 42
09	24 38
12	24 34
15	24 30
18	24 27
21	24 25
24	24 23
27	24 21
30	24 20

December 2017

Day	S. T.	☉	☽	☿	♀	♂	♃	♄	♅	♆	♇	☊ True
	h m s	° ' "	° ' "	° '	° '	° '	° '	° '	° '	° '	° '	° '
01 Fr	04 40 10	08♐ 58 36	02♉ 00 32	28♐ 53	29♏ 31	24♎ 44	11♏ 04	27♐ 46	25♈R01	11♓ 29	17♑ 48	17♌R41
02 Sa	04 44 06	09 59 24	16 39 09	29 10	00♐ 46	25 22	11 16	27 53	24 59	11 29	17 50	17 32
03 Su	04 48 03	11 00 13	01♊ 39 39	29 18	02 02	26 00	11 29	27 59	24 58	11 30	17 51	17 22
04 Mo	04 51 59	12 01 03	16 52 53	29R 16	03 17	26 37	11 41	28 06	24 56	11 30	17 53	17 12
05 Tu	04 55 56	13 01 54	02♋ 08 01	29 03	04 33	27 15	11 53	28 13	24 55	11 31	17 55	17 03
06 We	04 59 53	14 02 47	17 14 21	28 40	05 48	27 53	12 05	28 20	24 53	11 31	17 57	16 57
07 Th	05 03 49	15 03 40	02♌ 03 10	28 05	07 04	28 30	12 17	28 27	24 52	11 32	17 58	16 53
08 Fr	05 07 46	16 04 35	16 28 50	27 19	08 19	29 08	12 29	28 34	24 51	11 32	18 00	16 50
09 Sa	05 11 42	17 05 31	00♍ 29 01	26 22	09 35	29 46	12 41	28 41	24 50	11 33	18 02	16 49
10 Su	05 15 39	18 06 28	14 04 09	25 16	10 50	00♏ 24	12 53	28 48	24 48	11 33	18 04	16 49
11 Mo	05 19 35	19 07 26	27 16 28	24 02	12 06	01 01	13 05	28 55	24 47	11 34	18 06	16 48
12 Tu	05 23 32	20 08 26	10♎ 09 06	22 43	13 21	01 39	13 17	29 02	24 46	11 34	18 07	16 46
13 We	05 27 28	21 09 27	22 45 26	21 20	14 37	02 17	13 28	29 09	24 45	11 35	18 09	16 42
14 Th	05 31 25	22 10 28	05♏ 08 37	19 58	15 52	02 54	13 40	29 16	24 44	11 36	18 11	16 35
15 Fr	05 35 22	23 11 31	17 21 27	18 37	17 08	03 32	13 51	29 23	24 43	11 37	18 13	16 27
16 Sa	05 39 18	24 12 34	29 26 11	17 22	18 23	04 10	14 03	29 30	24 42	11 37	18 15	16 17
17 Su	05 43 15	25 13 39	11♐ 24 39	16 14	19 39	04 47	14 14	29 37	24 41	11 38	18 17	16 07
18 Mo	05 47 11	26 14 44	23 18 22	15 15	20 54	05 25	14 26	29 44	24 40	11 39	18 19	15 57
19 Tu	05 51 08	27 15 50	05♑ 08 45	14 27	22 10	06 02	14 37	29 51	24 40	11 40	18 21	15 49
20 We	05 55 04	28 16 56	16 57 23	13 49	23 25	06 40	14 48	29 59	24 39	11 41	18 23	15 43
21 Th	05 59 01	29 18 03	28 46 15	13 22	24 41	07 18	14 59	00♑ 06	24 38	11 42	18 25	15 39
22 Fr	06 02 57	00♑ 19 10	10♒ 37 57	13 06	25 56	07 55	15 10	00 13	24 38	11 43	18 27	15D 38
23 Sa	06 06 54	01 20 17	22 35 39	13 00	27 12	08 33	15 21	00 20	24 37	11 44	18 29	15 39
24 Su	06 10 51	02 21 25	04♓ 43 10	13D 04	28 27	09 10	15 32	00 27	24 37	11 45	18 31	15 41
25 Mo	06 14 47	03 22 33	17 04 49	13 17	29 43	09 48	15 43	00 34	24 36	11 46	18 32	15 43
26 Tu	06 18 44	04 23 40	29 45 09	13 38	00♑ 58	10 25	15 54	00 41	24 36	11 47	18 34	15 44
27 We	06 22 40	05 24 48	12♈ 48 36	14 07	02 14	11 03	16 04	00 48	24 35	11 48	18 36	15R 44
28 Th	06 26 37	06 25 56	26 18 51	14 42	03 29	11 40	16 15	00 55	24 35	11 49	18 38	15 42
29 Fr	06 30 33	07 27 04	10♉ 17 57	15 24	04 45	12 18	16 25	01 02	24 35	11 50	18 41	15 39
30 Sa	06 34 30	08 28 12	24 45 24	16 10	06 00	12 55	16 36	01 09	24 35	11 52	18 43	15 34
31 Su	06 38 26	09 29 19	09♊ 37 32	17 02	07 16	13 33	16 46	01 16	24 34	11 53	18 45	15 28

Data for	12-01-2017
Julian Day	2458088.50
Ayanamsa	24 06 12
SVP	05 ♓ 00 22
☽ ☊ Mean	18 ♌ 32 R

● ◐ PHASES ○ ◑

Day	Time	Phase	Position
03	15:48	○	11♊40
10	07:52	◑	18♍26
18	06:30	●	26♐31
26	09:21	◐	04♈48

LAST ASPECT Day	h m	☽ INGRESS Day	h m	Sign
02	01:54	02	21:22	♊
04	19:14	04	20:38	♋
06	17:57	06	20:38	♌
08	22:41	08	23:10	♍
11	03:03	11	05:02	♎
13	12:28	13	13:59	♏
15	01:42	16	01:08	♐
18	13:10	18	13:33	♑
20	15:37	21	02:30	♒
23	10:13	23	14:42	♓
25	02:48	26	00:28	♈
27	20:58	28	06:24	♉
29	14:02	30	08:32	♊

DECLINATION

Day	☉	☽	☿	♀	♂	♃	♄	♅	♆	♇
01 Fr	21S 48	07N36	25S 10	19S 19	08S 36	14S 11	22S 29	09N08	08S 07	21S 45
02 Sa	21 57	11 58	24 59	19 37	08 50	14 15	22 29	09 08	08 06	21 45
03 Su	22 05	15 41	24 46	19 55	09 04	14 18	22 29	09 07	08 06	21 45
04 Mo	22 14	18 24	24 32	20 12	09 18	14 22	22 30	09 06	08 06	21 45
05 Tu	22 21	19 50	24 16	20 28	09 31	14 26	22 30	09 06	08 06	21 45
06 We	22 29	19 49	23 59	20 44	09 45	14 29	22 30	09 06	08 06	21 44
07 Th	22 36	18 25	23 41	20 59	09 59	14 33	22 30	09 05	08 06	21 44
08 Fr	22 42	15 52	23 21	21 13	10 12	14 37	22 30	09 05	08 05	21 44
09 Sa	22 49	12 25	23 00	21 27	10 26	14 40	22 30	09 04	08 05	21 44
10 Su	22 54	08 25	22 38	21 40	10 39	14 44	22 31	09 04	08 05	21 44
11 Mo	22 59	04 07	22 14	21 53	10 53	14 47	22 31	09 03	08 05	21 44
12 Tu	23 04	00S 15	21 51	22 05	11 06	14 51	22 31	09 03	08 04	21 44
13 We	23 08	04 31	21 27	22 16	11 19	14 54	22 31	09 03	08 04	21 43
14 Th	23 12	08 31	21 05	22 27	11 32	14 57	22 31	09 02	08 04	21 43
15 Fr	23 16	12 07	20 43	22 37	11 45	15 01	22 31	09 02	08 03	21 43
16 Sa	23 19	15 10	20 23	22 46	11 58	15 04	22 31	09 02	08 03	21 43
17 Su	23 21	17 34	20 06	22 55	12 11	15 07	22 31	09 01	08 03	21 43
18 Mo	23 23	19 12	19 52	23 03	12 24	15 11	22 31	09 01	08 02	21 43
19 Tu	23 24	19 59	19 40	23 10	12 37	15 14	22 32	09 01	08 02	21 43
20 We	23 25	19 53	19 32	23 16	12 49	15 17	22 32	09 01	08 02	21 42
21 Th	23 26	18 55	19 27	23 22	13 02	15 20	22 32	09 00	08 01	21 42
22 Fr	23 26	17 07	19 25	23 27	13 14	15 24	22 32	09 00	08 01	21 42
23 Sa	23 26	14 34	19 26	23 32	13 27	15 27	22 32	09 00	08 01	21 42
24 Su	23 25	11 21	19 30	23 35	13 39	15 30	22 32	09 00	08 00	21 42
25 Mo	23 24	07 36	19 35	23 38	13 51	15 33	22 32	09 00	08 00	21 42
26 Tu	23 22	03 25	19 43	23 40	14 03	15 36	22 32	09 00	07 59	21 41
27 We	23 19	01N03	19 52	23 42	14 15	15 39	22 32	09 00	07 59	21 41
28 Th	23 17	05 35	20 03	23 42	14 27	15 42	22 32	08 59	07 58	21 41
29 Fr	23 14	10 00	20 15	23 42	14 39	15 45	22 32	08 59	07 58	21 41
30 Sa	23 10	13 59	20 27	23 42	14 51	15 47	22 32	08 59	07 57	21 41
31 Su	23 06	17 12	20 40	23 40	15 02	15 50	22 32	08 59	07 57	21 40

ASPECTARIAN

Day	Time	Aspect
01	02:41	☽ # ♆
	05:34	☽ # ♂
	08:08	☽ ∥ ♅
	09:15	♀ ♐
	10:05	♂ ☍ ♅
	15:08	☽ ☍ ♃
	15:37	☽ ✶ ♆
02	01:54	☽ △ ♆
	14:14	☽ # ♃
03	00:38	☽ ☍ ♀
	02:23	♃ △ ♆
	05:19	♂ # ♅
	07:35	☿ SR
	11:44	☉ □ ♆
	15:32	☽ □ ♆
04	12:39	☽ ✶ ♅
	15:58	☽ △ ♂
	17:47	☽ ☍ ♄
	19:14	☽ ☍ ☿
05	14:52	☽ △ ♆
	15:39	☽ △ ♃
06	01:08	☽ ☍ ♆
	03:21	☉ ∥ ♄
	12:06	☿ ☌ ♄
	12:18	☽ □ ♅
	15:58	☿ ✶ ♂
	17:57	☽ □ ♂
	21:22	♂ ✶ ♄
07	09:02	☽ △ ♀
	17:11	☽ □ ♃
	23:16	☽ △ ☉
08	09:11	☽ # ♃
	14:14	☽ △ ♅
	17:21	☽ △ ☿
	20:51	☽ △ ♄
	22:41	☽ ✶ ♂
09	09:00	♂ ♏
	09:47	☿ ∥ ☉
	11:37	☽ # ♂
	17:38	☽ □ ♀
	19:30	☽ ☍ ♆
	20:18	☽ ∥ ♅
	21:51	☽ ✶ ♃
10	01:57	☽ # ♆
	06:54	♀ ∥ ♇
	07:12	☽ △ ♇
	07:20	☿ ∥ ♄
	09:29	☿ △ ♅
	13:48	♀ □ ♆
	18:35	☽ □ ☿
11	03:03	☽ □ ♄
	14:32	☿ ∥ ♀
12	06:43	☽ ✶ ♀
	07:21	☿ ∥ ♇
	15:10	☽ □ ♇
	20:39	☽ ✶ ☉
	21:33	☽ ✶ ☿
13	01:49	☿ ☌ ☉
	03:50	☽ ☍ ♅
	12:28	☽ ✶ ♄
	19:24	☽ ☌ ♂
	21:08	☽ ∥ ♆
14	03:16	☽ # ♅
	10:13	♀ ∥ ♄
	12:39	☽ △ ♆
	16:59	☽ ☌ ♃
	21:15	☽ ∥ ♂
15	01:42	☽ ✶ ♇
	14:09	☿ ☌ ♀
	23:04	☽ ∥ ♃
16	11:28	☉ △ ♅
17	00:27	☽ □ ♆
	08:57	☽ ☌ ☿
	18:34	☽ ☌ ♀
18	02:46	☽ △ ♅
	13:10	☽ ☌ ♄
	13:20	☽ ∥ ☿
19	01:55	☽ ✶ ♂
	13:16	☽ ✶ ♆
	19:33	☽ ✶ ♃
20	02:54	☽ ☌ ♇
	04:49	♄ ♑
	13:00	☽ ∥ ☿
	15:37	☽ □ ♅
	23:13	♀ △ ♅
21	16:28	☉ ♑
	18:13	☽ □ ♂
	18:41	♀ ∥ ☉
	21:09	☉ ☌ ♄
22	04:54	☽ ✶ ☿
	09:17	☽ □ ♃
	16:41	☽ ∥ ♃
23	01:52	☿ SD
	04:02	☽ ✶ ♅
	08:24	☽ ∥ ♂
	10:13	☽ ✶ ♀
	15:31	☽ ✶ ♄
	18:56	☽ ✶ ☉
24	09:10	☽ △ ♂
	13:44	☽ ☌ ♆
	15:25	☽ # ♅
	16:32	☽ □ ☿
	21:20	☽ △ ♃
	21:35	☽ ∥ ♆
25	02:48	☽ ✶ ♇
	05:27	♀ ♑
	17:56	♀ ☌ ♄
26	01:45	☽ □ ♄
	02:31	☽ □ ♀
27	02:27	☽ △ ☿
	10:26	☽ □ ♇
	20:58	☽ ☌ ♅
28	06:01	♂ △ ♆
	08:04	☽ △ ♄
	12:47	☽ # ♇
	13:39	☽ △ ♀
	18:23	☽ ∥ ♅
	18:48	☽ △ ☉
29	02:36	☽ ✶ ♆
	03:31	☽ ☍ ♂
	10:23	☽ ☍ ♃
	14:02	☽ △ ♇
30	06:09	☽ # ♂
	12:48	☽ # ♃
31	03:36	☽ □ ♆
	12:31	☽ ☍ ☿
	23:39	☽ ✶ ♅

⚷ Chiron

01	Dec.	01 N 13
03		24♓19R
06		24 19
09		24 19
12		24 20
15		24 22
18		24 23
21		24 26
24		24 29
27		24 32
30		24 36

[illegible]

January 2018

31 13:31 11♌37 ✷ Total Lunar Eclipse (mag 1.321)

Day	S. T.	☉	☽	☿	♀	♂	♃	♄	♅	♆	♇	☊ True
	h m s	° ' "	° ' "	° '	° '	° '	° '	° '	° '	° '	° '	° '
01 Mo	06 42 23	10♑30 27	24♊47 19	17♐57	08♑31	14♏10	16♏56	01♑23	24♈R34	11♓54	18♑47	15♌R23
02 Tu	06 46 20	11 31 35	10♋05 11	18 56	09 47	14 47	17 06	01 30	24 34	11 56	18 49	15 18
03 We	06 50 16	12 32 43	25 20 22	19 59	11 02	15 25	17 16	01 37	24D34	11 57	18 51	15 15
04 Th	06 54 13	13 33 51	10♌22 44	21 05	12 18	16 02	17 26	01 44	24 34	11 58	18 53	15 13
05 Fr	06 58 09	14 34 59	25 04 14	22 13	13 33	16 40	17 36	01 51	24 34	12 00	18 55	15 12
06 Sa	07 02 06	15 36 07	09♍19 55	23 23	14 49	17 17	17 46	01 58	24 35	12 01	18 57	15D12
07 Su	07 06 02	16 37 15	23 07 55	24 36	16 04	17 54	17 55	02 05	24 35	12 03	18 59	15 13
08 Mo	07 09 59	17 38 24	06♎29 01	25 50	17 20	18 32	18 05	02 12	24 35	12 04	19 01	15 14
09 Tu	07 13 55	18 39 32	19 25 45	27 06	18 35	19 09	18 14	02 19	24 35	12 06	19 03	15R 14
10 We	07 17 52	19 40 41	02♏01 43	28 24	19 51	19 46	18 23	02 26	24 36	12 07	19 05	15 13
11 Th	07 21 49	20 41 49	14 20 58	29 43	21 06	20 24	18 32	02 33	24 36	12 09	19 07	15 11
12 Fr	07 25 45	21 42 58	26 27 30	01♑03	22 22	21 01	18 41	02 40	24 36	12 10	19 09	15 08
13 Sa	07 29 42	22 44 07	08♐25 08	02 24	23 37	21 38	18 50	02 47	24 37	12 12	19 11	15 05
14 Su	07 33 38	23 45 15	20 17 10	03 47	24 53	22 15	18 59	02 53	24 38	12 14	19 13	15 01
15 Mo	07 37 35	24 46 23	02♑06 29	05 10	26 08	22 52	19 08	03 00	24 38	12 15	19 15	14 58
16 Tu	07 41 31	25 47 31	13 55 26	06 35	27 24	23 30	19 16	03 07	24 39	12 17	19 17	14 55
17 We	07 45 28	26 48 39	25 46 06	08 00	28 39	24 07	19 25	03 14	24 40	12 19	19 19	14 54
18 Th	07 49 24	27 49 45	07♒40 24	09 26	29 55	24 44	19 33	03 20	24 40	12 20	19 21	14D 53
19 Fr	07 53 21	28 50 52	19 40 16	10 52	01♒10	25 21	19 41	03 27	24 41	12 22	19 23	14 54
20 Sa	07 57 18	29 51 57	01♓47 48	12 20	02 25	25 58	19 49	03 34	24 42	12 24	19 25	14 56
21 Su	08 01 14	00♒53 02	14 05 16	13 48	03 41	26 35	19 57	03 40	24 43	12 26	19 27	14 58
22 Mo	08 05 11	01 54 06	26 35 17	15 17	04 56	27 12	20 05	03 47	24 44	12 27	19 29	14 59
23 Tu	08 09 07	02 55 09	09♈20 39	16 46	06 12	27 49	20 13	03 53	24 45	12 29	19 31	15 01
24 We	08 13 04	03 56 11	22 24 17	18 16	07 27	28 26	20 20	04 00	24 46	12 31	19 33	15 02
25 Th	08 17 00	04 57 12	05♉48 47	19 47	08 42	29 03	20 27	04 06	24 47	12 33	19 35	15R 02
26 Fr	08 20 57	05 58 12	19 36 00	21 19	09 58	29 40	20 35	04 12	24 48	12 35	19 37	15 02
27 Sa	08 24 53	06 59 11	03♊46 21	22 51	11 13	00♐17	20 42	04 19	24 50	12 37	19 39	15 01
28 Su	08 28 50	08 00 09	18 18 16	24 24	12 28	00 54	20 49	04 25	24 51	12 39	19 41	15 00
29 Mo	08 32 47	09 01 05	03♋07 45	25 57	13 44	01 31	20 56	04 31	24 52	12 41	19 43	14 58
30 Tu	08 36 43	10 02 01	18 08 31	27 31	14 59	02 08	21 02	04 38	24 54	12 43	19 45	14 57
31 We	08 40 40	11 02 55	03♌12 24	29 06	16 14	02 44	21 09	04 44	24 55	12 45	19 47	14 56

Data for	01-01-2018
Julian Day	2458119.50
Ayanamsa	24 06 18
SVP	05 ♓ 00 18
☽ ☊ Mean	16 ♌ 54 R

● ◐ PHASES ○ ◑

Day	Time	Phase	Position
02	02:25	○	11♋38
08	22:25	◑	18♎36
17	02:18	●	26♑55
24	22:21	◐	04♉53
31	13:26	✷	11♌37

LAST ASPECT ☽ INGRESS

Day	h m	Day	h m	Sign
31	23:39	01	08:11	♋
02	22:47	03	07:23	♌
04	23:10	05	08:13	♍
07	02:51	07	12:15	♎
09	16:13	09	20:06	♏
11	14:54	12	07:05	♐
14	08:49	14	19:43	♑
17	06:31	17	08:33	♒
19	11:53	19	20:28	♓
22	01:14	22	06:28	♈
24	04:17	24	13:41	♉
26	03:17	26	17:40	♊
28	10:40	28	18:58	♋
30	16:41	30	18:53	♌

DECLINATION

Day	☉	☽	☿	♀	♂	♃	♄	♅	♆	♇
01 Mo	23S01	19N19	20S54	23S38	15S14	15S53	22S32	08N59	07S56	21S40
02 Tu	22 56	20 03	21 07	23 35	15 25	15 56	22 32	08 59	07 56	21 40
03 We	22 51	19 19	21 21	23 31	15 36	15 59	22 32	08 59	07 55	21 40
04 Th	22 45	17 13	21 34	23 27	15 48	16 01	22 32	08 59	07 55	21 40
05 Fr	22 38	14 00	21 47	23 22	15 59	16 04	22 32	08 59	07 54	21 39
06 Sa	22 31	10 03	21 59	23 16	16 10	16 06	22 32	09 00	07 54	21 39
07 Su	22 24	05 41	22 11	23 09	16 20	16 09	22 32	09 00	07 53	21 39
08 Mo	22 16	01 10	22 23	23 02	16 31	16 12	22 32	09 00	07 53	21 39
09 Tu	22 08	03S15	22 34	22 54	16 42	16 14	22 31	09 00	07 52	21 39
10 We	22 00	07 24	22 43	22 45	16 52	16 16	22 31	09 00	07 51	21 39
11 Th	21 51	11 08	22 53	22 36	17 03	16 19	22 31	09 00	07 51	21 38
12 Fr	21 41	14 22	23 01	22 26	17 13	16 21	22 31	09 00	07 50	21 38
13 Sa	21 31	16 57	23 08	22 15	17 23	16 24	22 31	09 01	07 49	21 38
14 Su	21 21	18 48	23 14	22 03	17 33	16 26	22 31	09 01	07 49	21 38
15 Mo	21 10	19 50	23 19	21 51	17 43	16 28	22 31	09 01	07 48	21 38
16 Tu	20 59	20 00	23 23	21 38	17 53	16 30	22 31	09 02	07 48	21 37
17 We	20 47	19 17	23 26	21 25	18 02	16 32	22 31	09 02	07 47	21 37
18 Th	20 36	17 42	23 28	21 11	18 12	16 35	22 31	09 02	07 46	21 37
19 Fr	20 23	15 20	23 29	20 56	18 21	16 37	22 30	09 02	07 45	21 37
20 Sa	20 11	12 16	23 28	20 41	18 30	16 39	22 30	09 03	07 45	21 37
21 Su	19 57	08 38	23 26	20 25	18 40	16 41	22 30	09 03	07 44	21 36
22 Mo	19 44	04 35	23 23	20 08	18 49	16 43	22 30	09 04	07 43	21 36
23 Tu	19 30	00 15	23 19	19 51	18 57	16 45	22 30	09 04	07 43	21 36
24 We	19 16	04N11	23 13	19 33	19 06	16 46	22 30	09 04	07 42	21 36
25 Th	19 01	08 32	23 06	19 15	19 15	16 48	22 29	09 05	07 41	21 36
26 Fr	18 47	12 33	22 58	18 56	19 23	16 50	22 29	09 05	07 40	21 35
27 Sa	18 31	15 59	22 48	18 37	19 32	16 52	22 29	09 06	07 40	21 35
28 Su	18 16	18 31	22 37	18 17	19 40	16 53	22 29	09 06	07 39	21 35
29 Mo	18 00	19 52	22 25	17 57	19 48	16 55	22 29	09 07	07 38	21 35
30 Tu	17 44	19 51	22 12	17 36	19 56	16 57	22 29	09 07	07 37	21 35
31 We	17 27	18 24	21 57	17 14	20 04	16 58	22 28	09 08	07 37	21 34

ASPECTARIAN

01 10:27 ☽ ☍ ♄
23:29 ☽ ☍ ♀
02 02:53 ☽ △ ♆
07:42 ☽ △ ♂
09:38 ☉ ⚹ ♆
11:08 ☽ △ ♃
13:44 ☽ ☍ ♇
14:12 ♅ SD
22:47 ☽ □ ♅
03 17:39 ♀ ⚹ ♆
04 09:34 ☽ □ ♂
09:39 ☽ # ♃
10:37 ☿ ∥ ♇
10:53 ☽ # ♂
11:34 ☽ □ ♃
18:52 ☽ △ ☿
23:10 ☽ △ ♅
05 11:25 ☽ △ ♄
15:08 ♂ ∥ ♃
22:59 ☉ ∥ ♄
06 04:37 ☽ ☍ ♆
05:59 ☽ ∥ ♅
10:23 ☽ △ ♀
11:40 ☽ △ ☉
12:05 ☽ # ♆
14:23 ☽ ⚹ ♂
14:44 ☽ ⚹ ♃
16:41 ☽ △ ♇
23:39 ☿ △ ♅
07 00:39 ♂ ☌ ♃
02:51 ☽ □ ☿
15:49 ☿ ∥ ☉
16:09 ☽ □ ♄
08 12:08 ☉ ⚹ ♃
16:14 ♀ ⚹ ♃
19:19 ☿ ∥ ♄
19:55 ♂ ⚹ ♇
22:15 ☽ □ ♀
23:17 ☽ □ ♇
09 07:02 ♀ ☌ ☉
09:02 ♀ ☌ ♇
09:31 ☉ ☌ ♇
09:45 ☽ ☍ ♅
16:13 ☽ ⚹ ☿
21:08 ♀ ⚹ ♂
10 00:47 ☽ ⚹ ♄
02:10 ☿ ∥ ♀
02:53 ☽ ∥ ♆
05:37 ☉ ⚹ ♂
10:02 ☽ # ♅
19:40 ☽ △ ♆
11 05:10 ☿ ♑
08:22 ☽ ☌ ♃
09:26 ☽ ⚹ ♇
10:42 ♀ ∥ ♄
12:35 ☽ ☌ ♂
13:41 ☽ ⚹ ☉
14:54 ☽ ⚹ ♀
12 07:10 ☉ ∥ ♇
18:12 ☽ ∥ ♃
13 05:17 ☽ ∥ ♂
07:04 ☿ ☌ ♄
07:39 ☽ □ ♆
19:09 ♀ □ ♅
14 08:49 ☽ △ ♅
20:45 ☉ □ ♅
15 01:50 ☽ ☌ ♄
07:03 ☽ ☌ ☿
20:39 ☽ ⚹ ♆
16 01:52 ♀ ∥ ♇
03:47 ♃ ⚹ ♇
10:54 ☽ ☌ ♇
10:59 ☽ ⚹ ♃
20:28 ☽ ⚹ ♂
21:45 ☽ □ ♅
17 06:31 ☽ ☌ ♀
18:15 ☽ ∥ ♂
18 01:44 ♀ ♒
12:11 ☽ ∥ ♃
19 00:02 ☽ □ ♃
09:58 ☽ ⚹ ♅
11:53 ☽ □ ♂
20 01:08 ☿ ⚹ ♆
03:10 ☉ ♒
03:30 ☽ ⚹ ♄
20:46 ☽ ☌ ♆
21:26 ☽ # ♅
23:22 ☽ ⚹ ☿
21 05:34 ☽ ∥ ♆
10:23 ☽ ⚹ ♇
11:26 ☽ △ ♃
22 01:14 ☽ △ ♂
10:56 ☽ ⚹ ☉
13:43 ☽ □ ♄
17:29 ☽ ⚹ ♀
23 15:30 ☽ □ ☿
18:48 ☽ □ ♇
24 04:17 ☽ ☌ ♅
10:09 ☉ ∥ ♂
19:18 ☽ # ♆
20:48 ☿ ☌ ♇
20:57 ☽ △ ♄
25 00:15 ♀ ∥ ♂
03:12 ☽ ∥ ♅
05:37 ☽ □ ♀
11:29 ☿ ⚹ ♃
11:50 ☽ ⚹ ♆
26 00:02 ☽ △ ♇
01:41 ☽ ☍ ♃
03:17 ☽ △ ☿
12:57 ♂ ♐
17:53 ☽ ☍ ♂
27 05:46 ☽ △ ☉
07:26 ☽ # ♃
13:32 ☽ △ ♀
14:42 ☽ □ ♆
21:26 ☽ # ☉
21:42 ☽ # ♀
28 06:05 ♀ ∥ ☉
07:08 ☿ □ ♅
10:40 ☽ ⚹ ♅
17:25 ☿ ∥ ♄
21:32 ☽ # ♂
29 02:15 ☽ ☍ ♄
15:19 ☽ △ ♆
21:04 ☽ # ♂
30 02:34 ☽ ☍ ♇
04:39 ☽ △ ♃
10:46 ☽ □ ♅
16:41 ☽ ☍ ☿
23:13 ☽ △ ♂
31 10:35 ☽ # ☉
13:11 ☽ # ♀
13:36 ☽ # ♃
13:40 ☿ ♒
16:14 ♀ ∥ ♃
22:48 ☽ ☍ ♀

⚷ Chiron

01 Dec. 01 N 13

Day	⚷
02	24♓40
05	24 45
08	24 50
11	24 55
14	25 01
17	25 08
20	25 15
23	25 22
26	25 29
29	25 37

February 2018

15 20:53 27♒08 ☀ Solar Eclipse (mag 0.599)

Day	S. T.	☉	☽	☿	♀	♂	♃	♄	♅	♆	♇	☊ True
	h m s	° ' "	° ' "	° '	° '	° '	° '	° '	° '	° '	° '	° '
01 Th	08 44 36	12♒ 03 49	18♌ 10 33	00♒ 41	17♒ 30	03♐ 21	21♏ 15	04♑ 50	24♈ 56	12♓ 47	19♑ 49	14♌ R56
02 Fr	08 48 33	13 04 41	02♍ 54 35	02 17	18 45	03 58	21 21	04 56	24 58	12 49	19 51	14 55
03 Sa	08 52 29	14 05 32	17 17 51	03 54	20 00	04 34	21 27	05 02	25 00	12 51	19 53	14 55
04 Su	08 56 26	15 06 23	01♎ 16 03	05 32	21 16	05 11	21 33	05 08	25 01	12 53	19 55	14 54
05 Mo	09 00 22	16 07 12	14 47 34	07 10	22 31	05 48	21 39	05 14	25 03	12 55	19 57	14 54
06 Tu	09 04 19	17 08 00	27 53 03	08 49	23 46	06 24	21 45	05 20	25 05	12 57	19 58	14 53
07 We	09 08 16	18 08 48	10♏ 35 02	10 29	25 01	07 01	21 50	05 26	25 06	12 59	20 00	14 53
08 Th	09 12 12	19 09 35	22 57 12	12 09	26 17	07 38	21 55	05 32	25 08	13 01	20 02	14D 53
09 Fr	09 16 09	20 10 21	05♐ 04 00	13 51	27 32	08 14	22 01	05 37	25 10	13 03	20 04	14 53
10 Sa	09 20 05	21 11 05	17 00 04	15 33	28 47	08 51	22 06	05 43	25 12	13 06	20 06	14 54
11 Su	09 24 02	22 11 49	28 49 59	17 16	00♓ 02	09 27	22 10	05 49	25 14	13 08	20 08	14 55
12 Mo	09 27 58	23 12 32	10♑ 37 58	18 59	01 17	10 03	22 15	05 54	25 16	13 10	20 09	14 56
13 Tu	09 31 55	24 13 13	22 27 43	20 44	02 32	10 40	22 19	06 00	25 18	13 12	20 11	14 57
14 We	09 35 51	25 13 53	04♒ 22 26	22 29	03 48	11 16	22 24	06 05	25 20	13 14	20 13	14 59
15 Th	09 39 48	26 14 32	16 24 41	24 16	05 03	11 52	22 28	06 11	25 22	13 16	20 15	14 59
16 Fr	09 43 45	27 15 10	28 36 32	26 03	06 18	12 29	22 32	06 16	25 24	13 19	20 16	15R 00
17 Sa	09 47 41	28 15 45	10♓ 59 30	27 51	07 33	13 05	22 36	06 21	25 26	13 21	20 18	14 59
18 Su	09 51 38	29 16 20	23 34 39	29 40	08 48	13 41	22 39	06 26	25 28	13 23	20 20	14 58
19 Mo	09 55 34	00♓ 16 52	06♈ 22 46	01♓ 29	10 03	14 17	22 43	06 31	25 31	13 25	20 21	14 57
20 Tu	09 59 31	01 17 23	19 24 25	03 20	11 18	14 53	22 46	06 36	25 33	13 27	20 23	14 55
21 We	10 03 27	02 17 53	02♉ 40 09	05 11	12 33	15 29	22 49	06 41	25 35	13 30	20 25	14 53
22 Th	10 07 24	03 18 20	16 10 24	07 02	13 48	16 05	22 52	06 46	25 38	13 32	20 26	14 52
23 Fr	10 11 20	04 18 45	29 55 27	08 55	15 03	16 41	22 55	06 51	25 40	13 34	20 28	14 51
24 Sa	10 15 17	05 19 09	13♊ 55 08	10 48	16 18	17 17	22 57	06 56	25 42	13 36	20 29	14D 51
25 Su	10 19 14	06 19 31	28 08 36	12 41	17 33	17 52	22 59	07 01	25 45	13 39	20 31	14 51
26 Mo	10 23 10	07 19 50	12♋ 33 55	14 35	18 48	18 28	23 02	07 05	25 47	13 41	20 33	14 52
27 Tu	10 27 07	08 20 08	27 07 57	16 28	20 03	19 04	23 04	07 10	25 50	13 43	20 34	14 52
28 We	10 31 03	09 20 24	11♌ 46 09	18 22	21 17	19 40	23 05	07 14	25 53	13 46	20 36	14R 52

Data for	02-01-2018
Julian Day	2458150.50
Ayanamsa	24 06 23
SVP	05 ♓ 00 15
☽ ☊ Mean	15 ♌ 15 R

LAST ASPECT ☽ INGRESS

Day	h m	Day	h m	
01	10:59	01	19:13	♍
03	07:07	03	21:47	♎
05	18:46	06	03:57	♏
08	07:17	08	13:55	♐
10	16:39	11	02:22	♑
13	05:44	13	15:13	♒
15	21:07	16	02:43	♓
17	22:15	18	12:06	♈
20	11:12	20	19:12	♉
22	11:46	23	00:08	♊
24	19:58	25	03:06	♋
26	21:51	27	04:42	♌

● ◐ PHASES ○ ◑

07	15:54	◑	18♏49
15	21:06	☀	27♒08
23	08:09	◐	04♊39

DECLINATION

Day	☉	☽	☿	♀	♂	♃	♄	♅	♆	♇
01 Th	17S 10	15N40	21S 40	16S 52	20S 11	17S 00	22S 28	09N08	07S 36	21S 34
02 Fr	16 53	11 58	21 23	16 30	20 19	17 01	22 28	09 09	07 35	21 34
03 Sa	16 36	07 38	21 03	16 07	20 26	17 03	22 28	09 10	07 34	21 34
04 Su	16 18	03 01	20 43	15 44	20 34	17 04	22 27	09 10	07 34	21 34
05 Mo	16 00	01S 36	20 21	15 20	20 41	17 05	22 27	09 11	07 33	21 33
06 Tu	15 42	06 00	19 57	14 56	20 48	17 07	22 27	09 12	07 32	21 33
07 We	15 23	09 59	19 33	14 31	20 54	17 08	22 27	09 12	07 31	21 33
08 Th	15 05	13 26	19 06	14 06	21 01	17 09	22 27	09 13	07 30	21 33
09 Fr	14 45	16 15	18 39	13 41	21 08	17 10	22 26	09 14	07 30	21 33
10 Sa	14 26	18 20	18 10	13 15	21 14	17 12	22 26	09 14	07 29	21 33
11 Su	14 07	19 37	17 39	12 49	21 20	17 13	22 26	09 15	07 28	21 32
12 Mo	13 47	20 02	17 07	12 23	21 26	17 14	22 26	09 16	07 27	21 32
13 Tu	13 27	19 34	16 34	11 56	21 32	17 15	22 25	09 17	07 26	21 32
14 We	13 07	18 13	15 59	11 29	21 38	17 16	22 25	09 17	07 25	21 32
15 Th	12 46	16 03	15 22	11 01	21 44	17 17	22 25	09 18	07 25	21 32
16 Fr	12 25	13 08	14 45	10 34	21 49	17 17	22 25	09 19	07 24	21 31
17 Sa	12 05	09 36	14 06	10 06	21 55	17 18	22 24	09 20	07 23	21 31
18 Su	11 43	05 35	13 25	09 37	22 00	17 19	22 24	09 21	07 22	21 31
19 Mo	11 22	01 16	12 43	09 09	22 05	17 20	22 24	09 21	07 21	21 31
20 Tu	11 01	03N11	12 00	08 40	22 10	17 20	22 24	09 22	07 20	21 31
21 We	10 39	07 33	11 16	08 11	22 15	17 21	22 23	09 23	07 19	21 31
22 Th	10 18	11 38	10 30	07 42	22 19	17 22	22 23	09 24	07 19	21 31
23 Fr	09 56	15 10	09 43	07 13	22 24	17 22	22 23	09 25	07 18	21 30
24 Sa	09 34	17 54	08 54	06 43	22 28	17 23	22 23	09 26	07 17	21 30
25 Su	09 11	19 36	08 05	06 13	22 32	17 23	22 22	09 27	07 16	21 30
26 Mo	08 49	20 03	07 15	05 44	22 36	17 23	22 22	09 28	07 15	21 30
27 Tu	08 27	19 10	06 24	05 14	22 40	17 24	22 22	09 29	07 14	21 30
28 We	08 04	16 59	05 32	04 43	22 44	17 24	22 22	09 30	07 13	21 30

ASPECTARIAN

01 05:00 ☽ □ ♃
08:31 ☿ ∥ ♇
10:59 ☽ △ ♅
13:41 ☉ ∥ ♃

02 01:49 ☽ □ ♂
03:22 ☽ △ ♄
15:50 ☽ ∥ ♅
16:29 ☽ ☍ ♆

03 00:20 ☽ # ♆
04:23 ☽ △ ♇
07:07 ☽ ⚹ ♃
15:58 ☿ ⚹ ♂

04 06:08 ♀ □ ♃
06:50 ☽ □ ♄
07:12 ☽ ⚹ ♂
07:47 ☿ ∥ ♂
08:30 ☽ △ ☿

05 02:36 ☽ △ ☉
09:22 ☽ □ ♇
15:33 ☽ △ ♀
18:46 ☽ ☍ ♅

06 08:58 ☽ ∥ ♆
14:06 ☽ ⚹ ♄
19:08 ☽ # ♅
23:46 ☽ □ ☿

07 01:38 ♀ ⚹ ♅
04:38 ☽ △ ♆
18:16 ☽ ⚹ ♇
21:58 ☽ ☌ ♃

08 04:37 ☽ ∥ ♀
07:17 ☽ □ ♀
11:58 ☽ ∥ ☉

09 06:41 ☽ ☌ ♂
09:45 ☽ ∥ ♃
16:05 ☽ □ ♆
20:34 ☽ ⚹ ☿
22:11 ☽ ∥ ☿

10 09:16 ☽ ⚹ ☉
16:39 ☽ △ ♅
23:20 ♀ ♓
23:21 ☉ □ ♃

11 02:44 ☽ ⚹ ♀
14:18 ☽ ☌ ♄
19:12 ☿ ∥ ♃

12 05:10 ☽ ⚹ ♆
19:23 ☽ ☌ ♇
22:55 ♂ ∥ ♇
23:43 ☽ ⚹ ♃

13 05:44 ☽ □ ♅
22:39 ☿ □ ♃

14 02:23 ☉ ⚹ ♅
11:32 ☽ ∥ ♃
14:30 ☽ ⚹ ♂

15 07:45 ☽ ∥ ☿
12:01 ☽ □ ♃
15:08 ☿ ⚹ ♅
17:42 ☽ ⚹ ♅
18:08 ☽ ☌ ☿
23:20 ♀ ⚹ ♄

16 05:39 ☽ ∥ ☉
14:59 ☽ ⚹ ♄
16:37 ☽ ☌ ♀
20:19 ☽ ∥ ♀

17 01:37 ☽ # ♅
04:13 ☽ □ ♂
04:32 ☽ ☌ ♆
11:21 ♂ □ ♆
12:28 ☿ ☌ ☉
13:33 ☽ ∥ ♆
17:50 ☽ ⚹ ♇
22:15 ☽ △ ♃

18 04:29 ☿ ♓
13:49 ♀ # ♅
17:19 ☉ ♓

19 00:16 ☽ □ ♄
15:19 ☽ △ ♂

20 01:47 ☽ □ ♇
11:12 ☽ ☌ ♅
22:42 ☽ # ♆
23:17 ☽ ⚹ ☉

21 03:13 ☽ # ♀
05:13 ☽ ⚹ ☿
07:14 ☽ △ ♄
10:32 ☽ ∥ ♅
16:32 ☽ # ☉
18:12 ☽ # ☿
18:42 ♀ ☌ ♆
19:20 ☽ ⚹ ♆
19:23 ☽ ⚹ ♀
20:24 ☿ ⚹ ♄

22 07:30 ☽ △ ♇
11:47 ☽ ☍ ♃
11:47 ☿ ∥ ☉
19:47 ♂ ∥ ♄
19:54 ♀ ∥ ♆

23 08:46 ☿ # ♅
17:51 ☽ □ ☿
18:43 ☽ # ♃
23:28 ☽ □ ♆

24 04:26 ☽ □ ♀
05:57 ☽ ☍ ♂
08:02 ☉ # ♅
19:58 ☽ ⚹ ♅

25 12:02 ♀ □ ♂
12:26 ☿ ☌ ♆
14:40 ☽ △ ☉
14:52 ☽ ☍ ♄
17:47 ☉ ⚹ ♄
23:54 ☿ ∥ ♆

26 01:51 ☽ △ ♆
03:50 ☽ △ ☿
11:15 ☽ △ ♀
13:11 ☽ ☍ ♇
17:18 ☽ △ ♃
21:51 ☽ □ ♅

27 10:19 ♀ ⚹ ♇
20:15 ☽ # ♃

28 13:29 ☽ △ ♂
18:37 ☽ □ ♃
23:14 ☽ △ ♅
23:57 ☿ □ ♂

⚷ Chiron

01	Dec.	01 N 33
01	25♓46	
04	25 54	
07	26 03	
10	26 12	
13	26 21	
16	26 31	
19	26 41	
22	26 51	
25	27 01	
28	27 11	

March 2018

Day	S. T. h m s	☉ ° ' "	☽ ° ' "	☿ ° '	♀ ° '	♂ ° '	♃ ° '	♄ ° '	♅ ° '	♆ ° '	♇ ° '	☊ True ° '
01 Th	10 34 60	10♓20 38	26♌22 49	20♓15	22♓32	20♐15	23♏07	07♑19	25♈55	13♓48	20♑37	14♌R50
02 Fr	10 38 56	11 20 50	10♍51 40	22 08	23 47	20 51	23 09	07 23	25 58	13 50	20 38	14 48
03 Sa	10 42 53	12 21 00	25 06 35	24 00	25 02	21 26	23 10	07 28	26 00	13 52	20 40	14 44
04 Su	10 46 49	13 21 08	09♎02 29	25 51	26 17	22 02	23 11	07 32	26 03	13 55	20 41	14 40
05 Mo	10 50 46	14 21 15	22 36 03	27 40	27 31	22 37	23 12	07 36	26 06	13 57	20 43	14 35
06 Tu	10 54 43	15 21 20	05♏45 56	29 27	28 46	23 12	23 12	07 40	26 09	13 59	20 44	14 30
07 We	10 58 39	16 21 23	18 32 53	01♈11	00♈01	23 47	23 13	07 44	26 11	14 01	20 45	14 26
08 Th	11 02 36	17 21 25	00♐59 21	02 53	01 15	24 23	23 13	07 48	26 14	14 04	20 47	14 23
09 Fr	11 06 32	18 21 25	13 09 04	04 31	02 30	24 58	23 13	07 51	26 17	14 06	20 48	14 22
10 Sa	11 10 29	19 21 24	25 06 39	06 05	03 45	25 33	23R 13	07 55	26 20	14 08	20 49	14D 23
11 Su	11 14 25	20 21 21	06♑57 09	07 35	04 59	26 08	23 13	07 59	26 23	14 11	20 50	14 24
12 Mo	11 18 22	21 21 17	18 45 39	08 59	06 14	26 43	23 13	08 02	26 26	14 13	20 52	14 26
13 Tu	11 22 18	22 21 10	00♒37 03	10 18	07 28	27 18	23 12	08 06	26 29	14 15	20 53	14 29
14 We	11 26 15	23 21 02	12 35 45	11 31	08 43	27 52	23 11	08 09	26 32	14 17	20 54	14 30
15 Th	11 30 12	24 20 52	24 45 30	12 37	09 57	28 27	23 10	08 13	26 35	14 20	20 55	14R 30
16 Fr	11 34 08	25 20 41	07♓09 11	13 36	11 12	29 02	23 09	08 16	26 38	14 22	20 56	14 28
17 Sa	11 38 05	26 20 27	19 48 36	14 28	12 26	29 36	23 08	08 19	26 41	14 24	20 57	14 23
18 Su	11 42 01	27 20 12	02♈44 25	15 13	13 41	00♑10	23 06	08 22	26 44	14 26	20 58	14 18
19 Mo	11 45 58	28 19 54	15 56 06	15 49	14 55	00 45	23 04	08 25	26 47	14 29	20 59	14 11
20 Tu	11 49 54	29 19 34	29 22 10	16 18	16 09	01 19	23 02	08 28	26 50	14 31	21 00	14 03
21 We	11 53 51	00♈19 13	13♉00 30	16 38	17 24	01 53	23 00	08 30	26 53	14 33	21 01	13 56
22 Th	11 57 47	01 18 49	26 48 47	16 50	18 38	02 27	22 58	08 33	26 57	14 35	21 02	13 51
23 Fr	12 01 44	02 18 23	10♊44 50	16 54	19 52	03 01	22 55	08 36	27 00	14 37	21 03	13 47
24 Sa	12 05 40	03 17 54	24 46 50	16R 51	21 06	03 35	22 53	08 38	27 03	14 40	21 04	13 45
25 Su	12 09 37	04 17 23	08♋53 15	16 39	22 21	04 09	22 50	08 40	27 06	14 42	21 05	13 44
26 Mo	12 13 34	05 16 50	23 02 51	16 21	23 35	04 43	22 47	08 43	27 09	14 44	21 06	13D 44
27 Tu	12 17 30	06 16 15	07♌14 14	15 55	24 49	05 16	22 44	08 45	27 13	14 46	21 07	13 44
28 We	12 21 27	07 15 37	21 25 38	15 24	26 03	05 50	22 40	08 47	27 16	14 48	21 07	13 43
29 Th	12 25 23	08 14 57	05♍34 30	14 47	27 17	06 23	22 37	08 49	27 19	14 50	21 08	13 40
30 Fr	12 29 20	09 14 15	19 37 34	14 06	28 31	06 56	22 33	08 51	27 23	14 53	21 09	13 34
31 Sa	12 33 16	10 13 30	03♎30 58	13 21	29 45	07 30	22 29	08 53	27 26	14 55	21 09	13 27

Data for 03-01-2018
Julian Day 2458178.50
Ayanamsa 24 06 26
SVP 05 ♓ 00 10
☽ ☊ Mean 13 ♌ 46 R

● ◐ PHASES ○ ◑

02	00:52	○	11♍23
09	11:21	◑	18♐50
17	13:11	●	26♓53
24	15:35	◐	03♋57
31	12:38	○	10♎45

LAST ASPECT ☽ Day	h m	INGRESS Day	h m	
28	23:14	01	05:58	♍
02	23:51	03	08:21	♎
05	06:20	05	13:24	♏
07	08:56	07	22:04	♐
10	02:29	10	09:53	♑
12	15:37	12	22:45	♒
15	07:33	15	10:12	♓
17	13:12	17	18:57	♈
19	19:29	20	01:07	♉
21	17:21	22	05:30	♊
24	03:53	24	08:54	♋
26	06:59	26	11:46	♌
28	09:56	28	14:32	♍
30	05:00	30	17:53	♎

DECLINATION

Day	☉	☽	☿	♀	♂	♃	♄	♅	♆	♇
01 Th	07S41	13N43	04S39	04S13	22S48	17S24	22S22	09N31	07S12	21S30
02 Fr	07 18	09 38	03 46	03 43	22 51	17 24	22 21	09 32	07 12	21 29
03 Sa	06 56	05 04	02 52	03 12	22 55	17 25	22 21	09 33	07 11	21 29
04 Su	06 33	00 20	01 58	02 42	22 58	17 25	22 21	09 34	07 10	21 29
05 Mo	06 09	04S17	01 04	02 11	23 01	17 25	22 21	09 35	07 09	21 29
06 Tu	05 46	08 34	00 11	01 40	23 04	17 25	22 20	09 36	07 08	21 29
07 We	05 23	12 20	00N42	01 09	23 06	17 25	22 20	09 37	07 07	21 29
08 Th	05 00	15 27	01 34	00 39	23 09	17 25	22 20	09 38	07 06	21 29
09 Fr	04 36	17 50	02 25	00 08	23 11	17 25	22 20	09 39	07 06	21 29
10 Sa	04 13	19 23	03 14	00N23	23 14	17 24	22 19	09 40	07 05	21 28
11 Su	03 49	20 05	04 02	00 54	23 16	17 24	22 19	09 41	07 04	21 28
12 Mo	03 26	19 53	04 48	01 25	23 18	17 24	22 19	09 42	07 03	21 28
13 Tu	03 02	18 48	05 31	01 56	23 20	17 24	22 19	09 43	07 02	21 28
14 We	02 38	16 52	06 12	02 27	23 22	17 23	22 18	09 44	07 01	21 28
15 Th	02 15	14 09	06 50	02 58	23 23	17 23	22 18	09 45	07 00	21 28
16 Fr	01 51	10 45	07 24	03 28	23 25	17 23	22 18	09 46	06 59	21 28
17 Sa	01 27	06 48	07 55	03 59	23 26	17 22	22 18	09 47	06 59	21 28
18 Su	01 04	02 28	08 23	04 29	23 28	17 22	22 18	09 49	06 58	21 28
19 Mo	00 40	02N04	08 47	05 00	23 29	17 21	22 17	09 50	06 57	21 28
20 Tu	00 16	06 35	09 07	05 30	23 30	17 20	22 17	09 51	06 56	21 28
21 We	00N08	10 50	09 22	06 01	23 31	17 20	22 17	09 52	06 55	21 27
22 Th	00 31	14 34	09 34	06 31	23 31	17 19	22 17	09 53	06 54	21 27
23 Fr	00 55	17 31	09 41	07 01	23 32	17 18	22 17	09 54	06 54	21 27
24 Sa	01 19	19 27	09 44	07 30	23 32	17 17	22 17	09 56	06 53	21 27
25 Su	01 42	20 12	09 42	08 00	23 33	17 17	22 16	09 57	06 52	21 27
26 Mo	02 06	19 41	09 36	08 30	23 33	17 16	22 16	09 58	06 51	21 27
27 Tu	02 29	17 54	09 26	08 59	23 33	17 15	22 16	09 59	06 50	21 27
28 We	02 53	15 01	09 12	09 28	23 33	17 14	22 16	10 00	06 49	21 27
29 Th	03 16	11 15	08 54	09 57	23 33	17 13	22 16	10 01	06 49	21 27
30 Fr	03 40	06 52	08 33	10 25	23 33	17 12	22 16	10 03	06 48	21 27
31 Sa	04 03	02 11	08 08	10 54	23 33	17 11	22 15	10 04	06 47	21 27

ASPECTARIAN

01 04:42 ☿ ✶ ♇
11:23 ♀ △ ♃
18:11 ☽ △ ♄

02 00:32 ☽ ∥ ♅
02:59 ☿ ∥ ♀
04:59 ☽ ☍ ♆
07:30 ☉ ∥ ♆
13:00 ☽ # ♆
13:06 ☿ △ ♃
13:29 ☽ # ☉
16:27 ☽ △ ♇
17:30 ☽ □ ♂
20:41 ☽ ✶ ♃
21:49 ☽ ☍ ☿
23:51 ☽ ☍ ♀

03 10:32 ☽ # ♀
13:42 ☽ # ☿
21:21 ☽ □ ♄
04 09:51 ☽ ∥ ☿
13:56 ☉ ☌ ♆
13:58 ☽ ∥ ♀
18:05 ☿ ☌ ♀
20:36 ☽ □ ♇

05 00:02 ☽ ✶ ♂
06:20 ☽ ☍ ♅
09:19 ☽ ∥ ☉
15:41 ☽ ∥ ♆

06 03:32 ☽ ✶ ♄
06:11 ☽ # ♅
07:35 ☿ ♈
15:24 ☽ △ ♆
19:29 ☽ △ ☉
23:46 ♀ ♈

07 04:13 ☽ ✶ ♇
07:56 ☿ # ♀
08:56 ☽ ☌ ♃
08 00:35 ☽ △ ♀
04:17 ☽ △ ☿
19:08 ☽ ∥ ♃

09 01:54 ☽ □ ♆
04:47 ♃ SR
10 00:55 ☽ ☌ ♂
02:29 ☽ △ ♅
19:32 ☽ □ ♀
19:41 ☿ # ☉

11 01:27 ☽ □ ☿
02:06 ☽ ☌ ♄
07:01 ☿ □ ♄
11:23 ♂ △ ♅
11:54 ☉ ✶ ♇
14:44 ☽ ✶ ♆
12 04:16 ☽ ☌ ♇
05:45 ☽ ✶ ☉
09:01 ☽ ✶ ♃
15:37 ☽ □ ♅

13 12:40 ♀ □ ♄
15:22 ☽ ✶ ♀
18:19 ☽ ∥ ♃
20:06 ☉ △ ♃
21:37 ☽ ✶ ☿
14 05:09 ♀ # ☉
20:54 ☽ □ ♃

15 03:34 ☽ ✶ ♅
07:07 ☿ # ♆
07:33 ☽ ✶ ♂

16 02:08 ☽ ✶ ♄
06:08 ☽ # ♅
13:47 ☽ ☌ ♆
18:07 ☽ # ☿
22:55 ☽ ∥ ♆

17 02:09 ☽ ✶ ♇
06:12 ☽ △ ♃
14:05 ☽ # ♀
16:42 ♂ ♑
19:04 ☽ □ ♂

18 08:09 ☽ ∥ ☉
10:19 ☽ □ ♄
17:09 ☽ # ☉
21:58 ☽ ☌ ♀
23:47 ☽ ☌ ☿

19 09:05 ☽ □ ♇
17:24 ☽ ∥ ♀
19:29 ☽ ☌ ♅
20 01:52 ☽ # ♆
03:36 ☽ △ ♂
04:03 ☿ ☌ ♀
14:56 ☽ ∥ ☿
16:05 ☽ △ ♄
16:16 ☉ ♈
18:17 ☽ ∥ ♅

21 02:42 ☽ ✶ ♆
13:59 ☽ △ ♇
17:21 ☽ ☍ ♃
22 08:22 ☽ ✶ ☉
18:30 ♀ # ♆
21:59 ☽ # ♃

23 00:20 ☿ SR
06:40 ☽ □ ♆
10:32 ☽ ✶ ☿
17:07 ☽ ✶ ♀
23:14 ♀ □ ♇

24 03:53 ☽ ✶ ♅
15:37 ☽ ☍ ♂
16:08 ☉ □ ♂
23:38 ☽ ☍ ♄

25 09:53 ☽ △ ♆
12:55 ☽ □ ☿
20:41 ☽ ☍ ♇
23:33 ☽ △ ♃

26 00:59 ☽ □ ♀
06:59 ☽ □ ♅
22:15 ☽ △ ☉

27 06:17 ☽ # ♃
14:11 ☽ △ ☿
15:12 ☿ ∥ ♀
28 02:06 ☽ □ ♃
08:34 ☽ △ ♀
09:56 ☽ △ ♅

29 00:48 ♀ ☌ ♅
01:26 ☽ △ ♂
04:06 ♀ ∥ ♅
05:32 ☽ △ ♄
06:38 ☽ ∥ ♀
06:55 ☽ ∥ ♅
14:17 ☽ ∥ ☿
14:17 ☉ □ ♄
15:50 ☽ ☍ ♆

30 00:22 ☽ # ♆
02:37 ☽ △ ♇
05:00 ☽ ✶ ♃
15:16 ☽ ∥ ☉

31 04:54 ♀ ♉
07:14 ☽ □ ♂
09:23 ☽ □ ♄
16:17 ☽ ☍ ☿

⚷ Chiron

01 Dec. 02 N 04

03	27♓21
06	27 32
09	27 42
12	27 53
15	28 04
18	28 15
21	28 25
24	28 36
27	28 47
30	28 57

April 2018

Day	S. T.	☉	☽	☿	♀	♂	♃	♄	♅	♆	♇	☊ True
	h m s	° ' "	° ' "	° '	° '	° '	° '	° '	° '	° '	° '	° '
01 Su	12 37 13	11♈12 44	17♎10 49	12♈R33	00♉59	08♑03	22♏R25	08♑55	27♈29	14♓57	21♑10	13♌R17
02 Mo	12 41 09	12 11 55	00♏33 51	11 44	02 13	08 36	22 21	08 56	27 33	14 59	21 11	13 07
03 Tu	12 45 06	13 11 05	13 37 58	10 55	03 27	09 08	22 17	08 58	27 36	15 01	21 11	12 56
04 We	12 49 03	14 10 12	26 22 37	10 05	04 40	09 41	22 12	08 59	27 39	15 03	21 12	12 47
05 Th	12 52 59	15 09 18	08♐48 51	09 17	05 54	10 14	22 07	09 00	27 43	15 05	21 12	12 39
06 Fr	12 56 56	16 08 22	20 59 16	08 31	07 08	10 46	22 03	09 02	27 46	15 07	21 13	12 34
07 Sa	13 00 52	17 07 24	02♑57 38	07 48	08 22	11 19	21 58	09 03	27 49	15 09	21 13	12 31
08 Su	13 04 49	18 06 25	14 48 37	07 09	09 35	11 51	21 52	09 04	27 53	15 11	21 14	12D 31
09 Mo	13 08 45	19 05 24	26 37 29	06 34	10 49	12 23	21 47	09 05	27 56	15 13	21 14	12 32
10 Tu	13 12 42	20 04 20	08♒29 39	06 03	12 03	12 55	21 42	09 06	28 00	15 15	21 15	12 33
11 We	13 16 38	21 03 16	20 30 23	05 38	13 16	13 27	21 36	09 06	28 03	15 17	21 15	12R 33
12 Th	13 20 35	22 02 09	02♓44 31	05 17	14 30	13 58	21 31	09 07	28 06	15 19	21 15	12 32
13 Fr	13 24 32	23 01 01	15 16 04	05 02	15 43	14 30	21 25	09 08	28 10	15 21	21 16	12 27
14 Sa	13 28 28	23 59 50	28 07 50	04 52	16 57	15 01	21 19	09 08	28 13	15 23	21 16	12 20
15 Su	13 32 25	24 58 38	11♈20 57	04 47	18 10	15 32	21 13	09 08	28 17	15 24	21 16	12 11
16 Mo	13 36 21	25 57 24	24 54 43	04D 48	19 24	16 03	21 06	09 09	28 20	15 26	21 16	12 00
17 Tu	13 40 18	26 56 08	08♉46 24	04 53	20 37	16 34	21 00	09 09	28 24	15 28	21 17	11 49
18 We	13 44 14	27 54 50	22 51 48	05 04	21 50	17 05	20 54	09 09	28 27	15 30	21 17	11 38
19 Th	13 48 11	28 53 30	07♊05 48	05 20	23 04	17 35	20 47	09R 09	28 31	15 32	21 17	11 29
20 Fr	13 52 07	29 52 08	21 23 22	05 40	24 17	18 06	20 41	09 09	28 34	15 34	21 17	11 22
21 Sa	13 56 04	00♉50 44	05♋40 12	06 05	25 30	18 36	20 34	09 09	28 37	15 35	21 17	11 18
22 Su	14 00 01	01 49 18	19 53 10	06 34	26 43	19 06	20 27	09 08	28 41	15 37	21 17	11 15
23 Mo	14 03 57	02 47 49	04♌00 24	07 08	27 56	19 36	20 20	09 08	28 44	15 39	21R 17	11 14
24 Tu	14 07 54	03 46 18	18 01 00	07 45	29 09	20 05	20 13	09 07	28 48	15 40	21 17	11 13
25 We	14 11 50	04 44 45	01♍54 30	08 26	00♊22	20 35	20 06	09 07	28 51	15 42	21 17	11 11
26 Th	14 15 47	05 43 10	15 40 31	09 11	01 35	21 04	19 59	09 06	28 55	15 44	21 17	11 08
27 Fr	14 19 43	06 41 33	29 18 15	09 59	02 48	21 33	19 52	09 05	28 58	15 45	21 17	11 01
28 Sa	14 23 40	07 39 53	12♎46 24	10 51	04 01	22 01	19 44	09 04	29 01	15 47	21 17	10 52
29 Su	14 27 36	08 38 12	26 03 15	11 46	05 14	22 30	19 37	09 03	29 05	15 48	21 16	10 40
30 Mo	14 31 33	09 36 29	09♏07 02	12 43	06 27	22 58	19 30	09 02	29 08	15 50	21 16	10 27

Data for	04-01-2018
Julian Day	2458209.50
Ayanamsa	24 06 29
SVP	05 ♓ 00 04
☽ ☊ Mean	12 ♌ 08 R

● ◐ PHASES ○ ◑

Day	h m	Phase	Position
08	07:18	◑	18♑24
16	01:57	●	26♈02
22	21:46	◐	02♌42
30	00:58	○	09♏39

LAST ASPECT ☽ Day	h m	INGRESS Day	h m	Sign
01	18:31	01	22:59	♏
03	16:07	04	06:56	♐
06	13:37	06	18:02	♑
09	02:40	09	06:51	♒
11	14:56	11	18:40	♓
13	11:27	14	03:26	♈
16	06:00	16	08:52	♉
17	22:06	18	12:03	♊
20	12:06	20	14:28	♋
22	14:59	22	17:10	♌
24	18:41	24	20:41	♍
26	09:50	27	01:14	♎
29	05:33	29	07:12	♏

DECLINATION

Day	☉	☽	☿	♀	♂	♃	♄	♅	♆	♇
01 Su	04N26	02S32	07N42	11N22	23S32	17S10	22S15	10N05	06S46	21S27
02 Mo	04 49	07 01	07 13	11 50	23 32	17 08	22 15	10 06	06 46	21 27
03 Tu	05 12	11 04	06 42	12 17	23 31	17 07	22 15	10 07	06 45	21 27
04 We	05 35	14 32	06 11	12 45	23 30	17 06	22 15	10 09	06 44	21 27
05 Th	05 58	17 15	05 39	13 12	23 29	17 05	22 15	10 10	06 43	21 27
06 Fr	06 21	19 08	05 07	13 38	23 28	17 03	22 15	10 11	06 42	21 27
07 Sa	06 43	20 09	04 36	14 04	23 27	17 02	22 15	10 12	06 42	21 27
08 Su	07 06	20 15	04 06	14 30	23 26	17 01	22 15	10 14	06 41	21 27
09 Mo	07 28	19 27	03 37	14 56	23 25	16 59	22 15	10 15	06 40	21 27
10 Tu	07 51	17 47	03 10	15 21	23 24	16 58	22 14	10 16	06 39	21 27
11 We	08 13	15 19	02 44	15 46	23 22	16 56	22 14	10 17	06 39	21 27
12 Th	08 35	12 08	02 22	16 11	23 21	16 55	22 14	10 19	06 38	21 27
13 Fr	08 57	08 20	02 01	16 35	23 19	16 53	22 14	10 20	06 37	21 27
14 Sa	09 19	04 04	01 43	16 58	23 18	16 52	22 14	10 21	06 37	21 27
15 Su	09 40	00N30	01 27	17 22	23 16	16 50	22 14	10 22	06 36	21 27
16 Mo	10 02	05 10	01 15	17 45	23 14	16 48	22 14	10 23	06 35	21 27
17 Tu	10 23	09 40	01 04	18 07	23 12	16 47	22 14	10 25	06 35	21 27
18 We	10 44	13 43	00 57	18 29	23 10	16 45	22 14	10 26	06 34	21 27
19 Th	11 05	17 01	00 52	18 50	23 08	16 43	22 14	10 27	06 33	21 28
20 Fr	11 25	19 18	00 49	19 11	23 06	16 41	22 14	10 28	06 33	21 28
21 Sa	11 46	20 22	00 49	19 31	23 04	16 40	22 14	10 30	06 32	21 28
22 Su	12 06	20 08	00 51	19 51	23 02	16 38	22 14	10 31	06 31	21 28
23 Mo	12 26	18 38	00 56	20 11	23 00	16 36	22 14	10 32	06 31	21 28
24 Tu	12 46	16 00	01 03	20 30	22 58	16 34	22 14	10 33	06 30	21 28
25 We	13 06	12 27	01 12	20 48	22 56	16 32	22 14	10 34	06 29	21 28
26 Th	13 26	08 16	01 23	21 06	22 53	16 30	22 14	10 36	06 29	21 28
27 Fr	13 45	03 43	01 36	21 23	22 51	16 28	22 14	10 37	06 28	21 28
28 Sa	14 04	00S59	01 51	21 39	22 49	16 26	22 14	10 38	06 28	21 28
29 Su	14 23	05 33	02 08	21 55	22 46	16 24	22 14	10 39	06 27	21 28
30 Mo	14 41	09 48	02 26	22 11	22 44	16 22	22 15	10 41	06 26	21 29

ASPECTARIAN

01 07:06 ☽ □ ♇
10:54 ☽ # ☉
17:54 ☿ ☌ ☉
18:31 ☽ ☍ ♅
22:32 ☽ ∥ ♆

02 00:57 ☽ # ☿
03:18 ☽ ☍ ♀
15:19 ☽ ✶ ♂
15:20 ☽ ✶ ♄
15:45 ♂ ☌ ♄
18:01 ☽ # ♅
22:02 ☿ # ♆

03 02:35 ☽ △ ♆
09:01 ☽ # ♀
14:10 ☽ ✶ ♇
16:07 ☽ ☌ ♃

04 07:06 ☿ □ ♂
15:37 ☿ ∥ ☉
22:12 ☽ ∥ ♃

05 00:52 ☽ △ ☿
08:23 ☿ □ ♄
12:20 ☽ □ ♆
13:32 ☽ △ ☉

06 13:37 ☽ △ ♅
22:09 ☉ # ♆

07 09:16 ☽ □ ☿
12:11 ☽ △ ♀
12:20 ☽ ☌ ♄
13:37 ♀ △ ♄
17:42 ☽ ☌ ♂

08 00:46 ☽ ✶ ♆
13:03 ☽ ☌ ♇
14:15 ☽ ✶ ♃

09 02:40 ☽ □ ♅
19:17 ☽ ✶ ☿

10 07:57 ☽ □ ♀
09:00 ☽ ∥ ♃
20:38 ☽ # ♀

11 01:11 ☽ ✶ ☉
02:09 ☽ □ ♃
04:52 ☉ □ ♇
06:03 ♀ △ ♂
14:56 ☽ ✶ ♅

12 11:55 ☽ # ♅
12:18 ☽ ✶ ♄
16:27 ♀ ✶ ♆
20:40 ☽ # ☉
22:29 ☽ ✶ ♂

13 00:09 ☽ ☌ ♆
00:57 ☽ ✶ ♀
09:56 ☽ ∥ ♆
11:16 ☽ ✶ ♇
11:27 ☽ △ ♃
17:23 ♀ # ♃

14 10:29 ♃ ✶ ♇
12:13 ☽ ☌ ☿
13:15 ☽ # ☿
17:29 ♂ ✶ ♆
20:02 ☽ □ ♄

15 04:40 ☽ ∥ ☿
07:47 ☽ □ ♂
09:22 ☿ SD
17:37 ☽ □ ♆

16 06:00 ☽ ☌ ♅
07:24 ☽ # ♆

17 00:39 ☽ △ ♄
02:19 ☉ ∥ ♅
04:12 ☽ ∥ ♅
04:22 ☽ ∥ ☉
07:00 ♀ ☍ ♃
11:28 ☽ ✶ ♆
13:03 ♀ △ ♇
13:49 ☽ △ ♂
20:42 ☽ ☍ ♃
21:19 ☽ △ ♇
22:06 ☽ ☌ ♀

18 01:48 ♄ SR
14:00 ☉ ☌ ♅
20:58 ☽ ✶ ☿
21:34 ☽ # ♃

19 14:11 ☽ □ ♆
21:57 ☽ ∥ ♀

20 03:13 ☉ ♉
12:06 ☽ ✶ ♅
15:17 ☽ ✶ ☉

21 00:43 ☽ □ ☿
05:51 ☽ ☍ ♄
16:46 ☽ △ ♆
22:37 ☽ ☍ ♂

22 00:57 ☽ △ ♃
02:22 ☽ ☍ ♇
05:03 ☽ ∥ ♀
12:41 ☽ ✶ ♀
14:59 ☽ □ ♅
15:24 ♇ SR

23 05:34 ☽ △ ☿
19:27 ☽ # ♃

24 03:46 ☽ □ ♃
05:13 ♂ ✶ ♃
16:40 ♀ ♊
18:41 ☽ △ ♅
20:18 ☽ ∥ ☉
21:05 ☽ □ ♀

25 05:18 ☽ △ ☉
11:07 ☽ ∥ ♅
12:31 ☽ △ ♄
21:29 ☿ □ ♄

26 00:05 ☽ ☍ ♆
07:29 ☽ ✶ ♃
09:39 ☽ # ♆
09:48 ☽ △ ♂
09:50 ☽ △ ♇
10:55 ♂ ☌ ♇

27 06:49 ☽ △ ♀
07:49 ♀ # ♇
10:20 ☽ ∥ ☿
17:23 ☽ □ ♄
20:18 ☽ ☍ ☿

28 04:48 ☽ # ☿
15:19 ☽ □ ♇
17:18 ☽ □ ♂

29 04:56 ☽ ∥ ♆
05:33 ☽ ☍ ♅
10:04 ☉ △ ♄
23:51 ☽ ✶ ♄

30 05:26 ☽ # ♅
06:06 ♀ # ♄
12:32 ☽ △ ♆
19:12 ☽ ☌ ♇
22:44 ☽ ✶ ♇

⚷ Chiron

01	Dec.	02 N 46
02	29♓08	
05	29 18	
08	29 29	
11	29 39	
14	29 49	
17	29 59	
20	00♈09	
23	00 18	
26	00 27	
29	00 36	

17 08:25 ♈

May 2018

Day	S. T. h m s	☉ ° ' ''	☽ ° ' ''	☿ ° '	♀ ° '	♂ ° '	♃ ° '	♄ ° '	♅ ° '	♆ ° '	♇ ° '	☊ True ° '
01 Tu	14 35 29	10♉ 34 44	21♏ 56 21	13♈ 44	07♊ 39	23♑ 26	19♏R22	09♑R01	29♈ 12	15♓ 51	21♑R16	10♌R13
02 We	14 39 26	11 32 57	04♐ 30 35	14 48	08 52	23 54	19 15	09 00	29 15	15 53	21 16	10 00
03 Th	14 43 23	12 31 09	16 50 09	15 54	10 05	24 22	19 07	08 58	29 18	15 54	21 15	09 50
04 Fr	14 47 19	13 29 19	28 56 41	17 03	11 17	24 49	19 00	08 57	29 22	15 56	21 15	09 42
05 Sa	14 51 16	14 27 28	10♑ 52 59	18 15	12 30	25 16	18 52	08 55	29 25	15 57	21 15	09 38
06 Su	14 55 12	15 25 35	22 42 52	19 29	13 42	25 43	18 45	08 53	29 28	15 59	21 14	09 36
07 Mo	14 59 09	16 23 41	04♒ 31 01	20 45	14 55	26 10	18 37	08 52	29 32	16 00	21 14	09D 36
08 Tu	15 03 05	17 21 45	16 22 39	22 04	16 07	26 36	18 29	08 50	29 35	16 01	21 14	09 37
09 We	15 07 02	18 19 48	28 23 14	23 25	17 20	27 02	18 22	08 48	29 38	16 03	21 13	09R 37
10 Th	15 10 58	19 17 50	10♓ 38 08	24 48	18 32	27 28	18 14	08 46	29 42	16 04	21 13	09 36
11 Fr	15 14 55	20 15 50	23 12 13	26 14	19 44	27 53	18 06	08 44	29 45	16 05	21 12	09 33
12 Sa	15 18 52	21 13 49	06♈ 09 22	27 42	20 56	28 18	17 59	08 42	29 48	16 06	21 12	09 27
13 Su	15 22 48	22 11 47	19 31 49	29 12	22 09	28 43	17 51	08 39	29 51	16 07	21 11	09 18
14 Mo	15 26 45	23 09 43	03♉ 19 37	00♉ 44	23 21	29 07	17 43	08 37	29 55	16 09	21 10	09 08
15 Tu	15 30 41	24 07 38	17 30 14	02 18	24 33	29 31	17 36	08 35	29 58	16 10	21 10	08 57
16 We	15 34 38	25 05 31	01♊ 58 36	03 54	25 45	29 55	17 28	08 32	00♉ 01	16 11	21 09	08 46
17 Th	15 38 34	26 03 23	16 37 52	05 33	26 57	00♒ 19	17 21	08 29	00 04	16 12	21 09	08 37
18 Fr	15 42 31	27 01 13	01♋ 20 28	07 13	28 09	00 42	17 13	08 27	00 07	16 13	21 08	08 30
19 Sa	15 46 27	27 59 02	15 59 21	08 56	29 21	01 04	17 06	08 24	00 11	16 14	21 07	08 25
20 Su	15 50 24	28 56 49	00♌ 29 01	10 40	00♋ 32	01 27	16 59	08 21	00 14	16 15	21 06	08 23
21 Mo	15 54 21	29 54 35	14 45 55	12 27	01 44	01 48	16 51	08 18	00 17	16 16	21 06	08 22
22 Tu	15 58 17	00♊ 52 18	28 48 21	14 16	02 56	02 10	16 44	08 15	00 20	16 17	21 05	08 22
23 We	16 02 14	01 50 01	12♍ 36 04	16 07	04 07	02 31	16 37	08 12	00 23	16 18	21 04	08 21
24 Th	16 06 10	02 47 41	26 09 34	18 00	05 19	02 52	16 30	08 09	00 26	16 19	21 03	08 18
25 Fr	16 10 07	03 45 20	09♎ 29 41	19 56	06 30	03 12	16 23	08 06	00 29	16 19	21 02	08 13
26 Sa	16 14 03	04 42 57	22 37 06	21 53	07 42	03 32	16 16	08 03	00 32	16 20	21 01	08 06
27 Su	16 17 60	05 40 33	05♏ 32 17	23 52	08 53	03 51	16 09	07 59	00 35	16 21	21 01	07 57
28 Mo	16 21 56	06 38 08	18 15 28	25 53	10 05	04 10	16 02	07 56	00 38	16 22	21 00	07 46
29 Tu	16 25 53	07 35 42	00♐ 46 46	27 56	11 16	04 29	15 55	07 53	00 41	16 22	20 59	07 34
30 We	16 29 50	08 33 14	13 06 33	00♊ 01	12 27	04 47	15 49	07 49	00 44	16 23	20 58	07 23
31 Th	16 33 46	09 30 45	25 15 31	02 07	13 38	05 05	15 42	07 45	00 47	16 24	20 57	07 15

Data for	05-01-2018
Julian Day	2458239.50
Ayanamsa	24 06 31
SVP	04 ♓ 59 59
☽ ☊ Mean	10 ♌ 33 R

● ◐ PHASES ○ ◑

Day	Time	Phase	Position
08	02:09	◑	17♒27
15	11:49	●	24♉36
22	03:49	◐	01♍02
29	14:20	○	08♐10

LAST ASPECT ☽ INGRESS

Last Aspect Day	h m	Ingress Day	h m	Sign
01	02:57	01	15:20	♐
04	00:50	04	02:07	♑
06	13:49	06	14:49	♒
09	02:29	09	03:11	♓
11	09:03	11	12:41	♈
13	18:06	13	18:17	♉
15	20:31	15	20:45	♊
17	18:19	17	21:49	♋
19	21:15	19	23:12	♌
21	03:31	22	02:04	♍
23	14:55	24	06:52	♎
25	21:04	26	13:40	♏
28	17:26	28	22:30	♐
30	06:26	31	09:27	♑

DECLINATION

Day	☉	☽	☿	♀	♂	♃	♄	♅	♆	♇
01 Tu	15N00	13S 31	02N46	22N26	22S 41	16S 20	22S 15	10N42	06S 26	21S 29
02 We	15 18	16 33	03 08	22 40	22 39	16 18	22 15	10 43	06 25	21 29
03 Th	15 36	18 47	03 32	22 53	22 37	16 16	22 15	10 44	06 25	21 29
04 Fr	15 53	20 08	03 56	23 06	22 34	16 14	22 15	10 45	06 24	21 29
05 Sa	16 10	20 34	04 23	23 19	22 32	16 12	22 15	10 46	06 24	21 29
06 Su	16 27	20 04	04 50	23 30	22 29	16 10	22 15	10 48	06 23	21 29
07 Mo	16 44	18 41	05 19	23 41	22 27	16 08	22 15	10 49	06 23	21 29
08 Tu	17 01	16 29	05 49	23 51	22 24	16 06	22 15	10 50	06 22	21 30
09 We	17 17	13 33	06 21	24 01	22 22	16 04	22 16	10 51	06 22	21 30
10 Th	17 33	09 59	06 53	24 10	22 20	16 02	22 16	10 52	06 21	21 30
11 Fr	17 49	05 54	07 26	24 18	22 17	16 00	22 16	10 53	06 21	21 30
12 Sa	18 04	01 25	08 01	24 26	22 15	15 58	22 16	10 55	06 21	21 30
13 Su	18 19	03N15	08 36	24 33	22 13	15 56	22 16	10 56	06 20	21 30
14 Mo	18 34	07 55	09 12	24 39	22 10	15 54	22 16	10 57	06 20	21 31
15 Tu	18 48	12 17	09 49	24 45	22 08	15 52	22 16	10 58	06 19	21 31
16 We	19 02	16 02	10 27	24 49	22 06	15 50	22 17	10 59	06 19	21 31
17 Th	19 16	18 49	11 05	24 53	22 04	15 48	22 17	11 00	06 18	21 31
18 Fr	19 29	20 23	11 44	24 57	22 02	15 46	22 17	11 01	06 18	21 31
19 Sa	19 42	20 34	12 24	24 59	22 00	15 44	22 17	11 02	06 18	21 32
20 Su	19 55	19 22	13 03	25 01	21 58	15 42	22 17	11 04	06 17	21 32
21 Mo	20 08	16 56	13 43	25 03	21 56	15 40	22 18	11 05	06 17	21 32
22 Tu	20 20	13 33	14 24	25 03	21 55	15 38	22 18	11 06	06 17	21 32
23 We	20 31	09 28	15 04	25 03	21 53	15 37	22 18	11 07	06 16	21 32
24 Th	20 43	04 59	15 45	25 02	21 51	15 35	22 18	11 08	06 16	21 33
25 Fr	20 54	00 19	16 25	25 01	21 50	15 33	22 18	11 09	06 16	21 33
26 Sa	21 05	04S 16	17 05	24 58	21 49	15 31	22 19	11 10	06 16	21 33
27 Su	21 15	08 36	17 45	24 55	21 47	15 29	22 19	11 11	06 15	21 33
28 Mo	21 25	12 29	18 24	24 52	21 46	15 27	22 19	11 12	06 15	21 34
29 Tu	21 34	15 45	19 03	24 47	21 45	15 26	22 19	11 13	06 15	21 34
30 We	21 44	18 17	19 40	24 42	21 44	15 24	22 20	11 14	06 15	21 34
31 Th	21 52	19 57	20 16	24 36	21 44	15 22	22 20	11 15	06 14	21 34

ASPECTARIAN

01 02:57 ☽ ⚹ ♂
12:12 ☽ # ☉
21:53 ☽ ∥ ♃
23:00 ♀ # ♂
02 09:21 ☽ ☍ ♀
21:59 ☽ △ ☿
22:10 ☽ □ ♆
04 00:50 ☽ △ ♅
20:02 ☽ ☌ ♄
05 02:33 ☉ # ♃
07:53 ☽ △ ☉
10:17 ☽ ⚹ ♆
16:01 ☽ ⚹ ♃
16:39 ☽ □ ☿
21:00 ☽ ☌ ♇
06 06:21 ☽ ☌ ♂
13:49 ☽ □ ♅
13:57 ☉ ⚹ ♆
07 08:51 ☿ □ ♆
19:31 ☽ # ☉
21:59 ♀ □ ♆
23:25 ☽ △ ♀
08 03:34 ☽ ∥ ♃
04:12 ☽ □ ♃
12:51 ☽ ⚹ ☿
09 00:40 ☉ ☍ ♃
00:58 ☿ # ♆
02:29 ☽ ⚹ ♅
18:27 ☽ # ♅
20:23 ☽ ⚹ ♄
10 10:28 ☽ ☌ ♆
14:27 ☽ △ ♃
16:19 ☽ # ☿
16:45 ☽ □ ♀
17:59 ☽ ⚹ ☉
20:14 ☽ ⚹ ♇
21:29 ☽ ∥ ♆
11 09:03 ☽ ⚹ ♂
14:18 ♂ ∥ ♄
23:08 ☉ △ ♇
12 04:36 ☽ □ ♄
13:31 ☿ □ ♂
13 02:55 ☽ □ ♇
05:03 ☽ ⚹ ♀
10:50 ☿ ☌ ♅
12:41 ☿ ♉
15:45 ☽ # ♆
16:32 ☽ □ ♂
18:06 ☽ ☌ ♅
18:58 ☽ ☌ ☿
14 07:56 ☽ ∥ ☿
09:00 ☽ △ ♄
16:29 ☽ ∥ ♅
21:45 ☽ ⚹ ♆
15 00:09 ☽ ☍ ♃
06:06 ☽ △ ♇
15:23 ♅ ♉
20:31 ☽ △ ♂
22:43 ☽ # ♃
16 04:56 ♂ ♒
07:04 ♂ □ ♅
20:52 ☿ ∥ ♅
23:18 ☽ □ ♆
17 05:58 ☽ ∥ ☉
18:19 ☽ ☌ ♀
22:00 ☽ ⚹ ♅
18 10:52 ☽ ⚹ ☿
11:35 ☽ ☍ ♄
16:49 ☿ △ ♄
19 00:24 ☽ △ ♆
01:49 ☽ △ ♃
08:27 ☽ ☍ ♇
13:11 ♀ ♋
16:48 ☽ ∥ ☉
17:34 ♀ ⚹ ♅
21:15 ☽ ⚹ ☉
23:34 ☽ □ ♅
20 01:39 ☽ ☍ ♂
19:32 ☽ □ ☿
21 02:15 ☉ ♊
03:31 ☽ □ ♃
09:48 ☽ # ♃
19:23 ☽ ∥ ☿
22 02:39 ☽ △ ♅
07:48 ☽ ⚹ ♀
14:46 ☽ ∥ ♅
16:20 ☽ △ ♄
23 02:14 ☿ ⚹ ♆
05:54 ☿ ☍ ♃
06:30 ☽ ☍ ♆
07:00 ☽ ⚹ ♃
07:11 ☽ △ ☿
14:55 ☽ △ ♆
17:16 ☽ # ♆
18:13 ☿ # ♃
24 02:40 ☉ △ ♂
12:20 ☽ △ ♂
12:49 ☽ △ ☉
18:03 ☽ □ ♀
21:29 ☽ □ ♄
25 09:50 ♃ △ ♆
13:36 ☿ △ ♆
21:04 ☽ □ ♆
26 06:41 ♀ ☍ ♄
10:50 ☽ ∥ ♆
14:43 ☽ ☍ ♅
20:46 ☽ □ ♂
27 04:35 ☽ ⚹ ♄
06:56 ☽ △ ♀
15:43 ☽ # ♅
19:49 ☽ ☌ ♃
20:24 ☽ △ ♆
28 05:12 ☽ ⚹ ♆
17:26 ☽ ☍ ☿
21:27 ☽ ∥ ♃
22:28 ☉ # ♆
29 07:21 ☽ ⚹ ♂
23:49 ☿ ♊
30 01:50 ☉ # ♂
06:26 ☽ □ ♆
31 11:03 ☽ △ ♅

⚷ Chiron

01 Dec.	03 N 25
02	00♈45
05	00 54
08	01 02
11	01 10
14	01 17
17	01 25
20	01 31
23	01 38
26	01 44
29	01 50

June 2018

Day	S. T.	☉	☽	☿	♀	♂	♃	♄	♅	♆	♇	☊ True
	h m s	° ' "	° ' "	° '	° '	° '	° '	° '	° '	° '	° '	° '
01 Fr	16 37 43	10♊ 28 16	07♑ 15 04	04♊ 15	14♋ 49	05♒ 22	15♏R36	07♑R42	00♉ 49	16♓ 24	20♑R56	07♌R08
02 Sa	16 41 39	11 25 45	19 07 20	06 24	16 00	05 38	15 29	07 38	00 52	16 25	20 55	07 05
03 Su	16 45 36	12 23 13	00♒ 55 17	08 34	17 11	05 54	15 23	07 34	00 55	16 25	20 54	07D 04
04 Mo	16 49 32	13 20 41	12 42 41	10 45	18 22	06 10	15 17	07 31	00 58	16 26	20 53	07 05
05 Tu	16 53 29	14 18 08	24 33 57	12 57	19 33	06 25	15 11	07 27	01 01	16 26	20 51	07 07
06 We	16 57 25	15 15 34	06♓ 34 02	15 09	20 43	06 39	15 06	07 23	01 03	16 27	20 50	07 09
07 Th	17 01 22	16 12 59	18 48 04	17 21	21 54	06 53	15 00	07 19	01 06	16 27	20 49	07 10
08 Fr	17 05 19	17 10 24	01♈ 21 05	19 33	23 04	07 06	14 54	07 15	01 09	16 28	20 48	07R 10
09 Sa	17 09 15	18 07 48	14 17 31	21 45	24 15	07 19	14 49	07 11	01 11	16 28	20 47	07 07
10 Su	17 13 12	19 05 12	27 40 36	23 56	25 25	07 31	14 44	07 07	01 14	16 28	20 46	07 03
11 Mo	17 17 08	20 02 35	11♉ 31 35	26 05	26 36	07 43	14 38	07 03	01 16	16 28	20 44	06 56
12 Tu	17 21 05	20 59 58	25 49 08	28 14	27 46	07 53	14 33	06 59	01 19	16 29	20 43	06 49
13 We	17 25 01	21 57 20	10♊ 28 59	00♋ 21	28 56	08 03	14 29	06 55	01 21	16 29	20 42	06 42
14 Th	17 28 58	22 54 41	25 24 15	02 27	00♌ 06	08 13	14 24	06 50	01 24	16 29	20 41	06 36
15 Fr	17 32 55	23 52 02	10♋ 26 16	04 31	01 16	08 22	14 19	06 46	01 26	16 29	20 40	06 32
16 Sa	17 36 51	24 49 22	25 26 01	06 32	02 26	08 30	14 15	06 42	01 28	16 29	20 38	06 29
17 Su	17 40 48	25 46 42	10♌ 15 32	08 32	03 36	08 37	14 11	06 38	01 31	16 29	20 37	06 28
18 Mo	17 44 44	26 44 00	24 48 50	10 30	04 46	08 44	14 07	06 33	01 33	16 30	20 36	06D 28
19 Tu	17 48 41	27 41 17	09♍ 02 22	12 26	05 55	08 50	14 03	06 29	01 35	16R 30	20 34	06 29
20 We	17 52 37	28 38 34	22 54 48	14 20	07 05	08 56	13 59	06 25	01 38	16 30	20 33	06 29
21 Th	17 56 34	29 35 50	06♎ 26 31	16 11	08 14	09 00	13 56	06 20	01 40	16 29	20 32	06R 29
22 Fr	18 00 30	00♋ 33 05	19 38 58	18 00	09 24	09 04	13 52	06 16	01 42	16 29	20 30	06 27
23 Sa	18 04 27	01 30 19	02♏ 34 07	19 47	10 33	09 08	13 49	06 11	01 44	16 29	20 29	06 24
24 Su	18 08 24	02 27 33	15 14 08	21 32	11 42	09 10	13 46	06 07	01 46	16 29	20 28	06 19
25 Mo	18 12 20	03 24 46	27 41 05	23 14	12 51	09 12	13 43	06 03	01 48	16 29	20 26	06 13
26 Tu	18 16 17	04 21 59	09♐ 56 53	24 54	14 00	09 13	13 40	05 58	01 50	16 29	20 25	06 07
27 We	18 20 13	05 19 12	22 03 16	26 32	15 09	09R 13	13 38	05 54	01 52	16 29	20 23	06 02
28 Th	18 24 10	06 16 24	04♑ 01 56	28 07	16 18	09 13	13 35	05 49	01 54	16 28	20 22	05 58
29 Fr	18 28 06	07 13 36	15 54 39	29 40	17 26	09 11	13 33	05 45	01 56	16 28	20 21	05 55
30 Sa	18 32 03	08 10 47	27 43 26	01♌ 10	18 35	09 09	13 31	05 41	01 58	16 28	20 19	05 54

Data for	06-01-2018
Julian Day	2458270.50
Ayanamsa	24 06 36
SVP	04 ♓ 59 55
☽ ☊ Mean	08 ♌ 54 R

● ◐ PHASES ○ ◑

06	18:33	◑	16♓00
13	19:44	●	22♊45
20	10:51	◐	29♍04
28	04:54	○	06♑28

LAST ASPECT Day	h m	☽ INGRESS Day	h m	
02	03:38	02	22:07	♒
04	05:11	05	10:55	♓
07	06:36	07	21:27	♈
09	19:38	10	04:05	♉
12	03:30	12	06:54	♊
13	19:44	14	07:21	♋
15	16:19	16	07:21	♌
18	03:26	18	08:41	♍
20	10:51	20	12:30	♎
22	01:34	22	19:11	♏
24	14:01	25	04:30	♐
26	12:54	27	15:54	♑
29	08:59	30	04:38	♒

DECLINATION

Day	☉	☽	☿	♀	♂	♃	♄	♅	♆	♇
01 Fr	22N01	20S41	20N51	24N30	21S43	15S21	22S20	11N16	06S14	21S35
02 Sa	22 09	20 30	21 25	24 23	21 42	15 19	22 20	11 17	06 14	21 35
03 Su	22 17	19 25	21 57	24 15	21 42	15 17	22 21	11 18	06 14	21 35
04 Mo	22 24	17 29	22 27	24 06	21 42	15 16	22 21	11 19	06 14	21 35
05 Tu	22 31	14 48	22 54	23 57	21 42	15 14	22 21	11 20	06 14	21 36
06 We	22 37	11 27	23 20	23 47	21 42	15 13	22 21	11 21	06 13	21 36
07 Th	22 43	07 35	23 43	23 37	21 42	15 11	22 22	11 22	06 13	21 36
08 Fr	22 49	03 18	24 03	23 26	21 43	15 10	22 22	11 22	06 13	21 36
09 Sa	22 54	01N16	24 21	23 14	21 43	15 08	22 22	11 23	06 13	21 37
10 Su	22 59	05 54	24 36	23 01	21 44	15 07	22 22	11 24	06 13	21 37
11 Mo	23 04	10 25	24 48	22 48	21 45	15 06	22 23	11 25	06 13	21 37
12 Tu	23 08	14 29	24 57	22 35	21 46	15 05	22 23	11 26	06 13	21 37
13 We	23 11	17 47	25 04	22 21	21 48	15 03	22 23	11 27	06 13	21 38
14 Th	23 15	19 58	25 08	22 06	21 49	15 02	22 23	11 28	06 13	21 38
15 Fr	23 18	20 46	25 09	21 51	21 51	15 01	22 24	11 28	06 13	21 38
16 Sa	23 20	20 05	25 08	21 35	21 53	15 00	22 24	11 29	06 13	21 39
17 Su	23 22	18 00	25 04	21 18	21 55	14 59	22 24	11 30	06 13	21 39
18 Mo	23 24	14 48	24 57	21 01	21 58	14 58	22 25	11 31	06 13	21 39
19 Tu	23 25	10 46	24 49	20 44	22 00	14 57	22 25	11 32	06 13	21 39
20 We	23 26	06 16	24 38	20 26	22 03	14 56	22 25	11 32	06 13	21 40
21 Th	23 26	01 35	24 25	20 07	22 06	14 55	22 25	11 33	06 13	21 40
22 Fr	23 26	03S04	24 10	19 48	22 10	14 55	22 26	11 34	06 13	21 40
23 Sa	23 26	07 29	23 54	19 28	22 13	14 54	22 26	11 35	06 13	21 41
24 Su	23 25	11 29	23 36	19 08	22 17	14 53	22 26	11 35	06 13	21 41
25 Mo	23 23	14 55	23 16	18 48	22 20	14 53	22 27	11 36	06 13	21 41
26 Tu	23 22	17 39	22 55	18 27	22 25	14 52	22 27	11 37	06 13	21 41
27 We	23 20	19 35	22 33	18 06	22 29	14 51	22 27	11 37	06 14	21 42
28 Th	23 17	20 37	22 10	17 44	22 33	14 51	22 27	11 38	06 14	21 42
29 Fr	23 14	20 42	21 45	17 21	22 38	14 51	22 28	11 39	06 14	21 42
30 Sa	23 11	19 53	21 20	16 59	22 43	14 50	22 28	11 39	06 14	21 43

ASPECTARIAN

01 00:54 ☽ ☌ ♄
14:14 ♀ △ ♂
14:29 ♀ △ ♃
16:42 ☽ ⚹ ♃
16:58 ☽ ☍ ♀
18:30 ☽ ⚹ ♆

02 03:38 ☽ ☌ ♇
07:22 ☿ # ♇
08:26 ♀ △ ♆
12:57 ☿ # ♂

03 00:00 ☽ □ ♅
10:23 ☽ ☌ ♂
13:31 ☉ # ♄
19:07 ☽ △ ☿
19:14 ☿ # ♄
21:03 ☿ ∥ ☉

04 01:24 ☽ △ ☉
05:11 ☽ □ ♃
20:20 ☽ ∥ ♃

05 12:59 ☽ ⚹ ♅

06 00:42 ☽ # ♅
01:36 ☽ ⚹ ♄
02:02 ☿ ☌ ☉
02:24 ♀ ☍ ♇
14:08 ☿ □ ♆
16:39 ☽ △ ♃
19:26 ☽ ☌ ♆
19:36 ☿ ∥ ♀
20:35 ☽ □ ☿

07 03:54 ☽ ⚹ ♇
06:00 ☉ □ ♆
06:36 ☽ [illegible] ♀
07:50 ☽ ∥ ♆

08 10:57 ☽ ⚹ ♂
10:59 ☽ □ ♄

09 07:30 ☽ ⚹ ☉
11:43 ☽ □ ♇
16:05 ☽ ⚹ ☿
19:38 ☽ □ ♀

10 01:38 ☽ # ♆
03:05 ♀ ∥ ☉
06:15 ☽ ☌ ♅
16:22 ☽ △ ♄
17:22 ☽ □ ♂

11 05:16 ☽ ☍ ♃
05:41 ☽ ∥ ♅
08:24 ☽ ⚹ ♆
15:32 ☽ △ ♇

12 03:30 ☽ ⚹ ♀
03:53 ☽ # ♃
19:52 ♀ # ♄
20:01 ☿ ♋
20:01 ☽ △ ♂

13 09:41 ☽ □ ♆
11:41 ☿ ⚹ ♅
21:55 ♀ ♌

14 09:36 ☽ ⚹ ♅
13:03 ☽ ☌ ☿
18:10 ☽ ☍ ♄
23:00 ♀ # ♂

15 03:33 ♀ □ ♅
06:10 ☽ △ ♃
09:40 ☽ △ ♆
16:19 ☽ ☍ ♇
18:17 ♀ # ♇

16 01:47 ☿ ☍ ♄
09:46 ☽ □ ♅
12:15 ☽ ☌ ♀
21:19 ☽ ☍ ♂

17 06:23 ☽ □ ♃
22:50 ☽ # ♃

18 03:26 ☽ ⚹ ☉
11:19 ☽ △ ♅
19:40 ☽ △ ♄
19:47 ☽ ∥ ♅
20:28 ♆ SR

19 06:45 ☽ ⚹ ☿
08:33 ☽ ⚹ ♃
12:49 ☽ ☍ ♆
19:43 ☿ △ ♃
19:53 ☽ △ ♇

20 00:18 ☽ # ♆
23:49 ☽ □ ♄

21 03:32 ☽ ⚹ ♀
03:59 ☿ △ ♆
04:38 ☽ △ ♂
10:08 ☉ ♋
16:55 ♀ ☍ ♂
20:30 ☽ □ ☿

22 01:34 ☽ □ ♇
16:54 ☽ ∥ ♆
21:51 ☽ △ ☉
22:26 ☽ ☍ ♅

23 05:58 ☉ ⚹ ♅
06:47 ☽ ⚹ ♄
09:26 ☿ ☍ ♇
12:25 ☽ □ ♂
16:35 ☽ □ ♀
21:12 ☽ ☌ ♃

24 00:39 ☽ # ♅
[illegible]
10:00 ☽ ⚹ ♇
14:01 ☽ △ ☿
14:47 ☿ ∥ ☉
23:41 ☽ ∥ ♃

25 17:20 ♀ □ ♃
22:33 ☽ ⚹ ♂

26 07:25 ☽ # ♀
08:50 ☽ △ ♀
12:54 ☽ □ ♆
14:10 ♂ ∥ ♄
21:06 ♂ SR

27 [illegible]
06:07 ☿ # ♄
13:29 ☉ ☍ ♄
19:42 ☽ △ ♅

28 03:35 ☽ ☌ ♄
19:14 ☽ ⚹ ♃

29 01:07 ☽ ⚹ ♆
02:38 ☿ # ♇
05:17 ☿ ♌
08:59 ☽ ☌ ♇

30 08:02 ☽ ☍ ☿
08:39 ☽ □ ♅
13:02 ☿ □ ♅
23:11 ☽ ☌ ♂

⚷ Chiron

01 Dec.	03 N 57
01	01♈55
04	02 00
07	02 05
10	02 09
13	02 12
16	02 16
19	02 18
22	02 21
25	02 23
28	02 24

13 03:02 20♋41 ☉ Solar Eclipse (mag 0.336)
27 20:22 04♒45 ✷ Total Lunar Eclipse (mag 1.613)

July 2018

Day	S. T. h m s	☉ ° ′ ″	☽ ° ′ ″	☿ ° ′	♀ ° ′	♂ ° ′	♃ ° ′	♄ ° ′	♅ ° ′	♆ ° ′	♇ ° ′	☊ True ° ′
01 Su	18 35 59	09♋07 59	09♒30 43	02♌39	19♌43	09♒R07	13♏R29	05♑R36	01♉59	16♓R27	20♑R18	05♌D55
02 Mo	18 39 56	10 05 10	21 19 23	04 05	20 51	09 03	13 28	05 32	02 01	16 27	20 16	05 57
03 Tu	18 43 53	11 02 22	03♓12 50	05 28	22 00	08 59	13 26	05 27	02 03	16 26	20 15	05 59
04 We	18 47 49	11 59 33	15 14 54	06 49	23 08	08 54	13 25	05 23	02 05	16 26	20 13	06 02
05 Th	18 51 46	12 56 45	27 29 49	08 07	24 15	08 48	13 24	05 19	02 06	16 25	20 12	06 04
06 Fr	18 55 42	13 53 57	10♈01 57	09 23	25 23	08 42	13 23	05 14	02 08	16 25	20 11	06 05
07 Sa	18 59 39	14 51 10	22 55 29	10 37	26 31	08 35	13 22	05 10	02 09	16 24	20 09	06R 06
08 Su	19 03 35	15 48 22	06♉13 55	11 47	27 38	08 27	13 21	05 06	02 11	16 24	20 08	06 05
09 Mo	19 07 32	16 45 36	19 59 25	12 55	28 46	08 18	13 21	05 01	02 12	16 23	20 06	06 03
10 Tu	19 11 28	17 42 49	04♊12 06	14 01	29 53	08 09	13 21	04 57	02 14	16 23	20 05	06 01
11 We	19 15 25	18 40 03	18 49 28	15 03	01♍00	07 59	13D 21	04 53	02 15	16 22	20 03	05 58
12 Th	19 19 22	19 37 18	03♋46 14	16 02	02 07	07 48	13 21	04 49	02 16	16 21	20 02	05 56
13 Fr	19 23 18	20 34 32	18 54 38	16 58	03 14	07 37	13 21	04 44	02 18	16 20	20 00	05 54
14 Sa	19 27 15	21 31 47	04♌05 26	17 51	04 20	07 25	13 22	04 40	02 19	16 20	19 59	05 53
15 Su	19 31 11	22 29 02	19 09 15	18 41	05 27	07 13	13 22	04 36	02 20	16 19	19 57	05 53
16 Mo	19 35 08	23 26 17	03♍57 55	19 27	06 33	07 00	13 23	04 32	02 21	16 18	19 56	05D 53
17 Tu	19 39 04	24 23 33	18 25 28	20 10	07 39	06 46	13 24	04 28	02 22	16 17	19 54	05 53
18 We	19 43 01	25 20 48	02♎28 33	20 49	08 45	06 32	13 25	04 24	02 23	16 16	19 53	05 53
19 Th	19 46 57	26 18 03	16 06 16	21 24	09 51	06 18	13 27	04 20	02 24	16 15	19 52	05R 53
20 Fr	19 50 54	27 15 19	29 19 42	21 55	10 57	06 03	13 28	04 16	02 25	16 15	19 50	05 52
21 Sa	19 54 51	28 12 35	12♏11 14	22 22	12 02	05 48	13 30	04 12	02 26	16 14	19 49	05 52
22 Su	19 58 47	29 09 51	24 44 01	22 44	13 08	05 33	13 32	04 09	02 27	16 13	19 47	05 51
23 Mo	20 02 44	00♌07 07	07♐01 33	23 02	14 13	05 17	13 34	04 05	02 28	16 12	19 46	05 50
24 Tu	20 06 40	01 04 24	19 07 18	23 15	15 18	05 01	13 37	04 01	02 28	16 11	19 44	05 50
25 We	20 10 37	02 01 41	01♑04 28	23 24	16 22	04 45	13 39	03 58	02 29	16 10	19 43	05D 50
26 Th	20 14 33	02 58 59	12 56 00	23 27	17 27	04 29	13 42	03 54	02 30	16 09	19 41	05 50
27 Fr	20 18 30	03 56 17	24 44 31	23R 26	18 31	04 12	13 45	03 50	02 30	16 07	19 40	05 50
28 Sa	20 22 26	04 53 36	06♒32 27	23 19	19 35	03 56	13 48	03 47	02 31	16 06	19 39	05 51
29 Su	20 26 23	05 50 56	18 22 05	23 08	20 39	03 40	13 51	03 44	02 31	16 05	19 37	05 52
30 Mo	20 30 20	06 48 16	00♓15 44	22 51	21 42	03 23	13 54	03 40	02 32	16 04	19 36	05 53
31 Tu	20 34 16	07 45 37	12 15 48	22 30	22 46	03 07	13 57	03 37	02 32	16 03	19 34	05 53

Data for	07-01-2018
Julian Day	2458300.50
Ayanamsa	24 06 41
SVP	04 ♓ 59 52
☽ ☊ Mean	07 ♌ 19 R

● ◐ PHASES ○ ◑

06	07:51	◑	14♈13
13	02:48	☉	20♋41
19	19:53	◐	27♎06
27	20:21	✷	04♒45

LAST ASPECT ☽ Day	h m	INGRESS Day	h m	
01	22:57	02	17:32	♓
04	09:47	05	04:50	♈
07	07:10	07	12:52	♉
09	16:10	09	16:59	♊
10	20:00	11	17:59	♋
13	02:48	13	17:31	♌
14	23:12	15	17:31	♍
17	10:50	17	19:43	♎
19	19:53	20	01:14	♏
22	09:19	22	10:14	♐
24	08:23	24	21:50	♑
26	13:42	27	10:42	♒
29	09:26	29	23:28	♓

DECLINATION

Day	☉	☽	☿	♀	♂	♃	♄	♅	♆	♇
01 Su	23N07	18S 11	20N54	16N36	22S 48	14S 50	22S 28	11N40	06S 14	21S 43
02 Mo	23 03	15 43	20 27	16 12	22 53	14 50	22 28	11 40	06 14	21 43
03 Tu	22 59	12 35	19 59	15 49	22 58	14 50	22 29	11 41	06 15	21 44
04 We	22 54	08 54	19 31	15 25	23 04	14 49	22 29	11 41	06 15	21 44
05 Th	22 48	04 47	19 03	15 00	23 09	14 49	22 29	11 42	06 15	21 44
06 Fr	22 43	00 23	18 34	14 35	23 15	14 49	22 30	11 43	06 15	21 45
07 Sa	22 36	04N09	18 05	14 10	23 21	14 49	22 30	11 43	06 16	21 45
08 Su	22 30	08 37	17 36	13 45	23 27	14 49	22 30	11 44	06 16	21 45
09 Mo	22 23	12 48	17 07	13 19	23 34	14 49	22 30	11 44	06 16	21 45
10 Tu	22 16	16 25	16 38	12 53	23 40	14 50	22 31	11 44	06 16	21 46
11 We	22 08	19 07	16 09	12 27	23 46	14 50	22 31	11 45	06 17	21 46
12 Th	22 00	20 34	15 40	12 00	23 53	14 50	22 31	11 45	06 17	21 46
13 Fr	21 52	20 34	15 12	11 33	24 00	14 51	22 31	11 46	06 17	21 47
14 Sa	21 43	19 04	14 44	11 06	24 06	14 51	22 32	11 46	06 18	21 47
15 Su	21 34	16 14	14 17	10 39	24 13	14 51	22 32	11 47	06 18	21 47
16 Mo	21 24	12 23	13 50	10 11	24 20	14 52	22 32	11 47	06 18	21 48
17 Tu	21 14	07 52	13 24	09 43	24 26	14 52	22 32	11 47	06 19	21 48
18 We	21 04	03 04	13 00	09 16	24 33	14 53	22 33	11 48	06 19	21 48
19 Th	20 53	01S 44	12 36	08 47	24 39	14 54	22 33	11 48	06 19	21 49
20 Fr	20 42	06 19	12 13	08 19	24 46	14 54	22 33	11 48	06 20	21 49
21 Sa	20 31	10 28	11 51	07 51	24 52	14 55	22 33	11 48	06 20	21 49
22 Su	20 19	14 04	11 31	07 22	24 59	14 56	22 34	11 49	06 21	21 50
23 Mo	20 07	17 00	11 12	06 53	25 05	14 57	22 34	11 49	06 21	21 50
24 Tu	19 55	19 08	10 55	06 25	25 11	14 58	22 34	11 49	06 21	21 50
25 We	19 42	20 24	10 40	05 56	25 17	14 59	22 34	11 49	06 22	21 50
26 Th	19 29	20 45	10 26	05 27	25 23	15 00	22 35	11 50	06 22	21 51
27 Fr	19 16	20 10	10 15	04 57	25 29	15 01	22 35	11 50	06 23	21 51
28 Sa	19 02	18 41	10 06	04 28	25 34	15 02	22 35	11 50	06 23	21 51
29 Su	18 48	16 25	09 59	03 59	25 39	15 03	22 35	11 50	06 24	21 52
30 Mo	18 34	13 26	09 54	03 30	25 44	15 05	22 35	11 50	06 24	21 52
31 Tu	18 20	09 52	09 52	03 00	25 49	15 06	22 36	11 50	06 25	21 52

ASPECTARIAN

01 08:04 ☽ □ ♃
18:57 ☽ # ♀
22:57 ☽ ☍ ♀

02 07:15 ☽ ∥ ♃
21:39 ☽ ⚹ ♅

03 00:45 ☉ # ♂
04:28 ☽ ⚹ ♄
06:05 ☽ # ♅
16:59 ☽ △ ☉
20:22 ☽ △ ♃

04 02:20 ☽ ☌ ♆
09:47 ☽ ⚹ ♇
15:37 ☽ ∥ ♆

05 10:27 ♀ # ♃
11:05 ☉ △ ♃
11:49 ☿ ☍ ♂
14:57 ☽ □ ♄
21:30 ☽ ⚹ ♂
22:39 ☽ △ ☿

06 18:54 ☽ □ ♇

07 07:10 ☽ △ ♀
11:15 ☽ # ♆
16:46 ☽ ☌ ♅
21:59 ☽ △ ♄
23:35 ☉ # ♄

08 03:52 ☽ □ ♂
10:41 ☽ □ ☿
12:31 ☽ ☍ ♃
14:42 ☉ △ ♆
17:34 ☽ ∥ ♅
17:48 ☽ ⚹ ♆
18:02 ☽ ⚹ ☉

09 00:12 ☽ △ ♇
02:48 ☽ ∥ ♀
09:14 ☿ □ ♃
12:49 ☽ # ♃
16:10 ☽ □ ♀

10 01:24 ☽ ∥ ☿
02:32 ♀ ♍
06:28 ☽ △ ♂
17:04 ♃ SD
17:24 ☽ ⚹ ☿
20:00 ☽ □ ♆

11 21:09 ☽ ⚹ ♀
21:36 ☽ ⚹ ♅

12 01:39 ☽ ☍ ♄
03:26 ♀ △ ♅
10:02 ☉ ☍ ♇
12:53 ♀ ∥ ♅
15:12 ☽ △ ♃
19:56 ☽ △ ♆

13 01:44 ☽ ☍ ♇
12:54 ☉ # ♇
18:13 ☿ # ♃
21:11 ☽ □ ♅

14 05:12 ☽ ☍ ♂
06:45 ♀ △ ♄
14:45 ☽ □ ♃
23:12 ☽ ☌ ☿

15 09:10 ☽ # ♃
14:25 ☽ ∥ ☿
21:21 ☽ △ ♅

16 00:56 ☽ △ ♄
03:19 ☽ ∥ ♅
04:36 ☽ ☌ ♀
13:20 ☽ ∥ ♀
15:35 ☽ ⚹ ♃
20:25 ☽ ☍ ♆

17 02:30 ☽ △ ♇
07:52 ☽ # ♆
10:50 ☽ ⚹ ☉

18 03:20 ☽ □ ♄
06:57 ☽ △ ♂

19 06:44 ☽ □ ♇
09:56 ☽ ⚹ ☿

20 00:06 ☽ ∥ ♆
05:43 ☽ ☍ ♅
09:06 ☽ ⚹ ♄
10:05 ☽ # ♀
12:14 ☽ □ ♂
23:42 ☽ ⚹ ♀

21 02:30 ☽ ☌ ♃
03:10 ☿ ∥ ♅
07:39 ☽ △ ♆
07:57 ☽ # ☿
08:23 ☽ # ♅
14:29 ☽ ⚹ ♇
20:02 ☽ □ ☿

22 06:25 ☽ ∥ ♃
09:19 ☽ △ ☉
09:21 ♀ ⚹ ♃
20:38 ☽ ⚹ ♂
21:01 ☉ ♌

23 15:37 ☽ □ ♀
18:08 ☽ □ ♆

24 02:37 ♀ # ♆
08:23 ☽ △ ☿
10:50 ☽ # ☉
19:23 ♀ ☍ ♆

25 02:51 ☽ △ ♅
05:48 ☽ ☌ ♄
11:36 ☉ □ ♅

26 01:33 ☽ ⚹ ♃
05:04 ☿ SR
06:30 ☽ ⚹ ♆
10:04 ☽ △ ♀
13:42 ☽ ☌ ♇

27 05:14 ☉ ☍ ♂
15:48 ☽ □ ♅
18:44 ☽ # ☉
18:49 ☽ ☌ ♂

28 01:24 ♀ △ ♇
14:47 ☽ □ ♃

29 09:26 ☽ ☍ ☿
11:29 ☽ ∥ ♃

30 04:33 ☽ ⚹ ♅
06:49 ☽ ⚹ ♄
11:08 ☽ # ♅

31 00:04 ☽ # ☿
03:23 ☽ △ ♃
07:30 ☽ ☌ ♆
14:27 ☽ ⚹ ♇
20:49 ☽ ∥ ♆
22:42 ☽ ☍ ♀

⚷ Chiron

01 Dec.	04 N 12
01	02♈25
04	02 25
07	02 25R
10	02 25
13	02 24
16	02 22
19	02 20
22	02 18
25	02 15
28	02 12
31	02 08

05 02:44 02♈25 R

August 2018

11 09:47 18♌42 ☉ Solar Eclipse (mag 0.736)

Day	S. T.	☉	☽	☿	♀	♂	♃	♄	♅	♆	♇	☊ True
	h m s	° ' "	° ' "	° '	° '	° '	° '	° '	° '	° '	° '	° '
01 We	20 38 13	08♌ 42 59	24♓ 24 51	22♌ R03	23♍ 49	02♒R51	14♏ 01	03♑R34	02♉ 33	16♓R02	19♑R33	05♌ 54
02 Th	20 42 09	09 40 23	06♈ 45 43	21 33	24 51	02 35	14 05	03 31	02 33	16 00	19 32	05R 54
03 Fr	20 46 06	10 37 47	19 21 26	20 58	25 54	02 19	14 09	03 27	02 33	15 59	19 30	05 54
04 Sa	20 50 02	11 35 12	02♉ 15 04	20 19	26 56	02 03	14 13	03 24	02 33	15 58	19 29	05 54
05 Su	20 53 59	12 32 39	15 29 28	19 37	27 58	01 48	14 17	03 22	02 33	15 57	19 28	05D 54
06 Mo	20 57 55	13 30 07	29 06 51	18 53	29 00	01 33	14 22	03 19	02 34	15 55	19 26	05 54
07 Tu	21 01 52	14 27 36	13♊ 08 14	18 07	00♎ 01	01 18	14 26	03 16	02 34	15 54	19 25	05 54
08 We	21 05 49	15 25 07	27 32 52	17 19	01 03	01 04	14 31	03 13	02R 34	15 53	19 24	05 55
09 Th	21 09 45	16 22 39	12♋ 17 46	16 32	02 03	00 51	14 36	03 11	02 34	15 51	19 23	05 55
10 Fr	21 13 42	17 20 12	27 17 31	15 45	03 04	00 37	14 41	03 08	02 34	15 50	19 21	05R 55
11 Sa	21 17 38	18 17 46	12♌ 24 32	15 00	04 04	00 25	14 47	03 06	02 33	15 48	19 20	05 55
12 Su	21 21 35	19 15 22	27 29 57	14 17	05 04	00 13	14 52	03 03	02 33	15 47	19 19	05 53
13 Mo	21 25 31	20 12 58	12♍ 24 47	13 38	06 04	00 01	14 57	03 01	02 33	15 46	19 18	05 51
14 Tu	21 29 28	21 10 36	27 01 17	13 03	07 03	29♑ 50	15 03	02 59	02 33	15 44	19 16	05 49
15 We	21 33 24	22 08 14	11♎ 13 55	12 32	08 02	29 40	15 09	02 57	02 32	15 43	19 15	05 46
16 Th	21 37 21	23 05 54	24 59 50	12 07	09 00	29 31	15 15	02 54	02 32	15 41	19 14	05 43
17 Fr	21 41 18	24 03 34	08♏ 18 45	11 49	09 58	29 22	15 21	02 53	02 31	15 40	19 13	05 41
18 Sa	21 45 14	25 01 16	21 12 32	11 37	10 56	29 14	15 28	02 51	02 31	15 38	19 12	05 40
19 Su	21 49 11	25 58 58	03♐ 44 31	11 32	11 53	29 06	15 34	02 49	02 30	15 37	19 11	05D 40
20 Mo	21 53 07	26 56 42	15 58 53	11D 34	12 50	29 00	15 41	02 47	02 30	15 35	19 10	05 41
21 Tu	21 57 04	27 54 27	28 00 07	11 44	13 46	28 54	15 47	02 46	02 29	15 33	19 08	05 43
22 We	22 01 00	28 52 13	09♑ 52 45	12 02	14 42	28 49	15 54	02 44	02 29	15 32	19 07	05 46
23 Th	22 04 57	29 50 00	21 40 57	12 27	15 37	28 45	16 01	02 43	02 28	15 30	19 06	05 48
24 Fr	22 08 53	00♍ 47 48	03♒ 28 28	13 00	16 32	28 42	16 08	02 41	02 27	15 29	19 05	05 50
25 Sa	22 12 50	01 45 38	15 18 31	13 40	17 26	28 39	16 16	02 40	02 26	15 27	19 04	05R 51
26 Su	22 16 47	02 43 29	27 13 47	14 28	18 20	28 38	16 23	02 39	02 25	15 26	19 03	05 50
27 Mo	22 20 43	03 41 22	09♓ 16 24	15 23	19 13	28 37	16 30	02 38	02 25	15 24	19 02	05 47
28 Tu	22 24 40	04 39 16	21 28 06	16 25	20 06	28D 37	16 38	02 37	02 24	15 22	19 02	05 44
29 We	22 28 36	05 37 12	03♈ 50 18	17 33	20 58	28 37	16 46	02 36	02 23	15 21	19 01	05 39
30 Th	22 32 33	06 35 09	16 24 14	18 48	21 49	28 39	16 54	02 35	02 22	15 19	19 00	05 34
31 Fr	22 36 29	07 33 08	29 11 09	20 08	22 40	28 41	17 02	02 35	02 20	15 18	18 59	05 30

Data for	08-01-2018
Julian Day	2458331.50
Ayanamsa	24 06 46
SVP	04 ♓ 59 48
☽ ☊ Mean	05 ♌ 40 R

● ◐ PHASES ○ ◑

04	18:18	◑	12♉19
11	09:58	☉	18♌42
18	07:49	◐	25♏20
26	11:56	○	03♓12

LAST ASPECT ☽ Day	h m	INGRESS Day	h m	
31	22:42	01	10:55	♈
03	02:53	03	19:51	♉
05	23:47	06	01:32	♊
07	07:55	08	04:02	♋
09	11:21	10	04:18	♌
11	09:59	12	04:00	♍
14	04:38	14	04:58	♎
16	07:57	16	08:55	♏
18	15:08	18	16:46	♐
20	23:48	21	04:01	♑
23	14:19	23	16:56	♒
25	04:39	26	05:33	♓
28	13:55	28	16:35	♈
30	23:04	31	01:31	♉

DECLINATION

Day	☉	☽	☿	♀	♂	♃	♄	♅	♆	♇
01 We	18N05	05S52	09N52	02N31	25S54	15S07	22S36	11N51	06S25	21S53
02 Th	17 50	01 35	09 54	02 01	25 58	15 08	22 36	11 51	06 26	21 53
03 Fr	17 34	02N52	09 59	01 32	26 02	15 10	22 36	11 51	06 26	21 53
04 Sa	17 18	07 17	10 07	01 02	26 06	15 11	22 37	11 51	06 27	21 53
05 Su	17 02	11 28	10 17	00 33	26 10	15 13	22 37	11 51	06 27	21 54
06 Mo	16 46	15 12	10 29	00 03	26 13	15 14	22 37	11 51	06 28	21 54
07 Tu	16 29	18 11	10 43	00S26	26 16	15 16	22 37	11 51	06 28	21 54
08 We	16 13	20 07	10 59	00 56	26 19	15 18	22 37	11 51	06 29	21 55
09 Th	15 56	20 45	11 17	01 25	26 21	15 19	22 38	11 51	06 29	21 55
10 Fr	15 38	19 55	11 36	01 55	26 24	15 21	22 38	11 51	06 30	21 55
11 Sa	15 21	17 39	11 57	02 24	26 25	15 23	22 38	11 51	06 31	21 55
12 Su	15 03	14 10	12 18	02 53	26 27	15 25	22 38	11 51	06 31	21 56
13 Mo	14 45	09 48	12 39	03 22	26 28	15 26	22 38	11 50	06 32	21 56
14 Tu	14 26	04 57	13 01	03 51	26 29	15 28	22 39	11 50	06 32	21 56
15 We	14 08	00S02	13 23	04 20	26 30	15 30	22 39	11 50	06 33	21 56
16 Th	13 49	04 51	13 44	04 49	26 30	15 32	22 39	11 50	06 34	21 57
17 Fr	13 30	09 17	14 04	05 18	26 30	15 34	22 39	11 50	06 34	21 57
18 Sa	13 11	13 09	14 23	05 47	26 30	15 36	22 39	11 50	06 35	21 57
19 Su	12 51	16 19	14 41	06 15	26 29	15 38	22 39	11 49	06 35	21 57
20 Mo	12 32	18 41	14 57	06 43	26 28	15 40	22 40	11 49	06 36	21 58
21 Tu	12 12	20 11	15 12	07 11	26 27	15 43	22 40	11 49	06 37	21 58
22 We	11 52	20 45	15 24	07 39	26 25	15 45	22 40	11 49	06 37	21 58
23 Th	11 32	20 24	15 34	08 07	26 24	15 47	22 40	11 49	06 38	21 58
24 Fr	11 11	19 09	15 42	08 35	26 22	15 49	22 40	11 48	06 38	21 59
25 Sa	10 51	17 04	15 47	09 02	26 20	15 52	22 40	11 48	06 39	21 59
26 Su	10 30	14 13	15 49	09 29	26 17	15 54	22 41	11 48	06 40	21 59
27 Mo	10 09	10 46	15 48	09 56	26 14	15 56	22 41	11 47	06 40	21 59
28 Tu	09 48	06 49	15 45	10 22	26 11	15 59	22 41	11 47	06 41	22 00
29 We	09 27	02 32	15 38	10 49	26 08	16 01	22 41	11 47	06 42	22 00
30 Th	09 06	01N54	15 28	11 15	26 05	16 03	22 41	11 46	06 42	22 00
31 Fr	08 44	06 20	15 15	11 41	26 01	16 06	22 41	11 46	06 43	22 00

ASPECTARIAN

01 16:05 ☽ ✶ ♂
17:45 ☽ □ ♄
21:14 ☽ # ♀

02 02:42 ♂ □ ♅
06:03 ☽ △ ☉
17:30 ☽ ∥ ♀

03 00:17 ☽ □ ♆
02:53 ☽ △ ☿
19:22 ☽ # ♆
23:39 ☽ □ ♂

04 00:34 ☽ ☌ ♅
02:07 ☽ △ ♄
16:37 ☽ ∥ ☿
21:50 ☽ ☍ ♃

05 00:48 ☽ ✶ ♆
02:15 ☽ ∥ ♅
06:58 ☽ □ ☿
07:03 ☽ △ ♆
23:47 ☽ △ ♀

06 00:17 ☽ # ♃
04:09 ☽ △ ♂
10:44 ☽ ∥ ☉
23:29 ♀ ♎
23:29 ☉ □ ♃

07 02:23 ☽ ✶ ☉
04:38 ☽ □ ♆
07:55 ☽ ✶ ☿
16:50 ♅ SR

08 00:33 ♀ △ ♂
06:09 ☽ □ ♀
08:13 ☽ ✶ ♅
09:16 ☽ ☍ ♄

09 02:06 ☿ ☌ ☉
03:44 ☽ △ ♃
05:43 ☽ △ ♆
11:21 ☽ ☍ ♇

10 01:35 ♀ □ ♄
05:13 ☽ ☍ ♂
08:22 ☽ □ ♅
09:50 ☽ ✶ ♀
16:56 ☿ ∥ ♅
21:15 ☉ # ♃

11 03:47 ☽ □ ♃
03:55 ☽ ☌ ☿
06:32 ☿ □ ♃
16:18 ☽ # ♃
18:07 ☽ ∥ ☉

12 08:05 ☽ △ ♅
08:52 ☽ △ ♄
09:57 ☽ ∥ ♅
13:17 ☽ ∥ ♅

13 02:15 ♂ ♑R
04:10 ☽ ✶ ♃
05:26 ☽ ☍ ♆
11:13 ☽ △ ♇
16:21 ☽ # ♆

14 04:38 ☽ △ ♂
04:50 ☽ # ♀
09:57 ☽ □ ♄
18:07 ☽ ☌ ♀

15 02:10 ☽ ✶ ☿

13:52 ☽ □ ♇
20:23 ☽ ✶ ☉
23:49 ☽ ∥ ♀

16 03:07 ☿ ∥ ☉
07:57 ☽ □ ♂
08:58 ☽ ∥ ♆
13:29 ☽ ☍ ♅
14:08 ☽ ✶ ♄

17 06:20 ☽ □ ☿
13:07 ☽ ☌ ♃
13:33 ☽ △ ♆
15:20 ☽ # ♅
20:13 ☽ ✶ ♇

18 00:11 ☽ # ♃
09:37 ☽ # ♅
15:08 ☽ ✶ ♂
15:35 ☿ ✶ ♀
18:17 ☽ ∥ ♃

19 04:26 ☿ SD
07:49 ♃ △ ♆
15:14 ☽ △ ☿
17:15 ☽ ✶ ♀
17:42 ♀ ∥ ♆
23:13 ☽ □ ♆

20 23:48 ☽ △ ☉

21 09:02 ☽ △ ♅
09:34 ☽ ☌ ♄

22 03:48 ☉ ∥ ♅

10:37 ☽ □ ♀
11:27 ☽ ✶ ♆
12:21 ☽ ✶ ♃
18:46 ☽ ☌ ♇

23 04:09 ☉ ♍
14:19 ☽ ☌ ♂
21:55 ☽ □ ♅

24 20:28 ☽ ☍ ☿

25 01:56 ☽ □ ♃
04:39 ☽ △ ♀
10:48 ☽ ∥ ♃
11:27 ☽ # ☿

16:38 ☉ △ ♅
22:08 ☉ △ ♄

26 10:22 ☽ ✶ ♅
10:49 ☽ ✶ ♄
17:19 ☽ # ♅
19:19 ♀ □ ♆

27 04:19 ☽ # ☉
04:49 ☽ ∥ ♀
06:43 ♀ # ☉
12:04 ☽ ☌ ♆
14:06 ♂ SD
14:?6 ☽ [illegible]
19:13 ☽ ✶ ♇

28 00:49 ☽ ∥ ♆
05:32 ☿ □ ♃
13:55 ☽ ✶ ♂
21:37 ☽ □ ♇

30 04:54 ☽ □ ♆
05:03 ☽ △ ☿
10:57 ☽ ☍ ♀
23:04 ☽ □ ♂

31 02:05 ☽ # ♆
04:40 ♀ # ♅
05:51 ☽ ☌ ♅
06:17 ☽ △ ♄
[illegible] ∥ [illegible]
16:43 ☽ △ ☉

⚷ Chiron

01 Dec. 04 N 08

03	02♈04R
06	02 00
09	01 55
12	01 49
15	01 44
18	01 38
21	01 32
24	01 25
27	01 18
30	01 11

September 2018

Day	S. T. h m s	☉ ° ' "	☽ ° ' "	☿ ° '	♀ ° '	♂ ° '	♃ ° '	♄ ° '	♅ ° '	♆ ° '	♇ ° '	☊ True ° '
01 Sa	22 40 26	08♍31 09	12♉12 21	21♌34	23♎30	28♑44	17♏10	02♑R34	02♉R19	15♓R16	18♑R58	05♌R26
02 Su	22 44 22	09 29 12	25 29 13	23 05	24 19	28 48	17 18	02 34	02 18	15 14	18 57	05 24
03 Mo	22 48 19	10 27 17	09♊03 01	24 41	25 08	28 53	17 27	02 33	02 17	15 13	18 56	05 23
04 Tu	22 52 15	11 25 24	22 54 39	26 20	25 56	28 59	17 35	02 33	02 16	15 11	18 56	05D 24
05 We	22 56 12	12 23 33	07♋04 16	28 03	26 43	29 05	17 44	02 33	02 14	15 09	18 55	05 24
06 Th	23 00 09	13 21 44	21 30 43	29 48	27 29	29 12	17 53	02 33	02 13	15 08	18 54	05 25
07 Fr	23 04 05	14 19 57	06♌11 10	01♍36	28 15	29 20	18 01	02D 33	02 12	15 06	18 54	05R 25
08 Sa	23 08 02	15 18 12	21 00 45	03 26	28 59	29 29	18 10	02 33	02 10	15 04	18 53	05 24
09 Su	23 11 58	16 16 28	05♍52 48	05 18	29 43	29 38	18 19	02 33	02 09	15 03	18 52	05 20
10 Mo	23 15 55	17 14 47	20 39 31	07 11	00 26	29 49	18 29	02 33	02 07	15 01	18 52	05 15
11 Tu	23 19 51	18 13 07	05♎13 00	09 04	01♏08	30 00	18 38	02 34	02 06	14 59	18 51	05 08
12 We	23 23 48	19 11 29	19 26 37	10 58	01 49	00♒11	18 47	02 34	02 04	14 58	18 50	05 00
13 Th	23 27 44	20 09 52	03♏15 49	12 53	02 28	00 24	18 57	02 35	02 03	14 56	18 50	04 53
14 Fr	23 31 41	21 08 18	16 38 39	14 47	03 07	00 37	19 07	02 35	02 01	14 54	18 49	04 46
15 Sa	23 35 38	22 06 45	29 35 44	16 41	03 44	00 51	19 16	02 36	01 59	14 53	18 49	04 41
16 Su	23 39 34	23 05 13	12♐09 39	18 35	04 21	01 05	19 26	02 37	01 58	14 51	18 49	04 39
17 Mo	23 43 31	24 03 43	24 24 26	20 28	04 56	01 20	19 36	02 38	01 56	14 50	18 48	04D 38
18 Tu	23 47 27	25 02 15	06♑24 58	22 21	05 30	01 36	19 46	02 39	01 54	14 48	18 48	04 39
19 We	23 51 24	26 00 49	18 16 25	24 13	06 02	01 53	19 56	02 40	01 52	14 46	18 47	04 41
20 Th	23 55 20	26 59 24	00♒03 59	26 04	06 33	02 10	20 06	02 42	01 50	14 45	18 47	04 43
21 Fr	23 59 17	27 58 01	11 52 31	27 54	07 02	02 28	20 17	02 43	01 48	14 43	18 47	04R 44
22 Sa	00 03 13	28 56 39	23 46 15	29 43	07 30	02 46	20 27	02 44	01 46	14 42	18 46	04 43
23 Su	00 07 10	29 55 20	05♓48 42	01♎32	07 57	03 05	20 38	02 46	01 44	14 40	18 46	04 39
24 Mo	00 11 07	00♎54 02	18 02 25	03 19	08 22	03 25	20 48	02 48	01 43	14 38	18 46	04 33
25 Tu	00 15 03	01 52 46	00♈29 03	05 06	08 45	03 45	20 59	02 49	01 40	14 37	18 46	04 24
26 We	00 18 60	02 51 32	13 09 15	06 52	09 06	04 06	21 10	02 51	01 38	14 35	18 46	04 14
27 Th	00 22 56	03 50 20	26 02 54	08 36	09 26	04 27	21 20	02 53	01 36	14 34	18 45	04 04
28 Fr	00 26 53	04 49 10	09♉09 20	10 20	09 43	04 49	21 31	02 55	01 34	14 32	18 45	03 54
29 Sa	00 30 49	05 48 02	22 27 40	12 03	09 59	05 11	21 42	02 57	01 32	14 31	18 45	03 46
30 Su	00 34 46	06 46 57	05♊57 08	13 45	10 13	05 34	21 53	03 00	01 30	14 29	18 45	03 40

Data for	09-01-2018
Julian Day	2458362.50
Ayanamsa	24 06 49
SVP	04 ♓ 59 43
☽ ☊ Mean	04 ♌ 02 R

● ◐ PHASES ○ ◑

Day	Time	Phase	Position
03	02:38	◑	10♊34
09	18:02	●	17♍00
16	23:15	◐	24♐02
25	02:53	○	02♈00

LAST ASPECT ☽ INGRESS

Day	h m	Day	h m	Sign
02	05:57	02	08:03	♊
04	06:38	04	12:05	♋
06	12:44	06	13:55	♌
08	13:32	08	14:30	♍
10	15:13	10	15:21	♎
11	22:58	12	18:16	♏
14	08:54	15	00:46	♐
16	23:15	17	11:08	♑
19	17:10	19	23:52	♒
21	17:14	22	12:27	♓
24	05:26	24	23:04	♈
26	10:29	27	07:17	♉
28	22:37	29	13:27	♊
30	15:39	01	18:02	♋

DECLINATION

Day	☉	☽	☿	♀	♂	♃	♄	♅	♆	♇
01 Sa	08N22	10N34	14N59	12S06	25S57	16S08	22S42	11N45	06S44	22S01
02 Su	08 01	14 23	14 40	12 32	25 53	16 11	22 42	11 45	06 44	22 01
03 Mo	07 39	17 31	14 17	12 56	25 49	16 14	22 42	11 45	06 45	22 01
04 Tu	07 17	19 44	13 52	13 21	25 44	16 16	22 42	11 44	06 46	22 01
05 We	06 55	20 47	13 24	13 45	25 39	16 19	22 42	11 44	06 46	22 01
06 Th	06 32	20 29	12 54	14 09	25 34	16 21	22 42	11 43	06 47	22 02
07 Fr	06 10	18 47	12 21	14 33	25 29	16 24	22 43	11 43	06 47	22 02
08 Sa	05 47	15 49	11 46	14 56	25 24	16 27	22 43	11 42	06 48	22 02
09 Su	05 25	11 47	11 08	15 19	25 18	16 29	22 43	11 42	06 49	22 02
10 Mo	05 02	07 04	10 29	15 41	25 13	16 32	22 43	11 41	06 49	22 02
11 Tu	04 39	02 01	09 49	16 03	25 07	16 35	22 43	11 41	06 50	22 02
12 We	04 17	03S01	09 07	16 25	25 01	16 38	22 43	11 40	06 51	22 03
13 Th	03 54	07 46	08 24	16 46	24 54	16 40	22 43	11 39	06 51	22 03
14 Fr	03 31	11 59	07 39	17 06	24 48	16 43	22 44	11 39	06 52	22 03
15 Sa	03 08	15 30	06 54	17 26	24 41	16 46	22 44	11 38	06 53	22 03
16 Su	02 45	18 11	06 08	17 46	24 34	16 49	22 44	11 38	06 53	22 03
17 Mo	02 21	19 59	05 22	18 05	24 27	16 52	22 44	11 37	06 54	22 03
18 Tu	01 58	20 50	04 35	18 24	24 20	16 54	22 44	11 36	06 55	22 04
19 We	01 35	20 44	03 48	18 42	24 13	16 57	22 44	11 36	06 55	22 04
20 Th	01 12	19 43	03 00	18 59	24 05	17 00	22 44	11 35	06 56	22 04
21 Fr	00S49	17 50	02 13	19 16	23 58	17 03	22 44	11 35	06 56	22 04
22 Sa	00 25	15 11	01 25	19 32	23 50	17 06	22 44	11 34	06 57	22 04
23 Su	00 02	11 50	00 38	19 48	23 42	17 09	22 45	11 33	06 58	22 04
24 Mo	00 21	07 58	00S09	20 02	23 34	17 12	22 45	11 32	06 58	22 04
25 Tu	00 45	03 41	00 57	20 16	23 26	17 15	22 45	11 32	06 59	22 05
26 We	01 08	00N50	01 43	20 30	23 17	17 18	22 45	11 31	06 59	22 05
27 Th	01 32	05 23	02 30	20 42	23 09	17 21	22 45	11 30	07 00	22 05
28 Fr	01 55	09 46	03 16	20 54	23 00	17 24	22 45	11 30	07 01	22 05
29 Sa	02 18	13 45	04 02	21 05	22 51	17 27	22 45	11 29	07 01	22 05
30 Su	02 42	17 05	04 48	21 15	22 42	17 30	22 45	11 28	07 02	22 05

ASPECTARIAN

01 05:34 ☽ ✶ ♆
07:07 ☽ ∥ ♅
09:07 ☽ ☍ ♃
10:24 ☽ # ♀
12:16 ☽ △ ♇
19:08 ☽ □ ☿
02 01:45 ☽ ∥ ☿
05:57 ☽ △ ♂
13:12 ☽ # ♃
03 10:43 ☽ □ ♆
12:59 ☿ ✶ ♀
04 05:28 ☽ △ ♀
06:38 ☽ ✶ ☿
14:27 ☿ # ♀
15:53 ☽ ✶ ♅
16:23 ☽ ☍ ♄
05 08:50 ☉ # ♆
09:32 ☽ ✶ ☉
13:28 ☽ △ ♆
17:56 ☽ △ ♃
19:42 ☽ ☍ ♇
06 02:39 ☿ ♍
10:21 ☽ □ ♀
11:10 ♄ SD
12:44 ☽ ☍ ♂
17:31 ☽ □ ♅
07 07:41 ☿ △ ♅
12:21 ☿ △ ♄
18:27 ☉ ☍ ♆
19:22 ☽ □ ♃
19:42 ☽ # ♃
08 02:18 ☿ ∥ ♅
05:15 ☽ # ♀
13:32 ☽ ✶ ♀
17:59 ☽ △ ♅
18:37 ☽ △ ♄
20:39 ♀ □ ♂
22:55 ☽ ☌ ☿
09 00:35 ☽ ∥ ♅
04:07 ☽ ∥ ☿
09:26 ♀ ♏
14:49 ☽ ☍ ♆
20:24 ☽ ✶ ♃
21:04 ☽ △ ♇
10 01:16 ☽ # ♆
10:37 ☽ ∥ ☉
15:13 ☽ △ ♂
19:35 ☽ □ ♄
11 00:57 ♂ ♒
12:11 ☉ ✶ ♃
15:29 ☉ △ ♇
22:58 ☽ □ ♇
12 05:46 ☽ # ☉
07:38 ♃ ✶ ♇
09:02 ♀ ☍ ♅
16:53 ♀ ∥ ♃
18:53 ☽ □ ♂
19:16 ☽ ∥ ♆
21:51 ☽ ☍ ♅
22:32 ☽ ☌ ♀
22:47 ☽ ✶ ♄
13 02:57 ☽ # ☿
03:58 ♀ ✶ ♄
20:03 ☽ ✶ ☿
20:51 ☽ △ ♆
22:02 ☽ # ♅
14 01:32 ☿ ☍ ♆
03:59 ☽ ✶ ♇
04:34 ☽ ☌ ♃
08:54 ☽ ✶ ☉
15 00:47 ☿ # ♆
02:24 ☽ ✶ ♂
10:36 ☽ ∥ ♃
19:04 ☽ ∥ ♀
16 02:52 ☿ △ ♇
05:13 ☽ □ ♆
11:52 ☿ ✶ ♃
14:48 ☽ □ ☿
17 14:57 ☽ △ ♅
16:25 ☽ ☌ ♄
22:03 ☽ ✶ ♀
18 16:54 ☽ ✶ ♆
23:02 ♂ □ ♅
19 01:03 ☽ ☌ ♇
03:26 ☽ ✶ ♃
14:20 ☽ △ ☿
17:10 ☽ △ ☉
20 03:36 ☽ □ ♅
04:23 ☽ ☌ ♂
09:16 ☽ ∥ ♀
13:46 ☽ □ ♀
21 01:53 ☿ ☌ ☉
07:40 ☽ ∥ ♃
17:14 ☽ □ ♃
22 03:40 ☿ ♎
15:57 ☽ ✶ ♅
17:57 ☽ ✶ ♄
23 01:54 ☽ # ♅
01:55 ☉ ♎
04:22 ☽ △ ♀
13:31 ☿ # ☉
16:47 ☿ □ ♄
17:23 ☽ ☌ ♆
24 01:25 ☽ ✶ ♇
01:30 ☿ △ ♂
05:26 ☽ △ ♃
05:42 ☽ ∥ ♆
12:11 ☿ ∥ ☉
25 04:28 ☽ □ ♄
06:24 ☽ ✶ ♂
10:13 ☽ ☍ ☿
12:30 ☽ ∥ ☿
14:27 ☽ ∥ ☉
23:51 ☉ □ ♄
26 01:47 ☽ # ☉
05:41 ☽ # ☿
10:29 ☽ □ ♇
27 08:45 ☽ # ♆
10:12 ☽ ☌ ♅
12:36 ☽ △ ♄
15:51 ☽ □ ♂
23:35 ☉ △ ♂
28 01:03 ☽ ☍ ♀
09:44 ☽ ✶ ♆
10:05 ☽ ∥ ♅
17:21 ☽ △ ♇
22:37 ☽ ☍ ♃
29 15:13 ♂ ∥ ♄
23:17 ☽ △ ♂
30 01:35 ☽ △ ☉
03:40 ☽ # ♃
15:00 ☽ □ ♆
15:39 ☽ △ ☿

⚷ Chiron

01 Dec.	03 N 46
02	01♈04R
05	00 56
08	00 48
11	00 41
14	00 33
17	00 24
20	00 16
23	00 08
26	30♓00
29	29 52

25 23:54 ♓R

October 2018

Day	S. T. h m s	☉ ° ' "	☽ ° ' "	☿ ° '	♀ ° '	♂ ° '	♃ ° '	♄ ° '	♅ ° '	♆ ° '	♇ ° '	☊ True ° '
01 Mo	00 38 42	07♎45 54	19♊37 13	15♎26	10♏24	05♒57	22♏04	03♑02	01♉R28	14♓R28	18♑45	03♌R36
02 Tu	00 42 39	08 44 53	03♋27 42	17 06	10 34	06 20	22 16	03 04	01 26	14 26	18D 45	03 34
03 We	00 46 36	09 43 55	17 28 27	18 45	10 41	06 45	22 27	03 07	01 23	14 25	18 45	03 34
04 Th	00 50 32	10 42 59	01♌39 06	20 24	10 47	07 09	22 38	03 09	01 21	14 23	18 45	03 33
05 Fr	00 54 29	11 42 05	15 58 28	22 01	10 50	07 34	22 50	03 12	01 19	14 22	18 45	03 32
06 Sa	00 58 25	12 41 14	00♍24 11	23 38	10R 50	08 00	23 01	03 15	01 17	14 20	18 46	03 29
07 Su	01 02 22	13 40 25	14 52 21	25 14	10 49	08 26	23 13	03 18	01 14	14 19	18 46	03 23
08 Mo	01 06 18	14 39 38	29 17 44	26 49	10 45	08 52	23 25	03 21	01 12	14 18	18 46	03 14
09 Tu	01 10 15	15 38 53	13♎34 14	28 24	10 38	09 19	23 36	03 24	01 10	14 16	18 46	03 04
10 We	01 14 11	16 38 10	27 35 55	29 57	10 29	09 46	23 48	03 27	01 07	14 15	18 46	02 52
11 Th	01 18 08	17 37 29	11♏17 57	01♏30	10 18	10 14	24 00	03 30	01 05	14 14	18 47	02 40
12 Fr	01 22 04	18 36 50	24 37 21	03 03	10 05	10 42	24 12	03 33	01 03	14 12	18 47	02 28
13 Sa	01 26 01	19 36 13	07♐33 26	04 34	09 49	11 11	24 24	03 37	01 00	14 11	18 47	02 19
14 Su	01 29 58	20 35 38	20 07 34	06 05	09 30	11 39	24 36	03 40	00 58	14 10	18 48	02 13
15 Mo	01 33 54	21 35 04	02♑22 53	07 35	09 10	12 08	24 48	03 44	00 55	14 09	18 48	02 09
16 Tu	01 37 51	22 34 33	14 23 47	09 04	08 47	12 38	25 00	03 47	00 53	14 07	18 49	02 08
17 We	01 41 47	23 34 03	26 15 24	10 33	08 22	13 08	25 13	03 51	00 51	14 06	18 49	02D 08
18 Th	01 45 44	24 33 35	08♒03 15	12 01	07 55	13 38	25 25	03 55	00 48	14 05	18 49	02 09
19 Fr	01 49 40	25 33 09	19 52 51	13 28	07 27	14 08	25 37	03 59	00 46	14 04	18 50	02R 08
20 Sa	01 53 37	26 32 44	01♓49 23	14 54	06 56	14 39	25 50	04 03	00 43	14 03	18 51	02 06
21 Su	01 57 33	27 32 21	13 57 23	16 20	06 24	15 10	26 02	04 07	00 41	14 02	18 51	02 01
22 Mo	02 01 30	28 32 00	26 20 23	17 45	05 51	15 42	26 14	04 11	00 38	14 01	18 52	01 54
23 Tu	02 05 27	29 31 41	09♈00 37	19 09	05 16	16 13	26 27	04 15	00 36	14 00	18 52	01 43
24 We	02 09 23	00♏31 24	21 58 55	20 32	04 41	16 45	26 40	04 19	00 33	13 58	18 53	01 31
25 Th	02 13 20	01 31 08	05♉14 33	21 55	04 05	17 18	26 52	04 24	00 31	13 57	18 54	01 18
26 Fr	02 17 16	02 30 55	18 45 25	23 16	03 28	17 50	27 05	04 28	00 29	13 57	18 54	01 06
27 Sa	02 21 13	03 30 44	02♊28 36	24 36	02 52	18 23	27 18	04 33	00 26	13 56	18 55	00 56
28 Su	02 25 09	04 30 35	16 20 46	25 55	02 15	18 56	27 30	04 37	00 24	13 55	18 56	00 47
29 Mo	02 29 06	05 30 28	00♋18 55	27 13	01 39	19 29	27 43	04 42	00 21	13 54	18 57	00 42
30 Tu	02 33 02	06 30 23	14 20 45	28 30	01 03	20 02	27 56	04 47	00 19	13 53	18 58	00 39
31 We	02 36 59	07 30 21	28 24 44	29 46	00 28	20 36	28 09	04 51	00 16	13 52	18 59	00 37

Data for	10-01-2018
Julian Day	2458392.50
Ayanamsa	24 06 52
SVP	04 ♓ 59 37
☽ ☊ Mean	02 ♌ 27 R

● ◐ PHASES ○ ◑

Day	Time	Phase	Position
02	09:46	◑	09♋09
09	03:47	●	15♎48
16	18:02	◐	23♑19
24	16:46	○	01♉13
31	16:40	◑	08♌12

LAST ASPECT ☽ INGRESS

Day	h m	Day	h m	Sign
03	08:34	03	21:13	♌
05	11:34	05	23:20	♍
07	14:03	08	01:11	♎
09	08:50	10	04:10	♏
11	23:13	12	09:53	♐
14	00:59	14	19:18	♑
16	21:50	17	07:37	♒
19	12:28	19	20:21	♓
21	23:48	22	06:59	♈
23	18:19	24	14:34	♉
26	14:50	26	19:41	♊
28	04:38	28	23:28	♋
31	02:31	31	02:42	♌

DECLINATION

Day	☉	☽	☿	♀	♂	♃	♄	♅	♆	♇
01 Mo	03S 05	19N31	05S 33	21S 24	22S 33	17S 33	22S 45	11N27	07S 02	22S 05
02 Tu	03 28	20 51	06 17	21 32	22 23	17 36	22 45	11 27	07 03	22 05
03 We	03 51	20 54	07 01	21 40	22 14	17 39	22 46	11 26	07 03	22 05
04 Th	04 14	19 38	07 45	21 46	22 04	17 42	22 46	11 25	07 04	22 05
05 Fr	04 38	17 06	08 28	21 51	21 55	17 45	22 46	11 24	07 05	22 06
06 Sa	05 01	13 29	09 10	21 55	21 45	17 48	22 46	11 23	07 05	22 06
07 Su	05 24	09 03	09 52	21 58	21 35	17 51	22 46	11 23	07 06	22 06
08 Mo	05 47	04 07	10 33	22 00	21 24	17 54	22 46	11 22	07 06	22 06
09 Tu	06 10	00S 59	11 14	22 00	21 14	17 57	22 46	11 21	07 07	22 06
10 We	06 32	05 56	11 53	21 59	21 03	18 00	22 46	11 20	07 07	22 06
11 Th	06 55	10 29	12 33	21 57	20 53	18 03	22 46	11 19	07 08	22 06
12 Fr	07 18	14 24	13 11	21 53	20 42	18 06	22 46	11 19	07 08	22 06
13 Sa	07 40	17 30	13 49	21 48	20 31	18 09	22 46	11 18	07 09	22 06
14 Su	08 03	19 41	14 26	21 42	20 20	18 12	22 46	11 17	07 09	22 06
15 Mo	08 25	20 54	15 02	21 34	20 09	18 16	22 46	11 16	07 10	22 06
16 Tu	08 47	21 07	15 37	21 25	19 57	18 19	22 46	11 15	07 10	22 06
17 We	09 09	20 23	16 11	21 15	19 46	18 22	22 46	11 14	07 11	22 06
18 Th	09 31	18 45	16 45	21 02	19 34	18 25	22 46	11 14	07 11	22 06
19 Fr	09 53	16 19	17 18	20 49	19 22	18 28	22 46	11 13	07 11	22 06
20 Sa	10 14	13 10	17 50	20 34	19 10	18 31	22 46	11 12	07 12	22 06
21 Su	10 36	09 25	18 21	20 18	18 58	18 34	22 46	11 11	07 12	22 06
22 Mo	10 57	05 13	18 51	20 00	18 46	18 37	22 46	11 10	07 13	22 06
23 Tu	11 18	00 41	19 20	19 41	18 34	18 40	22 46	11 09	07 13	22 06
24 We	11 39	03N58	19 48	19 21	18 21	18 43	22 46	11 08	07 13	22 06
25 Th	12 00	08 33	20 15	19 00	18 09	18 46	22 46	11 08	07 14	22 06
26 Fr	12 21	12 49	20 41	18 38	17 56	18 49	22 46	11 07	07 14	22 06
27 Sa	12 41	16 28	21 06	18 16	17 43	18 52	22 46	11 06	07 14	22 06
28 Su	13 01	19 14	21 30	17 52	17 30	18 55	22 46	11 05	07 15	22 06
29 Mo	13 21	20 53	21 53	17 28	17 17	18 58	22 46	11 04	07 15	22 06
30 Tu	13 41	21 15	22 15	17 04	17 04	19 01	22 46	11 03	07 15	22 06
31 We	14 01	20 17	22 35	16 39	16 51	19 04	22 46	11 03	07 16	22 06

ASPECTARIAN

01 02:02 ♆ SD
20:30 ☽ ⚹ ♅
23:19 ☽ ☍ ♄
02 12:19 ☽ △ ♀
18:47 ☽ △ ♆
03 00:03 ☿ □ ♆
01:07 ☿ ∥ ♆
02:11 ☽ ☍ ♆
02:28 ☽ □ ☿
08:34 ☽ △ ♃
21:10 ♂ ∥ ♆
23:30 ☽ □ ♅
04 09:32 ☽ ☍ ♂
15:23 ☽ □ ♀
16:20 ☽ ⚹ ☉
18:49 ☽ ⋕ ♃
05 05:59 ♀ ∥ ♂
11:21 ☽ ⚹ ☿
11:34 ☽ □ ♃
19:07 ♀ SR
06 01:27 ☽ △ ♅
04:44 ☽ △ ♄
11:46 ☽ ∥ ♅
17:17 ☽ ⚹ ♀
20:20 ☽ ⋕ ☿
23:05 ☽ ☍ ♆
07 06:28 ☽ △ ♆
09:42 ☽ ⋕ ♆
14:03 ☽ ⚹ ♃
16:39 ☽ ⋕ ☉
08 06:48 ☽ □ ♄
16:35 ☽ △ ♂
09 04:20 ☿ ⋕ ♅
08:50 ☽ □ ♆
10 00:41 ☿ ♏
03:17 ☽ ∥ ☉
04:36 ☽ ☌ ☿
06:00 ☽ ∥ ♆
06:06 ☽ ☍ ♅
10:12 ☽ ⚹ ♄
17:36 ☿ ☍ ♅
22:03 ☽ □ ♂
22:16 ☽ ☌ ♀
11 02:30 ♀ □ ♂
04:48 ☽ ⋕ ♅
05:12 ☽ △ ♆
13:23 ☽ ⚹ ♆
13:46 ☉ ∥ ♆
14:29 ☽ ∥ ☿
23:13 ☽ ☌ ♃
12 04:09 ☉ □ ♆
08:21 ☿ ⚹ ♄
13 06:22 ☽ ∥ ♃
07:07 ☽ ⚹ ♂
12:33 ☽ □ ♆
14 00:59 ☽ ⚹ ☉
09:07 ☽ ∥ ♂
21:08 ☽ △ ♅
15 02:41 ☽ ☌ ♄
11:48 ☽ ⚹ ☿
13:06 ☽ ⚹ ♀
20:21 ☿ ☌ ♀
23:27 ☽ ⚹ ♆
16 08:54 ☽ ☌ ♆
21:50 ☽ ⚹ ♃
17 09:18 ☽ □ ♅
12:06 ☽ ∥ ♂
23:45 ☽ □ ♀
18 03:46 ☽ ∥ ♃
09:11 ☽ □ ☿
11:51 ☽ ☌ ♂
16:41 ☽ ∥ ☿
19 09:48 ☿ △ ♆
11:46 ☽ □ ♃
12:28 ☽ △ ☉
17:24 ☿ □ ♂
21:48 ☽ ⚹ ♅
20 04:27 ☽ ⚹ ♄
09:45 ☽ △ ♀
13:03 ☽ ⋕ ♅
17:23 ☽ ∥ ☉
21 00:08 ☽ ☌ ♆
05:15 ☽ △ ☿
09:34 ☽ ⚹ ♆
11:26 ☿ ∥ ♃
12:52 ☽ ∥ ♆
21:17 ☿ ∥ ♂
23:48 ☽ △ ♃
22 14:14 ☉ ⋕ ♅
14:35 ♂ ∥ ♃
15:00 ☽ □ ♄
19:13 ☿ ⚹ ♆
23 10:40 ☿ ∥ ♀
11:23 ☉ ♏
13:59 ☽ ⚹ ♂
18:19 ☽ □ ♆
24 00:47 ☉ ☍ ♅
12:52 ♀ ⚹ ♄
15:32 ☽ ☌ ♅
16:55 ☽ ⋕ ♆
22:00 ☽ ☍ ♀
22:28 ☽ △ ♄
25 14:03 ♀ ∥ ♃
14:09 ☽ ∥ ♅
15:30 ☽ ⚹ ♆
20:59 ☽ ⋕ ☉
22:18 ☽ □ ♂
26 00:16 ☽ △ ♆
08:47 ☽ ☍ ☿
14:17 ♀ ☌ ☉
14:50 ☽ ☍ ♃
27 09:12 ☽ ⋕ ♂
12:44 ☽ ⋕ ♀
19:48 ☽ □ ♆
20:40 ☽ ⋕ ♃
28 02:53 ☉ ⚹ ♄
04:38 ☽ △ ♂
29 00:04 ☽ ⚹ ♅
02:11 ☽ △ ♀
07:33 ☽ ☍ ♄
09:34 ☽ △ ☉
11:05 ☿ ☌ ♃
14:34 ☿ ∥ ♆
23:13 ☽ △ ♆
23:42 ♀ ∥ ♂
30 07:53 ☽ ☍ ♆
23:32 ☽ △ ♃
31 02:31 ☽ △ ☿
03:10 ☽ □ ♅
03:22 ☽ □ ♀
04:39 ☿ ♐
08:46 ♀ ☍ ♅
13:42 ☿ ∥ ♄
14:20 ☽ ⋕ ♃
19:42 ♀ ♎R

⚷ Chiron

01 Dec.	03 N 13
02	29♓44R
05	29 36
08	29 28
11	29 20
14	29 13
17	29 05
20	28 58
23	28 51
26	28 45
29	28 39

November 2018

Day	S. T.	☉	☽	☿	♀	♂	♃	♄	♅	♆	♇	☊ True
	h m s	° ' "	° ' "	° '	° '	° '	° '	° '	° '	° '	° '	° '
01 Th	02 40 56	08♏ 30 21	12♌ 29 57	00♐ 59	29♎R54	21♒ 10	28♏ 22	04♑ 56	00♉R14	13♓R51	18♑ 59	00♌R36
02 Fr	02 44 52	09 30 22	26 35 41	02 12	29 21	21 44	28 35	05 01	00 11	13 51	19 00	00 34
03 Sa	02 48 49	10 30 26	10♍ 40 56	03 22	28 50	22 18	28 48	05 06	00 09	13 50	19 01	00 31
04 Su	02 52 45	11 30 32	24 44 04	04 30	28 21	22 53	29 01	05 11	00 07	13 49	19 02	00 25
05 Mo	02 56 42	12 30 41	08♎ 42 32	05 36	27 53	23 28	29 14	05 16	00 04	13 48	19 03	00 16
06 Tu	03 00 38	13 30 51	22 33 03	06 40	27 27	24 03	29 27	05 22	00 02	13 48	19 04	00 06
07 We	03 04 35	14 31 03	06♏ 11 57	07 40	27 03	24 38	29 40	05 27	29♈ 59	13 47	19 05	29♋ 53
08 Th	03 08 31	15 31 17	19 35 49	08 38	26 42	25 13	29 53	05 32	29 57	13 47	19 06	29 41
09 Fr	03 12 28	16 31 32	02♐ 42 06	09 31	26 22	25 49	00♐ 06	05 37	29 55	13 46	19 08	29 29
10 Sa	03 16 25	17 31 50	15 29 39	10 21	26 05	26 25	00 19	05 43	29 52	13 45	19 09	29 19
11 Su	03 20 21	18 32 09	27 58 54	11 07	25 51	27 00	00 33	05 48	29 50	13 45	19 10	29 12
12 Mo	03 24 18	19 32 30	10♑ 11 50	11 47	25 39	27 37	00 46	05 54	29 48	13 45	19 11	29 07
13 Tu	03 28 14	20 32 52	22 11 46	12 21	25 29	28 13	00 59	06 00	29 46	13 44	19 12	29 06
14 We	03 32 11	21 33 15	04♒ 03 03	12 50	25 22	28 49	01 12	06 05	29 43	13 44	19 14	29D 06
15 Th	03 36 07	22 33 40	15 50 41	13 11	25 17	29 26	01 26	06 11	29 41	13 43	19 15	29 07
16 Fr	03 40 04	23 34 06	27 40 08	13 24	25 15	00♓ 03	01 39	06 17	29 39	13 43	19 16	29 08
17 Sa	03 44 00	24 34 34	09♓ 36 53	13 29	25D 15	00 39	01 52	06 23	29 37	13 43	19 17	29R 07
18 Su	03 47 57	25 35 03	21 46 12	13R 25	25 17	01 16	02 06	06 28	29 35	13 42	19 19	29 04
19 Mo	03 51 54	26 35 33	04♈ 12 39	13 11	25 22	01 54	02 19	06 34	29 33	13 42	19 20	28 59
20 Tu	03 55 50	27 36 05	16 59 40	12 47	25 29	02 31	02 32	06 40	29 30	13 42	19 21	28 52
21 We	03 59 47	28 36 37	00♉ 09 10	12 13	25 39	03 08	02 46	06 46	29 28	13 42	19 23	28 43
22 Th	04 03 43	29 37 12	13 41 04	11 27	25 51	03 46	02 59	06 53	29 26	13 42	19 24	28 33
23 Fr	04 07 40	00♐ 37 47	27 33 16	10 32	26 05	04 23	03 12	06 59	29 24	13 42	19 26	28 23
24 Sa	04 11 36	01 38 24	11♊ 41 46	09 27	26 21	05 01	03 26	07 05	29 22	13 42	19 27	28 15
25 Su	04 15 33	02 39 03	26 01 18	08 15	26 39	05 39	03 39	07 11	29 20	13D 42	19 29	28 08
26 Mo	04 19 29	03 39 43	10♋ 26 11	06 57	26 59	06 17	03 53	07 17	29 19	13 42	19 30	28 04
27 Tu	04 23 26	04 40 25	24 51 12	05 36	27 21	06 55	04 06	07 24	29 17	13 42	19 32	28 02
28 We	04 27 23	05 41 08	09♌ 12 13	04 13	27 44	07 33	04 19	07 30	29 15	13 42	19 33	28 01
29 Th	04 31 19	06 41 53	23 26 23	02 53	28 10	08 12	04 33	07 36	29 13	13 42	19 35	28 00
30 Fr	04 35 16	07 42 39	07♍ 32 04	01 38	28 38	08 50	04 46	07 43	29 11	13 42	19 36	28 00

Data for	11-01-2018
Julian Day	2458423.50
Ayanamsa	24 06 55
SVP	04 ♓ 59 32
☽ ☊ Mean	00 ♌ 48 R

● ◐ PHASES ○ ◑

07	16:02	●	15♏11
15	14:55	◐	23♒11
23	05:39	○	00♊52
30	00:19	◑	07♍43

LAST ASPECT Day	h m	☽ INGRESS Day	h m	
02	04:32	02	05:48	♍
04	07:27	04	09:01	♎
06	08:19	06	13:03	♏
08	10:43	08	19:00	♐
11	03:36	11	03:56	♑
13	15:15	13	15:47	♒
16	03:59	16	04:42	♓
18	08:05	18	15:57	♈
20	22:47	20	23:44	♉
22	09:59	23	04:11	♊
25	05:31	25	06:38	♋
27	07:22	27	08:35	♌
29	09:47	29	11:08	♍

DECLINATION

Day	☉	☽	☿	♀	♂	♃	♄	♅	♆	♇
01 Th	14S 20	18N03	22S 54	16S 14	16S 37	19S 07	22S 46	11N02	07S 16	22S 06
02 Fr	14 39	14 44	23 12	15 50	16 24	19 10	22 46	11 01	07 16	22 06
03 Sa	14 58	10 34	23 28	15 25	16 10	19 13	22 46	11 00	07 17	22 06
04 Su	15 17	05 51	23 44	15 01	15 56	19 16	22 46	10 59	07 17	22 06
05 Mo	15 35	00 51	23 57	14 37	15 42	19 19	22 46	10 58	07 17	22 06
06 Tu	15 54	04S 09	24 10	14 13	15 28	19 22	22 46	10 58	07 17	22 06
07 We	16 12	08 52	24 20	13 51	15 14	19 24	22 45	10 57	07 17	22 06
08 Th	16 29	13 04	24 29	13 28	15 00	19 27	22 45	10 56	07 18	22 06
09 Fr	16 47	16 33	24 37	13 07	14 45	19 30	22 45	10 55	07 18	22 06
10 Sa	17 04	19 09	24 43	12 47	14 31	19 33	22 45	10 54	07 18	22 06
11 Su	17 20	20 47	24 47	12 28	14 16	19 36	22 45	10 54	07 18	22 06
12 Mo	17 37	21 24	24 49	12 09	14 02	19 39	22 45	10 53	07 18	22 05
13 Tu	17 53	21 00	24 49	11 52	13 47	19 41	22 45	10 52	07 19	22 05
14 We	18 09	19 40	24 47	11 36	13 32	19 44	22 45	10 51	07 19	22 05
15 Th	18 25	17 29	24 43	11 21	13 17	19 47	22 44	10 50	07 19	22 05
16 Fr	18 40	14 34	24 36	11 07	13 02	19 50	22 44	10 50	07 19	22 05
17 Sa	18 55	11 02	24 27	10 54	12 47	19 53	22 44	10 49	07 19	22 05
18 Su	19 09	06 59	24 16	10 43	12 31	19 55	22 44	10 48	07 19	22 05
19 Mo	19 23	02 34	24 01	10 32	12 16	19 58	22 44	10 47	07 19	22 05
20 Tu	19 37	02N04	23 44	10 23	12 00	20 01	22 43	10 47	07 19	22 05
21 We	19 51	06 46	23 24	10 15	11 45	20 03	22 43	10 46	07 19	22 05
22 Th	20 04	11 15	23 01	10 08	11 29	20 06	22 43	10 45	07 19	22 04
23 Fr	20 17	15 17	22 35	10 02	11 13	20 09	22 43	10 45	07 19	22 04
24 Sa	20 29	18 31	22 06	09 58	10 58	20 11	22 43	10 44	07 19	22 04
25 Su	20 41	20 39	21 35	09 54	10 42	20 14	22 42	10 43	07 19	22 04
26 Mo	20 53	21 28	21 02	09 51	10 26	20 16	22 42	10 43	07 19	22 04
27 Tu	21 04	20 52	20 28	09 50	10 10	20 19	22 42	10 42	07 19	22 04
28 We	21 15	18 54	19 54	09 49	09 54	20 22	22 42	10 41	07 19	22 04
29 Th	21 25	15 47	19 20	09 50	09 38	20 24	22 41	10 41	07 19	22 04
30 Fr	21 36	11 47	18 49	09 51	09 21	20 27	22 41	10 40	07 19	22 03

ASPECTARIAN

01 12:03 ☽ # ♂
15:23 ☽ ☍ ♂
15:51 ☽ # ♀
02 00:26 ☽ # ☉
03:26 ☽ □ ♃
04:32 ☽ ⚹ ♀
06:06 ☽ △ ♅
10:24 ☽ □ ☿
14:26 ☽ △ ♄
21:41 ☽ ∥ ♅
23:41 ☽ ⚹ ☉
03 05:22 ☽ ☍ ♆
14:15 ☽ △ ♇
14:51 ♀ ∥ ☉
16:56 ☽ # ♆
04 07:27 ☽ ⚹ ♃
18:03 ☽ □ ♄
18:12 ☽ ⚹ ☿
05 05:05 ☉ ∥ ♂
17:56 ☽ □ ♇
06 02:44 ☽ △ ♂
06:41 ☉ △ ♆
08:19 ☽ ☌ ♀
13:04 ☽ ☍ ♅
15:44 ☽ ∥ ♆
18:51 ♅ ♈R
22:39 ☽ ⚹ ♄
07 11:22 ☽ # ♅
13:31 ☽ △ ♆
23:07 ☽ ⚹ ♇
08 02:17 ☽ ∥ ♀
10:43 ☽ □ ♂
11:44 ☽ ∥ ♂
12:39 ♃ ♐
19:07 ☽ ☌ ♃
09 01:53 ☽ ∥ ☉
13:38 ☽ ☌ ☿
15:12 ♀ △ ♂
20:43 ☽ □ ♆
10 04:40 ☽ ∥ ♃
19:56 ☽ ⚹ ♀
22:01 ☽ ⚹ ♂
11 03:36 ☽ △ ♅
15:20 ☉ ⚹ ♇
15:26 ☽ ☌ ♄
12 07:03 ☽ ⚹ ♆
17:58 ☽ ☌ ♇
20:23 ☽ ⚹ ☉
13 06:34 ☽ □ ♀
15:15 ☽ □ ♅
18:07 ☽ ⚹ ♃
22:59 ☽ ∥ ♃
14 15:45 ☽ ∥ ☉
18:26 ☽ ⚹ ☿
15 09:31 ♂ ⚹ ♅
19:06 ☽ △ ♀
22:21 ♂ ♓
16 03:59 ☽ ⚹ ♅
05:03 ☽ ☌ ♂
08:11 ☽ □ ♃
10:52 ♀ SD
11:44 ☽ ∥ ♂
17:29 ☽ ⚹ ♄
17 00:49 ☽ ∥ ♀
01:20 ☽ # ♅
01:34 ☿ SR
07:42 ☽ □ ☿
08:09 ☽ ☌ ♆
11:28 ♀ # ♅
19:11 ☽ ⚹ ♇
22:06 ☽ ∥ ♆
18 08:05 ☽ △ ☉
20:19 ☽ △ ♃
19 04:31 ☽ □ ♄
16:27 ☽ △ ☿
20 01:33 ♂ □ ♃
04:22 ☽ □ ♇
15:45 ☽ ☍ ♀
22:47 ☽ ☌ ♅
21 02:52 ☽ # ♆
05:37 ☽ ⚹ ♂
11:55 ☽ △ ♄
17:58 ☽ # ♀
21:13 ☽ ∥ ♅
22 00:01 ☽ ⚹ ♆
01:13 ☽ # ♂
04:51 ☉ ∥ ♃
09:02 ☉ ♐
09:59 ☽ △ ♇
16:47 ☿ ∥ ♄
23 09:48 ☽ ☍ ♃
12:11 ☽ □ ♂
20:30 ☽ ☍ ☿
24 01:21 ☿ ∥ ♆
03:22 ☽ □ ♆
17:45 ☽ # ♃
21:35 ♂ # ♅
25 00:31 ☽ # ☉
01:04 ☽ △ ♀
01:10 ♆ SD
05:31 ☽ ⚹ ♅
13:09 ☽ # ☿
16:46 ☽ △ ♂
18:44 ☽ ☍ ♄
26 04:45 ☿ ∥ ☉
05:25 ☽ △ ♆
06:34 ☉ ☌ ♃
08:06 ☿ □ ♂
15:07 ☽ ☍ ♇
20:40 ☽ # ☉
27 04:16 ☽ □ ♀
05:43 ☿ ∥ ♃
07:22 ☽ □ ♅
08:26 ☽ # ♃
09:15 ☿ ☌ ☉
09:47 ☽ # ☿
15:42 ☽ △ ♃
16:23 ☽ △ ☿
17:39 ☽ △ ☉
21:31 ♂ ⚹ ♄
22:28 ☿ ☌ ♃
28 06:37 ♀ ∥ ♂
29 08:17 ☽ ⚹ ♀
09:47 ☽ △ ♅
14:43 ☽ □ ☿
19:12 ☽ □ ♃
30 00:19 ☽ △ ♄
02:20 ☽ ☍ ♂
06:02 ☽ ∥ ♅
10:18 ☽ # ♀
10:35 ☽ ☍ ♆
13:44 ☽ # ♂
20:49 ☽ △ ♇
23:19 ☽ # ♆

⚷ Chiron

01 Dec.	02 N 39
01	28♓33R
04	28 27
07	28 22
10	28 17
13	28 13
16	28 09
19	28 05
22	28 02
25	28 00
28	27 57

December 2018

Day	S. T.	☉	☽	☿	♀	♂	♃	♄	♅	♆	♇	☊ True
	h m s	° ' "	° ' "	° '	° '	° '	° '	° '	° '	° '	° '	° '
01 Sa	04 39 12	08♐43 27	21♍28 21	00♐R29	29♎07	09♓28	05♐00	07♑49	29♈R10	13♓42	19♑38	27♋R58
02 Su	04 43 09	09 44 16	05♎14 45	29♏29	29 37	10 07	05 13	07 56	29 08	13 42	19 40	27 55
03 Mo	04 47 05	10 45 07	18 50 45	28 40	00♏10	10 46	05 26	08 02	29 06	13 43	19 41	27 50
04 Tu	04 51 02	11 45 59	02♏15 41	28 02	00 43	11 24	05 40	08 09	29 05	13 43	19 43	27 43
05 We	04 54 58	12 46 53	15 28 41	27 36	01 18	12 03	05 53	08 16	29 03	13 43	19 45	27 35
06 Th	04 58 55	13 47 47	28 28 47	27 20	01 55	12 42	06 07	08 22	29 01	13 44	19 46	27 27
07 Fr	05 02 52	14 48 43	11♐15 12	27D 16	02 33	13 21	06 20	08 29	29 00	13 44	19 48	27 19
08 Sa	05 06 48	15 49 40	23 47 37	27 22	03 12	14 00	06 33	08 36	28 58	13 45	19 50	27 13
09 Su	05 10 45	16 50 38	06♑06 27	27 38	03 52	14 39	06 47	08 42	28 57	13 45	19 52	27 09
10 Mo	05 14 41	17 51 37	18 13 03	28 03	04 34	15 19	07 00	08 49	28 56	13 45	19 53	27 06
11 Tu	05 18 38	18 52 37	00♒09 40	28 35	05 16	15 58	07 13	08 56	28 54	13 46	19 55	27D 06
12 We	05 22 34	19 53 37	11 59 29	29 15	06 00	16 37	07 27	09 03	28 53	13 47	19 57	27 08
13 Th	05 26 31	20 54 38	23 46 25	00♐01	06 45	17 17	07 40	09 09	28 52	13 47	19 59	27 10
14 Fr	05 30 27	21 55 39	05♓35 01	00 52	07 31	17 56	07 53	09 16	28 50	13 48	20 01	27 13
15 Sa	05 34 24	22 56 41	17 30 16	01 48	08 18	18 36	08 06	09 23	28 49	13 48	20 02	27 15
16 Su	05 38 21	23 57 43	29 37 19	02 48	09 05	19 16	08 19	09 30	28 48	13 49	20 04	27R 16
17 Mo	05 42 17	24 58 46	12♈01 11	03 53	09 54	19 55	08 33	09 37	28 47	13 50	20 06	27 15
18 Tu	05 46 14	25 59 49	24 46 16	05 00	10 43	20 35	08 46	09 44	28 46	13 51	20 08	27 12
19 We	05 50 10	27 00 53	07♉55 53	06 10	11 33	21 15	08 59	09 51	28 45	13 51	20 10	27 09
20 Th	05 54 07	28 01 57	21 31 39	07 23	12 24	21 55	09 12	09 58	28 44	13 52	20 12	27 05
21 Fr	05 58 03	29 03 01	05♊32 56	08 38	13 16	22 35	09 25	10 05	28 43	13 53	20 14	27 00
22 Sa	06 01 60	00♑04 06	19 56 40	09 55	14 09	23 15	09 38	10 12	28 42	13 54	20 16	26 56
23 Su	06 05 56	01 05 11	04♋37 23	11 14	15 02	23 55	09 51	10 19	28 42	13 55	20 18	26 54
24 Mo	06 09 53	02 06 17	19 27 54	12 34	15 56	24 35	10 04	10 26	28 41	13 56	20 20	26 52
25 Tu	06 13 50	03 07 24	04♌20 24	13 55	16 51	25 15	10 17	10 33	28 40	13 57	20 22	26 51
26 We	06 17 46	04 08 30	19 07 29	15 18	17 46	25 55	10 30	10 40	28 40	13 58	20 24	26D 51
27 Th	06 21 43	05 09 38	03♍43 04	16 41	18 42	26 35	10 43	10 47	28 39	13 59	20 25	26 51
28 Fr	06 25 39	06 10 46	18 03 00	18 06	19 38	27 15	10 55	10 54	28 39	14 00	20 27	26 52
29 Sa	06 29 36	07 11 54	02♎04 59	19 31	20 35	27 55	11 08	11 01	28 38	14 01	20 29	26R 51
30 Su	06 33 32	08 13 03	15 48 18	20 57	21 33	28 36	11 21	11 08	28 38	14 02	20 31	26 51
31 Mo	06 37 29	09 14 12	29 13 27	22 24	22 31	29 16	11 34	11 16	28 37	14 04	20 33	26 49

Data for	12-01-2018
Julian Day	2458453.50
Ayanamsa	24 06 59
SVP	04 ♓ 59 28
☽ ☊ Mean	29 ♋ 13 R

● ◐ PHASES ○ ◑

Day	Time	Phase	Position
07	07:21	●	15♐07
15	11:49	◐	23♓27
22	17:49	○	00♋49
29	09:35	◑	07♎36

LAST ASPECT ☽ Day	h m	INGRESS Day	h m	Sign
01	14:35	01	14:49	♎
03	18:17	03	19:56	♏
05	21:55	06	02:50	♐
08	10:01	08	12:03	♑
10	21:28	10	23:40	♒
13	10:20	13	12:40	♓
15	11:50	16	00:44	♈
18	07:22	18	09:38	♉
20	00:42	20	14:35	♊
22	14:22	22	16:29	♋
24	14:51	24	16:59	♌
26	15:38	26	17:51	♍
28	16:28	28	20:24	♎
30	22:55	31	01:24	♏

DECLINATION

Day	☉	☽	☿	♀	♂	♃	♄	♅	♆	♇
01 Sa	21S 45	07N11	18S 21	09S 53	09S 05	20S 29	22S 41	10N40	07S 19	22S 03
02 Su	21 54	02 16	17 55	09 56	08 49	20 32	22 41	10 39	07 19	22 03
03 Mo	22 03	02S 41	17 34	10 00	08 32	20 34	22 40	10 38	07 19	22 03
04 Tu	22 12	07 26	17 18	10 04	08 16	20 36	22 40	10 38	07 18	22 03
05 We	22 20	11 47	17 06	10 10	07 59	20 39	22 40	10 37	07 18	22 03
06 Th	22 27	15 30	16 58	10 16	07 43	20 41	22 39	10 37	07 18	22 03
07 Fr	22 34	18 26	16 55	10 22	07 26	20 44	22 39	10 36	07 18	22 02
08 Sa	22 41	20 26	16 55	10 30	07 09	20 46	22 39	10 36	07 18	22 02
09 Su	22 47	21 25	16 59	10 38	06 52	20 48	22 38	10 35	07 18	22 02
10 Mo	22 53	21 23	17 07	10 46	06 36	20 50	22 38	10 35	07 17	22 02
11 Tu	22 58	20 22	17 17	10 55	06 19	20 53	22 38	10 34	07 17	22 02
12 We	23 03	18 28	17 29	11 05	06 02	20 55	22 37	10 34	07 17	22 02
13 Th	23 07	15 48	17 43	11 15	05 45	20 57	22 37	10 34	07 17	22 01
14 Fr	23 11	12 29	17 59	11 25	05 28	20 59	22 36	10 33	07 16	22 01
15 Sa	23 15	08 38	18 16	11 36	05 11	21 01	22 36	10 33	07 16	22 01
16 Su	23 18	04 24	18 34	11 47	04 54	21 04	22 36	10 32	07 16	22 01
17 Mo	23 20	00N07	18 53	11 59	04 37	21 06	22 35	10 32	07 16	22 01
18 Tu	23 22	04 44	19 13	12 11	04 19	21 08	22 35	10 32	07 15	22 00
19 We	23 24	09 17	19 32	12 23	04 02	21 10	22 34	10 31	07 15	22 00
20 Th	23 25	13 33	19 52	12 35	03 45	21 12	22 34	10 31	07 15	22 00
21 Fr	23 26	17 12	20 11	12 48	03 28	21 14	22 33	10 31	07 14	22 00
22 Sa	23 26	19 54	20 31	13 01	03 11	21 16	22 33	10 31	07 14	22 00
23 Su	23 26	21 22	20 50	13 15	02 53	21 18	22 33	10 30	07 13	21 59
24 Mo	23 25	21 21	21 08	13 28	02 36	21 20	22 32	10 30	07 13	21 59
25 Tu	23 24	19 50	21 26	13 41	02 19	21 22	22 32	10 30	07 13	21 59
26 We	23 22	16 58	21 43	13 55	02 02	21 24	22 31	10 30	07 12	21 59
27 Th	23 20	13 05	22 00	14 09	01 44	21 25	22 31	10 30	07 12	21 59
28 Fr	23 17	08 30	22 16	14 23	01 27	21 27	22 30	10 29	07 11	21 58
29 Sa	23 14	03 33	22 31	14 37	01 10	21 29	22 30	10 29	07 11	21 58
30 Su	23 11	01S 27	22 45	14 51	00 53	21 31	22 29	10 29	07 10	21 58
31 Mo	23 07	06 16	22 58	15 05	00 35	21 33	22 29	10 29	07 10	21 58

ASPECTARIAN

01 02:13 ♀ ☍ ♅
11:13 ☿ ♏R
14:35 ☽ ✶ ☿
23:57 ☽ ✶ ♃

02 04:45 ☽ □ ♄
08:32 ☽ ✶ ☉
17:02 ♀ ♏
23:24 ☉ ∥ ♇

03 00:35 ☉ □ ♂
01:30 ☽ □ ♇
18:17 ☽ ☍ ♅
21:06 ☽ ☌ ♀
23:17 ☽ ∥ ♆

04 04:04 ☽ ∥ ♂
10:44 ☽ ✶ ♄
14:28 ☽ ∥ ♀
17:16 ☽ # ♅
17:25 ☽ △ ♂
20:47 ☽ △ ♆

05 07:51 ☽ ✶ ♇
21:55 ☽ ☌ ☿
22:22 ☉ □ ♆

06 10:48 ☽ ∥ ☿
14:32 ☽ ☌ ♃
21:23 ☿ SD

07 04:12 ☽ □ ♂
04:43 ☽ □ ♆
11:21 ♂ ∥ ♆
14:12 ♂ ☌ ♆
15:59 ☉ ∥ ♄

08 05:51 ☽ ∥ ♃
10:01 ☽ △ ♅
17:51 ♀ # ♅
19:21 ☽ ✶ ♀

09 05:10 ☽ ☌ ♄
15:07 ☽ ✶ ♆
17:53 ☽ ✶ ♂

10 03:21 ☽ ☌ ♇
15:06 ☽ ∥ ♃
20:39 ☽ ✶ ☿
21:28 ☽ □ ♅

11 11:02 ☽ □ ♀
14:35 ☽ ✶ ♃

12 08:53 ☽ ∥ ☿
17:37 ☽ ✶ ☉
23:43 ☿ ♐

13 10:20 ☽ ✶ ♅
13:39 ☽ □ ☿

14 04:11 ☽ △ ♀
04:44 ☽ □ ♃
06:36 ☽ ∥ ♀
07:32 ☽ ✶ ♄
12:25 ☽ # ♅
16:35 ☽ ☌ ♆

15 02:19 ☽ ☌ ♂
05:05 ☽ ✶ ♇
07:59 ☽ ∥ ♆
21:06 ☽ ∥ ♂

16 06:48 ☽ △ ☿
14:27 ♀ ✶ ♄
17:13 ☽ △ ♃
19:21 ☽ □ ♄

17 06:54 ♂ ✶ ♇
15:20 ☽ □ ♇
22:01 ☽ # ♂

18 02:28 ☽ △ ☉
07:22 ☽ ☌ ♅
13:09 ☽ # ♆

19 03:28 ☽ △ ♄
06:43 ☽ ∥ ♅
06:54 ☽ ☍ ♀
10:34 ☽ ✶ ♆
18:06 ☽ # ♀
21:41 ☽ △ ♇

20 00:42 ☽ ✶ ♂
16:22 ☉ △ ♅

21 05:42 ☽ ☍ ☿
06:36 ☽ ☍ ♃
13:59 ☽ □ ♆
17:12 ♀ △ ♆
17:38 ☿ ☌ ♃
22:23 ☉ ♑

22 05:41 ☽ □ ♂
09:13 ☽ # ☿
14:22 ☽ ✶ ♅
21:52 ☽ # ♃

23 09:18 ☽ ☍ ♄
15:04 ☽ △ ♆
17:56 ☽ △ ♀

24 00:43 ☽ # ♃
01:24 ☽ ☍ ♇
04:20 ☽ # ☿
08:38 ☽ △ ♂
14:51 ☽ □ ♅
17:22 ☿ ∥ ♃

25 00:33 ☿ □ ♆
09:45 ☽ △ ♃
17:07 ☽ △ ☿
21:38 ☽ □ ♀

26 15:38 ☽ △ ♅
18:17 ☽ # ♀
22:02 ☿ ∥ ♇

27 02:35 ☽ △ ☉
11:49 ☽ □ ♃
11:52 ☽ △ ♄
13:55 ☽ ∥ ♅
17:10 ☽ ☍ ♆

28 00:05 ☽ □ ☿
02:53 ☽ ✶ ♀
04:05 ☽ △ ♇
06:30 ☽ # ♆
16:28 ☽ ☍ ♂
21:29 ♀ ✶ ♇
22:01 ☿ ∥ ♄

29 12:09 ☽ # ♂
15:43 ☽ □ ♄
16:01 ☽ ✶ ♃
21:24 ☽ ∥ ♂

30 08:24 ☽ □ ♇
10:14 ☽ ✶ ☿
22:55 ☽ ☍ ♅

31 04:39 ☽ ∥ ♆
12:37 ☿ ∥ ☉
19:48 ☽ ✶ ☉
22:10 ☽ ✶ ♄
22:44 ☽ # ♅

⚷ Chiron

01	Dec.	02 N 18
01	27♓56R	
04	27 55	
07	27 54	
10	27 54D	
13	27 54	
16	27 55	
19	27 56	
22	27 58	
25	28 01	
28	28 03	
31	28 07	
09	06:00	27♓54 D

06 01:43 15♑25 ☉ Solar Eclipse (mag 0.715)
21 05:13 00♌52 ✷ Total Lunar Eclipse (mag 1.201)

January 2019

Day	S. T.	☉	☽	☿	♀	♂	♃	♄	♅	♆	♇	☊ True
	h m s	° ' "	° ' "	° '	° '	° '	° '	° '	° '	° '	° '	° '
01 Tu	06 41 26	10♑15 22	12♏21 35	23♐51	23♏30	29♓56	11♐46	11♑23	28♈R37	14♓05	20♑35	26♋R47
02 We	06 45 22	11 16 32	25 14 14	25 19	24 29	00♈36	11 59	11 30	28 37	14 06	20 37	26 44
03 Th	06 49 19	12 17 43	07♐52 59	26 48	25 28	01 17	12 11	11 37	28 36	14 07	20 39	26 42
04 Fr	06 53 15	13 18 54	20 19 20	28 17	26 28	01 57	12 24	11 44	28 36	14 09	20 41	26 40
05 Sa	06 57 12	14 20 05	02♑34 44	29 46	27 29	02 38	12 36	11 51	28 36	14 10	20 43	26 39
06 Su	07 01 08	15 21 16	14 40 38	01♑16	28 30	03 18	12 49	11 58	28 36	14 11	20 45	26 38
07 Mo	07 05 05	16 22 27	26 38 36	02 47	29 31	03 58	13 01	12 05	28D36	14 13	20 48	26D38
08 Tu	07 09 01	17 23 37	08♒30 33	04 18	00♐33	04 39	13 13	12 12	28 36	14 14	20 50	26 39
09 We	07 12 58	18 24 48	20 18 43	05 49	01 35	05 19	13 25	12 19	28 36	14 16	20 52	26 41
10 Th	07 16 55	19 25 58	02♓05 51	07 21	02 37	06 00	13 37	12 26	28 36	14 17	20 54	26 43
11 Fr	07 20 51	20 27 07	13 55 10	08 53	03 40	06 41	13 50	12 33	28 36	14 19	20 56	26 44
12 Sa	07 24 48	21 28 16	25 50 26	10 26	04 43	07 21	14 02	12 40	28 37	14 20	20 58	26 46
13 Su	07 28 44	22 29 25	07♈55 50	11 59	05 46	08 02	14 14	12 47	28 37	14 22	21 00	26 47
14 Mo	07 32 41	23 30 33	20 15 49	13 32	06 50	08 42	14 25	12 55	28 37	14 23	21 02	26 48
15 Tu	07 36 37	24 31 40	02♉54 49	15 06	07 54	09 23	14 37	13 02	28 38	14 25	21 04	26 48
16 We	07 40 34	25 32 47	15 56 49	16 41	08 58	10 03	14 49	13 09	28 38	14 26	21 06	26 48
17 Th	07 44 30	26 33 53	29 24 55	18 16	10 03	10 44	15 01	13 15	28 39	14 28	21 08	26R48
18 Fr	07 48 27	27 34 58	13♊20 34	19 51	11 08	11 25	15 12	13 22	28 39	14 30	21 10	26 48
19 Sa	07 52 24	28 36 02	27 42 58	21 27	12 13	12 05	15 24	13 29	28 40	14 32	21 12	26 48
20 Su	07 56 20	29 37 06	12♋28 37	23 04	13 18	12 46	15 35	13 36	28 41	14 33	21 14	26 48
21 Mo	08 00 17	00♒38 09	27 31 13	24 41	14 24	13 27	15 47	13 43	28 41	14 35	21 16	26 47
22 Tu	08 04 13	01 39 11	12♌42 15	26 19	15 29	14 07	15 58	13 50	28 42	14 37	21 18	26 46
23 We	08 08 10	02 40 13	27 52 09	27 57	16 36	14 48	16 09	13 57	28 43	14 39	21 20	26 45
24 Th	08 12 06	03 41 14	12♍51 39	29 36	17 42	15 29	16 20	14 04	28 44	14 40	21 22	26 44
25 Fr	08 16 03	04 42 14	27 33 08	01♒15	18 48	16 09	16 31	14 11	28 45	14 42	21 24	26 42
26 Sa	08 19 59	05 43 14	11♎51 27	02 55	19 55	16 50	16 42	14 17	28 46	14 44	21 26	26 41
27 Su	08 23 56	06 44 14	25 44 12	04 36	21 02	17 30	16 53	14 24	28 47	14 46	21 28	26 39
28 Mo	08 27 53	07 45 13	09♏11 26	06 17	22 09	18 11	17 04	14 31	28 48	14 48	21 30	26 38
29 Tu	08 31 49	08 46 11	22 15 00	07 59	23 17	18 52	17 15	14 38	28 49	14 50	21 32	26D38
30 We	08 35 46	09 47 09	04♐57 56	09 42	24 24	19 32	17 25	14 44	28 50	14 52	21 34	26 39
31 Th	08 39 42	10 48 06	17 23 45	11 25	25 32	20 13	17 36	14 51	28 51	14 54	21 36	26 40

Data for 01-01-2019
Julian Day 2458484.50
Ayanamsa 24 07 05
SVP 04♓59 25
☽ ☊ Mean 27♋34 R

● ◐ PHASES ○ ◑

06 01:28 ☉ 15♑25
14 06:46 ◐ 23♈48
21 05:17 ✷ 00♌52
27 21:11 ◑ 07♏38

LAST ASPECT ☽ Day	h m	INGRESS Day	h m	
01	22:27	02	08:59	♐
04	17:42	04	18:55	♑
07	06:21	07	06:46	♒
09	16:53	09	19:44	♓
11	14:25	12	08:18	♈
14	15:56	14	18:32	♉
16	18:35	17	01:01	♊
19	01:34	19	03:45	♋
21	01:51	21	03:56	♌
23	01:21	23	03:23	♍
24	13:51	25	04:03	♎
27	05:22	27	07:32	♏
28	22:39	29	14:33	♐

DECLINATION

Day	☉	☽	☿	♀	♂	♃	♄	♅	♆	♇
01 Tu	23S02	10S42	23S10	15S19	00S18	21S34	22S28	10N29	07S09	21S58
02 We	22 57	14 34	23 21	15 33	00 01	21 36	22 27	10 29	07 09	21 57
03 Th	22 52	17 42	23 31	15 47	00N17	21 38	22 27	10 29	07 08	21 57
04 Fr	22 46	19 57	23 40	16 00	00 34	21 39	22 26	10 29	07 08	21 57
05 Sa	22 40	21 15	23 48	16 14	00 51	21 41	22 26	10 29	07 07	21 57
06 Su	22 33	21 32	23 54	16 28	01 08	21 43	22 25	10 29	07 07	21 57
07 Mo	22 26	20 49	24 00	16 41	01 26	21 44	22 25	10 29	07 06	21 56
08 Tu	22 18	19 10	24 04	16 55	01 43	21 46	22 24	10 29	07 06	21 56
09 We	22 10	16 43	24 07	17 08	02 00	21 47	22 24	10 29	07 05	21 56
10 Th	22 02	13 35	24 08	17 21	02 17	21 49	22 23	10 29	07 05	21 56
11 Fr	21 53	09 55	24 09	17 34	02 34	21 50	22 22	10 29	07 04	21 55
12 Sa	21 43	05 50	24 08	17 46	02 52	21 52	22 22	10 29	07 03	21 55
13 Su	21 34	01 28	24 06	17 59	03 09	21 53	22 21	10 29	07 03	21 55
14 Mo	21 23	03N03	24 02	18 11	03 26	21 54	22 21	10 29	07 02	21 55
15 Tu	21 13	07 32	23 57	18 23	03 43	21 56	22 20	10 30	07 02	21 55
16 We	21 02	11 49	23 50	18 34	04 00	21 57	22 19	10 30	07 01	21 54
17 Th	20 50	15 40	23 43	18 45	04 17	21 58	22 19	10 30	07 00	21 54
18 Fr	20 38	18 47	23 33	18 56	04 34	22 00	22 18	10 30	07 00	21 54
19 Sa	20 26	20 50	23 23	19 07	04 51	22 01	22 17	10 31	06 59	21 54
20 Su	20 14	21 33	23 11	19 17	05 07	22 02	22 17	10 31	06 58	21 53
21 Mo	20 01	20 43	22 57	19 27	05 24	22 03	22 16	10 31	06 58	21 53
22 Tu	19 47	18 24	22 42	19 37	05 41	22 05	22 16	10 31	06 57	21 53
23 We	19 34	14 47	22 26	19 46	05 58	22 06	22 15	10 32	06 56	21 53
24 Th	19 20	10 15	22 08	19 55	06 14	22 07	22 14	10 32	06 55	21 53
25 Fr	19 05	05 13	21 49	20 03	06 31	22 08	22 14	10 32	06 55	21 52
26 Sa	18 50	00 01	21 28	20 11	06 47	22 09	22 13	10 33	06 54	21 52
27 Su	18 35	05S01	21 06	20 18	07 04	22 10	22 12	10 33	06 53	21 52
28 Mo	18 20	09 39	20 42	20 26	07 20	22 11	22 12	10 34	06 52	21 52
29 Tu	18 04	13 43	20 16	20 32	07 37	22 12	22 11	10 34	06 52	21 51
30 We	17 48	17 02	19 50	20 38	07 53	22 13	22 10	10 34	06 51	21 51
31 Th	17 31	19 30	19 21	20 44	08 09	22 14	22 10	10 35	06 50	21 51

ASPECTARIAN

01 02:20 ♂ ♈
03:11 ☽ △ ♆
15:20 ☽ ⚹ ♇
22:27 ☽ ☌ ♀

02 05:50 ☉ ☌ ♄
07:25 ☽ ∥ ♀
10:42 ☽ △ ♂

03 08:24 ☽ ☌ ♃
12:01 ☽ □ ♆
04 05:13 ☿ △ ♅
16:11 ☽ △ ♅
17:42 ☽ ☌ ☿
19:59 ☉ ⚹ ♆

05 00:06 ☽ □ ♂
03:40 ☿ ♑
18:33 ☽ ☌ ♄
23:01 ☽ ⚹ ♆
06 12:12 ☽ ☌ ♇
20:27 ♅ SD

07 03:57 ☽ □ ♅
04:11 ☉ ∥ ♄
06:21 ☽ ⚹ ♀
11:19 ♀ ♐
15:42 ☽ ⚹ ♂

08 09:44 ☽ ⚹ ♃
10:05 ☿ □ ♂
20:46 ☽ ∥ ♀

09 16:53 ☽ ⚹ ♅
10 01:10 ☽ □ ♀
12:16 ☽ ⚹ ☿
16:41 ☉ ∥ ♇
20:29 ☽ # ♅
21:13 ☽ ⚹ ♄
23:48 ☽ □ ♃

11 00:47 ☽ ☌ ♆
05:44 ☉ ∥ ♃
11:37 ☉ ☌ ♇
14:11 ☽ ⚹ ♇
14:25 ☽ ⚹ ☉
17:00 ☽ ∥ ♆
12 15:28 ☽ # ♂
19:20 ☽ △ ♀

13 00:12 ☽ ☌ ♂
09:05 ☽ □ ☿
09:37 ☽ □ ♄
12:31 ☽ △ ♃
13:32 ☿ ☌ ♄
19:02 ♃ □ ♆

14 01:28 ☽ □ ♇
02:15 ☽ ∥ ♂
04:57 ♃ ∥ ♇
13:14 ☿ ⚹ ♆
15:56 ☽ ☌ ♅
21:20 ☽ # ♆

15 16:28 ☽ ∥ ♅
18:51 ☽ △ ♄
21:16 ☽ ⚹ ♆

16 01:31 ☽ △ ☿
09:18 ☽ △ ♇
18:35 ☽ △ ☉

17 19:55 ☽ ☍ ♀
20:33 ☽ ⚹ ♂

18 01:38 ☽ # ♀
01:57 ☽ □ ♆
03:12 ☽ ☍ ♃
16:50 ♀ △ ♂
18:39 ☽ # ☉
20:03 ☿ ☌ ♇

19 01:31 ☉ □ ♅
01:34 ☽ ⚹ ♅
20 00:29 ☽ □ ♂
01:50 ☽ ☍ ♄
03:21 ☽ △ ♆
09:00 ☉ ♒
14:02 ☽ ☍ ♇
18:58 ☽ ☍ ☿

21 01:51 ☽ □ ♅
04:16 ♀ □ ♆
10:04 ☽ # ☉
11:49 ♂ □ ♄
13:53 ☽ # ♀

22 02:20 ☽ △ ♂
04:45 ☽ △ ♀
05:13 ☽ △ ♃
11:06 ♀ ∥ ☉
12:27 ♀ ☌ ♃

23 01:21 ☽ △ ♅
11:13 ☿ □ ♅
15:42 ☿ ∥ ♄
22:38 ☽ ∥ ♅

24 01:19 ☿ ∥ ♃
01:57 ☽ △ ♄
02:56 ☽ ☍ ♆
05:42 ☽ □ ♃
05:50 ☿ ♒
08:29 ☽ □ ♀
13:51 ☽ △ ♇
16:04 ☽ # ♆
18:15 ☽ ∥ ♂
19:38 ☿ ∥ ♇

25 06:57 ☽ △ ☿
12:49 ☽ △ ☉
17:54 ♂ △ ♃

26 04:11 ☽ □ ♄
08:24 ☽ ⚹ ♃
08:57 ☽ ☍ ♂
08:59 ♂ # ♆
15:04 ☽ ⚹ ♀
16:31 ☽ □ ♇

27 05:22 ☽ ☍ ♅
09:23 ☽ ∥ ♆
10:58 ☽ # ♂
18:00 ☽ □ ☿

28 03:06 ♃ ∥ ♄
05:05 ☽ # ♅
09:47 ☽ ⚹ ♄
10:15 ☽ △ ♆
12:18 ☿ ∥ ♀
22:39 ☽ ⚹ ♇

30 02:52 ☿ ☌ ☉
06:01 ☽ ∥ ☉
10:04 ☽ ⚹ ☉
10:32 ☽ ⚹ ☿
19:07 ☽ □ ♆
22:41 ☽ ∥ ☿

31 00:24 ☽ ☌ ♃
05:50 ☽ △ ♂
14:22 ♄ ⚹ ♆
17:35 ☽ ☌ ♀
19:18 ☽ ∥ ♀
22:33 ☽ △ ♅

⚷ Chiron

01	Dec.	02 N 16
03	28♓11	
06	28 15	
09	28 20	
12	28 25	
15	28 30	
18	28 36	
21	28 43	
24	28 49	
27	28 57	
30	29 04	

February 2019

Day	S. T.	☉	☽	☿	♀	♂	♃	♄	♅	♆	♇	☊ True
	h m s	° ' "	° ' "	° '	° '	° '	° '	° '	° '	° '	° '	° '
01 Fr	08 43 39	11♒49 02	29♐36 01	13♒09	26♐40	20♈54	17♐46	14♑57	28♈52	14♓56	21♑38	26♋41
02 Sa	08 47 35	12 49 57	11♑38 06	14 54	27 48	21 34	17 56	15 04	28 54	14 58	21 39	26 43
03 Su	08 51 32	13 50 52	23 33 02	16 39	28 56	22 15	18 07	15 11	28 55	15 00	21 41	26 44
04 Mo	08 55 28	14 51 45	05♒23 26	18 25	00♑04	22 56	18 17	15 17	28 56	15 02	21 43	26 44
05 Tu	08 59 25	15 52 38	17 11 37	20 12	01 13	23 36	18 27	15 24	28 58	15 04	21 45	26R 44
06 We	09 03 22	16 53 29	28 59 40	21 59	02 21	24 17	18 37	15 30	28 59	15 06	21 47	26 42
07 Th	09 07 18	17 54 19	10♓49 34	23 46	03 30	24 57	18 47	15 36	29 01	15 08	21 49	26 40
08 Fr	09 11 15	18 55 08	22 43 23	25 34	04 39	25 38	18 56	15 43	29 03	15 10	21 51	26 37
09 Sa	09 15 11	19 55 55	04♈43 28	27 22	05 48	26 19	19 06	15 49	29 04	15 12	21 53	26 34
10 Su	09 19 08	20 56 41	16 52 28	29 11	06 57	26 59	19 15	15 55	29 06	15 14	21 54	26 31
11 Mo	09 23 04	21 57 25	29 13 29	01♓00	08 07	27 40	19 25	16 01	29 08	15 16	21 56	26 29
12 Tu	09 27 01	22 58 08	11♉49 56	02 48	09 16	28 20	19 34	16 08	29 09	15 18	21 58	26 28
13 We	09 30 57	23 58 49	24 45 21	04 36	10 25	29 01	19 43	16 14	29 11	15 20	22 00	26D 28
14 Th	09 34 54	24 59 28	08♊03 02	06 24	11 35	29 42	19 52	16 20	29 13	15 23	22 02	26 29
15 Fr	09 38 51	26 00 06	21 45 38	08 12	12 45	00♉22	20 01	16 26	29 15	15 25	22 03	26 30
16 Sa	09 42 47	27 00 42	05♋54 27	09 58	13 55	01 03	20 10	16 32	29 17	15 27	22 05	26 30
17 Su	09 46 44	28 01 16	20 28 37	11 43	15 04	01 43	20 19	16 38	29 19	15 29	22 07	26R 30
18 Mo	09 50 40	29 01 49	05♌24 25	13 26	16 14	02 24	20 27	16 44	29 21	15 31	22 09	26 29
19 Tu	09 54 37	00♓02 20	20 35 10	15 07	17 25	03 04	20 35	16 49	29 23	15 34	22 10	26 27
20 We	09 58 33	01 02 49	05♍51 38	16 45	18 35	03 45	20 44	16 55	29 25	15 36	22 12	26 22
21 Th	10 02 30	02 03 17	21 03 22	18 20	19 45	04 25	20 52	17 01	29 27	15 38	22 14	26 17
22 Fr	10 06 26	03 03 43	06♎00 20	19 52	20 55	05 06	21 00	17 06	29 30	15 40	22 15	26 11
23 Sa	10 10 23	04 04 08	20 34 31	21 19	22 06	05 46	21 08	17 12	29 32	15 42	22 17	26 05
24 Su	10 14 20	05 04 31	04♏40 53	22 41	23 16	06 26	21 16	17 18	29 34	15 45	22 18	26 00
25 Mo	10 18 16	06 04 53	18 17 46	23 57	24 27	07 07	21 23	17 23	29 37	15 47	22 20	25 56
26 Tu	10 22 13	07 05 14	01♐26 16	25 07	25 38	07 47	21 31	17 29	29 39	15 49	22 22	25 54
27 We	10 26 09	08 05 33	14 09 37	26 10	26 49	08 28	21 38	17 34	29 41	15 51	22 23	25D 54
28 Th	10 30 06	09 05 51	26 32 12	27 06	27 59	09 08	21 45	17 39	29 44	15 54	22 25	25 55

Data for	02-01-2019
Julian Day	2458515.50
Ayanamsa	24 07 10
SVP	04 ♓ 59 21
☽ ☊ Mean	25 ♋ 56 R

● ◐ PHASES ○ ◑

04	21:04	●	15♒45
12	22:27	◐	23♉55
19	15:54	○	00♍42
26	11:28	◑	07♐34

LAST ASPECT Day	h m	☽ INGRESS Day	h m	
31	22:33	01	00:48	♑
03	10:53	03	13:04	♒
06	00:00	06	02:03	♓
07	22:14	08	14:35	♈
10	23:49	11	01:29	♉
12	22:28	13	09:33	♊
15	12:50	15	14:04	♋
17	14:18	17	15:22	♌
19	13:52	19	14:48	♍
21	01:52	21	14:18	♎
23	15:11	23	15:57	♏
25	12:14	25	21:20	♐
28	06:18	28	06:49	♑

DECLINATION

Day	☉	☽	☿	♀	♂	♃	♄	♅	♆	♇
01 Fr	17S15	21S01	18S51	20S49	08N25	22S15	22S09	10N35	06S49	21S51
02 Sa	16 57	21 33	18 20	20 54	08 42	22 16	22 08	10 36	06 49	21 51
03 Su	16 40	21 06	17 47	20 58	08 58	22 17	22 08	10 36	06 48	21 50
04 Mo	16 22	19 42	17 13	21 01	09 14	22 18	22 07	10 37	06 47	21 50
05 Tu	16 04	17 26	16 37	21 04	09 29	22 19	22 06	10 37	06 46	21 50
06 We	15 46	14 28	16 00	21 07	09 45	22 20	22 05	10 38	06 46	21 50
07 Th	15 28	10 54	15 21	21 09	10 01	22 21	22 05	10 39	06 45	21 50
08 Fr	15 09	06 54	14 41	21 10	10 17	22 21	22 04	10 39	06 44	21 49
09 Sa	14 50	02 36	14 00	21 11	10 32	22 22	22 03	10 40	06 43	21 49
10 Su	14 31	01N51	13 17	21 12	10 48	22 23	22 03	10 40	06 42	21 49
11 Mo	14 11	06 18	12 33	21 12	11 03	22 24	22 02	10 41	06 41	21 49
12 Tu	13 52	10 35	11 48	21 11	11 18	22 24	22 01	10 42	06 41	21 49
13 We	13 32	14 29	11 02	21 10	11 33	22 25	22 01	10 42	06 40	21 48
14 Th	13 11	17 47	10 15	21 08	11 49	22 26	22 00	10 43	06 39	21 48
15 Fr	12 51	20 12	09 27	21 05	12 04	22 26	21 59	10 44	06 38	21 48
16 Sa	12 30	21 28	08 38	21 02	12 18	22 27	21 59	10 45	06 37	21 48
17 Su	12 10	21 20	07 49	20 59	12 33	22 28	21 58	10 45	06 36	21 48
18 Mo	11 49	19 43	07 00	20 54	12 48	22 28	21 57	10 46	06 36	21 47
19 Tu	11 27	16 39	06 11	20 50	13 03	22 29	21 57	10 47	06 35	21 47
20 We	11 06	12 27	05 22	20 44	13 17	22 29	21 56	10 48	06 34	21 47
21 Th	10 45	07 27	04 33	20 38	13 31	22 30	21 55	10 48	06 33	21 47
22 Fr	10 23	02 06	03 45	20 32	13 46	22 31	21 55	10 49	06 32	21 47
23 Sa	10 01	03S15	02 58	20 25	14 00	22 31	21 54	10 50	06 31	21 47
24 Su	09 39	08 14	02 13	20 17	14 14	22 32	21 53	10 51	06 30	21 46
25 Mo	09 17	12 39	01 30	20 09	14 28	22 32	21 53	10 52	06 30	21 46
26 Tu	08 54	16 18	00 49	20 00	14 42	22 33	21 52	10 53	06 29	21 46
27 We	08 32	19 03	00 10	19 51	14 55	22 33	21 51	10 53	06 28	21 46
28 Th	08 09	20 50	00N26	19 41	15 09	22 33	21 51	10 54	06 27	21 46

ASPECTARIAN

02 03:16 ♂ □ ♇
06:42 ☽ ⚹ ♆
06:57 ☽ ☌ ♄
20:14 ☽ ☌ ♇
21:13 ☽ □ ♂
23:42 ♀ △ ♅

03 02:49 ☽ ∥ ♀
10:53 ☽ □ ♅
21:55 ☿ ⚹ ♃
22:30 ♀ ♑

05 02:35 ☽ ⚹ ♃
07:11 ☽ ☌ ☿
09:06 ☽ ∥ ☿
12:55 ☽ ∥ ☉
13:50 ☽ ⚹ ♂

06 00:00 ☽ ⚹ ♅
07:34 ☽ ⚹ ♀
16:03 ☿ ∥ ☉

07 01:33 ☽ # ♅
05:07 ☽ # ♂
08:43 ☽ ☌ ♆
09:45 ☽ ⚹ ♄
16:17 ☽ □ ♃
22:14 ☽ ⚹ ♇

08 00:33 ☉ ⚹ ♃
00:54 ☽ ∥ ♆
01:25 ☿ ⚹ ♂

09 02:22 ☽ □ ♀
12:28 ♂ ∥ ♅
22:07 ☽ □ ♄
22:54 ☿ ⚹ ♅

10 04:43 ☽ △ ♃
08:40 ☽ ⚹ ☉
09:52 ☽ □ ♇
10:51 ☿ ♓
20:49 ☽ ☌ ♂
23:49 ☽ ☌ ♅

11 02:09 ☽ # ♆
03:58 ☽ ⚹ ☿
18:40 ☽ △ ♀

12 00:44 ☽ ∥ ♅
04:36 ☽ ∥ ♂
06:06 ☽ # ☿
06:32 ☽ ⚹ ♆
08:07 ☽ △ ♄
11:38 ☿ # ♂
18:23 ☽ # ☉
18:55 ☽ △ ♇

13 06:21 ♂ ☌ ♅
09:46 ☿ # ♅
20:37 ☽ □ ☿

14 10:52 ♂ ♉
12:57 ☽ □ ♆
20:57 ☽ ☍ ♃

15 07:50 ☽ △ ☉
12:50 ☽ ⚹ ♅
13:17 ☽ # ♀
15:26 ☽ ⚹ ♂

16 07:41 ☽ △ ☿
08:03 ☉ # ♂
14:25 ☽ ☍ ♀
15:50 ☽ △ ♆
17:41 ☽ ☍ ♄

17 02:40 ☽ ☍ ♇
08:03 ☽ # ♀
08:45 ♀ ⚹ ♆
14:18 ☽ □ ♅
18:58 ☽ □ ♂

18 07:55 ☉ ⚹ ♅
10:53 ♀ ☌ ♄
12:07 ☿ ∥ ♆
23:05 ☉ ♓

19 00:01 ☽ △ ♃
06:37 ☿ ☌ ♆
13:52 ☽ △ ♅
19:49 ☽ ∥ ♂
20:31 ☽ △ ♂

20 02:40 ☿ ⚹ ♄
07:16 ☽ # ☉
08:13 ☽ ∥ ♅
15:23 ☽ ☍ ♆
17:33 ☽ △ ♄
19:11 ☽ ☍ ☿
19:53 ☉ # ♅
21:45 ☽ △ ♀
23:42 ☽ □ ♃

21 01:52 ☽ △ ♇
04:16 ☽ # ♆
15:21 ☽ # ☿

22 18:20 ☽ □ ♄
20:40 ☿ □ ♃
22:54 ☽ ∥ ☿

23 00:56 ☽ ⚹ ♃
02:47 ☽ □ ♀
02:52 ☽ □ ♇
03:51 ♀ ☌ ♇
15:11 ☽ ☍ ♅
15:25 ☽ ∥ ♆
17:15 ☿ ⚹ ♇

24 00:44 ☽ △ ☉
03:13 ☽ ☍ ♂
06:11 [illegible] ∥ [illegible]
13:45 ☽ # ♅
19:29 ☽ △ ♆
22:21 ☽ ⚹ ♄

25 07:18 ☽ ⚹ ♇
11:15 ☽ △ ☿
11:59 ☽ # ♂
12:14 ☽ ⚹ ♀

27 03:16 ☽ □ ♆
08:21 ☽ ∥ ♀
14:34 ☽ ☌ ♃

28 01:11 ☽ □ ☿
02:33 [illegible]
06:18 ☽ △ ♅

⚷ Chiron

01	Dec.	02 N 35
02	29♓12	
05	29 20	
08	29 29	
11	29 37	
14	29 46	
17	29 56	
20	00♈05	
23	00 15	
26	00 25	
18	08:44	♈

March 2019

Day	S. T. h m s	☉ ° ' "	☽ ° ' "	☿ ° '	♀ ° '	♂ ° '	♃ ° '	♄ ° '	♅ ° '	♆ ° '	♇ ° '	☊ True ° '
01 Fr	10 34 02	10♓06 07	08♑38 55	27♓54	29♑10	09♉48	21♐52	17♑44	29♈46	15♓56	22♑26	25♋57
02 Sa	10 37 59	11 06 22	20 34 37	28 33	00♒21	10 29	21 59	17 50	29 49	15 58	22 28	25 58
03 Su	10 41 55	12 06 35	02♒23 47	29 03	01 32	11 09	22 06	17 55	29 51	16 01	22 29	25R 59
04 Mo	10 45 52	13 06 47	14 10 22	29 24	02 44	11 49	22 13	18 00	29 54	16 03	22 31	25 57
05 Tu	10 49 49	14 06 57	25 57 37	29 36	03 55	12 30	22 19	18 05	29 56	16 05	22 32	25 53
06 We	10 53 45	15 07 05	07♓48 09	29R 39	05 06	13 10	22 25	18 10	29 59	16 07	22 33	25 47
07 Th	10 57 42	16 07 11	19 43 55	29 32	06 17	13 50	22 32	18 14	00♉02	16 10	22 35	25 40
08 Fr	11 01 38	17 07 16	01♈46 25	29 16	07 29	14 30	22 38	18 19	00 04	16 12	22 36	25 30
09 Sa	11 05 35	18 07 18	13 56 54	28 52	08 40	15 10	22 44	18 24	00 07	16 14	22 37	25 21
10 Su	11 09 31	19 07 19	26 16 35	28 20	09 52	15 51	22 49	18 28	00 10	16 16	22 39	25 12
11 Mo	11 13 28	20 07 17	08♉46 55	27 41	11 03	16 31	22 55	18 33	00 13	16 19	22 40	25 05
12 Tu	11 17 24	21 07 13	21 29 37	26 56	12 15	17 11	23 00	18 38	00 16	16 21	22 41	25 00
13 We	11 21 21	22 07 07	04♊26 50	26 06	13 26	17 51	23 05	18 42	00 18	16 23	22 42	24 56
14 Th	11 25 17	23 06 59	17 40 56	25 12	14 38	18 31	23 10	18 46	00 21	16 26	22 44	24 55
15 Fr	11 29 14	24 06 49	01♋14 19	24 16	15 50	19 11	23 15	18 50	00 24	16 28	22 45	24 55
16 Sa	11 33 11	25 06 37	15 08 54	23 18	17 01	19 51	23 20	18 55	00 27	16 30	22 46	24 55
17 Su	11 37 07	26 06 22	29 25 25	22 21	18 13	20 31	23 25	18 59	00 30	16 32	22 47	24 54
18 Mo	11 41 04	27 06 05	14♌02 34	21 25	19 25	21 11	23 29	19 03	00 33	16 35	22 48	24 51
19 Tu	11 45 00	28 05 46	28 56 24	20 31	20 37	21 51	23 33	19 07	00 36	16 37	22 49	24 46
20 We	11 48 57	29 05 24	14♍00 06	19 41	21 49	22 31	23 37	19 11	00 39	16 39	22 50	24 38
21 Th	11 52 53	00♈05 00	29 04 40	18 55	23 01	23 11	23 41	19 14	00 42	16 41	22 51	24 29
22 Fr	11 56 50	01 04 35	14♎00 14	18 14	24 13	23 51	23 45	19 18	00 45	16 44	22 52	24 18
23 Sa	12 00 46	02 04 07	28 37 45	17 38	25 24	24 31	23 48	19 22	00 48	16 46	22 53	24 07
24 Su	12 04 43	03 03 38	12♏50 25	17 08	26 37	25 11	23 52	19 25	00 51	16 48	22 54	23 58
25 Mo	12 08 40	04 03 06	26 34 34	16 44	27 49	25 51	23 55	19 29	00 55	16 50	22 55	23 50
26 Tu	12 12 36	05 02 33	09♐49 43	16 25	29 01	26 31	23 58	19 32	00 58	16 52	22 56	23 45
27 We	12 16 33	06 01 59	22 38 03	16 13	00♓13	27 11	24 01	19 35	01 01	16 55	22 57	23 42
28 Th	12 20 29	07 01 22	05♑03 36	16 07	01 25	27 50	24 03	19 38	01 04	16 57	22 58	23D 41
29 Fr	12 24 26	08 00 44	17 11 25	16D 06	02 37	28 30	24 06	19 41	01 07	16 59	22 58	23 42
30 Sa	12 28 22	09 00 04	29 06 59	16 11	03 49	29 10	24 08	19 44	01 11	17 01	22 59	23 42
31 Su	12 32 19	09 59 22	10♒55 40	16 22	05 02	29 50	24 10	19 47	01 14	17 03	23 00	23R 41

Data for	03-01-2019
Julian Day	2458543.50
Ayanamsa	24 07 13
SVP	04 ♓ 59 17
☽ ☊ Mean	24 ♋ 27 R

LAST ASPECT ☽ Day	h m	INGRESS Day	h m	
02	18:48	02	19:07	♒
05	08:07	05	08:12	♓
07	19:09	07	20:29	♈
09	17:15	10	07:11	♉
12	09:32	12	15:49	♊
14	12:31	14	21:50	♋
16	18:04	17	00:57	♌
18	15:19	19	01:42	♍
20	15:22	21	01:28	♎
22	18:11	23	02:17	♏
25	02:25	25	06:07	♐
27	02:38	27	14:09	♑
30	00:06	30	01:47	♒

● ◐ PHASES ○ ◑

06	16:05	●	15♓47
14	10:27	◐	23♊33
21	01:43	○	00♎09
28	04:11	◑	07♑12

DECLINATION

Day	☉	☽	☿	♀	♂	♃	♄	♅	♆	♇
01 Fr	07S47	21S36	00N58	19S30	15N22	22S34	21S50	10N55	06S26	21S46
02 Sa	07 24	21 23	01 27	19 19	15 36	22 34	21 49	10 56	06 25	21 45
03 Su	07 01	20 11	01 51	19 07	15 49	22 35	21 49	10 57	06 24	21 45
04 Mo	06 38	18 08	02 12	18 55	16 02	22 35	21 48	10 58	06 23	21 45
05 Tu	06 15	15 18	02 28	18 43	16 15	22 35	21 47	10 59	06 22	21 45
06 We	05 52	11 50	02 39	18 29	16 28	22 36	21 47	11 00	06 22	21 45
07 Th	05 28	07 53	02 45	18 16	16 40	22 36	21 46	11 01	06 21	21 45
08 Fr	05 05	03 35	02 47	18 01	16 53	22 36	21 46	11 02	06 20	21 45
09 Sa	04 42	00N53	02 44	17 47	17 05	22 37	21 45	11 03	06 19	21 45
10 Su	04 18	05 23	02 36	17 31	17 18	22 37	21 44	11 04	06 18	21 44
11 Mo	03 55	09 43	02 23	17 15	17 30	22 37	21 44	11 05	06 17	21 44
12 Tu	03 31	13 43	02 07	16 59	17 42	22 37	21 43	11 06	06 16	21 44
13 We	03 08	17 09	01 46	16 42	17 53	22 38	21 43	11 07	06 15	21 44
14 Th	02 44	19 47	01 23	16 25	18 05	22 38	21 42	11 08	06 15	21 44
15 Fr	02 20	21 23	00 56	16 07	18 17	22 38	21 42	11 09	06 14	21 44
16 Sa	01 57	21 43	00 28	15 49	18 28	22 38	21 41	11 10	06 13	21 44
17 Su	01 33	20 40	00S03	15 31	18 39	22 39	21 41	11 11	06 12	21 44
18 Mo	01 09	18 13	00 34	15 12	18 50	22 39	21 40	11 12	06 11	21 44
19 Tu	00 45	14 31	01 05	14 52	19 01	22 39	21 40	11 13	06 10	21 44
20 We	00 22	09 51	01 36	14 32	19 12	22 39	21 39	11 14	06 09	21 43
21 Th	00N02	04 34	02 07	14 12	19 23	22 39	21 39	11 15	06 09	21 43
22 Fr	00 26	00S56	02 36	13 51	19 33	22 40	21 38	11 16	06 08	21 43
23 Sa	00 49	06 16	03 04	13 30	19 44	22 40	21 38	11 17	06 07	21 43
24 Su	01 13	11 07	03 29	13 09	19 54	22 40	21 37	11 18	06 06	21 43
25 Mo	01 37	15 14	03 53	12 47	20 04	22 40	21 37	11 20	06 05	21 43
26 Tu	02 00	18 26	04 14	12 24	20 14	22 40	21 36	11 21	06 04	21 43
27 We	02 24	20 36	04 33	12 02	20 23	22 40	21 36	11 22	06 04	21 43
28 Th	02 47	21 42	04 49	11 39	20 33	22 40	21 35	11 23	06 03	21 43
29 Fr	03 11	21 45	05 03	11 16	20 42	22 40	21 35	11 24	06 02	21 43
30 Sa	03 34	20 47	05 15	10 52	20 51	22 40	21 35	11 25	06 01	21 43
31 Su	03 57	18 56	05 23	10 28	21 00	22 41	21 34	11 26	06 00	21 43

ASPECTARIAN

01 02:27 ☽ △ ♂
03:10 ☽ ⚹ ☉
12:32 ♀ □ ♅
14:40 ☽ ⚹ ♆
16:46 ♀ ♒
18:24 ☽ ☌ ♄
02 03:49 ☽ ☌ ♇
16:56 ☽ ⚹ ☿
18:48 ☽ □ ♅
22:04 ☽ ☌ ♀
03 14:50 ☽ ∥ ♀
18:55 ☽ □ ♂
04 15:52 ☉ ∥ ♆
16:32 ☽ ⚹ ♃
16:59 ☽ # ♂
05 08:07 ☽ ⚹ ♅
18:20 ☿ SR
06 05:13 ☽ # ♅
08:36 ♅ ♉
11:27 ☽ ⚹ ♂
16:49 ☽ ☌ ♆
20:59 ☽ ⚹ ♄
07 01:02 ☉ ☌ ♆
05:38 ☽ □ ♃
05:42 ☽ ⚹ ♇
08:44 ☽ ∥ ♆
14:55 ☽ ∥ ☉
19:09 ☽ ☌ ☿
08 04:20 ☽ # ☿
12:30 ☽ ⚹ ♀
09 07:11 ☉ ⚹ ♄
08:45 ☽ □ ♄
09:32 ☽ ∥ ☿
16:57 ☽ □ ♇
17:15 ☽ △ ♃
18:39 ☽ # ☉
10 03:25 ♄ ∥ ♇
04:56 ☽ # ♆
07:32 ☽ ☌ ♅
11:51 ♀ # ♂
16:22 ♂ ⚹ ♆
11 04:46 ☽ □ ♀
07:52 ☽ ∥ ♅
14:19 ☽ ⚹ ♆
15:28 ☽ ☌ ♂
18:35 ☽ △ ♄
23:15 ☽ ⚹ ☉
12 02:14 ☽ △ ♇
09:32 ☽ ⚹ ☿
20:50 ☽ # ♀
13 06:23 ☽ ∥ ♂
14:27 ☉ ⚹ ♇
17:59 ☽ △ ♀
21:44 ☽ □ ♆
14 01:30 ☉ □ ♃
09:51 ☽ ☍ ♃
10:04 ♂ △ ♄
12:31 ☽ □ ☿
22:32 ☽ ⚹ ♅
15 01:48 ☿ ☌ ☉
10:20 ☽ # ♄
12:59 ☽ # ♇
23:10 ☽ # ♇
23:17 ☿ □ ♃
16 02:02 ☽ # ♄
02:18 ☽ △ ♆
06:25 ☽ ☍ ♄
08:23 ☽ ⚹ ♂
12:54 ☽ ☍ ♇
12:55 ☽ △ ☿
13:13 ☿ ⚹ ♇
18:04 ☽ △ ☉
17 01:48 ☽ □ ♅
19:16 ☽ ∥ ♂
18 03:24 ☿ ⚹ ♂
09:28 ☽ ☍ ♀
12:06 ☽ □ ♂
15:19 ☽ △ ♃
15:25 ☿ ∥ ☉
21:49 ☽ # ♀
19 02:40 ☽ △ ♅
17:16 ☽ ∥ ♅
20 04:13 ☽ ☍ ♆
08:16 ☽ △ ♄
08:35 ☽ ☍ ☿
11:38 ♂ △ ♇
14:04 ☽ △ ♇
14:11 ☽ △ ♂
14:27 ☿ ⚹ ♄
15:22 ☽ □ ♃
16:58 ☽ # ♆
21:59 ☉ ♈
21 08:08 ♀ □ ♂
09:49 ☽ # ☿
14:18 ♀ ⚹ ♃
18:26 ☽ ∥ ☉
21:36 ☽ # ☉
22 08:01 ☽ ∥ ☿
08:39 ☽ □ ♄
14:30 ☽ □ ♇
15:59 ☽ ⚹ ♃
18:11 ☽ △ ♀
23:15 ☽ ∥ ♆
23 03:38 ☽ ☍ ♅
24 00:58 ☽ # ♅
06:51 ☽ △ ♆
07:11 ☽ △ ☿
10:15 ☽ ∥ ♀
11:26 ☽ ⚹ ♄
17:28 ☿ ☌ ♆
17:31 ☽ ⚹ ♇
22:38 ☽ ☍ ♂
25 02:25 ☽ □ ♀
14:31 ☽ △ ☉
26 12:03 ☽ □ ☿
13:08 ☽ □ ♆
19:44 ♀ ♓
20:33 ☽ # ♂
27 02:38 ☽ ☌ ♃
16:08 ☽ ⚹ ♀
16:11 ☽ △ ♅
16:46 ♀ ⚹ ♅
19:38 ☽ ∥ ♄
28 00:13 ☽ ∥ ♇
14:00 ☿ SD
15:44 ♀ # ♅
21:49 ☽ ⚹ ☿
23:35 ☽ ⚹ ♇
29 01:49 ☽ ∥ ♇
05:01 ☽ ☌ ♄
06:48 ☽ ∥ ♄
11:37 ☽ ☌ ♇
23:08 ☽ # ♂
30 00:06 ☽ △ ♂
04:12 ☽ □ ♅
21:55 ☽ ⚹ ☉
31 06:13 ♂ ♊

⚷ Chiron

01 Dec.	03 N 05
01	00♈35
04	00 45
07	00 55
10	01 06
13	01 16
16	01 27
19	01 37
22	01 48
25	01 59
28	02 09
31	02 20

April 2019

Day	S. T.	☉	☽	☿	♀	♂	♃	♄	♅	♆	♇	☊ True
	h m s	° ' "	° ' "	° '	° '	° '	° '	° '	° '	° '	° '	° '
01 Mo	12 36 15	10♈ 58 38	22♒ 42 27	16♓ 38	06♓ 14	00♊ 29	24♐ 12	19♑ 50	01♉ 17	17♓ 05	23♑ 01	23♋R39
02 Tu	12 40 12	11 57 52	04♓ 31 39	16 59	07 26	01 09	24 14	19 53	01 20	17 07	23 01	23 33
03 We	12 44 09	12 57 05	16 26 42	17 24	08 39	01 49	24 15	19 56	01 24	17 10	23 02	23 24
04 Th	12 48 05	13 56 15	28 30 12	17 54	09 51	02 28	24 17	19 58	01 27	17 12	23 03	23 13
05 Fr	12 52 02	14 55 24	10♈ 43 46	18 28	11 03	03 08	24 18	20 01	01 30	17 14	23 03	23 01
06 Sa	12 55 58	15 54 31	23 08 17	19 07	12 16	03 48	24 19	20 03	01 34	17 16	23 04	22 47
07 Su	12 59 55	16 53 35	05♉ 44 04	19 49	13 28	04 27	24 20	20 05	01 37	17 18	23 04	22 35
08 Mo	13 03 51	17 52 37	18 31 05	20 35	14 41	05 07	24 20	20 07	01 40	17 20	23 05	22 24
09 Tu	13 07 48	18 51 38	01♊ 29 21	21 24	15 53	05 46	24 21	20 10	01 44	17 22	23 05	22 16
10 We	13 11 44	19 50 36	14 39 07	22 16	17 06	06 26	24 21	20 11	01 47	17 24	23 06	22 10
11 Th	13 15 41	20 49 32	28 01 02	23 12	18 18	07 06	24 R 21	20 13	01 51	17 26	23 06	22 07
12 Fr	13 19 37	21 48 25	11♋ 36 10	24 10	19 31	07 45	24 21	20 15	01 54	17 28	23 07	22 06
13 Sa	13 23 34	22 47 17	25 25 36	25 11	20 43	08 24	24 21	20 17	01 57	17 30	23 07	22 05
14 Su	13 27 31	23 46 06	09♌ 30 05	26 15	21 56	09 04	24 20	20 19	02 01	17 32	23 07	22 04
15 Mo	13 31 27	24 44 52	23 49 15	27 22	23 08	09 43	24 19	20 20	02 04	17 34	23 08	22 01
16 Tu	13 35 24	25 43 36	08♍ 21 02	28 31	24 21	10 23	24 18	20 21	02 08	17 35	23 08	21 55
17 We	13 39 20	26 42 19	23 01 14	29 42	25 33	11 02	24 17	20 23	02 11	17 37	23 08	21 47
18 Th	13 43 17	27 40 58	07♎ 43 40	00♈ 55	26 46	11 41	24 16	20 24	02 15	17 39	23 08	21 37
19 Fr	13 47 13	28 39 36	22 20 47	02 11	27 58	12 21	24 15	20 25	02 18	17 41	23 08	21 25
20 Sa	13 51 10	29 38 12	06♏ 44 56	03 29	29 11	13 00	24 13	20 26	02 21	17 43	23 09	21 13
21 Su	13 55 06	00♉ 36 46	20 49 36	04 49	00♈ 24	13 39	24 11	20 27	02 25	17 45	23 09	21 02
22 Mo	13 59 03	01 35 19	04♐ 30 26	06 11	01 36	14 19	24 09	20 28	02 28	17 46	23 09	20 52
23 Tu	14 02 60	02 33 49	17 45 40	07 35	02 49	14 58	24 07	20 29	02 32	17 48	23 09	20 46
24 We	14 06 56	03 32 18	00♑ 36 05	09 01	04 02	15 37	24 05	20 29	02 35	17 50	23 09	20 42
25 Th	14 10 53	04 30 45	13 04 32	10 28	05 14	16 16	24 02	20 30	02 39	17 52	23R 09	20 40
26 Fr	14 14 49	05 29 11	25 15 15	11 58	06 27	16 55	23 59	20 30	02 42	17 53	23 09	20D 41
27 Sa	14 18 46	06 27 35	07♒ 13 18	13 30	07 40	17 35	23 56	20 31	02 46	17 55	23 09	20 42
28 Su	14 22 42	07 25 57	19 04 07	15 03	08 52	18 14	23 53	20 31	02 49	17 57	23 09	20R 42
29 Mo	14 26 39	08 24 18	00♓ 53 06	16 38	10 05	18 53	23 50	20 31	02 52	17 58	23 09	20 41
30 Tu	14 30 35	09 22 37	12 45 15	18 15	11 18	19 32	23 47	20R 31	02 56	18 00	23 09	20 37

Data for	04-01-2019
Julian Day	2458574.50
Ayanamsa	24 07 16
SVP	04 ♓ 59 11
☽ ☊ Mean	22 ♋ 48 R

● ◐ PHASES ○ ◑

Day	Time	Phase	Position
05	08:51	●	15♈17
12	19:06	◐	22♋35
19	11:13	○	29♎07
26	22:19	◑	06♒24

LAST ASPECT Day	h m	☽ INGRESS Day	h m	Sign
01	03:03	01	14:49	♓
03	15:37	04	02:57	♈
06	02:16	06	13:07	♉
08	08:29	08	21:16	♊
10	17:27	11	03:32	♋
12	23:33	13	07:51	♌
15	01:39	15	10:15	♍
17	04:30	17	11:23	♎
19	11:14	19	12:42	♏
21	04:01	21	16:01	♐
23	11:45	23	22:52	♑
25	19:49	26	09:29	♒
28	09:45	28	22:12	♓

DECLINATION

Day	☉	☽	☿	♀	♂	♃	♄	♅	♆	♇
01 Mo	04N21	16S16	05S30	10S04	21N09	22S41	21S34	11N28	05S59	21S43
02 Tu	04 44	12 55	05 34	09 40	21 18	22 41	21 33	11 29	05 59	21 43
03 We	05 07	09 03	05 35	09 15	21 26	22 41	21 33	11 30	05 58	21 43
04 Th	05 30	04 46	05 34	08 50	21 34	22 41	21 33	11 31	05 57	21 43
05 Fr	05 53	00 15	05 31	08 25	21 42	22 41	21 32	11 32	05 56	21 43
06 Sa	06 16	04N21	05 26	07 59	21 50	22 41	21 32	11 33	05 55	21 43
07 Su	06 38	08 50	05 19	07 33	21 58	22 41	21 32	11 35	05 55	21 43
08 Mo	07 01	13 00	05 10	07 07	22 06	22 41	21 32	11 36	05 54	21 43
09 Tu	07 23	16 38	04 58	06 41	22 13	22 41	21 31	11 37	05 53	21 43
10 We	07 46	19 30	04 45	06 15	22 20	22 41	21 31	11 38	05 52	21 43
11 Th	08 08	21 21	04 30	05 48	22 27	22 41	21 31	11 39	05 52	21 43
12 Fr	08 30	22 01	04 14	05 22	22 34	22 41	21 31	11 40	05 51	21 43
13 Sa	08 52	21 20	03 55	04 55	22 41	22 41	21 30	11 42	05 50	21 43
14 Su	09 13	19 20	03 35	04 28	22 47	22 41	21 30	11 43	05 49	21 43
15 Mo	09 35	16 05	03 14	04 00	22 54	22 41	21 30	11 44	05 49	21 43
16 Tu	09 56	11 48	02 50	03 33	23 00	22 41	21 30	11 45	05 48	21 43
17 We	10 18	06 48	02 26	03 06	23 06	22 41	21 30	11 46	05 47	21 43
18 Th	10 39	01 24	02 00	02 38	23 12	22 41	21 29	11 48	05 47	21 43
19 Fr	11 00	04S03	01 32	02 10	23 17	22 41	21 29	11 49	05 46	21 43
20 Sa	11 21	09 12	01 04	01 43	23 22	22 41	21 29	11 50	05 45	21 43
21 Su	11 41	13 45	00 34	01 15	23 28	22 40	21 29	11 51	05 45	21 44
22 Mo	12 01	17 27	00 02	00 47	23 33	22 40	21 29	11 52	05 44	21 44
23 Tu	12 22	20 08	00N30	00 19	23 37	22 40	21 29	11 53	05 43	21 44
24 We	12 42	21 41	01 04	00N09	23 42	22 40	21 29	11 55	05 43	21 44
25 Th	13 01	22 07	01 38	00 37	23 46	22 40	21 29	11 56	05 42	21 44
26 Fr	13 21	21 28	02 14	01 05	23 51	22 40	21 29	11 57	05 41	21 44
27 Sa	13 40	19 51	02 51	01 33	23 55	22 40	21 29	11 58	05 41	21 44
28 Su	13 59	17 23	03 29	02 01	23 58	22 40	21 29	11 59	05 40	21 44
29 Mo	14 18	14 12	04 08	02 29	24 02	22 40	21 29	12 01	05 39	21 44
30 Tu	14 37	10 27	04 48	02 57	24 05	22 39	21 29	12 02	05 39	21 44

ASPECTARIAN

01 03:03 ☽ ✶ ♃
16:45 ☽ □ ♂
17:31 ☽ ✶ ♅
02 06:33 ☽ ☌ ♀
09:15 ☽ # ♅
09:39 ☿ ☌ ♆
22:41 ☽ ∥ ♀
03 01:26 ☽ ☌ ♆
02:00 ☽ ☌ ☿
06:59 ☽ ✶ ♄
13:10 ☽ ✶ ♇
15:37 ☽ □ ♃
17:32 ☽ ∥ ♆
19:36 ☽ ∥ ☿
19:37 ♂ # ♄
20:22 ☽ # ☉
04 04:16 ☿ # ☉
08:17 ☽ ✶ ♂
05 01:13 ♂ # ♆
03:34 ☉ # ♆
18:02 ☽ □ ♄
23:51 ☽ □ ♇
06 02:16 ☽ △ ♃
05:37 ☽ # ☿
08:19 ☽ # ♆
11:03 ☽ ∥ ☉
16:10 ☽ ☌ ♅
17:41 ☽ # ♀
07 09:18 ☿ ✶ ♄
15:34 ☽ ∥ ♅
16:05 ☽ ✶ ♀
21:47 ☽ ✶ ♆
08 03:00 ☽ △ ♄
03:14 ☿ # ☉
04:05 ☽ ✶ ☿
08:29 ☽ △ ♇
09 08:16 ☽ ☌ ♂
10 04:51 ☽ □ ♀
04:58 ☽ □ ♆
06:14 ♀ ☌ ♆
08:48 ☉ □ ♄
10:07 ☽ ✶ ☉
14:44 ☽ □ ☿
17:02 ♃ SR
17:27 ☽ ☍ ♃
20:53 ♀ ∥ ♆
21:43 ☿ ✶ ♇
11 02:56 ☽ # ♄
06:50 ☽ ✶ ♅
07:46 ☽ # ♆
12 04:19 ☿ □ ♃
10:15 ☽ △ ♆
15:06 ☽ △ ♀
15:07 ☽ ☍ ♄
15:09 ♀ ✶ ♇
16:04 ☽ # ♆
20:01 ☽ ☍ ♇
20:56 ☽ # ♄
23:33 ☽ △ ☿
23:40 ♂ # ♃
13 08:05 ☉ □ ♇
11:14 ☽ □ ♅
23:14 ☽ ✶ ♂
14 13:41 ☉ △ ♃
23:50 ♀ ✶ ♇
15 00:50 ☽ △ ♃
01:39 ☽ △ ☉
13:43 ☽ △ ♅
23:16 ♀ □ ♃
16 00:19 ☽ ∥ ♅
03:29 ☽ □ ♂
08:42 ☽ ∥ ☉
15:10 ☽ ☍ ♆
19:41 ☽ △ ♄
17 00:11 ☽ △ ♇
02:04 ☽ □ ♃
04:30 ☽ ☍ ♀
04:38 ☽ # ♆
06:01 ☿ ♈
11:52 ☽ ☍ ☿
18:07 ☽ # ♀
21:11 ☽ # ☿
18 06:47 ☽ △ ♂
13:46 ☽ ∥ ☿
16:20 ☽ ∥ ♀
20:49 ☽ □ ♄
19 01:19 ☽ □ ♇
03:08 ☽ ✶ ♃
07:46 ☽ ∥ ♆
16:36 ☽ ☍ ♅
20 08:56 ☉ ♉
11:42 ☽ # ☉
13:27 ☽ # ♅
16:11 ♀ ♈
18:41 ☽ △ ♆
23:21 ☽ ✶ ♄
21 04:01 ☽ ✶ ♇
12:29 ☉ ∥ ♅
18:20 ☽ △ ♀
22 03:20 ☽ △ ☿
18:36 ☽ ☍ ♂
19:41 ☿ # ♀
23:07 ☉ ☌ ♅
23 00:05 ☽ □ ♆
11:45 ☽ ☌ ♃
19:27 ☽ ∥ ♄
24 01:00 ☽ ∥ ♆
03:48 ☽ △ ♅
06:04 ☽ △ ☉
07:13 ☽ □ ♀
18:15 ☽ □ ☿
18:46 ♇ SR
25 09:23 ☽ ✶ ♆
14:35 ☽ ☌ ♄
17:58 ☽ ∥ ♇
19:49 ☽ ☌ ♇
23:52 ☽ ∥ ♄
26 14:58 ☽ □ ♅
27 [illegible]
13:04 ♂ □ ♆
14:36 ☽ ✶ ☿
22:12 ☽ △ ♂
28 09:45 ☽ ✶ ♃
23:26 ☽ # ☉
29 04:03 ☽ ✶ ♅
14:24 ☽ # ♅
16:35 ☽ ✶ ☉
30 00:55 ♄ SR
10:33 ☽ ☌ ♆
14:23 ☽ □ ♂
15:34 ☽ ✶ ♄
20:48 ☽ ✶ ♇
21:58 ☽ □ ♃

⚷ Chiron

01	Dec.	03 N 46
03	02♈30	
06	02 41	
09	02 51	
12	03 02	
15	03 12	
18	03 22	
21	03 31	
24	03 41	
27	03 51	
30	04 00	

Day	S. T. h m s	☉ ° ' "	☽ ° ' "	☿ ° '	♀ ° '	♂ ° '	♃ ° '	♄ ° '	♅ ° '	♆ ° '	♇ ° '	☊ True ° '
01 We	14 34 32	10♉20 55	24♓44 52	19♈54	12♈31	20♊11	23♐R43	20♑R31	02♉59	18♓01	23♑R08	20♋R30
02 Th	14 38 29	11 19 11	06♈55 25	21 35	13 44	20 50	23 39	20 31	03 03	18 03	23 08	20 21
03 Fr	14 42 25	12 17 25	19 19 16	23 17	14 56	21 29	23 35	20 31	03 06	18 04	23 08	20 10
04 Sa	14 46 22	13 15 38	01♉57 43	25 02	16 09	22 08	23 31	20 30	03 10	18 06	23 08	19 58
05 Su	14 50 18	14 13 49	14 50 56	26 48	17 22	22 47	23 27	20 30	03 13	18 07	23 07	19 47
06 Mo	14 54 15	15 11 59	27 58 12	28 36	18 35	23 26	23 22	20 29	03 16	18 09	23 07	19 37
07 Tu	14 58 11	16 10 07	11♊18 09	00♉26	19 48	24 05	23 18	20 29	03 20	18 10	23 07	19 30
08 We	15 02 08	17 08 13	24 49 10	02 17	21 00	24 44	23 13	20 28	03 23	18 12	23 06	19 25
09 Th	15 06 04	18 06 17	08♋29 48	04 11	22 13	25 23	23 08	20 27	03 26	18 13	23 06	19 23
10 Fr	15 10 01	19 04 19	22 18 56	06 06	23 26	26 02	23 03	20 26	03 30	18 14	23 06	19 22
11 Sa	15 13 58	20 02 20	06♌15 52	08 03	24 39	26 41	22 58	20 25	03 33	18 16	23 05	19D22
12 Su	15 17 54	21 00 18	20 20 02	10 02	25 52	27 20	22 53	20 24	03 37	18 17	23 05	19R22
13 Mo	15 21 51	21 58 15	04♍30 32	12 03	27 05	27 58	22 47	20 23	03 40	18 18	23 04	19 20
14 Tu	15 25 47	22 56 09	18 45 51	14 05	28 17	28 37	22 42	20 22	03 43	18 20	23 04	19 17
15 We	15 29 44	23 54 02	03♎03 25	16 09	29 30	29 16	22 36	20 20	03 46	18 21	23 03	19 11
16 Th	15 33 40	24 51 53	17 19 34	18 15	00♉43	29 55	22 30	20 19	03 50	18 22	23 02	19 04
17 Fr	15 37 37	25 49 43	01♏29 46	20 22	01 56	00♋34	22 24	20 17	03 53	18 23	23 02	18 55
18 Sa	15 41 33	26 47 31	15 29 13	22 30	03 09	01 12	22 18	20 16	03 56	18 24	23 01	18 46
19 Su	15 45 30	27 45 17	29 13 25	24 39	04 22	01 51	22 12	20 14	03 59	18 25	23 01	18 37
20 Mo	15 49 27	28 43 03	12♐39 03	26 50	05 35	02 30	22 05	20 12	04 03	18 26	23 00	18 30
21 Tu	15 53 23	29 40 46	25 44 22	29 01	06 48	03 09	21 59	20 10	04 06	18 27	22 59	18 26
22 We	15 57 20	00♊38 29	08♑29 23	01♊12	08 01	03 47	21 52	20 08	04 09	18 28	22 58	18 23
23 Th	16 01 16	01 36 10	20 55 48	03 24	09 14	04 26	21 46	20 06	04 12	18 29	22 58	18D23
24 Fr	16 05 13	02 33 51	03♒06 39	05 35	10 26	05 04	21 39	20 04	04 15	18 30	22 57	18 25
25 Sa	16 09 09	03 31 30	15 05 58	07 46	11 39	05 43	21 32	20 02	04 18	18 31	22 56	18 27
26 Su	16 13 06	04 29 08	26 58 27	09 57	12 52	06 22	21 25	19 59	04 22	18 32	22 55	18 30
27 Mo	16 17 02	05 26 45	08♓49 10	12 07	14 05	07 00	21 18	19 57	04 25	18 33	22 54	18 31
28 Tu	16 20 59	06 24 22	20 43 13	14 15	15 18	07 39	21 11	19 54	04 28	18 34	22 54	18R31
29 We	16 24 56	07 21 57	02♈45 24	16 22	16 31	08 17	21 04	19 52	04 31	18 34	22 53	18 28
30 Th	16 28 52	08 19 31	14 59 57	18 28	17 44	08 56	20 56	19 49	04 34	18 35	22 52	18 24
31 Fr	16 32 49	09 17 05	27 30 14	20 32	18 57	09 35	20 49	19 46	04 37	18 36	22 51	18 19

Data for	05-01-2019
Julian Day	2458604.50
Ayanamsa	24 07 19
SVP	04 ♓ 59 06
☽ ☊ Mean	21 ♋ 13 R

LAST ASPECT ☽ Day	h m	INGRESS Day	h m	
30	21:58	01	10:24	♈
03	08:47	03	20:18	♉
05	15:11	06	03:41	♊
07	23:50	08	09:07	♋
10	02:07	10	13:15	♌
12	12:26	12	16:23	♍
14	17:20	14	18:52	♎
16	09:39	16	21:27	♏
18	21:13	19	01:22	♐
20	17:06	21	07:57	♑
23	03:58	23	17:50	♒
25	12:51	26	06:08	♓
28	04:21	28	18:32	♈
30	15:08	31	04:43	♉

● ◐ PHASES ○ ◑

04	22:45	●	14♉11
12	01:13	◐	21♌03
18	21:12	○	27♏39
26	16:33	◑	05♓09

DECLINATION

Day	☉	☽	☿	♀	♂	♃	♄	♅	♆	♇
01 We	14N55	06S15	05N28	03N25	24N09	22S39	21S29	12N03	05S38	21S45
02 Th	15 13	01 45	06 10	03 53	24 12	22 39	21 29	12 04	05 38	21 45
03 Fr	15 31	02N54	06 52	04 21	24 14	22 39	21 29	12 05	05 37	21 45
04 Sa	15 49	07 32	07 35	04 49	24 17	22 39	21 29	12 06	05 37	21 45
05 Su	16 06	11 55	08 19	05 17	24 20	22 39	21 29	12 08	05 36	21 45
06 Mo	16 24	15 51	09 03	05 44	24 22	22 38	21 29	12 09	05 36	21 45
07 Tu	16 40	19 02	09 48	06 12	24 24	22 38	21 30	12 10	05 35	21 45
08 We	16 57	21 13	10 33	06 39	24 26	22 38	21 30	12 11	05 35	21 46
09 Th	17 13	22 12	11 18	07 07	24 27	22 38	21 30	12 12	05 34	21 46
10 Fr	17 29	21 51	12 04	07 34	24 29	22 38	21 30	12 13	05 34	21 46
11 Sa	17 45	20 08	12 50	08 01	24 30	22 37	21 30	12 14	05 33	21 46
12 Su	18 00	17 12	13 37	08 28	24 31	22 37	21 30	12 16	05 33	21 46
13 Mo	18 15	13 13	14 23	08 54	24 32	22 37	21 31	12 17	05 32	21 47
14 Tu	18 30	08 29	15 08	09 21	24 33	22 37	21 31	12 18	05 32	21 47
15 We	18 45	03 16	15 54	09 47	24 33	22 36	21 31	12 19	05 31	21 47
16 Th	18 59	02S06	16 39	10 13	24 33	22 36	21 31	12 20	05 31	21 47
17 Fr	19 13	07 20	17 23	10 39	24 34	22 36	21 32	12 21	05 30	21 47
18 Sa	19 26	12 07	18 07	11 05	24 33	22 36	21 32	12 22	05 30	21 48
19 Su	19 39	16 11	18 49	11 30	24 33	22 35	21 32	12 23	05 30	21 48
20 Mo	19 52	19 20	19 30	11 55	24 33	22 35	21 33	12 24	05 29	21 48
21 Tu	20 05	21 23	20 10	12 20	24 32	22 35	21 33	12 25	05 29	21 48
22 We	20 17	22 17	20 48	12 45	24 31	22 34	21 33	12 27	05 29	21 48
23 Th	20 29	22 02	21 24	13 09	24 30	22 34	21 34	12 28	05 28	21 49
24 Fr	20 40	20 44	21 58	13 33	24 29	22 34	21 34	12 29	05 28	21 49
25 Sa	20 51	18 32	22 30	13 57	24 27	22 33	21 34	12 30	05 28	21 49
26 Su	21 02	15 33	22 59	14 20	24 26	22 33	21 35	12 31	05 27	21 49
27 Mo	21 12	11 57	23 26	14 43	24 24	22 33	21 35	12 32	05 27	21 50
28 Tu	21 23	07 53	23 51	15 06	24 22	22 32	21 36	12 33	05 27	21 50
29 We	21 32	03 28	24 12	15 29	24 20	22 32	21 36	12 34	05 26	21 50
30 Th	21 41	01N09	24 31	15 51	24 17	22 31	21 37	12 35	05 26	21 50
31 Fr	21 50	05 49	24 47	16 12	24 15	22 31	21 37	12 36	05 26	21 51

ASPECTARIAN

01 03:25 ☽ ∥ ♆
03:44 ☽ # ☿
05:43 ☿ # ♆
06:38 ☿ ⚹ ♂
08:51 ☿ □ ♄
13:52 ☽ # ♀
02 14:39 ☽ ☌ ♀
21:49 ☿ □ ♇

03 02:17 ☽ □ ♄
04:00 ☿ △ ♃
04:22 ☽ ⚹ ♂
07:17 ☽ □ ♇
08:07 ☽ △ ♃
08:20 ☽ ∥ ♀
08:47 ☽ ☌ ☿
14:41 ☽ # ♆

04 00:24 ☽ ∥ ☿
02:16 ☽ ☌ ♅
05 01:14 ☽ ∥ ♅
06:02 ☽ ⚹ ♆
10:23 ☽ △ ♄
15:11 ☽ △ ♇
16:33 ♀ # ♆
21:58 ♂ ☍ ♃
06 04:06 ☽ ∥ ☉
18:26 ☿ ♉

07 12:15 ☽ □ ♆
13:27 ♀ □ ♄
16:36 ☽ ⚹ ♀
21:11 ☽ ☍ ♃
23:50 ☽ ☌ ♂
08 04:24 ☽ # ♄
09:27 ☽ # ♇
14:23 ☿ ☌ ♅
15:07 ☽ ⚹ ♅
15:14 ☽ ⚹ ☿

09 02:53 ☉ ⚹ ♆
16:56 ☽ △ ♆
16:57 ♀ △ ♃
17:19 ♀ □ ♇
17:58 ☽ ⚹ ☉
20:45 ☽ ☍ ♄

10 01:20 ☽ ☍ ♇
01:50 ☽ # ♇
02:07 ☽ □ ♀
04:48 ☿ ∥ ♅
06:47 ☽ # ♄
19:20 ☽ □ ♅
11 03:34 ☽ □ ☿
09:20 ☉ △ ♄
18:41 ☽ ∥ ☉
12 04:17 ☽ △ ♃
10:16 ☽ △ ♀
12:26 ☽ ⚹ ♂
18:34 ☽ ∥ ☿
22:34 ☽ △ ♅

13 05:03 ☽ ∥ ♅
14:49 ☽ △ ☿
20:11 ☽ ∥ ♀
23:16 ☽ ☍ ♆
14 02:41 ☽ △ ♄
03:05 ☉ △ ♇
06:33 ☽ □ ♃
07:13 ☽ △ ♇
07:31 ☽ △ ☉
13:49 ☽ # ♆
13:59 ♀ ⚹ ♂
17:20 ☽ □ ♂
15 09:46 ♀ ♉

16 01:21 ☿ ⚹ ♆
03:10 ♂ ♋
05:02 ☽ □ ♄
08:40 ☽ ⚹ ♃
09:39 ☽ □ ♇
15:31 ☽ ∥ ♆
22:20 ☽ △ ♂
23:10 ☿ △ ♄
17 00:49 ☽ ☍ ♀
04:05 ☽ ☍ ♅
18:00 ☽ # ♀
18 01:24 ☽ # ♅
05:04 ☽ △ ♆
05:47 ☿ △ ♇
08:16 ☽ ⚹ ♄
13:05 ☽ ⚹ ♇
14:27 ☽ ☍ ☿
16:17 ♀ ☌ ♅

20 02:18 ☽ # ☿
05:46 ☽ # ☉
10:33 ☽ □ ♆
17:06 ☽ ☌ ♃
19:16 ☿ ∥ ☉
21 02:58 ☽ ∥ ♄
05:21 ♀ ∥ ♅
08:00 ☽ ∥ ♇
08:00 ☉ ♊
10:52 ☿ ♊
13:07 ☿ ☌ ☉
14:36 ☽ ☍ ♂
15:44 ☽ △ ♅
22:59 ☽ △ ♀
22 14:46 ♂ ⚹ ♅
19:14 ☽ ⚹ ♆

22:23 ☽ ☌ ♄

23 03:58 ☽ ☌ ♇
05:39 ☽ ∥ ♇
06:42 ☿ # ♄
09:32 ☽ # ☿
10:50 ☽ ∥ ♄
17:15 ☿ # ♇
22:49 ☽ △ ☉
24 00:45 ☽ # ☉
02:17 ☽ □ ♅
06:01 ☽ △ ☿
16:18 ☽ □ ♀
25 02:36 ☿ # ♃
12:51 ☽ ⚹ ♃

26 07:38 ☽ # ♀
15:02 ☽ ⚹ ♅
20:07 ☽ △ ♂
20:23 ☽ # ♅

27 08:08 ☽ □ ☿
11:52 ☽ ⚹ ♀
19:39 ☽ ☌ ♆
22:22 ☽ ⚹ ♄
28 00:55 ☽ □ ♃
04:21 ☽ ⚹ ♇
13:28 ☽ ∥ ♆
29 07:50 ☿ ∥ ♂
09:52 ☽ ⚹ ☉
10:40 ☉ # ♄

11:31 ☽ □ ♂

30 01:23 ☿ □ ♆
08:03 ☽ ⚹ ☿
09:17 ☽ □ ♄
11:22 ☽ △ ♃
15:08 ☽ □ ♇
16:51 ♀ ⚹ ♆
21:01 ☽ # ♆

31 00:40 ☉ # ♇
03:12 ☿ ☍ ♃
13:27 ☽ ☌ ♅
15:27 ♀ △ ♄
23:49 ☽ ⚹ ♂

⚷ Chiron	
01 Dec.	04 N 25
03	04♈09
06	04 17
09	04 26
12	04 34
15	04 42
18	04 49
21	04 56
24	05 03
27	05 10
30	05 16

June 2019

Day	S. T. h m s	☉ ° ' "	☽ ° ' "	☿ ° '	♀ ° '	♂ ° '	♃ ° '	♄ ° '	♅ ° '	♆ ° '	♇ ° '	☊ True ° '
01 Sa	16 36 45	10♊14 38	10♉18 29	22♊34	20♉10	10♋13	20♐R42	19♑R43	04♉40	18♓37	22♑R50	18♋R13
02 Su	16 40 42	11 12 09	23 25 39	24 33	21 23	10 52	20 34	19 41	04 43	18 37	22 49	18 07
03 Mo	16 44 38	12 09 40	06♊51 14	26 31	22 37	11 30	20 27	19 38	04 46	18 38	22 48	18 01
04 Tu	16 48 35	13 07 10	20 33 24	28 26	23 50	12 09	20 19	19 35	04 48	18 38	22 47	17 57
05 We	16 52 31	14 04 39	04♋29 16	00♋18	25 03	12 47	20 12	19 31	04 51	18 39	22 46	17 55
06 Th	16 56 28	15 02 07	18 35 17	02 08	26 16	13 26	20 04	19 28	04 54	18 39	22 45	17 54
07 Fr	17 00 25	15 59 34	02♌47 50	03 56	27 29	14 04	19 57	19 25	04 57	18 40	22 44	17D 54
08 Sa	17 04 21	16 56 59	17 03 30	05 41	28 42	14 42	19 49	19 22	05 00	18 40	22 42	17 55
09 Su	17 08 18	17 54 24	01♍19 23	07 23	29 55	15 21	19 41	19 18	05 02	18 41	22 41	17 56
10 Mo	17 12 14	18 51 47	15 33 02	09 03	01♊08	15 59	19 34	19 15	05 05	18 41	22 40	17R 55
11 Tu	17 16 11	19 49 09	29 42 18	10 40	02 21	16 38	19 26	19 11	05 08	18 42	22 39	17 54
12 We	17 20 07	20 46 30	13♎45 13	12 14	03 34	17 16	19 18	19 08	05 10	18 42	22 38	17 52
13 Th	17 24 04	21 43 50	27 39 51	13 45	04 48	17 54	19 11	19 04	05 13	18 42	22 37	17 49
14 Fr	17 28 00	22 41 09	11♏24 18	15 14	06 01	18 33	19 03	19 00	05 16	18 43	22 35	17 45
15 Sa	17 31 57	23 38 27	24 56 41	16 40	07 14	19 11	18 55	18 57	05 18	18 43	22 34	17 40
16 Su	17 35 54	24 35 44	08♐15 19	18 03	08 27	19 49	18 48	18 53	05 21	18 43	22 33	17 37
17 Mo	17 39 50	25 33 01	21 18 57	19 24	09 40	20 28	18 40	18 49	05 23	18 43	22 32	17 34
18 Tu	17 43 47	26 30 17	04♑07 01	20 41	10 53	21 06	18 33	18 45	05 26	18 43	22 31	17 32
19 We	17 47 43	27 27 33	16 39 48	21 56	12 07	21 44	18 25	18 41	05 28	18 43	22 29	17D 32
20 Th	17 51 40	28 24 48	28 58 32	23 07	13 20	22 22	18 18	18 37	05 31	18 43	22 28	17 33
21 Fr	17 55 36	29 22 03	11♒05 20	24 16	14 33	23 01	18 10	18 33	05 33	18 43	22 27	17 34
22 Sa	17 59 33	00♋19 17	23 03 08	25 21	15 46	23 39	18 03	18 29	05 35	18R 43	22 25	17 37
23 Su	18 03 30	01 16 31	04♓55 29	26 24	17 00	24 17	17 56	18 25	05 37	18 43	22 24	17 39
24 Mo	18 07 26	02 13 45	16 46 33	27 23	18 13	24 55	17 49	18 21	05 40	18 43	22 23	17 42
25 Tu	18 11 23	03 10 59	28 40 52	28 18	19 26	25 34	17 41	18 17	05 42	18 43	22 21	17 43
26 We	18 15 19	04 08 13	10♈43 09	29 11	20 39	26 12	17 34	18 13	05 44	18 43	22 20	17 44
27 Th	18 19 16	05 05 26	22 57 58	29 59	21 53	26 50	17 27	18 08	05 46	18 43	22 19	17R 44
28 Fr	18 23 12	06 02 40	05♉29 29	00♌44	23 06	27 28	17 20	18 04	05 48	18 43	22 17	17 43
29 Sa	18 27 09	06 59 54	18 21 02	01 26	24 19	28 07	17 13	18 00	05 50	18 43	22 16	17 42
30 Su	18 31 05	07 57 08	01♊34 48	02 03	25 33	28 45	17 07	17 56	05 52	18 42	22 14	17 41

Data for 06-01-2019
Julian Day 2458635.50
Ayanamsa 24 07 23
SVP 04♓59 02
☽ ☊ Mean 19♋34 R

● ◐ PHASES ○ ◑

Day	Time	Phase	Position
03	10:03	●	12♊34
10	06:00	◐	19♍06
17	08:30	○	25♐53
25	09:47	◑	03♈34

LAST ASPECT ☽ INGRESS

Day	h m	Day	h m	Sign
01	22:53	02	11:49	♊
04	15:43	04	16:18	♋
06	14:12	06	19:17	♌
08	21:25	08	21:46	♍
10	12:02	11	00:30	♎
12	15:16	13	04:03	♏
14	19:46	15	09:04	♐
17	08:31	17	16:14	♑
19	11:19	20	02:01	♒
21	14:02	22	14:02	♓
24	23:11	25	02:39	♈
27	07:53	27	13:33	♉
29	18:40	29	21:10	♊

DECLINATION

Day	☉	☽	☿	♀	♂	♃	♄	♅	♆	♇
01 Sa	21N59	10N22	25N01	16N33	24N12	22S31	21S38	12N37	05S26	21S51
02 Su	22 07	14 33	25 12	16 54	24 09	22 30	21 38	12 38	05 25	21 51
03 Mo	22 15	18 06	25 20	17 15	24 06	22 30	21 38	12 39	05 25	21 51
04 Tu	22 22	20 43	25 26	17 34	24 02	22 29	21 39	12 40	05 25	21 52
05 We	22 29	22 10	25 29	17 54	23 59	22 29	21 39	12 41	05 25	21 52
06 Th	22 36	22 13	25 30	18 13	23 55	22 28	21 40	12 42	05 25	21 52
07 Fr	22 42	20 51	25 28	18 32	23 51	22 28	21 41	12 43	05 25	21 52
08 Sa	22 48	18 09	25 25	18 50	23 47	22 28	21 41	12 43	05 24	21 53
09 Su	22 53	14 22	25 19	19 07	23 43	22 27	21 42	12 44	05 24	21 53
10 Mo	22 58	09 46	25 12	19 25	23 38	22 27	21 42	12 45	05 24	21 53
11 Tu	23 03	04 41	25 03	19 41	23 34	22 26	21 43	12 46	05 24	21 54
12 We	23 07	00S37	24 52	19 57	23 29	22 26	21 43	12 47	05 24	21 54
13 Th	23 11	05 50	24 40	20 13	23 24	22 25	21 44	12 48	05 24	21 54
14 Fr	23 14	10 41	24 26	20 28	23 19	22 25	21 44	12 49	05 24	21 55
15 Sa	23 17	14 57	24 11	20 42	23 14	22 24	21 45	12 50	05 24	21 55
16 Su	23 20	18 23	23 54	20 56	23 08	22 24	21 46	12 50	05 24	21 55
17 Mo	23 22	20 50	23 37	21 10	23 02	22 23	21 46	12 51	05 24	21 55
18 Tu	23 23	22 09	23 19	21 23	22 57	22 23	21 47	12 52	05 24	21 56
19 We	23 25	22 19	22 59	21 35	22 51	22 22	21 47	12 53	05 24	21 56
20 Th	23 26	21 23	22 40	21 46	22 44	22 22	21 48	12 54	05 24	21 56
21 Fr	23 26	19 28	22 19	21 57	22 38	22 21	21 49	12 54	05 24	21 57
22 Sa	23 26	16 44	21 58	22 08	22 32	22 21	21 49	12 55	05 24	21 57
23 Su	23 26	13 19	21 36	22 18	22 25	22 20	21 50	12 56	05 24	21 57
24 Mo	23 25	09 24	21 15	22 27	22 18	22 20	21 51	12 57	05 24	21 58
25 Tu	23 24	05 06	20 52	22 35	22 11	22 19	21 51	12 57	05 24	21 58
26 We	23 22	00 35	20 30	22 43	22 04	22 19	21 52	12 58	05 24	21 58
27 Th	23 20	04N03	20 08	22 50	21 57	22 18	21 52	12 59	05 24	21 59
28 Fr	23 18	08 36	19 46	22 57	21 49	22 18	21 53	12 59	05 24	21 59
29 Sa	23 15	12 55	19 24	23 03	21 42	22 17	21 54	13 00	05 24	21 59
30 Su	23 12	16 45	19 02	23 08	21 34	22 17	21 54	13 01	05 25	22 00

⚷ Chiron

Day	Position
01 Dec.	04 N 57
02	05♈21
05	05 27
08	05 32
11	05 36
14	05 40
17	05 44
20	05 47
23	05 50
26	05 52
29	05 54

ASPECTARIAN

Day	Time	Aspect
01	12:41	☽ ∥ ♅
	15:16	☽ ⚹ ♆
	17:13	☽ △ ♄
	19:56	☽ ☌ ♀
	22:53	☽ △ ♇
02	17:06	☽ ∥ ♀
03	03:40	♀ △ ♇
	20:40	☽ □ ♆
	23:36	☽ ☍ ♃
04	12:58	☽ ⋕ ♄
	15:43	☽ ☌ ☿
	16:54	☽ ⋕ ♇
	20:05	☿ ♋
	23:03	☉ ⋕ ♃
05	00:38	☽ ⚹ ♅
	14:49	☽ ☌ ♂
06	00:07	☽ △ ♆
	01:29	☽ ☍ ♄
	07:02	☽ ☍ ♇
	08:48	☽ ⋕ ♇
	12:31	☽ ⋕ ♄
	14:12	☽ ⚹ ♀
07	03:38	☽ □ ♅
	14:17	☿ ⚹ ♅
	19:18	☽ ∥ ♀
	23:48	☽ ⚹ ☉
08	04:36	☽ △ ♃
	21:25	☽ □ ♀
09	01:38	♀ ♊
	06:17	☽ △ ♅
	08:51	☽ ∥ ♅
	11:34	☽ ⚹ ☿
	19:34	☉ □ ♆
10	00:46	☽ ⚹ ♂
	05:19	☽ ☍ ♆
	06:14	☽ △ ♄
	06:44	☽ □ ♃
	12:02	☽ △ ♇
	15:29	☉ ☍ ♃
	20:39	☽ ⋕ ♆
11	04:56	☽ △ ♀
	21:04	☽ □ ☿
12	06:20	☽ □ ♂
	09:12	☽ □ ♄
	09:28	☽ ⚹ ♃
	12:58	☽ △ ☉
	15:16	☽ □ ♇
	21:58	☽ ∥ ♆
13	13:12	☽ ☍ ♅
14	06:12	♂ △ ♆
	07:34	☽ △ ☿
	11:30	☽ ⋕ ♅
	12:54	☽ △ ♆
	13:14	☽ △ ♂
	13:22	☽ ⚹ ♄
	13:52	☉ ∥ ♂
	15:51	♂ ☍ ♄
	19:46	☽ ⚹ ♇
16	00:23	☽ ☍ ♀
	11:44	☿ △ ♆
	14:01	☿ ☍ ♄
	15:19	♃ □ ♆
	10:00	☽ [illegible]
	19:11	☽ □ ♆
17	05:14	☽ ⋕ ♀
	14:51	☽ ∥ ♄
	18:09	☽ ∥ ♇
	18:24	☿ ∥ ☉
18	02:30	☽ △ ♅
	11:41	♄ ⚹ ♆
	12:45	☽ ∥ ♃
	16:05	☿ ☌ ♂
	18:34	☽ ∥ ♃
19	03:54	☽ ☌ ♄
	03:59	☽ ⚹ ♆
	10:22	☽ ☍ ♂
	10:53	☿ ☍ ♇
	11:17	☽ ☌ ♇
	11:19	☽ ☍ ☿
	12:57	☽ ∥ ♇
	15:40	☿ ∥ ♂
	16:09	☽ ∥ ♄
	17:42	☽ ⋕ ♀
20	03:22	♂ ☍ ♇
	03:42	♀ ⋕ ♄
	12:57	☽ □ ♅
	21:26	☿ ⋕ ♃
	22:18	♀ ⋕ ♇
21	07:42	☽ △ ♀
	14:02	☽ ⚹ ♃
	14:37	♆ SR
	15:55	☉ ♋
	16:26	☿ ∥ ♀
22	01:00	☿ ⋕ ♇
	09:22	☿ ⋕ ♄
	15:58	☽ △ ☉
23	01:25	☽ ⚹ ♅
	02:26	☽ ⋕ ♅
	06:15	♀ ⋕ ♃
	10:54	♀ ∥ ♂
	16:46	♀ ☍ ♃
	18:02	♂ ⋕ ♃
24	02:04	☽ □ ♃
	03:10	☽ ⚹ ♄
	03:14	☽ □ ♀
	[illegible]	[illegible]
	09:59	♀ □ ♆
	11:18	☽ ⚹ ♇
	17:23	☽ △ ♂
	21:22	☽ ∥ ♆
	23:11	☽ △ ☿
26	13:22	☽ △ ♃
	14:39	☽ □ ♄
	17:54	♂ ⋕ ♇
	21:40	☽ ⚹ ♀
	22:44	☽ □ ♇
27	00:20	☿ ♌
	07:04	☽ ⋕ ♆
	07:53	☽ □ ♂
	12:23	♂ ⋕ ♄
	14:24	☽ □ ☿
	17:45	☉ ⚹ ♅
28	00:36	☽ ☌ ♅
	01:08	☽ ⚹ ☉
	23:21	☽ △ ♄
29	00:28	☽ ∥ ♅
	00:40	☽ ⚹ ♆
	07:09	☽ △ ♇
	18:40	☽ ⚹ ♂
30	[illegible]	☽ ⚹ ☿
	10:53	♀ ∥ ☉
	15:26	☽ ∥ ☿

02 19:23 10♋38 ☉ Total Solar Eclipse
16 21:32 24♑04 ✷ Total Lunar Eclipse (mag 0.658)

July 2019

Day	S. T.	☉	☽	☿	♀	♂	♃	♄	♅	♆	♇	☊ True
	h m s	° ' "	° ' "	° '	° '	° '	° '	° '	° '	° '	° '	° '
01 Mo	18 35 02	08♋54 21	15♊11 28	02♌36	26♊46	29♋23	17♐R00	17♑R51	05♉54	18♓R42	22♑R13	17♋R40
02 Tu	18 38 59	09 51 35	29 09 52	03 06	28 00	00♌01	16 53	17 47	05 56	18 42	22 12	17 39
03 We	18 42 55	10 48 49	13♋26 57	03 31	29 13	00 39	16 47	17 43	05 58	18 41	22 10	17 39
04 Th	18 46 52	11 46 03	27 57 58	03 51	00♋27	01 17	16 41	17 38	06 00	18 41	22 09	17 38
05 Fr	18 50 48	12 43 16	12♌36 58	04 07	01 40	01 56	16 34	17 34	06 02	18 41	22 07	17 38
06 Sa	18 54 45	13 40 29	27 17 36	04 19	02 54	02 34	16 28	17 29	06 04	18 40	22 06	17 37
07 Su	18 58 41	14 37 42	11♍53 52	04 26	04 07	03 12	16 22	17 25	06 05	18 40	22 04	17 37
08 Mo	19 02 38	15 34 55	26 20 44	04R 28	05 21	03 50	16 16	17 21	06 07	18 39	22 03	17 36
09 Tu	19 06 34	16 32 07	10♎34 34	04 25	06 34	04 28	16 11	17 16	06 09	18 39	22 02	17 35
10 We	19 10 31	17 29 20	24 33 07	04 18	07 48	05 06	16 05	17 12	06 10	18 38	22 00	17 34
11 Th	19 14 28	18 26 32	08♏15 22	04 06	09 01	05 44	16 00	17 07	06 12	18 37	21 59	17 33
12 Fr	19 18 24	19 23 44	21 41 18	03 50	10 15	06 22	15 54	17 03	06 13	18 37	21 57	17 33
13 Sa	19 22 21	20 20 56	04♐51 31	03 29	11 28	07 01	15 49	16 58	06 15	18 36	21 56	17D 33
14 Su	19 26 17	21 18 08	17 46 58	03 04	12 42	07 39	15 44	16 54	06 16	18 35	21 54	17 33
15 Mo	19 30 14	22 15 20	00♑28 43	02 36	13 55	08 17	15 39	16 50	06 18	18 35	21 53	17 33
16 Tu	19 34 10	23 12 33	12 57 57	02 04	15 09	08 55	15 34	16 45	06 19	18 34	21 51	17 34
17 We	19 38 07	24 09 46	25 15 56	01 29	16 23	09 33	15 30	16 41	06 20	18 33	21 50	17 34
18 Th	19 42 03	25 06 59	07♒24 13	00 51	17 36	10 11	15 25	16 37	06 21	18 32	21 48	17 35
19 Fr	19 46 00	26 04 12	19 24 34	00 12	18 50	10 49	15 21	16 32	06 23	18 32	21 47	17R 34
20 Sa	19 49 57	27 01 26	01♓19 06	29♋31	20 04	11 27	15 17	16 28	06 24	18 31	21 46	17 34
21 Su	19 53 53	27 58 41	13 10 21	28 50	21 18	12 05	15 13	16 24	06 25	18 30	21 44	17 33
22 Mo	19 57 50	28 55 56	25 01 20	28 09	22 31	12 43	15 09	16 19	06 26	18 29	21 43	17 32
23 Tu	20 01 46	29 53 12	06♈55 34	27 29	23 45	13 21	15 06	16 15	06 27	18 28	21 41	17 32
24 We	20 05 43	00♌50 29	18 57 02	26 51	24 59	13 59	15 02	16 11	06 28	18 27	21 40	17D 31
25 Th	20 09 39	01 47 46	01♉10 04	26 15	26 13	14 38	14 59	16 07	06 29	18 26	21 38	17 32
26 Fr	20 13 36	02 45 05	13 38 59	25 42	27 26	15 16	14 56	16 03	06 30	18 25	21 37	17 33
27 Sa	20 17 32	03 42 24	26 27 52	25 12	28 40	15 54	14 53	15 58	06 31	18 24	21 35	17 34
28 Su	20 21 29	04 39 45	09♊40 08	24 47	29 54	16 32	14 50	15 54	06 31	18 23	21 34	17 36
29 Mo	20 25 26	05 37 06	23 17 58	24 26	01♌08	17 10	14 47	15 50	06 32	18 22	21 33	17 37
30 Tu	20 29 22	06 34 29	07♋21 49	24 11	02 22	17 48	14 45	15 46	06 33	18 21	21 31	17 38
31 We	20 33 19	07 31 52	21 49 44	24 01	03 36	18 26	14 43	15 42	06 33	18 20	21 30	17R 37

Data for	07-01-2019
Julian Day	2458665.50
Ayanamsa	24 07 28
SVP	04♓58 59
☽ ☊ Mean	17♋59 R

● ◐ PHASES ○ ◑

02	19:17	☉ (eclipse)	10♋38
09	10:55	◐	16♎58
16	21:38	✷	24♑04
25	01:19	◑	01♉51

LAST ASPECT ☽ INGRESS

Day	h m	Day	h m	
01	21:49	02	01:25	♋
03	14:26	04	03:20	♌
05	06:25	06	04:26	♍
07	16:51	08	06:07	♎
09	19:36	10	09:29	♏
12	00:29	12	15:06	♐
14	01:31	14	23:05	♑
16	21:39	17	09:20	♒
18	15:55	19	21:20	♓
22	08:35	22	10:03	♈
24	14:49	24	21:44	♉
27	04:29	27	06:30	♊
28	15:25	29	11:31	♋
31	03:33	31	13:19	♌

DECLINATION

Day	☉	☽	☿	♀	♂	♃	♄	♅	♆	♇
01 Mo	23N08	19N48	18N41	23N13	21N26	22S16	21S55	13N01	05S25	22S00
02 Tu	23 04	21 46	18 20	23 17	21 18	22 16	21 56	13 02	05 25	22 00
03 We	23 00	22 23	18 00	23 20	21 10	22 15	21 56	13 03	05 25	22 01
04 Th	22 55	21 30	17 41	23 22	21 01	22 15	21 57	13 03	05 25	22 01
05 Fr	22 50	19 10	17 23	23 24	20 53	22 14	21 58	13 04	05 25	22 01
06 Sa	22 44	15 35	17 05	23 25	20 44	22 14	21 58	13 04	05 26	22 02
07 Su	22 38	11 04	16 48	23 26	20 35	22 14	21 59	13 05	05 26	22 02
08 Mo	22 32	05 59	16 33	23 26	20 26	22 13	22 00	13 05	05 26	22 02
09 Tu	22 25	00 39	16 19	23 25	20 17	22 13	22 00	13 06	05 26	22 02
10 We	22 18	04S37	16 06	23 23	20 08	22 12	22 01	13 06	05 27	22 03
11 Th	22 10	09 33	15 55	23 21	19 58	22 12	22 02	13 07	05 27	22 03
12 Fr	22 02	13 56	15 45	23 18	19 49	22 12	22 02	13 07	05 27	22 03
13 Sa	21 54	17 33	15 37	23 14	19 39	22 11	22 03	13 08	05 27	22 04
14 Su	21 45	20 15	15 31	23 10	19 29	22 11	22 03	13 08	05 28	22 04
15 Mo	21 36	21 52	15 26	23 05	19 20	22 10	22 04	13 09	05 28	22 04
16 Tu	21 26	22 22	15 23	22 59	19 10	22 10	22 05	13 09	05 28	22 05
17 We	21 17	21 46	15 22	22 53	18 59	22 10	22 05	13 10	05 29	22 05
18 Th	21 06	20 09	15 22	22 46	18 49	22 10	22 06	13 10	05 29	22 05
19 Fr	20 56	17 38	15 24	22 38	18 39	22 09	22 07	13 10	05 29	22 06
20 Sa	20 45	14 25	15 28	22 30	18 28	22 09	22 07	13 11	05 30	22 06
21 Su	20 34	10 38	15 33	22 21	18 17	22 09	22 08	13 11	05 30	22 06
22 Mo	20 22	06 27	15 40	22 11	18 07	22 09	22 08	13 11	05 31	22 07
23 Tu	20 10	02 01	15 48	22 01	17 56	22 08	22 09	13 12	05 31	22 07
24 We	19 58	02N32	15 57	21 50	17 45	22 08	22 10	13 12	05 31	22 07
25 Th	19 45	07 04	16 07	21 38	17 33	22 08	22 10	13 12	05 32	22 08
26 Fr	19 33	11 25	16 19	21 26	17 22	22 08	22 11	13 13	05 32	22 08
27 Sa	19 19	15 22	16 31	21 13	17 11	22 08	22 11	13 13	05 33	22 08
28 Su	19 06	18 41	16 44	20 59	16 59	22 08	22 12	13 13	05 33	22 09
29 Mo	18 52	21 05	16 57	20 45	16 48	22 07	22 13	13 13	05 34	22 09
30 Tu	18 38	22 17	17 10	20 30	16 36	22 07	22 13	13 14	05 34	22 09
31 We	18 23	22 02	17 24	20 15	16 24	22 07	22 14	13 14	05 35	22 10

ASPECTARIAN

01 03:07 ☽ ☍ ♃
06:05 ☽ □ ♆
17:17 ☽ ∥ ♂
21:49 ☽ ☌ ♀
23:20 ♂ ♌
02 03:20 ☽ # ♄
05:00 ☽ # ♇
11:28 ☽ ✱ ♅
12:28 ☽ # ♃
03 07:02 ☽ ☍ ♄
07:54 ☽ # ♃
08:42 ☽ △ ♆
14:26 ☽ ☍ ♇
14:48 ☽ # ♇
15:19 ♀ ♋
16:14 ☽ # ♄
04 05:42 ☽ ☌ ♂
06:48 ☽ ∥ ♂
09:51 ☽ ☌ ☿
13:12 ☽ □ ♅
05 06:25 ☽ △ ♃
14:03 ☽ ∥ ☿
06 10:01 ☽ ✱ ♀
13:49 ☽ ∥ ♅
14:25 ☽ △ ♅
07 04:50 ☽ ✱ ☉
07:21 ☽ □ ♃
09:05 ☽ △ ♄
11:11 ☽ ☍ ♆
16:51 ☽ △ ♇
23:16 ☿ SR
08 02:28 ☽ # ♆
13:10 ☽ ✱ ♂
13:37 ☽ ✱ ☿
15:32 ♀ ✱ ♅
16:33 ☽ □ ♀
22:28 ☿ ☌ ♂
09 09:30 ☽ ✱ ♃
11:23 ☽ □ ♄
17:08 ☉ ☍ ♄
19:36 ☽ □ ♇
10 03:51 ☽ ∥ ♆
16:48 ☽ □ ☿
17:42 ☉ # ♃
19:20 ☽ □ ♂
20:21 ☽ ☍ ♅
11 01:29 ☽ △ ♀
04:32 ☉ △ ♆
15:42 ☽ ✱ ♄
18:01 ♂ □ ♅
18:28 ☽ △ ♆
19:12 ☽ # ♅
19:33 ☽ △ ☉
19:51 ☉ # ♇
23:28 ☉ # ♄
12 00:29 ☽ ✱ ♇
10:53 ☽ # ☿
21:33 ☽ △ ☿
13 04:10 ☽ △ ♂
16:44 ☽ # ♂
20:11 ☽ ☌ ♃
14 01:31 ☽ □ ♆
14:49 ☉ ☍ ♇
19:07 ☽ # ☉
15 05:05 ☽ ∥ ♄
05:13 ☽ ∥ ♇
08:21 ☽ ∥ ♃
11:09 ☽ △ ♅
16 02:45 ♄ ∥ ♇
04:42 ☽ ☍ ♀
07:19 ☽ ☌ ♄
10:53 ☽ ✱ ♆
13:29 ☽ ∥ ♃
16:07 ☽ ∥ ♄
16:12 ☽ ∥ ♇
17:17 ☽ ☌ ♇
17 05:35 ♀ ☍ ♄
10:04 ☽ # ☉
11:40 ☽ ☍ ☿
21:55 ☽ □ ♅
18 05:51 ☽ ☍ ♂
14:39 ☽ # ♂
15:55 ☽ ✱ ♃
18:03 ♀ △ ♆
19 07:07 ☿ ♋R
16:52 ☽ # ☿
20 08:09 ☽ # ♅
10:17 ☽ ✱ ♅
21 04:07 ☽ □ ♃
06:29 ☽ ✱ ♄
08:31 ♀ ☍ ♇
10:47 ☽ ☌ ♆
12:35 ☿ ☌ ☉
17:19 ☽ ✱ ♇
18:21 ☽ △ ♀
22 02:13 ♃ ∥ ♄
05:09 ☽ ∥ ♆
05:29 ♀ # ♄
05:45 ♀ # ♃
05:59 ☽ △ ☿
08:35 ☽ △ ☉
09:34 ♀ # ♇
23 02:51 ☉ ♌
13:35 ☽ △ ♂
16:15 ☽ △ ♃
18:32 ☽ □ ♄
24 05:21 ☽ □ ♇
13:14 ☽ □ ♀
14:49 ☽ □ ☿
15:45 ☽ # ♆
25 00:27 ☿ ☌ ♀
10:18 ☽ ☌ ♅
12:23 ♂ △ ♃
13:30 ♃ ∥ ♇
26 03:13 ☽ □ ♂
04:30 ☽ △ ♄
09:00 ☽ ✱ ♆
10:34 ☽ ∥ ♅
14:58 ☽ △ ♇
21:45 ☽ ✱ ☿
27 04:29 ☽ ✱ ♀
08:11 ☽ ∥ ☿
11:43 ☽ ∥ ♂
14:17 ☽ ✱ ☉
28 01:54 ♀ ♌
03:08 ☽ ∥ ☉
09:09 ☽ ☍ ♃
12:46 ☽ ✱ ♂
15:09 ☿ ∥ ♂
15:25 ☽ □ ♆
20:11 ☽ ∥ ♀
29 17:59 ☽ # ♃
18:51 ☽ # ♇
20:54 ☽ # ♄
22:37 ☽ ✱ ♅
23:14 ☉ □ ♅
30 13:58 ☽ ☍ ♄
18:15 ☽ △ ♆
19:00 ☽ # ♄
21:09 ☽ # ♇
22:12 ☽ # ♃
23:27 ☽ ☍ ♇
31 03:33 ☽ ☌ ☿
20:51 ☽ ☌ ♀
23:55 ☽ □ ♅

⚷ Chiron

01 Dec. 05 N 14

02	05♈55
05	05 56
08	05 56
11	05 56R
14	05 56
17	05 55
20	05 53
23	05 51
26	05 49
29	05 46

08 21:44 05♈56 R

August 2019

Day	S. T. h m s	☉ ° ' "	☽ ° ' "	☿ ° '	♀ ° '	♂ ° '	♃ ° '	♄ ° '	♅ ° '	♆ ° '	♇ ° '	☊ True ° '
01 Th	20 37 15	08♌ 29 16	06♌ 37 11	23♋R57	04♌ 50	19♌ 04	14♐R41	15♑R38	06♉ 34	18♓R19	21♑R28	17♋R36
02 Fr	20 41 12	09 26 41	21 37 10	23D 59	06 04	19 42	14 39	15 35	06 34	18 17	21 27	17 33
03 Sa	20 45 08	10 24 07	06♍ 41 02	24 07	07 18	20 20	14 37	15 31	06 35	18 16	21 26	17 30
04 Su	20 49 05	11 21 33	21 39 44	24 22	08 32	20 58	14 36	15 27	06 35	18 15	21 24	17 26
05 Mo	20 53 01	12 19 00	06♎ 25 13	24 43	09 46	21 36	14 34	15 23	06 36	18 14	21 23	17 22
06 Tu	20 56 58	13 16 28	20 51 26	25 11	11 00	22 14	14 33	15 20	06 36	18 12	21 22	17 19
07 We	21 00 55	14 13 57	04♏ 54 56	25 45	12 14	22 53	14 32	15 16	06 36	18 11	21 20	17 17
08 Th	21 04 51	15 11 26	18 34 45	26 26	13 28	23 31	14 31	15 13	06 36	18 10	21 19	17 16
09 Fr	21 08 48	16 08 56	01♐ 51 55	27 13	14 42	24 09	14 31	15 09	06 37	18 09	21 18	17D 16
10 Sa	21 12 44	17 06 27	14 48 46	28 07	15 56	24 47	14 31	15 06	06 37	18 07	21 16	17 17
11 Su	21 16 41	18 03 58	27 28 15	29 06	17 10	25 25	14 30	15 02	06 37	18 06	21 15	17 19
12 Mo	21 20 37	19 01 31	09♑ 53 30	00♌ 12	18 24	26 03	14D 30	14 59	06R 37	18 04	21 14	17 21
13 Tu	21 24 34	19 59 05	22 07 24	01 24	19 38	26 41	14 31	14 56	06 37	18 03	21 13	17 22
14 We	21 28 30	20 56 39	04♒ 12 37	02 41	20 52	27 19	14 31	14 53	06 37	18 02	21 12	17R 21
15 Th	21 32 27	21 54 15	16 11 25	04 03	22 07	27 57	14 31	14 50	06 37	18 00	21 10	17 19
16 Fr	21 36 24	22 51 52	28 05 49	05 31	23 21	28 35	14 32	14 47	06 37	17 59	21 09	17 15
17 Sa	21 40 20	23 49 30	09♓ 57 40	07 03	24 35	29 13	14 33	14 44	06 36	17 57	21 08	17 09
18 Su	21 44 17	24 47 09	21 48 46	08 40	25 49	29 52	14 34	14 41	06 36	17 56	21 07	17 03
19 Mo	21 48 13	25 44 50	03♈ 41 08	10 21	27 03	00♍ 30	14 35	14 38	06 36	17 54	21 06	16 56
20 Tu	21 52 10	26 42 32	15 37 08	12 05	28 18	01 08	14 37	14 35	06 35	17 53	21 05	16 50
21 We	21 56 06	27 40 16	27 39 39	13 53	29 32	01 46	14 39	14 33	06 35	17 51	21 03	16 45
22 Th	22 00 03	28 38 01	09♉ 52 08	15 44	00♍ 46	02 24	14 40	14 30	06 34	17 50	21 02	16 42
23 Fr	22 03 59	29 35 49	22 18 25	17 36	02 00	03 02	14 42	14 28	06 34	17 48	21 01	16 41
24 Sa	22 07 56	00♍ 33 37	05♊ 02 33	19 31	03 15	03 40	14 45	14 25	06 33	17 47	21 00	16D 42
25 Su	22 11 53	01 31 28	18 08 32	21 28	04 29	04 19	14 47	14 23	06 33	17 45	20 59	16 43
26 Mo	22 15 49	02 29 21	01♋ 39 41	23 25	05 43	04 57	14 49	14 21	06 32	17 44	20 58	16 44
27 Tu	22 19 46	03 27 15	15 38 08	25 23	06 58	05 35	14 52	14 19	06 31	17 42	20 57	16R 45
28 We	22 23 42	04 25 11	00♌ 03 47	27 22	08 12	06 13	14 55	14 16	06 31	17 40	20 56	16 43
29 Th	22 27 39	05 23 09	14 53 38	29 21	09 26	06 51	14 58	14 14	06 30	17 39	20 55	16 40
30 Fr	22 31 35	06 21 08	00♍ 01 23	01♍ 20	10 41	07 29	15 01	14 13	06 29	17 37	20 54	16 34
31 Sa	22 35 32	07 19 09	15 17 51	03 19	11 55	08 08	15 05	14 11	06 28	17 36	20 53	16 27

Data for	08-01-2019
Julian Day	2458696.50
Ayanamsa	24 07 34
SVP	04 ♓ 58 55
☽ ☊ Mean	16 ♋ 21 R

● ◐ PHASES ○ ◑

01	03:12	●	08♌37
07	17:31	◐	14♏56
15	12:30	○	22♒24
23	14:56	◑	00♊12
30	10:37	●	06♍47

LAST ASPECT Day	h m	☽ INGRESS Day	h m	
01	20:48	02	13:21	♍
04	04:28	04	13:30	♎
06	07:36	06	15:32	♏
08	14:58	08	20:35	♐
10	19:52	11	04:51	♑
12	22:12	13	15:37	♒
16	01:03	16	03:51	♓
17	22:35	18	16:34	♈
21	04:07	21	04:38	♉
22	21:33	23	14:34	♊
25	06:59	25	21:06	♋
27	08:55	27	23:54	♌
29	00:07	29	23:58	♍
31	08:47	31	23:09	♎

DECLINATION

Day	☉	☽	☿	♀	♂	♃	♄	♅	♆	♇
01 Th	18N08	20N16	17N38	19N59	16N12	22S07	22S14	13N14	05S35	22S10
02 Fr	17 53	17 04	17 51	19 43	16 00	22 07	22 15	13 14	05 35	22 10
03 Sa	17 38	12 44	18 04	19 26	15 48	22 07	22 15	13 14	05 36	22 11
04 Su	17 22	07 38	18 17	19 08	15 36	22 07	22 16	13 14	05 36	22 11
05 Mo	17 06	02 10	18 29	18 50	15 23	22 07	22 16	13 14	05 37	22 11
06 Tu	16 50	03S18	18 39	18 31	15 11	22 07	22 17	13 14	05 38	22 11
07 We	16 33	08 27	18 49	18 12	14 58	22 07	22 17	13 15	05 38	22 12
08 Th	16 17	13 02	18 58	17 52	14 46	22 08	22 18	13 15	05 39	22 12
09 Fr	16 00	16 52	19 05	17 32	14 33	22 08	22 18	13 15	05 39	22 12
10 Sa	15 42	19 46	19 10	17 12	14 20	22 08	22 19	13 15	05 40	22 13
11 Su	15 25	21 37	19 14	16 50	14 07	22 08	22 19	13 15	05 40	22 13
12 Mo	15 07	22 23	19 15	16 29	13 54	22 08	22 20	13 15	05 41	22 13
13 Tu	14 49	22 02	19 15	16 07	13 41	22 08	22 20	13 15	05 41	22 13
14 We	14 31	20 40	19 12	15 44	13 28	22 09	22 21	13 15	05 42	22 14
15 Th	14 12	18 22	19 06	15 21	13 15	22 09	22 21	13 15	05 43	22 14
16 Fr	13 54	15 18	18 58	14 58	13 01	22 09	22 22	13 14	05 43	22 14
17 Sa	13 35	11 38	18 48	14 34	12 48	22 09	22 22	13 14	05 44	22 15
18 Su	13 16	07 32	18 34	14 10	12 34	22 10	22 22	13 14	05 44	22 15
19 Mo	12 56	03 08	18 18	13 45	12 21	22 10	22 23	13 14	05 45	22 15
20 Tu	12 37	01N23	18 00	13 20	12 07	22 10	22 23	13 14	05 46	22 15
21 We	12 17	05 55	17 38	12 55	11 53	22 11	22 24	13 14	05 46	22 16
22 Th	11 57	10 16	17 14	12 29	11 39	22 11	22 24	13 14	05 47	22 16
23 Fr	11 37	14 17	16 47	12 03	11 26	22 12	22 24	13 14	05 47	22 16
24 Sa	11 16	17 46	16 18	11 37	11 12	22 12	22 25	13 13	05 48	22 16
25 Su	10 56	20 27	15 46	11 10	10 58	22 12	22 25	13 13	05 49	22 17
26 Mo	10 35	22 05	15 12	10 43	10 43	22 13	22 25	13 13	05 49	22 17
27 Tu	10 14	22 25	14 36	10 16	10 29	22 13	22 26	13 13	05 50	22 17
28 We	09 53	21 17	13 59	09 48	10 15	22 14	22 26	13 12	05 51	22 17
29 Th	09 32	18 41	13 19	09 20	10 01	22 14	22 26	13 12	05 51	22 17
30 Fr	09 11	14 45	12 39	08 52	09 46	22 15	22 27	13 12	05 52	22 18
31 Sa	08 49	09 49	11 56	08 24	09 32	22 16	22 27	13 12	05 52	22 18

ASPECTARIAN

01 02:59 ☽ ∥ ♀
03:59 ☿ SD
12:54 ☽ △ ♃
18:22 ☽ ∥ ☉
19:17 ☽ ∥ ☿
20:48 ☽ ☌ ♂
02 01:41 ☿ ∥ ☉
06:52 ☽ ∥ ♂
10:01 ♀ □ ♅
21:31 ☽ ∥ ♅
23:50 ☽ △ ♅
03 12:40 ☽ □ ♃
14:03 ☽ △ ♄
18:31 ☽ ☍ ♆
23:35 ☽ △ ♇

04 04:28 ☽ ⚹ ☿
09:00 ☽ # ♆
05 06:01 ☽ ⚹ ♀
10:25 ☽ ⚹ ☉
13:28 ☽ ⚹ ♃
14:46 ☽ □ ♄
17:20 ☿ ∥ ♀
06 00:51 ☽ □ ♇
02:27 ☽ ⚹ ♂
07:36 ☽ □ ☿
10:37 ☽ ∥ ♆

07 02:56 ☽ ☍ ♅
07:32 ☉ △ ♃
14:02 ☽ □ ♀
18:02 ☽ ⚹ ♄
23:16 ☽ △ ♆
08 01:09 ☽ # ♅
04:53 ☽ ⚹ ♇
09:16 ☽ □ ♂
09:36 ☽ # ♂
14:58 ☽ △ ☿
18:27 ☽ # ☉
20:28 ♀ △ ♃
09 04:24 ☽ # ♀
18:06 ☽ # ☿
23:26 ☽ ☌ ♃

10 02:20 ☽ △ ♀
04:40 ☽ △ ☉
06:13 ☽ □ ♆
19:52 ☽ △ ♂
11 11:44 ☽ ∥ ♃
13:39 ♃ SD
14:33 ☽ ∥ ♇
17:38 ☽ △ ♅
19:44 ☽ # ♄
19:47 ☿ ♌
12 02:27 ♅ SR
09:54 ☽ ☌ ♄
12:47 ☽ ∥ ♄
15:59 ☽ ⚹ ♆
18:10 ☽ ∥ ♇
21:06 ☽ ∥ ♃
22:12 ☽ ☌ ♇

13 20:34 ☽ ☍ ☿
14 04:48 ☽ □ ♅
06:08 ♀ ☌ ☉
16:56 ☽ # ☿
20:39 ☽ ⚹ ♃
15 00:01 ♂ ∥ ♅
13:18 ☽ ☍ ♀

16 01:03 ☽ ☍ ♂
02:46 ☽ # ♀
10:37 ☽ # ☉
13:58 ☽ # ♅
16:22 ☽ # ♂
17:08 ☿ □ ♅
17:12 ☽ ⚹ ♅
17 09:18 ☽ □ ♃
09:37 ☽ ⚹ ♄
16:09 ☽ ☌ ♆
22:35 ☽ ⚹ ♇
18 01:31 ☉ ∥ ♅
05:19 ♂ ♍
09:57 ☽ ∥ ♆

19 15:42 ☽ △ ☿
21:56 ☽ □ ♄
21:59 ☽ △ ♃
20 05:53 ♀ ∥ ♅
10:53 ☽ □ ♇
23:16 ☽ # ♆
21 00:01 ☽ △ ☉
04:07 ☽ △ ♀
08:33 ☽ △ ♂
09:07 ♀ ♍
10:05 ☿ △ ♃
17:34 ☽ ☌ ♅

22 07:38 ☽ ∥ ♂
08:58 ☽ △ ♄
09:00 ☽ ∥ ☉
11:42 ☽ ∥ ♀
13:23 ☽ □ ☿
15:23 ☽ ⚹ ♆
17:27 ☽ ∥ ♅
21:33 ☽ △ ♇
23 10:03 ☉ ♍
14:42 ☽ ∥ ♅
20:17 ☽ □ ♀
21:19 ☽ □ ♂
24 17:05 ♀ ☌ ♂
17:54 ☽ ☍ ♃
17:58 ☉ ∥ ♂
23:18 ☽ □ ♆

25 06:59 ☽ ⚹ ☿
23:29 ♀ ∥ ♂
26 01:33 ☽ ⚹ ☉
03:25 ☽ # ♃
05:23 ☽ # ♇
05:59 ☽ ⚹ ♂
07:45 ☽ ⚹ ♀
08:27 ☽ ⚹ ♅
11:51 ☽ # ♄
15:39 ♀ △ ♅
21:46 ☽ ☍ ♄
23:37 ☽ # ♄
27 03:28 ☽ △ ♆
05:41 ♀ ∥ ☉
05:57 ☽ # ♇
07:43 ☽ # ♃
08:55 ☽ ☍ ♇

28 10:30 ☽ □ ♅
10:53 ♂ △ ♅
29 00:07 ☽ △ ♃
04:20 ☿ ∥ ♅
07:48 ☿ ♍
30 02:23 ☽ ☌ ☿
03:15 ☉ △ ♅
08:04 ☽ ∥ ♅
10:09 ☽ △ ♅
12:15 ☽ ☌ ♂
12:34 ☽ ∥ ☿
18:14 ☽ ☌ ♀
22:15 ☽ △ ♄
23:39 ☽ □ ♃

31 01:24 ☽ ∥ ♂
03:36 ☽ ☍ ♆
04:49 ☽ ∥ ☉
06:58 ☽ ∥ ♀
[illegible]
17:16 ☽ # ♆

⚷ Chiron

01 Dec.	05 N 12
01	05♈43R
04	05 39
07	05 35
10	05 31
13	05 26
16	05 20
19	05 15
22	05 09
25	05 03
28	04 56
31	04 49

September 2019

Day	S. T.	☉	☽	☿	♀	♂	♃	♄	♅	♆	♇	☊ True
	h m s	° ' "	° ' "	° '	° '	° '	° '	° '	° '	° '	° '	° '
01 Su	22 39 28	08♍ 17 11	00♎ 32 15	05♍ 17	13♍ 09	08♍ 46	15♐ 08	14♑R09	06♉R27	17♓R34	20♑R53	16♋R19
02 Mo	22 43 25	09 15 15	15 34 01	07 15	14 24	09 24	15 12	14 08	06 26	17 32	20 52	16 10
03 Tu	22 47 22	10 13 20	00♏ 14 34	09 11	15 38	10 02	15 16	14 06	06 25	17 31	20 51	16 03
04 We	22 51 18	11 11 27	14 28 27	11 07	16 53	10 41	15 20	14 05	06 24	17 29	20 50	15 57
05 Th	22 55 15	12 09 35	28 13 39	13 02	18 07	11 19	15 25	14 03	06 23	17 27	20 49	15 53
06 Fr	22 59 11	13 07 45	11♐ 31 10	14 56	19 22	11 57	15 29	14 02	06 22	17 26	20 48	15 52
07 Sa	23 03 08	14 05 56	24 24 03	16 49	20 36	12 35	15 34	14 01	06 21	17 24	20 48	15D 52
08 Su	23 07 04	15 04 09	06♑ 56 27	18 41	21 51	13 14	15 38	14 00	06 19	17 22	20 47	15 53
09 Mo	23 11 01	16 02 23	19 12 52	20 32	23 05	13 52	15 43	13 59	06 18	17 21	20 46	15 54
10 Tu	23 14 57	17 00 38	01♒ 17 39	22 22	24 19	14 30	15 49	13 58	06 17	17 19	20 46	15R 54
11 We	23 18 54	17 58 56	13 14 44	24 10	25 34	15 09	15 54	13 57	06 16	17 17	20 45	15 51
12 Th	23 22 51	18 57 15	25 07 26	25 57	26 48	15 47	15 59	13 57	06 14	17 16	20 44	15 46
13 Fr	23 26 47	19 55 35	06♓ 58 28	27 43	28 03	16 25	16 05	13 56	06 13	17 14	20 44	15 38
14 Sa	23 30 44	20 53 57	18 49 53	29 28	29 17	17 03	16 11	13 56	06 11	17 13	20 43	15 27
15 Su	23 34 40	21 52 21	00♈ 43 20	01♎ 12	00♎ 32	17 42	16 17	13 55	06 10	17 11	20 43	15 15
16 Mo	23 38 37	22 50 47	12 40 10	02 55	01 46	18 20	16 23	13 55	06 08	17 09	20 42	15 03
17 Tu	23 42 33	23 49 15	24 41 49	04 37	03 01	18 59	16 29	13 55	06 06	17 08	20 42	14 51
18 We	23 46 30	24 47 45	06♉ 50 00	06 18	04 15	19 37	16 35	13 55	06 05	17 06	20 41	14 42
19 Th	23 50 26	25 46 18	19 06 55	07 57	05 30	20 15	16 42	13D 55	06 03	17 04	20 41	14 34
20 Fr	23 54 23	26 44 52	01♊ 35 19	09 36	06 44	20 54	16 48	13 55	06 01	17 03	20 40	14 30
21 Sa	23 58 19	27 43 29	14 18 25	11 14	07 59	21 32	16 55	13 55	06 00	17 01	20 40	14 28
22 Su	00 02 16	28 42 07	27 19 47	12 50	09 14	22 11	17 02	13 55	05 58	16 59	20 40	14 27
23 Mo	00 06 13	29 40 48	10♋ 42 58	14 26	10 28	22 49	17 09	13 56	05 56	16 58	20 39	14 27
24 Tu	00 10 09	00♎ 39 32	24 30 51	16 01	11 43	23 27	17 16	13 56	05 54	16 56	20 39	14 26
25 We	00 14 06	01 38 17	08♌ 44 49	17 34	12 57	24 06	17 24	13 57	05 52	16 55	20 39	14 24
26 Th	00 18 02	02 37 05	23 23 45	19 07	14 12	24 44	17 31	13 57	05 50	16 53	20 39	14 19
27 Fr	00 21 59	03 35 55	08♍ 23 19	20 39	15 26	25 23	17 39	13 58	05 48	16 52	20 38	14 11
28 Sa	00 25 55	04 34 47	23 35 48	22 10	16 41	26 02	17 47	13 59	05 46	16 50	20 38	14 01
29 Su	00 29 52	05 33 41	08♎ 50 59	23 40	17 56	26 40	17 55	14 00	05 44	16 48	20 38	13 50
30 Mo	00 33 48	06 32 37	23 57 42	25 09	19 10	27 19	18 03	14 01	05 42	16 47	20 38	13 38

Data for	09-01-2019
Julian Day	2458727.50
Ayanamsa	24 07 37
SVP	04 ♓ 58 50
☽ ☊ Mean	14 ♋ 42 R

● ◐ PHASES ○ ◑

06	03:11	◐	13♐16
14	04:33	○	21♓05
22	02:41	◑	28♊49
28	18:27	●	05♎20

LAST ASPECT Day	h m	☽ INGRESS Day	h m	
02	08:34	02	23:36	♏
04	10:59	05	03:09	♐
06	16:05	07	10:39	♑
09	08:32	09	21:25	♒
11	05:23	12	09:52	♓
14	04:33	14	22:33	♈
16	16:02	17	10:31	♉
19	13:57	19	20:58	♊
22	02:41	22	04:51	♋
23	22:06	24	09:20	♌
25	16:15	26	10:38	♍
28	03:59	28	10:04	♎
30	02:07	30	09:43	♏

DECLINATION

Day	☉	☽	☿	♀	♂	♃	♄	♅	♆	♇
01 Su	08N28	04N17	11N13	07N56	09N17	22S 16	22S 27	13N11	05S 53	22S 18
02 Mo	08 06	01S 25	10 29	07 27	09 03	22 17	22 28	13 11	05 54	22 18
03 Tu	07 44	06 55	09 44	06 58	08 48	22 17	22 28	13 10	05 54	22 19
04 We	07 22	11 53	08 58	06 29	08 34	22 18	22 28	13 10	05 55	22 19
05 Th	07 00	16 05	08 12	05 59	08 19	22 19	22 28	13 10	05 56	22 19
06 Fr	06 38	19 18	07 26	05 30	08 04	22 19	22 29	13 09	05 56	22 19
07 Sa	06 15	21 26	06 39	05 00	07 49	22 20	22 29	13 09	05 57	22 19
08 Su	05 53	22 27	05 51	04 30	07 34	22 21	22 29	13 09	05 58	22 20
09 Mo	05 30	22 21	05 04	04 00	07 20	22 21	22 29	13 08	05 58	22 20
10 Tu	05 08	21 11	04 16	03 30	07 05	22 22	22 29	13 08	05 59	22 20
11 We	04 45	19 05	03 29	03 00	06 50	22 23	22 30	13 07	06 00	22 20
12 Th	04 22	16 10	02 42	02 30	06 35	22 23	22 30	13 07	06 00	22 20
13 Fr	03 59	12 36	01 54	02 00	06 20	22 24	22 30	13 06	06 01	22 20
14 Sa	03 36	08 33	01 07	01 29	06 04	22 25	22 30	13 06	06 02	22 21
15 Su	03 13	04 10	00 20	00 59	05 49	22 26	22 30	13 05	06 02	22 21
16 Mo	02 50	00N23	00S 26	00 28	05 34	22 26	22 30	13 05	06 03	22 21
17 Tu	02 27	04 57	01 13	00S 02	05 19	22 27	22 30	13 04	06 04	22 21
18 We	02 04	09 23	01 59	00 33	05 04	22 28	22 31	13 04	06 04	22 21
19 Th	01 41	13 30	02 44	01 04	04 48	22 29	22 31	13 03	06 05	22 21
20 Fr	01 18	17 07	03 29	01 34	04 33	22 30	22 31	13 02	06 05	22 22
21 Sa	00 54	20 00	04 14	02 05	04 18	22 30	22 31	13 02	06 06	22 22
22 Su	00 31	21 56	04 58	02 35	04 02	22 31	22 31	13 01	06 07	22 22
23 Mo	00 08	22 41	05 42	03 06	03 47	22 32	22 31	13 01	06 07	22 22
24 Tu	00S 16	22 05	06 25	03 36	03 31	22 33	22 31	13 00	06 08	22 22
25 We	00 39	20 05	07 08	04 07	03 16	22 34	22 31	12 59	06 09	22 22
26 Th	01 02	16 43	07 50	04 37	03 01	22 35	22 31	12 59	06 09	22 22
27 Fr	01 26	12 12	08 32	05 07	02 45	22 36	22 31	12 58	06 10	22 22
28 Sa	01 49	06 51	09 13	05 38	02 30	22 36	22 31	12 57	06 10	22 22
29 Su	02 13	01 05	09 53	06 08	02 14	22 37	22 31	12 57	06 11	22 23
30 Mo	02 36	04S 42	10 33	06 38	01 58	22 38	22 31	12 56	06 12	22 23

ASPECTARIAN

01 14:11 ☿ △ ♅
18:49 ♀ △ ♄
21:41 ☽ □ ♄
23:25 ☽ ⚹ ♃
02 08:34 ☽ □ ♇
10:43 ☉ ☌ ♂
16:27 ♀ □ ♃
19:18 ☽ ∥ ♆
03 00:13 ☽ # ♀
03:31 ☽ # ☉
08:22 ☽ # ♂
10:19 ☽ ☍ ♅
11:29 ☽ # ☿
15:40 ☿ ☌ ♂
17:13 ☽ ⚹ ♂
17:22 ☽ ⚹ ☿
17:59 ☽ ⚹ ☉
23:19 ☽ ⚹ ♄
04 01:41 ☿ ☌ ☉
04:33 ☽ ⚹ ♀
05:10 ☽ △ ♆
06:51 ☽ # ♅
10:59 ☽ ⚹ ♇
11:27 ♀ ☍ ♆
18:57 ☿ ∥ ♂
05 02:54 ♀ # ♆
12:38 ☿ △ ♄
06 00:17 ♃ ∥ ♇
00:50 ☽ □ ♂
07:12 ☿ □ ♃
07:21 ☽ ☌ ♃
07:23 ☽ □ ☿
10:54 ☽ □ ♆
16:05 ☽ □ ♀
21:57 ☉ △ ♄
07 03:45 ♀ △ ♇
07:19 ☿ ☍ ♆
18:42 ☽ ∥ ♇
19:03 ☉ # ♆
19:12 ☽ ∥ ♃
20:49 ☿ # ♆
22:28 ☿ ∥ ☉
22:49 ☽ △ ♅
08 01:52 ☽ # ♄
12:54 ☽ △ ♂
13:43 ☽ ☌ ♄
15:27 ☉ □ ♃
17:13 ☽ △ ☉
17:16 ☽ ∥ ♄
20:20 ☽ ⚹ ♆
23:42 ☽ ∥ ♃
09 00:35 ☽ ∥ ♇
03:04 ☽ △ ☿
03:04 ☽ ☌ ♇
03:07 ☿ △ ♇
04:15 ♂ △ ♄
08:32 ☽ △ ♀
10 07:24 ☉ ☍ ♆
09:58 ☽ □ ♅
11 05:23 ☽ ⚹ ♃
12 09:07 ♂ □ ♃
16:29 ☿ ∥ ♀
20:53 ☽ # ♅
22:27 ☽ ⚹ ♅
13 14:05 ☽ ⚹ ♄
15:11 ☿ ☌ ♀
18:35 ☽ □ ♃
19:40 ☉ △ ♇
20:13 ☽ ☍ ♂
20:44 ☽ ☌ ♆
14 03:49 ☽ ⚹ ♇
04:19 ♂ # ♆
05:26 ♂ ☍ ♆
07:15 ☿ ♎
13:44 ♀ ♎
14:02 ☽ ∥ ♆
14:39 ☽ # ♂
23:34 ☽ ☍ ♀
15 01:08 ☽ ☍ ☿
05:35 ☽ # ☉
19:03 ☽ # ♀
20:23 ☽ ∥ ☿
16 00:25 ☽ # ☿
00:28 ☽ ∥ ♀
00:36 ☿ # ♀
02:31 ☽ □ ♄
07:29 ☽ △ ♃
11:34 ☽ ∥ ☉
16:04 ☽ □ ♇
17 01:51 ☽ ∥ ♂
05:56 ☽ # ♆
22:31 ☽ ☌ ♅
18 01:55 ☿ # ☉
08:48 ♄ SD
13:52 ☽ △ ♄
20:02 ☽ ⚹ ♆
21:16 ☽ ∥ ♅
19 02:20 ☽ △ ♂
03:02 ☽ △ ♇
13:57 ☽ △ ☉
15:50 ♂ △ ♇
16:36 ♀ # ☉
20 10:51 ☽ △ ♀
17:24 ☽ △ ☿
21 01:28 ☿ # ♂
04:54 ☽ ☍ ♃
05:02 ☽ □ ♆
11:32 ♃ ∥ ♄
14:05 ☽ □ ♂
16:48 ♃ □ ♆
22 09:17 ☽ # ♇
13:54 ☽ # ♄
14:26 ☽ # ♃
15:32 ☽ ⚹ ♅
16:19 ☿ □ ♄
23:31 ☽ □ ♀
23 05:39 ☽ ☍ ♄
07:24 ☽ □ ☿
07:51 ☉ ♎
10:56 ☽ △ ♆
12:39 ☽ # ♃
13:38 ☽ # ♄
14:12 ☿ ∥ ♆
17:22 ☽ ☍ ♇
18:02 ☽ # ♇
20:26 ♀ # ♂
22:06 ☽ ⚹ ♂
24 11:13 ☽ ⚹ ☉
19:13 ☽ □ ♅
21:01 ☿ ⚹ ♃
25 07:36 ☽ ⚹ ♀
14:22 ☽ △ ♃
16:15 ☽ ⚹ ☿
19:21 ♀ □ ♄
26 19:54 ☽ △ ♅
20:15 ☽ ∥ ♅
23:50 ☿ □ ♇
27 08:51 ☽ △ ♄
13:22 ☽ ☍ ♆
14:46 ☽ □ ♃
14:54 ☽ # ☿
19:21 ☽ △ ♇
28 02:53 ☽ # ♆
03:59 ☽ ☌ ♂
04:47 ☽ # ♀
19:03 ☽ ∥ ♂
19:39 ☽ # ☉
23:41 ♀ ⚹ ♃
29 00:54 ☉ # ♂
02:45 ♀ ∥ ♆
08:09 ☽ □ ♄
13:03 ☽ # ♂
14:28 ☽ ⚹ ♃
14:30 ☽ ∥ ☉
15:40 ☽ ☌ ♀
18:41 ☽ □ ♇
30 02:07 ☽ ☌ ☿
06:22 ☽ ∥ ♆
09:03 ☽ ∥ ♀
18:57 ☽ ☍ ♅

⚷ Chiron

01 Dec.	04 N 51
03	04♈42R
06	04 35
09	04 27
12	04 19
15	04 12
18	04 04
21	03 56
24	03 48
27	03 39
30	03 31

October 2019

Day	S. T. h m s	☉ ° ' "	☽ ° ' "	☿ ° '	♀ ° '	♂ ° '	♃ ° '	♄ ° '	♅ ° '	♆ ° '	♇ ° '	☊ True ° '
01 Tu	00 37 45	07♎ 31 35	08♏ 45 55	26♎ 37	20♎ 25	27♍ 57	18♐ 11	14♑ 02	05♉R40	16♓R45	20♑R38	13♋R27
02 We	00 41 42	08 30 35	23 08 18	28 04	21 39	28 36	18 19	14 04	05 38	16 44	20 38	13 18
03 Th	00 45 38	09 29 37	07♐ 01 07	29 31	22 54	29 14	18 28	14 05	05 36	16 42	20 38	13 12
04 Fr	00 49 35	10 28 41	20 24 11	00♏ 56	24 08	29 53	18 36	14 07	05 34	16 41	20D 38	13 08
05 Sa	00 53 31	11 27 46	03♑ 20 00	02 20	25 23	00♎ 32	18 45	14 08	05 32	16 39	20 38	13 07
06 Su	00 57 28	12 26 53	15 52 48	03 43	26 38	01 10	18 54	14 10	05 30	16 38	20 38	13D 07
07 Mo	01 01 24	13 26 02	28 07 36	05 05	27 52	01 49	19 03	14 12	05 27	16 36	20 38	13R 07
08 Tu	01 05 21	14 25 13	10♒ 09 34	06 27	29 07	02 28	19 12	14 14	05 25	16 35	20 38	13 06
09 We	01 09 17	15 24 25	22 03 34	07 46	00♏ 21	03 06	19 21	14 16	05 23	16 34	20 38	13 03
10 Th	01 13 14	16 23 39	03♓ 53 59	09 05	01 36	03 45	19 30	14 18	05 21	16 32	20 39	12 57
11 Fr	01 17 11	17 22 55	15 44 28	10 23	02 51	04 24	19 40	14 20	05 18	16 31	20 39	12 48
12 Sa	01 21 07	18 22 13	27 37 50	11 39	04 05	05 02	19 49	14 22	05 16	16 30	20 39	12 37
13 Su	01 25 04	19 21 33	09♈ 36 07	12 53	05 20	05 41	19 59	14 24	05 14	16 28	20 39	12 24
14 Mo	01 29 00	20 20 55	21 40 40	14 07	06 34	06 20	20 09	14 27	05 11	16 27	20 40	12 10
15 Tu	01 32 57	21 20 19	03♉ 52 27	15 18	07 49	06 59	20 18	14 29	05 09	16 26	20 40	11 57
16 We	01 36 53	22 19 45	16 12 14	16 28	09 03	07 37	20 28	14 32	05 07	16 24	20 40	11 46
17 Th	01 40 50	23 19 13	28 40 58	17 36	10 18	08 16	20 38	14 35	05 04	16 23	20 41	11 37
18 Fr	01 44 46	24 18 44	11♊ 19 55	18 42	11 33	08 55	20 49	14 37	05 02	16 22	20 41	11 32
19 Sa	01 48 43	25 18 17	24 10 51	19 45	12 47	09 34	20 59	14 40	04 59	16 21	20 41	11 28
20 Su	01 52 39	26 17 52	07♋ 15 59	20 46	14 02	10 13	21 09	14 43	04 57	16 20	20 42	11 27
21 Mo	01 56 36	27 17 29	20 37 51	21 45	15 16	10 52	21 20	14 46	04 55	16 18	20 42	11 27
22 Tu	02 00 33	28 17 09	04♌ 18 50	22 40	16 31	11 31	21 30	14 50	04 52	16 17	20 43	11 26
23 We	02 04 29	29 16 51	18 20 29	23 32	17 45	12 09	21 41	14 53	04 50	16 16	20 44	11 23
24 Th	02 08 26	00♏ 16 35	02♍ 42 45	24 20	19 00	12 48	21 52	14 56	04 47	16 15	20 44	11 19
25 Fr	02 12 22	01 16 22	17 23 10	25 04	20 15	13 27	22 03	15 00	04 45	16 14	20 45	11 12
26 Sa	02 16 19	02 16 10	02♎ 16 32	25 44	21 29	14 06	22 14	15 03	04 42	16 13	20 45	11 03
27 Su	02 20 15	03 16 01	17 15 06	26 19	22 44	14 45	22 25	15 07	04 40	16 12	20 46	10 53
28 Mo	02 24 12	04 15 54	02♏ 09 39	26 48	23 58	15 24	22 36	15 10	04 37	16 11	20 47	10 42
29 Tu	02 28 08	05 15 48	16 50 56	27 11	25 13	16 03	22 47	15 14	04 35	16 10	20 48	10 32
30 We	02 32 05	06 15 45	01♐ 11 19	27 27	26 27	16 42	22 58	15 18	04 32	16 09	20 48	10 24
31 Th	02 36 02	07 15 44	15 05 51	27 37	27 42	17 22	23 10	15 22	04 30	16 08	20 49	10 18

Data for	10-01-2019
Julian Day	2458757.50
Ayanamsa	24 07 40
SVP	04 ♓ 58 45
☽ ☊ Mean	13 ♋ 07 R

● ◐ PHASES ○ ◑

Day	Time	Phase	Position
05	16:48	◐	12♑09
13	21:08	○	20♈14
21	12:40	◑	27♋49
28	03:39	●	04♏25

LAST ASPECT Day	h m	☽ INGRESS Day	h m	Sign
02	09:47	02	11:45	♐
04	07:35	04	17:44	♑
06	23:26	07	03:43	♒
08	18:27	09	16:05	♓
11	09:55	12	04:46	♈
13	21:59	14	16:24	♉
16	08:38	17	02:31	♊
19	02:15	19	10:44	♋
21	12:41	21	16:30	♌
23	09:16	23	19:31	♍
25	13:01	25	20:21	♎
27	08:23	27	20:30	♏
29	17:35	29	21:59	♐

DECLINATION

Day	☉	☽	☿	♀	♂	♃	♄	♅	♆	♇
01 Tu	02S 59	10S 07	11S 12	07S 07	01N43	22S 39	22S 31	12N55	06S 12	22S 23
02 We	03 22	14 49	11 50	07 37	01 27	22 40	22 31	12 55	06 13	22 23
03 Th	03 46	18 33	12 27	08 07	01 12	22 41	22 31	12 54	06 13	22 23
04 Fr	04 09	21 09	13 04	08 36	00 56	22 42	22 31	12 53	06 14	22 23
05 Sa	04 32	22 32	13 40	09 05	00 41	22 43	22 31	12 53	06 15	22 23
06 Su	04 55	22 44	14 15	09 34	00 25	22 43	22 31	12 52	06 15	22 23
07 Mo	05 18	21 49	14 50	10 03	00 09	22 44	22 31	12 51	06 16	22 23
08 Tu	05 41	19 54	15 23	10 31	00S 06	22 45	22 31	12 50	06 16	22 23
09 We	06 04	17 09	15 56	10 59	00 22	22 46	22 30	12 50	06 17	22 23
10 Th	06 27	13 43	16 27	11 27	00 38	22 47	22 30	12 49	06 17	22 23
11 Fr	06 49	09 44	16 58	11 55	00 53	22 48	22 30	12 48	06 18	22 23
12 Sa	07 12	05 22	17 28	12 22	01 09	22 49	22 30	12 47	06 18	22 23
13 Su	07 35	00 47	17 56	12 50	01 24	22 49	22 30	12 46	06 19	22 23
14 Mo	07 57	03N53	18 24	13 16	01 40	22 50	22 30	12 46	06 19	22 23
15 Tu	08 19	08 26	18 50	13 43	01 56	22 51	22 30	12 45	06 20	22 24
16 We	08 41	12 43	19 16	14 09	02 11	22 52	22 29	12 44	06 20	22 24
17 Th	09 04	16 31	19 40	14 35	02 27	22 53	22 29	12 43	06 21	22 24
18 Fr	09 25	19 38	20 03	15 01	02 42	22 54	22 29	12 43	06 21	22 24
19 Sa	09 47	21 49	20 24	15 26	02 58	22 55	22 29	12 42	06 22	22 24
20 Su	10 09	22 53	20 44	15 50	03 13	22 55	22 28	12 41	06 22	22 24
21 Mo	10 30	22 40	21 03	16 15	03 29	22 56	22 28	12 40	06 23	22 24
22 Tu	10 52	21 06	21 20	16 39	03 44	22 57	22 28	12 39	06 23	22 24
23 We	11 13	18 14	21 35	17 02	04 00	22 58	22 28	12 38	06 23	22 24
24 Th	11 34	14 12	21 48	17 25	04 15	22 59	22 27	12 38	06 24	22 24
25 Fr	11 55	09 14	22 00	17 48	04 31	22 59	22 27	12 37	06 24	22 24
26 Sa	12 16	03 40	22 10	18 10	04 46	23 00	22 27	12 36	06 25	22 23
27 Su	12 36	02S 08	22 17	18 32	05 01	23 01	22 26	12 35	06 25	22 23
28 Mo	12 56	07 48	22 22	18 53	05 17	23 02	22 26	12 34	06 25	22 23
29 Tu	13 16	12 57	22 25	19 14	05 32	23 02	22 26	12 34	06 26	22 23
30 We	13 36	17 15	22 25	19 34	05 47	23 03	22 25	12 33	06 26	22 23
31 Th	13 56	20 27	22 22	19 54	06 03	23 04	22 25	12 32	06 26	22 23

ASPECTARIAN

01 04:15 ♀ □ ♇
05:54 ☽ ∥ ☿
08:44 ☽ ⚹ ♄
13:13 ☽ △ ♆
13:46 ☽ # ♅
19:45 ☽ ⚹ ♇
02 09:47 ☽ ⚹ ♂
03 04:43 ☽ ⚹ ☉
06:38 ♇ SD
08:15 ☿ ♏
17:01 ☿ # ♅
17:15 ☽ □ ♆
20:41 ☽ ☌ ♃
04 04:23 ♂ ♎
07:35 ☽ ⚹ ♀
18:27 ☽ □ ♂
20:07 ☽ ∥ ♇
21:53 ☽ ⚹ ☿
23:35 ☽ ∥ ♄
05 04:09 ☽ △ ♅
07:23 ☽ ∥ ♃
20:40 ☽ ☌ ♄
06 00:09 ☽ ∥ ♃
01:27 ☽ ⚹ ♆
08:51 ☽ ∥ ♄
09:15 ☽ ☌ ♇
12:20 ☽ ∥ ♇
23:26 ☽ □ ♀
07 06:17 ☿ ☍ ♅
07:44 ☽ △ ♂
14:32 ☽ □ ♅
15:37 ☽ □ ☿
19:08 ☉ □ ♄
08 09:21 ☽ △ ☉
17:07 ♀ ♏
18:27 ☽ ⚹ ♃
09 07:46 ☽ ∥ ☿
13:46 ☉ ∥ ♆
18:47 ☽ △ ♀
10 02:55 ☽ ⚹ ♅
05:41 ☽ # ♅
11:49 ☽ △ ☿
12:30 ☽ ∥ ♀
21:08 ☽ ⚹ ♄
11 01:34 ☽ ☌ ♆
08:02 ☽ □ ♃
09:55 ☽ ⚹ ♇
14:55 ☽ ∥ ☉
19:01 ☽ ∥ ♆
12 15:43 ☽ ☍ ♂
20:58 ☽ ∥ ♂
21:18 ♀ # ♅
22:07 ♀ ☍ ♅
13 09:36 ☽ □ ♄
11:57 ☽ # ♂
18:02 ☉ ⚹ ♃
20:55 ☽ △ ♃
21:59 ☽ □ ♇
14 06:57 ☿ ⚹ ♄
07:36 ☉ □ ♇
12:47 ☽ # ♆
23:20 ☽ # ☉
15 02:29 ☽ ☌ ♅
08:34 ☽ ☍ ♀
20:45 ☽ △ ♄
22:45 ☿ △ ♆
16 00:04 ☽ ∥ ♅
00:23 ☽ ⚹ ♆
00:34 ☽ ☍ ☿
08:38 ☽ △ ♇
09:42 ☽ # ♀
17 19:12 ☽ △ ♂
18 04:29 ☽ # ☿
09:26 ☽ □ ♆
17:59 ☽ ☍ ♃
19 02:15 ☽ △ ☉
09:51 ☽ # ♇
11:37 ☽ # ♄
19:48 ☽ ⚹ ♅
22:19 ☿ ⚹ ♇
20 03:46 ☽ # ♃
05:37 ☽ □ ♂
11:59 ☽ ∥ ♃
13:28 ☽ △ ♀
13:30 ☽ ☍ ♄
13:59 ♀ ⚹ ♄
16:18 ☽ △ ♆
21 00:08 ☽ ☍ ♇
02:07 ☽ △ ☿
04:25 ☽ # ♄
05:58 ☽ # ♇
19:41 ♀ △ ♆
21:42 ☽ # ☿
22 00:57 ☽ □ ♅
12:59 ☽ ⚹ ♂
22:55 ☽ □ ♀
23 05:42 ☽ △ ♃
07:00 ☽ ∥ [illegible]
09:16 ☽ □ ☿
17:20 ☉ ♏
19:40 ☽ ⚹ ☉
24 03:25 ☽ △ ♅
07:59 ☽ ∥ ♅
12:17 ☽ # ☉
20:06 ☽ △ ♄
22:08 ☽ ☍ ♆
25 05:03 ☽ ⚹ ♀
05:27 ☽ △ ♇
07:38 ☽ □ ♃
09:51 ♀ ⚹ ♇
12:24 ☽ # ♆
[illegible]
19:34 ☽ # ♂
26 19:49 ☽ ☌ ♂
20:33 ☽ □ ♄
23:00 ☉ # ♅
27 05:39 ☽ □ ♇
08:23 ☽ ⚹ ♃
12:35 ☽ ∥ ♂
14:32 ♂ □ ♄
17:58 ☽ ∥ ♆
28 03:59 ☽ ☍ ♅
08:15 ☉ ☍ ♅
08:25 ☿ ∥ ♇
21:20 ☽ ⚹ ♄
22:01 ☽ # ♅
22:52 ☽ △ ♆
29 01:44 ☽ ∥ ☉
06:33 ☽ ⚹ ♇
15:15 ☽ ☌ ♀
17:35 ☽ ☌ ☿
30 14:06 ☿ ∥ ♇
18:29 ☽ ∥ ♀
22:06 ☿ ☌ ♀
31 01:49 ☽ □ ♆
04:11 ☽ ⚹ ♂
14:30 ☽ ☌ ♃
15:43 ☿ SR
21:53 ☽ ∥ ☿
23:59 ☽ ∥ ♇

⚷ Chiron

01	Dec. 04 N 19
03	03♈23R
06	03 15
09	03 07
12	02 59
15	02 52
18	02 44
21	02 37
24	02 30
27	02 23
30	02 16

November 2019

Day	S. T. h m s	☉ ° ' "	☽ ° ' "	☿ ° '	♀ ° '	♂ ° '	♃ ° '	♄ ° '	♅ ° '	♆ ° '	♇ ° '	☊ True ° '
01 Fr	02 39 58	08♏ 15 44	28♐ 32 43	27♏R38	28♏ 57	18♎ 01	23♐ 21	15♑ 26	04♉R28	16♓R07	20♑ 50	10♋R14
02 Sa	02 43 55	09 15 46	11♑ 32 55	27 31	00♐ 11	18 40	23 33	15 30	04 25	16 06	20 51	10 13
03 Su	02 47 51	10 15 50	24 09 37	27 14	01 26	19 19	23 44	15 34	04 23	16 06	20 52	10D 13
04 Mo	02 51 48	11 15 55	06♒ 27 14	26 49	02 40	19 58	23 56	15 38	04 20	16 05	20 53	10 14
05 Tu	02 55 44	12 16 02	18 30 48	26 13	03 55	20 37	24 08	15 43	04 18	16 04	20 54	10 15
06 We	02 59 41	13 16 10	00♓ 25 31	25 28	05 09	21 16	24 20	15 47	04 15	16 03	20 55	10R 14
07 Th	03 03 37	14 16 20	12 16 24	24 34	06 24	21 55	24 32	15 52	04 13	16 03	20 56	10 11
08 Fr	03 07 34	15 16 31	24 07 59	23 31	07 38	22 35	24 44	15 56	04 10	16 02	20 57	10 06
09 Sa	03 11 31	16 16 44	06♈ 04 07	22 21	08 53	23 14	24 56	16 01	04 08	16 01	20 58	09 59
10 Su	03 15 27	17 16 58	18 07 46	21 06	10 07	23 53	25 08	16 05	04 06	16 01	20 59	09 50
11 Mo	03 19 24	18 17 14	00♉ 21 01	19 47	11 22	24 32	25 20	16 10	04 03	16 00	21 00	09 41
12 Tu	03 23 20	19 17 32	12 45 08	18 27	12 36	25 12	25 32	16 15	04 01	16 00	21 01	09 32
13 We	03 27 17	20 17 52	25 20 39	17 09	13 51	25 51	25 45	16 20	03 59	15 59	21 02	09 24
14 Th	03 31 13	21 18 13	08♊ 07 38	15 54	15 05	26 30	25 57	16 25	03 56	15 59	21 03	09 18
15 Fr	03 35 10	22 18 36	21 05 57	14 47	16 20	27 10	26 10	16 30	03 54	15 58	21 04	09 14
16 Sa	03 39 06	23 19 01	04♋ 15 30	13 47	17 34	27 49	26 22	16 35	03 52	15 58	21 06	09 12
17 Su	03 43 03	24 19 28	17 36 28	12 58	18 49	28 29	26 35	16 40	03 49	15 57	21 07	09D 12
18 Mo	03 46 60	25 19 56	01♌ 09 16	12 20	20 03	29 08	26 47	16 45	03 47	15 57	21 08	09 12
19 Tu	03 50 56	26 20 26	14 54 30	11 54	21 18	29 47	27 00	16 51	03 45	15 57	21 10	09R 12
20 We	03 54 53	27 20 58	28 52 28	11 39	22 32	00♏ 27	27 13	16 56	03 43	15 57	21 11	09 12
21 Th	03 58 49	28 21 32	13♍ 02 44	11D 35	23 46	01 06	27 26	17 02	03 41	15 56	21 12	09 10
22 Fr	04 02 46	29 22 08	27 23 38	11 43	25 01	01 46	27 39	17 07	03 38	15 56	21 14	09 07
23 Sa	04 06 42	00♐ 22 46	11♎ 52 02	12 01	26 15	02 25	27 51	17 13	03 36	15 56	21 15	09 02
24 Su	04 10 39	01 23 25	26 23 12	12 28	27 30	03 05	28 04	17 18	03 34	15 56	21 16	08 56
25 Mo	04 14 35	02 24 05	10♏ 51 18	13 03	28 44	03 44	28 17	17 24	03 32	15 56	21 18	08 49
26 Tu	04 18 32	03 24 48	25 10 04	13 46	29 59	04 24	28 30	17 30	03 30	15 56	21 19	08 43
27 We	04 22 29	04 25 32	09♐ 13 41	14 36	01♑ 13	05 04	28 44	17 35	03 28	15 56	21 21	08 38
28 Th	04 26 25	05 26 17	22 57 43	15 31	02 27	05 43	28 57	17 41	03 26	15D 56	21 22	08 35
29 Fr	04 30 22	06 27 03	06♑ 19 40	16 32	03 42	06 23	29 10	17 47	03 24	15 56	21 24	08 34
30 Sa	04 34 18	07 27 51	19 19 07	17 37	04 56	07 03	29 23	17 53	03 22	15 56	21 25	08D 34

Data for	11-01-2019
Julian Day	2458788.50
Ayanamsa	24 07 43
SVP	04 ♓ 58 40
☽ ☊ Mean	11 ♋ 28 R

● ◐ PHASES ○ ◑

04	10:23	◐	11♒42
12	13:35	○	19♉52
19	21:11	◑	27♌14
26	15:06	●	04♐03

LAST ASPECT ☽ INGRESS

Day	h m	Day	h m	
31	14:30	01	02:39	♑
03	05:47	03	11:20	♒
05	14:37	05	23:08	♓
08	01:13	08	11:49	♈
10	14:02	10	23:19	♉
12	15:49	13	08:47	♊
15	11:41	15	16:16	♋
17	20:16	17	21:58	♌
19	21:12	20	01:55	♍
22	03:32	22	04:20	♎
24	02:50	24	05:59	♏
25	17:30	26	08:12	♐
28	10:50	28	12:33	♑
30	03:57	30	20:14	♒

DECLINATION

Day	☉	☽	☿	♀	♂	♃	♄	♅	♆	♇
01 Fr	14S 16	22S 23	22S 16	20S 13	06S 18	23S 05	22S 25	12N31	06S 27	22S 23
02 Sa	14 35	23 03	22 06	20 31	06 33	23 05	22 24	12 30	06 27	22 23
03 Su	14 54	22 29	21 53	20 49	06 48	23 06	22 24	12 29	06 27	22 23
04 Mo	15 12	20 50	21 36	21 07	07 03	23 07	22 24	12 29	06 28	22 23
05 Tu	15 31	18 17	21 15	21 24	07 18	23 07	22 23	12 28	06 28	22 23
06 We	15 49	14 59	20 50	21 40	07 33	23 08	22 23	12 27	06 28	22 23
07 Th	16 07	11 07	20 21	21 56	07 48	23 08	22 22	12 26	06 28	22 23
08 Fr	16 25	06 50	19 48	22 11	08 03	23 09	22 22	12 25	06 29	22 23
09 Sa	16 42	02 16	19 12	22 25	08 18	23 10	22 21	12 25	06 29	22 23
10 Su	16 59	02N26	18 33	22 39	08 33	23 10	22 21	12 24	06 29	22 23
11 Mo	17 16	07 07	17 52	22 52	08 48	23 11	22 20	12 23	06 29	22 22
12 Tu	17 33	11 35	17 11	23 04	09 02	23 11	22 20	12 22	06 29	22 22
13 We	17 49	15 38	16 29	23 16	09 17	23 12	22 19	12 22	06 30	22 22
14 Th	18 05	19 02	15 49	23 27	09 32	23 12	22 19	12 21	06 30	22 22
15 Fr	18 21	21 32	15 12	23 37	09 46	23 13	22 18	12 20	06 30	22 22
16 Sa	18 36	22 55	14 39	23 47	10 01	23 13	22 18	12 19	06 30	22 22
17 Su	18 51	23 01	14 10	23 56	10 15	23 14	22 17	12 18	06 30	22 22
18 Mo	19 06	21 47	13 46	24 04	10 29	23 14	22 17	12 18	06 30	22 22
19 Tu	19 20	19 15	13 28	24 12	10 44	23 15	22 16	12 17	06 30	22 22
20 We	19 34	15 34	13 15	24 19	10 58	23 15	22 16	12 16	06 30	22 21
21 Th	19 48	10 57	13 07	24 25	11 12	23 15	22 15	12 16	06 30	22 21
22 Fr	20 01	05 41	13 04	24 30	11 26	23 16	22 14	12 15	06 31	22 21
23 Sa	20 14	00 04	13 07	24 35	11 40	23 16	22 14	12 14	06 31	22 21
24 Su	20 26	05S 34	13 13	24 39	11 54	23 16	22 13	12 13	06 31	22 21
25 Mo	20 38	10 54	13 23	24 42	12 08	23 17	22 12	12 13	06 31	22 21
26 Tu	20 50	15 34	13 36	24 44	12 21	23 17	22 12	12 12	06 31	22 21
27 We	21 01	19 17	13 53	24 46	12 35	23 17	22 11	12 11	06 31	22 20
28 Th	21 12	21 50	14 11	24 47	12 49	23 17	22 10	12 11	06 31	22 20
29 Fr	21 23	23 04	14 32	24 47	13 02	23 17	22 10	12 10	06 31	22 20
30 Sa	21 33	23 00	14 54	24 47	13 16	23 18	22 09	12 09	06 30	22 20

ASPECTARIAN

01 00:27 ☽ ∥ ♄
10:47 ☽ △ ♅
14:11 ♂ ∥ ♆
19:22 ☽ ✶ ☉
20:25 ♀ ♐
02 07:29 ☽ ☌ ♄
08:35 ☽ ✶ ♆
14:11 ☽ □ ♂
17:39 ☽ ☌ ♇
03 01:43 ☽ ∥ ♄
02:00 ☽ ∥ ♇
05:47 ☽ ✶ ☿
12:39 ☽ ∥ ☿
15:43 ☽ ✶ ♀
19:50 ☽ □ ♅
21:14 ☽ ∥ ♀
04 18:42 ☿ ∥ ♀
05 04:28 ☽ △ ♂
10:24 ♂ □ ♇
10:55 ♄ ∥ ♇
11:29 ☽ ✶ ♃
14:37 ☽ □ ☿
18:48 ☽ ∥ ☉
06 07:43 ☽ ✶ ♅
10:42 ☽ □ ♀
16:07 ☽ # ♅
07 04:25 ☽ △ ☉
07:19 ☽ ✶ ♄
07:38 ☽ ☌ ♆
17:33 ☽ ✶ ♇
17:41 ☽ ∥ ♂
22:52 ☽ △ ☿
08 01:13 ☽ □ ♃
01:53 ☽ ∥ ♆
17:07 ☉ ✶ ♄
17:57 ☉ △ ♆
18:10 ♀ ∥ ♄
20:10 ♀ ∥ ♇
09 02:52 ♄ ✶ ♆
06:16 ☽ △ ♀
19:56 ☽ □ ♄
10 02:13 ☿ ✶ ♇
05:38 ☽ □ ♇
11:59 ☽ ☍ ♂
14:02 ☽ △ ♃
20:41 ☽ # ♆
11 07:11 ☽ ☌ ♅
09:17 ☽ # ♂
14:52 ☿ ∥ ☉
15:22 ☿ ☌ ☉
12 04:23 ☽ ∥ ♅
06:13 ☽ ✶ ♆
06:45 ☽ △ ♄
09:52 ☽ ☍ ☿
15:02 ♀ ∥ ♃
15:49 ☽ △ ♇
18:21 ♂ ✶ ♃
13 04:38 ☽ # ☿
14:35 ☿ ✶ ♄
16:01 ☽ # ☉
17:57 ☉ ✶ ♇
22:35 ☿ △ ♆
14 14:17 ☽ ☍ ♀
14:34 ☽ □ ♆
17:07 ♀ □ ♆
15 09:25 ☽ ☍ ♃
10:34 ☽ # ♄
11:39 ☽ # ♇
11:41 ☽ △ ♂
23:17 ☽ ✶ ♅
16 16:09 ☽ △ ☿
21:03 ☽ △ ♆
22:19 ☽ ☍ ♄
17 06:16 ☽ ☍ ♇
12:54 ☽ △ ☉
15:50 ☽ # ♇
17:09 ☽ # ♄
20:16 ☽ □ ♂
18 04:36 ☽ □ ♅
18:54 ☽ □ ☿
23:24 ☽ # ☉
19 07:41 ♂ ♏
12:05 ☽ △ ♀
21:07 ☽ △ ♃
20 02:49 ☽ ✶ ♂
08:13 ☽ △ ♅
12:59 ☽ # ☿
17:34 ☽ ∥ ♅
19:13 ☿ SD
21:33 ☽ ✶ ☿
22:51 ☽ # ♂
21 04:52 ☽ ☍ ♆
06:44 ☽ △ ♄
13:42 ☽ △ ♇
19:40 ☽ □ ♀
20:22 ☽ # ♆
22 00:25 ☽ □ ♃
03:32 ☽ ✶ ☉
14:59 ☉ ♐
23 08:53 ☽ □ ♄
15:32 ☽ □ ♇
24 02:00 ☽ ✶ ♀
02:50 ☽ ✶ ♃
04:04 ☽ ∥ ♆
11:37 ☽ ☌ ♂
11:52 ☽ ☍ ♅
13:34 ♀ ☌ ♃
16:51 ♂ ☍ ♅
25 03:50 ☽ ☌ ☿
06:13 ☽ ∥ ♂
06:19 ☽ # ♅
08:24 ♂ # ♅
08:28 ☽ △ ♆
11:00 ☽ ✶ ♄
12:47 ☽ ∥ ☿
17:30 ☽ ✶ ♇
26 00:29 ♀ ♑
27 11:38 ☽ □ ♆
12:34 ♆ SD
16:08 ☽ ∥ ☉
28 04:29 ☽ ∥ ♄
07:00 ☽ ∥ ♇
09:52 ☿ △ ♆
10:50 ☽ ☌ ♃
18:28 ♀ △ ♅
18:42 ☽ △ ♅
18:44 ☽ ☌ ♀
29 00:06 ☽ ✶ ♂
17:40 ☽ ✶ ♆
20:31 ☽ ✶ ☿
21:18 ☽ ☌ ♄
30 03:57 ☽ ☌ ♇
06:13 ☿ ✶ ♄
15:13 ☽ ∥ ♇
18:16 ☽ ∥ ♄

⚷ Chiron

01 Dec. 03 N 45

02	02♈10R
05	02 04
08	01 59
11	01 54
14	01 49
17	01 45
20	01 41
23	01 37
26	01 34
29	01 32

December 2019

26 05:16 04♑07 ☉ Annular Solar Eclipse

Day	S. T. h m s	☉ ° ' "	☽ ° ' "	☿ ° '	♀ ° '	♂ ° '	♃ ° '	♄ ° '	♅ ° '	♆ ° '	♇ ° '	☊ True ° '
01 Su	04 38 15	08♐ 28 39	01♒ 57 34	18♏ 46	06♑ 10	07♏ 42	29♐ 36	17♑ 59	03♉R20	15♓ 56	21♑ 27	08♋ 35
02 Mo	04 42 11	09 29 29	14 17 59	19 58	07 25	08 22	29 50	18 05	03 18	15 56	21 28	08 38
03 Tu	04 46 08	10 30 19	26 24 20	21 13	08 39	09 02	00♑ 03	18 11	03 17	15 56	21 30	08 40
04 We	04 50 05	11 31 10	08♓ 21 12	22 31	09 53	09 42	00 17	18 17	03 15	15 56	21 32	08 42
05 Th	04 54 01	12 32 02	20 13 29	23 51	11 08	10 21	00 30	18 23	03 13	15 57	21 33	08 42
06 Fr	04 57 58	13 32 55	02♈ 06 05	25 12	12 22	11 01	00 43	18 30	03 11	15 57	21 35	08R 42
07 Sa	05 01 54	14 33 49	14 03 38	26 35	13 36	11 41	00 57	18 36	03 10	15 57	21 37	08 40
08 Su	05 05 51	15 34 44	26 10 12	28 00	14 50	12 21	01 10	18 42	03 08	15 57	21 38	08 38
09 Mo	05 09 47	16 35 39	08♉ 29 08	29 25	16 05	13 01	01 24	18 49	03 06	15 58	21 40	08 35
10 Tu	05 13 44	17 36 35	21 02 49	00♐ 52	17 19	13 40	01 38	18 55	03 05	15 58	21 42	08 32
11 We	05 17 40	18 37 32	03♊ 52 34	02 19	18 33	14 20	01 51	19 01	03 03	15 59	21 44	08 29
12 Th	05 21 37	19 38 30	16 58 40	03 47	19 47	15 00	02 05	19 08	03 02	15 59	21 45	08 27
13 Fr	05 25 34	20 39 29	00♋ 20 18	05 16	21 01	15 40	02 18	19 14	03 00	16 00	21 47	08 26
14 Sa	05 29 30	21 40 28	13 55 50	06 45	22 15	16 20	02 32	19 21	02 59	16 00	21 49	08 26
15 Su	05 33 27	22 41 29	27 43 08	08 14	23 29	17 00	02 46	19 27	02 58	16 01	21 51	08D 26
16 Mo	05 37 23	23 42 30	11♌ 39 45	09 44	24 43	17 40	02 59	19 34	02 56	16 01	21 52	08 26
17 Tu	05 41 20	24 43 33	25 43 19	11 15	25 57	18 20	03 13	19 41	02 55	16 02	21 54	08 27
18 We	05 45 16	25 44 36	09♍ 51 36	12 46	27 11	19 00	03 27	19 47	02 54	16 03	21 56	08R 27
19 Th	05 49 13	26 45 40	24 02 24	14 17	28 25	19 40	03 41	19 54	02 53	16 04	21 58	08 26
20 Fr	05 53 09	27 46 45	08♎ 13 36	15 48	29 39	20 20	03 54	20 01	02 51	16 04	22 00	08 25
21 Sa	05 57 06	28 47 51	22 22 55	17 19	00♒ 53	21 00	04 08	20 08	02 50	16 05	22 02	08 23
22 Su	06 01 03	29 48 58	06♏ 27 59	18 51	02 07	21 41	04 22	20 14	02 49	16 06	22 04	08 21
23 Mo	06 04 59	00♑ 50 05	20 26 12	20 23	03 21	22 21	04 36	20 21	02 48	16 07	22 05	08 20
24 Tu	06 08 56	01 51 14	04♐ 14 55	21 56	04 35	23 01	04 50	20 28	02 47	16 08	22 07	08 18
25 We	06 12 52	02 52 23	17 51 33	23 28	05 49	23 41	05 03	20 35	02 47	16 09	22 09	08 18
26 Th	06 16 49	03 53 32	01♑ 13 52	25 01	07 03	24 21	05 17	20 42	02 46	16 09	22 11	08 17
27 Fr	06 20 45	04 54 42	14 20 17	26 34	08 16	25 02	05 31	20 49	02 45	16 10	22 13	08D 17
28 Sa	06 24 42	05 55 52	27 10 08	28 07	09 30	25 42	05 45	20 56	02 44	16 11	22 15	08 18
29 Su	06 28 38	06 57 02	09♒ 43 51	29 41	10 44	26 22	05 59	21 03	02 43	16 13	22 17	08 18
30 Mo	06 32 35	07 58 12	22 02 55	01♑ 15	11 57	27 02	06 13	21 10	02 43	16 14	22 19	08 19
31 Tu	06 36 32	08 59 23	04♓ 09 45	02 49	13 11	27 43	06 26	21 17	02 42	16 15	22 21	08 20

Data for	12-01-2019
Julian Day	2458818.50
Ayanamsa	24 07 48
SVP	04 ♓ 58 36
☽ ☊ Mean	09 ♋ 53 R

● ◐ PHASES ○ ◑

Day	Time	Phase	Position
04	06:59	◐	11♓49
12	05:13	○	19♊52
19	04:57	◑	26♍58
26	05:14	☉	04♑07

LAST ASPECT ☽ INGRESS

Day	h m	Day	h m	Sign
02	12:28	03	07:12	♓
05	08:16	05	19:46	♈
07	15:02	08	07:31	♉
10	01:14	10	16:48	♊
12	05:13	12	23:24	♋
14	15:57	15	03:57	♌
16	22:10	17	07:16	♍
19	08:07	19	10:05	♎
21	11:46	21	12:58	♏
23	03:28	23	16:35	♐
25	11:19	25	21:46	♑
27	21:04	28	05:22	♒
30	10:25	30	15:43	♓

DECLINATION

Day	☉	☽	☿	♀	♂	♃	♄	♅	♆	♇
01 Su	21S 43	21S 44	15S 18	24S 45	13S 29	23S 18	22S 08	12N09	06S 30	22S 20
02 Mo	21 52	19 27	15 42	24 43	13 42	23 18	22 08	12 08	06 30	22 20
03 Tu	22 01	16 21	16 07	24 40	13 55	23 18	22 07	12 08	06 30	22 19
04 We	22 10	12 37	16 33	24 37	14 08	23 18	22 06	12 07	06 30	22 19
05 Th	22 18	08 26	16 59	24 32	14 21	23 18	22 05	12 06	06 30	22 19
06 Fr	22 25	03 56	17 25	24 27	14 34	23 18	22 05	12 06	06 30	22 19
07 Sa	22 33	00N44	17 52	24 21	14 47	23 18	22 04	12 05	06 30	22 19
08 Su	22 39	05 27	18 18	24 15	14 59	23 18	22 03	12 05	06 30	22 18
09 Mo	22 46	10 01	18 43	24 08	15 12	23 18	22 02	12 04	06 29	22 18
10 Tu	22 52	14 17	19 08	24 00	15 24	23 18	22 01	12 04	06 29	22 18
11 We	22 57	17 59	19 33	23 51	15 37	23 18	22 01	12 03	06 29	22 18
12 Th	23 02	20 53	19 57	23 41	15 49	23 18	22 00	12 03	06 29	22 18
13 Fr	23 06	22 41	20 21	23 31	16 01	23 18	21 59	12 02	06 29	22 18
14 Sa	23 10	23 12	20 44	23 20	16 13	23 18	21 58	12 02	06 28	22 17
15 Su	23 14	22 19	21 06	23 09	16 25	23 18	21 57	12 01	06 28	22 17
16 Mo	23 17	20 03	21 27	22 57	16 37	23 17	21 56	12 01	06 28	22 17
17 Tu	23 20	16 35	21 47	22 44	16 48	23 17	21 55	12 01	06 28	22 17
18 We	23 22	12 09	22 06	22 30	17 00	23 17	21 55	12 00	06 27	22 16
19 Th	23 24	07 03	22 24	22 16	17 11	23 17	21 54	12 00	06 27	22 16
20 Fr	23 25	01 35	22 42	22 01	17 22	23 16	21 53	12 00	06 27	22 16
21 Sa	23 26	03S 57	22 58	21 45	17 33	23 16	21 52	11 59	06 26	22 16
22 Su	23 26	09 16	23 13	21 29	17 44	23 16	21 51	11 59	06 26	22 16
23 Mo	23 26	14 04	23 27	21 13	17 55	23 15	21 50	11 59	06 26	22 15
24 Tu	23 25	18 05	23 40	20 55	18 06	23 15	21 49	11 58	06 25	22 15
25 We	23 24	21 02	23 52	20 37	18 17	23 14	21 48	11 58	06 25	22 15
26 Th	23 23	22 46	24 02	20 19	18 27	23 14	21 47	11 58	06 24	22 15
27 Fr	23 21	23 13	24 11	20 00	18 37	23 13	21 46	11 57	06 24	22 14
28 Sa	23 18	22 24	24 19	19 40	18 47	23 13	21 45	11 57	06 24	22 14
29 Su	23 15	20 28	24 26	19 20	18 58	23 12	21 44	11 57	06 23	22 14
30 Mo	23 12	17 37	24 31	18 59	19 07	23 12	21 43	11 57	06 23	22 14
31 Tu	23 08	14 03	24 35	18 38	19 17	23 11	21 42	11 57	06 22	22 13

ASPECTARIAN

01 00:17 ☽ ∥ ☉
02:39 ☽ □ ♅
11:44 ☽ □ ♂
13:44 ☽ ✶ ☉
02 12:28 ☽ □ ☿
18:20 ♃ ♑
03 01:25 ☽ ∥ ☿
05:20 ☿ ✶ ♇
07:26 ☽ ✶ ♃
13:44 ☽ ✶ ♅
14:33 ☉ ∥ ♄
15:09 ☽ ∥ ♂
15:48 ♀ ✶ ♂
04 02:52 ☽ △ ♂
03:02 ☽ # ♅
03:28 ☽ ✶ ♀
15:20 ☽ ☌ ♆
20:15 ☽ ✶ ♄
05 02:42 ☽ ✶ ♇
03:57 ☉ ∥ ♇
08:16 ☽ △ ☿
10:28 ☽ ∥ ♆
21:10 ☽ □ ♃
06 22:59 ☽ □ ♀
07 01:06 ☽ △ ☉
09:07 ☽ □ ♄
15:02 ☽ □ ♇
08 05:23 ☽ # ♆
09:01 ☉ □ ♆
09:59 ☽ △ ♃
13:36 ☽ ☌ ♅
21:49 ♀ ✶ ♆
09 09:11 ☽ ☍ ♂
09:42 ☿ ♐
11:11 ☽ ∥ ♅
14:21 ☽ ✶ ♆
16:09 ☽ △ ♀
19:56 ☽ △ ♄
10 01:14 ☽ △ ♇
07:14 ☽ # ♂
20:44 ☽ ☍ ☿
11 10:05 ♀ ☌ ♄
13:56 ☽ # ☿
22:12 ☽ □ ♆
12 12:38 ☽ # ♄
16:59 ☽ # ♇
13 03:34 ☽ ☍ ♃
04:44 ☽ ✶ ♅
11:57 ♂ △ ♆
12:47 ☽ # ☉
15:15 ♀ ☌ ♇
14 02:57 ☽ # ☉
03:38 ☽ △ ♆
04:25 ☽ △ ♂
05:43 ♀ ∥ ♃
09:33 ☽ ☍ ♄
13:47 ☽ ☍ ♇
15:39 ♀ ∥ ☉
15:57 ☽ ☍ ♀
15 00:39 ☽ # ♇
05:13 ☽ # ♄
09:02 ☽ □ ♅
12:50 ☽ # ☿
19:00 ♃ △ ♅
20:18 ☽ △ ☿
16 00:54 ☉ ∥ ♃
10:47 ☽ □ ♂
22:10 ☽ △ ☉
22:48 ☽ # ♂
17 10:03 ☿ ∥ ♄
12:13 ☽ △ ♅
12:57 ☽ △ ♃
18 00:47 ☽ ∥ ♅
05:30 ☽ □ ☿
10:29 ☽ ☍ ♆
13:14 ☿ ∥ ♇
16:15 ☽ ✶ ♂
16:57 ☽ △ ♄
17:37 ☿ ∥ ♀
20:29 ☽ △ ♇
23:17 ♀ ∥ ♇
19 02:44 ☽ # ♆
08:07 ☽ △ ♀
10:01 ♂ ✶ ♄
16:34 ☽ □ ♃
20 04:20 ☿ □ ♆
06:42 ♀ ♒
13:40 ♀ ∥ ♄
14:23 ☽ ✶ ☿
20:08 ☽ □ ♄
23:24 ☽ □ ♇
21 11:01 ☽ ∥ ♆
11:46 ☽ ✶ ☉
15:52 ☽ □ ♀
17:47 ☽ ☍ ♅
20:21 ☽ ✶ ♃
22 04:10 ☿ ∥ ♃
04:20 ☉ ♑
13:09 ☽ # ♅
13:31 ♀ □ ♅
14:29 ♂ ✶ ♇
16:32 ☽ △ ♆
22:15 ☿ ∥ ☉
23:51 ☽ ✶ ♄
23 02:52 ☽ ✶ ♇
03:28 ☽ ☌ ♂
24 00:11 ☽ ∥ ♂
00:38 ☽ ✶ ♀
20:25 ☽ ∥ ♀
20:57 ☽ □ ♆
21:44 ☉ △ ♅
25 08:30 ☽ ∥ ♄
11:19 ☽ ☌ ☿
14:39 ☽ ∥ ♇
26 02:46 ☽ △ ♅
07:30 ☽ ☌ ♃
19:03 ☽ ∥ ♃
21:33 ☽ # ♃
27 03:24 ☽ ✶ ♆
12:09 ☽ ☌ ♄
14:43 ☽ ☌ ♇
18:26 ☉ ☌ ♃
21:04 ☽ ✶ ♂
28 02:35 ☽ ∥ ♇
[illegible] ∥ ♄
10:34 ☽ □ ♅
29 02:08 ☽ ☌ ♀
04:55 ☿ ♑
11:50 ☽ ∥ ♀
12:46 ☽ ∥ ♂
17:25 ♀ ∥ ♂
22:54 ☉ ∥ ♃
30 10:25 ☽ □ ♂
20:54 ☽ ✶ ☿
21:06 ☽ ✶ ♅
22:22 ☿ △ ♅
31 04:38 ☽ ✶ ♃
10:33 ☽ ✶ ☉
12:44 ☽ # ♅

⚷ Chiron

01	Dec.	03 N 23
02	01♈29R	
05	01 28	
08	01 27	
11	01 26	
14	01 26D	
17	01 26	
20	01 27	
23	01 29	
26	01 31	
29	01 33	
13	02:13	01♈26 D

10 19:11 20♋00 ✹ Penumbral Lunar Eclipse (mag 0.921)

January 2020

Day	S. T. h m s	☉ ° ′ ″	☽ ° ′ ″	☿ ° ′	♀ ° ′	♂ ° ′	♃ ° ′	♄ ° ′	♅ ° ′	♆ ° ′	♇ ° ′	☊ True ° ′
01 We	06 40 28	10♑00 33	16♓07 40	04♑23	14♒25	28♏23	06♑40	21♑24	02♉R42	16♓16	22♑23	08♋20
02 Th	06 44 25	11 01 43	28 00 34	05 58	15 38	29 03	06 54	21 31	02 41	16 17	22 25	08 21
03 Fr	06 48 21	12 02 52	09♈52 57	07 33	16 52	29 44	07 08	21 38	02 41	16 18	22 27	08 22
04 Sa	06 52 18	13 04 02	21 49 31	09 08	18 05	00♐24	07 22	21 45	02 40	16 19	22 29	08 23
05 Su	06 56 14	14 05 11	03♉55 00	10 43	19 18	01 05	07 35	21 52	02 40	16 21	22 31	08 24
06 Mo	07 00 11	15 06 20	16 13 52	12 19	20 32	01 45	07 49	21 59	02 40	16 22	22 33	08 25
07 Tu	07 04 07	16 07 29	28 49 57	13 56	21 45	02 25	08 03	22 06	02 39	16 23	22 35	08 26
08 We	07 08 04	17 08 38	11♊46 12	15 33	22 58	03 06	08 17	22 13	02 39	16 25	22 37	08 27
09 Th	07 12 01	18 09 46	25 04 15	17 10	24 11	03 46	08 31	22 20	02 39	16 26	22 39	08 28
10 Fr	07 15 57	19 10 54	08♋44 06	18 47	25 24	04 27	08 44	22 27	02 39	16 27	22 41	08R 28
11 Sa	07 19 54	20 12 02	22 43 54	20 25	26 37	05 07	08 58	22 34	02 39	16 29	22 43	08 27
12 Su	07 23 50	21 13 09	06♌59 57	22 04	27 50	05 48	09 12	22 42	02D 39	16 30	22 45	08 26
13 Mo	07 27 47	22 14 16	21 27 06	23 43	29 03	06 29	09 25	22 49	02 39	16 32	22 47	08 24
14 Tu	07 31 43	23 15 23	05♍59 30	25 22	00♓16	07 09	09 39	22 56	02 39	16 33	22 49	08 21
15 We	07 35 40	24 16 30	20 31 13	27 02	01 29	07 50	09 53	23 03	02 39	16 35	22 51	08 18
16 Th	07 39 36	25 17 36	04♎57 08	28 42	02 42	08 31	10 06	23 10	02 40	16 36	22 53	08 16
17 Fr	07 43 33	26 18 43	19 13 18	00♒23	03 54	09 11	10 20	23 17	02 40	16 38	22 55	08 14
18 Sa	07 47 30	27 19 49	03♏17 09	02 04	05 07	09 52	10 34	23 24	02 40	16 40	22 57	08 13
19 Su	07 51 26	28 20 55	17 07 25	03 46	06 19	10 33	10 47	23 31	02 41	16 41	22 59	08D 13
20 Mo	07 55 23	29 22 01	00♐43 47	05 28	07 32	11 13	11 01	23 38	02 41	16 43	23 01	08 13
21 Tu	07 59 19	00♒23 07	14 06 36	07 10	08 44	11 54	11 14	23 45	02 42	16 45	23 03	08 14
22 We	08 03 16	01 24 12	27 16 24	08 53	09 57	12 35	11 28	23 53	02 42	16 46	23 05	08 15
23 Th	08 07 12	02 25 16	10♑13 44	10 36	11 09	13 16	11 41	24 00	02 43	16 48	23 07	08R 16
24 Fr	08 11 09	03 26 21	22 59 03	12 20	12 21	13 56	11 54	24 07	02 43	16 50	23 09	08 15
25 Sa	08 15 06	04 27 24	05♒32 44	14 03	13 33	14 37	12 08	24 14	02 44	16 52	23 11	08 13
26 Su	08 19 02	05 28 26	17 55 22	15 46	14 45	15 18	12 21	24 21	02 45	16 53	23 13	08 10
27 Mo	08 22 59	06 29 28	00♓07 47	17 30	15 57	15 59	12 34	24 28	02 46	16 55	23 15	08 06
28 Tu	08 26 55	07 30 29	12 11 19	19 13	17 09	16 40	12 48	24 35	02 46	16 57	23 17	08 02
29 We	08 30 52	08 31 28	24 07 54	20 56	18 20	17 21	13 01	24 42	02 47	16 59	23 19	07 57
30 Th	08 34 48	09 32 27	06♈00 11	22 38	19 32	18 02	13 14	24 49	02 48	17 01	23 21	07 53
31 Fr	08 38 45	10 33 24	17 51 30	24 19	20 44	18 43	13 27	24 56	02 49	17 03	23 23	07 50

Data for	01-01-2020
Julian Day	2458849.50
Ayanamsa	24 07 53
SVP	04♓58 33
☽ ☊ Mean	08♋15 R

● ◐ PHASES ○ ◑

Day	Time	Phase	Position
03	04:46	◐	12♈15
10	19:21	✹	20♋00
17	12:59	◑	26♎52
24	21:43	●	04♒22

LAST ASPECT Day	h m	☽ INGRESS Day	h m	Sign
02	02:15	02	04:02	♈
04	01:19	04	16:16	♉
06	12:08	07	02:12	♊
08	22:16	09	08:44	♋
10	23:58	11	12:16	♌
13	13:42	13	14:07	♍
15	12:13	15	15:44	♎
17	13:00	17	18:22	♏
19	21:23	19	22:42	♐
21	04:47	22	05:01	♑
24	02:10	24	13:22	♒
25	19:08	26	23:45	♓
29	01:09	29	11:51	♈

DECLINATION

Day	☉	☽	☿	♀	♂	♃	♄	♅	♆	♇
01 We	23S04	09S58	24S38	18S16	19S27	23S11	21S41	11N57	06S22	22S13
02 Th	22 59	05 33	24 40	17 54	19 36	23 10	21 40	11 56	06 21	22 13
03 Fr	22 53	00 56	24 40	17 31	19 45	23 10	21 39	11 56	06 21	22 13
04 Sa	22 48	03N44	24 38	17 08	19 55	23 09	21 38	11 56	06 20	22 13
05 Su	22 41	08 20	24 35	16 44	20 03	23 08	21 37	11 56	06 20	22 12
06 Mo	22 35	12 41	24 31	16 20	20 12	23 07	21 36	11 56	06 19	22 12
07 Tu	22 28	16 36	24 26	15 55	20 21	23 07	21 35	11 56	06 19	22 12
08 We	22 20	19 51	24 18	15 30	20 29	23 06	21 34	11 56	06 18	22 12
09 Th	22 12	22 07	24 10	15 05	20 38	23 05	21 33	11 56	06 18	22 11
10 Fr	22 04	23 11	24 00	14 39	20 46	23 04	21 32	11 56	06 17	22 11
11 Sa	21 55	22 49	23 48	14 13	20 54	23 04	21 30	11 56	06 17	22 11
12 Su	21 46	20 58	23 35	13 47	21 02	23 03	21 29	11 56	06 16	22 11
13 Mo	21 36	17 45	23 21	13 20	21 09	23 02	21 28	11 56	06 15	22 10
14 Tu	21 26	13 26	23 04	12 53	21 17	23 01	21 27	11 56	06 15	22 10
15 We	21 15	08 21	22 47	12 25	21 24	23 00	21 26	11 56	06 14	22 10
16 Th	21 05	02 51	22 28	11 57	21 31	22 59	21 25	11 56	06 14	22 10
17 Fr	20 53	02S45	22 07	11 29	21 38	22 58	21 24	11 56	06 13	22 09
18 Sa	20 41	08 07	21 45	11 01	21 45	22 57	21 23	11 57	06 12	22 09
19 Su	20 29	13 01	21 21	10 32	21 52	22 56	21 21	11 57	06 12	22 09
20 Mo	20 17	17 10	20 55	10 03	21 58	22 55	21 20	11 57	06 11	22 09
21 Tu	20 04	20 21	20 28	09 34	22 04	22 54	21 19	11 57	06 10	22 08
22 We	19 51	22 24	20 00	09 05	22 10	22 53	21 18	11 57	06 10	22 08
23 Th	19 37	23 13	19 30	08 35	22 16	22 52	21 17	11 58	06 09	22 08
24 Fr	19 23	22 48	18 59	08 05	22 22	22 51	21 16	11 58	06 08	22 08
25 Sa	19 09	21 12	18 26	07 35	22 27	22 50	21 15	11 58	06 08	22 07
26 Su	18 54	18 38	17 51	07 05	22 32	22 49	21 13	11 58	06 07	22 07
27 Mo	18 39	15 16	17 16	06 35	22 38	22 48	21 12	11 59	06 06	22 07
28 Tu	18 23	11 19	16 39	06 04	22 42	22 46	21 11	11 59	06 05	22 07
29 We	18 08	06 58	16 00	05 33	22 47	22 45	21 10	11 59	06 05	22 06
30 Th	17 52	02 23	15 21	05 03	22 52	22 44	21 09	12 00	06 04	22 06
31 Fr	17 35	02N16	14 41	04 32	22 56	22 43	21 08	12 00	06 03	22 06

ASPECTARIAN

Day	Time	Aspect
01	00:17	☽ ☌ ♆
	10:44	☽ ⚹ ♄
	12:39	☽ ⚹ ♇
	19:46	☽ ∥ ♆
02	02:15	☽ △ ♂
	16:42	☿ ☌ ♃
	18:20	☽ □ ♃
	18:33	☽ □ ☿
03	09:38	♂ ♐
	15:39	☽ ⚹ ♀
	23:50	☽ □ ♄
04	01:19	☽ □ ♇
	13:30	☽ # ♆
	21:32	☽ ☌ ♅
05	07:21	☽ △ ♃
	15:19	☽ △ ☿
	19:43	☽ ∥ ♅
	21:38	☽ △ ☉
06	00:16	☽ ⚹ ♆
	09:08	☽ □ ♀
	11:08	☽ △ ♄
	12:08	☽ △ ♇
	20:01	☽ # ♀
07	06:22	☉ ⚹ ♆
	07:06	☽ ☍ ♂
08	06:08	☽ # ♂
	08:28	☽ □ ♆
	13:03	☿ ⚹ ♆
	16:45	☽ # ♄
	22:16	☽ △ ♀
09	00:58	☽ # ♇
	01:06	☽ # ☉
	02:50	☉ ∥ ♇
	13:23	☽ ⚹ ♅
10	00:01	☽ ☍ ♃
	13:20	☽ △ ♆
	15:20	☿ ☌ ☉
	17:09	☽ # ♃
	19:34	☽ ☍ ☿
	23:44	☽ ☍ ♄
	23:58	☽ ☍ ♇
11	01:49	♅ SD
	10:31	☽ # ♇
	15:02	☽ # ☉
	16:43	☽ □ ♅
	18:43	☽ # ♄
	21:54	☽ △ ♂
	23:25	☽ # ♂
12	09:52	☿ ☌ ♄
	10:12	☿ ☌ ♇
	16:23	♄ ☌ ♇
13	13:19	☉ ☌ ♇
	13:42	☽ ☍ ♀
	15:16	☉ ☌ ♄
	18:30	☽ △ ♅
	18:40	♀ ♓
	20:58	☉ ∥ ♄
14	02:01	☽ □ ♂
	03:08	☽ # ♀
	05:10	☿ ∥ ♃
	06:08	☽ △ ♃
	07:26	☽ ∥ ♅
	12:28	☉ ∥ ♂
	17:28	☽ ☍ ♆
15	03:52	☽ △ ♇
	04:13	☽ △ ♄
	05:39	♂ ∥ ♄
	06:42	☽ △ ☉
	09:24	☽ # ♆
	12:13	☽ △ ☿
	23:18	♀ ⚹ ♅
16	00:59	♀ # ♅
	06:15	☽ ⚹ ♂
	08:46	☽ □ ♃
	18:31	☿ ♒
	21:11	☿ ∥ ♇
17	06:17	☽ □ ♇
	06:57	☽ □ ♄
	15:17	☽ ∥ ♆
	21:37	☽ □ ☿
	22:56	☽ ☍ ♅
	23:45	☿ ∥ ♂
18	03:27	☽ △ ♀
	08:32	☿ □ ♅
	12:36	☽ ∥ ♀
	12:46	☽ ⚹ ♃
	18:29	☽ # ♅
	23:13	☿ ∥ ♄
	23:14	☽ △ ♆
19	10:19	☽ ⚹ ♇
	11:20	☽ ⚹ ♄
	21:23	☽ ⚹ ☉
20	09:41	☽ ⚹ ☿
	13:21	☽ □ ♀
	14:55	☉ ♒
	19:48	☽ ☌ ♂
	21:43	☽ ∥ ☉
21	01:01	☽ ∥ ☿
	04:47	☽ □ ♆
	09:39	☽ ∥ ♄
	16:07	♂ ∥ ♇
	19:58	☽ ∥ ♇
	20:14	☽ ∥ ♂
22	09:45	☽ ∥ ♃
	10:01	☽ △ ♅
	13:51	☿ ∥ ☉
23	01:54	☽ ⚹ ♀
	02:46	☽ ☌ ♃
	06:55	☉ □ ♅
	12:21	☽ ⚹ ♆
	13:08	♀ ⚹ ♃
	22:22	☽ ∥ ♃
24	00:19	☽ ☌ ♇
	02:10	☽ ☌ ♄
	07:46	☽ ∥ ♂
	12:03	☽ ∥ ♇
	18:35	☽ □ ♅
	23:32	☽ ∥ ♄
25	13:10	☿ ⚹ ♂
	18:36	☽ ⚹ ♂
	19:08	☽ ☌ ☿
	21:36	☽ ∥ ☉
26	07:19	☽ ∥ ☿
27	01:38	♀ □ ♂
	05:13	☽ ⚹ ♅
	20:01	♀ ☌ ♆
	20:07	☽ # ♅
	23:02	♀ ∥ ♆
28	01:14	☽ ⚹ ♃
	09:31	☽ □ ♂
	09:34	☽ ☌ ♆
	10:35	♂ □ ♆
	11:03	☽ ☌ ♀
	16:10	♂ ∥ ♃
	22:21	☽ ⚹ ♇
29	01:09	☽ ⚹ ♄
	04:44	☽ ∥ ♆
	08:25	☽ ∥ ♀
30	07:50	☽ ⚹ ☉
	14:55	☽ □ ♃
31	01:50	☽ △ ♂
	10:32	☽ # ♀
	11:11	☽ □ ♇
	14:25	☽ □ ♄
	15:10	☽ ⚹ ☿
	19:38	☽ # ♆

⚷ Chiron

01 Dec. 03 N 19

Day	° ′
01	01♈36
04	01 39
07	01 43
10	01 47
13	01 52
16	01 57
19	02 03
22	02 09
25	02 15
28	02 22
31	02 29

February 2020

Day	S. T.	☉	☽	☿	♀	♂	♃	♄	♅	♆	♇	☊ True
	h m s	° ' "	° ' "	° '	° '	° '	° '	° '	° '	° '	° '	° '
01 Sa	08 42 41	11♒ 34 20	29♈ 45 55	25♒ 59	21♓ 55	19♐ 24	13♑ 40	25♑ 03	02♉ 50	17♓ 05	23♑ 25	07♋R49
02 Su	08 46 38	12 35 15	11♉ 47 59	27 37	23 06	20 05	13 53	25 10	02 51	17 07	23 27	07D 49
03 Mo	08 50 34	13 36 09	24 02 31	29 14	24 17	20 46	14 06	25 16	02 53	17 09	23 29	07 50
04 Tu	08 54 31	14 37 01	06♊ 34 20	00♓ 48	25 29	21 27	14 19	25 23	02 54	17 11	23 30	07 52
05 We	08 58 28	15 37 52	19 27 50	02 20	26 39	22 08	14 32	25 30	02 55	17 13	23 32	07 53
06 Th	09 02 24	16 38 41	02♋ 46 31	03 47	27 50	22 49	14 45	25 37	02 56	17 15	23 34	07R 54
07 Fr	09 06 21	17 39 29	16 32 16	05 11	29 01	23 30	14 58	25 44	02 58	17 17	23 36	07 53
08 Sa	09 10 17	18 40 16	00♌ 44 35	06 31	00♈ 12	24 11	15 10	25 50	02 59	17 19	23 38	07 50
09 Su	09 14 14	19 41 01	15 20 08	07 44	01 22	24 52	15 23	25 57	03 01	17 21	23 40	07 45
10 Mo	09 18 10	20 41 46	00♍ 12 35	08 52	02 32	25 33	15 35	26 04	03 02	17 23	23 42	07 39
11 Tu	09 22 07	21 42 28	15 13 27	09 53	03 43	26 14	15 48	26 11	03 04	17 25	23 44	07 32
12 We	09 26 03	22 43 10	00♎ 13 19	10 46	04 53	26 55	16 00	26 17	03 05	17 27	23 45	07 24
13 Th	09 30 00	23 43 50	15 03 22	11 30	06 02	27 36	16 13	26 24	03 07	17 29	23 47	07 17
14 Fr	09 33 57	24 44 29	29 36 43	12 06	07 12	28 18	16 25	26 30	03 09	17 31	23 49	07 12
15 Sa	09 37 53	25 45 08	13♏ 49 10	12 32	08 22	28 59	16 37	26 37	03 10	17 33	23 51	07 09
16 Su	09 41 50	26 45 45	27 39 12	12 48	09 31	29 40	16 50	26 43	03 12	17 36	23 52	07 07
17 Mo	09 45 46	27 46 21	11♐ 07 32	12 53	10 41	00♑ 21	17 02	26 50	03 14	17 38	23 54	07D 07
18 Tu	09 49 43	28 46 56	24 16 16	12R 49	11 50	01 03	17 14	26 56	03 16	17 40	23 56	07 08
19 We	09 53 39	29 47 29	07♑ 08 13	12 33	12 59	01 44	17 26	27 03	03 18	17 42	23 58	07R 08
20 Th	09 57 36	00♓ 48 01	19 46 12	12 08	14 08	02 25	17 38	27 09	03 20	17 44	23 59	07 07
21 Fr	10 01 32	01 48 32	02♒ 12 48	11 34	15 16	03 07	17 50	27 15	03 22	17 46	24 01	07 04
22 Sa	10 05 29	02 49 02	14 30 04	10 51	16 25	03 48	18 01	27 22	03 24	17 49	24 03	06 59
23 Su	10 09 26	03 49 29	26 39 39	10 01	17 33	04 29	18 13	27 28	03 26	17 51	24 04	06 51
24 Mo	10 13 22	04 49 56	08♓ 42 50	09 04	18 41	05 11	18 25	27 34	03 28	17 53	24 06	06 42
25 Tu	10 17 19	05 50 20	20 40 45	08 04	19 49	05 52	18 36	27 40	03 30	17 55	24 08	06 31
26 We	10 21 15	06 50 43	02♈ 34 38	07 00	20 57	06 33	18 48	27 46	03 32	17 58	24 09	06 20
27 Th	10 25 12	07 51 04	14 26 04	05 55	22 05	07 15	18 59	27 52	03 34	18 00	24 11	06 10
28 Fr	10 29 08	08 51 23	26 17 15	04 50	23 12	07 56	19 10	27 58	03 37	18 02	24 12	06 01
29 Sa	10 33 05	09 51 40	08♉ 11 09	03 48	24 19	08 38	19 22	28 04	03 39	18 04	24 14	05 55

Data for	02-01-2020
Julian Day	2458880.50
Ayanamsa	24 07 58
SVP	04 ♓ 58 29
☽ ☊ Mean	06 ♋ 36 R

● ◐ PHASES ○ ◑

Day	Time	Phase	Position
02	01:41	◐	12♉40
09	07:34	○	20♌00
15	22:18	◑	26♏41
23	15:32	●	04♓29

LAST ASPECT Day	h m	☽ INGRESS Day	h m	
31	15:10	01	00:28	♉
03	11:28	03	11:29	♊
05	14:20	05	19:04	♋
07	15:44	07	22:46	♌
09	16:09	09	23:40	♍
11	18:27	11	23:39	♎
13	21:42	14	00:39	♏
15	22:21	16	04:08	♐
18	09:04	18	10:38	♑
20	14:19	20	19:42	♒
22	04:09	23	06:38	♓
25	14:12	25	18:48	♈
28	03:26	28	07:30	♉

DECLINATION

Day	☉	☽	☿	♀	♂	♃	♄	♅	♆	♇
01 Sa	17S 19	06N52	13S 59	04S 01	23S 00	22S 42	21S 06	12N00	06S 02	22S 06
02 Su	17 02	11 16	13 17	03 30	23 04	22 40	21 05	12 01	06 02	22 05
03 Mo	16 44	15 18	12 35	02 58	23 08	22 39	21 04	12 01	06 01	22 05
04 Tu	16 27	18 45	11 52	02 27	23 11	22 38	21 03	12 02	06 00	22 05
05 We	16 09	21 23	11 09	01 56	23 15	22 37	21 02	12 02	05 59	22 05
06 Th	15 51	22 57	10 26	01 24	23 18	22 35	21 00	12 03	05 58	22 05
07 Fr	15 32	23 12	09 43	00 53	23 21	22 34	20 59	12 03	05 58	22 04
08 Sa	15 14	21 58	09 02	00 22	23 23	22 33	20 58	12 04	05 57	22 04
09 Su	14 55	19 15	08 21	00N10	23 26	22 31	20 57	12 04	05 56	22 04
10 Mo	14 36	15 14	07 42	00 41	23 28	22 30	20 56	12 05	05 55	22 04
11 Tu	14 16	10 13	07 05	01 13	23 30	22 29	20 55	12 05	05 54	22 03
12 We	13 56	04 35	06 31	01 44	23 32	22 27	20 53	12 06	05 54	22 03
13 Th	13 37	01S 14	05 59	02 16	23 34	22 26	20 52	12 06	05 53	22 03
14 Fr	13 16	06 53	05 30	02 47	23 36	22 24	20 51	12 07	05 52	22 03
15 Sa	12 56	12 02	05 06	03 18	23 37	22 23	20 50	12 08	05 51	22 03
16 Su	12 36	16 26	04 45	03 49	23 38	22 22	20 49	12 08	05 50	22 02
17 Mo	12 15	19 52	04 29	04 20	23 39	22 20	20 48	12 09	05 49	22 02
18 Tu	11 54	22 10	04 17	04 51	23 39	22 19	20 46	12 10	05 49	22 02
19 We	11 33	23 14	04 10	05 22	23 40	22 17	20 45	12 10	05 48	22 02
20 Th	11 11	23 05	04 08	05 53	23 40	22 16	20 44	12 11	05 47	22 02
21 Fr	10 50	21 47	04 11	06 24	23 40	22 14	20 43	12 12	05 46	22 01
22 Sa	10 28	19 27	04 18	06 54	23 40	22 13	20 42	12 12	05 45	22 01
23 Su	10 06	16 16	04 30	07 25	23 40	22 12	20 41	12 13	05 44	22 01
24 Mo	09 44	12 27	04 46	07 55	23 39	22 10	20 39	12 14	05 43	22 01
25 Tu	09 22	08 10	05 06	08 25	23 38	22 09	20 38	12 15	05 42	22 01
26 We	09 00	03 36	05 29	08 55	23 37	22 07	20 37	12 15	05 42	22 01
27 Th	08 37	01N04	05 54	09 24	23 36	22 06	20 36	12 16	05 41	22 00
28 Fr	08 15	05 43	06 21	09 54	23 35	22 04	20 35	12 17	05 40	22 00
29 Sa	07 52	10 10	06 49	10 23	23 33	22 03	20 34	12 18	05 39	22 00

ASPECTARIAN

Day	Time	Aspect
01	06:10	☽ ☌ ♅
02	04:12	☽ △ ♃
	04:18	☽ ∥ ♅
	07:06	♀ ⚹ ♇
	10:00	☽ ⋕ ☿
	10:30	☽ ⚹ ♆
	22:54	☽ △ ♇
03	00:32	☽ ⚹ ♀
	02:24	☽ △ ♄
	08:49	☽ ⋕ ☉
	11:28	☽ □ ☿
	11:38	☿ ♓
	18:29	☿ ⋕ ♅
	22:02	♀ ⚹ ♄
04	16:21	☽ △ ☉
	19:51	☽ □ ♆
	20:14	☽ ⋕ ♄
05	05:08	☽ ☍ ♂
	08:35	☽ ⋕ ♇
	09:43	☿ ⚹ ♅
	14:20	☽ □ ♀
	16:38	☽ ⋕ ♃
06	00:18	☽ ⚹ ♅
	02:01	☽ △ ☿
	21:15	☽ ☍ ♃
07	01:16	☽ △ ♆
	12:03	☽ ☍ ♇
	15:42	☽ ⋕ ♃
	15:44	☽ ☍ ♄
	20:03	♀ ♈
	22:35	☽ ⋕ ♇
	23:00	☽ △ ♀
08	03:44	☽ □ ♅
	10:20	☽ ⋕ ♄
09	16:09	☽ △ ♂
10	03:28	☽ ⋕ ☉
	04:32	☽ △ ♅
	14:52	☽ ☍ ☿
	15:26	☽ ∥ ♅
11	00:56	☽ △ ♃
	03:30	☽ ☍ ♆
	13:37	☽ △ ♇
	15:04	☽ ⋕ ☿
	17:38	☽ △ ♄
	18:27	☽ □ ♂
	18:32	☽ ⋕ ♆
12	08:08	☽ ☍ ♀
	10:47	☽ ∥ ♀
13	01:55	☽ □ ♃
	04:41	☽ ⋕ ♀
	05:06	☿ ∥ ♆
	14:22	☽ □ ♇
	15:18	☽ △ ☉
	18:27	☽ ∥ ☿
	18:48	☽ □ ♄
	19:34	☽ ∥ ♆
	21:42	☽ ⚹ ♂
14	05:55	☽ ☍ ♅
	21:44	☽ △ ☿
15	00:26	☽ ⋕ ♅
	04:13	☽ ∥ [illegible]
	04:53	☽ ⚹ ♃
	06:26	☽ △ ♆
	17:22	☽ ⚹ ♇
	22:21	☽ ⚹ ♄
16	11:33	♂ ♑
	23:07	☽ △ ♀
17	00:56	☿ SR
	03:11	☽ □ ☿
	04:22	☿ ⋕ ♀
	06:30	☉ ⋕ ♅
	08:14	☽ ∥ ♄
	11:50	☽ □ ♆
	22:16	☽ ∥ ♇
18	02:14	☽ ∥ ♃
	09:04	☽ ⚹ ☉
	13:17	☽ ☌ ♂
	16:46	☽ △ ♅
19	04:58	☉ ♓
	09:57	☽ ⚹ ☿
	12:09	☽ □ ♀
	19:14	♀ ⋕ ♆
	19:51	☽ ☌ ♃
	20:06	☽ ⚹ ♆
20	08:07	☽ ☌ ♇
	14:19	☽ ☌ ♄
	16:00	♃ ⚹ ♆
	17:17	☽ ∥ ♃
	20:34	☽ ∥ ♇
21	00:11	[illegible]
	09:10	♂ △ ♅
	12:10	☽ ∥ ♄
22	04:09	☽ ⚹ ♀
	14:13	☉ ⚹ ♅
23	13:29	☽ ⚹ ♅
	16:30	☽ ⚹ ♂
	17:00	♀ □ ♃
24	00:40	☽ ☌ ☿
	01:15	☽ ⋕ ♅
	16:50	☽ ∥ ☉
	18:26	☽ ☌ ♆
	19:18	[illegible]
	22:47	☽ ⋕ ♀
25	02:06	☉ ⚹ ♂
	06:57	☽ ⚹ ♇
	13:05	☽ ∥ ♆
	14:12	☽ ⚹ ♄
	15:00	☽ ∥ ☿
26	01:45	☿ ☌ ☉
	02:20	♀ ⋕ ☉
	05:59	☿ ⚹ ♂
	08:33	☽ □ ♂
	11:54	☿ ∥ ♆
27	09:22	☽ □ ♃
	17:06	☽ ☌ ♆
	19:47	☽ □ ♇
	23:43	☽ ⋕ ♆
28	03:26	☽ □ ♄
	03:42	☽ ⋕ ☿
	12:24	☽ ⋕ ☉
	14:50	☽ ☌ ♅
	15:52	☽ ⚹ ☿
	22:07	♀ □ ♇
29	00:56	☽ △ ♂
	01:20	☽ ∥ ♀
	03:14	☿ ⚹ ♅
	03:40	☽ ⚹ ☉
	12:06	☽ ∥ ♅
	19:51	☽ ⚹ ♆
	22:42	☽ △ ♃

⚷ Chiron

01	Dec.	03 N 35
03	02♈36	
06	02 44	
09	02 52	
12	03 01	
15	03 09	
18	03 18	
21	03 28	
24	03 37	
27	03 47	

March 2020

Day	S. T. h m s	☉ ° ' "	☽ ° ' "	☿ ° '	♀ ° '	♂ ° '	♃ ° '	♄ ° '	♅ ° '	♆ ° '	♇ ° '	☊ True ° '
01 Su	10 37 01	10♓ 51 55	20♉ 11 25	02♓R48	25♈ 26	09♑ 19	19♑ 33	28♑ 10	03♉ 41	18♓ 07	24♑ 15	05♋R52
02 Mo	10 40 58	11 52 08	02♊ 22 23	01 53	26 33	10 01	19 44	28 16	03 44	18 09	24 17	05 50
03 Tu	10 44 55	12 52 19	14 48 48	01 03	27 39	10 42	19 55	28 21	03 46	18 11	24 18	05D 50
04 We	10 48 51	13 52 29	27 35 35	00 18	28 45	11 23	20 05	28 27	03 49	18 13	24 20	05 51
05 Th	10 52 48	14 52 36	10♋ 47 19	29♒ 41	29 51	12 05	20 16	28 33	03 51	18 16	24 21	05R 50
06 Fr	10 56 44	15 52 40	24 27 33	29 09	00♉ 57	12 46	20 27	28 38	03 54	18 18	24 23	05 47
07 Sa	11 00 41	16 52 43	08♌ 37 47	28 45	02 03	13 28	20 37	28 44	03 56	18 20	24 24	05 42
08 Su	11 04 37	17 52 44	23 16 24	28 28	03 08	14 09	20 48	28 49	03 59	18 23	24 25	05 35
09 Mo	11 08 34	18 52 42	08♍ 18 07	28 17	04 13	14 51	20 58	28 55	04 01	18 25	24 27	05 25
10 Tu	11 12 30	19 52 39	23 34 08	28 13	05 17	15 33	21 08	29 00	04 04	18 27	24 28	05 14
11 We	11 16 27	20 52 33	08♎ 53 21	28D 15	06 22	16 14	21 18	29 05	04 07	18 29	24 29	05 03
12 Th	11 20 24	21 52 26	24 04 21	28 23	07 26	16 56	21 28	29 11	04 10	18 32	24 30	04 53
13 Fr	11 24 20	22 52 17	08♏ 57 16	28 37	08 30	17 37	21 38	29 16	04 12	18 34	24 32	04 44
14 Sa	11 28 17	23 52 07	23 25 26	28 56	09 33	18 19	21 48	29 21	04 15	18 36	24 33	04 38
15 Su	11 32 13	24 51 54	07♐ 25 43	29 20	10 36	19 01	21 58	29 26	04 18	18 38	24 34	04 34
16 Mo	11 36 10	25 51 41	20 58 19	29 50	11 39	19 42	22 07	29 31	04 21	18 41	24 35	04 33
17 Tu	11 40 06	26 51 25	04♑ 05 43	00♓ 23	12 42	20 24	22 17	29 36	04 24	18 43	24 36	04 32
18 We	11 44 03	27 51 08	16 51 44	01 01	13 44	21 05	22 26	29 41	04 27	18 45	24 38	04 32
19 Th	11 47 59	28 50 49	29 20 33	01 43	14 45	21 47	22 35	29 45	04 30	18 47	24 39	04 30
20 Fr	11 51 56	29 50 28	11♒ 36 17	02 29	15 47	22 29	22 45	29 50	04 32	18 50	24 40	04 27
21 Sa	11 55 53	00♈ 50 05	23 42 30	03 18	16 48	23 10	22 54	29 55	04 35	18 52	24 41	04 20
22 Su	11 59 49	01 49 41	05♓ 42 12	04 10	17 49	23 52	23 02	29 59	04 38	18 54	24 42	04 10
23 Mo	12 03 46	02 49 14	17 37 43	05 05	18 49	24 34	23 11	00♒ 04	04 42	18 56	24 43	03 58
24 Tu	12 07 42	03 48 46	29 30 47	06 03	19 49	25 16	23 20	00 08	04 45	18 59	24 44	03 45
25 We	12 11 39	04 48 15	11♈ 22 47	07 04	20 48	25 57	23 28	00 12	04 48	19 01	24 45	03 30
26 Th	12 15 35	05 47 43	23 14 57	08 08	21 47	26 39	23 37	00 17	04 51	19 03	24 46	03 17
27 Fr	12 19 32	06 47 08	05♉ 08 46	09 14	22 46	27 21	23 45	00 21	04 54	19 05	24 46	03 06
28 Sa	12 23 28	07 46 31	17 06 10	10 22	23 44	28 02	23 53	00 25	04 57	19 07	24 47	02 57
29 Su	12 27 25	08 45 52	29 09 44	11 33	24 41	28 44	24 01	00 29	05 00	19 10	24 48	02 52
30 Mo	12 31 21	09 45 11	11♊ 22 41	12 46	25 38	29 26	24 09	00 33	05 03	19 12	24 49	02 49
31 Tu	12 35 18	10 44 28	23 48 53	14 00	26 35	00♒ 07	24 16	00 37	05 07	19 14	24 50	02 48

Data for	03-01-2020
Julian Day	2458909.50
Ayanamsa	24 08 02
SVP	04 ♓ 58 25
☽ ☊ Mean	05 ♋ 04 R

● ◐ PHASES ○ ◑

Day	Time	Phase	Position
02	19:58	◐	12♊42
09	17:48	○	19♍37
16	09:34	◑	26♐16
24	09:29	●	04♈12

LAST ASPECT ☽ INGRESS

Day	h m	Day	h m	Sign
01	15:53	01	19:22	♊
04	02:21	04	04:26	♋
06	07:13	06	09:29	♌
08	08:14	08	10:48	♍
10	08:33	10	10:04	♎
12	08:13	12	09:29	♏
14	10:07	14	11:10	♐
16	09:35	16	16:26	♑
19	00:49	19	01:17	♒
20	09:01	21	12:34	♓
23	14:52	24	00:59	♈
26	07:18	26	13:38	♉
28	23:06	29	01:39	♊
30	15:11	31	11:44	♋

DECLINATION

Day	☉	☽	☿	♀	♂	♃	♄	♅	♆	♇
01 Su	07S29	14N17	07S17	10N52	23S31	22S01	20S33	12N19	05S38	22S00
02 Mo	07 07	17 53	07 45	11 21	23 29	22 00	20 32	12 19	05 37	22 00
03 Tu	06 44	20 46	08 13	11 50	23 27	21 58	20 31	12 20	05 36	22 00
04 We	06 21	22 41	08 39	12 18	23 24	21 57	20 30	12 21	05 35	21 59
05 Th	05 57	23 26	09 04	12 46	23 22	21 55	20 28	12 22	05 35	21 59
06 Fr	05 34	22 49	09 27	13 14	23 19	21 54	20 27	12 23	05 34	21 59
07 Sa	05 11	20 45	09 49	13 41	23 16	21 52	20 26	12 24	05 33	21 59
08 Su	04 47	17 17	10 08	14 08	23 13	21 51	20 25	12 25	05 32	21 59
09 Mo	04 24	12 37	10 25	14 35	23 09	21 49	20 24	12 26	05 31	21 59
10 Tu	04 01	07 04	10 40	15 01	23 05	21 48	20 23	12 26	05 30	21 59
11 We	03 37	01 05	10 52	15 28	23 01	21 46	20 22	12 27	05 29	21 58
12 Th	03 13	04S55	11 03	15 53	22 57	21 45	20 21	12 28	05 28	21 58
13 Fr	02 50	10 32	11 11	16 19	22 53	21 43	20 20	12 29	05 27	21 58
14 Sa	02 26	15 24	11 17	16 44	22 49	21 42	20 19	12 30	05 27	21 58
15 Su	02 02	19 16	11 21	17 09	22 44	21 41	20 18	12 31	05 26	21 58
16 Mo	01 39	21 56	11 22	17 33	22 39	21 39	20 17	12 32	05 25	21 58
17 Tu	01 15	23 19	11 22	17 57	22 34	21 38	20 16	12 33	05 24	21 58
18 We	00 51	23 26	11 20	18 21	22 29	21 36	20 15	12 34	05 23	21 58
19 Th	00 28	22 21	11 15	18 44	22 23	21 35	20 15	12 35	05 22	21 58
20 Fr	00 04	20 13	11 09	19 07	22 18	21 34	20 14	12 36	05 21	21 58
21 Sa	00N20	17 12	11 01	19 29	22 12	21 32	20 13	12 37	05 21	21 57
22 Su	00 44	13 30	10 52	19 51	22 06	21 31	20 12	12 38	05 20	21 57
23 Mo	01 07	09 17	10 40	20 13	21 59	21 29	20 11	12 39	05 19	21 57
24 Tu	01 31	04 45	10 27	20 34	21 53	21 28	20 10	12 40	05 18	21 57
25 We	01 55	00 03	10 12	20 54	21 46	21 27	20 09	12 41	05 17	21 57
26 Th	02 18	04N40	09 56	21 15	21 40	21 25	20 08	12 42	05 16	21 57
27 Fr	02 42	09 14	09 38	21 34	21 33	21 24	20 08	12 44	05 15	21 57
28 Sa	03 05	13 29	09 19	21 54	21 26	21 23	20 07	12 45	05 15	21 57
29 Su	03 28	17 14	08 58	22 12	21 18	21 22	20 06	12 46	05 14	21 57
30 Mo	03 52	20 19	08 35	22 31	21 11	21 20	20 05	12 47	05 13	21 57
31 Tu	04 15	22 30	08 12	22 49	21 03	21 19	20 05	12 48	05 12	21 57

ASPECTARIAN

01 05:49 ☿ ∥ ☉
08:04 ☽ △ ♇
15:53 ☽ △ ♄
23:06 ☽ □ ☿
23:53 ♃ ∥ ♇
02 21:34 ☽ # ♄
03 06:25 ☽ □ ♆
13:12 ☽ # ♃
13:39 ☽ # ♇
16:45 ♀ □ ♄
04 02:21 ☽ ✱ ♀
02:56 ♀ ∥ ♅
04:45 ☽ △ ☿
11:08 ☿ ♒R
11:27 ☽ ✱ ♅
17:58 ☽ # ♂
21:25 ☿ ✱ ♀
05 02:26 ☽ ☍ ♂
03:08 ♀ ♉
07:51 ☽ △ ☉
10:42 ☽ ∥ ♂
13:16 ☽ △ ♆
16:57 ☽ ☍ ♃
23:51 ☽ ☍ ♇
06 00:30 ☉ ∥ ♆
07:13 ☽ ☍ ♄
11:50 ☽ # ♇
12:02 ☽ □ ♀
13:00 ☽ # ♃
16:07 ☽ □ ♅
07 02:37 ☽ # ♄
08 08:14 ☽ ☍ ☿
12:24 ☉ ☌ ♆
15:16 ☽ ∥ ♀
17:01 ☽ △ ♀
17:12 ☽ △ ♅
19:38 ♀ ☌ ♅
09 00:48 ☽ ∥ ♅
09:24 ☽ # ☿
10:49 ☽ △ ♂
15:58 ☽ ☍ ♆
20:09 ☽ △ ♃
10 01:24 ☽ △ ♇
03:50 ☿ SD
06:23 ☽ # ♆
08:33 ☽ △ ♄
13:14 ☽ # ☉
11 12:07 ☽ □ ♂
12:28 ☉ ✱ ♃
17:32 ☽ ∥ ☉
19:49 ☽ □ ♃
12 00:42 ☽ □ ♇
02:15 ☽ ∥ ♆
06:59 ☽ △ ☿
08:13 ☽ □ ♄
16:15 ☽ ☍ ♅
23:11 ☽ ☍ ♀
13 03:01 ☽ ∥ ☿
09:09 ☽ # ♅
15:00 ☽ ✱ ♂
15:54 ☽ △ ♆
21:14 ☽ ✱ ♃
14 00:48 ☽ △ ☉
01:54 ☽ ✱ ♇
08:22 ☽ # ♀
09:37 ☽ □ ☿
10:07 ☽ ✱ ♄
10:34 ♂ ✱ ♆
16:45 ☉ ✱ ♇
15 08:01 ☽ ∥ ♄
19:52 ☽ □ ♆
20:49 ☽ ∥ ♃
16 00:20 ☽ ∥ ♇
07:43 ☿ ♓
09:08 ☽ ∥ ♂
16:50 ☽ ✱ ☿
17 00:33 ☽ △ ♅
17:32 ☽ △ ♀
18 03:37 ☽ ✱ ♆
08:33 ☽ ☌ ♂
10:48 ☽ ☌ ♃
14:54 ☽ ☌ ♇
22:57 ☽ ✱ ☉
23:24 ☽ ∥ ♂
19 00:49 ☽ ☌ ♄
05:19 ☽ ∥ ♇
10:03 ☽ ∥ ♃
10:04 ☽ □ ♅
23:51 ☉ ✱ ♄
23:52 ☽ ∥ ♄
20 03:50 ☉ ♈
08:25 ☽ # ♀
09:01 ☽ □ ♀
11:36 ♂ ☌ ♃
21 20:40 ☽ ☌ ☿
21:52 ☽ ✱ ♅
22 03:58 ♄ ♒
05:06 ☽ # ♅
13:19 ☿ ✱ ♅
16:02 ☽ ∥ ☿
22:10 ♀ # ♄
23 02:37 ☽ ✱ ♀
02:39 ☽ ☌ ♆
03:09 ♀ ✱ ♆
05:17 ♂ ☌ ♇
08:02 ♂ ∥ ♇
11:21 ☽ ✱ ♃
14:19 ☽ ✱ ♇
14:52 ☽ ✱ ♂
21:10 ☽ ∥ ♆
24 01:16 ☽ ✱ ♄
15:17 ☽ # ☉
25 10:48 ☽ ∥ ☉
26 00:44 ☽ □ ♃
03:03 ☽ □ ♇
03:05 ☽ # ♆
07:18 ☽ □ ♂
12:23 ♀ # ♃
14:16 ☽ □ ♄
22:34 ♀ # ♂
23:30 ☽ ☌ ♅
27 02:01 ☽ # ☿
09:05 ☽ ✱ ☿
19:36 ☽ ∥ ♅
28 04:03 ☽ ✱ ♆
04:25 ♀ △ ♃
04:28 ♀ # ♇
10:34 ♂ ∥ ♃
13:40 ☽ △ ♃
14:22 ☽ ☌ ♀
15:20 ☽ △ ♇
23:06 ☽ △ ♂
29 02:37 ☽ △ ♄
02:55 ♀ △ ♇
20:33 ☽ ✱ ☉
21:56 ☽ # ♄
30 02:59 ☽ □ ☿
07:47 ☽ # ♂
09:47 ☽ # ♃
15:11 ☽ □ ♆
16:42 ☽ # ♇
19:44 ♂ ♒
31 05:43 ☽ ∥ ♀
18:32 ♂ ☌ ♄
21:25 ☽ ✱ ♅

⚷ Chiron

01 Dec.	04 N 05
01	03♈57
04	04 07
07	04 17
10	04 27
13	04 37
16	04 48
19	04 58
22	05 09
25	05 20
28	05 30
31	05 41

April 2020

Day	S. T.	☉	☽	☿	♀	♂	♃	♄	♅	♆	♇	☊ True
	h m s	° ' "	° ' "	° '	° '	° '	° '	° '	° '	° '	° '	° '
01 We	12 39 15	11♈43 42	06♋32 36	15♓17	27♉31	00♒49	24♑24	00♒40	05♉10	19♓16	24♑50	02♋R48
02 Th	12 43 11	12 42 54	19 38 16	16 36	28 26	01 31	24 31	00 44	05 13	19 18	24 51	02 48
03 Fr	12 47 08	13 42 03	03♌09 49	17 56	29 21	02 13	24 38	00 48	05 16	19 20	24 52	02 45
04 Sa	12 51 04	14 41 10	17 09 47	19 18	00♊15	02 54	24 46	00 51	05 20	19 23	24 52	02 41
05 Su	12 55 01	15 40 15	01♍38 18	20 42	01 09	03 36	24 52	00 55	05 23	19 25	24 53	02 34
06 Mo	12 58 57	16 39 18	16 32 03	22 08	02 02	04 18	24 59	00 58	05 26	19 27	24 54	02 25
07 Tu	13 02 54	17 38 18	01♎44 03	23 35	02 54	04 59	25 06	01 01	05 29	19 29	24 54	02 14
08 We	13 06 50	18 37 16	17 04 13	25 04	03 46	05 41	25 12	01 04	05 33	19 31	24 55	02 03
09 Th	13 10 47	19 36 12	02♏21 02	26 35	04 37	06 23	25 19	01 08	05 36	19 33	24 55	01 53
10 Fr	13 14 44	20 35 06	17 23 30	28 07	05 27	07 04	25 25	01 11	05 39	19 35	24 56	01 44
11 Sa	13 18 40	21 33 59	02♐03 02	29 41	06 16	07 46	25 31	01 14	05 43	19 37	24 56	01 38
12 Su	13 22 37	22 32 50	16 14 33	01♈16	07 04	08 28	25 37	01 16	05 46	19 39	24 57	01 34
13 Mo	13 26 33	23 31 38	29 56 33	02 53	07 52	09 10	25 43	01 19	05 50	19 41	24 57	01 33
14 Tu	13 30 30	24 30 26	13♑56 28	04 32	08 39	09 51	25 48	01 22	05 53	19 43	24 57	01D 33
15 We	13 34 26	25 29 11	25 59 43	06 12	09 25	10 33	25 54	01 24	05 56	19 45	24 58	01 33
16 Th	13 38 23	26 27 55	08♒28 42	07 54	10 10	11 15	25 59	01 27	06 00	19 47	24 58	01R 33
17 Fr	13 42 19	27 26 37	20 42 06	09 37	10 54	11 56	26 04	01 29	06 03	19 49	24 58	01 31
18 Sa	13 46 16	28 25 17	02♓44 27	11 22	11 37	12 38	26 09	01 31	06 07	19 50	24 58	01 27
19 Su	13 50 13	29 23 56	14 39 50	13 09	12 19	13 20	26 14	01 34	06 10	19 52	24 59	01 20
20 Mo	13 54 09	00♉22 33	26 31 44	14 57	13 00	14 01	26 18	01 36	06 14	19 54	24 59	01 11
21 Tu	13 58 06	01 21 07	08♈22 53	16 47	13 40	14 43	26 23	01 38	06 17	19 56	24 59	01 00
22 We	14 02 02	02 19 41	20 15 23	18 38	14 19	15 24	26 27	01 40	06 20	19 58	24 59	00 49
23 Th	14 05 59	03 18 12	02♉10 50	20 31	14 56	16 06	26 31	01 41	06 24	20 00	24 59	00 39
24 Fr	14 09 55	04 16 41	14 10 40	22 26	15 33	16 48	26 35	01 43	06 27	20 01	24 59	00 30
25 Sa	14 13 52	05 15 09	26 16 16	24 22	16 08	17 29	26 39	01 45	06 31	20 03	24 59	00 23
26 Su	14 17 48	06 13 35	08♊29 19	26 20	16 41	18 11	26 42	01 46	06 34	20 05	24R 59	00 19
27 Mo	14 21 45	07 11 58	20 51 50	28 20	17 13	18 52	26 45	01 48	06 38	20 07	24 59	00 17
28 Tu	14 25 42	08 10 20	03♋26 19	00♉21	17 44	19 34	26 49	01 49	06 41	20 08	24 59	00D 17
29 We	14 29 38	09 08 40	16 15 40	02 23	18 13	20 15	26 52	01 50	06 45	20 10	24 59	00 18
30 Th	14 33 35	10 06 57	29 23 00	04 27	18 41	20 57	26 54	01 51	06 48	20 12	24 59	00 19

Data for	04-01-2020
Julian Day	2458940.50
Ayanamsa	24 08 05
SVP	04 ♓ 58 19
☽ ☊ Mean	03 ♋ 25 R

● ◐ PHASES ○ ◑

01	10:22	◐	12♋09
08	02:35	○	18♎44
14	22:57	◑	25♑27
23	02:27	●	03♉24
30	20:38	◐	10♌57

LAST ASPECT Day	h m	☽ INGRESS Day	h m	
02	16:50	02	18:28	♌
03	19:30	04	21:19	♍
06	13:29	06	21:17	♎
08	12:50	08	20:17	♏
10	19:36	10	20:36	♐
12	11:47	13	00:06	♑
14	23:48	15	07:38	♒
17	14:35	17	18:31	♓
19	23:32	20	07:02	♈
22	12:33	22	19:37	♉
25	00:44	25	07:21	♊
27	17:01	27	17:29	♋
29	19:30	30	01:07	♌

DECLINATION

Day	☉	☽	☿	♀	♂	♃	♄	♅	♆	♇
01 We	04N38	23N36	07S 47	23N06	20S 55	21S 18	20S 04	12N49	05S 11	21S 57
02 Th	05 01	23 27	07 20	23 23	20 47	21 17	20 03	12 50	05 10	21 57
03 Fr	05 24	21 56	06 52	23 39	20 39	21 16	20 02	12 51	05 10	21 57
04 Sa	05 47	19 03	06 23	23 55	20 31	21 14	20 02	12 52	05 09	21 57
05 Su	06 10	14 55	05 53	24 10	20 22	21 13	20 01	12 53	05 08	21 57
06 Mo	06 33	09 46	05 22	24 25	20 14	21 12	20 00	12 54	05 07	21 57
07 Tu	06 55	03 55	04 49	24 39	20 05	21 11	20 00	12 56	05 06	21 57
08 We	07 18	02S 13	04 15	24 53	19 56	21 10	19 59	12 57	05 06	21 57
09 Th	07 40	08 13	03 40	25 06	19 47	21 09	19 59	12 58	05 05	21 57
10 Fr	08 02	13 38	03 04	25 19	19 37	21 08	19 58	12 59	05 04	21 57
11 Sa	08 24	18 07	02 26	25 31	19 28	21 07	19 57	13 00	05 03	21 57
12 Su	08 46	21 23	01 48	25 43	19 18	21 06	19 57	13 01	05 03	21 57
13 Mo	09 08	23 17	01 08	25 54	19 08	21 05	19 56	13 02	05 02	21 57
14 Tu	09 30	23 48	00 28	26 05	18 58	21 04	19 56	13 03	05 01	21 57
15 We	09 51	23 00	00N14	26 15	18 48	21 03	19 55	13 05	05 00	21 57
16 Th	10 13	21 04	00 56	26 25	18 38	21 03	19 55	13 06	05 00	21 57
17 Fr	10 34	18 13	01 40	26 34	18 28	21 02	19 55	13 07	04 59	21 57
18 Sa	10 55	14 38	02 24	26 42	18 17	21 01	19 54	13 08	04 58	21 57
19 Su	11 16	10 30	03 09	26 50	18 07	21 00	19 54	13 09	04 57	21 58
20 Mo	11 36	06 00	03 55	26 58	17 56	21 00	19 53	13 10	04 57	21 58
21 Tu	11 57	01 17	04 42	27 05	17 45	20 59	19 53	13 11	04 56	21 58
22 We	12 17	03N30	05 30	27 11	17 34	20 58	19 53	13 13	04 55	21 58
23 Th	12 37	08 11	06 18	27 17	17 23	20 58	19 52	13 14	04 55	21 58
24 Fr	12 57	12 35	07 07	27 23	17 12	20 57	19 52	13 15	04 54	21 58
25 Sa	13 16	16 32	07 56	27 28	17 00	20 56	19 52	13 16	04 53	21 58
26 Su	13 36	19 50	08 46	27 32	16 49	20 56	19 51	13 17	04 53	21 58
27 Mo	13 55	22 17	09 36	27 36	16 37	20 55	19 51	13 18	04 52	21 58
28 Tu	14 14	23 41	10 26	27 40	16 25	20 55	19 51	13 19	04 51	21 58
29 We	14 32	23 51	11 17	27 42	16 14	20 54	19 51	13 21	04 51	21 59
30 Th	14 51	22 44	12 08	27 45	16 02	20 54	19 51	13 22	04 50	21 59

ASPECTARIAN

01 17:52 ☽ △ ☿
23:24 ☽ △ ♆
02 01:35 ☽ ∥ ♀
08:50 ☽ ☍ ♃
09:08 ☉ # ♆
09:21 ☽ ☍ ♇
16:50 ☽ ⚹ ♀
19:50 ☽ ☍ ♄
22:15 ☽ ☍ ♂
23:51 ☽ # ♇
03 03:41 ☽ □ ♅
06:48 ☽ # ♃
12:41 ☽ # ♂
16:59 ☽ # ♄
17:11 ♀ ♊
19:30 ☽ △ ☉
04 01:15 ☿ ☌ ♆
16:25 ☿ # ☉
17:10 ♀ △ ♄
23:09 ☽ □ ♀
05 02:15 ♃ ☌ ♇
06:07 ☽ △ ♅
09:55 ☽ ∥ ♅
06 04:38 ☽ ☍ ♆
09:49 ☽ ☍ ☿
10:55 ☿ ∥ ♆
12:40 ☽ ∥ ☉
13:15 ☽ △ ♇
13:29 ☽ △ ♃
19:16 ☽ # ♆
20:05 ☽ # ☿
22:53 ☽ △ ♄
07 01:56 ☽ △ ♀
05:21 ☽ △ ♂
14:02 ♂ ∥ ♄
18:50 ♂ □ ♅
21:27 ☿ ⚹ ♇
08 02:21 ☿ ⚹ ♃
07:17 ☽ ∥ ☿
11:19 ☽ ∥ ♆
12:18 ☽ □ ♇
12:50 ☽ □ ♃
21:37 ☽ # ☉
22:04 ☽ □ ♄
09 05:10 ☽ ☍ ♅
06:41 ☽ □ ♂
20:54 ☽ # ♅
10 03:33 ☽ △ ♆
12:15 ☽ ⚹ ♇
13:08 ☽ ⚹ ♃
19:36 ☽ △ ☿
22:37 ☽ ⚹ ♄
11 04:49 ☿ ♈
07:28 ☽ ☍ ♀
08:21 ☽ ∥ ♂
10:04 ☽ ⚹ ♂
12:15 ☽ ∥ ♄
21:27 ☽ ∥ ♃
23:59 ☿ ⚹ ♄
12 05:33 ☽ ∥ ♇
05:54 ☽ □ ♆
11:47 ☽ △ ☉
13 06:01 ☽ □ ☿
10:37 ☽ △ ♅
14 11:05 ☉ □ ♇
12:11 ☽ ⚹ ♆
22:02 ☽ ☌ ♇
23:48 ☽ ☌ ♃
15 10:22 ☽ ☌ ♄
11:00 ☉ □ ♃
14:38 ☽ ∥ ♇
19:10 ☽ □ ♅
22:42 ☽ ⚹ ☿
16 00:23 ☽ ∥ ♃
03:30 ☽ △ ♀
05:43 ☽ ☌ ♂
10:41 ☽ ∥ ♄
22:07 ☽ ∥ ♂
17 14:35 ☽ ⚹ ☉
18 05:37 ☿ ⚹ ♀
06:48 ☽ ⚹ ♅
09:00 ☽ # ♅
18:59 ☽ □ ♀
20:06 ☽ # ☉
19 03:56 ☿ ⚹ ♂
10:33 ☽ ☌ ♆
14:46 ☉ ♉
20:52 ☽ ⚹ ♇
23:32 ☽ ⚹ ♃
20 05:27 ☽ ∥ ♆
09:10 ☽ # ☿
10:17 ☽ ⚹ ♄
21 07:01 ☉ □ ♄
07:01 ☿ # ♆
11:19 ☽ ⚹ ♀
13:36 [illegible]
20:07 ☽ ☌ ☿
22 07:14 ☽ # ♆
09:32 ☽ □ ♇
12:15 ☽ ∥ ☿
12:33 ☽ □ ♃
23:01 ☽ □ ♄
23 08:30 ☽ ☌ ♅
24 02:15 ☽ ∥ ☉
03:50 ☽ ∥ ♅
05:32 ☽ □ ♂
11:39 ☽ ⚹ ♆
21:28 ☽ △ ♇
23:36 ☉ ∥ ♅
25 00:44 ☽ △ ♃
02:57 ☽ # ♂
07:34 ☿ □ ♇
10:49 ☽ △ ♄
18:52 ♇ SR
26 00:08 ☽ # ♄
04:32 ☿ □ ♃
09:01 ☉ ☌ ♅
09:31 ☽ # ♃
16:40 ☽ ☌ ♀
19:56 ☽ △ ♂
20:16 ☽ # ♇
22:33 ☽ □ ♆
27 17:01 ☽ ⚹ ☿
19:54 ☿ ♉
28 06:09 ☽ ⚹ ♅
09:39 ☽ ⚹ ☉
17:28 ☿ □ ♄
29 07:13 ☽ △ ♆
16:01 ☽ ☍ ♇
19:30 ☽ ☍ ♃
30 04:28 ☽ ☍ ♄
08:50 ☽ # ♇
10:47 ☽ □ ☿
13:21 ☽ □ ♅
19:04 ☽ # ♃

⚷ Chiron

01 Dec.	04 N 46
03	05♈51
06	06 02
09	06 12
12	06 23
15	06 33
18	06 43
21	06 53
24	07 02
27	07 12
30	07 21

May 2020

Day	S. T. h m s	☉ ° ' "	☽ ° ' "	☿ ° '	♀ ° '	♂ ° '	♃ ° '	♄ ° '	♅ ° '	♆ ° '	♇ ° '	☊ True ° '
01 Fr	14 37 31	11♉05 13	12♌51 18	06♉33	19♊07	21♒38	26♑57	01♒52	06♉51	20♓13	24♑R59	00♋R19
02 Sa	14 41 28	12 03 26	26 42 42	08 39	19 31	22 19	26 59	01 53	06 55	20 15	24 59	00 17
03 Su	14 45 24	13 01 38	10♍57 47	10 47	19 54	23 01	27 02	01 54	06 58	20 16	24 59	00 13
04 Mo	14 49 21	13 59 47	25 34 42	12 55	20 15	23 42	27 04	01 55	07 02	20 18	24 58	00 08
05 Tu	14 53 17	14 57 54	10♎28 45	15 05	20 34	24 23	27 06	01 56	07 05	20 19	24 58	00 01
06 We	14 57 14	15 56 00	25 32 32	17 15	20 51	25 05	27 07	01 56	07 09	20 21	24 58	29♊54
07 Th	15 01 11	16 54 03	10♏36 49	19 25	21 06	25 46	27 09	01 57	07 12	20 22	24 58	29 47
08 Fr	15 05 07	17 52 05	25 32 00	21 35	21 19	26 27	27 10	01 57	07 15	20 24	24 57	29 41
09 Sa	15 09 04	18 50 06	10♐09 37	23 45	21 29	27 08	27 11	01 57	07 19	20 25	24 57	29 37
10 Su	15 13 00	19 48 05	24 23 38	25 54	21 38	27 49	27 12	01 57	07 22	20 26	24 56	29 35
11 Mo	15 16 57	20 46 02	08♑10 56	28 03	21 44	28 31	27 13	01 57	07 26	20 28	24 56	29D35
12 Tu	15 20 53	21 43 59	21 31 16	00♊11	21 48	29 12	27 14	01R57	07 29	20 29	24 56	29 36
13 We	15 24 50	22 41 54	04♒26 34	02 17	21 50	29 53	27 14	01 57	07 32	20 30	24 55	29 38
14 Th	15 28 46	23 39 47	17 00 16	04 22	21R50	00♓34	27 14	01 57	07 36	20 32	24 55	29 40
15 Fr	15 32 43	24 37 40	29 16 38	06 24	21 47	01 15	27R14	01 57	07 39	20 33	24 54	29 41
16 Sa	15 36 40	25 35 31	11♓20 20	08 25	21 42	01 56	27 14	01 56	07 42	20 34	24 54	29R41
17 Su	15 40 36	26 33 21	23 15 57	10 23	21 34	02 36	27 14	01 56	07 46	20 35	24 53	29 39
18 Mo	15 44 33	27 31 09	05♈07 48	12 19	21 24	03 17	27 13	01 55	07 49	20 37	24 52	29 35
19 Tu	15 48 29	28 28 57	16 59 40	14 12	21 11	03 58	27 13	01 54	07 52	20 38	24 52	29 31
20 We	15 52 26	29 26 43	28 54 37	16 02	20 56	04 39	27 12	01 54	07 56	20 39	24 51	29 26
21 Th	15 56 22	00♊24 28	10♉55 11	17 50	20 39	05 19	27 11	01 53	07 59	20 40	24 50	29 21
22 Fr	16 00 19	01 22 12	23 03 16	19 35	20 19	06 00	27 09	01 52	08 02	20 41	24 50	29 17
23 Sa	16 04 15	02 19 54	05♊20 26	21 16	19 58	06 40	27 08	01 51	08 05	20 42	24 49	29 14
24 Su	16 08 12	03 17 36	17 47 54	22 55	19 34	07 21	27 06	01 49	08 09	20 43	24 48	29 13
25 Mo	16 12 09	04 15 16	00♋26 47	24 30	19 07	08 01	27 04	01 48	08 12	20 44	24 47	29D13
26 Tu	16 16 05	05 12 55	13 18 06	26 02	18 39	08 42	27 02	01 47	08 15	20 45	24 47	29 14
27 We	16 20 02	06 10 32	26 22 55	27 31	18 09	09 22	27 00	01 45	08 18	20 46	24 46	29 15
28 Th	16 23 58	07 08 08	09♌42 22	28 57	17 38	10 02	26 57	01 44	08 21	20 46	24 45	29 16
29 Fr	16 27 55	08 05 42	23 17 27	00♋20	17 05	10 42	26 55	01 42	08 24	20 47	24 44	29 17
30 Sa	16 31 51	09 03 15	07♍08 44	01 39	16 30	11 22	26 52	01 40	08 27	20 48	24 43	29R17
31 Su	16 35 48	10 00 47	21 15 55	02 55	15 55	12 02	26 49	01 39	08 31	20 49	24 42	29 15

Data for	05-01-2020
Julian Day	2458970.50
Ayanamsa	24 08 08
SVP	04 ♓ 58 15
☽ ☊ Mean	01 ♋ 50 R

● ◐ PHASES ○ ◑

07	10:45	○	17♏20
14	14:03	◑	24♒14
22	17:39	●	02♊05
30	03:30	◐	09♍12

LAST ASPECT ☽ Day	h m	INGRESS Day	h m	
01	16:05	02	05:36	♍
04	02:25	04	07:10	♎
06	02:31	06	07:05	♏
08	02:40	08	07:16	♐
10	06:12	10	09:40	♑
12	10:31	12	15:40	♒
14	14:04	15	01:26	♓
17	08:01	17	13:37	♈
19	20:33	20	02:11	♉
22	08:02	22	13:36	♊
24	11:10	24	23:10	♋
27	01:07	27	06:34	♌
28	13:31	29	11:41	♍
31	09:17	31	14:39	♎

DECLINATION

Day	☉	☽	☿	♀	♂	♃	♄	♅	♆	♇
01 Fr	15N09	20N17	12N58	27N47	15S50	20S54	19S51	13N23	04S50	21S59
02 Sa	15 27	16 38	13 49	27 48	15 37	20 53	19 50	13 24	04 49	21 59
03 Su	15 45	11 56	14 38	27 49	15 25	20 53	19 50	13 25	04 48	21 59
04 Mo	16 02	06 27	15 28	27 49	15 13	20 53	19 50	13 26	04 48	21 59
05 Tu	16 19	00 29	16 16	27 49	15 00	20 53	19 50	13 27	04 47	21 59
06 We	16 36	05S34	17 04	27 48	14 48	20 52	19 50	13 28	04 47	22 00
07 Th	16 53	11 18	17 50	27 47	14 35	20 52	19 50	13 30	04 46	22 00
08 Fr	17 09	16 20	18 35	27 45	14 23	20 52	19 50	13 31	04 46	22 00
09 Sa	17 25	20 16	19 18	27 42	14 10	20 52	19 50	13 32	04 45	22 00
10 Su	17 41	22 51	19 59	27 39	13 57	20 52	19 50	13 33	04 45	22 00
11 Mo	17 57	23 58	20 39	27 36	13 44	20 52	19 50	13 34	04 44	22 01
12 Tu	18 12	23 38	21 16	27 31	13 31	20 52	19 50	13 35	04 44	22 01
13 We	18 27	22 02	21 51	27 26	13 18	20 52	19 51	13 36	04 43	22 01
14 Th	18 41	19 22	22 24	27 21	13 05	20 52	19 51	13 37	04 43	22 01
15 Fr	18 55	15 55	22 54	27 15	12 51	20 52	19 51	13 38	04 42	22 01
16 Sa	19 09	11 52	23 21	27 08	12 38	20 53	19 51	13 39	04 42	22 02
17 Su	19 23	07 24	23 46	27 01	12 25	20 53	19 51	13 41	04 41	22 02
18 Mo	19 36	02 42	24 08	26 53	12 11	20 53	19 51	13 42	04 41	22 02
19 Tu	19 49	02N06	24 28	26 44	11 58	20 53	19 52	13 43	04 41	22 02
20 We	20 02	06 51	24 45	26 34	11 44	20 54	19 52	13 44	04 40	22 02
21 Th	20 14	11 24	25 00	26 24	11 31	20 54	19 52	13 45	04 40	22 03
22 Fr	20 26	15 33	25 12	26 13	11 17	20 54	19 53	13 46	04 39	22 03
23 Sa	20 38	19 07	25 22	26 01	11 04	20 55	19 53	13 47	04 39	22 03
24 Su	20 49	21 51	25 30	25 49	10 50	20 55	19 53	13 48	04 39	22 03
25 Mo	21 00	23 33	25 35	25 36	10 36	20 56	19 54	13 49	04 38	22 04
26 Tu	21 10	24 03	25 38	25 22	10 22	20 56	19 54	13 50	04 38	22 04
27 We	21 20	23 14	25 40	25 07	10 09	20 57	19 55	13 51	04 38	22 04
28 Th	21 30	21 07	25 39	24 52	09 55	20 58	19 55	13 52	04 37	22 04
29 Fr	21 39	17 47	25 37	24 36	09 41	20 58	19 55	13 53	04 37	22 05
30 Sa	21 48	13 25	25 33	24 19	09 27	20 59	19 56	13 54	04 37	22 05
31 Su	21 57	08 15	25 27	24 02	09 13	21 00	19 56	13 55	04 37	22 05

ASPECTARIAN

01 03:25 ☽ # ♄
03:41 ☿ ☌ ♅
11:16 ☽ ✶ ♀
11:56 ☿ ∥ ♅
16:05 ☽ ☍ ♂
02 05:52 ☽ # ♂
06:06 ☽ ∥ ☉
08:15 ☉ # ♂
12:46 ☽ ∥ ☿
16:57 ☽ ∥ ♅
17:19 ☽ △ ♅
23:39 ☽ △ ☿

03 03:40 ☽ △ ☉
15:06 ☽ □ ♀
15:22 ☽ ☍ ♆
18:11 ☿ # ♂
23:01 ☽ △ ♇
04 02:25 ☽ △ ♃
03:55 ♀ □ ♆
06:50 ☽ # ♆
10:16 ☽ △ ♄
21:42 ☿ ☌ ☉
05 02:320 ☿ ∥ ☉
16:24 ☽ △ ♀
20:54 ☽ ∥ ♆
23:05 ☽ □ ♇
23:13 ☽ △ ♂

06 02:31 ☽ □ ♃
10:10 ☽ □ ♄
18:32 ☽ ☍ ♅
07 10:00 ☽ # ♅
10:42 ☿ ✶ ♆
14:34 ☽ ∥ ♂
15:41 ☽ △ ♆
16:31 ☽ ☍ ☿
23:04 ☽ ✶ ♇
08 01:34 ☽ □ ♂
02:40 ☽ ✶ ♃
04:49 ☽ # ☉
10:28 ☽ ✶ ♄
15:48 ☽ # ☿
20:59 ☽ ∥ ♄
09 04:36 ☽ ∥ ♃
13:16 ☿ △ ♇
14:37 ☽ ∥ ♇
17:15 ☽ □ ♆
18:36 ☿ # ♄
19:15 ☽ ☍ ♀

10 06:12 ☽ ✶ ♂
14:37 ☿ △ ♃
16:17 ☉ ✶ ♆
22:40 ☽ △ ♅
11 04:10 ♄ SR
07:34 ☿ □ ♂
08:19 ☿ # ♃
16:58 ♂ # ♅
21:58 ☿ ♊
22:06 ☽ ✶ ♆
12 00:25 ☽ △ ☉
06:15 ☽ ☌ ♇
10:31 ☽ ☌ ♃
19:09 ☽ △ ☿
19:19 ☽ ☌ ♄
20:15 ☿ △ ♄
13 00:06 ☽ ∥ ♇
01:32 ☽ # ☿
04:18 ♂ ♓
05:53 ☽ □ ♅
06:46 ♀ SR
07:07 ☿ # ♇
11:32 ☽ ∥ ♃
20:17 ☽ ∥ ♄
14 04:50 ☽ # ☉
09:21 ☽ △ ♀
14:33 ♃ SR
15 04:07 ☽ ☌ ♂
06:47 ☉ △ ♇
13:46 ☽ # ♅
16:42 ☽ ✶ ♅
16:59 ☽ □ ☿
19:25 ☽ ∥ ♂

16 18:35 ☽ ☌ ♆
20:37 ☽ □ ♀
17 03:16 ☽ ✶ ♇
07:14 ☽ ✶ ☉
08:01 ☽ ✶ ♃
14:00 ☽ ∥ ♆
16:41 ☉ △ ♃
17:30 ☽ ✶ ♄
18 17:18 ☽ ✶ ☿
19 04:56 ☉ # ♄
08:18 ☽ ✶ ♀
12:56 ☽ # ♆
15:51 ☽ □ ♇
20:33 ☽ □ ♃
20 05:58 ☽ □ ♄
12:10 ☽ ✶ ♂
13:50 ☉ ♊
18:07 ☽ ☌ ♅
23:03 ♀ □ ♆
21 00:37 ☽ # ♂
13:17 ☽ ∥ ♅
19:19 ☽ ✶ ♆
22 03:29 ☽ △ ♇
08:02 ☽ △ ♃
08:42 ☿ ☌ ♀
12:03 ☉ △ ♄
15:43 ☿ □ ♆
17:13 ☽ △ ♄
23 02:44 ☽ □ ♂
06:03 ☽ # ♄
13:12 ☽ ∥ ☉
14:55 ☽ # ♃

24 02:18 ☽ # ♇
03:15 ☽ ☌ ♀
05:34 ☽ □ ♆
11:10 ☽ ☌ ☿
15:24 ☉ # ♃
25 01:07 ☿ ∥ ♀
06:49 ♂ ✶ ♅
14:35 ☽ ✶ ♅
14:58 ☽ △ ♂
26 13:44 ☽ △ ♆
21:03 ☽ ☍ ♇

27 01:07 ☽ ☍ ♃
09:43 ☽ ☍ ♄
14:48 ☽ # ♇
19:03 ☽ ✶ ☉
20:40 ☽ ∥ ☉
21:35 ☽ □ ♅
28 01:15 ☽ # ♃
09:31 ☽ # ♄
13:31 ☽ ✶ ♀
18:10 ☿ ♋
29 13:33 ☽ ✶ ☿
21:31 ☽ ∥ ♅
30 02:15 ☽ △ ♅
07:35 ☽ ☍ ♂
15:19 ☽ □ ♀
19:29 ☽ # ♂
23:14 ☽ ☍ ♆

31 05:47 ☽ △ ♇
09:17 ☽ △ ♃
15:31 ☽ # ♆
17:20 ☽ △ ♄
21:17 ☽ □ ☿

⚷ Chiron	
01 Dec.	05 N 25
03	07♈31
06	07 39
09	07 48
12	07 56
15	08 04
18	08 12
21	08 20
24	08 27
27	08 34
30	08 40

June 2020

05 19:26 15♐34 ✹ Penumbral Lunar Eclipse (mag 0.593)
21 06:43 00♋21 ✪ Annular Solar Eclipse

Day	S. T.	☉	☽	☿	♀	♂	♃	♄	♅	♆	♇	☊ True
	h m s	° ' "	° ' "	° '	° '	° '	° '	° '	° '	° '	° '	° '
01 Mo	16 39 44	10♊ 58 17	05♎ 37 27	04♋ 07	15♊R18	12♓ 42	26♑R46	01♒R37	08♉ 34	20♓ 50	24♑R41	29♊R13
02 Tu	16 43 41	11 55 45	20 10 11	05 16	14 41	13 22	26 43	01 35	08 37	20 50	24 40	29 10
03 We	16 47 38	12 53 13	04♏ 49 24	06 21	14 04	14 01	26 39	01 33	08 40	20 51	24 39	29 07
04 Th	16 51 34	13 50 39	19 29 07	07 23	13 26	14 41	26 35	01 30	08 43	20 52	24 38	29 05
05 Fr	16 55 31	14 48 05	04♐ 02 46	08 21	12 48	15 20	26 32	01 28	08 46	20 52	24 37	29 03
06 Sa	16 59 27	15 45 29	18 24 00	09 16	12 11	16 00	26 28	01 26	08 48	20 53	24 36	29 01
07 Su	17 03 24	16 42 53	02♑ 27 36	10 06	11 34	16 39	26 23	01 23	08 51	20 53	24 35	29D 01
08 Mo	17 07 20	17 40 15	16 10 01	10 53	10 58	17 19	26 19	01 21	08 54	20 54	24 34	29 01
09 Tu	17 11 17	18 37 37	29 29 46	11 35	10 23	17 58	26 15	01 18	08 57	20 54	24 33	29 02
10 We	17 15 13	19 34 59	12♒ 27 10	12 14	09 49	18 37	26 10	01 16	09 00	20 55	24 32	29 04
11 Th	17 19 10	20 32 20	25 04 11	12 48	09 17	19 16	26 05	01 13	09 03	20 55	24 31	29 05
12 Fr	17 23 07	21 29 40	07♓ 23 53	13 18	08 46	19 54	26 00	01 10	09 05	20 56	24 30	29 06
13 Sa	17 27 03	22 27 00	19 30 14	13 44	08 17	20 33	25 55	01 07	09 08	20 56	24 29	29 08
14 Su	17 30 60	23 24 19	01♈ 27 37	14 06	07 50	21 12	25 50	01 04	09 11	20 56	24 27	29 08
15 Mo	17 34 56	24 21 38	13 20 39	14 22	07 25	21 50	25 44	01 01	09 14	20 57	24 26	29 09
16 Tu	17 38 53	25 18 56	25 13 50	14 35	07 02	22 29	25 39	00 58	09 16	20 57	24 25	29 09
17 We	17 42 49	26 16 15	07♉ 11 15	14 43	06 41	23 07	25 33	00 55	09 19	20 57	24 24	29 09
18 Th	17 46 46	27 13 32	19 16 31	14 46	06 23	23 45	25 27	00 52	09 21	20 57	24 23	29 10
19 Fr	17 50 42	28 10 50	01♊ 32 35	14R 44	06 07	24 23	25 21	00 48	09 24	20 57	24 21	29 10
20 Sa	17 54 39	29 08 07	14 01 42	14 39	05 53	25 01	25 15	00 45	09 26	20 57	24 20	29 10
21 Su	17 58 36	00♋ 05 24	26 45 23	14 29	05 42	25 38	25 09	00 41	09 29	20 58	24 19	29 11
22 Mo	18 02 32	01 02 40	09♋ 44 17	14 14	05 33	26 16	25 03	00 38	09 31	20 58	24 17	29R 11
23 Tu	18 06 29	01 59 56	22 58 17	13 56	05 26	26 53	24 56	00 34	09 34	20R 58	24 16	29 10
24 We	18 10 25	02 57 12	06♌ 26 36	13 34	05 22	27 30	24 50	00 31	09 36	20 58	24 15	29 10
25 Th	18 14 22	03 54 27	20 07 52	13 08	05 20	28 07	24 43	00 27	09 39	20 58	24 14	29 09
26 Fr	18 18 18	04 51 41	04♍ 00 24	12 40	05D 21	28 44	24 36	00 23	09 41	20 57	24 12	29 07
27 Sa	18 22 15	05 48 55	18 02 19	12 09	05 24	29 21	24 29	00 20	09 43	20 57	24 11	29 06
28 Su	18 26 12	06 46 08	02♎ 11 31	11 36	05 29	29 57	24 22	00 16	09 45	20 57	24 09	29 05
29 Mo	18 30 08	07 43 20	16 25 42	11 01	05 36	00♈ 34	24 15	00 12	09 48	20 57	24 08	29 04
30 Tu	18 34 05	08 40 33	00♏ 42 21	10 25	05 45	01 10	24 08	00 08	09 50	20 57	24 07	29 03

Data for	06-01-2020
Julian Day	2459001.50
Ayanamsa	24 08 13
SVP	04♓58 11
☽ ☊ Mean	00♋12 R

● ◐ PHASES ○ ◑

05	19:13	✹	15♐34
13	06:23	◑	22♓42
21	06:42	✪	00♋21
28	08:16	◐	07♎06

LAST ASPECT Day	h m	☽ INGRESS Day	h m	
02	10:41	02	16:07	♏
04	11:38	04	17:18	♐
06	04:12	06	19:46	♑
08	18:07	09	00:55	♒
10	14:36	11	09:32	♓
13	12:45	13	21:03	♈
16	00:50	16	09:36	♉
18	12:02	18	21:00	♊
20	21:48	21	06:02	♋
23	07:21	23	12:34	♌
24	05:35	25	17:06	♍
27	20:03	27	20:18	♎
29	13:03	29	22:49	♏

DECLINATION

Day	☉	☽	☿	♀	♂	♃	♄	♅	♆	♇
01 Mo	22N05	02N34	25N20	23N45	08S59	21S01	19S57	13N56	04S36	22S06
02 Tu	22 13	03S20	25 12	23 27	08 45	21 01	19 57	13 57	04 36	22 06
03 We	22 20	09 06	25 02	23 08	08 31	21 02	19 58	13 58	04 36	22 06
04 Th	22 28	14 22	24 52	22 50	08 18	21 03	19 59	13 59	04 36	22 06
05 Fr	22 34	18 46	24 40	22 31	08 04	21 04	19 59	14 00	04 36	22 07
06 Sa	22 41	21 58	24 27	22 12	07 50	21 05	20 00	14 01	04 35	22 07
07 Su	22 46	23 44	24 14	21 53	07 36	21 06	20 00	14 02	04 35	22 07
08 Mo	22 52	24 00	24 00	21 34	07 22	21 07	20 01	14 02	04 35	22 08
09 Tu	22 57	22 53	23 45	21 16	07 08	21 08	20 02	14 03	04 35	22 08
10 We	23 02	20 33	23 29	20 57	06 54	21 09	20 02	14 04	04 35	22 08
11 Th	23 06	17 17	23 14	20 39	06 40	21 10	20 03	14 05	04 35	22 09
12 Fr	23 10	13 21	22 57	20 22	06 26	21 11	20 04	14 06	04 35	22 09
13 Sa	23 13	08 56	22 41	20 05	06 12	21 12	20 05	14 07	04 34	22 09
14 Su	23 16	04 16	22 24	19 49	05 59	21 13	20 05	14 08	04 34	22 09
15 Mo	23 19	00N33	22 07	19 33	05 45	21 14	20 06	14 09	04 34	22 10
16 Tu	23 21	05 20	21 51	19 18	05 31	21 16	20 07	14 09	04 34	22 10
17 We	23 23	09 58	21 34	19 04	05 17	21 17	20 08	14 10	04 34	22 10
18 Th	23 24	14 16	21 17	18 51	05 04	21 18	20 08	14 11	04 34	22 11
19 Fr	23 25	18 04	21 01	18 38	04 50	21 19	20 09	14 12	04 34	22 11
20 Sa	23 26	21 07	20 45	18 26	04 36	21 21	20 10	14 13	04 34	22 11
21 Su	23 26	23 11	20 30	18 16	04 23	21 22	20 11	14 13	04 34	22 12
22 Mo	23 26	24 03	20 15	18 06	04 09	21 23	20 12	14 14	04 34	22 12
23 Tu	23 25	23 36	20 01	17 57	03 56	21 25	20 13	14 15	04 34	22 12
24 We	23 24	21 46	19 47	17 49	03 42	21 26	20 13	14 16	04 34	22 13
25 Th	23 23	18 40	19 35	17 41	03 29	21 27	20 14	14 16	04 34	22 13
26 Fr	23 21	14 30	19 23	17 35	03 16	21 29	20 15	14 17	04 34	22 13
27 Sa	23 19	09 30	19 12	17 29	03 03	21 30	20 16	14 18	04 34	22 14
28 Su	23 16	03 58	19 02	17 25	02 49	21 31	20 17	14 18	04 35	22 14
29 Mo	23 13	01S48	18 53	17 21	02 36	21 33	20 18	14 19	04 35	22 14
30 Tu	23 09	07 30	18 46	17 18	02 23	21 34	20 19	14 20	04 35	22 15

ASPECTARIAN

01 01:14 ☉ ⋕ ♇
09:29 ☽ △ ☉
15:21 ☽ △ ♀
02 05:11 ☽ ∥ ♆
07:23 ☽ □ ♇
10:41 ☽ □ ♃
18:39 ☽ □ ♄
21:38 ☽ ∥ ♂
03 00:41 ♀ □ ♂
02:42 ☽ △ ☿
06:18 ☽ ☍ ♅
15:45 ☽ △ ♂
17:44 ♀ ☌ ☉
22:06 ☽ ⋕ ♅
04 02:15 ☽ △ ♆
08:28 ☽ ⚹ ♇
11:38 ☽ ⚹ ♃
19:44 ☽ ⚹ ♄
20:57 ♀ ∥ ☉
05 08:06 ☽ ∥ ♄
11:05 ☿ ⚹ ♅
13:59 ☽ ☍ ♀
16:18 ☽ ∥ ♃
19:45 ☽ □ ♂
06 01:35 ☽ ∥ ♇
02:09 ☽ ⋕ ♀
04:12 ☽ □ ♆
06:19 ♀ ⋕ ♇
07:59 ☽ ⋕ ☉
19:12 ☉ □ ♂
07 11:09 ☽ △ ♅
14:07 ☽ ☍ ☿
16:26 ☽ ⋕ ☿
08 [illegible] ☽ ∥ [illegible]
02:08 ☽ ⚹ ♂
08:27 ☽ ⚹ ♆
15:02 ☽ ☌ ♇
18:07 ☽ ☌ ♃
23:56 ☽ ⋕ ☉
09 03:18 ☽ ☌ ♄
08:53 ☽ ∥ ♇
09:45 ♀ ⋕ ♃
17:31 ☽ □ ♅
18:43 ☽ ∥ ♃
19:16 ☽ △ ♀
19:57 ☽ ⋕ ♀
10 04:07 ☽ ∥ ♄
14:36 ☽ △ ☉
11 09:05 ☿ ∥ ☉
09:38 ☉ □ ♆
19:40 ☽ ⋕ ♅
12 02:35 ☽ □ ♀
03:21 ☽ ⚹ ♅
12:07 ☽ △ ☿
13 00:36 ♀ ⋕ ♄
02:13 ☽ ☌ ♂
02:51 ☽ ☌ ♆
09:56 ☽ ⚹ ♇
12:46 ☽ ⚹ ♃
14:14 ♂ ☌ ♆
14:52 ☽ ∥ ♂
21:26 ☽ ∥ ♆
23:13 ☽ ⚹ ♄
14 12:25 ☽ ⚹ ♀
20:36 ☿ ⋕ ♇
15 02:07 ☽ □ ☿
20:07 ☽ ⋕ ♆
22:22 ☽ □ ♇
16 00:11 ☽ ⚹ ☉
00:51 ☽ □ ♃
00:51 ☽ ⋕ ♂
11:29 ☽ □ ♄
17 04:15 ☽ ☌ ♅
15:03 ☽ ⚹ ☿
23:10 ☿ ⋕ ♃
23:28 ☽ ∥ ♅
18 03:18 ☽ ⚹ ♆
05:00 ☿ SR
09:17 ☽ ⚹ ♂
10:01 ☽ △ ♇
12:02 ☽ △ ♃
22:34 ☽ △ ♄
23:04 ♂ ⚹ ♇
19 03:48 ☽ ∥ ♀
08:40 ☽ ☌ ♀
15:45 ☽ ⋕ ♄
21:05 ☽ ∥ ☿
20 02:12 ☽ ⋕ ♃
03:51 ♂ ∥ ♆
07:58 ♂ ⚹ ♃
10:59 ☽ ⋕ ♇
13:08 ☽ □ ♆
21:44 ☉ ♋
21:48 ☽ □ ♂
21 04:28 ☽ ∥ ☉
13:30 ☽ [illegible] [illegible]
22 05:18 ☿ ⋕ ♄
08:02 ☽ ☌ ☿
20:23 ☽ △ ♆
23 02:20 ☽ ☍ ♇
03:18 ☽ ∥ ☉
03:30 ☽ ☍ ♃
04:33 ♆ SR
07:21 ☽ △ ♂
13:32 ☽ ☍ ♄
19:27 ☽ ⋕ ♇
22:06 ☽ ⚹ ♀
24 03:04 ☽ ⋕ ♃
05:35 ☽ □ ♅
13:00 ☽ ⋕ ♄
17:24 ☽ ∥ ☿
25 06:19 ☽ ∥ ♀
06:48 ♀ SD
26 01:03 ☽ ∥ ♅
01:35 ☽ ⚹ ☉
02:18 ☽ □ ♀
09:46 ☽ △ ♅
14:19 ☽ ⚹ ☿
27 04:58 ☽ ☍ ♆
10:25 ☽ △ ♇
10:52 ☽ △ ♃
20:45 ☽ △ ♄
21:25 ☽ ∥ ♆
28 01:46 ♂ ♈
04:58 ☽ ⋕ ♂
05:35 ☽ △ ♀
11:00 ♂ ⚹ ♄
15:15 ☽ □ ☿
29 03:12 ☽ ∥ ♂
11:34 ☽ ∥ ♆
12:56 ☽ □ ♇
13:03 ☽ □ ♃
23:02 ☽ □ ♄
30 06:23 ♃ ☌ ♇
14:21 ☽ △ ☉
[illegible]
15:40 ☽ △ ☿
22:14 ☿ ⚹ ♅

⚷ Chiron

01 Dec.	05 N 58
02	08♈46
05	08 52
08	08 57
11	09 02
14	09 06
17	09 10
20	09 14
23	09 17
26	09 20
29	09 [illegible]

July 2020

05 04:31 13♑38 ✹ Penumbral Lunar Eclipse (mag 0.379)

Day	S. T.	☉	☽	☿	♀	♂	♃	♄	♅	♆	♇	☊ True
	h m s	° ′ ″	° ′ ″	° ′	° ′	° ′	° ′	° ′	° ′	° ′	° ′	° ′
01 We	18 38 01	09♋37 44	14♏58 38	09♋R49	05♊56	01♈46	24♑R01	00♒R04	09♉52	20♓R57	24♑R05	29♊D03
02 Th	18 41 58	10 34 56	29 11 30	09 13	06 10	02 21	23 54	30♑00	09 54	20 56	24 04	29 03
03 Fr	18 45 54	11 32 07	13♐17 38	08 38	06 25	02 57	23 47	29 56	09 56	20 56	24 03	29 03
04 Sa	18 49 51	12 29 18	27 13 43	08 04	06 43	03 32	23 39	29 52	09 58	20 56	24 01	29R 03
05 Su	18 53 47	13 26 29	10♑56 37	07 33	07 02	04 07	23 32	29 48	10 00	20 55	24 00	29 03
06 Mo	18 57 44	14 23 40	24 23 49	07 04	07 23	04 42	23 24	29 43	10 02	20 55	23 58	29 02
07 Tu	19 01 41	15 20 51	07♒33 41	06 38	07 46	05 17	23 17	29 39	10 04	20 55	23 57	29 01
08 We	19 05 37	16 18 02	20 25 42	06 16	08 10	05 52	23 09	29 35	10 06	20 54	23 55	28 59
09 Th	19 09 34	17 15 13	03♓00 32	05 57	08 36	06 26	23 01	29 31	10 07	20 54	23 54	28 56
10 Fr	19 13 30	18 12 25	15 19 59	05 43	09 04	07 00	22 54	29 27	10 09	20 53	23 53	28 54
11 Sa	19 17 27	19 09 37	27 26 50	05 34	09 33	07 34	22 46	29 22	10 11	20 53	23 51	28 53
12 Su	19 21 23	20 06 49	09♈24 45	05 30	10 03	08 07	22 38	29 18	10 12	20 52	23 50	28 52
13 Mo	19 25 20	21 04 02	21 18 00	05D 31	10 35	08 41	22 31	29 14	10 14	20 51	23 48	28D 53
14 Tu	19 29 16	22 01 15	03♉11 16	05 37	11 09	09 14	22 23	29 09	10 16	20 51	23 47	28 54
15 We	19 33 13	22 58 29	15 09 18	05 48	11 43	09 46	22 15	29 05	10 17	20 50	23 45	28 56
16 Th	19 37 10	23 55 44	27 16 42	06 05	12 19	10 19	22 07	29 00	10 19	20 49	23 44	28 59
17 Fr	19 41 06	24 52 59	09♊37 37	06 27	12 56	10 51	22 00	28 56	10 20	20 49	23 42	29 01
18 Sa	19 45 03	25 50 15	22 15 27	06 55	13 35	11 23	21 52	28 51	10 22	20 48	23 41	29 03
19 Su	19 48 59	26 47 32	05♋12 36	07 28	14 14	11 54	21 44	28 47	10 23	20 47	23 39	29R 03
20 Mo	19 52 56	27 44 49	18 30 09	08 07	14 54	12 26	21 37	28 43	10 24	20 46	23 38	29 01
21 Tu	19 56 52	28 42 06	02♌07 30	08 52	15 36	12 57	21 29	28 38	10 25	20 45	23 37	28 58
22 We	20 00 49	29 39 24	16 02 26	09 41	16 18	13 27	21 21	28 34	10 27	20 45	23 35	28 54
23 Th	20 04 45	00♌36 43	00♍11 13	10 36	17 02	13 57	21 14	28 29	10 28	20 44	23 34	28 48
24 Fr	20 08 42	01 34 02	14 29 10	11 36	17 46	14 27	21 06	28 25	10 29	20 43	23 32	28 43
25 Sa	20 12 39	02 31 21	28 51 15	12 42	18 31	14 57	20 59	28 20	10 30	20 42	23 31	28 38
26 Su	20 16 35	03 28 40	13♎12 52	13 52	19 17	15 26	20 51	28 16	10 31	20 41	23 29	28 34
27 Mo	20 20 32	04 26 00	27 30 09	15 08	20 04	15 55	20 44	28 12	10 32	20 40	23 28	28 32
28 Tu	20 24 28	05 23 21	11♏40 22	16 28	20 51	16 23	20 37	28 07	10 33	20 39	23 27	28 31
29 We	20 28 25	06 20 42	25 41 44	17 53	21 40	16 51	20 30	28 03	10 34	20 38	23 25	28D 31
30 Th	20 32 21	07 18 03	09♐33 14	19 23	22 29	17 19	20 22	27 58	10 35	20 37	23 24	28 32
31 Fr	20 36 18	08 15 25	23 14 19	20 57	23 19	17 46	20 15	27 54	10 36	20 36	23 22	28 33

Data for	07-01-2020
Julian Day	2459031.50
Ayanamsa	24 08 18
SVP	04 ♓ 58 07
☽ ☊ Mean	28 ♊ 36 R

● ◐ PHASES ○ ◑

05	04:45	✹	13♑38
12	23:29	◑	21♈03
20	17:34	●	28♋27
27	12:33	◐	04♏56

LAST ASPECT ☽		INGRESS		
Day	h m	Day	h m	
02	01:22	02	01:22	♐
03	13:07	04	04:49	♑
06	09:36	06	10:09	♒
07	04:38	08	18:13	♓
11	03:49	11	05:06	♈
13	15:55	13	17:34	♉
16	03:22	16	05:20	♊
17	21:15	18	14:25	♋
20	17:56	20	20:18	♌
22	00:29	22	23:41	♍
24	23:09	25	01:55	♎
27	01:10	27	04:13	♏
29	04:02	29	07:25	♐
31	00:08	31	11:59	♑

DECLINATION

Day	☉	☽	☿	♀	♂	♃	♄	♅	♆	♇
01 We	23N05	12S 48	18N39	17N15	02S 10	21S 36	20S 20	14N20	04S 35	22S 15
02 Th	23 01	17 24	18 34	17 14	01 58	21 37	20 21	14 21	04 35	22 15
03 Fr	22 56	20 58	18 30	17 13	01 45	21 38	20 22	14 22	04 35	22 16
04 Sa	22 51	23 14	18 28	17 13	01 32	21 40	20 23	14 22	04 35	22 16
05 Su	22 45	24 03	18 26	17 13	01 20	21 41	20 24	14 23	04 36	22 17
06 Mo	22 40	23 27	18 27	17 14	01 07	21 43	20 25	14 24	04 36	22 17
07 Tu	22 33	21 32	18 28	17 15	00 55	21 44	20 26	14 24	04 36	22 17
08 We	22 27	18 34	18 31	17 17	00 42	21 46	20 27	14 25	04 36	22 18
09 Th	22 19	14 48	18 35	17 20	00 30	21 47	20 28	14 25	04 36	22 18
10 Fr	22 12	10 29	18 40	17 23	00 18	21 48	20 29	14 26	04 37	22 18
11 Sa	22 04	05 50	18 46	17 26	00 06	21 50	20 30	14 26	04 37	22 19
12 Su	21 56	01 01	18 53	17 30	00N06	21 51	20 31	14 27	04 37	22 19
13 Mo	21 47	03N48	19 02	17 34	00 18	21 53	20 32	14 27	04 37	22 19
14 Tu	21 38	08 29	19 11	17 38	00 29	21 54	20 33	14 28	04 38	22 20
15 We	21 29	12 54	19 20	17 43	00 41	21 55	20 34	14 28	04 38	22 20
16 Th	21 19	16 51	19 31	17 48	00 52	21 57	20 35	14 29	04 38	22 20
17 Fr	21 09	20 10	19 41	17 53	01 03	21 58	20 36	14 29	04 39	22 21
18 Sa	20 59	22 35	19 53	17 58	01 15	22 00	20 37	14 30	04 39	22 21
19 Su	20 48	23 54	20 04	18 03	01 26	22 01	20 38	14 30	04 39	22 21
20 Mo	20 37	23 53	20 15	18 09	01 36	22 02	20 39	14 30	04 40	22 22
21 Tu	20 25	22 28	20 26	18 15	01 47	22 04	20 40	14 31	04 40	22 22
22 We	20 13	19 40	20 37	18 20	01 58	22 05	20 41	14 31	04 40	22 22
23 Th	20 01	15 40	20 47	18 26	02 08	22 06	20 41	14 32	04 41	22 23
24 Fr	19 49	10 45	20 57	18 32	02 18	22 08	20 42	14 32	04 41	22 23
25 Sa	19 36	05 13	21 06	18 38	02 29	22 09	20 43	14 32	04 42	22 23
26 Su	19 22	00S 35	21 13	18 44	02 39	22 10	20 44	14 32	04 42	22 24
27 Mo	19 09	06 20	21 20	18 50	02 48	22 11	20 45	14 33	04 42	22 24
28 Tu	18 55	11 43	21 25	18 55	02 58	22 13	20 46	14 33	04 43	22 24
29 We	18 41	16 26	21 28	19 01	03 08	22 14	20 47	14 33	04 43	22 25
30 Th	18 27	20 12	21 30	19 06	03 17	22 15	20 48	14 34	04 44	22 25
31 Fr	18 12	22 47	21 29	19 12	03 26	22 16	20 49	14 34	04 44	22 25

ASPECTARIAN

01 02:53 ☿ ☌ ☉
06:07 ☉ ✶ ♅
07:33 ☽ # ♅
10:03 ☽ △ ♆
15:07 ☽ ✶ ♃
15:20 ☽ ✶ ♇
23:02 ☽ # ♀
23:40 ♄ ♑R
02 01:22 ☽ ✶ ♄
05:36 ☽ △ ♂
06:56 ☽ # ☿
12:03 ☽ ☍ ♀
19:19 ☽ ∥ ♄
03 05:54 ☽ ∥ ♃
11:59 ☽ ∥ ♇
13:07 ☽ □ ♆
18:56 ☽ # ☉
04 11:29 ☽ □ ♂
18:14 ☽ ☍ ☿
22:20 ☽ △ ♅
05 17:45 ☽ ✶ ♆
22:14 ☽ ☌ ♃
23:14 ☽ ☌ ♇
06 09:36 ☽ ☌ ♄
12:31 ☽ # ☉
16:17 ☽ ∥ ♇
19:37 ☽ ✶ ♂
22:07 ☽ ∥ ♃
07 00:23 ☽ △ ♀
04:38 ☽ □ ♅
09:51 ☽ ∥ ♄
08 00:23 ☽ # ☿
08:34 ☽ # ♀
10:42 ☿ □ ♂
09 02:12 ☽ # ♅
04:38 ☉ # ♇
05:35 ☽ △ ☿
11:15 ☽ □ ♀
13:50 ☽ ✶ ♅
10 06:10 ☽ △ ☉
10:57 ☽ ☌ ♆
14:47 ☽ ✶ ♃
16:51 ☽ ✶ ♇
11 03:49 ☽ ✶ ♄
06:06 ☽ ∥ ♆
16:09 ☽ □ ☿
21:16 ☽ ☌ ♂
12 01:22 ☽ ✶ ♀
04:24 ☽ # ♂
05:46 ☽ ∥ ♂
08:28 ☿ SD
11:02 ☉ # ♃
18:44 ☉ △ ♆
13 02:25 ☽ □ ♃
04:08 ☽ # ♆
05:03 ☽ □ ♇
15:55 ☽ □ ♄
14 04:57 ☽ ✶ ☿
08:00 ☉ ☍ ♃
14:15 ☽ ☌ ♅
15 09:09 ☽ ∥ ♅
11:17 ☽ ✶ ♆
13:57 ☽ △ ♃
16:51 ☽ ✶ ☉
17:02 ☽ △ ♇
19:10 ☉ ☍ ♇
16 03:22 ☽ △ ♄
06:23 ☽ ∥ ♀
19:52 ☽ ∥ ☿
17 02:27 ☽ ✶ ♂
03:35 ☽ # ♄
06:41 ☽ ☌ ♀
08:02 ☽ ∥ ☉
16:55 ☽ # ♃
21:00 ☽ # ♇
21:15 ☽ □ ♆
19 04:20 ☽ ☌ ☿
09:26 ☽ ✶ ♅
12:40 ☽ □ ♂
20:12 ☉ # ♄
20 04:02 ☽ △ ♆
05:28 ☽ ☍ ♃
09:05 ☽ ☍ ♇
17:56 ☽ ☍ ♄
22:28 ☉ ☍ ♄
22:54 ☿ ∥ ☉
21 01:04 ☽ # ♇
04:15 ☽ # ♃
14:23 ☽ □ ♅
16:34 ☽ # ♄
17:23 ☽ ∥ ☿
19:24 ☽ △ ♂
19:45 ☽ ∥ ☉
22 00:29 ☽ ✶ ♀
08:30 ☽ ∥ ♀
08:37 ☉ ♌
09:03 ☿ # ♄
20:25 ☿ ✶ ♅
23 05:56 ☽ ∥ ♅
17:17 ☽ △ ♅
18:48 ☽ ✶ ☿
24 05:47 ☽ □ ♀
10:24 ☽ ☍ ♆
10:58 ☽ △ ♃
15:06 ☽ △ ♇
23:09 ☽ △ ♄
25 02:11 ☽ # ♆
06:34 ☽ ✶ ☉
11:04 ☽ ∥ ♂
26 01:12 ☽ □ ☿
03:51 ☽ ☍ ♂
08:44 ☽ # ♂
10:46 ☽ △ ♀
12:42 ☽ □ ♃
17:06 ☽ ∥ ♆
17:13 ☽ □ ♇
27 01:10 ☽ □ ♄
16:03 ♃ ✶ ♆
17:50 ♀ □ ♆
21:47 ☿ □ ♂
22:05 ☽ ☍ ♅
23:51 ♀ ∥ ☉
28 09:05 ☽ △ ☿
13:55 ☽ # ♅
15:08 ☽ ✶ ♃
15:19 ☽ △ ♆
20:06 ☽ ✶ ♇
29 04:02 ☽ ✶ ♄
12:40 ☽ # ☉
16:04 ☽ # ♀
19:47 ☽ △ ☉
30 04:43 ☽ ∥ ♄
10:35 ☽ # ☿
14:02 ☽ △ ♂
14:18 ☿ ☍ ♃
18:06 ☽ ∥ ♃
18:45 ☿ △ ♆
19:21 ☽ □ ♆
19:49 ☽ ∥ ♇
31 00:08 ☽ ☍ ♀

⚷ Chiron

01 Dec. 06 N 15

02	09♈24
05	09 25
08	09 26
11	09 26
14	09 26R
17	09 26
20	09 25
23	09 23
26	09 21
29	09 19

11 19:04 09♈26 R

August 2020

Day	S. T. h m s	☉ ° ' "	☽ ° ' "	☿ ° '	♀ ° '	♂ ° '	♃ ° '	♄ ° '	♅ ° '	♆ ° '	♇ ° '	☊ True ° '
01 Sa	20 40 14	09♌12 47	06♑44 26	22♋35	24♊09	18♈13	20♑R08	27♑R50	10♉36	20♓R34	23♑R21	28♊R33
02 Su	20 44 11	10 10 11	20 03 01	24 17	25 00	18 39	20 02	27 46	10 37	20 33	23 20	28 31
03 Mo	20 48 08	11 07 35	03♒09 18	26 02	25 52	19 05	19 55	27 41	10 38	20 32	23 18	28 28
04 Tu	20 52 04	12 05 00	16 02 38	27 51	26 44	19 31	19 48	27 37	10 38	20 31	23 17	28 22
05 We	20 56 01	13 02 25	28 42 37	29 43	27 37	19 56	19 42	27 33	10 39	20 30	23 16	28 15
06 Th	20 59 57	13 59 52	11♓09 27	01♌38	28 31	20 21	19 35	27 29	10 39	20 29	23 14	28 07
07 Fr	21 03 54	14 57 20	23 24 01	03 35	29 25	20 45	19 29	27 25	10 40	20 27	23 13	27 59
08 Sa	21 07 50	15 54 49	05♈28 05	05 33	00♋20	21 08	19 23	27 20	10 40	20 26	23 12	27 52
09 Su	21 11 47	16 52 19	17 24 16	07 34	01 15	21 31	19 17	27 16	10 40	20 25	23 10	27 47
10 Mo	21 15 43	17 49 51	29 16 06	09 35	02 11	21 54	19 11	27 12	10 41	20 23	23 09	27 44
11 Tu	21 19 40	18 47 24	11♉07 51	11 38	03 07	22 16	19 05	27 08	10 41	20 22	23 08	27D43
12 We	21 23 37	19 44 58	23 04 20	13 41	04 04	22 38	18 59	27 04	10 41	20 21	23 06	27 43
13 Th	21 27 33	20 42 34	05♊10 40	15 44	05 01	22 59	18 54	27 01	10 41	20 19	23 05	27 45
14 Fr	21 31 30	21 40 12	17 31 52	17 47	05 58	23 19	18 48	26 57	10 41	20 18	23 04	27 47
15 Sa	21 35 26	22 37 50	00♋12 32	19 50	06 56	23 39	18 43	26 53	10 41	20 17	23 03	27R 47
16 Su	21 39 23	23 35 31	13 16 23	21 52	07 55	23 58	18 38	26 49	10R 41	20 15	23 02	27 46
17 Mo	21 43 19	24 33 13	26 45 32	23 53	08 54	24 17	18 33	26 46	10 41	20 14	23 00	27 42
18 Tu	21 47 16	25 30 56	10♌40 02	25 54	09 53	24 35	18 28	26 42	10 41	20 12	22 59	27 36
19 We	21 51 12	26 28 40	24 57 15	27 54	10 53	24 52	18 23	26 38	10 41	20 11	22 58	27 28
20 Th	21 55 09	27 26 26	09♍32 01	29 53	11 53	25 09	18 19	26 35	10 41	20 09	22 57	27 18
21 Fr	21 59 06	28 24 13	24 17 04	01♍50	12 53	25 25	18 14	26 31	10 41	20 08	22 56	27 08
22 Sa	22 03 02	29 22 02	09♎04 16	03 46	13 54	25 40	18 10	26 28	10 40	20 06	22 55	26 59
23 Su	22 06 59	00♍19 51	23 45 57	05 42	14 55	25 55	18 06	26 25	10 40	20 05	22 54	26 52
24 Mo	22 10 55	01 17 42	08♏16 00	07 35	15 56	26 09	18 02	26 22	10 40	20 03	22 52	26 47
25 Tu	22 14 52	02 15 34	22 30 30	09 28	16 58	26 22	17 59	26 18	10 39	20 02	22 51	26 44
26 We	22 18 48	03 13 28	06♐27 48	11 19	18 00	26 35	17 55	26 15	10 39	20 00	22 50	26 43
27 Th	22 22 45	04 11 22	20 08 01	13 09	19 02	26 47	17 52	26 12	10 38	19 59	22 49	26D 43
28 Fr	22 26 41	05 09 18	03♑32 21	14 58	20 04	26 58	17 49	26 09	10 38	19 57	22 48	26 43
29 Sa	22 30 38	06 07 15	16 42 26	16 45	21 07	27 08	17 46	26 06	10 37	19 55	22 47	26 41
30 Su	22 34 35	07 05 13	29 39 51	18 31	22 10	27 17	17 43	26 04	10 36	19 54	22 46	26 38
31 Mo	22 38 31	08 03 13	12♒25 48	20 15	23 14	27 26	17 41	26 01	10 36	19 52	22 46	26 32

Data for	08-01-2020
Julian Day	2459062.50
Ayanamsa	24 08 23
SVP	04 ♓ 58 04
☽ ☊ Mean	26 ♊ 58 R

PHASES ● ◐ ○ ◑

Day	Time	Phase	Position
03	15:58	○	11♒46
11	16:45	◑	19♉28
19	02:42	●	26♌35
25	17:57	◐	02♐59

LAST ASPECT ☽ Day	h m	INGRESS Day	h m	Sign
02	14:00	02	18:11	♒
04	21:46	05	02:28	♓
07	12:54	07	13:05	♈
09	19:51	10	1:29	♉
12	07:56	12	13:47	♊
14	11:20	14	23:37	♋
17	00:00	17	05:39	♌
19	05:39	19	08:21	♍
21	03:37	21	09:16	♎
23	04:20	23	10:16	♏
25	06:28	25	12:49	♐
27	12:00	27	17:37	♑
29	19:31	30	00:38	♒

DECLINATION

Day	☉	☽	☿	♀	♂	♃	♄	♅	♆	♇
01 Sa	17N57	24S00	21N27	19N17	03N35	22S17	20S50	14N34	04S45	22S26
02 Su	17 42	23 48	21 22	19 22	03 44	22 18	20 51	14 34	04 45	22 26
03 Mo	17 26	22 17	21 14	19 27	03 53	22 19	20 52	14 34	04 46	22 26
04 Tu	17 10	19 38	21 05	19 32	04 01	22 21	20 53	14 35	04 46	22 27
05 We	16 54	16 05	20 52	19 36	04 10	22 22	20 54	14 35	04 47	22 27
06 Th	16 38	11 54	20 37	19 40	04 18	22 23	20 55	14 35	04 47	22 27
07 Fr	16 21	07 18	20 19	19 44	04 26	22 24	20 56	14 35	04 48	22 28
08 Sa	16 04	02 29	19 59	19 48	04 34	22 25	20 57	14 35	04 48	22 28
09 Su	15 47	02N22	19 36	19 51	04 41	22 26	20 57	14 35	04 49	22 28
10 Mo	15 29	07 07	19 10	19 55	04 49	22 27	20 58	14 35	04 49	22 28
11 Tu	15 11	11 37	18 42	19 57	04 56	22 27	20 59	14 35	04 50	22 29
12 We	14 54	15 43	18 12	20 00	05 03	22 28	21 00	14 35	04 50	22 29
13 Th	14 35	19 13	17 40	20 02	05 10	22 29	21 01	14 35	04 51	22 29
14 Fr	14 17	21 56	17 06	20 04	05 16	22 30	21 02	14 35	04 52	22 30
15 Sa	13 58	23 39	16 30	20 05	05 23	22 31	21 02	14 35	04 52	22 30
16 Su	13 39	24 08	15 52	20 06	05 29	22 32	21 03	14 35	04 53	22 30
17 Mo	13 20	23 14	15 13	20 06	05 35	22 32	21 04	14 35	04 53	22 30
18 Tu	13 01	20 54	14 32	20 07	05 40	22 33	21 05	14 35	04 54	22 31
19 We	12 41	17 14	13 51	20 06	05 46	22 34	21 05	14 35	04 55	22 31
20 Th	12 22	12 27	13 08	20 06	05 51	22 35	21 06	14 35	04 55	22 31
21 Fr	12 02	06 54	12 25	20 05	05 56	22 35	21 07	14 35	04 56	22 32
22 Sa	11 42	00 57	11 41	20 03	06 01	22 36	21 08	14 35	04 56	22 32
23 Su	11 21	05S01	10 56	20 01	06 06	22 37	21 08	14 35	04 57	22 32
24 Mo	11 01	10 40	10 11	19 59	06 10	22 37	21 09	14 35	04 58	22 32
25 Tu	10 40	15 38	09 25	19 56	06 15	22 38	21 10	14 35	04 58	22 33
26 We	10 19	19 39	08 39	19 53	06 19	22 38	21 10	14 34	04 59	22 33
27 Th	09 58	22 29	07 53	19 49	06 22	22 39	21 11	14 34	05 00	22 33
28 Fr	09 37	23 59	07 07	19 44	06 26	22 39	21 12	14 34	05 00	22 33
29 Sa	09 16	24 07	06 20	19 40	06 29	22 40	21 12	14 34	05 01	22 33
30 Su	08 55	22 54	05 34	19 34	06 32	22 40	21 13	14 34	05 01	22 34
31 Mo	08 33	20 32	04 47	19 29	06 35	22 40	21 13	14 33	05 02	22 34

ASPECTARIAN

Day	Time	Aspect
01	06:56	☽ △ ♅
	10:52	☿ ☍ ♇
	21:23	☽ □ ♂
	23:57	☽ ☌ ♃
02	00:55	☽ ✶ ♆
	05:57	☽ ☌ ♇
	08:53	☽ ☍ ☿
	11:19	☉ □ ♅
	14:00	☽ ☌ ♄
	22:25	☽ ∥ ♇
	23:44	☽ ∥ ♃
03	11:26	☽ ⋕ ☿
	13:52	☽ □ ♅
	14:01	☽ ∥ ♄
	21:01	☿ ☍ ♄
04	00:54	☽ ⋕ ♀
	06:45	☽ ✶ ♂
	13:08	♂ □ ♃
	18:38	☽ ⋕ ☉
	20:59	☿ ⋕ ♄
	21:46	☽ △ ♀
05	03:33	☿ ♌
	09:05	☽ ⋕ ♅
	23:01	☽ ✶ ♅
06	16:21	☽ ✶ ♃
	18:12	☽ ☌ ♆
	23:38	☽ ✶ ♇
07	07:54	☽ ✶ ♄
	12:34	☽ ∥ ♆
	12:54	☽ □ ♀
	14:02	☽ ⋕ ♂
	15:22	♀ ♋
08	00:13	☽ △ ☿
	09:54	☿ ∥ ♀
	22:50	☽ △ ☉
09	03:45	☽ □ ♃
	08:36	☽ ☌ ♂
	11:38	☽ □ ♇
	11:54	☽ ∥ ♂
	12:16	☽ ⋕ ♆
	19:51	☽ □ ♄
10	03:05	♂ ⋕ ♆
	06:24	☽ ✶ ♀
	12:52	☿ □ ♅
	23:06	☽ ☌ ♅
11	01:13	☽ □ ☿
	15:53	☽ △ ♃
	17:07	☽ ∥ ♅
	18:33	☽ ✶ ♆
	19:19	☽ ∥ ☉
12	00:04	☽ △ ♇
	07:56	☽ △ ♄
	14:19	☽ ∥ ☿
	23:52	☉ ∥ ♅
13	04:58	♃ ∥ ♇
	06:26	☽ ∥ ♀
	07:06	♂ □ ♇
	14:55	☽ ⋕ ♄
14	00:34	☽ ✶ ☿
	05:17	☽ □ ♆
	06:13	☽ ⋕ ♇
	06:20	☽ ⋕ ♃
	08:34	☽ ✶ ☉
	11:20	☽ ✶ ♂
15	13:28	☽ ☌ ♀
	14:27	♅ SR
	19:19	☽ ✶ ♅
16	09:34	☽ ☍ ♃
	12:30	☽ △ ♆
	14:02	[illegible]
	17:24	☽ ☍ ♇
	19:33	☽ □ ♂
17	00:00	☽ ☍ ♄
	05:30	☿ △ ♂
	08:46	☽ ⋕ ♃
	09:10	☽ ⋕ ♇
	15:08	☿ ☌ ☉
	22:20	☿ ∥ ♅
	22:35	☽ ⋕ ♄
18	00:02	☽ □ ♅
	05:56	☽ ∥ ♀
	19:28	♀ ✶ ♅
	23:52	☽ △ ♂
19	05:39	☽ ☌ ☿
	13:53	☽ ∥ ♅
	20:19	☽ ∥ ☿
20	00:29	☽ ∥ ☉
	01:30	☿ ♍
	01:53	☽ △ ♅
	04:06	☽ ✶ ♀
	14:14	☽ △ ♃
	17:16	☽ ☍ ♆
	21:48	☽ △ ♇
21	02:22	☽ ∥ ♂
	03:37	☽ △ ♄
	06:07	☽ ⋕ ♆
	23:21	☿ ∥ ☉
22	08:26	☽ □ ♀
	14:46	☽ □ ♃
	15:46	☉ ♍
	22:34	☽ □ ♇
	23:44	☽ ∥ ♆
23	00:06	[illegible]
	04:20	☽ □ ♄
	04:32	☽ ⋕ ♂
	11:35	☽ ✶ ☉
	22:10	☽ ⋕ ☿
	22:42	☽ ✶ ☿
24	01:32	☽ ⋕ ☉
	04:00	☽ ☍ ♅
	13:52	☽ △ ♀
	16:21	☽ ✶ ♃
	18:21	♂ □ ♄
	18:36	☽ ⋕ ♅
	19:48	☽ △ ♆
25	00:36	☽ ✶ ♇
	00:18	[illegible]
	15:18	☿ △ ♅
	22:27	♀ ☍ ♃
26	01:36	☽ ⋕ ♀
	09:47	☽ □ ☿
	11:38	☽ ∥ ♄
	23:43	☽ □ ♆
27	00:47	☽ ∥ ♇
	01:51	☽ ∥ ♃
	12:00	☽ △ ♂
	21:13	♀ △ ♆
28	03:09	☽ △ ☉
	12:52	☽ △ ♅
	19:34	☿ ∥ ♂
29	00:05	☽ △ ♆
	01:56	☽ ☌ ♃
	05:55	☽ ✶ ♆
	08:51	☽ ☍ ♀
	11:12	☽ ☌ ♇
	13:28	☿ △ ♃
	17:19	☽ ☌ ♄
	19:31	☽ □ ♂
30	03:02	☽ ∥ ♃
	04:18	☽ ∥ ♇
	13:29	♀ ☍ ♇
	16:26	☿ ⋕ ♆
	18:05	☽ ∥ ♄
	18:44	☿ ☍ ♆
	20:32	☽ □ ♅
31	08:39	☽ ⋕ ♀

⚷ Chiron

01 Dec. 06 N 14

Day	Position
01	09♈16 R
04	09 13
07	09 09
10	09 05
13	09 00
16	08 55
19	08 50
22	08 44
25	08 38
28	08 32
31	08 26

September 2020

Day	S. T.	☉	☽	☿	♀	♂	♃	♄	♅	♆	♇	☊ True
	h m s	° ′ ″	° ′ ″	° ′	° ′	° ′	° ′	° ′	° ′	° ′	° ′	° ′
01 Tu	22 42 28	09♍ 01 14	25♒ 01 08	21♍ 59	24♋ 18	27♈ 34	17♑R38	25♑R58	10♉R35	19♓R51	22♑R45	26♊R23
02 We	22 46 24	09 59 17	07♓ 26 24	23 41	25 22	27 41	17 36	25 56	10 34	19 49	22 44	26 11
03 Th	22 50 21	10 57 22	19 42 05	25 22	26 26	27 48	17 34	25 53	10 33	19 47	22 43	25 59
04 Fr	22 54 17	11 55 28	01♈ 48 55	27 01	27 30	27 53	17 32	25 51	10 32	19 46	22 42	25 46
05 Sa	22 58 14	12 53 36	13 48 01	28 40	28 35	27 58	17 31	25 49	10 31	19 44	22 41	25 34
06 Su	23 02 10	13 51 45	25 41 11	00♎ 17	29 40	28 02	17 29	25 46	10 30	19 42	22 40	25 24
07 Mo	23 06 07	14 49 57	07♉ 31 02	01 53	00♌ 45	28 05	17 28	25 44	10 29	19 41	22 40	25 17
08 Tu	23 10 04	15 48 11	19 21 02	03 28	01 51	28 07	17 27	25 42	10 28	19 39	22 39	25 13
09 We	23 14 00	16 46 26	01♊ 15 28	05 02	02 56	28 08	17 26	25 40	10 27	19 37	22 38	25 11
10 Th	23 17 57	17 44 44	13 19 13	06 34	04 02	28R 09	17 25	25 38	10 26	19 36	22 38	25D 11
11 Fr	23 21 53	18 43 04	25 37 32	08 06	05 08	28 08	17 25	25 36	10 25	19 34	22 37	25 11
12 Sa	23 25 50	19 41 26	08♋ 15 39	09 36	06 15	28 07	17 25	25 35	10 23	19 33	22 36	25R 11
13 Su	23 29 46	20 39 50	21 18 16	11 05	07 21	28 04	17 24	25 33	10 22	19 31	22 36	25 09
14 Mo	23 33 43	21 38 16	04♌ 48 53	12 33	08 28	28 01	17D 24	25 32	10 21	19 29	22 35	25 04
15 Tu	23 37 39	22 36 44	18 48 50	14 00	09 35	27 57	17 25	25 30	10 19	19 28	22 34	24 57
16 We	23 41 36	23 35 15	03♍ 16 31	15 25	10 42	27 52	17 25	25 29	10 18	19 26	22 34	24 47
17 Th	23 45 33	24 33 47	18 06 50	16 49	11 49	27 46	17 26	25 28	10 16	19 24	22 33	24 35
18 Fr	23 49 29	25 32 21	03♎ 11 38	18 13	12 57	27 40	17 27	25 26	10 15	19 23	22 33	24 24
19 Sa	23 53 26	26 30 57	18 20 46	19 34	14 05	27 32	17 28	25 25	10 13	19 21	22 32	24 13
20 Su	23 57 22	27 29 35	03♏ 23 53	20 55	15 13	27 24	17 29	25 24	10 12	19 19	22 32	24 03
21 Mo	00 01 19	28 28 14	18 12 12	22 14	16 21	27 15	17 31	25 24	10 10	19 18	22 32	23 56
22 Tu	00 05 15	29 26 55	02♐ 39 44	23 32	17 29	27 05	17 32	25 23	10 08	19 16	22 31	23 52
23 We	00 09 12	00♎ 25 38	16 43 42	24 48	18 37	26 54	17 34	25 22	10 07	19 14	22 31	23 50
24 Th	00 13 08	01 24 23	00♑ 24 05	26 03	19 46	26 42	17 36	25 22	10 05	19 13	22 31	23 49
25 Fr	00 17 05	02 23 10	13 42 47	27 17	20 54	26 30	17 38	25 21	10 03	19 11	22 30	23 49
26 Sa	00 21 01	03 21 58	26 42 40	28 28	22 03	26 17	17 41	25 21	10 01	19 10	22 30	23 47
27 Su	00 24 58	04 20 47	09♒ 26 47	29 38	23 12	26 03	17 43	25 20	09 59	19 08	22 30	23 44
28 Mo	00 28 55	05 19 39	21 58 00	00♏ 46	24 21	25 49	17 46	25 20	09 58	19 06	22 30	23 38
29 Tu	00 32 51	06 18 32	04♓ 18 41	01 52	25 31	25 34	17 49	25 20	09 56	19 05	22 29	23 29
30 We	00 36 48	07 17 27	16 30 46	02 56	26 40	25 19	17 52	25D 20	09 54	19 03	22 29	23 18

Data for	09-01-2020
Julian Day	2459093.50
Ayanamsa	24 08 27
SVP	04 ♓ 58 00
☽ ☊ Mean	25 ♊ 19 R

● ◐ PHASES ○ ◑

02	05:23	○	10♓12
10	09:26	◑	18♊08
17	11:00	●	25♍01
24	01:55	◐	01♑29

LAST ASPECT ☽ Day	h m	INGRESS Day	h m	
01	04:57	01	09:35	♓
03	14:35	03	20:23	♈
06	04:46	06	08:45	♉
08	12:48	08	21:29	♊
11	04:49	11	08:23	♋
13	12:05	13	15:33	♌
15	15:10	15	18:38	♍
17	11:42	17	18:56	♎
19	14:29	19	18:33	♏
21	18:13	21	19:32	♐
23	17:32	23	23:17	♑
26	03:37	26	06:09	♒
28	07:19	28	15:35	♓

DECLINATION

Day	☉	☽	☿	♀	♂	♃	♄	♅	♆	♇
01 Tu	08N11	17S 13	04N01	19N22	06N38	22S 41	21S 14	14N33	05S 03	22S 34
02 We	07 49	13 10	03 15	19 16	06 40	22 41	21 14	14 33	05 03	22 34
03 Th	07 27	08 39	02 29	19 08	06 43	22 41	21 15	14 33	05 04	22 35
04 Fr	07 05	03 50	01 43	19 01	06 45	22 42	21 15	14 32	05 05	22 35
05 Sa	06 43	01N04	00 57	18 52	06 46	22 42	21 16	14 32	05 05	22 35
06 Su	06 21	05 53	00 12	18 44	06 48	22 42	21 16	14 32	05 06	22 35
07 Mo	05 58	10 30	00S 33	18 34	06 49	22 42	21 17	14 31	05 07	22 35
08 Tu	05 36	14 43	01 17	18 25	06 50	22 43	21 17	14 31	05 07	22 36
09 We	05 13	18 25	02 01	18 14	06 51	22 43	21 18	14 31	05 08	22 36
10 Th	04 51	21 22	02 45	18 04	06 51	22 43	21 18	14 30	05 09	22 36
11 Fr	04 28	23 25	03 28	17 52	06 51	22 43	21 19	14 30	05 09	22 36
12 Sa	04 05	24 20	04 11	17 41	06 52	22 43	21 19	14 29	05 10	22 36
13 Su	03 42	23 57	04 53	17 28	06 51	22 43	21 19	14 29	05 11	22 36
14 Mo	03 19	22 11	05 35	17 16	06 51	22 43	21 20	14 29	05 11	22 37
15 Tu	02 56	19 02	06 16	17 02	06 50	22 43	21 20	14 28	05 12	22 37
16 We	02 33	14 39	06 56	16 49	06 49	22 43	21 20	14 28	05 13	22 37
17 Th	02 10	09 16	07 36	16 34	06 48	22 43	21 21	14 27	05 13	22 37
18 Fr	01 46	03 16	08 15	16 20	06 47	22 43	21 21	14 27	05 14	22 37
19 Sa	01 23	02S 58	08 54	16 05	06 46	22 43	21 21	14 26	05 15	22 37
20 Su	01 00	09 00	09 32	15 49	06 44	22 43	21 21	14 26	05 15	22 37
21 Mo	00 36	14 26	10 08	15 33	06 42	22 43	21 21	14 25	05 16	22 38
22 Tu	00 13	18 54	10 45	15 16	06 40	22 43	21 22	14 25	05 16	22 38
23 We	00S 10	22 09	11 20	14 59	06 38	22 42	21 22	14 24	05 17	22 38
24 Th	00 34	24 01	11 54	14 42	06 35	22 42	21 22	14 23	05 18	22 38
25 Fr	00 57	24 26	12 28	14 24	06 33	22 42	21 22	14 23	05 18	22 38
26 Sa	01 20	23 30	13 00	14 06	06 30	22 42	21 22	14 22	05 19	22 38
27 Su	01 44	21 22	13 32	13 47	06 27	22 41	21 22	14 22	05 20	22 38
28 Mo	02 07	18 14	14 02	13 28	06 24	22 41	21 22	14 21	05 20	22 38
29 Tu	02 30	14 20	14 32	13 09	06 21	22 41	21 23	14 21	05 21	22 38
30 We	02 54	09 54	15 00	12 49	06 18	22 40	21 23	14 20	05 21	22 39

ASPECTARIAN

01 04:57 ☽ ✶ ♂
10:40 ☿ △ ♇
16:15 ☽ ⋕ ♅
02 06:05 ☽ ✶ ♅
12:18 ♀ ☍ ♄
14:09 ☉ △ ♅
19:49 ☽ ✶ ♃
03 00:10 ☽ ☌ ♆
05:56 ☽ ✶ ♇
06:35 ☽ ⋕ ☉
07:23 ☿ △ ♄
09:46 ☽ ⋕ ♂
12:11 ☽ ✶ ♄
12:58 ☽ ☍ ☿
14:35 ☽ △ ♀
17:57 ☽ ∥ ♆
04 09:13 ♀ □ ♂
12:22 ☽ ⋕ ☿
20:32 ☿ ✶ ♀
20:57 ☉ ∥ ♂
23:35 ☽ ∥ ☿
05 07:27 ☽ □ ♃
17:55 ☽ □ ♇
19:47 ☿ ♎
20:03 ☽ ⋕ ♆
06 00:10 ☽ □ ♄
02:11 ☽ ∥ ☉
04:39 ☽ ∥ ♂
04:46 ☽ ☌ ♂
07:22 ♀ ♌
08:53 ☽ □ ♀
07 06:01 ☽ ☌ ♅
16:10 ☽ △ ☉
20:09 ☽ △ ♃
22:47 ☽ ∥ ♅
08 00:37 ☽ ✶ ♆
06:40 ☽ △ ♇
12:48 ☽ △ ♄
22:53 ☽ ∥ ♀
09 03:42 ☽ ✶ ♀
05:27 ☉ ⋕ ♆
08:39 ☽ △ ☿
16:05 ☉ △ ♃
22:24 ♂ SR
23:25 ☽ ⋕ ♄
10 12:17 ☽ □ ♆
13:02 ☽ ⋕ ♇
14:28 ☽ ⋕ ♃
11 04:49 ☽ ✶ ♂
20:26 ☉ ☍ ♆
21:45 ☿ ⋕ ☉
12 02:49 ☽ □ ☿
03:58 ☽ ✶ ♅
16:55 ☽ ☍ ♃
20:46 ☽ △ ♆
22:45 ☽ ✶ ☉
13 00:42 ♃ SD
02:20 ☽ ☍ ♇
07:37 ☽ ☍ ♄
10:08 ☿ ∥ ♆
12:05 ☽ □ ♂
18:23 ☽ ⋕ ♃
19:37 ☽ ⋕ ♇
14 06:54 ☽ ☌ ♀
07:37 ☽ ⋕ ♄
09:34 ☽ □ ♅
14:54 ☽ ✶ ☿
23:07 ☉ △ ♇
15 12:20 ☽ ∥ ♀
15:10 ☽ △ ♂
15:30 ♀ □ ♅
20:00 ☿ ⋕ ♂
16 00:55 ☽ ∥ ♅
11:24 ☽ △ ♅
22:54 ☽ △ ♃
17 02:04 ☽ ☍ ♆
06:11 ☽ ⋕ ☿
07:06 ☽ △ ♇
10:07 ☽ ∥ ♂
10:35 ☿ □ ♃
11:43 ☽ △ ♄
16:22 ☽ ⋕ ♆
21:37 ☉ △ ♄
18 06:11 ☽ ∥ ☉
16:41 ☽ ✶ ♀
18:17 ☽ ⋕ ☉
22:36 ☽ □ ♃
19 02:08 ☽ ☌ ☿
06:40 ☽ □ ♇
08:56 ☽ ∥ ♆
11:14 ☽ □ ♄
14:29 ☽ ☍ ♂
14:52 ☽ ⋕ ♂
20 02:31 ☽ ∥ ☿
10:56 ☽ ☍ ♅
20:42 ☽ □ ♀
22:52 ☽ ✶ ♃
21 00:01 ☽ ⋕ ♅
01:47 ☽ △ ♆
05:16 ☽ ⋕ ♀
05:20 ☿ □ ♇
07:07 ☽ ✶ ♇
11:51 ☽ ✶ ♄
18:13 ☽ ✶ ☉
22 13:31 ☉ ♎
17:14 ☽ ∥ ♄
23 03:34 ☽ △ ♀
04:21 ☽ □ ♆
04:48 ☽ ∥ ♇
05:36 ☽ ∥ ♃
10:39 ☿ □ ♄
15:32 ☽ ✶ ☿
17:32 ☽ △ ♂
24 10:53 ☿ ☍ ♂
17:21 ☽ △ ♅
25 01:42 ♀ ∥ ♅
07:13 ☽ ☌ ♃
10:01 ☽ ✶ ♆
16:10 ☽ ☌ ♇
21:27 ☽ ☌ ♄
23:13 ☽ □ ♂
26 03:37 ☽ □ ☿
10:30 ☽ ∥ ♃
11:08 ☽ ∥ ♇
13:32 ☽ △ ☉
23:49 ☽ ∥ ♄
27 01:02 ☽ □ ♅
07:20 ☿ ⋕ ♀
07:41 ☿ ♏
28 05:06 ☽ ☍ ♀
07:19 ☽ ✶ ♂
14:55 ☿ ⋕ ♅
18:45 ☽ △ ☿
22:59 ☽ ∥ ☿
23:56 ☽ ⋕ ♅
29 01:02 ♀ △ ♂
05:13 ♄ SD
07:16 ☽ ⋕ ♀
10:59 ☽ ✶ ♅
21:50 ♂ □ ♄
30 02:42 ☽ ✶ ♃
05:01 ☽ ☌ ♆
11:50 ☽ ✶ ♇
17:31 ☽ ✶ ♄
18:28 ☽ ⋕ ♂
21:51 ☽ ∥ ♆

⚷ Chiron

01	Dec.	05 N 55
03	08♈19R	
06	08 12	
09	08 04	
12	07 57	
15	07 49	
18	07 41	
21	07 33	
24	07 25	
27	07 17	
30	07 09	

October 2020

Day	S. T.	☉	☽	☿	♀	♂	♃	♄	♅	♆	♇	☊ True
	h m s	° ' "	° ' "	° '	° '	° '	° '	° '	° '	° '	° '	° '
01 Th	00 40 44	08♎16 24	28♓35 43	03♏57	27♌50	25♈R03	17♑55	25♑20	09♉R52	19♓R02	22♑R29	23♊R06
02 Fr	00 44 41	09 15 23	10♈34 44	04 56	28 59	24 46	17 59	25 21	09 50	19 00	22 29	22 53
03 Sa	00 48 37	10 14 25	22 28 59	05 52	00♍09	24 29	18 03	25 21	09 48	18 59	22 29	22 40
04 Su	00 52 34	11 13 28	04♉19 51	06 45	01 19	24 12	18 07	25 21	09 46	18 57	22 29	22 30
05 Mo	00 56 30	12 12 33	16 09 14	07 35	02 29	23 54	18 11	25 22	09 44	18 56	22D 29	22 22
06 Tu	01 00 27	13 11 41	27 59 45	08 21	03 40	23 36	18 15	25 23	09 41	18 54	22 29	22 17
07 We	01 04 24	14 10 50	09♊54 41	09 04	04 50	23 18	18 19	25 23	09 39	18 53	22 29	22 15
08 Th	01 08 20	15 10 02	21 58 06	09 42	06 01	22 59	18 24	25 24	09 37	18 51	22 29	22D 15
09 Fr	01 12 17	16 09 17	04♋14 36	10 16	07 11	22 40	18 28	25 25	09 35	18 50	22 29	22 15
10 Sa	01 16 13	17 08 34	16 49 06	10 45	08 22	22 21	18 33	25 26	09 33	18 49	22 29	22R 16
11 Su	01 20 10	18 07 53	29 46 27	11 08	09 33	22 02	18 39	25 27	09 31	18 47	22 30	22 15
12 Mo	01 24 06	19 07 14	13♌10 49	11 25	10 44	21 43	18 44	25 28	09 28	18 46	22 30	22 12
13 Tu	01 28 03	20 06 38	27 04 45	11 36	11 55	21 23	18 49	25 30	09 26	18 44	22 30	22 06
14 We	01 31 59	21 06 04	11♍28 13	11 40	13 07	21 04	18 55	25 31	09 24	18 43	22 30	21 59
15 Th	01 35 56	22 05 32	26 17 47	11R 37	14 18	20 45	19 01	25 33	09 21	18 42	22 31	21 50
16 Fr	01 39 53	23 05 02	11♎26 31	11 25	15 30	20 26	19 07	25 34	09 19	18 40	22 31	21 40
17 Sa	01 43 49	24 04 34	26 44 31	11 06	16 41	20 08	19 13	25 36	09 17	18 39	22 31	21 31
18 Su	01 47 46	25 04 09	12♏00 37	10 38	17 53	19 49	19 19	25 38	09 14	18 38	22 32	21 24
19 Mo	01 51 42	26 03 45	27 04 12	10 01	19 05	19 31	19 25	25 40	09 12	18 37	22 32	21 18
20 Tu	01 55 39	27 03 23	11♐47 02	09 15	20 16	19 13	19 32	25 42	09 10	18 35	22 33	21 14
21 We	01 59 35	28 03 04	26 04 08	08 22	21 28	18 56	19 39	25 44	09 07	18 34	22 33	21 13
22 Th	02 03 32	29 02 45	09♑53 54	07 21	22 41	18 39	19 45	25 46	09 05	18 33	22 34	21D 13
23 Fr	02 07 28	00♏02 29	23 17 26	06 14	23 53	18 23	19 53	25 48	09 02	18 32	22 34	21 14
24 Sa	02 11 25	01 02 14	06♒17 34	05 01	25 05	18 07	20 00	25 51	09 00	18 31	22 35	21R 14
25 Su	02 15 22	02 02 01	18 57 55	03 46	26 17	17 51	20 07	25 53	08 57	18 30	22 35	21 13
26 Mo	02 19 18	03 01 50	01♓22 22	02 30	27 30	17 37	20 15	25 56	08 55	18 29	22 36	21 10
27 Tu	02 23 15	04 01 40	13 34 35	01 15	28 42	17 22	20 22	25 58	08 53	18 28	22 36	21 06
28 We	02 27 11	05 01 32	25 37 51	00 04	29 55	17 09	20 30	26 01	08 50	18 27	22 37	20 59
29 Th	02 31 08	06 01 25	07♈34 58	28♎59	01♎08	16 56	20 38	26 04	08 48	18 26	22 38	20 52
30 Fr	02 35 04	07 01 21	19 28 12	28 02	02 20	16 44	20 46	26 07	08 45	18 25	22 39	20 44
31 Sa	02 39 01	08 01 18	01♉19 26	27 14	03 33	16 33	20 54	26 10	08 43	18 24	22 39	20 36

Data for	10-01-2020
Julian Day	2459123.50
Ayanamsa	24 08 30
SVP	04 ♓ 57 54
☽ ☊ Mean	23 ♊ 44 R

● ◐ PHASES ○ ◑

01	21:06	○	09♈08
10	00:39	◑	17♋10
16	19:31	●	23♎54
23	13:24	◐	00♒36
31	14:49	○	08♉38

LAST ASPECT ☽ Day	h m	INGRESS Day	h m	
30	17:31	01	02:48	♈
03	05:48	03	15:13	♉
05	18:42	06	04:03	♊
08	01:57	08	15:46	♋
10	16:04	11	00:25	♌
12	14:30	13	04:56	♍
14	22:48	15	05:54	♎
16	22:12	17	05:06	♏
18	21:44	19	04:44	♐
21	03:39	21	06:45	♑
23	04:36	23	12:18	♒
24	21:55	25	21:19	♓
28	00:47	28	08:45	♈
30	16:13	30	21:19	♉

DECLINATION

Day	☉	☽	☿	♀	♂	♃	♄	♅	♆	♇
01 Th	03S 17	05S 08	15S 27	12N28	06N14	22S 40	21S 23	14N19	05S 22	22S 39
02 Fr	03 40	00 12	15 52	12 08	06 11	22 39	21 23	14 19	05 23	22 39
03 Sa	04 03	04N42	16 16	11 46	06 07	22 39	21 23	14 18	05 23	22 39
04 Su	04 26	09 25	16 39	11 25	06 04	22 39	21 23	14 17	05 24	22 39
05 Mo	04 49	13 48	17 00	11 03	06 00	22 38	21 23	14 17	05 24	22 39
06 Tu	05 13	17 41	17 19	10 41	05 56	22 37	21 22	14 16	05 25	22 39
07 We	05 35	20 52	17 36	10 18	05 53	22 37	21 22	14 15	05 26	22 39
08 Th	05 58	23 11	17 51	09 56	05 49	22 36	21 22	14 15	05 26	22 39
09 Fr	06 21	24 26	18 04	09 32	05 45	22 36	21 22	14 14	05 27	22 39
10 Sa	06 44	24 30	18 15	09 09	05 41	22 35	21 22	14 13	05 27	22 39
11 Su	07 07	23 14	18 23	08 45	05 38	22 34	21 22	14 13	05 28	22 39
12 Mo	07 29	20 39	18 29	08 21	05 34	22 34	21 22	14 12	05 28	22 39
13 Tu	07 52	16 49	18 31	07 57	05 30	22 33	21 21	14 11	05 29	22 39
14 We	08 14	11 53	18 30	07 32	05 27	22 32	21 21	14 10	05 29	22 39
15 Th	08 36	06 07	18 25	07 07	05 23	22 31	21 21	14 10	05 30	22 39
16 Fr	08 58	00S 07	18 17	06 42	05 20	22 31	21 21	14 09	05 30	22 39
17 Sa	09 20	06 25	18 04	06 17	05 16	22 30	21 20	14 08	05 31	22 39
18 Su	09 42	12 21	17 48	05 51	05 13	22 29	21 20	14 07	05 31	22 39
19 Mo	10 04	17 27	17 27	05 25	05 10	22 28	21 20	14 07	05 32	22 39
20 Tu	10 25	21 22	17 01	04 59	05 07	22 27	21 20	14 06	05 32	22 39
21 We	10 47	23 49	16 31	04 33	05 05	22 26	21 19	14 05	05 33	22 39
22 Th	11 08	24 42	15 56	04 07	05 02	22 25	21 19	14 04	05 33	22 39
23 Fr	11 29	24 07	15 18	03 40	05 00	22 24	21 18	14 04	05 34	22 39
24 Sa	11 50	22 13	14 37	03 13	04 58	22 23	21 18	14 03	05 34	22 39
25 Su	12 11	19 15	13 53	02 46	04 56	22 22	21 18	14 02	05 34	22 39
26 Mo	12 31	15 30	13 08	02 19	04 54	22 21	21 17	14 01	05 35	22 39
27 Tu	12 52	11 09	12 23	01 52	04 53	22 20	21 17	14 01	05 35	22 39
28 We	13 12	06 26	11 38	01 25	04 52	22 19	21 16	14 00	05 36	22 39
29 Th	13 32	01 31	10 57	00 57	04 51	22 18	21 16	13 59	05 36	22 39
30 Fr	13 51	03N26	10 18	00 30	04 50	22 16	21 15	13 58	05 36	22 39
31 Sa	14 11	08 16	09 44	00 02	04 50	22 15	21 15	13 57	05 37	22 39

ASPECTARIAN

01 08:24 ☽ ∥ ☉
02 14:59 ☽ □ ♃
20:33 ♀ ♍
20:49 ☽ # ☉
03 00:01 ☽ □ ♇
03:26 ☽ # ♆
03:58 ☽ ☌ ♂
05:48 ☽ □ ♄
07:01 ☽ ∥ ♂
11:30 ♃ ∥ ♇
17:14 ☽ △ ♀
04 05:17 ☽ ☍ ☿
09:49 ☽ ∥ ♀
10:59 ☽ ☌ ♅
13:31 ♇ SD
05 02:45 ☽ ∥ ♅
04:08 ☽ △ ♃
05:37 ☽ ⚹ ♆
12:50 ☽ △ ♇
18:42 ☽ △ ♄
21:22 ☽ # ☿
06 12:41 ☽ □ ♀
13:18 ☉ ∥ ♆
07 04:34 ☽ # ♄
09:18 ☽ △ ☉
15:20 ☉ # ♂
17:08 ☽ # ♃
17:37 ☽ # ♇
17:51 ☽ □ ♆
20:55 ☿ ☍ ♅
08 01:57 ☽ ⚹ ♂
09 06:16 ☽ ⚹ ♀
10:14 ☽ ⚹ ♅
12:03 ☽ △ ☿
13:19 ♂ □ ♇
10 03:17 ☽ ☍ ♃
03:44 [illegible]
10:05 ☽ □ ♂
10:36 ☽ ☍ ♇
16:04 ☽ ☍ ♄
23:09 ♀ △ ♅
11 06:36 ☽ # ♇
07:27 ☽ # ♃
13:35 ☉ □ ♃
17:28 ☽ □ ♅
18:27 ☽ # ♄
20:51 ☽ □ ☿
12 07:12 ♃ ⚹ ♆
11:10 ☽ ⚹ ☉
14:19 ☽ # ☿
14:30 ☽ △ ♂
16:39 ☿ ⚹ ♀
13 02:38 ♀ # ☉
08:30 ♂ # ♆
13:22 ☽ ∥ ♅
20:36 ☽ △ ♅
23:26 ☉ ☍ ♂
14 00:20 ☽ ⚹ ☿
01:06 ☿ SR
02:55 ☽ ☌ ♀
11:48 ☽ ☍ ♆
12:12 ☽ △ ♃
14:34 ☽ # ☉
17:56 ☽ △ ♇
19:41 ☽ ∥ ♀
22:48 ☽ △ ♄
15 02:26 ☽ # ♆
02:54 ☽ ∥ ♂
10:13 ☉ □ ♇
16 12:07 ☽ □ ♃
[illegible]
17:23 ☽ □ ♇
19:35 ☽ # ♂
20:27 ☽ ∥ ♆
22:12 ☽ □ ♄
23:27 ☽ # ♀
17 12:14 ☽ ∥ ☉
19:38 ☽ ☍ ♅
21:53 ☽ ☌ ☿
18 07:48 ☽ # ♅
10:06 ☽ ⚹ ♀
10:29 ☽ △ ♆
11:40 ☽ ⚹ ♃
13:58 ☉ □ ♄
14:49 ♀ ☍ ♆
16:43 ☽ ⚹ ♇
18:08 ♀ # ♆
21:44 ☽ ⚹ ♄
23:57 ☽ ∥ ☿
19 05:39 ♂ □ ♃
07:36 ♀ △ ♃
15:32 ♀ ∥ ♂
23:45 ☽ ∥ ♄
20 02:54 ☿ ☍ ♅
08:56 ☽ ∥ ♃
10:51 ☽ ∥ ♇
11:20 ☽ □ ♆
12:09 ☽ △ ♂
15:29 ☽ □ ♀
21 03:39 ☽ ⚹ ☉
19:50 ☽ ⚹ ☿
21:39 ♀ △ ♇
22:34 ☽ △ ♅
22 15:17 ☽ □ ♂
13:24 ☽ ⚹ ♆
17:45 ☽ ☌ ♃
22:41 ☽ ☌ ♇
23:00 ☉ ♏
23 01:11 ☽ △ ♀
04:36 ☽ ☌ ♄
19:35 ☽ ∥ ♇
21:50 ☽ □ ☿
22:17 ☽ ∥ ♃
24 05:04 ☽ □ ♅
08:18 ☽ ∥ ♄
15:41 ♀ △ ♄
19:04 ☿ # ♅
21:55 ☽ ⚹ ♂
25 18:24 ☿ ☌ ☉
26 02:00 ☽ △ ☿
03:32 ☽ △ ☉
08:26 ☽ # ♅
13:22 ☿ ∥ ☉
14:45 ☽ ⚹ ♅
15:30 ☽ ∥ ☉
16:06 ☽ ∥ ☿
27 09:41 ☽ ☌ ♆
13:39 ☽ ⚹ ♃
17:59 ☽ ⚹ ♇
28 00:47 ☽ ⚹ ♄
01:34 ☿ ♎R
01:42 ♀ ♎
04:07 ☽ ∥ ♆
07:44 ☽ # ♂
09:33 ☽ ☍ ♀
29 02:56 ☽ # ♀
10:55 ☽ ∥ ♀
18:34 ☽ ☌ ♂
30 02:39 ☽ □ ♃
06:26 ☽ □ ♇
06:48 ☽ ∥ ♂
08:00 ☉ # ♅
10:38 ☽ # ♆
13:30 ☽ □ ♄
16:13 ☽ ☍ ☿
31 06:10 ☽ ∥ ♆
14:56 ☽ ☌ ♅
15:53 ☉ ☍ ♅

⚷ Chiron

01	Dec.	05 N2 4
03	07♈01R	
06	06 53	
09	06 45	
12	06 37	
15	06 29	
18	06 21	
21	06 14	
24	06 07	
27	06 00	
30	05 53	

30 09:44 08♊38 ✸ Penumbral Lunar Eclipse (mag 0.854)

November 2020

Day	S. T. h m s	☉ ° ' "	☽ ° ' "	☿ ° '	♀ ° '	♂ ° '	♃ ° '	♄ ° '	♅ ° '	♆ ° '	♇ ° '	☊ True ° '
01 Su	02 42 57	09♏ 01 17	13♉ 10 19	26♎R37	04♎ 46	16♈R22	21♑ 03	26♑ 13	08♉R40	18♓R23	22♑ 40	20♊R30
02 Mo	02 46 54	10 01 18	25 02 31	26 11	05 59	16 12	21 11	26 16	08 38	18 22	22 41	20 25
03 Tu	02 50 51	11 01 22	06♊ 57 58	25 57	07 12	16 03	21 20	26 20	08 35	18 21	22 42	20 23
04 We	02 54 47	12 01 27	18 58 59	25D 54	08 25	15 54	21 28	26 23	08 33	18 20	22 43	20D 22
05 Th	02 58 44	13 01 34	01♋ 08 23	26 03	09 39	15 47	21 37	26 26	08 30	18 19	22 44	20 23
06 Fr	03 02 40	14 01 43	13 29 25	26 21	10 52	15 40	21 46	26 30	08 28	18 19	22 45	20 25
07 Sa	03 06 37	15 01 54	26 05 46	26 50	12 05	15 34	21 55	26 34	08 25	18 18	22 45	20 26
08 Su	03 10 33	16 02 07	09♌ 01 15	27 28	13 19	15 29	22 05	26 37	08 23	18 17	22 46	20 27
09 Mo	03 14 30	17 02 23	22 19 27	28 13	14 32	15 24	22 14	26 41	08 21	18 17	22 47	20R 26
10 Tu	03 18 26	18 02 40	06♍ 03 02	29 06	15 46	15 21	22 24	26 45	08 18	18 16	22 49	20 24
11 We	03 22 23	19 02 59	20 12 58	00♏ 05	16 59	15 18	22 33	26 49	08 16	18 15	22 50	20 21
12 Th	03 26 20	20 03 20	04♎ 47 44	01 10	18 13	15 16	22 43	26 53	08 13	18 15	22 51	20 16
13 Fr	03 30 16	21 03 43	19 42 52	02 19	19 27	15 14	22 53	26 57	08 11	18 14	22 52	20 11
14 Sa	03 34 13	22 04 08	04♏ 51 00	03 33	20 41	15 14	23 03	27 01	08 08	18 14	22 53	20 07
15 Su	03 38 09	23 04 35	20 02 45	04 50	21 54	15D 14	23 13	27 06	08 06	18 13	22 54	20 03
16 Mo	03 42 06	24 05 04	05♐ 08 03	06 10	23 08	15 16	23 23	27 10	08 04	18 13	22 55	20 00
17 Tu	03 46 02	25 05 34	19 57 46	07 32	24 22	15 18	23 33	27 15	08 01	18 12	22 56	19 59
18 We	03 49 59	26 06 05	04♑ 25 03	08 57	25 36	15 20	23 44	27 19	07 59	18 12	22 58	19D 59
19 Th	03 53 55	27 06 38	18 25 57	10 23	26 50	15 24	23 54	27 24	07 57	18 11	22 59	19 59
20 Fr	03 57 52	28 07 12	01♒ 59 28	11 51	28 05	15 28	24 05	27 28	07 55	18 11	23 00	20 01
21 Sa	04 01 49	29 07 48	15 06 56	13 20	29 19	15 33	24 16	27 33	07 52	18 11	23 02	20 02
22 Su	04 05 45	00♐ 08 24	27 51 19	14 50	00♏ 33	15 39	24 27	27 38	07 50	18 11	23 03	20 03
23 Mo	04 09 42	01 09 02	10♓ 16 30	16 21	01 47	15 46	24 38	27 43	07 48	18 10	23 04	20 04
24 Tu	04 13 38	02 09 41	22 26 47	17 52	03 01	15 53	24 49	27 48	07 46	18 10	23 06	20R 03
25 We	04 17 35	03 10 20	04♈ 26 27	19 24	04 16	16 01	25 00	27 53	07 43	18 10	23 07	20 02
26 Th	04 21 31	04 11 01	16 19 36	20 57	05 30	16 10	25 11	27 58	07 41	18 10	23 08	20 01
27 Fr	04 25 28	05 11 44	28 09 50	22 29	06 44	16 19	25 23	28 03	07 39	18 10	23 10	19 59
28 Sa	04 29 24	06 12 27	10♉ 00 17	24 03	07 59	16 29	25 34	28 08	07 37	18 10	23 11	19 58
29 Su	04 33 21	07 13 11	21 53 31	25 36	09 13	16 40	25 45	28 14	07 35	18D 10	23 13	19 57
30 Mo	04 37 18	08 13 57	03♊ 51 41	27 09	10 28	16 51	25 57	28 19	07 33	18 10	23 14	19 57

Data for	11-01-2020
Julian Day	2459154.50
Ayanamsa	24 08 33
SVP	04 ♓ 57 49
☽ ☊ Mean	22 ♊ 05 R

LAST ASPECT Day	h m	☽ INGRESS Day	h m	
02	02:29	02	10:00	♊
04	13:49	04	21:46	♋
07	01:27	07	07:19	♌
09	11:06	09	13:31	♍
11	11:00	11	16:10	♎
13	11:34	13	16:20	♏
15	11:14	15	15:48	♐
17	07:55	17	16:35	♑
19	16:30	19	20:25	♒
21	00:49	22	04:06	♓
24	10:45	24	15:05	♈
26	23:46	27	03:43	♉
29	12:49	29	16:17	♊

● ◐ PHASES ○ ◑

08	13:46	◑	16♌37
15	05:08	●	23♏18
22	04:45	◐	00♓20
30	09:30	✸	08♊38

DECLINATION

Day	☉	☽	☿	♀	♂	♃	♄	♅	♆	♇
01 Su	14S 30	12N48	09S 15	00S 25	04N49	22S 14	21S 14	13N57	05S 37	22S 39
02 Mo	14 49	16 52	08 52	00 53	04 49	22 13	21 14	13 56	05 37	22 38
03 Tu	15 08	20 16	08 35	01 21	04 50	22 11	21 13	13 55	05 38	22 38
04 We	15 27	22 51	08 23	01 49	04 50	22 10	21 12	13 54	05 38	22 38
05 Th	15 45	24 24	08 17	02 16	04 51	22 09	21 12	13 53	05 38	22 38
06 Fr	16 03	24 48	08 17	02 44	04 52	22 07	21 11	13 53	05 38	22 38
07 Sa	16 21	23 55	08 22	03 12	04 53	22 06	21 11	13 52	05 39	22 38
08 Su	16 38	21 47	08 31	03 40	04 55	22 04	21 10	13 51	05 39	22 38
09 Mo	16 55	18 25	08 44	04 08	04 57	22 03	21 09	13 50	05 39	22 38
10 Tu	17 12	13 59	09 01	04 35	04 59	22 01	21 09	13 50	05 39	22 38
11 We	17 29	08 40	09 21	05 03	05 01	22 00	21 08	13 49	05 40	22 38
12 Th	17 45	02 44	09 44	05 31	05 04	21 58	21 07	13 48	05 40	22 37
13 Fr	18 01	03S 30	10 10	05 58	05 07	21 57	21 06	13 47	05 40	22 37
14 Sa	18 17	09 38	10 37	06 26	05 10	21 55	21 06	13 47	05 40	22 37
15 Su	18 32	15 13	11 05	06 53	05 13	21 53	21 05	13 46	05 40	22 37
16 Mo	18 47	19 49	11 35	07 20	05 17	21 52	21 04	13 45	05 41	22 37
17 Tu	19 02	23 03	12 06	07 47	05 21	21 50	21 03	13 44	05 41	22 37
18 We	19 17	24 40	12 37	08 14	05 25	21 48	21 02	13 44	05 41	22 37
19 Th	19 31	24 39	13 09	08 41	05 29	21 46	21 01	13 43	05 41	22 36
20 Fr	19 44	23 08	13 42	09 08	05 34	21 45	21 01	13 42	05 41	22 36
21 Sa	19 58	20 25	14 14	09 34	05 39	21 43	21 00	13 41	05 41	22 36
22 Su	20 11	16 46	14 46	10 01	05 44	21 41	20 59	13 41	05 41	22 36
23 Mo	20 23	12 30	15 19	10 27	05 49	21 39	20 58	13 40	05 41	22 36
24 Tu	20 35	07 49	15 51	10 53	05 55	21 37	20 57	13 39	05 41	22 36
25 We	20 47	02 55	16 22	11 18	06 01	21 35	20 56	13 39	05 41	22 35
26 Th	20 59	02N03	16 53	11 44	06 06	21 33	20 55	13 38	05 41	22 35
27 Fr	21 10	06 57	17 24	12 09	06 13	21 31	20 54	13 37	05 41	22 35
28 Sa	21 21	11 36	17 54	12 34	06 19	21 29	20 53	13 37	05 41	22 35
29 Su	21 31	15 50	18 23	12 58	06 26	21 27	20 52	13 36	05 41	22 35
30 Mo	21 41	19 29	18 52	13 23	06 32	21 25	20 51	13 35	05 41	22 35

ASPECTARIAN

01 06:24 ☽ ∥ ♅
10:27 ☽ # ☉
10:33 ☽ ✶ ♆
16:07 ☽ △ ♃
19:06 ☿ □ ♄
19:14 ☽ △ ♇
02 02:31 ☽ △ ♄
03 00:32 ☽ △ ♀
07:44 ☽ # ♄
16:42 ☽ # ♃
17:51 ☿ SD
17:57 ☽ ✶ ♂
21:33 ☽ # ♇
22:43 ☽ □ ♆

04 13:49 ☽ △ ☿
05 14:19 ☽ ✶ ♅
18:23 ☽ □ ♀
06 01:08 ☽ △ ☉
04:09 ☽ □ ♂
09:13 ☿ □ ♄
09:14 ☽ △ ♆
16:01 ☽ ☍ ♃
17:41 ☽ ☍ ♇

07 00:53 ☽ ☍ ♄
01:27 ☽ □ ☿
15:58 ☽ # ♇
21:20 ☽ # ♃
22:50 ☽ □ ♅
08 05:06 ☽ # ♄
08:37 ☽ ✶ ♀
11:40 ☽ △ ♂
09 08:12 ☽ # ☉
11:06 ☽ ✶ ☿
16:09 ♀ ☍ ♂

10 00:46 ☽ ∥ ♅
03:51 ☽ △ ♅
05:12 ☉ △ ♆
20:43 ☽ ☍ ♆
21:14 ☽ # ☿
21:54 ☽ ✶ ☉
21:56 ☿ ♏
22:15 ♀ # ♂
11 03:56 ☽ △ ♃
04:21 ☽ △ ♇
11:00 ☽ △ ♄
12:23 ☽ # ♇
13:47 ☽ # ♀
14:51 ☽ ∥ ♂
12 08:02 ♀ ∥ ♆
16:51 ☽ ☍ ♂
21:18 ♃ ☌ ♇
23:32 ☽ ☌ ♀

13 05:01 ☽ □ ♇
05:06 ☽ □ ♃
06:15 ☽ # ♂
08:21 ☽ ∥ ♆
10:17 ☽ ∥ ♀
11:34 ☽ □ ♄
21:45 ☽ ☌ ☿

14 00:37 ♂ SD
04:19 ☽ ∥ ☿
05:11 ☽ ☍ ♅
17:25 ☽ # ♅
19:46 ☉ ✶ ♇
21:07 ☽ △ ♆
15 03:58 ☉ ✶ ♃
04:32 ☽ ✶ ♇
05:05 ☽ ✶ ♃
11:14 ☽ ✶ ♄
17:39 ☽ ∥ ☉
19:42 ♀ □ ♇

16 05:34 ♀ □ ♃
07:58 ☽ ∥ ♄
13:38 ☽ ∥ ♃
16:22 ☽ △ ♂
19:57 ☽ ∥ ♇
21:07 ☽ □ ♆
17 07:55 ☽ ✶ ♀
08:08 ☿ ☍ ♅
18 06:01 ☽ △ ♅
08:32 ☽ ✶ ☿
18:43 ☽ □ ♂
23:35 ☽ ✶ ♆

19 07:18 ☉ ✶ ♄
07:59 ☽ ☌ ♇
09:43 ☽ ☌ ♃
11:29 ♀ □ ♄
15:52 ☽ ☌ ♄
16:17 ☽ □ ♀
16:30 ☽ ✶ ☉

20 00:24 ☿ # ♅
05:35 ☽ ∥ ♇
10:42 ☽ □ ♅
13:38 ☽ ∥ ♃
19:36 ☽ ∥ ♄
20:16 ☽ □ ☿
21 00:49 ☽ ✶ ♂
03:08 ☽ ∥ ☉
11:30 ♂ # ♆
13:22 ♀ ♏
20:40 ☉ ♐

22 05:43 ☽ △ ♀
10:21 ☽ ∥ ☿
17:45 ☽ # ♅
19:11 ☽ ✶ ♅
23 09:51 ☽ ∥ ♀
13:38 ☽ △ ☿
15:31 ☽ ☌ ♆

24 01:17 ☽ ✶ ♇
04:40 ☿ △ ♆
04:47 ☽ ✶ ♃
09:16 ☽ # ♂
10:32 ☽ ∥ ♆
10:45 ☽ ✶ ♄
21:13 ☽ △ ☉

25 16:33 ☉ ∥ ♄
23:40 ☽ ☌ ♂
26 13:50 ☽ □ ♇
17:46 ☽ # ♆
18:15 ☽ □ ♃
20:16 ☽ ∥ ♂
23:46 ☽ □ ♄
27 10:36 ☿ ✶ ♇
17:11 ♀ ☍ ♅
19:11 ☽ ☌ ♅
19:25 ☽ ☍ ♀

28 05:49 ☽ # ♀
11:03 ☽ ∥ ♅
16:06 ☉ ∥ ♃
16:29 ☽ ✶ ♆

29 00:38 ♆ SD
02:40 ☽ △ ♇
02:52 ☿ ✶ ♃
07:54 ☽ △ ♃
08:34 ☽ ☍ ☿
12:49 ☽ △ ♄
18:52 ☽ # ☿

30 10:31 ☽ # ♄
12:21 ♀ # ♅
15:09 ☽ # ♃
18:53 ☽ # ☉
19:01 ☿ ✶ ♄

⚷ Chiron

01	Dec.	04 N 49
02	05♈46R	
05	05 40	
08	05 34	
11	05 29	
14	05 24	
17	05 19	
20	05 15	
23	05 11	
26	05 07	
29	05 04	

December 2020

14 16:19 23♐08 ☉ Total Solar Eclipse

Day	S. T.	☉	☽	☿	♀	♂	♃	♄	♅	♆	♇	☊ True
	h m s	° ' "	° ' "	° '	° '	° '	° '	° '	° '	° '	° '	° '
01 Tu	04 41 14	09♐ 14 44	15♊ 56 40	28♏ 43	11♏ 42	17♈ 03	26♑ 09	28♑ 24	07♉ R31	18♓ 10	23♑ 16	19♊ D57
02 We	04 45 11	10 15 32	28 10 06	00 16	12 57	17 16	26 21	28 30	07 29	18 10	23 17	19 57
03 Th	04 49 07	11 16 22	10♋ 33 34	01 50	14 11	17 29	26 32	28 35	07 27	18 10	23 19	19 58
04 Fr	04 53 04	12 17 13	23 08 43	03 23	15 26	17 43	26 44	28 41	07 25	18 10	23 21	19 58
05 Sa	04 57 00	13 18 05	05♌ 57 14	04 57	16 41	17 57	26 56	28 47	07 23	18 10	23 22	19 59
06 Su	05 00 57	14 18 58	19 00 57	06 31	17 55	18 12	27 09	28 52	07 21	18 11	23 24	19R 59
07 Mo	05 04 53	15 19 53	02♍ 21 37	08 05	19 10	18 27	27 21	28 58	07 20	18 11	23 25	19 58
08 Tu	05 08 50	16 20 48	16 00 38	09 38	20 25	18 43	27 33	29 04	07 18	18 11	23 27	19 58
09 We	05 12 47	17 21 46	29 58 36	11 12	21 40	19 00	27 45	29 10	07 16	18 11	23 29	19 57
10 Th	05 16 43	18 22 44	14♎ 14 51	12 46	22 54	19 17	27 58	29 16	07 14	18 12	23 30	19 56
11 Fr	05 20 40	19 23 43	28 47 06	14 20	24 09	19 35	28 10	29 22	07 13	18 12	23 32	19 55
12 Sa	05 24 36	20 24 44	13♏ 31 09	15 54	25 24	19 53	28 23	29 28	07 11	18 13	23 34	19 54
13 Su	05 28 33	21 25 46	28 21 05	17 28	26 39	20 11	28 36	29 34	07 10	18 13	23 36	19 53
14 Mo	05 32 29	22 26 49	13♐ 09 45	19 02	27 54	20 30	28 48	29 40	07 08	18 14	23 37	19 53
15 Tu	05 36 26	23 27 52	27 49 37	20 36	29 09	20 50	29 01	29 46	07 07	18 14	23 39	19 53
16 We	05 40 22	24 28 57	12♑ 13 51	22 10	00♐ 24	21 10	29 14	29 52	07 05	18 15	23 41	19 52
17 Th	05 44 19	25 30 01	26 17 10	23 45	01 39	21 30	29 27	29 59	07 04	18 15	23 43	19 52
18 Fr	05 48 16	26 31 07	09♒ 56 27	25 19	02 54	21 51	29 40	00♒ 05	07 02	18 16	23 45	19 51
19 Sa	05 52 12	27 32 12	23 10 51	26 54	04 09	22 12	29 53	00 11	07 01	18 17	23 46	19 50
20 Su	05 56 09	28 33 19	06♓ 01 31	28 28	05 24	22 34	00♒ 06	00 18	07 00	18 17	23 48	19 50
21 Mo	06 00 05	29 34 25	18 31 16	00♑ 03	06 39	22 56	00 19	00 24	06 58	18 18	23 50	19D 50
22 Tu	06 04 02	00♑ 35 31	00♈ 43 58	01 39	07 54	23 18	00 32	00 31	06 57	18 19	23 52	19 50
23 We	06 07 58	01 36 38	12 44 08	03 14	09 09	23 41	00 46	00 37	06 56	18 20	23 54	19 51
24 Th	06 11 55	02 37 45	24 36 34	04 50	10 24	24 04	00 59	00 44	06 55	18 21	23 56	19 53
25 Fr	06 15 51	03 38 52	06♉ 26 00	06 25	11 39	24 28	01 12	00 50	06 54	18 21	23 58	19 55
26 Sa	06 19 48	04 39 59	18 16 52	08 01	12 54	24 51	01 26	00 57	06 53	18 22	24 00	19 57
27 Su	06 23 45	05 41 06	00♊ 13 05	09 38	14 09	25 16	01 39	01 04	06 52	18 23	24 01	19 59
28 Mo	06 27 41	06 42 14	12 17 57	11 14	15 24	25 40	01 53	01 10	06 51	18 24	24 03	20 01
29 Tu	06 31 38	07 43 22	24 34 02	12 51	16 39	26 05	02 06	01 17	06 50	18 25	24 05	20R 01
30 We	06 35 34	08 44 29	07♋ 03 09	14 28	17 54	26 30	02 20	01 24	06 49	18 26	24 07	20 00
31 Th	06 39 31	09 45 37	19 46 14	16 05	19 10	26 56	02 33	01 31	06 49	18 27	24 09	19 59

Data for	12-01-2020
Julian Day	2459184.50
Ayanamsa	24 08 38
SVP	04 ♓ 57 45
☽ ☊ Mean	20 ♊ 30 R

● ◐ PHASES ○ ◑

08	00:37	◑	16♍22
14	16:17	☉	23♐08
21	23:41	◐	00♈35
30	03:29	○	08♋53

LAST ASPECT Day	h m	☽ INGRESS Day	h m	
01	04:23	02	03:34	♋
04	10:30	04	12:54	♌
05	22:29	06	19:48	♍
08	22:36	09	00:02	♎
11	00:57	11	01:59	♏
13	01:58	13	02:40	♐
14	16:17	15	03:35	♑
17	05:35	17	06:27	♒
19	08:45	19	12:39	♓
21	10:25	21	22:33	♈
23	22:52	24	10:56	♉
26	11:33	26	23:34	♊
29	03:02	29	10:30	♋

DECLINATION

Day	☉	☽	☿	♀	♂	♃	♄	♅	♆	♇
01 Tu	21S 50	22N19	19S 20	13S 47	06N39	21S 22	20S 50	13N35	05S 41	22S 34
02 We	21 59	24 11	19 47	14 10	06 46	21 20	20 49	13 34	05 41	22 34
03 Th	22 08	24 52	20 13	14 34	06 54	21 18	20 48	13 33	05 41	22 34
04 Fr	22 16	24 19	20 38	14 57	07 01	21 16	20 47	13 33	05 41	22 34
05 Sa	22 24	22 28	21 02	15 19	07 09	21 14	20 46	13 32	05 41	22 34
06 Su	22 31	19 26	21 25	15 41	07 17	21 11	20 44	13 32	05 41	22 33
07 Mo	22 38	15 20	21 48	16 03	07 24	21 09	20 43	13 31	05 41	22 33
08 Tu	22 44	10 23	22 09	16 25	07 33	21 07	20 42	13 31	05 41	22 33
09 We	22 50	04 47	22 29	16 46	07 41	21 04	20 41	13 30	05 40	22 33
10 Th	22 56	01S 10	22 48	17 06	07 49	21 02	20 40	13 30	05 40	22 32
11 Fr	23 01	07 11	23 06	17 26	07 58	20 59	20 39	13 29	05 40	22 32
12 Sa	23 05	12 54	23 23	17 46	08 06	20 57	20 37	13 29	05 40	22 32
13 Su	23 10	17 53	23 38	18 05	08 15	20 54	20 36	13 28	05 40	22 32
14 Mo	23 13	21 45	23 53	18 24	08 24	20 52	20 35	13 28	05 39	22 32
15 Tu	23 17	24 09	24 06	18 42	08 33	20 49	20 34	13 27	05 39	22 31
16 We	23 19	24 52	24 18	19 00	08 42	20 46	20 32	13 27	05 39	22 31
17 Th	23 22	23 58	24 28	19 17	08 51	20 44	20 31	13 26	05 39	22 31
18 Fr	23 23	21 39	24 37	19 34	09 01	20 41	20 30	13 26	05 38	22 31
19 Sa	23 25	18 14	24 45	19 50	09 10	20 38	20 29	13 25	05 38	22 30
20 Su	23 26	14 03	24 52	20 06	09 20	20 36	20 27	13 25	05 38	22 30
21 Mo	23 26	09 23	24 57	20 21	09 29	20 33	20 26	13 25	05 38	22 30
22 Tu	23 26	04 28	25 01	20 35	09 39	20 30	20 25	13 24	05 37	22 30
23 We	23 26	00N33	25 04	20 49	09 49	20 27	20 23	13 24	05 37	22 29
24 Th	23 25	05 29	25 05	21 02	09 59	20 24	20 22	13 24	05 36	22 29
25 Fr	23 23	10 13	25 05	21 15	10 09	20 22	20 20	13 23	05 36	22 29
26 Sa	23 21	14 36	25 03	21 27	10 19	20 19	20 19	13 23	05 36	22 29
27 Su	23 19	18 27	25 00	21 38	10 29	20 16	20 18	13 23	05 35	22 28
28 Mo	23 16	21 34	24 55	21 49	10 39	20 13	20 16	13 22	05 35	22 28
29 Tu	23 13	23 45	24 49	21 59	10 50	20 10	20 15	13 22	05 35	22 28
30 We	23 09	24 48	24 41	22 09	11 00	20 07	20 13	13 22	05 34	22 28
31 Th	23 05	24 34	24 32	22 18	11 10	20 04	20 12	13 22	05 34	22 27

ASPECTARIAN

01 02:14 ☽ ⚹ ♂
02:28 ☽ ∦ ♆
04:23 ☽ □ ♆
19:52 ☿ ♐
02 18:02 ☽ ⚹ ♅
03 07:44 ☽ △ ♀
13:30 ☽ □ ♂
14:33 ☽ △ ♆
04 00:23 ☽ ☍ ♆
06:53 ☽ ☍ ♃
08:09 ☿ ∥ ♄
10:30 ☽ ☍ ♄
21:53 ☽ △ ☿
23:10 ☽ ∦ ♆
05 00:45 ☽ ∦ ☉
02:39 ☽ □ ♅
10:33 ☿ ∥ ♃
11:03 ☽ ∦ ☿
11:06 ☽ ∦ ♃
14:40 ☽ ∦ ♄
14:42 ☽ △ ☉
21:48 ☽ □ ☿
22:29 ☽ △ ♂
06 04:54 ♀ △ ♆
08:03 ☉ ∥ ♆
20:27 ☽ ∦ ♀
07 08:47 ☽ △ ♅
09:15 ☽ ∥ ♅
11:26 ☽ □ ☿
08 03:47 ☽ ☍ ♆
08:23 ☽ ⚹ ♀
12:10 ☽ ∥ ♂
12:53 ☽ △ ♇
20:10 ☽ △ ♃
[illegible] ☽ ∥ ♆
22:36 ☽ △ ♄
09 04:32 ☿ ∥ ♆
19:41 ☉ □ ♆
21:14 ☽ ⚹ ☿
10 07:23 ☽ ⚹ ☉
08:32 ☽ ☍ ♂
11:51 ♀ ⚹ ♇
14:20 ☿ ∥ ☉
15:22 ☽ □ ♆
17:55 ☽ ∥ ♆
22:59 ☽ □ ♃
11 00:57 ☽ □ ♄
03:15 ☽ ∦ ♂
06:01 ☉ △ ♂
13:44 ☽ ☍ ♅
12 02:37 ☽ ∦ ♅
07:36 ☽ △ ♆
16:18 ☽ ⚹ ♇
21:00 ☽ ☌ ♀
13 00:24 ☽ ⚹ ♃
01:10 ☽ ∥ ♀
01:58 ☽ ⚹ ♄
11:39 ☿ □ ♆
15:50 ☽ ∥ ♄
17:43 ☽ ∥ ♃
14 06:13 ☽ ∥ ♇
08:15 ☽ □ ♆
10:42 ☽ ☌ ☿
12:14 ☽ △ ♂
13:03 ☽ ∥ ☉
20:59 ♀ ⚹ ♃
23:07 ☽ ∥ ☿
15 04:24 ☿ △ ♂
13:01 ♀ ⚹ ♄
15:23 ☽ △ ♂
16:22 ♀ ♐
16 10:12 ☽ ⚹ ♆
15:33 ☽ □ ♂
16:37 ☽ ∥ ☿
19:33 ☽ ☌ ♇
17 05:03 ♄ ♒
05:35 ☽ ☌ ♃
06:28 ☽ ☌ ♄
07:44 ☽ ∥ ☉
10:16 ☽ ⚹ ♀
16:34 ☽ ∥ ♆
18:51 ☽ □ ♅
18 07:41 ☽ ∥ ♃
09:01 ☽ ∥ ♄
14:23 ☽ ∥ ♀
22:09 ☽ ⚹ ♂
19 07:49 ☽ ⚹ ☿
08:45 ☽ ⚹ ☉
13:08 ♃ ♒
22:41 ☽ □ ♀
20 01:50 ☽ ⚹ ♅
03:27 ☽ ∦ ♅
03:29 ☿ ☌ ☉
23:08 ☿ ♑
23:34 ☽ ☌ ♆
23:35 ☽ ∦ ♂
21 07:45 ♀ ∥ ♄
10:03 ☉ ♑
10:25 ☽ ⚹ ♆
16:56 ♀ ∥ ♃
18:22 ♃ ☌ ♄
18:28 ☽ ∥ ♆
23:33 ☽ ⚹ ♄
23:36 ☽ ⚹ ♃
22 02:05 ☽ □ ☿
15:57 ☽ △ ♀
23 14:45 ♂ □ ♇
22:37 ☽ □ ♇
22:52 ☽ ☌ ♂
24 00:38 ☽ ∦ ♆
12:32 ☽ □ ♄
13:11 ☽ □ [illegible]
17:49 ☽ △ ☉
23:36 ☽ ∥ ♂
23:58 ☽ △ ☿
25 00:57 ☽ ☌ ♅
07:05 ☿ △ ♅
17:04 ☽ ∥ ♅
18:52 ♃ ∦ ♄
26 00:11 ☽ ⚹ ♆
11:33 ☽ △ ♇
27 01:42 ☽ △ ♄
02:55 ☽ △ ♃
12:57 ☽ ∦ ♃
13:18 ☽ ∦ ♄
28 [illegible] ☽ ∥ [illegible]
03:26 ☉ △ ♅
06:48 ☽ ☍ ♀
08:35 ☽ ∦ ♆
12:00 ☽ □ ♆
17:06 ☽ ∦ ☉
29 03:02 ☽ ⚹ ♂
20:00 ☽ ∦ ☿
23:34 ☽ ⚹ ♅
30 10:19 ♀ □ ♆
16:05 ☽ ☍ ☿
21:32 ☽ △ ♆
31 01:19 ☽ ∦ ☿
[illegible] ☽ ☍ ♆
13:46 ☽ □ ♂
21:58 ☽ ☍ ♄

⚷ Chiron

01 Dec.	04 N 26	
02	05♈02R	
05	05	00
08	04	58
11	04	57
14	04	57
17	04	56D
20	04	57
23	04	58
26	04	59
[illegible]	[illegible]	01
15	20:46	04♈56 D

Better books make better astrologers.
Here are some of our other titles:

AstroAmerica's Daily Ephemeris, 2000-2010
AstroAmerica's Daily Ephemeris, 2010-2020
- both for Midnight. Compiled & formatted by David R. Roell

Al Biruni
The Book of Instructions in the Elements of the Art of Astrology,
translated by R. Ramsay Wright

Derek Appleby
Horary Astrology: The Art of Astrological Divination

C.E.O. Carter
An Encyclopaedia of Psychological Astrology

Charubel & Sepharial
Degrees of the Zodiac Symbolized

H.L. Cornell, M.D.
Encyclopaedia of Medical Astrology

Nicholas Culpeper
Astrological Judgement of Diseases from the Decumbiture of the Sick,
and, **Urinalia**

Dorotheus of Sidon
Carmen Astrologicum, *translated by David Pingree*

Nicholas deVore
Encyclopedia of Astrology

Firmicus Maternus
Ancient Astrology Theory & Practice: Matheseos Libri VIII,
translated by Jean Rhys Bram

William Lilly
Christian Astrology, books 1 & 2
Christian Astrology, book 3

Claudius Ptolemy
Tetrabiblos, *translated by J.M. Ashmand*

Vivian Robson
Astrology and Sex
Electional Astrology
Fixed Stars & Constellations in Astrology

Richard Saunders
The Astrological Judgement and Practice of Physick

Sepharial
Primary Directions, a definitive study

James Wilson, Esq.
Dictionary of Astrology

H.S. Green, Raphael & C.E.O. Carter
Mundane Astrology: *3 Books*

If not available from your local bookseller, order directly from:
The Astrology Center of America
207 Victory Lane
Bel Air, MD 21014

on the web at:
http://www.astroamerica.com

Table of Logarithms

Minutes	0	1	2	3	4	5	6	7	8	9	10	11	Minutes
	HOURS OR DEGREES												
0	INFIN.	1.3802	1.0792	0.9031	0.7782	0.6812	0.6021	0.5351	0.4771	0.4260	0.3802	0.3388	0
1	3.1584	1.3730	1.0756	0.9007	0.7763	0.6798	0.6009	0.5341	0.4762	0.4252	0.3795	0.3382	1
2	2.8573	1.3660	1.0720	0.8983	0.7745	0.6784	0.5997	0.5331	0.4753	0.4244	0.3788	0.3375	2
3	2.6812	1.3590	1.0685	0.8959	0.7728	0.6769	0.5985	0.5320	0.4744	0.4236	0.3780	0.3368	3
4	2.5563	1.3522	1.0649	0.8935	0.7710	0.6755	0.5973	0.5310	0.4735	0.4228	0.3773	0.3362	4
5	2.4594	1.3454	1.0615	0.8912	0.7692	0.6741	0.5961	0.5300	0.4726	0.4220	0.3766	0.3355	5
6	2.3802	1.3388	1.0580	0.8888	0.7674	0.6726	0.5949	0.5290	0.4717	0.4212	0.3759	0.3349	6
7	2.3133	1.3323	1.0546	0.8865	0.7657	0.6712	0.5937	0.5279	0.4708	0.4204	0.3752	0.3342	7
8	2.2553	1.3259	1.0512	0.8842	0.7639	0.6698	0.5925	0.5269	0.4699	0.4196	0.3745	0.3336	8
9	2.2041	1.3195	1.0478	0.8819	0.7622	0.6684	0.5913	0.5259	0.4691	0.4188	0.3737	0.3329	9
10	2.1584	1.3133	1.0444	0.8796	0.7604	0.6670	0.5902	0.5249	0.4682	0.4180	0.3730	0.3323	10
11	2.1170	1.3071	1.0411	0.8773	0.7587	0.6656	0.5890	0.5239	0.4673	0.4172	0.3723	0.3316	11
12	2.0792	1.3010	1.0378	0.8751	0.7570	0.6642	0.5878	0.5229	0.4664	0.4164	0.3716	0.3310	12
13	2.0444	1.2950	1.0345	0.8728	0.7552	0.6628	0.5867	0.5219	0.4655	0.4156	0.3709	0.3303	13
14	2.0122	1.2891	1.0313	0.8706	0.7535	0.6614	0.5855	0.5209	0.4646	0.4149	0.3702	0.3297	14
15	1.9823	1.2833	1.0280	0.8683	0.7518	0.6601	0.5843	0.5199	0.4638	0.4141	0.3695	0.3291	15
16	1.9542	1.2775	1.0248	0.8661	0.7501	0.6587	0.5832	0.5189	0.4629	0.4133	0.3688	0.3284	16
17	1.9279	1.2719	1.0216	0.8639	0.7484	0.6573	0.5820	0.5179	0.4620	0.4125	0.3681	0.3278	17
18	1.9031	1.2663	1.0185	0.8617	0.7467	0.6559	0.5809	0.5169	0.4611	0.4117	0.3674	0.3271	18
19	1.8796	1.2607	1.0153	0.8595	0.7451	0.6546	0.5797	0.5159	0.4603	0.4110	0.3667	0.3265	19
20	1.8573	1.2553	1.0122	0.8573	0.7434	0.6532	0.5786	0.5149	0.4594	0.4102	0.3660	0.3259	20
21	1.8361	1.2499	1.0091	0.8552	0.7417	0.6519	0.5774	0.5139	0.4585	0.4094	0.3653	0.3252	21
22	1.8159	1.2445	1.0061	0.8530	0.7401	0.6505	0.5763	0.5129	0.4577	0.4086	0.3646	0.3246	22
23	1.7966	1.2393	1.0030	0.8509	0.7384	0.6492	0.5752	0.5120	0.4568	0.4079	0.3639	0.3239	23
24	1.7782	1.2341	1.0000	0.8487	0.7368	0.6478	0.5740	0.5110	0.4559	0.4071	0.3632	0.3233	24
25	1.7604	1.2289	0.9970	0.8466	0.7351	0.6465	0.5729	0.5100	0.4551	0.4063	0.3625	0.3227	25
26	1.7434	1.2239	0.9940	0.8445	0.7335	0.6451	0.5718	0.5090	0.4542	0.4055	0.3618	0.3220	26
27	1.7270	1.2188	0.9910	0.8424	0.7319	0.6438	0.5707	0.5081	0.4534	0.4048	0.3611	0.3214	27
28	1.7112	1.2139	0.9881	0.8403	0.7302	0.6425	0.5695	0.5071	0.4525	0.4040	0.3604	0.3208	28
29	1.6960	1.2090	0.9852	0.8382	0.7286	0.6412	0.5684	0.5061	0.4516	0.4033	0.3597	0.3201	29
30	1.6812	1.2041	0.9823	0.8361	0.7270	0.6398	0.5673	0.5051	0.4508	0.4025	0.3590	0.3195	30
31	1.6670	1.1993	0.9794	0.8341	0.7254	0.6385	0.5662	0.5042	0.4499	0.4017	0.3583	0.3189	31
32	1.6532	1.1946	0.9765	0.8320	0.7238	0.6372	0.5651	0.5032	0.4491	0.4010	0.3576	0.3183	32
33	1.6398	1.1899	0.9737	0.8300	0.7222	0.6359	0.5640	0.5023	0.4482	0.4002	0.3570	0.3176	33
34	1.6269	1.1852	0.9708	0.8279	0.7206	0.6346	0.5629	0.5013	0.4474	0.3995	0.3563	0.3170	34
35	1.6143	1.1806	0.9680	0.8259	0.7190	0.6333	0.5618	0.5004	0.4466	0.3987	0.3556	0.3164	35
36	1.6021	1.1761	0.9652	0.8239	0.7175	0.6320	0.5607	0.4994	0.4457	0.3979	0.3549	0.3158	36
37	1.5902	1.1716	0.9625	0.8219	0.7159	0.6307	0.5596	0.4984	0.4449	0.3972	0.3542	0.3151	37
38	1.5786	1.1671	0.9597	0.8199	0.7143	0.6294	0.5585	0.4975	0.4440	0.3964	0.3535	0.3145	38
39	1.5673	1.1627	0.9570	0.8179	0.7128	0.6282	0.5574	0.4965	0.4432	0.3957	0.3529	0.3139	39
40	1.5563	1.1584	0.9542	0.8159	0.7112	0.6269	0.5563	0.4956	0.4424	0.3949	0.3522	0.3133	40
41	1.5456	1.1540	0.9515	0.8140	0.7097	0.6256	0.5552	0.4947	0.4415	0.3942	0.3515	0.3126	41
42	1.5351	1.1498	0.9488	0.8120	0.7081	0.6243	0.5541	0.4937	0.4407	0.3934	0.3508	0.3120	42
43	1.5249	1.1455	0.9462	0.8101	0.7066	0.6231	0.5531	0.4928	0.4399	0.3927	0.3502	0.3114	43
44	1.5149	1.1413	0.9435	0.8081	0.7050	0.6218	0.5520	0.4918	0.4390	0.3919	0.3495	0.3108	44
45	1.5051	1.1372	0.9409	0.8062	0.7035	0.6205	0.5509	0.4909	0.4382	0.3912	0.3488	0.3102	45
46	1.4956	1.1331	0.9383	0.8043	0.7020	0.6193	0.5498	0.4900	0.4374	0.3905	0.3481	0.3096	46
47	1.4863	1.1290	0.9356	0.8023	0.7005	0.6180	0.5488	0.4890	0.4366	0.3897	0.3475	0.3089	47
48	1.4771	1.1249	0.9331	0.8004	0.6990	0.6168	0.5477	0.4881	0.4357	0.3890	0.3468	0.3083	48
49	1.4682	1.1209	0.9305	0.7985	0.6975	0.6155	0.5466	0.4872	0.4349	0.3882	0.3461	0.3077	49
50	1.4594	1.1170	0.9279	0.7966	0.6960	0.6143	0.5456	0.4863	0.4341	0.3875	0.3454	0.3071	50
51	1.4508	1.1130	0.9254	0.7948	0.6945	0.6131	0.5445	0.4853	0.4333	0.3868	0.3448	0.3065	51
52	1.4424	1.1091	0.9228	0.7929	0.6930	0.6118	0.5435	0.4844	0.4325	0.3860	0.3441	0.3059	52
53	1.4341	1.1053	0.9203	0.7910	0.6915	0.6106	0.5424	0.4835	0.4316	0.3853	0.3434	0.3053	53
54	1.4260	1.1015	0.9178	0.7891	0.6900	0.6094	0.5414	0.4826	0.4308	0.3846	0.3428	0.3047	54
55	1.4180	1.0977	0.9153	0.7873	0.6885	0.6081	0.5403	0.4817	0.4300	0.3838	0.3421	0.3041	55
56	1.4102	1.0939	0.9128	0.7855	0.6871	0.6069	0.5393	0.4808	0.4292	0.3831	0.3415	0.3034	56
57	1.4025	1.0902	0.9104	0.7836	0.6856	0.6057	0.5382	0.4798	0.4284	0.3824	0.3408	0.3028	57
58	1.3949	1.0865	0.9079	0.7818	0.6841	0.6045	0.5372	0.4789	0.4276	0.3817	0.3401	0.3022	58
59	1.3875	1.0828	0.9055	0.7800	0.6827	0.6033	0.5361	0.4780	0.4268	0.3809	0.3395	0.3016	59
	0	1	2	3	4	5	6	7	8	9	10	11	
	HOURS OR DEGREES												

Table of Logarithms

MINUTES	12	13	14	15	16	17	18	19	20	21	22	23	MINUTES
	HOURS OR DEGREES												
0	0.3010	0.2663	0.2341	0.2041	0.1761	0.1498	0.1249	0.1015	0.0792	0.0580	0.0378	0.0185	0
1	0.3004	0.2657	0.2336	0.2036	0.1756	0.1493	0.1245	0.1011	0.0788	0.0576	0.0375	0.0182	1
2	0.2998	0.2652	0.2331	0.2032	0.1752	0.1489	0.1241	0.1007	0.0785	0.0573	0.0371	0.0179	2
3	0.2992	0.2646	0.2325	0.2027	0.1747	0.1485	0.1237	0.1003	0.0781	0.0570	0.0368	0.0175	3
4	0.2986	0.2640	0.2320	0.2022	0.1743	0.1481	0.1233	0.0999	0.0777	0.0566	0.0365	0.0172	4
5	0.2980	0.2635	0.2315	0.2017	0.1738	0.1476	0.1229	0.0996	0.0774	0.0563	0.0361	0.0169	5
6	0.2974	0.2629	0.2310	0.2012	0.1734	0.1472	0.1225	0.0992	0.0770	0.0559	0.0358	0.0166	6
7	0.2968	0.2624	0.2305	0.2008	0.1729	0.1468	0.1221	0.0988	0.0767	0.0556	0.0355	0.0163	7
8	0.2962	0.2618	0.2300	0.2003	0.1725	0.1464	0.1217	0.0984	0.0763	0.0552	0.0352	0.0160	8
9	0.2956	0.2613	0.2295	0.1998	0.1720	0.1459	0.1213	0.0980	0.0759	0.0549	0.0348	0.0157	9
10	0.2950	0.2607	0.2289	0.1993	0.1716	0.1455	0.1209	0.0977	0.0756	0.0546	0.0345	0.0153	10
11	0.2944	0.2602	0.2284	0.1988	0.1711	0.1451	0.1205	0.0973	0.0752	0.0542	0.0342	0.0150	11
12	0.2939	0.2596	0.2279	0.1984	0.1707	0.1447	0.1201	0.0969	0.0749	0.0539	0.0339	0.0147	12
13	0.2933	0.2591	0.2274	0.1979	0.1702	0.1443	0.1197	0.0965	0.0745	0.0535	0.0335	0.0144	13
14	0.2927	0.2585	0.2269	0.1974	0.1698	0.1438	0.1193	0.0962	0.0741	0.0532	0.0332	0.0141	14
15	0.2921	0.2580	0.2264	0.1969	0.1694	0.1434	0.1189	0.0958	0.0738	0.0529	0.0329	0.0138	15
16	0.2915	0.2574	0.2259	0.1965	0.1689	0.1430	0.1186	0.0954	0.0734	0.0525	0.0326	0.0135	16
17	0.2909	0.2569	0.2254	0.1960	0.1685	0.1426	0.1182	0.0950	0.0731	0.0522	0.0322	0.0132	17
18	0.2903	0.2564	0.2249	0.1955	0.1680	0.1422	0.1178	0.0947	0.0727	0.0518	0.0319	0.0129	18
19	0.2897	0.2558	0.2244	0.1950	0.1676	0.1417	0.1174	0.0943	0.0724	0.0515	0.0316	0.0125	19
20	0.2891	0.2553	0.2239	0.1946	0.1671	0.1413	0.1170	0.0939	0.0720	0.0512	0.0313	0.0122	20
21	0.2885	0.2547	0.2234	0.1941	0.1667	0.1409	0.1166	0.0935	0.0716	0.0508	0.0309	0.0119	21
22	0.2880	0.2542	0.2229	0.1936	0.1663	0.1405	0.1162	0.0932	0.0713	0.0505	0.0306	0.0116	22
23	0.2874	0.2536	0.2224	0.1932	0.1658	0.1401	0.1158	0.0928	0.0709	0.0501	0.0303	0.0113	23
24	0.2868	0.2531	0.2218	0.1927	0.1654	0.1397	0.1154	0.0924	0.0706	0.0498	0.0300	0.0110	24
25	0.2862	0.2526	0.2213	0.1922	0.1649	0.1392	0.1150	0.0920	0.0702	0.0495	0.0296	0.0107	25
26	0.2856	0.2520	0.2208	0.1918	0.1645	0.1388	0.1146	0.0917	0.0699	0.0491	0.0293	0.0104	26
27	0.2850	0.2515	0.2203	0.1913	0.1640	0.1384	0.1142	0.0913	0.0695	0.0488	0.0290	0.0101	27
28	0.2845	0.2510	0.2198	0.1908	0.1636	0.1380	0.1138	0.0909	0.0692	0.0484	0.0287	0.0098	28
29	0.2839	0.2504	0.2193	0.1903	0.1632	0.1376	0.1134	0.0905	0.0688	0.0481	0.0284	0.0095	29
30	0.2833	0.2499	0.2188	0.1899	0.1627	0.1372	0.1130	0.0902	0.0685	0.0478	0.0280	0.0091	30
31	0.2827	0.2493	0.2183	0.1894	0.1623	0.1368	0.1126	0.0898	0.0681	0.0474	0.0277	0.0088	31
32	0.2821	0.2488	0.2178	0.1889	0.1619	0.1363	0.1123	0.0894	0.0678	0.0471	0.0274	0.0085	32
33	0.2816	0.2483	0.2173	0.1885	0.1614	0.1359	0.1119	0.0891	0.0674	0.0468	0.0271	0.0082	33
34	0.2810	0.2477	0.2169	0.1880	0.1610	0.1355	0.1115	0.0887	0.0670	0.0464	0.0267	0.0079	34
35	0.2804	0.2472	0.2164	0.1876	0.1605	0.1351	0.1111	0.0883	0.0667	0.0461	0.0264	0.0076	35
36	0.2798	0.2467	0.2159	0.1871	0.1601	0.1347	0.1107	0.0880	0.0663	0.0458	0.0261	0.0073	36
37	0.2793	0.2461	0.2154	0.1866	0.1597	0.1343	0.1103	0.0876	0.0660	0.0454	0.0258	0.0070	37
38	0.2787	0.2456	0.2149	0.1862	0.1592	0.1339	0.1099	0.0872	0.0656	0.0451	0.0255	0.0067	38
39	0.2781	0.2451	0.2144	0.1857	0.1588	0.1335	0.1095	0.0868	0.0653	0.0448	0.0251	0.0064	39
40	0.2775	0.2445	0.2139	0.1852	0.1584	0.1331	0.1091	0.0865	0.0649	0.0444	0.0248	0.0061	40
41	0.2770	0.2440	0.2134	0.1848	0.1579	0.1326	0.1088	0.0861	0.0646	0.0441	0.0245	0.0058	41
42	0.2764	0.2435	0.2129	0.1843	0.1575	0.1322	0.1084	0.0857	0.0642	0.0438	0.0242	0.0055	42
43	0.2758	0.2430	0.2124	0.1839	0.1571	0.1318	0.1080	0.0854	0.0639	0.0434	0.0239	0.0052	43
44	0.2753	0.2424	0.2119	0.1834	0.1566	0.1314	0.1076	0.0850	0.0635	0.0431	0.0235	0.0049	44
45	0.2747	0.2419	0.2114	0.1829	0.1562	0.1310	0.1072	0.0846	0.0632	0.0428	0.0232	0.0045	45
46	0.2741	0.2414	0.2109	0.1825	0.1558	0.1306	0.1068	0.0843	0.0628	0.0424	0.0229	0.0042	46
47	0.2736	0.2409	0.2104	0.1820	0.1553	0.1302	0.1064	0.0839	0.0625	0.0421	0.0226	0.0039	47
48	0.2730	0.2403	0.2099	0.1816	0.1549	0.1298	0.1061	0.0835	0.0621	0.0418	0.0223	0.0036	48
49	0.2724	0.2398	0.2095	0.1811	0.1545	0.1294	0.1057	0.0832	0.0618	0.0414	0.0220	0.0033	49
50	0.2719	0.2393	0.2090	0.1806	0.1540	0.1290	0.1053	0.0828	0.0615	0.0411	0.0216	0.0030	50
51	0.2713	0.2388	0.2085	0.1802	0.1536	0.1286	0.1049	0.0825	0.0611	0.0408	0.0213	0.0027	51
52	0.2707	0.2382	0.2080	0.1797	0.1532	0.1282	0.1045	0.0821	0.0608	0.0404	0.0210	0.0024	52
53	0.2702	0.2377	0.2075	0.1793	0.1528	0.1278	0.1041	0.0817	0.0604	0.0401	0.0207	0.0021	53
54	0.2696	0.2372	0.2070	0.1788	0.1523	0.1274	0.1037	0.0814	0.0601	0.0398	0.0204	0.0018	54
55	0.2691	0.2367	0.2065	0.1784	0.1519	0.1270	0.1034	0.0810	0.0597	0.0394	0.0201	0.0015	55
56	0.2685	0.2362	0.2061	0.1779	0.1515	0.1266	0.1030	0.0806	0.0594	0.0391	0.0197	0.0012	56
57	0.2679	0.2356	0.2056	0.1775	0.1510	0.1261	0.1026	0.0803	0.0590	0.0388	0.0194	0.0009	57
58	0.2674	0.2351	0.2051	0.1770	0.1506	0.1257	0.1022	0.0799	0.0587	0.0384	0.0191	0.0006	58
59	0.2668	0.2346	0.2046	0.1765	0.1502	0.1253	0.1018	0.0795	0.0583	0.0381	0.0188	0.0003	59
	12	13	14	15	16	17	18	19	20	21	22	23	
	HOURS OR DEGREES												

www.ingramcontent.com/pod-product-compliance
Lightning Source LLC
[illegible]
CBHW080525220326
41599CB00032B/6204

9781933303192